The First Global Integrated Marine Assessment
World Ocean Assessment I

The *World Ocean Assessment* is the culmination of the work of nearly 600 scientists and experts from many countries, representing various disciplines and steered by a 22-member Group of Experts. The *Assessment* examines the current state of knowledge of the world's oceans and the ways in which humans benefit from and affect them. The *Assessment* indicates that the oceans' carrying capacity is near or at its limit. It is clear that urgent action on a global scale is needed to protect the world's oceans from the many pressures they face.

The *World Ocean Assessment* – or, to give its full title, the *First Global Integrated Marine Assessment* – is the outcome of the first cycle of the United Nations' Regular Process for Global Reporting and Assessment of the State of the Marine Environment, including Socioeconomic Aspects. The *Assessment* provides vital, scientifically-grounded bases for the consideration of ocean issues, including climate change, by governments, intergovernmental agencies, non-governmental agencies and all other stakeholders and policy-makers involved in ocean affairs. Together with future assessments and related initiatives, it will support the implementation of the recently adopted 2030 *Agenda for Sustainable Development*, particularly its ocean related goals. Moreover, it will also form an important reference text for marine science courses.

The First Global Integrated Marine Assessment
World Ocean Assessment I

United Nations

CAMBRIDGE
UNIVERSITY PRESS

University Printing House, Cambridge CB2 8BS, United Kingdom

One Liberty Plaza, 20th Floor, New York, NY 10006, USA

477 Williamstown Road, Port Melbourne, VIC 3207, Australia

4843/24, 2nd Floor, Ansari Road, Daryaganj, Delhi - 110002, India

79 Anson Road, #06-04/06, Singapore 079906

Cambridge University Press is part of the University of Cambridge.

It furthers the University's mission by disseminating knowledge in the pursuit of education, learning, and research at the highest international levels of excellence.

www.cambridge.org
Information on this title: www.cambridge.org/9781316510018
DOI: 10.1017/9781108186148

© 2017 by the United Nations. All worldwide rights reserved.

This publication is in copyright. Subject to statutory exception and to the provisions of relevant collective licensing agreements, no reproduction of any part may take place without the written permission of Cambridge University Press.

First published 2017

Printed in the United States of America by Sheridan Books, Inc.

A catalogue record for this publication is available from the British Library.

Library of Congress Cataloging-in-Publication Data

ISBN 978-1-316-51001-8 Hardback
ISBN 978-1-316-64915-2 Paperback

Cambridge University Press has no responsibility for the persistence or accuracy of URLs for external or third-party internet websites referred to in this publication and does not guarantee that any content on such websites is, or will remain, accurate or appropriate.

The designations and the presentation of the materials used in this publication, including their respective citations, maps and bibliography, do not imply the expression of any opinion whatsoever on the part of the United Nations concerning the legal status of any country, territory, city or area or of its authorities, or concerning the delimitation of its frontiers or boundaries.

Also, the boundaries and names shown and the designations used in this publication do not imply official endorsement or acceptance by the United Nations.

Any information that may be contained in this publication emanating from actions and decisions taken by States does not imply recognition by the United Nations of the validity of the actions and decisions in question and is included without prejudice to the position of any Member State of the United Nations.

The contributions of the members of the Group of Experts and the Pool of Experts, who participated in the writing of the first global integrated marine assessment, were made in their personal capacity. The members of the Group and the Pool are not representatives of any Government or any other authority or organization.

The First Global Integrated Marine Assessment

World Ocean Assessment I

by

the Group of Experts of the Regular Process

Lorna Inniss and Alan Simcock (Joint Coordinators)

Amanuel Yoanes Ajawin, Angel C. Alcala (until July 2013), Patricio Bernal, Hilconida P. Calumpong (from July 2013), Peyman Eghtesadi Araghi, Sean O. Green, Peter Harris, Osman Keh Kamara, Kunio Kohata, Enrique Marschoff, Georg Martin, Beatrice Padovani Ferreira, Chul Park, Rolph Antoine Payet, Jake Rice, Andrew Rosenberg, Renison Ruwa, Joshua T. Tuhumwire, Saskia Van Gaever, Juying Wang, and Jan Marcin Węsławski

under the auspices of the United Nations General Assembly and its Regular Process for Global Reporting and Assessment of the State of the Marine Environment, including Socioeconomic Aspects

© 2016 United Nations

(Editorial Info)

Disclaimer

The designations and the presentation of the materials used in this publication, including their respective citations, maps and bibliography, do not imply the expression of any opinion whatsoever on the part of the United Nations concerning the legal status of any country, territory, city or area or of its authorities, or concerning the delimitation of its frontiers or boundaries.

Also, the boundaries and names shown and the designations used in this publication do not imply official endorsement or acceptance by the United Nations.

Any information that may be contained in this publication emanating from actions and decisions taken by States does not imply recognition by the United Nations of the validity of the actions and decisions in question and is included without prejudice to the position of any Member State of the United Nations.

The contributions of the members of the Group of Experts and the Pool of Experts, who participated in the writing of the first global integrated marine assessment, were made in their personal capacity. The members of the Group and the Pool are not representatives of any Government or any other authority or organization.

Table of Contents

Foreword and Preface 7

Summary of the first global integrated marine assessment 11

The context of the assessment 45

Chapter 1 – Introduction – Planet, oceans and life 47

Chapter 2 – Mandate, information sources and method of work 57

Assessment of Major Ecosystem Services from the Marine Environment (Other than Provisioning Services) 65

Chapter 3 – Scientific Understanding of Ecosystem Services 67

Chapter 4 – The Ocean's Role in the Hydrological Cycle 91

Chapter 5 – Sea-Air Interactions 105

Chapter 6 – Primary Production, Cycling of Nutrients, Surface Layer and Plankton 119

Chapter 7 – Calcium Carbonate Production and Contribution to Coastal Sediments 149

Chapter 8 – Aesthetic, Cultural, Religious and Spiritual Ecosystem Services Derived from the Marine Environment . 159

Chapter 9 – Conclusions on Major Ecosystem Services Other than Provisioning Services 171

Assessment of the Cross-cutting Issues: Food Security and Food Safety 183

Chapter 10 – The Oceans as a Source of Food 185

Chapter 11 – Capture Fisheries 191

Chapter 12 – Aquaculture 203

Chapter 13 – Fish Stock Propagation 213

Chapter 14 – Seaweeds 223

Chapter 15 – Social and Economic Aspects of Sea-Based Food and Fisheries 229

Chapter 16 – Synthesis of Part IV: Food Security and Safety 239

Assessment of Other Human Activities and the Marine Environment 243

Chapter 17 – Shipping 245

Chapter 18 – Ports 269

Chapter 19 – Submarine Cables and Pipelines 277

Chapter 20 – Coastal, Riverine and Atmospheric Inputs from Land 285

Chapter 21 – Offshore Hydrocarbon Industries 333

Chapter 22 – Other Marine-Based Energy Industries 353

Chapter 23 – Offshore Mining Industries 363

Chapter 24 – Solid Waste Disposal 379

Chapter 25 – Marine Debris 389

Chapter 26 – Land-Sea Physical Interaction 409

Chapter 27 – Tourism and Recreation 425

Chapter 28 – Desalinization 441

Chapter 29 – Use of Marine Genetic Resources 451

Chapter 30 – Marine Scientific Research 459

Chapter 31 – Conclusions on Other Human Activities 471

Chapter 32 – Capacity-Building in Relation to Human Activities Affecting the Marine Environment 479

Assessment of Marine Biological Diversity and Habitats 493

Chapter 33 – Introduction 495

Section A — Overview of Marine Biological Diversity 499

Chapter 34 – Global Patterns in Marine Biodiversity 501

Chapter 35 – Extent of Assessment of Marine Biological Diversity 525

Chapter 36 – Overview of Marine Biological Diversity 555

    Division 36.A – North Atlantic Ocean 557

    Division 36.B – South Atlantic Ocean 595

    Division 36.C – North Pacific Ocean 615

    Division 36.D – South Pacific Ocean 635

    Division 36.E – Indian Ocean 669

    Division 36.F – Open Ocean Deep Sea 685

    Division 36.G – Arctic Ocean 705

    Division 36.H – Southern Ocean 729

Section B — Marine Ecosystems, Species and Habitats Scientifically Identified as Threatened, Declining or Otherwise in need of Special Attention or Protection 749

I – Marine Species 751

Chapter 37 – Marine Mammals 753

Chapter 38 – Seabirds 763

Chapter 39 – Marine Reptiles 773

Chapter 40 – Sharks and Other Elasmobranchs 781

Chapter 41 – Tunas and Billfishes 789

II – Marine Ecosystems and Habitats 801

Chapter 42 – Cold-Water Corals 803

Chapter 43 – Tropical and Sub-Tropical Coral Reefs 817

Chapter 44 – Estuaries and Deltas 839

Chapter 45 – Hydrothermal Vents and Cold Seeps 853

Chapter 46 – High-Latitude Ice and the Biodiversity Dependent on it 863

Chapter 47 – Kelp Forests and Seagrass Meadows 869

Chapter 48 – Mangroves 877

Chapter 49 – Salt Marshes 887

Chapter 50 – Sargasso Sea 893

Chapter 51 – Biological Communities on Seamounts and Other Submarine Features Potentially Threatened by Disturbance 899

Section C — Environmental, economic and/or social aspects of the conservation of marine species and habitats and capacity-building needs 913

Chapter 52 – Synthesis of Part VI: Marine Biological Diversity and Habitats 915

Chapter 53 – Capacity-Building Needs in Relation to the Status of Species and Habitats 923

Overall Assessment 933

Chapter 54 – Overall Assessment of Human Impact on the Oceans 935

Chapter 55 – Overall Value of the Oceans to Humans 945

Annexes 953

Annex I – List of Contributors and Commentators 955

Annex II – Glossary 961

Annex III – Acronyms 969

FM Foreword and Preface

Foreword

The First Global Integrated Marine Assessment, also known as the "World Ocean Assessment I", is the outcome of the first cycle of the Regular Process for Global Reporting and Assessment of the State of the Marine Environment, including Socioeconomic Aspects.

Hundreds of scientists from many countries, representing various disciplines and steered by a 22-member Group of Experts, examined the state of knowledge of the world's oceans and the ways in which humans benefit from and affect them. Their findings indicate that the oceans' carrying capacity is near or at its limit. It is clear that urgent action on a global scale is needed to protect the world's oceans from the many pressures they face.

The first World Ocean Assessment provides an important scientific basis for the consideration of ocean issues by Governments, intergovernmental processes, and all policy-makers and others involved in ocean affairs. The Assessment reinforces the science-policy interface and establishes the basis for future assessments. Together with future assessments and related initiatives, it will help in the implementation of the recently adopted 2030 Agenda for Sustainable Development, particularly its ocean-related goals.

I thank the Group of Experts and all others whose dedication produced this Assessment, including the Co-Chairs, the Bureau of the Ad Hoc Working Group of the Whole of the Regular Process and its secretariat. I would also like to acknowledge the significant scientific, technical and financial assistance to the Regular Process provided by the European Union, the United Nations Environment Programme, the Intergovernmental Oceanographic Commission of the United Nations Educational, Scientific and Cultural Organization and other relevant intergovernmental organizations.

I look forward to working with all partners to ensure that the world's oceans, which are so essential for human well-being and prosperity, are healthy, productive and resilient for today's and future generations.

BAN Ki-moon
Secretary-General
United Nations

Preface

As we seek to pursue sustainable development, we all need an understanding of the ways – environmental, social and economic – in which we humans interact with the world around us. Globally, the drive towards sustainable development cannot ignore the seven-tenths of the planet covered by the ocean. Such thoughts led to the recommendation of the 2002 Johannesburg World Summit on Sustainable Development, that there should be a regular process for the global reporting and assessment of the state of the marine environment, including socioeconomic aspects. We need to understand the overall benefits of the ocean to us humans, and the overall impacts of humans on the ocean.

In September 2015, the United Nations adopted the Sustainable Development Goals, including SDG 14 ("Conserve and sustainably use oceans, seas and marine resources for sustainable development"). It is therefore timely that the work put in hand by the Johannesburg Summit has now produced The First Global Integrated Marine Assessment – World Ocean Assessment I under the auspices of the United Nations General Assembly. The General Assembly considered and endorsed not only its Outline, but also the terms of reference and working methods of the Group of Experts and the guidance to contributors. The approach to the assessment has therefore been carefully considered at the global level.

Implementing the approach has been a major task, relying essentially on voluntary efforts from hundreds of experts in many fields. We and our colleagues in the Group of Experts of the Regular Process have been privileged to organize, contribute to, and produce the final version of, this Assessment. Crucial support has been provided by the secretariat of the Regular Process in the United Nations Secretariat, the Division for Ocean Affairs and the Law of the Sea of the Office of Legal Affairs, by several international organizations and by a number of United Nations Member States, as detailed in Chapter 2 (Mandate, information sources and method of work). The full draft assessment was reviewed by United Nations Member States. Under the terms of reference and working methods, the Group of Experts is collectively responsible for the final text.

The Regular Process was tasked with providing a first Assessment that could serve as a baseline for future cycles of the process. The vast scale of the ocean and the complexities of its many facets are revealed again and again throughout this first Assessment. Likewise, the challenges that must be faced are presented. We have also sought to identify the main gaps in knowledge and in capacity-building that hinder the responses to these challenges. These elements are all summarized in Part I of the Assessment – Summary – under ten themes: climate change, over-exploitation of marine living resources, the significance of food security and food safety, patterns of biodiversity and the changes in them, the pressures from increased uses of ocean space, the threats from increased pollution, the effects of cumulative impacts, the inequalities in the distribution of benefits from the ocean, the importance of coherent management of human impacts on the ocean, and the problems of delay in implementing known solutions.

The General Assembly decided that the Assessment should not undertake policy analysis. We have therefore not ranked these themes in order of importance. Similarly, recommendations for action were not sought. It therefore remains for national governments and the competent international authorities to decide what action to take in the light of the Assessment.

Lorna Inis
Alan Simcock
Joint Coordinators of the Group of Experts of the Regular Process

I

Summary of the first global integrated marine assessment

1 Introduction[1]

Let us consider how dependent on the ocean we are. The ocean is vast: it covers seven tenths of the planet, is on average about 4,000 metres deep and contains 1.3 billion cubic kilometres of water (97 per cent of all the water on the surface of the Earth). There are, however, 7 billion people on Earth. This means that each one of us has just one fifth of a cubic kilometre of ocean as our portion to provide us with all the services that we get from the ocean. That small, one fifth of a cubic kilometre portion generates half of the annual production of the oxygen that each of us breathes, and all of the sea fish and other seafood that each of us eats. It is the ultimate source of all the freshwater that each of us will drink in our lifetimes.

The ocean is a highway for ships that carry the goods that we produce and consume. The seabed and the strata beneath it hold minerals and oil and gas deposits that we increasingly need to use. Submarine cables across the ocean floor carry 90 per cent of the electronic traffic of communications, financial transactions and information exchange. Our energy supply will increasingly rely on sea-based wind turbines and wave and tidal power from the ocean. Large numbers of us take our holidays by the sea. The seabed is a rich repository for archaeology.

That one fifth of a cubic kilometre also suffers from the sewage, garbage, spilled oil and industrial waste which we collectively allow to go into the ocean every day. Demands on the ocean continue to rise together with the world's population. By the year 2050, it is estimated that there will be 10 billion people on Earth. Our portion, or our children's portion, of the ocean will then have shrunk to one eighth of a cubic kilometre. That reduced portion will still have to provide each of us with oxygen, food and water, while still suffering from the pollution and waste that we allow to enter the ocean.

The ocean is also home to a rich diversity of animals, plants, seaweeds and microbes, from the largest animal on the planet (the blue whale) to plankton and bacteria that can only be seen with powerful microscopes. We use some of those directly, and many more contribute indirectly to the benefits that we derive from the ocean. Even those organisms without any apparent connection with humans are part of the biodiversity whose value we have belatedly recognized. However, our relationship with the ocean and its creatures works both ways. We intentionally exploit many components of that rich biodiversity and increase the mortality of other components, even though we are not deliberately harvesting them. Carelessly (for example, through the input of waste material) or because of an initial lack of knowledge (for example, through the ocean acidification from increased emissions of carbon dioxide), we are altering the environment in which those organisms live. All those actions are affecting their ability to thrive and, sometimes, even to survive.

The impacts of humanity on the ocean are parts of our inheritance and future. They have helped to shape our present and will shape not only the future of the ocean and its biodiversity as an integral physical and biological system, but also the ability of the ocean to provide the services that we use now, that we will increasingly need to use in the future and that are vital to each of us and to human well-being overall.

Managing our uses of the ocean is therefore vital. The successful management of any activity, however, requires an adequate understanding of the activity and of the context in which it takes place. Such an understanding is needed even more when management tasks are split among many players: unless each knows how the part they play fits into the overall pattern, there are risks of confusion, contradictory actions and failure to act. Managing the human uses of the ocean has inevitably to be divided among many players. In the course of their activities, individuals and commercial enterprises that use the ocean on a constant basis take decisions that affect the human impacts on the ocean.[2]

The United Nations Convention on the Law of the Sea[3] establishes the legal framework within which all activities in the oceans and seas must be carried out. National Governments and regional and global intergovernmental organizations all have their parts to play in regulating those activities. However, each of those many players tends to have a limited view of the ocean that is focused on their own sectoral interests. Without a sound framework in which to work, they may well fail to take into account the ways in which their decisions and actions interact with those of others. Such failures can add to the complexity of the manifold problems that exist.

It is therefore not surprising that, in 2002, the World Summit on Sustainable Development recommended that there be a regular process for global reporting and assessment of the state of the marine environment, including socioeconomic aspects, or that the General Assembly accepted that recommendation. In its resolution 64/71, the Assembly adopted the recommendation that the Regular Process for Global Reporting and Assessment of the State of the Marine Environment, including Socioeconomic Aspects should review the state of the marine environment, including socioeconomic aspects, on a continual and systematic basis by providing regular assessments at the global and supraregional levels and an integrated view of environmental, economic and social aspects.

Those regular reviews of the state of the ocean, the way in which the many dynamics of the ocean interact and the ways in which humans are using it should enable the many people and institutions involved in human uses to position their decisions more effectively in the overall context of the ocean. The first global integrated marine assessment, also known as the first world ocean assessment, is the first outcome of the Regular Process. It is divided into seven parts, which are described in detail below. The present part (part I, the summary) provides: (a) a summary of the organization of the

---

1 In the present summary, the chapters referred to in footnotes are chapters of parts II to VII of the first global integrated marine assessment. When placed at the end of a paragraph, such footnotes apply to all preceding paragraphs up to the previous such footnote.

2 See chaps. 1 and 3.

3 United Nations, Treaty Series, vol. 1833, No. 31363.

Process and the assessment; (b) a short description of the 10 main themes that have been identified; (c) a more detailed description of each of those themes, based on the content of parts II to VII; and (d) indications of the most serious gaps in our knowledge of the ocean and related human activities, as well as in the capacities to engage in some activities and to assess them all, drawing on the content of parts III to VII.[4]

2 Background to the assessment: the ocean around us

The starting point is the four main ocean basins of our planet: the Arctic Ocean, the Atlantic Ocean, the Indian Ocean and the Pacific Ocean.[5] Even though they have different names, they form one single interconnected ocean system. The basins have been created over geological times by the movement of the tectonic plates across the Earth's mantle. The tectonic plates have differing forms at their edges, giving broad or narrow continental shelves and varying profiles to the continental slopes leading down to the continental rises and the abyssal plains. Geomorphic activity in the abyssal plains between the continents gives rise to abyssal ridges, volcanic islands, seamounts, guyots (plateau-like seamounts), rift-valley segments and trenches. Erosion and sedimentation (either submarine or riverine, when the sea level was lower during the ice ages) have created submarine canyons, glacial troughs, sills, fans and escarpments. Around the ocean basins, there are marginal seas, more or less separated from the main ocean basins by islands, archipelagos or peninsulas, or bounded by submarine ridges and formed by various processes.[6]

The water of the ocean mixes and circulates within those geological structures. Although the proportion of the different chemical components dissolved in seawater is essentially constant over time, that water is not uniform: there are very important physical and chemical variations within the seawater. Salinity varies according to the relative balance between inputs of freshwater and evaporation. Differences in salinity and temperature of water masses can cause seawater to be stratified into separate layers. Such stratification can lead to variations in the distribution of both oxygen and nutrients, with an obvious variety of consequences in both cases for the biotas sensitive to those factors. A further variation is in the penetration of light, which controls where the photosynthesis on which nearly all ocean life depends can take place. Below a few tens of metres at the coastal level or a few hundred meters in the clearer open ocean, the ocean becomes dark and there is no photosynthesis.[7]

Superimposed on all this is a change in the acidity of the ocean. The ocean absorbs annually about 26 per cent of the anthropogenic carbon dioxide emitted into the atmosphere. That gas reacts with the seawater to form carbonic acid, which is making the ocean more acid.

The ocean is strongly coupled with the atmosphere, mutually transferring substances (mostly gases), heat and momentum at its surface, forming a single coupled system. That system is influenced by the seasonal changes caused by the Earth's tilted rotation with respect to the sun. Variations in sea-surface temperature among different parts of the ocean are important in creating winds, areas of high and low air pressure and storms (including the highly damaging hurricanes, typhoons and cyclones). In their turn, winds help to shape the surface currents of the ocean, which transport heat from the tropics towards the poles. The ocean surface water arriving in the cold polar regions partly freezes, rendering the remainder more saline and thus heavier. That more saline water sinks to the bottom and flows towards the equator, starting a return flow to the tropics: the meridional overturning circulation, also called the thermohaline circulation. A further overall forcing factor is the movements generated by the tidal system, predominantly driven by the gravitational effect of the moon and sun.[8]

The movements of seawater help to control the distribution of nutrients in the ocean. The ocean enjoys both a steady (and, in some places, excessive) input from land of inorganic nutrients needed for plant growth (especially nitrogen, phosphorus and their compounds, but also lesser amounts of other vital nutrients) and a continuous recycling of all the nutrients already in the ocean through biogeochemical processes, including bacterial action. Areas of upwelling, where nutrient-rich water is brought to the surface, are particularly important, because they result in a high level of primary production from photosynthesis by phytoplankton in the zone of light penetration, combining carbon from atmospheric carbon dioxide with the other nutrients, and releasing oxygen back into the atmosphere. Whether in the water column or when it sinks to the seabed, that primary production constitutes the basis on which the oceanic food web is built, through each successive layer up to the top predators (large fish, marine mammals, marine reptiles, seabirds and, through capture fisheries, humans).[9]

The distribution of living marine resources around the world is the outcome of that complex interplay of geological forms, ocean currents, nutrient fluxes, weather, seasons and sunlight. Not surprisingly, the resulting distribution of living resources reflects that complexity. Because some ocean areas have high levels of primary production, the density of living marine resources in those areas and the contiguous areas to

---

4   See chaps. 1 and 2.

5   The Southern Ocean is formed by the southernmost parts of the Atlantic, Indian and Pacific Ocean basins. The first world ocean assessment does not consider enclosed seas, such as the Caspian Sea or the Dead Sea.

6   See chap. 1.

7   See chaps. 1 and 4.

8   See chaps. 1 and 5.

9   See chaps. 1 and 6.

which currents carry that production is also high. Some of those areas of dense living marine resources are also areas of high biological diversity. The general level of biological diversity in the ocean is also high. For example, just under half of the world's animal phyla are found only in the ocean, compared to one single phylum found only on land.

Human uses of the ocean are shaped not only by the complex patterns of the physical characteristics of the ocean, of its currents and of the distribution of marine life, but also by the terrestrial conditions that have influenced the locations of human settlements, by economic pressures and by the social rules that have developed to control human activities — including national legislation, the law of the sea, international agreements on particular human uses of the sea and broader international agreements that apply to both land and sea.[10]

3 Carrying out the assessment

3.1 Organization

To carry out the complex task of assessing the environmental, social and economic aspects of the ocean, the General Assembly has established arrangements capable of bringing to bear the many different skills needed. After the holding of two international workshops to consider modalities for the Regular Process, the Assembly started the first phase in 2006, the assessment of assessments. This examined more than 1,200 ocean assessments — some regional, others global, some as thematically restricted as the status and trend of a single fish stock or pollutant in a specific area, others as broad as integrated assessments of entire marine ecosystems. The assessment of assessments resulted in conclusions on good practice in that field and in recommendations on how the task of carrying out fully integrated assessments might be approached.

The General Assembly set up an Ad Hoc Working Group of the Whole, which examined those conclusions and recommendations and put proposals to the Assembly. In 2009, the Assembly approved the framework for the Regular Process developed in that way. The framework consists of: (a) the overall objective for the Regular Process; (b) a description of the scope of the Regular Process; (c) a set of principles to guide its establishment and operation; and (d) best practices on key design features for the Regular Process, as identified in the assessment of assessments. The framework also provided that capacity-building, the sharing of data nd information and the transfer of technology would be crucial elements.

Between 2009 and 2011, the General Assembly set up, on the recommendation of the Ad Hoc Working Group of the Whole, the main institutional arrangements for the Regular Process, namely:

(a) The Ad Hoc Working Group of the Whole of the General Assembly on the Regular Process for Global Reporting and Assessment of the State of the Marine Environment, including Socioeconomic Aspects, which has overseen and guided the Process, meeting at least once a year. In 2011, the Working Group established a Bureau to put its decisions into practice during intersessional periods;

(b) The Group of Experts of the Regular Process, which has the task of carrying out assessments within the framework of the Regular Process at the request of the Assembly and under the supervision of the Working Group. The Group of Experts is collectively responsible for its work on the assessment. It consists of 22 members, for a maximum possible membership of 25, who are appointed through the regional groups within the Assembly. The work of the Group members has been either voluntary or supported by their parent institutions;

(c) The Pool of Experts, which provides a pool of skilled support to assist with the wide range of issues that an assessment of the ocean, integrated across ecosystem components, sectors and environmental, social and economic aspects, has to cover. The members of the Pool have been nominated by States through the chairs of the regional groups within the Assembly and are allocated tasks by the Bureau on the recommendations of the Group of Experts. The work of the Pool members has been either voluntary or supported by their parent institutions;

(d) The secretariat of the Regular Process, which has been provided by the Division for Ocean Affairs and the Law of the Sea of the United Nations. No additional staff were recruited specifically for this work, as it was to be carried out within the overall resource level of the Division;

(e) Technical and scientific support for the Regular Process, which has been available, as a result of invitations from the Assembly, from the Intergovernmental Oceanographic Commission of the United Nations Educational, Scientific and Cultural Organization (UNESCO), the United Nations Environment Programme (UNEP), the International Maritime Organization, the Food and Agriculture Organization of the United Nations (FAO), and the International Atomic Energy Agency;

(f) Workshops, which have been held as forums where experts could make an input to the planning and development of the assessment. Eight workshops have been held around the world to consider the scope and methods of the assessment, the information available in the region where each was held and capacity-building needs in that region;

(g) A website (www.worldoceanassessment.org), which has been established to make information about the assessment available and to provide a means of communication among members of the Group of Experts and of the Pool of Experts.

---

10 See chaps. 33 and 34.

In its resolution 68/70 adopted on 9 December 2013, the General Assembly took note of the guidance to contributors adopted by the Bureau of the Ad Hoc Working Group of the Whole (A/68/82 and Corr.1, annex II). In that guidance, it is stated that contributors are expected to act in their personal capacity as independent experts, and not as representatives of any Government or any other authority or organization. They should neither seek nor accept instructions from outside the Regular Process regarding their work on the preparation of the assessment, although they are free to consult widely with other experts and with government officials, in order to ensure that their contributions are credible, legitimate and relevant.

The Group of Experts proposed a draft outline for the first global integrated assessment of the marine environment. After detailed dialogue, revision and consideration by the Working Group, the outline was submitted in the report on the work of the Ad Hoc Working Group of the Whole (A/67/87, annex II)and adopted by the General Assembly on 11 December 2012 in its resolution 67/78. On 29 December 2014, the Assembly took note in its resolution 69/245 of the updated outline contained in annex II to A/69/77. The chapters have been prepared by writing teams of one or more members. Conveners from the Group of Experts or the Pool of Experts have led those teams. One or more lead members from the Group of Experts has overseen the preparation of (or, in some cases, prepared) each draft chapter. In some cases, the draft chapters have been reviewed by one or more commentators and, in all cases, by the Group of Experts as a whole. Synthesis chapters (drawing together the main points from each part) and the present summary have been prepared by members of the Group of Experts.

Notwithstanding the generous support of the hosts of the workshops and other support described in chapter 2, the production of the first world ocean assessment has been constrained by lack of resources. Apart from the costs of the workshops met by host States, support for the website from Australia and Norway and support by Australia, Belgium, Canada, China, the Republic of Korea, the United Kingdom of Great Britain and Northern Ireland and the United States of America for the travel costs of the members of the Group of Experts from those countries, outgoings have been met from a voluntary trust fund set up by the Secretary-General of the United Nations. Donations to that trust fund from Belgium, China, Côte d'Ivoire, Iceland, Ireland, Jamaica, New Zealand, Norway, Portugal, and the Republic of Korea have amounted to $315,000. Generous support to the Regular Process has also been provided, financially and technically, by the European Union, the Intergovernmental Oceanographic Commission and UNEP.[11]

3.2 Structure of the assessment

The assessment is divided into the seven parts described below.

Part I: summary
The summary describes how the assessment has been carried out, the overall assessment of the scale of human impact on the ocean, the overall value of the ocean to humans and the main pressures on the marine environment and human economic and social well-being. As guides for future action, it also sets out the gaps (general or partial) in knowledge and in capacity-building.

Part II: context of the assessment
Chapter 1 is a broad, introductory survey of the role played by the ocean in the life of the planet, the ways in which the ocean functions, and humans' relationships to the ocean. Chapter 2 explains in more detail the rationale for the assessment and how it has been produced.

Part III: assessment of major ecosystem services from the marine environment (other than provisioning services)
Ecosystem services are those processes, products and features of natural ecosystems that support human well-being. Some (fish, hydrocarbons or minerals) are part of the market economy. Others are not marketed. Part III looks at the non-marketed ecosystem services that the ocean provides to the planet. It considers, first, the scientific understanding of those ecosystem services and then the Earth's hydrological cycle, interactions between air and sea, primary production and ocean-based carbonate production. Finally, it looks at aesthetic, cultural, religious and spiritual ecosystem services (including some cultural objects that are in trade). Where relevant, it draws heavily on the work of the Intergovernmental Panel on Climate Change, with the aim of using the work of the Panel, not of duplicating or challenging it.

Part IV: assessment of the cross-cutting issues of food security and food safety
Part IV, which covers the one cross-cutting theme selected for examination, examines all aspects of the vital function of the ocean in providing food for humans. It draws substantially on information collected by FAO. The economic significance of employment in fisheries and aquaculture and the relationship those industries have with coastal communities are addressed, including gaps in capacity-building for developing countries.

Part V: assessment of other human activities and the marine environment
All other human activities that can impact on the ocean (other than those relating to food production) are covered in part V of the assessment. To the extent that the available information allows, each chapter describes the location and scale of the activity, the economic benefits, employment and social role, environmental consequences (where appropriate), links to other activities and gaps in knowledge and capacity-building.

Part VI: assessment of marine biological diversity and habitats
Part VI: (a) gives an overview of marine biological diversity and what is known about it; (b) reviews the status and trends of, and pressures on, marine ecosystems, species and habitats that have been scientifically

---

11 See chap. 2.

identified as threatened, declining or otherwise in need of special attention or protection; (c) examines the significant environmental, economic and social aspects of the conservation of marine species and habitats; and (d) identifies gaps in capacity to identify marine species and habitats that are recognized as threatened, declining or otherwise in need of special attention or protection, and to assess the environmental, social and economic aspects of the conservation of marine species and habitats.

Part VII: overall assessment
Finally, part VII considers the overall way in which the various human impacts cumulatively affect the ocean, and the overall benefits that humans draw from the ocean.[12]

4 Ten main themes

Ten main themes emerge from the detailed examination set out in parts III to VI of the first world ocean assessment. The order in which they are presented does not reflect any assessment of the order of importance for action. The present assessment has been prepared on the basis of the outline, in which it is stated that the first global integrated marine assessment will not include any analysis of policies. In the light of the dialogue in the Working Group, that limitation has been understood to include the prioritization of actions or the making of recommendations (A/69/77, annex II).

Theme A
Climate change and related changes in the atmosphere have serious implications for the ocean, including rises in sea level, higher levels of acidity in the ocean, the reduced mixing of ocean water and increasing deoxygenation. There are many uncertainties here, but the consensus is that increases in global temperature, in the amount of carbon dioxide in the atmosphere and in the radiation from the sun that reaches the ocean have already had an impact on some aspects of the ocean and will produce further significant incremental changes over time. The basic mechanisms of change are understood but the ability to predict the detail of changes is limited. In many cases, the direction of change is known, but uncertainty remains about the timing and rate of change, as well as its magnitude and spatial pattern.[13]

Theme B
The exploitation of living marine resources has exceeded sustainable levels in many regions. In some jurisdictions, various combinations of management measures, positive incentives and changes to governance have allowed those historical trends to be reversed, but they persist in others. Where fisheries have imposed levels of mortality on fish stocks and wildlife populations above sustainable levels for some considerable time, those stocks have become depleted. Overexploitation has also brought about changes to ecosystems (for example, overfishing of herbivorous fish in parts of the Caribbean has led to the smothering of corals by algae). Overexploitation can also make fish stocks less productive by reducing the numbers of spawning fish, with adverse effects often amplified by the removal of the larger, older fish, which produce disproportionately more eggs of higher quality than younger, smaller individuals. At the same time, reproductive success is also being reduced by pollution, loss of habitat and other forms of disturbance, including climate change. All those factors result, more generally, in declining biological resources with important implications for food security and biodiversity.[14]

Theme C
With regard to the cross-cutting issue of food security and food safety (part IV), fish products are the major source of animal protein for a significant fraction of the world's population, particularly in countries where hunger is widespread. Globally, the current mix of the global capture fisheries is near the ocean's productive capacity, with catches on the order of 80 million tons. Ending overfishing (including illegal, unreported and unregulated fishing) and rebuilding depleted resources could result in a potential increase of as much as 20 per cent in yield, but this would require addressing the transitional costs (especially the social and economic costs) of rebuilding depleted stocks. In some areas, pollution and dead zones are also depressing the production of food from the sea. Small-scale fisheries are often also a critical source of livelihoods, as well as of food, for many poor residents in coastal areas. Rebuilding the resources on which they depend and moving to sustainable exploitation will potentially have important benefits for food security. The contribution of aquaculture to food security is growing rapidly and has greater potential for growth than capture fisheries, but it brings with it new or increased pressures on marine ecosystems.[15]

Theme D
There are clear patterns in biodiversity around the world. The pressures on marine biodiversity are increasing, particularly near large population centres and in areas, such as the open ocean, that have so far suffered only limited impacts. Crucial areas for biodiversity, the so-called biodiversity hotspots, often overlap with the areas critical for the provision of ecosystem services by the ocean. In some of those hotspots, the ecosystem services create the conditions for high biodiversity, while in others, both the rich biodiversity and the ecosystem services result independently from the local physical and oceanographic conditions. In both cases, many of those hotspots have become magnets for human uses, in order to take advantage of the economic and social benefits that they offer. This creates enhanced potential for conflicting pressures.[16]

---

12 See chap. 1.

13 See also paras. 44-72 below.

14 See also paras. 73-87 below.

15 See also paras. 88-96 below.

16 See also paras. 97-108 below.

Theme E

Increased use of ocean space, especially in coastal areas, create conflicting demands for dedicated marine space. This arises both from the expansion of long-standing uses of the ocean (such as fishing and shipping) and from newly developing uses (such as hydrocarbon extraction, mining and the generation of renewable energy conducted offshore). In most cases, those various activities are increasing without any clear overarching management system or a thorough evaluation of their cumulative impacts on the ocean environment, thus increasing the potential for conflicting and cumulative pressures.[17]

Theme F

The current, and growing, levels of population and industrial and agricultural production result in increasing inputs of harmful material and excess nutrients into the ocean. Growing concentrations of population can impose, and in many areas are imposing, levels of sewage discharge that are beyond the local carrying capacity and which cause harm to human health. Even if discharges of industrial effluents and emissions were restrained to the lowest levels in proportion to production that are currently practicable, continuing growth in production would result in increased inputs to the ocean. The growing use of plastics that degrade very slowly result in increased quantities reaching the ocean and have many adverse effects, including the creation of large quantities of marine debris in the ocean, and negative impacts on marine life and on the aesthetic aspects of many ocean areas, and thus consequent socioeconomic effects.[18]

Theme G

Adverse impacts on marine ecosystems come from the cumulative impacts of a number of human activities. Ecosystems, and their biodiversity, that might be resilient to one form or intensity of impact can be much more severely affected by a combination of impacts: the total impact of several pressures on the same ecosystem often being much larger than the sum of the individual impacts. Where biodiversity has been altered, the resilience of ecosystems to other impacts, including climate change, is often reduced. Thus the cumulative impacts of activities that, in the past, seemed to be sustainable are resulting in major changes to some ecosystems and in a reduction in the ecosystem services that they provide.[19]

Theme H

The distribution around the world of the benefits drawn from the ocean is still very uneven. In some fields, this unevenness is due to the natural distribution of resources in areas under the jurisdiction of the various States (for example, hydrocarbons, minerals and some fish stocks). The distribution of some benefits is becoming less skewed: for example, the consumption of fish per capita in some developing countries is growing; the balance between cargoes loaded and unloaded in the ports of developing countries is moving closer to those in developed countries in tonnage terms. In many fields, however, including some forms of tourism and the general trade in fish, an imbalance remains between the developed and developing parts of the world. Significant differences in capacities to manage sewage, pollution and habitats also create inequities. Gaps in capacity-building hamper less developed countries in taking advantage of what the ocean can offer them, as well as reduce their capability to address the factors that degrade the ocean.[20]

Theme I

The sustainable use of the ocean cannot be achieved unless the management of all sectors of human activities affecting the ocean is coherent. Human impacts on the sea are no longer minor in relation to the overall scale of the ocean. A coherent overall approach is needed. This requires taking into account the effects on ecosystems of each of the many pressures, what is being done in other sectors and the way that they interact. As the brief summary above of the many processes at work in the ocean demonstrates, the ocean is a complex set of systems that are all interconnected. In all sectors, albeit unevenly, there has been a progressive, continuing development of management: from no regulation to the regulation of specific impacts, to the regulation of sector-wide impacts and finally to regulation taking account of aspects of all relevant sectors.

Such a coherent approach to management requires a wider range of knowledge about the ocean. Many of the gaps in the knowledge that such an integrated approach requires are identified in the present assessment. There are also widespread gaps in the skills needed to assess the ocean with respect to some aspects (for example, the integration of environmental, social and economic aspects). In many cases, there are gaps in the resources needed for the successful application of such knowledge and skills. Gaps in capacity-building are identified briefly at the end of the present summary, and in more detail in parts III to VI.[21]

Theme J

There is the delay in implementing known solutions to problems that have already been identified as threatening to degrade the ocean further. In many fields, it has been shown that there are practicable, known measures to address many of the pressures described above. Such pressures are continuously degrading the ocean, thereby causing social and economic problems. Delays in implementing such measures, even if they are only partial and will leave more to be done, mean that we are unnecessarily incurring those environmental, social and economic costs.[22]

Conclusion

The 10 themes are described in more detail in section V below. As explained above, the order in which the themes are presented does not represent any judgement on their priority. Elements in those themes

---

17 See also paras. 109-122 below.

18 See also paras. 123-151 below.

19 See also paras. 152-166 below.

20 See also paras. 167-186 below.

21 See also paras. 187-196 below.

22 See also paras. 197-202 below.

overlap, and the same issue may be relevant to more than one theme. The identification of knowledge gaps and capacity-building gaps follows in the final two sections of the summary.

5 Further details on the 10 main themes

5.1 Impacts of climate change and related changes in the atmosphere

Changes
Major features of the ocean are changing significantly as a result of climate change and related changes in the atmosphere. The work of the Intergovernmental Panel on Climate Change has been used, where climate is concerned, as the basis of the present assessment, as required in the outline (A/69/77, annex II).

Sea-surface temperature
The Intergovernmental Panel on Climate Change has reaffirmed in its fifth report its conclusion that global sea-surface temperatures have increased since the late nineteenth century. Upper-ocean temperature (and hence its heat content) varies over multiple time scales, including seasonal, inter-annual (for example, those associated with the El Niño-Southern Oscillation), decadal and centennial periods. Depth-averaged ocean-temperature trends from 1971 to 2010 are positive (that is, they show warming) over most of the globe. The warming is more prominent in the northern hemisphere, especially in the North Atlantic. Zonally averaged upper-ocean temperature trends show warming at nearly all latitudes and depths. However, the greater volume of the ocean in the southern hemisphere increases the contribution of its warming to the global heat content.

The ocean's large mass and high heat capacity enable it to store huge amounts of energy, more than 1,000 times than that found in the atmosphere for an equivalent increase in temperature. The earth is absorbing more heat than it is emitting back into space, and nearly all that excess heat is entering the ocean and being stored there. The ocean has absorbed about 93 per cent of the combined extra heat stored by warmed air, sea, land, and melted ice between 1971 and 2010. During the past three decades, approximately 70 per cent of the world's coastline has experienced significant increases in sea-surface temperature. This has been accompanied by an increase in the yearly number of extremely hot days along 38 per cent of the world's coastline. Warming has also been occurring at a significantly earlier date in the year along approximately 36 per cent of the world's temperate coastal areas (between 30° and 60° latitude in both hemispheres). That warming is resulting in an increasingly poleward distribution of many marine species.[23]

Sea-level rise
It is very likely that extreme sea-level maxima have already increased globally since the 1970s, mainly as a result of global mean sea-level rise. That rise is due in part to anthropogenic warming, causing ocean thermal expansion and the melting of glaciers and of the polar continental ice sheets. Globally averaged sea level has thus risen by 3.2 mm a year for the past two decades, of which about a third is derived from thermal expansion. Some of the remainder is due to fluxes of freshwater from the continents, which have increased as a result of the melting of continental glaciers and ice sheets.

Finally, regional and local sea-level changes are also influenced by natural factors, such as regional variability in winds and ocean currents, vertical movements of the land, isostatic adjustment of the levels of land in response to changes in physical pressures on it and coastal erosion, combined with human perturbations by change in land use and coastal development. As a result, sea levels will rise more than the global mean in some regions, and will actually fall in others. A 4°C warming by 2100 (which is predicted in the high-end emissions scenario in the report of the Intergovernmental Panel on Climate Change) would lead, by the end of that period, to a median sea-level rise of nearly 1 metre above the 1980 to 1999 levels.[24]

Ocean acidification
Rising concentrations of carbon dioxide in the atmosphere are resulting in increased uptake of that gas by the ocean. There is no doubt that the ocean is absorbing more and more of it: about 26 per cent of the increasing emissions of anthropogenic carbon dioxide is absorbed by the ocean, where it reacts with seawater to form carbonic acid. The resulting acidification of the ocean is occurring at different rates around the seas, but is generally decreasing the levels of calcium carbonate dissolved in seawater, thus lowering the availability of carbonate ions, which are needed for the formation by marine species of shells and skeletons. In some areas, this could affect species that are important for capture fisheries.[25]

Salinity
Alongside broad-scale ocean warming, shifts in ocean salinity (salt content) have also occurred. The variations in the salinity of the ocean around the world result from differences in the balance between freshwater inflows (from rivers and glacier and ice-cap melt), rainfall and evaporation, all of which are affected by climate change. The shifts in salinity, which are calculated from a sparse historical observing system, suggest that at the surface, high-salinity subtropical ocean regions and the entire Atlantic basin have become more saline, while low-salinity regions, such as the western Pacific Warm Pool, and high-latitude regions have become even less saline. Since variations in salinity are one of the drivers of ocean currents, those changes can have an effect on the circu-

---

23 See chap. 5.

24 See chap. 4.

25 See chaps. 5-7.

lation of seawater and on stratification, as well as having a direct effect on the lives of plants and animals by changing their environment.[26]

Stratification

Differences in salinity and temperature among different bodies of seawater result in stratification, in which the seawater forms layers, with limited exchanges between them. Increases in the degree of stratification have been noted around the world, particularly in the North Pacific and, more generally, north of 40°S. Increased stratification brings with it a decrease in vertical mixing in the ocean water column. This decreased mixing, in turn, reduces oxygen content and the extent to which the ocean is able to absorb heat and carbon dioxide, because less water from the lower layers is brought up to the surface, where such absorption takes place. Reductions in vertical mixing also impact the amount of nutrients brought up from lower levels into the zone that sunlight penetrates, with consequent reductions in ecosystem productivity.[27]

Ocean circulation

The intensified study of the ocean as part of the study of climate change has led to a much clearer understanding of the mechanisms of ocean circulation and its annual and decadal variations. As a result of changes in the heating of different parts of the ocean, patterns of variation in heat distribution across the ocean (such as the El Niño-Southern Oscillation) are also changing. Those changes in patterns result in significant changes in weather patterns on land. Water masses are also moving differently in areas over continental shelves, with consequent effects on the distribution of species. There is evidence that the global circulation through the open ocean may also be changing, which might lead, over time, to reductions in the transfer of heat from the equatorial regions to the poles and into the ocean depths.

Storms and other extreme weather events

Increasing seawater temperatures provide more energy for storms that develop at sea. The scientific consensus is that this will lead to fewer but more intense tropical cyclones globally. Evidence exists that the observed expansion of the tropics since approximately 1979 is accompanied by a pronounced poleward migration of the latitude at which the maximum intensities of storms occur. This will certainly affect coastal areas that have not been exposed previously to the dangers caused by tropical cyclones.[28]

Ultraviolet radiation and the ozone layer

The ultraviolet (UV) radiation emitted by the sun in the UV-B range (280-315 nanometres wavelength) has a wide range of potentially harmful effects, including the inhibition of primary production by phytoplankton and cyanobacteria, changes in the structure and function of plankton communities and alterations of the nitrogen cycle. The ozone layer in the Earth's stratosphere blocks most UV-B from reaching the ocean's surface. Consequently, stratospheric ozone depletion since the 1970s has been a concern. International action (under the Montreal Protocol on Substances that Deplete the Ozone Layer)[29] to address that depletion has been taken, and the situation appears to have stabilized, although with some variation from year to year. Given those developments and the variations in the water depths to which UV-B penetrates, a consensus on the magnitude of the ozone-depletion effect on net primary production and nutrient cycling has yet to be reached. There is, however, a potential effect of ultraviolet on nanoparticles.[30]

Implications for human well-being and biodiversity
Changes in seasonal life cycles in the ocean

It has been predicted under some climate change scenarios that up to 60 per cent of the current biomass in the ocean could be affected, either positively or negatively, resulting in disruptions to many existing ecosystem services. For example, modelling studies of species with strong temperature preferences, such as skipjack and bluefin tuna, predict major changes in range and/or decreases in productivity.[31]

The effects are found in all regions. For example, in the North-West Atlantic, the combination of changes in feeding patterns triggered by overfishing and changes in climate formed the primary pressures thought to have brought about shifts in species composition amounting to a full regime change, from one dominated by cod to one dominated by crustacea. Even in the open ocean, climate warming will increase ocean stratification in some broad areas, reduce primary production and/or result in a shift in productivity to smaller species (from diatoms of 2-200 microns to picoplankton of 0.2-2 microns) of phytoplankton. This has the effect of changing the efficiency of the transfer of energy to other parts of the food web, causing biotic changes over major regions of the open ocean, such as the equatorial Pacific.[32]

Loss of sea ice in high latitudes and associated ecosystems

The high-latitude ice-covered ecosystems host globally significant arrays of biodiversity, and the size and nature of those ecosystems make them critically important to the biological, chemical and physical balance of the biosphere. Biodiversity in those systems has developed remarkable adaptations to survive both extreme cold and highly variable climatic conditions.

High-latitude seas are relatively low in biological productivity, and ice algal communities, unique to those latitudes, play a particularly important role in system dynamics. Ice algae are estimated to contribute more than 50 per cent of the primary production in the permanently ice-covered central Arctic. As sea-ice cover declines, this productivity may decline and open water species may increase. The high-latitude ecosystems are undergoing change at a rate more rapid than in other places on earth. In the past

---

26 See chaps. 4 and 5.

27 See chaps. 1 and 4-6.

28 See chap. 5.

29 United Nations, *Treaty Series*, vol. 1522, No. 26369.

30 See theme F above and chap. 6.

31 See chaps. 42 and 52.

32 See chaps. 6 and 36A.

100 years, average Arctic temperatures have increased at almost twice the average global rate. Reduced sea ice, especially a shift towards less multi-year sea ice, will affect a wide range of species in those waters. For example, owing to low reproductive rates and long lifetimes, some iconic species (including the polar bear) will be challenged to adapt to the current fast warming of the Arctic and may be extirpated from portions of their range within the next 100 years.[33]

Plankton

Phytoplankton and marine bacteria carry out most of the primary production on which food webs depend. The climate-driven increases in the temperature of the upper ocean that had been predicted are now causing shifts in phytoplankton communities. This may have profound effects on net primary production and nutrient cycles over the next 100 years. In general, when smaller plankton account for most net primary production, as is typically the case in oligotrophic open-ocean waters (that is, areas where levels of nutrients are low), net primary production is lower and the microbial food web dominates energy flows and nutrient cycles. Under such conditions, the carrying capacity for currently harvestable fish stocks is lower and exports of organic carbon, nitrogen and phosphorus to the deep sea may be smaller.

On the other hand, as the upper ocean warms, the geographic range of nitrogen-fixing plankton (diazotrophs) will expand. This could enhance the fixation of nitrogen by as much as 35-65 per cent by 2100. This would lead to an increase in net primary production, and therefore an increase in carbon uptake, and some species of a higher trophic level may become more productive.

The balance between those two changes is unclear. A shift towards less primary production would have serious implications for human food security and the support of marine biodiversity.[34]

Fish stock distribution

As seawater temperatures increase, the distribution of many fish stocks and the fisheries that depend upon them is shifting. While the broad pattern is one of stocks moving poleward and deeper in order to stay within waters that meet their temperature preference, the picture is by no means uniform, nor are those shifts happening in concert for the various species. Increasing water temperatures will also increase metabolic rates and, in some cases, the range and productivity of some stocks. The result is changes in ecosystems occurring at various rates ranging from near zero to very rapid. Research on those effects is scattered, with diverse results, but as ocean climate continues to change, those considerations are of increasing concern for food production. Greater uncertainty for fisheries results in social, economic and food security impacts, complicating sustainable management.[35]

Seaweeds and seagrasses

Cold-water seaweeds, in particular kelps, have reproductive regimes that are temperature-sensitive. Increase in seawater temperature affects their reproduction and survival, which will consequently affect their population distribution and harvest. Kelp die-offs have already been reported along the coasts of Europe, and changes in species distribution have been noted in Northern Europe, Southern Africa and Southern Australia, with warm-water-tolerant species replacing those that are intolerant of warmer water. The diminished kelp harvest reduces what is available for human food and the supply of substances derived from kelp that are used in industry and pharmaceutical and food preparation.

Communities with kelp-based livelihoods and economies will be affected. For seagrasses, increased seawater temperatures have been implicated in the occurrence of a wasting disease that decimated seagrass meadows in the north-eastern and north-western parts of the United States. Changes in species distribution and the loss of kelp forest and seagrass beds have resulted in changes in the ways that those two ecosystems provide food, habitats and nursery areas for fish and shellfish, with repercussions on fishing yields and livelihoods.[36]

Shellfish productivity

Because of the acidification of the ocean, impacts on the production by shellfish of their calcium carbonate shells has already been observed periodically at aquaculture facilities, hindering production. As acidification intensifies, this problem will become more widespread, and occur in wild, as well as in cultured, stocks. However, like all other ocean properties, acidification is not evenly distributed, so that the effects will not be uniform across areas and there will be substantial variation over small spatial scales. In addition, temperature, salinity and other changes will also change shellfish distributions and productivity, positively or negatively in different areas. As with fishing, the course of those changes is highly uncertain and may be disruptive to existing shellfish fisheries and aquaculture.[37]

Low-lying coasts

Sea-level rise, due to ocean warming and the melting of land ice, poses a significant threat to coastal systems and low-lying areas around the world, through inundations, the erosion of coastlines and the contamination of freshwater reserves and food crops. To a large extent, such effects are inevitable, as they are the consequences of conditions already in place, but they could have devastating effects if mitigation options are not pursued. Entire communities on low-lying islands (including States such as Kiribati, Maldives and Tuvalu) have nowhere to retreat to within their islands and have therefore no alternative but to abandon their homes entirely, at a cost they are often ill-placed to bear. Coastal regions, particularly some low-lying river deltas, have very high population densities. Over 150 million people are estimated to live on land that is no more than 1 metre above today's high-tide levels, and 250 million at elevations within five metres of that level. Because of their

---

33 See chaps. 36G, 36H and 37.

34 See chap. 6.

35 See chaps. 36A-H and 52.

36 See chaps. 14 and 47.

37 See chaps. 5, 11 and 52.

high population densities, coastal cities are particularly vulnerable to sea-level rise in concert with other effects of climate change, such as changes in storm patterns.[38]

Coral reefs

Corals are subject to "bleaching" when the seawater temperature is too high: they lose the symbiotic algae that give coral its colour and part of its nutrients. Coral bleaching was a relatively unknown phenomenon until the early 1980s, when a series of local bleaching events occurred, principally in the eastern tropical Pacific and Wider Caribbean regions. Severe, prolonged or repeated bleaching can lead to the death of coral colonies. An increase of only 1°C to 2°C above the normal local seasonal maximum can induce bleaching. Although most coral species are susceptible to bleaching, their thermal tolerance varies. Many heat-stressed or bleached corals subsequently die from coral diseases.

Rising temperatures have accelerated bleaching and mass mortality during the past 25 years. The bleaching events in 1998 and 2005 caused high coral mortality at many reefs, with little sign of recovery. Global analysis shows that this widespread threat has significantly damaged most coral reefs around the world. Where recovery has taken place, it has been strongest on reefs that were highly protected from human pressures. However, a comparison of the recent and accelerating thermal stress events with the slow recovery rate of most reefs suggests that temperature increase is outpacing recovery.

Losses of coral reefs can have negative effects on fish production and fisheries, coastal protection, ecotourism and other community uses of coral reefs. Current scientific data and modelling predict that most of the world's tropical and subtropical coral reefs, particularly those in shallow waters, will suffer from annual bleaching by 2050, and will eventually become functionally extinct as sources of goods and services. This will have not only profound effects on small island developing States and subsistence fishermen in low-latitude coastal areas, but also locally significant effects even in major economies, such as that of the United States.[39]

Submarine cables

Submarine cables have always been at risk of breaks from submarine landslides, mainly at the edge of the continental shelf. As the pattern of cyclones, hurricanes and typhoons changes, submarine areas that have so far been stable may become less so and thus produce submarine landslides and consequent cable breaks. With the increasing dependence of world trade on the Internet, such breaks (in addition to breaks from other causes, such as ship anchors and bottom trawling) could delay or interrupt communications vital to that trade.[40]

Eutrophication problems

Where there are narrow continental shelves, some wind conditions can bring nutrient-rich, oxygen-poor water up into coastal waters, and produce hypoxic (low-oxygen) or even anoxic conditions (the implications of which are described under theme F). Changes in ocean circulation appear to be enhancing those effects. Examples of this can be found on the western coasts of the American continent immediately north and south of the equator, the western coast of sub-Saharan Africa and the western coast of the Indian subcontinent.[41]

Opening of Arctic shipping routes

Although the number of ships transiting Arctic waters is currently low, it has been escalating for the past decade, and the retreat of the polar sea ice as a result of planetary warming means that there are increasing possibilities for shipping traffic between the Atlantic and Pacific Oceans around the north of the American and Eurasian continents during the northern summer. The movement of species between the Pacific and the Atlantic demonstrates the scale of the potential impact. Those routes are shorter and may be more economic, but shipping brings with it increased risks of marine pollution both from acute disasters and chronic pollution and the potential introduction of invasive non-native species. The very low rate at which bacteria can break down spilled oil in polar conditions and the general low recovery rate of polar ecosystems mean that damage from such pollution would be very serious. Furthermore, the response and clear-up infrastructure found in other ocean basins is largely lacking today around the Arctic Ocean. Those factors would make such problems even worse. Over time, the increased commercial shipping traffic through the Arctic Ocean and the noise disturbance it creates may also displace marine mammals away from critical habitats.[42]

5.2 Higher mortality and less successful reproduction of marine biotas

Captures of fish stocks at levels above maximum sustainable yield

Globally, the levels of capture fisheries are near the ocean's productive capacity, with catches on the order of 80 million tons. Exploitation inevitably reduces total population biomass through removals. As long as the fish stock can compensate through increased productivity because the remaining individuals face less competition for access to food and therefore grow faster and produce more progeny, then fishing can be sustained. However, when the rate of exploitation becomes faster than the stock can compensate through increasing growth and reproduction, the removal level becomes unsustainable and the stock declines.

The concept of "maximum sustainable yield", entrenched in international legal instruments such as the United Nations Convention on the Law of the Sea and the Agreement for the Implementation of the Provisions of the United Nations Convention on the Law of the Sea of 10 December 1982 relating to the Conservation and Management of Straddling

---

38 See chap. 4.

39 See chaps. 34, 36D and 43.

40 See chap. 19.

41 See chaps. 6 and 20.

42 See chaps. 20 and 36G.

Fish Stocks and Highly Migratory Fish Stocks,[43] is based on the inherent trade-off between increasing harvests and decreasing the ability of a smaller resulting population to compensate for the removals.

At present, about one quarter of all assessed fish stocks are being overfished and more are still recovering from past overfishing. This is undermining the contribution that they could make to food security. Ending overfishing is a precondition for allowing stocks to rebuild. Other stocks may still be categorized as "fully exploited" despite being on the borderline of overfishing. Those could produce greater yields if effectively managed.

There are only a few means available to increase yields. Ending overfishing, eliminating illegal, unreported and unregulated fishing, bringing all fishery yields under effective management and rebuilding depleted resources may result in an increase of as much as 20 per cent in potential yield, provided that the transitional economic and social costs of rebuilding depleted stocks can be addressed.

Overfishing can also undermine the biodiversity needed to sustain marine ecosystems. Without careful management, such impacts on biodiversity will endanger some of the most vulnerable human populations and marine habitats around the world, as well as threaten food security and other important socioeconomic aspects (such as livelihoods).[44]

Impacts of changes in breeding and nursery areas

Changes in breeding and nursery areas are best documented for the larger marine predators. For seabirds, globally, the greatest pressure is caused by invasive species (mainly rats and other predators acting at breeding sites). That pressure potentially affects 73 threatened seabird species — 75 per cent of the total and nearly twice as many as any other single threat. The remaining most significant pressures are fairly evenly divided between those faced mainly at breeding sites, namely problematic native species, human disturbance and the loss of historical breeding and nursery sites to urban development (commercial, residential or infrastructural), and those faced mainly at sea, particularly by-catch in longlines, gillnets and trawl fisheries, when birds are foraging or moulting, migrating or in aggregations. The ingestion of marine plastic debris is also significant. For marine reptiles, decades of overharvesting of marine turtle eggs on nesting beaches have driven the long-term decline of some breeding populations. In some areas, tourist development has also affected reproductive success at historical turtle nesting beaches. All this has rendered them more vulnerable to fishery by-catch and other threats. Similar pressures apply to marine mammals.[45]

Levels of by-catch (non-target fish, marine mammals, reptiles and seabirds), discards and waste

Current estimates of the number of overfished stocks do not take into account the broader effects of fishing on marine ecosystems and their productivity. In the past, large numbers of dolphins drowned in fishing nets. This mortality greatly reduced the abundance of several dolphin species in the latter half of the twentieth century. Thanks to international efforts, fishing methods have changed and the by-catch has been reduced significantly. Commercial fisheries are the most serious pressure at sea that the world's seabirds face, although there is evidence of some reductions of by-catch in some key fisheries. Each year, incidental by-catch in longline fisheries is estimated to kill at least 160,000 albatrosses and petrels, mainly in the southern hemisphere. For marine reptiles, a threat assessment scored fishery by-catch as the highest threat across marine turtle subpopulations, followed by harvesting (that is, for human consumption) and coastal development.

The mitigation of those causes of mortality can be effective, even though the lack of reliable data can hamper the targeting of mitigation measures. Depending on the particular species and fishery methods, mitigation may include the use of acoustic deterrents, gear modifications, time or area closures and gear switching (for example, from gillnets to hooks and lines). In particular, the global moratorium on all large-scale pelagic drift-net fishing called for by the General Assembly in 1991 was a major step in limiting the by-catch of several marine mammal and seabird species that were especially vulnerable to entanglement.[46]

Impact of hazardous substances and eutrophication problems on reproduction and survival

Each of the reviews of regional biodiversity in part VI of the present assessment reported at least some instances of threats from hazardous substances. To give some examples, in the South Pacific, localized declines in species densities, assemblages and spatial distributions are being observed, particularly in areas close to population centres where overfishing, pollution from terrestrial run-off and sewage and damage from coastal developments are occurring. In the North Atlantic, impacts on the benthos have been particularly well documented, although their nature depends on the type, intensity and duration of the pollution or nutrient input. Persistent pressures of that type have been documented to alter greatly the species composition and biomass of the benthos directly and indirectly, through processes such as the formation of dead zones and hypoxic zones as a result of eutrophication problems and seawater circulation changes driven by climate change. Even in the open ocean, evidence is increasing for chemical contamination of deep-pelagic animals. Although the pathways for such contaminations are not well known, high concentrations of heavy metals and persistent organic pollutants have been reported.[47]

---

43 United Nations, *Treaty Series*, vol. 2167, No. 37924.

44 See chaps. 10, 11 and 15.

45 See chaps. 28 and 37-39.

46 See chaps. 11 and 37-39.

47 See chaps. 36A-H.

Impacts of disturbance from noise

Anthropogenic noise in the ocean increased in the last half of the past century. Commercial shipping is the main source, and the noise that it produces is often in frequency bands used by many marine mammals for communication. Many other types of marine biotas have also been shown to be affected by anthropogenic noise. Other significant sources of noise are seismic exploration for the offshore hydrocarbon industry and sonar. The impact of noise can be both to disrupt communication among animals and to displace them from their preferred breeding, nursery or feeding grounds, with consequent potential effects on their breeding success and survival.[48]

Impacts of recreational fishing

Recreational fishing is a popular activity in many industrialized countries, in which up to 10 per cent of the adult population may participate. The impact of that type of fishing is only sometimes taken into account in fishery management, although the quantities caught can be significant for the management of stocks experiencing overfishing. In several countries, there is a substantial industry supporting the recreational catching of sport fish (including trophy fish, such as marlins, swordfish and sailfish), but catch statistics are generally not available.[49]

Implications for human well-being and biodiversity
Food resources

The overfishing of some fish stocks is reducing the yield realized from those stocks. Such reductions in yield are likely to undermine food security. The role of fisheries in food security is further considered below.[50]

Species structure of highly productive sea areas

Many human activities have been documented to have impacts on marine life living on the seabed (benthic communities). The adverse effects of mobile bottom-contacting fishing gear on coastal and shelf benthic communities have been documented essentially everywhere that such gear has been used. Bottom trawling has caused the destruction of a number of long-lived cold-water coral and sponge communities that are unlikely to recover before at least a century. Many reviews show that, locally, the nature of those impacts and their duration depend on the type of substrate and frequency of trawling. Those effects have been found in all the regional assessments.[51]

With regard to fish and pelagic invertebrate communities, much effort has been devoted to teasing apart the influences of exploitation and of environmental conditions as drivers of change in fish populations and communities, but definitive answers are elusive. Most studies devote attention to explaining variation among coastal fish-community properties in terms of features of the physical and chemical habitats (including temperature, salinity, oxygen and nutrient levels, clarity of, and pollutants in, the water column) and of depth, sediment types, benthic communities, contaminant levels, oxygen levels and disturbance of the sea floor. All of those factors have been shown to influence fish-community composition and structure in at least some coastal areas of each ocean basin.

The scale at which a fish-community structure is determined and its variation is documented can be even more local, because some important drivers of change in coastal fish communities are themselves very local in scale, such as coastal infrastructure development. Other obvious patterns are recurrent, such as increasing mortality rates (whether from exploitation or coastal pollution) leading both to fish communities with fewer large fish and to an increase in species with naturally high turnover rates. However, some highly publicized projections of the loss of all commercial fisheries or of all large predatory fish by the middle of the current century have not withstood critical review.[52]

5.3 Food security and food safety

Seafood products, including finfish, invertebrates and seaweeds, are a major component of food security around the world. They are the major source of protein for a significant fraction of the global population, in particular in countries where hunger is widespread. Even in the most developed countries, the consumption of fish is increasing both per capita and in absolute terms, with implications for both global food security and trade.[53]

Fisheries and aquaculture are a major employer and source of livelihoods in coastal States. Significant economic and social benefits result from those activities, including the provision of a key source of subsistence food and much-needed cash for many of the world's poorest peoples. As a mainstay of many coastal communities, fisheries and aquaculture play an important role in the social fabric of many areas. Small-scale fisheries, particularly those that provide subsistence in many poor communities, are often particularly important. Many such coastal fisheries are under threat because of overexploitation, conflict with larger fishing operations and a loss of productivity in coastal ecosystems caused by a variety of other impacts. Those include habitat loss, pollution and climate change, as well as the loss of access to space as coastal economies and uses of the sea diversify.[54]

Capture fisheries

Globally, capture fisheries are near the ocean's productive capacity, with catches on the order of 80 million metric tons. Only a few means to increase yield are available. Addressing sustainability concerns more effectively (including ending overfishing, eliminating illegal, unreported and unregulated fishing, rebuilding depleted resources and reducing the broader ecosystem impacts of fisheries and the adverse impacts of pol-

---

48   See chaps. 17, 21 and 37.

49   See chaps. 28, 40 and 41.

50   See chap. 11.

51   See chaps. 36A-H, 42, 51 and 52.

52   See chaps. 10, 11, 15, 34, 36A-H and 52.

53   See chap. 10.

54   See chap. 15.

lution) is an important aspect of improving fishery yields and, therefore, food security. For example, ending overfishing and rebuilding depleted resources may result in an increase of as much as 20 per cent in potential yield, provided that the transitional costs of rebuilding depleted stocks can be addressed.[55]

In 2012, more than one quarter of fish stocks worldwide were classified by the Food and Agriculture Organization of the United Nations as overfished. Although those stocks will clearly benefit from rebuilding once overfishing has ended, other stocks may still be categorized as fully exploited despite being on the borderline of overfishing. Such stocks could yield more if effective governance mechanisms were in place.

Current estimates of the number of overfished stocks do not take into account the broader effects of fishing on marine ecosystems and their productivity. Those impacts, including by-catch, habitat modification and effects on the food web, significantly affect the ocean's capacity to continue to produce food sustainably and must be carefully managed. Fish stock propagation may provide a tool to help to rebuild depleted fishery resources in some instances.[56]

Fishing efforts are subsidized by many mechanisms around the world, and many of those subsidies undermine the net economic benefits to States. Subsidies that encourage overcapacity and overfishing result in losses for States, and those losses are often borne by communities dependent on fishery resources for their livelihood and food security.[57]

Aquaculture
Aquaculture production, including seaweed culture, is increasing more rapidly than any other source of food production in the world. Such growth is expected to continue. Aquaculture, not including the culture of seaweeds, now provides half of the fish products covered in global statistics. Aquaculture and capture fisheries are codependent in some ways, as feed for cultured fish is in part provided by capture fisheries, but they are competitors for space in coastal areas, markets and, potentially, other resources. Significant progress has been made in replacing feed sources from capture fisheries with agricultural production. Aquaculture itself poses some environmental challenges, including potential pollution, competition with wild fishery resources, potential contamination of gene pools, disease problems and loss of habitat. Examples of those challenges, and measures that can mitigate them, have been observed worldwide.[58]

Social issues
In both capture fisheries and aquaculture, gender and other equity issues arise. A significant number of women are employed in both types of activities, either directly or in related activities along the value chain.

Women are particularly prominent in product processing, but often their labour is not equitably compensated and working conditions do not meet basic standards. Poor communities are often subject to poorer market access, unsafe working conditions and other inequitable practices.[59]

Food safety
Food safety is a key worldwide challenge for all food production and delivery sectors, including all parts of the seafood industry, from capture or culture to retail marketing. That challenge is of course also faced by subsistence fisheries. In the food chain for fishery products, potential problems need to be assessed, managed and communicated to ensure that they can be addressed. The goal of most food safety systems is to avoid risk and prevent problems at the source. The risks come from contamination from pathogens (particularly from discharges of untreated sewage and animal waste) and toxins (often from algal blooms). The severity of the risk also depends on individual health, consumption levels and susceptibility. There are international guidelines to address those risks but substantial resources are required in order to continue to build the capacity to implement and monitor safety protocols from the water to the consumer.

5.4   Patterns of biodiversity

A basic, but key, conclusion of the present assessment is that there are clear patterns of biodiversity, both globally and regionally. A key question is whether there are consistent large-scale patterns of biodiversity, governed by underlying factors that constrain the distribution of the wide range of marine life across the wide variety of habitats. Global-scale studies to explore this question began long ago and have grown substantially in the past decade. The enormous amounts of data collected and compiled by the Census of Marine Life enable exploration and the mapping of patterns across more taxonomic groups than ever before, thus facilitating an understanding of the consistency of patterns of biodiversity.

Perhaps the most common large-scale biodiversity pattern on the planet is the "latitudinal gradient", typically expressed as a decline in the variety of species from the equator to the poles. Adherence to that pattern varies among marine taxa. Although coastal species generally peak in abundance near the equator and decline towards the poles, seals show the opposite pattern. Furthermore, strong longitudinal gradients (east-west) complicate patterns, with hotspots of biodiversity across multiple species groups in the coral triangle of the Indo-Pacific, in the Caribbean and elsewhere.

Oceanic organisms, such as whales, differ in pattern entirely, with species numbers consistently peaking at mid-latitudes between the equator and the poles. This pattern defies the common equator-pole gradient, suggesting that different factors are at play. Various processes may also control the difference in species richness between the oceanic and

---

55   See chaps. 11, 13, 36A-H and 52.

56   See chap. 13.

57   See chap. 15.

58   See chap. 12.

59   See chap. 15.

coastal environments (for example, in terms of dispersal, mobility or habitat structure), but general patterns appear to be reasonably consistent within each group.

However, across all groups studied, ocean temperature is consistently related to species diversity, making the effects of climate change likely to be felt as a restructuring factor of marine community diversity.

Although the patterns above hold for the species studied, numerous groups and regions have not yet been examined. For example, global-scale patterns of diversity in the deep sea remain largely unknown. Knowledge of diversity and distribution is biased towards large, charismatic species (for example, whales) or economically valuable species (for example, tuna). Our knowledge of patterns in microbial organisms remains particularly limited relative to the considerable biodiversity of those species. Enormous challenges remain even to measure this. Viruses remain another critical part of the oceanic system of which we lack any global-scale biodiversity knowledge.

Patterns of global marine biodiversity, other than species richness, are only just beginning to be explored. For example, investigations suggest that, globally, the higher the latitude at which a reef is located, the greater the evenness in the number of individuals of each species tend to be in that reef. Such a pattern, in turn, affects functional richness, which relates to the diversity of function in reef fish, a potentially important component of ecosystem productivity, resilience and provision of goods and services.[60]

Implications
Location of biodiversity hotspots and their relationship to the location of high levels of ecosystem services
Although marine life is found everywhere in the ocean, biodiversity hotspots exist where the number of species and the concentration of biotas are consistently high relative to adjacent areas. Some are subregional, such as the coral triangle in the Indo-Pacific, the coral reefs in the Caribbean, the cold-water corals in the Mediterranean and the Sargasso Sea. Some are more local and associated with specific physical conditions, such as biodiversity-rich habitat types. Key drivers of biodiversity are complex three-dimensional physical structures that create a diversity of physical habitats (associated with rocky sea floors), dynamic oceanographic conditions causing higher bottom-up productivity, effects of land-based inputs extending far out to sea (such as the inputs from the River Amazon) and special vegetation features creating unique and productive habitats near the shore. Those complex habitats, however, are often highly vulnerable to disturbance.

The high relative and absolute biodiversity of those hotspots often directly supports the extractive benefits of fishing and other harvests, providing a direct link between biodiversity and the provision of services by the ocean. The areas supporting high relative and absolute levels of biodiversity not only harbour unique species adapted to their special features, but also often serve as centres for essential life-history stages of species with wider distributions. For example, essentially all the biodiversity hotspots that have been identified have also been found to harbour juvenile fish, which are important for fisheries in adjacent areas.

105. Hotspots for primary productivity are necessarily also hotspots for production of oxygen as a direct result of photosynthesis. Furthermore, underlying the high biodiversity is often a high structural complexity of the habitats that support it. That structure often contributes other services, such as coastal protection and regeneration. In addition, it is the concentrated presence of iconic species in an area which adds to aesthetic services (supporting tourism and recreation) and spiritual and cultural services.[61]

Biodiversity and economic activity
Sometimes, because of the special physical features that contribute to high biodiversity, and sometimes because of the concentration of biodiversity itself, many societies and industries are most active in areas that are also biodiversity hotspots. As on land, humanity has found the greatest social and economic benefits in the places in the ocean that are highly productive and structurally complex. For example, 22 of the 32 largest cities in the world are located on estuaries; mangroves and coral reefs support small-scale (artisanal) fisheries in developing countries. Biodiversity hotspots tend to attract human uses and become socioeconomic hotspots. Hence biodiversity-rich areas have a disproportionately high representation of ports and coastal infrastructure, other intensive coastal land uses, fishing activities and aquaculture. This is one of the major challenges to the sustainable use of marine biodiversity.[62]

Some marine features, such as seamounts, often found in areas beyond national jurisdiction, have high levels of biodiversity, frequently characterized by the presence of many species not found elsewhere. Significant numbers of the species mature late, and therefore reproduce slowly. High levels of fishing have rapidly undermined the biodiversity of many such features, and risk continuing to do so in the absence of careful management.[63]

New forms of economic activity in the open ocean, such as seabed mining, and the expansion of existing forms of activity, such as hydrocarbon extraction, have the potential to have major impacts on its biodiversity, which is to date poorly known. Without careful management of those activities, there is a risk that the biodiversity of areas affected could be destroyed before it is properly understood.[64]

---

60 See chaps. 34, 35 and 36A-H.

61 See chaps. 8, 34, 36A-H and 52.

62 See chaps. 26, 34 and 36A-H.

63 See chaps. 36F and 51.

64 See chaps. 21-23 and 36F.

5.5 Increased use of ocean space

The world is seeing a greatly intensified use of ocean space. Since around the middle of the nineteenth century, there has been a great growth in the range of human activities in the ocean, each demanding its share of ocean space. At the same time, and in consequence, the regulation of activities in the ocean has increased. In a campaign to draw attention to this, the fishermen of the Netherlands coined the slogan "Fishing on a postage stamp", arguing that, by the time that all the other uses of the exclusive economic zone of the Netherlands (shipping lanes, offshore oil and gas extraction, sand and gravel extraction, dumping of dredged material, offshore wind-power installations, submarine cables and pipelines, etc.) had been allocated their spaces, not much space was left for their traditional fishing activities. Whether or not their activities were actually restricted, their slogan drew attention to a challenge faced all around the world as increasing demands are made for space for ocean-based activities.

Not all the uses of ocean space within national jurisdictions have the same implications. Some uses effectively exclude most other concurrent uses, for example where fishing rights for benthic species (such as oysters) in areas of national jurisdiction have been allocated to individual proprietors, where tourism would be hampered by other developments or where "no-take" marine protected areas have been created. Others may have a global distribution, but may have a lesser impact, such as shipping lanes and submarine cables. Yet others have, at least so far, only localized impacts, usually determined by the availability of some local resource. Those are likely to be intensive, limiting other uses in the areas where they occur, for example aquaculture, offshore oil and gas extraction, sand and gravel extraction and offshore wind-power installations.

Those differing implications of the developments in human uses of the ocean are important for policy decisions on how, and at what level (national, regional, global), activities should be best managed.[65]

Increased coastal population and urbanization (including tourism)

A large proportion of humans live in the coastal zone: 38 per cent of the world's population live within 100 km of the shore, 44 per cent within 150 km, 50 per cent within 200 km, and 67 per cent within 400 km. This proportion is steadily increasing. Consequently, there are growing demands for land in the coastal zone. Land reclamation has therefore been taking place on a large scale in many countries, particularly by reclaiming salt marshes, intertidal flats and mangroves. At the same time, where coastal land is threatened by erosion, large stretches of natural coastline have been replaced by "armoured", artificial coastal structures. Those can significantly affect coastal currents and the ability of marine biotas to use the coast as part of their habitat. Tourist developments have also significantly increased the lengths of artificial coastline. Changes in river management, such as the construction of dams, and the building of coastal infrastructures, such as ports, can significantly change the sedimentation pattern along coasts. Such changes can increase coastal erosion and promote other coastal changes, sometimes with the effect that coastal land is lost for its current use, producing demands for replacement space.[66]

Aquaculture and marine ranching

Increases in aquaculture, which is growing rapidly, and in marine ranching, which has substantial growth potential, require extensive ocean space as well as clean waters and, often, the dedicated use of an unpolluted seabed. Those requirements can result in conflicts with other uses, including, in some cases, the aesthetic or cultural values of sea areas. Similar demands for ocean space are also made by industries concerned with the production of cultural goods, such as pearls. Problems will result if management of such expansion is not integrated with that of other sectors.

Shipping routes and ports

World shipping has been growing consistently for the past three decades. Between 1980 and 2013, the annual tonnage carried in the five main shipping trades increased by 158 per cent. Although the use of ocean space by a ship is not continuous, on the more densely trafficked routes, shipping lanes cannot be used safely for other activities, even where those activities themselves are intermittent. Some of the ranges of the largest populations of seabirds in the northern hemisphere are intersected by major shipping routes, with consequent risk of disturbance to the wildlife and mortality from chronic or catastrophic oil and other spills.

The fundamental change in general cargo shipping (from loose bulk to containerized) has also produced a total change in the nature of the ports that act as terminals for that traffic, as large areas of flat land are needed for handling containers, both on departure and arrival. That land has, in many cases, been provided by means of land reclamation. As shipping traffic continues to grow, further substantial areas of land will be required. Dredging to create ports and to maintain navigation channels produces large amounts of dredged material that has to be disposed of. Most of that material is dumped at sea, where it smothers any biota on the seabed.[67]

Submarine cables and pipelines

The vital role that submarine cables now play in all forms of communication through the Internet — whether for academic, commercial, governmental or recreational purposes — means that there will continue to be a demand for more capacity, and hence for more submarine cables. Although submarine cables (and any protective corridors around them) cover only very narrow strips of seabed, they introduce a line break across the seabed that prevents other activities from spreading across it. Submarine cables will therefore continue to neutralize increasing segments of the seabed for any purpose that impinges on the seabed.

---

65 See chaps. 12, 17, 19, 21-24 and 28.

66 See chaps. 18, 26, 28, 48 and 49.

67 See chaps. 17 and 18.

Submarine pipelines are unlikely ever to venture into the open-ocean areas where many submarine cables have to be laid, but they have a growing role for transporting oil and gas through coastal zones and between continents and their adjacent islands. In some ways, therefore, their increased demand for seabed space is likely to be in areas where there are demands from other uses.[68]

Offshore hydrocarbon industries

The growth of the offshore oil and gas industry has increased the demand by that sector for access to ocean space within areas under national jurisdiction (including space for pipelines to bring the hydrocarbon products ashore). More than 620,000 km^2 (almost 9 per cent) of the exclusive economic zone (EEZ) of Australia is subject to oil and gas leases. In the United States, about 550,000 km^2 of the whole EEZ is subject to current oil and gas leases, including 470,000 km^2 in the Gulf of Mexico, representing 66 per cent of the EEZ of the United States in that area. When such significant proportions of the ocean areas under national jurisdiction are thus subject to such prior claims, overlaps in sectoral interests become inevitable.

Offshore mining

Offshore mining is currently confined to shallow-water coastal regions, although growing exploration activity is focused on deep-sea minerals. About 75 per cent of the world's tin, 11 per cent of gold, and 13 per cent of platinum are extracted from the placer deposits near the surface of the coastal seabed, where they have been concentrated by waves and currents. Diamonds are also an important mining target. Aggregates (sand, coral, gravel and seashells) are also important: the United Kingdom, the world's largest producer of marine aggregates, currently extracts approximately 20 million tons of marine aggregate per year, meeting around 20 per cent of its demand. Those activities are all concentrated in coastal waters, where other demands for space are high. Deep-water deposits that have generated continuing interest, but are not currently mined, include ferromanganese nodules and crusts, polymetallic sulphides, phosphorites, and methane hydrates. Demands for deep-sea space are likely to develop in the future.[69]

Offshore renewable energy

Offshore renewable energy generation is still in its early stages, although substantial offshore wind farms have been installed in some parts of the world. Most forms of marine-based renewable energy require ocean space, and wind farms already cover significant areas in the coastal North Sea. Wave and tidal energy will make equal, if not larger, demands. The location of wind, wave and tidal installations can have significant effects on marine biotas. Special care is needed in siting installations that can affect migration routes or feeding, breeding or nursery areas. This is therefore a field in which the requirements of the new energy sources for ocean space could be important competitors with other, longer-established uses or with the need to conserve marine biodiversity.[70]

Fishery management areas

Capture fisheries have a very long history, predating newer ocean uses, such as aquaculture, offshore energy infrastructure, submarine cables, pipelines or tourism. The fishermen exploiting those long-practised fisheries usually have a feeling of "ownership", even though they rarely have had any established legal rights to exclude others from their customary fishing grounds. There is a growing trend, however, as part of fishery management within national jurisdictions, for fishing enterprises or fishing communities (including indigenous fishing communities) to be recognized as having some form of rights to fish to a defined extent in a defined area. Those benefiting from such rights frequently see constraints on fishing from other activities in those defined areas as invasions of what they consider as entitlements. This is the "front line" of conflicts in uses. If it is not directly addressed, some ocean uses will find it difficult to thrive.[71]

Marine protected areas

The Plan of Implementation of the World Summit on Sustainable Development (Johannesburg Plan of Implementation),[72] adopted in 2002, called for the implementation of marine protected areas. Although a marine protected area does not necessarily imply an area in which all human activities are excluded, in many cases it does imply that some, or most, such activities will be at least controlled or regulated. The commitment made by many States to a target for such protected areas of at least 10 per cent of the areas under their jurisdiction[73] will be a factor in future use of ocean space, given that, at present, marine protected areas represent a much smaller part of the ocean area under national jurisdiction.

Implications of demands for ocean space

That long list of types of human activity shows there are simply too many demands for all to be accommodated in a way that will not constrain some aspect of their operation. The allocation of ocean space is a much more complex task than that of land-use planning onshore. In the first place, the ocean is three-dimensional. Some uses can be in the same area but vertically separated, thus ships, for example, can pass over submarine cables without any problem, except in shallow water. Secondly, some uses are transient: ships and fishing vessels in particular pass and repass, and other uses may take place in the intervals between them. Thirdly, there is no general tradition of permanent rights of private ownership, even in areas under national jurisdiction. However, the more

---

68 See chap. 19.

69 See chap. 22.

70 See chap. 23.

71 See chaps. 11 and 15.

72 *Report of the World Summit on Sustainable Development, Johannesburg, South Africa, 26 August-4 September 2002* (United Nations publication, Sales No. E.03. II.A.1 and corrigendum), chap. I, resolution 2, annex, para. 32 (c).

73 See United Nations Environment Programme, document UNEP/CBD/COP/10/27, annex, decision X/2, sect. IV, target 11.

intense the shipping or fishing, the more difficult it is for other uses to be accommodated. Developing effective ways of organizing the allocation of ocean space is not an easy task, given the wide range of interests that need to be considered and reconciled.

5.6 Increasing inputs of harmful material

Land-based inputs

The agricultural and industrial achievements of the past two centuries in feeding, clothing and housing the world's population have been at the price of seriously degrading important parts of the planet, including much of the marine environment, especially near the coast. Urban growth, unaccompanied in much of the world by adequate disposal of human bodily wastes, has also imposed major pressures on the ocean. Land-based inputs to the ocean have thus contributed much to the degradation of the marine environment. The Global Programme of Action for the Protection of the Marine Environment from Land-based Activities of 1995 highlighted the need for action to deal with sewage (including industrial wastes that are mixed with human bodily wastes) in developing countries. Although much has been done to implement national plans adopted under the Programme, particularly in South America, the lack of sewage systems and wastewater treatment plants is still a major threat to the ocean. This is particularly the case for very large urban settlements.[74]

Several aspects have to be considered in relation to the increasing inputs of harmful material from the land into the ocean.

Heavy metals and other hazardous substances

From the point of view of industrial development, many industrial processes have brought with them serious environmental damage, especially when the concentration of industries have led to intense levels of inputs to the sea of wastes which could not be assimilated. That damage is largely caused by heavy metals (especially lead, mercury, copper and zinc). With the development of organic chemistry, new substances have been created to provide important services in managing electricity (for example, polychlorinated biphenyls) and as pesticides. Chlorine has also been widely used in many industrial processes (such as pulp and paper production), producing hazardous by-products. Many of those chemical products and processes have proved to have a wide range of hazardous side-effects.

There are also problems from imperfectly controlled incineration, which can produce polycyclic aromatic hydrocarbons and, where plastics are involved, dioxins and furans. All those substances have adverse effects on the marine environment. As well as the long-known hazardous substances, there is evidence that some substances (often called endocrine disruptors), which do not reach the levels of toxicity, persistence and bioaccumulation[75] in the accepted definitions of hazardous substances, can disrupt the endocrine systems of humans and animals, with adverse effects on their reproductive success. Action is already being taken on several of those, but more testing is needed to clarify whether action is needed on others.

Over time, steps have been taken to reduce or, where possible, eliminate many of the impacts of heavy metals and hazardous substances. In some parts of the world, the efforts of the past 40 years have been successful, and concentrations in the ocean of many of the most seriously damaging heavy metals and other hazardous substances are now diminishing, for example in the North-East Atlantic, even though problems persist in some local areas. New technologies and processes have also been widely developed that have the ability to avoid those problems, but there are gaps in the capacities to apply those newer processes, often because of the costs involved.

The differential growth in industrial production between countries bordering the North Atlantic, on the one hand, and those bordering the South Atlantic, the Indian Ocean and the Pacific, on the other hand, means that much of that growth is now taking place in parts of the world that had not previously had to deal with industrial discharges on the current scale. In the past, industrial production had been dominated by the countries around the North Atlantic basin and its adjacent seas, as well as Japan. Over the past 25 years, the rapid growth of industries along the rest of the western Pacific rim and around the Indian Ocean has dramatically changed that situation. The world's industrial production and the associated waste discharges are rapidly growing in the South Atlantic, the Indian Ocean and the western Pacific. Even if the best practicable means are used to deal with heavy metals and hazardous substances in the waste streams from those growing industries, the growth in output and consequent discharges will increase the inputs of heavy metals and other hazardous substances into the ocean. It is therefore urgent to apply new less-polluting technologies, where they exist, and means of removing heavy metals and other hazardous substances from discharges, if the level of contamination of the ocean, particularly in coastal areas, is not to increase.

Frameworks have also emerged at the international level for addressing some of the problems caused by heavy metals and hazardous substances. In particular, the Stockholm Convention on Persistent Organic Pollutants[76] and the Minamata Convention on Mercury[77] provide agreed international frameworks for the States party to them to address the issues that they cover. Implementing them, however, will require much capacity-building.[78]

---

bodies.

76 United Nations, *Treaty Series*, vol. 2256, No. 40214.

77 United Nations Environment Programme, document UNEP(DTIE)/Hg/CONF/4, annex II.

78 See chap. 20.

---

74 See chap. 20.

75 Bioaccumulation is the process whereby substances are ingested by animals and other organisms, but not broken down or excreted, and thus build up in their

Oil

Although pollution from oil and other hydrocarbons is most obviously linked to offshore production and their maritime transport, substantial inputs of hydrocarbons occur from land-based sources, particularly oil refineries. In some parts of the world, it has proved possible to reduce such pressures on the marine environment substantially.[79]

Agricultural inputs

The agricultural revolution of the last part of the twentieth century, which has largely enabled the world to feed its rapidly growing population, has also brought with it problems for the ocean in the form of enhanced run-off of both agricultural nutrients and pesticides, as well as the airborne and waterborne inputs of nutrients from waste from agricultural stock. In the case of fertilizers, their use is rapidly growing in parts of the world where only limited use had occurred in the past. That growth has the potential to lead to increased nutrient run-off to the ocean if the increased use of fertilizers is not managed well. There are therefore challenges in educating farmers, promoting good husbandry practices that cause less nutrient run-off and monitoring what is happening to agricultural run-off alongside sewage discharges. In the case of pesticides, the issues are analogous to those of industrial development. Newer pesticides are less polluting than older ones, but there are gaps in the capacity to ensure that these less-polluting pesticides are used, in terms of educating farmers, enabling them to afford the newer pesticides, supervising the distribution systems and monitoring what is happening in the ocean.

Eutrophication

Eutrophication resulting from excess inputs of nutrients from both agriculture and sewage causes algal blooms. Those can generate toxins that can make fish and other seafood unfit for human consumption. Algal blooms can also lead to anoxic areas (i.e. dead zones) and hypoxic zones. Such zones have serious consequences from environmental, economic and social aspects. The anoxic and hypoxic zones drive fish away and kill the benthic wildlife. Where those zones are seasonal, any regeneration that happens is usually at a lower trophic level, and the ecosystems are therefore degraded. This seriously affects the maritime economy, both for fishermen and, where tourism depends on the attractiveness of the ecosystem (for example, around coral reefs), for the tourist industry. Social consequences are then easy to see, both through the economic effects on the fishing and tourist industries and in depriving the local human populations of food.[80]

Radioactive substances

In the case of radioactive discharges into the ocean, there have been, in the past, human activities that have given rise to concern, but responses to those concerns, and the actions taken, have largely removed the underlying problems, even though there is a continuing task to monitor what is happening to radioactivity in the ocean. In particular, the ending of atmospheric tests of nuclear weapons and, more recently, the improvements made in the controls on discharges from nuclear reprocessing plants have ended or reduced the main sources of concern. What remains is the risk voiced in the Global Programme of Action that public reaction to concerns about marine radioactivity could result in the rejection of fish as a food source, with consequent harm to countries that have a large fishery sector and damage to the world's ability to use the important food resources provided by the marine environment.[81]

Solid waste disposal

The dumping of waste at sea was the first activity capable of causing marine pollution to be brought under global regulation, in the form of the Convention on the Prevention of Marine Pollution by Dumping of Wastes and Other Matter, 1972[82] (the London Convention), regulating the dumping of wastes and other matter at sea from ships, aircraft and man-made structures. The controls under that agreement have been progressively strengthened, particularly in the 1996 Protocol to the Convention on the Prevention of Marine Pollution by Dumping of Wastes and Other Matter, 1972[83] which introduced the approach of a total ban on dumping, subject to limited exemptions. If the Convention or the Protocol were effectively and consistently implemented, that source of inputs of harmful substances would be satisfactorily controlled. However, there are gaps in knowledge about their implementation. Over half of the States party to the London Convention and the Protocol thereto do not submit reports on dumping under their control. This may mean that there is no such dumping, but it may also mean that the picture presented by the reports that are submitted is incomplete. Some of the world's largest economies have not become party to either agreement, and nothing is known of what is happening with respect to dumping under their control. The reported dumping is very largely of dredged material, most of it from the creation or maintenance of ports. Clear guidance under the London Convention lays down the conditions under which that material may be dumped. To the extent that that guidance is followed, there should be no significant impact on the marine environment, except for the smothering of the seabed, and to the extent that the dump sites are in areas with dynamic tidal activity, even that impact will be limited. There is also some evidence that illegal dumping is taking place, including that of radioactive waste, but complete proof of this has not been obtained.[84]

Marine debris

Marine debris is present in all marine habitats, from densely populated regions to remote points far from human activities, from beaches and shallow waters to the deepest ocean trenches. It has been estimated that the average density of marine debris varies between 13,000 and 18,000 pieces per square kilometre. However, data on plastic accumulation in the North Atlantic and Caribbean from 1986 to 2008 showed that the highest concentrations (more than 200,000 pieces per square

---

79 See chap. 20.

80 See chap. 20.

81 See chap. 20.

82 United Nations, *Treaty Series*, vol. 1046, No. 15749.

83 International Maritime Organization, document IMO/LC.2/Circ.380.

84 See chap. 24.

kilometre) occurred in the convergence zones between two or more ocean currents. Computer model simulations, based on data from about 12,000 satellite-tracked floats deployed since the early 1990s as part of the Global Ocean Drifter Program, confirm that debris will be transported by ocean currents and will tend to accumulate in a limited number of subtropical convergence zones or gyres.

Plastics are by far the most prevalent debris item recorded, contributing an estimated 60 to 80 per cent of all marine debris. Plastic debris continues to accumulate in the marine environment. The density of microplastics within the North Pacific Central Gyre has increased by two orders of magnitude in the past four decades. Marine debris commonly stems from shoreline and recreational activities, commercial shipping and fishing, and dumping at sea. The majority of marine debris (approximately 80 per cent) entering the sea is considered to originate from land-based sources.[85]

Nanoparticles are a form of marine debris, the significance of which is emerging only now. They are minuscule particles with dimensions of 1 to 100 nanometres (a nanometre is one millionth of a millimetre). A large proportion of the nanoparticles found in the ocean are of natural origin. It is the anthropogenic nanoparticles that are of concern. Those come from two sources: on the one hand, from the use of nanoparticles created for use in various industrial processes and cosmetics and, on the other hand, from the breakdown of plastics in marine debris, from fragments of artificial fabrics discharged in urban wastewater, and from leaching from land-based waste sites. Recent scientific research has highlighted the potential environmental impacts of plastic nanoparticles: they appear to reduce the primary production and the uptake of food by zooplankton and filter-feeders. Nanoparticles of titanium dioxide, which is widely used in paints and metal coatings and in cosmetics, are of particular concern. When nanoparticles of titanium dioxide are exposed to ultraviolet radiation from the sun, they transform into a disinfectant and have been shown to kill phytoplankton, which are the basis of primary production. The scale of the threats from nanoparticles is unknown, and further research is required.[86]

Shipping

Pollution from ships takes the form of both catastrophic events (shipwrecks, collisions and groundings) and chronic pollution from regular operational discharges. Good progress has been made over the past 40 years in reducing both. There have been large increases in the global tonnage of cargo carried by sea and in the distances over which those cargoes are carried. There have also been steady increases in the number of passengers carried on cruise ships and ferries. In spite of this, the absolute number of ship losses has steadily decreased. Between 2002 and 2013, the number of losses of ships of over 1,000 gross tonnage thus dropped by 45 per cent to 94. This is largely due to efforts under the three main international maritime safety conventions: the International Convention on the Safety of Life at Sea,[87] dealing with ship construction and navigation, the International Convention on Standards of Training, Certification and Watchkeeping for Seafarers, 1978,[88] dealing with crew, and the International Convention for the Prevention of Pollution from Ships (MARPOL).

Pollution from oil has been the most significant type of marine pollution from ships. The number of spills exceeding 7 tons has dropped steadily, in spite of the growth in the quantity carried and the length of voyages, from over 100 spills in 1974 to under five in 2012. The total quantity of oil released in those spills has also been reduced by an even greater factor. Progress has also been made in improving response capabilities, though much remains to be done, especially as coastal States have to bear the capital cost of acquiring the necessary equipment. Reductions in oil pollution have resulted from more effective enforcement of the MARPOL requirements, particularly in western Europe. The changes in arrangements for reparation for any damage caused by oil pollution from ships have improved the economic position of those affected.

In spite of all that progress, oil discharges from ships remain an environmental problem, for example, around the southern tip of Africa and in the North-West Atlantic. Off the coast of Argentina, however, a solution to the impact of those discharges on penguin colonies seems to have been found by rerouting coastal shipping. The likely opening of shipping routes through the Arctic between the Atlantic and the Pacific risks introducing that form of pollution into a sea area where response infrastructure is lacking, oil recovery in freezing conditions is difficult and the icy water temperature inhibits the microbial breakdown of the oil.[89]

Pollution from cargoes of hazardous and noxious substances appears to be a much smaller problem, even though there are clearly problems with misdescriptions of the contents of containers. Losses of containers, however, appear to be relatively small: in 2011, the losses were estimated at 650 containers out of about 100 million carried in that year.

Sewage pollution from ships is mainly a problem with cruise ships: with up to 7,000 passengers and crew, they are the equivalent of a small town and can contribute to local eutrophication problems. The local conditions around the ship are significant for the impact of any sewage discharges. The increased requirements under MARPOL on the discharges of ship sewage near the shore are likely to reduce the problems, but the identification of the cases where ships have contributed to eutrophication problems will remain difficult.

The dumping of garbage from ships is a serious element of the problem of marine debris. In 2013, new, more stringent controls under MARPOL came into force. Steps are being taken to improve the enforcement of those requirements. For example, the World Bank has helped several

---

85 See chap. 25.

86 See chaps. 6 and 25.

87 United Nations, Treaty Series, vol. 1184, No. 18961.

88 United Nations, Treaty Series, vol. 1361, No. 23001.

89 See chap. 17.

small Caribbean States to set up port waste-reception facilities, which has made it possible for the Wider Caribbean to be declared a special area under annex V of the Convention, under which stricter requirements apply. Other States (for example the Member States of the European Union) have introduced requirements for the delivery of waste ashore before a ship leaves port and have removed economic incentives to avoid doing so. It is, however, too early to judge how far those various developments have succeeded in reducing the problem.[90]

Offshore hydrocarbon industries

Major disasters in the offshore oil and gas industry have a global, historical recurrence of one about every 17 years. The most recent is the Deepwater Horizon blowout of 2010, which spilled 4.4 million barrels (about 600,000 tons) of oil into the Gulf of Mexico. The other main harmful inputs from that sector are drilling cuttings (contaminated with drilling muds) resulting from the drilling of exploration and production wells, "produced water" (the water contaminated with hydrocarbons that comes up from wells, either of natural origin or through having been injected to enhance hydrocarbon recovery), and various chemicals that are used and discharged offshore in the course of exploration and exploitation.

Those materials can be harmful to marine life under certain circumstances. However, it is possible to take precautions to avoid such harm, for example by prohibiting the use of the most harmful drilling muds, by limiting the proportion of oil in the produced water that is discharged or by controlling which chemicals can be used offshore. Such regulation has been successfully introduced in a number of jurisdictions. Nonetheless, given the growth in exploration and offshore production, there is no doubt that those inputs are increasing over time, even though exact figures are not available globally. Produced water, in particular, increases in quantity with the age of the field being exploited.[91]

Offshore mining

The environmental impacts of near-shore mining are similar to those of dredging operations. They include the destruction of the benthic environment, increased turbidity, changes in hydrodynamic processes, underwater noise and the potential for marine fauna to collide with vessels or become entangled in operating gear.[92]

Implications for human well-being and biodiversity
Human health, food security and food safety

Marine biotas are under many different pressures from hazardous substances on reproductive success. Dead zones and low-oxygen zones resulting from eutrophication and climate change can lead to systematic changes in the species structure at established fishing grounds. Either can reduce the extent to which fish and other species used as seafood will continue to reproduce at their historical rates. When those effects are combined with those of excessive fishing on specific stocks, there are risks that the traditional levels of the provision of food from the sea will not be maintained.

In addition, heavy metals and other hazardous substances represent a direct threat to human health, particularly through the ingestion of contaminated food from the sea. The episode of mercury poisoning at Minamata, in Japan, is probably the most widely known event of that kind, and the reason why the global convention to address such problems is named after the town. There are places around the world where local action has been taken to prevent or discourage the consumption of contaminated fish and other seafood. In other places, monitoring suggests that levels of contamination dangerous for human health are being reached. In yet other places, there are inadequate monitoring systems to check on risks of that kind. Ensuring linkages between adequate systems for controlling the discharge and emissions of hazardous substances and the systems for controlling the quality of fish and other seafood available for human consumption is therefore an important issue. In the case of subsistence fishing, the most effective approach is to ensure that contamination does not occur in the first place.

The lack of proper management of wastewater and human bodily wastes causes problems for human health, both directly through contact with water containing pathogens and through bacteriological contamination of food from the sea, and indirectly by creating the conditions in which algal blooms can produce toxins that infect seafood. Those problems are particularly significant in and near large and growing conurbations without proper sewage treatment systems, such as found in many places in developing countries.[93]

Impacts on marine biodiversity

Part of the standard definition of hazardous substances in the context of marine pollution is that they are bioaccumulative — that is, once they are taken into an organism, they are not broken down or expelled, and continue to accumulate in it. Because of that characteristic, they also are accumulated more in the higher levels of the food web. As creatures at the lower levels are eaten by those at higher levels, the hazardous substances in the former are retained and accumulated by the latter. Some of those substances affect the reproductive success of the biota in which they have accumulated. There are also some effects on immune systems, with the result that individuals and populations become less resistant to outbreaks of disease. The deaths of many seals in the North-East Atlantic in the 1990s from the phocine distemper virus have thus been linked to impaired immune systems. Likewise, improvements in a fish-health index in the same area in the 2000s have been attributed to reductions in the local concentrations of various hazardous substances.

The combined effects of hazardous substances, marine debris, oil and eutrophication (including the large and growing number of dead zones) resulting from the input of harmful material, waste and excessive

---

90 See chaps. 17 and 25.

91 See chap. 21.

92 See chap. 23.

93 See chaps. 4-6, 10-12, 15 and 20.

amounts of nutrients into the ocean therefore represent a significant pressure on marine biodiversity.[94]

5.7 Cumulative impacts of human activities on marine biodiversity

When the many pressures described above, from fishing and other types of marine harvesting to demand for ocean space and inputs of harmful materials, are brought together, the result is a complex but dangerous mix of threats to marine biodiversity. To those threats must be added several other significant factors. Those arise from a number of separate sources, including noise from ships and seismic exploration and the introduction of competing non-native species by aquaculture and long-distance shipping (and their further distribution by recreational boats). Taken altogether, those factors represent a massive set of pressures on marine biodiversity.[95]

Implications for marine biodiversity

Such cumulative impacts of human uses are reported in all the regional biodiversity assessments in part VI of the present assessment. There are indeed well-documented examples of cases where habitats, lower-trophic-level productivity, benthic communities, fish communities and seabird or marine mammal populations have been severely altered by pressures from a specific activity or factors (such as overfishing, pollution, nutrient loading, physical disturbance or the introduction of non-native species). However, many impacts on biodiversity, particularly at larger scales, are the result of the cumulative and interactive effects of multiple pressures from multiple drivers. It has repeatedly proved difficult to disentangle the effects of the individual pressures, which impedes the ability to address the individual causes.[96]

Even in the Arctic Ocean, where human settlements are relatively few and small, the potentially synergistic effects of multiple stressors come together. Furthermore, those stressors operate against a background of pressures from a changing climate and increasing human maritime activity, primarily related to hydrocarbon and mineral development and to the opening of shipping routes. Those changes bring risks of direct mortality, displacement from critical habitats, noise disturbance and increased exposure to hunting, which are superimposed on high levels of contaminants, notably organochlorines and heavy metals, as a result of the presence of those substances in the Arctic food web.[97]

In the open ocean (remote from land-based inputs), shifts in bottom-up forcing (that is, primary productivity) and competitive or top-down forcing (that is, by large predators) will also produce complex and indirect effects on ecosystem services. The stress imposed by low oxygen, low pH (that is, higher acidity) or elevated temperatures can reduce the resilience of individual species and ecosystems through shifts in organism tolerance and community interactions. Where this happens, it retards recovery from disturbances caused by human activities, such as oil spills, trawling and (potentially in the future) seabed mining. Slower growth of carbonate skeletons due to increased ocean acidification, delayed development under hypoxic conditions and increased respiratory demands with declining food availability illustrate how climate change could exacerbate anthropogenic impacts and compromise deep-sea ecosystem structures and functions, and ultimately its benefits to human welfare.[98]

Those multiple pressures interact in ways that are poorly understood but that can amplify the effects expected from each pressure separately. The North Atlantic has been, comparatively, the subject of much scientific research. It has many long-term ocean-monitoring programmes and a scientific organization that has functioned for over a century to promote and coordinate scientific and technical cooperation among the countries around the North Atlantic. Even there, however, experts are commonly unable to disentangle consistently the causation of unsustainable uses of, and impacts on, marine biodiversity. This may initially seem to be discouraging. Nevertheless, well-documented examples exist of the benefits that can follow from actions to address past unsustainable practices, even if other perturbations are also occurring in the same area.[99]

Marine mammals, marine reptiles, seabirds, sharks, tuna and billfish

Cumulative effects are comparatively well documented for species groups of the top predators in the ocean, including marine mammals, seabirds and marine reptiles. Many of those species tend to be highly mobile and some migrate across multiple ecosystems and even entire ocean basins, so that they can be exposed to many threats in their annual cycle. Some of those species are the subject of direct harvesting, particularly some pinnipeds (seals and related species) and seabirds, and by-catch in fisheries can be a significant mortality source for many species. However, in addition to having to sustain the impact of those direct deaths, all of those species suffer from varying levels of exposure to pollution from land-based sources and increasing levels of noise in the ocean. Land-nesting seabirds, marine turtles and pinnipeds also face habitat disturbance, such as through the introduction of invasive predators on isolated breeding islands, the disturbance of beaches where eggs are laid or direct human disturbance from tourism, including ecotourism.[100]

Some global measures have been helpful in addressing specific sources of mortality, such as the global moratorium on all large-scale pelagic drift-net fishing called for by the General Assembly in 1991, which was a major step in limiting the by-catch of several marine mammal and seabird species that were especially vulnerable to entanglement. However, for seabirds alone, at least 10 different pressures have been iden-

---

94 See chaps. 4-6, 20, 21, 25, 36A-H and 52.

95 See chaps. 11, 12, 17-23 and 25-27.

96 See chaps. 36A-H and 53.

97 See chap. 36G.

98 See chaps. 4-6, 11, 17, 20, 36F, 37-39 and 52.

99 See chap. 36A.

100 See chaps. 27, 37-39 and 52.

tified that can affect a single population throughout its annual cycle, with efforts to mitigate one pressure sometimes increasing vulnerability to others. Because of the complexity of those issues, conservation and management must therefore be approached with care and alertness to the nature of the interactions among the many human interests, the needs of the animals and their role in marine ecosystems.[101]

Ecosystems and habitats identified for special attention
Just as species can face the effects of multiple pressures over their annual cycle as they migrate (sometimes around an entire ocean basin), habitats can integrate the effects of multiple pressures across the interacting species that use them. Many cases are presented in the chapters on specialized habitats, which are often sites of concentrated human activities. For example, warm-water corals face major threats, such as extractive activities, sewage and other pollution, sedimentation, physical destruction and the effects of anthropogenic climate change, including increased coral bleaching. Such stressors often interact synergistically with one another and with natural stressors, such as storms. Likewise, cold-water corals are often challenged by the synergistic effects of low oxygen and increasing acidification, as well as by physical damage from fishing practices.[102]

All coastal habitats, including kelp forests, seagrass beds and mangroves, face multiple interacting threats from land-based sources, species invasions and direct anthropogenic pressures. For example, mangroves may face the aggregate effects of coastal and urban development, sewage and other pollutants, solid waste disposal, damage from extreme events, such as hurricanes, as well as conversion to aquaculture or agriculture and climate change. Each of the chapters on specific habitats presents similar lists of pressures, often present on the same sites. Although protection from direct human uses of areas where habitats occur (such as bans on converting mangroves to aquaculture or port facilities) can often produce immediate benefits, pressures such as land-based runoff, diseases and invasive species require coordinated efforts far beyond the specific habitats for which the protection is intended.[103]

Considering specific types of important marine and coastal habitats, estuaries and deltas are categorized globally as in poor overall condition, based on published assessments of them for 101 regions. In 66 per cent of cases, their condition has worsened in recent years. There are around 4,500 large estuaries and deltas worldwide, of which about 10 per cent benefit from some level of environmental protection. About 0.4 per cent is protected as strict nature reserves or wilderness areas (categories Ia and Ib of the categories of protected areas as defined by the International Union for Conservation of Nature).[104]

Mangroves are being lost at the mean global rate of 1-2 per cent a year, although losses can be as high as 8 per cent a year in some countries. While the primary threat to mangroves is overexploitation of resources and the conversion of mangrove areas to other land uses, climate-change-induced sea-level rise is now identified as a global threat to them, especially in areas of growing human settlements and coastal development.[105]

Kelp and seagrass habitats are declining worldwide for different reasons. The overfishing of dominant predators and climate change have reportedly caused changes in kelp community structures and distribution over time. Kelp forests are more affected by temperature changes owing to the narrow range in which their sexual reproduction can occur. Seagrass meadows are more affected by anthropogenic activities, such as siltation, pollution and reclamation.[106]

Fishing on seamounts has targeted fish aggregations to depths of 1,500 m. Aggregations on spatially limited topographic features are highly vulnerable, and many target species are slow-growing and long-lived, therefore exhibiting little resilience to disturbance. Furthermore, most fisheries use bottom trawls, gear that is highly destructive to benthic communities. Little recolonization is observed years after closure to fishing. Most sites of deep-water bottom fisheries have been overfished in the past, but there are now increased efforts to seek to regulate their use and to protect deep-water benthic habitats.[107]

Tourism and aesthetic, cultural, religious and spiritual marine ecosystem services
The changes in marine biodiversity can have consequential effects on the ecosystem services that humans obtain from the ocean. Particularly important is the link between the health of warm-water corals and tourism. Warm-water corals represent a major component of the attractiveness of many tourist resorts in the Caribbean, the Red Sea, the Indian Ocean and South-East Asia, and that attractiveness will be seriously undermined if tourists can no longer enjoy the corals. The same applies to other resorts (even in cold-water areas) where one of the attractions is scuba-diving to enjoy the marine wildlife. A different linkage is that to recreational fishing, where a significant industry relies on the availability of large sport fish such as marlins, swordfish and sailfish. In that case, there is a lack of information on which estimates of fish stocks and, consequently, judgements on the sustainable scale of the activity can be based.[108]

The disappearance or, more commonly, the reduction in numbers of iconic species can likewise adversely affect traditional practices. For example, native people on the North-East Pacific coast have seen their traditional whale-hunting halted because of the past overharvesting of

---

101 See chaps. 11 and 38.

102 See chaps. 42-51.

103 See chaps. 43, 44 and 47-49.

104 See chap. 44.

105 See chap. 48.

106 See chap. 47.

107 See chaps. 36F and 51.

108 See chaps. 27, 41 and 43.

grey whales carried out by other people. That hunting was an integral part of their cultural heritage and the affected tribes consider the cultural loss to be very serious. Pollution can have similar effects. For example, the Faroese authorities (Denmark) are taking measures to control the traditional food obtained in the islands from pilot whales because of the high levels of pollutants accumulated in their tissues.[109]

5.8 Distribution of ocean benefits and disbenefits

In assessing the social and economic aspects of the ocean, it is necessary to consider how different parts of the world, different States and different parts of society are gaining benefits (or suffering disbenefits) as a result of the ways in which human activities linked to the oceans are changing.

Changes in the universal ecosystem services from the ocean
The most obvious distributional effects of climate change relate to the rise in sea level. Some small island States are predicted to become submerged completely and some heavily populated deltas and other low-lying areas also risk inundation. Another important distributional effect is the poleward extension of major areas of storms, which is likely to lead to cyclones, hurricanes and typhoons in areas previously not seriously affected by them. Changes in patterns of variability of oscillations (such as the El Niño-Southern Oscillation) will bring climatic changes to many places and affect new areas, with consequent effects on agriculture and agricultural earnings.[110]

The changes in ocean conditions will affect many other ecosystem services indirectly. For example, some models predict that the warming ocean will increase the fish biomass available for harvesting in higher latitudes and decrease it in equatorial zones. This will shift provisioning services to benefit the middle and moderately high latitudes (which are often highly developed) at the expense of low latitudes, where small-scale (subsistence) fishing is often important for food security.[111]

Developments in fish and seafood consumption
The Food and Agriculture Organization of the United Nations (FAO) estimates that total fish consumption, including all aquaculture and inland and marine capture fisheries, has been rising from 9.9 kg per capita in the 1960s to 19.2 kg per capita in 2012 — an average increase of 3.2 per cent a year over half a century. The distribution of consumption per capita varies considerably, from Africa and Latin America and the Caribbean (9.7 kg) to Asia (21.6 kg), North America (21.8), Europe (22.0 kg) and Oceania (25.4 kg). Marine capture fisheries represent 51 per cent and marine aquaculture 13 per cent of the total production of fish (154 million tons), of which 85 per cent is used for food.

The annual consumption of fishery products per capita has grown steadily in developing regions (from 5.2 kg in 1961 to 17.0 kg in 2009) and low-income food-deficit countries (from 4.9 kg in 1961 to 10.1 kg in 2009). This is still considerably lower than in more developed regions, even though the gap is narrowing. A sizeable share of fish consumed in developed countries consists of imports and, owing to steady demand and declining domestic fishery production (down 22 per cent in the period 1992-2012), their dependence on imports, in particular from developing countries, is projected to grow.

FAO estimates indicate that small-scale fisheries contribute about half of global fish catches. When considering catches destined for direct human consumption, the share contributed by the subsector increases, as small-scale fisheries generally make broader direct and indirect contributions to food security (through affordable fish) and employment for populations in developing countries. As well as direct consumption, many small-scale fishermen sell or barter their catch. It is doubtful that much of that trade is covered by official statistics. However, studies have shown that selling or trading even a portion of their catch represents as much as one third of the total income of subsistence fishermen in some low-income countries. Thus an increase in imports of fish by more developed countries from less developed countries has the potential to increase inequities in food security and nutrition, unless those considerations are taken into account in global trade arrangements.[112]

Developments in employment and income from fisheries and aquaculture
The global harvest of marine capture fisheries has expanded rapidly since the early 1950s and is currently estimated to be about 80 million tons a year. That harvest is estimated to have a first (gross) value on the order of 113 billion dollars. Although it is difficult to produce accurate employment statistics, estimates using a fairly narrow definition of employment have put the figure of those employed in fisheries and aquaculture at 58.3 million people (4.4 per cent of the estimated total of economically active people), of which 84 per cent are in Asia and 10 per cent in Africa. Women are estimated to account for more than 15 per cent of people employed in the fishery sector. Other estimates, probably taking into account a wider definition of employment, suggest that capture fisheries provide direct and indirect employment for at least 120 million persons worldwide.

Small-scale fisheries employ more than 90 per cent of the world's capture fishermen and fish workers, about half of whom are women. When all dependants of those taking full- or part-time employment in the full value chain and support industries (boatbuilding, gear construction, etc.) of fisheries and aquaculture are included, one estimate concludes that between 660 and 820 million persons have some economic or livelihood dependence on fish capture and culture and the subsequent direct value chain. No sound information appears to be available on the levels of death and injury of those engaged in capture fishing or aquaculture, but capture fishing is commonly characterized as a dangerous occupation.

---

[109] See chaps. 8 and 20.

[110] See chaps. 4 and 5.

[111] See chaps. 11 and 15.

[112] See chaps. 10, 11 and 15.

Over time, a striking shift has occurred in the operation and location of capture fisheries. In the 1950s, capture fisheries were largely undertaken by developed fishing States. Since then, developing countries have increased their share. As a broad illustration, in the 1950s, the southern hemisphere accounted for no more than 8 per cent of landed values. By the last decade, the southern hemisphere's share had risen to 20 per cent. In 2012, international trade represented 37 per cent of the total fish production in value, with a total export value of 129 billion dollars, of which 70 billion dollars (58 per cent) was exports by developing countries.[113]

Aquaculture is responsible for the bulk of the production of seaweeds. Worldwide, reports show that 24.9 million tons was produced in 2012, valued at about 6 billion dollars. In addition, about 1 million tons of wild seaweed were harvested. Few data were found on international trade in seaweeds, but their culture is concentrated in countries where consumption of seaweeds is high.[114]

Developments in maritime transport

All sectors of maritime transport (cargo trades, passenger and vehicle ferries and cruise ships) are growing in line with the world economy. It is not possible to estimate the earnings from those activities, as the structure of the companies owning many of the ships involved is opaque. It seems likely that many of the major cargo-carrying operators were making a loss in 2012, as a result of overcapacity resulting from the general economic recession. On the other hand, cruise operators reported profits. According to estimates by the United Nations Conference on Trade and Development, owners from five countries (China, Germany, Greece, Japan and the Republic of Korea) together accounted for 53 per cent of the world tonnage in 2013. It seems likely that profits and losses are broadly proportional to ownership. Among the top 35 ship-owning countries and territories, 17 are in Asia, 14 in Europe and 4 in the Americas.

Worldwide, there are just over 1.25 million seafarers, only about 2 per cent of whom are women, mainly in the ferry and cruise-ship sectors. The crews are predominantly from countries members of the Organization for Economic Cooperation and Development and Eastern Europe (49 per cent of the officers and 34 per cent of the ratings) and from Eastern and Southern Asia (43 per cent of the officers and 51 per cent of the ratings). Africa and Latin America are noticeably underrepresented, providing only 8 per cent of the officers and 15 per cent of the ratings. Pay levels of officers differ noticeably according to their origin, with masters and chief officers from Western Europe receiving on average a fifth or a quarter, respectively, more than those from Eastern Europe or Asia, while pay levels for engineer officers are more in line with one another. The recent entry into force of the Maritime Labour Convention, 2006 should be noted in the context of the social conditions of seafarers.

statistics on the deaths of and injuries to seafarers are unreliable, and the Secretary-General of the International Maritime Organization has called for efforts to improve them. In general, it would appear that the levels of death and injury are worse than for many land-based industries. Over the past three decades, piracy and armed robbery have re-emerged as a serious risk to seafarers. Much attention has been focused on such attacks on ships in waters off Eastern Africa, but reports show that the problem is more widespread. In the past three years, action against attacks off Eastern Africa appears to have had some success, but attacks elsewhere are also of concern, especially in the South China Sea, the location of over half the incidents reported in 2013, and West Africa.[115]

Developments in offshore energy businesses

Global offshore oil production in mid-2014 was about 28 million barrels per day, which was worth about 3.2 billion dollars per day, and the industry directly employs about 200,000 people globally, mostly in the Gulf of Mexico (where about 60 per cent of the industry is located) and the North Sea. In the same year, the industry accounted for about 1.5 per cent of the gross domestic product (GDP) of the United States, 3.5 per cent of the GDP of the United Kingdom, 21 per cent of the GDP of Norway and 35 per cent of the GDP of Nigeria. The large majority of offshore hydrocarbon production is in the hands of international corporations or national companies usually working in partnership with them. This makes the tracking of the distribution of benefits from this sector, other than direct employment in extraction and processing, very difficult.[116]

Developments in offshore mining

There is limited information about the value of the offshore mining industry and the number of people it employs, but it is unlikely to be significant at present in comparison with terrestrial mining. For example, in the United Kingdom, which is the world's largest producer of marine aggregates, the industry directly employs approximately 400 people.[117]

Developments in tourism

Tourism has generally been increasing fairly steadily for the past 40 years (with occasional setbacks or slowing down during global recessions). In 2012, international tourism expenditure exceeded 1 billion dollars for the first time. Total expenditure on tourism, domestic as well as international, is several times that amount. The direct turnover of tourism contributed 2.9 per cent of gross world product in 2013, rising to 8.9 per cent when the multiplier effect on the rest of the economy is taken into account. The Middle East is the region where tourism plays the smallest part in the economy (6.4 per cent of GDP, including the multiplier effect), and the Caribbean is the region where it plays the largest part (13.9 per cent of GDP, including the multiplier effect).

Most reports of tourism revenues do not differentiate revenues from tourism directly related to the sea and the coast from other types of

---

113 See chaps. 11 and 15.

114 See chap. 14.

115 See chap. 17.

116 See chap. 21.

117 See chap. 23.

tourism. Even where tourism in the coastal zone can be separated from tourism inland, it may be generated by the attractions of the sea and coast or its maritime history, as it may be based on other attractions not linked to the marine environment. Consequently, the value of ocean-related tourism is a matter of inference. However, coastal tourism is a major component of tourism everywhere. In small island and coastal States, coastal tourism is usually predominant because it can only take place in the coastal zone in those countries. Particularly noteworthy is the way in which international tourism is increasing in Asia and the Pacific, both in absolute terms and as a proportion of world tourism. This implies that pressures from tourism are becoming of significantly more concern in those regions.

Tourism is also a significant component of employment. Globally, it is estimated that, in 2013, tourism provided 3.3 per cent of employment, when looking at the number of people directly employed in the tourism industry, and 8.9 per cent when the multiplier effect is taken into account. In the different regions, the proportion of employment supported by tourism is approximately the same as the share of GDP contributed by tourism, although, again, what proportion is based on the attractions of the sea and coast is not well known.[118]

Use of marine genetic material
The commercial exploitation of marine genetic resources had very modest beginnings in the twentieth century, particularly when measured against some estimates of the potential of the great diversity of species and biomolecules in the sea. Since 2000, the first drugs derived from marine organisms have been put into commerce (although, using the United States Food and Drug Administration approvals as a measure, only seven have so far received that approval). There has also been considerable growth in the use of marine natural products as food supplements and for other non-medical purposes. Economic and social aspects of the use of marine genetic material are therefore only just beginning to develop.[119]

Satellite national accounts
Information on the distribution of economic benefits from the ocean is hard to compile from current information sources. The work of the United Nations Statistics Division in developing a System of Environmental-Economic Accounting and an Experimental Ecosystem Accounting System seems likely to help to fill that information gap. In the same way, national satellite accounts dealing with tourism and fisheries should help to fill information gaps in those fields.[120]

5.9 Integrated management of human activities affecting the ocean

The Regular Process is to provide an assessment of all the aspects of the marine environment relevant to sustainable development: environmental, economic and social. Even though the marine environment covers seven tenths of the planet, it is still only one component of the overall Earth system. As far as environmental aspects are concerned, major drivers of the pressures producing change in the ocean are to be found outside the marine environment. In particular, most of the major drivers of anthropogenic climate change are land-based. Likewise, the main drivers of increased pressures on marine biodiversity and marine environmental quality include the demand for food for terrestrial populations, international trade in products from land-based agriculture and industries and coastal degradation from land-based development and land-based sources.

Thus, as far as social and economic aspects of the marine environment are concerned, many of the most significant drivers are outside the scope of the present assessment. For example, the levels of cargo shipping are driven mainly by world trade, which is determined by demand and supply for raw materials and finished products. The extent of cruising and other types of tourism is determined by the levels around the world of disposable income and leisure time. The patterns of trade in fish and other seafood and in cultural goods from the ocean are set by the location of supply and demand and the relative purchasing power of local markets as compared with international ones, modified by national and international rules on the exploitation of those resources. A wide range of factors outside the marine environment are thus relevant to policymaking for the marine environment.

The present assessment of the marine environment cannot therefore reach conclusions on some of the main drivers affecting the marine environment without stepping well outside the marine environment and the competences of those carrying out the assessment. It is essential to note, however, that the successful management of human activities affecting the marine environment will require the consideration of the full range of factors relating to human activities affecting the ocean.

Even within the scope of what has been requested, it has not proved possible to come to conclusions on one important aspect: a quantitative picture of the extent of many of the non-marketed ecosystem services provided by the ocean. Quantitative information is simply insufficient to enable an assessment of the way in which different regions of the world benefit from those services. Nor do current data-collection programmes appear to make robust regional assessments of ocean ecosystem services likely in the near future, especially for the less developed parts of the planet.[121]

---

118 See chap. 27.

119 See chap. 29.

120 See chaps. 3 and 9.

121 See chaps. 54 and 55.

The assessment of what is happening to aesthetic, cultural, religious and spiritual values is also very difficult. In essentially every coastal or island culture, the indigenous peoples have spiritual links to the sea. They often also have links with species or places, or both, that have high iconic values. The spiritual significance of those marine species and places may be part of their self-identification and reflects their beliefs about the origins of their culture. That is particularly true of island cultures, which are often intimately bound to the sea. Expressions of loss of, or threats to, such cultures and identities are readily found, but the marine component is not easily separated. Even populations that are economically fully developed with largely urbanized lifestyles still look to the ocean for spiritual and cultural benefits that have proven hard to value monetarily.[122]

Nevertheless, there is an overall message that the world has reached the end of the period when human impacts on the sea were minor in relation to the overall scale of the ocean. Human activities now have so many and such great impacts on the ocean that the limits of its carrying capacity are being (or, in some cases, have been) reached. It is instructive to look at the ways in which this has happened in one specific sector: fisheries. In the late nineteenth century, the regulation of fisheries was regarded by many as unnecessary: Thomas Huxley, the great defender of Charles Darwin's theory of natural selection and a leading marine biologist, speaking at the London Fisheries Exhibition, in 1883, said: "In relation to our present modes of fishing, a number of the most important sea fisheries … are inexhaustible. … [The] multitude of those fishes is so inconceivably great that the number that we catch is relatively insignificant; and secondly, … the magnitude of the destructive agencies at work on them is so prodigious, that the destruction effected by the fisherman cannot sensibly increase the death rate".

In less than 50 years, his qualification "in relation to our present modes of fishing" proved to be prophetic. Modes of fishing had changed to such an extent that international efforts were under way to regulate individual fisheries. We now know that those efforts were even then overdue. Furthermore, experience thereafter showed that the successful management of fisheries required a much broader approach. First to be acknowledged was the need for a multispecies approach: it was necessary to regulate the fisheries not only for each target species individually, but also to take into account the species on which the target species preyed and the species that preyed on it.

In the 1990s, it became clear that the effects of fisheries on other biotas made an ecosystem approach to fishery management necessary, taking into account how a fishery might directly kill other species through by-catches, alter habitats and change relationships in the food web. Since then, the increasing use of the ocean has shown how fisheries managers need to work with other sectors to manage their effects on each other and, collectively, on the ocean that they share.

When various conclusions in parts III to VI of the present assessment are linked together, they clearly show that a similar broadening of the context of management decisions will produce similar benefits in and among other sectors of human activities that affect the ocean. Examples of such interactions of pressures on the environment include:

(a) The lack of adequate sewage treatment in many large coastal conurbations, especially in developing countries, and other excessive inputs of nutrients (especially nitrogen) are producing direct adverse impacts on human health through microbial diseases as well as eutrophication problems. In many cases, they are creating harmful algal blooms, which are not only disrupting ecosystems, but also, as a consequence, damaging fisheries, especially small-scale fisheries and the related livelihoods and, in some cases, poisoning humans through algal toxins;[123]

(b) Plastic marine debris results from the poor management of waste streams on land and at sea. There is a clear impact of such debris in its original form on megafauna (fish caught in "ghost" nets, seabirds with plastic bags around their necks, etc.) and on the aesthetic appearance of coasts (with potential impacts on tourism). Less obviously, impacts on zooplankton and filter-feeding species have also been demonstrated from the nanoparticles into which those plastics break down, with potentially serious effects all the way up the food web. Likewise, nanoparticles from titanium dioxide (the base of white pigments found in many waste streams) have been shown to react with the ultraviolet component of sunlight and to kill phytoplankton;[124]

(c) Although much is being done to reduce pollution from ships, there is scope for more attention to the routes that ships choose and the effects of those routes in terms of noise, chronic oil pollution and operational discharges;[125]

(d) The cumulative effects of excessive nutrient inputs from sewage and agriculture and the removal of herbivorous fish by overfishing can lead to excessive algal growth on coral reefs. Where coral reefs are a tourist attraction, such damage can undermine the tourist business;[126]

(e) The ocean is acidifying rapidly and at an unprecedented rate in the Earth's history. The impact of ocean acidification on marine species and food webs will affect major economic interests and could increasingly put food security at risk, particularly in regions especially dependent on seafood protein.[127]

Better integrated management of human activities affecting the ocean can, in many cases, be achieved with existing knowledge. However, application of that knowledge in many countries requires improvements

---

122 See chap. 8.

123 See chap. 20.

124 See chaps. 6 and 25.

125 See chap. 17.

126 See chaps. 27 and 43.

127 See chaps. 4, 5, 10 and 52.

in the skills of those involved. The last section of the present summary deals with the gaps that have been identified in capacity-building. Furthermore, in many cases, better information is required. Significant knowledge gaps that would need to be filled in order to achieve more general improved and integrated management of human activities affecting the ocean are set out in the penultimate section of the summary.

5.10 Urgency of addressing threats to the ocean

The greatest threat to the ocean comes from a failure to deal quickly with the manifold problems that have been described above. Many parts of the ocean have been seriously degraded. If the problems are not addressed, there is a major risk that they will combine to produce a destructive cycle of degradation in which the ocean can no longer provide many of the benefits that humans currently enjoy from it.

In particular, the cumulative impact of many of the problems described in the present assessment must be considered. As always, addressing one aspect of a challenge without considering the other factors involved risks undermining what can be achieved. This means that addressing some challenges may require also addressing the problems of fragmented data collection, which makes it difficult to obtain a clear picture of the overall problem, and uncoordinated action in different fields (in either geographic or thematic terms).

On the other hand, the assessment contains many examples of efforts made to address individual problems that have resulted in improved ecosystems, economic benefits and improved livelihoods, even though other pressures could not be addressed at the same time. Feasible sectoral improvements do not need to be delayed until the benefits of integrated planning and management can be achieved. They can even facilitate action to address other pressures, either by demonstrating the gains from investing in improved management, or through bringing into clearer focus the costs imposed by other pressures.[128]

Some of the specific threats (such as the intensification of typhoons and hurricanes and changes in the stratification of seawater) are inextricably bound with the problems of climate change and acidification and can only be addressed as part of those issues.

However, many other threats derive from problems that are more local and constitute global problems simply because the same type of problem and threat occurs in many places. For most of those problems, techniques have been developed that can successfully address them. Implementing them successfully is then a question of building the capacities in infrastructure resources, organizational arrangements and technical skills.

Problems of that kind that can be addressed include:

(a) Reducing inputs of hazardous substances, waterborne pathogens and nutrients;[129]
(b) Preventing maritime disasters due to the collision, foundering and sinking of ships, and implementing and enforcing international agreements on preventing adverse environmental impacts from ships;[130]
(c) Improving fishery management;[131]
(d) Managing aquaculture;[132]
(e) Controlling tourism developments that will have adverse impacts on the future of the tourism industry in the locality where they occur;[133]
(f) Controlling solid waste disposal that can reach and affect the marine environment;[134]
(g) Improving the control of offshore hydrocarbon industries and offshore mining;[135]
(h) Establishing and maintaining marine protected areas.[136]

6 Knowledge gaps

Humans have been exploring the three tenths of the planet that is land for millennia. Serious scientific examination of the land and its plants and animals has been in progress for at least 500 years. Although humans have been using the ocean for millennia, it is only in the past 120 years or so that serious exploration of the seven tenths of the planet covered by the sea (other than charting coasts) has been in progress. It is therefore not surprising that our knowledge of the ocean is much more limited than our knowledge of the land. As the chapters of the present assessment demonstrate, much is known about much of the ocean, but nowhere do we have the detailed knowledge desirable for the effective future management of human use of the ocean. In some parts of the world, we do not even have sufficient knowledge to apply properly the techniques that have been successfully developed elsewhere. We have a basic framework of understanding, but there are many gaps to be filled in.

The information that we need to understand the ocean can be divided into four main categories: (a) the physical structure of the ocean; (b) the composition and movement of the ocean's waters; (c) the biotas of the ocean; and (d) the ways in which humans interact with the ocean. The

---

128 See, for example, chap. 36A.

129 See chap. 20.

130 See chap. 17.

131 See chap. 11.

132 See chap. 12.

133 See chap. 27.

134 See chaps. 24 and 25.

135 See chaps. 21 and 23.

136 See chap. 44.

identification of the gaps in that knowledge is best based on a survey of the gaps revealed in the chapters of the assessment. In general, we know least about the Arctic Ocean and the Indian Ocean. The parts of the Atlantic Ocean and the Pacific Ocean in the northern hemisphere are better studied than those in the southern hemisphere and, again in general, the North Atlantic and its adjacent seas are probably the most thoroughly studied — and even there major gaps remain.[137]

Physical structure of the ocean

Chapter 1 (Planet, ocean and life) of the assessment includes a map characterizing the geomorphic features of the ocean. The detail summarized in that map has been greatly enriched over the past quarter century by local and global studies. Although charting the oceans has been in progress for more than seven centuries in coastal waters and for 250 years along the main routes across the open ocean, many features still require more detailed examination. The designation of exclusive economic zones (EEZs) has led many countries to carry out more detailed surveys as a basis for managing their activities in those zones. Ideally, all coastal States would have such detailed surveys as a basis for their EEZ management.

Because of the significance of ocean acidification for carbonate formation, better information on the formation and fate of reef islands and shell beaches is desirable. It is possible to characterize the physical structure of the ocean in areas beyond national jurisdictions, but the reliability and detail of such characterizations varies considerably among different parts of the ocean: improvements in information of that kind are highly desirable to understand the interaction between the physical structure and the biotas, both in terms of conserving biodiversity and in terms of managing living marine resources.[138]

Waters of the ocean

Gaps persist in understanding sea temperature (both at the surface and at depth), sea-level rise, salinity distribution, carbon dioxide absorption, and nutrient distribution and cycling. The atmosphere and the ocean form a single linked system. Much of the information needed to understand the ocean is therefore also needed to understand climate change. Research promoted by the Intergovernmental Panel on Climate Change will look at many of those questions. It will thus be important to ensure that oceanic and atmospheric research is coordinated.

Ocean acidification is a consequence of carbon dioxide absorption, but understanding the implications for the ocean requires more than just a general understanding of how carbon dioxide is being absorbed, as the degree of acidification varies locally. The causes and implications of those variations are important for understanding the impact on the marine biotas.

In order to track primary production (on which the overwhelming majority of the ocean food web relies), routine and sustained measurements are highly desirable across all parts of the ocean of chlorophyll a (as an important marker of primary production), dissolved nitrogen and biologically active dissolved phosphorus (as the latter two are frequent limiting factors of primary production or causes of algal blooms).[139]

Biotas of the ocean

The Census of Marine Life has been an essential tool for ocean research in clarifying the biodiversity of the ocean and the number and distribution of species. Like all censuses, its value will decrease as time passes until it becomes a snapshot of a particular point in time, and less of an up-to-date picture of what is currently happening. It will be important for the Census to be regularly updated and improved. Improvement is particularly desirable for areas around and between Africa and Central and South America, across the Indian Ocean and in the South Pacific.[140]

Plankton are fundamental to life in the ocean. Information on their diversity and abundance is important for many purposes. Such information has been collected for over 70 years in some parts of the ocean (such as the North Atlantic) through continuous plankton recorder surveys. Nine organizations currently collaborate in extending such surveys, but the desirable comprehensive global coverage has not yet been achieved.

As well as information on biodiversity in the ocean and the number and distribution of the many marine species, information is also highly desirable on the health and reproductive success of separate populations. Many species contain separate populations that have limited interconnections. It is therefore important to understand how the local influences specific to each population are affecting them. As the regional surveys in part VI show, much is already known about the population health and reproductive success of many species, but there are also large gaps in knowledge, particularly in the southern hemisphere.[141]

Fish stock assessments are essential to the proper management of fisheries. A good proportion of the fish stocks fished in large-scale fisheries are the object of regular stock assessments. However, many important fish stocks of that kind are still not regularly assessed. More significantly, stocks important for small-scale fisheries are often not assessed, which has adverse effects in ensuring the continued availability of fish for such fisheries. This is an important knowledge gap to fill. Likewise, there are gaps in information about the interactions between large-scale and small-scale fisheries for stocks over which their interests overlap, and between recreational fishing and other fisheries for some species, such as some trophy fish (marlins, sailfish and others) and other smaller species.[142]

The present assessment sets out the main specific issues for which there are gaps in our knowledge of marine biotas, in particular of all the spe-

---

137 See chap. 30.

138 See chap. 9.

139 See chap. 9.

140 See chap. 35.

141 See chaps. 36A-H.

142 See chaps. 11 and 27.

cies and habitats that have been scientifically identified as threatened, declining or otherwise in need of special attention or protection. Those species include, with some indications of important issues identified in part VI: marine mammals, sea turtles, seabirds (particularly migration routes), sharks and other elasmobranchs (especially the lesser-known species and certain tropical areas), tuna and billfish (particularly the non-principally marketed species), cold-water corals (especially where they are found in the Indian Ocean), warm-water corals (particularly at locations in deeper water), estuaries and deltas (particularly integrated assessments of them), high-latitude ice, hydrothermal vents (especially the extent to which they are found in the Indian Ocean), kelp forests and seagrass beds (especially the degree of loss of kelp and the pathology of the diseases affecting them), mangroves (especially the taxonomy of associated species and their interactions with salt marshes), salt marshes (especially the ecosystem services that they provide) and the Sargasso Sea (especially the links with distant ecosystems).[143]

Ways in which humans interact with the ocean

Some of the issues relating to the ocean and to the ocean biotas (for example, ocean acidification and fish stock assessments) are linked to the way in which humans affect some aspects of the ocean (for example, through carbon-dioxide emissions or fisheries). However, there are many more areas in which we do not yet know enough about human activities that affect or interact with the ocean to enable us to manage those activities sustainably.

For shipping, much information is available about where ships go, their cargo and the economics of their operations. However, important gaps remain in our knowledge about how their routes and operations affect the marine environment. Those issues include primarily the noise that they make, chronic discharges of oil and the extent to which non-native invasive species are being transported. Other information gaps relate to the social aspects of shipping: in particular, little is known about the levels of death and injury of seafarers, an issue recently raised by the Secretary-General of the International Maritime Organization.[144]

Land-based inputs to the ocean have serious implications for both human health and the proper functioning of marine ecosystems. In some parts of the world, those have been studied carefully for over 40 years. In others, little systematic information is found. There are two important gaps in current knowledge. The first is how to link different ways of measuring discharges and emissions. Much information is available from local studies about inputs, but those are frequently measured and analysed in different ways, thereby making comparison difficult or impossible. There are sometimes good reasons for using different techniques, but ways of improving the ability to achieve standardized results and to make comparisons are essential to give a full global view. Secondly, different regions of the world have developed different systems for assessing the overall quality of their local waters. Again, good reasons for such differences almost certainly exist, but knowledge of how to compare the different results would be helpful, particularly in assessing priorities among different areas.[145]

Another area where there are important gaps in knowledge is the extent to which people are suffering from diseases that are either the direct result of inputs of waterborne pathogens or toxic substances, or the indirect result of toxins from algal blooms generated by excessive levels of nutrients. As well as gaps in information on the effects of such health hazards, there are also large gaps in knowledge of their economic effects.

The offshore hydrocarbon industries in some parts of the world collect and publish wide-ranging information on how their activities are affecting the local marine environment. In other parts of the world, little or no such information is found. Because the processes are very similar in most areas, filling the gaps in knowledge in what is happening around the world would be helpful.

The existing offshore mining industries are very diverse and, consequently, their impacts on the marine environment do not have much in common. Where they occur in the coastal zone, it is important that those responsible for integrated coastal zone management have good information on what is happening, particularly in relation to discharges of tailings and other disturbances of the marine environment. As offshore mining expands into deeper waters and areas beyond national jurisdiction, it will be important to ensure that information about their impacts on the marine environment is collected and published.[146]

Information on the disposal of solid waste at sea (dumping) is very patchy. Where reports under the London Convention and the Protocol thereto are not submitted, it is not clear whether dumping does not occur or occurs but is not reported. This represents an important gap in knowledge. The absence of information on dumping, if any, in other jurisdictions also impedes the understanding of the impact on the marine environment of that form of waste disposal.[147]

Our knowledge of marine debris has many gaps. Unless we understand better the sources, fates, and impacts of marine debris, we shall not be able to tackle the problems that it raises. Although the monitoring of marine debris is currently carried out in several countries around the world, the protocols used tend to be very different, preventing comparisons and the harmonization of data. Because marine debris is so mobile, the result is a significant gap in knowledge. There is also a gap in information for evaluating the impacts of marine debris on coastal and marine species, habitats, economic well-being, human health and safety, and social values. Because of their ability to enter into marine food chains, with a potential impact on human health, more information on the origin, fate and effects of plastic microparticles and nanoparticles

---

143 See chaps. 42 to 51.

144 See chap. 17.

145 See chap. 20.

146 See chap. 23.

147 See chap. 24.

is highly desirable. Likewise, because of their potential effects on phytoplankton, there is a gap in knowledge about titanium dioxide nanoparticles.[148]

Many aspects of integrated coastal zone management still present important knowledge gaps. Those responsible for managing coastal areas need information on, at least, coastal erosion, land reclamation from the sea, changes in sedimentation as a result of coastal works and changes in river regimes (such as damming rivers or increased water abstraction), the ways in which the local ports are working and dredging is taking place and the ways in which tourist activity is developing (and is planned to develop), and the impacts that those developments and plans are likely to have on the local marine ecosystem (and, for that matter, the local terrestrial ecosystems). It will help the development and effectiveness of integrated coastal management if recognized standards are set and followed for all such information, so that systematic best practices can be developed.[149]

The aesthetic, cultural, religious and spiritual ways in which humans relate to the ocean are also linked to some gaps in our knowledge. Over the centuries, many cultures have built up broad traditional knowledge of the ocean. Such knowledge is often under pressure and will be lost if it is not recorded. For example, Polynesian traditional navigational knowledge was disappearing fast and has been recorded only just in time. Cultural practices (such as traditional Chinese and Iranian boat-building) are also disappearing and risk being lost for future generations.[150]

Our knowledge of human interaction with the ocean is also very partial in terms of the ways in which we benefit from it. As has been noted above, it is not yet possible to place a value on the non-marketed ecosystem services derived from the ocean. There are many gaps in the information needed for such an exercise. Information on the effects of changes in the ways in which the planetary ecosystem works needs to be collected and evaluated, in order to permit an economic valuation of the choices for action that may have repercussions on non-marketed ecosystem services. The areas where such information seems particularly closely related to management decisions are integrated coastal zone management (including marine spatial management), offshore hydrocarbon exploitation, offshore mining, shipping routes, port development and waste disposal.[151]

Even with market-related ecosystem services and human activities, there are major information gaps. Such gaps include consistent definitions of what the ecosystem services and human activities cover, how to estimate the value of services and activities that are on the margins of the markets and, even more, the capture of the related data. Gaining a good understanding of the true overall economic situation of such activities as fishing, shipping and tourism would help to improve decision-making in those fields.[152]

Closing those gaps in our knowledge would amount to an ambitious programme of research. Research is already taking place on many more issues on which more information is desirable (for example, on how the genetic resources of the ocean can be used and what the practical possibilities are for seabed mining). Collaboration and sharing will be important for making the best uses of scarce research resources.[153]

7 Capacity-building gaps

The knowledge gaps identified in the present assessment all point to gaps in the capacities needed to fill them and to apply the resulting knowledge. On the basis of the information currently available, it is impossible to say what gaps currently exist in arrangements to build such capacities. Conclusions on where the capacity-building gaps exist could only be reached by conducting a survey, country by country, of the capacity-building arrangements that currently exist and of how suitable they are for each country's needs. The preliminary inventory of capacity-building for assessments[154] compiled by the Division for Ocean Affairs and the Law of the Sea as part of the Regular Process provides some initial information on which to base such a survey, but it would take a much more detailed study than has been possible in the first cycle of the Regular Process to match that information with the needs of each country. The present section therefore looks at the capacities that are desirable, rather than the gaps in capacities for building them.

The outline for the first global integrated marine assessment requires that capacities be identified to assess the status of the marine environment and to benefit from the various human activities that take place in the marine environment.

Certain capacities are desirable for multiple purposes. The most obvious of that kind of capacity is marine research vessels. Such vessels can provide multipurpose platforms capable of supporting geological and biota surveys, habitat mapping and similar tasks. The present assessment reviews the current distribution of research vessels around the world. Such vessels may be run by Governments, government institutes, universities, independent research institutes or commercial enterprises. Shared use, for example at a regional level, may be feasible.[155]

Turning from those points to the elements identified as knowledge gaps, the following are the main desirable capacity-building activities.

---

148 See chaps. 6 and 25.

149 See chaps. 4, 18 and 27.

150 See chap. 8.

151 See chap. 55.

152 See chaps. 3, 9 and 55.

153 See chap. 30.

154 See A/66/189, annex V, and A/67/87, annex V.

155 See chap. 30.

Physical structure of the ocean

Surveys of the physical structure of the ocean require both sea-going survey capacities and the laboratory and technical staff capabilities to analyse and interpret the resulting data. Both are essential to fill knowledge gaps about the physical structure of the ocean within and beyond national jurisdictions.

Waters of the ocean

Understanding the water column requires capacities to sample, analyse and interpret the ocean in terms of temperature, salinity, stratification, chemical composition and acidity. Much of that can be gathered by autonomous floating devices, such as the floats used by the Array for Real-time Geostrophic Oceanography, which are described in the present assessment.

Understanding primary production and the implications of sea-level rise requires information on sea levels and chlorophyll a. Such information is most effectively gathered from satellite sensors. Much of it is already available through the Internet, but the equipment and skills needed to access and interpret it are needed to be able to investigate local situations.

Ocean biotas

Better understanding of the ocean biotas demands capacities to organize the regular collection of sampling data on their number, distribution, health and reproductive success, to compile such data into databases (at the national or regional level), to analyse and interpret the data (for instance, taxonomic expertise is required to identify species) and to carry out assessments based on that information. Capacity to carry out marine scientific research is also highly desirable to improve the scientific understanding on which such monitoring is based.

The capacity to manage fisheries effectively requires ships, equipment and skills to monitor and assess fish stocks. Based on those assessments, capacities are then required to develop, apply and enforce appropriate fishery management policies. Such capacities are likely to include fishery protection vessels to monitor what is happening at sea, access to satellite data to monitor the movements of fishing vessels through transponders, institutional structures to regulate markets in fish and other seafood (including their freedom from contaminants and pathogens) and the necessary enforcement mechanisms at all stages from ocean to table.

Ways in which humans interact with the ocean

Many human activities affecting the oceans are carried out by commercial enterprises. Those can be expected to develop the capacities to generate the knowledge and infrastructure that they need to run their businesses and to comply with relevant regulations. For public authorities, however, capacities will be needed to ensure that they can create appropriate regulations to safeguard social and environmental interests and that they can deal effectively with such commercial enterprises (many of which are international companies). This may be particularly difficult when the public authority concerned is relatively local.

In developing ecosystem-based approaches to the management of human activities affecting the ocean (in parallel to those being developed for fisheries), capacities are necessary to gather and process information relating to the activity and to all the facets of the ocean ecosystems with which the activity in question interacts. The precise information required will vary from activity to activity. Examples of capacities likely to be needed in some specific human activities are those required in order to:

(a) Identify when ship-routing measures are needed to protect the marine environment, specify them and implement such measures;

(b) Plan and implement emergency response plans for maritime disasters. Such plans are likely to require significant capital investment in ships, aircraft, machinery and supplies;

(c) Develop and manage ports capable of handling international maritime traffic. Currently, many such port developments are being carried out and managed by commercial enterprises, in which case the proper regulation of those undertakings will be required;

(d) Ensure adequate port waste-reception facilities to enable ships to discharge their waste without being delayed;

(e) Carry out port State inspections of vessels and follow up any shortcomings detected;

(f) Sample, analyse and interpret land-based inputs to the ocean. Those capabilities need to be able to cover liquid and semi-liquid discharges by pipelines directly into the sea, discharges of liquids and suspended solids to rivers and the water quality of rivers at their mouths, and emissions into the air that may reach and affect the sea. In the case of emissions into the air, it is also desirable to be able to distinguish anthropogenic inputs from natural emissions;

(g) Ensure that new, cleaner technologies are applied to chemical and other production processes, so as to reduce the discharges and emissions of heavy metals and other hazardous substances;

(h) Manage solid waste placed in landfills, so as to prevent the leaching of heavy metals or other hazardous substances that can reach and affect the sea, and manage the incineration of waste to minimize emissions of heavy metals and other hazardous substances in the exhaust gases;

(i) Provide the necessary infrastructure and equipment for the proper handling of land-based industrial discharges, emissions and sewage, so as to minimize the content of heavy metals and other hazardous substances, to remove waterborne pathogens where they could pollute bathing waters and contaminate seafood and to prevent excessive nutrient discharges;

(j) Promote the proper handling of agricultural waste and slurry and the proper use of agricultural fertilizers and pesticides;

(k) Deliver the organization, equipment and skills to monitor and control other human activities that impact on the marine environment;

(l) Manage the coastal zone in an integrated way. Where tourism is significant, those capacities need to include the ability to monitor and regulate tourist developments and activities, so as to keep them within acceptable limits in relation to the carrying capacities of the local ecosystems.

A general gap exists in capacities for an integrated assessment of the marine environment. An integrated assessment needs to bring together: (a) environmental, social and economic aspects; (b) all the relevant sectors of human activities; and (c) all the components (fixed and living) of the relevant ecosystems. The idea of an integrated assessment in that sense is relatively recent. It presents a challenging requirement, which requires specialists in many different fields to work together.

In building capacities for integrated assessments, it is necessary to think further about the concept of an integrated marine assessment. The present assessment is the first global integrated assessment of the marine environment. The Group of Experts who are collectively responsible for it are convinced that the further development and refinement of techniques for making integrated assessments are needed.

The context of the assessment

1 Introduction – Planet, oceans and life

Contributors:
Peter Harris (Lead member and Convenor), Joshua Tuhumwire (Co-Lead member).

Introduction: Planet, oceans and life

1 Why the ocean matters

Consider how dependent upon the ocean we are. The ocean is vast – it covers seven-tenths of the planet. On average, it is about 4,000 metres deep. It contains 1.3 billion cubic kilometres of water (97 per cent of all water on Earth). But there are now about seven billion people on Earth. So we each have just one-fifth of a cubic kilometre of ocean to provide us with all the services that we get from the ocean. That small, one-fifth of a cubic kilometre share produces half of the oxygen each of us breathes, all of the sea fish and other seafood that each of us eats. It is the ultimate source of all the freshwater that each of us will drink in our lifetimes. The ocean is a highway for ships that carry across the globe the exports and imports that we produce and consume. It contains the oil and gas deposits and minerals on and beneath the seafloor that we increasingly need to use. The submarine cables across the ocean floor carry 90 per cent of the electronic traffic on which our communications rely. Our energy supply will increasingly rely on wind, wave and tide power from the ocean. Large numbers of us take our holidays by the sea. That one-fifth of a cubic kilometre will also suffer from the share of the sewage, garbage, spilled oil and industrial waste which we produce and which is put into the ocean every day. Demands on the ocean continue to rise: by the year 2050 it is estimated that there will be 10 billion people on Earth. So our share (or our children's share) of the ocean will have shrunk to one-eighth of a cubic kilometre. That reduced share will still have to provide each of us with sufficient amounts of oxygen, food and water, while still receiving the pollution and waste for which we are all responsible.

The ocean is also home to a rich diversity of plants and animals of all sizes – from the largest animals on the planet (the blue whales) to plankton that can only be seen with powerful microscopes. We use some of these directly, and many more contribute indirectly to our benefits from the ocean. Even those which have no connection whatever with us humans are part of the biodiversity whose value we have belatedly recognized. However, the relationships are reciprocal. We intentionally exploit many components of this biodiverse richness. Carelessly (for example, through inputs of waste) or unknowingly (for example, through ocean acidification from increased emissions of carbon dioxide), we are altering the circumstances in which these plants and animals live. All this is affecting their ability to thrive and, sometimes, even to survive. These impacts of humanity on the oceans are part of our legacy and our future. They will shape the future of the ocean and its biodiversity as an integral physical-biological system, and the ability of the ocean to provide the services which we use now and will increasingly need to use in the future. The ocean is vital to each of us and to human well-being overall.

Looking in more detail at the services that the ocean provides, we can break them down into three main categories. First, there are the economic activities in providing goods and services which are often marketed (fisheries, shipping, communications, tourism and recreation, and so on). Secondly, there are the other tangible ecosystem services which are not part of a market, but which are vital to human life. For example, marine plants (mainly tiny floating diatoms) produce about 50 per cent of atmospheric oxygen. Mangroves, salt marshes and sea grasses are also natural carbon sinks. Coastal habitats, including coral reefs, protect homes, communities and businesses from storm surges and wave attack. Thirdly, there are the intangible ecosystem services. We know that the ocean means far more to us than just merely the functional or practical services that it provides. Humans value the ocean in many other ways: for aesthetic, cultural or religious reasons, and for just being there in all its diversity – giving us a "sense of place" (Halpern et al., 2012). Not surprisingly, given the resources that the ocean provides, human settlements have grown up very much near the shore: 38 per cent of the world's population live within 100 km of the shore, 44 per cent within 150 km, 50 per cent within 200 km, and 67 per cent within 400 km (Small et al 2004).

All these marine ecosystem services have substantial economic value. While there is much debate about valuation methods (and whether some ecosystem services can be valued) and about exact figures, attempts to estimate the value of marine ecosystem services have found such values to be on the order of trillions of US dollars annually (Costanza, et al., 1997). Nearly three-quarters of this value resides in coastal zones (Martínez, et al., 2007). The point is not so much the monetary figure that can be estimated for non-marketed ecosystem services, but rather the fact that people do not need to pay anything for them – these services are nature's gift to humanity. But we take these services for granted at our peril, because the cost of replacing them, if it were possible to do so, would be immense and in many cases, incalculable.

There are therefore very many good reasons why we each need to take very good care of our-fifth of a cubic kilometre share of the ocean!

2 Structure of this Assessment

It is this significance of the ocean as a whole, and the relatively fragmented way in which it is studied and in which human activities impacting upon it are managed, that led in 2002 the World Summit on Sustainable Development to recommend (WSSD 2002), and the United Nations General Assembly to agree (UNGA 2002), that there should be a regular process for the global reporting and assessment of the marine environment, including socioeconomic aspects. Under the arrangements developed for this purpose, this Assessment is the first global integrated assessment of the marine environment (see further in Chapter 2).

Three possible focuses exist for structuring this Assessment: the ecosystem services (market and non-marketed, tangible and intangible) that the marine environment provides; the habitats that exist within the marine environment, and the pressures that human activities exert on the marine environment. All three have advantages and disadvantages.

Using ecosystem services as the basis for structuring the Assessment would follow the approach of the Millennium Ecosystem Assessment

(2005). This has the advantage of broad acceptance in environmental reporting. It would cover provisioning services (food, construction materials, renewable energy, coastal protection), while highlighting regulating services and quality-of-life services that are not captured using a pressures or habitats approach to structuring the Assessment. It would have the disadvantage that some important human activities using the ocean (for example, shipping, ports and minerals extraction) would be covered only incidentally.

Using marine habitats as the basis for structuring the Assessment would have the advantage that habitats are the property that inherently integrates many ecosystem features, including species at higher and lower trophic levels, water quality, oceanographic conditions and many types of anthropogenic pressures (AoA, 2009). The cumulative aspect of multiple pressures affecting the same habitat, that is often lost in sector-based environmental reporting (Halpern et al., 2008), is captured by using habitats as reporting units. It would have the disadvantage that consideration of human activities would be fragmented between the many different types of habitats.

Using pressures as the basis for structuring the Assessment would have the advantage that the associated human activities are commonly linked with data collection and reporting structures for regulatory compliance purposes. For instance, permits that are issued for offshore oil and gas development require specific monitoring and reporting obligations to be met by operators. It would have the disadvantage that many important ecosystem services would only be covered in relation to the impacts of the human activities.

Given that all three approaches have their own particular advantages and disadvantages, the United Nations General Assembly endorsed a structure for this Assessment that combined all three approaches, thereby structuring the World Ocean Assessment into seven main Parts, as follows.

Part I.   Summary
The Summary is intended to bring out the way in which the assessment has been carried out, the overall assessment of the scale of human impact on the oceans and the overall value of the oceans to humans, and the main threats to the marine environment and human economic and social well-being. As guides for future action it also describes the gaps in capacity-building and in knowledge.

Part II. The context of the Assessment
This chapter is intended as a broad, introductory survey of the role played by the ocean in the life of the planet, the way in which they function, and humans' relationships to them. Chapter 2 explains in more detail the rationale for the Assessment and how it has been produced.

Part III. Assessment of major ecosystem services from the marine environment (other than provisioning services)
Part III looks at the non-marketed ecosystem services provided to the planet by the ocean. It considers, first, the scientific understanding of such ecosystem services and then looks at the earth's hydrological cycle, air/sea interactions, primary production and ocean-based carbonate production. Finally it looks at aesthetic, cultural, religious and spiritual ecosystem services (including some cultural objects which are traded). Where relevant, it draws heavily on the work of Intergovernmental Panel on Climate Change (IPCC) – the aim is to use the work of the IPCC, not to duplicate or challenge it.

Part IV. Assessment of the cross-cutting issues: food security and food safety
The aim of Part IV is to look at all aspects of the vital function of the ocean in providing food for humans. It draws substantially on information collected by the Food and Agriculture Organization of the United Nations (FAO). The economic significance of employment in fisheries and aquaculture and the relationship these industries have with coastal communities are addressed, including gaps in capacity-building for developing countries.

Part V. Assessment of other human activities and the marine environment
All human activities that can impact on the oceans (other than those relating to food) are covered in Part V of the assessment. Each chapter describes the location and scale of activity, the economic benefits, employment and social role, environmental consequences, links to other activities and capacity-building gaps.

Part VI. Assessment of marine biological diversity and habitats
The aim of Part VI is: (a) to give an overview of marine biological diversity and what is known about it; (b) to review the status and trends of, and threats to, marine ecosystems, species and habitats that have been scientifically identified as threatened, declining or otherwise in need of special attention or protection; (c) to review the significant environmental, economic and/or social aspects in relation to the conservation of marine species and habitats; and (d) to find gaps in capacity to identify marine species and habitats that are viewed as threatened, declining or otherwise in need of special attention or protection and to assess the environmental, social and economic aspects of the conservation of marine species and habitats.

Part VII. Overall assessment
Part VII finally looks at the overall impact of humans on the ocean, and the overall benefit of the ocean for humans.

3   The physical structure of the ocean

Looking at a globe of the earth one thing that can be easily seen is that, although different names appear in different places for different ocean areas, these areas are all linked together: there is really only one world ocean. The seafloor beneath the ocean has long remained a mystery, but in recent decades our understanding of the ocean floor has improved. The publication of the first comprehensive, global map of seafloor physi-

Introduction: Planet, oceans and life　　Chapter 1

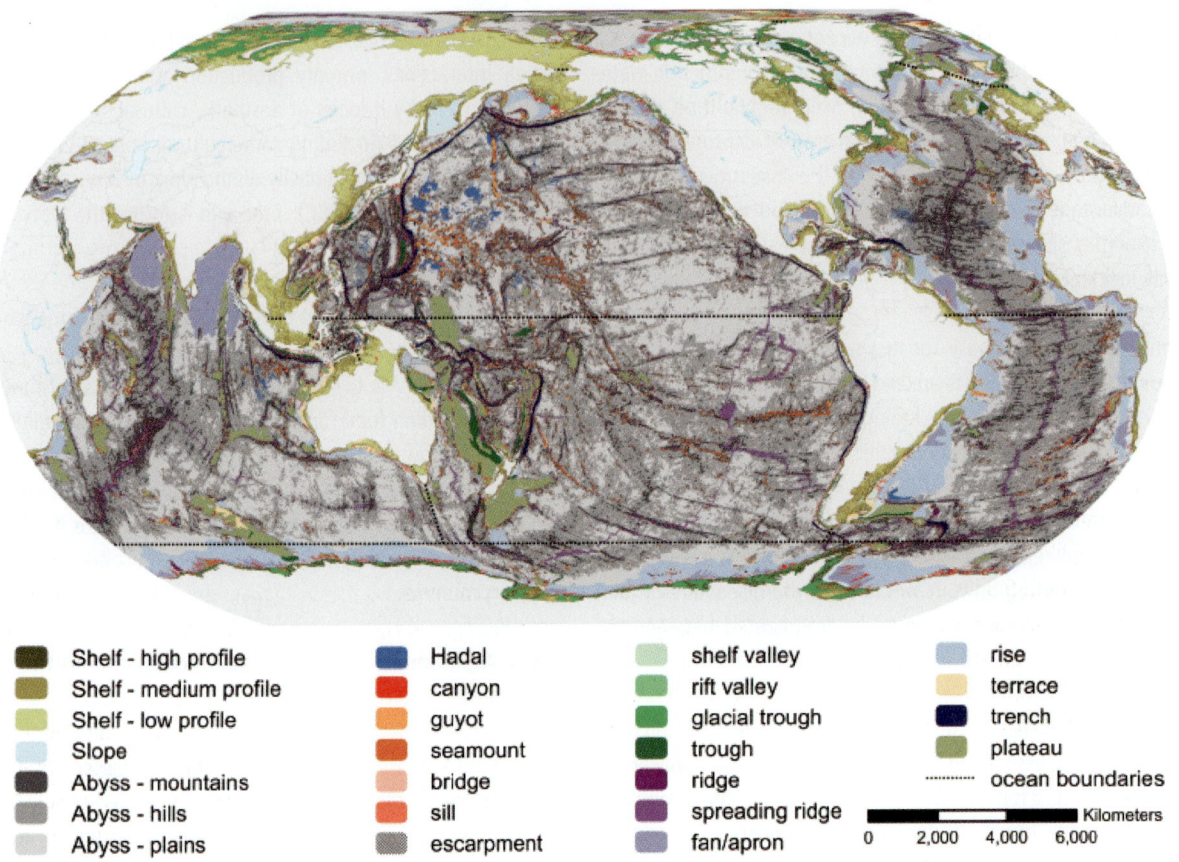

The boundaries and names shown and the designations used on this map do not imply official endorsement or acceptance by the United Nations.
Figure 1.1 | Geomorphic features map of the world's oceans (after Harris et al., 2014). Dotted black lines mark boundaries between major ocean regions. Basins are not shown.

ography by Bruce Heezen and Marie Tharp in 1977 provided a pseudo-three-dimensional image of the ocean that has influenced a long line of scholars. That image has been refined in recent years by new bathymetric maps (Smith and Sandwell, 1997) which are used to illustrate globes, web sites and the maps on many in-flight TV screens when flying over the ocean.

A new digital, global seafloor geomorphic features map has been built (especially to assist the World Ocean Assessment) using a combination of manual and ArcGIS methods based on the analysis and interpretation of the latest global bathymetry grid (Harris et al., 2014; Figure 1.1). The new map includes global spatial data layers for 29 categories of geomorphic features, defined by the International Hydrographic Organization and other authoritative sources.

The new map shows the way in which the ocean consists of four main basins (the Arctic Ocean, the Atlantic Ocean, the Indian Ocean and the Pacific Ocean) between the tectonic plates that form the continents. The tectonic plates have differing forms at their edges, giving broad or narrow continental shelves and varying profiles of the continental rises and continental slopes leading from the abyssal plain to the continental shelf. Geomorphic activity in the abyssal plains between the continents gives rise to abyssal ridges, volcanic islands, seamounts, guyots (plateau-like seamounts), rift valley segments and trenches. Erosion and sedimentation (either submarine or riverine when the sea level was lower during the ice ages) has created submarine canyons, glacial troughs, sills,

fans and escarpments. Around the ocean basins there are marginal seas, partially separated by islands, archipelagos or peninsulas, or bounded by submarine ridges. These marginal seas have sometimes been formed in many ways: for example, some result from the interaction between tectonic plates (for example the Mediterranean), others from the sinking of former dry land as a result of isostatic changes from the removal of the weight of the ice cover in the ice ages (for example, the North Sea).

The water of the ocean circulates within these geological structures. This water is not uniform: there are very important physical and chemical variations within the sea water. Salinity varies according to the relativity between inputs of freshwater and evaporation. Sea areas such as the Baltic Sea and the Black Sea, with large amounts of freshwater coming from rivers and relatively low evaporation have low salinity – 8 parts per thousand and 16 parts per thousand, respectively, as compared with the global average of 35 parts per thousand (HELCOM 2010, Black Sea Commission 2008). The Red Sea, in contrast, with low riverine input and high insolation, and therefore high evaporation, has a mean surface salinity as high as 42.5 parts per thousand (Heilman et al 2009). Seawater can also be stratified into separate layers, with different salinities and different temperatures. Such stratification can lead to variations in both the oxygen content and nutrient content, with critical consequences in both cases for the biota dependent on them. A further variation is in the penetration of light. Sunlight is essential for photosynthesis of inorganic carbon (mainly CO_2) into the organic carbon of plants and mixotrophic species[1]. Even clear water reduces the level of light that can penetrate by about 90 per cent for every 75 metres of depth. Below 200 metres depth, there is not enough light for photosynthesis (Widder 2014). The upper 200 metres of the ocean are therefore where most photosynthesis takes place (the euphotic zone). Variations in light level in the water column and on the sea bed are caused by seasonal fluctuation in sunlight, cloud cover, tidal variations in water depth and (most significantly, where it occurs) turbidity in the water, caused, for example, by resuspension of sediment by tides or storms or by coastal erosion. Where turbidity occurs, it can reduce the penetration of light by up to 95 per cent, and thus reduce the level of photosynthesis which can take place (Anthony 2004).

The new map provides the basis for global estimates of physiographic statistics (area, number, mean size, etc.): for example, it can be estimated that the global ocean covers 362 million square kilometres and the ocean floor contains: 9,951 seamounts covering 8.1 million square kilometres; 9,477 submarine canyons covering 4.4 million square kilometres; and the mid-ocean spreading ridges cover 6.7 million square kilometres with an additional 710,000 square kilometres of rift valleys where hydrothermal vent communities occur (Harris et al., 2014).

There is an important distinction to be made between the terminology used in scientific description of the ocean and the legal terminology used to describe States' rights and obligations in the ocean. Some important terms that will be used throughout this Assessment include the "continental shelf", "open ocean" and "deep sea".

Unless stated otherwise, "continental shelf" in this Assessment refers to the geomorphic continental shelf (as shown in Figure 1.1) and not to the continental shelf as defined by the United Nations Convention on the Law of the Sea. The geomorphic continental shelf is usually defined in terms of the submarine extension of a continent or island as far as the point where there is a marked discontinuity in the slope and the continental slope begins its fall down to the continental rise or the abyssal plain (Hobbs 2003). In total, continental shelves cover an area of 32 million square kilometres (out of a total ocean area of 362 million square kilometres).

The term "open ocean" in this Assessment refers to the water column of deep-water areas that are beyond (that is, seawards of) the geomorphic continental shelf. It is the pelagic zone that lies in deep water (generally >200 m water depth).

The term "deep sea" in this Assessment refers to the sea floor of deep-water areas that are beyond (that is, seawards of) the geomorphic continental shelf. It is the benthic zone that lies in deep water (generally >200 m water depth).

4 Seawater and the ocean/climate interaction

The Earth's ocean and atmosphere are parts of a single, interactive system that controls the global climate. The ocean plays a major role in this control, particularly in the dispersal of heat from the equator towards the poles through ocean currents. The heat transfer through the ocean is possible because of the larger heat-capacity of water compared with that of air: there is more heat stored in the upper 3 metres of the global ocean than in the entire atmosphere of the Earth. Put another way, the oceans hold more than 1,000 times more heat than the atmosphere. Heat transported by the major ocean currents dramatically affects regional climate: for example, Europe would be much colder than it is without the warmth brought by the Gulf Stream current. The great ocean boundary currents transport heat from the equator to the polar seas (and cold from the polar seas towards the equator), along the margins of the continents. Examples include: the Kuroshio Current in the northwest Pacific, the Humboldt (Peru) Current in the southeast Pacific, the Benguela Current in the southeast Atlantic and the Agulhas Current in the western Indian Ocean. The mightiest ocean current of all is the Circumpolar Current which flows from west to east encircling the continent of Antarctica and transporting more than 100 Sverdrups (100 million cubic meters per second) of ocean water (Rintoul and Sokolov, 2001). As well as the boundary currents, there are five major gyres of rotating currents: two in the Atlantic and two in the Pacific (in each case one north and one south of the equator) and one in the Indian Ocean.

The winds in the atmosphere are the main drivers of these ocean surface currents. The interface between the ocean and the atmosphere and the effect of the winds also allows for the ocean to absorb oxygen and, more importantly, carbon dioxide from the air. Annually, the ocean absorbs 2,300 gigatonnes of carbon dioxide (IPCC, 2005; see Chapter 5).

In addition to this vast surface ocean current system, there is the ocean thermohaline circulation (ocean conveyor) system (Figure 1.3). Instead of being driven by winds and the temperature difference between the equator and the poles (as are the surface ocean currents), this current system is driven by differences in water density. The most dense ocean water is cold and salty which sinks beneath warm and fresh seawater that stays near the surface. Cold-salty water is produced in sea ice "factories" of the polar seas: when seawater freezes, the salt is rejected (the ice is mostly fresh water), which makes the remaining liquid seawater saltier. This cold saltier water sinks into the deepest ocean basins, bringing oxygen into the deep ocean and thus enabling aerobic life to exist.

Wind-driven mixing affects only the surface of the ocean, mainly the upper 200 metres or so, and rarely deeper than about 1,000 metres. Without the ocean's thermohaline circulation system, the bottom waters of the ocean would soon be depleted of oxygen, and aerobic life there would cease to exist.

Superimposed on all these processes, there is the twice-daily ebb and flow of the tide. This is, of course, most significant in coastal seas. The

---

[1] That is, plankton species that both photosynthesize and consume other biota.

tidal range varies according to local geography: the largest mean tidal ranges (around 11.7 metres) are found in the Bay of Fundy, on the Atlantic coast of Canada, but ranges only slightly less are also found in the Bristol Channel in the United Kingdom, on the northern coast of France, and on the coasts of Alaska, Argentina and Chile (NOAA 2014).

Global warming is likely to affect many aspects of ocean processes. Changes in sea-surface temperature, sea level and other primary impacts will lead, among other things, to increases in the frequency of major tropical storms (cyclones, hurricanes and typhoons) bigger ocean swell waves and reduced polar ice formation. Each of these consequences has its own consequences, and so on (Harley et al., 2006; Occhipinti-Ambrogi, 2007). For example, reduced sea ice production in the polar seas will mean less bottom water is produced (Broecker, 1997) and hence less oxygen delivered to the deep ocean (Shaffer et al., 2009).

5 The ocean and life

The complex system of the atmosphere and ocean currents is also crucial to the distribution of life in the ocean, since it regulates, among other factors, (as said above) temperature, salinity, oxygen content, absorption of carbon dioxide and the penetration of light and (in addition to these) the distribution of nutrients.

The distribution of nutrients throughout the ocean is the result of the interaction of a number of different processes. Nutrients are introduced to the ocean from the land through riverine discharges, through inputs direct from pipelines and through airborne inputs (see Chapter 20). Within the ocean, these external inputs of nutrients suffer various fates and are cycled. Nutrients that are adsorbed onto the surface of particles are likely to fall into sediments, from where they may either be remobilised by water movement or settle permanently. Nutrients that are taken up by plants and mixotrophic biota for photosynthesis will also eventually sink towards the seabed as the plants or biota die; *en route* or when they reach the seabed, they will be broken up by bacteria and the nutrients released. As a result of these processes, the water in lower levels of the ocean is richer in nutrients.

The boundaries and names shown and the designations used on this map do not imply official endorsement or acceptance by the United Nations.
Figure 1.2 | The global ocean "conveyor" thermohaline circulation (Broecker, 1991). Bottom water is formed in the polar seas via sea-ice formation in winter, which rejects cold, salty (dense) water. This sinks to the ocean floor and flows into the Indian and North Pacific Oceans before returning to complete the loop in the North Atlantic. Numbers indicate estimated volumes of bottom water production in "Sverdrups" (1 Sverdrup = 1 million m3/s), which may be reduced by global warming because less sea ice will be formed during winter. Blue indicates cold currents and red indicates warm currents. The black question marks indicate sites long the Antarctic margin where bottom water may be formed but of unknown volumes. The question mark after the "5" indicates that this value is certain.

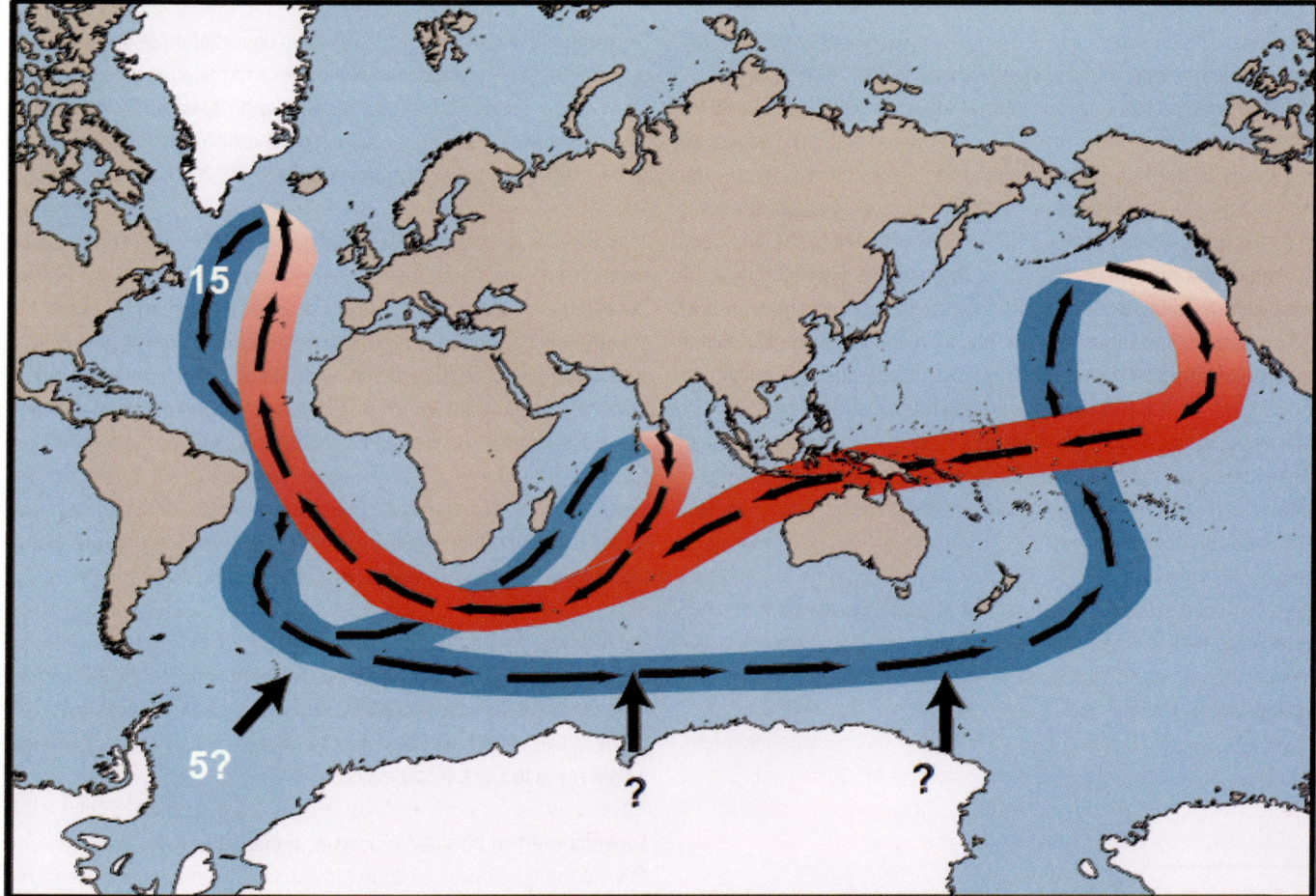

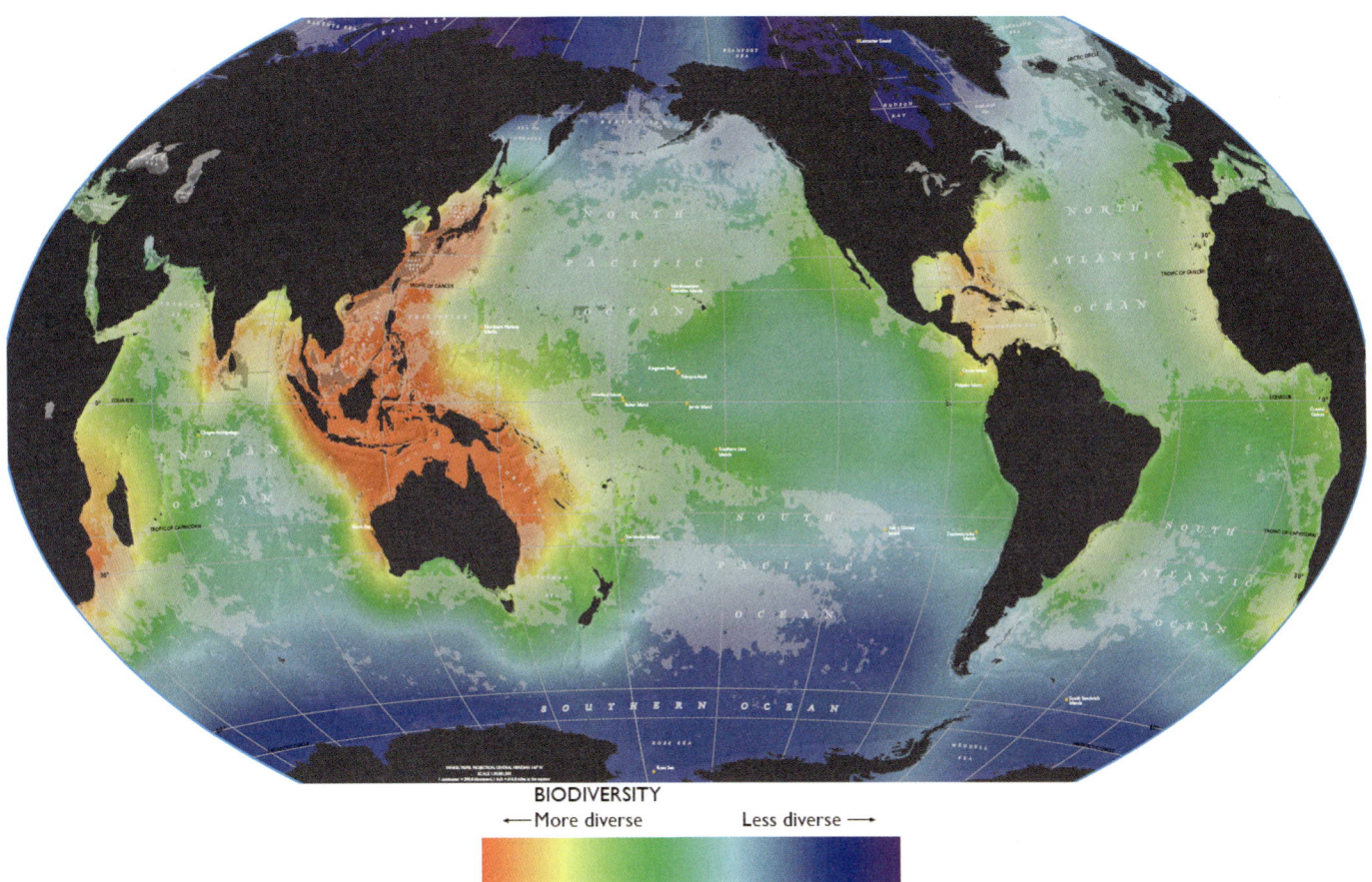

The boundaries and names shown and the designations used on this map do not imply official endorsement or acceptance by the United Nations.
Figure 1.3 | Distribution of biodiversity in the oceans. Biodiversity data: Tittensor et al., 2010. Human impact data: Halpern et al., 2008, Map: Census of Marine Life, 2010; Ausubel et al., 2010; National Geographic Society, 2010).

Upwelling of these nutrient-rich waters is caused by the interaction of currents and wind stress. In simple terms, along coasts (especially west-facing coasts with narrow continental shelves), coastal, longshore wind stress results in rapid upwelling; further out to sea, wind-stress produces a slower, but still significant, upwelling (Rykaczewski et al., 2008). Upwelled, nutrient-rich water is brought up to the euphotic zone (see previous section), where most photosynthesis takes place (see Chapter 6). The reality is far more complex, and upwelling is influenced by numerous other factors such as stratification of the water column and the influence of coastal and seafloor geomorphology, such as shelf-incising submarine canyons (Sobarzo et al., 2001). Other important factors are river plumes and whether the upwelling delivers the nutrient that is the local limiting factor for primary productivity (for example, nitrogen or iron; Kudela et al., 2008). Ocean upwelling zones commonly control primary productivity hotspots and their associated, highly productive fisheries, such as the anchoveta fishery off the coast of Peru. The Peruvian upwelling varies from year to year, resulting in significant fluctuations in productivity and fisheries yields. The major factor producing these variations is the El Niño Southern Oscillation, which is the best studied of the recurring variations in large-scale circulation, and its disruptive effects on coastal weather and fisheries are well-known (Barber and Chavez, 1983).

The major ocean currents connect geographic regions and also exert control on ocean life in other ways. Currents form natural boundaries that help define distinct habitats. Such boundaries may isolate different genetic strains of the same species as well as different species. Many marine animals (for example, salmon and squid) have migration patterns that rely upon transport in major ocean current systems, and other species rely on currents to distribute their larvae to new habitats. Populations of ocean species naturally fluctuate from year to year, and ocean currents often play a significant role. The survival of plankton, for example, is affected by where the currents carry them. Food supply varies as changing circulation and upwelling patterns lead to higher or lower nutrient concentrations.

The heterogeneity of the oceans, its water masses, currents, ecological processes, geological history and seafloor morphology, have resulted in great variations in the spatial distribution of life. In short, biodiversity is not uniformly distributed across the oceans: there are local and regional biodiversity "hotspots" (see Chapters 33 and 35). Figure 1.3 shows a way in which the diversity of species is consequently distributed around the world. Various classification systems have been devised to systematize this variety, including the European Nature Information System (EUNIS) (Davies and Moss, 1999; Connor et al., 2004) and the Global Open Ocean and Deep Sea-habitats (GOODS) classification and its refinements (Agnostini 2008; Rice et al 2011)).

Part VI (Assessment of marine biodiversity and habitats) describes in more detail the diversity that is found across the ocean, and the way in which it is being affected by human activities.

6   Human uses of the ocean

Humans depend upon the ocean in many ways and our ocean-based industries have had impacts on ocean ecosystems from local to global spatial scales. In the large majority of ocean ecosystems, humans play a major role in determining crucial features of the way in which the ecosystems are developing.

The impacts of climate change and acidification are pervasive through most ocean ecosystems. These, and related impacts, are discussed in Part III (Assessment of major ecosystem services from the marine environment (other than provisioning services)), together with the non-marketed ecosystem services that we enjoy from the ocean and the ways in which these may be affected by the pervasive impacts of human activities.

For wide swathes of the Earth's population, fish and other sea-derived food is a provisioning ecosystem of the highest importance. Part IV (Assessment of the cross-cutting issues: food security and food safety) examines the extent to which humans rely on the ocean for their food, the ways in which capturing, growing and marketing that food is impacting on ecosystems and the social and economic position of those engaged in these activities and the health risks to everyone who enjoys this food.

The wide range of other human activities is examined in Part V (Assessment of human activities and the marine environment): these activities include the growing importance of worldwide transport in the world economy; the major role of the seabed in providing oil and gas and other minerals; the non-consumptive uses of the ocean to provide renewable energy; the potential for non-consumptive use of marine genetic resources; the uses of seawater to supplement freshwater resources; and the vital role of the ocean in tourism and recreation. In addition, it is necessary to consider the way in which human activities that produce waste can affect the marine environment as the wastes are discharged, emitted or dumped into the marine environment, and the effects of reclaiming land from the sea and seeking to change the natural processes of erosion and sedimentation. Finally, we need to consider the marine scientific research that is the foundation of all our attempts to understand the ocean and to manage the human activities that affect it.

7   Conclusion

Our planet is seven-tenths ocean. From space, the blue of the ocean is the predominant colour. This Assessment is an attempt to produce a 360° review of where the ocean stands, what the range of natural variability underlies its future development and what are the pressures (and their drivers) that are likely to influence that development. As the description of the task set out in Chapter 2 (Mandate, information sources and method of work) shows, the Assessment does not attempt to make recommendations or analyse the success (or otherwise) of current policies. Its task is to provide a factual basis for the relevant authorities in reaching their decisions. The aim is that a comprehensive, consistent Assessment will provide a better basis for those decisions.

References

Agnostini, V., Escobar-Briones, E., Cresswell, I., Gjerde, K., Niewijk, D.J.A., Polacheck, A., Raymond, B., Rice, J., Roff, J.C., Scanlon, K.M., Spalding, M., Vierros, M., Watling, L. (2008). *Global Open Oceans and Deep Sea-habitats (GOODS) bioregional classification*, in: Vierros, M., Cresswell, I., Escobar-Briones, E., Rice, J., Ardron, J. (Eds.). United Nations Conference of the Parties to the Convention on Biological Diversity (CBD), p. 94.

Anthony, K.R.N., Ridd, P.V., Orpin, A.R., Larcombe, P. and Lough, J. (2004). Temporal Variation of Light Availability in Coastal Benthic Habitats: Effects of Clouds, Turbidity, and Tides. *Limnology and Oceanography*, Vol. 49, No. 6.

Ausubel, J.H., Crist, D.T., Waggoner, P.E. (Eds.) (2010). *First census of marine life 2010: highlights of a decade of discovery*. Census of Marine Life, Washington DC.

Barber, R.T., Chavez, F.P. (1983). Biological Consequences of El Niño. *Science* 222, 1203-1210.

Black Sea Commission (2008). Commission on the Protection of the Black Sea Against Pollution, State of Environment Report 2001 - 2006/7, Istanbul. (ISBN 978-9944-245-33-3).

Broecker, W.S. (1991). The great ocean conveyor. *Oceanography* 4, 79-89.

Broecker, W.S. (1997). Thermohaline circulation, the Achilles Heel of our climate system: will man-made CO2 upset the current balance? *Science* 278, 1582-1588.

Census of Marine Life (2010). *Ocean Life: Past, Present, and Future* http://comlmaps.org/oceanlifemap/past-present-future.

Connor, D.W., Allen, J.H., Golding, N., Howell, K.L., Lieberknecht, L.M., Northen, K.O., Reker, J.B. (2004). *Marine habitat classification for Britain and Ireland Version 04.05*. Joint Nature Conservation Committee, Peterborough UK.

Costanza, R., d'Arge, R., de Groot, R., Farber, S., Grasso, M., Hannon, B., Limburg, K., Naeem, S., O'Neill, R.V., Paruelo, J., Raskin, R.G., Sutton, P., van den Belt, M. (1997). The value of the world's ecosystem services and natural capital. *Nature* 387, 253-260.

Davies, C.E., Moss, D. (1999). *The EUNIS classification*. European Environment Agency, 124 pp.

Halpern, B.S., Walbridge, S., Selkoe, K.A., Kappel, C.V., Micheli, F., D'Agrosa, C., Bruno, J.F., Casey, K.S., Ebert, C., Fox, H.E., Fujita, R., Heinemann, D., Lenihan, H.S., Madin, E. M.P., Perry, M.T., Selig, E.R., Spalding, M., Steneck, R. and Watson, R. (2008). A Global Map of Human Impact on Marine Ecosystems. *Science*. 319, 948–952.

Halpern, B.S., Longo, C., Hardy, D., McLeod, K.L., Samhouri, J.F., Katona, S.K., Kleisner, K., Lester, S.E., O'Leary, J., Ranelletti, M., Rosenberg, A.A., Scarborough, C., Selig, E.R., Best, B.D., Brumbaugh, D.R., Chapin, F.S., Crowder, L.B., Daly, K.L., Doney, S.C., Elfes, C., Fogarty, M.J., Gaines, S.D., Jacobsen, K.I., Karrer, L.B., Leslie, H.M., Neeley, E., Pauly, D., Polasky, S., Ris, B., St Martin, K., Stone, G.S., Sumaila, U.R., Zeller, D. (2012). An index to assess the health and benefits of the global ocean. *Nature* 488, 615–620.

Harley, C.D.G., Hughes, A.R., Hultgren, K.M., Miner, B.G., Sorte, C.J.B., Thornber, C.S., Rodriguez, L.F., Tomanek, L., Williams, S.L. (2006). The impacts of climate change in coastal marine systems. *Ecology Letters* 9, 228–241.

Harris, P.T., MacMillan-Lawler, M., Rupp, J., Baker, E.K. (2014). Geomorphology of the oceans. *Marine Geology* 352, 4-24.

Heezen, B.C., Tharp, M. (1977). *World Ocean Floor Panorama*, New York, In full color, painted by H. Berann, Mercator Projection, scale 1:23,230,300, 1168 x 1930 mm.

Heilman et al (2009). S Heilman and N Mistafa, Red Sea, in United Nations Environment Programme, UNEP Large Marine Ecosystems Report, Nairobi 2009 (ISBN 978-92080702773-9).

HELCOM (2010). Helsinki Commission, Ecosystem Health of the Baltic Sea 2003–2007: HELCOM Initial Holistic Assessment, Helsinki (ISSN 0357 – 2994).

Hobbs, Carl III (2003). Article "Continental Shelf" in Encyclopedia of Geomorphology, ed Andrew Goudie, Routledge, London and New York.

IPCC (2005) Caldeira, K., Akai, M., Ocean Storage in *IPCC Special Report on Carbon dioxide Capture and Storage*, pp 277-318. https://www.ipcc.ch/pdf/special-reports/srccs/SRCCS_Chapter6.pdf

Kudela, R.M., Banas, N.S., Barth, J.A., Frame, E.R., Jay, D.A., Largier, J.L., Lessard, E.J., Peterson, T.D., Vander Woude, A.J. (2008). New Insights into the controls and mechanisms of plankton productivity in coastal upwelling waters of the northern California current system. *Oceanography* 21, 46-59.

Martínez, M.L., Intralawan, A., Vázquez, G., Pérez-Maqueo, O., Sutton, P., Landgrave, R. (2007). The coasts of our world: Ecological, economic and social importance. *Ecological Economics* 63, 254-272.

Millennium Ecosystem Assessment (2005). *Ecosystems and Human Well-being*: Synthesis. Island Press, Washington, DC., 155 p.

National Geographic Society (2010). *Ocean Life* (poster). National Geographic Society, Washington, D.C.

NOAA (2014). USA National Oceanic and Atmospheric Administration, Tide Predictions and Data (http://www.co-ops.nos.noaa.gov/faq2.html#26 accessed 15 October 2014).

Occhipinti-Ambrogi, A. (2007). Global change and marine communities: Alien species and climate change. *Marine Pollution Bulletin* 55, 342-352.

Rice, J., Gjerde, K.M., Ardron, J., Arico, S., Cresswell, I., Escobar, E., Grant, S., Vierros, M. (2011). Policy relevance of biogeographic classification for conservation and management of marine biodiversity beyond national jurisdiction, and the GOODS biogeographic classification. *Ocean & Coastal Management* 54, 110-122.

Rintoul, S.R., and Sokolov, S. (2001). Baroclinic transport variability of the Antarctic Circumpolar Current south of Australia (WOCE repeat section SR3). *Journal of Geophysical Research: Oceans* 106, 2815-2832.

Rykaczewski, Ryan R., and Checkley Jr., D.M. (2008). Influence of Ocean Winds on the Pelagic Ecosystem in Upwelling Regions, *Proceedings of the National Academy of Sciences of the United States of America*, Vol. 105, No. 6.

Shaffer, G., Olsen, S.M., Pedersen, J.O.P. (2009). Long-term ocean oxygen depletion in response to carbon dioxide emissions from fossil fuels. *Nature Geoscience*, 2, 105-109.

Small, Christopher and Cohen, J.E. (2004). Continental Physiography, Climate, and the Global Distribution of Human Population, *Current Anthropology* Vol. 45, No. 2.

Smith, W.H., Sandwell, D.T. (1997). Global Sea Floor Topography from Satellite Altimetry and Ship Depth Soundings. *Science Magazine* 277, 1956-1962.

Sobarzo, M., Figueroa, M., Djurfeldt, L. (2001). Upwelling of subsurface water into the rim of the Biobío submarine canyon as a response to surface winds. *Continental Shelf Research* 21, 279-299.

Tittensor, D.P., Mora, C., Jetz, W., et al. (2010). Global patterns and predictors of marine biodiversity across taxa. *Nature* 466:1098–1101. doi: 10.1038/nature09329.

UNEP, IOC-UNESCO (2009). *An Assessment of Assessments, findings of the Group of Experts. Start-up phase of the Regular Process for Global Reporting and Assessment of the State of the Marine Environment including Socio-economic aspects*. UNEP and IOC/UNESCO, Malta.

UNGA (2002). United Nations General Assembly, Resolution 57/141 (Oceans and the Law of the Sea), paragraph 45.

Widder (2014). Edith Widder, Deep Light in US National Oceanic and Atmospheric Administration, Ocean Explorer (http://oceanexplorer.noaa.gov/

explorations/04deepscope/background/deeplight/deeplight.htm accessed 15 October 2014).

WSSD (2002). Report of the World Summit on Sustainable Development, Johannesburg, South Africa, 26 August-4 September 2002 (United Nations publication, Sales No. E.03.II.A.1 and corrigendum), chap. I, resolution 2, annex, para. 36 (b).

2 Mandate, information sources and method of work

Contributors:
Alan Simcock (Convenor and Lead Member), Amanuel Ajawin, Beatrice Padovani Ferreira, Sean O. Green, Peter Harris, Jake Rice, Andrew Rosenberg and Juying Wang (Co-Lead Members)

The World Summit on Sustainable Development, held in Johannesburg, South Africa, in 2002, recommended that there should be established a Regular Process for the Global Reporting and Assessment of the Marine Environment, including Socioeconomic Aspects (WSSD, 2002). This recommendation was endorsed by the United Nations General Assembly (UNGA) in 2002 (UNGA, 2002).

After considerable preparatory work, including as a first phase the production of the assessment of assessments (AoA, 2009), the United Nations General Assembly approved in 2009 the framework for the Regular Process developed by its Ad Hoc Working Group of the Whole. This framework for the Regular Process consisted of: (a) the overall objective for the Regular Process, (b) a description of the scope of the Regular Process, (c) a set of principles to guide its establishment and operation and (d) the best practices on key design features for the Regular Process as identified by the group of experts established for the assessment of assessments (see below). The framework further provided that capacity-building, sharing of data, information and transfer of technology would be crucial elements of the framework. The following paragraphs set out these elements in the terms approved by the General Assembly (AH-WGW, 2009; UNGA, 2009).

1   Overall objective

The Regular Process, under the United Nations, would be recognized as the global mechanism for reviewing the state of the marine environment, including socioeconomic aspects, on a continual and systematic basis by providing regular assessments at the global and supraregional levels and an integrated view of environmental, economic and social aspects. Such assessments would support informed decision-making and thus contribute to managing in a sustainable manner human activities that affect the oceans and seas, in accordance with international law, including the United Nations Convention on the Law of the Sea[1] and other applicable international instruments and initiatives.

The Regular Process would facilitate the identification of trends and enable appropriate responses by States and competent regional and international organizations.

The Regular Process would promote and facilitate the full participation of developing countries in all of its activities.

Ecosystem approaches would be recognized as a useful framework for conducting fully integrated assessments.

2   Capacity-building and technology transfer

The Regular Process would promote, facilitate and ensure capacity-building and transfer of technology, including marine technology, in accordance with international law, including the United Nations Convention on the Law of the Sea and other applicable international instruments and initiatives, for developing and other States, taking into account the criteria and guidelines on the transfer of marine technology of the Intergovernmental Oceanographic Commission.

The Regular Process would promote technical cooperation, including South-South cooperation.

States and global and regional organizations would be invited to cooperate with each other to identify gaps and shared priorities as a basis for developing a coherent programme to support capacity-building in marine monitoring and assessment.

The value of large-scale and comprehensive assessments, notably in the Global Environment Facility's international waters large-marine ecosystems initiatives, in identifying and concentrating on capacity-building priorities would be recognized.

Opportunities for capacity-building would be identified, in particular on the basis of existing capacity-building arrangements and the identified capacity-building priorities, needs and requests of developing countries.

States and relevant international organizations, bodies and institutions would be invited to cooperate in building the capacity of developing countries in marine science, monitoring and assessment, including through workshops, training programmes and materials and fellowships.

Quality assurance procedures and guidance would be developed to assist Governments and international organizations to improve the quality and comparability of data.

3   Scope

The scope of the Regular Process is global and supraregional, encompassing the state of the marine environment, including socioeconomic aspects, both current and foreseeable.

In the first cycle, the scope of the Regular Process would focus on establishing a baseline. In subsequent cycles, the scope of the Regular Process would extend to evaluating trends.

The scope of individual assessments under the Regular Process would be identified by Member States in terms of, inter alia, geographic coverage, an appropriate analytical framework, considerations of sustainability,

---

1   United Nations, *Treaty Series*, vol. 1833, No. 31363.

issues of vulnerability and future scenarios that may have implications for policymakers.

4 Principles

The Regular Process would be guided by international law, including the United Nations Convention on the Law of the Sea and other applicable international instruments and initiatives, and would include reference to the following principles:

(a) Viewing the oceans as part of the whole Earth system;
(b) Regular evaluation by Member States of assessment products and the regular process itself to support adaptive management;
(c) Use of sound science and the promotion of scientific excellence;
(d) Regular analysis to ensure that emerging issues, significant changes and gaps in knowledge are detected at an early stage;
(e) Continual improvement in scientific and assessment capacity, including the promotion and development of capacity-building activities and transfer of technology;
(f) Effective links with policymakers and other users;
(g) Inclusiveness with respect to communication and engagement with all stakeholders through appropriate means for their participation, including appropriate representation and regional balance at all levels;
(h) Recognition and utilization of traditional and indigenous knowledge and principles;
(i) Transparency and accountability for the regular process and its products;
(j) Exchange of information at all levels;
(k) Effective links with, and building on, existing assessment processes, in particular at the regional and national levels;
(l) Adherence to equitable geographical representation in all activities of the regular process.

5 Reasons for these decisions

This framework largely reflected the recommendations of a group of experts, established by the General Assembly in 2005 (UNGA, 2005) and in place by the end of 2006, to carry out (under the guidance of an ad hoc steering group and with the assistance of the lead agencies, United Nations Environmental Programme (UNEP) and Intergovernmental Oceanographic Commission/United Nations Educational, Scientific and Cultural Organization (IOC-UNESCO)) an "assessment of assessments", reviewing the way in which past assessments, particularly of the marine environment at global and regional levels, had been carried out, in order to establish the approaches which could ensure that assessments under the Regular Process would be relevant, legitimate and credible – the three necessary conditions for an influential assessment.

The report of the assessment of assessments (AoA, 2009) summarised the justification for the Regular Process as follows:

"5.1 Marine ecosystems provide essential support to human well-being. However, they are undergoing unprecedented environmental changes, driven by human activities, and becoming depleted and disrupted... Keeping the world's oceans and seas under continuing review will help to improve the responses from national governments and the international community to the challenges posed by these changes. Reviews based on sound science can help the world as a whole understand better what is happening, what is causing it, [and] what the impacts are."

The report saw an urgent need for a more integrated approach, at the global level as well as at the regional and sub-regional levels. It indicated that such an integrated approach was feasible, and would help to develop a more coherent overview of the state of the global marine environment and its interactions with the world economy and human society. A better understanding is needed of how human activities themselves interact and cumulatively affect different parts of marine ecosystems. Baselines, reference points and reference values would also be needed as a basis for evaluating status and trends over time. More consistent information, both in coverage and quality, and integrated analyses would improve understanding of the rapid changes that are occurring in the oceans and their possible causes. The resulting knowledge would facilitate decisions to manage in a sustainable manner human activities affecting the oceans. Assessment is a necessary, integral part of the cycle of adaptive management of human activities that affect the oceans.

The report went on to explain the benefits from a Regular Process that could be a means for integrating existing information from different disciplines to show new and emerging patterns and to stimulate further development of the information base.

The elements relevant to the framework established by the General Assembly include actions to:

(a) Demonstrate the importance of oceans to human life and as a component of the planet;
(b) Integrate, analyze and assess environmental, social and economic aspects of all oceans components and interactions among all sectors of human activity affecting them; it could thus support sustainable, ecosystem-based management throughout the oceans;
(c) Promote well-designed assessment processes, conducted to the highest standards and fully documented by those responsible for them;
(d) Promote international collaboration to build capacity;
(e) Improve the quality, availability, accessibility, interoperability and usefulness of information for ocean assessment; it would also increase consistency in the selection and use of indicators, reference points and reference values;

(f) Support better policy and management at the appropriate scale by providing sound and integrated scientific analyses for decision-making by the relevant authorities;
(g) Build on existing assessment frameworks, processes and institutions and thus provide a base for cooperation among governments and at the level of international institutions.

The essential features which differentiate this assessment from earlier assessments are that it is global in scope, that it is to integrate the different sectors that are involved with the ocean and that it is to integrate environmental, social and economic aspects of the ocean. This is an ambitious project, and it has been clear from the outset that the first assessment of this kind would be breaking new ground, and that there would therefore be scope for improvement in future cycles of the Regular Process.

6   Timing

In 2009, the Ad Hoc Working Group of the Whole recommended that the Regular Process should involve a series of cycles and that the first cycle of the Regular Process should cover the five years from 2010 to 2014. This was endorsed by the General Assembly in 2009, on the basis that there would be two phases of the first cycle, the first phase up to the end of 2012 to agree the issues to be covered and the second phase from 2013 to 2014 to produce the first assessment (AHWGW, 2009; UNGA, 2009).

7   Modalities

In 2010, the General Assembly endorsed a series of recommendations from the Ad Hoc Working Group of the Whole on the modalities for the way in which the work of the Regular Process should be organized and implemented (AHWGW, 2009; AHWGW, 2010; UNGA, 2010). The modalities, consisting of key features, capacity-building and institutional arrangements, were developed further in a series of decisions of the General Assembly, on the basis of recommendation of the Ad Hoc Working Group of the Whole of the General Assembly (AHWGW, 2011a; UNGA, 2011a; AHWGW, 2011b; UNGA, 2011b; AHWGW, 2012; UNGA, 2012; AHWGW, 2013; UNGA, 2013; AHWGW, 2014; UNGA, 2014), informed, among other things, by material prepared by the initial group of experts appointed in 2009. The arrangements for the Group of Experts of the Regular Process were set out in the Terms of Reference and Working Methods (AHWGW, 2012; UNGA, 2012), and various paragraphs of the relevant General Assembly resolutions.

The main institutional arrangements thus established are as follows:

(a) *The Ad Hoc Working Group of the Whole on the Regular Process for Global Reporting and Assessment of the State of the Marine Environment, including Socioeconomic Aspects:* The Regular Process is to be overseen and guided by an Ad Hoc Working Group of the Whole of the General Assembly comprised of representatives of Member States. Relevant intergovernmental and non-governmental organizations with consultative status recognized by the Economic and Social Council are to be invited to participate in the meetings of the Ad Hoc Working Group. Relevant scientific institutions and major groups identified in Agenda 21 may request an invitation to participate in the meetings of the Ad Hoc Working Group. In 2011, the Ad Hoc Working Group agreed on the establishment of a Bureau to put in practice its decisions and guidance during the intersessional period (AHWGW, 2011b; UNGA, 2011b).

(b) *The Group of Experts of the Regular Process:* The general task of the Group of Experts, as set out in the Terms of Reference and Working Methods approved by the General Assembly, is "to carry out any assessments within the framework of the Regular Process at the request of the General Assembly under the supervision of the Ad Hoc Working Group of the Whole". It was noted that an assessment would only be carried out at the request of the General Assembly. Within this general task, the Group of Experts were to draw up a draft implementation plan and timetable, a draft outline of the assessment, proposals for writing teams for each chapter and proposals for independent peer review. Lead Members for each chapter, drawn from the Group of Experts, are to have a general task of managing each chapter, and a convenor of the writing team from the chapter (who might also be the Lead Member) is to be responsible for ensuring the proper development of the chapter. The Terms of Reference and Working Methods make clear that the Group of Experts is collectively responsible for the Assessment, and was to agree on a final text of any assessment for submission through the Bureau to the Ad Hoc Working Group of the Whole, and to present that text to the Ad Hoc Working Group of the Whole.

The Group of Experts, originally appointed in 2009 to develop thinking on the "basic building blocks" identified by the Assessment of Assessments, were invited to continue for the first cycle of the Regular Process pursuant to a series of decisions of the General Assembly.

The Group could be constituted of a maximum of 25 members, five appointed by each regional group within the General Assembly. One regional group only made two appointments, and therefore the full membership of the Group has been 22. In accordance with the Terms of Reference and Working Methods, the Group appointed two coordinators from within its membership, one from a developed country and one from a developing country. The members of the Group of Experts are volunteers or are supported by their parent institutions.

(c) *The Pool of Experts:* The General Assembly approved criteria for the appointment of experts to a Pool of Experts to assist in the preparation of the first assessment and to cover the wide range of issues that an assessment of the ocean integrated across sectors and across environmental, social and economic aspects would have to address. This assistance would include several distinct potential roles: convenors and members of the writing teams, commentators to enable expertise about parts of the world not covered by the writing teams to be brought in to the Assessment without making writing teams unmanageably large, and peer reviewers to review the complete draft of the Assessment. These experts have been nominated by States through the chairs of the regional groups of the United Nations. In addition, members of the Group of Experts and writing teams could consult widely with relevant experts.

(d) *Secretariat:* On the recommendation of the Ad Hoc Working Group of the Whole, the General Assembly requested the Secretary-General to designate the Division of Ocean Affairs and Law of the Sea as the secretariat of the Regular Process. Since no additional staff was allocated specifically for this work, the secretariat function has been provided by the existing staff.

(e) *Technical and Scientific Support:* Technical and scientific support for the Regular Process has been available from the IOC-UNESCO, UNEP, the International Maritime Organization (IMO) and the Food and Agriculture Organization of the United Nations (FAO), and the International Atomic Energy Agency (IAEA). These agencies were invited by the General Assembly, together with other competent United Nations specialized agencies, to provide such support as appropriate. A dedicated web-based platform was set up to make information about this Assessment available and to provide a means of communication between members of the Group of Experts and the members of the Pool of Experts. Agreement was reached between Australia, Norway and the United Nations Environment Programme to host such a website at GRID/Arendal in Norway.

(e) *Workshops:* In addition to the Pool of Experts, steps were taken to convene workshops as forums where experts (including government officials) could make an input to the planning and development of the Assessment. The General Assembly approved guidelines for these workshops, which were held in Santiago in September 2011 (at the invitation of the Government of Chile), in Sanya in February 2012 (at the invitation of the Government of China), in Brussels in June 2012 (at the invitation of the Government of Belgium, supported by the European Union), in Miami in November 2012 (at the invitation of the Government of the United States of America), in Maputo in December 2012 (at the invitation of the Government of Mozambique), in Brisbane in February 2013 (at the invitation of the Government of Australia), in Grand Bassam in October 2013 (at the invitation of the Government of Côte d'Ivoire) and in Chennai in January 2014 (at the invitation of the Government of India). The workshops were open to representatives of all States, although participation was mainly from experts in the respective regions. Each workshop aimed to consider the scope and methods of this Assessment, the information available in the region where it was held, and capacity-building needs in that region. Reports of each workshop were made available on the website of the Division of Ocean Affairs and Law of the Sea and on the website of the first Assessment.

8 Finance

The General Assembly decided that the costs of the first cycle of the Regular Process should be financed from a voluntary trust fund, and invited the Secretary-General to establish such a fund for the purpose of supporting the operations of the first five-year cycle of the Regular Process, including for the provision of assistance to members of the Group of Experts from developing countries. The Trust Fund is managed and administered by the Division of Ocean Affairs and Law of the Sea. Contributions to this fund have been made by Belgium, China, Côte d'Ivoire, Iceland, Ireland, Jamaica, New Zealand, Norway, Portugal and the Republic of Korea. In addition, Australia, Belgium, Canada, Chile, China, Côte d'Ivoire, India, Mozambique, the Republic of Korea, the United Kingdom of Great Britain and Northern Ireland and the United States of America have supported workshops in the region and/or the travel and accommodation costs of members of the Group of Experts from their countries. Generous support to the Regular Process has also been provided, financially and technically, by the European Union, IOC-UNESCO and UNEP.

9 Guidance

On the advice of the Group of Experts, the Ad Hoc Working Group decided that there should be comprehensive guidance for the Regular Process. Accordingly it prepared such guidance, covering the responsibilities of the Group of Experts, the members of the Pool of Experts, the writing teams and their convenors, the commentators and the peer reviewers, the approaches to achieve integration and to deal with uncertainty, risk, ethical questions and style. This was approved by the General Assembly (UNGA, 2012), and can be found in AHWGW, 2012.

10 Collection of information

When the methods of work were being developed, it was thought that there would be time for a number of working papers to bring together detailed information and thus to serve as the basis for the preparation of this Assessment. In practice, the time available has not proved sufficient to adopt this approach generally. In some cases, detailed background information has been included in appendices to the relevant chapter.

11 Development of the first World Ocean Assessment

The starting point for each substantive chapter has been the outline developed by the Ad Hoc Working Group of the Whole, on the basis of proposals from the Group of Experts, approved by the General Assembly (AHWGW, 2012; UNGA, 2012) and slightly amended by the Ad Hoc Working Group of the Whole in 2014 (AHWGW, 2014). The writing teams, constituted as described above, elaborated this outline and, in some cases, assigned drafting duties within the Group. A draft chapter was prepared, reviewed by the Lead Member (where not part of the writing team), by other members of the Group of Experts to ensure consistency among chapters, and (in some cases) by a panel of commentators chosen from the Pool of Experts, but not otherwise part of the writing team. The writing teams responded as necessary to comments from these reviews and prepared a consensus draft chapter. The consensus draft was submitted to the Group of Experts and secretariat. The Group of Experts collectively reviewed all these consensus draft chapters, in order to ensure consistency and to prepare the synthesis chapters for each Part of this Assessment and Part I (the summary). An editor overseen by the secretariat reviewed each chapter for format and consistency, raising questions for clarification with the writing team where necessary. After any concerns raised by the copy editor had been addressed, the secretariat circulated the entire draft of the first Assessment for review by States, by a team of peer reviewers assigned by the Bureau of the Ad Hoc Working Group of the Whole, on a proposal from the Group of Experts and by intergovernmental organizations. In March 2015, close to 5000 comments were received. The Group of Experts and the writing teams then proceeded to respond to the comments and revise the draft chapters accordingly. At the end of April 2015, the Group of Experts met again in New York to discuss the finalization of the responses and the revision of the chapters. Following a review by the secretariat of the responses and revisions, all chapters of the Assessment were ready for submission to the Bureau by mid-July. The Assessment, including its summary[2] is to be considered by the Ad Hoc Working Group of the Whole in September 2015.

---

2 See A/70/112.

References

AHWGW (2009). *Report on the work of the Ad Hoc Working Group of the Whole to recommend a course of action to the General Assembly on the regular process for global reporting and assessment of the state of the marine environment, including socio-economic aspects*, United Nations General Assembly document A/64/347.

AHWGW (2010). *Report on the work of the Ad Hoc Working Group of the Whole on the Regular Process for Global Reporting and Assessment of the State of the Marine Environment, including Socio-Economic Aspects*, United Nations General Assembly document A/65/358.

AHWGW (2011a). *Report on the work of the Ad Hoc Working Group of the Whole on the Regular Process for Global Reporting and Assessment of the State of the Marine Environment, including Socio-Economic Aspects*, United Nations General Assembly document A/65/759.

AHWGW (2011b). *Report on the work of the Ad Hoc Working Group of the Whole on the Regular Process for Global Reporting and Assessment of the State of the Marine Environment, including Socio-Economic Aspects*, United Nations General Assembly document A/66/189.

AHWGW (2012). *Report on the work of the Ad Hoc Working Group of the Whole on the Regular Process for Global Reporting and Assessment of the State of the Marine Environment, including Socio-Economic Aspects*, United Nations General Assembly document A/67/87.

AHWGW 2013). *Report on the work of the Ad Hoc Working Group of the Whole on the Regular Process for Global Reporting and Assessment of the State of the Marine Environment, including Socioeconomic Aspects*, United Nations General Assembly document A/68/82.

AHWGW (2014). *Report on the work of the Ad Hoc Working Group of the Whole on the Regular Process for Global Reporting and Assessment of the State of the Marine Environment, including Socioeconomic Aspects*, United Nations General Assembly document A/69/77.

AoA (2009). UNEP and IOC-UNESCO, *An Assessment of Assessments, Findings of the Group of Experts. Start-up Phase of a Regular Process for Global Reporting and Assessment of the State of the Marine Environment including Socio-economic Aspects*. (ISBN 978-92-807-2976-4).

UNGA (2002). United Nations General Assembly, Resolution 57/141 (Oceans and the Law of the Sea), paragraph 45.

UNGA (2005). United Nations General Assembly, Resolution 60/30 (Oceans and the Law of the Sea), paragraph 91.

UNGA (2009). United Nations General Assembly, Resolution 64/71 (Oceans and the Law of the Sea).

UNGA (2010). United Nations General Assembly, Resolution 65/37 A (Oceans and the Law of the Sea).

UNGA (2011a). United Nations General Assembly, Resolution 65/37 B (Oceans and the Law of the Sea).

UNGA (2011b). United Nations General Assembly, Resolution 66/231 (Oceans and the Law of the Sea).

UNGA (2012). United Nations General Assembly, Resolution 67/78 (Oceans and the Law of the Sea).

UNGA (2013). United Nations General Assembly, Resolution 68/70 (Oceans and the Law of the Sea).

UNGA (2014). United Nations General Assembly, Resolution 69/245 (Oceans and the Law of the Sea).

WSSD (2002). *Report of the World Summit on Sustainable Development, Johannesburg, South Africa, 26 August-4 September 2002* (United Nations publication, Sales No. E.03.II.A.1 and corrigendum), chap. I, resolution 2, annex, para. 36 (b).

III | Assessment of Major Ecosystem Services from the Marine Environment (Other than Provisioning Services)

3 Scientific Understanding of Ecosystem Services

Contributors:
Marjan van den Belt (Convenor), Patricio Bernal (Lead Member), Elise Granek, Françoise Gaill, Benjamin Halpern and Michael Thorndyke

1 Introduction to the concept of ecosystem services from oceans

Humanity has always drawn sustenance from the ocean through fishing, harvesting and trade. Today 44 per cent of the world's population lives on or within 150 kilometres from the coast (United Nations Atlas of Oceans). However this fundamental connection between nature and people has only very recently been incorporated into trans-disciplinary thinking on how we manage and account for the human benefits we get from nature. Today, when a product taken from an ecosystem[1], for example, fibres, timber or fish, enters the economic cycle (i.e., a part of the human system), it receives a monetary value that accounts at least for the costs associated with its extraction and mobilization. If that natural product is the result of cultivation, as in the case of agriculture, forestry and aquaculture, the monetary value also includes the production costs. However, the extraction of natural products and other human benefits from ecosystems has implicit costs of production and other ancillary costs associated with preserving the integrity of the natural production system itself. Traditionally these benefits and costs have been hidden within the "natural system," and are not accounted for financially; such hidden costs and benefits are considered "externalities" by neoclassical economists. While the neoclassical economic toolbox includes non-market valuation approaches, an ecosystem services approach emphasizes that 'price' is not equal to "value" and highlights human well-being, as a normative goal. The emergence and evolution of the ecosystem services concept offers an explicit attempt to better capture and reflect these hidden or unaccounted benefits and associated costs when the natural "production" system is negatively affected by human activities. The ecosystem services approach has proven to be very useful in the management of multi-sector processes and already informs many management and regulatory processes around the world (e.g. United Kingdom National Ecosystem Assessment, 2011).

Ecosystems, including marine ecosystems, provide services to people, which are life-sustaining and contribute to human health and wellbeing (Millennium Ecosystem Assessment, 2005; de Groot, 2011). The Millennium Ecosystem Assessment defines an ecosystem as "a dynamic complex of plant, animal and micro-organism communities and their non-living environment interacting as a functional unit" and goes on to define ecosystem services as "the benefits that humans obtain from ecosystems" (p. 27). This definition encompasses both the benefits people perceive and those benefits that are not perceived (van den Belt et al., 2011b). In other words, a benefit from ecosystems does not need to be explicitly perceived (or empirically quantified) to be considered relevant in an ecosystem services approach. Similarly, ecosystems and their processes and functions can be described in biophysical (and other) relationships whether or not humans benefit from them. Ecosystem services reflect the influence of these processes on society's wellbeing; including people's physical and mental well-being. While ecosystems provide services not only to people, the evaluations of services are, by definition anthropocentric.

The deliberate interlinking between human and natural systems is not new, but over the past few decades interest in "ecosystem services" as a concept has surged, with research and activities involving natural and social scientists, governments and businesses alike (Costanza et al., 1997; Daily, 1997; Braat and de Groot, 2012). This interest is in part driven by the growing recognition that the collective impact of humans on the Earth is pushing against the biophysical limits of many ecosystems to sustain the well-being of humankind. Such pressures are well recognized (e.g., Halpern et al., 2008; Rockstrom et al., 2009) and are felt by pelagic, coastal, and intertidal ecosystems.

The human system – comprising built, human and social capital[2] – ultimately is fully dependent on natural capital. Ecosystems can exist without humans in them, but humans cannot survive without ecosystems. Therefore, the human system can usefully be considered as a sub-system of natural capital. An ecosystem services approach then becomes an organizing principle to make visible the relative contribution of natural capital toward the goal of human well-being. The use of such an organizing principle can be the basis for investments to maintain and enhance natural capital to ensure a flow of ecosystem services (Costanza et al., 2014).

Natural capital is the natural equivalent of the human-made agricultural and aquaculture production systems mentioned above (Daly and Cobb, 1989). In essence, natural capital refers to ecosystems (i.e., coastal shelves, kelp forests, mangroves, coral reefs and wetlands) as a network of natural production systems in the most fundamental sense. Humans with our many production systems are part of this natural capital and collectively have much to gain or lose from maintaining or neglecting, respectively, its sustainability.

The normative goal underpinning the ecosystem services concept is to maintain long-term sustainability, as well as local and immediate enhancement of human well-being within the carrying capacity of the biophysical system. To continue receiving a sustainable flow of ecosystem services, it is crucial to manage the scale of the human system relative to its natural capital base (Rockstrom et al., 2009). The ecosystem services approach acknowledges natural capital as the paradigm in which the human subsystem exists, highlighting (but not limiting to) the anthropocentric aspect of this concept (Costanza et al., 2014). At the same time the ecosystem services approach draws into decision-making the less visible aspects of sustainable development, such as supporting, regulating and cultural services. Through an ecosystem services approach, people, governments and businesses are increasingly using this approach as

---

[1] Synonyms for 'ecosystems' in the literature are: natural systems, natural capital, nature, natural assets, ecological resources, natural resources, ecological infrastructure.

[2] Built Capital refers to human-made infrastructure. Human Capital refers to the ability to deal with complex societal challenges, including education, institutions and health. Social Capital refers to the networks of relationships among people who live and work in a particular society, enabling that society to function effectively.

an organizing principle for finding new ways to invest their human, social and built capital in this common goal (Döring and Egelkraut, 2008).

The magnitude of human pressures on the Earth's natural systems and acknowledgement of the interconnectedness between ecosystems and human sub-systems has revealed a need to transition from an emphasis on single-species or single-sector management to multi-sector, ecosystem-based management (TEEB, 2010a; Kelble et al., 2013) across multiple geographic (Costanza, 2008) and temporal (Shaw and Wlodarz, 2013) dimensions. Intensification of use of natural capital increases interactions between sectors and production systems that in turn increase the number of mutual impacts (i.e., externalities). This requires accountability among tradeoffs in a way that was, perhaps, not as necessary when the use of natural capital was less intense. On land, negative impacts can be partially managed or contained in space. However, in the ocean, due to its fluid nature, impacts may broadcast far from their site of origin and are more difficult to contain and manage. For example, there is only one Ocean when considering its role in climate change through the ecosystem service of "gas regulation".

An ecosystem services approach supports assessment and decision-making across land and seascapes; i.e., to consider benefits from ecosystems in natural, urban, rural, agricultural, coastal and marine environments in an integrated way, and ultimately to understand the potential and nature of tradeoffs among services given different management actions. An example derived from Food and Agriculture Organization (FAO) states that 50 billion United States dollars is lost annually from global income derived from marine fisheries, compared to a more sustainable fishing, due to fish stocks over-exploitation, when viewed through an ecosystem services lens (FAO, 2012).

Principles for sustainable governance of oceans[3] are straightforward (Costanza et al., 1998; Crowder et al., 2008,), but use of an ecosystem services approach has the potential to provide a basis for collaborative investments (in monetary or governance efforts), based on common ground and shared values. In other words, the ecosystem services approach has the potential to provide a new "currency" or organizing principle to consider multi-scale and cross-sectoral synergies and tradeoffs.

Several recently developed and evolving frameworks outline an ecosystem services approach and its underlying connection between natural and human systems. Although the essence of the ecosystem services concept is the dependence of human well-being on ecosystems, there are diverse definitions of the concept, reflecting differing worldviews on how human systems relate to ecosystems. For example, ecological economists emphasize that human societies are a sub-set of ecosystems and as a consequence assume limited substitutability between built/manufactured and natural capital (Daly and Farley, 2004; van den Belt 2011a; Braat and de Groot, 2012; Farley, 2012). Some definitions of ecosystem services emphasize the functional aspects of ecosystems from which people derive benefits (Costanza et al., 1997; Daily, 1997) and others put more emphasis on their utilitarian aspects and seek conformity with economic accounting (Boyd and Banzhaf, 2007; United Nations Statistics Division, 2013). Still others emphasize human health and well-being (Fisher et al., 2009) and values (TEEB, 2010a).

The ecosystem services approach aims to address and make explicit the inherent complexity of the coupling between biophysical and human systems. For example, it allows regulating ecosystem services at a global scale, such as climate regulation and sea level rise, to be integrated into local decision-making (Berry and Bendor, 2015). An important point here is that though climate change is perceived as a broadly global phenomenon, its impacts will be local, depending on a host of local/regional drivers that will interact with global climate changes. This means that assessments of natural capital and ecosystem services are best done at multiple scales. At the same time, integration across and between regions is essential to ensure shared best practices, agreed protocols and data-access policies, etc. This is an important function for governance at the global level.

The ecosystem services approach has been embraced by different fields and perspectives. For example, those concerned with biodiversity (e.g., TEEB, 2009; TEEB, 2010a; TEEB, 2010b; TEEB, 2010c; Intergovernmental Panel for Biodiversity and Ecosystem Services-IPBES) and climate change (e.g., Intergovernmental Panel for Climate Change-IPCC) have generally aligned themselves with this approach. Many international organizations (e.g., United Nations, World Bank, the Organization for Economic Cooperation and Development (OECD), The Nature Conservancy, International Union for the Conservation of Nature(IUCN), FAO), governments (e.g., European Union, United Kingdom, United States of America), and increasingly companies (e.g., Dow Chemical and potentially those connected to the World Oceans Council) are collaborating to explore the potential for efficient and effective decision-making offered by an ecosystem services approach. An example of intergovernmental collaboration on ecosystem services is the Group on Earth Observations (GEO)[4] and particularly GEO's Biodiversity Network (GEO BON), a voluntary partnership among intergovernmental, non-governmental and governmental organizations (www.earthobservations.org/geobon). The Intergovernmental Science-Policy Platform on Biodiversity and Ecosystem Services (IPBES) enhances this integration effort at sub-regional, regional and global levels (Larigauderie and Mooney, 2010; www.ipbes.net).

---

[3] 'Lisbon' Principles for Sustainable Development of Oceans: 1) Responsibility: ability to respond to social and ecological goals. 2) Scale-matching: ensuring flow of ecological and social information allows for timely and appropriate action across scales. 3) Precaution: in the face of uncertainty about potentially irreversible ecological impacts, decisions about natural capital err on the side of precaution. The burden of proof shifts to those whose activities potentially damage natural capital. 4) Adaptive management: decision-makers collect and integrate socio-cultural-economic-ecological information, adapting their decisions accordingly. 5) Full-cost accounting: where appropriate, external costs allow markets to reflect full costs.6) Participation: foster stakeholder awareness and collaboration.

[4] GEO, the Group on Earth Observations has today 89 member states and the European Commission.

Scientific Understanding of Ecosystem Services — Chapter 3

Although the concept has achieved broad acceptance, caution is needed in implementing ecosystem services approaches to avoid a simplistic or biased commodification of ecosystems that prioritizes some elements of nature that are economically useful to the detriment of overall ongoing preservation of those ecosystems for their intrinsic value. An unbalanced approach focused primarily on assigning monetary values can exacerbate power asymmetries and increase socio-ecological conflicts (e.g., Beymer-Farris and Bassett, 2012). Giving equal focus to non-market/non-use services within the ecosystem services framework is both a desirable approach and a strength of this method for decision-making (Chan et al., 2012). When ecosystem services are approached as an organizing principle, this includes the development of common units of measurement for decision support, beyond application of existing tools in the natural and social science toolboxes. It needs to be acknowledged that we don't, and may never, fully understand social-ecological systems to the point that people can confidently predict changes and impact or 'optimize' these systems. A precautionary stance regarding management and governance for maintenance of resilience of social-ecological systems is highlighted (Bigagli, 2015).

The ecosystem services approach gained momentum in the late 1990s, when monetary values associated with ecosystem services from natural capital were conservatively estimated (at a rate double that of global Gross Domestic Product (GDP) to highlight the potential economic and societal value of previously unvalued ecosystem services (Costanza et al., 1997). These values were globally expressed with a single spatial dimension, a snapshot of which is shown in Figure 3.1. These values only provided a starting point of a necessary debate, as they relied on many and generally conservative assumptions about how to, in a broader sense, value services globally. Although they expressed these services in monetary values, the authors did not claim that these services were suitable for exchange in the market system (Costanza et al., 1997). A recent re-assessment of these global values indicated that the values of global ecosystem services have increased with additional studies on ecosystem services, but these values simultaneously have decreased where natural capital has been converted to other types of capital (Costanza et al., 2014).

An ecosystem services approach certainly isn't without controversy and critique is offered by neoclassical economists and ecologists (McCauley, 2006), albeit for different reasons. Some critiques of an ecosystem services approach are highlighting the utilitarian manner in which this approach has been implemented (Wegner and Pascual, 2011; Bscher et al., 2012). Ecosystem services, or "nature's benefits" provide a strengths-based, organizing principle to more deliberately and systematically consider the contributions biophysical communities (including biodiversity

The boundaries and names shown and the designations used on this map do not imply official endorsement or acceptance by the United Nations.

Figure 3.1 | Global map of values of estimated ecosystem services in 1997. Source: Costanza et al., 1997.

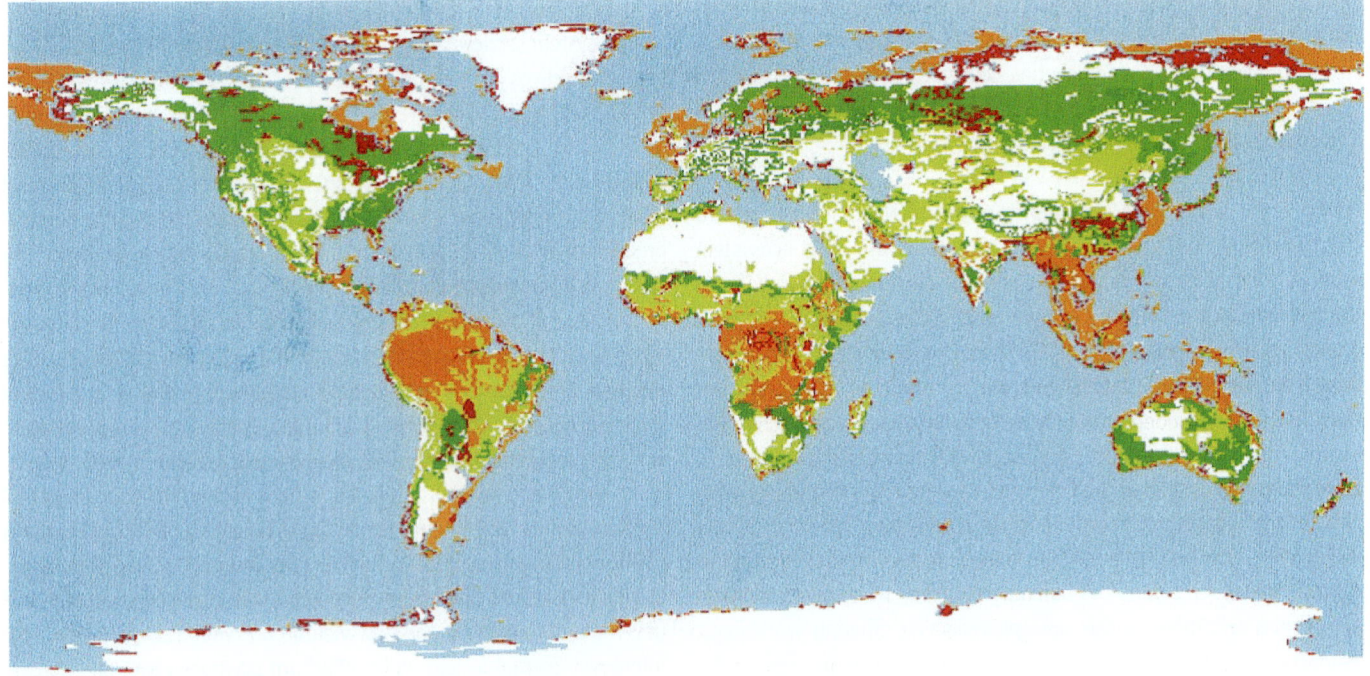

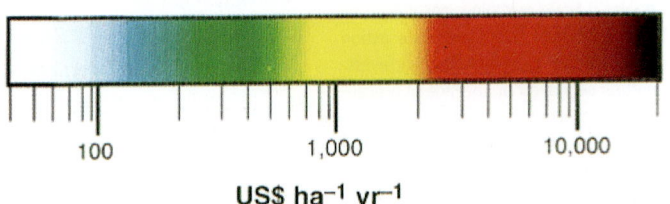

US$ ha⁻¹ yr⁻¹

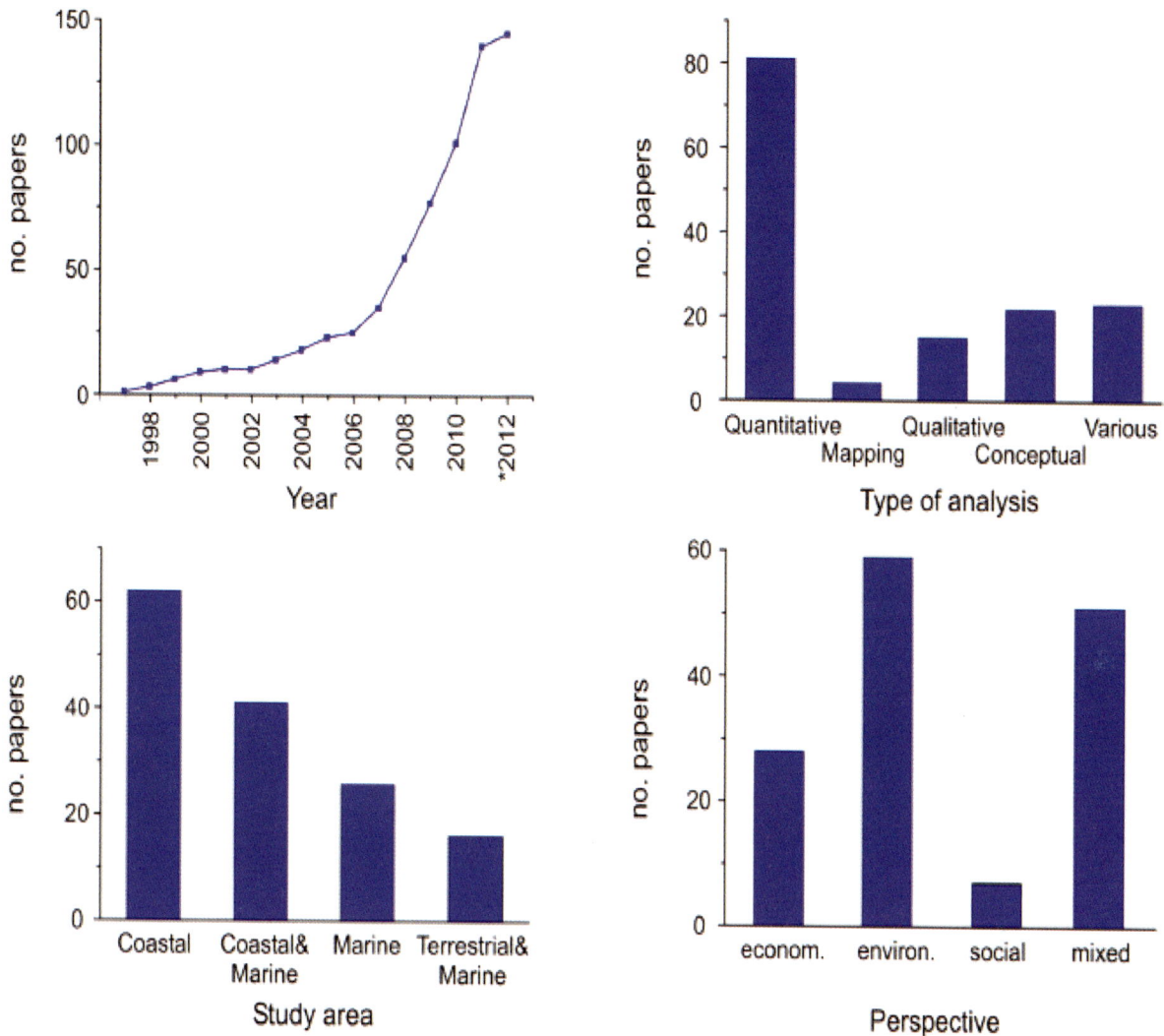

Figure 3.2 | Data and analysis from 145 MCES assessments by Liquete et al. (2013). A. Number of publications per year. *The year 2012 covers 1 January to 4 April. B. Number of studies per type of analysis. C. Number of papers per type of environment analyzed. D. Number of publications per scientific discipline.

and habitat) provide to human well-being (including health). A weak application of an ecosystem services approach builds on traditional natural resource management tools by considering a broader appreciation of the advantages provided by natural systems to include social, economic, health and ecological benefits. This approach is then used to analyze, in more detail, aspects of ecosystem services currently considered externalities and builds upon natural resource management strategies of the 20th century. This may incrementally expand the quality and quantity of relevant indicators considered when making decisions about tradeoffs. In a strong application of an ecosystem services approach, it can be used to synthesize systemic aspects of managing the human sub-system within an ecosystem. A strong application of an ecosystem services approach requires the design of tools and skill sets suitable to support multi-faceted management and governance strategies fit for the 21st century.

The Millennium Ecosystem Assessment (2005) classified ecosystem services as: provisioning services (e.g., food – including food traded in formal markets and subsistence trade and barter -, pharmaceutical compounds, building material); regulating services (e.g., climate regulation, moderation of extreme events, waste treatment, erosion protection, maintaining populations of species); supporting services (e.g., nutrient cycling, primary production) and cultural services (e.g., spiritual experience, recreation, information for cognitive development, aesthetics).

Supporting services are often considered at an 'intermediate' level as support functions toward "final ecosystem services" (Landers and Nahlik, 2013). While the intermediate nature of supporting services makes accounting more challenging, i.e. avoiding double counting, it is also important to acknowledge the "unaccountable"' characteristics of ecosystems for three reasons. First, the complexity of ecosystems is such that applying accounting practices modelled in accordance with traditional economic accounting is often both impossible and inappropriate. In other words, while economic activities can be aggregated to a certain extent[5], attributes of ecosystems and their functions do not lend themselves well to aggregation. Second, supporting services or support func-

---

[5] The System of National Accounts does not account for everything either.

Table 3.1 | Overview of thematic working groups of the Ecosystem Service Partnership (ESP), which would be useful to complete for a subsequent World Oceans Assessment.

Thematic working groups of ESP	Biomes	Scale
1. Ecosystem services assessment frameworks and typologies		
2. Biodiversity and ecosystem services		
3. Ecosystem service indicators		
4. Mapping ecosystem services		
5. Modeling ecosystem services		
6. Valuation of ecosystem services -6A. Cultural services and values -6B. Ecosystem services and public health -6C. Economic and monetary valuation -6D. Value integration		
7. Ecosystem services in trade-off analysis and project evaluation		
8. Ecosystem services and disaster-risk reduction		
9. Application of ecosystem services in planning and management -9A. Restoring ecosystems and their services		
10. Co-investment and reward mechanisms for ecosystem services -10A. Ecosystem services and poverty alleviation		
11. Ecosystem service accounting and greening the economy		
12. Governance and institutional aspects		

tions underlie all other services (e.g., provisioning and cultural services are made available in part by supporting services). Third, supporting services are often considered to be most important from cultural and spiritual perspectives, which have their own specific value (Chan et al 2012).

Scientific publications concerning ecosystem services have grown exponentially since the late 1990s. As shown in Figure 3.2, the marine and coastal ecosystem services (MCES) literature is no exception. Liquete et al. (2013) recently categorized 145 articles on the current status of MCES.

The analysis by Liquete et al. (2013) found that most of the MCES case studies they reviewed: 1) were concentrated in Europe and North America; 2) did not cover the area beyond the continental shelf edge, with benthic habitats generally lacking, and 3) focused on mangroves for supporting and provisioning services and on coastal wetlands for regulating and supporting services. A primary focus on local or regional geographic location raises a concern for MCES, as biophysical events and conditions are generated further afield. For example, patterns of upwelling and migratory species will be influenced by benthic and oceanic conditions that might occur at some distance from the affected region and thus will be difficult to predict. As in other domains, decision-makers have to make decisions under conditions of high uncertainty with limited ability to conclusively consider all risks. An ecosystem services approach has the advantage of making visible the non-linear behaviour[6] of ecosystems and draw attention in decision-making to fundamentally different alternatives (Barbier et al., 2008). Such alternatives may lead to synergies (i.e., shared values across sectors as a basis for social-ecological enterprises and poverty alleviation) or to difficult trade-offs between different uses or user groups. A valuation spectrum should include "all that is important to people", whether the people themselves perceive this or not (van den Belt et al., 2011b) and regardless of whether the value is monetary, spiritual, cultural, or otherwise.

2 Evolving ecosystem services frameworks, principles and methods

An overview follows of accepted typologies, principles and methods currently used for assessing and measuring ecosystem services in the rapidly growing international literature. Although concepts and methodologies show a consistent pattern in local applications, no generally accepted classification of ecosystem goods and services for global accounting purposes exists (Haines-Young and Potschin, 2010; Böhnke-Henrichs et al., 2013). The complexity of such a task requires a pluralistic approach across temporal and spatial scales to make ecosystem services visible in decision-making processes and to decision-makers. Capabilities for temporal and spatial analyses are evolving rapidly (e.g. Altman et al., 2014). These now enable decision support and the use of an ecosystem services approach at local, regional, national and global scales (e.g. Zurlini et al., 2014). However, consistency across scales and across terrestrial and marine environments has not been achieved. This is often highlighted as a research, policy and management priority (Braat and de Groot, 2012). For example, the Ecosystem Service Partnership (ESP) (www.es-partnership.org) attracts scientists and practitioners working with the ecosystem services concept in a self-organizing manner. The ESP website allows the assessment of ecosystem services through the various themes, geographic locations and biomes. The themes (Table 3.1) provide a good overview of the variety of methods and tools and required skills through which the ecosystem services concept can be

---

6 Non-linear behaviour refers to the characteristic of complex systems where effects are not proportional to their causes.

Figure 3.3 | Process of ecosystem service assessments based on TEEB, redrawn after Hendriks et al., 2012.

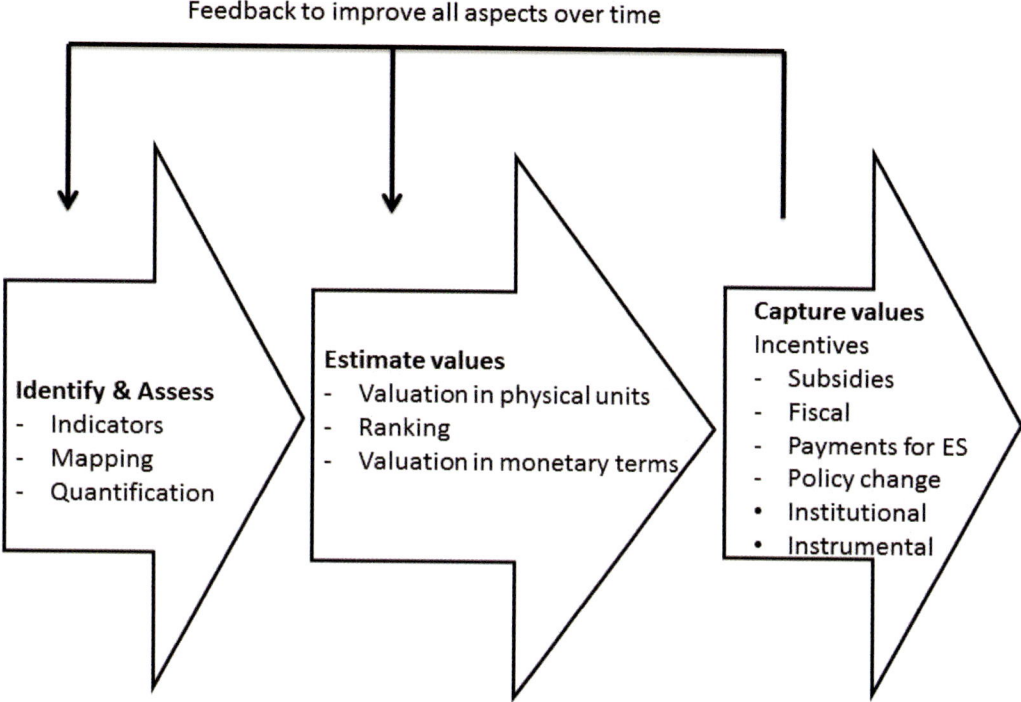

viewed. Associated with ESP, the Marine Ecosystem Services Partnership (http://marineecosystemservices.org/) features a library of valuation-oriented literature, organized by ecosystem, on the delivery of ecosystem services and offering interconnection with other databases (see Appendix 2 for an overview of relevant databases). Currently organized by country, further analyses of scale addressed by the valuation studies included may help progress toward a multi-scale approach. For example, completion of Table 3.1 for marine ecosystem services could be very useful for a future second United Nations World Ocean Assessment.

The Economics of Ecosystems and Biodiversity (TEEB) started as a UNEP project (2007 – 2010) initiated by the G8. This resulted in the promotion of steps toward the management of values that people derive from ecosystems (Figure 3.3). In essence, the TEEB framework clusters and links the ESP themes into a process suitable for decision support for projects, governments and businesses (TEEB, 2010b). This process is then ideally implemented systemically, with appropriate feedback mechanisms for on-going assessments of all aspects involved at multiple scales.

2.1 The flow of ecosystem services

For this introductory chapter on ecosystem services, however, we elaborate on the cascading Haines-Young and Potschin (2010) framework. This framework is relevant because of its close alignment with the evolving United Nations System of Environmental-Economic Accounting (United Nations Statistics Division, 2013) and its effort to seek a consistent classification system and set of accounting principles (Boyd and Banzhaf, 2007; Landers and Nahlik, 2013).

Conceptual models, such as the Common International Classification of Ecosystem Goods and Services (CICES) (Haines-Young and Potschin, 2010), enable practitioners to differentiate between natural capital, i.e., the natural resources or ecological infrastructure, and the services that are derived from that infrastructure. This is presented in a framework cascading from biome to function/process, service, benefit and value (Figure 3.4). This framework is influenced by two perspectives: 1) the desire to account for ecosystem services and avoid double counting by economists and 2) an opportunity for natural scientists to rapidly communicate the value of particular ecological structures and processes. When applying this framework, supporting and cultural ecosystem services are easily ignored, as non-market[7] values are at best considered at the end of the cascade and more often are not considered at all; and the flow of ecosystem services is portrayed as linear or unidirectional, mimicking a production chain, and implies a "trickling down" from natural capital to value for people, whose task it is to perceive this value. Appreciated for its simplicity, this framework relies, in theory, on coherent and collective policy action to correct cumulative pressures when values are perceived. This feedback requires active management to allow natural capital to function and provide essential services and benefits, whether people perceive such values or not. This framework shows similarities to the DPSIR (Driver-Pressure-State-Impact-Response[8]) framework. In

---

[7] In a weak application of an ecosystem services approach, cultural services are often limited to a monetary equivalent of 'recreation'. In a stronger application of this approach spiritual connections, sense of place and mental well-being are recognized. Social sciences contribute a myriad of tools to appreciate such values (e.g. (Pike et al.,, 2014).

[8] DPSIR: Drivers-Pressures-State-Impact-Response generally focusses on impacts as in costs rather than on the benefits people derive from ecosystems. Another difference is that the 'State' in DPSIR has a biophysical focus, whereas in the ES

comparison, the U.S. EPA draft classification system for Final Ecosystem Goods and Services (FEGS-CS) attempts to provide a categorization of beneficiaries and assist in tracking changes in ecosystem services upon those beneficiaries (Landers and Nahlik, 2013).

Economists often use the term 'ecosystem goods and services', in part to seek comparability and consistency with the System of National Accounting (United Nations Statistics Division, 2013). It is important to recognize that the provision of ecosystem goods and services relies on the integrity of ecosystem processes and functions, referred to as regulating and supporting ecosystem services, with characteristics that make them less than suitable for rigorous accounting (Farley, 2012). Disparate disciplinary perspectives occur in the context of applying an ecosystem services approach; e.g., economists appreciate an ability to account for outputs and optimization of the 'production process', whether it is human- or nature-made, whereas ecologists tend to resist such a linear accounting of ecosystems as inaccurate because ecosystems are 'complex systems', with highly non-linear behaviours, and simplifying these complexities can lead to misrepresentation of management needs required to maintain valued services.

Following the steps of this cascading framework, marine ecological infrastructure includes (but are not limited to) biophysical structures, e.g., the open ocean, continental shelves, coral reefs, kelp forests, seagrass beds, mangroves, salt marshes, rocky intertidal and subtidal zones, sand dunes and beaches. These are ecological systems and the associated structures created by biological and physical processes, e.g., primary production, wave generation, and decomposition of organic matter. Ecosystem functions and processes emphasize the potential capacity of natural capital to deliver an ecosystem service, which includes resource functions (e.g., mineral deposits and deep-sea fish), sink capacity (e.g., the ability to absorb, dilute or keep out of sight unwanted by-products) and service functions (e.g., habitat to support biodiversity, wave attenuation, degradation of organic matter). 9

This flow from biophysical structures to functions and processes to ecosystem services is labelled the "supply of ecosystem services" (Figure 3.4). Ecosystem services also provide benefits (such as, air to breathe, water to drink, fish to eat, sustenance of marine life, energy to harness from wave/wind/tidal/thermal power, health, safety and increased human well-being). Because these benefits are essential for human well-being, a market or non-market value10 can, in some cases, be placed on these ecosystem services. This is part of the cascade labelled 'demand for ecosystem services'.

---

framework, the 'State' of the human dimension is equally important. (Kelble et al., 2013).

9   Some scholars (e.g., Aronson et al., 2007) separate natural capital into renewable natural capital (living species and ecosystems); non-renewable natural capital (subsoil assets, e.g., petroleum, coal, diamonds); replenishable natural capital (e.g., the atmosphere); and cultivated natural capital (e.g., aquaculture).

10   Market and non-market values are sometimes also referred to as use or non-use values or as instrumental and intrinsic values.

In essence, the flow diagram has two fundamental purposes: (1) identifying the ecological processes required to attain ecosystem services; and (2) developing the ability to account more rigorously for this natural 'production system', particularly at a global level. At this analytical level, the ecosystem services concept effectively reveals and communicates the 'invisible' biophysical processes and functions and thereby broadens, guides and informs local decision alternatives and scenarios. This is not a uni-directional flow - the 'cascading production chain' (as shown in Figure 3.4) also requires attention for reverse processes taking 'values' in a broad pluralistic sense, as a starting point, to collectively develop solutions (Haines-Young and Potschin, 2010; van den Belt, 2014; Maes et al., 2012; Tallis et al., 2012). Understanding this flow of ecosystem services at multiple scales, top-down and bottom-up, facilitates practical local solution-oriented responses, enabled by global guidance.

Sometimes a limited set of ecosystem services can be locally managed for short-term benefits, whereas other ecosystem services have globalized characteristics and/or have longer-term benefits. Therefore, this approach has the potential to effectively connect mutual or competing interests at local to global scales and facilitate cohesive decision support. Given that the ecosystem services approach is an inherently anthropocentric concept and is context-dependent, any value attributed to ecosystem services is not absolute and depends on the supply of (i.e., how much of a service is available, if it is limiting) and demand for the service (i.e., how much people need or want a service). A 'gap' between supply and demand of ecosystem services indicates a shortage or abundance (Figure 3.4). The gap varies temporally and spatially, per societal sector, and by the political scale of the perspective (i.e., local, regional, or global). When an abundant supply of ecosystem services exists relative to demand, the governance or management requirement is primarily one of monitoring. A shortage of supply of ecosystem services, relative to demand, makes the necessity of effective governance and management more acute (see also 'time preference' below) - quality and efficiency of delivery of ecosystem service need to be considered. Supply and demand are dynamically interconnected and therefore employment of methodologies beyond market-based theories is crucial.

2.2   Biophysical supply of ecosystem services

Any assessment of ecosystem services must begin with natural capital. The natural system encompasses species present, the flows of matter and energy to which these species contribute, their functional attributes, and the interactions with the physical environment that serve to enhance or dampen the functional attributes and processes. This may require principles and practical guidelines codifying simplification schemes (e.g., Townsend et al., 2011), as science will not be able to provide all of the answers in the time needed to develop management responses. An assessment of natural capital in marine systems should include the distribution and level of ecosystem services in relation to space and time, so that changes in ecosystem services may be better understood following different management practices and proximity to tipping points of marine ecosystems (MacDiarmid et al, 2013; Townsend and Thrush, 2010).

Assessing the supply of ecosystem services in practice requires a process similar to the generic TEEB approach highlighted in Figure 3.3. First, one must define, as specifically as possible, how an ecosystem function or process of interest connects to specific human benefits of interest and exactly which aspects of a species or ecosystem structure are connected to that function. Developing such a conceptual model following ecological principles (Foley et al., 2010) is important because, for example, a single species can provide more than one function, and different attributes or processes of the species may be more or less important for (a) particular service(s) of interest. For example, mangrove forests provide coastal protection, carbon storage, nursery habitat, and wood, among other services, and these services are provided primarily by the density of above-ground biomass, below-ground biomass, submerged root structures, and the absolute amount of above-water/ground biomass, respectively. Mangroves can provide bundles of ecosystem services, which are inter-related to each other. Measurements require knowledge of such bundles and how they occur at multiple spatial scales over which their benefits are conferred (Costanza, 2008).

The second step is to develop a model describing how the biophysical system produces or inhibits production of the metric of interest, and which key drivers modify that production. This step corresponds to step 1 in Figure 3.3. In the mangrove example above, if we are interested in the coastal protection function of mangrove forests and thus the above-ground density of the woody biomass, we ideally would have or develop a mangrove growth model that could predict how wave height and intensity, sunlight, rainfall, sedimentation, etc., affect production, and especially the inter-plant density, of the woody biomass. In order to do this modelling, for all potential functions (and services) of interest, one can draw on or develop species-specific population models coupled with ecosystem dynamics models, although the parameters of the model may vary spatially and temporally. Once in place, these models then permit relatively simple sensitivity analyses that identify key drivers of change in the metric of interest.

Such models are always challenged by the availability of data, particularly in many developing countries. Thus, model development must proceed hand-in-hand with data discovery and, where possible, data-gap filling, so that models are tailored to the scale, resolution, and complexity of the data available for a region (Figure 3.5). Typically useful data include physical data on sea level, pH, temperature and wave height and intensity, and biological data on the demographics, densities, dispersal, and trophic dynamics of species. Although the data needs are similar at a global level across the major oceans, these data will vary by locale and temporally (sometimes seasonally). Availability of data and scientific understanding to properly paramatize such models in particular, depends on scale and differs between regions. Local/regional data for marine ecosystem services assessments are generally much more available for counties including, but not limited to Europe, North America, Australia/ New Zealand, and Japan, and are very poor in most of Africa, Asia, and Latin America. A complete world assessment of ecosystem services is beyond the scope of this Assessment, but would ideally be undertaken for a future assessment.

The final step in the process of assessing the supply of ecosystem services is to map and monitor the modelled or empirically derived values for the metrics of interest (step 2 in Figure 3.3) and the communication thereof (step 3 in Figure 3.3). Mapping and modelling are inherently constrained by the spatial resolution of the input data for the models described above. Without such maps, one cannot say from where within a region of interest the supply of and demand for the service is actually coming, and thus managers are left to make decisions about how to maintain or improve the supply, in order to meet demand, at the coarsest scale of assessment (for example, for an entire country). Such coarse-scale decision-making may be appropriate, and in fact is often all that is needed for many decision contexts that occur at a scoping level. Scoping is the process used to identify the key issues of concern at an early stage in any planning process. Scoping should be carried out at an early stage to facilitate strategic planning and reporting. However, when management is using an ecosystem services framework to make smaller-scale decisions, such as designation of Marine Protected Areas, issuing permits for offshore mining, oil or wind-energy installations, and offshore aquaculture installations, then more detailed maps of service supply are critical.

Numerous examples of both types of decision-making exist. On the one hand is the more general, coarse-scale, often data-poor heuristic assessment, where decision-makers are primarily interested in whether service supply will go up, stay constant, or decline under a given management action. For example, model-building, including indigenous stakeholders, can be used to scope for changes over time in ecosystem service values in a non-spatial manner (van den Belt et al., 2012). On the other hand, more specific, finer-scale, often data-rich quantitative scenario development requires detailed assessments of who wins and loses under a given management action, and by how much, when and where. Examples include decisions on wave energy (Kim et al., 2012) and offshore aquaculture facility locations (Buck et al., 2004), considering specific tradeoffs.

At local and regional scales, often considerable but incomplete data are available, to make visible the biophysical supply of ecosystem services. Fundamental to such efforts are sufficient data to map the location and interaction of key biophysical attributes (such as wave energy, ocean temperature, species density and composition, quality and health of those species, etc.), and for some places around the world such data exist. However, for many regions of the world such data do not exist or are extremely limited, constraining the ability to produce precise global, regional and local estimates of the supply of and demand for ecosystem services. A detailed assessment of the most limiting data gaps between regions is a highly desirable study to be conducted before a second United Nations World Ocean Assessment. The ability to map and monitor key areas for ecosystem service supply is crucial for the development of scenarios and strategies to ensure future supply (Burkhard et al., 2012; Maes et al., 2012a; Maes et al., 2012b; Martinez-Harms and Balvanera, 2012). Furthermore, more complete data sets can be achieved through complementary strategies including baseline assessments in key ecosystems and/or in-depth pilot research efforts that can support model development for extrapolation to similar habitats/ecosystems.

Scientific Understanding of Ecosystem Services   Chapter 3

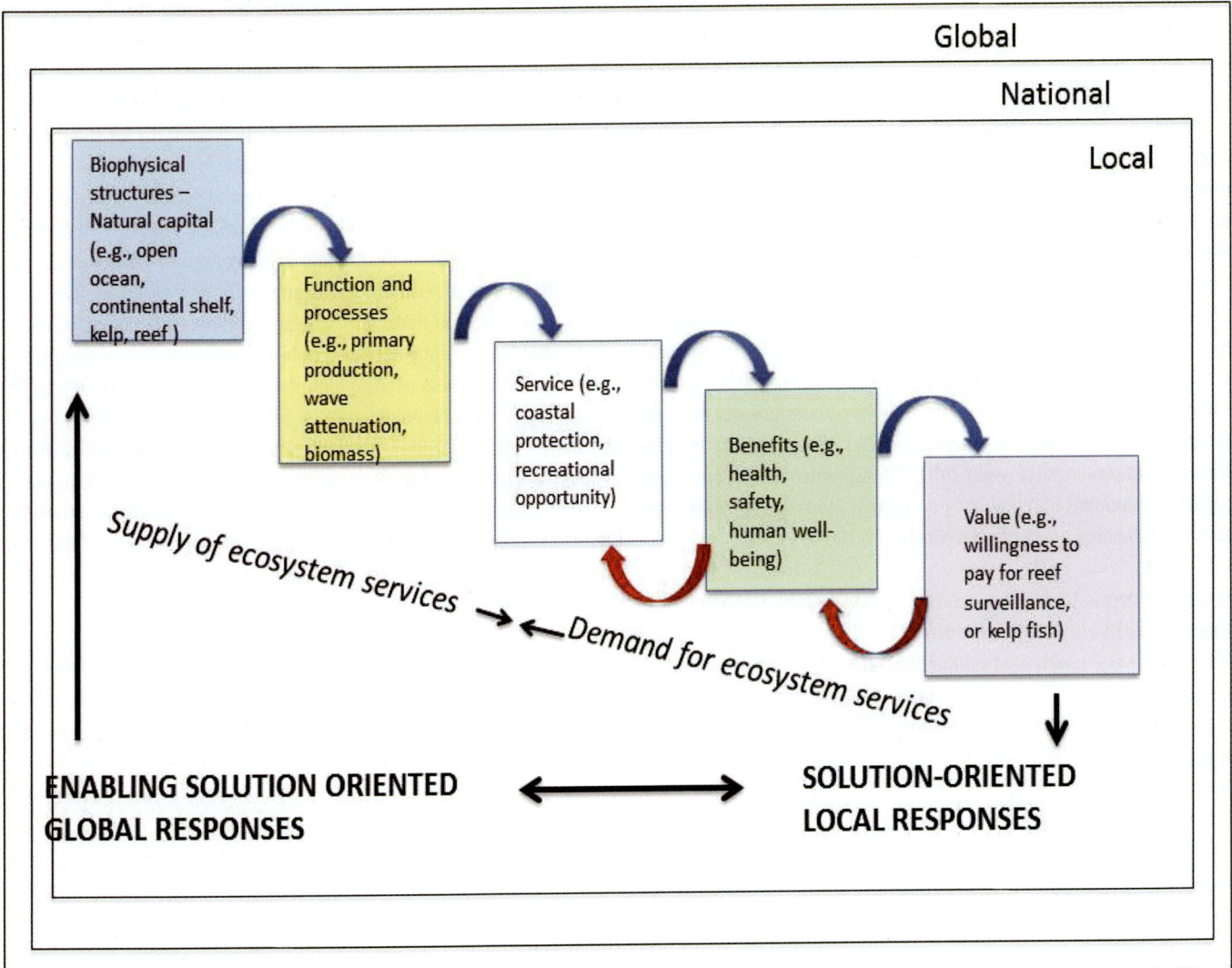

Figure 3.4 | The flow of ecosystem services at multiple scales. Adapted from Haines-Young and Potschin (2010. While not a part of the original model, we added and highlight the 'supply of and demand for ecosystem services' and the gap between 'supply and demand', signalling a shortage or abundance of ecosystem services. This is one basis for establishing 'value' in a broader sense.

The provisioning of ecosystem services depends not only on the presence of biophysical structure and processes, but the condition (intact vs. degraded) and, in some cases, temporal variability (e.g., seasonal variability in the density or height of seagrasses or kelps, or variability in storm-driven waves). To determine the quantity of an ecosystem service, one must identify the spatial scale (local, regional, global) and temporal scale (short- to long-term) of both supply and demand (also illustrated in Figure 3.4). A mismatch often exists between the data available on supply versus demand due to the variability in spatial provisioning and jurisdictional disconnects between supply and demand and the corresponding data available. For example, global studies often draw on low-resolution, remotely sensed data on a global scale, whereas local studies draw on higher-resolution data on a smaller spatial scale. This difference in data quality and spatial extent can lead to different conclusions on the quantity and quality of service provisioning available and the need to handle differences and uncertainty with care. Nevertheless, considering this 'mismatch' of data and information available to assess a gap between supply and demand of ecosystem services is an important move toward broadening the notion of value away from narrow commodification of ecosystem services.

Of particular importance is the multi-scale aspect of the ecosystem services approach, as it provides an invitation to consider a connection between local and global scales at different temporal/seasonal intervals (Costanza, 2008). Some ecosystem services are produced and consumed in situ (e.g., coastal protection), whereas others have clear global aspects (e.g., carbon sequestration, climate regulation, biodiversity, global fisheries and mineral extraction). Certain services are primarily seasonal (e.g., coastal protection), and others are provided or utilized year-round (e.g., food provision).

2.3 Demand for ecosystem services

The 'Benefits' and "Value" steps in the cascading framework (Figure 3.4) represent the 'demand for ecosystem services' and indicate where drivers of management and decision-making can be incorporated. The

Chapter 3 Scientific Understanding of Ecosystem Services

perception of values and benefits sets the context when determining the 'supply of ecosystem services'. Therefore, it is important to consider demand for ecosystem services through at least two lenses: (1) demand, as identified by market-based, economic sectors (as defined in the United Nations System of National Accounts); and (2) demand from non-market sectors or societal groups, including 'needs' and 'wants', whether perceived by people or not. Therefore, value statements, if perceived, are bi-directional and can be viewed as "trickling down" through Total Economic Values and/or "trickling up" through participatory involvement of local communities.

Although the biophysical knowledge of the supply of ecosystems services is progressing, the understanding and visibility of socio-cultural-health-economic benefits from ecosystems (i.e., the understanding of the demand for ecosystem benefits) remain fragmented and are lagging behind, especially for oceans. One difficulty in profiling demand is partly due to the vast geographic scope and overall invisibility of supporting and regulating ecosystem services. Demand for ecosystem services is frequently assessed based on diverse rationales, such as risk reduction, revealed preferences, direct use or consumption of goods and services (Wolff et al, 2015). Also, the relative importance of these ecosystem services is often locally perceived by non-market sectors, especially through diverse cultural perspectives. As a result, management and decision-making frequently prioritize quantifiable ecosystem services (e.g., provisioning services). This prioritization of provisioning services often occurs to the exclusion or detriment of supporting and regulating services. On the other hand, cultural services are frequently highlighted together with provisioning services, as indigenous livelihoods are often tightly coupled to provisioning services as part of cultural services.

As a consequence, in any comprehensive process of ecosystem services valuation, it will be necessary to utilize both monetary and non-monetary valuations, as befits the spatial and temporal characteristics of each ecosystem service. When classical economic theory addresses "market failures", it resorts to the following distinctions:

- A rival good declines in abundance as it is consumed or used, e.g., when one fishing boat catches a fish, the same fish cannot be caught be another boat.
- Non-rival goods can be used by many without being 'used up', e.g., one and the same fish can be admired by multiple divers, or clean coastal waters can be available.
- A good is excludable if the use of it can be prevented, e.g., one needs permission to drill for minerals in the Exclusive Economic Zone.
- A non-excludable good is freely accessible to all, e.g. Storm protection provided by mangroves, seagrasses and reefs and dunes.

Most provisioning goods are 'rival and excludable' and therefore more suitable for valuation through markets, (e.g., fisheries in an Exclusive Economic Zone). However, some provisioning services are 'rival but non-excludable' (e.g., fisheries outside of Exclusive Economic Zones). Depending on place, some non-rival, excludable goods can be enjoyed by those who can afford them; these include some recreational and research services. Most regulatory and cultural services are non-rival and non-excludable, such as the existence of diverse marine life or practically, whale-watching from shores. Based on these characteristics, it is generally inappropriate and unconventional to value non-rival and/or non-excludable ecosystem services using market mechanisms. Even non-market valuation approaches have severe limitations in this realm, which requires socio-political and institutional considerations. Hence, processes to support "trickling up" of local demand for ecosystem ser-

Table 3.2 | Each marine ecosystem provides a suite of ecosystem services, a subset of which are identified; policy and management decisions result in tradeoffs among ecosystem services. * Open ocean may include benthic and pelagic systems. Grey boxes indicate services provided by the ecosystem on the left. Numbers are examples of studies of the ecosystem service in that particular ecosystem. The numbers in table 2 correspond to the case studies listed in Appendix 1. (expanded from Granek et al. 2010).

Marine ecosystems \ Selected ecosystems services	Aquaculture production	Carbon sequestration, climate regulation	Fisheries production	Pharmaceuticals	Pollution buffering and water quality	Protection against storm surges, wind damage	Recreation and Tourism	Shoreline stabilization, Erosion control
Rocky intertidal							13,45, 50	
Salt marshes		12,36, 37						22,29,62
Mangrove forests	15,39,48	3,20,33	16,17, 41		4,23, 47	6,30, 61		10,49,61
Seagrass beds		16,17, 19, 27, 41	16,17, 41		1,34, 52	6,30		
Coral reefs		21,28, 42	9,16, 17,41	9,61		6,30, 61	9,11, 13,25	61
Kelp forests		32,54	24,43, 55,56			30	2,38, 62	
Sand dunes							13,51, 57	5,35, 40
Open ocean*	7,8,26	18,31, 59	44,53, 60				14,46, 58	

vices become increasingly important, preferably supported by appropriate data and an ability to integrate and make these data visible.

Some basic global data is available that can be used for the socio-economic component of assessments based on ecosystem services, such as revenue from coastal and marine related economic sectors. Jobs related to coastal and marine related economic sectors - and cultural values related to culturally important species - may be available at regional level in some places, but are less available in other places. Until the multiple ecosystem services, their interconnections and tradeoffs between different sectors are more accurately recognized and at least semi-quantified in the decision-making sphere, full inclusion of all available global databases is beyond the scope of this first assessment. However, the distinction between markets and other interests, resolution, geographic spread and ease of access are important characteristics of any evolving framework of data sets. 'Scale' sets the direct context for any situation where an ecosystem services approach is envisioned, used and under improvement. The ecosystem services approach has the ability to effectively communicate land-sea connectivity and tradeoffs associated with a variety of ocean- and land-based human uses, economic sectors, stakeholders and governance (Butler et al., 2013). In such an analysis, costs (e.g., due to a loss of ecosystem services, often expressed in indirect values) and benefits (e.g., due to a monetary or non-monetary gain in direct or indirect values) are incurred by different groups over different time scales.

Data on ecosystem services and their valuation for specific case studies are often re-used for similar case studies in different locations, because local data collection and analysis are expensive and require specific skills in non-market analysis. Such 'benefits transfer' approaches to valuation can be controversial because they require assumptions about similarities among regions that are often inaccurate, but they remain a powerful and necessary approach to filling data gaps, when used with caution. Table 3.2 provides a sample of references to local case studies of ecosystem services and their values associated with a sample of particular marine ecosystems. The development of such matrices is often referred to as a 'rapid ecosystem service assessment (RESA)' to identify where ecosystem services and valuation data are available and where data gaps exist. The 17 per cent of boxes that are grey and have no studies referenced represent ecosystem services provided by a particular ecosystem for which insufficient studies have been conducted.

Because it is both essential and expensive to initiate studies of local ecosystem services, various databases have been developed to extract relevant information from site-specific case studies and 'transfer' such knowledge to similar sites. The 'benefit transfer' approach also comes with severe limitations and risk of propagation of errors (Liu et al., 2011). Appendix 2 provides a limited overview of publicly searchable databases that can assist decision-makers in populating matrices suitable to their region, following the exemplified structure of Table 2. The selection of data bases in Appendix 2 was based on explicit reference to an 'ecosystem services' approach, and does not provide an exhaustive list of databases that could be used when applying an ecosystem services approach.

2.4 Managing gaps, tradeoffs, and values across multiple spatial scales

Managing tradeoffs, for example between prioritizing fish-protein production from coastal waters versus coastal protection (Maes et al., 2012b), recreational use (Ghermandi et al., 2011) or cultural considerations (Chan et al., 2012), can lead to difficult decisions for managers and policy-makers. Fairness of distribution and environmental justice beyond direct costs and benefits for user groups need to be considered. The supply of ecosystem services is affected by decision-making that may favour production or provisioning of one service over others. For example, if kelp harvest is a favoured service that is managed, associated "costs" may be a reduction in fish protein, as fish habitat is reduced, and/or a reduction in recreational diving, as the kelp forest is extracted from the ocean (Menzel et al., 2013). Poor decision-making often results in benefits to some users (i.e., those who harvest kelp) and costs to other users (i.e., those who fish for animals that live in kelp, recreational divers, etc.). To achieve equitable distributions via policy-making, it is necessary to consider who wins (i.e., gains, benefits) and who loses (i.e., suffers a cost or loss), directly and indirectly as well as now and in the future. In the absence of regulation or when decision-making fails to consider the suite of services provided by an ecosystem and the range of users of those services, decisions on how best to manage a marine ecosystem may lead to unintended consequences (e.g., costs to recreational divers and fishing communities).

In decision making, stakeholders or managers often choose a set of possible actions to take and then assess the tradeoffs that exist among the identified options. One strength of an inclusive ecosystem services assessment is that it allows exploration of a broader set of possible actions and outcomes and distributive impacts, often identifying and highlighting true 'win-win' solutions (e.g., Lester et al., 2012; White et al., 2012).

Decision-makers are faced with the challenge of considering the spatial and temporal distribution of these services, which directly affects the flow of services. Certain services may be provisioned in close proximity to local communities, but utilized by both local users and others that live far from the location of provisioning. For example, coral reefs may provide protein and coastal protection to local community members on an island, and recreational opportunities, as well as some protein, to outsiders who visit the location as tourists. Even within the local community, individuals residing along the coast may prioritize the coastal protection service of the reefs or mangroves, whereas residents who live inland or upland may prioritize the provisioning of marine protein. The ecosystem services framework, when systematically applied, allows for considerations of multiple ecosystems services over time and space and thus, in this example, highlighting regulating and supporting services, such as habitat needed for spawning to ensure long term provisioning of protein.

Decisions on how best to manage marine resources frequently require consideration of the tradeoffs among a suite of possible scenarios. These tradeoffs generally entail values gained or lost with each scenario. Most commonly such values assigned are monetary. Historically, this has led to consideration of values that can be given a monetary worth, whereas services that are difficult to measure and value are often excluded from the decision-making process (TEEB, 2010a). Rodriguez et al. (2006) found that provisioning, regulating, cultural and supporting services are generally traded off in this respective order. This approach results in a focus on one or a few ecosystem services and in decisions that have an unequal distribution of costs and benefits across sectors of the population. Failure to include supporting and cultural services, specifically on par with provisioning services, may have unintended consequences.

In other words, understanding the flow of production (i.e., supply) and consumption (i.e., demand) of ecosystem services is complex, leaves room for cultural interpretation (Chan et al., 2012), and has distributive implications (Rodríguez et al., 2006; Halpern et al., 2011). However, tools are available - ranging from simple (for scoping purposes or in the face of poor data) to complex (for management purposes and when adequate data are available) - to assist in the development of scenarios and decision-support for this purpose.

2.5 Time preferences

Just as spatial analysis at multiple scales is crucial in understanding the supply of ecosystem services, the understanding of time scales and time preferences are important in assessing tradeoffs, especially with regard to the demand for ecosystem services. The perception of time is often culturally defined. Indigenous peoples often think in terms of multiple generations and time can have a spiritual element. For a market-oriented investor or government, time is captured in a 'discount rate'. In essence, a high discount rate reflects a desire to consume resources now rather than later. From an economic perspective, this choice also determines how quickly an investment returns a profit. Long-term planning to safeguard the benefits of less visible, non-provisioning ecosystem services requires low or even negative discount rates (Carpenter et al., 2007). For investments in natural capital and for people to receive ecosystem services and benefits, multiple discount rates are required. Such ecological discount rates may be place-based (e.g., when considering in situ ecosystem services) or universal (e.g., when ecological infrastructure is providing global ecosystem services) and should also reflect the (often slow) recovery time of ecosystems. This would apply to most supporting, regulatory and cultural services, as they are 'non-rival, non-excludable' services. In addition, certain ecosystem services may be provisioned (e.g., coastal protection when seagrass beds are dense enough to attenuate waves) or utilized (intertidal or inshore fisheries during seasons when ocean conditions do not permit offshore fishery) seasonally, highlighting the importance of managing for time frames hat reflect seasonal availability of or access to a service (TEEB, 2010a).

2.6 The challenge of multi-scale integrated assessments for ecosystem services

There are indicators that allow us to reflect on the health of oceans, e.g., the Ocean Health Index (Halpern et al., 2012) and retrospectively how ocean health is changing. A general indicator for ecosystem services from oceans is not available, nor may it be desirable as one indicator. Such an indicator would require integration across biophysical and human dimensions, with relevance across multiple scales and developing a transparent ability to consider tradeoffs with a forward perspective. This requires the gathering of data at local, regional, national and global scales, and in principle with three dimensions: space, time and values. Although not unique to the ecosystem services concept, the need to connect local to global scales through bottom-up and top-down governance is paramount.

Database management and modeling capacity are increasingly important to support decision-making at multiple levels of scale. This capacity needs to be 'fit for purpose' (i.e., it needs to answer specific questions by decision-makers in a timely fashion), as well as contribute to the development of knowledge across scales (i.e., be relevant beyond the boundary of an individual decision-maker). Currently several tools are available, each emphasizing particular strengths, such as the ability to: (1) communicate effectively with local stakeholders (e.g., Rapid Ecosystem Service Assessments (RESA), Seasketch (McClintock et al., 2012);

Table 3.3 | A subset of tools that can be included in an ecosystem services valuation 'toolbox'. The tools range from crude conversation starters (e.g. RESA) to spatially dynamic decision support frameworks (e.g. MIMES).

	Dimension	Rapid Ecosystem Service Assessment (RESA)	SeaSketch	InVEST	Mediated Modeling	MIMES
Context	Social / values	Possible	Yes	Yes	Yes	Yes
Content	Spatial	Limited	Yes	Yes	No	Yes
	Dynamic/ changes over time	No	No	No	Yes	Yes
	Ecological	Yes	Yes	Yes	Yes	Yes
	Economic	Yes	Limited	Yes, where benefits are perceived	Yes, where benefits are not perceived	Yes, where benefits are not perceived
Process	Adaptive	Scoping	Scoping	Research	Scoping	Management

(2) illustrate spatial aspects (e.g., InVEST (Lester et al., 2012; White et al., 2012); and (3) consider scenarios and changes over time, e.g., Mediated Modeling at the scoping (van den Belt et al., 2012), research, and MIMES/MIDAS (Altman et al., 2014) at management levels. Table 3 illustrates some tools with differing strengths and weaknesses. A comprehensive overview of all tools is beyond the scope of this assessment.

These tools draw on local 'small data' and global 'big data' to various extents. Each case study has the potential to be used in education and add to the collective building of knowledge on ecosystem services. As discussed, multiple databases on ecosystem services and their values are already available (Appendix 1), many of which feature ecosystem-based management tools (e.g., http://ebmtoolsdatabase.org). Newly initiated local case studies, as well as the output from modelling tools and applications of TEEB-like processes, add to this body of knowledge, and draw on 'big data' sets. Bringing together the various databases, tools and knowledge gained from various applications is a top priority for multiple stakeholders, such as policy makers, industry and non-governmental organizations. The iMarine infrastructure is one example of an emerging "Community Cloud" platform which offers Virtual Research Environments that integrate a broad range of data services with scientific data and advanced analysis. Such scenarios then result in new datasets. This could be expanded to include protocols for an ecosystem services approach. Figure 3.5 illustrates a connection between: (1) 'big data', primarily spatial information relevant to the supply of ecosystem service and (2) 'small data', the transferable insights that can be gained from local case studies. These data are brought together in (modeling) tools, evolving (1) from scoping to management level and (2) from static to dynamic tools. In the same way, but with a much more "bottom-up" and integrated emphasis, the European Marine Biodiversity Observation System (EMBOS: http://www.embos.eu/) offers the advantages of scale and expert identification of relevant organisms (taxonomy). This holistic approach is important since marine biodiversity provides many ecosystem services. However, biodiversity is undergoing profound changes, due to anthropogenic pressures, climatic warming and natural variation. Proper understanding of biodiversity patterns and ongoing changes is needed to assess consequences for ecosystem integrity, in order to be in a position to manage the natural resources.

Appropriate application of an ecosystem services approach as an organizing principle in a consistent manner across multiple scales (space, time and values), requires capacity development.

3 Capacity-building and knowledge gaps

This section highlights knowledge gaps regarding the application of ecosystem services and discusses opportunities for capacity development. This concerns 'human capital', often interpreted as the 'ability to deal with complex societal challenges'. In the context of marine ecosystem services, this is reflected in the capacity to collect and use available data to make visible 'the benefits that people derive from ecosystems' relevant for effective decision-making at multiple scales. This includes effective global policies and agreements, education and awareness programmes. Assessing governance and institutional changes that are required at multiple scales is beyond the scope of this chapter, although it should be noted that a feedback to this effect is included in all of the ecosystem services frameworks.

There is a gap in social sciences and economics' ability to support ecosystem-based science. Application of an ecosystem services approach emphasizes the need for human dimensions of well-being, bridging natural and social sciences. Such integrative approach requires capability building in skills beyond existing disciplines. Generic skills that are needed to work within an ES framework, include: technical (e.g. modellers) and specialists (including scientists in specific disciplines), integrators (to make links between the parts), translators (to change policy questions into assumptions) and interpreters (who can communicate complex issues in simple terms).

The multi-scale and process-oriented aspects of an ecosystem services approach provide both a challenge and an opportunity for capacity development in understanding and capturing value regarding the supply of and demand for ecosystem services. Table 4 attempts to relate the scale of the demand for and supply of ecosystem services with data gaps and capacity to interlink/disseminate data for decision-support.

Figure 3.5 | Evolution of ecosystem services knowledge. Adapted from van den Belt et al., 2013.

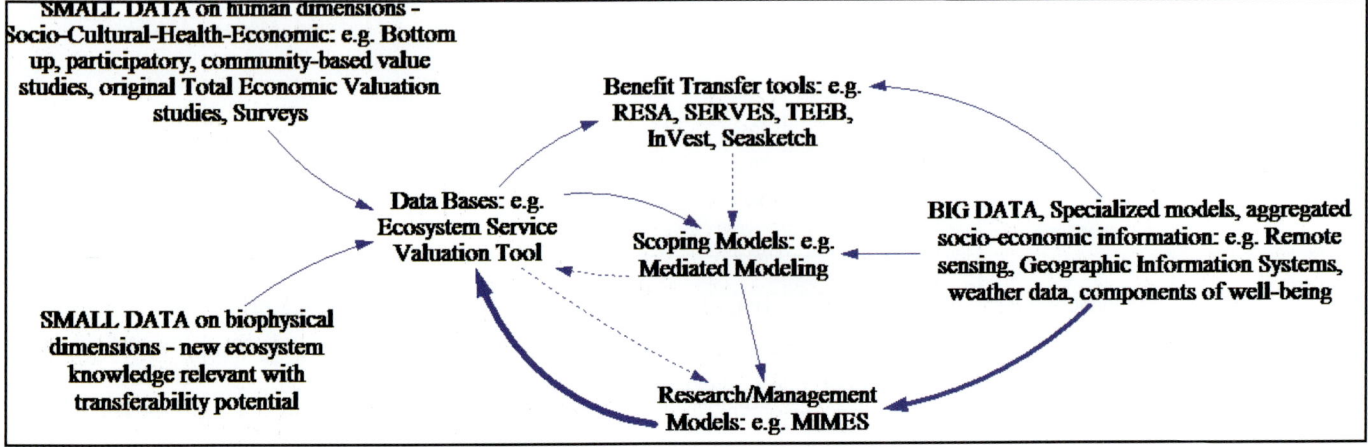

The following are important capacity-development needs:

Data availability and resolution at different scales and geographic spread: Here the most important action item will be to map key areas, identify existing gaps and put in place mechanisms for filling those gaps in a coordinated and strategic way. For example, in the developing world data gaps complicate even rapid ecosystem service assessments at the scoping level. Although other areas have access to data for scoping purposes, crucial knowledge is lacking to use such data through an ecosystem services approach for management purposes.

The ability to use data in an integrated manner, both for 'trickling down' accounting, as well as for "trickling up" community empowerment and participatory purposes: This is exacerbated by the severe lack of local empowerment and understanding of the ecosystem services concept and by the fact that it is a multi-factoral and trans-regional, trans-national issue. This can be addressed by coordinated knowledge transfer and information exchange at the global level, for example, in coordination with IPBES.

Capabilities to undertake heuristic/participatory processes: Once again, this should be approached in a regional to global dimension, albeit for enhancement of specific purposes at each level. Heuristics approaches to problem solving can be used in the domains of natural science and social science and refers to 'operating under less than perfect circumstances to arrive at a way forward'. Perhaps most important will be to encourage, facilitate, collate and promote understanding of regional differences in valuation of ecosystem services according to culture and history. The first step in capacity-building and filling in knowledge gaps will be to empower local stakeholder communities and enable them to understand the impact that ecosystem services have on their lives and well-being. Empowerment and enablement are key concepts in the social sciences and if we are to improve and develop an ecosystem services approach, it will be vital to equip communities, from the bottom up, to develop a stronger sense of ownership and responsibility for the protection and sustainability of their local and global ecosystems and resultant services. However, collectively, it is crucial for people to understand that ecosystem services do not respect national and international boundaries, necessitating an integrated approach and a trading off with adjacent regions. If not accomplished in a transparent manner, this approach is likely to exacerbate regional conflicts. A simple example is the need for an understanding of ecosystem life-processes by the community at large and the interdependence and cascading links between individual ecosystem services. Furthermore, it is vital to understand how this varies region-to-region and culture-to-culture.

Relevance and capacity for different regions, specifically for marine ecosystem services: Human capacity-building (e.g., technology training/education) and the associated physical infrastructure (e.g., coastal marine laboratories and institutes, marine observatories/observations, oceanographic fleets, together with appropriate and robust technology/instrumentation) are important to understand marine biomes as natural capital. This is expensive infrastructure and it is often lacking or operating at a low level in developing countries. Marine research stations are scattered worldwide, are often long established, and can act as important focal points for community-wide understanding and appreciation of marine/coastal ecosystem services. However, they lack capacity to recognize and value ecosystem services or use this approach as an organizing principle. Yet, these infrastructures have the potential to underpin the ecosystem services approach and facilitate gap-filling, e.g., by collecting data relevant to different sea-users and providing avenues to educate local communities. Improvement in these domains requires appropriate national policies in science and significant institutional strengthening. Education and training are vital to share best practices, data and experience and to create a truly global approach. Good examples of the human capital that is available but is, as yet, frag-

Table 3.4 | Evolution of ecosystem services knowledge. Adapted from van den Belt et al., 2013.

	Local	National	Global
Supply of ecosystem services	Need = high resolution data and ability to interlink data for decision-support in the short term. Available = Mixed data and multiple tools; sufficient for scoping purposes in developed countries. Insufficient for management in developed countries. Insufficient for scoping or management in developing countries.	Need = mixed resolution data and ability to interlink data for decision-support in the short and long term. Available = Multiple databases often organized per country and multiple tools.	Need = low resolution data and high ability to interlink and disseminate data for decision-support in the long term. Available = Sufficient data for scoping, insufficient ability to interlink.
Demand for ecosystem services	Need = high ability for recognizing market and non-market sectors in managing tradeoffs. Available = Market-based information often available through the system of national accounting. Non-market-based information depends on local governance and community involvement.	Need = ability for recognizing market and supporting non-market sectors in managing tradeoffs in the short and long term. Available = market-based information and some socio-cultural information depending on country.	Need = ability to support all sectors with understanding of global ecosystem services and humanity's long-term, collective needs. Available = market-based information and some socio-cultural information.
Gap	Matching data between supply and demand of ecosystem services and ability to interconnect with regional/global scales.	Examples of ecosystem services supply; demand-side lagging. Interconnections among ecosystem services and between local and global scales elusive.	Shortage in some global ecosystem services. Interlinkages among global ecosystem services elusive.

mented, in terms of supporting the development and understanding of ecosystem services, are the various networks of marine infrastructures exemplified by MARS (http://www.marsnetwork.org) in Europe and NAML (http://www.naml.org/) in the USA, together with smaller Japanese and Australian counterparts. Recently a global initiative has been launched with the help of the Intergovernmental Oceanographic Commission, i.e., the World Association of Marine Stations (WAMS) (http://www.marsnetwork.org/world-association-marine-stations-wams), with the mission to unite and integrate their strategies from training, education, and outreach to best practice and shared research agendas. New initiatives emerging from the EMBRC consortium (http://www.embrc.eu) and Euromarine (www.euromarinenetwork.eu) are acting as vibrant platforms bringing together all actors in the marine sphere. An important development recently available is The European Marine Training Portal (http://www.marinetraining.eu/). The European Marine Training portal is a centralized access point for education and training in the field of marine sciences. It will help European scientists, technicians and other stakeholders to navigate in the jungle of courses and training opportunities. Marinetraining.eu offers a variety of services to both training organizers and trainees.

Databases and tools available to Marine Stations and Meteorological Centres need to integrate and share data/tools/strategy. Time series are vital for biological/chemical/physical/geological datasets.

As original local studies of ecosystem services are expensive, guidance is needed for local stakeholders and decision-makers to progress from scoping to management tools. This includes a continuum of multiple discount rates relevant to the various ecosystem services (TEEB, 2010a). The network of existing marine research stations and institutes can play a central and coordinating role in providing relevant information and assist in preparation of options to consider bundles of ecosystem services. Many marine stations have historical data sets that, if properly digitized and shared, could help to fill gaps. Many are still locally collecting biogeochemical, biophysical and biodiversity data and recording their changes. These are powerful tools but tend to be restricted to local or regional databases. Although generally not private, they are often not widely known; this is where the United Nations Member States could come together to identify all sources and repositories of knowledge and data and bring them together to benefit the global community. Indeed this is one of the key missions of WAMS, supported by UNESCO-IOC. Whereas this is well recognized in Europe, North America and Australia, for example, an urgent need exists to embrace and empower other less well-supported regions, including but not limited to Africa, South America, the Caribbean, and the Polar regions.

4   Conclusion

Many fundamental Earth system processes are approaching or have crossed safe boundaries for their continued sustainability. Oceans play a crucial role in these Earth systems. After two decades of development, the ecosystem services approach has made good progress in making more visible the benefits people derive from ecosystems, which are often taken for granted. The ecosystem services approach outlined above provides an organizing methodology to assess and analyze the supply of and demand for ecosystem services and to connect across multiple geographic and temporal scales. However, this chapter does not fully outline the necessary steps to determine the potential supply and tradeoffs of ecosystem services for a region. The trans-disciplinary nature of an ecosystem services approach is complex and goes well beyond a mechanical application of both natural and social science, including decision making. The definitions of ecosystem services are multiple and broad and leave room for interpretation. A strong application of the ecosystem services concept can have a transformational impact, shift paradigms and provide new organizing principles advancing sustainability. A weak application of this concept may provide justification for business-as-usual. For example, a robust and strategic application has the potential to create a collaborative space to address fundamental challenges facing humanity; a weak application may address scattered local challenges at best or justify undesirable outcomes at worst.

Establishing principles, approaches and consistent terminology and guidelines for use of marine ecosystem services are needed. Linkages to people are often missing and more data and knowledge on attitudes, perceptions and beliefs of resource users and resource dependents is key. Several networks (e.g., MEA, GEO-BON, IPBES, TEEB, Lisbon Principles) have developed and are further developing such principles and guides. A significant development in Europe is EMBOS (http://www.embos.eu/). This has a focus on observation systems for marine biodiversity. This represents a significant challenge since biodiversity varies over large scales of time and space, and requires research strategies beyond the tradition and capabilities of classic research. Research that covers these scales requires a permanent international network of observation stations with an optimized and standardized methodology. In this way, we recognize that it is increasingly important to develop 'frameworks of frameworks' and understand the underlying purpose and worldview of each contributing framework in order to unify instead of divide the potential support for an ecosystem services approach, especially for oceans. Developing overarching principles, creating consistency in reporting, and generating relevant shared data and information, as well as the capacity to use such information, are creating an exciting opportunity for the United Nations and its members.

The ecosystem services approach has the potential to support a variety of management frameworks, including Marine Spatial Planning and tools for coordinating national and international sustainable marine resource management. Marine laboratories and fleets provide much of the needed data and human capital to better understand the supply of ecosystem services. Opportunities to fill data gaps exist (especially in developing countries), as well as developing capability to make available data suitable for use in ecosystem services approaches. These opportunities should be identified and acted upon with some urgency.

An increasing amount of spatial data/information is readily available/accessible. However, global data are often too coarse in resolution to make accurate estimates for certain regions and the capacity to access and use global data is limited and often lacking in developing countries and even developed countries. In addition, local or fine resolution spatial data and information are often unavailable and expensive. Also, nomenclatures and protocols should be standardized to enable meaningful integration, comparison and shared analyses.

Perhaps the most important gap in knowledge is understanding and integration of an ecosystem services ethos. This can be remedied by initiating a global approach with coordinated knowledge and education transfer amongst both developed and developing nations. Marine ecosystems exist regardless of the status of development of nations, but their integrity is certainly dependent on anthropogenic effects of all kinds under the influence of cultures around the globe. Thus, the ecosystem services approach must be multi-scalar in all facets. A thematic link with IPBES for oceans could address this.

Numerous methodologies have been developed to guide the ecosystem services approach; these range from scoping to highly advanced research and management approaches. Some methodologies provide static 'snapshots,' and others provide a spatially dynamic framework highlighting inter-linkages between bundles of ecosystem services and their changes over time.

The top-down progression of the cascading model (Figure 3.4) reflects steps involved in scoping the provision and value of ecosystem services. Inclusive, participatory approaches are important if we are to enhance ecosystem service models with bottom-up considerations to incorporate non-market and monetary values. The incorporation of local or bottom-up perspectives provides the opportunity to better integrate the distribution of costs and benefits and thereby enhance the fairness of decision-making.

When a participatory, bottom-up approach to ecosystem service valuation is taken, the 'gap' between 'supply of' and 'demand for' ecosystem services can more accurately define and measure 'value'; either there is an abundance, a sufficiency, or a shortage in time and space, applying both market and non-market perspectives. Mapping such gaps and how they change over time and space can be used to identify 'hotspots' for prioritization of management actions at multiple scales. Increasingly, marine ecosystem services are used in marine spatial planning (White et al., 2012; Altman et al., 2014).

It is important that the ecosystem services approach is used to influence beyond the immediate jurisdiction of those undertaking or sponsoring an ecosystem services assessment. Marine ecosystems function independently of national boundaries and Exclusive Economic Zones and so require an integrated global approach, if humanity wants to receive ecosystem services. When local biophysical data are not available, more heuristic methods can still guide conversations among multiple stakeholders to consider options to govern, manage and sustain the 'benefits people derive from ecosystems'.

At a global level, assessment of slow-moving biophysical processes (e.g., climate regulation, ocean acidification) need to be interpreted in terms of ecosystem services for their relevance to and impact on bundles of local ecosystem service in case studies.

In order to facilitate and enable the use of an ecosystem services approach, agreement on a global nomenclature and resulting classification would be useful. However, such a classification ought to be flexible enough to allow for local variability in applications. Therefore, the design of nomenclature, principles, and data management needs to be transparent and display characteristics appropriate to scale and purpose.

In addition to multiple scales, comparability between locations and case studies and over time is important. Some databases go to great lengths to encourage long-term comparability, e.g., Marine Ecosystem Service Partnership and Ecosystem Valuation Tool at Earth Economics. Comparability and transferability apply not only to data-gathering and -formatting, but also to the human component of socializing; using/interlinking such data is equally important (e.g., exchanges and collaborative opportunities).

The available ecosystem services frameworks emphasize that this is an iterative, evolving process and therefore needs an adaptive programme of strategic assessment.

References

Altman, I., Boumans, R., Roman, J., Gopal, S. and Kaufman, L. (2014). An Ecosystem Accounting Framework for Marine Ecosystem-Based Management. *In:* Fogarty, M. J. and McCarthy, J. J. (eds.) *Marine Ecosystem-based Management.* Harvard University Press.

Aronson, J., Milton, S.J., and Blignaut, J.N. (eds.). (2007). *Restoring natural capital: science, business, and practice.* Island Press, Washington, DC. DOI 10.3368/er.24.1.22.

Barbier, E.B., Koch, E.W., Silliman, B.R., Hacker, S.D., Wolanski, E., Primavera, J., Granek, E.F., Polasky, S., Aswani, S., Cramer, L. A., Stoms, D.M., Kennedy, C. J., Bael, D., Kappel, C.V., Perillo, G.M.E. and Reed, D.J. (2008). Coastal ecosystem-based management with nonlinear ecological functions and values. *Science,* 319, 321-323. DOI 10.1126/science.1150349.

Berry, M. and BenDor, T.K. 2015. Integrating sea level rise into development suitability analysis. *Computers, Environment and Urban Systems,* 51, 13-24. Available: DOI 10.1016/j.compenvurbsys.2014.12.004

Beymer-Farris, B.A. and Bassett, T.J. (2012). The REDD menace: Resurgent protectionism in Tanzania's mangrove forests. *Global Environmental Change,* 22, 332-341. DOI: 10.1016/j.gloenvcha.2011.11.006.

Bigagli, E. (2015). The EU legal framework for the management of marine complex social-ecological systems. *Marine Policy,* 54, 44-51. Available: DOI 10.1016/j.marpol.2014.11.025.

Böhnke-Henrichs, A., Baulcomb, C., Koss, R., Hussain, S.S. and de Groot, R.S. (2013). Typology and indicators of ecosystem services for marine spatial planning and management. *Journal of Environmental Management,* 130, 135-145. Available: DOI 10.1016/j.jenvman.2013.08.027.

Boyd, J. and Banzhaf, S. (2007). What are ecosystem services? The need for standardized environmental accounting units. *Ecological Economics,* 63, 616-626.

Braat, L.C. and de Groot, R. (2012). The ecosystem services agenda: bridging the worlds of natural science and economics, conservation and development, and public and private policy. *Ecosystem Services,* 1, 4-15.

Bscher, B., Sullivan, S., Neves, K., Igoe, J., and Brockington, D. (2012). Towards a synthesized critique of neoliberal biodiversity conservation. *Capitalism, Nature, Socialism, 23*(2), 4-30.

Buck, B.H., Krause, G. and Rosenthal, H. (2004). Extensive open ocean aquaculture development within wind farms in Germany: the prospect of offshore co-management and legal constraints. *Ocean & Coastal Management,* 47, 95-122.

Burkhard, F., Kroll, Nedkov, S. and Müller, F. (2012). Mapping ecosystem service supply, demand and budgets. *Ecological Indicators,* 21, 17-29.

Butler, J.R.A., Wong, G.Y., Metcalfe, D.J., Honzak, M., Pert, P.L., Rao, N., van Grieken, M.E., Lawson, T., Bruce, C., Kroon, F.J. and Brodie, J.E. (2013). An analysis of trade-offs between multiple ecosystem services and stakeholders linked to land use and water quality management in the Great Barrier Reef, Australia. *Agriculture Ecosystems & Environment,* 180, 176-191.

Carpenter, S. R., Brock, W. A. and Ludwig, D. (2007). Appropriate discounting leads to forward-looking ecosystem management. *Ecological Research,* 22, 10-11.

Chan, K.M.A., Guerry, A.D., Balvanera, P., Klain, S., Satterfield, T., Basurto, X., Bostrom, A., Chuenpagdee, R., Gould, R., Halpern, B.S., Hannahs, N., Levine, J., Norton, B., Ruckelshaus, M., Russell, R., Tam, J. and Woodside, U. (2012). Where are cultural and social in ecosystem services? A framework for constructive engagement. *BioScience,* 62, 744-756.

Costanza, R., Andrade, F., Antunes, P., van den Belt, M., Boersma, D., Boesch, D.F., Catarino, F., Hanna, S., Limburg, K., Low, B., Molitor, M., Pereira, J. G., Rayner, S., Santos, R., Wilson, J. and Young, M. (1998). Principles for sustainable governance of the oceans. *Science,* 281, 198-199.

Costanza, R. (2008). Ecosystem services: Multiple classification systems are needed. *Biological Conservation,* 141, 350-352. DOI 10.1016/j.biocon.2007.12.020.

Costanza, R., Darge, R., deGroot, R., Farber, S., Grasso, M., Hannon, B., Limburg, K., Naeem, S., Oneill, R. V., Paruelo, J., Raskin, R. G., Sutton, P. and van den Belt, M. (1997). The value of the world's ecosystem services and natural capital. *Nature,* 387, 253-260. DOI 10.1038/387253a0.

Costanza, R., de Groot, R., Sutton, P.C., van der Ploeg, S., Anderson, S, Kubiszewski, I., Farber, S. Turner, K. (2014). "Changes in the glogal value of ecosystem services." *Global Environmental Change,* vol 26: 152-158.

Daily, G. (Ed.). (1997). *Nature's Services: Societal Dependence on Natural Ecosystems.* Washington DC: Island Press.

Daly, H. and Cobb, J. (1989). *For the Common Good: Redirecting the Economy Toward Community, the Environment and a Sustainable Future.* Boston, Beacon Press.

Daly, H.E. and Farley, J. (2004). *Ecological economics: principles and applications.* Washington: Island Press. 454 pages.

de Groot, R. (2011). What are Ecosystem Services. *In:* Van den Belt, M. and Costanza, R. (eds.) *Ecological Economics of Estuaries and Coasts.* Elsevier Academic Press

Döring, R., and Egelkraut, T.M. (2008). Investing in natural capital as management strategy in fisheries: The case of the Baltic Sea cod fishery. *Ecological Economics, 64*(3), 634-642.

FAO. (2012). Payments for Ecosystem Services,(accessed 7 May 2015) http://www.fao.org/fileadmin/templates/nr/sustainability_pathways/docs/Factsheet_PES.pdf.

Fisher, B., Turner, R. K., and Morling, P. (2009). Defining and classifying ecosystem services for decision making. *Ecological Economics, 68*(3), 643-653. doi: 10.1016/j.ecolecon.2008.09.014.

Farley, J. (2012). Ecosystem Services: The Economic Debate. *Ecosystem Services,* 1, 9.

Foley, M.M., Halpern, B.S., Micheli, F., Armsby, M.H., Caldwell, M.R., Crain, C.M., Prahler, E., Rohr, N., Sivas, D., Beck, M.W., Carr, M.H., Crowder, L.B., Emmett Duffy, J., Hacker, S.D., McLeod, K.L., Palumbi, S.R., Peterson, C.H., Regan, H.M., Ruckelshaus, M.H., Sandifer, P.A. and Steneck, R.S. (2010). Guiding ecological principles for marine spatial planning. *Marine Policy,* 24, 11. DOI http://dx.doi.org/10.1016/j.marpol.2010.02.001.

Ghermandi, A., Nunes, P., Portela, R., Rao, N. and Teelucksingh, S. (eds.) (2011). *Recreational, Cultural, and Aesthetic Services from Estuarine and Coastal Ecosystems.* Burlington, MA: Elsevier.

Granek, E.F., Polasky, S., Kappel, C.V., Stoms, D.M., Reed, D.J., Primavera, J., Koch, E.W., Kennedy, C., Cramer, L.A., Hacker, S.D., Perillo, G.M.E., Aswani, S., Silliman, B., Bael, D., Muthiga, N., Barbier, E.B., Wolanski, E. (2010). Ecosystem services as a common language for coastal ecosystem-based management. *Conservation Biology* 24, 207-216.

Haines-Young, R. and Potschin, M. (2010). *Proposal for a common international classification of ecosystem goods and services (CICES) for integrated environmental and economic accounting.* Department of Economic and Social Affairs Statistical division, United Nations (ESA/STAT/AC.217-UNCEEA/5/7/Bk).

Halpern, B.S., Walbridge, S., Selkoe, K.A., Kappel, C.V., Micheli, F., D'Agrosa, C., Bruno, J.F., Casey, K.S., Ebert, C., Fox, H.E., Fujita, R., Heinemann, D., Lenihan, H.S., Madin, E.M.P., Perry, M.T., Selig, E.R., Spalding, M., Steneck, R. and Watson, R. (2008). A global map of human impact on marine ecosystems. *Science,* 319, 948-952. DOI 10.1126/science.1149345.

Halpern, B.S., Longo, C., Hardy, D., McLeod, K.L., Samhouri, J.F., Katona, S.K., Kleisner, K., Lester, S.E., O'Leary, J., Ranelletti, M., Rosenberg, A.A., Scarborough, C., Selig, E.R., Best, B.D., Brumbaugh, D.R., Chapin, F.S., Crowder, L.B., Daly, K.L., Doney, S.C., Elfes, C., Fogarty, M.J., Gaines, S.D., Jacobsen, K.I., Karrer, L.B., Leslie, H.M., Neeley, E., Pauly, D., Polasky, S., Ris, B., St. Martin, K., Stone, G.S., Sumalia, U.R., Zeller, D., (2012). An index to assess the health and benefits of the global ocean. *Nature,* 488, 615. DOI 10.1038/nature11397.

Hendriks, K. Braat, L. Ruijs, A., van Egmond, P., Melman, D., van der Heide, M., Klok, C., Gaaff, Dietz, F. (2012). *TEEB voor Fysiek Nederland*, Alterra, Wageningen, ISSN 1566-7197 (page 32).

Kelble, C.R., Loomis, D.K., Lovelace, S., Nuttle, W.K., Ortner, P.B., Fletcher, P., Cook, G.S., Lorenz, J.J. and Boyer, J.N. (2013). The EBM-DPSER Conceptual Model: Integrating Ecosystem Services into the DPSIR Framework. *PLoS ONE,* 8(8): DOI:10.1371/journal.pone.0070766.

Kim, C.-K., Toft, J.E., Papenfus, M., Verutes, G., Guerry, A.D., Ruckelshaus, M.H., Arekema, K.K., Guannel, G., Wood, S.A., Bernhardt, J.R., Tallis, H., Plummer, M.L., Halpern, B.S., Pinsky, M.L., Beck, M.W., Chan, F., Chan, K.M.A. and Polasky, S. (2012). Catching the right wave: evaluating wave energy resources and potental compatibility with existing marine and coastal uses. *PLoS ONE,* 7 (11). DOI: 10.1371/journal.pone.0047598.

Landers, D. H. and Nahlik, A.M. (2013). *Final ecosystem goods and services classification system (FEGS-CS)*. Corvallis, Oregon: US Environmental Protection Agency.

Larigauderie, A., Mooney, H.A. (2010). The Intergovernmental science-policy Platform on Biodiversity and Ecosystem Services: moving a step closer to an IPCC-like mechanism for biodiversity. *Current Opinion in Environmental Sustainability,* 2(1-2), 9-14. doi:10.1016/j.cosust.2010.02.006.

Lester, S.E., Costello, C., Halpern, B.S., Gaines, S.D., White, C. and Barth, J.A. (2012). Evaluating tradeoffs among ecosystem services to inform marine spatial planning. *Marine Policy* 38, 80-89. DOI 10.1016/j.marpol.2012.05.022.

Liquete, C., Piroddi, C., Drakou, E. G., Gurney, L., Katsanevakis, S., Charef, A. and Egoh, B. (2013). Current Status and Future Prospects for the Assessment of Marine and Coastal Ecosystem Services: A Systematic Review. *PLoS ONE,* 8(7). DOI 10.1371/journal.pone.0067737.

Lui, S., Portela, R., Rao, N., Ghermandi, A., Wang, X. (2011). Environmental benefit transfers of ecosystem service valuation. *In:* Wolanski, E. and McClusky, D.S. (eds.) *Treatise on Estuarine and Coastal Science.* Burlington MA: Academic Press.

MacDiarmid A.B., Law C.S., Pinkerton M., Zeldis J. (2013). New Zealand marine ecosystem services. In Dymond J.R. (ed.) *Ecosystem services in New Zealand: conditions and trends.* Manaaki Whenua Press, Lincoln, New Zealand

Maes, J., Egoh, B., Willemen, L., Liquete, C., Vihervaara, P., Schägner, J.P., Grizzetti, B., Drakou, E.G., Notte, A. L., Zulian, G., Bouraoui, F., Luisa Paracchini, M., Braat, L. and Bidoglio, G. (2012a). Mapping ecosystem services for policy support and decision making in the European Union. *Ecosystem Services,* 1, 31-39.

Maes, J., Paracchini, M.L., Zulian, G., Dunbar, M.B. and Alkemade, R. (2012b). Synergies and trade-offs between ecosystem service supply, biodiversity, and habitat conservation status in Europe. *Biological Conservation,* 155, 1-12. DOI http://dx.doi.org/10.1016/j.biocon.(2012).06.016.

Martinez-Harms, M.J. and Balvanera, P. (2012). Methods for mapping ecosystem service supply: A review. *International Journal of Biodiversity Science, Ecosystems Services and Management,* 8, 17-25.

McClintock, W., Paul, E., Burt, C. and Bryan, T. (2012). *McClintock Lab: seasketch*. UC Santa Barbara, Santa Barbara, CA 93106-6150 Marine Science Institute. http://mcclintock.msi.ucsb.edu/projects/seasketch [Accessed 13th September 2012].

McCauley, D.J. (2006). Nature: McCauley replies [4]. *Nature, 443*(7113), 750.

Menzel, S., Kappel, C.V., Broitman, B.R., Micheli, F. and Rosenberg, A.A. (2013). Linking human activity and ecosystem condition to inform marine ecosystem based management. Aquatic Conservation: *Marine and Freshwater Ecosystems,* 506.

Millennium Ecosystem Assessment (2005). *Ecosystems and human well-being.* Island Press, Washington, D.C.

Pike, K., Wright, P., Wink, B., and Fletcher, S. (2014). The assessment of cultural ecosystem services in the marine environment using Q methodology. *Journal of Coastal Conservation*.

Rockstrom, J., Steffen, W., Noone, K., Persson, A., Chapin, F.S., Lambin, E.F., Lenton, T.M., Scheffer, M., Folke, C., Schellnhuber, H.J., Nykvist, B., de Wit, C. A., Hughes, T., van der Leeuw, S., Rodhe, H., Sorlin, S., Snyder, P.K., Costanza, R., Svedin, U., Falkenmark, M., Karlberg, L., Corell, R.W., Fabry, V.J., Hansen, J., Walker, B., Liverman, D., Richardson, K., Crutzen, P. and Foley, J. A. (2009). A safe operating space for humanity. *Nature,* 461, 472-475. DOI Doi 10.1038/461472a.

Rodríguez, J.P., Beard Jr, T.D., Bennett, E.M., Cumming, G.S., Cork, S.J., Agard, J., Dobson, A.P. and Peterson, G.D. (2006). Trade-offs across space, time, and ecosystem services. *Ecology and Society,* 11.

Shaw, W. D. and Wlodarz, M. (2013). Ecosystems, ecological restoration, and economics: Does habitat or resource equivalency analysis mean other economic valuation methods are not needed? *Ambio,* 42, 628-643.

Tallis, H., Mooney, H., Andelman, S., Balvanera, P., Cramer, W., Karp, D., Polasky, S., Reyers, B., Ricketts, T., Running, S., Thonicke, K., Tietjen, B. and Walz, A. (2012). A Global System for Monitoring Ecosystem Service Change. *BioScience,* 62, 977-986. DOI 10.1525/bio.2012.62.11.7.

TEEB. (2009). *The Economics of Ecosystems and Biodiversity for National and International Policy Makers*. The Economics of Ecosystems and Biodiversity (TEEB).

TEEB. (2010a). *The Economics of Ecosystems and Biodiversity Ecological and Economic Foundations* Earthscan, London and Washington.

TEEB. (2010b). *Mainstreaming the Economics of Nature: A synthesis of the approach, conclusions and recommendations of TEEB*. The Economics of Ecosystems and Biodiversity.

TEEB. (2010c). *A quick guide to TEEB for Local and Regional Policy Makers*. The Economics of Ecosystems and Biodiversity (TEEB).

Townsend, M. Thrush, S. (2010). Ecosystem functioning, goods and services in the coastal environment. Prepared by the National Institute of Water and Atmospheric Research for Auckland Regional Council. Auckland Regional Council Technical Report 2010/033.

Townsend, M., Thrush, S. F. and Carbines, M. J. (2011). Simplifying the complex: An 'Ecosystem principles approach' to goods and services management in marine coastal ecosystems. *Marine Ecology Progress Series,* 434, 291-301.

United Nations Atlas of Oceans, www.oceansatlas.org, accessed on 16 April, 2015

United Nations Statistics Division. (2013). *System of Environmental-Economic Accounting 2012: Experimental Ecosystem Accounting*. European Commission, Organisation for Economic Co-operation and Development, United Nations, World Bank.

UK National Ecosystem Assessment. (2011). *The UK National Ecosystem Assessment: Synthesis of the Key Findings*. Cambridge: UNEP-WCMC.

van den Belt, M. (2011a). Ecological Economics of Estuaries and Coasts. *In:* Wolanski, E. and McClusky, D.S. (eds.) *Treatise on Estuarine and Coastal Science.* Burlington MA: Academic Press.

van den Belt, M., Forgie, V.E. and Farley, J. (2011b). Valuation of Coastal Ecosystem Services. *In:* Wolanski, E. and D.S., M. (eds.) *Ecological Economics of Estuaries and*

Coasts. Burlington MA: Elsevier.

van den Belt, M., McCallion, A., Wairepo, S., Hardy, D., Hale, L. and Berry, M. (2012). *Mediated Modelling of Coastal Ecosystem Services: A case study of Te Awanui Tauranga Harbour*, Manaaki Taho Moana project.

van den Belt, M. (2013, August 26-30). Integrating Ecosystem Services Valuation Case Studies, Valuation Databases and Ecosystem Services Modeling, *Proceedings of the* 6th Annual International Ecosystem Services Partnership Conference, Bali, Indonesia.

van den Belt, M. and Cole, A.O. (2014). *Ecosystem Services of Marine Protected Areas*. Wellington, NZ: Department of Conservation.

Wegner, G., and Pascual, U. (2011). Cost-benefit analysis in the context of ecosystem services for human well-being: A multidisciplinary critique. *Global Environmental Change, 21*(2), 492-504.

White, C., Halpern, B.S. and Kappel, C.V. (2012). Ecosystem service tradeoff analysis reveals the value of marine spatial planning for multiple ocean uses. *Proceedings of the National Academy of Sciences of the United States of America,* 109, 4696-4701. DOI 10.1073/pnas.1114215109.

Wolff, S., Schulp, C.J.E. and Verburg, P.H. (2015). Mapping ecosystem services demand: A review of current research and future perspectives. *Ecological Indicators*, 55, 159-171. Available: DOI 10.1016/j.ecolind.2015.03.016.

Zurlini, G., Petrosillo, I., Aretano, R., Castorini, I., D'Arpa, S., De Marco, A., Pasimeni, M.R., Semeraro, T. and Zaccarelli, N. (2014). Key fundamental aspects for mapping and assessing ecosystem services: Predictability of ecosystem service providers at scales from local to global. *Annali di Botanica*, 4, 53-63. Available: DOI 10.4462/annbotrm-11754.

Appendix 1. Case studies and references related to Tables 2 and 3

1. Abu-Hilal, A. H. 1994. Effect of Depositional Environment and Sources of Pollution on Uranium Concentration in Sediment, Coral, Algae and Seagrass Species from the Gulf of Aqaba (Red Sea). *Marine Pollution Bulletin,* 28, 81-88.
2. Airoldi, L., Balata, D. and Beck, M. W. 2008. The Gray Zone: Relationships between habitat loss and marine diversity and their applications in conservation. *Journal of Experimental Marine Biology and Ecology*, 8. Available: DOI 10.1016/j.jembe.2008.07.034.
3. Alongi, D. M. 2012. Carbon sequestration in mangrove forests. *Carbon Management,* 3, 313-322.
4. Alongi, D. M., Chong, V. C., Dixon, P., Sasekumar, A. & Tirendi, F. 2003. The influence of fish cage aquaculture on pelagic carbon flow and water chemistry in tidally dominated mangrove estuaries of peninsular Malaysia. *Marine Environmental Research,* 55, 313-333.
5. Avis, A. M. A review of coastal dune stabilization in the Cape Province of South Africa. *Landscape and Urban Planning,* 18, 55-68. Available: DOI 10.1016/0169-2046(89)90055-8.
6. Barbier, E. B., Hacker, S. D., Kennedy, C., Koch, E. W., Stier, A. C. & Silliman, B. R. 2011. The value of estuarine and coastal ecosystem services. *Ecological Monographs*, 169.
7. Bridger, C. J. & Costa-Pierce, B. A. (eds.) 2003. *Open Ocean Aquaculture: From Research to Commercial Reality.* Baton Rouge, LA: The World Aquaculture Society.
8. Buck, B. H., Krause, G. & Rosenthal, H. 2004. Extensive open ocean aquaculture development within wind farms in Germany: the prospect of offshore co-management and legal constraints. *Ocean & Coastal Management,* 47, 95.
9. Burke, L. & Maidens, J. 2004. *Reefs at risk in the Caribbean*. World Resources Institute, Washington, D.C. Report number: 9781569735671156973567 0.
10. Carleton, J. M. 1974. Land-building and Stabilization by Mangroves. *Environmental Conservation,* 1, 285.
11. Cesar, H., Burke, L. & Pet-Soede, L. 2003. The economics of worldwide coral reef degradation.
12. Chmura, G. L., Anisfeld, S. C., Cahoon, D. R. & Lynch, J. C. 2003. Global carbon sequestration in tidal, saline wetland soils. *Global Biogeochemical Cycles,* 17, n/a.
13. Davenport, J. & Davenport, J. L. 2006. The impact of tourism and personal leisure transport on coastal environments: A review. *Estuarine, Coastal and Shelf Science*, 280. Available: DOI 10.1016/j.ecss.2005.11.026.
14. Davis, D., Banks, S., Birtles, A., Valentine, P. & Cuthill, M. 1997. Whale sharks in Ningaloo Marine Park: managing tourism in an Australian marine protected area. *Tourism Management (United Kingdom)*.
15. DeWalt, B. R., Vergne, P. & Hardin, M. 1996. Shrimp aquaculture development and the environment : people, mangroves and fisheries on the Gulf of Fonseca, Honduras. *World development : the multi-disciplinary international journal devoted to the study and promotion of world development,* 24, 1193-1208.
16. Dorenbosch, M., Grol, M. G. G., Christianen, M. J. A., Nagelkerken, I. & van der Velde, G. 2005. Indo-Pacific seagrass beds and mangroves contribute to fish density and diversity on adjacent coral reefs. *Marine Ecology - Progress Series,* 302, 63-76.
17. Dorenbosch, M., van Riel, M. C., Nagelkerken, I. & van der Velde, G. 2004. The relationship of reef fish densities to the proximity of mangrove and seagrass nurseries. *Estuarine Coastal & Shelf Science,* 60, 37. Available: DOI 10.1016/j.ecss.2003.11.018.
18. Falkowski, P., Scholes, R. J., Boyle, E., Canadell, J., Canfield, D., Elser, J., Gruber, N., Hibbard, K., Hogberg, P., Linder, S., Mackenzie, F. T., Moore, B., III, Pedersen, T., Rosenthal, Y., Seitzinger, S., Smetacek, V. & Steffen, W. 2000. The Global Carbon Cycle: A Test of Our Knowledge of Earth as a System. *Science*.
19. Fourqurean, J. W., Duarte, C. M., Kennedy, H., Marbà, N., Holmer, M., Mateo, M. A., Apostolaki, E. T., Kendrick, G. A., Krause-Jensen, D., McGlathery, K. J. & Serrano, O. 2012. Seagrass ecosystems as a globally significant carbon stock.
20. Fujimoto, K. 2004. Below-ground carbon sequestration of mangrove forests in the Asia-Pacific region. *In:* Vannucci, M. (ed.) *Mangrove management and conservation workshop.* Okinawa, Japan.
21. Gattuso, J. P., Frankignoulle, M. & Wollast, R. 1998. Carbon and Carbonate Metabolism in Coastal Aquatic Ecosystems. *Annual Review of Ecology, Evolution & Systematics,* 29, 405.
22. Gedan, K. B., Kirwan, M. L., Wolanski, E., Barbier, E. B. & Silliman, B. R. 2011. The present and future role of coastal wetland vegetation in protecting shorelines: answering recent challenges to the paradigm. *Climatic Change*, 7.
23. Harbison, P. 1986. Mangrove muds--a sink and a source for trace metals. *Marine Pollution Bulletin,* 17, 246.
24. Harrold, C. & Reed, D. C. 1985. Food Availability, Sea Urchin Grazing, and Kelp Forest Community Structure. *Ecology,* 1160. Available: DOI 10.2307/1939168.
25. Hawkins, J. P. & Roberts, C. M. 1994. The Growth of Coastal Tourism in the Red Sea: Present and Future Effects on Coral Reefs. *Ambio*, 503. Available: DOI 10.2307/4314268.
26. Hoagland, P., Jin, D. & Kite-Powell, H. 2003. The optimal allocation of ocean space: aquaculture and wild-harvest fisheries. *Marine Resource Economics*, 129.
27. Kennedy, H., Beggins, J., Duane, C. M., Fourqurean, J. W., Holmer, M., Marbà, N. & Middelburg, J. J. 2010. Seagrass sediments as a global carbon sink: Isotopic constraints. *Global Biogeochemical Cycles,* 24, 1.
28. Kinsey, D. W. & Hopley, D. 1991. Research paper: The significance of coral reefs as global carbon sinks— response to Greenhouse. *Palaeogeography, Palaeoclimatology, Palaeoecology,* 89, 363-377. Available: DOI 10.1016/0031-0182(91)90172-n.
29. Knutson, P., Brochu, R., Seelig, W. & Inskeep, M. 1982. Wave damping in Spartina alterniflora marshes. *Wetlands,* 2, 87.
30. Koch, E. W., Barbier, E. B., Silliman, B. R., Reed, D. J., Perillo, G. M. E., Hacker, S. D., Granek, E. F., Primavera, J. H., Muthiga, N., Polasky, S.,

Halpern, B. S., Kennedy, C. J., Kappel, C. V. & Wolanski, E. 2009. Non-Linearity in Ecosystem Services: Temporal and Spatial Variability in Coastal Protection. *Frontiers in Ecology and the Environment*, 29. Available: DOI 10.2307/25595035.

31. Kuhlbusch, T. A. J. 1998. Enhanced: Black Carbon and the Carbon Cycle *Science*, 280, 1903-1904. Available: DOI DOI:10.1126/science.280.5371.1903.

32. Laffoley, D. d. A. & Grimsditch, G. (eds.) 2009. *The management of natural coastal carbon sinks.* Switzerland: International Union for Conservation of Nature and Natural Resources.

33. Lal, R. 2005. Forest soils and carbon sequestration. *Forest Ecology and Management.* Elsevier.

34. Lemmens, J. W. T. J., Clapin, G., Lavery, P. & Cary, J. 1996. Filtering capacity of seagrass meadows and other habitats of Cockburn Sound, Western Australia. *MARINE ECOLOGY- PROGRESS SERIES,* 143, 187-200.

35. Levin, N. & Ben-Dor, E. Monitoring sand dune stabilization along the coastal dunes of Ashdod-Nizanim, Israel, 1945–1999. *Journal of Arid Environments,* 58, 335-355. Available: DOI 10.1016/j.jaridenv.2003.08.007.

36. Li, Y.-L., Wang, L., Zhang, W.-Q., Zhang, S.-P., Wang, H.-L., Fu, X.-H. & Le, Y.-Q. 2010. Variability of soil carbon sequestration capability and microbial activity of different types of salt marsh soils at Chongming Dongtan. *Ecological Engineering*, 1754. Available: DOI 10.1016/j.ecoleng.2010.07.029.

37. McLeod, E., Chmura, G. L., Bouillon, S., Salm, R., Björk, M., Duarte, C. M., Lovelock, C. E., Schlesinger, W. H. & Silliman, B. R. 2011. A blueprint for blue carbon: toward an improved understanding of the role of vegetated coastal habitats in sequestering CO2. *Frontiers in Ecology & the Environment,* 9, 552.

38. Menzel, S., Kappel, C. V., Broitman, B. R., Micheli, F. & Rosenberg, A. A. 2013. Linking human activity and ecosystem condition to inform marine ecosystem based management. *Aquatic Conservation: Marine and Freshwater Ecosystems*, 506.

39. Minh, T., Yakuitiyage, A., and Macintosh, D.J. 2001. *Management of the integrated mangrove-aquaculture farming systems in the Mekong Delta of Vietnam.* Pathumthani, Thailand Integrated Tropical Coastal Zone Management, School of Environment, Resources, and Development, Asian Institute of Technology, .

40. Moreno-Casasola, P. 1986. Sand Movement as a Factor in the Distribution of Plant Communities in a Coastal Dune System. *Vegetatio*, 67. Available: DOI 10.2307/20037269.

41. Nagelkerken, I., Velde, G. v. d., Gorissen, M. W., Meijer, G. J., Hof, T. V. t. & Hartog, C. d. Regular Article: Importance of Mangroves, Seagrass Beds and the Shallow Coral Reef as a Nursery for Important Coral Reef Fishes, Using a Visual Census Technique. *Estuarine, Coastal and Shelf Science*, 51, 31-44. Available: DOI 10.1006/ecss.2000.0617.

42. Ohde, S. & van Woesik, R. 1999. Carbon dioxide flux and metabolic processes of a coral reef, Okinawa. *Bulletin of Marine Science*, 65, 559-576.

43. Paddack, M. J. & Estes, J. A. 2000. Kelp forest fish populations in marine reserves and adjacent exploited areas of central California. *Ecological Applications*, 855.

44. Pauly, D. & Christensen, V. 1995. Primary production required to sustain global fisheries. *Nature*, 255.

45. Pinn, E. H. & Rodgers, M. 2005. The influence of visitors on intertidal biodiversity. *Journal of the Marine Biological Association of the United Kingdom,* 85, 263.

46. Quiors, A. 2005. Whale shark "ecotourism" in the Philippines and Belize: evaluating conservation and community benefits. *Tropical Resources: Bulletin of the Yale Tropical Resources Institute,* 24, 42-48.

47. Robertson, A. I. & Phillips, M. J. 1995. Mangroves as filters of shrimp pond effluent: predictions and biogeochemical research needs. *Hydrobiologia,* 295, 311.

48. Ronnback, P. 1999. The Ecological Basis for Economic Value of Seafood Production Supported by Mangrove Ecosystems. *Ecological Economics,* 29, 235-252. Available: DOI http://www.sciencedirect.com/science/journal/09218009.

49. Sathirathai, S. & Barbier, E. B. 2001. Valuing Mangrove Conservation in Southern Thailand. *Contemporary Economic Policy*, 19, 109-122. Available: DOI http://onlinelibrary.wiley.com/journal/10.1111/%28ISSN%291465-7287.

50. Schiel, D. R. & Taylor, D. I. Effects of trampling on a rocky intertidal algal assemblage in southern New Zealand. *Journal of Experimental Marine Biology and Ecology,* 235, 213-235. Available: DOI 10.1016/s0022-0981(98)00170-1.

51. Schlacher, T. A., de Jager, R. & Nielsen, T. 2011. Vegetation and ghost crabs in coastal dunes as indicators of putative stressors from tourism. *Ecological Indicators,* 11, 284-294. Available: DOI 10.1016/j.ecolind.2010.05.006.

52. Short, F. T. & Short, C. A. 1984. The seagrass filter: purification of estuarine and coastal waters. *In:* Kennedy, V. S. (ed.) *The estuary as a filter.* Orlando: Academic Press.

53. Siegfried, W. R., Crawford, R. J. M., Shannon, L. V., Pollock, D. E., Payne, A. I. L. & Krohn, R. G. 1990. Scenarios for global-warming induced change in the open-ocean environment and selected fisheries of the west coast of Southern Africa. *South African Journal of Science*, 281-285.

54. Smith, S. V. 1981. Marine macrophytes as global carbon sink. *Science*, 838.

55. Steneck, R. S., Graham, M. H., Bourque, B. J., Corbett, D., Erlandson, J. M., Estes, J. A. & Tegner, M. J. 2002. Kelp forest ecosystems: biodiversity, stability, resilience and the future. *Environmental Conservation,* 29, 436-459.

56. Tegner, M. J. & Dayton, P. K. 2000. Ecosystem effects of fishing in kelp forest communities. *ICES Journal of Marine Science / Journal du Conseil,* 57, 579.

57. van der Meulen, F. & Salman, A. H. P. M. Management of Mediterranean coastal dunes. *Ocean and Coastal Management,* 30, 177-195. Available: DOI 10.1016/0964-5691(95)00060-7.

58. Vianna, G. M. S., Meekan, M. G., Pannell, D. J., Marsh, S. P. & Meeuwig, J. J. 2012. Socio-economic value and community benefits from shark-diving tourism in Palau: A sustainable use of reef shark populations. *Biological Conservation*, 267. Available: DOI 10.1016/j.biocon.2011.11.022.

59. Walsh, J. J., Rowe, G. T., Iverson, R. L. & McRoy, C. P. 1981. Biological export of shelf carbon is a sink of the global CO2 cycle. *Nature,* 291, 196.
60. Ward, P. & Myers, R. A. 2005. Shifts in Open-Ocean Fish Communities Coinciding with the Commencement of Commercial Fishing. *Ecology*, 835. Available: DOI 10.2307/3450838.
61. Wells, S., Ravilious, C. & Corcoran, E. 2006. *In the front line. Shoreline protection and other ecosystem services from mangroves and coral reefs. UNEP-WCMC Biodiversity Series 24.* UNEP-WCMC 2006. [Accessed 20130927].
62. Woodhouse, W. W., Broome, S. W., Seneca, E. D., Center, C. E. R. & University., N. C. S. 1974. *Propagation of Spartina alterniflora for substrate stabilization and salt marsh development / by W.W. Woodhouse, Jr., E.D. Seneca, and S.W. Broome.* Fort Belvoir, Va. :, Coastal Engineering Research Center. Available: DOI 10.5962/bhl.title.47587.

Appendix 2. Overview of databases available to support ecosystem services assessments.

1 Sources of information for ecosystem services approaches: from biomes to case studies and tradeoffs

CBD: Convention on Biological Diversity
, www.cbd.int/ebsa
Conservation International
http://www.conservation.org
CICES: Common International Classification of ES
http://unstats.un.org/unsd/envaccounting/
UNEP: United Nations Environment Programme
www.unep.org
EMBOS: European Marine Biodiversity Observation System :
http://www.embos.eu
EEA: European environment agency
www.eea.europa.eu
GEF: Global Envirinment Facility International Waters Learning Exchange & Resource Network
http://iwlearn.net
TEEB: The Economics of Ecosystems & Biodiversity
www.teebweb.org/
Earth Economics
www.eartheconomics.org
GEOBON
www.earthobservations.org/geobon.shtml
NCEAS: National Center for Ecological Analysis and Synthesis
www.nceas.ucsb.edu/ebm
Ocean Health Index
http://www.oceanhealthindex.org/
MESP: marine ecosystem services partnership
http://marineecosystemservices.org/explore

2 Additional databases /Biomes

Ocean
http://www.oceanhealthindex.org/
all biomes
www.teebweb.org
NatureServe: Tools for EBM of Coastal and Marine Environments
www.natureserve.org
http://ebmtoolsdatabase.org/

3 Case studies

MAES: Mapping & Assessment of Ecosystems and their Services (EU)
http://ec.europa.eu/environment/nature/knowledge/
GIFS: The Geography of Inshore Fishing and Sustainability (North Sea)
http://www.gifsproject.eu/en/toolkit

Global Socioeconomic Monitoring Initiative for Coastal Management (SocMon)
www.socmon.org

4 Regional cases

EU Integrated Maritime Policy *Marine protected areas (*Mediterranean Sea)
http://planbleu.org/en/ressources-donnees/simedd
MPAs+MCZs+SACs (UK)
http://uknea.unep-wcmc.org/
MPA (California, US)
http://www.dfg.ca.gov/mlpa
HELCOM (Baltic)
www.helcom.fi
Rhode Island (US)
www.seagrant.gso.uri.edu
Oregon (US)
http://www.oregon.gov/LCD/OCMP/
ESP: Ecosystem Services Partnership EBM (west pacific/Philiples)
http://www.es-partnership.org

5 Arctic

Arctic
http://www.abds.is/

6 Tradeoffs and decision support

InVEST: Integration Valuation of Environmental Services and Tradeoffs
www.naturalcapitalproject.org/InVEST.html
MIMES: Multi-scale Integrated Model of Ecosystem Services & Natural Capital
http://www.afordablefutures.com
SoLVES: Social Values for Ecosystem Services
http://solves.cr.usgs.gov
iMarine
https://portal.i-marine.d4science.org
FSD: Foundation for Sustainable Development
http://www.fsd.nl

4 The Ocean's Role in the Hydrological Cycle

Contributors:
Deirdre Byrne and Carlos Garcia-Soto (Convenors), Gordon Hamilton, Eric Leuliette, LisanYu, Edmo Campos, Paul J. Durack, Giuseppe M.R. Manzella, Kazuaki Tadokoro, Raymond W. Schmitt, Phillip Arkin, Harry Bryden, Leonard Nurse, John Milliman, Lorna Inniss (Lead Member), Patricio Bernal (Co-Lead Member)

1 The interactions between the seawater and freshwater segments of the hydrological cycle

The global ocean covers 71 per cent of the Earth's surface, and contains 97 per cent of all the surface water on Earth (Costello et al., 2010). Freshwater fluxes into the ocean include: direct runoff from continental rivers and lakes; seepage from groundwater; runoff, submarine melting and iceberg calving from the polar ice sheets; melting of sea ice; and direct precipitation that is mostly rainfall but also includes snowfall. Evaporation removes freshwater from the ocean. Of these processes, evaporation, precipitation and runoff are the most significant at the present time.

Using current best estimates, 85 per cent of surface evaporation and 77 per cent of surface rainfall occur over the oceans (Trenberth et al., 2007; Schanze et al., 2010). Consequently, the ocean dominates the global hydrological cycle. Water leaving the ocean by evaporation condenses in the atmosphere and falls as precipitation, completing the cycle. Hydrological processes can also vary in time, and these temporal variations can manifest themselves as changes in global sea level if the net freshwater content of the ocean is altered.

Precipitation results from the condensation of atmospheric water vapour, and is the single largest source of freshwater entering the ocean (~530,000 km3/yr). The source of water vapour is surface evaporation, which has a maximum over the subtropical oceans in the trade wind regions (Yu, 2007). The equatorward trade winds carry the water vapour evaporated in the subtropics to the Intertropical Convergence Zone (ITCZ) near the equator, where the intense surface heating by the sun causes the warm moist air to rise, producing frequent convective thunderstorms and copious rain (Xie and Arkin, 1997). The high rainfall and the high temperature support and affect life in the tropical rainforest (Malhi and Wright, 2011).

Evaporation is enhanced as global mean temperature rises (Yu, 2007). The water-holding capacity of the atmosphere increases by 7 per cent for every degree Celsius of warming, as per the Clausius-Clapeyron relationship. The increased atmospheric moisture content causes precipitation events to change in intensity, frequency, and duration (Trenberth, 1999) and causes the global precipitation to increase by 2-3 per cent for every degree Celsius of warming (Held and Soden, 2006).

Direct runoff from the continents supplies about 40,000 km3/yr of freshwater to the ocean. Runoff is the sum of all upstream sources of water, including continental precipitation, fluxes from lakes and aquifers, seasonal snow melt, and melting of mountain glaciers and ice caps. River discharge also carries a tremendous amount of solid sediments and dissolved nutrients to the continental shelves.

The polar ice sheets of Greenland and Antarctica are the largest reservoirs of freshwater on the planet, holding 7 m and 58 m of the sea-level equivalent, respectively (Vaughan et al., 2013). The net growth or shrinkage of such an ice sheet is a balance between the net accumulation of snow at the surface, the loss from meltwater runoff, and the calving of icebergs and submarine melting at tidewater margins, collectively known as marine ice loss. There is some debate about the relative importance of these in the case of Greenland. Van den Broeke et al. (2009), show the volume transport to the ocean is almost evenly split between runoff of surface meltwater and marine ice loss. In a more recent work, Box and Colgan (2013) estimate marine ice loss at about twice the volume of meltwater (see Figure 5 in that article), with both marine ice loss and particularly runoff increasing rapidly since the late 1990s. According to the Arctic Monitoring and Assessment Programme (AMAP, 2011), the annual mass of freshwater being added at the surface of the Greenland Ice Sheet (the surface mass balance) has decreased since 1990. Model reconstructions suggest a 40% decrease from 350 Gt/y (1970 - 2000) to 200 Gt/y in 2007. Accelerating ice discharge from outlet glaciers since 1995 - 2002 is widespread and has gradually moved further northward along the west coast of Greenland with global warming. According to AMAP (2011), the ice discharge has increased from the pre-1990 value of 300 Gt/y to 400 Gt/y in 2005.

Antarctica's climate is much colder, hence surface meltwater contributions are negligible and mass loss is dominated by submarine melting and ice flow across the grounding line where this ice meets the ocean floor (Rignot and Thomas, 2002). Freshwater fluxes from ice sheets differ from continental river runoff in two important respects. First, large fractions of both Antarctic ice sheets are grounded well below sea level in deep fjords or continental shelf embayments; therefore freshwater is injected not at the surface of the ocean but at several hundred meters water depth. This deep injection of freshwater enhances ocean stratification which, in turn, plays a role in ecosystem structure. Second, unlike rivers, which act as a point source for freshwater entering the ocean, icebergs calved at the grounding line constitute a distributed source of freshwater as they drift and melt in adjacent ocean basins (Bigg et al., 1997; Enderlin and Hamilton, 2014).

Sea ice is one of the smallest reservoirs of freshwater by volume, but it exhibits enormous seasonal variability in spatial extent as it waxes and wanes over the polar oceans. By acting as a rigid cap, sea ice modulates the fluxes of heat, moisture and momentum between the atmosphere and the ocean. Summertime melting of Arctic sea ice is an important source of freshwater flux into the North Atlantic, and episodes of enhanced sea ice export to warmer latitudes farther south give rise to rapid freshening episodes, such as the Great Salinity Anomaly of the late 1960s (Gelderloos et al., 2012). The spatial distributions of these freshwater fluxes drive important patterns in regional and global ocean circulation, which are discussed in Chapter 5.

The Southern Ocean (defined as all ocean area south of 60°S) deserves special mention due to its role in the storage of heat (and carbon) for the entire planet. The Antarctic Circumpolar Current (ACC) connects the three major southern ocean basins (South Atlantic, South Pacific and Indian) and is the largest current by volume in the world. The ACC flows eastward, circling the globe in a clockwise direction as viewed from the

South Pole. In addition to providing a lateral connection between the major ocean basins (Atlantic, Indian, Pacific), the Southern Ocean also connects the shallow and deep parts of the ocean through a mechanism known as the meridional overturning circulation (MOC) (Gordon, 1986; Schmitz, 1996, see Figures I-90 and I-91). Because of its capacity to bring deep water closer to the surface, and surface water to depths, the Southern Ocean forms an important pathway in the global transport of heat. Although there is no observational evidence at present, (WG II AR5, 30.3.1, Hoegh-Guldberg, 2014) model studies indicate with a high degree of confidence that the Southern Ocean will become more stratified, weakening the surface-to-bottom connection that is the hallmark of present-day Southern Ocean circulation (WG I AR5 12.7.4.3, Collins et al., 2013). A similar change is anticipated in the Arctic Ocean and subarctic seas (WG I AR5 12.7.4.3, Collins et al., 2013), another region with this type of vertical connection between ocean levels (Wüst, 1928). These changes will result in fresher, warmer surface ocean waters in the polar and subpolar regions (WGII AR5 30.3.1, Hoegh-Guldberg, 2014; WG I AR5 12.7.4.3, Collins et al., 2013), significantly altering their chemistry and ecosystems.

Imbalances in the freshwater cycle manifest themselves as changes in global sea level. Changes in global mean sea level are largely caused by a combination of changes in ocean heat content and exchanges of freshwater between the ocean and continents. When water is added to the ocean, global sea level adjusts, rapidly resulting in a relatively uniform spatial pattern for the seasonal ocean mass balance, as compared to the seasonal steric signal, which has very large regional amplitudes (Chambers, 2006). 'Steric' refers to density changes in seawater due to changes in heat content and salinity. On annual scales, the maximum exchange of freshwater from land to ocean occurs in the late Northern Hemisphere summer, and therefore the seasonal ocean mass signal is in phase with total sea level with an amplitude of about 7 mm (Chambers et al., 2004). Because most of the ocean is in the Southern Hemisphere, the seasonal maximum in the steric component occurs in the late Southern Hemisphere summer, when heat storage in the majority of the ocean peaks (Leuliette and Willis, 2011). Because globally averaged sea level variations due to heat content changes largely cancel out between the Northern and Southern Hemispheres, the size of the steric signal, globally averaged, is only 4 mm.

Globally averaged sea level has risen at 3.2 mm/yr for the past two decades (Church et al., 2011), of which about a third comes from thermal expansion. The remainder is due to fluxes of freshwater from the continents, which have increased as the melting of continental glaciers and ice sheets responds to higher temperatures. Multi-decadal fluctuations in equatorial and mid-latitude winds (Merrifield et al., 2012; Moon et al., 2013) cause regional patterns in sea-level trends which are reflected in the El Niño/Southern Oscillation (ENSO) and the Pacific decadal oscillation (PDO) indices in the Pacific (Merrifield et al., 2012; Zhang and Church, 2012) and northern Australia (White et al., 2014). Interannual changes in global mean sea level relative to the observed trend are largely linked to exchanges of water with the continents due to changes in precipitation patterns associated largely with the ENSO; this includes a drop of 5 mm during 2010-11 and rapid rebound in 2012-13 (Boening et al., 2012; Fasullo et al., 2013).

Some key alterations are anticipated in the hydrological cycle due to global warming and climate change. Changes that have been identified include shifts in the seasonal distribution and amount of precipitation, an increase in extreme precipitation events, changes in the balance between snow and rain, accelerated melting of glacial ice, and of course sea-level rise. Although a global phenomenon, it is the impact of sea-level rise along the world's coastlines that has major societal implications. The impacts of these changes are discussed in the next Section.

Changes in the rates of freshwater exchange between the ocean, atmosphere and continents have additional significant impacts. For example, spatial variations in the distribution of evaporation and precipitation create gradients in salinity and heat that in turn drive ocean circulation; ocean freshening also affects ecosystem structure. These aspects and their impacts are discussed in Sections 3 and 4.

Another factor potentially contributing to regional changes in the hydrological cycle are changes in ocean surface currents. For example, the warm surface temperatures of the large surface currents flowing at the western boundaries of the ocean basins (the Agulhas, Brazil, East Australian, Gulf Stream, and Kuroshio Currents) provide significant amounts of heat and moisture to the atmosphere, with a profound impact on the regional hydrological cycle (e.g., Rouault et al., 2002). Ocean surface currents like these are forced by atmospheric winds and sensitive to changes in them - stronger winds can mean stronger currents and an intensification of their effects (WGII AR5 30.3.1, Hoegh-Guldberg, 2014), as well as faster evaporation rates. Shifts in the location of winds can also alter these currents, for example causing the transport of anomalously warm waters (e.g., Rouault, 2009). However, despite a well-documented increase in global wind speeds in the 1990s (Yu, 2007), the overall effect of climate change on winds is complex, and difficult to differentiate observationally from decadal-scale variability, and thus the ultimate effects of these currents on the hydrological cycle are difficult to predict with any high degree of confidence (WGII AR5 30.3.1, Hoegh-Guldberg, 2014).

2 Environmental, economic and social implications of ocean warming

As a consequence of changes in the hydrological cycle, increases in runoff, flooding, and sea-level rise are expected, and their potential impacts on society and natural environment are among the most serious issues confronting humankind, according to the Fifth Assessment Report (AR5) of the United Nations Intergovernmental Panel on Climate Change (IPCC). This report indicates that it is very likely that extreme sea levels have increased globally since the 1970s, mainly as a result of global mean sea-level rise due in part to anthropogenic warming causing ocean thermal expansion and glacier melting (WGI AR5 3.7.5, 3.7.6;

WGI AR5 10.4.3). In addition, local sea-level changes are also influenced by several natural factors, such as regional variability in oceanic and atmospheric circulation, subsidence, isostatic adjustment, and coastal erosion, among others; combined with human perturbations by land-use change and coastal development (WGI AR5 5.3.2). A 4°C warming by 2100 (Betts et al., 2011; predicted by the high-end emissions scenario RPC8.5 in WGI AR5 FAQ12.1) leads to a median sea-level rise of nearly 1 m above 1980-1999 levels (Schaeffer et al., 2012).

The vulnerability of human systems to sea-level rise is strongly influenced by economic, social, political, environmental, institutional and cultural factors; such factors in turn will vary significantly in each specific region of the world, making quantification a challenging task (Nicholls et al., 2007; 2009; Mimura, 2013). Three classes of vulnerability are identified: (i) early impacts (low-lying island states, e.g., Kiribati, Maldives, Tuvalu, etc.); (ii) physically and economically vulnerable coastal communities (e.g., Bangladesh); and (iii) physically vulnerable but economically "rich" coastal communities (e.g., Sydney, New York). Table 4.1 outlines the main effects of relative sea-level rise on the natural system and provides examples of socio-economic system adaptations.

It is widely accepted that relative trends in sea-level rise pose a significant threat to coastal systems and low-lying areas around the world, due to inundation and erosion of coastlines and contamination of freshwater reserves and food crops (Nicholls, 2010); it is also likely that sea-level effects will be most pronounced during extreme episodes, such as coastal flooding arising from severe storm-induced surges, wave overtopping and rainfall runoff, and increases in sea level during ENSO events. An increase in global temperature of 4°C is anticipated to have significant socio-economic effects as sea-level rise, in combination with increasingly frequent severe storms, will displace populations (Field et al., 2012). These processes will also place pressure on existing freshwater resources through saltwater contamination (Nicholls and Cazenave, 2010). Figure 4.1 outlines in more detail the effects of sea-level rise on water resources of low-lying coastal areas.

Small island countries, such as Kiribati, Maldives and Tuvalu, are particularly vulnerable. Beyond this, entire identifiable coherent communities also face risk (e.g., Torres Strait Islanders; Green, 2006). These populations have nowhere to retreat to within their country and thus have no alternative other than to abandon their country entirely. The low level of economic activity also makes it difficult for these communities to bear the costs of adaptation. A shortage of data and local expertise required to assess risks related to sea-level rise further complicate their situation. Indeed the response of the island structure to sea-level rise is likely to be complex (Webb and Kench, 2010). Traditional customs are likely to be at risk and poorly understood by outside agencies. Yet traditional knowledge is an additional resource that may aid adaptation in such settings and should be carefully evaluated within adaptation planning. A significant part of the economy of many island nations is based on tourism;

Table 4.1 | The main effects of relative sea-level rise on the natural system, interacting factors, and examples of socio-economic system adaptations. Some interacting factors (for example, sediment supply) appear twice as they can be influenced both by climate and non-climate factors. Adaptation strategies: P = Protection; A = Accommodation; R = Retreat. Source: based on Nicholls and Tol, 2006.

Natural System Effects		Interacting Factors		Socio-economic System Adaptations
		Climate	Non-climate	
1. Inundation, flood and storm damage	a. Surge (sea)	– wave/stormclimate – erosion – sediment supply	– sediment supply – floodmanagement – erosion – land use	– dykes/surge barriers [P] – building codes/floodwise buildings [A] – land use planning/hazard delineation [A/R]
	b. Backwater effect (river)	– run-off	– catchment management – land use	
2. Wetland loss (and change)		– CO_2 fertilization – sediment supply	– sediment supply – migration space – direct destruction	– land-use planning [A/R] – managed realignment/forbid hard defences [R] – nourishment/sediment management [P]
3. Erosion (direct and indirect morphological change)		– sediment supply – wave/stormclimate	– sediment supply	– coast defences [P] – nourishment [P] – building setbacks [R]
4. Saltwater Intrusion	a. Surface Waters	– run-off	– catchment management – land use	– saltwater intrusion barriers [P] – change water abstraction [A/R]
	b. Ground-water	– rainfall	– land use – aquifer use	– freshwater injection [P] – change water abstraction [A/R]
5. Rising water tables/impeded drainage		– rainfall – run-off	– land use – aquifer use – catchment management	– upgrade drainage systems [P] – polders [P] – change land use [A] – land use planning/hazard delineation [A/R]

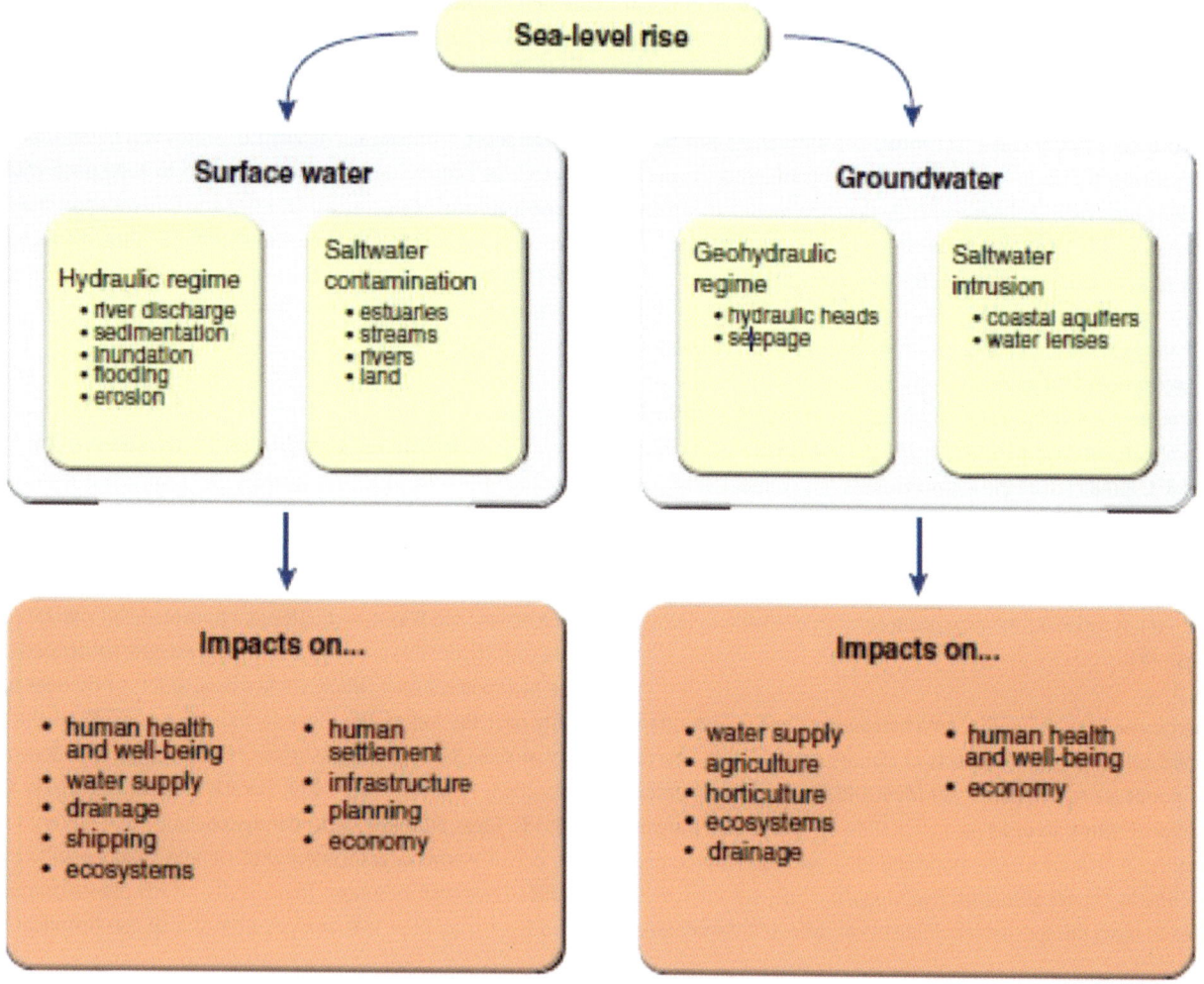

Figure 4.1 | Effects of sea-level rise on water resources of small islands and low-lying coastal areas. Source: Based on Oude Essink et al. (1993); Hay and Mimura (2006).

this too will be affected by sea-level rise through its direct effects on infrastructure and possibly also indirectly by the reduced availability of financial resources in the market (Scott et al., 2012).

Coastal regions, particularly some low-lying river deltas, have very high population densities. It is estimated that over 150 million people live within 1 metre of the high-tide level, and 250 million within 5 metres of high tide. Because of these high population densities (often combined with a lack of long-range urban planning), coastal cities in developing regions are particularly vulnerable to sea-level rise in concert with other effects of climate change (World Bank, 2012).

Effects of sea-level rise are projected to be asymmetrical even within regions and countries. Nicholls and Tol (2006), extending the global vulnerability analysis of Hoozemans et al. (1993) on the impacts of and responses to sea-level rise with storm surges over the 21st century, show East Africa (including small island States and countries with extensive coastal deltas) as one of the problematic regions that could experience major land loss. Dasgupta et al. (2009) undertook a comparative study on the impacts of sea-level rise with intensified storm surges on developing countries globally in terms of its impacts on land area, population, agriculture, urban extent, major cities, wetlands, and local economies. They based their work on a 10 per cent future intensification of storm surges with respect to current 1-in-100-year storm-surge predictions. They found that Sub-Saharan African countries will suffer considerably from the impacts. The study estimated that Mozambique, along with Madagascar, Mauritania and Nigeria account for more than half (9,600 km2) of the total increase in the region's storm-surge zones.

Of the impacts projected for 31 developing countries, just ten cities account for two-thirds of the total exposure to extreme floods. Highly vulnerable cities are found in Bangladesh, India, Indonesia, Madagascar, Mexico, Mozambique, the Philippines, Venezuela and Viet Nam (Brecht et al., 2012). Because of the small population of small islands and potential problems with implementing adaptations, Nicholls et al. (2011) conclude that forced abandonment of these islands seems to be a possible outcome even for small changes in sea level. Similarly, Barnett and Adger (2003) point out that physical impact might breach a threshold that pushes social systems into complete abandonment, as institutions that could facilitate adaptation collapse.

The Ocean's Role in the Hydrological Cycle — Chapter 4

Impacts of climate change on the hydrological cycle, and notably on the availability of freshwater resources, have been observed on all continents and many islands. Glaciers continue to shrink worldwide, affecting runoff and water resources downstream. Figure 4.2 shows the changes anticipated by the late 21st century in water runoff into rivers and streams. Climate change is the main driver of permafrost warming and thawing in both high-latitude and high-elevation mountain regions (IPCC WGII AR5 18.3.1, 18.5). This thawing has negative implications for the stability of infrastructure in areas now covered with permafrost.

Projected heat extremes and changes in the hydrological cycle will in turn affect ecosystems and agriculture (World Bank, 2012). Tropical and subtropical ecoregions in Sub-Saharan Africa are particularly vulnerable to ecosystem damage (Beaumont et al., 2011). For example, with global warming of 4°C (predicted by the high-end emissions scenario RPC8.5 in WGI AR5 FAQ 12.1), between 25 per cent and 42 per cent of 5,197 African plant species studied are projected to lose all their suitable range by 2085 (Midgley and Thuiller, 2011). Ecosystem damage would have the follow-on effect of reducing the ecosystem services available to human populations.

The Mediterranean basin is another area that has received a lot of attention in regard to the potential impacts of climate change on it. Several modelling groups are taking part in the MedCORDEX (www.medcordex.eu) international effort, in order to better simulate the Mediterranean hydrological cycle, to improve the modelling tools available, and to produce new climate impact scenarios. Hydrological model schemes must be improved to meet the specific requirements of semi-arid climates, accounting in particular for the related seasonal soil water dynamics and the complex surface-subsurface interactions in such regions (European Climate Research Alliance, 2011).

Even the most economically resilient of States will be affected by sea-level rise, as adaptation measures will need to keep pace with ongoing sea-level rise (Kates et al., 2012). As a consequence, the impacts of sea-level rise will also be redistributed through the global economic markets as insurance rates increase or become unviable and these costs are passed on to other sectors of the economy (Abel et al., 2011).

3   Chemical composition of seawater

3.1   Salinity

Surface salinity integrates the signals of freshwater sources and sinks for the ocean, and if long-term (decadal to centennial) changes in salinity are considered, this provides a way to investigate associated changes in the hydrological cycle. Many studies have assessed changes to ocean salinity over the long term; of these, four have considered changes on a global scale from the near-surface to the sub-surface ocean (Boyer et al., 2005; Hosoda et al., 2009; Durack and Wijffels, 2010; Good et al., 2013). These studies independently concluded that alongside broad-scale ocean warming associated with climate change, shifts in ocean salinities have also occurred. These shifts, which are calculated using methods such as objective analysis from the sparse historical observ-

The boundaries and names shown and the designations used on this map do not imply official endorsement or acceptance by the United Nations.

Figure 4.2 | Changes in water runoff into rivers and streams are another anticipated consequence of climate change by the late 21st Century. This map shows predicted increases in runoff in blue, and decreases in brown and red. (Map by Robert Simmon, using data from Milly et al., 2005; Graham et al., 2010; NASA Geophysical Fluid Dynamics Laboratory.)

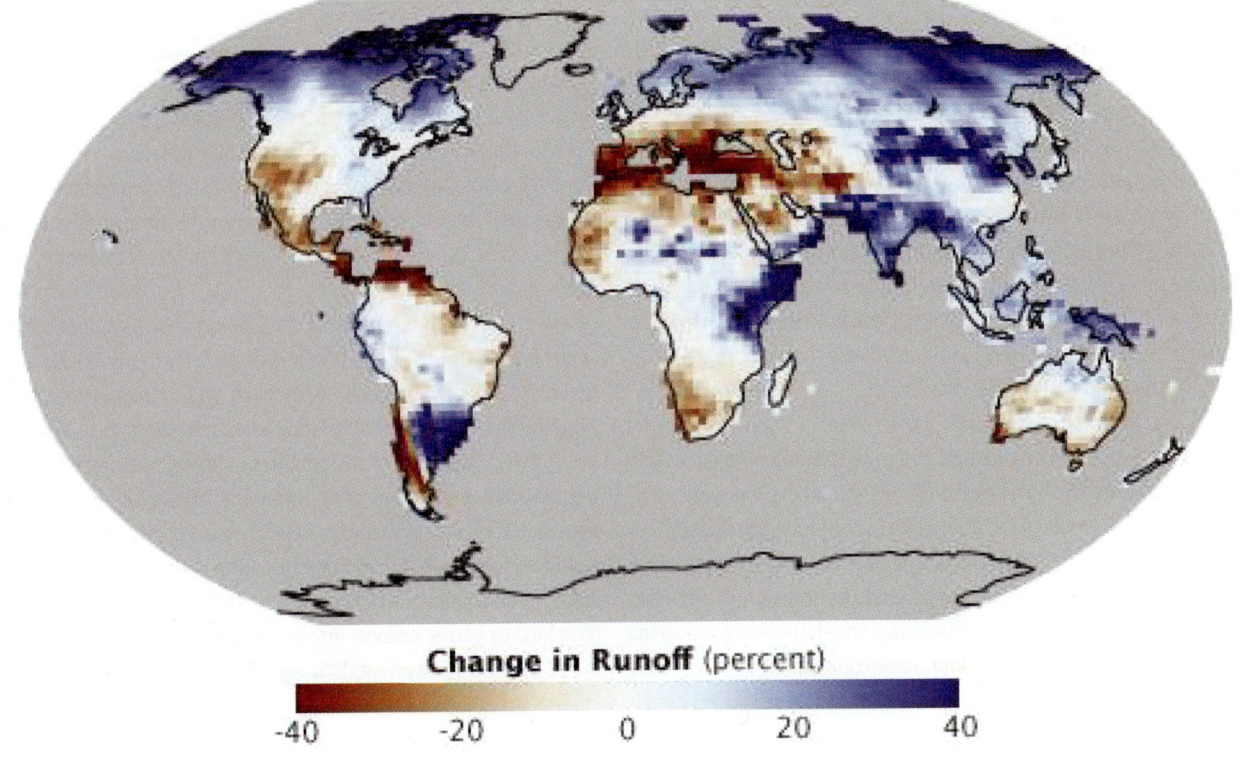

Chapter 4

The Ocean's Role in the Hydrological Cycle

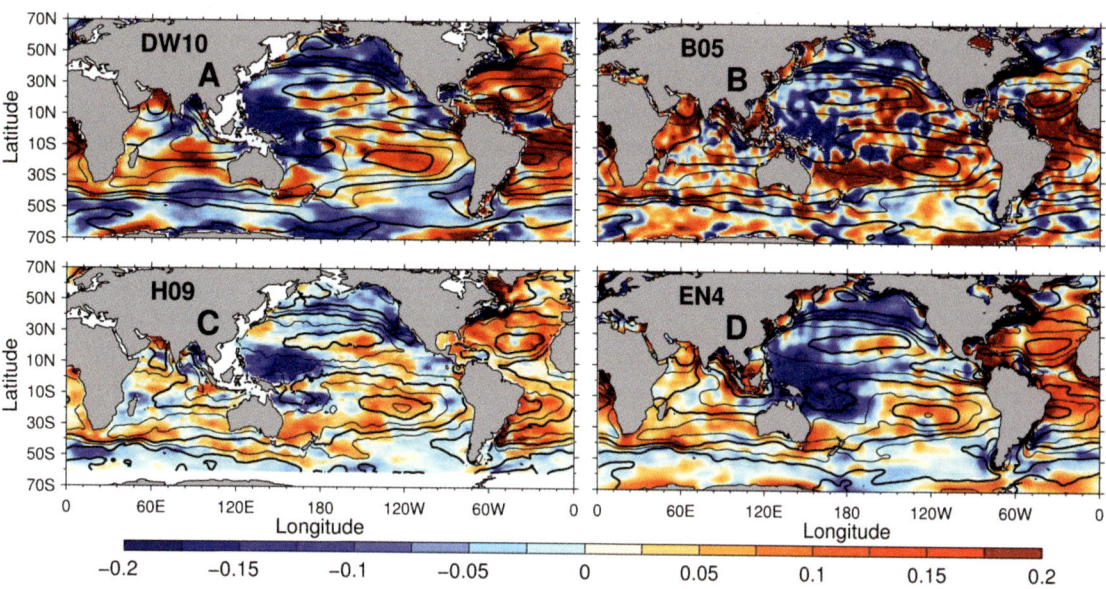

The boundaries and names shown and the designations used on this map do not imply official endorsement or acceptance by the United Nations.

Figure 4.3 | Four long-term estimates of global sea-surface salinity (SSS) change according to (A) Durack and Wijffels (2010; ©American Meteorological Society. Used with permission.), analysis period 1950-2008; (B) Boyer et al. (2005), analysis period 1955-1998; (C) Hosoda et al. (2009), analysis period 1975-2005; and (D) Good et al. (2013), analysis period 1950-2012; all are scaled to represent equivalent magnitude changes over a 50-year period (PSS-78 50-year-1). Black contours show the associated climatological mean SSS for the analysis period. Broad-scale similarities exist between each independent analysis of long-term change, and suggest an increase in spatial gradients of salinity has occurred over the period of analysis. However, regional-scale differences are due to differences in data sources, temporal periods of analysis, and analytical methodologies.

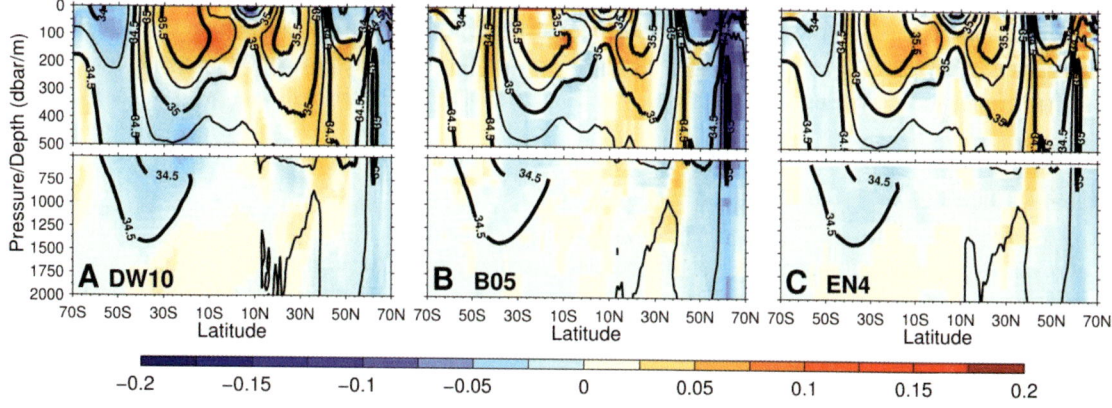

The boundaries and names shown and the designations used on this map do not imply official endorsement or acceptance by the United Nations.

Figure 4.4 | Three long-term estimates of global zonal mean subsurface salinity changes according to (A) Durack and Wijffels (2010; ©American Meteorological Society. Used with permission.), analysis period 1950-2008; (B) Boyer et al. (2005), analysis period 1955-1998; and (C) Good et al. (2013), analysis period 1950-2012; all scaled to represent equivalent magnitude changes over a 50-year period (PSS-78 50-year-1). Black contours show the associated climatological mean subsurface salinity for the analysis period. Broad-scale similarities also exist in the subsurface salinity changes, which suggest a decreasing salinity in ocean waters fresher than the global average, and an increasing salinity in waters saltier than the global average. However, regional differences, particularly in the high-latitude regions, are due to limited data sources, different temporal periods of analysis and different analytical methodologies.

ing system, suggest that at the surface, high-salinity subtropical ocean regions and the entire Atlantic basin have become more saline, and low-salinity regions, such as the western Pacific Warm Pool, and high-latitude regions have become even fresher over the period of analysis (Figure 4.3). Significant regional-scale differences may be ascribed to the paucity of observational data, particularly in the pre-Argo era, the difference in temporal period over which each analysis was conducted, and differences in methodology and data selection criteria.

Despite regional differences, the broad-scale patterns of change suggest that long-term, coherent changes in salinity have occurred over the observed record, and this conclusion is also supported by shifts in salinity apparent in the subsurface ocean (Figure 4.4). These subsurface changes also show that spatial gradients of salinity within the ocean interior have intensified, and that at depth, salinity-minimum (intermediate) waters have become fresher, and salinity-maximum waters have become saltier (Durack and Wijffels, 2010; Helm et al., 2010; Skliris et al., 2014). Taken together, this evidence suggests intensification of the global hydrological cycle; this is consistent with what is expected from global warming (see Section 1). Actual changes in the hydrological cycle may be even more intense than indicated by patterns of surface salinity anomalies, as these may be spread out and reduced in intensity by being transported (advected) by ocean currents. For example, the work of Hosoda et al. (2009) and Nagano et al. (2014) indicates that large (ENSO-scale) salinity anomalies are rapidly transported from the central Pacific to the northwestern North Pacific (the Kuroshio Extension region).

3.2 Nutrients

Many different nutrients are required as essential chemical elements that organisms need to survive and reproduce in the ocean. Macronutrients, needed in large quantities, include calcium, carbon, nitrogen, magnesium, phosphorus, potassium, silicon and sulphur; micronutrients like iron, copper and zinc are needed in lesser quantities (Smith and Smith, 1998). Macronutrients provide the bulk energy for an organism's metabolic system to function, and micronutrients provide the necessary co-factors for metabolism to be carried out. In aquatic systems, nitrogen and phosphorus are the two nutrients that most commonly limit the maximum biomass, or growth, of algae and aquatic plants (United Nations Environment Programme (UNEP) Global Environment Monitoring System (GEMS) Water Programme, 2008). Nitrate is the most common form of nitrogen and phosphate is the most common form of phosphorus found in natural waters. On the other hand, one of arguably the most important groups of marine phytoplankton is the diatom. Recent studies, for example, Brzezinski et al., (2011), show that marine diatoms are significantly limited by iron and silicic acid.

About 40 per cent of the world's population lives within a narrow fringe of coastal land (about 7.6 per cent of the Earth's total land area; United Nations Environment Programme, 2006). Land-based activities are the dominant source of marine nutrients, especially for fixed nitrogen, and include: agricultural runoff (fertilizer), atmospheric releases from fossil-fuel combustion, and, to a lesser extent, from agricultural fertilizers, manure, sewage and industrial discharges (Group of Experts on the Scientific Aspects of Marine Environmental Protection, 2001; Figure 4.5).

An imbalance in the nutrient input and uptake of an aquatic ecosystem changes its structure and functions (e.g., Arrigo, 2005). Excessive nutrient input can seriously impact the productivity and biodiversity of a marine area (e.g., Tilman et al., 2001); conversely, a large reduction in natural inputs of nutrients (caused by, e.g., damming rivers) can also adversely affect the productivity of coastal waters. Nutrient enrichment between 1960-1980 in the developed regions of Europe, North America, Asia and Oceania has resulted in major changes in adjacent coastal ecosystems.

Nitrogen flow into the ocean is a good illustration of the magnitude of changes in anthropogenic nutrient inputs since the industrial revolution. These flows have increased 15-fold in North Sea watersheds, 11-fold in the North Eastern USA, 10-fold in the Yellow River basin, 5.7-fold in the Mississippi River basin, 5-fold in the Baltic Sea watersheds, 4.1-fold in the Great Lakes/St Lawrence River basin, and 3.7-fold in South-Western Europe (Millennium Ecosystem Assessment, 2005). It is expected that global nitrogen exports by rivers to the oceans will continue to rise.

Box 4.1 | Example – Nutrients in the Pacific region

The Pacific Ocean basins form the largest of the mid-latitude oceans. In addition, the subarctic North Pacific Ocean is one of the most nutrient-rich areas of the world ocean; in 2013, the most recent year for which statistics have been compiled, the North Pacific (north of 40° N) provided 30% of the world's capture, by weight, of ocean fish (FAO, 2015). Many oceanographic experiments have been carried out over the last half century in the North Pacific Ocean; studies based on these datasets reveal the decadal-scale variation of nutrient concentrations in the surface and subsurface (intermediate) layers, as seen in Figure 4.6.

A linearly increasing trend of nutrient concentrations (nitrate and phosphate) has been observed in the intermediate waters in a broad area of the North Pacific (Figure 4.6b); Ono et al., 2001; Watanabe et al., 2003; 2008; Tadokoro et al., 2009; Guo et al., 2012; Whitney et al., 2013). Conversely, the concentration of nutrients in the surface layer has decreased (Figure 4.6a; Freeland, 1997; Ono et al., 2002; 2008; Watanabe et al., 2005; 2008; Aoyama et al., 2008, Tadokoro et al., 2009; Whitney, 2011). Surface nutrients are primarily supplied by the subsurface ocean through a process known as "vertical mixing", an exchange between surface and subsurface waters. Vertical mixing is partly dependent on the differences in density between adjacent ocean layers: layers closer to one another in density mix more easily.

A significant increase in temperature and a corresponding decrease in salinity (see above) have been observed during the last half-century in the upper layer of the North Pacific (IPCC, 2013, WG1 AR5). These changes are in the direction of increased stratification in the upper ocean and thus it is possible that this increased stratification has caused a corresponding decrease in the vertical mixing rate.

Superimposed on the linear trends, nutrient concentrations in the ocean have also exhibited decadal-scale variability, which is evident in both surface and subsurface waters (Figure 4.6c). Unlike the linear trends, the decadal-scale variability appeared synchronized between the surface and subsurface layers in the western North Pacific (Tadokoro et al., 2009). These relationships suggest that the mechanisms producing the trends and more cyclical variability are different.

Figure 4.5(a) | Trends in annual rates of application of nitrogenous fertilizer (N) expressed as mass of N, and of phosphate fertilizer (P) expressed as mass of P2O5, for all States of the world except for many of the countries belonging to the United Nations regional group of Eastern European States and the former USSR (scale on the left in 106 metric tons), and trends in global total area of irrigated crop land (H2O) (scale on the right in 109 hectares). Source: Tilman et al., 2001.

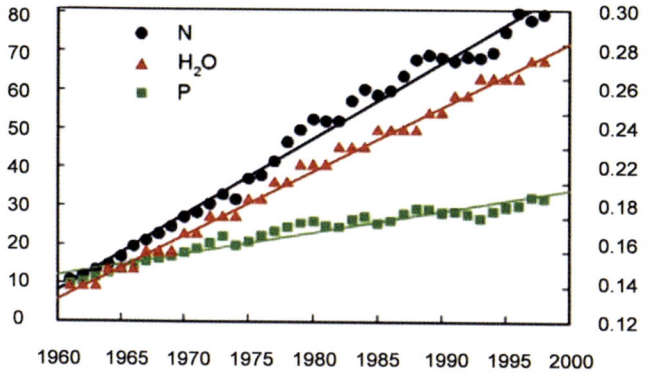

Figure 4.5(b) | Estimated growth in fertilizer use, 1960-2020. From GESAMP (2001). Source: Bumb and Baanante, 1996.

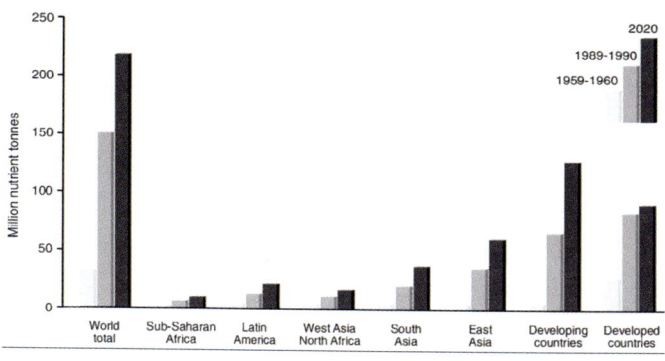

Projections for 2030 show an increase of 14 per cent compared to 1995. By 2030, global nitrogen exports by rivers are projected to be 49.7 Tg/yr; natural sources will contribute 57 per cent of the total, agriculture 34 per cent, and sewage 9 per cent (Bouwman et al., 2005). An example of this is discussed in Box 4.1.

4 Environmental, economic and social implications of changes in salinity and nutrient content

4.1 Salinity

Although changes to ocean salinity do not directly affect humanity, changes in the hydrological cycle that are recorded in the changing patterns of ocean salinity certainly do. Due to the scarcity of hydrological cycle observations over the ocean, and the uncertainties associated with these measurements, numerous studies have linked salinity changes to the global hydrological cycle by using climate models (Durack et al., 2012; 2013; Terray et al., 2012) or reanalysis products (Skliris et al., 2014). However, these studies only considered long-term salinity changes, and not changes that occur on interannual to decadal timescales. These latter scales are strongly affected by climatic variability (Yu, 2011; Vinogradova and Ponte, 2013). As mentioned in Section 3, these studies collectively conclude that changes to the patterns of ocean salinity are likely due to the intensification of the hydrological cycle, in particular patterns of evaporation and rainfall at the ocean surface. This result concurs well with the "rich-get-richer" mechanism proposed in earlier studies, suggesting that terrestrial "dry" zones will become dryer and terrestrial "wet" zones will become wetter due to ongoing climate change (Chou and Neelin, 2004; Held and Soden, 2006).

4.2 Nutrients

Marine environments are unsteady systems, whose response to climate-induced or anthropogenic changes is difficult to predict. As a result, no published studies quantify long-term trends in ocean nutrient concentrations. However, it is well understood that imbalances in nutrient concentration cause widespread changes in the structure and functioning of ecosystems, which, in turn, have generally negative impacts on habitats, food webs and species diversity, including economically important ones; such adverse effects include: general degradation of habitats, destruction of coral reefs and sea-grass beds; alteration of marine food-webs, including damage to larval or other life stages; mass mortality of wild and/or farmed fish and shellfish, and of mammals, seabirds and other organisms.

Among the effects of nutrient inputs into the marine environment it is important to mention the link with marine pH. The production of excess algae from increased nutrients has the effect, inter alia, to release CO_2 from decaying organic matter deriving from eutrophication (Hutchins et al., 2009; Sunda and Cai, 2012). The effects of these acidification processes, combined with those deriving from increasing atmospheric CO_2, can reduce the time available to coastal managers to adopt approaches to avoid or minimize harmful effects on critical ecosystem services, such as fisheries and tourism. Globally, the manufacture of nitrogen fertilizers has continued to increase (Galloway et al., 2008) accompanied by increasing eutrophication of coastal waters and degradation of coastal ecosystems (Diaz and Rosenberg, 2008; Seitzinger et al., 2010; Kim et al., 2011), and amplification of CO_2 drawdown (Borges and Gypens, 2010; Provoost et al., 2010). In addition, atmospheric deposition of anthropogenic fixed nitrogen may now account for up to about 3 per cent of oceanic new production, and this nutrient source is projected to increase (Duce et al., 2008).

The Ocean's Role in the Hydrological Cycle Chapter 4

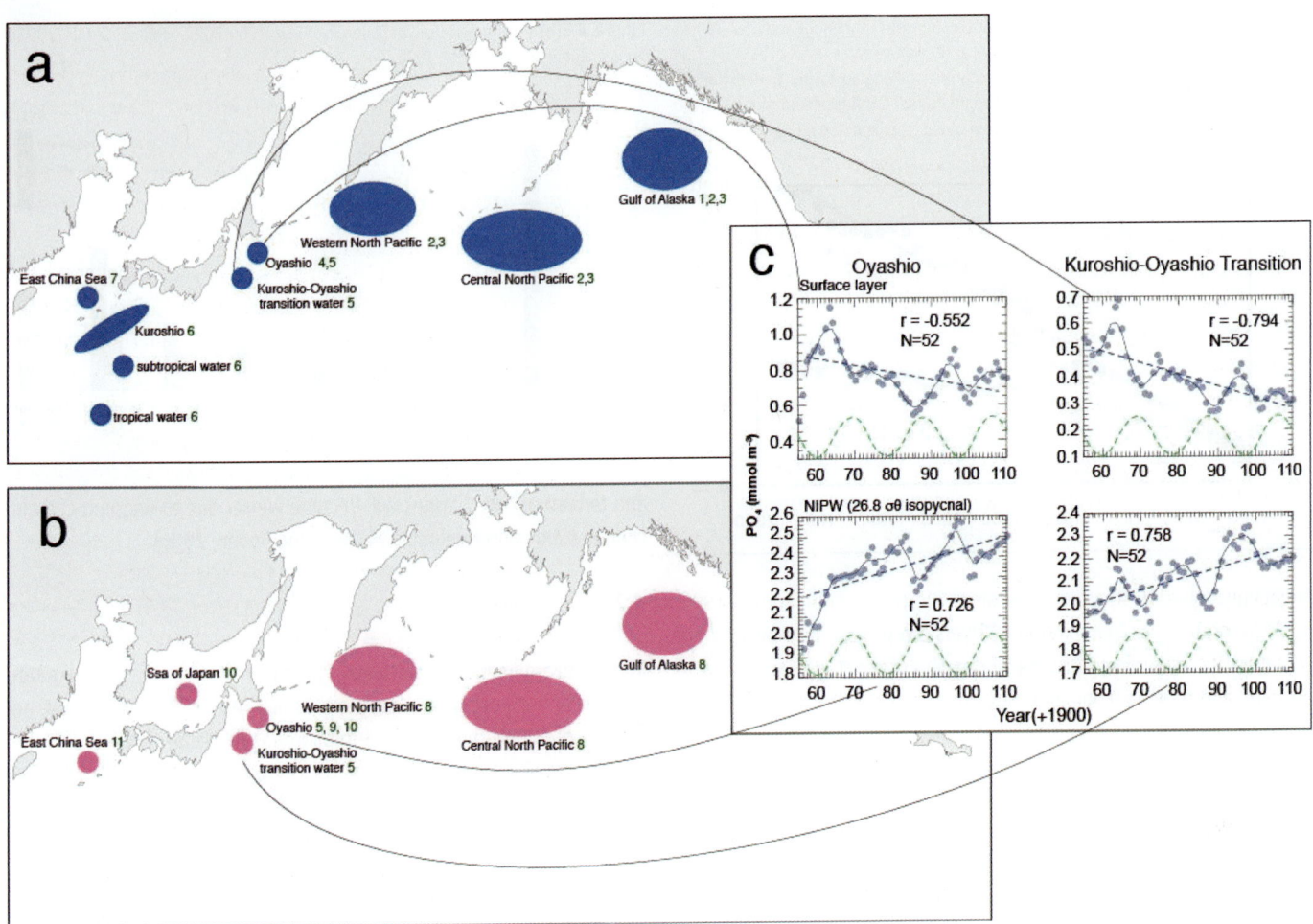

The boundaries and names shown and the designations used on this map do not imply official endorsement or acceptance by the United Nations.

Figure 4.6 | Synthesis of the decadal-scale change in nutrient concentrations in the North Pacific Ocean in the last fifty years. (a) The blue area shows the waters for which a decreasing trend in nutrient concentrations was reported in the surface layer. (b) The pink area shows the waters for which an increasing trend in nutrient concentrations was reported in the subsurface. (c) Example of the nutrient change in the North Pacific Ocean. Five-year running mean of the annual mean concentration (mmol m-3) of Phosphate concentration in the surface and North Pacific Intermediate Water (NPIW) of the Oyashio and Kuroshio-Oyashio transition waters from the mid-1950s to early 2010. (Time series from Tadokoro et al., 2009). Blue broken lines indicate statistically significant trends of PO4. Thin green broken lines represent the index of diurnal tidal strength represented by the sine curve of the 18.6-yr cycle.*The numbers following each area name indicate the referenced literature: (1) Freeland et al., (1997); (2) Ono et al., (2008); (3) Whitney (2011); (4) Ono et al., (2002); (5) Tadokoro et al., (2009); (6) Watanabe et al., (2005); (7) Aoyama et al., (2008); (8) Watanabe et al., (2008); (9) Ono et al., (2001); (10) Watanabe et al., (2003); (11) Guo et al., (2012).

References

Abel, N., Gorddard, R., Harman, B., Leitch, A., Langridge, J., Ryan, A. and Heyenga, S. (2011). Sea level rise, coastal development and planned retreat: analytical framework, governance principles and an Australian case study, *Environmental Science & Policy*, 14:279-288.

Aoyama, M., et al. (2008). Marine biogeochemical response to a rapid warming in the main stream of the Kuroshio in the western North Pacific, *Fisheries Oceanography*, 17, 206–218, doi:10.1111/j.1365-2419.2008.00473.x.

Arrigo, K. (2005). Marine microorganisms and global nutrient cycles. *Nature*, 437, 15 doi:10.1038/nature04158.

Barnett, J., and Adger, W.N. (2003). Climate Dangers and Atoll Countries. *Climatic Change*, 61(3), 321–337. doi:10.1023/B:CLIM.0000004559.08755.88.

Beaumont, L.J., Pitman, A., Perkins, S., Zimmermann, N.E., Yoccoz, N.G., and Thuiller, W. (2011). Impacts of climate change on the world's most exceptional ecoregions. *Proceedings of the National Academy of Sciences of the United States of America*, 108(6), 2306–11. doi:10.1073/pnas. 1007217 108.

Betts, R.A., Collins, M., Hemming, D.L., Jones, C.D., Lowe, J.A. and Sanderson, M.G. (2011). When could global warming reach 4°C? *Philosophical transactions. Series A, Mathematical, physical, and engineering sciences*, 369(1934), 67–84.

Bigg G.R., Wadley, M.R., Stevens, D.P. and Johnson, J.A. (1997). Modelling the dynamics and thermodynamics of icebergs. *Cold Regions Science and Technology*, 26, 113–135.

Boening, C., Willis, J.K., Landerer, J.K., Nerem, R.S. and Fasullo, J. (2012). The 2011 La Niña: So strong, the oceans fell. *Geophysical Research Letters*, 39, L18607, doi:10.1029/2012GL052885.

Borges, A.V., and Gypens, N., (2010). Carbonate chemistry in the coastal zone responds more strongly to eutrophication than to ocean acidification. *Limnology and Oceanography*, 55, 346–353.

Bouwman, A.F., Van Drecht, G. and Van der Hoek, K.V. (2005). Global and regional surface balances in intensive agricultural production systems for the period 1970 - 2030. *Pedosphere*, 15, 2, 137-155.

Box, J.E. and Colgan, W. (2013). Greenland Ice Sheet Mass Balance Reconstruction. Part III: Marine Ice Loss and Total Mass Balance (1840–2010). *Journal of Climate*, 26, 6990–7002. doi: http://dx.doi.org/10.1175/JCLI-D-12-00546.1 Boyer, T.P., Levitus, S., Antonov, J.I., Locarnini, R.A., and Garcia, H.E. (2005) Linear trends in salinity for the World Ocean, 1955-1998. *Geophysical Research Letters*, 32, L01604. doi: 10.1029/2004GL021791.

Brecht, H., Dasgupta, S., Laplante, B., Murray, S., and Wheeler, D. (2012). Sea-Level Rise and Storm Surges: High Stakes for a Small Number of Developing Countries. *The Journal of Environment & Development*, 21(1), 120–138. doi:10.1177/1070496511433601.

Brzezinski, M.A., Baines, S.B., Balch, W.M., Beucher, C.P., Chai, F., Dugdale, R.C., Krause, J.W., Landry, M.R., Marchi, A., Measures, C.I., Nelson, D.M., Parker, A.E., Poulton, A.J., Selph, K.E., Strutton, P.G., Taylor, A.G. and Twining, B.S. (2011). Co-limitation of diatoms by iron and silicic acid in the equatorial Pacific. *Deep-Sea Research II*, 58, 493-511. 10.1016/j.dsr2.2010.08.005.

Bumb, B and Baanante, C. A., (1996). World trends in fertilizer use and projections to 2020. *2020 Brief* 38 (International Food Policy Research Institute, Washington, DC, USA).

Chambers, D., (2006). Observing seasonal steric sea level variations with GRACE and satellite altimetry. *Journal of Geophysical Research*, 111(3):C03010, doi:10.1029/2005JC002914.

Chambers, D.P., Wahr, J. and Nerem, R.S. (2004) Preliminary observations of global ocean mass variations with GRACE. *Geophysical Research Letters*,, 31, L13310, doi:10.1029/2004GL020461.

Chou, C. and Neelin, J.D. (2004). Mechanisms of Global Warming Impacts on Regional Tropical Precipitation. *Journal of Climate*, 17 (13), pp 2688-2701. doi:10.1175/1520-0442(2004)017<2688:MOGWIO>2.0.CO;2

Church, J.A., et al., (2011). Revisiting the Earth's sea-level and energy budgets from 1961 to 2008. *Geophysical Research Letters*, 38, L18601, doi:10.1029/2011GL048794.

Collins, M., Knutti, R., Arblaster, J., Dufresne, J.-L., Fichefet, T., Friedlingstein, P., Gao, X., Gutowski, W.J., Johns, T., Krinner, G., Shongwe, M., Tebaldi, C., Weaver, A.J. and Wehner, M. (2013). *Long-term Climate Change: Projections, Commitments and Irreversibility. In: Climate Change 2013: The Physical Science Basis. Contribution of Working Group I to the Fifth Assessment Report of the Intergovernmental Panel on Climate Change* [Stocker, T.F., Qin, D., Plattner, G.-K., Tignor, M.,

Allen, S.K., Boschung, J., Nauels, A., Xia, Y., Bex, V. and Midgley, P.M. (eds.)]. Cambridge University Press, Cambridge, United Kingdom and New York, NY, USA.

Costello M.J., Cheung A., De Hauwere, N. (2010). Topography statistics for the surface and seabed area, volume, depth and slope, of the world's seas, oceans and countries. *Environmental Science and Technology* 44(23), 8821-8828. doi:10.1021/es1012752.

Dasgupta, S., Laplante, B., Murry, S., and Wheeler, D. (2009). Sea-Level Rise and Storm Surges: A Comparative Analysis of Impacts in Developing Countries. *World Bank Policy Research Working Paper* 4901. Washington, DC: World Bank.

Diaz, R. J., and Rosenberg, R., (2008). Spreading dead zones and consequences for marine ecosystems. *Science*, 321, 926–929.

Duce, R.A., LaRoche, J., Altieri, K., Arrigo, K.R., Baker, A.R., Capone, D.G., Cornell, S., Dentener, F., Galloway, J., Ganeshram, R.S., Geider, R.J., Jickells, T., Kuypers, M.M., Langlois, R., Liss, P.S., Liu, S. M., Middelburg, J.J., Moore, C.M., Nickovic, S., Oschlies, A., Pedersen, T., Prospero, J., Schlitzer, R., Seitzinger, S., Sorensen, L.L., Uematsu, M., Ulloa, O., Voss, M., Ward, B., Zamora, L. (2008). Impacts of atmospheric anthropogenic nitrogen on the open ocean. *Science*, 320, 893–897.

Durack, P. J. and Wijffels, S.E., (2010). Fifty-Year Trends in Global Ocean Salinities and Their Relationship to Broadscale Warming. *Journal of Climate*, 23, pp 4342-4362. doi: 10.1175/2010JCLI3377.1.

Durack, P.J., Wijffels, S.E. and Matear, R.J. (2012). Ocean Salinities Reveal Strong Global Water Cycle Intensification During 1950 to 2000. *Science*, 336 (6080), pp 455-458. doi: 10.1126/science.1212222.

Durack, P.J., Wijffels, S.E., and Boyer, T.P. (2013). Long-term salinity changes and implications for the global water cycle. In: *Ocean Circulation and Climate: a 21st Century Perspective*. Siedler, G., S.M. Griffies, S.M., Gould, J. and Church, J.A. (Eds.). pp 727-757, International Geophysics series, vol. 103. Academic Press, Elsevier, Oxford, doi: 10.1016/B978-0-12-391851-2.00028-3.

Enderlin, E.M. and Hamilton, G.S. (2014). Estimates of Iceberg Submarine Melting from High-Resolution Digital Elevation Models: Application to Sermilik Fjord, East Greenland. *Journal of Glaciology*, in press.

European Climate Research Alliance, (2011). *Collaborative Programs: Changes in the Hydrological Cycle* (full text), http://www.ecra-climate.eu/index.php/collaborative-programmes/hydrocycle/11-cllaborative-programmes/20-changes-in-the-hydrological-cyclefull, last accessed on 2014-07-18.

FAO (2015) (Food and Agriculture Organization of the United Nations), on-line Global Capture Production Database 1950-2013, http://www.fao.org/figis/servlet/TabLandArea?tb_ds=Capture&tb_mode=TABLE&tb_act=SELECT&tb_

grp=COUNTRY, last accessed on 2015-05-15.

Fasullo, J.T., Boening, C., Landerer, F.W., and Nerem, R.S., (2013). Australia's unique influence on global sea level in 2010–2011. *Geophysical Research Letters*, 40, 4368–4373, doi:10.1002/grl.50834.

Field, C. B., Barros, V., Stocker, T. F., Qin, D., Dokken, D. J., Ebi, K. L., Mastrandrea, M. D., et al. (2012). IPCC: *Managing the risks of extreme events and disasters to advance climate change adaptation*. A Special Report of Working Groups I and II of the Fourth Assessment Report of the Intergovernmental Panel on Climate Change.

Freeland, H., Denman, K., Wong, C.S., Whitney, F., and Jacques, R. (1997). Evidence of change in the winter mixed layer in the Northeast Pacific Ocean, *Deep Sea Research Part I*, 44, 2117–2129.

Galloway, J. N., et al., (2008): Transformation of the nitrogen cycle: Recent trends, questions, and potential solutions. *Science*, 320, 889–892.

Gelderloos R., Straneo, F., and Katsman, C. (2012). Mechanisms behind the temporary shutdown of deep convection in the Labrador Sea: Lessons from the Great Salinity Anomaly years 1968-1971. *Journal of Climate*, 25, 6745-6755.

GESAMP (2001) (Group of Experts on the Scientific Aspects of Marine Environmental Protection: a IMO/FAO/UNESCO-IOC/WMO/WHO/IAEA/UN/UNEP Joint Advisory Committee on Protection of the Sea). *Protecting the oceans from land-based activities - Land-based sources and activities affecting the quality and uses of the marine, coastal and associated freshwater environment*. Rep. Stud. GESAMP No. 71, 162 pp.

Giannini, A., Saravanan, R., and Chang, P. (2003). Oceanic forcing of Sahel rainfall on interannual to interdecadal time scales. *Science*, 302, 1027-1030.

Good, S.A., Martin, M.J. and Rayner, N.A. (2013). EN4: Quality controlled ocean temperature and salinity profiles and monthly objective analyses with uncertainty estimates. *Journal of Geophysical Research: Oceans*, 118 (12), pp. 6704-6716. doi: 10.1002/2013JC009067.

Gordon, A.L. (1986). Inter-Ocean Exchange of Thermocline Water. *Journal of Geophysical Research*, 91(C4), 5037-5046.

Graham, S., Parkinson, C., and Chahine, M. (2010). The Water Cycle. http://earthobservatory.nasa.gov/Features/Water/page3.php

Green, D. (2006) *How Might Climate Change Affect Island Culture in the Torres Strait?* CSIRO Marine and Atmospheric Research Paper 011.

Guo, X., Zhu, X.-H., Wu, Q.-S., and Huang, D. (2012). The Kuroshio nutrient stream and its temporal variation in the East China Sea, *Journal of Geophysical Research*, 117, C01026, doi:10.1029/2011JC007292.

Hartmann, D.L., Klein Tank, A.M.G., Rusticucci, M., Alexander, L.V., Brönnimann, S., Charabi, Y., Dentener, F.J., Dlugokencky, E.J., Easterling, D.R., Kaplan, A., Soden, B.J., Thorne, B.W., Wild, M., and Zhai, P.M. (2013). Observations: Atmosphere and Surface. In: *Climate Change 2013: The Physical Science Basis*. Contribution of Working Group I to the Fifth Assessment Report of the Intergovernmental Panel on Climate Change, Stocker, T.F., Qin, D., Plattner, G.-K., Tignor, M., Allen, S.K., Boschung, J., Nauels, A., Xia, Y., Bex, V., and Midgley, P.M. (eds.). Cambridge University Press, Cambridge, UK, 1535 pp.

Hay, J.E. and Mimura, N. (2006). Sea-level rise: Implications for water resources management. *Mitigation and Adaptation Strategies for Global Change*, 10, 717-737.

Held, I. M., and Soden, B.J. (2006). Robust responses of the hydrological cycle to global warming. *Journal of Climate*, 19(21), 5686-5699, doi: 10.1175/JCLI3990.1.

Helm, K.P., Bindoff, N.L. and Church, J.A. (2010). Changes in the global hydrological-cycle inferred from ocean salinity. *Geophysical Research Letters*, 37 (18), L18701. doi: 10.1029/2010GL044222.

Hoegh-Guldberg, O., Cai, R., Poloczanska, E.S., Brewer, P.G., Sundby, S., Hilmi, K., Fabry, V.J. and Jung, S. (2014). The Ocean. In: *Climate Change 2014: Impacts, Adaptation, and Vulnerability. Part B: Regional Aspects. Contribution of Working Group II to the Fifth Assessment Report of the Intergovernmental Panel on Climate Change* [Barros, V.R., Field, C.B., Dokken, D.J., Mastrandrea, M.D., Mach, K.J., Bilir, T.E., Chatterjee, M., Ebi, K.L., Estrada, Y.O., Genova, R.C., Girma, B., Kissel, E.S., Levy, A.N., MacCracken, S., Mastrandrea, P.R., and White, L.L. (eds.)]. Cambridge University Press, Cambridge, United Kingdom and New York, NY, USA, pp. 1655-1731.

Hoozemans, F.M.J., Marchand, M., and Pennekamp, H.A. (1993). *A Global Vulnerability Analysis: Vulnerability Assessment for Population, Coastal Wetlands and Rice Production on a Global Scale.* 2nd edition. Delft, Netherlands: Delft Hydraulics Software.

Hosoda, S., Suga, T., Shikama, N., and Mizuno, K. (2009). Global Surface Layer Salinity Change Detected by Argo and Its Implication for Hydrological Cycle Intensification. *Journal of Oceanography*, 65, pp 579-596. doi: 10.1007/s10872-009-0049-1

Hutchins, D.A., Mulholland, M.R., and Fu, F. (2009). Nutrient Cycles and Marine Microbes in a CO2-Enriched Ocean. *Oceanography* 22(4):128–145

IPCC (2013), *Climate Change 2013: The Physical Science Basis*. Contribution of Working Group I to the Fifth Assessment Report of the Intergovernmental Panel on Climate Change. Stocker, T.F., Qin, D., Plattner, G.-K., Tignor, M., Allen, S.K., Boschung, J., Nauels, A., Xia, Y., Bex, V., and Midgley, P.M. (eds.). Cambridge University Press, Cambridge, UK, 1535 pp.

Kates, R.W., Travis, W.R. and Wilbanks, T.J. (2012). Transformational adaptation when incremental adaptations to climate change are insufficient. *Proceedings of the National Academy of Sciences of the United States of America* 109 (19):7156-7161. doi:10.1073/pnas.1115521109

Kim, T.W., Lee, K., Najjar, R.G., Jeong, H.D. and Jeong, H.J. (2011). Increasing N abundance in the northwestern Pacific Ocean due to atmospheric nitrogen deposition. *Science*, 334, 505–509.

Kelly, K.A., Thompson, L., Lyman, J. (2014). The coherence and impact of meridional heat transport anomalies in the Atlantic Ocean inferred from observations. *Journal of Climate*, 27, 1469–1487, doi: 10.1175/JCLI-D-12-00131.1.

Leuliette, E.W., and Willis, J.K. (2011). Balancing the sea level budget. *Oceanography*, 24 (2), 122–129, doi:10.5670/oceanog.2011.32.

Malhi, Y., and Wright, J. (2011). Spatial patterns and recent trends in the climate of tropical rainforest regions. *Philosophical Transactions of the Royal Society*, London B 359(1443), 311-29. DOI: 10.1098/rstb.2003.1433.

Merrifield, M.A., Thompson, P.R., and Lander, M. (2012). Multidecadal sea level anomalies and trends in the western tropical Pacific. *Geophysical Research Letters*, 39, L13602, doi:10.1029/2012GL052032.

Midgley, G. and Thuiller, W. (2011). Potential responses of terrestrial biodiversity in Southern Africa to anthropogenic climate change. *Regional Environmental Change*, 11, 127–135.

Millennium Assessment (2005). *Ecosystems and Human Well-being: A Framework for Assessment*, Island Press, The Center for Resource Economics, Washington DC, 245 pp., ISBN 1-55963-402-2.

Milly, P.C.D., Dunne, K., and Vecchia, A.V. (2005). Global pattern of trends in streamflow and water availability in a changing climate, *Nature*, 438 (7066), 347–350, http://dx.doi.org/10.1038/nature04312.

Mimura N. (2013). Sea-level rise caused by climate change and its implications for society. *Proceedings of the Japan Academy, Series B Physical and Biological Sciences* 89(7): 281–301. doi: 10.2183/pjab.89.281.

Moon, J.-H., Song, Y.T., Bromirski, P.D., and Miller, A.J. (2013). Multidecadal regional sea level shifts in the Pacific over 1958-2008. *Journal of Geophysical Research*, 118, 7024–7035, doi:10.1002/2013JC009297.

Nagano, A., Uehara, K., Suga, T., Kawai, Y., Ichikawa, H. and Cronin, M.F. (2014). Origin of near-surface high-salinity water observed in the Kuroshio Extension region, *Journal of Oceanography*, 70, 389-403, doi: 10.1007/s10872-014-0237-5.

Nakamura, T., Toyoda, T., Ishikawa, Y., and Awaji, T. (2006). Effects of tidal mixing at the Kuril Straits on North Pacific ventilation: Adjustment of the intermediate layer revealed from numerical experiments, *Journal of Geophysical Research*, 111, C04003, doi:10.1029/2005JC003142.

Nicholls, R.J. and Cazenave, A. (2010). Sea-level rise and its impact on coastal zones, *Science*, 328, 1517-20.

Nicholls, R.J., Marinova, N., Lowe, J.A., Brown, S., Vellinga, P., de Gusmão, D., Hinkel, J. et al. (2011). Sea-level rise and its possible impacts given a "beyond 4°C world" in the twenty-first century. *Philosophical transactions. Series A, Mathematical, physical, and engineering sciences*, 369(1934), 161–81. doi:10.1098/rsta.2010.0291.

Nicholls, R.J., and Tol, R.S.J. (2006). Impacts and responses to sea-level rise: a global analysis of the SRES scenarios over the twenty-first century. *Philosophical transactions. Series A, Mathematical, physical, and engineering sciences* 364:1073–1095.

Nicholls, R.J., Wong, P.P., Burkett, V.R., Codignotto, J.O., Hay, J.E., McLean, R.F., Ragoonaden, S., and Woodroffe, C.D. (2007). Coastal systems and low-lying areas. In: *Climate Change 2007: Impacts, Adaptation and Vulnerability*. Contribution of Working Group II to the Fourth Assessment Report of the Intergovernmental Panel on Climate Change. In: Parry, M.L., Canziani, O.F., Palutikof, J.P., van der Linden, P.J., and Hanson, C.E. (eds.). Cambridge University Press, Cambridge, UK, pp. 315-356.

Nicholls, R.J., Woodroffe, C.D. and Burkett, V. (2009). Coastline degradation as an indicator of global change. In: *Climate Change: Observed Impacts on Planet Earth*. Letcher, T.M. (ed.). Elsevier, 409-24.

Nicholls, R.J. (2010). Impacts of and responses to sea-level rise. In: *Understanding Sea-level rise and variability*. Church, J.A., Woodworth, P.L., Aarup, T., Wilson, W.S. (eds.). Wiley-Blackwell. ISBN 978-4443-3451-7, 17-43.

Ono, T., Midorikawa, T., Watanabe, Y.W., Tadokoro, K., and Saino, T. (2001). Temporal increases of phosphate and apparent oxygen utilization in the subsurface waters of the western subarctic Pacific from 1968 to 1998, *Geophysical Research Letters*, 28(17), 3285–88.

Ono, T., Tadokoro, K., Midorikawa, T., Nishioka, J., and Saino, T. (2002). Multi-decadal decrease of net community production in western subarctic North Pacific, *Geophysical Research Letters*, 29(8), 1186, doi:10.1029/2001GL014332.

Ono T., Shiomoto, A., and Saino, T. (2008). Recent decrease of summer nutrients concentrations and future possible shrinkage of the subarctic North Pacific high-nutrient low-chlorophyll region, *Global Biogeochemical Cycles*, 22, GB3027, doi:10.1029/2007GB003092.

Osafune, S., and Yasuda, I. (2006). Bidecadal variability in the intermediate waters of the northwestern subarctic Pacific and the Okhotsk Sea in relation to 18.6-year period nodal tidal cycle, *Journal of Geophysical Research*, 111, C05007, doi:10.1029/2005JC003277

Oude Essink, G.H.P., Boekelman, R.H. and Bosters, M.C.J. (1993). Physical impacts of sea level change. In: Jelgersma, S., Tooley, M.J., Oerlemans, J., van Dam, J.C., Liebscher, H.J., Oude Essink, G.H.P., Boekelman, R.H., Bosters, M.C.J., Wolff, W.J., Dijksma, K.S., et al. *State of the art report: Sea level changes and their consequences for hydrology and water management: Fundamental Aspects, Policy and Protection in Low Lying Coastal Regions and Deltaic Areas.* Netherlands Institute for Marine and Coastal Zone Management (formerly Tidal Waters Division), The Hague, The Netherlands.

Provoost, P., van Heuven, S., Soetaert, K., Laane, R.W.P.M., and Middelburg, J.J. (2010). Seasonal and long-term changes in pH in the Dutch coastal zone. *Biogeosciences*, 7, 3869–78.

Rignot, E. and Thomas, R.H. (2002). Mass Balance of Polar Ice Sheets. *Science*, 297, 1502–06.

Rouault, M., Penven, P., and Pohl, B. (2009). Warming in the Agulhas Current system since the 1980's. *Geophysical Research Letters*, 36(12), L12602, doi: 10.1029/2009GL037987.

Rouault, M., White, S.A., Reason, C.J.C. Lutjeharms, J.R.E., and Jobard, I. (2002). Ocean–Atmosphere Interaction in the Agulhas Current Region and a South African Extreme Weather Event. *Weather Forecasting*, 17, 655–669. doi: 10.1175/1520-0434(2002)017<0655:OAIITA>2.0.CO;2

Sabine, C.L., Feely, R.A., Gruber, N., Key, R.M., Lee, K., Bullister, J.L., Wanninkhof, R., Wong, C.S., Wallace, D.W.R., Tilbrook, B., Millero, F.J., Peng, T.-H., Kozyr, A., Ono, T., and Rios, A.F. (2004). The Oceanic Sink for Anthropogenic CO2, *Science*, 305 (5682) pp. 367-71, doi: 10.1126/science.1097403.

Schaeffer, M., Hare, W., Rahmstorf, S., and Vermeer, M. (2012). Long term sea-level rise implied by 1.5 °C and 2 °C warming levels. *Nature Climate Change*, 2: 867-870. . doi:10.1038/nclimate1584.

Schanze, J.J., R. Schmitt, W. and Yu, L.L.(2010) The global oceanic freshwater cycle: A state-of-the-art quantification. *Journal of Marine Research*, 68, pp 569-595. doi: 10.1357/002224010794657164

Schmitz, W.J. Jr. (1996). *On the World Ocean Circulation*, Volume I. Some Global Features/North Atlantic Circulation. Woods Hole Oceanographic Institution, Technical Report WHOI-96- 03, 140 pp.

Scott, D., Simpson, M.C., and Sim, R. (2012). The vulnerability of Caribbean coastal tourism to scenarios of climate change related sea level rise, *Journal of Sustainable Tourism*, 20:883-898. Doi: 10.1080/09669582.2012.699063.

Seitzinger, S. P., et al. (2010). Global river nutrient export: A scenario analysis of past and future trends. *Global Biogeochemical Cycles*, 24, DOI: 10.1029/2009GB003587.

Skliris, N., Marsh, R., Josey, S.A., Good, S.A., Liu, C., and Allan, R.P. (2014). Salinity changes in the World Ocean since 1950 in relation to changing surface freshwater fluxes. *Climate Dynamics*,43:709-736. doi: 10.1007/s00382-014-2131-7.

Smith R.L., and Smith, T.M., (1998). *Elements of Ecology*, Pearson, Benjamin Cummings, San Francisco, USA.

Sunda, W.G. and Cai, W.-J. (2012). Eutrophication Induced CO2 Acidification of Subsurface Coastal Waters: Interactive Effects of Temperature, Salinity, and Atmospheric PCO2, *Journal of Environmental Science & Technology*, October.

Tadokoro, K., Ono, T., Yasuda, I., Osafune, S., Shiomoto, A., and Sugisaki, H. (2009) Possible mechanisms of decadal-scale variation in PO4 concentration in the western North Pacific, *Geophysical Research Letters*, 36, L08606, doi:10.1029/2009GL037327.

Tilman, D., Fargione, Wolff, B., D'Antonio, C., Dobson, A., Howarth, R., Schindler, D., Schlesinger, W.H., Simberloff, D., and Swackhamer, D. (2001) Forecasting Agriculturally Driven Global Environmental Change, *Science*, 292, 13 April 2001.

Terray, L., Corre, L., Cravatte, S., Delcroix, T., Reverdin, G., and Ribes, A. (2012) Near-Surface Salinity as Nature's Rain Gauge to Detect Human Influence on the Tropical Water Cycle. *Journal of Climate*, 25 (3), pp 958-977. doi: 10.1175/JCLI-D-10-05025.1

Trenberth, K.E., (1999). Conceptual framework for changes of extremes of the hydrological cycle with climate change. *Climatic Change*, 42, 327-339.

Trenberth, K.E., Smith, L., Qian, T., Dai, A., and Fasullo, J. (2007). Estimates of the Global Water Budget and Its Annual Cycle Using Observational and Model Data. *Journal of Hydrometeorology*, 8, 758-769. doi: 10.1175/JHM600.1.

United Nations Environment Programme (2006). *The State of the Marine Environment: A regional assessment*. Global Programme of Action for the Protection of the Marine Environment from Land-based Activities, United Nations Environment Programme, The Hague.

UNEP Global Environment Monitoring System Water Programme (2008). *Water Quality for Ecosystem and Human Health*, 120 pp., Ontario, Canada.

Van den Broeke, M.R., Bamber, J.L., Ettema, J., Rignot, E. Schrama, E., van de Berg, W.J., Velicogna, I., and Wouters, B. (2009). Partitioning recent Greenland mass loss. *Science*, 326(5955), 984-986.

Vaughan, D.G., Comiso, J.C., Allison, I., Carrasco, J., Kaser, G., Kwok, R., Mote, P., Murray, T., Paul, F., Ren, J., Rignot, E., Solomina, O., Steffen, K., and Zhang, T. (2013). Observations: Cryosphere. In: *Climate Change 2013: The Physical Science Basis*. Contribution of Working Group I to the Fifth Assessment Report of the Intergovernmental Panel on Climate Change, Stocker, T.F., Qin, D., Plattner, G.-K., Tignor, M., Allen, S.K., Boschung, J., Nauels, A., Xia, Y., Bex, V., and Midgley, P.M. (eds.). Cambridge University Press, Cambridge, UK, 1535 pp.

Vinogradova, N.T. and Ponte, R.M. (2013). Clarifying the link between surface salinity and freshwater fluxes on monthly to inter-annual timescales. *Journal of Geophysical Research*, 108 (6), pp 3190-3201, doi: 10.1002/jgrc.20200.

Watanabe, Y.W., Wakita, M., Maeda, N., Ono, T., and Gamo, T. (2003). Synchronous bidecadal periodic changes of oxygen, phosphate and temperature between the Japan Sea deep water and the North Pacific intermediate water, *Geophysical Research Letters*, 30(24), 2273, doi:10.1029/2003GL018338.

Watanabe, Y.W., Ishida, H., Nakano, T., and Nagai, N. (2005) Spatiotemporal Decreases of Nutrients and Chlorophyll-a in the Surface Mixed Layer of the Western North Pacific from 1971 to 2000. *Journal of Oceanography*, 61, 1011-16.

Watanabe, Y.W., Shigemitsu, M., and Tadokoro, K. (2008). Evidence of a change in oceanic fixed nitrogen with decadal climate change in the North Pacific subpolar region, *Geophysical Research Letters*, 35(1), L01602, doi:10.1029/2007GL032188.

Webb, A.P. and Kench, P.S. (2010). The dynamic response of reef islands to sea-level rise: Evidence from multi-decadal analysis of island change in the Central Pacific, *Global and Planetary Change*, 72: 234-246. dx.doi.org/10.1016/j.gloplacha.2010.05.003.

White, N.J., Haigh, I.D., Church, J.A., Koen, T., Watson, C.S., Pritchard, T.R., Watson, P.J., Burgette, R.J., McInnes, K.L., You, Z.-J., Zhang, X., Tregoning, P. (2014). Australian sea levels—Trends, regional variability and influencing factors. *Earth-Science Reviews*, 136, 155–74, doi: 10.1016/j.earscirev.2014.05.011.

Whitney, F.A. (2011) Nutrient variability in the mixed layer of the subarctic Pacific Ocean, 1987–2010, *Journal of Oceanography*., 67, 481–92, doi:10.1007/s10872-011-0051-2.

Whitney, F.A., Bograd, S.J., and Ono, T. (2013) Nutrient enrichment of the subarctic Pacific Ocean pycnocline, *Geophysical Research Letters*, 40, 2200–2205, doi:10.1002/grl.50439.

World Bank. (2012). *Turn Down the Heat: Why a 4 °C Warmer World must be Avoided* (Potsdam Institute for Climate Impact Research and Climate Analytics). http://www.worldbank.org/content/dam/Worldbank/document/Full_Report_Vol_2_Turn_Down_The_Heat_%20Climate_Extremes_Regional_Impacts_Case_for_Resilience_Print%20version_FINAL.pdf, last accessed on 2014-07-18.

Wüst, G. (1928). Der Ursprung der Atlantischen Tiefenwassar. Jubiläums' Sonderband. *Zeitschrift der Gesellschaft für Erdkunde*.

Xie, P., and Arkin, P.A. (1997). Global precipitation: A 17-year monthly analysis based on gauge observations, satellite estimates, and numerical model outputs. *Bulletin of the American Meteorological Society*, 78(11), 2539-2558.

Yu, L., (2007). Global variations in oceanic evaporation (1958-2005): The role of the changing wind speed. *Journal of Climate*, 20(21), 5376-90.

Yu, L. (2011). A global relationship between the ocean water cycle and near-surface salinity. *Journal of Geophysical Research*, 116 (C10), C10025. doi: 10.1029/2010JC006937

Zhang, X., and J.A. Church, (2012). Sea level trends, interannual and decadal variability in the Pacific Ocean. *Geophysical Research Letters*, 39(21), L21701. doi:10.1029/2012GL053240.

5 Sea-Air Interactions

Contributors:
Jeremy T. Mathis (Convenor), Patricio Bernal (Co-Lead Member), Lorna Inniss (Lead Member), Alberto Mavume, Renzo Mosetti, Alberto Piola, Regina Rodrigues, Chris Reason, Jose Santos and Craig Stevens

Sea-Air Interactions Chapter 5

1 Introduction

From the physical point of view, the interaction between these two turbulent fluids, the ocean and the atmosphere, is a complex, highly nonlinear process, fundamental to the motions of both. The winds blowing over the surface of the ocean transfer momentum and mechanical energy to the water, generating waves and currents. The ocean in turn gives off energy as heat, by the emission of electromagnetic radiation, by conduction, and, in latent form, by evaporation.

The heat flux from the ocean provides one of the main energy sources for atmospheric motions. This source of energy for the atmosphere is affected by the turbulence at the air/sea interface, and by the spatial distribution of the centres of high and low energy transfer affected by the ocean currents. This coupling takes place through processes that fundamentally occur at small scales. The strength of this coupling depends on air-sea differences in several factors and therefore has geographic and temporal scales over a broad range. At these small scales on the sea-surface interface itself, waves, winds, water temperature and salinity, bubbles, spray and variations in the amount of solar radiation that reaches the ocean surface, and other factors, affect the transfer of properties and energy.

In the long term, the convergence and divergence of oceanic heat transport provide sources and sinks of heat for the atmosphere and partly shape the mean climate of the earth. Analyzing whether these processes are changing due to anthropogenic influences and the potential impact of these changes is the subject of this chapter. Following guidance from the Ad Hoc Working Group of the Whole, much of the information presented here is based on or derives from the very thorough analysis conducted by the Intergovernmental Panel on Climate Change (IPCC) for its recent Fifth Assessment Report (AR5).

The atmosphere and the ocean form a coupled system, exchanging at the air-sea interface gases, water (and water vapour), particles, momentum and energy. These exchanges affect the biology, the chemistry and the physics of the ocean and influence its biogeochemical processes, weather and climate (exchanges affecting the water cycle are addressed in Chapter 4).

From a biogeochemical point of view, gas and chemical exchanges between the oceans and the atmosphere are important to life processes. Half of the Global Net Primary Production of the world is by phytoplankton and other marine plants, uptaking CO_2 and releasing oxygen (Field et al., 1998; Falkowski and Raven, 1997). Phytoplankton is therefore also responsible for half of the annual production of oxygen by plants and, through the generation of organic matter, is at the basis of most marine food webs in the ocean. Oxygen production by plants is a critical ecosystem service that keeps atmospheric oxygen from otherwise declining. However, in many regions of the ocean, phytoplankton growth is limited by a deficit of iron in seawater. Most of the iron alleviating this limitation reaches the ocean through wind-borne dust from the deserts of the world.

Gas and chemical exchanges between the atmosphere and ocean are also important to climate change processes. For example, marine phytoplankton produces dimethyl sulphide (DMS), the most abundant biological sulphur compound emitted to the atmosphere (Kiene et al., 1996). DMS is oxidized in the marine atmosphere to form various sulphur-containing compounds, including sulphuric acid, which influence the formation of clouds. Through this interaction with cloud formation, the massive production of atmospheric DMS over the ocean may have an impact on the earth's climate. The absorption of CO_2 from the atmosphere at the sea surface is responsible for the fundamental role of the ocean as a carbon sink (see section 3 below).

2 Heat flux and temperature

2.1 Sea-Surface Temperature

Sea-surface temperature (SST) has been measured in surface waters by a variety of methods that have changed significantly over time. Furthermore the spatial patterns of SST change are difficult to interpret. Nevertheless a robust trend emerges from these historical series after careful inspection and analysis of the datasets. Figure 5.1 shows the historical SST trend instrumentally observed using the best datasets of spatially interpolated products, contrasted against the 1961 – 1990 climatology. Changes in SST are reported in this section and in Chapter 2 of the IPCC (Hartmann et al., 2013).

The IPCC in AR5 concluded that 'recent' warming (since the 1950s) is strongly evident in SST at all latitudes of each ocean. Prominent spatiotemporal structures, including the El Niño Southern Oscillation (ENSO), decadal variability patterns in the Pacific Ocean, and a hemispheric asymmetry in the Atlantic Ocean, were highlighted as contributors to the regional differences in surface warming rates, which in turn affect atmospheric circulation (Hartmann et al., 2013).

"It is certain that global average sea surface temperatures (SSTs) have increased since the beginning of the 20th century. (…) Intercompari-

Figure 5.1 | Global annual average sea surface temperature (SST) and Night Marine Air Temperature (NMAT) relative to a 1961–1990 climatology from state of the art data sets. Spatially interpolated products are shown by solid lines; non-interpolated products by dashed lines. From Hartmann et al. 2013, Fig. 2.18.

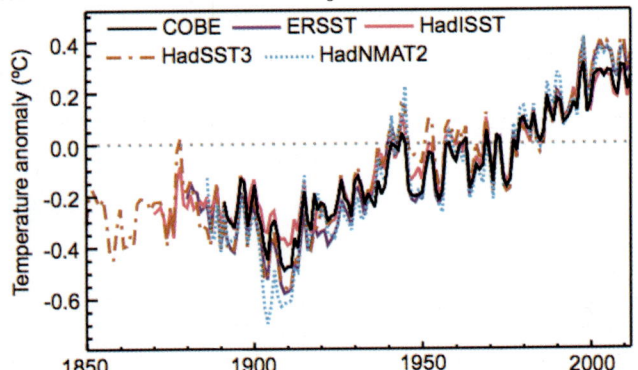

sons of new SST data records obtained by different measurement methods, including satellite data, have resulted in better understanding of uncertainties and biases in the records. Although these innovations have helped highlight and quantify uncertainties and affect our understanding of the character of changes since the mid-20th century, they do not alter the conclusion that global SSTs have increased both since the 1950s and since the late 19th century." (Hartmann et al., 2013).

2.2 Changes in sea-surface temperature (SST) as inferred from subsurface measurements.

Upper ocean temperature (hence heat content) varies over multiple time scales, including seasonal, interannual (e.g., associated with El Niño), decadal and centennial (Rhein et al., 2013). Depth-averaged (0 to 700 m) ocean-temperature trends from 1971 to 2010 are positive over most of the globe. The warming is more prominent in the Northern Hemisphere, especially in the North Atlantic. This result holds true in different analyses, using different time periods, bias corrections and data sources (e.g., with or without XBT or MBT data1) (Rhein et al. 2013). Zonally averaged upper-ocean temperature trends show warming at nearly all latitudes and depths (Figure 5.2a). However, the greater volume of the Southern Hemisphere ocean increases the contribution of its warming to the global heat content (Rhein et al., 2013). Strongest warming is found closest to the sea surface, and the near-surface trends are consistent with independently measured SST (Hartmann et al., 2013). The global average warming over this period is 0.11 [0.09 to 0.13] °C per decade in the upper 75 m, decreasing to 0.015°C per decade by 700 m (Figure 5.2c) (Rhein et al 2013).

The globally averaged temperature difference between the ocean surface and 200 m increased by about 0.25°C from 1971 to 2010. This change, which corresponds to a 4 per cent increase in density stratification, is widespread in all the oceans north of about 40°S. Increased stratification will potentially diminish the exchanges between the interior and the surface layers of the ocean; this will limit, for example, the input of nutrients from below into the illuminated surface layer and of oxygen from above into the deeper layers. These changes might in turn result in reduced productivity and increased anoxic waters in many regions of the world ocean (Capotondi et al., 2012).

2.3 Upper Ocean Heat Content (UOHC)

The ocean's large mass and high heat capacity allow it to store huge amounts of energy: more than 1000 times that found in the atmosphere for an equivalent increase in temperature. The earth is absorbing more heat than it is emitting back into space, and nearly all this excess heat is entering the ocean and being stored there.

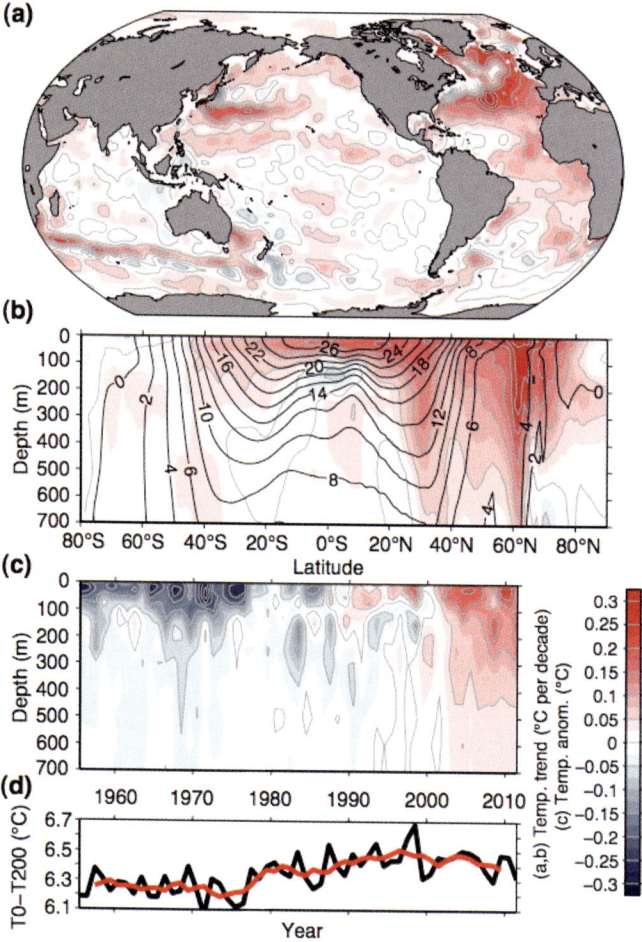

The boundaries and names shown and the designations used on this map do not imply official endorsement or acceptance by the United Nations.

Figure 5.2 | (a) Depth-averaged (0 to 700) m ocean-temperature trend for 1971-2010 (longitude vs. latitude, colours and grey contours in degrees Celsius per decade); (b) Zonally averaged temperature trends (latitude vs. depth, colours and grey contours in degree Celsius per decade) for 1971-2010 with zonally averaged mean temperature over-plotted (black contours in degrees Celsius). Both North (25-65°N) and South (south of 30°S), the zonally averaged warming signals extend to 700 m and are consistent with poleward displacement of the mean temperature field. Zonally averaged upper-ocean temperature trends show warming at nearly all latitudes and depths (Figure 2 (b). A relative maximum in warming appears south of 30°S. (c) Globally averaged temperature anomaly (time vs. depth, colours and grey contours in degrees Celsius) relative to the 1971–2010 mean; (d) Globally averaged temperature difference between the ocean surface and 200 m depth (black: annual values, red: 5-year running mean). All panels are constructed from an update of the annual analysis of Levitus et al. (2009). From Rhein et al. (2013) Fig 3.1

The upper ocean (0 to 700 m) heat content increased during the 40-year period from 1971 to 2010. Published rates range from 74 TW to 137 TW (1 TW = 1012 watts), while an estimate of global upper (0 to 700 m depth) ocean heat content change, using ocean statistics to extrapolate to sparsely sampled regions and estimate uncertainties (Domingues et al., 2008), gives a rate of increase of global upper ocean heat content of 137 TW (Rhein, et al. 2013). Warming of the ocean accounts for about 93 per cent of the increase in the Earth's energy inventory between 1971 and 2010 (high confidence), Melting ice (including Arctic sea ice, ice sheets and glaciers) and warming of the continents and atmosphere account for the remainder of the change in energy (Rhein et al. 2013).

---

1   XBT are expendable bathythermographs, probes that using electronic solid-state transducers register temperature and pressure while they free fall through the water column. MBT are their mechanical predecessors, that lowered on a wire suspended from a ship, used a metallic thermocouple as transducer.

Sea-Air Interactions

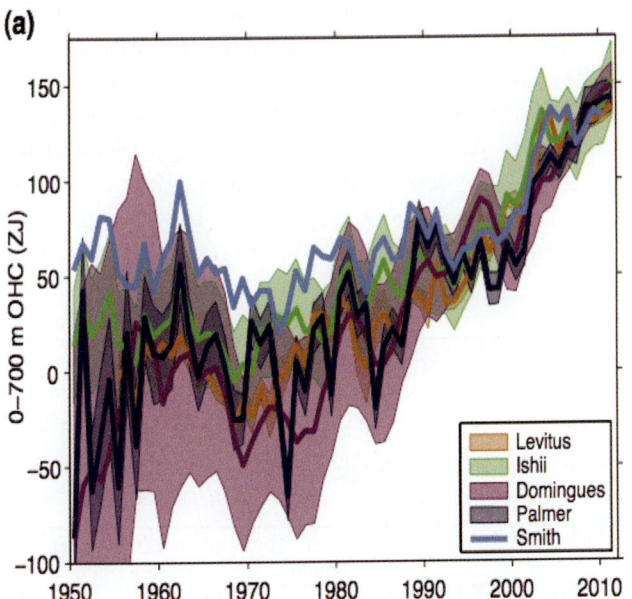

Figure 5.3 | Observation-based estimates of annual global mean upper (0 to 700 m) ocean heat content in ZJ (1 ZJ = 1021 Joules) updated from (see legend): Levitus et al. (2012), Ishii and Kimoto (2009), Domingues et al. (2008), Palmer et al. (2009; ©American Meteorological Society. Used with permission.) and Smith and Murphy (2007). Uncertainties are shaded and plotted as published (at the one standard error level, other than one standard deviation for Levitus, with no uncertainties provided for Smith). Estimates are shifted to align for 2006-2010, 5 years that are well measured by the ARGO Program of autonomous profiling floats, and then plotted relative to the resulting mean of all curves for 1971, the starting year for trend calculations

Global integrals of 0 to 700 m upper ocean heat content (UOHC) (Figure 5.3.) estimated from ocean temperature measurements all show a gain from 1971 to 2010 (Rhein et al. 2013).

2.4 The ocean's role in heat transport

Solar energy is unevenly distributed over the earth's surface, leading to excess heat reaching the tropics and a heat deficit in latitudes poleward of about 40o in each hemisphere. The heat balance, and therefore a relatively stable climate, is maintained through the meridional redistribution, or flux, of heat by the atmosphere and the ocean. Quantification and understanding of this heat content and its redistribution have been achieved through diverse methods, including international programmes maintaining instrumented moorings, transoceanic lines of XBTs, satellite observations, numerical modelling and, more recently, the ARGO Program of autonomous profiling instruments (Abraham et al., 2013; von Schuckmann and Le Traon, 2011).

In the latitude band between 25°N and 25°S, the atmospheric and oceanic contributions to the meridional heat fluxes are similar, and the atmosphere dominates at higher latitudes. In the ocean, the heat flux is accomplished by contributions from the wind-driven circulation in the upper ocean, by turbulent eddies, and by the Meridional Overturning Circulation (MOC). The MOC is a component of ocean circulation that is driven by density contrasts, rather than by winds or tides, and one which exhibits a pronounced vertical component, with dense water sinking at high latitudes, offset by broadly distributed upwelling at lower ones. As

distinct circulation patterns characterize each of the ocean basins, their individual contributions to the meridional heat flux differ significantly. Estimates indicate that, on a yearly average, the global oceans carry 1-2 PW (1PW=1015W) of heat from the tropics to higher latitudes, with somewhat higher transports to the northern hemisphere (Fasullo and Trenberth, 2008).

Most of the heat excess due to increases in atmospheric greenhouse gases goes into the ocean (IPCC, 2013). Although all ocean basins have warmed during the last decades, the increase in heat content is not uniform; the increase in heat content in the Atlantic during the last four decades exceeds that of the Pacific and Indian Oceans combined (Levitus et al., 2009; Palmer and Haines, 2009). Enhanced northward heat flux in the subtropical South Atlantic, which includes heat driven from the subtropical Indian Ocean through the Agulhas Retroflection, may have contributed to the larger increase in heat content in the Atlantic Ocean compared with other basins (Abraham et al., 2013; Lee et al., 2011).

Numerical simulations also indicate that changes in ocean heat fluxes are the main mechanism responsible for the observed temperature fluctuations in the subtropical and subpolar North Atlantic (Grist et al., 2010).

Meridional heat flux estimates inferred from the residual of heat content variations suggest that the heat transferred northward throughout the Atlantic is transferred to the atmosphere in the subtropical North Atlantic (Kelly et al., 2014). Observations from the Rapid/Mocha instrument array at 26°N in the North Atlantic indicate that the mean Atlantic meridional heat flux at this latitude is 1.33 PW, with substantial variability due to changes in the strength of the MOC (Cunningham et al., 2007; Kanzow et al., 2007; Johns et al., 2011; McCarthy et al., 2012). Moreover, recent studies show that interannual changes in the MOC (and the associated heat flux measured at 26°N) lead to temperature anomalies in the subtropical North Atlantic which, in turn, can have a strong impact on the northern hemisphere climate (Cunningham et al., 2013; Buchan et al., 2014).

2.5 Air-sea Heat fluxes

Heat uptake by the ocean can be substantially altered by natural oscillations in the earth's ocean and atmosphere. The effects of these large-scale climate oscillations are often felt around the world, leading to the rearrangement of wind and precipitation patterns, which in turn substantially affect regional weather, sometimes with devastating consequences.

The ENSO is the most prominent of these oscillations and is characterized by an anomalous warming and cooling of the central-eastern equatorial Pacific. The warm phase is called El Niño and the cold, La Niña. During El Niño events, a weakening of the Pacific trade winds decreases the upwelling of cold waters in the eastern equatorial Pacific and allows warm surface water that generally accumulates in the western Pacific to flow east.

As a consequence, El Niños release heat into the atmosphere, causing an increase in globally averaged air temperature. However, the "recharge oscillator theory" (Ren and Jin, 2013) indicates that a buildup of upper-ocean heat content is a necessary precondition for the development of El Niño events. La Niñas are associated with a strengthening of the trade winds, which leads to a strong upwelling of cold subsurface water in the eastern Pacific. In this case, the ocean uptake of heat from the atmosphere is enhanced, causing the global average surface temperature to decrease (Roemmich and Gilson, 2011).

The cycling of ENSO between El Niño and La Niña is irregular. In some decades El Niño has dominated and in other decades La Niña has been more frequent, also seen in phase shifts of the Interdecadal Pacific Oscillation (Meehl et al., 2013), which is related to build up and release of heat. A strengthening of the Pacific trade winds in the past two decades has led to a more frequent occurrence of La Niñas (England et al., 2014). Consequently, the heat uptake by the subsurface ocean was enhanced, leading to a slowdown of the surface warming (Kosaka and Xie, 2013). This is one of the factors affecting the global mean temperature, expected to increase by 0.21°C per decade from 1998 to 2012, but which instead warmed by just 0.04°C (the so-called recent warming hiatus, IPCC, 2013). Although there are several hypotheses on the cause of the global warming hiatus, the role of ocean circulation in this negative feedback is certain. Drijfhout et al. (2014) have shown that the North Atlantic, Southern Ocean and Tropical Pacific all play significant roles in the ocean heat uptake associated with the warming hiatus.

Chen and Tung (2014) analyzed the historical and recent record of sea surface temperature and Ocean Heat Content (OHC), and found distinct patterns at the surface and in deeper layers. On the surface, the patterns conform to the El Niño/La Niña patterns, with the Pacific Ocean playing a dominant role by releasing heat during an El Niño (or capturing heat during La Niña). At depth, the dominant pattern shows heating taking place in the Atlantic Ocean and in the Circumpolar Current region. Coinciding in time, changes in OHC could help to explain the observed slowdown in global warming. It is anticipated that the mechanisms involved may at some point reverse, releasing large amounts of heat to the atmosphere and accelerating global warming (e.g., Levermann, et al., 2012).

Many other naturally occurring ocean-atmosphere oscillations in the Pacific, Atlantic, and Indian Oceans have also been recognized and named. The ENSO as a global phenomenon, has an expression in the Atlantic basin called the Atlantic Niño. In the last six decades, this mode has weakened, leading to a warming of the equatorial eastern Atlantic of up to 1.5°C (Tokinaga and Xie, 2011). Although the role of the Atlantic Niño on the global heat budget is not significant, this Atlantic warming trend has led to an increase in precipitation over the equatorial Amazon, Northeast South America, Equatorial West Africa and the Guinea coast, and a decrease in rainfall over the Sahel (Gianinni et al., 2003; Tokinaga and Xie, 2011; Marengo et al., 2011; Rodrigues et al., 2011). Moreover, recent studies have shown that the Atlantic Niño can have an effect on ENSO (Rodriguez-Fonseca et al., 2009; Keenlyside et al., 2013).

In the Indian Ocean, the dominant basin-wide oscillation is the Indian Dipole Mode (Saji et al., 1999). A positive phase is characterized by cool surface-temperature anomalies in the eastern Indian Ocean, warm-temperature anomalies in the western Indian Ocean, and easterly wind-stress anomalies along the equator. Similarly to ENSO, meridional heat transport and the associated buildup of upper-ocean heat content are a possible precondition for the development of the Indian Ocean Dipole event (McPhaden and Nagura, 2014). The warm surface temperatures in the western Indian Ocean are associated with an increase in subsurface heat content and vice-versa for the east (Feng et al. 2001; Rao et al., 2002). This zonal contrast of ocean heat content is induced by anomalies of zonal wind along the equator and the resulting variability in zonal mass and heat transport (Nagura and McPhaden 2010). The warm surface temperatures in the western Indian Ocean are associated with an increase in subsurface heat content and vice-versa for the east; the positive dipole causes above-average rainfall in eastern Africa and droughts in Indonesia and Australia (Behera et al., 2005; Yamagata et al., 2004; Ummenhofer et al., 2009; Cai et al., 2011; Section 5 below). Although the phenomena discussed here are global, many of the most significant impacts are on the coastal environment (see following Section).

2.6 Environmental, economic and social impacts of changes in ocean temperature and of major ocean temperature events

Coastal waters are valuable both ecologically and economically because they support a high level of biodiversity. They act as nursery areas for many commercially important fish species, and are the marine areas most accessible to the public. Because inshore habitats are shallow, water temperatures in coastal areas are closely linked to the regional climate and its seasonal and long-term fluctuations. Coastal waters also host some of the most vulnerable marine habitats, because they are intensively exploited by (including, but not limited to) the fishing industry and recreational craft, and because of their proximity to outlets of pollution, such as rivers and sewage outfalls. Coastal development and the threat of rising sea level may also impinge upon these valuable habitats (Halpern et al., 2008). Ecological degradation can lower the socio-economic value of coastal regions, with negative impacts on commercial fisheries, aquaculture facilities, damage to coastal infrastructure, problems with power-station cooling, and exert a dampening effect on coastal tourism from degraded ecological services.

It has been recently shown that when compared with estimates for the global ocean, decadal rates of SST change are higher at the coast. During the last three decades, approximately 70 per cent of the world's coastline has experienced significant increases in SST (Lima and Wethey, 2012). This has been accompanied by an increase in the number of yearly extremely hot days along 38 per cent of the world's coastline, and warming has been occurring significantly earlier in the year along approximately 36 per cent of the world's temperate coastal areas (defined as those between latitudes 30° and 60° in both hemispheres) at an average rate of 6.1 ± 3.2 days per decade (Lima and Wethey, 2012).

The warming of coastal waters can have many serious consequences for the ecological system (Harley et al., 2006). This can include changes in the distribution of important commercial fish and shellfish species, particularly the movement of species to higher latitudes due to thermal stress (Perry et al., 2005). Warming of coastal waters also can lead to more favourable conditions for many organisms, among them marine invasive species that can devastate commercial fisheries and destroy marine ecosystem dynamics (Occhipinti-Ambrogi, 2007). Water quality might also be impacted by higher temperatures that can increase the severity of local outbreaks by pathogenic bacteria or the occurrence of Harmful Algal Blooms (HABs). These in turn would cause harm to seafood, consumers and marine organisms (Bresnan et al., 2013). Increased coral reef bleaching and mortality from warming seas (combined with ocean acidification, see next sections) will lead to the loss of important marine habitats and associated biodiversity.

Changes in ocean temperatures have global impacts. As ocean temperatures warm, species that prefer specific temperature ranges may relocate – as has been observed, for instance, in copepod assemblages in the North Atlantic (Hays et al., 2005). Some organisms, like corals, are sedentary and cannot relocate with changing temperatures. If the water becomes too warm, they may experience a bleaching event. Higher sea level and warmer ocean temperatures can alter ocean circulation and current flow and increase the frequency and intensity of storms, leading to changes in the habitat of many species worldwide.

Changes in ocean temperatures affect not only marine ecosystems, but also the climate over land, with devastating economic and social implications. Many natural oceanic oscillations are known to have an impact on (terrestrial) climate, but these oscillations and the response of the climate to them are also changing during recent decades. For instance, an El Niño phase of ENSO (see previous Section for more details on ENSO) displaces great amounts of warm water from the western to the eastern Pacific, leading to more evaporation over the latter. As a consequence, western and southern South America and parts of North America experience wetter conditions. At the same time, Australia, Brazil, India, Indonesia, the Philippines, parts of Africa and the United States of America suffer droughts. La Niña events usually cause the opposite patterns. However, in the last several decades, ENSO events have changed their spatial and temporal characteristics (Yeh et al., 2009; McPhaden, 2012).

During recent decades, the warm waters of El Niño events have been displaced to the central Pacific instead of to the eastern Pacific. It is not clear yet whether these changes are linked to anthropogenic climate change or natural variability (Yeh et al., 2011). In any case, the effects on climate of an ENSO event centred in the central Pacific (a central Pacific ENSO) are in sharp contrast to that associated with one centred in the eastern Pacific.

For instance, northeastern and southeastern Australia experience a reduction in rainfall during the eastern Pacific El Niños and there is a decrease in rainfall over northwestern and northern Australia during central Pacific events (Taschetto and England, 2009; Taschetto et al., 2009). The Indian monsoon fails during eastern Pacific El Niños, but is enhanced during central Pacific El Niños (Kumar et al., 2006). Over the semi-arid region of northeast Brazil, eastern Pacific El Niños/La Niñas cause dry/wet conditions; central Pacific El Niños have the opposite effect, with the worst drought in the last 50 years associated with the strong 2011/12 La Niña and not with El Niños as in the past (Rodrigues et al., 2011; Rodrigues and McPhaden, 2014). This drought caused the displacement of 10 million people and economic losses on the order of 3 billion United States dollars in relation to agriculture and cattle raising alone. In contrast to drought in Brazil, the 2011/12 La Niña caused floods across southeastern Australia.

In other ocean basins, changes in oceanic oscillations and temperatures have also had an impact on climate. For instance, in the Indian Ocean, a positive phase of the Indian Dipole Mode (warm/cold temperatures in the western/eastern equatorial Indian Ocean) leads to flooding in east Africa and droughts in Indonesia, Australia, and India (Saji et al., 1999; Ashok et al., 2001; Gadgil et al., 2004; Yamagata et al., 2004; Behera et al., 2005; Ummenhofer et al., 2009; Cai et al., 2011). The counterpart of ENSO in the Atlantic (Atlantic Niño) has weakened during the last six decades, leading to an increase in SST in the eastern equatorial Atlantic. As a consequence, rainfall has been enhanced over the equatorial Amazon and West Africa (Tokinaga and Xie, 2011). On the other hand, an unusual warming of the tropical North Atlantic in 2005 was responsible for one of the worst droughts in the Amazon River basin and a record Atlantic hurricane season. Hurricanes Rita and Katrina caused the loss of almost 2000 lives and an estimated economic toll of 150 billion —135 billion US dollars from Katrina and 15 billion US dollars from Rita. (http://www.datacenterresearch.org/data-resources/katrina/facts-for-impact/). Anomalous warm conditions also occurred in the tropical North Atlantic in 2010 leading to two once-in-a-century droughts in less than five years in the Amazon River basin (Marengo et al., 2011).

Ocean warming will stress species both through thermic changes in their environmental envelope and through increased interspecies competition. These shifts become all the more important in shelf seas once they reach terrestrial boundaries, i.e., the shifting species runs out of shelf. For example, changes in the coastal currents in south-eastern Australia cause changes to primary production through to fisheries productivity. This then feeds through to local and regional socio-economic impacts (Suthers et al., 2011).

The IPCC AR5 concluded that "it is unlikely that annual numbers of tropical storms, hurricanes and major hurricanes counts have increased over the past 100 years in the North Atlantic basin. Evidence, however, is for a virtually certain increase in the frequency and intensity of the strongest tropical cyclones since the 1970s in that region" (Hartmann et al. 2013, Section 2.6.3). Moreover, the IPCC AR5 states that "it is difficult to draw firm conclusions with respect to the confidence levels associated with observed trends prior to the satellite era and in ocean basins outside of the North Atlantic" (Hartmann et al. 2013, Section 2.6.3). Although a strong scientific consensus on the matter does not exist, there is some evidence supporting the hypothesis that global warming might lead to fewer but more intense tropical cyclones globally (Knutson et al., 2010). Evidence exists that the observed expansion of the tropics since approximately 1979 is accompanied by a pronounced poleward migration of the latitude at which the maximum intensities of storms occur at a rate of 1° of latitude per decade (Kossin et al., 2014; Hartmann et al., 2013; Seidel et al., 2008). If this trend is confirmed, it would increase the frequency of events in coastal areas that are not exposed regularly to the dangers caused by cyclones. Hurricane Sandy in 2012 may be an example of this (Woodruff et al., 2013).

3   Water flux and salinity

3.1   Regional patterns of salinity, and changes in salinity[2] and freshwater content

The ocean plays a pivotal role in the global water cycle: about 85 per cent of the evaporation and 77 per cent of the precipitation occur over the ocean (Schmitt, 2008). The horizontal salinity distribution of the upper ocean largely reflects this exchange of freshwater: high surface salinity is generally found in regions where evaporation exceeds precipitation, and low salinity is found in regions of excess precipitation and runoff. Ocean circulation also affects the regional distribution of surface salinity.

The Earth's water cycle involves evaporation and precipitation of moisture at the Earth's surface. Changes in the atmosphere's water vapour content provide strong evidence that the water cycle is already responding to a warming climate. Further evidence comes from changes in the distribution of ocean salinity (Rhein et al. 2013; FAQ. 3.2). Diagnosis and understanding of ocean salinity trends are also important, because salinity changes, like temperature changes, affect circulation and stratification, and therefore the ocean's capacity to store heat and carbon as well as to change biological productivity.

Seawater contains both salt and fresh water, and its salinity is a function of the weight of dissolved salts it contains. Because the total amount of salt does not change over human time scales, seawater's salinity can only be altered—over days or centuries—by the addition or removal of fresh water.

The water cycle is expected to intensify in a warmer climate. Observations since the 1970s show increases in surface and lower atmospheric water vapour (Figure 5.4a), at a rate consistent with observed warming. Moreover, evaporation and precipitation are projected to intensify in a warmer climate. Recorded changes in ocean salinity in the last 50 years support that projection (Rhein et al. 2013; FAQ. 3.2).

The atmosphere connects the ocean's regions of net fresh water loss to those of fresh water gain by moving evaporated water vapour from one place to another. The distribution of salinity at the ocean surface largely reflects the spatial pattern of evaporation minus precipitation (Figure 5.4b), runoff from land, and sea ice processes. There is some shifting of the patterns relative to each other, because of the ocean's currents. Ocean salinity acts as a sensitive and effective rain gauge over the ocean. It naturally reflects and smoothes out the difference between water gained by the ocean from precipitation, and water lost by the ocean through evaporation, both of which are very patchy and episodic (Rhein et al. 2013; FAQ. 3.2). Data from the past 50 years show widespread salinity changes in the upper ocean, which are indicative of systematic changes in precipitation and runoff minus evaporation.

(Figure 5.4b). Subtropical waters are highly saline, because evaporation exceeds rainfall, whereas seawater at high latitudes and in the tropics—where more rain falls than evaporates—is less so. The Atlantic, the saltiest ocean basin, loses more freshwater through evaporation than it gains from precipitation, while the Pacific is nearly neutral, i.e., precipitation gain nearly balances evaporation loss, and the Southern Ocean is dominated by precipitation. (Figure 5.4b; Rhein et al. 2013; FAQ. 3.2).

The boundaries and names shown and the designations used on this map do not imply official endorsement or acceptance by the United Nations.

Figure 5.4 | Changes in sea surface salinity are related to the atmospheric patterns of evaporation minus precipitation (E − P) and trends in total precipitable water: (a) Linear trend (1988–2010) in total precipitable water (water vapour integrated from the Earth's surface up through the entire atmosphere) (kg m−2 per decade) from satellite observations (Special Sensor Microwave Imager) (after Wentz et al., 2007) (blues: wetter; yellows: drier). (b) The 1979–2005 climatological mean net E −P (cm yr−1) from meteorological reanalysis (National Centers for Environmental Prediction/National Center for Atmospheric Research; Kalnay et al., 1996) (reds: net evaporation; blues: net precipitation). (c) Trend (1950–2000) in surface salinity (PSS78 per 50 years) (after Durack and Wijffels, 2010) (blues freshening; yellows-reds saltier). (d) The climatological-mean surface salinity (PSS78) (blues: <35; yellows–reds: >35). From Rhein et al. 2013; FAQ. 3.2. Fig 1.

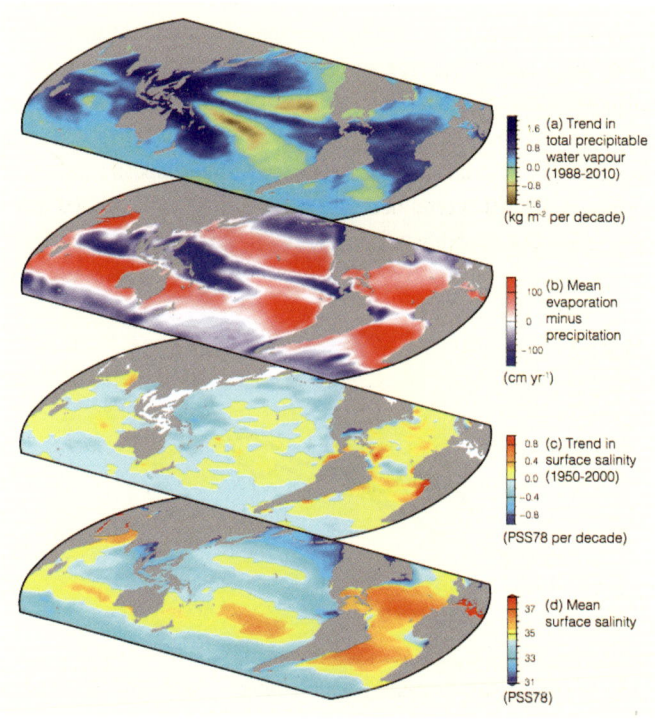

---

[2]   'Salinity' refers to the weight of dissolved salts in a kilogram of seawater. Because the total amount of salt in the ocean does not change, the salinity of seawater can be changed only by addition or removal of fresh water.

Changes in surface salinity and in the upper ocean have reinforced the mean salinity pattern (4c). The evaporation-dominated subtropical regions have become saltier, while the precipitation-dominated subpolar and tropical regions have become fresher. When changes over the top 500 m are considered, the evaporation-dominated Atlantic has become saltier, while the nearly neutral Pacific and precipitation-dominated Southern Ocean have become fresher (Figure 5.4d; Rhein et al. 2013; FAQ. 3.2).

Observed surface salinity changes also suggest a change in the global water cycle has occurred (Chapter 4). The long-term trends show a strong positive correlation between the mean climate of the surface salinity and the temporal changes in surface salinity from 1950 to 2000. This correlation shows an enhancement of the climatological salinity pattern: fresh areas have become fresher and salty areas saltier.

Ocean salinity is also affected by water runoff from the continents, and by the melting and freezing of sea ice or floating glacial ice. Fresh water added by melting ice on land will change global-averaged salinity, but changes to date are too small to observe (Rhein et al. 2013; FAQ. 3.2).

In conclusion, according to the last IPCC AR5, "It is very likely that regional trends have enhanced the mean geographical contrasts in sea surface salinity since the 1950s: saline surface waters in the evaporation-dominated mid-latitudes have become more saline, while relatively fresh surface waters in rainfall-dominated tropical and polar regions have become fresher" (Stocker et al., 2013). "The mean contrast between high- and low-salinity regions increased by 0.13 [0.08 to 0.17] from 1950 to 2008. It is very likely that the inter-basin contrast in freshwater content has increased: the Atlantic has become saltier and the Pacific and Southern Oceans have freshened. Although similar conclusions were reached in AR4, recent studies based on expanded data sets and new analysis approaches provide high confidence in this assessment" (Stocker et al., 2013). "The spatial patterns of the salinity trends, mean salinity and the mean distribution of evaporation minus precipitation are all similar. These similarities provide indirect evidence that the pattern of evaporation minus precipitation over the oceans has been enhanced since the 1950s (medium confidence)" Stocker et al., (2013). "Uncertainties in currently available surface fluxes prevent the flux products from being reliably used to identify trends in the regional or global distribution of evaporation or precipitation over the oceans on the time scale of the observed salinity changes since the 1950s" (Stocker et al., 2013).

4 Carbon dioxide flux and ocean acidification

4.1 Carbon dioxide emissions from anthropogenic activities

Since the start of Industrial Revolution, human activities have been releasing large amounts of carbon dioxide into the atmosphere. As a result, atmospheric CO_2 has increased from a glacial to interglacial cycle of 180-280 ppm to about 395 ppm in 2013 (Dlugokencky and Tans, 2014). Until around 1920, the primary source of carbon dioxide to the atmosphere was from deforestation and other land-use change activities (Ciais et al., 2013). Since the end of World War II, anthropogenic emissions of CO_2 have been increasing steadily. Data from 2004 to 2013 show that human activities (fossil fuel combustion and cement production) are now responsible for about 91 per cent of the total CO_2 emissions (Le Quéré et al. 2014).

CO_2 emissions from fossil fuel consumption can be estimated from the energy data that are available from the United Nations Statistics Division and the BP Annual Energy Review. Data in 2013 suggests that about 43 per cent of the anthropogenic CO_2 emissions were produced from coal, 33 per cent from oil and 18 per cent from gas, and 6 per cent from cement production (Figure 5.5).

Coal is an important and, recently, growing proportion of CO_2 emissions from fossil fuel combustion. From 2012 to 2013, CO_2 emissions from coal increased 3.0 per cent, compared to the increase rate of 1.4 per cent for oil and gas (Le Quéré et al. 2014). Coal accounted for about 60 per cent of the CO_2 emission growth in the same period. This is largely because many large economies of the world have recently resorted to using coal as an energy source for a wide variety of industrial processes, instead of oil, gas and other energy sources.

4.2 The ocean as a sink for atmospheric CO_2

The global oceans serve as a major sink of atmospheric CO_2. The oceans take up carbon dioxide through mainly two processes: physical air-sea flux of atmospheric CO_2 at the ocean surface, the so called "solubility pump" and through the active biological uptake of CO_2 into the biomass and skeletons of plankters the so-called "biological pump". Colder water can take up CO_2 more than warm water, and if this cold, denser water sinks to form intermediate, deep, or bottom water, there is transport

Figure 5.5 | CO_2 emissions from different sources from 1958 to 2013 (Le Quéré et al. 2014).

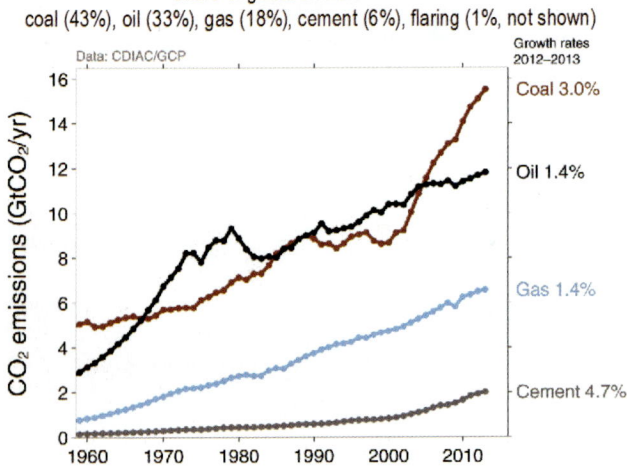

Chapter 5 — Sea-Air Interactions

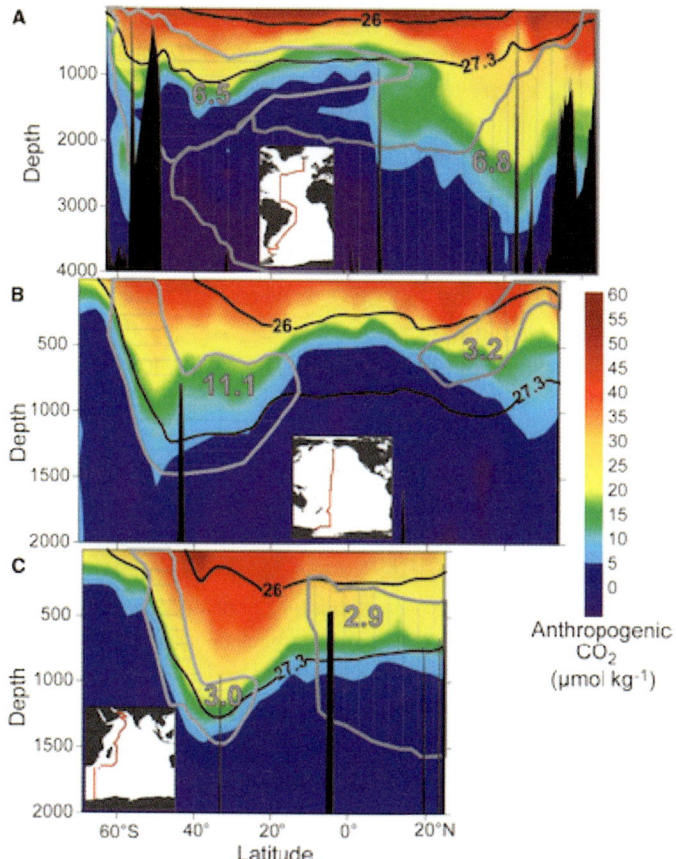

The boundaries and names shown and the designations used on this map do not imply official endorsement or acceptance by the United Nations.

Figure 5.6 | Anthropogenic CO_2 distributions along representative meridional sections in the Atlantic, Pacific, and Indian oceans for the mid-1990s (Sabine et al. 2004).

mixed into the deeper waters as dissolved organic or inorganic carbon. Part of this carbon is permanently buried in the sediments and other part enters into the slower circulation of the deep ocean. This "biological pump" serves to maintain the gradient in CO_2 concentration between the surface and deep waters.

Because the ocean mixes slowly, about half of the anthropogenic CO_2 (Cant) stored in the ocean is found in the upper 10 per cent of the ocean (Figure 5.6.). On average, the penetration depth is about 1000 meters and about 50 per cent of the anthropogenic CO_2 in the ocean is shallower than 400 meters.

Globally, the ocean shows large spatial variations in terms of its role as a sink of atmospheric CO_2 (Takahashi et al. 2009). Over the past 200 years the oceans have absorbed 525 billion tons of CO_2 from the atmosphere, or nearly half of the fossil fuel emissions over the period (Feely et al. 2009). The oceanic sink of atmospheric CO_2 has increased from 4.0 ± 1.8 $GtCO_2$ (1 $GtCO_2$ = 109 tons of carbon dioxide) per year in the 1960s to 9.5 ± 1.8 $GtCO_2$ per year during 2004-2013. During the same period, the estimated annual atmospheric CO_2 captured by the ocean was 2.6 ±0.5 Gt of CO_2 compared with around 1.9 Gt of CO_2 during the sixties (Le Queré et al., 2014). However, due to the decreased buffering capacity, caused by this CO_2 uptake, the proportion of anthropogenic carbon dioxide that goes into the ocean has been decreasing.

Estimates of the global inventory of anthropogenic carbon (Cant), including marginal seas, have a mean value of 118 PgC and a range of 93 to 137 PgC in 1994 and a mean of 160 PgC and range of 134 to 186 PgC in 2010 (Rhein et al 2013). When combined with model results Khatiwala et al. (2013) arrive at a "best" estimate of the global ocean inventory (including marginal seas) of anthropogenic carbon from 1750 to 2010 of 155 PgC with an uncertainty of ±20 per cent (Rhein et al 2013).

The storage rate of anthropogenic CO_2 is assessed by calculating the change in Cant concentrations between two time periods. Regional observations of the storage rate are in general agreement with that expected from the increase in atmospheric CO_2 concentrations and with the tracer-based estimates. However, there are significant spatial and

of carbon away from the surface ocean and thus from the atmosphere into the ocean interior. This "solubility pump" helps to keep the surface waters of the ocean on average lower in CO_2 than the deep water, a condition that promotes the flux of the gas from the atmosphere into the ocean.

Phytoplankton take up CO_2 from the water in the process of photosynthesis, some of which sinks to the bottom in the form of particles or is

The boundaries and names shown and the designations used on this map do not imply official endorsement or acceptance by the United Nations.
Figure 5.7 | Maps of storage rate distribution of C_{ant} in (mol m^{-2} yr^{-1}) averaged over 1980-2005 for the three ocean basins (left to right: Atlantic, Pacific and Indian Ocean). From Khatiwala et al 2009, a slightly different colour scale is used for each basin.

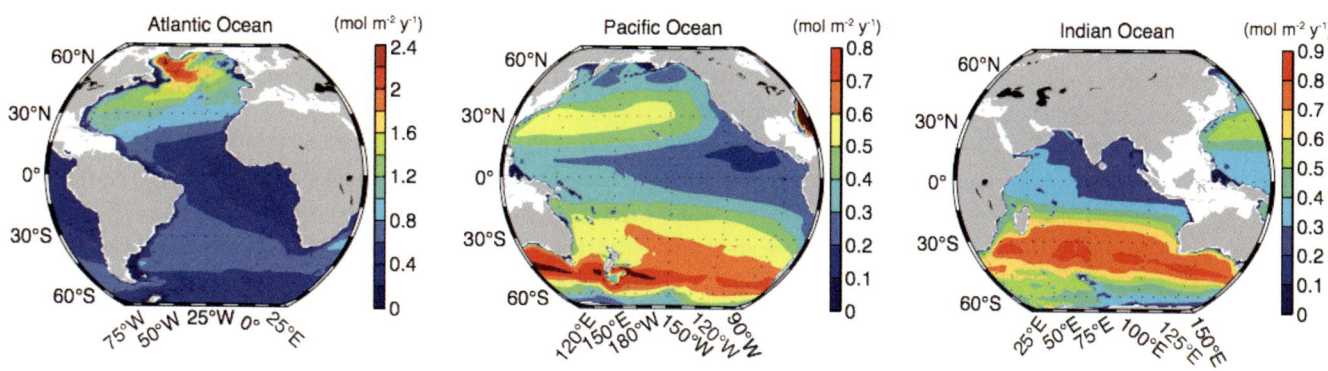

Sea-Air Interactions

temporal variations in the degree to which the inventory of Cant tracks changes in the atmosphere (Figure 5.7, Rhein et al 2013).

Comprehensive evaluation of available data shows that in the context of the global carbon cycle, it is only the ocean that has acted as a net sink of carbon from the atmosphere. The land was a source early in the industrial age, and since about 1950 has trended toward a sink, but it is not yet clearly a net sink. (Ciais et al. 2013 and Khatiwala et al. 2009, Khatiwala et al. 2013). Latest data from 2004 to 2013 show that the global oceans take up about one-fourth (26 per cent, Le Quéré, 2014) of the total annual anthropogenic emissions of CO_2. This is a very important physical and ecological service that the ocean has performed in the past and performs today, that underpins all strategies to mitigate the negative impacts of global warming.

4.3 Ocean acidification

As already seen in the previous section, the global oceans serve as an important sink of atmospheric CO_2, effectively slowing down global climate change. However, this benefit comes with a steep bio-ecological cost. When CO_2 reacts with water, it forms carbonic acid, which then dissociates and produces hydrogen ions. The extra hydrogen ions consume carbonate ions (CO_3^{2-}) to form bicarbonate (HCO_3^-). In this process, the pH and concentrations of carbonate ions (CO_3^{2-}) are decreasing. As a result, the carbonate mineral saturation states are also decreasing. Due to the increasing acidity, this process is commonly referred to as "ocean acidification (OA)". According to the IPCC AR 4 and 5, "Ocean acidification refers to a reduction in pH of the ocean over an extended period, typically decades or longer, caused primarily by the uptake of carbon dioxide (CO_2) from the atmosphere." (…)" Anthropogenic ocean acidification refers to the component of pH reduction that is caused by human activity" (Rhein et al. 2013).

Although the average oceanic pH can vary on interglacial time scales, the changes are usually on the order of ~0.002 units per 100 years; however, the current observed rate of change is ~0.1 units per 100 years, or roughly 50 times faster. Regional factors, such as coastal upwelling, changes in riverine and glacial discharge rates, and sea-ice loss have created "OA hotspots" where changes are occurring at even faster rates. Although OA is a global phenomenon that will likely have far-reaching implications for many marine organisms, some areas will be affected sooner and to a greater degree.

Recent observations show that one such area in particular is the cold, highly productive region of the sub-arctic Pacific and western Arctic Ocean, where unique biogeochemical processes create an environment that is both sensitive and particularly susceptible to accelerated reductions in pH and carbonate mineral concentrations. The OA phenomenon can cause waters to become undersaturated in carbonate minerals and thereby affect extensive and diverse populations of marine calcifiers.

4.4 The CO_2 problem

Based on the most recent data of 2004 to 2013, 35.7 $GtCO_2$ (1 $GtCO_2$ = 109 tons of carbon dioxide) of anthropogenic CO_2 are released into the atmosphere every year (Le Quéré et al. 2014). Of this, approximately 32.4 $GtCO_2$ come directly from the burning of fossil fuels and other industrial processes that emit CO_2. The remaining 3.3 $GtCO_2$ are due to changes in land-use practices, such as deforestation and urbanization. Of this 35.7 $GtCO_2$ of anthropogenically produced CO_2 emitted annually, approximately 10.6 $GtCO_2$ (or 29 per cent) are incorporated into terrestrial plant matter. Another 15.8 $GtCO_2$ (or 46 per cent) are retained in the atmosphere, which has led to some planetary warming. The remaining 9.5 $GtCO_2$ (or 26 per cent) are absorbed by the world's oceans (Le Quéré et al. 2014).

As the hydrogen ions produced by the increased CO_2 dissolution take carbonate ions out of seawater, the rate of calcification of shell-building organisms is affected; they are confronted with additional physiological challenges to maintain their shells. Although alteration of the carbonate equilibrium system in the ocean reducing carbonate ion concentration, and saturation states of calcium carbonate minerals will play a role imposing an additional energy cost to calcifier organisms, such as corals and shellbearing plankton, this is by no means the sole impact of OA.

4.5 What are the impacts of a more acidic ocean?

Throughout the last 25 million years, the average pH of the ocean has remained fairly constant between 8.0 and 8.2. However, in the last three decades, a fast drop has begun to occur, and if CO_2 emissions are left unchecked, the average pH could fall below 7.8 by the end of this century (Rhein, et al. 2013).

This is well outside the range of pH change of any other time in recent geological history. Calcifying organisms in particular, such as corals, crabs, clams, oysters and the tiny free-swimming pteropods that form calcium carbonate shells, could be particularly vulnerable, especially during the larval stage. Many of the processes that cause OA have long been recognized, but the ecological implications of the associated chemical changes have only recently been investigated. OA may have important ecological and socioeconomic consequences by impacting directly the physiology of all organisms in the ocean.

The altered environment is imposing an extra energy cost for the acid-base regulation of their internal body milieu. Through biological and evolutionary adaptation this process might have a huge variation of expression among different types of organisms, a subject that only recently has become the focus of intense scientific research.

Calcification is an internal process that in its vast majority does not depends directly on seawater carbonate content, since most organism use bicarbonate, that is increasing under acidification scenarios, or CO_2 originating in their internal metabolism. It has been demonstrated in the laboratory and in the field that some calcifiers can compensate and thrive in acidification conditions.

OA is not a simple phenomenon nor will it have a simple unidirectional effect on organisms. The abundance and composition of species may be changed, due to OA with the potential to affect ecosystem function at all trophic levels, and consequential changes in ocean chemistry could occur as well. Some species may also be better able than others to adapt to changing pH levels due to their exposure to environments where pH naturally varies over a wide range. However, at this point, it is still very uncertain what the ecological and societal consequences will be from any potential losses of keystone species.

4.6 Socioeconomic impacts of ocean acidification

Some examples of economic disruptions due to OA have been reported. The most visible case is the harvest failure in the oyster hatcheries along the Pacific Northwest coast of the USA. Hatcheries that supply the majority of the oyster spat to farms nearly went out of business as they unknowingly pumped low pH water, apparently corrosive to oyster larvae, into their operation. Although intense upwelling that could have brought low oxygen water to hatcheries might also be a factor in these massive mortalities, low pH, "corrosive water" tends to recur seasonally in this region. Innovations and interactions with scientists allowed these hatcheries to monitor the presence of corrosive incoming waters and adopt preventive measures.

Economic studies have shown that potential losses at local and regional scales may have negative impacts for communities and national economies that depend on fisheries. For example, Cooley and Doney (2009) using data from 2007, found that of the 4 billion dollars in annual domestic sales, Alaska and the New England states likely to be affected by hotspots of OA, contributed the most at 1.5 billion dollars and 750 million dollars, respectively. These numbers clearly show that any disruption in the commercial fisheries in these regions due to OA could have a cascading effect on the local as well as on the national economy.

References

Abraham, J.P., Baringer, M., Bindoff, N.L., Boyer, T., Cheng, L.J., Church, J.A., Conroy, J.L., Domingues, C.M., Fasullo, J.T., Gilson, J., Goni, G., Good, S.A., Gorman, J.M., Gouretski, V., Ishii, M., Johnson, G.C., Kizu, S., Lyman, J.M., Macdonald, A.M., Minkowycz, W.J., Moffitt, S. E., Palmer, M.D., Piola, A.R., Reseghetti, F., Schuckmann, K., Trenberth, K.E., Velicogna, I., & Willis, J.K. (2013). A review of global ocean temperature observations: Implications for ocean heat content estimates and climate change, *Review of Geophysics*, 51, 450–483, doi:10.1002/rog.20022.

Ashok K., Guan, Z., Yamagata, T. (2001), Impact of the Indian Ocean dipole on the relationship between the Indian monsoon rainfall and ENSO, *Geophysical Research Letters*, 28, 4499–4502.

Behera S. K., Luo, J.-J., Masson, S., Delecluse, P., Gualdi, S., Navarra, A., Yamagata, T. (2005), Paramount impact of the Indian Ocean dipole on the east African short rains: a CGCM study. *Journal of Climate*, 18, 4514–4530.

Bresnan, E., Davidson, K., Edwards, M., Fernand, L., Gowen, R., Hall, A., Kennington, K., McKinney, A., Milligan, S., Raine, R., Silke, J. (2013). Impacts of climate change on harmful algal blooms, *MCCIP Science Review*, 236-243, doi:10.14465/2013.arc24.236-243.

Buchan, J., Hirschi, J. J.-M., Blaker, A.T., and Sinha, B. (2014). North Atlantic SST anomalies and the cold north European weather events of winter 2009/10 and December 2010. *Monthly Weather Review*, 142, 922–932, doi: 10.1175/MWR-D-13-00104.1.

Cai W., van Rensch, P., Cowan, T., Hendon, H.H. (2011), Teleconnection pathways of ENSO and the IOD and the mechanisms for impacts on Australian rainfall, *Journal of Climate*, 24, 3910–3923.

Capotondi, A., Alexander, M.A., Bond, N.A., Curchitser, E.N., and Scott, J.D. (2012): Enhanced upper ocean stratification with climate change in the CMIP3 models. *Journal of Geophysical Research*, 117, C04031, doi:10.1029/2011JC007409.

Chen, X. and Tung, K.-K. (2014). Varying planetary heat sink led to global-warming slowdown and acceleration. *Science* 345, 897-903.

Ciais, P., Sabine, C., Govindasamy, B., Bopp, L., Brovkin, V., Canadell, J., Chhabra, A., DeFries, R., Galloway, J., Heimann, M., Jones, C., Le Quéré, C., Myneni, R., Piao, S., and Thornton, P.(2013). Chapter 6: Carbon and Other Biogeochemical Cycles, in: *Climate Change 2013 The Physical Science Basis*, edited by: Stocker, T., Qin, D., and Platner, G.-K., Cambridge University Press, Cambridge.

Cooley. S. R., and Doney, S.C. (2009). Anticipating ocean acidification's economic consequences for commercial fisheries. *Environmental Research Letters* 4.

Cunningham, S. A., Kanzow, T., Rayner, D., Baringer, M.O., Johns, w.E., Marotzke, J., Longworth, H.R., Grant, E.M., Hirschi, J.J.-M., Beal, L.M., Meinen, C.S., Bryden, H.L. (2007), Temporal variability of the Atlantic meridional overturning circulation at 26.5 °N, *Science*, 317, 935–938, doi: 10.1126/science.1141304.

Cunningham, S. A., Roberts, C.D., Frajka-Williams, E., Johns, W.E., Hobbs, W., Palmer, M.D., Rayner, D., Smeed, D.A., and McCarthy, G. (2013), Atlantic Meridional Overturning Circulation slowdown cooled the subtropical ocean, *Geophysical Research Letters*, 40, 6202–6207, doi:10.1002/2013GL058464.

Doney S.C., Ruckelshaus M., Duffy J.E., Barry J.P., Chan F., English C.A., Galindo H.M., Grebmeier J.M., Hollowed, A.B., Knowlton, N., Polovina, J., Rabalais, N.N., Sydeman, W.J., Talley, L.D. (2012). Climate change impacts on marine ecosystems. *Annual Review of Marine Science* 4:11-37.

Doney, S.C., Mahowald, N., Lima, I., Feely, R.A., Mackenzie, F.T., Lamarque, J.-F. and Rasch, P.J. (2007). Impact of anthropogenic atmospheric nitrogen and sulfur deposition on ocean acidification and the inorganic carbon system. *Proceedings of the National Academy of Sciences of the United States of America*, 104, 14580-14585.

Domingues, C.M., Church, J. A., White, N.J., Gleckler, P.J., Wijffels, S.E., Barker, P.M. and Dunn, J.A. (2008). Improved estimates of upper-ocean warming and multi-decadal sea-level rise. *Nature*, 453, 1090–1093.

Dlugokencky, E. and Tans, P. (2014). http://www.esrl.noaa.gov/gmd/ccgg/ trends, last access: 25 May 2015.

Drijfhout, S.S., Blaker, A.T., Josey, S.A., Nurser, A.J.G., Sinha, B. and Balmaseda, M.A. (2014), Surface warming hiatus caused by increased heat uptake across multiple ocean basins, *Geophysical Research Letters*, 41, 7868–7874.

England, M.H., McGregor, S., Spence, P., Meehl, G.A., Timmermann, A., Cai, W., Gupta, A.S., McPhaden, M.J., Purich, A., Santoso, A. (2014). Recent intensification of wind-driven circulation in the Pacific and the ongoing warming hiatus. *Nature Climate Change*, 4, 403–407.

Fabry, V.J., McClintock, J.B., Mathis, J.T., and Grebmeier, J.M., (2009). Ocean Acidification at high latitudes: the Bellwether. *Oceanography*, 22(4), 160–171.

Falkowski, P.G. and Raven, J.A. (1997). *Aquatic photosynthesis*. Blackwell Science, Malden, MA.

Fasullo, J.T., and Trenberth, K.E. (2008). The annual cycle of the energy budget. Part II: Meridional structures and poleward transports, *Journal of Climate*, 21, 2313–2325, doi:10.1175/2007JCLI1936.1.

Feng, M., Meyers, G. and Wijffels, S. (2001). Interannual upper ocean variability in the tropical Indian Ocean *Geophysical Research Letters*, 28(21), 4151-4154.

Field, C.B., Behrenfeld, M.J., Randerson, J.T. and Falkowski, P. (1998). Primary production of the biosphere: integrating terrestrial and oceanic components, *Science* 281, 237–240.

Gadgil, S., Vinayachandran, P.N., Francis, P.A., and Gadgil, S. (2004). Extremes of the Indian summer monsoon rainfall, ENSO and equatorial Indian Ocean oscillation, *Geophysical Research Letters*, 31, L12213, doi: 10.1029/2004GL019733.

Grist, J.P., Josey, S.A., Marsh, R., Good, S.A., Coward, A.C., de Cuevas, B.A., Alderson, S.G., New, A.L., and Madec, G. (2010). The roles of surface heat flux and ocean heat transport convergence in determining Atlantic Ocean temperature variability, *Ocean Dynamics*, 60, 771–790, doi:10.1007/s10236-010-0292-4.

Halpern, B.S., Walbridge, S., Selkoe, K.A., Kappel, C.V., Micheli, F., D'Agrosa, C., Bruno, J.F., Casey, K.S., Ebert, C., Fox, H.E., Fujita, R., Heinemann, D., Lenihan, H.S., Madin, E.M.P., Perry, M.T., Selig, E.R., Spalding, M., Steneck, R., and Watson, R. (2008). A global map of human impact on marine ecosystems, *Science*, 319, 948-952.

Harley, C.D.G., Hughes, R.A., Hultgren, K.M., Miner, B.G., Sorte, C.J., Thornber, C.S., Rodriguez, L.F., Tomanek, L., Williams, S.L. (2006). The impacts of climate change in coastal marine systems, *Ecology Letters*, 9, 228–241.

Hartmann, D.L., Klein Tank, A.M.G., Rusticucci, M., Alexander, L.V., Brönnimann, S., Charabi, Y., Dentener, F.J., Dlugokencky, E.J., Easterling, D.R., Kaplan, A., Soden, B.J., Thorne, P.W., Wild, M. and Zhai, P.M. (2013). Observations: Atmosphere and Surface. In: *Climate Change 2013: The Physical Science Basis. Contribution of Working Group I to the Fifth Assessment Report of the Intergovernmental Panel on Climate Change* [Stocker, T.F., Qin, D., Plattner, G.-K., Tignor, M., Allen, S.K., Boschung, J., Nauels, A., Xia, Y., Bex, V. and Midgley, P.M. (eds.)]. Cambridge University Press, Cambridge, United Kingdom and New York, NY, USA.

Hays, G.C., Richardson, A.J., and Robinson, C. (2005). *Trends in Ecology and Evolution*. 20, 337-344.

IOC-UNESCO, (2011). *Methodology for the GEF Transboundary Waters Assessment Programme*. Volume 6. Methodology for the Assessment of the Open Ocean,

UNEP, vi + 71 pp.

IPCC (2013), *Climate Change 2013: The Physical Science Basis.* Contribution of Working Group I to the Fifth Assessment Report of the Intergovernmental Panel on Climate Change. Stocker, T.F., Qin, D., Plattner, G.-K., Tignor, M., Allen, S.K., Boschung, J., Nauels, A., Xia, Y., Bex, V., and Midgley, P.M. (eds.). Cambridge University Press, Cambridge, UK, 1535 pp.

Ishii, M., and Kimoto, M. (2009). Reevaluation of Historical Ocean Heat Content Variations with Time-Varying XBT and MBT Depth Bias Corrections. *Journal of Oceanography*, Vol. 65: pp. 287-299.

Johns, W.E., Baringer, M.O. Beal, L.M., Cunningham, S.A., Kanzow, T., Bryden, H.L., Hirschi, J.J.M., Marotzke, J., Meinen, C.S., Shaw, B. and Curry, R. (2011). Continuous, array-based estimates of Atlantic Ocean heat transport at 26.5°N. *Journal of Climate*, 24, 2429–2449, doi: 10.1175/2010JCLI3997.1.

Kanzow, T., Cunningham, S.A., Rayner, D., Hirschi, J.J.-M., Johns, W.E., Baringer, M.O., Bryden, H.L., Beal, L.M., Meinen, Ch. S., Marotzke J. (2007). Observed flow compensation associated with the MOC at 26.5 °N in the Atlantic, *Science*, 317,938–941, doi: 10.1126/science.1141304.

Keenlyside, N.S., Ding, H. and Latif, M. (2013). Potential of equatorial Atlantic variability to enhance El Niño prediction, *Geophysical Research Letters*, 40, 2278–2283.

Kelly, K.A., Thompson, L-A. and Lyman, J. (2014). The Coherence and Impact of Meridional Heat Transport Anomalies in the Atlantic Ocean Inferred from Observations. *Journal of Climate*, 27, 1469–1487.

Kiene, R.P., Visscher, P.T., Keller, M.D. and Kirst, G.O. (eds.). (1996). *Biological and environmental chemistry of DMSP and related sulfonium compounds*. Plenum Press, New York (ISBN 0-306-45306-1).

Knutson, T.R., McBride, J.L., Chan, J., Emanuel, K., Holland, G., Landsea, C., Held, I., Sugi, M. (2010). Tropical cyclones and climate change. *Nature Geoscience*, 3(3), 157-163.

Kosaka, Y., and Xie, S.-P. (2013). Recent global-warming hiatus tied to equatorial Pacific surface cooling. *Nature*, 501, 403–407.

Kossin, J.P., Emanuel, K.A., Vecchi, G.A. (2014).The poleward migration of the location of tropical cyclone maximum intensity. *Nature*, 509(7500), 349-352.

Kumar, K.K., Rajagopalan, B., Hoerling, M., Bates, G. and Cane, M. (2006). Unraveling the mystery of Indian monsoon failure during El Niño, *Science*, 314, 115-119.

Le Quéré, C., Moriarty, R. Andrew, R.M., Peters, G.P., Ciais, P., Friedlingstein, P., Jones, S.D., Sitch, S., Tans, P., Arneth, A., Boden,T.A., Bopp, L., Bozec, Y., Canadell, J.G., Chevallier, F., Cosca, C.E., Harris, I., Hoppema, M., Houghton, R.A., House, J.I., Johannessen, T., Kato, E., Jain, A.K., Keeling, R.F., Kitidis, V., Klein Goldewijk, K., Koven, C., Landa, C., Landschützer, P., Lenton, A., Lima, I., Marland, G., Mathis, J.T., Metzl, N., Nojiri, Y., Olsen, A., Peters, W., Ono, T., Pfeil, B., Poulter, B., Raupach, M.R., Regnier, P., Rödenbeck, C., Saito, S., Salisbury, J.E., Schuster, U., Schwinger, J., Séférian, R., Segschneider, J., Steinhoff, T., Stocker, B.D., Sutton, A.J., Takahashi, T., Tilbrook, B., Viovy,N., Wang, Y.-P., Wanninkhof, R., Van der Werf, G., Wiltshire, A. and Zeng, N. (2014). Global Carbon Budget 2014. *Earth System Science Data Discuss.*, 7, 521-610. DOI:10.5194/essdd-7-521-2014.

Lee, S.-K., W. Park, W., van Sebille, E., Baringer, M.O., Wang, C., Enfield, D.B., Yeager, S.G., and B.P. Kirtman (2011). What caused the significant increase in Atlantic Ocean heat content since the mid-20th century? *Geophysical Research Letters*, 38, L17607, doi:10.1029/2011GL048856.

Levermann, A., Bamber, J.L., Drijfhout, S., Ganopolski, A., Haeberli, W., Harris, N.R.P., Huss, M., Krüger, K., Lenton, T.M., Lindsay, R.W., Notz, D., Wadhams, P. and Weber, S.(2012). Potential climatic transitions with profound impact on Europe: Review of the current state of six 'tipping elements of the climate system'. *Climatic Change*, 110, 845–878, DOI 10.1007/s10584-011-0126-5.

Levitus, S., Antonov, J.I., Boyer, T.P., Locarnini, R.A., Garcia, H.E., and Mishonov, A.V. (2009). Global ocean heat content 1955–2008 in light of recently revealed instrumentation problems, *Geophysical Research Letters*, 36, L07608, doi:10.1029/2008GL037155.

Lima, F.P., and Wethey, D.S. (2012). Three decades of high-resolution coastal sea surface temperatures reveal more than warming, *Nature communications*, 3.

Marengo, J. A., Tomasella, J., Alves, L.M., Soares, W.R., and Rodriguez, D.A. (2011), The drought of 2010 in the context of historical droughts in the Amazon region, *Geophysical Research Letters*, 38, L12703, doi:10.1029/2011GL047436.

McCarthy, G., Frajka-Williams, E., Johns, W.E., Baringer, M.O., Meinen, C.S., Bryden, H.L., Rayner, D., Duchez, A., Roberts. C., and Cunningham, S.A. (2012). Observed interannual variability of the Atlantic meridional overturning circulation at 26.5°N, *Geophysical Research Letters*, 39, L19609, doi:10.1029/2012GL052933.

McPhaden, M. J., (2012). A 21st century shift in the relationship between ENSO SST and warm water volume anomalies, *Geophysical Research Letters*, 39, L09706, doi:10.1029/2012GL051826.

McPhaden, M.J., and Nagura, M. (2014). Indian Ocean dipole interpreted in terms of recharge oscillator theory. *Climate Dynamics*, 42, 1569-1586.

Meehl, G. A., Hu, A., Arblaster, J.M., Fasullo, J.Y., and Trenberth, K.E. (2013). Externally forced and internally generated decadal climate variability associated with the Interdecadal Pacific Oscillation, *Journal of Climate*, 26, 7298–7310.

Nagura, M., and McPhaden, M.J. (2010). Dynamics of zonal current variations associated with the Indian Ocean dipole. *Journal of Physical Research*, 115, C 11026,

Occhipinti-Ambrogi, A. (2007). Global change and marine communities: Alien species and climate change, *Marine Pollution Bulletin.*, 55, 342-52.

Palmer, M. D., and Haines, K. (2009). Estimating oceanic heat content change using isotherms, *Journal of Climate*, 22, 4953–4969, doi: 10.1175/2009JCLI2823.1.

Perry, A., Low, P. J., Ellis, J.R., Reynolds, J.D. (2005). Climate change and distribution shifts in marine fishes, *Science*, 308, 1912–15.

Ren, H.-L., and Jin, F.-F. (2013). Recharge oscillator mechanisms in two types of ENSO. *Journal of Climate*, 26, 6506–6523. doi: http://dx.doi.org/10.1175/JCLI-D-12-00601.1.

Rhein, M., Rintoul, S.R., Aoki, S., Campos, E., Chambers, D., Feely, R.A., Gulev, S., Johnson, G.C., Josey, S.A., Kostianoy, A., Mauritzen, C., Roemmich, D., Talley, L.D. and Wang, F. (2013). Observations: Ocean. In: *Climate Change 2013: The Physical Science Basis. Contribution of Working Group I to the Fifth Assessment Report of the Intergovernmental Panel on Climate Change* [Stocker, T.F., Qin, D., Plattner, G.-K, Tignor, M., Allen, S.K., Boschung, J., Nauels, A., Xia, Y., Bex V. and Midgley, P.M. (eds.)]. Cambridge University Press, Cambridge, United Kingdom and New York, NY, USA.

Rodrigues, R. R., Haarsma, R.J., Campos, E.J.D., and Ambrizzi, T. (2011). The impacts of inter-El Niño variability on the Tropical Atlantic and Northeast Brazil climate, *Journal of Climate*, 24, 3402–22.

Rodriguez-Fonseca, B., Polo, I., Garcia-Serrano, J., Losada, T., Mohino, E., Mechoso, C.R., and Kucharski, F. (2009). Are Atlantic Niños enhancing Pacific ENSO events in recent decades? *Geophysical Research Letters*, 36, L20705, doi:10.1029/2009GL040048.

Roemmich, D., and Gilson, J. (2011), The global ocean imprint of ENSO, *Geophysical Research Letters*, 38, L13606

Sabine C. L., Feely, R. A., Gruber, N., Key, R. M., Lee, K., Bullister, J. L., Wanninkhof, R., Wong, C. S., Wallace, D. W. R., Tilbrook, B., Millero, F. J., Peng, T. H., Kozyr, A., Ono, T., and Rios, A. F., (2004). The Oceanic sink for anthropogenic CO_2. *Science*

305: 367-371.

Sabine, C.L. and Feely, R.A. (2007). The oceanic sink for carbon dioxide. pp. 31–49. In Reay, D., Hewitt, N., Grace, J., and Smith, K. (Eds.), *Greenhouse Gas Sinks*, CABI Publishing, Oxfordshire, UK.

Saji, N.H., Goswami, B.N., Vinayachandran, P.N., and Yamagata, T. (1999). A dipole mode in the tropical Indian Ocean. *Nature*, 401, 360–63.

Seidel, D.J., Fu, Q., Randel, W.J., Reichler, T.J. (2008), Widening of the tropical belt in a changing climate. *Nature Geoscience*, 1(1), 21-24.

Smith, D.M., Murphy, J.M. (2007). An objective ocean temperature and salinity analysis using covariances from a global climate model. *Journal of Geophysical Research*. Vol. 112, Issue C2.

Stocker, T.F., Qin, D., Plattner, G.-K., Alexander, L.V., Allen, S.K., Bindoff, N.L., Bréon, F.-M., Church, J.A., Cubasch, U., Emori, S., Forster, P., Friedlingstein, P., Gillett, N., Gregory, J.M., Hartmann, D.L., Jansen, E., Kirtman, B., Knutti, R., Krishna Kumar, K., Lemke, P., Marotzke, J., Masson-Delmotte, V., Meehl, G.A., Mokhov, I.I., Piao, S., Ramaswamy, V., Randall, D., Rhein, M., Rojas, M., Sabine, C., Shindell, D., Talley, L.D., Vaughan, D.G., and Xie, S.-P. (2013). Technical Summary. In: *Climate Change 2013: The Physical Science Basis. Contribution of Working Group I to the Fifth Assessment Report of the Intergovernmental Panel on Climate Change*. Stocker, T.F., Qin, D., Plattner, G-K., Tignor, M., Allen, S.K., Boschung, J., Nauels, A., Xia, Y., Bex, V. and Midgley, P.M. (eds.). Cambridge University Press, Cambridge, United Kingdom and New York, NY, USA.

Suthers, I.M., Young, J.W., Baird, M.E., Roughan, M., Everett, J.D., Brassington, G.B., Byrne, M., Condie, S.A., Hartog, J.R., Hassler, C.S., Hobday, A.J., Holbrook, N.J., Malcolm, H.A., Oke, P.R., Thompson, P.A., Ridgway, K. (2011). The strengthening East Australian Current, its eddies and biological effects - An introduction and overview. Deep Sea Research Part II: *Topical Studies in Oceanography*. 58 (2011) 538–546.

Takahashi, T., Sutherland, S.C., Wanninkhof, R., Sweeney, C., Feely, R.A., Chipman, D.W., Hales, B., Friederich, G., Chavez, F., Watson, A., Bakker, D.C.E., Schuster, U., Metzl, N., Yoshikawa-Inoue, H., Ishii, M., Midorikawa, T., Nojiri, Y., Sabine, C., Olafsson, J., Arnarson, Th. S., Tilbrook, B., Johannessen, T., Olsen, A., Richard Bellerby, Körtzinger, A., Steinhoff, T., Hoppema, M., de Baar, H.J.W., Wong, C.S., Bruno Delille and Bates N.R. (2009). Climatological mean and decadal changes in surface ocean pCO2, and net sea-air CO2 flux over the global oceans. *Deep-Sea Research II*, 56, 554-57.

Taschetto, A.S., and England, M.H. (2009). El Niño Modoki impacts on Australian rainfall, *Journal of Climate*, 22, 3167–3174, doi:10.1175/2008JCLI2589.1.

Tokinaga, H., and Xie, S.-P. (2011). Weakening of the equatorial Atlantic cold tongue over the past six decades. *Nature Geosciences*, 4, 222–226.

Ummenhofer, C.C., Sen Gupta, A., England, M.H., and Reason, C.J.C. (2009). Contributions of Indian Ocean sea surface temperatures to enhanced East African rainfall. *Journal of Climate*, 22, 993-1013.

von Schuckmann, K. and Le Traon, P.-Y. (2011). How well can we derive Global Ocean Indicators from Argo data? *Ocean Science*, 7, 783-91, doi:10.5194/os-7-783-2011.

Woodruff, J. D., Irish, J.L., and Camargo, S.J. (2013). Coastal flooding by tropical cyclones and sea-level rise. *Nature*, 504(7478), 44-52.

Yamagata T., Behera, S.K., Luo, J.–J., Masson, S., Jury, M., Rao, S.A. (2004). *Coupled ocean–atmosphere variability in the tropical Indian Ocean. Earth Climate: Ocean-Atmosphere Interaction*, Geophysical Monograph Series, 147, 189–211.

Yeh, S.-W., Kug, J.-S., Dewitte, B., Kwon, M.-H., Kirtman, B., and Jin, F.-F. (2009). El Niño in a changing climate, *Nature*, 461, 511–514, doi:10.1038/nature08316.

Yeh, S.-W., Kirtman, B.P. Kug, J.-S. Park, W. and Latif, M. (2011). Natural variability of the central Pacific El Niño event on multi-centennial timescales, *Geophysical Research Letters*, 38, L02704, doi:10.1029/2010GL045886.

6 Primary Production, Cycling of Nutrients, Surface Layer and Plankton

Writing team:
Thomas Malone (Convenor), Maurizio Azzaro, Antonio Bode, Euan Brown, Hilconida Calumpong (Co-Lead Member), Robert Duce, Peyman Eghtesadi Araghi (Co-Lead Member), Dan Kamykowski, Sung Ho Kang, Yin Kedong, Chul Park (Lead Member), Michael Thorndyke and Jinhui Wang

1 Primary Production[1]

1.1 Definition and ecological significance

Gross primary production (GPP) is the rate at which photosynthetic plants and bacteria use sunlight to convert carbon dioxide (CO_2) and water to the high-energy organic carbon compounds used to fuel growth. Free oxygen (O_2) is released during the process. Net primary production (NPP) is GPP less the respiratory release of CO_2 by photosynthetic organisms, i.e., the net photosynthetic fixation of inorganic carbon into autotrophic biomass. NPP supports most life on Earth; it fuels global cycles of carbon, nitrogen, phosphorus and other nutrients and is an important parameter of atmospheric CO_2 and O_2 levels (and, therefore, of anthropogenic climate change).

Global NPP is estimated to be ~105 Pg C yr-1, about half of which is by marine plants (Field et al., 1998; Falkowski and Raven, 1997; Westberry et al., 2008).[2] Within the euphotic zone of the upper ocean,[3] phytoplankton and macrophytes[4] respectively account for ~94 per cent (~50 ± 28 Pg C yr-1) and ~6 per cent (~3.0 Pg C yr-1) of NPP (Falkowski et al., 2004; Duarte et al., 2005; Carr et al., 2006; Schneider et al., 2008; Chavez et al., 2011; Ma et al., 2014; Rousseaux and Gregg, 2014). All NPP is not equal in terms of its fate. Marine macrophytes play an important role as carbon sinks in the global carbon cycle, provide habitat for a diversity of animal species, and food for marine and terrestrial consumers (Smith, 1981; Twilley et al., 1992; Duarte et al., 2005; Duarte et al., 2010; Heck et al., 2008; Nellemann et al., 2009; McLeod et al., 2011, Fourqurean et al., 2012). Phytoplankton NPP fuels the marine food webs upon which marine fisheries depend (Pauly and Christensen, 1995; Chassot et al., 2010) and the "biological pump" which transports 2-12 Pg C yr-1 of organic carbon to the deep sea (Falkowski et al., 1998; Muller-Karger et al., 2005; Emerson and Hedges, 2008; Doney, 2010; Passow and Carlson, 2012), where it is sequestered from the atmospheric pool of carbon for 200-1500 years (Craig, 1957; Schlitzer et al., 2003; Primeau and Holzer, 2006; Buesseler, et al., 2007).

Changes in the size structure of phytoplankton communities influence the fate of NPP (Malone, 1980; Legendre and Rassoulzadegan, 1996; Pomeroy et al., 2007; Marañón, 2009). In general, small cells (picophytoplankton with equivalent spherical diameters < 2 µm) account for most NPP in subtropical, oligotrophic (< 0.3 mg chlorophyll-a m-3), nutrient-poor (nitrate + nitrite < 1 µM), warm (> 20°C) waters. Under these conditions, the flow of organic carbon to harvestable fisheries and the biological pump are relatively small. In contrast, larger cells (microphytoplankton > 20 µm) account for > 90 per cent of NPP in more eutrophic (> 5 mg chlorophyll-a m-3), nutrient-rich (nitrate + nitrite >10 µM), cold (< 15°C) waters (Kamykowski, 1987; Agawin et al., 2000; Buitenhuis et al., 2012). Under these conditions, diatoms[5] account for most NPP during spring blooms at high latitudes and periods of coastal upwelling when NPP is high and nutrients are not limiting (Malone, 1980). The flow of organic carbon to fisheries and the biological pump is higher when larger cells account for most NPP (Laws et al., 2000; Finkel et al., 2010).

1.2 Methods of measuring net primary production (NPP)

1.2.1 Phytoplankton Net Primary Production

Phytoplankton (NPP) has been estimated using a variety of in situ and remote sensing methods (Platt and Sathyendranath, 1993; Geider et al., 2001; Marra, 2002; Carr et al., 2006; Vernet and Smith, 2007; Cullen, 2008a; Cloern et al., 2013). Multiplatform (e.g., ships, moorings, drifters, gliders, aircraft, and satellites) sampling strategies that utilize both approaches are needed to effectively detect changes in NPP on ecosystem to global scales (UNESCO-IOC, 2012).

On small spatial and temporal scales (meters-kilometres, hours-days), several techniques have been used including oxygen production and the incorporation of ^{13}C and ^{14}C labelled bicarbonate (Cullen, 2008a). The most widely used and standard method against which other methods are compared or calibrated is based on the incorporation of 14C-bicarbonate into phytoplankton biomass (Steeman-Nielsen, 1963; Marra, 1995; Marra, 2002; Vernet and Smith, 2007; Cullen, 2008a). On large spatial scales (Large Marine Ecosystems[6] to the global ocean), the most effective way to detect space-time variability is via satellite-based measurements of water-leaving radiance combined with diagnostic models of depth-integrated NPP as a function of depth-integrated chlorophyll-a concentration (Ψ Chl), photosynthetically active solar radiation, and temperature (Antoine and Morel, 1996; Perry, 1986; Morel and Berthon, 1989; Platt and Sathyendranath, 1993; Behrenfeld and Falkowski, 1997; Sathyendranath, 2000; Gregg et al., 2003; Behrenfeld et al., 2006; Carr et al., 2006; Arrigo et al., 2008; Bissinger et al., 2008; McClain, 2009; Westberry et al., 2008; Cullen et al., 2012; Siegel et al., 2013). An overview of the latest satellite based models may be found at the Ocean Productivity website.[7] Satellite ocean-colour radiometry (OCR)

---

1 Microbenthic, epiphytic and symbiotic algae can be locally important in shallow waters and corals, but are not addressed here. Chemosynthetic primary production is addressed elsewhere.

2 1 Pg = 1015 g

3 Defined here to include the epipelagic (0-200 m) and mesopelagic (200 – -1000 m) zones. The euphotic zone lies within the epipelagic zone.

4 Macrophytes include sea grasses, macroalgae, salt marsh plants and mangroves. Phytoplankton are single-celled, photosynthetic prokaryotic and eukaryotic microorganisms growing in the euphotic zone (the layer between the ocean's surface and the depth at which photosynthetically active radiation [PAR] is 1 per cent of surface intensity). Most phytoplankton species are > 1 µm and < 1 mm in equivalent spherical diameter (cf. Ward et al., 2012).

5 Diatom growth accounts for roughly half of marine NPP and therefore for about a quarter of global photosynthetic production (Smetacek, 1999).

6 Large marine ecosystems (200,000 km2 or larger) are coastal ecosystems characterized by their distinct bathymetry, hydrography, productivity and food webs (Sherman et al., 1993).

7 http://www.science.oregonstate.edu/ocean.productivity/.

data have been used to estimate surface chlorophyll-a fields and NPP since the Coastal Zone Color Scanner (CZCS) mission (1978-1986). Uninterrupted OCR measurements began with the Sea-viewing Wide Field-of-view Sensor (SeaWiFS) mission (1997-2010) (Hu et al., 2012). A full accounting of current polar orbiting and geostationary ocean-colour sensors with their capabilities (swath width, spatial resolution, spectral coverage) can be found on the web site of the International Ocean-Colour Coordinating Group.[8]

The skill of model-based estimates of NPP has been improving (O'Reilly et al., 1998; Lee, 2006; Friedrichs et al., 2009; Saba et al., 2010; Saba et al., 2011; Mustapha et al., 2012), but further improvements are needed through more accurate estimates of Ψ Chl. Chlorophyll-a fields can be estimated more accurately by blending data from remote sensing and in situ measurements, especially in regions where in situ measurements are sparse and in turbid, coastal ecosystems (Conkright and Gregg, 2003; Gregg et al., 2003; Onabid, 2011). An empirical approach has been developed for ocean-colour remote sensing called Empirical Satellite Radiance-In situ Data (ESRID) algorithm (Gregg et al., 2009).

1.2.2 Macrophyte Net Primary Production

The NPP of macroalgae, sea grasses, salt marsh plants and mangroves can be estimated by sequentially (e.g., monthly during the growing season) measuring increases in biomass (including leaf litter in salt marshes and mangrove forests) using a combination of in situ techniques (e.g., Mann, 1972; Cousens, 1984; Dame and Kenny, 1986; Amarasinghe and Balasubramaniam, 1992; Long et al., 1992; Day et al., 1996; Ross et al., 2001; Curcó et al., 2002; Morris, 2007) and satellite-based multispectral imagery (e.g., Gross et al., 1990; Zhang et al., 1997; Kovacs et al., 2001; Gitelson, 2004; Liu et al., 2008; Kovacs et al., 2009; Heumann, 2011; Mishra et al., 2012; Son and Chen, 2013). For remote sensing, accurate in situ measurements are critical for validating models used to map these habitats and estimate NPP (Gross et al., 1990; Kovacs et al., 2009; Roelfsema et al., 2009; Mishra et al., 2012; Jia et al., 2013; Trilla et al., 2013). These include shoot- or leaf-tagging techniques, measurements of 14C incorporation into leaves, and measurements of dissolved O2 production during the growing season (Bittaker and Iverson, 1976; Kemp et al., 1986; Duarte, 1989; Kaldy and Dunton, 2000; Duarte and Kirkman, 2001; Plus et al., 2001, Silva et al., 2009).

1.2.3 The Phenology[9] of Phytoplankton Annual Cycles

The timing of seasonal increases in phytoplankton NPP is determined by environmental parameters, including day length, temperature, changes in vertical stratification, and the timing of seasonal sea-ice retreat in polar waters. All but day length are affected by climate change. Thus, phytoplankton phenology provides an important tool for detecting climate-driven decadal variability and secular trends. Phenological metrics to be monitored are the time of bloom initiation, bloom duration and time of maximum amplitude (Siegel et al., 2002; Platt et al., 2009).

1.3 Spatial patterns and temporal trends

Marine NPP varies over a broad spectrum of time scales from tidal, daily and seasonal cycles to low-frequency basin-scale oscillations and multi-decade secular trends (Malone, 1971; Pingree et al., 1975; Steele, 1985; Cloern, 1987; Cloern, 2001; Cloern et al., 2013; Duarte, 1989; Powell, 1989; Malone et al., 1996; Henson and Thomas, 2007; Vantrepotte and Mélin, 2009; Cloern and Jassby, 2010; Bode et al., 2011; Chavez et al.,

The boundaries and names shown and the designations used on this map do not imply official endorsement or acceptance by the United Nations.

Figure 6.1 | Climatological map Distribution of annual marine NPP for (a) NASA Ocean Biogeochemical Model and (b) Vertically-Integrated Production Model (VGPM) for the period from September 1998 to 2011 (Rousseaux – August 1999 (Blue < 100 g C m^{-2}, Green > 110 g C m^{-2} and < 400 g C m^{-2}, Red > 400 g C m^{-2}) (Rutgers Institute of Marine and Gregg, 2014). Globally, diatoms accounted for about 50 per cent of NPP while coccolithophores, chlorophytes and cyanobacteria accounted for about 20 per cent, 20 per cent and 10 per cent, respectively. Diatom NPP was highest at high latitudes and in equatorial and eastern boundary upwelling systems. Coastal Sciences, http://marine.rutgers.edu/opp/). Coastal ecosystems (red – green) and the permanently stratified subtropical waters of the central gyres (blue) each account for ~30 per cent of the ocean's NPP, whereas the former accounts for only ~8 per cent of the ocean's surface area compared to ~60 per cent for the open ocean waters of the subtropics (Geider et al., 2001; Marañón et al., 2003; Muller-Karger et al., 2005).

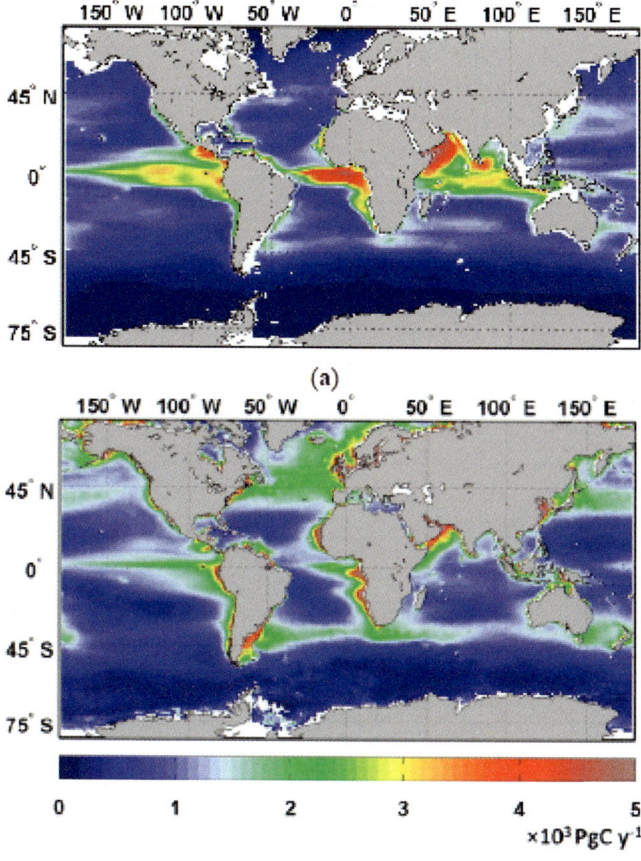

---

8 http://www.ioccg.org/sensors/current.html.

9 Phenology is the study of the timing and duration of cyclic and seasonal natural phenomena (e.g., spring phytoplankton blooms, seasonal cycles of zooplankton reproduction), especially in relation to climate and plant and animal life cycles.

2011). Our focus here is on low-frequency cycles and multi-decade trends.

1.3.1 Phytoplankton NPP

For the most part, the global pattern of phytoplankton NPP (Figure 6.1) reflects the pattern of deep-water nutrient inputs to the euphotic zone associated with winter mixing and thermocline erosion at higher latitudes, thermocline shoaling at lower latitudes, and upwelling along the eastern boundaries of the ocean basins and the equator (Wollast, 1998; Pennington et al., 2006; Chavez et al., 2011; Ward et al., 2012). The global distribution of phytoplankton NPP is also influenced by iron limitation and grazing by microzooplankton in so-called High Nutrient Low Chlorophyll (HNLC) zones which account for ~20 per cent of the global ocean, e.g., oceanic waters of the subarctic north Pacific, subtropical equatorial Pacific, and Southern Ocean (Martin et al., 1994; Landry et al., 1997; Edwards et al., 2004). Nutrient inputs associated with river runoff enhance NPP in coastal waters during the growing season (Seitzinger et al., 2005; Seitzinger et al., 2010). Annual cycles of NPP associated with patterns of nutrient supply and seasonal variations in sunlight tend to increase in amplitude and decrease in duration with increasing latitude. Seasonal increases in NPP generally follow winter mixing when nutrient concentrations are high, the seasonal thermocline sets up, and day length increases. Annual cycles are also more pronounced in coastal waters subject to seasonal upwelling.

The global distribution of annual NPP in the ocean can be partitioned into broad provinces with eastern boundary upwelling systems and estuaries exhibiting the highest rates and subtropical central gyres the lowest rates (Figure 6.1, Table 6.1).

Province	mg C m-2 d-1	g C m-2 yr-1
Subtropical Central Gyres	20 – 1,040	150 – 170
Western Boundaries	10 – 3,500	200 – 470
Eastern Boundaries	30 – 7,300	460 – 1,250
Equatorial Upwelling	640 – 900	240
Arctic Ocean	3 – 1,100	5 – 400
Southern Ocean	290 – 370	50 – 450
Coastal Seas	100 – 1,400	40 – 600
Estuaries & Coastal Plumes	100 – 8,000	70 – 1,890

Table 6.1 | Ranges of phytoplankton mean daily NPP and annual NPP reported for different marine provinces. Estimates are based on in situ measurements and models using satellite-based observations of chlorophyll-a fields. Western boundaries of the ocean basins generally feature broad continental shelves and eastern boundaries tend to have narrow shelves with coastal upwelling. (Data sources: Malone et al., 1983; O'Reilly and Busch, 1983; Pennock and Sharp, 1986; Cloern, 1987; Malone, 1991; Barber et al., 1996; Karl et al., 1996; Malone et al., 1996; Pilskaln et 173 al., 1996; Smith and DeMaster, 1996; Lohrenz et al., 1997; Cloern, 2001; Smith et al., 2001; Steinberg et al., 2001; Marañón et al., 2003; Sakshaug, 2004; PICES, 2004; Teira et al., 2005; Tian et al., 2005; Pennington et al., 2006; Subramanian et al., 2008; Vernet et al., 2008; Bidigare et al., 2009; Sherman and Hempel, 2009; Chavez et al., 2010; 176 Saba et al., 2011; Brown and Arrigo, 2012; Cloern et al., 2013; Lomas et al., 2013).

Interannual variability and multi-decadal trends in phytoplankton NPP on regional to global scales are primarily driven by: (1) climate change (e.g., basin-scale oscillations and decadal trends, including loss of polar ice cover, upper ocean warming, and changes in the hydrological cycle); (2) land-based, anthropogenic nutrient loading; and (3) pelagic and benthic primary consumers. Global-scale trends in phytoplankton NPP remain controversial (Boyce et al., 2010; Boyce et al., 2014; Mackas, 2011; Rykaczewski and Dunne, 2011; McQuatters-Gollop et al., 2011; Dave and Lozier, 2013; Wernand et al., 2013).). Remote sensing (sea-surface temperature and chlorophyll fields), model simulations and marine sediment records suggest that global phytoplankton NPP may have increased over the last century as a consequence of basin-scale climate forcing that promotes episodic and seasonal nutrient enrichment of the euphotic zone through vertical mixing and upwelling (McGregor et al., 2007; Bidigare et al., 2009; Chavez et al., 2011; Zhai et al., 2013). In contrast, global analyses of changes in chlorophyll distribution over time suggest that annual NPP in the global ocean has declined over the last 100 years (Gregg et al., 2003; Boyce et al., 2014). A decadal scale decline is consistent with model simulations indicating that both NPP and the biological pump have decreased by ~7 per cent and 8 per cent, respectively, over the last five decades (Laufkötter et al., 2013), trends that are likely to continue through the end of this century (Steinacher et al., 2010).

Given uncertainties concerning global trends, long-term impacts of secular changes in phytoplankton NPP on food security and climate change cannot be assessed at this time with any certainty. Resolving this controversy and predicting future trends will require sustained, multi-decadal observations and modelling of phytoplankton NPP and key environmental parameters (e.g., upper ocean temperature, pCO2, pH, depth of the aragonite saturation horizon, vertical stratification and nutrient concentrations) on regional and global scales – observations that may have to be sustained for at least another 40-50 years (Henson et al., 2010).

1.3.2 Macrophyte NPP

Marine macrophyte NPP, which is limited to tidal and relatively shallow waters in coastal ecosystems, varies from 30-1,200 g C m-2 yr-1 (Smith, 1981; Charpy-Roubaoud and Sournia, 1990; Geider et al., 2001; Duarte et al., 2005; Duarte et al., 2010; Fourqurean et al., 2012; Ducklow et al., 2013). In contrast to the uncertainty of decadal trends in phytoplankton NPP, decadal declines in the spatial extent and biomass of macrophytes (a proxy for NPP) over the last 50-100 years are relatively well documented. Macrophyte habitats are being lost and modified (e.g., fragmented) at alarming rates (Duke et al., 2007; Valiela et al., 2009; Waycott et al., 2009; Wernberg et al., 2011), i.e., 2 per cent for macrophytes as a group, with total areal losses to date of 29 per cent for seagrasses, 50 per cent for salt marshes and 35 per cent for mangrove forests (Valiela et al., 2001; Hassan et al., 2005; Orth et al., 2006; Waycott et al., 2009; Fourqurean et al., 2012). As a whole, the world is losing its macrophyte ecosystems in coastal waters four times faster than its rain forests (Duarte et al., 2008), and the rate of loss is accelerating (Waycott et al., 2009).

2 Nutrient Cycles

Nitrogen (N) and phosphorus (P) are major nutrients required for the growth of all organisms, and NPP is the primary engine that drives the cycles of N and P in the oceans. The cycles of C, N, P and O2 are coupled in the marine environment (Gruber, 2008). As discussed in section 6.1.3, the global pattern of phytoplankton NPP reflects the pattern of dissolved inorganic N and P inputs to the euphotic zone from the deep ocean (Figure 6.1). Superimposed on this pattern are nutrient inputs associated with N fixation, atmospheric deposition, river discharge and submarine ground water discharge. In regard to the latter, ground water discharge may be a significant source of N locally in some parts of Southeast Asia, North and Central America, and Europe, but on the scale of ocean basins and the global ocean, ground water discharge of N has been estimated to be on the order of 2-4 per cent of river discharge (Beusen et al., 2013). Given this, and challenges of quantifying ground water inputs on ocean basin to global scales (NRC, 2004), this source is not considered herein.

2.1 Nitrogen

The ocean's nitrogen cycle is driven by complex microbial transformations, including N fixation, assimilation, nitrification, anammox and denitrification (Voss et al., 2013) (Figure 6.2). NPP depends on the supply of reactive N (Nr)[10] to the euphotic zone. Although most dissolved chemical forms of Nr can be assimilated by primary producers, the most abundant chemical form, dissolved dinitrogen gas (N2), can only be assimilated by marine diazotrophs.[11] Nr inputs to the euphotic zone occur via fluxes of nitrate from deep water (vertical mixing and upwelling), marine N2 fixation, river discharge, and atmospheric deposition.[12] Nr is removed from the marine N inventory through denitrification and anammox[13] with subsequent efflux of N2 and N2O to the atmosphere (Thamdrup et al., 2006; Capone, 2008; Naqvi et al., 2008; Ward et al., 2009; Ward, 2013). Although there is no agreement concerning the oxygen threshold that defines the geographic extent of denitrification and anammox (Paulmier and Ruiz-Pino, 2009), these processes are limited to suboxic waters with very low oxygen concentrations (< 22 μM).

Variations in the ocean's inventory of Nr have driven changes in marine NPP and atmospheric CO2 throughout the Earth's geological his-

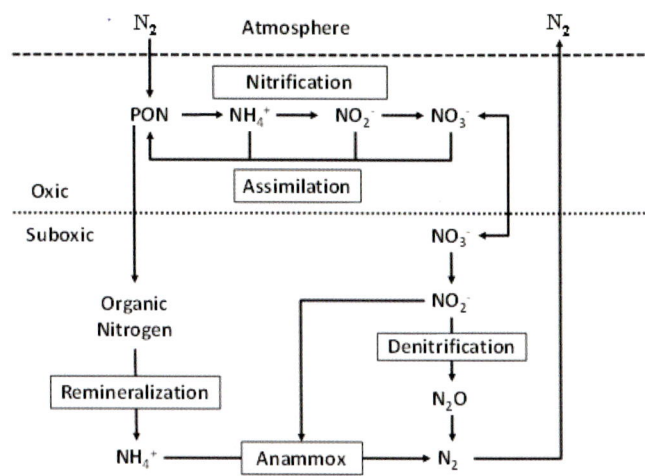

Figure 6.2 | The biological nitrogen cycle showing the main inorganic forms in which nitrogen occurs in the ocean (PON-pariculate organic nitrogen) (adapted from Ward, 2012).

tory (Falkowski, 1997; Gruber, 2004; Arrigo, 2005). Marine N2 fixation provides a source of "new" N and NPP that fuel marine food webs and the biological pump. Thus, the rate of N2 fixation can affect atmospheric levels of CO2 on time-scales of decades (variability in upper ocean nutrient cycles) to millennia (changes in the Nr inventory of the deep sea). This makes the balance between the conversion of N2 to biomass (N2 fixation) and the production of N2 (reduction of nitrate and nitrite by denitrification and anammox) particularly important processes in the N cycle that govern the marine inventory of Nr and sustain life in the oceans (Karl et al., 2002; Ward et al., 2007; Gruber, 2008; Ward, 2012).

2.1.1 The Marine Nitrogen Budget

Estimates of global sources and sinks of Nr vary widely (Table 6.2). Marine biological N2 fixation accounts for ~50 per cent of N2 fixation globally (Ward, 2012). Most marine N2 fixation occurs in the euphotic zone of warm (> 20°C), oligotrophic waters between 30° N and 30° S (Karl et al., 2002; Mahaffey et al., 2005; Stal, 2009; Sohm et al., 2011). Denitrification and anammox in benthic sediments and mid-water oxygen minimum zones (OMZs) account for most losses of N from the marine Nr inventory (Ulloa et al., 2012; Ward, 2013).

Table 6.2 | Summary of estimated sources and sinks (Tg N yr-1) in the global marine nitrogen budget. (Data sources: Codispoti et al., 2001; Gruber and Sarmiento, 2002; Karl et al., 2002; Galloway et al., 2004; Mahaffey et al., 2005; Seitzinger et al., 2005; Boyer et al., 2006; Moore et al., 2006; Deutsch et al., 2007; Duce et al., 2008; DeVries et al., 2012; Grosskopf et al., 2012; Luo et al., 2012; Naqvi, 2012.)

Sources		
	N fixation	60-200
	Rivers	35-80
	Atmosphere	38-96
	TOTAL	133-376
Sinks	Denitrification & anammox	120-450
	Sedimentation	25
	N2O loss	4-7
	TOTAL	149-482

---

10 Reactive or fixed N forms include dissolved inorganic nitrate, nitrite, ammonium and organic N compounds, such as urea and free amino acids.

11 Prokaryotic, free-living and symbiotic bacteria, cyanobacteria and archaea.

12 River discharge and atmospheric deposition include nitrate from fossil fuel burning and fixed N in synthetic fertilizer produced by the Haber-Bosch process for industrial nitrogen fixation.

13 Anaerobic ammonium oxidation.

Assuming a C:N:P ratio of 106:16:1 (the Redfield Ratio, Redfield et al., 1963), the quantity of N_r needed to support NPP globally is ~8800 Tg N yr^{-1}. Given current estimates, inputs of N_r from river discharge and atmospheric deposition support 2-4 per cent of NPP annually, i.e., most NPP is supported by recycled nitrate from deep waters (cf. Okin et al., 2011).

Although the N_2O flux is a small term in the marine N budget (Table 6.2), it is a significant input to the global atmospheric N_2O pool. Given a total input of 17.7 Tg N yr^{-1} (Freing et al., 2012), marine sources may account for 20-40 per cent of N_2O inputs to the atmosphere. As N_2O is 200-300 times more effective than CO_2 as a greenhouse gas, increases in N_2O from the ocean may contribute to both global warming and the destruction of stratospheric ozone. We note that although global estimates for anammox have yet to be made, this anaerobic process may be responsible for most N_2 production in some oxygen minimum zones (OMZs) (Strous et al., 2006; Hamersley et al., 2007; Lama et al., 2009; Koeve and Kähler, 2010; Ulloa et al., 2012).

The accounting in Table 6.2 suggests that total sinks may exceed total sources, but the difference is not significant. Many scientists believe that biological N_2 fixation is underestimated or the combined rates of denitrification and anammox are overestimated (Capone, 2008). On average, the Redfield Ratio approximates the C:N:P ratio of phytoplankton biomass, and the distribution of deviations from the Redfield Ratio (Martiny et al., 2013) suggests that: sources exceed sinks in the subtropical gyres; sources and sinks are roughly equal in upwelling systems (including their OMZs); and sources tend to be less than sinks at high latitudes. This pattern is consistent with the known distribution of marine diazotrophs and the observation that most marine N_2 fixation occurs in warm, oligotrophic waters between 30° N and 30° S (Mahaffey et al., 2005; Stal, 2009; Sohm et al., 2011). However, given the wide and overlapping ranges of current estimates of N_r sources and sinks (Table 6.2), the extent to which the two are in steady state remains controversial.

Atmospheric deposition of iron to the oceans via airborne dust may ultimately control the rate of N_2 fixation in the global ocean and may account for the relatively high rate of N_2 fixation in the subtropical central gyres (Karl et al., 2002). Fe II is required for photosynthetic and respiratory electron transport, nitrate and nitrite reduction, and N_2 fixation. The large dust plume that extends from North Africa over the subtropical North Atlantic Ocean dominates the global dust field (Stier et al., 2005). Consequently, iron deposition is particularly high in this region (Mahowald et al., 2005) where it may increase phytoplankton NPP by stimulating N_2 fixation (Mahowald et al., 2005; Krishnamurthy et al., 2009; Okin et al., 2011). Model simulations indicate that the distribution and rate of N_2 fixation may also be influenced by non-Redfield uptake of N and P by non-N_2 fixing phytoplankton (Mills and Arrigo, 2010). In these simulations, N_2 fixation in ecosystems dominated by phytoplankton with N:P ratios < Redfield is lower than expected when estimated rates are based on Redfield stoichiometry. In contrast, in systems dominated by phytoplankton with N:P ratios > Redfield, N2 fixation is higher than expected based on Redfield stoichiometry.

2.1.2 Time-Space Coupling of N_2 Fixation and Denitrification/Anammox

Early measurements of N_2 fixation and the geographic distribution of *in situ* deviations from the Redfield Ratio suggest that the dominant sites of N_2 fixation and denitrification are geographically separated and coupled on the time scales of ocean circulation (Capone et al., 2008 and references therein). In this scenario, the ocean oscillates between being a net source and a net sink of N_r on time scales of hundreds to thousands of years (Naqvi, 2012). However, there is also evidence that N_2 fixation is closely coupled with denitrification/anammox in upwelling-OMZ systems[14], i.e., rates of N_2 fixation are high downstream from OMZs where denitrification/anammox is high (Deutsch et al., 2007). Their findings indicate that N_2 fixation and denitrification are in steady state on a global scale. Results from 3-D inverse modelling (DeVries et al., 2013) and observations that the marine N_r inventory has been relatively stable over the last several thousand years (Gruber, 2004; Altabet, 2007) support the hypothesis that rates of N_2 fixation and denitrification/anammox are closely coupled in time and space.

At the same time, global biogeochemical modelling suggests that the negative feedbacks stabilizing the N_r inventory cannot persist in an ocean where N_2 fixation and denitrification/anammox are closely coupled, i.e., spatial separation, rather than spatial proximity, promoted negative feedbacks that stabilized the marine N inventory and sustained a balanced N budget (Landolfi et al., 2013). If the coupling is close as argued above, the budget may not be in steady state. In this scenario, increases in vertical stratification of the upper ocean and expansion of OMZs associated with ocean warming (Keeling et al., 2010) could lead to closer spatial coupling of N_2 fixation and denitrification, a net loss of N from the marine N_r inventory, and declines in NPP and CO_2 sequestration during this century.

2.2 Phosphorus

Phosphorus (P) is an essential nutrient utilized by all organisms for energy transport and growth. The primary inputs of P occur via river discharge and atmospheric deposition (Table 6.3). Biologically active P (BAP) in natural waters usually occurs as phosphate (PO_4^{-3}), which may be in dissolved inorganic forms (including orthophosphates and polyphosphates) or organic forms (organically bound phosphates). Natural inputs of BAP begin with chemical weathering of rocks followed by complex biogeochemical interactions, whose time scales are much longer than anthropogenic P inputs (Benitez-Nelson, 2000). Primary anthropogenic sources of BAP are industrial fertilizer, sewage and animal wastes.

The Marine Phosphorus Budget: River discharge of P into the coastal ocean accounts for most P input to the ocean (Table 6.3). However, most riverine P is sequestered in continental shelf sediments (Paytan and McLaughlin, 2007) so that only ~25 per cent of the riverine input

---

14 Oxygen minimum zones (OMZs) are oxygen-deficient layers in the ocean's water column (Paulmier and Ruiz-Pino, 2009).

Table 6.3 | Summary of estimated sources and sinks (Tg P yr-1) in the global marine phosphorus budget. (Data sources: Filippelli and Delaney, 1996; Howarth et al., 1996; Benitez-Nelson, 2000; Compton et al., 2000; Ruttenberg, 2004; Seitzinger et al., 2005; Paytan and McLaughlin, 2007; Mahowald et al., 2008; Harrison et al., 2010; Krishnamurthy et al., 2010.)

Sources	River discharge	10.79 – 31.00
	Atmospheric deposition	0.54 – 1.05
	TOTAL	11.33 – 32.05
Sinks	Open ocean sedimentation	1.30 – 10.57

enters the NPP-driven marine P cycle. Estimates of BAP reaching the open ocean from rivers range from a few tenths to perhaps 1 Tg P yr-1 (Seitzinger et al., 2005; Meybeck, 1982; Sharpies et al., 2013). Mahowald et al., (2008) estimated that atmospheric inputs of BAP are ~0.1 Tg P yr-1. Together these inputs would support ~0.1 per cent of NPP annually. Thus, like Nr, virtually all NPP is supported by BAP recycled within the ocean on a global scale.

The primary source of P in the atmosphere is mineral dust, accounting for approximately 80 per cent of atmospheric P. Other important sources include biogenic particles, biomass burning, fossil-fuel combustion, and biofuels. The P in mineral particles is not very soluble, and most of it is found downwind of desert and arid regions. Only ~0.1 Tg P yr^{-1} of BAP appears to enter the oceans via atmospheric deposition (Mahowald et al., 2008). Although a small term in the P budget (Table 6.3), atmospheric deposition appears to be the main external source of BAP in the oligotrophic waters of the subtropical gyres and the Mediterranean Sea (Paytan and McLaughlin, 2007; Krishnamurthy et al., 2010).

Burial in continental shelf and deep-sea sediments is the primary sink, with most riverine input being removed from the marine P cycle by rapid sedimentation of particulate inorganic (non-reactive mineral lattices) P in coastal waters (Paytan and McLaughlin, 2007). Burial in deep-sea sediments occurs after transformations from dissolved to particulate forms in the water column. Of the riverine input, 60-85 per cent is buried in continental shelf sediments (Slomp, 2011). Assuming that inputs from river discharge and atmospheric deposition are, respectively, ~15 Tg P yr^{-1} and 1 Tg P yr^{-1}, and that 11 Tg P yr^{-1} and 5 Tg P yr^{-1}, respectively, are buried in shelf and open-ocean sediments, the P budget appears to be roughly balanced on the scale of P turnover times in the ocean (~1500 years, Paytan and McLaughlin, 2007).

3 Variability and Resilience of Marine Ecosystems

3.1 Phytoplankton species diversity and resilience

Biodiversity enhances resilience by increasing the range of possible responses to perturbations and the likelihood that species will functionally compensate for one another following disturbance (functional redundancy) (McCann, 2000; Walker et al., 2004; Hooper et al., 2005; Haddad et al., 2011; Appeltans et al., 2012; Cleland, 2011). Annually averaged phytoplankton species diversity of the upper ocean tends to be lowest in polar and subpolar waters, where fast-growing (opportunistic) species account for most NPP, and highest in tropical and subtropical waters, where small phytoplankton (< 10 µm) account for most NPP (Barton et al., 2010). Phytoplankton species diversity is also a unimodal function of phytoplankton NPP, with maximum diversity at intermediate levels of NPP and minimum diversity associated with blooms of diatoms, dinoflagellates, *Phaeocystis* sp., and coccolithophores (Irigoien et al., 2004). This suggests that pelagic marine food webs may be most resilient to climate and anthropogenic forcings at intermediate levels of annual phytoplankton NPP.

3.2 Events, phenomena and processes of special interest

Zooplankton grazing: Zooplankton populations play key roles in both microbial food webs[15] supported by small phytoplankton (< 10 µm) and metazoan food webs[16] supported by large phytoplankton (> 20 µm). As such, they are critical links in nutrient cycles and the transfer of NPP to higher trophic levels of metazoan consumers. They fuel the biological pump and they limit excessive increases in NPP (e.g., Corten and Linley, 2003; Greene and Pershing, 2004; Steinberg et al., 2012). Microbial food webs dominate the biological cycles of C, N and P in the upper ocean and feed into metazoan food webs involving zooplankton, planktivorous fish, and their predators (Pomeroy et al., 2007; Moloney et al., 2011; Ward et al., 2012). Zooplankton in microbial food webs are typically dominated by heterotrophic and mixotrophic flagellates and ciliates. Metazoan food webs dominate the flow of energy and nutrients to harvestable fish stocks and to the deep sea (carbon sequestration). Zooplankton in metazoan food webs are typically dominated by crustaceans (e.g., copepods, krill and shrimp) and are part of relatively short, efficient, and nutritionally rich food webs supporting large numbers of planktivorous and piscivorous fish, seabirds, and marine mammals (Richardson, 2008; Barnes et al., 2010; Barnes et al., 2011).

Microbial food webs support less zooplankton biomass than do metazoan food webs, and a recent analysis suggests that zooplankton/phytoplankton ratios range from a low of ~0.1 in the oligotrophic subtropical gyres to a high of ~10 in upwelling systems and subpolar regions (Ward et al., 2012). Such a gradient is consistent with a shift from "bottom-up", nutrient-limited NPP in the oligotrophic gyres, where microflagellates are the primary consumers of NPP (Calbet, 2008), to "top-down", grazing control of NPP by zooplankton in more productive high-latitude and upwelling ecosystems, where planktonic crustaceans are the primary grazers of NPP (Ward et al., 2012). Thus, zooplankton grazing on phytoplankton is an important parameter of spatial patterns and tempo-

---

[15] The microbial food web (or microbial loop) consists of small phytoplankton (mean spherical diameter < 10 µm), heterotrophic bacteria, archaea and protozoa (flagellates and ciliates).

[16] The so-called "classical" food web is dominated by larger phytoplankton, metazoan zooplankton and nekton.

ral trends in NPP, particularly at high latitudes and in coastal upwelling systems (section 6.1.4).

3.2.1 NPP and Fisheries

Fish production depends to a large extent on NPP but the relationship between NPP and fish landings is complex. For instance, Large Marine Ecosystems (LMEs) of the coastal ocean account for ~30 per cent of marine phytoplankton NPP and ~80 per cent of marine fish landings globally (Sherman and Hempel, 2009). They are also "proving grounds" for the development of ecosystem-based approaches (EBAs) to fisheries management (McLeod and Leslie, 2009; Sherman and Hempel, 2009; Malone et al., 2014b). EBAs are guided in part by the recognition that the flow of energy and nutrients from NPP through marine food webs ultimately limits annual fish landings (Pauly and Christensen, 1995; Pikitch et al., 2004).

Both mean annual and maximum fish landings have been shown to be related to NPP on regional scales, with increases in potential landings at high latitudes (> 30 per cent) and decreases at low latitudes (up to 40 per cent) (Pauly and Christensen, 1995; Ware, 2000; Ware and Thomson, 2005; Frank et al., 2006; Chassot et al., 2007; Sherman and Hempel, 2009; Blanchard et al., 2012). However, the NPP required to support annual fish landings (PPR) varies among LMEs, e.g., fisheries relying on NPP at the Eastern Boundary Upwelling Systems require substantially higher levels of NPP than elsewhere (Chassot et al., 2010). Variations in PPR/NPP are related to a number of factors, including the relative importance of microbial and metazoan food webs and differences in the efficiencies of growth and transfer efficiencies among trophic levels. The level of exploitation (PPR/NPP) increased by over 10 per cent from 2000 to 2004, and the NPP appropriated by current global fisheries is 17-112 per cent higher than that appropriated by sustainable fisheries. Temporal and spatial variations in PPR/NPP call into question the usefulness of global NPP *per se* as a predictor of global fish landings (Friedland et al., 2012). Friedland et al. (2012) found that NPP is a poor predictor of fish landings across 52 LMEs, with most variability in fish landings across LMEs accounted for by chlorophyll-a concentration, the fraction of NPP exported to deep water, and the ratio of secondary production to NPP. Given these considerations and uncertainties concerning the effects of climate change on fluxes of nutrients to the euphotic zone, it is not surprising that there is considerable uncertainty associated with projections of how changes in NPP will affect fish landings over the next few decades.

3.2.2 NPP Fisheries and zooplankton

Zooplankton is a critical link between NPP and fish production (Cushing, 1990; Richardson, 2008). Efficient transfer of phytoplankton NPP to higher trophic levels ultimately depends on the relative importance of microbial and metazoan foods webs and the coherence between the timing of phytoplankton blooms (initiation, amplitude, duration) and the reproductive cycles of zooplankton and planktivorous fish (Cushing, 1990; Platt et al., 2003; Koeller et al., 2009; Jansen et al., 2012).

Energy transfer to higher trophic levels via microbial food webs is less efficient than for metazoan food webs (e.g., Barnes et al., 2010; Barnes et al., 2011; Suikkanen et al., 2013). Coherence in time and space is especially important in higher-latitude ecosystems (Sherman et al., 1984; Edwards and Richardson, 2004; Richardson, 2008; Ohashia et al., 2013), where seasonal variations in NPP are most pronounced and successful fish recruitment is most dependent on synchronized production across trophic levels (Cushing, 1990; Beaugrand et al., 2003). The phenological response to ocean warming differs among functional groups of plankton, resulting in predator-prey mismatches that may influence PPR/NPP in marine ecosystems. For example, phytoplankton blooms in the North Atlantic begin earlier south of 40°N (autumn – winter) and in spring north of 40°N (Siegel et al., 2002; Ueyama and Monger, 2005; Vargas et al., 2009). Likewise, a 44-year time series (1958-2002) revealed progressively earlier peaks in abundance of dinoflagellates (23 days), diatoms (22 days) and copepods (10 days) under stratified summer conditions in the North Sea (Edwards and Richardson, 2004). Such differential responses in phytoplankton and zooplankton phenology lead to mismatches between successive trophic levels and, therefore, a decline in PPR/NPP, i.e., a decrease in carrying capacity for harvestable fish stocks.

3.2.3 Coastal Eutrophication and "Dead Zones"

Excess phytoplankton NPP in coastal ecosystems can lead to accumulations of phytoplankton biomass and eutrophication. Anthropogenic N and P loading to estuarine and coastal marine ecosystems has more than doubled in the last 100 years (Seitzinger et al., 2010; Howarth et al., 2012),[17] leading to a global spread of coastal eutrophication and associated increases in the number of oxygen-depleted "dead zones" (Duarte, 1995; Malone et al., 1999; Diaz and Rosenberg, 2008; Kemp et al., 2009), loss of sea grass beds (Dennison et al., 1993; Kemp et al., 2004; Schmidt et al., 2012), and increases in the occurrence of toxic phytoplankton blooms (see below). Current global trends in coastal eutrophication and the occurrence of "dead zones" and toxic algal events indicate that phytoplankton NPP is increasing in many coastal ecosystems, a trend that is also likely to exacerbate future impacts of over-fishing, sea-level rise, and coastal development on ecosystem services (Dayton et al., 2005; Koch et al., 2009; Waycott et al., 2009).

3.2.4 Oxygen minimum zones (OMZs)

OMZs, which occur at midwater depths (200-1000 m) in association with eastern boundary upwelling systems, are expanding globally as the solubility of dissolved O_2 decreases and vertical stratification increases due to upper ocean warming (Chan et al., 2008; Capotondi et al., 2012; Bijma et al., 2013). Currently, the total surface area of OMZs is estimated to be ~30 x 10^6 km² (~8 per cent of the ocean's surface area) with a volume of ~10 x 10^6 km³ (~0.1 per cent of the ocean's volume). It is expected that the spatial extent of OMZs will continue to increase (Oschlies et al., 2008), a trend that is likely to affect nutrient cycles and

---

[17] Primarily due to the rapid rise in fertilizer use in agriculture, production of manure from farm animals, domestic sewage, and atmospheric deposition associated with fossil-fuel combustion.

fisheries – especially when combined with the spread of coastal dead zones associated with coastal eutrophication.

3.2.5 Toxic Algal Blooms

Toxin-producing algae are a diverse group of phytoplankton species with only two characteristics in common: (1) they harm people and ecosystems; and (2) their initiation, development and dissipation are governed by species-specific population dynamics and oceanographic conditions (Cullen, 2008b). Negative impacts of algal toxins include illness and death in humans who consume contaminated fish and shellfish or are exposed to toxins via direct contact (swimming, inhaling noxious aerosols); mass mortalities of wild and farmed fish, marine mammals and birds; and declines in the capacity of ecosystems to support goods and services (Cullen, 2008b; Walsh et al., 2008). Impacts associated with toxic algal blooms are global and appear to be increasing in severity and extent in coastal ecosystems as a consequence of anthropogenic nutrients, introductions of non-native toxic species with ballast water from ships, and climate-driven increases in water temperature and vertical stratification of the upper ocean (Glibert et al., 2005; Glibert and Bouwman, 2012; Cullen, 2008b; Franks, 2008; Malone, 2008; Hallegraeff, 2010; Moore et al., 2008, Babin et al., 2008).

3.2.6 Nanoparticles

Nanoparticles have dimensions of 1-100 nm and are produced both naturally and anthropogenically. Of concern here are anthropogenic nanoparticles, such as titanium dioxide (TiO_2)[18] and nanoplastics[19]. Nanoparticulate TiO_2 is highly photoactive and generates reactive oxygen species (ROS) when exposed to ultraviolet radiation (UV). Consequently, TiO_2 has been used for antibacterial applications, such as wastewater treatment. It also has the potential to affect NPP. For example, it has been found that ambient levels of UV from the sun can cause TiO_2 nanoparticles suspended in seawater to kill phytoplankton, perhaps through the generation of ROS (Miller et al., 2012). Recent work has also highlighted the potential environmental impacts of microplastics (cf. Depledge et al., 2013; Wright et al., 2013). Experimental evidence suggests that nanoplastics may reduce grazing pressure on phytoplankton and perturb nutrient cycles. For example, Wegner et al., (2012) found that mussels (*Mytilus edulis*) exposed to nanoplastics consume less phytoplankton and grow slower than mussels that have not been exposed. In addition, microplastics contain persistent organic pollutants, and both mathematical models and experimental data have demonstrated the transfer of pollutants from plastic to organisms (Teuten et al., 2009).

Understanding the ecotoxicology of anthropogenic nanoparticles in the marine environment is an important challenge, but as of this writing there is no clear consensus on environmental impacts *in situ* (cf. Handy et al., 2008). We know so little about the persistence and physical behaviour of anthropogenic nanoparticles *in situ* that extrapolating experimental results, such as those given above, to the natural marine environment would be premature. We urgently need to develop the means to reliably and routinely monitor nanoparticles of anthropogenic origin and their impacts on production and fate of phytoplankton biomass. A first step towards risk assessment would be to establish and set limits based on their intrinsic toxicity to phytoplankton and the consumers of plankton biomass. The provision of such information is part of the mission of Working Group 40 of the Joint Group of Experts on the Scientific Aspects of Marine Environmental Protection (GESAMP). WG 40 was established to assess the sources, fate and effects of micro-plastics in the marine environment globally.[20]

3.2.7 Ultraviolet Radiation and the Ozone Layer

The Sun emits ultraviolet radiation (UV, 400-700 nanometers), with UV-B (280-315 nm) having a wide range of potentially harmful effects, including inhibition of primary production by phytoplankton and cyanobacteria (Häder et al., 2007; Villar-Argaiz et al., 2009; Ha et al., 2012), changes in the structure and function of plankton communities (Ferreyra et al., 2006; Häder et al., 2007; Fricke et al., 2011; Guidi et al., 2011; Santos et al., 2012a; Ha et al., 2014), and alterations of the N cycle (Goes et al., 1995; Jiang and Qiu, 2011). The ozone layer in the Earth's stratosphere blocks most UV-B from reaching the ocean's surface. Consequently, stratospheric ozone depletion since the 1970s has been a concern, especially over the South Pole, where a so-called ozone hole has developed.[21] However, the average size of the ozone hole declined by ~2 per cent between 2006 and 2013 and appears to have stabilized, with variation from year to year driven by changing meteorological conditions.[22] It has even been predicted that there will be a gradual recovery of ozone concentrations by ~2050 (Taalas et al., 2000). Given these observations and variations in the depths to which UV-B penetrates in the ocean (~1-10 m), a consensus on the magnitude of the ozone-depletion effect on NPP and nutrient cycling has yet to be reached.

4 Socioeconomic importance

Marine NPP supports a broad range of ecosystem services valued by society and required for sustainable development (Millennium Ecosystem Assessment, 2005; Worm et al., 2006; Conservation International, 2008;

---

[18] The world production of nanoparticulate TiO_2 is an order of magnitude greater than the next most widely produced nanomaterial, ZnO. About 70 per cent of all pigments use TiO_2, and it is a common ingredient in products such as sunscreen and food colouring. Titanium dioxide is therefore likely to enter estuaries and oceans, for example, from industrial discharge.

[19] Plastic nanoparticles are released when plastic debris decomposes in seawater. Nanoparticles are also released from cosmetics and from clothes in the wash, and enter sewage systems where they are discharged into the sea.

[20] http://www.gesamp.org/work-programme/workgroups/working-group-40.

[21] Ozone can be destroyed by reactions with by-products of man-made chemicals, such as chlorine from chlorofluorocarbons (CFCs). Increases in the concentrations of these chemicals have led to ozone depletion.

[22] http://www.nasa.gov/content/.

Perrings et al., 2010; Schlitzer et al., 2012; Malone et al., 2014b; Chapter 3 in this assessment). These include:

(a) food security through the production of harvestable fish, shellfish and macroalgae (Sherman and Hempel, 2009; Chassot et al., 2010; Barbier et al., 2011);
(b) climate regulation through carbon sequestration (Twilley et al., 1992; Cebrian, 2002; Schlitzer et al., 2003; Duarte et al., 2005; Bouillon et al., 2008; Mitsch and Gosselink, 2008; Schneider et al., 2008; Subramaniam et al., 2008; Laffoley and Grimsditch, 2009; Nellemann et al., 2009; Chavez et al., 2011; Crooks et al., 2011; Henson et al., 2012);
(c) maintenance of water quality through nutrient recycling and water filtration (Falkowski et al., 1998; Geider et al., 2001; Dayton et al., 2005; Howarth et al., 2011);
(d) protection from coastal erosion and flooding through the growth of macrophyte habitats (Danielsen et al., 2005; UNEP-WCMC, 2006; Davidson and Malone, 2006/2007; Braatz et al., 2007; Koch et al., 2009; Titus et al., 2009; Barbier et al., 2011), and
(e) development of biofuels and discovery of pharmaceuticals through the maintenance of biodiversity (Chynoweth et al., 2001; Orhan et al., 2006; Han et al., 2006; Yusuf, 2007; Negreanu-Pîrjol et al., 2011; Vonthron-Sénécheau et al., 2011; Pereira et al., 2012; Sharma et al., 2012).

On a global scale, the value of these services in coastal marine and estuarine ecosystems has been estimated to be > 25 trillion United States dollars annually, making the coastal zone among the most economically valuable regions on Earth (Costanza et al., 1997; Martínez et al., 2007). The global loss of macrophyte ecosystems threatens the ocean's capacity to sequester carbon from the atmosphere (climate control), support biodiversity (Part V of this Assessment) and living marine resources (Part IV of this Assessment), maintain water quality, and protect against coastal erosion and flooding (Boesch and Turner, 1984; Dennison et al., 1993; Duarte, 1995; CENR, 2003; Scavia and Bricker, 2006; Davidson and Malone, 2006/07; Diaz and Rosenberg, 2008; MacKenzie and Dionne, 2008; Nellemann et al., 2009). Estimates of the value of these services by Koch et al., (2009) and Barbier et al., (2011) suggest that the socio-economic impact of the degradation of marine macrophyte ecosystems is on the order of billions of US dollars per year.

5 Anthropogenic Impacts on Upper Ocean Plankton and Nutrient Cycles

5.1 Nitrogen loading

The rate of industrial Nitrogen gas (N_2) fixation increased rapidly during the 20th century and is now about equal to the rate of biological N_2 fixation, resulting in a two- to threefold increase in the global inventory of Reactive nitrogen (N_r) (Galloway et al., 2004; Howarth, 2008), a trend that has accelerated the global N cycle (Gruber and Galloway 2008). Today, anthropogenic N_r inputs to surface waters via atmospheric deposition and river discharge are now roughly equivalent to marine N_2 fixation (Table 6.2) and are expected to exceed marine N_2 fixation in the near future as a result of increases in emissions from combustion of fossil fuels and use of synthetic fertilizers. This trend is expected to continue (Duce et al., 2008; Seitzinger et al., 2010).

Atmospheric deposition of anthropogenic N_r increased by an order of magnitude during the 20th century to ~54 Tg N y^{-1} (80 per cent of total deposition), an amount that could increase NPP by ~0.06 per cent. Estimates of anthropogenic emissions for 2030 indicate a 4-fold increase in total atmospheric N_r deposition to the ocean and an 11-fold increase in AAN deposition (Duce et al., 2008). However, Lamarque et al., (2013) suggest that oxidized Nr may decrease later this century because of increased control of the emission of oxidized N compounds. At the same time, the geographic distribution of atmospheric deposition has also changed (Suntharalingam et al., 2012). In the late 1800s, atmospheric deposition over most of the ocean is estimated to have been < 50 mg N m^{-2} y^{-1}. By 2000, deposition over large ocean areas exceeded 200 mg N m^{-2} y^{-1} with intense deposition plumes (> 700 mg N m^{-2} y^{-1}) extending downwind from Asia, India, North and South America, Europe and West Africa. Predictions for 2030 indicate similar patterns, but with higher deposition rates extending farther offshore into the oligotrophic, subtropical central gyres. Likewise, marine N_2O production has increased compared to pre-industrial times downwind of continental population centres (in coastal and inland seas by 15-30 per cent, in oligotrophic regions of the North Atlantic and Pacific by 5-20 per cent, and in the northern Indian Ocean by up to 50 per cent). These regional patterns reflect a combination of high N_r deposition and enhanced N_2O production in suboxic zones.

The major pathway of anthropogenic Nr loading to the oceans is river runoff. Anthropogenic Nr input to the coastal ocean via river discharge more than doubled during the 20th century due to increases in fossil-fuel combustion, discharges of human and animal wastes, and the use of industrial fertilizers in coastal watersheds (Peierls et al., 1991; Galloway et al., 2004; Seitzinger et al., 2010). Riverine input of Nr to the coastal ocean is correlated with human population density in and net anthropogenic inputs (NANI)[23] to coastal watersheds (Howarth et al., 2012). NPP in coastal marine and estuarine ecosystems increases with increasing riverine inputs of Nr (Nixon, 1992). Given predicted increases in population density in coastal watersheds and climate-driven changes in the hydrological cycle, global nutrient-export models predict that riverine inputs of Nr to coastal waters will double again by 2050 (Seitzinger et al., 2010). In this context, it is noteworthy that anthropogenic perturbations of the N-cycle caused by NANI already exceed the estimated "planetary boundary" (35 x 103 kg yr-1) within which sustainable development is possible (Rockstram et al., 2009).

---

[23] Net anthropogenic nitrogen input (NANI) is the sum of synthetic N fertilizer used, N fixation associated with agricultural crops, atmospheric deposition of oxidized N, and the net movement of N into or out of the region in human food and animal feed.

Ocean warming and associated increases in vertical stratification are likely to exacerbate the effects of increases in NANI on phytoplankton NPP in coastal waters (Rabalais et al., 2009). As a consequence, excess NPP and the global extent of coastal eutrophication are likely to continue increasing, especially in coastal waters near large watersheds, population centres and areas of industrial agriculture (Kroeze and Seitzinger, 1998; Dayton et al., 2005; Seitzinger et al., 2005; UNESCO, 2008; Kemp et al., 2009; Rabalais et al., 2009; Sherman and Hempel, 2009).

5.2 Ocean warming

5.2.1 Global impacts on NPP

Henson et al., (2013) used the results of six global biogeochemical models to project the effects of upper ocean warming on the amplitude and timing of seasonal peaks in phytoplankton NPP. Amplitude decreased by 1-2 per cent over most of the ocean, except in the Arctic, where an increase of 1 per cent by 2100 is projected. These results are supported by the response of phytoplankton and zooplankton to global climate-change projections carried out with the IPSL Earth System Model (Chust et al., 2014). Projected upper ocean warming by the turn of the century led to reductions in phytoplankton and zooplankton biomass of 6 per cent and 11 per cent, respectively. Simulations suggest such declines are the predominant response over nearly 50 per cent of the ocean and prevail in the tropical and subtropical oceans while increasing trends prevail in the Arctic and Antarctic oceans. These results suggest that the capacity of the oceans to regulate climate through the biological carbon pump may decrease over the course of this century. The model runs also indicate that, on average, a 30-40 year time series of observations will be needed to validate model results.

Regardless of the direction of global trends in NPP, climate change may be causing shifts in phytoplankton community size spectra toward smaller cells which, if confirmed, will have profound effects on the fate of NPP and nutrient cycling during this century (Polovina and Woodworth, 2012). The size spectrum of phytoplankton communities in the upper ocean's euphotic zone largely determines the trophic organization of pelagic ecosystems and, therefore, the efficiency with which NPP is channelled to higher trophic levels, is exported to the deep ocean, or is metabolized in the upper ocean (Malone, 1980; Azam et al., 1983; Cushing, 1990; Kiørboe, 1993; Legendre and Rassoulzadegan, 1996; Shurin et al., 2006; Pomeroy et al., 2007; Marañón, 2009; Barnes et al., 2010; Finkel et al., 2010; Suikkanen et al., 2013; and section 6.3.2).

In today's ocean, the proportion of NPP accounted for by small phytoplankton (cells with an equivalent spherical diameter < 10 μm) generally increases with increasing water temperature in the ocean (Atkinson et al., 2003; Daufresne et al., 2009; Marañón, 2009; Huete-Ortega et al., 2010; Morán et al., 2010; Hilligsøe et al., 2011) and with increasing vertical stratification of the euphotic zone (Margalef, 1978; Malone, 1980; Kiørboe, 1993). Small cells also have a competitive advantage over large cells in nutrient-poor environments (Malone, 1980a; Chisholm, 1992; Kiørboe, 1993; Raven, 1998; Marañón, 2009). Thus, as the upper ocean warms and becomes more stratified, it is likely that the small phytoplankton species will account for an increasingly large fraction of NPP (Morán et al., 2010) resulting in increases in energy flow through microbial food webs and decreases in fish stocks and organic carbon export to the deep sea (see section 6.1.1 and references therein).

This trend may be exacerbated by increases in temperature that are likely to stimulate plankton metabolism, enhancing both NPP and microbial respiration. Recent studies (Montoya and Raffaelli, 2010; Sarmento et al., 2010; Behrenfeld, 2011; Taucher and Oschlies, 2011; Taucher et al., 2012) suggest that predicted climate-driven increases in the temperature of the upper ocean will stimulate the NPP of smaller picophytoplankton cells (equivalent spherical diameter < 2μm), despite predicted decreases in nutrient inputs to the euphotic zone from the deep sea in permanently stratified regions of the ocean (e.g., the oligotrophic, subtropical central gyres). However, if this does occur, it will not result in an increase in fishery production or in the ocean's uptake of atmospheric CO_2, because increases in picophytoplankton NPP will be accompanied by equivalent increases in the respiratory release of CO_2 by bacterioplankton and other heterotrophic microbial consumers in the upper ocean (Sarmento et al., 2010; Behrenfeld, 2011).

5.2.2 Regional impacts on NPP

Regional trends in phytoplankton NPP are less controversial. The area of low NPP in the subtropical central gyres increased by 1-4 per cent yr^{-1} from 1998 through 2006 (Polovina et al., 2008; Vantrepotte and Mélin, 2009), a trend that is likely to continue through this century (Polovina et al., 2011). Decreasing NPP has been attributed to climate-driven (ocean warming) increases in vertical stratification and associated decreases in nutrient fluxes from deep water to the euphotic zone in the permanently stratified subtropical gyres (Rost et al., 2008; Jang et al., 2011; Polovina et al., 2011; Capotondi et al., 2012; Moore et al., 2013). In the North Atlantic, upper ocean warming and increases in stratification have been accompanied by decreasing NPP in waters south of ~50°N, whereas warming and increases in stratification to the north have been accompanied by increasing NPP (Richardson and Shoeman, 2004; Bode et al., 2011). These divergent responses to stratification reflect increases in the availability of sunlight in nutrient-rich, well-mixed subpolar waters and increases in nutrient limitation in nutrient-poor, permanently stratified[24] subtropical waters (Richardson and Shoeman, 2004; Steinacher et al., 2010; Bode et al., 2011; Capotondi et al., 2012).

Polar ecosystems are particularly sensitive to climate change (Smith et al., 2001; Anisimov et al., 2007; Bode et al., 2011; Doney et al., 2012; Engel et al., 2013), and the impacts of shrinking ice cover on NPP are expected to be especially significant in the Arctic Ocean (Wang and Overland, 2009). Loss of Arctic sea ice has accelerated in recent years

---

[24] The permanent or main thermocline extends from ~50° N to ~50° S. North Atlantic Deep Water and Antarctic Bottom Water formation take place at higher latitudes.

(with a record low in 2012),[25] a trend that is correlated with an increase in annual NPP by an average of 27.5 Tg C yr^{-1} since 2003, with an overall increase of 20 per cent from 1998 to 2010 (Arrigo et al., 2008; Arrigo and van Dijken, 2011; Brown and Arrigo, 2012). Of this increase, 30 per cent has been attributed to a decrease in the spatial extent of summer ice and 70 per cent to a longer growing season (the spring bloom is occurring earlier). The change in NPP is not spatially homogeneous. Positive trends are most pronounced in seasonally ice-free regions, including the eastern Barents shelf, Siberian shelves (Kara and east Siberian seas), western Mackenzie shelf, and the Bering Strait. NPP is expected to continue increasing during this century due to continued sea-ice retreat and the associated increase in available sunlight. However, this trend may be short-lived if nitrate supplies from deep water are insufficient (Tremblay and Gagnon, 2009). Neglecting the latter, Arrigo and van Dijken (2011) project a > 60 per cent increase in NPP for a summer ice-free Arctic using a linear extrapolation of the historical trend. Should these trends continue, additional loss of ice during Arctic spring could boost NPP more than three-fold above 1998-2002 levels and alter marine ecosystem structure and the degree of pelagic-benthic coupling. However, predictions of future trends in Arctic NPP are uncertain, given the possibility of nitrate limitation (Vancoppenolle et al., 2013). Reducing uncertainty for both nitrate fields and rates of biogeochemical processes in the sea-ice zone should improve the skill of projected changes in NPP needed to anticipate the impact of climate change on Arctic food webs and the carbon cycle.

The coastal marine ecosystem of the West Antarctic Peninsula supports massive spring-summer phytoplankton blooms upon which the production of Antarctic krill depends. NPP associated with these blooms is correlated with the spatial and temporal extent of ice cover during the previous winter. Air temperatures over the West Antarctic Peninsula have warmed by 7°C since the 1970s, resulting in a 40 per cent decline in winter sea-ice cover and a decrease in phytoplankton NPP (Flores et al., 2012; Ducklow et al., 2013; Henley, 2013). Continued declines in the extent of winter sea-ice cover is likely to drive decadal-scale reductions in NPP and the production of krill, reducing the food supply for their predators (marine mammals, seabirds and people).

5.2.3 Distribution and abundance of toxic phytoplankton species

The socioeconomic impacts of toxic dinoflagellate species are increasing globally (Van Dolah, 2000; Glibert et al., 2005; Hoagland and Scatasta, 2006; Babin et al., 2008; UNESCO, 2012), and their distribution and abundance are sensitive indicators of the impacts of anthropogenic nutrient inputs and climate-driven increases in water temperature and vertical stratification on ecosystem services (see section 6.3.2).

Alexandrium tamarense represents a group of species that cause paralytic shellfish poisoning (PSP) (*Alexandrium catenella*, *A. fundyense*, *Pyrodinium bahamense* and *Gymnodinium catenatum*) globally (Boesch et al., 1997). Since the 1970s, PSP episodes have spread from coastal waters of Europe, North America and Japan to coastal waters of South America, South Africa, Australia, the Pacific Islands, India, all of Asia and the Mediterranean (Lilly et al., 2007). Climate-driven shifts in the geographic ranges of *Ceratium furca* and *Dinophysis spp.* in the NE Atlantic have also occurred (Edwards et al., 2006), and the abundance of dinoflagellates in the North Sea have been positively correlated with the North Atlantic Oscillation (NAO) and sea surface temperature (Edwards et al., 2001).

5.2.4 Distribution and abundance of indicator zooplankton species

The distribution and abundance of calanoid copepods are also sensitive indicators of climate-driven increases in upper ocean temperature and basin-scale oscillations (Hays et al., 2005; Burkill and Reid, 2010; Edwards et al., 2010) including poleward shifts in species distributions (Beaugrand et al., 2002; Beaugrand et al., 2003; Cheung et al., 2010; Chust et al., 2014), decreases in size, and higher growth rates (e.g., Beaugrand et al., 2002; Richardson, 2008; Mackas and Beaugrand, 2010). There have also been phenological changes, with the seasonal peak in abundance advancing to earlier in the year for some species and being delayed for others (Edwards and Richardson, 2004, section 6.3.2). In the North Pacific, there is a strong correlation between sea-surface temperature in the spring and the latitude at which subtropical species reach their seasonal peak in abundance.[26] Water temperature also influences the annual cycle of *Neocalanus plumchrus* biomass in the Northeast Pacific, where decadal-scale variations include a shift to an earlier occurrence of the seasonal biomass peak, as well as a decrease in the duration of the bloom under warm ocean conditions (Mackas et al., 2007; Batten and Mackas, 2009).

The frequency and magnitude of gelatinous zooplankton blooms may be important indicators of the status and performance of marine ecosystems (Hay, 2006; Graham et al., 2014). Both predators (medusa and ctenophores) and herbivores (tunicates) can affect the fate of NPP (Pitt et al., 2009; Lebrato and Jones, 2011). Predators disrupt metazoan food webs by consuming copepods and small fish (Richardson et al., 2009). Tunicates reduce the transfer of NPP to upper trophic levels and to the deep sea as their gelatinous remains are degraded via microbial food webs (Lebrato and Jones, 2011).

Although, there is no evidence for an increase in the frequency and magnitude of gelatinous zooplankton on a global scale (Condon et al., 2012), decadal scale increases have been reported in several coastal marine ecosystems (Brodeur et al., 2002; Kideys, 2002; Lynam et al., 2006; Uye, 2008; Licandro et al., 2010). A rigorous analysis of multidecadal (using a 1950 baseline) abundance data for 45 Large Marine Ecosystems, Brotz et al., 2012 found that 28 (62 per cent) exhibited increasing trends while 3 (7 per cent) exhibited decreasing trends. Thus,

---

25 http://nsidc.org/arcticseaicenews//.

26 http://www.pices.int/publications/pices_press/volume16/v16_n2/pp_19-21_CPR_f.pdf.

while increases of jellyfish populations may not be globally universal, they are both numerous and widespread. The most likely causes of these trends include ocean warming, overfishing, coastal eutrophication, habitat modification, aquaculture, and introductions of non-indigenous gelatinous species (Brotz et al., 2012; Purcell, 2012). While direct evidence is lacking for most of these pressures, jellyfish tend to be most abundant in warm waters with low forage fish populations, and it is likely that ocean warming will provide a rising baseline of abundance leading to increases in the magnitude of jellyfish blooms and associated impacts on ecosystem services (Graham et al., 2014).

5.3 Ocean acidification

The oceans are becoming more acidic due to increases in uptake of atmospheric CO_2 (Calderia and Wickett, 2003; Calderia and Wickett, 2005; Doney et al., 2009; Beardall et al., 2009), and most of the upper ocean is projected to be undersaturated with respect to aragonite within 4-7 decades (Orr et al., 2005) with undersaturation expected to occur earliest at high latitudes (> 40°) and in upwelling systems where the aragonite saturation horizon is expected to shoal most rapidly (Feely et al., 2009, Gruber et al., 2009). These chemical changes in turn affect marine plankton via several mechanisms including the following: (1) decreases in the degree of aragonite saturation makes it harder for calcifying organisms (e.g., coccolithophores, foraminifera, and pteropods) to precipitate their mineral structures; (2) decreases in pH alters the bioavailability of essential algal nutrients such as iron and zinc; and (3) increases in CO_2 decrease the energy requirements for photosynthetic organisms to synthesize biomass. Such biological effects are likely to perturb marine biogeochemical cycles including carbon export to the deep sea via the biological pump which may have a positive feedback on the buildup of CO_2 in the upper ocean and atmosphere. However, assessments of the impacts of ocean acidification on NPP and nutrient cycling remain controversial and are a subject of much research (cf., Delille et al., 2005; Doney et al., 2009; Shi et al., 2009; Shi et al.,2010; Shi et al., 2012; Moy et al., 2009; Kristy et al., 2010). For example, increases in CO_2 may stimulate N_2 and carbon fixation by colonial cyanobacterial diazotrophs (Barcelos e Ramos et al., 2007). In addition, as the upper ocean warms, the geographic range of diazotrophs will expand. These effects may combine to enhance N_2 fixation by as much as 35-65 per cent by the end of this century (Hutchins et al., 2009). It is noteworthy interesting that projected increases in N_2 fixation are about the same magnitude as increases in denitrification projected by Oschlies et al., (2008). Although both of these estimates have large uncertainties, if input and output fluxes accelerate at about the same rate, the ocean's global inventory of N_r would not change, whereas NPP could increase (Sarmento et al., 2010; Behrenfeld, 2011).

In regard to macrophytes, photosynthetic rates of calcifying macroalgae do not appear to be stimulated by elevated CO_2 conditions, i.e., the majority of studies to date have shown a decrease or no change in photosynthetic rates under elevated CO_2 conditions (Hofmann and Bischof, 2014). On the other hand, there is clear evidence that ocean acidification (higher pCO_2) stimulates seagrass NPP resulting in increases in above- and below-ground biomass suggesting that the capacity of seagrasses to sequester carbon may be significantly increased (Garrard and Beaumont, 2014).

5.4 Sea-level rise, coastal development and macrophyte NPP

Sea levels have increased globally since the 1970s, mainly as a result of global mean sea-level rise due in part to anthropogenic warming causing ocean thermal expansion and glacier melting (Chapter 4 of this Assessment). Sea-level rise will not be uniform globally. Regional differences in sea-level trends will be related to changes in prevailing winds, ocean circulation, gravitational pull of polar ice sheets, and subsidence, so that sea-level rise will exceed the global mean in some regions and will actually fall in others.[27]

To date, the global decline in macrophyte habitats has been primarily due to coastal development, artificially hardened shorelines, aquaculture operations, dredging and eutrophication. This will change with sea-level rise (Short and Neckles, 1999; Nicholls and Cazenave, 2010). Macrophyte habitats are projected to be negatively affected by sea-level rise and subsidence, especially where distributions are constrained on their landward side by geomorphology and human activities along the shoreline (Pernetta, 1993; Short and Neckles, 1999; Orth et al., 2006; Alongi, 2008; Gilman et al., 2008; Silliman et al., 2009; Waycott et al., 2009; Donato et al., 2011). Together, sea-level rise, subsidence, coastal development and aquaculture operations are destroying mangrove forests, tidal marshes and seagrass beds at an alarming rate. The combination of sea-level rise and the loss of these coastal habitats will decrease the capacity of coastal ecosystems to provide services, including climate regulation (carbon sequestration), protection against coastal flooding and erosion, and the capacity to support biodiversity and living marine resources.

5.5 Regions of special interest

5.5.1 Coastal river plumes

Increases in land-based anthropogenic inputs of N and P to coastal waters is driving increases in annual phytoplankton NPP in estuaries and coastal marine ecosystems near population centres and areas of industrial agriculture in large river basins (sections 6.2.1 and 6.2.2). This may lead to further increases in the spatial extent and/or number of coastal ecosystems experiencing eutrophication and oxygen-depleted dead zones associated with the coastal plumes of major river-coastal systems, including the Yangtze (E. China Sea), Mekong (S. China Sea), Niger (Gulf of Guinea), Nile (Mediterranean Sea), Parana (Atlantic Ocean), Mississippi (Gulf of Mexico), and Rhine (North Sea) (UNESCO, 2012).

5.5.2 Polar waters and subtropical gyres

---

27 http://tidesandcurrents.noaa.gov/sltrends//.

Ocean warming appears to be driving opposing trends in phytoplankton NPP in polar waters (interannual increases in NPP) and subtropical gyres (interannual decreases in NPP) and a global expansion of oxygen minimum zones associated with upwelling systems. Regions of special interest include the Arctic Ocean, coastal waters of the western Antarctic Peninsula, permanently stratified subtropical gyres of the North Pacific and North Atlantic, and major coastal upwelling centers (Cariaco Basin and California, Humboldt, Canary, Benguela and Somali Currents).

5.5.3 Subpolar waters

Early expressions of the impacts of ocean acidification on upper ocean plankton are most likely to occur at high latitudes. Pteropods and foraminifera (dominated by *Globigerina bulloides*) are most abundant at high latitudes (> 40°N) in surface waters of the North Atlantic (Barnard et al., 2004; Fraile et al., 2008; Bednaršek et al., 2012), whereas the coccolithophore *E. huxleyi* is most abundant in the "Great Southern Coccolithophore Belt" of the Southern Ocean[28] and at high latitudes in the NE Atlantic (Barnard et al., 2004; Balch et al., 2011; Sadeghi et al., 2012). If the abundance of these functional groups declines in these regions, likely impacts will be to reduce the capacity of the oceans to take up CO_2, export carbon to the deep sea, and support fisheries (Cooley et al., 2009).

6 Information needs

As shown above, anthropogenic nutrient-loading of coastal waters and climate-change pressures on marine ecosystems (ocean warming and acidification, sea-level rise) are driving changes in NPP and nutrient cycles that are affecting the provision of ecosystem services and, therefore, sustainable development. However, although changes in macrophyte NPP and their impacts are relatively well documented (and must continue to be), a consensus on the magnitude of changes and even the direction of change in phytoplankton NPP and upper ocean nutrient cycles has yet to be reached.

Documenting spatial patterns and temporal trends in NPP and nutrient cycles (and their causes and socioeconomic consequences) will rely heavily on the accuracy and frequency with which changes in NPP and nutrient cycling can be detected over a broad range of scales (cf. deYoung et al., 2004; UNESCO, 2012; Mathis and Feeley, 2013). Given the importance of marine NPP and the species diversity of primary producers to sustaining ecosystem services, rapid detection of changes in time-space patterns of marine NPP and in the diversity of primary producers that contribute to NPP is an important dimension of the Regular Process[29] for global reporting and assessment of the state of the marine environment, including socioeconomic aspects.

Data requirements for the Regular Process have been used to help guide the development of the Global Ocean Observing System and an implementation strategy for its coastal module (UNESCO, 2012; Malone et al., 2014a; Malone et al., 2014b). The essential variables required to compute key indicators of ecosystem health include species richness, chlorophyll-*a*, dissolved Nr, and dissolved BAP (UNESCO, 2012). Routine and sustained measurements of these variables over a range of temporal and spatial scales are required for rapid and timely detection of changes in NPP and nutrient cycles and the impacts of these changes on ecosystem services on regional (e.g., Large Marine Ecosystems) to global scales. Although satellite imagery, limited *in situ* measurements and numerical models are making it possible to detect interannual and decadal changes in NPP on these scales, the same cannot be said for observations of species richness and nutrient distributions (UNESCO, 2012).

6.1 Net primary production

Sustained observations of chlorophyll, irradiance and temperature fields are required for model-based estimates of phytoplankton NPP (see section 6.1.2). An integrated approach using long term data streams from both remote sensing and frequent *in situ* observations is needed to capture the dynamics of marine phytoplankton NPP and to detect decadal trends. Remote sensing provides a cost-effective means to observe physical and biological variables synoptically in time and space with sufficient resolution to elucidate linkages between climate-driven changes in the NPP of ecosystems and the dynamic relationship between phytoplankton NPP and the provision of ecosystem services (Platt et al., 2008; Forget et al., 2009). For details on requirements, advantages and limitations of satellite-based remote sensing of ocean colour, see IOCCG (1998), Sathyendranath (2000), and UNESCO (2006, 2012).

Two related activities, both contributions to the Global Ocean Observing System, provide the core of an integrated observing system needed to provide data required to assess the state of the marine environment in terms of both time-space variations in phytoplankton NPP and the impacts of these variations on ecosystem services: the Chlorophyll Global Integrated Network (ChloroGIN)[30] (Sathyendranath et al., 2010) and Societal Applications in Fisheries and Aquaculture using Remotely-Sensed Imagery (SAFARI) (Forget et al., 2010). FARO (Fisheries Applications of Remotely Sensed Ocean Colour) has recently been initiated to coordinate the development of ChloroGIN and SAFARI for the provision of ocean-colour data and data products for use in fisheries research and ecosystem-based management of living marine resources.[31] Likewise, the GEO Biodiversity Observation Network, the Global Biodiversity Information Facility (GBIF), and the Ocean Biogeographical Information System (UNESCO, 2012) provide data and information on the species richness of marine primary producers.

---

28 The belt is centered around the sub-Antarctic front and has a spatial extent of 88 x 106 km2 (~25 per cent of the global ocean).

29 http://www.un.org/Depts/los/global_reporting/global_reporting.htm.

30 http://www.chlorogin.org/.

31 http://www.faro-project.org/index.html.

6.2 Nitrogen and phosphorus cycles

The N cycle is more dynamic[32] and less well understood than previously thought (Codispoti et al., 2001; Capone and Knapp, 2007; Zehr and Kudela, 2011; Landolfi et al., 2013; Voss et al., 2013). Major impediments to detecting and understanding decadal changes in the marine N cycle are: current uncertainties about the rates (undersampling); distribution and coupling of sources and sinks; sensitivity of N_2 fixation, denitrification, and anammox to anthropogenic inputs of N_r; and changes in the marine environment associated with climate change (warming and increases in stratification of the upper ocean, ocean acidification, oxygen depletion, and sea-level rise).

Quantifying inputs of N and P to coastal ecosystems and the open ocean requires a network of coordinated and sustained observations on local to global scales. For atmospheric deposition, monitoring should focus on regions that have intense deposition plumes downwind of major population centres in West Africa, East Asia, Europe, India, North and South America (section 6.2.1 and Schulz et al., 2012). This is a major goal of the SOLAS programme[33]. Shipboard time-series observations of biogeochemical variables that are being established globally[34] should provide deposition data for these plumes. For riverine inputs, rivers that are part of the Global Terrestrial Network for River Discharge (GTN-R)[35] and that represent a broad range of volume discharges and catchment-basin population densities are high priorities for monitoring land-based inputs and associated land-cover/land-use practices in their watersheds (UNESCO, 2012).

All global ocean biogeochemistry models require oceanographic data on physical and chemical variables, including temperature, salinity, mixed-layer depth, and the concentration of macro-nutrients (N, P, Si) (Le Quéré et al., 2010). Over the last decade, autonomous technologies for measuring essential physical variables (including temperature, salinity and mixed-layer depth) have revolutionized our ability to observe the sea surface and the ocean's interior. By integrating data from both remote sensing (satellite-based sensors and land-based HF radar) and *in situ* measurements (from ships of opportunity, research vessels and automated moorings, profiling floats, gliders, surface drifters and large pelagic predators), observations of atmospheric and upper ocean geophysics are now made continuously in four dimensions; data are transmitted to data assembly centers in near-real time via satellites, fiber-optic cables, and the internet; and predictions (nowcasts and forecasts) of atmospheric and upper ocean "weather" are made routinely using data assimilation techniques and coupled atmospheric-hydrodynamic models (Hall et al., 2010).

Over the last decade, autonomous technologies have revolutionized our ability to measure nitrate, nitrite, ammonium and reactive phosphate *in situ* (Johnson and Coletti, 2002; ACT, 2003; Sakamoto et al., 2004; Adornato et al., 2010). Efforts are also underway to expand sampling programmes such Repeat Hyrdrography (Hood 2009), Argo (Rudnick et al., 2004; Testor et al., 2010), and OceanSites[36] to incorporate *in situ* nutrient sensors.

6.3 Plankton species diversity

Sustaining marine species richness[37] is the single most important indicator of the capacity of ecosystems to support services valued by society (Worm et al., 2006). A biodiversity observation network (GEO BON)[38] has been established to document changes in species biodiversity, and the Ocean Biogeographic Information System (OBIS)[39] documents the species diversity, distribution and abundance of life in the oceans. Both are contributions to GEOSS.[40] A set of sentinel sites should be targeted for sustained observations of species richness including Large Marine Ecosystems and the emerging network of marine protected areas that is nested within them (Malone et al., 2014a). As a group, these sites represent a broad range of species diversity, climate-related changes in the marine environment, and anthropogenic nutrient inputs. Here we underscore the importance of rapid detection of changes in plankton diversity and early warnings of impacts on marine ecosystem services.

---

32 Estimates of turnover times of Nr have decreased from 10,000 years to 1,500 years (Codispoti et al., 2001).

33 http://www.solas-int.org/.

34 e.g., For example, http://www.unesco.org/new/en/natural-sciences/ioc-oceans/sections-and-programmes/ocean-sciences/biogeochemical-time-series/.

35 http://www.fao.org/gtos/gt-netRIV.html; http://gtn-r.bafg.de, http://www.bafg.de/GRDC/EN/Home/homepage_node.html.

36 http://www.whoi.edu/virtual/oceansites/

37 Species richness is an unweighted list of species present in an ecosystem that is especially important to monitor because it is the simplest indicator of species diversity and it does not discount rare species that are often the primary concern.

38 https://www.earthobservations.org/geobon.shtml

39 http://www.iobis.org/

40 www.earthobservations.org/geobon.shtml

References

Adornato, L., Cardenas-Valencia, A., Kaltenbacher, E., Byrne, R.H., Daly, K., and others (2010). In situ nutrient sensors for ocean observing systems. In *Proceedings of OceanObs'09: Sustained Ocean Observations and Information for Society* (Vol. 2), Venice, Italy, 21-25 September 2009. Hall, J., Harrison, D.E., & Stammer, D., eds. *ESA Publication* WPP-306, doi:10.5270/OceanObs09.cwp.01.

ACT (Alliance for Coastal Technologies) (2003). *State of Technology in the Development and Application of Nutrient Sensors*; ACT (2007). Recent Developments in *In Situ* Nutrient Sensors: Applications and Future Directions, ACT 06-08 (UMCES CBL 07-048).

Agawin, N.S.R., Duarte, C.M., and Agustí, S. (2000). Nutrient and temperature control of the contribution of picoplankton to phytoplankton biomass and production. *Limnology and Oceanography*, 45(3): 591-600.

Alongi, D. (2008). Mangrove forests: Resilience, protection from tsunamis, and responses to global climate change. *Estuarine Coastal and Shelf Science*, 76, 1-13.

Altabet, M.A. (2007). Constraints on oceanic N balance/imbalance from sedimentary 15N records, *Biogeosciences*, 4: 75–86.

Amarasinghe, M.D., and Balasubramaniam, S. (1992). Net primary productivity of two mangrove forests stands on the northwestern coast of Sri Lanka. *Hydrobiologia*, 247: 37–47.

Anisimov, O.A., Vaughan, D.G., Callaghan, T.V., Furgal, C., Marchant, H., Prowse, T.D., Vilhjálmsson, H., and Walsh, J.E. (2007). Polar regions (Arctic and Antarctic). *Climate Change 2007: Impacts, Adaptation and Vulnerability. Contribution of Working Group II to the Fourth Assessment Report of the Intergovernmental Panel on Climate Change.* Parry, M.L., Canziani, O.F., Palutikof, J.P., van der Linden, P.J., and Hanson, C.E., eds. Cambridge University Press, Cambridge, 653-685.

Antoine, D., and Morel, A. (1996). Oceanic primary production. 1. Adaptation of a spectral light-photosynthesis model in view of application to satellite chlorophyll observations. *Global Biogeochemical Cycles*, 10 (1): 43-55.

Appeltans, W., Ahyong, S.T., Anderson, G., Angel, M.V., Artois, T., et al. (2012). The magnitude of global marine species diversity. *Current Biology*, 22: 2189 – 2202.

Arrigo, K.R. (2005). Marine microorganisms and global nutrient cycles. *Nature*, 437, 349-355.

Arrigo, K.R. and van Dijken, G.L. (2011). Secular trends in Arctic Ocean net primary production, *Journal of Geophysical Research*, 116, C09011, doi:10.1029/2011JC007151.

Arrigo, K.R., van Dijken, G.L., and Pabi, S. (2008). Impact of a shrinking Arctic ice cover on marine primary production, *Geophysical Research Letters*, 35(19), L19603.

Atkinson, D., Ciotti, B.J., and Montagnes, D.J.S. (2003). Protists decrease in size linearly with temperature. *Proceedings of the Royal Society of London B*, 270: 2605-2611.

Azam, F., Fenchel, T., Field, J. G., Gray, J. S., Meyer-Reil, L. A., Thingstad, F. (1983). The Ecological Role of Water-Column Microbes in the Sea. *Marine Ecology Progress Series*, 10: 257-263.

Babin, M., Roesler, C.S., and Cullen, J.J., eds. (2008). *Real-Time Coastal Observing Systems for Marine Ecosystem Dynamics and Harmful Algal Blooms: Theory, Instrumentation and Modelling*. Monographs on Oceanographic Methodology Series, UNESCO.

Balch, W.M., Drapeau, D.T., Bowler, B.C., Lyczskowski, E., Booth, S., and Alley, D. (2011). The contribution of coccolithophores to the optical and inorganic carbon budgets during the Southern Ocean Gas Exchange Experiment: New evidence in support of the "Great Calcite Belt" hypothesis, *Journal of Geophysical Research*, 116, C00F06, doi:10.1029/2011JC006941.

Balch, W.M., and Utgoff, P.E. (2009). Potential interactions among ocean acidification, coccolithophores, and the optical properties of seawater. *Oceanography*, 22 (4): 146-159.

Barber, R.T., Sanderson, M.P., Lindley, S.T., Chai, F., Newton, J., Trees, C.C., Foley, D.G., and Chavez, F.P. (1996). Primary productivity and its regulation in the equatorial Pacific during and following the 1991–92 El Niño. *Deep-Sea Research II*, 43: 933–969.

Barbier, E.B., Hacker, S.D., Kennedy, C., Koch, E.W., Stier, A.C., and Silliman, B.R. (2011). The value of estuarine and coastal ecosystem services. *Ecological Monographs*, 81: 169–193

Barcelos Ramos, J., Biswas, H., Schulz, K.G., LaRoche, J., and Riebesel, U. (2007). Effect of rising atmospheric carbon dioxide on the marine nitrogen fixer *Trichodesmium*. *Global Biogeochemical Cycles*, 21: GB2028, doi:10.1029/2006GB002898.

Barnard, R., Batten, S.D., Beaugrand, G. et al. (2004). Continuous Plankton Records: Plankton atlas of the North Atlantic Ocean (1958 – 1999). II. Biogeograhical charts. *Marine Ecology Progress Series*, Supplement: 11-75.

Barnes, C., Maxwell, D., Reuman, D.C., and Jennings, S. (2010). Global patterns in predator–prey size relationships reveal size dependency of trophic transfer efficiency. *Ecology*, 91:222–232.

Barnes, C., Irigoien, X., De Oliveira, J.A.A., Maxwell, D., and Jennings, S. (2011). Predicting marine phytoplankton community size structure from empirical relationships with remotely sensed variables. *Journal of Plankton Research*, 33: 13–24.

Barton, A.D., Dutkiewicz, S., Flierl, G., Bragg, J., Follows, M.J. (2010). Patterns of diversity in marine phytoplankton. *Science*, 327, 1509-1511.

Batten, S.D., and Mackas, D.L. (2009). Shortened duration of the annual *Neocalanus plumchrus* biomass peak in the Northeast Pacific. *Marine Ecology Progress Series*, 393: 189–198.

Beardall, J., Stojkovic, S., and Larsen S. (2009). Living in a high CO2 world: impacts of global climate change on marine phytoplankton. *Plant Ecology & Diversity*, 2: 191-205.

Beaugrand, G., K.M. Brander, J.A. Lindley, S. Souissi, and P.C. Reid. (2003). Plankton effect on cod recruitment in the North Sea. *Nature*, 426: 661–664.

Beaugrand, G., Reid, P.C., Ibañez, F., Lindley, J.A., and Edwards, M. (2002). Reorganization of the North Atlantic Marine Copepod Biodiversity and Climate. *Science*, 296: 1692-1694.

Bednaršek, N., Tarling, G.A., Bakker, D.C.E., Fielding, S., Jones, E.M., Venables, H.J., Ward, P., Kuzirian, A., Lézé, B., Feely, R.A., Murphy, E.J. (2012). Extensive dissolution of live pteropods in the Southern Ocean. *Nature,* 5: 881-885.

Behrenfeld, M.J. (2011). Biology: Uncertain future for marine algae. *Nature Climate Change*, 1: 33-34.

Behrenfeld, M.J., O'Malley, R.T., Siegel, D.A., McClain, C.R., et al. (2006). Climate-driven trends in contemporary ocean productivity. *Nature* 444: 752–755.

Behrenfeld, M.J., and Falkowski, P.G. (1997). Photosynthetic Rates Derived from Satellite-Based Chlorophyll Concentration. *Limnology and Oceanography*, 42: 1–20.

Benitez-Nelson, C.R. (2000). The biogeochemical cycling of phosphorus in marine systems. *Earth-Science Review*, 51(1-4): 109-135.

Beusen, A.H.W., Slomp, C.P., and Bouwman, A.F. (2013). Global land–ocean linkage: direct inputs of nitrogen to coastal waters via submarine groundwater discharge. *Environmental Research Letters*, 8: 034035 (doi:10.1088/1748-9326/8/3/034035).

Bidigare, R.R., Chai, F., Landry, M.R., Lukas, R., Hannides, C.C.S., Christensen, S.J., Karl, D.M., Shi, L., and Chao. Y. (2009). Subtropical ocean ecosystem structure changes forced by North Pacific climate variations. *Journal of Plankton Research*,

31 (10): 1131-39.

Bijma, J., Pörtner, H-O. Yesson, C., and Rogers, A.D. (2013). Climate change and the oceans – What does the future hold? *Marine Pollution Bulletin*. 74 (2): 495-505.

Bissinger, J.E., Montagnes, D.J.S., Sharples, J., and Atkinson, D. (2008). Predicting marine phytoplankton maximum growth rates from temperature: Improving on the Eppley curve using quantile regression. *Limnology and Oceanography*, 53: 487–93.

Bittaker, H.F., and Iverson R.L. (1976). Thalassia testudinum productivity: A field comparison of measurement methods. *Marine Biology*, 37: 39-46.

Blanchard, J.L., Jennings, S., Holmes, R., Harle, J., Merino, G., Allen, J. I., Holt, J., Dulvy N. K., and Barange, M. (2012). Potential consequences of climate change for primary production and fish production in large marine ecosystems. *Philosophical transactions of the Royal Society of London, Series B, Biological Sciences*, 367: 2979–89.

Bode, A., Hare, J., Li, W.K.W., Morán, X.A.G., and Valdés L. (2011). Chlorophyll and primary production in the North Atlantic. In: Reid, P.C., and Valdés, L., eds. *ICES Status Report on Climate Change in the North Atlantic*. International Council for the Exploration of the Sea, Copenhagen, pp. 77-102.

Boesch, D.F., and Turner, R.E. (1984). Dependence of Fishery Species on Salt Marshes: The Role of Food and Refuge. *Estuaries and Coasts*, 7:460–468.

Boesch, D.F., Anderson, D.M., Horner, R.A., Shumway, S.E., Tester, P.A., and Whitledge, T.E. (1997). Harmful Algal Blooms in Coastal Waters: Options for Prevention, Control and Mitigation. *Science for Solutions*, NOAA Coastal Ocean Program Decision and Analysis Series, No. 10, NOAA Coastal Ocean Office, Silver Spring, MD, 46 pp. + appendix.

Bouillon, S., Borges, A.V., Castañeda-Moya, E., Diele, K., Dittmar, T., Duke, N.C., Kristensen, E., Lee, S.Y, Marchand, C., Middelburg, J.J., Rivera-Monroy, V.H., Smith III, T.J, and Twilley, R.R. (2008). Mangrove production and carbon sinks: A revision of global budget estimates. *Global Biogeochemical Cycles*, 22(2), doi:10.1029/2007GB003052.

Boyce, D.G., Dowd, M., Lewis, M.R., Worm, B. (2014). Estimating global chlorophyll changes over the past century. *Progress in Oceanography*, 122: 163-177.

Boyce D.G., Lewis, M.R., and Worm, B. (2010). Global phytoplankton decline over the past century. *Nature*, 466: 591-596.

Boyer, E.W., Howarth, R.W., Galloway, J.N., et al. (2006). Riverine nitrogen export from the continents to the coasts. *Global Biogeochemical Cycle*, 20: GB1S91; doi:10.1029/2005GB002537.

Braatz, S., Fortuna, S., Broadhead, J., and Leslie, R. (2007). Coastal protection in the aftermath of the Indian Ocean tsunami: What role for forests and trees? *Rap publication*, 2007/07, Proceedings of the Regional Technical Workshop, Khao Lak, Thailand, 28-31 August, 2006.

Brodeur, R.D., Sugisaki, H., Hunt, G.L. Jr., (2002). Increases in jellyfish biomass in the Bering Sea: implications for the ecosystem. *Marine Ecology Progress Series*, 233, 89–103.

Brotz, L., Cheung, W.W.L., Kleisner, K., Pakhomov, E., Pauly, D. (2012). Increasing jellyfish populations: trends in Large Marine Ecosystems. *Hydrobiologia*, 690: 3-20.

Brown, Z.W., and Arrigo, K.R. (2012). Contrasting trends in sea ice and primary production in the Bering Sea and Arctic Ocean *ICES Journal of Marine Science*, 69: 1180-1193.

Buitenhuis, E.T., Li, W.K.W., Vaulot, D., Lomas, M.W., Landry, M.R., Partensky, F., Karl, D.M., Ulloa, O., Campbell, L., Jacquet, S., Lantoine, F., Chavez, F., Macias, D., Gosselin, M., and McManus, G.B., (2012). Picophytoplankton biomass distribution in the global ocean. *Earth System Science Data*, 4: 37–46.

Buesseler, K.O., Antia, A.N., Chen, M., Fowler, S.W., Gardner, W.D., Gustafsson, O., Harada, K., Michaels, A.F., van der Loeff, M.R., Sarin, M., Steinberg, D.K., and Trull, T. (2007). An assessment of the use of sediment traps for estimating upper ocean particle fluxes. *Journal of Marine Research*, 65: 345–416.

Burkill, P.H., and Reid, P.C. (2010). Plankton biodiversity of the North Atlantic: changing patterns revealed by the Continuous Plankton Recorder survey. In: *Proceedings of OceanObs'09: Sustained Ocean Observations and Information for Society*, (Vol. 1). Hall, J., Harrison, D.E., and Stammer, D., eds. ESA Publication WPP-306, doi:10.5270/OceanObs09.pp.09.

Calbet, A. (2008). The trophic roles of microzooplankton in marine systems. *ICES Journal of Marine Science*, 65: 325–331.

Caldeira, K., and Wickett, M.E. (2003). Oceanography: Anthropogenic carbon and ocean pH. *Nature*, 425: 365-365.

Caldeira K., and Wickett, M.E. (2005). Ocean model predictions of chemistry changes from carbon dioxide emissions to the atmosphere and ocean. *Journal of Geophysical Research*, 110:C09S04, doi:10.1029/2004JC002671.

Capone, D.G. (2008). The Marine Nitrogen Cycle. *Microbe*, 3 (4): 186 – 192.

Capone, D.G., and Knapp, A.N. (2007). Oceanography: A marine nitrogen cycle fix? *Nature*, 445: 159-160.

Capone, D.G., Bronk, D.A., Mulholland, M.R., and Carpenter, E.J. (2008). Nitrogen in the Marine Environment. *Elsevier Inc.*, Amsterdam, 1729 pp. pp.

Capotondi, A., Alexander, M.A., Bond, N.A., Curchitser, E.N., and Scott, J.D. (2012). Enhanced upper ocean stratification with climate change in the CMIP3 models. *Journal of Geophysical Research*, 117, C04031, doi:10.1029/2011JC007409.

Carr, M-E., Friedrichs, M.A.M., et al. (2006). A comparison of global estimates of marine primary production from ocean color. *Deep-Sea Research Part II*, 53: 741-770.

Cebrian, J. (2002). Variability and control of carbon consumption, export, and accumulation in marine communities. *Limnology and Oceanography, Inc.*, 47(1): 11–22.

CENR. (2003). An Assessment of Coastal Hypoxia and Eutrophication in U.S. Waters. Washington DC: *National Science and Technology Council, Committee on Environment and Natural Resources.*

Chan, F., Barth, J.A., Lubchenco, J., et al. (2008). Emergence of Anoxia in the California Current Large Marine Ecosystem. *Science*, 319: 920.

Charpy-Roubaud, C., and Sournia, A. (1990). The comparative estimation of phytoplanktonic, microphytobenthic and macrophytobenthic primary production in the oceans. *Marine Microbial Food Webs*, 4 (1): 31-57.

Chassot, E., Melin, F., Le Pape, O., and Gascuel, D. (2007). Bottom-up control regulates fisheries production at the scale of eco-regions in European seas. *Marine Ecology Progress Series*, 343: 45–55.

Chassot, E., Bonhommeau, S., Dulvy, N.K., Mélin, F., Watson, R., Gascuel, D., and Le Pape. O. (2010). Global marine primary production constrains fisheries catches. *Ecology Letters*, 13(4): 495 – 505.

Chavez, F.P., Messié, M., and Pennington, J.T. (2011). Marine Primary Production in Relation to Climate Variability and Change. *Annual Reviews Marine Science*, 3: 227-260.

Chaalali A., Beaugrand, G., Raybaud, V., Goberville, E., David, V., et al. (2013). Climatic Facilitation of the Colonization of an Estuary by *Acartia tonsa*. *PLoS ONE* 8(9): e74531. doi:10.1371/journal.pone.0074531

Cheung, W., Lam, V., Sarmieno, J.L., Kearney, K., Watson, R., Zeller, D., Pauly, D. (2010) . Large scale redistribution of maximum fisheries catch potential in the global ocean unser climate change. *Global Change Biology.*, 16: 24-35. DOI: 10.1111/j.1365-2486.2009.01995.x

Chisholm, S.W. (1992). Phytoplankton size, p. 213–237. In: Falkowski, P.G., and Woodhead, A.D., (eds.), *Primary Productivity and Biogeochemical Cycles in the Sea*. Plenum Press.

Church, J.A., and White, N.J. (2006). A 20th century acceleration in global sea-level rise. *Geophysical Research Letters*, 33: L01602.

Chust, G., Allen, J. I., Bopp, L., Schrum, C., Holt, J., Tsiaras, K., Zavatarelli, M., Chifflet, M., Cannaby, H., Dadou, I., Daewel, U., Wakelin, S. L., Machu, E., Pushpadas, D., Butenschon, M., Artioli, Y., Petihakis, G., Smith, C., Garçon, V., Goubanova, K., Le Vu, B., Fach, B. A., Salihoglu, B., Clementi, E. and Irigoien, X. (2014). Biomass changes and trophic amplification of plankton in a warmer ocean. *Global Change Biology*, 20: 2124–2139. doi: 10.1111/gcb.12562.

Chynoweth, D.P., Owens, J.M., and Legrand, R. (2001). Renewable methane from anaerobic digestion of biomass. *Renewable Energy*, 22:1-8.

Cleland, E.E. (2011). Biodiversity and Ecosystem Stability. *Nature Education Knowledge*, 3(10):14

Cloern, J.E. (1987). Turbidity as a control on phytoplankton biomass and productivity in estuaries. *Continental Shelf Research*, 7: 1367-1381.

Cloern, J.E. (2001). Review. Our evolving conceptual model of the coastal eutrophication problem. *Marine Ecology Progress Series*, 210: 223–253.

Cloern, J.E., and Jassby, A.D. (2010). Patterns and Scales of Phytoplankton Variability in Estuarine–Coastal Ecosystems. *Estuaries and Coasts*, 33:230–241.

Cloern, J.E., Foster, S.Q., and Kleckner, A.E. (2013). Review: phytoplankton primary production in the world's estuarine-coastal ecosystems. *Biogeosciences Discussion*, 10: 17725–17783.

Codispoti, L.A., Brandes, J.A., Christensen, J.P., Devol, A.H., Naqvi, S.W.A., Paerl, H.W., and Yoshinari, T. (2001). The oceanic fixed nitrogen and nitrous oxide budgets: Moving targets as we enter the anthropocene? *Scientia Marina*, 65(Suppl. 2): 85–105.

Compton, J.S., Mallinson, D., Glenn, C.R., Filippelli, G.M., Föllmi, K.B., Shields-Zhou, G.A., and Zanin, Y. (2000). Variations in the global phosphorus cycle, in Marine Authigenesis: From Global to Microbial, Glenn, C.R., Prévol-Lucas, L., and Lucas, J., eds. *SEPM Special Publication*, 66, 21-33.

Condon, R. H., Graham, W.M., Duarte, C.M., Pitt, K.A., Lucas, C.H., Haddock, S.H.D., Sutherland, K.R., Robinson, K.L., Dawson, M.N., Decker, M.B., and others. (2012). Questioning the Rise of Gelatinous Zooplankton in the World's Oceans. *BioScience*, 62: 160-169.

Conkright, M.E., and Gregg, W.W. (2003). Comparison of global chlorophyll climatologies: In situ, CZCS, blended in situ-CZCS and SeaWiFS. *International Journal of Remote Sensing*, 24 (5): 969–991.

Conservation International. (2008). *Economic Values of Coral Reefs, Mangroves, and Seagrasses: A Global Compilation*. Center for Applied Biodiversity Science, Arlington, VA.

Cooley, S.R., Kite-Powell, H.L., and Doney, S.C. (2009). Ocean acidification's potential to alter global marine ecosystem services. *Oceanography*, 22: 172-181.

Corten, A., and Lindley, J.A. (2003). The use of CPR data in fisheries research. *Progress in Oceanography*, 58: 285-300.

Costanza, R., d'Arge, R., de Groot, R., Farber, S., Grasso, M., Hannon, B., Limburg, K., Naeem, S., O'Neill, R.V., Paruelo, J., Raskin, R.G., Suttonk, P., and van den Belt, M. (1997). The value of the world's ecosystem services and natural capital. *Nature*, 387: 253-260.

Cousens, R. (1984). Estimation of Annual Production by the Intertidal Brown Alga *Ascophyllum nodosum* (L.) Le Jolis. *Botanica Marina*, 27: 217-227.

Craig, H. (1957). The Natural Distribution of Radiocarbon and the Exchange Time of Carbon Dioxide between Atmosphere and Sea. *Tellus*, 9(1): 1-17.

Crooks, S., Herr, D., Tamelander, J., Laffoley, D., and Vandever, J. (2011). Mitigating Climate Change through Restoration and Management of Coastal Wetlands and Near-shore Marine Ecosystems: Challenges and Opportunities. *Environment Department Paper* 121, World Bank, Washington DC.

Cullen, J.J. (2008a). Primary production methods. pp. 578-584, In Steele, J.H., Turekian, K.K., and Thorpe, S.A., eds. *Encyclopedia of Ocean Science*, Elsevier, ISBN: 978-0-12-374473-9.

Cullen, J.J. (2008b). Observation and prediction of harmful algal blooms, In Babin, M., Roesler, C.S., and Cullen, J.J., eds. *Real-Time Coastal Observing Systems for Marine Ecosystem Dynamics and Harmful Algal Blooms: Theory, Instrumentation and Modelling*. Monographs on Oceanographic Methodology Series, UNESCO, pp. 1-41.

Cullen, J.J., Davis, R.F., and Huot, Y. (2012). Spectral model of depth-integrated water column photosynthesis and its inhibition by ultraviolet radiation. *Global Biolgeochemical Cycles*, 26, GB1011, doi:10.1029/2010GB003914.

Curcó, A., Ibàñez, C., Day, J.W., and Prat, N. (2002). Net primary production and decomposition of salt marshes of the Ebre delta (Catalonia, Spain). *Estuaries and Coasts*, 25(3): 309-324.

Cushing, D.H. (1990). Plankton production and year-class strength in fish populations: an update of the match/mismatch hypothesis. *Advances in Marine Biology*, 26: 249-293.

Dame, R.F., and Kenny, D. (1986). Variability of Spartina alterniflora primary production in the euhaline North Inlet estuary. *Marine Ecology - Progress Series*, 32: 71-80.

Danielsen, F., Sørensen, M.K., Olwig, M.F., Selvam, V., Parish, F., Burgess, N.D., Hiraishi, T., Karunagaran, V.M., Rasmussen, M.S., Hansen, L.B., Quarto, A., and Suryadiputra, N. (2005). The Asian tsunami: A protective role for coastal vegetation. *Science*, 310: 643.

Daufresne, M., Lengfellner, K., and Sommer, U. (2009). Global warming benefits the small in aquatic ecosystems. *Proceedings of the National Academy of Sciences, USA* 106: 12788–12793

Dave, A.C., and Lozier, M.S. (2013). Examining the global record of interannual variability in stratification and marine productivity in the low-latitude and mid-latitude ocean. *Journal of Geophysical Research. Oceans*, 118, 3114–3127, doi:10.1002/jgrc.20224.

Davidson, M., and Malone, T.C., eds. (2006/2007). Stemming the tide of coastal disasters. *Marine Technology Society Journal*, 40: 1-125.

Day, J.W., Jr., Coronado-Molina, C., Vera-Herrera, F.R., Twilley, R., Rivera-Monroy, V.H., Alvarez-Guillen, H., Day, R., and Conner, W. (1996). A 7 year record of aboveground net primary production in a southeastern Mexican mangrove forest. *Aquatic Botany*, 55: 39–60.

Dayton, P., Curran, S., Kitchingman, A., Wilson, M., Catenazzi, A., Restrepo, J., Birkeland, C., Blaber, S., Saifullah, S., Branch, G., Boersma, D., Nixon, S., Dugan, P., Davidson, N., and Vörösmarty, C. (2005). Coastal ecosystems In: *Ecosystems and Human Well-being: Current State and Trends: Findings of the Condition and Trends Working Group*, Hassan, R., Scholes, R., and Ash, N., eds. Island Press, p. 513-550.

Delille, B., Harlay, J., Zondervan, I., Jacquet, S., Chou, L., et al. (2005). Response of primary production and calcification to changes of pCO2 during experimental blooms of the coccolithophorid Emiliania huxleyi. *Global Biogeochemical Cycles*, 19, GB2023, doi:10.1029/2004GB002318.

Dennison, W.C., Orth, R.J., Moore, K.A., Stevenson, J.C., Carter, V., Kollar, S., Bergstrom, P.W., and Batiuk, R.A. (1993). Assessing Water Quality with Submersed Aquatic Vegetation. Habitat requirements as barometers of Chesapeake Bay health. *BioScience*, 43: 86–94.

Depledge, M.H., Galgani, F., Panti, C., Caliani, I., Casini, S., and Fossi, M.C. (2013). Plastic litter in the sea. *Marine Environmental Research*, 92: 279-281.

Deutsch, C., Sarmiento, J.L., Sigman, D.M., Gruber, N., and Dunne, J.P. (2007). Spatial coupling of nitrogen inputs and losses in the ocean. *Nature*, 445: 163-167.

DeVries, T., Deutsch, C., Primeau, F., Chang, B., and Devol, A. (2012). Global rates of water-column denitrification derived from nitrogen gas measurements. *Nature Geosci*ence, 5: 547–550.

DeVries, T., Deutsch, C., Rafter, P.A., and Primeau, F. (2013). Marine denitrification rates determined from a global 3-dimensional inverse model. *Biogeosciences*, 10: 2481–2496.

deYoung, B., Heath, M., Werner, F., Chai, F., Megrey, B., and Monfray, P. (2004). Challenges of Modeling Ocean Basin Ecosystems. *Science*, 304: 1463-1466.

Diaz, R.J., and Rosenberg, R. (2008). Spreading Dead Zones and Consequences for Marine Ecosystems. *Science*. 321: 926-929.

Donato, D.C., Kauffman, J.B., Murdiyarso, D., Kurnianto, S., Stidham, M., and Kanninen, M. (2011). Mangroves among the most carbon-rich forests in the tropics. *Nature Geosci*ence, 4: 293-297.

Doney, S.C. (2010). The Growing Human Footprint on Coastal and Open-Ocean Biogeochemistry. *Science*, 328: 1512-1516.

Doney, S.C., Ruckelshaus, M., Duffy, J.E., Barry, J.P., Chan, F., English, C.A., Galindo, H.M., Grebmeier, J.M., Hollowed, A.B., Knowlton, N., Polovina, J., Rabalais, N.N., Sydeman, W.J., and Talley, L.D. (2012). Climate change impacts on marine ecosystems. *Annual Reviews - Marine Science*, 4: 11-37.

Doney, S.C., Balch, W.M., Fabry, V.J., and Feely, R.A. (2009). Ocean Acidification: A Critical Emerging Oroblem for the Ocean Sciences. *Oceanography*, 22 (4): 16-25.

Duarte, C.M. (1989). Temporal biomass variability and production/biomass relationships of seagrass communities. *Marine Ecology Progress Series*, 51: 269-276.

Duarte, C.M. (1995). Submerged aquatic vegetation in relation to different nutrient regimes. *Ophelia*, 41: 87-112.

Duarte, C.M., and Kirkman, H. (2001). Methods for the measurement of seagrass abundance and depth distribution. In *Global Seagrass Research Methods*, Short, F.T., and Coles, R.G., eds. 141–153. Elsevier Science B.V.

Duarte, C.M., Borum, J., Short, F.T., and Walker, D.I. (2008). Seagrass ecosystems: their global status and prospects, p. 281-306. In *Aquatic Ecosystems*, Polunin, N., ed. Cambridge University Press, Cambridge, U.K.

Duarte, C.M., Middelburg, J.J., and Caraco, N.F. (2005). Major role of marine vegetation on the oceanic carbon cycle. *Biogeosciences*, 2: 1-8.

Duarte, C.M., Marbà, N., Gacia, E., Fourqurean, J.W., Beggins, J., Barrón, C., and Apostolaki, E.T. (2010). Seagrass community metabolism: Assessing the carbon sink capacity of seagrass meadows. *Global Biogeochemical Cycles*, 24 (4), GB4032, doi:10.1029/2010GB003793

Duce, R.A., LaRoche, J., Altieri, K., Arrigo, K.R., Baker, A.R., et al. (2008). Impacts of Atmospheric Anthropogenic Nitrogen on the Open Ocean. *Science*, 320: 893-897.

Ducklow, H.W., Fraser, W.R., Meredith, M.P., et al. (2013). West Antarctic Peninsula: An Ice-Dependent Coastal Marine Ecosystem in Transition. *Oceanography*, 26 (3): 190 – 203.

Duke, N.C., Meynecke, J-O., Dittmann, S., Ellison, A.M., Anger, K., et al. (2007). A world without mangroves? *Science*, 317: 41-42.

Edwards, M., Reid, P.C., and Planque, B. (2001). Long-term and regional variability of phytoplankton biomass in the north-east Atlantic (1960-1995). *ICES Journal of Marine Science* 58: 39-49.

Edwards, M., and Richardson, A.J. (2004). Impact of climate change on marine pelagic phenology and trophic mismatch. *Nature*, 430: 881-884.

Edwards, M., Johns, D.G., Leterme, S.C., Svendsen, E., and Richardson, A.J. (2006). Regional climate change and harmful algal blooms in the northeast Atlantic. *Limnology and Oceanography, Inc.*, 51 (2): 820-829.

Edwards, M., Beaugrand, G., Johns, D.G., Licandro, P., McQuatters-Gollop, A., and Wootton, M. (2010). Ecological status report: results from the CPR survey 2009. *SAHFOS Technical Report*, 7: 1-8. Plymouth, U.K. ISSN 1744-0750.

Emerson, S., and Hedges, J.I. (2008). *Chemical Oceanography and the Marine Carbon Cycle*, Cambridge University Press, Cambridge, U.K.

Engel, I., Luo, B.P., Pitts, M.C., Poole, L.R., Hoyle, C.R., Grooß, J.-U., Dörnbrack, A., and Peter, T. (2013). Heterogeneous formation of polar stratospheric clouds – Part 2: Nucleation of ice on synoptic scales. *Atmospheric Chemistry and Physics Discussions*, 13: 8831-8872.

Falkowski, P.G. (1997), Evolution of the nitrogen cycle and its influence on the biological sequestration of CO2 in the ocean, *Nature*, 387: 272–275.

Falkowksi P.G., Katz, M.E., Knoll, A.H., Quigg, A., Raven, J.A., Schofield, O., and Taylor, F.J.R. (2004). The Evolution of Modern Eukaryotic Phytoplankton. *Science*, 305 (5682): 354-360.

Falkowski, P.G., and Raven, J.A. (1997). *Aquatic photosynthesis*. Blackwell Science, Malden, MA.

Falkowski, P.G. Barber, R.T., and Smetacek, V. (1998). Biogeochemical Controls and Feedbacks on Ocean Primary Production. *Science*, 281: 200–206.

Feely, R.A., Doney, S.C., and Cooley, S.R. (2009). Ocean Acidification: Present Conditions and Future Changes in a High-CO2 World. *Oceanography*, 22 (4): 36-47.

Ferreyra, G.A., Mostajir, B., Schloss, I.R., Chatila, K., Ferrario, M.E., Sargian, P., Roy, S., Prod'homme, J., and Demers. S. (2006). Ultraviolet-B radiation effects on the structure and function of lower trophic levels of the marine planktonic food web. *Photochem Phtobiol*, 82, 887-897.

Field, C.B., Behrenfeld, M.J., Randerson, J.T., and Falkowski, P. (1998). Primary Production of the Biosphere: Integrating Terrestrial and Oceanic Components. *Science*, 281, 237–240.

Filippelli, G.M., and Delaney, M.L. (1996). Phosphorus geochemistry of equatorial Pacific sediments. *Geochimica et Cosmochimica Acta*, 60:1479-1495.

Finkel, Z.V., Beardall, J., Flynn, K.J., Quigg, A., Rees, T.A.V., and Raven J.A. (2010). Phytoplankton in a changing world: cell size and elemental stoichiometry. *Journal of Plankton Research*, 32: 119–137.

Flores, H., Atkinson, A., Kawaguchi, S., Krafft, B.A., Milinevsky, G., Nicol, S., et al. (2012). Impact of climate change on Antarctic krill. *Marine Ecology Progress Series*, 458: 1–19.

Forget, M-H., Stuart, V., and Platt, T., eds. (2009). Remote Sensing in Fisheries and Aquaculture. *IOCCG Report*. No. 8, 120 pp.

Forget, M-H., Platt, T., Sathyendranath, S., Stuart, V., and Delaney, L. (2010). "Societal Applications in Fisheries and Aquaculture Using Remotely-Sensed Imagery - The SAFARI Project" in *Proceedings of OceanObs'09: Sustained Ocean Observations and Information for Society (Vol. 2)*, Venice, Italy, 21-25 September 2009, Hall, J., Harrison, D.E., and Stammer, D., eds., ESA Publication WPP-306. doi:10.5270/OceanObs09.cwp.30.

Fourqurean, J.W., Duarte, C.M., Kennedy, H., Marbà, N., Holmer, M., Mateo, M.A., Apostolaki, E.T., Kendrick, G.A., Krause-Jensen, D., McGlathery, K.J., and Serrano, O. (2012). Seagrass ecosystems as a globally significant carbon stock. *Nature Geoscience*, 5: 505–509.

Fraile, I., Schulz, M., Mulitza, S., and Kucera, M. (2008). Predicting the global distribution of planktonic foraminifera using a dynamic ecosystem model. *Biogeosciences*, 5: 891–911.

Frank, K.T., Petrie, B., Shackell, N.L., and Choi, J.S. (2006). Reconciling differences in trophic control in mid-latitude marine ecosystems. *Ecology Letters*, 9: 1096–1105.

Franks, P.J.S. (2008). Physics and physical modeling of harmful algal blooms. In Babin, M., Roesler, C.S., and Cullen, J.J., eds. *Real-Time Coastal Observing Systems for Marine Ecosystem Dynamics and Harmful Algal Blooms: Theory, Instrumentation and Modelling*. Monographs on Oceanographic Methodology Series, UNESCO., pp. 561-598.

Freing, A., Wallace, D.W.R., and Bange, H.W. (2012). Global oceanic production of nitrous oxide. *Philosophical Transactions of the Royal Society B*, 367: 1245–1255.

Fricke, A., Molis, M., Wiencke, C., Valdivia, N., and Chapman, A.S. (2011). Effects of UV radiation on the structure of Arctic macrobenthic communities. *Polar Biology*, 34, 995-1009.

Friedland, K.D., Stock, C., Drinkwater, K.F., Link, J.S., Leaf, R.T., et al. (2012). Pathways between Primary Production and Fisheries Yields of Large Marine Ecosystems. *PLoS ONE*, 7(1): e28945, doi: 10.1371/journal.pone.0028945.

Friedrichs M.A.M, Carr, M-E., Barber, R.T., Scardi, M., Antoine, D., et al. (2009). Assessing the uncertainties of model estimates of primary productivity in the tropical Pacific Ocean. *Journal of Marine Systems,* 76: 113–133.

Galloway, J.N., Dentener, F.J., Capone, D.G., Boyer, E.W., Howarth, R.W., Seitzinger, S.P., Asner, G.P., Cleveland, C.C., Green, P.A., Holland, E.A., Karl, D.M., Michaels, A.F., Porter, J.H., Townsend, A.R., and Vörösmarty, C.J. (2004). Nitrogen cycles: past, present and future. *Biogeochemistry*, 70:153-226.

Garrard S. L., and N. J. Beaumont. (2014). The effect of ocean acidification on carbon storage and sequestration in seagrass beds; a global and UK context. *Marine Pollution Bulletin*, 86: 138-146.

Geider, R.J., Delucia, E.H., Falkowski, P.G., Finzi, A.C., Grime, J.P., Grace, J., Kana, T.M., La Roche, J., Long, S.P., Osborne, B.A., Platt, T., Prentice, I.C., Raven, J.A., Schlesinger, W.H., Smetacek, V., Stuart, V., Sathyendranath, S., Thomas, R.B., Vogelmann, T.C., Williams, P., and Woodward, F.I. (2001). Primary productivity of planet earth: biological determinants and physical constraints in terrestrial and aquatic habitats. *Global Change Biology*, 7: 849-882.

Gilman, E.L., Ellison, J., Duke, N.C., and Field, C. (2008). Threats to mangroves from climate change and adaptation options: a review. *Aquatic Botany*, 89: 237–250.

Gitelson, A.A. (2004). Wide dynamic rage vegetation index for remote quantification of biophysical characteristics of vegetation. *Journal of plant physiology*, 161: 165-173.

Glibert, P.M., Anderson, D.M., Gentien, P., Granéli, E., and Sellner, K.G. (2005). The global, complex phenomena of harmful algal blooms. *Oceanography*, 18(2): 136-147.

Glibert, P.M., and Bouwman, L. (2012). Land-based Nutrient Pollution and the Relationship to Harmful Algal Blooms in Coastal Marine Systems. *Inprint*, 2: 5-7 (www.loicz.org).

Goes, J.I., Handa, N., Taguchi, S., and Hama, T. (1995). Changes in the patterns of biosynthesis and composition of amino acids in a marine phytoplankton exposed to ultraviolet-B radiation: nitrogen limitation implicated. *Photochemistry and Photobiology*, 62, 703-710.

Graham, W.M., Gelcich, S., Robinson, K.L., Duarte, C.M., Brotz, L., Purcell, J.E., Madin, L.P., Mianzan, H., Sutherland, K.R., Uye, S.-I., and others. (2014). Linking human well-being and jellyfish: Ecosystem services, impacts and societal responses. *Frontiers in the Ecology and Environment*, 12: 515–523.

Gregg, W.W., Conkright, M.E., Ginoux, P., O'Reilly, J.E., and Casey, N.W. (2003). Ocean primary production and climate: global decadal changes. *Geophysical Research Letters*, 30 (15): 1809, doi:10.1029/2003GL016889.

Gregg, W.W., Casey, N.W., O'Reilly, J.E., and Esaias, W.E. (2009). An empirical approach to ocean color data: Reducing bias and the need for post-launch radiometric re-calibration. *Remote Sensing of Environment*, 113: 1598–1612.

Greene, C.H., and Pershing, A.J. (2004). Climate and the Conservation Biology of North Atlantic Right Whales: The Right Whale at the Wrong Time? *Frontiers in Ecology and the Environment*, 2: 29-34.

Gross, M.F., Klemas, V., and Hardisky, M.A. (1990). Long-term remote monitoring of salt marsh biomass. Proceedings SPIE 1300, *Remote Sensing of the Biosphere*, 59, doi:10.1117/12.21390.

Groβkopf, T., Mohr, W., Baustian, T., Schunck, H., Gill, D., Kuypers, M.M.M., Lavik, G., Schmitz, R.A., Wallace, D.W.R., and LaRoche, J. (2012). Doubling of marine dinitrogen-fixation rates based on direct measurements. *Nature*, 488, 361-364.

Gruber, N. (2004). The Dynamics of the Marine Nitrogen Cycle and its Influence on Atmospheric CO_2 Variations. In *The Ocean Carbon Cycle and Climate*, NATO Science Series, Follows, M., and Oguz. T., eds. Dordrecht: Kluwer Academic, pp. 97–148.

Gruber, N. (2008). The Marine Nitrogen Cycle: Overview and Challenges. In *Nitrogen in the marine environment*. Capone, D.G., Bronk, D.A., Mulholland, M.R., and Carpenter, E.J., eds. Chapter 1, 2nd Edition, San Diego, CA, Academic Press.

Gruber, N., and Sarmiento, J.L. (2002). Large-Scale Biogeochemical–Physical Interactions in Elemental Cycles. In *The Sea: Biological-Physical Interactions*. Robinson, A.R., McCarthy, J.F., Rothschild, B., eds. (Wiley, New York), vol. 12, pp. 337-399.

Gruber, N., and Galloway, J.N. (2008). An Earth-system perspective of the global nitrogen cycle. *Nature*, 451: 293 -296.

Gruber, N., C. Hauri, and G-K. Plattner. (2009). High vulnerability of eastern boundary upwelling systems to ocean acidification. *Global Change Newsletter (IGBP)*, Issue No. 73.

Guidi, L., Degl'Innocenti, E., Remorini, D., Biricolti, S., Fini, A., Ferrini, F., Nicese, F.P., and Tattini, M. (2011). The impact of UV-radiation on the physiology and biochemistry of *Ligustrum vulgare* exposed to different visible-light irradiance. *Environmental and Experimental Botany*, 70, 88-95.

Ha, S-Y., Joo, H-M., Kang, S-H., Ahn, I-Y., and Shin, K-H. (2014). Effect of ultraviolet irradiation on the production and composition of fatty acids in plankton in a sub-Antarctic environment. *Journal of Oceanography,* 70: 1-10. DOI 10.1007/s10872-013-0207-3.

Ha, S-Y., Kim, Y-N., Park, M-O., Kang, S-H., Kim, H-C. and Shin, K-H. (2012). Production of mycosporine-like amino acids of in situ phytoplankton community in Kongsfjorden, Svalbard, Arctic. *Journal of Photochemistry and Photobiology, B: Biology*, 114, 1-14.

Haddad, N.M., Crutsinger, G.M., Gross, K., Haarstad, J., and Tilman, D. (2011). Plant diversity and the stability of foodwebs. *Ecology Letters*, 14: 42–46.

Häder, D.-P., Kumar, H.D., Smith, R.C., and Worrest, R.C. (2007). Effects of solar UV radiation on aquatic ecosystems and interactions with climate change. *Photochemical and Photobiological Sciences*, 6, 267-285.

Hall, J., Harrison, D.E., and Stammer, D., eds. (2010). *Proceedings of OceanObs'09: Sustained Ocean Observations and Information for Society* (Volumes 1 and 2), ESA Publication WPP-306, doi:10.5270/OceanObs09.

Hallegraeff, G.M. (2010). Ocean climate change, phytoplankton community responses, and harmful algal blooms: a formidable predictive challenge. *Journal of the Phycology*, 46: 220–235.

Hamersley, M.R., Lavik, G., Woebken, D., Rattray, J.E., Lam, P., Hopmans, E.C., Damsté, J.S.S., Krüger, S., Graco, M., Gutiérrez, D., and Kuypers, M.M.M. (2007). Anaerobic ammonium oxidation in the Peruvian oxygen minimum zone. *Limnology and Oceanography*, 52 (3): 923-933.

Xu, H., Miao, X., and Wu, Q. (2006). High quality biodiesel production from a macroalga Chlorella prototheocoides by heterotrophic growth in fermenters. *Journal of Biotechnology* 126:499-507.

Handy, R.D., Owen, R., and Valsami-Jones, E. (2008). The ecotoxicology of nanoparticles and nanomaterials: current status, knowledge gaps, challenges, and future needs. *Ecotoxicology* 17:315–325.

Harrison, J.A., Bouwman, A.F., Mayorga, E., and Seitzinger, S. (2010). Magnitudes and sources of dissolved inorganic phosphorus inputs to surface fresh waters and the coastal zone: a new global model. *Global Biogeochemical Cycles*, 24, GB1003. doi: 10.1029/2009GB003590.

Hassan, R., Scholes, R., and Ash, N., eds. (2005). Coastal ecosystems In *Ecosystems and Human Well-being: Current State and Trends*, Island Press.

Hay, S. (2006). Marine ecology: gelatinous bells may ring change in marine ecosystems. *Current Biology* 16(17): R679-82.

Hays G.C., Richardson, A.J., and Robinson, C. (2005). Climate change and marine plankton. *Trends in Ecology & Evolution*, 20 (6): 337-344.

Hay, S. (2006). Marine Ecology: Gelatinous Bells May Ring Change in Marine Ecosystems. *Current Biology*, 16: 679-682.

Heck, Jr., K.L., Carruthers, T.J.B., Duarte, C.M., Hughes, A.R., Kendrick, G., Orth, R.J., and Williams, S.W. (2008). Trophic Transfers from Seagrass Meadows Subsidize Diverse Marine and Terrestrial Consumers. *Ecosystems*, 11: 1198–1210.

Henley, S.F. (2013). Climate-induced changes in carbon and nitrogen cycling in the rapidly warming Antarctic coastal ocean. (http://hdl.handle.net/1842/7626).

Henson, S.A., and Thomas, A.C. (2007). Phytoplankton Scales of Variability in the California Current System: 1. Interannual and Cross-Shelf Variability, *Journal of Geophysical Research-Oceans*, 112, C07017, doi:10.1029/2006JC004039.

Henson, S.A., Sarmiento, J.L., Dunne, J.P., Bopp, L., Lima, I., Doney, S.C., John, J., and Beaulieu, C. (2010). Detection of anthropogenic climate change in satellite records of ocean chlorophyll and productivity. *Biogeosciences*, 7: 621–640.

Henson, S.,Lampitt, R., and Johns, D. (2012). Variability in phytoplankton community structure in response to the North Atlantic Oscillation and implications for organic carbon flux. *Limnology and Oceanography*, 57(6): 1591-1601.

Henson, S., Cole, H., Beaulieu, C., and Yool, A. (2013). The impact of global warming on seasonality of ocean primary production. *Biogeosciences*, 10: 4357–4369

Heumann, B.W. (2011). Satellite remote sensing of mangrove forests: Recent advances and future opportunities. *Progress in Physical Geography*, 35 (1): 87-108.

Hilligsøe, K.M., Richardson, K., Bendtsen, J., Sørensen, L-L., Nielsen, T.G., and Lyngsgaard, M.M. (2011). Linking phytoplankton community size composition with temperature, plankton food web structure and sea-air CO2 flux. *Deep-Sea Research Part I: Oceanographic Research Papers*, 58: 826–838.

Hoagland P., and Scatasta, S. (2006). The Economic Effects of Harmful Algal Blooms. In Granéli, E., and Turner, J.T., eds. *Ecology of Harmful Algae.* Ecology Studies Series: Springer-Verlag Berlin Heidelber. 189: 391-402.

Hofmann, L.C., Bischof, K. (2014). Ocean acidification effects on calcifying macroalgae. *Aquatic Biology*, 22: 261-279.

Hood, M., ed. (2009). *Ship-based Repeat Hydrography: A Strategy for a Sustained Global Program.* UNESCO-IOC Technical Series, 89. IOCCP Reports, 17. ICPO Publication 142. (www.go-ship.org/Docs/IOCTS89_GOSHIP.pdf).

Hooper, D.U., Chapin III, F.S., Ewel, J.J., Hector, A., Inchausti, P., Lavorel, S., Lawton, J.H., Lodge, D.M., Loreau, M., Naeem, S., Schmid, B., Setälä, H., Symstad, A.J., Vandermeer, J., and Wardle, D.A. (2005). Effects of biodiversity on ecosystem functioning: A consensus of current knowledge. *Ecological Monographs*, 75: 3–35.

Howarth, R.W., et al. (1996). Regional nitrogen budgets and riverine N and P fluxes for the drainages to the North Atlantic Ocean: Natural and human influences. *Biogeochemistry*, 35, 75– 139.

Howarth, R.W. (2008). Coastal nitrogen pollution: A review of sources and trends globally and regionally. Harmful Algae, 8: 14–20.

Howarth, R.W., Chan, F., Conley, D.J., Garnier, J., Doney, S.C., Marino, R., and Billen, G. (2011). Coupled biogeochemical cycles: eutrophication and hypoxia in temperate estuaries and coastal marine ecosystems. *Frontiers in Ecology and the Environment*, 9(1): 18–26.

Howarth, R.W., Swaney, D.P., Billen, G., Garnier, J., Hong, B., Humborg, C., Johnes, P.J., Mörth, C-M., and Marino, R. (2012). Nitrogen fluxes from large watersheds to coastal ecosystems controlled by net anthropogenic nitrogen inputs and climate. *Frontiers in Ecology and the Environment*. 10(1): 37–43.

Hu, C., Feng, L., Lee, Z., Davis, C.O., Mannino, A., McClain, C.R., and Franz, B.A. (2012). Dynamic range and sensitivity requirements of satellite ocean color sensors: learning from the past. *Applied Optics*, 51 (25): 6045-6062.

Huete-Ortega, M., Marañón, E., Varela, M., Bode, A. (2010). General patterns in the size scaling of phytoplankton abundance in coastal waters during a 10-year time series. *Journal of Plankton Research*, 32: 1-14.

Hutchins, D.A., Mulholland, M.R., and Fu, F-X. (2009). Nutrient Cycles and Marine Microbes in a CO2-Enriched Ocean. *Oceanography*, 22: 128–145.

IOCCG. (1998). *Minimum Requirements for an Operational Ocean-Colour Sensor for the Open Ocean.* Report 1, Morel, A., ed. International Ocean-Colour Coordinating Group (IOCCG) Report 1, pp.48.

Irigoien, X., Huisman, J., and Harris, R.P. (2004). Global biodiversity patterns of marine phytoplankton and zooplankton. *Nature*, 429: 863-867.

Jang, C.J., Park, J., Park, T., and Yoo, S. (2011) Response of the ocean mixed layer depth to global warming and its impact on primary production: a case for the North Pacific Ocean. *ICES Journal of Marine Science*. doi:10.1093/icesjms/fsr064.

Jansen, T., Campbell, A., Brunel, T., Clausen, L.M.. (2013). Spatial Segregation within the Spawning Migration of North Eastern Atlantic Mackerel (*Scomber scombrus*) as Indicated by Juvenile Growth Patterns. *PloS One*, doi: 10.1371/journal.pone.0058114.

Jia, M., Wang, Z., Liu, D., Ren, C., Tang, X., and Dong, Z. (2013). Monitoring Loss and Recovery of Salt Marshes in the Liao River Delta, China. *Journal of Coastal Research* (in-press). 31: 371-377. Doi: http://dx.doi.org/10.2112/JCOASTRES-D-13-00056.1.

Jiang, H., and Qiu, B. (2011). Inhibition of photosynthesis by UV-B exposure and its repair in the bloom-forming cyanobacterium *Microcystis aeruginosa*. *Journal of Applied Phycology*, 23 (4): 691-696.

Johnson, K.S., and Coletti, L.J. (2002). *In situ* ultraviolet spectrophotometry for high resolution and long-term monitoring of nitrate, bromide and bisulfide in the ocean. *Deep-Sea Research Part I: Oceanography Research Papers*, 49: 1291-1305.

Kaldy, J.E., and Dunton, K.H. (2000). Above- and below-ground production, biomass and reproductive ecology of *Thalassia testudinum* (turtle grass) in a subtropical coastal lagoon. *Marine Ecology Progress Series*, 193: 271-283.

Kamykowski, D. (1987). A preliminary biophysical model of the relationship between temperature and plant nutrients in the upper ocean. *Deep Sea Research Part A. Oceanographic Research Papers*, 34 (7): 1067–1079.

Karl, D.M., Christian, J.R., Dore, J.E., Hebel, D.V., Letelier, R.M., Tupas, L.M., and Winn, C.D. (1996). Seasonal and interannual variability in primary production and particle flux at Station ALOHA. *Deep Sea Research Part II: Topical Studies in Oceanography*, 43: 539-568.

Karl, D., Michaels, A., Bergman, B., Capone, D., Carpenter, E., Letelier, R., Lipschultz, F., Paerl, H., Sigman, D., and Stal, L. (2002). Dinitrogen fixation in the world's oceans. *Biogeochemistry*, 57/58: 47–98.

Keeling, R.F., Körtzinger, A., and Gruber, N. (2010). Ocean deoxygenation in a warming world. *Annual Reviews Marine Science*, 2: 199–229.

Kemp, W.M., Lewis, M.R., Jones, T.W. (1986). Comparison of methods for measuring productlon by the submersed macrophyte, *Potamogeton perfoliatus*. *Limnology and Oceanography*, 31: 1322-1334.

Kemp W.M., Testa, J.M., Conley, D.J., Gilbert, D., and Hagy, J.D. (2009). Temporal responses of coastal hypoxia to nutrient loading and physical controls, *Biogeosciences*, 6: 2985–3008.

Kemp, W.M., Batiuk, R., Bartleson, R., Bergstrom, P., Carter, V., Gallegos, G., Hunley, W., Karrh, L., Koch, E., Landwehr, J., Moore, K., Murray, L., Naylor, M., Rybicki, N., Stevenson, J.C., and Wilcox, D. (2004). Habitat Requirements for Submerged Aquatic Vegetation in Chesapeake Bay: Water Quality, Light Regime, and Physical-Chemical Factors. *Estuaries*, 27: 363–377.

Kideys, A. E. (2002). Fall and Rise of the Black Sea Ecosystem. *Science*, 297: 1482-1484.

Kiørboe T. (1993). Turbulence, phytoplankton cell size, and the structure of pelagic food webs. *Advances in Marine Biology* 29:72.

Koch, E.W., Barbier, E.B., Silliman, B.R., Reed, D.J., Perillo, G.M.E., Hacker, S.D., Granek, E.F., Primavera, J.H., Muthiga, N., Polasky, S., Halpern, B.S., Kennedy, C.J., Kappel, C.V., and Wolanski, E. (2009). Non-linearity in ecosystem services: temporal and spatial variability in coastal protection. *Frontiers in Ecology and the Environ*ment, 7(1): 29–37, doi:10.1890/080126.

Koeller, P., Fuentes-Yaco, C., Platt, T., Sathyendranath, S., Richards, A., Ouellet, P., Orr, D., Skúladóttir, U., Wieland, K., Savard, L., and Aschan, M. (2009). Basin-Scale Coherence in Phenology of Shrimps and Phytoplankton in the North Atlantic Ocean. *Science*, 324:791-793.

Koeve, W., and Kähler, P.K. (2010). Heterotrophic denitrification vs. autotrophic anammox – quantifying collateral effects on the oceanic carbon cycle. *Biogeosciences*, 7: 2327–2337.

Kovacs, J.M., King, J.M.L., de Santiago, F.F., Flores-Verdugo. F. (2009). Evaluating the condition of a mangrove forest of the Mexican Pacific based on an estimated leaf area index mapping approach. *Environmental Monitoring and Assessment*, 157: 137-149.

Kovacs, J., Wang, M.J., Blanco-Correa, M.. (2001). Mapping mangrove disturbances using multi-date Landsat TM imagery. *Environmental Management*, 27: 763–776.

Krishnamurthy, A., Moore, J. K., Mahowald, N., Luo, C., Doney, S.C., Lindsay, K., Zender, C.S. (2009). Impacts of increasing anthropogenic soluble iron and nitrogen deposition on ocean biogeochemistry. *Global Biogeochemical Cycles*, 23, GB3016, doi:10.1029/2008GB003440.

Krishnamurthy, A., Moore, J.K., Mahowald, N., Luo, C., and Zender. C.S. (2010). Impacts of atmospheric nutrient inputs on marine biogeochemistry. *Journal of Geophysical Research*, 115: 13 pp., G01006, doi:10.1029/2009JG001115.

Kroeker, K.J., Kordas, R.L., Crim, R.N., and Singh, G.G. (2010). Meta-analysis reveals negative yet variable effects of ocean acidification on marine organisms. *Ecology Letters*, 13: 1419–1434.

Kroeze, C., and Seitzinger, S.P. (1998). Nitrogen inputs to rivers, estuaries and continental shelves and related nitrous oxide emissions in 1990 and 2050: a global model. *Nutrient Cycling in Agroecosystems*, 52: 195-212.

Laffoley, D., and Grimsditch, G. (2009). The management of natural coastal carbon sinks. Gland: IUCN.

Lam, P., Lavik, G., Jensen, M.M., van de Vossenberg, J., Schmid, M., Woebken, D., Gutiérrez, D., Amann, R., Jetten, M.S.M., and Kuypers, M.M.M. (2009). Revising the nitrogen cycle in the Peruvian oxygen minimum zone. *Proceedings of the National Academy of Sciences* (USA), 106: 4752-4757.

Lamarque, J.-F., Dentener, F., McConnell, J., et al. (2013). Multi-model mean nitrogen and sulfur deposition from the Atmospheric Chemistry and Climate Model Intercomparison roject (ACCMIP): evaluation of historical and projected future changes. *Atmospheric Chemistry and Physics*, 13: 7997-8018.

Landolfi, A., Dietze, H., Koeve, W., and Oschlies, A. (2013). Overlooked runaway feedback in the marine nitrogen cycle: the vicious cycle. *Biogeosciences*, 10: 1351–1363.

Landry, M.R., Barber, R.T., Bidigare, R.R., Chai, F., Coale, K.H., et al. (1997). Iron and grazing constraints on primary production in the central equatorial Pacific: An EqPac synthesis. *Limnology and Oceanography*, 42: 405-418.

Laufkötter, C., Vogt, M., and Gruber, N. (2013). Long-term trends in ocean plankton production and particle export between 1960-2006. *Biogeosciences*, 10: 7373–7393.

Laws, E.A., Falkowski, P.G., Smith, Jr., W.O., Ducklow, H., and McCarthy, J.J. (2000). Temperature effects on export production in the open ocean. *Global Biogeochemical Cycles*, 14: 1231–1246.

Lebrato, M., and Jones, D.O.B. (2011). Jellyfish biomass in the biological pump: Expanding the oceanic carbon cycle. *The Biochemical Society Journal*, p. 35-39.

Lee, Z.-P., ed. (2006). Remote sensing of inherent optical properties: Fundamentals, Tests of algorithms, and applications. *IOCCG Report* No. 5, 126 pp., Dartmouth, Canada.

Legendre, L., and Rassoulzadegan, F. (1996). Food-web mediated export of biogenic carbon in oceans: hydrodinamic control. *Marine Ecology Progress Series*, 145: 179–193.

Le Quéré, C., Takahashi, T., Buitenhuis, E.T., Rödenbeck, C., and Sutherland, S.C. (2010). Impact of climate change and variability on the global oceanic sink of CO2. *Global Biogeochemical Cycles*, 24 (4): GB4007, doi: 10.1029/2009GB003599.

Licandro, P., Conway, D.V.P., Daly Yahia, M.N., Fernandez de Puelles, M.L., Gasparini, S., Hecq, J.H., Tranter, P., Kirby, R.R. (2010). A blooming jellyfish in the northeast Atlantic and Mediterranean. *Biology Letters*, doi:10.1098/rsbl.2010.0150.

Lilly, E.L., Halanych, K.M., and Anderson, D.M. (2007). Species boundaries and global biogeography of the *Alexandrium tamarense* complex (Dinophyceae). *Journal of Phycology*, 43: 1329–1338.

Liu, J., Liu, S., Loveland, T.R., Tieszen, L.L. (2008). Integrating remotely sensed land cover observations and a biogeochemical model for estimating forest ecosystem carbon dynamics. *Ecological Modelling*, 219: 361-372.

Lohrenz, S.E., Fahnenstiel, G.L., Redalje, D.G., Lang, G.A., Chen, X., and Dagg, M.J. (1997). Variations in primary production of northern Gulf of Mexico continental shelf waters linked to nutrient inputs from the Mississippi River. *Marine Ecology Progress Series*, 155: 45-54.

Lomas, M.W., Bates, N.R., Johnson, R.J., Knap, A.H., Steinberg, D.K., and Carlson, C.A. (2013). Two decades and counting: 24-years of sustained open ocean biogeochemical measurements in the Sargasso Sea. *Deep Sea Research Part II: Topical Studies in Oceanography*. 93: 16-32, doi: 10.1016/j.dsr2.2013.01.008.

Long, S.P, Jones, M.B., and Roberts, M.J., eds. (1992). *Primary Productivity of Grass Ecosystems of the Tropics and Sub-tropics*. Chapman and Hall, London, 267 pp.

Luo, Y.-W., Doney, S.C., Anderson, L.A., Benavides, M., Bode, A., et al. (2012). Database of diazotrophs in global ocean: abundances, biomass and nitrogen fixation rates. *Earth System Science Data*, 4: 47-73.

Lynam C. P., Gibbons, M.J., Axelsen, B.E., Sparks, C.A.J., Coetzee, J., Heywood, B.G., Brierley, A.S. (2006). Jellyfish overtake fish in a heavily fished ecosystem. *Curr. Biol.*, 16: 492–493.

Ma, S., Tao, Z., Tang, X., Yu, Y., Zhou, X., Ma, W., and Li, Z. (2014). Estimation of Marine Primary Productivity From Satellite-Derived Phytoplankton Absorption Data. *IEEE Journal of Selected Topics in Applied Earth Observations and Remote Sensing*, 7 (7): 3084-3092.

Mackas, D.L. (2011). Does blending of chlorophyll data bias temporal trend? *Nature*, 472: E4-E5.

Mackas, D.L., and Beaugrand, G. (2010). Comparisons of zooplankton time series. *Journal of Marine Systems*, 79: 286-304.

Mackas, D.L., Batten, S.D., and Trudel, M. (2007). Effects on zooplankton of a warming ocean: recent evidence from the Northeast Pacific. *Progress in Oceanography*, 75(2): 223–252.

MacKenzie, R.A., and Dionne, M. (2008). Habitat heterogeneity: importance of salt marsh pools and high marsh surfaces to fish production in two Gulf of Maine salt marshes. *Marine Ecology Progress Series*, 368: 217–230.

Mahaffey, C., Michaels A.F., Capone, D.G. (2005). The conundrum of marine N2 fixation. *American Journal of Science*, 305 (6-8): 546–595. doi: 10.2475/ajs.305.6-8.546.

Mahowald, N.M., Baker, A.R., Bergametti, G., Brooks, N., Duce, R.A., Jickells, T.D., Kubilay, N., Prospero, J.M., and Tegen, I. (2005). Atmospheric global dust cycle and iron inputs to the ocean, *Global Biogeochemical Cycles*, 19, GB4025, doi:10.1029/2004GB002402.

Mahowald, N., Jickells, T.D., Baker, A., Artaxo, P., Benitez-Nelson, C.R., et al. (2008). Global distribution of atmospheric phosphorus sources, concentrations and deposition rates, and anthropogenic impacts, *Global Biogeochemical Cycles*, 22, GB4026, doi:10.1029/2008GB003240.

Malone, T.C. (1971). Diurnal rhythms in netplankton and nanoplankton assimilation ratios. *Marine Biology*, 10: 285-289.

Malone, T.C. (1980). Algal size and phytoplankton ecology. pp. 433-464. In: Morris, I., ed. *The Physiological Ecology of Phytoplankton*, Blackwell Scientific Publications, London.

Malone, T.C. (1991). River flow, phytoplankton production and oxygen depletion in Chesapeake Bay, In *Modern and Ancient Continental Shelf Anoxia*, Tyson, R.V., and Pearson, T.H., eds. Geological Society Publication No. 58, p. 83-94.

Malone, T.C. (2008). Ecosystem dynamics, harmful algal blooms and operational oceanography. In Babin, M., Roesler, C.S., and Cullen, J.J., eds. *Real-Time Coastal Observing Systems for Marine Ecosystem Dynamics and Harmful Algal Blooms*, Monographs on Oceanographic Methodology Series, UNESCO, pp. 527-560.

Malone, T.C., Conley, D.J., Fisher, T.R., Glibert, P.M., Harding, L.W., Sellner, K. (1996). Scales of Nutrient-Limited Phytoplankton Productivity in Chesapeake Bay. *Estuaries*, 19 (2B): 371-385.

Malone, T.C., DiGiacomo, P.M., Gonçalves, E., Knap, A.H., Talaue-McManus, L., and de Mora, S. (2014a). A Global Ocean Observing System Framework for Sustainable Development. *Marine Policy*, 43:262-272.

Malone, T.C., DiGiacomo, P.M., Gonçalves, E.J., Knap, A.H., Talaue-McManus, L., de Mora, S., and Muelbert. J. (2014b). Enhancing the Global Ocean Observing System to meet evidence based needs for ecosystem-based management of coastal ecosystem services. *Natural Resources Forum*, 38: 168–181.

Malone, T.C., Hopkins, T.S., Falkowski, P.G., and Whitledge, T.E. (1983). Production and transport of phytoplankton biomass over the continental shelf of the New York Bight. *Continental Shelf Research*, 1: 305-337.

Malone, T.C., Malej, A., Harding, L.W., Smodlaka, N., eds. (1999). *Ecosystems at the Land-Sea Margin: Drainage Basin to the Coastal Sea*. AGU, *Coastal and Estuarine Studies*, No. 55: 377 p.

Mann, K.H. (1972). Ecological energetics of the sea-weed zone in a marine bay on the Atlantic coast of Canada II. Productivity of the seaweeds. *Marine Biology*, 14, 199-209.

Marañón, E. (2009). Phytoplankton Size Structure. In *Encyclopedia of Ocean Science*, 2nd Edition, Steele, J.H., Turekian, K.K., and Thorpe, S.A., eds. Academic Press.

Marañón, E., Behrenfeld, M.J., González, N., Mouriño, B., and Zubkov, M.V. (2003). High variability of primary production in oligotrophic waters of the Atlantic Ocean: uncoupling from phytoplankton biomass and size structure. *Marine Ecology Progress Series*, 257: 1-11

Margalef, R. (1978). Life-forms of phytoplankton as survival alternatives in an unstable environment. *Oceanologica Acta*, 1: 493-509.

Marra, J., and Ducklow, H.W. (1995). Primary Production in the North Atlantic: Measurements, Scaling, and Optical Determinants. *Philosophical Transactions: Biological Sciences*, *Royal Society London B*, 348: 153-160.

Marra, J. (2002). Approaches to the Measurement of Plankton Production. pp. 78-108, In Williams, P.J.L., Thomas, D.N., and Reynolds, C.S., eds. *Phytoplankton Productivity: Carbon Assimilation in Marine and Freshwater Ecosystems*, Blackwell Science, Oxford.

Martin, J.H., Coale, K.H., Johnson, K.S., Fitzwater, S.E., et al. (1994). Testing the iron hypothesis in ecosystems of the equatorial Pacific Ocean. *Nature*, 371: 123-129.

Martínez, M.L., Intralawan, A., Vázquez, G., Pérez-Maqueo, O., Sutton, P., and Landgrave, R. (2007). The coasts of our world: Ecological, economic and social importance. *Ecological Economics* 63: 254-272.

Martiny, A.C., Pham, C.T.A., Primeau, F.W., Vrugt, J.A., Moore, J.K., Levin, S.A., Lomas, M.W. (2013), Strong latitudinal patterns in the elemental ratios of marine plankton and organic matter. *Nature geosciences*, 6: 279-283, doi: 10.1038/ngeo1757.

Mathis, J.T., and Feely, R.A. (2013). Building an integrated coastal ocean acidification monitoring network in the U.S. *Elementa: Science of the Anthropocene*, 1: 000007 doi: 10.12952/journal.elementa.000007.

McCann, K.S. (2000). The diversity–stability debate. *Nature*, 405: 228-233.

McClain, C.R. (2009). A Decade of Satellite Ocean Color Observations. *Annual Reviews Marine Science*, 1: 19-42.

McGregor, H.V., Dima, M., Fischer, H.W., and Mulitza, S. (2007). Rapid 20th-century increase in coastal upwelling off northwest Africa. *Science*, 315: 637-639.

McLeod, K., and Leslie, H., eds. (2009). *Ecosystem-Based Management for the*

Oceans. Washington, DC: Island Press. 392 pp.

McLeod, E., Chmura, G.L., Bouillon, S., Salm, R., Björk, M., Duarte, C.M., Lovelock, C.E., Schlesinger, W.H., and Silliman, B.R. (2011). A blueprint for blue carbon: toward an improved understanding of the role of vegetated coastal habitats in sequestering CO2. *Frontiers in Ecology and the Environment* 9: 552-560.

McQuatters-Gollop, A., Reid, P.C., Edwards, M., Burkill, P.H., Castellani, C., Batten, S., Gieskes, W., et al. (2011). Is there a decline in marine phytoplankton? *Nature*, 472: E6-E7.

Meier, M.F., Dyurgerov, M.B., Rick, U.K., O'Neel, S., Pfeffer, W.T., Anderson, R.S., Anderson, S.P., and Glazovsky, A.F. (2007). Glaciers Dominate Eustatic Sea-Level Rise in the 21st Century. *Science*, 317 (5841): 1064-1067.

Meybeck, M. (1982). Carbon, Nitrogen, and Phosphorus Transport by World Rivers. *American Journal of Science*, 282: 401-450.

Millennium Ecosystem Assessment. (2005). *Ecosystems and Human Well-being: Synthesis*. Island Press.

Miller, R.J., Bennett, S., Keller, A.A., Pease, S., and Lenihan, H.S. (2012). TiO2 Nanoparticles are phototoxic to marine phytoplankton. *PLoS ONE* 7 (1): e30321. doi:10.1371/journal.pone.0030321.

Mills, M.M., and Arrigo, K.R. (2010). Magnitude of oceanic nitrogen fixation influenced by the nutrient uptake ratio of phytoplankton. *Nature Geoscience*, 3: 412-416.

Mishra, D.R., Cho, H.J., Ghosh, S., Fox, A., Downs, C., Merani, P.B.T., Kirui, P., et al. (2012). Post-spill state of the marsh: Remote estimation of the ecological impact of the Gulf of Mexico oil spill on Louisiana Salt Marshes. *Remote Sensing of Environment*, 118: 176-185.

Mitsch, W.J., and Gosselink, J.G. (2008). *Wetlands*. Van Nostrand Reinhold, New York, New York, USA.

Moloney, C.L., St John, M.A., Denman, K.L., Karl, D.M., Köster, F.W., Sundby, S., and Wilson, R.P. (2011). Weaving marine food webs from end to end under global change. *Journal of Marine Systems*, 84: 106-116.

Montoya, J.M., and Raffaelli, D. (2010). Climate change, biotic interactions and ecosystem services. *Philosophical Transactions of the Royal Society B*, 365, doi: 10.1098/rstb.2010.0114.

Moore, C.M., Mills, M.M., Arrigo, K.R., Berman-Frank, I., Bopp, L., Boyd, P.W., Galbraith, E.D., Geider, R.J., Guieu, C., Jaccard, S.L., Jickells, T.D., La Roche, J., Lenton, T.M., Mahowald, N.M., Marañón, E., Marinov, I., Moore, J.K., Nakatsuka, T., Oschlies, A., Saito, M.A., Thingstad, T.F., Tsuda A., Ulloa O. (2013). Processes and patterns of oceanic nutrient limitation. *Nature Geoscience*, 6: 701-710, doi: 10.1038/ngeo/1765.

Moore, J.K., Doney, S.C., Lindsay, K., Mahowald, N., Michaels A.F. (2006). Nitrogen fixation amplifies the ocean biogeochemical response to decadal timescale variations in mineral dust deposition. *Tellus, Series B*. 58: 560-572, doi:10.1111/j.1600-0889.2006.00209.x.

Moore, S.K., Trainer, V.L., Mantua, N.J., Parker, M.S., Laws, E.A., Backer, L.C., and Fleming, L.E. (2008). Impacts of climate variability and future climate change on harmful algal blooms and human health. *Environmental Health*, 7(Suppl 2): S4.

Morán, X.A.G., López-Urrutia, A., Calvo-Díaz, A., and Li, W.K. (2010). Increasing importance of small phytoplankton in a warmer ocean. *Global Change Biology*, 16: 1137–1144.

Morel, A., and Berthon, J-F. (1989). Surface pigments, algal biomass profiles, and potential production of the euphotic layer: Relationships reinvestigated in view of remote-sensing applications. *Limnology and Oceanography*, 34: 1545-1562.

Morris, J. T. (2007). Ecological engineering in intertidal saltmarshes. *Hydrobiologia*, 577: 161-168.

Moy, A.D., Howard, W.R., Bray, S.G., and Trull, T.W. (2009). Reduced calcification in modern Southern Ocean planktonic foraminifera. *Nature Geoscience*, 2: 276-280.

Muller-Karger, F.E., Varela, R., Thunell, R.C., Luerssen, R., Hu, C., Walsh, J.J. (2005). The importance of continental margins in the global carbon cycle. *Geophysical Research Letters*, 32, L01602, doi:10.1029/2004GL021346.

Mustapha, S.B., Bélanger, S., and Larouche, P. (2012). Evaluation of ocean color algorithms in the southeastern Beaufort Sea, Canadian Arctic: New parameterization using SeaWiFS, MODIS, and MERIS spectral bands. *Canadian Journal of Remote Sensing*, 38(05): 535-556.

Naqvi, W. (2012). Marine nitrogen cycle. *The Encyclopedia of Earth*. (http://www.eoearth.org/view/article/154479).

Naqvi, S.W.A., Voss, M., and Montoya, J.P. (2008). Recent advances in the biogeochemistry of nitrogen in the ocean. *Biogeosciences*, 5: 1033–1041.

National Research Council. (2004). *Groundwater Fluxes across Interfaces*. Washington, D.C.: National Academies Press.

Negreanu-Pîrjol, B., Negreanu-Pîrjol, T., Paraschiv, G., Bratu, M., Sîrbu, R., Roncea, F., and Meghea, A. (2011). Physical-chemical characterization of some green and red macrophyte algae from the Romanian Black Sea littoral. *St. Cerc. St. CICBIA*, 12 (2): 173-184.

Nellemann, C., Corcoran, E., Duarte, C.M., Valdés, L., De Young, C., Fonseca, L., Grimsditch, G., eds. (2009). *Blue Carbon. A Rapid Response Assessment.* United Nations Environment Programme, GRID-Arendal, www.grida.no.

Nicholls, R.J., and Cazenave, A. (2010). Sea-Level Rise and its Impact on Coastal Zones. *Science*, 328 (5985): 1517–1520.

Nixon, S.W. (1992). Quantifying the relationship between nitrogen input and the productivity of marine ecosystems. In Takahashi, M., Nakata, K., and Parsons, T.R., eds. *Proceedings of Advanced Marine Technology Conference (AMTEC)*, Vol. 5: 57-83.

Ohashi, R., Yamaguchi, A., Matsuno, K., Saito, R., Yamada, N., Iijima, A., Shiga, N., and Imai, I. (2013). Interannual changes in the zooplankton community structure on the southeastern Bering Sea shelf during summers of 1994–2009. *Deep Sea Research Part II: Topical Studies in Oceanography*, 94: 44-56.

Okin, G.S., Baker, A.R., Tegen, I., Mahowald, N.M., Dentener, F.J., Duce, R.A., et al. (2011). Impacts of atmospheric nutrient deposition on marine productivity: Roles of nitrogen, phosphorus, and iron. *Global Biogeochemical Cycles*, 25, GB2022, doi:10.1029/2010GB003858.

Onabid, M.A. (2011). Improved ocean chlorophyll estimate from remote sensed data: The modified blending technique. *African Journal of Environmental Science and Technology*, 5(9): 732-747.

O'Reilly, J.E., Busch, D.A. (1983). Phytoplankton primary production for the Northwestern Atlantic shelf. *Rapports et procès-verbaux des reunions, Conseil permanent international pour l'exploration de la mer*, 183: 255-268.

O'Reilly, J.E., Maritorena, S., Mitchell, B.G., Siegel, D.A., Carder, K.L., Garver, S.A., Kahru, M., and McClain, C. (1998). Ocean color chlorophyll algorithms for SeaWiFS. *Journal of Geophysical Research*, 103 (C11): 24937-24953.

Orhan, I., Sener, B., Atici, T., Brun, R., Perozzo, R., and Tasdemir, D. (2006). Turkish freshwater and marine macrophyte extracts show in vitro antiprotozoal activity and inhibit Fabl, a key enzyme of Plasmodium falciparum fatty acid biosynthesis. *Phytomedicine*, 13(6): 388-393.

Orr, J. C., V. J. Fabry, O. Aumont, L. Bopp, S. C. Doney, R. A. Feely, A. Gnanadesikan,

N. Gruber, A. Ishida, F. Joos, R. M. Key, K. Lindsay, E. Maier-Reimer, and others. (2005). Anthropogenic ocean acidification over the twenty-first century and its impact on calcifying organisms. *Nature*, 437: 681-686.

Orth, R.J., Carruthers, T.J.B., Dennison, W.C., Duarte, C.M., Fourqurean, J.W., Heck, Jr., K.L., Hughes, A.R., Kendrick, G.A., Kenworthy, W.J., Olyarnik, S., Short, F.T., Waycott, M., and Williams, S.L. (2006). A global crisis for seagrass ecosystems. *BioScience* 56:987-996.

Oschlies, A., Schulz, K.G., Riebesell, U., and Schmittner, A. (2008). Simulated 21st century's increase in oceanic suboxia by CO2-enhanced biotic carbon export. *Global Biogeochemical Cycles*, 22, GB4008, doi:10.1029/2007GB003147.

Passow, U., and Carlson, C.A. (2012). The biological pump in a high CO2 world. *Marine Ecology Progress Series*, 470: 249-271.

Paulmier, A., and Ruiz-Pino, D. (2009). Oxygen minimum zones (OMZs) in the modern ocean. *Progress in Oceanography*, 80: 113-28.

Pauly, D., and Christensen, V. (1995). Primary production required to sustain global fisheries. *Nature*, (374): 255-257.

Paytan, A., and McLaughlin, K. (2007). The oceanic phosphorus cycle. *Chemical Reviews*, 107, 563–576.

Peierls, B.L., Caraco, N.F., Pace, M.L., and Cole, J.J. (1991). Human influence on river Nitrogen. *Nature,* 350: 386–87.

Pennington, J.T., Mahoney, K.L., Kuwahara, V.S., Kolber, D.D., Calienes, R., and Chavez, F.P. (2006). Primary production in the eastern tropical Pacific: A review. *Progress in Oceanography*, 69: 285–317.

Pennock, J.R., and Sharp, J.H. (1986). Phytoplankton production in the Delaware Estuary: temporal and spatial variability. *Marine Ecology Progress Series*, 34: 143-155.

Pereira H., Barreira, L., Figueiredo, F., Custódio, L., Vizetto-Duarte, C., Polo, C., Rešek, E., Engelen, A., and Varela, J. (2012). Polyunsaturated Fatty Acids of Marine Macroalgae: Potential for Nutritional and Pharmaceutical Applications. *Marine Drugs*, 10(9): 1920-1935.

Pernetta, J.C. (1993). *Mangrove forests, climate change and sea level rise: hydrological influences on community structure and survival, with examples from the Indo-West Pacific.* Marine Conservation Development Report, International Union for Conservation of Nature, Gland, Switzerland.

Perrings, C., Naeem, S., Ahrestani, F., Bunker, D.E., Burkill, P., et al. (2010). Ecosystem services for 2020. *Science*, 330: 323-324.

Perry, M.J. (1986). Assessing marine primary production from space. *BioScience*, 36 (7): 461-467.

PICES. 2004. Marine Ecosystems of the North Pacific. PICES Special Publication 1, 280 p.

Pikitch, E.K., Santora, C., Babcock, E.A., Bakun, A., Bonfil, R., et al. (2004). Ecosystem-based fishery management. *Science*, 305: 346–347.

Pilskaln, C.H., Paduan, J.B., Chavez, F.P., Anderson, R.Y., and Berelson, W.M. (1996). Carbon export and regeneration in the coastal upwelling system of Monterey Bay, central California. *Journal of Marine Research*, 54: 1149-1178.

Pingree, R.D., Pugh, P.R., Holligan, P.M., and Forster, G.R. (1975). Summer phytoplankton blooms and red tides along tidal fronts in the approaches to the English Channel. *Nature*, 258: 672-677.

Pitt, K.A., Welsh, D.T., Condon, R.H. (2009). Influence of jellyfish blooms on carbon, nitrogen and phosphorus cycling and plankton production. *Hydrobiologia*, 616 (1): 133-149.

Platt, T., and Sathyendranath, S. (1993). Estimators of primary production for the interpretation of remotely sensed data on ocean color. *Journal of Geophysical Research*, 98: 14561-14576.

Platt, T., Fuentes-Yaco, C., and Frank, K. (2003). Marine ecology: Spring algal bloom and larval fish survival. *Nature*, 423: 398-399.

Platt, T., Hoepffner, N., Stuart, V., and Brown, C., eds. (2008). Why ocean colour? The Societal benefits of ocean-colour technology. *IOCCG Report*, No. 7, 141 pp.

Platt, T., White, III, G.N., Zhai, L., Sathyendranath, S., and Roy, S. (2009). The phenology of phytoplankton blooms: Ecosystem indicators from remote sensing. *Ecological Modelling*, 220 (21): 3057-3069.

Polovina, J.J., Howell, E.A., and Abecassis, M. (2008). The ocean's least productive waters are expanding. *Geophysical Research Letters*, 35, L03618, http://dx.doi.org/10.1029/2007GL031745

Polovina, J.J., Dunne, J.P., Woodworth, P.A., Howell, E.A. (2011). Projected expansion of the subtropical biome and contraction of the temperate and equatorial upwelling biomes in the North Pacific under global warming. *ICES Journal of Marine Science*, 68 (6), 986–995.

Polovina, J.J., and Woodworth, P.A. (2012). Declines in phytoplankton cell size in the subtropical oceans estimated from satellite remotely-sensed temperature and chlorophyll, 1998–2007. *Deep Sea Research Part II: Topical Studies in Oceanography*, 77–80, 82-88.

Pomeroy, L.R., Williams, P.J.leB., Azam, F., and Hobbie, J.E. (2007). The microbial loop. *Oceanography*, 20 (2): 28-33.

Powell, T.M., (1989). Physical and biological scales of variability in lakes, estuaries, and the coastal ocean, p. 157-180. In Roughgarden, J., May, R.M., and Levin, S.A., eds. *Perspectives in Ecological Theory*, Princeton University Press, Princeton, N.J.

Plus, M., Deslous-Paoli, J-M., Auby, I., and Dagault, F. (2001). Factors influencing primary production of seagrass beds (*Zostera noltii* Hornem.) in the Thau Lagoon (French Mediterranean coast). *Journal of Experimental Marine Biology and Ecology*, 259:63–84.

Primeau, F.W., and Holzer, M. (2006). The Ocean's Memory of the Atmosphere: Residence-Time and Ventilation-Rate Distributions of Water Masses. *Journal of Physical Oceanography*, 36: 1439-1456.

Purcell, J.E., (2012). Jellyfish and ctenophore blooms coincide with human proliferations and environmental perturbations. *Annual Reviews, Marine Science*, 4: 209-235.

Rabalais, N.N., Turner, R.E., Díaz, R.J., and Justíc, D. (2009). Global change and eutrophication of coastal waters. *ICES Journal of Marine Science*, 66: 1528–1537.

Rahmstorf, S., Foster, G., and Cazenave, A. (2012). Comparing climate projections to observations up to 2011. *Environmental Research Letters*, 7: 1-5.

Raven, J.A. (1998). Small is beautiful: The picophytoplankton. *Functional Ecology*, 12: 503–513.

Redfield, A.C., Ketchum, B.H., and Richards, F.A. (1963). *The influence of organisms on the composition of sea-water*. In The Sea, vol. 2, M.N. Hill (ed.), pp. 26-77, Interscience, New York.

Richardson, A.J. (2008). In hot water: zooplankton and climate change. *ICES Journal of Marine Science*, 65, 279-295.

Richardson, A.J., Bakun, A., Hays, G.C., Gibbons, M.J. (2009). The jellyfish joyride: causes, consequences and management responses to a more gelatinous future. *Trends in Ecology and Evolution* 24(6), 312-322.

Richardson, A.J., and Shoeman, D.S. (2004). Climate impact on plankton ecosystems in the Northeast Atlantic. *Science*, 305: 1609-1612.

Rockstram, J., Steffen, W., Noone, K., Persson, A., Chapin, F.S. III., Lambin, E., Lenton, T.M., Scheffer, M., Folke, C., Schellnhuber, H.J., Nykvist, B., de Wit, B.A., Hughes, T., van der Leeuw, S., Rodhe, H., Sorlin, S., Snyder, P.K., Coastanza, R., Svedin, U., Falkenmark, M., Karlberg, L., Corell, R.W., Fabry, V.J., Hansen, J., Walker, B., Liverman, D., Richardson, K., Crutzen, P., Foley, J. (2009). Planetary Boundaries: Exploring the Safe Operating Space for Humanity. *Ecology and Society*, 14: 32. [online] URL: http://www.ecologyandsociety.org/vol14/iss2/art32/.

Rockström, J., Steffen, W. Noone, K., Persson, Å., Chapin, F.S. III, Lambin, E.F., Lenton, T.M., Scheffer, M., Folke, C., Schellnhuber, H.J., Nykvist, B., de Wit, C.A., Hughes, T., van der Leeuw, S., Rodhe, H., Sörlin, S., Snyder, P.K., Costanza, R., Svedin, U., Falkenmark, M., Karlberg, L., Corell, R.W., Fabry, V.J., Hansen, J., Walker, B.H., Liverman, D., Richardson, K., Crutzen, P., and Foley, J.A. (2009). A safe operating space for humanity. *Nature* 461:472-475.

Roelfsema, C.M., Phinn, S.R., Udy, N., and Maxwell, P. (2009). An Integrated Field and Remote Sensing Approach for Mapping Seagrass Cover, Moreton Bay, Australia. *Spatial Science*, 54 (1): 45-62.

Ross, M.S., Ruiz, P.L., Telesnicki, G.J., and Meeder, J.F. (2001). Estimating above-ground biomass and production in mangrove communities of Biscayne National Park, Florida (U.S.A.). *Wetlands Ecology and Management*, 9: 27–37.

Rost, B., Zondervan, I., and Wolf-Gladrow, D. (2008). Sensitivity of phytoplankton to future changes in ocean carbonate chemistry: current knowledge, contradictions and research directions. *Marine Ecology Progress Series*, 373: 227–237.

Rousseaux, C.S., and Gregg, W.W. (2014). Interannual variations in phytoplankton primary production at a global scale. *Remote Sensing*, 6: 1-19.

Rudnick, D.L., Davis, R.E., Eriksen, C.C., Fratantoni, D.M., and Perry, M.J. (2004). Underwater gliders for ocean research. *Marine Technology Society Journal*, 38(1): 48-59.

Ruttenberg, K.C. (2004). The global phosphorus cycle. In: Holland, H.D. and Turekian, K.K., eds. *Treatise on Geochemistry*. Elsevier, New York, NY, pp. 585–643.

Rykaczewski, R.R., and Dunne, J.P. (2011). A measured look at ocean chlorophyll trends. *Nature*, 472: E5-E6.

Saba, V.S., Friedrichs, M.A.M., Carr, M-E., Antoine, D., Armstrong, R.A., et al. (2010). Challenges of modeling depth-integrated marine primary productivity over multiple decades: A case study at BATS and HOT. *Global Biogeochemical Cycles*, 24, GB3020, doi:10.1029/2009GB003655.

Saba, V.S., Friedrichs, M.A.M., Antoine, D., Armstrong, R.A., Asanuma, I., Behrenfeld, M.J., Ciotti, A.M., Dowell, M., Hoepffner, N., Hyde, K.J.W., Ishizaka, J., Kameda, T., Marra, J., Mélin, F., Morel, A., O'Reilly, J., Scardi, M., Smith W.O., Jr., Smyth, T.J., Tang, S., Uitz, J., Waters, K., and Westberry, T.K. (2011). An evaluation of ocean color model estimates of marine primary productivity in coastal and pelagic regions across the globe. *Biogeosciences*, 8, 489–503.

Sadeghi, A., Dinter, T., Vountas, M., Taylor, B., Altenburg-Soppa, M., and Bracher, A. (2012). Remote sensing of coccolithophore blooms in selected oceanic regions using the PhytoDOAS method applied to hyper-spectral satellite data. *Biogeosciences*, 9: 2127–2143.

Sakamoto, C.M., et al. (2004). Influence of Rossby waves on nutrient dynamics and the plankton community structure in the North Pacific subtropical gyre. *Journal of Geophysical Research*, 109, doi:10.1029/2003JC001976.

Sakshaug, E. (2004). Primary and secondary production in the Arctic Seas. In *The Organic Carbon Cycle in the Arctic Ocean*, Stein, R., and MacDonald, R.W., eds. pp. 57-81, Springer, Berlin.

Santos, A.L., Oliveira, V., Baptista, I., Henriques, I., Gomes, N.C.M., Almeida, A., Correia, A., and Cunha, A. (2012a). Effects of UV-B radiation on the structureal and physiological diversity of bacterioneuston and bacterioplankton. *Applied and Environmental Microbiology*, 78: 2066-2069.

Sarmento, H., Montoya, J.M., Vázquez-Domínguez, E., Vaqué, D., and Gasol, J.M. (2010). Warming effects on marine microbial food web processes: how far can we go when it comes to predictions? *Philosophical Transactions of the Royal Society B*, 365: 2137-2149.

Sathyendranath, S., ed. (2000). Remote sensing of ocean colour in coastal and other optically complex waters. *Reports of the International Ocean-Colour Coordinating Group*, No. 3, 140 p., Dartmouth, Canada.

Sathyendranath, S., Ahanhanzo, J., Bernard, S., Byfield, V., Delaney, L., and others. (2010). ChloroGIN: Use of satellite and in situ data in support of ecosystem-based management of marine resources in *Proceedings of OceanObs'09: Sustained Ocean Observations and Information for Society* (Vol. 2), Venice, Italy, 21-25 September 2009, Hall, J., Harrison, D.E., & Stammer, D., eds. ESA Publication WPP-306.

Scavia, D., and Bricker, S.B. (2006). Coastal eutrophication assessment in the United States. *Biogeochemistry*, 79: 187–208.

Schlitzer, R., Usbeck, R., and Fischer, G. (2003). Inverse modeling of particulate organic carbon fluxes in the South Atlantic. In: *The South Atlantic in the Late Quaternary: Reconstruction of material budget and current systems.* Wefer, G., Mulitza, S., and Ratmeyer, V., eds. Springer, Berlin, Heidelberg, New York, pp. 1-19.

Schlitzer, R., Monfray, P., Hoepffner, N., Fischer, G., Gruber, N., Lampitt, R., Lévy, M., Schmidt, A.L., Wysmyk, J.K.C., Craig, S.E., and Lotze, H.K. (2012). Regional-scale effects of eutrophication on ecosystem structure and services of seagrass beds. *Limnology and Oceanography*. 57(5): 1389–1402.

Schmidt, A.L., Wysmyk, J.K.C., Craig, S.E., and Lotze, H.K. (2012). Regional-scale effects of eutrophication on ecosystem structure and services of seagrass beds. *Limnology and Oceanography*. 57(5): 1389–1402.

Schneider, B., Bopp, L., Gehlen, M., Segschneider, J., Frolicher, T.L., Cadule, P., Friedlingstein, P., Doney, S.C., Behrenfeld, M.J., and Joos, F. (2008). Climate-induced interannual variability of marine primary and export production in three global coupled climate carbon cycle models. *Biogeosciences*, 5: 597–614.

Schulz, M., Prospero, J.M., Baker, A.R., Dentener, F., et al. (2012). Atmospheric transport and deposition of mineral dust to the ocean: implications for research needs. *Environmental Science & Technology*, 46: 10,390-10,404.

Seitzinger, S.P., Harrison, J.A., Dumont, E., Beusen, A.H.W., Bouwman A.F. (2005). Sources and delivery of carbon, nitrogen, and phosphorus to the coastal zone: An overview of the Global Nutrient Export from Watersheds (NEWS) models and their application, *Global Biogeochemical Cycles*, 19, GB4S01, doi:10.1029/2005GB002606.

Seitzinger, S.P., Mayorga, E., Bouwman, A.F., et al. (2010). Global nutrient export: A scenario analysis of past and future trends. *Global Biogeochemical Cycles*, 23, GB0A08, doi: 10.1029/2009GB003587.

Sharma, K.K., Schuhmann, H., and Schenk, P.M. (2012). High lipid induction in microalgae for biodiesel production. *Energies*, 5: 1532-1553.

Sharpies, J., Middelburg, J.J., Fennel, K., Jickells, T.D. (2013), Riverine delivery of nutrients and carbon to the oceans, *Nature*, 504:61-70. doi:10.1038/nature12857.

Sherman, K., and Hempel, G. (2009). *The UNEP Large Marine Ecosystem Report: A perspective on changing conditions in LMEs of the world's Regional Seas*. UNEP Regional Seas Reports and Studies, 182. United Nations Environmental Programme, Nairobi, 872 pp.

Sherman, K., Alexander, L.M., and Gold, B.D., eds. (1993). *Large Marine Ecosystems: Stress, Mitigation, and Sustainability*, AAAS Press, Washington, D.C., 376 pp.

Sherman, K., Smith, W., Morse, W., Berman, M., Green, J., and Ejsymont, L. (1984). Spawning strategies of fishes in relation to circulation, phytoplankton production, and pulses in zooplankton off the northeastern United States. *Marine Ecol-*

ogy Progress Series, 18: 1-19.

Shi, D., Xu, Y., and Morel, F.M.M. (2009). Effects of the pH/pCO2 control method on medium chemistry and phytoplankton growth. *Biogeosciences*, 6: 1199-1207.

Shi, D., Xu, Y., Hopkinson, B.M., Morel, F.M.M. (2010). Effect of ocean acidification on iron availability to marine phytoplankton. *Science*, 327(5966): 676-679.

Shi, D., Kranz, S.A., Kim, J-M., and Morel, F.M.M. (2012). Ocean acidification slows nitrogen fixation and growth in the dominant diazotroph Trichodesmium under low-iron conditions. *Proceedings of the National Academy of Sciences*, 109 (45): E3094-100.

Short, F.T., and Neckles, H.H. (1999). The effects of global climate change on seagrasses. *Aquatic Botany*, 63: 169-196.

Shurin, J.G., Gruner, D.S., and Hillebrand, H. (2006). All wet or dried up? Real differences between aquatic and terrestrial food webs. *Proceedings of the Royal Society B*, 273: 1-9.

Siegel, D.A., Behrenfeld, M.J., Maritorena, S., McClain, C.R., Antoine, D., and others. (2013). Regional to global assessments of phytoplankton dynamics from the Sea-WiFS mission. *Remote Sensing of Environment* 135: 77–91.

Siegel, D.A., Doney, S.C., and Yoder, J.A. (2002). The North Atlantic spring phytoplankton bloom and Sverdrup's critical depth hypothesis. *Science*, 296: 730–733.

Silliman, B.R., Grosholz, T., and Bertness, M.D. (2009). Salt marshes under global siege. Pages 103–114 in Silliman, B.R., Grosholz, T., and Bertness, M.D., eds. *Human impacts on salt marshes: a global perspective*. University of California Press, Berkeley, California, USA.

Silva, J., Sharon, Y., Santos, R., and Beer, S. (2009). Measuring seagrass photosynthesis: methods and applications. *Aquatic Biology*, 7: 127–141.

Slomp, C.P. (2011). Phosphorus cycling in the estuarine and coastal zones: sources, sinks, and transformations. In: Wolanski, E., and McLusky, D.S., eds. *Treatise on Estuarine and Coastal Science*, Vol 5, pp. 201–229. Waltham: Academic Press.

Smetacek, V. (1999). Diatoms and the ocean carbon cycle. *Protist*, 150 (1): 25-32.

Smith, R.C., Baker, K.S., Dierssen, H.M., Stammerjohn, S.E., and Vernet, M. (2001). Variability of primary production in an Antarctic marine ecosystem as estimated using a multi-scale sampling strategy. *American Zoology*, 41: 40–56.

Smith, S.V. (1981). Marine macrophytes as a global carbon sink. *Science*, 211: 838-840.

Smith, Jr., W.O., and Demaster, D.J. (1996). Phytoplankton biomass and productivity in the Amazon River plume: correlation with seasonal river discharge. *Continental Shelf Research*, 16 (3): 291-319.

Sohm, J.A., Webb, E.A., and Capone, D.G. (2011). Emerging patterns of marine nitrogen fixation. *Nature Reviews: Microbiology*, 9: 499-508.

Son, N-T., and Chen, C-F. (2013). Remote sensing of mangrove forests in Central America. *SPIE Newsroom*, doi: 10.1117/2.1201304.004771 (http://spie.org/x93599.xml)

Stal, L.J. (2009). Is the distribution of nitrogen-fixing cyanobacteria in the oceans related to temperature? *Environmental Microbiology*, 11 (7): 1632-45.

Steele, J.H. (1985). A comparison of terrestrial and marine ecological systems. *Nature*, 313, 355-358.

Steeman-Nielsen, E. (1963). *Productivity, definition and measurement*. In Hill, M.W. ed. The Sea, vol. 1, pp 129-164, New York, John Wiley.

Steinacher, M., Joos, F., Frölicher, T.L., Bopp, L., Cadule, P., Cocco, V., Doney, S.C., Gehlen, M., Lindsay, K., Moore, J.K., Schneider, B., and Segschneider, J. (2010). Projected 21st century decrease in marine productivity: a multi-model analysis, *Biogeosciences*, 7: 979-1005.

Steinberg, D.K., Carlson, C.A., Bates, N.R., Johnson, R.J., Michaels, A.F., and Knap, A.H. (2001). Overview of the US JGOFS Bermuda Atlantic Time-series Study (BATS): a decade-scale look at ocean biology and biogeochemistry. *Deep-Sea Research II*, 48: 1405-1447.

Steinberg, D.K., Lomas, M.W., Cope, J.S., (2012). Long-term increase in mesozooplankton biomass in the Sargasso Sea: Linkage to climate and implications for food web dynamics and biogeochemical cycling. *Global Biogeochemical Cycles* 26 (1), GB1004.

Stier, P., Feichter, J., Kinne, S., Kloster, S., Vignati, E., Wilson, J., Ganzeveld, L., Tegen, I., Werner, M., Balkanski, Y., Schulz, M., Boucher, O., Minikin, A., and Petzold, A. (2005). The aerosol-climate model ECHAM5-HAM. *Atmospheric Chemistry and Physics*, 5: 1125–1156.

Strous, M., Pelletier, E., Mangenot, S., et al. (2006). Deciphering the evolution and metabolism of an anammox bacterium from a community genome. *Nature*, 440, 790–794.

Subramaniam, A., Yager, P.L., Carpenter, E.J., Mahaffey, C., Björkman, K., Cooley, S., Kustka, A.B., Montoya, J.P., Sañudo-Wilhelmy, S.A., Shipe, R., and Capone, D.G. (2008). Amazon River enhances diazotrophy and carbon sequestration in the tropical North Atlantic Ocean. *PNAS*, 105 (3): 10460-10465.

Suntharalingam, P., Buitenhuis, E., Le Quéré, C., Dentener, F., Nevison, C., Butler, J.H., Bange, H.W., and Forster, G. (2012). Quantifying the impact of anthropogenic nitrogen deposition on oceanic nitrous oxide, *Geophysical Research Letters*, 39: L07605, doi:10.1029/2011GL050778.

Suikkanen, S., Pulina, S., Engström-Öst, J., Lehtiniemi, M., Lehtinen, S., et al. (2013) Climate Change and Eutrophication Induced Shifts in Northern Summer Plankton Communities. *PLoS ONE* 8 (6): e66475. doi:10.1371/journal.pone.0066475.

Taalas, P., Kaurola, J., Kylling, A., Shindell, D., Sausen, R., Dameris, M., Grewe, V., Herman, J., Damski, J. and Steil, B. (2000). The impact of greenhouse gases and halogenated species on future solar UV radiation doses. *Geophysical Research Letters*, 27: 1127-1130.

Taucher, J. and Oschlies, A. (2011). Can we predict the direction of marine primary production change under global warming? *Geophysical Research Letters*, 38: L02603.

Taucher, J., Schulz, K.G., Dittmar, T., Sommer, U., Oschlies, A., and Riebesell, U. (2012) Enhanced carbon overconsumption in response to increasing temperatures during a mesocosm experiment. *Biogeosciences*, 9: 3479–3514.

Teira, E., Mouriño, B., Marañón, E., Pérez, V., Pazó, M.J., Serret, P., de Armas, D., Escánez, J., Woodward, E.M.S., and Fernández, E. (2005). Variability of chlorophyll and primary production in the eastern north Atlantic subtropical gyre: potential factors affecting phytoplankton activity. *Deep-Sea Research*, 52: 569–588.

Testor, P., Meyers, G., Pattiaratchi, C., Bachmayer, R., Hayes, D., Pouliquen, S., et al. (2010). *Gliders as a component of future observing systems* in Proceedings of OceanObs'09: Sustained Ocean Observations and Information for Society (Vol. 2), Hall, J., Harrison, D.E., and Stammer, D., eds. ESA Publication WPP-306, doi:10.5270/OceanObs09.

Teuten, E.L., Saquing, J.M., Knappe, D.R.U., Barlaz, M.A., Jonsson, S., Bjorn, A., Rowland, S.J., Thompson, R.C., Galloway, T.S., Yamashita, R., Ochi, D., Watanuki, Y., Moore, C., Viet, P.H., Tana, T.S., Prudente, M., Boonyatumanond, R., Zakaria, M.P., Akkhavong, K., Ogata, Y., Hirai, H., Iwasa, S., Mizukawa, K., Hagino, Y., Imamura, A., Saha, M., Takada, H. (2009). Transport and release of chemicals from plastics to the environment and to wildlife. *Philosophical Transactions of the Royal Society of London B: Biological Science*, 364: 2027-2045.

Thamdrup, B., Dalsgaard, T., Jensen, M.M., et al. (2006). Anaerobic ammonium oxidation in the oxygen-deficient waters off northern Chile. *Limnology and Oceanography*, 51: 2145–2156.

Tian, T., Wei, H., Su, J., and Chung, C. (2005). Simulations of annual cycle of phytoplankton production and the utilization of nitrogen in the Yellow Sea. *Journal of Oceanography*, 61: 343-357.

Titus, J.G., Anderson, K.E., Cahoon, D.R., Gesch, D.B., Gill, S.K, Gutierrez, B.T., Thieler, E.R., and Williams, S.J. (2009). *Coastal sensitivity to sea level rise: A Focus on the Mid-Atlantic Region*. U.S. Climate Change Science Program, Synthesis and Assessment Product 4.1.

Tremblay, J.E., and Gagnon, J. (2009). The effect of irradiance and nutrient supply on the productivity of Arctic waters: A perspective on climate change, pp. 73–94, In *Influence of Climate Change on the Changing Arctic and Sub-Arctic Conditions*, Springer, New York.

Trilla, G.G., Pratolongo, P., Beget, M.E., Kandus, P, Marcovecchio, J., and Di Bella, C. (2013). Relating biophysical parameters of coastal marshes to hyperspectral reflectance data in the Bahia Blanca Estuary, Argentina. *Journal of Coastal Research*, 29 (1): 231-238.

Twilley, R.R., Chen, R.H., and Hargis, T. (1992). Carbon sinks in mangroves and their implications to carbon budget of tropical coastal ecosystems. *Water Air and Soil Pollution*, 64 (1): 265-265.

Ueyama, R., and Monger, B.C. 2005. Wind-induced modulation of seasonal phytoplankton blooms in the North Atlantic derived from satellite observations, *Limnology and Oceanography*, 50: 1820–1829.

Ulloa, O., Canfieldb, D.E., DeLong, E.F., Letelier, R.M., and Stewart, F.J. (2012). Microbial oceanography of anoxic oxygen minimum zones. *Proceedings of the National Academy of Sciences* (USA), 109 (40): 15996-16003.

UNEP-WCMC. (2006). *In the front line: Shoreline protection and other ecosystem services from mangroves and coral reefs*. UNEP-WCMC, Cambridge, UK 33 pp.

UNESCO. (2006). *A Coastal Theme for the IGOS Partnership for the Monitoring of our Environment from Space and from Earth*. Paris, 60 pp.

UNESCO. (2008). *Filling Gaps in LME Nitrogen Loadings Forecast for 64 LMEs*. GEF/LME

UNESCO-IOC. (2012). *Requirements for Global Implementation of the Strategic Plan for Coastal GOOS*. GOOS Report 193. Paris: Intergovernmental Oceanographic Commission. (www.ioc-goos.org/index.php?option=com_oe&task=viewDocumentRecord&docID=7702&lang=en)

Uye, S. (2008). Blooms of the giant jellyfish *Nemopilema nomurai*: A threat to the fisheries sustainability of the East Asian Marginal Seas. *Plankton & Benthos Research*, 3: 125–131.

Valiela, I., Bowen, J.L., and York, J.K. (2001). Mangrove forests: One of the world's threatened major tropical environments. *BioScience*, 51 (10): 807–815.

Valiela, I., Kinney, E., Culbertson, J., Peacock, E., and Smith, S. (2009). Global losses of mangroves and salt marshes: Magnitudes, causes and consequences. Pp. 107-138 in *Global Loss of Coastal Habitats: Rates, Causes, and Consequences*, Duarte, C. ed. Fundación BBVA. Bilbao.

Vancoppenolle, M., Bopp, L., Madec, G., Dunne, J., Ilyina, T., Halloran, P.R., and Steiner, N. (2013), Future Arctic Ocean primary productivity from CMIP5 simulations: Uncertain outcome, but consistent mechanisms, *Global Biogeochemical Cycles*, 27, doi:10.1002/gbc.20055.

Van Dolah, F.M. (2000). Marine algal toxins: origins, health effects, and their increased occurrence. *Environmental Health Perspectives*, 108 (suppl 1): 133-141.

Vantrepotte, V., and Mélin, F. (2009). *Temporal variability of 10-year global SeaWiFS time-series of phytoplankton chlorophyll a concentration*. ICES Journal of Marine Science, 66: 1547–1556.

Vargas, M., Brown, C.W., and Sapiano, M.R.P. (2009). Phenology of marine phytoplankton from satellite ocean color measurements. *Geophysical Research Letters*, 36: L01608, doi: 10.1029/2008GL036006.

Vernet, M., and Smith, R.C. (2007) Measuring and Modeling Primary Production in Marine Pelagic Ecosystems, in *Principles and Standards for measuring Primary Production*, Fahey, J.T., and Knapp, A.K., eds. Oxford University Press, Ch.9, pp.142-174.

Vernet, M., Martinson, D., Iannuzzi, R., Stammerjohn, S., Kozlowski, W., Sines, K., Smith, R.C., and Garibotti, I. (2008). Primary production within the sea-ice zone west of the Antarctic Peninsula: I – Sea ice, summer mixed layer, and irradiance. *Deep Sea Research Part II: Topical Studies in Oceanography*, 55: 2068-2085.

Villar-Argaiz, M., Median-Sánchez, J.M., Bullejos, F.J., Delgado-Molina, J.A., Ruíz Pérez, O., Navarro, J.C., and Carrillo, P. (2009). UV radiation and phosphorus interact to influence the biochemical composition of phytoplankton. *Freshwater Biology*, 54, 1233-1245.

Vonthron-Sénécheau, C., Kaiser, M., Devambez, L., Vastel, A., Mussio, I., Rusig, A.M. (2011). Antiprotozoal activities of organic extracts from French marine seaweeds. *Marine Drugs*, 9 (6):922-33.

Voss, M., Bange, H.W., Dippner, J.W., Middelburg, J.J., Montoya, J.P., and Ward, B.B. (2013). The marine nitrogen cycle: recent discoveries, uncertainties and the potential relevance of climate change. *Philosophical Transactions of the Royal Society B, Biological Sciences*, 368 (1621): 20130121, doi: 10.1098/rstb.2013.0121.

Walker, B., Holling, C.S., Carpenter, S.R., Kinzig, A. (2004). "Resilience, adaptability and transformability in social–ecological systems". *Ecology and Society*, 9 (2): 5.

Walsh, P.J., Smith, S.L., Fleming, L.E., Solo-Gabriele, H.M., and Gerwick, W.H., eds. (2008). *Oceans and Human Health: Risk and Remedies from the Sea*. Elsevier, 644 p.

Wang, M., and Overland, J. (2009). An sea ice free summer Arctic within 30 years. *Geophysical Research* Letters, 36, L07502. http://dx.doi.org/10.1029/2009GL037820.

Ward, B.A., Dutkiewicz, S., Jahn, O., and Follows, M.J. (2012). A size-structured food-web model for the global ocean. *Limnology and Oceanography*, 57(6): 1877–1891.

Ward, B.B. (2012). The Global Nitrogen Cycle. In: Knoll, A.H., Canfield, D.E., and Konhauser, K.O., eds, *Fundamentals of Geomicrobiology*, Wiley-Blackwell, Chichester, UK, pp. 36-48.

Ward, B.B. (2013). How is nitrogen lost? *Science*, 341: 352 – 352.

Ward, B.B., Capone, D.G., and Zehr, J.P. (2007). What's New in the Nitrogen Cycle? *Oceanography*, 20 (2): 101-109.

Ward, B.B., Devol, A.H., Rich, J.J., Chang, B.X., Bulow, S.E., Naik, H., Pratihary, A., and Jayakumar, A. (2009). Denitrification as the dominant nitrogen loss process in the Arabian Sea. *Nature*, 461:78–81.

Ware, D.M. (2000). Aquatic ecosystems: Properties and models. In: Harrison P.J., Parsons T.R., eds. *Fisheries oceanography: An integrative approach to fisheries ecology and management*. Oxford: Balckwell Science. pp. 161–200.

Ware, D.M. and Thomson, R.E. (2005). Bottom-up ecosystem trophic dynamics determine fish production in the northeast Pacific. *Science*, 308: 1280–1284.

Waycott, M., Duarte, C.M., Carruthers, T.J.B., Orth, R.J., Dennison, W.C., Olyarnik, S., Calladine, A., Fourqurean, J.W., Heck, K.L., Hughes, A.R., Kendrick, G.A., Kenworthy, W.J., Short, F.T., and Williams, S.L. (2009). Accelerating loss of seagrasses across the globe threatens coastal ecosystems. *Proceedings of the National Academy of Sciences* USA 106:12377–12381.

Wegner, I.A., Besseling, E., Foekema, E.M., Kamermans, P., and Koelmans, A.A. (2012). Effects of nanopolystyrene on the feeding behavior of the blue mussel (*Mytilus edulis* L.). *Environmental Toxicology and Chemistry*, 31 (11): 2490-2497.

Wernand, M.R., van der Woerd, H.J., Gieskes, W.W.C. (2013). Trends in ocean colour and chlorophyll concentration from 1889 to 2000, worldwide. *PLoS One*, 8(6).

Wernberg, T., Russell, B.D., Thomsen, M.S., Gurgel, C.F.D., Bradshaw, C.J.A., Poloczanska, E.S., Connell, S.D. (2011). Seaweed communities in retreat from ocean warming. *Current Biology*, 21: 1828-1832.

Westberry, T., Behrenfeld, M.J., Siegel, D.A., and Boss, E. (2008). Carbon based primary productivity modeling with vertically resolved photoacclimation. *Global Biogeochemical Cycles*, 22, GB2024, doi:10.1029/2007GB003078.

Wollast, R. (1998). Evaluation and comparison of the global carbon cycle in the coastal zone and the open ocean. In *The Sea*, Brink, K.H. and A.R. Robinson, eds. 10: 213-253.

Worm, B., Barbier, E.B., Beaumont, N., Duffy, J.E., Folke, C., Halpern, B.S., Jackson, J.B.C., Lotze, H.K., Micheli, F., and Palumbi, S.R. (2006). Impacts of biodiversity loss on ocean ecosystem services. *Science*, 314: 787-790.

Wright, S.L., Thompson, R.C., and Galloway, T.S. (2013). The physical impacts of microplastics on marine organisms: A review. *Environmental Pollution*, 178, 483e492.

Yusuf, C. (2007). Biodiesel from microalgae. *Biotechnology Advances*, 25:294-306.

Zehr, J.P., Kudela, R.M. (2011), Nitrogen Cycle of the Open Ocean: From Genes to Ecosystems. *Annual Reviews Marine Science*, 3: 197–225.

Zhang, M., Ustin, S.L., Rejmankova, E., and Sanderson, E.W. (1997). Monitoring Pacific coast salt marshes using remote sensing. *Ecological Applications*, 7 (3): 1039–1053.

Zhai, L., Platt, T., Tang, C., Sathyendranath, S., and Walne, A. (2013). The response of phytoplankton to climate variability associated with the North Atlantic Oscillation. *Oceanography*, 93: 159-168.

7 Calcium Carbonate Production and Contribution to Coastal Sediments

Contributors:
Colin D. Woodroffe (Convenor), John W. Farrell, Frank R. Hall and Peter Harris (Lead Member)

1 Calcium carbonate production in coastal environments

Biological production of calcium carbonate in the oceans is an important process. Although carbonate is produced in the open ocean (pelagic, see Chapter 5), this chapter concentrates on production in coastal waters (neritic) because this contributes sediment to the coast through skeletal breakdown producing sand and gravel deposits on beaches, across continental shelves, and within reefs. Marine organisms with hard body parts precipitate calcium carbonate as the minerals calcite or aragonite. Corals, molluscs, foraminifera, bryozoans, red algae (for example the algal rims that characterize reef crests on Indo-Pacific reefs) are particularly productive, as well as some species of green algae (especially *Halimeda*). Upon death, these calcareous organisms break down by physical, chemical, and biological erosion processes through a series of discrete sediment sizes (Perry et al., 2011). Neritic carbonate production has been estimated to be approximately 2.5 Gt year^{-1} (Milliman and Droxler, 1995; Heap et al., 2009). The greatest contributors are coral reefs that form complex structures covering a total area of more than 250,000 km^2 (Spalding and Grenfell, 1997; Vecsei, 2004), but other organisms, such as oysters, may also form smaller reef structures.

Global climate change will affect carbonate production and breakdown in the ocean, which will have implications for coastal sediment budgets. Rising sea level will displace many beaches landwards (Nicholls et al., 2007). Low-lying reef islands called sand cays, formed over the past few millennia on the rim of atolls, are particularly vulnerable, together with the communities that live on them. Rising sea level can also result in further reef growth and sediment production where there are healthy coral reefs (Buddemeier and Hopley, 1988). In areas where corals have already been killed or damaged by human activities, however, reefs may not be able to keep pace with the rising sea level in which case wave energy will be able to propagate more freely across the reef crest thereby exposing shorelines to higher levels of wave energy (Storlazzi et al., 2011; see also Chapter 43).

Reefs have experienced episodes of coral bleaching and mortality in recent years caused by unusually warm waters. Increased carbon dioxide concentrations are also causing ocean waters to become more acidic, which may affect the biological production and supply of carbonate sand. Bleaching and acidification can reduce coral growth and limit the ability of reef-building corals and other organisms to produce calcium carbonate (Kroeker et al., 2010). In some cases, ocean acidification may lead to a reduced supply of carbonate sand to beaches, increasing the potential for erosion (Hamylton, 2014).

1.1 Global distribution of carbonate beaches

Beaches are accumulations of sediment on the shoreline. Carbonate organisms, particularly shells that lived in the sand, together with dead shells reworked from shallow marine or adjacent rocky shores, can contribute to beach sediments. Dissolution and re-precipitation of carbonate can cement sediments forming beachrock, or shelly deposits called coquina. On many arid coasts and islands lacking river input of sediment to the coast, biological production of carbonate is the dominant source of sand and gravel. Over geological time (thousands of years) this biological source of carbonate sediment may have formed beaches that are composed entirely, or nearly entirely, of calcium carbonate. Where large rivers discharge sediment to the coast, or along coasts covered in deposits of glacial till deposited during the last ice age, beaches are dominated by sediment derived from terrigenous (derived from continental rocks) sources. Carbonate sediments comprise a smaller proportion of these beach sediments (Pilkey et al., 2011).

Sand blown inland from carbonate beaches forms dunes and these may be extensive and can become lithified into substantial deposits of carbonate eolianite (wind-blown) deposits. Significant deposits of eolianite are found in the Mediterranean, Africa, Australia, and some parts of the Caribbean (for example most of the islands of the Bahamas). The occurrence of carbonate eolianites is therefore a useful proxy for mapping the occurrence of carbonate beaches (Brooke, 2001).

Carbonate beaches may be composed of shells produced by tropical to sub-polar species, so their occurrence is not limited by latitude, although carbonate production on polar shelves has received little attention (Frank et al., 2014). For example, Ritchie and Mather (1984) reported that over 50 beaches in Scotland are composed almost entirely of shelly carbonate sand. There is an increase in carbonate content towards the south along the east coast of Florida (Houston and Dean, 2014). Carbonate beaches, comprising 60-80 per cent carbonate on average, extend for over 6000 km along the temperate southern coast of Australia, derived from organisms that lived in adjacent shallow-marine environments (James et al., 1999; Short, 2006). Calcareous biota have also contributed along much of the western coast of Australia; carbonate contents average 50-70 per cent, backed by substantial eolianite cliffs composed of similar sediments along this arid coast (Short, 2010). Similar non-tropical carbonate production occurs off the northern coast of New Zealand (Nelson, 1988) and eastern Brazil (Carannante et al., 1988), as well as around the Mediterranean Sea, Gulf of California, North-West Europe, Canada, Japan and around the northern South China Sea (James and Bone, 2011).

On large carbonate banks, biogenic carbonate is supplemented by precipitation of inorganic carbonate, including pellets and grapestone deposits (Scoffin, 1987). Ball (1967) identified marine sand belts, tidal bars, eolian ridges, and platform interior sand blankets comprising carbonate sand bodies present in Florida and the Bahamas. This is also one of the locations where ooids (oolites) form through the concentric precipitation of carbonate on spherical grains. Inorganic precipitation in the Persian Gulf, including the shallow waters of the Trucial Coast, reflects higher water temperature and salinity (Purser, 1973; Brewer and Dyrssen, 1985).

1.2 Global distribution of atolls

The most significant social and economic impact of a possible reduction in carbonate sand production is the potential decrease in supply of sand to currently inhabited, low-lying sand islands on remote reefs, particularly atolls. Atolls occur in the warm waters of the tropics and subtropics. These low-lying and vulnerable landforms owe their origin to reef-building corals (see Chapter 43 which discusses warm-water corals in contrast to cold-water corals dealt with in Chapter 42). The origin of atolls was explained by Charles Darwin as the result of subsidence (sinking) of a volcanic island. Following an initial eruptive phase, volcanic islands are eroded by waves and by slumping, and gradually subside, as the underlying lithosphere cools and contracts. In tropical waters, fringing coral reefs grow around the volcanic peak. As the volcano subsides the reef grows vertically upwards until eventually the summit of the volcano becomes submerged and only the ring of coral reef (i.e., an atoll) is left behind. The gradual subsidence can be understood in the context of plate tectonics and mantle "hot spots". Many oceanic volcanoes occur in linear chains (such as the Hawaiian Islands and Society Islands) with successive islands being older along the chain and moving into deeper water as the plate cools and contracts (Ramalho et al., 2013).

Most atolls are in the Pacific Ocean (in archipelagoes in the Tuamotu Islands, Caroline Islands, Marshall Islands, and the island groups of Kiribati, Tuvalu and Tokelau) and Indian Ocean (the Maldives, the Laccadive Islands, the Chagos Archipelago and the Outer Islands of the Seychelles). The Atlantic Ocean has fewer atolls than either the Pacific or Indian Oceans, with several in the Caribbean (Vecsei, 2003; 2004). The northernmost atoll in the world is Kure Atoll at 28°24′ N, along with other atolls of the northwestern Hawaiian Islands in the North Pacific Ocean. The southernmost are the atoll-like Elizabeth (29°58′ S) and Middleton (29°29′ S) Reefs in the Tasman Sea, South Pacific Ocean (Woodroffe et al., 2004). The occurrence of seamounts (submarine volcanoes) is two times higher in the Pacific than in the Atlantic or Indian Oceans, explaining the greater frequency of atolls.

Corals, which produce aragonite, are the principal reef-builders that shape and vertically raise the reef deposit, and there are secondary contributions from other aragonitic organisms, particularly molluscs, as well as coralline algae, bryozoans and foraminifera which are predominantly made of calcite. Carbonate sand and gravel is derived from the breakdown of the reef. Bioerosion is an important process in reefs, with bioeroders, such as algae, sponges, polychaete worms, crustaceans, sea urchins, and boring molluscs (e.g., *Lithophaga*) reducing the strength of the framework and producing sediment that infiltrates and accumulates in the porous reef limestone (Perry et al., 2012). Erosion rates by sea urchins have been reported to exceed 20 kg $CaCO_3$ m^{-2} $year^{-1}$ in some reefs, and parrotfish may produce 9 kg $CaCO_3$ m^{-2} $year^{-1}$ (Glynn, 1996). Over time, cementation lithifies the reef. Whereas the reef itself is the main feature produced by these calcifying reef organisms, loose carbonate sediment is also transported from its site of production. Transported sediment can be deposited, building sand cays. Broken coral or larger boulders eroded from the reef by storms form coarser islands (termed *motu* in the Pacific). Sand and gravel can be carried across the reef and deposited together with finer mud filling in the lagoon (Purdy and Gischler, 2005).

Carbonate production on reefs has been measured by at least three different approaches; hydrochemical analysis of changes in alkalinity of water moving across a section of reef, radiometric dating of accretion rates in reef cores, and census-based approaches that quantify relative contributions made by different biota (including destruction by bioeroders). These approaches indicate relatively consistent rates of ~10 kg $CaCO_3$ m^{-2} $year^{-1}$ on flourishing reef fronts, ~4 kg $CaCO_3$ m^{-2} $year^{-1}$ on reef crests, and <1 kg $CaCO_3$ m^{-2} $year^{-1}$ in lagoonal areas (Hopley et al., 2007; Montaggioni and Braithwaite, 2009; Perry et al., 2012; Leon and Woodroffe, 2013). These rates have been described in greater detail in specific studies (Harney and Fletcher, 2003; Hart and Kench, 2007), and have been used to produce regional extrapolations of net production (Vecsei, 2001, 2004).

2 Changes known and foreseen –sea-level rise and ocean acidification.

Several climate change and oceanographic drivers threaten the integrity of fragile carbonate coastal ecosystems. Anticipated sea-level rise will have an impact on the majority of coasts around the world. In addition, carbonate production is likely to be affected by changes in other climate drivers, including warming and acidification. Tropical and subtropical reefs would appear to be some of the worst affected systems. However, it is also apparent that already many degraded systems can be attributed to impacts from social and economic drivers of change; pollution, overfishing and coastal development have deteriorated reef systems and many severely eroded beaches can be attributed to poor coastal management practices.

2.1 Potential impacts of sea-level rise on beaches

Sea-level rise poses threats to many coasts. Between 1950 and 2010, global sea level has risen at an average rate of 1.8 ± 0.3 mm $year^{-1}$; approximately 10 cm of anthropogenic global sea-level rise is therefore inferred since 1950. Over the next century, the mean projected sea-level rise for 2081-2100 is in the range 0.26-0.54 m relative to 1986-2005, for the low-emission scenario (RCP 2.6). The rate of rise is anticipated to increase from ~3.1 mm $year^{-1}$ indicated by satellite altimetry to 7-15 mm $year^{-1}$ by the end of the century (Church et al., 2013). The rate experienced on any particular coast is likely to differ from the global mean trend as a result of local and regional factors, such as rates of vertical land movement or subsidence. Beach systems can be expected to respond to this gradual change in sea level, and the low-lying reef islands on atolls appear to be some of the most vulnerable coastal systems (Nicholls et al., 2007).

Based on predictions from the Bruun Rule, a simple heuristic that uses slope of the foreshore and conservation of mass, sea-level rise will cause erosion and net recession landwards for many beaches (Bruun, 1962). Although this approach has been widely applied, it has been criticized as unrealistic for many reasons, including that it does not adequately incorporate consideration of site-specific sediment budgets (Cooper and Pilkey, 2004). Few analyses consider the contribution of biogenic carbonate and none foreshadow the consequences of any reduction in supply of carbonate sand. This is partly because of time lags between production of carbonate and its incorporation into beach deposits, which is poorly constrained in process studies and which is subject to great variability between different coastal settings, ranging from years to centuries (Anderson et al., 2015). In view of uncertainties in rates of sediment supply and transport, probabilistic modeling of shoreline behavior may be a more effective way of simulating possible responses, including potential accretion where sediment supply is sufficient (Cowell et al., 2006).

2.2 Potential impacts of sea-level rise on reef islands

Small reef islands on the rim of atolls appear to be some of the most vulnerable of coastal environments; they are threatened by exacerbated coastal erosion, inundation of low-lying island interiors, and saline intrusion into freshwater lenses upon which production of crops, such as taro, depends (Mimura, 1999). Sand cays, on atolls as well as on other reefs, have accumulated incrementally over recent millennia because reefs attenuate wave energy sufficiently to create physically favourable conditions for deposition of sand islands (Woodroffe et al., 2007), as well as enabling growth of sediment-stabilizing seagrasses and mangrove ecosystems (Birkeland, 1996). Sand cays are particularly low-lying, rarely rising more than a few metres above sea level; for example, <8 per cent of the land area of Tuvalu and Kiribati is above 3 m above mean sea level, and in the Maldives only around 1 per cent, reaches this elevation (Woodroffe, 2008). This has led to dramatic warnings in popular media and inferences in the scientific literature that anthropogenic climate change may lead to reef islands on atolls submerging beneath the rising ocean, with catastrophic social and economic implications for populations of these atoll nations (Barnett and Adger, 2003; Farbotko and Lazrus, 2012).

However, reef islands may be more resilient than implied in these dire warnings (Webb and Kench, 2010). Unlike the majority of temperate beaches that have a finite volume of sediment available, biogenic production of carbonate sediments means that there may be an ongoing supply of sediment to these islands. Although coral is a major contributor, it is not necessarily the principal constituent of beaches; large benthic foraminifera (particularly *Calcarina*, *Amphistegina* and *Baculogypsina*) contribute more than 50 per cent of sediment volume on many islands on Pacific atolls (Woodroffe and Morrison, 2001; Fujita et al., 2009). One survey of Pacific coral islands (Webb and Kench, 2010) reported that 86 per cent of islands had remained stable or increased in area over recent decades, and only 14 per cent of islands exhibited a net reduction in area; however, the greatest increases in area resulted from artificial reclamation (Biribo and Woodroffe, 2013). Further studies of shoreline changes on atoll reef islands using multi-temporal aerial photography and satellite imagery indicate accretion on some shorelines and erosion on others, but with the most pronounced changes associated with human occupation and impacts (Rankey, 2011; Ford, 2012; Ford 2013; Hamylton and East, 2012; Yates et al., 2013).

The impacts of future sea-level rise on individual atolls remain unclear (Donner, 2012). Healthy reef systems may be capable of keeping pace with rates of sea-level rise. There is evidence that reefs have coped with much more rapid rates of rise during postglacial melt of major ice sheets than are occurring now or anticipated in this century. Reefs have responded by keeping up, catching up, or in cases of very rapid rise giving up, often to backstep and occupy more landward locations (Neumann and Macintyre, 1985; Woodroffe and Webster, 2014). Geological evidence suggests that healthy coral reefs have exhibited accretion rates in the Holocene of 3 to 9 mm year^{-1} (e.g., Perry and Smithers, 2011), comparable to projected rates of sea-level rise for the 21st century. However, reef growth is likely to lag behind sea-level rise in many cases resulting in larger waves occurring over the reef flat and affecting the shoreline (Storlazzi et al., 2011; Grady et al., 2013). It is unclear whether these larger waves, and the increased wave run-up that is likely, will erode reef-island beaches, overtopping some and inundating island interiors, or whether they will more effectively move sediments shoreward and build ridge crests higher (Gourlay and Hacker, 1991; Smithers et al., 2007). Dickinson (2009) inferred that reef islands on atolls will ultimately be unable to survive because once sea level rises above their solid reef-limestone foundations, which formed during the mid-Holocene sea-level highstand 4,000 to 2,000 years ago, formerly stable reef islands will be subject to erosion by waves.

2.3 Impact of climate change and ocean acidification on production

The impact of climate change on the rate of biogenic production of carbonate sediment is also little understood, but it seems likely to have negative consequences. Although increased temperatures may lead to greater productivity in some cases, for example by extending the latitudinal limit to coral-reef formation, ocean warming has already been recognised to have caused widespread bleaching and death of corals (Hoegh-Guldberg, 1999; Hoegh-Guldberg, 2004; Hoegh-Guldberg et al., 2007). Ocean acidification will have further impacts, and may inhibit some organisms from secreting carbonate shells; for example reduction in production of the Pacific oyster has been linked to acidification (Barton et al., 2012). Decreased seawater pH increases the sensitivity of reef calcifiers to thermal stress and bleaching (Anthony et al., 2008). Based on the density of coral skeleton in >300 long-lived *Porites* corals from across the Great Barrier Reef, De'ath et al. (2009) inferred that a decline in calcification of ~14 per cent had occurred since 1990 manifested as a reduction in the extension rate at which coral grows, which they attributed to temperature stress and declining saturation state of seawater aragonite (which is related to a decrease in pH). However, this extent of the apparent decline has been questioned because of inclusion of many young corals (Ridd et al., 2013); it is not observed in corals collected more recently from inshore (D'Olivo et al., 2013).

There has been some debate about the role of carbonate sediments acting as a chemical buffer against ocean acidification; in this scenario, dissolution of metastable carbonate mineral phases produces sufficient alkalinity to buffer pH and carbonate saturation state of shallow-water environments. However, it is apparent that dissolution rates are slow compared with shelf water-mass mixing processes, such that carbonate dissolution has no discernable impact on pH in shallow waters that are connected to deep-water, oceanic environments (Andersson and Mckenzie, 2012). The seawater chemistry within a reef system can be significantly different from that in the open ocean, perhaps partially offsetting the more extreme effects (Andersson et al., 2013; Andersson and Gledhill, 2013). Corals have the ability to modulate pH at the site of calcification (Trotter et al. 2011; Venn et al. 2011; Falter et al., 2013). Internal pH in both tropical and temperate coral is generally 0.4 to 1.0 units higher than in the ambient seawater, whereas foraminifera exhibit no elevation in internal pH (McCulloch et al., 2012).

Changes in the severity of storms will affect coral reefs; storms erode some island shorelines, but also provide inputs of broken coral to extend other islands (Maragos et al., 1973; Woodroffe, 2008). Alterations in ultra-violet radiation may also have an impact, as UV has been linked to coral bleaching. Furthermore, if reefs are not in a healthy condition due to thermal stress (bleaching) coupled with acidification and other anthropogenic stresses (pollution, overfishing, etc.), then reef growth and carbonate production may not keep pace with sea-level rise. This could, in the long-term, reduce carbonate sand supply to reef islands causing further erosion, although ongoing erosion of cemented reef substrate is also a source of sediment on reefs, indicating that supply of carbonate sand to beaches is dependent upon several interrelated environmental processes. Disruption of any one (or combination) of the controlling processes (carbonate production, reef growth, biological stabilization, bioerosion, physical erosion and transport) may result in reduction of carbonate sand supply to beaches.

3   Economic and social implications of carbonate sand production.

More than 90 per cent of the population of atolls in the Maldives, Marshall Islands, and Tuvalu, as well many in the Cayman Islands and Turks and Caicos (which all have populations of less than 100,000), live at an elevation <10 m above sea level and appear vulnerable to rising sea level, coastal erosion and inundation (McGranahan et al., 2007). The social disruption caused by relocating displaced people to different islands or even to other countries is a problem of major concern to many countries (Farbotko and Lazrus, 2012, see also Chapter 26). Beach aggregate mining is a small-scale industry on many Pacific and Caribbean islands employing local peoples (McKenzie et al., 2006), but mining causes environmental damage when practised on an industrial scale (Charlier, 2002; Pilkey et al., 2011, see also Chapter 23). In the Caribbean, illegal beach mining is widespread but there is little information on what proportion is carbonate (Cambers, 2009). Beach erosion reduces the potential opportunities associated with tourism (see Chapter 27), and decreases habitat for shorebirds and turtles (Fish et al., 2005; Mazaris et al., 2009).

Without coral reefs producing sand and gravel for beach nourishment and protecting the shoreline from currents, waves, and storms, erosion and loss of land are more likely (see also Chapter 39). In Indonesia, Cesar (1996) estimated that the loss due to decreased coastal protection was between 820 United States dollars (for remote areas) and 1,000,000 dollars per kilometre of coastline (in areas of major tourist infrastructure) as a consequence of coral destruction (based on lateral erosion rates of 0.2 m year^{-1}, and a 10 per cent discount rate [similar to an interest rate] over a 25-year period). In the Maldives, mining of coral for construction has had severe impacts (Brown and Dunne, 1988), resulting in the need for an artificial substitute breakwater around Malé at a construction cost of around 12,000,000 dollars (Moberg and Folke, 1999).

4   Conclusions, Synthesis and Knowledge Gaps

There has been relatively little study of rates of carbonate production, and further research is needed on the supply of biogenic sand and gravel to coastal ecosystems. Most beaches have some calcareous biogenic material within them; carbonate is an important component of the shoreline behind coral-reef systems, with reef islands on atolls entirely composed of skeletal carbonate.

The sediment budgets of these systems need to be better understood; direct observations and monitoring of key variables, such as rates of calcification, would be very useful. Not only is little known about the variability in carbonate production in shallow-marine systems, but their response to changing climate and oceanographic drivers is also poorly understood. In the case of reef systems, bleaching as a result of elevated sea temperatures and reduced calcification as a consequence of ocean acidification seem likely to reduce coral cover and production of skeletal material. Longer-term implications for the sustainability of reefs and supply of sediment to reef islands would appear to decrease resilience of these shorelines, although alternative interpretations suggest an increased supply of sediment, either because reef flats that are currently exposed at low tide and therefore devoid of coral, may be re-colonized by coral under higher sea level, or because the disintegration of dead stands of coral may augment the supply of sediment.

Determining the trend in shoreline change, on beaches in temperate settings and on reef islands on atolls or other reef systems, requires monitoring of beach volumes at representative sites. This has rarely been undertaken over long enough time periods, or with sufficient attention to other relevant environmental factors, to discern a pattern or assign causes to inferred trends. Although climate and oceanographic drivers threaten such systems, the most drastic erosion appears to be the result of more direct anthropogenic stressors, such as beach mining, or the construction of infrastructure or coastal protection works that interrupt sediment pathways and disrupt natural patterns of erosion and deposition.

References

Anderson, T.R., Fletcher, C.H., Barbee, M.M., Frazer, L.N., Romine, B.M., (2015). Doubling of coastal erosion under rising sea level by mid-century in Hawaii. *Natural Hazards*, doi 10.1007/s11069-015-1698-6.

Andersson, A.J., Mackenzie, F.T., (2012). Revisiting four scientific debates in ocean acidification research. *Biogeosciences* 9: 893–905.

Andersson, A.J., Gledhill, D., (2013). Ocean acidification and coral reefs: effects on breakdown, dissolution, and net ecosystem calcification. *Annual Reviews of Marine Science* 5, 321–48.

Andersson, A.J., Yeakel, K.L., Bates, N.R., de Putron, S.J., (2013). Partial offsets in ocean acidification from changing coral reef biogeochemistry. *Nature Climate Change* 4, 56-61.

Anthony, K., Kline, D., Diaz-Pulido, G., Dove, S., Hoegh-Guldberg, O., (2008). Ocean acidification causes bleaching and productivity loss in coral reef builders. *Proceedings of the National Academy of Science* 105, 17442-17446.

Ball, M.M., (1967). Carbonate sand bodies of Florida and the Bahamas. *Journal of Sedimentary Petrology* 37, 556-591.

Barnett, J., Adger, N., (2003). Climate dangers and atoll countries. *Climatic Change*. 61, 321-337.

Barton, A., Hales, B., Waldbusser, G.G., Langdon, C., Feely, R.A., (2012). The Pacific oyster, *Crassostrea gigas*, shows negative correlation to naturally elevated carbon dioxide levels: implications for near-term ocean acidification effects. *Chinese Journal of Limnology and Oceanography* 57, 698-710.

Biribo, N., Woodroffe, C.D., (2013). Historical area and shoreline change of reef islands around Tarawa Atoll, Kiribati. *Sustainability Science* 8, 345–362.

Birkeland, C. (ed.) (1996). Life and Death of Coral Reefs. (New York, Chapman & Hall).

Brewer, P.G., Dyrssen, D., (1985). Chemical oceanography of the Persian Gulf. *Progress in Oceanography* 14, 41-55.

Brooke, B., (2001). The distribution of carbonate eolianite. *Earth-Science Reviews* 55, 135-164.

Brown, B.E., Dunne, R.P., (1988). The environmental impact of coral mining in the Maldives. *Environmental Conservation* 15, 159-166.

Bruun, P., (1962). Sea-level rise as a cause of shore erosion. American Society of Civil Engineering Proceedings, *Journal of Waterways and Harbors Division* 88, 117-130.

Buddemeier, R.W., Hopley, D., (1988). Turn-ons and turn-offs: causes and mechanisms of the initiation and termination of coral reef growth. *Proceedings of the 6th International Coral Reef Congress* 1, 253-261.

Cambers, G. (2009). Caribbean beach changes and climate change adaptation. *Aquatic Ecosystem Health & Management*, 12, 168-176.

Carannante, G., Esteban, M., Milliman, J.D., Simone, L. (1988). Carbonate lithofacies as paleolatitude indicators: problems and limitations. *Sedimentary Geology* 60, 333-346.

Cesar, H., (1996). *Economic analysis of Indonesian coral reefs*. World Bank Environment Department, Washington DC, USA., p. 103.

Charlier, R.H., (2002). Impact on the coastal environment of marine aggregates mining. *International Journal of Environmental Studies* 59, 297-322.

Church, J.A., Clark, P.U., Cazenave, A., Gregory, J.M., Jevrejeva, S., Levermann, A., Merrifield, M.A., Milne, G.A., Nerem, R.S., Nunn, P.D., Payne, A.J., Pfeffer, W.T., Stammer, D., Unnikrishnan, A.S., 2013. Sea Level Change, in: Stocker, T.F., Plattner, G.-K., Tignor, M., Allen, S.K., Boschung, J., Nauels, A., Xia, Y., Bex, V., Midgley, P.M. (Eds.), *Climate Change 2013: The Physical Science Basis. Contribution of Working Group I to the Fifth Assessment Report of the Intergovernmental Panel on Climate Change*. Cambridge University Press, Cambridge, United Kingdom and New York, NY, USA, pp. 1137-1216.

Cooper, J.A.G., Pilkey, O.H., (2004). Sea-level rise and shoreline retreat: time to abandon the Bruun Rule. *Global and Planetary Change* 43, 157-171.

Cowell, P.J., Thom, B.G., Jones, R.A., Everts, C.H., Simanovic, D., (2006). Management of uncertainty in predicting climate-change impacts on beaches. *Journal of Coastal Research* 22, 232-245.

De'ath, G., Lough, J.M., Fabricus, K.E., (2009). Declining coral calcification on the Great Barrier Reef. *Science* 323, 116-119.

Dickinson, W.R., (2009). Pacific atoll living: How long already and until when? *GSA Today* 19, 4-10.

D'Olivo, J.P., McCulloch, M.T., Judd, K., (2013). Long-term records of coral calcification across the central Great Barrier Reef: assessing the impacts of river runoff and climate change. *Coral Reefs* 32, 999-1012.

Donner, S., (2012). Sea level rise and the ongoing battle of Tarawa. EOS, *Transactions of the American Geophysical Union* 93, 169-176.

Falter, J., Lowe, R., Zhang, Z., McCulloch, M., (2013). Physical and biological controls on the carbonate chemistry of coral reef waters: effects of metabolism, wave forcing, sea level, and geomorphology. *PLoS One* 8, e53303.

Farbotko, C., Lazrus, H. (2012). The first climate refugees? Contesting global narratives of climate change in Tuvalu. *Global Environmental Change* 22, 382-390.

Fish, M.R., Cote, I.M., Gill, J.A., Jones, A.P., Renshoff, S., Watkinson, A. (2005). Predicting the impact of sea-Level rise on Caribbean sea turtle nesting habitat. *Conservation Biology* 19, 482-491.

Ford, M., (2012). Shoreline changes on an urban atoll in the central Pacific Ocean: Majuro Atoll, Marshall Islands. *Journal of Coastal Research* 28, 11-22.

Ford, M., (2013). Shoreline changes interpreted from multi-temporal aerial photographs and high resolution satellite images: Wotje Atoll, Marshall Islands. *Remote Sensing of Environment* 135, 130-140.

Frank, T.D., James, N.P., Bone, Y., Malcolm, I., Bobak, L.E., (2014). Late Quaternary carbonate deposition at the bottom of the world. *Sedimentary Geology*, 306, 1-16.

Fujita K., Osawa, Y., Kayanne, H., Ide, Y., Yamano, H. (2009). Distribution and sediment production of large benthic foraminifers on reef flats of the Majuro Atoll, Marshall Islands. *Coral Reefs* 28, 29-45.

Glynn, P.W., (1996). Coral reef bleaching: facts, hypotheses and implications. *Global Change Biology* 2, 495-509.

Gourlay, M.R., Hacker, J.L.F. (1991). *Raine Island: coastal processes and sedimentology*. CH40/91, Department of Civil Engineering, University of Queensland, Brisbane.

Grady, A.E., Reidenbach, M.A., Moore, L.J., Storlazzi, C.D., Elias, E., (2013). The influence of sea level rise and changes in fringing reef morphology on gradients in alongshore sediment transport. *Geophysical Research Letters* 40, 3096–3101.

Hamylton, S.M., East, H., (2012). A geospatial appraisal of ecological and geomorphic change on Diego Garcia Atoll, Chagos Islands (British Indian Ocean Territory). *Remote Sensing* 4, 3444-3461.

Hamylton, S., (2014). Will coral islands maintain their growth over the next century? A deterministic model of sediment availability at Lady Elliot Island, Great Barrier Reef. *PLoS ONE* 9, e94067.

Harney, J.N., Fletcher, C.H., (2003). A budget of carbonate framework and sediment production, Kailua Bay, Oahu, Hawaii. *Journal of Sedimentary Research* 73, 856-868.

Hart, D.E., Kench, P.S., (2007). Carbonate production of an emergent reef platform, Warraber Island, Torres Strait, Australia. *Coral Reefs* 26, 53-68.

Heap, A.D., P.T. Harris, L. Fountain, (2009). Neritic carbonate for six submerged coral reefs from northern Australia: Implications for Holocene global carbon dioxide. *Palaeogeography, Palaeoclimatology, Palaeoecology* 283, 77-90.

Hoegh-Guldberg, O., (1999). Climate change, coral bleaching and the future of the world's coral reefs. *Marine and Freshwater Research* 50, 839-866.

Hoegh-Guldberg, O., (2004). Coral reefs in a century of rapid environmental change. *Symbiosis* 37, 1-31.

Hoegh-Guldberg, O., Mumby, P.J., Hooten, A.J., Steneck, R.S., Greenfield, P., Gomez, E., Harvell, C.D., Sale, P.F., Edwards, A.J., Caldeira, K., Knowlton, N., Eakin, C.M., Glesias-Prieto, R., Muthiga, N., Bradbury, R.H., Dubi, A., Hatziolos, M.E. (2007). Coral reefs under rapid climate change and ocean acidification. *Science* 318, 1737-1742.

Houston, J.R., Dean, R.G. (2014). Shoreline change on the east coast of Florida. *Journal of Coastal Research* 30, 647-660.

Hopley, D., Smithers, S.G. and Parnell, K., (2007). *Geomorphology of the Great Barrier Reef: development, diversity and change*. Cambridge University Press.

James, N.P., Collins, L.B., Bone, Y., Hallock, P., (1999). Subtropical carbonates in a temperate realm: modern sediments on the southwest Australian shelf. *Journal of Sedimentary Research* 69, 1297-1321.

James, N.P., Bone, Y., (2011). Neritic carbonate sediments in a temperate realm. Springer, Dordrecht.

Kroeker, K.J., Kordas, R.L., Crim, R.N., Singh, G.G. (2010). Meta-analysis reveals negative yet variable effects of ocean acidification on marine organisms. *Ecology Letters* 13, 1419-1434.

Leon, J.X., Woodroffe, C.D., (2013). Morphological characterisation of reef types in Torres Strait and an assessment of their carbonate production, *Marine Geology* 338, 64-75.

Maragos, J.E., Baines, G.B.K. and Beveridge, P.J. (1973). Tropical cyclone creates a new land formation on Funafuti atoll. *Science* 181: 1161-1164.

Mazaris, A.D., Matsinos, G., Pantis, J.D. (2009). Evaluating the impacts of coastal squeeze on sea turtle nesting. *Ocean & Coastal Management* 52, 139-145.

McCulloch, M., Falter, J.L., Trotter, J., Montagna, P., (2012). Coral resilience to ocean acidification and global warming through pH up-regulation. *Nature Climate Change* 2, 1–5.

McGranahan, G., Balk, D., Anderson, B., (2007). The rising tide: assessing the risks of climate change and human settlements in low elevation costal zones. *Environment and Urbanization* 19, 17-37.

McKenzie, E., Woodruff, A., McClennen, C., (2006). "Economic assessment of the true costs of aggregate mining in Majuro Atoll, Republic of the Marshall Islands'. SOPAC Technical Report 383, p. 74.

Milliman, J.D., and Droxler, A.W. (1995). Calcium carbonate sedimentation in the global ocean: Linkages between the neritic and pelagic environments. *Oceanography* 8(3):92–94, http://dx.doi.org/10.5670/oceanog.1995.04.

Mimura, N. (1999). Vulnerability of island countries in the South Pacific to sea level rise and climate change. *Climate Research* 12, 137-143.

Moberg, F. Folke, C., (1999). Ecological goods and services of coral reef ecosystems. *Ecological Economics* 29, 215–233.

Montaggioni, L.F., Braithwaite, C.J.R., (2009). Quaternary Coral Reef Systems: history, development processes and controlling factors. Elsevier, Amsterdam.

Nelson, C.S., (1988). An introductory perspective on non-tropical shelf carbonates. *Sedimentary Geology* 60, 3-12.

Neumann, A.C., Macintyre, I., (1985). Reef response to sea level rise: keep-up, catch-up or give-up. *Proceedings of the 5th International Coral Reef Congress* 3, 105-110.

Nicholls, R.J., Wong P.P., Burkett V.R., Codignotto J.O., Hay J.E., McLean R.F., Ragoonaden, S. and Woodroffe, C.D., et al., Coastal systems and low-lying areas. In: Parry, M.L., Canziani, O.F., Palutikof, J.P., van der Linden, P.J., Hanson, C.E., (Editors) (2007), *Climate Change 2007: impacts, adaptation and vulnerability. Contribution of Working Group II to the Fourth Assessment Report of the Intergovernmental Panel on Climate Change (IPCC).* Cambridge University Press, pp. 315-357.

Perry, C.T., Smithers, S.G., (2011). Cycles of coral reef 'turn-on', rapid growth and 'turn-off' over the past 8500 years: a context for understanding modern ecological states and trajectories. *Global Change Biology* 17, 76–86.

Perry, C.T., Kench, P.S., Smithers, S.G., Riegl, B., Yamano, H., O'Leary, M.J., (2011). Implications of reef ecosystem change for the stability and maintenance of coral reef islands. *Global Change Biology* 17, 3679-3696.

Perry, C., Edinger, E., Kench, P., Murphy, G., Smithers, S., Steneck, R., Mumby, P., (2012). Estimating rates of biologically driven coral reef framework production and erosion: a new census-based carbonate budget methodology and applications to the reefs of Bonaire. *Coral Reefs* 31, 853–868.

Pilkey, O.H., Neal, W.J., Cooper, J.A.G., Kelley, J.T., (2011). *The World's Beaches: A global guide to the science of the shoreline*. University of California Press.

Purdy, E.G., Gischler, E., (2005). The transient nature of the empty bucket model of reef sedimentation. *Sedimentary Geology* 175, 35-47.

Purser, B.H. (Ed) (1973). *The Persian Gulf: Holocene carbonate sedimentation and diagenesis in a shallow epicontinental sea.* Springer-Verlag.

Ramalho, R.S., Quartau, R., Trenhaile, A.S., Mitchell, N.C., Woodroffe, C.D., Ávila, S.P. (2013) Coastal evolution on volcanic oceanic islands: a complex interplay between volcanism, erosion, sedimentation, sea-level change and biogenic production. *Earth-Science Reviews*, 127: 140-170.Rankey, E.C., (2011). Nature and stability of atoll island shorelines: Gilbert Island chain, Kiribati, equatorial Pacific. *Sedimentology* 58, 1831-1859.

Rankey, E.C. (2011) Nature and stability of atoll island shorelines: Gilbert Island chain, Kiribati, equatorial Pacific. *Sedimentology* 58, 1831-1859.

Ridd, P.V., Teixeira da Silva, E., Stieglitz, T., (2013). Have coral calcification rates slowed in the last twenty years? *Marine Geology* 346, 392-399.

Ritchie, W., Mather, A.S., (1984). "The beaches of Scotland". Commissioned by the Countryside Commission for Scotland 1984, Report No. 109. http://www.snh.org.uk/pdfs/publications/commissioned_reports/ReportNo109.pdf

Scoffin, T.P., *An Introduction to Carbonate Sediments and Rocks*. (1987). Chapman & Hall, New York, 274 pp.

Short, A.D., (2006). Australian beach systems, nature and distribution. *Journal of Coastal Research* 22, 11-27.

Short, A.D., (2010). Sediment transport around Australia - sources, mechanisms, rates and barrier forms. *Journal of Coastal Research* 26, 395-402.

Smithers, S.G., Harvey, N., Hopley, D. and Woodroffe, C.D., (2007). Vulnerability of geomorphological features in the Great Barrier Reef to climate change. In Johnson J.E., Marshall, P.A. (Editors) in *Climate Change and the Great Barrier Reef. Great Barrier Reef Marine Park Authority and Australian Greenhouse Office, Australia*, pp. 667-716.

Spalding, M.D. and Grenfell, A.M., (1997). New estimates of global and regional coral reef areas. *Coral Reefs* 16, 225-230.

Storlazzi, C.D., Elias, E., Field, M.E. and Presto, M.K., (2011). Numerical modeling of the impact of sea-level rise on fringing coral reef hydrodynamics and sediment transport. *Coral Reefs* 30, 83-96.

Trotter, J., Montagna, P., McCulloch, M., Silenzi, S., Reynaud, S., Mortimer, G., Martin, S., Ferrier-Pages, C., Gattuso, J-P., Rodolfo-Metalpa, R., (2011). Quantifying the pH 'vital effect' in the temperate zooxanthellate coral *Cladocora caespitosa*: Validation of the boron seawater pH proxy. *Earth and Planetary Science Letters*, 303, 163–173.

Vecsei, A., (2001). Fore-reef carbonate production: development of a regional census-based method and first estimates. *Palaeogeography Palaeoclimatology Palaeoecology* 175, 185-200.

Vecsei, A., (2003). Systematic yet enigmatic depth distribution of the world's modern warm-water carbonate platforms: the 'depth window'. *Terra Nova* 15, 170-175.

Vecsei, A., (2004). A new estimate of global reefal carbonate production including the fore-reefs. *Global and Planetary Change* 43, 1-18.

Venn, A., Tambutté, E., Holcomb, M., Allemand, D., Tambutté, S., (2011). Live tissue imaging shows reef corals elevate pH under their calcifying tissue relative to seawater. *PLoS One* 6, e20013.

Webb, A.P., Kench, P., (2010). The Dynamic Response of Reef Islands to Sea Level Rise: Evidence from Multi-Decadal Analysis of Island Change in the Central Pacific. *Global and Planetary Change* 72, 234-246

Woodroffe, C.D., (2008). Reef-island topography and the vulnerability of atolls to sea-level rise. *Global and Planetary Change* 62, 77-96.

Woodroffe, C.D., Morrison, R.J., (2001). Reef-island accretion and soil development, Makin Island, Kiribati, central Pacific. *Catena* 44, 245-261.

Woodroffe, C.D., Kennedy, D.M., Jones, B.G., Phipps, C.V.G. (2004). Geomorphology and Late Quaternary development of Middleton and Elizabeth Reefs. *Coral Reefs* 23, 249-262.

Woodroffe, C.D., Samosorn, B., Hua, Q., Hart, D.E., (2007). Incremental accretion of a sandy reef island over the past 3000 years indicated by component-specific radiocarbon dating, *Geophysical Research Letters* 34, L03602, doi:10.1029/2006GL028875.

Woodroffe, C.D., Webster, J.M., (2014). Coral reefs and sea-level change. *Marine Geology* doi 10.1016/j.margeo.2013.12.006.

Yates, M.L., Le Cozannet, G., Garcin, M., Salai, E., Walker, P., (2013). Multidecadal atoll shoreline change on Manihi and Manuae, French Polynesia. *Journal of Coastal Research* 29 870-882.

8 Aesthetic, Cultural, Religious and Spiritual Ecosystem Services Derived from the Marine Environment

Contributors:
Alan Simcock (Lead Member)

Aesthetic, Cultural, Religious and Spiritual Ecosystem Services Derived from the Marine Environment

1 Introduction

At least since the ancestors of the Australian aboriginal people crossed what are now the Timor and Arafura Seas to reach Australia about 40,000 years ago (Lourandos, 1997), the ocean has been part of the development of human society. It is not surprising that human interaction with the ocean over this long period profoundly influenced the development of culture. Within "culture" it is convenient to include the other elements – aesthetic, religious and spiritual – that are regarded as aspects of the non-physical ecosystem services that humans derive from the environment around them. This is not to decry the difference between all these aspects, but rather to define a convenient umbrella term to encompass them all. On this basis, this chapter looks at the present-day implications of the interactions between human culture and the ocean under the headings of cultural products, cultural practices and cultural influences.

2 Cultural products

No clear-cut distinction exists between objects which have a utilitarian value (because they are put to a use) and objects which have a cultural value (because they are seen as beautiful or sacred or prized for some other non-utilitarian reason). The two categories can easily overlap. Furthermore, the value assigned to an object may change: something produced primarily for the use to which it can be put may become prized, either by the society that produces it or by some other society, for other reasons (Hawkes, 1955). In looking at products from the ocean as cultural ecosystem services, the focus is upon objects valued for non-utilitarian reasons. The value assigned to them will be affected by many factors: primarily their aesthetic or religious significance, their rarity and the difficulty of obtaining them from the ocean. The example of large numbers of beads made from marine shells found in the burial mounds dating from the first half of the first millennium CE of the Mound People in Iowa, United States of America, 1,650 kilometres from the sea, shows how exotic marine products can be given a cultural value (Alex, 2010).

Another good – albeit now purely historical – example is the purple dye derived from marine shellfish of the family *Muricidae*, often known as Tyrian purple. In the Mediterranean area, this purple dye was very highly valued, and from an early date (around 1800-1500 BCE) it was produced in semi-industrial fashion in Crete and later elsewhere. Its cost was high because large numbers of shellfish were required to produce small amounts of the dye. Because of this, its use became restricted to the elite. Under the Roman republic, the togas of members of the Senate were distinguished by a border of this colour, and under the Roman empire it became the mark of the emperors (Stieglitz, 1994). This usage has produced a whole cultural structure revolving around the colour purple and spreading out into a range of metaphors and ideas: for example, the concept of the "purple patch," an elaborate passage in writing, first used by the Roman poet Horace (Horatius).

Goods derived from marine ecosystems that are given a cultural value because of their appearance and/or rarity include pearls, mother-of-pearl, coral and tortoiseshell. In the case of coral, as well as its long-standing uses as a semi-precious item of jewellery and inlay on other items, a more recent use in aquariums has developed.

2.1 Pearls and mother-of-pearl

Pearls and mother-of-pearl are a primary example of a marine product used for cultural purposes. Many species of molluscs line their shells with nacre – a lustrous material consisting of platelets of aragonite (a form of calcium carbonate (see Chapter 7)) in a matrix of various organic substances (Nudelman et al., 2006). The shells with this lining give mother-of-pearl. Pearls themselves are formed of layers of nacre secreted by various species of oyster and mussel around some foreign body which has worked its way into the shell (Bondad-Reantaso et al., 2007).

Archaeological evidence shows that pearls were already being used as jewellery in the 6th millennium BCE (Charpentier et al., 2012). By the time of the Romans, they could be described as "holding the first place among things of value" (Pliny). For the ancient world, the main source was the shellfish beds along the southern coast of the Persian Gulf, with Bahrain as the main centre. The pearl fishery in the Persian Gulf maintained itself as the major source of pearls throughout most of the first two millennia CE, and by the 18th century was sufficiently profitable to support the founding of many of the present Gulf States. It developed further in the 19th century, and by the start of the 20th century the Persian Gulf pearl trade reached a short-lived peak in value at about 160 million United States dollars a year, and was the mainstay of the economies of the Gulf States (Carter, 2005).

During the 20th century, however, the Persian Gulf pearl trade declined steadily, due substantially to competition from the Japanese cultured pearl industry and general economic conditions. With the emergence of the Gulf States as important oil producers, the economic significance of the pearl trade for the area declined. The Kuwait pearl market closed in 2000, and with its closure the Persian Gulf pearl fishery ceased to be of economic importance (Al-Shamlan, 2000). However, some pearling still continues as a tourist attraction and, with Japanese support, an attempt has been made to establish a cultivated pearl farm in Ras Al Kaimah (OBG, 2013). Other traditional areas for the harvesting of natural pearls include the Gulf of Cutch and the Gulf of Mannar in India, Halong Bay in Viet Nam and the Islas de las Perlas in Panama (CMFRI, 1991; Southgate, 2007).

The great transformation of the pearl industry came with the success of Japanese firms in applying the technique developed in Australia by an Englishman, William Saville-Kent. The technique required the insertion of a nucleus into the pearl oyster in order to provoke the formation of a pearl. Using the oyster species from the Persian Gulf, this meant that, instead of the three or four pearls that could be found in a thousand wild oysters, a high percentage of the farmed oysters would deliver pearls. The Japanese industry started in about 1916. By 1938, there were about

360 pearl farms in Japanese waters, producing more than 10 million pearls a year (15 tons). Production continued to increase after World War II and reached a peak of 230 tons in 1966, from 4,700 farms. Pollution and disease in the oyster, however, rapidly caused the industry to contract. By 1977, only about 1,000 farms remained, producing about 35 tons of pearls. Competition from Chinese cultured freshwater pearls and an oyster epidemic in 1996 reduced the Japanese industry to the production of less than 25 tons a year. Nevertheless, this industry was still worth about 130 million dollars a year. From the 1970s, other Indian Ocean and Pacific Ocean areas were developing cultured pearl industries based on the traditional pearl oyster species: in India and in Viet Nam in the traditional pearling regions, and in Australia, China, the Republic of Korea and Venezuela. Apart from China, where production had reached 9-10 tons a year, these are relatively small; the largest is apparently in Viet Nam, which produces about 1 ton a year (Southgate, 2007).

At the same time, new forms of the industry developed, based on other oyster species. The two main branches are the "white South Sea" and "black South Sea" pearl industries, based on *Pinctada maxima* and *Pinctada margaritefera*, respectively. "Black" pearls are a range of colours from pale purple to true black. Australia (from 1950) and Indonesia (from the 1970s) developed substantial industries for "white South Sea" pearls, earning around 100 million dollars a year each. Malaysia, Myanmar, Papua New Guinea and the Philippines have smaller industries. The black "South Sea" pearl industry is centred in French Polynesia, particularly in the Gambier and Tuamotu archipelagos. The industry in French Polynesia was worth 173 million dollars in 2007 (SPC, 2011). The Cook Islands, building on a long-standing mother-of-pearl industry, started a cultured-pearl industry in 1972, which grew to a value of 9 million dollars by 2000. However, in that year poor farm hygiene and consequent mass mortality of the oysters led to a collapse to less than a quarter of that value by 2005. The trade has recovered somewhat since then, largely due to increased sales to tourists in the islands. Small "black South Sea" pearl industries also exist in the Federated States of Micronesia, Fiji, the Marshall Islands and Tonga. Small pearl industries based on the oyster species *Pterea penguin* and *Pterea sterna* exist in Australia, China, Japan, Mexico and Thailand (SPC, 2011; Southgate, 2007).

Reliable information on the cultured pearl industries is not easy to obtain: for example, significant divergences exist between the statistics for the *Pintada margaritifera* industry in the FAO Fisheries Global Information System database and those reported by the South Pacific Secretariat in their newsletters (SPC, 2011). The FAO itself noted the lack of global statistics on pearls (FAO, 2012). However, all sources suggest that the various industries suffered severe set-backs in 2009-2012 from a combination of the global economic crisis and overproduction. It is also clear that, apart from local sales to tourists, the bulk of all production passes through auctions in Hong Kong, China, and Japan.

Mother-of-pearl is produced mainly from the shells of pearl oysters, but other molluscs, such as abalone, may also be used. In the 19th century it was much used as a material for buttons and for decorating small metal objects and furniture. In many of these uses it has been superseded by plastics. It developed as an important industry in the islands around the Sulu Sea and the Celebes Sea, but substantial industries also existed in western Australia (now overtaken by the cultured-pearl industry), the Cook Islands and elsewhere (Southgate 2007). It remains important in the Philippines, which still produces several thousand tons a year (FAO, 2012).

2.2 Tortoiseshell

For several centuries, material from the shells of sea turtles was used both as a decorative inlay on high-quality wooden furniture and for the manufacture of small items such as combs, spectacle frames and so on. The lavish use of tortoiseshell was a particular feature of the work of André Charles Boulle, cabinetmaker to successive 18th century French kings. This established a pattern which was widely imitated (Penderel-Brodhurst, 1910). The shells of hawksbills turtles (*Eretmochelys imbricata*), in particular, were used for this purpose. The demand for the shells of hawksbill turtles produced an enormous and enduring effect on hawksbill populations around the world. Within the last 100 years, millions of hawksbills were killed for the tortoiseshell markets of Asia, Europe and the United States (NMFS, 2013). The species has been included in the most threatened category of the IUCN's Red List since the creation of the list in 1968, and since 1977 in the listing of all hawksbill populations on Appendix I of the Convention on International Trade in Endangered Species of Wild Fauna and Flora[1] (CITES) (trade prohibited unless not detrimental to the survival of the species). Some production of objects with tortoiseshell continues (particularly in Japan), but on a very much reduced scale.

2.3 Coral (and reef fish)

The Mediterranean red coral (*Corallium rubrum*), was used from a very early date for decoration and as a protective charm. In the 1st century, Pliny the Elder records both its use a charm to protect children and its scarcity as a result of its export to India (Pliny). As late as the second half of the 19th century, teething-rings were still being made with coral (Denhams, 2014). It is now principally used for jewellery. The Mediterranean red coral is still harvested. Similar genera/species from the western Pacific near Japan, Hawaii, and some Pacific seamounts are also harvested. The global harvest reached a short-lived peak at about 450 tons a year in 1986, as a result of the exploitation of some recently discovered beds on the Emperor Seamounts in the Pacific. It has fallen back to around 50 tons a year, primarily from the Mediterranean and adjoining parts of the Atlantic (CITES, 2010). This trade in the hard coral stone is estimated to be worth around 200 million dollars a year (FT, 2012), although another estimate places it at nearer 300 million dollars) (Tsounis, 2010). Despite proposals in 2007 and 2010, these corals are not listed under the CITES.

Other corals of cultural interest, on the other hand, have been listed under CITES. The cultural use made of these genera and species is very

---

[1] United Nations, *Treaty Series*, vol.. 993, No. 14537.

different. The main use is inclusion in aquariums. Some experimental evidence exists that the ability to watch fish in aquariums has a soothing effect on humans (especially when suffering from dementia) (for example, Edwards et al., 2002). For similar reasons, many homes, offices, surgeries and hospitals have installed such aquariums. Suitable pieces of coral, either alive or dead, are seen as attractive parts of such aquarium scenes. The demand for coral for this purpose is substantial. International trade in coral skeletons for decorative purposes began in the 1950s. Until 1977 the source was largely the Philippines. In that year a national ban on export was introduced, and by 1993 the ban was fully effective. The main source then became Indonesia. Until the 1990s, the trade was mainly in dead corals for curios and aquarium decoration. Developments in the technology of handling live coral led to a big increase in the trade in live coral. CITES lists 60 genera of hard corals in Appendix II; hence their export is permitted only if the specimens have been legally acquired and export will not be detrimental to the survival of the species or its role in the ecosystem. For coral rock, the trade averaged about 2,000 tons a year in the decade 2000-2010, although declining slightly towards the end of the decade. Fiji (with 60 per cent) and Indonesia (with 11 per cent) were the major suppliers over this decade. Other countries supplying coral rock included Haiti, the Marshall Islands, Mozambique, Tonga, Vanuatu and Viet Nam, although the last five introduced bans towards the end of this period. The major importers were the United States (78 per cent) and the European Union (12 per cent). For live coral, the picture was slightly different: over the same decade, the number of pieces of live coral traded rose from some 700,000 to some 1,200,000. Of these, Indonesia supplied an average of about 70 per cent, with other important suppliers including Fiji (10 per cent), Tonga (5 per cent), Australia (5 per cent) and the Solomon Islands (4 per cent). The United States accounted for an average of 61 per cent of the imports, and the European Union took 31 per cent. For some species of coral, mariculture is possible, and by 2010 pieces produced by mariculture accounted for 20 per cent of the trade (Wood et al., 2012).

An aquarium would not be complete without fish, and this need has produced another major global trade: in reef fish. Because few marine ornamental fish species have been listed under CITES, a dearth of accurate information on the precise details of the trade exists. The FAO noted the lack of global statistics on the catches of, and trade in, ornamental fish in its 2012 Report on the State of the World's Fisheries and Aquaculture (FAO, 2012). The late Director of the trade association Ornamental Fish International, Dr. Ploeg, likewise lamented the lack of data (Ploeg, 2004). One estimate puts the scale of the trade in ornamental fish (freshwater and marine) at 15 billion dollars. In 2000 to 2004 an attempt was made to set up in UNEP/WCMC a Global Marine Aquarium Database (GMAD), drawing not only on official trade records, but also on information supplied by trade associations. This provides some interesting, albeit now dated, information, but it has not been kept up-to-date because of lack of funding. One of the most notable features was that the number of fish reported as imported was some 22 per cent more than the number reported as exported (Wabnitz et al., 2003). The need for better information is a matter of on-going debate; the European Union has conducted a consultation exercise in 2008-2010 (EC, 2008).

The GMAD data suggested that some 3.5-4.3 million fish a year, from nearly 1,500 different species, were being traded worldwide. The main sources of fish (in order of size of exports) were the Philippines, Indonesia, the Solomon Islands, Sri Lanka, Australia, Fiji, the Maldives and Palau. These countries accounted for 98 per cent of the recorded trade, with the Philippines and Indonesia together accounting for nearly 70 per cent. The main destinations of the fish were the United States, the United Kingdom, the Netherlands, France and Germany, which accounted for 99 per cent of the recorded trade; the United States accounted for nearly 70 per cent. These figures probably do not include re-exports to other countries. It was estimated that the value of the trade in 2003 was 1 million to 300 million dollars (Wabnitz et al., 2003).

From the social perspective, the number of people depending on the trade is relatively small. A workshop organized by the Secretariat of the Pacific Community in 2008 showed that some 1,472 people in 12 Pacific island countries and territories depended on the trade in ornamental fish for their livelihoods (Kinch et al., 2010). GMAD reported an estimate of 7,000 collectors providing marine ornamental fish in the Philippines (Wabnitz et al., 2003). It also reported a much higher estimate of some 50,000 people in Sri Lanka being involved with the export of marine ornamentals, but this probably reflects the large, long-standing trade based on the aquaculture of ornamental freshwater fish.

2.4 Culinary and medicinal cultural products

Items of food, and specific ways of preparing dishes from them, can be very distinctive features of cultures. Products derived from marine ecosystems often play a significant role. One almost universal feature is salt. For millennia, salt was vital in much of the world for the preservation of meat and fish through the winter months. Although nowadays salt is mainly obtained from rock-salt and brine deposits in the ground, salt is still widely prepared by the evaporation of seawater, especially in those coastal areas where the heat of the sun can be used to drive the evaporation. Although statistics for the production of salt often do not differentiate between the sources for salt production, countries such as Brazil, India and Spain are recorded as producing many millions of tons of salt from the sea (BGS, 2014).

A further common preparation used in many forms of cooking is a sauce derived from fermenting or otherwise processing small fish and shellfish. Such sauces are recorded as *garum* and *liquamen* among the Romans from as long ago as the 1st century (Pliny). They are also crucial ingredients in the cuisines of many east Asian countries – China, Republic of Korea, Thailand, Viet Nam – and other fish-based sauces are found in many western cuisines, for example, *colatura de alici* (anchovy sauce) and Worcestershire sauce.

Cultural pressures can interact with the sustainable use of products derived from marine ecosystem services. Just as the demand for tortoiseshell inlay and objects was driven by desire to emulate the élite in both Asia and Europe, and affected the hawksbill turtle, other species of marine turtle were also affected by the status of turtle soup as a prestige dish. In Europe, soup made from green turtles (*Chelonia mydas*) became a prestige dish when the turtles were brought back by European trading ships passing through the tropics. It was served lavishly at formal dinners – in the mid-19th century, a report of a routine large dinner refers to "four hundred tureens of turtle, each containing five pints" – that is, 1,136 litres in total (Thackeray, 1869). Large amounts were also commercialized in tins. In spite of growing conservation concerns, it was still seen as appropriate for inclusion in the dinner to welcome the victorious General Eisenhower back to the United States in 1945 (WAA, 1945). The dish has disappeared from menus since the green turtle was listed under Appendix I to CITES in 1981, except in areas where turtles are farmed or where freshwater species are used.

Another group of species where cultural forces create pressures for excessive harvesting is the sharks (see also chapter 40). Shark's fin soup is a prestige dish in much of eastern Asia, especially among Chinese-speaking communities. Prices for shark's fins are very high (hundreds of dollars per kilogramme). As shown in Figure 8.1, the trade in shark fins peaked in 2003-2004 and has subsequently levelled out at quantities 17-18 per cent lower (2008-2011). The statistics are subject to many qualifications, but trade in shark fins through Hong Kong, China (generally regarded as the largest trade centre in the world) rose by 10 per cent in 2011, but fell by 22 per cent in 2012. The FAO report from which the figure is drawn suggests that a number of factors, including new regulations by China on government officials' expenditures, consumer backlash against artificial shark fin products, increased regulation of finning (the practice of cutting fins of shark carcasses and discarding the rest), other trade bans and curbs, and a growing conservation awareness, may have contributed to the downturn. At the same time, new figures suggest the shark fin markets in Japan, Malaysia and Thailand, though focused on small, low-value fins, may be among the world's largest (FAO, 2014a).

Similar cultural pressures exist in relation to other aspects of marine ecosystems. Traditional medicine in eastern Asia, for example, uses dried seahorses for a range of illnesses. Most dried seahorses (caught when they are about 12-16 cm in size) are exported to China. The value in 2008 was 100-300 dollars per kilogramme, depending on the size and species; the larger animals are the most valuable. Production is said to be more than 20 million sea horses (70 tons) a year. Viet Nam and China are the major producers; Viet Nam has developed its seahorse aquaculture since 2006. This trade is seen as a significant pressure on the conservation status of several species of seahorse (FAO, 2014b).

Not all consequences of the cultural uses of the ocean's ecosystem services in relation to food are necessarily negative. The Mediterranean diet, with its substantial component of fish and shellfish, was inscribed in 2013 on the UNESCO Representative List of the Intangible Cultural Heritage of Humanity (UNESCO, 2014).

3 Cultural practices

3.1 Cultural practices that enable use of the sea

Humans interact with the ocean in a large number of ways, and many of these lead to cultural practices which enrich human life in aesthetic,

Figure 8.1 | Source: FAO, 2014a.

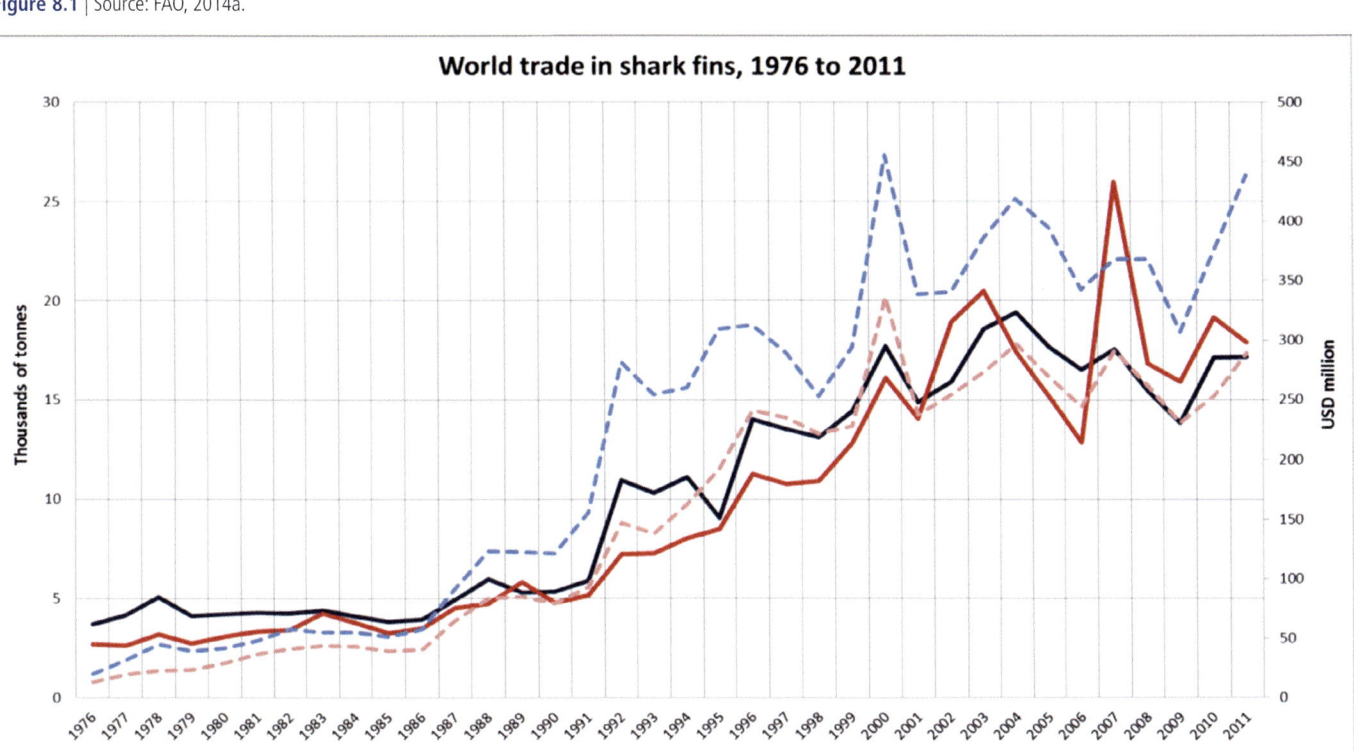

religious or spiritual ways, as well as in purely practical matters. Such practices are beginning to be inscribed in the UNESCO Representative List of the Intangible Cultural Heritage of Humanity. Those listed so far include a practice in Belgium of fishing for shrimp on horse-back: twice a week, except in winter months, riders on strong Brabant horses walk breast-deep in the surf, parallel to the coastline, pulling funnel-shaped nets held open by two wooden boards. A chain dragged over the sand creates vibrations, causing the shrimp to jump into the net. Shrimpers place the catch (which is later cooked and eaten) in baskets hanging at the horses' sides. In approving the inscription, the Intergovernmental Committee for the Safeguarding of the Intangible Cultural Heritage (ICSICH) noted that it would promote awareness of the importance of small, very local traditions, underline the close relations between humans, animals and nature, and promote respect for sustainable development and human creativity (UNESCO, 2014).c

Similarly, the Chinese tradition of building junks with separate watertight bulkheads has been recognized as a cultural heritage that urgently needs protection. The ICSICH noted that, despite the historical importance of this shipbuilding technology, its continuity and viability are today at great risk because wooden ships are replaced by steel-hulled vessels, and the timber for their construction is in increasingly short supply; apprentices are reluctant to devote the time necessary to master the trade and craftspeople have not managed to find supplementary uses for their carpentry skills. Furthermore, the ICSICH noted that safeguarding measures designed to sustain the shipbuilding tradition are underway, including State financial assistance to master builders, educational programmes to make it possible for them to transmit their traditional knowledge to young people, and the reconstruction of historical junks as a means to stimulate public awareness and provide employment (UNESCO, 2014).

Another cultural tradition linked to the sea is that of the lenj boats in the Islamic Republic of Iran. Lenj vessels are traditionally hand-built and are used by inhabitants of the northern coast of the Persian Gulf for sea journeys, trading, fishing and pearl diving. The traditional knowledge surrounding lenjes includes oral literature, performing arts and festivals, in addition to the sailing and navigation techniques, terminology and weather forecasting that are closely associated with sailing, and the skills of wooden boat-building itself. This tradition is also under threat, and the Islamic Republic of Iran has proposed a wide range of measures to safeguard it (UNESCO, 2014).

Along the north-east Pacific coast, sea-going canoes were one of the three major forms of monumental art among the Canadian First Nations and United States Native Americans, along with plank houses and totem poles. These canoes came to represent whole clans and communities and were a valuable trade item in the past, especially for the Haida, Tlingit and Nuu-Chah-Nulth. Recently, there has been a revival in the craft of making and sailing them, and they are capable of bringing prestige to communities (SFU, 2015).

Similar important navigational traditions survive in Melanesia, Micronesia and Polynesia. Using a combination of observations of stars, the shape of the waves, the interference patterns of sea swells, phosphorescence and wildlife, the Pacific Islanders have been able to cross vast distances at sea and make landfall on small islands. Although now largely being replaced by modern navigational aids, the Pacific navigational tradition shows how many aspects of the marine ecosystems can be welded together to provide results that at first sight seem impossible. Since the 1970s the tradition has been undergoing a renaissance (Lewis, 1994).

Apart from the practical cultural practices linked to the sea that support navigation, cultural practices in many parts of the world reflect the dangers of the ocean and the hope of seafarers to gain whatever supernatural help might be available. The fishing fleet is blessed throughout the Roman Catholic world, usually on 15 August, the Feast of the Assumption. This dates back to at least the 17th century in Liguria in Italy (Acta Sanctae Sedis, 1891). It spread generally around the Mediterranean, and was then taken by Italian, Portuguese and Spanish fishermen when they emigrated, and has been adopted in many countries, even those without a Roman Catholic tradition.

In many places in China and in the cultural zone influenced by China, a comparable festival is held on the festival of Mazu, also known (especially in Hong Kong, China) as Tian Hou (Queen of Heaven). According to legend, she was a fisherman's daughter from Fujian who intervened miraculously to save her father and/or her brothers and consequently became revered by fishermen, and was promoted by the Chinese Empire as part of their policy of unifying devotions. The main festival takes place on the 23rd day of the 3rd lunar month (late April/early May). A tradition of visiting a local shrine before a fishing voyage also continues in some places (Liu, 2003).

Miura, on the approaches to Tokyo Bay in Japan, developed as a military port and a harbour providing shelter to passing ships. Drawing on dances from other cities demonstrated to them by visiting sailors, the people of Miura began the tradition of Chakkirako to celebrate the New Year and bring fortune and a bountiful catch of fish in the months to come. By the mid-eighteenth century, the ceremony had taken its current form as a showcase for the talent of local girls. The dancers perform face-to-face in two lines or in a circle, holding fans before their faces in some pieces and clapping thin bamboo sticks together in others, whose sound gives its name to the ceremony. Now included in the UNESCO Representative List of the Intangible Cultural Heritage of Humanity, the ceremony is intended to demonstrate cultural continuity (UNESCO, 2014).

A specific cultural practice that acknowledges the importance of sea trade is the "Marriage of the Sea" (*Sposalizio del Mare*) in Venice, Italy. This takes the form of a boat procession from the centre of city to the open water, where the civic head (originally the Doge, now the Sindaco) throws a wedding ring into the sea. In 1177, Venice had successfully established its independence from the Emperor and Patriarch in Constantinople (Istanbul), from the Pope in Rome and from the Holy Roman

Emperor, by using its leverage to reconcile the two latter powers, and had become the great entrepôt between the eastern and western Mediterranean. Pope Alexander III acknowledged this by giving the Doge a ring. Henceforth, annually on Ascension Day, the Doge would "wed" the sea to demonstrate Venice's control of the Adriatic (Myers et al., 1971). Abolished when Napoleon dissolved the Venetian Republic, the ritual has been revived since 1965 as a tourist attraction (Veneziaunica, 2015).

3.2 Cultural practices that react to the sea

A verse in the Hebrew psalms speaks of the people "that go down to the sea in ships and...see...the wonders of the deep" (Psalm 107(106)/23, 24). A similar sense of awe at the sea appears in the Quran (Sura 2:164). This sense of awe at the ocean is widespread throughout the world. In many places it leads to a special sense of place with religious or spiritual connotations, which lead to special ways of behaving: in other words, to religious or spiritual ecosystem services from the ocean. A reductionist approach can see no more in such ways of behaviour than bases for prudential conduct: for example, fishing may be halted in some area at a specific time of year, which coincides with the spawning of a particular fish population, thus promoting the fish stock recruitment. But such a reductionist approach is not necessary, and can undermine a genuine sense of religious or spiritual reaction to the sea.

The risk exists that such reductionist approaches will be seen as the natural interpretation of ritual or religious practices. In a survey of the environmental history of the Pacific Islands, McNeill writes that "Lagoons and reefs probably felt the human touch even less [than the islands], although they made a large contribution to island sustenance...human cultural constraints often operated to preserve them. Pacific islanders moderated their impact on many ecosystems through restraints and restrictions on resource use. In many societies taboos or other prohibitions limited the exploitation of reefs, lagoons, and the sea. These taboos often had social or political purposes, but among their effects was a reduction in pressures on local ecosystems. Decisions about when and where harvesting might take place were made by men who had encyclopaedic knowledge of the local marine biota" (McNeill, 1994).

This clearly sets out the external ("etic") view of the system of taboos and beliefs, i.e., the view that can be taken by an outside, dispassionate observer. It does not allow for the internal ("emic") view as seen by someone who is born, brought up and educated within that system. It is important to understand this distinction and allow for the way in which the insider will have a different frame of reference from the outsider.

Good examples of the way in which such an insider's religious or spiritual reactions can underpin a whole system of community feeling can be found among the First Nations of the Pacific seaboard of Canada. A member of the Huu-ay-aht First Nation, a tribe within the Nuu-chah-nulth Tribal Group in this area, describes their traditional approach to whaling as follows:

"Whaling within Nuu chah nulth society was the foundation of our economic structure. It provided valuable products to sell, trade and barter. In essence it was our national bank... Whaling [however, also] strengthened, maintained and preserved our cultural practices, unwritten tribal laws, ceremonies, principles and teachings. All of these elements were practiced throughout the preparations, the hunt and the following celebrations. Whaling strengthened and preserved our spirituality and is clearly illustrated through the discipline that the Nuu chah nulth hereditary whaling chiefs exemplified in their months of bathing, praying and fasting in preparation for the hunt. The whale strengthened our relationships with other nations and communities. People came from great distances and often resulted in intertribal alliances, relationships and marriages. The whale strengthened the relationships between families because everyone was involved in the processing of the whale, the celebrations, the feasting, and the carving of the artefacts that can still be seen today in many museums around the world. The whale strengthened the relationships between family members since everyone shared in the bounty of the whale. And the whale strengthened our people spiritually, psychologically and physically" (Happynook, 2001).

Because of the restrictions imposed to respond to the crises in the whale population caused by commercial whaling, the Nuu-chah-nulth are not permitted to undertake whaling, and the related peoples further south in Washington State, United States, need to obtain special authorization (a request for which has been under consideration since 2005), and feel that part of their cultural heritage has been taken away from them. As the draft evaluation of the Makah request to resume whale-hunting puts it, with no authorization this element of their culture would remain a connection to the past without any present reinforcement. In effect, a cultural ecosystem service would be lost (NOAA, 2015).

3.3 Cultural practices tied to a specific sea area

Not all interactions between communities with traditions based on their long-standing uses of the ocean result in such clashes between opposing points of view. In Brazil, for example, the concept has been introduced of the Marine Extractive Reserve (*Reserva Extrativista Marinha*). These are defined areas of coast and coastal sea which aim to allow the long-standing inhabitants to continue to benefit from the resources of the reserve, applying their traditional knowledge and practices, while protecting the area against non-traditional, new exploitation, and protecting the environment (Chamy, 2002). Six such reserves have been created, and a further 12 are in the process of designation and organization (IBAMA, 2014).

In Australia, before colonization, the coastal clans of indigenous peoples regarded their territories as including both land and sea. The ocean, or "saltwater country", was not additional to a clan estate on land: it was inseparable from it. As on land, saltwater country contained evidence of the Dreamtime events by which all geographic features, animals, plants and people were created. It contained sacred sites, often related to these creation events, and it contained tracks, or Songlines, along which mythological beings travelled during the Dreamtime. Mountains, rivers, waterholes, animal and plant species, and other cultural resources came into being as a result of events which took place during these Dreamtime journeys. The sea, like the land, was integral to the identity of each clan, and clan members had a kin relationship to the important marine animals, plants, tides and currents. Many of these land features and heritage sites of cultural significance found within landscapes today have associations marked by physical, historical, ceremonial, religious and ritual manifestations located within the indigenous people's cultural beliefs and customary law. The Commonwealth and State Governments in Australia are now developing ways in which the groups of indigenous people can take a full part in managing the large marine reserves which have been, or are being, created, in line with their traditional culture. The techniques being used must vary, because they must take account of other vested rights and Australia's obligations under international law (AIATSIS, 2006).

Madagascar provides an interesting example of the way in which traditional beliefs can influence decisions on sea use. On the west coast of the northern tip of the island, a well-established shrimp-fishing industry is largely, but not entirely, undertaken by a local tribal group, the Antankarana. This group has a traditional set of beliefs, including in the existence of a set of spirits – the *antandrano* – who represent ancestors drowned in the sea centuries ago in an attempt to escape a local opposing tribal group, the Merina. These spirits are honoured by an annual ceremony focused on a particular rock in the sea in the shrimp fishery area. A proposal was made to create a shrimp aquaculture farm, which would have severely reduced the scope of the shrimp fishery. The Antankarana leader successfully invoked against this proposal reports from local mediums participating in the annual ceremony that the *antandrano* spirits would oppose the aquaculture proposal (which might well have been under Merina control). Thus a religious ecosystem service from the sea was deployed to defend a provisioning ecosystem service (Gezon, 1999).

At a global level, specific marine sites were inscribed by UNESCO in the World Heritage List, and thus brought under certain commitments and controls to safeguard them. So far 42 marine or coastal sites have been designated on the basis of their natural interest:

(a) 22 "contain superlative natural phenomena or areas of exceptional natural beauty and aesthetic importance";
(b) 12 are "outstanding examples representing major stages of earth's history, including the record of life, significant ongoing geological processes in the development of landforms, or significant geomorphic or physiographic features";
(c) 14 are "outstanding examples representing significant ongoing ecological and biological processes in the evolution and development of terrestrial, fresh water, coastal and marine ecosystems and communities of plants and animals"; and
(d) 29 "contain the most important and significant natural habitats for in-situ conservation of biological diversity, including those containing threatened species of outstanding universal value from the point of view of science or conservation".

(Sites can qualify under more than one criterion.)

Fifteen are islands. Three have been declared to be in danger: the Belize barrier reef (the largest in the northern hemisphere), which is threatened by mangrove cutting and excessive development (2009); the Florida Everglades in the United States, which have suffered a 60 per cent reduction in water flow and are threatened by eutrophication (2010); and East Rennell in the Solomon Islands, which is threatened by logging (2013). In addition, four marine or coastal sites have been inscribed in the World Heritage List because of their mixed cultural and natural interest – the island of St Kilda in the United Kingdom (for centuries a very remote inhabited settlement, featuring some of the highest cliffs in Europe); the island of Ibiza in Spain (a combination of prehistoric archaeological sites, fortifications influential in fortress design and the interaction of marine and coastal ecosystems); the Rock Islands Southern Lagoon (Ngerukewid Islands National Wildlife Preserve) in Palau (a combination of neolithic villages and the largest group of saltwater lakes in the world); and Papahānaumokuākea (a chain of low-lying islands and atolls with deep cosmological and traditional significance for living native Hawaiian culture, as an ancestral environment, as an embodiment of the Hawaiian concept of kinship between people and the natural world, and as the place where it is believed that life originates and where the spirits return after death) (UNESCO, 2014).

Other marine sites of cultural interest are those which offer the possibility of learning more about their past through underwater archaeology. Underwater archaeology draws on submerged sites, artefacts, human remains and landscapes to explain the origin and development of civilizations, and to help understand culture, history and climate change. Three million shipwrecks and sunken ruins and cities, like the remains of the Pharos of Alexandria, Egypt – one of the Seven Wonders of the Ancient World - and thousands of submerged prehistoric sites, including ports and methods of marine exploitation, such as fish traps, are estimated to exist worldwide. Material here is often better preserved than on land because of the different environmental conditions. In addition, shipwrecks can throw important light on ancient trade patterns; for example, the Uluburun shipwreck off the southern coast of Turkey, which illuminated the whole pattern of trade in the Middle East in the Bronze Age in the second millennium BCE (Aruz et al., 2008). Shipwrecks can also yield valuable information about the sociocultural, historical, economic, and political contexts at various scales of reference (local, regional, global) between the date of the vessel's construction (e.g. hull design, rig, materials used, its purpose, etc.) and its eventual demise in the sea (e.g. due to warfare, piracy/privateering, intentional abandonment, natural weather events, etc.) (Gould, 1983). Many national administrations pursue policies to ensure that underwater archaeological sites within their jurisdictions are properly treated. At the global level, the UNESCO Convention on the Protection of the Underwater Cultural Heritage (2001)[2] entered into force in 2009, and provides a framework for cooperation in this field and a widely recognized set of practical rules for the treatment and research of underwater cultural heritage.

---

2   United Nations, *Treaty Series*, vol. 2562. No. 45694.

Where such approaches are not applied, there are risks that irreplaceable sources of knowledge about the past will be destroyed. Bottom-trawling is a specific threat to underwater archaeological sites, with implications for the coordination of fisheries and marine archaeological site management. Questions also arise over archaeological sites outside national jurisdictions (mainly those of shipwrecks).

Cultural practices related to the sea, coastal sites of cultural interest (such as the UNESCO World Heritage Sites) and underwater archaeological sites form important elements for ocean-related tourism, which is discussed in Chapter 27 (Tourism and recreation). In particular, shipwrecks provide attractions for divers.

Special problems arise over recent shipwrecks where close relatives of people who died in the shipwreck are still living, particularly where the wreck occurred in wartime. Where the wrecks are in waters within national jurisdiction, many States have declared such sites to be protected, and (where appropriate) as war graves. As underwater exploration techniques improve, the possibility of exploring such wrecks in water beyond national jurisdiction increases, and this gives rise to sharp controversies.

Even without special remains or outstanding features, the ocean can provide an ecosystem service by giving onlookers a sense of place. The sense of openness and exposure to the elements that is given by the ocean can be very important to those who live by the sea, or visit it as tourists (see also Chapter 27). Even where the landward view has been spoiled by development, the seaward view may still be important. This is well demonstrated by a recent legal case in England, seeking to quash an approval for an offshore wind-farm at Redcar. Redcar is a seaside town with a large steel plant and much industrialization visible in its immediate hinterland. The beach and its view to the south-east are, however, described as spectacular. The court had to decide whether construction of the wind-farm about 1.5 kilometres offshore would introduce such a major new industrial element into the seascape/landscape as to undermine efforts to regenerate the seaside part of the town. The court decided that the ministry was justified in its approval, but the case underlines the importance of the aesthetic ecosystem service that the sea can provide (Redcar, 2008).

As described in Chapter 27 (Tourism and recreation), over the past 200 years there has been a growing cultural practice worldwide of taking recreation in coastal areas and at sea. Some evidence is emerging of positive links between human health and the enjoyment of the coastal and marine environment (Depledge et al., 2009; Wyles et al., 2014; Sandifer et al., 2015).

4 Cultural influences

Art reflects the society in which it is produced, and is influenced by that society's interests. The relationship between a society and the ocean is therefore likely to be reflected in its art. Much visual art therefore reflects the sense of place that is predominant in the society that generates it. The sense of place in societies that are much concerned with the sea reflects the aesthetic ecosystem services provided by the sea, hence the visual arts are also likely to reflect the same service. Examples of the way in which this occurs are not difficult to find. The Dutch painting school of the 17th century developed the seascape – ships battling the elements at sea – just at the period when the Dutch merchant ships and Dutch naval vessels were the dominant forces on the local ocean. The French impressionists of the second half of the 19th century took to painting coastal and beach scenes in Normandy just at the period when the railways had enabled the Parisian élite – their most likely patrons – to escape to the newly developed seaside resorts on the coast of the English Channel. Similarly, Hokusai's *The Great Wave at Kanagawa* is focused on a distant view of Mount Fuji rather than on the ocean – not surprising given that it was painted at a time when shipping in Japan was predominantly coastal. Today, the advances in cameras capable of operating under water, and the availability of easily managed breathing gear and protective clothing, result in the most stunning pictures of submarine life.

This reflection of the aspects of the aesthetic ecosystem services from the ocean that preoccupy the society contemporaneously with the work of the artist can also be found in literature and music. Camões's great epic *The Lusiads* appears just at the time when Portugal was leading the world in navigation and exploration. In the same period, Chinese literature saw the emergence of both fictional and non-fictional works based on the seven voyages of Admiral Zheng He in the south-east Asian seas and the Indian Ocean. It is with the emergence in the 19th century of widespread trading voyages by American and British ships that authors like Conrad, Kipling and Melville bring nautical novels into favour. Likewise, the impressionist seascapes in visual art are paralleled by impressionist music such as Debussy's *La mer*.

5 The ultimate ecosystem service for humans

Burial at sea has long been practiced as a matter of necessity during long voyages. It was specifically provided for in 1662 in the English Book of Common Prayer (BCP, 1662). Both the London Convention on the Prevention of Marine Pollution by Dumping of Wastes and Other Matter, 1972[3] and its Protocol[4] (see chapter 24), which regulate the dumping of waste and other matter at sea, are careful to leave open the possibility of the burial of human remains at sea. Western European States regularly authorize a small number of such disposals every year (LC-LP, 2014). The United States authorities have issued a general permit for burial at sea of human remains, including cremated and non-cremated remains, under certain conditions (USA-ECFR, 2015). In Japan, increasing prices for burial plots and concerns about the expanding use of land

---

[3] United Nations, *Treaty Series*, vol. 1046, No. 15749.

[4] 36 *International Legal Materials* 1 (1997).

for cemeteries have led to a growing pattern of cremation followed by the scattering of the cremated remains, often at sea. The practice started in 1991, when the law on the disposal of corpses was relaxed, and has become more popular following such funeral arrangements for a number of prominent people (Kawano, 2004).

6 Conclusions and identification of knowledge and capacity-building gaps

This chapter set out to review the ways in which ecosystem services from the sea interrelate with human aesthetic, cultural, religious and spiritual desires and needs. Five main conclusions emerge:

(a) Several goods produced by the ocean have been taken up as élite goods, that is, goods that can be used for conspicuous consumption or to demonstrate status in some other way. When that happens, a high risk exists that the pressures generated to acquire such élite goods, whether for display or consumption, will disrupt marine ecosystems, especially when the demand comes from relatively well-off consumers and the supply is provided by relatively poor producers. The development of the market in shark's fin is a good example of this (although signs exist that that particular situation has stopped getting worse).

(b) Some producers could be helped by a better understanding of the techniques and precautions needed to avoid ruining the production. As well as better knowledge, they may also need improved skills, equipment and/or machinery to implement that better understanding. The production of cultured pearls in the Cook Islands is a good example.

(c) Some élite goods pass through a number of hands between the original producer and the ultimate consumer. There appears to be a gap in capacity-building to safeguard producers and ensure more equitable profit-sharing in the supply chain. The case of small producers of cultured pearls is an example.

(d) Very different perceptions of marine ecosystem services and how humans relate to them can exist between different groups in society, even when such groups are co-located. Understanding on all sides of the reasons for those differences is a prerequisite for effective management of the ecosystem services.

(e) Aspects of the marine environment that are valued as cultural assets of humanity need constant consideration; they cannot just be left to fend for themselves. Where technology or social change has overtaken human skills that are still seen as valuable to preserve, conditions need to be created in which people want to learn those skills and are able to deploy them. Where an area of coast or sea is seen as a cultural asset of humanity, the knowledge is needed of how it can be maintained in the condition which gives it that value.

References

Acta Sanctae Sedis (1891). In: *Compendium Opportune Redacta et Illustrata Studio et Cura Victorii Piazzesi*. Congregatio de Propaganda Fide, Rome.

AIATSIS (2006). Australian Institute of Aboriginal and Torres Straits Islanders Studies. *Sea Countries of the South: Indigenous Interests and Connections within the South-west Marine Region of Australia*. http://www.environment.gov.au/indigenous/publications/pubs/sea-country-report.pdf (accessed 3 June 2014).

Alex, L.M. (2010). *Iowa's Archaeological Past*. University of Iowa Press, Iowa City (ISBN 0-87745-680-1).

Al-Shamlan, S.M. (2000). *Pearling in the Arabian Gulf: A Kuwaiti Memoir*. English Edition, trans. P. Clark, London Centre for Arab Studies.

Aruz, J., Benzel, K., and Evans, J.M. (2008). *Beyond Babylon: Art, Trade and Diplomacy in the Second Millennium BC*. New York (ISBN 978-1-58839-295-4).

Bartley, D. (2014). Fisheries and Aquaculture topics. Ornamental fish. Topics Fact Sheets. Text by Devin Bartley. In: *FAO Fisheries and Aquaculture Department* [online]. Rome. Updated 27 May 2005. (accessed 17 August 2014).

BCP - Book of Common Prayer (1662). *The Book of Common Prayer, and Administration of the Sacraments, and other Rites and Ceremonies of the Church According to the Use of the Church of England (Forms of Prayer to be used at Sea)*. Oxford.

BGS - British Geological Survey (2014). *World Mineral Production 2008 – 2012*. Keyworth, Nottingham (ISBN 978-0-85272-767-6).

Bondad-Reantaso, M.G., McGladdery, S.E., and Berthe, F.C.J. (2007). *Pearl Oyster Health Management - A Manual*. Food and Agriculture Organization of the United Nations, Fisheries Technical Paper No. 503, Rome.

Carter, R. (2005). The History and Prehistory of Pearling in the Persian Gulf. *Journal of the Economic and Social History of the Orient*, Vol. 48, No. 2.

Chamy, P. (2002). Reservas Extrativistas Marinhas: Um Estudo sobre Posse Tradicional e Sustentabilidade. First National Meeting of the Associação Nacional de Pós-Graduação e Pesquisa em Ambiente e Sociedade. http://www.anppas.org.br/encontro_anual/encontro1/gt/conhecimento_local/Paula%20Chamy.pdf (accessed 23 May 2014).

Charpentier, V., Phillips, C.S., and Méry, S. (2012). Pearl fishing in the ancient world: 7500 BP. *Arabian Archaeology and Epigraphy* 23: 1-6.

CITES - Conference of the Parties of the Convention on the International Trade in Endangered Species (2010). 15th Meeting, Proposition 21 (Inclusion of all species in the family in Appendix II) CITES Document AC27 Inf. 14. www.traffic.org/cites-cop-papers/CoP15_Prop21_Rec.pdf (accessed 15 June 2015).

CMFRI - Indian Central Marine Fisheries Research Institute (1991). *Training Manual on Pearl Oyster Farming and Pearl Culture in India*. Tuticorin.

Denhams Auction Catalogue (2014). Lot 960 (embossed silver gilt rattle with coral teething bar and 5 bells, London 1857). http://www.denhams.com/search.php?searchtext=rattle (accessed 12 August 2014).

Depledge, M.H., Bird, W.J., (2009). The Blue Gym: Health and wellbeing from our coasts. *Marine Pollution Bulletin* 58, 947-948.

EC - European Commission (2008). *Monitoring of International Trade in Ornamental Fish*. Consultation Paper prepared by the United Nations Environment Programme – World Conservation Monitoring Centre. http://old.unep-wcmc.org/medialibrary/2011/11/02/5fbf9a43/Monitoring%20of%20international%20trade%20in%20ornamental%20fish%20-%20Consultation%20Paper.pdf (accessed 15 June 2015).

Edwards N.E, and Beck, A.M. (2002). Animal-assisted therapy and nutrition in Alzheimer's disease. *Western Journal of Nursing Research*, Vol. 24.

FAO (2012). Food and Agriculture Organization of the United Nations. *The State of World Fisheries and Aquaculture*. Rome (ISBN 978-92-5-107225-7).

FAO (2014a) Food and Agriculture Organization of the United Nations. *State of the Global Market for Shark Commodities - Summary of the Draft FAO Technical Paper*.

FAO (2014b). Food and Agriculture Organization of the United Nations. *Aquaculture Fact Sheet: Seahorses*. http://www.fao.org/fishery/culturedspecies/Hippocampus_comes/en (accessed 15 June 2015).

FT (2012). Financial Times, London, 7 March 2012: Doulton, M. How to regulate harvest of precious pink bounty?

Gezon, L. (1999). Of Shrimps and Spirit Possession: Toward a Political Ecology of Resource Management in Northern Madagascar. *American Anthropologist, New Series*, Vol. 101, No. 1.

Gould, R.A. (1983). *Shipwreck Anthropology*. University of New Mexico Press, Albuquerque, New Mexico, USA (ISBN 9780826306876).

Happynook, T.M. *(Mauk-sis-a-nook) (2001).* Securing Nuu chah nulth Food, Health and Traditional Values through the Sustainable Use of Marine Mammals. Presentation at "Whaling and the Nuu chah nulth People", A Symposium at the Autry Museum of Western Heritage Griffith Park, Los Angeles, 24 March 2001. http://www.turtleisland.org/news/news-Nuuchahnulth.htm (accessed 31 May 2014).

Hawkes J., and Priestley J.B. (1955). *Journey down a rainbow*. Heinemann-Cresset, London.

Horatius - Quintus Horatius Flaccus. *De Arte Poetica*. Lines 14-21.

IBAMA (2014). Instituto Brasileiro do Meio Ambiente e dos Recursos Naturais Renováveis, *Reservas Extractivistas*. (http://www.ibama.gov.br/resex/resex.htm accessed 31 July 2014).

Kawano, S. (2004). Scattering Ashes of the Family Dead: Memorial Activity among the Bereaved in Contemporary Japan. *Ethnology*, Vol. 43, No. 3.

Kinch, J., and Teitelbaum, A. (2010). *Proceedings of the Sub-regional Workshop on the Marine Ornamental Trade in the Pacific, 2-5 December 2008, Noumea, New Caledonia*. Secretariat of the Pacific Community, Aquaculture Technical Papers (ISBN: 978-982-00-0373-6).

LC-LP - London Convention and London Protocol (2014). 36th Consultative Meeting of Contracting Parties (1972 London Convention) and 9th Meeting of Contracting Parties (1996 London Protocol), 3-7 November 2014. http://www.imo.org/MediaCentre/MeetingSummaries/LCLP/Pages/LC-36-LP-9.aspx (accessed 20 November 2014).

Lewis, D. (1994). *We the Navigators – The Ancient Art of Landfinding in the Pacific*. Honolulu (ISBN 978-0-8248-1582-0).

Liu, T.-S. (2003). A Nameless but Active Religion: An Anthropologist's View of Local Religion in Hong Kong and Macau. *The China Quarterly*, No. 174.

Lourandos H. (1997). *Continent of Hunter-Gatherers: New Perspectives in Australian Prehistory*. Cambridge University Press, Cambridge (ISBN: 978-052135946-7).

McNeill, J.R. (1994). Of Rats and Men: A Synoptic Environmental History of the Island Pacific. *Journal of World History*, Vol. 5, No. 2. University of Hawai'i Press.

Myers, M. L. and S. Boorsch (1971), Grand Occasions, *The Metropolitan Museum of Art Bulletin, New Series*, Vol. 29, No. 5.

NMFS - USA National Marine Fisheries Service (2013). *Hawksbill Sea Turtle (Eretmochelys Imbricata) 5-Year Review: Summary and Evaluation*. http://www.nmfs.noaa.gov/pr/pdfs/species/hawksbillseaturtle2013_5yearreview.pdf (accessed 15 June 2015).

NOAA - USA National Oceanic and Atmospheric Administration (2015). *Draft*

Environmental Impact Statement on the Makah Tribe Request to Hunt Gray Whales. http://www.westcoast.fisheries.noaa.gov/publications/protected_species/marine_mammals/cetaceans/gray_whales/makah_deis_feb_2015.pdf (accessed 20 April 2015).

Nudelman , F., Gotliv, B.A., and Addali, L. (2006). Mollusk Shell Formation: Mapping the Distribution of Organic Matrix Components Underlying a Single Aragonitic Tablet in Nacre. *Journal of Structural Biology*, Volume 153, Issue 2.

OBG (Oxford Business Group) (2013). *The Report : Ras al Khaimah* (page 135). (ISBN 978-1907065835), Oxford Business Group, Oxford, United Kingdom.

Penderel-Brodhurst, J. (1910). Boulle, André Charles. *Encyclopedia Britannica*. Cambridge University Press.

Pliny – Gaius Plinius Secundus (Pliny the Elder). *Naturalis Historia*. Book IX, chapter 54 (pearls), and Book XII, chapter 11 (coral).

Ploeg, A. (2004). *The Volume of the Ornamental Fish Trade*. http://www.ornamental-fish-int.org/files/files/volume-of-the-trade.pdf (accessed 12 July 2014).

Redcar (2008). The Queen on the Application of Redcar and Cleveland Borough Council *versus* the Secretary of State for Business, Enterprise and Regulatory Reform EDF Energy (Northern Offshore Wind) Limited, [2008] EWHC 1847 (Admin).

Sandifer, P.A., A.E. Sutton-Grier, and B.P. Ward. 2015. Exploring connections among nature, biodiversity, ecosystem services, and human health and well-being: Opportunities to enhance health and biodiversity conservation. *Ecosystem Services*. 12:1-15 (http://www.sciencedirect.com/science/article/pii/S2212041614001648)

SFU (2015). Simon Fraser University, Bill Reid Centre for North-West Coast Studies (2015). *North-West Coast Canoes*. https://www.sfu.ca/brc/art_architecture/canoes.html (accessed 24 April 2015).

Southgate, P.C. (2007). Overview of the cultured marine pearl industry. In: *Pearl Oyster Health Management*. Ed. Bondad-Reantaso et al. Food and Agriculture Organization of the United Nations, Fisheries Technical Paper No. 503, Rome.

SPC - Secretariat of the Pacific Community (2011). *Pearl Oyster Information Bulletin*, Issue 19 (ISSN 1021-1861).

Stieglitz R. (1994). The Minoan Origin of Tyrian Purple. *The Biblical Archaeologist*, Vol. 57, No. 1.

Thackeray, W.M. (1869). A Dinner in the City. In: *Sketches and Travels in London*. Smith & Elder, London.

Tsounis, G., Rossi, G., Grigg, R., Santangelo, G., Bramanti, L., and Gili, J.-M. (2010). The Exploitation and Conservation of Precious Corals. In: *Oceanography and Marine Biology: An Annual Review*. Ed. R. N. Gibson, R. J. A. Atkinson, and J. D. M. Gordon.

UNESCO - United Nations Educational, Scientific and Cultural Organization (2014). Lists of Intangible Cultural Heritage and Register of Best Safeguarding Practices. http://www.unesco.org/culture/ich/index.php?lg=en&pg=00011#tabs (accessed 15 June 2015).

UNESCO - United Nations Educational, Scientific and Cultural Organization (2015). World Heritage List. http://whc.unesco.org/en/list/ (accessed 16 June 2015).USA-ECFR - Electronic Code of Federal Regulations (2015). Title 40, Part 229, §229-1 (Burial at Sea). http://www.ecfr.gov/cgibin/retrieveECFR?gp=1&SID=f280c25ad55688c1cf1ec8ff69c3885f&ty=HTML&h=L&mc=true&r=SECTION&n=se40.25.229_11 (accessed 24 April 2015)

Veneziaunica (2015). Festa Della Sensa. http://www.veneziaunica.it/it/content/festa-della-sensa (accessed 15 June 2015).

WAA - Waldorf Astoria Hotel Archive (1945). *Menu for Dinner on 19 June 1945 for General Dwight D. Eisenhower*. http://www.hosttotheworld.com/omeka/files/original/4182ef6f0e6c0a7947f4533f0b8e93f3.jpg (accessed 12 July 2014).

Wabnitz, C., Taylor, M., Green, E. and Razak, T. (2003). The Global Marine Aquarium Database (GMAD). *From Ocean to Aquarium*. United Nations Environment Programme – World Conservation Monitoring Centre, Cambridge (ISBN: 92-807-2363-4).

White, M., Smith, A., Humphryes, K., Pahl, S., Snelling, D., Depledge, M. (2010). Blue space: the importance of water for preference, affect, and restorativeness ratings of natural and built scenes. Journal of Environmental Psychology 30, 482-493.

Wood, E., Malsch, K., and Miller, J. (2012). *International trade in hard corals: review of management, sustainability and trends*. Proceedings of the 12th International Coral Reef Symposium, Cairns, Australia, 9-13 July 2012. http://www.icrs2012.com/proceedings/manuscripts/ICRS2012_19C_1.pdf (accessed 14 June 2014).

Wyles, K.J., Pahl, S., Thompson, R.C. (2014). Perceived risks and benefits of recreational visits to the marine environment: Integrating impacts on the environment and impacts on the visitor. *Ocean & Coastal Management* 88, 53-63.

9 Conclusions on Major Ecosystem Services Other than Provisioning Services

Contributors:
Patricio A. Bernal (Lead Member)

Chapter 9
Conclusions on Major Ecosystem Services Other than Provisioning Services

1   Introduction

The ecosystem services assessed in Part III are large-scale; some of them are planetary in nature and provide human benefits through the normal functioning of the natural systems in the ocean, without human intervention. This makes them intrinsically difficult to value. However, some of these same ecosystem services in turn sustain provisioning services that generate human benefits through the active intervention of humans. This is the case, for example, for the global ecosystem service provided by primary production by marine plants, which by synthesizing organic matter from CO_2 and water, provide the base of nearly all food chains in the ocean (except the chemosynthetic ones), and provide the food for animal consumers that in turn sustain important provisioning ecosystem services from which humans benefit, such as fisheries.

The services in Part III are not the only ones provided by the ocean. Many other ecosystem services are directly or indirectly referred to in Parts IV to VI of this Assessment. The provisioning ecosystem services related to food security are addressed in Part IV, *Assessment of Cross-Cutting Issues: Food Security And Food Safety* (Chapters 10 through 16); those related to coastal protection are referred to in Part VI, *Assessment of Marine Biological Diversity and Habitats*, in *Warm Water Corals* (Chapter 43), *Mangroves* (Chapter 48), and in *Aquaculture* (Chapter 12), *Estuaries and Deltas* (Chapter 44), *Kelp Forests and Seagrass Meadows* (Chapter 47) and *Salt Marshes* (Chapter 49); the maintenance of special habitats are addressed in Chapters on *Open Ocean Deep-sea Biomass* (Chapter 36F); *Cold Water Corals* (Chapter 42) and *Warm Water Corals* (Chapter 43), *Hydrothermal Vents and Cold Seeps* (Chapter 45), *High-Latitude Ice* (46) and *Seamounts and Other Submarine Geological Features Potentially Threatened by Disturbance* (Chapter 51); the sequestration of carbon in coastal sediments, the so-called blue carbon, is addressed in Chapters on *Mangroves* (Chapter 48), *Estuaries and Deltas* (Chapter 44) and *Salt Marshes* (Chapter 49); the cycling of nutrients is covered in *Estuaries and Deltas* (Chapter 44) and *Salt Marshes* (Chapter 49, but also Chapter 6).

Because of the very large scale of the services analysed in Part III, although they are influenced by human activities, they cannot be easily managed, and in certain cases they cannot be managed at all. The uptake of atmospheric CO_2 (Chapter 5) and the role of the ocean in the hydrological cycle (Chapter 4) are two examples of regulatory ecosystem services that cannot be managed or valued easily.

2   Accounting for the human benefits obtained from nature

Ecosystems can exist without humans in them, but humans cannot survive without ecosystems. Throughout history, humanity has made use of nature for food, shelter, protection and engaging in cultural activities. The intensity of humanity's use of nature has changed with the evolution of society and reached high levels with the introduction of modern technologies and industrial systems. Today, at a planetary scale, including the deepest ocean, no natural or pristine systems are found without people or unaffected by the impact of human activities; nor do social systems exist that can thrive without the support of nature. Social and ecological systems are truly interdependent and constantly co-evolving.

This fundamental connection between humans and nature has received different levels of recognition with regard to how we deal with the benefits humans extract from nature in economic terms. Extractive activities, e.g., of minerals, or of living natural resources, such as fibre, timber and fish, raise the issues of irreplaceability and sustainability. The use of nature is multifaceted and, as a norm, a given ecosystem can provide many goods and several services at the same time. For example, a mangrove ecosystem provides wood fibre, fuel, and nursery habitat for numerous species (provisioning services); it detoxifies and sequesters pollutants coming from upstream sources, stores carbon, traps sediment, and thus protects downstream coral reefs, and buffers shores from tsunamis and storms (regulating services); it provides beautiful places to fish or snorkel (cultural services); and it recycles nutrients and fixes carbon (supporting services; Lubchenco and Petes, 2010). When humans convert a natural ecosystem to another use, some ecosystem services may be lost and others services gained. Such a process gives rise to trade-offs between natural services and between these and services not derived from natural capital. For example, when mangroves are converted to shrimp ponds, airports, shopping malls, agricultural lands, or residential areas, new services are obtained: food production, space for commerce, transportation, and housing, but the original natural services are lost (Lubchenco and Petes, 2010).

Therefore, human benefits can be derived from a series of different activities that simultaneously affect the same ecosystem, but that are not necessarily connected with their harvesting or production processes. Sustainability requires that users take only a fraction of the resources, preserving in this way the natural capability of the ecosystem to regenerate the same resources, making them available for use by future generations. The appropriate spatial scale and the time sufficient to recover are part of the sustainability requirement. To extract anchovies (*Engraulis spp.*) or sardines (*Sardinops spp.*) that can regenerate their populations in three to eight years has different implications than to extract orange roughy (*Hoplostethus atlanticus*) from the top of seamounts that needs 100 to 150 years or more to recover.

3   The evolution of management tools

The increase in "*the magnitude of human pressures on the natural system has caused a transition from single-species or single-sector management to multi-sector, ecosystem-based management across multiple geographic and temporal dimensions. (…) Intensification of use of ecosystems increases interactions between sectors and production systems that in turn increase the number of mutual negative impacts (i.e., externalities)*" (Chapter 3).

Chapter 9   Conclusions on Major Ecosystem Services Other than Provisioning Services

Because all these processes take place in an integrated socio-ecological system, we have seen an expansion of scope in the decision-making process, incorporating the simultaneous consideration of several uses or industries at the same time and the livelihoods and other social aspects connected with this ensemble of activities. These approaches enable the consideration of tradeoffs among different uses and beneficiaries, enlarging the range of policy options. Only recently have regulatory instruments for better accounting for the indirect and cumulative impacts on natural systems of these multiple uses been incorporated into the management and regulation of human activities. Mobilized by a series of high-level World Conferences addressing these issues, in Stockholm 1972, Rio de Janeiro 1992, Johannesburg 2002, and Rio de Janeiro 2012, the international community has acted to advance and implement this enlarged scope of decision-making across all societies.

Assigning value to the human benefits obtained from nature is more easily done when the goods and services obtained are traded, thereby becoming part of commerce. Prices in different markets are readily available and comparisons are possible. It is not that simple, however, for certain types of benefits, for example, when subsistence livelihoods that do not enter into trade are concerned, or other intangible cultural, recreational, religious or spiritual benefits are involved.

However, the extraction of natural products and other human benefits from wild ecosystems can affect other processes inside the ecosystems that provide valuable permanent services to humanity that are not part of commerce. Examples include the production of organic matter and oxygen through primary production in the ocean, the protection of the coast by mangrove forests, the re-mineralization of decaying organic matter at the coastal fringes, and the absorption of heat and CO_2 by the ocean that has delayed the impacts of global warming. Furthermore, fluctuations in the provision of these natural services can have significant impacts on those natural products that are in commerce.

Climate patterns drive the magnitude and variability of the circulation and heat storage capacity of the surface layers of the ocean, as described in Chapters 4 and 5. The displacement of warm and cold water pools on the surface of the ocean feeds into the dynamics of the atmosphere, generating enormous transient fluctuations in weather patterns, such as the El Niño and La Niña cycles, that cascade down affecting the production of a series of goods and services, not only in the ocean, but most notably also on land. For example, El Niño adversely affects the availability and price of fishmeal and fish oil, also key components of the diet of carnivorous species in aquaculture (see Chapter 12).

4   Scientific understanding of ecosystem services

The fundamental connection between humans and nature has received uneven levels of recognition in how we deal in economic terms with the benefits humans extract from nature.

Humans derive many benefits from all aspects of the natural world. Some of these benefits are provided by nature without human intervention and some require human inputs, often with substantial labour and economic investment. The features and functions of nature which provide these services can be regarded as "natural capital", and the way in which this natural capital is organized and how it functions in delivering benefits to humans, has led to these types of benefits being described as "ecosystem services". The Millennium Ecosystem Assessment characterizes ecosystem services as: provisioning services (e.g., food, pharmaceutical compounds, building material); regulating services (e.g., climate regulation, moderation of extreme events, waste treatment, erosion protection, maintaining populations of species); supporting services (e.g., nutrient cycling, primary production); and cultural services (e.g., spiritual experience, recreation, information for cognitive development, aesthetics).

The rent for land or the royalties on mineral extraction are examples of long-established approaches adopted to account for uses of nature. They are based on a one-to-one relationship between one activity or industry and one natural source of the goods or services. The effect of other industries on the same ecosystem is not considered; neither are the impacts on other members of the social system affected by these industries.

To comprehensively account for human benefits and costs, "natural capital" needs to be considered alongside the assets that humans have themselves developed, whether in the form of individual skills ("human capital"), the social structures they have created ("social capital") or the physical assets that they have developed ("built capital"). Managing the scale of the human efforts in using natural capital is crucial. The ecosystem-services approach allows decision-making to be integrated across land, sea and the atmosphere and enables an understanding of the potential and nature of trade-offs among services given different management actions.

The increasing magnitude of human pressures on the natural system has caused a transition from an emphasis on single-species or single-sector management to multi-sector, ecosystem-based management and hence to the explicit incorporation of human actions in socio-ecological systems management.

A number of variants of the ecosystem-services approach exist. Some emphasize the functional aspects of ecosystems from which people derive benefits. Others put more emphasis on their utilitarian aspects and seek to apply mainstream economic accounting methods, assigning them monetary values obtained in the market or using non-market methodologies. Yet others emphasize human well-being and ethical values. Looking at ecosystem services requires consideration of a wide range of scales, from the completely global (for example, the role of the oceans in distributing heat around the world; Chapter 5) to the very local (for example, the protection offered by coral reefs to low-lying islands; Chapter 42).

Most studies conducted on marine and coastal ecosystem services have been focused locally and, in general, have not taken into account benefits generated further afield. An ecosystem-services approach can help with decision-making under conditions of uncertainty, and can bring to light important synergies and trade-offs between different uses of the ocean. However, attempting to assess the relationship between the operation of ecosystem services and the interests of humans requires a much broader management approach and an understanding that many aspects of the ocean have non-linear behaviour and responses. One difficulty is that to date no generally agreed classification of goods and services derived from natural capital exists that could facilitate the task. Another obstacle is that the range of factors involved at all levels might require the consideration of their interaction at the relevant scale, making their treatment very complex.

Some ecosystem services are more visible and easily understood than others. There is a risk that the less visible an ecosystem service is, the less it will be taken into account in decision-making. There is also a risk that ecosystem services that can be valued in monetary terms will be understood more easily than others, thus distorting decisions. Likewise, the time scales over which some ecosystem services will be affected by decisions will be much longer than others, which an approach with traditionally used discount rates would completely dismiss.

4.1 Information gaps

Describing and mapping the full range of ecosystem services at different scales requires much data on the underlying functions and structure of the way in which ecosystems operate. Although much work has been done, mostly on terrestrial ecosystems, this assessment draws attention in its various chapters to the gaps in the information needed to understand the way in which individual ecosystem services operate in the ocean and along the coasts. In addition to these specific gaps, a more general, overarching gap exists in understanding how all the individual ecosystem services fit together.

4.2 Capacity-building gaps

Many skills have yet to be developed that should integrate an understanding of the operation of ecosystems: knowledge is currently too fragmented between different specialisms.

Even in cases where the appropriate skills exist, some parts of the world lack institutions with the status, resources and commitment to make the necessary inputs into decision-making that will affect a range of ecosystem services from the oceans.

The many institutions that already exist to study the ocean also require new or enhanced abilities and connections, but can be expected to work together. Some international networks already exist to facilitate this. More are needed.

5 The ocean's role in the hydrological cycle

Water is essential for life and the existence of water in a liquid state on the surface of the earth is probably a critical reason why life is found on this planet and not on others. The presence of water on the surface of the earth is the combined result of the cosmic and geological history of the planet during its 4.5 billion years of existence. These processes, at human time-scales of thousands of years, can be considered as quasi-stable.

The ocean dominates the hydrological cycle. The great majority of the water on the surface of the planet, 97 per cent, is stored in the ocean. Only 2.5 per cent of the global balance of water is fresh water, of which approximately 69 per cent is permanent ice or snow and 30 per cent is ground water. The remainder 1 per cent is available in soil, lakes, rivers, swamps, etc. (Trenberth et al., 2007).

Water evaporates from the planet's surface, is transported through the atmosphere and falls as rain or snow. Rain is the largest source of fresh water entering the ocean (~530,000 km^3 yr^{-1}). At the ocean-atmosphere interface, 85 per cent of surface evaporation and 77 per cent of surface rainfall occur (Trenberth et al., 2007; Schanze et al., 2010). The residence time of all water in the atmosphere is only seven days. It is fair to say that "*all atmospheric water is on a short-term loan from the ocean*".

However, as with many other cycles, the water cycle is a dynamic system in a quasi steady-state condition. This steady-state condition can be altered if the factors controlling the cycle change. The great glaciations, or ice ages, are processes in which, due to interplanetary and planetary changes, huge amounts of water pass from the liquid to the solid state, altering the availability of liquid water on the surface of the earth and dramatically changing the sea level. As a consequence of the change in sea level, the shape of world coastlines and the amount of emerged (or submerged) land is drastically changed.

Changes are occurring today at an unprecedentedly fast but still uncertain rate. A warmer ocean expands and, being contained by rigid basins, the only surface that can move is the free surface in contact with the atmosphere, raising sea level.

In addition, the melting of ice due to a warmer atmosphere and a warmer ocean is increasing the volume of the ocean and in the long run will dominate the amount of total change in sea level.

As the ocean warms, evaporation will increase, and global precipitation patterns will change. The IPCC assess (AR 3, 2001; AR 4, 2007; and AR 5, 2013 and 2014) that the dynamic system of water on earth, driven by global warming, is changing sea level at a mean rate of 1.7 [1.5 to 1.9] mm yr^{-1} between 1901 and 2010, increased to 3.2 [2.8 to 3.6] mm yr^{-1} between 1993 and 2010 and, due to the changes in climate, will also change the patterns of rain on land.

Warming affects the polar ice caps, and changes in their melting affect the salinity of the ocean. This in turn can affect the ocean circulation, especially the thermohaline vertical circulation, also known as the "conveyor belt" (Chapter 5) and the operation of associated ecosystem services.

Due to the concentration of human population and built infrastructure in the coastal zone, sea level rise will seriously affect the way in which humans operate. The effects of sea-level rise will vary widely between regions and areas, with some of the regions most affected least able to manage a response.

Changes in water run-off will affect both land and sea. Salinity in the different parts of the ocean has changed over time, but is now changing more rapidly. Gradients in salinity are becoming more marked. Because the distribution of marine biota is affected by the salinity of the water that they inhabit, changes in the distribution of salinity are likely to result in changes in distribution of the biota.

Changes in run-off from land are affecting the input of nutrients to the ocean. These changes will also affect marine ecosystems, due to increases in the acidity of the ocean.

The warming of the ocean is not uniform. It is modulated by oscillations such as El Niño. These oscillations cause significant transient changes to the climate and ecosystems, both on land and at sea, and have serious economic effects. The variations in ocean warming will affect the interaction with the atmosphere and affect the intensity and distribution of tropical storms.

5.1 Information gaps

The sheer scale of the changes that are happening to the ocean makes present knowledge inadequate to understand all the implications. There are gaps in understanding sea-level rise, and interior temperature, salinity, nutrient and carbon distributions. Many of these gaps are being addressed as part of the world climate change agenda, but more detail about ocean conditions is needed for regional and local management decisions. The information gaps are particularly serious in some of the areas most seriously affected (e.g., the inter-tropical band).

5.2 Capacity-building gaps

Parts of the water cycle are still subject to uncertainties due to insufficient *in situ* measurements and observations. This is particularly true for surface water fluxes at the regional meso-scale. Because changes in the ocean's role on the hydrological cycle will be pervasive, all parts of the world need to have access to, and the ability to interpret, these changes. People and institutions with the necessary skills exist in many countries, but in many others they lack the status and resources to make the necessary input to decision-making.

6 Sea-Air Interaction

Most of the heat excess due to increases in atmospheric greenhouse gases is absorbed by the ocean. All ocean basins have warmed during recent decades, but the increase in heat content is not uniform; the increase in heat content in the Atlantic during the last four decades exceeds that of the Pacific and Indian Oceans combined.

'Recent' warming (since the 1950s) is strongly evident in sea surface temperatures at all latitudes in all part of the ocean. Prominent structures that change over time and space, including the El Nino Southern Oscillation (ENSO), decadal variability patterns in the Pacific Ocean, and a hemispheric asymmetry in the Atlantic Ocean, have been highlighted as contributors to the regional differences in surface warming rates, which in turn affect atmospheric circulation.

The effects of these large-scale climate oscillations are often felt around the world, leading to the rearrangement of wind and precipitation patterns, which in turn substantially affect regional weather, sometimes with devastating consequences.

Compared with estimates for the global ocean, coastal waters are warming faster: during the last three decades, approximately 70 per cent of the world's coastline has experienced significant increases in sea surface temperature. Such coastal warming can have many serious consequences for the ecological system, including species relocation.

These changes are also affecting the salinity of the ocean: saline surface waters in the evaporation-dominated mid-latitudes have become more saline, while relatively fresh surface waters in rainfall-dominated tropical and polar regions have become fresher.

Approximately 83 per cent of global CO_2 increase is currently generated from the burning of fossil fuels and industrial activity. Forests and grasslands that usually absorb CO_2 from the atmosphere are being removed, causing even more CO_2 to be absorbed by the ocean. The ocean thus serves as an important sink of atmospheric CO_2, effectively slowing down global climate change. However, this benefit comes with a steep bio-ecological cost. When CO_2 reacts with water, it forms carbonic acid, leading to the ocean becoming more acid – referred to as "ocean acidification". Through various routes, this imposes an additional energy cost to calcifier organisms, such as corals and shell-bearing plankton, although this is by no means the sole impact of ocean acidification (OA).

OA is not a simple phenomenon nor will it have a simple unidirectional effect on organisms. Calcification is an internal process that in the vast majority of cases does not depend directly on seawater carbonate content, since most organism use bicarbonate, which is increasing under acidification scenarios, or CO_2 originating in their internal metabolism. It has been demonstrated in the laboratory and in the field that some calcifiers can compensate and thrive in acidification conditions.

Without significant intervention to reduce CO_2 emissions, by the end of this century, average surface ocean pH is expected to be below 7.8, an unprecedented level in recent geological history. These changes will be recorded first at the ocean's surface, where the highest biodiversity and productivity occur.

The abundance and composition of species may be changed, due to OA, with the potential to affect ecosystem function at all trophic levels. Consequential changes in ocean chemistry could occur as well. Economic studies have shown that potential losses at local and regional scales may be catastrophic for communities and national economies that depend on fisheries.

6.1 Information Gaps

Regular monitoring of relevant fluxes across the ocean-atmosphere interface needs to be maintained, including the regular assessment of the accumulation of heat and CO_2 (changes in alkalinity) in the surface layers of the ocean. The state of knowledge regarding OA is only currently moving beyond the nascent stage. Therefore several major information gaps are yet to be filled. Neither all areas of the globe nor all potentially affected animal and plant groups have yet been covered in terms of research. The full range of response and adaptation of organisms, although an active field of research, is very seldom known.

Additional information is required around the effects of mean changes in OA versus changes in variability and extremes; as well as multi-generational effects and adaptive potential of different organisms (Riebesell and Gattuso, 2015; Sunday et al., 2014). The tolerance level by individual organisms to changes in pH must be understood in situ rather than exclusively in laboratory conditions, along with the possible consequential changes in competition by organisms for resources. The effects of potential loss of keystone species within ecosystems are not yet clear, neither are the chemical changes due to pH.

The future agenda of research in OA should include integrating knowledge on multiple stressors, competitive and trophic interactions, and adaptation through evolution and moving from single-species to community assessments (Sunday et al., 2014). Future economic impacts of OA are being studied but much more needs to be done. Monitoring and management strategies for maintaining the marine economy will also be needed. In this regard, it has been suggested that future OA research could focus on species related to ecosystem services in anticipation that these case studies might be most useful for modellers and managers (Sunday et al. 2014).

6.2 Capacity-Building Gaps

OA was put in evidence only through long-term observational programmes coordinated by the international research community. To monitor OA on a regular basis, these international efforts need to be continued and institutionally consolidated. OA adaptation is a demanding field of research that requires significant infrastructure (i.e. mesocosm experimental facilities) and highly qualified human resources. These are not readily available in all regions of the world. Further understanding of the scope of adaptation capabilities to OA by plants and animals, the application of mitigation strategies, or the successful management of productive systems cultivating organisms with calcareous exo-skeletons that are regularly exposed to corrosive waters sources, require the existence of these capabilities in place.

7 Primary Production, Cycling of Nutrients, Surface Layer and Plankton

"Marine primary production" is the photosynthesis of plant life in the ocean to produce organic matter, using the energy from sunlight, and carbon dioxide and nutrients dissolved in seawater. Carbon dioxide dissolved in seawater is drawn from the atmosphere. Oxygen is produced as a by-product of photosynthesis both on land and in the ocean. Of the total annual oxygen production from photosynthesis on land and ocean, approximately half originates in marine plants. The plants involved in this process range from the microscopic phytoplankton to giant seaweeds. On land, the other 50 per cent of the world's oxygen originates in the photosynthesis from all plants and forests. Present-day animals and bacteria rely on present-day oxygen production by plants on land and in the ocean as a critical ecosystem service that keeps atmospheric oxygen from otherwise declining.

Marine primary production, as the primary source of organic matter in the ocean, is the basis of nearly all life in the oceans, playing an important role in the global cycling of carbon. Phytoplankton absorbs about 50 billion tons of carbon a year, and large seaweeds and other marine plants (macrophytes) about 3 billion tons. At a planetary level, this ecological function plays an important role in removing CO_2 – one of the significant greenhouse gases – from the atmosphere. Total annual anthropogenic emissions of CO_2 are estimated at 49.5 billion tons, one-third of which is taken up by the ocean. This ecological service provided by the ocean has so far prevented warming of the planet above 2º C.

Marine primary production also plays a major role in the cycling of nitrogen around the world. A moderate level of uncertainty exists about the extent to which the ocean is currently a net absorber or releaser of nitrogen.

Anthropogenic nutrient loading in coastal waters, ocean warming, ocean acidification, and sea-level rise are driving changes in the phenology (see below) and spatial pattern of phytoplankton and in net primary production as well as in nutrient cycles. These changes are threatening the provision of several ecosystem services at different scales (local, regional).

Changes in macrophyte net primary production and their impacts (losses of habitat and of carbon sinks) are also well documented.

Changes in net primary production by phytoplankton or in the nutrient cycles in the upper levels of the world ocean have been the subject of recent debate. Some analyses suggested a diminishing trend in world primary production (Boyce et al., 2010). However, this result was challenged by Rykaczewski and Dunne (2011) and finally reviewed by the original authors (Boyce, et al., 2014). After recalibration of the datasets, despite an overall decline of chlorophyll concentration over 62 per cent of the global ocean with sufficient observations, they describe a more balanced picture between regional increases and decreases of primary production, without an overall globally pervasive negative trend. The wider consequences of this long-term trend are presently unresolved.

The efficient use of primary production to support animals at higher levels in the food web depends on a good relationship between the timing of bursts of primary production and breeding periods of zooplankton and planktivorous fish. The phenology of species, i.e., the timing of events in the life cycle of species, plays a significant role here. Changes in the phenology of plankton species and planktivorous fish due to ocean warming is starting to produce significant mismatches and could produce many more, affecting the local level of production in the ocean.

Warming of the upper ocean and associated increases in vertical stratification may lead to a major decrease in the proportion of primary production going to zooplankton and planktivorous fish, and an increase in the proportion of phytoplankton being broken down by microbes without first entering the higher levels of the food web. Such a trend would reduce the carrying capacity of the oceans for fisheries and the capacity of the oceans to mitigate the impacts of anthropogenic climate change.

Coastal eutrophication (see Chapter 20) is likely to lead to an increase in the numbers and area of dead zones and toxic phytoplankton blooms. Both can have serious effects on the supply to humans of food from the sea.

Nanoparticles (microscopic fragments of plastic and other anthropogenic substances) pose a potential serious threat to plankton and the vast numbers of marine biota which depend on them.

Increases in nitrogen inputs to the ocean and in sea temperatures may have serious impacts on the type (species, size) and amount of marine primary production, although much debate still occurs about the scale of this phenomenon. Different responses are likely to be found in different regions. This may be particularly significant in the Southern Ocean, where a major drop in primary production has been forecast.

7.1   Information gaps

Completing a worldwide observing system for the biology and water quality of the ocean that could provide cost-effectively improved information to future assessments under the Regular Process is seen as an important gap. Routine and sustained measurements across all parts of the ocean are needed on planktonic species diversity, chlorophyll *a*, dissolved nitrogen and dissolved biologically active phosphorus. Due to their ability to enter into marine food chains, with a potential impact on both marine organism and human health, plastic microparticles need to be systematically monitored.

Without this additional information, it will not be possible to understand or predict the changes that will occur due to the accumulated and combined effect of several drivers. Some information can be derived from satellite remote sensing, but *in situ* observations at the surface and especially at sub-surface levels are irreplaceable, given the fact that the ocean is essentially opaque to electromagnetic radiation, the medium *par excellence* of remote sensing.

7.2   Capacity-building gaps

The gathering of such information requires a worldwide network for data collection. Both a Global Ocean Observing System (GOOS) and a Global Biodiversity Observation Network (GEO BON) are currently being developed to collect biological and ecologically relevant information that, if completed, would fill in some of the information gaps described above. The capacities to participate in these systems need to be extended worldwide. It is also important to develop the skills to explain to decision-makers and the general public the importance of plankton and its significance for the ecosystem services provided by the ocean.

8   Ocean-sourced carbonate production

Many marine organisms secrete calcium carbonate to produce a hard skeleton. These vary in size from the microscopic plankton, through corals, to large mollusc shells. Carbonate production by corals is particularly important, because the reefs that they form are fundamental to the existence of many islands and some entire States.

Sand beaches are also often formed by the fragmented shells of marine biota. Beaches are dynamic structures, under constant change from the effects of the oceans; hence a constant supply of new sand of this kind is needed to sustain them.

Sea-level rise will particularly affect beaches, causing them to move inland. In the case of small islands, this may diminish the already limited inhabitable area. Such changes may also be affected by the availability of new supplies of sand. Such changes could be very serious for States or parts of States comprised of atolls.

The impact of climate change on the rate of biogenic production of carbonate sediment is also little understood, but it seems likely to have negative consequences. Ocean warming has already led to the death (bleaching) of some coral reefs or parts of them.

The acidification of the ocean may lead to significant changes in the biogenic production of calcium carbonate, with implications for the future of coral reefs and shell beaches.

8.1 Knowledge gaps

Long-term data are lacking on the formation and fate of reef islands and shell beaches. Information is particularly lacking on the links between the physical structures and the environmental circumstances that may affect changes in these structures.

8.2 Capacity-building gaps

A gap often exists in the capacities of people living on atolls and in countries depending heavily on tourism from shell beaches to understand the drivers that shape the development of these structures. Without such capacities it is impossible to bring the factors affecting the future of these structures into the making of decisions which can fundamentally affect them.

9 Aesthetic, cultural, religious and spiritual ecosystem services derived from the marine environment

The development of human culture over the centuries has been influenced by the ocean, through transport of cultural aspects across the seas, the acquisition of cultural objects from the sea, the development of culture to manage human activities at sea, and the interaction of cultural activities with the sea.

The ocean has been and continues to be the source of prized materials for cultural use, for example: pearls, mother-of-pearl, coral, and tortoiseshell. Some marine foodstuffs are also ingredients in culturally significant dishes. The high value given to many of these culturally significant objects can lead to their over-exploitation and the long-term damage of the ecosystems that supply them. The sea is an important element in many sites of cultural significance. Forty-six marine and coastal sites are included in the UNESCO World Heritage List.

Even where sites do not reach this high threshold of cultural significance, great aesthetic and economic importance may be attached to preserving the seascape and coastal views. This concern should also be extended to the tangible cultural heritage of past human interaction with the sea and of past human activities in periods of low sea levels. This can be particularly significant in making decisions about the location of new offshore installations.

The intangible cultural heritage of human interaction with the sea is also important. Ten per cent of the items on the UNESCO List of Intangible Cultural Heritage in Need of Urgent Safeguarding involve the ocean. Important cultural areas are derived from the need of humans to operate on the ocean, leading to skills such as navigation, hydrography, naval architecture, chronometry and many other techniques.

The need is increasingly being recognised to understand the knowledge systems developed by many indigenous peoples to understand and manage their interactions with the marine environment.

9.1 Knowledge gaps

In the current fast-changing world, it is important to record much traditional knowledge before it is lost.

9.2 Capacity-building gaps

For cultural goods derived from the sea, a gap exists in many parts of the world for the skills needed to identify and implement potential opportunities to turn local marine objects to good account. Means to support traditional cultural practices so that they endure for future generations need to be provided, and anthropological skills to record and interpret them are also important.

10 The ecosystem services concept and the United Nations and other systems of environmental-economic accounting

As can be seen from the foregoing, there is a general need to bring together information about ecosystem services, in a way which allows judgments to be made about trade-offs. In 1992, the United Nations Conference on Environment and Development (UNCED) called for the implementation of integrated environmental-economic accounting in countries, to complement national accounts by accounting for environmental losses and gains. As a consequence, the United Nations Statistics Division (in charge of maintaining the framework as well as the world standards for the System of National Accounts) led the development of the System of Environmental-Economic Accounting (SEEA) under the auspices of the United Nations Committee of Experts on Environmental-Economic Accounting (UNCEEA). The SEEA Central Framework was adopted as an international standard by the United Nations Statistical Commission at its 43rd Session in 2012 and the SEEA Experimental Ecosystem Accounting was endorsed by the same commission at its 44th session in 2013 (http://unstats.un.org/unsd/envaccounting/seea.asp). Both the SEEA Central Framework and SEEA Experimental Ecosystem Accounting use the accounting concepts, structures and principles of the System of National Accounts (SNA).

The SSEA provides a way of organizing information in both "physical terms" and "monetary terms" using consistent definitions, concepts and classifications. Physical measures and valuation of ecosystem services and ecosystem assets are discussed in the SEEA Experimental Ecosystem Accounting. Those ecosystem services and ecosystem assets are not typically traded on markets, as explained in Chapter 3, and are difficult to measure because observed prices cannot be used to measure these assets and services as in standard economic accounting (Statistics Division of the United Nations, 2012). Although concepts and methods are

Chapter 9 — Conclusions on Major Ecosystem Services Other than Provisioning Services

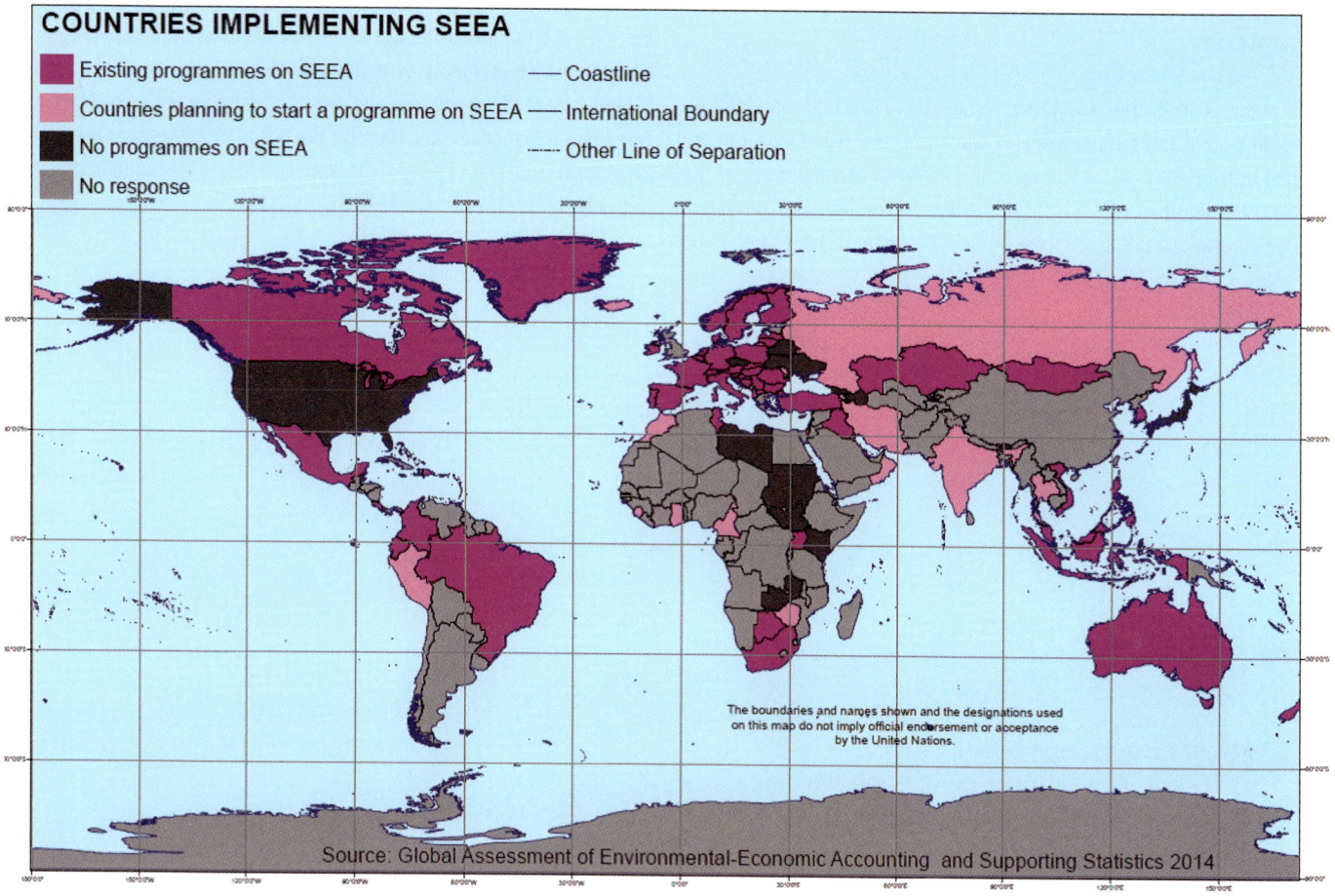

Figure 9.1 | Countries of the world implementing natural capital accounting programmes. The map is provided by the United Nations Statistics Division.

still experimental, many United Nations Member States have engaged in ambitious programmes to apply these new concepts (Figure 9.1).

Simplified ecosystem capital accounts are currently being implemented in Europe by the European Environment Agency, in cooperation with Eurostat, as one of the responses to recurrent policy demands in Europe for accounting for ecosystems and biodiversity (The Economics of Ecosystems and Biodiversity:TEEB). The European Union has developed the MAES programme, for Mapping and Assessment of Ecological Services in the 27 countries of the European Union (http://www.eea.europa.eu/publications/an-experimental-framework-for-ecosystem).

The United Kingdom Government has recently completed a national ecosystem assessment (UK National Ecosystem Assessment, 2011) and, building on this report, has made a commitment to include the value of natural capital and ecosystems fully into the United Kingdom environmental accounts by 2020.

11   Conclusions

Long-established approaches adopted to account for uses of nature, like rent or royalties, are based on a one-to-one relationship between one activity or industry and one natural source of the goods or services. The effect of other industries on the same natural source is not considered; neither are the impacts on other members of the social system affected by these industries.

Traditionally these invisible benefits and costs are mostly hidden in the "natural system", and usually are not accounted for at all in economic terms. The emergence and evolution of the concept of ecosystem services is an explicit attempt to better capture and reflect these hidden or unaccounted benefits and costs, expanding the scope of policy options already available in integrated management approaches through the consideration of the trade-offs among different uses and beneficiaries.

The ecosystem services approach has proven to be very useful in the management of multi-sector processes and is informing today many management and regulatory processes around the world, especially on land. However, the methodologies and different approaches to assess and measure ecosystem services, or to assign them value, as presented in Chapter 3, are far from benefiting from a common set of standards and a consensual framework for their application. Furthermore, as stated in Chapter 3, "*On land, negative impacts can be partially managed or contained in space. However, in the ocean, due to its fluid nature, impacts may broadcast far from their site of origin and are more difficult to contain and manage. For example, there is only one Ocean when considering its role in climate change through the ecosystem service of 'gas regulation'."*

In Chapters 4 to 8, the ecosystems services analyzed are provided on different spatial and temporal scales. Most of those ecosystems are described only on large to very large scales, many of them, indeed, at the planetary scale. In contrast, much of the work on valuing ecosystem services in the ocean has focused on smaller systems, for example, on assessing and valuing the services provided by marine parks, marine reserves and marine protected areas, in order to enable local judgments (including such elements as making local planning decisions or establishing fees to charge to visitors). Despite a significant amount of work on ecosystem-services valuation for the ocean and coasts, as reported in Chapter 3, evidence of its broader application in decision-making at the larger scales is still very limited.

References

Boyce, D.G., Lewis, M. R. and Worm, B., (2010). Global phytoplankton decline over the past century. *Nature*, 466: 591-596.

Boyce D.G., Dowd M, Lewis MR, Worm, B., (2014). Estimating global chlorophyll changes over the past century. *Progress in Oceanography* 122:163–173

Lubchenco, J., Petes, L.E., (2010). The interconnected biosphere: science at the ocean's tipping points. *Oceanography* 23 (2), 115–129.

Riebesell, U. and Gattuso J.-P., (2015). Lessons learned from ocean acidification research. *Nature Climate Change* 5:12-14.

Rykaczewski, R.R. and Dunne, J.P. (2011). A measured look at ocean chlorophyll trends. *Nature* 472, E5-6.

Schanze, J. J., Schmitt, R. W. and Yu, L.L. (2010). The global oceanic freshwater cycle: A state-of-the-art quantification. *Journal of Marine Research* 68, 569–595.

Sunday, J.M., P. Calosi, T. Reusch, S. Dupont, P. Munday, J. Stillman, (2014). Evolution in an acidifying ocean. *Trends in Ecology and Evolution* 29(2): 117-125.

Trenberth, K.E., Smith, L., Qian, T., Dai, A., and Fasullo, J. (2007). Estimates of the Global Water Budget and Its Annual Cycle Using Observational and Model Data. *Journal of Hydrometeorology*, 8: 758 – 769.

United Nations Statistics Division (2012). The System of Environmental-Economic Accounting (SEEA) Experimental Ecosystem Accounting. (http://unstats.un.org/unsd/envaccounting/eea_white_cover.pdf).

United Kingdom National Ecosystem Assessment (2011). *The UK National Ecosystem Assessment: Synthesis of the Key Findings.* UNEP-WCMC, Cambridge.

IV Assessment of the Cross-cutting Issues: Food Security and Food Safety

Introduction

One of the main services provided by the oceans is food for human consumption, resulting in benefits for human health and nutrition, economic returns, and employment. These benefits can be enjoyed sustainably, but only if the intensity and nature of harvesting and culture are appropriately planned and managed, and access to the potential benefits is made available.

Part IV of the WOA reviews these issues under the headings of the Ocean as a source of food (Chapter 10), Capture fisheries (Chapter 11), Aquaculture (12), Fish stock propagation (13), Specialized marine food sources (14), and Social and economic aspects of fisheries (15). Chapter 10 summarizes the contributions of seafood[1] to human nutrition and alleviation of hunger, discussing both patterns at regional and sub-regional scales and their trends over time. Chapter 11 looks in more detail at capture fisheries, presenting trends over time both globally and regionally in overall harvest levels and fishing gear used. It also looks at major species harvested at these scales, and the sustainability of use of the harvested species. It also looks at the ecosystem effects of fishing, considering the nature, levels, and, where information is available, trends, in effects on bycatch species, marine food webs, and habitats. Chapter 12 reviews the same types of information for aquaculture, considering overall production and production of key species at global and regional scales, and, with regard to ecosystem effects, considers issues such as introduction of alien species, local degradation and conversion of habitats, use of antibodies, genetic manipulations, and other similar factors in this form of production. Chapters 13 and 14 address focused issues of artificial propagation of fish and use of marine plants and species other than fish and invertebrates as food. Chapter 15 then assesses the magnitude of economic and social benefits from fisheries and aquaculture. The assessment again looks at trends both globally and regionally, and in addition at differences in the nature, scales, and distribution of social and economic benefits of large-scale and small-scale fisheries. The role of trade, hunger, poverty, worker safety and related issues are all addressed, with particular attention to the interactions of trade, hunger, and poverty alleviation in how benefits may be taken and distributed.

The synthesis in Chapter 16 brings these aspects of the ocean as a source of food together. It integrates the perspectives of the sustainability of harvested and cultured stocks and the impacts on marine ecosystems from fishing and aquaculture, with the perspectives of economic benefits and social / livelihoods benefits.

---

1 Both the terms "seafood" and "fish" are used to include a variety of marine sources of food, depending on the source being consulted. In Part IV both terms are used generically to refer to all types of fish (including both bony and cartilaginous species) and invertebrates consumed as food. When information is presented on a subset of these taxa, the text is explicit about the intended group of species.

10 The Oceans as a Source of Food

Contributors:
Jake Rice (Lead Member), Beatrice Padovani Ferreira (Co-Lead Member) and Andrew Rosenberg (Co-Lead Member)

1 Introduction

One of the main services provided by the oceans to human societies is the provisioning service of food from capture fisheries and culturing operations. This includes fish, invertebrates, plants, and for some cultures, marine mammals and seabirds for direct consumption or as feed for aquaculture or agriculture. These ocean-based sources of food have large-scale benefits for human health and nutrition, economic returns, and employment.

A major challenge around the globe is to obtain these benefits without compromising the ability of the ocean to continue to provide such benefits for future generations, that is, to manage human use of the ocean for sustainability. In effect, this means that capture fisheries and aquaculture facilities must ensure that the supporting stocks are not overharvested and the ecosystem impacts of the harvesting or aquaculture facilities do not undermine the capacity of a given ocean area to continue to provide food and other benefits to society (see Chapter 3). Further, the social and economic goals of the fisheries and aquaculture should fully consider sustainable use in order to safeguard future benefits.

2 Dimensionality of the oceans as a source of food

Capture fisheries and aquaculture operate at many geographical scales, and vary in how they use marine resources for food production. Here, "small-scale" refers to operations that are generally low capital investment but high labour activities, relatively low production, and often family or community-based with a part of the catch being consumed by the producers (Béné et al., 2007; Garcia et al., 2008). Large-scale operations require significantly more capital equipment and expenditure, are more highly mechanized and their businesses are more vertically integrated, with generally global market access rather than focused on local consumption. These descriptions are at the ends of a spectrum continuum of scales with enormous variation in between.

The geography of harvesting and food production from the sea is also important. Williams (1996) documents that until the mid-1980s, developed countries dominated both harvesting and aquaculture, but thereafter developing countries became dominant, first in capture fisheries and later in mariculture. A general division of large-scale fisheries and mariculture in the developed world and small-scale operations in the developing world was never absolute. Small-scale operations were present in all areas, but highly mobile large-scale fisheries are increasingly operating around the globe (Beddington et al., 2007; World Bank/FAO, 2012), and large aquaculture facilities for export products are increasing in the developing world (Beveridge et al., 2010; Hall et al., 2011).

3 Trends in capture fisheries and aquaculture

According to FAO statistics reported by member States, production of fish from capture fisheries and aquaculture for human consumption and industrial purposes has grown at an annual rate of 3.4 per cent for the past half century from about 20 to above 162 mmt by 2013 (FAO, 2014a; FAO, 2015). Over the last two decades though, almost all of this growth has come from increases in aquaculture production. Chapters 11 and 12 of this Assessment describe the time course of capture fisheries and aquaculture development over the last several decades.

Globally aquaculture production has increased at approximately 8.6 per cent per year since 1980, to reach an estimated 67 mmt in 2012, although the rate of growth has slowed slightly in recent years. Of that total, however, more than 60 per cent is from freshwater aquaculture. In addition nearly 24 million tons of aquatic plants (mostly seaweeds) were cultured on 2012. Total marine aquaculture production is growing slightly faster than freshwater aquaculture in all regions, but, like freshwater aquaculture, over 80 per cent of production is concentrated in a few countries, particularly China, as well as some other east and south Asian countries (FAO, 2014a).

Some of the fish taken in capture fisheries are used as feed in aquaculture, fishmeal, fish oil and other non-human consumption uses. Thus the total harvest from capture fisheries and production from mariculture is not all available for human consumption. This use of fish is debated with regard to the best use of production from capture fisheries (Naylor et al., 2009; Pikitch et al., 2012). The total amount of fish used for purposes other than direct consumption has been declining slowly since the early

Table 10.1 | Total and per capita food fish supply by continent and economic grouping in 2011[1]. Source: FAO Information and Statistics Branch, Fisheries and Aquaculture Department, 2015.

	Total food supply (million tonnes live weight equivalent)	Per capita food supply (kg/year)
World	**132.2**	**18.9**
World (excluding China)	86.3	15.3
Africa	11.0	10.4
North America	7.6	21.7
Latin America and the Caribbean	5.9	9.8
Asia	90.3	21.5
Europe	16.4	22.1
Oceania	0.9	25.0
Industrialized countries	26.4	27.0
Other developed countries	5.6	13.7
Least-developed countries	10.3	12.1
Other developing countries	89.9	18.9
LIFDCs[2]	21.2	8.6

1   Preliminary data
2   Low-income food-deficit countries.

2000s from about 30 per cent to just over 20 per cent of total capture fishery harvest in 2012 (FAO 2014a). Consequently, fish for human consumption has been increasing slightly faster than the human population, increasing the importance of fish in meeting food security needs (HLPE, 2014).

Finally, fishing is also undertaken for recreational, cultural and spiritual reasons. Even though fish taken for these purposes may be consumed, they are addressed in chapters 8 and 27, and will not be considered further here.

4 Value of marine fisheries and mariculture

Fish harvested or cultured from the sea provide three classes of benefits to humanity: food and nutrition, commerce and trade, and employment and livelihoods (see Chapter 15 for additional detail). All three classes of benefits are significant for the world.

4.1 Food and nutrition

According to FAO (2014a) estimates, fish and marine invertebrates provide 17 per cent of animal protein to the world population, and provide more than 20 per cent of the animal protein to over 3 billion people, predominantly in parts of the world where hunger is most widespread. Asia accounts for 2/3 of the total consumption of fish. However, when population is taken into account, Oceania has the highest per capita consumption (approximately 25 kg per year), with North America, Europe, South America and Asia all consuming over 20 kg per capita, and Africa, Latin America and the Caribbean are around 10 kg per capita. Per capita consumption does not capture the full importance of the marine food sources to food security, however. Many of the 29 countries where these sources constitute more than a third of animal protein consumed are in Africa and Asia. Of these, the United Nations has identified 18 as low-income, food deficient economies (Karawazuka Béné, 2011, FAO, 2014b). Thus fish and invertebrates, usually from the ocean, are most important where food is needed most.

Not only are marine food sources important for overall food security, fish are rich in essential micronutrients, particularly when compared to micronutrients available when meeting human protein needs from consumption of grains (WHO 1985). Compared to protein from livestock and poultry, fish protein is much richer in poly-unsaturated fatty acids and several vitamins and minerals (Roos et al., 2007, Bonhan et al., 2007). Correspondingly, direct health benefits relative to reducing risk of obesity, heart disease, and high blood pressure have been linked to diets rich in fish (Allison et al., 2013).

It should be noted, however, that there are also potential health risks from consumption of seafood, particularly as fish at higher trophic levels may concentrate environmental contaminants, and there are occasional outbreaks of toxins in shellfish. Substantial effort is invested in monitoring for these risks, and avoiding the conditions where probability of toxin outbreaks may increase. More broadly, food safety is a key worldwide challenge facing all food production and delivery sectors including all parts of the seafood industry from capture or culture to retail marketing. This challenge of course faces subsistence fisheries as well. In the food chain for fishery products, risk of problems needs to be assessed, managed and communicated to ensure problems can be addressed. The goal of most food safety systems is to avoid risk and prevent problems at the source. The risks come from contamination from toxins or pathogens and the severity of the risk also depends on individual health, consumption levels and susceptibility. There are international guidelines to address these risks but substantial resources are required in order to continue to build the capacity to implement and monitor safety protocols from the water to the consumer.

Because of the several limiting factors affecting wild fish catch today (see Chapter 11), it is forecasted that aquaculture production will supply all of the increase in fish consumption in the immediate future. Production is projected to rise to 100 million tons by 2030 (Hall et al., 2011) and to 140 million tons by 2050, if growth continues at the same rate.

Estimates by the World Resources Institute (Waite et al., 2014), assuming (a) the same mix of fish species, (b) that all aquaculture will go to human consumption and (c) that there will be a 10 per cent decrease in wild fish capture for food, indicate that the growth in aquaculture production cited above would boost fish protein supply to 20.2 million tons, or 8.7 million tons above 2006 levels. This increase would meet 17 per cent of the increase in global animal protein consumption required by 9.6 billion people for 2050 (Waite et al. 2014).

4.2 Commerce and trade

The total value of world fish production from capture fisheries and marine and freshwater aquaculture was estimated to be 252 billion USD in 2012, with the "first-sale" value of fish from capture fisheries at approximately 45 per cent of that value (FAO 2014a). Consistent accounting for "value" has been elusive, providing alternative value estimates that are as much as 15-20 per cent greater (e.g., Dyck and Sumalia, 2010). The different possible accounting schemes make it correspondingly difficult to estimate the growth rate of economic value of fisheries, but all approaches project the value to have increased consistently for decades and likely to continue to increase. This increase in economic value is attributable to several factors, including increased production (primarily from aquaculture), an increasing proportion of catches directed to human consumption, improvements in processing and transportation technologies that add to the product's value, and changing consumer demand (Delgado et al 2003). Several factors contribute to increasing consumer demand. The factors include increasing awareness of health benefits of eating fish, increasing economic consumer power in developed and developing economies, and market measures such a certification of sustainably harvested fish and aquaculture products (FAO 2014a).

Just as total per capita consumption of fish underestimates the importance of fish to food security in many food-deficit countries, the total economic value of fish sales underrepresents the value of fish sales to low-income parts of the world. There is a "cash crop" value to fish catches of even small-scale subsistence fishers. Most of this "value" is not captured in the formal economic statistics of countries, and probably varies locally and seasonally (Dey et al., 2005). However studies have shown that the selling or trading of even a portion of their catch represents as much as a third of the total income of subsistence fishers in some low income countries (Béné et al., 2009).

4.3 Employment and livelihoods

These differences between large-scale and small scale fishers are particularly important in considering employment benefits from food from the ocean. Estimates of full-time or part-time jobs derived from fishing, vary widely, with numbers ranging from 58 million to over 120 million jobs being available (BNP 2009, FAO 2014a). All sources agree that over 90 per cent are employed in small-scale fisheries. This includes jobs in the processing and trading sectors, where opportunities for employment of women are particularly important (BNP 2009). The value-chain jobs are considered to nearly triple the employment benefits from fishing and mariculture, compared to direct employment from harvesting (World Bank 2012). All sources report that more than 85 per cent of the employment opportunities are in Asia and a further 8 per cent in Africa, largely in income-deficit countries or areas. It is even harder to track direct and value-chain employment from small-scale aquaculture production and break out the portion that is derived from marine aquaculture (Beveridge et al., 2010), but recent estimates for employment from aquaculture exceed 38 million persons (Phillips et al., 2013).

Of the 58.3 million people estimated to be employed in fisheries and aquaculture (4.4 per cent of total estimated economically active people), 84 per cent were in Asia and 10 per cent in Africa. Women are estimated to account for more than 15 per cent of people employed in the fisheries sector (FAO, 2014).

When full- or part-time participants in the full value-chain and support industries (boat-building, gear construction, etc.) of fisheries and aquaculture and their dependents are included, FAO estimated that between 660 and 820 million persons derive some economic and/or livelihood benefits (FAO 2012, Allison 2013). Direct employment in fishing is also growing over 2 per cent per year, generally faster than human population growth (Allison, 2013). However, there has been a shift from 87 per cent in capture fisheries and the rest in aquaculture (primarily freshwater) in 1990, to approximately a 70:30 division in 2010, with slightly faster growth in employment in mariculture than in freshwater aquaculture (FAO, 2012).

Trade in fishery products further complicates efforts to evaluate trends in the contribution of the oceans to human well-being. Fish is one of the most heavily traded food commodities on the planet, with an estimated 38 per cent of fishery production by 2010, up from 25 per cent in 1976 (FAO, 2012). This represents about 10 per cent of international agricultural exports. The direct value of international exports was over 136 billion USD in 2012, up 102 per cent in just 10 years (FAO, 2014a. http://www.fao.org/3/a-i4136e.pdf); European Union (EU) countries alone imported more than 514 billion in fish products in 2013, although slightly over half of that was from trade among EU Member States (http://www.fao.org/3/a-i4136e.pdf). Fish trade is truly global, with FAO recording fish and fishery products exported by 197 countries, led by China, which contributes 14 per cent of the total exports.

Developing countries contribute over 60 per cent by volume and over 50 per cent by value of exports of fish and fish products. Although this trade generates significant revenues for developing countries, through sales, taxation, license fees, and payment for access to fish by distant water fleets, there is a growing debate about the true benefits to the inhabitants of these countries from these revenue sources (Bostock et al., 2004; World Bank 2012). The debate centres on whether poor fishers would benefit more from personal or community consumption of the fish than from sales of the fish to obtain cash or credit. The issue is complicated by the leasing of access rights for foreign vessels which may compete for resources with coastal small scale fishers. With small-scale and large-scale fisheries each harvesting about half of the world's fish, resolving the relative importance of large-scale and small-scale fisheries to food security, in an increasingly globalized economy, is complex. Reviews found the issue to be polarized in the early 2000s (FAO 2003; Kurien, 2004), and there has been little convergence of views over the ensuing decade (HLPE, 2014).

5 Impacts of fisheries and mariculture, on marine ecosystems

Harvesting or culturing marine fish, invertebrates or plants necessarily has at least direct and immediate, and often indirect and longer-term impacts on marine ecosystems. For over a century fisheries experts have sought ways to evaluate the short-term and long-term sustainability of varying levels of fish harvests (Smith 1994), and to manage fisheries to keep these harvests within sustainable bounds (Garcia et al., 2014). Assessing and managing the wider ecosystem impacts of fisheries and aquaculture is even more challenging (Garcia et al., 2014). These impacts may range from loss of habitat due to destructive fishing practices to impacts on the structure of marine food webs by selectively harvesting some species that play a key role in the integrity of a given ecosystem. The fact that these effects may be difficult to quantify in no way diminishes their importance in sustaining the capacity of the oceans to provide food and other benefits to human society. Moreover, the scope of assessments of impacts continues to expand, as life cycle analyses are introduced into fisheries (Avadí and Fréon, 2013). Results indicate that, for example, the carbon footprint of a kg of fish at market depends greatly on modes of capture and transport. However, the carbon footprint is often substantially lower than the footprint of a kg of poultry or livestock (Mogensen et al., 2012). Other chapters in this

Assessment, primarily in Part VI, consider a broad range of impacts on the ocean of human activities. Since food production from the ocean is such an important benefit, particular care must be taken to ensure that sustained capacity to produce food from fisheries and aquaculture is not diminished.

6   Conclusions

This chapter sets the stage for assessing the role of the oceans as a source of food. The chapters to follow will assess in depth the ways that food is taken from the sea. Each chapter will consider the trends in yields, resources, economic benefits, employment, and livelihoods, the interactions among the trends, and their main drivers, on global and regional scales as appropriate. They will also look at the main impacts of the various food-related uses of the ocean on biodiversity – both species and habitats. Some of these interactions will also be considered, from the perspective of the affected components of biodiversity, in Part VI of the World Ocean Assessment. Each chapter will also consider the main factors that affect the trends in benefits, resources used and impacts. Together a picture will emerge of the importance of the ocean as a source of food, and of fisheries and mariculture as sources of commerce, wealth, and livelihoods for humankind, with a particular focus on the world's coastal peoples.

References

Allison, E.H., Delaporte, A., and Hellebrandt de Silva, D. (2013). *Integrating fisheries management and aquaculture development with food security and livelihoods for the poor.* Report submitted to the Rockefeller Foundation, School of International Development, University of East Anglia Norwich, 124 p.

Avadí, A., and Fréon, P. (2013) Life cycle assessment of fisheries: A review for fisheries scientists and managers. *Fisheries Research* 143: 21-38.

Beddington, J.R., Agnew, D.J., and Clark, C.W. (2007). Current problems in the management of marine fisheries. *Science* 316(5832): 1713–1716.

Béné, C., Macfadyen, G., and Allison, E. (2007). Increasing the contribution of small-scale fisheries to poverty alleviation and food security. *FAO Fisheries Technical Paper* No. 481. Food and Agriculture Organization of the United Nations, Rome, 125 p.

Béné, C., Belal, E., Baba, M.O., Ovie, S., Raji, A., Malasha, I., Njaya, F., Na Andi, M., Russell, A., and Neiland, A. (2009). Power Struggle, Dispute and Alliance over Local Resources: Analyzing 'Democratic' Decentralization of Natural Resources through the Lenses of Africa Inland Fisheries. *World Development* 37: 1935–1950.

Beveridge M., Phillips, M., Dugan, P., and Brummett, R. (2010). *Barriers to Aquaculture Development as a Pathway to Poverty Alleviation and Food Security: Policy Coherence and the Roles and Responsibilities of Development Agencies.* OECD Workshop, Paris, France, 12–16 April 2010.

BNP (2009). Big Number Program. Intermediate report. Rome: Food and Agriculture Organization and WorldFish Center.

Bonham, M.P., Duffy, E.M., Robson, P.J., Wallace, J.M., Myers, G.J., Davidson, P.W., Clarkson, T.W., Shamlaye, C.F., Strain, J., and Livingstone, M.B.E. (2009). Contribution of fish to intakes of micronutrients important for foetal development: a dietary survey of pregnant women in the Republic of Seychelles. *Public Health Nutrition* 12(9):1312–1320.

Bostock, T., Greenhalgh, P., and Kleih, U. (2004). *Policy Research: Implications of Liberalization of Fish Trade for Developing Countries.* Synthesis report. Natural Resources Institute, University of Greenwich, Chatham, UK, 68 p.

Delgado, C., Wada, N., Rosegrant, M.W., Meijer, S., and Ahmed, M. (2003). *Fish to 2020: Supply and Demand in Changing Global Markets*. International Food Policy Research Institute. Washington, DC and WorldFish Center, Penang, Malaysia.

Dey, M.M., Mohammed, R.A., Paraguas, F.J., Somying, P., Bhatta, R., Ferdous, M.A., and Ahmed, M. (2005). Fish consumption and food security: a disaggregated analysis by types of fish and classes of consumers in selected Asian countries. *Aquaculture Economics and Management* 9: 89–111.

Dyck, A.J., and Sumaila, U.R. (2010). Economic impact of ocean fish populations in the global fishery. *Journal of Bioeconomics* 12: 227–243.

FAO (2003). *Report of the expert consultation on international fish trade and food security*. FAO Fisheries Report. No.708. Rome.

FAO (2012). *The State of the World Fisheries and Aquaculture*. FAO Rome. 209 pp.

FAO (2014a). *The State of the World Fisheries and Aquaculture*. FAO Rome. 239 pp.

FAO (2014b). Low-Income Food-Deficit Countries (LIFDC) – List for 2014. http://www.fao.org/countryprofiles/lifdc/en/.

Garcia S., Allison, E.H, Andrew, N., Béné, C., Bianchi, G., de Graaf, G., Kalikoski, D., Mahon, R., and Orensanz, L.. (2008). Towards integrated assessment and advice in small-scale fisheries: Principles and Processes. *FAO Fisheries and Aquaculture Technical Paper* No.515. Food and Agriculture Organization of the United Nations, Rome, 84 p.

Garcia, S.M., Rice, J., and Charles, A.T. (eds). (2014). *Governance of Marine Fisheries and Biodiversity Conservation: Interaction and Co-evolution*. Wiley Interscience. London. 486 pp.

Hall, S.J., Delaporte A., Phillips M.J., Beveridge M., O'Keefe M. (2011). *Blue Frontiers: Managing the Environmental Costs of Aquaculture*. The WorldFish Center, Penang, Malaysia.

HLPE, (2014). *Sustainable fisheries and aquaculture for food security and nutrition*. A report by the High Level Panel of Experts on Food Security and Nutrition of the Committee on World Food Security, Rome.

Kawarazuka, N., and Béné, C. (2011). The potential role of small fish species in improving micronutrient deficiencies in developing countries: building evidence. *Public Health Nutrition* 14: 1927–1938.

Kawarazuka, N., and Béné C. (2010). Linking small-scale fisheries and aquaculture to household nutritional security: a review of the literature. *Food Security* 2: 343–357.

Kurien, J. (2004). *Fish trade for the people: Toward Understanding the Relationship between International Fish Trade and Food Security*. Report of the Study on the impact of international trade in fishery products on food security, Food and Agriculture Organization of the United Nations and the Royal Norwegian Ministry of Foreign Affairs.

Mogensen, L., Hermansen, J.E., Halberg, N., Dalgaard, R., Vis, R.C., and Smith, B.G. (2012). Life Cycle Assessment Across the Food Supply Chain. In: Baldwin, C., editor. *Sustainability in the Food Industry*. Wiley, London. pp. 115-144.

Naylor, R.L., Hardy, R.W., Bureau, D.P., Chiu, A., Elliott, M., Farrell, A.P., Forster, I., Gatlin, D.M., Goldburg, R.J., Hua, K., and Nichols, P.D. (2009). Feeding aquaculture in an era of finite resources. *Proceedings of the National Academy of Sciences of the United States of America* 106: 18040.

Phillips, M., Van, N.T., and Subasinghe, R. (2013). *Aquaculture Big Numbers*. Working Paper. 12 June 2012. WorldFish and FAO.

Pikitch, E., Boersma, P.D., Boyd, I.L., Conover, D.O., Cury, P., Essington, T., Heppell, S.S., Houde, E.D., Mangel, M., Pauly, D., Plagányi, É., Sainsbury, K., and Steneck, R.S. (2012). *Little Fish, Big Impact: Managing a Crucial Link in Ocean Food Webs*. p. 108. Lenfest Ocean Program. Washington, DC.

Roos, N., Wahab, M.A., Chamnan, C, and Thilsted, S.H. (2007). The role of fish in food-based strategies to combat Vitamin A and mineral deficiencies in developing countries. *Journal of Nutrition* 137: 1106–1109.

Smith, T.D. (1994). Scaling Fisheries: *The Science of Measuring the Effects of Fishing 1855-1955*. Cambridge Studies in Applied Ecology and Resource Management. Cambridge, 384 pp.

WHO (1985). *Energy and protein requirements*. World Health Organization, Geneva.

Williams, M.J. (1996). The transition in the contribution of living aquatic resources to food security. International Food Policy Research Institute: *Food Agriculture and the Environment Discussion Paper* No. 13: 41pp.

Waite, R. et al. 2014. *Improving Productivity and Environmental Performance of Aquaculture*.

Working Paper, Installment 5 of Creating a Sustainable Food Future. Washington, DC: World Resources Institute. Accessible at http://www.worldresourcesreport.org.

World Bank/FAO/WorldFish (2012). *Hidden Harvest: The Global Contribution of Capture Fisheries*. World Bank, Report No. 66469-GLB, Washington, DC. 69 pp.

Capture Fisheries

Writing team:
Enrique Marschoff (Convenor and Lead Member), Beatrice Padovani Ferreira (Co-Lead Member), Fábio Hazin, Jake Rice (Co-Lead Member) and Andrew Rosenberg (Co-Lead Member)

Capture Fisheries

1 Present status and trends of commercially exploited fish and shellfish stocks

Production of fish from capture fisheries (Figure 11.1) and aquaculture for human consumption and industrial purposes has grown at the rate of 3.2 per cent for the past half century from about 20 to nearly 160 million mt by 2012 (FAO 2014).

Globally, marine capture fisheries produced 82.6 million mt in 2011 and 79.7 million mt in 2012. The relatively small year-to-year variations largely reflect changes in the catch of Peruvian anchoveta, which can vary from about 4 to 8 million tons *per annum*.

In 2011 and 2012, 18 countries accounted for more than 76 per cent of global marine harvests in marine capture fisheries (Table 11.1). Eleven of these countries are in Asia.

In 2011-2012, the top ten species (by tonnage) in marine global landings were Peruvian anchoveta, Alaska pollock, skipjack tuna, various sardine species, Atlantic herring, chub mackerel, scads, yellowfin tuna, Japanese anchovy and largehead hairtail. In 2012, 20 species had landings over a half a million tons and this represented 38 per cent of the total global marine capture production. Many of these top species are small pelagic fishes (e.g. sardines, chub mackerels) and shellfish (squids and shrimp) whose abundance is highly sensitive to changing climatic conditions, resulting in significant interannual variability in production.

Tuna harvests in 2012 were a record high, exceeding more than seven million tons. Sharks, rays and chimaera catches have been stable during the last decade at about 760,000 tons annually. Shrimp production from marine capture fisheries reached a record high in 2012 at 3.4 million tons; much of this catch was from the Northwest and Western Central Pacific, although catches also increased in the Indian Ocean and the Western Atlantic. Cephalopod catches exceeded 4 million tons in 2012.

1.1 Regional Status

Significant growth in marine capture fisheries has occurred in the eastern Indian Ocean, the eastern central Atlantic and the northwest, western central and eastern central Pacific over the last decade, but landings

Figure 11.1 | Evolution of world's capture of marine species. From SOFIA (FAO 2014).

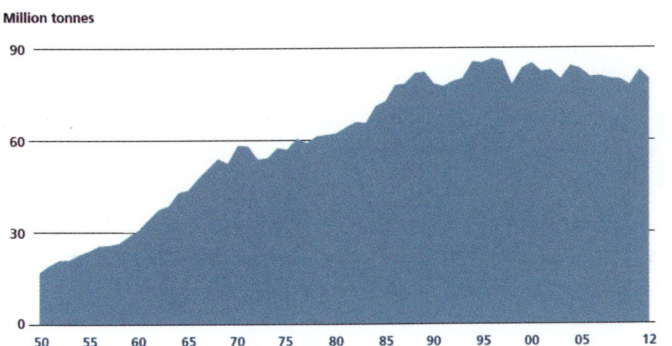

Table 11.1 | Marine capture fisheries production per country. From SOFIA (FAO, 2014).

2012 Ranking	Country	Continent	2003	2011	2012	Variation 2003–2012	Variation 2011–2012
			(Tonnes)			(Percentage)	
1	China	Asia	12 212 188	13 536 409	13 869 604	13.6	2.4
2	Indonesia	Asia	4 275 115	5 332 862	5 420 247	27.0	1.7
3	United States of America	Americas	4 912 627	5 131 087	5 107 559	4.0	−0.5
4	Peru	Americas	6 053 120	8 211 716	4 807 923	−20.6	−41.5
5	Russian Federation	Asia/Europe	3 090 798	4 005 737	4 068 850	31.6	1.6
6	Japan	Asia	4 626 904	3 741 222	3 611 384	−21.9	−3.5
7	India	Asia	2 954 796	3 250 099	3 402 405	15.1	4.7
8	Chile	Americas	3 612 048	3 063 467	2 572 881	−28.8	−16.0
9	Viet Nam	Asia	1 647 133	2 308 200	2 418 700	46.8	4.8
10	Myanmar	Asia	1 053 720	2 169 820	2 332 790	121.4	7.5
11	Norway	Europe	2 548 353	2 281 856	2 149 802	−15.6	−5.8
12	Philippines	Asia	2 033 325	2 171 327	2 127 046	4.6	−2.0
13	Republic of Korea	Asia	1 649 061	1 737 870	1 660 165	0.7	−4.5
14	Thailand	Asia	2 651 223	1 610 418	1 612 073	−39.2	0.1
15	Malaysia	Asia	1 283 256	1 373 105	1 472 239	14.7	7.2
16	Mexico	Americas	1 257 699	1 452 970	1 467 790	16.7	1.0
17	Iceland	Europe	1 986 314	1 138 274	1 449 452	−27.0	27.3
18	Morocco	Africa	916 988	949 881	1 158 474	26.3	22.0
	Total 18 major countries		58 764 668	63 466 320	60 709 384	3.3	−4.3
	World total		79 674 875	82 609 926	79 705 910	0.0	−3.5
	Share 18 major countries (percentage)		73.8	76.8	76.2		

in many other regions have declined. Thus, even though overall landings have been quite stable, the global pattern is continuing to adjust to changing conditions and regional development of fishing capacity (Table 11.2).

An estimated 3.7 million fishing vessels operate in marine waters globally; 68 per cent of these operate from Asia and 16 per cent from Africa. Seventy per cent are motorized, but in Africa only 36 per cent are motorized. Of the 58.3 million people estimated to be employed in fisheries and aquaculture (4.4 per cent of total estimated economically active people), 84 per cent are in Asia and 10 per cent in Africa. Women are estimated to account for more than 15 per cent of people employed in the fisheries sector (FAO 2014).

Table 11.2 | Fishing areas and captures (from SOFIA, FAO, 2014)

Fishing area code	Fishing area name	2003	2011	2012	Variation 2003–2012	Variation 2011–2012
		(Tonnes)			(Percentage)	
21	Atlantic, Northwest	2 293 460	2 002 323	1 977 710	−13.8	−1.2
27	Atlantic, Northeast	10 271 103	8 048 436	8 103 189	−21.1	0.7
31	Atlantic, Western Central	1 770 746	1 472 538	1 463 347	−17.4	−0.6
34	Atlantic, Eastern Central	3 549 945	4 303 664	4 056 529	14.3	−5.7
37	Mediterranean and Black Sea	1 478 694	1 436 743	1 282 090	−13.3	−10.8
41	Atlantic, Southwest	1 987 296	1 763 319	1 878 166	−5.5	6.5
47	Atlantic, Southeast	1 736 867	1 263 140	1 562 943	−10.0	23.7
51	Indian Ocean, Western	4 433 699	4 206 888	4 518 075	1.9	7.4
57	Indian Ocean, Eastern	5 333 553	7 128 047	7 395 588	38.7	3.8
61	Pacific, Northwest	19 875 552	21 429 083	21 461 956	8.0	0.2
67	Pacific, Northeast	2 915 275	2 950 858	2 915 594	0.0	−1.2
71	Pacific, Western Central	10 831 454	11 614 143	12 078 487	11.5	4.0
77	Pacific, Eastern Central	1 769 177	1 923 433	1 940 202	9.7	0.9
81	Pacific, Southwest	731 027	581 760	601 393	−17.7	3.4
87	Pacific, Southeast	10 554 479	12 287 713	8 291 844	−21.4	−32.5
18, 48, 58, 88	Arctic and Antarctic areas	142 548	197 838	178 797	25.4	−9.6
	World total	79 674 875	82 609 926	79 705 910		

2 Present status of small-scale artisanal or subsistence fishing

The FAO defines small-scale, artisanal fisheries as those that are household based, use relatively small amounts of capital and remain close to shore. Their catch is primarily for local consumption. Around the world there is substantial variation as to which fisheries are considered small-scale and artisanal. The United Nations Conference on Sustainable Development (Rio+20) emphasized the role of small-scale fisheries in poverty alleviation and sustainable development. In some developing countries, including small island States, small-scale fisheries provide more than 60 per cent of protein intake. Its addition to the diets of low-income populations (including pregnant and breastfeeding mothers and young children) offers an important means for improving food security and nutrition. Small-scale fisheries make significant contributions to food security by making fish available to poor populations, and are critical to maintain the livelihoods of vulnerable populations in developing countries. Their role in production and its contribution to food security and nutrition is often underestimated or ignored; subsistence fishing is rarely included in national catch statistics (HLPE, 2014). Anyhow, the key issues in artisanal fisheries are their access both to stocks and to markets (HLPE, 2014).

Significant numbers of women work in small-scale fisheries and many indigenous peoples and their communities rely on these fisheries. The "Voluntary Guidelines on the Responsible Governance of Tenure of Land, Fisheries and Forests in the Context of National Food Security" (FAO 2012) are important in consideration of access issues. FAO also notes the linkage to international human rights law, including the right to food. Most of the people involved in small-scale fisheries live in developing countries, earn low incomes, depend on informal work, are exposed to the absence of work regulations and lack access to social protection schemes. Although the International Labour Organization adopted the Work in Fishing Convention, 2007 (No.188), progress towards ratification of the Convention has been slow.

FAO continues to encourage the establishment of fishers' organizations and cooperatives as a means of empowerment for small-scale fishers in the management process to establish responsible fisheries policy. They have also highlighted the need to reduce post-harvest losses in small-scale fisheries as a means of improving production. Two special sections discuss these issues in SOFIA. Besides the "Voluntary Guidelines on the Responsible Governance of Tenure of Land, Fisheries and Forests in the Context of National Food Security", FAO also adopted the "Voluntary Guidelines for Securing Sustainable Small-Scale Fisheries in the Context of Food Security and Poverty Eradication" in June 2014.

3 Impacts of capture fisheries on marine ecosystems

The effects of exploitation of marine wildlife were first perceived as a direct impact primarily on the exploited populations themselves. These concerns were recognized in the 19th and early 20th centuries (e.g., Michelet, 1875; Garstang, 1900; Charcot, 1911) and began to receive policy attention in the Stockholm Fisheries Conference of 1899 (Rozwadowski, 2002). In 1925, an attempt to globally manage "marine industries" and their impact on the ecosystems was presented before the League of Nations (Suarez, 1927), but little action was taken. Only following WWII, with rapid increases in fishing technology, was substantial overfishing in both the Atlantic and Pacific Oceans (Gulland and Carroz, 1968) acknowledged. Establishment in 1946 of FAO, with a section for fisheries, provided an initial forum for global discussions of the need for regulation of fisheries.

Capture fisheries affect marine ecosystems through a number of different mechanisms. These have been summarized many times, for example by Jennings and Kaiser (1998) who categorized effects as:

(a) The effects of fishing on predator-prey relationships, which can lead to shifts in community structure that do not revert to the original condition upon the cessation of fishing pressure (known as alternative stable states);
(b) Fishing can alter the population size and body-size composition of species, leading to fauna composed of primarily small individual organisms (this can include the whole spectrum of organisms, from worms to whales);
(c) Fishing can lead to genetic selection for different body and reproductive traits and can extirpate distinct local stocks;
(d) Fishing can affect populations of non-target species (e.g., cetaceans, birds, reptiles and elasmobranch fishes) as a result of by-catches or ghost fishing;
(e) Fishing can reduce habitat complexity and perturb seabed (benthic) communities.

Here these impacts are discussed first for the species and food webs being exploited directly, and then for the other ecosystem effects on by-catches and habitats of fishing. Part VI of this Assessment provides additional detail regarding impacts on biodiversity and habitats.

3.1 Target species and communities

The removal of a substantial number of individuals of the target species affects the population structure of the target species, other species taken by the gear, and the food web. The magnitude of these effects is highly variable and depends on the species considered and the type and intensity of fishing. In general, policies and management measures were instituted first to manage the impact of fisheries on the target species, with ecosystem considerations being added to target species management primarily in the past two to three decades.

If the exploited fish stock can compensate through increased productivity because the remaining individuals grow faster and produce more larvae, with the increase in productivity extracted by the fishery, then fishing can be sustained. However, if the rate of exploitation is faster than the stock can compensate for by increasing growth and reproduction, then the removals will not be sustained and the stock will decline. At the level of the target species, sustainable exploitation rates will result in the total population biomass being reduced roughly by half, compared to unexploited conditions.

The ability of a given population of fish to compensate for increased mortality due to fishing depends in large part on the biological characteristics of the population such as growth and maturation rates, natural mortality rates and lifespan, spawning patterns and reproduction dynamics. In general, slow growing long-lived species can compensate for and therefore sustain lower exploitation rates (the proportion of the stock removed by fishing each year) than fast growing shorter lived species (Jennings et al. 1998). In addition, increased exploitation rates inherently truncate the age composition of the population unless only certain ages are targeted. This truncation results in both greater variability in population abundance through time (Hsieh et al. 2006) and greater vulnerability to changing environmental conditions, including climate impacts. Very long-lived species with low rates of reproduction may not be able to truly compensate for increased mortality, and therefore any significant fishing pressure may not be sustainable on such species. Of course there are many complicating factors, but this general pattern is important for understanding sustainable exploitation of marine species.

The concept of "maximum sustainable yield" (MSY), adopted as the goal of many national and international regulatory bodies, is based on this inherent trade-off between increasing harvests and the decreasing ability of a population to compensate for removals. Using stock size and exploitation rates that would produce MSY, or other management reference points, FAO has concluded that around 29 per cent of assessed stocks are presently overfished (biomass below the level that can produce MSY on a continuing basis; Figure 11.2 below). That percentage may be declining in the more recent years, but has shown little overall trend since the early 1990s. FAO estimates that if overfished stocks were rebuilt, they would yield an additional 16.5 million mt of fish worth 32 billion United States dollars in the long term (Ye et al., 2013). However, significant social and economic costs may be incurred during the transition, as many fisheries would need to reduce exploitation in the short term to allow this rebuilding.

Anyhow, for many ecological reasons, the MSY is an over-simplified reference point for fisheries (Larkin, 1997; Pauly, 1994). For example, declines in productivity can result as fewer fish live to grow to a large size, because larger, older fish produce disproportionately more eggs of higher quality than younger, smaller individuals (Hixon et al. 2013). Long-term overfishing may even change the genetic pool of the species concerned, because the larger and faster-growing specimens have a greater probability of being removed, thereby reducing overall productivity (Hard et al., 2008; Ricker, 1981). Interactions between species may also mean that all stocks cannot be maintained at or above the biomass that will produce MSY. Strategies for taking these interactions into account have been developed (Polovina 1984, Townsend et al. 2008, Fulton et al. 2011; Farcas and Rossberg 2014, http://arxiv.org/abs/1412.0199), but are not yet in routine practice.

Figure 11.2 | State of world marine fish stocks (from SOFIA, FAO 2014)

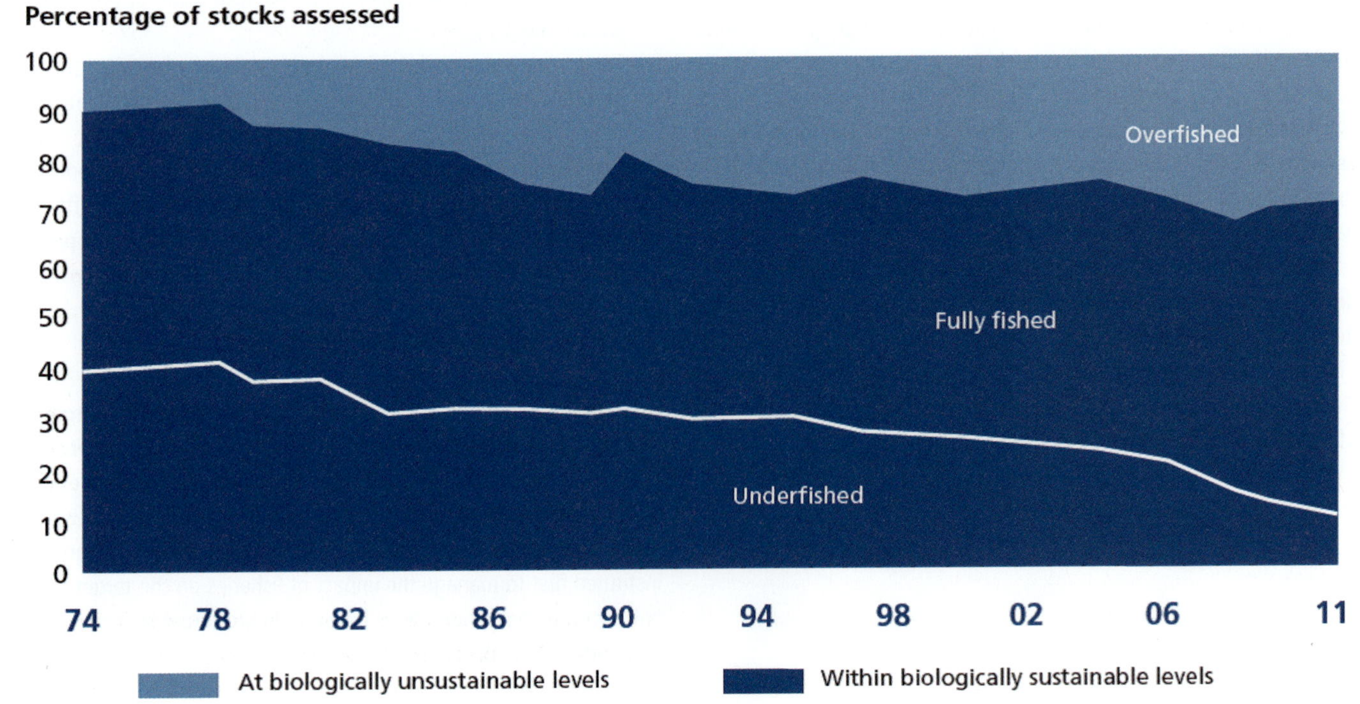

3.2 Ecosystem effects of fishing

The FAO Ecosystem approach to Fisheries (FAO 2003) has detailed guidelines describing an ecosystem approach to fisheries. The goal of such an approach is to conserve the structure, diversity and functioning of ecosystems while satisfying societal and human needs for food and the social and economic benefits of fishing (FAO 2003). There are ongoing efforts around the world to implement an ecosystem approach to fisheries that encompasses the aspects considered below, among others.

3.3 Ecosystem effects of fishing – food webs

Marine food webs are complex and exploiting commercially important species can have a wide range of effects that propagate through the food web. These include a cascading effect along trophic levels, affecting the whole food web (Casini et al., 2008; Sieben et al., 2011). The removal of top predators may result in changes in the abundance and composition of lower trophic levels. These changes might even reach other and apparently unrelated fisheries, as has been documented, for example, for sharks and scallops (Myers et al., 2007) and sea otters, kelp, and sea urchins (Szpak et al., 2013). Because of these complexities in both population and community responses to exploitation, it is now widely argued that target harvesting rates should be less than MSY. No consensus exists on how much less, but as information about harvest amounts and stock biology is more uncertain, it is agreed that exploitation should be reduced correspondingly (FAO, 1995).

The controversial concept of "balanced harvesting" refers to a strategy that considers the sustainability of the harvest at the level of the entire food web (see, for example, Bundy, A., et al. 2005; Garcia et al., 2011; FAO 2014). Rather than harvesting a relatively small number of species at their single-species MSYs, balanced harvesting suggests there are benefits to be gained by exploiting all parts of the marine ecosystem in direct proportion to their respective productivities. It is argued that balanced harvesting gives the highest possible yield for any level of perturbation of the food web, On the other hand, the economics of the fishery may be adversely affected by requiring the harvest of larger amounts of low-value but highly productive stocks.

Table 11.3 | Discards of fish in major fisheries by gear type. From Kelleher, 2005.

Fishery	Landings	Discards[1]	Weighted average discard rate (%)	Range of discard rates (%)
Shrimp trawl	1 126 267	1 865 064	62.3	0–96
Demersal finfish trawl	16 050 978	1 704 107	9.6	0.5–83
Tuna and HMS longline	1 403 591	560 481	28.5	0–40
Midwater (pelagic) trawl	4 133 203	147 126	3.4	0–56
Tuna purse seine	2 673 378	144 152	5.1	0.4–10
Multigear and multispecies	6 023 146	85 436	1.4	n.a.
Mobile trap/pot	240 551	72 472	23.2	0–61
Dredge	165 660	65 373	28.3	9–60
Small pelagics purse seine	3 882 885	48 852	1.2	0–27
Demersal longline	581 560	47 257	7.5	0.5–57
Gillnet (surface/bottom/trammel)[2]	3 350 299	29 004	0.5	0–66
Handline	155 211	3 149	2.0	0–7
Tuna pole and line	818 505	3 121	0.4	0–1
Hand collection	1 134 432	1 671	0.1	0–1
Squid jig	960 432	1 601	0.1	0–1

[1] The sum of the discards presented in this table is less than the global estimate, as a number of discard database records could not be assigned to particular fisheries.

3.4 Other ecosystem effects of fishing by-catches

Fisheries do not catch the target species alone. All species caught or damaged that are not the target are known as by-catch; these include, *inter alia*, marine mammals, seabirds, fish, kelp, sharks, mollusks, etc. Part of the by-catch might be used, consumed or processed (incidental catch) but a significant amount is simply discarded (discards) at sea. Global discard levels are estimated to have declined since the early 1990s, but at 7.3 million tons are still high (Kelleher, 2005).

Fisheries differ greatly in their discard rates, with shrimp trawls producing by far the greatest discard ratios relative to landed catches of target species (Table 11.3).

Very few time series have been found that document trends in by-catch levels for marine fisheries in general, or even for particular fisheries or species groups over longer periods. Although both Alverson et al. (1994) and Kelleher (2005) provide global estimates of discards in fisheries that differ by a factor of three, the latter source (with the lower estimate) stresses that the methodological differences between the two estimates were so large that two estimates should not be compared (a warning confirmed in the Kelleher report by the authors of the earlier report).

When even rough trend information is available, it is for particular species of concern in particular fisheries, and is usually intended to document the effectiveness of mitigation measures that have been implemented already. As an illustration, in the supplemental information to Anderson et al. (2011), which reports a global examination of longline fisheries, of the 67 fisheries for which data could be found, two estimates of seabird by-catches were available for only 17 of them. Of those, the more recent seabird by-catch estimates were at least 50 per cent lower than the earlier estimates in 15 of the fisheries, and reduced to 5 per cent or less of the earlier estimates in 10 of the fisheries. Several reasons were given, depending on the fishery; they included reduction in effort and the use of a variety of technical and occasionally temporal and/or spatial mitigation measures. These can be taken as illustrative of the potential effectiveness of mitigation efforts, but should not be extrapolated to other longline fisheries.

The more typical case is reflected in FAO Fisheries and Aquaculture Department (2009) and the report of a FAO Expert Consultation (FAO, 2010), which call for efforts to monitor by-catches and discards more consistently, in order to provide the data needed to document trends. Even the large initiative by the United States to document by-catches in fisheries (National Marine Fisheries Service, 2011) considers the reported estimates to be a starting point for gaining insight into trends in by-catch and discards. It documents the very great differences among fisheries within and among the United States fisheries management regions, but has neither tables nor figures depicting trends for any fishery.

By-catch rates may result in overfishing of species with less ability to cope with fishing pressures. The biological impact of by-catches varies greatly with the species being taken, and depends on the same life-

Capture Fisheries

Chapter 11

history characteristics that were presented above for the target species of fisheries. By-catch mortality is a particular concern for small cetaceans, sea turtles and some species of seabirds and sharks and rays. These issues are discussed in the corresponding chapters in Part VI on marine mammals (Chapter 37), seabirds (38), marine reptiles (39) and elasmobranchs (40). In general, long-lived and slow-growing species are the most affected (Hall et al., 2000). Thus, the benchmarks set for a given fishery also consider by-catch species.

The geographic distribution of discard rates is shown in Figure 11.3 (from Kelleher, 2005).

The numbers in bold are the FAO Statistical Areas and the tonnages are of by-catch. By-catches are clearly a global issue, and can be addressed from local to global scales. The review by Kelleher (2005) reports a very large number of cases where measures have been implemented by States, by international organizations, or proactively by the fishing industry (especially when the industry is seeking independent certification for sustainability), and by-catch and discard rates have decreased and in a few cases been even eliminated.

A recent global review of practices by regional fisheries management organizations and arrangements (RFMO/As) for deep sea fisheries found that all RFMO/As have adopted some policies and measures to address by-catch issues in fisheries in their regulatory areas. However, almost nowhere was full monitoring in place to document effectiveness of these policies (UNEP/CBD/FAO, 2011). Nevertheless, extensive evidence exists that by-catches can be mitigated by changes in fishing gear, times, and places, and the incremental cost is often, but not always, small.

At the global level, calls for action on by-catch and discards have been raised at the United Nations General Assembly, including in UNGA resolutions on sustainable fisheries and at the Committee on Fisheries. In response, FAO developed International Guidelines on Bycatch Management and Reduction of Discards; these were accepted in 2011 (FAO, 2011).

3.5 Ecosystem effects of fishing – benthic and demersal habitats

Fishing gear impacts on the seafloor and other habitats depend on the gear design and use, as well as on the particular environmental features. For example, in benthic habitats, substrate type and the natural disturbance regime are particularly important (Collie et al., 2000). Mobile bottom-contacting gear (including bottom trawls) also can resuspend sediments, mobilizing contaminants and particles with unknown ecological effects on both benthic and demersal habitats (Kaiser et al., 2001).

The boundaries and names shown and the designations used on this map do not imply official endorsement or acceptance by the United Nations.

Figure 11.3 | Distribution of discards by FAO statistical areas (numbers in bold are FAO statistical areas, catches in tons). * Note: the high discard rate in FAO Area 81 is a data artefact. Source: Kelleher, 2005.

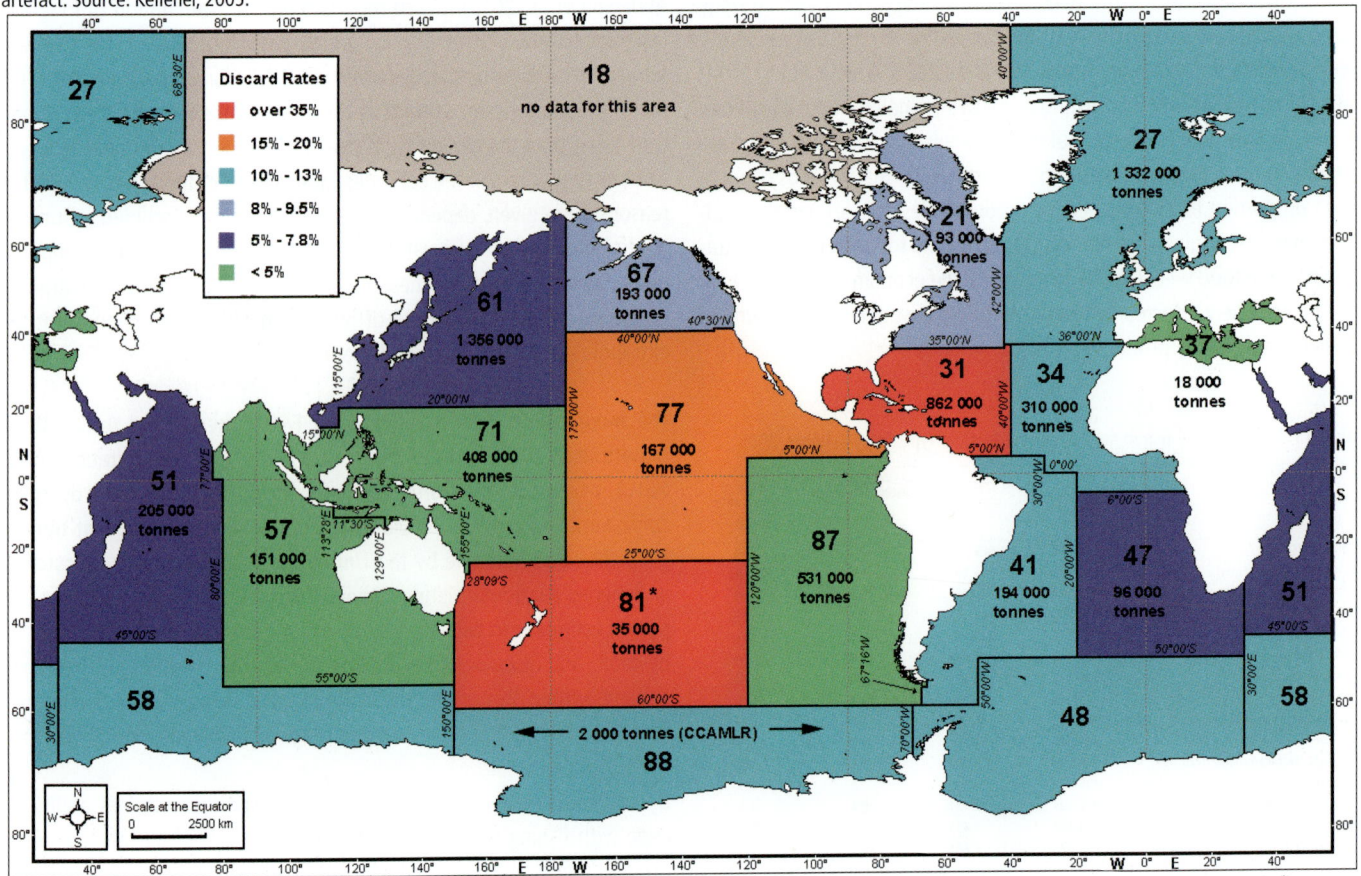

Chapter 11

Capture Fisheries

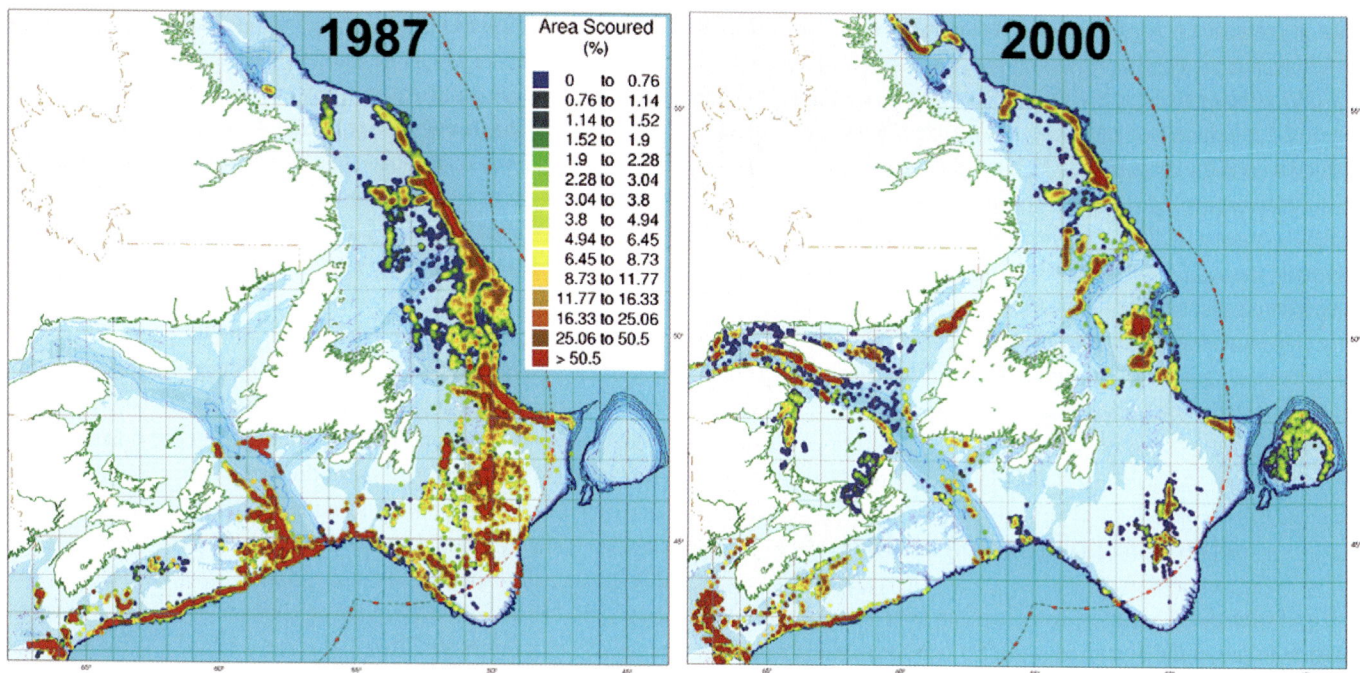

The boundaries and names shown and the designations used on this map do not imply official endorsement or acceptance by the United Nations.

Figure 11.4 | Distribution of trawling effort in Atlantic Canadian waters in 1987 and 2000, based on data of bottom-trawl activity adjusted to total effort for <150 t. From Gilkinson et al., 2006.

A very large literature exists on habitat impacts of fishing gear; experts disagree on both the magnitude of the issue and the effectiveness of management measures and policies to address the impacts. In the late 2000s, several expert reviews were conducted by FAO and the Convention on Biological Diversity in cooperation with UNEP. These reports (FAO, 2007; 2009) provide a recent summary of the types of impacts that various types of fishing gear can have on the seafloor. Most conclusions are straightforward:

All types of gear that contact the bottom may alter habitat features, with impacts larger as the gear becomes heavier.

Mobile bottom-contacting gear generally has a larger area of impact on the seabed than static gear, and consequently the impacts may be correspondingly larger.

The nature of the impact depends on the features of the habitat. Structurally complex and fragile habitats are most vulnerable to impacts, with biogenic features, such as corals and glass sponges, easily damaged and sometimes requiring centuries to recover. On the other hand, impacts of trawls on soft substrates, like mud and sand, may not be detectable after even a few days.

The nature of the impacts also depends on the natural disturbance regime, with high-energy (strong current and/or wave action) habitats often showing little incremental impacts of fishing gear, whereas areas of very low natural disturbance may be more severely affected by fishing gears.

Impacts of fishing gears can occur at all scales of fishery operations; some of the most destructive practices, such as drive netting, dynamite and poisons, although uncommon, are used only in very small-scale fisheries (Kaiser 2001).

All gear might be lost or discarded at sea, in particular pieces of netting. These give rise to what is known as "ghost fishing", that is fishing gear continuing to capture and kill marine animals even after it is lost by fishermen. Assessment of their impacts at either a global or local level is difficult, but the limited number of studies available on its incidence and prevalence indicate that ghost fishing can be a significant problem (Laist et al., 1999, Bilkovic et al. 2012).

Quantitative trend information on habitat impacts is generally not available. Many reports provide maps of how the geographical extent and intensity of bottom-contacting fishing gear have changed over time (e.g. Figure 11.4 from Gilkinson et al., 2006; Greenstreet et al., 2006). These maps show large changes in the patterns of the pressure, and accompanying graphs show the percentage of area fished over a series of years. However, these are individual studies, and broad-scale monitoring of benthic communities is not available. Insights from individual studies need to be considered along with information on the substrate types in the areas being fished to know how increases in effort may be increasing benthic impacts. Furthermore, the recovery potential of the benthic biota has been studied in some specific geographies and circumstances but broadly applicable patterns are not yet clear (e.g., Steele et al. 2002, Claudet et al. 2008).

Even without quantitative data on trends in benthic communities, however, marine areas closed to fishing have increased. Views differ on what level of protection is actually given to areas that are labelled as closed to fishing, but the trend in increasing area protection is not challenged (c.f. CBD, 2012; Spalding et al., 2013). Moreover, the size of the areas being closed to fishing that are not already affected by historical fishing is unknown, as is the recovery rate for such areas, and high-seas fisheries continue to expand into new areas, although probably at a slower rate as RFMO/As increase their actions to implement United Nations General Assembly Resolution 61/105 (FAO, 2014). Hence the pressure on seafloor habitats and benthic communities from bottom-contacting fishing gear may be decreasing slightly, but has been very high for decades on all continental shelves and in many offshore areas at depths of less than several hundred meters (FAO, 2007).

4 Effects of pollution on seafood safety

Fish and particularly predatory fish are prone to be contaminated with toxic chemicals in the marine environment (e.g., organochlorines, mercury, cadmium, lead); these are found mostly in their liver and lipids. Because many sources of marine contamination are land-based (Chapter 20), freshwater fish may contain higher concentrations of contaminants than marine species (Yamada et al., 2014). Furthermore, contamination of the organisms found there is highly variable at the regional and local levels.

Processing methods might significantly reduce the lead and cadmium contents of fish (Ganjavi et al., 2010) and presumably those of other contaminants, whose concentrations generally increase with size (age) of fish (Storelli et al., 2010).

Some species of fish might be toxic (venomous) on their own, such as species of the genus *Siganus* and *Plotosus* in Singapore, which are being culled to reduce their presence on beaches (Kwik, 2012) and *Takifugu rubripes* (fugu), whose properties are relatively well known, such that it is processed accordingly (Yongxiang et al., 2011). However, in extreme situations, human consumption of the remains of fugu processing resulted in severe episodes (Saiful Islam et al., 2011).

Fish, mussels, shrimp and other invertebrates might become toxic through their consumption of harmful algae, whose blooms increased due to climate change, pollution, the spreading of dead (hypoxic/anoxic) zones, and other causes.

Harmful algal blooms are often colloquially known as red tides. These blooms are most common in coastal marine ecosystems but also the open ocean might be affected and are caused by blooms of microscopic algae (including cyanobacteria). Toxins produced by these organisms are accumulated by filtrators that become toxic for species at higher trophic levels, including man. Climate change and eutrophication are considered as part of a complex of environmental stressors resulting in harmful blooms (Anderson et al., 2012). The problem has prompted research to develop models to predict the behaviour of these blooms (Zhao and Ghedira, 2014). Since the 1970s, the phenomenon has spread from the northern hemisphere temperate waters to the southern hemisphere and has now been well documented at least in Argentina, Australia, Brunei Darussalam, China, Malaysia, Papua New Guinea, the Philippines, Republic of Korea and South Africa, but the expansion might also be due to increased awareness of the phenomenon (Anderson et al., 2012) The impact of toxic algal blooms is mostly economic, but episodes of severe illness, even with high mortality rates, might occur, which prompt regulations closing the affected fisheries.

One of the best-known risks in this category is ciguatera, a well-known toxin ingested by human consumption of predatory fish in some regions of the world. The toxin comes from a dinoflagellate and is passed along and concentrated up the food chain (Hamilton et al., 2010). Processed foods are usually safer from the standpoint of contamination. Thus, processing results in added value to the raw food (Satyanarayana et al., 2012). However, inadequate harvest and postharvest handling and processing of the catches might result in contamination with pathogenic organisms (Boziaris et al., 2013).

The general trend expected is an increase in the frequency of harmful algal blooms, in the bioaccumulation of chemical contaminants and in the prevalence of common food-borne pathogenic microorganisms (Marques et al., 2014), although the occurrence of catastrophic events seems to be diminishing.

5 Illegal, unreported and unregulated (IUU) fishing

The FAO International Plan of Action for IUU fishing (FAO 2001) defines IUU fishing as:

- Illegal fishing refers to activities conducted by national or foreign vessels in waters under the jurisdiction of a State, without the permission of that State, or in contravention of its laws and regulations; conducted by vessels flying the flag of States that are parties to a relevant regional fisheries management organization but operate in contravention of the conservation and management measures adopted by that organization and by which the States are bound, or relevant provisions of the applicable international law; or in violation of national laws or international obligations, including those undertaken by cooperating States to a relevant regional fisheries management organization;
- Unreported fishing refers to fishing activities which have not been reported, or have been misreported, to the relevant national authority, in contravention of national laws and regulations; or undertaken in the area of competence of a relevant regional fisheries management organization which have not been reported or have been misreported, in contravention of the reporting procedures of that organization;

- Unregulated fishing refers to fishing activities in the area of application of a relevant regional fisheries management organization that are conducted by vessels without nationality, or by those flying the flag of a State not party to that organization, or by a fishing entity, in a manner that is not consistent with or contravenes the conservation and management measures of that organization; or in areas or for fish stocks in relation to which there are no applicable conservation or management measures and where such fishing activities are conducted in a manner inconsistent with State responsibilities for the conservation of living marine resources under international law.

Notwithstanding the definitions above, certain forms of unregulated fishing may not always be in violation of applicable international law, and may not require the application of measures envisaged under the International Plan of Action (IPOA). FAO considers IUU fishing to be a major global threat to sustainable management of fisheries and to stable socio-economic conditions for many small-scale fishing communities. This illegal fishing not only undermines responsible fisheries management, but also typically raises concerns about working conditions and safety. Illegal fishing also raises concerns about connections to other criminal actions, such as drugs and human trafficking. IUU fishing activity has escalated over the last two decades and is estimated to take 11-26 million mt of fish per annum with a value of 10-23 billion United States dollars. In other words, IUU fishing is responsible for about the same amount of global harvest as would be gained by ending overfishing and rebuilding fish stocks. It is an issue of equal concern on a global scale.

International efforts by RFMO/As, States and the European Union are aimed at eliminating IUU fishing. FAO notes that progress has been slow and suggested (FAO 2014) that better information-sharing regarding fishing vessels engaged in illegal activities, traceability of vessels and fishery products, and other additional measures might improve the situation.

6 Significant economic and/or social aspects of capture fisheries

Capture fisheries are a key source of nutrition and employment for millions of people around the world. FAO (2014) estimates that 800 million people are still malnourished and small-scale fisheries in particular are an important component of efforts to alleviate both hunger and poverty.

Growth in production of fish for food (3.2 per cent *per annum*) has exceeded human population growth (1.6 *per annum*) over the last half century. Recently the growth of aquaculture, which is among the fastest-growing food-producing sectors globally, has formed a major part of meeting rising demand and now accounts for half of the fish produced for human consumption. By 2030 this figure will rise to two-thirds of fish production.

Per capita consumption of fish has risen from 9.9 kg *per annum* to 19.2 kg in 2012. In developing countries this rise is from 5.2 kg to 17.8 kg. In 2010, fish accounted for 16.7 per cent of the global population's consumption of animal protein and 4.3 billion people obtained 15 per cent of their animal protein from fishery products.

Employment in the fisheries sector has also grown faster than the world population and faster than in agriculture. However, of the 58.3 million people employed in the fishery sector, 83 per cent were employed in capture fisheries in 1990. But employment in capture fisheries has decreased to 68 per cent of total fishery sector employment in 2012 according to FAO (2014) statistics.

7 The future status of fish and shellfish stocks over the next decade

World population growth, together with urbanization, increasing development, income and living standards, all point to an increasing demand for seafood. Capture fisheries provide high-quality food that is high in protein, essential amino acids, and long-chain poly-unsaturated fatty acids, with many benefits for human health. The rate of increase in demand for fish was more than 2.5 per cent since 1950 and is likely to continue (HPLE, 2014).

Climate change is expected to have substantial and unexpected effects on the marine environment as detailed throughout this Assessment. Some of these impacts may not negatively impact fisheries and indeed may result in increased availability for capture fisheries in some areas. Nevertheless, there will certainly be an increase in uncertainty with regard to effects on stock productivities and distributions, habitat stability, ecosystem interactions, and the configuration of ecosystems around the globe. Whether these effects on the resources will be "mild" or "severe" will require prudent fisheries management that is precautionary enough to be prepared to assist fishers, their communities and, in general, stakeholders in adapting to the social and economic consequences of climate change (Grafton, 2009).

Small-scale, artisanal fisheries are likely to be more vulnerable to the impacts of climate change and increasing uncertainty than large-scale fisheries (Roessig et al. 2004). While small-scale fisheries may be able to economically harvest a changing mix of species, varying distribution patterns and productivity of stocks may have severe consequences for subsistence fishing. Further, the value of small-scale fisheries as providers not only of food, but also of livelihoods and for poverty alleviation will be compromised by direct competition with large-scale operations with access to global markets (Alder and Sumaila, 2004).

The data clearly indicate that the amount of fish that can be extracted from historically exploited wild stocks is unlikely to increase substantially. Some increase is possible through the rebuilding of depleted stocks, a central goal of fisheries management. Current trends diverge between well-assessed regions showing stabilization of fish biomass and other regions continuing to decline (Worm and Branch, 2012).

In Europe, North America and Oceania, major commercially exploited fish stocks are currently stable, with the prospect that reduced exploitation rates should achieve rebuilding of the biomass in the long term. In the rest of the world, fish biomass is, on average, declining due to lower management capacity. Many fisheries may still be productive, but prospects are poor (Worm et al., 2009).

The growing demand for fish products cannot be met from sustainable capture fisheries in the next decade. On the other hand, the potential for sustainable exploitation of non-traditional stocks is not well known. Particularly in light of the growth of the aquaculture sector with a need for fishmeal for feed, the pressure to exploit non-traditional resources will increase even if the impacts on marine ecosystems are not well understood.

8   Identify gaps in capacity to engage in capture fisheries and to assess the environmental, social and economic aspects of capture fisheries and the status and trends of living marine resources

Rebuilding overfished stocks is a major challenge for capture fisheries management. Another key challenge is making better, more sustainable use of existing marine resources while conserving the ecosystem upon which they depend. From a global perspective this will require filling a number of gaps, both scientific and in management capacity (Worm et al., 2009):

- The transfer of fishing effort from developed to developing countries is a process that has been accelerating since the 1960s. Almost all of the fish caught by foreign fleets is consumed in industrialized countries and will have important implications for food security (Alder and Sumaila, 2004) and biodiversity in the developing world. In many regions there is insufficient capacity to assess and manage marine resources in the context of this pressure;

- The increase in IUU fishing operations is a major challenge for management that will require increased management capacity if it is to be controlled;
- Recovery of depleted stocks is still a poorly understood process, particularly for demersal species. It is potentially constrained by the magnitude of the previous decline, the loss of biodiversity, species' life histories, species interactions, and other factors. In other words, the basic principle for recovery is straightforward – fishing pressure needs to be reduced. But the specific application of plans to promote recovery of the stock once fishing pressure is reduced requires significant scientific and management capacity;
- Addressing the challenges of spatial management of the ocean for fisheries, conservation and many other purposes, and the overall competition for ocean space, will depend upon greater scientific and management capacity in most regions.

The average performance of stock-assessed fisheries indicates that most are slowly approaching the fully fished status (*sensu* FAO). On the other hand, recent analyses of unassessed fish stocks indicate that they are mostly in poorer condition (Costello et al., 2012). The problem is severe because most of these stocks sustain small-scale fisheries critical for the food security in developing countries. Better information and the capacity to manage many of these stocks will be needed to improve the situation.

Debates among fisheries specialists have been more concerned about biological sustainability and economic efficiencies than about reducing hunger and malnutrition and supporting livelihoods (HLPE, 2014). It is necessary to develop the tools for managing small-scale fisheries efficiently, particularly in view of the competing long-distance fleets. The fishing agreements allowing long-distance fleets to operate in developing countries had not yielded the expected results in terms of building the capacity to administer or sustainably fish their resources. IUU fishing becoming more prominent has exacerbated the situation (Gagern and van den Bergh, 2013). It is necessary for developing countries to build the capacity to develop sustainable industrial fisheries and to develop stock assessment capabilities for small-scale fisheries balancing food security and conservation objectives (Allison and Horemans, 2006).

References

Alder, J., and Sumaila, U.R. (2004). Western Africa: A Fish Basket of Europe Past and Present. *The Journal of Environment Development*, June, 13 (2): 156-178.

Allison, E.H., Horemans, B. (2006). Putting the principles of the Sustainable Livelihoods Approach into fisheries development policy and practice. *Marine Policy*, 30: 757-766.

Alverson, D.L., Freeberg, M.H., Murawski, S.A., and Pope, J.G. (1994). A global assessment of fisheries bycatch and discards. *FAO Fisheries Technical Paper*, No. 339: 235 p.

Anderson, O.R.J., Small, C.J., Croxall, J.P., Dunn, E.K., Sullivan, B.J., Yates, O., Black, A. (2011). Global seabird bycatch in longline fisheries. *Endangered Species Research* 14: 91–106.

Bilkovic, D.M., Havens, K.J., Stanhope, D.M., Angstadt, K.T. (2012). Use of fully biodegradable panels to reduce derelict pot threats to marine fauna. *Conservation Biology* 26(6): 957-966.

Boziaris, I.S., Stmatiou, A.P., and Nychas, G.J.E. (2013). Microbiological aspects and shelf life of processed seafood products. *Journal of the Science of Food and Agriculture*, 93 (5): 1184–1190.

Bundy, A., Fanning, P., Zwanenburg, K.C. (2005). Balancing exploitation and conservation of the eastern Scotian Shelf ecosystem: application of a 4D ecosystem exploitation index. *ICES Journal of Marine Science: Journal du Conseil* 62 (3): 503-510.

Casini, M., Lövgren, J., Hjelm, J., Cardinale, M., Molinero, J.C., and Kornilovs, G. (2008). Multilevel trophic cascades in a heavily exploited open marine ecosystem. *Proc. R. Soc.* B. 275, doi: 10.1098/rspb.2007.1752.

CBD (2012). *Review of Progress in Implementation of the Strategic Plan for Biodiversity 2011-2020, Including the Establishment of National Targets and the Updating of National Biodiversity Strategies and Action Plans*, UNEP/CBD/COP/11/12, paragraph 26 https://www.cbd.int/doc/meetings/cop/cop-11/official/cop-11-12-en.pdf

Charcot, J. (1911). *The Voyage of the 'Why Not?' in the Antarctic*. Philip Walsh (trans.). Hodder and Stoughton. New York and London.

Claudet, J., Osenberg, C.W., Benedetti-Cecchi, L., Domenici, P., Garcia-Charton, J.-A., Pérez-Ruzafa, A., Badalamenti, F., Bayle-Smpere, J., Brito, A., Bulleri, F., Culioli, J.-M., Dimech, M., Falcón, J.M., Guala, I., Milazzo, M., Sánchez-Meca, J., Somerfield, P.J., Stobart, B., Vandeperre, F., Valle, C., and Planet, S. (2008). Marine reserves: size and age do matter. *Ecology Letters*, 11, 481-489.

Collie, J.S., Hall, S.J., Kaiser, M.J., and Poiner, I.R. (2000). A quantitative analysis of fishing impacts on shelf-sea benthos. *Journal of Animal Ecology* 69: 785-798.

Costello, C., Ovando, D., Hilborn, R., Gaines, S.D., Deschenes, O., Lester, S.E. (2012). Status and Solutions for the World's Unassessed Fisheries. *Science* 338: 517-520.

FAO (1995). *Code of Conduct for Responsible Fisheries*. Rome, FAO. 48 p.

FAO (2001). *International Plan of Action to Prevent, Deter and Eliminate Illegal, Unreported and Unregulated Fishing*, Rome, FAO. 24p.

FAO (2003). The ecosystem approach to fisheries. *FAO Technical Guidelines for Responsible Fisheries* No. 4, Suppl. 2. Rome, FAO. 112 p.

FAO (2007). Report of the Expert Consultation on International Guidelines for the Management of Deep-sea Fisheries in the High Seas, Bangkok, 11-14 September 2007 *FAO Fisheries Report* No. 855. http://www.fao.org/docrep/011/i0003e/i0003e00.HTM

FAO (2009). FAO/UNEP Expert Meeting on Impacts of Destructive Fishing Practices, Unsustainable Fishing, and Illegal, Unreported and Unregulated (IUU) Fishing on Marine Biodiversity and Habitats. *FAO Fisheries and Aquaculture Report* No. 932. http://www.fao.org/docrep/012/i1490e/i1490e00.pdf

FAO (2010). Report of the Expert Consultation on International Guidelines for Bycatch Management and Reduction of Discards. Rome, 30 November–3 December 2009. Fisheries and Aquaculture Report. No. 934. Rome, FAO: 28p.

FAO (2011). *International Guidelines on Bycatch Management and Reduction of Discards*. FAO. 73 pp. www.fao.org/docrep/015/ba0022t/ba0022t00.pdf

FAO (2014). *The State of World Fisheries and Aquaculture*. FAO, Rome.

FAO Fisheries and Aquaculture Department (2009). Guidelines to reduce sea turtle mortality in fishing gears. FAO, Rome: 128 pages.

Farcas, A., and Rossberg, A.G. (2014). Maximum sustainable yield from interacting fish stocks in an uncertain world: two policy decisions and underlying trade-offs. *arXiv preprint arXiv:1412.0199*.

Fulton, E.A., et al. (2011). Lessons in modelling and management of marine ecosystems: the Atlantis experience. *Fish and Fisheries* 12.2: 171-188.

Gagern, A., and van den Bergh, J. (2013). A critical review of fishing agreements with tropical developing countries. *Marine Policy*. 38: 375-386.

Ganjavi, M., Ezzatpanah, H., Givianrad, H.M., Shams, A. (2010). Effect of canned tuna fish processing steps on lead and cadmium contents of Iranian tuna fish. *Food Chemistry*, 118 (3): 525-528.

Garcia, S.M. (Ed.), Kolding, J., Rice, J., Rochet, M.J., Zhou, S., Arimoto, T., Beyer, J., Borges, L., Bundy, A., Dunn, D., Graham, N., Hall, M., Heino, M., Law, R., Makino, M., Rijnsdorp, A.D., Simard, F., Smith, A.D.M. and Symons, D. (2011). Selective Fishing and Balanced Harvest in Relation to Fisheries and Ecosystem Sustainability. Report of a scientific workshop organized by the IUCN-CEM Fisheries Expert Group (FEG) and the European Bureau for Conservation and Development (EBCD) in Nagoya (Japan), 14–16 October 2010. Gland, Switzerland and Brussels, Belgium: IUCN and EBCD. iv + 33pp.

Garstang, W. (1900). The Impoverishment of the Sea. A Critical Summary of the Experimental and Statistical Evidence bearing upon the Alleged Depletion of the Trawling Grounds. *Journal of the Marine Biological Association of the United Kingdom* 6(1): 1–69.

Gilkinson, K., Dawe, E., Forward, B., Hickey, B., Kulka, D., and Walsh, S., (2006). *A Review of Newfoundland and Labrador Region Research on the Effects of Mobile Fishing Gear on Benthic Habitat and Communities*. DFO Can. Sci. Advis. Sec. Res. Doc. 2006/055: 30p. http://www.dfo-mpo.gc.ca/csas/Csas/DocREC/2006/RES2006_055_e.pdf

Grafton, R.Q. (2009). Adaptation to Climate Change in Marine Capture Fisheries. *Environmental Economics Research Hub Research Reports*, 37: 33pp.

Greenstreet, S.P.R., Shanks, A.M., and Buckett, B.E. (2006). *Trends in fishing activity in the North Sea by UK registered vessels landing in Scotland over the period 1960 to 1998*. Fisheries Research Services Collaborative Report 2.06.

Gulland, J.A. and Carroz, J.E. (1968) Management of fishery resources. *Advances in Marine Biology* 6: 1–71.

Hall, M.A., Alverson, D.L., and Metuzals, K.I. "By-catch: Problems and Solutions." *Marine Pollution Bulletin* 41.1 (2000): 204-219.

Hamilton, B., Whittle, N., Shaw, G., Eaglesham, G., Moore, M.R., Lewis, R.J. (2010). Human fatality associated with Pacific ciguatoxin contaminated fish. *Toxicon*, 56 (5): 668–673.

Hard, J.J., Gross, M.R., Heino, M., Hilborn, R., Kope, R.G., Law, R., and Reynold, J.D. (2008). Evolutionary consequences of fishing and their implications for salmon. *Evolutionary Applications*, vol. 1, no. 2: 388-408.

Hixon, M.A., Johnson, D.W., and Sogard, S.M. (2013). BOFFFFs: on the importance of conserving old-growth age structure in fishery populations. – *ICES Journal of Marine Science*, doi:10.1093/icesjms/fst200.

HLPE (2014). *Sustainable fisheries and aquaculture for food security and nutrition*. A report by the High level Panel of Experts on Food Security and Nutrition of the Committee on World Food Security, Rome.

Hsieh, C., Reiss, C.S., Hunter, J.R., Beddington, J.R., May, R.M., and Sugihara, G. (2006). Fishing elevates variability in the abundance of exploited species. *Nature* 443: 859-862.

Jennings, S., and Kaiser, M. (1998). The effects of fishing on marine ecosystems. *Advances in Marine Biology*, 34: 201-352.

Jennings, S., Reynolds, J.D., and Mills, S.C. (1998). Life history correlates of responses to fisheries exploitation. *Proceedings of the Royal Society London B*: 265:333-339.

Kaiser, M.J., Collie, J.S., Hall, S.J., Jennings, S., and Poiner, I.R. (2001). *Impacts of Fishing Gear on Marine Benthic Habitats*. Reykjavik Conference on Responsible Fisheries in the Marine Ecosystem. Reykjavik, Iceland, 1-4 October 2001.

Kelleher, K. (2005). Discards in the World's Marine Fisheries. An Update. Rome, FAO, *FAO Fisheries Technical Paper* 470. http://www.fao.org/docrep/008/y5936e/y5936e00.HTM

Kwik, J.T.B. (2012). Controlled Culling of Venomous Marine Fishes Along Sentosa Island Beaches: A Case Study of Public Safety Management in the Marine Environment of Singapore. *The Raffles Bulletin of Zoology*. Supplement No. 25: 93–99.

Laist, D.W., Coe, J.M., and O'Hara, K.J. (1999). Marine Debris Pollution. In: Twiss, Jr., J.R., and Reeves, R.R. (eds.) *Conservation and Management of Marine Mammals*: 342-366. Smithsonian Institution Press. Washington, D.C.

Larkin, P.A. (1997). An epitaph for the concept of maximum sustained yield. *Transactions of the American Fisheries Society*, 106(1): 1-11.

Marques, A., Rosa, R. (2014). Seafood Safety and Human Health Implications. In: Goffredo, S., Dubinsky, Z. (eds.), *The Mediterranean Sea: its history and present challenges*: 589-603. Springer Netherlands.

Michelet, J. (1875). *La Mer*. Paris, Michel Lévy Frères: 428 pp.

Myers, R.A., Baum, J.K., Shepherd, T.D., Powers, S.P., and Peterson, C.H. (2007). Cascading Effects of the Loss of Apex Predatory Sharks from a Coastal Ocean. *Science* 315: 1846-1850.

National Marine Fisheries Service (2011). *U.S. National Bycatch Report*, Karp, W.A., Desfosse, L.L., Brooke, S.G. (eds.) U.S. Dep. Commer., NOAA Tech. Memo. NMFS-F/SPO-117C, 508 p.

Pauly, D. (1994) *On the sex of fish and the gender of scientists: A collection of essays in fisheries science*, Chapman and Hall, London.

Polovina, J.J. (1984). Model of a coral reef ecosystem. *Coral reefs* 3.1: 1-11.

Ricker, W.E. (1981). Changes in the average size and age of Pacific salmon. *Canadian Journal of Fisheries and Aquatic Sciences* 38: 1636–1656.

Roessig, J.M., Woodley, C.M., Cech, J.J., Hansen, L.J. (2004). Effects of global climate change on marine and estuarine fishes and fisheries. *Reviews in Fish Biology and Fisheries* 14: 251-275.

Rozwadowski, H.M. (2002). *The Sea Knows No Boundaries: A Century of Marine Science under ICES*. ICES, University of Washington Press, Copenhagen, Seattle, London.

Saiful Islam, M., Luby, S.P., Rahman, M., Parveen, S., Homaira, N., Begum, N.H., Dawlat Khan, A.K.M., Sultana, R., Akhter, S., and Gurley, E.S. (2011). Social Ecological Analysis of an Outbreak of Pufferfish Egg Poisoning in a Coastal Area of Bangladesh. *American Journal of Tropical Medicine and Hygiene* 85 (3): 498-503.

Satyanarayana S.D.V., Pavan Kumar, P., Amit, S., Dattatreya, A., Aditya, G. (2012). Potential Impacts of Food and it's Processing on Global Sustainable Health. *Journal of Food Processing & Technology* 3: 143.

Sieben, K., Rippen, A.D., and Eriksson, B.K. (2011). Cascading effects from predator removal depend on resource availability in a benthic food web. *Marine Biology* 158:391-400.

Spalding, M.D., Meliane, I., Milam, A., Fitzgerald, C., and Hale, L.Z. (2013). Protecting Marine Spaces: global targets and changing approaches, *Ocean Yearbook* 27: 213-248.

Steele, M.A., Malone, J.C., Findlay, A.M., Carr, M. and Forrester, G. (2002). A simple method for estimating larval supply in reef fishes and a preliminary test of population limitation by larval delivery in the kelp bass *Paralabrax clathratus*. *Marine Ecology Progress Series* 235:195–203.

Storelli, M.M., Barone, G., Cuttone, G., Giungato, D. (2010). Occurrence of toxic metals (Hg, Cd and Pb) in fresh and canned tuna: public health implications. *Food and Chemical Toxicology* (48), 11: 3167–3170.

Suarez, J.L. (1927). Rapport au Conseil de la Société des Nations. Exploitation des Richesses de la Mer. Publications de la Société des Nations V. *Questions Juridiques*. V.1. 120: 125.

Szpak, P., & Orchard, T.J., Salomon, A.K., and Gröcke, D.R. (2013). Regional ecological variability and impact of the maritime fur trade on nearshore ecosystems in southern Haida Gwaii (British Columbia, Canada): evidence from stable isotope analysis of rockfish (Sebastes spp.) bone collagen. *Archaeol Anthropol Sci* DOI 10.1007/s12520-013-0122-y.

Townsend, H.M., Link, J.S., Osgood, K.E., Gedamke, T., Watters, G.M., Polovina, J.J., Levin, P.S., Cyr, N., and Aydin, K.Y. (eds) (2008). *Report of the National Ecosystem Modeling Workshop (NEMoW)*. NOAA Technical Memorandum NMFS-F/SPO-87.

UNEP/CBD/FAO (2011). *Report of Joint Expert Meeting on Addressing Biodiversity Concerns in Sustainable Fisheries* http://www.cbd.int/doc/meetings/mar/jem-bcsf-01/official/jem-bcsf-01-sbstta-16-inf-13-en.pdf

Worm, B., Hilborn, R., Baum, J.K., Branch, T.A., Collie, J.S., Costello, C., Fogarty, M.J., Fulton, E.A., Hutchings, J.A., Jennings, S., Jensen, O.P., Lotze, H.K., Mace, P.M., McClanahan, T.R., Minto, C., Palumbi, S.R., Parma, A.M., Ricard, D., Rosenberg, A.A., Watson, R., Zeller D. (2009). Rebuilding Global Fisheries. *Science* 325: 578-584.

Yamada, A., Bemrah, N., Veyrand, B., Pollono, C., Merlo, M., Desvignes, V., Sirot, V., Oseredczuk, M., Marchand, P., Cariou, R., Antignac, J.P., Le Bizec, B., and Leblanc, J.C. (2014). Perfluoroalkyl Acid Contamination and Polyunsaturated Fatty Acid Composition of French Freshwater and Marine Fishes. *Journal of Agricultural and Food Chemistry*, 62 (30): 7593–7603.

Ye, Y., Cochrane, K., Bianchi, G., Willmann, R., Majkowski, J., Tandstad, M., Carocci, F. (2013). Rebuilding global fisheries, the World Summit Goal, costs and benefits. *Fish and Fisheries*, 14: 174-185.

Yongxiang, F., Rong, J., Ning, L., Weixing, Y. (2011). Study on a management system for safely utilizing puffer fish resources. *Chinese Journal of Food Hygiene*: 2011-03.

Zhao, J., Ghedira, H. (2014). Monitoring red tide with satellite imagery and numerical models: A case study in the Arabian Gulf. *Marine Pollution Bulletin* (79): 305-313.

12 Aquaculture

Writing team:
Patricio Bernal (Convenor and Lead Member), Doris Oliva

Aquaculture   Chapter 12

1 Scale and distribution of aquaculture

Aquaculture is providing an increasing contribution to world food security. At an average annual growth rate of 6.2 per cent between 2000 and 2012 (9.5 per cent between 1990 and 2000), aquaculture is the world's fastest growing animal food producing sector (FAO, 2012; FAO 2014). In 2012, farmed food fish contributed a record 66.6 million tons, equivalent to 42.2 per cent of the total 158 million tons of fish produced by capture fisheries and aquaculture combined (including non-food uses, see Figure 12.1). Just 13.4 per cent of fish production came from aquaculture in 1990 and 25.7 per cent in 2000 (FAO, 2014).

In Asia, since 2008 farmed fish production has exceeded wild catch (freshwater and marine), reaching 54 per cent of total fish production in 2012; in Europe aquaculture production is 18 per cent of the total and in other continents is less than 15 per cent. Nearly half (49 per cent) of all fish consumed globally by people in 2012 came from aquaculture (FAO, 2014).

In 2012, world aquaculture production, for all cultivated species combined, was 90.4 million tons (live weight equivalent and 144.4 billion dollars in value). This includes 44.2 million tons of finfish (87.5 billion dollars), 21.6 million tons of shellfish (crustacea and molluscs with 46.7 billion dollars in value) and 23.8 million tons of aquatic algae (mostly seaweeds, 6.4 billion dollars in value). Seaweeds and other algae are harvested for use as food, in cosmetics and fertilizers, and are processed to extract thickening agents used as additives in the food and animal feed industries. Finally 22,400 tons of non-food products are also farmed (with a value of 222.4 million dollars), such as pearls and seashells for ornamental and decorative uses (FAO, 2014).

According to the latest (but incomplete) information for 2013, FAO estimates that world food fish aquaculture production rose by 5.8 per cent to 70.5 million tons, with production of farmed aquatic plants (including mostly seaweeds) being estimated at 26.1 million tons.

2 Composition of world aquaculture production: inland aquaculture and mariculture

Although this Chapter is part of an assessment of food security and food safety from the ocean, to understand the trends in the development of world aquaculture and its impact on food security it is relevant to compare inland aquaculture, conducted in freshwater and saline estuarine waters in inland areas, versus true mariculture, conducted in the coastal areas of the world ocean.

Of the 66.6 million tons of farmed food fish[1] produced in 2012, two-thirds (44.2 million tons) were finfish species: 38.6 million tons grown from inland aquaculture and 5.6 million tons from mariculture. Inland aquaculture of finfish now accounts for 57.9 percent of all farmed food fish production globally.

Although finfish species grown from mariculture represent only 12.6 percent of the total farmed finfish production by volume, their value (23.5 billion United States dollars) represents 26.9 percent of the total value of all farmed finfish species. This is because mariculture includes a large proportion of carnivorous species, such as salmon, trouts and groupers, "cash-crops" higher in unit value and destined to more affluent markets.

FAO (2014) concludes that freshwater fish farming makes the greatest direct contribution to food security, providing affordable protein food, particularly for poor people in developing countries in Asia, Africa and Latin America. Inland aquaculture also provides an important new source of livelihoods in less developed regions and can be an important contributor to poverty alleviation.

3 Main producers of aquaculture products

In 2013, China produced 43.5 million tons of food fish and 13.5 million tons of aquatic algae (FAO, 2014, p 18), making it by far the largest producer of aquaculture products in the world. Aquaculture production is still concentrated in few countries of the world. Considering national total production, the top five countries (all in Asia: China, India, Viet Nam, Indonesia, Bangladesh) account for 79.8 per cent of world production while the top five countries in finfish mariculture (Norway, China, Chile, Indonesia, and Philippines) account for 72.9 per cent of world production (Table 12.1, Figure 12.2).

4 Species cultivated

It is estimated that more than 600 aquatic species are cultured worldwide[2] in a variety of farming systems and facilities of varying technologi-

Figure 12.1 | World capture fisheries and aquaculture production between 1950 and 2012 (HLPE, 2014).

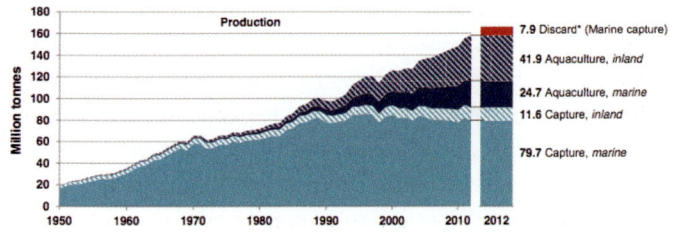

---

1 The generic term "farmed food fish" used here and by FAO, includes finfishes, crustaceans, molluscs, amphibians, freshwater turtles and other aquatic animals (such as sea cucumbers, sea urchins, sea squirts and edible jellyfish) produced for intended use as food for human consumption.

2 Up to 2012, the number of species registered in FAO statistics was 567, including finfishes (354 species, with 5 hybrids), molluscs (102), crustaceans (59), amphibians and reptiles (6), aquatic invertebrates (9), and marine and freshwater algae

Table 12.1 | Farmed food fish production by 15 top producers and main groups of farmed species in 2012 (FAO, 2014).

Producer	Finfish Inland aquaculture	Finfish Mariculture	Crustaceans	Molluscs	Other species	National total	Share in world total
	(Tonnes)			(Tonnes)			(Percentage)
China	23 341 134	1 028 399	3 592 588	12 343 169	803 016	41 108 306	61.7
India	3 812 420	84 164	299 926	12 905	...	4 209 415	6.3
Viet Nam	2 091 200	51 000	513 100	400 000	30 200	3 085 500	4.6
Indonesia	2 097 407	582 077	387 698	...	477	3 067 660	4.6
Bangladesh	1 525 672	63 220	137 174	...	...	1 726 066	2.6
Norway	85	1 319 033	...	2 001	...	1 321 119	2.0
Thailand	380 986	19 994	623 660	205 192	4 045	1 233 877	1.9
Chile	59 527	758 587	...	253 307	...	1 071 421	1.6
Egypt	1 016 629	...	1 109	...	...	1 017 738	1.5
Myanmar	822 589	1 868	58 981	...	1 731	885 169	1.3
Philippines	310 042	361 722	72 822	46 308	...	790 894	1.2
Brazil	611 343	...	74 415	20 699	1 005	707 461	1.1
Japan	33 957	250 472	1 596	345 914	1 108	633 047	1.0
Republic of Korea	14 099	76 307	2 838	373 488	17 672	484 404	0.7
United States of America	185 598	21 169	44 928	168 329	...	420 024	0.6
Top 15 subtotal	36 302 688	4 618 012	5 810 835	14 171 312	859 254	61 762 101	92.7
Rest of world	2 296 562	933 893	635 983	999 426	5 288	4 871 152	7.3
World	38 599 250	5 551 905	6 446 818	15 170 738	864 542	66 633 253	100

Note: The symbol "..." means the production data are not available or the production volume is regarded as negligibly low.

cal sophistication, using freshwater, brackish water and marine water (FAO, 2014). In 2006, the top 25 species being farmed accounted for over 90 percent of world production (FAO, 2006a). Of the more than 200 species of fish and crustaceans currently estimated to be cultivated and fed on externally supplied feeds, just 9 species account for 62.2 percent of total global-fed species production, including grass carp (*Ctenopharyngodon idellus*), common carp (*Cyprinus carpio*), Nile tilapia (*Oreochromis niloticus*), catla (*Catla catla*), whiteleg shrimp (*Litopenaeus vannamei*), crucian carp (*Carassius carassius*), Atlantic salmon (*Salmo solar*), pangasiid catfishes (striped/tra catfish [*Pangasianodon hypophthalmus*] and basa catfish [*Pangasius bocourti*]), and rohu (*Labeo rohita*; Tacon et al., 2011). The farming of freshwater tilapias, including Nile tilapia and some other cichlid species, is the most widespread type of aquaculture in the world. FAO has recorded farmed tilapia production statistics for 135 countries and territories on all continents (FAO, 2014). In this respect, aquaculture is no different from animal husbandry, in that global livestock production is concentrated in a few species (Tacon et al. 2011).[3] Among molluscs only 6 species account for the 64.5 per cent of the aquaculture production (15.5 million tons in 2013) and all of them are bivalves: the cupped oyster (*Crassostrea spp*), Japanese carpet shell *(Ruditapes philippinarum)*, constricted Tagelus (*Sinnovacula constricta*), blood cocked *Anadara granosa*, Chilean mussel (*Mytilus chilensis*) and Pacific cupped oyster (*Crassostrea gigas*).

5 Aquaculture systems development

The cultivation of farmed food fish is the aquatic version of animal husbandry, where full control of the life cycle enables the domestication of wild species, their growth in large-scale farming systems and the application of well-known and well-established techniques of animal artificial selection of desirable traits, such as resistance to diseases, fast growth and size.

For most farmed aquatic species, hatchery and nursery technologies have been developed and well established, enabling the artificial control of the life cycle of the species. However wild seed is still used in many farming operations. For a few species, such as eels (*Anguilla* spp.), farming still relies entirely on wild seed (FAO, 2014).

Aquaculture can be based on traditional, low technology farming systems or on highly industrialized, capital-intensive processes. In between there is a whole range of aquaculture systems with different efficiencies that can be adapted to local socioeconomic contexts.

Physically, inland aquaculture and coastal shrimp mariculture uses fixed ponds and raceways on land that put a premium on the use of land. Finfish mariculture and some farming of molluscs such as oysters and mussels tend to use floating net pens, cages and other suspended systems in the water column of shallow coastal waters, enabling these systems to be fixed by being anchored to the bottom.

Direct land use needs for fish and shrimp ponds can be substantial. Current aquaculture production occupies a significant quantity of land, both in inland and coastal areas. Aquaculture land use efficiency, however, differs widely by production system. While fish ponds use relatively high amounts of land (Costa-Pierce et al., 2012, cited in WRI, 2014), flow-through systems (raceways) use less land, while cages and pens suspended in water bodies use very little (if any) land (WRI, 2014).

The handling of monocultures with high densities of individuals in confinement replicates the risks typical to monocultures in land-based animal husbandry, such as the spread and proliferation of parasites, and the contagion of bacterial and viral infections producing mass mortalities, and the accumulation of waste products. If on land these risks can be partially contained, in mariculture, the use of semi-enclosed systems open to the natural flow of seawater and sedimentation to the bottom, propagate these risks to the surrounding environment affecting the health of the ecosystems in which aquaculture operations are implanted.

The introduction of these risks to the coastal zones puts a premium in the application of good management practices and effective regulations for zoning, site selection and maximum loads per area.

In 1999 during the early development of shrimp culture, a White Spot Syndrome Virus (WSSV) epizootic quickly spread through nine Pacific coast countries in Latin America, costing billions of dollars (McClennen, 2004). Disease outbreaks in recent years have affected Chile's Atlantic salmon production with losses of almost 50 percent to the virus of "infectious salmon anaemia" (ISA). Oyster cultures in Europe were attacked by herpes virus Os HV-1 or OsHV-1 μvar, and marine shrimp farming in

---

(37).

3 On land, the top eight livestock species are pig, chicken, cattle, sheep, turkey, goat, duck and buffalo (Tacon et al. 2011)

several countries in Asia, South America and Africa have experienced bacterial and viral infections, resulting in partial or sometimes total loss of production. In 2010, aquaculture in China suffered production losses of 1.7 million tons caused by natural disasters, diseases and pollution. Disease outbreaks virtually wiped out marine shrimp farming production in Mozambique in 2011 (FAO, 2010, 2012).

New diseases also appear. The early mortality syndrome (EMS) is an emerging disease of cultured shrimp caused by a strain of *Vibrio parahaemolyticus*, a marine micro-organism native in estuarine waters worldwide. Three species of cultured shrimp are affected (*Penaeus monodon*, *P. vannamei* and *P. chinensis*). In Viet Nam, about 39 000 hectares were affected in 2011. Malaysia estimated production losses of 0.1 billion dollars (2011). In Thailand, reports indicated annual output declines of 30–70 percent. The disease has been reported in China, Malaysia, Mexico, Thailand and Viet Nam (FAO, 2014).

It is apparent that intensive aquaculture systems are likely to create conditions that expose them to disease outbreaks. When semi-enclosed systems are used, as in mariculture, pathogens in their resting or reproductive stages propagate directly to the environment, where they can persist for long periods of time as a potential source of recurring outbreaks.

Optimization of industrial systems selects for few or a single preferred species. This is the case in the oyster culture with the widespread culture of *Crassostrea gigas* and in the shrimp industry by the dominance of *Penaeus vanamei*, the white shrimp as the preferred species. This can be also an additional source of risk, if evolving pathogens develop resistance to antibiotics or other treatments used to control well-known diseases.

6 Fed and non-fed aquaculture

Animal aquaculture production can be divided among those species that feed from natural sources in the environment in which they are grown, and species that are artificially fed. The output of naturally-fed aquaculture represents a net increase of world animal protein stock, while the contribution of fed aquaculture, consuming plant or animal protein and fat, depends on conversion rates controlled by the physiology of the species and the effectiveness of the farming system.

In 2012, global production of non-fed species from aquaculture was 20.5 million tons, including 7.1 million tons of filter-feeding carps and 13.4 million tons of bivalves and other species. Accordingly, 46.09 million tons or 69.2 per cent of total farmed food fish (FAO, 2014) was dependent upon the supply of external nutrient inputs provided in the form of (i) fresh feed items, (ii) farm-made feeds or (iii) commercially manufactured feeds (Tacon et al., 2011).

The share of non-fed species in total farmed food fish production continued to decrease to 30.8 percent in 2012 compared with about 50 percent in 1982, reflecting stronger growth in the farming of fed species, especially of high value carnivores (FAO, 2014).

In Europe, after much publicly and privately sponsored research, the technology to farm cod was fully developed and supported by large amounts of venture capital, and industrial production of cod started. In the early 2000s this industrial development suffered from the financial crisis of 2008, and further growth and development almost stopped. Although the participation of risk capital in the development of aquaculture might be an option in particular places and circumstances, it is far from being the preferred option. Development of aquaculture systems, supplying domestic and international markets, has a better chance to succeed if supported by a mix of long-term public support systems (credit, technical assistance) for small and rural producers coupled with entrepreneurial initiatives well implanted in the markets.

Marine finfish aquaculture is rapidly growing in the Asia-Pacific region, where high-value carnivorous fish species (e.g. groupers, barramundi, snappers and pompano) are typically raised in small cages in inshore environments. In China this development has led to experiments in offshore mariculture using larger and stronger cages. (FAO, 2014).

These examples show that at least to the present, decision-making for the development of mariculture, particularly finfish mariculture, tends to be dominated by economic growth and not by food security considerations. To balance this trend, the intergovernmental High Level Panel of Experts on Food Security has recently advocated the need to define specific policies to support current targets on food security in view of the projected growth of human population (HLPE, 2014).

The potential for non-fed mariculture development is far from being fully explored particularly that of marine bivalves in Africa and in Latin America and in the Caribbean. Limited capacity in mollusc seed production is regarded as a constraint in some countries (FAO, 2014).

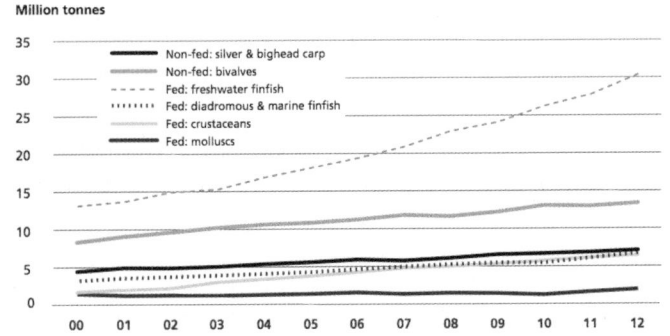

Figure 12.2 | World aquaculture production, fed and non-fed between years 2000 and 2012 (FAO, 2014)

Figure 12.3 | The aquaculture industry has reduced the share of fishmeal in farmed fish diets (percent) (FAO, 2014).

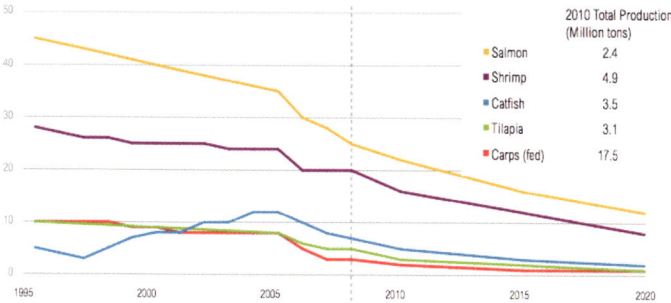

Note: Fishmeal use varies within and between countries; the figures presented are global means. Data represent observations between 1995-2008, and projections for 2009-2020.
Source: Tacon and Metian (2008), Tacon et al. (2011).

7 Aquafeed production

Total industrial compound aquafeed production increased, from 7.6 million tons in 1995 to 29.2 million tons in 2008 (last estimate available, Tacon et al., 2011). These are estimates because there is no comprehensive information on the global production of farm-made aquafeeds (estimated by FAO at between 18.7 and 30.7 million tons in 2006) and/or on the use of low-value fish/trash fish as fresh feed.

Fishmeal is used as high-protein feed and fish oil as a feed additive in aquaculture (FAO, 2014). Fishmeal and fish oil are produced mainly from harvesting stocks of small, fastreproducing fish (e.g., anchovies, small sardines and menhaden) and for which there is some, but limited, demand for human consumption. This use, promoted in the 1950s by FAO as a means to add value to the massive harvesting of small pelagic fish, raises the question of the alternative use of this significant fish biomass for direct human consumption (HLPE 2014).

In 2012 about 35 per cent of world fishmeal production was obtained from fisheries by-products (frames, off-cuts and offal) from the industrial processing of both wild caught and farmed fish. Commercial operations harvesting myctophids[4] for fishmeal and oil are being piloted in some regions, though the ecological consequences of exploiting these previously untapped resources have not been evaluated. In 2007 the largest producer of fishmeal was Peru (1.4 million tons) followed by China (1.0 million tons) and Chile (0.7 million tons). Other important producers were Thailand, the United States of America, Japan, Denmark, Norway and Iceland (Tacon et al., 2011).

Estimates of total usage of terrestrial animal by-product meals and oils in compound aquafeeds ranges between 0.15 and 0.30 million tons, or less than 1 percent of total global production.

Patterns in the use of fishmeal and fish oil have changed in time due to the growth and evolution of the world aquaculture industry. On a global basis, in 2008 (the most recent published estimate), the aquaculture sector consumed 60.8 percent of global fishmeal production (3.72 million tons) and 73.8 percent of global fish oil production (0.78 million ons, Tacon et al., 2011). In contrast, the poultry and pork industries each used nearly 26 per cent and 22 per cent respectively of the available fishmeal in 2002 while aquaculture consumed only 46 percent of the global fishmeal supply and 81 percent of the global fish oil supply (Pike, 2005; Tacon et al., 2006)

Fish oil has become also a product for direct human consumption for health reasons. Long-chain Omega-3 fatty acids, specifically EPA and DHA, have been shown to play a critical role in human health: EPA in the health of the cardiovascular system and DHA in the proper functioning of the nervous system, most notably brain function. In 2010 fish oil for direct human consumption was estimated at 24 per cent of the total world production, compared with 5 per cent in 1990. (Shepherd and Jackson, 2012).

The total use of fishmeal by the aquaculture sector is expected to decrease in the long term in favour of plant-based materials (Figure 12.3). It has gone down from 4.23 million tons in 2005 to 3.72 million tons in 2008 (or 12.8 percent of total aquafeeds by weight), and is expected to decrease to 3.49 million tons by 2020 (at an estimated 4.9 per cent of total aquafeeds by weight) (Tacon et al., 2011).

These trends reflect that fishmeal is being used by industry as a strategic ingredient fed in stages of the growth cycle where its unique nutritional properties can give the best results or in places where price is less critical (Jackson, 2012). The most commonly used alternative to fishmeal is that of soymeal. Time series of the price of both products show that use of fishmeal is being reduced in less critical areas such as grower feeds, but remains in the more critical and less price-sensitive areas of hatchery and brood-stock feeds. (Jackson and Shepherd, 2012)

The use of fish oil by the aquaculture sector will probably increase in the long run albeit slowly. It is estimated that total usage will increase by more than 16 percent, from 782,000 tons (2.7 percent of total feeds by weight) in 2008 to the estimated 908.000 tons (1.3 percent of total feeds for that year) by 2020. It is forecast that increased usage will shift from salmonids, to marine finfishes and crustaceans because of the current absence of cost-effective alternative lipid sources that are rich in long-chain polyunsaturated fatty acids. (Tacon et al., 2011)

8 Economic and social significance

At the global level, the number of people engaged in fish farming has, since 1990, increased at higher annual rates than that of those engaged in capture fisheries. The most recent estimates (FAO 2014, Table 12.2) indicate that about 18.9 million people were engaged in fish farming, 96 per cent concentrated primarily in Asia, followed by Africa (1.57 percent), Latin America and the Caribbean (1.42 percent), Europe (0.54 per cent), North America (0.04 per cent) and Oceania (0.03 per cent). The

---

4 Myctophids are small-size mesopelagic fish inhabiting between 200 and 1000 metres that vertically migrate on a daily basis. Biomass of myctophids is estimated to be considerable worldwide.

Table 12.2 | FAO (2014) estimates that the total number of fish farmers in the world has grown from 8 million in 1995 to close to 19 million today, representing an increasing source of livelihoods. Not all these jobs are permanent and year-around, since many are seasonal.

	1995	2000	2005	2010	2011	2012
	(Thousands)					
Of which, fish farmers						
Africa	65	91	140	231	257	298
Asia	7 762	12 211	14 630	17 915	18 373	18 175
Europe	56	103	91	102	103	103
Latin America and the Caribbean	155	214	239	248	265	269
North America	6	6	10	9	9	9
Oceania	4	5	5	5	6	6
World	8 049	12 632	15 115	18 512	19 015	18 861

18,175 million fish farmers in 2012 represented 1.45 per cent percent of the 1.3 billion people economically active in the broad agriculture sector worldwide. (FAO, 2014).

Out of the 18.8 million of fish farmers in the world (Table 12.2), China alone employs 5.2 million, representing 27.6 per cent of the total, while Indonesia employs 3.3 million farmers, representing 17.7 per cent of the total. Employment at farm level includes full-time, part-time and occasional jobs in hatcheries, nurseries, grow-out production facilities, and labourers. Employment at other stages along aquaculture value-chains includes jobs in input supply, middle trade and domestic fish distribution, processing, exporting and vending (HLPE, 2014). More than 80 percent of global aquaculture production may be contributed by small- to medium-scale fish farmers, nearly 90 per cent of whom live in Asia (HLPE, 2014). Farmed fish are expected to contribute to improved nutritional status of households directly through self-consumption, and indirectly by selling farmed fish for cash to enhance household purchasing power (HLPE, 2014).

The regional distribution of jobs in the aquaculture sector reflects widely disparate levels of productivity strongly linked to the degree of industrialization of the dominant culture systems in each region. In Asia, low technology is used in non-fed and inland-fed aquaculture, using extensive ponds, which is labour intensive compared with mariculture in floating systems. In 2011, the annual average production of fish farmers in Norway was 195 tons per person, compared with 55 tons in Chile, 25 tons in Turkey, 10 tons in Malaysia, about 7 tons in China, about 4 tons in Thailand, and only about 1 ton in India and Indonesia (FAO, 2014).

Extrapolating from a ten-country case study representing just under 20 percent of the global aquaculture production, Phillips and Subasinghe (2014, personal communication, cited in HLPE, 2014) estimated that "*total employment in global aquaculture value chains could be close to 38 million full-time people.*"

Fish is among the most traded food commodities worldwide. Fish can be produced in one country, processed in a second and consumed in a third. There were 129 billion dollars of exports of fish and fishery products in 2012 (FAO, 2014)

In the last two decades, in line with the impressive growth in aquaculture production, there has been a substantial increase in trade of many aquaculture products based on both low- and high-value species, with new markets opening up in developed and developing countries as well as economies in transition.

Aquaculture is contributing to a growing share of international trade in fishery commodities, with high-value species such as salmon, seabass, seabream, shrimp and prawns, bivalves and other molluscs, but also relatively low-value species such as tilapia, catfish (including *Pangasius*) and carps (FAO 2014). Pangasius is a freshwater fish native to the Mekong Delta in Viet Nam, new to international trade. However, with production of about 1.3 million tons, mainly in Viet Nam and all going to international markets, this species is an important source of low-priced traded fish. The European Union and the United States of America are the main importers of Pangasius. (FAO, 2014)

9 Environmental impacts of aquaculture

Environmental effects from aquaculture include land use and special natural habitats destruction, pollution of water and sediments from wastes, the introduction of non-native, competitive species to the natural environment through escapes from farms, genetic effects on wild populations (of fish and shellfish) from escapes of farmed animals or their gametes, and concerns about the use of wild forage fish for aquaculture feeds.

Table 12.3 | Per capita average outputs per fish farmer by region (in FAO, 2014).

	Production[1] per person				
	2000	2005	2010	2011	2012
	(Tonnes/year)				
Aquaculture					
Africa	4.4	4.6	5.6	5.4	5.1
Asia	2.3	2.7	2.9	3.0	3.2
Europe	19.8	23.5	24.9	26.0	27.8
Latin America and the Caribbean	3.9	6.3	7.8	9.0	9.7
North America	91.5	68.2	70.0	59.5	59.3
Oceania	23.1	29.5	33.8	30.4	32.7
World	**2.6**	**2.9**	**3.2**	**3.3**	**3.5**

[1] Production excludes aquatic plants.

9.1 Land use

WRI (2014) estimate that inland aquaculture ponds occupied between 12.7 million ha and 14.0 million ha of land in 2010, and that brackish water or coastal ponds occupied approximately 4.4 million ha—for a combined area of roughly 18 million hectares, overwhelmingly in Asia. Many of these ponds were converted from rice paddies and other existing cropland rather than newly converted natural lands—but even so, aquaculture adds to world land use demands.

In 2008, global land use efficiencies of inland and brackish water ponds averaged 2.3 tons of fish per hectare per year (t/ha/yr). Expanding aquaculture to 140 million tons by 2050 without increases in that average efficiency would imply an additional area of roughly 24 million ha directly for ponds—about the size of the United Kingdom. (WRI, 2014)

9.2 Interaction with mangroves

Land conversion for aquaculture can lead to severe ecosystem degradation, as in the case of the proliferation of extensive low-yield shrimp farms that destroyed large extensions of mangrove forests in Asia and Latin America (Lewis et al., 2002, cited in WRI, 2014). Since the 1990s, non-governmental organizations and policy-makers have focused on curbing the expansion of extensive, shrimp farms into mangrove forests in Asia and Latin America (FAO et al., 2006b). As a result, mangrove clearance for shrimp farms has greatly decreased, thanks to mangrove protection policies in affected countries and the siting of new, more high-yield shrimp farms away from mangrove areas. (Lewis et al., 2002).

9.3 Pollution of water and sediments

Wastes from mariculture generally include dissolved (inorganic) nutrients, particulate (organic) wastes (feces, uneaten food and animal carcasses), and chemicals for maintaining infrastructure (anti-biofouling agents) and animal health products (antiparasitics, disinfectants and antibiotics). These wastes impose an additional oxygen demand on the environment, usually creating anoxic conditions under pens and cages.

Research in Norway has shown that benthic effects decline rapidly with increasing depth of water under salmon nets, but situating farms as close to shore as possible may be a prerequisite for economic viability of the industry. Fallowing periods of several years have been found necessary in Norway to allow benthic recovery. Research elsewhere indicates that benthic recovery may be quicker under some conditions (WHOI, 2007)

9.4 Impact of escapes

With the use of floating semi-enclosed systems, escapes are inevitable in mariculture and inland aquaculture. Catastrophic events (e.g., hurricanes or other storms), human error, seal and sea lion predation and vandalism will remain potential paths for farmed fish to escape into the wild. Advancements in technology are likely to continue to reduce the frequency and severity of escape events but it is unlikely that this ecological and economic threat will ever disappear entirely. There is considerable evidence of damage to the genetic integrity of wild fish populations when escaped farmed fish can interbreed with local stocks. Furthermore, in semi-enclosed systems, cultured organisms release viable gametes into the water. Mariculture industry has undertaken a significant effort to produce and use variants of cultivated species that are infertile, diminishing the risk of gene-flow from cultivated/domesticated species to their wild counterparts when escapes occur.

9.5 Non-native species.

Aquaculture has been a significant source of intentional and unintentional introductions of non-native species into local ecosystems worldwide. The harm caused by invasive species is well documented.

Intensive fish culture, particularly of non-native species, can be and has been involved in the introduction and/or amplification of pathogens and disease in wild populations (Blazer and LaPatra, 2002, cited in WHOI, 2007).

Non-native oysters have been introduced in many regions to improve failing harvests of native varieties due to diseases or overexploitation. The eastern oyster, *Crassostrea virginica*, was introduced to the West Coast of the United States in 1875. The Pacific or Japanese oyster *Crassostrea gigas*, native to the Pacific coast of Asia, has been introduced in North and South America, Africa, Australia, Europe, and New Zealand and has also spread through accidental introductions either through larvae in ballast water or on the hulls of ships (Helm, 2006).

9.6 Genetically modified organisms

Although the use of transgenic, or genetically modified organisms (GMOs), is not common practice in aquaculture (WHOI, 2007), nevertheless the potential use of GMOs would pose severe risks. The production and commercialization of aquatic GMOs should be analyzed considering economic issues, environmental protection, food safety and social and health well-being (Muir, et al., 1999; Le Curieux-Belfond et al., 2009).

9.7 Use of chemicals as pesticides and for antifouling

A wide variety of chemicals are currently used in aquaculture production. As the industry expands, it requires the use of more drugs, disinfectants and antifouling compounds (biocides)[5] to eliminate the microorganisms in the aquaculture facilities. Among the most common disinfectants are hydrogen peroxide and malachite green. Pyrethroid insecticides and avermectins are used as anthelmintic agents (Romero et al., 2012). Organic booster biocides were recently introduced as alternatives to the organotin compounds found in antifouling products after restrictions were imposed on the use of tributyltin (TBT). The replacement products are generally based on copper metal oxides and organic biocides. The biocides that are most commonly used in antifouling paints include chlorothalonil, dichlofluanid, DCOIT (4,5-dichloro-2-n-octyl-4-isothiazolin-3-one, Sea-nine 211®), Diuron, Irgarol 1051, TCMS pyridine (2,3,3,6-tetrachloro-4-methylsulfonyl pyridine), zinc pyrithione and Zineb. (Guardiola et al., 2012). The use of biocides is not as well-regulated as drug use in aquaculture because the information available on their effects on ecosystems is still limited.

9.8 Use of antibiotics

Antibiotic drugs used in aquaculture may have substantial environmental effects. The use of antibiotics in aquaculture can be categorized as therapeutic, prophylactic or metaphylactic. Therapeutic use is the treatment of established infections. Metaphylaxis are group-medication procedures, aimed at treating sick animals while also medicating others in the group to prevent disease. Prophylaxis means the precautionary use of antimicrobials in either individuals or groups to prevent the development of infections (Romero et al., 2012).

In aquaculture, antibiotics at therapeutic levels are frequently administered for short periods of time via the oral route to groups of fish that share tanks or cages. Fish do not effectively metabolize antibiotics and will pass them largely unused back into the environment in feces. 70 to 80 per cent of the antibiotics administered to fish as medicated pelleted feed are released into the aquatic environment via urinary and fecal excretion and/or as unused medicated food (Romero et al., 2012). For this among other reasons, antibiotic use in net, pen or cage mariculture is a concern because it can contribute to the development of resistant strains of bacteria in the wild. The spread of antimicrobial resistance due to exposure to antimicrobial agents is well documented in both human and veterinary medicine. It is also well documented that fish pathogens and other aquatic bacteria can develop resistance as a result of antimicrobial exposure. Examples include *Aeromonas salmonicida, Aeromonas hydrophila, Edwardsiella tarda, Yersinia ruckeri, Photobacterium damselae* and *Vibrio anguillarum*. Research has shown that antibiotics excreted tend to degrade faster in sea-water, while they persist more in sediments. (Romero et al., 2012)

The public health hazards related to antimicrobial use in aquaculture are twofold: the development and spread of antimicrobial-resistant bacteria and resistance genes and the presence of antimicrobial residues in aquaculture products and the environment (Romero et al., 2012). The high proportions of antibiotic-resistant bacteria that persist in sediments and farm environments may provide a threat to fish farms because they can act as sources of antibiotic-resistance genes for fish pathogens in the vicinity of the farms. Because resistant bacteria may transfer their resistance elements to bacterial pathogens, the implementation of efficient strategies to contain and manage resistance-gene emergence and spread is critical for the development of sustainable aquaculture practices.

Industry faced with uncertainties created by the limited knowledge of infectious diseases and their prevalence in a particular environment tends to abuse the use of antibiotics. Defoirdt et al. (2011, cited by Romero et al., 2012) estimated that approximately 500–600 metric tons of antibiotics were used in shrimp farm production in Thailand in 1994; he also emphasized the large variation between different countries, with antibiotic use ranging from 1 g per metric ton of production in Norway to 700 g per metric ton in Viet Nam. In the aftermath of the ISA infection in the salmon culture in Chile, SERNAPESCA, the Chilean National Fisheries and Aquaculture Service, recently released data reporting unprecedentedly high amounts of antibiotics used by the salmon industry.[6] Inefficiencies in the antibiotic treatment of fish illnesses now may lead to significant economic losses in the future (Romero et al., 2012).

---

[5] Biocides are chemical substances that can deter or kill the microorganisms responsible for biofouling.

[6] According to SERNAPESCA, the industry used an estimated 450,700 kilos of antibiotics in 2013.

Antimicrobial-resistant bacteria in aquaculture also present a risk to public health. The appearance of acquired resistance in fish pathogens and other aquatic bacteria means that such resistant bacteria can act as a reservoir of resistance genes from which genes can be further disseminated and may ultimately end up in human pathogens. Plasmid-borne resistance genes have been transferred by conjugation from the fish pathogen *A. salmonicida* to *Escherichia coli*, a bacterium of human origin, some strains of which are pathogenic for humans (Romero et al., 2012).

9.9 Diseases and parasites

Farming marine organisms in dense populations results in outbreaks of viral, bacterial, fungi and parasite diseases. Diseases and parasites constitute a strong constraint on the culture of aquatic species and disease and parasite translocation by host movements in different spatial scales is common. In molluscs the main parasites are protozoans of the genus *Bonamia*, *Perkinsus* and *Marteilia*. The pathogens *Haplosporidium*, bacteria (rickettsial and vibriosis) and herpes-type virus have a great impact on the rates of mortality. In shrimps the most relevant diseases are viral (white spot disease, WPS, yellow head disease, YHD, taura syndrome disease, TSD) (Bondad-Reantaso et al., 2005).

The "*Sea lice (*Copepoda, Caligidae*) have been the most widespread pathogenic marine parasite*" in Salmon farming, affecting also other cultured fishes and wild species (Ernst et al., 2001; Costello, 2006). The global economic cost of sea lice control was estimated at over 480 million dollars in 2006 (Costello, 2009); however, there are other impacts such as the decrease in conversion efficiency (Sinnott, 1998) and the depression of immune systems, which allow the outbreak of bacterial (vibriosis and furuncolosis) and viral diseases (infectious salmon anaemia virus, ISA, infectious pancreatic necrosis, IPN and pancreas disease, PD) (Robertson, 2011).

References

Blazer, V.S., and LaPatra, S.E. (2002). Pathogens of cultured fishes: potential risks to wild fish populations. pp. 197-224. In *Aquaculture and the Environment in the United States*, Tomasso, J., ed. U.S. Aquaculture Society, A Chapter of the World Aquaculture Society, Baton Rouge, LA.

Bondad-Reantaso, M.G., Subasinghe, R. P., Arthur, J. R., Ogawa, K., Chinabut, S., Adlard, R., Tan, Z and Shariff, M., (2005). Disease and health management in Asian aquaculture. *Veterinary Parasitology* vol 132, pp. 249–272.

Costa-Pierce, B.A., Bartley, D.M., Hasan, M., Yusoff, F., Kaushik,S.J., Rana, K., Lemos, D., Bueno, P. and Yakupitiyage, A. (2012). Responsible use of resources for sustainable aquaculture, In *Global Conference on Aquaculture 2010*, Subasinghe, R., ed. Sept. 22-25, 2010, Phuket, Thailand. Rome: FAO. Available from http://ecologicalaquaculture.org/Costa-PierceFAO(2011).pdf.

Costello, M.J. (2006). Ecology of sea lice parasitic on farmed and wild fish. *Trends in Parasitology* vol 22, No 10, pp 475-483

Costello, M.J. (2009). The global economic cost of sea lice to the salmonid farming industry. *Journal of Fish Diseases*, vol 32. pp 115-118.

Defoirdt, T., Sorgeloos, P., Bossier, P. (2011). Alternatives to antibiotics for the control of bacterial disease in aquaculture. *Current opinion in microbiology*, vol. 14, No. 3, pp. 251-58.

Ernst, W., Jackman, P., Doe, K., Page, F., Julien, G., Mackay, K., Sutherland, T. (2001). Dispersion and toxicity to non-target aquatic organisms of pesticides used to treat sea lice on salmon in net pen enclosures. *Marine Pollution Bulletin*, vol. 42, No. 6, pp. 433-44.

FAO (2006a). State of world aquaculture 2006. *FAO Fisheries Technical Paper*, No. 500. Rome: FAO. 134 pp.

FAO, NACA, UNEP, WB, WWF (2006b). *International Principles for Responsible Shrimp Framing*. Network of Aquaculture Centres in Asia-Pacific (NACA). Bangkok, Thailand. 20 pp. Available from http://www.enaca.org/uploads/international-shrimp-principles-06.pdf.

FAO (2010). *The State of World Fisheries and Aquaculture* 2010. Rome: FAO. 197 pp.

FAO (2012). *The State of World Fisheries and Aquaculture* 2012. Rome: FAO. 209 pp.

FAO (2014). *The State of World Fisheries and Aquaculture* 2014. Rome: FAO. 223 pp.

Guardiola, F.A., Cuesta, A., Meseguer, J. and Esteban, M.A. (2012). Risks of Using Antifouling Biocides in Aquaculture. *International Journal of Molecular Sciences* 13(2): 1541–1560.

Helm, M.M. (2006). *Crassostrea gigas*. Cultured Aquatic Species Information Programme, FAO Fisheries and Aquaculture Department, Rome: FAO. Available from http://www.fao.org/fishery/culturedspecies/Crassostrea_gigas/en.

HLPE (2014). *Sustainable fisheries and aquaculture for food security and nutrition*. The High Level Panel of Experts on Food Security and Nutrition of the Committee on World Food Security. Rome: FAO.

Jackson, A.J., and Shepherd, J. (2012). The future of fish meal and oil. In *Second International Conference on Seafood Technology on Sustainable, Innovative and Healthy Seafood*. Ryder, R., Ababouch, L., and Balaban, M. (eds) FAO/The University of Alaska, 10–13 May 2010, Anchorage, the United States of America. pp. 189–208. FAO Fisheries and Aquaculture Proceedings, No. 22. Rome: FAO. 238 pp. Available from www.fao.org/docrep/015/i2534e/i2534e.pdf.

Jackson, A.J. (2012). Fishmeal and Fish Oil and its role in Sustainable Aquaculture. *International Aquafeed*, September/October, pp.18 – 21.

Le Curieux-Belfond, O., Vandelac, L., Caron, J., Séralini, G.-E., (2009). Factors to consider before production and commercialization of aquatic genetically modified organisms: the case of transgenic organisms: the case of transgenic salmon. *Environmental Science & Policy*, 12: 170-189.

Lewis, R. R., Phillips, M. J., Clough, B., Macintosh, D.J. (2002). *Thematic Review on Coastal Wetland Habitats and Shrimp Aquaculture*. Washington, DC: World Bank, Network of Aquaculture Centres in Asia-Pacific, World Wildlife Fund, and FAO.

McClennen, C. (2004). *The Economic, Environmental and Technical Implications of the Development of Latin American Shrimp Farming*. Master of Arts in Law and Diplomacy Thesis, The Fletcher School. Available from http://dl.tufts.edu/bookreader/tufts:UA015.012.DO.00040#page/1/mode/2up.

Muir, W.M., and Howard, R.D. (1999). Possible ecological risks of transgenic organism release when transgenes affect mating success: Sexual selection and the Trojan gene hypothesis. *Proceedings of the National Academy of Sciences, USA*, vol. 96, No. 24, pp. 13853-56.

Pike, I.H. (2005). Eco-efficiency in aquaculture: global catch of wild fish used in aquaculture. *International Aquafeed*, 8 (1): 38–40.

Robertson, B. (2011). Can we get the upper hand on viral diseases in aquaculture of Atlantic salmon? *Aquaculture Research* 2011, vol. 42, pp 125-131.

Romero, J., Feijoo, C.F., Navarrete, P. (2012). Antibiotics in Aquaculture – Use, Abuse and Alternatives. In *Health and Environment in Aquaculture*, Carvalho, E.D., David, G.S., Silva, R.J., (eds.) InTech. Available from http://www.intechopen.com/books/health-and-environment-in-aquaculture/antibiotics-in-aquaculture-use-abuse-and-alternatives.

Shepherd, C.J., and Jackson, A.J. (2012). *Global fishmeal and fish oil supply - inputs, outputs, and markets*. International Fishmeal & Fish Oil Organisation, World Fisheries Congress, Edinburgh. Available from http://www.seafish.org/media/594329/wfc_shepherd_fishmealtrends.pdf.

Sinnot, R. (1998). Sea lice – watch out for the hidden costs. *Fish Farmer*, vol 21 No 3, pp 45-46.

Tacon, A.G.J.; Hasan, M.R.; Subasinghe, R.P. (2006). Use of fishery resources as feed inputs for aquaculture development: trends and policy implications. *FAO Fisheries Circular*. No.1018. Rome, FAO. 99p.

Tacon, A. G. J., Hasan, M. R., Metian, M. (2011). Demand and supply of feed ingredients for farmed fish and crustaceans -Trends and prospects. *FAO Fisheries Technical Paper*, No. 564. Rome, FAO.

WHOI (2007). *Sustainable Marine Aquaculture: Fulfilling the Promise; Managing the Risks*. Marine Aquaculture Task Force, Marine Finfish Aquaculture Standards Project, 128 pp.

WRI (2014). Creating a SuStainable Food Future: A menu of solutions to sustainably feed more than 9 billion people by 2050. World Resources Report 2013–14: Interim Findings. World Resources Institute, Washington D.C., USA, 144 pp.

13 Fish Stock Propagation

Contributors:
Kai Lorenzen (Convenor), Michael Banks, Andrew Rosenberg (Lead Member), V. N. Sanjeevan, Stephen Smith, Zacharie Sohou and Chang Ik Zhang

1 Definition

Fish stock propagation, more commonly known as fisheries enhancement, is a set of management approaches involving the use of aquaculture technologies to enhance or restore fisheries in natural ecosystems (Lorenzen, 2008). "Aquaculture technologies" include culture under controlled conditions and subsequent release of aquatic organisms, provision of artificial habitat, feeding, fertilization, and predator control. "Fisheries" refers to the harvesting of aquatic organisms as a common pool resource, and "natural ecosystems" are ecosystems not primarily controlled by humans, whether truly natural or modified by human activity. This places enhancements in an intermediate position between capture fisheries and aquaculture in terms of technical and management control (Anderson, 2002).

The present chapter focuses primarily on enhancements involving releases of cultured organisms, the most common form of enhancements often described by terms such as 'propagation', 'stock enhancement', 'sea ranching' or 'aquaculture-based enhancement'.

2 Enhancements in marine resource management

Enhancements are developed when fisheries management stakeholders or agencies take a proactive, interventionist approach towards achieving management objectives by employing aquaculture technologies instead of relying solely on the protection of natural resources and processes. Enhancement approaches may be used effectively or ineffectively in resource management. To understand how enhancement initiatives can give rise to such different outcomes, it is important to consider not only the technical intervention but the management context in which the initiative has arisen, including ecological and socioeconomic factors as well as the governance arrangements (Lorenzen, 2008).

2.1 Effective enhancements

Enhancement approaches may be employed towards different ends commonly referred to as sea ranching, stock enhancement and restocking (Bell et al. 2008). Sea ranching entails releasing cultured organisms to maintain stocks that do not recruit naturally in the focal ecosystem. This may involve stocks that once recruited naturally but no longer do so due to loss of critical habitat, or it may involve creation of fisheries for desired "new" species for which the focal system provides a habitat suitable for adult stages but not for spawning or for juveniles. Stock enhancement is the practice of releasing cultured organisms into natural stocks of the same species on a regular basis, with the aim of increasing abundance or harvest beyond the level supported by natural recruitment. Restocking entails temporary releases of cultured organisms into wild stocks that have been depleted by overfishing or extreme environmental events, with the aim of accelerating recovery or enabling recovery of stocks "trapped" in a depleted or declining state. The use of enhancement approaches represents a spectrum from strongly production/catch-oriented applications to strongly conservation/restoration-oriented ones, and entails quite different management practices (Section 13.5; Table 13.1).

The technical intervention of enhancements interacts synergistically with governance arrangements. Stakeholders or management agencies invest in enhancements when they have incentives to do so, either because they stand to gain material benefits (e.g. increase in harvests) or because engaging in enhancement activities increases the perceived legitimacy of management arrangements or agencies (for example, stakeholders may be more supportive of a management agency that engages in fisheries enhancement activities than of one that only regulates fishing). Enhancements require a reasonable level of governance control to emerge at all (they are unlikely to emerge under unregulated open access), and they tend to further strengthen governance control when implemented (Anderson, 2002; Drummond, 2004; Lorenzen, 2008). By helping to strengthen and transform governance arrangements, enhancement initiatives can sometimes generate fisheries management benefits beyond those directly attributable to the technical intervention.

Economic and social benefits of enhancements may arise from biological outcomes such as increased catches or maintenance of fisheries and other ecosystem services in highly modified environments. Successful enhancements often have further, more derived benefits. Pinkerton (1994), for example, describes economic benefits of Alaska salmon enhancements that result from greater consistency and quality of harvests, as well as greater volume. Enhancements can make economic and social benefits from aquaculture technologies available to stakeholders, such as traditional fishers who may lack the assets, skills or interest to engage in conventional aquaculture.

In addition to direct management benefits, enhancements provide opportunities for advancing basic knowledge of ecology, evolution and exploitation dynamics of marine resources (Lorenzen 2014).

2.2 Ineffective enhancements

Often, enhancements are initiated under conditions that are fundamentally unsuitable for their effective use, or designed inappropriately. Such ineffective enhancements can nonetheless persist for a considerable time and sometimes do considerable ecological and economic damage. Incentives for stakeholders or management agencies to engage in enhancement activities can exist even in the absence of evidence of their technical effectiveness, and once investments have been made and stakeholders have become vested, it becomes increasingly difficult to discontinue such initiatives. These issues point to the need for constructive science and management engagement with the development of new, and the reform of existing, enhancements (Section 13.4).

2.3 Examples of enhancement efforts

The following examples illustrate the potential for well-managed enhancements to contribute to fisheries management goals and the interactions between the technical and governance dimensions of such initiatives.

Very large-scale enhancement efforts are undertaken in the Pacific Northwest of the United States of America (Naish et al., 2007). These efforts include enhancements to support commercial and recreational fisheries (Knapp et al., 2007), enhancement and restocking initiatives to meet tribal treaty obligations (Smith, 2014), and restoration efforts for endangered populations (Kline and Flagg, 2014). Pacific Northwestern habitats once hosted a tremendous biomass of salmon that comprised a significant component of food and nutrient webs linking ocean and freshwater biomes. For example, it is estimated that the Columbia River once hosted returns of 10-16 million wild salmon (Johnson et al., 1997). Historical overharvest, irrigation withdrawals, hydropower dams and other factors have reduced returns. Of the current returns of around 1 million, hatchery fish make up around 80 per cent (95 per cent of the coho, 70 to 80 per cent of the spring and summer chinook, 50 per cent of the fall chinook, and 70 per cent of the steelhead) (NMFS, 2000)). In Oregon, Nicholas and Hankin (1989) estimated that 21 of 36 coastal stocks of spring and fall chinook salmon were almost entirely comprised of wild fish. In the remaining stocks, the percentage of hatchery fish in the runs ranged from 10 to 75 per cent. Oregon's hatchery programme annually releases 74 million salmonids: 60.4 million salmon, 6.4 million steelhead and 7.6 million trout (ODFW, 1998). Such hatchery programmes can maintain fisheries when essential habitats are degraded or inaccessible and help conserve or restore endangered populations, but they also pose ecological and genetic risks to wild populations. A major scientific review of Columbia River hatchery programmes successfully used population modelling to identify hatchery operation and harvest policies that simultaneously improve the conservation status of wild populations and provide moderate increases in harvest (Paquet et al., 2011). In Alaska, large-scale salmon enhancements are run by community-based Aquaculture Associations. Since the mid-1970s, Aquaculture Associations produce and release juvenile salmon and, in return, gain exclusive rights to a share of the harvest in the form of "cost-recovery fish". The associations have since become engaged in many aspects of salmon fisheries management, effectively creating a co-management system with the State of Alaska.

The world's largest marine invertebrate fisheries enhancement is the scallop enhancement operation run by fishing cooperatives in Hokkaido, Japan (Uki, 2006). Development of an effective spat collecting, on-growing and releasing technology in the mid-1960s created the opportunity to seed scallop grounds with high densities of juveniles. Fishing cooperatives adopted rotational seeding and harvesting of fishing grounds, combined with predator control, and increased regional production from an average of 40,000 tons to around 300,000 tons per year. The success of this enhancement has been attributed to a combination of factors including suitable habitat, the species' biology (young optimal harvest age, low post-release dispersal), integration of spat releasing with predator control and rotational harvesting, and devolution of management to a fishing cooperative with exclusive rights over the resource (Uki, 2006).

In New Zealand, the Japanese scallop enhancement technology was adapted to revive the Southern Scallop Fishery in what became a restocking initiative combined with far-reaching changes in governance. Adoption of aquaculture technology allowed the fishery to opt out of the fisheries management framework of the time and transition to an individual quota-based regime and rotational seeding and harvesting. Cultured juveniles contributed strongly to initial recovery but natural recruitment became dominant as the fishery was rebuilt (Drummond, 2004). More recently, low spat survival has led to a sharp reduction in catches and to the closure of some of the main grounds (Williams et al. 2014). This decline in survival may be related due to changes in productivity due to increasing sedimentation in the area.

In the Republic of Korea, the National Fisheries Research and Development Institute (NFRDI) developed seed production technology to release healthy juveniles of rockfish and sea bream. Since 1998, seed production and fish release have successfully enhanced fishery resources and increased the income of fishermen. In the early stages of seed production, national facilities took the lead to develop techniques, but currently private companies produce the seed. Between 1986 and 2012, 46 marine species including abalone, various flatfish, sea bream and sea slugs were targets for production and 1,410 million juveniles of fish and shellfish species were stocked in the sea in the Republic of Korea. In the Republic of Korea, habitat restoration tools are also widely applied together with fish release in situations where habitat has been identified as the primary factor limiting production. These tools refer to the increase in available habitat and/or access to key habitat for at least some stages of the life history of a target species. Although artificial habitats are currently popular in some areas and widely used, scientific evaluation of the effectiveness of habitat restoration is incomplete. In the Republic of Korea, construction of artificial reefs is aimed at improving productivity of devastated fishing grounds by providing fish resources with habitats, and spawning and nursery grounds. Since 1971, about 3,000 fishing grounds have been augmented, with artificial reefs covering a total area of 216 kha as of 2012. Fifty-five per cent of the area with artificial reefs is utilized as fishing grounds and the other 45 per cent is preserved as spawning and nursery grounds of fish resources. Enhanced fisheries are managed cooperatively with fishing communities and marine enhancement in the Republic of Korea is becoming integrated into a comprehensive ecosystem-based fisheries management approach (Zhang et al., 2009).

In India, efforts with regard to stock enhancement of Penaeid prawns along the Kerala coast have not met with the desired success. This probably reflects heavy mortality of hatchery grown post larvae on their release to the sea, as they are neither acclimatized to the stress conditions of the sea nor have they acquired adequate predator avoidance skills. An additional effort in India is intended to revive depleted marine snail species along the coast of Tamil Nadu; *Xancus pyrum* (sacred chank), *Babylonia spirata* (whelk), *Hemifusus pugilinus* (spindle shells), *Chicoreus ramosus* (murex) and *C. virgineus*. Wild stocks of all of these species are heavily exploited for their meat (India exports 700 to 900 tons of frozen whelk meat every year), shells (used as a trumpet in temples and for the manufacture of ornaments) and opercula (which have medicinal value and are exported to Australia, France, Germany, Italy, Japan). About 10,000 juveniles and 0.5 million larvae of the above species were sea-ranched in the Gulf of Mannar in October 2010. It is premature to comment on the success of this experiment, but regular surveys of the grow out site show only a few dead organisms.

2.4 Global extent of enhancements

Marine fisheries enhancement is a widespread activity. Between 1984 and 1997, 64 countries reported stocking over 30 billion individuals of over 180 species in marine environments (Born et al., 2004). The global contribution of enhancements to marine fish production is difficult to quantify exactly, but is unlikely to exceed one to two million tons per year (around 1-2 per cent of global marine fisheries and aquaculture production) (Lorenzen 2014). This modest contribution to global production should not distract from the fact that considerable efforts and monetary investments are expended on enhancement initiatives, and that enhancements contribute substantially to several high-value fisheries as well as to restoration efforts for various species of conservation concern.

2.5 Developing or reforming enhancements

According to the reviewed assessments, enhancements are often initiated or promoted by fisheries stakeholders, but require scientific and management engagement in order to assess the potential of such initiatives, to develop effective enhancement systems where the potential exists, and to discontinue initiatives that are likely to be ineffective or harmful. Constructive science and management engagement with enhancements may be guided by the widely used and recently updated "responsible approach" (Blankenship and Leber, 1995; Lorenzen et al., 2010). The updated responsible approach consists of 15 recommended actions, divided into three stages of development or reform (Box 13.1). A staged approach ensures that the basic potential of enhancements is assessed (Stage I) prior to investment in technology development and pilot studies (Stage II), which in turn precede operational-scale implementation(Stage III). Qualitative and quantitative modelling are crucial in Stage I, and experimental (adaptive) management is central to assessing enhancement capacity and ecological impacts in later stages. This requires monitoring of temporal and spatial controls where fisheries are not enhanced and possibly not exploited (Caddy and Defeo 2003; Leleu et al., 2012; Costello, 2014). The most systematic and rigorous application of many ideas summarized in the responsible approach can be found in the Hatchery Reform process being applied to Pacific salmon hatchery programmes (Mobrand et al., 2005; Paquet et al., 2011).

3 Management considerations

3.1 The fisheries system and management context

Enhancements enter into existing fisheries systems and it is crucial to gain a broad-based understanding of the system prior to defining management objectives and assessing possible courses of action. At a minimum the following should be considered: the biology and status of the target fish stock (biological resource), the supporting habitat and ecosystem, the aquaculture operation, stakeholder characteristics (of fishers, aquaculture producers and resource managers), markets for inputs and outputs, governance arrangements, and the linkages between these components. A framework for enhancement-fisheries system analysis is outlined in Lorenzen (2008).

3.2 Stakeholder involvement

Stakeholder involvement is central to effective scientific and management engagement with enhancement initiatives because stakeholders tend to have a large influence on the initiation and development of such initiatives. Only when stakeholders are constructively involved in the assessment and decision-making process is the enhancement initiative likely to develop towards a beneficial conclusion (which may be an effective enhancement or the discontinuation of an ineffective or damaging initiative). Stakeholder involvement also makes the often considerable knowledge and experience of stakeholders accessible to the scientific and management process.

3.3 Identifying appropriate biological and technical system designs

Different enhancement strategies, such as sea ranching, stock enhancement and restocking, involve quite different management approaches and considerations (Utter and Epifanio, 2002; Naish et al., 2007 and Lorenzen et al., 2010; Lorenzen et al., 2012). Table 13.1 outlines the different practices involved with regards to aquaculture, stock and genetic management (based on Lorenzen et al., 2012).

3.4 Stock dynamics and management

Quantitative assessment of stock dynamics and the potential of enhancement as well as alternative management options, such as harvest restrictions to contribute to stock management objectives, is important at all stages of enhancement initiatives (Caddy and Defeo, 2003; Walters and Martell, 2004; Lorenzen, 2005). Different considerations apply to ranching, stock enhancement and restocking systems (Table 13.1). In ranching systems where maintaining natural recruitment is not a management goal, stock structure could be manipulated to maximize biomass production in food fisheries or to maximize abundance of 'catchable' size fish in put-and-take recreational fisheries. In stock enhancements where cultured fish are released into wild populations, it would be desirable to manage stocking and harvesting activities so as to limit negative impacts on naturally recruiting stock components which may arise from compensatory ecological responses to stocking or from overfishing of the natural spawning stock (Hilborn and Eggers 2000; Lorenzen, 2005). Such effects may reduce or eliminate net benefits from enhancement and pose conservation threats to wild stocks. Impacts of enhancements on wild stocks could be reduced by separating the cultured and wild population components as far as technically possible at the point of stocking, and through differential harvesting and possibly induced sterility of cultured fish (Lorenzen, 2005; Naish et al., 2007; Mobrand et al., 2005). According to these authors, restocking is likely to be advantageous over natural recovery only for populations that have been depleted to a very low fraction of their carrying capacity and requires concomitant reductions in fishing effort (Lorenzen 2005). Fisheries models and assessment tools are now available to conduct such quantitative assessment at all stages in the development or reform of enhancements (Lorenzen, 2005; Michael et al. 2009).

3.5 Aquaculture production for enhancements

Rearing of marine organisms in culture facilities subjects them to domestication processes that have strong and almost always negative impacts on their capacity to survive, grow, and reproduce in the wild (Le Vay et al., 2007; Lorenzen et al., 2012). A variety of measures, such as rearing in near-natural environments, environmental enrichment, life-skills training and soft release strategies, can counteract such domestication effects, but none are likely to be wholly effective (Olla et al., 1998; Brown and Day, 2002). Aquaculture production for release into natural ecosystems may benefit from culture practices that differ from those normally employed in facilities producing organisms for on-growing in aquaculture facilities and may also require different genetic management.

3.6 Genetic management

Genetic management is important for maximizing post-release fitness and enhancement effectiveness, and for minimizing risks to the genetic integrity of wild stocks. Three main sets of issues need to be considered: (1) potential disruption of neutral and adaptive spatial population structure due to translocation; (2) impacts of hatchery spawning and rearing on the genetic diversity of stocked fish and the enhanced, mixed stock; (3) impacts of hatchery rearing on the fitness of released fish and their naturally recruiting offspring; and (4) hybridization between stocked and wild species (Utter and Epifanio, 2002; Tringali et al., 2007; Araki et al., 2008). Appropriate sourcing and management of brood stock, possibly combined with rearing practices that minimize domestication selection are key genetic management actions and it may also be necessary to limit the contribution of cultured fish to the naturally spawning population (Miller and Kapuscinski, 2003; Tringali et al., 2007; Baskett and Waples, 2013). Different genetic management approaches may apply in sea ranching systems or "separated" stock enhancement programmes where direct genetic interactions between stocked and wild fish are absent and where, for example, selective breeding may be used to improve the post-release performance of hatchery fish (Table 13.1; Jonasson et al., 1997).

3.7 Pathogen interactions

Impacts on wild stocks from pathogen and parasite interactions that may cause disease may occur via three mechanisms: (1) introduction of alien pathogens, (2) transfer of pathogens that have evolved increased virulence in culture, (3) changes in host population density, age/size structure, or immune status that affect the dynamics of established pathogens. It is therefore important to implement an epidemiological, risk-based approach to managing disease interactions that accounts for ecological and evolutionary dynamics of transmission and host population impacts (Bartley et al., 2006).

3.8 Governance

Enhancements require governance systems that are effective at restricting exploitation and ensuring that those who invest in the resource through stocking can reap at least a sufficient share of the benefits. Depending on the wider governance framework, such arrangements can be based on individual or communal use rights (e.g., individual quotas or territorial use rights) or on government regulation (and taxation to recoup costs). A second important requirement of governance systems for enhanced fisheries is coordination of the fisheries and aquaculture components in terms of stock, genetic and health management.

3.9 Impacts on marine ecosystems

Potential impacts of enhancements on marine ecosystems differ between types of enhancement system. Impacts on non-target species are of the most concern in ranching systems where organisms that do not recruit naturally in the receiving ecosystem may be released in high numbers and harvested intensively. Species introduced outside their native range pose particular risks (many have minimal impacts, but a small proportion become invasive and inflict massive ecological and economic damage). In stock enhancement systems, ecological and genetic impacts on the wild stock component tend to be of the most concern. Restocking initiatives will have broadly positive impacts on marine ecosystems as long as good stock and genetic management approaches are in place. Although potential impacts of marine enhancement activities are well understood, empirical evidence for such impacts is limited except for the large-scale salmon enhancements in the Pacific Northwest and the Laurentian Great Lakes of North America (Naish et al., 2007; Crawford, 2001). This paucity of information likely reflects the limited scale of marine enhancements to date.

3.10 Interactions with other sectors

Aquaculture technologies enable enhancements in the first place and availability of cultured organisms from the commercial aquaculture sector can greatly reduce the barriers for fisheries stakeholders to engage in enhancements. Interactions with fisheries may occur in terms of access conflicts or impacts on wild target or non-target species and such interactions may increase as marine enhancements become more common. Market interactions between products from enhancements and from aquaculture and capture fisheries can be significant where enhancements account for substantial market share as in the case of salmon (Knapp et al., 2007). However, the market share of enhancements is small for most species and products, so that enhancements are more often impacted through the market by developments in the aquaculture and capture fisheries sectors than vice versa.

3.11 Technical and economic performance

As discussed previously, the technical and economic performance of marine enhancements is highly variable. Reviews by Hilborn (1998) and Arnason (2001) concluded that only a small proportion of documented enhancements are demonstrably economically successful, but for many information is insufficient to assess economic viability, and some are demonstrably unsuccessful. Further assessments and comparative analyses are urgently required.

Box 13.1 | Elements of the updated "responsible approach" to fisheries enhancement (Lorenzen et al., 2010).

Stage I: Initial appraisal and goal setting
(1) Understand the role of enhancement within the fishery system
(2) Engage stakeholders and develop a rigorous and accountable decision-making process
(3) Quantitatively assess contributions of enhancement to fisheries management goals
(4) Prioritize and select target species and stocks for enhancement
(5) Assess economic and social benefits and costs of enhancement

Stage II: Research and technology development including pilot studies
(6) Define enhancement system designs suitable for the fishery and management objectives
(7) Design appropriate aquaculture systems
(8) Use genetic resource management to avoid deleterious genetic effects
(9) Use disease and health management
(10) Ensure that released hatchery fish can be identified
(11) Use an empirical process for defining optimal release strategies

Stage III: Operational implementation and adaptive management
(12) Devise effective governance arrangements
(13) Define a stock management plan with clear goals, measures of success and decision rules
(14) Assess and manage ecological impacts
(15) Use adaptive management

---

4 International agreements and guidelines

There are currently no international agreements pertaining directly to fisheries enhancements. Some FAO instruments, including the FAO Technical Guidelines for Responsible Fisheries, deal with issues associated with fisheries enhancements (e.g., FAO, 2008). In addition, eco-labelling of products from enhanced fisheries has been considered at the Expert Consultation on the Development of Guidelines for the Ecolabelling of Fish and Fishery Products from Inland Capture Fisheries held in 2010 (FAO, 2010). The FAO Committee on Fisheries adopted these Guidelines in 2011 (FAO, 2011). The *ICES Code of Practice on the Introductions and Transfers of Marine Organisms* (ICES, 2005) is widely accepted and applies to introductions carried out for the purpose of fisheries enhancements.

5 Future trends

Enhancements are likely to become more widespread as burgeoning demand for seafood and increasingly severe human impacts on the coastal oceans create greater demand for proactive management, aquaculture technologies become available for an ever-increasing number of marine species, and governance arrangements for many fisheries move towards rights-based systems that provide strong incentives for investment in resources (Lorenzen et al., 2013). Greater scientific and management attention to enhancements is required to aid the development of potentially effective initiatives and to avoid widespread investment in ineffective or damaging enhancements (Lorenzen, 2014).

6 State of scientific knowledge, application and recommendations

Rapid progress has been made in the scientific understanding of marine enhancements over the past 20 years (Leber, 2013). Unfortunately, the scientific knowledge and tools now available to aid the development or reform of enhancements are not widely applied (Lorenzen 2014). Reasons may include that mainstream fisheries and aquaculture scientists are often unaware of developments in this interdisciplinary area or not adequately trained to conduct the necessary assessments. Research providers and management agencies need to build capacity for engaging with enhancement initiatives using current science. Improved reporting on enhancement initiatives and outcomes at national and international level is also important. Currently, harvests from enhanced fisheries tend to be lumped into either capture fisheries or aquaculture production figures in national and international statistics (Born et al., 2004; Klinger et al., 2012).

Table 13.1 | Design criteria for biological-technical components of marine enhancement fisheries systems serving different objectives (adapted from Lorenzen et al., 2012).

	Sea ranching	Stock enhancement	Re-stocking
Aim of enhancement	Increase fisheries catch	Increase fisheries catch while conserving or increasing naturally recruiting stock	Rebuild depleted wild stock to higher abundance
Wild population status	Absent or insignificant	Numerically large Possibly depleted relative to carrying capacity	Numerically large or small Depleted relative to carrying capacity
Aquaculture management	Production-oriented Partial domestication Conditioning for release Possibly induced sterility	Integrated programmes: as for re-stocking Separated programmes: as for sea ranching	Conservation-oriented Minimize domestication Conditioning for release
Genetic management	Maintain genetic diversity Selection for high return	Integrated programmes: as for re-stocking Separated programmes: as for sea ranching; also selection to promote separation	Preserve all wild population genetic characteristics
Population management	Stocking and harvesting to create desired population structure	Integrated programmes: restricted stocking and harvesting to increase catch while conserving naturally recruiting stock Separated programmes: as for sea ranching; also measures to promote separation	High stocking density over short period; temporarily restricted harvesting or moratorium

References

Anderson, J.L. (2002). Aquaculture and the future: why fisheries economists should care. *Marine Resource Economics* 17: 133-151.

Arnason, R. (2001). *The economics of ocean ranching: experiences, outlook and theory*. FAO Fisheries Technical Paper 413. Rome: Food and Agriculture Organization of the United Nations.

Araki, H., Berejikian, B.A., Ford, M.J., and Blouin, M.S. (2008). Fitness of hatchery-reared salmonids in the wild. *Evolutionary Applications* 1: 342-355.

Bartley, D.M., Bondad-Reantaso, M.G., and Subasinghe, R.P. (2006). A risk analysis framework for aquatic animal health management in marine stock enhancement programmes. *Fisheries Research* 80: 28-36.

Baskett, M.L., and Waples, R.S. (2013). Evaluating alternative strategies for minimizing unintended fitness consequences of cultured individuals on wild populations. *Conservation Biology* 27: 83-94.

Bell, J.D., Leber, K.M., Blankenship, H.L., Loneragan, N.R., and Masuda, R. (2008). A new era for restocking, stock enhancement and sea ranching of coastal fisheries resources. *Reviews in Fisheries Science* 16: 1-9.

Blankenship, H.L. and Leber, K.M. (1995). A responsible approach to marine stock enhancement. *American Fisheries Society Symposium* 15: 67-175.

Born, A.F., Immink, A.J., and Bartley, D.M. (2004). *Marine and coastal stocking: global status and information needs*. FAO Fisheries Technical Paper 429. Rome: Food and Agriculture Organization of the United Nations. pp. 1-18.

Brown, C., and Day, R.L. (2002). The future of enhancements: lessons for hatchery practice from conservation biology. *Fish & Fisheries* 3: 79-94.

Caddy, J.F., and Defeo, O. (2003). *Enhancing or restoring the productivity of natural populations of shellfish and other marine invertebrate resources*. FAO Fisheries Technical Paper 448. Rome: Food and Agriculture Organization of the United Nations. pp. 159.

Costello, M.J. (2014). Long live Marine Reserves: A review of experiences and benefits. *Biological Conservation*, 176: 289-296.

Crawford, S.S. (2001). Salmonine introductions to the Laurentain Great Lakes: an historical review and evaluation of ecological effects. Canadian Special Publication of Fisheries and Aquatic Science 132: 205 pp.

Drummond, K. (2004). The role of stock enhancement in the management framework for New Zealand's southern scallop fishery. In: Leber, K.M, Kitada, S., Blankenship H.L., and Svåsand, T., editors. *Stock Enhancement and Sea Ranching: Developments, Pitfalls and Opportunities*. Oxford: Blackwell Publishing. pp. 397-411.

FAO (1995). *Code of Conduct for Responsible Fisheries*. Rome: FAO. 41 p.

FAO (2008). *Technical Guidelines for Responsible Fisheries*, No. 6, Inland fisheries, Suppl. 1 (Rehabilitation of inland waters for fisheries). Rome: FAO.

FAO (2010). Report of the Expert Consultation on the Development of Guidelines for the Ecolabelling of Fish and Fishery Products from Inland Capture Fisheries. Rome, 25–27 May 2010. FAO Fisheries and Aquaculture Report No. 943. Rome, FAO. 37p.

FAO (2011). *Guidelines for the Ecolabelling of Fish and Fishery Products from Inland Capture Fisheries*. Rome: FAO.

Hilborn, R. (1998). The economic performance of marine stock enhancement projects. *Bulletin of Marine Science* 62: 661–674.

Hilborn, R., and Eggers, D. (2000). A review of the hatchery programmes for pink salmon in Prince William Sound and Kodiak Island, Alaska. *Transactions of the American Fisheries Society* 129: 333-350.

ICES (2005). *ICES Code of Practice on the Introductions and Transfers of Marine Organisms 2005*. Copenhagen: International Council for the Exploration of the Sea. 30 pp.

Johnson, T. H., Lincoln, R., Graves, G. R. and Gibbons, R. G. (1997). Status of wild salmon and steelhead stocks in Washington State. In Stouder, D.J., Bisson, P.A. and Naiman, R. (editors). *Pacific Salmon & their Ecosystems*. New York: Springer. pp. 127-144.

Jonasson, J., Gjedre, B., and Gjedrem, T. (1997). Genetic parameters for return rate and body weight in sea-ranched Atlantic salmon. *Aquaculture* 154: 219-231.

Kline, P. A., & Flagg, T. A. (2014). Putting the red back in Redfish Lake, 20 years of progress toward saving the Pacific Northwest's most endangered salmon population. *Fisheries* 39: 488-500.

Klinger, D.H., Turnipseed, M., Anderson, J.L., Asche, F., Crowder, L.B., Guttormsen, A.G., Halpern, B.S., O'Connor, M.I., Sagarin, R., Selkoe, K.A., Shester, G.G., Smith, M.D., and Tyedmers, P. (2012). Moving beyond the fished or farmed dichotomy. *Marine Policy* 38: 369-374.

Knapp, G., Roheim, C. & Anderson, J. (2007). *The Great Salmon Run: Competition Between Wild and Farmed Salmon*. TRAFFIC North America. Washington D.C.: World Wildlife Fund

Leber, K.M. (2013). Marine fisheries enhancement: Coming of age in the new millennium. In: Christou, P., Savin, R., Costa-Pierce, B.A., Misztal, I., and Whitelaw, C.B.A., editors. *Sustainable Food Production*. New York, NY: Springer Science. pp. 1139-1157.

Leleu, K., Remy-Zephir, B., Grace, R., and Costello, M.J. (2012). Mapping habitat change after 30 years in a marine reserve shows how fishing can alter ecosystem structure. *Biological Conservation* 155: 193–201.

Le Vay, L., Carvalho, G.R., Quinitio, E.T., Lebata, J.H., Ut, V.N., and Fushimi, H. (2007). Quality of hatchery-reared juveniles for marine fisheries stock enhancement. *Aquaculture* 268: 169-180.

Lorenzen, K. (2005). Population dynamics and potential of fisheries stock enhancement: practical theory for assessment and policy analysis. *Philosophical Transactions of the Royal Society* B 360: 171-189.

Lorenzen, K. (2008). Understanding and managing enhancement fisheries systems. *Reviews in Fisheries Science* 16: 10-23.

Lorenzen, K. (2014) Understanding and managing enhancements: why fisheries scientists should care. *Journal of Fish Biology 85*: 1807-1829.

Lorenzen, K., Leber, K.M., and Blankenship, H.L. (2010). Responsible approach to marine stock enhancement: an update. *Reviews in Fisheries Science* 18: 189-210.

Lorenzen, K., Beveridge, M.C.M., and Mangel, M. (2012). Cultured fish: integrative biology and management of domestication and interactions with wild fish. *Biological Reviews* 87: 639-660.

Lorenzen, K., Agnalt, A.L. Blankenship, H.L. Hines, A.H., Leber, L.M., Loneragan, N.R., and Taylor, M.D. (2013). Evolving context and maturing science: aquaculture-based enhancement and restoration enter the marine fisheries management toolbox. *Reviews in Fisheries Science* 21: 213-221.

Michael, J.H., Appleby, A., and Barr, J. (2009). Use of the AHA model in Pacific salmon recovery, hatchery, and fishery planning. *American Fisheries Society Symposium* 71: 455-464.

Miller, L.M., Kapuscinski, A.R. (2003). Genetic guidelines for hatchery supplementation programmes. In: Hallerman, E.M., editor. *Population Genetics: Principles and Applications for Fisheries Scientists*. Bethesda, MD: American Fisheries Society. pp. 329-355.

Mobrand, L.E., Barr, J., Blankenship, L., Campton, D.E., Evelyn, T.T., Flagg, T.A., Mahnken, C.V.W, Seeb, L.W., Seidel, P.R., and Smoker, W.W. (2005). Hatchery reform in

Washington State: principles and emerging issues. *Fisheries* 30: 11-23.

Naish, K.A., Taylor, J.E., Levin, P.S., Quinn, T.P., Winton, J.R., Huppert, D., and Hilborn, R. (2007). An evaluation of the effects of conservation and fishery enhancement hatcheries on wild populations of salmon. *Advances in Marine Biology* 53: 61-194.

Nicholas, J.W. and Hankins, D.G. (1989) Chinook salmon populations in Oregon coastal river basins: description of life histories and assessment of recent trends in run strengths. Corvallis: Oregon State University Extension Service. 359 pp.

NMFS 2000. Viable Salmonid Populations and the Recovery of Evolutionarily Significant Units. U.S. Department of Commerce, NOAA Technical Memorandum NMFS-NWFSC-42. Seattle: Northwest Fisheries Center.

ODFW (1998). Fish Propagation Annual Report for 1997. Salem, OR: Oregon Department of Fish and Wildlife.

Olla, B.L., Davis, M.W., and Ryer, C.H. (1998). Understanding how the hatchery environment represses or promotes the development of behavioral survival skills. *Bulletin of Marine Science* 62: 531-550.

Paquet, P.J., Flagg, T., Appleby, A., Barr, J., Blankenship, L., Campton, D., Delarm, M., Evelyn, T., Fast, D., Gislason, J. Kline, P., Maynard, D., Mobrand, L., Nandor, G., Seidel, P., and Smith, S. (2011). Hatcheries, conservation, and sustainable fisheries—achieving multiple goals: results of the Hatchery Scientific Review Group's Columbia River basin review. *Fisheries* 36: 547-561.

Pinkerton, E. (1994). Economic and management benefits from the coordination of capture and culture fisheries: the case of Prince William Sound pink salmon. *North American Journal of Fisheries Management* 14: 262-277.

Smith, C. (2014). Hatcheries and harvest: meeting treaty obligations through artificial propagation. *Fisheries* 39: 541-542.

Tringali, M.D., Bert, T.M., Cross, F., Dodrill, J.W., Gregg, L.M., Halstead, W.G., Krause, R.A., Leber, K.M., Mesner, K., Porak, W., Roberts, D., Stout, R., and Yeager, D. (2007). *Genetic Policy for the Release of Finfishes in Florida*. Florida Fish and Wildlife Research Institute Publication Number IHR-2007-001. St. Petersburg: Florida Fish and Wildlife Research Institute.

Uki, N. (2006). Stock enhancement of the Japanese scallop *Patinopecten yessoensis* in Hokkaido. *Fisheries Research* 80: 62-66.

Utter, F., and Epifanio, J. (2002). Marine aquaculture: Genetic potentials and pitfalls. *Reviews in Fish Biology and Fisheries* 12: 59-77.

Walters, C.J., and Martell, S.J.D. (2004). *Fisheries Ecology and Management*. Princeton, NJ: Princeton University Press. 399 pp.

Williams, J.R., Hartill, B., Bian, R. and Williams, C.L. (2014). Review of the Southern scallop fishery (SCA 7). New Zealand Fisheries Assessment Report 2014/07, 71 pp.

Zhang, C.I., Kim, S., Gunderson, D., Marasco, R., Lee, J.B., Park, H.W., and Lee, J.H. (2009). An ecosystem-based fisheries assessment approach for Korean fisheries. *Fisheries Research* 100: 26-41.

14 Seaweeds

Contributors:
John West (Convenor), Hilconida Calumpong (Co-Lead Member) and Georg Martin (Lead Member)

Seaweeds — Chapter 14

1 Introduction

Seaweeds are a group of photosynthetic non-flowering plant-like organisms (called macroalgae) that live in the sea. They belong to three major groups based on their dominant pigmentation: red (Rhodophyta), brown (Phaeophyta) and green (Chlorophyta). Seaweeds were traditionally and are currently still used as food in China, Japan and the Republic of Korea. About 33 genera of seaweeds, mostly red and brown, are harvested and farmed commercially (McHugh, 2003), although close to 500 species in about 100 genera are collected and utilized locally (Mouritsen, 2013). Currently about 80 per cent of total seaweed production is for direct human consumption, eaten dried or fresh for its nutritional value or for flavouring (see Kilinc et al., 2013 for a comprehensive listing of nutrients and compounds) in the form of sushi, salad, soup, dessert and condiments, and the remaining 20 per cent is used as a source of the phycocolloids extracted for use in the food, industrial, cosmetic, and medical industry (Browdy et al., 2012, Critchly et al., 2006, Lahaye, 2001, McHugh, 2003, Mouritsen, 2013, Ohno and Critchley, 1993), as well as for animal feed additive, fertilizer, water purifier, and probiotics in aquaculture (Abreu et al., 2011, Chopin, 2012, Chopin et al., 2001, Chopin et al., 2012, Fleurence et al., 2012, Kim et al., (2014), Neori et al., 2004, Pereira and Yarish, 2008, 2010, Rose et al., 2010). Carrageenan and agar are extracted from red seaweeds, and alginates and fucoidan are extracted from brown seaweeds, generally from kelp species. Recently, the kelp species Saccharina lattisima was considered for bioethanol production (Adams et al., 2009).

2 Production

World production of seaweeds comes from two sources: harvesting from wild stocks and from aquaculture (including land-based culture, mariculture and farming). Production from harvesting of wild stocks has been stable at over 1 million tons (wet weight) in the last 10 years (2003 to 2012) according to FAO (2014) statistics (see Figure 14.1). Top producers in 2012 were Chile (436,035 tons representing 39 per cent of total world production), China (257,640 tons or 23 per cent), Norway (140,336 or 13 per cent), Japan (98,514 or 9 per cent), France (41,229 tons or 4 per cent), Ireland (29,500 tons or 2.73 per cent), Iceland (18,079 tons or 2 per cent), South Africa (14,509 tons or 1 per cent) and Canada (13,833 tons or 1 per cent). Contributing less than 1 per cent each were 24 other countries. Chile has consistently been the number one top producer since 2003, except in 2007 when China exceeded Chile's production by 1 per cent. Norway and Japan have maintained their position as third and fourth top producers, respectively, since 2003.

Three countries posted only one year's production in 10 years (Namibia in 2003 with 408 tons, Samoa in 2004 with 478 tons, Senegal in 2012 with 1,028 tons. India posted 1 ton of production in 2004 to 2008, except in 2005 when it posted 2 tons of production).

The bulk of seaweeds produced worldwide come from aquaculture. The FAO (2014) reported that the production of aquatic seaweeds from mariculture, reached 24.9 million tons in 2012, valued at about $6 billion United States dollars. The red, brown and green seaweeds constitute about 88 per cent (21 million tons). About 96 per cent (23.8 million tons) of the total production were produced from aquaculture (see Figure 14.2). Data from FAO showed a steady increase of about 8 per cent per year over the last 10 years (range of 4-12 per cent), specifically for red seaweeds (Figure 14.3) with the brown seaweeds showing stable production. The cultured seaweeds are mainly those that produce carrageenan (*Kappaphycus alvarezii* and *Eucheuma spp.* - 8.3 million tons), followed by the alginate-producing brown seaweeds (kelps - 5.7 million tons). China is the consistent top supplier, although showing a

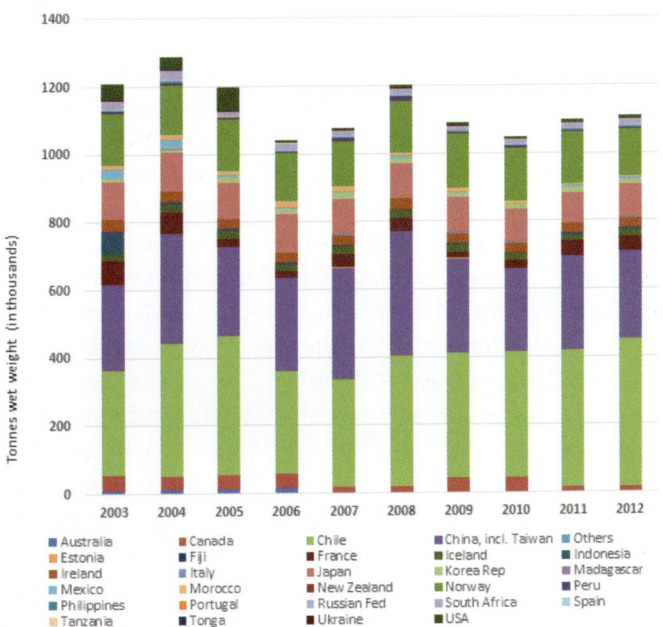

Figure 14.1 | World seaweed production from wild stocks in 2003-2012 by country/territory in tons wet weight. Data from FAO, 2014. Four countries with production in 10 years of less than 1000 tons or with only one production within 10 years are lumped under Others (see text). tp://www.fao.org/fishery/statistics/software/fishstatj/en.

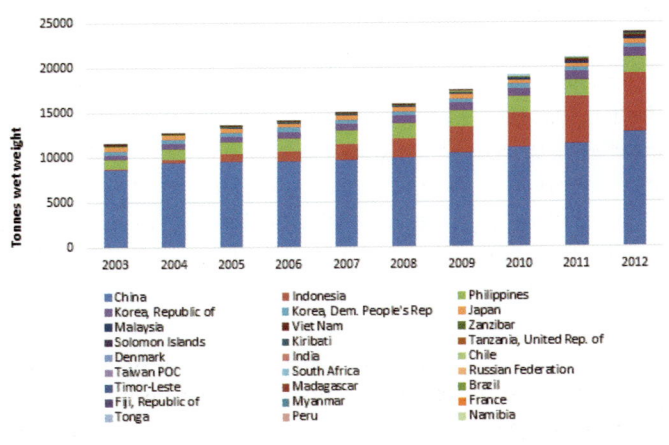

Figure 14.2 | World seaweed production from aquaculture in 2003-2012 by country/territory in tons wet weight. Data from FAO 2014. http://www.fao.org/fishery/statistics/software/fishstatj/en.

Figure 14.3 | World aquaculture production from 2003-2012 by species groups in tons wet weight and total value in United States dollars per group. (Unidentified aquatic plants excluded.) Green algae production is minimal, as shown in this graph. Data from FAO 2014.

decreasing trend, with 50 per cent of the world production over a 10-year period (2003-2012). The Philippines ranked second in 2003 to 2006, producing 9-10 per cent, after which it was overtaken by Indonesia. The Democratic People's Republic of Korea, the Republic of Korea, and Japan produced between 2-5 per cent of the annual total, and 31 other countries produced less than 1 per cent of the annual total, except for Malaysia, which showed an increasing production equivalent to 1.09-1.39 per cent of the annual global quantity during 2010 to 2012.

3   Social and economic impacts and challenges

Harvests from wild populations are affected by overexploitation and climatic changes. In Northern Ireland, for example, which is listed as one of the top 10 producers of wild stocks globally (FAO, 2014), McLaughlin et al., (2006) described in detail the adverse impacts of seaweed harvesting at small, artisanal and commercial scales on areas of conservation importance, protected and priority habitats and species, including disturbance of birds and wildlife, disruption of food webs, damage to substrata, habitat destruction, localized biodiversity changes, and changes in particle-size distribution in sediments. Direct effects on the seaweed population include mortalities due to increased growth rate and cover of other algae which are not harvested, such as filamentous green algae and the brown seaweed, *Fucus vesiculosus*, which outcompete the desired species, and die-back due to increased predation. In several areas of Norway, the kelp *Saccharina lattisima* has been reported by Moy and Christie (2012) to have suffered dieback by 40-80 per cent due to sea urchin predation.

The brown seaweed kelps are most affected by rising water temperature, because sexual reproduction (gamete formation) in most kelps will not occur above 20°C (Dayton 1985, Dayton et al., 1999). Already along the European coasts and especially in Brittany, France, the brown kelp, *Laminaria digitata*, which is heavily harvested for commercial uses, is reported to be on the verge of local extinction. The already reduced reproductive potential of the kelp due to dwindling population and harvesting-induced ecosystem changes may be exacerbated by climate-caused increase in sea temperature (Brodie et al., 2014, Raybaud et al., 2013). Two other kelp species, *Laminaria ochroleuca*, a warm-temperate perennial, and *Saccorhiza polyschides*, a wide-ranging cool- to warm-temperate annual, have somewhat higher temperature tolerances for sexual reproduction than other kelps (Pereira et al., 2011); however, *Saccorhiza* outcompetes *L. ochroleuca* in shared habitats. Brittany is the northern limit of *L. ochroleuca*'s range. Since 1940, *L. ochroleuca* has been found on the coasts of southern England, which is apparently indicative of a slow northward extension of warmer waters. Anticipated increasing ocean temperatures in the future in the boreal region may result in *L. ochroleuca* possibly replacing *L. hyperborea* (Brodie et al., 2014). On the other hand, the kelp *Ecklonia maxima* is extending eastward on the tip of South Africa because of a northward intrusion by cooler inshore water (Bolton et al., 2012); this could greatly benefit the whole ecosystem and provide more food for the abalone industry there. All this is quite a contrast from southward intrusion patterns by warm water on the east and west coasts of Australia, causing extensive retreat of kelps and fucoids (another group of brown algae) southward from their previous northern-most limits (Wernberg et al., 2011, Millar, 2007).

Seaweed farming and culture are seriously affected by diseases. *Ice-ice* disease has impacted the farming of the kappa-carrageenan-producing *Kappaphycus alvarezii*, commercially called "cottonii". Another species, *Eucheuma denticulatum*, commercially called "spinosum," is *ice-ice*-resistant, but contains iota-carrageenan which fetches a much lower price on the world market (Valderrama, 2012). This problem may be a result of the low genetic variation in *K. alvarezii*, all of whose cultured stocks around the world have a similar mitochondrial haplotype, which is not the case for *E. denticulatum* (Halling et al., 2013; Zuccarello et al., 2006). Significant diseases affecting cultivated kelps (e.g., *Saccharina japonica*) include green-rot, white-rot, blister disease, which may be environmentally induced, and malformation disease of summer sporelings and swollen stipe or "frond twist disease" which are caused by bacteria (Brinkhaus et al., 1987, Tseng, 1986). Parasites such as *Pythium*, an oomycete fungus, causes "red rot" or "red wasting" disease in the red seaweed *Pyropia* commonly used in making sushi (Hurd et al., 2014). However, based on case studies from six countries, Valderrama (2012) reported that the socioeconomic impacts of seaweed farming have been positive. He attributed this mainly to small-scale, family operations resulting in the generation of substantial employment as compared to other forms of aquaculture. He added that seaweed farming is often undertaken in remote areas where coastal communities face fewer economic alternatives and where many of these communities have traditionally relied on coastal fisheries which are currently being affected by overexploitation. Valderrama stated that the impact of seaweed farming in these cases goes beyond its immediate economic benefits to communities as it also reduces the incentives for overfishing. However, one challenge faced by farmers in these remote areas is low profits due to high shipping costs. This disadvantage is exacerbated by the dependence of farmers on processors for the procurement of their farming materials and their lack of farm-management skills. In addition, food safety issues can sometimes affect markets and prices. This is because seaweeds are efficient nutrient extractors (Kim et al., 2014) and may accumulate compounds that pose harm to human health (Mouritsen 2013; see also Chapter 10).

4 Information and Knowledge Gaps

Despite the long history of utilization, it is reported that kelp-dominated habitats along much of the NE Atlantic coastline have been chronically understudied over recent decades in comparison with other regions such as Australasia and North America. For example, McLaughlin et al. (2006) noted that information on the distribution and biomass of commercial seaweeds in Northern Ireland is lacking. Smale and Wernberg (2013) highlight the changing structure of kelp forests in the North-East Atlantic in response to climate- and non-climate-related stressors, which will have major implications for the structure and functioning of coastal ecosystems. This paucity of field-based research is impeding ability to conserve and manage this important resource.

References

Abreu, M.H., Pereira, R., Yarish, C., Buschmann, A.H., Sousa-Pinto, I. (2011). IMTA with Gracilaria vermiculophylla: productivity and nutrient removal performance of the seaweed in a Land-based pilot scale system. *Aquaculture* 312 (1-4): 77-87.

Adams, J., Gallagher, J., Donnison, I. (2009). Fermentation study on Saccharina lattisima for bioethanol production considering variable pre-treatments. *Journal of Applied Phycology* 21: 569-574.

Bolton, J., Anderson, R., Smit, A., Rothman, M. (2012). South African kelp moving eastwards: the discovery of Ecklonia maxima (Osbeck) Papenfuss at De Hoop Nature Reserve on the South Coast of South Africa, *African Journal of Marine Science* 34: 147-151.

Brinkhuis, B.H., Levine, H.G.,Schlenk, C.G., Tobin, S. (1987). Laminaria cultivation in the far-east and North America. In: *Seaweed Cultivation for Renewable Resources*. (Bird, K.T. Benson, P.H., eds.). Developments in Aquaculture and Fisheries Science 16: 107-146.

Brodie, J., Williamson, C.J., Smale, D.A., Kamenos, N.A., Mieszkowska, N., Santos, R., Cunliffe, M., Steinke, M., Yesson, C., Anderson, K.M., Asnaghi, V., Brownlee, C., Burdett, H.L., Burrows, M.T., Collins, S., Donohue, P.J.C., Harvey, B., Noisette, F., Nunes, J., Ragazzola, F., Raven, J.A., Foggo, A., Schmidt, D.N., Suggett, D., Teichberg, M., Jason M. Hall-Spencer, J.M. (2014). The future of the northeast Atlantic benthic flora in a high CO2 World. *Ecology and Evolution* 1-12. doi:10.1002/ece3.1105.

Browdy, C.L., Hulata, G., Liu, Z., Allan, G.L., Sommerville, C., Passos de Andrade, T., Pereira, R., Yarish, C., Shpigel, M., Chopin, T., Robinson, S., Avnimelech, Y., Lovatelli, A. (2012). Novel and emerging technologies: can they contribute to improving aquaculture Sustainability? In Subasinghe, R.P., Arthur, J.R., Bartley, D.M., De Silva, S.S., Halwart, M., Hishamunda, N., Mohan, C.V., Sorgeloos, P. (eds.), Farming the Waters for People and Food. *Proceedings of the Global Conference on Aquaculture 2010*, Phuket, Thailand. 22–25 September 2010. pp. 149–191. FAO, Rome and NACA, Bangkok.

Chopin, T. (2012). Aquaculture, Integrated Multi-Trophic (IMTA). In: Meyers, R.A. (ed.), *Encyclopedia of Sustainability Science and Technology*. Springer, Dordrecht, The Netherlands. pp. 542–64.

Chopin, T., Buschmann, A. H., Halling, C., Troell, M., Kautsky, N., Neori, A., Kraemer, G.P., Zertuche-Gonzales, J.A., Yarish, C., Neefus, C. (2001). Integrating seaweeds into marine aquaculture systems: a key toward sustainability. *Journal of Phycology* 37: 975–986.

Chopin, T., Cooper, J. A. ,Reid, G., Cross, S., Moore, C. (2012). Open-water integrated multi-trophic aquaculture: environmental biomitigation and economic diversification of fed aquaculture by extractive aquaculture. *Reviews in Aquaculture* 4: 209–220.

Critchley, A.T., Ohno, M,. Largo, D.B. (2006). *World Seaweed Resources: An Authoritative Reference System*. DVD–ROM. Wokingham, UK: ETI Information Services.

Dayton, P.K. (1985). Ecology of kelp communities. *Annual Review of Ecology and Systematics* 16: 215–245.

Dayton, P.K., Tegner, M.J., Edwards, P.B., Riser, K.L. (1999). Temporal and spatial scales of kelp demography: the role of oceanography and climate. *Ecological Monographs* 69: 219–250.

FAO. (2014). *Fishery and Aquaculture Statistics. Aquaculture production 1950-2012 (FishstatJ)*. In: FAO Fisheries and Aquaculture Department [online or CD-ROM]. Rome. Updated 2014. http://www.fao.org/fishery/statistics/software/fishstatj/en.

Fleurence, J., Morançais, M., Dumay, J., Decottignies, P., Turpin, V., Munier, M., Garcia Bueno, N., Jaouen, P. (2012). What are the prospects for using seaweed in human nutrition and for marine animals raised through aquaculture? *Trends in Food Science & Technology* 27:57-61.

Halling, C., Wikström, S.A., Lilliesköld-Sjöö, G., Mörk, E., Lundsør, E., Zuccarello, G.C. (2013). Introduction of Asian strains and low genetic variation in farmed seaweeds: indications for new management practices. *Journal of Applied Phycology* 25:89–95, doi: 10.1007/s10811-012-9842-0.

Hurd, C.L., Harrison, P.J., Bischof, K., Lobban, C.S. (2014). *Seaweed Ecology and Physiology*, (2nd ed.). Cambridge University Press.

Kılınç, B. Cirik, S., Turan, G., Tekogul, H., Koru, E. (2013). *Seaweeds for Food and Industrial Applications*. http://dx.doi.org/10.5772/53172. In: *Food Industry* http://cdn.intechopen.com/pdfs/41694/InTech-Seaweeds_for_food_and_industrial_applications.pdf

Kim, J.K., Kraemer, G.P., Yarish, C. (2014). Field scale evaluation of seaweed aquaculture as a nutrient bioextraction strategy in Long Island Sound and the Bronx River Estuary. *Aquaculture* 433: 148-156.

Lahaye, M. (2001). Chemistry and physico-chemistry of phycocolloids, *Cahiers de Biologie Marine*. 42: 137-157.

McHugh, D.J. (2003). A Guide to the Seaweed Industry. *FAO Fisheries Technical Paper* 441.

McLaughlin, E., Kelly, J., Birkett, D., Maggs, C., Dring, M. (2006). *Assessment of the Effects of Commercial Seaweed Harvesting on Intertidal and Subtidal Ecology in Northern Ireland*. Environment and Heritage Service Research and Development Series. No. 06/26.

Millar, A.J.K. (2007). The Flindersian and Peronian Provinces. In: McCarthy, P., Orchard, A., (eds.), *Algae of Australia. An Introduction*. CSIRO Publishing, Melbourne, pp. 554-559.

Mouritsen, O.G. (2013). *Seaweeds Edible, Available & Sustainable*. The University of Chicago Press, Chicago & London, 287 pp.

Moy, F., Christie, H. (2012). Large-scale shift from sugar kelp (Saccharina latissima) to ephemeral algae along the south and west coast of Norway. *Marine Biology Research* 8: 309-321.

Neori, A., Chopin, T., Troell, M., Buschmann, A.H., Kraemer, G. Halling, C., Shpigel, M., Yarish, C. (2004). Integrated aquaculture: rationale, evolution and state of the art emphasizing seaweed biofiltration in modern aquaculture. *Aquaculture* 231: 361-391.

Ohno, M., Critchley, A., (eds.). (1993). *Seaweed Cultivation and Marine Ranching*. JICA, Yokosuka, Japan, i-xvii, 431, i-xii pp.

Pereira, T., Engelen, A., Pearson, G., Serrão, E., Destombe, C., Valero, M. (2011). Temperature effects on the microscopic haploid stage development of Laminaria ochroleuca and Saccoriza polyschides, kelps with contrasting life histories. *Cahiers de Biologie Marine* 52: 395-403.

Pereira, R., Yarish, C. (2008). Mass production of Marine Macroalgae. In: Jørgensen, S.E., Fath, B.D., (eds.), *Ecological Engineering*. Vol. [3] of Encyclopedia of Ecology, 5 vols. pp. 2236-2247. Elsevier: Oxford.

Pereira, R., Yarish, C. (2010). The role of Porphyra in sustainable culture systems: Physiology and Applications. In: Israel, A., Einav, R., (eds.), *Role of Seaweeds in a Globally Changing Environment*. Springer Publishers, pp. 339-354.

Pereira, R., Yarish, C., Critchley, A. (2012). Seaweed Aquaculture for Human Foods in Land Based and IMTA Systems. In: Meyers, R. (eds.), *Encyclopedia of Sustainability Science and Technology*. Springer Science, N.Y. pp. 9109-9128.

Raybaud,V., Beaugrand, G., Goberville, E., Delebecq, G., Destombe, C. Valero, M., Davoult, D., Morin, P., Gevaert, F. (2013). Decline in Kelp in West Europe and Climate.

PLoS ONE 8(6): e66044. doi:10.1371/journal.pone.0066044.

Rose, J.M., Tedesco, M., Wikfors, G.H., Yarish, C. (2010). *International Workshop on Bioextractive Technologies for Nutrient Remediation Summary Report*. US Department of Commerce, Northeast Fisheries Science Center Reference Document 10-19; Available from: National Marine Fisheries Service, 166 Water Street, Woods Hole, MA 02543-1026, or online at http://www.nefsc.noaa.gov/nefsc/publications/12 p.

Smale, D.A., Wernberg, T. (2013). Extreme climatic event drives range contraction of a habitat-forming species. *Proceedings of the Royal Society B* 280: 20122829.

Tseng, C.K. (1986). Laminaria mariculture in China. In: Doty, M.S., Caddy, J.F., Santelices, B. (eds.), Case studies of seven commercial seaweed resources. *FAO Fisheries Technical Papers*, (281): 311 p.

Valderrama, D. (2012). *Social and economic dimensions of seaweed farming: a global review*. IIFET Tanzania Proceedings. https://ir.library.oregonstate.edu/xmlui/handle/1957/33886

Wernberg, T., Russell, B., Thomsen, M., Gurgel, F., Bradshaw, C., Poloczanska, E., Connell, S. (2011). Seaweed communities in retreat from Ocean Warming. *Current Biology* 21: 1828-1832.

Zuccarello G.C., Critchley, A.T., Smith, J., Sieber, V., Lhonneur, G.B. (2006). Systematics and genetic variation in commercial Kappaphycus and Eucheuma (Solieriaceae, Rhodophyta). *Journal of Applied Phycology* (2006) 18: 643-651 doi: 10.1007/s10811-006-9066-2.

15 Social and Economic Aspects of Sea-Based Food and Fisheries

Contributors:
Beatrice Ferreira (Convenor and Lead Member), Ratana Chuenpagdee, Patrick McConney, Gordon Munro and Enrique Marschoff, Jake Rice and Andrew Rosenberg (Co-Lead Members)

Social and Economic Aspects of Sea-Based Food and Fisheries

1 Introduction

Fish are one of the most internationally traded foods, and the value of global fish trade exceeds the value of international trade of all other animal proteins combined (World Bank, 2011). In 2012, international trade represented 37 per cent of the total fish production in value, with a total export value of 129 billion United States dollars, of which 70 billion dollars constituted developing countries' exports (FAO, 2014). Estimates indicate that small-scale fisheries contribute about half of global fish catches (FAO, 2014; HLPE, 2014). When considering catches destined for direct human consumption, the share contributed by the subsector increases, as small-scale fisheries generally make broader direct and indirect contributions to food security through affordable fish and employment to populations in developing countries.

This chapter, in addressing the economic and social aspects of marine fisheries, examines both macro and micro issues. The macro issues considered are some aspects of the economics of marine capture fishery. Among the micro issues explored are local to regional socioeconomic effects, competition for space between various ocean activities and user groups, the relationship between capture fisheries and aquaculture, and gender issues in fisheries and aquaculture.

The contribution of small-scale fisheries has been increasingly recognized as a major factor for food security and livelihoods at household and community levels, particularly for poor communities around the world. Information on small-scale fisheries is often not captured in national statistics as a result of difficulties due to many factors, including their socioeconomic complexity and the highly dynamic nature of their operation (Chuenpagdee, 2011). Numerous initiatives around the world reflect their importance, including those led by FAO in the development

The boundaries and names shown and the designations used on this map do not imply official endorsement or acceptance by the United Nations.

Figure 15.1 | Spatial distribution of average annual landed values (2005 United States dollars per square kilometre per year) by decade (from Swartz et al 2013; with permission of Springer).

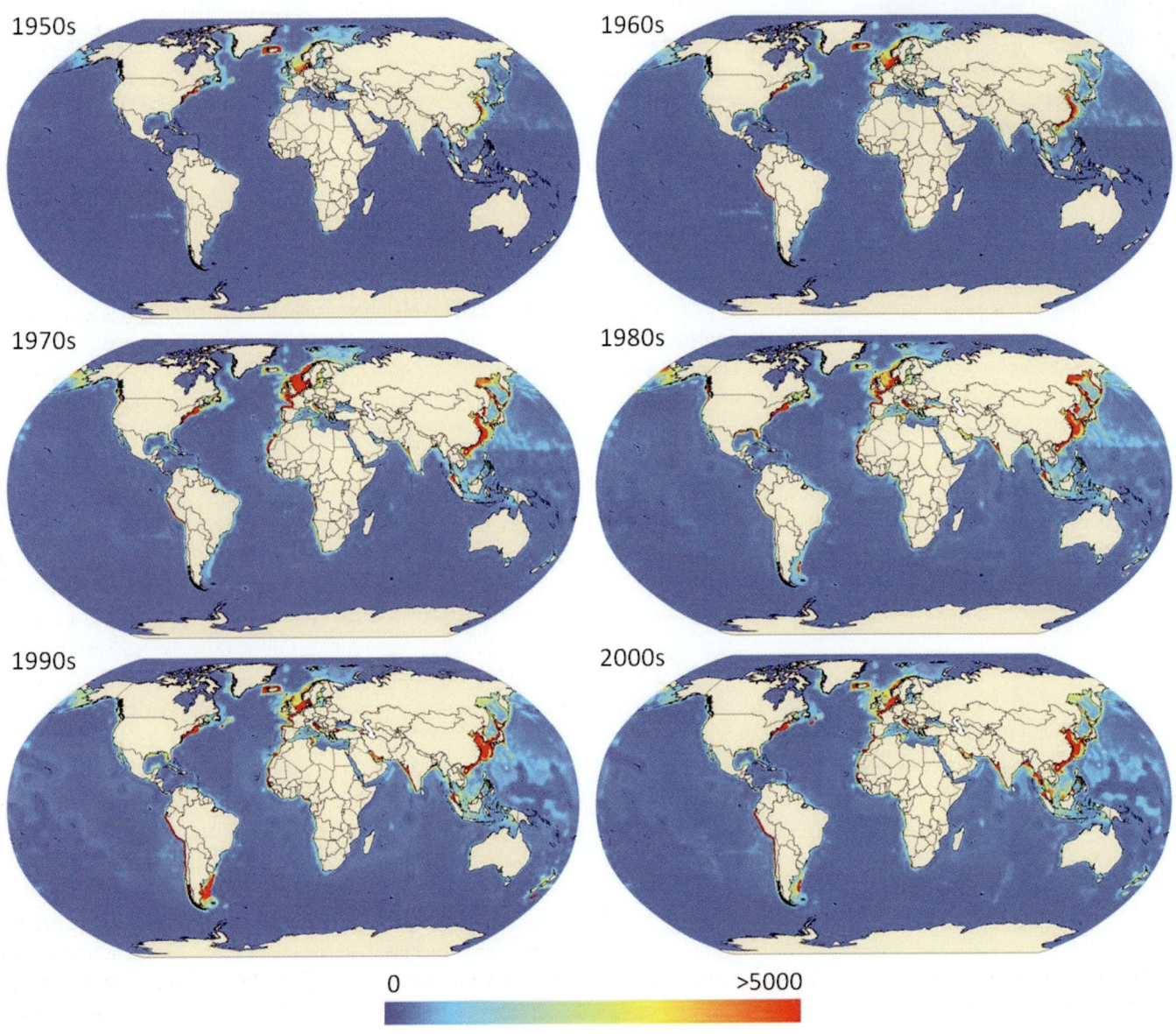

of the Voluntary Guidelines for Securing Sustainable Small-Scale Fisheries.[1]

2 Marine Capture Fisheries Social and Economic Value

The global marine capture fisheries harvest expanded rapidly from the early 1950s, and is currently estimated to be about 80 million tons per annum (see Chapter 11 and FAO, 2014). This harvest is estimated to have a first value (gross) in the order of 80 billion US dollars (World Bank and FAO, 2009). Although it is difficult to produce accurate employment statistics, capture fisheries provide, direct and indirect employment, for at least 120 million persons worldwide (ibid.).

Global and regional fishery catch statistics in most cases do not distinguish between large scale and small-scale fisheries, so the small-scale sector is often poorly covered in official statistics and chronically underevaluated in general. The Big Numbers Project (BNP)[2] carried out case studies in populous developing countries and the results from these case studies, together with other available information, formed the basis for a first disaggregated review of the fisheries sector as a whole (WorldFish Center, 2008). Tentative estimates were calculated for developing countries at 28-30 million MT/year for marine fisheries. This represents half of the catch in those countries, of which 90-95 per cent is destined for domestic human consumption. Those figures highlight the importance of small-scale fisheries for food security in developing countries.

Small-scale fisheries employ more than 90 per cent of the world's capture fishers and fish workers, about half of whom are women. In addition to employment as full- or part-time fishers and fish workers, seasonal or occasional fishing and related activities provide vital supplements to the livelihoods of millions. These activities may be a recurrent sideline activity or become especially important in times of difficulty. Many small-scale fishers and fish workers are self-employed and engaged in directly providing food for their household and communities as well as working in commercial fishing, processing and marketing (FAO, 2014).

The quality of such employment is increasingly seen as an important social and economic aspect of fisheries as attested to by the attention to decent work in the FAO Voluntary Guidelines on Securing Small-Scale Fisheries (SSF Guidelines) that draws from several international instruments concerning, gender, child labour, workers' rights and the like. Much of this labour is linked directly, through short value chains, to providing critical income along with food and nutrition security, especially in rural coastal communities.

---

1 The Guidelines have recently been adopted at the 31st Session of the Committee on Fisheries, June 2014. The final text is available at www.fao.org.

2 This is a joint activity of FAO and the WorldFish Center and funded through the World Bank's PROFISH1 Partnership.

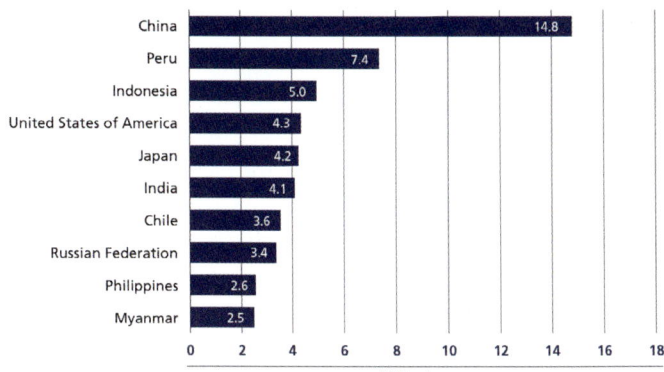

Figure 15.2 | Marine and inland capture fisheries: top ten producer countries in 2008. From FAO, 2010.

Over time, there has been a shift in the relative scale and geography of capture fisheries. In the 1950s, capture fisheries were largely undertaken by developed fishing States in the northern hemisphere. Since then, developing countries increased their share of the total. Consider Figure 15.1, which presents geo-referenced distributions of decadal averages of annual landed values of the world's fisheries and highlights the southward and offshore expansion of the fishing grounds over time (Swartz et al., 2013). Although the two hemispheres do not reflect developed vs. developing fishing States precisely, the figures are, nonetheless, indicative. In the 1950s, the Southern hemisphere accounted for no more than 8 per cent of landed values. By the last decade, the Southern hemisphere's share had risen to 20 per cent of the total. This change likely resulted from a combination of factors including transfer of fishing effort from north to south, overall increases in fisheries in the south and improvement in reporting systems. Nevertheless, the relative contribution to global landings from the two hemispheres has changed.

In terms of volume, the shift seen in Figure 15.1 is even more striking; as shown in Figure 15.2, the top ten capture fisheries producers include seven developing countries.

Indeed, net exports of fish and fishery products from developing countries have grown significantly in recent decades, rising from 3.7 billion dollars in 1980 to 18.3 billion dollars in 2000, 27.7 billion dollars in 2010, and reaching 35.1 billion dollars in 2012. For Low-Income Food-Deficit Countries (LIFDCs) net export revenues amounted to 4.7 billion dollars in 2010, compared with 2.0 billion dollars in 1990 (HLPE, 2014). The share of exports from developing countries is close to 50 per cent (value) and 60 per cent (in volume of live weight equivalent) of global fish exports (FAO, 2012).

This also reflects the impacts of globalization of fish markets, which have grown at an accelerating rate in the last decades. This has been viewed either as positive or negative, depending on the value systems used (Taylor et al., 2007). Although fish trade contributes to food security through the generation of revenues, adverse effects by international trade on the environment, small-scale fisheries culture, livelihoods and special needs related to food security are a matter of concern. Articulation with global demand may provide incentives to overexploit or waste resources, endanger the lives of fisherfolk, change cultural traditions and more – much of which can be unintended – shark finning, spiny lobster dive fisheries, and sea cucumber fisheries are examples. Small-scale fisheries stakeholders cannot often adapt to, and benefit equitably from, opportunities of global market trends (FAO, 2014-consultation). Also, there have been evidences that when global figures are considered, although there is quantity equivalence in trade, a quality exchange also takes place, with developing countries exporting high-quality seafood in exchange for lower quality seafood (Asche et al., 2015).

Regarding the trends in world marine capture fisheries, production has levelled off as the capacity of the ocean to produce ongoing harvest is approached (FAO, 2014- SOFIA). Overall production might be increased however, if overfished stocks are rebuilt and fisheries and ecosystems are used more sustainably. This requires overall reductions in exploitation rates, achievable through a range of context dependent management tools (Worm et al., 2009).

As noted in Chapter 11, global fisheries agreements and the FAO generally utilize the concept of Maximum Sustainable Yield (MSY) as a reference point for gauging whether a fishery resource is fully exploited, overexploited, and less than fully exploited. According to this reference point, FAO classifies the status of marine capture fishery resources (Table 15.1).

In the beginning of the 1950s, fully exploited and overexploited fishery resources *combined* accounted for less than 5 per cent of the total. Over 95 per cent fell into the less than fully exploited category (FAO, 1997, p. 7).

Over the following 25 years, the percentage of overexploited marine capture fish stocks rose to 10 per cent of the total. The percentage of these overexploited stocks then increased alarmingly from 10 to 26 per cent between the mid-1970s and the end of the 1980s. That percentage has continued to increase, but at a much slower pace (FAO, 2014).

The FAO states that:
> "[…] the declining global marine catch over the last few years together with the increased percentage of overexploited fish stocks […] convey the strong message that the state of world marine fisheries is worsening […] which leads to negative social and economic consequences" (FAO, 2012, p.12).

Further, these analyses of individual stocks do not fully account for the broader, ecosystem-level effects of fisheries exploitation that may be hindering future productivity in various ways, such as loss of habitat, or impacts on food webs and ecological functions needed to continue to produce desirable fish for harvest. There are two inter-related general considerations regarding management of these ecosystem-level effects: 1) the potential impacts of fisheries themselves on the ecosystems, in order to maintain overall ecosystem function including productivity, usually referred to as ecosystem-based fishery management (FAO, 2003); 2) the interaction of fisheries with other sectors of human activity and consideration of the cumulative impact of all sectors on marine ecosystems, usually referred to as ecosystem-based management (McLeod and Leslie, 2009).

The discussion here and in Chapter 11 on full exploitation and overexploitation of capture fishery resources was essentially cast in biological terms. When examined in economic terms, the situation portrayed in Table 15.1 implies a loss in the potential of economic returns accruing to society from capture fisheries compared to the situation where all fisheries were managed to maximize economic benefits. The maximum economic yield (MEY), when adopted as a reference point, is more conservative and reached at lower fishing effort levels than the MSY, the latter argued to be used as an upper limit rather than a management target (Worm et al., 2009; Froese and Proelß, 2010).

Translated into monetary terms, the figures in Table 15.1 have been estimated in some analyses to cost to the world economy in the order of 50 billion dollars per year in lost resource rent (World Bank and FAO, 2009). This implies that, the economic return from marine capture fisheries could be improved compared to the current situation. If other incentives such as subsidies of the fisheries sector are taken into account, there are some estimates that this global economic return amounts to *minus* 5 -12 billion dollars per year (World Bank and FAO, 2009; Munro, 2010; Sumaila et al., 2012). Some estimates of world fishery subsidies are in the order of 25-30 billion dollars per year (Sumaila, et al., 2010). Other estimates are of lower levels of subsidies (Cox and Schmidt, 2002). The differences may be largely due to definitional issues with regard to what is considered to be a subsidy in the different analyses.

This is not to say that all world capture fisheries are yielding negative economic returns. Clearly several capture fisheries are yielding positive, and in some cases large positive, net economic returns. From a global perspective, however, the positive returns from these fisheries are more than offset by those yielding negative net economic returns. No clear divide between developed and developing fishing States is observed. (Sumaila et al., 2012, p.3).

From an economic standpoint, the extent of the capture fishery's resource depletion shown in Table 15.1, which was due to the rapid expansion of the world capture fishing industry over several decades, involved the running down of world's stock of the capture fishery's natural capital.

Rebuilding capture fishery resources requires reducing harvests below the net growth rates of the fish stock. As the resources grow, potential

Table 15.1 | Status of World Marine Capture Fishery Resources 2011. Source: FAO, 2014, p.7.

Status	Percentage
Less than fully exploited	10
Fully exploited	61
Overexploited	29

resource rent can be expected to emerge, which must go unrealized in all or in part, if the resource investment is to continue – hence the cost. Using a 50-year time horizon, Sumaila et al. (2012) estimate that after 12 years of resource investment, the net economic returns from the investment would begin to outweigh the costs. Over the 50-year period, the returns would far outweigh the costs (Sumaila, et al., 2012). Economic and technical considerations that arise in rebuilding fisheries were explored in additional detail in an Organisation for Economic Co-operation and Development workshop (OECD, 2012).

3 Issues in Regulation of Marine Capture Fisheries

It has now long been recognized that the inherent difficulties in regulating marine capture fishery resources are a problem of scope and management objectives in the decision-making process, and are often framed as the well-known "Tragedy of the Commons" (Hardin, 1968). When access is open to all for exploitation, incentives are created that promote inefficiencies, including: (1) loss of economic "rent" because of the "race to fish", (2) high transaction and enforcement costs incurred to reduce overuse and (3) low productivity, because no one has an incentive to work hard in order to increase their private returns (Ostrom, 2000). All of these factors reduce the net economic return from fisheries. The management of common property requires a minimum set of rules, defining access conditions and conservation measures to ensure sustainability and economic returns.

Where social, economic, and governance circumstances allow effective management of entry into a fishery and effort by those allowed to participate, substantial progress can be made at improving both the ecological and economic performance of a fishery, but often at the cost of few people receiving employment. On the west coast of Canada, for example, a move to Individual Transferrable Quotas in a complex, multispecies fishery for rockfish (*Sebastes* spp) resulted in improved stock status for the entire complex, and particularly reduced catches of the stocks most in need of reduced fishing mortality, while improving economic returns to the fishery. However, the fleet size and employment dropped by nearly half from the period before the programme was introduced (Rice, 2003; Branch, 2006; Branch and Hilborn, 2008).

In the context of fisheries, management efforts also need to take into consideration how the legitimacy of rules and regulations may be perceived differently when applied to large- vs. small-scale. The majority of the world's fisheries comprise small-scale, multi-species, multi-gear, commercial fishing vessels, operating in all bodies of water (inland, brackish and marine), both near urban centres and in remote areas. Their operation involves family members, in pre-harvest, harvest and post-harvest parts of the fish chain. Women and children often participate in the fisheries. Small-scale fisheries catches are landed relatively close to where fishing occurs and are distributed through various channels. A certain portion is generally sold to local markets or to intermediaries by family members and some remains for household consumption. These characteristics of the fisheries imply that they require different management approaches than large-scale, industrialized fisheries. As at least half of the world's fish catches derive from small-scale fisheries, success in fisheries management needs to be demonstrated, not only where large-scale fisheries dominate, but also in the small-scale sector, with its high potential to address global food security.

Community-based resource management has been shown to be effective in establishing fishery rules (Berkes, 2005). Cinner and Aswani (2007), however, found that customary management was effective in smaller, remote communities with high levels of equality, but it is susceptible to economic pressures and by fishermen who do not practice customary fishing traditions.

4 Impacts of Illegal, Unreported, and Unregulated (IUU) fishing

There are additional economic and social considerations related to IUU fishing (see also Chapter 11). It is a complex phenomenon involving vessel owners, vessels, crew, flag State authorities and logistics. Often IUU vessels are related, through ownership, to authorized vessels obtaining cover to sell their catches.

Marine Resources Assessment Group (2005) states that the most obvious impact of IUU fishing is direct loss of the value of the catches that could be taken by the coastal State if the IUU fishing was not occurring. This is mostly from vessels operating without licences and licensed vessels misreporting catches (quantity, species, fishing area, etc.) and illegal trans-shipment of catches. Secondary economic impacts from the loss of fish to IUU vessels may include reduced revenue from seafood exports and reduced employment in the harvest and postharvest sectors. Reduced fishing port activity has a ripple or multiplier effect across economies, adversely affecting labour and transportation as well as the manufacturing sector.

IUU fishing may also increase poverty and reduce food security and food sovereignty. Conflict between authorized, compliant vessels and IUU vessels is common in some fisheries and can become violent with threats to both life and livelihoods on a large scale. Armed resistance to surveillance and enforcement is increasing in some locations with the potential to undermine all monitoring, control and surveillance (MCS) as resources are allocated to address what may be seen as a threat to national security rather than fisheries management. It can be noted that conflicts and IUU fishing generally occur between vessels of any size. There may also be gender and socio-cultural effects, depending upon the composition of the harvest and post-harvest labour forces.

5 Space-use conflicts: industrial capture fisheries vs. artisanal capture fisheries; aquaculture vs. artisanal capture fisheries

Due to recent improvements in technology and affordability, vessel monitoring systems (VMS) are increasingly available for both large- and small-scale fishing vessels, and thus can provide geo-referenced data that accurately describe fishing areas on geographic scales applicable to MSP. Combined with validated logbook data, rich time-series data are potentially available from intensely fished and monitored sea areas in developed countries. The data situation is slowly improving in developing countries. Land tenure systems that extend to parcels of seabed and water for aquaculture also provide clear boundaries. Superimposed on these spaces are increasingly sophisticated layers of information on the interactions among fisheries, and between aquaculture and fisheries. Although not all fisheries conflicts concern spatial use, or can be managed through MSP, many are potential candidates for spatial conflict management.

Sources of conflict between large and small-scale fisheries are a well-reported concern (FAO, 2014). Spatial components of conflict concern:

- Sea tenure and territorial use rights
- Fishery resource allocations by site
- Fishing gear and method interactions
- Ecosystem (species) interactions
- IUU fishing (several aspects)
- Port access and market transactions
- Management jurisdiction and governance

Sources of conflict between fisheries and aquaculture with spatial components concern:

- Sea tenure and territorial use rights
- Natural resource allocations by site
- Fishing interactions with infrastructure
- Ecosystem (species) interactions
- Area access and market transactions
- Management jurisdiction and governance

The lists are quite similar, although the specific nature of the conflicts varies greatly between the lists and site-specific situations. The next section looks more closely at fisheries-aquaculture conflicts (see also Chapter 12).

Cataudella et al. (2005) note that the FAO (1995) Code of Conduct for Responsible Fisheries (CCRF) defines the global framework in which marine aquaculture and capture fisheries are to be considered as interactive parts of the same system. The assessment of such interactions is crucial for implementing the CCRF, especially in areas where the use of the coastal zone results in conflicts between many resource users competing for space (e.g. fisheries, aquaculture, tourism, shipping, energy). The CCRF treats aquaculture as an important part of the fisheries system to be responsibly developed and managed for sustainability (FAO, 1999), but in the nearly two decades that have intervened, this has proven to be challenging.

The relationships between marine aquaculture and capture fisheries can be complex, operating at multiple levels of governance and crossing several spatial and temporal scales, affecting different points along value chains, as well as ecosystems or target and culture species in a variety of ways. Cataudella et al. (2005) categorize the conflict interactions as old and new, somewhat based arbitrarily on the currency of the topic.

Old interactions are issues generated by the:

- Allocation of public financial resources
- Likelihood of disease spreading and new outbreaks
- Environmental pollution
- Employment threats and opportunities
- Introduction of exotic or invasive species
- Need for stocking programmes
- Ownership of resources and of confined environments
- Use of wild seed to supply aquaculture
- Use of fishery products to supply the fish-feed farming industry.

New interactions are issues concerning the:

- Stocking and restocking models
- Genetic origin of cultured organisms
- Biodiversity conservation and value
- Genetic improvement through breeding programmes and genetic engineering
- Development of aquaculture in sensitive environments
- Direct impact of farmed products on markets and prices
- Growing role of aquaculture in meeting the demand for fishery products
- Product quality and labelling
- Feasibility of capture fisheries and aquaculture within a sustainable system.

The above interactions are most in need of conflict management through legislation and policy related to planning for integrated coastal zone management and marine spatial planning. However, considerable guidance is available on appropriate approaches that include conflict management (e.g. Ehler and Douvere, 2009) as well as enabling policy (e.g. EU Marine Strategy Framework Directive).

Marine spatial planning (MSP) is the public process of analyzing and allocating the spatial and temporal distribution of human activities in marine areas to achieve ecological, economic, and social objectives that are usually specified through a political process (Ehler and Douvere, 2006). It is linked to ecosystem-based management (EBM) (see McLeod and Leslie, 2009), the ecosystem approach to fisheries (EAF) (see FAO, 2003), marine protected areas (MPAs) (FAO report on MPAs and Fisheries, 2011) and similar endeavours that have the potential to assist in managing conflicts through participation among diverse stakeholders (Ehler and Douvere, 2009). Managing space use conflicts between large- and small-scale fisheries and with other sectors is an increasingly important issue in many parts of the world.

6 Gender in fisheries

On a global level, fisheries are often perceived as male-dominated, laden with culturally stereotypical images of fishermen. The term "fishing industry", for example, conjures an image that focuses attention on harvest and men's work more than the term "seafood industry" which is more equitable (Aslin et al., 2000). The involvement of women is now reflected by the increasing use of gender-neutral terms such as "fisher" and "fisherfolk", and more international discussion of gender (Williams et al., 2005). Yet recent global investigation has shown that if post-harvest (e.g., fish processing and trade) and ancillary activities (e.g., fishing inputs and financing) are taken into account, then the gendered image is quite different. Overall, women may be in the majority in fisheries, or nearly so (FAO et al., 2008). This does not take into account the growing number of women engaged worldwide in fisheries policy, planning, management, science, education, civil society advocacy and other activities related to fisheries that were previously more male-dominated.

The post-harvest situation is particularly inequitable. Women outnumber men in fish processing and trading across the world, but their informal sector activities are often not recorded, and they are invisible in national labour and economic statistics. Thus the socioeconomic contribution of women to fisheries is underestimated at national and global levels. Only a few countries in the developing world collect and use gender-disaggregated statistical data and other information data for fisheries policy and planning (Weeratunge and Snyder, 2009). Without comparative data for women and men, it is difficult in most places to determine the disparity between female and male socioeconomic activities and well-being. This scarcity of gender-disaggregated fisheries data constrains gender-sensitive policies and mainstreaming, with little action taken to address the disadvantageous position of women (Sharma, 2003).

It is widely accepted in the developing world that women strongly influence the social, economic and cultural aspects of fishing households and the industry as a whole. There are increasing numbers of women in technical, scientific and managerial fisheries jobs around the world, but this varies markedly by region. In some societies where men engage in the most conspicuous fisheries-related socioeconomic and political activities, the women are labelled "fisher wives", but the implied subordination is misleading (Weeratunge and Snyder, 2009). In Ghana, "fisher wives" or "fish mammies" support the entire small-scale fishing industry as they invest in fishing boats and gear, and provide loans to husbands and other fishers while running small socioeconomic empires without formal political power (Walker, 2001). Although addressing gender-inequity is critical, interventions need to be carefully designed. 'Women in development' projects have contributed to reducing the real power that women held, for example, by introducing poorly designed credit and fish marketing schemes that exacerbate unsustainable fishing for short-term monetary gain or loan servicing.

Small-scale fisheries in developed and developing countries have striking similarities. In both, gender issues are often overlooked or misunderstood because of an analytical focus that looks at the fisheries sector in isolation from the broader society, and is concerned primarily with narrow ecological and economic factors such as maintaining fish stocks to ensure a viable long-term harvest. Interventions have been directed more at men harvesting at sea, rather than at women engaged in post-harvest on shore, or at the interconnections between harvest and post-harvest (Weeratunge and Snyder, 2009). Although this narrow, male sectoral perspective is changing as the EAF becomes more widely adopted (FAO 2003), gender is not yet mainstreamed into this approach despite advances in incorporating other social, cultural and institutional dimensions (De Young et al, 2008). EAF is just one facet of the changing face of fisheries governance. Gender issues are more appropriately considered in the wider context of fisheries governance than fisheries management.

Gender remains a key governance issue in both developed and developing countries. Its many interconnected dimensions relate to vulnerabilities, assets, opportunities, capabilities, coping strategies, outcomes, food security, empowerment and more. With new attention to sustainable development goals based on blue and green economies, gender in fisheries should feature more prominently. State and civil society agencies realize that well-being will not be improved and poverty will not be reduced if gender is not adequately addressed. Gender mainstreaming should be an integral part of fisheries, but this is not occurring, because gender research to support fisheries policy is insufficient. As the links between gender in fisheries and poverty, climate, health and other major developmental issues become apparent (Bene and Merten, 2008; Bennett, 2005; FAO, 2006; Neis et al., 2005), more attention will need to be paid to gender in fisheries in the context of the development post-2015 agenda.

Certain issues, particularly at the micro level, demand additional research. The state of small-scale fisheries throughout the world, and gender issues in fisheries are particularly prominent. A further issue that has been seriously under-researched is that of the relationship between capture fisheries and aquaculture.

7 Climate change and small-scale fisheries

Pollution, environmental degradation, climate change impacts and natural and human-induced disasters pose serious challenges to fisheries sustainability. Because of the heavy reliance on fisheries for food security, employment and livelihoods, these factors become additional threats facing small-scale fishing communities (FAO, 2011-2015).

Expected impacts of climate change include increase in the severity and intensity of natural disasters and changes in the local distribution and abundance of harvested fish and shellfish populations (Barange et al., 2014), with consequences on the post-harvest and trade (FAO, 2011-2015; HPLE, 2014). Impacts of climate change are predicted to be more severe where the relative importance of fisheries to national economies and diets is higher and there is limited societal capacity to adapt to potential impacts and opportunities (Allison et al., 2009). The severity of threats increases due to combined effects of climate change and ecosystem degradation and overfishing, highlighting the importance of appropriate co-management measures (HPLE, 2014).

A comprehensive understanding of how communities respond to these threats and other global change, in their environmental, social and political contexts, is required (Bundy et al., 2015). These issues are also treated in the Summary (under Impacts of the Climate Changes).

8 Specific additional issues raised in regional workshops for the World Ocean Assessment

Fisheries management requires time-consuming and dedicated human resources and failure to meet or prioritize these efforts is a widespread problem, leading to poor fisheries management. During the regional workshops for this World Ocean Assessment it became apparent that lack of data, including difficulties in maintaining data collection and conducting stock assessments, as well as obtaining fishery-independent data, was an issue for all developing countries. Problems with databases and data integration, due to different methods of data collection and lack of long time-series, were raised in all regions. Lack of data on the small-scale, as well as recreational fisheries, was a problem in developed and developing States. In particular, catches from subsistence fishing are often missing from national catch statistics, leaving a gap in the ecological, social and economic aspects of fisheries. Ecosystem-based management is seldom applied due to the lack of practical examples and applications, and difficulties in assessing ecosystem impacts.

Fish is one of the most internationally-traded foods. This has an impact on the infrastructure needed to commercialize the product, especially given the fact that fish is a perishable commodity. The difficulties to adapt to international-market requirements - including means to abide by regulations - and the lack of fish preserving and processing facilities was a recurring issue, especially in developing countries that are near, or trade often with, developed countries.

Contamination of fish products as well as the effects on catches caused by pollution and habitat degradation were raised at the workshops. Developing countries reported difficulties in assessing those risks and monitoring those impacts. The main focus of fish certification has been eco-labelling that addresses environmental sustainability issues. With limited exceptions, certification concerns predominantly developed countries and large-scale fisheries. Fish certification is progressively moving to include social responsibility and labour considerations, but it is unclear whether food security and nutrition considerations can or will be included in future.

9 Conclusion

Fisheries around the world are deeply embedded in the issues of food and economic security, livelihoods for large numbers of people, gender equity and poverty alleviation. Both large and small-scale fishery operations provide essential economic and social benefits to society. Small-scale fisheries, in particular, constitute half of the world's total catches and involve more than 90 per cent of total fishing population (in harvest and post-harvest activities). The significant contribution to food security, livelihoods and local economic development means that small-scale fisheries can no longer be overlooked. Instead, management and governance of fisheries needs to incorporate key features distinguishing small-scale fisheries from their large-scale counterpart. This implies changes in information systems, fisheries assessment, monitoring and surveillance, and research and development. Importantly, issues related to fishing rights, tenure and access to resources, health and safety, gender and social justice, among others, deserve special attention in policy and decision-making. Finally, it is worth noting that small-scale fisheries governance would have different priorities, focusing for instance on stakeholder participation and subsidiarity principles. Tension and conflicts between different scales of operations, and with other marine activities, will continue to challenge policy-makers in many areas. They can be overcome, however, with an attempt to create policy coherence through a holistic and integrated approach to fisheries governance. During the regional workshops the need to improve the capacity of States to more effectively manage these critical resources, and in particular in regions where sustainability of fisheries needs to be improved, was recognized. The need to build capacity is also essential to address issues of equity and broader sustainable development efforts.

References

Allison, E.H., Perry, A. L., Badjeck, M.-C., Adger, W. N., Brown, K., Conway, D., Halls, A.S., Pilling, G.M., Reynolds, J.D., Andrew, N.L., and Dulvy, N.K. (2009). *Vulnerability of national economies to the impacts of climate change on fisheries*. Blackwell Publishing Ltd, FISH and FISHERIES. Available from http://www.uba.ar/cambioclimatico/download/Allison%20et%20al%202009.pdf. Accessed on: 14 July, 2015.

Asche, F., Bellemare, M.F., Roheim, C., Smith, M.D., & Tveteras, S. (2015). Fair Enough? Food Security and the International Trade of Seafood. *World Development*, 67, 151-160.

Aslin, H.J., Webb, T. and Fisher, M. (2000). *Fishing for Women: Understanding Women's Roles in the Fishing Industry*. Canberra: Bureau of Rural Sciences.

Barange, M., Merino, G., Blanchard, J. L., Scholtens, J., Harle, J., Allison, E.H., Allen, J.I., Holt, J., and Jennings, S. (2014). Impacts of climate change on marine ecosystem production in societies dependent on fisheries. *Nature Climate Change* 4: 211-216. Available from http://www.nature.com/nclimate/journal/v4/n3/full/nclimate2119.html?message-global=remove. Accessed on: 14 July, 2015.

Bene, C. and Merten, S. (2008). Women and Fish-for-Sex: Transactional Sex, HIV/AIDS and Gender in African Fisheries. *World Development* 36: 875-899.

Bennett, E. (2005). Gender, Fisheries and Development. *Marine Policy* 29: 451-459.

Berkes, F. (2005). Why keep a community-based focus in times of global interactions? Keynotes of the 5th International Congress of Arctic Social Sciences (ICASS), University of Alaska Fairbanks, May 2004. Topics in *Arctic Social Sciences* 5: 33-43.

Branch, T.A. (2006). Discards and revenues in multispecies groundfish trawl fisheries managed by trip limits on the US West Coast and by ITQs in British Columbia. *Bulletin of Marine Science* 78: 669-689.

Branch, T.A. and R. Hilborn (2008) Matching catches to quotas in a multispecies trawl fishery: targeting and avoidance behavior under individual transferable quotas. *Canadian Journal of Fisheries and Aquatic Sciences* 65: 1435-1446.

Bundy, A., Chuenpagdee, R., Cooley, S. R., Defeo, O., Glaeser, B., Guillotreau, P., Isaacs, M., Mitsutaku, M. and Perry, I. (2015). A decision support tool for response to global change in marine systems: the IMBER-ADApT framework. *Fish and Fisheries*. DOI: 10.1111/faf.12110.

Cataudella, S., Massa, F., and Crosetti, D., eds. (2005). *Interactions between Aquaculture and Capture Fisheries: A Methodological Perspective*. FAO, Studies and Reviews. General Fisheries Commission for the Mediterranean; 78. Rome.

Chuenpagdee, R. (2011). A matter of scale: prospects in small-scale fisheries. In Chuenpagdee (ed.). *Contemporary Visions for World Small-Scale Fisheries*. Eburon, Delft.

Cinner, J. and Aswani, S. (2007). Integrating Customary Management into Marine Conservation. *Biological Conservation* 140 (3/4): 201-216.

Cox, A. and Schmidt, C.-C. (2002). *Subsidies in the OECD Fisheries Sector: A review of recent analysis and future directions*. Background paper for the FAO expert consultation on identifying, assessing and reporting on subsidies in the fishing industry; 3.6.

De Young, C., Charles, A., and Hjort, A. (2008). Human Dimensions of the Ecosystem Approach to Fisheries: An Overview of Context, Concepts, Tools and Methods. *FAO Fisheries Technical Paper*. No. 489. Rome.

Ehler, C., and Douvere., F. (2006). *Visions for a Sea Change*. UNESCO, Intergovernmental Oceanographic Commission. Paris.

Ehler, C., and Douvere., F. (2009). *Marine Spatial Planning: A Step-by-Step Approach Toward Ecosystem-Based Management*. UNESCO, Intergovernmental Oceanographic Commission, Paris.

FAO (1995). *Code of Conduct for Responsible Fisheries*. FAO, Rome.

FAO (1997). *Review of the State of World Fishery Resources: Marine Fisheries*. FAO Fisheries Circular No. 920, Rome.

FAO (1999). *Indicators for Sustainable Development of Marine Capture Fisheries*. FAO Technical Guidelines for Responsible Fisheries, Rome.

FAO (2003). *The Ecosystem Approach to Fisheries*. FAO Technical Guidelines for Responsible Fisheries. No. 4, Suppl. 2, Rome.

FAO (2006). *Gender Policies for Responsible Fisheries*. FAO, Rome. (*See also* links from http://www.fao.org/fishery/topic/16605/en).

FAO (2010). *The State of World Fisheries and Aquaculture 2010*. FAO, Rome.

FAO (2011). *Marine protected areas and fisheries*. FAO Technical Guidelines for Responsible Fisheries, No. 4, Suppl. 4, Rome.

FAO (2012). *The State of World Fisheries and Aquaculture 2012*. FAO, Rome.

FAO (2014). *The State of World Fisheries and Aquaculture 2014*. FAO, Rome.

FAO (2011-2015). Small-scale fisheries - Web Site. Voluntary Guidelines on Securing Sustainable Small-Scale Fisheries [SSF Guidelines]. FI Institutional Websites. In: *FAO Fisheries and Aquaculture Department* [online]. Rome. Updated 15 April 2014. http://www.fao.org/fishery/ssf/guidelines/en

FAO, WorldFish, and World Bank (2008). *Small-scale Capture Fisheries - A Global Overview with Emphasis on Developing Countries: A Preliminary Report of the Big Numbers Project*, Rome and Penang.

Froese, R., and Proelß, A. (2010). Rebuilding fish stocks no later than 2015: will Europe meet the deadline? *Fish and fisheries* 11.2: 194-202.

Hardin, G. (1968). The Tragedy of the Commons. *Science*, Vol. 162 no. 3859: 1243-1248.

HLPE, 2014. Sustainable fisheries and aquaculture for food security and nutrition. A report by the High Level Panel of Experts on Food Security and Nutrition of the Committee on World Food Security, Rome 2014.

Kronbak, L.G., and Lindroos, M. (2006). An Enforcement-Coalition Model: Fishermen and Authorities Forming Coalitions. *Environmental and Resource Economics* 35: 169-194.

Marine Resources Assessment Group Ltd (2005). *Review of Impacts of Illegal, Unreported and Unregulated Fishing on Developing Countries*, Synthesis Report, prepared by MRAG for the UK's Department for International Development (DFID), with support from the Norwegian Agency for Development Cooperation (NORAD).

McLeod, K.O. and Leslie, H. (eds). (2009). *Ecosystem-based management for the oceans*. Island Press, Washington, D.C. USA. 392p.

Munro, G. (2010). *From Drain to Gain in Capture Fishery Rents: A Synthesis Study*. FAO Fisheries and Aquaculture Technical Paper No. 538, Rome.

Munro, G. (2011). On the Management of Shared Living Marine Resources, *Proceedings of the Danish Conference on Environmental Economics 2011*. Available from: http://www.dors.dk/graphics/SynkronLibrary/Konference%20201/Abstracts/Munro_paper.pdf. Accessed on: 6 August, 2014.

Munro, G., Van Houtte, A., and Willmann, R. (2004). *The Conservation and Management of Shared Fish Stocks: Legal and Economic Aspects*. FAO Fisheries Technical Paper No. 465, Rome.

Munro, G., and Sumaila, U.R. (2011). *On the Curbing of Illegal, Unreported and Unregulated (IUU) Fishing*. In: Tubiana, L., Jacquet, P., and Pachauri, R., editors. *A Planet for Life 2011 – Oceans*. IDDRI, Paris.

Munro, G., Sumaila, U.R., and Turris, B. (2012). Catch Shares, the Theory of Cooperative Games and the Spirit of Elinor Ostrom: A Research Agenda. *IIFET 2012 Conference Proceedings*. Available at: https://ir.library.oregonstate.edu/xmlui/bitstream/

handle/1957/33889/Paper%20for%20Dar%20es%20Salaam.pdf?sequence=1. Accessed on: 6 August, 2014.

Munro, G., Turris, B., Kronbak, L., Lindroos, M., and Sumaila, U.R. (2013). *Catch Share Schemes, the Theory of Dynamic Coalition Games and the Groundfish Trawl Fishery of British Columbia.* Paper presented to the North American Association of Fisheries Economists Conference, 2013.

Neis, B., Binkley, M., Gerrard, S. and M. Maneschy (eds.). 2005. *Changing Tides: Gender, Fisheries and Globalisation.* Halifax: Fernwood

OECD (2012). *Rebuilding Fisheries: The Way Forward.* OECD, Paris.

Ostrom, E. (2000). Private and common property rights. In Brouckaert, B., and De Geest, G., editors. *Encyclopedia of Law and Economics, Vol.II: The History and Methodology of Law and Economics.* Cheltenham, UK: Edward Elgar: pp. 332-379.

Rice, J. (2003) The British Columbia rockfish trawl fishery. In *Report and documentation of the International Workshop on Factors of Unsustainability and Overexploitation in Fisheries, Mauritius, 3–7 February 2003.* Edited by J. Swan and D. Gréboval. FAO, Rome, Italy.

Sharma, C. (2003). The Impact of Fisheries Development and Globalization Processes on Women of Fishing Communities in the Asian Region. *APRN Journal* 8:1-12. ICSF: Chennai.

Sumaila, U.R. (2013). *Game Theory and Fisheries: Essays on the Tragedy of Free For All Fishing.* London, Routledge.

Sumaila, U.R., Marsden, D., Watson, R., and Pauly, D. (2007). Global Ex-vessel Fish Price Database: Construction and Applications. *Journal of Bioeconomics* 9: 39-51.

Sumaila, U.R., Teh, L., Watson, R., and Munro, R. (2010). A Bottom-up Re-estimation of Global Fisheries Subsidies. *Journal of Bioeconomics* 12: 201-225.

Sumaila, U.R., Cheung, W., Dyck, A., Gueye, K., Huang, L., Lam, V., Pauly, D., Srinivasan, T., Swartz, W., Watson, R., and Zeller, D. (2012). Benefits of Rebuilding Global Marine Fisheries Outweigh Costs. *PLoS ONE* 7: e40542. DOI:10.1371/journal.pone.0040542.

Swartz, W., Sumaila, U.R., and Watson, R. (2013). Global Ex-vessel Price Database Revisited: A New Approach for Estimating "Missing" Prices. *Environmental and Resource Economics* 56: 467-480. DOI: 10.1007/s10640-012-9611-1.

Taylor, W., Schechter, M.G., and Wolfson, L.G. (2007). *Globalization: Effects on Fisheries Resources.* Cambridge University Press.

Walker, B.L.E. (2001). Sisterhood and Seine-nets: Engendering Development and Conservation in Ghana's Marine Fishery. *Professional Geographer* 53: 160-177.

Weeratunge, N., and Snyder., K. (2009). *Gleaner, Fisher, Trader, Processor: Understanding Gendered Employment in the Fisheries and Aquaculture Sector.* Paper presented at the FAO-IFAD-ILO Workshop on Gaps, Trends and Current Research in Gender Dimensions of Agricultural and Rural Employment: Differentiated Pathways out of Poverty, Rome, 31 March - 2 April 2009.

Williams, M.J., Nandeesha, M.C., and Choo, P.S. (2005). Changing Traditions: First Global Look at the Gender Dimensions of Fisheries. *NAGA, Worldfish Center Newsletter;* vol. 28, No. 1 & 2 (January and June).

World Bank (2005). Hamilton, K., Ruta, G., Bolt, K., Markandya, A., Pedroso-Galinato, S., Silva, P., Ordoubadi, M.S., Lange, G., and Tajibaeva, L. *Where Is the Wealth of Nations? Measuring Capital for the 21st Century.* World Bank, Washington.

World Bank (2011). *The Global Program on Fisheries: Strategic Vision for Fisheries and Aquaculture.* World Bank, Washington.

World Bank and FAO (2009). Kelleher, K., Willmann, R., and Arnason, R., eds. *The Sunken Billions: The Economic Justification for Fisheries Reform.* World Bank and FAO, Washington.

WorldFish Center (2008). "Small-scale capture fisheries: a global overview with emphasis on developing countries: a preliminary report of the Big Numbers Project." *The WorldFish Center Working Papers.*

Worm, B., Hilborn, R., Baum, J. K., Branch, T. A., Collie, J. S., Costello, C., Fogerty, M.J., Fulton, E.A., Hutchings, J.A., Jennings, S., Jensen, O.P., Lotze, H.K., Mace, P.M., McClanahan, T.R. Minto, C., Palumbi, S.R., Parma, A.M., Ricard, D., Rosenberg, A.A., Watson, R., Zeller, D. (2009). Rebuilding global fisheries. *Science, 325*(5940), 578-585

16 Synthesis of Part IV: Food Security and Safety

Group of Experts:
Andrew Rosenberg (Lead Member)

Fish products, including finfish, invertebrates and seaweeds, are a major component of food security around the world. In addition to providing a source of high-quality protein and critical long chain omega-3 fatty acids with well-known nutritional benefits in many countries, fish and fishery products are the major source of animal protein for a significant fraction of the global population, and in particular in countries where hunger is widespread. Even in the most developed countries, consumption of fish is increasing both *per capita* and in absolute terms, with implications for both global food security and trade.

Fisheries and aquaculture are a major employer and source of livelihoods in coastal States. Significant economic and social benefits result, including providing a key source of both subsistence food and much-needed cash for many of the world's poorest peoples. As a mainstay of many coastal communities, fisheries and aquaculture play an important role in the social fabric of many areas.

Small-scale fisheries, particularly those that provide subsistence in many poor communities, are often a key source of employment, cash, and food in coastal areas. Many such coastal fisheries are under threat due to over-exploitation, conflict with larger fishing operations, and loss of productivity in coastal ecosystems due to a variety of other impacts. These include habitat loss, pollution and climate change, as well as loss of access to space as coastal economies and uses of the sea diversify.

Globally, world capture fisheries are near the ocean's productive capacity with catches in the order of 80 million metric tons. Only a few means to increase yields are available. More effectively addressing sustainability concerns including ending overfishing, eliminating illegal, unreported and unregulated (IUU) fishing, rebuilding depleted resources, reducing broader ecosystem impacts of fisheries, and adverse impacts on them from pollution, are important aspects of improving fishery yields and thereby food security. For example, ending overfishing and rebuilding depleted resources may result in an increase of as much as 20 per cent in potential yield, if the transitional costs of rebuilding depleted stocks can be addressed.

In 2012, more than one-quarter of fish stocks worldwide were classified by FAO as overfished. Although these stocks clearly will benefit from rebuilding once overfishing has ended, other stocks may still be classed as fully fished despite being on the borderline of overfishing; these stocks could yield more if effective governance mechanisms were in place.

Current estimates of the number of overfished stocks do not take into account broader effects of fishing on marine ecosystems and their productivity. These impacts, including by-catch, habitat modification, and food web effects, are important elements in the sustainability of the ocean's capacity to continue to produce food and must be carefully managed. These very real threats endanger some of the most vulnerable populations and marine habitats around the world and need to be directly addressed to improve food security and answer other social needs.

Fish stock propagation may provide a tool to help rebuild depleted fishery resources in some instances. Propagation programs must be carefully designed and maintained in order to really benefit resource sustainability.

Fishing effort is subsidized by many mechanisms around the world and many of these subsidies undermine the net economic benefits to States. Subsidies that encourage over-capacity and overfishing result in losses for States and these losses are often borne by communities dependent upon fishery resources for livelihoods and food security.

Aquaculture production, including seaweed culture, is increasing more rapidly than any other source of food production in the world and is expected to continue to increase. Aquaculture, not including the culture of seaweeds, now provides half of the fish products covered in the global statistics.

Aquaculture and capture fisheries are co-dependent in some ways, as feed for cultured fish is in part provided from capture fisheries. They are also competitors for space in coastal areas, for markets, and potentially for other resources (labour, governmental support and attention, etc.). Significant progress has been made in replacing feed sources from capture fisheries with agricultural production (e.g., soybeans), although more work is certainly needed.

Aquaculture itself poses some environmental challenges, including potential pollution, competition with wild fishery resources, potential contamination of gene pools, disease problems, and loss of habitat (e.g., from the construction of shrimp ponds). Examples of these challenges, and measures that can mitigate them, have been observed worldwide and need to be directly addressed by management action.

In both capture fisheries and aquaculture, gender and other equity issues arise. A significant number of women are employed in both types of activities, either directly or in related activities along the value chain. Women are particularly prominent in product processing, but often their labour is not equitably compensated, and working conditions do not meet basic standards. Poor communities are often subject to poorer market access, unsafe conditions for labour, and other inequitable practices that need to be remedied.

The ongoing impacts of a changing climate, including ocean acidification, pose great challenges for fisheries and aquaculture. Climate change is already resulting in shifts in the distribution and productivity of fishery resources and marine ecosystems more generally. This impacts fishing businesses and communities, yields and food security. Changes in availability and yields for individual resources may be positive or negative but in any case result in greater uncertainty for fishers, communities, businesses and fishery governance frameworks.

There are major capacity-building needs with regard to food security and food safety.

Chapter 16 — Synthesis of Part IV: Food Security and Safety

- The complexity of the issues concerning food provisioning from the sea requires a multidisciplinary approach to research. While the fields of fishery and aquaculture science are well developed, there are critical needs for research on small-scale subsistence uses of the marine environment as well as recreational, cultural and spiritual aspects of marine resources. In addition, greater understanding must be developed of the structure, function and dynamics of marine ecosystems and of the economic and social aspects of human society that depend upon these resources.
- It is necessary to improve understanding of the role of fisheries and aquaculture in commerce, employment and the support livelihoods. Therefore advanced capacity building is necessary for appropriate skills to be able to use advanced technologies to create wealth from capture fisheries and aquaculture in a sustainable way.

1 Capture fisheries

- Efforts have been made to create awareness to reduce post-harvest losses, especially in small-scale fisheries, as a means of increasing production. However, little is known about what new methods are being implemented and to what extent they impact on production. There is a gap in capacities needed to develop, deploy, and evaluate approaches to reduce waste and post-harvest losses and ensure that new technology is transferred to those that need it most.
- Efforts have been made to reduce by-catch and other broader ecosystem impacts of fishing and to increase awareness of these problems. For example, globally it is still poorly known whether by-catch excluder devices have been successfully adopted in terms of the relative ratio of the target catch landed and the by-catch either landed or discarded. It is necessary to build capacity to monitor and ensure compliance with measures such as these that are intended to reduce ecosystem level impacts.
- If ecosystem-based approaches to management are to be implemented, integrating fisheries governance with governance of other marine sectors, greater scientific and technical capacity will be needed to inform the process.
- If further depletion of fishery stocks due to overfishing, climate impacts or other pressures is to be avoided, trends in fishing effort, landings, geographical scope, species composition and other key attributes must be ascertained and consistently monitored, and data must be made broadly available. It is necessary to build enough capacity with appropriate technological and scientific skills and the necessary equipment to provide adequate information and data to facilitate regional and global management.
- Technical capacity to monitor and control seafood safety is urgently needed. Methodologies must be shared and deployed and greater training in procedures that safeguard seafood supplies is necessary.

- Certain issues, particularly at the micro level, demand additional research and therefore need capacity-building to address them. The state of small-scale fisheries throughout the world, and gender issues in fisheries, are particularly prominent and are poorly studied. A further issue that has been seriously under-researched is the relationship between capture fisheries and aquaculture.

2 Aquaculture

- Much better data and analysis of the trends, character and factors influencing aquaculture production are needed. In principle these data should be more accessible than capture fisheries data but in practice this is not the case. Understanding this rapidly growing sector is vital to the understanding of food security patterns and needs.
- Disease and product safety are a key challenge for aquaculture. Greater scientific and technical capacity is needed to address these challenges in many countries and data and scientific information must be shared in order to exchange lessons learned.
- Aquaculture technology crosses the spectrum from relatively simple small-scale operations to larger-scale enterprises. It includes breeding, feeding, health and safety aspects. Sharing both technology and approaches to improve efficiency and sustainability is an important aspect of improving food security and safety.

3 Fish stock propagation

- For propagation efforts to be successful, capacity must be developed that will promote efficient and effective approaches and comprehensive monitoring of these efforts. These must be well designed experiments that rely on lessons learned from other efforts around the globe.
- Proposed propagation efforts will benefit from a comprehensive, integrated, ecosystem-based fisheries-management approach. Capacity is needed in terms of individuals, infrastructure and institutions to deliver effective stock propagation.

4 Seaweeds as a resource

- Seaweed aquaculture is seriously affected by disease, as with other forms of intensive aquaculture. Research on seaweed diseases and new techniques for combating the diseases are needed along with the technical capacity to deploy new methods.
- Undertaking and building capacity for biochemical research on seaweed extracts from various species will enable them to be harnessed for their wide variety of nutrient, medicinal and food values.

V Assessment of Other Human Activities and the Marine Environment

Shipping

Contributors:
Alan Simcock (Lead member) and Osman Keh Kamara (Co-Lead member)

Shipping   Chapter 17

1 Introduction

For at least the past 4,000 years, shipping has been fundamental to the development of civilization. On the sea or by inland waterways, it has provided the dominant way of moving large quantities of goods, and it continues to do so over long distances. From at least as early as 2000 BCE, the spice routes through the Indian Ocean and its adjacent seas provided not merely for the first long-distance trading, but also for the transport of ideas and beliefs. From 1000 BCE to the 13th century CE, the Polynesian voyages across the Pacific completed human settlement of the globe. From the 15th century, the development of trade routes across and between the Atlantic and Pacific Oceans transformed the world. The introduction of the steamship in the early 19th century produced an increase of several orders of magnitude in the amount of world trade, and started the process of globalization. The demands of the shipping trade generated modern business methods from insurance to international finance, led to advances in mechanical and civil engineering, and created new sciences to meet the needs of navigation.

The last half-century has seen developments as significant as anything before in the history of shipping. Between 1970 and 2012, seaborne carriage of oil and gas nearly doubled (98 per cent), that of general cargo quadrupled (411 per cent), and that of grain and minerals nearly quintupled (495 per cent) (UNCTAD, 2013). Conventionally, around 90 per cent of international trade by volume is said to be carried by sea (IMO, 2012), but one study suggests that the true figure in 2006 was more likely around 75 per cent in terms of tons carried and 59 per cent by value (Mandryk, 2009). Not only has the quantity of cargo increased, the average length of voyages has also increased: between 2000 and 2013 the estimated amount of international seaborne shipments measured in ton miles increased by 65 per cent from 30,648 to 50,506 billion ton miles, while the total amount of international cargo rose by only about 50 per cent (UNCTAD, 2013). This growth in the average length of voyages has been largely in the carriage of coal, grain and ores.

2 Nature and Magnitude of World Shipping Movements

2.1 Cargo traffic

Global shipping movements naturally mirror the world economy. The modern period up to 2008 therefore generally showed a steady increase. The economic crisis of 2008, not surprisingly, produced a drop in activity, but this was less than the drop in the world's Gross Domestic Product, largely because of the continuing demand in eastern Asia for bulk movement of iron ore and coal (UNCTAD, 2013). Figure 17.1 shows the way in which world cargo movements are increasing. Figure 17.2 shows the distribution of total shipping movements around the world. The different main trades have substantially different distributions and patterns of sailings: the container routes are concentrated in the East/West belt around the southern part of the northern hemisphere and are very regular in their sailings, while both the five main bulk dry cargoes (iron ore, coal, grain, bauxite/alumina and phosphate rock) and the oil and gas trade are focused on the sources of these cargoes. Their sailings are also affected by changes in the market prices for these commodities. The carriage of bulk dry cargoes and oil and gas tends to have a higher pro-

Figure 17.1 | International Seaborne Trade: selected years 1980 – 2013. Millions of tons loaded. The "Five Major Bulks" are iron ore, grain, coal, bauxite/alumina and phosphate rock. "Other Dry Cargo" includes agricultural produce, metals, and forest products). Source: UNCTAD, 2013.

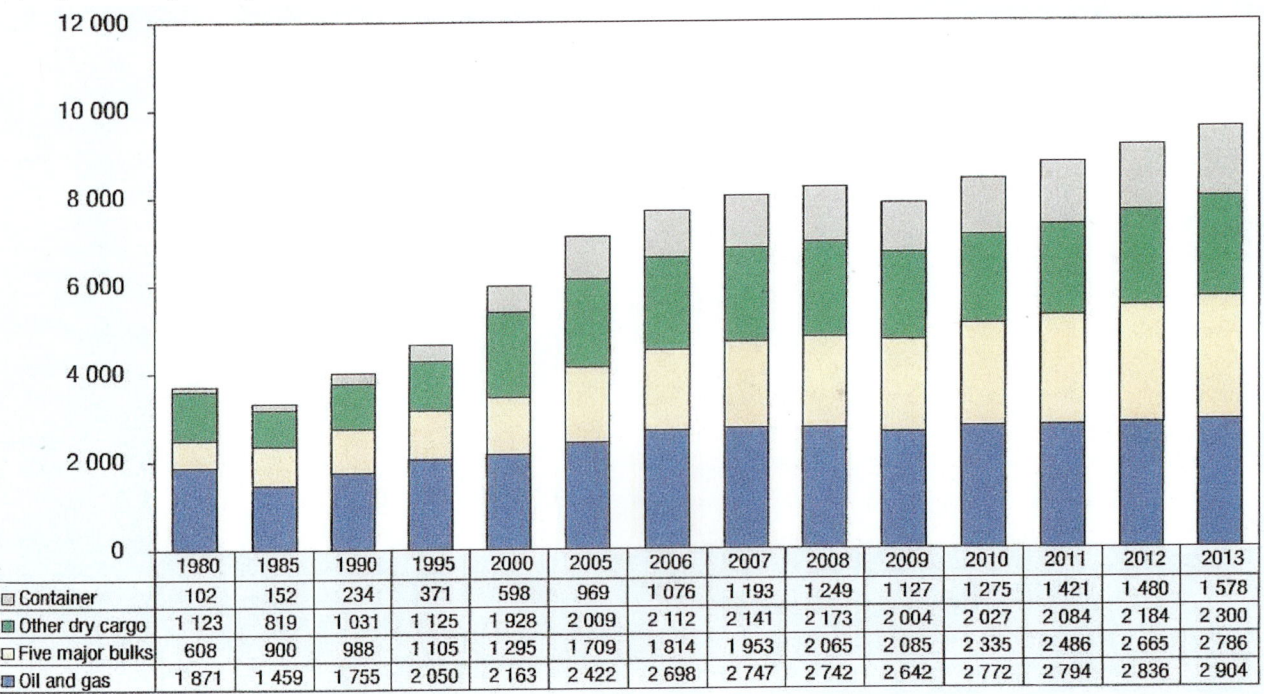

246

Chapter 17 — Shipping

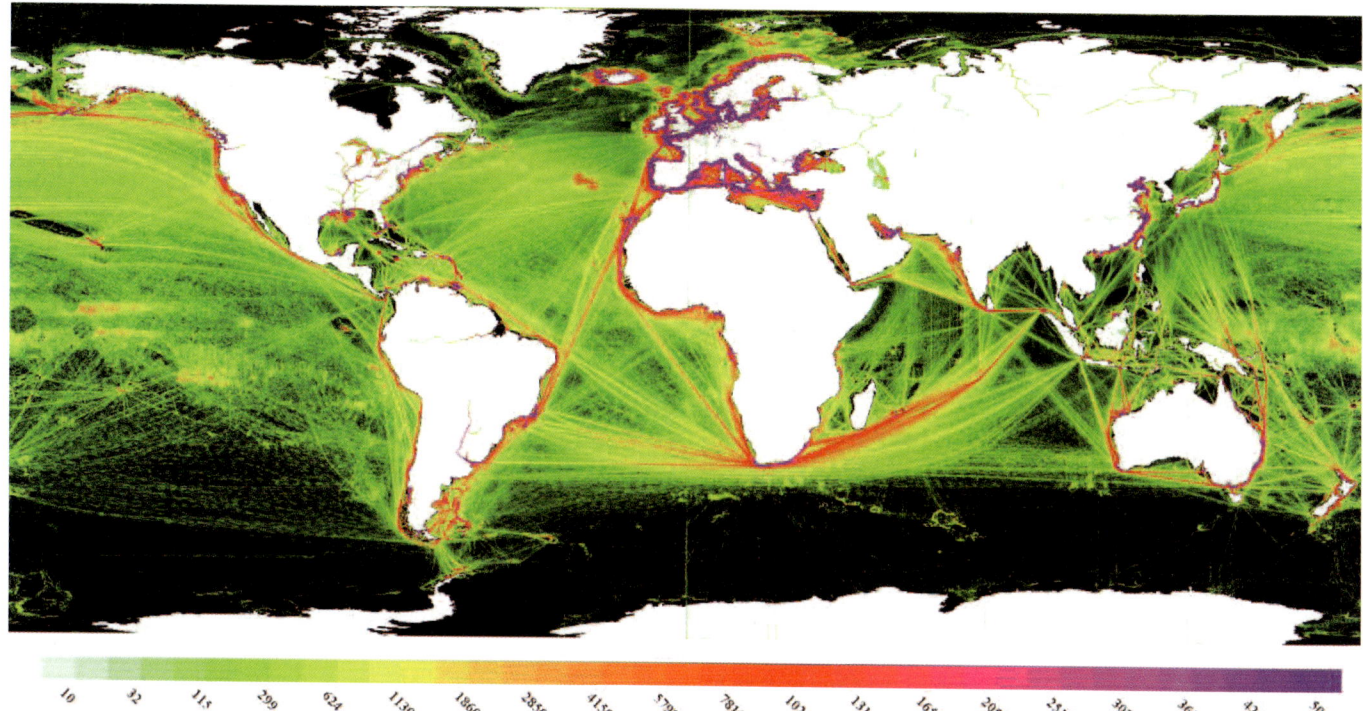

The boundaries and names shown and the designations used on this map do not imply official endorsement or acceptance by the United Nations.

Figure 17.2 | Global Network of Ship Movements (data 2012). Data derived from daily Automatic Identification System (AIS) messages recorded for each 0.2° X 0.2° grid square. The coloured scale shows the number of messages recorded over the year for the grid squares. Source: IMO, 2014o.

portion of return journeys in ballast. The mineral cargoes, in particular, have strong emphases on routes from Africa, South America, Australia and Indonesia to eastern Asia (Kaluza et al., 2010). Significant changes in maritime traffic routes could result from developments in extracting hydrocarbons from the earth: the growth of the shale gas industry of the United States of America, for example, is leading to major falls in United States imports, and growth of United States exports, with consequent changes in trade routes (EIA, 2014a).

For a long time there was an imbalance in cargo movements between developed and developing countries: cargo volumes loaded in the ports of developing countries far exceeded the volumes of goods unloaded. This reflected the difference in volume of exports from developing countries (dominated by raw materials) and their imports (substantially finished goods). As Figure 17.3 shows, over the past four decades a steady change has occurred: loadings and unloadings in the ports of developing countries reached near parity in 2012, driven by the fast-growing import demand in developing regions, fuelled by their industrialization and rapidly rising consumer demand.

General cargo transport has been transformed by the introduction of container shipping. Before 1957, when the first container shipment was made from Houston to New York in the United States, general cargo had to be loaded and unloaded package by package, with relatively long times needed to turn ships around, and high labour costs. The introduction of standardized containers (the Twenty-foot Equivalent Unit (TEU) and the Forty-foot Equivalent Unit (FEU)) enabled ships and ports to be constructed so that compatibility was not an issue. (Ninety per cent of current shipments are of FEU, but the TEU is widely used for statistical purposes) (Levinson, 2007). The convenience of being able to handle practically all forms of general cargo in this way is a major factor in producing the massive expansion of long-distance maritime transport. For a long time, growth in the volume of container traffic was three to four times the growth in world GDP (the average was 3.4 times over the period 1990–2005). A variety of factors now seem to be changing and some analysts suggest that the multiplier has fallen to only 1.5 times in 2012, and may continue at this level. This would imply that in future the global shipping industry would grow more slowly (UNCTAD, 2013). Table 17.1 shows how trade levels and the consequent distribution of container movements between parts of the world vary widely.

General cargo transport has been transformed by the introduction of container shipping. Before 1957, when the first container shipment was

Figure 17.3 | Cargoes loaded and unloaded in the ports of developing countries 1970 – 2012. Percentage share in tonnage of global loadings and unloadings. Source: UNCTAD, 2013.

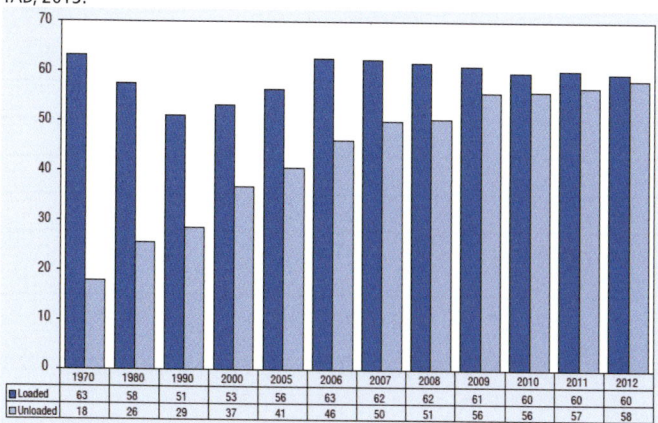

247

Shipping   Chapter 17

Table 17.1 | Container movements on the 10 most heavily trafficked routes 2006.

Route	Number of TEU Movements ('000)
Far East to Far East	21,750
Far East to North America	13,764
Far East to North-West Europe	8,951
North America to Far East	3,950
Far East to Mediterranean	3,750
North-West Europe to Far East	3,571
Far East to the Persian Gulf and Indian Subcontinent	3,322
North-West Europe to North America	3,191
Latin America to North America	2,791
North-West Europe to North-West Europe	2,518

made from Houston to New York in the United States, general cargo had to be loaded and unloaded package by package, with relatively long times needed to turn ships around, and high labour costs. The introduction of standardized containers (the Twenty-foot Equivalent Unit (TEU) and the Forty-foot Equivalent Unit (FEU)) enabled ships and ports to be constructed so that compatibility was not an issue. (Ninety per cent of current shipments are of FEU, but the TEU is widely used for statistical purposes) (Levinson, 2007). The convenience of being able to handle practically all forms of general cargo in this way is a major factor in producing the massive expansion of long-distance maritime transport. For a long time, growth in the volume of container traffic was three to four times the growth in world GDP (the average was 3.4 times over the period 1990–2005). A variety of factors now seem to be changing and some analysts suggest that the multiplier has fallen to only 1.5 times in 2012, and may continue at this level. This would imply that in future the global shipping industry would grow more slowly (UNCTAD, 2013). Table 17.1 shows how trade levels and the consequent distribution of container movements between parts of the world vary widely.

Increases in the size of cargo vessels, and consequent efficiency gains, have been a major factor in enabling unit freight costs for containers, for oil and gas and for bulk traffic to be contained, thus encouraging growth in trade. It seems likely that the trend of increases in the size of vessels will continue. This trend is, however, constrained by the limitations on the size of vessels that ports can handle and by navigational choke-points. At present, three main choke-points constrain the size of vessels. These are the Panama Canal, the Suez Canal and the Straits of Malacca. The approximate maximum dimensions of vessels that can navigate these three passages are shown in Table 17.2. Vessels larger than these dimensions must seek alternative routes: around Cape Horn, around the Cape of Good Hope and around or through the Indonesian archipelago, respectively. These alternative routes add significantly to the costs of some of the main shipping routes, but may be offset by economies of scale in using larger vessels. Work is in progress to provide a new set of locks on the Panama Canal, which are expected to open in 2016, enabling ships within the "New Panamax" classification to pass through the canal (ACP, 2014). This is expected to result in significant changes in patterns of shipping between the Atlantic and Pacific. Proposals exist for a further canal through Nicaragua, which might (if completed) have even larger effects. Work has already started on expanding the Suez Canal. Subject to such changes and the emergence of alternative routes, the scope for efficiency savings from increasing the size of ships, and thus for containing costs, is likely to diminish, as the limits at the choke-points restrict further growth in the size of vessels.

The other source of potential increases in the deployment of larger vessels is the effects of climate change. As a result of the warming of the Arctic, it is becoming possible (at least in summer) to navigate between the Pacific and the Atlantic through both the North-West Passage (through the Canadian Arctic archipelago) and the Northern Sea Route (NSR - along the Arctic coast of the Russian Federation). These possibilities are currently only open to ice-class vessels. The extent to which larger vessels can be deployed depends on the routes that are feasible: Arctic shipping routes, especially the NSR, are subject to significant draft and beam restrictions (Humpert et al., 2012). Increases in the frequency and severity of northern hemisphere blizzards and Arctic cyclones may also limit the use these routes (Wassman, 2011). The *Nordic Orion* (75,600 dead-weight tons (dwt)) became the first commercial vessel to pass through the North-West Passage in October 2013 (G&M, 2014). The NSR has been used for Russian internal traffic since the 1930s. Some international transit traffic took place (with the aid of

Table 17.2 | Choke-points in international shipping: maximum sizes of vessels.

Classification	Length	Beam	Draft	Air-draft (overhead clearance above water)	Approximate dead-weight tonnage[1]	Approximate Twenty-foot Unit (TEU) container capacity
Malaccamax	333-400 metres	59 metres	25 metres	Unlimited[2]	300,000	15,000 – 18,000
Panamax	294.13 metres	32.31 metres	12.04 metres	57.91 metres	65,000 – 85,000	5,000
New Panamax	366 metres	49 metres	15.2 metres	57.91 metres	120,000	13,000
Suezmax	Unlimited	50 metres	20.1 metres	68 metres	120,000 – 200,000	14,500
Largest current crude-oil tankers	415 metres	65 metres	35 metres		320,000 – 500,000	
Largest current container ships	400 metres	59 metres	16 metres		184,600	19,100

1   Dead-weight tonnage (DWT) is a measure of how much weight a ship can safely carry. It is the aggregate of the weights of cargo, fuel, fresh water, ballast water, provisions, passengers, and crew.
2   There are now proposals for a bridge across the Malacca Strait, which would introduce a limit.

icebreakers) in the early 1990s and the number of ships using the transit passage rose from four in 2010 to 71 in 2013 (Liu, 2010 and Economist, 2014). The route between Shanghai and Rotterdam via the NSR is approximately 4,600 km (about 40 per cent) shorter than the route via the Suez Canal, and would take 18–20 days compared to 28–30 days via the Suez Canal (Verny and Grigentin, 2009). Some estimates suggest that, in the longer term, up to 20 per cent - 25 per cent of global shipping movements could be affected by possible Arctic routes, which could offer up to 35 per cent savings in movement time and, hence, costs (Laulajainen, 2009). Others are more pessimistic, but can see some possibilities (Liu et al., 2010). The International Maritime Organization (IMO) has developed a new International Code for Ships Operating in Polar Waters (the Polar Code), covering both Arctic and Antarctic waters. The Code has been made mandatory under the International Convention on the Safety of Life at Sea (SOLAS) and the International Convention for the Prevention of Pollution from Ships (MARPOL) through the adoption of relevant amendments to those Conventions, respectively in November 2014 and May 2015. The expected date of entry into force for the Code is 1 January 2017. Nevertheless, the requirements in the Code will need support through infrastructure such as improved charts and emergency response plans and waste-reception and other facilities capable of dealing with activities on a much larger scale than at present exists (COMNAP 2005, TRB 2012).

As well as the global, long-haul traffic, sea transport also carries much freight on shorter routes. Comparable statistics on this are difficult to find. Within Europe, a study for the European Commission in 1999 showed that 43 per cent of the total freight ton-miles within Europe (including both international and national traffic) were carried on short-sea journeys – an amount about the same as the ton-miles of road haulage. This high proportion was due to the fact that the average movement length by sea was much greater: the average sea movement was nearly 14 times that of the average road movement. Efforts are being made to increase the amount of freight carried on short-sea movements, in order to reduce both the pressure on roads and air pollution emissions (EC, 1999). Similar motives underlie the "America's Marine Highway Program", under which the United States is investing to increase the amount of short-sea freight movements along the Atlantic and Pacific coasts and from the Gulf of Mexico to the east coast (MARAD, 2014). Elsewhere, containerization is leading to rapid growth in short-sea coastal freight movements: for example, in Brazil, the volume of containers carried in coastwise traffic has grown between 1999 and 2008 from 20,000 TEU to 630,000 TEU (+3,050 per cent) (Dias, 2009). To a large extent, the scale of coastwise freight transport reflects the need to distribute more locally the large number of containers arriving in global movements in very large ships. Roll-on/roll-off ferries also play an important role in the more local movement of containers and other cargo, often combined with passenger traffic.

One specialised form of maritime transport that attracts concern in some quarters is the transport of radioactive materials. A wide range of materials need to be transported, from supplies for nuclear medicine to the components in the nuclear fuel cycle. Since 1961 the International Atomic Energy Agency (IAEA) has published advisory regulations on the safe transport of radioactive material, which are generally adopted. Particular concern has been expressed about the shipment of used nuclear fuel for recycling. Since 1971, some 7,000 civil shipments of over 80,000 tons of used nuclear fuel have been reported, mostly to the reprocessing plants at Cap la Hague (France) and Sellafield (United Kingdom of Great Britain and Northern Ireland). These include 160 shipments (totalling 7,140 tons) from Japan to Europe (WNA, 2014). A 2011 survey of the transport of radioactive material in northern Europe confirmed that there had been no maritime transport accidents involving a release of radioactive materials (KIMO, 2011), and none have been reported since then (European Union, 2013).

2.2 Passenger traffic

Since the advent of large aircraft, maritime passenger traffic has effectively been confined to short-sea ferries and cruise ships. Every State with inhabited offshore islands too far offshore for the strait to be bridged has ferry services. States consisting of, or containing, archipelagos rely heavily on ferries for internal passenger transport. International passenger ferries are particularly important in the Baltic Sea, the North Sea and the Caribbean, where several States face each other across relatively short sea-crossings. Roll-on/roll-off ferries (where passenger vehicles and their passengers can make the journey together) have substantially aided the growth of short-sea passenger transport. Roll-on/roll-off ferries are also important for local freight movements, especially in Europe. Growth in passenger transport by ferries is governed mainly by improvements in the facilities and general economic growth in the countries concerned. Over the past decade, for example, the traffic on Greek passenger ferries has stagnated in the light of the Greek economic crisis, while traffic on passenger ferries in Indonesia and the Philippines has continued to grow substantially. Total ferry passengers worldwide in 2008 and its regional components are shown in Table 17.3.

Table 17.3 | World ferry traffic volume and distribution 2008.

	Passengers	Cars	Buses	Freight vehicles
World traffic volumes (millions of journeys)	2,052	252	677	32
Percentages of world total in each region				
America and Caribbean	14.6	29.7	11.9	2.7
Baltic	10.9	33.7	38.6	24.3
Mediterranean	21.2	14.3	14.9	26.7
North Sea	4.4	7.5	32.4	31.7
Pacific	1.5	0.4	<0.1	1.5
Red Sea and Persian Gulf	3.7	0.5	0.9	0.2
South-East Asia	43.7	13.9	1.3	12.9

Shipping

2.3 Cruise ships

The other major sector of passenger maritime transport is cruise ships. Although maritime tourist travel can be traced back to 1837, and a substantial business developed, especially in the Mediterranean Sea in the 19th century (P&O, 2014), the modern cruise ship industry emerged in the 1960s and 1970s as a means of employing ocean-going passenger liners at a time when mass long-distance passenger air-travel was emerging and coming to dominate the long-distance passenger market. When the market demand became clear, specialized cruise ships began to be built, with less emphasis on speed than passenger liners, and more on space for entertainment and relaxation. The market has grown steadily and rapidly since then: the estimated numbers were 3,774,000 journeys in 1990 and 21,556,000 journeys in 2013 (Figure 17.4). The total turnover of the cruise market was estimated at 37.1 billion United States dollars (CMW, 2014). Growth has slowed somewhat since 2008, but has continued.

Over half the market demand in 2013 was from the United States (51.7 per cent). The remaining demand is reported as 26.6 per cent from Europe, 3.6 per cent from Australia and New Zealand, 3.4 per cent from Brazil, 3.4 per cent from Canada and 11.3 per cent from the rest of the world. The main target areas in terms of itineraries and ship deployment for cruises in 2013 are reported as: the Caribbean (34 per cent), the Mediterranean (22 per cent), the rest of Europe (11 per cent), Australia (5 per cent), Alaska (5 per cent), South America (4 per cent), Asia (3 per cent) and other areas (16 per cent) (CLIA, 2014).

There appears to be a trend towards larger vessels. In the Baltic Sea, the Helsinki Commission has calculated that the average number of passengers on the cruise ships calling at Baltic ports rose between 2006 and 2012 by 49 per cent, from 1,099 to 1,635 (HELCOM, 2014a).

2.4 The world fleet of ships

The size of the world's fleet of ships has been increasing rapidly in the period from 2000 – 2013 (Figure 17.5). The persistence of a high rate of growth after the economic crisis of 2008 is accounted for by the lead time between the ordering and delivery of vessels: 2012 was the first year since 2001 in which the tonnage of ships delivered fell below the tonnage delivered in the previous year (UNCTAD, 2013).

Figure 17.4 | Growth in numbers of passenger cruise journeys 2007 – 2014 (millions of journeys). Note: ROW is "Rest of the World" Source: CLIA, 2014.

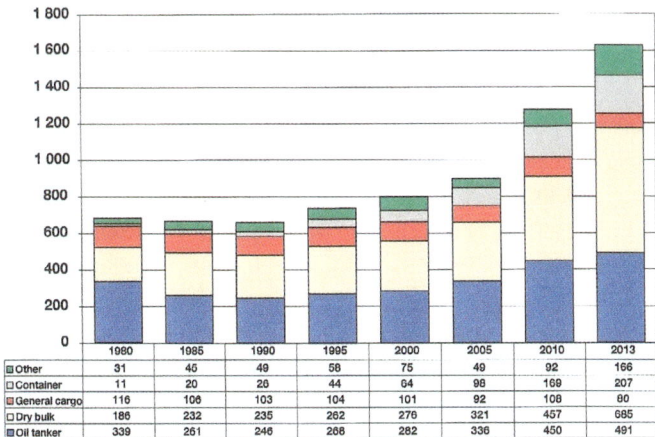

Figure 17.5 | Total size of the world fleet of ships 1980 – 2013. Source: UNCTAD, 2013. Note: Figure 5. includes all propelled seagoing merchant vessels of 100 GT and above, excluding inland waterway vessels, fishing vessels, military vessels, yachts, and offshore fixed and mobile platforms and barges (with the exception of floating production storage and offloading units (FPSOs) and drillships).

The age profile of the world fleet has also been changing: by January 2013, 20 per cent of all seagoing merchant ships were less than five years old, representing 40 per cent of the world's deadweight tonnage. At that point, the average age (per ship) in January 2013 was 9.9 years for dry-bulk carriers, 10.8 years for container ships, 16.7 years for oil tankers, and highest for general-cargo ships (25 years) and the miscellaneous ships (22.6 years). The average ages of oil tankers and dry bulk carriers are lower because of the rapid increases in their numbers. The figure for oil tankers also reflects the phasing out of single-hulled oil tankers (UNCTAD, 2013).

The practice of registering ships in flag States other than that of the owner's nationality has grown, particularly under what are called "open registries", which (among other things) may not impose requirements on the nationality of officers or other crew. The proportion of vessels over 1,000 gross tonnage[1] flying a flag different from that of their owner's nationality has increased steadily from less than 41.5 per cent in 1988 to 73 per cent in 2013. In 2013, more than half the tonnage of the world's ships was registered with four registries – Panama (21.52 per cent), Liberia (12.16 per cent), the Marshall Islands (8.60 per cent) and Hong Kong, China (7.97 per cent) (UNCTAD, 2013) (Figure 17.6). Because of the attractions of "open registries", a number of States have created international shipping registers. These usually have less stringent requirements on the nationality of crew, but may not be open to ships trading solely within national waters.

The pattern of ownership of vessels varies widely between the registries. For example, among the 12 largest registries, some have negligible proportions owned or controlled by nationals. For others (China and Greece), the tonnage is predominantly controlled by nationals. Yet others have substantial, but not predominant, proportions controlled by na-

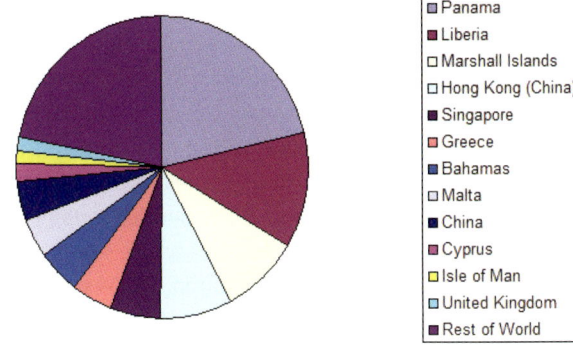

Figure 17.6 | Share of the World's Gross Tonnage by Registry 2013. (12 Registries with the largest gross tonnage on the register and the total gross tonnage of all other Registries). Source: compiled from UNCTAD, 2013).

tionals. This variable pattern is also found among all the other registries. Figure 17.7 shows the estimated spread of ownership and/or control for ships of over 1,000 gross tonnage between the 12 largest ship-owning States and the rest of the world. Owners from five countries (China, Germany, Greece, Japan, and the Republic of Korea) together account for 53 per cent of the world tonnage. Among the top 35 ship-owning countries and territories, 17 are in Asia, 14 in Europe, and 4 in the Americas (UNCTAD, 2013).

2.5 Ship safety

Whether ships are carrying passengers or cargo, the main aim is that the ship should reach port safely at the end of the voyage. In the 1960s, concern about the numbers of collisions of ships, the damage that they could inflict on the environment, the risks to the lives of those on board and the economic effects of losses led to the development of various recommendations on methods of navigation in areas with high levels of shipping activity. In 1971, the IMO Assembly adopted the principle of compulsory measures for ships' routing under the SOLAS Convention, of which the scheme in the Dover Straits was the first (IMO, 2014c).

The IMO has now established some 152 ships' routing measures around the world. These include Traffic Separation Schemes (which require ships going in opposite directions to stay in designated lanes), Inshore Traffic Zones (designated areas landward of a traffic separation scheme for

Figure 17.7 | Spread of the control/ownership of vessels of over 1,000 gross tonnage between the 12 largest fleets and the rest of the world 2013. Source: compiled from UNCTAD, 2013.

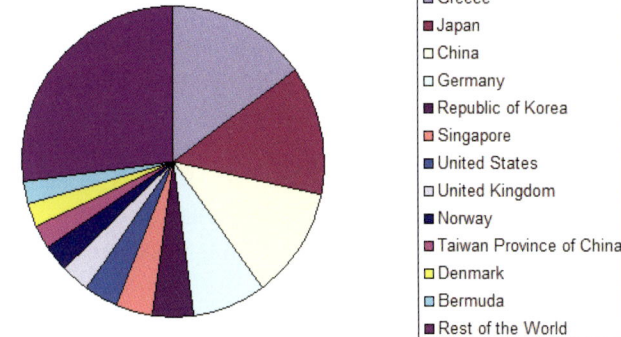

---

[1] "Gross tonnage" is a measure of "the moulded volume of all enclosed spaces of the ship" (International Convention on Tonnage Measurement of Ships, 1969) and is calculated from the volume of the ship multiplied by a reduction factor which increases with the size of the ship.

coastal traffic), Deep-Water Routes (routes which have been accurately surveyed for clearance of sea-bottom and submerged objects), and Areas To Be Avoided (areas in which either navigation is particularly hazardous or it is exceptionally important to avoid casualties and which should be avoided by all ships, or by certain classes of ships). In addition, a number of Governments and port authorities have established similar schemes, particularly in the approaches to major ports (UKHO, 2014). The importance of ships' routing measures can be seen from the straits linking the Black Sea and the Mediterranean. A 2004 study showed that the majority of the accidents in the period 1953 to 1994 were collisions between two or more ships; after the introduction of a traffic separation scheme in 1994, the majority of the accidents were groundings or strandings (Akten, 2004).

In addition to the ships' routing measures adopted by IMO, the IMO has designated, on the proposal of coastal States, Particularly Sensitive Sea Areas (PSSAs) in their exclusive economic zones, where associated measures to protect the environment can be approved. Fourteen such areas have so far been designated. Eight establish Areas to be Avoided (ATBA), six impose mandatory reporting requirements on some or all ships (MSR), four include Traffic Separation Schemes (TSS), two impose a ban on any ship anchoring in a specified area, and one imposes a mandatory deep-water route (DWR). In addition, two recommend the use of pilotage and two recommend the use of an established two-way route. Three of the areas (the Great Barrier Reef, Malpelo Island and the Paracas National Reserve) are linked to areas designated as World Heritage Sites (see Chapter 8 – Cultural Ecosystem Services from the Ocean). The 14 areas, the States proposing them, and the additional protective measures adopted are (in the order of their designation):

(a) Great Barrier Reef (Australia, 1990) (for measures see (h) below);
(b) Sabana-Camagüey Archipelago (Cuba, 1997, ATBA);
(c) Malpelo Island (Colombia, 2002 – ATBA);
(d) Marine area around the Florida Keys (United States, 2002 – ATBA, mandatory no-anchoring areas);
(e) Wadden Sea (Denmark, Germany, Netherlands, 2002 – DWR);
(f) Paracas National Reserve (Peru, 2003 – ATBA);
(g) Western European Waters (Belgium, France, Ireland, Portugal, Spain and the United Kingdom – MSR);
(h) Extension of the Great Barrier Reef PSSA to include the Torres Strait (Australia and Papua New Guinea, 2003 – MSR, two-way route, recommended pilotage);
(i) Canary Islands (Spain, 2004 – ATBA, TSS, MSR, recommended routes);
(j) Galapagos Archipelago (ATBA, MSR, recommended routes);
(k) Baltic Sea (Denmark, Estonia, Finland, Germany, Latvia, Lithuania, Poland and Sweden, 2004 – TSS, DWR, ATBA, MSR);
(l) Papahānaumokuākea Marine National Monument (USA, 2007 – ATBA, MSR for all USA ships and all other ships over 300 gross tonnage, recommended reporting for other ships);
(m) Strait of Bonifacio (France and Italy, 2011 – MSR, recommendation on navigation);
(n) Saba Bank (Caribbean Netherlands, 2012 – ATBA, mandatory no-anchoring area) (IMO, 2014d).

Work continues to improve the safety of navigation, including through on-going improvements to charts, electronic aids to navigation and other navigation services. Such improvements have played a significant part in achieving the reductions in shipping casualties described in this chapter (IMO, 2014e).

The combined effect of all these measures has been to achieve a steady reduction in the number of ships lost at sea, with environmental, social and economic benefits: less pollution of the sea, fewer lives lost and less disruption of trade. Over the long term, in 1912, the casualty rate was about 1 in 100 ships from a fleet of about 33,000 ships; in 2009, the casualty rate was about 1 in 670 ships from a fleet of about 100,000 (Allianz, 2012).

As for the locations of these events, they are naturally concentrated on the main shipping lanes.

In the late 1990s a series of disasters involving ships not covered by the IMO Conventions and their requirements prompted IMO to undertake action as part of their Integrated Technical Cooperation Programme to help States in various regions to develop codes to improve the standard of shipping in those regions. By 1999, States in Asia, the Caribbean and the Pacific agreed codes for non-Convention ships, which many States have incorporated into their legislation. Draft regulations have been developed for Africa, and the IMO has been assisting some States to use these as a basis for improving the safety of ships operating in their waters. Much work still remains to be done, because the infrastructure and skilled personnel needed to implement the regulations are often not available (Williams, 2001). Even where regional codes have been adopted, they do not apply to vessels under 15 – 24 metres (depending on the region). Concern therefore remains in many parts of the world (especially Africa) about improving the safety of small vessels – for example, a 2012 study showed that 70 per cent of the shipping incidents between 1998 and 2011 reported in South Africa involved small ships or fishing vessels (Mearns et al., 2012).

3 Socioeconomic aspects of shipping

3.1 Profitability of the world fleet

The profitability of ferries and the carriage of cargoes within a State ("cabotage") will depend very heavily on local circumstances. Many States subsidize ferries to offshore islands from general taxation in order to remove the disadvantages which those living on such islands would otherwise suffer. Restrictions on which ships may carry out cabotage (which can be restricted to ships flying the national flag of the State within which the carriage takes place) can significantly affect the profitability of routes. States may also intervene to make cabotage more profitable in order to encourage cargo traffic to use sea routes rather than land routes, in order to reduce pressure on roads or the need to build or improve them.

The profitability of international traffic, particularly on intercontinental routes, is, on the other hand, very much a matter of the global shipping market. Long-distance cargo capacity is largely traded on a global market, which is focused on certain cities with well-established local shipbroking networks. Among these are Amsterdam, the Netherlands; Athens (Piraeus), Greece; Copenhagen, Denmark; Hong Kong, China; London, United Kingdom of Great Britain and Northern Ireland; New York, United States; Oslo, Norway; Shanghai, China; and Tokyo, Japan. This market covers both ships operated principally by those who own them, and ships whose owners generally expect to charter them out to other firms to operate. Ships can easily switch between these categories, depending on the levels of supply and demand in the market. Since the level of activity in global shipping is closely linked to the level of global trade within the four main markets of oil and gas, the main bulk cargoes, containers and general cargo, the levels of supply and demand of shipping in the market will likewise fluctuate. Extended periods of growth in global trade encourage ship-owners to invest in new capacity – even more so if growth causes demand to outstrip supply significantly, as happened at some periods over the past two decades. 2001 to 2012 saw the longest sustained growth in the size of the world fleet in history, with record deliveries of new vessels year after year. Additions to capacity will also be caused by investment in new vessels which can operate more efficiently, usually because they are larger. The entry into service of such improved vessels will cause other vessels to be cascaded down the markets. Mismatches between increases in capacity and growth in global trade will lead to overcapacity and falls in freight rates and consequent drops in profitability. Likewise, if global demand drops because of economic recessions, the long lead-time for new shipping means that capacity will continue to grow from deliveries of vessels ordered at a time of growth, thus enhancing the effects of falls in demand (UNCTAD, 2013).

The drop in global trade from the 2008 global recession led to serious drops in freight rates, especially for container ships. For container ships, recovery in trade in 2011 did not immediately lead to better freight rates, in view of the large increases in new capacity. Measures such as slowing voyages (which both saves fuel and uses more capacity) and transferring capacity from east-west routes (where trade remained low) to north-south routes (where trade was growing) meant that in 2012 container freight rates recovered. For tankers carrying oil, gas and chemicals, the position has been similarly poor, with rates over 2009 – 2013 roughly half those of the peak of early 2008. For bulk carriage of coal, grain and minerals, rates over the same period have been only 10 per cent - 30 per cent of the peak 2008 rates. In 2012, a survey revealed that 21 carriers of the top 30 that publish financial results reported an overall operating loss of 239 million dollars in 2012, with only seven carriers turning in positive results (UNCTAD, 2013).

Overall, the pattern has been one of bigger ships offered by fewer companies. Although, in general, the level of service for cargo carriage by regular sailings, as shown by the UNCTAD Liner Shipping Connectivity Index, has improved over the past decade, the result of concentrating cargo in bigger vessels owned by bigger companies has been to reduce the level of competition. In 2004, 22 countries were served by three or fewer carriers: in 2013, 31 countries were facing this situation (UNCTAD, 2013). In 2013, three of the largest container shipping lines proposed collaboration in the scheduling of sailings and allocation of cargo to sailings, while retaining separate sales and pricing systems. These proposals were not accepted by all States and have now been dropped. However, in 2014, the shipping companies involved in those proposals and other companies formed two alliances. Some in the shipping industry believe that further arrangements may be proposed (SCD, 2014, Lloyds List, 2014).

In recent years, passenger cruise ships have suffered some bad publicity, with the loss of the *Costa Concordia* and a number of other adverse events. However, all the major cruise lines have been reporting profits. Across the industry, profits per passenger were reported to grow by nearly 18 per cent between 2011 and 2013, from 157 dollars to 185 dollars (Statista, 2014).

4 Seafarers

4.1 Numbers of seafarers

Worldwide, there are just over 1.25 million seafarers. Only about 2 per cent of these are women, mainly in the ferry and cruise-ship sectors (ITF, 2014). Their origins are shown in Table 17.5.

Although there are many uncertainties, a recent survey by the Baltic International Maritime Council Organization and the International Shipping Federation indicates that the industry will most probably face a continuing shortage of qualified crew (and particularly of officers) when shipping markets recover. There is also a high wastage rate of qualified

Table 17.5 | Broad Geographical Origins of Seafarers 2010. Source: BIMCO/ISF, 2010.

Origin	Officers	per cent	Ratings	per cent
Africa / Latin America	50,000	8	112,000	15
Eastern Europe	127,000	20	109,000	15
Far East	184,000	30	275,000	37
Indian Sub-Continent	80,000	13	108,000	14
OECD Countries	184,000	29	143,000	19
Total - All National Groups	624,000	100	747,000	**100**

crew leaving the industry, and this contributes to potential shortages (BIMCO/ISF, 2010).

4.2 Conditions of work for seafarers

Because ships and those who work on them are operating in a world market, and are frequently (and, in international shipping, usually) not operating within the State under whose flag they fall, international action to regulate the pay and conditions of seafarers has been a major concern of the International Labour Organization (ILO) since its foundation in 1919. In 2006, the ILO adopted the Maritime Labour Convention (MLC) as the "fourth pillar" of international maritime law (alongside SOLAS, the International Convention on Standards of Training, Certification and Watchkeeping for Seafarers, as amended (STCW), and MARPOL). The MLC embodies "all up-to-date standards of existing international maritime labour Conventions and Recommendations, as well as the fundamental principles to be found in other international labour Conventions". The MLC entered into force on 20 August 2013 and, by June 2014, had been ratified by 61 States representing 80 per cent of global shipping tonnage (ILO, 2014).

The MLC continues and restates a unique system for setting recommended minimum wages for seafarers from all countries. A Subcommittee of the ILO Joint Maritime Commission (with representation of Governments, seafarers and ship-owners) has agreed on an increase in the minimum monthly basic wage figure for able seafarers to 592 dollars (7,104 dollars a year) from 1 January 2015 and to 614 dollars (7,368 dollars a year) from 1 January 2016 (ILO 2014)

The pay of officers is determined by the market. However, there are noticeable differences between pay rates, depending on the national origin of the officers concerned. A global survey in 2012 showed the following pattern of salaries (Table 17.6).

Because seafarers can find themselves in foreign ports without many of the support services available to land-based workers, the rights under the MLC on such issues as enforcing arrears of pay are very important. In addition, a survey based on 3,480 cases presented to the Legal Committee of the IMO strongly suggested that the rights of seafarers, as set out in the IMO/ILO "Guidelines on fair treatment of seafarers in the event of a maritime accident", are often subject to violation. Among the views expressed were that the survey showed a need to keep up pressure for better implementation of the Guidelines, and that seafarers were more exposed to criminal proceedings than many other workers (UNCTAD, 2013).

4.3 Safety of seafarers

There are difficulties in obtaining a clear picture of the deaths and injuries suffered by seafarers. In this context, it seems necessary also to consider deaths and injuries suffered by those working at sea in the fishing industry, since these have similar causes. In 1999, a study looking at 19 major shipping administrations over the period 1990 – 1994 concluded that casualties arising from disasters involving merchant vessels were grossly underreported, and in addition failed to address mortality from all other causes of death at sea (Nielsen, 1999). In 2013, when the

Table 17.7 | Pattern of Deaths of Seafarers 2008 – 2012. Source: Sekimizu, 2013.

Year	Deaths of seafarers
2008	1,942
2009	2,395
2010	1,501
2011	1,095
2012	1,051

Secretary-General of IMO launched the Accident Zero Campaign, he noted that the available statistics are neither accurate nor comprehensive, and suggested that there is a need for an official global statistical base (Sekimizu, 2013). The statistics quoted by the IMO Secretary-General showed the following pattern of deaths (Table 17.7) as far as they could be ascertained:

Of these, about 10 per cent were in the fishing sector, 40 per cent in domestic shipping and 50 per cent in other categories, including international shipping. Statistics on serious injuries to seafarers are even less easily established.

Over the past three decades, acts of piracy and armed robbery have re-emerged as serious risks to seafarers. Much attention has been focused on such attacks on shipping in waters off eastern Africa, but reports show that the problem is more widespread. In the last three years, action against attacks off eastern Africa appears to have had some success, but attacks elsewhere are also of concern - especially in the South China Sea area, the location of over half the incidents reported in 2013. The statistics cover reports of alleged piracy (outside the territorial sea) and armed robbery at sea in the territorial sea and port areas (Figure 17.8). Of the 132 attacks reported in 2013 in the South China Sea area, 70 per cent allegedly occurred while the ship was in port. Worldwide, 17 per cent of the attacks were reported to involve actual violence against the crew (IMO, 2014a).

4.4 Safety of Passengers

There are several aspects to the safety of passengers on board passenger ships. The aspect on which most attention has been focused by the

Table 17.6 | Pay of ships' officers in United States dollars a year. Source: compiled from Faststream, 2012.

Origin	Job type			
	Master Mariner	Chief Engineer	Chief Officer	Second Engineer
Asia	111,422	102,740	74,319	72,996
Eastern Europe	109,627	104,448	74,653	81,125
Western Europe	138,320	104,628	90,273	81,871

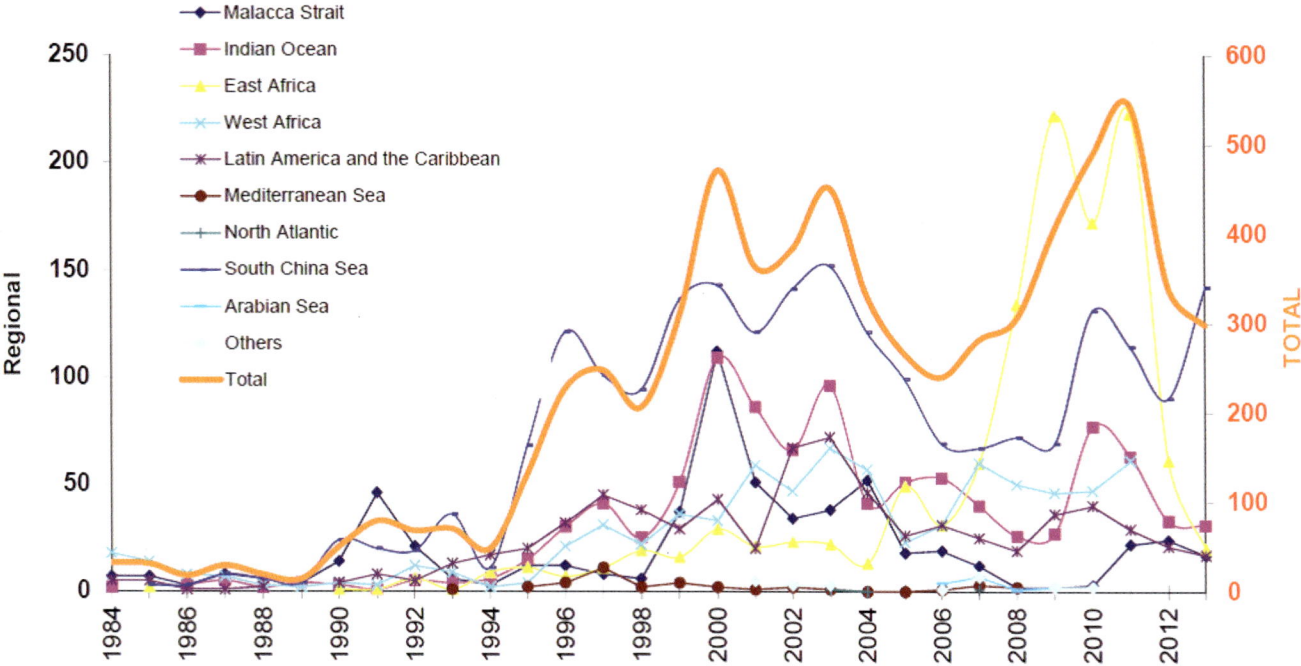

Figure 17.8 | Reports of Alleged Acts of Piracy and Armed Robbery Committed or Attempted 1984 – 2013. Source: IMO, 2014a.

international community, since at least the *Titanic* disaster in 1912, is protection against incidents involving ships on international voyages. Although (as with deaths and injuries to seafarers) there are problems in establishing the relevant statistics for deaths of, and injuries to, passengers, there seems to be little doubt that the number of incidents involving passenger ships on international voyages, and the consequent harm to passengers, is small and has fallen steadily: 13 passenger vessels were lost in 2002 as compared to six in 2013 (Allianz, 2014). A second aspect of passenger safety is that of passenger ferries on domestic voyages. Although, again, statistics are limited, the IMO Secretary-General has drawn attention to the fact that in the 2½ years from January 2012 to June 2014, 2,932 lives were lost in domestic passenger ship accidents around the world. The third main aspect of passenger safety is the accidents and other events of a kind that have nothing to do with the failure of vessels: these cover not only medical emergencies, but also crimes committed by crew or other passengers, and people falling overboard. The size of modern cruise ships (with up to about 5,000 passengers and 2,500 crew) means that this kind of event is as likely to happen on a ship as in a small town. However, investigation and follow-up are a matter for the flag State of the vessel. Statistics on these aspects of passenger safety are not consistently collected. However, because of the large proportion of cruise trips that start in its territory or involve its citizens, the United States since 2010 has required reporting these incidents to the Federal Bureau of Investigation (FBI) for, among others, cruises starting or ending in its ports. The reports show 130 reports of alleged serious crimes in 2011 – 2012 (Rockefeller, 2013).

5 Links to other industries

5.1 Shipbuilding

The steady growth in the numbers and size of vessels of all kinds resulted in record levels of ships being on order in 2008: the dead-weight tonnage (dwt) on order grew by between 50 per cent (general cargo) and 1000 per cent (bulk carriers) to a total of around 600 million dwt. The economic crisis of 2008 resulted in a rapid decline in new orders, and as a consequence many shipyards are thought likely to have to reduce employment. Over the decade from 2002 to 2012, there have been major changes in the location of shipbuilding: in 2012, China (41 per cent), Republic of Korea (33 per cent) and Japan (18 per cent) produced about 92 per cent of all new dwt tonnage; six years earlier in 2006, China had been producing only about 15 per cent, about the same as the European shipyards taken together. The Philippines also has a growing shipbuilding industry: 3 per cent of global dwt completed in 2012. Eastern Asia has thus become dominant in the global shipbuilding market (UNCTAD, 2013).

5.2 Ship-breaking

The ship-breaking industry has likewise become more concentrated. In 2012, 70 per cent of all gross tonnage reported as sold for demolition was sold to ship-breaking yards in Bangladesh, India and Pakistan; 22 per cent was sold to ship-breaking yards in China and 3 per cent to yards in Turkey, leaving only 5 per cent being sold to the rest of the world (UNCTAD, 2013). The ship-breaking industry has given rise to concerns about both the impact on the workers employed and the effects on the marine environment (see Chapter 20). An ILO expert group described it as very hazardous for the workers and presenting many threats to the environment, with major difficulties in enforcing regulations (ILO, 2003). The Hong Kong International Convention for the Safe and Environmentally Sound Recycling of Ships was adopted in 2009. To enter into force, this requires ratification by 15 States representing 40 per cent of the gross tonnage of the world fleet and the combined maximum annual ship recycling volume of these States constituting not less than 3 per cent of the gross tonnage of their combined fleet. However, at the end of 2014, only three States representing 1.98 per cent of world tonnage had ratified it (IMO, 2014b). From the start of 2015, the IMO will implement, in coordination with the Government of Bangladesh, a project aimed at enhancing the standards of health, safety and environmental compliance of ship recycling in Bangladesh.

5.3 Bunkers

Ships need to burn fuel to move, and there is therefore a substantial world-wide industry delivering ships' bunkers (as ships' fuel is named). Estimates of total worldwide fuel consumption by ships vary: calculations in the IMO Greenhouse Gas study based on a bottom-up approach using data on ship movements from Automatic Identification Systems (AIS) show a total consumption of about 327 million tons per year in 2011, compared with a figure of 254 million tons a year based on top-down data on sales of bunkers (IMO 2014o). This compares with an estimate by the United States Energy Information Agency of around 164 million tons a year (EIA, 2014b). In 2010, this represented about 1.5 per cent of the world's total primary energy supply (OECD, 2014). Most of the bunkers are residual fuel oil – that is, the fuel oil that remains after lighter fractions have been removed for other uses. As a consequence, it often has high sulphur content and presents other problems (such as the need to heat it before it can be pumped to the engines). Restrictions are being introduced on bunkers, in order to combat air pollution from ships (see below). Much of the delivery of bunkers takes place in the larger ports, especially those situated near navigational choke-points. Singapore is the world's leading port for the supply of ships' bunkers (MPA, 2014).

5.4 Marine insurance

Alongside the maritime transport industry, a major industry has grown up to insure ships and their cargoes while they are in transit. This is an important component of maritime transport, since cargo owners, ship owners, crew and the rest of the world (including the marine environment) can easily be damaged by ship accidents. A means of ensuring compensation is essential. Many States require a valid insurance certificate as a precondition of entry to their ports: for example, this is a requirement in all European Economic Area States (EC, 2009) The annual premium income worldwide of the marine insurance industry was estimated at 28,930 million United States dollars (excluding the offshore energy industries, whose insurance is often included in marine insurance figures, because it is provided by the same firms). The premiums on cargoes and freight costs represented 18,139 million dollars (62.7 per cent) of this business, the ships themselves 8,563 million dollars (29.6 per cent), and 2,228 million dollars (7.7 per cent) the cost of insurance against causing damage to the marine environment. In addition to the commercial insurance of ships, cargoes, freight costs and environmental risks, many ships are entered into Protection and Indemnity Clubs (P & I Clubs). These are non-profit associations of ship owners, which cover their members against other risks not covered by the marine insurance policy. These clubs are financed by calls on members. In 2013, the total of calls from P & I Clubs was estimated at 3,630 million dollars (Seltmann, 2014).

An important element of these insurance arrangements is the inspection of ships by independent surveyors. These inspections are organized by Classification Societies, which also lay down construction standards for ships that are consistent with the legal requirements of the States with which the Classification Society works. Registration by a Flag State, as well as obtaining insurance, is normally conditional on a Classification Society issuing a certificate that the ship meets the standards laid down for its class. There are over 100 Classification Societies in the various parts of the world. The major Classification Societies formed, in 1961, the International Association of Classification Societies (IACS), which currently consists of 12 member societies and has adopted common approaches to the task of classification through the development of unified rules, requirements and interpretations– for example, the Common Structural Rules for Tankers and Bulk Carriers 2006 (IACS, 2014).

6 Pathways by which shipping impacts on the marine environment and the nature of those impacts

Shipping's impacts on the marine environment can be divided into the catastrophic and the chronic. Catastrophic impacts on the marine environment result from disasters involving the ship, and may lead to its total loss: for example, collisions, fires, foundering and wrecks. Chronic impacts are those that result from the day-to-day operation of ships, without calling into question the ship's integrity or continued functioning (Donaldson, 1994). Both are important, and both are addressed by very similar methods, including by ensuring the safe construction of vessels and their safe operation through construction standards, safe navigation methods, and the proper training and deployment of the crew.

Figure 17.2 highlights the way in which major shipping routes pass through certain choke-points: among the more significant are the Malacca and Singapore Straits, the Strait of Hormuz, the Bab al Mandab at the entrance to the Red Sea, the Suez Canal, the straits linking the Black Sea and the Mediterranean, the Sound and the Belts at the entrance to the Baltic Sea, the English Channel and the Straits of Dover, and the Panama Canal. Concerns about chronic effects are therefore greatest in these areas, because it is there that those effects are most concentrated.

Catastrophic events produce the most serious impacts on the marine environment, as well as being very serious from the point of view of the crew and any passengers and in their economic impact. As explained above, the combined effect of efforts under a number of international conventions has been to reduce steadily the number of ship losses and other catastrophic events.

6.1 Combined impacts of catastrophic events and chronic inputs to the ocean from ships

For most of the major threats to the ocean from shipping, MARPOL provides the technical specifications for preventing and reducing the threats. It was adopted in 1973, adapted in 1978 to facilitate its entry into force, and entered into force for the provisions relating to oil and noxious liquids in bulk in 1983. Since then, it has developed over time (as explained below), both by strengthening the requirements and by bringing into force regulations relating to additional fields. Other international conventions also address threats to the marine environment arising from ships (see also below).

Effective implementation and enforcement of the requirements of these international conventions are crucial.

6.2 Oil

Oil spills from shipping have a wide range of impacts. Catastrophic discharges of large amounts of hydrocarbons will produce large oil slicks with consequentially massive impacts. Smaller slicks will have lesser impacts, but may be equally serious if they are repeated frequently. The impacts range from covering seabirds with oil (which can lead to death), through killing and tainting fish and shellfish and making the stock of fish farms unusable to covering beaches and rocky shores with oil (which can adversely affect tourism). In specific cases, problems can be caused for industries that rely on an intake of seawater (such as marine salt production, desalinization plants and coastal power stations) and coastal installations (such as marinas, ports and harbours) (ITOPF, 2014a). In summarising general experience with oil spills, the study on the environmental impact of the spill of 85,000 tons of crude oil in the 1993 *Braer* catastrophe (Ritchie et al., 1994) drew attention to three important features of major oil spills:

(a) There is an initial, very serious impact, usually with extensive mortality of seabirds, marine mammals, fish and benthic biota and coastal pollution;
(b) In many circumstances, however, marine ecosystems will recover relatively quickly from oil spills: crude oil loses most of its toxicity within a few days of being spilled at sea, mortality of marine biota declines rapidly thereafter, sub-lethal effects are of limited long-term significance and marine ecosystems recover well where there are nearby sources of replacement biota;
(c) Nevertheless, the local circumstances of an oil spill will be very significant. The impact on seabirds, marine mammals and sessile biota will obviously be worse if the spill occurs in areas where they are present in large numbers at the season when the spill occurs – the location of breeding and nursery areas and migration routes and other regular concentrations being particularly important.

The ambient temperature is one of the local circumstances that are most significant for the duration of the impact and the timing of recovery. In warmer areas, the bacteria that break down hydrocarbons are more active, and the effects will disappear more quickly. In spite of the size (about 1 million tons) of the discharges (not including the airborne deposits from the burning of a further 67 million tons), the effect on the coasts of Kuwait and Saudi Arabia of the discharges from oil wells during the Gulf War in 1991 was largely disappearing within 18 months. These coasts had largely recovered within five years. However, oil appears to have persisted in salt marshes and at lower depths in the lower sediments as a result of their anaerobic condition (Readman et al., 1992; Jones et al., 1994; Otsuki et al., 1998; Barth, 2001). In colder areas, on the other hand, bacterial activity is much lower, and the effects of oil spills persist much longer. The impact of the *Exxon Valdez* disaster, in which 35,000 tons of oil were spilt in 1989, was still measurable 20 years later (EVOSTC, 2010).

Local circumstances will also determine the appropriate response to an oil spill. In relatively calm water, it is often appropriate to contain an oil spill with floating booms and use skimmers to retrieve as much oil as possible. With such equipment, it is possible to recover a large proportion of the spill – two-thirds of the 934 tons spilt from the *Fu Shan Hai* in the Baltic in 2003 were recovered (HELCOM, 2010). The other major approach is the use of chemical dispersants. Opinion is divided on the appropriateness of using them: some States regard them as appropriate in many cases, depending on the meteorological circumstances, the local environment and the nature of the oil spill; other States regard them as unacceptable (for examples, see the different opinions in BONN, 2014).

The problem of pollution from oil was the starting point of MARPOL, and the rules to prevent it are in its Annex I. The Annex covers the construction of oil tankers, their operation, what discharges of oily water are permitted, the equipment that must be used and the record-keeping required about any discharges. These requirements have been strengthened over time. In particular, it requires the phasing out of single-hulled oil tankers by, at the latest, 2015.

MARPOL Annex I not only prohibits any discharge into the sea of oil or oily mixtures from any ships in the waters around Antarctica, but also provides for the designation of Special Areas, in which more stringent limits on the discharge of oily water apply. As a counterpart to the designation of Special Areas, coastal States in a Special Area must be parties to MARPOL and must provide appropriate reception facilities for oily waste (see also Chapter 18 - Ports). An important feature of Special Areas is that the maximum permitted level of oil in water discharged is 15 parts per million. In a number of States, the legal system considers that any visible slick on the sea surface must have been caused by a discharge above this level (for examples, see NSN, 2012). Special Areas have been designated, and are in force, in the Mediterranean Sea, the Baltic Sea, the Black Sea, the "Gulfs Area"[2], the Antarctic Area (south of

---

2 The "Gulfs Area" is the sea area between the Arabian Peninsula and the Asian mainland.

Shipping Chapter 17

Figure 17.9 | Seaborne oil trade and number of tanker spills of more than 7 tons 1970 - 2012. Source: ITOPF, 2014b

60°S), North-West European Waters and Southern South African waters. Three further areas have been designated, but are not yet in force because the coastal States have not all notified IMO that adequate reception facilities are in place: the Red Sea, the Gulf of Aden and the Oman area of the Arabian Sea (IMO, 2014f).

In some parts of the world, special measures have been introduced to reduce oil pollution. Aerial surveillance, supplemented and guided more recently by the use of satellite surveillance, has been used in North-West Europe. Coupled with an effective programme of prosecutions of owners and masters of ships observed unlawfully discharging oil, this has led to decreases over the last two decades in the numbers of oil spills, both in absolute terms and in terms of numbers of oil spills observed per hour flown (BONN, 2013; HELCOM, 2014). In the Mediterranean, pilot projects of this kind have been undertaken (REMPEC, 2014). Canada also has set up similar surveillance programmes, using both aerial and satellite surveillance (Canada, 2011).

Over the past forty years, there have been substantial reductions in the scale of marine environmental problems from oil pollution. As Figure 17.9 shows, after the amount of oil transported by sea started to recover from the effects of the 1974 price increases, the amount transported (measured in ton/miles) has steadily increased. At the same time, the number of recorded spills of more than 7 tons has steadily decreased. Forty-six per cent of the spills between 7 and 700 tons between 1970 and 2013 occurred as a result of collision or grounding and 26 per cent as a result of hull or equipment failure or fire or explosion. For spills of over 700 tons in that period, 63 per cent were the result of collision or grounding and 28 per cent of hull or equipment failure or fire or explosion. In both cases the remaining causes were unidentified (ITOPF, 2014a, UNCTAD, 2012).

As Figure 17.10 shows, a similar decrease is observed in the amount of oil involved in these oil.

Nevertheless, a significant problem remains, especially near major shipping routes. A study has shown that even low levels of oil fouling in Magellanic penguins appear to be sufficient to interfere with reproduction (Fowler et al., 1995). One way in which the extent of the remaining problem can be seen is from observations on shorelines of the proportion of the dead seabirds found there which have been contaminated by oil. Diving seabirds are very sensitive to oil pollution: once such a bird is polluted with oil, it is likely to die from hypothermia and/or inability to forage. In the MARPOL North-Western Europe Special Area, the proportion of common guillemots (*Uria aalge*) stranded near the major shipping routes in the southern North Sea was about 40 per cent in 2010, compared with about 4 per cent around the Orkney Islands (OSPAR, 2010). Similar reports have been made about the oiling of seabirds in other areas with high levels of shipping: in the MARPOL Southern South Africa Waters Special Area, studies note that, on the basis of the proportion of the population that has been affected, the African penguin is considered to have suffered more from oiling than any other seabird species globally (Wolfaardt, 2009; Garcia-Borboroglu et al., 2013). In the Straits of Malacca, there is a serious problem with illegal discharges of oil: during the five-year period from 2000 to 2005, there were 144 cases of oil spills into the sea; of this number, 108 cases were due to illegal discharges from ships (BOBLME, Malaysia, 2011). In the waters around south-eastern South America, used both by coastwise local shipping and large vessels travelling between the Atlantic and Pacific Oceans, a study showed that between 1980 and 1994 some 22,000 adult and some 20,000 juvenile Magellanic penguins (*Spheniscus magellanicus*) were being killed each year by oil from discharges from ships passing through the foraging areas for their colonies on the coast (Grandini, 1994). Happily, the solution adopted in 1997 of requiring coastal shipping to follow routes further out to sea may have reduced this problem: over the years 2001 – 2007, the number of oiled penguins observed annually was around 100 (Argentina, 1998; Boersma, 2008). However, other reports are less optimistic (see Chapter 36B). Further north, on the Atlantic coast of Canada, there are also reports of substantial numbers of seabirds being killed by oil. A conservative estimate is put at 300,000 birds a year, with appreciable effects on the populations of species commonly suffering this fate (Canada, 2011).

Figure 17.10 | Quantities of oil in spills of more than 7 tons in the years 1970 – 2013. (with notes of the major recent oil spills and their sizes). Source: ITOPF, 2014b.

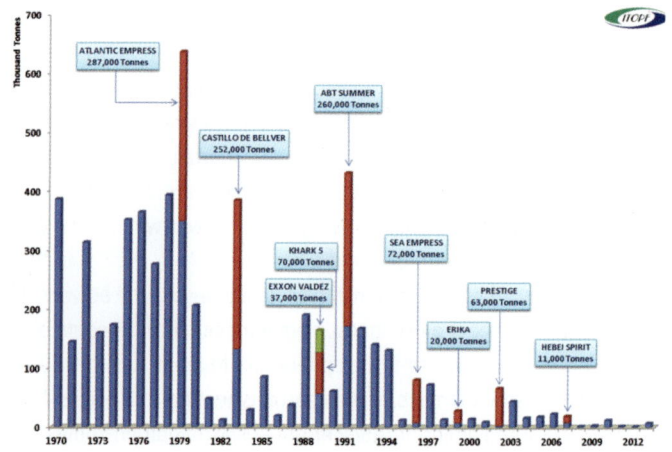

Effective response to oil spills requires a good deal of organization and equipment. The international framework for this is provided by the 1990 International Convention on Oil Pollution Preparedness, Response and Co-operation (OPRC Convention). This entered into force in 1995, and 107 States are now parties. The IMO plays an important role in coordination and in providing training (IMO, 2014g). Coastal States have to bear the capital cost of establishing adequate response capability, but may be able to recover operational costs if and when that capacity is deployed to deal with an oil spill. Developing countries can have difficulties in mobilising the resources for investment in the necessary facilities (Moller et al., 2003).

Major oil spills can cause serious economic damage to a wide range of people and enterprises. After the 1967 *Torrey Canyon* disaster, many States sought to make it easier for those suffering economic damage to obtain reparation. The 1969 International Convention on Civil Liability for Oil Pollution Damage and the 1971 International Convention on the Establishment of an International Fund for Compensation for Oil Pollution aimed to achieve this. These Conventions were revised in 1992 and the revisions came into force in 1996. By July 2014, 115 States were parties to both the 1992 Conventions, and 24 States have become parties to a supplementary protocol providing for additional compensation if the damage exceeds the limits of the 1992 Convention. The economic effect of the Conventions is basically to transfer the economic consequences of an oil spill from the coastal State to the States in which undertakings receive cargoes of oil. This is done either through the insurance costs which the cargo carriers have to incur and include in the costs of the voyages or (to the extent that the damage exceeds the amount insured and the coastal State participates in the funds) through the contributions paid to the funds by those that receive oil cargoes and are located in the States parties.

6.3 Hazardous and noxious substances and other cargoes capable of causing harm

Oil is not the only ship's cargo capable of causing damage. Much depends on the quantities involved – large quantities of nearly any cargo can have an adverse impact, at least on the local environment. SOLAS and MARPOL require precautions against damage from a range of other cargoes, including through requiring compliance with the International Maritime Solid Bulk Cargoes Code, the International Maritime Dangerous Goods Code, the International Code for the Safe Carriage of Grain in Bulk and the International Code for the Construction and Equipment of Ships carrying Dangerous Chemicals in Bulk.

Data on marine pollution incidents involving hazardous and noxious substances are scarce (FSI, 2012). A 2010 study looking at 312 reported incidents of this kind between 1965 and 2009, mainly in the North Atlantic, concluded that reports had become much more frequent since about 2000, with the advent of the internet. It found that about 33 per cent of the cases involved bad weather or structural damage, 30 per cent collision or grounding, 11 per cent fire or explosion and only 6 per cent failures in loading or unloading. Only about half the cases involved discharges into the sea. The three most common substances involved were iron ore, sulphuric acid and caustic soda (Cedre, 2010).

The increased use of containers means that a substantial amount of hazardous or noxious substances is being carried in containers. In 2010 a group of container owners set up a voluntary system to report incidents involving containers, such as fires and spillages, with a view to analysing the data to see if any patterns emerged which could be useful for risk reduction. The Container Notification Information System now covers about 60 per cent of all container slot capacity. Data on the number of incidents have not yet been published, but some preliminary conclusions have been announced: nearly 50 per cent of incidents involved containers where the contents had been mis-declared; 75 per cent of incidents involved hazardous or noxious cargos; no particular global pattern of loading ports emerged from the incidents, but incidents appeared to be higher with containers packed in June, July and August (CINS, 2014).

Containers lost overboard are another source of potential pollution from hazardous and noxious substances. Some estimates have suggested that the numbers of such containers could be in the thousands annually. However, the World Shipping Council, based on a survey to which 70 per cent of the global container shipping capacity responded, estimated in 2011 that about 350 containers are lost overboard each year, excluding mass losses of 50 or more containers as a result of a major ship disaster. If those mass losses are included, the number of containers lost rises to about 650 a year out of about 100 million carried annually (WSC, 2011). On the other hand, it must be remembered that even one container lost overboard can have a lasting and widespread effect on the marine environment: a container holding 28,800 plastic yellow ducks, red beavers, blue turtles and green frogs was lost in 1992 in the middle of the Pacific. The toys have been washed up not only all around the Pacific, but also as far away as the Hebrides in the United Kingdom in 2003 (Ebbesmeyer, 2009).

Following on from the International Convention on Oil Pollution Preparedness, Response and Cooperation (OPRC), a protocol dealing with preparedness and response to incidents involving hazardous and noxious substances was adopted in 2000. This follows the same model as the OPRC Convention. It came into force in 2007, but so far only 33 States have become parties. Efforts to set up an international agreement to deal with compensation for liability and damage from hazardous and noxious ships' cargoes were started as long ago as 1984. A convention was agreed in 1996 but, despite further efforts, no scheme is yet in force to provide international support where a hazardous or noxious cargo causes economic damage.

6.4 Sewage

The problems from the discharge of sewage (in the narrow sense of human and animal urine and fæcal waste) from ships are the same as those for similar discharges from land, which are discussed in Chapter 20. Basically, the problems are the introduction of nutrients into the sea, and the introduction of waterborne pathogens. Away from land, the oceans are capable of assimilating and dealing with raw sewage through natural bacterial action. Therefore, the regulations in Annex IV to MARPOL prohibit the discharge of sewage into the sea within a specified distance of the nearest land, unless ships have in operation an approved sewage treatment plant. (IMO, 2014j).

In summary, discharge of sewage into the sea outside a Special Area is permitted:

(a) When a ship has in operation an approved sewage treatment plant to meet the relevant operational requirements (these are broadly similar to the performance of an effective secondary sewage-treatment plant on land);
(b) When a ship is discharging comminuted and disinfected sewage using an approved system at a distance of more than three nautical miles from the nearest land;
(c) When a ship is discharging sewage which is not comminuted or disinfected at a distance of more than 12 nautical miles from the nearest land (MARPOL Annex IV as in force from 2005).

Because of the problems of eutrophication described in Chapter 20, the amendments to MARPOL Annex IV by IMO in 2011 introduces the Baltic Sea as a special area under Annex IV and adds new discharge requirements for passenger ships while in a special area. In effect, when adequate reception facilities are in place, passenger ships capable of carrying more than 12 passengers may only discharge sewage if nitrogen and phosphorus have been removed to specified standards. (MEPC, 2012).

"Grey water" (that is, waste water from baths, showers, sinks, laundries and kitchens) is not covered by MARPOL Annex IV. Some States (for example, the United States in respect of Alaska) have introduced controls over the discharge of sewage and grey water from larger passenger ships putting into their ports because the local conditions (in Alaska, particularly the water temperature) make the breakdown of any contaminants it may contain quite slow (EPA, 2014a). Furthermore, some States, particularly small island developing States, have difficulties in managing sewage discharged ashore from cruise ships and from the large numbers of such ships visiting their ports. These challenges for small island developing States are discussed further in Chapter 25.

6.5   Garbage

There is no doubt that a substantial part of the marine debris considered in Chapter 25 originates from ships. The damage to the environment from this marine debris is described in that chapter. This debris is constituted by waste from the normal operations of the ship that is thrown overboard. All the serious (and not entirely understood) consequences of marine debris described in that chapter therefore apply to this chronic form of discharge from ships. Because of the large numbers of passengers that they carry, cruise ships generate a high proportion of the garbage generated at sea – in 1995, the United States National Research Council estimated that cruise ships produced 24 per cent of the solid waste generated on board ships, although they represented only 1 per cent of the world fleet (NRC, 1995). Because of the scale of the challenge, most large cruise ships now incinerate on board each day a high proportion of the waste that they generate (75 to 85 per cent of garbage is generally incinerated on board on large ships (EPA, 2008)).

Annex V to MARPOL seeks to eliminate and reduce the amount of garbage being discharged into the sea from ships. Although the Annex is not a compulsory part of the requirements of MARPOL, 15 States, with combined merchant fleets constituting no less than 50% of the gross tonnage of the world's merchant shipping became parties to enable its entry into force on 31 December 1988. Experience showed that the requirements in the original version of Annex V were not adequately preventing ships' garbage from polluting the sea. United Nations General Assembly resolution 60/30 invited IMO to review the Annex. This was done and a revised version entered into force in 2013. Alongside this, IMO adopted guidelines to promote effective implementation. The revised Annex V prohibits generally the discharge of all garbage into the sea, with exceptions related to food waste, cargo residues, cleaning agents and additives and animal carcasses. It also provides for Special Areas where the exceptions are much more restricted. The Special Areas comprise the Mediterranean Sea, the Baltic Sea, the Black Sea, the Red Sea, the "Gulfs" area[3], the North Sea, the Antarctic area (south of 60°S) and the Wider Caribbean Region (including the Gulf of Mexico and the Caribbean Sea) (IMO, 2014h).

Providing adequate waste reception facilities in ports and ensuring that those facilities are used is important. The provision of waste-reception facilities in ports is considered in Chapter 18. However, it should be noted here that small island developing States face major problems in establishing adequate port waste-reception facilities (Corbin, 2011). The greatest effort to promote use of waste-reception facilities has been in Europe, by requiring ships to deliver garbage on shore before leaving port, and removing any economic incentive to avoid doing so. Under this approach, with a few exceptions, all ships are required to deliver their garbage to the port waste-reception facility before leaving port, and the cost of such facilities is to be recovered from ships using the ports, with all ships (again with some exceptions) contributing substantially towards the cost of those facilities, whether or not they made use of them (European Union, 2000). This substantially removes any economic advantage from not using them. This has resulted in a significant (about 50 per cent) increase between 2005 and 2008 in the amount of garbage delivered on shore in European Union ports (EMSA, 2010).

As the OECD pointed out in its 2002 report: "Illegal discharge of wastes at sea often takes place away from shorelines and under cover of night. These two factors make it difficult for port and coastal States to detect acts of pollution, and/or positively identify the polluting vessel" (OECD, 2002). As said in Chapter 25, more information is needed.

---

3   The sea area between the Arabian Peninsula and the mainland of Asia.

6.6 Air pollution

Since the replacement of sail by steam and then diesel, ships have been making emissions to the air. By the early 1990s it was becoming apparent that, in some parts of the world, emissions of nitrogen oxides (NOx) and sulphur oxides (SOx) from ships were becoming a serious element in air pollution for coastal States with heavy shipping traffic in their coastal waters (OSPAR, 2000). Even short-term exposure to NOx produces adverse respiratory effects, including airway inflammation, in healthy people and increased respiratory symptoms in people with asthma. It also reduces resistance to respiratory infections (Knelson et al., 1977; Lee, 1980; EPA, 2014b). Airborne NOx is also a substantial source of nitrogen inputs into coastal waters, and can thus contribute to excessive levels of nutrients (OSPAR, 2010: see also Chapter 20). Exposure to SOx likewise weakens resistance to respiratory infections, and is linked to higher rates of mortality in humans. It is also a contributor (with land-based emissions) to acid rain, which can harm forests and fresh waters (Greaver et al., 2012). SOx emissions from ships have been worsening for decades, as a result of the increasing restrictions on the levels of sulphur in hydrocarbon fuels used on land: as restrictions have reduced the extent to which fuel oils with higher sulphur content can be used on land, so such fuel oils have become more attractive for use at sea, because there were no restrictions and the reduced demand on land lowered the price. NOx and SOx, together with volatile organic compounds (VOCs), can also react in sunlight to produce smog, which affects many major cities: for coastal cities, emissions from ships can contribute to this problem (EPA, 2014c). In addition, shipping was seen as a further source of chlorofluorocarbons and other substances which were contributing to the depletion of the ozone layer, and thus increasing ultraviolet radiation on the earth's surface (GESAMP, 2001). Estimates in 1997 of total global NOx emissions from shipping suggested that they were equivalent to 42 per cent of such emissions in North America and 74 per cent of those in European OECD countries, and that total global SOx emissions from shipping were equivalent to 35 per cent of such emissions in North America and 53 per cent of such emissions in European OECD countries. The global emissions of both NOx and SOx were concentrated in the northern hemisphere (Corbett et al., 1997). Emissions from shipping have therefore been seen as a significant contributory source of air pollution in many parts of the world.

In 1997 a new annex to MARPOL (Annex VI) was adopted to limit the main air pollutants contained in ships' exhausts, including NOx and SOx. It also prohibits deliberate emissions of ozone-depleting substances and regulates shipboard incineration and emissions of VOCs from tankers. Following its entry into force in 2005, it was revised in 2008 to reduce progressively up to 2020 (or, in the light of a review, 2025) global emissions of NOx, SOx and particulate matter, and to introduce emission control areas (ECAs) to reduce emissions of those air pollutants further in designated sea areas (IMO, 2014n). These requirements can be achieved either by using bunkers with lower sulphur content (which may have higher prices) or by installing exhaust scrubbers. Some shipping companies have announced fuel surcharges to meet extra costs which they attribute to the new requirements (Container Management, 2014).

6.7 Anti-fouling treatments

Ships have always been at risk of marine organisms (such as barnacles) taking up residence on their hulls. This increases the resistance of the hull in its passage through water, and thus slows its speed and increases the fuel requirement. With fuel being around half the operating cost of a vessel, this can be a significant extra cost. Historically, the response involved taking the ship out of water and scraping the hull. Because of the inconvenience and cost of this, various treatments developed, mostly involving the application copper sheeting or copper-based paints. In the 1960s, organic compounds of tin were developed, which were shown to be very effective when applied as paints to ships' hulls, with the tin compounds leaching into the water. The most effective was tributyl tin (TBT) (Santillo et al., 2001). By the late 1970s they were commonly used on commercial and recreational craft from developed countries. In the late 1970s and early 1980s, oyster (*Crassostrea gigas*) harvests in Arcachon Bay, France, failed. Subsequent research identified that TBT was the cause. At the same time, research in the United Kingdom showed that TBT was an endocrine disruptor in a marine whelk species (*Nucella lapillus*) causing masculinisation (imposex) in females and widespread population decline. Bans on TBT on boats less than 25 metres long first started in the 1980s. In 1990, the IMO recommended that Governments should eliminate the use of antifouling paints containing TBT. This resolution was intended as a temporary restriction until the IMO could implement a more far-reaching measure. The International Convention on the Control of Harmful Anti-fouling Systems on Ships was adopted in 2001. This prohibited the use of organotin compounds as biocides in anti-fouling paints. This Convention came into force in 2008, and has been ratified by 69 States, representing 84.41 per cent of the gross tonnage of the world's merchant fleet (IMO, 2014j). There are the typical enforcement problems with this Convention. There is also a legacy problem in that dry docks and port berths may have deposits of old anti-fouling paint in the sediments on their bottoms. As and when this sediment has to be removed, disposal into the sea will be a problem, since it may remobilise the TBT remains.

6.8 Wrecks

The seabed is littered with the remains of shipwrecks, some dating as far back as the second millennium BCE. The main impact on the marine environment comes from more recent wrecks, since the introduction of fuel oil as the source of the motive force. Such more recent wrecks will usually contain bunkers, which will eventually leak, and become a new source of oil pollution of the sea. Likewise, cargoes may present dangers of pollution from oil or hazardous substances. There are a number of other problems: first, and depending on its location, a wreck may constitute a hazard to navigation. Secondly, substantial costs are likely to be involved in the location, marking and removal of hazardous wrecks. The Nairobi International Convention on the Removal of Wrecks, 2007, aims to resolve these and related issues. It sets out rules on how to determine whether a wreck presents a hazard, makes the owner of the ship liable for costs of removal and marking (subject to the rules on limits for liability for marine damage) and requires compulsory insurance to cover such costs for ships registered in, or other ships entering or leaving, States parties to the Convention. The Convention will enter into force in 2015. So far there are 12 contracting States, representing 13.84 per cent of the gross tonnage of the world's merchant fleet (Bray et al, 2007; IMO, 2014).

6.9 Invasive species

Invasive non-native species are a major and growing cause of biodiversity loss. They can cause health problems, damage infrastructure and facilities, disrupt capture fisheries and aquaculture and destroy habitats and ecosystems. In some cases, the transport by shipping is clear. For example, in 1991 and 1992, the bacterium that causes cholera (*Vibrio cholerae*) was found in ballast water from five cargo ships in ports in the United States along the Gulf of Mexico (McCarthy, 1994). In other cases, it can be inferred. There are two main ways in which ocean shipping transports invasive species: as attachments to hulls, and in ballast water that has been taken up and discharged by ballasting operations during the stages of a voyage.

So far, the International Union for the Conservation of Nature (IUCN) has identified 84 non-native invasive marine species which have appeared in marine habitats outside their natural distribution (GISD, 2014). A separate review (Molnar et al, 2008), using a wide range of reports (not all peer-reviewed), identified 329 marine invasive species, with at least one species documented in 194 marine eco-regions (84 per cent of the 232 marine eco-regions worldwide used in the review). The main groups of species listed were crustaceans (59 species), molluscs (54), algae (46), fish (38), annelids (worms) (31), plants (19), and cnidarians (sea-anemones, jellyfish, etc) (17). The review found 205 species with detailed shipping pathway information. Of these, 39 per cent are thought to have been, or likely to have been, transported only by fouling of ships' hulls, 31 per cent in ballast water, and 31 per cent by one or other of these routes. Some regional reviews have also identified high numbers of non-native species: for example, 120 in the Baltic Sea and over 300 in the Mediterranean (Zaiko et al., 2011).

Another review of reports of cases of invasive species (Williams et al, 2008) found several estimates of economic damage in the range of millions of US dollars to the localities where the invasive species had been studied. Figure 17.11 shows a summary of the scale of transfers between origins and destinations (Nelleman, 2008).

The scale of these problems led to international efforts to address the pathways through ships' ballast water and biofouling. In 1991 the IMO Marine Environment Protection Committee (MEPC) adopted guidelines for preventing the introduction of unwanted organisms and pathogens from ships' ballast water and sediment discharges. In 1993, the IMO Assembly followed this up by asking the MEPC to review the guidelines with a view to developing an international convention and, in 1997 invited States to use the guidelines to address this problem. More than fourteen years of negotiations were needed to develop the 2004 International Convention for the Control and Management of Ships' Ballast Water and Sediments (the BWM Convention). The Convention will require all ships to implement a Ballast Water and Sediments Management Plan. All ships will have to carry a Ballast Water Record Book and will be required to carry out ballast water management procedures to a given standard. States parties can take additional measures subject to specified criteria and guidelines. The MEPC completed the work of developing the guidelines in 2008. The basic requirements of the BWM Convention are the ballast water exchange standard and the ballast water performance standard. Ships performing ballast water exchange must do so with an efficiency of 95 per cent volumetric exchange of ballast water. Ships using a ballast water management system must meet a performance standard based on agreed numbers of organisms per unit of volume. The BWM Convention requires acceptance by 30 States rep-

The boundaries and names shown and the designations used on this map do not imply official endorsement or acceptance by the United Nations.
Figure 17.11 | Major pathways and origins of invasive species infestations in the marine environment. Source: Nelleman et al, 2008.

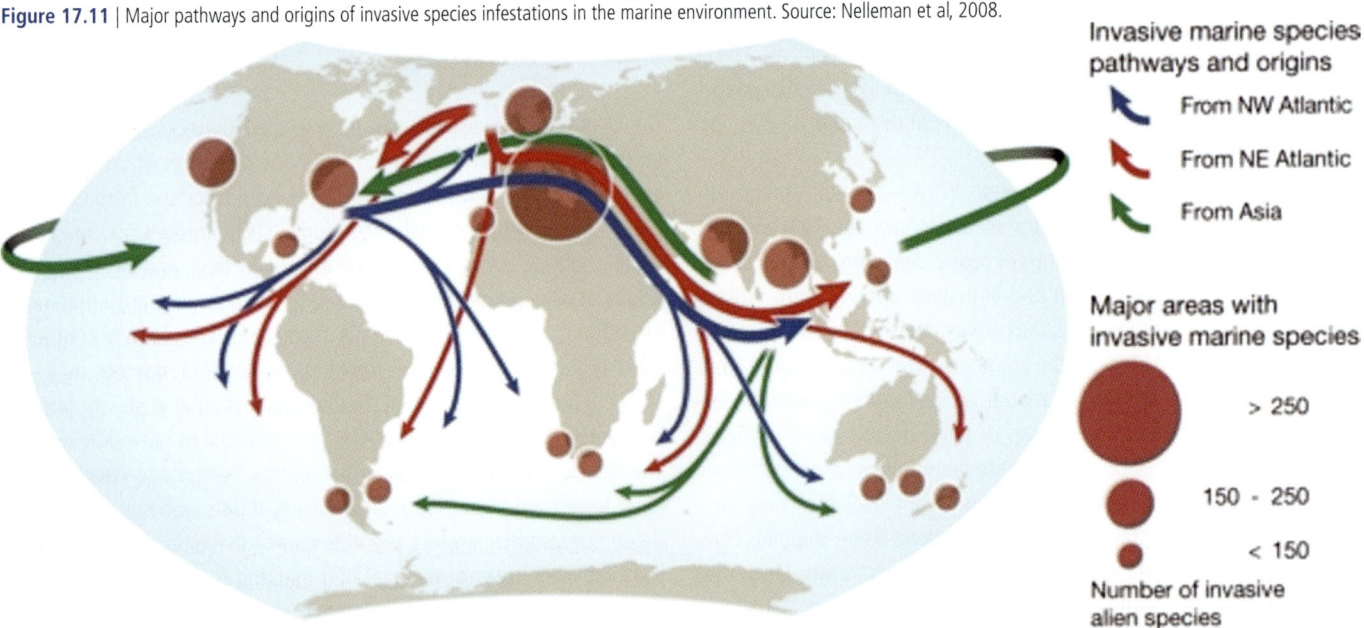

resenting 35 per cent of the gross tonnage of the world merchant fleet before it can enter into force. By November 2014, 43 States representing 32.54 per cent of the tonnage of the world merchant fleet had accepted it. It therefore seems likely to enter into force fairly soon (IMO, 2014j, IMO, 2014k).

The Guidelines for the control and management of ships' biofouling to minimize the transfer of invasive aquatic species (Biofouling Guidelines) were adopted in July 2011. The Biofouling Guidelines are intended to provide a globally consistent approach to the management of aquatic organisms on ships' hulls, and represent a decisive step towards reducing the transfer of invasive aquatic species by ships. In addition, biofouling management can improve a ship's hydrodynamic performance and, therefore, be an effective tool in enhancing energy efficiency and reducing air emissions. In October 2012, the IMO supplemented the Biofouling Guidelines with Guidance for minimizing the transfer of invasive aquatic species as biofouling (hull fouling) for recreational craft, less than 24 metres in length.

6.10   Noise

The marine environment is subject to a wide array of human-made noise from activities such as commercial shipping, oil and gas exploration and the use of various types of sonar. This human activity is an important component of oceanic background noise and can dominate in coastal waters and shallow seas. Long-term measurements of ocean ambient sound indicate that low frequency anthropogenic noise has been increased, primarily due to commercial shipping, both as a result of increases in the amount of shipping and as a result of developments in vessel design (particularly of propellers), which have not prioritised noise reduction. Shipping noise is centred in the 20 to 200 Hz frequency band. Noise at these low frequencies propagates efficiently in the sea, and can therefore affect marine biota over long distances. Baleen whales use the same frequency band for some of their communication signals. A variety of other marine animals are known to be affected by anthropogenic noise in the ocean. Negative impacts for least 55 marine species (cetaceans, fish, marine turtles and invertebrates) have been reported in scientific studies. The effects can range from mild behavioural responses to complete avoidance of the affected area. A 1993 study concluded that "low-frequency noise levels increased by more than 10 dB in many parts of the world between 1950 and 1975," corresponding to about 0.55 dB per year. A 2002 study indicated an increase of approximately 10 dB over 33 years (about 0.3 dB per year). Subsequent measurements up to 2007 confirmed this but suggest that, in some places at least, the subsequent rate of increase has slowed or stopped. It is generally agreed that anthropogenic noise can be an important stressor for marine life and is widely regarded as a global issue that needs addressing (NRC, 2003, Tyack 2008, Andrew et al., 2011, UNEP 2012).

6.11   Enforcement

The effectiveness of the internationally agreed rules to protect the marine environment from the adverse effects of shipping depends to a large extent on the extent of enforcement. There are important economic aspects to the ways in which this enforcement is carried out. The United Nations Convention on the Law of the Sea (UNCLOS) gives flag States, coastal States and port States a range of powers to enforce internationally agreed rules and standards. For port-State control, ports are often competing with their neighbours. This makes it economically important for the port-States to be certain that their enforcement actions are not disadvantaging the competitive positions of their ports. Port-State inspection is, therefore, carried out in many regions in accordance with memorandums of understanding between the States of the region. Memorandums of understanding (MoU) have been set up covering most ocean regions: Europe and the north Atlantic (Paris MoU – 27 States); Asia and the Pacific (Tokyo MoU – 19 States and territories); Latin America (Acuerdo de Viña del Mar – 15 States); Caribbean (Caribbean MoU – 14 States and territories); West and Central Africa (Abuja MoU – 14 States); the Black Sea region (Black Sea MoU – 6 States); the Mediterranean (Mediterranean MoU – 10 States); the Indian Ocean (Indian Ocean MoU – 17 States); and the Riyadh MoU (part of the Persian Gulf – 6 States). These port-State inspection organizations publish details of the results of their inspections, which can have economic significance for ship operators, since cargo consignors tend not to want to use shipping lines which have a poor performance.

Attention to the implementation of international rules and standards demands qualified inspectors to undertake the various controls. Most port-State control inspectors have qualified through service on board ships. A demand for more inspectors may well be in competition for the same pool of staff, which (as noted above) may be itself inadequate for the primary task of crewing ships if the steady growth in shipping activity continues.

7   Significant environmental, economic and/or social aspects in relation to shipping (including information gaps and capacity-building gaps)

This section summarises the most significant elements from the foregoing sections.

Shipping is a vital component of the world economy. As the world economy has become increasingly globalized, the role of shipping has become more important. The economic crisis of 2008 produced some reductions in the levels of shipping, but those have recovered and growth has resumed, though not at quite the previous rate. Shipping has provided means for many States rich in primary resources to export those resources, and for many States that are developing their economies to export their products. Gradually, the balance of the tonnage of goods loaded in developed and in developing countries is becoming more equal. Increasing human wealth will therefore continue to be a driver in increasing the scale of shipping that is needed.

The pressures that shipping imposes on the environment are significant and widespread. In total they represent a significant contribution to the cumulative pressures that humans are imposing on the rest of the marine environment, and that is affecting the harvest from the sea and the maintenance of biodiversity. The pressures are particularly concentrated at certain choke points where shipping routes crowd through narrow sea-passages, e.g., straits or canals. Those pressures are also diverse – some result from shipping disasters, and some are chronic (oil discharges, loss of containers, garbage, sewage, air pollution, noise, antifouling treatments of hulls, transport of invasive species). Over the past 40 years, global rules and standards have been developed to regulate most of these. Steps are now being taken to make the enforcement of these rules and standards more uniform throughout the world.

However, there is still a significant number of States and territories that have not been able to become parties to the various international conventions and agreements that embody these rules and standards. These States and territories need to build the capacities which will enable them to commit themselves to implementing these rules and standards. The necessary skilled staff and facilities will also be needed to implement those rules and standards. Although the IMO and other international organizations have programmes to support such capacity-building, there are still gaps.

There are signs that, at least in some regions, the implementation of these global rules and standards and other local measures is helping to improve the status of the marine environment. The overall number of ships lost at sea has continued to decrease, and in some areas oil discharges seem to have declined. But there is a continuing growth in shipping. Unless the everyday pressures generated by shipping can be steadily reduced, the continuing growth in shipping will lead to increased pressure on the ocean. Even if all ships can meet the standards of the best, increased numbers and tonnage of shipping will eventually increase the pressures on the environment.

However, in many parts of the world, coastal States do not have adequate plans to respond to maritime casualties. Such plans often require substantial investment in plant and equipment and the training of personnel. The resources for such investment are sometimes lacking.

From the social aspect of shipping, it is noteworthy that Africa and South America are underrepresented in the labour force. The divergence between the genders is even more noteworthy, with only an estimated 2 per cent of the maritime labour force being female. It is also noteworthy that there are significant differences in the levels of pay of ships' officers depending on their national origin.

Current reports suggest that there is an adequate labour force to provide the crews of the current levels of shipping. If (as is expected) the amount of world shipping continues to grow, crew shortages may develop. Skilled staff will also be needed to enforce the internationally agreed rules and standards, and this demand may be competing with those for ships' crews.

The entry into force of the 2006 Maritime Labour Convention in 2013 was a major step forward in ensuring the provision of decent working and living conditions for seafarers. Capacities will need to be built for its enforcement.

There is a gap in the information available on deaths of, and injuries to, seafarers. This information is essential for ensuring that they have decent working conditions and for reducing the numbers of seafarers' deaths and injuries. Likewise, the information available on the deaths of, and injuries to, passengers does not appear to be adequate to support policy development in this field, although such information as is available does not suggest that this is a major problem.

References

ACP (2014). Autoridad del Canal de Panamá – https://www.pancanal.com/eng/pr/press-releases/2014/06/25/pr514.html (accessed 4 July 2014).

Akten, Nekmettin (2004). Analysis of Shipping Casualties in the Bosphorus, *Journal of Navigation*, vol. 57, issue 03.

Allianz Global Corporate & Specialty (2012). SE with Cardiff University, Safety and Shipping 1912 to 2012. http://www.agcs.allianz.com/assets/PDFs/Reports/AGCS_safety_and_shipping_report.pdf (accessed 22 July 2014).

Allianz Global Corporate & Specialty (2014) SE, Safety and Shipping Review. http://www.agcs.allianz.com/insights/white-papers-and-case-studies/shipping-review-2014/ (accessed 22 July 2014).

Andrew, R.K., Howe, B.M., and Mercer, J.A. (2011). Long-time trends in ship traffic noise for four sites off the North American West Coast, *Acoustical Society of America*, 129, 642.

Argentina 1998 – Prefectura Naval Argentina, Ordenanza No 13/98 (Rutas de los buques que transportan hidrocarburos y sustancias nocivas líquidas en granel en navegación marítima nacional), 8 January.

Barth, H. (2001). The coastal ecosystems 10 years after the 1991 Gulf War oil spill (unpublished) http://www.uni-regensburg.de/Fakultaeten/phil_Fak_III/Geographie/phygeo/downloads/barthcoast.pdf (accessed 21 June 2014).

BIMCO/ISF (Baltic International Marine Council and International Shipping Federation) (2010). Manpower 2010 Update: The Worldwide Demand for and Supply of Seafarers. https://www.bimco.org/en/News/2010/11/~/media/About/Press/2010/Manpower_Study_handout_2010.ashx (accessed 20 June 2014).

BOBLME Malaysia (2011). Bay of Bengal Large Marine Ecosystem Project, Country report on pollution – Malaysia - http://www.boblme.org/documentRepository/BOBLME-2011-Ecology-07.pdf (accessed 17 March 2014).

Boersma, P. Dee (2008). Penguins as Marine Sentinels, *BioScience*, vol. 58, No. 7.

BONN (2013). Bonn Agreement, Annual report on aerial surveillance for 2012 http://www.bonnagreement.org/eng/html/welcome.html (accessed 31 May 2014).

BONN (2014). Bonn Agreement, Counter-Pollution Manual http://www.bonnagreement.org/eng/html/welcome.html (accessed 31 May 2014).

Bray, S. and Langston, W.J., *Tributyltin pollution on a global scale - An overview of relevant and recent research: impacts and issues*, WWF, Godalming.

Brazil (2013). Marinha do Brasil - Diretoria de Hidrografia e Navegação, Manual de Utilização do Sistema de Tarifa de Utilização de Faróis. https://sistuf.dhn.mar.mil.br/arquivos/manual.pdf (accessed 20 June 2014).

Canada (2011). Environment Canada, Birds oiled at Sea. http://www.ec.gc.ca/mbc-com/default.asp?lang=En&n=C6E52970-1 (accessed 20 April 2014).

Cedre (2010). Centre de documentation, de recherche et d'expérimentation sur les pollutions accidentelles des eaux. *Bulletin de information du Cedre No 27*, 2010 (ISSN 1247-603X).

CINS (Container Incident Notification System Organization) (2014). *Liner 'Cargo Incident Notification System' growing up.* http://www.containerownersassociation.org/news/370-liner-%E2%80%98cargo-incident-notification-system%E2%80%99-growing-up.html (accessed 31 May 2014).

CLIA (Cruise Lines International Association) (2014). *State of the Cruise Industry Report.* http://www.cruising.org/docs/default-source/market-research/pressconferencepresentation.pdf?sfvrsn=0 (accessed 5 July 2014).

CMW (Cruise Market Watch) (2014). *Growth.* http://www.cruisemarketwatch.com/growth/ (accessed 8 July 2014).

COMNAP (Council of Managers of National Antarctic Programmes) (2005). *International Coordination of Hydrography in Antarctica: Significance to Safety of Antarctic Ship Operations*, XXX Antarctic Consultative Meeting, New Delhi, 2007 (HCA7-INF5).

Corbett J.J., and Fischbeck, P.S. (1997). Emissions from Ships, *Science*, vol 278, 1997.

Container Management (2014): "Maersk announces Low-Sulphur Surcharge". (http://container-mag.com/2014/10/09/maersk-announces-low-sulphur-surcharge/ accessed 10 December 2014).

Corbin, C. (2011). Protecting the Caribbean Sea from marine-based pollution: lessons from the MARPOL Annex V Special Area Designation, In *Technical Proceedings of the Fifth International Marine Debris Conference March 20–25, Honolulu, Hawai'i, USA* (Ben Carswell, Kris McElwee, and Sarah Morison, eds), National Oceanic and Atmospheric Administration, Technical Memorandum NOS-OR&R-38, 2011.

Dias, M. (2009). *O Desenvolvimento do Transporte de Contêineres na Cabotagem Brasileira*, in www.antaq.gov.br/portal/SeminarioCabotagem.asp (accessed 4 July 2014).

Lord Donaldson of Lymington (1994): Safer Ships, Cleaner Seas: Report of Lord Donaldson's Inquiry into the Prevention of Pollution from Merchant Shipping (Command Paper Cm 2560), Her Majesty's Stationery Office, London, United Kingdom.

Ebbesmeyer, C. and Scigliano, E. (2009). *Flotsametrics and the Floating World: How One Man's Obsession with Runaway Sneakers and Rubber Ducks Revolutionized Ocean Science*, Collins, London (ISBN. London Collins (ISBN: 9780061971150).

EC (European Commission). (1999). *The Development of Short Sea Shipping in Europe: A Dynamic Alternative in a Sustainable Transport Chain*, (COM(1999) 317 final).

EC (European Parliament and the Council of the European Union) (2000). Directive of 27 November 2000 on port reception facilities for ship-generated waste and cargo residues. (Directive 2000/59/EC as amended). EC (European Parliament and the Council of the European Union) (2009). Directive of 23 April 2009 on the insurance of shipowners for maritime claims (Directive 2009/20/EC).

Economist (2014). Polar bearings, *The Economist*, 12 July 2014.

EIA (USA Energy Information Administration) (2014a). U.S. *Natural Gas Imports and Exports 2013* - http://www.eia.gov/naturalgas/importsexports/annual/ (accessed 10 July 2014).

EIA (USA Energy Information Agency (2014b), *International Energy Satistics – Consumption of Residual Fuel Oil for Bunkering* (http://www.eia.gov/cfapps/ipdbproject/IEDIndex3.cfm?tid=5&pid=66&aid=13 (accessed 20 October 2014).

EMSA (European Maritime Safety Agency) (2010). Horizontal Assessment Report – Port Reception Facilities (*Directive 2000/59/EC*), 2010. http://ec.europa.eu/transport/modes/maritime/consultations/doc/prf/emsa-report.pdf (accessed 6 June 2014).

EPA (USA Environmental Protection Agency) (2008). Cruise Ship Discharge Assessment Report, (*document EPA842-R-07-005*).

EPA (USA Environmental Protection Agency) (2014a). *Sewage and Graywater Standards Development* http://water.epa.gov/polwaste/vwd/sewage_gray.cfm (accessed 14 August 2014).

EPA (USA Environmental Protection Agency) (2014b). *Nitrogen Oxides – Health* http://www.epa.gov/oaqps001/nitrogenoxides/health.html (accessed 21 July 2014).

EPA (USA Environmental Protection Agency) (2014c). Southern California Air Quality http://www.epa.gov/Region9/socal/air/index.html (accessed 21 July 2014).

EU (2013). European Union, *Report from the Commission to the European Parliament, the Council and the European Economic and Social Committee on the implementation by the Member States of Council Directive 2006/117 EURATOM on* the supervision and control of shipments of radioactive waste and spent fuels (COM(2013) 240 final).

EVOSTC (Exxon Valdez Oil Spill Trustee Council) (2010). Status of Injured Resources and Services. http://www.evostc.state.ak.us/index.cfm?FA=status.injured (accessed 20 June 2014).

Faststream (Faststream Recruitment Ltd) (2012). *Maritime Salary Review*. http://www.faststream.com/faststream_recruitment_news/2012_maritime_salary_review/ (accessed 20 May 2014).

Fowler, G.S., et al. (1995). Hormonal and Reproductive Effects of Low Levels of Petroleum Fouling in Magellanic Penguins (*Spheniscus Magellanicus*), *The Auk*, vol. 112, section 2.

FSI (IMO Sub-committee on Flag State Implementation) (2012). Casualty Statistics and Investigations – report by IMO Secretariat, IMODOCS FSI 20/5/4.

G&M (2014). A reality check on the North-West Passage "boom". *The Globe and Mail*, Toronto, 7 January.

Garcia-Borboroglu, P., Boersma, P.D., Reyes, L.M. and Skewgar, E. (2013). *Petroleum Pollution and Penguins: Marine Conservation Tools to Reduce the Problem*. Paper presented at the annual meeting of the International Marine Conservation Congress, George Madison University, Fairfax, Virginia. http://citation.allacademic.com/meta/p296486_index.html (accessed 30 June 2014).

GESAMP (2001). Joint Group of Experts on the Scientific Aspects of Marine Pollution, *A Sea of Troubles*. Arendal, Norway *(ISBN 82-7701-019-9)*.

GISD (2014). International Union for the Conservation of Nature, *Global Invasive Species Database*. http://www.issg.org/database/species/search.asp?sts=sss&st=sss&fr=1&x=12&y=9&sn=&rn=&hci=8&ei=-1&lang=EN (accessed 25 November 2014).

Gandini, P., Boersma, D., Frere, E., Gandini, M., Holik T., and Lichtschein, V. (1994). *Magellanic Penguins (Spheniscus magellanicus) Affected by Chronic Petroleum Pollution along the Coast of Chubut, Argentina*. The Auk, Vol. 111(1), 20-27.

Greaver, T.L., Sullivan, T.J., Herrick, J.D., Barber, M.C., Baron, J.S., Cosby, B.J., Deerhake, M.E., Dennis, R.L., Dubois, J-J.B., Goodale, C.L., Herlihy, A.T., Lawrence, G.B., Liu, L., Lynch, J.A., and Novak, K.J. (2012). Ecological effects of nitrogen and sulfur air pollution in the US: what do we know? *Frontiers in Ecology and the Environment*, Vol. 10, No. 7. 365-372.

HELCOM (Helsinki Commission) (2010). Ecosystem Health of the Baltic Sea 2003–2007: HELCOM Initial Holistic Assessment, *Baltic Sea Environment Proceedings No. 122*. Helsinki (ISSN 0357 – 2994).

HELCOM (Helsinki Commission) (2014a). *2013 HELCOM overview on port reception facilities for sewage in the Baltic Sea area and related trends in passenger traffic* (HELCOM document 4/16/Rev 1).

HELCOM (Helsinki Commission) (2014b). Annual report on illegal discharges observed during aerial surveillance 2013. http://helcom.fi/Lists/Publications/HELCOM%20report%20on%20Illegal%20discharges%20observed%20during%20aerial%20surveillance%20in%202013.pdf (accessed 12 July 2014).

Hildebrand, J.A. (2009). Anthropogenic and natural sources of ambient noise in the ocean *Marine Ecology Progress Series*, Vol 395.

Humpert M., and Raspotnik, A. (2012). *The Future of Arctic Shipping*, The Arctic Institute. http://www.thearcticinstitute.org/2012/10/the-future-of-arctic-shipping.html (accessed 25 July 2014).

IACS, (International Association of Classification Societies) (2014). *Classification Societies - What, Why And How?* http://www.iacs.org.uk/document/public/explained/Class_WhatWhy&How.PDF (accessed 20 October 2014).

IAEA (International Atomic Energy Agency) (2008). Security in the transport of radioactive material, Vienna, *IAEA Nuclear Security Series No. 9* (ISBN 978–92–0–107908–4).

ICS (International Chamber of Shipping) (2014). *World Seaborne Trade* http://www.ics-shipping.org/shipping-facts/shipping-and-world-trade/world-seaborne-trade (accessed 3 July 2014)

IEA (International Energy Agency) (2014). *Key World Energy Statistics 2014*, Paris.

ILO (International Labour Organization) (2003). Safety and health in shipbreaking: Guidelines for Asian countries and Turkey, *ILO 2003* (MESHS/2003/1).

ILO (International Labour Organization) (2014). Press Release – "ILO body adopts new minimum monthly wage for seafarers" 28 February 2014 - http://www.ilo.org/global/about-the-ilo/media-centre/press-releases/WCMS_236644/lang--en/index.htm (accessed 31 May 2014).

IMO (International Maritime Organization) (2012). International Shipping Facts and Figures – Information Resources on Trade, Safety, Security, Environment. http://www.imo.org/KnowledgeCentre/ShipsAndShippingFactsAndFigures/TheRoleandImportanceofInternationalShipping/Documents/International per cent20Shipping per cent20- per cent20Facts per cent20and per cent20Figures.pdf (accessed 17 April 2014).

IMO (International Maritime Organization) (2013). IMO Instruments Implementation Code (III Code) (IMO document A 28/Res 1070).

IMO (International Maritime Organisation) (2014). Special areas under MARPOL. http://www.imo.org/OurWork/Environment/Pollution/Prevention/SpecialAreasUnderMARPOL/Pages/Default.aspx (accessed 21 June 2014).

IMO (International Maritime Organization) (2014a). Reports on Acts of Piracy and Armed Robbery against Ships - Annual Report – 2013 (MSC.4/Circ.208 1 March 2013[1]) http://www.imo.org/OurWork/Security/SecDocs/Documents/PiracyReports/208_Annual_2013.pdf (accessed 31 May 2014).

IMO (International Maritime Organization) (2014b). Ships' Routing http://www.imo.org/OurWork/Safety/Navigation/Pages/ShipsRouting.aspx (accessed 8 July 2014).

IMO (International Maritime Organization) (2014b). Hong Kong Convention http://www.imo.org/about/conventions/listofconventions/pages/the-hong-kong-international-convention-for-the-safe-and-environmentally-sound-recycling-of-ships.aspx (accessed 10 July 2014).

IMO (International Maritime Organization) (2014c). Summary of Status of Conventions http://www.imo.org/About/Conventions/StatusOfConventions/Pages/Default.aspx, (accessed 12 July 2014).

IMO (International Maritime Organization) (2014d). Particularly Sensitive Sea Areas http://pssa.imo.org/#/intro (accessed 31 May 2014).

IMO (International Maritime Organization) (2014e). International Convention on Oil Pollution Preparedness, Response and Co-operation 1990 http://www.imo.org/About/Conventions/ListOfConventions/Pages/International-Convention-on-Oil-Pollution-Preparedness,-Response-and-Co-operation-(OPRC).aspx (accessed on 31 May 2014).

IMO (International Maritime Organization) (2014f). Carriage of Chemicals by Ship. http://www.imo.org/OurWork/Environment/PollutionPrevention/ChemicalPollution/Pages/Default.aSpx (accessed 31 May 2014).

IMO (International Maritime Organization) (2014g). Prevention of Pollution by Garbage from Ships. http://www.imo.org/OurWork/Environment/PollutionPrevention/Garbage/Pages/Default.aspx (accessed 31 May 2014).

IMO (International Maritime Organization) (2014h). Prevention of Pollution by Sew-

---

[1] The document is dated 1 March 2013, but states that it contains the information for the year ending 31 December 2013.

age from Ships. http://www.imo.org/OurWork/Environment/PollutionPrevention/Sewage/Pages/Default.aspx (accessed 7 June 2014).

IMO (International Maritime Organization) (2014j). Prevention of Pollution by Sewage from Ships. http://www.imo.org/OurWork/Environment/PollutionPrevention/Sewage/Pages/Default.aspx (accessed 7 June 2014).

IMO (International Maritime Organization) (2014k). Water Management. http://www.imo.org/OurWork/Environment/BallastWaterManagement/Pages/Default.aspx (accessed 20 November 2014).

IMO (International Maritime Organization) (2014l). Status of Conventions. http://www.imo.org/About/Conventions/StatusOfConventions/Documents/Status%20-%202014.pdf (accessed 28 November 2014).

IMO (International Maritime Organization) (2014m). Special areas under MARPOL. http://www.imo.org/OurWork/Environment/Pollution/Prevention/SpecialAreasUnderMARPOL/Pages/Default.aspx (accessed 21June 2014).

IMO, (2014) – International Maritime Organization, Prevention of Air Pollution from Ships. (http://www.imo.org/OurWork/Environment/PollutionPrevention/AirPollution/Pages/Air-Pollution.aspx accessed 3 December 2014).

IMO, (2014) – International Maritime Organization, *Third IMO Greenhouse Gas Study 2014*.

IOPCF (International Oil Pollution Compensation Funds) (2013). Annual Report. http://www.iopcfunds.org/uploads/tx_iopcpublications/annualreport2013_e.pdf (accessed on 31 May 2014).

Irons, David B., et al (2000). Nine Years after the «Exxon Valdez» Oil Spill: Effects on Marine Bird Populations in Prince William Sound, Alaska, *The Condor*, vol. 102, No. 4.

ITF (International Transport Workers Federation) (2014). Women Seafarers http://www.itfseafarers.org/ITI-women-seafarers.cfm (accessed 1 July 2014).

ITOPF (International Tanker Owners Federation) (2014a). Information and Statistics. http://www.itopf.com/knowledge-resources/data-statistics/statistics/ (accessed 30 May 2014).

ITOPF (International Tanker Owners Federation) (2014b). Oil Tanker Spill Statistics 2013. http://www.itopf.com/fileadmin/data/Documents/Company_Lit/OilSpillstats_2013.pdf (accessed 30 May 2014).

Jones, D.A., et al (1994). Intertidal recovery in the Dawhat ad-Dafi and Dawhat al-Musallamiya region (Saudi Arabia) after the Gulf War oil spill. *Courier der Forschungs-Institut Senckenberg*, vol 166.

Kaluza, P., Kölzsch, A., Gastner, M.T. and Blasius, B. (2010). The complex network of global cargo ship movements. *Journal of the Royal Society: Interface*, (7).

Kendall, Steven J., Erickson, W.P., Lance, B.K., Irons, D.B. and McDonald, L.L. (2001). Twelve Years after the Exxon Valdez Oil Spill. *The Condor*, Vol. 103, No. 4.

KIMO (Kommunenes Internasjonale Miljøorganisasjon) (2011). *The Transport of Nuclear Materials by Sea in Northern Europe*. Lerwick (United Kingdom), 2011.

Knelson, J.H. and Lee, R.E. (1977). Oxides of Nitrogen in the Atmosphere: Origin, Fate and Public Health Implications. *Ambio*, vol. 6, No. 2/3.

Laulajainen, R. (2009). The Arctic Sea Route. *International Journal of Shipping and Transport Logistics*, vol. 1(1).

Lee, S.D. (ed.) (1980). *Nitrogen Oxides and their Effects on Health*. Ann Arbor Science Publishers, Michigan.

Levinson, Marc. (2008). *The Box: How the Shipping Container Made the World Smaller and the World Economy Bigger*. Princeton (ISBN 9781400828586).

Liu, Miaojia and Kronbak, Jacob. (2010). The potential economic viability of using the Northern Sea Route (NSR) as an alternative route between Asia and Europe. *Journal of Transport Geography, vol.* 18, Issue 3.

Lloyds List, 2014. CMA CGM, UASC and China Shipping unveil rival to 2M, G6 and CKYHE, Tuesday 09 September 2014.

Mandryk, W. (2009) (Lloyd's Marine Intelligence Unit), *Measuring global seaborne trade*. International Maritime Statistics Forum, New Orleans. http://www.imsf.info/papers/NewOrleans2009/Wally_Mandryk_LMIU_IMSF09.pdf (accessed 10 July 2014).

MARAD (2014) USA Department of Transportation, Maritime Administration, America's Marine Highway Program. http://www.marad.dot.gov/ships_shipping_landing_page/mhi_home/mhi_home.htm (accessed 20 June 2014).

MARPOL (1973/78). The International Convention for the Prevention of Pollution from Ships. United Nations *Treaty Series*, vol. 1340, I-22484.

McCarthy, S.A., Khambaty, F.M., (1994). International dissemination of epidemic *Vibrio cholerae* by cargo ship ballast and other nonpotable waters. *Applied Environmental Microbiology*, Vol 60, 7.

Mearns, K., Olivier J. and Jordaan M. (2012). Shipping Accidents along the South African Coastline. http://www.academia.edu/537251/Shipping_Accidents_Along_the_South_African_Coastline (accessed 20 June 2014).

MEPC (IMO Marine Environment Protection Committee) (2012). Resolution MEPC.227(64), IMO Document MEPC 64/23/Add.1, Annex 22.

MLC (International Labour Organization) (2006). The Maritime Labour Convention. http://www.ilo.org/dyn/normlex/en/f?p=NORMLEXPUB:91:0::::P91_SECTION:MLC_A5 (accessed on 2 July 2014).

Moller, T.H., Molloy, F.C. and Thomas, H.M. (2003): Oil Spill Risks and the State of Preparedness in the Regional Seas. *International Oil Spill Conference Proceedings*: Vol. 2003, No. 1, pp. 919-922.

Molnar, J.L., Gamboa, R.L., Revenga, C., and Spalding , M.D. (2008). Assessing the Global Threat of Invasive Species to Marine Biodiversity. *Frontiers in Ecology and the Environment*, Vol. 6, 9.

MPA (Maritime and Port Authority of Singapore) (2014) *Bunkering*. http://www.mpa.gov.sg/sites/port_and_shipping/port/bunkering/bunkering.page (accessed 20 October 2014).

Nelleman, C., Hain, S., and Alder, J. (Eds) (2008). *In Dead Water – Merging of climate change with pollution, over-harvest, and infestations in the world's fishing grounds*. United Nations Environment Programme, GRID-Arendal, Norway.

Nielsen, D., and Roberts S. (1999). Fatalities among the world's merchant seafarers (1990–1994). *Marine Policy*, Volume 23, 1.

NRC (USA National Research Council (1995)). *Clean Ships, Clean Ports, Clean Oceans: Controlling Garbage and Plastic Wastes at Sea*. National Academy Press: Washington, DC, 1995.

NRC (USA National Research Council - Committee on Potential Impacts of Ambient Noise in the Ocean on Marine Mammals (2003)). *Ocean Noise and Marine Mammals*. National Academies Press. Washington (DC) USA.

NSN (North Sea Network) (2012). *North Sea Manual on Maritime Oil Pollution Offences*. London, (ISBN 978-1-906840-45-7).

OECD (Organization for Economic Cooperation and Development) (2003). *Cost Savings Stemming from Non-Compliance with International Environmental Regulations in the Maritime Sector*. http://www.oecd.org/officialdocuments/publicdisplaydocumentpdf/?cote=DSTI/DOT/MTC(2002)8/FINAL&docLanguage=En (accessed 7 June 2014).

OSPAR (OSPAR Commission) (2000). *Quality Status Report on the North-East Atlantic*. London, 2000 (ISBN 0 946956 52 9).

OSPAR (OSPAR Commission) (2010). *Quality Status Report 2010*, London (ISBN 978-1-907390-38-8).

Otsuki, A., Abdulraheem, M.Y., and Reynolds, R.M. (eds) (1998). *Offshore Environment of the ROPME Sea Area after the War-Related Oil Spill – Results of the 1993-94 Umitaka-Maru Cruises*, Tokyo 1998 (ISBN 4-88704-123-3).

P&O (2014). *Peninsular and Oriental Steam Navigation Company, Our History*. http://www.poheritage.com/our-history (accessed 6 July 2014).

Rall, David P. (1974). Review of the Health Effects of Sulfur Oxides. *Environmental Health Perspectives*, Vol. 8.

Readman, J.W., Fowler, S.W., Villeneuve, J.-P., Cattini, C., Oregioni B., Mee, L.D. (1992). Oil and combustion-product contamination of the Gulf marine environment following the war. *Nature*, 358.

REMPEC (2014). Regional Marine Pollution Emergency Response Centre for the Mediterranean, Illicit Discharges. http://www.rempec.org/rempec.asp?theIDS=2_161&theName=PREVENTION&theID=8&daChk=2&pgType=1 (accessed 12 July 2014).

Ritchie, W., and O'Sullivan, M. (eds). (1994). *The Environmental Impact of the Wreck of the Braer, Scottish Office Environment Department*. Edinburgh (ISBN 07480 0900 0).

Rockefeller (2013). Staff Report Prepared For Senator John D. Rockefeller IV, Chairman, USA Senate Committee On Commerce, Science, And Transportation, Cruise Ship Crime. http://www.lipcon.com/files/cruise-ship-crime-consumers-have-incomplete-access-to-cruise-crime-data.pdf (accessed 20 July 2014).

Santillo, D., Johnston, P., Langston, W.J. (2001). Tributyltin (TBT) antifoulants: a tale of ships, snails and imposex. In: Harremoes P, Gee D, MacGarvin M, Stirling A, Keys J, Wynne B, et al., eds *Late Lessons from Early Warnings: The Precautionary Principle*. Environment Issue Report, no. 22., European Environment Agency, Copenhagen.

SCD (Supply Chain Digest) (2014). http://www.scdigest.com/ontarget/14-07-23-3.php?cid=8310 (accessed 26 July 2014).

Sekimizu, K. (2013). Address of the Secretary-General of the International Maritime Organization at the Opening of the Fifty-Sixth Session of the Sub-Committee on Fire Protection, 7 January 2013 http://www.imo.org/MediaCentre/SecretaryGeneral/Secretary-GeneralsSpeechesToMeetings/Pages/FP-56-opening.aspx (accessed 31 May 2014).

Sekimizu, K. (2014). Address of the Secretary-General of the International Maritime Organization at the Opening of the Ninety-Third Session of the Maritime Safety Committee, 14 May 2014. http://www.imo.org/MediaCentre/SecretaryGeneral/Secretary-GeneralsSpeechesToMeetings/Pages/MSC93opening.aspx (accessed 30 June 2014).

Seltmann, A. (2014). *Global Marine Insurance Report 2014*. International Union for Marine Insurance http://www.iumi.com/images/gillian/HKfromHH/20140922_1200_Seltmann_Astrid_FactsFigures_corr.pdf (accessed 23 May 2014).

Statista Inc. (2014) http://www.statista.com/statistics/204563/cruise-revenue-expenses-and-profit-per-passenger-worldwide/ (accessed 21 July 2014).

STH (2014). Ship Trade House – Types of Vessel. (http://shiptradehouse.com/vsltypes accessed 22 May 2015).

Stebbing, A.R.D. (1985). Organotins and water quality – some lessons to be learned, *Marine Pollution Bulletin*. 16 (10).

TRB (Transportation Research Board) (2012). *Safe Navigation in the U.S. Arctic* - Summary of a Conference, Seattle. http://onlinepubs.trb.org/onlinepubs/conf/CPW11.pdf (accessed 3 July 2014).

Tyack, P.L., 2008. Implications for Marine Mammals of Large-Scale Changes in the Marine Acoustic Environment. *Journal of Mammalogy*, 89(3), 549–558.

UKHO (United Kingdom Hydrographic Office) (2014). *Annual Summary of Notices to Mariners 2014*. Taunton (ISBN 978-0-70-773-2138).

UNCLOS (1982). United Nations Convention on the Law of the Sea, United Nations, *Treaty Series*, vol. 1833, No. 31363.

UNCTAD (2012). United Nations Conference on Trade and Development, *Liability and Compensation for Ship-Source Oil Pollution: An Overview of the International Legal Framework for Oil Pollution Damage from Tankers - Studies in Transport Law and Policy - 2012 No. 1*, New York and Geneva.

UNCTAD (2013). Secretariat of the United Nations Conference on Trade and Development (UNCTAD), *Review of Maritime Transport 2013*, New York and Geneva (ISBN 978-92-1-112872-7). http://unctad.org/en/PublicationsLibrary/rmt2013_en.pdf

UNEP (United Nations Environment Programme) (2012). Convention on Biological Diversity, Subsidiary Body on Scientific, Technical and Technological Advice, Scientific Synthesis on the Impacts of Underwater Noise on Marine and Coastal Biodiversity and Habitats (SBSTTA document 16/INF/12).

USMA (USA Maritime Administration) (2014). *Marine Highway Fact sheet*. http://www.marad.dot.gov/documents/AMH_Fact_Sheet_V11.pdf (accessed 7 July).

Verny J. and Grigentin, C. (2009). Container shipping on the Northern Sea Route. *International Journal of Production Economics*, 122.

Wergeland, T. (2012). Ferry Passenger Markets. In *The Blackwell Companion to Maritime Economics* (Wayne K. Talley, ed.). http://onlinelibrary.wiley.com/doi/10.1002/9781444345667.ch9/summary (accessed 2 July 2014).

Williams, I. and Hoppe, H. (2001). Safety Regulations for Non-Convention Vessels – The IMO Approach. http://www.imo.org/blast/blastDataHelper.asp?data_id=18002&filename=Safety.pdf (accessed 20 June 2014).

Williams, S.L. and Grosholz, E.D. (2008). The Invasive Species Challenge in Estuarine and Coastal Environments: Marrying Management and Science, *Estuaries and Coasts*, Vol. 31 (1).

WNA (World Nuclear Association) (2014). Transport of Radioactive Materials. http://www.world-nuclear.org/info/Nuclear-Fuel-Cycle/Transport/Transport-of-Radioactive-Materials/ (accessed 1 July 2014).

WNA (World Nuclear Association) (2014). Transport of Nuclear Fuels. http://www.world-nuclear.org/info/Nuclear-Fuel-Cycle/Transport/Transport-of-Radioactive-Materials/ (accessed 7 July 2014).

Wolfaardt, A.C. et al. (2009). Review of the rescue, rehabilitation and restoration of oiled seabirds in South Africa, especially African penguins *Spheniscus demersus* and Cape gannets *Morus capensis*, 1983–2005, *African Journal of Marine Science*, Volume 31, Issue 1.

WSC (World Shipping Council). (2011). *Containers Lost at Sea*. http://www.worldshipping.org/industry-issues/safety/Containers_Overboard__Final.pdf (accessed 31 May 2014).

Zaiko, A., M. Lehtiniemi, A. Narščius, and S. Olenin (2011). Assessment of bioinvasion impacts on a regional scale: a comparative approach. *Biological Invasions*, Volume 13, Issue 8, 1739-1765.

18 Ports

Group of Experts:
Alan Simcock (Lead member)

1 Introduction

Ports are the nodes of the world's maritime transport system. Every voyage of a ship must begin and end at a port. Their size and distribution will therefore both reflect and contribute to the pattern of maritime transport described in Chapter 17 (Shipping). Since the maritime transport system is part of a much larger global transport system, including road, rail, river and canal transport and the interchanges between all the modes, the factors that determine the location and growth (and decline) of ports are manifold, and go well beyond an assessment of the marine environment. These non-marine factors (such as land and river transport connections, location of population and industry and size of domestic markets) will determine, to a large extent, the development of ports and, therefore, the way in which they affect the marine environment. Nodes, however, can become bottlenecks, restricting the free flow of trade. Before the economic crisis of 2008, there were fears that port capacity could limit the development of world trade (UNCTAD, 2008). That problem has receded with the widespread economic slow-down, but could easily re-appear. This would lead to increased pressure for new port developments.

Just as containerization has transformed general cargo shipping from the mid-20th century onwards, so it has also transformed the nature of the ports that container ships use. In the past, ports relied on large numbers of relatively unskilled dockworkers to do the physical work of loading and unloading general cargo, often on a basis of casual labour, with no security of regular work. Containerization and parallel improvements in the handling of bulk cargoes have transformed this situation. Ports now require smaller numbers of much more skilled workers, and even more investment in handling equipment.

2 Scale and magnitude of port activity

Ports can be classified in several different ways. Some ports are dedicated to a single function (such as the handling of oil). Others are general, handling a variety of trades. Some are private, used for the traffic of one trader (or a small number of traders). Others are general, open to shipping in general. Some are designed for bulk traffic, both dry and liquid. Others are for general cargo, which today usually implies containers. And some ports are a mix of these various categories. (This chapter does not deal with marinas and other harbours for recreational vessels: those are covered in Chapter 27 (Tourism and recreation)).

Dry bulk traffic covers the five major bulk trades (iron ore, coal, grain, bauxite/alumina and phosphate rock), together amounting to 2,786 million tons in 2013, and the minor bulk trades (soymeal, oilseed/meal, rice, fertilizers, metals, minerals, steel and forest products), together amounting to 2,300 million tons in 2013. The main tanker bulk traffic (crude oil, petroleum products, and liquefied natural gas) amounted to 2,904 million tons. There is also a much smaller market in bulk tanker carriage of chemicals (UNCTAD, 2013).

The location of ports for handling bulk traffic is usually determined by the location of their sources of supply and demand. A new oil field may well demand the creation of a completely new port, as happened with the creation of Sullom Voe in the Shetland Islands in the United Kingdom in the 1970s at the beginning of the exploitation of North Sea oil and gas (Zetland, 1974). A large iron and steel works may be linked to the creation of new port facilities to receive imports of iron ore, as is happening at Zhanjiang in China (Baosteel, 2008). As a result of geographical or historical factors, some ports for bulk traffic can have awkward conjunctions in their location. For example, in Australia, the coal mines in Queensland need more port outlets, but the likely locations for ports are near the Great Barrier Reef, which gives rise to difficult decisions (Saturday Paper, 2014). In the United Kingdom, the Milford Haven oil terminal grew up gradually over many years in the safe natural harbour of Milford Haven. It is currently the United Kingdom's largest oil port, with a throughput of hydrocarbons in bulk of 40 million tons a year. However, the United Kingdom's first marine nature reserve, Skomer Island, is near the mouth of the harbour (Donaldson, 1994; DfT, 2014).

The containerization of general cargo, the consequent reduction of trans-shipment costs and the use of ever larger ships has changed the nature of the demand for general cargo ports over the past half century. Instead of relatively small ships moving directly from the origin to the destination of the cargo, thus minimising the then expensive trans-shipment costs, there is now a hierarchy of ports, with cargoes passing through entrepôts where they are trans-shipped. Rotterdam, in the Netherlands, is a good example of such an entrepôt, with many other North Sea ports receiving the trans-shipped goods. (Haralambides, 2002). The proportion of worldwide total container movements that involve trans-shipment is gradually increasing (25 per cent in 2000: 28 per cent in 2012 (Notteboom et al., 2014)). The nature of this hierarchy shows that there is a major equatorial shipping route linking major ports, with supporting north-south and transoceanic routes. The "trans-shipment markets" identified are the zones within which ports are competing with each other for the long-haul business, which will be trans-shipped for delivery to its final destination by ship, road or rail (Rodrigue, 2010, Figure 13). Containerized general cargo amounted to 1.6 billion tons in 2012 – an estimated 52 per cent of global seaborne trade in terms of value (UNCTAD, 2013). The imbalances in containerized exports and imports, the liberalization of trade regulation and transit facilitation are resulting in a growth of containerization of trades previously handled as bulk. Since more containerized imports arrive in some ports than there are exports from those ports to fill the containers, the shipping costs for the return or onward journey using the surplus containers are low. This acts as a form of subsidy on the use of such containers, and thus attracts business from the bulk trades. For example, between 2008, when grain trading was deregulated in Australia, and 2013, the country's containerized wheat export shipments increased tenfold (UNCTAD, 2013).

The world's busiest container port is Shanghai in China, with 33.62 million TEU movements in 2013. Table 18.1 sets out the numbers of container movements for each of the further five container ports with the heaviest traffic. Outside these areas, there are of course other very large and busy ports – for example (with millions of TEU movements in 2013): Los Angeles, California, USA (7.87), Long Beach, California, USA (6.73) and New York/New Jersey, USA (5.47). In total, the world's 50 busiest container ports in 2013 were spread as follows:

(a) Twenty-four in the west Pacific (ten in China; three in Japan; two each in Indonesia and Malaysia; and one each in Hong Kong, China, the Philippines, the Republic of Korea, Singapore, Taiwan Province of China, Thailand and Viet Nam);
(b) Four in the eastern Pacific (two in the United States of America and one each in Canada and Panama);
(c) Seven in the Indian Ocean (two in the United Arab Emirates and one each in India, Oman, Saudi Arabia, Sri Lanka and South Africa);
(d) Eleven in the eastern Atlantic and adjacent seas (two each in Germany and Spain and one each in Belgium, Egypt, Italy, Malta, the Netherlands, Turkey and the United Kingdom); and
(e) Four in the western Atlantic (two in the United States and one each in Brazil and Panama) (WSC, 2014).

3 Socioeconomic aspects of ports

The arrival of containerization of general cargo and the increased mechanization of the handling of bulk cargoes has transformed employment in the dock industry. It has reduced the amount of human physical effort, increased the amount of work done by machinery and reduced substantially the risks of death and injury to dockworkers. As a result, it has also decreased substantially the number of dockworkers required. Negotiations over the change have therefore often been difficult, particularly in the early years of the introduction of containerization. The change has now spread worldwide, and few ports still rely on the handling of general cargo parcel by parcel. However, statistics at global level on the effects of the change are not available (ILO, 2002).

The economic effects on port operations have been no less thoroughgoing. Three main strands of change have been noticeable:

(a) As the economics of ship operation have created pressures for ever larger ships, both for bulk carriage of cargoes and for containers (see Chapter 17 – Shipping), so pressures have developed on ports to create the facilities to handle these larger ships. These pressures have required ports to invest in deeper-water facilities, bigger cranes and navigational improvements in order to accommodate the larger ships. These have all required substantial investment;

(b) The general liberalization of the terms of world trade and consequent growth in shipping have led to ports being placed more and more in competition with each other. Coupled with the development of hierarchies among ports in container traffic, where large ships are used for long voyages between hubs, and the containers are then re-distributed in smaller

Table 18.1 | The world's busiest container ports in the five major transhipment markets – 2013. Source: WSC, 2014: http://www.worldshipping.org/about-the-industry/global-trade/top-50-world-container-ports.
* Not among the world's 50 busiest container ports.

Port	Country	TEU movements 2013 (Millions)
World's busiest container port		
Shanghai	China	33.62
North-East Asia		
Busan	Republic of Korea	17.69
Qingdao	China	15.52
Tianjin	China	13.01
Dalian	China	10.86
Keihin ports (Kawasaki, Tokyo, Yokohama)	Japan	8.37
Central East Asia		
Hong Kong	China	22.35
Ningbo-Zhoushan	China	17.33
Guangzhou	China	15.31
Kaohsiung	Taiwan Province of China	9.94
Xiamen *(formerly known as Amoy)*	China	8.01
South-East Asia		
Singapore	Singapore	32.60
Port Kelang	Malaysia	10.35
Tanjung Pelepas	Malaysia	7.63
Tanjung Priok	Indonesia	6.59
Laem Chang	Thailand	6.04
Middle East and Indian Sub-Continent		
Jebel Ali, Dubai	United Arab Emirates	13.64
Jeddah	Saudi Arabia	4.56
Colombo	Sri Lanka	4.31
Jawaharlal Nehru Port *(near Mumbai)*	India	4.12
Sharjah	United Arab Emirates	4.12
Mediterranean		
Algeciras Bay	Spain	4.50
Valencia	Spain	4.33
Ambarli *(near Istanbul)*	Turkey	3.38
Port Said	Egypt	3.12
Marsaxlokk	Malta	2.75
North-West Europe		
Rotterdam	Netherlands	11.62
Hamburg	Germany	9.30
Antwerp	Belgium	8.59
Bremen and Bremerhaven	Germany	5.84
Felixstowe	United Kingdom	3.74
South-East USA and Central America		
Colon	Panama	3.36
Balboa	Panama	3.19
Georgia Ports (Savannah, Brunswick)	United States	3.03
Hampton Roads (Newport News, Norfolk, Virginia Beach)*	United States	2.22
Houston*	United States	1.47

ships on shorter voyages, this has led to the need for ports to work together to offer shipping lines and (through them) shippers a comprehensive service. At the same time, in many parts of the world there has been a substantial transfer of the operation of ports (and, in some cases, the ownership of the land and equipment of the ports) from the public sector to the private sector. The combined effect of these various trends has been the creation of large commercial groupings of ports around the world. Some of these groupings have sprung from a successful operator of a specific port: the Port of Singapore Authority is the leading example of this type of development, with interests in 25 terminals around the world. Others have sprung from major shipping lines: APM Terminals is controlled by the major Danish maritime shipping enterprise A P Møller Mærsk, and has interests in 71 ports around the world. Another starting point for assembling a chain of ports has been sovereign wealth funds: for example, Dubai Ports World has interests in more than 65 terminals around the world. The final major type of port grouping is represented by Hutchison Port Holdings, part of the Hutchison Whampoa group, which developed from a dock-operating company in Hong Kong; it has interests in 52 ports. These four groups alone therefore have major interests in over 200 ports worldwide. There are a number of smaller similar chains, largely with a regional focus: these include SSA Marine in North America and Eurogate in Europe (privately-owned companies), Hanjin and Evergreen (linked to ocean carriers) and Ports America (owned by financial holding companies) (Rodrigue, 2010). In many countries, however, ports remain under the control of government agencies or chambers of commerce, or are independent public agencies;

(c) The larger sizes of ships have intensified the pressures to handle them in port in the shortest possible time. Ship owners want their capital to be earning money on voyages as much as possible, and therefore dislike the ships being tied up in port – or, even more, waiting at sea until they can get into a port berth. This, coupled with the more stringent requirements arising from growing trade volumes, global value chains, increasingly time-sensitive trade and lean supply chains, has led to increased competition between ports, intensified the pressure on ports to service ships and handle their cargo the shortest possible time and produced an intense focus on the efficiency of ports.

One important aspect of the economics of port operation is security against theft and disruption. In 2002, the International Maritime Organization adopted a new chapter in the International Convention on the Safety of Life at Sea (SOLAS) and promulgated the International Ship and Port Facility Security (ISPS) Code to improve ship and port security. This is supported by the joint IMO/International Labour Organization code of practice on security in ports. These instruments provide a consistent baseline worldwide, by clarifying the desirable division of responsibilities for issues such as access control, cargo and ship stores control, and facility monitoring to prevent unauthorized persons and materials from gaining access to the port. The ISPS Code came into force in 2004. Gaps still remain in some areas to implement these arrangements (IMO, 2015).

3.1 Efficiency

In 2012, the Organization for Economic Cooperation and Development (OECD) published a study on port efficiency that it had commissioned (Merk and Dang, 2012). This study sought to compare the efficiency of ports around the world, in the different fields of containers, grain, iron ore and oil, looking at proxies for the inputs of each type of port to the handling of cargoes and the throughput achieved, measured in terms of the dead-weight tonnage (dwt) passing through the port. For container ports, the study concluded that, with the exception of Rotterdam in the Netherlands, the most efficient ports were mostly located in Asia. The most efficient container ports were not necessarily the largest ports. Among most efficient ports are some of the largest global container ports (for example, Hong Kong, China; Singapore; and Shenzhen and Shanghai in China) (handling from 20 to 60 million dwt per port per month), but also medium to small size ports. For bulk oil ports, it concluded that, with the exception of Galveston, Texas, in the United States and (again) Rotterdam in the Netherlands, the most efficient oil ports are mostly located in the ROPME/RECOFI area[1], but not all ports in that region are operating efficiently. In this case, size does matter: the most efficient terminals are largely those with the largest throughput. In the case of bulk coal ports, the study concluded that a group of coal ports in Australia and China were clearly more efficient than nearly all the rest of the sample, although Velsen/IJmuiden in the Netherlands, Banjamarsin in India and Puerto Bolivar in Colombia were equally good. In the case of iron-ore and grain ports, the study concluded that, in both cases, larger ports were more efficient. It also concluded that, for grain ports, the least efficient terminals tend to be found in developed OECD countries. It should be noted, however, that the methodology of the study inevitably tends to rate a port as less efficient if, for historical reasons, its past investment has provided more facilities than is required for current levels of traffic.

It is instructive to compare the results of this study with the ranking published by the World Bank of the quality of the infrastructure of ports in different countries. This is based on a questionnaire to members of the World Economic Forum, which has been running for some 30 years. Recent rounds of the survey have included around 13,000 respondents from around 130 countries. Although subjective, the views expressed are likely to influence trade and investment decisions. The classification runs from 7 (efficient by international standards) to 1 (extremely underdeveloped). In 2012, the best-regarded ports were those in the Netherlands

---

1 Regional Organization for the Protection of the Marine Environment (ROPME) Members: Bahrain, Iran (Islamic Republic of), Iraq, Kuwait, Oman, Qatar, Saudi Arabia, and the United Arab Emirates. Regional Commission for Fisheries (RECOFI) Members: Bahrain, Iran (Islamic Republic of), Iraq, Kuwait, Oman, Qatar, Saudi Arabia, United Arab Emirates.

> **Box 18.1 | Quality of Port Infrastructure**
>
> **Category 6:** Bahrain, Belgium, Finland ↓, Germany, Hong Kong, China ↓, Iceland, Netherlands ↑, Panama ↑, Singapore, United Arab Emirates ↑.
>
> **Category 5:** Australia ↑, Barbados, Canada ↑, Chile ↓, Cyprus ↓, Denmark ↓, Estonia, France ↓, Ireland ↑, Jamaica ↓, Japan, Lithuania ↑, Malaysia, Malta ↑, Namibia, New Zealand, Norway ↓, Oman ↑, Portugal, Qatar ↑, Republic of Korea ↑, Saudi Arabia ↑, Seychelles, Slovenia, Spain ↑, Suriname ↑, Sweden, United Kingdom ↑, United States of America ↓.
>
> *Those countries marked ↑ had a higher ranking, and those marked ↓ a lower ranking, in 2012 than in 2009.*

and Singapore, both being ranked at 6.8. Box 18.1 shows the countries whose ports are regarded as being in categories 6 and 5.

The message from both these sources is that well-equipped and well-managed ports can be found in all parts of the world – as can less well-equipped and less well-managed ports. Given the importance of port effectiveness for world trade, improving capacities both in the planning and construction of ports and in their management could have beneficial effects. The facilities for the provision of accurate and timely navigational information to ships using ports is an important element of the equipment for the efficiency and effectiveness of ports, particularly in view of the adverse impacts on the marine environment from ships' casualties.

3.2 Charging

Charges for the use of ports raise some important issues. First, there is how to charge for services rendered. The normal recommendation of economists is that charges should only be levied if a service is delivered: economic theory argues against cross-subsidization between services. In the case of ports, however, there is a strong argument that ships' operators should not normally be able to opt out of paying for port waste-reception facilities. If they can opt out, they have an economic incentive not to pay for the disposal of their waste and to retain it on board until they can throw it into the sea, thereby aggravating the problem of marine debris. The European Union has adopted legislation requiring its ports generally to apply the rule of no separate charge for waste-reception facilities (EU, 2000). Whatever form a charge takes, it is important that the money is applied towards the environmentally sustainable disposal of the waste (see Chapter 17).

Secondly, there is the question of how far the port operator should be expected to cover the costs of providing the port. This applies both landward and seaward. In the landward direction, it is important that ports have adequate road, rail or inland-waterway connections to the port's hinterland. Otherwise, any efficiency gains in the port are cancelled out by the inefficiencies of transport into the hinterland. This can be very important for the economic viability of the port, since competitors may be able to offer a better deal overall. There is then the question of how far the costs of such adequate connections should be financed from the port charges rather than from government revenues or charges on the users of the connections. Decisions on this can only be taken for each port in the light of the policies of its possible competitors.

A parallel situation arises in the seaward direction, where there is often a need for dredging to maintain the access channels. In some countries, port operators have pressed governments to fund all or part of the costs of deepening and widening navigation channels, since they find themselves faced with competition from neighbouring ports which have natural deep-water harbours.

3.3 Landlocked countries

Because of the large proportion of international trade that is transported by sea (see Chapter 17 – Shipping), landlocked countries have particular difficulties from their lack of seaports. The 31 landlocked developing countries (LLDCs), 16 of which are among the least-developed countries (LDCs), face serious challenges to their growth and development, derived in substantial part from their problems in accessing maritime transport. In general, LLDCs face a 45 per cent higher ratio of freight charges to total value of exports and imports than the average of the developing countries through which their exports and imports must transit (LLDCs, 2011). This is a further aspect of capacity-building gaps to improve the efficiency of ports in the transit countries.

4 Impacts on the marine environment from port operations

The direct impacts on the marine ecosystem from ports take three main forms: first, the concentration of shipping, secondly, the demand for coastal space and, thirdly, the need for deep water. Chapter 26 (Land/sea interaction) considers other impacts that result from the transformations caused to the shoreline by the creation of ports and harbours.

4.1   Concentration of shipping

The concentration of shipping is generally an inevitable result of a successful port. Where a port takes part in a general market for port services, the more successful the port is, the greater are the size and number of the ships that it will serve. This means that discharges and emissions from the ships will be higher and have a more concentrated effect on the marine environment around the port. Even if each individual ship maintains the best practicable level of control over its impact, increasing levels of shipping to and from a port will result in increasing overall impacts, unless the best practicable means of control can be improved. Chapter 17 (Shipping) discusses the impacts from ships, particularly chronic oil discharges, garbage, sewage, anti-fouling treatments, air pollution and noise. All these can be controlled, but that control is more in the hands of the ships' masters and owners than in the hands of the port authority. Port authorities and governments can, however, influence these aspects through their charging policies and their enforcement of international standards. Because many ports have competition from their neighbours, effective action is likely to require agreement at a regional level. For this reason, the regional memorandums of understanding on port-state control have an important role in managing the impact of ports on the marine environment. Other effects, such as the turbidity caused by ships' propellers disturbing sediments, are more site-specific, and can to some extent be controlled by port navigation rules. Nevertheless, such turbidity (and the subsequent re-settlement of sediment) can have adverse impacts on sensitive habitats, such as corals and sea-grass beds (Jones, 2011).

In all these cases, port authorities and port operators have some important roles to play in managing the impacts of ships. Adequate waste-reception (and especially for cruise ships) sewage-reception facilities are important for preventing marine debris and eutrophication problems. Likewise, adequate land-based electricity supplies ("cold ironing") for ships that need to run equipment while in port (especially refrigerator ships) are essential to reduce air pollution, since otherwise they must run the ships' generators while they are in port.

The IMO has set up a system whereby ships' operators can report inadequacies in port reception facilities. This can be found at https://gisis.imo.org/Public/PRF/ReportedCases.aspx. It enables ships to report the problems that they have encountered and port authorities to offer (if they wish) explanations for such shortcomings and information on steps that are being taken to resolve them. Since the beginning of 2005, 279 inadequacies have been reported. States have responded in only 76 cases (although there are several where the port State had not been notified).

4.2   Coastal space

The demand for coastal space in ports is tied up with the growth in container traffic. Space is needed next to the berths for the containers to be off-loaded. In step with the development of container traffic, there has therefore been a substantial growth in the land needed for container ports. Rodrigue (2010, in Figure 3) shows the current scale of coastal space occupied by container ports. These are particularly demanding of coastal space because they have to have level space to hold the containers until they can be forwarded into the hinterland: bulk cargoes are normally transferred directly to less space-demanding storage.

Further growth in port throughput will inevitably result in further demand for container storage space at ports. This demand is rarely going to be able to be met from land that is not part of the coast, because around most ports this land is already committed to other forms of development (such as housing or industry) which are also essential for the growth of the port. As discussed in Chapter 26 (Land/sea physical interaction), this demand has therefore often been met by land reclamation – often from mangroves or salt marshes (for the pressures on which see Chapters 48 (Mangroves) and 49 (Salt marshes). These pressures are likely to continue. There is therefore a need for further investigation on how ports can handle increasing numbers of containers without increasing their demands for coastal space.

4.3   Deep water

The third pressure generated by ports is for deep water access channels. This normally means that dredging is used to deepen and widen the channels through sedimentary deposits, although in some cases it can involve blasting a channel through rock or (in rare cases) through coral reefs. Lack of available dredging services may constrain what can be done to provide deep-water access, and thus affect a port's competitiveness. Dredging can also affect the hydrodynamics of an estuary with consequences for adjacent beaches and seabed stability over broad areas (Pattiaratchi and Harris, 2002). Where dredging is used on areas not previously dredged, the impact on the bottom-dwelling flora and fauna may have to be balanced against the advantages of the improved access for ships. Where blasting is the only method available for providing the necessary deep-water access, the judgement is even more difficult, because it may mean the destruction of ecosystems based on a rocky or coral reef substrate. The quantities of material to be lifted by dredging can be immense (see Chapter 24 – Disposal of solid waste) and difficult judgements may have to be made about where the disposal should take place (Brodie, 2014). Where the dredging has to be done in the estuary of a river with a history of heavy industrial development, even more difficult judgements may have to be made about whether the dredged material should be re-introduced to the sea at all, given the risk of remobilising hazardous substances that have been sequestered in the sediments (see again Chapter 24 – Disposal of solid waste). The effects of elevated turbidity from dredging operations can have negative impacts on seagrasses (Erftemeijer and Lewis, 2006) and other benthic communities (Newell et al., 1998).

5 Integrating environmental, social and economic aspects

Port development is a special case of the issues raised by integrated coastal-zone management. Economically, it is always of high importance for the coastal State (and for the landlocked States that depend on transit through the coastal State). The pressures from ports will grow in step with the growth in international trade between coastal States, except to the extent that it is possible to improve the performance of ships and port installations. Port development also focuses together a large bundle of difficult trade-offs: increased benefits from trade, increased impacts from shipping, increased demand for coastal space and increased demand for creating or maintaining access channels. The growth in port throughput will therefore nearly always be accompanied by increased pressures on the environment. Social effects will be less pressing, because the changes needed as a result of the changeover to containerization are now largely in the past, and the social adjustments have been made. They will, however, need to be taken into consideration for those ports that have not yet joined the global consensus on containerization. A careful review of the different interests will therefore always be essential if port development is to be sustainable.

6 Information and capacity-building gaps

6.1 Knowledge gaps

Since ports constitute a significant economic sector, much information is available about them and their operations. What seems to be lacking is systematic information bringing together worldwide the operational aspects of ports and their impacts on the local marine environment, and their contribution to economic activity.

6.2 Capacity-building

Since the operation of a port can significantly affect both the successful operation of ships and the economic performance of the countries it serves, some ports need capacity-building in the operational skills needed for successful port operation. This is particularly important for ports that are serving as transit ports for landlocked countries, since the landlocked countries rely on the quality of port management in the transit country or countries, and are not in a position to insist on improvements.

It is important to develop (and then maintain) the capacities of port States both to implement the International Ship and Port Facility Security Code and related instruments and to carry out port-State inspections of ships, so as to enforce the internationally agreed standards for ships. Capacities to provide ships with good, real-time information on local navigational issues are also important.

Since the delivery to shore of garbage from ships in general is an important element of combating marine debris problems, ports which do not have adequate and easily used port waste-reception facilities need to have their capacities in this field improved. The same applies to sewage-reception facilities for cruise ships in relation to eutrophication problems.

Where ports which need dredging to maintain or improve navigation adjoin bays, rivers or estuaries with a history of industrial discharges, there is a need for them to have the capacity to examine the dredged material to decide whether it can safely be re-deposited in the sea.

References

Baosteel (2008). *Baosteel Bought Shares of Zhanjiang Port Group*, http://www.baosteel.com/group_en/contents/2863/38876.html (accessed 16 June 2014).

Brodie, J. (2014). Dredging the Great Barrier Reef: Use and misuse of science. *Estuarine, Coastal and Shelf Science* 142.

DfT (United Kingdom Department for Transport) (2014). *UK Port Freight Statistics 2013*. https://www.gov.uk/government/uploads/system/uploads/attachment_data/file/347745/port-freight-statistics-2013.pdf (accessed 20 October 2014).

Donaldson of Lymington, Lord (1994). *Cleaner Seas, Safer Ships: Report of Lord Donaldson's Inquiry into the Prevention of Pollution from Merchant Shipping*, Her Majesty's Stationery Office, London (ISBN 978-0101256025).

Erftemeijer, P.L.A., Lewis III, R.R.R. (2006). Environmental impacts of dredging on seagrasses: A review. *Marine Pollution Bulletin* 52.

EU (European Union) (2000). Directive on port reception facilities (Directive 2000/59/EC).

Haralambides, H.E. (2002). Competition, Excess Capacity, and the Pricing of Port Infrastructure, *International Journal of Maritime Economics*, Vol. 4 (4).

ILO (International Labour Organization) (2002). *General Survey of the reports concerning the Dock Work Convention (No. 137) and Recommendation (No. 145), 1973*. (ISBN 92-2-112420-7).

IMO (International Maritime Organization) (2015). *The International Ship and Port Facility Security Code (ISPS Code)* (http://www.imo.org/OurWork/Security/Instruments/Pages/ISPSCode.aspx accessed 20 April 2015).

Jones, R.J. (2011). Environmental Effects of the Cruise Tourism Boom: Sediment Resuspension from Cruise Ships and the Possible Effects of Increased Turbidity and Sediment Deposition on Corals (Bermuda). *Bulletin of Marine Science*, Volume 87, Number 3, 2011.

LLDCs (Group of Landlocked developing Countries) (2011). *Position Paper on the draft outcome document for UNCTAD XIII*, Geneva (UNCTAD Document TD/450).

Merk, O., Dang, T.T. (2012). Efficiency of World Ports in Container and Bulk Cargo (oil, coal, ores and grain), *OECD Regional Development Working Papers*, 2012/09, OECD Publishing, Paris.

Newell, R.C., Seiderer, L.J., Hitchcock, D.R., (1998). The impact of dredging works in coastal waters: a review of the sensitivity to disturbance and subsequent recovery of biological resources on the sea bed. *Oceanography and Marine Biology Annual Review* 36.

Notteboom, T., Parola, F. and Satta, G. (2014). *Progress Report on EU Research Project: Synthesis of the information regarding the container transshipment volumes* (http://www.portopia.eu/wp-content/uploads/2015/01/Transshipment.pdf accessed on 20 April 2015).

Pattiaratchi, C.B., Harris, P.T. (2002). Hydrodynamic and sand transport controls on en echelon sandbank formation: an example from Moreton Bay, eastern Australia. *Journal of Marine Research* 53.

Rodrigue, J. (2010). Maritime Transportation: Drivers for the Shipping and Port Industries, in *International Transport Forum 2010 "Transport and Innovation: Unleashing the Potential"*. http://www.internationaltransportforum.org/Proceedings/Genoa2010/Rodrigue.pdf (accessed 29 November 2013).

Saturday Paper (2014). Great Barrier Reef dredging goes to federal court, 29 March. http://www.thesaturdaypaper.com.au/news/environment/2014/03/29/great-barrier-reef-dredging-goes-federal-court/1396011600 (accessed 3 December 2014).

UNCTAD (United Nations Conference on Trade and Development) (2008). *Outcome of the meeting "Globalization of port logistics: opportunities and challenges for developing countries"* (UNCTAD document TD/419).

UNCTAD (United Nations Conference on Trade and Development) (2013). *Review of Maritime Transport*, Geneva (ISBN 978-92-1-112872-7).

World Bank (2012). *Quality of Port Infrastructure*. http://data.worldbank.org/indicator/IQ.WEF.PORT.XQ (accessed 14 January 2014).

WSC (World Shipping Council) (2014). *Top 50 World Container Ports*. http://www.worldshipping.org/about-the-industry/global-trade/top-50-world-container-ports (accessed 20 October 2014).

Zetland (1974). *Zetland County Council Act* (1974 c. viii).

19 Submarine Cables and Pipelines

Group of Experts:
Alan Simcock (Lead member)

Submarine Cables and Pipelines — Chapter 19

1 Submarine communications cables

1.1 Introduction to submarine communications cables

In the last 25 years, submarine cables have become a dominant element in the world's economy. It is not too much to say that, without them, it is hard to see how the present world economy could function. The Internet is essential to nearly all forms of international trade: 95 per cent of intercontinental, and a large proportion of other international, internet traffic travels by means of submarine cables. This is particularly significant in the financial sphere: for example, the SWIFT (Society for Worldwide Interbank Financial Telecommunication) system was transmitting financial data between 208 countries via submarine cables in 2010. As long ago as 2004, up to 7.4 trillion United States dollars were transferred or traded on a daily basis by cables (Rauscher, 2010). The last segment of international internet traffic that depended mainly on satellite communications was along the East coast of Africa: that was transferred to submarine cable with the opening of three submarine cables along the East coast of Africa in 2009-2012 (Terabit, 2014). Submarine cables have advantages over satellite links in reliability, signal speed, capacity and cost: the average unit cost per Mb/s capacity based on 2008 prices was 740,000 dollars for satellite transmission, but only 14,500 dollars for submarine cable transmission (Detecon, 2013).

Submarine telegraph traffic by cable began between England and France in 1850-1851. The first long-term successful transatlantic cable was laid between Newfoundland, Canada, and Ireland in 1866. The early cables consisted of copper wire insulated by gutta percha, and protected by an armoured outer casing. The crucial development that enabled the modern systems was the development of fibre-optic cables: glass fibres conveying signals by light rather than electric current. The first submarine fibre-optic cable was laid in 1986 between England and Belgium; the first transatlantic fibre-optic cable was laid in 1988 between France, the United Kingdom and the United States. It was just at that time that the Internet was beginning to take shape, and the development of the global fibre-optic network and the Internet proceeded hand in hand. The modern Internet would not have been possible without the vastly greater communications possibilities offered by fibre-optic cables (Carter et al., 2009). Over the 25 years from 1988 to 2013, an average of 2,250 million dollars a year was invested in laying 50,000 kilometres of cable a year. However, this includes a great burst in the development of the global fibre-optic network that took place in 2000-2002, in conjunction with the massive interest in investment in companies based on the Internet: the so-called dot-com bubble. At the peak, in 2001, 12,000 million dollars were invested in submarine cables in one year. After the dot-com bubble burst in 2002, the cable-laying industry contracted severely, but by 2008 had recovered to what has since been a steady growth (Terabit, 2014). Figures 1 and 2 show diagrammatically the transatlantic and transpacific submarine communications cables that exist. More detailed diagrammatic maps showing submarine cables in the Caribbean, the Mediterranean, North-West Europe, South and East Asia, and Sub-Saharan Africa can be found here: http://submarine-cable-map-2014.telegeography.com/.

Two Arctic submarine communications cables are reported to be planned, linking Tokyo and London: one will go around the north of the Eurasian continent, the other around the north of the American continent through the North-West passage; both would service Arctic communities *en route*. In 2012, both were planned to be in service by 2016. The link by the American route is said to be under construction but is not now expected to be complete until 2016. The link around the Eurasian route is reported to be stalled (Hecht, 2012; Arctic Fibre, 2014; Telegeography, 2013; APM, 2015).

Deployed international bandwidth (in other words, the total capacity of the world's international cables) increased at a compound annual growth rate of 57 per cent between 2007 and 2011. It reached 67 Terabits per second (Tbps) in 2011, which was six times the bandwidth in use in 2007 (11.1 Tbps). It has increased steadily since then and was estimated to be increasing to about 145 Tbps in 2014 (Detecon, 2013). Submarine cable bandwidth is somewhat lower, as shown in Table 19.1. The investment necessary to support this steady stream of investment is provided through consortia. The precise balance of the different interests varies from case to case, but the major players are nearly always national telecommunications operators, internet service providers and private-sector equity investors. Governments are rarely involved, except through government-owned national telecommunications operators (Terabit, 2014; Detecon, 2013).

1.2 Magnitude of the impact of submarine cables on the marine environment

In 2007, the total route length of submarine fibre-optic cables was about 1 million route kilometres (Carter et al., 2009). This has now extended to

Table 19.1 | Activated Capacity on Major Undersea Routes (Tbps), 2007-2013. Source: Terabit, 2014.

	2007	2008	2009	2010	2011	2012	2013	CAGR, 2007-2013
Transatlantic	6	8	11	13	15	19	23	25%
Transpacific	3	7	8	12	12	14	20	35%
Pan-East Asian	2	2	6	8	10	12	17	46%
South Asia & Middle East Intercontinental	1	2	3	3	4	8	12	42%
North America-South America	1	1	3	4	6	7	9	52%
Australia & New Zealand Intercontinental	1	1	2	2	2	3	5	40%
Sub-Saharan African Intercontinental	0	0	0	1	1	2	2	57%
Global Transoceanic Bandwidth (Tbps)	14	22	33	43	51	65	87	36%
Percent Change		57%	49%	32%	19%	26%	35%	

about 1.3 million route kilometres, given the extensions reported in the 2014 Submarine Cable Report (Terabit, 2014). Although these are great lengths, the breadth of the impact on the marine environment is much, much less: the diameter of the fibre-optic cables on the abyssal plain is about 17-20 millimetres – that is, the width of a typical garden hose. On the continental shelf, the width of the cable has to be greater – about 28-50 millimetres – to allow for the extra armour to protect it from impacts and abrasion in these more dynamic waters and the greater threats from shipping and bottom trawling (Carter et al., 2009).

The cable is normally buried in the seabed if the water depth is less than 1,000-1,500 metres and the seabed is not rocky or composed of highly mobile sand. This is to protect the cable against other users of the sea, such as bottom trawling. Known areas where mineral extraction or other uses are likely to disturb the seabed are avoided. In greater water depths, the cable is normally simply laid on the seabed (Carter et al., 2009). Where a cable is buried, this is normally done by a plough towed by the cable ship that cuts a furrow into which the cable is fed. In a soft to firm sedimentary seabed, the furrow will usually be about 300 millimetres wide and completely covered over after the plough has passed. On other substrates, the furrow may not completely refill. The plough is supported on skids, and the total width of the strip disturbed may be between two and eight metres, depending on the type of plough used. Various techniques have been used to minimise disturbance in specially sensitive areas: on the Frisian coast in Germany, a specially designed vibrating plough was used to bury a cable through salt marshes (recovery was monitored and the salt-marsh vegetation was re-established in one to two years and fully recovered within five years); in Australia, cables crossing seagrass beds were placed in narrow slit trenches (400 millimetres wide), which were later replanted with seagrass removed from the route prior to installation; in the Puget Sound in Washington State in the USA, cables were installed in conduits drilled under a seagrass bed. Mangroves are reported to have recovered within two to seven months, and physical disturbance of sandy coasts subject to high-energy wave and tide action is reported to be removed within days or weeks. Where burial has not been possible, it has sometimes been necessary to impose exclusion zones and to monitor such zones (as between the North and South Islands of New Zealand (Carter et al., 2009)).

The boundaries and names shown and the designations used on this map do not imply official endorsement or acceptance by the United Nations.
Figure 19.1 | Diagrammatic map of transatlantic submarine cables. Source: Telegeography, 2014.

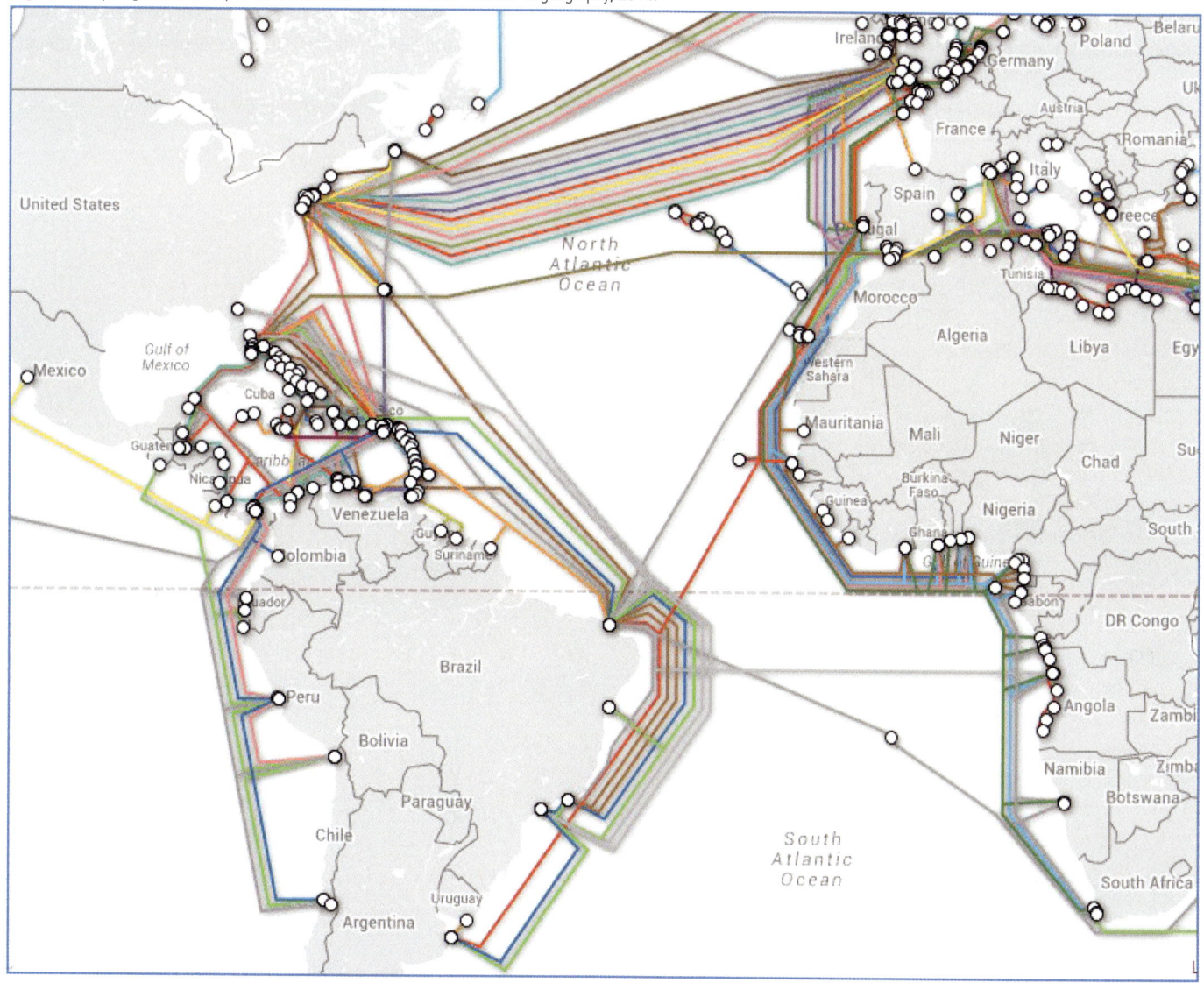

Submarine Cables and Pipelines

Further disturbance will occur if a cable failure occurs. Areas of cable failure are likely to have already been disturbed by the activity that caused the cable failure. Normally, the cable will have to be brought to the surface for repair. This will involve the use of a grapnel dragged across the seabed, unless a remotely operated robot submarine can be used. Reburial of the cable may involve agitating the sediment in which it has been buried. This disturbance will mobilise the sediment over a strip up to 5 metres wide. Fibre-optic cables have a design life of 20-25 years, after which the cable will need to be lifted and replaced, with a recurrence of the disturbance, although there is also the possibility of leaving them in place for use for purposes of scientific research (Carter et al., 2009; Burnett et al., 2014).

Evaluating the impact on marine animals and plants of this disturbance is not easy, since the area affected, though long, is narrow. In general, the verdict is that the seabed around a buried cable will have returned to its normal situation within at most four years. In waters over 1,000-1,500 metres deep (where burial is unusual), no significant disturbance of the marine environment has been noted, although any repairs will disturb the plants and animals that may grow on the cable. Such growth is common on exposed cables in shallow calm water, but is limited in water depths greater than 2000 metres, where biodiversity and macrofaunal abundance are much reduced (Carter et al., 2009). Some noise disturbance may be caused by the process of laying cables, but this is not significantly more than would be caused by ordinary shipping (OSPAR, 2008).

1.3 Threats to communications cables from the marine environment

Soon after transoceanic communications cables were laid, problems were experienced from impacts of the marine environment on the cables: specifically, submarine earthquakes and landslides breaking the cables (Milne, 1897). However, around 70 per cent of all cable failures are associated with external impacts caused by fishing and shipping in water depths of less than 200 metres (Carter et al., 2009).

Nevertheless, the risks of damage through catastrophic geological events (including those triggered by storms) are real, and some aspects of such risks are probably growing (see the discussion of the effects of climate change on storms in Chapter 5). The most recent major events have been near the Taiwan Province of China. On 26 December 2006, an earthquake occurred at the south end of the island. This triggered multiple submarine landslides. The landslides and subsequent turbidity currents travelled over 330 kilometres and caused 19 breaks in seven cable systems. Damage was located in water depths to 4,000 metres.

The boundaries and names shown and the designations used on this map do not imply official endorsement or acceptance by the United Nations.

Figure 19.2 | Diagrammatic map of transatlantic submarine cables. Source: Telegeography, 2014.

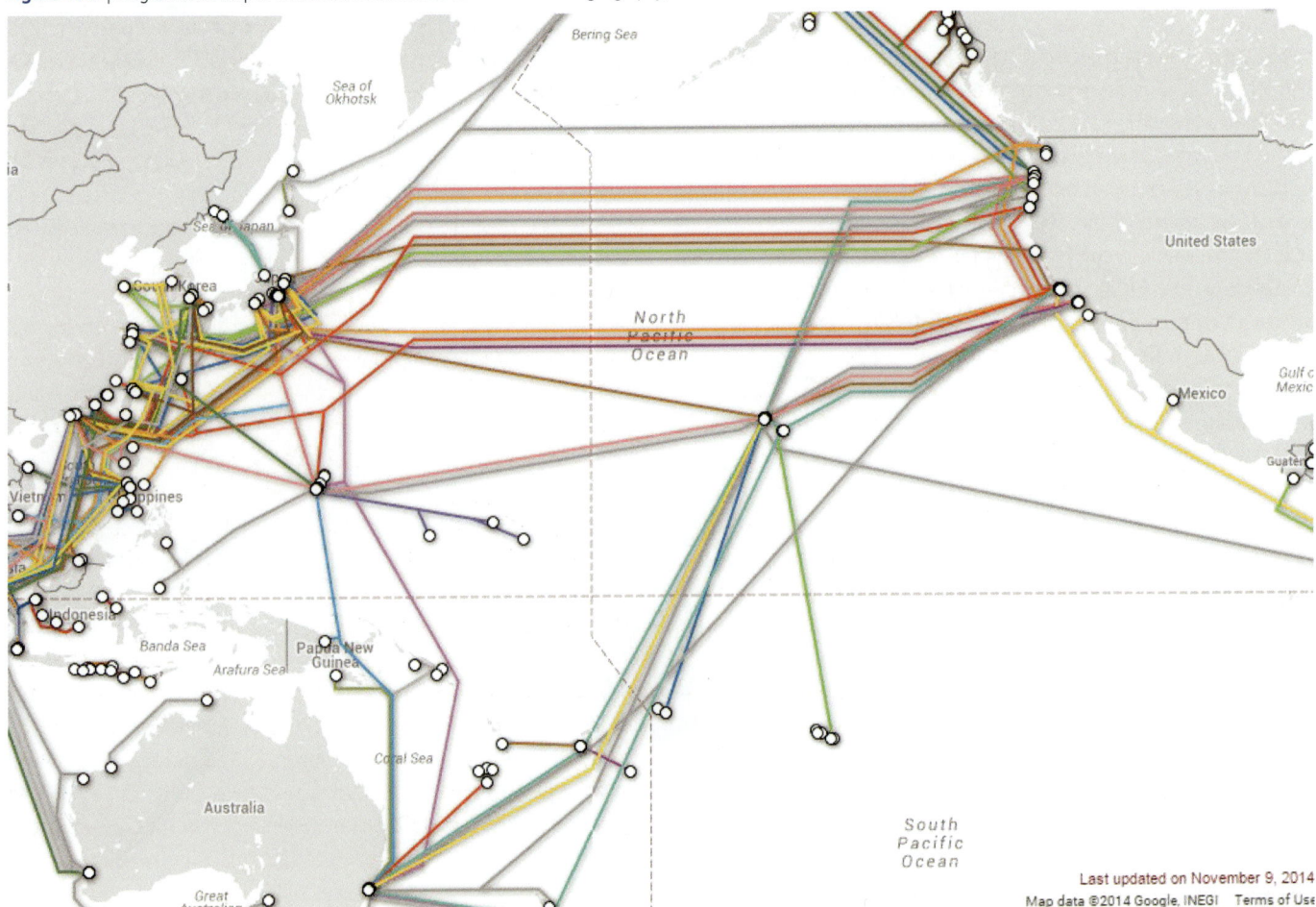

The cable repair works involved 11 repair vessels and took 49 days. The result was a major disruption of services in the whole region: the internet connections for China, Japan, Philippines, Singapore and Viet Nam were seriously impaired. Banking, airline bookings, email and other services were either stopped or delayed and financial markets and general commerce were disrupted (Detecon, 2013; Carter et al., 2014).

Three years later, Typhoon Morakot hit the island of Taiwan Province of China, on 7 August 2009. Three metres of rain fell on the central mountains, causing much erosion. The sediment carried into sea caused several submarine landslides which broke a number of cables. The level of disruption was shorter and less serious than in 2006. This case is particularly significant, however, because it was the result of an extreme weather event. Given the consensus that climate change is causing the poleward migration of storms, areas that have previously been spared this kind of event are more likely in future to suffer from such storms. This is likely to increase the chances of submarine landslides, since an instability will be introduced into areas where it has not previously been generated (Carter et al., 2012).

The seas off East Asia present a combination of a very dense network of submarine communications cables (see the diagrammatic map in http://submarine-cable-map-2014.telegeography.com/) and an area of unstable geology. The scale of disruption that might be caused, either by a geological incident or by a vessel, can be envisaged by considering the Straits of Malacca. Fourteen of the 37 main submarine cables in the Western Pacific run through this narrow strait. These cables represent virtually the entire data connection between Asia, India, the Middle East and Europe. In addition, it is one of the busiest shipping routes worldwide. This drastically increases the likelihood of disruptions by anchors and other manmade hazards. Such disruptions unfortunately do happen regularly (Detecon, 2013). This, and the situation on the Isthmus of Suez, is one of the main attractions in a submarine cable route from the Pacific to the Atlantic around the north of either the American or the Eurasian continent. There is further a risk from deliberate human interference, but statistically this is a rare event (Burnett et al., 2014).

The International Cable Protection Committee Ltd. (ICPC) is a non-profit organization that facilitates the exchange of technical, legal and environmental information concerning submarine cable installation, maintenance and protection. It has over 150 members representing telecommunication and power companies, government agencies and scientific organizations from more than 50 countries, and encourages cooperation with other users of the seabed. It is thus the main forum in which issues about the protection of these submarine cable connections, vital to global commerce, are being discussed.

1.4 Information and capacity-building gaps

A large body of knowledge already exists about the construction and operation of submarine communication cables, including how to survey environmentally acceptable routes and allow for the submarine geology. Coastal States need access to these skills to decide on safe locations and to take account of areas of potential geological change and disruption, or (at least) to negotiate successfully with commercial undertakings planning to install cables.

As with many other uses of the marine environment that involve uses of the seabed within their jurisdictions that may prevent or limit other legitimate uses of the sea, States need to have the capacities, in taking decisions on submarine cables, for resolving the conflicting demands of these uses with the other parties involved.

2 Submarine power cables

2.1 The nature and magnitude of submarine power cables

The number and extent of submarine cables carrying power rather than communications are much less significant, both in terms of their impact on the marine environment and in their importance to the world economy. They are essentially of only local interest.

Most of the world's submarine power cables are found in the waters around Europe. The cables fall into one of two classes, depending on whether the electricity is carried as direct current (DC) or alternating current (AC). The choice depends on several factors, including the length of the submarine cable and the transmission capacity needed: DC cables are preferred for longer distances and higher transmission capacities. DC cables can be either monopolar (when the current returns through the sea water) or bipolar (when the cable has two components with opposite polarities). Because monopolar DC cables tend to produce electrolysis, they are now rarely used for major projects.

The AC cables include those between the mainland of Germany and its island of Heligoland, between Italy and its island of Sicily, between Spain and Morocco, between Sweden and the Danish island of Bornholm and, outside Europe, between the islands of Cebu, Negros and Panay in the Philippines. The DC cables include cables linking the Danish islands of Lolland, Falster and Zealand to Germany, Denmark to Norway, Denmark to Sweden, Estonia to Finland, Finland to Sweden, France to the United Kingdom, Germany to Sweden, the Italian mainland to its island of Sardinia and to the French island of Corsica, the Netherlands to Norway (at 580 kilometres, this is the longest submarine power cable in the world), the Netherlands to the United Kingdom, Northern Ireland to Scotland in the United Kingdom of Great Britain and Northern Ireland and the mainland of Sweden to its island of Gotland. Outside Europe, there are DC cables linking the mainland of Australia to its island of Tasmania, the mainland of Canada to its Vancouver Island, Honshu to Shikoku in Japan, the North Island to the South Island in New Zealand and Leyte to Luzon in the Philippines.[1] As can be seen, all these cables (with the exception of the Netherlands/Norway cable) cross fairly narrow stretches of water. They play a locally important part in managing electricity supply, enabling surpluses in one country or area to be transferred to another, or to enable an island to benefit from the economies of scale in power generation through a link to power stations serving a much bigger area. The links between Denmark, Norway and Sweden play an important role in the common power policy of those three States.

2.2 Environmental impacts of submarine power cables

The disturbance of the marine environment caused by the installation of a power cable will usually be larger than that for a communications cable, simply because the cable will be larger, in order to carry the current. However, neither the physical disturbance nor the associated noise is likely to have more than a temporary effect.

The other two aspects that have given rise to concern are heat and electromagnetic fields. There are few empirical studies of heat emitted from submarine power cables. AC cables are theoretically likely to emit more heat than DC cables. Calculations for the cable from the Australian mainland through the Bass Strait to Tasmania suggested that the external surface temperature of the cable would reach about 30°C-35°C. The seabed surface temperature directly overlying the cable was expected to rise by a few degrees Celsius at a burial depth of 1.2 metres. Readings taken at a Danish wind farm in 2005 showed that, for a 132 kV cable, the highest temperature recorded closest to the cable between March and September was 17.7°C. German authorities have set a precautionary standard for new cables such that the cables should not raise the temperature at a depth of 20 cm in the seabed by more than 2°C. This can be achieved by burying the cables at an appropriate depth (OSPAR, 2008).

Concerns have been raised about the effects of the electromagnetic fields generated by the electric current flowing along submarine power cables, since some fish and marine mammals have been shown to be sensitive to either electric fields or magnetic fields. The World Health Organization, however, concluded in 2005 that "…none of the studies performed to date to assess the impact of undersea cables on migratory fish (e.g. salmon and eels) and [on] all the relatively immobile fauna inhabiting the sea floor (e.g. molluscs), have found any substantial behavioural or biological impact" (WHO, 2005). A literature survey in 2006 reached a similar conclusion (Acres, 2006), and nothing had emerged by the 2010 European Union report on the implementation of the EU Marine Strategy Directive to cast doubt on those conclusions (Tasker et al., 2010).

2.3 Knowledge and capacity-building gaps

As with communications cables, coastal States need to have access to the skills to locate submarine power cables in a safe and environmentally acceptable way, and to evaluate the economic and social benefits of introducing such links.

3 Submarine Pipelines

3.1 The nature and magnitude of submarine pipelines

Submarine pipelines are used for transporting three main substances: gas, oil and water. Submarine gas and oil pipelines fall into three groups: intra-field pipelines, which are used to bring the oil or gas from wellheads to a point within the operating field for collection, processing and onward transport; export pipelines, which transport the gas and oil to land; and transport pipelines, which have no necessary connection with the operating field, but transport gas or oil between two places on land. The last category is often included with the export pipelines. The intra-field and export pipelines are discussed in Chapter 21 as part of the processes of extracting the oil and gas. This section is concerned only with the transport pipelines. In general, what is said about submarine pipelines in Chapter 21 applies to gas and oil transport pipelines.

Submarine transport pipelines are used mainly for the transport of gas and are located predominantly around the Mediterranean and the Baltic and North Seas. Many have been created since 2000. In the Mediterranean, the earliest gas pipeline was the Trans-Mediterranean Pipeline, built in 1983 to link Algeria and the Italian mainland, via Sicily. This was followed in 1996 by the Maghreb-Europe Pipeline to link Morocco and Spain across the Strait of Gibraltar. Subsequent Mediterranean pipelines are: the Greenstream Pipeline, built in 2004 between Libya and Sicily, the interconnector built in 2007 between Greece and Turkey, the link completed in 2008 between Arish in Egypt and Ashkelon in Israel (which has been out of service since 2012), and the Medgaz Pipeline built in 2011 between Algeria and Spain. Further north, a link was built between Scotland and Northern Ireland in the United Kingdom in 1996. An interconnector was built between Belgium and the United Kingdom in 1998. The Balgazand/Bacton Line (BBL) connected the Netherlands and the United Kingdom in 2006. Finally, the Nord Stream Pipeline was completed in 2011 and 2012 through the Baltic, between Vyborg in the Russian Federation and Kiel in Germany. This is the longest gas transport pipeline in the world (1,222 kilometres in length). Issues about its environmental impact bulked large in the negotiations leading to its construction, and particular problems were encountered over munitions dumped in the Baltic at the end of the Second World War (see Chapter 24 (Solid waste disposal)).[2] There are also a number of gas pipelines linking Norwegian gas production to its export markets. The Norwegian upstream gas transportation system has been developed from

---

[1] This list has been compiled from a variety of sources.

[2] This list has been compiled from a variety of sources.

the 1970s, and continues to develop, to cater for the transportation of natural gas produced on the Norwegian continental shelf. Norwegian domestic consumption of natural gas is limited. Almost all the gas produced is exported (101,000 million standard cubic metres) to European gas markets through landing terminals in Belgium, France, Germany and the United Kingdom. The pipeline network in 2014 forms a 7,980-kilometre integrated transportation system, transporting gas from nearly 60 offshore fields and three large gas processing plans on the Norwegian mainland, to European gas markets. The latest main addition to the system is the Langeled Pipeline, opened in 2007, which goes from the onshore processing plant in Norway for the Ormen Lange gas field to the United Kingdom, via a riser platform at the Sleipner field.

Outside Western Europe and the Mediterranean, there is a gas pipeline linking the Russian Federation and Turkey across the South-Eastern corner of the Black Sea, and one linking the island of Sakhalin to the mainland of the Russian Federation in the North-West Pacific. Oil transport pipelines exist between Indonesia and Singapore across the Strait of Malacca, and in China, linking the island of Hainan to Hong Kong.[3] Generally, these submarine transport pipelines have been built and financed by oil and gas operators (including national oil and gas companies), sometimes in consortiums with national gas distribution undertakings.

3.2 Environmental impacts of oil and gas pipelines

The environmental impacts of intra-field and export submarine pipelines are discussed in Chapter 21 (Offshore hydrocarbon industries). The impacts of oil and gas submarine transport pipelines are essentially the same.

3.3 Submarine water pipelines

Because of the high cost and maintenance difficulties, submarine pipelines are only used to supply small islands close to continents or larger islands where the natural water supplies of the islands are insufficient for their needs. The supply of water to Singapore from Malaysia is the only significant international example (PUB, 2014). Domestic examples include: China (where Xiamen Island receives some of its water from the mainland through 2.3 kilometres of submarine pipelines), Fiji (where several small islands with tourism resorts are supplied through submarine pipelines), Malaysia (where Penang receives some of its water supply from the Malaysian mainland through 3.5 kilometres of submarine pipelines), the Seychelles (where five small islands are supplied through submarine pipelines of up to 5 kilometres in length) and, most significantly, in Hong Kong, China (where water is supplied to some of the islands, including the densely populated Hong Kong Island, from the Chinese mainland, through 1.3 kilometres of submarine pipelines) (UNESCO, 1991).

3.4 Knowledge and capacity-building gaps

For oil and gas transport pipelines, the requirements are likely to arise from the overall planning of the exploitation of hydrocarbon reserves and the provision of gas services. The comments in Chapter 21 on this subject are therefore relevant.

For submarine water pipelines, the essential questions will be linked to the planning and implementation of freshwater supply services. Questions of access to information and the necessary skills need to be addressed in that context. As with the laying of submarine communication cables, in taking decisions on submarine water pipelines within their jurisdictions, States need to have the capacities for resolving the conflicting demands of these uses.

---

3 This information has also been compiled from a variety of sources.

References

Acres, H. (2006). Literature Review: *Potential electromagnetic field (EMF) effects on aquatic fauna associated with submerged electrical cables*. Supplement to the Environmental Assessment Certificate (EAC) Application for the Vancouver Island Transmission Reinforcement (VITR) Project. Prepared for BC Hydro Environment & Sustainability Engineering, Victoria BC.

Arctic Fibre (2014). www.arcticfibre.com (accessed 10 November 2014).

APM (Alaska Public Media). (2015). "Arctic Fiber Project Delayed Into 2016" (http://www.alaskapublic.org/2014/12/23/arctic-fiber-project-delayed-into-2016/ accessed 10 June 2015).

Burnett, D.R., Beckman, R.C. and Davenport, T.M. (eds.), (2014). *Submarine Cables: The Handbook of Law and Policy*, Nijhoff, Leiden (Netherlands) and Boston (USA) (ISBN 978-90-04-26032-0).

Carter, L., Burnett, D. Drew, S. Marle, G. Hagadorn, L. Bartlett-McNeil, D., and Irvine, N. (2009). *Submarine Cables and the Oceans – Connecting the World*. UNEP-WCMC Biodiversity Series No. 31. ICPC/UNEP/UNEP-WCMC, Cambridge (England.

Carter, L., Milliman, J.D., Talling, P.J., Gavey, R., and Wynn, R.B. (2012). Near-synchronous and delayed initiation of long run-out submarine sediment flows from a record-breaking river flood, offshore Taiwan, *Geophysical Research Letters*, Volume 39, 12, doi:10.1029/2012GL051172.

Carter, L., Gavey, R. Talling, P.J. and Liu, J.T. (2014). Insights into submarine geohazards from breaks in subsea telecommunication cables. *Oceanography* 27(2).

Detecon (2013). Detecon Asia-Pacific Ltd, *Economic Impact of Submarine Cable Disruptions*, prepared for Asia-Pacific Economic Cooperation Policy Support Unit (Document APEC#213-SE-01.2).

Hecht, J. (2012). Fibre optics to connect Japan to the UK – via the Arctic, *New Scientist*, 2856.

Milne, J. (1897). Sub-Oceanic Changes: Section III, *The Geographical Journal*, Vol. 10(3).

OSPAR (2008). OSPAR Commission, *Background Document on potential problems associated with power cables other than those for oil and gas activities*. London.

PUB (Singapore Public Utilities Board) (2014). *The Singapore Water Story Water: From Vulnerability to Strength.* http://www.pub.gov.sg/water/Pages/singaporewaterstory.aspx (accessed 25 October 2014).

Rauscher, K. F. (2010). ROGUCCI – *Reliability of Global Undersea Cable Communications Infrastructure – Report*. IEEE Communications Society, New York, USA.

Tasker, M.L., Amundin, M., Andre, M., Hawkins, A., Lang, W., Merck, T., Scholik-Schlomer, A., Teilmann, J., Thomsen, F., Werner, S. and Zakharia, M. (2010). Marine Strategy Framework Directive Task Group 11 Report, *Underwater noise and other forms of energy*, Luxembourg.

Telegeography (2013). Is dormant Polarnet project back on the agenda? *Telegeography*, (https://www.telegeography.com/products/commsupdate/articles/2013/01/28/is-dormant-polarnet-project-back-on-the-agenda/ accessed 10 October 2014).

Telegeography (2014). Submarine Cable Map 2014. *Telegeography* (http://submarine-cable-map-2014.telegeography.com/ accessed 30 September 2014).

Terabit (2014). Terabit Ltd/Submarine Telecoms Forum Inc, *Submarine Cables Industry Report*, Issue 3. (http://www.terabitconsulting.com/downloads/2014-submarine-cable-market-industry-report.pdf accessed 20 August 2014).

UNESCO (1991). United Nations Education Scientific and Cultural Organization, *Hydrology and Water Resources of Small Islands, A Practical Guide*. Studies and Reports on Hydrology No. 49, UNESCO, Paris.

WHO (2005). World Health Organization, *Electromagnetic Fields and Public Health – Effects of EMF on the Environment*, (http://www.who.int/peh-emf/publications/facts/envimpactemf_infosheet.pdf accessed on 21 November 2014). Geneva.

20 Coastal, Riverine and Atmospheric Inputs from Land

Contributors:
Alan Simcock (Convenor and Lead Member), Benjamin Halpern, Ramalingam Kirubagaran, Md. M. Maruf Hossain, Marcos Polette, Emma Smith and Juying Wang (Co-Lead Member)

Commentators:
Arsonina Bera, Mark Costello, Robert Duce, Ralf Ebinghaus, Jim Kelley, Thomas Malone and Jacquis Rasoanaina.

Some material originally prepared for Chapter 36F (Open Ocean Deep Sea) and Chapter 51 (Seamounts and other submarine features potentially threatened by disturbance) has been incorporated into parts of this Chapter. The contributors to those chapters were Jeroen Ingels, Malcolm R. Clark, Michael Vecchione, Jose Angel A. Perez, Lisa A. Levin, Imants G. Priede, Tracey Sutton, Ashley A. Rowden, Craig R. Smith, Moriaki Yasuhara, Andrew K. Sweetman, Thomas Soltwedel, Ricardo Santos, Bhavani E. Narayanaswamy, Henry A. Ruhl, Katsunori Fujikura, Linda Amaral Zettler, Daniel O. B. Jones, Andrew R. Gates, Paul Snelgrove, J. Anthony Koslow, Peter Auster, Odd Aksel Bergstad, J. Murray Roberts, Alex Rogers, Michael Vecchione.

1 Introduction

The movement of materials from land to sea is an inevitable part of the hydrological cycle and of all geological processes. Nevertheless, human activities have both concentrated and increased these flows as a result of the creation of large human settlements, the development of industrial processes and the intensification of agriculture. Until the 1960s, many took the view that the oceans were capable of assimilating everything that humans wanted to discharge into the oceans. In the 1960s, this view came to be seen as out-dated (UNESCO, 1968). Following the 1972 Stockholm Conference on the Human Environment, many steps were taken to address issues of marine pollution. During the preparations for the United Nations Conference on Environment and Development ("the first Earth Summit"), held in Rio de Janeiro, Brazil, in 1992, there was general agreement that, in spite of what had been done, a major initiative was needed to address the problems of land-based inputs to the oceans. As a result, Agenda 21 (the non-binding action plan from the 1992 Earth Summit) invited the United Nations Environment Programme ("UNEP") to convene an intergovernmental meeting on protection of the marine environment from land-based activities (Agenda 21, 1992). In October 1995, the Global Programme of Action for the Protection of the Marine Environment from Land-Based Activities (GPA) was adopted in Washington, DC. First among the priorities of this Programme was improving the management of waste-water: this concerned not only waste-water containing human wastes, but also waste-water from industrial processes. In addition, a wide range of other source categories also creating problems for the marine environment was identified (UNEP, 1995). The Programme reflected the experience of over twenty years' work by governments, both individually and through regional seas organizations, to address these problems. Subsequent intergovernmental reviews of the implementation of the GPA show that progress is being made in many parts of the world, but only slowly.

In evaluating the impacts of contamination on the marine environment, there are significant difficulties in comparing the situations in different areas. For many aspects of contamination, evaluating the levels of contamination requires chemical analysis of the amounts of the contaminants in samples of water, biota and/or sediments. Unless there is careful control of the sampling methods and analytical techniques in all the cases to be compared, it is difficult to achieve scientifically and statistically valid comparisons. Lack of clear and practical comparisons creates problems in setting priorities. For example, a fairly recent review of available evidence in the Wider Caribbean concluded that, among the 30 pollution studies examined, analyses varied in terms of sampling schemes, parameters and analytical techniques, and data were presented in ways which made comparison all but impossible – whether data were in terms of dry weight or wet weight, what sediment fraction was analyzed, and whether data were presented in absolute terms or as a percentage of lipids (Fernandez et al., 2007). Such differences make meaningful comparisons of the available data very difficult. For this reason, this chapter does not attempt to give detailed figures on concentrations of contaminants. The Global Environment Facility is supporting the Transboundary Waters Assessment Programme (TWAP), to enable priorities between different areas to be established.

The main issues relating to inputs to the oceans and seas from land-based sources can be categorized for the purposes of this chapter under the headings of: hazardous substances (including the effects of desalinization plants), endocrine disruptors, oil, nutrients and waterborne pathogens, and radioactive substances.

In all cases, consideration has to be given to the variety of means by which the movement of the substances from land to water takes place. The main distinction is between waterborne and airborne inputs. Waterborne inputs can either be direct (through a pipeline from the source directly into the sea, by run-off from land directly into the sea or by seepage of groundwater directly into the sea) or riverine (through runoff or leaching from land to a watercourse or by a direct discharge into a watercourse, and the subsequent flow from such watercourses into the sea). Waterborne inputs are much more readily measured, and for that reason have so far attracted more attention. There is increasing evidence that airborne inputs are more significant than has hitherto been thought, not only for heavy metals and other hazardous substances but also for nitrogen (GESAMP, 2009; Duce et al., 2008).

2 Hazardous Substances

2.1 Which substances are hazardous?

A wide range of substances can adversely affect marine ecosystems and people. The adverse effects can range from straightforward fatal poisoning to inducing cancers, weakening immune systems so that diseases develop more easily, reducing reproductive performance and inducing mutations in offspring. A first requirement for controlling the input of hazardous substances into the marine environment, whether from point or diffuse sources or through the atmosphere is therefore to establish what substances show sufficient grounds for concern that regulatory action is needed. Lists of substances identified as hazardous can never be closed: new substances are constantly being developed, and new uses are likewise constantly being found for a wide range of elements and compounds.

International effort to define substances hazardous to the marine environment began in relation to dumping of waste at sea (see Chapter 24). In this context, the Convention for the Prevention of Marine Pollution by Dumping from Ships and Aircraft of 15 February 1972 (Oslo Convention) and the Convention on the Prevention of Marine Pollution by Dumping of Wastes and Other Matter of 13 November 1972 (London Convention) established the first internationally agreed lists of substances whose introduction into the marine environment should be controlled. In both Conventions, a ban on dumping was agreed for similar "black lists". These included, among other items, substances such as toxic organohalogen compounds, agreed carcinogenic substances and mercury and cadmium and their compounds. Controls were agreed on dumping for similar "grey lists", which included, among other items, arsenic, lead, copper and zinc and their compounds, organosilicon compounds, cyanides, fluorides and pesticides not in the "black list" (Oslo Convention 1972; IMO, 1972). When attention was thereafter given to dealing with discharges and emissions from land, these "black" and "grey" lists were adapted and used by many national and international authorities concerned with the marine environment for the initial work in the field of regulation of land-based inputs of hazardous substances.

Over the past 40 years, regulatory authorities have added further categories to be controlled. In 1976 the United States Environment Protection Agency produced a list of "toxic pollutants" and an explanatory list of "priority pollutants" (EPA, 1976; EPA, 2003). An important contribution to the more general debate on the approach to control of hazardous substances was made in 1993, when the Great Lakes Commission proposed the virtual elimination of discharges of substances which are toxic and persistent (IJC, 1993). The most extensive exercise that has focused specifically on the marine environment was undertaken from 1998 by the OSPAR Commission for the Protection of the Marine Environment of the North-East Atlantic to implement its long-term strategy of eliminating discharges, emissions and losses of hazardous substances which could reach and affect the marine environment. For this purpose, "hazardous substances" were defined as substances that are toxic, persistent and liable to bioaccumulate (bioaccumulation occurs when a substance taken in by an organism is not excreted, but builds up in the organism), or which give rise to an equivalent level of concern (OSPAR, 1998). This required a definition of thresholds of toxicity, persistence and bioaccumulativity. These agreed levels were applied to the more than 11,000 substances listed in the Nordic Substances Database with experimental data. The resulting list of substances of possible concern was then analysed in 2001-2004 to see which substances were only found as intermediates in closed systems, or were not being produced or used, and were therefore unlikely to affect marine ecosystems. After these had been discounted, the resulting list of chemicals for priority action was used to see what action was needed to meet the cessation target (OSPAR, 2010). The European Union, through its Regulation, Evaluation, Authorization and Restriction of Chemicals (REACH) Regulation (EU, 2006), is addressing all "persistent, bioaccumulative and toxic" (PBT) and "very persistent and very bioaccumulative" (vPvB) substances that are in substantial (more than 1 ton/year) use in its area, or proposed to be introduced. China has developed its Catalogue of Toxic Chemicals Prohibited or Strictly Controlled (China, 2014). Other organizations, such as the Arctic Monitoring and Assessment Programme, have developed similar lists (Macdonald et al., 1996).

Although there is substantial overlap between the various lists of substances where action is considered to be needed to protect the marine environment, there are variations. These result from differences in evaluation of the level of risk. Different methods of evaluation, and different choices of cut-off levels for toxicity, persistence and bioaccumulativity can lead to differing views. Different judgements are made on the extent to which precautions by users can sufficiently guard against the risks to the environment. Different views are taken on the reasonable practicability of the use of acceptable substitutes: what is regarded in some jurisdictions as acceptable (because, for example, its use can be managed acceptably) is regarded in others as unacceptable. Sometimes (as with chlordane) international action can help change what is regarded as reasonably practical. The result is that there is no single agreed list of hazardous substances that are of concern: substances that are regarded as acceptable in one area are banned in another.

Table 20.1 shows the principal substances which the range of authorities mentioned in the previous paragraph have regarded as hazardous to the marine environment, and on which action is being taken in all or some parts of the world to control inputs of them to the sea from land.

Particularly important are:

(a) *Heavy metals:* All heavy metals occur naturally and, because of natural weathering processes and the immunity of natural elements to destruction, are found at measurable levels even in waters generally regarded generally as pristine. Some heavy metals (such as cadmium, mercury and lead) are always highly toxic. Others (especially copper and zinc) are essential trace elements in diet or intake for many biota. Some heavy metals, especially copper and arsenic, have been used extensively in the past for plant protection purposes, resulting in widespread additional dispersal and higher concentrations in some areas. In excessive amounts, however, even these can interfere with the absorption of other essential trace elements and, at high levels, become toxic. At lower levels, they also appear capable of affecting the immune systems of biota (Coles et al., 1995; Kakuschka et al., 2007) or their reproductive success (Leland et al., 1978);

(b) *Persistent organic pollutants (POPs):* In contrast, POPs are man-made. They are organic compounds (that is, compounds involving carbon, most often combined with hydrogen and/or with chlorine, bromine or other halogens) that resist degradation in the environment through chemical, biological or other processes. Many were developed as biocides (insecticides, herbicides, etc.) since about 1910-1930. Others are used in manufacturing processes or in electrical appliances. From the 1960s, concerns developed about their effects on immune systems and reproductive success, and about their carcinogenic effects. As a consequence of the call in the GPA in 1995, subsequently endorsed by the UNEP Governing Council, the Stockholm Convention on Persistent Organic Pollutants was adopted on 22 May 2001[1] and now provides a global mechanism for controlling the production and use of POPs. Initially, agreement was reached in 2004 that production and use of 12 POPs should be banned or strictly controlled. Since then a

---

1   2256 United Nations *Treaty Series* 119.

further 10 POPs have been brought under the Convention's controls;

(c) **Polycyclic aromatic hydrocarbons (PAHs):** PAHs are complex compounds of hydrogen and carbon (and, in some cases, other elements such as nitrogen, oxygen or sulphur). They occur naturally, and are also typically created by imperfect combustion processes. Many, but not all, are carcinogenic and/or affect reproductive success;

It is important to note that the category of hazardous substances is not closed. New substances are constantly being developed, and new uses are constantly being found for a wide range of elements and compounds. The questions whether these substances and elements are toxic, persistent and bioaccumulative and whether their uses present risks to the marine environment need to be kept under continual review. Substances where such questions arise are sometimes referred to as "contaminants of emerging concern" (see, for example, Yuan et al., 2013).

Knowledge of the extent of the presence of hazardous substances in the marine environment is patchy. Some issues, such as the presence in the marine environment of contamination from heavy metals and lindane, have been studied for over 30 years in some areas and, to a lesser extent, have also been studied quite widely around the world. Other issues have only been looked at more recently, and a number have only been examined from the point of view of laboratory tests of substances on marine biota, without monitoring for the presence of the substances in the sea itself or its biota and sediments.

Table 20.1 | Background information on substances classified by the authorities mentioned in the text as presenting hazardous characteristics and therefore justifying action.

Substances	Sources and Main Uses[1]	Production and Related Developments
† = Persistent Organic Pollutant (POP) under the Stockholm Convention on Persistent Organic Pollutants 2001; †? = Substance under consideration for listing as a POP) under the Stockholm Convention	* = diffuse sources, where the pathways will be mainly from leaching (especially from land-fill waste disposal), emissions to air and/or runoff	
Heavy metals		
Cadmium	Large combustion plants; electro-plating; incinerators; paints*; batteries*	World production of cadmium is fairly stable (around 20,000 tons/year) between 2001 and 2011[2]
Copper	Mining; electric wiring and machinery*; pesticides*	World production of copper increased 15% to 16.2 million tons/year during 2001-2011[2]
Lead and organic lead compounds	Roofing*; fuel for internal combustion engines*; paint*; PVC stabilizer*	The phasing out of lead in vehicle fuel has significantly reduced inputs of lead to the seas. Emissions in Europe have decreased by 92% during 1990-2003, with similar decreases in North America[3]. World production of lead has, however, risen 53% to 4.75 million tons/year during 2001-2011[2]. Over half of this is in Australia and China.
Mercury and organic mercury compounds	Large combustion plants; electrolysis chlor-alkali plants; primitive gold-refining*	World production of mercury is relatively stable, fluctuating between 1,120 and 2,280 tons/year during 2001-2011[2]. A substantial stockpile is, however, emerging as mercury-cell chlor-alkali plants change technology. A global Convention was adopted at Minamata, Japan, in 2013 to control trade in, and the use of, and plants discharging or emitting it. The Convention is not yet in force.
Zinc	Large combustion plants; surface-treatment of sheet metal*; cosmetics*	World production of zinc has risen by 38 % to 12.6 million tons during 2001-2011, over half in Australia, China and Peru.
Organohalogens		
Brominated diphenyls (BDPs) (hexa-BDP†) and BDP ethers (BDEs)(tetra-BDE†, penta-BDE†, hexa-BDE†, hepta-BDE and deca-BDEs)	Fire retardants in automobiles; plastics and textiles*	World production is about 40,000 tons/year. All BDEs are now controlled in a number of countries. Production and use of hexa-BDP and tetra-, penta-, hexa- and hepta-BDEs are to be eliminated under the Stockholm Convention[4].
Hexabromobiphenyl†	Fire retardant*	No current production or use is known.
Hexabromocyclododecane (HBCDD) †?	Fire retardant in plastic foam*	At its peak in the 1970s, production was about 6,000 tons /year. No production is now reported.
Hexachlorobutadiene†?	Fumigant*; transformer, hydraulic or heat transfer liquid*; viticulture pesticide*	Production and use have ceased in Europe.
Perfluorooctanyl sulphonic acid and its salts (PFOS)† and perfluorooctanesulfonyl fluoride (POSF-F)†	Electronic components*; fire-fighting foams*; insecticide*; stain repellent for carpets*; fat repellent in food-packaging	Production and use to be eliminated under Stockholm Convention, subject to specific exemptions.
Polychlorinated biphenyls (PCBs)†	Heat exchange fluids*; electric transformers and capacitors*; paint additives*; carbonless copy paper*; plastics*	Production and use to be eliminated under the Stockholm Convention. Such a prohibition has been in force since about 1990 in many States, but residues often remain.

Substances	Sources and Main Uses[1]	Production and Related Developments
Polychlorinated dibenzodioxins (PCDDs)[†] and polychlorinated dibenzofurans (PCDFs)[†]	Incomplete combustion of material containing organic substances and chlorine; emissions from polyvinyl-chloride (PVC) plants.	Emissions to be minimised under the Stockholm Convention
Polychlorinated naphthalenes[†?]	Wood preservatives*, additives to paints and engine oils*, cable insulation*; in capacitors*	
Short chained chlorinated paraffins (SCCPs)[†?]	Lubricants in metal working; leather treatment; production of rubber and plastics	SCCPs are produced in Brazil, China, India, Japan, Russia, Slovakia and the United States. Use in Europe and North America has dropped by about 75% since peaks in the 1990s.
Vinyl chloride	Mainly used in the production of polyvinylchloride (PVC);	
Pesticides/biocides		
Aldrin[†], Dieldrin[†], Endrin[†] (Aldrin rapidly converts to Dieldrin)	Insecticides*. Endrin also used in rodent control*.	Production and use to be eliminated under the Stockholm Convention, subject to some transitional exemptions.
Atrazine and Simazine	Herbicide (used extensively in maize and sugarcane agriculture to control weeds)*	Production and use have been phased out in some countries (where it has largely been replaced by less persistent herbicides. Still produced and used in some other countries, where controls on use are seen as sufficient to keep it out of the water environment.
Chlordane[†]	Insecticide, particularly for termites	Production and use to be eliminated under the Stockholm Convention. The Global Environment Facility in 2006 provided USD 14 million for a programme to enable China to achieve this.
Chlordecone[†]	Insecticide, particularly used in banana culture	No current production or use is known.
Dichlorodiphenyltrichloroethane (DDT)[†]	Originally use widely as a broad-spectrum insecticide, now almost exclusively for controlling insect disease-vectors*	Production and use controlled under the Stockholm Convention. 18 Convention Parties have registered to continue to use DDT for disease-vector control, of which 5 reported no use in their last report. One Party (the Gambia) reported use, but had not registered.
Dicofol	Pesticide, especially for mites on tomatoes and melons*	Some countries have phased out the use of dicofol. It is still used in Brazil, China, India and Israel. Produced by chemically modifying DDT.
Endosulfan[†]	Pesticide*	Production and use to be eliminated under the Stockholm Convention, subject to specific exemptions.
Heptachlor	Insecticide, especially for soil insects and termites*	Production and use to be eliminated under the Stockholm Convention.
Hexachlorobenzene	Fungicide*	Production and use to be eliminated under the Stockholm Convention.
Lindane (γ-hexachlorocyclohexane (HCH)[†], including α-HCH[†] and β-HCH[†] isomers (produced in large quantities as by-products to γ-HCH)	Insecticide*	Production and use to be eliminated under the Stockholm Convention, subject to exception for use against head-lice and scabies. Production and use have largely already ceased, but stockpiles of α- and β-HCH exist.
Methoxychlor	Insecticide for use on both animals and plants*	Phased out in the European Union and the United States. Information is lacking on production and use elsewhere.
Mirex[†]	Insecticide, particularly for termites*; fire retardant*	Production and use to be eliminated under the Stockholm Convention. The Global Environment Facility in 2006 provided US$14 million for a programme to enable China to achieve this.
Pentachlorophenol (PCP) and its salts and esters[†?]	General pesticide, now widely restricted to use as a fungicide and wood preservative*	PCP is being considered under the Stockholm Convention because it transforms into pentachloroanisole (PCA) which is seen as a problem. PCP has been phased out in the European Union.
Pentachlorobenzene[†]	Used to make PCBs less viscous*; in dyestuff carriers*; as a fungicide*; as a flame retardant*	No current intentional production or use is known; probably still produced as a by-product in imperfect incineration.
Toxaphene[†]	Insecticide, particularly used for cotton and soya-bean culture*	Production and use is to be eliminated under the Stockholm Convention.
Aromatics		
Polycyclic aromatic hydrocarbons (PAHs)	Incinerators; large combustion plants; Söderberg-process aluminium-smelting plants; coke plants; imperfect combustion of wood and fossil fuels (including vehicles)*; coal-tar surface treatments (including creosote)*	Reductions in PAH emissions are being achieved by tighter regulation of vehicles, combustion plants and incinerators, technology changes in aluminium-smelting plants and (in some areas) elimination of the use of some surface-treatment processes.

[1] http://chm.pops.int/TheConvention/ThePOPs/tabid/673/Default.aspx and associated risk assessments, together with the relevant OSPAR Background Papers (http://www.ospar.org/content/content.asp?menu=00200304000000_000000_000000)
[2] British Geological Survey, World Minerals Statistics Archive (http://www.bgs.ac.uk/mineralsuk/statistics/wms.cfc?method=searchWMS
[3] United Nations Environment Programme, Final review of scientific information on lead, Nairobi, 2010.
[4] Stockholm Convention on Persistent Organic Pollutants, United Nations Treaty Series, vol. 2256, No. 40214.

Some hazardous substances reach the marine environment in inflows of water, others are airborne. Waterborne contaminants tend to be found mainly near the inflows, and thus concentrated in estuarial and coastal waters, particularly where they are adsorbed onto particles in the water and settle as sediments. Airborne contaminants are carried much further out to sea, and therefore are found more generally. For some hazardous substances, sampling around the world's continents has shown that they are present in all continents (for example, dioxins and furans (which are most often airborne) have been found in butter samples from all continents, though to a lesser extent in the southern hemisphere (Weiss et al., 2005). Where hazardous substances have been spread worldwide largely by air transport, it can be assumed that they have also reached the ocean. It is known that some POPs have been concentrated on the higher latitudes of northern hemisphere land-masses by a process of volatilization from land and redeposition – sometimes described a "multi-hop" process, as compared with "one-hop" contaminants that are carried in one step to their final destination.

3 Point Sources

The most obvious threats to the marine environment from hazardous substances come from point sources. Such point sources can be either discharges into rivers which ultimately reach the sea, or direct discharges through pipelines into the sea. There can be cases (usually volcanic eruptions) in which natural processes result in the introduction of naturally occurring hazardous substances into the ocean. However, many point sources are large industrial plants which provide a concentrated source from which the hazardous substances pass into the marine environment. Waste-water treatment plants can also be regarded as point sources, since they can concentrate hazardous substances from a substantial area and funnel them to a single discharge point. Historically, it was the impact of such point sources on inland waters that first gave rise to concern. In England, effective legislation was introduced as early as 1875 (Rivers (Prevention of Pollution) Act 1875). Similar legislation followed in other industrialized countries. Because of the then current belief in the almost infinite absorptive capacity of the sea, general measures on discharges and emissions reaching and affecting the sea were not adopted until the 1970s. Initially the measures were "end of the pipe" methods of removing contaminants from discharges and emissions. Gradually, the emphasis has moved more to "clean technology", where the contaminants are not used in, or not generated by, the process. Among the most significant point sources in respect of hazardous substances are the following:

   (a) **Large combustion plants:** Since fossil fuels naturally contain other minerals, such as heavy metals, their combustion releases those elements. Since the gases from combustion are released to the air, large combustion plants are a significant source for airborne transport of contaminants to the ocean. Many large combustion plants do not have sufficient scrubbers to clean the flue gases. Such plants are particularly sig-

nificant for emissions of mercury: all forms of coal-burning account for 24 per cent of the total global estimated annual anthropogenic releases of 1,960 tons (estimate range: 304 to 678 tons) (UNEP, 2013a based on a 2010 inventory). This estimate differs in absolute amount and relative proportion of the total emissions from an earlier one in 2008 based on a 2005 inventory: in 2008 all forms of coal-burning were estimated to be in the range of 1,230 to 2,890 tons, and to constitute the largest sector emitting mercury; the change is due to revised estimates of emissions from domestic heating (revised downwards from 2008 to 2013), and emissions from artisanal gold refining (revised upwards from 2008 to 2013, and thus estimated in 2013 to be the largest mercury-emitting sector). If the 2005 inventory figures are compared with the 2010 inventory figures and the same methodology used is considered, and the estimates employ the same 2010 methodology, the emissions in 2010 from coal combustion in power generation and industrial uses combined are the same as, or perhaps slightly higher than, in 2005. The fact that emissions from this sector are not higher, even though new coal-fired power plants are being built, rests on the improving combustion efficiency and emissions controls in most parts of the world (UNEP, 2008; UNEP, 2013a). Emissions of mercury from large combustion plants should eventually be controlled by the actions required under the Minamata Convention on Mercury of 10 October 2013 (Minamata Convention). Coal-fired power stations are also significant sources of cadmium, zinc and PAHs. Cement production is another form of large combustion plant which can emit heavy metals both from the fuel and from the raw materials: in 2013, mercury emissions from this sector were estimated on the basis of the 2010 inventory at 173 tons (estimate range 65.5 – 646 tons) (EU BREF, 2013; UNEP, 2013a).

Between 2001 and 2012, the proportion of the total amount of electricity generated by coal-fired power stations declined

Figure 20.1 | World Cement Production 2001-12. Source: European Cement Association, 2014.

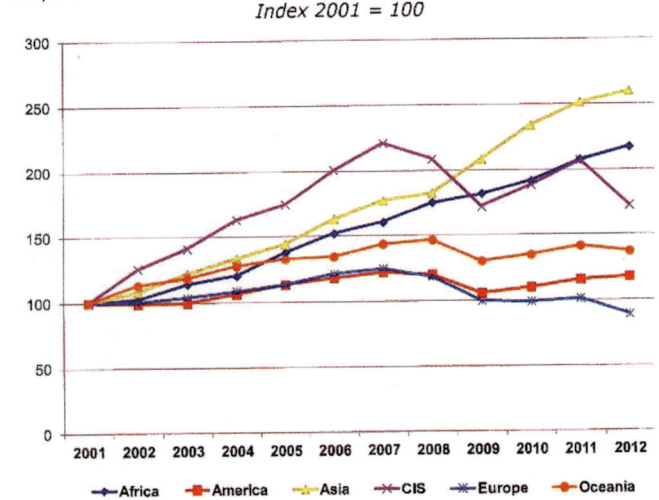

or remained stable in much of the world (Africa, Europe and Central Asia, North and South America, South Asia). It can therefore be expected that emissions of mercury from such power stations reaching the ocean will stabilise or decline. The proportion, however, grew steadily in East Asia – from 51 per cent to 63 per cent (although China's proportion of coal-derived electricity remained stable at around 80 per cent). Unless even greater efforts are made to control emissions of hazardous substances from coal-fired power stations, the levels of contaminants reaching the ocean from this source in that part of the world are likely to increase (World Bank, 2014).

The pattern of development in cement production is different from that of coal-fired power generation: except in Europe, there has been significant growth over the past decade: 33 per cent in the Americas, 66 per cent in Oceania, over 200 per cent in Africa and over 250 per cent in Asia. This increase in production appears to have been accompanied by marked improvements in the quality of control of emissions: for one of the most significant, mercury, the UNEP 2013 estimate of mercury emissions from this sector was lower than the 2008 estimate (173 tons as against 189 tons) (UNEP, 2008; UNEP, 2013a).

(b) **Chemical industries:** Chemical industries can give rise to a wide range of contaminant emissions and discharges: the products themselves may present problems – as can be seen from the list of substances in Table 20.1 – and other hazardous substances can be released either in the production process or as part of the waste stream. Where efforts have been focused on combating pollutant discharges and emissions, chemical plants have usually been high on the list of targets. For traditional technologies, the focus has to be on removing pollutants from the waste streams and preventing leaks during the process. Increasingly, however, the focus is on new technologies which do not present the same pollution problems as the traditional technologies (for example, the membrane process in chlor-alkali production and the "no-chlorine" processes in paper and pulp production).

The world of chemical production is, moreover, changing fast. Measuring the overall situation is not easy, because of the wide range of products that come under the umbrella of the chemical industries. One measure that can be used to indicate the scale of change in the chemical industries, however, is the value of the goods produced. In real terms, the statistics of such product values will show changes in the level of activity of the chemical industries in different countries. Such statistics will, of course, hide changes in chemical industries where bulk production of basic chemicals is replaced by production of specialist chemicals of higher intrinsic value. Nevertheless, they can give an overall view of the way in which the world's chemical industries are changing (see Appendix to this chapter):

(i) Between 2003 and 2012, the value of the total world output of chemicals rose by 12 per cent in real terms;
(ii) In 2003, 60 per cent by value of the world output of chemicals was in North America and Europe. By 2012, this had dropped to 40 per cent;
(iii) In contrast, the proportion by value of world chemical production in Asia and the Pacific rose from 29 per cent to 49 per cent, in spite of a reduction of 24 per cent in the value of Japanese chemical products. The value of Chinese chemical products in real terms rose by 293 per cent between 2003 and 2012 (to 29 per cent of total world production), that of Singapore by 74 per cent, that of India by 56 per cent, and that of the Republic of Korea by 32 per cent.

There has therefore been a significant change in the potential for the impact of chemical industries on the marine environment, with a change of focus from the Atlantic Ocean basin to the Pacific Ocean basin.

Figure 20.2 | World Chemical Production by value in 2003 and 2012. Source: Appendix to this chapter.

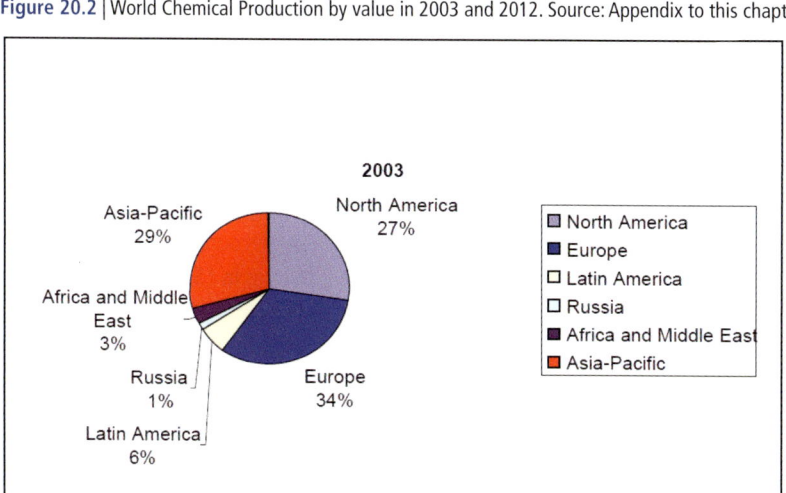
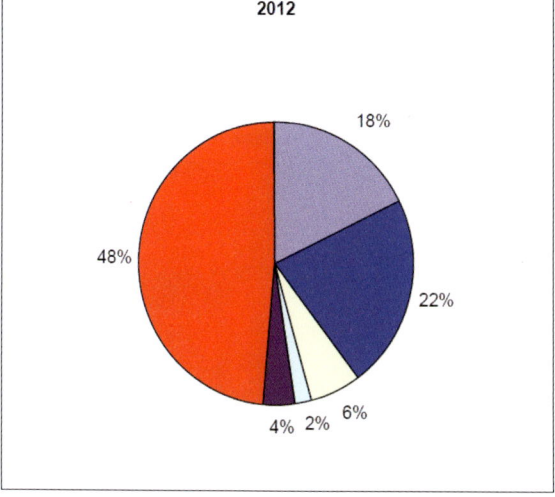

Certain types of chemical plants merit specific mention: chlor-alkali plants, polyvinyl chloride (PVC) plants and titanium-dioxide plants.

(c) **Chlor-alkali plants:** Chlorine and caustic soda are basic requirements for many chemical industries. Since 1892, they have been produced by electrolysis of brine. The (original) mercury-cell process uses a layer of mercury as the cathode, which is constantly withdrawn and reacted with water. The resulting water discharges, unless purified, have a high mercury content. This original process is increasingly being replaced, initially by the diaphragm process (which used an asbestos diaphragm) and now by the membrane process, neither of which use mercury. One hundred mercury-cell process plants still exist in 44 countries. Existing plans will result in this number diminishing to 55 plants in 25 countries by 2020 (UNEP, 2013b).

(d) **Polyvinyl chloride plants:** PVC plants use various processes to convert vinyl chloride monomer into the plastic PVC which has manifold uses throughout the world. Global production in 2009 was around 30 million tons, representing a growth of 50 per cent since 1995. The world's production capacity is significantly higher (around 48 million tons/year), but the economic recession reduced use of this capacity (Deloitte, 2011). Production levels are expected to recover quickly and to continue to grow. Capacity in China has grown particularly rapidly, from about 375,000 tons in 2001 to nearly 16,000,000 tons in 2008.

The adverse environmental impact from PVC plants consists mainly of the emission of dioxins and furans and the risks from the emission of vinyl chloride monomer (VCM), a known carcinogen. Although the immediate threats are to the vicinity of the plants, there is evidence that all these emissions can reach the marine environment (OSPAR, 2000).

In China, there is an additional problem in that the large majority of PVC plants in that country develop the PVC by an acetylene-based process starting from coal, in contrast to plants in the rest of the world which mainly use an ethylene-based process starting from oil. The production of the acetylene from coal requires a catalyst, which is currently mercury chloride (although research is in hand to develop a mercury-free alternative). About 574-803 tons/year of mercury are used (2009 figures), of which about 368-514 tons are lost in waste (China, 2010). It is not clear how much of this reaches the sea.

(e) **Titanium dioxide (TiO2) plants:** TiO2 is used as a very white pigment, mainly in paint, plastics and paper. Different production processes are used for the two main mineral sources (ilmenite and rutile), but both produce large amounts of acid waste. In Europe in the 1970s, much of this was disposed of into the sea, either by pipeline or dumping. This gave rise to concern about effects on fish (Vethaak et al., 1991). Improved waste management methods, mainly through recycling the acid or its use in other products, have now largely removed these problems in Europe. Estimated production of TiO2 is about 1.4 million tons/year each in Europe and the United States, and about 2.3 million tons/year in China, where production is growing rapidly (USGS, 2013).

(f) **Mining:** Mining is a significant part of the economy in a number of States, and everywhere is a basic source of supply to manufacturing industry. In 2010, eight States were responsible for over 70 per cent by value of global production from mining: Australia (15.6 per cent), China (15.0 per cent), Brazil (10.2 per cent), Chile (6.8 per cent), the Russian Federation (6.2 per cent), South Africa (5.9 per cent), India (5.6 per cent) and the United States (5.0 per cent). Mining formed over a tenth of Gross Domestic Product (GDP) in Papua New Guinea (33.4 per cent), Zambia (23.8 per cent), Chile (14.7 per cent), Ghana (12.7 per cent) and Peru (12.0 per cent) (ICMM, 2012).

From the point of view of the aquatic environment, the main concern about mining is the disposal of waste. Large amounts of pulverized rock mixed with water ("tailings") are produced, which have to be stored or disposed of. Except that some mines remove cyanide (used in extracting metals), tailings are not treated before disposal. They therefore contain a large range of potentially hazardous substances. They can also cause problems through siltation and smothering of biota, particularly in the sea. Concerns about the consequent problems for the sea go back 600 years in south-west England (Worth, 1953). Tailings are most often stored on land behind dams. Catastrophic collapses of tailings dams can release toxic materials into watercourses and thence to the sea. Twenty-six tailing-dam failures in 15 countries have been noted between 2000 and 2014 (WISE, 2014). Not all of these will have affected the sea, but the 1996 event at Marinduque, in the Philippines, clearly had effects on the marine environment, making the sea much more acid, with elevated heavy-metal levels, as well as smothering a substantial area of the seabed (USGS, 2000). In some cases, tailings are also disposed

Figure 20.3 | China: PVC production capacity and output (in 10,000 tons). Source: Chlor-Alkali Industry Association in China, 2010.

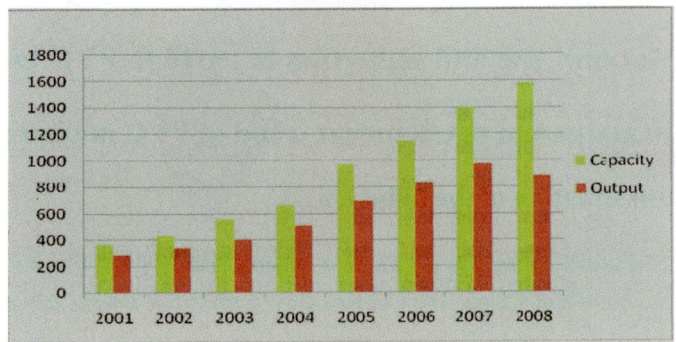

of directly into rivers and into the sea. In 2012, there were 12 mines (1 in Indonesia, 5 in Norway, 4 in Papua-New Guinea, 1 in Turkey and 1 in the United Kingdom) practising disposal direct into the sea (IMO, 2012). In all cases the aim was to have a pipeline taking the waste well below the bulk of marine life in the water column. However, the tailings smother a large area of seabed, and are capable (depending on the local geology) of introducing substantial amounts of heavy metals into the marine environment.

(g) **Smelting:** The smelting of metals, both ferrous and non-ferrous, can result in the emission of heavy metals to the atmosphere, which may then be deposited in coastal catchments and transported to the sea by watercourses, or deposited direct onto the sea by wet or dry deposition. For example, around 70 per cent of the emissions of lead to the atmosphere in Australia in 2003/04 were from non-ferrous metal processing (though not generally adversely affecting the marine environment). Ferrous-metal production also leads to emissions of lead to air: in Europe in 2000, it was estimated that lead emissions from iron and steel production was about half as much again as that from non-ferrous metal production. There is no recent estimate of the amount of global emissions of lead to air, but in 1983 it was estimated at 87,000-113,000 tons (this will have reduced substantially since with the reduction in use of leaded petrol/gasoline) (UNEP, 2010). When properly managed, metal smelting can have very limited adverse effects on the marine environment. However, production of many metals is increasing rapidly (see, for example, the figures in Table 20.1 for heavy metals). Likewise, the production of iron and steel is also increasing rapidly: pig-iron production increased by 85 per cent between 2001 and 2011 to 1,158 million tons/year (BGS, 2012). In particular, pig-iron production in China over the same period rose by over 300 per cent, so that it is now over half the annual world production. Even if the levels of emissions per ton of production are kept steady at the present level, the total load will increase in proportion.

Aluminium presents a special case. The primary form of production is by electrolysis. The Söderberg process became the predominant method. The carbon anode used in this process is consumed at a rate of about 0.5 ton for every ton of aluminium produced (ALCOA, 2014). Much of this carbon used to be emitted as polycyclic aromatic hydrocarbons (PAHs). Over time, better controls on PAH emission have been introduced, and more importantly, the Söderberg-process aluminium-smelting plants are being phased out – only about 7 per cent of global aluminium production is now by that process. From 2005 to 2010, world primary aluminium production increased by almost 30 per cent to 41.6 million tons (over one-third of which is in China, which has no Söderberg plants), but PAH emissions to air were reduced by 50 per cent per ton of aluminium produced (IAI, 2013);

(h) **Paper industry:** Paper mills can give rise to a variety of environmental concerns. In relation to hazardous substances, the problems arise mainly from bleaching the pulp, a process needed for the production of most paper. During the period before the 1970s-1980s, the pulp and paper industry was the source of inputs giving rise to concern: polychlorinated dioxins and furans (PCDDs and PCDFs) were detected in effluents of pulp mills, resulting from the long-established use of chlorine in bleaching. It has proved possible to reduce this problem substantially by a mix of measures: principally by replacing elemental chlorine with chlorine dioxide and other oxygen-containing substances and by introducing closed systems and recycling the bleach-plant effluent. New processes have also been introduced: the Elemental-Chlorine-Free (ECF) and the Totally-Chlorine-Free (TCF) processes, which avoid the by-products of the chlorine bleaching process (EU BREF, 2001).

The paper industry has seen substantial growth in the period 2001-2012: worldwide production has increased by 23 per cent to just over 400 million tons. This growth has not been uniform: production in Canada and the United States has declined, while production levels in Africa, Europe, Oceania and the Russian Federation have remained more or less stable. The growth has been in Asia and Latin America, where production has increased over this period by 76 per cent and 34 per cent, respectively; production in China alone has grown by nearly 220 per cent to 103 million tons in 2012 (see Table 1 in Appendix to this chapter). Even if levels of contaminants per ton of production are kept at previous levels, growth on this scale will substantially increase the total load of contaminants finding its way to the sea. There is evidence (Zhuang, 2005) that the expansion of Chinese paper-making capacity has been accompanied by improved environmental management, but data to show the total effect do not seem to have been collated.

(i) **Incinerators:** Increasingly, significant amounts of domestic and municipal waste consist of plastics containing chlorine. Much of this waste is disposed of through incineration. Where this happens in uncontrolled open-air burning, there is a substantial risk of the formation of dioxins and furans: almost any combination of carbon, hydrogen, oxygen and chlorine can yield some polychlorinated dioxins/furans under the wrong conditions (Altwicker et al., 1990). Even where the incineration takes place in purpose-built incinerators, a risk of such formation remains, especially where controls do not ensure that appropriate temperatures are reached during combustion or where devices to scrub or filter the flue gases are not installed or not properly maintained and operated. The same problem arises where incineration is used to dispose of wastes from industries that produce waste containing hazardous substances: if incineration is not properly done, both the hazardous substances in the waste and other newly created hazardous substances may be emitted.

(j) **Fertilizer production:** The production of phosphate fertilizer produces substantial amounts of waste from the rock that has to be processed. Heavy metals, especially cadmium, are found in this waste, and reach the sea either from direct discharges or, in some cases, by leaching from land-based waste storage. Total world fertilizer production has risen by 23 per cent between 2002 and 2011, rising even more in South America (89 per cent) and East Asia (78 per cent). Production in Africa represents a fifth of the total world production, and concern has been raised about the impacts of some of the discharges (Gnandi et al., 2006).

(k) **Desalination:** Desalination is very important in some parts of the world where fresh water is in short supply (see chapter 28). Desalination plants require massive intakes from the sea (capacity in the north and central Red Sea, for example, is over 1,750 megalitres[2] a day (PERSGA, 2006), and in the Persian Gulf, it is over 10,900 megalitres a day (Sale et al., 2011)) and produce substantial discharges. The potential contaminants are found in discharges of heated, concentrated brine and of chemicals added to improve performance and to prevent corrosion (chlorine, copper and antiscalants). The effects of the brine discharge are mostly local (within tens of metres of the discharge), and are quickly diluted and dispersed, but in extreme cases they can be traced for several kilometres (Roberts et al., 2010). They are particularly significant in areas with high tidal ranges where the discharge is above the high-tide mark, where they can affect biota in the inter-tidal zone. Chlorine concentrations in discharges in the Red Sea average 0.25 ppm (standard swimming-pool chlorination is 1.0-3.0 ppm), and so local biocidal effects are possible. Copper concentrations in the discharges of a typical desalination plant are around 15 ppb, significantly above generally accepted criteria for satisfactory water-quality. In Red Sea desalination plants, about 9 tons of antiscalants a day are used and discharged. They have a relatively low toxicity and are diluted rapidly, and are therefore judged unlikely to pose a significant threat, but there is limited information on them. In general, the conclusion of a review of articles studying these problems was that discharge site selection is the primary factor that determines the extent of ecological impacts of desalination plants (Roberts et al., 2010). Overall, the Regional Organization for the Conservation of the Environment of the Red Sea and Gulf of Aden (PERSGA) determined in 2006 that desalination was not a threat to the Red Sea (PERSGA, 2006). No overall assessment of effects in the Persian Gulf appears to have been made (Sale et al., 2011).

4 Diffuse Sources

There are manifold diffuse sources of hazardous substances that can reach and affect the ocean. The main pathways are through surface water runoff in watercourses (both from liquid discharges and from leaching), groundwater discharges, and wet and dry deposition of emissions to the atmosphere. The most significant processes are waste disposal, routine combustion processes, abrasion, use of biocides and accidents. All of these affect both land and sea, and there is nothing special about the methods to control these processes for the purpose of protecting the marine environment. It is, however, necessary to ensure that marine aspects of the impact of all hazardous substances are specifically considered in decision-making on control measures, because the effects of some hazardous substances are significantly greater (or different) than in freshwater or land environments. Other compounds released from diffuse sources that have been suggested for consideration include pharmaceuticals (both human and veterinary) and cosmetic ingredients (such as musk xylene). Evaluation of such substances has not yet shown general agreement that there are significant problems which need action, although some regulatory bodies are keeping some of these substances under observation.

4.1 Waste disposal

Adverse effects on the marine environment from waste disposal can arise from a wide range of processes. Leaching from land-fills into which waste has been deposited is probably the major source. This can be significant for brominated flame retardants (PBDEs and related substances (see entry in Table 20.1)). Industrial liquid waste will often enter into municipal waste water treatment systems – these can be regarded as point sources, but at the same time they usually collect waste water from a large area. The waste entering municipal waste water treatment systems also includes runoff from accidents involving the spilling of hazardous substances. A large number of hazardous substances will form part of materials in waste streams. Among the heavy metals, lead and cadmium are particularly significant given their widespread use in batteries: 80 per cent of all lead used in OECD countries is used in batteries (ILZG, 2014). Although there is a strong economic interest in recycling such lead (and lead is the most recycled non-ferrous metal), there is a substantial risk that it will eventually leach to the ocean from badly managed waste streams. The same applies to other heavy metals (such as cadmium) which are also used in batteries and electronic equipment.

Plastics containing chlorine compounds (such as PVC) form a significant part of waste streams in most countries. These therefore also present problems for the marine environment if disposal is not properly managed, because inadequately controlled combustion can result in the release of hazardous substances to the marine environment.

The Global Alliance on Health and Pollution (which includes, among others, UNEP, UNDP, UNIDO and the World Bank) has developed an international register of over 2,000 sites in middle- and low-income countries (as defined by the World Bank) where pollution problems are occur-

---

2   A megalitre is equivalent to one million litres or one thousand cubic metres.

ring (http://www.pollutionproject.org/about-tsip/). A large number of these sites are in the immediate coastal zone. Although this exercise has been focused on implications for human health, the extent to which the problems at these sites consist of uncontrolled releases of hazardous substances gives an indication of the extent to which badly managed waste-disposal sites and other sites with toxic deposits can present a problem for the ocean. In more developed countries, there are also problems from sites with toxic deposits, but remediation efforts appear to have been implemented in many of these cases (Ericson et al., 2013).

4.2 Use of Pesticides

The purpose of pesticides is that they are spread into the environment in order to control the pest against which they are aimed. If the pesticides are applied improperly, if surplus pesticides are not adequately disposed of, or if the chemicals involved have a sufficiently high degree of persistence before they degrade, they will eventually reach the marine environment. As shown above, action has been taken to remove from use many of the pesticides that give rise to most concern about their impact on the marine environment because of their toxicity, persistence and bioaccumulativity. However, even where such pesticides have been removed from the market, stocks often remain, and residues from past use persist in the soil and watercourse sediments that can make their way to the sea. In some cases, the judgement has been made that controls on the use of the pesticide will be sufficient to guard against harm to the oceans (see above on atrazine). In all these cases, therefore, there is a strong case for continued monitoring to check that bans are working and that usage conditions are being observed.

4.3 Routine combustion processes

Some hazardous substances, especially polycyclic aromatic hydrocarbons (PAHs), can be created by relatively common combustion processes, such as wood-burning stoves (Oanh et al., 1999). Uncontrolled burning of waste, such as rubber tyres, is another such source. Such emissions can be limited by better design of stoves and by better management of waste disposal. However, effective control of all such sources is unlikely to be practicable.

4.4 Abrasion

Some hazardous substances are used in products such as vehicle tyres and paint, where eventual abrasion is likely to free them into the environment, as the tyres are worn down or the paint peels off. Significant progress was made in reducing this kind of contaminant with the replacement of white lead paint by paint based on titanium dioxide (Waters, 2011). Substitution of this kind is the most effective way of resolving this kind of problem.

4.5 Small-scale gold-mining

A traditional, but crude, refining process for recovering gold from ore uses mercury to create an amalgam with the gold and subsequently vaporizes the mercury to leave high-quality gold. The vaporized mercury becomes an airborne contaminant, and can reach and affect the ocean. Artisanal gold-mining has been estimated to account for about 25 per cent of global gold production (Donkor et al., 2006). The predominant refining process in artisanal gold-mining is the mercury-amalgam process. It is judged to be the sector with the largest source of mercury emissions to the air (UNEP, 2008). The Minamata Convention on Mercury (2013) requires States bound by the Convention which have artisanal and small-scale gold mining to reduce and, where possible eliminate the use and environmental releases of mercury from such mining and processing.

4.6 Accidents

Wherever hazardous substances are produced, stored or transported, there is scope for accidental releases. There is no effective global source for statistics of accidents involving hazardous substances (ILO, 2007). In several countries, systems have been established to provide for the location, design and inspection of premises where hazardous substances are produced or stored and of vehicles carrying them, and for response to, and investigation of, significant accidents that do occur (for example, the European Union Seveso Directive (EU, 1996)).

5 Regional View of the Impact of Hazardous Substances on the Ocean

The lack of data makes it impossible to develop a general assessment of the relative impacts of hazardous substances on the ocean in the different parts of the world. In some areas, regional or national efforts have produced time-series of observations that enable trends to be established. But even here, the need to work through a number of institutions often means that clear comparisons between the absolute situation in different areas is not possible: different measuring techniques may be used; significantly different ranges of varieties of chemicals may be observed; and there is often an absence of any ring-testing to validate the accuracy of different institutions.

5.1 Open ocean generally

Observations of the presence of heavy metals and other hazardous substances in the open ocean[3] are very limited, including areas around islands and archipelagos in the open ocean. Few specific studies of pollution in the open ocean have been conducted. What information is available is concentrated on the north Atlantic. The Indian Ocean and the southern parts of the Atlantic and Pacific Oceans have hardly been assessed.

---

[3] As explained in Chapter 1, "open ocean" in this Assessment refers to the water column of deep-water areas that are beyond (that is, seawards of) the geomorphic continental shelf. It is the pelagic zone that lies in deep water (generally >200 m water depth).

For hazardous substances, the most significant route for impacts on the open ocean is transport through the atmosphere: hazardous substances can be carried either as aerosols (that is, microscopically fine particles of solids or liquids suspended in the air) or as gases (particularly in the case of mercury). The substances can remain suspended for long periods, and thus travel long distances. However, available evidence does not show that heavy metals in the open ocean are at levels causing adverse effects on humans or biota – with the exception of mercury. The load of mercury in the atmosphere has approximately tripled in the last two centuries. This has led to a probable doubling of inputs to the ocean. However, evidence also exists that, in some open-ocean areas such as near Bermuda, levels of mercury in the sea have decreased from the early 1970s to 2000. Nevertheless, there is good evidence that some fish concentrate mercury in their flesh to levels which give rise to risks for humans who eat a lot of such fish. Mercury concentrations in midwater fishes are several-fold higher than in epipelagic fishes at the same trophic level. Mercury levels in deep-sea fishes, such as morids and grenadiers, are substantially higher than in shelf-dwelling fishes, such as cod; notably long-lived fishes on seamounts, such as orange roughy and black cardinalfish, have mercury levels near or at the levels normally regarded as permissible for human consumption (0.5 ppm). Human activities have also led to higher levels of airborne inputs of lead and cadmium, but in these cases there is no evidence yet of toxic effects (Monteiro et al., 1996; Koslow, 2007; GESAMP, 2009).

For persistent organic pollutants (POPs), there is no doubt about their ability to be carried long distances through the atmosphere – this was one of the major reasons for the concerns that led to the Stockholm Convention. Although the effects of deposition of POPs on land have been extensively studied, information specifically on the levels of deposition of these substances in the open ocean and their possible effects is very limited (GESAMP, 2009). Estimates suggest that concentration of POPs may be an order of magnitude higher in deep-sea than in near-surface-dwelling fishes, and the deep sea has been referred to as their ultimate global sink (Froescheis et al., 2000; Mormede and Davies, 2003).

5.2 Arctic Ocean

In the Arctic, downward trends are reported in concentrations of the POPs controlled by the Stockholm Convention. Levels in marine mammals, some seabirds and polar bears are still high enough to cause adverse effects on their immune systems and reproductive success, but this is not the case for fish. Of the heavy metals, lead concentrations in biota were assessed as low in 1997 and since then they have been found to be decreasing. Mercury has been found at relatively high levels in whales, but the presence of selenium is also high enough to neutralize any detrimental effects. Parts of northern Canada have substantial natural levels of cadmium. The runoff from these deposits is reflected in the marine biota. Local pollution from heavy metals and some POPs is found around some mines, especially on the Kola Peninsula (Russian Federation) and some military installations, such as the Distant Early Warning System stations in northern Canada. In addition, one report suggests that 12 million drums of unknown, but potentially polluting, contents have been left in the Russian Federation: remediation is under way (AMAP, 1997; AMAP, 2009). Nevertheless, atmospheric transport and transport by ocean currents of pollutants are still significant issues for the Arctic (Stemmler et al., 2010; Ma et al., 2015).

5.3 Atlantic Ocean and Adjoining Seas

5.3.1 North-East Atlantic Ocean, North Sea, Celtic Seas

The North-East Atlantic is one of the most thoroughly assessed areas of the ocean: two comprehensive assessments were carried out in 2000 and 2010 (OSPAR, 2010). It is also an area where major efforts have been made since 1975 to reduce inputs of hazardous substances. Assessments are made of each of the contaminants studied, rather than attempting to combine them in a single indicator.

Statistically robust results show major reductions in the amounts of heavy metals being introduced into the marine environment in this area (Green et al., 2003). This is also demonstrated from monitoring by the OSPAR Commission (see Table 20.2).

A large part of these reductions was achieved in the 1990s: progress since 1998 has been slower. Concentrations in some areas, such as around the industrial estuaries of the Rhine (the Netherlands), the Seine (France) and the Tyne, Tees and Thames (United Kingdom), as well as in certain industrialized estuaries in Norway (Inner Sørfjord) and Spain (Ría de Pontevedra) and the inner German Bight, are still at levels giving rise to risk of pollution effects. High concentrations of cadmium found in fish and shellfish around Iceland seem to be linked to volcanic activity, such as the eruption of the Eyjafjellajökull volcano in Iceland in 2010 (OSPAR, 2010).

Trends in concentrations of PAHs in fish and shellfish are predominantly downward, especially in the Celtic Sea but, in many estuaries and urbanized and industrialized locations, they are still at levels which pose pollution risks. In many locations in coastal waters, concentrations of at least one polychlorinated biphenyl (PCB) congener pose a risk of causing pollution effects. Similar concern has arisen over the exposure to perfluorinated compounds, particularly perfluorinated octanoic sulfonate (PFOS). Over 25 years after being banned, PCBs are thought to be possibly causing adverse biological impacts in some areas: the Faroese

Table 20.2 | Percentage change in inputs of some heavy metals into North Sea and Celtic Seas 1990-2006. Source: OSPAR, 2010

Area	Cadmium - riverine input	Cadmium - direct discharges	Lead - riverine inputs	Lead – direct discharges	Mercury –riverine inputs	Mercury - direct discharges
North Sea	-20%	-75%	-50%	-80%	-75%	-70%
Celtic Seas	-60%	-95%	No trend	-90%	-85%	-95%

authorities (Denmark) have initiated a risk management process for the human consumption of pilot-whale meat (a traditional food source in the Faroe Islands) because of the presence of POPs.

Observations show that concentrations in fish and shellfish of the pesticide lindane (which has been banned since the early 1980s) are decreasing generally. However, concentrations in some localities are still of concern. These probably represent past use on nearby land. The more recent cessation of the use of other pesticides classed as hazardous substances is seen as likely to achieve similar results.

5.3.2 Baltic Sea

The Baltic Sea is an enclosed water-body with very limited water exchange with the North Sea and the North-East Atlantic. Periodically major inflows occur, bringing in substantial amounts of new water with high salinity from the North Sea. These inflows were fairly frequent until about 1980, but thereafter became infrequent, occurring in 1993, 1997, 2003, 2011 and 2014. The large quantity of freshwater from the Baltic catchments, together with the limited exchange with the North Sea, allows the build-up of hazardous substances in the basins of the Baltic Sea. Like the North-East Atlantic, the Baltic has a long-standing practice of assessment of the state of the marine environment. The Helsinki Commission has developed a multimetric indicator-based assessment tool. This has been used to integrate the status of contamination by individual chemicals and biological effects at specific sites or areas into a single status value termed the "Contamination Ratio" (CR). This CR is the ratio of the current status (the measurement of the concentration of a substance or biological effect) and a threshold level or quality criterion for that particular substance or biological effect. The CRs of all substances or indicators are grouped under four different ecological objectives (contaminant concentrations in the environment generally, contaminant concentrations in fish, biological effects on wildlife and levels of radioactivity) and integrated to yield a status classification ("high", "good", "moderate", "poor" or "bad") for each ecological objective. The ecological objective receiving the lowest status classification serves as the overall classification of the assessed site or area, giving the classification of the "hazardous substances status" of that site or area. The criteria used are not all uniform, but may include nationally set criteria. Therefore the results are not strictly comparable between assessment units. The overall picture is shown in the adjacent map, based on assessments at 144 sites, where "high" indicates good conditions and "bad" bad conditions of the marine environment with respect to hazardous substances (HELCOM, 2010a; HELCOM, 2010b).

Overall, there has been a steady and substantial improvement in the quality of the Baltic in respect of hazardous substances over the past two decades. This is due partly to the focus of the Baltic States in tackling the hotspots of pollution that were identified, and partly to the closure during this period of a number of the more polluting plants in countries in economic transition, as a result of economic circumstances. In the countries in economic transition, the former large installations have been superseded by a larger number of small and medium-sized enterprises, which makes the task of adequate regulation more difficult. There is, nevertheless, much further progress to be made before the goals set by the Helsinki Commission are reached.

5.3.3 Mediterranean Sea

The Barcelona Commission has carried out assessments of many aspects of the state of the Mediterranean over the last four decades. Nevertheless, there are major gaps in the data available for assessment: much more is known about contaminants in the sea off the northern coasts of the Mediterranean than about those off the southern and eastern coasts. Major sources of discharges and emissions of heavy metals are seen as the cement industry, electricity generation, metal mining and smelting, and fertilizer production. Many waste-water treatment plants are also seen as a problem. Based upon the available information, high concentrations of heavy metals (especially lead and/or mercury) in sediments and shellfish (blue mussel (*Mytilus galloprovincialis*)) are found around Barcelona, Cartagena and Malaga (Spain), Marseilles/Fos and Toulon (France), the Gulf of Genoa, the Po delta, the Gulf of Trieste and around Naples (Italy), the coast of Croatia, Vlora Bay (Albania), around Athens, Thessaloniki and Kavala (Greece), around Izmir (Turkey) (though subsequent Turkish Government tests have found nothing that would require action to be taken), Haifa Bay (Israel), the Nile delta (Egypt) and the coastal lagoons of Bizerta and Tunis (Tunisia). Insufficient data were available for robust trend analysis, but the limited analysis possible showed a general pattern of stable to declining trends, although in some places there were slightly increasing trends (UNEP, 2012).

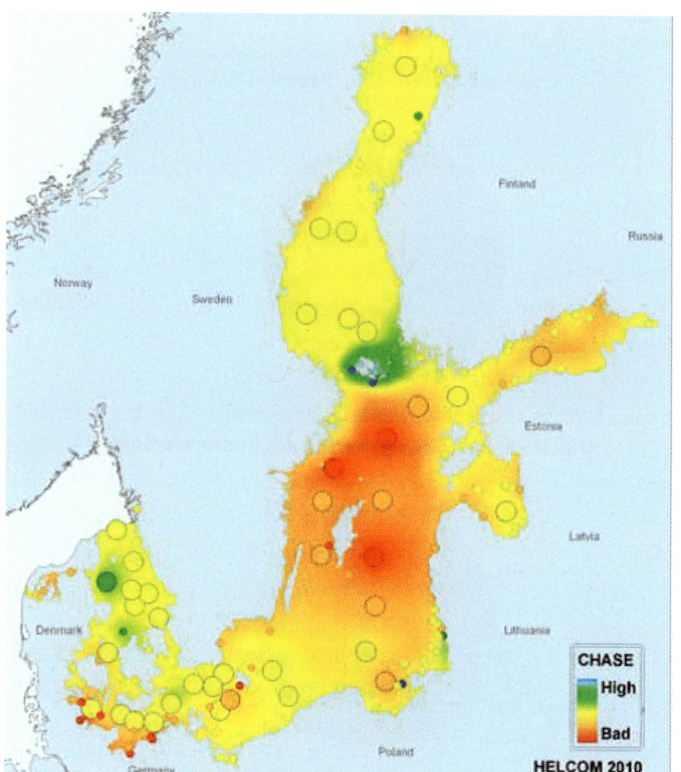

Figure 20.4 | Baltic Sea: Combined Hazardous Substances Contamination Index. Source: HELCOM, 2010a.

In the past, high levels of POPs have been measured in top predators in the Mediterranean. More recently, a study of data from 1971-2005 has concluded that the contamination of sediments by POPs is mainly associated with major urban areas, the mouths of major rivers, major ports and coastal lagoons, and that there has been a general decline in such concentrations. A 2011 study identified the areas of the mouth of the River Ebro and Barcelona (Spain), the mouth of the Rhône and Marseilles (France), the coast from Nice (France) to Livorno (Italy), the area around Genoa (Italy), the coast of Croatia and the port of Piraeus (Greece) as showing elevated levels of PCBs. Most of these locations, together with the Bay of Naples, the coast of the Marche and the Gulf of Trieste (Italy), the area around Dürres and Vloa Bay (Albania), the Ambracian, Saronic and Thermaic Gulfs (Greece), the area around Izmir (Turkey), the Bay of Tunis (Tunisia) and the Bay of Algiers (Algeria) also showed moderate to high levels of chlorinated pesticides. Again, data were insufficient for trend analysis (UNEP, 2012). Turkish authorities have subsequently indicated that there have been no findings which would have required measures to be taken.

5.3.4 Black Sea

Contamination by pesticides and heavy metals has not been judged to be a basin-wide problem by the Black Sea Commission. Elevated concentrations of heavy metals in bottom sediments and biota near river mouths, hot-spots and ports are decreasing. Pesticides are mostly introduced through rivers and streams discharging from agricultural areas. However, as a result of economic change, the use of these substances has decreased considerably and no longer presents a major hazard, except where their use was very intensive in the past. Elevated concentrations of HCH (mainly lindane) have been found along the coastal areas influenced by the Danube River: some sites near the Danube Delta were found to be among the highest levels of HCH recorded globally. In 2002, evidence was found of DDT and its breakdown products, probably from inappropriate storage of expired pesticides (Black Sea Commission, 2008; Heileman et al., 2008c).

5.3.5 North-West Atlantic

As in the Arctic, the problems of airborne transport of POPs found in the Arctic are also of concern in Labrador and Newfoundland. The main influence further south in Canada is the outflow of the St Lawrence River, which drains a large part of the heavily populated interior of Canada and the United States. The work derived from the efforts of the Canada/United States International Joint Commission (on shared water bodies) has done much to reduce the hazardous-substance content of this outflow. Similarly much has been done in Canada to address the problems posed by coastal industries, especially paper and pulp mills. As a result, hazardous substances are not seen as a priority for the Canadian Atlantic (Janowicz et al., 2006). Nevertheless, some problems remain, particularly in the Saguenay Fjord, where mercury and other metals were found in beluga whales at levels sufficient to cause concern. Cultured and wild scallops have been found to contain cadmium above the levels acceptable for human consumption, although its main source seems to be of natural origin (Dufour et al., 2007).

In the United States, the National Coastal Condition Reports (NCCR) (of which the latest NCCR IV was completed at the end of 2012, though based on data from 2003-2006 (EPA, 2012)) have been prepared regularly since 2000. They consider indices of water quality, sediment quality, benthic quality, coastal habitats and fish-tissue contaminants. They examine the coastal waters (estuaries and embayments) and also look at some of the waters further offshore. The sediment-quality index and the fish-tissue contaminants are the most relevant to the question of contamination by hazardous substances, although the benthic index (which looks at the structure of the benthos and the extent to which it is affected by pollution) can also be illuminating. The sediment-quality index is based on measurements of toxicity, amounts of contaminants (heavy metals, PAHs and PCBs) and total organic carbon content in samples taken from a range of stations, which number thousands across the country. The fish-tissue contaminants index is based on samples of fish for human consumption of species appropriate to the region. The indices for the sampling stations within a region are used to classify the region as "good", "fair" or "poor", according to the proportions of sampling stations within different bands of the indices.

The United States divides its Atlantic coast into two regions: the North-East region (Maine to Virginia, including Chesapeake Bay), and the South-East region (North Carolina to Florida). The North-East region is the most heavily populated part of the United States, and the overall condition of its coastal waters is judged to be "fair". The positions on the sediment-quality index (overall "fair") and the fish-tissue contaminant index (overall "fair to poor") are shown in the pie charts below.

Figure 20.5 | United States North-East Region Sediment and Fish-Tissue Contaminants Indices. Source: EPA, 2012.

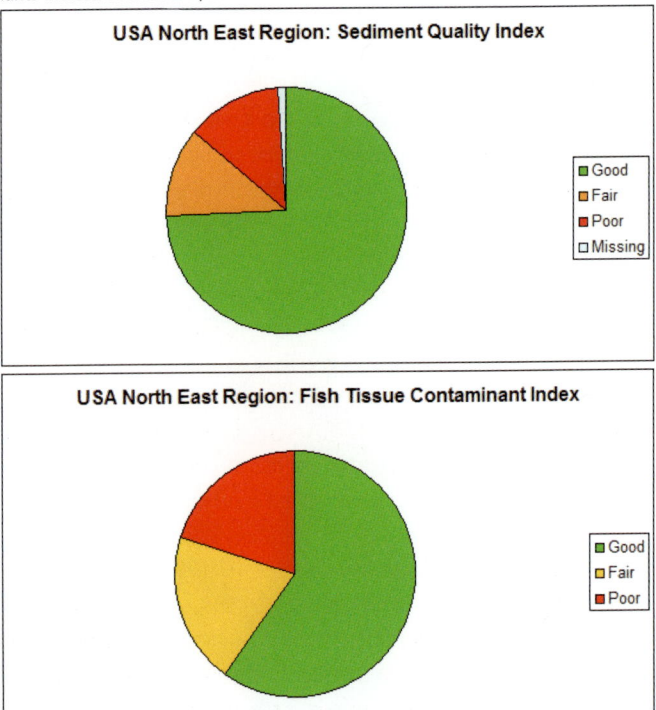

The problem zones for both sediments and fish-tissue contaminants are principally in Great Bay (New Hampshire)[DDT, Hg], Narragansett Bay (Rhode Island), Long Island Sound, the New York/New Jersey harbour area [DDT, Hg], the Upper Delaware Estuary and the western tributaries of Chesapeake Bay. The impaired ratings for the large majority of these sites were due to the presence of PCBs. Advice was also issued at various dates during 2006 against eating fish caught along about 84 per cent of the length of the coast of the North-East region – mainly because of the presence of PCBs. Those marked [DDT] also showed, above the thresholds, moderate to high levels of DDT and those marked [Hg] showed moderate levels of mercury. The NCCR also considered whether trends could be detected over the period from 2001 to 2006. No overall statistically valid trends were noted, but a significant reduction was observed in the areas judged as "poor" on the element of the presence of contaminants from Narrangassett Bay to the Delaware River.

In Chesapeake Bay, where a combination of problems of toxic contaminants and eutrophication resulted in 2009 in a special programme involving the Federal Government and the five States, the most recent report shows that for PCBs and mercury, many locations in the catchment have an impaired ecological status, largely stemming from concentrations in sediments and in fish tissue (where human consumption often has to be discouraged). A limited number of locations have severe problems from dioxins/furans, PAHs, some chlorinated pesticides (aldrin, chlordane, dieldrin, DDT/DDE, heptachlor epoxide, mirex), and some metals (aluminium, chromium, iron, lead, manganese, zinc). For other products (atrazine, some pharmaceuticals, some household and cosmetic products, some brominated flame retardants and biogenic hormones), it was not possible to assess where severe impacts were occurring, but it is known that that the substances have potential for adverse, sub-lethal ecological effects (EPA et al., 2012).

The NCCR also looked at the condition of the Mid-Atlantic Bight: that is, the seas between Cape Cod and Cape Hatteras out to the edge of the continental shelf. None of the contaminants for which tests are made for the sediment-quality index or the fish-tissue contaminants index was found in excess of their corresponding Effects Range Medium (ERM) values (values probably causing harmful effects). Only three chemicals (arsenic, nickel, and total DDT) exceeded their corresponding Effects Range Low (ERL) values (values possibly causing harmful effects), and these lower-threshold exceedances occurred at only a few sites. This implies that, on the same basis as for waters closer to the shore, this sea area should be regarded as in a "good" condition.

The overall condition of the South-East region (North Carolina to Florida) was judged to be fair. The sediment-quality index was judged to be "fair to poor" and the fish-tissue contaminants index was rated "good". No statistically significant trends were observed for the period 2001-2006. Conditions in the bight between Cape Hatteras and the south of Florida were also examined. Three metals (arsenic, cadmium, and silver) were found at concentrations between ERL and ERM values at 9 of the 50 offshore sampling sites, but no sites had more than one ERM value exceeded. Nevertheless, advice was in force in the whole area against eating king mackerel (*Scomberomorus cavalla*), because of mercury contamination.

A separate study – the National Mussel Watch – looked over a 20-year period (1986 – 2005) at levels of contamination by hazardous substances in mussels and oysters along the coast. For the Atlantic coast, this has shown in general no significant trends in contamination by heavy metals, but some locations show a decreasing trend. It has shown no significant trend in cadmium contamination in Chesapeake Bay, in spite of major efforts at reduction, and has shown significant increasing trends in mercury and lead at a few locations in the zones where problems were identified by the NCCR. Nevertheless, it has shown significant decreasing trends in contamination by POPs all along the Atlantic coast of the United States (Kimbrough et al., 2009; Mussel Watch, 2011).

5.3.6 Wider Caribbean

Information on hazardous substances in the Wider Caribbean (that is, the Gulf of Mexico, the Caribbean Sea and the Atlantic immediately east of the Leeward and Windward Islands) is mixed: for the United States and its dependencies, the same type of information is available as for the Atlantic coast; elsewhere, there is no systematic record.

The overall condition of the United States Gulf coast is judged to be "fair". The sediment-quality index was judged to be "fair to poor" and the fish-tissue contaminants index was rated "good". The areas rated "poor" on the sediment-quality index lay mainly around the Florida Keys, the coasts of Alabama and Mississippi, Galveston (Texas) and the Texas coast south of Corpus Christi. No statistically significant trends were observed for the period 2001-2006, but a substantial reduction in areas failing the test for the presence of contaminants was found. Nevertheless, in 2006 advice was in force along the whole of the United States Gulf Coast against the eating of king mackerel (*Scomberomorus cavalla*), because of mercury contamination (EPA, 2012). The Mussel Watch has likewise shown some locations with decreasing trends in heavy-metal contamination and a general decreasing trend in contamination by POPs (Mussel Watch, 2011).

There is no recent, comprehensive compilation and analysis of inputs of hazardous substances to the remainder of the Wider Caribbean (Fernandez et al., 2007), although specific areas are known where problems of this kind are found (Cartagena Bay, Colombia; Puerto Limon, Costa Rica; Havana Bay, Cuba; Kingston Harbour, Jamaica; and some locations in Puerto Rico). These largely result from the discharge of untreated waste-water from local industries. Mining also presents significant problems, particularly mining of bauxite in Guyana, Jamaica and Suriname and (to a lesser extent) in the Dominican Republic and Haiti (GEF, 1998; EPA, 2012). Heavy and increasing usage of agricultural pesticides is reported from the mainland countries of Central America, and from Jamaica and Cuba (UNEP-UCR/CEP, 2010). In addition, chlordecone (an agricultural pesticide, the use of which was prohibited from 1993) gave rise to concern in fish and seafood from Martinique in 2003 (Bocquenéa and Franco, 2005). The presence of the same chemical in seafood was still preventing it being marketed in 2013 (*Le Monde*, 2013).

5.3.7 North Atlantic open ocean

The OSPAR Quality Status Report 2000 (OSPAR, 2000) examined the situation in the open ocean of the Atlantic (beyond the 200-m isobath) east of 42°W longitude. This showed limited information about the state of the marine environment in this area. No later comprehensive survey has been made. The conclusions on contamination by hazardous substances were that:

(a) Airborne inputs of hazardous substances from land were very significant: probably equal to the effects of waterborne inputs reaching the deep Atlantic;
(b) Anthropogenic inputs to the North Atlantic were higher than those to other deep ocean areas, representing up to 25 per cent of the total estimated global deposition rates for a range of substances;
(c) Nevertheless, the level of concern for the area about contamination by hazardous substances was rated as low.

5.3.8 South-East Atlantic

The coastal waters of the South-East Atlantic are dominated by three currents: from north to south, the Canaries Current, the Guinea Current and the Benguela Current. These three areas have been adopted for the tasks of addressing the problems of the marine environment. Little detailed information is available about land-based sources of pollution in these areas. What is clear is that the main problems are "hot spots" in the proximity of the principal coastal cities: Abidjan (Côte d'Ivoire), Accra (Ghana), Cape Town (South Africa), Casablanca and Rabat (Morocco), Dakar (Senegal), Douala (Cameroon), Lagos and Port Harcourt (Nigeria), Luanda (Angola) and Walvis Bay (Namibia). Most of the industries operating in the region are located in or around these coastal areas and discharge untreated effluents directly into sewers, canals, streams and rivers that end up in the ocean. Outside the immediate areas of discharge, however, the effects are limited by the strong marine current (Heileman, 2008b).

Nevertheless, some specific problems are more than local. Mercury emissions from artisanal gold-mining in West Africa are a general problem. This gold production is an important part of the national economies of several States in the area, but significant levels of mercury have been found in many West African rivers, and therefore present risks to the marine environment (Donkor et al., 2006). Other mining activities also present significant threats. For example, phosphate mining at Hahatoé-Kpogamé in Togo results in discharges of tailings and other waste with high levels of cadmium and lead being found in fish and crustaceans (Gnandi et al., 2006).

Samples from the Korle (Accra, Ghana), Ebrié (Abidjan, Côte d'Ivoire) and Lagos (Nigeria) lagoons, show heavy metals in the sediments up to three (Cd), six (Hg) and eight (Pb) times more than those from uncontaminated areas, and in shellfish at or above WHO standards for Cu, Pb and Zn (GCLMEP, 2003).

5.3.9 South-West Atlantic

Although there are studies of several locations along the coast of Brazil, there does not appear to be a comprehensive study of the levels of heavy metals or POPs for the coastal sea of Brazil as a whole. It is clear that there are many untreated direct and riverine discharges from coastal cities which produce significant local effects: São Paulo, with a population of over 11 million and a concentration of petrochemical and fertilizer industries, and Rio de Janeiro, with over 6 million inhabitants, are the most significant, but there are other examples, such as Rio Grande. The River Amazon also has a major effect on the northern part of the area. The diffuse sources contributing to this effect include agricultural pesticides and mercury from small-scale gold mining (Heileman, 2008a, Heileman, 2008c; Heileman and Gasalla, 2008; Niencheski et al., 2006).

The situation is much the same further south: there are hot-spots associated with major coastal cities, but no overall survey. The River Plate is a major influence, since it drains areas with a high concentration of potentially polluting industries, and is assessed as highly polluted. Apart from that, the most serious area is around Bahia Blanca, where the general level of contamination has been assessed as moderate (Marchovecchio, 2009). However, the San Matias Gulf has also been identified as having relatively high levels of cadmium and lead (Heileman, 2008e).

5.3.10 South Atlantic open ocean

Very limited information is available on levels of contaminants in the central South Atlantic. Nevertheless, samples of skipjack tuna, tested for brominated flame retardants as a marker for widely dispersed POPs, show levels that are lower than in the open ocean of the Pacific (Ueno et al., 2004).

5.4 Indian Ocean

No comprehensive studies or time series of the incidence of hazardous substances in the Indian Ocean exist, although there are a number of local, one-off studies.

5.4.1 Western Indian Ocean

In general, the areas of both the Agulhas Current (the waters off the coasts of eastern South Africa and Mozambique and around the Comoros, Madagascar, Mauritius, Réunion (France) and the Seychelles) and the Somali Current (the waters off the coasts of the Federal Republic of Somalia, Kenya and the United Republic of Tanzania) are not heavily polluted with heavy metals or POPs (Heileman et al., 2008b). Nevertheless, some top predators (yellowfin tuna) are reported to show high concentrations of HCB and lindane by comparison with the same species elsewhere, although levels of PCBs and DDTs are not so high. The residues of

PCBs, DDTs, lindane and HCB were higher than those measured in 1999 (Machado Torres et al., 2009).

However, relatively severe localized problems are found near major cities and industrialized areas in all the countries. The main industries that contribute towards chemical contamination in this region include: manufacturing, textiles, tanneries, paper and pulp mills, breweries, chemical, cement, and sugar and fertilizer factories. Coastal solid-waste dumps add to the problems. The intensive use of agro-chemicals, such as DDT, aldrin and toxaphene, has been common throughout the region. Inappropriate utilization, storage and dumping of agrochemicals are a growing concern. Direct discharge of wastes from fertilizer factories is a severe problem in the region (Heileman et al., 2008a). Mozambique has instituted a legal and institutional framework for the management and treatment of municipal and industrial waste, including the development of sanitation infrastructure (landfills, industrial and wastewater treatment plants.

5.4.2 Red Sea, including the Gulf of Suez and the Gulf of Aqaba, and the Gulf of Aden

Slow water turnover makes the Red Sea particularly vulnerable to pollution build-up. Pollution is severe in localized areas around industrial zones and facilities, including especially the Gulfs of Suez and Aqaba and near the port of Aden. The installations include phosphate mines, desalination plants, chemical industrial installations and oil production and transportation facilities. In 2003, elevated levels of some heavy metals were found near Suez, in the Sharm al Maya Bay in Egypt (PERSGA, 2006; EEAA, 2003).

5.4.3 Persian Gulf

Major manufacturing industries operate in the coastal States of the Persian Gulf, based largely on the raw materials from oil and gas extraction, producing fertilizers, chemicals, petrochemicals, minerals and plastics. The demand for fresh food has also led to intensive agriculture and the use of pesticides. All these activities have resulted in waste-water and runoff taking heavy metals and other hazardous substances into the semi-enclosed sea of this area (Sale et al., 2011).

5.4.4 Arabian Sea and waters west of India, the Maldives and Sri Lanka

Overall, pollution from hazardous substances in the northern Arabian Sea has been assessed as severe in several coastal hotspots, but in general it has been evaluated as moderate. The major issues in these hotspots are heavy metals from industrial installations. Other hotspots are found at the mouths of some major rivers (for example, the Tigris, Euphrates, Karun, Hileh and Mand rivers). Other hotspots involving PAHs have been recorded in coastal areas receiving effluents from highly industrialized zones. In waters off Pakistan, chlorinated pesticides are more prominent. Since persistent organic pesticides are not to be marketed in the States bordering the Arabian Sea, these findings likely result from the remains of historic use (Heileman et al., 2008a).

Along the coasts of India the picture is mixed. High concentrations have been noted, for example, off the Maharashtra coast. Near Mumbai, sample fish have also been shown to have concentrations of lead, cadmium and mercury above levels that are generally regarded as fit for human consumption (Heileman et al., 2008a; Deshpande et al., 2009). Around the Alang-Sosiya ship-breaking yards in Gujarat, India (which employ 40,000 people), on the basis of samples taken in 2001, particularly high levels, above approved limits of heavy metals in sediments, have been reported (Janil et al., 2011). On the other hand, along the Gujarat coast, the concentrations of mercury in sediments have decreased to below the limits of detection, reflecting the decrease in its concentrations in land-based effluents. Along the Maharashtra coast, mercury levels in sediments have declined and are currently about 0.1μg/g. Along the Karnataka coast, observations have been made off Mangalore and Karwar. At both locations, concentrations of mercury in sediments have shown decreasing trends. Along the Kerala coast, concentrations of mercury were low at all sampling locations, and exhibited declining trends (NIOT, 2014).

In Sri Lanka, most industrial plants are concentrated near Colombo. They lack waste-water treatment capacity, and textile and metal-finishing plants are discharging significant quantities of some heavy metals. The Lunawa coastal lagoon has been ruined by such discharges (BOBLME, Sri Lanka, 2011).

5.4.5 Waters east of India, the Maldives and Sri Lanka (Bay of Bengal, Andaman Sea, Malacca Strait)

The dominant influence in the north of this area is the River Ganges, the second-largest hydrological system on the planet. The Bay of Bengal Large Marine Ecosystem Project (BOBLME) has organized recent surveys of marine pollution in all the coastal States. These show that much information is available, but there are no time series or sufficient metadata for comparisons.

In the waters off India, along most of the coast, concentrations of mercury in sediments have declined: off Tamil Nadu, concentrations were observed at <0.1 μg/g; off Andhra Pradesh, there has been a substantial decline to a similar level; and the coast off Orissa exhibited a decline to concentrations of 0.1-0.2 μg/g. In contrast, off West Bengal mercury content of sediments showed a marginally increasing trend both inshore and near-shore: recent values of up to 0.3 μg/g suggest continued release of industrial waste containing mercury (NIOT, 2014). Hot-spots for heavy-metal pollution are found in the Ganges estuary (the Hooghly River, Diamond Harbour, Sagar Island and Haldia). Further south, hot-spots have been reported at Bhitarkarnik, Visakhapatnam, Ennore, Cuddalore, and Tuticorin. At all these places, pollution from heavy metals results from direct industrial discharges and the inputs of rivers carrying industrial discharges. Pollution from POPs stems mainly from pesticide-manufacturing plants and ineffective storage of withdrawn pesticides, as well as the leaching of pesticides and past uses of PCBs. Recent surveys indicate significant levels of DDT, PCBs and dieldrin in both nearshore and off-shore fish in the Bay of Bengal (BOBLME, India, 2011).

For Bangladesh, the picture is very similar. In addition, heavy metal pollution is particularly noticeable near the Sitakunado ship-breaking area of Chittagong. Reports from 2004 suggest that banned organo-chlorine pesticides are being sold on the black market and that some are being used in fish-processing plants (BOBLME, Bangladesh, 2011).

In the waters off Myanmar, significant levels of heavy metals have been reported in fish samples, but at levels below those at which human consumption is not advised. Organochlorine pesticides are not regarded as a problem because of lack of availability for use (BOBLME Myanmar, 2011).

In the Andaman Sea off Thailand, levels in sediments of lead and cadmium were reported in 2009 that were well above levels regarded elsewhere as likely to cause harm, and in 2007 levels of mercury in sample fish were reported that were above Thai and many other national standards for human consumption (BOBLME, Thailand, 2011).

On the western coast of peninsular Malaysia, levels of contamination of hazardous substances observed in 2009 and 2010 were in general within national standards, which are consistent with generally recognized standards. Exceptions were off the coast of Perak, where significant numbers of samples showed lead and cadmium levels in excess of these standards. This is attributed to major historic mining activities in that State (BOBLME, Malaysia, 2011).

5.4.6 Waters north and west of Australia

In general, the waters north and west of Australia are in very good condition. The large-scale mining in the catchments has not generally caused problems with hazardous substances because of the low rainfall and consequent absence of major watercourses. There are some localized problems around the Gulf of Carpentaria, such as Darwin Harbour and Melville Bay (Nhulunbuy, Northern Territory), where a localized, biologically dead area has been created by mining wastes (SE2011 Committee, 2011).

5.4.7 Indian Ocean open ocean

As with other open-ocean areas, information on levels of contamination from hazardous substances is limited. Studies in 1996 around the Chagos Archipelago (over 500 km from the nearest continental land) showed that only some PCBs and lindane were above the limits of detection, and then only just. The conclusions therefore were that atmospheric transport was the main source, and that the area was amongst the least affected coastal areas (Everaarts et al., 1999). Air sampling around the Chagos Archipelago has also concluded that the atmosphere over the Indian Ocean in 2006 was substantially less contaminated from atmospheric POPs than it was according to the available data from the 1970s and 1990s (Wurl et al., 2006). Samples of skipjack tuna from the mid-Indian Ocean were studied as a way of examining the distribution of airborne contaminants. They showed (like all tuna studied from around the world) detectable levels of brominated diphenyl ethers, but at lower levels than in the north Pacific (Ueno et al., 2004; Tanabe et al., 2008).

5.5 Pacific Ocean

5.5.1 North-East Pacific

Waters west of Canada and the mainland of the United States
The United States applies the general principles of the National Coastal Condition Report to its Pacific coasts, although the sheer scale of the Alaskan coast makes some aspects inappropriate. For Alaska, the results show an overwhelmingly good condition: as with the Alaskan Arctic coast, there are some local natural sources of heavy metals, but the presence of other hazardous substances is mainly from long-range transport through the seas or the atmosphere (EPA, 2012).

In Canada, the situation is much the same for the northern and central coast of British Columbia, where population density and levels of industrial development are low. Further south, however, some hot-spots with severe adverse effects from high concentrations of chlorinated hydrocarbons and heavy metals have been found, for example, in the Port Moody arm of the strait separating Vancouver Island from the mainland (Belan, 2004).

For the area of the Alaska current as a whole, samples of biota obtained during 2003-08 have generally not detected concentrations of PCBs, pesticides or mercury at levels of concern (PICES, 2009).

The waters off the western coasts of the contiguous United States are also assessed as being "good", with 86-89 per cent of the sampling stations being put in this class on the individual indices. The areas where the sampling stations fail to achieve "good" status are mainly around San Francisco, Los Angeles and San Diego in California (EPA, 2012). Nonetheless, the Mussel Watch shows significant decreasing trends of levels of heavy metals and other hazardous substances at sampling stations all along the Pacific coast of the contiguous United States, except for Puget Sound, where increases in the levels of lead contamination have been found (Mussel Watch, 2011).

Waters west of Mexico, Guatemala, El Salvador, Honduras, Nicaragua and Costa Rica
From the point of view of contamination by hazardous substances, the waters west of the countries of Central America are affected mainly by the discharges from local, relatively small-scale industries and agricultural runoff. The overwhelming majority of industrial effluent is discharged to the sea without treatment, and usage of pesticides is one of the highest in Latin America (Heileman, 2008e).

5.5.2 East Asian Coastal Seas - General

At the regional level, no information is regularly collected on inputs of hazardous substances and their effects. Levels of heavy metal contamination in the East Asian coastal seas are, however, known to have been rising over the last two decades, largely due to untreated municipal waste-water and industrial effluents. The rise was rapid in some areas, particularly in coastal waters of China, over the twenty years to 2000. For most of the area, there is no general evidence that this has ceased. In the Gulf of Thailand, lead and cadmium have been found at high levels near the mouths of all major rivers. In some areas, the levels of heavy metals in fish and shellfish have made them unsuitable for human consumption. Depth profiles of sediment samples suggest in many cases that these inputs of heavy metals are linked to recent rapid growth of electronics, ship-painting and chemical industries.

Persistent organic pollutants are measurably present in most coastal areas of the East Asian Seas at levels higher than many other parts of the world, but studies have shown decreasing levels of those which had previously been banned. Endosulfan has been found in most coastal sediments, especially off Malaysia, suggesting recent use. Both DDT and PCBs are found at levels above limits generally recognised as tolerable, but in some areas (for example, the Macau estuary in China), studies have shown that such levels peaked as much as twenty years ago (UNEP/COBSEA, 2010).

5.5.3 Coastal waters of China

At the national level, the Chinese authorities have been carrying out systematic monitoring of the coastal waters and coastal sediments as part of their national environmental monitoring programme. For water quality, their system uses an assessment classification based on a range of parameters covering not only hazardous substances but also problems caused by non-hazardous waste elements. As far as hazardous substances are concerned, levels are reported as satisfying national standards. The combined assessment is shown below in the section on nutrients. In addition, the Chinese authorities study sediment quality in relation to hazardous substances. Decreases in concentrations have been observed generally since 1997. However, in 2010 areas around 36 coastal pollutant discharge outfalls did not meet the sediment quality requirements, mainly because of the levels of copper and cadmium, and some of the area showed a worsening (China, 2012). In Hong Kong, China, waters, monitoring focused specifically on hazardous substances has been undertaken to evaluate the effects of the stringent measures to combat pollution from these substances. These monitoring programmes have demonstrated a steady decline between 1991 and 2012 of, in total, about 30 per cent in the levels of heavy metals in sediments at the west end of Hong Kong Island. Even there, however, levels in enclosed bays used as typhoon shelters can be much higher. Sediments around Hong Kong, China, show a slight decline in levels of PCBs (Hong Kong, China, 2013). Other evidence (from sea-bird feathers) shows that there are high levels of mercury in the South China Sea (Watanuki et al., 2013).

5.5.4 Yellow Sea and waters between Japan and the Korean Peninsula

The absence of data, and (even where they are available) their incompleteness and lack of consistency, make any assessment of the impact of hazardous substances on the Yellow Sea as a whole very difficult. There seems to be little doubt, however, that inputs of hazardous substances are at levels which give rise to concern, largely because of the discharges of untreated industrial waste-water (NOWPAP, 2007). Where detailed information is available about specific areas, however, there is good evidence of improvement: in the waters around Japan, for example, levels of PCB concentrations in both fish and shellfish have decreased by over 75 per cent between 1979 and 2005. Nevertheless, concentrations in both fish and sediments remain at levels sufficient to cause concern, particularly in enclosed sea areas surrounded by big cities (Japan MOE, 2009). Local sources of pollution, mainly from mining, are also significant along the northern coast of the Russian Asia-Pacific Region (Kachur et al., 2000).

5.5.5 North-West Pacific (Kuroshio and Oyashio Currents)

The areas east of the Japanese main island and the islands north and south of it are significantly less affected by industrial, mining and agricultural activities than the seas along the coast of the Asian mainland, except close to the Japanese main islands. In enclosed sea areas surrounded by big cities on the Japanese main island, there are levels of contamination from POPs much (up to five times) higher than on the western coast. This is particularly the case in Ise Bay and Tokyo Bay (I. Belkin et al., 2008; Heileman and Belkin, 2008; Japan MOE, 2009).

5.5.6 North Pacific open ocean

Apart from some major island ports, such as Pearl Harbor, Hawaii (which shows evidence of PCB contamination (EPA, 2012)), contamination from heavy metals and other hazardous substances is not a major concern in the central north Pacific. However, mercury concentrations in the North-East Pacific increased between 2002 and 2006 and modelling suggests an average increase of 3 per cent per year between 1995 and 2006 (Sunderland et al., 2009). Samples of skipjack tuna from the western part of the central north Pacific, studied to assess airborne transport of contaminants, showed higher levels of brominated diphenyl ethers than samples from the central Indian Ocean and off Brazil. The suggestion is made that this might be the result of atmospheric transport from locations in south-east Asia where waste goods containing these fire retardants were being dismantled (Ueno et al., 2004).

5.5.7 South-East Pacific (Waters west of Panama, Colombia, Ecuador, Peru and Chile)

Little new information on the situation in this area as a whole is available since the survey conducted in 2000 by the Permanent Commission of the South Pacific (CPPS, 2000). In respect of discharges of hazardous substances, this survey showed a major problem with mining waste, particularly in the south of Peru and the north of Chile. The mining industry in these countries is mainly in the coastal areas. A substantial number of mines disposed of their waste directly into the sea, and others indirectly through rivers: none of these wastes were treated. The areas said to be highly polluted were at the mouth of the River Rímac (into which a number of mines discharged) and between Pisco and Ite in Peru, and around Chañaral in Chile. In the north of Chile, as well, there were beaches which had been used in the past for the disposal of mine waste, and from which heavy metals (especially copper) were leaching into the sea. In addition, a heavy load of agricultural pesticides was thought to be present: nearly 5,000 tons of active ingredient were thought to be used annually in the 1990s, resulting in what was judged to be serious pollution in the coastal areas of the province of Chiriqui in Panama, in the extreme south of Colombia, around Guayaquil in Ecuador, around Pisco in Peru, and in regions VI (Rancagua), VII (Talca), VIII Concepción), X (Puerto Montt) and XI (Cohaique) in Chile.

5.5.8 South Pacific open ocean

Even less is known about the contamination of the open ocean by hazardous substances in the South Pacific than in the other open-ocean areas. The island States of the area have neither major industrial sites, nor major mines. A wide range of pesticides has been used for local agriculture, although the most hazardous are no longer used. A result of this use, however, is that residues have been found in the soil, as well as stocks of persistent organo-chlorine pesticides and contaminated sites where the pesticides were stored (Samoa, for example, has three such sites). There were also a number of electrical devices containing PCBs (the Federated States of Micronesia had 13.5 tons). With Australian and GEF assistance, programmes are in place to dispose of stocks and remediate contaminated sites. The States have, however, recorded their concern about lack of capacity to prevent the accidental creation of dioxins and furans from imperfect incineration (Samoa, 2004; FSM, 2004).

5.5.9 Waters east of Australia and around New Zealand

The waters to the east of Australia are renowned for the Great Barrier Reef – the world's largest coral reef system. Although the catchments draining into this part of the sea are not heavily industrialized, they contain intensive agriculture, especially for sugar cane. The pesticides (and other runoff) from these catchments are judged likely to cause environmental harm, particularly to the central and southern parts of the Great Barrier Reef. Models of the mean annual loads of a range of common pesticides (ametryn, atrazine, diuron, hexazinone, tebuthiuron and simazine) show that inputs are in the range of 16-17 tons of active ingredient. The total pesticide load to the Great Barrier Reef lagoon is likely to be considerably larger, given that another 28 pesticides have been detected in the rivers (Lewis et al., 2009; RWQPP, 2013).

Further south, the coast of the south-eastern part of Australia is the most heavily populated area in the country: nearly one-third of the total population is in central New South Wales. Port Jackson (Sydney Harbour) is locally contaminated with heavy metals, especially lead and zinc, and a large proportion of the estuary has sedimentary metals at concentrations where some adverse biological effects can be expected. Most of the contamination is a legacy of past poor industrial practice, but some evidence exists for continuing entry of contaminants, probably from leaching (Birch, 2000). Further south, in the State of Victoria, the Government acknowledges that data on the condition of marine and coastal ecosystems are not gathered in a comprehensive manner, making assessment of the condition of coastal and marine systems difficult. Estuarine and bay systems, such as Port Phillip Bay (Melbourne), Western Port and the Gippsland Lakes, have impaired water quality, partly from industrial and agricultural sources (Victoria, 2013).

In New Zealand, a study was made of dioxins, furans, PCBs, organochlorine pesticides, estuarine sediments and shellfish. The catchments covered ranged from highly urbanized to areas relatively remote from anthropogenic influences. The results showed that concentrations of these substances in New Zealand estuaries are low, and in most cases markedly lower than concentrations reported for estuaries in other countries, although concentrations in some estuaries are approaching those reported elsewhere for urbanized estuaries (NZMOE, 1999). Examination of sediment cores from Tamaki Creek, near Auckland (New Zealand's largest city) has shown a four-fold increase in levels of heavy metals since the European settlement of the area, with most of the increase in the last 50 years. Tamaki Estuary is classified as a polluted area (Abrahim et al., 2008). The estuaries around Auckland and near other large urban areas seem likely to be subject to the same pressures.

5.6 Southern Ocean

Levels of contamination by heavy metals and other hazardous substances are low. Long-distance transport through marine currents and the atmosphere means that measurable levels of contamination are found, but not at levels that give rise to concern. Some of the research stations have accumulated waste contaminated with heavy metals and other hazardous substances. Australia has removed a quantity from the Thala Valley base (Australian Government, 2011). Recently, brominated flame retardants are reported to have escaped from McMurdo Sound base (NGN, 2014).

6 Endocrine Disruptors

As discussed above, hazardous substances are usually defined in relation to the qualities of being toxic, persistent and liable to bioaccumulate. Toxicity is usually defined in relation to being fatal when ingested, to causing cancers (carcinogens), to causing birth defects (teratogens) or to causing mutations (mutagens). Many of the substances regarded as hazardous substances within these accepted definitions were also shown to affect endocrine systems and thus to interfere with the reproductive success of individuals and populations, and were therefore described as "endocrine disruptors".

In the 1990s, a consensus emerged that certain substances outside the accepted definitions of hazardous substances could also disrupt endocrine systems, and thus affect the ability of individuals and populations to reproduce successfully. In the marine context, the issue was highlighted by the discovery that tributyl tin, which had been adopted widely as a component in anti-fouling treatments for ships' hulls, had a severe effect at very low concentrations on molluscs: initially, the effects were observed at concentrations so low that they could not be detected by the then available methods of chemical analysis. The effects were referred to as "imposex", and took the form of females developing male sexual characteristics and thus becoming infertile. In some harbours, whole populations of molluscs disappeared. Where such substances were not within the accepted definitions of hazardous substances, new initiatives were needed. The question of "endocrine disruptors" for those concerned with the marine environment therefore became more focused on substances which are not within the accepted definitions of hazardous substances, but which may nonetheless have serious effects on the health of the marine environment (OSPAR, 2003).

The case of tributyl tin is discussed further in Chapter 17 (Shipping). Systems have been developed, principally by the Organization for Economic Cooperation and Development (OECD), to test substances to see whether they have the potential to disrupt endocrine systems (OECD, 2012).

In the application of these testing procedures, some substances not otherwise identified as hazardous substances have been identified that are, or may be, of particular concern to the marine environment. These include:

(a) *Nonyl phenyl ethoxylates:* These are used as emulsifiers, dispersive agents, surfactants and/or wetting agents. These degrade quickly to nonyl phenyls and short-chained nonyl phenyl ethoxylates, which are toxic to aquatic organisms and are thought to have endocrine-disrupting properties. The main users are the industrial, institutional and domestic cleaning sectors (30 per cent of use in Europe; other significant sectors in Europe are emulsion polymerisation (12 per cent), textiles (10 per cent), chemical synthesis (9 per cent) and leather (8 per cent)). Estimated use in Western Europe in 1997 was 76,600 tons. Action has been taken within the European Union and is proposed in the United States (OSPAR, 2009; EPA, 2010);

(b) *Estrogenic contaminants:* There is some evidence that human-derived steroids, such as estradiol and ethinyl estradiol (the active ingredient of the contraceptive pill) can affect aquatic biota. In fresh water, intersex conditions induced in male fish (trout) in rivers in England were attributed to ethinyl estradiol from sewage (Desbrow et al., 1998; Routledge et al., 1998; Tyler and Jobling, 2008). In contrast, androgenic effects have been found in female fish in rivers carrying pulp and paper mill effluents (mosquito fish) and feedlot effluent (fathead minnows) (Orlando et al., 2004). Whether such substances persist enough to continue to cause such effects after a lapse of time, and how the substances might operate in more dynamic or more dilute environments (such as the sea) is not clear;

(c) *Phthalates:* Phthalates are a class of chemicals most commonly used to make rigid plastics (especially PVC) soft and pliable. They can leach from PVC products, particularly when they enter waste streams. Phthalates can affect reproduction and development in a wide range of wildlife species. Reproductive and developmental disturbances include changes in the number of offspring and/or reduced hatching success and disruption of larval development. They generally do not persist in the environment over the long term, but there is evidence that environmental concentrations are above levels that give rise to concern in some aquatic environments. (Engler, 2012; Oehlmann et al., 2009).

7 Oil

The United States National Research Council has carried out two major studies, in 1985 and 2003, on the amounts of oil that enter the marine environment, both for United States waters and for the world as a whole (NRC, 2003). The studies concluded that the global estimates of hydrocarbons from land-based sources were particularly uncertain. The 2003 study placed the best estimate of runoff from land globally at 140,000 tons/year, but recognized that this could be as much as 5 million tons, or as little as 6,800 tons. This compares with its estimate of:

(a) The amounts of oil spilled (on average) in the process of extracting hydrocarbons from the seabed. Here the best estimate was 38,000 tons, within a range of 20,000 tons to 62,000 tons;
(b) The amounts of oil seeping naturally into the sea from submarine seeps, such as those off south California. Here the best estimate was 600,000 tons/year, within a range of 200,000 tons to 2 million tons;
(c) The total amount of hydrocarbons entering the sea from all sources. Here the best estimate was 1.3 million tons/year, within a range of 470,000 tons to 8.3 million tons.

Land runoff is therefore a significant component of the impact of hydrocarbons on the sea. As discussed in chapter 17 (Shipping), however, the significance for the marine environment depends crucially on the location: warm, sunny zones will result in much more rapid breakdown of the hydrocarbons by bacteria into harmless substances. Likewise, in areas with high levels of natural seepage of hydrocarbons, oleophilic bacteria will often be abundant and thus the breakdown of anthropogenic inputs will be quicker than in areas with little or no natural seepage. Moreover, within the land-based sources, much of the runoff is the result of relatively large numbers of relatively small accidents and mishaps, which are difficult to prevent. Mitigation, in the form of well designed drainage systems, accident-response systems and public education, has to be the main response.

Table 20.3 | Levels of oil content of aqueous discharges from European oil refineries. Source: Baldoni-Andrey et al., 2012.

Year	Throughput (million tons per year)	Oil content of water discharges per ton of throughput (g per ton of throughput)	Year	Throughput (million tons per year)	Oil content of water discharges per ton of throughput (g per ton of throughput)
1981	440	24.0	1997	627	1.86
1984	422	12.1	2000	524	1.42
1987	449	10.3	2005	670	1.57
1990	511	6.54	2008	748	1.33
1993	557	3.62	2010	605	1.30

Oil refineries, however, can represent significant point-sources of hydrocarbon discharges that can reach and affect the sea. No global information seems to be available on losses and discharges from oil refineries. In some areas the impact on the marine environment is serious. In the Persian Gulf, heavy (>200 µg/g) contamination of sediments in the central offshore basin is reported, and attributed to industrial effluents from onshore industries (Elshorbagy, 2005). Efforts in Europe and North America have shown, however, that it is possible to reduce this impact substantially. In Europe, CONCAWE (the oil companies' environmental organization) reports that, under pressure from regulators, the oil companies have diminished substantially the amounts of oil discharged in process water from refineries in relation to throughput:

8 Nutrients and Waterborne Pathogens

8.1 General

The second main aspect of land-based inputs that cause marine pollution involves the input of organic matter and nutrients. Organic matter and nutrients are not in themselves harmful, but can cause pollution problems when the inputs are excessive. There are a number of sources from which they enter the ocean. One of these is sewage, in the narrow sense of the waterborne disposal of human faeces and urine. Given the origin of sewage in this narrow sense, inputs of human pathogens are unavoidably bound up with sewage inputs. It is convenient therefore to consider issues of waterborne pathogens alongside nutrients.

8.2 The effects of organic matter

Sewage, in the narrow sense described above, contains high levels of organic matter, both particulate and dissolved. In a broader sense, the terms "sewage" and "municipal waste water" are used to describe the mix of waterborne disposal of human waste and discharges from artisanal and industrial undertakings when these are processed together. Organic matter also enters riverine discharges from natural sources, from direct or riverine inputs of industry and from aquaculture. Many artisanal or industrial discharges also contain high levels of organic matter, both particulate and dissolved. All this particulate and dissolved organic matter tends either to be oxidised or broken down by bacteria. Both processes require oxygen. The need for oxygen for chemical oxidisation is the Chemical Oxygen Demand (COD). The oxygen needed by the bacteria is the Biological Oxygen Demand (BOD). When the COD and BOD in a body of water exceed the oxygen available, the body of water can become hypoxic or anoxic, with a reduced ability to support aquatic life (Metcalfe & Eddy, 2004).

8.3 The effects of nutrients

Several nutrients are significant for the marine environment: mainly compounds of nitrogen, phosphorus, silicon and iron. In much of the ocean, primary production is limited by the availability of nitrogen. The inputs of nitrogen compounds are therefore of greatest significance. However, other aspects of nutrient input are also important: changes in the balance between available nitrogen and phosphorus can be the cause of blooms of various species of algae. Some species of algae produce toxins which can lead to amnesic shellfish poisoning, neurotoxic shellfish poisoning and paralytic shellfish poisoning (which can have death rates of 10 per cent-20 per cent) (GESAMP, 2001).

The boundaries and names shown and the designations used on this map do not imply official endorsement or acceptance by the United Nations.

Figure 20.6 | Atmospheric Transport of Nitrogen to the North-East Atlantic from North-West Europe. Source: EMEP model in OSPAR, 2010.

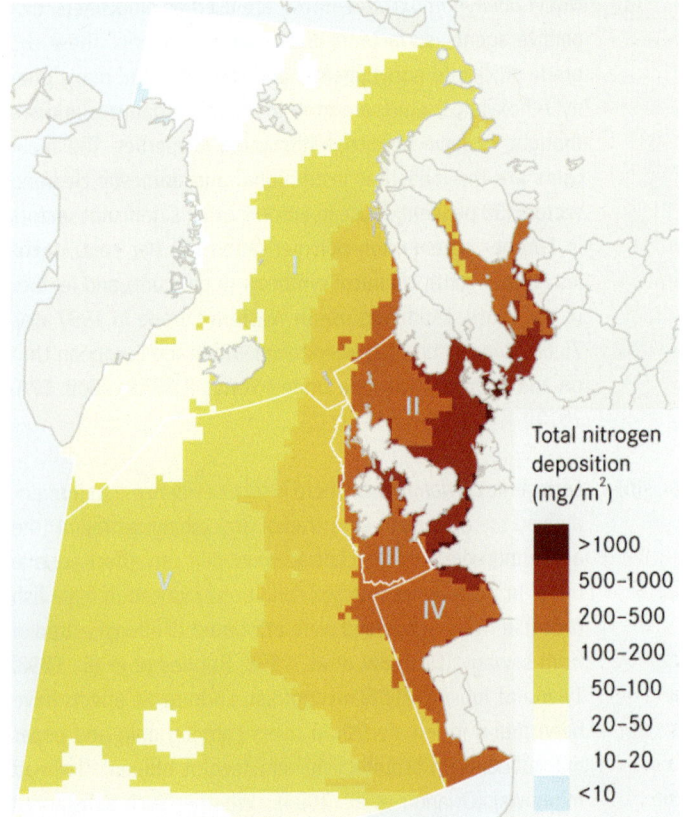

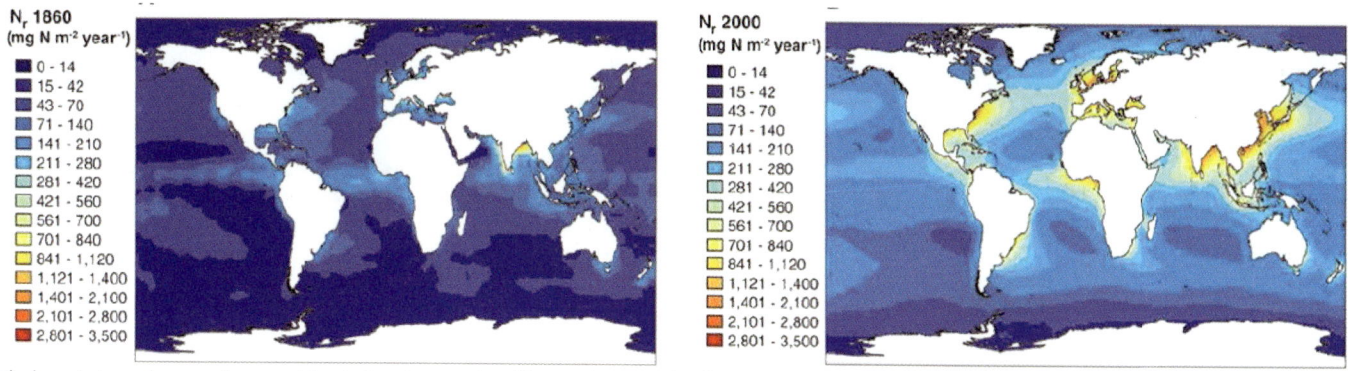

The boundaries and names shown and the designations used on this map do not imply official endorsement or acceptance by the United Nations.

Figure 20.7 | Airborne Reactive Nitrogen Inputs 1860 and 2000. Source: Duce et al., 2008. Total atmospheric reactive nitrogen (Nr) deposition to the ocean in mg per square metre per year in 1860 and 2000. Both organic and inorganic forms of nitrogen are included.

Anthropogenic inputs of nitrogen and phosphorus into estuarial and coastal ecosystems have more than doubled in the last century. These inputs are made through both the waterborne routes described above, but also significantly through airborne inputs, particularly in the forms of oxidized nitrogen, ammonia (especially from livestock), and water-soluble organic nitrogen. The importance of this airborne route for problems of the marine environment can be seen from the statistics for the North Sea and the Celtic Sea in the North-East Atlantic. In 2005 the total amounts of nitrogen estimated to be discharged in liquid discharges (riverine and direct) into these areas was 1,205 kilotons. These discharges came from 11 of the 15 States in the North-East Atlantic catchments. If we look at the airborne emissions of nitrogen from these 15 States, we find that these are estimated at 4,600 kilotons – 47 per cent from agriculture, 28 per cent from transport (including shipping), 21 per cent from combustion, and 3 per cent from other sources (OSPAR, 2010). These 4,600 kilotons of emissions are from a larger area (the total area of 15 States rather than the catchments in those States that discharge into the North-East Atlantic). Some will therefore be carried to sea areas other than the North Sea and the Celtic Sea, and some are already included in the 1,205 kilotons of riverine and direct inputs, since they will fall on land in the catchments of the North Sea and Celtic Sea. Nevertheless, it is clear that consideration of managing excessive inputs of nutrients must take into account the nutrients that reach the sea through the atmospheric transport of nutrients.

Atmospheric transport of nutrients is also important for the range over which the nutrients can be carried. As an example, the adjacent map shows the spread of nitrogen inputs to the North-East Atlantic from North-Western Europe in 2006: the inputs reach well into the open ocean (the area marked V on the map).

At the global level, the scale of the problem can be grasped from studies aimed at showing the implications of the secular trends in airborne inputs of nitrogen to the world ocean. Drawing on work by Galloway and others (Galloway et al., 1995) and Ducklow (Ducklow, 1996), Duce and others have demonstrated the increases in the inputs of total atmospheric reactive nitrogen (N_r) over the last 140 years (Figure 20.7). This brings out the significance of urbanization and industrialization and of intensified agriculture.

Inputs of nitrogen and phosphorus to the ocean provide nutrients to marine plants – especially to phytoplankton. Increased inputs stimulate growth (unless there is some countervailing factor, such as turbidity to reduce the availability of the light needed for photosynthesis). Excessive net phytoplankton production in coastal and shelf ecosystems can lead to an accumulation of phytoplankton biomass and to eutrophication problems. Among other phenomena, excessive net production of phytoplankton can result in *marées vertes* ("green tides") and red tides, when large areas of sea are infested with algae. Eventually, primary production will tail off as nutrients are exhausted and again growth is limited. The masses of algae (phytoplankton) will decay under the action of bacteria. This process will use up the oxygen dissolved in the seawater, and the resulting hypoxic (where oxygen is below 2 mg per litre) or anoxic (absence of oxygen) conditions will result in the death of the animals on the seabed and of fish that cannot leave the area. In the worst cases, these conditions will lead to "dead zones" (Diaz et al., 2008), loss of sea grass beds (Orth et al., 2006)), and increases in the occurrence of toxic phytoplankton blooms (Heisler et al., 2008). These dead zones reduce the amount of habitat available to aerobic animals upon which fisheries depend. The number of low-oxygen zones in coastal waters has increased exponentially to over 400 systems since the 1960s and has reached an area of about 245,000 km² worldwide (Figure 20.8; Rabalais et al., 2001; Diaz et al., 2008).

The occurrence of stratification, where different layers in the sea do not mix, can be significant for problems from nutrients, since concentrations of nutrients may be much higher in one layer as a result of the water inputs having a different density. Since stratification is often seasonal, problems from nutrients can often also be seasonal. Significant seasonal meteorological changes (such as the monsoons, rainy seasons or changes in insolation (the amount of sunlight)) can also create seasonal problems from nutrients through changes in runoff and primary production.

At the same time, even high levels of nutrients in discharges to the seas do not necessarily create problems: in estuaries and coastal lagoons, depending on local circumstances, bacterial action can result in a net conversion of nitrates from land runoff into nitrogen gas released to the air, thus reducing the load to the sea. Also, the turbidity of coastal water, resulting from tidal disturbance of sediments and/or coastal erosion and

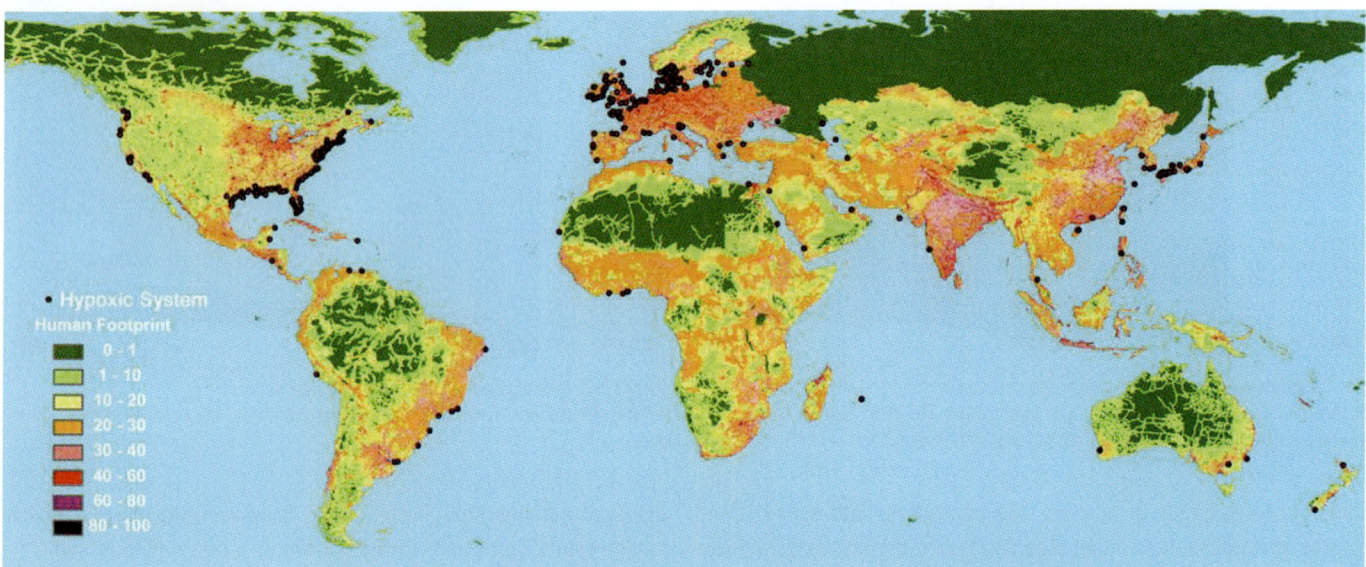

The boundaries and names shown and the designations used on this map do not imply official endorsement or acceptance by the United Nations.
Figure 20.8 | Global Map of Dead and Hypoxic Zones. Source: http://www.scientificamerican.com/media/inline/2008-08-15_bigMap.jpg.

other causes, can limit the depth to which sunlight can penetrate and thus inhibit photosynthesis and the growth of phytoplankton. The precise consequences of heavy loads of nutrients in discharges to the sea will therefore often depend on local circumstances, including the rate at which semi-enclosed areas are flushed by tides and currents.

In certain circumstances, anoxic zones occur naturally (Helly and Levin, 2004). In the Black Sea, the large inflows of fresh water from rivers such as the Danube, Don and Dnieper result in a high degree of stratification, with little mixing between the layers. The result is that a large part of the central deep water of the Black Sea (estimated at about 90 per cent of its volume) is naturally anoxic (Heileman et al., 2008c). Likewise, where narrow continental shelves and currents flowing from the open ocean against the continental slopes are found, nutrient-rich, oxygen-poor water can be brought up into coastal waters, and produce hypoxic or even anoxic conditions. Examples of this are found on the western coasts of America immediately north and south of the equator, the western coast of sub-Saharan Africa, and the western coast of the Indian subcontinent (Chan et al., 2008).

8.4 Sources of nutrients

8.4.1 Municipal waste-water

Urban settlements have, of course, always produced waste-water, but a steep change in the quantities and their effect on the environment occurred from the middle of the 19th century, with the introduction of waterborne methods of disposing of human excrement and their connection to collective drains. Until then, the main system of disposal had been cess-pits and the collection of "night soil" in carts and its disposal on land for use as agricultural fertilizer. The first major changeover came in England in 1848, when legislation made the use of sewers for disposal of human excrement compulsory as a measure against cholera. Within ten years, some of the problems that the new approach could cause were shown by the Great Thames Stink of 1858, which, among other things, made the newly built Houses of Parliament almost unusable: the decomposition of the waterborne excrement produced intolerable levels of stinking gas. Sewage treatment for waste-water discharges to inland waters was adopted as the solution, and sewage treatment processes gradually improved in effectiveness and spread to more and more parts of the world.

The idea that treatment of waste-water discharges direct to the sea was also essential took much longer to be accepted: as late as 1990, some large towns in Western Europe still discharged major parts of their municipal waste-water untreated direct to the sea. The belief that the ocean had an almost unlimited absorptive capacity of the ocean was difficult to eradicate. However, the problems resulting from municipal waste-water discharges to relatively enclosed bays and harbours were acknowledged early - in the United Kingdom, the problems of the semi-enclosed Belfast Lough were one of the reasons for the investigations of the 1896 Royal Commission on Sewage Disposal, which established ground-breaking standards for sewage treatment, in terms of suspended solids and BOD per unit volume of discharge. However, it took the recognition of the significance of the problems from nutrients in relatively open sea areas such as Chesapeake Bay or the German Bight in the 1980s to create more general acceptance that action was needed, and to draw attention more generally to the issue of nutrient inputs to the sea.

The main routes of nutrient input are through rivers and direct discharges through pipelines of waste-water from sewage-collection systems and factories. When the Global Programme of Action for the Protection of the Marine Environment (GPA) was adopted in 1995, there was general agreement that the most important need for protecting the marine environment and improving human well-being was improved management of sewage, especially that from large conurbations. Where sewage treatment is applied to sewage discharges, three levels of treatment are typically recognized: primary (removal of solids and floating oil and grease), secondary (breaking-down of biological substances by microbes or protozoa) and tertiary (disinfection and removal of nutrients). It is not always essential for discharges from sewage-collection systems to be treated before discharge. In some circumstances, very long discharge pipelines can take untreated sewage sufficiently far out to sea, and into sufficiently dynamically active waters, that the nutrients and microbes in the sewage are adequately dispersed and assimilated and problems of eutrophication avoided. For this to be the case, the pipelines must take the sewage well beyond immediate coastal waters and strong currents must be present to provide the dispersal. Even then, in most cases, at least primary treatment of the sewage is preferable. Progress has been made in many parts of the world but, overall, untreated sewage inputs remain a major threat to the marine environment.

Increasingly, in addition, inputs of water across the coastline through underground aquifers are being recognized as a significant pollution route, although statistical estimates of the amounts of water, nutrients and contaminants through this route are rarely available.

8.4.2 Food and related industries

The preparation of human food inevitably results in the generation of organic waste: milling grain produces chaff; brewing and distilling produce the spent malt or other grain used; wine-making leaves the pressed grapes; fish preparation leaves guts, heads and tails, and so on. Some of these wastes are liquid or semi-liquid and can be discharged to the sea. Others can conveniently be disposed of into the sea (especially the waste from fish-processing), directly or through a watercourse. As explained in chapter 12, aquaculture is also a source of nutrients to the marine environment. All these elements will create BOD or COD, and will release nutrients as they are decomposed or oxidized.

8.4.3 Land runoff

The world has been able to produce more and more food from land, through a combination of improvements in strains of crops, agricultural techniques and pesticides, increased use of fertilizers, as well as bringing new areas into cultivation. The scale of this increase in agricultural production can be seen from FAO statistics on cereal production: an increase of over 25 per cent in the tonnage of cereals produced worldwide between 2002 and 2012. This increase in overall tonnage is also reflected in increased yield per hectare: over the period 2002 to 2012, yields increased by over 7 per cent in southern Asia, by over 9 per cent in eastern and south-eastern Asia, by over 18 per cent in Africa and by over 20 per cent in western Asia.

The substantial increases in total crops and in yield, while essential to feed the world's growing population, carry with them some environmental problems for the marine environment. As discussed above, some of the pesticides used on land have had impacts on the marine environment as a result of runoff. Likewise, the increased use of fertilizers has resulted in increased runoff of nutrients to the seas. These nutrients, intended to promote photosynthesis in land plants, also encourage primary production in the seas, and result in eutrophication problems. The runoff not only takes the obvious form of surface water entering the sea through rivers and watercourses, it can also enter the sea through groundwater seepage through aquifers. Estimates suggest that direct subterranean/submarine discharges of fresh water to the oceans around the world deliver up to 12 per cent of total surface water runoff, with the most accepted values between 5 per cent and 10 per cent (Basterretxea et al., 2010).

The use of nitrogen-based fertilizers has grown enormously in recent decades. This growth continues, as Table 20.4 shows: world consumption has increased by 42 per cent between 2002 and 2012, including more than doubling in Latin America, southern Asia, eastern Asia and Oceania.

Increased use of agricultural fertilizers does not necessarily result in increased nutrient inputs to the seas: good agricultural practices can help avoid this. Adjusting amounts of fertilizer applied to likely take-up by crops, applying fertilizers at the time of year when take-up by crops will be greatest, ploughing so that the furrows do not promote runoff, and leaving buffer zones along watercourses can all help reduce the leaching of nutrients to the watercourses and the seas.

Table 20.4 | World Nitrogen Fertilizer Consumption. Source: FAOSTAT.

											Million tons
	2002	2003	2004	2005	2006	2007	2008	2009	2010	2011	2012
Europe and Central Asia	5,330	5,090	5,743	5,798	5,705	6,699	6,902	6,711	6,559	6,997	7,174
North America	2,736	2,620	2,715	3,150	2,763	3,151	2,884	2,466	2,685	2,868	2,959
Latin America	691	722	865	880	1,043	1,253	1,106	1,091	1,277	1,455	1,459
Africa	800	920	1,112	1,160	1,022	990	1,013	1,017	1,054	1,067	1,142
West Asia	409	462	547	533	441	440	456	595	550	455	417
South Asia	99	96	137	164	141	138	167	197	210	232	238
East Asia	908	1,199	1,275	1,315	1,391	1,490	1,671	1,428	1,692	1,737	1,962
Oceania	161	372	378	462	389	422	468	531	544	556	679
World	11,295	11,779	12,775	13,473	12,901	14,583	14,665	14,045	14,578	15,374	16,030

The type of crop cultivated can also be very significant. Legumes, such as soya beans, naturally fix nitrogen from the air into soil, from where it can then run off. The vast increases in soya bean cultivation in some tropical countries (such as Brazil) can increase nitrogen fluxes in the same way as the use of nitrogen fertilizers (Filoso et al., 2006).

Intensive raising of livestock is another major source of nutrients: both solid and liquid wastes are involved, as well as gaseous emissions of ammonia and methane, all of which can find their way to the seas through runoff from the land or deposition from the air.

8.4.4 Other sources of nutrients

The processing of many food products by food and drink industries for consumption also frequently results in waste-water containing nutrients in various forms. These waste-water streams are a further factor affecting nutrient inputs to the seas.

The combustion of petrol/gasoline and other liquid fuels also produces nitrogen compounds, which can be carried through the air to the seas. Vehicles powered by internal-combustion engines are obvious sources of such compounds (especially of ammonia). Near major shipping routes, the contribution from ships can also be significant. In north-western Europe, for example, over 25 per cent of nitrogen emissions to the atmosphere are from these sources (OSPAR, 2010).

8.5 Waterborne pathogens

Untreated municipal waste-water inevitably contains infectious microbes from humans. If these microbes reach the seas, they can infect humans both when they immerse themselves in sea water (sea-bathing) and through the consumption of fish and shellfish (especially the latter, since shellfish filter large quantities of seawater in the course of obtaining their food). Similar contamination also arises from animal excrement. In bathing waters, the probability of respiratory and intestinal diseases and infections rises for bathers rises in an almost direct relationship with the sewage pollution in the water. GESAMP and the World Health Organization estimated in 1999 that bathing in polluted seas causes some 250 million cases a year of gastroenteritis and upper respiratory disease. The same study estimated that eating contaminated shellfish is responsible for the loss every year of 3,500,000-7,000,000 disability-adjusted life-years (a standard measure of time lost due to premature death and time spent disabled by disease), putting it in the same bracket as stomach cancer, intestinal nematodes and upper respiratory tract infections (GESAMP, 2001).

9 Regional view of impacts of Nutrients and Waterborne Pathogens

The foregoing review of the sources of oxygen demand (both COD and BOD), nutrients and waterborne pathogens and the ways in which they can affect the ocean sets out the general mechanisms. It is necessary then to see to what extent the various parts of the ocean have in fact been affected. Because this kind of problem is confined to coastal waters (since distribution and dilution remove the detrimental effects), it is not necessary to examine the open ocean area.

9.1 Arctic Ocean

No problems are reported from elevated levels of nutrients in the Arctic Ocean because there are no major concentrations of population or agriculture.

9.2 Atlantic Ocean and Adjacent Seas

9.2.1 North-East Atlantic, North Sea and Celtic Seas

Serious problems from eutrophication became apparent in the North Sea in the 1980s, as dead zones appeared, particularly in the German Bight. As a result, the coastal States committed themselves to a 50 per cent reduction in inputs of nitrogen and phosphorus compounds by 1995. The 1998 OSPAR Eutrophication Strategy extended the goals of combating eutrophication to the whole of the North-East Atlantic. In 1991, the European Union adopted legislation requiring improved treatment of urban waste-water and reduction in inputs of nitrates from agriculture. Assessing the impacts of anthropogenic nutrient inputs (especially inputs from diffuse sources) is complicated by the delivery of nutrient-rich water from the deep Atlantic. Most North Sea countries achieved the target reduction in phosphorus inputs by 1995, and some countries have now reduced phosphorus inputs to 15 per cent or less of their 1985 level. Although the target for 50 per cent reductions in nitrogen inputs between 1985 and 1995 was not achieved (and still has not been achieved, except in Denmark), the resulting major programmes have achieved substantial reductions in inputs. Germany and the Netherlands have almost achieved the 50 per cent reduction. Even after these reductions, eutrophication problem areas, with enhanced levels of nutrients, are still found along the coasts of Belgium, Demark, Germany and the Netherlands, while a number of estuaries and fjords in Ireland, Norway, Portugal, Spain and the United Kingdom also show such levels and are therefore regarded as eutrophication problem areas. In France, the estuaries of the Loire and Seine and much of the coast of Brittany (where beaches covered in sea-lettuce create serious health impacts on both locals and tourists) are still eutrophication problem areas. Mass mortality of benthic and pelagic animals has, however, been limited to a few estuaries and fjords in Denmark, the Netherlands, Norway and Sweden (OSPAR, 2010).

Since 1976, the European Union has had programmes to reduce the inputs of waterborne pathogens to the waters of its member States. This has required major investment in sewage treatment and the management of storm-water runoff. The results have been a steady improvement in water quality, both for bathing and for shell-fish production. By 2012 (which was a very wet summer in Europe) with consequential high levels of storm runoff), only 1.7 per cent of the monitored coastal bathing sites failed to meet the European Union's mandatory standards. Most of these were in the North Sea (EEA, 2013).

9.2.2 Baltic Sea

The Baltic Sea is sensitive to eutrophication because of the strong halocline, the limited water exchange with the North Sea and the consequent long residence time of water in it. High nutrient loads and a long residence time mean that nutrients discharged to the sea will remain in the basin for a long time. In addition, the stratification of the water masses increases the vulnerability of the Baltic Sea to eutrophication, because it hinders or prevents ventilation and oxygenation of the bottom waters and sediments. Furthermore, absence, or low levels, of oxygen worsen the situation by affecting nutrient transformation processes by bacteria, such as nitrification and denitrification and the capacity of sediments to bind phosphorus, and lead to release of significant quantities of it.

As a result, most of the Baltic is regarded as suffering from problems of eutrophication. Only the Gulf of Bothnia (the northernmost part of the waters between Finland and Sweden) is generally free of these problems, although even here, there are small coastal sites with pronounced eutrophication problems. The worst affected areas are the Gulf of Finland, the Gulf of Riga, the Baltic Proper (the area between Sweden and Estonia and Latvia), the area east of the island of Bornholm, the Belts and the Kattegat. Smaller sites on the coasts of Sweden and the Gdansk Bight are also classified as suffering eutrophication (HELCOM, 2010a). In general, the anoxic and hypoxic areas of the Baltic Sea are regarded as one of the largest areas of dead zones in the world.

9.2.3 Mediterranean Sea

Eutrophication is assessed as being a localized problem in the Mediterranean basin. The main causes, as elsewhere, are inadequately treated sewage and runoff and emissions from animal husbandry and high usage levels of agricultural fertilizers. 37 per cent of coastal towns with a population of more than 2,000 have no sewage treatment at all, and a further 12 per cent have only primary treatment. These towns are concentrated on the southern shore of the western Mediterranean, in coastal Sicily, on the eastern coast of the Adriatic and in the Aegean and the north-eastern corner of the Levantine basin. Fertilizer usage reaches over 200 kg per hectare of arable land in Croatia, Egypt, Israel and Slovenia, and is over 100 kg per hectare in France, Greece, Italy and Spain. Since the eastern Mediterranean is naturally oligotrophic, locally enhanced levels of nutrient input can produce marked results. The main hypoxic area is along the delta of the River Nile (Egypt) and there are areas at high risk of hypoxia at the mouth of the River Po (Italy) and the River Rhône (France). Medium risks of hypoxia have been reported at the mouth of the River Ebro (Spain) and in the Gulf of Gabès (Tunisia), the Gulf of Sidra (Libya), some bays and estuaries around the Aegean Sea and the Gulf of Iskenderun (Turkey), although Turkish authorities indicated that risks of hypoxia have not been confirmed in the latter area (UNEP, 2012).

9.2.4 Black Sea

As noted above, the Black Sea has historically had an anoxic zone in deep waters below 200 m. However, a major hypoxic (and, at times, anoxic) zone developed in the shallower north-western shelf from the 1950s. The inputs of nutrients by 1990 were estimated at approximately 80 per cent from agriculture and 15 per cent from municipal wastewater (most large towns having at least secondary sewage treatment). Between 1960 and 1990 the nutrient input into the catchments of the Rivers Danube, Dnipro and Don increased approximately 10-fold. The resulting anoxic or hypoxic zones at their peak in 1983-1990 extended to between 18,000 and 40,000 square kilometres, with consequential effects on fisheries and benthic biodiversity.

Three causes reduced the massive agricultural inputs: the economic problems of Eastern Europe from 1990 onwards, the adoption of stringent standards for nitrate emissions by the European Union (which required changes in the practices of States in the upper Danube basin) and the preparation for the entry of some States in the lower Danube basin into the European Union (which required the adoption of those standards). The very substantial reductions in the nutrient inputs meant that the worst effects of the hypoxia had disappeared by 1995, although the effects of changes in benthic biodiversity are still being felt (Borysova et al., 2005; Diaz et al., 2008).

9.2.5 North-West Atlantic

Nitrogen releases to air and water are low in most of Canada, but southern areas where rapid development is taking place show signs of emerging problems. At present, there is little sign of estuarine eutrophication on the Atlantic coast of Canada, but hypoxic conditions have been found in the lower St Lawrence estuary areas since the mid-1980s. These are at depths below 275 m. About a third of the problem is attributed to land-based inputs. The other two-thirds seem to be the result of changed oceanic circulation, resulting in larger amounts of Atlantic water from south of the Gulf Stream entering the estuary. This water has lower oxygen levels and a higher temperature (resulting in more bacteriological activity and consequent consumption of oxygen) than the previously dominant Labrador Current water (Schindler et al., 2006; DFO-MPO, 2013).

The United States National Coastal Condition Report (NCCR) uses a measure of water quality relevant to the occurrence of eutrophication based on a combination of levels of dissolved inorganic nitrogen (DIN), dissolved inorganic phosphorus (DIP), chlorophyll a, dissolved oxygen and the degree of water clarity. Cut-off points (varying between regions) are used to classify these indicators into good, fair and poor categories, and an algorithm gives an overall value in the light of these classifications (EPA, 2012). The United States has also carried out a National Estuarine Eutrophication Assessment (NEEA), looking at 141 estuaries in the contiguous 48 states. An update was published in 2007. Although full conclusions could be reached on only 64 of the estuaries, 29 showed moderately high to high eutrophic conditions (NEEA, 2007).

For the North-East Region (Maine to Virginia, including Chesapeake Bay), there is a marked gradient from north to south: the overall evaluation is that the region has fair water quality, but this ranges from very good quality in the open estuaries of the north, to poor in many of the southern estuaries, which have poor levels of water exchange and drain densely populated catchments. Particular problem areas are Great Bay (New Hampshire), Narragansett Bay (Rhode Island), Long Island Sound (between Connecticut and Long Island, New York), New York/New Jersey (NY/NJ) harbour, the Delaware Estuary, and the western tributaries of Chesapeake Bay. High levels of enteric bacteria resulted in advice for short periods in the mid-2000s against bathing at about 17 per cent of the beaches monitored. Further out to sea, the water quality in the Mid-Atlantic Bight was generally rated as "good" (EPA, 2012). The NEEA showed a similar division between the estuaries in the northern and southern parts of this region, with the former being generally good and the latter generally showing the worst conditions nationally. It also noted a worsening between 1999 and 2004 in the status of 8 of the 22 estuaries assessed in the southern part (NEEA, 2007).

Chesapeake Bay presents a complex of problems, in part because of its large catchment basin with extensive industrial agriculture and a large and rapidly growing population, and in part because of the long residence time of water in the estuarine system (measured in months) (Kemp et al., 2005). Efforts to address the problems began in 1983 with the Environmental Protection Agency's Chesapeake Bay Program. By the mid-1990s, this seemed to be bearing fruit, but by 2005 it was clear that it was not reaching its goals (GAO, 2005). New efforts began in 2009, focused on a range of measures to address the multiple causes of the problems, including measures to achieve by 2025 a specified total maximum daily load from all sources of nitrogen and phosphorus (Chesapeake Bay Program, 2014).

For the South-East Region (North Carolina to Florida), the overall rating was that the water quality was "fair", with only 22 per cent of the sampling points rated "good" and 13 per cent "poor". The main problem areas were the large estuaries of Albemarle and Pamlico Sounds (North Carolina) and the major ports of Charleston (South Carolina) and Savannah (Georgia). Away from the immediate coast, the South Atlantic Bight was regarded as having overwhelmingly "good" water quality (EPA, 2012). This picture is generally consistent with the NEEA, but that assessment was unable to classify the Albemarle and Pamlico Sounds, while judging the Charleston and Savannah port areas as presenting lesser problems (NEEA, 2007).

9.2.6 Wider Caribbean

The water quality of the waters of the Gulf Coast of the United States along the immediate shoreline is judged by the National Coastal Condition Report to be "fair", with 30 per cent of the sampling stations "good" and 10 per cent "poor" (and 7 per cent not evaluated because of lack of data) (EPA, 2012). This picture was consistent with that presented by the NEEA (NEEA, 2007).

However, a little further out into the shelf waters, there is a major eutrophication problem area. Since 2000 this dead zone has fluctuated annually in size from about 8,500 sq km to about 21,750 sq km (NOAA, 2013). This is regarded as the second-largest dead zone in the world. The reasons for the fluctuation are not fully understood, but are largely connected with the levels of flow in the Mississippi River. This drains about 3.1 million sq km (about 40 per cent of the contiguous United States), with a very high level of arable and livestock agriculture and a correspondingly high level of nitrogen and phosphorus runoff. The first problems were noted by shrimp fishermen in the 1950s. Studies of the sediments show that algal growth (and hence eutrophication problems) in the area of the dead zone increased significantly in the second half of the 20th century. The dead zone has had a significant effect on the shrimp fishery (Turner et al., 2008; Rabalais et al., 2001; Diaz et al., 2008).

In other parts of the wider Caribbean, significant progress has been made in addressing the problems of sewage and other nutrient discharges. In 2010, the Caribbean Environment Programme conducted a comprehensive survey of the problems. In spite of some uncertainties, this showed major progress in Colombia and Venezuela in reducing inputs from municipal waste-water since an earlier survey in 1994: total nitrogen inputs had reduced by more than 80 per cent, and also substantial reductions in organic matter (BOD). Elsewhere, smaller reductions had been achieved in mainland States, but large increases were found in the island States. Much more general progress had been made in reducing organic matter (BOD) and nutrients from industrial sites in coastal areas: reductions of 50 per cent-90 per cent in the former in all parts, and 90 per cent or more in the latter everywhere except in the United States and Mexico (UNEP-UCR/CEP, 2010).

Nevertheless, there are major issues with sewage, both in terms of health and eutrophication. The Caribbean relies heavily on the tourist industry for its economic well-being. Clean bathing waters and coral reefs are two important supports for that tourist industry. Eutrophication leads to excessive algal growth which can smother and kill corals – especially if the herbivore fish (such as groupers) have been reduced by over-fishing. Untreated sewage harms the health of both local populations and visiting tourists. Both effects have serious implications for the tourist industry.

9.2.7 South-East Atlantic

Detailed studies and analysis conducted in the Guinea Current region, and more generally in West and Central Africa, show that sewage constitutes the main source of pollution in that area from land-based activities. A similar situation applied in the regions of the Benguela Current (where harmful algal blooms also appear to be on the increase) and the Canaries Current (where the waters off the cities of Dakar in Senegal and Casablanca and Rabat in Morocco appear to be specially affected). All the countries assessed reflect high urban, domestic loads, sometimes from industrial origin, which create problems from BOD, suspended sediments, nutrients, bacteria and pathogens. For example, the mean annual amount of oxygen required to meet BOD for the entire West and Central Africa region, including the countries adjoining the Guinea Current, has been estimated to be 288,961 tons for BOD from municipal sewage and 47,269 tons for BOD from industrial discharges: a total of 336,230 tons (For comparison, the mean annual amount of oxygen required to meet BOD for the River Rhine at the border between Germany and the Netherlands is about 60,000 tons.). Again, the rapid growth of urban populations is far beyond the capacity of relevant authorities and municipalities to provide adequate basic services of sewage and waste-water-treatment facilities (GCLMEP, 2003; Heileman, 2008b).

9.2.8 South-West Atlantic

The waters off the northern coasts of Brazil have naturally relatively low levels of nutrients. During most of the year, therefore, there are no problems from eutrophication. However, during the rainy season, runoff from land brings sudden increases in the levels of nutrients, and consequently algal blooms then occur (de Lacerda et al., 2002). Estuaries, bays and lagoons close to the larger conurbations tend to show eutrophication from sewage and other nutrient inputs, often enhanced by the effects of limited water circulation (Costa, 2007).

Further south, in the heavily populated areas of south-eastern Brazil, high levels of nutrients and consequent eutrophication problems are common. Guanabara Bay (on which the city of Rio de Janeiro (population 6.3 million) is located) is the most heavily affected area, with very high nutrient levels and high levels of microbial pollution (de Lacerda et al., 2002). In the south of Brazil, in the State of Santa Catarina, in urban estuaries, the dissolved inorganic nitrogen (DIN) was generally three times greater than in non-urban ones (Pagliosa et al., 2006).

In Brazil, the majority of households and industries generally do not have access to sewerage. The national average of those with connections to a sewer in 2008 was 44 per cent, ranging from 1.7 per cent in the State of Pará in the north, to 82 per cent in the State of São Paulo in the south. Supply of piped water was much more common than sewerage connections, and sewerage connections were more common than sewage treatment: only 28 per cent of the volume of water supplied passed into the sewer system and only 68 per cent of the sewage was treated, only a little over half of that treated receiving secondary or higher treatment. This situation in 2008, however, represented a big improvement (for example, an increase of 40 per cent of households connected) on that at the time of the previous survey in 2000. Brazil currently has a major programme of investment (equivalent to 4.2 billion United States dollars) for the improvement of sanitation. So it is reasonable to hope that the situation will improve (IBGA, 2008; PAC2, 2014). Further south again, Uruguay and Argentina, which contain the large conurbations of Montevideo and Buenos Aires, have serious microbial pollution in some localized areas of their coastal waters, where pathogens have been detected which in some cases have exceeded international recommended levels for recreational water. Toxic red tides are becoming more frequent and of longer duration (Heileman, 2008e).

9.3 Indian Ocean

9.3.1 Western Indian Ocean

Throughout this area, there is a tendency for high nutrient levels to encourage ecosystem change, leading to dominance by algal communities. On the coasts of the Agulhas Current, the growing coastal populations and increasing tourism, for which sewage treatment facilities are inadequate, result in the increasing discharge of raw sewage directly into rivers or the sea, leading to eutrophication in localized areas. Untreated effluents from fish processing plants and abattoirs are also frequently discharged into the sea, causing varying degrees of localized pollution.

On the coasts of the States bordering the Somali current, most of the coastal municipalities do not have the capacity to handle the vast quantities of sewage and solid wastes generated daily. Raw sewage containing organic materials, nutrients, suspended solids, parasitic worms and benign and pathogenic bacteria and viruses is discharged into coastal areas. High microbial levels are observed in areas near sewage outfalls (Heileman and Scott, 2008).

In the Comoros, there is no sewerage, drainage or waste-water treatment. In Kenya, microbial water quality studies have been completed in a number of locations and microbial pollution levels near urban centres, such as Mombasa, were several orders of magnitude higher than in coastal waters in rural areas. In Mozambique, faecal coliform counts in the channel adjacent to the Infulene River in Maputo were found to be worryingly high. In Madagascar, similar high counts of bacteria from human excrement have been measured in coastal waters. Microbial pollution is an ongoing problem in several Mascarene coastal areas. Periodic draining of waste-water ponds on fish farms adds to nutrient discharges. At present, in Mauritius, 73 per cent of households use cesspits or septic tanks whilst 2 per cent use pit latrines; so most of the effluents are discharged directly to the sea or are carried to the sea by runoff and rivers with higher potential for microbial pollution, particularly after heavy rains. Agricultural practices in Mauritius (both intensive agriculture and small-scale market gardening, and livestock rearing) also pose a serious threat to coastal ecosystems and give rise to algal blooms and red tides (ASCLME/SWIOFP, 2012).

9.3.2 Red Sea, including the Gulf of Aden, the Gulf of Aqaba and the Gulf of Suez

Although its effects are usually limited to a small area around urban areas and large tourist developments, sewage is a major source of coastal contamination throughout the Red Sea and the Gulfs of Aden, Aqaba and Suez. Because of rapid population growth and inadequate treatment and disposal facilities, poorly treated or untreated sewage is dumped in coastal areas. Inputs of sewage also results in eutrophication of the coastal waters around some population centres, major ports and tourist facilities (Gerges, 2002).

9.3.3 Persian Gulf

The shortage of freshwater resources and the availability of financial resources resulted in an extensive investment in sewage treatment in the Gulf States on the southern shore of the Persian Gulf, in order to permit re-use of the treated water for irrigation and other purposes. The treatment applied is generally secondary or tertiary. This re-use also reduces the demand for water from desalinization. Hence there has not been the same pressure from discharge of nutrients as in other parts of the world from urban growth and consequent increases in urban waste-water. As long ago as 1999, 252 million cubic metres of water were being produced annually in this way (Alsharhan et al., 2001). The latest FAO figures show that this has risen to 551 million cubic metres/year. Elsewhere in the area, coastal water quality at the Iraq-Kuwait border has declined as a result of increased agricultural pollution due to the draining and subsequent loss of the filtering role of the Mesopotamian marshlands (Heileman et al., 2008b). On the northern shore, moreover, some cities, such as Bushehr, are discharging treated sewage effluent, which is giving rise to enhanced levels of nutrients, although it is not clear that this results in eutrophication problems (Rabbaniha et al., 2013).

9.3.4 Arabian Sea, including waters west of India, the Maldives and Sri Lanka

This area is affected by natural nutrient enrichment, at the time of the south-western monsoon, as deep-level nutrient-rich water is brought up onto the narrow continental shelf (Naqvi et al., 2009).

In the north of the area, sewage, fertilizers and other effluents have resulted in eutrophication in coastal areas such as Karachi. Fish kills in some localities, such as off the Karachi coast and Gawadar Bay, have been attributed to harmful algal blooms caused by the growing pollution (Heileman et al.,2008b).

Further south, the Indian Central Pollution Control Board (CPCB) estimates that the 644 cities and towns of over 50,000 population across the country (coastal and inland) discharge 5,500 megalitres a day of sewage, of which only 522 megalitres a day – less than a tenth – receives any treatment before discharge. Of this, the 120 cities and towns of populations of over 50,000 in the coastal area generate about 6,835 megalitres a day of waste-water, out of which only 1,492 megalitres (22 per cent) receive any treatment. The rest is discharged without any kind of treatment to the coastal waters. This represents an increase of about 150 per cent over the levels of discharge twenty years ago, although the rate of increase has recently slowed (CPCB, 2014). There have also been large increases in the amounts of artificial fertilizers used. However, it is argued that much of this usage is in relative dry areas from which there is little runoff (NIOT, 2014). Considering the west coast of India separately, the state of Maharashtra, in the middle, accounts for the majority of the 3,220 megalitres discharged daily into the Arabian Sea (CPCB, 2014). In spite of this heavy nutrient load, which produces some hypoxic zones, few eutrophication problems (such as harmful algal blooms) are reported, probably because of the very dynamic tidal action which produces rapid dispersal. The algal mass, measured as chlorophyll-a, is lower in this area than in the Bay of Bengal (BOBLME, India 2011).

Given the statistics on sewage, it is not surprising that high levels of pathogenic bacteria are reported all along the coast (except in the Karwar (Karnataka) region), with increasing levels on the coasts of Goa, the rest of Karnataka and Kerala. These increasing trends in levels of nutrients and waterborne pathogens point to the significant influence of sewage inputs (NIOT, 2014).

9.3.5 Waters east of India, the Maldives and Sri Lanka (Bay of Bengal, Andaman Sea, Malacca Strait)

In the waters to the east of the Indian subcontinent, hypoxic areas regularly occur along the coast, although severe eutrophication problems appear to be rare. These hypoxic areas are partly a natural situation brought about by enhancement of nutrient levels by the monsoon winds bringing nutrient-rich water to the surface (Vinayachandran, 2003), and partly by high levels of nutrient input from the land. The major inputs are from West Bengal in India (which provides well over 50 per cent of the inputs from the Indian coast) and from Bangladesh. The Indian input of sewage is around 2,330 megalitres/day into the Bay of Bengal, 80 per cent of which has had no treatment (CPCB, 2014). The hypoxic areas are also associated with frequent harmful algal blooms, for which seven hotspots have been identified (Gopalpur (Orissa), Visakhapatnam and Coringa (Andra Pradesh) and Ennore, Kalpakkam, Porto Nova, and the Gulf of Mannar (Tamil Nadu)) (BOBLEME, India 2011; Satpathy et al., 2013; NIOT, 2014). High levels of pathogenic bacteria are found all along the Indian coast of the Bay of Bengal (NIOT, 2014).

In Bangladesh, sewage collection and treatment exists only for one-third of Dacca (the capital), although investment is taking place to extend this and develop a sewerage system for the port city of Chittagong. Human wastes from most of the 150 million population are therefore liable, eventually, to find their way into the Bay of Bengal. Increasing amounts of artificial fertilizers are being used – imports rose by 2.3 times between 2003 and 2006 – but no data are available for inputs to the sea. Harmful algal blooms are frequent, and have been linked to mass mortalities in shrimp farms. Information is lacking on illnesses linked to food from the sea, but is thought to be increasing in parallel to increasing marine pollution (BOBLME, Bangladesh, 2011).

In Myanmar, there seems to be no evidence of hypoxic zones linked to major population centres. Generally, seawater samples showed acceptable levels of nutrients and dissolved oxygen, although samples from the mouth of the Yangon (Rangoon) river showed increased levels of suspended solids and COD (BOBLME, Myanmar, 2011).

On the Andaman Sea coast of Thailand, little provision is made for sewage treatment of the human wastes from the massive tourist industry. In particular, at Patong (the main town of the tourist island of Phuket), sewage discharges are leading to elevated nutrient levels and algal blooms in December-February of most years. However, the Thai authorities have established a comprehensive marine water-quality monitoring system, which shows that around 90 per cent of the sampling stations on this coast are "fair" or better. The only station with badly deteriorated water quality is at the mouth of the Ranong River, on the border with Myanmar. A major algal bloom and fish kill took place on this coast in 2007, but it seems likely that this was due to unusual upwelling of nutrient-rich water from the deep ocean (BOBLME, Thailand, 2011).

Malaysia also has a long-standing marine water-quality monitoring system. On the basis of this Malaysia more recently developed a marine water-quality index. This brings together parameters for suspended solids, oxygen demand and microbes, together with those for heavy metals. For the coasts facing the Andaman Sea and the Straits of Malacca, this index shows that in 2012, of the 62 coastal monitoring stations in this area, 3 per cent were rated "excellent", 10 per cent "good", 79 per cent "moderate" and 8 per cent "poor". Three of the five "poor" monitoring stations were near the port of Malacca and the other two were beaches apparently badly affected by oil pollution. Similar results were reported for estuarine and island monitoring stations (BOBLME, Malaysia, 2011).

9.3.6 Waters west of Australia

In general, the waters around Australia have naturally low levels of nutrients, since they are not affected by any marine current that can bring water with a high nutrient content to the coastal waters, and since much of the coast has limited land runoff because of the low rainfall. Blooms of toxic and nuisance algae, however, continue to be a problem in a number of the estuaries and in inshore waters along the western coast, with adverse impacts that include major events of fish mortality. When they occur, algal blooms in this region can cover large areas. In Western Australia, major nutrient and algal bloom problems have a long history in the Peel–Harvey Estuary, caused principally by nutrient pollution from upstream agricultural lands. Major works were undertaken to improve flushing of the estuary, but they seem to have brought only temporary relief (SE2011 Committee, 2011).

9.4 Pacific Ocean

9.4.1 Waters west of Canada and the mainland of the United States

The low population density and the small areas that are devoted to arable and livestock farming of Alaska as compared with the rest of the United States mean that problems of enhanced nutrient and microbiological inputs do not exist. The handful of sampling sites classified as "fair" rather than "good" from the point of view of water quality by the National Coast Condition Report are thought probably to be the result of so-far-unidentified natural factors rather than of human impact (EPA, 2012).

The west coast of Canada also does not show any problems of eutrophication or microbiological disease. However, there appear to be risks that such problems may develop near the border with the United States, where the main population centres and agriculture are found. This is the possible result of expanding human populations and intensifying agriculture in the lower Fraser Valley and Puget Sound (Schindler et al., 2006).

9.4.2 Waters west of Mexico, Guatemala, El Salvador, Honduras, Nicaragua and Costa Rica

In the coastal waters of these countries, waste-water discharges and agriculture runoff are the main sources of anthropogenic nutrient enrichment. Very little urban waste-water is treated: for example, in El Salvador, less than 3 per cent is treated (Romero Deras, 2013). Fertilizer consumption increased from 76 kilogrammes per hectare to about 131 kilogrammes per hectare between 1990 and 2001, and has continued to rise. Deforestation and associated increases in erosion and runoff also contribute to enhanced nutrient runoff. Eutrophication problems have been noted in the Gulf of Nicoya (Costa Rica), Jiquilisco Bay (El Salvador) and Corinto and El Realejo (Nicaragua). Harmful algal blooms associated with eutrophication have also been observed (Heileman, 2008d).

9.4.3 East Asian Coastal Seas – General

Both municipal waste-water and agricultural runoff present problems for the East Asian Seas. No consistent assessment is possible across the area as a whole, but it is clear that both these major sources are causing problems, particularly in the areas near the major conurbations. In the Philippines, Manila Harbour is a clear example of this. In Thailand, the national marine water-quality index shows that the main problem areas are in the inner Gulf of Thailand, around the mouths of the Chao Phraya, Thachin, Mae Klong, and Bangpakong Rivers. In Malaysia, the overwhelming majority of sampling stations on the east coast of the peninsula and in Sarawak, Sabah and Labuan were put into the "moderate" quality classification. The best areas are in the north of Sabah. Harmful algal blooms have become much more frequent in recent years in all parts of the region (UNEP/COBSEA 2009).

9.4.4 Coastal waters of China

The Chinese authorities have developed a water-quality assessment system which looks at the parameters related to (a) oxygen and nutrients (dissolved oxygen, COD, pH, inorganic nitrogen and phosphates), (b) heavy metals and (c) oil. Microbiological parameters are also monitored. Norms have been established for each of four categories (Category I: Clean water, Category II: Relatively clean water, Category III: Slightly polluted water, Category IV: Medium polluted water). Water that is worse than Category IV is classed as "Heavily polluted water". Classifying waters, works on the "one out, all out" principle: if the samples from an area fail to meet the level specified for a category for any one of the parameters, then the area is demoted to a lower category. In practice, the determinant parameter for all areas is the parameter for inorganic nitrogen, except for Liaodong Bay (the north-eastern gulf of the Bohai Sea), where the determinant parameter is inorganic phosphate. Figure 20.9 shows the results in 2014 for studies in major bays along the coast of China: many are heavily polluted (the absence of indications seaward of the lines enclosing the bays, of course, does not mean that the water

Coastal, Riverine and Atmospheric Inputs from Land — Chapter 20

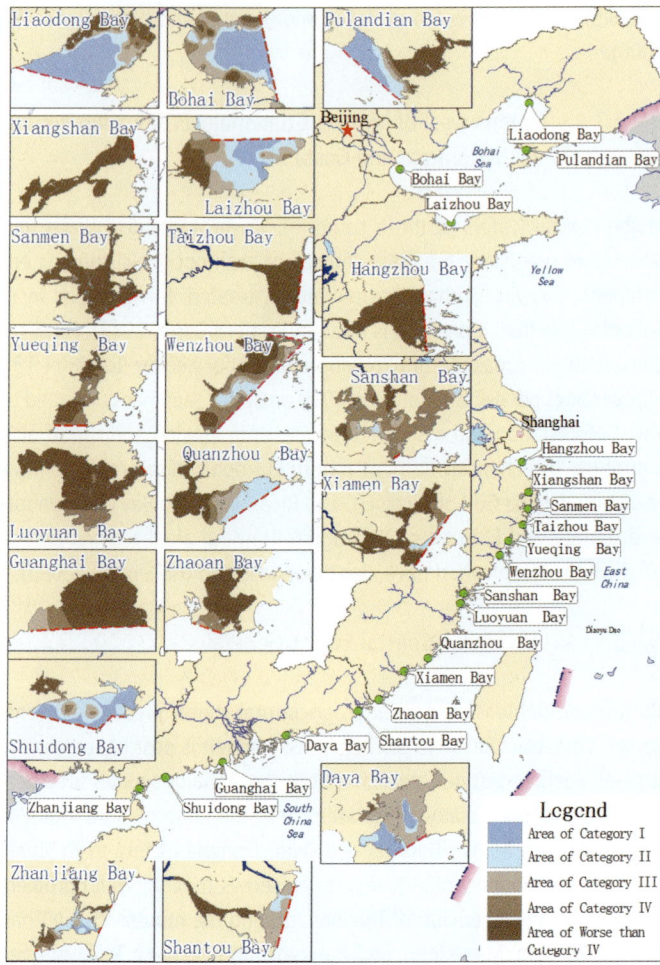

The boundaries and names shown and the designations used on this map do not imply official endorsement or acceptance by the United Nations.

Figure 20.9 | Water Quality in 2014 in Major Bays along the Coast of China. Source: China NMEMC, 2015.

there is clean; merely that the data for such areas is not included in this map). The total area of waters that could not be classified as Category I (clean water) increased steeply, at about 20,000 square kilometres/year, from 1990 to 2000. Since then, the amount of water that is classified as other than clean has remained more stable, although the areas within the different categories below Category I have fluctuated. In particular, the total area classified as heavily polluted water (worse than Category IV) has remained more or less stable over the decade from 2000 to 2009. The fluctuations have, however, been different in the different areas. In the Bohai Sea, although the area of clean water has increased, the other areas have deteriorated in status. It should be remembered that about 10 per cent of the planet's population live in the catchments of the Bohai Sea. In the Yellow Sea, the area in category II and worse increased by about 40 per cent between 2003 and 2004, but by 2009 had recovered its pre-2003 level. In the East China Sea, the area in Category I (clean water) increased until 2005, but after that point remained constant. In the South China Sea, the area of water in Categories II and worse increased by about 75 per cent from 2000 to 2004, but then fell back again in 2005; it then worsened again by 2009 to a level worse than in 2004. These figures show that the extent of marine pollution measured in this way is probably significantly related to changing levels of runoff from land, since it is the levels of nutrients that are determinative (Wang et al., 2011).

Harmful algal blooms in Chinese coastal waters increased massively in number and extent since the 1990s, affecting areas up to 30,000 square kilometres (Wang et al., 2011). Since 2006, the areas affected by "red tides" have decreased, now being less than 20,000 square kilometres. The areas affected by "green tides" have, however, increased since 2008 (China, 2012).

9.4.5 Bohai Sea, Yellow Sea, and the NOWPAP region

Assessment of relative inputs of nutrients to the Yellow Sea from China and the Korean Peninsula is not possible because comparable data are lacking. The same applies to discharges into the parts of the NOWPAP region[4]. From the pattern of harmful algal blooms, however, it is clear that real problems exist here. Harmful algal blooms have been observed along all the coasts, particularly concentrated in the Bohai Sea, on the south of the Korean Peninsula and on the north-west of the island of Kyushu (Japan). Harmful algal blooms off the Chinese coast are usually judged to be much larger than those off the Korean Peninsula and Japan. This may be due to the means of observation: China uses aircraft more than the Republic of Korea and Japan, which rely on ships. In Russian Federation waters, the blooms are confined to Peter the Great Bay, and appear to be the result of the size of the local urban population: no serious harm is attributed to them (NOWPAP, 2007). A UNDP-GEF Strategic Action Programme has committed China and the Republic of Korea to reduce nutrient discharges to the Yellow Sea by 10 per cent every 5 years up until 2020 (UNDP, 2011).

9.4.6 North-West Pacific (Kuroshio and Oyashio Currents)

Japan has a long history of sewage collection and treatment. The sewage from about 85 per cent of the population is treated: about 60 per cent by sewerage networks and about 25 per cent by small local plants. Nutrient removal during sewage treatment is, however, much less common (JSWA, 2014). On the eastern coasts of Japan, there are problems of high levels of nutrients, but these appear to be confined to the more enclosed waters near major conurbations, such as Tokyo Bay and Osaka Bay (Japan MOE, 2009).

9.4.7 South-East Pacific Ocean

As with hazardous substances, the information available on a consistent basis is relatively old. What it showed is that major conurbations lack effective sewage treatment works: the Tumaco estuary in Colombia, the

---

4 The NOWPAP (Northwest Pacific Action Plan) was established by China, Japan, the Republic of Korea and the Russian Federation in 1994 as an integral part of the Regional Seas Programme of the United Nations Environment Programme (UNEP). As stated in the Northwest Pacific Action Plan, it covers the "marine environment and coastal zones of the following States: [Democratic People's Republic of Korea]; Japan; People's Republic of China; Republic of Korea; and Russian Federation from about 121°E to 143°E longitude and from approximately 52°N to 33°N latitude, without prejudice to the sovereign right of any State".

Gulf of Guayaquil (especially the inner area, where oxygen levels were so low that fish were absent) in Ecuador, areas near Ferrol, Callao and Ilo-Ite in Peru, and the bays of San Vicente, Valparaiso and Concepción in Chile showed high levels of nutrients, and consequent eutrophication problems (CPPS, 2000). In spite of substantial programmes of investment in sewage collection and treatment (Peru has increased the proportion of the population served from 9 per cent to 37 per cent between 1985 to 2010), problems remain. Likewise, high levels of fertilizer use add to the problems.

Darwin was one of the first to record red tides (algal blooms) in this area, but they remained rare until the 1980s. Since then, they have become frequent (several a year) along the whole length of the coast from Colombia to Chile (ISP, 2010).

9.4.8 South-West Pacific

The east coast of Australia suffers from enhanced levels of nutrient runoff. These are particularly serious for the Great Barrier Reef. Compared to pre-European conditions (before 1850), modelled mean annual river loads to the Great Barrier Reef lagoon have increased 3.2 to 5.5-fold for total suspended solids, 2.0 to 5.7-fold for total nitrogen and 2.5 to 8.9-fold for total phosphorus. However, the effects vary widely depending on the level of agriculture in the catchment. Almost no change in loading for most pollutants has been observed in the rivers capable of affecting the northern Great Barrier Reef, but there have been much greater changes in rivers capable of affecting the central and southern Great Barrier Reef. Given the sensitivity of the corals of the Great Barrier Reef, the risk of adverse effects is high. Recent work suggests that a substantial part of the decline in hard coral is due to the high nutrient levels in the southern areas (Bell et al., 2014).

Further south, more than half the estuaries in New South Wales are subject to double the natural levels of sediment and nutrient inputs, and around one-third of these estuaries have been cleared of more than 50 per cent of their natural marine vegetation. These and other pressures are directly linked to the poor water quality found in a high proportion of New South Wales estuaries: only 11 per cent of the estuaries have been found to comply more than 90 per cent of the time with the guidelines for chlorophyll-a. Many of the estuaries are under pressure from excessive inputs of sediments and nutrients, and altered freshwater inputs and hydrological regimes (SE2011 Committee, 2011).

In New Zealand, significant eutrophication problems generally only occur in shallow estuaries and bays with restricted circulation. Guidelines for nutrient discharge have been adopted to deal with these problems (ESNZ, 2014). However, delivery of suspended sediment to the sea around New Zealand is 1.7 percent of the world total delivery, when the New Zealand land area is only 0.2% of the global land area (Hicks et al., 2011).

10 Inputs of Radioactive Substances

The waters, biota and sediments of the ocean all contain radioactivity. Some of this is entirely natural, representing the dispersion of naturally radioactive isotopes throughout the earth and the effects of cosmic radiation. Some, however, is the product of relatively recent human activities: the use of atomic bombs during World War II, the testing of further nuclear weapons, discharges and emissions from nuclear power plants and nuclear reprocessing plants, dumping of radioactive waste, accidents involving nuclear material and other less significant activities. Some human activities that concentrate naturally occurring radioactive material (NORM) have a longer history.

In considering radioactivity in the marine environment, it is essential to distinguish between:

(a) The occurrence of ionizing radiation, emitted through the decay of radionuclides, with the level of activity measured in becquerels (one becquerel being the activity of a quantity of radioactive material in which one nuclide decays every second); and

(b) The impact of such radiation on living organisms, where the energy deposited in the tissues of the organism (the absorbed dose) is measured in grays, and the sum of the effects of that dose on the different parts of the body (the effective dose) is measured, for humans, in sieverts. The biological effects of the absorbed dose vary according to the nature of the radiation (α-radiation can have a much more significant effect than β- or γ-radiation) and the part of the body affected. When the radioactive substance is incorporated into the body (for example, by being eaten), its effects integrated over a period of 50 years (70 years for children) are estimated through the committed effective dose, expressed in sieverts.

10.1 Naturally occurring radioactivity in the oceans

The natural background radioactivity in the ocean varies considerably. A study conducted under the auspices of the International Atomic Energy Agency (IAEA) in 1995 examined the variations between the FAO major fishing areas. This looked at the distribution of polonium-210 (^{210}Po), based on the view that, on a global scale, this isotope was radiologically the most important representative of naturally occurring radioactive material. The study concluded that insufficient evidence was then available to estimate the levels of polonium radioactivity in seawater in the different areas of the world. However, data for levels of polonium radioactivity in fish (shown in Figure 20.10), crustacea and molluscs for those areas for which data were available showed variations by factors of 58, 250 and 71, respectively, between the highest and lowest levels (MARDOS, 1995).

Coastal, Riverine and Atmospheric Inputs from Land — Chapter 20

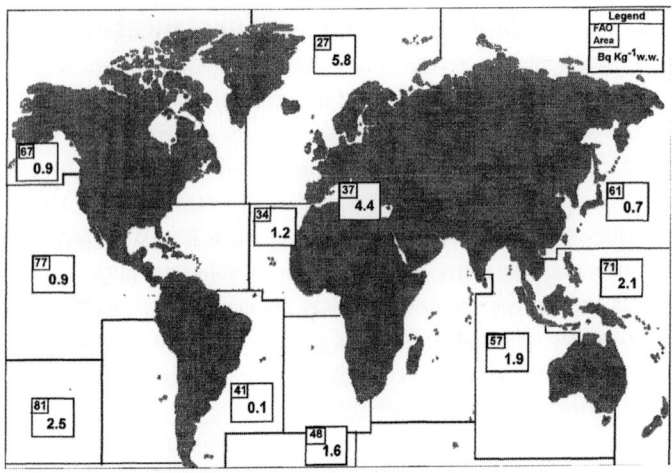

The boundaries and names shown and the designations used on this map do not imply official endorsement or acceptance by the United Nations.
Figure 20.10 | Concentrations of ^{210}Po in fish for FAO major fishing areas. Becquerels per kilogramme of wet weight. Source: IAEA, MARDOS, 1995.

10.2 Anthropogenic radioactivity in the oceans

Anthropogenic releases of radionuclides into the ocean have had a measurable effect on the levels of radioactivity in the oceans and its distribution. The distribution in space and time can be quite complex, but is always related to four general processes: the type and location of the input, radioactive decay, biogeochemistry and oceanic processes, such as transport by ocean currents and sedimentation. The complex interaction of these processes over time means that all parts of the ocean are affected by anthropogenic releases of radionuclides, but that the distribution is quite varied. The 1995 IAEA study, using caesium-137 (^{137}Cs) as typical of anthropogenic radionuclides, estimated that radioactivity levels of ^{137}Cs in seawater and fish vary by factors of around 40-60 between the Southern Ocean (the lowest) and the North-East Atlantic (the highest) (MARDOS, 1995). Although the ocean contains most of the anthropogenic radionuclides released into the environment, the radiological impact of this contamination is low. Radiation doses from naturally occurring radionuclides in the marine environment (for example, ^{210}Po), are on average two orders of magnitude higher (WOMARS, 2005).

10.2.1 Testing of nuclear weapons

Much of the anthropogenic radioactivity in the ocean derives from global fall-out from the atmospheric testing of nuclear weapons between 1945 and 1963. Most of this global fall-out resulted from the input of radioactive material from the explosions into the stratosphere. There was also additional local fall-out from material which did not reach the stratosphere. The United Nations Scientific Committee on the Effects of Atomic Radiation (UNSCEAR) estimates that this global fall-out totalled around 2,500 million terabecquerels (TBq) (UNSCEAR, 2008). Using ^{90}Sr and ^{137}Cs as indicators, an IAEA study estimated that about 64 per cent of this fell on the oceans, of which 1 per cent fell on the Arctic Ocean, 26 per cent on the North Atlantic Ocean, 7 per cent on the South Atlantic Ocean, 14 per cent on the Indian Ocean, 35 per cent on the North Pacific Ocean and 17 per cent on the South Pacific Ocean. The IAEA study further estimated that the inventory of radioactivity from this source in the oceans had decreased (through natural decay) by 2000 to about 13,850,000 TBq. Much of this reduction was, of course, the result of the decay of short-lived isotopes (WOMARS, 2005). There have been no atmospheric tests of nuclear weapons since 1980, and so this major source of anthropogenic radioactivity appears to be purely historic.

10.2.2 Nuclear reprocessing

Overall, the second largest source of anthropogenic inputs of radioactive material into the ocean has been nuclear re-processing plants. In this sector, the major sources are the plants at Cap de la Hague (France: current capacity 1,700 tons/year of waste reprocessed) and Sellafield (United Kingdom: current capacity 2,100 tons/year). When the plants at these sites started work in the 1970s, relatively high levels of radioactive materials were discharged to the sea, reaching a peak in 1975 of 5,230 TBq of ^{137}Cs and 466 TBq of ^{90}Sr from Sellafield. Over the period 1970 – 1983, discharges from Cap de la Hague were much lower, rep-

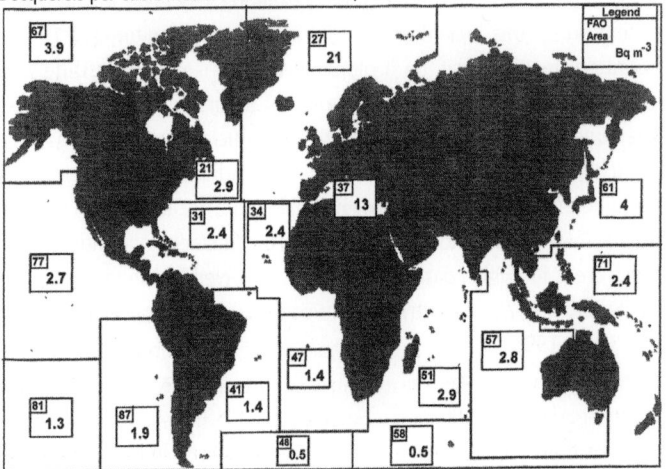

The boundaries and names shown and the designations used on this map do not imply official endorsement or acceptance by the United Nations.
Figure 20.11 | Concentrations of ^{137}Cs in seawater for FAO major fishing areas. Unit: Becquerels per cubic metre. Source: MARDOS, 1995.

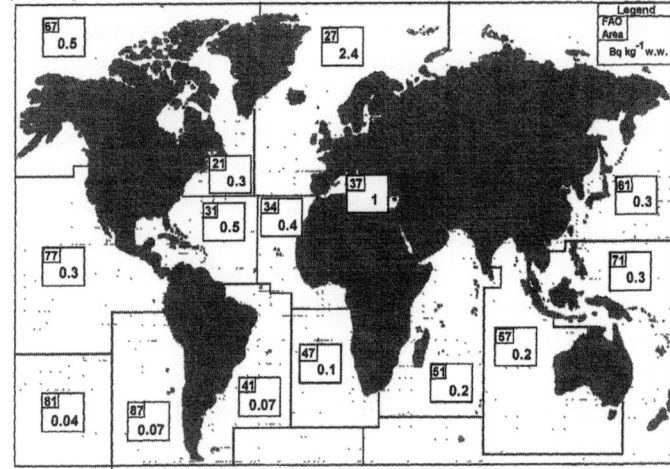

The boundaries and names shown and the designations used on this map do not imply official endorsement or acceptance by the United Nations.
Figure 20.12 | Concentrations of ^{137}Cs in fish for FAO major fishing areas. Becquerels per kilogramme of net weight. Source: MARDOS, 1995.

resenting about 2 per cent and 16 per cent respectively, of the levels of ^{137}Cs and ^{90}Sr discharges from Sellafield. In both cases, steps were taken to reduce discharge levels drastically, and new technology was developed and installed. The result was that aggregate annual discharges (other than tritium) from the two sites were reduced, by 2000, to around 98.2 TBq (0.2 TBq of α-emitting substances and 98 TBq of β-emitting substances other than tritium). Since then efforts at reductions have continued: by 2011, discharges were down to 18.2 TBq (0.1 TBq of α-emitting substances and 18.1 TBq of β-emitting substances other than tritium). This represents a reduction of 99.7 per cent from the peak of annual discharges (WOMARS, 2005; OSPAR, 2013; NEA, 2013). Although some States remain concerned at these discharges, the major impact is now only historic. It has been announced that one of the Sellafield plants (the Thermal Oxide Reprocessing Plant (THORP)) will close in 2018, when the currently programmed reprocessing has been completed, although this programme is currently reported to be behind schedule (NDA, 2014).

During the implementation of these reductions, some European States raised concerns about discharges of technetium-99 (^{99}Tc) from a new plant at Sellafield. Technetium has chemical properties close to those of manganese, which is naturally concentrated by many crustacea, especially lobsters. The ^{99}Tc discharges from the new plant at Sellafield were initially high: just over 180 TBq in 1994. This was substantially due to treating a backlog accumulated while the new plant was built, but ^{99}Tc discharges were still at around 40 TBq/year in 2003. In response to continued pressure from these European States, the United Kingdom has now implemented a chemical process to remove much of the ^{99}Tc from the discharge stream, and levels are now below 5TBq/year (OSPAR, 2010).

Other civilian reprocessing plants on much smaller scales were operational in Belgium, Germany and Italy, but have been closed since 1991 or earlier. China has a small nuclear reprocessing plant (capacity 50 tons/year) in operation in the inland province of Gansu. A larger plant (capacity 800 tons/year) is reported to be planned to start operation in the same province in 2017, and plans for a further plant exist. India has small reprocessing plants at coastal sites at Trombay (near Mumbai: capacity 60 tons/year), Tarapur (in Maharashtra: capacity 100 tons/year) and Kalpakkam (in Tamil Nadu: capacity 100 tons/year). A further plant (capacity 100 tons/year) was opened in 2011 at Tarapur, and further capacity is being built at Kalpakkam. Japan has a pilot reprocessing plant at Tokai on the coast north of Tokyo (capacity 40 tons/year) and is in the process of opening a large plant (capacity 800 tons/year) at Rokkasho (on the coast at the northern end of Honshu). No data on discharges from any of these plants seem to be available. The Russian Federation has operated the Mayak reprocessing plant (capacity 400 tons/year) near Ozyersk in the Ural Mountains since 1971. The nearby Lake Karachay has been used for the discharge of large quantities of radioactive waste. The IAEA 2005 study noted that this lake represents a potential source for future contamination of the Ob River system, and thus of the Arctic Ocean. New reprocessing facilities are also under construction at Zhelenogorsk, near the border with the Ukraine. Apart from the risk from Lake Karachay, there is no evidence to suggest that these other civilian reprocessing plants have led, or might lead, to significant contamination of the ocean (NEA, 2013; WNA, 2013; WOMARS, 2005).

10.2.3 Nuclear accidents

There have been two nuclear accidents that reached level 7 (the highest level) on the IAEA's International Nuclear and Radiological Event Scale: Chernobyl and Fukushima. These have resulted in substantial amounts of radioactive material reaching the ocean.

10.3 Chernobyl

On 26 April 1986, the Number Four reactor at the nuclear power plant at Chernobyl, Ukraine, went out of control during a test at low power, leading to an explosion and fire that demolished the reactor building and released large amounts of radioactive material into the atmosphere. Around 100,000 TBq of ^{137}Cs were released to the atmosphere. Although most of this activity was deposited on land, a significant part went to the sea, particularly the Baltic Sea. The total input of ^{137}Cs from Chernobyl to the Baltic Sea has been estimated at 4,700 TBq, including post-Chernobyl river discharges of ^{137}Cs to the Baltic Sea estimated at 300 TBq. Inputs from Chernobyl to the Black Sea have been estimated at 2,000-3000 TBq ^{137}Cs. The North Sea and the Mediterranean Sea also received inputs of radioactive material, and continue to do so through outflows from the Baltic Sea and Black Sea, respectively.

Because of the Chernobyl input, the Baltic Sea has the highest concentrations of ^{137}Cs of any sea region. Average concentrations of ^{137}Cs in fish from the Baltic Sea in 1990 were similar to those in the Irish Sea (which were affected by the Sellafield discharges), about 4 times higher than in the Black Sea, and about 30 times higher than in the Mediterranean Sea. However, radiation doses to humans in the Baltic Sea area from marine pathways (including those from ^{137}Cs in fish) during 1999-2006 have not exceeded an annual value of 0.02 mSv, and the dose for a person eating 90 kilogrammes a year of fish was estimated at 0.014 mSv over the period 2007 – 2010 – both well below the limit of 1 mSv per year for the general public set in the IAEA Basic Safety Standards. HELCOM assessments in 2009 and 2013 concluded that concentrations of radioactive substances in the Baltic Sea are not expected to cause harmful effects to wildlife in the foreseeable future (WOMARS, 2005; HELCOM, 2009; HELCOM, 2013). Likewise, a 2006 IAEA report concluded that radioactivity concentrations in marine fish resulting from the inputs from the Chernobyl disaster to the marine environment are not of concern (IAEA, 2006).

10.4 Fukushima

On 11 March 2011, a 9.0-magnitude earthquake occurred near Honshu, Japan, creating a devastating tsunami that left a trail of death and destruction in its wake. The earthquake and the subsequent tsunami, which flooded over 500 square kilometres of land, resulted in the loss of more than 20,000 lives and destroyed property, infrastructure and natural resources. They also led to the worst civil nuclear disaster since Chernobyl. Three of the six nuclear reactors at Fukushima Daiichi nuclear power station suffered severe core damage. This resulted in the release, over a prolonged period, of very large amounts of radioactive material into the environment. UNSCEAR concluded that the information available to it implied atmospheric releases of iodine-131 (^{131}I) and caesium-137 (^{137}Cs) in the ranges of 100,000-500,000 TBq and 6,000-20,000 TBq, respectively. (^{131}I and ^{137}Cs are two of the most significant radionuclides from the point of view of exposures of people and the environment). Winds transported a large portion of the atmospheric releases onto the Pacific Ocean. In addition, liquids containing radioactivity were discharged directly into the surrounding sea. The direct discharges amounted to around 10 per cent (for ^{131}I) and 50 per cent (for ^{137}Cs) of the corresponding atmospheric discharges. Low-level releases into the ocean were still ongoing in May 2013. The estimated releases are about 10 per cent (^{131}I) and 20 per cent (^{137}Cs) of the corresponding estimated atmospheric releases from the Chernobyl accident (UNSCEAR, 2013), but because of the sea-side site and the effects of the winds, the Fukushima event was the largest-ever accidental release of radioactive material to the ocean, being slightly more than the amount reaching the sea from the intrinsically much larger Chernobyl event (Japan, 2011; Pacchioli, 2013).

UNSCEAR further concluded that exposures of marine biota to radioactivity following the accident were, in general, too low for acute effects to be observed, though there may have been some exceptions because of local variability. Effects on biota in the marine environment would have been confined to areas close to where the highly radioactive water was released into the ocean (UNSCEAR, 2013).

Within a few weeks of the disaster, traces of ^{134}Cs were found over 1,900 km across the Pacific from Fukushima. By August 2011, bluefin tuna caught off California were found to contain ^{134}Cs which could only have come from Fukushima. ^{134}Cs has a half-life of only two years, and so material from pre-Fukushima sources (such as weapons testing) would have decayed long before. Further sampling suggested that the strong Kurushio current acted as a barrier preventing significant amounts of radioactive material moving south in the Pacific, and confining it to around the latitude of Fukushima (Pacchioli, 2013; Fisher et al., 2013).

In December 2013, the IAEA confirmed that a comprehensive Sea Area Monitoring Plan had been established, noting that radionuclide concentrations remain within the WHO guidelines for drinking water and that the public is safe. The IAEA assessment also addressed monitoring of food products, adding that the joint FAO/IAEA Division concluded that measures taken to monitor and rapidly respond to any issues regarding radionuclide contamination in the food system are appropriate and that the public food supply (including food from the sea) is safe (IAEA, 2013a; IAEA, 2014).

10.5 Other nuclear accidents

The 2005 IAEA study reviewed the full range of accidents involving radioactive material resulting in inputs to the ocean, but did not consider that the amounts were significant, beyond noting that the six sunken nuclear submarines which remain in the ocean with their reactors may be considered as potential sources of radioactive contamination of the ocean, and that nuclear-powered satellites burning up in the atmosphere on re-entry can affect radioactivity in the ocean (a 1964 incident over the southern hemisphere resulted in a measurable increase in the ratio between ^{238}Pu and ^{239,240}Pu between the northern and southern hemispheres (WOMARS, 2005).

10.5.1 Nuclear power plants

There were 434 commercial nuclear power reactors in 30 countries in operation at the end of 2013. The plants containing them have a total capacity of over 370,000 megawatts (MW). A little over 300,000 MW of this capacity is in OECD countries. About 72 more reactors are under construction. These plants produce over 11 per cent of the world's electricity: from nearly 75 per cent of the national supply in France to 1.5 per cent in the Islamic Republic of Iran (see Table 20.5). Other States which do not have nuclear power plants, such as Denmark and Italy, import substantial amounts of their electricity from neighbouring States which

Table 20.5 | Proportion of electricity generated from nuclear power 2013. Source: IAEA PRIS Database, IAEA, 2013b.

State	per cent of Electricity from Nuclear Power	State	per cent of Electricity from Nuclear Power	State	per cent of Electricity from Nuclear Power
France	73.3	Bulgaria	30.7	South Africa	5.7
Belgium	52.1	Armenia	29.2	Mexico	4.6
Slovakia	51.7	Korea, Republic of	27.6	Argentina	4.4
Hungary	50.7	United States of America	19.4	Pakistan	4.4
Ukraine	43.6	United Kingdom	18.3	India	3.5
Sweden	42.7	Russia	17.5	Brazil	2.8
Switzerland	36.4	Romania	19.8	Netherlands	2.8
Czech Republic	35.9	Spain	19.7	China	2.1
Slovenia	33.6	Canada	16.0	Japan	1.7
Finland	33.3	Germany	15.4	Iran, Islamic Republic of	1.5

rely substantially on nuclear power (IAEA, 2013b). Electricity generated from nuclear power is therefore a significant source of energy.

Emissions and discharges are inevitable from the operation of these plants. For the purposes of the World Ocean Assessment, the crucial question is the extent of the impact of these emissions and discharges on the marine environment. The 2005 IAEA survey of sources of anthropogenic inputs of radioactive materials to the ocean commented that routine discharges from nuclear power plants contribute orders of magnitude less to the radioactive contamination of the world ocean than nuclear-weapons testing, nuclear reprocessing plants and nuclear accidents (WOMARS, 2005). The supporting material for the 2008 UNSCEAR report to the United Nations General Assembly gives a figure of approximately 1.3 TBq as the worldwide aggregate level of radioactivity from radionuclides other than tritium released from nuclear power plants in liquid effluents in 2002 (UNSCEAR, 2008). Data from some plants is not included but, as can be seen from comparison with the figures quoted above for other sources, this is consistent with the WOMARS conclusion. The 2008 UNSCEAR report further comments that radiation doses from nuclear power reactors decrease over time because of lower discharge levels. This is consistent with the observations recorded by the OSPAR Commission, which noted a statistically significant reduction of 38 per cent in total β-activity (other than tritium) from nuclear industries discharging into the North-East Atlantic (OSPAR, 2010). At the same time, aggregate discharges from nuclear power plants are likely to increase somewhat as the nuclear power stations under construction and planned come on stream.

Discharges of tritium are, however, rather different. The production of tritium by nuclear power plants is normally related to the level of electricity generated. No accepted abatement technology exists, and the amount of radioactivity in discharges can be many times that from other radionuclides. However, tritium is a natural product produced by cosmic rays. This source accounts for a significant amount of the radionuclide found in the sea. It also has a very low dose coefficient and therefore exhibits a very low radiotoxicity to humans and inherently low radiotoxicity to biota (OSPAR, 2007).

10.5.2 Human activities concentrating naturally occurring radioactive material (NORM)

A wide range of materials used in an even wider range of human activities contain natural radioactivity. The effects of the human activities can be to concentrate this naturally occurring radioactive material (NORM) from these materials, usually in the form of waste. Recent studies by the OSPAR Commission (summarized in OSPAR, 2010) conclude that the major source of NORM reaching the North-East Atlantic is the offshore oil and gas industry, where produced water (water coming from the reservoir with the oil and gas) and the scale that it deposits in pipelines (which has to be cleared periodically) contains low levels of radionuclides (mainly ^{210}Pb, ^{210}Po, and $^{226/228}$Ra). Although the proportion of total-α activity is higher than for discharges from the nuclear industries, the overall concentrations are not so far thought to be significant, although work to assess the levels is continuing. Apart from phosphate rock processing (see below), other anthropogenic sources of NORM in the marine environment are not thought to be significant.

One form of NORM reaching the marine environment that has been thought to be significant in some States is the ^{210}Po found in phosphogypsum, a by-product of the treatment of rock containing phosphate to produce phosphate fertilizers. In many cases, this phosphogypsum has been discharged directly into the sea as slurry. At Workington, England, in the area affected by the Whitehaven phosphate-processing plant releases, it was found that molluscs were concentrating this ^{210}Po to an extent that those who consumed substantial quantities of the molluscs might be ingesting ^{210}Po at potentially dangerous levels. The closure of the plant in 1992 resolved the problem. Similar problems were also found at a plant at Rotterdam in the Netherlands, which was also closed. For these and other reasons, this method of disposing of phosphogypsum has been phased out in most countries. It continues in Lebanon, Morocco (where it is under review) and South Africa (IAEA, 2013c).

10.6 Impact of radioactivity in the marine environment

Two issues need to be differentiated: the impact of radioactivity from the marine environment on humans, and the impact of such radioactivity on marine biota.

As far as concerns the radiation impact on humans through food from the marine environment, the IAEA MARDOS study in 1995 reported on the exposure of humans to ^{137}Cs and ^{210}Po, as the anthropogenic and natural radionuclides, respectively, of most radiological significance. This study concluded that, worldwide, the mean individual committed effective doses in 1990 were:

In another way of considering the data, the study identified the critical group of humans (the group most at risk) as those eating seafood from the North-East Atlantic, the FAO major fishing area with the highest

Table 20.6 | Estimated mean individual committed effective doses in 1990 from ^{137}Cs and ^{210}Po. Source, MARDOS, 1995.

Radionuclide	Food source	Mean individual effective dose commitment worldwide (microsieverts)	Uncertainty factor as a result of limited data
137Cs (anthropogenic)	Fish	0.03 µSv	0.5
137Cs (anthropogenic)	Shellfish	0.002 µSv	0.5
210Po (natural)	Fish	1.9 - 2.3 µSv	5
210Po (natural)	Shellfish	2.8 – 7.2 µSv	5

Coastal, Riverine and Atmospheric Inputs from Land — Chapter 20

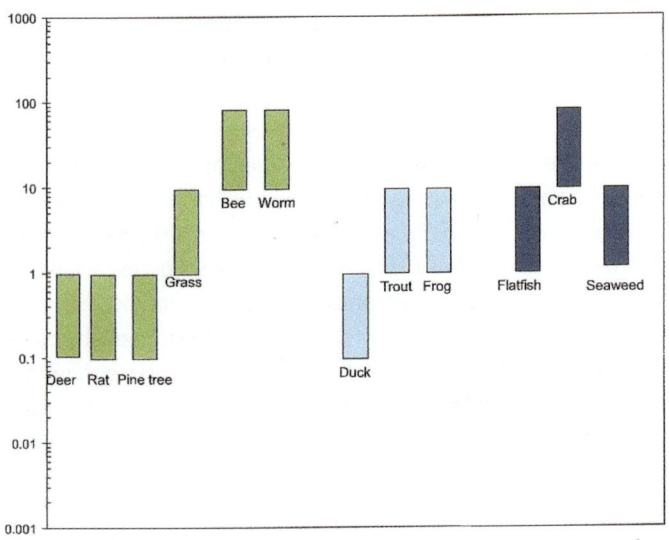

Figure 20.13 | Derived Consideration Reference Levels for Reference Animals and Plants. In milligrays (mGy)/day. Source: IPRC, 2014.

levels of radioactivity. Taking as the definition of the critical group those consuming 100 kg of fish and 10 kg of shellfish per year (a daily consumption of about 300 g (about 10½ oz) of seafood), the total individual committed effective doses were estimated for 1990 at 3.1 µSv from ^{137}Cs and 160 µSv from ^{210}Po. There is no reason to consider that current levels would be significantly higher. All these figures must be considered in relation to the IAEA's recommended annual limit for exposure of the general public to radiation of 1mSv (1,000 µSv).

For a long time, the International Commission on Radiation Protection (ICRP – the international body of experts that agrees standards of radiation protection) considered that the precautions necessary to protect humans will be adequate to protect other species: "The Commission believes that…if man is adequately protected, then other living things are also likely to be sufficiently protected" (ICRP, 1977). In the 1990s, this approach was questioned, particularly for habitats where humans do not go – which covers much of the marine environment. It was not suggested that there were any obvious cases where the approach was failing to protect non-human species, but rather that it was desirable to adopt an approach which would explicitly demonstrate the proper protection of the whole environment. The ICRP debated this extensively from 2000 and in 2005 set up a new standing committee to consider the radiological protection of the environment. This debate resulted in the inclusion of an approach for developing a framework to demonstrate radiological protection of the environment, as part of the general 2007 revision of the ICRP recommendations (ICRP, 2007). The ICRP considered that this approach needed to be based on a sound scientific system similar to that developed for human protection, and that this could best be achieved by the creation of a set of Reference Animals and Plants. Descriptions of 12 "reference animals and plants" have been developed, of which three – a flatfish, a crab and a seaweed – are relevant to the marine environment. These are generic biological descriptions of the types of animal and plant, accompanied by consideration of their vulnerability to radiation and the relationship between environmental

levels of radionuclides and the corresponding levels in such animals and plants. Most recently, in 2014, the ICRP has published guidance on the application of their recommendations to different exposure situations with respect to the animals and plants living in different types of natural environments. Central to this approach is the "Derived Consideration Reference Level" (DCRL): a band of dose rates within which there is some chance of a deleterious effect from ionizing radiation occurring to individuals of that type of Reference Animal or Plant. The recommended DCRLs are shown in Figure 20.13 (ICRP, 2009; ICRP, 2014). This work is being taken forward through the IAEA International Plan of Activities on the Radiation Protection of the Environment.

11 Significant environmental, economic and social aspects of land-based Inputs and related information and capacity-building gaps

The world needs to feed, clothe, house and keep happy its 7¼ billion people. The agricultural and industrial developments of the past two centuries have substantially enabled this to be done and, to a significant extent, have assisted in improving human well-being. But these achievements have been obtained at a price: these agricultural and industrial developments have seriously degraded important parts of the planet, including much of the marine environment. Land-based inputs to the ocean have contributed substantially to this degradation of the marine environment.

The GPA highlighted the need for action to deal with sewage. Although much has been done to implement national plans adopted under the GPA, particularly in South America, this chapter shows that lack of sewage systems and waste-water treatment plants is still a major threat to the ocean. This is particularly the case in respect of very large urban settlements. The lack of proper management of waste-water and human wastes leads to excessive inputs of both nutrients and hazardous substances, which damage the marine environment. It also causes problems for human health, both directly and through bacteriological contamination of food from the sea.

From the point of view of industrial development, many of the earlier industrial processes brought with them serious environmental damage, especially when concentrations of industry led to intense levels of waste inputs to the sea, beyond its carrying capacity. New technologies and processes have largely been developed which have the ability to avoid these problems, but there can be gaps in the capacity to apply these newer processes, often because of the costs involved.

This is particularly significant because of the major transfer in the growth of industrial production demonstrated in this chapter. In the past, industrial production has been dominated by the countries around the North Atlantic basin and its adjacent seas, together with Japan. Over the past 15 years, the rapid growth of industries along the rest of the western Pacific Rim and around the Indian Ocean has changed this. Rapidly growing proportions of the world's industrial production – and the associated waste discharges – are focused on the South Atlantic, the Indian Ocean and the western Pacific.

The survey in this chapter shows that some major information gaps need to be filled before this process of industrial growth can be managed in a way that can avoid reproducing, in the new areas of growth, the many problems that have been discovered in the areas that have been industrialized longer. For long stretches of the coastal zones, information is lacking on what is happening with heavy metals and other hazardous substances. Information is also lacking on the extent to which developing industries are able to apply the newer, cleaner technologies. Moreover, information is very scarce on how problems in the coastal zones are affecting the open ocean.

The agricultural revolution of the last part of the 20th century, which has largely enabled the world to feed its rapidly growing population, has also brought with it problems for the oceans, in the form of enhanced runoff of both agricultural nutrients and pesticides, as well as the airborne and waterborne inputs of nutrients from wastes from agricultural stock. In the case of fertilizers, there is a rapid growth in their use in parts of the world where only limited use has occurred in the past. This has the potential to lead to increased nutrient runoff to the ocean, if the increased use of fertilizers is not managed well. There are therefore problems in educating farmers, promoting good husbandry that causes less nutrient runoff and monitoring what is happening to agricultural runoff alongside sewage discharges. In the case of pesticides, the issues are analogous to those of industrial development. Newer pesticides are less polluting than older ones, but gaps remain in the capacity to ensure that these less-polluting pesticides are used, in terms of educating farmers, enabling them to afford the newer pesticides, supervising the distribution systems, and monitoring what is happening in the oceans.

The growth of dead zones, resulting from excessive nutrient runoff and the consequent eutrophication problems, is serious in terms of all three of environment, economics and society. The dead zones drive fish away and kill the benthic animals. Where the dead zone is seasonal, such regeneration as happens is usually at a lower trophic level, and the ecosystems are therefore degraded. This affects the maritime economy seriously, both for fishers and (where tourism has some dependence on the attractiveness of the ecosystem (for example, where there are coral reefs)) for the tourist industry. Social consequences are then easy to see, both through the direct economic effects on the fishing and tourist industries and in depriving the local human populations of the benefits of an attractive environment.

In respect of heavy metals and hazardous substances, frameworks have emerged at the international level for addressing some of these problems. In particular, the Stockholm Convention on Persistent Organic Pollutants and the Minamata Convention on Mercury provide agreed international frameworks for the States that are party to them to address the issues that they cover. Implementing them, however, will require many capacities and, as the organizations involved with these Conventions have noted, there are important gaps in those capacities around the world.

In the case of radioactive discharges into the ocean, the survey shows that, in the past, there have been human activities that have given rise to concern, but that reactions to these concerns have largely removed the underlying problems, although there is a continuing need to monitor what is happening to radioactivity in the ocean. What remains is the concern voiced in the GPA that public reaction to concerns about marine radioactivity could result in rejection of fish as a food source, with consequent harm to countries that have a large fisheries sector and damage to the world's ability to use the important food resources provided by the marine environment.

Underlying all these issues is the major information gap in the information needed to see what is happening around the world to the ocean as a result of land-based inputs. This chapter has noted a range of differing systems for assessing the state of the ocean in respect of both hazardous substances and eutrophication. These systems usually differ for good reasons: conditions vary widely around the world. There is a lack, however, of methods to compare explicitly the information that each assessment system produces. This does not imply a need for a single global system of monitoring: as has been said, good reasons for the differences often exist. But an important gap in information results from the lack of any means of comparing the answers given by the different assessment systems. Comparison between monitoring systems also presupposes good quality-assurance of monitoring data.

An even more important gap in information is the absence of any form of regular, systematic assessment of the impact of land-based inputs in many parts of the world. In some parts of the world, such as the Caribbean, many one-off, independent examinations of several aspects of the marine environment have been undertaken, but they are not in forms which enable them to be assembled into a wider, continuous assessment. Given the potential significance of transboundary effects from land-based inputs, this is a very serious information gap. In at least some parts of the world where this is the case, universities and marine research institutes have the capacity to carry out the monitoring and analysis that is needed: what is lacking is more the capacity to organize these existing capacities to fill the wider information gap.

In summary, therefore, important changes are under way in the location around the world of industrial activity and agriculture, which have the potential to cause serious problems if past errors are reproduced. Worrying gaps exist in the capacities needed: to provide sewerage systems and waste-water treatment plants, to implement international conventions regulating which substances can be put into the sea from the land, and to monitor what is happening in the marine environment as a result. Finally, overall, major gaps remain in knowledge about land-based inputs and what knowledge about them is available in different parts of the world.

References

Abraham, G.M.S. and Parker, R.J. (2008). Assessment of heavy metal enrichment factors and the degree of contamination in marine sediments from Tamaki Estuary, Auckland, New Zealand. *Environmental Monitoring Assessment*, 136, 227–238.

ADB (2012). *Promoting beneficial sewage sludge utilization in the People's Republic of China*. Asian Development Bank (ADB), Mandaluyong City, Philippines.

Agenda 21, (1992). United Nations Conference on Environment and Development, Document A/CONF.151/26/Rev.l (Vol. I), 1992.

ALCOA (2014). *Aluminium Smelting*. ALCOA, http://www.alcoa.com/global/en/about_alcoa/pdf/Smeltingpaper.pdf.

Alsharhan, A.S., Rizk, Z.A., Nairn, A.E.M., Bakhit, D.W., and Alhajari, S.A. (2001). *The Hydrogeology of an Arid Region: The Arabian Gulf and Adjoining Areas*. Elsevier, Amsterdam, the Netherlands.

Altwicker, E.R., Schonberg, J.S., Konduri, R.K.N.V., and Milligan, M.S. (1990). Polychlorinated Dioxin/Furan Formation in Incinerators. *Hazardous Waste and Hazardous Materials* 7(1).

AMAP (1997). *Arctic Pollution Issues: A State of the Arctic Environment Report*. AMAP, Oslo, Norway.

AMAP (2009). Arctic Pollution 2009. AMAP, Oslo, Norway.

ASCLME/SWIOFP (2012). Transboundary Diagnostic Analysis for the western Indian Ocean. Volume 2: Diagnostic Analysis. UNEP, South Africa.

Australian Government (2011). This week at Casey: 21 January 2011. Department of the Environment, http://www.antarctica.gov.au/living-and-working/stations/casey/this-week-at-casey/2011/21-january-2011/3.

Baldoni-Andrey, P., Girling, A., Bakker, A., Muller, A, Struijk, K., Fotiadou, I., Andrés Huertas, A., Negroni, J., Neal, G., and den Haan, K. (2012). *Trends in oil discharged with aqueous effluents from oil refineries in Europe: 2010 survey data*. CONCAWE Report no.6/12, Brussels, Belgium, https://www.concawe.eu/DocShareNoFrame/Common/GetFile.asp?PortalSource=1856&DocID=37556&mfd=off&pdoc=1.

Basterretxea, G., Tovar-Sanchez, A., Beck, A.J., Masqué, P., Bokuniewicz, H.J., Coffey, R., Duarte, C.M., Garcia-Orellana, J., Garcia-Solsona, E., Martinez-Ribes, L., Vaquer-Sunyer, R. (2010). Submarine Groundwater Discharge to the Coastal Environment of a Mediterranean Island (Majorca, Spain). Ecosystem and Biogeochemical Significance. *Ecosystems*, 13(5), 629-643.

Belan, T.A. (2004). Marine environmental quality assessment using polychaete taxocene characteristics in Vancouver Harbour. *Marine Environmental Research*, 57(1–2), 89-101.

Belkin, I., Aquarone, M.C., and Adams, S. (2008). Kuroshio Current Large Marine Ecosystem. In: K. Sherman, Hempel G., eds. *The UNEP Large Marine Ecosystem Report: A perspective on changing condi-tions in LMEs of the world's Regional Seas*. UNEP, Rep. Studi., 182: 393-402.

Bell, P., Elmetri, I., Lapointe, B.E. (2014). Evidence of large-scale chronic eutrophication in the Great Barrier Reef: quantification of chlorophyll a thresholds for sustaining coral reef communities. *Ambio*, 43(3), 361-376.

Birch, G.F. (2000). Marine pollution in Australia, with special emphasis on central New South Wales estuaries and adjacent continental margin. *International Journal of Environment and Pollution*, 13(1), 573-607.

Black Sea Commission (2008). *State of the Environment of the Black Sea (2001-2006/7)*. Temel Oguz, Istanbul, Turkey, http://www.blacksea-commission.org/_publ-SOE2009.asp.

BOBLME, Bangladesh (2011). *Bay of Bengal Large Marine Ecosystem Project, Country report on pollution – Bangladesh*. BOBLME-2011-Ecology-01, http://www.boblme.org/documentRepository/BOBLME-2011-Ecology-01.pdf.

BOBLME, India (2011). *Bay of Bengal Large Marine Ecosystem Project, Country report on pollution – India*. BOBLME-2011-Ecology-07, http://www.boblme.org/documentRepository/BOBLME-2011-Ecology-07.pdf.

BOBLME, Malaysia (2011). *Bay of Bengal Large Marine Ecosystem Project, Country report on pollution – Malaysia*. BOBLME-2011-Ecology-11, http://www.boblme.org/documentRepository/BOBLME-2011-Ecology-11.pdf.

BOBLME, Myanmar (2011). *Bay of Bengal Large Marine Ecosystem Project, Country report on pollution – Myanmar*. BOBLME-2011-Ecology-13, http://www.boblme.org/documentRepository/BOBLME-2011-Ecology-13.pdf.

BOBLME, Thailand (2011). *Bay of Bengal Large Marine Ecosystem Project, Country report on pollution – Thailand*. BOBLME-2011-Ecology-08, http://www.boblme.org/documentRepository/BOBLME-2011-Ecology-08.pdf.

BOBLME, Sri Lanka (2011). *Bay of Bengal Large Marine Ecosystem Project, Country report on pollution – Sri Lanka*. BOBLME-2011-Ecology-14, http://www.boblme.org/documentRepository/BOBLME-2011-Ecology-14.pdf.

Bocquenéa, G. and Franco, A. (2005). Pesticide contamination of the coastline of Martinique. *Marine Pollution Bulletin*, 51(5), 612-619.

Borysova, O., Kondakov, A., Paleari, S., Rautalahti-Miettinen, E., Stolberg F., and Daler, D. (2005). *Eutrophication in the Black Sea region; Impact assessment and Causal chain analysis*. University of Kalmar, Kalmar, Sweden, http://www.unep.org/dewa/giwa/areas/reports/r22/giwa_eutrophication_in_blacksea.pdf.

BGS, 2012: British Geological Survey, *World Mineral Production 2008-2012*, British Geological Survey, Keyworth, Nottingham.

Chan, F., Barth, J.A., Lubchenco, J., Kirincich, A., Weeks, H., Peterson, W.T., and Menge, B.A. (2008). Emergence of Anoxia in the California Current Large Marine Ecosystem. Science, 319 (5865), 920.

Chesapeake Bay Program (2014). *Reducing Nitrogen Pollution*. Annapolis, United States, http://www.chesapeakebay.net/indicators/indicator/reducing_nitrogen_pollution.

China (2010). *Project Report on the Reduction of Mercury Use and Emission in Carbide PVC Production*. People's Republic of China, Ministry of Environmental Protection, http://www.unep.org/chemicalsandwaste/Portals/9/Mercury/VCM Production/Phase I Final Report - PVC Project Report for China.pdf.

China (2012). China Marine Environment Status Bulletin (http://wenku.baidu.com/view/512b78245901020207409cea accessed 13 April 2015).

China (2014). *Catalogue of Toxic Chemicals Prohibited or Strictly Controlled*. People's Republic of China, Ministry of Environmental Protection, http://english.mep.gov.cn/inventory/toxic_chemicals/200712/t20071212_114161.htm.

China NMEMC (National Marine Environment Monitoring Centre) (2015). Direct communication from the organization.

Coles, J.A., Farley, S.R., Pipe, R.K. (1995). Alteration of the immune response of the common marine mussel *Mytilus edulis* resulting from exposure to cadmium. *Diseases of Aquatic Organisms*, 22, 59-65.

Costa, O.S. (2007). Anthropogenic nutrient pollution of coral reefs in Southern Bahia, Brazil. *Brazilian Journal of Oceanography*, 55(4).

CPCB (2009). *Wastewater Generation and Treatment In Class-I Cities & Class-II Towns Of India*. Indian Central Pollution Control Board, Status of Water Supply, http://www.cpcb.nic.in/upload/NewItems/NewItem_153_Foreword.pdf.

CPCB (2014). *Coastal Ocean Monitoring and Prediction System (COMAPS) at a Glance*. Indian Central Pollution Control Board, http://www.cpcb.nic.in/comaps.php.

CPPS (2000). *Estado del Medio Ambiente marino y costero del Pacífico Sudeste*. Comisión Permanente del Pacífico Sur, Quito, Ecuador.

de Lacerda, L.D., Kremer, H.H., Kjerfve, B., Salomons, W., Marshall, J.I. and Crossland, C.J. (2002). *South American Basins: LOICZ Global Change Assessment and Synthesis of River Catchment - Coastal Sea Interaction and Human Dimensions*. LOICZ, Texel, the Netherlands, http://www.loicz.org/imperia/md/content/loicz/print/rsreports/report_21.pdf.

Deloitte (2011). *PVC production profitability*. Deloitte, http://www.deloitte.com/assets/Dcom-Russia/Local%20Assets/Documents/PDF_2011/dttl_PVC-production-profitability_08042011.pdf.

Desbrow C., Routledge, E.J., Brighty, G.C., Sumpter, J.P., and Waldock, M. (1998). Identification of estrogenic chemicals in STW effluent. I: Chemical fractionation and in vitro biological screening. *Environmental Science and Technology*, 32, 1549-1558.

Deshpande, A., Bhendigeri, S., Shirsekar, T., Dhaware, D., and Khandekar, R.N. (2009). Analysis of heavy metals in marine fish from Mumbai Docks. *Environmental Monitoring and Assessment*, 159(1-4), 493-500.

DFO-MPO (2013). *Canadian Department of Fisheries and Oceans/Ministère de la Pêche et des Océans, "Will "Dead Zones" Spread in the St. Lawrence River?"*. Fisheries and Oceans Canada, http://www.dfo-mpo.gc.ca/science/publications/article/2005/01-12-2005-eng.htm.

Diaz, R.J. and Rosenberg, R. (2008). Spreading Dead Zones and Consequences for Marine Ecosystems. *Science*, 321(5891), 926-929.

Donkor, A.K., Nartey, V.K. Bonzongo, J.C., Adotey, D.K. (2006). Artisanal Mining of Gold with Mercury in Ghana. *West African Journal of Applied Ecology*, 9(1).

Duce, R.A., LaRoche, J., Altieri, K., et al. (2008). "Impacts of atmospheric anthropogenic nitrogen on the open ocean", *Science*, 320, 893-897.

Dufour, R., and Ouellet, P. (2007). *Estuary and Gulf of St. Lawrence marine ecosystem overview and assessment report*. Canadian Technical Report on Fisheries and Aquatic Science, 2744.

EEA (2013). *European bathing water quality in 2012*. European Environment Agency, Copenhagen, Denmark.

EEAA (2003). *Marine Pollution in the Gulf of Aqaba and Gulf of Suez and its Effects on South Sinai: A Comprehensive Review*. Egyptian Environmental Affairs Agency, http://st-katherine.net/downloads/Marine%20Pollution.pdf.

Elshorbagy, W. (2005). *Overview of Marine Pollution in the Arabian Gulf, Keynote Address Arabian Gulf Conference*. Academia.edu, https://www.academia.edu/2595870/Overview_of_marine_pollution_in_the_Arabian_Gulf_with_emphasis_on_pollutant_transport_modeling.

Engler, R.E. (2012). The Complex Interaction between Marine Debris and Toxic Chemicals in the Ocean. *Environmental Science and Technology*, 2012, 46 (22), 12302–12315

EPA (1976). *List of Toxic pollutants (as amended from time to time) and Appendix A to Part 423-126 (List of Priority Pollutants)*. US Environmental Protection Agency, Code of Federal Regulations, §401.15.

EPA (2003). *Guide for industrial waste management*. US Environmental Protection Agency, Washington D.C, EPA-530-R-03-001.

EPA (2010). *Nonylphenol (NP) and Nonylphenol Ethoxylates (NPEs) Action Plan*. US Environmental Protection Agency, http://www.epa.gov/oppt/existingchemicals/pubs/actionplans/RIN2070-ZA09_NP-NPEs%20Action%20Plan_Final_2010-08-09.pdf.

EPA, 2012: *National Coastal Condition Report* IV, U.S. Environmental Protection Agency, Washington D.C., EPA-842-R-10-003.

EPA, USGS, FWS (2012). *Toxic Contaminants in the Chesapeake Bay and its Watershed: Extent and Severity of Occurrence and Potential Biological Effects*. US Environmental Protection Agency - Chesapeake Bay Program Office, Annapolis, United States.

EPA (2013). *Atrazine Up-date January 2013*. US Environmental Protection Agency, http://www.epa.gov/oppsrrd1/reregistration/atrazine/atrazine_update.htm.

Ericson, B., Caravanos, J., Chatham-Stephens, K., Landrigan, P. and Fuller, R. (2013). Approaches to systematic assessment if environmental exposures posed at hazardous waste sites in the developing world: The Toxics Sites Identification Program. *Environmental Monitoring and Assessment*, 185(2), 1755–1766. http://link.springer.com/article/10.1007%2Fs10661-012-2665-2#page-1.

ESNZ (2014). *Guidance Document: Nutrient Load Criteria to Limit Eutrophication in Three Typical New Zealand Estuary Types - ICOLL's, Tidal Lagoon, and Tidal River Estuaries*. Environment Southland New Zealand, http://www.es.govt.nz/media/26710/nutrient_criteria_nz_shallow_estuaries_leigh_copy.pdf.

EU (1996). *Council Directive on the control of major-accident hazards involving dangerous substances*. EU Council Directive 96/82/EC of 9 December 1996.

EU (2004). *European Commission Decision concerning the non-inclusion of atrazine in Annex I to Council Directive 91/414/EEC and the withdrawal of authorisations for plant protection products containing this active substance*. European Commission Decision 2004/248/EC of 10 March 2004.

EU (2006). *Regulation of the European Parliament and of the Council concerning the Registration, Evaluation, Authorisation and Restriction of Chemicals (REACH), establishing a European Chemicals Agency, amending Directive 1999/45/EC and repealing Council Regulation (EEC) No 793/93 and Commission Regulation (EC) No 1488/94 as well as Council Directive 76/769/EEC and Commission Directives 91/155/EEC, 93/67/EEC, 93/105/EC and 2000/21/EC*. EU Council and Parliament Regulation (EC) No 1907/2006 of 18 December 2006.

EU BREF (2001). *Reference Document on Best Available Techniques in the Pulp and Paper Industry*. European Integrated Pollution Prevention and Control Bureau, http://eippcb.jrc.ec.europa.eu/reference/BREF/ppm_bref_1201.pdf.

EU BREF (2013). *Best Available Techniques (BAT) Reference Document for the Production of Cement, Lime and Magnesium Oxide*. European IPC Bureau, Luxembourg.

European Cement Association (2014). *Key facts & figures*. European Cement Association, http://www.cembureau.be/about-cement/key-facts-figures.

Everaarts, J.M., Booij, K., Fischer, C.V., Maas, Y.E.M., Nieuwenhuize, J. (1999). Assessment of the environmental health of the Chagos Archipelago. In *Ecology of the Chagos Archipelago*, Sheppard, C.R.C., and Seaward, M.R.D., eds., Linnean Society, Westbury Academic and Scientific Publishing, West Yorkshire, United Kingdom, pp. 305-326.

Fernandez, A., Singh, A., and Jaffé, R. (2007). A literature review on trace metals and organic compounds of anthropogenic origin in the Wider Caribbean Region. *Marine Pollution Bulletin*, 54(11), 1681-91.

Fisher, N.S., Madigan, D.J., and Baumann, Z. (2013), Radioactive Cesium from Fukushima Japan Detected in Bluefin Tuna off California: Implications for Public Health and for Tracking Migration, *Proceedings of the 16th International Conference on Heavy Metals in the Environment*, Volume 1, 2013 (doi.org/10.1051/e3sconf/20130132001) Filoso, S., Martinelli, L.A., Howarth, R.W., Boyer, E.W., and Dentener, F. (2006). Human Activities Changing the Nitrogen Cycle in Brazil. *Biogeochemistry*, 71(1-2), 68-89.

Froescheis, O., Looser, R., Cailliet, G.M., Jarman, W.M., Ballschmiter, K. (2000) The deep-sea as a final global sink of semivolatile persistent organic pollutants? Part I: PCBs in surface and deep-sea dwelling fish of the North and South Atlantic and the Monterey Bay Canyon (California). *Chemosphere* 40: 651–660.

FSM (2004). *Federated States of Micronesia POPs Project Country Plan*. Secretariat of the Pacific Regional Environmental Programme, http://www.sprep.org/att/IRC/eCOPIES/Countries/FSM/29.pdf.

GAO (2005). United States Government Accountability Office. Report to Congressional Requesters, *Chesapeake Bay Program: Improved Strategies Are Needed to Better Assess, Report, and Manage Restoration Progress*, GAO-06-96.

GCLMEP (2003). *Guinea Current Large Marine Ecosystem, Transboundary Diagnostic Analysis*. Guinea Current Large Marine Ecosystem Project, http://gclme.iwlearn.org/publications/our-publications/transboundary-diagnostic-analysis/view.

GEF (1998). *Planning and Management of Heavily Contaminated Bays and Coastal Areas in the Wider Caribbean*. Report on Project RLA/93/G41, Global Environment Facility.

GESAMP (2001a). *A Sea of Troubles*. Joint Group of Experts on the Scientific Aspects of Marine Protection (GESAMP), Arendal.

GESAMP (2001b). *Protecting the oceans from land-based activities*. Joint Group of Experts on the Scientific Aspects of Marine Protection (GESAMP), Arendal.

GESAMP (2009). *Pollution in the Open Ocean: A Review of Assessments and Related Studies*. Joint Group of Experts on the Scientific Aspects of Marine Protection (GESAMP), UNEP and UNESCO-IOC.

Gerges, M.A. (2002). The Red Sea and Gulf of Aden Action Plan - Facing the challenges of an ocean gateway. *Ocean & Coastal Management*, 45 (11-12), 885–903.

Gnandi, K., Tchangbedji, T., Killi, K., Baba, G., and Abbe, K. (2006). Impact of Phosphate Mining on the Bioaccumulation of Heravy Metals in Marine Fish and Crustacea from the Coastal Zonbe of Togo. *Mine Water and the Environment*, 25(1), 56-62.

Green, N., Bjerkeng, B., Hylland, K., Ruus, A., Rygg, B. (2003). *Hazardous substances in the European marine environment: Trends in metals and persistent organic pollutants*. European Environment Agency, Copenhagen.

Haggarty, D.R., McCorquodale, B., Johannessen, D.I., Levings, C.D., and Ross, P.S. (2003). *Marine environmental quality in the central coast of British Columbia, Canada: A review of contaminant sources, types and risks*. Canadian Technical Reports on Fisheries and Aquatic Science, 2507, Sydney, Canada.

Hammond, A.L. (1972). Chemical Pollution: Polychlorinated Biphenyls. *Science*, 175 (4018), 155–156.

Hara A., Hirano, K., Shimizu, M., Fukada, H., Fujita, T., Itoh, F., Takada, H., Nakamura M., and Iguchi, T. (2007). Carp (Cyprinus caprio) vitellogenin: Characterization of yolk proteins, development of immunoassays and utility as a biomarker of exposure to environmental estrogens. *Environmental Sciences* 14: 95-108.

Heileman, S. (2008a). East Brazil Shelf Large Marine Ecosystem. In: K. Sherman, Hempel, G. (eds.), *The UNEP Large Marine Ecosystem Report: A perspective on changing conditions in LMEs of the world's Regional Sea*s. UNEP, Rep. Studi., 182: 711-722.

Heileman, S. (2008b). Guinea Current Large Marine Ecosystem. In: Sherman, K., Hempel, G. (eds.), *The UNEP Large Marine Ecosystem Report: A perspective on changing conditions in LMEs of the world's Regional Sea*s. UNEP Regional Seas Report and Studies, 182: 117-130.

Heileman, S. (2008c). North Brazil Shelf Large Marine Ecosystem. In: Sherman, K., Hempel G. (eds.), *The UNEP Large Marine Ecosystem Report: A perspective on changing conditions in LMEs of the world's Regional Sea*s. UNEP Regional Seas Report and Studies,182: 701-710.

Heileman, S. (2008d). Pacific Central-American Coastal Large Marine Ecosystem. In: Sherman, K., Hempel, G.(eds.), *The UNEP Large Marine Ecosystem Report: A perspective on changing conditions in LMEs of the world's Regional Sea*s. UNEP Regional Seas Report and Studies, 182: 735-746.

Heileman, S. (2008e). Patagonian Shelf Large Marine Ecosystem. In: Sherman, K., Hempel, G. (eds.), *The UNEP Large Marine Ecosystem Report: A perspective on changing conditions in LMEs of the world's Regional Sea*s. UNEP Regional Seas Report and Studies, 182: 735-746.

Heileman, S. and Belkin, I. (2008). Oyashio Current Large Marine Ecosystem. In: K. Sherman, Hempel, G. (eds.), *The UNEP Large Marine Ecosystem Report: A perspective on changing conditions in LMEs of the world's Regional Sea*s. UNEP Regional Seas Report and Studies, 182: 403-412.

Heileman, S., Bianchi, G. and Funge-Smith, S. (2008a). Bay of Bengal Large Marine Ecosystem. In: Sherman, K., Hempel, G. (eds.), *The UNEP Large Marine Ecosystem Report: A perspective on changing conditions in LMEs of the world's Regional Sea*s. UNEP Regional Seas Report and Studies, 182: 237-254.

Heileman, S., Eghtesadi-Araghi, P., and Mistafa, N. (2008b). Arabian Sea Large Marine Ecosystem. In: Sherman, K., Hempel, G., eds. *The UNEP Large Marine Ecosystem Report: A perspective on changing conditions in LMEs of the world's Regional Sea*s. UNEP Regional Seas Report and Studies, 182: 221-234.

Heileman, S., and M. Gasalla (2008). South Brazil Shelf Large Marine Ecosystem. In: K. Sherman, Hempel G., eds. *The UNEP Large Marine Ecosystem Report: A perspective on changing conditions in LMEs of the world's Regional Sea*s. UNEP Regional Seas Report and Studies, 182: 723-734.

Heileman, S., W. Parr, and G. Volovik (2008c). Black Sea Large Marine Ecosystem. In: K. Sherman, Hempel G., eds. *The UNEP Large Marine Ecosystem Report: A perspective on changing conditions in LMEs of the world's Regional Sea*s. UNEP Regional Seas Report and Studies,182: 203-218.

Heileman, S., and L. E. P. Scott (2008). Somali Current Large Marine Ecosystem. In: K. Sherman, Hempel G., eds. *The UNEP Large Marine Ecosystem Report: A perspective on changing conditions in LMEs of the world's Regional Sea*s. UNEP Regional Seas Report and Studies, 182: 159-172.

Heisler, J., Glibert, P.M., Burkholder, J.M., Anderson, D.M., Cochlane, W., Dennison, W.C., Dortch, Q., Gobler, C.J., Heil, C.A., Humphries, E., Lewitus, A., Magnien, R., Marshall, H.G., Sellner, K., Stockwell, D.A., Stoecker, D.K., and Suddleson, M. (2008). *Eutrophication and harmful algal blooms: A scientific consensus*. Harmful Algae, Vol. 8(1), 3–13.

HELCOM (2009). *Radioactivity in the Baltic Sea, 1999-2006: HELCOM thematic assessment*. Baltic Marine Environment Protection Commission, Baltic Sea Environment Proceedings No. 117, Helsinki, Finland.

HELCOM (2010a). *Ecosystem Health of the Baltic Sea 2003–2007: HELCOM Initial Holistic Assessment*. Baltic Marine Environment Protection Commission, Helsinki, Finland.

HELCOM (2010b). *Hazardous substances in the Baltic Sea: An integrated thematic assessment of hazardous substances in the Baltic Sea*. Baltic Marine Environment Protection Commission, Helsinki, Finland.

HELCOM, 2013. Thematic assessment of long-term changes in radioactivity in the Baltic Sea, 2007-2010. Baltic Sea Environment Proceedings, No. 135.

Helly, J.J. and Levin, L.A. (2004). Global Distribution of Naturally Occurring Marine Hypoxia on Continental Margins. *Deep-Sea Research Part I*, 51(9), 1159-1168.

Hicks, D.M., Shankar, U., McKerchar, A.I., Basher, L., Lynn, I., Page, M., Jessen, M. (2011). Suspended sediment yields from New Zealand rivers, *Journal of Hydrology (New Zealand)*, 50 (1), 81-142.

Hong Kong, China (2013). *Marine Water Quality in Hong Kong in 2012*. Government of the China, Hong Kong Special Administrative Region, http://www.epd.gov.hk/epd/english/environmentinhk/water/marine_quality/files/Report2012eng.pdf.

IAEA (2006). *Environmental consequences of the Chernobyl accident and their remediation : twenty years of experience: a report of the Chernobyl Forum Expert Group 'Environment'*. International Atomic Energy Agenc, Vienna, Austria.

IAEA (2013a). Events and highlights on the progress related to recovery operations at Fukushima Daiichi Nuclear Power Station, International Atomic Energy Agency, http://www.iaea.org/newscenter/news/2014/eventshighlightsjune2014.pdf.

IAEA (2013b). Power Reactor Information System (PRIS), data for 2013, International Atomic Energy Agency, http://www.iaea.org/PRIS/WorldStatistics/NuclearShareofElectricityGeneration.aspx.

IAEA (2013c). Radiation protection and management of NORM residues in the phosphate industry. International Atomic Energy Agency, Vienna, Austria.

IAEA (2014). IAEA International Peer Review Mission on Mid-And-Long-Term Roadmap towards the Decommissioning Of Tepco's Fukushima Daiichi Nuclear Power Station Units 1-4, http://www.iaea.org/newscenter/focus/fukushima/final_report120214.pdf.

IAI (2013). *Global Life Cycle Inventory Data for the Primary Aluminium Industry*. International Aluminium Institute, London, UK.

IBGA (2008). *Pesquisa Nacional de Saneamento Básica*. Instituto Brasileiro de Geografia e Estatística, Rio de Janeiro, Brazil.

ICMM (2012). *The role of mining in national economies*. International Council on Mining and Metals, https://www.icmm.com/document/4440.

ICRP (1977). 1977 Recommendations of the International Commission on Radiological Protection (ICRP Publication 26). *Annals of the International Commission on Radiological Protection*, 1(3).

ICRP (2007). 2007 Recommendations of the International Commission on Radiological Protection (ICRP Publication 103). *Annals of the International Commission on Radiological Protection*, 37(2-4).

ICRP (2009). Environmental protection: transfer parameters for reference animals and plants. (ICRP Publication 114). *Annals of the International Commission for Radiological Protection*, 39(6).

ICRP (2014). Protection of the Environment under Different Exposure Situations (ICRP Publication 124). *Annals of the International Commission for Radiological Protection*, 43(1).

IJC (1993). *Seventh Biennial Report on Great Lakes Water Quality*. International Joint Commission, Washington D.C. and Ottawa.

ILO (2007). *Background information for developing an ILO policy framework for hazardous substances*. International Labour Organization, Geneva, Switzerland.

ILZG (2014). *End Uses if Lead*. International Zinc and Lead Study Group, http://www.ilzsg.org/static/enduses.aspx?from=2.

IMO (1972). Convention on the prevention of marine pollution by dumping of wastes and other matter. London, 13 November 1972, United Nations, *Treaty Series*, vol. 1046, No.15749.

IMO (2012). *International Assessment of Marine and Riverine Disposal of Mine Tailings*. International Maritime Organization, London, UK.

ISP (2010). *Informe Programa De Vigilancia De Fenómenos Algales Nocivos (Fan) En Chile 2009*. Instituto de Salud Publica de Chile (http://www.ispch.cl/informes-programa-marea-roja-2002-2009).

Janil, H.J., Kurup, S., Trivedi, R., Chatrabhuji, P.M., and Bhatt, P.N. (2011). Seawater quality and trends in heavy metal distribution in marine sediment along Alang Sosiya Ship Breaking Yard (ASSBY) region. *Analytical Chemistry*, 10, 374-380.

Janowicz, M. and D. Tremblay, eds (2006). *Implementing Canada's National Programme of Action for the Protection of the Marine Environment from Land-Based Activities: Second Atlantic Team Status Report 2004-2006*. Environment Canada, Dartmouth, NS.

Japan MOE (2009). *Present Status of Marine Pollution in the Sea around Japan*. Japanese Ministry of the Environment, Tokyo.

Japan (2011). *Additional Report of the Japanese Government to the IAEA - The Accident at TEPCO's Fukushima Nuclear Power Stations - (Second Report) (Summary) September 2011*. International Atomic Energy Agency, http://www.iaea.org/newscenter/focus/fukushima/japan-report2/japanreport120911.pdf.

JSWA (2014). Japan Sewage Works Association, *Making Great Breakthroughs: All About the Sewage Works in Japan* (http://www.jswa.jp/en/jswa-en/allabout.html accessed 20 October 2014).

Kachur, A.N., and Tkalin, A.V. (2000). Sea of Japan. In: C. Sheppard (ed), *Seas at the Millennium*. Elsevier, Oxford, UK.

Kakuschke, A. and Prange, A. (2007). The Influence of Metal Pollution on the Immune System – A Potential Stressor for Marine Mammals in the North Sea. *Int. J. Comp. Psy.*, 20, 179-193.

Kemp, W.M., Boynton, W.R., Adolf, J., Boesch, D., Boicourt, W., Brush, G., Cornwell, J., Fisher, T., Glibert, P., Hagy, J., Harding, L., Houde, E., Kimmel, D., Miller, W.D., Newell, R.I.E., Roman, M., Smith, E., and Stevenson, J.C. (2005). Eutrophication of Chesapeake Bay: Historical trends and ecological interactions. *Marine Ecology Progress Series*, 303, 1–29.

Kimbrough, K. L., Johnson, W.E., Lauenstein, G.G., Christensen, J.D., and Apeti, D.A. (2008). *An Assessment of Two Decades of Contaminant Monitoring in the Nation's Coastal Zone*. United States National Ocean and Atmospheric Administration Technical Memorandum NOS NCCOS 74, Silver Spring, Maryland.

Le Monde 16 April 2013: *Guadeloupe : monstre chimique*.

Leland, H.V., Luoma, S.N. and Wilkes, D.J. (1977). Heavy Metals and Related Trace Elements. *Journal (Water Pollution Control Federation)*, 49(6), 1340-1369.

Lewis, S.E., Brodie, J.E. Bainbridge, Z.T., Rohde, K.W., Davis, A.M., Masters, B.L., Maughan, M, Devlin, M.J., Mueller, J.F. and Schaffelke, B. (2009). Herbicides: A new threat to the Great Barrier Reef. *Environmental Pollution* 157(8-9), 2470-84.

Ma, Y., Xie, Z., Halsall, C., Möller, A., Yang, H., Zhong, H., Cai, G. (2015). The spatial distribution of organochlorine pesticides and halogenated flame retardants in the surface sediments of an Arctic fjord: The influence of ocean currents vs. glacial runoff, *Chemosphere*, Vol. 119, 953–960.

Macdonald, R.W. and Bewers, J.M. (1996). Contaminants in the arctic marine environment: priorities for protection. *ICES Journal of Marine Science*, 53(3), 537–563.

Marchovecchio, J.E. (2009). *Land Sources of Pollutants Impacting Coastal Marine Ecosystems From Argentina*. Instituto do Milenio, http://www.institutomilenioestuarios.com.br/pdfs/Paticipacao_Eventos/10_CBO2008/apresentacoes/04_J_E_Marcovecchio_PollutionArgentina.pdf.

MARDOS (1995). *Sources of radioactivity in the marine environment and their relative contributions to overall dose assessment from marine radioactivity (MARDOS)*, International Atomic Energy Agency, IAEA-TECDOC-838, Vienna, Austria.

Metcalf & Eddy (2004). Tchobanoglous, G., Burton, F.L., and Stensel, H.D. (eds), *Wastewater engineering: Treatment and reuse* (4th ed.). Boston: McGraw-Hill.

Minamata Convention on Mercury of 10 October 2013. http://treaties.un.org/doc/Treaties/2013/10/20131010%2011-16%20AM/CTC-XXVII-. 17.pdf.

Mormede, S., and Davies, I.M., (2003) Horizontal and vertical distribution of organic contaminants in deep-sea fish species. *Chemosphere* 50: 563–574.

Mussel Watch (2011). *Mussel Watch Program 2011*. National Centers for Coastal Ocean Science, http://ccma.nos.noaa.gov/about/coast/nsandt/download.aspx.

Naqvi, S.W.A., Naik, H., Jayakumar, A., Pratihary, A.K., Narvenkar, G., Kurian, S., Agnihotri, R., Shailaja, M.S. and Narvekar, P.V, (2009). Seasonal Anoxia Over the

Western Indian Continental Shelf. *Geophysical Monograph Series,* 185.

NDA (2014). *Annual Report and Accounts 2013-2014.* United Kingdom Nuclear Decommissioning Authority, House of Commons Paper 187 of Session 2013/2014.

NEA (2013). *Nuclear Energy Data/Données sur l'énergie nucléaire 2013.* Nuclear Energy Agency of the Organization for Economic Cooperation and Development, NEA publication 7162, Paris, France.

NGN (2014). National Geographic News 4 March 2014. National Geographic, http://news.nationalgeographic.com/news/2014/03/140304-antarctica-research-toxic-adelie-penguins-mcmurdo-station-science/#close-modal.

Niencheski, L.F.H., Baraj, B., Windom, H.L. and Franca, R.G. (2006). Natural background assessment and its anthropogenic contamination of Cd, Pb, Cu, Cr, Zn, Al and Fe in the sediments of the Southern area of Patos Lagoon. *Journal of Coastal Research,* SI39(II),1040-1043.

NIOT (2014). *National Report for the Workshop, under the Auspices of the United Nations, in Support of the Regular Process for Global Reporting and Assessment of the State of the Marine Environment, Including Socio-economic Aspects, Chennai.* DOALOS, http://www.un.org/depts/los/global_reporting/Chennai_2013/National%20Report.pdf).

NOAA (2013). (National Oceanic and Atmospheric Administration). NOAA-supported scientists find large Gulf dead zone, but smaller than predicted. http://www.noaanews.noaa.gov/stories2013/2013029_deadzone.html.

NOWPAP (2007). *State of the Marine Environment in the NOWPAP Region.* UNEP/NOWPAP.

NRC (National Research Council, Marine Board and Ocean Studies Board) (2003). *Oil in the Sea III: Inputs, Fates, and Effects.* National Academies Press, Washington, DC.

NZMOE (1999). *Ambient concentrations of selected organochlorines in estuaries* (Revised edition), New Zealand Ministry of the Environment, Wellington, New Zealand.

Oanh, N.T.K., Reutergårdh, L.B., Dung, N.T. (1999). Emission of Polycyclic Aromatic Hydrocarbons and Particulate Matter from Domestic Combustion of Selected Fuels. *Environmental Science and Technology,* 33(16), 2703-2709.

OECD (2012). *Information on OECD Work Related to Endocrine Disrupters.* Organization for Economic Cooperation and Development, http://www.oecd.org/chemicalsafety/testing/50067203.pdf.

Oehlmann J., Schulte-Oehlmann, U, . Kloas, W. Jagnytsch, O., Lutz, I, Kusk, K.O., . Wollenberger, L., Santos, E,M., Paull, G.C., Van Look, K.J.W. and Tyler, C.R. (2009). A critical analysis of the biological impacts of plasticizers on wildlife. *Philosophical Transactions of the Royal Society B: Biological Sciences* 364.1526, 2047-2062.

Orlando E. F., Kolok, A.S., Binzcik, G.A., Gates, J.L., Horton, M.K. and Lambright, C.S. (2004). Endocrine-disrupting effects of cattle feedlot effluent on an aquatic sentinel species, the fathead minnow. *Environmental Health Perspectives* 112, 353–358.

Orth, R. J., Carruthers, T.J.B., Dennison, W.C., Duarte, C.M., Fourqurean, J.W., Heck Jr., K.L., Hughes, A.R., Kendrick, G.A., Kenworthy, W.J., Olyarnik, S., Short, F.T., Waycott, M. and Williams, S.L. (2006). A Global Crisis for Seagrass Ecosystems. *BioScience* Volume 56 (12), 987-996.

Oslo Convention 1972 – Convention for the Prevention of Marine Pollution by Dumping from Ships and Aircraft. Oslo, 15 February 1972, United Nations, *Treaty Series,* vol. 932 No. 13269.

OSPAR (1998). *Ministerial Meeting of the OSPAR Commission: Sintra, 22-23 July 1998 - The Main Results.* OSPAR Commission, London, UK.

OSPAR (2000). *Quality Status Report 2000.* OSPAR Commission, London, UK.

OSPAR (2003). *Survey of the use of effect related methods to assess and monitor wastewater discharges - Testing of endocrine effects.* OSPAR Commission, London, UK.

OSPAR (2007). *Second Periodic Evaluation of progress towards the objective of the Radioactive Substance Strategy 2007.* OSPAR Commission, London, UK.

OSPAR (2009). *Background Document on Nonyl Phenyl Ethoxylates.* OSPAR Commission, London, UK.

OSPAR (2010). *Quality Status Report 2010.* OSPAR Commission, London, UK.

OSPAR (2013). *Liquid discharges from nuclear installations 2011.* OSPAR Commission, London, UK.

PAC2. (2014). *Programma de Açeleracão de Crescimento 2.* Ministerio do Planejamento, http://www.pac.gov.br/cidade-melhor/saneamento.

Pacchioli, D. (2013). Radioisotopes in the Ocean. *Oceanus,* 50(1).

Pagliosa, P.R., Fonseca, A., Barbosa, F.A.R., Braga, E. (2006). Urbanization impact on Subtropical Estuaries: a Comparative Study of Water Properties in Urban Areas and in Protected Areas. *Journal of Coastal Research,* II, 731-735.

PERSGA (2006). *State of the Marine Environment Report for the Red Sea and Gulf of Aden.* Regional Organization for the Conservation of the Environment of the Red Sea and Gulf of Aden, Jeddah, Saudi Arabia.

PICES (2010). *Marine Ecosystems of the North Pacific Ocean, 2003-2008.* North Pacific Marine Science Organisation (PICES) Special Publication 4, https://www.pices.int/publications/special_publications/NPESR/2010/NPESR_2010.aspx.

Rabalais, N.N., and Turner, R.E., eds. (2001). *Coastal Hypoxia:Consequences for Living Resources and Ecosystems.* American Geophysical Union, Washington D.C.

Rabbaniha, M., Ghasemzadeh, J., Owfi, F. (2013). Spatial and Temporal Patterns of Fish Larvae Assemblage in the Northern Coastal Waters of Persian Gulf along the Bushehr Province Shoreline. *Journal of Fisheries Science,* 7.2 (2013), 141-151.

Roberts, D.A., Johnston, E.L., Knott, N.A. (2010). Impacts of desalination plant discharges on the marine environment: A critical review of published studies. *Water Research,* 44(18), 5117-5128.

Romero Deras, M.A.R. (2013). *Proyecto de Desarrollo de Capacidades para el Uso Seguro de Aguas Servidas en Agricultura (FAO, WHO, UNEP, UNU-INWEH, UNW-DPC, IWMI e ICID). Producción de aguas servidas, tratamiento y uso en El Salvado.* UNWater, http://www.ais.unwater.org/ais/pluginfile.php/378/mod_page/content/144/EL_SALVADOR.pdf accessed 24 April 2014.

Routledge, E.J., Sheahan, D., Desbrow, C., Brighty, G.C., Waldock, M., Sumpter, J.P. (1998). Identification of estrogenic chemicals in STW effluent. 2. In vivo responses in fish and roach. *Environmental Science and Technology,* 32, 1559-1565.

RWQPP (2013). *Reef Water Quality Protection Plan, 2013 Scientific Consensus Statement.* Government of Australia and Government of Queensland, http://www.reefplan.qld.gov.au/about/assets/scientific-consensus-statement-2013.pdf.

Sale, P.F., Feary, D.A., Burt, J.A., Bauman, A.G., Cavalcante, G.H., Drouillard, K.G., Kjerfve, B., Marquis, E., Trick, C.G., Usseglio, P., Van Lavieren, H. (2011). The Growing Need for Sustainable Ecological Management of Marine Communities of the Persian Gulf. *Ambio,* 40(1), 4-17.

Samoa (2004). Government of Samoa, POPS assessment for Samoa. Secretariat of the Pacific Regional Environmental Programme, http://www.sprep.org/att/IRC/eCOPIES/Countries/Samoa/4.pdf.

Satpathy, K.K., Panigrahi, S., Mohanty, A.K., Sahu, G., Achary, M.S., Bramha, S.N., Padhi, R.K., Samantara, M.K., Selvanayagam, M. and Sarkar, S.K. (2013). Severe oxygen depletion in the shallow regions of the Bay of Bengal off Tamil Nadu coast. *Current Science* India, 104(11), 1467.

Schindler, D.W., Dillon, P.J. and Schreier, H. (2006). A Review of Anthropogenic Sources of Nitrogen and Their Effects on Canadian Aquatic Ecosystems. *Biogeochemistry*, 79, 25-44.

SE2011 Committee (2011). *Australia state of the environment 2011: Independent report to the Australian Government Minister for Sustainability, Environment, Water, Population and Communities*. DSEWPaC, Canberra, Australia, http://www.environment.gov.au/science/soe/2011-report/download.

SOA (2010). *Marine Environment Quality Communiqué of China (1989–2009)*. State Oceanic Administration of China, http://www.mep.gov.cn/plan/zkgb.2010.8.13.

Stemmler, I. and Lammel, G. (2010). Pathways of PFOA to the Arctic: variabilities and contributions of oceanic currents and atmospheric transport and chemistry sources, *Atmospheric Chemistry and Physics*, 10.

Stockholm Convention on Persistent Organic Pollutants, United Nations *Treaty Series*, vol. 2256, No. 40214.

Sunderland, E.M., Krabbenhoft, D.P., Moreau, J.W., Strode, S.A., Landing, W.M. (2009) Mercury sources, distribution, and bioavailability in the North Pacific Ocean: Insights from data and models. *Global Biogeochemical Cycles* 23(2): GB2010.

Tanabe, S., Ramu, K., Isobe, T. and Takahashi, S. (2008). Brominated flame retardants in the environment of Asia-Pacific: an overview of spatial and temporal trends. *Journal of Environmental Monitoring*, 10, 188-197.

Machado Torres, J.P., Munschy, C., Héas-Moisan, K., Potier, M., Ménard F. and Bodin, N. (2009). *Organohalogen Compounds in Yellowfin Tuna (Thunnus Albacares) from the Western Indian Ocean*. IFREMER, http://archimer.ifremer.fr/doc/00077/18866/16445.pdf accessed 20 April 2014.

Turner, R. E., Rabalais, N.N. and Justic, D. (2008). Gulf of Mexico Hypoxia: Alternate States and a Legacy. *Environmental Science and Technology* 42, 2323–2327.

Tyler C. R. and Jobling, S. (2008). Roach, sex, and gender-bending chemicals: The feminization of wild fish in English rivers. *Bioscience* 58 (11), 1051-1059.

Ueno, D., N. Kajiwara, N. Tanaka, H., Subramanian, A. Fillmann, G., Lam, P.K.S., Zheng,G.J. Muchitar, M. Razak, M., Prudente, M. Chung, K-H. and Tanabe, S. (2004). Global Pollution Monitoring of Polybrominated Diphenyl Ethers Using Skipjack Tuna as a Bioindicator. *Environmental Science and Technology*, 38(8), 2312-2316.

UNDP (2011). *Historic deal to safeguard Yellow Sea*. UNDP, http://www.undp.org/content/undp/en/home/presscenter/articles/2011/06/07/historic-deal-to-safeguard-yellow-sea-is-made.html.

UNEP (1995). *Global Programme of Actions for thwe Protection of the Marine Environment against Land-Based Activities*. UNEP Document UNEP(OCA)/LBA/IG.2/7.

UNEP (2008). *The Global Atmospheric Mercury Assessment: Sources, Emissions and Transport*. UNEP, Nairobi, Kenya.

UNEP (2010). *Final review of scientific information on lead*. UNEP, Nairobi, Kenya.

UNEP (2012). *State of the Mediterranean Marine and Coastal Environment*. UNEP/MAP – Barcelona Convention, Athens.

UNEP (2013a). *The Global Atmospheric Mercury Assessment: Sources, Emissions, Releases and Environmental Transport*. UNEP, Geneva.

UNEP (2013b). *Up-dated Global Inventory of Mercury-Cell Chlor-Alkali Facilities*. UNEP, Nairobi, Kenya.

UNEP/COBSEA (2010). *State of the Marine Environment Report for the East Asian Seas 2009*. COBSEA Secretariat, Bangkok, Thailand.

UNEP-UCR/CEP, (2010). *Domestic and Industrial Pollutant Loads and Watershed Inflows in the Wider Caribbean Region: Updated CEP Technical Report No. 33 Land-based Sources and Activities in the Wider Caribbean Region*. UNEP, Kingston, Jamaica, http://www.cep.unep.org/publications-and-resources/technical-reports/technical-reports.

UNESCO (1968). *Inter-Governmental Conference of Experts on the Scientific Bases for the Rational Utilization and Conservation of Biospheric Resources*. UNESCO, Paris, France.

UNSCEAR (2008). *Sources and Effects of Ionizing Radiation – Report to the General Assembly*. United Nations Scientific Committee on the Effects of Atomic Radiation, New York.

UNSCEAR (2013). *Sources and Effects of Ionizing Radiation – Report to the General Assembly*. United Nations Scientific Committee on the Effects of Atomic Radiation, New York.

USGS (2000). *An Overview of Mining-Related Environmental and Human Health Issues, Marinduque Island, Philippines: Observations from a Joint U.S. Geological Survey*. Armed Forces Institute of Pathology Reconnaissance Field Evaluation, May 12-19, 2000 (U. S. Geological Survey Open-File Report 00-397).

USGS (2013). *Titanium and Titanium Dioxide*. USGS, http://minerals.usgs.gov/minerals/pubs/commodity/titanium/mcs-2013-titan.pdf.

Vethaak, D. and Meer, J.V.D. (1991). Fish Disease Monitoring in the Dutch Part of the North Sea in Relation to the Dumping of Waste from Titanium Dioxide Production. *Chemistry and Ecology*, 5(3), 149-170.

Victoria (2013). *Victoria: State of the Environment – Science, Policy, People*. Victoria Commissioner for Environmental Sustainability, http://www.ces.vic.gov.au/__data/assets/pdf_file/0019/230770/Introduction.pdf.

Vinayachandran, P.N. and Mathew, S. (2003). Phytoplankton bloom in the Bay of Bengal during the northeast monsoon and its intensification by cyclones. *Geophysical Researearch Letters*, 30(11), 1572.

Wang, B., Xie, L., and Sun, X. (2011). Water quality in marginal seas off China in the last two decades. *International Journal of Oceanography*, Jan2011, 1-6.

Watanuki, Y., Yamashita, A., Ishizuka, M., Ikenaka, Y., Nakayama, S.M.M., Ishii, C., Yamamoto, T., Ito, M., Kuwae, T. and Trathan, P.N. *Feathers of tracked seabirds reveal a spatial pattern of marine pollution*. PICES, http://www.pices.int/publications/presentations/PICES-2013/2013-S3/S3-0955-Watanuki.pdf.

Waters, P. (2011). White Hiding Pigments, *Brushstrokes,* May 2011 (http://www.scanz.org.nz/pm/may-white-hiding-pigments accessed 13 April 2015).

Weiss, J., Päpke, O., and Bergman, Å. (2005). A Worldwide Survey of Polychlorinated Dibenzo-p-dioxins, Dibenzofurans, and Related Contaminants in Butter. *Ambio*, 34(8), 589-597.

WISE (2014). *Chronology of major tailings dam failures*. World Information System on Energy, http://www.wise-uranium.org/mdaf.html.

WNA (2013). *Information Library*. World Nuclear Association, http://www.world-nuclear.org/info/countryprofiles.

WOMARS (2005). *Worldwide marine radioactivity studies (WOMARS). Radionuclide levels in oceans and seas*.International Atomic Energy Agency, Vienna, Austria.

World Bank (2009). *World Bank, Project Appraisal Document on a Proposed Adaptable Program Loan in the Amount Of US$840 Million to the Argentine Republic for the Matanza-Riachuelo Basin Sustainable Development Project Phase 1 (Apli) in support of the First Phase of the Matanza-Riachuelfo Basin Sustainable Development Program*. World Bank, http://www-wds.worldbank.org/external/default/WDSContentServer/WDSP/IB/2009/05/19/000333037_20090519030356/Rendered/PDF/484430PAD0P105101Official0Use0Only1.pdf.

World Bank (2013). *Catalina Marulanda, Implementation Status & Results – Argentina – Matanza-Riachuelo Basin (MRB) Sustainable Development Adaptable Lending Program (P105680)*. World Bank, http://www-wds.worldbank.org/external/default/WDSContentServer/WDSP/LCR/2013/11/25/090224b0820beb2a/1_0/Rendered/PDF/Argentina000Ma0Report000Sequence008.pdf.

World Bank (2014). *World Bank Data*. World Bank, http://data.worldbank.org/indicator/EG.ELC.COAL.ZS.

Worth, R.H. (1953). *Worth's Dartmoor*, David & Charles, Newton Abbot 1967.

Wurl, O., Potter, J.R., Obbard, J.P. and Durville, C. (2006). Persistent organic pollutants in the equatorial atmosphere over the open Indian Ocean. *Environmental Science and Technology*, 40(5), 1454-1461.

Yuan, S., Gou, N., Akram, N., Alshawabkeh, A., Gu, Z. (2013). Efficient degradation of contaminants of emerging concerns by a new electro-Fenton process with Ti/MMO cathode, *Chemosphere* 93(11), 2796-2804.

Zhuang Z., Ding, L., and Li, H. (2005). *China's Pulp and Paper Industry: A Review*. Georgia Tech, http://www.cpbis.gatech.edu/files/papers/CPBIS-FR-08-03%20Zhuang_Ding_Li%20FinalReport-China_Pulp_and_Paper_Industry.pdf

Chapter 20 — Coastal, Riverine and Atmospheric Inputs from Land

Appendix. World Paper Production 2001 - 2012

Regions	2001	2002	2003	2004	2005	2006	2007	2008	2009	2010	2011	2012
Africa	3,655,820	3,706,273	3,765,252	4,104,096	3,778,037	3,932,437	4,147,837	4,139,737	3,507,477	3,815,477	3,475,477	3,721,963
Asia	102,951,021	102,951,021	109,526,900	117,813,100	124,423,962	138,241,795	148,491,773	154,706,283	158,414,796	170,121,999	176,988,475	180,801,105
of which												
China	32,140,000	37,800,000	43,000,000	49,500,000	56,000,000	65,000,000	73,500,000	79,800,000	86,400,000	92,700,000	99,300,000	102,500,000
Japan	30,717,000	30,686,000	30,457,000	30,891,000	30,953,000	31,097,000	31,268,000	30,628,000	26,268,000	27,364,000	26,609,000	26,370,000
Republic of Korea	9,332,000	9,812,000	10,148,000	10,511,000	10,549,000	10,703,000	10,932,000	9,890,000	9,726,000	11,022,000	11,368,000	11,368,000
Latin America and the Caribbean	15,105,539	15,519,009	16,152,297	17,500,580	18,534,937	18,635,566	18,783,352	19,208,364	19,385,196	19,773,267	20,176,139	20,176,139
of which												
Brazil	7,354,000	7,661,000	7,811,000	8,221,000	8,597,000	8,738,000	9,008,000	9,154,000	9,428,000	9,844,000	10,150,000	10,213,000
Europe	93,411,157	96,188,052	98,480,020	102,994,612	103,737,147	107,161,246	107,396,340	103,715,241	93,396,655	100,368,756	99,757,224	97,708,687
of which												
Finland	12,502,000	12,789,000	13,058,000	14,036,000	12,391,140	14,189,360	14,334,000	13,126,000	10,602,000	11,758,000	11,329,000	10,696,000
France	9,625,000	9,809,000	9,939,000	10,255,000	10,332,000	10,006,200	9,870,500	9,404,000	8,331,500	8,829,800	8,527,213	8,099,583
Germany	17,879,000	18,526,000	19,310,000	20,391,000	21,679,000	22,656,000	23,317,000	22,828,000	20,870,000	23,072,000	22,706,000	22,630,000
Sweden	10,534,000	10,724,000	11,061,600	11,589,000	11,775,000	12,066,000	11,511,100	11,663,000	10,932,000	11,410,000	11,298,000	11,416,000
Canada	19,834,000	20,073,000	19,964,000	20,462,000	19,498,000	18,189,000	17,367,000	15,789,000	12,823,000	12,755,000	12,057,000	10,755,000
Russian Federation	5,624,800	5,978,000	6,377,000	6,830,000	7,126,000	7,434,000	7,581,000	7,700,000	7,373,000	5,606,000	7,549,120	7,661,145
United States of America	81,248,828	81,879,072	80,712,166	82,084,368	83,697,335	84,316,937	83,915,965	80,178,382	71,355,455	77,689,080	76,430,822	75,533,121
Oceania	3,511,000	3,519,000	3,900,000	4,014,000	4,195,000	4,162,000	4,064,000	4,152,000	4,148,000	4,100,000	3,955,000	4,066,825
	325,342,165	329,813,427	338,877,635	355,802,756	364,990,418	382,072,981	391,747,267	389,589,007	370,403,579	394,229,579	400,389,257	400,423,985

NOTE: These figures are taken from the FAOSTAT Forest statistics (http://faostat.fao.org/DesktopDefault.aspx?PageID=626&lang=en#ancor). They show the sum for each area of the figures for production of newsprint, printing and writing paper other paper and paperboard. They were accessed on 8 and 9 January 2014.

Value of Chemical Output by Country/Region - billions of US $

At current prices

Country	2003	2004	2005	2006	2007	2008	2009	2010	2011	2012
United States	487.7	540.9	610.9	657.7	716.2	738.7	624.4	697.8	776.8	769.4
Canada	32.7	38.2	40.2	44.0	44.9	46.5	35.1	41.8	56.7	46.5
Mexico	36.7	41.5	48.3	51.5	54.8	61.0	47.7	52.9	58.3	60.2
North America	**557.1**	**620.5**	**699.3**	**753.2**	**815.8**	**846.2**	**707.2**	**792.5**	**891.7**	**876.2**
Brazil	45.5	60.3	72.3	82.6	103.5	112.2	100.9	128.5	157.3	153.0
Brazil Other	69.2	76.0	83.6	95.7	103.8	116.8	113.5	120.0	138.0	141.0
Latin America	**114.8**	**136.3**	**160.8**	**178.3**	**207.3**	**229.0**	**214.4**	**248.5**	**295.3**	**293.9**
France	99.5	111.6	116.0	123.4	139.6	157.2	129.2	131.6	148.0	140.0
Germany	148.0	167.8	175.9	190.4	225.0	247.3	204.9	225.3	259.9	237.4
Italy	77.6	88.0	90.0	96.3	106.9	113.9	96.2	98.6	107.3	99.4
United Kingdom	77.9	87.0	92.3	104.7	116.9	116.0	85.7	86.8	95.8	92.4
Belgium	35.8	40.3	41.8	45.5	51.7	60.0	46.2	51.9	61.6	54.8
Ireland	34.6	36.0	36.1	34.9	44.5	49.1	52.0	59.2	66.0	62.4
The Netherlands	42.9	53.5	58.0	63.6	73.4	80.7	61.1	71.3	85.0	81.1
Spain	42.5	49.2	52.6	56.9	66.4	73.9	64.1	68.3	78.1	73.2
Sweden	15.6	17.4	18.4	20.2	21.6	22.0	18.4	20.2	24.8	25.4
Switzerland	30.8	35.1	37.7	41.6	48.3	55.7	57.9	65.4	76.6	74.6
Other	66.6	78.3	91.3	102.1	115.4	143.2	120.9	115.4	154.8	154.8
Europe	**671.8**	**764.2**	**810.1**	**879.6**	**1,009.7**	**1,119.0**	**936.6**	**994.0**	**1,157.9**	**1,095.5**
Russia	30.2	34.4	40.6	47.3	55.6	69.5	50.8	66.2	78.8	76.1
Africa and Middle East	71.3	85.4	95.4	100.9	115.2	142.6	126.6	155.5	183.9	183.8
Japan	216.9	241.7	249.5	249.5	258.3	299.5	277.3	323.1	368.0	357.3
China	168.2	218.8	289.4	364.1	494.2	667.0	746.5	978.7	1,303.3	1,431.9
India	38.8	47.1	55.3	62.6	78.1	92.5	92.4	115.5	134.5	130.8
Australia	14.0	15.7	16.9	17.3	20.8	22.7	22.7	28.7	33.9	35.7
Rep of Korea	61.9	75.9	89.1	100.7	113.9	122.5	108.5	138.5	169.4	176.3
Singapore	16.3	23.3	25.5	33.1	36.6	37.8	31.7	43.9	58.0	61.4
Chinese province of Taiwan	36.9	49.1	54.7	59.9	70.8	72.5	59.2	81.0	89.8	90.5
Other	46.2	54.9	62.4	76.4	93.1	108.3	99.3	123.6	156.4	163.7
Asia-Pacific	**599.1**	**726.5**	**842.8**	**963.5**	**1,165.8**	**1,422.8**	**1,437.8**	**1,833.0**	**2,313.2**	**2,447.4**
Global Total	**2,044.2**	**2,370.1**	**2,648.9**	**2,922.6**	**3,373.5**	**3,829.1**	**3,462.6**	**4,107.5**	**4,920.9**	**4,973.2**

21 Offshore Hydrocarbon Industries

Contributors:
Peter Harris (Convenor and Lead Member), Babajide Alo, Arsonina Bera, Marita Bradshaw, Bernard Coakley, Bjorn Einar Grosvik, Nuno Lourenço, Julian Renya Moreno, Mark Shrimpton, Alan Simcock (Co-Lead Member) and Asha Singh

Commentators:
Ana Paula Falcão, Jim Kelley and Nathan Young

Chapter 21

Offshore Hydrocarbon Industries

1 Scale and significance of the offshore hydrocarbon industries and their social and economic benefits.[1]

1.1 Location of offshore exploration and production activities

Offshore oil and gas exploration and development is focused in specific geographic areas where important oil fields have been discovered. Notable offshore fields are found in: the Gulf of Mexico (Fig. 1); the North Sea (Fig. 2); California (in the Santa Barbara basin); the Campos and Santos Basins off the coast of Brazil; Nova Scotia and Newfoundland in Atlantic Canada; West Africa, mainly west of Nigeria and Angola; the Gulf of Thailand; off Sakhalin Island on the Russian Pacific coast; in the ROPME/RECOFI area[2] and on the Australia's North-West Shelf.

1.2 Production

According to the United States of America National Research Council (2003) in a snapshot of the global offshore oil and gas industry, there were (in 2003) more than 6,500 offshore oil and gas installations worldwide in 53 countries, 4,000 of which were in the United States Gulf of Mexico, 950 in Asia, 700 in the Middle East and 400 in Europe. These numbers are constantly changing as the industry expands and contracts in different places in response to numerous factors involved in the global energy market. An indicator of this volatility is that by 2014 there were only 2,410 rigs in the United States Gulf of Mexico, for example.

Global crude oil production is currently 84 million barrels per day (BPD; CIA Factbook, 2012 figures) of which about 33 per cent is from the offshore (Fig. 3). Data compiled by Infield (2014) indicate that onshore crude production plateaued at around 65 million BPD as early as the 1990s and growth in offshore production has accounted for most of the increased global productivity since then. Production from deep water[3] (>100 m water depth and as deep as 2,900 m at Shell Oil's "*Stones*" field in the Gulf of Mexico) platforms accounted for about 1 per cent of production in 2000 but this figure had increased to 7 per cent by 2010 and is anticipated to reach 11 per cent of total global production by 2015 (Infield, 2014; Fig. 3).

Over 1,000 offshore oilfields are forecast to be developed between 2011 and 2015, about 80 per cent of which will be in shallow water depths (<100 m). Capital spending on shallow water platforms and pipelines is forecast to grow from an estimated 50 billion United States dollars in 2011 to nearly 60 billion United States dollars by 2015, whereas spending on deep water infrastructure is forecast to rise from 45 billion dollars in 2011 to nearly 80 billion dollars by 2015 (Infield, 2014).

Offshore natural gas production is geographically dispersed: key areas include the North Sea, Gulf of Mexico, Southeast Asia, Australia, New Zealand, Qatar, West Africa and South America. The geographic areas of major investment are (in order of decreasing numbers of projects): the North Sea; Southeast Asian seas; the Gulf of Mexico; Eastern Indian Ocean; and Gulf of Guinea.

In 2001, the United States received 23 per cent of its domestic natural gas from the Gulf of Mexico but by 2013 federal waters of the Gulf of Mexico provided only 5 per cent of United States production (EIA, 2014). The reason is the fracking (hydraulic fracturing) revolution combined with horizontal drilling into tight (i.e. low permeability) geological formations, which has led to a significant increase in the United States' production of onshore shale gas and shale oil such that domestic production will meet US requirements in the short to medium term (until the mid 2020s). Fracking is also employed in some offshore locations (e.g. off southern California).

The southernmost offshore petroleum facilities in production in the world are in gas fields located 70 km offshore Tierra del Fuego, Argentina. These fields are currently producing 15 million cubic metres of gas per day. The offshore platforms are designed to resist the roughest sea conditions and wind speeds of up to 180 km/hr. The northernmost facilities in production are the Prirazlomnoye oil fields, located off the coast of Russia in the Pechora Sea (adjacent to the Barents Sea). The field is estimated to hold 72 million tons of recoverable oil and production is expected to reach six million tons annually. In 2014, some 300,000 tons will be shipped out from waters that are ice-covered for 7-8 months a year (Barents Observer, 2014).

1.3 Exploration

Oil and gas explorers rely on seismic reflection surveys to produce images of the stratigraphy and structure of subsurface rocks. They use this information to determine the location and size of oil and gas reservoirs. Globally, there were about 142 specialized seismic vessels in operation in 2013 (Offshore Magazine, 2013) and each year the Bureau of Ocean Energy Management[4] gives permits for about 20 3-D seismic surveys in

---

[1] Regional Organization for the Protection of the Marine Environment (ROPME) Members: Bahrain, Iran (Islamic Republic of), Iraq, Kuwait, Oman, Qatar, Saudi Arabia, and the United Arab Emirates. Regional Commission for Fisheries (RECOFI) Members: Bahrain, Iran (Islamic Republic of), Iraq, Kuwait, Oman, Qatar, Saudi Arabia, United Arab Emirates.

[2] Although Infield (2014) uses 500 ft (152 m) as the divide between shallow and deep water, there is no agreed definition of "deep water". The geomorphic continental shelf break is typically around 100 to 200 m in depth (it is around 120 m deep in the Gulf of Mexico between Texas and Florida; Harris et al., 2014), so any rig in water deeper than this is located on the continental slope (deep water).

[3] On October 1, 2011, the Bureau of Ocean Energy Management, Regulation and Enforcement (BOEMRE), formerly the Minerals Management Service (MMS), was replaced by the Bureau of Ocean Energy Management (BOEM) and the Bureau of Safety and Environmental Enforcement (BSEE) as part of a major reorganization (http://www.boemre.gov).

[4] The estimate of over 200,000 offshore workers was derived by adding the number for the North Sea (about 24,000) to the number in the Gulf of Mexico (about 121,000) and multiplying by 1.5 to account for the remainder of the global workforce. It does not include shore-based staff. Another way to estimate the numbers working offshore is to use the average crew numbers (from www.oilpro.com) as follows: (A) 540 jack-up exploration rigs (crew = 55 per rig) total of ~30,000; plus (B) 142 seismic vessels (crew = 80 per vessel) total of ~11,000; plus (C) 80 drill ships (crew = 80 per ship) total of ~6,500. For each crew at sea there is

the United States Gulf of Mexico. Every seismic vessel tows a seismic source comprising a number of compressed air-guns, and an array of hydrophones in a "streamer" to capture the sound waves reflected from sedimentary layers below. Multi-streamer marine seismic surveys can image the subsurface in 3 dimensions (3-D seismic), acquired by a vessel equipped with between 8 and 16 streamers towed 50 to 100m apart, each 3 to 8 km long. Seismic reflection surveys are not restricted to the exploration phase but may be periodically repeated during the production phase of an offshore field. Once in production, some oil fields are re-surveyed to assess how well the reservoir is drained over time (4-D seismic).

Exploratory drilling is carried out mainly by jack-up rigs, semi-submersibles, or drillships. A jack-up rig consists of a buoyant hull fitted with extendable legs that, resting on the sea floor, are capable of raising its hull over the surface of the sea. There are about 540 jack-up rigs currently in operation globally, normally limited to shallow water (<100 m) drilling.

In order to explore in deep water, either drillships or semi-submersible vessels (also referred to as Mobile Offshore Drilling Units, or MODUs) are used. These vessels must maintain their position over the well to within a percent or so of water depth, requiring either dynamic position capability (using powerful propeller "thrusters") or anchoring to the seabed. The world fleet of offshore drilling ships currently comprises about 80 vessels of various sizes and capabilities. Semi-submersible vessels are built with ballasted pontoons with tall legs that support a platform. Once on location the pontoons are partly flooded so that they sink below the ocean surface and wave action, while holding the platform at a safe height above the waves. When oil fields were first developed in deep water offshore locations, drilling semi-submersibles were converted for use as combined drilling and production platforms. As the oil industry has progressed into deeper water, purpose-built production semi-submersible platforms were designed. There are currently about 40 semi-submersible, deep-water exploration vessels in operation globally (Offshore Magazine, 2013).

1.4 Social aspects of the offshore oil and gas industry

Over 200,000 people work on offshore rigs and platforms globally, although the exact number is difficult to estimate. In 2011 there were 23,758 core offshore workers spending over 100 nights a year offshore on the United Kingdom's North Sea oil rigs. According to the United States Bureau of Labor Statistics, there were 120,676 people employed in the Gulf of Mexico oil and gas industry in 2009 earning 15.6 billion United States dollars (average income of 129,000 dollars pa). Salaries of offshore oil and gas industry workers have a broad range (Hays, 2013). In Nigeria, expatriate workers are the highest paid in Africa with an average annual salary of 140,800 dollars whereas local workers in Nigeria's oil and gas sector have an average salary of 55,100 dollars (43 times higher than the average annual income in Nigeria, which is 1,280

another crew ashore (on leave) so multiply by 2, making over 100,000 personnel in total (oilpro.com), not including shore-based workers or the crews of fixed production platforms.

dollars (World Bank, 2010)). Local oil and gas sector workers have average salaries ranging from 31,100 dollars in Sudan to 163,600 dollars in Australia (Hays, 2013). The average salaries paid to foreign workers are generally higher than local remuneration rates, ranging from 59,800 dollars in Sudan to 171,000 dollars in Australia (Hays, 2013).

Most offshore oil workers spend extended periods, often one or more weeks, at their workplace – usually a production platform or a MODU. They then leave to live at home onshore for a non-work period that is also commonly one or more weeks. The offshore accommodations, recreational facilities and food are provided by their employer, which also provides transportation between the workplace and some onshore "pick-up point", commonly a heliport. This work system is variously called "fly-in", "fly-in/fly-out", "FIFO" or "long-distance commute" employment (Shrimpton and Storey, 2001).

Like other employment systems, this work pattern offers advantages and disadvantages for offshore workers, their families and the communities and regions in which they live. However, it is important to note that there are important limitations to the understanding of offshore employment effects; for example, the research to date has focused on developed countries (mostly Australia, Canada, Norway, United Kingdom and of Great Britain and Northern Ireland and the United States), large operations and companies, fixed work schedules and married male workers.

This work system has implications for various interrelated work and family life issues, the most important of which are Health and Safety and Family Life:

Health and Safety: This includes issues relating to working in a hazardous environment, the remoteness of the operations, the hazardous and stressful nature of the commute, the use of extended shifts and rotations and in some instances the possible risk of abduction by pirates or militants.

Family Life: Offshore commute work presents challenges to family relationships, but these must be assessed in the context of an understanding of the range of advantages and disadvantages that the system can present. As identified by workers and their family members, these are: income from offshore work;

- secondary and family employment;
- separation of work and family life;
- access to services and facilities;
- independence and decision-making;
- inappropriate worker behaviour;
- family separation; and
- isolation from and within the community.

However, these advantages and disadvantages are not always experienced in the same ways and to the same degree, with the main factors underlying such variations being differences in the availability of alternative employment, the work environment and workers' experience of it, the regularity and security of employment, family members' experience and expectations of family life, and workers' and spouses' perceptions of the effects on the family.

Various responses and interventions may be appropriate in addressing family life challenges, including those that improve the compatibility of the work organization and family life, improve the compatibility of the work culture and home life, improve self-selection during hiring, help newhires and their families get used to a new work pattern, and provide counselling or other support to employees and family members.

Overall, while research has shown that commute operations have somewhat higher proportions of separated and divorced workers than do conventional ones, it is not clear that this is a direct consequence of the work system, because these workplaces seem to attract separated and divorced employees.

Piracy and abductions: (Kashubsky, 2011) compiled a database of 60 known attacks against maritime and petroleum infrastructure between 1975 and 2010. Out of these incidents, 41 have occurred since 2004, the majority of which have taken place in Nigeria (Kashubsky, 2011). The majority of incidents involved violence (whether actual use of violence or threat of violence), but 15 of 60 incidents (25 per cent) were non-violent.

Since 2006 there have been about 200 abductions in the Niger Delta of foreign workers from offshore platforms, survey vessels and pipe-laying barges. Such abductions are carried out by militant groups, especially the Movement for the Emancipation of the Niger Delta (MEND), whose stated goals are to localize control of Nigeria's oil and to secure reparations from the federal government for pollution caused by the oil industry.

1.5 Communities wholly or mostly dependent upon the offshore hydrocarbon industries

Offshore petroleum activity has had significant, and sometimes dramatic, effects on infrastructure development, education and training, and research and development (Stantec, 2012), primarily focused on such major centres of activity as Aberdeen (United Kingdom of Great Britain and Northern Ireland), Stavanger (Norway), Houston (United States), New Orleans (United States) and St John's (Canada). Industry activity has also increased the entrepreneurship and competitiveness of local individuals and companies at the local level, and generated population growth, commonly reversing previous demographic trends.

Aberdeen is a good example of the way in which these effects can affect a port city. It receives many benefits from the petroleum industry but there are also "displacement and deterrence effects" on traditional industries (Harris et al., 1988). Displacement sees existing activity being crowded out by new activity, while deterrence sees new activity preventing other activity by making a region unattractive for it. The offshore petroleum industry generated upward pressure on wages and increased the price of housing and office space, although industrial property shortages were avoided due to an increase in warehouse and factory space. The high housing prices deterred outside workers from entering the region for employment.

As a result of these forces, several industries had local growth rates below their national averages, while ones that were already declining saw that decline accelerate. The industries that declined faster than average in the 1970s included: fishing, food and drink, clothing and footwear, building materials, and timber and furniture. Harris et al. (1988) conclude that "for every 100 jobs created by the oil industry in Aberdeen, at least eight jobs have been lost in traditional industries. By 1981, displaced and deterred employment amounted to more than 3,000 jobs. Of this, only about 25 percent has been absorbed by the oil sector."

This decline of other industries resulted in a higher dependence on the oil industry. Newlands (2000) estimates that, in 1985, 40 percent of Aberdeen's workforce relied upon oil. He also notes that, in the 1960s, "most businesses in Aberdeen were locally owned and controlled with only a few examples of external ownership [but] a survey conducted in 1984 suggested that the figure had fallen to as low as 11 percent".

By contrast with the potential negative impacts on traditional sectors of the economy, there may also be benefits for them. Harris et al. (1988), note that "better communications... benefit individuals as well as firms. Indeed, they are just one example of the improvement in the range and quality of services available to people in Aberdeen which has taken place in recent years. There has been a marked increase in the number, variety, and quality of shops and restaurants. There are more entertainment spots such as wine bars, discos and nightclubs. These developments cannot be attributed wholly to the establishment of the oil industry in Aberdeen, but oil developments have undoubtedly influenced the extent and pace of change". However, such benefits are concentrated in and around major centres of offshore petroleum activity.

Such changes have had significant positive consequences for tourism in Aberdeen, Stavanger, St. John's and other activity centres. Various studies have shown the industry making a major contribution to tourism through improved air links, meetings, conferences, trade shows, corporate hospitality and the personal expenditures of petroleum industry personnel. Newlands (2000) notes that new hotels opened, and others expanded, in Aberdeen in the 1970s. This increased the number of hotel rooms by 27 percent between 1970 and 1975 and by 58 percent between 1975 and 1980. The number of restaurants rose from 17 to 36. A similar expansion and "cosmopolitanization" of the hospitality, accommodations and hence tourism sectors has been seen in St. John's (Shrimpton 2002).

Generally speaking, management strategies have limited the negative biophysical environmental and other effects of offshore petroleum activity on Norway, the Shetland Islands, Nova Scotia, Newfoundland and

Labrador, which continue to experience rapid growth in tourism, including eco-tourism and adventure tourism.

Harris et al. (1988) also note that: "the maintenance or improvement of services applies also to the public sector. There are better hospital facilities and a larger and more comprehensive educational system than would have been the case had oil developments not reversed the trend of economic decline and emigration from Aberdeen".

Given the nature of the offshore employment system (Section 1.4) it has a number of effects on the communities and regions where the workers live (Shrimpton and Storey, 2001), including those on:

- *Residential Patterns*: The commute system can give workers and their families considerable flexibility as to where they live. Depending largely on the schedule, transportation systems and employee preferences, they may live close to, or distant from, the workplace.
- *Expenditures*: Offshore work wage rates are often high, they are commonly combined with long hours of work and considerable amounts of overtime, and workers generally have few expenses or spending opportunities at the workplace. As a result, these workers generally also have high disposable incomes. Their expenditure patterns, including payment of taxes, are largely dependent on where they live, and can make a significant contribution to the economy of those communities and regions.
- *Non-Commute Employment*: Some offshore employees have secondary paid work while in their home communities. This can involve the use of oil industry work skills and/or involvement in traditional local farming or fishing activity. In the latter case, offshore oil labour and incomes can help sustain the local primary sector.
- *Community Life and Social and Recreational Services*: Offshore work removes some citizens from communities on a part-time basis, affecting their ability to participate in formal and informal social events and networks, including local service groups, sports teams and elected government.

1.6 Description of economic benefits to States

Daily global offshore oil production is currently about 28 million barrels (Fig. 3), which is worth between 1.4 billion dollars and 2.8 billion dollars per day (assuming 50 dollars and 100 dollars per barrel). Oil and gas production from the United Kingdom continental shelf (for example) has contributed 271 billion United Kingdom pounds (2008 money) in tax revenues over the last forty years. In 2008, tax rates on United Kingdom continental shelf production ranged from 50 – 75 per cent, depending on the field. The industry paid 12.9 billion pounds in corporate taxes in 2008-9, the largest since the mid-1980s, because of high oil and gas prices. This represented 28 per cent of total corporation tax paid in the UK. In addition to production taxes, the supply chain contributes another 5-6 billion pounds per year in corporation and payroll taxes (UK National Archives, 2013).

In Australia, from 1999 to 2013, the offshore petroleum industry has contributed over 21.9 billion Australian dollars in petroleum resource rent tax in addition to corporate taxes (data.gov.au, 2014).

The offshore oil and gas industry accounts for about 1.5 per cent of United States GDP, 3.5 per cent of the United Kingdom's GDP, 12 per cent of Malaysia's GDP, 24 per cent of Norway's GDP and 35 per cent of Nigeria's GDP (OPEC, 2013; EIA, 2014). In Norway, crude oil, natural gas and pipeline transport services accounted for about 100 billion dollars in 2010, nearly half of the value of Norway's total exports and 10 times higher than the export value of fish (Norwegian Petroleum Directorate, 2012). In Nigeria crude oil export was valued at around 94 billion dollars pa and accounts for about 70 per cent of total exports revenue (OPEC, 2013 figures). Thus the overall value of the offshore oil and gas industry

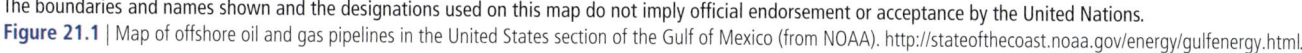
The boundaries and names shown and the designations used on this map do not imply official endorsement or acceptance by the United Nations.
Figure 21.1 | Map of offshore oil and gas pipelines in the United States section of the Gulf of Mexico (from NOAA). http://stateofthecoast.noaa.gov/energy/gulfenergy.html.

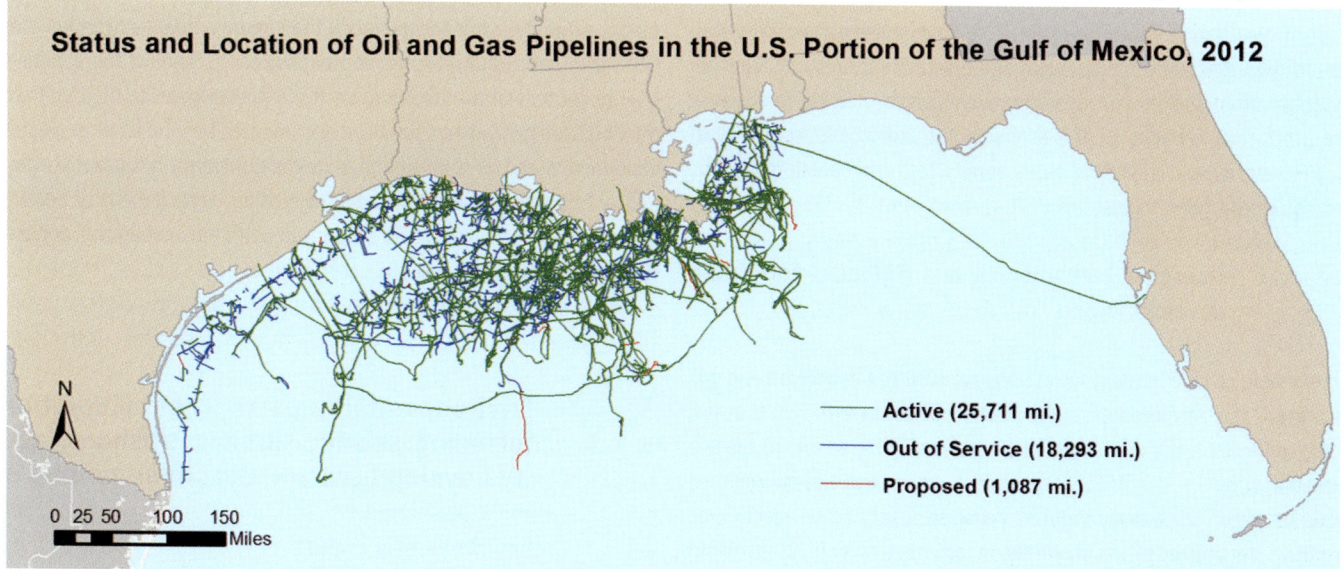

accounts for a significant part of GDP but varies dramatically among countries in terms of its overall importance.

Offshore petroleum activity can have a range of other impacts, some of them negative, on the local economy. Some early literature described negative effects on the local economy. For example, Galenson (1986) argues that Norway's performance in curbing inflation was less than it might have been without oil revenues, and that they allowed the government to pursue policies that harmed manufacturing: "Oil money was used to preserve the existing pattern of industry. The restructuring necessary to meet changing market demands was slowed, if not stopped. New initiatives were not encouraged". Mallakh et al. (1984) and Noreng (1980) argue that one of problems was that Norwegian government policy prevented labour from moving to more productive firms and sectors. It was not only prevented from moving to and from manufacturing, but from less to more productive uses within manufacturing.

However, petroleum taxes and royalties can help address the challenges posed by the fact that the sector is cyclical and involves a non-renewable resource, meaning that the state's revenue from it can be highly volatile. In 1990, Norway established a government pension fund to transfer capital from the state's petroleum revenue. The fund was designed to be invested for the long term, but in a way that made it possible to draw on when required. Its purpose is to support the government's long-term management of the petroleum revenues. The fund gives the government room for manoeuvring in fiscal policy should oil prices drop or the mainland economy contract. This facilitates economic stability and predictability. The fund also serves as a tool to manage the financial challenges of an ageing population and an expected drop in petroleum revenues.

There is growing interest, globally, nationally and locally, in creating sustainable economic development. Notwithstanding the fact that the offshore petroleum industry activity involves large technologically-complex projects and the exploitation of a non-renewable resource, the evidence from Canada, Norway, the United Kingdom, and other States indicates that it has been the engine for significant and sustainable (over many decades) economic development in a number of jurisdictions on both sides of the Atlantic. This is partly because it can make a major contribution to output, income, employment and government finances. Such activity has often also had a transformative effect, by helping to enhance the productive capacity of the economy through stimulating growth in the quantity and quality of factor inputs to the production process, thereby contributing to sustainable long-term economic development.

1.7 Emerging technologies and potential for future developments

An example of an emerging technology relevant to offshore oil and gas facilities is the design of structures that could be deployed on the seafloor (rather than floating). Equipment that can be fixed directly to the sea floor, where it is relatively protected from ice and violent weather, could be used for subsea produced water removal and re-injection or disposal, single-phase and multi-phase boosting of well fluids, sand and solid separation, gas/liquid separation and boosting, and gas treatment and compression (Sorenson, 2013). Re-injection of produced gas, water and waste increases pressure within the reservoir that has been depleted by production. Also, re-injection helps to decrease unwanted waste, such as flaring (because the gas that would have been flared is re-injected), by using the separated components to boost recovery.

Disadvantages of robotic (unmanned) seafloor mounted facilities include difficulties in monitoring their operation and implementing any necessary repairs. The 2010 Deepwater Horizon (DWH) underwater spill in the Gulf of Mexico took months to bring under control, partly because of the challenges imposed by the water depth of the structure. The added complications that would arise in an Arctic setting, where repairs may have to be undertaken in winter months beneath floating sea ice, are factors that any such operations would need to address.

Another new technology is Floating Liquefied Natural Gas (FLNG) production where the processing of gas from offshore fields takes place at sea. The world's first FLNG vessel, Shell's Prelude, is currently under construction in the Republic of Korea for deployment on Australia's North-West Shelf. The LNG plant will be located on a large vessel that will be moored above the gas field – several hundred kilometres from the coast. The successful deployment of FLNG may allow for the production of gas from smaller or more remote offshore fields (Geoscience Australia and BREE, 2012).

A potential future development in the offshore energy sector is the possibility of mining methane gas hydrates from seabed deposits. Methane clathrate, also called methane hydrate, is composed of methane trapped within the crystal structure of water, forming a solid similar to ice, found in seabed sediments in water depths of greater than 300 to 500 m (Ruppel, 2011). When brought to the earth's surface, one cubic metre of gas hydrate releases 164 cubic metres of natural gas. Methane that forms hydrate can be both biogenic, created by biological activity in sediments, and thermogenic, created by geological processes deeper within the earth. A conservative estimate (Boswell & Collett 2011) for the global gas hydrate inventory is ~1,800 gigatons of carbon. While global estimates vary considerably, the energy content of methane occurring in hydrate form is immense, possibly exceeding the combined energy content of all other known fossil fuels. However, methane production from hydrate has not been documented beyond small-scale field experiments and its contribution to global gas supply has probably been delayed by several decades by the increasing development of onshore gas resources from shale, coal seams and other unconventional deposits (Geoscience Australia and BREE, 2012).

2 Environmental impacts from exploration, including seismic surveys, offshore facility development and decommissioning

2.1 Environmental impacts

Environmental impacts arise throughout petroleum exploration-drilling-production development operations as well as in the decommissioning of facilities once the oil field is no longer economic, although the nature and degree of impact varies (Swan et al., 1994). Seismic surveys, oil and gas production, transportation and decommissioning all have associated environmental impacts; these are described briefly below.

The risks of impact from oil spills are greatest during transport, from pipeline rupture and vessel loss or spillage, when large volumes of oil can be released suddenly. It is important to realize that accidental (anthropogenic) spills occur against a background of continuous leakage from the seafloor. Crude oil is a naturally-occurring substance. An amount of oil approximately equal to that spilled accidentally by humans enters the oceans each year through natural seepage (Kvenvolden and Cooper, 2003; National Research Council, 2003). Natural seepage is a gradual, ongoing process and ecosystems have evolved that use it as a food source. Spills are ecologically damaging because they result in unnatural concentrations of oil at a particular site that are incompatible with local marine life.

Also it should be noted that accidental oil spills account for only a small percentage of the total volume of oil that enters the oceans due to humans. Most oil enters the ocean mixed with sewage and urban stormwater runoff (GESAMP, 2007) but such diffuse sources do not have the same dramatic impact on ecosystems as a spill because the oil is delivered continuously in low concentrations over a broad area. The long-term effects of low-level oil pollution from diffuse sources are unknown (GESAMP, 2007).

2.2 Drilling and production activities

Drilling activities are carried out from ships or fixed platforms during exploration and to extract oil once it has been found. Direct damage to the seafloor is caused by the anchors used to hold the rig in place as well as by the impact of the well emplacement itself. Drilling requires the use of lubricant (drilling mud) and the disposal of drill cuttings onto the seabed at the drill site. Drilling mud and some of the drill-cuttings contain crude oil residues, polycyclic aromatic hydrocarbons (PAHs) and heavy metals that can be toxic. The environmental impacts of drilling may include smothering the seabed by blanketing it with dense drilling mud and cuttings and toxicity effects (Swan et al., 1994). The initial impact is generally confined to the immediate surrounds, typically within 150 m of the drill site. However, Olsgard and Gray (1995) reported that barium, total hydrocarbons, zinc, copper, cadmium and lead contamination sourced from production platforms on the Norwegian shelf had spread considerably after a period of 6 to 9 years, so that evidence of contamination was found 2 to 6 km away from the platforms. This led to a ban in the North Sea on discharging oil-based muds or cuttings contaminated with them from 1993. The ban gradually decreased the affected zone around drilling installations from several km to approximately 500 m (Bakke et al., 2013).

During the production of oil and gas, water from the hydrocarbon reservoir is also brought to the surface. This is known as "produced water" (PW), is a by-product of oil production and it is either disposed of into the ocean or may be re-injected into the well to promote oil recovery (Swan et al., 1994). Produced water can also include sea-water injected into the well to promote recovery. Compared with ambient seawater, PW may contain elevated concentrations of heavy metals (e.g. arsenic, mercury, barium, copper, lead and zinc), radium isotopes, as well as hydrocarbons. The proportion of oil/water varies between locations but generally the proportion of water increases over time as the oil deposit is depleted (i.e. older wells discharge more PW than new wells). The proportion of water produced per barrel of oil typically ranges from around 3:1 to 7:1 although the relative amount of PW increases over time, such that in extreme cases the fluid pumped from a well might be 98 per cent water and only 2 per cent oil (Holdway and Heggie, 2000). At the production platform, most of the PW is separated from the oil, treated (typically to around 30 mg/L hydrocarbon) and disposed of into the ocean. Because the increase in the absolute amount of PW discharged leads to an increase in the absolute amount of oil discharged, unless the proportion of oil in the PW discharged is decreased, regulatory measures have

The boundaries and names shown and the designations used on this map do not imply official endorsement or acceptance by the United Nations.

Figure 21.2 | Offshore oil and gas fields under exploitation, new discoveries not yet in production and pipelines in the North Sea in 2009. Figure taken from OSPAR, 2010. (http://qsr2010.ospar.org/en/ch07_01.html).

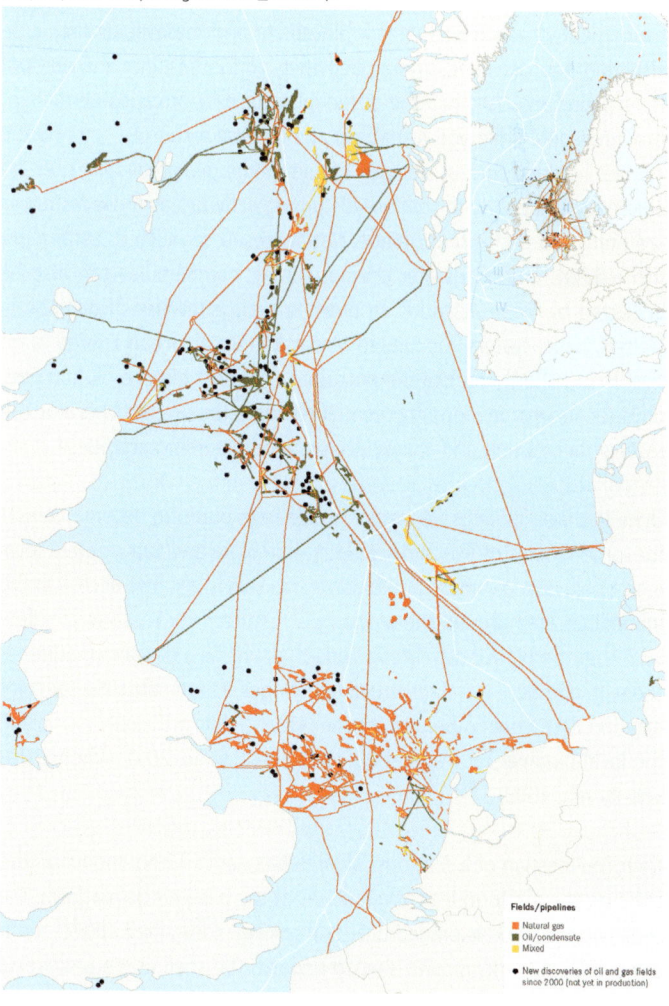

been adopted in some areas, such as the North Sea, which have resulted in reductions between 2001 and 2006 of around 33 per cent in the amount of oil discharged in PW (OSPAR Decision 2001/1; OSPAR 2010).

PW forms a buoyant plume because it is typically 40° to 80°C warmer (and therefore less dense) than ambient seawater and thus it will be dispersed by wind and currents away from the production platform. Mixing and dilution with seawater results in toxic effects of PW being generally confined to within 1 km of production platforms, although PW plumes may be detected in surface waters for distances exceeding 10 km from the point source (Jones and Hayward, 2003). Cases of coral discolouration (coral bleaching) have been attributed to dilute (~12 per cent) PW concentrations (ITOPF, 2007). Hence, the situation of production platforms in relation to prevailing winds and currents and to the proximity of sensitive habitats is a consideration for offshore petroleum development.

There are additional environmental consequences of the emplacement of rigs and floating platforms into the marine environment, which include effects on migratory birds and artificial habitat. Electric lights used on the rigs have been shown to interfere with the natural migration pathways of some species of birds, causing them to accidently collide with platform structures. Although the actual numbers of birds killed due to having collided with platforms is unknown, modelling has shown that the numbers could be significant (OSPAR, 2012b). Any seafloor disturbing activities such as anchor placement and retrieval, drilling, construction and decommissioning activities, and jetting into the seafloor for pipeline trenches has the potential to disturb or cause permanent and irreversible damage to natural and cultural resources. Underwater cultural heritage such as shipwrecks and submerged prehistoric sites are especially vulnerable to seafloor disturbing activities as these resources are finite and each site is unique. Natural resources such as coral reefs, fish habitat, and deepwater chemosynthetic communities can also be impacted by these activities. In order to reduce the risk of damage in United States waters, the United States Bureau of Ocean Energy Management (BOEM) requires the operator to conduct high-resolution geophysical surveys to identify potential resources before the operator can receive their permit and commence seafloor disturbing activities.

Once in place, the legs (jacket) of an offshore platform become habitat for some species of fish and sessile marine biota and can create a local area of elevated biomass and biodiversity. This is because access to the immediate area around the platform is restricted for reasons of safety such that the platform creates a zone that acts as a de facto marine reserve. In addition, the steel structure provides a hard substrate for colonization that would otherwise not be present, thus artificially increasing the local biodiversity (Page et al., 1999; Shaw et al., 2001; Whomersley and Picken, 2003).

Over the lifespan of a platform, shell debris derived from molluscs that colonize the platform legs accumulates at the base of the platform. The shell accumulation is draped over drill cuttings (described above) forming a characteristic, mound-shaped deposit. The shell drape provides a new habitat that has different properties from the surrounding seabed and thus offers habitat to different species. Disturbance of the shell drape will expose the (potentially toxic) drill cuttings that are a factor for consideration for rig decommissioning.

Because of the biological colonization of MODUs, the relocation of the vessel between drilling locations has been identified as a vector for the introduction of non-native species (Paula and Creed, 2005; Sammarco et al., 2010).

2.3 Seismic surveys and their impact on marine mammals and other ocean life

Marine acoustic survey equipment is used by the oil and gas industry as well as by the military, marine industries and academic researchers to map the seafloor, study the sediments beneath the seafloor and image the water column. Depending on the purpose, sonars differ in frequency, source level and beam pattern. Sonar signals diminish as they propagate, affecting different parts of the water column with different biological consequences. Towed, low frequency systems (such as seismic air guns) are omni-directional, radiating sound in all directions. Hull-mounted systems are higher frequency (kHz or greater) and utilize beam forming to obtain higher resolution images at lower source levels.

Airguns employed to acquire seismic reflection data are far more powerful (225 to 255 decibels re 1 micro-Pascal peak; Richardson et al., 1995) than equipment used for marine research or normal ship navigation. Airguns used in seismic reflection surveys emit sound at a frequency of typically ~100 Hz which overlaps with the range of marine mammals' hearing and is therefore most likely to affect marine mammals and other marine life (McCauley et al., 2000; O'Brien et al., 2002; NRC, 2003; Boebel et al., 2005; Nowacek et al., 2007; 2013; CBD, 2014).

Cetaceans have been observed avoiding powerful, low frequency sound sources. A study by McCauley et al (2000) has shown that migrating humpback whales will leave a minimum 3 km gap between themselves and an operating seismic vessel, with resting humpback whale pods (groups) containing cows exhibiting increased sensitivity and leaving an increased gap of 7–12 km. Conversely, the study found that male humpback whales were attracted to a single operating airgun as they were believed to have confused the low-frequency sound with that of whale breaching behaviour. In addition to whales, sea turtles, fish and squid all showed alarm and avoidance behaviour in the presence of an approaching seismic source. While there has not been any documented direct linkage between seismic surveys and the beaching of marine mammals, Gordon et al., (2004) noted that concerns over the stranding of beaked whales in two separate incidents was sufficient for United States courts to agree to a restraining order on seismic operations by the RV *Maurice Ewing*. Nowacek et al. (2007) noted that displacement (relocation to an un-impacted area) is a common response of mammals, which may cause harm if the impacted site is an important feeding ground (see also Cerchio et al., 2014). Lucke et al (2009) found that harbour porpoises consistently showed aversive behavioural reactions at received sound pressure levels above a certain threshold level produced using a seismic airgun.

The historical record of cetaceans stranding themselves prior to the industrial age includes the English Crown holding rights on stranded cetaceans from at least 1324, when they were known as "fishes royal" since the Crown had first claim on them (Fraser, 1977). Thus cetacean stranding occurs under natural environmental conditions. Additional evidence in support of this conclusion is the occurrence of marine mammal strandings within the geological record (Pyenson et al., 2014).

It is not known whether there has been any increase in whale strandings that can be attributed directly to seismic surveys. Nowacek et al. (2007; 2013) conclude that major data deficiencies are: the lack of studies linking animal responses to received acoustic level data; gaps in species representation; and poor understanding of habituation, sensitization and tolerance. In the case of marine animals other than cetaceans, there is some evidence for short-term displacement of seals and fish by seismic surveys, but there is little literature available (Thompson et al., 2013; see also Chapter 37).

2.4 Pipeline construction, main causes of leaks and decommissioning

In order to bring oil and gas ashore to refineries and transport systems, pipelines are laid across the seabed, in places forming complex networks (Fig. 2).[5] Before a pipeline can be laid, the proposed route is surveyed using acoustic seabed mapping technology and underwater cameras to identify any obstacles such as natural bedrock formations, boulders, seabed valleys or migrating sand dunes as well as shipwrecks, other cables and pipelines and dumped ammunition. The pipe is then laid along the surveyed route using either a specialised pipe-laying vessel or else a pull/tow system where the pipeline is built onshore and then towed by ship to its desired location. The first process usually involves one pipe-laying vessel supported by barges that supply the ship with pipe sections plus other support ships that monitor the seabed. Pipe sections are welded together on board, the joints tested (by ultrasound) and the pipe coated with an anticorrosion application.

The diameter of the pipe varies between 0.15 to 1.4 m, but is ~0.3 m in most cases. A distinction is made between a flowline and a pipeline. Flowlines are intrafield pipelines used to connect subsea wellheads, manifolds and the platform within a particular development field. Pipelines (export pipelines, also called trunk lines) are used to bring the resource to shore.

Problems that affect the stability and integrity of submarine pipelines are: (1) expansion/contraction of the steel pipe causing lateral or upheaval buckling; (2) erosion of the seabed around the pipeline by storms, waves and currents leaving unsupported spans of pipeline vulnerable to fracture; and (3) corrosion. Solutions to the expansion/contraction problem are: adding expansion joints; burial; anchoring; applying a concrete or rock cover; or laying the pipe in an "S"- shaped configuration (S-lay). Problems of seabed erosion can be overcome by burial (Xu et al., 2009; Yang et al., 2012). Burial of the pipeline is also required in areas where vessels might anchor, where bottom fishing activities occur or where the pipeline crosses the shoreline. However, burial is more expensive than simply laying the pipeline along the seabed, so it is not used unless necessary (or a legal requirement).

Pipelines are maintained and inspected using a "pig," a tool that can be inserted in one end of the pipeline and pushed by the fluid to the other end. The most basic pigs are used to clean the inside of the pipes; highly-complex "smart pigs" can inspect the condition and thickness of the pipeline and detect points of corrosion or fracturing. Smart pigs are used more in the North Sea than in the Gulf of Mexico because the majority of existing Gulf of Mexico lines were not originally designed or built to accommodate smart tool pigs (MSL, 2000).

Pipelines are monitored for leakage by analog and computer-assisted systems (Stafford and Williams, 1996). The mass balance approach simply measures the amount of oil going in the pipeline and the amount coming out. Real-time transient modelling compares actual measured data with a computer model. If the results are outside normal operating limits, an alarm alerts the operator to take appropriate action. Other methods of monitoring pipelines include chemical and radioactive tracers, acoustic emission, neural networks, fibre-optic sensors and pressure point analysis (Stafford and Williams, 1996).

Woodson (1990) compiled a database of 1,047 submarine pipeline failures reported in the Gulf of Mexico between 1967 and 1990. The results indicate that a pipeline failure occurred in the Gulf of Mexico on average once every 5 days over the period of data collection. The source of failure was reported in 916 incidents in which the main causes were attributed as follows: 50 per cent (456 out of 916 incidents) due to internal or external pipeline corrosion; 12 per cent (106 incidents) due to storms and hurricanes; 14 per cent (124 incidents) due to damage from ship's anchors and fishing gear; 10 per cent (94 incidents) due to material failure of valves, gaskets or other joints; and 15 per cent (136 incidents) due to other or unknown causes. Pipelines over 10 years in age had a greater number of failures due to corrosion than younger pipelines; there was a trend of an increasing rate of failure due to corrosion in the last 5 years of the database (1986-1990), presumably attributable to the aging of the pipeline network. Pipelines damaged by storms (in which pipes are excavated by waves and currents exposing an unsupported span that is liable to fracture) were not buried in 40 out of 52 cases (Woodson, 1990).

In its review of offshore pipeline safety, the United States Marine Board Committee on the Safety of Marine Pipelines (Marine Board, 1994) cited the work of Woodson (1990) and noted that during the 1990s, transmission and production pipeline leakage and accidents accounted for about 98 per cent of accidental releases by offshore production activi-

---

[5] The following summary of pipeline construction is based primarily on Palmer and King (2004).

ties. However, although corrosion is the most commonly cited cause of pipeline failure, corrosion-related ruptures do not result in significant release of oil into the environment. Rather, damage caused by a few major incidents involving ship's anchors caused pipe leakage that is attributed to 95 per cent of the 250,000 barrels that leaked from pipelines in the Gulf of Mexico from 1967 to 1990 (Marine Board, 1994).

The National Research Council (NRC) (1997) noted that, in the United States Gulf of Mexico "no agency coordinates the collection of information and the available data on offshore pipeline failures are correspondingly "incomplete". Data on pipeline failures from state waters are collected by states (when collected) and may not be readily accessible. This is in spite of the fact that pipelines in state waters are commonly the oldest and most exposed to collision with ships (Marine Board, 1994). Since 2006, data on offshore pipelines (outside of state waters) has been the responsibility of the Bureau of Safety and Environmental Enforcement (BSEE).

In Europe, a pipeline data base complied by PARLOC (2001) includes records from 1971 to 2000 during which time there were 542 pipeline incidents including 96 pipeline leaks into the environment plus 92 leaks of pipeline fittings. The cause of failure of pipeline fittings is not recorded (PARLOC, 2001); for pipelines the database shows that corrosion is the major cause of failure (51 per cent) followed by maritime actions such as anchoring (23 per cent), material failure and other or unknown causes (26 per cent).

Less than 2 per cent of the North Sea pipeline inventory has been decommissioned as of 2013 (Oil and Gas UK, 2013). Of North Sea pipelines that have been decommissioned, 80 per cent are less than 16 inches (40.64 cm) in diameter. Half of the larger diameter pipelines (16 inches (40.64 cm) or greater) decommissioned to date were removed (i.e. 10 per cent of decommissioned pipelines were removed). Cleaning and purging is carried out following cessation of production, pipeline system depressurisation and removal of bulk hydrocarbons. Cleaning involves chemical cleaning to detach hydrocarbon residue from the pipe wall and bi-directional magnetic, disc and brush cleaning to remove ferrous and other loose debris using a specialist pig (Oil and Gas UK, 2013; see also Chapter 20).

The Gulf of Mexico offshore oil and gas has been operating since 1936 (Owen, 1975) and more of the pipeline infrastructure has been retired than in other parts of the world. Rach (2013) reported, based on data from the United States Bureau of Safety and Environmental Enforcement, that the inventory of Gulf of Mexico pipelines (as of mid-2013) includes 24,126 miles (38,827 km) that are in active use, 2,409 miles (3,877 km) proposed for installation, 12,628 miles (20,323 km) that have been abandoned, 2,264 miles (3,643 km) proposed to be abandoned and 2,425 miles (3,902 km) of pipeline that are out of service. Thus 42 per cent of existing (66,695 km) pipelines in the Gulf of Mexico are either abandoned, proposed to be abandoned or are no longer in service.

Apart from leakage of oil, other environmental and economic consequences of abandoned pipelines are their impacts on fishing (inhibiting bottom-trawling), other pipe and cable-laying activities and creating artificial habitats.

2.5 Rig decommissioning, dismantling and disposal, "Rigs to Reefs" programme

In its assessment of environmental governance in 27 developing countries, the World Bank (2010) found that governments lack a policy and process for decommissioning and abandonment and do not routinely assess, determine, or assign the future liability costs of decommissioning and abandonment. Only about 50 per cent of countries have an established process for managing the decommissioning and abandonment of oil and gas projects. Disposal of man-made structures (including platforms) at sea and abandonment or toppling on site of man-made structures falls under the scope of the Convention on the Prevention of Marine Pollution by Dumping of Wastes and Other Matter, 1972 (1972 London Dumping Convention) (LDC) and the 1996 London Protocol (LP). These treaties are global agreements and provide relevant dumping management policies, provisions, and assessment guidelines (London Convention and Protocol/UNEP, 2009). The 1989 International Maritime Organization (IMO) guidelines provide for the removal of offshore installations in order to leave 55 metres of clear water over any remains left in place (IMO, 1989).

In the United States, the BSEE has jurisdiction over decommissioning of wells and structures, pipelines, and the so-called "Rigs-to-Reefs" program. The Rigs-to-Reefs programme is the practice of converting decommissioned offshore oil and gas rigs into artificial reefs, which has occurred mostly in the Gulf of Mexico. However, less than 10 per cent of rigs decommissioned in the Gulf of Mexico have so far been converted to reefs; 90 per cent are removed (The Economist, 2014; BSEE data). Apart from a few platforms converted to reefs in Brunei Darussalam and Malaysia, opposition to the practice has meant that none have been allowed off California or in the North Sea to date (Day, 2008; Macreadie et al., 2011, 2012; Jørgensen, 2012). In the North Sea, 122 decommissioned installations were brought ashore between 1999 and 2010, and four large concrete installations and the footings of one large steel installation have been left in place (OSPAR 2010). One estimated cost of removing the existing North Sea production platforms, in the United Kingdom sector alone, was over 14 billion dollars (Prince, 2004). About 60 rigs are nearing the end of their working life in Australia (Macreadie et al., 2011).

The BSEE has granted permits for about 420 platforms to be converted to artificial reefs in the Gulf of Mexico. There are three methods for converting a non-producing platform into an artificial reef: (1) partially remove the platform; (2) topple the platform in place; or (3) tow-and-place the platform to one of about 28 sites designated as an artificial reef area. Partial removal typically relies on non-explosive means to cut the platform below the sea surface, leaving the legs in their vertical position. Toppling in place uses non-explosive or explosive severance to cut the platform from its legs and lay the platform legs on their side (the platform itself is removed). The tow-and-place method entails removing the platform and detaching the structure from the seafloor before towing it to a designated artificial reef area for disposal.

In the United States Gulf of Mexico there were about 2,996 production platforms as of March 2013, of which 813 (27 per cent) are no longer producing. In 2010, the United States Government issued notices to companies requiring them to set permanent plugs in nearly 3,500 nonproducing wells and dismantle about 650 unused oil and gas production platforms (Rach, 2013). Decommissioning means ending operations and returning the lease or pipeline right-of-way to a condition that meets the requirements of regulations of BSEE and other agencies that have jurisdiction over decommissioning activities. The regulations apply to any installation, other than a pipeline, that is permanently or temporarily attached to the seabed. Very few deep sea (>200 m depth) oil and gas fields have as yet been depleted, hence decommissioning of infrastructure has not yet become an issue (Macreadie et al., 2011).

Regulations for the decommissioning of a platform vary between countries. The United States requires plugging the well(s) supported by the platform and severing the well casings 15 feet (5 m) below the seabed; cleaning and removing all production and pipeline risers supported by the platform; removing the platform from its foundation by severing all bottom-founded components at least 15 feet (5 m) below the seabed; disposing of the platform onshore or placing the platform at an artificial reef site; and cleaning the platform site to ensure that no debris or potential obstructions remain. Over the lifespan of a platform a mound of debris accumulates beneath the platform that may contain toxic chemicals; removal of such mounds is also a requirement for decommissioning.

In the North Sea, OSPAR Decision 98/3 requires the topsides of all decommissioned installations to be removed to shore and all sub-structures or jackets weighing less than 10,000 tons to be completely removed. The regulations stipulate that, where there may be practical difficulty in removing installations (i.e. the footings of large steel platforms weighing over 10,000 tons, the concrete gravity based platform sub-structures, or concrete anchor bases and other structures with significant damage or deterioration, which would prevent removal), a decision may be taken to leave parts of the structure on the seabed, on a case-by-case basis.

3 Offshore installation disasters and their impacts, including longer-term effects.

3.1 Impacts of offshore installation disasters

Impacts of offshore oil and gas installation disasters include loss of human life, loss of revenue and environmental impacts. Offshore disasters have resulted in loss of life on several occasions in the history of the offshore oil and gas industry. In March 1980, the "flotel" (floating hotel) platform Alexander L. Kielland capsized in a storm in the North Sea with the loss of 123 lives. In February 1982, the Ocean Ranger semi-submersible mobile offshore drilling unit sank on the Grand Banks of Newfoundland; none of the 84 crew members survived. In July 1988, 167 people died when Occidental Petroleum's Piper Alpha offshore production platform exploded in the United Kingdom sector of the North Sea after a gas leak. In 2001 Petrobras-36 in Brazil exploded and sank five days later killing 11 people. In April 2010, the Deepwater Horizon platform exploded, killing 11 people. The Kolskaya floating oil rig capsized and sank in the Sea of Okhotsk in December 2011, killing 53 crew members. In December, 2013, a rig owned by Saudi Arabia's state-run petroleum company, Aramco, sank in the ROPME/RECOFI area, killing three crew members. Such disasters have resulted in the imposition of new regulations on industry (Turner, 2013); for example, the Piper Alpha disaster resulted in the United Kingdom Government passing the 1992 Offshore Installations (Safety Case) Regulations. However, there have been subsequent disasters around the world and the industry, as a whole, has continued to make the changes needed to improve its safety record (Harris, 2013).

From 2001 to 2010, the United States Minerals Management Service reported 69 offshore deaths, 1,349 injuries, and 858 fires and explosions on offshore rigs in the Gulf of Mexico. During 2003–2010, the United States oil and gas extraction industry (onshore and offshore, combined) had a collective fatality rate seven times higher than that for all United States workers (27.1 versus 3.8 deaths per 100,000 workers; Centers for Disease Control and Prevention, 2013). Catastrophic events attract intense media attention but do not account for the majority of work-related fatalities during offshore operations. A report by Baker et al. (2011) found that helicopter crashes were the most frequent fatal event in this industry.

Economic impacts stemming from offshore oil and gas installation disasters include the direct loss of income for the period that the facility remains offline, the costs to repair the facility, the costs to other industries (e.g. fishing and tourism) affected by the disaster and other compensation. As a result of the Deepwater Horizon oil spill, BP established a Trust Fund of 20 billion dollars for natural resource damages, state and local response costs and individual compensation. Other industry-wide consequences may follow such disasters. For example, exploration drilling in the Gulf of Mexico was slow to recover from the moratorium that followed the Deepwater Horizon oil spill in 2010. By 2012 36 rigs were back working off Louisiana and 4 off Texas, compared with 21 and 2, respectively, in late 2010. On 3 June, 2008, a high-pressure 12 inch export sales gas pipeline (SGL), critically weakened by a region of external corrosion, ruptured and exploded on the beach of Varanus Island off the coast of Western Australia. There was approximately 60 million Australian dollars in damage to the plant. Plant closure led to up to 3 billion dollars of losses to the West Australian economy, which lost 30 per cent of its gas supply for two months.

A blowout of the Montara wellhead platform on 21 August 2009, on Australia's remote North-West Shelf, leaked an estimated 30,000 barrels of crude plus an unknown quantity of gas until 3 November 2009 (total of 74 days), when the leak was finally stopped. The rig later caught fire but all 69 workers on the rig were safely evacuated with no injuries or fatalities. The company spent about 5.3 million Australian dollars on clean-up, about 300 million dollars in lost revenue and repair bills and was fined 510,000 dollars in August 2012 by the Australian Government (ABC News, 2012); the well finally went into production in June, 2013. However, fishermen and seaweed farmers in Indonesia are seeking compensation with support from the Australian Lawyers Alliance (ALA), claiming that environmental damage caused by the spill has cost them more than 1.5 billion dollars per year in lost earnings (ALA, 2014).

In 1980, within the ROPME/RECOFI area, the Hasbah Platform Well 6 blew out for 8 days, spilling 100,000 barrels of oil and costing the lives of 19 men. However, the worst disaster in the region occurred during the 1991 war between Iraq and Kuwait, when there were 22 incidents that spilled amounts of oil variously estimated at between 2 and 11 million barrels into the ROPME/RECOFI area (Khordagui and Al-Ajmi, 1993; Elshorbagy, 2005). The coast of Saudi Arabia was the most heavily impacted by the spill. Initial assessments of the environmental damage caused by the spilled oil were optimistic of rapid recovery (e.g. Fowler et al., 1993), but more recent studies have documented lingering effects of oil trapped in intertidal sediments and salt marshes over broad spatial scales (Michel et al., 2005; Barth, 2007). As of 2011, the Government of Saudi Arabia had invested 180 million United States dollars and the United Nations had spent U45 million dollars in rehabilitating impacted areas.

Natural disasters also take a toll on offshore oil and gas facilities. In August 2005, Hurricane Katrina affected 19 per cent of United States oil production by destroying 113 offshore oil and gas platforms, damaging 457 oil and gas pipelines, and spilling an unquantified amount of oil (http://www.bsee.gov/Hurricanes/2005/katrina/).

Environmental consequences of offshore oil and gas installation disasters are perhaps the most widely publicized aspect of such events. In the 2010 Deepwater Horizon (DWH) Gulf of Mexico oil spill, it is estimated that around 4.9 million barrels (about 670,000 tons, assuming a specific gravity of 0.88) was discharged into the sea before the well was capped, approximately 16 times more oil than was spilled by the Exxon Valdez in 1989 (about 37,000 tons of crude oil; Crone and Tolstoy, 2010; Oil Spill Commission, 2011). The impact of this huge volume of oil on deep water habitats in the Gulf of Mexico is unknown at the time of this writing (April, 2014). Prior to the DWH incident, the total volume of oil spilled in the Gulf of Mexico between 1964 and 2009 is estimated to be 517,847 barrels (Mufson, 2010).

Numerous smaller-sized spills have also occurred in recent years. Examples include a North Sea spill of 200 tons in August 2011 that occurred at Shell's platform Gannet Alpha; in 2012 Chevron suspended activities after two oil leaks (of around 5,000 barrels) occurred in a space of four months off the Brazilian coast; in March 2012 the platform Elgin-Franklin, operated by the Total group in the North Sea, was evacuated after an uncontrollable gas leak of an estimated 300 million cubic feet over a 45 day period (Beall and Ferreti, 2012).

The number of accidents reflects the massive scale of the offshore drilling and production enterprise. For every accident there are environmental consequences. Before the Deepwater Horizon accident, such major incidents were anticipated to occur with such extreme rarity that they were not considered relevant. One explanation could be that risk assessments are performed for single wells and not for whole areas (or on an industry-wide basis). However, a study of accidental oil spills based on global historical data has shown that the DWH accident was not an outlier, but an accident that can happen every 17 years with an uncertainty interval from 8 to 91 years (5–95 per cent). When the DWH accident was excluded from the data set, the resulting frequency was 23 years with an uncertainty interval from 10 to 177 years (Eckle et al., 2012).

Accidents that occur in coastal waters have the most severe environmental impact. Most oil floats on the sea surface where it can be readily delivered to the shoreline, where the concentrated consequences are evident. The coast is also a habitat for a diversity of species of birds, mammals, invertebrates and marine plants. For this reason spills that impact the coast, such as the Exxon Valdez spill that occurred in Alaska in 1989, have the greatest impact on the ecosystem (Shaw, 1992). The speed of ecosystem recovery is generally slower for colder and deeper habitats than it is for warmer and shallower habitats (Harris, 2014). However, every ecosystem is different and recovery times are difficult to estimate. For example, oil spilled in the Niger Delta over the last 50 years has penetrated up to 5 m into the soil profile and caused groundwater contamination in 8 out of 15 sites investigated that could take up to 30 years to clean up (UNEP, 2011).

3.2 Impact of oil spills on the marine ecosystem

The impacts of oil spills range from the immediate effects of oiling to longer term consequences of habitats being modified by the presence of oil and tar balls. Traces of hydrocarbons can remain in coastal sediments for many years after an oil spill (Hester and Mendelssohn, 2000). For example, for some of the rocky shores where oil stranded after the Exxon Valdez spill in 1989, oil is still found subsurface, only slightly weathered (Irvine et al., 2014). Similarly, oil from the 1991 Gulf War is still apparent in intertidal sediments and in salt marshes along the coast of Saudi Arabia (Michel et al., 2005; Barth, 2007).

There is no clear relationship between the amount of oil spilled in the marine environment and the likely impact on wildlife. A smaller spill at a particular season of the year and in a sensitive environment may prove much more harmful than a larger spill at another time of the year in another or even the same environment. Even small spills can have very large effects. Species that use the sea surface are most vulnerable (birds and mammals) and the eggs and larvae of many other species can be damaged by oil (Alford et al., 2014).

Some species may exhibit reduced abundance due to spills (Sánchez et al., 2006) although direct causal evidence is not always available (Carls et al., 2002). Some opportunistic species are able to take advantage of the changed habitat conditions and the attendant reduced abundance of impacted species, giving rise to a short-term increase in local biodiversity (Edgar et al., 2003; Yamamoto et al., 2003); this is an example of why biodiversity statistics alone are not a reliable indicator of environmental health. Recovery time for sites varies as a function of the type of oil spilled, the biological assemblage impacted, substrate type, climate, wave/current regime and coastal geomorphology and ranges from years to decades depending on these and other factors (Ritchie, 1993; Jewett et al., 1999; French-McCay, 2004).

- There are a number of pathways for oil to reach the oceans, namely:
- Land-based sources (urban runoff, coastal refineries);
- Oil transporting and shipping (operational discharges, tanker accidents);
- Offshore oil and gas facilities (operational discharges, accidents, blow-outs);
- Atmospheric fallout;
- Natural seeps.

Figures published by National Research Council (2003) range from an average of 470,000 tons to a possible 8.4 million tons per year for the sum of all of these sources. It is generally agreed that the largest single source is the land-based (urban runoff, coastal refineries) input, although there is little agreement on the absolute values for any source terms.

When oil enters the sea, it reacts according to physical, chemical and biological processes that change the properties of the oil and consequently, its behaviour. Factors include:

- The quantity and duration of the discharge/spill;
- The time of the year at which it occurs;
- The temperature of the air and the receiving water body;
- The weather and sea (e.g. waves and currents) conditions;
- The species composition in the area affected;
- The properties of the shoreline (rocky, sandy, mud flats, mangroves, etc.);
- The presence and abundance of oil-degrading micro-organisms;
- The concentration of dissolved oxygen in the water.

Different types of oil have different physical properties, which affect the way the oil will react in the environment.

Oil, when spilled at sea, will normally break up and be dissipated or scattered into the marine environment over time. This dissipation is a result of a number of chemical and physical processes that change the compounds that make up oil when it is spilled. The key natural processes are evaporation, dispersion, dissolution, oxidation, emulsification, biodegradation and sedimentation. The addition of chemical dispersants (also surfactants) can accelerate this process of natural dispersion.

Lighter components of the oil will evaporate to the atmosphere. The amount of evaporation and the speed at which it occurs depend upon the volatility of the oil. Evaporation of oil with a large percentage of light and volatile compounds occurs more quickly than one with a larger amount of heavier compounds.

Waves, currents and turbulence at the sea surface can cause all or part of a slick to break-up into fragments and droplets of varying sizes. These become mixed into the upper levels of the water column.

Water soluble compounds in oil may dissolve into the surrounding water. This depends on the composition and state of the oil, and occurs most quickly when the oil is finely dispersed in the water column. Components that are most soluble in sea water are the light aromatic hydrocarbons compounds, such as benzene and toluene. However, these compounds are also those first to be lost through evaporation, a process which is 10-100 times faster than dissolution.

Oils react chemically with oxygen either breaking down into soluble products or forming persistent compounds called tars. This process is promoted by sunlight although it is very slow even in strong sunlight such that thin films of oil break down at no more than 0.1 per cent per day. The formation of tars is caused by the oxidation of thick layers of high viscosity oils or emulsions. This process forms an outer protective coating of heavy compounds that results in the increased persistence of the oil as a whole. Tar balls, which are often found on shorelines and have a solid outer crust surrounding a softer, less weathered interior, are a typical example of this process.

Emulsification occurs when two liquids combine, one suspended in the other. Emulsification of crude oils refers to the process whereby sea water droplets become suspended in the oil. Oils with an asphaltene content greater than 0.5 per cent tend to form stable emulsions which may persist for many months after the initial spill has occurred. Emulsions may separate into oil and water again if heated by sunlight under calm conditions or when stranded on shorelines.

Sea water contains a range of micro-organisms or microbes that can partially or completely degrade oil to water soluble compounds and eventually to carbon dioxide and water. Many types of microbe exist and each tends to degrade a particular group of compounds in crude oil (Hazen et al., 2010).

Sinking usually occurs due to the adhesion of particles of sediment or organic matter to the oil, in which case the oil accumulates in the seabed sediments. In the ROPME/RECOFI area for example, Elshorbagy (2005) reported that oil-contaminated seabed sediments occur, particularly in coastal areas. The highest levels of hydrocarbons were 1600 µg l^{-1} found near Bahrain compared with background levels of 10 to 15 µg l^{-1}. Oil washed ashore at Pensacola Beach, Florida, from the DWH spill resulted in weathered oil petroleum hydrocarbon concentrations in beach sands ranging from 3.1 to 4,500 mg kg^{-1} (Kostka et al., 2011).

4 Significant environmental aspects in relation to offshore hydrocarbon installations.

Drilling operations may require the use of many chemicals. In the OSPAR region, chemicals are categorised in four colour classes depending on degradability, octanol-water coefficient and toxicity. The green category means it "shall pose little or no risk to the environment". Chemicals in the black category should be prohibited and chemicals in the red category should be substituted. Chemicals in the yellow category have characteristics between the red and the green class and are considered to be environmentally acceptable.

Other operational discharges are drill cuttings and small spills of oil and chemicals connected to exploration, production or transport.

For the offshore industry in the North East Atlantic, OSPAR has agreed on several decisions and recommendations to reduce discharges of oil and chemicals. These include Recommendation 2001/1: Management of PW and 15 per cent reduction target for oil discharged with PW, in addition to agreements on decisions and recommendations for use of chemicals offshore, decommissioning and environmental management (OSPAR, 2010).

OSPAR reports discharges, spills and emissions from oil and gas installations. Between 2001 and 2007 an average of between 400 and 450 million m^3 yr^{-1} PW were discharged (OSPAR 2009). For 2010 a sum of 361 million m^3 PW was discharged in this area. The main contributing countries were Denmark (25 million m^3), Norway (131 million m^3) and United Kingdom (196 million m^3; OSPAR, 2010). The numbers account for the whole OSPAR region, but the main region of activities is the North Sea. Oil content in PW was reported as dissolved and dispersed oil, and annual average oil content was reported to be 12 and 13 mg/l, respectively. Annual average oil content in dissolved and dispersed oil was 4,227 tons and 4,746 tons, respectively, giving a total of 8,972 tons in 2010. Annual quantity of injected PW was 81 million m^3 (OSPAR, 2012a). Yearly discharge of approximately 400 million m^3 PW has been relatively constant since 2001 (OSPAR, 2010).

Most of the concern regarding negative effects on ecosystems due to operational discharges from offshore oil and gas activities has been directed to the oil fraction of PW discharges, and less towards the added chemicals. This is due to the content of polycyclic aromatic hydrocarbons (PAHs) and alkylated phenols from the oil fraction. PAHs have received focus because they can be metabolically activated in fish and bound to DNA as DNA adducts. PAHs are also shown to damage early developmental stages of fish creating several effects at low doses, including effects on heart development (Incardona et al., 2004; Brette et al., 2014). Alkylated phenols have received focus because some of these compounds have hormone-mimicking effects (Heemken et al, 2001).

Norway monitors the levels of contaminants in sediments, deployed fish and mussels and in wild caught fish, in addition to effect studies in fish and mussels. Monitoring of sediments shows very small areas with increased total hydrocarbon content or disturbed fauna. Mussels and fish deployed in cages for 6 weeks around different platforms did not show effects in distances further than 1 km (Brooks et al., 2011). No increased levels of contaminants were found in fillets of wild caught fish, although in some cases wild caught fish had increased levels of PAH metabolites in bile. The most surprising findings have been levels of DNA adducts in haddock liver from the North Sea, at levels giving rise to environmental concern in 2002 (Balk et al., 2011) and in 2011 (Grøsvik et al., 2012) because the levels were above the environmental assessment criteria (EAC) for DNA adduct in haddock liver (ICES, 2011).

5 Gaps in capacity to engage in offshore hydrocarbon industries and to assess the environmental, social and economic aspects.

The hydrocarbon industry is an extremely technical endeavour and it has evolved over a period covering more than 70 years. The offshore industry must deal with the fact that the hydrocarbon resource lies hidden, often several kilometres beneath the seafloor. Therefore, the industry employs highly skilled specialists and advanced technologies to image and sample the seafloor to find the hidden hydrocarbons. In general, oil and gas resource development requires the deployment of considerable technology to access, control, and transport the hydrocarbons.

In many parts of the world, massive oil revenues have not overcome high levels of poverty. Indeed, in some cases, they have led to significant social problems. Many oil and gas companies do not publish information what royalties, taxes and fees they pay country by country, and there is thus often a lack of transparency about these transactions. (FESS, 2006; Ross, 2008).

Only 80 out of the known 974 sedimentary basins on earth contain exploitable hydrocarbons (Li, 2011). Therefore, huge risks and uncertainty are inherently associated with hydrocarbon exploration activities. Many "dry" exploration wells are drilled for every winner (for example, see BSEE web site). The industry invests expertise, money and time to reduce risk and uncertainty so that the maximum amount of hydrocarbons is found with minimum effort and investment.

Exploration and production companies are international and operate with sophisticated technologies to make discoveries, across international boundaries wherever hydrocarbons may be found. Individual fields may cross international boundaries, further complicating their development and adding risk to investors; the Timor Gap, an area of disputed seafloor located on the border between Australia and Timor-Leste is a good example (Nevins, 2004).

In the initial stages of their development, many oil-endowed countries lacked the highly specialized knowledge or the substantial funds required to successfully find and produce offshore hydrocarbons. So, offshore hydrocarbon exploration became a primarily private-sector activity worldwide, dominated by international oil companies with the relevant skills, experience and finances needed to take on significant risk.

Over the past few decades there has been a significant shift in ownership of the offshore oil and gas global enterprise. In the 1970s, 85 per cent of all offshore oil reserves were owned by seven international oil companies (IOCs). By 2012, 18 of the top 25 oil and gas producers were National Oil Companies (NOCs), controlling 75 per cent of oil production and holding 90 per cent of the world's oil reserves (Wagner and Johnson, 2012). NOCs have competed with IOCs developing new technologies and productive resource capacity to potentially overtake the largest IOCs in size and scope. NOC's like Brazil's Petrobras, Malaysia's Petronas and Norway's Statoil have specialized in deepwater drilling technologies, once monopolized by the IOCs (Wagner and Johnson, 2012). This is an example of technology transfer that has benefitted developing countries by allowing them to participate in (if not dominate) the offshore oil and gas industry.

Assessing the environmental impact of offshore oil and gas development in developing countries has not progressed at the same pace as the capacity to develop and exploit the resource. In its assessment of environmental governance in 27 oil-producing developing countries, the World Bank (2010) found there was a "lack of a sufficiently organized administrative structure that enables efficient regulatory compliance and enforcement. Additionally, the human and financial resources needed for effective environmental governance are generally lacking." A case study from Trinidad and Tobago published by Chandool (2011) illustrates the key issues: paucity of accessible data, lack of public participation, lack of post-approval enforcement and lack of quality control in environmental impact assessment (EIA) practice.

In Malaysia, companies are required to complete an EIA that is prepared by a registered consultant (Mustafa, 2011). The industry was found to have breached its license to operate in 28 cases that went to court in 2009, with fines totalling Ringgit Malaysia (RM) 250,000 (about 76,000 dollars or an average fine of 2,700 dollars). Given the overall value of the industry (Petronas alone had a net worth in 2012 of over 157 billion dollars and had nearly 40,000 employees), there is a *prima facie* lack of proportion between the potential damage from the offences and the deterrent level of such fines.

More broadly, there are gaps in monitoring the offshore oil and gas industry by the responsible government departments. It has already been mentioned above (Section 2.4) that the collection of information on offshore pipeline failures is not coordinated in the Gulf of Mexico between state and federal jurisdictions (NRC, 1997). The data that are collected by government departments charged with monitoring the offshore oil and gas industry are often incomplete or not strategic. For example, PARLOC (2001) notes that whilst corrosion has been identified as one of the major causes of leaking pipes in the North Sea, corrosion protection data are not currently recorded in the pipeline database, so it is not possible to derive failure rates for specific types of corrosion prevention. The lack of coordination between different agencies having a share of responsibility in managing the offshore oil and gas sector was identified as a key issue by the Australian Government's Montara Commission of Inquiry Report (2010), whose conclusion was: "A single, independent regulatory body should be created, looking after safety as a primary objective, well integrity and environmental approvals. Industry policy and resource development and promotion activities should reside in government departments and not with the regulatory agency. The regulatory agency should be empowered (if that is necessary) to pass relevant petroleum information to government departments to assist them to perform the policy roles."[6] Thus, there is a gap in the capability of the responsible government departments to collect and share relevant information among different departments and authorities.

Another gap is with respect to the capacity for local communities to engage with the offshore oil and gas industry in decision-making. As has been already noted above for Trinidad and Tobago, lack of public participation in environmental impact assessments is a clear capacity gap (Chandool, 2011). The Australian company, Santos, engaged with local communities in Indonesia's Jawa Timur Province, who were concerned with the impact of offshore oil activities on their coastal habitats (Anggraeni, 2013). Although the community expressed concerns over Santos' air quality risk management and employment opportunities for local people, a dialogue was established allowing for a more satisfactory outcome for the local community (Anggraeni, 2013).

---

[6] Partly in response to this conclusion, the Australian Government established, on 1 January 2012, the National Offshore Petroleum Safety and Environmental Management Authority (NOPSEMA; http://www.nopsema.gov.au), to be the national regulator for health and safety, well integrity and environmental management for offshore oil and gas operations.

Offshore Hydrocarbon Industries

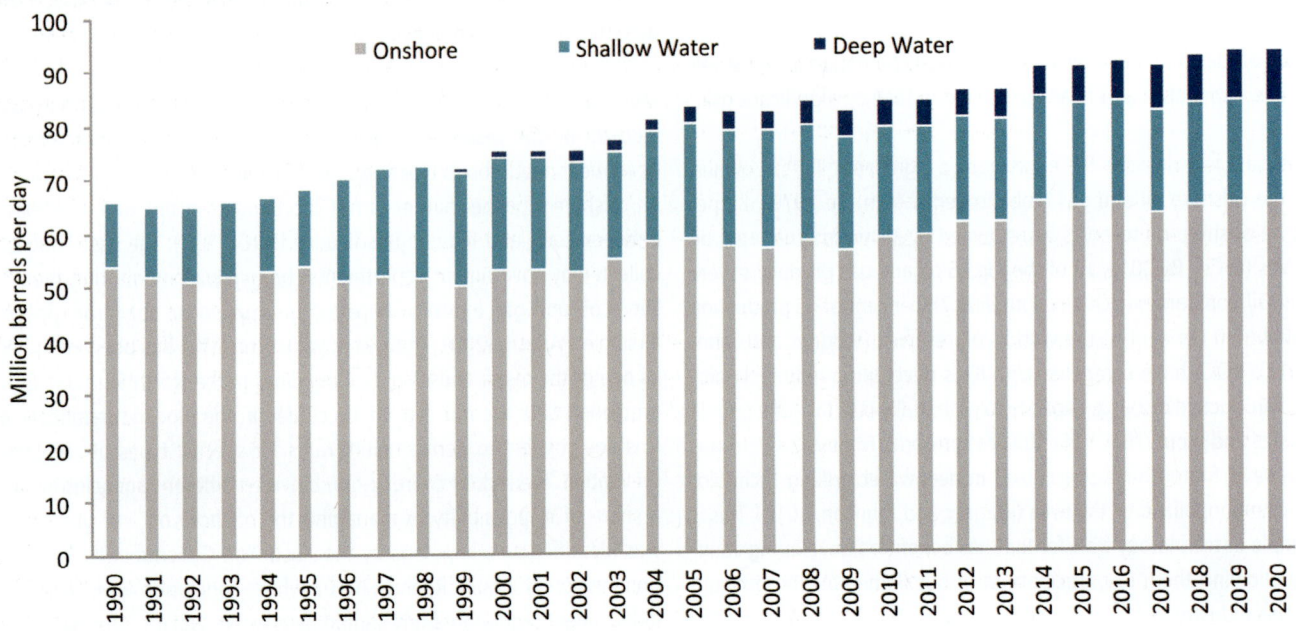

Sources: Infield Systems, BP Statistical Review 2014

Figure 21.3 | Global crude oil production, comparing onshore, shallow offshore (<100 m water depth) and offshore deep (>100 m water depth) production (from Infield, 2014). Whereas onshore production has remained stable at around 50-60 million barrels per day since the early 1970s, offshore production has grown steadily over the last four decades. Deep water production has accounted for nearly all growth since about 2005.

References

ABC News (2012) Indonesian fishermen want Montara oil spill investigated, Felicity James and Matt Brann, http://www.abc.net.au/news/2014-07-15/indonesian-fishermen-want-montara-oil-spill-investigation/5598650.

ALA, (2014). Australia's overseas development assistance program and the Montara oil spill - Environmental degradation in East Nusa Tenggara, Indonesia. Submission to Senate Standing Committee on Foreign Affairs, Defence and Trade, Inquiry into Australia's overseas aid and development assistance program, Australia's overseas aid and development assistance program Submission 31. Australian Lawyers Alliance, Sydney, Australia.

Alford, J.B., Peterson, M.S., Green, C.C., (2014). *Impacts of Oil Spill Disasters on Marine Habitats and Fisheries in North America.* CRC Press, New York, p. 340.

Anderson, C.M., Mayes, M., LaBelle, R., (2012). *Update of Occurrence Rates for Offshore Oil Spills.* Bureau of Safety and Environmental Enforcement and Bureau of Ocean Energy Management, Herndon, VA, USA, p. 87.

Anggraeni, R., (2013). *Corporate Social Responsibility Performance Assessment, Case Study Santos (Madura Offshore) Pty Ltd, Jawa Timur Province.* Unpublished Masters Thesis, IPB - Bogor Agricultural University, Kota Bogor, Indonesia.

Australian Government (2010). Report of the Montara Commission of Inquiry, in: Borthwick, D.C. (Ed.). Australian Government Department of Canberra, p. 395. http://www.industry.gov.au/AboutUs/CorporatePublications/MontaraInquiryResponse/Documents/Montara-Report.pdf

Baker, S.P., Shanahan, D.F., Haaland, W., Brady, J.E., Li, G., (2011). Helicopter crashes related to oil and gas operations in the Gulf of Mexico. *Aviation and Space Environmental Medicine*; 82:885–9.

Bakke, T.H., Klungsøyr, J., Sanni, S. (2013). Environmental impacts of produced water and drilling waste discharges from the Norwegian offshore petroleum industry. *Marine Environmental Research* 92:154-169.

Balcomb, K. C. and Claridge, D. E., (2001). A mass stranding of cetaceans caused by naval sonar in the Bahamas. *Bahamas Journal of Science*. 5, 2-12.

Balk, L., Hylland, K., Hansson, T., Berntssen, M.G.H., Beyer, J., Jonsson, G., Melbye, A., Grung, M., Torstensen, B.E., Børseth, J.F., Skarphedinsdottir, H., Klungsøyr, J. (2011). Biomarkers in Natural Fish Populations Indicate Adverse Biological Effects of Offshore Oil Production. *PLoS ONE*, 6(5) DOI: 10.1371/journal.pone.0019735.

Barents Observer, (2014). World's northernmost offshore oil in transport, April 2014, http://barentsobserver.com/en/energy/2014/04/worlds-northernmost-offshore-oil-transport-22-04

Barth, H.-J., (2007). Crab induced salt marsh regeneration after the 1991 Gulf War oil spill. *Aquatic Ecosystem Health & Management* 10, 327-334.

Beall, J. and Ferreti, A., (2012). Forage Pétrolier Offshore: La nécessaire prévention des risques. *Marine & Océans* 35, 62-63.

Boebel, O., Clarkson, P., Coates, R., Larter, R., O'Brien, P. E., Ploetz, J., Summerhayes, C., Tyack, T., Walton, D. W. H. and Wartzok, D., (2005). Risks posed to the Antarctic marine environment by acoustic instruments: A structured analysis - SCAR action group on the impacts of acoustic technology on the Antarctic marine environment. *Antarctic Science*. 17, 533-540.

Boswell, R. and Collett, T. S., (2011). Current perspectives on gas hydrate resources. *Energy and Environmental Science* 4, 1206-1215.

Brette, F., Machado, B., Cros, C., Incardona, J.P., Scholz, N.L., Block, B.A. (2014). Crude Oil Impairs Cardiac Excitation-Contraction Coupling in Fish. *Science* 343:772-776.

Brooks, S.J., Harman, C., Grung, M., Farmen, E., Ruus, A., Vingen, S., Godal, B.F., Barsiene, J., Andreikenaite, L., Skarpheoinsdottir, H., (2011). Water Column Monitoring of the Biological Effects of Produced Water from the Ekofisk Offshore Oil Installation from 2006 to 2009. *Journal of Toxicology and Environmental Health-Part A-Current Issues*. 74: 582-684.

Carls, M. G., Marty, G. D. and Hose, J. E., (2002). Synthesis of the toxicological impacts of the Exxon Valdez oil spill on Pacific herring (Clupea pallasi) in Prince William Sound, Alaska, U.S.A. *Canadian Journal of Fisheries and Aquatic Sciences.* 59, 153-172.

CBD. Convention on Biological Diversity (2014). *Report on the expert workshop on underwater noise and its impacts on marine and coastal biodiversity*, London, 25–27 February 2014, London, p. 22.

Centers for Disease Control and Prevention (2013). Fatal Injuries in Offshore Oil and Gas Operations — United States, 2003–2010. *Morbidity and Mortality Weekly Report*, April 26, 2013, 62(16); 301-304. http://www.cdc.gov/mmwr/preview/mmwrhtml/mm6216a2.htm

Cerchio, S., Strindberg, S., Collins, T., Bennett, C., Rosenbaum, H., (2014). Seismic Surveys Negatively Affect Humpback Whale Singing Activity off Northern Angola. *PLoS ONE* 9, e86464.

Chandool, C., (2011). Oil and Gas and the Environment in Trinidad and Tobago: Experience and Challenges. Impact Assessment and Responsible Development for Infrastructure, Business and Industry 31st Annual Meeting of the International Association for Impact Assessment. *IAIA-11 Conference Proceedings*, Puebla, Mexico.

Crone, T. J. and Tolstoy, M., (2010). Magnitude of the 2010 Gulf of Mexico Oil Leak. *Science*. 330, 634

Day, M.D., (2008). Decommissioning of Offshore Oil and Gas Installations, in: Orszulik, S. (Ed.), *Environmental Technology in the Oil Industry*. Springer Netherlands, pp. 189-213.

Eckle, P., Burgherr, P., and Michaux, E. (2012). Risk of large oil spills: a statistical analysis in the aftermath of Deepwater Horizon. *Environmental Science and Technology* 46:13002-13008.

The Economist (2014). In Deep Water, *The Economist*, 22 February, 2014 http://www.economist.com/node/21596990/print.

Edgar, G.J., Kerrison, L., Shepherd, S.A. and Toral-Granda, M.V., (2003). Impacts of the Jessica oil spill on intertidal and shallow subtidal plants and animals. *Marine Pollution Bulletin*. 47, 276-283.

EIA. (2014) United States Energy Information Administration, *U.S. Crude Oil and Natural Gas Proved Reserves 2013*, downloaded March, 2014, http://www.eia.gov/naturalgas/crudeoilreserves/.

Elshorbagy, W. (2005). *Overview of marine pollution in the Arabian Gulf with emphasis on pollutant transport modeling*, Arabian Coast 2005 Conference, Dubai, 20 pp.

FESS. (2006). *Oil and Gas and Conflict - Development Challenges and Policy Approaches.* Foundation for Environmental Security and Sustainability (FESS), Washington DC, p. 44. http://www.fess-global.org/files/OilandGas.pdf

Fowler, S.W., Readman, J.W., Oregioni, B., Villeneuve, J.-P., Mckay, K., (1993). Petroleum hydrocarbons and trace metals in nearshore Gulf sediments and biota before and after the 1991 war: an assessment of temporal and spatial trends. *Marine Pollution Bulletin* 27, 171-182.

Fraser, F. C. (1977). Royal fishes: the importance of the dolphin, in *Functional Anatomy of Marine Mammals, Volume 3* (ed. R. J. Harrison), Academic Press, London and New York.

French-McCay, D. P., (2004). Oil spill impact modelling: Development and validation. *Environmental Toxicology and Chemistry*. 23, 2441-2456

Galenson, W., (1986). *A Welfare State Strikes Oil: The Norwegian Experience*, University Press of America Ltd., Lantham, MD.

Geoscience Australia and Bureau of Resource and Energy Economics, (2012). *Australian Gas Resource Assessment 2012*, Canberra. https://www.ga.gov.au/products/servlet/controller?event=GEOCAT_DETAILS&catno=74032.

GESAMP. (2007). Estimates of oil entering the marine environment from sea-based activities. IMO/FAO/UNESCO-IOC/UNIDO/ WMO/IAEA/UN/UNEP Joint Group of Experts on the Scientific Aspects of Marine Environmental Protection, *Reports and Studies* 75. p. 96.

Gordon, J., Gillespie, D., Potter, J., Frantzis, A., Simmonds, M.P., Swift, R., Thompson, D., (2004). A Review of the Effects of Seismic Survey on Marine Mammals *Marine Technology Society Journal* 37, 16-34.

Grøsvik, B.E., Kalstveit, E., Liu, L., Nesje, G., Westrheim, K., Bertnssen, M.H.G., Le Goff, J., Meier, S. (2012). Condition monitoring in the water column 2011: Oil hydrocarbons in fish from Norwegian waters. *IMR Report* No. 19-2012.

Heemken, O.P., Reincke, H., Stachel, B., Theobald, N., (2001). The occurrence of xenoestrogens in the Elbe river and the North Sea. *Chemosphere*. 45: 245-259.

Harris, A., (2013). Piper Alpha 25 years on - have we learned the lessons? *Engineering and Technology Magazine*. The Institution of Engineering and Technology, London. http://eandt.theiet.org/magazine/2013/07/never-stop-learning.cfm

Harris, A.H., Lloyd, M.G. and Newlands, D.A., (1988). *The impact of oil on the Aberdeen economy*, Avebury, Aldershot.

Harris, P.T., (2014). Shelf and deep-sea sedimentary environments and physical benthic disturbance regimes: a review and synthesis. *Marine Geology* 353, 169–184.

Harris, P.T., MacMillan-Lawler, M., Rupp, J., Baker, E.K., (2014). Geomorphology of the oceans. *Marine Geology* 352, 4-24.

Hays, (2013). *Oil & Gas global salary guide 2013 - global salaries and recruiting trends*, Hays Recruiting, London, p. 32.

Hazen, T.C., Dubinsky, E.A., DeSantis, T.Z., Andersen, G.L., Piceno, Y.M., Singh, N., Jansson, J.K., Probst, A., Borglin, S.E., Fortney, J.L., Stringfellow, W.T., Bill, M., Conrad, M.E., Tom, L.M., Chavarria, K.L., Alusi, T.R., Lamendella, R., Joyner, D.C., Spier, C., Baelum, J., Auer, M., Zemla, M.L., Chakraborty, R., Sonnenthal, E.L., D'haeseleer, P., Holman, H.-Y.N., Osman, S., Lu, Z., Van Nostrand, J.D., Deng, Y., Zhou, J., Mason, O.U., (2010). Deep-Sea Oil Plume Enriches Indigenous Oil-Degrading Bacteria. *Science* 330, 204-208.

Hester, M.W. and Mendelssohn, I.A., (2000). Long-term recovery of a Louisiana brackish marsh plant community from oil-spill impact: Vegetation response and mitigating effects of marsh surface elevation. *Marine Environmental Research*. 49, 233-254.

Holdway, D. and Heggie, D.T., (2000). Direct hydrocarbon detection of produced formation water discharge on the Northwest Shelf, Australia. *Estuarine, Coastal and Shelf Science* 50, 387-402.

ICES. (2011). Report of the Study Group on Integrated Monitoring of Contaminants and Biological Effects (SGIMC). *ICES/OSPAR report*. Pp 265.

IMO. (International Maritime Organization) (1989). *Guidelines and Standards for the Removal of Offshore Installations and Structures on the Continental Shelf and in the Exclusive Economic Zone* (IMO Resolution A.672 (16)).

Incardona, J.P., Collier, T.K., Scholz, N.L,. (2004). Defects in cardiac function precede morphological abnormalities in fish embryos exposed to polycyclic aromatic hydrocarbons. *Toxicology and Applied Pharmacology*. 196(2): 191-205.

Infield, (2014). *Offshore outlook*. Data provided to the authors via e-mail.

Irvine, G.V., Mann, D.H., Carls M.G., Reddy C., Nelson R.K. (2014), *Exxon Valdez oil after 23 years on rocky shores in the Gulf of Alaska: Boulder armor stability and persistence of slightly weathered oil.* Abstract for 2014 Ocean Sciences Meeting, 23-28 February 2014, Hawaii, USA.

ITOPF. (2007). International Tanker Owners Pollution Federation, http://www.itopf.com/index.html.

Jewett, S.C., Dean, T.A., Smith, R.O. and Blanchard, A., (1999). 'Exxon Valdez' oil spill: Impacts and recovery in the soft-bottom benthic community in and adjacent to eelgrass beds. *Marine Ecology Progress Series*. 185, 59-83.

Jones, R.J. and Hayward, A.J., (2003). The effects of Produced Formation Water (PFW) on coral and isolated symbiotic dinoflagellates of coral. *Marine and Freshwater Research*. 54, 153-162.

Jørgensen, D., (2012). Rigs-to-reefs is more than rigs and reefs. *Frontiers in Ecology and the Environment* 10, 178-179.

Kashubsky, M., (2011). A Chronology of Attacks on and Unlawful Interferences with, Offshore Oil and Gas Installations, 1975 to 2010. *Perspectives on Terrorism* 5. http://www.terrorismanalysts.com/pt/index.php/pot/article/view/offshore-gas-and-oil-attacks.

Khatib, Z. and Verbeek, P., (2003). Water to value – produced water management for sustainable field development of mature and green fields. *Journal of Petroleum Technology*. 55, 26-28.

Khordagui, H., Al-Ajmi, D., (1993). Environmental impact of the Gulf War: An integrated preliminary assessment. *Environmental Management* 17, 557-562.

Kostka, J.E., Prakash, O., Overholt, W.A., Green, S.J., Freyer, G., Canion, A., Delgardio, J., Norton, N., Hazen, T.C., Huettel, M., (2011). Hydrocarbon-Degrading Bacteria and the Bacterial Community Response in Gulf of Mexico Beach Sands Impacted by the Deepwater Horizon Oil Spill. *Applied Environmental Microbiology* 77, 7962-7974.

Kvenvolden, K.A. and Cooper, C.K., (2003). Natural seepage of crude oil into the marine environment. *Geo-Marine Letters*. 23, 140-146.

Li, G., (2011). *World atlas of oil and gas basins*. John Wiley and Sons, 496 pp.

London Convention and Protocol/UNEP (2009) *Guidelines for the Placement of Artificial Reefs*, Adopted at the Thirtieth Consultative Meeting of Contracting Parties to the 1972 Convention on the Prevention of Marine Pollution by Dumping of Wastes and Other Matter and Third Meeting of Contracting Parties to the 1996 Protocol to the 1972 Convention on the Prevention of Marine Pollution by Dumping of Wastes and Other Matter, 27-31 October, 2008.

Macreadie, P.I., Fowler, A.M., Booth, D.J., (2011). Rigs-to-reefs: will the deep sea benefit from artificial habitat? *Frontiers in Ecology and the Environment* 9, 455-461.

Macreadie, P.I., Fowler, A.M., Booth, D.J., (2012). Rigs-to-reefs policy: can science trump public sentiment? Frontiers in Ecology and the Environment 10, 179-180.

Mallakh, R.E., Noreng, Ø. and Poulson, B.W., (1984). *Petroleum and Economic Development*, Lexington Books, Lexington, MA.

Marine Board, (1994). *Improving the Safety of Marine Pipelines* - Committee on the Safety of Marine Pipelines, Marine Board, Commission on Engineering and Technical Systems, Division on Engineering and Physical Sciences, National Research Council. National Academies Press, Washington DC. 121 pp.

McCauley, R.D., Fewtrell, J., Duncan, A.J., Jenner, C., Jenner, M.-N., Penrose, J.D., Prince, R.I.T., Adhitya, A., Murdoch, J., McCabe, K., (2000). Marine seismic surveys - a study of environmental implications. *APEA Journal* 40, 692-708.

Michel, J., Hayes, M.O., Getter, C.D., Cotsapas, L., (2005). The Gulf War oil spill twelve years later: consequences of eco-terrorism. *International Oil Spill Conference Proceedings*, pp. 957-961.

MSL, (2000). *Appraisal and development of pipeline defect assessment methodologies* - Report prepared by MSL Engineering Limited for Minerals Management

Service, U.S. Department of the Interior. MSL Engineering Limited, Ascot, Berkshire, UK, p. 93.

Mufson, S., (2010). Federal records show steady stream of oil spills in gulf since 1964, *Washington Post*, Washington DC. http://www.washingtonpost.com/wp-dyn/content/article/2010/07/23/AR2010072305603.html.

Mustafa, M., (2011). The role of environmental impact assessment in addressing marine environmental issues arising from oil and gas activities: examples from Malaysia, 2011 *International Conference on Environment and BioScience IPCBEE*. IACSIT Press, Singapore, Cairo, Egypt, pp. 58-62.

National Research Council, (2003). *Oil in the Sea III: Inputs, Fates, and Effects*. Washington, DC: The National Academies Press, 280 pp. http://www.nap.edu/catalog.php?record_id=10388

Nevins, J., (2004). Contesting the Boundaries of International Justice: State Countermapping and Offshore Resource Struggles between East Timor and Australia. *Economic Geography* 80, 1-22.

Newlands, D., (2000). 'The Oil Economy,' pp 126-152 in Fraser and Lee (eds.), *Aberdeen 1800 – 2000: a New History*, Tuckwell Press Ltd., East Lothian.

Noreng, Ø .,(1980), *The Oil Industry and Government Strategy in the North Sea*, Croom Helm Ltd., London.

Norsk olje & gass. *Environmental Report* (2013) (https://www.norskoljeoggass.no/en/Publica/Environmental-reports/Environmental-report-2013/)

Norwegian Petroleum Directorate (2012) *The petroleum sector – Norway's largest industry*. http://www.npd.no/en/Publications/Facts/Facts-2012/Chapter-3/

Nowacek, D.P., Bröker, K., Donovan, G., Gailey, G., Racca, R., Reeves, R.R., Vedenev, A.I., Weller, D.W., Southall, B.L., (2013). Responsible Practices for Minimizing and Monitoring Environmental Impacts of Marine Seismic Surveys with an Emphasis on Marine Mammals. *Aquatic Mammals* 39, 356-377.

Nowacek, D.P., Thorne, L.H., Johnston, D.W., Tyack, P.L., (2007). Responses of cetaceans to anthropogenic noise. *Mammal Review* 37, 81-115.

NRC, (1997). *A Focus on Offshore Safety: Recent Reports by the Marine Board of the National Research Council*. Marine Board of the National Research Council, Washington DC, 20 pp.

O'Brien, P E., Arnt, W., Everson, I., Gohl, K., Gordon, J. C. D., Goss, K., Kremser, U., Laws, R., Martin, T., Ploetz, J., Rodhouse, P. G., Rogers, T., Thompson, D., Sexton, M., Wilson, R. and Woehler, E., (2002). *Impacts of marine acoustic technology on the Antarctic environment*. SCAR Ad Hoc Group on marine acoustic technology and the environment.

Oil & Gas UK, (2013). *Decommissioning of pipelines in the North Sea region*. Oil and Gas UK, London, p. 52. www.oilandgasuk.co.uk

Oil Spill Commission (2011). National Commission on the BP Deepwater Horizon Oil Spill and Offshore Drilling staff working papers, *The amount and fate of the oil*, http://oscaction.org/resource-center/staff-papers/.

Offshore Magazine, (2013). Seismic survey vessel capabilities go up while count goes down. *Offshore Magazine*, vol. 73, issue 3.

Olsgard, F. and Gray, J. S., (1995). A comprehensive analysis of the effects of offshore oil and gas exploration and production on the benthic communities of the Norwegian continental shelf. Marine Ecology Progress Series. 122, 277-306.

OPEC. (2013). *Annual statistical bulletin*. Organization of the Petroleum Exporting Countries, Vienna, Austria.

OSPAR. (2009). Assessment of impacts of offshore oil and gas activities in the North-East Atlantic, *Offshore Industry Series*. The Convention for the Protection of the Marine Environment of the North-East Atlantic, London.

OSPAR. (2010). *Quality Status Report 2010* (http://qsr2010.ospar.org/en/index.html)

OSPAR. (2012a), Draft OSPAR report on discharges, spills and emissions from offshore oil and gas installations in 2010. OIC 12/6/4-E.

OSPAR. (2012b). Report of the OSPAR Workshop on research into possible effects of regular platform lighting on specific bird populations, *Offshore Industry Series*. OSPAR Commission, London.

Owen, E.W. (1975) *Trek of the Oil Finders*, American Association of Petroleum Geologists, Memoir 6, 800 pp.

Page, H.M., Dugan, J.E., Dugan, D.S., Richards, J.B., Hubbard, D.M. (1999). Effects of an offshore oil platform on the distribution and abundance of commercially important crab species. *Marine Ecological Progress Series* 185: 47-57.

Palmer, A.C., and King, R.A. (2004). *Subsea Pipeline Engineering*. PennWell Corporation, Tulsa, Oklahoma, 570 pp.

PARLOC, (2001). *The Update of Loss of Containment Data for Offshore Pipelines - Prepared by Mott MacDonald Ltd. for: The Health and Safety Executive, The UK Offshore Operators Association and The Institute of Petroleum*. Mott MacDonald Ltd., Croydon, p. 161.

Paula, A.F., Creed, J.C., (2005). Spatial distribution and abundance of nonindigenous coral genus Tubastraea (Cnidaria, Scleractinia) around Ilha Grande, Brazil. *Brazil Journal of Biology* 65, 661-673.

Pyenson, N.D., Gutstein, C.S., Parham, J.F., Le Roux, J.P., Chavarria, C.C., Little, H., Metallo, A., Rossi, V., Valenzuela-Toro, A.M., Velez-Jurabe, J., Santellie, C.M., Rogers, D.R., Cozzuol, M.A., and Suarez, M.E., (2014). Repeated mass strandings of Miocene marine mammals from Atacama Region of Chile point to sudden death at sea. *Proceedings of the Royal Society B: Biological Sciences*, 22 April 2014 vol. 281 no. 1781. http://rspb.royalsocietypublishing.org/content/281/1781/20133316.full

Pinder, D., (2001). Offshore oil and gas: global resource knowledge and technological change. *Ocean & Coastal Management* 44, 579-600.

Prince, I., (2004). North Sea abandonment heats up, *E&P Magazine*. Hart Energy, Houston, Texas, November, 2004.

Rach, N., (2013). BSEE: Offshore decommissioning accelerates in Gulf of Mexico, *Offshore Engineer*, August 2013.

Richardson, W.J., Greene, C.R.J., Malme, C.I., Thomson, D.H., (1995). *Marine mammals and noise*. Academic Press, San Diego.

Ritchie, W., 1993. The short-term impact of the Braer oil spill in Shetland and the significance of coastal geomorphology. Scottish Geographical Magazine. 109, 50-56.

Ross, M.L., (2008). *Blood Barrels - Why Oil Wealth Fuels Conflict*, Foreign Affairs. Council on Foreign Relations, Tampa, Florida.

Ruppel, C.D., (2011). Methane hydrates and contemporary climate change. *Nature Education Knowledge* 3, 29.

Sammarco, P.W., Porter, S.A., Cairns, S.D., (2010). A new coral species introduced into the Atlantic Ocean - Tubastraea micranthus (Ehrenberg 1834) (Cnidaria, Anthozoa, Scleractinia): An invasive threat? *Aquatic Invasions* 5, 131-140.

Sánchez, F., Velasco, F., Cartes, J.E., Olaso, I., Preciado, I., Fanelli, E., Serrano, A. and Gutierrez-Zabala, J.L., (2006). Monitoring the Prestige oil spill impacts on some key species of the Northern Iberian shelf. *Marine Pollution Bulletin*. 53, 332-349

Shaw, D. G., 1992. The Exxon Valdez oil-spill: ecological and social consequences. *Environmental Conservation*. 19, 253-258

Shaw, R.F., D.C. Lindquist, M.C. Benfield, T. Farooqi and J.T. Plunket. (2001). *Offshore Petroleum Platforms: Functional Significance for Larval Fish Across Longitudinal and Latitudinal Gradients*. Prepared by the Coastal Fisheries Institute, Louisiana State University. U.S. Department of the Interior, Minerals Management Service,

Gulf of Mexico OCS Region, New Orleans, LA. OCS Study MMS 2002-077. 107 pp.

Shrimpton, Mark and Storey, Keith, (2001). *The Effects of Offshore Employment in the Petroleum Industry: a Cross-National Perspective*, Minerals Management Service, US Department of the Interior, Herndon, VA.

Shrimpton, M., (2002). *'Benefiting Communities: Lessons from Around the Atlantic'*, paper SPE 74057 presented at the Society of Petroleum Engineers International Conference on Health, Safety and Environment in Exploration and Production, Kuala Lumpur, Malaysia, March 2002.

S.L. Ross Environmental Research Ltd., (2009). *Assessing Risk and Modelling a Sudden Gas Release Due to Gas Pipeline Ruptures*. A Report to: U.S. Department of the Interior Minerals Management Service. S.L. Ross Environmental Research Ltd., Ottawa, p. 93.

Sorensen, C., (2013). *The latest move in offshore drilling: oil rigs on the ocean floor*. Macleans. http://www.macleans.ca/economy/business/maybe-its-better-down-where-its-wetter/.

Stafford, M., Williams, N. (1996). *Pipeline leak detection study*, UK Health and Safety Executive - Offshore Technology Report. Bechtel Ltd., London, p. 58.

Stantec, (2012). *Socio-economic Benefits from Oil and Gas Activity in Newfoundland and Labrador: 2008-2010*, Petroleum Research NL, St. John's, NL.

Swan, J.M., Neff, J.M. and Young, P.C., eds., (1994). *Environmental Implications of Offshore Oil and Gas Development in Australia - the findings of an independent scientific review*. Australian Petroleum Exploration Association, Sydney.

Thompson, P.M., Brookes, K.L., Graham, I.M., Barton, T.R., Needham, K., Bradbury, G., Merchant, N.D., (2013). Short-term disturbance by a commercial two-dimensional seismic survey does not lead to long-term displacement of harbour porpoises. *Proceedings of the Royal Society (B)* 280, 1-8.

Turner, J., (2013). *Sea change: offshore safety and the legacy of Piper Alpha*, Offshore-technology.com. Kable Intelligence Limited. http://www.offshore-technology.com/features/feature-piper-alpha-disaster-anniversary-offshore-safety/

UNEP (2011). *Environmental Assessment of Ogoniland*, http://www.unep.org/nigeria/.

UK National Archives (2013). http://webarchive.nationalarchives.gov.uk/+/http://hmrc.gov.uk/statistics/

Wagner, D., and Johnson, B., (2012). The Rise of National Oil Companies, *Huff Post*, 28 April, 2012, http://www.huffingtonpost.com/daniel-wagner/the-rise-of-national-oil-_b_2138965.html.

Whomersley P. and Picken G.B. (2003). Long-term dynamics of fouling communities found on offshore installations in the North Sea. *Journal of the Marine Biological Association of the United Kingdom* 83(5): 897-901.

Woodson, R.D., (1990). *Offshore pipeline failures*, MSc Thesis, Department of Civil Engineering. University of California, Berkley, Berkley, California, p. 74.

World Bank, (2010). Environmental governance in oil-producing developing countries – Findings from a survey of 32 countries. *Extractive Industries for Development Series #17*. World Bank.

Xu, Y., Wu, J., Zhu, L., Wang, N., (2009). Study on the Unstable Geological Factors of Oil Gas Submarine Pipeline in Shallow Sea Shelf Area, *Proceedings ICPTT 2009*, pp. 281-295.

Yamamoto, T., Nakaoka, M., Komatsu, T., Kawai, H. and Ohwada, K., (2003). Impacts by heavy-oil spill from the Russian tanker Nakhodka on intertidal ecosystems: Recovery of animal community. *Marine Pollution Bulletin*. 47, 91-98.

Yang, L., Guo, Y., Shi, B., Kuang, C., Xu, W., Cao, S., (2012). Study of Scour around Submarine Pipeline with a Rubber Plate or Rigid Spoiler in Wave Conditions. Journal of Waterway, Port, Coastal, and Ocean Engineering 138, 484-490.

22 Other Marine-Based Energy Industries

Contributors:
Peyman Eghtesadi Araghi (Convenor and Lead Member), Amardeep Dhanju, Lars Golmen, Peter Harris (Co-Lead Member)

1 Marine Renewable Energy Resources: Background

This chapter concerns ocean processes that are viable sources of renewable energy in various forms, such as offshore wind, waves, tides, ocean currents, marine biomass, and energy from ocean thermal differences among different layers (Appiott et al., 2014). Most of these energy forms are maintained by the incoming heat from the sun, so they represent indirect solar energy. Tidal energy is an exception, driven by the varying gravitational forces that the moon and sun exert on both the earth and its oceans (Butikov 2002). Marine renewable energy offers the potential to meet the increasing global energy demand, while reducing long-term carbon emissions. Although some marine renewable energy resources are still in a conceptual stage, other sources have been operational with varying degrees of technical and commercial success. The following section briefly discusses various forms of marine renewable energy sources that are currently in operation or in a demonstration phase.

1.1 Offshore Wind Power: Background

Offshore wind power relates to the installation of wind turbines in large water bodies. On average, winds blow faster and more uniformly at sea than on land, and a faster and steadier wind means less wear on the turbine components and more electricity generated per turbine (Musial et al., 2006). The potential energy produced from wind is proportional to roughly the cube of the wind speed. As a result, a marginal increase in wind speed results in a significantly larger amount of energy generation. For instance, a turbine at a site with an average wind speed of 25 km/h would provide roughly 50 per cent more electricity than the same turbine at a site with average wind speeds of 22 km/h.

Offshore wind power is also the most developed form of marine renewable energy in terms of technology development, policy frameworks, and installed capacity. Turbine design and other project elements for offshore wind have benefited significantly from research on and experience with land-based wind energy projects and offshore oil and gas development (Steen and Hansen, 2014). It is already a viable source of renewable energy in many regions and is attracting global attention because of its large-scale resource potential, also often close to major electrical load centers in coastal areas. In light of these factors, offshore wind energy appears to have the greatest immediate potential for energy production, grid integration, and climate change mitigation.

1.2 Ocean Wave Energy: Background

As the wind flows over the ocean, air-sea interface processes transfer some of the wind energy to the water, forming waves which store this energy as potential energy and kinetic energy (Special Report on Renewable Energy Sources and Climate Change, 2011). The immense power of waves can be observed at the coast, where this energy can have considerable impacts on coastal landscapes, shoreline topography, and infrastructure. Efforts are now underway to tap this resource for electric generation using wave energy conversion (WEC) devices. WECs transform mechanical energy from the surface motion of ocean waves or from velocity fluctuations below the surface into electrical current.

1.3 Tidal Power: Background

Tides are regular and predictable changes in the height of the ocean, driven by gravitational and rotational forces between the Earth, Moon, and Sun, combined with centrifugal and inertial forces. Many coastal areas experience roughly two high tides and two low tides per day (called "semi-diurnal" tides); however in some locations there is only one tidal cycle per day (these are "diurnal" tides Special Report on Renewable Energy Sources and Climate Change, 2011). Tidal power can be harnessed either through a barrage or through submerged tidal turbines in straits or sounds. Tidal barrages involve the use of a dam across an inlet. Sluice gates on the barrage allow the tidal basin to fill on the incoming high tides and to empty through the turbine system on the outgoing tide (U.S. EIA, 2013).

Energy extraction using tidal barrages is derived from the potential energy created when elevation differences develop between two water bodies separated by a dam or barrage, which is analogous to the way a turbine operates in a hydroelectric plant on a dammed river. Conversely, submerged tidal turbines that operate without a barrage only rely on the kinetic energy of the freely moving water. Because water is about 800 times denser than air and is more corrosive, tidal turbines must be much sturdier than wind turbines (U.S. EIA, 2013).

1.4 Ocean Current Energy: Background

Ocean currents are the continuous flow of ocean waters in certain directions, driven or controlled by wind flows, salinity, temperature gradients, gravity, and the Earth's rotation or coriolis; (U.S DOE, 2013). Although surface ocean currents are generally wind driven, most deep ocean currents are a result of thermohaline circulation – a process driven by density differences in water due to temperature (thermo) and salinity (haline) in different parts of the ocean. Currents driven by thermohaline circulation move much slower than surface currents (NOAA, 2007). Many large and powerful ocean currents, such as the Gulf Stream off the east coast of the United States and the Kuroshio Current off the east coast of Japan, represent an enormous source of untapped energy that can be harnessed through large underwater turbines.

1.5 Ocean Thermal Energy Conversion (OTEC): Background

The OTEC system produces electricity from the natural thermal gradient of the ocean between the surface and subsurface ocean waters in tropical and subtropical regions. Heat stored in warm surface water is used to create vapor to drive a turbine and generator, and cold, deep water pumped to the surface recondenses the vapor (Avery and Wu, 1994). The OTEC heat engine is usually configured to operate as a thermodynamic Rankine vapor cycle in which a low-boiling-point working fluid (such as ammonia) is evaporated by heat transfer from the surface seawater and produces electricity by expanding through a turbine connected to a generator. The vapor exiting the turbine is condensed with the cold deep water. This is the Closed-Cycle process. The Open-Cycle process uses expendable water/seawater as the driving fluid, under low (<0.1 Atmosphere) pressure. After the flash evaporation of the fluid, fresh, potable water may be collected on the condensers, while new seawater is being evaporated upstream. Additional freshwater may be collected in a second exchanger utilizing the remaining Delta-T after the first stage. Thus, the Open-Cycle process generates both electricity and potable water.

The OTEC systems have been demonstrated to work successfully, but no large-scale plant has been built yet. A benefit of OTEC is that it produces constant base-load electricity, in contrast to other forms of ocean energy sources that fluctuate according to varying winds and currents.

1.6 Osmotic Power: Background

Salinity gradient energy is an often-overlooked potential source of renewable energy from the ocean. The mixing of freshwater and seawater that occurs where rivers and streams flow into the salty ocean releases large amounts of energy. Various concepts on how to make use of this salinity gradient have existed for over twenty years. One such concept is Pressure-Retarded Osmosis (PRO), in which seawater is pumped into a pressure chamber where the pressure is less than the osmotic pressure difference between fresh river water (low salinity water) and seawater (higher salinity water). Freshwater then flows through a semi-permeable membrane and increases the volume (or pressure) within the chamber; a turbine is spun as the pressure is relieved. Early technologies were not considered to be promising, primarily because they relied on expensive membranes. Membrane technologies have advanced, but they remain the main technical barrier to economical osmotic energy production (Appiott et al., 2014). Also, water on both sides must be low in particulates and other solids, eliminating many rivers from being a potential freshwater source.

1.7 Marine Biomass Energy: Background

Some researchers are looking towards marine biomass, including seaweeds and marine algae as a viable source of biofuel. Interest in marine biomass is driven both by the potential productivity of microalgae, which is tenfold greater than that of agricultural crops, and because, unlike first-generation biofuels, microalgae do not require arable land or freshwater, nor do they compete with food production (NERC, 2014). Marine ecosystems are highly productive because they cycle energy and nutrients much more rapidly and efficiently than terrestrial ecosystems. Marine algae are photosynthetic aquatic plants that use light as the energy source and seawater as a growth medium. Algae can be harvested and processed into biofuels, including biodiesel and bioethanol.

Biodiesel is a non-toxic and biodegradable fuel that is being used in existing diesel engines without requiring significant modification. Bioethanol can also be used as fuel when mixed with gasoline. Algae grown for biofuels can also provide a sink for carbon dioxide, thereby contributing to climate change mitigation (replacing fossil CO_2 with biogenic CO_2 emissions). Algae are an economical choice for biodiesel production because of their wide availability and low cost. Despite these advantages, however, offshore production of algae is still developing and most algae production takes place onshore.

Figure 22.1 | Photo credit: Principle Power. The WindFloat Prototype (WF1) floating wind turbine, deployed by Principle Power in 2011, 5km off the coast of Aguçadoura, Portugal. The WF1 is outfitted with a Vestas v80 2.0 MW offshore wind turbine. As of December 2015, the system has produced in excess of 16 GWh of renewable energy delivered to the local grid.

2 Resource Assessment and Installed Capacity (Global and Regional Scales)

2.1 Offshore Wind Capacity

An assessment of the world's exploitable offshore wind resources has placed the estimates around 22 TWa[1] (Arent et al., 2012) which is approximately nine times greater than the International Energy Agency's (IEA) 2010 estimate of average global electricity generation (IEA, 2012). According to a report by the United States National Renewable Energy Laboratory (NREL), offshore wind resource potential for contiguous United States and Hawaii for annual average wind speeds greater than 7.0 m/sec and at 90 m above the surface is 4,150 GWa (Schwartz et al., 2010). Similarly, a report by the European Environmental Agency (EEA) calculated the technical offshore wind power potential, based on the

---

[1] TWa: The average number of terawatt-hours, not terawatts, over a specified time period. For example, over the course of one year, an average terawatt is equal to 8,760 terawatt-hours, or 24 hours x 365 days x 1 terawatt.

forecasted costs of developing and running wind power projects in 2020 at 2,850 GWa (EEA, 2009). This figure does not account for spatial use conflicts in developing the wind resource offshore.

As of 2012, large-scale commercial offshore wind projects and demonstration-scale or pilot projects are already operational in the Belgium, China, Denmark, Finland, Germany, Ireland, Italy, Japan, Netherlands, Norway, Portugal, Republic of Korea, Spain, Sweden, United Kingdom of Great Britain and Northern Ireland and United States of America (EWEA, 2008; RenewableUK, 2010; 4C Offshore, 2013; WWEA, 2014). Currently, the North Sea region is considered to be the global leader in offshore wind, both in installed and planned capacity and in technical capability. By the end of 2013, the offshore wind industry has achieved a cumulative global installed capacity of 7,357 MW (WWEA, 2014). Offshore wind turbines can be either bottom-mounted or floating. Currently, most offshore wind projects are bottom-mounted; floating wind turbines are still in the demonstration phase. Moreover, most installations are near the shore in relatively shallow waters due to the higher cost of transmission cabling further offshore, and due to the technical and economic challenges of installing turbines in deeper waters.

2.2 Wave Energy Capacity

The global exploitable wave energy resource is estimated at around 3,700 GWa (Mørk et al., 2010), which is large enough to meet the average global electric generation (IEA, 2012). Wave energy can be harnessed using either floating or fixed conversion devices. Floating devices convert the wave energy by coupling it to a hydraulic system as the device is lifted up and down by the movements of the waves. Fixed devices generally use the oscillating water column generated by the wave to push air (or water) through a turbine. Other types of wave energy technology, such as overtopping devices and attenuators, are also undergoing testing and demonstration.

The world's first grid-connected wave energy device, a 500 kW unit in Scotland (United Kingdom) called the Islay Limpet, has been operating successfully since 2000 (UKDTI, 2004). The Aguçadoura Wave Farm, the world's first utility-scale wave energy project, was launched off the coast of Portugal in September 2008. This installation, developed by PelamisWave Power, utilized three 750 kW devices with a total capacity of 2.25 MW (RenewableUK, 2010). The project operated for two months before technical problems forced the developers to abandon it.

2.3 Tidal Energy Capacity

The global tidal energy resource is estimated to be 3,000 GWa by the World Offshore Renewable Energy Report 2004-2008 (UK DTI, 2004), however, less than 3 per cent of this energy is located in areas suitable for power generation. Tidal energy is feasible only where strong tidal flows are amplified by factors such as funneling in estuaries, making it highly site-specific (UK DTI, 2004). Traditionally, tidal energy has been harnessed using large barrages in areas of high tidal ranges. Many countries, such as Canada, China, France, Republic of Korea, Russian Federation and the United Kingdom have sites with large tidal ranges that are viable for tidal energy capture facilities. The Sihwa Lake Tidal Power Station in the Republic of Korea, which has been operational since August 2011, is the world's largest tidal power barrage with a capacity of 254 MW, surpassing the 240 MW Rance Tidal Power Station in France, which has been generating power since 1967. Numerous projects have also been proposed in other areas, including in the Severn Estuary in the United Kingdom (Hall, 2012).

As part of its technology development initiative, the United States Department of Energy (US DOE) has funded research into several new types of technology, including a turbine under development by Verdant Power Inc. Verdant Power was the first company in the United States of America to be granted a license for a commercial tidal energy project, and looks to build upon an earlier demonstration project in New York's East River with an installation of up to 30 turbines along the strait that connects Long Island Sound and the Atlantic Ocean in the New York harbour.

2.4 Ocean Current Energy Capacity

There are no commercial grid-connected turbines currently operating, although a number of prototypes and demonstration units are under development. In 2014, the United States Bureau of Ocean Energy Management (BOEM) issued a lease to Florida Atlantic University (FAU) for testing ocean current turbines. FAU's Southeast National Marine Renewable Energy Center (SNMREC) plans to deploy experimental demonstration devices in areas located 10 to 12 nautical miles offshore Florida (Bureau of Ocean Energy Management, 2014).

2.5 Ocean Thermal Energy Conversion (OTEC) Capacity

OTEC technologies have been tested as early as the 1930s. However, OTEC has been limited to small-scale pilot projects, and has yet to encourage much investment and commercial development (US DOE, 2008). Research initiatives in the France, India, Japan and United States of America and elsewhere are currently examining and testing differ-

Figure 22.2 | Waves4power OWC plant in operation offshore Sweden (Reprinted with permission from Waves4power AB).

ent types of OTEC technologies (Lockheed Martin Corporation, 2012). A modern-type, but very small OTEC plant was constructed in Hawaii, United States, in 1979 (Kullenberg et al., 2008), and similar demonstration projects have been proposed by other nations (IOES, 2015). Most experience is derived from land-based plants. For floating installations, reference material is mostly derived from design studies, but with successful demonstrations in Hawaii (MiniOTEC, OTEC-1) and Okinawa, Japan (OTEC Okinawa, 2014).

2.6 Osmotic Power Capacity

The world's first complete prototype osmotic power plant was launched in Norway in 2009 by Statkraft. This plant is located in Tofte, southwest of Oslo. According to the company's assessment, osmotic-power technologies remain several years away from commercial viability (Kho, 2010). Statkraft recently (2014) decided to shelve development plans.

2.7 Marine Biomass Energy Capacity

Research into algae production has largely been guided down three tracks: open and covered ponds, photobioreactors, and fermenters; the first two are the most widely pursued. Siting algae farms in ocean areas has also been investigated (Lane, 2010). In the United States, the National Aeronautics and Space Administration (NASA) is investigating the feasibility of growing algae in floating photobioreactors on the outer (geomorphic) continental shelf or in the open ocean. The Offshore Membrane Enclosures for Growing Algae (OMEGA) system would use freshwater algae and wastewater in the photobioreactors to produce biofuel while also cleaning wastewater, creating oxygen, and providing a sink for carbon dioxide (NASA, 2012). As the technology is developed further and States with favourable growing conditions begin to look towards marine renewable energy, this option may become commercially viable in the future.

3 Environmental Benefits and Impacts from Offshore Renewable Energy Development

Marine renewable energy installation and generation invariably has environmental impacts, both positive and negative. These impacts depend on the installation size and footprint, location, and the use of specific technology. A major positive impact of ocean renewable energy is the provision of low-carbon electricity. Analysis suggests that the carbon intensity of offshore wind and marine hydrokinetic resources, such as wave and tidal power, is more than an order of magnitude lower than fossil fuel generation. In a life cycle analysis, greenhouse gas emissions from wave energy projects average between 13-50 gCO_2eq/kWh[2], tidal current projects emit approximately 15 gCO_2eq/kWh, and offshore wind projects between 4-6 gCO_2eq/kWh (Raventós et al., 2010); these are to be compared with emissions closer to 800-1000 gCO_2eq/kWh for coal power plants and 400-600 gCO_2eq/kWh for natural gas power plants (POST, 2006 and 2011).

Other environmental benefits include no emissions of toxic air or water pollutants (such as NO_x [Nitrogen dioxide], SO_2 [sulfur dioxide], mercury, particulate matter, and thermal pollution from cooling water discharge) and minimal land-use disturbance with the exception of land-use changes related to assembly, equipment loading/offloading and cable landfalls along the coast. In addition, there are potential biodiversity benefits from the installation of offshore turbines or marine energy conversion devices. Offshore renewable energy (ORE) structures increase the amount of hard substrate for colonization and provide marine organisms with artificial reefs. Such structures also create increased heterogeneity in the area: this is important for maintaining species diversity and density (Langhamer, 2012). Investigations have found greater fish abundance in the vicinity of offshore turbines compared to the surrounding areas (Wilhelmsson et al., 2006). One negative impact is that invasive species can find new habitats in these artificial reefs and possibly adversely influence the native habitats and associated environment (Langhamer, 2012).

The broader environmental impacts from marine renewable energy can be understood in the context of an ecological risk assessment framework developed by the United States Environmental Protection Agency (EPA). This approach provides a conceptual model for developing a systematic view of possible ecological effects (McMurray, 2008). The terminology needed for this model requires defining stressors and receptors. Stressors are features of the environment that may change due to installation, operation, or decommissioning of the facilities, and receptors are ecosystem elements with a potential for some form of response to the stressor(s) (Boehlert and Gill, 2010). Stressors can be considered in terms of different stages of development (survey, construction, operation, and decommissioning) as well as the duration, frequency, and intensity of the disturbance. Project size and scale are also determining factors in the magnitude of stressors and receptors. Within this framework, we discuss various ecological impacts in the following sections.

Potential negative impacts on flora and fauna including marine mammals, birds, and benthic organisms, as well as impacts on the larger ecosystem may occur from offshore renewable energy (OREI) development. Some of these impacts are limited to the construction phase, while other impacts span operation and decommissioning phases (Linley et. al., 2009). Potential impacts include habitat loss or degradation at various stages of a project life cycle; injurious noise and displacement of marine mammals from pile driving of wind and tidal-stream generators. Tide power turbines may also induce local seabed scouring and/or changes to the current regime, with unintended consequences for biota. Turbine construction may induce mortality due to physical collision with the OREI structures; effects of operational noise; and electromagnetic field (EMF) impacts from submerged cables (U.S. Department of Interior, 2011).

---

2   gCO_2eq/kWh : grams of CO_2 equivalent per kilowatt hour of generation.

Noise created during pile-driving operations involves sound pressure levels that are high enough to impair hearing in marine mammals and disrupt their behaviour at a considerable distance from the construction site (Thomsen et. al., 2006). During pile driving for the Horns Rev II offshore wind project in Denmark, a negative effect was detected out to a distance of 17.8 km (Brandt et. al, 2011). Although it has been observed that marine mammals temporarily abandon the construction area, they tend to return once pile driving operations cease. Acoustic impact on marine mammals is a major concern and an important topic of assessment and mitigation strategies in many States. More information is required almost everywhere to understand impacts on and responses by marine organisms to such stresses.

Fixed and moving parts of Ocean Renewable Energy (ORE) devices can lead to fatal strikes or collisions with birds and aquatic fauna. Blades used in marine turbines, such as those in ocean current or tidal energy devices, are relatively slow-moving and therefore not considered to pose a significant threat to wildlife (Scott and Downie, 2003). However the speed of the tip of some horizontal axis rotors could be an issue for cetaceans, fish, or diving strike birds (Boehlert and Gill, 2010). Operation of the SeaGen tidal energy device in Strangford Lough, United Kingdom, considered the presence of seals and porpoises and the potential threat of blade strikes; to minimize strike risk, the turbine was shut down when the presence of seals was observed within 30 meters (Copping et al., 2013). Similarly, investigation of long-tailed geese and ducks in and around the Nysted offshore wind project in Denmark suggests that flocks employ an avoidance strategy. Research suggests that the percentage of flocks entering the wind project area decreased significantly from pre-construction to initial operation. Overall, less than 1 per cent of ducks and geese migrated close enough to be at any risk of collision (Desholm and Kahlert, 2005). This avoidance strategy or adjustment of flight paths, a form of receptor, has also been observed in other projects, such as Horns Rev (NERI, 2006). It is important to highlight that the additional distances travelled by migratory birds to avoid these wind farms were relatively trivial (around 500 m) compared to their total migratory trajectory of 1,400 km. However, construction of further utility-scale projects could have a cumulative impact on the population, especially when considered in combination with other human actions (Masden et. al., 2009).

Submerged cables carrying electricity from ORE devices to onshore substations emit low-frequency Electric and Magnetic Fields (EMF). Marine and avian species are sensitive and responsive to naturally occurring magnetic fields; these are commonly used for direction-finding using the Earth's geomagnetic field. Anthropogenic sources of EMFs are an overlay to naturally occurring sources, and as these sources become increasingly common, there are potential impacts on marine organisms. Industry standards for the design of submarine cables require shielding, which restricts directly emitted electric fields, but cannot shield the magnetic field component of EMFs (Boehlert and Gill, 2010). Moreover, an alternating current (AC) magnetic field has a rotational component that induces an additional electric field in the surrounding environment.

There is evidence that EMF's from wind farms can cause disturbance to air traffic control radar systems (De la Vega et al., 2013).

Magnetic fields are strongest over the cables, decreasing rapidly with vertical and horizontal distance from the cables. In projects where the electric current is delivered along two sets of cables that were separated by at least several meters, the magnetic field appeared as a bimodal peak (Normandeau et al., 2011). Studies suggest that behavioural effects of EMF on species occur, although the impacts vary significantly among species.

Thermal aspects of electricity-transmission cables should also be considered. When electric energy is transported, a certain amount is lost as heat, leading to an increased temperature in the cable surface and subsequent warming of the surrounding marine environment (Merck and Wasserthal, 2009). Temperature changes can affect benthic organisms, although data on measureable impacts are sparse. Increased temperatures can also attract marine organisms, exposing them to a higher amount of EMF radiation.

In addition, there are other potential impacts such as chemical effects from potential spills or leaching of anti-fouling paints from ocean renewable energy devices (Boehlert and Gill, 2010), or impacts on benthic creatures and certain fish species which have not yet been fully investigated and assessed. Many such effects are localized, depend on marine and avian biodiversity in a region and can be understood only through comprehensive site-specific environmental impact assessment. Substantial work is proceeding to gather and disseminate available information and data on ecological impacts of ocean energy devices. The International Energy Agency-Implementing Agreement on Ocean Energy Systems (OES) Annex IV[3] is maintaining the Tethys database, an important reference for both developers and policy makers (http://tethys.pnnl.gov/).

4   Socioeconomic Benefits and Impacts from Offshore Renewable Energy Deployment

Socioeconomic impacts cover a range of issues, including access to the ocean, visual impacts amenity, impact on coastal and offshore cultural heritage sites, and other uses of the ocean, including recreational tourism and fisheries, related to offshore renewable energy sites. In many regions, these issues have been examined within the context of comprehensive marine spatial planning. Marine spatial planning provides an understanding of the extent to which certain activities take place in an area identified for offshore renewable energy development and provides a baseline assessment of critical ecological and cultural sites.

---

[3]   Annex IV. Assessment of Environmental Effects and Monitoring Efforts for Ocean Wave, Tidal and Current Energy Systems, Tethys.pnnl.gov.

Sociological surveys of coastal residents, ocean users and other stakeholders have been widely used to assess perceived and experienced impacts from offshore renewable energy facilities. Survey results from two operational offshore wind projects in Denmark, Hons Rev and Nysted, indicate a generally positive attitude among coastal residents (Ladenburg et al, 2006). Relative open access to the projects for marine resource extraction could be a reason for high levels of public approval of the projects. Also, the visual impact of the installations may not have affected those people who were surveyed, since wind farms have caused a loss of amenity in other areas (see below). Both the projects provide access to sailing and fishing within their waters. The Nysted offshore wind project provides access for fishing with net and line, Horns Rev allows only line fishing, and bottom-trawling fishing is prohibited in both projects. Fishing can be further restricted as setting of lobster traps may be limited near cables and turbines. Wave, tidal power, and ocean current energy sources have not yet been commercially deployed at a large enough scale to enable an assessment of potential socio-economic impacts.

The effect of impaired visual amenity from ORE deployment can affect property prices, result in loss of recreational value, and reduce demand for tourism in coastal areas. It can also affect historic and culturally significant resources. These impacts are expected to be more prominent for an offshore wind project than for a wave, tidal, or ocean current installation, as the latter will be underwater and of smaller scale, for similar capacity. Research in the United Kingdom, England and Wales, which have extensive offshore wind capacity, concludes that offshore wind projects have a measureable impact on property prices. The impact is more pronounced in areas which are closest to the wind projects. On the other hand, a small increase in housing value is also seen in areas where wind projects are not visible, indicating a potential economic benefit to landowners near non-visible wind project operations due to an increased rental rate (Gibbons, 2014). This latest assessment is consistent with earlier published studies, which strongly suggest that, given a choice, consumers prefer offshore wind projects sited away from the coast and, in some cases, completely out of sight (Ladenburg and Lutzeyer, 2012).

Utility-scale offshore renewable energy projects can use significant ocean space and impose restrictions for navigational purposes. A large project, if placed along an existing navigational route, can increase the distance that ships and boats would be required to travel. The extent of transit through offshore renewable energy projects depends on safety issues and ease of access. In Denmark, transit through offshore wind energy projects is possible via certain routes; in Germany, navigation is allowed as close as 500 meters (Albrecht et al., 2013). The International Association of Marine Aids to Navigation and Lighthouse Authorities has promulgated recommendations on how to mark different types of offshore renewable energy installations so that they are conspicuous under different meteorological conditions. This, along with proper charting of installations and associated cables, can limit navigational risks (Detweiler, 2011).

Planning, construction, and maintenance of offshore renewable energy operations have the potential to create direct and indirect employment in various sectors including manufacturing, construction, operation, and maintenance. In 2013, offshore wind represented 3.6 per cent of the United Kingdom's electricity supply, contributed close to 1 billion United Kingdom pounds to the economy and supported 20,000 jobs, including 5,000 direct jobs (Offshore Renewable Energy Catapult, 2014). As the industry continues to grow, it has the potential to add thousands of new jobs, not just in the United Kingdom, but around Europe and other parts of the world that form a critical link to the supply chain. In Europe, offshore wind energy and ocean energy create 7-9 job-years/MW[4] during construction and installation. In addition, offshore wind projects can generate up to 11 job-years/MW in manufacturing, and 0.2 jobs/MW for operation and maintenance during the operational years of a project. These figures are comparable to those for conventional sources like coal, although offshore wind and other marine renewable sources have no job-creation potential related to fuel extraction, processing, and transportation for the life of a project (Energy [r]evolution, 2012).

Above and beyond the employment generation factor, offshore renewable energy also offers other intrinsic economic and electrical system integration benefits. For instance, many offshore renewable energy projects are sited, or proposed, close to densely populated coastal areas. Proximity to major electrical load centres can significantly reduce the cost of transmission and offset transmission congestion. Moreover, ocean renewable sources, particularly offshore wind power, offers the additional value proposition of load coincidence in many regions (Bailey and Wilson, 2014).

5   Offshore Renewable Energy Assessment Capacity Gaps

A capacity gap is a lack of information that, if available, would or could identify whether environmental effects [of a project] will have substantial negative impacts (McMurray, 2012). In many regions, sufficient knowledge exists in the near-shore and offshore waters to provide an initial baseline assessment, although it is often insufficient to provide a site-specific impact assessment. Significant capacity gaps exit in assessing environmental, social, and economic impacts from deploying devices in the marine environment. Most forms of ocean renewable energies have still not reached commercial scale, although some are at a high Technological Readiness Level (TRL).

Certain marine environments and species present additional challenges in addressing capacity gaps. For instance, it is technically difficult to obtain information on benthic biota, as compared to species in the pelagic zone. Due to the lack of high-quality benthic information, resource managers and developers are often required to conduct time-consuming and

---

4   Job-years per MW denote the total amount of labour needed to manufacture equipment or construct a power plant that will deliver a peak output of one megawatt of power (The Energy Policy Institute, 2013).

resource-extensive surveys before siting decisions can be finalized. In the pelagic zone, marine migratory species pose additional challenges for site characterization. Further site-specific research is required to understand migratory species such as whales to ensure that project siting has minimal impact on migratory routes or traditional foraging grounds. Similarly, impacts on avian species in the offshore environment, particularly migratory bird species and bats, have not been fully understood and require further research and assessment.

In the absence of operational utility-scale projects, it is often difficult to determine the socioeconomic impacts of an emerging renewable energy technology. One way to address such capacity gaps is to make long-term monitoring an integral part of the construction and operation phase, though if long-term monitoring regimes are too costly developers may be dissuaded from pursuing commercial projects. Studies and surveys assessing impacts before and during the operation of a project can provide valuable information on impacts, and can suggest substantive mitigation measures to address those impacts.

The knowledge and capacity gaps should be addressed within a comprehensive framework that considers all ecological resources and human uses in an area. This framework, also referred to as marine spatial planning, provides a process for analysing and allocating spatial and temporal distribution of human activities in marine areas to achieve ecological, economic, and social objectives that are usually specified through a political process (UNESCO, 2014). States are increasingly using marine spatial planning as the tool for identifying and siting offshore renewable energy projects. More importantly, the collaborative processes at the heart of marine spatial planning foster relationships and linkages among ocean uses, stakeholders and resources managers to enhance the quality of scientific information and traditional knowledge available. This collaboration and information exchange can lead to better-informed siting decisions and can minimize social and environmental impacts.

6 Conclusion

Offshore renewable energy is an immense resource awaiting efficient usage. Technological progress to harness the resource is steadily increasing around the world. When fully developed and implemented, ocean renewable energy can enhance the diversity of low-carbon energy options and provide viable alternatives to fossil fuel sources. For developing countries and new growing economies, installing renewable energy systems represents a viable path towards a low-carbon future.

To achieve a commercial break-through such that ocean renewable energy becomes cost-competitive, many governments have funded Research and Development (R&D) projects and provided financial support for technological developments and demonstrations within this sector. Traditional commercial funding sources are often insufficient to achieve this goal in the long-term, so innovative strategies are required. In addition, higher education courses on ocean renewable energies must be promoted, and research to understand and mitigate potential environmental and socio-economic impacts of these new technologies must be conducted. Given its immense potential, offshore renewable energy is well positioned to be part of a carbon-constrained energy future.

References

4C Offshore (2013). *Global Offshore Wind Farms Database*. Retrieved 23 July 2013, from http://www.4coffshore.com/offshorewind/.

Albrecht, C., Wagner, A., Wesselmann, K. and Korb, M. (2013). *The impact of offshore wind energy on tourism: Good practices and perspectives for the South Baltic region*. Retrieved from http://www.offshore-stiftung.com/60005/Uploaded/Offshore_Stiftung%7C2013_04SBO_SOW_tourism_study_final_web.pdf.

Appiott, J., Dhanju, A., and Cicin-Sain, B. (2014). Encouraging renewable energy in the offshore environment. *Ocean and Coastal Management*, 90, 58-64.

Arent, D., Sullivan, P., Heimiller, D., Lopez, A., Eurek, K., Badger, J., Jørgensen, H., and Kelly, M. (2012). *Improved Offshore Wind Resource Assessment in Global Climate Stabilization Scenarios*. National Renewable Energy Laboratory (NREL). Technical Report: NREL/TP-6A20e55049. Retrieved from http://www.nrel.gov/docs/fy13osti/55049.pdf.

Avery, W. and Wu, C. (1994). *Renewable Energy from the Ocean: A guide to OTEC*. Oxford University Press.

Bailey, B. and Wilson, W. (2014). *The value proposition of load coincidence and offshore wind. North American Windpower*. Retrieved from http://www.nawindpower.com/issues/NAW1401/Cover_NAW1401.html.

Boehlert G.W. and Gill, A. B. (2010). Environmental and ecological effects of ocean renewable energy development: Current synthesis. *Oceanography*, 23 (2): 68-81.

Brandt, M. J., Diederichs, A., Betke, K. and Nehls, G. (2011). Responses of harbour porpoises to pile driving at the Horns Rev II offshore wind farm in the Danish North Sea. *Marine Ecology Progress Series*, 421:201-216. Doi:10.3354/meps08888.

Bureau of Ocean Energy Management (BOEM) (2014). *BOEM issues first renewable energy lease for marine hydrokinetic technology testing*. Retrieved from http://www.boem.gov/press06032014/.

Butikov, E.I. (2002). A dynamic picture of the oceanic tides. *American Journal of Physics*, 70, 1001-101. doi: 10.1119/1.1498858.

Copping A., Hanna, L., Whiting, J., Geerlofs, S., Grear, M., Blake, K., Coffey, A., Massaua, M., Brown-Saracino, J., and Battey, H. (2013). *Environmental Effects of Marine Energy Development around the World for the OES Annex IV*, [Online], Available: www.ocean-energy-systems.org.

De la Vega, D., Matthews, J.C.G., Norin, L., Angulo, I. (2013). Mitigation techniques to reduce the impact of wind turbines on radar services. *Energies* 6, 2859-2873.

Desholm, M. and Kahlert, J. (2005). Avian collision risk at an offshore wind farm. *Biological letters*, 1:296-298. doi:10.1098/rsbl.2005.0336.

Detweiler, G.H. (2011). *Offshore renewable energy installations: impact on navigation and marine safety*. The Coast Guard Proceedings of the Marine Safety and Security Council. Retrieved from http://www.uscg.mil/proceedings/spring2011/articles/19_Detweiler.pdf.

Energy [r]evolution: A sustainable EU 27 Energy Outlook (2012). Retrieved from http://www.greenpeace.org/eu-unit/Global/eu-unit/reports-briefings/2012%20pubs/Pubs%203%20Jul-Sep/E%5bR%5d%202012%20lr.pdf.

European Environmental Agency (EEA) (2009). Europe's onshore and offshore wind energy potential: An assessment of environmental and economic constraints. *EEA Technical report* No 6/2009. Retrieved from http://www.energy.eu/publications/a07.pdf.

European Wind Energy Association (EWEA) (2008). *Pure Power: Wind Energy Scenarios up to 2030*. European Wind Energy Association. Retrieved from http://www.ewea.org/fileadmin/ewea_documents/documents/00_POLICY_document/PP.pdf.

Gibbons, S. (April 2014). *Gone with the Wind: Valuing the visual impacts of wind turbines through house prices*. Spatial Economic Research Center (SERC) Discussion Paper 159. Retrieved from http://docs.wind-watch.org/Gone-with-wind-SERC-April-2014.pdf.

Hall, C. (2012). *Largest renewable energy projects in the World. Energy Digital*. Retrieved from http://www.energydigital.com/top_ten/top-10-business/largest-renewable-energy-projects-in-theworld.

Institute of Ocean Energy (IOES) (website accessed May, 2015) *Study of OTEC technology in the world*. Retrieved from http://www.ioes.saga-u.ac.jp/en/about_oetc_03.html

International Energy Agency (IEA) (2012). *Key World Energy Statistics, 2012*. Retrieved from http://www.iea.org/publications/freepublications/publication/kwes.pdf.

Kho, J. (2010). *Osmotic Power: a Primer*. Retrieved from http://www.statkraft.com/Images/Osmotic_Power_report_KACHAN_061010%5B1%5D_tcm9-19279.pdf.

Kullenberg, G., Mendler de Suarez, J., Wowk, K., McCole, K. and Cicin-Sain, B. (2008). *Policy brief on climate, oceans, and security*. In: Presented at the 4th Global Conference on Oceans, Coasts, and Islands, April 7-11, 2008, Hanoi, Vietnam. Retrieved from http://globaloceanforumdotcom.files.wordpress.com/2013/03/climate-and-oceans-pb-april2.pdf.

Ladeburg, J. and Lutzeyer, S. (2012). The properties of visual disamenity costs of offshore wind farms – the impacts on wind farm planning and cost of generation. International Association for Energy Economics: Energy Forum. Retrieved from http://www.iaee.org/en/publications/fullnewsletter.aspx?id=23.

Ladenburg, J., Tranberg, J. and Dubgaard, A. (2006). Chapter 8: Socioeconomic effects, in *Danish offshore wind: Key environmental issues*. Retrieved from http://188.64.159.37/graphics/Publikationer/Havvindmoeller/danish_offshore_wind.pdf.

Lane, J. (2010). *Salt water: The tangy taste of energy freedom, Biofuels Dig*. Retrieved from http://www.biofuelsdigest.com/bdigest/2010/04/09/salt-water-the-tangy-taste-of-energy-freedom/.

Langhamer, O. (2012). Artificial reef effect in relation to offshore renewable energy conversion: State of the art. *The Scientific World Journal*, 2012. Doi:10.1100/2012/386713.

Linley, A., Laffont, K., Wilson, B., Elliott, M., Perez-Dominguez, R. and Burdon, D. (2009). *Offshore and coastal renewable energy: Potential ecological benefits and impacts of large-scale offshore and coastal renewable energy projects*. Retrieved from http://web2.uconn.edu/seagrantnybight/documents/Energy%20Docs/nercmarinerenewables.1.pdf.

Lockheed Martin Corporation (2012). *Ocean Thermal Extractable Energy Visualization* http://energy.gov/sites/prod/files/2013/12/f5/1055457.pdf.

Masden, E.A., Haydon, D.T., Fox, A.D., Furness, R.W., Bullman, R. and Desholm, M. (2009). Barriers to movement: impacts of wind farms on migrating birds. *ICES Journal of Marine Science*, 66 (4): 746-753. doi: 10.1093/icesjms/fsp031.

McMurray, G. R. (2008). Wave energy ecological effects workshop: Ecological assessment briefing paper. Pp. 25-66 in *Ecological Effects of Wave Energy Development in the Pacific Northwest: A Scientific Workshop. October 11-12, 2007*, G.W. Boehlert, G.R. McMurray, and C.E. Tortorici, eds, NOAA Technical Memorandum NMFS-F/SPO-92.

McMurray, G.R. (2012). *Gap analysis: Marine renewable energy environmental effects on the U.S. West Coast*. Oregon Marine Renewable Energy Environmental Science Conference, Corvallis, Oregon, November 28-29, 2012. Retrieved from http://hmsc.oregonstate.edu/rec/gap-analysis.

Merck, T. and Wasserthal, R. (2009). *Assessment of environmental impacts of cables*. OSPAR Commission. Retrieved from http://qsr2010.ospar.org/media/assessments/p00437_Cables.pdf.

Mørk, G., Barstow, S., Kabuth, A., and Pontes, M.T. (2010). Assessing the global wave energy potential. In: *Proceedings of OMAE2010 29th International Conference on Ocean, Offshore Mechanics and Arctic Engineering*, June 6e11, Shanghai, China. Retrieved from http://www.oceanor.no/related/59149/paper_OMAW_2010_20473_final.pdf.

Musial, W., Butterfield, S. and Ram, B. (2006). *Energy from Offshore Wind*. Conference paper presented at Offshore Technology Conference, Houston, TX, May 1-4, 2006. Retrieved from http://www.nrel.gov/docs/fy06osti/39450.pdf.

National Aeronautics and Space Administration (NASA) (2012). *OMEGA: Offshore Membrane Enclosure for Growing Algae*. National Aeronautics and Space Administration (NASA), 17 April 2012. Web. 30 August 2012. Retrieved from http://www.nasa.gov/centers/ames/research/OMEGA/index.html.

National Environmental Research Institute (NERI) (2006). *Final result of bird studies at the offshore wind farms at Nysted and Horns Rev*, Denmark. Retrieved from http://www.vattenfall.dk/da/file/69662-Horns-Rev--Nysted-birds_7842547.pdf.

National Oceanic and Atmospheric Administration (NOAA) (2007). *Welcome to currents*. Retrieved from http://oceanservice.noaa.gov/education/tutorial_currents/lessons/currents_tutorial.pdf.

National Environment Research Council (NERC) (2014). *Algal bioenergy special interest group*. Retrieved from http://www.nerc.ac.uk/research/programmes/algal/background.asp?cookieConsent=A.

Normandeau, Exponent, Tricas, T. and Gill, A. (2011). *Effects of EMFs from Undersea Power Cables on Elasmobranchs and Other Marine Species*. U.S. Dept. of the Interior, Bureau of Ocean Energy Management, Regulation, and Enforcement, Pacific OCS Region, Camarillo, CA. OCS Study BOEMRE 2011-09. Retrieved from http://www.data.boem.gov/PI/PDFImages/ESPIS/4/5115.pdf.

Offshore Renewable Energy Catapult (2014). *Generation energy and prosperity: Economic impact study of the offshore renewable energy industry in the UK*. Retrieved from https://ore.catapult.org.uk/documents/2157989/0/ORE+Catapult+UK+economic+impact+report/2c49a781-ff1e-462f-a0c7-b25eb9478b0f?version=1.0.

OTEC Okinawa (2014). Retrieved from http://otecokinawa.com/en/.

Parliamentary Office of Science and Technology (POST) (October 2006). Carbon footprint of electricity generation. (Report Number 268). Retrieved from http://www.parliament.uk/documents/post/postpn268.pdf

Parliamentary Office of Science and Technology (POST) (June 2011). Carbon footprint of electricity generation. (Report number 383). Retrieved from http://www.parliament.uk/documents/post/postpn_383-carbon-footprint-electricity-generation.pdf

Raventós, A., Simas, T., Moura, A., Harrison, G., Thomson, C. and Dhedin, J. (2010). *Life cycle assessment for marine renewables*. (Deliverable D6.4.2). Retrieved from http://mhk.pnl.gov/wiki/images/e/eb/EquiMar_D6.4.2.pdf

RenewableUK (2010). *Marine Renewable Energy*. RenewableUK, London. Retrieved from http://www.bwea.com/marine/resource.html.

Schwartz, M., Heimiller, D., Haymes, S., and Musial, W. (2010). *Assessment of offshore wind energy resources for the United States*. National Renewable Energy Laboratory (Technical Report NREL/TP-500-45889). Retrieved from http://energy.gov/sites/prod/files/2013/12/f5/45889.pdf

Scott, W. and Downie, A.J. (2003). *A review of possible marine renewable energy development projects and their natural heritage impacts from a Scottish perspective*. Scottish Natural Heritage Commissioned Report F02AA414.

SRREN (2011). *Special report on Renewable Energy Sources and Climate Change Mitigation*). Retrieved from http://srren.ipcc-wg3.de/report

Steen, M., and Hansen, G.H. (2014). Same sea, different ponds: Cross-sectional knowledge spillovers in the North Sea. *European Planning Studies*, 22 (10), 2030-2049. doi: 10.1080/09654313.2013.814622.

The Energy Policy Institute (March 2013). *Employment estimates in the energy sector: Concepts, methods, and results*. Retrieved from http://epi.boisestate.edu/media/16370/employment%20estimates%20in%20the%20energy%20sector;%20concepts%20methods%20and%20results.pdf.

Thomsen, F., Lüdemann, K., Kafemann, R. and Piper, W. (2006). *Effects of offshore wind farm noise on marine mammals and fish*. A report funded by COWRIE Ltd. Retrieved from http://users.ece.utexas.edu/~ling/2A_EU3.pdf.

UK Department of Trade and Industry (UK DTI) (2004). *The World Offshore Renewable*

Energy Report 2004-2008. UK Department of Trade and Industry, London. Retrieved from http://www.ppaenergy.co.uk/Insights/d,czoxMToiMzU2ODY2ZGYyZDIiOw==.html.

United Nations Educational, Scientific and Cultural Organization (UNESCO) (2014). *Marine Spatial Planning*. Retrieved from http://www.unesco-ioc-marine-sp.be/marine_spatial_planning_msp.

U.S. Department of Interior (U.S. DOI) (2011). Effects of EMFs from undersea power cables on Elasmobranchs and other marine species. OCS Study BOEMRE 2011-09. Retrieved from http://www.data.boem.gov/PI/PDFImages/ESPIS/4/5115.pdf.

U.S. Department of Energy (US DOE) (2008). *Ocean Thermal Energy Conversion* (Updated 30 Dec 2008). U.S. Department of Energy, Washington, DC. Available at: http://www.energysavers.gov/renewable_energy/ocean/index.cfm/mytopic¼50010?print.

U.S. Department of Energy (US DOE) (2012). *Turbines off NYC East River will provide power to 9,500 residents*. Retrieved from http://energy.gov/articles/turbines-nyc-east-river-will-provide-power-9500-residents.

U.S Department of Energy (DOE) (2013). *Assessment of energy production potential from ocean currents along the United States coastline*. Retrieved from http://www1.eere.energy.gov/water/pdfs/energy_production_ocean_currents_us.pdf

U.S. Energy Information Administration (EIA) (2013). *Hydropower explained: Tidal power*. Retrieved from http://www.eia.gov/energyexplained/index.cfm?page=hydropower_tidal.

World Wind Energy Association (WWEA) (2014). *Key statistics of World Wind Energy Report 2013*. Retrieved from http://www.wwindea.org/webimages/WWEA_WorldWindReportKeyFigures_2013.pdf

Wilhelmsson, D., Malm, T. and Öhman, M.C. (2006). The influence of offshore windpower on demersal fish. *ICES Journal of Marine Science*, 63: 775-784. doi:10.1016/j.icesjms.2006.02.001.

23 Offshore Mining Industries

Contributors:
Elaine Baker (Convenor), Françoise Gaill, Aristomenis P. Karageorgis, Geoffroy Lamarche, Bhavani Narayanaswamy, Joanna Parr, Clodette Raharimananirina, Ricardo Santos, Rahul Sharma and Joshua Tuhumwire (Lead Member)

Consultors:
James Kelley, Nadine Le Bris, Eddy Rasolomanana, Alex Rogers and Mark Shrimpton

Offshore Mining Industries — Chapter 23

1 Introduction

Marine mining has occurred for many years, with most commercial ventures focusing on aggregates, diamonds, tin, magnesium, salt, sulphur, gold, and heavy minerals. Activities have generally been confined to the shallow near shore (less than 50 m water depth), but the industry is evolving and mining in deeper water looks set to proceed, with phosphate, massive sulphide deposits, manganese nodules and cobalt-rich crusts regarded as potential future prospects.

Seabed mining is a relatively small industry with only a fraction of the known deposits of marine minerals (Figure 23.1) currently being exploited. In comparison, terrestrial mining is a major industry in many countries (estimated to be worth in excess of 700 billion United States dollars per year, PWC, 2013). Pressure on land-based resources may spur marine mining, especially deep seabed mining. However, global concerns about the impacts of deep seabed mining have been escalating and may influence the development of the industry (Roche and Bice, 2013).

The exploitation of marine mineral resources is regulated on a number of levels: global, regional and national. At the global level, the most important applicable instrument is the United Nations Convention on the Law of the Sea (UNCLOS). It is complemented by other global and regional instruments. At the national level, legislation governing the main marine extractive industries (i.e. aggregate mining) may be extremely complex and governed in part by national or subnational authorities (Radzevicius et al., 2010). As regards national legislation to regulate deep-sea mining, terrestrial mining legislation often applies to the continental shelf or EEZ, rather than specific deep-sea mining legislation (EU, 2014). However many Pacific Islands States, that are gearing up for deep seabed mining have made significant efforts to adopt concise and comprehensive domestic laws (SPC, 2014).

2 Scale and significance of seabed mining

2.1 Sand and gravel extraction

Aggregates are currently the most mined materials in the marine environment and demand for them is growing (Bide and Mankelow, 2014). Due to the low value of the product, most marine aggregate extractions are carried out at short distances from landing ports close to the consumer base and at water depths of less than 50 m (UNEP, 2014).

In Europe, offshore sand and gravel mining is an established industry in Denmark, France, Germany, the Netherlands and the United Kingdom of Great Britain and Northern Ireland (Earney, 2005). Marine aggregates

The boundaries and names shown and the designations used on this map do not imply official endorsement or acceptance by the United Nations.
Figure 23.1 | Global distribution of known marine mineral resources (from Rona, 2008).

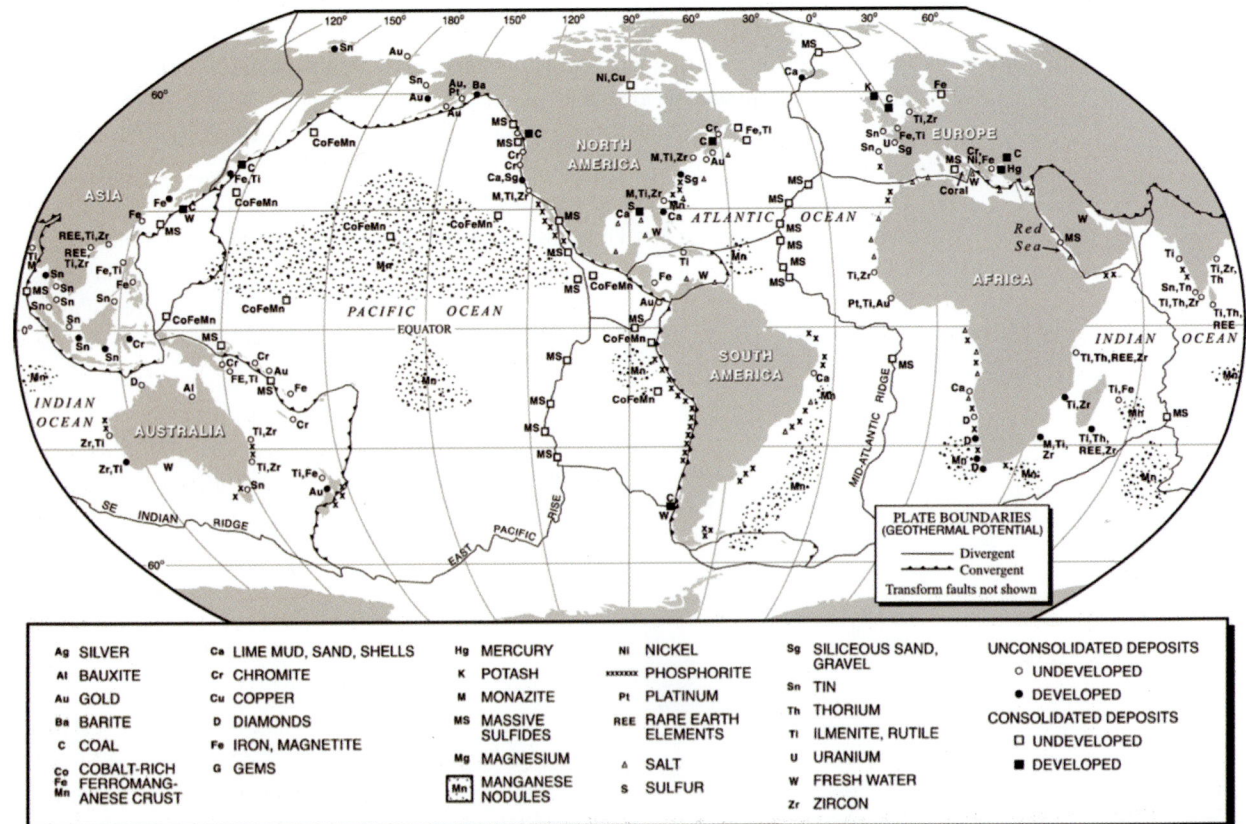

are also mined in the tidal channels of the Yellow River China, the west coast of the Republic of Korea, tidal channels between the islands south of Singapore and in a range of settings in the waters surrounding Hong Kong, China (James et al 1999). In many of the Pacific Islands States, aggregates for building are in short supply and the mining of terrestrial sources, principally beaches, has been associated with major increases in coastal vulnerability (e.g. impacts of beach mining in Kiribati and the Marshall Islands are well documented (Webb 2005, McKenzie et al 2006). Therefore, marine sources of aggregates are considered as a preferred source. The Secretariat of the Pacific Islands Applied Geoscience Commission (SOPAC), now part of the Secretariat of the Pacific Community, has been involved in assisting Pacific Island States in the planning, development and management of sand and gravel resources, (SOPAC, 2007).

Although globally the majority of the demand for aggregates is met by aggregates extracted from land-based sources, the marine-based industry is expanding (JNCC, 2014). However, no figures are available on the global scale of marine aggregate mining.

2.1.1 Case Study: North-East Atlantic

The Working Group on the Effects of Extraction of Marine Sediments (WGEXT) of the International Council for the Exploration of the Sea (ICES) has provided yearly statistics since 1986 on marine aggregate production (ICES 2007, 2008, 2009, 2010, 2011, 2012, 2013; Sutton and Boyd, 2009; Velegrakis et al., 2010). Since 1995, an average of 56 million m^3 y^{-1} has been extracted from the seabed of the North-East Atlantic (Figure 23.2). Five countries account for 93 per cent of the total marine aggregate extraction (Denmark, France, Germany, the Netherlands, and the United Kingdom; OSPAR, 2009). The Netherlands is the largest producer (average 27.3 million m^3 y^{-1}). There are thirteen landing ports and 17 specialist wharves in Europe (Belgium, France and the Netherlands; Highley et al., 2007).

The United Kingdom, one of the largest producers of marine aggregates in the region, currently extracts approximately 20 million tons of marine aggregate (sand and gravel) per year from offshore sites (Figure 23.3). Production meets around 20 per cent of the demand in England and Wales (Crown Estate, 2013). Around 85 per cent of the mined aggregate

Figure 23.2 | Total marine aggregate extraction in the OSPAR maritime area (in million m3). Data from: ICES, 2005, 2006, 2007, 2008, 2009 (OSPAR 2009).

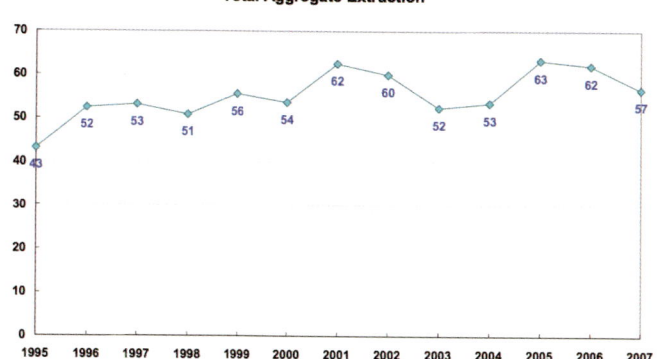

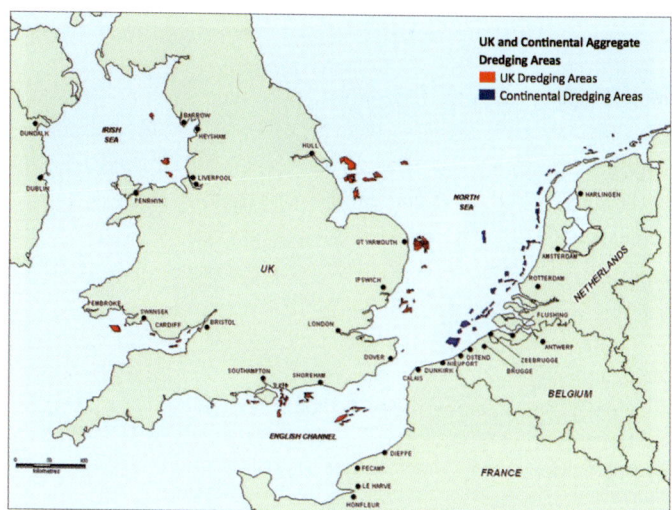

The boundaries and names shown and the designations used on this map do not imply official endorsement or acceptance by the United Nations.

Figure 23.3 | Map of the coastline showing the location of aggregate license areas in the United Kingdom and the adjacent coast of continental Europe (Newell and Woodcock, 2013).

is used for concrete, with the remainder used for beach nourishment and reclamation. In 2010, the area of seabed dredged was 105.4 km^2, although 90 per cent of dredging effort was confined to just 37.63 km^2. Between 1998 and 2007, aggregate extraction produced a dredge footprint of 620 km^2 (BMAPA, 2014). In 2012, 23 dredging vessels were operating (BMAPA, 2014) and aggregates were landed at 68 wharves in 45 ports in England and Wales. Wharves are mainly located in specific regions where a shortfall in land-derived supplies exists and/or there are economic advantages because of river access and proximity to the market (Highley et al., 2007).

The European Union Marine Strategy Framework Directive (MSFD: 2008/56/EC) requires that its Member States take measures to achieve or maintain Good Environmental Status (GES) by 2020. The Descriptor 6 of the MSFD, referred to as "Sea-floor integrity", is closely linked to marine aggregate extraction from the seabed – seafloor integrity is defined as a level that ensures that the structure and functions of the ecosystems are safeguarded and benthic ecosystems, in particular, are not adversely affected (Rice et al., 2010). Descriptor 6 requires immediate actions from Member States to develop suitable pressure indicators (calculated from several parameters such as the species diversity, the number of species and the proportion of different types of species in benthic invertebrate samples) and launch continuous monitoring schemes to contribute to GES achievement.

2.1.2 Case Study: Pacific Islands - Kiribati

The adverse effects of sand mining on the beaches (above the high water mark) of South Tarawa, the main island of Kiribati, were recognized in the 1980s. Removal of the beach sand changes the shape of the beach, increasing erosion and the island's vulnerability to flooding from storm surges and rising sea level. As a consequence of ongoing beach mining, the EU-funded Environmentally Safe Aggregate Project

Offshore Mining Industries Chapter 23

The boundaries and names shown and the designations used on this map do not imply official endorsement or acceptance by the United Nations.

Figure 23.4 | Tarawa Atoll. ESAT resource area in yellow (50-70 year supply). The dot is larger than the absolute maximum surface area that could be mined in any given year (SOPAC, 2013, Figure courtesy Dr. Arthur Webb).

Table 23.1 | Principal marine placer mining activities (from Murton, 2000)

Placer Minerals	Mined locations
Rutile and ilmenite	South-east and south-west Australia
	Eastern South Africa
	South India
	Mozambique
	Senegal
	Brazil
	Florida
Titanium-rich magnetite	North Island, New Zealand
	Java, Indonesia
	Luzon, Philippines
	Hokkaido, Japan
Tin	Indonesian Sunda shelf, extending from the islands of Bangka, Belitung, and Kundur
	Malaysia
	Thailand
Diamonds	West Coast, South Africa
	Namibia
	Northern Australia

for Tarawa (ESAT) was started in 2008. A purpose-built dredge vessel, the "*MV Tekimarawa*" was commissioned and a State-owned dredging company was developed to provide marine aggregates for urban construction. The mined material is processed by local people at a processing facility, used on the island for building material and also sold to other islands. The resource area in Tarawa Lagoon (Figure 23.4), which is currently being mined for coarse sand and gravel, is expected to provide aggregates for 50 to 70 years. ESAT also has a license to excavate access channels on the intertidal reef flats in Beito and Bonriki. This provides fine intertidal silt suitable for road base.

The introduction of marine mining in Tarawa Lagoon has not stopped illegal beach mining. Reviews have found that controlling beach mining by communities is difficult, and that trying to regulate this practice in the absence of a suitable alternative source of revenue is next to impossible (Babinard et al., 2014).

The shoreline and beach profile in South Tarawa has been severely altered, with the almost complete removal of the high protective berm. Mining has now moved on to other untouched beaches. It is estimated that natural recovery of damaged areas will take decades (SOPAC, 2013).

2.2 Placer mining

Placer deposits include minerals that have been concentrated by physical processes, such as waves, wind and currents. Globally, diamonds dominate this sector, but placer deposits also contain valuable minerals. Harben and Bates (1990) identify the most economically important of these minerals (and their associated elements) as: cassiterite (tin), ilmenite (titanium), rutile (titanium), zircon (zirconium), chromite (chromium), monazite (thorium), magnetite (iron), gold and diamonds. About 75 per cent of the world's tin, 11 per cent of gold, and 13 per cent of platinum are extracted from placers (Daesslé and Fischer, 2013).

Diamond placer deposits exist in two distinct areas: a 700-km stretch along the coastal borders of Namibia and South Africa, and an area off the northern coast of Australia (Rona, 2005). Deposits off the coast of South Africa have not been actively mined since 2010 (De Beers, 2012) and Australian operations have not progressed since discovery. Offshore of Namibia, five vessels operated by NAMDEB (a joint partnership between the Namibian government and De Beers) currently extract approximately 1 million carats/year (De Beers, 2007; 2012). In addition there are diver operated mining activities conducted from smaller vessels. A report from The World Wide Fund for Nature (WWF) South Africa (Currie et al., 2008) identified a number of environmental concerns associated with offshore diamond mining. These included destruction of kelp beds, which provide important habitat for juvenile rock lobsters and the destruction of healthy reefs during the removal of diamondiferous gravels. The authors also suggested that the dumping of tailings back into the ocean or onto the beach (after processing) could also potentially result in the formation of land bridges from some islands to the mainland in the vicinity of islands.

Dredging of tin placers is the largest marine metal mining operation in the world (Scott, 2011). The tin belt, as it is called, stretches from Myanmar, down through Thailand, Malaysia, Singapore and Indonesia. The largest operations are offshore of Indonesia, where submerged and buried fluvial and alluvial fan deposits are mined up to 70 meters below sea level, using large dredgers. P.T. TIMAH, a state-owned enterprise, operates the official tin mine offshore of Bangka and Belitung islands. Their dredges can recover more than 3.5 million cubic meters of material per month (Timah, 2014). Numerous "informal miners" also dredge in the shallow coastal area (see Figure 23.5). These operations use divers to suck sediment from the seafloor using plastic tubing connected to a diesel pump (which also pumps air to the divers). The Indonesian islands produce 90 per cent of Indonesia's tin, and Indonesia is the world's second-largest exporter of the metal.

Commercial production of tin began in Thailand in the late 1800s. Most of the offshore tin is located off the Malay Peninsula. The major offshore mining operations ceased in 1985 when the tin price collapsed. Prior to that, large-scale operations were located in the Andaman Sea and the Gulf of Siam (now Gulf of Thailand). The Thaisarco tin smelter in Phuket processes tin from inside and outside Thailand. While most of the Thai-sourced tin originates from land-based deposits, a number of privately owned suction boats still work the near shore during the dry season; a typical boat can recover about 15 kg of cassiterite ore per day.

Gold placer deposits along the Gulf of Alaska of the United States of America coast have been worked since 1898. The gold is recovered from sands exposed at low tide, but the gold-bearing sands extend for approximately 5 km offshore to water depths of 20 m (Jewett et al., 1999). The deposit was most recently actively mined from 1987 to 1990, when the lease was terminated. During that period, 3,673 kg of gold were recovered (Garnett, 2000). The Placer Marine Mining Company purchased an offshore lease at Nome from the Alaska Department of Natural Resources in 2011. The AngloGold-De Beers partnership also has an offshore lease and has invested several million US dollars in exploration and baseline studies. They are hoping to have the required permits in place to begin mining by 2017. There are also a number of individual leases, and due to interest from the general public in shallow water gold mining, the Alaska Department of Natural Resources has also established two recreational mining areas offshore of Nome.

2.2.1   Case Study: New Zealand

Iron sands constitute a very large potential resource in New Zealand. Iron sands occur extensively in the coastal zone, and exploration off the west coast of the North Island of New Zealand's exclusive economic zone has identified potential resources concentrated on the continental shelf. In 2014, following an exploration phase, Trans-Tasman Resources Limited (TTR) was granted a 20-year mineral mining permit by the New Zealand Ministry of Business, Innovation and Employment for the extraction of iron sand from the South Taranaki Bight (Figure 23.6). This permit is the first step in a regulatory process that may allow the company to extract iron sand over a 66-km^2 area of seabed located in water

Figure 23.5 | Homemade dredges operating offshore Bangka Island Indonesia (Photo Rachel Kent, The Forest Trust).

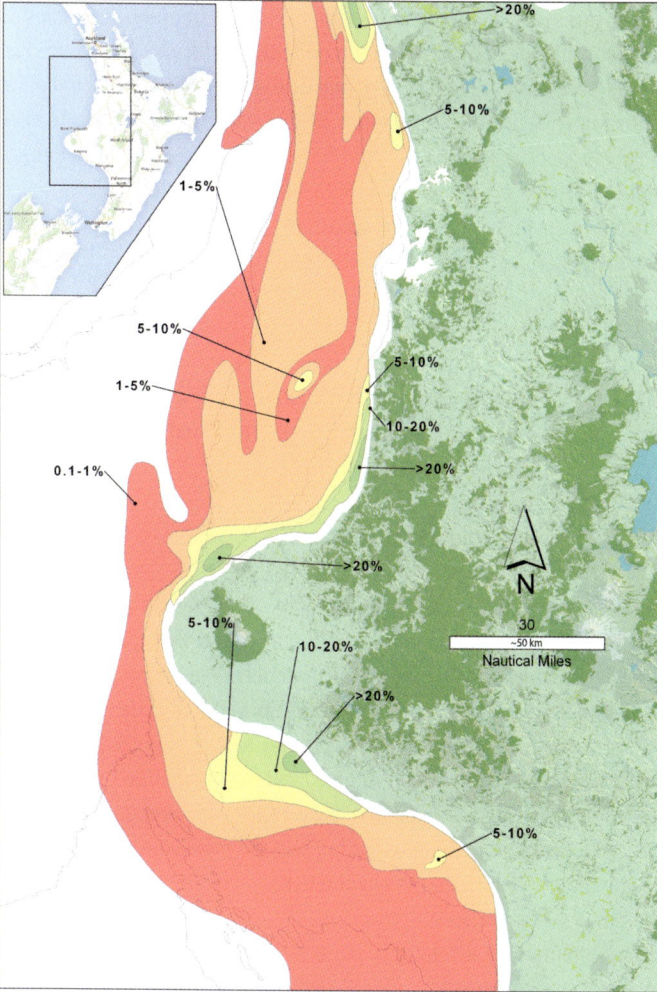

The boundaries and names shown and the designations used on this map do not imply official endorsement or acceptance by the United Nations.
Figure 23.6 | Surficial concentrations of iron sand along the west coast of the North Island of New Zealand (Taranaki region) (modified from Carter, 1980, Taylor & Francis, Ltd., www.tandfonline.com).

depths of between 20-42 m, up to 36 km offshore. It is estimated that 50 million tons per year of sand could be extracted from the seabed (TTR, 2015). It may still take several years before mining commences and, in addition, the company also needs to obtain consent from the New Zealand Petroleum and Minerals branch of the Environmental Protection Authority (EPA) before any mining can begin (NZ Petroleum and Minerals, 2014). At the time of publication of this report, the decision-making Committee appointed by the EPA has refused to grant the mining consent to TTR (NZ EPA, 2015). The reason for this decision is related in part to the uncertainties about the scope and significance of the potential adverse environmental effects.

2.3   Sulphur mining

Sulphur is used in manufacturing and agriculture. Most is produced onshore, but native sulphur is associated with offshore salt domes in the Gulf of Mexico. One offshore mine, the Main Pass 299 facility, located in shallow water off central Louisiana, United States, was operational until 2000 (Kyle, 2002). The sulphur was extracted by the Frasch system, which uses the injection of superheated water through boreholes to melt the sulphur, which is then forced to the surface by compressed air (Ober, 1995). The mine facility is one of the largest platform configurations in the Gulf, with 18 platforms. However, it is unlikely that the mine will resume operations in the near future, due to a glut in the supply of sulphur. This over-supply stems from the fact that sulphur is now extracted in environmental control systems and petroleum refining, which account for 55 per cent of the world sulphur production.

3 Significant environmental, economic and/or social aspects in relation to offshore mining industries

3.1 Environmental Impacts

The current shallow-water seabed mining activities all employ dredging systems to excavate material from the seabed. Dredging techniques vary depending on the nature of the material being mined. They include: a plain suction dredge, which vacuums up unconsolidated material; a rotary cutter dredge, which has a cutting tool at the suction inlet to dislodge more consolidated material; and bucket dredges, which drag a bucket along the sea floor. In marine mining, the dredged material is generally placed into an onboard hopper and excess water and tailings are discharged back into the environment.

Environmental impacts include physical alteration of the benthic environment and underwater cultural heritage. Table 23.2 summarises the environmental impacts associated with aggregate mining, which are potentially applicable to all types of shallow water marine mining. Examples of documented impacts are listed in Table 23.3. The most immediate impacts relate to sediment removal resulting in loss of benthic communities. The removal of the sediment may also affect (re) colonization and recovery rates of impacted communities (Tillin et al., 2011). Most studies on the impact of dredging on marine benthos show that dredging can result in a 30–70 per cent reduction in species variety, a 40–95 per cent reduction in the number of individuals, and a similar reduction in biomass in dredged areas (Newell et al., 1998).

In addition to removal, sediment disturbance can expose marine organisms to increased turbidity and elevated suspended sediment concentrations. This can reduce light availability, which can impact photosynthetic organisms like phytoplankton. Tides and currents can spread turbidity plumes and sediment beyond the mine area. This can be accompanied by changes in water chemistry and contamination (such as algal spores, and from formerly buried substances).

Changes in hydrodynamic processes and seabed geomorphology can also occur. For example, trailer suction dredging, a common form of aggregate dredging, involves dragging the dredge slowly along the seabed, resulting in furrows that are up to 2-3 m wide and 0.5 m deep. These furrows can persist, depending on the local current regime and mobility of the sediments (Newell and Woodcock, 2013). Static suction dredges are employed at sites where deposits are thick and can result in the formation of large pits. Hitchcock and Bell (2004 and references therein) reported that pits within gravelly substrates may fill very slowly and persist after several years, whereas pits in channels with high current velocities have been observed to fill within one year, and those in intertidal watersheds can take 5–10 years to fill.

The European SANDPIT project (Van Rijn et al 2005) aimed to develop reliable techniques to predict the morphological behaviour of large-scale sand mining pits/areas and to understand associated sediment transport processes (Idier et al., 2010). In the study, a baseline pit, based on an actual Dutch pit, was defined as an inverted truncated pyramid 10 m below the seabed, with dimensions at the seabed of 500m x1300m, an excavated volume of $3.5Mm^3$, and located 1.5km from shore at a water depth of 10m (Soulsby et al., 2005). Modelling results using this baseline pit indicate that, for example, there could be a reduction of current speed of up to 10 per cent in the pit; an increase in wave height in the centre of the pit of 1-5 per cent, increasing to 10-15 per cent in the areas surrounding the pit; a reduction of sediment transport in the centre of the pit by 40-90 per cent and an increase of 70-200 per cent outside the pit (Soulsby et al., 2005).

Changes in sediment grain size composition can also occur. For example, diamond mining on the continental shelf of Namibia in 130 m depth was shown to have altered the surficial sediments in a mined area, from previously predominantly homogenous well-sorted sediment, to a more heterogeneous mud, coarse sand and gravel. This is because, as part of the on-board processing, cobbles, pebbles and tailings are discarded over the side (Rogers and Li, 2002). Long-term or permanent changes in grain size characteristics of sediments will affect other factors such as organic content, pore-water chemistry, and microbe abundance and composition (Anderson, 2008).

Less well-documented potential impacts include underwater noise. A review by Thomsen et al. (2009) summarized information on the potential risks from dredging noise. They noted that dredging produces broadband and continuous low frequency sound, that studies indicate that dredging can trigger avoidance reaction in marine mammals, and that marine fish can detect dredging noise over considerable distances. They report that the sparse data available indicates that dredging is not as noisy as seismic surveys, pile driving and sonar; but it is louder than most shipping, operating offshore wind turbines and drilling, and should be considered as a medium impact activity. Marine fauna and birds may collide with or become entangled in operating vessels, but this potential impact is also not well studied. Todd et al (2015) noted that collisions with marine mammals are possible, but unlikely, given the slow speed of dredgers.

Because most marine mining currently occurs close to shore there has been considerable concern regarding the potential impact of mining on archaeological sites. Mining activities, particularly aggregate dredging, has been shown to irreversibly damage underwater cultural heritage,

Table 23.2 | Spatial and temporal scale of the main effects arising from aggregate extraction activities and the confidence associated with the evidence (from Tillin et al 2011).

Effects arising from aggregate extraction activities	Spatial Scale of Effect	Temporal Scale of Effect	Confidence in Evidence
Direct Impacts: Removal of aggregates:	Impacts on benthic marine organisms and seabed morphology. Confined to footprint of extraction: the active dredge zone.	Recovery may begin after cessation of activity.	Good evidence for impacts on seabed habitats and biological assemblages (Newell et al 2004).
Direct Impacts: Removal of aggregates:	Impacts on cultural heritage and archaeology	May be permanent and irreversible	Good evidence for impacts (Michel et al., 2004)
Direct Impacts: Formation of sediment plumes	From 300-500m for sand particle deposition to 3km where particles are remobilised by local hydrodynamic conditions	Longevity of sediment plumes, up to 4-5 tidal excursions for fine particles (MALSF 2009)	Confidence in understanding of sediment plume has been assessed as high (MALSF 2009)
Indirect Impacts: Visual Disturbance	May affect seabirds and marine mammals, spatial extent of effect depends on visual acuity of organism and response.	Confined to period of extraction activities	Little evidence, unlikely to be different from other forms of shipping.
Indirect Impacts: Noise Disturbance	Changes in noise levels detectable up to several km. Behavioural responses likely to occur over much more limited distances and little risk of hearing damage.	Confined to period of extraction activities	Evidence of hearing thresholds only available for a few species (Cefas 2009).
Indirect Impacts: Collision Risk	Confined to activity footprint	Confined to period of extraction activities	Little evidence, unlikely to be different from other forms of shipping.
Indirect Impacts: Sediment deposition	From 300-500m for sand particle deposition to 3km where particles are remobilised by local hydrodynamic conditions.	Heaviest particles settle almost immediately, lightest particles will settle within 1 tidal excursion (a tidal cycle of ebb and flood) (Cefas 2009).	High (from modelling studies and direct observations at a number of sites).

including shipwrecks, airplane crash sites and submerged prehistoric sites (Firth, 2006). Individual States, such as the United States have prepared recommendations and guidelines to avoid dredging impacts on cultural sites (Michel et al., 2004). These include improved location of cultural sites using remote sensing technology, the establishment of buffer zones around known sites, and preparation of plans to preserve resources and subsequent monitoring of dredging activity. Government policies in the United Kingdom on marine mineral extraction from the seabed off the coast of England are set out in Marine Minerals Guidance Note 1 (MMG 1; Wenban-Smith, 2002). The MMG 1 states that all applications for dredging in previously undredged areas require an environmental impact assessment. The Office of the Deputy Prime Minister, which approves applications, can request the applicant to provide information relating to potential impacts to archaeological heritage and landscape and provide information on the measures envisaged to prevent, reduce and where possible offset any significant adverse effects. A review by Firth (2013) of marine archaeology in the United Kingdom recommends that thorough exploration of cultural sites, to constrain their area, may be more cost effective than blanket buffer zones, which can disrupt dredging activity.

The scale of impacts will vary depending on the method and intensity of dredging, level of screening (for example in aggregate mining screening may be employed to alter the sand to gravel ratio, in which case significant quantities of sediment, typically unwanted fine sediment particles, can be returned to the seabed), sediment type and local hydrodynamics (Newell and Woodcock, 2013).

Physical and biological impacts (e.g. smothering leading to death or impaired function) may persist well after the mining finishes. Recovery times are likely to vary greatly and be species dependent (Foden et al., 2009). Cumulative impacts such as climate change and other anthropogenic activities may also affect recovery timing.

Some of the mitigation measures now used with dredging operations include:

- The use of silt curtains to contain dredge plumes;
- The return of overflow waste to the seabed rather than in the water column;
- Locating mining activities away from known migratory pathways and calving or feeding grounds;
- Limiting the number of vessels or operations in given areas;
- Requiring reduced boat speeds in areas likely to support marine mammals;
- Engineering to reduce the noise of the primary recovery and ore-lift operations;
- Limiting unnecessary use of platform and vessel flood lights at night and ensuring that those that are required are directed approximately vertically onto work surfaces to avoid or mitigate seabird strikes;
- Leaving patches within a mining site un-mined to increase the rate of recolonization and recovery of benthic fauna;
- Excluding areas from mining if they support unique populations of marine life;

Table 23.3 | Documented environmental impacts of offshore mining.

Mining activity	Location	Impact	Reference
Shell and sand extraction	Owen Anchorage, south-west of Fremantle, Western Australia	Dredging in shallow near-shore waters associated with significant conservation values, e.g., seagrass, coral communities; adverse effects on marine habitats due to direct seabed disturbance and indirect effects, such as elevated turbidity levels. Other concerns include changes in near-shore wave and current conditions, which could affect shipping movements and seabed/shoreline stability	Walker et al., 2001
Sand and gravel extraction	European Union	Loss of abundance, species diversity and biomass of the benthic community in the dredged area. Similar effects from turbidity and resuspension of sediment over a wide area. Benthic impact is a key concern where dredging activities may impinge on habitats or species classified as threatened or in decline (such as Maerl or Sabellaria reefs).	OSPAR, 2009
Sand and gravel extraction	Dieppe, France	10-year monitoring programme revealed a change in substrate from gravel and coarse sand to fine sand in the dredged area. The maximum impact on benthic macrofauna was a reduction by 80 per cent in species richness and 90 per cent in both abundance and biomass. In the surrounding area, the impact was almost as severe. Following cessation of dredging, species richness was fully restored after 16 months, but densities and biomass were still 40 per cent and 25 per cent, respectively, lower than in reference stations after 28 months. The community structure differed from the initial one, corresponding to the new type of sediment.	Desprez, 2000
Sand and gravel extraction	United States of America	Comprehensive review of impacts from dredging operations identifying the most severe effects: entrainment of benthic organisms; destruction of essential habitat; increased turbidity affecting sensitive fauna like corals and suspension-feeding organisms.	Michel et al., 2013
Sand and gravel extraction	Moreton Bay, Australia	Alteration of the existing tidal delta morphology by the removal of a small area of shallow banks. In most cases, the prevailing sediment transport processes would result in a gradual infill of extraction sites.	Fesl, 2005
Sand and gravel extraction	Puck Bay, Southern Baltic Sea	Benthic re-colonization at a site formed by sand extraction was investigated some 10 years after the cessation of dredging. The examined post-dredging pit is one of five deep (up to 14 m) pits created with a static suction hopper on the sandy, flat and shallow (1–2 m) part of the inner Puck Bay (the southern Baltic Sea). Organic matter was found to accumulate in the pit, resulting in anaerobic conditions and hydrogen sulfide formation. Macrofauna was absent from the deepest part of the pit and re-colonization by pre-mining benthic fauna was considered unlikely.	Szymelfenig et al., 2006
Diamond mining	Benguela Region, Africa (offshore of Namibia and South Africa)	Cumulative impacts of seabed diamond mining assessed over time and as a combination of numerous operations. Four to 15 years for benthic recovery, biodiversity altered in favour of filter feeders and algae, resulting in decreased biodiversity but increased biomass.	Pulfrich et al., 2003; Pulfrich and Branch, 2014
Diamond mining	Offshore Namibia, Orange Delta	Changes in surficial sediment grain size composition from unimodal to polymodal, with increased coarse sand and gravel.	Rogers and Li, 2002
Tin mining	Bangka-Belitung Province Indonesia	Hundreds of makeshift pontoons operate alongside a fleet of 52 dredgers belonging to P.T. TIMAH. The island coastline has been altered by tailing dumps, and up to 70 per cent of coastal ecosystems, particularly coral, sea-grass and mangroves, are degraded.	IDH, 2013
Gold mining	Norton Sound, northeastern Bering Sea, United States.	Mining with a bucket-line dredge occurred near shore in 9 to 20 m during June to November 1986 to 1990. Sampling a year after mining ceased indicated that benthic macrofaunal community parameters (total abundance, bio- mass, diversity) and abundance of dominant families were significantly reduced at mined stations	Jewett et al., 1999

- Excluding areas of mining if they are potential sites of cultural heritage;
- Depositing tailings within as small an area as possible surrounding the mining block, or onshore; and
- Avoiding the need for re-mining areas by mining target areas to completion during initial mining.

Several studies have looked at the restoration of seabed habitat after mining activity (e.g., Cooper et al., 2013, Kilbride et al., 2006, Boyd et al., 2004). In the OSPAR region, where damage to protected species and habitat occurs, restoration is identified within the obligations of the Convention for the Protection of the Marine Environment of the North-East Atlantic, various European directives, and in various United Kingdom marine policy documents, (Cooper et al., 2013). A study on seabed restoration identified three issues central to decisions about whether to attempt restoration following marine aggregate dredging. They include: (i) necessity (e.g. a clear scientific rationale for intervention and/or a policy/legislative requirement), (ii) technical feasibility (i.e. whether it is possible to restore the impacts), and (iii) whether is it affordable (Cooper et al., 2013).

A recent study of the Thames Estuary, United Kingdom, an area of aggregate extraction, used the estimated value of ecosystem goods and services to determine if seabed restoration was justifiable in terms of costs and benefits; they concluded that in this case it was not (Cooper et al., 2013). The proposed restoration involved levelling the seabed and restoring the sediment character for an estimated cost of over 1 million British pounds. In order to determine if this expenditure could be justified, the authors assessed the significance of the persistent impacts on the ecosystem goods and services and the cost and likelihood of successful restoration. While the site-specific cost benefit analysis precluded restoration, they suggest that the approach taken could be used at other sites to determine if restoration is practical and effective.

In the United Kingdom a research fund, (the Aggregate Levy Sustainability Fund), was established in 2002 and ran until March 2011, using revenue from the Aggregates Levy introduced in 2002 - a tax of 2.00 British pounds per ton on primary aggregate sales (including land- and marine-derived aggregates; Newell and Woodcock, 2013). There was intense public criticism when the Fund was discontinued in 2011, as previously 7 per cent of the Fund had been directed to communities, non-governmental organizations and other stakeholders to fund projects delivering conservation, local community and other sustainability benefits (e.g., BBC 2011; MPA 2011). Cooper et al., 2013 also suggest that the fund could have been used to finance seabed restoration projects.

3.2 Social impacts

Social impacts of offshore mining are likely to be complex and different and generally less than that for terrestrial mining (Roche and Bice, 2013). Table 23.4 details potential social impacts from offshore mining. In countries where offshore mining is relatively new and untested (like Australia), societal expectations set higher standards for its acceptance, particularly with regard to environmental protection and strengthening of the national economy (Mason et al., 2014).

Regional initiatives, targeted at developing a holistic approach to decision-making, that incorporate social, environmental and economic

Table 23.4 | Positive and negative potential social impacts identified (after Tillin et al, 2011; Roche and Bice, 2013)

Impact	Effect
Environmental degradation	Loss of ecosystem services that negatively affects livelihoods.
Provision of material	For coastal defence and beach replenishment.
Revenue	Revenue to industry, government and community; Foreign exchange earner.
Reduced pressure on land based resources	Avoidance of social impacts for resource extraction on land, including competing resources, community relocations.
Employment	Employment for local community, accompanied by influx of people to new industry; particularly for small island communities.
Cultural impacts	Loss of cultural sites; changes/loss in resource distribution (food, territory, etc.); ignoring of/loss of traditional knowledge.
Governance and policy	New regulatory regimes; implementation of policy; social and environmental degradation can lead to conflict.

Table 23.5 | Relevant regional and national initiatives

	Initiative
European Union	MSFD (2008): "Directive 2008/56/EC on establishing a framework for community action in the field of marine environmental policy" This directive provides a transparent legislative framework for an ecosystem-based approach to the management of human activities; supports the sustainable use of marine goods and services; and integrates the value of marine ecosystem services into decision making.
United Kingdom	Marine Environment Protection Fund 2010: Framework to allow marine aggregates extraction options to be analysed using socio-economic information. The framework analyses the interactions between different uses of the marine environment at both local and regional levels (Dickie et al., 2010)
Pacific Islands	SPC-EU DSM Project (2011-2016): Technical assistance and advisory service for Pacific Island countries choosing to engage in deep sea mining to help them improve governance and management in accordance with international law, with particular attention to the protection of the marine environment and securing equitable financial arrangements for their people.
United States	Executive Order 13547- Stewardship of the Ocean, Our Coasts, and the Great Lakes. The Order adopts the recommendations of the Interagency Ocean Policy Task Force, except where otherwise provided in this Order, and directs executive agencies to implement those recommendations under the guidance of a National Ocean Council. Based on those recommendations, this Order establishes a national policy to ensure, amongst other things, the protection, maintenance, and restoration of the health of ocean and coastal ecosystems and resources.

evaluation and stakeholder engagement, are outlined in Table 23.5. In some areas, such as the Pacific Islands region, emphasis is on making informed decisions about deep-sea mining. Countries which decide to engage in deep sea mining can obtain assistance from the Secretariat of the Pacific Community to develop national regulatory frameworks (offshore national policy, legislation and regulations) in close collaboration with all key stakeholders and in particular, local communities (SPC-EU, 2012). Elsewhere, the framework is focused more on the sustainable management of the marine environment, including non-living resources, and includes ecosystem-based approaches and valuation of ecosystem services affected by human activity. For example the European Union Marine Strategy Framework Directive (2008) advocates a transition from a sector-specific policy landscape to a system-based one, in which activities are regulated in concert, based on shared space and time across boundaries. Uncertainty remains, however, about how to value coastal assets and quantitatively measure social impact (Beaumont et al., 2007).

Awareness is increasing of the potential social impacts of marine and coastal extractive mineral industries, such as coastal dredging for aggregates and beach re-nourishment schemes (e.g., Austen et al., 2009; Drucker et al., 2004). Strong public sentiments about environmental and social issues already exist around land-based mining (e.g., Mudd, 2010). However, there is currently not the same level of understanding and informed debate around offshore mining (Mason et al., 2014). As offshore mining becomes more commonplace, information and data on the marine environment and impacts will be collected, and it is important that this information is disseminated to stakeholders. It is worth noting that the value of stakeholder participation in developing and implementing policy was included in Principle 10 of the Rio Declaration, which states that: "environmental issues are best handled with the participation of all concerned citizens, at the relevant level…"

Studies suggest that for an informed society to accept a nascent offshore mining industry, stakeholders require: better information (particularly rigorous scientific analysis of potential impacts, costs and benefits); a transparent and socially responsive management process within a consistent and efficient regulatory regime; and meaningful engagement with stakeholders (Boughen et al., 2010; Mason et al., 2010).

3.2.1 Case Study: Kiribati

A recent study by Babinard et al. (2014) examined the potential social impacts of offshore aggregate mining in South Tarawa (see section 2.1.3). The authors determined that as the ESAT (Environmentally Safe Aggregates for Tarawa) dredging operation develops, it could have adverse consequences for the welfare of those Kiribati residents who are either sellers or users of aggregates. Sellers of aggregates rely on beach mining for their livelihood (they currently receive 1 Australian dollar per bag). A 2006 household survey found that 206 out of 280 households surveyed were involved in some form of beach mining (Pelesikoti, 2007). There is widespread belief that they are acting within their rights as customary owners of the land, and they will likely lose economic opportunities as a result of the offshore dredging operations. For users of aggregates on the island, the main issue is whether they will be legally able to continue to mine aggregates from their own beaches. Residents argue that the customary rights to mine are included in the Foreshore Amendment Act of 2006 (Pelesikoti, 2007).

3.3 Economic benefits from marine mining

The economic benefits from near-shore mining are difficult to estimate. Marine aggregates are often sourced locally and reporting is scattered, but the marine sector is often distinguished from the land sector, so the value of the resource can be estimated. In contrast, commodities like tin and diamonds are part of a global market, which does not distinguish between land-derived and marine-derived materials. Table 23.6 gives estimated values where reported.

4 Developments in other forms of seabed mining: current state and potential scale

4.1 Phosphate mining

Phosphorites are natural compounds containing phosphate in the form of cement-binding sediments in tropical to sub-tropical regions (Murton, 2002). They are widely distributed on the continental shelves and upper slopes, oceanic islands, seamounts and flanks of atolls. Deposits have been found off the west coast of Tasmania, Australia; Congo, Ecuador, Gabon, Mexico, Morocco, Namibia, New Zealand, Peru, South Africa, and the United States. They are usually located in less than 1,000 m of water and their formation is linked to zones of coastal upwelling, divergence and biological productivity.

Currently proposals to mine phosphate are under consideration in New Zealand, Namibia and Mexico. In New Zealand, the Ministry of Business, Innovation and Employment has granted a 20-year mining permit

Figure 23.6 | Estimates of marine aggregates and minerals

Locations	Resource	Quantity	Revenue	Employment	References
European Union, United Kingdom, Japan, United States (minor)	Aggregate	~ 50-150+ million m3 / year (can vary strongly year to year depending on demand)	1-3+ billion US dollars)	5,000–15,000 (estimate)	Ifremer, 2014 Herbich, 2000 Marinet, 2012 Newell and Woodcock, 2013
South Africa, Namibia, Australia (Inactive)	Diamond Placers	1.1 million carats (2012).	3.5 billion US dollars	~1,600	NAMDEB, 2010 NAMDEB, 2014
Indonesia, Malaysia, Thailand; Australia (inactive)	Tin	19,000 tons /yr tin	Indonesia 500 million US dollars	Indonesia ~3,500 Malaysia & Thailand N/A	Timah, 2012
New Zealand (inactive)	Iron Sands	0	0	0	
United States, South America, Australia, New Zealand, Africa, Portugal, India (all inactive)	Phosphates	0	0	N/A	
Mexico (inactive)	Phosphates	Total of 327.2 million ore tons at 18.5% P2O5	0	N/A	Don Deigo (2015)
United States (now inactive)	Sulphur	0	0	0	

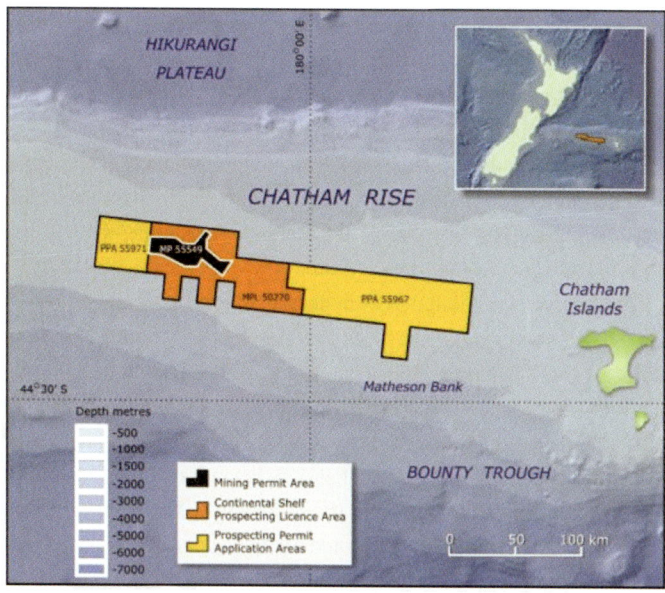

The boundaries and names shown and the designations used on this map do not imply official endorsement or acceptance by the United Nations.
Figure 23.7 | Location of Chatham Rise phosphate project area (RSC, 2014).

to Chatham Rock Phosphate Ltd. for the extraction of rock phosphate nodules from an 820-km2 area of the Chatham Rise (Figure 23.7). Before mining can commence, the company still needs to obtain consent from the Environmental Protection Authority. At the time of publication of this report the Authority had refused an application by Chatham Rise Phosphate limited for a marine consent to mine phosphorite nodules on the Chatham Rise (NZ EPA, 2015). The decision-making committee found that that "the destructive effects of the extraction process, coupled with the potentially significant impact of the deposition of sediment on areas adjacent to the mining blocks and on the wider marine ecosystem, could not be mitigated by any set of conditions or adaptive management regime that might be reasonably imposed." They also concluded that the economic benefit to New Zealand of the proposal would be modest at best.

In Namibia, an Environmental Impact Assessment Report and an Environmental Management Plan were submitted in March 2012 for the Sandpiper Phosphate Project (Figure 23.8), which proposed to dredge phosphate-enriched sediments south of Walvis Bay, Namibia, in depths of 180-300 m (Midgley, 2012). The company planned to extract 5.5 Mt of phosphate-enriched marine sediments on an annual basis, for over 20 years. The environmental impact assessment (EIA) identified low-level potential adverse impacts including biogeochemical changes, benthic habitat loss, loss of biodiversity and cumulative impacts (Namibian Marine Phosphates, 2012; Midgley, 2012; McClune, 2012). No official decision has been issued on the Sandpiper Phosphate Project application as yet, however in September 2013, an 18-month moratorium on environmental clearances for bulk seabed mining activities for industrial minerals, base and/or rare metals (including phosporites) was declared by the Government of Namibia. During this period the Ministry of Fisheries and Marine Resources is required to make a strategic impact assessment on the potential impacts of the proposed phosphate mining on the fish-

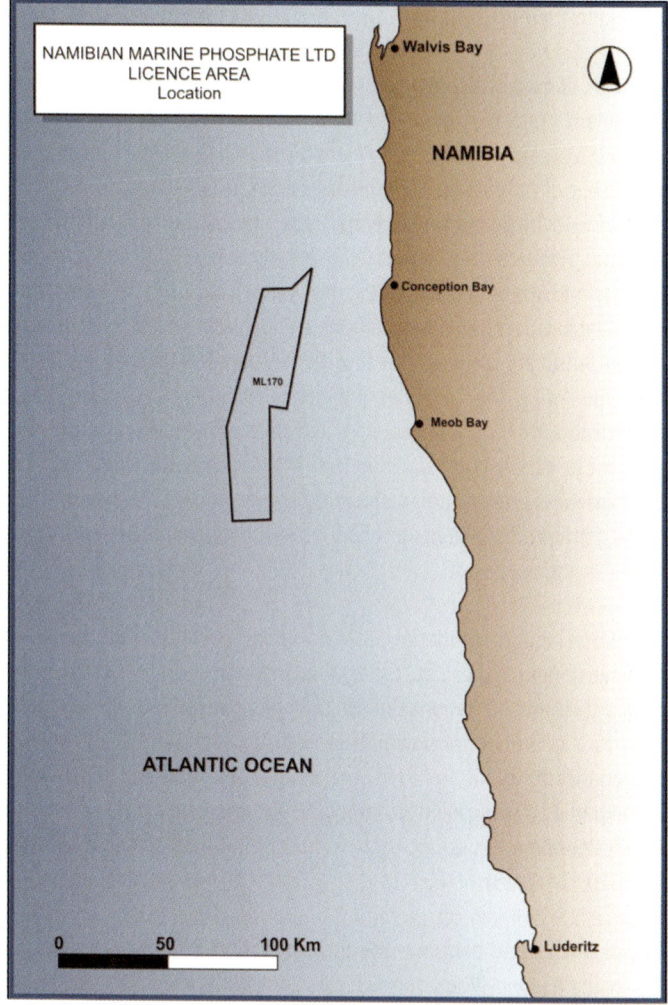

The boundaries and names shown and the designations used on this map do not imply official endorsement or acceptance by the United Nations.
Figure 23.8 | The Sandpiper Project (license area shown) includes the zone of highest regional phosphate concentration (Namibian Marine Phosphate, 2012).

ing industry. While the Ministry of Mines and Energy is allowing marine phosphate exploration activities to continue during the moratorium period, such activities are not currently being undertaken in areas within the national jurisdiction of Namibia.

A proposed Mexican underwater phosphate mine, the Don Diego project, is located in 60-90m water depth, approximately 40 km off the cost of the Bay of Ulloa, on the west coast of Baja California. The permit area is 912 km^2 and it is estimated that if the project proceeds the area dredged annually would be around 1 per cent (1.7 km^2; Don Diego, 2015). Phosphorite resources at the Don Diego deposit have been estimated to total 327.2 million ore tons at 18.5 per cent P$_2$O$_5$. Odyssey Marine Exploration has lodged an environmental impact assessment for the recovery of the phosphate sands with the Mexican Secretary of Environment and Natural Resources and is awaiting a decision (Odyssey Marine Exploration, 2014). Local non-governmental organizations including WildCoast, Centro Mexicano Derecho Ambiental (CEMDA), Grupo Tortuguero, Vigilantes de Bahia Magdalena and Medio Ambiente Sociedad have been vocal in their opposition to the project (Pier, 2014).

4.2 Deep-Sea Mining

Although commercial deep-sea mining has not yet commenced, the three main deep-sea mineral deposit types – sea-floor massive sulphides (SMS), polymetallic nodules and cobalt-rich crusts – have been the subject of interest for some time (see SPC 2013a,b,c,d). Recent announcements make it seem likely that SMS mining will begin in the Manus Basin of Papua New Guinea (Nautilus Minerals, 2014a and b). Other Pacific Island States (e.g., Fiji, Solomon Islands, Tonga and Vanuatu) have issued exploration licenses to various companies to evaluate the commercial feasibility of mineral resources development in their exclusive economic zones. The economic interest in SMS deposits is their high concentrations of copper, zinc, gold, and silver; polymetallic nodules for manganese, nickel, copper, molybdenum and rare earth elements; and ferromanganese crusts for manganese, cobalt, nickel, rare earth elements, yttrium, molybdenum, tellurium, niobium, zirconium, and platinum.

In addition, the International Seabed Authority (ISA), which regulates deep-sea mining in the Area (the seabed, ocean floor and subsoil thereof beyond the limits of national jurisdiction) has entered into 15-year contracts for exploration for polymetallic nodules, SMS and cobalt-rich ferromanganese crusts in the deep seabed with 26 contractors (composed of companies, research institutions and government agencies) plus 1 contract pending ISA Council action in July 2015 (ISA, 2000; ISA 2001; ISA 2010; ISA 2013).

Seventeen of these contracts are for exploration for polymetallic nodules in the Clarion-Clipperton Fracture Zone (CCZ, 16) and Central Indian Ocean Basin (1). There are six contracts for exploration for SMS in the South West Indian Ridge, Central Indian Ridge and the Mid-Atlantic Ridge and four contracts for exploration for cobalt-rich crusts in the Western Pacific Ocean (3) and Atlantic (1) (ISA 2015a). These licences allow contractors to explore for seabed minerals in designated areas of the Area.

The ISA has called for comments on draft regulations for exploitation licensing in the Area (ISA 2015b). The decision to commence deep-sea mining in the Area will depend in part on the availability of metals from terrestrial sources and their prices in the world market, as well as technological and economic considerations based on capital and operating costs of the deep-sea mining system.

5 Gaps in capacity to engage in offshore minerals industries and to assess the environmental, social and economic aspects.

Despite the importance of marine extractive industries in many developing countries, the environmental, social and economic aspects are often not adequately understood. Therefore it is necessary to strengthen the approach to planning and managing these activities. This includes implementing the precautionary principle and adaptive management, as well as transparent monitoring. There is also a lack of consensus on what is an acceptable condition in which to leave the seabed post mining. Increasing public awareness and engendering a custodial and stewardship attitude to the environment may help curb the most damaging practices.

Unregulated mining often occurs in parallel to regulated mining activities. For example, numerous small operators participate in the marine sector of the tin mining industry in Bangka and Belitung, Indonesia. Many of the practices associated with these workers are unsafe and miners are killed or injured every year; local news reports refer to over 100 fatalities per year (Jakarta Post, 2010). The lack of regulation or the lack of enforcement of regulations, allows mining to take place in critical marine habitats and extensive damage has been done to coral reefs and mangrove environments (IDH, 2013). Improved licensing, regulation, enforcement and monitoring, in conjunction with social programmes to find alternative sources of revenue, would be needed. How the industry is being regulated would also need to be considered. The export data, published by the Bangka Belitung regional administration, showed that P.T. Timah, which owns 473,800 hectares of concession areas, exported 8,899 tons of tin in 2009, and privately owned smelters, which operate concession areas of 16,884 hectares, exported 13,867 tons. These discrepancies highlight the magnitude of the problem. The penalties provided by mining and/or environmental legislation may need to be strengthened to stop these practices.

For any State or company planning resource development, integrating coastal and marine ecosystem services into the development process is important; however, information on the services provided or the value of these services is often scarce. In many developing countries the interface between governments and offshore minerals industries needs to be strengthened. Deficiencies exist in the information available and in the institutional capacity to manage non-living marine resources. In summary, the following gaps can be identified:

- Increased capacity in coastal and marine geosciences information systems (including social, cultural, economic, ecological, biophysical and geophysical information) to improve geoscientific advice for management and monitoring of coastal environments to meet the requirements of ecosystem-based management and sustainable development;
- Development and implementation of robust regulatory frameworks for marine mineral extraction industries, which include environmental impact assessments, environmental quality and social laws, environmental liability, and monitoring capacity;
- Increased public awareness of the vulnerability of coastal environments, the benthic habitats and the fishery nursery grounds that may be affected by marine mining; and
- Technology transfer and skills development to ensure best practice in marine mineral extraction.

References

Austen, M.C., Hattam, C., Lowe, S., Mangi, C., Richardson, K. (2009). *Quantifying and Valuing the Impacts of Marine Aggregate Extraction on Ecosystem Goods and Services*. MALSF funded project MEPF 08-P77. www.cefas.co.uk/media/462458/mepf-08-p77-final-report.pdf. Accessed June 2014.

Babinard, J., Bennett, C.R., Hatziolos, M.E., Faiz, A., Somani, A. (2014). Sustainably managing natural resources and the need for construction materials in Pacific island countries: The example of South Tarawa, Kiribati. *National Resources Forum*, 38, 58-66.

Beaumont, N.J., Austen, M.C., Atkins, J.P., Burdon, D., Degraer, S., Dentinho, T.P., Derous, S., Holm, P., Horton, T., van Ierland, E., Marboe, A.H., Starkey, D.J., Townsend, M., Zarzycki, T. (2007). Identification, definition and quantification of goods and services provided by marine biodiversity: Implications for the ecosystem approach. *Marine Pollution Bulletin*, 54 (3), 253-265.

BBC (2011). *Dismay in Cumbria at quarrying tax fund end.* 10 April 2011. http://www.bbc.co.uk/news/uk-england-cumbria-13025923. Accessed June 2014.

Bide, T. and Mankelow, J. (2014). Mapping marine sand and gravel. *Planet Earth*, Spring 2014, pp. 14-15. www.planetearth.nerc.ac.uk. Accessed June 2014.

BMAPA (2014). British Marine Aggregate Producers Association. http://www.bmapa.org/. Accessed June 2014.

Boughen, N., Mason, C., Paxton, G., Parsons, R., Johns, S., Parr, J., Moffat, K. (2010). Seafloor exploration and mining in Australia: Stakeholder reactions, expectations and desired level of engagement. *CSIRO Wealth from Oceans Flagship, Report EP111562*, Australia: 20pp.

Boyd, S.E., Cooper, K.M., Limpenny, D.S., Kilbride, R., Rees, H.L., Dearnaley, M.P., Stevenson, J., Meadows, W.J., Morris, C.D. (2004). Assessment of the re-habilitation of the seabed following marine aggregate dredging. *Science Series Technical Report*. CEFAS Lowestoft, 121, 154pp.

Cefas (Centre for Environment, Fisheries & Aquaculture Science) (2009). A Generic Investigation into Noise Profiles of Marine Dredging in relation to the Acoustic Sensitivity of the Marine Fauna in UK waters with particular emphasis on Aggregate Dredging: Phase 1 Scoping and Review of Key Issues MEPF Ref No: MEPF 08/P21 Project.

Cooper, K., Burdon, D., Atkins, J.P., Weiss, L., Somerfield, P., Elliott, M., Turner, K., Ware, S., Vivian, C. (2013). Can the benefits of physical seabed restoration justify the costs? An assessment of a disused aggregate extraction site off the Thames Estuary, UK. *Marine Pollution Bulletin*, 75(1), 33-45.

Crown Estate (2013). *Marine Aggregates, Capability & Portfolio 2013*. http://www.thecrownestate.co.uk/media/495658/ei-marine-aggregates-capability-and-portfolio-2013.pdf. Accessed June 2014.

Currie, H., Grobler, K., Kemper, J., Roux, J.P., Currie, B., Moroff, N., Ludynia, K., Jones, R., James, J., Pillay, P., Cadot, N., Peard, K., de Couwer, V. and Holtzhausen, H. (2008). *Namibian islands' marine protected area*. Ministry of Fisheries and Marine. Resources, Windhoek.

Daesslé, L.W., Fischer, D.W. (2001). Marine Minerals in the Mexican Pacific: Toward Efficient Resource Management. *Marine Georesources & Geotechnology*, 19(3), 197-206.

De Beers (2007) *Operating and Finance Review*. http://www.debeersgroup.com/content/dam/de-beers/corporate/documents/Archive%20Reports/Operating_and_Financial_Review_2007_March_2008.PDF Accessed June 2014.

De Beers (2012) *Operating and Finance Review*. http://www.debeersgroup.com/content/dam/de-beers/corporate/documents/articles/reports/2013/OFR_2012.PDF Accessed June 2014.

Desprez, M. (2000). Physical and biological impact of marine aggregate extraction along the French coast of the Eastern English Channel: short-and long-term post-dredging restoration. *ICES Journal of Marine Science*, 57(5), 1428-1438.

Dickie, S., Hime, E., Lockhart-Mummery, E., Ozdemiroglu, R., Tinch (2010). Including the Socio-Economic Impacts of Marine Aggregate Dredging in Decision-Making. Published February 2011.

Don Diego (2015). http://www.dondiego.mx/ Accessed March 2015.

Drucker, B.S., Waskes, W., Byrnes, M.R. (2004). The U.S Minerals Management Service Outer Continental Shelf sand and gravel program: Environmental studies to assess the potential effects of offshore dredging operations in Federal Waters. *Journal of Coastal Research*, 20(1), 1-5.

Earney, F.C. (2005). Marine mineral resources. Routledge.

Fesl, E. (2005). Moreton Bay sand extraction study. Queensland Environment Protection Agency. 14pp.

Firth, A. (2006). Marine Aggregates and Prehistory. HERITAGE AT RISK, 8.

Firth, A. (2013). Marine Archaeology, in Newell, R.C., Woodcock, T.A., (2013). *Aggregate Dredging and the Marine Environment: an overview of recent research and current industry practice*. The Crown Estate, pp. 165.

Foden, J., Rogers, S.I., Jones, A.P. (2009). Recovery rates of UK seabed habitats after cessation of aggregate extraction. *Marine Ecology Progress Series*, 390, 15–26.

Garnett, R.H.T. (2000). Marine placer gold, with particular reference to Nome, Alaska, in: Cronan, D.S. Ed. 2000. *Handbook of marine mineral deposits*. pp. 67-10.

Harben, P.W. and Bates, R.L. (1990). *Industrial minerals: geology and world deposits*. Metal Bulletin Plc.

Herbich, J.B. (2000). *Handbook of coastal engineering*. New York: McGraw-Hill.

Highley, D.E., Hetherington, L.E., Brown, T.J., Harrison, D.J., Jenkins, G.O. (2007). *The strategic importance of the marine aggregate industry to the UK*. British Geological Survey Research Report, 0R/07/019, 44 pp.

ICES (2007). Report of the Working Group on the Effects of Extraction of Marine Sediments on the Marine Ecosystem (WGEXT). 17–20 April 2007, Helsinki, Finland, *ICES CM 2007/MCH:08*, 92pp.

ICES (2008). Report of the Working Group on the Effects of Extraction of Marine Sediments on the Marine Ecosystem (WGEXT). 8–11 April 2008, Burnham-on-Crouch, UK, *ICES CM 2008/MHC09*, 86pp.

ICES (2009). Report of the Working Group on the Effects of Extraction of Marine Sediments on the Marine Ecosystem (WGEXT). 14-17 April 2009, New York, USA, *ICES CM 2009/MHC:09*, 98pp.

ICES (2010). Report of the Working Group on the Effects of Extraction of Marine Sediments on the Marine Ecosystem (WGEXT). 31 May-4 June 2010, Djurönäset, Sweden, *ICES CM 2010/SSGHIE*, 108pp.

ICES (2011). Report of the Working Group on the Effects of Extraction of Marine Sediments on the Marine Ecosystem (WGEXT). 11–15 April 2011, Delft, the Netherlands, *ICES CM 2011/SSGHIE*, 89pp.

ICES (2012). Report of the Working Group on the Effects of Extraction of Marine Sediments on the Marine Ecosystem (WGEXT). 16–20 April 2012, Rouen, France, *ICES CM 2012/SSGHIE*, 104pp.

ICES (2013). Report of the Working Group on the Effects of Extraction of Marine Sediments on the Marine Ecosystem (WGEXT). 22-25 April 2013, Faial, Portugal, *ICES CM 2013/SSGHIE*, 54pp.

IDH (2013). *IDH Tin working group communiqué*. http://www.foe.co.uk/sites/default/files/downloads/idh-tin-working-group-communiqu-18070.pdf. Accessed June 2014.

Idier, D., Hommes, S., Brière, C., Roos, P.C., Walstra, D.J.R., Knaapen, M.A. and Hulscher, S.J. (2010). Morphodynamic models used to study the impact of offshore aggregate extraction: a review. *Journal of Coastal Research*, 39-52.

Ifremer (2014). *Marine aggregate extraction*. http://www.ifremer.fr/demf/aggregate_eng.html. Accessed June 2014.

ISA (2000). Decision of the assembly relating to the regulations on prospecting and exploration for polymetallic nodules in the Area. International Seabed Authority, Jamaica, ISBA/6/A/18, pp.48.

ISA (2001). Recommendations for guidance of contractors for the assessment of the possible environmental impacts arising from exploration for polymetallic nodules in the Area. International Seabed Authority, Jamaica, ISBA/7/LTC/1 2001, pp.11.

ISA (2010). The International Marine Minerals Society's Code for Environmental Management of Marine Mining. Note by the Secretariat; ISBA/16/LTC/2 http://www.isa.org.jm/files/documents/EN/16Sess/LTC/ISBA-16LTC-2.pdf. Accessed June 2014.

ISA (2013). Towards the Development of a Regulatory Framework for Polymetallic Nodule Exploitation in the Area. *ISA Technical Study, No. 11*, pp. 90.

ISA (2015a). Deep sea mineral contractors. http://www.isa.org.jm/deep-seabed-minerals-contractors Accessed March 2015.

ISA (2015b). Developing a Regulatory Framework for Mineral Exploitation in the Area. http://www.isa.org.jm/files/documents/EN/Survey/Report-2015.pdf Accessed April 2015.

Jakarta Post (2010). Hundreds of illegal miners killed in last four years. *The Jakarta Post*, http://www.thejakartapost.com/news/2010/02/08/hundreds-illegal-miners-killed-last-four-years.html. Accessed June 2014.

James, J., Evans, C., Harrison, D., Ooms, K., Vivan, C. & Boyd, S.E. (1999).The effective development of offshore aggregates in south-east Asia. Technical Report WC/99/9. Notthingam, British Geological Survey.

Jewett, S.C., Feder, H.M. and Blanchard, A. (1999). Assessment of the benthic environment following offshore placer gold mining in the northeastern Bering Sea. Marine Environmental Research, 48(2), 91-122.

JNCC (2014). *Offshore Marine Aggregates*. Joint Nature Conservation Committee. http://jncc.defra.gov.uk/page-4278. Accessed June 2014.

Kilbride, R., Boyd, S.E., Rees, H.L., Dearnaley, M.P. and Stevenson, J. (2006). Effects of dredging activity on epifaunal communities: surveys following cessation of dredging. *Estuarine Coastal and Shelf Science*, 70 (1-2), pp207-223.

Kyle, J.R. (2002). A Century of fire and brimstone: the rise and fall of the Frasch sulphur industry of the Gulf of Mexico Basin. *Industrial Minerals and Extractive Industry Geology*. Geological Society of London, Special Publication, 189-198.

MALSF (2009). Marine aggregate extraction: helping to determine good practice. Summary Report. Marine Aggregate Levy Sustainability Fund.

Marinet (2012). *Aggregate dredging and the Crown Estate – facts and figures*. http://www.marinet.org.uk/campaign-article/aggregate-dredging-and-the-crown-estate-facts-and-figures. Accessed June 2014.

Mason, C., Paxton, G., Parr, J., Boughen, N. (2010). Charting the territory: Exploring stakeholder reactions to the prospect of seafloor exploration and mining in Australia. *Marine Policy*, 34(6), 1374-1380.

Mason, C., Paxton, G., Parsons, R., Parr, J., Moffat, K. (2014). "For the Benefit of Australians": Exploring expectations for the mining industry from a national perspective, *Resources Policy*, 41, 1-8.

McClune, J. (2012). Marine phosphate mining generates global concern. *Namib Times*. http://www.namibtimes.net/forum/topics/marine-phosphate-mining-generates-global-concern. Accessed June 2014.

McKenzie, E., Woodruff, A. & McClennen, C. (2006). Economic Assessment of the True Costs of Aggregate Mining in Majuro Atoll Republic of the Marshall Islands. SOPAC.

Michel, J., Bejarano, A.C., Peterson, C.H. and Voss, C. (2013). Review of Biological and Biophysical Impacts from Dredging and Handling of Offshore Sand. U.S. Department of the Interior, Bureau of Ocean Energy Management, Herndon, VA. *OCS Study BOEM 2013-0119*. 258 pp.

Midgley, J. (2012). *Environmental impact assessment report for the marine component - Sandpiper project*. http://www.envirod.com/draft_environmental_impact_report2.html. Accessed June 2014.

MPA (2011). *Mineral Products Association press release*. MPA Says Scrapping of Aggregates Levy Sustainability Fund Works Against 'Localism'. 12 January 2011. http://www.mineralproducts.org/11-release001.htm. Accessed June 2014.

MSFD (2008). Directive 2008/56/EC of the European Parliament and of the Council – 17 June 2008 establishing a framework for community action in the field of marine environmental policy (Marine Strategy Framework Directive) http://eur-lex.europa.eu/LexUriServ/LexUriServ.do?uri=OJ:L:2008:164:0019:0040:EN:PDF. Accessed June 2014.

Mudd, G.M. (2010). The environmental sustainability of mining in Australia: key mega-trends and looming constraints. *Resources Policy*, 35 (2), 98-115.

Murton, B.J. (2002). A Global review of non-living resources on the extended continental shelf. *Brazilian Journal of Geophysics*, 18(3), 281-306.

NAMDEB (2010). *Financial performance for the year ended 31 December 2010*. http://www.namdeb.com/pdf/NamdebFinancialAdvert.pdf. Accessed June 2014.

NAMDEB (2014). Corporate website. http://www.namdeb.com/about_org_profile.php. Accessed June 2014.

Namibian Marine Phosphate (2012). *Sandpiper marine phosphates project*. http://www.namphos.com/ Accessed June 2014.

Nautilus Minerals (2014a). *Press release* Nautilus Minerals and State of PNG Resolve Issues and Sign Agreement. http://www.nautilusminerals.com/s/Media-NewsReleases.asp?ReportID=649293. Accessed June 2014.

Nautilus Minerals (2014b). *Community consultation in Papua New Guinea*. http://www.cares.nautilusminerals.com/PapuaConsultationProcess.aspx. Accessed May 2014.

Newell, R.C., Woodcock, T.A. (2013). *Aggregate Dredging and the Marine Environment: an overview of recent research and current industry practice*. The Crown Estate, pp. 165.

Newell, R.C., Seiderer, L.J., Simpson, N.M. and Robinson, J.E. (2004). Impacts of marine aggregate dredging on benthic macrofauna off the south coast of the United Kingdom. *Journal of Coastal Research*, 20 (1): 115–125.

Newell, R.C., Seiderer, L.J. and Hitchcock, D.R. (1998). The impact of dredging works in coastal waters: A review of the sensitivity to disturbance and subsequent recovery of biological resources on the seabed. *Oceanography and Marine Biology: An Annual Review*, 36: 127-78.

NZ Petroleum and Minerals (2014). *Minerals mining permit granted to Trans-Tasman Resources Limited*. http://www.nzpam.govt.nz/cms/news_media/2014/trans-tasman-resources-ltd-mining-permit-granted. Accessed June 2014.

NZ EPA (2015). New Zealand Environmental Protection Authority http://www.epa.govt.nz/news/epa-media-releases/Pages/EPA-refuses-marine-consent-application-by-CRP.aspx Accessed March 2015.

Ober, J. (1995). Sulfur, from Mineral Commodity Summaries, U.S. Bureau of Mines, January 1995, pp. 166-167. http://www.epa.gov/epawaste/nonhaz/industrial/special/mining/minedock/id/id4-sulf.pdf Accessed March 2015.

Odessey Marine Exploration Inc. (2014). http://ir.odysseymarine.com/releasedetail.cfm?ReleaseID=869839 Accessed March 2015.

OSPAR (2009). *Summary assessment of sand and gravel extraction in the OSPAR maritime area.* OSPAR Commission. http://qsr2010.ospar.org/media/assessments/p00434_Sand_and_Gravel_Summary_Assessment.pdf. Accessed June 2014.

Pelesikoti, N. (2007). Reducing vulnerability of Pacific ACP states. Kiribati Technical Report. 1. Extent of Household Aggregate Mining in South Tarawa 2. Proposed Integrated Monitoring Framework for Tarawa Lagoon. EU EDF 8 —Project Report 72. South Pacific Applied Geoscience Commission.

Pier, D. (2014). A Project of this Magnitude has Never Been Undertaken on Earth." A Gigantic Ocean Mine Threatens Baja California. The Scuttlefish. http://thescuttlefish.com/2014/12/a-project-of-this-magnitude-has-never-been-undertaken-on-earth-a-gigantic-ocean-mine-threatens-baja-california/ .

Pulfrich, A. and Branch, G.M. (2014). Effects of sediment discharge from Namibian diamond mines on intertidal and subtidal rocky-reef communities and the rock lobster Jasus lalandii. *Estuarine, Coastal and Shelf Science*, 150, 179-191.

Pulfrich, A., Parkins, C.A., Branch, G.M., Bustamante, R.H. and Velásquez, C.R. (2003). The effects of sediment deposits from Namibian diamond mines on intertidal and subtidal reefs and rock lobster populations. Aquatic Conservation: *Marine and Freshwater Ecosystems*, 13(3), 257-278.

PWC (2013). *Review of global trends in the mining industry.* http://www.pwc.com/en_GX/gx/mining/publications/assets/pwc-mine-a-confidence-crisis.pdf. Accessed June 2014.

Radzevicius, R., Velegrakis, A.F., Bonne, W., Kortekaas, S., Garel, E., Blazauskas, N. and Asariotis, R. (2010). Marine aggregate extraction regulation in EU member states. *Journal of Coastal Research*, SI 51: 15-37.

RCS (2014). Independent JORC (2012) Technical Report and Mineral Resource Estimate on the Chatham Rise Project in New Zealand.

Rice, J., Arvanitidis, C., Borja, A., Frid, C., Hiddink, J., Krause, J., Lorance, P., Ragnarsson, S.Á., Sköld, M., Trabucco, B. (2010). Marine Strategy Framework Directive – Task Group 6 Report Seafloor integrity. *EUR 24334 EN* – JRC, Luxemburg, pp. 73.

Roche, C., Bice, S. (2013). Anticipating Social and Community Impacts of Deep Sea Mining, in: Baker, E., Beaudoin, Y. (Eds.), *Deep Sea Minerals: Deep Sea Minerals and the Green Economy.* Vol. 2, Secretariat of the Pacific Community, pp. 59-80.

Rona, P.A. (2005). TAG hydrothermal vent field: A key to modern and ancient seafloor hydrothermal VMS ore-forming systems, in: Mao, J., Bierlein, F.P. (Eds), *Mineral Deposit Research: Meeting the Global Challenge*, Vols 1 & 2, 687-690.

Rona, P.A., (2008). The changing vision of marine minerals. *Ore Geology Reviews*, 33 (3-4) 618-666.

Scott, S.D., (2011). Marine minerals: their occurrences, exploration and exploitation. *Oceans 2011*. IEEE, pp. 1-8.

SOPAC (2007). *EU EDF 8 – SOPAC Project Report 74.* Reducing Vulnerability of Pacific ACP States. Samoa. Technical report on aggregate sources assessment in selected parts of Upolu and Savai'i Islands. 72pp. http://ict.sopac.org/VirLib/ER0074.pdf. Accessed June 2014.

SOPAC (2013). *Annual report*. http://www.sopac.org/sopac/sopac-3/SOPAC-3_2-1-4_OIP%202013%20report.pdf Accessed June 2014.

SPC (2013a). Deep Sea Minerals and the Green Economy. Baker, E., Beaudoin, Y. (eds). Vol 2 Secretariat of the Pacific Community, 210pp.

SPC (2013b). Sea-floor Massive Sulphides: A physical, biological, environmental, and technical review. Baker, E., Beaudoin, Y. (eds) Vol 1A, Secretariat of the Pacific Community, 65 pp.

SPC (2013c). Manganese Nodules: A physical, biological, environmental, and technical review. Baker, E., Beaudoin, Y. (eds). Vol. 1B, Secretariat of the Pacific Community 70 pp.

SPC (2013d). Cobalt-rich Ferromanganese Crusts: A physical, biological, environmental, and technical review. Baker, E., Beaudoin, Y. (eds), Vol. 1C Secretariat of the Pacific Community 60 pp.

SPC-EU (2012). *ACP states regional legislative and regulatory framework for deep sea minerals exploration and exploitation*. Prepared under the SPC-EU EDF10 Deep Sea Minerals Project. - 1st ed., pp70. http://www.smenet.org/docs/public/FinalDeepSeaMineralsProjectReport.pdf Accessed March 2015.

Sutton, G., Boyd, S. (2009). Effects of Extraction of Marine Sediments on the Marine Environment 1998 – 2004. *ICES Cooperative Research Report No. 297*. 180pp.

Szymelfenig, M., Kotwicki, L. & Graca, B. (2006). Benthic re-colonization in post-dredging pits in the Puck Bay (Southern Baltic Sea). Estuarine, Coastal and Shelf Science, 68(3), 489-498.

Tillin, H.M., Houghton, A.J., Saunders, J.E., Drabble, R., Hull, S.C. (2011). Direct and Indirect Impacts of Aggregate Dredging, Marine Aggregate Levy Sustainability Fund (MALSF). *Science Monograph Series: No. 1; MEPF 10/P144*, 46pp.

TIMAH (2012). *Press release.* http://www.idx.co.id/Portals/0/StaticData/NewsAndAnnouncement/ANNOUNCEMENTSTOCK/From_EREP/201303/805e679600_5ffbd03093.pdf. Accessed June 2014.

TIMAH (2014). http://www.timah.com/v2/eng/our-business/4110052012111846/tin-product/. Accessed June 2014.

Thomsen, F., McCully, S., Wood, D., Pace, F. and White, P. (2009). A generic investigation into noise profiles of marine dredging in relation to the acoustic sensitivity of the marine fauna in UK waters with particular emphasis on aggregate dredging: phase 1 scoping and review of key issues. Cefas MEPF Ref No. MEPF/08/P21. 59 pp.

Todd, V.L., Todd, I.B., Gardiner, J.C., Morrin, E.C., MacPherson, N.A., DiMarzio, N.A. and Thomsen, F. (2015). A review of impacts of marine dredging activities on marine mammals. ICES Journal of Marine Science: Journal du Conseil, 72(2), 328-340.

TTR (2015) Trans Tasman Resources http://www.ttrl.co.nz/south-taranaki-bight-project/overview/ Accessed march 2015.

UNEP (2014). Sand, rarer than one thinks. *UNEP Global Environment Alert Service (GEAS)*. http://na.unep.net/geas/getUNEPPageWithArticleIDScript.php?article_id=110. Accessed June 2014.

United Nations (1992a) Convention on Biological Diversity. Rio de Janeiro, 5 June 1992.

United Nations (1992b) Agenda 21. United Nations, Rio de Janeiro, 13 June 1992.

Van Rijn, L.C., Soulsby, R.L., Hoekstra, P. & Davies, A.G. (2005). SANDPIT, *Sand Transport and Morphology of Offshore Mining Pits*. Aqua Publications.

Velegrakis, A.F., Ballay, A., Poulos, S., Radzevičius, R., Bellec, V., Manso, F. (2010). European marine aggregates resources: Origins, usage, prospecting and dredging techniques. *Journal of Coastal Research*, SI 51, 1–14.

Walker, D.I., Hillman, K.A., Kendrick, G.A., Lavery, P. (2001). Ecological significance of seagrasses: Assessment for management of environmental impact in Western Australia. *Ecological Engineering*, 16(3), 323-330.

Wenban-Smith, F.F. (2002). Marine Aggregate Dredging and the Historic Environment: Palaeolithic and Mesolithic archaeology on the seabed. BMAPA and EH, London.

Webb, A. (2005). Technical Report--An assessment of coastal processes, impacts, erosion mitigation options and beach mining. (Bairiki/Naikai causeway, Tungaru Central Hospital coastline and Bonriki runway--South Tarawa, Kiribati). *EU-SOPAC Project Report, 46.*

24 Solid Waste Disposal

Contributors:
Alan Simcock (Lead member), Juying Wang (Co-lead member)

1 Introduction – the regulatory system

The disposal at sea of waste generated on land and loaded on board vessels for dumping is the object of long-standing global, and (in many areas) regional, systems of regulation. (These systems also cover, for completeness, dumping from aircraft and waste (other than operational discharges) from fixed installations in the sea). Such dumping must be distinguished from discharges into rivers and directly from land into the sea and emissions to air from land-based activities discussed in Chapter 20 (Land-based inputs).

When concerns about the environment developed in the 1960s, growing constraints on the land disposal of waste and discharges into rivers led to pressures to find new routes for waste disposal. Concerns about these pressures led to action in several forums. Several United Nations specialized agencies set up the Group of Experts on the Scientific Aspects of Marine Pollution (GESAMP[1] – later altered to "Marine Environmental Protection").

The preparatory committee for the 1972 Stockholm Conference on the Human Environment, set up by the United Nations General Assembly, established an intergovernmental working group on marine pollution. At the national level, several countries started developing approaches to control such dumping. The United States of America put forward proposals for an international agreement on the subject. Spurred from the national level by an attempt by the vessel *Stella Maris* to dump 650 tons of chlorinated waste, several countries started developing approaches to control such dumping. States adjoining the North-East Atlantic adopted an international convention regulating dumping in that area in Oslo, Norway, on 15 February 1972 (OSPAR, 1982; IMO, 1991).

Later that year, the Stockholm Conference adopted a set of principles for international environmental law and called, among other things, for an international instrument to control dumping of waste at sea. The United Kingdom, in consultation with the United Nations Secretariat, organized a further conference in London, and the Convention on the Prevention of Marine Pollution by Dumping of Wastes and Other Matter 1972 (the 1972 London Convention) was signed on 13 November 1972 in London, Mexico City and Moscow (ICG, 1982, IMO, 2014f).[2]

1.1 The 1972 London Convention

The main provisions of the 1972 London Convention can be summarized as follows:

(a) A definition of "dumping" to cover the deliberate disposal of waste and other matter at sea from ships, aircraft, platforms or other man-made structures in the sea;

(b) A ban on dumping at sea of any of the substances on the "black list" (Annex I to the Convention): toxic organohalogen compounds, agreed carcinogenic substances, mercury and cadmium and their compounds, crude oil and petroleum products[3] taken on board for the purpose of dumping them, high-level radioactive substances as defined by the International Atomic Energy Agency and persistent synthetic substances (including plastics) liable to float or remain in suspension. Exceptions were allowed for *force majeure* and for trace amounts not added for disposal purposes;

(c) A requirement for a special prior permit for any dumping of any substances on the "grey list" (Annex II to the Convention) – arsenic, lead, copper and zinc and their compounds, organosilicon compounds, cyanides, fluorides and pesticides not in Annex I, bulky objects and tar likely to obstruct fishing or navigation, medium-level and low-level radioactive waste and substances to be dumped in such quantities as to cause harm;

(d) A requirement for at least a general prior permit for all other dumping. Such permits were required to follow an approach set out in Annex III to the Convention, which required consideration of alternative land-based disposal and the avoidance of harm to legitimate uses of the sea;

(e) A requirement to appraise the effectiveness of the regulatory assessment process through compliance monitoring and field monitoring of effects;

(f) An obligation to report to the Secretariat of the Convention (which is hosted by the International Maritime Organization (IMO) in London) on dumping permits issued and amounts permitted to be dumped (IGC, 1982; LC-LP, 2014a).

When the 1972 London Convention entered into force in 1975, dumping at sea was still a major disposal route for many kinds of waste. Over the years, the meetings of the Contracting Parties have tightened the requirements of the Convention, with the result that the amounts of waste that may be dumped were reduced significantly:

(a) Guidance was adopted on the approaches to the grant of special and general permits for dumping. In many respects this guidance was gradually made more precise and restrictive (IMO, 2014a);

(b) In 1972 incineration of hazardous waste at sea was just beginning to be practised. In 1978 an amendment was adopted clarifying that the incineration at sea of oily wastes and organohalogen compounds was permitted as an interim solution, but requiring a special prior permit in accordance with agreed guidelines for this practice. This amendment came into force in 1979 (IGC, 1982). In 1988, the Consultative Meeting of the States parties called for such incineration to be minimized and for a re-evaluation of the practice (LDC, 1988). In 1993 an amendment to prohibit this practice was adopted and entered into force from 1994 (IMO, 2012);

(c) In 1990, the Contracting Parties adopted a resolution calling for the phasing out of the dumping of industrial waste (LDC, 43(13)).

---

1 At present, it is jointly sponsored by IMO, FAO, IAEA, WMO, UNESCO-IOC, UN, UNDP, UNEP and UNIDO.

2 United Nations, *Treaty Series*, vol. 1046, No. 15749.

3 "Petroleum products" includes wastes from crude oil, refined petroleum products, petroleum distillate products, and any mixtures containing these substances.

Following this, an amendment to Annex I of the Convention was adopted in 1993, which entered into force in 1994, to prohibit the dumping of industrial waste from the end of 1995 (IMO, 2012; IMO, 2014c).

(d) Even though the 1972 London Convention, as adopted, prohibited the dumping of high-level radioactive waste, many Contracting Parties remained unhappy with any dumping of radioactive waste of any kind. In 1983, a voluntary moratorium on such dumping was agreed. In 1993 an amendment was adopted to prohibit all dumping of radioactive waste, subject to a review before February 2019, and every twenty-five years thereafter. The Consultative Meeting of the Contracting Parties is beginning preparations for this review (IMO, 2012; LC-LP, 2014).

1.2 The 1996 London Protocol[4]

The generally restrictive policy of the Contracting Parties to the 1972 London Convention towards the dumping of waste and other matter at sea resulted in a further development in 1996, when a protocol to the convention was adopted. This Protocol is intended gradually to replace the 1972 London Convention. The London Protocol entered into force in 2006. Among a number of other changes, the fundamental difference between the 1972 Convention and the 1996 London Protocol is that the Protocol adopts a "reverse list" approach. All dumping of waste is prohibited, except for a limited number of categories where dumping could be permitted, in contrast to the 1972 Convention approach, which prohibited dumping only of a specified list of substances, while requiring a permit (general or special) for everything else. The limited number of categories where dumping can still be permitted under the Protocol as originally adopted are:

(a) Dredged material;
(b) Sewage sludge;
(c) Fish waste, or material resulting from industrial fish processing operations;
(d) Vessels and platforms or other man-made structures at sea;
(e) Inert, inorganic geological material;
(f) Organic material of natural origin;
(g) Bulky items primarily comprising iron, steel, concrete and similar unharmful materials for which the concern is physical impact and limited to those circumstances, where such wastes are generated at locations, such as small islands with isolated communities, having no practicable access to disposal options other than dumping.

Shortly after the Protocol entered into force in 2006, the Meeting of Contracting Parties to the London Protocol adopted an amendment to add "sub-seabed carbon-dioxide (CO_2) streams from CO_2 capture processes for sequestration" to the list of permitted forms of disposal (LP.1(1)). States Parties may therefore issue permits to allow the injection into a sub-seabed geological formation of CO_2 streams from CO_2 capture processes. This amendment entered into force in 2007. In 2012, specific guidelines were adopted to for such disposal activities and the potential effects on the marine environment in the proximity of the receiving formations. In 2009, a further amendment was adopted, allowing the export of CO_2 from CO_2 capture processes for sequestration in sub-seabed geological formations (LP.3(4)). This amendment is not yet in force. Guidance on the implementation of the export of CO_2 streams for disposal in sub-seabed geological formations for the purposes of sequestration was adopted in 2013. The intention of carbon dioxide sequestration in sub-seabed geological formations is to prevent release into the biosphere of substantial quantities of carbon dioxide derived from human activities, by retaining the carbon dioxide permanently within such geological formations.

In 2008, the Contracting States to both the 1972 London Convention and the 1996 London Protocol adopted a resolution agreeing that the scope of the London Convention and Protocol includes ocean fertilization activities, that is, any activity undertaken by humans with the principal intention of stimulating primary productivity in the oceans. (Ocean fertilization does not include ordinary aquaculture, or mariculture, or the creation of artificial reefs). It was further agreed that:

(a) In order to provide for legitimate scientific research, such research should be regarded as placement of matter for a purpose other than the mere disposal thereof under Article III.1(b) (ii) of the London Convention and Article 1.4.2.2 of the London Protocol;
(b) Scientific research proposals should be assessed on a case-by-case basis using an assessment framework to be developed by the Scientific Groups under the London Convention and Protocol;
(c) Such an assessment framework should include, *inter alia*, tools for determining whether the proposed activity is contrary to the aims of the Convention and Protocol;
(d) Until specific guidance is available, Contracting Parties should be urged to use utmost caution and the best available guidance to evaluate the scientific research proposals to ensure protection of the marine environment consistent with the Convention and Protocol;
(e) For the purposes of the resolution, legitimate scientific research should be defined as those proposals that have been assessed and found acceptable under the assessment framework;
(f) Given the present state of knowledge, ocean fertilization activities other than legitimate scientific research should not be allowed. To this end, such other activities should be considered as contrary to the aims of the Convention and Protocol and should not currently qualify for any exemption from the definition of dumping in the Convention and the Protocol (LC-LP, 2008).

In 2010, the Contracting Parties to the 1972 London Convention and the 1996 London Protocol adopted the Assessment Framework for Scientific Research Involving Ocean Fertilization (LC-LP, 2010). In 2013, the Contracting Parties to the London Protocol adopted amendments to incorporate into the Protocol provisions regulating the placement of matter for ocean fertilization and other marine geo-engineering activities (LP.4(8)). These amendments are not yet in force (LC-LP, 2013). Guidance on implementing the provisions was adopted in 2014 (LC-LP, 2014).

---

4   36 *International Legal Materials* 1 (1997).

1.3 Acceptance of the system of regulation

As of October 2014, there are 87 parties to the 1972 London Convention, and 45 parties to the 1996 London Protocol. Thirty-four States are parties to both the Convention and the Protocol (IMO, 2014b). There are, however, many regional conventions on marine environmental protection that have specific references to, or contain provisions relating to, the regulation of disposal of wastes into the sea. Most regional conventions (the Abidjan, Antigua, Barcelona, Bucharest, Cartagena, Helsinki, Jeddah, Kuwait, Lima, Nairobi, Noumea, OSPAR Conventions[5]) have specific provisions that regulate sea dumping. The dumping clauses are largely based on, or are more stringent than, the London Convention or London Protocol. (An overview of Contracting Parties to the London Protocol, London Convention and Regional Agreements that include management of sea dumping issues is set out in IMO 2014e). Most States are therefore Contracting Parties to an international agreement that relates to the management of sea dumping of solid waste or other matter. However, there remain some States, including some of the world's 20 largest economies, which are not party to any of these agreements. It is not known how far such States apply policies along the lines of those required by the 1972 London Convention or the 1996 London Protocol.

2 Amounts and nature of current dumping

Agreements in, and under, the 1972 London Convention and the 1996 London Protocol provide for annual reporting of the number of permits and the quantity and nature of the waste dumped under them. However, reporting under the Convention and the Protocol is not consistent. Figure 24.1 shows, for 1976 to 2010, the number of States that are Contracting States of the 1972 London Convention, the number submitting reports and the proportion that the latter are of the former.

When the Meeting of Contracting Parties to the 1996 London Protocol set up a compliance mechanism in 2007, the worrying decline in reporting led it to include the issue of reporting in the terms of reference of the Compliance Group, which formed part of that mechanism (LC-LP, 2007). Reports under the London Convention and Protocol take some time to be compiled and submitted. It is usually only in the fourth year after the year being reported on that it is possible to take a final view on the reporting for that year. It is worth noting that non-reporting is the highest amongst London Convention parties, while reporting from London Protocol parties is above 75per cent. It may well be that some or all of the 59 per cent of Contracting States that did not submit reports had not authorized any dumping –like eight of the States in 2010 that did submit reports – but the absence of reports makes it impossible to draw clear conclusions. Also, several non-reporting States are land-locked, and therefore may also not have had any dumping to report. There is also a substantial degree of variation from year to year in which States submit reports.

The Meetings of the Contracting Parties have made efforts to try to improve the level of reporting on the dumping of waste at sea, but so far with limited success. The steps taken include reviews and simplifications of the reporting forms and more recently the introduction of on-line reporting. Improved outreach to Parties and contact with the industrial organizations (such as the International Association of Ports and Harbours) involved in dumping is beginning to produce some results. Some States (such as Nigeria and South Africa) have also sought to assist neighbours to set up reporting systems (LC-LP, 2013).

In spite of these efforts, it is therefore difficult to derive a clear picture of the quantity and nature of wastes and other matter being dumped at sea from the reports under the 1972 London Convention and 1996 London Protocol.

Nevertheless, it is clear that the overwhelming type of dumping is of dredged material. For the last year for which a summary of the national reports is available (2010), 35 of the 38 reports submitted recorded the dumping of dredged material. Most, if not all, of this is derived from dredging for navigational purposes. Some is "capital dredging" for the creation of new berths or shipping channels, but most is "maintenance dredging" for the maintenance of existing harbours and shipping channels. The quantity of material involved is considerable. For example, Belgium reported dumping 52 million tons in 2010: over 200,000 tons per working day. It is not, however, possible to give an overall picture

---

[5] Convention for Co-operation in the Protection and Development of the Marine and Coastal Environment of the West and Central African Region (Abidjan Convention). http://abidjanconvention.org/index.php?option=com_content&view=article&id=100&Itemid=200&lang=en

The Convention for Cooperation in the Protection and Sustainable Development of the Marine and Coastal Environment of the Northeast Pacific (Antigua Convention). http://www.unep.org/regionalseas/programmes/nonunep/nepacific/instruments/nep_convention.pdf

Convention for the Protection of the Marine Environment and the Coastal Region of the Mediterranean (Barcelona Convention). United Nations Treaty Series. vol. 1102, No. 16908.

Convention on the Protection of the Black Sea Against Pollution (Bucharest Convention). United Nations Treaty Series. vol. 1764, No. 30674.

Convention for the Protection and Development of the Marine Environment of the Wider Caribbean Region (Cartagena Convention). United Nations Treaty Series, vol. 1506, No. 25974.

Convention on the protection of the marine environment of the Baltic sea Area, 1992 (Helsinki Convention). United Nations Treaty Series, vol. 2099, No. 36495.

Regional Convention for the Conservation of the Red Sea and Gulf of Aden Environment (Jeddah Convention). http://www.persga.org/Documents/Doc_62_20090211112825.pdf.

Kuwait Regional Convention for Co-operation on the Protection of the Marine Environment from Pollution (Kuwait Convention). United Nations Treaty Series, vol. 1140, No. 17898.

Agreement on the Protection of the Marine Environment and Coastal Area of the South-East Pacific (Lima Convention). United Nations Treaty Series, vol. 1648, No. 28325.

The Convention for the Protection, Management and Development of the Marine and Coastal Environment of the Eastern African Region (Nairobi Convention). http://www.unep.org/NairobiConvention/The_Convention/index.asp.

Convention for the Protection of Natural Resources and Environment of the South Pacific Region (Noumea Convention). https://www.sprep.org/attachments/Legal/Files_updated_at_2014/NoumeaConvProtocols.pdf

Convention for the protection of the marine environment of the north-east Atlantic (the 'OSPAR Convention'). United Nations Treaty Series, vol. 2354, No. 42279.

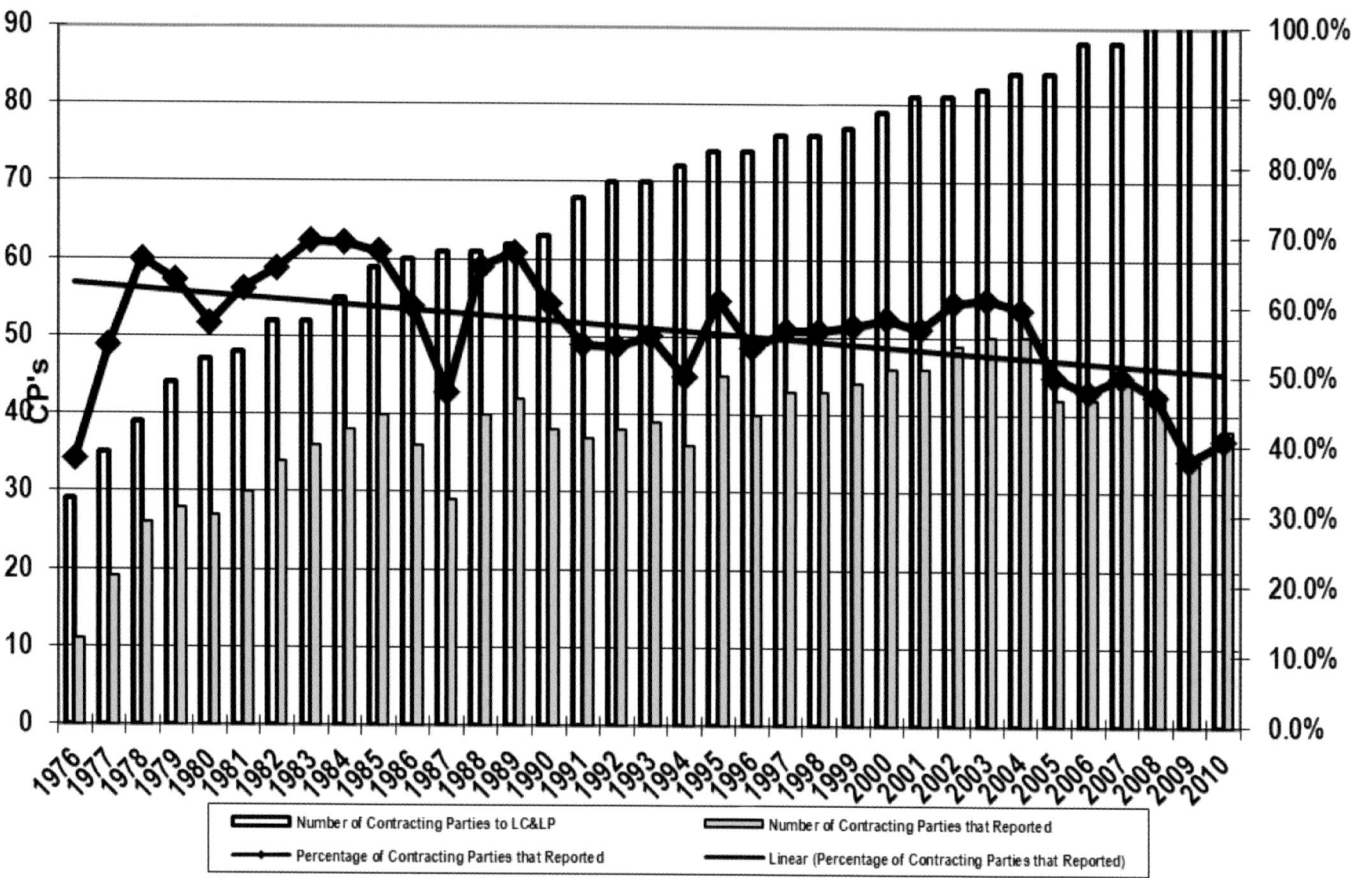

Figure 24.1 | Contracting Parties to the 1972 London Convention, Contracting Parties submitting reports to the Convention Secretariat and the latter as a proportion of the former, 1976 – 2010. Source: IMO, 2014g.

of how much is the result of regular dredging and how much is new construction, because many reports do not differentiate between capital dredging and maintenance dredging.

The impacts of this dumping of dredged material are essentially twofold (although there can be other effects): the smothering of the seabed by the dredged material, and the remobilization of hazardous substances contained in the dredged material. The effects of smothering depend essentially on the nature of the dump area. If the dumpsite were to have a biodiverse benthic life, such smothering would be catastrophic. Where tidal action is very dynamic and there is a sedimentary bottom, effects are limited, because much of the seabed material will be kept in motion by the tidal action. The choice of dumpsite is therefore important. The regular use of the same dumpsites (which is reported to be common) limits adverse effects. The remobilization of hazardous substances is a different matter. The Guidance under the London Convention and Protocol sets out procedures and criteria for deciding whether it is safe to dump contaminated dredged material. Where the harbour from which the dredged material comes is on the estuary of a river with a history of heavy industry (for example, the Rhine), it is frequently contrary to this Guidance (or, in the example quoted, parallel guidance from OSPAR, the local regional organization) to dump the material at sea, and it should be returned to land.

In the past, a substantial number of States dumped sewage sludge or animal slurry at sea. Where this was done, of course, it was an addition to the nutrient input. In many areas, this has now been stopped because it was a potential contributor to eutrophication problems. In 2010, only Australia (up to 20,000 litres) and the Republic of Korea (556,534 tons) reported dumping of this kind (IMO, 2014b). The Republic of Korea has also reported that dumping of sewage sludge will end by the end of 2015 (LC-LP, 2013).

The other substances reported as dumped cover a miscellaneous range. Dumping of fish waste was reported in 2010 by six countries. The total amount dumped was around 100,000 tons (not all reporting was in terms of tonnage). The other categories of material dumped included rock, sand and gravel, spoilt cargoes (for example, wheat, rice and fertilizer), molasses waste and a handful of ships and platforms (some of the latter being intended to create artificial reefs). In addition, permits were granted for a few burials at sea (see Chapter 8 Cultural ecosystem services). The overall impression is that, for the countries submitting reports, disposal of waste at sea is now a minor impact on the marine environment and human uses of the sea, except for the dumping of dredged material.

3 Dumping of radioactive material

As noted above, the dumping of high-level radioactive waste has been prohibited under the 1972 London Convention since 1975, and dumping of medium- and low-level radioactive waste has been prohibited also under the 1996 London Protocol (subject to a review every 25 years) since 1994. The first reported sea disposal of radioactive waste took place in 1946 and the last authorized disposal appears to have been in 1993. During the 48-year history of sea disposal, 14 countries have used more than 80 sites to dispose of approximately 85,000 terabecquerels of radioactive waste. Some countries used this waste management option only for small quantities of radioactive waste. Two countries conducted only one disposal each and one country conducted only two disposals (IAEA, 1999).

In 1992, reports that the former Soviet Union had dumped large amounts of high-level radioactive wastes for over three decades in shallow waters in the Arctic Ocean caused widespread concern, especially in countries with Arctic coastlines. In 1992, a joint Norwegian-Russian Expert Group was established to investigate radioactive contamination due to dumped nuclear waste in the Barents and Kara Seas. The Russian Federation provided information on the dumping, some of which had taken place before 1975. It arranged exploratory cruises to the dumping areas, with the participation of the International Atomic Energy Agency. The results obtained during the cruises did not indicate any significant radioactive contamination at the dumping sites, although the levels near some dumped objects are slightly elevated compared with elsewhere (IAEA, 1995).

Norway undertook further radiological monitoring of the Barents Sea in 2007, 2008 and 2009. Activity concentrations of the anthropogenic radionuclides usually used to trace the impact of radioactive waste were reported as low, and up to an order of magnitude lower than in previous decades, including in marine biota. Weighted absorbed dose rates to biota from anthropogenic radionuclides were low, and orders of magnitude below a predicted no-effect screening level of 10 micrograys per hour (μGy/hr). Dose rates to man from consumption of seafood and dose rates to biota in the marine environment were found to be dominated by the contribution from naturally occurring radionuclides (Gwynn et al., 2012). In 2012, a further joint Norwegian/Russian project examined radioactive pollution in the Kara Sea (Stråleverninfo, 2012). It concluded that the situation gave rise to no immediate cause for concern, but that further monitoring of the situation is warranted (JNREG, 2014). A further joint Norwegian/Russian study of radioactive contamination in the Barents Sea has been launched.

4 Dumped explosives and military chemicals

After both World Wars, States were faced with the problem of how to dispose of the residues of explosive materials and other warlike stores ("munitions"), including a number of containers of poisonous gases. The solution adopted for substantial quantities was to dump them in the sea. During peacetime, some States have also adopted this method of disposal for unwanted explosives and military chemicals. The dump sites were usually chosen to avoid seabed areas then being used by people, but over time some of these areas have come into use as a result of improved technologies and pressures from other uses of the sea.

In 2010, the United Nations General Assembly adopted a resolution noting the importance of raising awareness of the environmental effects related to waste originating from chemical munitions dumped at sea, and invited relevant international organizations to keep the issue under review (UNGA, 2010).

Munitions dumped at sea present a risk to several classes of users of the sea. Fishers in the location of the dump sites can bring the munitions up in their nets, especially bottom-trawling nets. Construction of offshore installations, submarine cables and submarine pipelines can interact with dumped munitions. Some munitions based on phosphorus can break out from the (often wooden) boxes in which they were stored at the time of disposal, float to the surface, be stranded on beaches and then (as the tide recedes and they dry out) spontaneously burst into flame, and burn at temperatures around 1,000 degrees centigrade. These present potential risks to users of beaches, especially tourists (HELCOM, 2013).

Exercises have been carried out in several parts of the world to map the dump sites and to establish what was dumped there. The Baltic Marine Environment Protection Commission (HELCOM) estimated that 40,000 tons of munitions were dumped in the Baltic at the end of World War II. Some of these munitions are contained in ships onto which they were loaded and which were then scuttled. Others were thrown overboard piece by piece, a process which means that the munitions can end up scattered over a wide area. Similar conclusions about dispersed dumping have been reached in other areas. The four main dumping areas in the Baltic were south-east of the Swedish island of Gotland and south-west of the Latvian city of Liepaja, east of the Danish island of Bornholm and south of the Little Belt between the main Danish islands and Schleswig-Holstein in Germany. There is also evidence that munitions were thrown overboard as the ships left port (HELCOM, 2013). The OSPAR Commission has carried out a similar exercise, resulting in an "Overview of Past Dumping at Sea of Chemical Weapons and Munitions", together with a database on encounters with dumped conventional and chemical munitions, which it is intended to keep up-to-date. Best estimates suggest that over one million tons of munitions were dumped in Beaufort's Dyke (a trough in the United Kingdom of Great Britain and Northern Ireland between Scotland and Northern Ireland), some 168,000 tons of ammunition were dumped in the Skagerrak, some 300,000 tons of munitions of various types, such as bombs, grenades, torpedoes and mines, were dumped in the North Sea and an estimated 35,000 tons were dumped off Knokke-Heist, Belgium (OSPAR, 2010).

In other parts of the world, problems have arisen with dumped munitions. For example, in 2006 New Zealand had problems with munitions that had been dumped improperly at the end of the Second World War. An estimated 1,500 tons of munitions had ended up in relatively shallow water and were posing threats to fisheries and recreational uses of the sea. The New Zealand authorities concluded that the best solution was to lift them and re-dump them in much deeper water before they dried out: if they were brought ashore and allowed to dry, there was a high risk that they would become unstable (LC-LP, 2006).

A non-governmental organization, the James Martin Center for Nonproliferation Studies, conducted a general survey of dumped chemical warfare munitions and published an interactive map of 168 munitions dump-sites, with the publicly available information about them, on the internet (https://www.google.com/maps/d/viewer?mid=zwm9Gb8KEKxI.kMpXo9rjqLZM&hl=en).

In 2010, the Research and Technology Organization of the North Atlantic Treaty Organization (NATO) reviewed the environmental aspects of the disposal of unwanted munitions. The overall conclusion was that that the technology and expertise existed to deal with immediate problems and with the current generation of munitions, including the legacy of munitions dumped at sea, but that the expertise and technology was often lodged in countries where there was no significant problem, and that a mechanism was required to assist in the transfer of the technology and expertise to the places where it was needed. It was noted that this could be significant in measures to control terrorism (NATO, 2010).

5 Illegal dumping

If there are problems in obtaining an overall global picture of dumping authorized under the London Convention and London Protocol, trying to gain an overview of the potential effects of illegal dumping presents much greater problems. While the 1972 London Convention and the 1996 London Protocol have a mechanism for reporting illegal dumping[6], no report has been received in the recent past. An alleged case of illegal dumping in Canadian waters is currently under investigation with a report expected to be provided to the governing bodies of the London Convention and Protocol in the near future.

Several cases have been reported of illegal export of waste from industrialized countries for disposal in States in Africa. Most of these have concerned disposal on land. There have also been persistent informal reports of dumping of radioactive or toxic waste in the sea off the coast of the Federal Republic of Somalia. Informal information given to INTERPOL suggested that the naval force present off the coast of the Federal Republic of Somalia to combat piracy may have detected vessels suspected of illegal dumping of waste. Following the tsunami on 26 December 2004, UNEP responded to an urgent request from the authorities in the Puntland region of the Federal Republic of Somalia for help in assessing potential environmental damage. After an initial UNEP report, an inter-agency mission, which included FAO, UNDP, UNEP and WHO, went to Puntland in March 2005. It investigated three sample sites along a 500-kilometre coastal stretch between the three main populated coastal locations of Xaafuun, Bandarbeyla and Eyl where toxic waste had reportedly been uncovered by the tsunami. No evidence of toxic waste was found by the mission. In June 2010, Greenpeace International claimed to have proof of the dumping of toxic waste in the Federal Republic of Somalia by European and American companies in the period from 1990 to 1997, citing testimony from an Italian parliamentary commission, evidence uncovered by an Italian prosecutor (including wiretapped conversations with alleged offenders) and warnings by the Special Representative of the Secretary-General for Somalia in 2008 of possible illegal dumping in the Federal Republic of Somalia. While INTERPOL and some of the entities cited in the Greenpeace International report have uncovered fragmentary evidence and signs of the dumping of toxins, no international investigation has ever been able to verify the dumping of illegal waste in the Federal Republic of Somalia, largely because of the security situation (UNSC, 2011).

Other evidence of illegal dumping appears from time to time as a result of ocean monitoring. For example, the authorities in Japan have detected within areas under its jurisdiction high levels of polychlorinated biphenyls (PCBs) and butyl tin and phenyl tin compounds. The origins of such pollution could not be identified (Japan MOE, 2009).

6 Conclusions on knowledge gaps and capacity-building gaps

The disposal of solid waste at sea has been regulated under international agreements for the past 40 years. The majority of coastal States have accepted this regime. If the 1972 London Convention and the 1996 London Protocol were effectively and consistently applied, this source of inputs of harmful substances would be satisfactorily controlled. The problem is basically that we do not know whether this regime is generally being fully implemented, since there is substantial under-reporting of what is happening.

There is therefore a major knowledge gap about the implementation of the 1972 London Convention and the 1996 London Protocol, as has been acknowledged by the Meetings of the Contracting Parties to the two agreements. Some capacity-building is available from the International Maritime Organization and some of the Contracting Parties, to promote better implementation of the agreements and better reporting of what is being done. However, a significant capacity-building gap remains.

The information gap about the scale and nature of dumping of waste and other matter that is taking place is further compounded by the ab-

---

6 See http://www.imo.org/OurWork/Environment/LCLP/Reporting/incidents/Pages/default.aspx

sence of information about dumping under the control of States which are subject to any formal reporting system under the 1972 London Convention, the 1996 London Protocol or regional dumping agreements and which do not publish any national data. This category includes some of the world's largest economies.

Much work has been done to identify the locations where munitions have been dumped. However, some gaps in the knowledge remain on this subject. There are gaps in building capacities to help fishers and other users of the sea to draw on this knowledge, in order to reduce the risks to which they are subjected and to know how they should respond if they bring up dumped munitions in their nets.

References

Gwynn, J.P., Heldal, H.E., Gäfvert, T., Blinova, O., Eriksson, M., Sværen, I., Brungot, A.L., Strålberg, E., Møller, B., Rudjord, A.L. (2012). Radiological status of the marine environment in the Barents Sea, *Journal of Environmental Radioactivity*, 113.

HELCOM (Baltic Marine Environment Protection Commission) (2013). *Chemical Munitions Dumped in the Baltic Sea.* Report of the ad hoc Expert Group to update and Review the Existing Information on Dumped Chemical Munitions in the Baltic Sea, Baltic Sea Environment Proceeding (BSEP) No. 142, Helsinki.

IAEA (International Atomic Energy Agency) (1995). Special Report: Marine scientists on the Arctic Seas: Documenting the radiological record by Pavel Povinec, Iolanda Osvath, and Murdoch Baxter, in *IAEA Bulletin* 2/1995.

IAEA (International Atomic Energy Agency) (1999). *Inventory of Radioactive Waste Disposals at Sea*, IAEA-TECDOC-1105.

IGC (Inter-Governmental Conference on the Convention on the Dumping of Wastes at Sea (1982). *Final Act of the Conference,* International Maritime Organization, London.

IMO (International Maritime Organization) (1991). *The London Dumping Convention: The First Decade and Beyond*. International Maritime Organization, London.

IMO (International Maritime Organization) (2012). International Maritime Organization, Status of the London Convention and Protocol (IMO Document LC 34/2), 2012.

LC-LP (International Maritime Organization) (2014a). Convention on the Prevention of Marine Pollution by Dumping of Wastes and Other Matter. (http://www.imo.org/About/Conventions/ListOfConventions/Pages/Convention-on-the-Prevention-of-Marine-Pollution-by-Dumping-of-Wastes-and-Other-Matter.aspx accessed 9 April 2014).

IMO (International Maritime Organization) (2014b). Final report on permits issued in 2010 (IMO Document LC-LP.1/Circ.63).

IMO (International Maritime Organization) (2014c). Status of multilateral Conventions and instruments in respect of which the International Maritime Organization or its Secretary-General performs depositary or other functions, 2014. (http://www.imo.org/About/Conventions/StatusOfConventions/Documents/Status%20-%20 2014.pdfaccessed 28 October 2014).

IMO (International Maritime Organization) (2014e). *The London Protocol – What is it and how to implement it*, IMO I533E.

IMO (International Maritime Organization) (2014f). *Origins of the London Convention.* (http://www.imo.org/KnowledgeCentre/ReferencesAndArchives/IMO_Conferences_and_Meetings/London_Convention/VariousArticlesAndDocumentsAboutTheLondonConvention/Documents/Origins%20of%20the%20London%20Convention%20-%20Historic%20events%20and%20documents%20%20M.%20Harvey%20September%202012.pdf accessed 12 October 2014).

IMO International Maritime Organization (2014g). Direct Communication from the IMO Secretariat in 2014.

Japan MOE (Ministry of the Environment) (2009). *Present Status of Marine Pollution in the Sea around Japan*, Ministry of Environment, Tokyo.

JNREG (Joint Norwegian-Russian Expert Group) (2014). Investigation into the Radioecological status of Stepovogo Fjord. The dumping site of the nuclear submarine K-27 and solid radioactive waste. Result from the 2012 research cruise. Norwegian Radiation Protection Authority. ISBN: 978-82-90362-33-6.

LC-LP (1972 London Convention and 1996 London Protocol) (2006). *Notification under Article 8.2 of the 1996 London Protocol regarding a case of emergency*. London Convention document LC-LP.1/Circ.2.

LC-LP (1972 London Convention and 1996 London Protocol) (2007). Compliance Procedures and Mechanisms pursuant to Article 11 of the 1996 Protocol to the 1972 London Convention (Report of the Twenty-Ninth Consultative Meeting Annex 7 (London Convention document LC 29/1 7, annex 7).

LC-LP (1972 London Convention and 1996 London Protocol) (2008). Resolution LC-LP.1 on the Regulation of Ocean Fertilization (LC-LP document 30/16, Annex 6).

LC-LP (1972 London Convention and 1996 London Protocol) (2010). Resolution LC-LP.2 on the Assessment Framework for Scientific Research (LC-LP document 32/15, Annex 5).

LC-LP (1972 London Convention and 1996 London Protocol) (2013). Report of the Thirty-Fifth Consultative Meeting of Contracting Parties to the London Convention & Eighth Meeting of Contracting Parties to the London Protocol (London Convention document LC 35/15).

LC-LP (1972 London Convention and 1996 London Protocol) (2014). (36th Consultative Meeting of Contracting Parties (1972 London Convention) and 9th Meeting of Contracting Parties (1996 London Protocol), 3-7 November 2014. (http://www.imo.org/MediaCentre/MeetingSummaries/LCLP/Pages/LC-36-LP-9.aspx accessed 20 November 2014).

LDC (London Convention) (1988). Resolution LDC.35 (11) Status of Incineration of Noxious Liquid Wastes at Sea.

NATO (North Atlantic Treaty Organization) (2010). *Environmental Impact of Munition and Propellant Disposal*. RTO Technical Report Tr-Avt-115.

OSPAR (Oslo and Paris Commissions) (1982). *The Oslo and Paris Commissions – the first ten years.* London.

OSPAR (Oslo and Paris Commissions) (2010). OSPAR Commission for the Protection of the North-East Atlantic, Overview of Past Dumping at Sea of Chemical Weapons and Munitions, London 2010 (ISBN 978-1-907390-60-9).

Stråleverninfo (2012). Statens Strålevern, Felles norsk-russisk tokt til dumpet atomavfall I Kara havet (http://www.nrpa.no/dav/6ced2cea4b.pdf accessed 19 April 2014).

UNGA (United Nations General Assembly) (2010). Cooperative measures to assess and increase awareness of environmental effects related to waste originating from chemical munitions dumped at sea (A/RES/65/149).

UNSC (United Nations Security Council) (2011). Report of the Secretary-General on the protection of Somali natural resources and waters (S/2011/661).

25 Marine Debris

Contributors:
Juying Wang (Convenor and Lead Member), Maria Clare Baker, Arsonina Bera, Kiho Kim, Rainer Lohmann, Douglas Ofiara and Yuhui Zhao

1 Overview

1.1 Definition of marine debris

Litter disposal and accumulation in the marine environment is one of the fastest-growing threats to the health of the world's oceans (Pham et al., 2014). Marine debris, also known as marine litter, has been defined by UNEP (2009) as "any persistent, manufactured or processed solid material discarded, disposed of or abandoned in the marine and coastal environment". Marine debris consists of items that have been made or used by people and deliberately discarded into the sea or rivers or on beaches; brought indirectly to the sea with rivers, sewage, storm water or winds; accidentally lost, including material lost at sea in bad weather (fishing gear, cargo); or deliberately left by people on beaches and shores (UNEP, 2005). In 1997, the United States of America Academy of Sciences estimated the total input of marine litter into the oceans, worldwide, at approximately 6.4 million tons per year (UNEP, 2005). Jambeck et al (2015) recently calculated that 275 million metric tons (MT) of plastic waste was generated in 192 coastal countries in 2010, with 4.8 to 12.7 million MT entering the ocean.

Marine debris is present in all marine habitats, from densely populated regions to remote points far from human activities (UNEP, 2009) from beaches and shallow waters to the deep-ocean trenches (Miyake et al. 2011). The density of marine debris varies greatly among locations, influenced by anthropogenic activities, hydrological and meteorological conditions, geomorphology, entry point, and the physical characteristics of debris items. However, a recent study presented data on detectable floating plastic accumulation with visual observation in the North Atlantic and Caribbean from 1986 to 2008, the highest concentrations (> 200,000 pieces per square kilometre) occurred in the convergence zones (Law et al., 2010). Computer model simulations, based on data from about 12,000 satellite-tracked floats deployed since the early 1990s as part of the Global Ocean Drifter Program (GODP, 2011), confirm that debris will be subject to transport by ocean currents and will tend to accumulate in a limited number of sub-tropical convergence zones or gyres (IPRC, 2008; UNEP and NOAA, (2011)) (Figure 25.1).

1.2 Types of marine debris

Marine debris comprises of various material types, and can be classified into several distinct categories (ANZECC, 1996; Edyvane et al., 2004; Ribic et al., 1992; Galgani et al., 2010):

(a) *Plastics*, covering a wide range of synthetic polymeric materials, including fishing nets, ropes, buoys and other fisheries-related equipment; consumer goods, such as plastic bags, plastic packaging, plastic toys; tampon applicators; nappies; smoking-related items, such as cigarette butts, lighters and cigar tips; plastic resin pellets; microplastic particles;

(b) *Metal*, including drink cans, aerosol cans, foil wrappers and disposable barbeques;

(c) *Glass*, including bottles, bulbs;

(d) *Processed timber*, including pallets, crates and particle boards;

The boundaries and names shown and the designations used on this map do not imply official endorsement or acceptance by the United Nations.

Figure 25.1 | A model simulation of the distribution of marine litter in the ocean after ten years shows plastic converging in the five gyres: the Indian Ocean gyre, the North and South Pacific gyres, and the North and South Atlantic gyres. The simulation, derived from a uniform initial distribution and based on real drifter movements, shows the influence of the five main gyres over time. Source: IPRC, 2008.

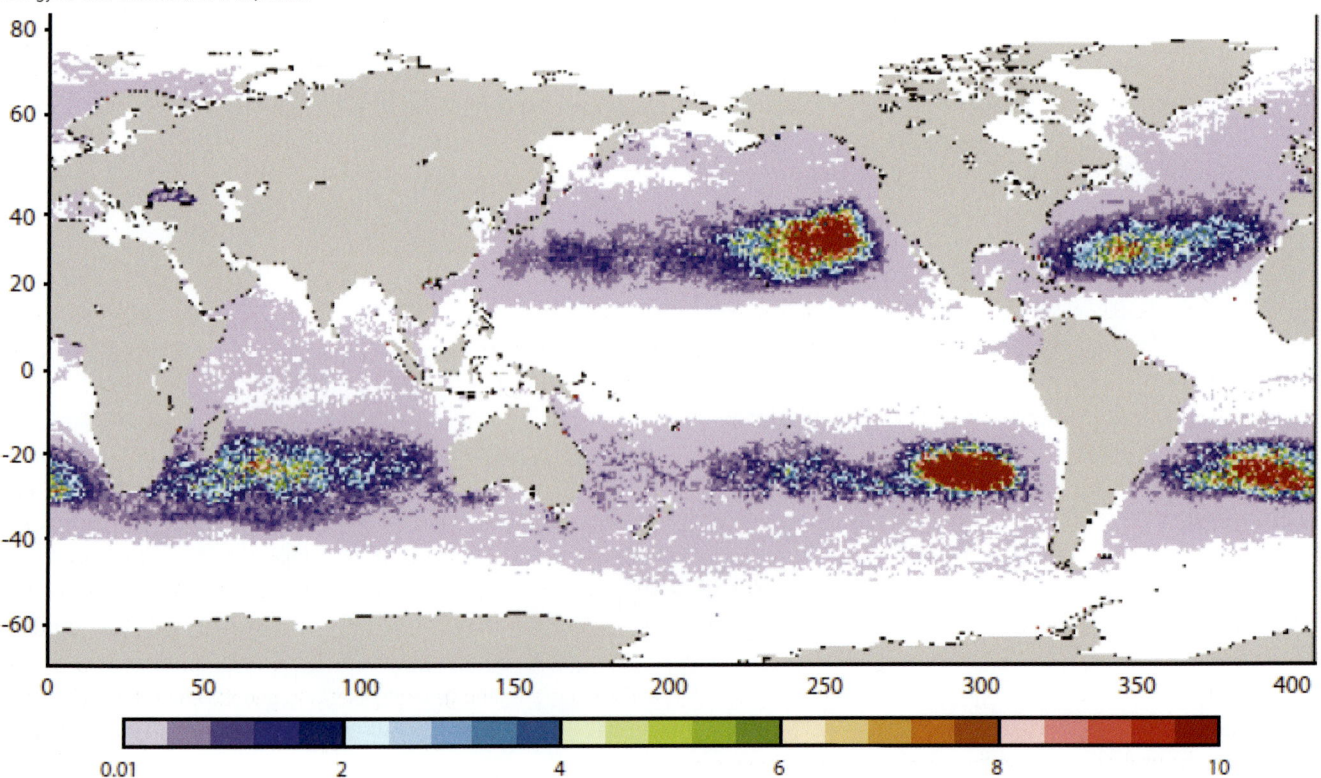

(e) *Paper and cardboard*, including cartons, cups and bags;
(f) *Rubber*, including tyres, balloons and gloves;
(g) *Clothing and textiles*, including shoes, furnishings and towels.

1.3 Sources of marine debris

Marine debris originates from a wide and diverse range of sources. The majority of marine debris (approximately 80 per cent) entering the seas and oceans is considered to originate from land-based sources (Allsopp, et al., 2006), including sewage treatment, combined sewer overflows, people using the coast for recreation or shore fishing, shore-based solid waste disposal, inappropriate or illegal dumping of domestic and industrial rubbish, poorly managed waste dumps, street litter which is washed, blown or discharged into nearby waterways by rain, snowmelt, and wind, etc. The remaining can be attributed to maritime transport, industrial exploration and offshore oil platforms, fishing and aquaculture (UNEP, 2009) and loss and purposeful disposal (e.g. ballast weights made of steel, lead or cement) of scientific equipment.

2 Environmental Impacts

The incidence of debris in the marine environment is a cause for concern. It is known to be harmful to biota, it presents a hazard to shipping (propeller fouling), it is aesthetically detrimental, and it may also have the potential to transport contaminants over long distances (STAP, 2011). Marine debris, and in particular the accumulation of plastic debris, has been identified as a global problem alongside other contemporary key issues, such as climate change, ocean acidification and loss of biodiversity (CBD and STAP-GEF, 2012).

2.1 Entanglement and Ingestion

Marine debris results in entanglement of and ingestion by organisms, and poses a direct threat to marine biota. Adverse effects of marine debris have been reported for 663 species by reviewing available publications (CBD and STAP-GEF, 2012). Over half of these reports documented entanglement in, and ingestion of, marine debris, representing almost a 40 per cent increase since a review in 1997, which reported 247 species (Laist, 1997). Reports revealed that all known species of sea turtles, about half of all species of marine mammals, and one-fifth of all species of sea birds were affected by entanglement in, or ingestion of, marine debris. Species with the greatest number of individuals affected by entanglement or ingestion were the Northern fur seal, *Callorhinus ursinus*, the California sea lion, *Zalophus californianus*, and the seabird *Fulmarus glacialis*; the most frequently reported species are all either birds or marine mammals. About 15 per cent of the species affected through entanglement and ingestion are on the IUCN Red List (CBD and STAP-GEF, 2012).

Abandoned, lost or discarded fishing gear (including monofilament line, nets and ropes), as well as ropes, netting and plastic packaging, can be a cause of entanglement for pinnipeds (seals and related genera), cetaceans, turtles, sharks, sirenia (dugongs and related genera) and birds (WSPA, 2012). The effects range from immediate mortality through drowning to progressive debilitation over a period of months or years (Laist, 1997). Pinniped entanglement usually involves plastic collar-like debris which is often referred to as "neck collars", where the plastic forms a collar around the neck. The animal cannot remove it and it hampers normal feeding or breathing (Allen et al., 2012; Waluda and Staniland, 2013). As the animal grows, the collar effectively tightens and cuts into tissues becoming firmly embedded in skin, muscle and fat (WSPA, 2012) and may cause death. "Ghost fishing" as it is known, can affect many species of fish and invertebrates such as crabs, corals and sponges. For example, several dead and moribund Geryon crabs were found associated with discarded nets in the deep Mediterranean (Ramirez-Llodra et al., 2013). In addition, lost and abandoned traps and the associated by-catch are a global issue with annual trap loss rates approaching 90 per cent in some fisheries (Al-Masroori et al. 2009; Bilkovic et al. 2012).

Marine debris can be mistaken for food items and be ingested by a wide variety of marine biota (Pham et al., 2014). Many species of seabirds, marine mammals and sea turtles have been reported to eat marine debris. Ingestion of sharp debris may damage their guts and result in infection, pain or death. Plastic polymer mass may irritate the stomach tissue, cause abdominal discomfort, and stimulate the animal to feel full and cease eating (Derraik, 2002; Galgani et al., 2010). Two sperm whales (*Physeter macrocephalus*) were found off the coast of northern California in 2008 with a large amount of fishing gear in their gastrointestinal tracts (Jacobsen et al., 2010). A total of 141 mesopelagic fishes from 27 species in the North Pacific Subtropical Gyre, were dissected to examine whether their stomach contents contained plastic particles. The incidence of plastic in fish stomachs was 9.2 per cent (Davison and Ash, 2011). The study of planktivorous fish from the North Pacific gyre found an average of 2.1 plastic items per fish (Boerger et al., 2010). However, the consequences of ingestion are not fully understood, because effects associated with ingestion can mostly be determined by necropsy (CBD and STAP-GEF, 2012; Hong et al., 2013).

2.2 Transport of chemicals

Plastics have a wide variety of chemicals, including those from manufacturing and those that accumulate from the marine environment (i.e. ambient seawater).

Plastics contain a wide variety of potentially toxic chemicals incorporated during manufacture which could be released into the environment (Lithner et al, 2011). Research has established that chemicals used in some plastics, such as phthalates and flame retardants, can have toxicological effects on fish, mammals and molluscs (STAP, 2011). Experimental studies show that phthalates and bisphenol-A (BPA) affect reproduction in all the species studied, impairing development in crustaceans and amphibians, and generally inducing genetic aberrations (Teuten et al., 2009). There is recent evidence that large concentrations of microplastic and additives can harm ecophysiological functions performed by organisms (Browne et al., 2013; Wright et al., 2013).

Because of their small size, microplastics (<1 mm) have a large ratio of surface area to volume that promotes adsorption of chemical contaminants to their surface, and therefore have a high capacity to facilitate the transport of contaminants. An estimated amount of about 35,000 tons, of microplastics are floating in the world's oceans (Cozar et al. 2014; Eriksen et al. 2014). Boerger et al. (2010) found that 35 per cent of the fish sampled in the North Pacific central gyre revealed microplastics in the gut. A range of marine biota are reported to have ingested microplastics, including zooplankton (Cole et al., 2013), amphipods, lugworms and barnacles (Thompson et al., 2004), mussels (Browne et al., 2008), decapod crustaceans (Murray and Cowie, 2011), fish (Boerger et al., 2010; Rochman et al., 2013) and seabirds (Tanaka et al., 2013; van Franeker, 2011). Ingestion of microplastics has caused more and more concern in recent years, as it can provide a pathway for long-distant transport and bioaccumulation of contaminants, and may be compounded by plastic microbead additives in many personal care products (Fendall and Sewell 2009, Kershaw and Leslie 2012).

Plastic debris can accumulate persistent, bio-accumulative and toxic substances (PBTs) that are present in the oceans from other sources, such as PCBs, PAHs, DDTs and HCHs (Mato et al., 2001; Ogata et al., 2009). Within a few weeks these substances can become concentrated on the surface of or in plastic debris by orders of magnitude more than in the surrounding water column (Mato et al., 2001; Teuten et al., 2009; Hirai et al., 2011; Rios et al., 2010). Japanese medaka (*Oryzias latipes*) exposed to a mixture of polyethylene with chemical pollutants absorbed from the marine environment, bioaccumulate these chemical pollutants and suffer liver toxicity and pathology (Rochman et al., 2013). Plastics may provide a mechanism to facilitate the transport of chemicals to remote, pristine locations where they are ingested by biota (Teuten et al., 2007; Hirai et al.,2011). However, it is not yet clear whether chemicals accumulated on plastic debris are effectively transferred to marine biota (Gouin et al., 2011; Koelmans et al., 2013a and b).

2.3 Habitat Destruction

Marine debris can cause destruction of habitats in a number of ways, including smothering, entanglement, and abrasion. The extent of the impact depends on the nature of the debris (i.e., size, quantity, composition, persistence) and the susceptibility of the affected environment (i.e., habitat vulnerability and resilience).

In spite of the growing number of studies documenting the distribution and abundance of marine debris, the ecological impacts, including effects on habitats, are not well documented (NRC, 2009). The few studies that do exist looked at the impacts of derelict fishing gear (that is, gear that has been abandoned, lost or discarded) on coral reefs and other structurally complex benthic communities (Bauer et al., 2008). For example, in the Florida Keys, USA, Chiappone et al. (2005) found that 87 per cent of all debris was recreational hook-and-line fishing gear, but because of low debris density, less than 0.2 per cent of the sessile species were affected. However, Lewis et al. (2009) noted that lost lobster traps, upwards of 100,000 of which are lost each year, represent a significant threat to seagrass beds and coral reefs in the Florida Keys, especially during storms. Also, when gear and other marine debris wash up on shore, especially during storms, they can cause shoreline destruction and smother the underlying substrate where the debris comes to rest.

Although studies of the effects of marine debris on habitat have focused mainly on benthic environments, the presence of floating debris can similarly undermine the quality of pelagic habitats by: (i) affecting the mobility of species, either through entanglement or ghost fishing (that is, entangling fish in lost, abandoned or discarded fishing nets, traps or pots); (ii) reducing the quality of food available in the environment through accidental ingestion of the debris, which may have accumulated toxins on its surface and interfere with digestion and excretion; and (iii) altering the behaviour and fitness of species, as in the case of debris acting as a fish-aggregating device (Hallier and Gaertner, 2008; Hammer et al., 2012; NRC, 2009).

Abandoned and derelict vessels are a widespread problem for the marine environment. Besides the fact that sunken, stranded, and decrepit vessels can be an eyesore and become hazards to navigation, these vessels can pose significant threats to natural resources. They can physically destroy sensitive marine and coastal habitats, sink or move during coastal storms, disperse oil and toxic chemicals still on board, become a source of marine debris, and spread derelict nets and fishing gear that entangle and endanger marine life.[1]

2.4 Introduction and Spread of Alien Species

Marine debris can serve as a vector for numerous species. Hence, floating debris can potentially transport and introduce species to new environments (Barnes, 2002; Winston et al., 1997). Donohue et al. (2001) recorded 13 invertebrate and 10 vertebrate species living on or within a tangle of debris comprising mostly derelict fishing gear in the Northwestern Hawaiian Islands. Similarly, Barnes and Fraser (2003) documented 10 species from 5 different phyla on a single plastic packing band floating in the Southern Ocean. Although none of the species documented in these studies were non-native, the results nonetheless point to the potential for marine debris to serve as vectors for alien species.

To date, the establishment of an alien species via marine debris has yet to be documented (Lewis et al., 2005; Barnes, 2002; Barnes and Milner, 2005; Masó et al., 2003). The absence of such evidence probably reflects the paucity of research rather than the unlikelihood of such events. However, examples of non-native species arriving in new habitat have been well documented. For example, a 180-ton concrete dock cast adrift from Misawa, Japan, by the March 2011 tsunami was carried across the Pacific where it washed ashore in Oregon in the United States in June

---

1 (http://response.restoration.noaa.gov/oil-and-chemical-spills/oil-spills/abandoned-and-derelict-vessels.html).

2012 carrying at least 90 Japanese species including 6 species of non-native algae, crustaceans, and molluscs known to be invasive species in other parts of the world (Lam et al., 2013; Portland State University 2012). Removal of the dock and its burden of non-native species cost 85,000 United States dollars (Barnea et al., 2014).

A recent study by Goldstein et al. (2013) hints at the possibility of marine debris contributing to habitat expansion for the sea skater *Halobates sericeus* (of the *Hemiptera* order). They showed that abundance of *H. sericeus* was related to the availability of floating marine debris, and that such debris was used by the sea skater to attach its egg masses. This suggests that, in principle, *H. sericeus* and similar species could spread across ocean basins with the aid of marine debris.

Because marine debris is subject to surface and deep-water currents, the geographic spread of alien species by such debris is not expected to be random. For instance, the North Pacific convergence zone, which tends to concentrate marine debris, regularly occurs around the north-western Hawaiian Islands. Thus, the islands are subject to unusually high loads of marine debris, and perhaps associated invasive species.

Marine debris can also support the growth and transport of microbes (e.g., cyanobacteria, fungi, algae) to new habitats (Masó et al., 2003; Thiel and Gutow, 2005a and b; Zettler et al., 2013). Masó et al. (2003) found dinoflagellates, including those responsible for harmful algal blooms, growing on plastic debris, and raised the possibility that the increase in harmful algal blooms may be facilitated by the increasing abundance of marine debris.

2.5 Socioeconomics Impacts

The socioeconomic impacts of marine debris are a difficult problem to quantify, because many pollution problems and biological and environmental effects have taken a long time to identify and quantify, partly because of the diverse sources (lack of awareness, inadequate waste management, etc.), and because data on volume/mass, occurrence and distribution are seldom recorded. Furthermore, the literature is sparse for economic analyses addressing elements of potential effects. The Kommunenes Internasjonale Miljøorganisasjon (KIMO) studies (Hall, 2000; Mouat et al., 2010) are the most thorough, but inconsistencies, missing data, and absence of detail have been noted. In such cases, verifiable data were used for point estimates using a Benefits Transfer Approach (Ofiara and Brown, 1999; Unsworth and Petersen, 1995).

2.6 Impacts on Beach Communities, Beach Use, Coastal Tourism

2.6.1 Beach cleaning

Several references in the literature cite anecdotal information related to costs of beach cleaning. NRC (1995) reports the 1993 cost of beach cleaning at Virginia Beach, VA, United States of America, was 43,646 euros per km/yr (60,724 United States dollars per km/yr) and for Atlantic City, NJ, United States, was 215,225 euros per km/yr (299,439 US dollars per km/yr) (2011 Euro values given in parentheses; for all the conversions see Appendix). OSPAR Commission (2009) reports this cost for 2004 for the coast of the United Kingdom at 14 million British pounds per year (19.7 million euros per yr), for the Skagerrak coast, Sweden at 5.1 million euros per year (1.87 million euros per yr) for 2006, and Naturvardsverket (2009) reports the cost of cleaning marine debris on five beaches and in two ports in Poland for 2009 at 570,000 euros per year (632,120 euros per yr). Lane et al., 2007, estimated it would cost 286 million dollars per year to remove debris from the wastewater stream in South Africa (311 million dollars per year, 224 million euros per year -2011 values). A recent study by the Natural Resources Defense Council (NRDC) reports beach cleaning costs and waterway debris removal for 43 communities from South San Francisco to San Diego, California, as 10,993,010 dollars spent (Stickel et al., 2013).

2.6.2 Damage to beach use

Studies in the United States examined damage to beach use from marine debris and medical waste (see Appendix). A major wash-up of marine debris on the shore in 1976 closed New York beaches and caused 15-25 million dollars in lost revenues (43-71 million euros, 59-99 million dollars, in 2011 values; Swanson et al., 1978). ERA (1979) found that clean beaches in an adjacent state suffered piggyback effects from the 1976 event; the public avoided going to an "open-clean" beach in an adjacent state (Seaside Heights, New Jersey, United States) as if it too had marine debris and was closed, an example of avoidance behaviour resulting in lost revenues (943,638 euros per year, 2011 values). Extensive pollution and medical waste wash-ups occurred in 1987-1988 on New Jersey and New York beaches, with losses estimated at of 201-749 million euros at 2011 values for marine debris and medical waste; an average of 475 million euros (Ofiara and Brown, 1989 and 1999; Kahn et al., 1989; Swanson et. al., 1991) in 2011 values.

2.6.3 Losses to tourism

Ofiara and Brown (1989, 1999) found that marine debris wash-ups in New Jersey, United States, decreased beach attendance by 8.9 per cent -18.7 per cent in 1987 and by 7.9 to 32.9 per cent in 1988 (Appendix). A study in South Africa found that a decrease in beach cleanliness could decrease tourism spending by up to 52 per cent (Balance et al., 2000). In Sweden, research found that marine debris on beaches reduced tourism by between one and five per cent (OSPAR, 2009). Hence, even a limited presence of marine debris can decrease coastal tourism by between one to five per cent, and severe events can decrease beach visits by 8.4 per cent to 25.8 per cent (averaged limits).

2.7 Impacts on Commercial Fishing

The Marine Pollution Monitoring Management Group (MPMMG, 2002) reported the cost of marine debris removal in the United Kingdom fisheries at 33 million euros, and Watson and Bryson (Macfayden, 2009) reported a cost for one trap fisherman in the Scottish Clyde fishery of

Table 25.1 | Summary of impacts of lesser magnitude, point estimates

Ghost Fishing:			
Brown et al. (2005)	Cantabrian Sea, Spain	1.46% loss-landings	Monkfishery
NRC (2008)	Not Available	up to 5% EU landings	
Allsopp et al. (2006)	United States	$250mill/yr loss-landings	Lobster fishery
Macfayden et al. (2009)	Louisiana, United States	4-10mill. Crabs/yr lost	Blue Crab fishery
Hall (2000)	United Kingdom	Avg. Cost cleanup = £L2355/hbr	
Mouat et al. (2010)	United Kingdom	Avg. Cost cleanup = €8034/hbr-harbours, €9492/hbr-marinas; €8253.hbr-composite	
Hall (2000)	Shetland Is. Livestock crofts, United Kingdom	96% reported marine debris caught in fences, 36% reported animals entangled, 20% reported animals ill	
Mouat et al. (2010)	Shetland Is. Livestock crofts, United Kingdom	71% reported marine debris caught in fences, 42% reported animals entangled or ill	
Hall (2000)	United Kingdom	11% reported cleanup costs of €20,199/yr, rest €0	

21,000 dollars in lost gear and 38,000 dollars in lost time. Without more information, it is hard to give these estimates their proper context. Studies for the Kommunenes Internasjonale Miljøorganisasjon (KIMO) have estimated average losses per vessel from marine debris as follows: cleaning marine debris from nets GBP 4,065 or Euro 12,007; contaminated catch GBP 1,686 or Euro 2,183, snagged nets GBP 3,392 or Euro 3,820; fouled propellers Euro 182 euros (Hall, 2000; Mouat et al., 2010 - Appendix; GBP at 1998 values, Euro at 2008 values).

A recent study that examined blue crab ghost fishing from lost/abandoned traps/pots found an average mortality rate of 18 crabs/trap/year were harvested in Virginia-Chesapeake Bay, United States waters (sampled in the winter) (Bilkovic et al. 2014), compared to earlier mortality rate estimates of 20 crabs/trap/year in Maryland-Chesapeake Bay waters (Giordano et al. 2011), and 26 crabs/trap/year in Gulf of Mexico waters (Guillory, 1993). An earlier study examined ghost fishing catch rates during the crabbing season of 50 crabs/trap/year (live catch rate-capture rate) in Virginia-Chesapeake Bay waters (Havens et al. 2006). Bilkovic et al. (2014) further estimated an overall loss of 900,000 crabs or 300,000 United States dollars for Virginia-Chesapeake Bay, United States waters. Impacts of lesser magnitude are summarized in Table 25.1.

2.8 Impacts from Invasive Species

The literature pertaining to economic impacts of invasive species is silent regarding marine debris, but it does contain some evidence about the dimensions of the impacts from invasive species. The Swedish Naturvardsverket (2009) cites the collapse of the anchovy fishery in the Black Sea due to the introduction of the American comb jellyfish at an estimated 240 million euros per year. Holt (2009) examined control and eradication costs associated with the Carpet sea squirt in Holyhead Harbour, Wales, and estimated those costs at 525,000 pounds over a 10-yr period (2009-2019); the costs of inaction were estimated at 6.87 million pounds for the same 10-yr period.

3 Assessment of the status of marine litter

3.1 Floating Marine Debris

Floating marine debris in the water column has been documented in the open ocean and in coastal waters. Results for densities of floating marine debris in different regions of the world's oceans are shown in Table 25.3. However, comparisons between studies or even systematic status and trend analyses are challenging because of differences in the collection and measurement methodology used.

In coastal waters, the type, composition and density of floating debris vary greatly among locations. The spatial distribution is influenced by anthropogenic activities, hydrographic and geomorphological factors, prevailing winds, and entry point (Barnes et al., 2009; Derraik, 2002). Generally, the distribution and composition of marine debris floating at sea depends largely on near-shore circulation patterns (Aliani et al.,

Table 25.2 | Summary - Projections (2011 values). Note: KIMO (2000, 2009) = Hall (2000), Mouat et al. (2010), S-O = Swanson et al., 1991, Ofiara and Brown, 1999, NA: not available.

Beach Cleaning Costs (KIMO, 2000,2009)	United Kingdom	€14.301mill/yr - €14.487mill/yr	(avg. €14.394mill/yr)
Damage to Beach Use (S-O),	New York, New Jersey, United States	All causes: €1,403mill - €5,236mill MD, Medical Waste: €201mill - €749mill	(avg. €3,319mill) (avg. €475mill)
Commercial Fishing (KIMO, 2000,2009),	United Kingdom	€8.308mill/yr - €8.935mill/yr	(avg. €8.6215mill/yr)
Aquaculture (KIMO, 2000,2009),	United Kingdom	€94,338/yr	
Harbors, Marinas (KIMO, 2000,2009),	United Kingdom	€491,641 - €944,510/yr	(avg. €718,076/yr)
Damages to Vessels (S-O),	New York Harbour, United States	€749mill	
Coastal Agriculture (KIMO, 2000,2009),	United Kingdom	€486,270 - €614,461/yr	(avg. €550,366/yr)

Table 25.3 | Densities of floating marine debris in different regions. Note: Ship-based trawling surveys and visual observations are used for small and large debris, respectively.

Location	Method	Density	Reference
Coastal North Atlantic Ocean	0.333mm mesh net	3537 items/km2, 286.8 kg/km2	Carpenter and Smith, 1972
North Atlantic Ocean Caribbean	0.947mm mesh net	1.023 g/cm3	Colton et al., 1974
Northwest Pacific	0.50mm mesh net	up to 37.6 items/km2	Day et al., 1990
North Pacific central gyre	0.333mm mesh net	334,271 items/km2, 5,114 g/km2	Moore et al., 2001
Southern California's coastal waters	0.333mm mesh net	7.25 items/m3, 0.02g/m3	Moore et al., 2002
California Current	0.333mm mesh net	3.29 items/m3, 0.003g/m3	Lattin et al., 2004
North Pacific Ocean, Kuroshio Current	0.333mm mesh net	174,000 items/km2, 3600 g/km2	Yamashita and Tanimura, 2007
California Current	0.505mm mesh net	0.011-0.033 items/m3 (Median)	Gilfillan et al., 2009
Caribbean Sea	0.335mm mesh net	1414 ± 112 items/km2	Law et al., 2010
Gulf of Maine	0.335mm mesh net	1534 ± 200 items/km2	Law et al., 2010
North Atlantic Subtropical Gyre (near 30°N).	0.335mm mesh net	20,328±2324 items/km2	Law et al., 2010
Cape Cod, Massachusetts, United States to Caribbean Sea	0.335mm mesh net	0.80~1.24 g/ ml, 0.97~1.04 g/ml	Moret-Ferguson et al., 2010
North Atlantic Ocean	0.335mm mesh net	0.808-1.24 g/ml	Moret-Ferguson et al., 2010
Southeast Bering Sea and United States west coast	0.505mm mesh net	0.004-0.19 items/m3, 0.014-0.209 mg/m3	Doyle et al., 2011
Northeast Pacific Ocean	0.333mm mesh net 0.202mm mesh net	Summer 2009: 0.448 items/m2 (Median), Fall 2010: 0.021 items/m2 (Median)	Goldstein et al., 2013
South Pacific subtropical gyre	0.333mm mesh net	Mean: 26,898 items/km2, 70.96 g/km2	Eriksen et al., 2013
Australia	0.333mm mesh net	4256.4-8966.3 items/km2	Reisser et al., 2013
Bay of Calvi (Mediterranean-Corsica)	0.2 mm mesh net	6.2 particles/100 m2	Collignon et al., 2014
South-East Pacific (Chile)	visual observations	40°S and 50°S : <1 items/km2; nearshore waters: >20 items/km2	Thiel et al., 2003
Ligurian Sea, north-western Mediterranean	visual observations	1997:15-25 items/km2, 2000: 3-1.5 items/km2	Aliani et al., 2003
Floating marine debris in fjords, gulfs and channels of southern Chile	visual observations	1- 250 items/km2	Hinojosa and Thiel, 2009
North East Pacific Ocean	visual observations	0-15,222 items/km2	Titmus and Hyrenbach et al., 2011
Northeast Pacific Ocean	visual observations	0.0014-0.0032 items/m2	Goldstein et al., 2013
Straits of Malacca	visual observations	578 ± 219 items/m2	Ryan, 2013
Bay of Bengal	visual observations	8.8 ± 1.4 items/m2	Ryan, 2013

2003; Lattin et al., 2004; Ribic et al., 2010; Thiel et al., 2003). Prevailing winds also affect the pattern of debris abundance. Greater quantities of plastics were observed at downwind sites (Browne et al., 2010; Collignon et al., 2012). Collignon et al. (2014) observed that the density of floating debris was five times higher before a strong wind event than afterwards. This was explained by the wind stress increasing the mixing and vertical redistribution of the plastic particles in the upper layers of the water column. However, most land-based litter is carried by water currents through rivers and storm-water (Ryan et al., 2009). The density of the debris in the southern California, United States coast water, after the storm was seven times higher than prior to the storm (Moore et al., 2002). The weight of plastic increased by more than 200 times after a storm in Santa Monica Bay, California, United States (Lattin et al., 2004). Higher densities of debris in coastal waters are also associated with human population density (Lebreton et al., 2012; Thiel et al., 2003).

In the open ocean, spatial patterns of debris are influenced by the interaction of large-scale atmospheric and oceanic circulation patterns, leading to particularly high accumulations of floating debris in the sub-tropical gyres (Howell et al., 2012; Goldstein et al., 2013; Martinez et al., 2009). A high profile publication in the *Science* journal presented over 20 years of data clearly demonstrating that some of the most substantial accumulations of debris are now in oceanic gyres far from land (Law et al., 2010). The models developed by Martinez et al. (2009) suggest that marine debris deposited in coastal zones tends to accumulate in the central oceanic gyres within two years after deposition. The persistent floating debris will accumulate in mid-ocean sub-tropical gyres, forming so-called garbage patches (Kaiser, 2010; Lebreton et al., 2012) (See Figure 25.1).

Although the type of litter found in the world's oceans is highly diverse, plastics are by far the most abundant material recorded. Plastic debris was first reported in the oceans in the early 1970s (Carpenter and Smith, 1972; Colton et al., 1974). Plastics are estimated to represent between 60 per cent and 80 per cent of the total marine debris (Derraik, 2002; Gregory and Ryan, 1997). Almost all aspects of daily life involve plastics, and consequently the production of plastics has increased substantially in the last 60 years and this trend continues. The fragmentation of plastics generates microplastics. For example, in sampling the South Pacific subtropical gyre, 1.0mm - 4.7mm particles accounted for 55 per cent of the total count and 72 per cent of the total weight (Eriksen et al., 2013).

Table 25.4 | Density of beach debris in different beaches. GM: geometric mean; M: surveying results in May; S: surveying results in September.

Location	Density	Reference
Dominica	1.9-6.2 items/m, 51.5-153.7 g/m	Corbin and Singh, 1993
St. Lucia	4.5-11.2 items/m, 8.2-109.2 g/m	Corbin and Singh, 1993
Panama	3.6 items/m2 (180/50 m2)	Garrity and Levings, 1993
Persian Gulf, United Arab Emirates	0.84 items/ m2	Khordagui and Abu-Hilal, 1994
Tasmania, Australia	300 items/km, 0.09-0.35 items/m	Jones, 1995
Marmion Marine Park, Australia	2.74 items/m, 0.54 g/m	Jones, 1995
West Australia, Marmion Marine Park, Australia	3.66 items/m, 0.12 g/m	Jones, 1995
Northern New South Wales, Australia	10.9 items/km2	Frost and Cullen, 1997
Transkei Coast, South Africa	19.6-72.5 items/m, 42.8-164.1 g/m	Madzena and Lasiak, 1997
Bird Island, South Georgia	0.014-0.21 items/m	Walker et al., 1997
New Jersey, United States	0.36-6.4 items/m	Ribic, 1998
Cliffwood Beach, New Jersey, United States	2.7-3.7 items/m2	Thornton and Jackson, 1998
Caribbean Sea: Curaçao	60 items/m, 4.5 kg/m	Debrot et al., 1999
Orange County, California, United States	1709 items/m	Moore et al., 2001
Ensenada, Baja California, Mexico	1.525 items/m2 (including natural litter)	Silva-Iñiguez and Fischer, 2003
Japanese beaches	2144 g/100 m2, 341 items/100 m2	Kusui and Noda, 2003
Russian beaches	1344 g/100 m2, 20.7 items/100 m2	Kusui and Noda, 2003
Volunteer Beach, Playa Voluntario, Falkland Islands (Malvinas)	accumulation rate:77±25 items/km/month	Otley and Ingham, 2003
Gulf of Aqaba, Red Sea	1.64-7.38 items/m	Abu-Hilal and Al-Najjar, 2004
Gulf of Oman, Oman	1.79 items/m; 27.02g/m	Claerboudt, 2004
Anxious Bay, Australia	1.9-15.0 kg/km	Edyvane et al. 2004
Point Pleasant Park, Halifax Harbour, Canada	accumulation rate: 355±68 items/month	Walker et al., 2006
Rio de Janeiro, Brazil	13.76 items /100 m2	Oigman-Pszczol and Creed, 2007
NOWPAP region	570 items/100 m2, 3864 g/100 m2	UNEP/NOWPAP, 2008
Gulf of Aqaba, Red Sea	2.8 items/m2, 0.31 kg/m2	Abu-Hilal and Al-Najjar, 2009
Chile	1.8 items/m2	Bravo et al., 2009
OSPAR region	712 items/100m	OSPAR Commission, 2009
Belgium	6429 ± 6767 items/100m	Van Cauwenberghe et al., 2013
Caribbean Sea, Bonaire	115 ± 58 items/m, 3408 ± 1704 g/m (GM)	Debrot, et al., 2013
Chile	27 items/m2 (small plastic)	Hidalgo-Ruz and Thiel, 2013
Mumbai, India	68.83 items/m2, 7.49 g/m2 (Plastic debris)	Jayasiri et al., 2013
Nakdong River Estuary, Republic of Korea	large plastics: 8205(M), 27,606 items/m2(S); mesoplastics:238 (M), 237 particles/m2(S) macroplastics : 0.97(M), 1.03 particles/m2(S)	Lee et al., 2013
Monterey Bay, CA, United States	0.03-17.1 items/m2 1±2.1 items/m2	Rosevelt et al., 2013
Turkish Western Black Sea coast	0.085-5.058 items/m2	Topçu et al., 2013
20 beaches, Republic of Korea	480.9 (±267.7) count · 100 m−1 for number, 86.5 (±78.6) kg · 100 m−1 for weight, 0.48 (±0.38) m3 · 100 m−1 for volume	Hong et al., 2013

Research on the amount, distribution, composition and potential impact of microparticles has received increasing attention.

Plastic debris continues to accumulate in the marine environment. Goldstein et al. (2013) show that the density of microplastics within the North Pacific Central Gyre has increased by two orders of magnitude in the past four decades. In contrast, there is no significant trend in the density of surface water plastics in the North Atlantic from 1986 to 2008, despite increases in plastic production during this time (Law et al., 2010). Some form of loss must be taking place to offset the presumed increase in input of plastics to the ocean. Possible sinks for floating plastic debris include fragmentation, sedimentation, shore deposition, and ingestion by marine organisms (Law et al., 2011).

3.2 Beach debris

Millions of volunteers in more than 150 countries are involved in beach-cleanup activities on International Coastal Cleanup Day every year

(Ocean Conservancy, 2011). The volunteers' participation contributes to extensive sampling and helps to obtain more information from a wider range of sites (Rees and Pond, 1995). The density of debris reported from the beaches in different regions of the world is listed in Table 25.4. For most of the beaches, the major debris is plastic. The spatial distribution of plastic debris is affected by multiple factors, including land uses, human population, fishing activity, and oceanic current systems (Ribic et al., 2010).

Beach debris density may be linked to the number of tourists and the cleaning frequency (Bravo et al., 2009; Kuo and Huang, 2014). For example, beach debris densities in central Chile were lower than in northern and southern Chile, which could be due to different attitudes of beach users or intensive beach cleaning in central regions (Bravo et al., 2009). Rodriguez-Santos et al. (2005) found that the quantity of litter depends on beach visitor density. Ocean current patterns, sand types, wave action, and wind exposure have further effects on litter abundance. For example, in Monterey Bay, California, United States, the seasonal variability in debris abundance may be a function of oceanic winds, as well as the possibility that seasonal current patterns may drive debris deposition (Rosevelt et al., 2013).

Although marine debris density is usually associated with population density, a few studies contradict this. Ribic et al. (2010) show no trends over several decades in beach-debris densities along the Eastern Atlantic seaboard of the United States, although large percentage increases in coastal population occurred in the south-east Atlantic region and a smaller percentage increase in coastal population occurred in the north-east region.

3.3   Benthic marine debris

The occurrence of litter on the seafloor has been far less investigated than in surface waters or on beaches, principally because of the high cost and the technical difficulties involved in sampling the seafloor. Nevertheless, a few investigations of benthic debris have been recorded, including on the continental shelves, on raised seabed features, such as seamounts, ridges and banks, in canyons and in polar regions. The surveying methods for the density and composition of benthic marine debris include bottom trawling, coring, scuba diving, the use of submersibles, snorkelling, manta tows and sonar (Spengler and Costa, 2008) and more recently, towed camera systems and remotely operated vehicles (ROVs).

Abundances of benthic debris range from dozens to more than hundreds of thousands items per square kilometre. As more areas of Europe's seafloor are being explored, benthic litter is progressively being revealed to be more widespread than previously assumed. Pham et al. (2014) reported data on litter distribution and density collected during 588 video and trawl surveys across 32 sites in European waters (35-4500 m depth). Debris was found to be present in the deepest areas and at locations as remote from land as the Charlie-Gibbs Fracture Zone across the Mid-Atlantic Ridge. The highest litter density occurred in submarine canyons, reaching an average (± SE) of 9.3±2.9 items ha^{-1}. The lowest density was found on continental shelves and on ocean ridges; mean (± SE) litter density of 2.2±0.8 and 3.9±1.3 items ha^{-1}, respectively. As for most other marine environments studied, plastic was the most prevalent litter item found on the seafloor. Woodall et al (2015) showed the litter was ubiquitous on deep-sea raised benthic features, such as seamounts, banks and ridges, A total of 56 items was found in the Atlantic Ocean over a survey area of 11.6 ha, and 31 items in the Indian Ocean over 5.6 ha, with a significant difference in the type of litter between areas sampled in the Indian Ocean (where the dominant litter type was fishing gear) and sites in the Atlantic Ocean (which had mixed refuse).

Litter from fishing activities (derelict fishing lines and nets) was particularly common on seamounts, banks, mounds and ocean ridges. A significant source of benthic debris is lost and discarded fishing gear, which is of particular concern due to ghost fishing effects that can kill both commercial and non-commercial species. Laist (1996) reports annual gear loss rates of about one percent for gillnet fisheries and between 5 - 30 percent for trap fisheries in United States fisheries. Whereas trap loss rates in the American lobster fishery are relatively low (5-10 percent), because the fishery involves more than 3 million deployed traps, the lobster fishery alone may account for the loss of more than 150,000 traps per year.

Hydrography, geomorphology, and anthropogenic activities all affect the abundance, type, and location of debris reaching the seafloor (Barnes et al., 2009; Galgani et al., 2000; Schlining et al., 2013). Because they facilitate the transport and deposition of debris, submarine canyons act as conduits for debris, transporting it from the coast to the deep sea (Ramirez-Llodra et al., 2013; Schlining et al., 2013). Ramirez-Llodra et al. (2013) suggest that debris in a canyon mainly originates from coastal areas, that plastic debris can be transported easily by canyon-enhanced currents, whereas heavy debris is usually discarded from ships. Wei et al. (2012) indicate that the debris density was higher in the eastern than that in the western Gulf of Mexico, primarily because of shipping lanes, offshore oil- and gas-installation platforms, as well as fishing activities. The litter density and diversity were independent of depth of water and of distance from land. Galgani et al. (2000) report that only small amounts of debris were collected on the continental shelf, mostly in canyons descending from the continental slope. Ramirez-Llodra et al. (2013) report accumulation of litter with increasing depth, but the mean weight at different depths, or between the open slope and canyons, showed no significant variation. Schlining et al. (2013) found debris clustered just below the edge of canyon walls or on the outside of canyon meanders. Wei et al. (2012) indicated that the total density of anthropogenic waste was significantly different between parallel depth transects. Woodall et al (2015) concluded that the pattern of accumulation and composition of the litter was determined by a complex range of factors both environmental and anthropogenic.

Debris continuously accumulates on the deep seabed; some research shows a significant increasing trend. Watters et al. (2010) reported a significant increase in the amount of litter at some of shelf locations off

Table 25.5 | Density of benthic debris in different regions. ROV: Remotely Operated Vehicle; SCUBA: Self-Contained Underwater Breathing Apparatus.

Location	Method	Density	Depth range	Reference
Bay of Biscay, France	trawl	0.263-4.94 items/ha	0-100m	Galgani et al., 1995a
Northwestern Mediterranean	trawl	19.35 items/ha	750m	Galgani et al., 1995b
French Mediterranean coast	trawl	0-78 items/ha	100-1600m	Galgani et al., 1996
European coast	trawl	0-1010 items /ha	<2200m	Galgani et al., 2000
Eastern China Sea and the south coast of the Republic of Korea	trawl	30.6-109.8 kg/km2	—	Lee et al., 2006
Greek Gulfs	trawl	72-437 items/km2, 7-47.4 kg/km2	—	Koutsodendris et al., 2008
Gulf of Aqaba, Red Sea	SCUBA	2.8 items/m2; 0.31 kg/m2	—	Abu-Hilal et al., 2009
submarine canyons and the continental shelf off California, United States	submersible	1.7 items/100m	20-365 m	Watters et al., 2010
West coast of the United States	trawl	67.1 items/km2	55-1280m	Keller et al., 2010
West coast of Portugal	ROV	1100 items/km2	850-7400 m	Mordecai et al., 2011
Eastern Fram Strait west of Svalbard	Image observation	3635-7710 items/km2	2500m	Bergmann et al., 2012
Gulf of Mexico	trawl	<28.4 items/ha	359-3724m	Wei et al., 2012
Antalya Bay, Eastern Mediterranean	trawl	18.5-2,186 kg/km2, 115-2,762 items/km2	200-800m	Güven et al., 2013
Belgium	trawl	3125 ± 2830 items/km2	—	Van Cauwenberghe et al., 2013
Mediterranean Sea	trawl	0.02-3264.6 kg/km2	900-2700m	Ramirez-Llodra et al., 2013
Monterey Bay, California, United States	ROV	---	25-3971m	Schlining et al., 2013
Atlantic Ocean, Mediterranean Sea and Indian Ocean	Core (microplastic)	1.4-40 pieces/50ml sediment 13.4±3.5 pieces/50ml sediment;	1000-3500m	Woodall et al,. 2014
Atlantic Ocean	ROV	12.23-0.59 items/ha	200-2800m	Woodall et al,. 2015
Indian Ocean	ROV	17.39-0.75 items/ha	1320-1610m	Woodall et al,. 2015

California, United States, between 1993 and 2007. The debris density has continued increasing, and has doubled during the last decade in the Arctic deep sea (Bergmann and Klages, 2012). The density of microplastics in sediments has been increasing along the Belgian coast (Claessens et al., 2011). However, some studies did not observe significant temporal increases, for example, in litter abundance between 1989 and 2010 in Monterey Canyon, central California, United States (Schlining et al., 2013).

4 Prevention and Clean-up of Marine Debris

Numerous policies, global, international, national and local, address various aspects of marine debris. Some countries have banned outright the use of certain plastic derivative products.

5 Gaps, Needs, Priorities

Marine debris is a complex cultural and multi-sectoral problem that imposes tremendous ecological, economic, and social costs around the world. One of the substantial barriers to addressing marine debris is the absence of adequate scientific research, assessment, and monitoring. There is a gap in scientific research to better understand the sources, fates, and impacts of marine debris (NOAA and EPA, 2011; NRC, 2008).

Scalable, statistically rigorous and, where possible, standardized monitoring protocols are needed to monitor changes in conditions as a result of efforts to prevent and reduce the impacts of marine debris. Although monitoring of marine debris is currently carried out within several countries around the world (often on the basis of voluntary efforts by nongovernmental organizations), the protocols used tend to be very different, preventing comparisons and harmonization of data across regions or timescales (NOAA and EPA, 2011; Cheshire et al., 2009).

There is a gap in information needed to evaluate impacts of marine debris on coastal and marine species, habitats, economic health, human health and safety, and social values. More information is also needed to understand the status and trends in amounts, distribution and types of marine debris. There is also a gap in capacity in the form of new technologies and methods to detect and remove accumulations of marine debris (NOAA and EPA, 2011), as well as in means of bringing home to the public in all countries the significance of marine debris and the important part that the public can play in combating it.

Besides, the ways in which waste management is conducted are often a barrier. This is a global problem, but waste is managed on a very local level. Truly biodegradable, naturally occurring, biopolymers are becoming more wide spread and commercially available. There is a need to pursue truly biodegradable biopolymer alternatives to plastic (Chanprateep, 2010).

References

Abu-Hilal, A.H., Al-Najjar, T. (2004). Litter pollution on the Jordanian shores of the Gulf of Aqaba (Red Sea). *Marine Environmental Research* 58, 39–63.

Abu-Hilal, A.H., Al-Najjar, T. (2009). Marine litter in coral reef areas along the Jordan Gulf of Aqaba, Red Sea. *Journal of Environmental Management* 90, 1043–1049.

Aliani, S., Griffa, A., Molcard, A. (2003). Floating debris in the Ligurian Sea, northwestern Mediterranean. *Marine Pollution Bulletin* 46, 1142-1149.

Al-Masroori, H., Al-oufi, H., McShane, P. (2009). Causes and mitigations on trap ghost fishing in Oman: Scientific approach to local fisher' perception. *Journal of Fisheries and Aquatic Science* 4: 129-135.

Allen, R., Jarvis, D., Sayer, S., Mills, C. (2012). Entanglement of grey seals Halichoerus grypus at a haul out site in Cornwall, UK. *Marine Pollution Bulletin* 64, 2815-2819.

Allsopp, M., Walters, A., Santillo, D. and Johnston, P. (2006). Plastic Debris in the World's Oceans. *Greenpeace*. Netherlands.

ANZECC (1996). *ANZECC Strategy to Protect the Marine Environment. Working together to reduce impacts from shipping operations: The Australian Marine Debris Status Review*. Australian and New Zealand Environment and Conservation Council.

Balance, A., Ryan, P.G. and Turpie, J.K. (2000). How Much is a Clean Beach Worth? The Impact of Litter on Beach Users in the Cape Peninsula, South Africa. *South African Journal of Science*. 96 (5): 210-213.

Barnes, D.K.A. (2002). Invasions by marine life on plastic debris. *Nature* 416, 808-809.

Barnes, D.K.A. (2005). Remote Islands reveal rapid rise of Southern Hemisphere sea debris. *The Scientific World Journal* 5, 915-921.

Barnes, D.K.A., Milner, P. (2005). Drifting plastic and its consequences for sessile organism dispersal in the Atlantic Ocean. *Marine Biology* 146, 815–825 (doi:10.1007/s00227-004-1474-8).

Barnes, D.K.A., Fraser, K.P.P. (2003). Rafting by five phyla on man-made flotsam in the Southern Ocean. *Marine Ecology-Progress Series* 262, 289-291.

Barnes, D.K.A., Galgani, F., Thompson, R.C., Barlaz, M. (2009). Accumulation and fragmentation of plastic debris in global environments. *Philosophical Transactions of the Royal Society*, B 364, 1985-1998.

Barnea, N., Albins, K., Cialino, K., Koyanagi, K., Lippiatt, S., Murphy, P., Parker, D. (2014). Proceedings of the Japan Tsunami Marine Debris Summary Meeting. NOAA Marine Debris Program. U.S. Dept. of Commerce, NOAA Technical Memorandum NOS OR&R 50. Silver Spring, MD. 53 pp. http://marinedebris.noaa.gov/sites/default/files/Proceedings%20of%20the%20JTMD%20Summary%20Meeting.pdf.

Bauer, L., Kendall, M., Jeffery, C. (2008). Incidence of marine debris and its relationships with benthic features in Gray's Reef National Marine Sanctuary, Southeast USA. *Marine Pollution Bulletin* 56, 402-413.

Bergmann, M., Klages, M. (2012). Increase of litter at the Arctic deep-sea observatory HAUSGARTEN. *Marine Pollution Bulletin* 64 (12), 2734-2741.

Bilkovic, D.M., Havens, K.J., Stanhope, D.M., Angstadt, K.T. (2012). The use of fully biodegradable panels to reduce derelict pot threats to marine fauna. *Conservation Biology* 26(6): 957-966.

Bilkovic, D.M., Havens, K.J., Stanhope, D., Angstadt, K. (2014). Derelict fishing gear in Chesapeake Bay, Virginia: Spatial patterns and implications for marine fauna. *Marine Pollution Bulletin* 80: 114-123.

Boerger, C.M., Lattin, G.L., Moore, S.L., Moore, C.J. (2010). Plastic ingestion by planktivorous fishes in the North Pacific Central Gyre. *Marine Pollutant Bulletin* 60 (12), 2275–2278.

Bravo, M., de los Ángeles Gallardo, M., Luna-Jorquera, G., Núñez, P., Vásquez, N., Thiel, M. (2009). Anthropogenic debris on beaches in the SE Pacific (Chile): Results from a national survey supported by volunteers. *Marine Pollution Bulletin* 58, 1718-1726.

Brown, J., Macfadyen, G., Huntington, T., Magnus, J. and Tumilty, J. (2005). *Ghost fishing by Lost Fishing Gear*. Final Report to DG Fisheries and Maritime Affairs of the European Commission. Institute for European Environmental Policy/Poseidon Aquatic Resource Management Ltd.

Browne, M.A., Dissanayake, A., Galloway, T.S., Lowe, D.M., Thompson, R.C. (2008). Ingested microscopic plastic translocates to the circulatory system of the mussel, Mytilus edulis (L.). *Environmental Science & Technology* 42, 5026-5031.

Browne, M.A., Galloway, T.S., Thompson, R.C. (2010). Spatial Patterns of Plastic Debris along Estuarine Shorelines. *Environmental Science & Technology* 44, 3404-409.

Browne, M.A., Niven, S.J., Galloway, T.S., Rowland, S.J., Thompson, R.C. (2013). Microplastic Moves Pollutants and Additives to Worms, Reducing Functions Linked to Health and Biodiversity. *Current Biology*, 23, (23), 2388 – 2392.

Carpenter, E.J., Smith, K.L J. (1972). Plastics on the Sargasso Sea Surface. *Science* 175, 1240-1241.

CBD, STAP-GEF (2012). Impacts of Marine Debris on Biodiversity: Current Status and Potential Solutions. *Technical Series No. 67*. Secretariat of the Convention on Biological Diversity (CBD) and the Scientific and Technical Advisory Panel (STAP) – GEF.

Chanprateep, S. (2010). Current trends in biodegradable polyhydroxyalkanoate. *Journal of Bioscience and Bioengineering* 110(6): 621-632.

Cheshire, A.C., Adler, E., Barbière, J., Cohen, Y., Evans, S., Jarayabhand, S., Jeftic, L., Jung, R.T., Kinsey, S., Kusui, E.T., Lavine, I., Manyara, P., Oosterbaan, L., Pereira, M.A., Sheavly, S., Tkalin, A., Varadarajan, S., Wenneker, B., Westphalen, G. (2009). UNEP/IOC Guidelines on Survey and Monitoring of Marine Litter. *UNEP Regional Seas Reports and Studies*, No. 186; IOC Technical Series No. 83.

Chiappone, M., White, A., Swanson, D. and Miller, S. (2005). Occurrence and biological impacts of fishing gear and other marine debris in the Florida Keys. *Marine Pollution Bulletin*, 44, 597-604.

Claerboudt, R.C. (2004). Shore litter along sandy beaches of the Gulf of Oman. *Marine Pollution Bulletin* 49, 770–777.

Claessens, M., De Meester, S., Van Landuyt, L., De Clerc, K., Janssen, C.K. (2011). Occurrence and distribution of microplastics in marine sediments along the Belgian coast. *Marine Pollution Bulletin* 62, 2199-2204.

Cole, M., Lindeque, P., Fileman, E., Halsband, C., Goodhead, R., Moger, J., Galloway, T.S. (2013). Microplastic ingestion by zooplankton. *Environmental Science & Technology* 47, 6646-6655.

Collignon, A., Hecq, J., Galgani, F., Collard, F., Goffart, A. (2014). Annual variation in neustonic micro- and meso-plastic particles and zooplankton in the Bay of Calvi (Mediterranean-Corsica). *Marine Pollution Bulletin* 79, 293-298.

Collignon, A., Hecq, J.H., Galgani, F., Voisin, P., Collard, F., Goffart, A. (2012). Neustonic microplastic and zooplankton in the North Western Mediterranean Sea. *Marine Pollution Bulletin* 64, 861-864.

Colton, J.B., Knapp, F.D., Burns, B.R. (1974). Plastic particles in surface waters of the Northwestern Atlantic. *Science* 185, 491-497.

Corbin, C.J., Singh, J.G. (1993). Marine debris contamination of beaches in St. Lucia and Dominica. *Marine Pollution Bulletin* 26, 325–328.

Cozar, A., Echevarria, F., Gonzalez-Gordillo, J.I., Irigoien, X., Ubeda, B., Hernandez-

Leon, S., et al. (2014). Plastic debris in the open ocean. *Proceedings of the National Academy of Sciences of the United States of America* 111, 10239–10244. doi: 10.1073/pnas.1314705111.

Day, R.H., Shaw, D.G., Ignell, S.E. (1990). The quantitative distribution and characteristics of neuston plastic in the North Pacific Ocean, 1984-1988. Proceedings of the Second International Conference on Marine Debris, 2-7 April, 1989, Honolulu, Hawaii. *NOAA Technical Memorandum*, NOAA-TM-NMFS-SWFSC-154. 182-211.

Davison, P., Asch, R. (2011). Plastic ingestion by mesopelagic fishes in the North Pacific Subtropical Gyre. Marine Ecology Progress Series; 432:173–180.

Debrot, A.O., Van Rijn, J., Bron, P.S. and de León, R. (2013). A baseline assessment of beach debris and tar contamination in Bonaire, Southeastern Caribbean. *Marine Pollution Bulletin* 71(1-2), 325-329.

Debrot, A.O., Tiel, A.B., Bradshaw, J.E. (1999). Beach Debris in Curaçao. *Marine Pollution Bulletin* 38(9), 795-801.

Derraik, J.G.B. (2002). The pollution of the marine environment by plastic debris: a review. *Marine Pollution Bulletin* 44, 842-852.

Donohue, M.J., Boland, R.C., Sramek, C.M., Antonelis, G.A. (2001). Derelict fishing gear in the Northwestern Hawaiian Islands: diving surveys and debris removal in 1999 confirm threat to coral reef ecosystems. *Marine Pollution Bulletin* 42, 1301-1312.

Doyle, M.J., Watson, W., Bowlin, N.M., Sheavly, S.B. (2011). Plastic particles in coastal pelagic ecosystems of the Northeast Pacific ocean. *Marine Environmental Research* 71, 41-52.

Edyvane, K.S., Dalgetty, A., Hone, P.W., Higham, J.S., Wace, N.M. (2004). Long-term Marine Litter Monitoring in the Remote Great Australian Bight, South Australia. *Marine Pollution Bulletin* 48, 1060-1075.

ERA (Economic Research Associates) (1979). Cost Impact of Marine Pollution on Recreational Travel Patterns. PB–290655. National Technical Information Service: Springfield, VA.

Eriksen, M., Lebreton, L.C.M., Carson, H.S., et al. Plastic Pollution in the World's Oceans: More than 5 Trillion Plastic Pieces Weighing over 250,000 Tons Afloat at Sea. Dam HG, ed. PLoS ONE. (2014); 9(12): e111913. doi:10.1371/journal.pone.0111913.

Eriksen, M., Maximenko, N., Thiel, M., Cummins, A., Lattin, G., Wilson, S., Hafner, J., Zellers, A., Rifman, S. (2013). Plastic pollution in the South Pacific subtropical gyre. *Marine Pollution Bulletin* 68(1-2), 71-76.

Fendall, L.S., Sewell, M.A. (2009). Contributing to marine pollution by washing your face: microplastics in facial cleansers. *Marine Pollution Bulletin* 58: 1225-1228.

Frost, A., Cullen, M. (1997). Marine debris on Northern New South Wales beaches (Australia): sources and the role of beach usage. *Marine Pollution Bulletin* 34, 348–352.

Galgani, F., Burgeot, T., Bocquéné, G., Vincent, F., Leauté, J.P., Labastie, J., Forest, A., Guichet, R. (1995a). Distribution and abundance of debris on the continental shelf of the Bay of Biscay and in Seine Bay. *Marine Pollution Bulletin* 30, 58-62.

Galgani, F., Jaunet, S., Campillot, A., Guenegen, X., His, E. (1995b). Distribution and abundance of Debris on the continental shelf of the North-Western Mediterranean Sea. *Marine Pollution Bulletin* 31, 713–717.

Galgani, F., Fleet, D., Van Franeker, J., Katsavenakis, S., Maes, T., Mouat, J., Oosterbaan, L., Poitou, I., Hanke, G., Thompson, R., Amato, E., Birkun, A. & Janssen, C. (2010). Marine Strategy Framework Directive Task Group 10 Report Marine litter, *JRC Scientific and technical report*, ICES/JRC/IFREMER Joint Report (no 31210 – 2009/2010), Editor: Zampoukas, N., pp. 57.

Galgani, F., Leaute, J.P., Moguedet, P., Souplet, A., Verin, Y., Carpentier, A., Goraguer, H., Latrouite, D., Andral, B., Cadiou, Y., Mahe, J.C., Poulard, J.C., Nerisson, P. (2000). Litter on the sea floor along European coasts. *Marine Pollution Bulletin* 40, 516-527.

Galgani, F., Souplet, A., Cadiou, Y. (1996). Accumulation of debris on the deep sea floor off the French Mediterranean coast. *Marine Ecology Progress Series* 142, 225-234.

Garrity, S.D., Levings, S.C. (1993). Marine debris along the Caribbean coast of Panama. *Marine Pollution Bulletin* 26, 317–324.

Gilfillan, L.R., Ohman, M.D., Doyle, M.J., Watson, W. (2009). Occurrence of Plastic Micro-debris in the Southern California Current System. *California Cooperative Oceanic Fisheries Investigations Reports* 50,123-133.

Giordano, S., Lazar, J., Bruce, D., Little, C., Levin, D., Slacum Jr., H.W., Dew-Baxter, J., Methratta, L., Wong, D. and Corbin, R. (2011). Quantifying the impacts of derelict blue crab traps in Chesapeake Bay. In: Carswell, B., McElwee, K. and Morison, S., eds. Proceedings of the Fifth International Marine Debris Conference. Technical Memorandum NOS OR&R-38. NOAA: Silver Spring, MD.

GODP (Global Ocean Drifter Program) (2011). *The Global Ocean Drifter Program. Satellite-tracked surface drifting buoys*. http://www.aoml.noaa.gov/phod/dac/gdp.

Goldstein, M.C., Titmus, A.J., Ford, M. (2013). Scales of spatial heterogeneity of plastic marine debris in the northeast Pacific Ocean. *PloS ONE*, 8(11): e80020. doi:10.1371/journal.pone.0080020.

Gouin, T., Roche, N., Lohmann, R., Hodges, G.A. (2011). Thermodynamic approach for assessing the environmental exposure of chemicals absorbed to microplastic. *Environmental Science & Technology*. 45, 1466-1472.

Gregory, M.R., Ryan, P.G. (1997). Pelagic plastic and other seaborne persistent synthetic debris: A review of Southern Hemisphere perspectives, in: Coe, J.M., Rogers, D.B. (Eds.), *Marine debris: Sources, impacts, and solution*, New York, Springer, pp. 49-66.

Guillory, V. (1993). Ghost fishing in blue crab traps. *North American Journal of Fisheries Management* 13(3): 459-466.

Güven, O., Gülyavuz, H., Deva, M.C. (2013). Benthic Debris Accumulation in Bathyal Grounds in the Antalya Bay.Eastern Mediterranean. *Turkish Journal of Fisheries and Aquatic Sciences* 13, 43-49.

Hall, K. (2000). Impacts of Marine Debris and Oil. *Kommunenes Internasjonale Miljoorganisasjon (KIMO)* (Local Authorities International Organisation) Shetland.

Hallier, J.P., Gaertner, D. (2008). Drifting fish aggregation devices could act as an ecological trap for tropical tuna species. *Marine Ecology-Progress Series* 353, 255-264.

Hammer, J., Kraak, M.S., Parsons, J. (2012). Plastics in the Marine Environment: The Dark Side of a Modern Gift, in: Whitacre, D.M. (Ed.), *Reviews of Environmental Contamination and Toxicology*. Springer New York, pp. 1-44.

Havens, K. J., Bilkovic, D.M., Stanhope, D., Angstadt, K. and Hershner, C. (2006). Derelict Blue Crab Trap impacts on marine fisheries in the lower York River, Virginia. Marine Debris Survey in Virginia. Final Report to NOAA Chesapeake Bay Program Office. Center for Coastal Resources Management, Virginia Institute of Marine Science, College of William & Mary, Williamsburg, Virginia, USA. 12 pp.

Hidalgo-Ruz, V., Thiel, M. (2013). Distribution and abundance of small plastic debris on beaches in the SE Pacific (Chile): A study supported by a citizen science project. *Marine Environmental Research* 87-88, 12-18.

Hinojosa, I.A., Thiel, M. (2009). Floating marine debris in fjords, gulfs and channels of southern Chile. *Marine Pollution Bulletin* 58, 341-350.

Hirai, H., Takada, H., Ogata, Y., Yamashita, R., Mizukawa, K., Saha, M., Kwan, C., Moore, C., Gray, H., Laursen, D., Zettler, E., Farrington, J., Reddy, C., Peacock, E. &

Ward, M. (2011). Organic micropollutants in marine plastics debris from the open ocean and remote and urban beaches. *Marine Pollution Bulletin* 62, 1683-1692.

Holt, R. (2009). *The Carpet Sea Squirt Didemnum vexillum: Eradication from Holyhead Marina*. Presentation to the Scottish Natural Heritage Conference "Marine Non-invasive Species: Responding to the Threat. 27 October, 2009. Battleby, Scotland.

Hong, S., Lee, J., Jang, Y.C., Kim, Y.J., Kim, H.J., Han, D., Hong, S.H., Kang, D., Shim, W.J. (2013). Impacts of marine debris on wild animals in the coastal area of Korea, *Marine Pollution Bulletin*. 66: 117-124.

Howell, E.A., Bograd, S.J., Morishige, C., Seki, M.P., Polovina, J.J. (2012). On North Pacific circulation and associated marine debris concentration. *Marine Pollution Bulletin* 65, 16-22.

IPRC (International Pacific Research Center) (2008). Tracking Ocean Debris. *IPRC Climate*, 8, 2.

Jacobsen, J.K., Massey, L. & Gulland, F. (2010). Fatal ingestion of floating net debris by two sperm whales (Physeter macrocephalus). *Marine Pollution Bulletin*, 60(5), 765–767.

Jayasiri, H.B., Purushothaman, C.S., Vennila, A. (2013). Quantitative analysis of plastic debris on recreational beaches in Mumbai, India. *Marine Pollution Bulletin* 77(1-2), 107-112.

Jambeck, J.R., Geyer, R., Wilcox, C., Siegler, T.R., Perryman, M., Andrady, A., Narayan, R., Law, K.L. (2015). Plastic waste inputs from land into the ocean. Science, vol. 347(6223):768-771, DOI: 10.1126/science.1260352.

Jones, M.M. (1995). Fishing debris in the Australian marine environment. *Marine Pollution Bulletin* 30, 25–33.

Kaiser, J. (2010). The dirt on ocean garbage patches. *Science* 328, 1506.

Kahn, J., Ofiara, D. and McCay, B. (1989). "Economic Measures of Beach Closures," "Economic Measures of Toxic Seafoods," "Economic Measures of Pathogens in Shellfish," "Economic Measures of Commercial Navigation and Recreational Boating-Floatable Hazards." In WMI, SUNY. Use Impairments and Ecosystem impacts of the New York Bight. SUNY: Stony Brook, NY.

Keller, A.A., Fruh, E.L., Johnson, M.M., Simon, V., McGourty, C. (2010). Distribution and abundance of anthropogenic marine debris along the shelf and slope of the US West Coast. *Marine Pollution Bulletin* 60(5), 692-700.

Kershaw, P.J., Leslie, H. (2012). Sources, fate, & effects of micro-plastics in the marine environment - a global assessment. Report of the Inception Meeting. 13-15th March 2012. UNESCO-IOC, Paris, 45pp.

Khordagui, H.K., Abu-Hilal, A.H. (1994). Man-made litter on the shores of the United Arab Emirates on the Arabian Gulf and the Gulf of Oman. *Water, Air, and Soil Pollution* 76, 343–352.

Koelmans, A.A., Besseling, E., Wegner, A., Foekema, E.M. (2013a). Plastic as a carrier of POPs to aquatic organisms. A model analysis. *Environmental Science and Technology* 47, 7812e7820.

Koelmans, A.A., Besseling, E., Wegner, A., Foekema, E.M. (2013b). Correction to plastic as a carrier of POPs to aquatic organisms. A model analysis. *Environmental Science and Technology* 47, 8992e8993.

Koutsodendris, A., Papatheodorou, G., Kougiourouki, O., Georgiadis, M. (2008). Benthic marine litter in four Gulfs in Greece, Eastern Mediterranean; abundance, composition and source identification *Estuarine, Coastal and Shelf Science* 77 (3), 501-512.

Kuo, F.J., Huang, H.W. (2014). Strategy for mitigation of marine debris: Analysis of sources and composition of marine debris in northern Taiwan. *Marine Pollution Bulletin*, http://dx.doi.org/10.1016/j.marpolbul.2014.04.019.

Kusui, T., Noda, M. (2003). International survey on the distribution of stranded and buried litter on beaches along the Sea of Japan. *Marine Pollution Bulletin* 47, 175-179.

Laist, D.W. (1996). Marine debris entanglement and ghost fishing: A cryptic and significant type of bycatch in anonymous. Proceedings of the solving bycatch workshop. September 25-27 1995. Seattle, Washington. University of Alaska Sea Grant College program. Fairbanks, Alaska. USA. pp 33-39.

Laist, D.W. (1997). Impacts of marine debris: entanglement of marine life in marine debris, including a comprehensive list of species with entanglement, in: Coe, J.M., Rogers, D.B. (Eds.), *Marine Debris – Sources, Impacts and Solutions*. Springer-Verlag, New York, pp. 99–139. Lattin, G.L., Moore, C.J., Zellers, A.F., Moore, S.L., Weisberg, S.B. (2004). A comparison of neustonic plastic and zooplankton at different depths near the southern California shore. *Marine Pollution Bulletin* 49(4), 291-294.

Lam, J., Chan, S., Hansen, G., Chapman, J., Miller, J., Carlton, J., Boatner, R., Cooper, R. and Kight, P. (2013). Japanese tsunami marine debris: Key aquatic invasive species watch. ORESU-G-13. Oregon State University Sea Grant Program. (Oregon Sea Grant). 8 p. http://www.adfg.alaska.gov/static/species/nonnative/invasive/pdfs/tsunami_debris_species_watch_osu.pdf.

Lane, S.B., Gonzalves, C., Lukambusi, L., Ochiewo, J., Pereira, M., Rasolofojaona, H., Ryan, P. (2007). *Regional Overview and Assessment of Marine Litter Related Activities in the WIO Region*. Nairobi Convention and GEF WIO-LaB Project Countries. Prepared on behalf of UNEP (GPA and Regional Seas Programme).

Lattin, G.L., Moore, C.J., Zellers, A.F., Moore, S.L., Weisberg, S.B. (2004). A comparison of neustonic plastic and zooplankton at different depths near the southern Californian shore. *Marine Pollution Bulletin* 49, 291-294 (doi:10.1016/j.marpolbul.2004.01.020) [PubMed].

Law, K.L., Morét-Ferguson, S., Maximenko, N.A., Proskurowski, G., Peacock, E.E., Hafner, J., Reddy, C.M. (2010). Plastic Accumulation in the North Atlantic Subtropical Gyre. *Science* 329(5996), 1185-1188.

Law, K.L., Morét-Ferguson, S., Proskurowski, G., Maximenko, N.A., Reddy, C.M., Peacock, E., Hafner, J. (2011). Plastic Accumulation in the North Atlantic Subtropical Gyre. *Science* 329, 1185-1188.

Lebreton, L.C. M., Greer, S.D., Borrero, J.C. (2012). Numerical modelling of floating debris in the world's oceans Original Research Article. *Marine Pollution Bulletin* 64(3), 653-661.

Lee, D., Cho, H., Jeong, S. (2006). Distribution characteristics of marine litter on the sea bed of the East China Sea and the South Sea of Korea. *Estuarine, Coastal and Shelf Science* 70, 187-194.

Lee, J., Hong, S., Song, Y., Hong, S., Jang, Y., Jang, M., Heo, N., Han, G., Lee, M., Kang, D., Shim, W. (2013). Relationships among the abundances of plastic debris in different size classes on beaches in South Korea. *Marine Pollution Bulletin* 77, 349-354.

Lewis, C.F., Slade, S.L., Maxwell, K.E., Matthews, T.R. (2009). Lobster trap impact on coral reefs: effects of wind-driven trap movement. *New Zealand Journal of Marine and Freshwater Research* 43, 271-282.

Lewis, P.N., Riddle, M.J., Smith, S.D.A. (2005). Assisted passage or passive drift: a comparison of alternative transport mechanisms for non-indigenous coastal species into the Southern Ocean. *Antarctic Science* 17, 183-191.

Lithner, D., Larsson, A. & Dave, G. (2011). Environmental and health hazard ranking and assessment of plastic polymers based on chemical composition. *Science of the Total Environment.* 409: 3309–3324.

Macfadyen, G., Huntington, T. and Cappell, R. (2009). Abandoned, Lost or Otherwise discarded Fishing Gear. *UNEP Regional Seas Reports and Studies No. 185. FAO*

Fisheries and Aquaculture Technical Paper No. 523. Rome: UNEP/FAO.

Madzena, A., Lasiak, T. (1997). Spatial and temporal variations in beach litter on the Transkei coast of South Africa. *Marine Pollution* Bulletin 34 (11), 900–907.

Martinez, E., Maamaatuaiahutapu, K., Taillandier, V. (2009). Floating marine debris surface drift: Convergence and accumulation toward the South Pacific subtropical gyre. *Marine Pollution Bulletin* 58 ,1347-1355.

Masó, M., Garcés, E., Pagès, F., Camp, J. (2003). Drifting plastic debris as a potential vector for dispersing Harmful Algal Bloom (HAB) species. *Scientia Marina* 67, 107–111.

Mato, Y., Isobe, T., Takada, H., Kanehiro, H., Ohtake, C., Kaminuma, T. (2001). Plastic resin pellets as a transport medium for toxic chemicals in the marine environment. *Environmental Science & Technology* 35, 318-324.

Miyake, H., Shibata, H., Furushima, Y. (2011). Deep-sea litter study using deep-sea observation tools. In: Omori K., Guo X., Yoshie N., Fujii N., Handoh I.C., Isobe A., Tanabe S. (Eds.), *Interdisciplinary Studies on Environmental Chemistry–Marine Environmental Modeling and Analysis*. Terrapub, pp.261–269.

Moore, C.J., Moore, S.L., Leecaster, M.K., Weisberg, S.B. (2001). A Comparison of Plastic and Plankton in the North Pacific Central Gyre. *Marine Pollution Bulletin* 42(12), 1297-1300.

Moore, C.J., Moore, S.L., Weisberg, S.B., Lattin, G.L., Zellers, A.F. (2002). A comparison of neustonic plastic and zooplankton abundance in southern California's coastal waters. *Marine Pollution Bulletin* 44(10), 1035-1038.

Moore, E., Lyday, S., Roletto, J., Litle, K., Parrish, J.K., Nevins, H., Harvey, J., Mortenson, J., Greig, D., Piazza, M., Hermance, A., Lee, D., Adams, D., Allen, S., Kell, S. (2009). Entanglements of marine mammals and seabirds 2001–2005. *Marine Pollution Bulletin* 58, 1045-1051.

Moore, S.L., Gregorio, D., Carreon, M., Weisberg, S.B., Leecaster, M.K. (2001). Composition and distribution of beach debris in Orange County, California. *Marine Pollution Bulletin* 42, 241–245.

Mordecai, G., Tyler, P., Masson, D.G., Huvenne, V.A.I. (2011). Litter insubmarine canyons off the west coast of Portugal Gideon Mordecai. *Deep-Sea Research II* 58, 2489-2496.

Morét-Ferguson, S., Law, K.L., Proskurowski, G., Murphy, E.K., Peacock, E.E., Reddy, C.M. (2010). The size, mass, and composition of plastic debris in the western North Atlantic Ocean. *Marine Pollution Bulletin* 60(10), 1873-1878.

Mouat, J., Lozano, R.L. and Bateson, H. (2010). *Economic Impacts of Marine Litter*. Kommunenes Internasjonale Miljøorganisasjon (KIMO), Shetland.

MPMMG (Marine Pollution Monitoring Management Group) (2002). *The Impacts of Marine Litter*, Report of the Marine Litter Task Team (MaLiTT).

Murray, F., Cowie, P.R. (2011). Plastic contamination in the decapod crustacean Nephrops norvegicus (Linnaeus, 1758). *Marine Pollution Bulletin*. 62 (6):1207–1217.

Naturvårdsverket (2009). What's in the Sea for Me? Ecosystem Services Provided by the Baltic Sea and Skaggerak. Report 5872. Swedish Department of Environmental Protection, Stockholm, Sweden. Available from: http://www.naturvardsverket.se/Documents/publikationer/978-91-620-5872-2.pdf.

(NOAA). U.S. Department of Commerce, National Oceanic and Atmospheric Administration. *Abandoned and Derelict Vessels*. http://response.restoration.noaa.gov/oil-and-chemical-spills/oil-spills/abandoned-and-derelict-vessels.html.

NRC (National Research Council) (1995). *Clean Ships, Clean Ports, Clean Oceans*. National Academy Press, Washington, DC.

NRC (National Research Council) (2008). *Tackling Marine Debris in the 21st Century*. Committee of the Effectiveness of International and National Measures to Prevent and Reduce Marine Debris and its Impacts. National Academy Press: Washington, DC.

NRC (National Research Council) (2009). *Tackling Marine Debris in the 21st Century*. National Academy Press, Washington DC, p. 206.

Ocean Conservancy (2011). Tracking Trash: 25 Years of Action for the Ocean (*ICC report*). Available from: http://act.oceanconservancy.org/pdf/Marine_Debris_2011_Report_OC.pdf.

Ofiara, D.D. and Brown, B. (1989). "Marine Pollution Events of 1988 and Their Effect on Travel, Tourism and Recreational Activities in New Jersey." In *Proceedings of the Conference on Floatable Waste in the Ocean: Social, Economic, and Public Health Implications*. SUNY University Press: Stony Brook, NY.

Ofiara, D.D. and Brown, B. (1999). "Assessment of Economic Losses to Recreational Activities from 1988 Marine Pollution Events and Assessment of Economic Losses from Long-term Contamination of Fish Within the New York Bight to New Jersey." *Marine Pollution Bulletin* 38(11): 990-1004.

Ogata, Y., Takada, H., Mizukawa, K., Hiraia, H., Iwasaa, S., Endo, S., Mato, Y., Saha, M., Okuda, K., Nakashima, A., Murakami, M., Zurcher, N., Booyatumanondo, R., Zakaria, M.P., Dung, L.Q., Gordon, M., Miguez, C., Suzuki, S., Moore, C., Karapanagiotik, H.K., Weerts, S., McClurg, T., Burres, E., Smith, W., Van Velkenburg, M., Lang, J.S., Lang, R.C., Laursen, D., Danner, B., Stewardson, N., Thompson, R.C. (2009). International Pellet Watch: global monitoring of persistent organic pollutants (POPs) in coastal waters. 1. Initial phase data on PCBs, DDTs, and HCHs. *Marine Pollution Bulletin* 58, 1437-1446.

Oigman-Pszczol, S.S., Creed, J.C., (2007). Quantification and Classification of Marine Litter on Beaches along Armação dos Búzios, Rio de Janeiro, Brazil. *Coastal Education and Research Foundation* 23, 421-428.

OSPAR Commission. (2009). "*Marine Litter in the North-East Atlantic region: Assessment and Priorities for Response*." London, UK.

Otley, H., Ingham, R., (2003). Marine debris surveys at Volunteer Beach, Falkland Islands, during the summer of 2001/02. *Marine Pollution Bulletin* 46, 1534–1539.

Pham, C.K., Ramirez-Llodra, E., Alt, C.H.S., Amaro, T., Bergmann, M., et al., (2014). Marine Litter Distribution and Density in European Seas, from the Shelves to Deep Basins. *PLoS ONE* 9(4): e95839. doi: 10.1371/journal.pone.0095839

Portland State University. (2012). *Response protocols for biofouled debris and invasive species generated by the 2011 Japan tsunami*. July 31-August 1 2012 Final Workshop report. Portland, Oregon, USA. 60p. http://www.anstaskforce.gov/Tsunami/FINAL%20JTMD%20Biofouling%20Response%20Protocol_19%20Oct%202012.pdf

Ramirez-Llodra, E., Mol, B.D., Company, J.B. (2013). Effects of natural and anthropogenic processes in the distribution of marine litter in the deep Mediterranean Sea. *Progress in Oceanography* 118, 273–287.

Rees, G., Pond, K. (1995). Marine litter monitoring programmes – a review of methods with special reference to national surveys. *Marine Pollution Bulletin* 30, 103–108.

Reisser, J., Shaw, J., Wilcox, C., Hardesty, B.D., Proietti, M., Thums, M., Pattiaratchi, C., (2013). Marine Plastic Pollution in Waters around Australia: Characteristics, Concentrations, and Pathways. *PLoS ONE* 8(11), e80466. doi:10.1371/journal.pone.0080466.

Ribic, C.A., Sheavly, S.B., Rugg, D.J., Erdmann, E.S., (2010). Trends and drivers of marine debris on the Atlantic coast of the United States 1997-2007. *Marine Pollution Bulletin* 60, 1231–1242.

Ribic, C.A., Sheavly, S.B., Rugg, D.J., Erdmann, E.S., (2012). Trends in marine debris along the U.S. Pacific Coast and Hawaii 1998–2007. *Marine Pollution Bulletin* 64, 994–1004.

Ribic, C.A., (1998). Use of indicator items to monitor marine debris on a New Jersey beach from 1991–1996. *Marine Pollution Bulletin* 36, 887–891.

Ribic, C.A., Dixon, T.R., Vining, I., (1992). Marine Debris Survey Manual. *NOAA Technical Report NMFS* 108.

Rios, L., Jones, P., Moore, C. & Narayan, U., (2010). Quantitation of persistent organic pollutants adsorbed on plastic debris from the Northern Pacific Gyre's "eastern garbage patch". *Journal of Environmental Monitoring* 12, 2226-2236.

Rochman, C.M., Hoh, E., Hentschel, B.T., Kaye, S., (2013). Long-Term Field Measurement of Sorption of Organic Contaminants to Five Types of Plastic Pellets: Implications for Plastic Marine Debris. *Environmental Science & Technology* 47, 1646-1654.

Rochman, C.M., Hoh, E., Kurobe, T., Teh, S.J., (2013). Ingested plastic transfers hazardous chemicals to fish and induces hepatic stress. *Nature/ Scientific Reports* 3, 3263.

Rodriguez-Santos, I., Friedrich, A.C., Wallner-Kersanach, M., Fillmann, G., (2005). Influence of socio-economic characteristics of beach users on litter generation. *Ocean Coast. Manage.* 48, 742–752.

Rosevelt, C., Los Huertos, M., Garza, C., Nevins, H.M., (2013). Marine debris in central California: Quantifying type and abundance of beach litter in Monterey Bay, CA. *Marine Pollution Bulletin* 71, 299-306.

Ryan, P.G., (2013). A simple technique for counting marine debris at sea reveals steep litter gradients between the Straits of Malacca and the Bay of Bengal. *Marine Pollution Bulletin* 69, 128–136.

Ryan, P.G., Moore, C.J., van Franeker, J.A., Moloney, C.L., (2009). Monitoring the abundance of plastic debris in the marine environment. *Philosophical Transactions of the Royal Society of London. Series B. Biological Sciences* 364, 1999–2012.

Schlining, K., von Thun, S., Kuhnz, L., Schlining, B., Lundsten, L., Stout, N.J., Chaney, L., Connor, J., (2013). "Debris in the deep: Using a 22-year video annotation database to survey marine litter in Monterey Canyon, central California, USA." *Deep Sea Research Part I. Oceanographic Research Papers* 79, 96-105.

Silva-Iñiguez, L., Fischer, D.W., (2003). Quantification and classification of marine litter on the municipal beach of Ensenada, Baja California, Mexico. *Marine Pollution Bulletin* 46, 132–138.

Spengler, A., Costa, M.F., (2008). Methods applied in studies of benthic marine debris. *Marine Pollution Bulletin* 56, 226–230.

STAP (2011). Marine Debris as a Global Environmental Problem: Introducing a solutions based framework focused on plastic. A STAP Information Document. Global Environment Facility, Washington, DC.

Stickel, B.H., Jahn, A. and Kier, B. (2013). Waste in our water: The Annual Cost to California Communities of Reducing Litter That Pollutes our waterways. Kier Associates: San Rafael, CA.

Swanson, R.L., Stanford, H.M. and O'Connor, J.S. (1978). "June 1976 Pollution of Long Island Ocean Beaches." *Journal of the Environmental Engineering Division, ASCE.* 104(EE6): 1067-1085.

Swanson, R.L., Bell, T.M., Kahn, J. and Olga, J. (1991). "Use Impairments and Ecosystem Impacts of the New York Bight." *Chemistry and Ecology* 5: 99-127.

Tanaka, K., Takada, H., Yamashita, R., Mizukawa, K., Fukuwaka, M., Watanuki, Y., (2013). Accumulation of plastic-derived chemicals in tissues of seabirds ingesting marine plastics. *Marine Pollution Bulletin* 69, 219-222.

Teuten, E.L., Rowland, S.J., Galloway, T.S., Thompson, R.C., (2007). Potential for plastics to transport hydrophobic contaminants. *Environmental Science & Technology* 41, 7759-7764.

Teuten, E.L., Saquing, J.M., Knappe, D.R.U., Barlaz, M.A., Jonsson, S., Björn, A., Rowland, S.J., Thompson, R.C., Galloway, T.S., Yamashita, R., Ochi, D., Watanuki, Y., Moore, C., Viet, P.H., Tana, T.S., Prudente, M., Boonyatumanond, R., Zakaria, M.P., Akkhavong, K., Ogata, Y., Hirai, H., Iwasa, S., Mizukawa, K., Hagino, Y., Imamura, A., Saha, M., Takada, H. (2009). Transport and release of chemicals from plastics to the environment and to wildlife. *Philosophical Transactions of the Royal Society of London Biology.* 364, 2027-2045.

Ten Brink, P., Lutchman, I., Bassi, S., Speck, S., Sheavly, S., Register, K. and Woolaway, C. (2009). *Guidelines on the Use of Market-based Instruments to Address the Problem of Marine Litter.* Institute for European Environmental Policy, Brussels, Belgium and Sheavly Consultants, VA Beach, VA, USA.

Thiel, M., Bravo, M., Hinojosa, I.A., Luna, G., Miranda, L., Nunez, P., Pacheco, A.S., Vasquez, N. (2011). Anthropogenic litter in the SE Pacific: an overview of the problem and possible solutions. *Journal of Integrated Coastal Zone Manage* 11, 115-134.

Thiel, M., Gutow, L. (2005a) The ecology of rafting in the marine environment I: the floating substrata. *Oceanogr Mar Biol* 42:181–263.

Thiel M, Gutow L (2005b) The ecology of rafting in the marine environment II: the rafting organisms and community. *Oceanography and Marine Biology* 43:279–418.

Thiel, M., Hinojosa, I., Vasquez, N., Macaya, E. (2003). Floating marine debris in coastal waters of the SE-Pacific (Chile). *Marine Pollution Bulletin* 46, (2), 224-231.

Thompson, R.C., Olsen, Y., Mitchell, R.P., Davis, A., Rowland, S.J., John, A.W.G., McGonigle, D., Russell, A. E. (2004). Lost at sea: Where is all the plastic? *Science* 304 (5672), 838.

Thornton, L., Jackson, N.L., (1998). Spatial and temporal variations in debris accumulation and composition on an estuarine shoreline, Cliffwood beach, New Jersey, USA. *Marine Pollution Bulletin* 36, 705–711.

Titmus, A.J., Hyrenbach, K.D., (2011). Habitat associations of floating debris and marine birds in the North East Pacific Ocean at coarse and meso spatial scales. *Marine Pollution Bulletin* 62, (11), 2496-2506.

Topçu, E.N., Tonay, A.M., Dede, A., Öztürk, A.A., Öztürk, B., (2013). Origin and abundance of marine litter along sandy beaches of the Turkish Western Black Sea Coast. *Marine Environmental Research* 85, 21-28.

UNEP (2005). Marine Litter. *An analytical overview*. Available from: http://www.unep.org/regionalseas/marinelitter/publications/docs/anl_oview.pdf

UNEP (2006). Ecosystems and Biodiversity in Deep Waters and High Seas. UNEP Regional Seas *Reports and Studies No. 178*. UNEP/ IUCN, Switzerland 2006. ISBN: 92-807-2734-6.

UNEP (2009). *Marine Litter: A Global Challenge.* United Nations Environment Programme, Nairobi.

UNEP and NOAA (2011). The Honolulu Strategy: A Global Framework for Prevention and Management of Marine Debris.

UNEP-MAP-WHO-MEDPOL (2011). *Assessment of the status of marine litter in the Mediterranean*. UNEP-MAP, p. 89.

UNEP (United Nations Environment Programme), NOWPAP (Northwest Pacific Action Plan) , (2008). *Marine Litter in the Northwest Pacific Region, A report by the Northwest Pacific Action Plan (NOWPAP)*, United Nations Environment Programme.

Unsworth, R.E. and Petersen, T.B. (1995). *A Manual for Conducting Natural Resource Damage Assessments: The Role of Economics*. Division of Economics, Fish and Wildlife Service. US Department of Interior, Washington, DC.

Van Cauwenberghe, L., Claessens, M., Vandegehuchtea, M.B., Mees, J., Janssen, C.R. (2013). Assessment of marine debris on the Belgian Continental Shelf. *Marine*

Pollution Bulletin 73, 161-69.

van Franeker, J.A., Blaize, C., Danielsen, J., Fairclough, K., Gollan, J., Guse, N., Hansen, P.L., Heubeck, M., Jensen, J.K., Le Guillou, G., Olsen, B., Olsen, K.O., Pedersen, J., Stienen, E.W.M., Turner, D.M. (2011). Monitoring plastic ingestion by the northern fulmar Fulmarus glacialis in the North Sea. *Environmental Pollution* 159 (10), 2609–2615.

Walker, T.R., Grant, J., Archambault, M., (2006). Accumulation of Marine Debris on an Intertidal Beach in an Urban Park (Halifax Harbour, Nova Scotia). *Water Quality Research Journal of Canada* 41, 256-262.

Walker, T.R., Reid, K., Arnould, J.P.Y., Croxall, J.P., (1997). Marine debris surveys at Bird Island, South Georgia 1990–1995. *Marine Pollution Bulletin* 34, 61–65.

Waluda, C.M., Staniland, I.J., (2013). Entanglement of Antarctic fur seals at Bird Island, South Georgia. *Marine Pollution Bulletin* 74, 244–252.

Watters, D.L., Yoklavich, M.M., Love, M.S., Schroeder, D.M., (2010). Assessing marine debris in deep seafloor habitats off California. *Marine Pollution Bulletin* 60(1), 131-138.

Wei, C.L., Rowe, G.T., Nunnally, C., Wicksten, M.K., (2012). Anthropogenic "litter" and macrophyte detritus in the deep northern Gulf of Mexico. *Marine Pollution Bulletin* 64, 966-973.

Winston, J.E., Gregory, M.R., Stevens, L.M., (1997). Encrusters, epibionts, and other biota associated with pelgaic plastics: a review of biogeographical, environmental, and conservation issues, In: Coe, J.M., Rogers, D.B. (Eds.), *Marine Debris: sources impact, and solutions*, New York, Springer-Verlag, pp. 81-97.

WMI. (1989). *Use Impairments and Ecosystem Impacts of the New York Bight*: SUNY: Stony Brook, NY.

Woodall L. C., Sanchez-Vidal, A., Canals, M., Paterson, G.L.J., Coppock, R., Sleight, V., Calafat, A., Rogers, A. D., Narayanaswamy, B. E., Thompson, R. C., (2014). The deep sea is a major sink for microplastic debris. *Royal Society Open Science*.1: 140317. DOI: 10.1098/rsos.140317.

Woodall, L.C., Robinson, L.F., Rogers, A.D., Narayanaswamy, B.E. and Paterson, G.L.J., (2015). Deep-sea litter: a comparison of seamounts, banks and a ridge in the Atlantic and Indian Oceans reveals both environmental and anthropogenic factors impact accumulation and composition. *Frontiers in Marine Science.*, 02 February 2015. doi: 10.3389/fmars.2015.00003.

World Society for the Protection of Animals (WSPA), (2012). *Untangled-Marine debris: a global picture of the impact on animal welfare and of animal-focused solutions.* London: World Society for the Protection of Animals.

Wright, S. Thompson, R.C., & Galloway, T.S. (2013). The physical impacts of microplastics in marine organisms: a review. *Environmental Pollution* 178, 483-492. http://dx.doi.org/10.1016/j.envpol. 2013. 02.031.

Yamashita, R., Tanimura, A., (2007). Floating plastic in the Kuroshio Current area, western North Pacific Ocean. *Marine Pollution Bulletin* 54(4), 485-488.

Zettler, E.R., Mincer, T.J., Amaral-Zettler, L.A., (2013). Life in the "Plastisphere": Microbial Communities on Plastic Marine Debris. *Environmental Science & Technology* 47, 7137-7146.

Appendix. Economic Impacts of Marine Debris

Category-Resource Affected, Study, Location	Date-Sample size	Method/Approach-Data	Findings: Estimated Effects/Losses	
Beaches-				
Beach cleaning:				
NRC (1995) VA Beach, VA, USA	1993$	Data contact	$24,240/mi/yr ($39,009/km/yr)	€43,646/km/yr $60,724/km/yr
NRC (1995) Atlantic City, NJ, USA	1993$	Data contact	$119,530/mi/yr ($192,359/km/yr)	€215,225/km/yr $299,439/km/yr
Hall (2000): UK	1998£ n=69	Survey UK	77%bc, 897.1km, 10079.7tns, tc=£2,330,719/yr (n=36) £2598/km, €3931/km	€14.301mill/yr $19.905mill/yr 157 x .931 x £64,742/munic.
Hall (2000): Denmark, Germany, Norway, Sweden	1998£ n=13	Survey KIMO members	84.6%bc, 3983km, 655tns, tc=£716,657/yr	
Hall (2000): all countries	1998£ n=82	Survey UK, KIMO members	77%bc, 897.1km, 10079.7tns, tc=£2,913,795/yr	
Mouat et al (2010): UK	2008€ n=61	Survey UK	93%bc, 839km, 21,757tns, tc=€4,513,189.28/yr (n=31) €5379/km, €5804/km, (tc=€3,964,152-2011) P= €17,936,000-18,780,000/yr, midpt=(P=€15.574mill/yr)	€14.487mill/yr $21.688mill/yr 129 x .93 x €120,735/munic
Mouat et al (2010): Belgium, Netherlands	2008€ n=12	Survey KIMO members	92%bc, 68.8km, 724tns, tc=€2,265,415.3/yr P=€10.4mill/yr	
Mouat et al (2010): Denmark, Sweden, Ireland, Portugal, Spain	2008€ n=9	Survey KIMO members	100%bc, 210km, NA-tns, tc=€1,236,453/yr	
Mouat et al (2010): all countries	2008€ n=82	Survey UK,KIMO members	95%bc, 1117.8km, 22,481tns, tc= €7,913,057.82/yr, (tc=£6,950,421-2011)	
OSPAR (2009):all UK coast (from Environmental Agency, 2004)	2004£	Data contact	tc=-£14mill/yr; 2011 values: £17.1mill/yr, €19.7mill/yr	
OSPAR (2009): Skagerrak coast, Sweden	2006€	Data contact	tc=€1.5mill; 2011 value €1.87mill	
Naturvardsverket (2009): Poland 5 municipalities, 2 ports	NA, 2009€ likely	Data contact	tc=€570,000; 2011 value €5,346,207	
Lane (2007): So. Africa To remove litter from waste stream	NA, 2007 likely	NA	cost to remove litter from wastewater stream= R2bill/yr ($279mill/yr)	(2011: $303mill/yr, € 218mill/yr)

Marine Debris   Chapter 25

Study	Year	Source	Findings	Costs
Damage to Beach Use/Attendance:				
ERA (1979) Beach closures, NY, NJ, USA MD washups	1976$	3 beaches contacted	NY: Jones Beach, Robert Moses Beach: lost revenues= $8.88mill/yr NY: Smith Point Beach: lost revenues=$734,100/yr NJ: Seaside Heights, NJ lost revenues=$332,100/yr, avoidance clean beach total= $9,946,200/yr	€943,638/yr $1,312,869/yr €28.261mill/yr $39.320mill/yr
NYDEC (1977), Swanson et al (1978) Beach closures from floatable MD, trash Washups, NY, USA	1976$	City-State data	NY shore cleanup by Peace Corps=$100,000; lost tourist revenue=$15-25mill.	€43-71mill $59-98.9mill
NJDEP & USEPA (1987): NJ, USA Beach cleaning	1987$	State data	NJ beaches cleaned, 127mi, 25,000cu yd, $3mill/yr 204km, $14,706/km,	€4.27mill/yr $5.9mill/yr €20,930/km $29,119/km
Ofiara & Brown (1989,1999) Beach closures, NJ, USA MD washups & bacteria	1988 (1987$)	Data contact	lost NEV: $132-644mill, midpt=$388mill lost revenues: $251-1227mill, midpt=$739mill Gross EV= $383-1871mill	€545-2662mill $758-3704mill
Kahn et al (1989), WMI(1989), Swanson et al (1991) Beach closures, NY & NJ, USA	1988 (1987$) MD washups & bacteria	Data contact	lost NEV: $447-1515mill, midpt=$981mill lost revenues: $539-2165mill, midpt=$1352mill Gross EV= $986-3689mill	€1403-5236mill $1952-7286mill
Losses to Tourism:				
Balance et al (2000): S. Africa from decrease in beach cleanliness	NA, 2000 likely	NA	Decrease in beach cleanliness could decrease tourism revenue up to 52%	
OSPAR (2007): Sweden from MD on beach	NA, 2007€ likely	NA	MD decreases tourism 1-5%/yr, loss revenues= £15mill/yr	
Ofiara & Brown (1989, 1999): NJ, USA MD washups, NJ beaches	1987 1988, 1987$	Data contact Data contact	8.9%-18.7% decrease in beach attendance from MD washups in 1987, NJ 7.9%-32.9% decrease in beach attendance from MD washups in 1988, NJ	€545-2662mill $758-3704mill -see above
Damages to Fishing:				
Hall (2000): UK (Shetland Is Fisheries) Cost of MD removal fr nets, contaminated catch, damage to nets fr snagging	1998£ n=25	Survey	92%caught MD, 69%catch contaminated, 92%snag nets alv=£6,000-30,000/vessel, P=£885,400-4,428,000/yr alv: cMD=£4065/boat, cc=£1686/boat, sn=£3392/boat	€1.4-6.7mill/yr $11.564mill/yr €8.308mill/yr
Mouat et al (2010): UK (Shetland Is Fisheries) Cost of MD removal fr nets, contaminated catch, damage to nets fr snagging	2008€ n=22	Survey	86%caught MD, 82%catch contaminated, 95%snag nets, 82%fouled prop alv=€17,219-19,165/vessel, P=€11.7mill - 13mill/yr alv: cMD=€12,007/boat, cc=€2183/boat, sn=€3820/boat, fp=€182/boat	€9.7-10.8mill/yr €8.935mill/yr $12.444mill/yr
Mouat et al (2010): Portugal, Spain	Portugal(n=21) 2008€ Spain (n=6) 2008€	Survey Survey	29%caught MD, 38%catch contaminate, 0-snag nets, 57%fouled propellers, 19%blocked intakes, ac=€2930/boat, 81%covered insurance 100%caught MD, 50%catch contaminate, 83.33%snag nets, 100%fouled propellers & blocked intakes	

Chapter 25 — Marine Debris

Study	Year/Currency	Method	Details	Costs
MPMMG (2002): UK fishery	NA, 2002€ likely	NA	cost of MD removal=€33mill/yr	
Watson & Bryson (2003): Scotland Clyde fishery	2002$	NA	avl=$21,000 lost gear, $38,000 lost time to single trap fisherman (per vessel, n=1)	
Damage from Ghost Fishing:				
NRC (2008): tangle & gillnet fishery, EU	2008€ likely	NA	loss=<5% of European Union commercial landings	
Brown et al (2005): Cantabrian Sea, Spain Monkfishery	2005€ likely	NA	loss=1.46% of landings, .0146(768)=11,213tns	
Allsopp et al (2006): US lobster fishery	2006$ likely	NA	lost value of landings=$250mill/yr	
Macfadyen et al (2009): Blue crab Fishery, LA, USA	2009 likely	NA	loss=4-10mill blue crabs/yr	
Damages to Aquaculture:				
Hall (2000): UK Shetland fishery	1998£ n=15	Survey	40%caught MD, 20%net contaminated, 1hr/mo-1.80/mo some fouled propeller, ac=£150-1200/incident, (avg-ac=£675/incident) KIMO mthd: P=196x £675/incident, P=£132,300/yr	€173,683/yr $212,222/yr
Mouat et al (2010): UK Scotland fishery	2008€ n=11	Survey	73%caught MD, N=268(.73)=196, al=€52.24/farm, P=€10,239/yr foul prop, N=196, al=€528.17/farm, P=€103,521/yr TotalP=€113,760/yr	€8491 $11,824 €85,847/yr $119,500/yr €94,338 $131,374
Damages to Shipping, Harbors/Marinas:				
Hall (2000): UK Harbors, Marinas Removal of floatables and MD in harbor	1998£ n=42	Survey	31%cleanup MD, ac=£100-15,000, mean ac=£2354.67/hbr KIMO mthd: 300(.4615)=138, ac=£2354.67/hbr, P=138xac=£324,944/yr fouled propellers, 180incidents, ac=£3947/incident, P=£710,406/yr	€491,641/yr $684,287/yr €1.075mill/yr $1.496mill/yr
Mouat et al (2010): UK Harbors, Marinas Removal of floatables and MD in harbor	2008€ n=48	Survey	39.56%cleanup floatable MD, 6.59%dredge MD, 46.15%cleanup MD 69%reported fouled propellers, 28.6% report blocked intakes, no costs cleanup MD, ac=€8253/hbr, N=300(.4615), P=138x ac=€1,138,924/yr	€944,510/yr $1.315mill/yr
Mouat et al (2010): Denmark, Norway, Portugal, Spain, all 2008€ Removal of floatables & MD in harbor	Denmark (n=5) Norway (n=4) Portugal (n=5) Spain (n=21)		80%cleanup floatables, 20%dredge MD, 60%fouled propellers, 20%fouled propellers, ac=€10,760.21/hbr 75%cleanup floatables, 25%dredge MD, 75%fouled propellers, ac=€10,052.07/hbr 20%cleanup floatables, 0-dredge, 69%fouled propellers, 20%blocked intakes 95%cleanup floatables, 0-dredge, 48%fouled propellers, 14%blocked intakes Spain tc=€563,917.33/yr (tc split as follows 97.38% - harbors, 2.62% - marinas)	
Kahn et al (1989), Swanson et al (1991) Damage to vessels (Commercial, Pleasure) NY, USA	1988 (1987$)	Data contacts	MD floatables in NY Harbor	Commer. Boats: added repair costs= $500mill Pleasure boats: lost NEV= $26mill Gross EV= $526mill €749mill $1041mill

Rescues-Vessels Disabled from MD:

Study	Year	Method	Data	Economic Values
Hall (2000): UK rescues	1998£	Log records	230 rescues, ac=£4000/rescue, P=£506,000-1,334,000/yr	€765,579-2,018,345/yr, €1.392mill/yr $1,065,566-2,809,221/yr, $1.937mill/yr
Mouat et al (2010): UK rescues	2008€	Log records	286 rescues from fouled prop in 2008, ac=€2902-7653/incident N=286, P=£830,000-2,189,000/yr	€688,293-1.815mill/yr, €1.252mill/yr $959,517-2.528mill/yr, $1.743mill/yr
Moore (2008): US rescues	2005	Log records	269 rescues; 116 injuries, 15 deaths, $3mill property damages	

Damage to Coastal Agriculture:

Study	Year	Method	Data	Economic Values
Hall (2000): UK (Shetland Is.)	1998£ n=30	Survey	96%MD in fences, 36%animals entangled in MD, 20%animals ingest-ill ac=£400/croft, N=1500crofts, P=£600,000/yr clear MD: 1440x £213/farm, animal entangle: 540x£10.5/farm, ill: 300x£30/farm (£321,390/yr)	€486,270/yr $676,826/yr
Mouat et al (2010): UK (Shetland Is)	2008€ n=31	Survey	71.4%MD in fences, 41.9%animals entangled in MD or ingest-ill ac=€841.10/farm, N=25% of 1200crofts, P=€252,331/yr clear MD: 1200x .714x €840/farm=€719,712/yr, entangled 1200x .419x€17.663/farm=€8884/yr (TotalP=€728,596/yr)	€614,461/yr $855,698/yr

Damage to Coastal Power Plants:

Study	Year	Method	Data	Economic Values
Hall (2000): UK n=9	1998£	Survey	Clean corse, fine screens, 78% rpt seaweed/organic main prob, 11% rpt litter-£26,000 1 rpt 20-25% human litter (barrels of liquid, sewage, plastic bottles, general litter), <5% MD, tc=(.05)14,000=£700, 33% rpt must clean screens regardless of type of debris 11% MD problem, ac=(£26,000+700)/2=£13,350/yr, 33% rpt must clean regardless, hence ac=£0/yr	for 11%: ac=€20,199/yr, $28,113/yr for 89%: ac=£0/yr
Mouat et al (2010): UK n=3	2008€	Survey	100% seaweed/organic main prob, 67% rpt some MD collected, tc=€16,516 (n=1) 1 rpt 5% is MD, 1 rpt 1% is MD, 1 rpt not affected; ac=1-5% (tc)= €165-826/yr 33% rpt must clean regardless of type of debris, hence ac=€0/yr	for 67%: ac=€137-685/yr, mean=€411/yr ac=$191-954/yr, mean=$573/yr for 33%: ac=€0/yr

Damage from Invasive Species:

Study	Year	Method	Data	Economic Values
Naturvardsverket (2009): Black Sea Intro of American comb jellyfish	2009€ likely	NA	intro of Amer. comb jellyfish caused collapse of anchovy fishery in Black Sea, €240mill/yr	
Holt (2009): Wales, UK Holyhead Harbour	2009£ likely	NA	control & eradication of invasive species, Carpet sea squirt over 2009-2019=£525,000 cost of inaction=up to £6.876mill 10-yr period	

Notes: Data contact refers to data obtained from authoritative agencies, USEPA, US Coast Guard, state/municipal/beach park data and/or representative-officials responsible. Abbreviations where not obvious: MD=marine debris, mill.=million, bc=beach clean, ac=average cost, tc=total cost, P=projection, munic=municipality or local authority, avl=average loss per vessel, hbr=harbor, al=average loss, N=universe projections based on, Nbc=no. municipalities beach cleaned for projections, TotalP=total projection, midpt=midpoint, NEV=net economic value, EV=economic value. Totals may not add due to rounding. Mouat et al is abbreviated as KIMO in places for shorthand. One date appears if the study date and year of monetary value were the same, a monetary symbol appears after the date. All conversions: 1987$ to 2011$: 1.9801from US CPI-U, 1976$ to 2011£: 3.9532 using US CPI-U, 1998£ to 2011£: 1.3128 from UK CPI, 2011$ to 2011€: .71876 exchange rate, 2011£ to 2011€: 1.1525 exchange rate, 2011£ to 2011$: 1.6041 exchange rate, 2011£ to 2011$: 1.3926 exchange rate, 1993$ to 2011$: 1.5567 from US CPI-U, 1998€ to 2011€: 0.8293 using historical inflation rates for € currency.

26 Land-Sea Physical Interaction

Contributors:
Julián Reyna (Convenor), Arsonina Bera, Hong-Yeon Cho, William Douglas Wilson, Regina Folorunsho, Kazimierz Furmanczyk, Sean Green (Co-lead member), Frank Hall, Peter Harris (Co-lead member), Lorna Inniss (lead member), Sung Yong Kim, Teruhisa Komatsu, Renzo Mosetti, Kareem Sabir, Wilford Schmidt, Hannes Tõnisson, Joshua Tuhumwire (Co-lead member)

1 Introduction

This chapter deals with how human activities have changed the physical interaction between the sea and the land. This physical interaction is important because about 60 per cent of the world's population live in the coastal zone (Nicholls et al., 2007). The "coastal zone" is defined in a World Bank publication as "the interface where the land meets the ocean, encompassing shoreline environments as well as adjacent coastal waters. Its components can include river deltas, coastal plains, wetlands, beaches and dunes, reefs, mangrove forests, lagoons and other coastal features." (Post et al., 1996) In some places, natural coastal erosion processes cause damage to property, harm to economic activities and even loss of life. In other places, human activities have modified natural processes of erosion of the coast and its replenishment, through: (1) coastal development such as land reclamation, sand mining and the construction of sea defences that change the coastal alongshore sediment transport system; (2) modification of river catchments to either increase or decrease natural sediment delivery to the coast; and (3) through global climate change and attendant sea level rise changes to surface wave height and period and the intensity and frequency of storm events.

2 Natural coastal erosion and property damage

Coastal erosion is a natural, long-term process that contributes to the shaping of present coastlines, but it can also pose a threat to life and property (Rangel-Butrago and Anfuso, 2009). For example, the total coastal area (including houses and buildings) currently being lost in Europe through marine erosion is estimated to be about 15 km^2 per year (Van Rijn, 2011). Over 70 per cent of the world's beaches experience coastal erosion, some portion of which is a natural process (Dar and Dar, 2009). Other natural processes influencing coastal sediment dynamics include the supply of biogenic carbonate sand and gravel (see Chapter 7) and volcanism which can provide an important sediment source to some coastal areas, including some continental coasts, such as in Italy (de Rita et al., 2002) and to volcanic islands, such as in Polynesia, Indonesia, the Caribbean, the Azores and sub-Antarctic islands (e.g., Dey and Smith, 1989; Ross and Wall, 1999). Volcanic activity may supply sediments to the coast directly in the form of ash deposited as atmospheric fallout, or as lava flows or debris flows down the flanks of volcanoes that are adjacent to the coast (Fisher and Smith, 1991).

3 Impacts of anthropogenic climate change

The impacts on coastal ecosystems of anthropogenic global climate change (Jaagus, 2006; Jylhä et al., 2004) are associated with sea-level rise (Johansson et al., 2004 and increased storminess (Alexandersson et al., 1998) (Lowe et al., 2001; Masselink and Russell, 2006; Meier et al., 2004; Morton et al., 2005; Tõnisson et al., 2011; Suursaar et al., 2006; Wang et al., 1998). Coastal sedimentation and morphology are influenced directly by regional land morphology and composition and anthropogenic activities that affect the amounts and locations of precipitation, run-off from both point and non-point sources, sea level, and storm activity. In addition, aeolian (i.e., wind-blown) dust, especially from deserts in Africa and Asia, affects some coastal communities. Aeolian processes are discussed in Chapter 5 and will not be considered further here.

The Intergovernmental Panel on Climate Change (IPCC) (2013) has shown that the rate of global sea-level rise throughout the 20th century has increased, due to melting ice caps and glaciers and the thermal expansion of the oceans, both resulting from increased global temperatures. Local sea level is further affected by processes including sediment discharge and subsidence (the natural, gravitational sinking of land over time), hydrological management, fluid withdrawal, and tectonic activity (Millman et al., 1989; Reed and Yuill, 2009; Boon et al., 2010). The human response to sea-level rise will include armouring coastlines to protect real estate, thus cutting off the natural (landward) retreat path of coastal and intertidal organisms. Coastal development that has occurred on low-gradient, sandy coastlines is the most vulnerable, since the natural response of such systems is to retreat landward as sea level rises.

Another apparent response to the warming of the Earth's atmosphere is a change in the ocean wave climate, manifested as increased wave heights associated with more intense storm events (Carter and Draper, 1988; WASA Group, 1998; Gulev and Hasse, 1999; Allan and Komar, 2000). Changes in wave regime may affect the stability of sandy shorelines and potentially dramatic changes in coastal geomorphology may occur locally. For example, the transformation from tide-dominated to wave-dominated coastal systems is possible in some locations (Harris et al., 2002). Increased wave height and period translates into an increase in the water depth at which sediments may be mobilized, thereby fundamentally changing the character of the seabed habitat. For example, areas of sandy seafloor previously stable under the prevailing wave and current regime may become transformed into a different habitat type, subject to mobilizing forces of episodic storms (Hughes et al., 2010).

4 Impacts of coastal development

Increasing human encroachment, land reclamation, coastal development and economic activity (e.g., shipping, recreation, mining) are considered to be among the major anthropogenic impacts on the coastal environment. These impacts have both direct and indirect influences on the physical interaction of the ocean with the coast.

4.1 Land Reclamation

Land reclamation is a significant component of economic growth and development for many countries around the world. The need for space to accommodate an increasing world population, which is projected to exceed 8.1 billion by 2025 (United Nations, 2013), has been a contributing factor to the growing trend of large-scale reclamation projects in many coastal areas to provide suitable land for housing and recreation, industry, commerce, agriculture and, in some cases, to provide coastal protection for the adjacent coastline. Many land-reclamation projects are also found in coastal cities that are short of space for expansion, particularly for port activities.

Two methods of land reclamation are generally used: infilling and draining of tidal and submerged wetlands. Infilling is most common in coastal areas associated with dredging activities, either indirectly by utilizing dredged material form port and harbour development or directly from offshore sources. Improved dredging technologies over time have increased the scale efficiency, cost-effectiveness and value of the land created in reclamation projects (Kolman, 2012). Previously land reclamation was restricted to shallow near-shore environments; however, with improvements in dredging methods and technologies, land reclamation has progressed to deeper water.

Large-scale land reclamation was pioneered in the Netherlands by the building of "polders" (areas of former swamp or intertidal land enclosed by embankments known as dikes). Since the 11th century, the Dutch have reclaimed marshes and fenland, resulting in some 3,000 polders nationwide enclosing about 7,000 km^2. About half of the total surface area of polders in northwest Europe is in the Netherlands. With the advent of modern machinery, such reclamation of wetlands is achieved much more readily. In West Bengal, India, 1,500 km^2 of the coastal Sundarbans wetlands have been reclaimed over the last 100 years (UNEP, 2009). In China, 9,200 km^2 (16 per cent) of the wetlands present in the 1970s had disappeared by 2007 (Zuo et al., 2013). At a more local level, 28 per cent of the tidal flats around the coast of the Yellow Sea were reclaimed between the 1980s and the late 2000s (Murray et al., 2014). Around the world, polder-type reclamation has been undertaken in Egypt, Morocco, Senegal and Tunisia in Africa, in Bangladesh, China, India, Indonesia, Iraq, Japan, Malaysia, Myanmar, the Republic of Korea, Pakistan, Thailand and Viet Nam in Asia; in Belgium, Denmark, Germany, Italy, Poland, Portugal, Romania, Spain and the United Kingdom of Great Britain and Northern Ireland in Europe; in The Russian Federation; in the south-eastern states of the United State of America in North America; and in Argentina, Colombia, Suriname and Venezuela in South America (MVenW, 1983). This reclamation work has particularly affected mangroves and salt marshes (see Chapters 48 and 49).

In recent times significant large-scale dredging projects have been undertaken in several countries in Asia and the Middle East. Rapidly emergent economies, such as China, Japan, Singapore, and United Arab Emirates, have all undertaken large-scale reclamation projects as a solution to finding land for economic development (Glaser et al., 1991; Suzuki, 2003). Reclamation is no longer limited to near-shore coastal environments. Technological advances enabled the creation of the international airport of Hong Kong, China, Japan's Kansai airport and resort developments, such as the Palm Jumeirah, Palm Jebel Ali, the Deira Islands and other similar facilities that are now a prominent feature of the coastline of Dubai, United Arab Emirates.

Major projects, such as the creation of Hulhumalé Island in the Maldives between 1997 and 2002, have been used to relieve overpopulated urban areas and to enable urban expansion. More than 10 per cent of the developed land area of Hong Kong, China, is reclaimed from the sea (Jiao et al., 2001). Other examples of such urban land reclamation projects are: Rotterdam in Europe, New York/New Jersey Port Authority and San Francisco in North America, Rio de Janeiro and Rio Grande in South America, Shanghai, Singapore and Tokyo in East Asia, Chennai and Kolkata in South Asia, Bahrain and Dubai in West Asia, and Cape Town and Lagos in Africa. In special cases, such as the Principality of Monaco, reclaimed land forms 20 per cent of the land area of the State (Anthony, 1994).

Several examples of reclamation are found in China, where the primary objective was to provide suitable living space for a growing population and to promote economic development. Between 2003 and 2006, the Shanghai government spent 40 billion yuan (6.5 billion United States dollars) on the so-called Lingang New City Project to reclaim 133.3 km^2 of artificial land from the sea. From 1949 to 2000, China reclaimed about 12,000 km^2 of land. Land reclamation has also been used to tackle long-term coastal erosion problems from storm surges and extreme climatic events. For example, significant financial investments have been allocated to reclaim beaches along Florida's eastern coastline in the United States and support a major tourism industry. The Sand Engine[1] in the Netherlands is also an innovative example of an effort to alleviate long-term coastal erosion problems while at the same time boosting local biological diversity and the economy (Stive et al., 2013).

4.2   Environmental impacts of land reclamation

Land reclamation causes significant negative impacts on coastal habitats and the ecosystem services they provide (Wang et al., 2014; Wang et al., 2010). Degradation of wetlands, seagrass beds and coastal water quality is commonly associated with large-scale reclamation projects. Examples of studies examining the impacts of coastal wetlands reclamation are available from, e.g., Hong Kong, China (Jiao et al., 2001), the Republic of Korea (Lee, 1998) and the Netherlands (Waterman et al., 1998).

Large-scale reclamation can also affect the regional groundwater regime, causing changes of the groundwater level and the modification of natural groundwater discharge to the coast. For example, Mahamood and Twigg (1995) conducted a statistical analysis of water table data to document a rise in the water table in areas of Bahrain associated

---

[1] The Sand Motor is an innovative method for coastal protection. The Sand Motor (also known as Sand Engine) is a huge volume of sand that has been applied along the coast of Zuid-Holland (the Netherlands) at Ter Heijde in 2011. Wind, waves and currents will spread the sand naturally along the coast of Zuid-Holland. This is called 'Building with Nature'. The Sand Motor will gradually change in shape and will eventually be fully incorporated into the dunes and the beach. The coast will be broader and safer. From: http://www.dezandmotor.nl/en-GB/.

with land reclaimed from the sea. Some of the problems related to land reclamation in coastal areas are reported by Jiao (2000) as including:

- Rises in water level will lead to reduction in the bearing capacity of foundations and in the stability of slopes.
- Ground water may also penetrate underground concrete and cause corrosion of steel reinforcements.
- An increase in water level may cause damp surfaces and superficial damage to the floors of residential buildings (Mahamood and Twigg, 1995).

Furthermore, aquifers behave like underground reservoirs and can receive only a certain amount of rainfall before they overflow. If the water level is increased as a result of land reclamation, the additional water storage capacity will be significantly decreased, resulting in increased rainwater run-off. Thus, aquifers that are full will increase the chance of flooding during heavy rainfall periods. As submarine groundwater input to the sea depends on the groundwater level relative to sea level, a change in the groundwater regime may modify the coastal marine environment and ecology (Jiao, 2000).

4.3 Socioeconomic impacts of land reclamation

The significance of land reclamation to the development of local and national economies cannot be overstated. Land development is inextricably linked to rapid economic growth in many countries. For example, Small Island Developing States depend on coastal economies for their growth and development. Several islands rely heavily on the provision of suitable tourism infrastructure and seaports to maintain their development objectives. More generally, growth in international trade entails growth in shipping (see Chapter 17) both in volume and in the size of ships, which makes port expansion necessary (see chapter 18). In many cities, such port expansion can only be achieved by land reclamation.

As well as benefits, interference with other uses of the sea can occur For example, in Japan, part of Isahaya Bay on the Ariake Sea was cut off from the rest of the bay by a dike to reclaim land for agriculture and to create freshwater storage. Fishers and seaweed growers complained that the effects, especially on tidal flows, have harmed fish, shellfish and seaweed production, a view that a Japanese court upheld in 2010 (JT, 2014; AS, 2013).

4.4 Habitats, coastal development and coastal squeeze

Measured normal to the coast, the assemblage of coastal habitats ranges in width from only a few metres to tens of kilometres, offering very diverse environmental conditions to numerous plant and animal species. These habitats formed as a result of long-term stability in ecological conditions influenced by the constant distance from the shoreline. The influence of the sea decreases rapidly towards the land and biodiversity decreases rapidly just a few hundred metres from the shoreline as well. Therefore, it is noted that the loss of the land on the coastal zone often leads to the rapid decrease of biodiversity, especially on the uplifting regions (Kont et al., 2011).

Coastal vegetation (e.g., dune vegetation, reed-beds, mangroves, salt marshes) inhibits erosion, reducing the effect of storm surges and wave attack (Kathiresan and Rajendran, 2005), and serves as natural protection. Removing the vegetation for coastal development can increase coastal erosion (Waycott et al., 2009) and may add to the amount of sediment moving along the shore, causing siltation of harbours and burial of coastal and coastal sea habitats.

One of the most recent and most notable coastal defence structures is the 25-km-long Saint Petersburg Flood Prevention Facility Complex. The estimated cost of this structure was approximately 3.85 billion United States dollars (Reuters, 2011). Recent winter storms have caused extreme erosion events on the surrounding sandy beaches (Tõnisson et al., 2012).

Several countries in the world have legislatively defined shoreline locations, and major efforts are made to implement the laws. The most notable examples are Poland, the Netherlands and the United States of America. Today, over one-fourth of the 1032-km-long Polish coast is more or less engineered (www.climateadaptation.eu). For the period 2004-2023 a Long-Term Coastal Protection Strategy was developed for which 249 million euros has been secured (approximately 7.5 million euros per year). Approximately 3.5 million inhabitants live in the region that might be potentially affected by the sea in Poland (ec.europa.eu). This cost is rather big compared to what the Baltic States (which have a shoreline that is four times longer) altogether are spending on coastal protection. The only known cost is the Lithuanian Palanga nourishment project reaching 1.65 million euros per year (Gulbinskas et al., 2009).

Most of the effort in the Netherlands is focused on the flood protection, because over three-fourths of the Netherlands is less than 5 m above sea level. Approximately 9 million people live in the region at high risk from coastal flooding and over 50 per cent of GDP is generated in this region. In principle, the Netherlands addresses coastal protection activities in a comprehensive (national) approach to counteract flood risk (coast and river flooding). The total amount spent on flood-risk protection is estimated at 550 million euros per year. Over the period 1998-2015, measures to protect the Dutch coast against flooding and erosion and adapt to increased storminess and sea-level rise amount to 3.4 billion euros. The yearly expenditure on beach nourishments under the Sand Nourishment Programme has increased from 22 million euros in 2000 to 70 million euros in 2008 (ec.europa.eu).

Over 50 per cent of the population in the United States live in the coastal region and are more or less affected by changes in coastal processes. In the United States, coastal erosion is responsible for approximately 500 million United States dollars per year in coastal property loss, including damage to structures and loss of land. To mitigate coastal erosion, the federal government spends an average of 150 million dollars every year on beach nourishment and other shoreline erosion control measures.

However, it is estimated that despite the measures, every fourth house within an approximately 500-m-wide coastal zone will be destroyed by erosion by 2050 (NOAA, 2013).

Economic losses related to coastal erosion are more complex to assess. Detailed analyses carried out in Louisiana in the United States show that major disruptions in economic activities caused by coastal erosion and flooding might result in over 70,000 jobs lost in the United States and a long-term loss for the United States economy of approximately 10 billion dollars (dnr.louisiana.gov). Loss of coastal habitat in Louisiana results from three primary processes: control of the Mississippi River, which reduces the amount of replenishing sediment that rebuilds the coast, subsidence, and industrial activity.

The United States Army Corps of Engineers will be spending upward of 5 billion dollars on shore protection projects following Hurricane Sandy. The vast majority of these funds will be spent on pumping sand onto beaches from Delaware to Connecticut, an amount approaching 25 to 35 million m³. This is an adaptation model that cannot be exported to most developing countries due to its cost, and it is doubtful that it can be maintained for an extended period of time in the United States of America (Young et al., 2014).

The use of initially less expensive hard protection measures (which are extremely destructive to surrounding shores and natural habitats) has decreased, whereas the share of soft coastal protection measures has significantly increased in recent decades. However, in light of global sea-level rise, researchers and coastal communities have started the debate on managed realignment. In the United Kingdom, realignment of private property and infrastructure occurred in a 66-km-long stretch of shoreline during 1991-2013. The United Kingdom government plans to reach 550 km (10 per cent of the United Kingdom coastline) by 2030 (Esteves and Thomas, 2014). These zones will function as natural buffer zones and reduce the risk of erosion and coastal defence costs to the surrounding coasts. Furthermore, most of the local communities are in favour of this measure and the strongest opposition is from those whose property will be most severely affected. This topic is also recognized by European Commission, and a new Horizon 2020 call (open until 2014 autumn) was issued on the topic: "Science and innovation for adaptation to climate change: from assessing costs, risks and opportunities to demonstration of options and practices", where realignment is considered as a key solution (http://ec.europa.eu/research/participants/portal/desktop/en/opportunities/h2020/).

Many countries do not have the economic resources to protect their coast actively through engineering solutions, which are very expensive and often not technologically successful. Effective non-engineering solutions are available such as restoring the processes that were in place before human intervention, improved urban planning, and orderly retreat strategies. The cost of not dealing with coastal erosion can be even more costly over the long-term. For example, in 2006, Mozambique requested 3 million US dollars from the international community to carry out engineering coastal protection countermeasures for potential flooding. This request was turned down in 2007, and since then, the international community has spent over 90 million dollars to address extreme flooding in the region (Ashdown, 2011).

5 Impacts of catchment disturbance

5.1 Decline in marine sedimentation as a result of water management

On a global basis, the natural (pre-Anthropocene) sediment load to the oceans from rivers is estimated to be about 15×10^9 tons; the majority is supplied by mountain rivers with steep catchments (Milliman and Syvitski, 1992; Ludwig and Probst, 1998; Syvitski et al., 2005). Over the last few centuries, the global annual sediment flux into the coastal zone has increased by 2.3×10^9 tons due to anthropogenic soil erosion and decreased by 3.7×10^9 tons due to retention in reservoirs: the net effect is a reduction of sediment input by 1.4×10^9 tons (Syvitski et al., 2005). A major environmental consequence of river sediment starvation is erosion of the coast and attendant loss of habitat; conversely, the consequence of increased sediment input is elevated coastal and estuarine turbidity and smothering and burial of biota. Many examples of the two opposite consequences are documented from different places around the world (Fig. 1).

A well-documented example of the first problem (sediment starvation) is the Mississippi River, which is controlled by the United States Army Corps of Engineers and has one of the most diverse and largest watersheds in the world. Over the Holocene period (the last 10,000 years) the Mississippi deposited the largest deltaic coastal complex in North America. Prior to European settlement of North America, the sediment load of the Mississippi was around 400 million tons per year, but at present it is around 100 million tons per year as a result of the construction of around 45,000 water reservoirs in the catchment (Syvitski, 2008). The decreased sediment load contributes to loss of wetlands in the delta; this loss peaked between 1955 and 1978 at 11,114 ha/year and declined to 2,591 ha per year from 1990 to 2000. Reductions in sedimentation, compounded by sea-level rise, water abstraction (causing subsidence), oil and gas extraction (also causing subsidence), tidal erosion and storm surges, have resulted in an approximate total land loss of 113,300 acres of the coastal Mississippi delta over the past 60 years (Morton et al., 2005).

Other examples of river systems that have undergone a similar pattern of anthropogenic sediment retention that has reduced the sediment discharge of the river to below pristine conditions include the Yellow, Indus, Colorado, Nile and Danube Rivers (Syvitski, 2008). In extreme cases the sediment load is reduced to near zero (e.g., the Nile, Indus and Colorado Rivers). The Volta River in West Africa had a sediment load of 11 million tons per year prior to the construction of the Akosombo Dam, but after the dam's completion in 1963, this load was reduced to nearly zero (Milliman and Meade, 1983). With the recent construction of another dam

Land-Sea Physical Interaction — Chapter 26

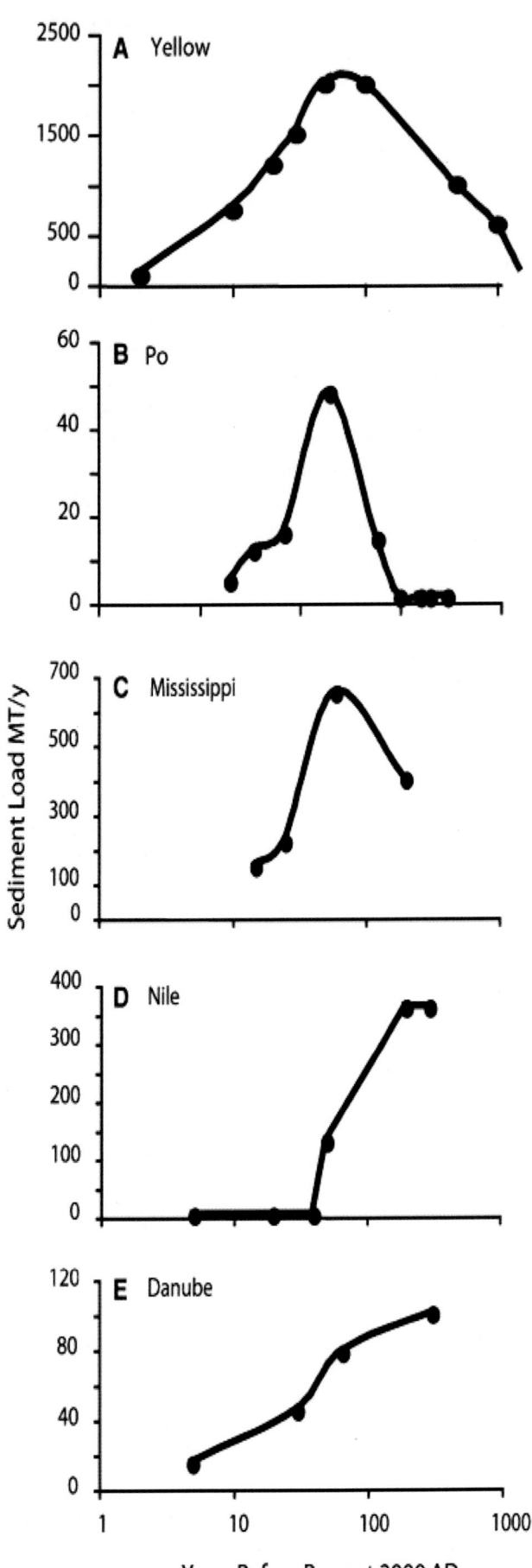

Figure 26.1 | Change in the sediment load in millions of tons per year (MT/year) delivered to the deltas of five rivers over a period of 1,000 years: (A) Yellow; (B) Po; (C) Mississippi; (D) Nile; and (E) Danube (from Syvitski, 2008). In some cases an initial phase of catchment clearing and disturbance, characterized by increased sediment load, is followed by a reduced sediment load phase attributed to dam building (A, B and C). In other cases the curves indicate only a steady decrease in sediment load attributed to dam building (D and E) (Source: Syvitski, 2008).

on the Volta River (the Kpong Dam), its sediment yield has diminished further, starving the delta and coast of sediment input.

5.2 Increase in marine sedimentation as a result of catchment disturbance

In many instances, dam building and sediment entrapment in reservoirs does not offset the impacts of land clearing and soil disturbance in the catchment, such that river sediment loads exceed pristine levels in spite of dam building. The data presented in Figure 26.1 for the Yellow River 10 years ago and for the Po and Mississippi 100 years ago all showed this effect. Another example is the coast of northeast Queensland in Australia adjacent to the Great Barrier Reef; modelling work by Brodie et al. (2003) indicates that the total suspended sediment load of rivers draining into the Great Barrier Reef region is currently about 16 million tons per year, compared with pre-European settlement estimates of around 2 million tons per year. This is in spite of the fact that nearly all of the rivers in the region have been dammed. Consequently, increased sedimentation (with attendant inputs of nutrients and pesticides) to the ecosystem are affecting inshore areas of the Great Barrier Reef, causing increased algal growth, accumulation of pollutants in sediments and marine species, reducing light and smothering corals (GBRMPA, 2009).

One other key point noted by Milliman and Meade (1983, p. 19) is that "much of the sediment probably accumulates on the subaerial parts of subsiding deltas, never really reaching salt water. Along coastlines that include a number of large estuaries, most of the inflowing river sediment is trapped in embayments." Hence a significant proportion of sediment discharged by most rivers is trapped within the coastal biome. Nevertheless, more widespread impacts of increased sediment loads may occur during flood events, when buoyant river plumes extend across the continental shelf. The input of fine-grained sediment in historical times may also have increased the turbidity of the coastal zone above pristine environmental conditions, affecting some light-dependent species.

5.3 Modelling Coastal Sedimentary Processes

Computer models have assisted in improving our understanding of sedimentation processes in the near shore (Syvitski and Alcott, 1993; Syvitski and Alcott, 1995; Mulder et al., 1997; Syvitski and Milliman, 2007; Chu et al., 2013). Computer algorithms provide process-related information to understand flow dynamics, particulate and chemical distributions, and the overall oceanography of coastal environments.

5.4 Impacts of increased sediment input on coastal habitats

The impact of sediments has been recognized by both global reviews and the United Nations Environment Programme (UNEP/GEMS, 2008). These impacts include:

- Smothering of marine communities and, in severe cases, complete burial leading to suffocation of corals, mangrove stands and seagrass beds.
- Decreases in the amount of available sunlight, which may in turn limit the production of algae and macrophytes, increase water temperatures and reduce growth of natural vegetation.
- Injuring fish by irritating or scouring their gills and degrading fish habitats as gravel containing fish eggs becomes filled with fine particles, thus reducing available oxygen
- Reducing the success of visual predators, and may also harm some benthic macroinvertebrates.
- Infilling watercourses, storm drains and reservoirs, leading to costly dredging and an increased risk of flooding.
- Many toxic organic chemicals, heavy metals and nutrients are physically and/or chemically adsorbed by sediments, so that an increase of sediment loading to the marine environment can also lead to increased deposition of these toxic substances that result in further negative impacts such as eutrophication.

It is important to recall that human impacts on marine habitats rarely (if ever) act in isolation (Lotze et al., 2006; Harris, 2012). Therefore it is difficult, if not impossible, to attribute specific changes in species composition or abundance to single, specific impacts, such as increased sediment loads of rivers caused by human activities. Nevertheless, increased sediment loads of rivers have evidently affected the marine environment, and this section reviews these effects as far as possible.

Coastal habitats and species that are most affected by increased sediment loads of rivers are those which are least tolerant of rapid sediment deposition and high turbidity (in relation to natural background levels), and changes in sediment grain size and composition. Among such ecosystems, coral reefs and estuaries with their associated mangrove, salt marshes and sea grasses are among the most vulnerable habitats.

In the Caribbean, Burke et al. (2004) reported on the assessment of more than 3,000 watersheds across the region that discharged into an area where coral reefs were present. They found that 20 per cent of coral reefs in the region are at high threat, and about 15 per cent at medium threat from damage caused by increased sediment and pollution from agricultural lands and other land modification.

Sediment input to temperate ecosystems also affects habitats and associated species. For example, eelgrass meadows, which provide important near-shore marine habitat for fish, shellfish, and invertebrates, as well as food for waterfowl and detritus feeders, can be buried or fragmented by increased sediment delivery associated with river-delta channelization (Lotze et al., 2006; Grossman et al., 2011). Increased turbidity is a particular problem because seagrasses require some of the highest light levels of any plant group in the world, approaching 25 per cent of incident radiation in some seagrass species, compared with 1 per cent or less for other angiosperm species (Dennison et al., 1993).

Therefore, even small increases in turbidity can cause significant reductions in seagrass cover over a short time span (Orth et al., 2006).

Changes in sediment grain-size composition can also affect ecosystems. For example, many shellfish beds and forage-fish spawning beaches depend on a specific sediment grain-size composition that is linked to land-use activities and hydrological conditions that release and carry sediment into coastal waters (Gelfenbaum et al., 2009). Water quality, near-shore and offshore habitats, and aquatic ecosystem health are affected by contaminants and nutrients that preferentially adsorb to fine sediments. Specialized organisms that have adapted to fine sediments, high sedimentation rates, and mobile substrate utilize estuaries. The macroinvertebrates that are found at the bottom of estuaries are much smaller than those found in streambeds with larger particle sizes, and they tend to be opportunistic species (Schaffner et al., 1987). Within the estuary, the density of fauna is commonly greater in the freshwater tidal areas than in parts of the estuary having (tidally) varying salinity (Schaffner et al., 1987). The species diversity of macroinvertebrates is usually lower in fine sediments than that in sediments with coarser particles. The diversity and evenness of species decline with an increasing percentage of silt/clay and organic matter (Junoy and Vieitez, 1990). However, fine-sediment beds are important for burrowing tube-building invertebrates and other burrowing species (Minshall, 1984).

The impacts of increased sedimentation from river input will differ between deltas and estuaries (Harris and Heap, 2003). Estuaries and deltas formed along coasts at the end of the last Ice Age when sea level reached its present position, around 6,500 years ago. Deltas evolved in places where river sediment discharge filled the palaeo-river valley to form a deltaic protrusion from the coast. The formation of a delta relies on the river supplying sediment to the coast more rapidly than can be redistributed by waves and tides, causing seaward progradation. For estuaries and lagoons, slow rates of sediment delivery to the coast results in the palaeo-valley remaining partially un-filled. Waves and tides produce deltas and estuaries with distinctively different shapes in plain view (e.g., Harris and Heap, 2003; Syvitski, 2008).

Prograding coastlines, characterized by deltas, strand plains and tidal flats, export most of their sediment load to the sea and generally have a naturally low sediment-trapping efficiency. They contain a suite of habitats that will not be significantly affected by sedimentation. In contrast, transgressive coastlines, characterized by estuaries and lagoons, have a high sediment-trapping efficiency (Fig. 2). They are, therefore, more susceptible to the accumulation of particle-associated contaminants such as heavy metals. They also contain a suite of habitats that will change (evolve) as they infill with sediments (Roy et al., 2001), and are therefore more susceptible to catchment perturbations that affect river sediment

Land-Sea Physical Interaction — Chapter 26

Type of Coastal Environment	Sediment Trapping Efficiency	Turbidity	Circulation	Habitat Change due to Sedimentation
 Tide-dominated Delta	Low	Naturally High	Well Mixed	Low Risk
 Wave-dominated Delta	Low	Naturally Low	Salt Wedge/ Partially Mixed	Low Risk
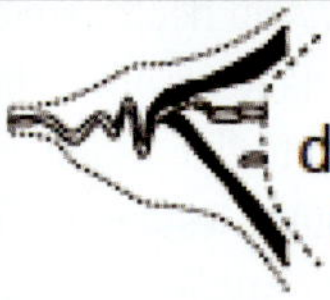 Tide-dominated Estuary	Moderate	Naturally High	Well Mixed	Some Risk
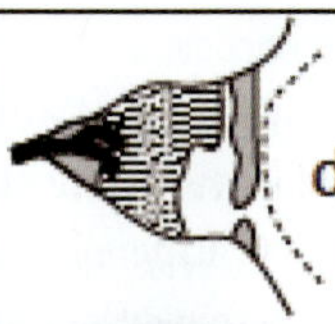 Wave-dominated Estuary	High	Naturally Low	Salt Wedge/ Partially Mixed	High Risk
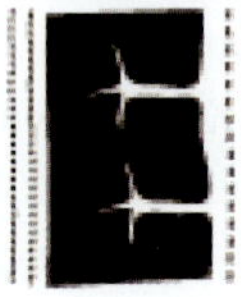 Tidal Flats	Low	Naturally High	Well Mixed	Low Risk
 Strand Plains	Low	Naturally Low	Negative/ Salt Wedge/ Partially Mixed	Low Risk
 Lagoon	High	Naturally Low	Negative/ Well Mixed	High Risk

Figure 26.2 | Diagram showing different types of coastal depositional environments in relation to four common management issues: sediment trapping efficiency, turbidity, water circulation and habitat change due to sedimentation (from Harris and Heap, 2003).

loads. The water contained in tide-dominated deltaic distributary channels, estuaries, and creeks that drain intertidal flats, is naturally turbid and generally well mixed. In contrast, water contained in wave-dominated deltaic distributary channels, estuaries, and lagoons is naturally clear (low turbidity) and exhibits mainly stratified (estuarine) circulation patterns (Fig. 2).

In arid regions of the Earth, this circulation is often of the inverse (negative) type, driven by high evaporation rates. From a management perspective, therefore, human activities that give rise to higher turbidity levels are likely to have a greater impact in wave-dominated systems (that have naturally low-turbidity, clear water) than in tide-dominated systems that are naturally turbid (Harris and Heap, 2003).

5.5 Coastal erosion caused by sediment starvation of coasts attributed to dam building and impacts on habitats

When sediment delivery to the coast is reduced by dam building in the catchment, critical near-shore habitat and beaches can be eroded by natural coastal processes and lost (Malini and Rao, 2004; Warrick et al., 2009). Erosion along pristine sedimentary coastlines will cause a landward retreat of habitats, such that an ecological succession is observed at a fixed observation point. For example, supratidal vegetated dunes give way to intertidal salt marsh and tidal flats. Thus the entire suite of habitats is preserved, but laterally displaced.

Several factors can complicate this simple view. In some cases the space that could allow habitats to retreat is simply not available (so-called coastal squeeze; Doody, 2013; Pontee, 2013) For example, in a modelling study of continental United States coastal habitats, Feagin et al. (2005) found that beach erosion resulted in the disappearance of sand dune plants or in their confinement to a narrow belt because existing coastal development (or coastal defences like seawalls) occupies the available space landward of the dune zone. This results in a breakdown of the successional process in built-up areas along the coastal zone.

An example of loss of habitat attributed to coastal erosion induced by dam building was reported for mangrove habitat of the Godavari Delta, India (Malini and Rao, 2004). Elsewhere, coastal erosion caused by a combination of factors has been attributed to loss of habitat (see Box 26.1). For example, loggerhead turtle nesting areas in southern Italy have been affected by coastal erosion (Mingozzi et al., 2008). Although numerous examples exist from around the world where coastal erosion and habitat loss have occurred, it is often not possible to single out dam building as the single (or even the dominant) cause of coastal erosion. This is because of complicating, often related factors, such as subsidence, diversion of freshwater input, coastal development and land-use changes, (e.g., Syvitski, 2008). For example, the loss of wetlands in the Mississippi Delta region is a famous example of habitat destruction caused by human interference with a major river system (see Box 26.2), but the loss of habitat has not been caused so much by dam building as by diversion of the river channel by the construction of artificial levees and dredging activities (Shaffer et al., 2009).

Box 26.1 | Case Study – West African Coastal Erosion

The West African coastal zone stretches from Mauritania to Cape of Good Hope in South Africa and consists of a narrow coastal zone backed by a gradually rising Precambrian landmass that is drained by several rivers. The rivers carry sediments, nutrients and water to the coastal areas. The major rivers responsible for most of the sediment load are the Congo, Niger, Volta, Benue, Gambia, and the Orange. Of these rivers, the Congo, Niger and Orange are classified among the 10 largest rivers in the world in terms of sediment and water yield (Milliman et al., 1983). Minor rivers, such as Senegal, Ouémé, Mono, Bandama and Calvally, collectively contribute significant amounts of sediments and water to the coast because of their basin size, length, and flood plains as compared to rivers such as the Niger, Congo and Orange (Fig. 1).

The ability of these rivers to effectively nourish the coast is hindered by climatic, environmental and anthropogenic factors that determine the amount of sediments, nutrients and water delivered, and include different origins of sediments, and climate variability in the catchment areas. (Folorunsho et al., 1998; Awosika et al., 2013; Awosika and Folorunsho, 2014).

Sediments carried to the coast by the major rivers and ephemeral rivers are derived predominantly from the weathered Precambrian rock complex comprised of schist, gneiss, and granite. The rock complex has now been reduced by erosion to piedmonts that dot most of the West African region. Younger sedimentary rock types in the lower Niger River and flood plains have also been eroded along the river courses and carried to the coast. Suspended sediments, comprising fine silt and fine sand blown from the Sahara desert through the savannah and the Equatorial region, constitute another major sediment type in the region.

The large-scale impoundment of upstream reservoirs has reduced downstream flow volumes and velocities. As a consequence, the strength of the Niger River in its downstream segment is reduced. The reduction in flow velocity encourages sediment deposition within river channels, and is reflected in shallower river cross-sections and enlarged sand bars. Assessment of the spatial distribution and relative sizes of sand bars indicates increased deposition over the years of impoundment. The sand bar just north of Patani on the Forcados River increased from 0.1701 km2 in 1963 to 0.486 km2 in 1988. The sand bar at Anibeze, east of Patani on the same river, increased from 0.3645 km2 to 0.9315 km2 in the same period. A similar trend is observed on the River Nun, where the sand bar just south of Odoni increased its size by 100 per cent between 1963 and 1988. The sand bar opposite Odi not only increased in size, it also expanded in the upstream direction.

At Sampor, a major river channel completely silted up. Several other sand bars on the river between Onitsha and the bifurcation of the Niger into the Forcados and Nun Rivers have evolved into larger sand bars, thereby significantly modifying the river morphology. The NDES environmental change atlas for this area (NDES, 2000) confirms that changes have largely been confined to the flood plain. This channel modification has adversely affected the economy of the river channel and the Nigerian Federal Government's plan for large-scale dredging of the River Niger.

> **Box 26.2 | Coastal Louisiana and the Mississippi River Delta, USA, and the Impact of Hurricane Katrina**
>
> The Mississippi River Delta provides the primary buffer for inland communities within the State of Louisiana and is a major resource for fisheries. Land loss resulting from subsidence has historically been replenished by the deposition of silt from the Mississippi River. However, human activity, including resource exploitation, such as petroleum extraction and controlling the flow of the Mississippi River by the U.S. Army Corps of Engineers, has resulted in reduced sediment flow and significant loss of land. It is important to note that whereas southeast Louisiana contains 37 per cent of marsh habitat in the United States, it has the highest rate of land loss of any region of the country (Glick et al., 2013).
>
> This land loss resulted in the increased landward incursion of seawater during Hurricane Katrina in 2005, flooding coastal communities and eroding and displacing habitat. In addition, the Mississippi River Gulf Outlet provided a channel through which seawater flowed, overtopping the levees that protected the City of New Orleans, causing billions of dollars of damage to property and economic activities and loss of life (Shaffer et al., 2009).

Other responses to sediment starvation of the coast from dam building also occur (see Box 26.1). Erosion of a once stable or prograding coast can transform the composition and character of the subtidal habitat such that it no longer supports the original community. For example, erosion of the Elwa River delta in Washington State, USA, transformed a sedimentary terrace in front of the delta from a sandy habitat colonized by molluscs into a cobbled substrate hosting a different ecosystem (Warrick et al., 2009).

5.6 Significant environmental, economic and/or social aspects in relation to changes in sediment input

The social and economic aspects of loss of specific habitats include loss of livelihood (where important fish-food sources and access to croplands have been displaced or where tourism assets are lost), homes and communities (for example, where coastal erosion and shoreline retreat have damaged or destroyed buildings), and habitats of cultural or amenity values. The concept of waterfront property value is also linked to tourism, as it is often the case that the aesthetic and cultural aspects driving property values and tourism are the same (Phillips and Jones, 2006). Many of these issues are described in relation to the specific habitats involved in other parts of this Assessment (coral reefs, see Chapter 43 and 44; estuaries and deltas, see Chapter 44; kelp and seagrass, see Chapter 47; mangroves, see Chapter 48; and salt marshes, see Chapter 49).

Shaffer et al. (2009) note in relation to the cost of restoring wetlands on the Mississippi Delta (estimated to be about 5,300 US dollars per hectare) that "the most significant twenty-first century public works projects will be those undertaken to correct environmental damage caused by twentieth-century projects." In other words, the cost of restoring damaged habitats is often much greater than the cost of the projects that caused the damage in the first place. The cost of maintenance of coastal protection infrastructure is estimated to be about 300,000 dollars per mile (1.8 km) of coast in the United States (Dunn et al., 2000). Along coasts that are eroding, the cost of allowing the coastline to retreat is immense. If the coast of Delaware in the United States were allowed to recede at its present pace (without shoreline defence engineering works), the cost of lost property by 2050 has been estimated by Parsons and Powell (2001) to be about 291 million dollars (year 2000 dollars), which the authors argue justifies the dollar cost of defence.

6 Gaps in Capacity to Assess Land/Sea Physical Interactions

The assessment of land/sea physical interactions is multidisciplinary, requiring expertise in all aspects of oceanography (physical, geological, chemical, and biological) and concomitant terrestrial counterparts. Local, regional, and basin-wide scales must be considered, along with implications of climate change and sea-level rise. Ultimately, the natural sciences must inform and serve the social realities of economic and infrastructural needs. This section is arranged according to gaps in knowledge of coastal processes, gaps in knowledge of ocean processes, and gaps in capacity to apply existing science to effective decision-making. Some examples are taken from the Caribbean; however, the discussion is global, with many areas of extrapolation.

Globally, coastal engineering science has evolved dramatically during the last century, from the application of crude coastal protection measures such as the *ad hoc* placement of rocks and other natural materials, to the well-modelled and sophisticated coastal structures built today. These changes are mainly due to some improvement in understanding of coastal processes, such as the interaction of nearshore waves with littoral sediment transport paths. However, methodologies are still developing, linking changes in sea level to changes in shape, sediment volume, and beach width. New models have been developed, and produce better results when validated than the Bruun model (DECCW, 2010). Since the 1950s, the Bruun model has been used to model coastal retreat caused by increases in sea level, in spite of its apparent limitations as a two-dimensional representation of a multi-dimensional process (Cooper and Pilkey, 2004).

In developed countries, such as the European Union Member States, the United States and Canada, permanent, extensive coastal and nearshore monitoring ensures that the body of data required for modelling coastal processes is readily available. An abundance of information exists on the sources and sinks of sediment, and on sediment transport processes. However, the situation is different in many African and Asian coastalstates, and in Small Island Developing States, where capacity and technologies for widespread monitoring are limited. The UNESCO Sandwatch programme has enabled establishment of beach profile monitoring in many developing countries by non-governmental organizations and school students (UNESCO, 2010).

Development of a specific global framework for land/sea physical interaction assessment needs to be initiated. This could include improving capacity of persons who collect and analyze existing and new data at local, regional, and basin-wide levels. Ultimately, a standardized training programme, through an inter-institutional network, should be established. Support is needed for ongoing *in-situ* measurements and for the re-establishment of discontinued data collection programmes, and for initiating new studies. Above all, the scientific questions should be clearly articulated. Forecasting ocean processes is a necessary capability for addressing climate change and sea-level rise.

Many countries, including small island States, are susceptible to climate change impacts, many of which will directly affect land-sea interaction (Carter et al., 2014). Inundation of coastal lowlands and small islands will lead to the loss of mangroves, seagrasses, and coral reef habitats and ecosystems. It is also postulated that rainfall patterns over the region will be altered, changing river discharge and non-point-source sedimentation. It is also thought that hurricanes will become less frequent but more intense, posing a particular threat to the coasts of island States. The global Integrated Coastal Area Management (ICAM) programme of IOC-UNESCO is designed to assist countries in their efforts to build marine scientific and technological capabilities as a follow-up to Chapter 17 of Agenda 21, and to Chapter IV of the Mauritius Strategy. The main objectives of ICAM are to increase capacity to respond to change and challenges in coastal and marine environments through further development of such science-based management tools as marine spatial planning, ecosystem-based management, and the Large Marine Ecosystem (LME) approach.

Activities include syntheses of scientific information and preparation of methodological manuals, a strategic alliance with the International Geosphere-Biosphere Programme (IGBP) and its core project on Land-Ocean Interaction in the Coastal Zone (LOICZ), and a project on the development and application of indicators for integrated coastal and ocean management.

In the same way the SPINCAM (Southeast Pacific Data and Information Network In Support to Integrated Coastal Area Management) project was designed to establish an ICAM indicator framework at national and regional levels in the countries of the Southeast Pacific region (Chile, Colombia, Ecuador, Panama and Peru), focusing on the state of the coastal and marine environment and socioeconomic conditions, to provide stakeholders with information and atlases on the sustainability of existing and future coastal management practices and development. The SPINCAM model is proposed for replication in other regions.

With respect to gaps in knowledge of ocean processes, it is noted that anthropogenic climate change will affect wave climate, sea level and ocean acidification (OA). Impacts of the latter are discussed in chapter 5, and the production of beach sand is discussed in chapter 7. However, given that the science on OA is now developing, significant gaps exist in knowledge of this phenomenon and its effects. Thus far only certain regions of the world's oceans are being studied, e.g. off North America, Canada and New Zealand. The Global Ocean Acidification Observing Network (GOA-ON) is striving to assist in the development of standards, and of capacity for researchers in other countries.

Sea-level monitoring and research is quite mature; almost all coastal States participate in the collection of sea-level data. Instrumentation is easily accessible and States utilize data for early warning of coastal hazards. The Global Sea Level Observing System has improved monitoring in regions not previously monitored. The main gaps in knowledge are in the use of sea-level data in models to determine changes in coastal processes and changes in shorelines.

Other gaps in capacity relate to the application of existing knowledge to inform the development of coastal areas. An understanding of coastal dynamics is vital in design and construction of coastal infrastructure. Diversion of sediment pathways through dam building, port/harbour construction, and infrastructure development constructed on dunes and sand spits may result in losses following coastal erosion. Land reclamation may destroy habitat and ecosystem services that protect the coast from sea-level-related hazards such as storm surge and tsunamis. One significant gap in capacity is in coastal engineering, especially in Small Island Developing States. Once capacities in physical oceanography and coastal engineering are linked, ocean and coastal processes may be better understood for application by coastal developers.

Even where capacity in Small Island Developing States exists, such as in Barbados, the challenge is to develop a comprehensive succession planning framework that maintains that capacity over time. Additionally, the costs of appropriately located, designed and constructed coastal infrastructure may be prohibitive. Currently, Barbados is undertaking a Coastal Risk Assessment and Management Programme (CRMP) of 42 million US dollars through concessional loan financing. The aim of this project is to conduct diagnostic studies of ocean and coastal processes, in order to determine the impact of sea-level rise and other coastal hazards on the country's coastal assets (IDB, 2011). Thus, the costs of monitoring, including maintenance and calibration of instrumentation, and training, are a factor in the application of existing science for effective coastal planning.

References

Agardy, T., Alder, J., Dayton, P.K., Curran, S., Kitchingman, A., Wilson, M., Catenazzi, A., Restrepo, J., Birkeland, C., Blaber, S., Saifullah, S., Branch, G., Boersma, D., Nixon, S., Dugan, P., Davidson, N., Vorosmarty, C., 2005. Coastal Systems, in: Hassan, R., Scholes, R., Ash, N. (Eds.), *Ecosystems and Human Well-being: Current State and Trends - Findings of the Condition and Trends Working Group of the Millennium Ecosystem Assessment.* Island Press, Washington, pp. 513-549.

Alexandersson, H., Schmith, T., Iden, K. & Tuomenvirta, H. (1998). Long-term variations of the storm climate over NW Europe. *Global Atmosphere and Ocean System*, Vol. 6, 97–120.

Allan, J. and Komar, P. (2000). Are ocean wave heights increasing in the eastern North Pacific? *EOS Transactions American Geophysical Union* 81, 561.

Anthony, E.J. (1994). Natural and Artificial Shores of the French Riviera: An Analysis of Their Interrelationship. *Journal of Coastal Research*, Vol. 10(1).

AS (Asahi Shimbun) (2013). Floodgates need to be reopened in Isahaya reclamation project, 30 December 2013.

Ashdown, P. (Chair), (2011). *Humanitarian Emergency Response Review.* Available from http://www.dfid.gov.uk/Documents/publications1/HERR.pdf

Awosika, L.F., Folorunsho, R., and Imovbore, V. (2013). Morphodynamics and features of littoral cell circulation observed from sequential aerial photographs and Davies drifter along a section of the strand coast East of the Niger Delta, Nigeria. *Journal of Oceanography and Marine Science*, Vol. 4(1), 12-18.

Awosika, L. F. and Folorunsho, R., (2014). Estuarine and ocean circulation dynamics in the Niger Delta, Nigeria: Implications for oil spill and pollution management. In: *The Land/Ocean Interactions in the Coastal Zone of West and Central Africa*: Estuaries of the World -. Springer International Publishing Switzerland.

Boon, John D., Brubaker, John M. and Forrest, David R. (2010). Chesapeake Bay land subsidence and sea level change: an evaluation of past and present trends and future outlook. Virginia Institute of Marine Science Special Report No. 425 in Applied Marine Science and Ocean Engineering, A report to the U.S. Army Corps of Engineers Norfolk District, http://www.vims.edu/GreyLit/VIMS/sramsoe425.pdf

Brodie, J., McKergow, L.A., Prosser, I.P., Furnas, M., Hughes, A.O., Hunter, H. (2003). *Sources of sediment and nutrient exports to the Great Barrier Reef World Heritage Area.* Australian Centre for Tropical Freshwater Research, Report No. 03/11. James Cook University, Townsville.

Bruun, P. (1988). The Bruun rule of erosion by sea-level rise. *Journal of Coastal Research*, Vol. 4(4), 627-648

Bruun, P., (1962). Sea-level rise a cause of shore erosion. *Journal of Waterways and Habours Division*, Vol. 88, 117-130 ASCE

Burke, L., Maidens, J., Spalding, M., Kramer, P., Green, E., Greenhalgh, S., Nobles, H., Kool, J. (2004). *Reefs at Risk in the Caribbean*. World Resources Institute, Washington DC, p. 84.

Carter, L.M., Jones, J.W., Berry, L., Burkett, V., Murley, J.F., Obeysekera, J., Schramm, P.J. and Wear, D. (2014). Ch. 17: Southeast and the Caribbean. In: *Climate Change Impacts in the United States: The Third National Climate Assessment*, J. M. Melillo, Terese (T.C.) Richmond, and G. W. Yohe, (eds.), U.S. Global Change Research Program, 396-417. doi:10.7930/J0NP22CB.

Carter, D.J.T. and Draper, L. (1988). Has the northeast Atlantic become rougher? *Nature* 332, 494.

Chu, M.L, Knouft, J.H., Ghulam, A., Guzman, J.A., and Pan, Z., (2013). Impacts of urbanization on river flow frequency: A controlled experimental modeling-based evaluation approach, *Journal of Hydrology*, Vol. 495, 1-12.

Cloern, J.E. (1987). Turbidity as a control on phytoplankton biomass and productivity in estuaries, *Continental Shelf Research*, Vol. 7, Nos. 11-12, 367-381.

Cooper, J.A.G., Pilkey, O.H. (2004). Sea-level rise and shoreline retreat: time to abandon the Bruun Rule. *Global and Planetary Change* 43: 157-171.

Crowell, M., Coulton, K., Johnson, C., Westcott, J., Bellomo, D., Edelman, S., Hirsh, E. (2010). An estimate of the U.S. population living in 100-year coastal flood hazard areas. *Journal of Coast Research*, Vol. 26(2), 201-211.

Dar I.A., Dar, M.A., (2009). Prediction of shoreline recession using geospatial technology; a case study of Chennai Coast, Tamil Nadu, India. *Journal of Coastal Research*, Vol. 25(6), 1276-1286.

Dennison, W.C., Orth, R.J., Moore, K.A., Stevenson, J.C., Carter, V., Kollar, S., Bergstrom, P.W., Batiuk, R.A. (1993). Assessing water quality with submersed aquatic vegetation. *BioScience*, Vol. 43, 86–94.

de Rita, D., Fabbri, M., Mazzini, I., Paccara, P., Sposato, A., Trigari, A. (2002). Volcaniclastic sedimentation in coastal environments: the interplay between volcanism and Quaternary sea level changes (central Italy). *Quaternary International* 95/96, 141-154.

Dey, S and Smith, L. (1989). Carbonate and Volcanic Sediment Distribution Patterns on the Grenadines Bank, Lesser Antilles Island Arc, East Caribbean. *Bulletin of Canadian Petroleum Geology*, Vol. 37 (1989), No. 1. (March), Pages 18-30

Doody, J.P., (2013). Coastal squeeze and managed realignment in southeast England, does it tell us anything about the future? *Ocean & Coastal Management*, Vol. 79, 34-41.

Dunn, S., Friedman, R., Baish, S., (2000). Coastal erosion: Evaluating the risk. *Environment*, Vol. 42, 36-45.

Esteves, L.S., Thomas, K. (2014). Managed realignment in practice in the UK: results from two independent surveys. In: Green, A.N. and Cooper, J.A.G. (eds.), Proceedings 13th International Coastal Symposium Durban, South Africa, *Journal of Coastal Research*, Special Issue No. 70, 407-413.

Feagin, R.A., Sherman, D.J., Grant, W.E., (2005). Coastal erosion, global sea-level rise, and the loss of sand dune plant habitats. *Frontiers in Ecology and the Environment* Vol. 3, 359-364.

Fisher, R.V., Smith, G.A., 1991. *Sedimentation in Volcanic Settings*. Society for Sedimentary Geology Special Publication 45, Tulsa, Oklahoma. 255pp.

Folorunsho, R., Awosika, L.F. and Dublin Green, C.O. (1998). An assessment of river inputs into the Gulf of Guinea. In: Ibe, A.C., Awosika, L.F. and Aka, K. (eds): *Nearshore dynamics and sedimentology of the Gulf of Guinea*. Centre for Environment and Development in Africa, 163-172.

GBRMPA (2009). Great Barrier Reef outlook report 2009 in brief. Great Barrier Reef Marine Park Authority, Australia. http://elibrary.gbrmpa.gov.au/jspui/bitstream/11017/429/1/Great-Barrier-Reef-outlook-report-2009-in-brief.pdf

Gelfenbaum, G., Fuentes, T.L., Duda, J.J., Grossman, E.E., and Takesue, R.K., eds., (2009). Extended abstracts from the Coastal Habitats in Puget Sound (CHIPS) 2006 Workshop, Port Townsend, Washington, November 14–16, 2006: *U.S. Geological Survey Open-File Report 2009–1218*, 136 p.

Glaser, R., Haberzettl, P., & Walsh, R.P.D. (1991). Land reclamation in Singapore, Hong Kong and Macau. *GeoJournal*, Vol. 24, 365–373. doi:10.1007/BF00578258.

Glick, P., Clough, J., Polaczyk, A., Couvillion, B. and Nunley, B. (2013). Potential effects of sea-level rise on coastal wetlands in southeastern Louisiana. In: Brock, J.C.; Barras, J.A., and Williams, S.J. (eds.), Understanding and Predicting Change in the Coastal Ecosystems of the Northern Gulf of Mexico. *Journal of Coastal Research*, Special Issue No. 63, 211–233, Coconut Creek (Florida), ISSN 0749-0208.

Grossman, E.E., George, D.A., and Lam, A., (2011). *Shallow stratigraphy of the Skagit*

River Delta, Washington, derived from sediment cores. U.S. Geological Survey Open-File Report 2011–1194.

Gulbinskas, S., Trimonis, E., Blažauskas, N., Michelevicius, D. (2009). Sandy deposits study offshore Lithuania, SE Baltic Sea. *Baltica,* Vol. 22(1), 1-9.

Gulev, S.K. and Hasse, L. (1999). Changes of wind-waves in the North Atlantic over the last 30 years, *International Journal of Climatology* 19, 1091–1117.

Harris, P.T., Heap, A.D., (2003). Environmental management of coastal depositional environments: inferences from an Australian geomorphic database. *Ocean and Coastal Management,* Vol. 46, 457-478.

Harris, P.T., Heap, A., Bryce, S., Smith, R., Ryan, D., Heggie, D. (2002). Classification of Australian clastic coastal depositional environments based upon a quantitative analysis of wave, tidal and fluvial power. *Journal of Sedimentary Research* 72, 858-870.

Harris, P.T., (2012). Anthropogenic threats to benthic habitats. In: Harris, P.T., Baker, E.K. (Eds.), *Seafloor Geomorphology as Benthic Habitat: GeoHab Atlas of seafloor geomorphic features and benthic habitats.* Elsevier, Amsterdam, p. 39-60.

Hughes, M.G., Harris, P.T., Brooke, B.P. (2010). *Seabed exposure and ecological disturbance on Australia's North-West shelf: Potential surrogates for marine biodiversity*. Geoscience Australia Record 2010/43, Canberra, p. 76.

IPCC (Intergovernmental Panel on Climate Change) (2007). *Climate Change 2007: The physical science basis.* Contribution of Working Group I to the Fourth Assessment report of the Intergovernmental Panel on Climate Change, S. Solomon, D. Qin, M. Manning, Z. Chen, M. Marquis, K.B. Averyt, M. Tignor, H.L. Miller (eds.), Cambridge Univ. Press, New York. 996 pp.

IPCC (Intergovernmental Panel on Climate Change) (2013). *Fifth Assessment Report: Climate Change*, Cambridge University Press, New York.

Jaagus, J. (2006). Climatic changes in Estonia during the second half of the 20th century in relationship with changes in large-scale atmospheric circulation. *Theoretical Applied Climatology*, Vol. 83, 77–88.

Jiao J.J., Subhas Nandy, and Hailong, L.P., (2001). Analytical Studies on the Impact of Land Reclamation on Ground Water Flow. *Ground Water,* Vol. 39(6),912-920.

Jiao, J.J. (2000). Modification of regional groundwater regimes by land reclamation. Hong Kong *Geologist* 6, 29-36.

Johansson, M.M., Kahma, K.K., Boman, H. & Launiainen, J. (2004). Scenarios for sea level on the Finnish coast. *Boreal Environment Research*, Vol. 9, 153–166.

JT (Japan Times) (2014). Court orders state to pay for not opening Isahaya Bay gates, 11 April 2014.

Junoy, J. and Vieitez, J.M. (1990). Macrozoobenthic community structure in the Ria de Foz, an intertidal estuary (Galicia, Northwest Spain). *Marine Biology*, Vol. 107, 329-339.

Jylhä, K., Tuomenvirta, H. and Ruosteenoja, K. (2004). Climate change projections for Finland during the 21st century. *Boreal Environment Research,* Vol. 9, 127–152.

Kalnejais, L.H., Martin, W.R. and Bothner, M.H. (2010). The release of dissolved nutrients and metals from coastal sediments due to resuspension. *Marine Chemistry.* Vol 121(1-4), 224-235.

Kathiresan, K. and Rajendran, N., (2005). Coastal mangrove forests mitigated tsunami. *Estuarine, Coastal and Shelf Science*, Vol. 65, 601-606.

Kolman, R., (2012). New Land by the Sea : Economically and Socially, Land Reclamation Pays. https://www.iadc-dredging.com/ul/cms/fck-uploaded/documents/PDF%20Articles/article-new-land-by-the-sea.pdf, Accessed 18 August 2014.

<u>Kont, A., Jaagus, J., Orviku, K., Palginõmm, V., Ratas, U., Rivis, R., Suursaar, Ü., Tõnisson, H. (2011). Natural development and human activities on Saaremaa Island (Estonia) in the context of climate change and integrated coastal zone management. In: Schernewski, G.; Hofstede, J.; Neumann, T. (eds.). *Global Change and Baltic Coastal Zones*, Coastal Research Library, Vol. 1117 – 134, Springer B.V.</u>

Lambeck, K., Chappell, J., (2001). Sea level change through the last glacial cycle. *Science,* Vol. 292, 679-686.

Larson, M. and Kraus, N.C. (1989). S-beach: numerical model simulating storm-induced beach change. Report 1: empirical foundation and model development. Technical Report cerc-89-9, USA Army Engineer Waterways experiment Station, Vicksburg USA.

Lee, H.-D. (1998). Economic value comparison between preservation and agricultural use of coastal wetlands. *Ocean Research* 20, 145-152.

Reed, D., and Yuill, B. (2009). Understanding Subsidence in Coastal Louisiana, Pontchartrain Institute for Environmental Sciences, *Journal of Coastal Research,* Special Issue 54, 23-36.

Lotze, H.K., Lenihan, H.S., Bourque, B.J., Bradbury, R.H., Cooke, R.G., Kay, M.C., Kidwell, S.M., Kirby, M.X., Peterson, C.H., Jackson, J.B.C., (2006). Depletion, Degradation, and Recovery Potential of Estuaries and Coastal Seas. *Science,* Vol. 312, 1806-1809.

Lowe, J.A., Gregory, J.M. & Flather, R.A. (2001). Changes in the occurrence of storm surges around the United Kingdom under a future climate scenario using dynamic storm surge model driven by the Hadley Centre climate models. *Climate Dynamics*, Vol. 18, 179–188.

Ludwig, W., Probst, J.-L., (1998). River sediment discharge to the oceans; present-day controls and global budgets. *American Journal of Science,* Vol. 298, 265-295.

Mahamood, H.R., Twigg, D.R. (1995). Statistical analysis of water table variations in Bharain. *Quarterly Journal of Engineering Geology & Hydrogeology* 28:s63–s64.

Malini, B.H., Rao, K.N., (2004). Coastal erosion and habitat loss along the Godavari delta front - a fallout of dam construction (?). *Current Science,* Vol. 87, 1232-1236.

Masselink, G. and Russell, P. (2006). Flow velocities, sediment transport and morphological change in the swash zone of two contrasting beaches. *Marine Geology*, Vol. 227, 227-240.

Meier, H.E.M., Broman, B. and Kjellström, E. (2004). Simulated sea level in past and future climates of the Baltic Sea. *Climate Research*, Vol. 27: 59–75.

Milliman, J.D., Meade, R.H., (1983). World-wide delivery of river sediment to the oceans. *The Journal of Geology,* Vol. 91, 1-21.

Milliman, J.D., Broadus, J.M., Gable, F. (1989). Environmental and Economic Implications of Rising Sea Level and Subsiding Deltas: The Nile and Bengal Examples, *Ambio,* Vol. 18(6), 340-345

Milliman, J.D., Syvitski, J.P.M., (1992). Geomorphic/tectonic control of sediment discharge to the ocean: the importance of small mountainous rivers. *The Journal of Geology,* Vol. 100, 525-544.

Mingozzi, T., Masciari, G., Paolillo, G., Pisani, B., Russo, M., Massolo, A., (2008). Discovery of a regular nesting area of loggerhead turtle Caretta caretta in southern Italy: a new perspective for national conservation.In: Hawksworth, D., Bull, A. (eds.), *Biodiversity and Conservation in Europe*. Springer Netherlands, 277-299.

Minshall, G.W. (1984). Aquatic insect-substratum relationships. In: V. H. Resh and D.M. Rosenberg, (eds.), *The ecology of aquatic insects*. VPraeger, New York.

Morton, R.A., Miller, T., and Moore, L. (2005). Historical shoreline changes along the US Gulf of Mexico: A summary of recent shoreline comparisons and analyses. *Journal of Coastal Research*, Vol. 21(4), 704–709.

Mulder, T., Savoye, B. and Syvitski, J.P.M. (1997). Numerical modelling of the sediment budget for a mid-sized gravity flow: the 1979 Nice turbidity current. *Sedimentology*, Vol. 44, 305-326

Murray, N.J., Clemens, R.S., Phinn, S.R., Possingham, H.P. and Fuller, R.A. (2014). Tracking the rapid loss of tidal wetlands in the Yellow Sea, *Frontiers in Ecology and the Environment*, Vol. 12(5), 267-272.

MVenW (Netherlands Ministerie van Verkeer en Waterstaat) (1983). *Polders of the World*, The Hague.

Nicholls, R.J., Wong, P.P., Burkett, V.R., Codignotto, S.L., Hay, J.E., Mclean, R.F., Ragoonaden, S., Woodroffe, C.D., (2007). Coastal systems and low-lying areas. In: Parry, M.L., Canziani, O.F., Palutikof, J.P., Van der Linden, P.J., Hanson, C.E. (eds) *Climate Change 2007: Impacts, adaptation and vulnerability. Contribution of the Working Group II to the Fourth Assessment report of the Intergovernmental Panel on Climate Change*, Cambridge University Press, Cambridge, pp315-356

NOAA (2013. (United States, National Oceanic and Atmostpheric Association). *National Coastal Population Report: Population Trends from 1970 to 2020*, 22p. State of the Coast Report Series. Available from http://stateofthecoast.noaa.gov.

Orth, R.J., Carruthers, T.J.B., Dennison, W.C., Duarte, C.M., Fourqurean, J.W., Heck, K.L., Hughes, A.R., Kendrick, G.A., Kenworthy, W.J., Olyarnik, S., Short, F.T., Waycott, M., Williams, S.L., (2006). A Global Crisis for Seagrass Ecosystems. *Bioscience*, Vol. 56, 987-996.

Parsons, G.R., Powell, M., (2001). Measuring the Cost of Beach Retreat. *Coastal Management*, Vol. 29, 91-103.

Phillips, M.R., Jones, A.L., (2006). Erosion and tourism infrastructure in the coastal zone: Problems, consequences and management. *Tourism Management*, Vol. 27, 517-524.

Post, J.C. and Lundin, C.G, eds. (1996). *Guidelines for integrated coastal zone management*. Environmentally Sustainable Development Studies and Monographs Series No. 9 (World Bank).

Pontee, N., 2013. Defining coastal squeeze: A discussion. *Ocean & Coastal Management*, Vol. 84, 204-207.

Rangel-Butrago, N., Anfuso, G., (2009). Assessment of coastal vulnerability in La Guajra Peninsula, Columbian Carribean Sea. *Journal of Coastal Research*, SI56, 792-796.

Roy, P.S., Williams, R.J., Jones, A.R., Yassini, I., Gibbs, P.J., Coates, B., West, R.J., Scanes, P.R., Hudson, J.P., Nichol, S., (2001). Structure and function of southeast Australian estuaries. *Estuarine, Coastal and Shelf Science*, Vol. 53, 351-384.

Schaffner, L.C., Diaz, R.J. Olsen, C.R. and Larsen, I.L. (1987). Faunal characteristics and sediment accumulation processes in the James River Estuary, Virginia. *Estuarine, Coastal, and Shelf Science*, Vol. 25, 211-226.

Schubel, J.R., (1977). Sediment and the quality of the Estuarine Environment: Some Observations. In: Y,H. Suffet (ed), Fate of Pollutants in the Air and Water Environments, Part 1, Vol. 8, John Wiley & Sons, New York, 399-423.

Shaffer, G.P., Day, J.W., Mack, S., Kemp, G.P., Heerden, I.V., Poirrier, M.A., Westphal, K.A., FitzGerald, D., Milanes, A., Morris, C.A., Bea, R., Penland, P.S., (2009). The MRGO Navigation Project: A Massive Human-Induced Environmental, Economic, and Storm Disaster. *Journal of Coastal Research*, Vol. 54, 206-224.

Steetzel, H., (1993). Cross-shore transport during storm surges. Doctoral Thesis, Delft University of Technology, Delft, the Netherlands.

Stive, M., de Schipper, M., Luijendijk, A., Ranasinghe, R., van Thiel de Vries, J., Aarninkhof, S., van Gelder-Maas, C., de Vries, S., Henriquez, M., Marx, S., (2013). The sand engine: a solution for vulnerable deltas in the 21st century? *7th International Conference on Coastal Dynamics*, Bordeaux, France.

Suursaar, Ü., Jaagus, J. and Kullas, T. (2006). Past and future changes in sea level near the Estonian coast in relation to changes in wind climate. *Boreal Environment. Research*, Vol. 11 (2): 123–142.

Suzuki, T., (2003). Economic and geographic backgrounds of land reclamation in Japanese ports. *Marine Pollution Bulletin*, Vol. 47(1-6), 226–9. doi:10.1016/S0025-326X(02)00405-8

Syvitski, J.P.M., (2008). Deltas at risk. *Sustainability Science*, Vol. 3, 23-32.

Syvitski, J.P.M., Alcott, J.M. (1995). RIVER3: Simulation of water and sediment river discharge from climate and drainage basin variables. *Computers and Geoscience*, Vol. 21(1), 89-151

Syvitski, J.P.M., Alcott, J.M. (1993). GRAIN2: Predictions of particle size seaward of river mouths. *Computers and Geoscience*, Vol. 19(3), 399-446

Syvitski, J.P.M., Milliman, John, D., (2007). Geology, Geography, and Humans Battle for Dominance over the Delivery of Fluvial Sediment to the Coastal Ocean, *The Journal of Geology*, Vol. 115(1), 1-19

Syvitski, J.P.M., Vörösmarty, C.J., Kettner, A.J., Green, P., (2005). Impact of humans on the flux of terrestrial sediment to the global coastal ocean. *Science*, Vol. 308, 376-380.

Tõnisson H., Suursaar Ü., Orviku K., Jaagus.J., Kont A., Willis D.A. and Rivis, R. (2011). Changes in coastal processes in relation to changes in large-scale atmospheric circulation, wave parameters and sea levels in Estonia. *Journal of Coastal Research*, SI64-1, 701 - 705.

Tõnisson, H., Suursaar, Ü., Suuroja, S., Ryabchuk, D., Orviku, K., Kont, A., Sergeev, Y., Rivis, R. (2012). Changes on coasts of western Estonia and Russian Gulf of Finland, caused by extreme storm Berit in November 2011. In: IEEE/OES Baltic International Symposium, May 8-11, 2012, Klaipeda, Lithuania, Proceedings: IEEE, 2012, 1 - 7.

United Nations, Department of Economic and Social Affairs, Population Division (2013). *World Population Prospects: The 2012 Revision*.

UNEP/GEMS (United Nations Environment Programme, Global Environment Monitoring System). (2008). Water Quality for Ecosystem and Human Health, 2nd Edition (ISBN 92-95039-51-7).

UNEP (United Nations Environment Programme) (2009). Sherman, K. and Hempel, G., (eds.), The UNEP Large Marine Ecosystem Report: A perspective on changing conditions in LMEs of the world's Regional Seas. UNEP Regional Seas Report and Studies No. 182. Nairobi, Kenya.

UNESCO (United Nations Educational, Social and Cultural Organization) (2010). Sandwatch: adapting to climate change and educating for sustainable development. Paris, 136 pp.

Van Rijn, L.C. (1998). *Principles of Coastal Morphology*. Aqua Publications, The Netherlands Van Rijn, L.C., (2009). Prediction of dune erosion due to storms. *Coastal Engineering*, Vol. 56, 441-457

Van Rijn, L.C. (2011). Coastal erosion and control. Ocean & Coastal Management, Vol. 54(12) (December 2011), 867-887

Van Thiel de Vries, J.S.M., Van de Graaff, J., Raubenheimer, B., Reiners, A.J.H.M., and Strive, M.J.F., (2006). Modeling inner surf hydrodynamics during storm surges. *Proceedings of the 30th International Conference*, San Diego, USA, 896-908

Wang, W., Liu, H., Li, Y., and Su, J. (2014). Development and management of land reclamation in China. *Ocean & Coastal Management*. In press. doi:10.1016/j.ocecoaman.2014.03.009

Wang, X., Chen, W., Zhang, L., Jin, D., & Lu, C. (2010). Estimating the ecosystem service losses from proposed land reclamation projects: A case study in Xiamen. *Ecological Economics*, Vol. 69(12), 2549–2556. doi:10.1016/j.ecolecon.2010.07.031

Wang, P., Kraus, N.C. and Davis, R.A. Jr., (1998). Total longshore sediment transport rate in the surf zone: Field measurements and empirical predictions. *Journal of Coastal Research*, Vol.14, 1, 269-282.

Warrick, J.A., George, D.A., Gelfenbaum, G., Ruggiero, P., Kaminsky, G.M., and Beirne, M., (2009). Beach morphology and change along the mixed grain-size delta of the dammed Elwha River, Washington. *Geomorphology*, Vol. 111, p. 136-148.

WASA Group (1998). Changing waves and storms in the northeast Atlantic? The WASA group, *Bulletin of the American Meteorological Society* 79 (1998) 741–760.

Waterman, R., Misdorp, R., Mol, A., (1998). Interactions between water and land in The Netherlands. *Journal of Coastal Conservation* 4, 115-126.

Waycott, M., Duarte, C.M., Carruthers, T.J.P., Orth, R.J., Dennison, W.C., Olyarnik, S., Calladine, A., Fourquerean, J. W., Heck, K. L., Hughes, A. R., Kendrick, G. A., Kenworthy, W. J., Short, F. T. and Williams, S. L. (2009). Accelerating loss of seagrasses across the globe threatens coastal ecosystems. *Proceedings of the National Academy of Sciences of the United States of America*. Vol. 106(30), 12377-12381.

Withers, P.J.A. and Lord, E.I., (2002). Agricultural nutrient inputs to rivers and groundwaters in the UK: policy, environmental management and research needs, *Science of The Total Environment*, Vols. 282–283, 9-24.

Young. R., Stancheva, M., Stanchev, H., Palazov, A., Peek, K., Coburn, A. and Griffith, A. (2014). Adapting to Sea Level Rise and Storms: Missed Opportunities and Continuing Development (case studies from USA and Bulgaria). Geophysical Research Abstracts, EGU General Assembly, Vol. 16, EGU 2014-3108.

Zuo, P., Li, Y., Liu, C.A., Zhao S.H. and Guan D.M. (2013). Coastal Wetlands of China: Changes from the 1970s to 2007 Based on a New Wetland Classification System. *Estuaries and Coasts*, Vol. 36, 390-400.

27 Tourism and Recreation

Group of Experts:
Alan Simcock (Lead Member) and Lorna Inniss (Co-Lead Member)

1 Introduction

Seaside holidays have a long history. They were popular for several hundred years (100 BCE – 400) among the ruling classes of the Roman Empire: these visited the coast of Campania, the Bay of Naples, Capri and Sicily for swimming, boating, recreational fishing and generally lounging about (Balsdon, 1969). But thereafter seaside holidaying largely fell out of fashion. In the mid-18th century, the leisured classes again began frequenting seaside resorts, largely as a result of the health benefits proclaimed by Dr. Richard Russell of Brighton, England, in 1755 (Russell, 1755). Seaside resorts such as Brighton and Weymouth developed in England, substantially helped by the royal patronage of Kings George III and George IV of Great Britain (Brandon, 1974). After the end of the Napoleonic wars, similar developments took place across Europe, for example at Putbus, on the island of Rügen in Germany (Lichtnau, 1996). The development of railway and steamship networks led both to the development of long-distance tourism for the wealthy, with the rich of northern Europe going to the French Riviera, and to more local mass tourism, with new seaside resorts growing up to serve the working classes of industrialised towns in all countries where industrialisation took place. In England, whole towns would close down for a "wakes week", and a large part of the population would move to seaside resorts to take a holiday: for example, in 1860 in north-west England, 23,000 travelled from the one town of Oldham alone for a week in the seaside resort of Blackpool (Walton, 1983). Between 1840 and 1969, the population of Blackpool (based almost entirely on the tourist industry) grew from 500 to 150,000 (Pevsner, 1969).

This relatively local mass tourism industry gave way to the modern mass tourist industry from the 1960s onwards. This was facilitated mainly by the introduction of, first, large passenger jet aircraft in the 1960s and, then, large-bodied jet aircraft in the 1970s, which (like the railways a century earlier) enabled relatively cheap mass transit over long distances that were not previously feasible (Sezgin et al., 2012).

2 Present nature and magnitude of tourism

International tourism has grown immensely over the last half century. In 1965, the number of international tourist arrivals worldwide was estimated at 112.9 million. Thirty-five years later, in 2000, this figure had grown to 687.3 million – an increase of 509 per cent, equivalent to an average annual compound growth rate of 5.3 per cent (WTO, 2014). A significant feature of these statistics is the increase in both absolute numbers and as a proportion of world tourist traffic of Asia and the Pacific: in absolute terms, the numbers of international tourist arrivals in that region has more than doubled, and the share of world traffic has increased by 6 percentage points, from 17 per cent to 23 per cent of the global total. Likewise, tourist numbers in Africa have also risen both in absolute terms and as a proportion, although from a much lower base. Tourist arrivals in Sub-Saharan Africa rose by over a quarter between 2007 and 2012, from 3.5 percent to 5 percent of the global total. Nevertheless, Europe continues to dominate international tourism, with 51 per cent of all international tourist arrivals in 2012. Fuller details are in the Appendix to this chapter.

When the origins of the tourists represented by these arrivals are considered, the pattern shown in Figure 27.1 is not markedly different: European tourists dominate the departures as much as the arrivals; Asian and Pacific tourism is growing strongly, and African tourism is also growing significantly, although from a low base. This is not surprising since most tourists tend to visit countries in their own region (Orams, 2003). It is for small States that the growth in long-distance tourism is most important: taking, for example, the 25 States and territories that cooperate in the Caribbean Tourist Organization, 35 per cent of their 24 million arrivals[1] in 2012 were from the United States of America, 14 per cent from Europe and 12 per cent from Canada, meaning that at least 61 per cent of arrivals were from outside their immediate area (CTO, 2013).

Although the figures for international tourist arrivals are the standard measure for looking at the tourism industry, they are somewhat misleading. They relate to international tourism. In global regions where there are many States (as, for example, in Europe), journeys will count as international when, in other parts of the world, they would be classed as domestic. This means that, for example, that a 1,400 km journey from the Ruhr, Germany to the Costa Brava, Spain in Europe will count towards international tourist statistics, while a 3,000 km journey from Beijing to Hainan Island in China will not. As a measure of global tourism, the statistics for international tourism will therefore exaggerate the proportion of world tourist activity in those global regions where there are relatively numerous States.

Statistics on total tourism (both international and domestic) are difficult to produce, because there is not the opportunity to capture information that arises when tourists cross national boundaries. What is clear, however, is that the numbers of domestic tourists are substantially more in large States than those of international tourists. In Brazil, it has been estimated that, in 2011, 49 million of the inhabitants made one or more visits within the country for the purpose of tourism (FIPE, 2012). This compares with 5.4 million tourists arriving from outside the country (AET, 2012). In China, in 2013, domestic tourism in mainland China was estimated to involve 3,260 million domestic tourists, compared with 129 million from Hong Kong, China, Macau, China and Taiwan Province of China, and 29 million from the rest of the world (NBSC, 2014). In the United States, domestic tourism accounted for 1,600 million person-trips for leisure purposes in 2013, compared with international arrivals of 70 million (US Travel, 2014). On the other hand, in smaller States (particularly Small Island Developing States), international tourist arrivals will be more closely aligned with total levels of tourism.

Even when allowances can be made for domestic tourism, the available statistics tend to be too broad-brush to allow a clear analysis of the im-

---

1   This figure differs from that given for the Caribbean in Table 27.1 because the Caribbean Tourist Organization includes Belize and Cancun, Mexico in Central America and Guyana in South America.

Chapter 27 — Tourism and Recreation

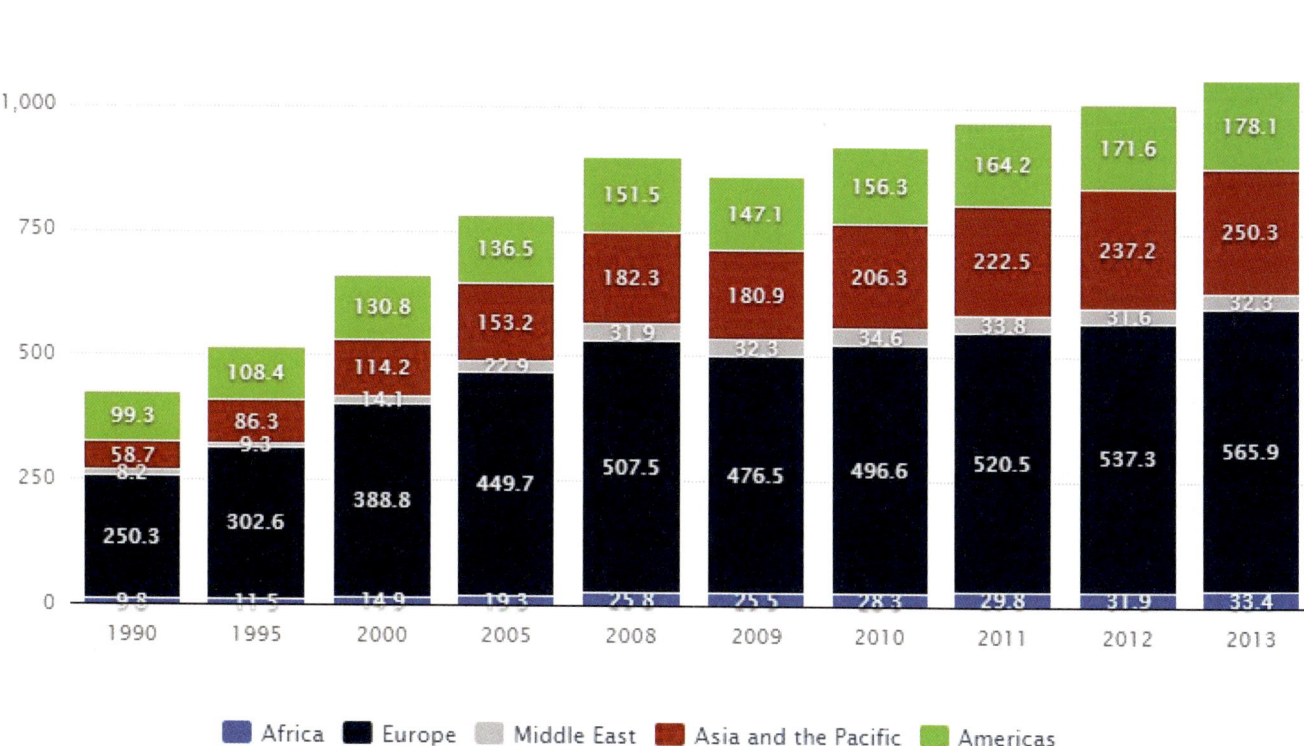

Figure 27.1 | Origins of tourists by WTO region. Source: WTO, 2014.

pact of tourism on the ocean, since they include tourism of all kinds. The statistics quoted above give the total number of tourists, irrespective of whether they are visiting a country for a beach holiday, to view ancient monuments or to climb mountains. Again, in smaller coastal States (particularly Small Island Developing States), the total number of tourists will be close to the number of tourists who will have an impact on the ocean, since there is only the coastal zone to accommodate them. But the available global figures are not sufficiently differentiated to allow conclusions focused precisely on coastal tourism.

In Europe, efforts have been made to determine the proportion of tourists that are staying in the coastal zone. For 27 countries of the European Union, there were in total in 2009 about 28.1 million "bed-places" (hotels, hostels, camp sites, etc.). Of these, about 60 per cent were in coastal regions (coastal regions being defined as the 447 third-level statistical units (34 per cent out of a total of 1,294) that have either a coastline or more than half their population within 50 km of the sea) (Eurostat, 2014a). Looking at use, rather than supply, surveys showed that, in 2012, for the 28 European Environment Agency countries for which data are available, 599 million tourist person/nights were spent in coastal regions out of the total of 1,416 million tourist person/nights spent in those countries – that is, 42 per cent of all tourism in those countries was in coastal regions, which (as said above) represent only 34 per cent of the total number of regions (Eurostat, 2014b – extracted in Appendix)[2]. Surveys of the European population have also confirmed

a strong wish for seaside holidays: 46 per cent of people in the European Union give a beach holiday as their reason for holiday travel; to this must be added the proportion of the 14 per cent giving a sporting holiday, since this covers scuba diving among other activities (EU, 2014). This factor is made more important by the high proportion of international holiday travel originating in Europe. In Brazil, it was estimated that, in 2011, 78 per cent of domestic tourism destinations were in the coastal Federal Units (although, of course, several of these stretch far inland) (FIPE, 2012). In the United States, surveys of the reasons for domestic travel in 2013 showed that visiting beaches was one of the five main reasons for travel, after visiting relatives, shopping, visiting friends and fine dining (US Travel, 2014). Also in the United States, in 2008, it was noted that Miami Beach attracted more than twice as many visitors than the Grand Canyon, Yellowstone National Park and Yosemite National Park combined, and that California beaches attract more visitors than all 388 National Park Service properties combined (Houston, 2008). Coastal tourism therefore appears to represent a dominant form of tourism generally.

The statistics quoted above do not include Antarctica. Since 1966, a trade has developed both by cruise ships and (to a lesser extent) for airborne tourists. This has grown steadily (with a dip in the 2010/11 season) over the last decade from a total of 27,537 in the 2003/04 season to 37,405 in the 2013/14 season. In 2013/04, 74 per cent of the tourists landed on Antarctica. Four-fifths of the tourists come from the USA (30.1 per cent), Australia (12.6 per cent), China (11.3 per cent), Ger-

---

[2] The difference between the 60 per cent for bed-spaces and the 42 per cent for overnight stays in coastal regions is probably due to the fact that much coastal tourism in Europe is highly seasonal, with many bed-spaces being unoccupied during the winter months.

many (8.4 per cent), the United Kingdom (7.3 per cent), Canada (4.4 per cent), France (3.4 per cent) and Switzerland (2.4 per cent) (IAATO, 2014). There are obvious concerns about the potential impact (from, among other things, waste, accidents, accidental introduction of organisms and exhausts and oil spills), although the authorities and tour operators attempt to minimise these.

Land-based tourism in the Arctic is included in the statistics for the different continents above. In addition, there is also a very significant component of cruise ships which do not land their passengers, who therefore are not counted in the statistics. The limited statistics available on Arctic land-based tourism suggest that it is growing quite quickly, but is still only counted in the 100,000s. Cruising is probably growing more quickly, with Arctic seas becoming ice-free in parts during the summer (Lück et al., 2010). The challenges posed for the marine environment are similar to those for the Antarctic.

In spite of the limitations of the available statistics, it is clear that the total amount of tourism has generally been increasing fairly steadily for the last 40 years (with occasional set-backs or slowing down in times of global recession), that the domestic component of tourism is very important in large countries, that international tourism is important in small States, and that coastal tourism is a major component of tourism, if not everywhere the predominant one. Particularly noteworthy is the way in which international tourism is increasing in Asia and the Pacific, both in absolute terms and as a proportion of world tourism, with the implication that pressures from tourism are becoming of significantly more concern in those regions.

3. Socioeconomic aspects of the human activities

Movements of people on the scale of the tourism described above require substantial inputs in transportation, accommodation, feeding and recreation. As a study of foreign direct investment in tourism by the United Nations Conference on Trade and Development (UNCTAD) puts it: "A significant part of tourism's development potential stems from the fact that it links together a series of cross-cutting activities involving the provision of goods and services such as accommodation, transport, entertainment, construction, and agricultural and fisheries production" (UNCTAD, 2007). Tourism has therefore become a major economic activity. (Since it is often difficult to distinguish travel for business purposes from travel for recreational purposes, it is often necessary to describe this economic activity as "tourism and travel"; in the rest of this section, tourism must be understood in this wider sense.)

Even though international tourism is only a part of the picture, it is worthwhile examining the statistics on expenditure from international tourism to see the situation in the different regions of the world. The World Bank World Development Indicator 6.14 (Inbound tourism expenditure) gives details of inbound international tourism expenditure for 2006 and 2012 for 114 coastal States and territories. Table 27.1 shows an analysis of this data by global regions, showing also the proportion that the inbound tourism expenditure forms of total exports. Fuller details are in the appendix to this chapter.

This shows that, on the basis of this sample of 114 States and territories, tourism and travel accounts for about 6 per cent of total exports globally. However, some regions of the world (particularly the Caribbean) are economically very dependent on international tourism in terms of

Table 27.1 | Inbound international tourism expenditure by global region, ranked by regional average percentage of total exports. Source: Compiled from World Bank, 2014.

Region (and number of States and territories covered)	Inbound tourism expenditure (million USD$)	Regional average % of total exports	Inbound tourism expenditure (million USD$)	Regional average % of total exports	State or territory with highest % of total exports in region in 2012	State or territory with lowest % of total exports in region in 2012
	2006	2006	2012	2012		
Caribbean Islands (11)	10,467	40.3	12,008	44.2	Aruba (Netherlands) (65.7%)	Haïti (16.3%)
Oceania (7)	26,453	13.7	41,108	11.3	Fiji (61.1%)	Solomon Islands (10.5%)
Sub-Saharan Africa (16)	14,981	7.8	20,740	5.9	Cabo Verde (60.6%)	Democratic Republic of the Congo (0.1%)
Western and Central Europe* (18)	378,794	6.9	440,661	6.1	Cyprus (27.8%)	Germany (3.0%)
Central and South America (17)	22,245	4.9	36,606	4.5	Belize (28.9%)	Brazil (2.4%)
North America (3)	163,599	7.4	234,108	7.4	USA (9.0%)	Mexico (3.4%)
Middle East & North Africa (12)	38,092	6.7	53,889	5.3	Jordan (33.0%)	Algeria (0.4%)
East Asia (12)	132,024	4.0	273,708	4.7	Macau, China (94.2%)	Japan (1.8%)
South Asia (5)	11,882	5.0	23,093	4.4	Maldives (79.9%)	Bangladesh (0.4%)
Eastern Europe (13)	17,488	4.0	28,624	3.7	Albania (45.9%)	Russian Federation (3.0%)

* including Cyprus and Turkey

Table 27.2 | Estimated contribution of tourism to GDP and employment 2013, ranked by total contribution to GDP. Source: Compiled from WTTC, 2014.

Region	Direct contribution to GDP US$ million	% share of total GDP	Total contribution to GDP US$ million, including the multiplier effect	% share of total GDP	% share of direct employment	% share of total employment, including multiplier effect
World	2,155,500	2.9	6,990,540	9.5	3.3	8.9
Caribbean	15,299	4.3	48,994	13.9	3.6	11.3
South East Asia	121,166	5.0	294,376	12.3	3.7	9.7
North Africa	34,951	5.6	74,998	12.1	5.2	11.6
Oceania	49,606	2.8	188,018	10.8	4.4	12.4
European Union (27)	552,148	3.2	1,512,360	9.0	4.0	9.9
Central and South America	142,476	3.2	387,609	8.8	2.8	7.9
North East Asia	431,742	2.6	1,389,330	8.5	2.9	8.2
North America	544,342	2.7	1,665,850	8.3	4.2	10.4
Remainder of Europe and Central Asia	111,596	2.3	362,120	7.2	NA	NA
Sub-Saharan Africa	36,623	2.6	95,713	6.9	2.3	5.8
Middle East	63,988	2.4	167,598	6.4	2.5	6.4
South Asia	NA	NA	NA	NA	NA	NA

foreign-currency earnings. It also shows that most small coastal States and territories are more dependent on such earnings than larger countries with more diversified and larger industries or resources of raw materials – although it is not unimportant in countries such as Australia or the United States.

Expenditure by international tourists, however, is not the only important aspect of coastal tourism. As shown above, domestic tourism is also very important, particularly in larger States. Although there are no global estimates of the total expenditure in coastal regions by domestic and foreign tourists combined, it is helpful to look at estimates of this total expenditure for countries as a whole, given the evidence (see above) that coastal tourism can be nearly as much as a half or more of total tourism.

In assessing the importance of an economic activity such as tourism for a country, it is important to consider not only the direct expenditure on that activity, but also the "indirect" expenditure on that activity and the resulting "induced" economic activity. The indirect expenditure is that which those active in the economic activity have to spend to buy assets and supplies that they need to carry it out. In the case of tourism, this includes the construction of hotels and other necessary buildings and the purchase of food, power and services, etc. The induced economic activity (sometimes called the multiplier effect) is that which is generated by those supported by the economic activity in question. In the case of tourism, this includes the spending of those who are directly or indirectly employed in tourism. The World Travel and Tourism Council (an industry body) has commissioned research to estimate the scale of the contribution of the tourism sector (in the wider sense explained above) to national economies. Table 27.2 summarizes the conclusions of this research (unlike Table 27.1, information on land-locked States cannot be separated out from that for coastal States). The Table also shows estimates of the proportion of employment in the different regions supported directly and in total.

These statistics show that tourism is a significant component of many economies. As a result, many international organizations promote tourism development as a valuable way forward in improving national economies. However, three important factors need to be borne in mind in evaluating its importance:

First, the direct employment provided by tourism in many countries has a very large proportion of female workers. Studies by the World Travel and Tourism Council showed that in four (Australia, France, Germany and South Africa) of the five countries studied, the proportion of women employed in tourism is over 60 per cent of the work force. The exception was Turkey, where the proportion was 27 per cent, no doubt as a result of cultural differences (WTTC, 2014). It has been noted that this predominance of female workers makes tourism significant in giving economic status to women (Wilson, 2008).

Secondly, there will be a "leakage" from the earnings generated by a country's tourism activity to the rest of the world economy. This leakage will have four main components:

(a) Goods and services (such as wine or entertainers) must be purchased from abroad to meet demands from tourists (especially international tourists) that cannot be met from indigenous sources;
(b) Expatriate staff (especially managers) will remit at least part of their earnings to their home countries);
(c) International hotel companies will remit earnings to their non-resident owners. The terms on which they are able to do so, and the taxation regime to which any such earnings are subject will, in many cases, be the subject of negotiations between the local authorities

and the hotel companies, especially where a large investment is concerned. Small States may be at a disadvantage when negotiating such terms with large international companies, especially where there is a credible threat of choosing a site in another country as an alternative; and

(d) Commissions will be payable to tourism organizers for directing tourists to tourist establishments.

This "leakage" is usually a higher proportion of earnings in developing countries than in developed countries, although it is not easy to quantify (Yu, 2012).

Thirdly, there is a risk that the employment in tourism will be relatively low-grade and/or seasonal. The risk of the employment being low-grade comes largely from the fact that tourists frequently expect routine tasks (such as cooking, cleaning and making beds) to be done for them, although they would commonly perform these tasks themselves in their own homes. In some areas, when managerial staff are expatriates, the grade of work for the local population can be even lower. The extent to which this is the case varies according to the quality of local trained staff that is available. An important factor is therefore the extent to which related training and education are provided: for example, the University of the West Indies has undertaken specific programmes for this purpose, including a specialist training centre in the Bahamas (UWI, 2002; UWI, 2014).

Tourism has a further socioeconomic significance, going beyond its macroeconomic importance and the effects on those involved in providing services. Where tourist resorts are created, the circumstances of those already living in the area are affected – sometimes adversely. For example, the literature notes cases where (a) the local residents lost access to beaches that they have previously enjoyed, even where the beaches are public property, because hotels or other tourist developments block access to those beaches; (b) local residents lost access to other areas that they have enjoyed for recreation because they are taken for resort building; (c) local residents had their property expropriated without compensation for the erection of hotels; (d) large increases in land values as a result of the erection of tourist establishments effectively prevented local residents from acquiring land; (e) local residents have seen land to which they attach religious or cultural significance diverted to tourism use (Bartolo et al., 2008; Cater, 1995; Wilson, 2008). On the other hand, cases (mentioned in the same literature) are also found where careful planning and collaboration with the local people produced "win-win" situations, in which successful tourist resorts have been created and the local people have benefited substantially. Information is lacking, and would almost certainly be impossible to collect, to make an assessment of the balance of adverse and beneficial effects, even in one country or region, let alone globally.

4 Major impacts on the marine environment

4.1 Coastal built environment

Coastal tourism needs coastal infrastructure. In the first place, transport is needed to get the tourists to the coast. This requires airports, roads, car-parks and (in some cases) railways. All this tends to change the coastal landscape. In addition, tourism demands accommodation. Hotels and restaurants are therefore built in large numbers, with many completely new resorts being developed. These commonly include marine promenades, bathing places and other hard landscape features, which completely change the shoreline (Davenport et al., 2006).

Globally, there are few statistics on the extent to which coastal areas have been built up to meet tourism needs. Many studies of specific areas are available, most using satellite-based photographs or sensing, but a comprehensive overview is lacking. Particular efforts, however, have been made in Europe, making a more general overview possible. Studies by the European Environment Agency have shown that, for the coastal zone up to 1 km from the shoreline, more than 10 per cent was built up in Bulgaria, Germany, Latvia, Lithuania, the Netherlands, Poland, Portugal and Romania, more than 20 per cent in France, Italy, Spain, more than 30 per cent in Slovenia and nearly 50 per cent in Belgium (the last two countries having very short coastlines). Information was not available for the United Kingdom of Great Britain and Northern Ireland. The proportion of the area close to the shoreline covered with urban development has also been growing rapidly: between 1990 and 2000, nearly 8 per cent of the area within 10 km of the shoreline in the States mentioned (together with Denmark, Estonia, Finland, Greece and Ireland) was changed from agricultural or natural uses to artificial land cover (EEA, 2006). Some regional studies in the United States have shown a similar picture: more than 10 per cent of the estuarine coastlines of Delaware, Maryland, Virginia, and North Carolina now have artificial shorelines (Currin, 2013).

One significant factor is the extent to which built development for tourism is linked to more general urban development. In many parts of the world (for example, Cyprus, Rousillon in France, southern Spain, Costa Rica, the Algarve in Portugal, and California and Florida in the United States), tourist development is linked with the development of residential property. This has often been targeted at the retirement market from colder industrialised areas. The tourist market and the retirement market overlap, and support a variety of land-use demands – in particular golf courses, which create specific pressures from high levels of fertilizer, pesticide and water use and the consequent run-off (see chapter 20) (Honey et al., 2007).

This change from agricultural or natural uses to hard, artificial land cover has been happening wherever coastal tourism has been developed. The fundamental (and normally irreversible) changes that it brings about have significant implications for coastal ecosystems. These changes are most obvious for species that use both land and sea, such as seabirds, marine reptiles and some marine mammals, and for habitats such as mangroves and salt marshes which combine both land and sea (see Chapters 48 and 49). The changes usually introduce a barrier of artificial land cover between the sea and the natural or agricultural land cover in the hinterland, thus preventing animals moving between one and the other, and affecting the plant cover in the marginal zone. The changes also usually introduce night-time illumination, which also affects the way in which animals (particularly nocturnal animals such as bats) can use the terrain.

The impact of these changes is most obvious for sea turtles, which need to come ashore onto sandy beaches to lay their eggs, which are usually deposited near the vegetation fringe at the top of the beach. Such areas are obviously most affected by coastal development. In the Mediterranean, at the beginning of the 19th century, there were significant breeding populations of green turtles (*Chelonia mydas*), loggerhead turtles (*Caretta caretta*) and leather-back turtles (*Dermochelys coriacea*). Because of the transformation of so many Mediterranean sandy beaches into tourist resorts, these breeding areas are now reduced to Cyprus (for the green turtle) and small areas of Greece and Turkey (for loggerhead turtles); breeding by leather-back turtles is now virtually unknown, except for occasional reports from Israel and Syria (Davenport, 1998, and see chapter 39). Night-time lighting of tourist developments is also a significant problem at turtle-hatching time: turtle hatchlings, which emerge at night, are programmed to make for the lightest part of the horizon, which in natural conditions will be the sea; they are confused by street lighting and fail to reach the sea (Tuxbury et al., 2005; Arianoutsou, 1988).

However, the change from natural to artificial shorelines also affects purely marine species. The difference between a naturally sloping beach and a more vertical seawall produces a significantly different environment. There is growing evidence that the biota living on breakwaters, seawall, groynes and similar structures, and the fish assemblages associated with them, differ from those on natural shorelines. Even where the natural shoreline is rocky, the replacement artificial shoreline will have different effects; for example, replacing natural rock with concrete may provide a different acid/alkali balance as a result of leaching (Bulleri et al., 2010).

The introduction of artificial hard coastal constructions can also affect the longshore movement of sediments, changing the patterns of sand transport and sedimentation. This can result in changes to beaches. The exact pattern will depend on local circumstances: for example, at Nouakchott, Mauritania, the construction of port facilities is resulting in erosion of dune systems, with increased risks of sea flooding of coastal settlements, reduction of beach area and threats of siltation of the harbour (Elmoustaphat, 2007). Even though sophisticated computer modelling of the possible effects of coastal constructions can be used to reduce the risks, a study of the Herzliya marina in Israel has shown that the effects in practice diverged extensively from those predicted by a meticulous prior environmental planning exercise (Klein et al., 2001).

4.2   Waste and sewage

The influx of tourists to coastal resorts inevitably results in problems in the treatment and disposal of the large amounts of solid waste and urban waste water (sewage) that result. Inadequate handling of solid waste often results in marine debris. Indeed, litter dropped on beaches by tourists is itself a significant source of marine debris (see chapter 24 for both these problems). As described in chapter 20, achieving adequate disposal of sewage is a problem in many areas. Not only do the nutrients contained in this sewage add to the enhanced levels of nutrients in the seawater, leading to eutrophication problems in many areas, but inadequate management can also easily result in health risks to tourists bathing or boating in the sea, a problem that is more directly linked to the success of tourist resorts. Such health hazards for tourists can be self-defeating in attracting business in a highly competitive market.

A special case of these problems of waste and sewage is presented by cruise ships, particularly in the Caribbean (as described in chapter 17, one of the major cruising markets), where large cruise ships put into relatively small ports which have limited facilities for handling waste and sewage. Islands with populations in the range of 20,000 to 100,000 are faced with handling the waste and/or sewage from ships with combined passengers and crew of up to 7,000 people. The resulting difficulties were the main reason why it took so long for the Caribbean to be declared a Special Area for garbage under the International Convention for the Prevention of Marine Pollution from Ships (MARPOL), because adequate reception facilities were a precondition of such a designation. A World Bank project in the member States of the Organization of Eastern Caribbean States (OECS), costing about 50 million United States dollars, enabled much progress to be made in tackling both problems. However, problems remain. During the implementation of this project, cruise lines are reported by the World Bank to have warned the individual OECS governments that any island that imposed waste disposal charges would lose cruise tourism because the cruise lines would merely make a substitute call at ports in less demanding States. The OECS managed to agree a common levy on cruise passengers entering ports, but the World Bank has reported problems in ensuring that these resources are devoted to the waste and sewage management tasks (World Bank, 2003; ECLAC, 2005).

4.3   Beach and shore usage

For many people, the main point of a seaside holiday is the use of a sandy beach for a mixture of sun-bathing, lounging, swimming and surfing. In general, such usage does not require any change to the natural state of the beach. In many places, however, steps have been taken to try to improve the beach. Often this has taken the form of erecting groynes (wood or stone structures perpendicular to the shore) to try to prevent longshore movement of sand and thus maintain a more sandy beach. "Beach feeding" has also frequently taken place, involving the dredging of sand from further out to sea and its deposition on the beach. These efforts are one form of human intervention in the land/sea interaction, which is considered more generally in Chapter 26. In addition, more recently, attempts have been made to use the creation of artificial reefs to improve the size of surf breakers, and thus the attractiveness of beaches to surfers. The first attempt of this kind was the Cables Reef at Mossman, West Australia. This involved dumping large amounts of natural rock to build up the existing reef. The local municipality reports that it has universally been judged a success in improving the surfing (Mossman, 2003). Later attempts, mainly using a technology based on large sand-filled containers, have been less successful. At El Segundo, California, USA, an artificial reef was created in 2000 but did not achieve its purpose, and its removal began in 2008 (CCT, 2008). At Bournemouth, in the United Kingdom, an artificial surfing reef was created in 2009, but the structure has failed and hoped-for economic benefits have not materialized. (Rendle et al., 2012; Rendle, 2014; Bailey, 2012). Other examples at the Bay of Plenty in New Zealand and at Tuvalom, in India, have also not produced the hoped-for improvements in surfing (Mull, 2014).

Protection of bathers from attacks by sharks has been seen as necessary in some parts of the world, notably Australia and the United States. This has had some adverse effects on local populations of rays, dolphins and turtles, because they have become entangled in this netting (Davenport, 2006).

The need to keep beaches attractive to tourists often leads to the local beach managers (either the communal authority or a hotel which has a concession on the beach) to clean up the debris left by the beach users. However, such clean-up operations usually also include the removal of the natural deposits along the high-tide line of seaweed and other marine material (including dead seabirds and other biota). Such removal of natural material has been shown to reduce substantially the biodiversity of sandy shore shorelines, especially seabirds (Llewellyn et al., 1996; Mann, 2000). Mediterranean and Baltic beaches used substantially by tourists have been shown to have lower densities and diversity of marine invertebrates than neighbouring beaches with less use of this kind. This has been attributed to the combination of cleaning and trampling pressure from the tourists (Gheskiere et al., 2005). Nevertheless, such beach cleaning may often be necessary to maintain tourism, especially where large amounts of seaweed are brought up onto beaches by the sea. A special problem of this kind has recently emerged in the Eastern Caribbean: since 2011, high numbers of large mats of *Sargassum* species have been washed up on beaches. The same problem, which appears to be emerging from north equatorial recirculation region, has been encountered on the island of Fernando de Noronha, Brazil, and in Sierra Leone (Johnson et al.,2012).

Similar usage impacts can be found on rocky shores. Here even a relatively low number of humans walking over rocks where seaweed and barnacles are found can reduce the coverage of these biota significantly, and heavy (more than 200 visitors a day during the tourist season) usage can take more than a year to recover. Such effects have been shown for New Zealand (Schiel et al., 1999), Italy (Milazzo et al., 2002) and the United Kingdom (Pinn et al., 2005).

Dunes are also vulnerable to heavy usage by tourists, since the footfall can disturb the vegetation cover on which the dunes' stability relies. Because of the importance of dunes in coastal protection, this has been studied extensively in Europe, where it has been shown that 200 passages a day over a dune was enough to reduce vegetation cover by 50 per cent (Hylgaard et al., 1981). Fixed dunes also have a relatively low resilience to damage from vegetation removal (Lemauviel et al., 2003).

Use of the near shore for anchoring ships can also result in damage to the seabed. This is particularly important for shores where the immediate underwater habitat is coral reefs or seagrass beds. Damage has been noted from small pleasure vessels, which often anchor over coral areas so that those on board can dive to see the corals. But more serious damage is caused by cruise ships anchoring in such areas. Destruction of corals of up to 300 square metres has been observed from one anchoring of one cruise ship. Recovery from such damage can take a long time (Allen, 1992).

4.4 Enjoyment of wildlife

Over the past few decades, coastal tourism has come to include creating opportunities for the public to enjoy the local wildlife. This has generated a large number of businesses serving tourists. Six major categories of business are involved, though others do occur. Five are non-consumptive (general marine diving, viewing corals, watching seabirds, watching whales and other marine mammals and watching sharks), and one (recreational fishing and hunting) has a direct impact on the marine biota.

4.5 General marine diving

All around the world, tourists (both domestic and international) engage in diving (usually using self-contained underwater breathing apparatus (SCUBA)). The attractions of this activity are both the sense of freedom conveyed by being in the water and the interesting rock and coral formations and biota that can be seen. The scale of this tourist activity can be judged from the activities of the Professional Association of Diving Instructors, a global organization of experts training people in scuba diving: between 2000 and 2013, the number of firms in its membership grew by 24 per cent to 6,197, and the number of individual trainers by 26 per cent to 135,615. The annual number of people trained in this period has been around 900,000 (PADI, 2014). At low levels of usage, diving sites do not appear to be adversely affected by recreational diving. There are, however, thresholds above which both the divers' experience is affected by over-crowding and the marine environment is adversely affected (by physical damage and disturbance of fish and other biota). The problem lies in establishing where those thresholds lie, particularly in the absence of long-term monitoring (Davis, 1996).

To enhance the experience of recreational divers, artificial reefs have been created in several locations, including, Australia, Canada, Japan, New Zealand, the Cayman Islands (United Kingdom) and the United States. Many of these used former naval ships as the basis of the new reef. These ships were cleaned of potentially polluting material and then sunk at the desired location. Studies have shown that these have brought substantial economic benefits to the areas from increased visits by tourists for the experience of diving around them (SWEC, 2003; Morgan et al., 2009).

4.6 Coral viewing

The sheer splendour and variety of tropical and sub-tropical coral reefs has made them a very popular tourist attraction: people are prepared to travel great distances and pay substantial costs to see coral reefs in their native state. This has therefore generated a large component of the tourist trade. The scale of this component can be judged from what is said on tourism and recreation in Chapter 43.

The specific pressures on corals generated by such viewing can be seen from an assessment of the tourism pressures on the Great Barrier Reef of Australia. These cover (in addition to what is said about anchor damage above):

(a) Damage (particularly to branching corals) by untrained scuba divers – damage by qualified scuba divers is not seen as a problem;
(b) Damage by trampling at landing points where large concentrations of tourists were landed from boats to walk on the reef – more generally, even in heavily trafficked areas, damage by tourists walking on the reef was not seen as a problem;
(c) Some reduction in growth caused by shading from pontoons moored to provide facilities (lecture theatres, restaurants, etc.) for tourists – this could be avoided by careful choice of mooring sites, so that the pontoons were moored over sand rather than corals. Likewise, problems from the anchor points could be avoided by correct design and choice of site;
(d) Fish feeding by tourists: inappropriate types of food can adversely affect the health of fish, and frequent feeding of large volumes of food could promote unduly large and aggressive fish aggregations. Again, such problems can be avoided by proper management;
(e) Shell collecting: this was not seen as a major problem, provided that operators gave guidance to tourists;
(f) Glass-bottomed boats and semi-submersible vessels were seen as potentially capable of causing damage through collisions with corals. However, a survey of one heavily used site could find no overt damage caused by operations of semi-submersible vessels over a five-year period.

The conclusion therefore was that coral viewing, even on a major and locally intensive scale, was compatible with sustaining the reef in a good condition, provided that appropriate management steps were taken (Dinesen et al., 1995). Other studies, however, suggest that: (1) diving can, through abrasion, make large massive coral communities more susceptible to other pressures (Hawkins et al., 1999); (2) damage is virtually impossible to avoid (based on studies in St Lucia and the Cayman Islands); (3) substantial damage can still occur even when restrictive and highly-policed management is in place and (4) camera-users do more damage than divers not undertaking photography (Rouphael et al., 2001; Tratalos et al., 2001; Barker et al., 2004). In some places (for example, Eilat in Israel), artificial reefs have been created to reduce pressure on natural coral reefs (Wilhelmson, 1998).

4.7 Bird-watching

There are no global statistics to show the extent of coastal tourism based on bird-watching (Balmforth, 2009). This is largely due to two facts. First, it is not easy to identify bird-watching tourism as a distinct activity: many people may spend a day or two bird-watching out of a longer holiday, although many others will go to destinations where they intend to spend much of their time bird-watching. The latter is particularly the case for destinations where the main attraction is the presence of birds, particularly during migration seasons. Secondly, the resources demanded for bird-watching are not elaborate. Although some sites may provide hides to enable closer observation, much bird-watching is done in the open with no more equipment than binoculars. The resource demand is therefore not easy to capture. Nevertheless, bird-watching is a substantial and growing part of the tourism market. As a major source of tourists of this type, the United States' market is worth noting. In the 2012 National Survey on Recreation and the Environment (NSRE, 2012), it is estimated that 19.9 million people in the United States took trips away from home to watch birds, although not all of these will have visited coastal areas. This group is reported to have both higher educational qualifications and higher incomes than the national average. However, the previous rapid increase in the numbers watching birds (a 332 per cent increase between 1983 and 2002) has slowed down or stopped. Some coastal resorts in the United States can rely very substantially on bird-watching: in 1991, Cape May on the Atlantic Coast of New Jersey was estimated to be attracting 100,000 bird-watching visitors a year, who were spending about 10 million dollars a year (Kerlinger et al., 1991). The Caribbean Tourist Organization has accepted an estimate that, worldwide, three million tourist arrivals a year may be primarily for the purpose of bird-watching (CTO, 2014).

The adverse impact of bird-watching arises from the interaction of the tourist and bird populations. On land, tourists entering nesting areas during the breeding season can disturb breeding birds, potentially leading to the abandonment of nests. On water, boats carrying bird-watchers can disrupt seabird feeding. This is particularly significant at staging-post sites where migrant birds congregate, since the energy balance of migrating birds is often delicate. Such sites are particularly attractive to bird-watchers because of the numbers of birds (often of many different species) passing through them. On both land and water, bird-watchers can cause birds to flush into the air, making them use energy which (particularly during migration) can be in a tight balance. Careful management of bird-watching sites can minimize this kind of problem (Green et al., 2010; Parsons et al., 2006). One survey of the literature has, however, commented on the lack of research in this field (Steven et al., 2014).

Table 27.3 | Whale-watching numbers and expenditure. Source: Compiled from IFAW, 2009.

Region	Whale-watchers 1998	Whale-watchers 2008	Average annual growth rate 1998 - 2008	Number of countries 1998	Number of countries 2008	Jobs supported	Direct expenditure* USD millions	Total expenditure* USD millions
Africa and Middle East	1,552,250	1,361,330	-1.3%	13	22	1,065	31.7	163.5
Europe	418,332	828,115	7.1%	18	22	794	32.3	97.6
Asia	215,465	1,055,781	17.2%	13	20	2,191	21.6	65.9
Oceania, Pacific Islands and Antarctica	976,063	2,477,200	9.8%	12	17	1,868	117.2	327.9
North America	5,500,654	6,256,277	1.3%	4	4	6,278	566.2	1,192.6
Central America and Caribbean	90,720	301,616	12.8%	19	23	393	19.5	53.8
South America	266,712	696,900	10.1%	8	11	615	84.2	211.8
GLOBAL TOTAL:	9,020,196	12,977,218	3.7%	87	119	13,205	872.7	2,113.1

* In this table, "direct expenditure" is the expenditure on whale-watching trips, and "indirect expenditure" covers the other costs of the tourist trip (travel, hotels, food, etc).

4.8 Whale, seal and dolphin watching

As a tourist activity, whale watching dates back to about 1950, when part of Point Loma in San Diego, California, United States, was declared a public venue for observing grey whales and the spectacle attracted 10,000 visitors in its first year. Within a few years, boat trips to see whales from the sea were added to the land-based opportunities for watching whales (Hoyt, 2009). The attraction spread to other areas and countries. A survey in 2008 showed that the activity was by then taking place in 119 countries, all around the world, involved about 13 million people a year taking part in whale-watching, supported about 13,000 jobs and generated expenditure by tourists of about 2.1 billion dollars (IFAW, 2009 – see Table 27.3). Whale watching involves not just whales in the strict sense, but also dolphins and other marine mammals: in total around 40 species.

Other marine mammals also support tourism based on watching them. Dolphin-watching has developed as a tourism activity since the 1980s, and is now practised around the world (Constantin et al., 1996). Seal-watching has also developed within the ranges of the various species of seals and other pinnipeds. Since seals and other pinnipeds regularly haul themselves out of the water onto rocks and beaches, seal-watching can offer more reliable viewing of the animals to both operators and tourists, and therefore has enhanced popularity where it is feasible (Newsom, 1996; Bosetti et al., 2002).

Whale-watching involves risks to both humans and the animals. For humans, the risks come from their presence, often in relatively small boats, in the vicinity of large marine animals. The risks are enhanced where the activity involves being in the water – "swimming with dolphins". The threats to the animals are various. The most obvious are those of collisions between whale-watching boats and cetaceans. With quite large boats, often travelling at high speeds (in order to minimize the "blank" time to get from the shore to where the cetaceans are), such collisions can often be fatal to whales (IWC, 2007).

In addition, the literature documents many responses by cetaceans to less extreme pressures from whale-watching traffic (whether on the surface or underwater): surfacing or diving (sometimes to considerable depths), slapping the tail on the water, breaching (that is, leaping out of the water), making noises, changing the size of the group or the way in which the members of a group interact, changing their swimming patterns, changing their patterns of feeding and/or resting (Senigaglia, 2012; Parsons, 2012). The difficult issue to resolve is whether such behavioural changes are having long-term harmful effects. The result of increased demands for energy and/or increases in stress levels and/or changes in patterns of feeding and resting may affect overall health. One study of bottle-nose dolphins (*Tursiops*) suggests that, in the long term, such pressures may lead to reduced reproductive rates (Bejder, 2006). Pressuring cetaceans to move from their chosen feeding grounds may result in them settling in areas providing less (or less appropriate) food, with obvious deleterious effects. Noise from whale-watching boats may disrupt communication between individuals (which may be important for promoting mating or avoiding harm). The cumulative effect of these various pressures may worsen the situation. These were the kinds of considerations that led the whale-watching subcommittee of the International Whaling Commission to state in 2006 that "… there is new compelling evidence that the fitness of individual odontocetes [that is, the toothed whales (such as the sperm whale (*Physeter macrocephalus*), the killer whale (*Orcinus orca*), beaked whales (*Ziphiidae*) and dolphins (*Delphinidae*)] repeatedly exposed to whale-watching vessel traffic can be compromised and that this can lead to population-level effects" (IWC, 2006). The effect of whale-watching on the life-patterns of the majority of species of baleen (plankton-feeding) whales, with feeding and breeding grounds separated by long migrations, is still under study. Nevertheless, action has been taken in some areas, such as South Africa, to prevent whale-watching in nursery areas (Workshop, 2004; IWCSC, 2013)

As a result, the International Whaling Commission has instituted a five-year strategic plan (2011 – 2016) on whale-watching (IWC, 2014). This aims to provide a framework for research, monitoring, capacity-building, development and management by national authorities. The work includes analysis of the methods adopted by various national administrations to control or regulate whale-watching and coordination of scientific research (IWCSC, 2013).

The impact of watching on seals and other pinnipeds is rather different because of their habit of hauling themselves out of the water. This means that they can be observed both on foot and from boats without any interaction in the water. Furthermore a distinction exists between pinniped species which are inherently "tame" and readily allow very close human approach often to less than 20m with little overt response (most fur seals, sea lions and southern phocid seals) and those which are generally wary of human approach and flush to the water when boats may be at a distance of 200m or more (grey and harbour seals). Some seal species can also become habituated to human presence without any adverse reaction (Wilson, 2015).

4.9 Shark watching

Chapter 40 describes the growth of the tourist activity in shark watching and shark diving, resulting in an industry that, on one estimate, exceeds 300 million dollars a year. The activity in many cases involves placing tourists wearing scuba gear in metal cages and lowering them into the water, and then attracting sharks by throwing "chum" (fish waste and offal) into the water. It therefore has considerable potential both for injury to the tourists and for disturbing the local ecology. On the other hand, strong arguments are made that the potential economic gains for developing economies are large and the environmental risks are low and can be kept within acceptable bounds by suitable management and monitoring (Martin, 2006).

4.10 Recreational fishing

In most countries, marine recreational fishing is less significant than inland recreational fishing. Nevertheless, estimates suggest that recreational fishing is important in 76 per cent of the world's exclusive economic zones (Mora, 2009). Some coastal marine stocks in more industrialized nations are exclusively exploited for recreation, or intensive co-exploitation for commercial and recreational purposes occurs. Overall, there is a growing recognition of the immense economic, sociocultural and ecological importance of recreational fishing as a significant component of global capture fisheries (FAO, 2012). One estimate puts the global level of expenditure in 2003 on recreational fishing at 40 billion dollars a year, supporting 954,000 jobs (Cisneros-Montemayor et al., 2009). This includes fishing by people in the localities around their homes, and the proportion that is attributable to tourists (whether international or domestic) is uncertain. Recreational fisheries are most developed in economically developed countries, but they are emerging as a social and economic factor in many other economies (for example, Argentina, Brazil, China, India) and some other developing countries. Where statistics are available, some 4 per cent to 16 per cent of the populations engage in recreational fishing (FAO, 2012). For example, in Brazil in 2007, about 200,000 fishers have amateur angling permits and it was estimated that there were an additional one million unregistered recreational fishers. In addition, sport fishing in Brazil has grown at a rate of up to 30 per cent a year, with a corresponding growth in tourist numbers. This is reflected, among other things, in the growing success of the sport fishing trade that draws thousands of visitors (FAO, 2010).

In more detail for one developed economy where recreational fishing is popular, in Great Britain (that is, the United Kingdom less Northern Ireland) in 2012, about 2.2 per cent of the adult population (1.08 million people) went fishing in the sea at least once. Total resident sea-angler spending in that year was estimated at 1,230 million pounds (1,685 million dollars). This directly supported 10,400 jobs (Armstrong et al., 2013).

The environmental impact of this recreational fishing activity is twofold. First, it is a driver increasing the demand for small boats in coastal waters: most marine recreational fishing is carried out from boats, rather than from the shore. This demand is one of the factors underlying the development of coastal marinas (see below). Secondly, the catch from recreational fishing is a component of the total fishing mortality caused by capture fisheries. Traditionally, it has been regarded as of marginal importance in this regard. However, figures are beginning to emerge that show that it can be a significant component and needs to be taken into account in the general management of fish stocks. For example, in the United States, recreational landings in 2002 accounted for 4 per cent of total marine fish landed in the country. When large industrial fisheries (such as menhaden (*Brevoortia spp*) and pollock (*Theragra chalcogramma*)) are excluded, the recreational component was 10 per cent; when only the fish populations are considered where there are concerns about sustainability, recreational landings in that year accounted for 23 per cent of the total nationwide, rising to 38 per cent in the waters of the States on the southern Atlantic coast of the United States and to 64 per cent in the Gulf of Mexico (Coleman, 2004).

The extent to which the effects of recreational fisheries are taken into account in managing fish stocks varies around the world. Of the authorities responsible for managing the world's exclusive economic zones (EEZs), for recreational fishing, 29 per cent impose regulations on the size of fish caught, 15 per cent regulate the number of fish caught, 13 per cent collect data on what is happening and 3 per cent impose a limit on the number of recreational fishers (Mora et al., 2009).

It is therefore likely that the impacts of recreational fisheries are not being taken into account in managing fish stocks in much of the world. The acquisition of information on local impacts of recreational fishing and the skills to incorporate this information into fisheries management (especially since those undertaking the fishing will in most cases be very different from the usual populations of fisherfolk) will therefore represent significant gaps in much of the world.

Large sport fish (marlin (*Makaira nigricans*, *Istiompax indica* and *Tetrapturus spp*), sailfish (*Istiophorus platypterus*), swordfish (*Xiphias gladius*) and similar species) are a special case. Fishing by rod for these large fish requires relatively large and powerful boats. The tourist market for these species is therefore focused on the more wealthy tourists, especially from the USA. It is particularly significant in the American tropics and sub-tropics, but it is also found, for example, off Mozambique. The economic value of the total of the various recreational fisheries of this kind has been estimated at about 143 million dollars (2003 prices) (Ditton et al., 2003; IOTC, 2013). Although some data on recreational fishing around Mexico have been collected, no reliable data are available for catches of sailfish for the other recreational fisheries of Central and South America, one of the main areas for recreational fishing for large sport fish (for which we know of no reliable data on catches of sailfish (Hinton et al., 2013). Similarly, the data for the Indian Ocean are only partial (IOTC, 2013).

Waste discarded from recreational fishing boats can cause problems. For example, discarded monofilament fishing lines have been found on 65 per cent of coral colonies at Oahu, Hawaii, United States, apparently causing substantial mortality by abrading polyps when moved by wave surge (Yoshikawa et al., 2004).

Recreational hunting for seabirds and some marine mammals and reptiles also takes place. In many countries, such hunting is prohibited or strictly controlled, especially for species regarded as threatened or endangered. Nevertheless, such recreational hunting can be of some economic significance for local communities. For example, Canada is the only one of the five jurisdictions in which polar bears are found that allows recreational (trophy) hunting for them; in the two other jurisdictions where such hunting is permitted it is restricted to indigenous peoples (Lunn et al., 2002). An average of about 100 bears per year is taken by recreational hunters, representing about 20 per cent of the total number taken in Canada. This has been estimated to bring an income of about 1.3 million dollars per year (2010 prices) to the 30 or so communities where such hunting is permitted (Écoressources, 2010).

4.11 Boating and personal leisure transport

In North America and Europe a massive growth has occurred over the last fifty years in the numbers of small vessels used for pleasure boating. For example, in the United States (including the Great Lakes and internal waterways), in 2013 just under 12 million such craft were notified to the authorities (USCG, 2014), a slight reduction on the previous year, suggesting that the market may be becoming saturated. A high proportion (82 per cent) is motorized, with consequent pollution problems from oil and noise. This activity is economically significant, with the turnover in the United States estimated at 121,500 million dollars a year. It is estimated that 36 per cent of the adult population take part in recreational boating at least once a year (NMMA, 2013). Such widespread activities are not without their risks; global figures are not available but, for example, in the United States in 2013 4,062 boating accidents occurred, involving 560 deaths. This shows that safety measures and instruction can be effective, because these represent reductions of 50 per cent (accidents) and 31 per cent (deaths) over the last 15 years (USCG, 2014). Although the current level of participation in the rest of the world is much lower, it is expected to grow rapidly over the next few years in the fast-growing economies: in Brazil, sales of leisure boats have been growing at a rate of over 10 per cent per year since 2005 (except in 2009) (FT, 2011); in China, it is forecast that the number of pleasure yachts will increase to over 100,000 by 2020 (CCYIA, 2013).

All these boats require moorings when they are not being used for recreational sailing. There has therefore been a parallel growth in marinas and specialized harbours for small boats. These installations form a significant part of the hard coastal constructions discussed above, and therefore present the problems analyzed there.

The other main environmental problems from yachts and small boats are parallel to those from larger ships (see chapter 17). Apart from the inevitable impact of oil from motor engines, the most significant are the residual problems from anti-fouling paints (especially tributyltin (TBT)), the role of small boats as vectors of non-indigenous species, waste disposal and anchoring and movement impacts.

The use of TBT has been banned since the 1980s for small vessels (under 25 metres) in many parts of the world and, more generally, under the International Convention on Control of Harmful Anti-Fouling Systems on Ships since 2003 for new applications and from 2008 for vessels already treated with TBT (see further in chapter 17). However, some States have still not accepted this prohibition: 16 per cent of the tonnage of the world's shipping is registered in States that have not become parties to this Convention (IMO, 2014). Even where States are parties to the Convention, areas still remain where TBT is being found in small-boat harbours and associated areas – for example, in Brazil, the Ilha Grande Bay, Rio de Janeiro (described as one of the most heavily protected tourist areas in the country) was shown in 2009 to be still heavily affected by TBT (Pessoa, 2009).

As concern has grown over the transport of non-indigenous organisms by ships, the role of small boats as vectors of such biota has been shown to be significant – not so much in long-distance transport, as in the more local distribution of species once they have been introduced into a region. Problems with the transport of non-indigenous organisms by recreational boats are being found in locations as far apart as British Columbia on the Pacific coast of Canada and Cornwall on the Atlantic coast of England. The problem species include shellfish, seaweeds and bryozoa (Davenport, 2006; Murray, 2011; ERCCIS, 2014). In 2012, the International Maritime Organization issued guidance for minimizing the transfer of invasive aquatic species by recreational craft (IMO, 2012).

As with cruise ships, although on a smaller scale, recreational boat anchors can cause damage to coral reefs. Their anchors can likewise cause problems to seagrass beds (Backhurst et al., 2000). Recreational motor boats can cause further damage to seagrass beds from the action of their propellers in shallow water; re-growth after such damage can take up to four years (Sargent et al., 1994; Dawes et al., 1997). Powerboats (high-speed motor-boats) cause disturbance through noise and wake to seabirds, marine mammals and sea turtles, particularly to slower-swimming species that are unable to get away, and by disturbing foraging. They can also affect the enjoyment of beaches and inshore waters by other human users. Other devices can cause similar disturbances. Such devices include Jet-Skis® and Wave-Runners®, sand-yachts and kites and paragliders towed by all-terrain vehicles and motor-boats (Davenport, 2006). In effect, these newer forms of recreation on shore and in inshore waters are competing for marine space with non-motorized uses and the natural ecology.

5 Integration of environmental and socio-economic trends

Successful management of sustainable tourism requires a complex balancing exercise. Many factors have to be taken into account: the means of access, the urban development of hotels, other accommodation, restaurants, and other support facilities, sewage and waste disposal, the many kinds of recreational activity, the interests of the local inhabitants, the profitability of resorts and, last but by no means least, the maintenance of the local natural ecology. The levels at which such varied interests can be managed will vary, hence a successful balancing exercise will usually require the involvement of a wide range of authorities, residents and commercial interests. Examples show, however, that it is possible for such successful balances to be achieved and – even more difficult – maintained over long periods.

A further factor in this balancing exercise is the cross-effects between different recreational activities and different environmental compartments and species. The impact of bird-watching on nesting and feeding areas for seabirds is explained above. But similar impacts on birds are also caused by other recreational activities: boats used in birds' feeding areas for sea-fishing, whale-watching, shark-watching or simply for sailing will cause the same type of disruption as boats used for bird-watching; coastal walking through nesting areas for simple enjoyment will cause the same kind of disruption as access for bird-watching. Likewise, all kinds of uses of boats in areas frequented by cetaceans will have similar impacts to the use of boats for whale-watching. For wildlife which is disturbed, it does not matter what the purpose of the recreational use is: it is the cumulative impact which gives rise to a threat. Secondly, the balancing exercise is further complicated where marine protected areas are created, since these add a further set of factors which have to be taken into account.

Tourism generally, however, suffers from the problem that success risks undermining itself. For many tourists, the attraction of a tourist location relies on a combination of a relatively high level of service provision and a feeling of a relatively low level of pressure from other tourists. As a tourist location becomes recognised as providing this desirable combination, the pressure to intensify the provision of tourist services increases, thus undermining the balance. If the balance is maintained, prices are likely to rise as the service providers take advantage of the demand in the market, and the location will become less available to the less well-off. If the balance is not maintained, those who can afford to will look elsewhere, creating pressures for the development of new resorts. The world's supply of good sandy beaches is fixed. As disposable wealth in an increasing number of economies increases, the pressure to open up more and more areas for tourism will also increase.

At the same time, within any specific resort, there will be conflicting demands for marine space: among others, sun-bathing *versus* beach volley-ball, swimming *versus* Jet-Skis® and similar devices, wildlife watchers *versus* water-skiers. All this will require management – alongside the many other demands on coastal marine space from fishing (both commercial and recreational), shipping, ports, sand and gravel dredging and all the other human activities that affect the coastal zone. The management of ecotourism is particularly important, because it is essential to ensure that the pressures on the wildlife from it are not more than can be accepted: once a natural ecology is damaged, it is often impossible to restore. Since maintaining a good quality of local environment will be essential for the success of a tourist location, tourism management will be one of the first areas to feel these combined pressures.

6 Capacity-building gaps

Any successful management process will rely on a combination of good information about what is happening and the skills needed to integrate and apply that information. Successful management of sustainable tourism has therefore to be based on a wide-ranging collection of all the relevant information and the development of the necessary skills. The information and skills should include: (a) knowledge about the main features of the local marine environment and their vulnerability to the tourist and recreational activities that affect the marine environment; (b) information about the location, scale and economic significance of those tourist and recreational activities; (c) the relationships between those tourist and recreational activities and the other uses of the marine environment in the locality; and (d) skills to evaluate what would be the most appropriate balance between the various interests involved (including the conservation of the local marine environment and any formally protected coastal or marine areas) and to broker or settle an acceptable agreement between all those interests.

In many parts of the world, much progress has been made in developing the skills necessary to monitor local ecosystems, and in training local residents in managing hotels and the many other trades necessary for the proper functioning of tourist services, although there is scope for improvement. Less progress has been made in integrating these two sides of a successful sustainable tourist operation. Since so many branches of administration are relevant, a focus on ensuring that general administrators understand the importance of an integrated approach to tourism and are willing to implement it is necessary.

References

AET (2012). Secretaria Nacional de Políticas de Turismo, *Anuário Estatístico de Turismo – 2012*. (http://www.dadosefatos.turismo.gov.br/export/sites/default/dadosefatos/anuario/downloads_anuario/Anuxrio_Estatxstico_de_Turismo_2012_-_Ano_base_2011.pdf Accessed 15 August 2014)

Allen, W.H. (1992). Increased Dangers to Caribbean Marine Ecosystems, *BioScience* Vol 42, No. 5.

Arianoutsou, M. (1988). Assessing the impacts of human activities on nesting of loggerhead sea-turtles (*Caretta caretta* L.) on Zakynthos Island, Western Greece. *Environmental Conservation* 15, 1988.

Armstrong, M., Brown, A., Hargreaves, J., Hyder, K., Pilgrim-Morrison, S., Munday, M. (2013). Proctor, S., Roberts, A. and Williamson, K. (2013). *Sea Angling 2012 – a survey of recreational sea angling activity and economic value in England*, Department of Environment, Food and Rural Affairs, London.

Backhurst, M.K. and Cole, R.G. (2000). Biological impacts of boating at Kawau Island, north-eastern New Zealand. *Journal of Environmental Management* 60, 2000.

Bailey, S. (2012). Closed - but Boscombe bodyboarders still love the surf reef, *Bournmouth Daily Echo*, 2 April.

Balmford, A., Beresford, J., Green, J., Naidoo, R., Walpole, M., Manica, A. (2009). A Global Perspective on Trends in Nature-Based Tourism, *PLoS Biol* 7(6): e1000144. doi:10.1371/journal.pbio.1000144.

Balsdon, J.P.V.D. (1969). *Life and Leisure in Ancient Rome*, Bodley Head, London.

Barker, H.H.L., Robert, C.M. (2004). Scuba diver behaviour and the management of diving impacts on coral reefs. *Biological Conservation* 120.

Bartolo, R., Delamaro, M., Bursztyn, I. and Hallewell, L. (2008).Tourism for Whom? Different Paths to Development and Alternative Experiments in Brazil, *Latin American Perspectives*, Vol. 35, (3).

Bejder, L., Samuels, A., Whitehead, H. and Gales, N. (2006). Interpreting short-term behavioural responses to disturbance within a longitudinal perspective, *Animal Behaviour*, vol. 72 (5).

Bosetti, V. and Pearce, D. *A Study of Environmental Conflict: The Economic Value of Grey Seals in South West England*, Centre for Social and Economic Research on the Global Environment Working Paper GEC 01-02 (ISSN 0967-8875).

Brandon, P. (1974). *The Sussex Landscape*, Hodder and Stoughton, London.

Bulleri, F., and Chapman, M.G., (2010). The introduction of coastal infrastructure as a driver of change in marine environments, *Journal of Applied Ecology* vol 47.

Cater, E. (1995). Environmental Contradictions in Sustainable Tourism, *The Geographical Journal*, Vol. 161(1).

CCT (Contra Costa Times) (2008). Surf's not up, so reef's coming out, 29 Sepember. (http://www.contracostatimes.com/california/ci_10586139 Accessed 10 October 2014.

CCYIA (China Cruise and Yacht Industry Association) (2013). Report 2012/13, Beijing.

Cisneros-Montemayor, A.M. and Sumaila, U.R. (2010). A global estimate of benefits from ecosystem-based marine recreation: potential impacts and implications for management. *Journal of Bioeconomics*, Vol 12(3).

Coleman, F., Figueira, W.F., Ueland, J.S. and Crowder, L.B. (2004). The Impact of United States Recreational Fisheries on Marine Fish Populations, *Science* 305.

Constantin, R., and Bejder, L. (1996). Managing the Whale- and Dolphin-Watching Industry, in *Marine Wildlife and Tourism Management: Insights from the Natural and Social Sciences* (ed. E. S. Higham and M. Lück), CABI, Wallingford, UK.

CTO (Caribbean Tourist Organization) (2013). Latest Statistics 2012. (http://www.onecaribbean.org/content/files/13MARCH2013Lattab12.pdf. Accessed 14 August 2014).

CTO (Caribbean Tourist Organization) (2014). Birdwatching. (http://www.onecaribbean.org/content/files/BirdwatchingCaribbeanNicheMarkets-2.pdf. Accessed 20 October 2014).

Currin, C. (2013). Carolyn Currin in Living Shoreline Summit Committee, *2013 Mid-Atlantic Living Shorelines Summit Proceedings*. (http://www.dnr.maryland.gov/ccs/pdfs/ls/2013summit/SummitProceedings.pdf. Accessed 17 October 2014).

Davenport, J. (1998). Temperature and the life history strategies of sea-turtles, *Journal of Thermal Biology* 22.

Davenport, J. and Davenport, J.L. (2006). The impact of tourism and personal leisure transport on coastal environments, *Estuarine, Coastal and Shelf Science* 67.

Davis, D. and Tisdell, C. (1996). Economic Management of Recreational Scuba Diving and the Environment, *Journal of Environmental Management*, 4.

Dawes, C.J., Andorfer, J., Rose, C., Uranowski, C., and Ehringer, N. (1997). Regrowth of the seagrass *Thalassia testudinum* into propeller scars, *Aquatic Botany* 59.

Dinesen, Z. and Oliver, J. (1995). Tourism Impact, in *State of the Great Barrier Reef World Heritage Area Workshop* (eds. D. R. Wachenfeld, J. Oliver and K. Davis), Great Barrier Reef Marine Park Authority, Townsville, Queensland.

Ditton, R.B. and Stoll, J.R. (2003). Social and Economic Perspectives on Recreational Billfish Fisheries, *Marine and Freshwater Research*, Vol 54.

EC (European Commission) (2014). *Flash Eurobarometer 392, Preferences of Europeans Towards Tourism* (http://ec.europa.eu/public_opinion/flash/fl_392_en.pdf Accessed 12 December 2014).

Écoressources Consultants (2010). *Evidence of the Socio-Economic Importance of Polar Bears for Canada*. ISBN: 978-1-100-18970-3.

ECLAC (United Nations Economic Commission for Latin America and the Caribbean) (2005). Issues and Challenges in Caribbean Cruise Ship Tourism, Santiago, Chile, 2005 (Document LC/CAR/L.75).

EEA (European Environment Agency) (2006). *The changing faces of Europe's coastal areas*, Copenhagen.

Elmoustaphat, A.O, Franck Levoy, Olivier Monfort, and Vladimir G. Koutitonsky (2007). A Numerical Forecast of Shoreline Evolution after Harbour Construction in Nouakchott, Mauritania, *Journal of Coastal Research*, Vol. 23(6).

ERCCIS – (Environment Records Centre for Cornwall) – Invasive Species (http://www.erccis.org.uk/invasivespecies/Investigate_Invasives_Marine/Marina+and+Boat+Owners accessed 10 October 2014).

Eurostat (Directorate-General for Statistics of the European Commission)(2014a). Coastal Regional Statistics (http://ec.europa.eu/eurostat/statistics-explained/index.php/Coastal_region_statistics Accessed 15 August 2014).

Eurostat (Directorate-General for Statistics of the European Commission) (2014b). Coastal Regional Statistics. (http://appsso.eurostat.ec.europa.eu/nui/show.do?dataset=tour_occ_nin2c&lang=en accessed 15 August 2014).

FAO (Food and Agriculture Organization) (2010). Fishery and Aquaculture Country Profiles: Brazil , (http://www.fao.org/fishery/facp/BRA/en#CountrySector-ProductionSector accessed 20 July 2014).

FAO (Food and Agriculture Organization) (2012). FAO Technical Guidelines for Responsible Fisheries 13: Recreational Fisheries, FAO, Rome.

FIPE (Fundação Instituto de Pesquisas Econômicas)(2012). Caracterização e Dimensionamento do Turismo Doméstico no Brasil – 2010/2011. (http://www.dadosefatos.turismo.gov.br/export/sites/default/dadosefatos/demanda_turistica/domestica/downloads_domestica/Demanda_domestica_-_2012_-_Relatorio_Executivo_nov.pdf. Accessed 17 August 2014).

FT (2011). Boatbuilders ride wave of economic prosperity. *Financial Times*, Brazil, 20

September.

Gheskiere, T., Vincx, M., Weslawski, J.M., Scapini, F., Degraer, S. (2005). Meiofauna descriptor of tourism-induced changes at sandy beaches, *Marine Environmental Research* 60.

Green, R. and Darryl, N. Jones (2010). *Practices, needs and attitudes of bird-watching tourists in Australia*. Cooperative Research Centre for Sustainable Tourism, Gold Coast, Queensland.

Hawkins, J.P., Roberts, C.M., Van't Hof, T., De Meyer, K., Tratalos, J., and Aldam, C. (1999). Effects of Recreational Scuba Diving on Caribbean Coral and Fish Communities, *Conservation Biology*, Vol. 13(4).

Hinton, M.G. and Maunder, M.N. (2013). *Status of Sailfish in the Eastern Pacific Ocean in 2011 and Outlook for the Future*, Inter-American Tropical Tuna Commission Scientific Advisory Committee document SAC-04-07c.

Honey, M. and Krantz, D. (2007). *Global Trends in Coastal Tourism*, Centre on Ecotourism and Sustainable Development, Stanford and Washington DC.

Houston, J.R. (2008). The economic value of beaches – a 2008 Update, *Shore & Beach*, Vol. 76(3).

Hoyt, E. (2009). Whale watching, in *Encyclopedia of Marine Mammals* (eds. W.F. Perrinan, B. Würsig and J.G.M. Thewissen) Academic Press, San Diego, California.

Hylgaard, T. and Liddle, M.J. (1981). The effect of human trampling on a sand dune ecosystem dominated by *Empetrum nigrum*, *Journal of Applied Ecology* 18.

IAATO (International Association of Antarctic Tourism Operators) (2014). Tourism Statistics (www. http://iaato.org/tourism-statistics accessed 20 October 2014).

IFAW (International Fund for Animal Welfare) (2009). *Whale Watching Worldwide – Tourism numbers, expenditures and expanding economic benefits*, Yarmouth Port, Massachusetts.

IMO (International Maritme Organization) (2012). *Guidance for Minimizing the Transfer of Invasive Aquatic Species as Biofouling (Hull Fouling) for Recreational Craft* (MEPC.1/Circ.792).IMO (International Maritime Organization) (2014). *Status of Multilateral Conventions*. (http://www.imo.org/About/Conventions/StatusOfConventions/Documents/Status%20-%202014.pdf)

IOTC (Indian Ocean Tuna Commission)(2013). *Billfish Caught in the Recreational and Sport Fishing of South Coast of Mozambique: Results of the First Census of Recreational and Sport Fishing in 2007 and the Sampling Program in 2012* (Document IOTC-2013-WPB11-17).

IWC (International Whaling Commission) (2007). Report of the scientific committee, *Journal of Cetacean Research and Management*, vol. 9, supplement.

IWC (International Whaling Commission) (2014). *Five Year Strategic Plan for Whalewatching 2011–2016*. (http://iwc.int/private/downloads/61pp7v1qdn4sss40ow88kgso4/AC-002s3%20IWC%20Whale%20Booklet_HR.pdf accessed 12 October 2014).

IWCSC (International Whaling Commission) (2013). *Report of the Scientific Committee Annual Meeting 2013*. (https://archive.iwc.int/pages/view.php?ref=3285&search=%21collection73&order_by=relevance&sort=DESC&offset=0&archive=0&k=&curpos=1 accessed 12 October 2014)

Johnson, D.R., Ko, D.R., Franks, J.S., Mareno, P, and Snachez-Rubio, G. (2012). The Sargassum Invasion of the Eastern Caribbean and Dynamics of the Equatorial North Atlantic. *Proceedings of the 65th Gulf and Caribbean Fisheries Institute* 65.

Kerlinger, P. and Wiedner, D.S. (1991). "The economics of birding at Cape May, New Jersey" in *Ecotourism and resource conservation, a collection of papers,* ed. J. A. Kusler. Ecotourism and Resource Conservation.

Klein, M. and Zviely, D. (2001). The environmental impact of marina development on adjacent beaches: a case study of the Herzliya marina, Israel. *Applied Geography* Vol 21.

Lemauviel, S., and Rose, F. (2003). Response of three plant communities to trampling in a sand dune system in Brittany (France). *Environmental Management* 31.

Lichtnau, B. (1996). *Architektur in Mecklenburg und Vorpommern, 1800-1950*, Steinbecker Verlag, Ernst-Moritz-Arndt-Universität Greifswald.

Llewellyn, P.J. and Shackley, S.E. (1996). The effects of mechanical beachcleaning on invertebrate populations. *British Wildlife* 7.

Lück, M., Maher, P.T. and Stewart, E.J. (2010). *Cruise Tourism in Polar Regions*, Earthscan, London.

Lunn, N.J., Schliebe, S. and Born, E.W. (eds.) (2002). *Proceedings of the 13th Working Meeting of the IUCN/SSC Polar Bear Specialist Group, 23–28 June 2001, Nuuk, Greenland,* . IUCN, Gland, Switzerland and Cambridge, UK.Mann, K.H. (2000). *Ecology of Coastal Waters: with Implications for Management*, Blackwell Science, Massachusetts, USA.

Mann, K.H. (2000). *Ecology of Coastal Waters: with Implications for Management*. Blackwell Science, Massachusetts, USA.

Martin, R.and Abdul Hakeem, A.A. (2006). Development of a Sustainable Shark Diving Ecotourism Industry in the Maldives: Challenges and Opportunities, *Maldives Marine Research Bulletin*, No 8.

Milazzo, M., Chemello, R., Badalementi, F. and Riggio, S. (2002). Short-term effect of human trampling on the upper infralittoral macroalgae of Ustica Island MPA (western Mediterranean, Italy). *Journal of the Marine Biological Association of the United Kingdom* 82.

Mora C., Myers, R.A., Coll, M., Libralato, S., Pitcher, T.J. Sumaila, R.U., Zeller, D., Watson, R., Gaston, K.J., Worm, B. (2009) Management Effectiveness of the World's Marine Fisheries. *PLoS Biology* 7(6).

Morgan, O.A., Massey, D.M. and Huth, W. (2009). Diving Demand for Large Ship Artificial Reefs, *Marine Resource Economics*, Vol. 24(1).

Mossman (2003). Town of Mossman Park, Mossman Beach Management Plan, adopted 26 August 2003 (www.mosmanpark.wa.gov.au/council/public-documents/download/28/).

Mull, J. (2014). Pipe Dreams – the reality of artificial reefs, *The Surfer*, 2 June 2014 (http://www.surfermag.com/features/pipe-dreams-artificial-reef/ accessed 10 October 2014).

Murray, C.C., Pakhomov, E.A., and Therriault, T.W. (2011). Recreational boating: a large unregulated vector transporting marine invasive species, *Diversity and Distributions*, Vol. 17(6). DOI: 10.1111/j.1472-4642.2011.00798.x.

NBSC (National Bureau of Statistic of China) (2014). *Statistical Communiqué of the People's Republic of China on the 2013 National Economic and Social Development* http://www.stats.gov.cn/english/PressRelease/201402/t20140224_515103.html Accessed on 20 December 2014.

Newsom, D. and Rodger, K. (1996). *Impact of Tourism on Pinnipeds and Implications for Toursim Management* in *Marine Wildlife and Tourism Management: Insights from the Natural and Social Sciences* (ed. E. S. Higham and M. Lück), CABI, Wallingford, UK.

NMMA (National Marine Manufacturers Association) (2013). *2013 U.S. Recreational Boating Statistical Abstract*, Chicago.

Orams, M.B. (2003). Sandy Beaches as a Tourism Attraction: A Management Challenge for the 21st Century, *Journal of Coastal Research*, Special Issue No. 35.

PADI (Professional Association of Diving Instructors) (2014). Statistics (http://www.padi.com/scuba-diving/about-padi/ accessed 10 October 2014).

Parsons, E.C.M. (2012). The Negative Impacts of Whale-Watching, *Journal of Marine Biology*, Volume 2012.

Parsons, M., Mavor, R.A. and Mitchell, P.I. (2006). The UK Seabird Monitoring Programme, in Waterbirds around the world (ed. G C Boere, C A Galbraith and D A Stroud). 2006, The Stationery Office, Edinburgh, UK, 2006.

Pessoa, I., Fernandez, M., Toste, R., Dore, M. and Parahyba, M. (2009). Imposex in a touristic area in Southeastern Brazilian coast. *Journal of Coastal Research*, Special Issue 56.

Pevsner, N. (1969). *The Buildings of England – North Lancashire*, Penguin Books, Harmondsworth, England.

Pinn, E.H. and Rodgers, M. (1996). The influence of visitors on intertidal biodiversity, *Journal of the Marine Biological Association of the United Kingdom*, 85.

Rendle, E. and Davidson, M. (2012). An Evaluation of the Physical Impact and Structural Integrity of a Geotextile Surf Reef, *Coastal Engineering*, Vol. 33.

Rendle E. and Rodwell, L. (2014). Artificial surf reefs: A preliminary assessment of the potential to enhance a coastal economy, *Marine Policy*, vol. 45.

Rouphael, A.B. and Inglis, G.J. (2001). "Take only photographs and leave only footprints"?: An experimental study of the impacts of underwater photographers on coral reef dive sites, *Biological Conservation*, Vol. 100(3).

Russell, R. (1755). De tabe glandulari sive de usu aquae marinae (Oeconomia naturalis in morbis acutis et chronicis glandularum), London.

Sargent, F.J., Leary, T.J., Crewz, D.W. and Kruer, C.R. (1994). *Scarring of Florida's Seagrasses: Assessment and Management Options.* Technical Report FMRI 1/94, Florida Marine Research Institute, St Petersburg, Florida.

Schiel, D.R. and Taylor, D.I. (1999). Effects of trampling on a rocky intertidal assemblage in southern New Zealand. *Journal of Experimental Marine Biology & Ecology* 235.

Senigaglia, V., Bejder, L., Christiansen, F., Gendron, D., Lundquist, D., Noren, D., Schaffar, A., Smith, J.C., Williams, R. and Lusseau, D. (2012). *Meta-analyses of whale-watching impact studies: differences and similarities in disturbance responses among species*, International Whaling Commission Scientific Committee Paper SC/64/WW6. (http://iwc.int/private/downloads/burvztaqbdsg8sk8gwoc4g4o0/SC-64-WW6.pdf).

Sezgin, E. and Medet, Y. (2012). Golden Age of Mass Tourism: Its History and Development, in *Visions for Global Tourism Industry - Creating and Sustaining Competitive Strategies*, Dr. Murat Kasimoglu (Ed.).

Steven. R., Morrison, C. and Castley, J.G. (2014). Bird watching and avitourism: A global review of research into its participant markets, distribution and impacts, highlighting future research priorities to inform sustainable avitourism management, *Journal of Sustainable Tourism* (published at www.academia.edu/7812374 26 June 2014).

SWEC (South-West Economy Centre, University of Plymouth) (2003). *An Assessment of the Socio-Economic Impact of the Sinking of HMS Scylla*, Plymouth. http://www.wrecktoreef.co.uk/wp-content/uploads/2012/12/scylla-eco-soc.pdf Accessed 10 October 2014.

Tratalos, J.A., and Austin, T.J. (2001). Impacts of recreational SCUBA diving on coral communities of the Caribbean island of Grand Cayman. *Biological Conservation* 102.

Tuxbury, S.M. and Salmon, M. (2005). Competitive interactions between artificial lighting and natural cues during seafinding by hatchling marine turtles, *Biological Conservation* 121.

UNCTAD (United Nations Conference on Trade and Development) (2007). *FDI and Tourism, the Development Dimension*, Geneva.

USA Travel (US Travel Association) (2014). *US Travel Answer Sheet June 2014* https://www.ustravel.org/sites/default/files/page/2009/09/US_Travel_AnswerSheet_June_2014.pdf Accessed 15 August 2014.

USCG (United States Coast Guard) (2014). *Recreational Boating Statistics 2013*, Washington DC.

UWI (University of the West Indies) (2002). Kenneth O. Hall, Jean S. Holder and Chandana Jayawardena. Caribbean Tourism and the Role of UWI in Tourism and Hospitality Education, *Social and Economic Studies*, Vol. 51(1).

UWI (University of the West Indies) (2014). Centre for Hotel and Tourism Management (http://myspot.mona.uwi.edu/chtm/about-center accessed 20 October 2014).

Walton. J. (1983). *Leisure in Britain (1780-1939).* Basingstoke: Palgrave Macmillan.

Wilhelmson, D., Marcus C. Öhman, Henrik Ståhl and Yechiam Shlesinger (1998). Artificial Reefs and Dive Tourism in Eilat, Israel, *Ambio*, Vol. 27(8).

Wilson, S.C. (2015) *The impact of human disturbance at seal haul-outs: A literature review for the Seal Conservation Society* (http://www.pinnipeds.org/seal-information/human-interaction accessed 17 April 2015).

Wilson, T.D. (2008). Economic and Social Impacts of Tourism in Mexico, *Latin American Perspectives*, Vol. 35(3).

Workshop (2004). *Report of the Workshop on the Science for Sustainable Whale-watching* http://iwc.int/private/downloads/356zlh078ds0oww04w4o00gcw/ww_workshop.pdf Accessed 12 October 2014.

World Bank (2003). *OECS Ship-Generated Waste Management Project and the Solid Waste Management Project* (Report 27270) (http://www-wds.worldbank.org/external/default/WDSContentServer/WDSP/IB/2003/12/05/000012009_20031205130005/Rendered/PDF/272700OECS.pdf accessed 2 September 2014)

World Bank (2014). *World Development Indicator 6.14* (http://wdi.worldbank.org/table/6.14. Accessed 10 October 2014.

WTO (World Tourism Organization) (2014). *Compendium of Tourist Statistics*, Madrid.

WTTC (World Travel and Tourism Council) (2014). Economic Data (http://www.wttc.org/focus/research-for-action/economic-data-search-tool/ accessed 18 September 2014).

Yoshikawa, T. and Asoh, K. (2004). Entanglement of monofilament fishing lines and coral death, *Biological Conservation*, 117 (5).

Yu, L. (2012). *The International Hospitality Business: Management and Operations*, Routledge, London 2012.

28 Desalinization

Contributors:
Abdul-Rahman Ali Al-Naama (Convenor), Alan Simcock (Lead member)

1 Introduction

Desalinization of seawater is an essential process for the support of human communities in many places around the world. Seawater has a salt content of around 35,000 parts per million (ppm) depending on the location and circumstances: to produce the equivalent of freshwater (with around 1000 ppm (AMS, 2014) therefore requires the removal of over 97 per cent of the salt content. The main purpose of desalinization is to produce water for drinking, sanitation and irrigation. The process can also be used to generate ultra-pure water for certain industrial processes. This chapter reviews the scale of desalinization, the processes involved and its social and economic benefits. Issues relating to discharges from desalinization plants are considered in Chapter 20 (Coastal, riverine and atmospheric inputs from land).

2 Nature, location and magnitude of desalinization

As Figure 28.1 shows, desalinization capacity has grown rapidly over the past half-century. About 16,000 desalinization plants were built worldwide between 1965 and 2011. About 3,800 of these plants are thought to be currently out of service or decommissioned. The current operational capacity is estimated to be about 65,200 megalitres per day (65,200,000 cubic metres per day (m³/d) – in comparison, the public water supply of New York City, United States of America, delivers in total about 3,800 megalitres per day) (GWI, 2015; NYCEP, 2014).

Historically, human settlements have tended to grow up where freshwater was available, and their growth has been conditioned by freshwater availability and the possibilities of bringing it to serve the settlement. As long ago as 312 BCE, the Romans had had to build a 16.4-kilometre aqueduct to bring water to Rome in order to avoid this constraint (Frontinus). Desalinization represents an alternative technology for avoiding this constraint on the growth of human settlements in areas with very limited availability of freshwater. That capability, however, comes at the price of considerable capital investment and energy consumption. Gleick et al. (2009) give an overview of the worldwide distribution of desalinization capacity in 2009.

The nature of the industry, however, varies in many ways between the different regions, particularly in respect of the technology used: the Middle East has relied more on thermal processes, while the United States has relied more on membrane processes. Thermal processes (mainly Multi-Stage-Flash (MSF) and Multiple-Effect-Distillation (MED)) evaporate the water and then re-condense it. At peak performance these distillation processes produce a freshwater output of about 30-40 per cent of the seawater taken in. The residue has to be discharged as brine. Membrane-based processes (such as reverse osmosis (RO), electro-deionization (EDI) and electro-dialysis (ED)) force feed-water through a semi-permeable membrane that blocks various particulates and dissolved ions, leaving the feed-water behind as an enhanced brine, with or without further refinements. (Details of these processes can be found in WHO, 2007 and in GCC, 2014). The energy needed for all forms of desalinization is usually obtained from fossil fuels. However, combined plants for nuclear power generation and water desalinization have been developed in a number of places (for example, Argentina, India, Japan and Pakistan), and the International Atomic Energy Agency has conducted studies on how far this might be developed (IAEA, 2007). At present, very little desalinization is powered by solar energy. One estimate puts it as low as 1 per cent (Kalogirou, 2009). Projects are emerging, however, to develop this form of desalinization. For example, in Abu Dhabi, United Arab Emirates, the Environment Agency completed 22 small (25 m³/day) solar desalinization plants for brackish groundwater in 2012 (The National, 2012). In Chile, Fundación Chile started a small pilot project partly powered by solar energy in Arica in 2013 (Arica, 2013).

Many countries have installed major amounts of desalinization capacity over the past 70 years: the largest amounts of capacity have been installed in Saudi Arabia (capable of producing over 10 million m³ per day), the United States of America (over 8 million m³ per day (but see

Figure 28.1 | Global desalinization capacity 1965 – 2015. Source: GWI, 2015. "Contracted" covers plant that is complete or under construction; "commissioned" covers plant that is in operation or is available for operation.

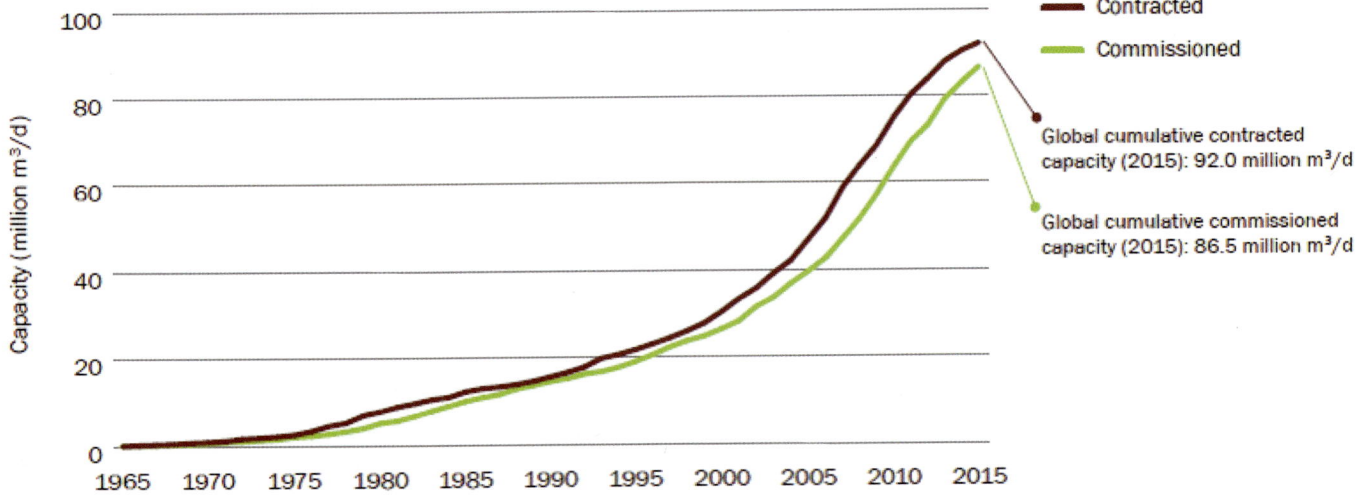

Chapter 28 — Desalinization

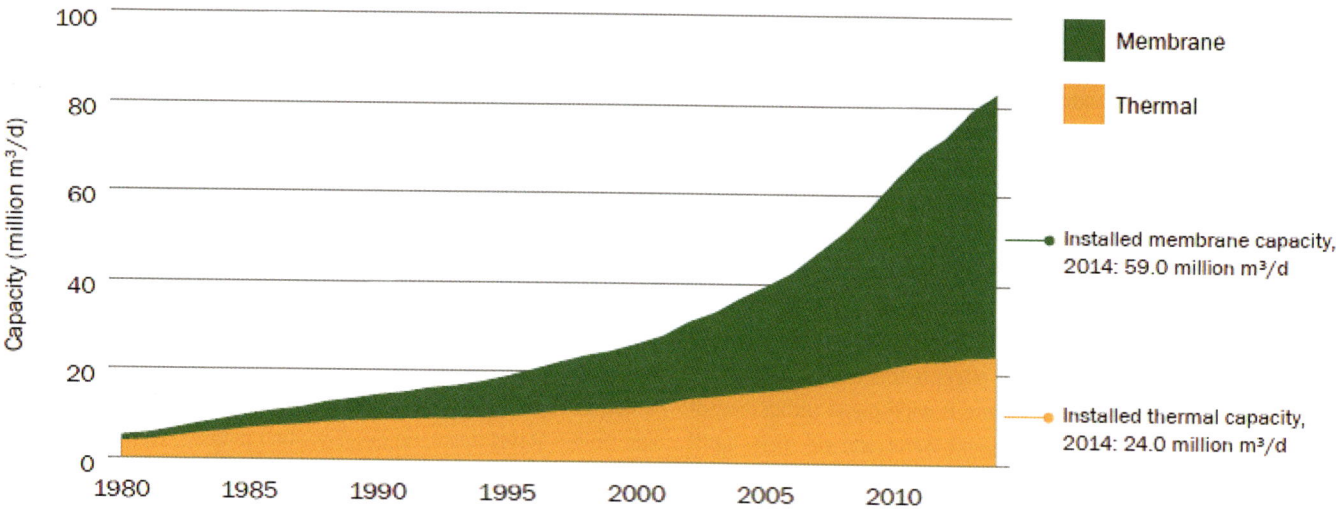

Figure 28.2 | Proportion of thermal and membrane technologies installed 1980 – 2014. Source: GWI, 2015.

below), the United Arab Emirates (about 7 million m³ per day) and Spain (about 5 million m³ per day). . More recently, Algeria, Australia, India and Israel also substantially increased their capacities (GWI, 2014).

2.1 The Persian Gulf area

The Persian Gulf area has the biggest concentration of installed desalinization capacity in the world. In total, the desalinization capacity of the area is around 9.2 million megalitres a year. Ninety-six per cent of this capacity is located in the six countries that form the Gulf Cooperation Council (GCC - Bahrain, Kuwait, Oman, Qatar, Saudi Arabia and the United Arab Emirates).

2.2 The GCC States

These six States have a common approach to desalinization. They are located in an arid, hot desert region, characterized by an average rainfall ranging between 75 and 140 mm a year and by limited, non-renewable groundwater resources. Surface freshwater resources comprise only 0.6 per cent of their total area. Natural freshwater resources range from 60 to 370 cubic meters a year per head across the GCC countries (World Bank, 2005). The resources *per capita* are expected to decrease in the future by up to 20 per cent, due largely to population growth. The total GCC population in 2012 was 44,643,654, of which Saudi Arabia constituted 62 per cent. This population is increasing at an average rate of 14 per cent annually. The discovery of oil and gas resulted in the GCC countries being the world's top fossil-fuel exporters with the highest *per capita* incomes and the fastest-growing economies in the world, which underlies the population growth. From 1998 to 2008, real GDP for GCC countries grew at an average rate of 5.2 per cent annually.

Bridging the gap between demand for, and supply of, freshwater has remained a major issue. To meet the need for freshwater, desalinization of seawater has been one of the main water-supply alternatives that the GCC countries have adopted. Desalinization has become the backbone of many GCC States. For example, Qatar draws as much as 99 per cent of its drinking water from this source. In Saudi Arabia, 50 per cent of municipal water supplies are obtained from desalinization: in 2012 this represented deliveries of 955,000 megalitres per year (SWCC, 2014). The total desalinization capacity installed in the GCC States in 2012 for water production was 8.9 million megalitres a year. This production capacity was divided: Saudi Arabia (KSA) 39 per cent, United Arab Emirates (UAE) 18 per cent, Kuwait 18 per cent, Qatar 15 per cent, Bahrain 6 per cent, and Oman 4 per cent (GCC, 2012). This is shown in Figure 28.4.

Figure 28.3 | Proportion of different technologies in use 2014. RO: Reverse Osmosis; MSF: Multi-stage flash; MED: Multi-effect distillation; ED/EDR: Electrodialysis/Electrodialysis Reversal; NF/SR: Nanofiltration/Sulfate Removal. Source: GWI, 2015.

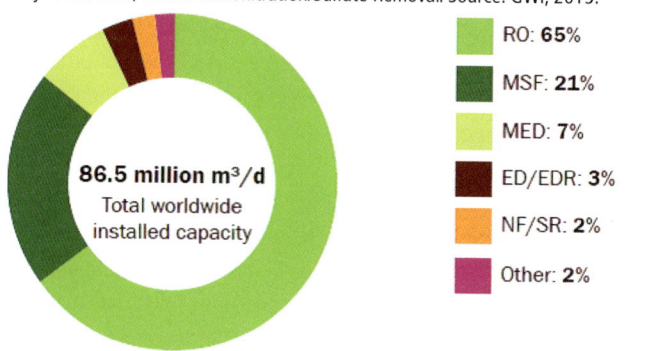

Figure 28.4 | Desalinization capacity in the GCC States, 2012. Source GCC, 2012.

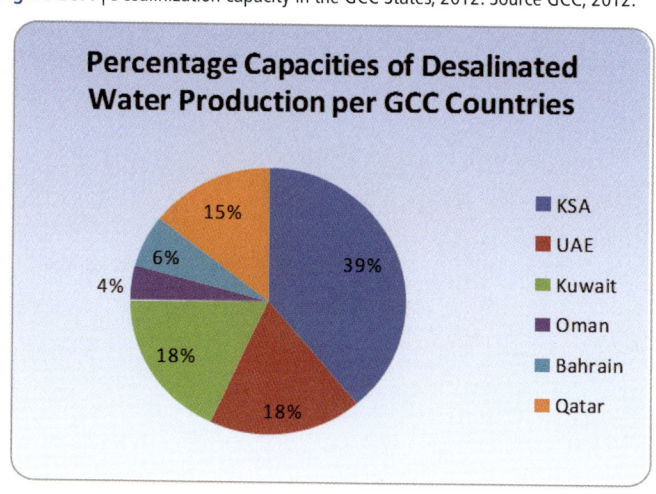

The practice of desalinization in the GCC States is heavily influenced by the high local level of electricity consumption, which is largely due to the demand for air-conditioning and cooling, necessary in the hot desert climate, and to the energy-intensive petro-chemical industries. The demand for electricity and water is also influenced by the pricing policy. Water and electricity subsidies are a commonplace practice among the GCC countries. The shared rationale behind energy and water subsidization includes: cultural considerations, expanding access to energy and water, protecting the poor, consumption smoothing, fostering industrial development, avoiding inflationary pressures, and political considerations. The result of the lower prices is to increase demand for both electricity and water. However, there is widespread recognition of the harmful effects caused by the current water and electricity tariff rates (Saif, 2012).

The high level of use of thermal technologies for desalinization in the GCC States is mainly due to the predominant method of electricity generation, which is through gas-fired power plants. A by-product of the electricity generation process is steam, which can be utilized by MSF and MED thermal desalinization plants for their energy needs. The two plants need to be co-located in order for the desalinization plant to capitalize on the power stations' by-product of steam. This co-location of power and plants is referred to as co-generation. Roughly 60 per cent of the MSF plants in the United Arab Emirates are co-generation, while that percentage stands at 70 per cent in Qatar. The quality of the water available for desalinization also plays a role. It has the 4 Hs: high salinity, high turbidity, high temperature and high marine life. These factors have in the past made it less suitable for membrane technology (Al Hashemi et al. 2014).

The thermal technology most used in the GCC States is the MSF, which is characterized by a high consumption of energy. Reverse osmosis (RO) is the next most used, and the least used is MED: see Figure 28.5. The GCC States constitute around 88 per cent of the world's use of desalinization by thermal processes.

Although this was the balance between thermal and membrane technologies in the GCC States in 2012, the situation is changing quickly, because the GCC States will in future be adopting more RO projects, as a step towards minimizing energy consumption and decreasing environmental impacts. Most of the desalinization plants under construction in 2012 were RO or combined RO/MSF, and the balance is expected to change even more in the future (GCC, 2012).

The GCC States are continuing to invest heavily in their water and energy sectors as shown by many independent water and power plant (IWPP) projects. For example, in 2009, Qatar initiated a 30-year water and electricity master plan that will see major investments in desalinization, water infrastructure and wastewater treatment (GWI, 2015). Between 2010 and 2015, Qatar plans to invest approximately 5,470 million United States dollars in desalinization projects, with an additional 1,100 million dollars investment in IWPP production facilities between 2013 and 2017. Likewise, the UAE plans to invest 13,890 million dollars in new desalinization plants and distribution networks between 2012 and 2016.

Generally, this investment in further desalinization is counterbalanced by a new interest in adopting an integrated water policy that uses wastewater and drainage water as a valuable source of water and to augment the water supply by enforcing recycling and re-use in agricultural and industrial activities. To this are added an interest in increasing water storage, particularly through groundwater recharge, and attempts to educate the public on the need for water conservation (Darwish and Mohtara, 2013; Al Hashemi et al. 2014). For example, Qatar has also created a National Food Security Program (QNFSP), with a mandate to manage water resources efficiently in agriculture and food production through the use of technologies to minimize water consumption. As well as supplying the agricultural sector with freshwater, a core objective of the QNFSP is to use the solar desalinization of water to replenish the country's aquifers (QNFSP, 2012). Similarly, Abu Dhabi, United Arab Emirates, has already invested to increase water storage capacity (EAD, 2009).

2.3   Other States in the Persian Gulf area

The other States in the Persian Gulf area (the Islamic Republic of Iran and Iraq) make significantly less use of desalinization than the GCC States, although it still plays a part in their water supply arrangements.

It is assumed that the Islamic Republic of Iran has a desalinization capacity of about 400 megalitres per day (this is about 4.5 per cent of that installed in the GCC States). In terms of technology, the Islamic Republic of Iran's existing desalinization plants use a mix of thermal processes and RO. MSF is the most widely used thermal technology, although MED and vapour compression (VC) also feature (GWI, 2014).

Although Iraq is believed to have about the same amount of installed desalinization capacity as the Islamic Republic of Iran (Iraq is reported to have a capacity of 430 megalitres per day), it is used in a quite different way. Much of the water of the Euphrates and Tigris Rivers has a salinity above 1,000 parts per million. The Iraqi authorities therefore use desali-

Figure 28.5 | Use of the different Desalinization Technologies in GCC States. Source: GCC, 2012.

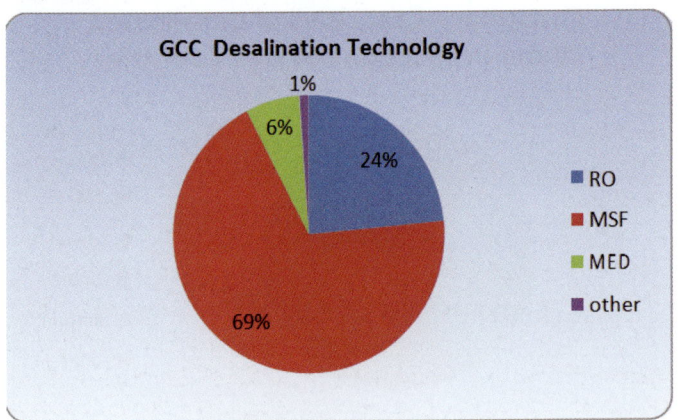

nization mainly to improve the poor quality of the river water, and only undertake a modest amount of seawater desalinization (ESCWA, 2009).

2.4 United States of America

Outside the Persian Gulf area, the United States has the largest installed desalinization capacity in the world. This is concentrated in California, Florida and Texas. However, desalinization of seawater is only a small part of the desalinization carried out in the United States. In 2010, seawater desalinization represented only 10 per cent of the desalinization capacity – 82 per cent was for brackish-water desalinization (largely from brackish groundwater, but also from rivers) and 8 per cent for waste-water re-use (Shea, 2010).

In California, however, the situation is changing. In November 2002, California voters adopted by the initiative procedure (by-passing the State legislature) Proposition 50, the "Water Security, Clean Drinking Water, Coastal and Beach Protection Act, 2002". This legislation allocated the sum of 50 million dollars for grants for brackish-water and ocean-water desalinization projects. This grant programme - administered by the State Department of Water Resources - aimed to assist local public agencies to develop new local water supplies through the construction of brackish water and ocean water desalinization projects and help advance water desalinization technology. Two rounds of funding were conducted in 2004 – 2006. A third funding round announced eight further grants in August 2014, totalling nearly 9 million dollars for a mix of plant construction, pilot projects and research (DWR, 2014).

Statistics on desalinization in California show that there are 10 seawater desalinization plants in California, with a daily capacity of about 23 megalitres. Not all these plants are in regular operation, but are used only when other water supplies need to be supplemented. Currently, there are proposals for a further 15 seawater desalinization plants. If all these plants were built, they would have capacity to provide some 946 – 1,400 megalitres per day – about 5 - 7 per cent of California's water demand in the early 2000s. One of these projects (Carlsbad) is expected to become operational in 2016 and will then be the largest seawater-desalinization plant in the United States, capable of delivering 190 megalitres per day of drinkable water (SWRCB, 2015).

3 Other countries with large desalinization capacities

A review of the countries other than those in the ROPME/RECOFI area[1] and the United States that have recently installed major desalinization capacitiesshows the following picture.

---
[1] Regional Organization for the Protection of the Marine Environment (ROPME) Members: Bahrain, Iran, Iraq, Kuwait, Oman, Qatar, Saudi Arabia, and the United Arab Emirates. Regional Commission for Fisheries (RECOFI) Members: Bahrain, Iran (Islamic Rep. of), Iraq, Kuwait, Oman, Qatar, Saudi Arabia, United Arab Emirates.

3.1 Algeria

Algeria is invested heavily in seawater desalinization capacity during the 2000s. By 2013, ten major plants had been put into service, with a capacity of 1,410 megalitres per day. In addition, 21 smaller plants with an aggregate capacity of 60 megalitres per day have been created. Further plants, with a further aggregate capacity of 850 megalitres per day, are expected to come into service in the near future, including that at Al Mactaa (500 megalitres per day), which will be one of the largest in the world (ADE, 2014)

3.2 Australia

At the start of the 2000s, the only operational seawater desalinization plants in Australia were on small islands. A prolonged drought in the middle of the 2000s, affecting most of the most heavily populated areas, led in 2007 to the creation of a National Urban Water and Desalination Plan, estimated to cost 840 million dollars, together with a National Centre of Excellence in Desalination, based in Perth, Western Australia. Desalinization is one of the options to be regularly considered under this plan. A major desalinization plant, with a capacity of 144 megalitres per day, has been in operation near Perth since 2006. Another major desalinization plant was built, partly financed by this plan, at a cost of 1,500 million dollars, for the city of Adelaide, South Australia, with a capacity of 280 megalitres per day. This is said to give the city a climate-independent source of water. In the light of the operating costs and the recovery of the local river system, the plant has, however, been placed on stand-by (ANAO, 2013). Similarly, major desalinization plants have been built to service Brisbane, Melbourne and Sydney, but have been placed in stand-by because of the recovery of other sources of supply (BT, 2010; ABC, 2012; SDP, 2015).

3.3 China

In 2012, the Chinese National Development and Reform Commission (NDRC) said that, under a first plan covering 2011-2015, China aims by 2015 to produce 2,200 megalitres per day of freshwater from desalinization. This compares with 660 megalitres per day in 2011. The NDRC said it will encourage innovation and upgrade desalinization facilities, as well as cultivate a number of desalinization facility manufacturers with international competitiveness. The NRDC will also encourage the use of desalinated seawater. More than half of freshwater provided in the islands of China, and more than 15 per cent of water delivered to coastal factories, is to come from the sea by 2015, according to the plan (PD (E), 2012).

3.4 Israel

Israel relies substantially on desalinization of seawater for its water supply. In 2008, desalinization represented 17 per cent (383 megalitres per day) of its water supply. By 2013, this was expected to rise to 32 per cent (4,950 megalitres per day) of the supply (Tenne, 2011).

3.5 Japan

Although Japan has relatively limited natural freshwater resources (it has about half the world average of natural freshwater resources per head), desalination has not so far played a major role in meeting general demand for water: a 2006 World Bank review of water management in Japan does not mention desalination (World Bank, 2006). The main focus in Japan on desalination has been in providing suitable cooling water for nuclear power plants – at least 10 such plants have associated desalination plants (IAEA, 2002). Nevertheless, desalination is used locally to supplement natural freshwater supplies for domestic and industrial use. For example, the authorities on Okinawa, the main island of the Ryukyu archipelago, installed in 1997 a desalination plant with a capacity of 40 megalitres per day (about 10 per cent of the island's daily demand for water) (Yamazato, 2006), and the city of Fukuoka on the southern Japanese island of Kyushu, after some major water shortages, installed in 2005 a desalination plant capable of supplying 50 megalitres per day (Shimokawa, 2008).

3.6 Spain

Spain has long had difficulties in providing adequate water supplies in some parts of the country. This is particularly the case along the Mediterranean coast, which has undergone massive development for tourism. In 2001, the Spanish authorities adopted legislation for a National Hydrological Plan. Among other things, this legislation declared a substantial number of desalination plants as projects being in the public interest (Spain, 2001). In 2005, this National Hydrological Plan was amended, and a new list of projects along the Mediterranean coastline was added, which the Ministry of the Environment and its associated bodies were to implement as a matter of urgency. This list included about 20 desalination projects (Spain, 2005). The desalination component of the Plan is reported to have had an estimated cost of about 3,000 million dollars. By 2013, 27 of the 51 approved plants had been built at a cost of about 2,200 million dollars. However, the economic recession starting in 2008 is reported to have reduced the demand for water to such an extent that many of the plants are standing idle or working at well below their planned capacity (Cala, 2013).

3.7 Other States

Many small islands have very limited natural freshwater resources, and have decided to supplement these with desalination. In the Caribbean, the following use desalination: Antigua and Barbuda, Aruba (the Netherlands), the Cayman Islands (United Kingdom of Great Britain and Northern Ireland), Curaçao (the Netherlands), Cuba and Trinidad and Tobago (Scalley, 2012; CWCL, 2015). Elsewhere, Malta (46 megalitres per day, 57 per cent of supply; (NSO, 2014)) and Singapore (capacity of 455 megalitres per day, 25 per cent of supply; (PUB, 2014)) are examples of island States which derive high proportions of their public water supplies from desalination.

In temperate zones, where natural freshwater supplies are usually adequate, authorities in some places are creating desalination plants as an insurance against long droughts and other disruptions of supply. For example, Thames Water in the United Kingdom has built a plant on the Thames estuary with a capacity of 150 megalitres per day (WTN, 2014).

In Chile, the northern provinces are some of the most arid areas in the world, yet it is here that the main minerals deposits are found that enabled mining to contribute 12.1 per cent of Chilean GDP in 2013 (BdeC, 2014). Since the extraction of metals from ore requires substantial quantities of water, there is a growing pressure in these northern provinces on freshwater resources. Many mines rely on freshwater from local rivers, but such abstractions (also called withdrawals) compete with growing demand from the local population. Some mines use seawater, but this imposes extra costs of safeguards against the corrosion that seawater causes. More recently, some mines have installed desalination plants to provide them with freshwater. There has been debate in the Chilean Chamber of Deputies about making the use of desalinized water compulsory if freshwater is to be used (Moreno et al., 2011; CdD, 2013). It seems likely that further desalination plants will be constructed.

4 Social and economic aspects of desalinization

Freshwater is essential to all life on land. Yet 97 per cent of all the water on earth is in the ocean (USGS, 2014). According to the Intergovernmental Panel on Climate Change (IPCC) (IPCC, 2014), about 80 per cent of the world's population already suffers serious threats to its water security, as measured by indicators including water availability, water demand, and pollution.

As the description of the nature, location and magnitude of desalinization shows, there are parts of the world where desalination is essential to human populations at present, or greater, levels. The largest area of this kind is the six GCC States, but island States, such as Malta and Singapore, are also in this category. Such States are likely to continue to generate significant growth in population over the coming years, together with the associated economic development. The only source of additional freshwater for such growing populations is likely to remain desalination.

Climate change is likely to add to the number of States that will wish to explore the use of desalination. The IPCC Fifth Report (IPCC, 2014) concludes that:

(a) The spatial distribution of the impacts of climate change on freshwater resource availability varies considerably between climate models, and depends strongly on the projected pattern of rainfall change. There is strong consistency in projections of reduced availability around the Mediterranean and parts of southern Africa, but much greater variation in projections for south and East Asia;

(b) Some water-stressed areas are likely to see increased runoff in the future, and therefore less exposure to water resources stress;
(c) Over the next few decades, and for increases in global mean temperature of less than around 2°C above preindustrial levels, changes in population will generally have a greater effect on changes in the freshwater available per head than will climate change. Climate change would, however, regionally exacerbate or offset the effects of population pressures;
(d) Estimates of future water availability are sensitive not only to climate and population projections and assumptions on usage per head, but also to the choice of hydrological impact model and to the measure of stress or scarcity that is adopted.

As an indication of the potential magnitude of the impact of climate change, one estimate quoted by the IPCC forecasts that

(a) A 1°C rise in global mean temperature (compared to the 1990s) will meant that about 8 per cent of the global population will see a severe reduction in water resources (that is, a reduction in runoff either greater than 20 per cent or more than the standard deviation of current annual runoff);
(b) A 2°C rise (on the same basis) will increase that proportion to 14 per cent; and
(c) A 3°C rise (on the same basis) will increase that proportion to 17 per cent.

The spread across climate and hydrological models was, however, large. The IPCC report includes desalinization as one of the range of adaptive measures that may prove particularly effective but notes that desalinization will increase green-house gas emissions to the extent that it relies on fossil fuel for its energy requirements.

It therefore seems likely that desalinization will increasingly be considered as a future adaptation measure for communities suffering increased water stress. Given that;

(a) Desalinization is, at least at present, significantly more expensive than most other forms of water supply when other options are available, and
(b) Most current forms of desalinization are using fossil fuels as an energy source, it is more likely that, outside areas where adequate alternative sources of water are simply not available, desalinization plants will be built as a fall-back provision, rather than as a primary source of freshwater.

5 Environmental impacts of desalinization

A major environmental impact of the majority of the present desalinization plants is the emission of greenhouse gases to generate the required energy. In some cases, especially in the GCC States, this is to some extent reduced through co-generation, by which the waste heat from electricity generation is used to desalinate water, without further major demands for energy. In those States, some 60 – 70 per cent of desalinization is done in this way. Where solar or nuclear energy can be used to power the desalinization, this impact is reduced or eliminated.

The other main forms of environmental impact arise from the intake of feedwater and the discharge of brine. The discharges are discussed in Chapter 20. Intake pipes create a risk to marine biota. The risk is highly variable, and is dependent on the technology employed for the seawater intake. In particular, it depends on how far the intake pipe is from the shore, as well as how the intake pipes are located with reference to the water column or the seabed. Biota living in the vicinity of a desalination plant's intake pipe can collide with, or be held against, the intake screens (impingement), or be sucked in with the feedwater into the plant (entrainment). Careful planning of the intake arrangements for each desalinization plant is needed to minimize this form of impact. For example, the intake arrangement Fukuoka in Japan has the intake pipes buried in a sandy seabed, which acts as a form of sand-filter to prevent non-microscopic biota entering the pipes (Shimokawa, 2008). However, such an arrangement can require substantial disruption of the seabed during construction and also lead to maintenance problems.

6 Significant environmental, social and economic aspects, knowledge gaps and capacity-building gaps

Desalinization has become essential to the functioning of many communities around the world. This is most evident in the GCC States and a number of small island States and territories. The impacts of climate change on freshwater supply are likely to increase the usefulness of desalinization as one of the effective forms of adaptation to these impacts, at least as a fall-back provision for periods when natural freshwater supplies are deficient.

There are many commercial firms specializing in the design and construction of desalinization plants. The technology is therefore available on the market. States and communities, however, need to have the capacities to negotiate in this market and to obtain the technologies that they need at a fair price.

In several cases, desalinization plants have been built which have proved to be inefficient or too large for the eventual requirement. There is therefore a case for more efficient sharing of knowledge on the planning, construction and operation of these plants.

Given their continuing importance, the need exists for better knowledge of how to operate desalinization plants with the lowest possible inputs of energy. Considerable progress seems to be possible in this direction: Malta, for example, reports having reduced the energy demand for its desalinization by 33 per cent in a decade (NSO, 2014).

References

ABC (Australian Broadcasting Corporation) News 18 December 2012. http://www.abc.net.au/worldtoday/content/2012/s3656791.htm

ADE (L'algérienne des eaux) (2014). *Déssalement*. http://www.ade.dz/index.php/projets-2/dessalement

Al Hashemi, R., Zareen, S., Al Raisi, A., Al Marzooqi, F.A., Hasan, S.W. (2014). A Review of Desalination Trends in the Gulf Cooperation Council Countries, *International Interdisciplinary Journal of Scientific Research*, Vol 1, No. 2.

AMS (American Meteorological Society) (2014). *Meteorological Glossary under the word "Freshwater"*. http://glossary.ametsoc.org/wiki/Freshwater .

ANAO (Australian National Audit Office) (2013). *Grants for the Construction of the Adelaide Desalination Plant (Audit Report 32/2012/13)*, Canberra, (ISBN 0 642 81327 2).

Arica, G.R. (2013). *Inauguran la primera planta desalinizadora de agua que funciona con energía solar en Arica*, 12 de Marzo. http://www.gorearicayparinacota.cl/w2/index.php/2013/03/12/inauguran-la-primera-planta-desalinizadora-de-agua-que-funciona-con-energia-solar-en-arica/.

BdeC (Banco de Chile) (2014). *Serie actividad económica sectorial mensual 2008 – 2013*. http://www.bcentral.cl/estadisticas-economicas/series-indicadores/index_aeg.htm.

BT (Brisbane Telegraph) 5 December 2010 http://www.brisbanetimes.com.au/queensland/tugun-desalination-plant-to-be-mothballed-20101205-18l30.html.

Cala, A. (2013). Spain's Desalination Ambitions Unravel, *New York Times*, October 9, 2013. http://www.nytimes.com/2013/10/10/business/energy-environment/spains-desalination-ambitions-unravel.html?pagewanted=all&_r=0 .

CdD (Chile Cámara de Diputados) (2013). *Redacción de Sesiones*, Martes 10 de Diciembre de 2013. http://www.camara.cl/pdf.aspx?prmID=10338%20&prmTIPO=TEXTOSESION Cooley, H., Donnelly, K. (2012). *Key Issues in Seawater Desalination in California – Proposed Seawater Desalination Facilities*, Pacific Institute, Oakland, California. http://pacinst.org/wp-content/uploads/sites/21/2014/04/desalination-facilities.pdf.

Cooley, H., Gleick, P.H., Wolff, G., (2006). *Desalination, with a Grain of Salt, a California Perspective*, Pacific Institute, Oakland, California. http://www.pacinst.org/wp-content/uploads/sites/21/2013/02/desalination_report3.pdf.

CWCL (Cayman Water Company Ltd) (2015). *Frequently Asked Questions* (http://www.caymanwater.com/html/FAQs.html).

Darwish, M.A., Mohtara, R. (2013). Qatar water challenges, *Desalination and Water Treatment*, Volume 51, Issue 1-3.

DWR (California Department of Water Resources) (2014). *Desalination*. http://www.water.ca.gov/desalination/ .

EAD (Abu Dhabi Environment Agency) (2009). *Annual Report 2009*.

ESCWA (United Nations Economic and Social Commission for Western Asia) (2009). *Water Development Report 3: Role Of Desalination In Addressing Water Scarcity*, United Nations, New York (ISBN. 978-92-1-128329-7).

Frontinus - Sextus Julius Frontinus, *De Aquaeductu (Lib I.5)*, edited by R H Rodgers (2004). Cambridge University Press, Cambridge, England.

GCC (Gulf Cooperation Council) (2012). *Water Statistical Report*, Riyadh.

GCC (Gulf Cooperation Council) (2014). *Desalination In The GCC – The History, The Present & The Future*, Riyadh.

Gleick, P.H., Allen, L., Christian-Smith, J., Cohen, M.J., Cooley, H., Herberger, M., Morrison, J., Palaniappan, M., Schulte, P. (2009). *The World's Water - The Biennial Report On Freshwater Resources*, Volume 7, Island Press, Washington, DC.

GWI (Global Water Information) (2014). *Water Desalinization Report – Market Data*, citing Desal.Data. http://www.desalinization.com/market/desal-markets (accessed 20 October 2014).

GWI (Global Water Intelligence) (2015). Section 1: Market profile. In: *IDA Desalination Yearbook 2015-2016*.

IAEA (International Atomic Energy Agency) (2002). *Status of Design Concepts of Nuclear Desalination Plants*, IAEA, Vienna (IAEA-TECDOC-1326, ISBN 92–0–117602–3).

IAEA (International Atomic Energy Agency) (2007). *Economics of Nuclear Desalination — New Developments and Site Specific Studies*, Vienna (IAEA-TECDOC-1561, ISBN 978–92–0–105607–8).

IPCC (2014). Jiménez Cisneros, B.E, Oki, T., Arnell, N.W., Benito, G., Cogley, J.G., Döll, P., Jiang, T. and Mwakalila, S.S., Freshwater resources, in: *Climate Change 2014: Impacts, Adaptation, and Vulnerability. Part A: Global and Sectoral Aspects*. Contribution of Working Group II to the Fifth Assessment Report of the Intergovernmental Panel on Climate Change, Cambridge University Press, Cambridge, United Kingdom.

Jenkins, S., Paduan, J., Roberts, P., Schlenk, D., Weis, J. (2012). *Management of Brine Discharges to Coastal Waters, recommendations of a Science Advisory Panel*. Technical Report 694, Southern California Coastal Water Research Project, Costa Mesa, California, United States of America.

Kalogirou, S.A. (2009). *Solar Energy Engineering - Processes and Systems*, Academic Press, Burlington, Massachusetts (ISBN: 978-0-12-374501-9).

Moreno, P.A., Aral, H., Cuevas, J., Monardes, A., Adaro, M., Norgate, T., Bruckard, W. (2011). The use of seawater as process water at Las Luces copper–molybdenumbeneficiation plant in Taltal (Chile), *Minerals Engineering*, Vol 24.

NSO (Malta National Statistics Office) (2014). *World Water Day 2014: Water and Energy*. http://www.nso.gov.mt/statdoc/document_file.aspx?id=3974.

NYCEP (New York Environmental Protection Department) (2014). *History of Drought and Water Consumption*. http://www.nyc.gov/html/dep/html/drinking_water/droughthist.shtml .

PD(E) (People's Daily (English version)) (2012). *China unveils plan to boost seawater desalination*, 27 December. http://en.people.cn/90778/8072171.html .

PUB (Singapore Public Utilities Board) (2014). *The Singapore Water Story Water: From Vulnerability to Strength*. http://www.pub.gov.sg/water/Pages/singaporewaterstory.aspx

QNFSP (Qatar National Food Security Programme) (2014). *The Qatar National Food Security Plan*. (www.qnfsp.gov.qa).

Saif, O. (2012). *The Future Outlook of Desalination in the Gulf: Challenges & opportunities faced by Qatar & the UAE*. http://inweh.unu.edu/wp-content/uploads/2013/11/The-Future-Outlook-of-Desalination-in-the-Gulf.pdf.

Scalley, T.H. (2012). Freshwater Resources in The Insular Caribbean: An Environmental Perspective, *Caribbean Studies*, Vol. 40, No. 2.

Shea, A.L. (2010). *Status and Challenges for Desalination in the United States*. https://www.watereuse.org/sites/default/files/u8/Status_Challenges_US.pdf.

Shimokawa, A. (2008) *Desalination plant with Unique Methods in Fukuoka*. (http://www.niph.go.jp/soshiki/suido/pdf/h21JPUS/abstract/r9-2.pdf.

Spain (2001). Ley 10/2001, de 5 de julio, del Plan Hidrológico Nacional (Boletín Oficial del Estado, 161/2001 of 6 July).

Spain (2005). Ley 11/2005, de 22 de junio, por la que se modifica la Ley 10/2001, de 5 de julio, del Plan Hidrológico Nacional (Boletín Oficial del Estado, 149/2005 of 23 June).

SDP (Sydney Desalination Plant) (2015), *Our History* http://www.sydneydesal.com.au/

who-we-are/our-history/ .

SWCC (Saline Water Conversion Corporation). (2014). التقرير السنوي ١٤٣٥/١٤٣٦ (*Annual Report 1435-1436 AH – 2014 CE*) http://www.swcc.gov.sa/files/assets/Reports/annual2014.pdf

SWRCB (California State Water Resources Control Board) (2015). *Draft Staff Report: Desalination Facility Intakes, Brine Discharges, and the Incorporation of Other Non-Substantive Changes (March 20, 2015).* (http://www.waterboards.ca.gov/water_issues/programs/ocean/desalination/docs/amendment/150320_sr_sed.pdf.

Tenne, A. (2011). *The Master Plan for Desalinization in Israel 2020*. http://www.water.gov.il/Hebrew/ProfessionalInfoAndData/2012/07-Israel-Water-Sector-Desalination.pdf.

The National (2012). *Twenty-two solar desalination plants completed, agency says*, 18 January. http://www.thenational.ae/news/uae-news/environment/twenty-two-solar-desalination-plants-completed-agency-saysUSGS (United States Geological Survey) (2014). *The Water Cycle – The Oceans.* http://water.usgs.gov/edu/watercycleoceans.html.

WHO (World Health Organization) (2007). *Desalination for Safe Water Supply: Guidance for the Health and Environmental Aspects Applicable to Desalination*, Geneva. http://www.who.int/water_sanitation_health/gdwqrevision/desalination.pdf

World Bank (2005). *Annual Report 2005.* http://siteresources.worldbank.org/INTANREP2K5/Resources/51563_English.pdf.

World Bank (2006) *Water Resources Management in Japan Policy, Institutional and Legal Issues*. http://siteresources.worldbank.org/INTEAPREGTOPENVIRONMENT/Resources/WRM_Japan_experience_EN.pdf.

WTN (Water-Technology.net) (2014). *Thames Water Desalination Plant, London, United Kingdom*. http://www.water-technology.net/projects/water-desalination/.

Yamazato, T. (2006) *Seawater Desalination Facility on Okinawa.* http://www.niph.go.jp/soshiki/suido/pdf/h19JPUS/abstract/r27.pdf .

29 Use of Marine Genetic Resources

Writing team:
Sophie Arnaud-Haond, Caroline Bissada, Peyman Eghtesadi Araghi (Lead Author and Lead member), Elva Escobar-Briones, Françoise Gaill, S. Kim Juniper, Ahmed Kawser, Ellen Kenchington, Nigel Preston, Gabriele Procaccini, Nagappa Ramaiah, Jake Rice (Co-Lead member) Alex Rogers, Wouter Rommens, Zheng Senlin and Michael Thorndyke

1 Introduction

The natural environment has long been a source of inspiration for new drugs and other products of biotechnology. Until relatively recently, the terrestrial environment, in particular, has been the primary source of genetic material and natural products at the centre of major new developments in biotechnology, including new drugs. Examples of natural products used in drug development include the anti-malarial drug quinine isolated from the bark of the *Chinchona*, the analgesics codeine and morphine from *Papaver somnifetum* latex, and antibiotics such as penicillins and tertracyclines from strains of *Penicillium* sp. and *Streptomyces* sp. The terrestrial environment contains far more known species of plants and animals than are at present known in the oceans (Hendricks et al., 2006; Mora et al. 2011), and has contributed greatly to the development of new biotechnologies, and new drugs in particular (Molinski et al., 2009; Arrieta et al., 2010; Leal et al., 2012). Yet there are many reasons to expect that the marine environment should represent a rich reservoir of novel genetic material and natural products, particularly those derived from animals and their microbiomes. Covering more than 70 per cent of the planet, and constituting 95 per cent of the volume of the biosphere, the oceans are home to a greater diversity of major animal groups (phyla) than the terrestrial environment (34 of 36 known phyla are found in the oceans *versus* 17 found on land). Most marine organisms have a large dispersal potential, either through the movement of adults, or through the dispersal of larvae by ocean circulation, potentially crossing hundreds to thousands of kilometres during their development. It is thus likely for many species that the same genomic background could be sampled both within several exclusive economic zones (EEZs) and in areas beyond national jurisdiction (ABNJ).

The study and utilization of marine genetic resources is a fairly recent human activity and, compared to the terrestrial environment, examples are relatively few and scattered throughout the world ocean. This chapter will therefore provide a general review of marine genetic resources (MGRs) rather than providing a regionally comprehensive and inclusive assessment. We will use a fairly broad definition of marine genetic resources that includes nucleic acid sequences, chemical compounds produced by marine organisms and unrefined materials extracted from marine biomass. Within areas under national jurisdiction, where marine organisms are most abundant and most accessible to researchers, MGRs and marine biodiversity are best known. MGRs in the vast, offshore oceanic areas beyond national jurisdiction (ABNJ) are, by comparison, poorly documented. The growing appreciation of the diversity and novelty of life in the oceans that has emerged from the results of programs such as the Census of Marine Life, and MicroB3 (Marine Microbial Biodiversity, Bioinformatics and Biotechnology - a European Union 7th Framework Program), have fuelled interest in the commercial possibilities of MGRs. Currently MGRs are important to the economies and sustainability of many sectors including the pharmaceutical industry (new medicines), cosmetics, the emerging nutraceutical industry, and aquaculture (new high-value high nutrition, healthy foods), biomedicine and many other economically and culturally important sectors.

Molinski et al. (2009) noted a decline since the mid-1990s in the interest of large pharmaceutical companies in the development of 'drugs from the sea'. This is likely related to a general decline in all natural product research (Dias et al. 2012; Lahlou, 2012). Some hints of a recent resurgence exist, but it will be several years before it can be determined if this is a sustainable trend. In parallel with large commercial pharmaceutical interests reducing their activity in this sector, activity increased in smaller academic-industry partnerships in developed countries, resulting in the emergence of small-scale, start-up companies. New and affordable developments in analytical technologies (gene sequencing, biomolecule characterization) have helped drive this new trend. Further evidence of increased interest in MGR is found in an analysis by Arrieta et al. (2010), who noted growth over the past decade in the accumulation of patent claims related to marine organism genes (currently increasing by 12 per cent per year) and identified marine natural products. In the context of its work on the conservation and sustainable use of marine biological diversity of areas beyond national jurisdiction (ABNJ), the United Nations General Assembly stressed the need for a comprehensive regime to better address the conservation and sustainable use of marine biological diversity of areas beyond national jurisdiction. Marine genetic resources, including questions on the sharing benefits, are part of the package of issues to be addressed by the preparatory committee established to make substantive recommendations to the Assembly on the elements of a draft text of an international legally-binding instrument under the United Nations Convention on the Law of the Sea (resolution 69/292).

2 Marine Pharmaceuticals

Marine biodiversity in theory has enormous biotechnological potential. Yet, to date, despite "repeated waves of enthusiasm and much early promise," examples of successful development of commercial products are very few. In the early 1950s, the first marine bioactive compounds, spongouridine and spongothymidine, were isolated from the Caribbean sponge *Cryptotheca crypta*. The 1970s saw the beginning of basic scientific research in chemistry and pharmacology of marine natural products and directed efforts in drug development (Molinski et al., 2009; Mayer et al., 2010). However, only in the last decade have these research efforts resulted in the production of a first generation of drugs from the sea into clinical trials. Not until 2004 was the first drug from the sea (developed from a neurotoxin produced by a tropical marine cone snail, *Conus magnus*) finally approved for sale on the market under the name Prialt; it is used in the treatment of chronic pain associated with spinal cord injuries. More recently, in 2007, a second drug, the anti-tumour compound trabectedin (known as Yondelis, discovered from the colonial tunicate *Ecteinascidia turbinata*), was approved for the treatment of soft-tissue sarcoma in the European Union. Seven drugs derived from marine organisms are currently approved by the United States Food and Drug Administration (FDA) and on the market (Mayer et al., 2010; http://marinepharmacology.midwestern.edu/clinPipeline.htm). An example of a medical but non-pharmaceutical usage of compounds produced by marine organisms are wound treatment 'dressings' made of marine diatom polymers (Marine Polymer Technologies Inc.).

3   Marine Nutraceuticals

Related to the biopharma industry, attention also increasingly focuses on so-called marine nutraceuticals. These compounds (or 'substances') are found to be beneficial to human health and "delivered" via food or food products (Kim, 2013). Commercial efforts are also underway to synthetically produce some of these substances. Compounds of interest comprise a wide variety, including polysaccharides, polyphenols, bioactive peptides, polyunsaturated fatty acids, and carotenoids; with identified anticancer, anti-inflammatory, anti-oxidant, and antimicrobial activities (Guerard et al., 2010; Ngo et al., 2011; Vidanarachchi et al., 2012). Many of the products derive from marine algae and include, for example, Fucoidan, a sulfonated polysaccharide found in brown algae and shown to benefit immune system and gastrointestinal health by neutralizing free radicals. Other examples are: Vitamin C and carotenoids, anti-oxidants obtained from *Chlorella* sp.; Spirulina (cyanobacteria of the genus *Arthrospira*), used as a natural blue food colorant and contains a wide variety of nutrients, such as B vitamins, minerals, proteins, linolenic acid, and anti-oxidants such as beta carotene and Vitamin E.

In contrast to the more limited development of new drugs from marine organisms, the nutraceutical industry in Europe and Asia is experiencing significant growth. Marine and algal omega-3 products alone accounted for 1.5 billion United States dollars in sales in 2009, and the global nutraceutical industry as a whole is estimated to grow to a 180 billion US dollars business by 2017 (Kim, 2013). This rapidly growing sector often uses marine biomass, such as fish waste or harvested algae, to produce health food products and restorative cosmetics.

4   Marine organism-derived anti-foulants and adhesives

The antifoulant and marine adhesives industries have a long history. The costs of biofouling to the fleets around the world are estimated to run into 100 million US dollars each year. Many biocidal antifoulants are now banned (e.g., copper-based paints) because of their direct toxicity to marine organisms. Tributyl tin-based products, once used extensively, are now banned because of their now well-known impact on sex determination in marine molluscs and other organisms. Naturally derived antifoulants include enzymes, antimicrobials, biomimetics such as novel topographies, and natural chemical signals (Callow and Callow, 2011; Kirschner and Brennan, 2012; Gittens et al., 2013)

Marine algae are an important source of novel antifouling compounds and mangroves and sponges also feature high on the list of exploited marine organisms. Many algal species produce compounds that inhibit growth of marine bacteria. A red algal metabolite has been shown to affect the composition and density of bacterial colonies, and compounds from the mangrove plant reduce larval barnacle settlement. Such identified active natural compounds inspire the commercial synthesis of mimetics that are more stable and easily applied to surfaces, such as ship hulls, docks etc. (Callow and Callow, 2011; Kirschner and Brennan, 2012; Gittens et al., 2013).

Although the antifoulant industry seeks to prevent undesirable attachments, the bio-adhesive industry focuses on marine glues, materials that enable stable and robust adhesion under water. Perhaps not surprisingly, target organisms used as potential sources of novel marine glues are precisely those that the antifoulant community seeks to deter, for example, barnacles, mussels, tube-building worms and echinoderms. All successfully attach to wet surfaces with tenacity and therefore provide a rich source of ideas for novel adhesives. Some (starfish) uniquely have strong yet temporary adhesives used to anchor their feet during locomotion (Kamino, 2010; Stewart et al., 2011; Petrone, 2013). The mechanisms employed often involve complex surface chemistries and physicochemical three-dimensional properties, but nevertheless often have a simple biochemical basis. For example, one of the most widely employed core substances is L-Dopa, whilst extracellular DNA is also thought to be an important component in some bio-adhesives (Kamino, 2010; Stewart et al., 2011; Petrone, 2013). Bulk harvesting of natural glues from marine organisms is impractical and any economic exploitation will require an extensive research effort directed at the chemical structure and genetic basis of the natural glues produced by marine organisms.

5   Environmental, economic and social aspects of MGR research

Marine scientific research related to MGR involves the collection of ocean data and the collection of biological samples from the water column and the seabed. Most scientific instruments and sampling devices so far cause negligible harm to marine habitats. Nonetheless, in some parts of the seabed scientific sampling is concentrated in relatively small areas, such as at well-studied hydrothermal vent fields. There, competing uses have arisen between researchers requiring physical samples (biological or geological) and researchers wishing to observe the natural evolution of local ecosystems (Godet et al., 2011). Such use conflicts led to the adoption of codes of best practices by international organizations such as InterRidge (http://www.interridge.org/publications) (see Chapter 30 on Marine scientific research).

Although the possibility still exists of direct extraction of new natural products from marine biomass, such as the extraction of compounds from seaweeds and oils from Sargassum or krill, future exploitation of MGR for purposes of developing the genetic resource is unlikely to involve substantial large-scale harvesting of marine organisms and the resultant destruction of habitat and depletion of species populations as caused by fisheries. Indeed, if recent trends can be used as a guide, future MGR products will come from laboratories and larger facilities where genes and biomolecules will be discovered, and then incorporated into industrial processes that may involve the artificial synthesis of biomolecules (e.g. Mizuki et al., 2014; Wilde et al., 2012; Newman and Cragg, 2012), or the insertion of desired genes into microbes that then produce the desired molecule (Newman and Cragg, 2012). However, potential for habitat damage exists even from scientific fieldwork, particularly in vulnerable marine ecosystems and areas of intensive study and sample collection. In addition, the success of nutraceutical extraction from fish waste could lead to the large-scale harvesting of stocks with little or no food value to humans but attractive as sources of biomolecules for commercial products.

Benefits arising from MGR research include improved knowledge of marine biodiversity and the importance of biodiversity to the provision of ecosystem services and the maintenance of ocean health. Related marine natural products research will also result in more efficient, value-added exploitation of marine resources, as described above. Any accelerated growth in the MGR research sector will not only profit the marine pharmaceutical industry, but also result in health and other social benefits from access to the pharmaceutical or other products themselves (Leary and Juniper, 2014; Broggiato et al., 2014).

6 Capacity to engage in MGR research

6.1 Research Vessels

The first step in the process of accessing MGRs is the collection of samples in the field. In inter-tidal and coastal waters, the capacity to collect biological samples does not require advanced technology. Smaller vessels, even sailing vessels, can be used to sample water column organisms. For example, two recent expeditions used the privately owned sailing yachts; Tara (http://oceans.taraexpeditions.org/index-o.php/en) and Sorcerer II (www.jcvi.org/cms/research/projects/gos/overview/) to conduct ocean basin-scale surveys of planktonic microorganisms. Further offshore sampling of deep-water organisms and the seabed requires an offshore-capable ship (defined here as greater than 60 metres in length)

and in most cases a specialized research vessel. Operating costs for these larger vessels are typically greater than 25 000 United States dollars per day. At first glance, basic capability to engage in MGRs research appears to be fairly widespread among nations. The International Research Vessel database (www.researchvessels.org) lists 271 vessels greater than 60 metres in length available in more than 40 countries. However, the majority of these vessels belong to a few developed countries, as shown in Figure 29.1 (Juniper, 2013).

At sea, most biological specimens are collected by lowering or towing sampling devices from the vessel (Juniper, 2013). Heavier gear, e.g., winches, cables, dredges and corers, is required to sample the seabed, where the majority of known marine species are found. This latter requirement reinforces the importance of larger, motorized research vessels for accessing biodiversity at depths of several kilometres. Some sampling devices are highly specialized and not all research vessels have the same collecting capacity. This is especially true for remotely operated vehicles (ROVs), which are used to locate and sample hydrothermal vent ecosystems, a source of novel species adapted to extreme conditions. Hydrothermal vent sites are very small in area, usually tens to hundreds of square meters, and cannot be easily sampled or even located by lowering standard sampling devices from ships. They require precisely navigated diving with ROVs or human-occupied submersibles, although the latter are used less frequently than ROVs. These vehicles are capable of diving to much deeper depths than military submarines, and are equipped with robotic arms, video cameras and various sampling devices. The list of States possessing and operating deep-diving scientific submersibles is limited to a subset of the developed countries currently leading marine scientific research efforts globally, such as the USA, Canada, the UK, France, the Russian Federation, Japan and most recently China, India and South Korea (Juniper, 2013). ROVs can also be chartered from commercial firms that provide support for the offshore petroleum industry, but most of these vehicles are limited to depths of less than 2,000 metres.

6.2 Biodiversity Expertise

Table 29.1 | Geographic distribution of patent claims for genes of marine origin. Source: Arnaud-Haond et al. (2011)

Country	Marine Organism Patent Claims
USA	199
Germany	149
Japan	128
France	34
United Kingdom	33
Denmark	24
Belgium	17
Netherland	13
Switzerland	11
Norway	9

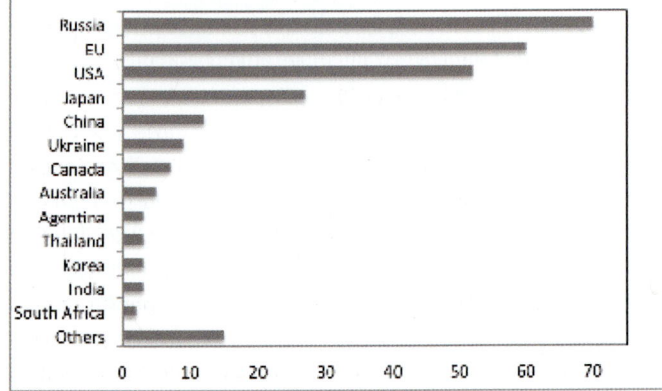

Figure 29.1 | Geographic distribution of offshore research vessels (60 m or greater in length). Source: International Research Vessel Database.

Another important capacity related to accessing and deriving value from marine biodiversity is the specialist knowledge required to identify species, both known species and those new to science (Hendriks & Duarte, 2008; Juniper, 2013). Marine biodiversity specialists are mostly trained in developed countries with a long history of botanical and zoological scholarship in universities and museums. One way to measure the worldwide distribution of marine biodiversity expertise is to examine the level of publication activity in the scientific literature in relation to the country of affiliation of the lead author on these publications. A recent review of the literature revealed that the majority of publications in the field of marine biodiversity come from relatively few developed countries (Hendriks and Duarte, 2008).

These specialists are experts in the morphological identification of specimens and, increasingly, the interpretation of DNA sequence information that is used to identify marine plants, animals and microbes. It could be argued that this scientific expertise is of greater importance to the usage and application of MGR than is access to laboratories that can produce gene sequence and biochemical composition information from field samples. Many countries have commercial sequencing and biochemical analysis facilities that serve national and international organizations/institutions, and academic and private clients. The raw data produced by these facilities require expert interpretation before they can be used to identify species, genes and biochemical compounds of interest. Table 29.1 illustrates the dominance by applicants from a few developed countries in recent patent claims associated with genes of marine origin according to a database undertaken by Arnaud-Haond et al. (2011). The top ten countries account for 90 per cent of filed gene patents, with 70 per cent from the top three (Arnaud-Haond et al. 2011). Relatively new approaches, such as microbial metagenomics, also require sophisticated bio-informatics tools and training and these are most accessible in developed countries. Nevertheless, some (growing?) capabilities in bio-informatics and genomics exist in developing countries, particularly in the health and agricultural science sectors, and these skills could be adapted and applied to the exploitation of MGRs.

6.3 Deriving Economic Value from MGR

Few examples exist of a direct developmental path from field collections of marine organisms through to the commercialization of marine natural products or genes from marine organisms. The Vent Polymerase enzyme is one example of a rapid transition from the discovery of a new microorganism in the field to the commercialization of a biomolecule (Mattila et al, 1991). Field research in the marine environment is primarily led by academic or government scientists and is aimed at increasing our knowledge of the ocean and the organisms that it supports. It is this knowledge base that may be later used by laboratory-based scientists in academia, industry and government in research more directly related to the eventual use of MGR for commercial purposes (Brock, 1997; Juniper, 2013). Surprisingly few direct connections are found between these basic and applied research sectors.

A review by the journal *Nature* (Macilwain, 1998) concluded that tens to hundreds of thousands of failed prospects exist for every example of a commercially successful natural product. One measure of success at extracting drugs from the sea is the number of drugs from marine organisms approved for commercialization by the FDA. Currently (June 2014) only seven drugs from marine organisms have received FDA approval (see also above), and approximately twice that number are in clinical trials (http://marinepharmacology.midwestern.edu/clinPipeline.htm). Mayer et al. (2010) note that the pre-clinical pipeline "continues to supply several hundred novel marine compounds every year and those continue to feed the clinical pipeline with potentially valuable compounds."

Among the myriad marine life forms, the greatest unknown potential source of novel bioproducts is marine microorganisms (Gerwick et al., 2001). The advent and applications of genome, proteome, and transcriptome analyses are helping to recognize the enormous extent of marine microbial diversity and deepening our understanding of how the chemistry of the ocean and its interaction with the atmosphere are also mediated by microbial metabolism. From the bioprospecting perspective, the biomolecules produced by marine microbes remain virtually unstudied for potential commercial applications.

Microbial bioproducts will remain undeveloped unless there is a feasible alternative to mass culturing and efficient harvesting. In general, molecular approaches offer particularly promising alternatives, not only to the supply of known natural products (e.g., through the identification, isolation, cloning, and expression of genes involved in the production of the chemicals), but also to the discovery of novel sources of molecular diversity (e.g., through the identification of genes and biosynthetic pathways from uncultured microorganisms; Bull et al., 2000).

6.4 "Omics" Tools

Recent breakthroughs in marine metagenomics are paving the way for a new era of molecular marine research. Metagenomic studies of marine life are yielding new insights into ocean biodiversity and the functioning of marine ecosystems; for the first time we can explore ecological interrelationships at the gene level. The first study of this type, led by the J. Craig Venter Institute, used these tools to survey marine microbial diversity, discovering thousands of new species, millions of new genes and thousands of new protein families. A seminal 2004 paper described how the analysis of 200 litres of surface water from the Sargasso Sea enabled the identification of about 1,800 genomic microbial species and 1.2 million unknown genes using an environmental metagenomic shotgun approach (Venter et al., 2004). This work led to other marine metagenomic studies, such as those reviewed by Gilbert and Dupont (2011) that show how massive sequencing of environmental samples can lead to the discovery of extraordinary microbial biodiversity and to the unravelling of important components of the pathways of phosphorus, sulphur, and nitrogen cycling. These powerful molecular tools are enabling a new "study it all approach" to discovering organismal, genetic, biomolecular and metabolic diversity in the oceans at an accelerated pace, as exemplified by the recent round-the-world Tara expedition

(Ainsworth, 2013) that returned over 25,000 samples of water column organisms for intensive molecular analysis with so-called 'omics' tools.

7 The importance of understanding life histories, plasticity and the impact of climate change

Climate change and ocean acidification are widely recognized as increasing threats to marine ecosystems impacting growth, survival, reproduction, and many other phenotypic features of all marine organisms, leading to changes in species abundances and distribution. Many key marine invertebrates, including crabs, echinoderms and molluscs often have equally if not more complex life histories with several, vulnerable, free-swimming planktonic stages before they metamorphose and settle to their adult benthic form. These sophisticated morphological and physiological processes are underpinned by complex gene regulatory networks and genomic pathways (Gilbert, 2013). This is significant because although it is now increasingly clear that predicted climate change events will affect marine biota, it is also now clear that several key organisms exhibit a valuable plasticity in the face of environmental stresses and challenges (Byrne and Przeslawski, 2013; Chan et al., 2015a, b; Dorey et al., 2013; Harley et al., 2006; Merila and Hendry, 2014; Reusch, 2014; Stumpp et al., 2011a, b; 2012; Thor and Dupont, 2015). Understanding and exploring this potential will be vital if we are to identify resilient and potentially phenotypically plastic populations.

Identification of the genetic bases of this plasticity will be an important resource to the future of exploited marine species in a changing ocean.

8 Conclusion

The commercial utilization of MGRs had very modest beginnings in the 20th century, particularly when measured against the estimated potential of the great diversity of species and biomolecules in the sea. More promisingly, the past decade has seen the commercialization of the first drugs derived from marine organisms, and considerable growth in nutraceutical and other non-medical uses of marine natural products. This past decade has also seen an astounding increase in our capacity to discover novel marine organisms and biomolecules and understand the genomic basis of life in the oceans. New technologies are fostering a new wave of optimism about the commercial potential of MGRs that is influencing funding priorities for marine research and has led to the emergence of futuristic terms, such as 'blue growth' and the 'knowledge-based blue economy of tomorrow'. Much of the capacity for discovery and commercialization of MGRs remains in the hands of a few developed countries. Much of the genetic diversity in our seas and oceans remains unknown and relatively unexplored; yet more potential is to be realized, particularly in the context of climate change. While this chapter has emphasized the commercial utilization of marine genetic resources, there are strong arguments to be made for the value to societies and ecosystems of simply protecting and conserving marine genetic resources (e.g. Pearce and Moran, 1994).

References

Ainsworth, C. (2013). Systems ecology: Biology on the high seas. *Nature* 501, 20–23 doi:10.1038/501020a.

Arnaud-Haond, S., Arrieta J.M., and Duarte, C.M. (2011). Marine biodiversity and gene patents. *Science* 331, 1521-1522, doi: 10.1126/science.1200783.

Arnaud-Haond, S., Arrieta J.M., and Duarte, C.M. (2010). What lies beneath: Conserving the oceans' genetic resources. *Proceedings of the National Academy of Sciences* 107, 18318-18324. doi/10.1073/pnas.0911897107.

Brock, T.D. (1997). The value of basic research: Discovery of *Thermus aquaticus* and other extreme thermophiles. *Genetics* 146, 1207-1210.

Broggiato, S., Arnaud-Haond, S., Chiarolla, C., and Greiber, T. (2014). Fair and equitable sharing of benefits from the utilization of marine genetic resources in areas beyond national jurisdiction: Bridging the gaps between science and policy. *Marine Policy* 49, 176-185. doi.org/10.1016/j.marpol.2014.02.012.

Bull, A.T., Ward, A.C., Goodfellow, M. (2000). Search and discovery strategies for biotechnology: the paradigm shift. *Microbiology and Molecular Biology Reviews*: 64:573–606.

Byrne M. and Przeslawski, R. (2013). Multistressor Impacts of Warming and Acidification of the Ocean on Marine Invertebrates' Life Histories. *Integrative and Comparative biology* 53, *582-596*.

Callow, J.A. and Callow, M.E. (2011). Trends in the development of environmentally friendly fouling-resistant marine coatings. *Nature Communications* 2, 244. doi: 10.1038/ncomms1251.

Chan, K.Y.K., García E., and Dupont S. (2015a). Acidification reduced growth rate but not swimming speed of larval sea urchins. *Nature Scientific Reports* 5, 9764 doi:10.1038/srep09764.

Chan, K.Y.K., Grünbaum, D., Arnberg, M. and Dupont S. (2015b). Impacts of ocean acidification on survival, growth, and swimming behaviours differ between larval urchins and brittlestars. *ICES Journal of Marine Science* . doi: 10.1093/icesjms/fsv073.

Dias, D.A., Urban, S. and Roessner, U. (2012) A historical overview of natural products in drug discovery. *Metabolites* 2012, 2, 303-336; doi:10.3390/metabo2020303.

Dorey, N., Lancon, P., Thorndyke, M.C. and Dupont, S. (2013). Assessing physiological tipping point of sea urchin larvae exposed to a broad range of pH. *Global Change Biology* 19, 3355-3367. doi: 10.1111/gcb.12276.

Gerwick, W.H., Tan, L.T. and Sitachitta, N. (2001). *Nitrogen-containing metabolites from marine cyanobacteria*. P. 75-184 in The Alkaloids, G. Cordell, editor. (Ed.), Academic Press, San Diego.

Gilbert, J. and Dupont, C. (2011). Microbial Metagenomics: Beyond the Genome. *Annual Review of Marine Science* 3, 347-371.

Gilbert, S.F. (2013). Developmental Biology. Sinauer Associates, Inc. Sunderland, M.A., USA, pp 719.

Gittens, J.E., Smith, T.J., Suleiman R. and Akid, R. (2013). Current and emerging environmentally friendly systems for fouling control in the marine environment. *Biotechnology Advances* 31, 1738–1753.

Godet, L., Zelnio, K.A. and van Dover, C.L. (2011) Scientists as stakeholders in the conservation of hydrothermal vents. *Conservation Biology* 25, 214-222, doi: 10.1111/j.1523-1739.2010.01642.x.

Guerard, F., Decourcelle, N., Sabourin, C., Floch-Laizet C., Le Grel, L., Le Floch, P., Gourlay, F., Le Delezir, R., Jaouen, P. and Bourseau, P. (2010). Recent developments of marine ingredients for food and nutraceutical applications: a review. *Journal des Sciences Halieutique et Aquatique* 2, 21-27.

Harley, C.D.G., Randall Hughes, A., Hultgren, K.M., Miner, B.G., Sorte, C.J.B., Thornber, C.S., Laura, F. Rodriguez, C.S., Lars, Tomanek L., and Williams, S.L., (2006). The impacts of climate change in coastal marine systems. *Ecology Letters* 9, 228-41.

Hendriks, E., Duarte, C.M. and Heip, C.H.R. (2006). Biodiversity research still grounded. *Science* 312, 1715.

Hendriks, I.E. and Duarte, C.M. (2008). Allocation of effort and balances in biodiversity research. *Journal of Experimental Marine Biology and Ecology* 360, 15-29, doi:10.1016/j.jembe.2008.03.004.

Juniper, S. Kim (2013). *Information Paper 3 - Technological, Environmental, Social and Economic Aspects of Marine Genetic Resources*. In IUCN Informational Papers for the United Nations Inter-sessional Workshop on Marine Genetic Resources 2-3 May 2013. Prepared by International Union for Conservation of Nature (IUCN) Environmental Law Centre, Bonn, German (www.iucn.org/law).

Kamino, K. (2010). Biofouling. *The Journal of Bioadhesion and Biofilm Research*, 29, 735-749, DOI: 10.1080/08927014.2013.800863.

Kim, S.-K. (Editor) *Marine Nutraceuticals*. CRC Press, Boca Raton, 2013. 464 pp.

Kirschner, C.M. and Brennan, A.B. (2012). Bio-Inspired Antifouling Strategies. *Annual Review of Materials Research* 42, 211–29.

Lahlou, M. (2012). The success of natural products in drug discovery. *Pharmacology & Pharmacy* 4, 17-31. DOI: 10.4236/pp.2013.43A003.

Leal, M.C., Puga, J., Serôdio, J., Gomes, N.C.M. and Calado, R. (2012). Trends in the Discovery of New Marine Natural Products from Invertebrates over the Last Two Decades – Where and What Are We Bioprospecting? *PloS One* 7, e30580.

Leary, D. and Juniper, S.K. (2014). Addressing the marine genetic resources issue: is the debate heading in the wrong direction? Chapter 34 (p. 768-785) in Clive Schofield, Seokwoo Lee, and Moon-Sang Kwon (eds.), *The Limits of Maritime Jurisdiction*, Martinus Nijhoff Publishers, The Netherlands, 794pp.

Macilwain, C. (1998). When rhetoric hits reality in debate on bioprospecting. *Nature* 392, 535-540.

Mattila, P., Korpela, J., Tenkanen, T., and Pitkanen, K. (1991) Fidelity of DNA synthesis by the *Thermococcus litoralis* DNA polymerase - An extremely heat-stable enzyme with proof-reading activity. *Nucleic Acids Research* 19: 4967-4973.

Mayer, A.M.S. et al. (2010). The odyssey of marine pharmaceuticals: a current pipeline perspective. *Trends in Pharmacological Sciences* 31, 255–265. doi:10.1016/j.tips.2010.02.005

Merila, J. and Hendry, A.P. (2013). Climate change, adaptation, and phenotypic plasticity: the problem and the evidence. *Evolutionary Applications* 7: 1-14.

Mizuki, K., Iwahashi, K., Murata, N., Ikeda, M., Nakai, Y., Yoneyama, H., Harusawa, S., and Usami, Y. (2014). Synthesis of Marine Natural Product (−)-Pericosine E. *Organic Letters* 2014 16 (14), 3760-3763. doi: 10.1021/ol501631r.

Molinski, T.F., Dalisay, D.S., Lievens, S.L. and Saludes, J.P. (2009). Drug development from marine natural products. *Nature Reviews Drug Discovery* 8, 69-85, doi:10.1038/nrd2487.

Mora, C., Tittensor, D.P., Adl, S., Simpson, A.G.B., Worm, B. (2011). How Many Species Are There on Earth and in the Ocean? *PLoS Biol* 9(8): e1001127. doi:10.1371/journal.pbio.1001127.

Newman, D.J. and Cragg, G.M. (2012) Meeting the supply needs of marine natural products. pp. 1285-1313 in E. Fattorusso, W. H. Gerwick, O. Taglialatela-Scafati (eds.) *Handbook of Marine Natural Products*. Springer Dordrecht, Heidelberg, New York, London. DOI: 10.1007/978-90-481-3834-0.

Ngo, D.-H., Wijesekara, I., Vo, T-S., Ta Q.V., and Kim, S-K. (2011). Marine food-derived functional ingredients as potential antioxidants in the food industry: An overview. *Food Research International* 44 523–529.

Pearce, D.W., & Moran, D. (Eds.). (1994). *The economic value of biodiversity*. Earthscan. p. 167. Accessible at https://www.cbd.int/financial/values/g-economicvalue-iucn.pdf.

Petrone L. (2013). Molecular surface chemistry in marine bioadhesion. *Advances in Colloid and Interface Science* 195–196: 1-18.

Reusch, T.B.H. (2014). Climate change in the oceans: evolutionary versus phenotypically plastic responses of marine animals and plants. *Evolutionary Applications* 7, 104-122. doi:10.1111/eva.12109.

Stewart R.J., Todd, C. Ransom, T.C. and Hlady, V. (2011). Natural Underwater Adhesives. *Journal of Polymer Science Part B: Polymer Physics*, 49(11): 757-771. DOI: 10.1002/polb.22256.

Stumpp, M., Dupont, S., Thorndyke, M.C., Melzner, F. (2011a). CO_2-induced seawater acidification impacts sea urchin larval development II: Gene expression patterns in pluteus larvae. *Comparative Biochemistry and Physiology. A, Molecular & integrative physiology* 160, 320-330.

Stumpp, M., Wren, J., Melzner, F., Thorndyke, M.C. and Dupont, S. (2011b). CO_2-induced seawater acidification impacts sea urchin larval development I: Elevated metabolic rates decrease scope for growth and induce developmental delay. *Comparative Biochemistry and Physiology. A, Molecular & Integrative Physiology* 160, 331–340.

Stumpp, M., Hu, M.Y., Melzner, F., Gutowska, M., Dorey, N., Himmerkusa, N., Holtmann, W.C., Dupont, S.T., Thorndyke, M.C. and M. Bleich. (2012). Acidified seawater impacts sea urchin larvae pH regulatory systems relevant for calcification. *Proceedings of the National Academy of Sciences of the United States of America* 109, 18192–18197.

Thor, P. and Dupont, S., (2015). Transgenerational effects alleviate severe fecundity loss during ocean acidification in a ubiquitous planktonic Copepod. *Global Change Biology*, doi: 10.1111/gcb.12815.

Venter, J.C. et al. (2004). Environmental genome shotgun sequencing of the Sargasso Sea. *Science*, 304: 66-74.

Vidanarachchi, J.K., Kurukulasuriya, M.S., Malshani Samaraweera A. and Silva, K.F. (2012). Applications of marine nutraceuticals in dairy products. Adv. Food Nutr. Res. 65: 457-478.

Wilde, V.L., Morris, J.C. and Phillips, A.J. (2012). Marine Natural Products Synthesis. Pp. 601-673 in E. Fattorusso, W. H. Gerwick, O. Taglialatela-Scafati (eds.) Handbook of Marine Natural Products. Springer Dordrecht, Heidelberg, New York, London. DOI: 10.1007/978-90-481-3834-0.

Additional Reading

The European Marine Board (http://www.marineboard.eu/) has been responsible for generating policy advice, position papers, vision documents, etc., that review the importance of "blue biotechnology;" most are available as free downloads, for example "Marine Biotechnology: a Vision and Strategy for Europe" (http://www.marineboard.eu/publications/full-list). This is a review of the state of the art in marine biotechnology and its significant potential to contribute to scientific, societal and economic needs (2010). In addition, the European Commission has hosted several projects that summarize and highlight the needs and advantages of developing coordinated programmes for harnessing and sustainably exploiting products from our seas and oceans; often generically referred to as "Blue Growth," it includes all aspects of marine and maritime technology (http://ec.europa.eu/maritimeaffairs/policy/blue_growth/index_en.htm).

Other organizations, for example the National Association of Marine Laboratories (NAML) (http://www.naml.org/) in North America also produce regular position papers from marine expert groups.

Marine Scientific Research

Contributors:
Patricio Bernal (Lead member), Alan Simcock (Co-Lead member)

Chapter 30

1 Introduction

A scientific understanding of the ocean is fundamental to carry out an effective management of the human activities that affect the marine environment and the biota that it contains. This scientific understanding is also essential to predict or forecast, mitigate and guide the adaptation of societies to cope with the many ways the ocean affects human lives and infrastructures at different spatial and temporal scales.

Ideally, in order to manage human activities so as to achieve sustainable use of the marine environment and its resources, we need to know the geology and geophysics of ocean basins, the physical processes at work as the waters of the world's different oceans and seas move around, the input, distribution and fate of substances (both natural and artificial), the occurrence and distribution of flora and fauna (including the assemblages and habitat dependencies that control the different ecosystems), the biological processes that regulate and sustain the productivity of ecosystems and the way in which all these elements interact. Marine scientific research is the main way in which we can move towards this goal.

From a more fundamental perspective, the ocean is still one of the least known areas of the world. Humanity in its search of understanding has reached beyond our solar system and seeks fundamental answers in the infinitely distant and in the infinitely small. It has been said that we know more about the morphological features on the surface of other planets than of our own ocean. A significant effort of ocean exploration, using the most advanced techniques available today, is still probably one of the most rewarding collective efforts for humanity, as is attested to by the series of achievements of major international scientific programmes of the past..

Sustainability has to do with the mode by which humanity make use of nature. The increasing pressures that we impose on natural systems leave no room for complacency. At any point in time it is possible to extract the best advice that science can provide to completely or partially remove uncertainties around a phenomenon. From a scientific point of view, the need for better information always exists, therefore unresolved uncertainties are not a valid ground for delaying action. There are many improvements that can be made to managing human impacts on the ocean on the basis of current scientific knowledge. However, it is not long since the need to effectively communicate scientific results to policy makers has been recognized and is being systematically addressed through internationally validated efforts. At the national level, it is becoming common practice among institutions funding research to request those receiving their grants to undertake explicit initiatives of outreach towards the general public or to summarize the result of publicly funded research for policy makers. From a more basic perspective, publicly funded projects in data intensive sciences, like earth sciences, geophysics, and genomics are requested to deposit and disseminate the raw data collected through open access repositories.

The traditional knowledge of those who work with the sea has, in many cases, built up over millennia an understanding of many of these elements. It is essential that this traditional knowledge also be incorporated in our overall understanding of the ocean. Marine scientific research has an important role in validating traditional knowledge and identifying emerging issues. Marine scientific research is therefore fundamental to achieving sustainable use of the oceans.

2 The scale and extent of marine scientific research

The scale and extent of marine scientific research are as wide as the scope of the World Ocean Assessment: every field that needs to be covered in an assessment of the state of the world's marine environment needs to be explored scientifically. This Assessment therefore shows the results of the work that is being done in all these fields and assesses the major gaps in information, thus pointing the way to judgements on priorities for further scientific research.

In order to obtain a full picture, it is necessary to consider where, by whom and how the scientific research is being done. This is not an easy task, because until now no systematic collection of this information has occurred, although the Intergovernmental Oceanographic Commission of UNESCO (IOC/UNESCO) has initiated a process to produce regularly a Global Ocean Science Report (GOSR) aiming to conduct a global and regional assessment of capacity development needs in the field of marine science research and ocean observations.

Table 30.1 | Regions of study of IOC experts. Source: Analysis of IOC, 2014.

Area of Study declared by experts	Experts located in a coastal State of Area of Study	Experts located elsewhere	Total number of experts declaring an interest in the Area of Study
Arctic Ocean	59	78	137
North Atlantic Ocean	519	208	807
Baltic Sea	91	7	98
Black Sea	135	11	146
Mediterranean Sea	393	71	464
North Sea	117	4	121
Wider Caribbean	314	12	326
South Atlantic Ocean	169	562	731
Indian Ocean	588	137	625
Red Sea	61	18	79
The Persian Gulf	49	16	65
North Pacific Ocean	375	102	477
West Pacific Ocean and fringing seas	100	34	134
South Pacific Ocean	157	102	259
Southern Ocean	142[1]	6	148

[1] For the Southern Ocean, the coastal States have been taken to be those States that maintain research stations on Antarctica.

One starting point is the question of who is doing this research. The IOC/UNESCO maintains a database of "ocean and freshwater experts", which can be analysed to help answer this question. The IOC/UNESCO database is compiled on the basis of voluntary self-recording by experts without any independent validation procedure (http://www.oceanexpert.net). There is therefore no reason to think that it is comprehensive, and the status and experience of the experts listed may vary. Examination suggests that it contains practically no experts whose expertise is solely in fresh water, and that nearly all experts have chosen to declare a geographic area of study. It therefore enables an initial understanding of research demographics for the various parts of the ocean. Table 30.1 shows the information on geographic areas of study derived from an analysis of this database. As this information is provided individually by the experts without any independent validation procedure, any analysis based on it may well be affected by biases or incompleteness in the database, and as much of the detail of the information provided is determined by the experts themselves, any analysis is bound to be fairly broad-brush, but this database is the best comprehensive basis available for examining the question of the spread of interests of marine scientists. This appears as a gap of information that needs to be addressed.

Subject to the reservations explained above about the nature of the evidence, the main conclusions are that:

(a) Significantly more marine scientific researchers regard themselves as experts on the Atlantic Ocean than on the Pacific Ocean
(b) The Indian Ocean is relatively well served by marine scientific researchers; and
(c) The two main Southern hemisphere ocean basins (South Atlantic and Pacific and the Southern Ocean) attract relatively fewer marine scientific researchers.

3 Status and trends of scientific output by regions

3.1 Status and trends relating to personnel

An alternative approach to assessing the capacities for marine scientific research by region is to review the number of scientific papers published about each region. This approach also suffers from limitations, and care must be exerted in using it. Some of the issues in assessing the numbers of scientific papers published about the different regions lie with the language of publication and the cost per publication. Although most of the international scientific literature is published today in English, a large potential bias remains with regard to scientific papers and reports published in other languages, or in local journals not reported to international data bases. Although some data bases containing full-text articles address these issues, for example, SciELO (Scientific Electronic Library Online; http://www.scielo.org) for several geographic areas and CNKI (China Knowledge Resource Integrated Database; http://www.cnki.net) for China, the risk of under-reporting of certain countries, languages and regions exists. Another issue is the attribution of the origin of papers. Usually, the attribution is assigned according to the location of the principal, first or corresponding author. This means that the attribution of where work has been done will exclude the location of junior authors in the multi-authored papers, which are now very common.

The analysis below shows the global results analyzed by region of scientific papers published on oceanography from one database for scientific papers which has a wide coverage – ScImago (http://www.scimagojr.com/). Although the papers used are classified *sensu lato* as "oceanography", the data base lists articles in 122 journals with a broader scope than what is usually understood as "oceanography" in the strict sense (Appendix). The journals covered are those that were regularly published in 209 distinct jurisdictions, national or otherwise (described as "countries and territories" in the data base) over the 17 years between 1996 and 2013.

During the 19 years of the data base, a healthy, clear, positive trend of increasing scientific contributions on oceanography each year from around 7,000 to nearly 14,000 altogether is shown (Figure 30.1). On the other hand, the number of countries from which contributors are drawn has shown only a slight increase of about 20 new countries in 17 years, probably due to the fact that the database is reaching the upper limit of the number of countries.

When the countries and territories are grouped into eight regions, the following breakdown emerges of the origins of the 213,760 articles published between 1996 and 2013 (Figure 30.2).

These proportions per region, with North America and Western Europe having the highest number, do not differ significantly from those obtained when analyzing papers from other scientific disciplines. This suggests that they accurately reflect the level of scientific activity in general, not merely a situation specific to the marine sciences, and therefore that this analysis may reflect common, broad issues on available research infrastructure, investment and institutional development that, together

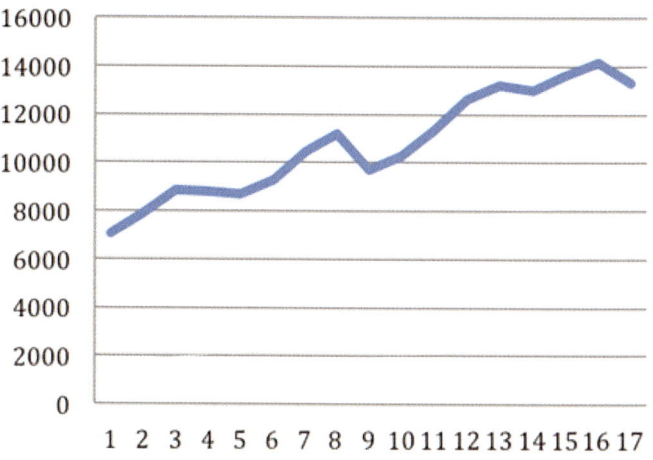

Figure 30.1 | Increase in number of scientific papers on oceanography. Adapted from Analysis of ScImago, 2014.

Marine Scientific Research   Chapter 30

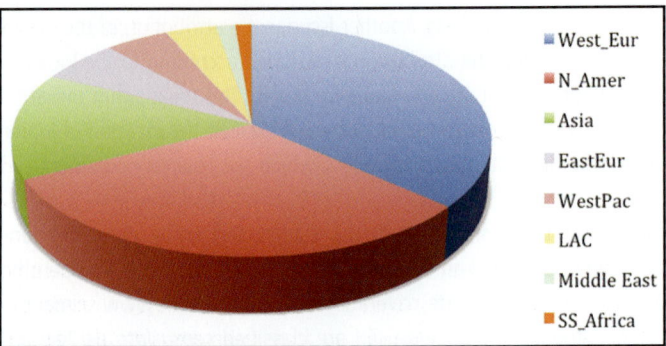

Figure 30.2 | Geographic areas of origin of scientific papers on oceanography 1996 – 2013. Source: Analysis of SciImago 2014.

with appropriate national policies, do control the development of scientific research in general.

3.2 Status and trends relating to equipment

Almost as important as the personnel involved in marine scientific research are the facilities available to them. It is even more difficult than with the personnel to gain an overall view of how far researchers studying the marine environment have adequate equipment. Nevertheless, one indication can be gained from the available information about research vessels. The University of Delaware in the USA maintains an online catalogue of research vessels, including both surface and submersible vessels, and their cruise schedules (www.researchvessels.org). This covers 836 research vessels based in 59 countries, including both publicly owned and commercial research vessels. Again, given that it relies on voluntary recording, it is not comprehensive, but gives a general impression of the distribution of research vessels. Judging by their size, many of these vessels are for coastal operations: 224 are less than 20 m length, while only 138 larger than 80 m. Of those with ocean going capabilities 179 are clustered between 40 and 60 m and 139 between 60 and 80 m. The different capabilities of the vessels can be roughly assessed by the type of equipment they have. All vessels in the database have echo-sounding capabilities, while only 187 are equipped with Conductivity, Temperature and Depth probes (CTDs), 124 with Acoustic Doppler Current Profilers (ADCP), 116 with multi-beam mapping systems and 57 with dynamic positioning systems. Of all the reported fleet

129 have icebreaker capabilities and 103 can berth and deploy remotely operated vehicles (ROVs), autonomous underwater vehicles (AUVs) or submersibles. It is likely that many members of the modal classes (40-80 m length) are fisheries R/V or multipurpose platforms capable of fisheries survey capabilities (acoustic or standard trawling). Table 30.2 shows an analysis of the areas in which these research vessels are based.

Even with the limited information available, this analysis shows a preponderance of research vessels based in the northern hemisphere. Closer analysis suggests that the vessel capacities for research in the Indian Ocean, in other parts of the waters around Africa and in much of the Pacific Ocean are also limited. Anecdotal information suggests that this imbalance is also applicable to other equipment needed for marine scientific research.

4 Collaboration in Marine Scientific Research

One way of overcoming imbalances in national capabilities to undertake marine scientific research is through international joint activities.

Oceanography has always been considered as an international endeavour. The organizers of the Challenger Expedition, that conventionally marks the origin of modern oceanography, took every step necessary to secure the contribution of the best international specialists of the time to produce the fifty volumes of the Challenger Report, containing the results of the Expedition. The first efforts in the study of the North Sea, North Atlantic and the Arctic were also international, and gave rise to the creation in 1902 of the International Council for the Exploration of the Sea (ICES), which plays a fundamental role in codifying the methodologies that enable progress in physical and chemical oceanography.

After the Second World War, the main event that brought together international scientific cooperation was the International Geophysical Year (IGY) of 1957-58. Although the IGY included some oceanographic research, this was not its main focus and the oceanographic community reacted to this situation by planning a major international expedition to the least-known ocean basin at the time: the Indian Ocean.

These initiatives gave rise to two international institutions: first, the Scientific Committee of Ocean Research (SCOR) under the International Council of Scientific Unions (ICSU) in 1957 to coordinate ocean research in the IGY, and then the IOC/UNESCO in December 1960, following a recommendation of the First Oceanographic Congress held in July 1960 in the Danish Parliament. During the International Indian Ocean Expedition, the IOC coordinated the efforts of 27 nations employing over 40 oceanographic research vessels in more than 70 cruises in the Indian Ocean during 1962-1965.

Later SCOR and IOC, through the regular organization of the Joint Oceanographic Assemblies, kept the focus of the community on the

Table 30.2 | Marine Research Vessels. Source: Analysis of IRVSI 2014.

Geographic Area of the World	Number of Research Vessels reported	Largest number recorded in the Geographic Area by one State
Africa	6	4 (South Africa)
Asia	179	108 (Japan)
Eastern Europe	153	116 (Russian Federation)
Western Europe	184	39 (United Kingdom)
North America	288	230 (USA)
Oceania	10	7 (Australia)
Latin America and Caribbean	29	7 (Argentina)
Total	849	

design of international research programmes. For example, the Committee on Climatic Changes and the Ocean, sponsored by SCOR and IOC, is at the origin of two global research projects of primarily physical studies: the World Ocean Circulation Experiment (WOCE) and the Tropical Ocean-Global Atmosphere Study (TOGA) also co-sponsored by the World Climate Research Programme (WCRP). WCRP was established in 1980 under the joint sponsorship of ICSU and the World Meteorological Organization (WMO) and since 1993 the IOC has also sponsored it.

In the 1980s two major international programmes requiring ocean going capabilities were co-sponsored by SCOR, the International Geosphere-Biosphere Programme (IGBP) and jointly with IOC: first the Joint Global Ocean Flux Study (JGOFS), focusing on the role of the ocean in the global carbon cycle, and second the Global Ocean Ecosystem Dynamics (GLOBEC) programme. Over ten years GLOBEC developed seven regional comparative studies to understand marine ecosystem responses to global changes, including both environmental and human pressures, and produced over 3,500 publications, including 30 special issues of primary journals. The Integrated Marine Biogeochemistry and Ecosystem Research (IMBER) programme has followed GLOBEC.

SCOR and IOC/UNESCO have also developed the Global Ecology and Oceanography of Harmful Algal Blooms (GEOHAB) Programme with a focus on obtaining an understanding of the ecological and oceanographic conditions that cause harmful algal blooms and promote their development. Other international programmes are the Global Coral Reef Monitoring Network and the Census of Marine Life, a ten year effort focusing on the biology of the ocean that mobilized more than 2,700 scientists, published 3,100 scientific papers and described 1,200 new species for science, leaving as one of its legacies the Ocean Biodiversity Information System (OBIS), the largest repository of marine biodiversity to date.

In the domain of marine geology and geophysics, the Integrated Ocean Drilling Programme, that initially built and operated the R/V "*Glomar Challenger*" in the seventies, was followed after October 2013 by the International Ocean Discovery Programme (IODP) currently operating the R/V "*JOIDES Resolution*". These international programmes were instrumental in developing the technology to drill the sea floor and to obtain the long cores that provide a wealth of research activities expanding our knowledge in different areas, including plate tectonics and seismology.

In geochemistry, the Geochemical Ocean Sections Study, GEOSECS, obtained very accurate sections and profiles of the distribution of chemical, isotopic, and radiochemical tracers in the ocean, building a global three-dimensional view of the chemical composition, including alkalinity, of the ocean, enabling the establishment of a solid baseline to measure acidification worldwide. GEOSECS is now being followed by GEOTRACES, which is measuring the distributions of trace elements in the sea.

4.1 The development of a permanent infrastructure to observe the Ocean

Perhaps one of the most fundamental changes in marine scientific research was the realization that what was needed to underpin many of the more focused or local research efforts was a common infrastructure to observe the oceans at the global but also at other relevant temporal and spatial scales. In the late 1980s, oceanographers had come to realise that the ocean played a tremendously important role in the climate system through its ability to store large amounts of heat and to move this source of energy for the atmosphere slowly around the globe. Accordingly, understanding and forecasting climate change was seen to require observations over much longer periods of time, than the time-limited experiments such as the ocean observations done during the First GARP[1] Global Experiment (FGGE) during 1978-79 or the TOGA study, which ran from January 1985 to December 1994.

In 1989 IOC's Technical Committee for Ocean Processes and Climate (TC/OPC) recommended the design and implementation of a global operational observing system. The WMO Executive Council endorsed that call in June 1989, as did the 15th IOC Assembly in July 1989. Finally, in June of 1990, the Intergovernmental Panel for Climate Change (IPCC) called for a Global Ocean Observing System (GOOS) which was endorsed by the Second World Climate Conference in September 1990, that saw GOOS as a major component of the proposed Global Climate Observing System (GCOS). In February 1991, the TC/OPC agreed that the concept of GOOS should be broadened to include physical, chemical and biological coastal ocean monitoring; climate was no longer to be the sole focus. In May 1991, WMO's 11th Congress accepted to co-sponsor the GOOS.

Existing physical oceanographic observing systems developed over the years by UNESCO/IOC became fundamental building blocks of GOOS. For example, the IOC's global sea-level observing system (GLOSS) and the joint IOC/WMO Integrated Global Ocean Services System (IGOSS), which included the Ship-of-Opportunity Programme and the drifting and other buoys of the Data Buoy Cooperation Panel.

The creation of GOOS reflected the desire of many nations to establish systems of ocean observations dealing with environmental, biological and pollution aspects of the ocean and coastal seas, to raise the capacity of developing nations to acquire and use ocean data effectively and to integrate existing systems of observation and data management within a coherent framework.

That desire was reflected in the call made in Rio de Janeiro from the United Nations Conference on Environment and Development in June 1992 to develop GOOS as one of the mechanisms required to support sustainable development. This required that the initial focus on climate research had to be enlarged to include other aspects, like the impact of pollution and the status of marine living resources. The Health of the

---

1 GARP is the Global Atmosphere Research Programme

Ocean (HOTO) Panel was established as an *ad hoc* group in 1993, and became a formal advisory group to J-GOOS in 1994. An *ad hoc* Living Marine Resources (LMR) Panel met in 1993 and in 1996, and an *ad hoc* Coastal Panel met in 1997.

IOC/UNESCO and WMO gave first priority to the implementation of the physical oceanographic component of GOOS, as the ocean component of the climate observing system GCOS. This part of GOOS has been successfully in operation since 2005; however the development of the other parts of GOOS has continued as new technologies emerge and mature, enabling the automatic long-term measurements of chemical and biological variables.

4.2 Operating Systems of GOOS

Although fundamentally underpinning most of the research conducted to understand the role of the ocean in climate change, strictly speaking GOOS is not a research project. GOOS should be better recognized as a *large and distributed scientific facility* or infrastructure, equivalent to the large observatories of astrophysics or the big accelerators of particles of physics. This section describes its components (IOC/GOOS, 2015).

4.2.1 Surface moorings

Surface moorings are large fixed buoys, moored to the bottom of the ocean, mostly deployed in the Equatorial region. They measure surface winds, air temperature, relative humidity, sea-surface temperature and subsurface temperatures from a 500-m-long thermistor chain hanging below the buoy. Daily data are broadcast to shore through satellite links (TAO, 2015).

4.2.2 Argo Profiling Floats Programme

The Argo floats are autonomous observation systems which drift with ocean currents making detailed physical measurements of the upper 2 kilometres of the water column. Floating along at a depth of 2,000 metres, every 10 days an Argo float awakens and increases its buoyancy by pumping fluid into an external bladder. During its journey upward through the water column, it records the conductivity (salinity) of the seawater, its temperature, and pressure. Once at the surface, the Argo float finds its geographical position via global positioning systems (GPS) and transmits its data by satellite to Argo data centres. After completing the upward profile it decreases its buoyancy and sinks again, collecting a similar record on the trip down to 2000 m. The information is joined with data from over 3,000 other Argo floats to form a synoptic 3-D view of the ocean in near real time (http://www.argo.ucsd.edu/index.html).

About 800 profiling floats are deployed on a yearly basis by a number of States. Between 2004 and 2009, 26 States deployed at least one float to maintain a global array of 3,200 units, spaced 3° by 3° of latitude and longitude. Profiling float technology has evolved to reach the initial desired five-year lifetime, and a float deployed today will probably last between 5 and 10 years. Argo floats spend 90 per cent of their time at 2,000-m depth and on average rise to the surface every ten days to transmit their data.

This system has revolutionized oceanography since its inception in 1998 through the Climate Variability and Predictability (CLIVAR) programme and the Global Ocean Data Assimilation Experiment (GODAE).

Argo floats take more than 100,000 salinity and temperature profiles each year, more than 20 times the number of annual hydrography profiles taken from research ships. The Argo array is maintained by the active engagement of 30 countries that contribute floats and ship-time for the deployments. The original engineering specifications of the floats were made available to many research institutions around the world and floats are now made in several countries. The International Argo Steering Team oversees technically the project and operations are monitored at the Argo Information Centre, a part of the IOC – WMO, Joint Technical Commission of Oceanography and Marine Meteorology - operational centre (JCOMMOPS).

Argo data have transformed ocean circulation studies. Today Argo data are routinely assimilated into global circulation models, giving accurate and timely global views of the circulation patterns and heat distribution of the ocean. This product has become an essential element of atmospheric forecast models and greatly improves seasonal climate, monsoon, El Niño forecasts, as well as tropical cyclone simulations. The value of subsurface heat content measurements to the study of global warming and climate change has made the Argo an invaluable component of 21st-century environmental observation systems.

4.2.3 The Ship-of-Opportunity Programme (SOOP).

Ships of opportunity are usually ordinary cargo ships on regular routes, whose owners and crew agree to carry and, where necessary, operate oceanographic equipment during their regular voyages. Other types of vessel are also used. The Ship-of-Opportunity Programme (SOOP) and its Implementation Panel (SOOPIP) is an operational programme under the intergovernmental governance of the Joint WMO-IOC Technical Commission for Oceanography and Marine Meteorology (JCOMM). The primary goal of SOOP is to satisfy upper-ocean data requirements which have been established by GOOS and GCOS, and which can be met at present by measurements from ships of opportunity (SOO).

SOOP operates a global network of Expendable Bathythermograph (XBT) and ThermoSalinoGraphs (TSG) systems on board of merchant ships, from which data are transmitted in real time and made available to the oceanographic and meteorological communities for operational use in ocean models and for other scientific purposes. Around 14,000 XBT probes are launched every year and more than 30,000 TSG observations are collected annually. Other types of measurements are also made. The following devices are commonly used:

(a) *XBT (Expendable BathyThermograph)* is an expendable (disposable) temperature- and depth-profiling system;

(b) *TSG (ThermoSalinoGraph)* is an automated sea-surface temperature and salinity measurement system for making continuous underway measurements from the ship's water intake;

(c) *CTD* is an electronic set of instruments to make precise conductivity, temperature, and depth measurements. The instrument is connected to the ship by a conducting cable; Accuracies better than 0.005 mS/cm are usually achieved for conductivity, better than 0.002° C for temperature, and better than 0.1 per cent of full-scale range for depth;

(d) *XCTD* is an expendable (disposable) conductivity, temperature and depth profiling system.

(e) *ADCP (Acoustic Doppler Current Profiler).* A beam of sound of known frequency is reflected from small particles moving with the water. Adequate sampling of this backscattering beam allows current measurements by the Doppler effect at different depths. ADCPs can, for example, be installed on the hull of the ship "looking downwards" or lowered from a ship to different depths to measure a wider range of current profiles. An accurate GPS positioning system can then be used on a moving ship to subtract the ship's speed from the measured current vector;

(f) *pCO2*. Measurements of the "partial pressure of CO_2" (pCO_2) on the ocean surface indicate whether the local ocean is acting as a source or a sink of CO_2. Measurements use a standardized infrared analyzer or a gas chromatograph to determine the concentration of CO_2. The probe is installed in the hull of a ship, and measurements can be made while the ship is under way. Partial pressure of CO_2 in the air can also be measured. Accuracies in the order of 0.2 parts per million can be achieved.

4.2.4 Hydrography

The direct sampling of ocean water by lowering bottles from a ship and bringing water samples up on board ship for analysis remains one of the fundamental tools of ocean observations. A CTD rosette, equipped with Niskin bottles, is lowered to its deepest point and then as it is winched up to the ship the bottles are closed, one at a time, capturing a CTD profile of the water column along the way. The water can be sampled for CO_2, chlorophyll, microorganisms, biogeochemistry, and a wide variety of other uses. The International Ocean Carbon Coordinating Programme and the CLIVAR Programme organize and coordinate major hydrography cruises and maintain databases of tens of thousands of hydrography profiles taken throughout the world's ocean. These programmes provide essential data streams for GOOS, as they provide precise and accurate *in situ* measurements that benchmark observations measuring the penetration of heat in the ocean or the changes in alkalinity, monitoring the ocean's uptake of CO_2 that is changing the ocean acidity levels.

4.2.5 Surface drifting buoys

The Global Drifter Programme manages the deployment of surface drifting buoys around the world. These simple buoys take measurements of seawater-surface temperature, salinity and marine meteorological variables that are telemetered in real time through the WMO's Global Telecommunications System (GTS) to support global meteorological services, climate research and monitoring. The surface drifters are a flexible component of GOOS and can be deployed quickly for such tasks as monitoring an approaching typhoon. The global array is designed to use 1,250 buoys to cover the oceans at a resolution of one per 5° x 5° of latitude and longitude. This array provides over 630,000 sea-surface observations per year. The surface temperature data are used to calibrate satellite temperature imagery, bringing bias errors down from 0.7° Celsius to less than 0.3° C, allowing accurate climate-change monitoring. Along with the Argo profilers, the surface drifter programme has contributed to the success of a real-time monitoring system of the oceans, enabling much more accurate weather and climate forecasts.

4.2.6 Continuous Plankton Recorder

Launched over the side of a research vessel, merchant ship, or other vessel of opportunity, the Continuous Plankton Recorder (CPR) captures plankton from the near-surface waters as the ship tows the instrument during its normal sailing. Since 1946, the CPR has been regularly deployed in the North Atlantic and North Sea on several routes. The CPR is a critical component of GOOS and monitors the near-surface plankton in the North Atlantic and North Sea on a monthly basis from a network of shipping routes. Many other tracks around the world are now covered by the CPR programme. The amounts and types of phytoplankton and zooplankton captured by the CPR are analyzed in a laboratory. After analysis, the counts are checked and added to the CPR database, which contains details of the plankton found in over 170,000 samples taken since 1946 in the North Sea and North Atlantic Ocean, and increasingly elsewhere.

4.2.7 Global Sea Level Observing System (GLOSS)

The Global Sea Level Observing System (GLOSS) is an international programme conducted under the auspices of the JCOMM of the WMO and the IOC. It coordinates a network of sea-level monitoring gauges installed along the coasts of over 70 countries. The main component of GLOSS is the "Global Core Network" (GCN) of 290 sea-level stations around the world for long-term climate-change and oceanographic sea-level monitoring. Each station is capable of accurately monitoring sea-level changes with high accuracy, and many are able to transmit information in real time via satellite links. The GLOSS sub-network that transmits in real time is part of the global tsunami warning systems.

Real-time measurements of water-level changes can provide tsunami warnings for locations surrounding the affected sea basins. Sea-level observations are also useful for local navigation and continual refinement of tide-table predictions. Tide gauges measure rising water levels from storms and extreme tides, which can be responsible for billions of United States dollars in damage and lost productivity every year.

4.2.8 Ocean Biological Data

As a result of the ten-year-long effort by the Census of Marine Life, a significant increase in biological data took place. This new data was integrated to pre-existing data into the Ocean Biodiversity Information System (OBIS http://www.coml.org/global-marine-life-database-obis). Several of these new data streams are associated to the tagging and tracking of live animals, for example the Tagging of Pacific Pelagics (TOPP) programme in Western North America and the Australian Animal Tracking and Monitoring System (AATAMS). The tagging of marine animals, fish, birds, turtles, sharks, mammals, with electronic sensors is increasingly being undertaken by scientists worldwide to track their movements. Electronic tags such as archival, pop-up archival and satellite positioning tags are revealing when, where and how marine animals travel, and how these movements relate to the ocean environment. (http://www.scor-int.org/observations.htm). An Ocean Tracking Network is being developed. The network will track thousands of marine animals around the world using acoustic tags safely attached to the animals. At the same time, the network will be building a record of data relevant to climate change, through observation of changes in the animals' patterns of movement.

5 Socioeconomic aspects of marine scientific research

Three major points emerge from the foregoing analyses and the material in other chapters on the results of marine scientific research in the fields they cover.

First, the success of the management of human activities that affect the marine environment is conditional upon having reliable information about that environment. If adequate information is not being collected, then management decisions will be less than optimal. Parts of the world that do not have adequate infrastructure for an adequate collection of information about their local marine environment are disadvantaged. Although research based in other parts of the world may provide a good understanding of how the marine ecosystems operate, and of the pressures to which they are subject, such a general understanding must be supplemented by adequate local information. Such collection of local information is always likely to be more efficient, effective and economical.

Second, as the world's marine environment is very much interconnected, sub-optimal management in one part of the world is likely to affect the quality of the marine environment in other parts of the world. This is the case of land-based point-sources of pollution that, depending on circulation, can broadcast their negative impacts across maritime borders; or if stocks of marine living resources are not well managed in one part of the world, diminishing the landings of a certain target-species, this may increase fishing pressure on the same or similar species in other parts of the world.

Third, even though universities and other educational establishments produce good-quality marine experts throughout the world, graduates will experience pressure to move to those parts of the world where they can hope to have access to the best equipment for their further research. It is only in that way that they can hope to develop their careers most successfully. Such a "brain drain" will undermine efforts to establish adequate marine research in all parts of the world until appropriate local conditions for the development of scientific research exist.

6 Environmental impacts of marine scientific research

Any observation of a natural system has the risk that it will disturb that system. Proper design of marine scientific research can reduce, or even eliminate, this risk. It is particularly important that efforts that aim at improving the understanding of marine ecosystems should not damage those ecosystems.

The IOC has an important role in establishing safeguards for marine research projects that risk adversely affecting the marine environment. Efforts have been increasingly made to address this task. The International Ship Operators forum, answering to concerns of the impact of both ship operations and marine scientific research operations, developed a Code of Conduct for Marine Scientific Research Vessels that was approved at the 21st International Ship Operators Meeting (ISUM) in Qingdao, China. The code calls for *"the utilisation of environmentally responsible practices"* (…) and to *"adopt the precautionary approach as the basis for the proposed mitigation measures"*. *"Every vessel conducting marine science should develop a marine environmental management plan"* which *"should be designed to employ the most appropriate tool(s) to collect the scientific information while minimizing the environmental impact."* Among the activities addressed by the code are: dredging, grab & core sampling, lander operations, trawling, mooring deployments, remotely operated vehicle (ROV) sampling, jetting system operations for cable burial, high intensity lighting for camera operations.

Other example can be taken from the Argo floats programme. Every year, about 1 per cent of floats are beached or trapped in fishing nets. These are recovered, secured and redeployed when possible, or recycled through a procedure coordinated by JCOMMOPS. All other floats finish their mission at depth, which is the best compromise found to date to limit the impact on the environment: (1) to avoid energy consumption to recover the instruments at sea by the use of motor vessels, and (2) to avoid having floats drifting at the surface for a long time (after a set of predefined cycles) and becoming a potential issue for navigation. The total mass of float hardware reaching the sea floor every year (less than 30 tons), and more precisely the small fraction of polluting material inside, can be more than fairly compared to old metro trains sunk to provide structure for artificial reefs, merchant ships, fishing vessels, off-shore stations, lost containers and decommissioned offshore oil platforms, that sink to or stay on the sea floor.

Technological improvement now allows the use of a bi-directional telecommunication system, which can "control" the behaviour of the platform by sending new configuration parameters and receiving data. About 30 per cent of the Argo array is now equipped with this system. A float can then be asked to stay at surface to await its imminent retrieval. This is already being done today in pilot projects, and is used in particular to recover biogeochemical floats, which are equipped with expensive sensors and require some post-calibration. The involvement of civil society (for example, the yachting community, non-governmental organizations and foundations) and industry in offering deployment opportunities to cover large ocean areas can be also a way to improve retrieval capacity. This requires a large networking capacity and is encouraged by IOC through its operational Centre JCOMMOPS. The manufacturers of floats are also encouraged, with rest of the world industry, to use environmentally friendly materials, whenever possible. As Argo is the main pillar of the ocean climate warning system, the ratio between advantages and disadvantages for the environment is judged to be more than satisfactory by the marine scientific research community, and at the same time that community continues to develop strategies to mitigate its impact.

Another example of the development of precautions against damage to the marine environment from marine scientific research concerns hydrothermal vents. In the 1990s, an international organization called Interridge was established and is today supported by China, France, Germany, Japan, the United Kingdom and the United States of America, together with Canada, India, Norway, Portugal and the Republic of Korea as associates, to pool resources for the investigation of oceanic ridges. Within this framework, recommendations have been developed on how to protect hydrothermal vents during research (Interridge, 2001). This provides a helpful model for developing protocols to ensure that marine scientific research does not harm the very objects that it wants to study.

7 Conclusions and capacity-building and information gaps

Major disparities exist in the capacities around the world to undertake the marine scientific research necessary for proper management of human activities that can affect the marine environment. The other chapters of this Assessment demonstrate how these disparities constrain the tasks of managing these human impacts. Capacities to undertake marine scientific research exist in most parts of the world.

Although a full assessment of all the existing programmes of capacity development is beyond the scope of this chapter, several long-standing international programmes have addressed these disparities. For example, the Train-Sea-Coast Programme, established in 1993 by the United Nations Division for Ocean Affairs and the Law of the Sea (DOALOS) with initial funding from the United Nations Development Programme and then by the Global Environment Facility, although now closed, aimed to build capabilities to enhance national/regional capabilities on key transboundary topics/problems in coastal and ocean-related matters. Topics addressed were quite wide, and ranged from coastal zone management, marine pollution control to marine protected areas and responsible fisheries. On geophysics, the IOC/UNESCO has maintained an annual Training Through Research ocean-going programme for young students to acquire hands-on experience in the operation, use and interpretation of data from current equipment used in marine geology and geophysics. In the area of living marine resources, the Food and Agriculture Organization of the United Nations (FAO) and Norway have developed for the last 40 years the ocean-going Nansen Programme funded by the Norwegian Agency for Development Cooperation (Norad) and executed in a partnership between the Norwegian Institute of Marine Research (IMR) and FAO. The first R/V *Dr Fridtjof Nansen* was commissioned in October 1974. The third version of the R/V is currently being built and expected to be commissioned in 2016. The International Seabed Authority (ISA) has three active training streams, the *Endowment Fund* supporting the participation of qualified researchers from developing countries in cooperative research on the seabed; the *ISA/Contractors Training programme* aimed at training developing countries' scientists and managers and the *ISA Internship Programme* that, in a twofold approach, receives young scientists and managers from developing countries at ISA headquarters to learn about the goals and functions of ISA, but also receives young, highly qualified personnel to reside and contribute for short periods to ISA activities.

Many other international training initiatives on marine sciences, bilateral or multilateral, do exist, especially in the academic/education domain, but no comprehensive global reporting or cataloguing of these important efforts exists to date.

Gaps remain in the abilities to integrate the results of scientific research into the development of policy: capacity-building gaps thus exist in creating an effective science/policy interface first and foremost at the national level, but also at the regional and global levels.

Furthermore, efforts to fill the capacity-building and information gaps identified in other chapters will be much less productive if they are not made against a background of developing a global coverage of systems that can provide adequate integrated management information to global, regional and national authorities. This will be both more efficient and more economical, because a coherent body of scientific information will ensure that unexpected results of human activities and efforts to manage them will not go undetected, and will avoid duplication and overlap.

As this chapter has suggested, systematic information and knowledge about the progress of marine science is lacking. This therefore strengthens the case for supporting within the UN System the IOC's efforts to develop a World Ocean Science Report (see Decision EC-XLVII/6.2) that would eventually complement the existing World Science Report of UNESCO.

References

Interridge (2001). *Management and Conservation of Hydrothermal Vent Ecosystems* (http://www.interridge.org/files/interridge/Management_Vents_May01.pdf accessed 29 November 2014).

IOC (Intergovernmental Oceanographic Commission) (2014). *Ocean Expert – Directory of Marine and Freshwater Professionals* (www.oceanexpert.net accessed 9 October 2014).

IOC/GOOS (Intergovernmental Oceanographic Commission – Global Ocean Observation System) (2015). http://www.ioc-goos.org/ accessed 9 October 2015.

IRVSI (2014). University of Delaware, *International Research Vessels Schedules and Information* (http://www.researchvessels.org/ accessed 9 October 2014).

ScImago (2014). ScImago Journal and Country Rank Portal (http://www.scimagojr.com/ accessed 6 August 2014).

TAO (Tropical Atmosphere Ocean project) (2015) http://www.pmel.noaa.gov/tao/ accessed 15 July 2015.

Appendix. Alphabetic List of Journals Indexed in http://www.scimagojr.com/

1. 28th International North Sea Flow Measurement Workshop 2010
2. Acta Adriatica
3. Acta Oceanologica Sinica
4. Advanced Series on Ocean Engineering
5. Annual Review of Marine Science
6. Antarctic Science
7. Aquatic Biology
8. Atmosphere - Ocean
9. Atoll Research Bulletin
10. Australian Meteorological and Oceanographic Journal
11. Bollettino di Geofisica Teorica ed Applicata
12. Brazilian Journal of Oceanography
13. Bulletin of Marine Science
14. Bulletin of the Plankton Society of Japan
15. China Ocean Engineering
16. Chinese Journal of Oceanology and Limnology
17. Coastal Engineering Practice - Proceedings of the 2011 Conference on Coastal Engineering Practice
18. Continental Shelf Research
19. Deep-Sea Research Part II: Topical Studies in Oceanography
20. Deep-Sea Research Part I: Oceanographic Research Papers
21. Developments in Sedimentology
22. Dynamics of Atmospheres and Oceans
23. Earth System Monitor
24. Estuarine, Coastal and Shelf Science
25. Fish and Fisheries
26. Fisheries and Aquatic Science
27. Fisheries Oceanography
28. GeoArabia
29. Geo-Marine Letters
30. Global and Planetary Change
31. Helgoland Marine Research
32. Hidrobiologica
33. Hydrographic Journal
34. Hydro International
35. IAHS-AISH Publication
36. ICES Journal of Marine Science
37. Indian Journal of Marine Sciences
38. Integrated Ocean Drilling Program: Preliminary Reports
39. International Journal of Marine and Coastal Law
40. International Journal of Nautical Archaeology
41. International Journal of Oceans and Oceanography
42. International Ocean Systems
43. IWMI Research Report
44. Izvestiya - Atmospheric and Oceanic Physics
45. Journal of Coastal Conservation
46. Journal of Earthquake and Tsunami
47. Journal of Geophysical Research
48. Journal of Island and Coastal Archaeology
49. Journal of King Abdulaziz University, Marine Science
50. Journal of Marine Research
51. Journal of Marine Science and Technology
52. Journal of Marine Science and Technology
53. Journal of Marine Science and Technology (Japan)
54. Journal of Marine Systems
55. Journal of Navigation
56. Journal of Northwest Atlantic Fishery Science
57. Journal of Oceanography
58. Journal of Ocean University of China
59. Journal of Operational Oceanography
60. Journal of Physical Oceanography
61. Journal of Sea Research
62. Journal of the Indian Ocean Region
63. Kuste
64. La Mer
65. Latin American Journal of Aquatic Research
66. Limnology and Oceanography
67. Limnology and Oceanography Bulletin
68. Marine and Freshwater Behaviour and Physiology
69. Marine and Freshwater Research
70. Marine and Petroleum Geology
71. Marine Biodiversity
72. Marine Biodiversity Records
73. Marine Biology Research
74. Marine Chemistry
75. Marine Environmental Research
76. Marine Geodesy
77. Marine Geology
78. Marine Geophysical Researches
79. Marine Georesources and Geotechnology
80. Marine Micropaleontology
81. Marine Ornithology
82. Marine Pollution Bulletin
83. Marine Resource Economics
84. Mariner's Mirror
85. Marine Technology Society Journal
86. Mediterranean Marine Science
87. Memoirs of Museum Victoria
88. Methods in Oceanography
89. NAFO Scientific Council Studies
90. NIWA Biodiversity Memoirs
91. Ocean and Coastal Management
92. Ocean Dynamics
93. Ocean Modelling
94. Ocean Modelling
95. Oceanography
96. Oceanography and Marine Biology
97. Oceanologia
98. Oceanological and Hydrobiological Studies
99. Oceanology

100. Ocean Science
101. Ocean Science Journal
102. Oceanus
103. Offshore
104. Palaeogeography, Palaeoclimatology, Palaeoecology
105. Palaeontologia Electronica
106. Paleoceanography
107. Phuket Marine Biological Center Research Bulletin
108. Physical Oceanography
109. Plankton and Benthos Research
110. Polarforschung
111. Polar Research
112. Proceedings of the 9th (2010) ISOPE Pacific/Asia Offshore Mechanics Symposium, PACOMS-2010
113. Proceedings of the Coastal Engineering Conference
114. Revista de Biologia Marina y Oceanografia
115. Russian Journal of Marine Biology
116. Russian Journal of Pacific Geology
117. Science of Tsunami Hazards
118. Scientia Marina
119. Tellus, Series A: Dynamic Meteorology and Oceanography
120. Terrestrial, Atmospheric and Oceanic Sciences
121. Thalassas
122. World Dredging, Mining and Constructions

31 Conclusions on Other Human Activities

Contributor:
Alan Simcock (Lead member)

Chapter 31
Conclusions on Other Human Activities

1 The nature and magnitude of the human activities

1.1 Communications and transport

The network of shipping routes covers the whole ocean. There are particular choke points, where large numbers of ships pass through relatively limited areas, with consequent increases in the risks of both disasters and chronic pollution problems. The impending opening of the Panama Canal to larger ships will tend to modify the pattern of ship movements. Global warming is likely to lead to more use of the routes between the Atlantic and Pacific Oceans through Arctic waters, with increased risks to ecosystems that have slow recovery times, and where infrastructure for response to disasters does not currently exist. Shipping traffic grows in relation to world trade, and considerable further growth is therefore likely. Cargo ships have been steadily increasing in size, but limits are probably being reached because of the draught limitations of some of the world's choke points. More emphasis is being placed in many areas on coastwise movement of goods by ship to reduce pressures on roads. Passenger shipping is largely divided into cruise ships and ferries. The cruise-ship market is growing steadily and is also moving to larger vessels. Ferries are most important around the Baltic Sea, the North Sea and the Mediterranean Sea (where there are large international cargo movements over relatively short crossings) and in States with a large scatter of islands (such as Greece, Indonesia and the Philippines).

Ports form the nodes of the network of shipping routes. General cargo ports have changed completely over the past 50 years with the introduction of containerization. A hierarchy of these ports is developing, with transhipment as cargoes are cascaded to the ports nearest to their final destinations. Specialized oil and gas ports are naturally located near the sources of supply and the major centres of demand. This pattern is likely to change as a result of changes in the oil and gas markets. Other bulk terminals respond to the same drivers. In some cases, there are challenges because of the location of sources of supply or established delivery centres near important sites for marine ecosystems and biodiversity.

Submarine cables likewise cross nearly all ocean basins. The development of fibre-optic cables in the 1980s permitted the parallel development of the internet. Submarine pipelines linking land-terminals are relatively limited in their coverage, being mainly found in the Mediterranean Sea, the Baltic Sea and the waters around the United Kingdom. The environmental challenges that they present are the same as those for pipelines linked to offshore oil and gas operations.

1.2 Waste

All parts of the ocean are affected by waste materials arriving by a variety of routes. Waste can take the form of discharges of liquid waste from land and emissions to the air, of dumping solid waste and other matter at sea, and of marine debris resulting from poor management of waste on land, discharges of garbage by ships and loss of fishing gear. Areas of particular concern are large conurbations, where large amounts of human bodily waste have to be disposed of, and areas of heavy industrial concentrations. Sewage from areas of high human population does not inevitably cause problems, since it can be treated to remove the potential to cause problems. However, in many parts of the world, particularly in developing countries, there is a lack of adequate sewage collection and treatment systems, and large amounts of untreated sewage are discharged to the sea. Much progress is being made in some places (particularly South America), but there is still a vast amount of further installation needed. Likewise, methods exist to avoid discharges of hazardous substances from industries, or to control them to acceptable levels, but these are not being applied everywhere. In this context, the massive growth of chemical industries in East Asia over the past decade-and-a-half presents particular difficulties. High levels of use of agricultural fertilizers and pesticides are also leading to discharges and emissions of nutrients and hazardous substances. Recent studies suggest that windborne transport of all these kinds of emissions is causing problems in the open ocean.

1.3 Extractive industries

Offshore oil and gas exploration and development is focussed in specific geographic areas where important oil fields have been discovered. Notable offshore fields are found: in the Gulf of Mexico; in the North Sea; off the coast of California (in the Santa Barbara basin), United States; in the Campos and Santos Basins off the coast of Brazil; off the coast of Newfoundland, Canada; off West Africa, mainly west of Nigeria and Angola; in the Gulf of Thailand; off Sakhalin Island on the Russian Pacific coast; in the Persian Gulf, on the Australian North West Shelf and off the west coast of New Zealand.

Offshore mining is a localised activity and is currently limited to relatively close to the shore. The activity on the largest scale is sand and gravel dredging, which takes place in Canada, Denmark, Japan, the Netherlands, the United Kingdom, and the United States. Other minerals extracted from the seabed include tin, titanium ores and diamonds. Countries where such activities are currently active include Australia, Brazil, India, Indonesia, Malaysia, Myanmar, Namibia, New Zealand, South Africa and Thailand. Other minerals currently being considered for extraction include gold (Alaska, United States) and iron (New Zealand – where currently a permit has been refused by the Environmental Protection Authority). The impacts are largely in the form of smothering the seabed. As regards deep-sea mining in the Area (the seabed and ocean floor and their subsoil beyond national jurisdiction) no actual extraction has yet started. Without careful management of such activities, there is a risk that the biodiversity of areas affected could be destroyed before it is properly understood.

1.4 Coastal zone

All around the world, there is a constant interaction between sea and land. On rocky coasts, changes usually take place over geological time. On softer coasts, the changes can happen within a human lifetime. The changes happen both by the sea eroding land and by sedimentation creating new land. In many places, humans assist these processes, either by undertaking works to protect the land against erosion, or by reclaiming land from the sea. Significant areas of land reclamation have occurred in Europe (about 15,000 km^2) and in Asia (about 12,000 km^2) over the past century. This has lead to losses of much coastal habitat (especially of mangroves, salt marshes and coastal flood plains) and to significant adverse effects from changes in the form of the coast (in particular, the creation of "armoured", artificial coastlines). In comparison, about 15 km^2 is eroded from the coast of Europe every year. An important driver in reclamation processes is the increasing proportion of the human population who live in coastal areas.

Tourism and recreation is a major use of the coastal zone. It is also a significant element of many national economies, and (especially in the case of many small island States) it may be the main support of the local economy. In most cases, the attractions of a coastal tourist resort will lie in beaches, dramatic scenery or interesting flora and fauna (either on land or in the sea), or some combination of these. But the natural attractions have to be linked to the provision of adequate facilities for the tourists. In many cases, therefore, there is an inherent tension between conserving the natural attractions and developing the necessary facilities.

1.5 Other activities

The other activities discussed in Part V have very distinct profiles. Desalinization is crucial for the States on the southern shore of the Persian Gulf: without it, the present populations of that area could not be supported. The same applies to some island States, such as Malta and Singapore. Elsewhere, it is more of a fall-back for situations when natural water resources are insufficient, but may have an important role in avoiding constraints on urban development and in facilitating mining in desert areas. Renewable energy production from the sea (through offshore wind-farms and from wave and tidal power) is still in its infancy in much of the world, but clearly has great potential for growth. The use of marine genetic resources is also in its infancy. Finally, marine scientific research is fundamental for improving the management of all other human activities that affect the ocean. Without adequate coverage of all parts of the ocean, it will be difficult to make progress.

2 Socioeconomic aspects of the human activities

2.1 Communications and transport

Shipping is fundamental to the world economy, both for the supply of raw materials to those who will process them as well as for the delivery of agricultural and industrial products to consumers. The lead time for the delivery of new ships and the fluctuations in the world economy mean that there are substantial variations in the demand and supply sides of shipping, with consequential effects on the profitability of the industry. Ship-building has become concentrated in East Asia, and ship-breaking in South Asia. The personnel of the industry are drawn mainly from North America, Europe and South and East Asia, with Africa and South America being under-represented. At present supply and demand for qualified officers and crew are more or less in balance, but there is a risk of a lack of qualified officers and crew if and when the world economy expands rapidly. Women form only about two per cent of those employed in shipping, and are mainly on cruise ships and ferries. There is a lack of information about deaths and injuries to seafarers.

The move in general cargo shipping to containerization has revolutionized ports: investment in machinery that requires skilled operators has ended the need for large numbers of dockworkers capable of handling heavy loads by hand. The efficiency of port operations is very important for ship-operators. Increasingly, the ports in a region are in competition with each other, and the charges are affected by the extent to which port charges are expected to cover the costs of construction and maintenance of ports and their road links to the hinterland. The quality of port infrastructure and operations varies widely, although many developing countries achieve the highest standards in both.

The provision of submarine cables is driven by demand for internet bandwidth. This shows no signs of slowing its growth. Submarine cables to carry the traffic are mainly provided by consortiums of telecommunications companies, internet service providers and private-sector investors. The market appears to be working smoothly.

2.2 Waste

The socioeconomic aspects of waste reaching the sea have both direct and indirect features. Waterborne pathogens are commonly carried with sewage. These can seriously affect human health, both through bathing in contaminated water, and through eating fish and seafood contaminated with them. Other wastes have socioeconomic implications by affecting food quality and through their effects on fish and other species used for food by, for example, adverse effects on reproduction. More indirectly, effects of waste on water quality can damage tourism and reduce the aesthetic, cultural, religious and spiritual ecosystem services that humans get from the sea.

2.3 Extractive industries

The offshore oil and gas industries are significant for the economies of the countries that have started them: the industry accounts for about 21 per cent of Norway's gross domestic product (GDP), 35 per cent of Nigeria's GDP, 3.5 per cent of the United Kingdom's GDP and 1.5 per cent of United States GDP. The number of people employed is relatively small: estimated at 200,000 worldwide.

Compared with land-based mining, the extraction of minerals from the seabed is a very small-scale activity. The United Kingdom industry for extracting sand and gravel seems to be the largest, with 400 employees.

2.4 Coastal zone

Since a high proportion of humans live in the coastal zone, there is a preoccupation with making sure that: (1) land used for housing, industry or agriculture is not lost or flooded, (2) the demand for land suitable for urban development and ports (and in some cases agriculture) is met, and (3) existing homes and infrastructure are not destroyed. This leads to a readiness to invest substantial amounts in both coastal protection and land reclamation. The long-term effectiveness of hard engineering approaches to these tasks has been called into question, and in many parts of the world the approach tends more towards adjusting the natural process of erosion and sedimentation to achieve the desired ends.

As has been said, in many parts of the world, tourism and recreation is a major economic activity in the coastal zone. It requires a relatively high proportion of labour in preparing and serving food and in cleaning and maintaining accommodations, providing jobs that in many regions are strongly seasonal. A high proportion of these jobs are filled by women.

2.5 Other activities

Desalinization is essential for the continued existence of many States. It may likewise be important for avoiding constraints on future economic development in other places. Renewable energy from maritime sources, which is beginning to be implemented in some parts of the world, has a significant potential role to play in mitigating climate change,. The use of marine genetic resources offers possibilities of finding and applying new marine ecosystem services. Marine scientific research is an essential underpinning of managing the sustainable use of the ocean.

3 Pathways from the human activity to its environmental impacts

3.1 Communications and transport

There are three main pathways by which shipping impacts on the environment: loss of ships, chronic discharges and emissions, and noise.

Ports also impact on the environment in three main ways: the demand for coastal land (which often leads to reclamation of the necessary land from the sea), changes in the form of the coast (with hard coastlines replacing softer ones) and dredging to maintain navigation channels (and the consequent need to dispose of the dredged material). Ports also inevitably lead to concentrations of shipping, and therefore represent areas where the impacts of shipping are equally concentrated.

Submarine cables have very limited environmental impacts, since they are very slim (typically 25 – 40 millimetres wide in the deep sea), and since their routes are usually chosen to avoid, where possible, areas that may cause problems from bottom trawling and ships' anchors. In soft substrates on continental shelves the cables are usually buried by ploughing, but again the zone affected is narrow.

3.2 Waste

Waste products reach and affect the marine environment through a variety of routes. Liquid discharges may reach the sea either though discharges to rivers or directly through pipelines. Waste emissions to air can be carried to the sea directly, or through run-off from the land on which they are originally deposited. Substances applied to land may volatilize and be re-deposited, either directly to the sea, or through successive re-volatilizations and re-depositions. Solid waste and other matter may be deliberately dumped into the sea, or may reach it from badly-managed waste disposal on land.

3.3 Extractive industries

The offshore oil and gas industries affect the marine environment through six main pathways: the effects of seismic exploration during the exploration phase; the drill cuttings (and the drilling muds used to lubricate the drills which are mixed with them) that are discarded on the seabed; the chemicals that are used and discharged during operation; the produced water (and its admixture of oil) that is discharged during production, and which increases in quantity as the wells age; the gas that may be flared off during production; and the oil spills that may occur. In addition, there is the question of the disposal of offshore production platforms when they are decommissioned at the end of the field's life.

Since offshore mining is currently based on dredging, the impacts on the marine environment come from disturbance of the seabed and (except with sand and gravel extraction) the discharge of the dredged material that is rejected as not containing the minerals that are sought. Disturbance of the bed of the deep sea by future mining has considerable potential to harm benthic biodiversity, about which there is as yet limited knowledge.

3.4 Coastal zone

There are two main ways in which anthropogenic change to the natural processes of erosion and sedimentation along the coast affects ecosystems. First, wherever and however the anthropogenic change happens, there are likely to be consequential changes in the processes elsewhere in the general neighbourhood. Secondly, such change usually involves moving from soft shore forms (gravel, sand or mud) to hard shore forms (stone or concrete). In addition to these two processes, changes in land affecting river regimes (for example, through the building of dams) tend to reduce the flow of sediment from land to sea, weakening beach replenishment, as well as weakening the force of rivers, which can lead to the extension of sand bars at their mouths.

Significant provision of tourist and recreational facilities usually leads to major changes in the shore environment. Hard shorelines can replace soft ones. Urban development can destroy the natural hinterland of the shore. Night-time lighting can significantly affect habitats. Increased discharges of sewage (especially if not properly treated) can produce the problems associated with excessive nutrient discharges. Regular cleaning of beaches removes the natural detritus which is important for many shore-living birds and animals.

3.5 Other human activities

The pathways by which desalinization can affect the marine environment are mainly through its intakes of seawater and discharges of brine. The energy required for desalinization can also have important implications when it is provided by the burning of fossil fuels. Most forms of renewable energy generation make substantial demands for ocean space which cannot be used for other purposes. The structures required for wind-farms may have effects on migrating birds, but with proper planning these seem to be limited. Although potential pathways for environmental effects exist for other forms of human activity, they are currently judged to be minimal. Provided collection of specimens is done with care, there is no reason why the development of the use of marine genetic resources should adversely affect the marine environment. Marine scientific research has to be involved with all forms of marine biota and habitats, but provided proper protocols are followed, no significant adverse effects should result.

4 Major ecosystem impacts

4.1 Communications and transport

Steady progress has been made in reducing the numbers of ships lost at sea, thus damage to the environment from shipping disasters has dropped significantly, particularly in respect of the amounts of oil spilled in such disasters. Regimes have been established to control chronic discharges from ships, in particular in the form of oil, sewage, garbage and air pollutants. Measures have also been taken to deal with invasive species carried in ballast water and with wrecks. The challenge now is to improve enforcement of these regimes. Noise from ships may well be a source of significant human impact on marine life. Losses of containers overboard are low.

Ports have an important role to play in enforcing the control regime over ships. Since they are often in competition with other ports in the region, the regional memorandums of understanding on port state control have an important function. The quality of port reception facilities for waste oil, sewage and garbage is also important in reducing environmental impacts of ports and their users. The disposal of dredged material needs proper management. Even where the material is harmless, it can damage bottom-living plants and animals by smothering them. Where the material contains contaminants (usually from historic industrial activities), disposal at sea risks remobilizing them and again requires proper management.

Submarine cables have always been at risk of breaks from submarine landslides, mainly at the edge of the continental shelf. As the pattern of cyclones, hurricanes and typhoons changes, submarine areas that have so far been stable may become less so, and thus produce submarine landslides and consequent cable breaks. With the increasing dependence of world trade on the internet, such breaks (in addition to breaks from other causes, such as ships' anchors and bottom-trawling) could delay or interrupt communications vital to that trade.

4.2 Waste

Waste material introduced into the ocean can cause problems for the marine environment in a variety of ways. Hazardous substances (heavy metals, persistent organic pollutants, polycyclic aromatic hydrocarbons (PAHs), pesticides and endocrine disruptors) can be toxic to marine biota, reduce their reproductive success or weaken their immune systems so that they succumb more easily to disease. Excessive discharges and emissions of nutrients (particularly nitrogen compounds) from human bodily wastes, animal excreta, food-processing plants, agricultural fertilizers and traffic can produce hypoxic (low oxygen) dead zones in the sea, which can kill bottom-living plants and animals and reduce fish stocks. They can also cause algal blooms which can smother beaches and which can consist of algae species that generate toxins harmful both to humans and marine life. Pathogens in human and animal waste can cause illness from contact with the water into which they are discharged and from food from the sea that they contaminate. All these impacts undermine human health and ecosystems and make them much less resilient to other pressures.

4.3 Extractive industries

It is possible to regulate all the pathways by which the offshore oil and gas industries affect the marine environment to keep the impacts at an acceptable level. The success of such regulation depends on the regulatory methods chosen and the degree to which they are enforced. The impact of noise disturbance from seismic exploration depends very much on the overlap of the area to be explored with the habitats of marine life that is sensitive to noise. These animals often have seasonal patterns of migration, and seismic exploration can be timed so that overlaps do not happen. Information on the impact is limited. The impact of drill cuttings is mainly from the drilling muds with which they are mixed, although some release of metals can occur form the rock cuttings themselves. Regulation of the drilling muds used can control this problem. The same approach can be used to control the chemicals used on, and discharged from, offshore installation. Produced water, because of its quantity, has to be discharged to the sea. The problem which it poses is the oil content. This oil content can be largely removed by centrifuges, and an acceptable level (usually 30 parts per million or less) can be achieved. In the North Sea, steps have been taken to tighten limits as the amount of produced water increases. Spills from offshore installations or breaks in pipelines occur. The main safeguard against such spills is the proper environmental management of the whole operation. In at least one region, steps have been taken to encourage high standards of such management. There are two main approaches to the removal of decommissioned installations. Under 1989 IMO Guidelines, installations should be removed so that there are 55 metres of clear water over any remains. In the North Sea, removal of the rest of the installation is the norm, although exceptions can be made. In the Gulf of Mexico, it is much more usual to allow installations to be placed to form artificial reefs.

For current offshore mining, the main issue are the turbidity created by the dredging and the management of the disposal of unwanted dredged material. In most cases, both are likely to mean that the area mined will be effectively cleared of marine biota. Recovery and the scale and speed of re-colonization after mining has ceased in an area will very much depend on local circumstances.

4.4 Coastal zone

The form of consequential changes resulting from anthropogenic changes to the land/sea boundary will depend entirely on local circumstances: it may take the form of promoting erosion that would not otherwise have happened, because the longshore water movements are re-focused, or it may take the form of causing siltation where it has not previously happened, or it may be a mixture of these in different places. Only very careful modelling during the planning process can hope to identify the consequences, and even then the best models can prove not to have been adequate. Where soft shorelines are replaced with hard shorelines, the local biota will be affected. It will become more difficult for animals to move from sea to land and back, which may disturb their foraging or breeding patterns. It may also affect local plant communities. Such changes are likely to make the shoreline less resilient. Finally, it will offer a new hard substrate to biota that arrive, which is likely to change the local fish and shellfish communities. The natural consequence of beach erosion is the landward retreat of coastal habitats, but this natural process is hindered by coastal development, which causes so-called "coastal squeeze"; there is no space for the habitats to retreat. Coastal squeeze results in the fragmentation and removal of some habitats and the species they support. Finally, in areas where sediment loads have increased above natural levels, impacts include burial of habitat, reduced light levels caused by turbidity, changes to substrate (e.g., mud draping over once rocky habitat), and smothering of coral reefs and other sessile fauna.

Tourism and recreational development is likely to produce all the problems of coastal development described above. In addition, it is likely to lead to large numbers of people walking on the shore (and thus compacting sand and disturbing breeding sites) and using the water (and thus disturbing larger animals and fish (not least with noisy motorized devices), creating oil films on the water surface from sun-tan preparations and leaving litter that becomes marine debris).

4.5 Other activities

Any effects of desalinization intakes and discharges are very local – a matter of tens of metres – and even these can be reduced by proper design. The remaining other activities should not create major environmental impacts, but the use of autonomous floats in marine scientific research needs – and is receiving – care over their eventual fate.

5 Integration of environmental and socio-economic trends

5.1 Communications and transport

The levels of activities in shipping and in the ports that it uses, and the demand for submarine cables, all respond fairly directly to the world economic situation. The growth of industry in the Pacific basin, especially in the west, means that the pressures from all forms of communications and transport will particularly increase there. Growth in communications and transport activity means that, even if all ships, ports and cables achieve the best currently practicable level of environmental protection, the pressures from them will nevertheless continue to increase. Since the environment is finite, it will only be possible to limit the pressure on the environment to no more than current levels, if improved performance in safeguarding the marine environment against those pressures is achieved.

5.2 Waste

Waste generation has tended to increase at least in step with economic growth. In some ways, economic growth has generated additional forms of waste, particularly through the development of packaging to protect agricultural and industrial products. It is only relatively recently that efforts have been devoted to waste minimization. Population and industrial growth means that, even if the best currently practicable levels of waste reduction and control are generally achieved, there will be increasing pressure on the finite marine environment. As with shipping, this implies that continuous improvement in environmental protection is needed.

5.3 Extractive industries

The creation of an extractive enterprise is the result of a decision balancing the economic benefits to be gained against the environmental and other impacts. It is therefore important that, when a decision is initially taken, there should be a clear view of what measures will be required to protect the marine environment. There is now widespread understanding of what is required in most environments. Problems arise, however, when the circumstances of a new development are very different from past experience. This is particularly the case in Arctic conditions, where natural processes to break down contamination will be slower because of the cold.

5.4 Coastal zone

Socioeconomic pressures are likely to be strongly in favour of intervention to reclaim land and prevent erosion. With economic growth, there will be pressure for ports to improve their capacities, which is likely to require more land. Short of complete relocation of the port, reclamation of land from the sea may be the only option. Equally, where erosion or flooding is threatened, there will always be pressure to safeguard existing investments in housing, industrial buildings and infrastructure.

Because of the economic importance of tourism and recreation in so many places, there is usually pressure to extend tourist facilities. The problem is that too great an extension is likely to lead to devaluing the attractions that enabled a tourist industry to develop in the first place, and thus to undermine the justification for any extension. Without integrated coastal zone management to balance economic gain against environmental change, there is a risk that the basis of the economic gain can be eroded.

5.5 Other activities

Pressures to increase the amount of desalinization that is carried out will result from pressures to expand populations in water-poor locations. The economic and social drivers of such developments are outside the scope of this Assessment. The need for mitigation of climate change is undeniable, and the economics of expanding the amount of energy produced from marine renewable sources will depend on judgments about how far this form of mitigation is good value. Much of the genetic diversity in our seas and oceans remains unknown and relatively unexplored. Without marine scientific research, it will be difficult (if not impossible) to achieve and maintain sustainable management of all human activities affecting the sea.

6 Environmental, economic and social influences

6.1 Communications and transport

The global shipping industry has recognized the need for improvement of its environmental performance. States have thus adopted, through the International Maritime Organization (IMO), a range of measures to try to improve the industry's performance. Enforcement of these measures is crucial. The IMO has now started a collective process to improve the performance of its members in this enforcement.

For submarine cables, the International Cable Protection Committee brings together cable owners, maintenance authorities, cable system manufacturers, cable ship operators, cable route survey companies and governments to consider all aspects of ensuring the safety of submarine cables and reducing their impact on the environment.

6.2 Waste

Global systems have been put in place to control persistent organic pollutants and mercury. These offer the possibility that the impacts of some forms of hazardous substances on the marine environment will be brought under control. In some regions (for example, the European Union), more general systems have been set up for controlling present and future chemicals that are placed on the market. There is general recognition in most international investment programmes of the need to improve sewage collection and treatment in many developing countries. The Global Programme of Action to Protect the Marine Environment from Land-Based Activities provides a framework within which States can consider their overall approach to these problems. The challenge is to make all these various steps operational.

6.3 Extractive industries

Offshore oil and gas extraction is usually significantly more expensive than extraction on land by normal drilling processes. Placer mining from the coastal seabed, however, may well not be much more expensive than winning the same minerals on land. Commitment to offshore developments will therefore depend on judgements about future demand, costs and prices. The hydrocarbon market is currently changing significantly with the emergence of new sources such as "fracking" and the Athabasca oil sands. The global market in hydrocarbons, and the other minerals mined from the seabed, will therefore determine the pressures to develop marine extractive industries further. However, in the longer term, the increasing difficulties of sourcing minerals on land are likely to increase the interest in mining the bed of the deep sea. The International Seabed Authority is charged with regulating such activity in the Area.

6.4 Coastal zone

In the absence of integrated coastal zone management, with its capacity to bring together in one decision-making process all the various factors that can affect the sustainability of a coastal zone, there are bound to be problems in balancing the different factors involved in the land/sea interface and in maintaining a successful tourist industry.

6.5 Other activities

The factors that will affect developments in the other human activities discussed in Part V are those sketched out above.

7 Capacity building gaps

The capacity-building gaps identified in this Part are summarised in Chapter 32 (Capacity-building in relation to human activities affecting the marine environment).

32 Capacity-Building in Relation to Human Activities Affecting the Marine Environment

Group of Experts:
Renison Ruwa (Lead Member), Amanuel Yoanes Ajawin, Sean O. Green, Lorna Inniss, Osman Keh Kamara and Alan Simcock (Co-Lead Members)

1 Introduction

The oceans provide various ecosystem services or what are also referred to as the "benefits that people desire from ecosystems" (Millennium Ecosystem Assessment, 2005). It is therefore necessary to know the types or nature of services that humans receive from the oceans and the scale or level of human activities that can be exerted without causing imbalances that could affect sustainability. Achieving sustainability requires strong public understanding of the importance of the ocean. This therefore calls for enhanced outreach and communication efforts through the development of mechanisms and partnerships to build capacity for outreach and awareness programmes. The major types of ecosystem services are described in Chapter 3. For sustainability the following are needed: scientific understanding of the services; assessment of the level of food production which results from various ecological processes, in order to address food security and safety; assessment of aesthetic uses of the ocean environment; and the level and type of capacity for studying and managing human activities and their impacts arising from exploitation of the ecosystem services. The level of capacity-building reflects, among other things, the efforts at identifying knowledge gaps in science, technological advances, human skills development and infrastructure.

To fulfil the overall objective of the Regular Process, all States need to address the overall objectives of the Regular Process as set out in the reports of the Ad Hoc Working Group of the Whole (AHWGW) to the United Nations General Assembly (UNGA) (A/64/347, 65/358), and the United Nations Secretary-General's Report (A65/69/Add.1) (UNGA 2010, UNGA/AHWGW 2009 and 2010). This outcome can only be achieved with significant efforts at capacity-building. The Regular Process itself therefore promotes, facilitates, and, within its capabilities, ensures that capacity-building and technology transfer are undertaken through promoting technical cooperation, including South-South cooperation amongst developing countries and taking gender and equitable geographical distribution into account. Over the long-term (i.e., beyond this first Assessment), the Regular Process will support and promote capacity-building through identifying opportunities and facilitate linkages for international cooperation that includes technical cooperation and technology transfer with regard to developing countries (in particular the least developed countries, African coastal States and Small Island Developing States), in order to improve the capacity in these geographical areas to undertake integrated assessments. Substantial capacity-building efforts are being undertaken by United Nations agencies through technical cooperation programmes. It is also important that gaps are identified and priorities shared so that a coherent programme to support capacity-building in marine monitoring and assessment, including socioeconomic aspects, is achieved, The approach for this first baseline Assessment was to conduct integrated assessments using the "Driver Pressure State Impact Response" (DPSIR) methodology commonly used to represent human-environmental /economic interactions , including scaling up assessments (national, sub-regional, regional and global). The workshops were also used as fora to explain the processes for conducting integrated assessments. The workshops were participatory and helped to promote ownership of the Regular Process outcomes at various scales. Furthermore, the workshops not only added further value in creating and promoting awareness of the Regular Process, but also promoted institutional capacity linkages. Various regional and international reviews of capacity-building have been conducted by various agencies. These also provide sources of information for a critical analysis of this subject, in particular for identification of gaps; therefore this chapter also includes an overview of these regional and international initiatives as per the chapters authored in this section.

This first Assessment has two chapters on capacity-building: one each in Parts V and VI. Part V deals with "Assessment of other human activities affecting the marine environment" and includes this chapter on "Capacity-building in relation to human activities affecting the marine environment." Part VI, entitled: "Assessment of marine biological diversity and habitats", includes Chapter 53 on "Capacity-building needs in relation to the status of species and habitats". The topics addressed in the chapters are based on the DPSIR Methodology as approved by the AHWGW. Furthermore, pursuant to the guidance of the AHWGW, the regional workshops will also contribute to identification of capacity-building strategies to address the approved themes in the two chapters on capacity-building for regional needs.

2 Outcomes based on regional workshops on capacity-building needs

The analysis showed that for most regions the main capacity needs were cross-cutting issues among the regions; these are summarized as follows: (i) Data accessibility and data sharing; (ii) The provisions for mentoring and training opportunities for less experienced scientists and practitioners; (iii) Data collection and marine habitat mapping to inform management of ecosystems, biodiversity and fisheries; (iv) Need to improve professional capacities to assess socioeconomic issues; and (v) Capacity to conduct integrated and ecosystem-services assessments.

The regional workshops were undertaken in the following regions: south-west Pacific region (UNGA 2013a), Wider Caribbean region (UNGA 2013b), eastern and southeastern Asian Seas (UNGA 2012a), South-East Pacific region (UNGA 2011), the joint North Atlantic, Baltic Sea, Mediterranean and the Black Sea region (UNGA 2012b), the Western Indian Ocean (UNGA 2013c), South Atlantic Ocean (UNGA 2013d) and Northern Indian Ocean (UNGA 2014). The regional outcomes in terms of knowledge gaps and capacity needs were as follows:

2.1 Capacity needs for marine assessments in the south-west Pacific Region

This workshop was held in Brisbane, Australia, 25-27 February 2013 (UNGA 2013a). The focus was on linkages and upscaling from national to regional and global scales to promote synergies for building capacity which will include mentoring, learning and cooperation in communication, data and information transfer, as follows:

- The production of global marine assessments should be linked to ongoing efforts to support regional (led by the Secretariat of the Pacific Regional Environment Programme) and national state-of-the-environment reporting and streamlining of reporting arrangements (led by the Pacific Islands Forum Secretariat/Secretariat of the Pacific Regional Environment Programme). By providing capacity development and other support to these initiatives, the region will be better placed to contribute to and benefit from the Regular Process. The production of global marine assessments should be done in a way that provides mentoring and learning opportunities for less experienced scientists and practitioners.
- Active facilitation of involvement of practitioners from Pacific Island countries and territories in producing global marine assessments, including improved communication efforts to ensure awareness of the opportunity to be involved, assistance in registering for the Pool of Experts and resourcing support for and formal recognition of work done will all contribute to capacity-building in those countries.
- A large quantity of data and information exists, but it is often not readily identifiable or accessible. Enhanced regional and national capacities to store, access, share and interrogate data and information would assist the production of global marine assessments and facilitate the meeting of regional and national objectives.
- Resourcing is a substantial constraint on the capacity of the region to contribute to the production of global marine assessments. This can in part be addressed by the nature, scope and process for the development of assessments that more deliberately support national and regional objectives, as well as the objective of producing a global report. For example, the global marine assessment could provide region-specific information and access to the underlying data and information.
- Because of the limited capacity of the region to engage in the drafting of this Assessment, the review stage might be an efficient point for the region to ensure that regional information and perspectives are appropriately reflected therein. A second workshop or network among involved practitioners may provide mechanisms for doing this. Similarly, providing support to an appropriate Pacific regional organization to facilitate and coordinate ongoing regional engagement may be useful.

2.2 Capacity needs for marine assessments in the Wider Caribbean Region

This workshop was held in Miami, United States of America, 13-15 November 2012 (UNGA 2013b). The emphasis was placed on: needs for projects to include capacity-building and have specialized research institutions and research vessels offer opportunities for training, including the use of ships of opportunity; specialized research institutions to offer learning and mentoring opportunities, especially data and information analysis and synthesis; building collaboration and networks across experts, institutions and a variety of stakeholders, and promoting a culture of manpower retention for sustaining research in institutions. Other points included:

- Previous or ongoing regional marine assessments, specifically the Caribbean Coastal Marine Productivity Programme, the Caribbean Planning for Adaptation to Climate Change Project and the Caribbean Large Marine Ecosystem Project, were highlighted as successful cases of capacity-building.
- In some disciplines, such as physical oceanography and remote sensing of the ocean environment, capacity is highly concentrated in a few institutions. In other disciplines, such as social sciences, it is highly dispersed.
- Access to research vessels (e.g., NOAA ships) and ships of opportunity (e.g., those used in relation to the Living Oceans Foundation) offer opportunities and synergies on a wider scale with advanced technology for enhanced marine assessments.
- Data are often abundant, including data collected by ships of opportunity; the limitation is in the capacity to manage the data, including how to organize, store, synthesize and analyse them. Participants discussed the need for nationals to study at institutions where data are already being used and then to bring the expertise home.
- Building collaboration among scientists, resource managers and other stakeholders is central to capacity-building, especially as it includes building a willingness to share and communicate. With this in mind, capacity-building in the region would benefit from establishing and promoting networks of practitioners, experts, institutions and countries and promoting regional programmes.
- A fundamental shortfall exists in capacity to integrate the key insights of existing research into policy and management agendas, and this is a core area where capacity-building would yield benefits.
- There would be great costs in capacity from failure to retain the knowledge that is invested in training employees and management leadership. Such retention requires fiscal incentives to retain individuals in positions. The constant cycle of promotion at all levels results in an export of knowledge out of the field. Often, the bulk of expert individuals will be lost from policy and management to narrow academic research fields.

2.3 Capacity-building needs for the eastern and south-eastern Asian Seas

This workshop was held in Sanya, China, 21-23 February 2012 (UNGA 2012a). The focus was on building skills in integrated assessments, methodologies and quality assurance of data through effective creation of synergies and communication for data and information sharing. Creating awareness of the Regular Process within the scientific community of the region was emphasized. A successful WOA would require the ability to understand the implications of what we know about the status of biodiversity and link this with the state of the environment, as well as with ecosystem-based fisheries assessments in order to produce accurate fisheries status reports. In addition to assessing capture fisheries correctly, there is insufficient capacity for assessing impacts of aquaculture on the surrounding marine ecosystems and more generally for assessing environmental impacts that are anthropogenic, and/or due to climate change and invasion of alien species, as well as for socio-economic assessments of human well-being. All these are candidates for capacity-building that would improve capacity to conduct integrated assessments. Other points included the following.

1. At the highest level, the workshop participants identified as the first priority the need for improved skills in and knowledge on the conduct of integrated assessments (i.e., including environmental, economic and social aspects). Such experience/skills were lacking throughout the region and training in methodologies for conducting integrated assessments would be of direct benefit to the Regular Process.
2. Additional short-term capacity-building needs (i.e., that could deliver results within the next 18 months) identified by the workshop included the following:
(a) Building awareness of the need for interoperability between States and regions regarding several areas, including: an international classification standard for marine economic activities; quality assurance/quality control for data collection and analysis; enhancing comparability and compatibility of data from different sources; and biological information management, including taxonomy;
(b) Improved international networking and resource sharing, including a network to facilitate international communication and cooperative platform-building related to marine environmental, social and economic data;
(c) Following the kind offer from UNEP, IOC-UNESCO and the Asia-Pacific Network for Global Change Research (APN), the organization of a regional workshop focusing on capacity-building and the technical and scientific aspects of the Regular Process would aim to share information about available assessments, data and knowledge of methodologies to be used in compiling and developing the first global integrated marine assessment.
3. This regional workshop would aim at gathering scientists and relevant national authorities to raise awareness of the Regular Process within the scientific community of the region. The workshop would also aim at facilitating the appointment by States of individual scientists from the region to the pool of experts. The workshop would be co-organized by UNEP, IOC-UNESCO, GRID-Arendal, the North-West Pacific Action Plan (NOWPAP) and the Coordinating Body on the Seas of East Asia (COBSEA), with the support of APN.
4. Long-term capacity-building needs (i.e., that should be started quickly but which would only deliver results in the next three to five years) identified by the workshop included the following:
(a) Conduct of marine habitat mapping to inform management of ecosystems, biodiversity and fisheries. This included the development of skills in areas such as collection and analysis of remote sensing data, acoustic seafloor mapping, underwater video analysis and statistical analysis of biophysical environmental data;
(b) Long-term and well-planned biodiversity assessments were needed on both commercial and non-commercial marine species, including using genetic information to trace and determine stocks and species;

(c) Ecosystem-based fisheries assessment for capture fisheries and forecasting the status of fish and shellfish stocks;
(d) Assessing impacts of capture fisheries on the marine ecosystem;
(e) Assessing impacts of aquaculture on the surrounding marine ecosystem;
(f) Assessing impacts of habitat degradation (e.g., using ecological modelling and forecasting) on projected fish and shellfish stocks and aquaculture;
(g) Monitoring anthropogenic contamination of water, sediment and biota, to ensure maintenance of food security;
(h) Assessing impacts of climate change on marine biota and ecosystems, including the effects of ocean temperature change, ocean acidification, changes in coastal sediment and water discharge, changes in tidal and other currents, swell and wave patterns and coastal habitat changes due to sea-level rise;
(i) Assessing impacts of alien species;
(j) Assessing socioeconomic aspects.

2.4 Capacity needs for marine assessments in the south-east Pacific Region

This workshop was held in Santiago, Chile, 13-15 September 2011 (UNGA 2011). The focus was on addressing institutional and individual capacity-building, especially with regard to technical support and joint development and implementation of partnership projects. It called attention to the insufficient capacity to monitor harmful and alien species using remote sensing capabilities, as well as creating capacity to organize databases using standardized tools and formats. It was also important to build capacity to assess the effects on biodiversity of human activities and to address biophysical and socioeconomic issues for human well-being. Other needs included:

2.4.1 Information

- Information on this vast ocean region of the South Pacific is scattered and has not been summarized and collated, although it exists in the form of reports of scientific expeditions, historical records of fishing activities (fishing fleets) and a large number of scientific publications. The large increase in databases on biodiversity from 5 million entries in 2005 to over 32 million geo-referenced records in 2011 was noted;
- The South-East Pacific Group of Experts considered it essential in the short term to strengthen the capacities of the competent technical bodies with regard to integrated assessment methods. The DPSIR methodology adopted by the UNGA as the conceptual basis for carrying out this first integrated assessment of the marine environment, although known in the region and widely used in the terrestrial environment, has thus far not been regularly used in marine environmental assessments. The fruitful exchange of information between experts from the west coast of the Americas, from Mexico to Chile is noted;

- Incorporate geo-referencing information systems for ecosystem-focused analysis;
- Improve information and monitoring systems;
- Compile base-line data, which is difficult and costly;
- Improve information systems that can be shared.

2.4.2 Capacity-building

- The South-East Pacific Group of Experts acknowledges the shortfall in ability to generate capacity to analyse the ocean environment in areas beyond national jurisdiction;
- More experts able to conduct research on climate change with reference to oceans;
- Capacity to organize databases using standardized formats and tools for access by the public;
- Strengthen methodology for economic assessment;
- Pilot project in Chile to harmonize economic assessment methodologies.

2.4.3 Knowledge gaps

- Technical support for the maintenance of equipment and sensors;
- Development of projects and research capacity on palaeoclimatology at the regional level, including effects on marine coastal areas (corals, sediments, ice cores, etc.);
- Monitoring of harmful algal blooms by remote sensing;
- Assessment of wide-scale processes at the level of the entire South Pacific Basin is of great importance in understanding and predicting the behaviour of living marine resources, particularly those exhibiting migratory behaviour (birds, turtles, mammals and pelagic fish species) in the south-east Pacific region;

2.5 *Capacity needs for marine assessments in the North Atlantic, the Baltic Sea, the Mediterranean Sea and the Black Sea*

This workshop was held in Brussels, Belgium, 27-29 June 2012 (UNGA 2012b). The meeting determined that transfers of skills within the region were needed and that the region can provide a source of knowledge and skills for other regions through creation of partnerships. It is necessary to address food security, marine biodiversity and habitats and information on anthropogenic impacts on the marine environment. Other points included:

1. It was agreed that capacity shortfalls did exist within the area covered by the workshop, and that the region could serve as a source of knowledge for other regions. Transfers of skills within the region were needed both from north to south (particularly within the Mediterranean) and from west to east.
2. Knowledge gaps at national and regional scales were identified in the report entitled "Analysis of the existing marine assessment in Europe", prepared in June 2012, including information on:
(a) Food security;
(b) Marine biological diversity and habitats;
(c) Human activity affecting the marine environment.

2.6 Capacity needs for marine assessments in the western Indian Ocean

This workshop was held in Maputo, Mozambique, 6-7 December 2012 (UNGA 2013c). The focus was on capacity needs to address biophysical issues, which are important for alteration of biodiversity, and socio-economic impacts due to anthropogenic impacts which consequently influence human well-being. Further emphasis was placed on building institutional and individual capacity to address biodiversity, fisheries, tourism, aquaculture, information and data, mining, and economic valuation of natural resources and the environment for human well-being.

The experts assembled at this workshop clearly endorsed the Regular Process; however, capacity-building needs were not highlighted in the December 2012 workshop report. In the assessment workshop of August 2012 the following capacity needs and gaps were identified:

- Information on environmental flows for major rivers;
- Information on ocean acidification: degree and extent of ocean acidification resulting from human activities (including coral bleaching) and socioeconomic implications;
- Regional perspective on ocean-source carbonate production;
- Information on pollution determination from aquaculture use and modification of habitats;
- Environmental flow assessments of coastal, riverine and atmospheric inputs from land;
- Lack of capacity for assessing offshore hydrocarbon industries;
- Lack of capacity to assess offshore mining industries;
- Carrying-capacity studies need for tourism and recreation;
- Economic valuation of resources/environment.

2.7 Capacity needs for the marine assessments of the South Atlantic Ocean

This workshop was held in Grand-Bassam, Ivory Coast, 28 to 30 October 2013 (UNGA 2013d). The focus was on identifying knowledge gaps with regard to the biophysical, food security and safety, socioeconomic, and biodiversity aspects, based on which the capacity needs were identified.

2.7.1 Biophysical aspects

The principal gaps identified by the experts are: (i) Absence of continuous long time-series on sea-level rise and its impact on the coastal and marine environment; (ii) Absence of information on the knock-on effect of El Niño in the sub-region, especially in West Africa; (iii) Poor links between meteorological and oceanographic institutes; (iv) Lack of continuous, long time-series on acidification, especially *in situ* measurements at tropical latitudes; (v) Scarcity of studies on the factors influencing surface-layer and *species variation, notably studies based on in situ measurements of surface layers and plankton.*

2.7.2 Food security and safety aspects

In the South Atlantic region, many national institutions and regional organizations conduct assessments of the status of fish and shellfish stocks and fisheries. Although fisheries statistics are available, continuous time series are lacking in many areas. In fact, many assessments are project-related, so when financing stops, the data collection is discontinued. This happens in all countries; the only exceptions are Argentina and Uruguay, where fairly complete time series are available for the most economically important fish stocks. Vessel availability for independent fishery surveys is a constraint for the whole region.

2.7.3 Socioeconomic aspects related to fishing

The principal gaps identified by the experts in the economic evaluation of fishing activities are: (i) Scarcity of evaluations of economic consequences (risk assessment) of disasters and impact of other activities on fisheries and the living standards of fishers; (ii) Scarcity of studies on the impacts of the global economy on fisheries; (iii) Lack of data on post-fishing losses (during processing, marketing, etc.); (iv) Absence of studies on the impact of harmful algal blooms on fisheries in West Africa; (v) Lack of information on the contribution of artisanal fisheries.

The principal gaps identified by the experts on fishing practices and health and safety are: (i) Stock assessments of species caught by both the industrial and artisanal sectors (they are frequently pooled together, although some countries have good reporting systems); (ii) Scarcity of information on illegal, unreported and unregulated (IUU) fisheries, although the Food and Agriculture Organization of the United Nations (FAO) evaluates the implementation of the Code of Conduct for Responsible Fisheries country by country; (iii) Scarcity of assessments of incidental catches of marine mammals, turtles and birds, especially in the African countries; (iv) Scarcity of information on the number of people employed by the sector; (v) Ineffective implementation of health and safety control systems (poor reporting mechanisms).

2.7.4 Socioeconomic aspects related to environment

The principal gaps identified by the experts on environmental pollution affecting human health and their socioeconomic impacts are: (i) Poor reporting mechanisms and/or difficulty in accessing existing documentation (reports) on oil leakages and spills; (ii) Lack of information on the types and amounts of oil dumped into the sea and trends for the next decade; (iii) Poor capacity in the region to assess the disposal of solid waste in the ocean; (iv) Impacts of exploration and exploitation activities and the lack of regulation of offshore oil and gas exploration and exploitation as well as of sand and gravel mining; (v) Scarcity of studies on land reclamation and habitat modification; (vi) Lack of socio-economic data and technological skills; (vii) Scarcity of studies on the tourism industry and poor capacity to assess tourism and all associated (i.e., economic, environmental and social) aspects.

2.7.5 Biodiversity aspects

The principal gaps identified by the experts regarding coastal areas, continental shelf and deep sea habitats are: (i) Scarcity of information on deep sea and continental shelf habitats; (ii) Lack of information on the current status of the mangrove species (in this regard, surveys and geographic information system (GIS) mapping projects need to be conducted); (iii) Scarcity of seagrass mapping programmes; (iv) Lack of research on vulnerability and adaptation in response to climate change; (v) Scarcity of close monitoring programmes of cetaceans, especially in West Africa; (vi) Absence of monitoring programmes for certain estuarine areas, especially in West Africa; and (vii) Scarcity of knowledge with regard to deep-water corals and plankton.

2.7.6 Capacity needs

A major capacity shortage facing many countries in the South Atlantic region is the ability to conduct assessments of the state of the marine environment at national to regional spatial scales. This need is mainly due to the lack of funding, but also due to the lack of resources and capability to conduct such studies, especially at the local and national levels. It is important to note, however, that capacity needs are unevenly distributed and that South-South cooperation also represents an opportunity to fill existing gaps. The experts therefore suggested that more capacity-building activities be organized under the umbrella of the Regular Process.

Another important gap concerns the geographical discontinuity of information in the South Atlantic region, and in particular the scarcity of studies on biophysical and socioeconomic dynamics in the region. This was deemed to be an important gap that hinders the development of an integrated regional assessment. Optimizing the coordination of marine environmental data-collection activities within countries and within the region should contribute to the production of an integrated regional assessment.

2.8 Capacity needs for marine assessments in the northern Indian Ocean

This workshop was held in Chennai, India, 27-29 January 2014 (UNGA 2014). The meeting focused on identifying short-term and long-term capacity-building needs that were determined through gap analyses. The capacity-building should concentrate on developing methodologies for integrated assessments and standardization of data and information generation for national, sub-regional, regional and global assessments. It is also a priority to create regional partnerships for undertaking joint research and to mobilize funds for capacity-building. Capacity-building to address biodiversity, critical habitats, microbial assessments, shipping, and environmental monitoring using satellite technology is also highlighted as a priority. Other points included:

1. The immediate action plan recommendation includes identification of the needs for capacity-building (including the acquisition of necessary

technology) for marine monitoring and assessment (including integrated assessments). The capacity-building activities need to concentrate on the following issues:

(a) methodologies to obtain the information from various sources on a regular basis;
(b) standardization of the information content for assessments at various levels;
(c) developing common methodologies to carry out the assessment and to train data collectors: this is very important for uniform data collection. The procedure, data collection, formatting and preparation of reports should be standardized for all the member countries.
(d) developing methodologies for scaling up national, sub-regional, regional and global assessments; and
(e) developing reporting forms to assist the integration process, with the aim of securing coherence, consistency and comparability as far as possible.

2. Development of a short-term capacity-building plan to mobilize the information and knowledge that is known to exist but has not yet been systematically organized in a way that would allow its use for the Regular Process. However, for this purpose, it may be necessary to identify the gap areas, and make efforts in capacity-building for those areas. India can help other States in capacity-building at various levels.
3. To identify and fill gaps like microbial assessment, seagrass mapping, etc. Satellite-based techniques can be used to identify mangroves, seagrasses, etc., and create an ecosystem report card.
4. Undertake assessments on the open ocean and activities related to shipping.
5. To enhance cooperation between member States of the region, a template / matrix will be developed for circulation to neighbouring countries to complete; the questionnaire will include information for identifying gaps and capacity needs.
6. It is stressed that improving communication among the countries of the region is the most important first step.
7. Shortfalls in continuous monitoring of the environment using satellite technology.
8. Insufficient involvement of regional organizations, undertaking joint research programmes, and securing funds for capacity-building activities.

3 Outcomes based on chapters focussing on knowledge gaps to inform capacity-building needs

3.1 Assessment of major ecosystem services from the marine environment (other than provisioning services)

This section deals with three types of ecosystem services: regulating, cultural and supporting services (Chapters 3-9). The identified gaps and capacity-building needs to address them are as follows:

3.1.1 Ecosystem services other than provisioning services

- Data availability and resolution at different scales and geographical ranges: the developing world especially has massive gaps.
- Capacity to undertake heuristic/participatory processes at regional and global levels: this should involve training and empowering local stakeholders to enable them to understand the impact of ecosystem services on their well-being.
- Human capacity and infrastructure (research laboratories and institutes, observatories and oceanographic fleets) should be developed on a continual basis.

3.1.2 Oceans and the hydrological cycle

- Skills to quantify potential impacts on society and natural environment due to flooding and sea-level rise: the latter are acknowledged as being among the most serious issues confronting humankind.
- Capacity is inadequate to determine local sea-level changes which are also influenced by several natural factors, such as regional variability in the ocean and atmospheric circulation, subsidence, isostatic adjustment, and coastal erosion. It is necessary to study the latter too.
- Regional capacity is not sufficient to study changes in the rates of freshwater exchange between the ocean, atmosphere and continents because of their significant impacts. There is also inadequate ability to determine spatial variations in the distribution of evaporation and precipitation that create gradients in salinity and heat that in turn help drive ocean circulation.
- Capacity is insufficient to utilize traditional knowledge as an additional resource to address adaptation in given impact settings; this knowledge should be carefully evaluated within adaptation planning.
- Capacity is insufficient for standardizing methodologies to address regional differences which are due to differing data sources, temporal periods of analysis, and analysis methodologies.
- Capacity is insufficient for disaster preparedness to address high-intensity cyclones, because the scientific consensus shows that global warming will lead to fewer but more intense tropical cyclones globally. This will certainly affect coastal areas that have not been exposed previously to the dangers caused by tropical cyclones.

3.1.3 Sea-air interface

- Regional capacity is not adequate to determine levels of rising carbon dioxide (CO_2) in the atmosphere and increased absorption of CO_2 by the oceans, which has created an unprecedented ocean acidification (OA) phenomenon that is altering pH levels and threatening a number of marine ecosystems. It is necessary to map OA hotspots, which have now become a global problem.
- Capacity is insufficient to study the impact of shellfish farming due to acidification and to establish indicators for OA to facilitate determination of OA hot spots.

3.1.4   Plankton productivity and nutrients

- There are important shortfalls in regional capacity in terms of both infrastructure and human skills to enable measurement of primary production *in situ* and through remote sensing. The infrastructure includes multiplatform infrastructure, e.g., laboratories, oceanographic ships, moorings, drifters, gliders, aircraft, and satellites that can enable continuous measurements for both short-term and long-term monitoring.
- Various regions lack long-term measurements of primary production and therefore lack long-term data to construct predictive models to estimate trends.
- Phytoplankton can play a significant role in climate regulation to undertake continuous regional measurements of phytoplankton production through carbon sequestration, which is an order magnitude higher than that provided by grasslands and forest vegetation, and also form a basis for prediction of fisheries production to address food security. For both reasons it is important to undertake continuous regional measurements of phytoplankton production, and these measurements will require improved capacity for plankton monitoring.
- There is insufficient ability to identify which species of phytoplankton are most suitable for development of bio-fuels and pharmaceuticals.
- There is insufficient ability to identify which species of phytoplankton engage in nutrient recycling or nutrient stripping from seawater, culture them and use them for management of water quality in aquaculture.

3.1.5   Ocean-sourced carbonate production

- There is a shortfall in capacity to deal with the impacts of global warming and sea-level rise.
- There are gaps in our knowledge of the impacts of future rises in sea level on individual atolls; determining shoreline changes has rarely been undertaken, and long-term studies are especially lacking.
- Drastic effects from loss of sand dunes to beach mining and interruption of sediment pathways, especially as caused by coastal protection works.
- There is shortfall in capacity to deal with the impacts of acidification, which inhibit organisms from secreting carbonate shells or skeletons. Furthermore, reduction in sand-carbonate production leads to a decrease in supply to sand beaches. Relatively few studies exist of rates of carbonate production and transport of marine sand and gravel to contribute to coastal ecosystems.

3.1.6   Aesthetic, cultural, religious and spiritual ecosystem services derived from the marine environment

- It is necessary to identify the priority concerns in terms of the nature of the aesthetic, cultural, religious hand spiritual ecosystem services derived from the marine environment in relation to the various geographical areas, developed and developing countries, and find out how humans have adapted for their own well-being.

3.2   Assessment of the cross-cutting issues: Food security and food safety

Food security and food safety are important activities which play a crucial role in human well-being in the provisioning services category of the ecosystem services panoply. The major activities covered are capture fisheries and aquaculture, as well as scientific and socioeconomic aspects. From the gap analyses, the capacity-building needs to address are as follows:

3.2.1   Oceans and seas as sources of food

- Covering 71 per cent of the earth's surface, the oceans offer a variety of habitats for various fisheries species which are used for various competing needs: these are both consumptive and non-consumptive but of varying socioeconomic value. To maximize benefits, to address these competing needs would require multidisciplinary research teams. Fisheries must address food security as well as recreational, cultural and spiritual aspects.
- To enhance the traditional subsistence type of fishing commonly practised in the developing world will require addressing fishing in terms of commerce and profit and thereby creating employment and supporting livelihoods. Advanced capacity-building for appropriate skills will be required to be able to use advanced technologies to create wealth from capture fisheries and mariculture in a sustainable way.

3.2.2   Capture fisheries

- Efforts have been made to create awareness to reduce post-harvest losses, especially in small-scale fisheries, as a means of increasing production. However, little is known to what extent this is implemented and to what extent it has increased production, although this would greatly improve the socioeconomic benefits to small-scale fishers. Enhanced capacity-building for appropriate research and innovative technology and its transfer would address these issues.
- Efforts have been made to reduce by-catch and increase awareness of this problem, including efforts to make by-catch excluder devices. Globally it is still poorly known whether this has been successfully achieved in terms of the relative ratio of the target catch landed and the by-catch caught and either landed or discarded. To address these issues would require building capacity to monitor and ensure compliance and promote observer programmes effectively.
- To improve the ecosystem approach to fisheries management to address not only ecological issues and governance but also socioeconomic issues for human well-being will require increased efforts to promote ecosystem-based management.
- To avoid fisheries depletion requires controlling fishing effort for stocks. For most important fisheries, historical fishing trends are unknown and their recovery rates are also poorly known, but their

fisheries continue to expand into new areas. These issues can only be addressed with increased efforts to build enough capacity with appropriate technological and scientific skills to provide adequate information and data to facilitate regional and global management.
- There is insufficient capacity to address fish diseases from capture fisheries and illnesses caused by ingestion of toxic fish. Globally the phytosanitary issues are not well known, especially in developing countries.

3.2.3 Aquaculture

- To obtain a clear understanding of the trends and contribution of mariculture globally in terms of aquatic farming will require building capacity to address the relative ratio of freshwater aquaculture production and mariculture. Mariculture includes marine plant cultivation, which mostly consists of seaweeds.
- There is insufficient knowledge of mariculture diseases and how to combat them because they are poorly known, especially in the developing world. Filling this knowledge gap would require greater capacity in fish health in mariculture contexts.
- There is insufficient capacity to categorize mariculture for addressing food security, ornamental and decorative uses and clearly document their socioeconomic benefits.
- There is insufficient capacity to map cultivated species, where they are farmed regionally and globally, and share information and data to facilitate world production.
- To promote sustainability of mariculture will require building capacity to improve mariculture technologies that are environmentally friendly.
- There is insufficient capacity for improving industrial production of fish feed using low-value or trash fish, including by-catch that would otherwise be discarded. However, this should not compete with fish for direct human consumption or deliberate fishing that would be undesirable for biodiversity conservation.

3.2.4 Fish stock propagation

- There is insufficient capacity in aquaculture technologies which will promote efficient and effective stock propagation; this includes culture techniques under controlled conditions, provision of artificial habitat, feeding, fertilization, predator control and subsequent release of the aquatic organisms into the sea.
- Improved sustainability of fish stock propagation requires applying a comprehensive integrated ecosystem-based fisheries-management approach and therefore it is necessary to build capacity in terms of individuals, infrastructure and institutions that can deliver effective stock propagation.

3.2.5 Seaweeds and other benthic food

- Seaweed farming and aquaculture are seriously affected by disease and there is insufficient capacity to research seaweed diseases and build techniques for combating the diseases.

- To harness their wide variety of nutrients, medicinal and food values would require undertaking and building capacity for biochemical research on seaweed extracts from various species.

3.2.6 Social and economic aspects of fisheries and other marine food

- Certain issues, particularly at the micro level, demand additional research and therefore need capacity-building to address them. The state of small-scale fisheries throughout the world, and gender issues in fisheries, are particularly prominent and are poorly studied. A further issue that has been seriously under-researched is the relationship between capture fisheries and aquaculture.

3.3 Assessment of other human activities and the marine environment

The activities addressed in this section are basically centred on *in situ* use of the ocean, e.g., in shipping, ports, tourism, waste disposal and extractive uses, e.g., mining, desalination, etc. The gaps and the needed capacity-building are as follows:

3.3.1 Shipping

Knowledge gaps
- The IMO has emphasized the need for better information on the health and well-being of ships' crews. The death rate is unacceptably high, and little is known about causes of death, injuries and illnesses, with the result that it is difficult to formulate policies to address the problems.
- The potential development of Arctic shipping routes between the Atlantic and the Pacific highlights the inadequacy of charts of these waters: some date back to surveys in the mid-19th century. Similar shortcomings exist in Antarctic waters.
- As new anti-fouling systems for ships are developed, the resolution of the parties to the IMO Convention on the Control of Harmful Anti-fouling Systems on Ships calling for the harmonization of test methods and performance standards for anti-fouling systems containing biocides presents a necessity to investigate and evaluate such methods and standards.

Capacity-building
- Potential shortages exist in adequately trained ships' officers and crew, and both Africa and South America are proportionally underrepresented in the global pool of such officers and crew. Capacity-building to develop training institutions of high quality and to use such institutions to meet the demand is therefore desirable.
- Increased navigation in the Arctic Ocean and (in spite of the emergency response plans of the International Association of Antarctic Tour Operators) the presence of large passenger cruise ships in the Southern Ocean mean that there are gaps in adequate emergency response systems in both areas.

- In coastal areas where large numbers of very small vessels (especially with wooden hulls) operate, to ensure that the operators of such vessels have the knowledge and equipment to make them safe would require capacity-building. This could include capacity-building to ensure that maritime administrations can apply regional safety codes where they exist, or develop them where they have not yet been prepared.
- Improved port-state control is very important for ensuring the safety of shipping and the protection of the marine environment from accidents and unacceptable practices involving ships. There are gaps in the technical skills and equipment in some States for implementing effective port-state control.

3.3.2 Ports

- Because the operation of a port can significantly affect both the successful operation of ships and the economic performance of the countries it serves, some ports need capacity-building in the operational skills needed for successful port operation.
- The delivery to shore of garbage from ships is an important element of combating marine debris. Building capacity in this field for ports which do not have adequate and easily used port waste-reception facilities would improve their ability to combat marine debris.
- Many ports that need dredging to maintain or improve navigation adjoin bays, rivers or estuaries with a history of industrial discharges. Decisions on whether such material can safely be re-deposited in the sea, guided by international standards, requires the capacity to examine the dredged material relative to such standards.

3.3.3 Submarine cables and pipelines

- If coastal States wish to safely locate submarine cables and pipelines that cross areas of potential geological change and disruption, or (at least) to negotiate successfully with commercial undertakings planning to install cables in such locations, they need access to the skills in marine geology needed.
- In taking decisions on submarine cables and pipelines, States need to have the capacities to address possible competing uses of the seabed on which the cables and pipeline are laid.

3.3.4 Coastal, riverine and atmospheric inputs from land

- Shortfalls were found the skills and capacities for several important disciplines, including:
- Skills and infrastructure to monitor wastes and waste water (municipal, cruise ships and degree of treatment, industrial discharges, agricultural runoff, atmospheric emissions).
- Skills and infrastructure to treat waste and wastewater.
- Gaps in capacity to assess the environmental, social and economic aspects related to coastal, riverine and atmospheric inputs from land.
- Capacity to identify hazardous substances, which also includes ability to establish: thresholds of toxicity, persistence and bio-accumulation, a substance database with experimental data, monitoring and assessment programmes.
- Ability to monitor and assess atmospheric circulation and detect airborne inputs.

3.3.5 Offshore hydrocarbon industries

- A major capacity gap is the ability to manage environmental impact assessments and monitor compliance, mainly within (but not confined to) developing countries.

3.3.6 Other marine energy-oriented industries

The other sources of marine energy production industries are: offshore wind, waves, tides, ocean currents, marine biomass and energy from ocean thermal differences between different water layers. The capacity gaps to assess the environmental, social and economic aspects of offshore renewable energy deployment/generation are:

- Lack of information and data for full evaluation of Environmental Impact Assessments (EIAs). Data gaps are very common due to remoteness, or the level of technology not being available for long-term data and information gathering (especially for developing countries).
- Capacity in terms of enabling infrastructure to exploit these sources of energy.
- Skills or knowledge capacity lacking in most developing countries.
- High organizational capacity to foster relationships and linkages among ocean users, stakeholders and resource managers required to enable proper planning for use of these sources with minimal conflict and environmental impact.

More awareness campaigns would enhance appreciation of the fact that these renewable sources of energy, given their immense potential, can reduce use of the fossil-fuel carbon-based energy sources and reduce CO_2 emissions.

3.3.7 Offshore mining industries

- As in oil and gas, the major gap for this activity is the ability to undertake EIAs and monitor compliance, especially because of their remoteness; this is mainly so in developing countries.
- The offshore mining technology and management are still nascent and in mostly shallow water (<50m depth). Where such mining affects various stakeholder activities, social and economic conflicts can arise. Enhanced capacity for meaningful engagement with stakeholders will contribute to avoiding and resolving such conflicts.

3.3.8 Solid waste disposal

Information gaps

- Serious information gaps exist on the nature and volume of dumping. These gaps exist with regard to waters under the jurisdiction or control of both parties and non-parties to the London Convention and Protocol. The understanding of the potential effect of the dumping of solid waste on the marine environment is directly affected by these gaps.

- In areas where the possibility exists that explosives or containers of harmful substances, such as chemical weapons, have been dumped in the past, especially in areas where fishing vessels operate or where it is planned to locate submarine cables or pipelines, information on the location of such dumping must be available to the authorities, fishers, and others involved in activities in those areas.

Capacity-building

- Where States are still authorizing the dumping of solid waste, they need access to the skills and equipment needed to analyze the chemical constituents of potential hazardous waste to see whether it may be acceptable to be dumped in the sea.

3.3.9   Marine debris

- One of the major barriers to addressing marine debris is the absence of adequate scientific research, assessment, and monitoring. Scientific research is needed to better understand the sources, fates, and impacts of marine debris.
- Research is inadequate to qualify and evaluate the effects of plastic polymer masses that cause irritation in the stomach tissue and abdominal discomfort, and stimulate the organism to feel full and cease eating.
- Scientific evidence is insufficient to test for direct links between the chemical characteristics of marine debris and adverse effects on marine life.
- In spite of the growing number of studies documenting the distribution and abundance of marine debris, the ecological impacts, including effects on habitats, are not well documented.
- Research is insufficient to qualify and evaluate the presence of floating debris which can similarly undermine the quality of pelagic habitats; as is information on the impacts of marine debris in benthic habitats which are comparatively well studied.
- Scientific evidence and assessment efforts have not been adequate to evaluate the impacts of microplastics in the water column of the ocean.
- To date, the introduction of an alien species via marine debris has yet to be documented and there are important shortfalls in the scientific evidence of the role of marine debris in introducing alien species, especially in developing countries.
- Research, assessment, and monitoring are not sufficient to evaluate impacts of marine debris on coastal and marine species, habitats, economic health, human health and safety, and social values. Research and monitoring are insufficient to understand and in many parts of the sea, to qualify, the status and trends of marine debris. Development of new technologies and methods for detecting and removing accumulations of marine debris will also require additional research.

- The capacity to raise awareness about the problems posed by marine debris needs to be strengthened, especially in developing countries.

3.3.10   Land/sea physical alteration

- Capacity for data acquisition, especially in developing countries which suffer data-poor conditions.
- Capacity to undertake integrated assessments by multidisciplinary teams in the framework of ecosystem-based management in order to assess and understand the impacts on coastal and shoreline changes caused by a multiplicity of factors which include both anthropogenic and natural causes; capacity for modelling coastal processes, and to collect quality data based on defined standard techniques for use in developing such models.
- Due to the transboundary nature of large coastal water flows and sediment dispersal, undertaking to meet identified research needs can only be done with an improved regional capacity of individuals from various disciplines and a network of institutions.

3.3.11   Tourism and recreation

- Information is inadequate in many parts of the world on the extent of coastal tourism and its contribution to the local economy.
- Authorities concerned with the management of coastal areas where tourism is or could be occurring as an important activity need access to the skills necessary for integrated coastal management.

3.3.12   Desalinization

- Many areas suffering from shortages of freshwater could be helped by the creation of installations for desalinization and the skills needed to maintain and manage them. This is likely to become increasingly important with changes in rainfall as a result of climate change.

3.3.13   Use of marine genetic resources

- Marine biodiversity is best known in areas within national jurisdiction, and it is least known in the vast offshore oceanic areas beyond national jurisdiction.
- Biotechnology of marine biodiversity for commercial products is similarly at its infancy at the global level, and it is almost non-existent in developing countries.
- If marine genetic resources are to be explored and where appropriate developed, there are currently insufficient analytical technologies, especially for developing countries.
- There is current insufficient knowledge and skills to ensure application of environmentally friendly harvesting techniques in poorly known habitats and vulnerable marine ecosystems, such as cold water coral and sponge systems or hydrothermal vents; any exploitation in such areas requires a precautionary approach.

- There is inadequate capacity to study and collect marine genetic resources: this will require suitable vessels, both for deep sea and shallower waters, and appropriate research laboratories; the absence of this spectrum of needed resources is usually an important constraint in developing countries.

References

Millennium Ecosystem Assessment (2005). *Ecosystems and human well-being.* Washington, D.C., Island Press.

UNGA (2010). *Report of the Secretary-General* (A/65/69/Add.1).

UNGA (2011). *Final report of the workshop held under the auspices of the United Nations in support of the regular process for global reporting and assessment of the state of the marine environment, including Socioeconomic aspects.* Santiago, Chile, 13-15 September 2011 (A/66/587).

UNGA (2012a). *Final report of the Workshop held under the auspices of the United Nations in support of the Regular Process for Global Reporting and Assessment of the state of the Marine Environment, including Socioeconomic Aspects.* Sanya, China, 21-23 February 2012 (A/66/799).

UNGA (2012b). *Final Report of the workshop held under the auspices of the United Nations in support of the Regular Process for Global Reporting and Assessment of the State of the Marine Environment, including Socioeconomic Aspects.* Brussels, 27 to 29 June 2012 (A/67/679).

UNGA (2013a). *Final Report of the sixth workshop held under the auspices of the United Nations in support of the Regular Process for Global Reporting and Assessment of the State of the Marine Environment including, Socioeconomic Aspects.* Brisbane Australia, 25-27 February 2013 (A/67/885).

UNGA (2013b). *Final report of the fourth workshop held under the auspices of the United Nations in support of the Regular Process for Global Reporting and Assessment of the state of the Marine Environment, including Socioeconomic Aspects.* Miami, United States of America, 13-15 November 2012 (A/67/687).

UNGA (2013c). *Final report of the fifth workshop held under the auspices of the United Nations in support of the Regular Process for the Global Reporting and Assessment of the State of the Marine Environment, including Socioeconomic Aspects.* Maputo, Mozambique, 6 and 7 December 2012 (A/67/896).

UNGA (2013d). *Final report of the fourth workshop held under the auspices of the United Nations in support of the Regular Process for Global Reporting and Assessment of the state of the Marine Environment, including Socioeconomic Aspects.* Grand-Bassam, Côte d' Ivoire, 28-30 October 2013 (A/68/766).

UNGA (2014). *Report of the eighth workshop held under the auspices of the United Nations in support of the Regular Process for Global Reporting and Assessment of the State of the Marine Environment, including Socioeconomic Aspects.* Chennai, India, 27-29 January 2014 (A/68/812).

UNGA/AHWGW (2009). *Report on the work of the Ad Hoc Working Group of the Whole to recommend a course of action to the General Assembly on the Regular Process for Global Reporting and Assessment of the state of the Marine Environment, including Socioeconomic Aspects. (A/64/347).*

UNGA/AHWGW (2010). *Report on the work of the Ad Hoc Working Group of the Whole to recommend a course of action to the General Assembly on the Regular Process for Global Reporting and Assessment of the state of the Marine Environment, including Socioeconomic Aspects.* (A65/358).

VI Assessment of Marine Biological Diversity and Habitats

33 Introduction

Group of Experts:
Jake Rice (Lead Member and Editor of Part VI) and Alan Simcock (Co-Lead Member)

Introduction

Chapter 33

The biodiversity of the world's oceans directly supports many of the services and industries reviewed in Parts III, IV, and V, and may be affected by how the various social and economic benefits are used. To ensure the ongoing availability of those benefits to current and future generations, and to maintain healthy oceans, it is essential that the uses made of the ocean are sustainable, both individually and in the aggregate. In Part VI we examine ocean biodiversity from several perspectives, and when trends are apparent, link those trends to their main drivers. From this multi-perspective investigation of biodiversity trends, we obtain the third part of the information to be integrated in this first Assessment. This information may contribute importantly to improving global ocean literacy worldwide, and informing policies and selection of management measures from local to global scales.

The Convention on Biological Diversity[1] (CBD 1992) emphasizes that "biodiversity" exists on many scales: from genetic diversity within populations, through diversity of populations of the same species, the diversity of species in ecosystems, to the diversity of habitats within geographic areas. The diversity at all of these scales reveals patterns and structures that are crucial to the functioning of ecosystem processes and the delivery of ecosystem functions. However, in-depth analyses of patterns and trends, linking them to all drivers that underlie them and their ecological, social, and economic consequences are not feasible for the entire ocean, at even one of these scales.

Therefore Part VI presents overviews of these biodiversity features first spatially, and then followed by more focused examinations of key species groups and habitats. From these overviews, it is possible to present an analysis that integrates how global ocean biodiversity is changing as a result of the impacts of humanity's uses of the ocean, with the ability of the ocean to sustain itself, and humanity's uses of it into the future.

Chapters 34 and 35 present the main global patterns of diversity of populations, species, and habitats. Chapter 34 summarizes what has been learned about the nature and scales of those global patterns of diversity in species, and the dominant natural gradients that underlie those patterns, including features of the seabed such as depth, topography and types of substrate, of the water column such as temperature, salinity, nutrients and currents, and planet-scale features such as latitude, seasonality, and proximity to coasts. These patterns and their natural drivers underlie natural variation in biodiversity and provide the foundation for evaluating the potential for supporting human uses and sustaining perturbations from those uses. The ocean is vast and complex, hence these patterns in global biodiversity are incompletely quantified and their natural drivers are not fully understood. Chapter 35 summarizes the degree to which the biodiversity of the world's oceans has been sampled at even a descriptive level, and the subset of places and taxa for which sufficient sampling exists to quantify trends in biodiversity components in space and time, and to provide strong evidence linking those trends to specific natural and anthropogenic drivers.

Two contrasting messages emerge:

(a) An immense amount remains to be learned about the ocean's biodiversity. Sampling has been insufficient to fully quantify patterns and relationships with potential drivers in most of the ocean, and even to describe the biodiversity present in many parts of the ocean. This presents major challenges to fulfilling a core task of this Assessment – setting baselines against which to measure future changes.

(b) Nevertheless, with the current levels of sampling, much can be concluded about how the ocean has changed in the past decades and centuries and trends that may continue into the future. These past changes and current trends provide information about the sustainability of human interactions with marine biodiversity, whether those interactions take the form of direct use or indirect impacts. Although relationships between biodiversity variation and its drivers need to be better quantified almost in all parts of the world, we have sufficient knowledge to indicate which outcomes are likely to be more sustainable or less sustainable, and thus inform our choices. Nevertheless, we must acknowledge that uncertainties will remain and surprises will be encountered.

Chapter 36 comprises the largest part of Part VI. That chapter presents the major temporal trends in biodiversity from the primary and secondary producers (phytoplankton and zooplankton), through the benthos and fish to the top predators. To the extent that they are known in each case, it also links the major trends to their key natural and anthropogenic drivers. Where important changes in trends are documented, the knowledge of the degree to which the changes are due to natural causes to impacts of human uses or to responses to efforts to mitigate past impacts, is summarized. This provides the raw material for integration of biodiversity trends with sustainability of human uses of marine biodiversity. Chapter 36 is subdivided into eight divisions, presenting similar assessments of trends in biodiversity for the northern and southern sub-basins of the Atlantic and the Pacific Oceans, for the Indian Ocean, the Arctic Ocean and the Southern Ocean, and for the open-ocean and deep-sea areas of all the oceans together.

For all but the open-ocean division, the trends are generally reported separately for coastal and offshore/shelf areas, because the species and habitats, and the main drivers of trends, often differ in the near-shore and the offshore/shelf areas. Even at this level of subdivision, only major trends in biodiversity can be reported, and often it is necessary to report the patterns and trends at finer geographic scales, because these sub-basins and oceans contain many partially independent marine ecosystems, with differing species and habitats, different temporal trends, and possibly different drivers of trends.

None of these regional assessments can be exhaustive of all trends in biodiversity at the regional scale. They focus on trends that are well documented, and likely to be representative of what is happening to biodiversity at those regional scales. Illustrative examples of the trends and their relationships to drivers are presented consistently, and some chapters are supported by technical annexes with additional data and

---

[1] United Nations, *Treaty Series*, vol. 1760, No. 30619.

references to enable more in-depth documentation of the conclusions in the sub-chapters. Although several divisions of Chapter 36 have made some use of traditional knowledge in assessing patterns and drivers, in all divisions, institutional assessments and scientific publications were the predominant source of information on patterns, trends, and drivers. That creates an unavoidable (but unintentional) bias towards more extensive assessment of biodiversity trends and drivers in the parts of the world with larger investments in scientific research and monitoring, and where such investments have occurred over longer periods. If the divisions of Chapter 36 report fewer trends in biodiversity in parts of the world with less research and monitoring, this reflects the mandate of this Assessment to use only published sources. Hence at least some of the differences among the regional assessments arise from differences in the quantity of information that informs this Assessment, and not necessarily differences in what is happening to marine biodiversity at these large regional scales. When strong trends are present consistently in regional assessments where appropriate data are available, it would be precautionary and sound scientific practice to consider them as relevant information for less data-rich areas as well.

Regional assessments are only one perspective for documenting, and to the extent possible explaining, trends in biodiversity. Other perspectives are legitimate, and for some aspects of biodiversity the regional assessments may not be the most powerful assessment approach, for two reasons. First, some species groups include wide-spread and highly migratory species; hence no one region captures the trends in the populations, and some habitat types important to biodiversity often are found in relatively small individual occurrences (as well as, rarely, larger ones) but in many of the regions, and a piecemeal regional approach may be inefficient. Second, some groups of species and certain types of habitats are considered particularly vulnerable to natural and anthropogenic drivers. Hence assessments of trends and drivers are most powerful when all the relevant information is viewed together, regardless of the specific regions from which the parts of the whole story were documented.

Therefore Part VI concludes with short chapters on five key species groups with the above characteristics: marine mammals, seabirds, marine reptiles, elasmobranchs (sharks, rays, and their relatives), and tuna and billfish. Chapters then follow on a number of key habitats. Many are near coastal habitats of known high biodiversity value and subject to multiple human pressures, such as estuaries, kelp forests, and similar habitats. Several are offshore or deep-sea habitats known often to be biodiversity hotspots, and to be attracting increasing pressure from direct or indirect human uses. Examples include seamounts, tropical and sub-tropical corals, cold-water corals, cold seeps and hydrothermal vents. These shorter and more focused assessments of species groups and habitats that are both highly valued economically, culturally, or both, and highly vulnerable to impacts from unsustainable use, provide flagship cases to integrate ecological, social, and economic aspects of marine biodiversity and its contributions to human well-being.

Collectively, across Part VI, this Assessment evaluates the patterns of occurrence of marine biodiversity (Chapter 34) and the extent to which it is known (Chapter 35), the major trends in coastal, offshore, and deep-sea biodiversity, usually at regional scales and sometimes sub-regional scales, as well as the main known drivers of those trends (Chapter 36A-36H), and examines issues of particular concern for species (Chapters 37-41) and habitats (Chapters 42-51). This provides the third main pillar of this Assessment, for integration with Parts II-V in Part VII (Overall Assessment).

A Overview of Marine Biological Diversity

Chapter 34. Global Patterns in Marine Biodiversity

Contributors:
Paul Snelgrove (Convenor), Edward Vanden Berghe, Patricia Miloslavich, Phil Archambault, Nicolas Bailly, Angelika Brandt, Ann Bucklin, Malcolm Clark, Farid Dahdouh-Guebas, Pat Halpin, Russell Hopcroft, Kristin Kaschner, Ben Lascelles, Lisa A. Levin, Susanne Menden-Deuer, Anna Metaxas, David Obura, Randall R. Reeves, Tatiana Rynearson, Mayumi Sato, Karen Stocks, Marguerite Tarzia, Derek Tittensor, Verena Tunnicliffe, Bryan Wallace, Ross Wanless, Tom Webb, Patricio Bernal (Co-Lead member), Jake Rice (Co-Lead member), Andrew Rosenberg (Co-Lead member)[1]

---

[1] The writing team thanks Roberto Danovaro, Esteban Frere, Ron O'Dor and Chih-Lin Wei for their valuable contributions to this chapter.

Global Patterns in Marine Biodiversity

1 Introduction

Marine environments encompass some of the most diverse ecosystems on Earth. For example, marine habitats harbour 28 animal phyla and 13 of these are endemic to marine systems. In contrast, terrestrial environments contain 11 animal phyla, of which only one is endemic. The relative strength and importance of drivers of broad-scale diversity patterns vary among taxa and habitats, though in the upper ocean the temperature appears to be consistently linked to biodiversity across taxa (Tittensor et al. 2010). These drivers of pattern have inspired efforts to describe biogeographical provinces (e.g. the recent effort by Spalding et al., 2013)) that divide the ocean into distinct regions characterized by distinct biogeochemical and physical combinations). Biogeographers such as Briggs (1974) examined broad-scale pattern in marine environments in historical treatises and although many of the patterns described therein hold true today, the volume and diversity of data available to address the question have increased substantially in recent decades. We therefore focus our chapter on more recent analyses that build on those early perspectives. The International Census of Marine Life programme that ran from 2000-2010 provided significant new data and analyses of such patterns that continue to emerge today (McIntyre, 2011; Snelgrove, 2010). Indeed, many of our co-authors were part of that initiative and that influence is evident in the summary below. In the few years since that programme ended, some new perspectives have emerged which we include where space permits, noting that we cannot be exhaustive in coverage and also that the large data sets necessary to infer broad-scale patterns do not accumulate quickly.

Not surprisingly, the different scales, at which many organisms live, from ambits of microns for microbes to ocean basins for migratory fishes and marine mammals, along with variation in the drivers and patterns of diversity, render a single analysis impossible. In order to assess gradients in marine biodiversity we use a taxonomic framework for some groups of organisms and a habitat framework for others. Comparatively well-sampled groups taxonomically (primarily pelagic (water column) vertebrates and cephalopods) could be treated at a phylum or class level globally, whereas for taxa in which taxonomic or geographic knowledge is highly uneven, we followed a habitat framework, noting that a group by group treatment of benthic invertebrates would encompass more than 30 phyla and would render the chapter unwieldy. We therefore organized the chapter into an Introduction, a series of summaries on biodiversity patterns in pelagic taxa, and then summaries of knowledge on biodiversity in contrasting benthic ecosystems. Although this strategy is imperfect (e.g. many fishes occupy primarily benthic environments), it nonetheless creates a framework in which to evaluate current knowledge of biodiversity patterns within a relatively short chapter. Space limitations also preclude comprehensive coverage of all habitats and taxa, and we therefore present a broad but incomplete summary that omits kelps, seagrasses, and salt marshes, for example. We therefore encourage readers to also review the more detailed chapters within the World Ocean Assessment that focus on the biology and status of specific taxa and ecosystems. Our goal in this chapter is to identify the key environmental drivers of global diversity patterns based on current knowledge, while acknowledging many data gaps that will necessitate revising these patterns as new data become available. Specifically, we address how depth, latitude, productivity, temperature and substrate influence broad-scale distributions and diversity patterns, and identify the knowledge gaps (taxonomic, geographic) that constrain our ability to assess such patterns. Below we summarize knowledge on biodiversity gradients with a few key references, but we also include a more extensive reading list for those seeking more detailed information (Appendix).

2 Pelagic ecosystems

2.1 Marine Mammals

Marine mammals include cetaceans (baleen whales and toothed whales, dolphins and porpoises), pinnipeds (seals, sea lions, walrus), sirenians (manatees and dugongs), the marine otter (*Lontra felina*), and sea otters (*Enhydra lutris* and subspecies) and the polar bear. Excluding the seven extant freshwater species, about 120 wholly or partly marine species are currently recognized (www.marinemammalscience.org). However, ongoing taxonomic revision will keep this number in flux. Marine mammal species occupy almost all marine habitats: from fast ice to the tropics, on shorelines where pinnipeds haul out during their moulting, mating or pupping season, in shallow coastal waters where some dolphins and baleen whales spend much of their time, and in the open ocean where many pelagic pinnipeds, baleen whales, and toothed cetaceans occur. However, other than sperm whales and perhaps some of the beaked whales capable of diving beyond 2,000 m, air-breathing limits marine mammals to bathypelagic depths at most. In contrast to highly restricted distributions in some smaller cetaceans and pinnipeds, many species exhibit circumglobal or circumpolar distributions, with some (baleen whales in particular) undertaking long annual migrations.

Many marine mammals spend most of their time offshore, but sirenians, marine and sea otters, and some small cetaceans and pinnipeds, as well as benthic feeders such as walruses and grey whales, rarely venture beyond the continental shelf except when migrating. Global marine mammal species richness peaks at mid-latitude (around 60°) in both hemispheres (Kaschner et al., 2011; Figure 34.1a). Pinnipeds mostly drive the peaks, which despite some highly endemic species in the subtropics, largely concentrate in polar to temperate waters (Figure 34.1b). Thick blubber or fur layers insulate marine mammals; hence they occupy all climate zones, although temperature nevertheless influences distributions (Tittensor et al., 2010), often through feeding and breeding constraints. As with other marine predators, food availability drives patterns in many marine mammals, but so does availability of breeding habitat for many species. Dugongs, as highly specialized herbivores, associate with sea grass beds in warm, shallow coastal waters and estuaries. Baleen whales, which feed at low trophic levels, require dense prey aggregations to sustain their metabolic needs, forcing most species to migrate to high latitudes during peak feeding seasons to utilize high summer productivity (except for some resident tropical populations associated

Chapter 34 Global Patterns in Marine Biodiversity

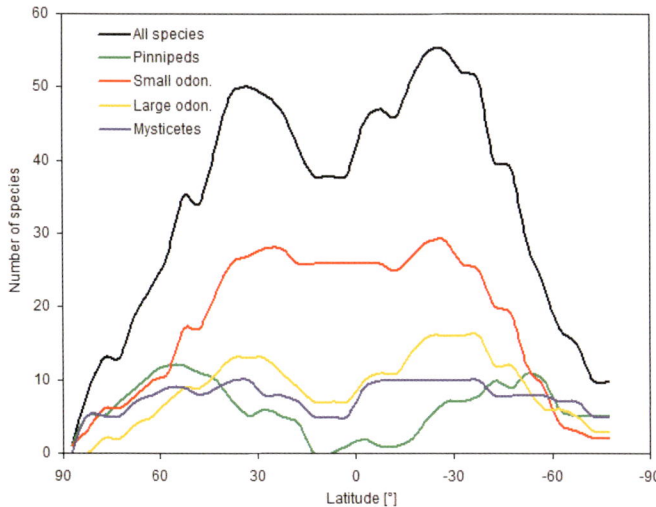

Figure 34.1a | Global marine mammal species richness peaks between 30° and 60° in both hemispheres. Number of species (as predicted by a relative environmental suitability model, (Kaschner, 2006) was summed over 5° latitudinal bands for all species, mysticetes, small odontocetes, large odontocetes (beaked whales and sperm whales), and pinnipeds (from Kaschner et al., 2011).

with productive upwelling waters). Some species concentrate at specific breeding or calving grounds in winter. Toothed cetaceans generally feed at higher trophic levels and are not linked, as are some baleen whales, to zooplankton aggregations along the polar ice edges, and this helps drive higher species diversity at mid-latitudes (Figure 34.1a) (Kaschner et al., 2011). Nevertheless, the finer-scale distribution of most species links with ocean features that aggregate prey such as eddies, fronts or upwelling areas, or with specific breeding grounds. Many pinniped distributions correlate with prey aggregations; however, availability of suitable haul-out sites, on either land or sea ice, for moulting, breeding and pupping, limits most pinniped species, as does maximum length of foraging trips, as well as temperature (Tittensor et al., 2010).

Despite their often impressive body size, new discoveries of whale species still occur by re-evaluation of molecular and morphological evidence. Despite well-established broad-scale distributions of most small cetaceans, pinnipeds, sirenians and otters (Reeves et al., 2002), range maps or predictions of environmental suitability for the beaked whales (ziphiids) largely represent guesswork; their actual distributions potentially span entire ocean basins (Kaschner et al., 2011). Similarly, even after centuries of intense whaling in all oceans, the breeding grounds and migration patterns of some of the large baleen whales, such as North Pacific right whales, are still not well known.

Relatively low densities with often very large ranges, low detectability, and inconspicuous behaviour of many species limit studies on marine mammal distributions. To date, most dedicated sampling has centred on Northern hemisphere continental shelves and slopes (Halpin et al., 2009). Large knowledge gaps remain on species occurrence throughout much of the tropics and Southern hemisphere temperate waters north of 60° (Kaschner et al., 2011). Although range maps and opportunistic sightings document species presence in an area and facilitate larger-scale biodiversity inventories, extending these types of sources to estimate density or abundance or to assess relative ecological importance often proves problematic. Dedicated sighting surveys of cetaceans conducted annually in offshore areas of the North Pacific under sponsorship of the International Whaling Commission (IWC) (IWC-POWER: IWC) can help address these gaps and evaluate North Pacific whale recovery trajectories (Halpin et al., 2009)

2.2 Seabirds

The boundaries and names shown and the designations used on this map do not imply official endorsement or acceptance by the United Nations.

Figure 34.1b | Predicted patterns of marine mammal species richness. A. All species included in the analysis (n=115), B. Odontocetes (n=69), C. Mysticetes (n=14), D. Pinnipeds (n=32). Colours indicate the number of species predicted to occur in each 0.5°x0.5° grid cell from a relative environmental suitability (RES) model, using environmental data from 1990-1999, and assuming a presence threshold of RES>0.6. (from Kaschner et al., 2011).

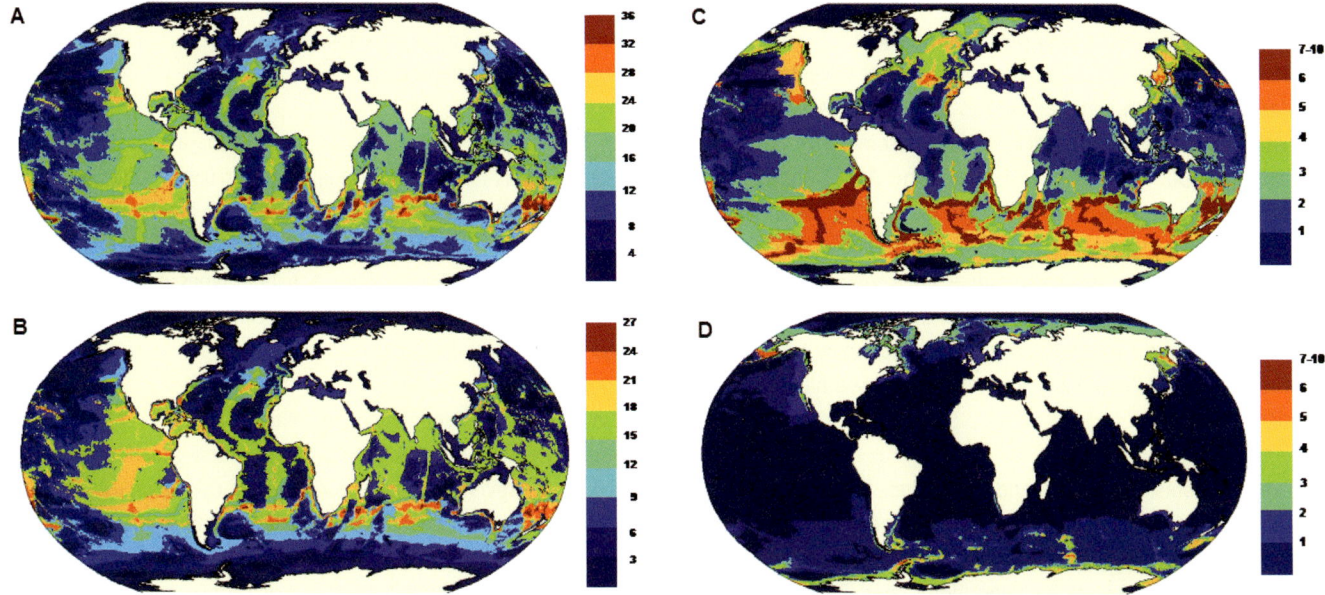

503

"Seabirds" denotes species that rely on the marine environment for at least part of the year, and include many spectacularly mobile species that travel thousands of kilometres, returning to land only to breed. Seabirds as a group occur in all seas and oceans worldwide, exploiting surface waters in all habitats from the intertidal zone to the open ocean. Globally, seabird density, diversity and endemism are highest in the highly productive temperate latitudes and in upwelling areas (Croxall et al., 2012; Chown et al., 1998).

Seabirds are central-place foragers (foragers that return to a particular place to consume food) during the breeding season, with many adapted to exploit highly clumped prey. Therefore largest aggregations occur where food availability is high within a restricted foraging range from a suitable nesting habitat (Lascelles et al., 2012). Foraging ranges vary from a few kilometres from shore (e.g., seaduck and small terns) to several thousand (e.g., larger albatross). Seabirds adopt a range of behaviours to capture prey, from surface-seizing to plunge or pursuit diving. Feeding generally occurs at or immediately below the water's surface, although the Emperor Penguin reaches depths over 500 m.

Seabirds can be roughly subdivided into three groups. "Pelagic seabirds", such as porcellariiformes, pelecaniformes, alcids and penguins, often travel far from land, primarily use oceanic pelagic water (seas above the open ocean, typically >200 m in depth). In contrast, "coastal seabirds (year-round)", including most larids, are those that primarily use coastal inshore water (seas along coasts, typically <8 km from the shoreline) throughout the year. "Coastal seabirds (nonbreeding season)", such as seaduck, grebes and divers, are those that primarily use coastal inshore water during the non-breeding season.

For much of the year coastal species tend to be relatively static, often tied to particular habitats or topographic features. Pelagic species distributions, however, often link to dynamic processes and variables and require complex analyses to define. BirdLife International recognizes around 350 species as seabirds (i.e., 3.5 per cent of all bird species), of which over 280 meet a stricter definition (excluding ducks, loons, etc.) used in some earlier reviews. However, ongoing taxonomic revision will keep this number in flux. In recent years new species have been found, as well as rediscovery of some thought to be extinct. Re-evaluation of molecular and morphological evidence has split some taxa, adding an additional eight species since 2000 with a further 15-20 under review in the coming years. Knowledge of the at-sea distribution of species remains patchy. Many species are relatively well studied at specific sites, but at-sea movements across entire ranges are known for only a few species, as are areas used during non-breeding periods and those areas visited by juvenile birds. The at-sea distributions for many tropical species, particularly in the Central and South Pacific and South East Asia are also under-studied.

Seabird distribution may vary depending on their breeding site (e.g., tropical vs. temperate zones), age, sex, whether it is day or night and the time of year (Lascelles et al., 2012). In addition, many species, particularly procellariiforms, alternate between "long" and "short" foraging trips during the breeding season. Areas most important for their survival have rarely been defined in any systematic way, although recent studies, such as the BirdLife Marine Important Bird Area Atlas, have helped fill this gap and show distribution patterns at multiple scales.

2.3 Turtles

Marine turtles have inhabited the world's oceans for more than 100 million years, having survived the dinosaurs and numerous major global shifts in climate. Today there are seven recognized marine turtle species, six belonging to Chelonidae, green turtles *Chelonia mydas*, hawksbills *Eretmochelys imbricata*, loggerheads *Caretta caretta*, olive ridleys *Lepidochelys olivacea*, Kemp's ridleys *Lepidochelys kempii*, and flatbacks *Natator depressus*, and one extant member of Dermochelyidae, the leatherback *Dermochelys coriacea*. Despite few species, marine turtles occur circumglobally, inhabit nearly all oceans, occupy unique ecological niches, and exhibit variations in abundance and trends, as well as reproduction and morphology among populations of the same species (Wallace et al., 2010).

Marine turtles have evolved several adaptations to marine habitats (e.g., maintaining water balance in saltwater, hydrodynamic body shape and swimming efficiency) that are unique compared to other turtle species, but because they are reptiles, temperature fundamentally constrains their distributions and life history (Spotila, 2004). For example, the development, and survival of marine turtle embryos means successful hatchling production requires the consistently warm temperature (28-33°C) of sandy beach environments. Because these habitats are limited to the tropics and subtropics, most major marine turtle nesting sites occur between the equator and 30° latitude (Wallace et al., 2010) (see Figure 39.1).

Temperate also limits marine distributions, as most population ranges only reach 45° latitude (see Figure 39.1), extending only seasonally into northern and southern extremes of their ranges (Spotila, 2004). Leatherbacks defy this pattern, with core migratory and foraging habitats into temperate and even sub-arctic regions and average water temperatures between 10-20°C (Eckert et al., 2012).

Within ocean basin-scale distributions, adult marine turtles generally migrate hundreds to thousands of kilometres from nesting beaches for foraging, often showing high site fidelity to both breeding and feeding areas. Immature turtles also show site fidelity to areas used for foraging and growth. For some species, primary habitat types, e.g., coral reefs for hawksbills, seagrass beds for green turtles, constrain foraging to tropical regions (Spotila, 2004).

Many marine turtle populations demonstrate ontogenetic variation in habitat use that is related to geography and oceanography (Bolten, 2003). In several places around the world, hatchlings disperse from nesting beaches and orient toward persistent, offshore current systems (e.g., Gulf Stream in the Atlantic Ocean), where they associate with ephemeral habitats in convergence zones, such as *Sargassum* communities. After spending the first few years of life in these oceanic areas and growing to larger body sizes, juvenile turtles tend to recruit to neritic habitats where they remain—for the most part—until reaching sexual maturity. Although this description provides a useful heuristic for understanding sea turtle life history distribution patterns, significant within-population variation exists in timing and duration of recruitment by individuals from one life stage—and habitat type—to another; these variations have implications for overall population dynamics and management (Bolten, 2003).

The wide distributions of marine turtles can vary greatly among populations, which are subject to multiple threats that operate on different spatial and temporal scales. Effectively prioritizing limited management resources requires understanding which threats will most strongly influence distribution patterns in space and time (Wallace et al., 2011).

2.4 Fishes

In English, 'fish' designates any aquatic multicellular animal (jellyfish, cuttlefish, starfish, etc.). The term 'finfish' designates those with a central spine comprised of vertebrae (Chordata/Vertebrata – vertebrates), whether or not present in adults, ossified, or with paired and/or impair fins supported by rays. Although no longer recognized as a valid taxonomic group, 'finfish' (hereafter "fish") offers a practical descriptor of a group exclusive to aquatic life that constrains many adaptations and defines a similar body plan, while acknowledging diverse body forms.

By the end of 2013, more than 33,000 valid species of extant species were described (Eschmeyer, 2014), constituting more than half of all vertebrates; ca. 17,500 occur in marine environments for at least part of their life cycle. Surprisingly, species descriptions have accelerated since World War II: between 1999 and 2013 (15 years), a new fish species was described every day, a rate that is still increasing (Figure 34.2). This increase comes in spite of decreasing fish taxonomists around the world, and cannot be attributed to recent molecular advances, given the low proportion of species discovered by genetics methods. Different editions of Fishes of the World (e.g., Nelson, 2006) report new discoveries (Table 34.1), and although the deep seas were expected to deliver many

Table 34.1 | Number of species by life zone (Saltwater including diadromous, Freshwater) from the successive editions of Fishes of the World by J.S. Nelson (1976, 1984, 1994, 2006). The last line gives the current counts from the Catalog of Fishes (Eschmeyer, 2014).

Year	Salt Water		Freshwater		Total
Fishes of the World (Nelson) successive editions					
1976	11967	64%	6851	36%	18818
1984	13312	61%	8411	39%	21723
1994	14652	60%	9966	40%	24618
2006	16025	57%	11952	43%	27977
Catalog of Fishes					
2013	17535	53%	15467	47%	33002

new species, given the assumed high rate of exploration of continental shelves <200 m, species richness per surface area in shallow waters remains much higher, noting that waters below 2,000 m deep remain largely unexplored. Coral reefs still deliver most newly described marine fishes each year, especially for cryptic species such as gobies and small labrids. Coral reefs are not the only source of new species: for instance, populations of many fish species previously thought to be distributed widely in the Indo-Pacific region are now recognized as different species between the Indian and the Pacific Ocean, and even more recently between the Red Sea and the Indian Ocean.

Fishes are ubiquitous throughout the world ocean, in locations as small as tidal pools that may dry up daily, and from the poles down to the base of the Marianas Trench in the West Central Pacific (11,782 m). They live in caves, on the shoreline, sometimes out of the water for some periods in mangroves or intertidal areas, over, on or in soft or hard bottoms, in crevices in rocky or coral reefs, and a few are even found in poorly oxygenated water. As in many groups, species richness (as well as species per family or genus, and number of genera) generally increases from

Figure 34.3 | Number of species of fish per genus, in each latitudinal band of 1 degree, as calculated from distribution data available in OBIS on 26 September 2011.

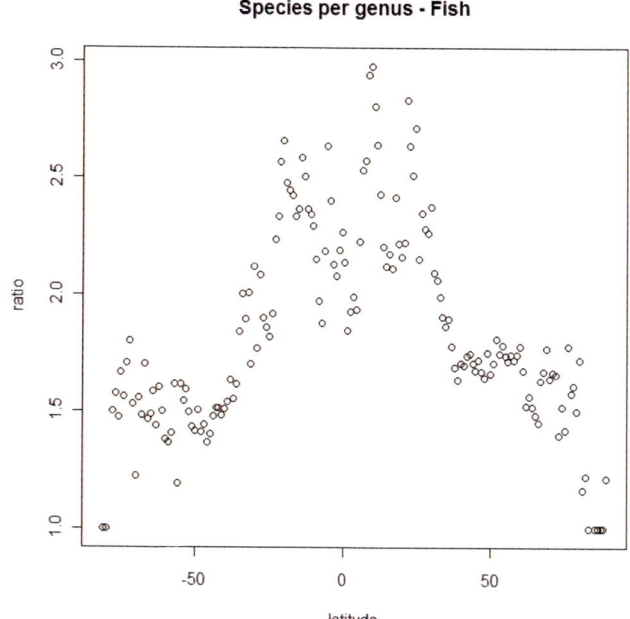

Figure 34.2 | Cumulated number of species described per year between 1758 and 2013. The rate of descriptions accelerates after 1950, mainly due to an increasing number of freshwater species descriptions; the rate of marine species descriptions has been linear since the early 1980s.

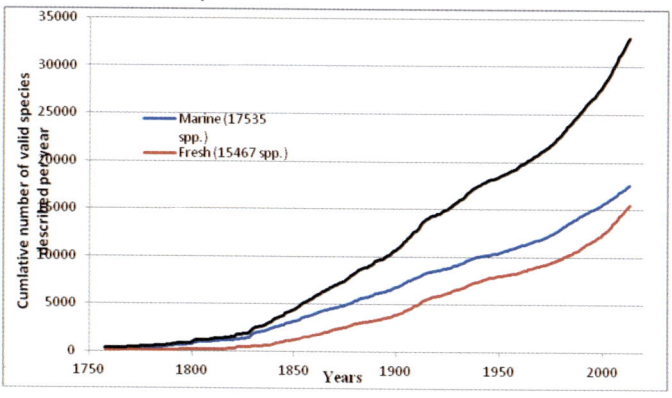

high to low latitudes (Figure 34.3), more so than the number of families, but local climate, oceanography or phylogenetic history may alter this pattern (Tables 2 and 3). Shallow coral reefs with a high diversity of habitats and high biomass productivity support the most species; deep-sea corals are less species-rich. Several studies show highest species richness in the Coral Triangle (Indonesia, Philippines, Papua New Guinea region), thought to be the centre of tropical marine diversity. More than 300 widespread species (2 per cent) occur in 10 or more FAO areas (from over 19 possible areas), and 8,000 (47 per cent) occur in just one area (12,000 – 65 per cent in 1 or 2). Geographic and/or hydrological conditions restrict the majority of species distributions. Amphitropical distributions occur only in large pelagic species.

Although fishes occupy all depths, species diversity drops dramatically below the continental shelves. Depth ranges are incomplete for many species, but about 6,800 (58 per cent) of the 11,000+ species with recorded ranges in FishBase (www.fishbase.org; Froese & Pauly, 2014) are restricted to the upper 200 m, with only ca. 620 spp. (4 per cent) below 2,000 m. Lack of data on deep-sea species, except for a few targeted around submarine seamounts in the high seas by non-sustainable fisheries, illustrates a serious sampling bias that demands cautious interpretation.

Only a few species of shark and rays have been caught below 3,000 m and none below 4,000 m. Gobies that constitute the most speciose family in marine life zones in tropical and temperate waters in general, are barely present in the North Pacific above 40° N. The life cycles of more than 700 hundred species (including salmonids, eels), alternate between marine and fresh waters (amphidromy, diadromy, catadromy, anadromy).

Few herbivorous species occupy high latitudes as compared to tropical areas. Although anti-freeze blood proteins prevent ice formation in the blood of some cold water fishes, digesting plants requires a higher metabolic activity than most cold water fishes can maintain. Based on recorded information for about 6,400 species in Fishbase, about 1,000 species are top predators and carnivores, 4,400 are predators or omnivores, and 1,000 are herbivores or omnivores. The commercial large species that are most studied predominantly occupy the upper trophic levels.

The high diversity of forms, behaviour, ecology and biology based on one body plan enables great success in the marine environment. However, the populations of many exploited species are threatened by fisheries that now access stocks in almost the entire water volume between 0 and 1,500 m depth. Despite some local extirpations, no marine fishes are reported to be globally extinct; however, large species with few offspring, such as some sharks and manta rays, are endangered, often because of threats along migration pathways. Populations of some shark species targeted for their fins have decreased by 90 per cent, but although the populations are no longer economically exploitable, no sign of extirpation has been noted so far (Ferretti et al., 2010).

2.5 Cephalopods

Shell-less coleoid cephalopods occur from pole to pole, and from the ocean's surface to depths of many thousands of metres; many can even fly above the ocean's surface. They range from surface-dwelling tropical forms with adults the size of a grain of rice to 30-m giants in the deep oceans.

Squid compete with fishes in nearly all marine niches, although there are only one tenth as many species, perhaps reflecting their relatively recent radiation since the disappearance of the dinosaurs. The same event killed all of the Ammonites, a highly diverse group of cephalopods that lived near the sea surface. The deep ocean remains sparsely sampled for

Table 34.2 | Number of marine fish species per FAO area.

FAO area	Spp. count
Arctic Ocean	147
Atlantic, Northwest	1129
Atlantic, Northeast	1115
Mediterranean and Black Sea	811
Atlantic, Western Central	2428
Atlantic, Eastern Central	1699
Atlantic, Southwest	1777
Atlantic, Southeast	1779
Atlantic, Antarctic	250
Indian Ocean, Western	4432
Indian Ocean, Eastern	4757
Indian Ocean, Antarctic	250
Pacific, Northwest	5299
Pacific, Northeast	717
Pacific, Western Central	6490
Pacific, Eastern Central	4138
Pacific, Southwest	2249
Pacific, Southeast	1916
Pacific, Antarctic	170

Table 34.3 | Number of marine fish species per Ocean and FAO area. E: East; N: North; S: South; W: West: indicates which part of the ocean. Note: The second eastern central line for Atlantic represents the Mediterranean and Black Seas. The Northwestern Pacific includes some coral reef areas in its southern part which explains the high number of species compared to the Northeastern part.

Ocean	Atlantic				Indian				Pacific			
Latitude												
Arctic	N	147										
North	W	1129	E	1115					W	5299	E	717
Central	W		E	811					W	6490	E	4138
Central	W	2428	E	1699	W	4432	E	4757				
South	W	1779	E	1777					W	1916	E	2249
Antarctic	S	250			S	250			S	170		

cephalopods, raising questions about their total biomass and global patterns. For those areas that have been sampled, recent evidence suggests that primarily oceanic squid peak in diversity in the northern hemisphere at temperate latitudes, a pattern reflected by (primarily coastal) non-squid cephalopods in the Pacific Ocean at least; temperature strongly drives these patterns (Tittensor et al., 2010).

More information on the biology, biogeography and diversity of cephalopods is available from CephBase (cephbase.eol.org), which is now available through the pages of the Encyclopedia of Life.

2.6 Marine Microbes

Marine microbes, defined as single-celled or chain forming microorganisms, span a very broad size range, from microscopic cells that are <1/50th the diameter of a human hair to forms visible to the naked eye. They are found throughout all the oceans, from the tropics to the poles and from the surface to the deepest depths. These single-celled organisms divide asexually, up to several times per day, leading to high biomass that fuels nearly all marine productivity, including all important fisheries around the globe, and drives global biogeochemical cycles, including carbon, oxygen, and many others. Marine microbes also represent the most phylogenetically diverse organisms on Earth. A single litre of seawater can contain representatives of all major branches of the tree of life: Archaea, Bacteria and all major kingdoms of Eukaryotes. Microbial diversity within the plankton far exceeds that in terrestrial habitats. For example, planktonic photoautotrophs represent deep phylogenetic diversity, including 20 diverse clades. In contrast, autotrophic diversity in terrestrial environments is dominated by just one clade (Falkowski et al., 2004). Planktonic heterotrophs are equally diverse.

Eukaryotic plankton includes purely autotrophic species (phytoplankton) that convert inorganic to organic carbon, fuelled by light energy through photosynthesis. Primary production supplied by phytoplankton forms the basis of the food web and ultimately feeds all marine organisms, up to the largest whales. Eukaryotic plankton also includes heterotrophic microbes that ingest organic carbon through a myriad of feeding strategies, and so-called mixotrophic species, which include species either simultaneously or sequentially alternating between phototrophic and heterotrophic modes. Feeding by heterotrophic and mixotrophic plankton is the single largest factor in reducing primary production; it can control the abundance and biogeochemical activity of phytoplankton, and it is essential for the transfer of matter and energy to higher levels in the food web, and for the recycling of nutrients (Sherr et al., 2007). Bacteria are also essential for recycling and remineralizing organic matter and contribute substantially to primary production.

Latitude, proximity to land, and season primarily delimit global large-scale distribution patterns of plankton. Abundance declines from high nutrient coastal areas to the vast areas of the generally low-nutrient (oligotrophic) waters of the open ocean. But exceptions exist. For example, Charles Darwin on his '*Beagle*' voyage noted that nitrogen-fixing phytoplankton can become very abundant in the open ocean and form surface mats and filaments. Latitude interacts with season in forcing plankton abundance patterns. In lower latitudes, seasonal variations in irradiance and temperature, including ice cover, result in highly variable plankton abundances seasonally, with spring and fall peaks. Organism physiology, nutrient availability, susceptibility to grazing, and viral attack, as well as the fluid flow regime, further define distributions.

2.7 Zooplankton

Zooplankton occur from pole to pole, and from tide-pools to the deepest trenches in the ocean. They span the size range from single cells and multicellular organisms that are smaller than 0.05 mm, to gelatinous colonies that are longer than the largest whales. In the vast scale of the oceans, they are united by their inability to control their movement in the horizontal scale, but many perform vertical migration of hundreds of metres per day.

As an assemblage, the ~7,000 described species of multicellular zooplankton (Wiebe et al., 2010) encompass species from every major animal phylum, and the majority of the minor phyla; some of these 15 phyla are almost exclusively planktonic (Bucklin et al., 2010). Many additional phyla are classically considered as non-planktonic, but they do in fact live within the plankton for their earliest life stages, and are referred to as meroplankton, in contrast to the holozooplankton generally considered. This means that the zooplankton encompass an exceedingly wide range of body plans, and modes of life, ranging from relatively passive herbivorous species, to blindingly fast attack carnivores. It also includes some of the world's most passive predators that literally rely on prey blundering into them. Finally, some zooplankton taxa have developed symbiosis with internally housed algae so successfully that they no longer rely on other organisms as prey.

The majority of zooplankters range from <1 mm to 1 cm in length. With ~2,000 described planktonic species typically representing 80-90 per cent of total zooplankton abundance and living biomass in most marine ecosystems, copepods represent the most successful body plan. These small, robust crustaceans are easily collected with simple nets and manipulated for experimental purposes, making them the central focus of ecological research on plankton for the past century. Different species of copepods play almost every imaginable ecological role: the majority are suspension-feeding grazers on smaller single-celled plankton, some are scavengers and detrital feeders, and others range from active attack to passive ambush predators. Several other diverse crustacean groups illustrate a wide range of feeding strategies: ostracods (detritivores), euphausiids (filter-feeders), amphipods (predators, or commensalists), mysids (scavengers) and decapods (predators); note that the latter two groups may be considered either planktonic or benthic, given their tight association with the seafloor.

Lacking the arthropod skeleton, most other planktonic groups are considered "gelatinous" zooplankton, which are generally not well collected in nets because of their fragility and often lower abundances. With the exception of the nearly 140 species of pelagic tunicates (larvaceans, pyrosomes, doliolids and salps) and about 80 species of shelled pteropods, all other groups are clearly predatory. Two of the three classes of medusae or "jellyfish" within the phylum Cnidaria (hydrozoans and scyphozoans), are clearly the most speciose gelatinous groups, followed by the phylum Mollusca with its three functional groups (shell-less pteropods, heteropods, and cephalopods – although the latter are considered by many as nekton, as are the fishes). Once grouped with the cnidarians, the Ctenophora, or comb-jellies, are probably the most seriously underestimated group in terms of their biodiversity: their extremely fragile body construction confounds specimen collection. Of the extant worm-like groups, only Chaetognatha (arrow worms) occur in plankton samples in high abundance; primarily benthic polychaetes and nemertines usually occur in the plankton in modest diversity and abundance.

The >100-year quest to find patterns in zooplankton distribution shows that each species has its own environmental preferences and tolerances, with some species confined to specific regional habitats and others that are relatively wide-spread. Over time, several broad patterns have emerged for zooplankton that exhibit consistency across multiple taxa (Dolan et al., 2007). Diversity in offshore habitats exceeds that in coastal regions, although coastal abundance and biomass may be higher. Diversity increases from the poles to the tropics, often with an equatorial dip (Boltovskoy, 1999) (again in contrast with abundance and biomass). Diversity increases with increasing depth in polar systems (Kosobokova et al., 2011), has a mid-depth peak in temperate/subarctic systems, but may peak in surface waters of tropical oceans. Although these trends hold for the overall zooplankton community, they vary among every taxonomic group within the assemblage.

Zooplankton experts seek to create global maps for every major taxonomic grouping, or even for entire communities or ecosystems, particularly using observational data, in conjunction with environmental data, to predict biodiversity distribution (e.g., Rombouts et al., 2009). Obtaining sufficient data for all taxa under consideration, across the full spectrum of habitats, remains a primary hurdle. One of the greatest accomplishments of the Census of Marine Life (CoML) was to build the Ocean Biogeographic Information System (OBIS), a system that can address such questions by pulling together the disparate datasets to allow such synthetic tasks to be undertaken (e.g., Tittensor et al., 2010; Vanden Berghe et al., 2010).

New insights derived from DNA-based approaches represent the biggest current challenges in understanding zooplankton biodiversity. Initially, these tools offered great promise in tackling simple issues, such as phenotypic variation or rates of hybridization. In practice they are revealing numerous cryptic species not previously recognized based on morphology alone, that force rethinking on what represents a species, and the geographic/environmental boundaries between them. Many species believed to span several ocean basins may in fact represent species assemblages, suggesting that current estimates may severely underestimate the overall diversity of marine zooplankton in all groups.

3 Benthic ecosystems

3.1 Rocky Shore Ecosystems

The ease of access and suitability for experimental work of rocky shore habitats have attracted a long history of scientific study and engaged a broad audience. Biodiversity assessments in this habitat typically use quadrat and transect assessment with visual or photographic identification, and do not require ships or complex technology. In many rocky intertidal environments, and particularly in regions with large tidal ranges, aerial exposure at upper tidal levels and predation at low tidal levels create distinct bands of species, or zonations, that represent one of the most striking and well-documented gradients in the ocean (first described by Stephenson and Stephenson 1949). The global distribution of rocky intertidal habitats creates opportunities to compare latitudinal trends and to detect large-scale patterns and changes. Despite fairly well-understood local patterns and processes, large-scale patterns (regional to global) are difficult to discern; however, many drivers, such as temperature and exposure, act on large scales, as do human influences, such as invasive species and pollution. Examples from around the world (e.g. Thompson et al., 2002) demonstrate the past and current effects on rocky shores of pollution (e.g. oil, eutrophication), overfishing, introduced exotic species, modification of coastal processes (e.g. coastal defences, sedimentation) and global change (e.g. temperature, sea level). The relative magnitudes of some anthropogenic pressures differ among industrialized countries and developing countries.

Very few long-term studies have addressed temporal trends in rocky shore biodiversity and most focus locally and regionally. Barry et al. (1995) and Sagarin et al. (1999) observed changes in the abundance of macroinvertebrate species in a rocky intertidal community in California between surveys in 1931-33 and 1993-96. These changes are consistent with recent climate warming that shifts species northward. Eight of nine southern species increased in abundance and five of eight northern species decreased; however, cosmopolitan species displayed no trend. Blanchette et al. (2008) described the spatial pattern of distribution of species abundance for rocky intertidal communities along the Pacific coast of North America from Alaska to Mexico (more than 4000 km).

This biogeographic study represents one on the larger-scale analyses of this habitat, and reported strong spatial structure in the rocky intertidal communities of the north-east Pacific. Breaks in similarity among clusters generally linked with known biogeographical and oceanographic discontinuities. Sea surface temperature and species similarity both correlated strongly, coinciding with long-term temporal trends along the California coast that point to both geography and oceanographic conditions as primary determinants of patterns of intertidal community structure.

Recent efforts through the NaGISA project of the Census of Marine Life (e.g., Iken et al., 2010) and others demonstrate large-scale patterns. The NaGISA project gathered information on rocky shore systems globally, and compared diversity and abundance of key benthic groups from intertidal and shallow subtidal rocky shore sites in order to identify latitudinal trends and their environmental drivers. Global analyses were constrained by differences in sampling efforts (numbers, years, and strata sampled), timing of sampling, under-sampling of some ecoregions, and unbalanced representation of the northern and southern hemispheres. Results indicate that distribution patterns of diversity and biomass of various taxonomic groups (e.g., macroalgae, gastropods, decapods and echinoderms) are very complex and sometimes defy the expected latitudinal gradient of species decreasing towards the poles. Regional diversity hotspots often complicate any simple broad-scale pattern. Despite differences in sampling effort, timing, and coverage, this effort identified likely drivers of diversity of specific taxa and communities, several of which were tightly linked to human activities. For example, pollution indices correlate significantly with diversity in several phyla (e.g., Iken et al., 2010). Although the natural heterogeneity of these systems complicates unequivocal establishment of cause-consequence relationships, a larger data base for the analysis of global diversity trends and their drivers will provide more substantive evidence for the identification of likely drivers.

However, although much progress has been made recently, our understanding of rocky shore biodiversity patterns remains incomplete, especially beyond local or small regional scales. Similarly, the complexity of these systems constrains efforts to assign environmental or human-induced drivers to rocky shore diversity, because such drivers can act on different scales (Benedetti-Cecchi et al., 2010), and may act cumulatively, synergistically, or antagonistically. That 40 per cent of the world's population currently lives within 100 km of the coast enhances the urgency of this issue, particularly given that as population density and economic activities increase, so will pressures on rocky shores, as well as other coastal ecosystems from tropical to temperate systems, and even some polar regions.

3.2   Tropical coral reef ecosystems

Tropical coral reefs span the Indo-Pacific and Atlantic Oceans, although cool upwelling associated with boundary currents limits distributions along the west coasts of Africa and South/Central America. We refer readers to Chapter 43 for a more detailed discussion on tropical and subtropical coral reef habitats that extends beyond our focus on biodiversity gradients. The diversity and productivity of coral reefs and associated ecosystems (mangroves, seagrasses and pelagic habitats) are among the highest globally, providing essential ecosystem services to tropical countries. The taxonomic richness of corals reefs is second to none, with tropical coral reefs housing 25per cent of all known marine life on the planet including sea fans, sponges, worms, starfish, brittle stars, sea urchins, crustaceans, and fish. In fact the variety of life supported by coral reefs rivals that of the tropical forests of the Amazon or New Guinea. Temperature and habitat complexity, particularly for stony corals and bony fish, have been shown quantitatively to drive tropical reef diversity on global scales (Tittensor et al., 2010), with other features such as habitat area (rocky substrates within the photic zone), and historical factors (historical speciation/extinction patterns) being secondary drivers of regional to global patterns of coral reef biodiversity.

Tropical coral reefs are restricted to warm waters with average annual temperatures typically above 18 °C though with annual mean temperatures between 20-27 °C, which enable stony corals, via photosynthetically enhanced calcification, to lay down a skeleton fast enough to build up reefs over multiple generations. The symbiosis between stony corals and their intra-cellular symbionts (zooxanthellae) limit coral reefs to sunlit substrates, optimized in the top 10-15 m of the water column but with reef build-up possible down to 30-50 m depth. Thus coral reefs are restricted to island and continental fringes, and shallow oceanic banks that reach the photic zone. Classic coral reef descriptions emphasize oligotrophic low-sediment oceanic conditions being ideal for coral reefs, however vibrant coral and reef growth, with high-diversity communities, can occur in relatively high turbidity and sedimentation conditions in eutrophic waters near highly productive major estuaries (e.g., along the Andaman Sea). This ability for coral reefs to flourish in both sets of conditions is due to tight cycling of carbon and nitrogen between the coral symbionts.

A total of 836 tropical reef-associated coral species have been described globally, with 759 from the Indo-Pacific and just 77 from the Atlantic, with no species in common across these two ocean systems. The broad swath of equatorial currents that cross the Pacific and Indian Oceans, joined through the 'leaky' Indonesian region, dominates global patterns of coral reef diversity connecting East Africa to French Polynesia - of 4,000 species of tropical fishes, 492 are shared between the Western Indian Ocean and French Polynesia (Randall, 1998). Westwards from here, the large deep-water barrier in the East Pacific isolates the Eastern Tropical Pacific from the broader Indo-Pacific.

At the centre of the Indo-Pacific, the Indo-Australian Arc (IAA) provides optimal temperature conditions and the highest habitat area for corals; within this 'Coral Triangle' (Roberts et al., 2002; Hoeksema, 2007) sub-regions show peaks of over 600 species of hard corals, 550 species of reef fishes and 50 species of stomatopods (Reaka et al., 2008). Diversity declines east and west from here into the Central Indian and Pacific Oceans (<300 coral and fish species, 5-15 stomatopods), though new evidence suggests a second peak of diversity in the Western Indian Ocean/Eastern African region, in both stomatopods (30 species, Reaka et al., 2008) and hard corals (350-400 species, Obura, 2012). The lowest diversity in tropical Indo-Pacific coral reefs is in the Eastern Pacific due to isolation, with <150 coral and fish species, and <5 stomatopod species. Coral reefs harbour high levels of cryptic diversity: for instance, populations of many coral and fish species previously thought to be distributed widely in the Indo-Pacific region are now recognized as different species between the Indian and the Pacific Ocean, and even more recently between the Red Sea and the Indian Ocean.

High levels of endemism are shown in remote island coral reef systems, in low-dispersive groups such as fishes. In Hawaii and the Easter Island group levels of endemism in fish exceed 20 per cent (Randall, 1998), and only 12 shore fish species are shared between the Red Sea, Easter Island, and the Hawaiian Island archipelago. The small size and lower diversity of the tropical Atlantic, with subregions in the Caribbean, Gulf of Mexico and around Brazil, may increase vulnerability to anthropogenic and climate threats compared to the larger, more diverse Indo-Pacific.

Deep-water corals also form reef structures, which we describe briefly in the deep-sea section below; we also refer readers to Chapter 42, which presents a more detailed description of these environments.

3.3 Mangrove forest ecosystems

Mangroves are woody plants that grow normally in tropical and subtropical latitudes along the land–sea interface, bays, estuaries, lagoons, and backwaters (Mukherjee et al., 2014). These plants and their associated organisms constitute the 'mangrove forest community' or 'mangal'. Although mangrove ecosystems occur in more than 120 countries worldwide, encompassing just over 80 plant species, subspecies and varieties globally (Massó i Alemán et al., 2010), they are generally species-poor vegetation formations. Nonetheless, they support a complex community of animals and micro-organisms, the numbers of which have never been reliably estimated. Mangrove decapods and, to a lesser extent, insects are better studied than most taxa but the scientific community is only just beginning to understand how mangrove ecosystems work and what they contribute to ecosystem functions, goods and services, including biodiversity support, storm protection, fisheries production, effects on water quality and providing significant carbon sinks (Dahdouh-Guebas, 2013). Several recent papers document immense carbon sequestration and suggest they may represent the most carbon-rich tropical forests (Donato et al., 2011).

Mangrove tree species may be divided into two distinct floristic groups, an Atlantic-East-Pacific and Indo-West-Pacific, the latter of which represents the mangrove species richness peak between 90 and 135 degrees East. Latitudinal richness peaks globally near the equator (Figure 34.4). Local species richness links significantly to regional richness (Ellison, 2002), with recent recognition that high variability temperature interacts with aridity in defining upper latitudinal limits of *Avicennia* and *Rhizophora* (Quisthoudt et al., 2012).

Figure 34.4 | Latitudinal species richness of mangrove plant species.

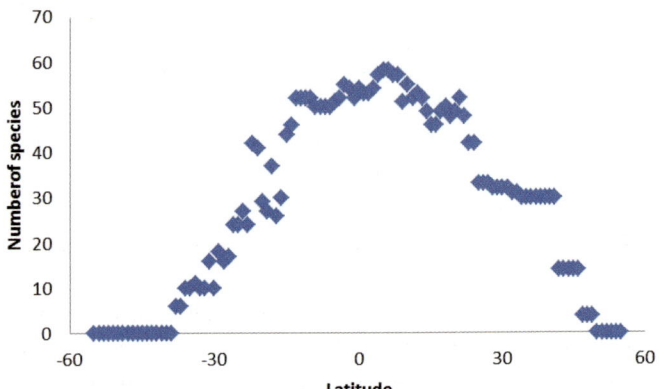

The International Union for Conservation of Nature (IUCN) rarely lists mangrove species as threatened because they are often widely distributed, creating conservation challenges. Yet, reports of local extirpations, sometimes hidden as cryptic ecological degradation, may affect local and regional fisheries or other coastal functions (Dahdouh-Guebas et al., 2005). These losses point to an urgent need to re-assess mangrove ecosystems nationally and regionally to identify regions most at risk of losing mangrove ecosystems and associated functions, goods, and services (Mukherjee et al., 2014).

3.4 Coastal sedimentary ecosystems

The coastal zone denotes the relatively narrow transition zone between land and ocean where strong interactions occur with humans. Sediment covers much of the continental shelf from the poles to the equator and supports a wide diversity of invertebrates spanning almost all animal phyla. The shallowest depths along the shoreline support seagrasses, mangroves, and salt marshes, but seabed primary production is otherwise limited to benthic photosynthetic microbes that quickly disappear as light attenuates with depth. Resuspended material and phytoplankton sinking from the photic zone add significantly to benthic production.

For organisms ranging from large megafaunal clams and crabs to meiofaunal nematodes and copepods, temperature primarily defines broad biogeographic provinces, but within regions substrate composition plays a major role in defining composition and diversity of sedimentary fauna. Sand and coarser substrates typically occur in high-energy exposed environments, with muds characterizing quiescent areas. Depth and productivity also strongly influence faunal patterns, with peak diversity at mid-shelf depths and locations with moderate organic input (Renaud et al., 2007). A hump-shaped pattern of maximum diversity linking with productivity has been reported elsewhere, including the Arctic (Witman et al., 2008) and in fossil marine invertebrates (brachiopods) (Lockley, 1983). Intertidal sediments typically exhibit very low diversity, irrespective of substrate composition; this is likely to be due to the harsh, dynamic nature of that environment.

In a broad sense, evidence suggests high species richness in tropical sediments relative to temperate and polar regions, but because sampling effort is strongly biased towards temperate seafloor environments, this complicates broad-scale comparisons. For example, total species number for sedimentary invertebrates in the Canadian Arctic compares favourably with the Canadian Pacific and Atlantic (Archambault et al., 2010) and on the eastern and western coasts of the United States, gastropod molluscs increase in species richness from the Arctic towards the equator, but peak in the subtropics (Roy et al., 1998). Multiple studies testing latitudinal gradients over varying scales find few consistent patterns, and suggest that complex differences in local environments play a much greater role than latitude in defining diversity and composition; indeed, landscape heterogeneity may obliterate a simple broad-scale pattern, except where broad physical drivers dominate. The strong influence of the Gulf Stream on coastal European waters produces high species richness in high latitude sediments (Ellingsen and Gray, 2002).

Algal genera generally exhibit an inverse latitudinal gradient, with biodiversity hotspots in temperate regions, but bryopsidalean algae peak in diversity in the tropical Indo-Pacific region (Kerswell, 2006).

Key data gaps in species richness data for coastal sediments of Africa, South America, the western tropical Pacific and polar regions (Costello et al., 2010) constrain latitudinal comparisons, but we do know that Australian coastal sediments support very high species richness (Butler et al., 2010), and that Antarctica diversity (Griffiths, 2010) only modestly exceeds Arctic diversity (Piepenburg et al., 2011).

However, it is recognized that even with less attention than European waters, the Indo-Pacific area is the most diverse area of our oceans. Molluscs have the largest diversity of all phyla in the marine environment (Bouchet, 2006), and mollusc diversity is exceedingly high in the tropical waters of the Indo-Pacific, particularly in coral reef environments (Crame, 2000). How this pattern translates to sedimentary fauna remains unknown and leads us to conclude that we need far more studies in many regions to drawn firm conclusions about general patterns of marine biodiversity in the coastal zone.

Very few studies compare sedimentary fauna across sharply contrasting sediments because differences in sampling gear complicate such comparisons. Sampling cobbles and gravel, which support a wide diversity of encrusting epifauna and flora, requires very different tools than sampling muds or even coarse sand. Nonetheless, a qualitative comparison of species lists indicates strongly different faunas within these substrata, linking to a wide range of variables spanning physical disturbance, larval supply, and food quality, to name just a few.

3.5 Deep-sea benthic ecosystems

The deep sea spans depths from 200 m to almost 11,000 m, encompassing more than 90 percent of the global ocean area, and representing the largest ecosystem on Earth (Watling et al., 2013). However, less than 5 percent of its area has been explored, and less than 0.001 per cent, the equivalent of a few football fields, has been sampled quantitatively, making it among the least known environments on Earth. Decades ago researchers projected 1 to 10 million total species in the deep sea (Grassle and Maciolek, 1992). More recently, Mora et al. (2011) predicted that 91 per cent of marine species remain unknown, largely due to undersampling of the deep sea (other works downsize that estimate – see Appendix). For example, 585 of 674 isopods species collected in recent expeditions to the deep Weddell Sea were new to science (Brandt et al., 2007). Sampling is uneven, with very limited understanding of hard substrate biota outside of chemosynthetic ecosystems, and undersampling of metazoan meiofauna and protozoa.

Deep-sea sediments cover much of the deep-sea floor, with little variation in temperature and salinity. The absence of photosynthesis below ~200 m means that most deep-sea life depends exclusively on sinking food from surface waters. These sediments support a highly diverse fauna spanning most phyla; the first comprehensive sampling covered just 21 m^2 of seafloor, yielding 1,597 species from 13 different phyla of invertebrates (Grassle and Maciolek, 1992). The best known broad-scale diversity patterns for deep-sea invertebrates and fishes are the unimodal diversity-depth relationship, bathymetric zonation, and a general decline in species richness towards the poles. Macrofauna in total and as individual taxa provide the strongest evidence for unimodal diversity-depth relationships (Rex and Etter, 2010) in which, despite some contradictory patterns in some locations, highest diversity occurs at depths of ~1,500-2,000 m. Significant unimodal diversity-depth relationships have also been reported for nematodes, ostracods, and foraminifers in the Arctic Ocean, and for megafaunal invertebrates and fishes in the western North Atlantic. Reduced population densities under extreme food limitation may suppress species diversity in the deep oligotrophic abyss (>3,000-4,000 m depth), whereas elevated carbon fluxes at shallow depths may suppress diversity by driving competitive exclusion or creating physiological stress. On upwelling along continental margins, low oxygen at upper bathyal depths (100-1000 m) suppresses diversity. Declining food supply and thermal energy with depth (and distance from continents) are likely to drive these patterns, which are complicated by regional variation in food availability. Topographic isolation or complexity, boundary effects, sediment characteristics, currents, oxygen concentration, physical disturbance, biological interactions and patch dynamics, as well as evolutionary history, also influence diversity and distribution patterns at regional and local scales (Rex and Etter, 2010).

Many fish and invertebrate taxa occur at a similar depth range in "bands" of distinct assemblages or "zones". Zonation describes sequences where few changes in species composition occur within a band, but abrupt faunal boundaries occur. However, major oceanographic features, such as oxygen minimum zones, strong bottom currents and abrupt shifts of water masses, can obscure, alter, or create zonation patterns. Globally, clear faunal differences are observed between upper bathyal depths compared to mid and lower bathyal depths (Rex and Etter, 2010). The preponderance of rare species complicates these analyses, particularly in abyssal environments where many species occur only once in samples (Grassle and Maciolek, 1992).

Multiple biological and physical factors, such as larval dispersal, competition, predation, temperature, oxygen concentration, hydrostatic pressure, and food supply, all potentially drive zonation. Temperature, oxygen, and food supply vary most in the upper bathyal region, where pronounced species turnover occurs, creating a ubiquitous shelf-slope transition zone. In lower bathyal (3,000-4,000 m) and abyssal depths, declining food supply homogenizes fauna and reduces species turnover.

Multiple studies report latitudinal diversity gradients in many deep-sea invertebrate groups, although patterns vary in sparse sampling of limited taxa (Kaiser et al., 2013). For several macrofaunal groups, and for meiofaunal foraminifers, diversity decreases with increasing latitude in the North Atlantic. In the South Atlantic, isopod diversity increases, whereas gastropods and bivalves decline poleward (although there are exceptions to this trend). Strong seasonality and pulses of phytodetritus at high latitudes may depress diversity, much like physical disturbance. Other work suggests increasing diversity in the northern hemisphere for nematodes.

Thousands of topographic features, such as submarine canyons and fjords, incise continental and island margins and increase complexity of bottom topography, modify abundance and diversity by intensifying mixing, amplifying currents, enhancing productivity, sediment and food deposition, and channel cascading shelf waters. Enhanced food deposition drives species aggregations and consequent increased diversity in these habitats compared to adjacent, topographically simpler areas. Steep topography and amplified currents often expose bedrock and boulders within canyons, supporting additional fauna, such as megafaunal and macrofaunal suspension feeders that require hard surfaces for attachment and strong currents for food delivery.

Biogenic reefs (formed mainly by deep-water coral and sponges) occupy hard substrata with high currents, often within basins or along the continental margins. These reefs form from skeletons of dense aggregations of one or a few species. The skeletons create surfaces for colonization, extend higher in the water than the surrounding seafloor (thus reaching faster currents and greater food delivery) and add spaces for protection from predators and other disturbances. These reefs therefore enhance local species diversity, and provide nursery areas for many macro- and megafaunal invertebrates and fishes.

Despite their remoteness, human activities affect deep-sea diversity, resulting in declines in deep-sea fishes, loss of habitat-forming invertebrates (e.g., deep-water corals), and increased contaminants in deep-sea biota. Although these impacts illustrate local to regional effects, the cascading effects of warming surface layers portend substantial changes in the food supply to deep-sea ecosystems. Manifestations of climate change including ocean warming, acidification and deoxygenation, may reduce the bathyal habitats available, with concomitant broad-scale changes in patterns of species distribution and diversity. Deep-sea biodiversity loss could adversely affect ecosystem functions of the Earth's largest environment.

Hydrothermal vents and cold seeps occur where dissolved chemical compounds emerge at the seafloor at rates and concentrations high enough to sustain chemosynthesis. Chemosynthesis is the process that some microbes use to transform CO_2 into organic molecules. The emerging fluid is often associated with active tectonic features such as spreading centres, subduction zones, and volcanoes, but seeps may also be linked to methane escape via mass wasting, brine pools, turbidity flows, diapirs, and pockmarks, canyons and faults. The resulting habitat distribution tends to be linear, following mid-ocean and back-arc spreading centres, as well as volcanic arcs in the case of vents, and along continental margins in the case of seeps. The fluid emissions at most regions of high carbon accumulation – oil, gas and clathrate-rich deposits – support chemosynthesis. Thus, the microbes form the basis of a food web for a metazoan community that is mostly endemic to these systems. Habitats supporting chemosynthetic production and communities occur in every ocean.

No overall assessment of diversity patterns and drivers exists, although alphadiversity at vents and seeps is often lower than in the surrounding non-chemosynthetic ecosystems. The considerable work on biogeographic patterns includes exploration of faunal relationships and origins. Despite similarity in many taxa at vents and seeps, they usually differ at species or genus levels; however, both habitats harbour many endemic taxa. Some taxa at vents and seeps are new to science at higher taxonomic levels, especially those housing microbial symbionts.

Overall, species diversity at seeps exceeds that at hydrothermal vents, driven by high variability in the geological settings of methane and sulphide release and within-site heterogeneity (Levin and Sibuet, 2012). Depth may describe both the biogeographic similarity of seeps across the Atlantic and, possibly, the decrease of symbiont-hosting species with depth in general at seeps. However, depth may reflect more direct drivers, such as greater production and predation at shallower depths or behaviour of fluid flux sustaining chemosynthesis. Local site longevity and stability of the fluid source will influence any pattern analysis, as will depth; vents in the photosynthetic zone above 200 m differ notably in taxa and structure. These habitats exhibit low diversity within taxa. Habitat and depth drive a variety of patterns in vesicomyid clams hosting symbionts, but better systematics are needed. The East Pacific Rise represents a diversity hotspot for the most speciose family, the vent-endemic dirivultid copepods.

Biogeographic patterns at vents are likely to share controlling factors with all ocean fauna: continental barriers, oceanographic barriers, and pressure gradients with depth. However, similarity analyses with growing datasets indicate strong control by the history of spreading ridges from the mid-Mesozoic to the present. Thus, diversity analysis is likely to identify connectivity, geological longevity and ridge stability as important, reflecting the smaller-scale drivers currently known. Recent discovery of vent communities in the Antarctic and Arctic reveal a unique community composition and suggest that dispersal barriers are also important drivers of diversity (Rogers et al., 2012).

Increasing evidence suggests that the character of the venting fluids fundamentally drives taxonomic composition, overlaid on geographic separation, particularly in the complex settings of the Atlantic and the western Pacific back-arc rifting and volcanism (Desbruyères, 2000); relevant factors may include reduced compound composition, water temperature and metal content. Similarly, the high diversity of animals recently recognized from mud volcanoes relates to the nature of the chemical substrates in emerging fluids and the adaptations of associated microbes and symbiont hosts both across and within sites (Rodrigues et al., 2013). Where geochemical drivers characteristic of vents and seeps come together, an intermediate ecosystem with biodiversity elements from both vents and seeps emerges (Levin et al. 2012). Decay of large organic falls also supports microbial processes and species reliant on chemosynthesis (Smith et al., 2015). Thus vents and seeps also hold many taxa in common with organic remains, such as wood falls and whale carcasses.

Seamounts are undersea mountains historically defined by an elevation of 1 km or more, but more recently by a more ecological definition, that includes knolls and hills with an elevation of 100 m or more. They occur in all oceans of the world, from the tropics to the poles, and cover depth ranges from near the surface to the abyss. The total number remains uncertain because so little of the deep ocean has been surveyed, but estimates range from 14,000-50,000 large seamounts and tens to hundreds of thousands of smaller ones (Stocks, 2010).

Three important characteristics distinguish seamounts from the surrounding deep-sea habitat (Clark, 2009). First, as "islands" of shallower sea floor, they provide a range of depths for different communities. Second, their typical hard and often bare rock surfaces contrast with the fine, unconsolidated sediments that cover the majority of the sea floor. Third, the physical structure of some seamounts alters local hydrography and currents so as to concentrate species and productivity over the seamount, thus increasing their importance as oceanic ecosystems but also attracting commercial exploitation. The very low proportion of seamounts sampled globally limits understanding of their diversity, and the composition, structure, function, and connectivity of seamount ecosystems remain unexplored and unknown except in a few locations (Stocks and Hart, 2007).

Seamount benthic communities are rich and varied; sandy or muddy sediments dominate where currents are slow, with mostly deposit-feeding species of polychaetes, echinoderms, various crustaceans, sipunculids, nemertean worms, molluscs, sponges, and nematodes utilizing sinking particulate matter. Suspension feeders, including corals, crinoids, hydroids, ophiuroids, and sponges, dominate where faster currents expose rocky areas. The large corals and sponges can form extensive and complex reef-like or thicket structures, which add habitat for smaller mobile fauna. Seamount biodiversity research has rapidly increased the number of known species in recent decades. A global review in 1987 (Wilson and Kaufmann, 1987) recorded 449 species of fish and 596 species of invertebrates from 100 seamounts, but more recent surveys suggest much higher numbers (Stocks, 2010). The Census of Marine Life on Seamounts amalgamated data on over 5,400 taxa (although not all to species) from 258 seamounts into the public database SeamountsOnline (Stocks, 2010), which can currently be accessed through the Ocean Biogeographic Information System portal (www.iobis,org) by selecting the Seamounts Online database. However, gear selectivity and generally few samples per seamount limit biodiversity knowledge for any one seamount.

Depth-related environmental parameters strongly influence seamount species composition, together with seafloor type and character (e.g., substratum, hardness, composition, mobility) (see Clark et al., 2010a). Habitat complexity on seamounts largely determines benthic species occurrence, distribution and diversity. Volcanic activity, lava flows and areas of hydrothermal venting add to habitat diversity on seamounts, creating unique environmental conditions that support specialized species and assemblages (see preceding section). Water column stratification and oceanic flow conditions also add local dynamic responses that can regulate the spatial scale of faunal distributions.

Many early studies suggested high seamount endemism given their geographic isolation, often separated from other seamounts by deep water and considerable distance. Although seamount assemblages can differ in species abundance or frequency, similarity in deep-sea fish assemblages between seamounts and adjacent continental slopes or islands (scales of km), as well as across oceans (1000s of km), contradicts the idea of ecological islands (Clark et al., 2010b). In the latter case, the global-scale circulation of deep-sea water masses presumably influences fish distribution. Regional-scale similarities in faunal composition between seamounts and other habitats in the South Pacific demonstrate that seamounts share a common regional pool of species with non-seamount communities. Schlacher et al. (2014) found high species turnover with depth and distance in seamount assemblages off Hawaii at the scale of individual seamounts, but geographic separation was a poor predictor of ecological separation for the region as a whole. These studies emphasize that the spatial scales over which faunal assemblages of seamounts are structured cannot be generalized. Nevertheless, recent biogeographic classifications for the deep ocean suggest that benthic

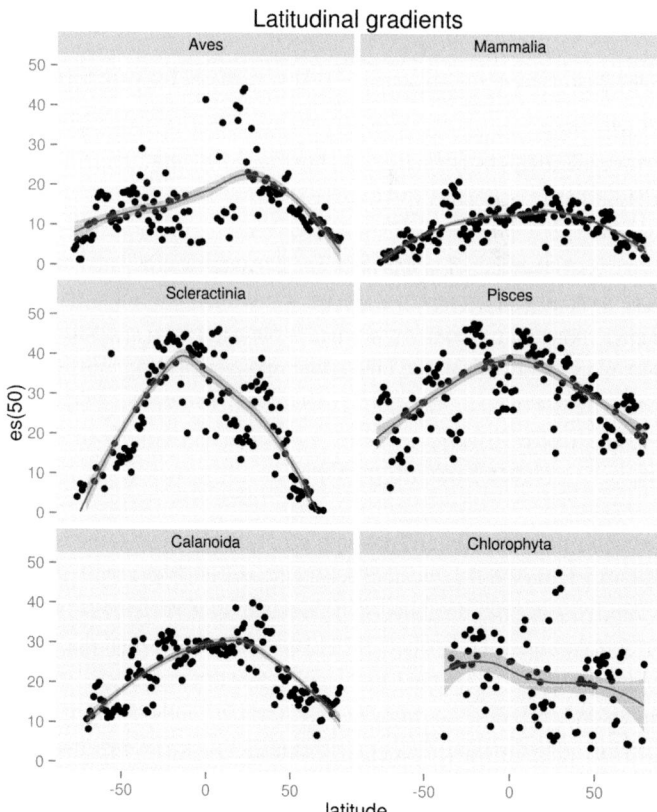

Figure 34.5 | ES(50) calculated for various groups, from the data as available in OBIS as of the end of 2012. ES(50) (or Hurlbert's index) represents the numbers of species expected to be present in a random sample of 50 individuals; this metric measures the diversity (not species richness, as its name might suggest), independent of sample size. Points in the graphs above represent calculation of ES(50) for bands of 1 degree of latitude. The blue line is the LOESS (LOcal regrESSion) prediction/smoothing; the darker grey bands are the 95% confidence intervals around the LOESS estimate. Most groups, but not all (e.g. Chlorophyta) show a clear unimodal pattern. All calculations were made with R (R Development Core Team, 2014), using package ggplot2 for LOESS and plotting (Wickham, 2009).

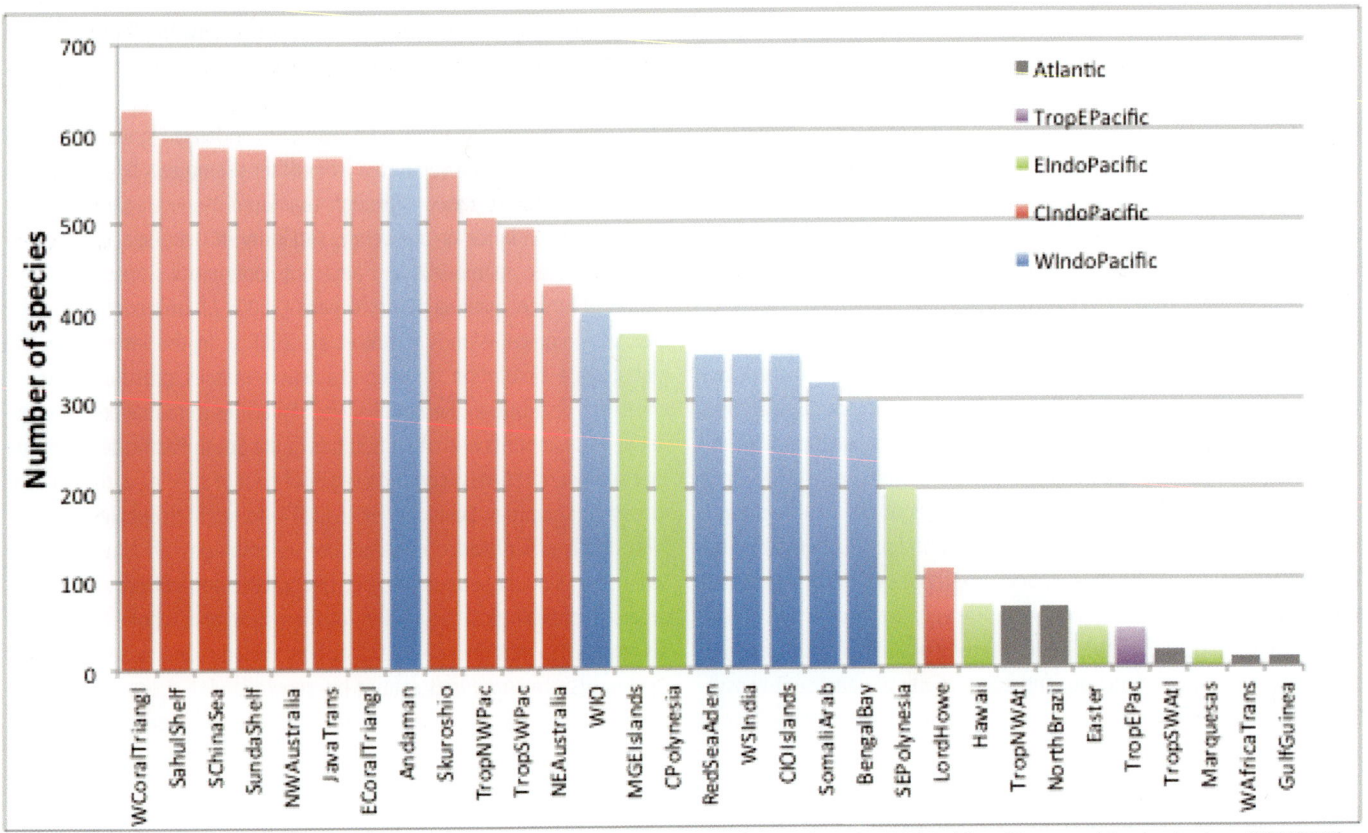

Figure 34.6 | Coral species richness by province in the Marine Ecosystems of the World Classification, from the IUCN Red List of Threatened Species database (IUCN 2013).

community composition will vary markedly among basins (e.g., Watling et al., 2013).

Better understanding of global deep-sea biodiversity gradients requires more sampling, but predictive species distribution modelling and use of environmental surrogates can improve our short-term understanding (e.g., Clark et al., 2012) and help inform management options for the deep sea.

3.6 Cross-taxa integration

The global ocean houses an enormous variety of life. In total, the oceans support an estimated 2.2 million eukaryotic species (Mora et al., 2011), of which science has described some 220,000 (WoRMS Editorial Board, 2013). A key question is whether consistent 'rules' constrain the distribution of this life across the variety of different organisms and habitats examined here, and if so, whether they result in consistent large-scale patterns of biodiversity. Global-scale studies to explore this question began long ago and especially in the last decade (e.g., Rutherford et al., 1999; Roberts et al., 2002), but the enormous amounts of data collected and compiled by the Census of Marine Life enable exploration and mapping patterns across more taxonomic groups than ever before (Tittensor et al., 2010) to understand the consistency of diversity patterns.

Perhaps the most common large-scale biodiversity pattern on the planet is the 'latitudinal gradient', typically expressed as a decline in species from the equator to the poles (Figure 34.5). Adherence to this pattern varies among marine taxa; Chlorophyta and other macroalgae, for example (Figure 34.5, lower right panel), do not exhibit the same latitudinal gradient, as noted earlier above in this chapter). Although coastal species generally peak in abundance near the equator and decline towards the poles, seals show the opposite pattern; indeed pinnipeds peak at high latitudes (Fig. 1a). Furthermore, strong longitudinal (east-west) gradients, complicate patterns, with 'hotspots' of richness across multiple species groups in the 'Coral Triangle' in the Indo-Pacific and the Caribbean (Figure 34.6).

Oceanic organisms, such as whales, differ in pattern entirely, with species numbers consistently peaking at mid-latitudes between the equator and poles. This pattern defies the common equator-pole gradient, suggesting that different factors are involved. Different processes may control species richness among oceanic and coastal species (for example, in terms of dispersal, mobility, or habitat structure), but general patterns appear to be reasonably consistent within each group. However, across all groups studied, ocean temperature is consistently related to species diversity (Tittensor et al., 2010), hence the effects of climate change are likely to be observed as a restructuring of marine community diversity (Worm and Lotze, 2009).

Although the above patterns hold for the 11,000 species studied (Tittensor et al., 2010), numerous groups and regions were not represented. For example, global-scale patterns of diversity in the deep ocean remain largely unknown (Rex and Etter, 2010). Our diversity and distribution knowledge is taxonomically biased towards large, charismatic (e.g., whales) or economically valuable (e.g., tunas) species. Our knowledge of patterns in microbial organisms remains particularly limited relative to the enormous diversity therein, and enormous challenges to even measure biodiversity remain. Viruses remain a critical part of the oceanic system for which we lack any global-scale biodiversity knowledge.

Other than species richness, we are just beginning to explore other patterns of global marine biodiversity. Patterns of 'evenness' in reef fishes, which relates to relative proportions of individual species in the community, apparently show an inverse gradient (Stuart-Smith et al., 2013). This pattern, in turn, affects 'functional richness', which relates to the diversity of functions in reef fishes, a potentially important component of ecosystem productivity, resilience, and goods and services provision. The importance of these patterns depends on their robustness and consistency across different species groups, but opening such new insights into other facets of biodiversity provides additional information with which to manage it, particularly in support of human welfare.

References

Archambault, P., Snelgrove, P.V.R., Fisher, J.A.D., Gagnon, J.M., Garbary, D.J., Harvey, M., Kenchington, E.L., Lesage, V., Lévesque, M., Lovejoy, C., Mackas, D.L., McKindsey, C.W., Nelson, J.R., Pepin, P., Piché, L., Poulin, M. (2010). From Sea to Sea: Canada's Three Oceans of Biodiversity. *PLoS ONE*, vol. 5, No. 8, e12182. doi:10.1371/journal.pone.0012182.

Barry, J.P., Baxter, C.H., Sagarin, R.D., Gilman, S.E. (1995). Climate-related, long-term faunal changes in a California rocky intertidal community. *Science* vol. 267, 672-675.

Benedetti-Cecchi, L., Iken, K., Konar, B., Cruz-Motta, J.J., Knowlton, A., Pohle, G., Castelli, A., Tamburello, L., Mead, A., Trott, T., Miloslavich, P., Wong, M., Shirayama, Y., Lardicci, C., Palomo, G., Maggi, E. (2010). Spatial relationships between polychaete assemblages and environmental variables over broad geographical scales. *PLoS ONE*, vol. 5, e12946. doi:10.1371/journal.pone.0012946.

Blanchette, C.A., Miner, C.M., Raimondi, P.T., Lohse, D. Heady, K.E.K., Broitman, B.R. (2008) Biogeographical patterns of rocky intertidal communities along the Pacific coast of North America. *Journal of Biogeography* vol. 35, 1593–1607.

Bolten, A.B. (2003). Variation in sea turtle life history patterns: neritic vs. oceanic development stage, in: Lutz, P.L., Musick, J.A., and Wyneken, J., eds., *Biology of Sea Turtles*, Vol. II, CRC Press, pp. 243-258.

Boltovskoy, D., ed. (1999). *Radiolaria Polycystina. "South Atlantic Zooplankton"*. Backhuys Publishers, Leiden, pp. 149-212.

Bouchet, P. (2006). The magnitude of marine biodiversity. In: Duarte, C.M., ed. *The Exploration of Marine Biodiversity: Scientific and Technological Challenges*. pp. 31-62.

Brandt, A., Gooday, A.J., Brix, S.B., Brökeland, W., Cedhagen, T., Choudhury, M., Cornelius, N., Danis, B., De Mesel, I., Diaz, R.J., Gillan, D.C., Ebbe, B., Howe, J., Janussen, D., Kaiser, S., Linse, K., Malyutina, M., Brandao, S., Pawlowski, J., Raupach, M. (2007). The Southern Ocean deep sea: first insights into biodiversity and biogeography. *Nature*, vol. 447, pp. 307-11.

Briggs, J.C. (1974). *Marine zoogeography*, McGraw-Hill.

Bucklin, A., Nishida, S., Schnack-Schiel, S., Wiebe, P.H., Lindsay, D., Machida, R.J., and Copley, N.J. (2010). A Census of Zooplankton of the Global Ocean, Chapter 13. In: McIntyre, A.D., eds. *Life in the World's Oceans: Diversity, Distribution, and Abundance*. Blackwell Publishing Ltd., Oxford. pp. 247-265.

Butler, A.J., Rees, T., Beesley, P., Bax, N.J. (2010). Marine Biodiversity in the Australian Region. *PLoS ONE*, vol. 5, No. 8, e11831. doi:10.1371/journal.pone.0011831.

Chown, S.L., Gaston, K.J., Williams, P.H. (1998). Global patterns in species richness of pelagic seabirds: the Procellariiformes. *Ecography* vol. 21, pp. 342-350.

Clark, M.R. (2009). Seamounts: Biology. In: Gillespie, R.G., and Clague, D.A., eds., *Encyclopaedia of Islands*. University of California Press. pp. 818–821.

Clark, M.R., Rowden, A.A., Schlacher, T., Williams, A., Consalvey, M. (2010a). The ecology of seamounts: structure, function, and human impacts. *Annual Review of Marine Science*, vol. 2, pp. 253–78.

Clark, M.R., Althaus, F., Williams, A., Niklitschek E, and Menezes, G. (2010b). Are deep-sea fish assemblages globally homogenous? Insights from seamounts. *Marine Ecology*, vol. 31, pp. 39–51.

Clark, M.R., Schlacher, T.A., Rowden, A.A., Stocks, K.I., and Consalvey, M. (2012). Science priorities for seamounts: research links to conservation and management. *PLoS ONE*, vol. 7, No. 1, e29232. doi:10.1371/journal.pone.0029232.

Costello, M.J., Coll, M., Danovaro, R., Halpin, P., Ojaveer, H., Miloslavich, P. (2010). A Census of Marine Biodiversity Knowledge, Resources, and Future Challenges. *PLoS ONE*, vol. 5, No. 8, e12110. doi:10.1371/journal.pone.0012110.

Crame, J.A. (2000). Evolution of taxonomic diversity gradients in the marine realm: evidence from the composition of recent bivalve faunas. *Paleobiology*, vol. 26, pp. 188–214.

Croxall, J.P., Butchart, S.H.M., Lascelles, B., Stattersfield, A.J., Sullivan, B., Symes, A., Taylor, P. (2012). Seabird conservation status, threats and priority actions: a global assessment. *Bird Conservation International*, vol. 22, pp. 1-34.

Dahdouh-Guebas, F., Hettiarachchi, S., Sooriyarachchi, S., Lo Seen, D., Batelaan, O., Jayatissa, L.P., Koedam, N. (2005). Transitions in ancient inland freshwater resource management in Sri Lanka affect biota and human populations in and around coastal lagoons. *Current Biology* vol. 15, No. 6, pp. 579-86.

Dahdouh-Guebas, F. (2013). Les biens et services écosystémiques: l'exemple des mangroves. In: P. Meerts, ed., *Vers une nouvelle synthèse écologique: de l'écologie scientifique au développement durable*. Centre Paul Duvigneaud de Documentation Ecologique, Bruxelles, Belgique. pp.182-93.

Desbruyères., D., Almeida., A., Biscoito., M., Comtet., T., Khripounoff., A., Le Bris, N., Sarradin, P.M., Segonzac., M. (2000). A review of the distribution of hydrothermal vent communities along the northern Mid-Atlantic Ridge: dispersal vs. environmental controls. *Hydrobiologia*, vol. 440, No. 1-3, p. 201-16.

Dolan, J.R., Ritchie, M.R. and Ras, J. (2007). The "neutral" community structure of planktonic herbivores, tintinnid ciliates of the microzooplankton, across the SE Tropical Pacific Ocean. *Biogeosciences*, 4, 297-310.

Donato, D.C., Kauffma, J.B., Murdiyarso, D., Kurnianto, S., Stidham, M., Kanninen, M. (2011). Mangroves among the most carbon-rich forests in the tropics. *Nature Geoscience*, vol. 4, pp. 293-97.

Eckert, K.L., Wallace, B.P., Frazier, J.G., Eckert, S.A., Pritchard, P.C.H. (2012). *Synopsis of the biological data on the leatherback sea turtle (Dermochelys coriacea)*. US Fish and Wildlife Service, Biological Technical Publication BTP-R4015-2012.

Ellingsen, K.E., and Gray, J.S. (2002). Spatial patterns of benthic diversity: is there a latitudinal gradient along the Norwegian continental shelf? *Journal of Animal Ecology*, vol. 71, pp. 373-89

Ellison, A.M. (2002). Macroecology of mangroves: large-scale patterns and processes in tropical coastal forests. *Trees - Structure and Function*, vol. 16, issue 2-3, pp. 181-94.

Eschmeyer, W. N. (ed) (2014). Catalog of Fishes: Genera, Species, References. (http://research.calacademy.org/research/ichthyology/catalog/fishcatmain.asp). Electronic version 19 May 2014.

Falkowski, P.G., Katz, M.E., Knoll, A.H., Quigg, A., Raven, J.A., Schofield, O., Taylor, F.J.R. (2004). The Evolution of Modern Eukaryotic Phytoplankton. *Science*, vol. 305, no. 5682, pp. 354-60.

Ferretti, F., Worm, B., Britten, G.L., Heithaus, M.R., Lotze, H.K. (2010). Patterns and ecosystem consequences of shark declines in the ocean. Ecology *Letters* vol. 13, 1055-1071.

Froese, R. and Pauly, D. (eds.) (2014) FishBase. World Wide Web electronic publication. www.fishbase.org, version (02/2014).

Grassle, J.F., and Maciolek, N.J. (1992). Deep-Sea Species Richness: Regional and Local Diversity Estimates from Quantitative Bottom Samples. *The American Naturalist*, vol. 139, pp. 313-41.

Griffiths H.J. (2010). Antarctic marine biodiversity – what do we know about the distribution of life in the Southern Ocean? *PLoS ONE*, vol. 5, No. 8, e11683. doi:10.1371/journal.pone.0011683

Halpin, P.N., Read, A.J., Fujioka, E., Best, B.D., Donnelly, B., Hazen, L.J., Kot, C., Urian, K., LaBrecque, E., Dimatteo, A., Cleary, J., Good, C., Crowder, L.B., and Hyrenbach, K.D. (2009). OBIS-SEAMAP The World Data Center for Marine Mammal, Sea Bird, and Sea Turtle Distributions. *Oceanography*, vol. 22, pp. 104-15.

Hoeksema, B. (2007). Delineation of the Indo-Malayan Centre of Maximum Marine Biodiversity: The Coral Triangle. *Biogeography, Time and Place: distributions, Barriers and Islands*, vol. 29, pp. 117–78.

Iken, K., Konar, B., Benedetti-Cecchi, L., Cruz-Motta, J.J., and Knowlton, A. (2010). Large-scale spatial distribution patterns of echinoderms in nearshore rocky habitats. *PLoS ONE*, vol. 5, e13845. doi:10.1371/journal.pone.0013845

IUCN 2013. The IUCN Red List of Threatened Species. Version 2013.2. <http://www.iucnredlist.org>. Downloaded on 21 November 2013.

Kaiser, S., Brandão, S.N., Brix, S., Barnes, D.K.A., Bowden, D., Ingels, J., Leese, F., Linse, K., Schiaparelli, S.,Arango, C., Bax, N., Blazewicz-Paszkowycz, M., Brandt, A., Catarino, A.I., Danis B., David, B., De Ridder, C., Dubois, P., Ellingsen, K.E., Glover, A., Griffiths, H.J., Gutt, J., Halanych, K., Havermans, C., Held, C., Janussen, D., Lörz, A.-N., Pearce, D., Pierrat, B., Riehl, T., Rose, A., Sands, C.J., SoleriMembrives, A., Schüller, M., Strugnell, J., Vanreusel, A., Veit-Köhler, G., Wilson, N., Yasuhara, M. (2013). Pattern, process and vulnerability of Southern Ocean benthos - a decadal leap in knowledge and understanding. *Marine Biology*, vol. 160, pp. 2295-2317. doi 10.1007/s00227-013-2232-6.

Kaschner, K., Tittensor, D.P., Ready, J., Gerrodette, T., and Worm, B. (2011). Current and future patterns of global marine mammal biodiversity. *PLoS ONE*, vol. 6, e19653.

Kaschner, K., Watson, R., Trites, A.W., Pauly, D. (2006). Mapping worldwide distributions of marine mammals using a Relative Environmental Suitability (RES) model. *Marine Ecology Progress Series*, vol. 316, pp. 285-310.

Kerswell, A.P. (2006). Global biodiversity patterns of benthic marine algae. *Ecology*, vol. 87, pp. 2479–88.

Kosobokova, K.N., Hopcroft, R.R., Hirche, H.J. (2011). Patterns of zooplankton diversity through the depths of the Arctic's central basins. *Marine Biodiversity*, vol. 4, pp. 29-50.

Lascelles, B.G., Langham, G.M., Ronconi, R.A., Reid, J.B. (2012). From hotspots to site protection: identifying Marine Protected Areas for seabirds around the globe. *Biological Conservation*, vol. 156, pp. 5–14.

Levin, L.A., Orphan, V.J., Rouse, G.W., Ussler, W., Rathburn, A.E., Cook, G.S., Goffredi, S., Perez, E., Waren, A., Grupe, B., Chadwick, G., Strickrott, B. (2012). A hydrothermal seep on the Costa Rica margin: Middle ground in a continuum of reducing ecosystems. *Proceedings of the Royal Society B* vol. 279, 2580-88. doi: 10.1098/rspb.2012.0205.

Levin, L.A., and Sibuet, M. (2012). Understanding Continental Margin Biodiversity: A New Imperative. *Annual Review of Marine Science*, vol. 4, pp. 79-112.

Lockley, M.G. (1983). A review of brachiopod dominated palaeocommunities from the type Ordovician. *Palaeontology*, vol. 26, pp. 111-45.

McIntyre, A., (2011). *Life in the World's Oceans: Diversity, Distribution, and Abundance*, Wiley.

Massó i Alemán, S., Bourgeois, C., Appeltans, W., Vanhoorne, B., De Hauwere, N., Stoffelen, P., Heughebaert, A., Dahdouh-Guebas, F. (2010). The 'Mangrove Reference Database and Herbarium'. *Plant Ecology and Evolution*, vol. 143, No. 2, pp. 225-32.

Mora, M., Tittensor, D.P., Adl, S., Simpson, A.G.B., and Worm, B. (2011). How many species are there on earth and in the ocean? *PLoS Biology*, vol. 9, e1001127, pp. 1-8.

Mukherjee, N., Sutherland, W.J., Khan, M.N.I., Berger, U., Schmitz, N., Dahdouh-Guebas, F., Koedam, N. (2014). Using expert knowledge and modelling to define mangrove composition, functioning and threats and estimate time-frame for recovery. *Ecology and Evolution*, vol. 4, No. 11, pp. 2247-62.

Nelson, J.S. (1976). *Fishes of the world*. Wiley-Interscience, New York 416 pp. ISBN 0-471-01497-4.

Nelson, J.S. (1984). *Fishes of the world*. John Wiley and Sons, New York. 2nd edition. 523 pp. ISBN: 0-471-86475-7.

Nelson, J.S. (1994). *Fishes of the world*. John Wiley and Sons, Inc. New York. 3rd edition. 600 pp. ISBN: 0-471-54713-1.

Nelson, J.S. (2006). *Fishes of the world*. John Wiley and Sons, Inc. New York. 4th edition. pp. 601.

Obura, D. (2012). The Diversity and Biogeography of Western Indian Ocean Reef-Building Corals. *PLoS ONE*, vol. 7, e45013. doi:10.1371/journal.pone.0045013.

Piepenburg, D., Archambault, P., Ambrose, W.G., Blanchard, A., Bluhm, B.A., Carroll, M.L., Conlan, K.E., Cusson, M., Feder, H.M., Grebmeier, J.M., Jewett, S.C., Lévesque, M., Petryashev, V.V., Sejr, M.K., Sirenko, B.I., Włodarska-Kowalczuk, M. (2011). Towards a pan-Arctic inventory of the species diversity of the macro- and megabenthic fauna of the Arctic shelf seas. *Marine Biodiversity*, vol. 41, pp. 51-70. doi:10.1007/s12526-010-0059-7.

Quisthoudt, K., Schmitz, N., Randin, C.F., Dahdouh-Guebas, F., Robert, E.M.R., Koedam. N. (2012). Temperature variation among mangrove latitudinal range limits worldwide. *TREES: Structure and Function*, vol. 26, pp. 1919-31.

R Development Core Team (2014). R: A language and environment for statistical computing. R Foundation for Statistical Computing, Vienna, Austria. URL http://www.R-project.org/.

Randall, J.E. (1998). Zoogeography of shore fishes of the Indo- Pacific region. *Zoological Studies*, vol. 37, pp. 227–68.

Reaka, M.L., Rodgers, P.J., Kudla, A.U. (2008). Patterns of biodiversity and endemism on Indo-West Pacific coral reefs. *Proceedings of the National Academy of Sciences of the United States of America*, vol. 105, pp. 11474–81.

Reeves, R.R., Stewart, B.S., Clapham, P.J., Powell, J.A. (2002). *Guide to Marine Mammals of the World*. Vol. National Audubon Society, Alfred A. Knopf, New York, 527 pp.

Renaud, P.E., Webb, T.J., Bjørgesæter, A., Karakassis, I., Kedra, M., Kendall, M.A., Labrune, C., Lampadariou, N., Somerfield, P.J , Wlodarska-Kowalczuk, M., Vanden Berghe, E., Claus, S., Aleffi, I.F., Amouroux, J.M., Bryne, K.H., Cochrane, S.J., Dahle, S., Degraer, S., Denisenko, S.G., Deprez, T., Dounas, C., Fleischer, D., Gil J., Grémare, A., Janas, U., Mackie, A.S.Y., Palerud, R., Rumohr, Heye, Sardá, R., Speybroeck, J., Taboada, S., Van Hoey, G., Weslawski, J.M, Whomersley, P., Zettler, M.L. (2007). Continental-scale patterns in benthic invertebrate diversity: insights from the MacroBen database. *Marine Ecology Progress Series*, vol. 382, pp. 239–52.

Rex, M.A., and Etter, R.J. (2010). *Deep-Sea Biodiversity: Pattern and Scale*. Harvard University Press, 366 pp.

Roberts, C. M., McClean, C.J., Veron, J. E.N., Hawkins, J. P., Allen, G. R., McAllister, D.E., Mittermeier, C. G., Schueler, F. W., Spalding, M., Wells, F., Vynne, C., Werner, T.B. (2002). Marine biodiversity hotspots and conservation priorities for tropical reefs. *Science*, vol. 295, pp. 1280–84.

Rodrigues, C. F., Hilário, A., Cunha, M. R. (2013). Chemosymbiotic species from the Gulf of Cadiz (NE Atlantic): distribution, life styles and nutritional patterns. *Biogeosciences*, vol. 10, pp. 2569-81.

Rogers, A.D., Tyler, P.A., Connelly D.P, Copley, J.T, James, R., Larter, R.D., Lins,e K., Mills, R.A., Garabato, A.N., Pancost, R.D., Pearce, D.A., Polunin, N.V.C., German, C.R., Shank, T., Boersch-Supan, P.H., Alker, B.J., Aquilina, A., Bennett, S.A., Clarke, A., Dinley, R.J.J., Graham, A.G.C., Green, D.R.H., Hawkes, J.A., Hepburn, L, Hilario, A., Huvenne, V.A.I., Marsh, L., Ramirez-Llodra E, Reid, W.D.K., Roterman, C.N., Sweeting, C.J., Thatje, S., and Zwirglmaier, K (2012). The Discovery of New Deep-

Sea Hydrothermal Vent Communities in the Southern Ocean and Implications for Biogeography. *PLoS Biology*, vol. 10, No. 1, e1001234.

Rombouts, I., Beaugrand, G., Ibanez, F., Gasparini, S., Chiba, S. and Legendre, L. (2009). Global latitudinal variations in marine copepod diversity and environmental factors. *Proceedings of the Royal Society*, vol. 276, pp.3053-62.

Roy, K., Jablonski, D., Valentine, J.W., Rosenberg, G. (1998). Marine latitudinal diversity gradients: Tests of causal hypotheses. *Proceedings of the National Academy of Sciences of the United States of America*, vol. 95, pp. 3699–3702.

Rutherford S., D'Hondt, S., Prell, W. (1999). Environmental controls on the geographic distribution of zooplankton diversity. *Nature*, vol. 400, pp. 749-53.

Sagarin, R.D., Barry, J.P., Gilman, S.E., Baxter, C.H. (1999). Climate-related change in an intertidal community over short and long time scales. *Ecological Monographs* vol. 69, 465-490.

Schlacher, T.A., Baco, A.R., Rowden, A.A., O'Hara, T.D., Clark, M.R., Kelley, C., Dower, J.F. (2014). Seamount benthos in a cobalt-rich crust region of the central Pacific: conservation challenges for future seabed mining. *Diversity and Distributions*, vol. 20. pp. 491-502.

Sherr, B.F., Sherr, E.B., Caron, D.A., Vaulot, D., Worden, A.Z. (2007).Oceanic Protists. *Oceanography*, vol. 20, No. 2, pp. 130–34.

Smith, C.R., Glover, A.G., Treude, T., Higgs, N.D., Amon, D.J. (2015). Whale-fall ecosystems: recent insights into ecology, paleoecology, and evolution. *Annual Review of Marine Science* vol. 7, 571-96.

Snelgrove, P. V. R., (2010). *Discoveries of the Census of Marine Life: Making Ocean Life Count*, Cambridge University Press.

Spalding, M. D., Melanie, I., Milam, A., Fitzgerald, C., and Hale, L. Z., (2013). Protecting marine spaces: global targets and changing approaches, *Ocean Yearbook*, Volume 237, p. 213-248.

Spotila, J.R. (2004). *Sea Turtles: A Complete Guide to Their Biology, Behavior, and Conservation*. Johns Hopkins University Press, 227 pp.

Stephenson, T.A., Stephenson, A. (1949). The universal features of zonation between the tidemarks on rocky coasts. *Journal of Ecology* vol. 38, 289-305

Stocks, K.I., Hart, P.J.B. (2007). Biogeography and biodiversity of seamounts. In: Pitcher, T.J., Morato, T., Hart, P.J.B., Clark, M.R., Haggan, N., Santos, R.S., eds. *Seamounts: ecology, fisheries, and conservation*. Blackwell Fisheries and Aquatic Resources Series 12. Oxford, UK: Blackwell Publishing. pp 255–81.

Stocks, K.I. (2010). SeamountsOnline: an online information system for seamount biology. Available from http://seamounts.sdsc.edu. Accessed 15 July 2014.

Stuart-Smith R.D., Bates A.E., Lefcheck, J.S., Duffy, J.E., Baker, S.C., Thomson, R.J., Stuart-Smith, J.F., Hill, N.A., Kininmonth, S.J., Airoldi, L., Becerro, M.A., Campbell, S.J., Dawson, T.P., Navarette, S.A., Soler, G.A., Strain, E.M.A., Willis, T.J., Edgar, G.J. (2013). Integrating abundance and functional traits reveals new hotspots of fish diversity. *Nature*, vol. 501, pp. 539-54.

Thompson, R.C., Crowe, T.P., Hawkins, S.J. (2002). Rocky intertidal communities: past environmental changes, present status and predictions for the next 25 years. *Environmental Conservation* vol. 29, 168-191.

Tittensor, D.P., Mora, C., Jetz, W., Lotze, H.K., Ricard, D., Vanden Berghe, E., Worm, B. (2010). Global patterns and predictors of marine biodiversity across taxa. *Nature*, vol. 466, pp.1098–1101. doi:10.1038/nature09329.

Vanden Berghe, E., Stocks, K.I., and Grassle, J.F. (2010). Data integration: The Ocean Biogeographic Information System. In: Mcintyre, A.D., ed. *Life in the world's oceans: diversity, distribution, and abundance*, pp. 333-353.

Wallace, B.P., DiMatteo, A.D., Hurley, B.J., Finkbeiner, E.M., Bolten, A.B., Chaloupka, M.Y., Hutchinson, B.J., Abreu-Grobois, F.A., Amorocho, D., Bjorndal, K.A., Bourjea, J., Bowen, B.W., Dueñas, R.B., Casale, P., Choudhury, B.C., Costa, A., Dutton, P.H., Fallabrino, A., Girard, A., Girondot, M., Godfrey, M.H., Hamann, M., López-Mendilaharsu, M., Marcovaldi, M.A., Mortimer, J.A., Musick, J.A., Nel, R., Pilcher, N.J., Seminoff, J.A., Troëng, S., Witherington, B., Mast, R.B. (2010). Regional Management Units for marine turtles: A novel framework for prioritizing conservation and research across multiple scales. *PLoS ONE*, vol. 5, No. 12, e15465. doi:10.1371/journal.pone.0015465.

Wallace, B.P., DiMatteo, A.D., Bolten, A.B., Chaloupka, M.Y., Hutchinson, B.J., Abreu-Grobois, F.A., Mortimer, J.A., Seminoff, J.A., Amorocho, D., Bjorndal, K.A., Bourjea, J., Bowen, B.W., Dueñas, R.B., Casale, P., Choudhury, B.C., Costa, A., Dutton, P.H., Fallabrino, A., Finkbeiner, E.M., Girard, A., Girondot, M., Hamann, M., Hurley, B.J., López-Mendilaharsu, M., Marcovaldi, M.A., Musick, J.A., Nel, R., Pilcher, N.J., Troëng, S., Witherington, B., Mast, R.B.(2011). Global Conservation Priorities for Marine Turtles. *PLoS ONE*, vol. 6, No. 9, e24510. doi:10.1371/journal.pone.0024510.

Watling, L., Guinotte, J., Clark, M.R., Smith, C.F. (2013). A proposed biogeography of the deep sea. *Progress in Oceanography*, vol. 111, pp. 91-112.

Wickham, H (2009). ggplot2: Elegant graphics for data analysis. Springer New York.

Wiebe, P.H., Bucklin, A., Madin, L.P., Angel, M.V., Sutton, T., Pagés, F., Hopcroft, R.R., Lindsay, D. (2010). Deep-sea sampling on CMarZ cruises in the Atlantic Ocean – an Introduction. *Deep-Sea Research II*, vol. 57, pp. 2157-2166.

Wilson, R.R., and Kaufmann, R.S. (1987). Seamount biota and biogeography. In: Keating, B.H., Fryer, P., Batiza, R., Boehlert, G.W., eds., *Seamounts, islands and atolls*. Washington DC, USA: American Geophysical Union. pp. 319–334.

Witman, J.D., Cusson, M., Archambault, P., Pershing, A.J., Mieskowska, N. (2008). The relation between productivity and species diversity in temperate- arctic marine ecosystems. *Ecology*, vol. 89, pp. S66-S80.

Worm, B., and Lotze, H.K. (2009). Changes in marine biodiversity as an indicator of climate change. In: Letcher, T., ed., *Climate change: observed impacts on planet Earth*. Elsevier.

WoRMS Editorial Board (2013). *World Register of Marine Species*. VLIZ. Available from http://www.marinespecies.org. Accessed 21 November 2013.

Appendix. Additional reading on marine biodiversity patterns

Mammals

Anderson, P.K. (1995). Competition, predation, and evolution and extinction of Steller's sea cow, *Hydrodamalis gigas*. *Marine Mammal Science* 11:391-394

Bailleul, F., Charrassin, J.-B., Monestiez, P., Roquet, F., Biuw, M., Guinet, C. (2007). Successful foraging zones of southern elephant seals from the Kerguelen Islands in relation to oceanographic conditions. *Philosophical Transactions of the Royal Society of London B: Biological Sciences* 362:2169-2181

Balance, L.T., Pitman, R.L., Fiedler, P.C. (2006). Oceanographic influences on seabirds and cetaceans of the eastern tropical Pacific: A review. *Progress in Oceanography* 69:360-390

Biuw, M., Boehme, L., Guinet, C., Hindell, M., Costa, D., Charrassin, J.B., Roquet, F., Bailleul, F., Meredith, M., Thorpe, S., Tremblay, Y., McDonald, B., Park, Y.H., Rintoul, S.R., Bindoff, N., Goebel, M., Crocker, D., Lovell, P., Nicholson, J., Monks, F., Fedak, M.A. (2007). Variations in behavior and condition of a Southern Ocean top predator in relation to in situ oceanographic conditions. *Proceedings of the National Academy of Sciences*. USA 104:13705-13710

Branch, T.A., Stafford, K.M., Palacios, D.M., Allison, C., Bannister, J.L., Burton, C.L.K., Cabrera, E., Carlson, C.A., Vernazzani, B.G., Gill, P.C., Hucke-Gaete, R., Jenner, K.C.S., Jenner, M.N.M., Matsuoka, K., Mikhalev, Y.A., Miyashita, T., Morrice, M.G., Nishiwaki, S., Sturrock, V.J., Tormosov, D., Anderson, R.C., Baker, A.N., Best, P.B., Borsa, P., Brownell, R.L., Jr., Childerhouse, S., Findlay, K.P., Gerrodette, T., Ilangakoon, A.D., Joergensen, M., Kahn, B., Ljungblad, D.K., Maughan, B., McCauley, R.D., McKay, S., Norris, T.F., Group OWADR, Rankin, S., Samaran, F., Thiele, D., Van Waerebeek, K., Warneke, R.M. (2007). Past and present distribution, densities and movements of blue whales *Balaenoptera musculus* in the Southern Hemisphere and northern Indian Ocean. *Mammal Review* 37:116-175

Dalebout, M.L., Mead, J.G., Baker, C.S., Baker, A.N., van Helden, A.L. (2002). A new species of beaked whale *Mesoplodon perrini* sp. n. (Cetacea: Ziphiidae) discovered through phylogenetic analysis of mitochondrial DNA sequences. *Marine Mammal Science* 18:577-608

Davis, R.W., Ortega-Ortiz, J.G., Ribic, C.A., Evans, W.E., Biggs, D.C., Ressler, P.H., Cady, R.B., Leben, R.R., Mullin, K.D., Würsig, B. (2002). Cetacean habitat in the northern oceanic Gulf of Mexico. *Deep Sea Research (Part I): Oceanographic Research Papers* 49:121-142

Gilles, A., Adler, S., Kaschner, K., Scheidat, M., Siebert, U. (2011). Modelling harbour porpoise seasonal density as a function of the German Bight environment: implications for management. *Endangered Species Research* 14:157-169

Gilles, A., Scheidat, M., Siebert, U. (2009). Seasonal distribution of harbour porpoises and possible interference of offshore wind farms in the German North Sea. *Marine Ecology Progress Series* 383:295-307

Gregr, E.J., Coyle, K.O. (2009). The biogeography of the North Pacific right whale (*Eubalaena japonica*). *Progress in Oceanography* 80:188-198

Kaschner, K., Quick, N., Jewell, R., Williams, R., Harris, C.M. (2012). Global coverage of cetacean line-transect surveys: status quo, gaps and future challenges. *Plos One* 7:13

Kot, C.Y., Fujioka, E., Hazen, L.J., Best, B.D., Read, A.J., Halpin, P.N. (2010). Spatio-temporal gap analysis of OBIS-SEAMAP project data: assessment and way forward. Plos One 5:e12990

Kraus, S.D., Rolland ,R.M. (2007). The Urban Whale: North Atlantic Right Whales at the Crossroads, Vol. Harvard University Press, Cambridge, MA

MacLeod, C.D., Mitchell, G. (2006). Key areas for beaked whales worldwide. *Journal of Cetacean Research & Management* 7:309-322

Mattila, D.K., Clapham, P.J., Vasquez, O., Bowman, R.S. (1994). Occurrence, population composition, and habitat use of humpback whales in Samana Bay, Dominican Republic. *Canadian Journal of Zoology* 72:1898-1907

Schick, R.S., Halpin, P.N., Read, A.J., Slay, C.K., Kraus, S.D., Mate, B.R., Baumgartner, M.F., Roberts, J.J., Best, B.D., Good, C.P., Loarie, S.R., Clark, J.S. (2009). Striking the right balance in right whale conservation. *Canadian Journal of Fisheries and Aquatic Sciences* 66:1399-1403

Schipper, J., Chanson, J.S., Chiozza, F., Cox, N.A., Hoffmann, M., Katariya, V., Lamoreux, J., Rodrigues, A.S.L., Stuart, S.N., Temple, H.J., Baillie, J., Boitani, L., Jr. T.E.L., Mittermeier, R.A., Smith, A.T., Absolon, D., Aguiar, J.M., Amori, G., Bakkour, N., Baldi, R., Berridge, R.J., Bielby, J., Black, P.A., Blanc, J.J., Brooks, T.M., Burton, J.A., Butynski, T.M., Catullo, G., Chapman, R., Cokeliss, Z., Collen, B., Conroy, J., Cooke, J.G., Fonseca, G.A.B., Derocher, A.E., Dublin, H.T., Duckworth, J.W., Emmons, L., Emslie, R.H., Festa-Bianchet, M., Foster, M., Foster, S., Garshelis, D.L., Gates, C., Gimenez-Dixon, M., Gonzalez, S., Gonzalez-Maya, J.F., Good, T.C., Hammerson, G., Hammond, P.S., Happold, D., Happold, M., Hare, J., Harris, R.B., Hawkins, C.E., Haywood, M., Heaney, L.R., Hedges, S., Helgen, K.M., Hilton-Taylor, C., Hussain, S.A., Ishii, N., Jefferson, T.A., Jenkins, R.K.B., Johnston, C.H., Keith, M., Kingdon, J., Knox, D.H., Kovacs, K.M., Langhammer, P., Leus, K., Lewison, R., Lichtenstein, G., Lowry, L.F., Macavoy, Z., Mace, G.M., Mallon, D.P., Masi, M., McKnight, M.W., Medellín, R.A., Medici, P., Mills, G., Moehlman, P.D., Molur, S., Mora, A., Nowell, K., Oates, J.F., Olech, W., Oliver, W.R.L., Oprea, M., Patterson, B.D., Perrin, W.F., Polidoro, B.A., Pollock, C., Powel, A., Protas, Y., Racey, P., Ragle, J., Ramani, P., Rathbun, G., Reeves, R.R., Reilly, S.B., III J.E.R., Rondinini, C., Rosell-Ambal, R.G., Rulli, M., Rylands, A.B., Savini, S., Schank, C.J., Sechrest, W., Self-Sullivan, C., Shoemaker, A., Sillero-Zubiri, C., Silva, N.D., Smith, D.E., Srinivasulu, C., Stephenson, P.J., Strien, N.V., Talukdar, B.K., Taylor, B.L., Timmins, R., Tirira, D.G., Tognelli, M.F., Tsytsulina, K., Veiga, L.M., Vié, J.-C., Williamson, E.A., Wyatt, S.A., Xie, Y., Young, B.E. (2008). The status of the world's land and marine mammals: Diversity, threat, and knowledge. *Science*:225-230

Smith, T.D., Reeves, R.R., Josephson, E.A., Lund, J.N. (2012). Spatial and seasonal distribution of American whaling and whales in the age of sail. *Plos One* 7:e34905

SMM Committee on Taxonomy (2012). List of marine mammal species and subspecies, Society for Marine Mammalogy, www.marinemammalscience.org, Consulted on: 28/08/2013.

Stevick, P.T., McConnell, B.J., Hammond, P.S. (2002). Patterns of Movement. In: Hoelzel AR (ed) Marine Mammal Biology - An Evolutionary Approach. Blackwell Science Ltd, Oxford, Malden, MA, p 185-216

Stockin, K.A., Pierce, G.J., Binedell, V., Wiseman, N., Orams, M.B. (2008). Factors affecting the occurrence and demographics of common dolphins (*Delphinus sp.*) in the Hauraki Gulf, New Zealand. *Aquatic Mammals* 34:200-211

Wada S, Oishi M, Yamada TK (2003). A newly discovered species of living baleen whale. *Nature* 426:278-281

Cephalopods

O'Dor, R.K., (2012). The incredible flying squid. *New Scientist* 2865: 39-41)

Microbes

Chisholm, S.W., Falkowski, P.G. and Cullen, J.J. (2001). Dis-crediting ocean fertilization. *Science* 294: 309-310

Doney, S.C., Fabry, V.J., Feely, R.A., Kleypas, J.A. (2009). Ocean acidification: The other CO(2) problem. *Annual Review of Marine Science* 1, 169-192.

Field, C.B., Behrenfeld, M.J., Randerson, J.T., and Falkowski, P. (1998). Primary production of the biosphere: Integrating terrestrial and oceanic components. *Science* 281:237-240.

Hoegh-Guldberg, O., Bruno, J.F. (2010). Marine ecosystems. The impact of climate change on the world's marine ecosystems. *Science* 328, 1523-1524

Jessup, D.A. et al. (2009). Mass stranding of marine birds caused by surfactant-producing red tide. *PLoS One*. 4: e4550.

Landsberg, J.H. (2002) The effects of harmful algal blooms on aquatic organisms. *Reviews in Fisheries Science*. 10: 113-390.

Pollard, R.T. et al. 2009. Southern Ocean deep-water carbon export enhanced by natural iron fertilization. *Nature* pg: 577-580.

Ray, G.C., Grassle, J.F. (1991) Marine Biological Diversity Program. *BioScience* 41 (7), 453-457.

Rost, B., Zondervan, I., Wolf-Gladrow, D. (2008) Sensitivity of phytoplankton to future changes in ocean carbonate chemistry: current knowledge, contradictions and research directions. *Marine Ecology Progress Series* 373, 227-237.

Sherr, E. and Sherr, B. 2009. Capacity of herbivorous protists to control initiation and development of mass phytoplankton blooms. *Aquatic Microbial Ecology* 57:253-262.

Shumway, S.E., Allen, S.M., Boersma, P.D. (2003) Marine birds and harmful algal blooms: sporadic victims or under-reported events? *Harm. Algae*. 2: 1-17.

Strom, S.L. (2008). Microbial ecology of ocean biogeochemistry: A community perspective. *Science* 320:1043-1045.

Van Dolah, F.M., Roelke, D., Greene, R.M. (2001) Health and ecological impacts of harmful algal blooms: risk assessment needs. *Human and Ecological Risk Assessment: An International Journal*. 7: 1329-1345.

Winder, M., Sommer, U., 2012. Phytoplankton response to a changing climate. *Hydrobiologia* 698 (1), 5-16.

Wolf-Gladrow, D.A., Riebesell, U., Burkhardt, S., Bijma, J., 1999. Direct effects of CO2 concentration on growth and isotopic composition of marine plankton. *Tellus B* 51 (2), 461-476.

Rock shores

Blanchette, C.A., Wieters, E.A., Broitman, B.R., Kinlan, B.P., Schiel, D.R. (2009) Trophic structure and diversity in rocky intertidal upwelling ecosystems: A comparison of community patterns across California, Chile, South Africa and New Zealand. *Progress in Oceanography* 83: 107-116.

Crain, C.M., Kroeker, K., Halpern, B.S. (2008) Interactive and cumulative effects of multiple human stressors in marine systems. *Ecology Letters* 11: 1304-1315.

Cruz-Motta, J.J., Miloslavich, P., Palomo, G., Iken, K., Konar, B., et al. (2010). Patterns of spatial variation of assemblages associated with intertidal rocky shores: A global perspective. PLoS ONE 5(12): e14354. doi:10.1371/journal.pone.0014354

Darling, E.S., Côté, I.M. (2008) Quantifying the evidence for ecological synergies. *Ecology Letters* 11: 1278-1286.

Helmuth, B., Mieszkowska, N., Moore, P., Hawkins, S.J. (2006) Living on the edge of two changing worlds: Forecasting the responses of rocky intertidal ecosystems to climate change. *Annual Review of Ecology, Evolution, and Systematics* 37: 373-404.

Konar, B., Iken, K., Cruz-Motta, J.J., Benedetti-Cecchi, L., Knowlton, A., et al. (2010) Current patterns of macroalgal diversity and biomass in northern hemisphere rocky shores. *PLoS ONE* 5(10): e13195. doi:10.1371/journal.pone.0013195

Peterson, C.H., Lubchenco, J. (1997). Marine ecosystem services. In: Daily G (ed) Nature's services: societal dependence on natural ecosystems. Island Press, Washington, DC. Pp 177-194.

Pohle, G., Iken, K., Clarke, K.R., Trott, T., Konar, B., et al. (2011) Aspects of benthic decapod diversity and distribution from rocky nearshore habitat at geographically widely dispersed sites. *PLoS ONE* 6(4): e18606. doi:10.1371/journal.pone.0018606

Thompson, R.C., Crowe, T.P., Hawkins, S.J. (2002) Rocky intertidal communities: past environmental changes, present status and predictions for the next 25 years. *Environment Conservation* 29: 168-191.

Underwood, A.J. (1992) Beyond BACI: the detection of environmental impacts on populations in the real, but variable, world. *Journal of Experimental marine biology and Ecology* 161: 145-178.

Underwood, A.J. (1994) On beyond BACI - sampling designs that might reliably detect environmental disturbances. *Ecological Applications* 4: 3-15.

Coral Reefs

Barott, K.L., Rodriguez-Brito, B., Janouškovec, J., Marhaver, K.L., Smith, J.E., Keeling, P., Rohwer, F.L. (2011) Microbial diversity associated with four functional groups of benthic reef algae and the reef-building coral Montastraea annularis. *Environmental Microbiology*, 13: 1192–1204. doi: 10.1111/j.1462-2920.2010.02419.x

Bellwood, D.R., Hughes, T.P. (2001). Regional-scale assembly rules and biodiversity of coral reefs. *Science* 292: 1532–1534.

Bruno, J.F., Selig, E.R. (2007) Regional decline of coral cover in the Indo-Pacific: timing, extent, and subregional compari- sons. *PLoS ONE*, 2, e711.

Budd, A.F., Romano, S.L., Smith, N., Barbeitos, M.S. (2010). Rethinking the Phylogeny of Scleractinian Corals: A Review of Morphological and Molecular Data. Integrative and Comparative Biology 50: 411–427. Available: http://icb.oxfordjournals.org/cgi/doi/10.1093/icb/icq062.

Burke, L., Reytar, K., Spalding, M.D., Perry, A. (2011). Reefs at Risk-hires. World Resources Institute. 130 pp.

Darwin, C.R. (1842). The structure and distribution of coral reefs. Being the first part of the geology of the voyage of the Beagle, under the command of Capt. Fitzroy, R.N. during the years 1832 to 1836. London: Smith Elder and Co.

Fukami, H., Chen, C.A., Budd, A.F., Collins, A., Wallace, C., et al. (2008). Mitochondrial and Nuclear Genes Suggest that Stony Corals Are Monophyletic but Most Families of Stony Corals Are Not (Order Scleractinia, Class Anthozoa, Phylum Cnidaria). *PLoS ONE* 3: e3222. doi:10.1371/journal.pone.0003222.t002.

Hopley, D. (2009). Encyclopedia of Modern Coral Reefs - Structure, Form and Process. James Cook University, Townsville, Australia. Springer. 1250 pp.

IPCC (2007). Synthesis Report. Contribution of Working Groups I, II and III to the Fourth Assessment Report of the Intergovernmental Panel on Climate Change. Geneva: Intergovernmental Panel on Climate Change.

IUCN (2013). The IUCN Red List of Threatened Species. Version 2013.2. <http://www.iucnredlist.org>. Downloaded on 21 November 2013.

Mora, C., Aburto-Oropeza, O., Ayala Bocos, A., Ayotte, P.M., Banks, S., et al. (2011). Global human footprint on the linkage between biodiversity and ecosystem functioning in reef fishes. *PLOS Biology* 9: e1000606. doi:10.1371/journal.

pbio.1000606.

Renema, W., Bellwood, D.R., Braga, J.C., Bromfield, K., Hall, R., et al. (2008). Hopping Hotspots: Global Shifts in Marine Biodiversity. *Science* 321: 654–657. doi:10.1126/science.1155674.

Sheppard, C.R.C. (1987). Coral species of the Indian Ocean and adjacent seas: a synonymized compilation and some regional distributional patterns. *Atoll Research Bulletin*, Vol. 307, pp. 1–33.

Spalding, M.D., Fox, H.E., Allen, G.R., Davidson, N., Ferdana, Z.A., Finlayson, M., Halpern, B.S., Jorge, M.A., Lombana, A.L., Lourie, S.A. (2007). Marine ecoregions of the world: a bioregionalization of coastal and shelf areas. *BioScience*, Vol. 57m, pp. 573–583.

Veron, J.E.N. (2000). Corals of the World. Townsville, Australia: Australian Institute of Marine Science.

Wafar, M., Venkataraman, K., Ingole, B., Khan, S.A., Loka Bharathi, P. (2011). State of Knowledge of Coastal and Marine Biodiversity of Indian Ocean Countries. *PLoS ONE*, Vol. 6(1).

Wilkinson, C. (2004). Status of Coral Reefs of the World. Global Coral Reef Monitoring Network/International Coral Reef Initiative.

Wilson, M., Rosen, B. (1998). Implications of paucity of corals in the Paleogene of SE Asia: plate tectonics or Centre of Origin? In: Hall R, Holloway J, editors. Backhuys Publishers, Leiden, The Netherlands: Biogeography and Geological Evolution of SE Asia. pp. 165–195.

Mangrove Forests

Barbier, E.B., Koch, E.W., Silliman, B.R., Hacker, S.D., Wolanski, E., Primavera, J., Granek, E.F., Polasky, S., Aswani, S., Cramer, L.A., Stoms, D.M., Kennedy, C.J., Bael, D., Kappel, C.V., Perillo, G.M.E., Reed, D.J. (2008). Coastal ecosystem–based management with nonlinear ecological functions and values. *Science* 319: 321-323.

Cannicci, S., Burrows, D., Fratini, S., Lee, S.Y., Smith, T.J. III, Offenberg, J., Dahdouh-Guebas, F. (2008). Faunal impact on vegetation structure and ecosystem function in mangrove forests: a review. *Aquatic Botany* 89: 186-200.

Duke, N.C., Meynecke, J.-O., Dittmann, S., Ellison, A.M., Anger, K., Berger, U., Cannicci, S., Diele, K., Ewel, K.C., Field, C.D., Koedam, N., Lee, S.Y., Marchand, C., Nordhaus, I., Dahdouh-Guebas, F. (2007). A world without mangroves? *Science* 317: 41-42.

Hutchison, J., Manica, A., Swetnam, R., Balmford, A., Spalding, M. (2013) Predicting global patterns in mangrove forest biomass. *Conservation Letters* 7(3): 233–240.

Lee, S.Y., Primavera, J.H., Dahdouh-Guebas, F., McKee, K., Bosire, J.O., Cannicci, S., Diele, K., Fromard, F., Koedam, N., Marchand, C., Mendelssohn, I., Mukherjee, N., Record, S. (2014). Ecological role and services of tropical mangrove ecosystems: A reassessment. *Global Ecology and Biogeography* 23: 726–743.

Nagelkerken, I., Blaber, S., Bouillon, S., Green, P., Haywood, M., Kirton, L.G., Meynecke, J.-O., Pawlik, J., Penrose, H.M., Sasekumar, A., Somerfield, P.J. (2008). The habitat function of mangroves for terrestrial and marina fauna: a review. *Aquatic Botany* 89(2): 155-185.

Coastal Sediments

Griffiths, C.L., Robinson, T.B., Lange, L., Mead, A. (2010). Marine Biodiversity in South Africa: An Evaluation of Current States of Knowledge. *PLoS ONE* 5: e12008. doi:10.1371/journal.pone.0012008

Hedges, J.I., Keil, R.G. (1995). Sedimentary organic matter preservation: an assessment and speculative synthesis. *Marine Chemistry* 49:81-115

Miloslavich, P., Klein, E., Díaz, J.M., Hernández, C.E., Bigatti, G., et al. (2011). Marine Biodiversity in the Atlantic and Pacific Coasts of South America: Knowledge and Gaps. *PLoS ONE* 6: e14631. doi:10.1371/journal.pone.0014631

Snelgrove, P.V.R., Butman, C.A. (1994). Animal-sediment relationships revisited: Cause versus effect Oceanogr. *Mar. Biol. Ann. Rev.* 32: 111-177.

Deep-Sea Sediments

Bailey, D.M., Collins, M.A., Gordon, J.D.M., Zuur, A.F., Priede, I.G. (2009). Long-term changes in deep-water fish populations in the northeast Atlantic: a deeper reaching effect of fisheries? *Proc. R. Soc. B Biol. Sci.* 276, 1965–1969.

Brandt, A., De Broyer, C., Ebbe, B., Ellingsen, K.E., Gooday, A.J., Janussen, D., Kaiser, S., Linse, K., Schueller, M., Thomson, M.R.A., Tyler, P.A., Vanreusel, A. (2012). Southern Ocean Deep Benthic Biodiversity, in: Rogers, A.D., Johnston, N.M., Murphy, E.J., Clarke, A. (Eds.), Antarctic Ecosystems. John Wiley & Sons, Ltd, pp. 291–334.

Brandt, A., Linse, K., Schüller, M. (2009). Bathymetric distribution patterns of Southern Ocean macrofaunal taxa: Bivalvia, Gastropoda, Isopoda and Polychaeta. *Deep Sea Res.* Part Ocean. Res. Pap. 56, 2013–2025.

Carney, R.S. (2005). Zonation of deep-sea biota on continental margins. Ocean. *Mar. Biol. Annu. Rev.* 43, 211–279.

Danovaro, R., Gambi, C., Dell'Anno, A., Corinaldesi, C., Fraschetti, S., Vanreusel, A., Vincx, M., Gooday A. J. (2008). Exponential decline of deep-sea ecosystem functioning linked to benthic biodiversity loss. *Current Biology* 18: 1-8.

Danovaro, R., Canals, M., Gambi, C., Heussner, S., Lampadariou, N., Vanreusel, A. (2009) Exploring Benthic Biodiversity Patterns and Hotspots on European Margin Slopes *Oceanography* 22: 16-25.

De Leo, F.C., Smith, C.R., Rowden, A.A., Bowden, D.A., Clark, M.R. (2010). Submarine canyons: hotspots of benthic biomass and productivity in the deep sea. *Proc. R. Soc. B* 277: 2783-92

Devine, J.A., Baker, K.D., Haedrich, R.L. (2006). Fisheries: Deep-sea fishes qualify as endangered. *Nature* 439, 29–29.

Fosså, J.H., Mortensen, P.B., Furevik, D.M. (2002). The deep-water coral Lophelia pertusa in Norwegian waters: distribution and fishery impacts. *Hydrobiologia* 471, 1–12.

Helly, J., Levin, L.A. (2004). Global distribution of naturally occurring marine hypoxia on continental margins. *Deep Sea Research*. 51: 1159-1168.

Herndl, G.M.J., Reinthaler, T., Teira, E., van Aken, H., Veth, C., Pernthaler, A., Pernthaler, J. (2005). 2309 Contribution of Archaea to Total Prokaryotic Production in the Deep Atlantic Ocean Appl. *Environ. Microbiol.* 71:2303-

Levin, L.A. (2003). Oxygen minimum zone benthos: Adaptation and community response to hypoxia. Oceanography and Marine Biology: An Annual Review *41*: 1-45. (2003)

Levin, L.A., Ron, J.E., Rex, M.A., Gooday, A.J., Smith, C.R., Pineda, J., Stuart, C.T., Hessler, R.R., Pawson, D. (2001). Environmental influences on regional deep-sea species diversity. *Annu. Rev. Ecol. Syst.* 32, 51–93.

McClain, C.R., Allen, A.P., Tittensor, D.P., Rex, M.A. (2012). Energetics of life on the deep seafloor. *Proceedings of the National Academy of Sciences* 109, 15366–15371.

Menot, L., Sibuet, M., Carney, R.S., Levin, L.A., Rowe, G.T., Billett, D.S.M., Poore, G., Kitazato, H., Vanreusel, A., Galéron, J., Lavrado, H.P., Sellanes, J., Ingole, B., Krylova, E. (2010). New Perceptions of Continental Margin Biodiversity, in: McIntyre, A.D. (Ed.), Life in the World's Oceans. Wiley-Blackwell, pp. 79–102.

Ramirez-Llodra, E., Brandt, A., Danovaro, R., De Mol, B., Escobar, E., German, C.R.,

Levin, L.A., Martinez Arbizu, P., Menot, L., Buhl-Mortensen, P., Narayanaswamy, B.E., Smith, C.R., Tittensor, D.P., Tyler, P.A., Vanreusel, A., Vecchione, M. (2010). Deep, diverse and definitely different: unique attributes of the world's largest ecosystem. *Biogeosciences* 7, 2851–2899.

Ruhl, H.A., Ellena, J.A., Smith, K.L. (2008). Connections between climate, food limitation, and carbon cycling in abyssal sediment communities. *Proceedings of the National Academy of Sciences* 105, 17006–17011.

Smith, C.R., De Leo, F.C., Bernardino, A.F., Sweetman, A.K., Arbizu, P.M. (2008). Abyssal food limitation, ecosystem structure and climate change. Trends Ecol. Evol. 23, 518–528.

Smith, K.L., Ruhl, H.A., Bett, B.J., Billett, D.S.M., Lampitt, R.S., Kaufmann, R.S. (2009). Climate, carbon cycling, and deep-ocean ecosystems. *Proceedings of the National Academy of Sciences* U. S. A. 106, 19211–19218.

Snelgrove, P.V.R., Smith, C.R. (2002). A riot of species in an environmental calm: The paradox of the species-rich deep-sea floor. Ocean. *Marine Biology*. 40, 311–342.

Somero, G.N. (1992). Biochemical ecology of deep-sea animals. *Experientia* 48, 537–543.

Stegeman, J.J., Kloepper-Sams, P.J., Farrington, J.W. (1986). Monooxygenase induction and chlorobiphenyls in the deep-sea fish *Coryphaenoides armatus*. *Science* 231, 1287–1289.

Stramma, L., Schmidt, S., Levin, L.A., Johnson, G.C. (2010). Ocean oxygen minima expansions and their biological impacts. *Deep-Sea Research* I 210: 587-595. (2010)

Tittensor, D.P., Rex, M.A., Stuart, C.T., McClain, C.R., Smith, C.R. (2011). Species–energy relationships in deep-sea molluscs. *Biology Letters*. 7, 718–722.

Tyler, P., Amaro, T., Arzola, R., Cunha, M.R., de Stigter, H., Gooday, A., Huvenne, V., Ingels, J., Kiriakoulakis, K., Lastras, G., Masson, D., Oliveira, A., Pattenden, A., Vanreusel, A., Van Weering, T., Vitorino, J., Witte, U., Wolff, G. (2009). Europe's Grand Canyon: Nazaré Submarine Canyon. *Oceanography*, 22, (1), 46-57.

Yakimov, M.M., Cono, V.L., Smedile, F., DeLuca, T.H., Juárez, S., Ciordia, S., Fernández, M., Albar, J.P., Ferrer, M., Golyshin, P.N., Giuliano, L. (2011). Contribution of crenarchaeal autotrophic ammonia oxidizers to the dark primary production in Tyrrhenian deep waters (Central Mediterranean Sea). ISME J. 2011 Jun;5(6):945-61.

Yasuhara, M., Hunt, G., van Dijken, G., Arrigo, K.R., Cronin, T.M., Wollenburg, J.E. (2012). Patterns and controlling factors of species diversity in the Arctic Ocean. *Journal of Biogeography*. 39, 2081–2088.

Seamounts

Castelin, M., Puillandre, N., Lozouet, P., Sysoev, A., Richer de Forges, B. et al. (2011). Molluskan species richness and endemism on New Caledonian seamounts: are they enhanced compared to adjacent slopes? *Deep Sea Research* I 58: 637–646.

Cho, W., Shank, T.M. (2010). Incongruent patterns of genetic connectivity among four ophiuroid species with differing coral host specificity on North Atlantic seamounts. *Marine Ecology* 31 (Suppl. 1): 121–143.

Clark, M.R. (2009). Seamounts: Biology. Pp. 818–821 In: Gillespie, R.G. & Clague, D.A. (eds). *Encyclopaedia of Islands* University of California Press. 1074p.

Clark, M.R., Watling, L., Rowden, A.A., Guinotte, J.M., Smith, C.R. (2011). A global seamount classification to aid the scientific design of marine protected area networks. *Ocean and Coastal Management* 54: 19–36.

Howell, K.L., Mowles, S.L., Foggo, A. (2010). Mounting evidence: near-slope seamounts are faunally indistinct from an adjacent bank. *Marine Ecology* 31 (suppl. 1): 52–62.

Kim, S.-S., Wessel, P. (2011). New global seamount census from altimetry-derived gravity data. *Geophysical Journal International* 186: 615–631.

Kitchingman, A., Lai, S. (2004). Inferences on potential seamount locations from mid resolution bathymetric data. In: Morato T, Pauly D, editors. Seamounts: Biodiversity and Fisheries. Vancouver, BC, Canada: Fisheries Centre, University of British Columbia. pp 7-12.

McClain, C.R., Lundsten L, Barry J, DeVogelaere A. (2010). Bathymetric patterns in diversity, abundance and assemblage structure on a northeast Pacific seamount. *Marine Ecology* 31 (suppl 1): 14–25.

McClain, C.R., Lundsten L, Ream M, Barry J, De Vogelaere A (2009). Endemicity, biogeography, composition and community structure on a northeast Pacific seamount. *PLoS ONE* 4:e4141, 8pp.

Miller, K., Williams A, Rowden AA, Knowles C, Dunshea G (2010). Conflicting estimates of connectivity among deep-sea coral populations. *Marine Ecology* 31 (suppl 1):144–157.

Morato, T., Clark, M.R. (2007). Seamount fishes: ecology and life histories. In: Pitcher, T.J., Morato, T., Hart, P.J.B., Clark, M.R., Haggan, N., Santos, R.S., editors. Seamounts: ecology, fisheries, and conservation. Blackwell Fisheries and Aquatic Resources Series 12. Oxford, UK: Blackwell Publishing. pp. 170–188.

O'Hara, T.D., Consalvey, M., Lavrado, H.P., Stocks, K.I. (2010). Environmental predictors and turnover of biota along a seamount chain. *Marine Ecology* 31 (suppl 1): 84–94.

Parin, N.V., Mironov, A.N., Nesis, K.N. (1997). Biology of the Nazca and Sala y Gomez submarine ridges, and outpost of the Indo-West Pacific fauna in the Eastern Pacific Ocean: Composition and distribution of the fauna, its communities and history. *Advances in Marine Biology* 32:145–242.

Raymore, P.A. (1982). Photographic investigations on three seamounts in the Gulf of Alaska. *Pacific Science* 361: 15–34.

Richer de Forges, B., Koslow, J.A., Poore, G.C.B. (2000). Diversity and endemism of the benthic seamount fauna in the southwest Pacific. *Nature* 405: 944-947.

Rowden, A.A., Dower, J.F., Schlacher, T.A., Consalvey M, Clark MR (2010). Paradigms in seamount ecology: fact, fiction, and future. *Marine Ecology* 31 (suppl 1): 226–239.

Rowden, A.A., Schlacher, T.A., Williams, A., Clark, M.R., Stewart, R. et al. (2010). A test of the seamount oasis hypothesis: seamounts support higher epibenthic megafaunal biomass than adjacent slopes. *Marine Ecology* 31 (suppl 1): 95–106.

Rowden, A.A., Schnabel, K.E., Schlacher, T.A., Macpherson, E., Ahyong, S.T. et al. (2010). Squat lobster assemblages on seamounts differ from some, but not all, deep-sea habitats of comparable depth. *Marine Ecology* 31 (suppl 1): 63–83.

Stocks, K.I., Clark, M.R., Rowden, A.A., Consalvey, M. (2012). CenSeam, and international program on seamounts within the Census of Marine Life: achievements and lessons learned. *PLoS ONE 7(2):* doi:10.1371/journal.pone.0032031

Thoma, J.N., Pante, E., Brugler, M.R., France, S.C. (2009). Deep-sea octocorals and antipatharians show no evidence of seamount scale endemism in the NW Atlantic. *Marine Ecology Progress Series* 397: 25–35.

Tracey, D.M., Bull, B., Clark, M.R., MacKay, K.A. (2004). Fish species composition on seamounts and adjacent slope in New Zealand waters. *New Zealand Journal of Marine and Freshwater Research* 38: 163–182.

UNESCO (2009). Global Oceans and Deep Seabed--Biogeographic Classification. Paris, France: UNESCO-IOC. IOC Technical Series 84. 87 p.

White, M., Bashmachnikov, I., Aristegui, J., Martins, A. (2007). Physical processes and seamount productivity. In: Pitcher TJ, Morato T, Hart PJB, Clark MR, Haggan N, Santos RS, editors. Seamounts: ecology, fisheries, and conservation. Blackwell Fisheries and Aquatic Resources Series 12. Oxford, UK: Blackwell Publishing. pp. 65–84.

Yesson, C., Clark, M.R., Taylor, M., Rogers, A.D. (2011). The global distribution of seamounts based on 30-second bathymetry data. *Deep Sea Research* I 58: 442–453.

Hydrothermal Vents

DeWitt, T.J., Sih, A., Wilson, D.S. (1998). Costs and limits of phenotypic plasticity: Trends in *Ecology & Evolution*, v. 13:. 77-81.

Embley, R.W., Baker, E.T., Butterfield, D.A., Chadwick, Jr. W.W., Lupton, J.E., Resing, J.A., de Ronde, C.E.J., Nakamura, K.-I., Tunnicliffe, V., Dower, J.F., Merle, S.G. (2007). Exploring the submarine ring of fire: *Mariana Arc - Western Pacific: Oceanography* 20: 68-79.

Kawecki, T. (1995). Demography of source—sink populations and the evolution of ecological niches: Evolutionary Ecology 9: 38-44.

Levin, L.A. (2005). Ecology of cold seep sediments: interactions of fauna with flow, chemistry and microbes.: *Oceanography and Marine Biology Annual Reviews*, v. 43: 1 - 46.

Olu, K., Cordes, E.E., Fisher, C.R., Brooks, J.M., Sibuet, M., Desbruyères, D. (2010). Biogeography and potential exchanges among the Atlantic equatorial belt cold-seep faunas: *PLoS ONE* 5: e11967.

Sibuet, M., Olu, K. (1998). Biogeography, biodiversity and fluid dependence of deep-sea cold-seep communities at active and passive margins: Deep Sea Research Part II: Topical Studies in Oceanography, v. 45: 517-567.

Tarasov, V.G., Gebruk, A.V., Mironov, A.N., Moskalev, L.I. (2005). Deep-sea and shallow-water hydrothermal vent communities: Two different phenomena?: *Chemical Geology*, v. 224: 5-39.

Treude, T., Kiel, S., Linke, P., Peckmann, J., Goedert, J. (2011). Elasmobranch egg capsules associated with modern and ancient cold seeps: a nursery for marine deep-water predators: *Marine Ecology Progress Series* 437: 175-181.

Vrijenhoek, R.C., (2013). On the instability and evolutionary age of deep-sea chemosynthetic communities. *Deep Sea Research Part II: Topical Studies in Oceanography*, vol. 92. pp. 189-200.

35 Extent of Assessment of Marine Biological Diversity

Contributors:
Patricia Miloslavich (Convenor), Philippe Archambault, Nicholas J. Bax, Edward Vanden Berghe, Andre Boustany, Angelika Brandt, Alan J. Butler, Malcolm Clark, Bruce B. Collette, Nicholas Dulvy, Martin Edwards, Elva Escobar-Briones, Katsunori Fujikura, Judith Gobin, John E. Graves, Charles L. Griffiths, Huw J. Griffiths, Patrick N. Halpin, Russell Hopcroft, Kristin Kaschner, Ben Lascelles, Xinzheng Li, Telmo Morato, Fabio Moretzsohn, Massa Nakaoka, David Obura, Henn Ojaveer, Pierre Pepin, C. Raghunathan, Randall R. Reeves, Jake Rice (Lead Member and Editor of Part VI), Mayumi Sato, George Schillinger, Colin A. Simpfendorfer, John W. Tunnell, Jr., Verena Tunnicliffe, Paul Snelgrove, Karen Stocks, Marguerite Tarzia, K. Venkataraman, Bryan P. Wallace, Ross McLeod Wanless, Tom Webb and Hiroya Yamano[1]

---

[1] The writing team thanks Esteban Frere for his contributions to this chapter.

Chapter 35

1 Introduction

This chapter provides a summary of currently assessed marine biodiversity in terms of its coverage for the most conspicuous and well known taxonomic groups, particular ecosystems, and large geographic regions. Assessments will be focused on the evaluation of the state of knowledge of marine biodiversity; however, for some groups, such evaluations are provided indirectly by studies aimed to establish threat and or risk status. The groups that have been summarized globally are the sea mammals (cetaceans and pinnipeds), seabirds, sea turtles, sharks, tunas, billfish, corals, and plankton. The special ecosystems are seamounts, vents, and seeps. Regional summaries of coverage of assessments are provided whenever possible for large basins, such as North Atlantic, South Atlantic, North Pacific, South Pacific, Indian Ocean, Arctic Ocean, and Southern Ocean. However, in some cases, information is compiled by countries (e.g., Canada) when these have more than one basin, or by large continents (e.g., South America) which share a history of surveys and exploration. After each of the sections, a global analysis of the status of knowledge of marine biodiversity is summarized within a few synthesis graphs. About 40 scientists contributed to this effort, each within their area of expertise and specified for each subsection. Supplementary material providing a list of assessments with date, special area, habitat, taxonomic groups, and web information has also been compiled for a few of the regions (Caribbean, Europe, Gulf of Mexico, the Southern Ocean and Sub-Saharan Africa) and States (China, India and Japan), as well as for vents and seeps ecosystems and for turtles (Appendix I). In addition, a complete reference list for further reading for each of the taxonomic groups and regions is provided (Appendix II).

2 Groups summarized globally: Cetaceans, pinnipeds, seabirds, sea turtles, sharks, tunas, billfish, corals, seamounts, vents and seeps.

2.1 Marine Mammals

Global assessments of marine mammal distributions are limited by geographic and seasonal biases in data collection, as well as by biases in taxonomic representation due to rarity and detectability. In addition, not all data collected have been published in open-access repositories, thus further constraining our ability to develop comprehensive assessments. Given the financial, logistical and methodological challenges of mounting surveys, especially for animals that spend most of their time underwater, assessments have been most extensive and intensive on the coastal shelves and continental slopes along the coastlines of developed countries (Kaschner et al., 2012 & Figure 35.1A). Ship-board surveys of large ocean areas have been and continue to be carried out in the Southern Ocean and North Pacific under among others, the auspices of the International Whaling Commission, focusing on those whales previously subject to commercial whaling. Advances in satellite telemetry are helping to fill in some gaps in offshore areas for both cetaceans and pinnipeds (Block et al., 2011).

However, the remaining geographic biases in sampling coverage are very apparent, for example, in the map portal Ocean Biogeographic Information System Spatial Ecological Analysis of Megavertebrate Populations OBIS-SEAMAP (http://seamap.env.duke.edu), which is an online data portal compiling occurrence records of higher vertebrates living in the marine environment (Halpin et al., 2009), where the majority of species observation records fall within the coastal shelves and continental slopes of the Northern Hemisphere. Around 95 per cent of the marine mammal observations published on the portal are from inside the 200-nautical-mile (nm) exclusive economic zone (EEZ), while ~5 per cent are in areas beyond national jurisdiction (ABNJ). A recent analysis of global coverage of cetacean visual line-transect survey coverage showed that only ~ 25 per cent of the world's ocean area had been covered by systematic surveys by the year 2006, and only 6 per cent had been covered frequently enough to be able to detect population trends (Kaschner et al., 2012). Pinniped and cetacean populations are monitored fairly frequently in the United States of America, European and Southern Ocean waters; more than half of the total global line-transect effort from 1978 – 2006 was in areas within the national jurisdiction of the United States (Kaschner et al., 2012) and ~35 per cent of all marine mammal observations held in OBIS-SEAMAP are from within the 200-nm EEZ of the United States.

Geographically, the largest gaps in sampling coverage are in the Indian Ocean and the temperate South Atlantic and South Pacific, where comparatively few dedicated surveys have been conducted. In the Southern Hemisphere, surveys have been carried out mostly in the EEZs of Australia, New Zealand, Chile, Argentina, and South Africa where more than 50 per cent of the world's species are found, and endemism is relatively high. Seasonally, most data collection using standard visual monitoring methods is concentrated in the summer as poor weather conditions seriously lower detectability, but again, satellite telemetry and passive acoustic monitoring are helping to fill in some of the temporal gaps. Although passive acoustic monitoring can be very useful in detecting the calls of certain species, and thus help determine their presence in a region, during seasons of poor visibility or low survey effort, such monitoring cannot yet be used routinely for the development of abundance or density estimates.

Sampling effort and reporting is also highly variable for different species. For example, although OBIS-SEAMAP currently contains a total of >560,000 marine mammal occurrence records covering 106 species of the roughly 120 marine mammal species (~88 per cent), the data sets are uneven, with no records available of some uncommon species (~12 per cent) and fewer than 10 records available for others (~14 per cent). Overall, the distributions of some well-known species, such as the humpback whale (*Megaptera novaeangliae*) and the harbour porpoise (*Phocoena phocoena*), have been studied extensively and are well established; they are based on sightings and strandings or analyses of catch data. Relatively large proportions of the known ranges of these species

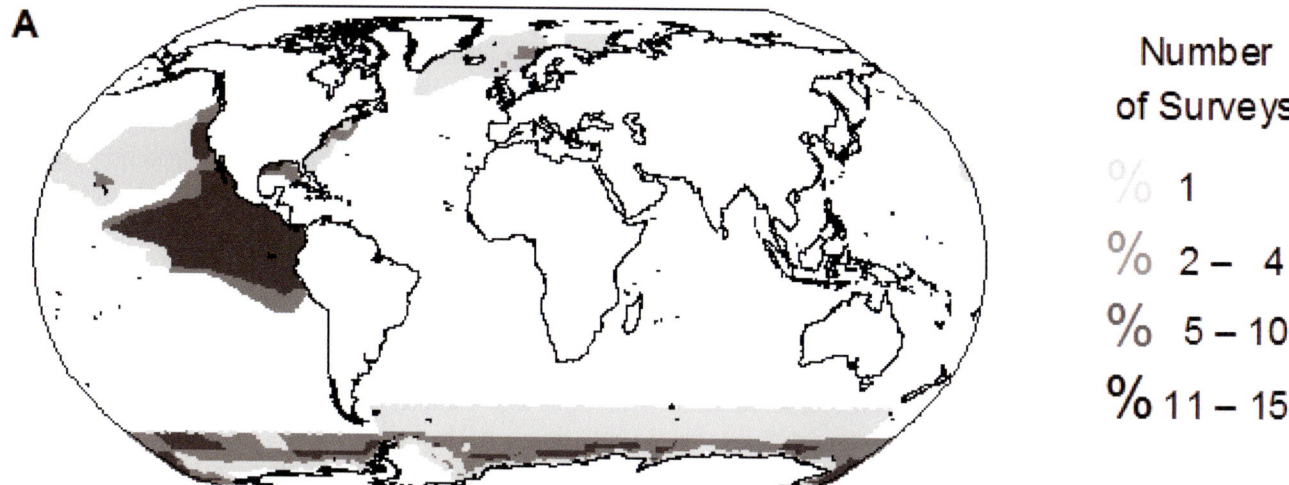

Figure 35.1A | Global coverage of visual cetacean line-transect survey effort (Frequency of coverage) (from Kaschner et al., 2012). See also the OBIS SEAMAP data available through http://seamap.env.duke.edu.

are being monitored frequently, using different survey techniques (Kaschner et al., 2012). Similarly, at-sea movements and occurrence of other species, such as the southern and northern elephant seals (*Mirounga leonina*, *M. angustirostris*), have been investigated in considerable detail, using data loggers and satellite tracking (Block et al., 2011). In contrast, the information on some species is limited and patchy due to their rarity and/or cryptic behaviour. Some have rarely, if ever, been seen alive and are known only from a few stranding records (e.g., Perrin's beaked whale, *Mesoplodon perrini*).

Assessments of marine mammal species distribution and status derived from available data sets must be viewed in comparison to survey effort to control for unsurveyed regions or areas where observation data have not been shared with open-access information systems (Figure 35.1A). Accumulated data sets and understanding of marine mammal species distributions are improving, but any interpretation of the state of knowledge needs to take account of the significant biases, as noted. In summary, assessment of marine mammals globally is far from comprehensive, with abundance estimation and trend analysis at the population level limited to relatively few species in particular geographic regions, and for some species even such basic information as their actual range of occurrence is still lacking.

2.2 Seabirds

BirdLife International is the International Union for the Conservation of Nature (IUCN) Red List authority for birds, and assesses the status, trends and threats of all Critically Endangered seabirds each year, as well as species thought to warrant immediate uplisting. In addition BirdLife International carries out a comprehensive assessment of all 350+ species of seabird every four years. Some seabird populations and/or species lack population monitoring altogether, resulting in unknown population trends for 53 seabird species on the IUCN Red List (Croxall et al., 2012). The International Waterbird Census (IWC) includes certain seabird species. It has run since 1967 and today covers over 25,000 sites in more than 100 countries. Results are reviewed and published in Waterbird Population Estimates, which assess the trends and 1 per cent thresholds for over 800 species and 2,300 biogeographic populations worldwide.

At the global and regional levels many Multilateral Environmental Agreements (MEAs) include priority species lists for which aspects of status, trends and threats are supposed to be assessed. Seabirds are included on some of these lists and are used as indicators for assessing the state of the marine environment. Those currently most actively undertaking work include the Agreement on the Conservation of Albatrosses and Petrels (ACAP) (29 species), the Directive 2009/147/EC of the European Parliament and of the Council of 30 November 2009 on the conservation of wild birds (the European Union (EU) Birds Directive (all seabirds in the EU), the Convention for the Protection of the Marine Environment of the North-East Atlantic (OSPAR Convention) (9 species), the Agreement on the Conservation of African-Eurasian Migratory Waterbirds (AEWA) (82 species), the Convention for Protection against Pollution in the Mediterranean Sea (Barcelona Convention) (14 species), the Convention on the Conservation of Migratory Species of Wild Animals (CMS) (20 seabird species on Annex I; 50 on Annex II), the Convention on the Conservation of European Wildlife and natural habitats (Bern Convention) (over 30 species), the Convention on the Protection of the Marine Environment of the Baltic Sea (HELCOM) (11 species), the Convention on the Protection of the Black Sea against Pollution (Bucharest Convention) (2 species), the Commission for the Conservation of Antarctic Marine Living Resources (CCAMLR) (7 species), the Conservation of Arctic Flora and Fauna (CAFF) (3 species), the North American Agreement on Environmental Cooperation (1 species), and the Convention on International Trade in Endangered Species of Wild Flora and Fauna (CITES) (6 species). Other MEAs that have this remit but are not yet active include the Nairobi Convention for the Protection, Management and Development of Marine and Coastal and Environment of the Western Indian Ocean Region (the Nairobi Convention) (47 species), the Regional Convention for the Conservation of the Red Sea and Gulf of Aden Environment (Jeddah

Convention) (lists not yet provided by contracting parties), Convention for Cooperation in the Protection, Management and Development of the Marine and Coastal Environment of the Atlantic Coast of West, Central and Southern Africa Region (Abidjan Convention) (considering adding a species list), and the Convention for the Protection and Development of the Marine Environment in the Wider Caribbean Region (Cartagena Convention) (5 species). These MEAs (and other processes) have led to the development of individual species management plans, which often outline how (and by whom) monitoring of status, trends and threats can be addressed.

Numerous online databases have pooled seabird data at regional or global scales (as well as national programmes); these include data for:

- Colonies – Sea Around Us Project, BirdLife International World Bird Database and Marine E-atlas, Global Seabird Colony Register, Circumpolar Seabird Data portal, New Zealand National Aquatic Biodiversity Information System (NABIS)
- Productivity - Pacific Seabird Monitoring Database.
- Tracking - Seabirdtracking.org, OBIS-SEAMAP, Movebank, seaturtle.org, Tagging of Pacific Predators (TOPP), British Antarctic Survey, CNRS-Chize
- At-sea surveys - European Seabirds at Sea, North Pacific Seabird Portal, Australia Antarctic Division, Royal Navy Birdwatching Society, OBIS-SEAMAP, Atlas of Seabirds @ Sea, eBird, North Pacific Pelagic Seabird Database, Global Biodiversity Information Facility.
- Threats - the New Zealand Threat Classification System: conservation status of 473 taxa assessed. For seabirds, perhaps the world second largest in terms of number of species assessed (Robertson et al., 2013).

The BirdLife International Marine Important Bird Areas (IBA) e-Atlas provides a site-based information portal for seabird conservation. This first global network of over 3,000 sites covers 6.2 per cent of the world's oceans and was compiled by BirdLife International drawing on work from 1,000 seabird scientists, government ministries and secretariats of conventions. The World Seabird Union (comprised of 22 seabird organizations) has established the Seabird Information Network aiming to showcase, and link, different global seabird databases.

2.3 Marine Turtles

The primary global assessment framework for marine turtle species is the IUCN Red List of Threatened SpeciesTM (www.iucnredlist.org). The IUCN Marine Turtle Specialist Group (MTSG), one of the IUCN/Species Survival Commission's specialist groups, is responsible for conducting regular Red List assessments of each marine turtle species on a global scale. Currently the Red List identifies the olive ridley (*Lepidochelys olivacea*) as Vulnerable, the loggerhead (*Caretta caretta*) and green (*Chelonia mydas*) turtles as Endangered, the Kemp's ridley (*Lepidochelys kempii*), hawksbill (*Eretmochelys imbricata*), and leatherback (*Dermochelys coriacea*) turtles as Critically Endangered, and the flatback turtle (*Dermochelys coriacea*) as Data Deficient.

To address the critical issue of geographically variable population traits in marine turtle species, the MTSG developed an alternative assessment framework and a new approach to Red List assessments that better characterize variation in status and trends of individual populations (Wallace et al., 2010), which results in official Red List categories for subpopulations in addition to the single global listing.

To address the challenges presented by the mismatched scales of global Red List assessments and regional/population-level variation in status, the IUCN Marine Turtle Specialist Group (MTSG) convened the Burning Issues Working Group of marine turtle experts who developed (1) regional management units (RMUs) (i.e., spatially explicit population segments defined by biogeographical data for marine turtle species) as the framework for defining population segments for assessments (Wallace et al., 2010). These RMUs are functionally equivalent to IUCN subpopulations, thus providing the appropriate demographic unit for Red List assessments. The Group also developed (2) a flexible yet robust framework for assessing population viability and the degree of threats that could be applied to any population in any region, and (3) a "conservation priorities portfolio" for all RMUs, with globally included identification of critical data needs by RMUs, as well as individual population risk and threats criteria, and that reflects the wide variety of conservation objectives held by different stakeholders depending on institutional or regional priorities. South Asia had the highest proportion of RMUs categorized as requiring critical data needs (~40 per cent), followed by the West Indian Ocean (25 per cent) and Australasia (20 per cent). Similarly, population risk and threats scores for RMUs in the Indian Ocean were associated with the lowest availability and quality of data among ocean basins.

Among population risk criteria, insufficient information was available to assess recent and long-term trends for roughly 25-30 per cent of all RMUs. Among threats, climate change was scored "data deficient" in two-thirds of all RMUs, while pollution and pathogens were scored "data deficient" in more than half of all RMUs. These results demonstrate the need to enhance data collection and reporting on population trends, as well as current and future impacts of climate change and pollution/pathogens on marine turtles.

In addition to the two primary global assessment frameworks described above, several other global status assessments exist for marine turtles. The Convention on International Trade in Endangered Species of Wild Fauna and Flora (CITES) and the Convention on the Conservation of Migratory Species of Wild Animals (CMS, or Bonn Convention) include all marine turtle species in their lists, meaning that international trade in any products of any marine turtle species is prohibited and marine turtles are categorized as being in danger of extinction throughout all or a significant proportion of their range.

National laws to assess and protect endangered species can also result in global assessments. For example, all marine turtle species (except the flatback, *Natator depressus* which does not appear in the United States territorial waters) are listed globally as either Endangered or Threatened under the United States Endangered Species Act. Recently, the United States designated "distinct population segments"—which are similar to the RMUs and IUCN subpopulations described above—for loggerhead turtles (*Caretta caretta*) and listed all populations as either Threatened or Endangered. In addition, global status reviews are performed every five years for all marine turtle species listed under this act in the United States (Wallace, 2010).

Regional assessments offer more detailed views of marine turtle status, significant threats, and data gaps. Three noteworthy examples of regional marine turtle status are highlighted here. First, the Wider Caribbean Sea Turtle Conservation Network (WIDECAST, www.widecast.org) generated an "atlas" of marine turtle nesting sites, legal protection, and other relevant information for more than 40 countries and territories in the Wider Caribbean. Second, regional members of the IUCN Marine Turtle Specialist Group conducted a comprehensive assessment of the distribution, threats, and conservation priorities with regard to marine turtles in the Mediterranean. Third, the Indian Ocean-Southeast Asia Marine Turtle Memorandum of Understanding (IOSEA, www.ioseaturtles.org) has produced status and threats assessments of two species (loggerheads and leatherbacks) across more than 30 countries in the Indian Ocean and Southeast Asia. Assessments of marine turtle status at national and local levels occur around the globe, but a complete review is beyond the scope of this section (see Appendix I – Turtles: Summary of existing assessment frameworks and resources for marine turtles at global and regional scales).

In general, an urgent need remains for enhanced monitoring and reporting of marine turtle population status and trends, as well as of threats to marine turtles globally. Although much information exists in some regions (e.g., Wider Caribbean, Mediterranean, North America), significant data needs are apparent in other regions (e.g., West and East Africa, North Indian Ocean, Southeast Asia). Regional and global efforts to compile all available information in such regions are vital to filling these data gaps. Nonetheless, significant efforts to quantify fundamental marine turtle demographic rates and processes (NRC, 2010) are still required to improve assessments of marine turtle status at global, regional and local scales (Wallace, 2011).

2.4 Sharks, Rays, and Chimaeras

Sharks, rays and chimaeras comprise the Class Chondrichthyes. This group is highly diverse (at least 1,200 valid species) and occur in a broad range of habitats, so a wide range of approaches has been taken to assess the status of individual populations. The most publicly available assessments for chondrichthyans are available from the IUCN Red List. The IUCN Species Survival Commission's Shark Specialist Group (SSG), is a global network of experts in the biology, taxonomy, and conservation of sharks, rays, and chimaeras which continuously conducts global and regional assessments of the Red List Status of chondrichthyans. Established in 1991, the SSG currently has more than 123 members from 33 countries collaborating to assess threat level, collate knowledge, highlight species at risk, and advise decision-makers on effective, science-based policies for sustainable use and long-term conservation (www.iucnssg.org). In 2011, using the 2001 IUCN Red List Categories and Criteria (version 3.1; http://www.iucnredlist.org/technical-documents/categories-and-criteria), a total of 1,041 chondrichthyan species had been assessed and their extent of occurrence mapped, highlighting considerable gaps in knowledge (Dulvy et al., 2014). A total of 487 out of the 1,041 species were categorized as Data Deficient, particularly in four regions: (1) Caribbean Sea and Western Central Atlantic Ocean, (2) Eastern Central Atlantic Ocean, (3) Southwest Indian Ocean, and (4) the South and East China Seas (Dulvy et al., 2014), and in 2014, the assessed number of species was raised to 1,088. Since assessments are considered out of date after ten years, a concerted effort has been initiated to reassess all species in support of the 2020 Aichi targets of the Convention on Biological Diversity's Strategic Plan for Biodiversity (e.g. North-East Pacific and Europe regions in 2014; Australia and Oceania planned for 2015).

The Red List Assessments are complemented by data from catch landings, fishery catch rates, fisheries stock assessments, fishery-independent surveys, transect surveys (divers, boats, and aerial), as well as increasing quantities of individual photographic identification data, satellite tracking data and population genetics data, which vary in availability, quality, and geographic and taxonomic coverage. These data collection programmes and research projects are also combined with historical ecological information and traditional knowledge-based assessments of the change in species distributions.

National catch landings data are reported annually to the Food and Agriculture Organization of the United Nations (FAO). From 2000 to 2009, 143 countries/entities reported shark, ray and chimaera catches to FAO. The taxonomic resolution of the global landed catch has improved, but remains poor. By 2010, only a small proportion (29 per cent) of the catch was reported to species level, the remaining bulk of the global catch reported at much coarser taxonomic levels, and around one-third of global catches reported at the taxonomic Class level (i.e. "Sharks, rays, skates, etc"). Among the top shark fishing nations (Indonesia, India, Spain, Taiwan Province of China, Argentina, Mexico, United States of America, Malaysia, Pakistan, Brazil, Japan, France, New Zealand, Thailand, Portugal, Nigeria, Islamic Republic of Iran, Sri Lanka, Republic of Korea, Yemen), half (11) report 50 per cent or more of their catch at the species and genus level.

The taxonomic and geographic distribution of fisheries assessments of stock biomass and fishing mortality is very sparse. To date, we are aware of 41 stock assessments for 28 chondrichthyan species. The United States and Australia conduct most stock assessments; the majority conducted in the Atlantic Ocean (21), followed by the Indian Ocean (11) and 9 in the Pacific Ocean. Research surveys and shark control programmes are increasingly being used to assess the trajectory and status of shark and ray populations, particularly in the coastal waters of the United States, Europe, South Africa, New Zealand, and Australia. Many of these time series are ongoing and the specific assessment of the status of chondrichthyans is periodic and dependent on research funding.

Emerging technologies, such as satellite tags and acoustic tracking arrays, as well as the widespread availability of digital underwater photography, web-based database capability and photo identification systems, are providing information for better population estimates and refined geographic distributions. The miniaturization and longevity of electronic tags have revealed complex sex-biased migrations, migratory routes and infrequent but biologically important ocean transits connecting populations that were previously thought to be separate. The development of pattern-matching algorithms has transformed collections of photographs into mark-recapture methods for estimating local abundance and spatial dynamics of larger, more charismatic species, such as: White Shark (*Carcharodon carcharias*), Whale Shark (*Rhincodon typus*), and manta rays (*Manta birostris* and *M. alfredi*). Assessment approaches have been complemented by the rapid emergence of worldwide tissue-sampling and population genetics work that has led to an increasing understanding of the variation in gene flow and connectedness of populations within species, and increasingly the degree to which their ecology and life histories shape patterns of genetic relatedness. Genetic information is also increasingly used to assess the scale of the trade in shark fins and other valuable traded items, including the species composition of trade, and occurrences of illegal sale and trade (Abercrombie et al., 2005).

Assessments of long-term changes in distribution and population trajectories are increasingly being compiled from less formal sources of historical ecological information, including newspaper reports, trade records, and sightings. Compilations of museum records, newspaper reports and sightings have been used to reconstruct the former distribution of sawfishes, prompting conservation action. Assessments of historical landings and the traditional ecological knowledge of fishers have revealed a massive reduction in the diversity of chondrichthyans landed in Southeast Asian markets.

Key challenges that remain include continuing ongoing assessment activities, research surveys, and expanding assessments to include other species, not just the larger and more charismatic species. Assessments would also need to be expanded to include lesser known species, which are often more threatened, particularly the rays and ray-like sharks, and the 90 obligate and euryhaline freshwater species. Geographically, greater attention would need to be paid to Central and South America, Africa, and South East Asia.

2.5 Tuna

As many tunas are commercially important fisheries species, most assessments are based on fisheries-dependent catch data, although these are occasionally augmented by fisheries independent datasets, such as larval trawls, aerial surveys and scientific catch surveys. Fisheries catch data have the potential for bias due to extrinsic factors, such as those that may influence fishing effort (e.g., fuel and fish prices; regulations on fishing times and areas; changes to gear that influences fishing efficiency over time), as well as lack of reporting of catch and/or effort, and changes in the distribution of tuna species that may cause changes in the interaction rates with individual fisheries (Collette et al., 2011). In addition to limitations in data on catch rates, data on basic biological parameters necessary for accurate stock assessments (e.g., growth rates, stock structure, size/age of maturity, natural mortality rates) are often poorly known, thus also affecting the accuracy of the assessments. These limitations have begun to be addressed through advances in scientific methodologies. Electronic and conventional tagging studies have shed light on all these biological parameters, and population genetic and micro-constituent studies have facilitated delineation of stock structure in many tuna species. In addition, traditional sampling methodologies, such as histological sampling of gonads, counting rings on hard parts, such as otoliths and fin spines, and cohort analysis have all provided information on growth rates and reproductive maturity schedules. However, collecting these data costs money, hence data, and therefore the assessments, are usually better for the tuna species that are more economically important.

Most tuna assessments are conducted through regional fisheries management organizations (RFMOs), a collection of national and other fishing parties that jointly manage shared fish stocks (Aranda et al., 2010). Five tuna RFMOs currently exist that regulate fisheries for their member States: the International Commission for the Conservation of Atlantic Tunas (ICCAT), regulates tuna in the Atlantic Ocean, the Inter-American Tropical Tuna Commission (IATTC), regulates tuna in the eastern tropical Pacific, the Western and Central Pacific Fisheries Commission (WCPFC), regulates tuna in the Western and Central Pacific Ocean, the Indian Ocean Tuna Commission (IOTC), regulates tuna in the Indian Ocean and the Commission for the Conservation of Southern Bluefin Tuna (CC-SBT), regulates southern Bluefin tuna (*Thunnus maccoyii*) throughout its range. In addition to assessments conducted through RFMOs, other international organizations such as the IUCN, the CITES, national governments and independent scientists also occasionally conduct assessments of various tuna species. Unlike RFMO assessments, which generally seek to assess stock health in relation to optimal fisheries yield, most other organizations conducting assessments on tunas attempt to estimate extinction risk. As a result, RFMOs' and others' assessments may differ greatly in their evaluations of the health of tuna stocks.

The most commercially important tuna species have been assessed recently, either regionally or throughout their range by the above-mentioned tuna Commissions. The Bluefin tunas (Pacific [*Thunnus orientalis*], Southern [*T. maccoyii*] and Atlantic [*T. thynnus*]) have all had a full stock assessment within the last four years through their respective RFMOs. Likewise, Bigeye tuna (*T. obesus*), Yellowfin tuna (*T. albacares*), Albacore tuna (*T. alalunga*) and Skipjack tuna (*Katsuwonus pelamis*) have all been assessed regionally through the RFMO assessment process, as well as globally through the IUCN. Other species, such as Blackfin tuna (*T. atlanticus*) and Longtail tuna (*T. tonggol*) have not had full assessments conducted through their respective RFMOs, although localized assessments in part of their range may have been undertaken. RFMOs for these are the ICCAT for Blackfin tuna, and the IOTC and WCPFC for Longtail tuna.

Less is known on the stock status of tuna species for which there are only small, directed fisheries or for which most of the catch occurs as by-catch. Slender tuna (*Allothunnus fallai*), frigate tuna (*Auxis thazard*), and bullet tuna (*Auxis rochei*) all range widely, but formal assessments have not been conducted by RFMOs in each ocean basin. Black skipjack (*Euthynnus lineatus*), Kawakawa (*Euthynnus affinis*), and little tunny (*Euthynnus alletteratus*) are all regionally distributed (Eastern Pacific, Western Pacific and tropical Atlantic, respectively), and few data are available on range-wide catches over time; this is necessary for a full population assessment. However, the wide ranges of these six species, coupled with relatively low and localized exploitation, caused these species to be classified under "Least Concern" by the IUCN.

Although formal stock assessments have been completed for almost half of the tuna species (7 out of 15), few standardized data sets exist on catch rates over time for the remainder of the species. Improvements in the collection of fishery-dependent data or initiation of fisheries-independent data collection would be necessary to obtain accurate estimates of stock health. In the meantime, relatively stable catches over time for the unassessed species suggest that there is little immediate threat to the viability of any of these species.

2.6　Billfish

Billfish are epipelagic marine fishes distinguished by elongated spears or swords on their snouts. Most of the species have very large, ocean-wide or cosmopolitan ranges in tropical and subtropical waters and all are tied to the tropics for reproduction. However, the Swordfish extends into temperate waters. All are of commercial or recreational importance; hence our knowledge of their distribution comes largely from fisheries. Three species are restricted to the Indo-West Pacific: *Istiompax indica*, Black Marlin; *Tetrapturus angustirostris*, Shortbill Spearfish; and *Kajikia audax*, Striped Marlin. The other three species of spearfish and the White Marlin, *Kajikia albida*, are restricted to the Atlantic Ocean.

Most of the species are well known and easily distinguished; fisheries records document their distribution. However, this is not the case for Black *versus* Blue Marlin or for the spearfish. The Atlantic Longbill Spearfish, *Tetrapturus pfluegeri*, was not described until 1963 and a second species, *T. georgii*, Roundscale Spearfish, although originally described in 1841, was not validated as a species until 2006. This species is easily confused with White Marlin (*Kajikia albida*), hence the exact distributions of these two species are not yet completely clear, but appear to include much of the North and South Atlantic. Due to overfishing, two billfish meet the IUCN Red List criteria for a threatened category, Vulnerable (Collette et al., 2011): *Makaira nigricans*, the Blue Marlin, and *Kajikia albida*, the White Marlin.

The ICCAT, operating since 1969, is the organization responsible for the conservation of tunas and tuna-like species in the Atlantic Ocean and adjacent seas including several species of billfishes including the White Marlin, Blue Marlin, Sailfish (*Istiophorus albicans*) and Longbill Spearfish. Studies carried out by the ICCAT are mostly focused on the effects of fishing on stock abundance and include data on biometry, ecology, and oceanography. In the tropical Atlantic, this responsibility is held by the Inter-American Tropical Tuna Commission (IATTC) operating since 1950.

2.7　Coral and coral reef assessments

Coral reefs are among the most charismatic of tropical marine ecosystems (Knowlton et al., 2010) and have been assessed globally under several frameworks. Interestingly, however, because they occur in complex shallow seas, the application of large scale oceanographic tools and observation systems on major vessels is impossible; at the same time, they are accessible to small-scale, small-vessel direct observation methods. Their visual attractiveness, ecological complexity and the growth of observational science due to the invention of SCUBA technology in the 1960s have made them a focus for direct observational methods by researchers. As a result, even in the most accessible of coral reef systems in the Caribbean, recent synthesis has found diversity of methods and incompatibility of datasets to be the norm (Jackson et al. 2014).

Coral reefs bear among the highest taxonomic diversity of any ecosystem, and at the same time reef science is relatively youthful. This has resulted in high fluidity in the taxonomy of reef species, in particular complex symbiotic organisms, such as corals, though some well-sampled groups such as, bony fish and molluscs have benefited from taxonomic work inherited from other ecosystems. At the same time, molecular techniques, such as barcoding, are showing high levels of un-described diversity in microbial and invertebrate communities, both major components of coral reef biota.

Nevertheless, the broad global patterns in marine biodiversity are well-described through patterns in indicator groups including corals, stomatopods, and fish. The Indo-Malayan region is a clear centre of diversity for coral reef taxonomic groups, resulting from a broad range of biodiversity-generating and -maintaining processes from short to long time scales (Roberts et al. 2002). Diversity assessments in other regions have, however, been less complete, but resulting in emerging evidence of elevated diversity in other regions, such as in Eastern Africa and the South China Sea. The Caribbean or tropical Atlantic region is highly depauperate in terms of species diversity compared to the Indo-Pacific, resulting from isolation mechanisms over tens of millions of years, as well as since the formation of the isthmus of Panama (Veron, 2000). Coral species have been assessed for the IUCN Red List of Threatened Species, resulting in over one-third of species being identified as Threatened, among the highest proportions of all taxonomic groups globally (See also chapter 43).

Most of the global level assessments of coral reefs are ecological in nature, or use higher taxonomic levels, such as genera for recording absolute or relative abundance. The principal variables used in reporting coral reef health include proportional cover for attached benthic taxa (e.g. hard coral cover, in percent), abundance or density per unit area for key mobile taxa (invertebrates and vertebrates) and biomass, particularly for fish. The establishment of the Global Coral Reef Monitoring Network (GCRMN) was in response to the largest global reef impact ever recorded: the 1998 El Niño event. The resulting series of GCRMN reports (Wilkinson 1999, 2000, 2002, 2004 and 2008) exemplify the challenges in continuing such a large effort. At the same time, remote sensing technologies and global threat datasets have been used to prepare global assessments of the health of coral reefs in the Reef at Risk publications of the World Resources Institute (Burke et al. 2011). The most recent reporting under the GCRMN has focused on regional level reporting, with the first major regional assessment focusing on the Caribbean (Jackson et al. 2014). In this regard, it matches the scope of regional assessment frameworks, which have included those for: (1) the Caribbean, such as the Atlantic and Gulf Rapid Reef Assessment (AGGRA), the Caribbean Coastal Marine Productivity (CARICOMP); (2) parts of the Pacific such as the Coral Reef Monitoring Activities in Polynesia Mana Node, Coral Triangle Initiative on Coral Reefs, Fisheries and Food Security (CTI-CFF) and (3) parts of the Indian Ocean such as the Coral Reef Degradation in the Indian Ocean (CORDIO), all building on the governance frameworks that are the product of the United Nations Environment Programme (UNEP) Regional Seas programme established in 1974. Due to the popularity of coral reefs for SCUBA diving, volunteer and participatory monitoring are popular alternatives, with the most comprehensive one being that of Reef Check (Hodgson, 1999). In these, assessments are based on indicator species and estimates of variables such as benthic cover. Though variable in quality and coverage, the resulting data can be invaluable in broad scale scientific assessments of reef status. Coral reef areas with the least investment in assessments are those in poor developing countries with generally low national dependence on the sea (though they may have large sectors of society with high levels of livelihood/subsistence dependence on the sea, e.g. the Indian Ocean) and of middle/low biodiversity and interest for international science. Coral reef areas with the most investment in assessments are dispersed throughout the Pacific and tropical Atlantic.

Cold-water or deep-water corals are found globally, but have been most extensively mapped in the North Atlantic, due to extension of fishing and exploration for seabed resources in that region, and New Zealand has undertaken significant coral mapping. With greatest development from 200-1,000 m, and on topographic promontories such as seamounts, they can form large reefs of several 100s of m across and 10s of m above the substrate, but are highly vulnerable to damage and changing chemical oceanographic conditions. Further discoveries on the distribution of cold water corals are continuing to be made, such as in the southern Indian Ocean (Cairns, 2007).

Coral reefs are mentioned as a model ecosystem under Aichi Target 10 of the Convention on Biological Diversity (CBD) (to reverse impacts on climate-sensitive ecosystems). The search for "Essential Biodiversity Variables" (EBVs) to support monitoring for such targets and commitments is gaining momentum, and there is recognition that coral reefs may provide one of the ten globally-consistent sources to support this process, and not only with respect to biodiversity - greater recognition of the ecosystem services contributed by coral reefs (to communities, global tourism, and national/global economies and trade) should secure resources for monitoring of the ecosystem processes/indicators that underpin those services and goods, incentivizing monitoring and assessment to manage them for future benefit. In parallel with the CBD, the Intergovernmental Science-Policy Platform on Biodiversity and Ecosystems services (IPBES), and the Sustainable Development Goals may generate increased justification for upscaling coral reef assessments globally.

2.8 Plankton

At the global level, the seasonal pattern of Chlorophyll a is the best known and most studied phytoplankton-related variable in most marine ecosystems. Long-term studies on seasonal changes in phytoplankton diversity and abundance have been more localized geographically. The Western Channel Observatory (WCO) run by the Plymouth Marine Laboratory and the Marine Biological Association (United Kingdom) holds a marine biodiversity reference dataset for the Western English Channel with some of the longest time-series in the world for zooplankton and phytoplankton. At the Chesapeake Bay, a monthly, continuous 20-year phytoplankton database exists. In the Baltic Sea, historical phytoplankton data on community composition shows long-term changes in comparison to the early 1900s (Hällfors et al., 2013). However, much less is known about how changing diversity affects the productivity and functioning of marine food webs as well as the drivers behind these changes. Some global changes in diversity have been addressed based on the Continuous Plankton Recorder (CPR) data.

Perhaps the longest time-series sets exist for planktonic organisms (zooplankton and fish larvae) from the North-eastern Atlantic (North and Baltic Seas, English Channel and Bay of Biscay). It is important to note that plankton monitoring has extended to practically all the regions of the European coast due to the implementation of relatively recent legislation (e.g., the European Water Framework Directive). In this way, the study of plankton taxonomic composition and dynamics is being conducted in many areas that have been poorly studied or not studied at all. Surveys on micro-, nano-, picoplankton also exist, but these are spatially more restricted and substantially more comprehensive.

Some of the early formal accounts on zoobenthos date back to the famous Michael Sars expedition in the Atlantic in 1910, amended with several major contributions from Census of Marine Life activities (such as "Patterns and processes of the ecosystems of the northern mid-Atlantic", the Mid-Atlantic Ridge Ecosystem Project (MAR-ECO), and the global project Census of Marine Zooplankton) about a century later and introducing innovative identification techniques involving molecular biology.

2.8.1 The Continuous Plankton Recorder (CPR) survey

The Continuous Plankton Recorder (CPR) survey is recognized as the longest sustained (operating since 1931) and geographically most extensive marine biological survey in the world. The dataset comprises a uniquely large record of marine biodiversity covering ~1,000 taxa over multi-decadal periods. The survey determines the abundance and distribution of microscopic plants (phytoplankton) and animals (zooplankton, including fish larvae) in our oceans and shelf seas. Using ships of opportunity from ~30 different shipping companies, it obtains samples at monthly intervals on ~50 trans-ocean routes. In this way the survey autonomously collects biological and physical data from ships covering ~20,000 km of the ocean per month, ranging from the Arctic to the Southern Ocean. The survey is an internationally funded charity with a wide consortium of stakeholders.

Plankton are collected on a band of silk and subsequently visually identified (~1,000 taxa) by experts from around the world. Additionally, over the last decade the CPRs have been equipped with modern chemical and physical sensors, as well as molecular probes, to provide an array of additional information about our changing oceans. The final stages in the operation of the survey are the archiving of the resulting data and samples and interpreting the results at its central hub in Plymouth, United Kingdom. Strict quality control procedures are maintained for all CPR activities to ensure the integrity and long-term value of the programme. The database and sample archive together provide a resource that can be utilized in a wide range of environmental, ecological and fisheries-related research, e.g., molecular analyses of marine pathogens, modelling for forecasting and data for incorporation in new approaches to ecosystem and fishery management. Since the first tow of a CPR on a "ship of opportunity" in 1931, more than 6 million nautical miles of sea have been sampled and over 100 million data entries have been recorded.

Over the last eight decades the purpose of the survey has also co-evolved, with changing environmental policy, from purely monitoring plankton distributions to addressing and providing indicators for major marine management issues ranging from fisheries, harmful algal blooms, biodiversity, pollution, eutrophication, ocean acidification and climate change. For example, the CPR survey has documented a northerly shift of 1,000 km of marine organisms around Europe associated with climate change over the last four decades, with large ramifications for the European fishing industry. In the Arctic, the CPR survey recently recorded the first modern trans-Arctic migration of a diatom species (*Neodenticula seminae*) related to declining ice coverage (Reid et al., 2007; Edwards et al., 2012).

This diatom species, normally found in the Pacific Ocean, has been absent from the North Atlantic for over 800,000 years; perhaps it signifies the rapidity and unprecedented nature of climate change in the Arctic over recent geological history (Reid et al., 2008). In 2011, the Sir Alister Hardy Foundation for Ocean Science (SAHFOS), along with 12 other research organizations using the CPR from around the world, formed a Global Alliance of CPR surveys (GAC) with the aim of training new surveyors, producing a global ocean status report, capacity-building and developing a global plankton database. This global network of CPR surveys now routinely monitors the North Sea, North Atlantic, Arctic, North Pacific and Southern Ocean.

New surveys are underway in Australian, New Zealand, Japanese and South African waters; a Brazilian and an Indian Ocean survey are under development (Figure 35.2). These surveys provide coverage of large parts of the world's oceans, but many gaps still exist, particularly in

The boundaries and names shown and the designations used on this map do not imply official endorsement or acceptance by the United Nations.

Figure 35.2 | Start of sustained open-ocean biological time series and records (temporally broken and coastal time series are not included) plotted along the global mean SST time series from 1900 (Hadley Centre). Note that the majority of the time series are less than 30 years long. Station P (North Pacific); VICM (Vancouver Island Continental Margin time series); NMFS (National Marine Fisheries Service collection); BATS (Bermuda Atlantic Time Series Study); HOT (Hawaii Ocean Time Series program); Iberian coastal time series (North Coast of Spain); SO CPR survey (Southern Ocean CPR survey); AMT (Atlantic Meridional Transect); AZMP (Atlantic Zone Monitoring Program) (Edwards et al., 2010).

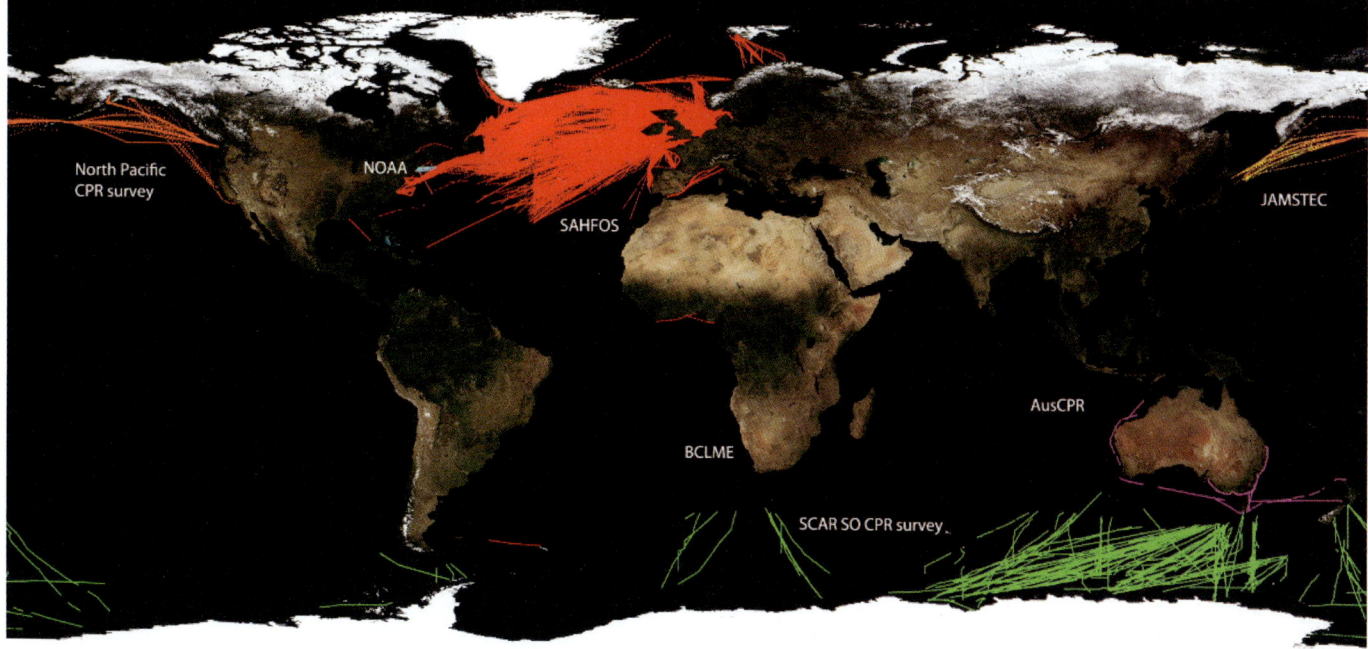

the South Atlantic, Indian and Pacific Oceans. This global network also brings together the expertise of approximately 50 plankton specialists, scientists and technicians from 12 laboratories around the world. Working together, centralizing the database and working in close partnership with the marine shipping industry, this global network of CPR surveys with its low costs and new technologies makes the CPR an ideal tool for an expanded and comprehensive marine biological sampling programme. The database and website can be accessed via www.sahfos.ac.uk (Edwards et al., 2012).

2.9 Seamounts

Several global seamount databases have been compiled, including the Seamount Catalogue (mainly geological), Seamounts Online (SMOL) (biological) and the Seamount Ecosystem Evaluation Framework (SEEF) (ecological) (Kvile et al., 2014). There are also detailed national datasets on seamount location and faunal composition, such as off New Zealand, the Azores, and for the western South Pacific (Allain et al., 2008). These databases and knowledge of seamounts have benefited from increased research on seamounts in the early 2000s by New Zealand, the United Kingdom, the United States, Japan, Australia, Portugal (Azores), among other countries, including the international CenSeam project of the Census of Marine Life (Clark et al., 2010).

A total of 684 seamounts have data recorded in the SEEF and SMOL up to the end of 2012. Their spatial distribution is: 458 in the Pacific Ocean, 164 in the Atlantic Ocean, 22 in the Mediterranean Sea, 12 in the Indian Ocean and 28 in the Southern Ocean. Their distribution is shown in Figure 35.3.

The seamounts have a mixture of data types: all have geological information, and 54 per cent have had some level of biological investigation (Kvile et al., 2014). Overall, the seamounts in the North Atlantic Ocean and Mediterranean Sea have been the most studied; other oceans are typically patchy. For example, in the Pacific Ocean, over 60 per cent of seamounts in the database had biological data, but extensive sampling was focused on a few areas: on the Nazca and Sala y Gomez chains in the eastern South Pacific, around New Zealand and southern Australia in the southwest Pacific, and off parts of Hawaii, Alaska and the west coast of the United States in the North Pacific.

The last decade has seen a dramatic increase in the number of seamounts being surveyed. This has in part been due to efforts by the fishing industry to find new fish stocks, but also by major national or international (such as the Census of Marine Life project CenSeam) research programmes carrying out biodiversity surveys (see Figure 35.4, from Kvile et al., 2014). The CenSeam data can be accessed through the OBIS portal (www.iobis.org) by selecting the Seamounts Online database.

Biological surveys of seamounts vary considerably in the methods and equipment used. In the North Atlantic, off the west coast of the United States, and in the North Pacific, remotely operated underwater vehicles (ROVs) and manned submersibles have been used to carry out extensive survey work. However, these tools tend to focus sampling on the mega-fauna, the large taxa that are clearly visible to the eye or camera equipment. Fish trawls have been used on many seamounts, and although these sample fish very effectively, they are poor at retaining fragile or small invertebrates. Off New Zealand and Australia, epibenthic sledges have been used on seamounts, which tend to catch a wide size range down to macroinvertebrates. Typically, a combination of sampling gear is necessary to adequately describe the benthic biodiversity on seamounts.

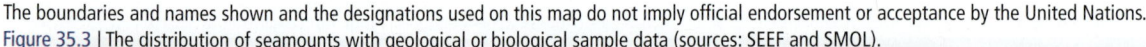

The boundaries and names shown and the designations used on this map do not imply official endorsement or acceptance by the United Nations.
Figure 35.3 | The distribution of seamounts with geological or biological sample data (sources: SEEF and SMOL).

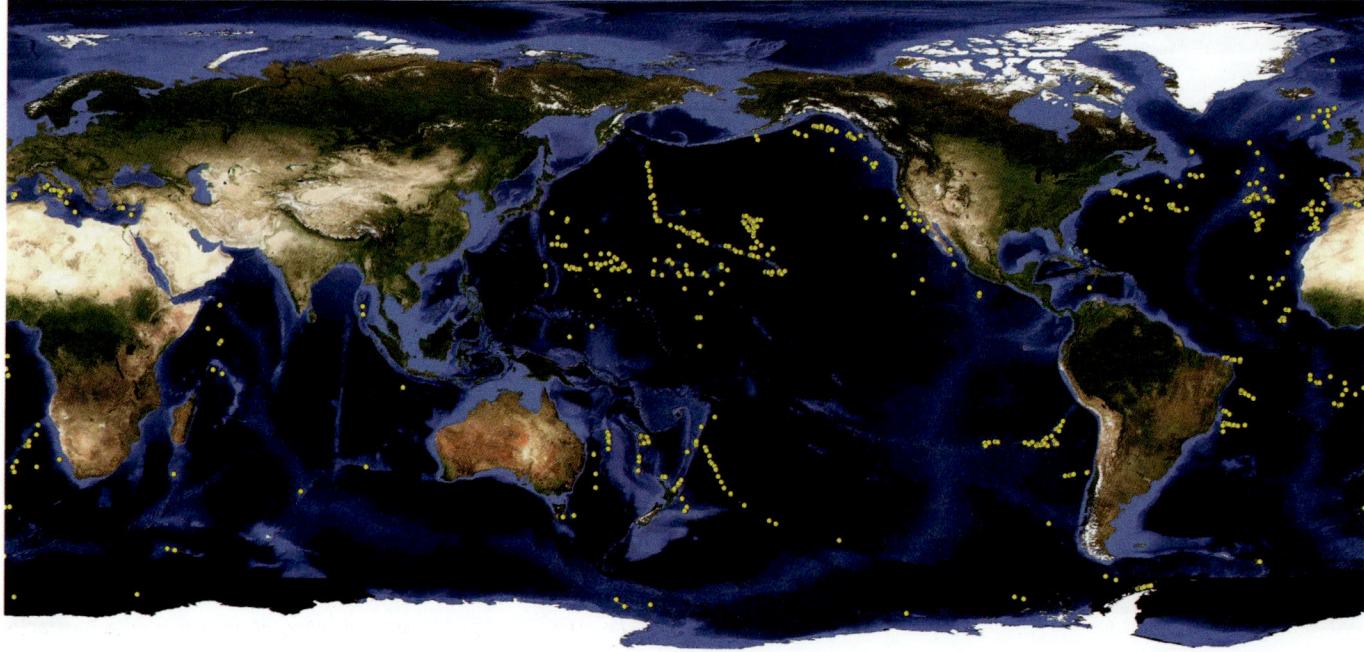

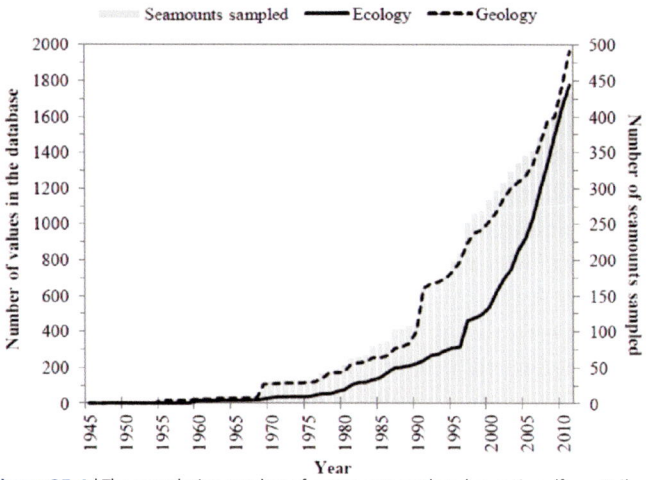

Figure 35.4 | The cumulative number of seamounts explored over time (from Kvile et al., 2014).

However, globally, a very low proportion of the large number of seamounts have been sampled. Of the nearly 450 seamounts sampled, relatively good data exist only for 300, and few of these are in equatorial latitudes, or deeper than about 2,000 m. Therefore, much about the structure, function and connectivity of seamount ecosystems remains unexplored and unknown. Continual, and potentially increasing, threats to seamount resources from fishing and seabed mining are creating a pressing demand for research to inform conservation and management strategies. To meet this need, intensive scientific effort in the following areas would be needed: (1) Improved physical and biological data; of particular importance is information on seamount location, physical characteristics (e.g., habitat heterogeneity and complexity), more complete and intensive biodiversity inventories, and increased understanding of seamount connectivity and faunal dispersal; (2) New human impact data; these should encompass better studies on the effects of human activities on seamount ecosystems, as well as monitor long-term changes in seamount assemblages following impacts (e.g., recovery); (3) Global data repositories; there is a pressing need for more comprehensive fisheries catch and effort data, especially on the high seas, and compilation or maintenance of geological and biodiversity databases that underpin regional and global analyses; (4) Application of support tools in a data-poor environment; conservation and management will have to increasingly rely on predictive modelling techniques, critical evaluation of environmental surrogates as faunal "proxies", and ecological risk assessment.

2.10 Vents and seeps

Since the discovery of hydrothermal vents and continental margin seeps in 1977, many animals found there were recognized as new to science, often at higher taxonomic levels. A recent assessment suggested that about 70 per cent of vent animals and 40 per cent of seep species now known are endemic to these habitats (German et al., 2011). In addition, many animals displayed unusual adaptations to habitats of reduced chemical compounds and physiological stresses. Systematic studies have generated many dispersed publications and some compilations have examined evolutionary patterns in a larger taxonomic group, ecological phenomena within a seep or vent region, and biogeographic pattern analyses. In this Assessment, only the most recent of these compilations are presented.

The InterRidge Vents Database is an important tool for metadata on hydrothermal vent sites that is currently maintained in an open-source content-management system in which updates depend on researcher input (Beaulieu et al., 2013). Over 500 confirmed and inferred referenced locations are listed. In the last decade, vent discoveries in arc and back-arc settings increased the known vent fields within exclusive economic zones. No biological or collection information is listed, hence it is a site for geographic information only. No similar information for seeps and other chemosynthesis-based habitats exists. The most recent map for locations of seeps globally appears to be that of Seuss (2010).

The Census of Marine Life (CoML) sponsored the eight-year project called Chemosynthetic Ecosystem Science (ChEss), with a primary focus to document the distribution of diversity in deep-water chemosynthetic ecosystems. The development of the ChEss database remains ongoing, but it is currently linked to OBIS; however, they are faunal lists and must be manipulated for site or taxonomic comparisons. Assessments emerging from use of the database are underway, but many are specific to faunal groups or regions.

To facilitate work in the field and laboratory, a handbook of vent fauna that assimilated most of the taxonomic papers and locality information related to hydrothermal vent fauna is available. The format of the book is a page for each species with a brief description, drawings or photographs and distribution information; it covers over 500 species in 12 phyla. No overall assessment of diversity is presented, but managers will find useful information on the species in their area of interest.

The ChEss Database has been used to formulate an assessment of the drivers of biogeographic patterns in vent faunas. The assessment focuses on historical relationships and centres of diversity. The approach uses network analysis and is a high-quality assessment of patterns. Another similar analysis for seep faunas of the equatorial Atlantic highlights discoveries on the western African margins and points out regions that still need investigation. Both datasets are available through IFREMER, France.

An overall assessment of vent and seep faunal diversity distribution and major drivers has yet to be executed. Exploration of these habitats is still underway, as biogeographic analysis shows major gaps in knowledge. Because of the close relationship of seeps on continental margins to areas of resource exploration interest (oil and gas, methane hydrates), assessment of the nature of diversity and its role in ecosystem function is important before potential alteration and/or destructions takes place. Similarly, growing interest in the metals associated with hydrothermal vent deposits may drive more work to define diversity patterns at vent habitats, as well as their relationships with other chemosynthesis-based habitats (See Appendix I-Vents and seeps: Major Inventories Available for Vent and Seep Habitats and Faunas). At present, many deep sea research initiatives have teamed up in INDEEP (International Network for Scientific investigation of deep-sea ecosystems) with the goal of providing a framework that will allow coordinated efforts in deep sea research across all its habitats while reaching out scientific results to managers and society.

3 Regional summaries of coverage of assessments: Arctic, North Atlantic, North Pacific, Indian, South Atlantic, South Pacific, and Southern Ocean

The following regional summaries are largely based on the work carried out by the network of National and Regional Committees of the Census of Marine Life (O'Dor et al., 2010; Costello et al., 2010). In this summary, the Caribbean and Mediterranean Seas as well as the Gulf of Mexico are included in the North Atlantic section. There are a few gaps in terms of geographic coverage (e.g. the Pacific coast of the Russian Federation, the Atlantic coast of Africa) due to lack of information or difficulties obtaining the data. However, a complete global analysis including all of the world's oceans was carried out using data from OBIS and presented in part 3 of the chapter.

3.1 The Arctic

Despite more than a century of observations on the Arctic's marine life, information on basic species inventories, as well as a quantitative synthesis, has remained fragmented until recently; however, some areas are still poorly known. Renewed interest in marine biodiversity generated by activities, such as those of the CBD, the Census of Marine Life and the International Polar Year, has begun to change this situation, but it is a slow process. At best, we are now armed with relatively complete lists of the species present, and have begun the process of assembling various datasets throughout the Arctic with the ultimate goal of establishing pan-Arctic patterns and trends over time (see Gradinger et al., 2010) for the taxonomic groups. The most comprehensive assessment in the Arctic is the Arctic Biodiversity Assessment (ABA), which includes chapters on marine mammals, birds, fishes, invertebrates, parasites, and ecosystems, contributed and reviewed by more than 100 scientists (Caff, 2013).

There are about 24 species of large unique vertebrates, such as the Polar Bear, Walrus, seals and several species of whales, found in the Arctic. As for seabirds, 64 species are recognized as breeding in the Arctic, and an even greater number of species is known to exploit its productivity seasonally (Archambault et al., 2010). Many of these air-breathing species are displaced southward during the winter's ice cover. As for fishes, nearly 250 species are known from the Arctic. About 30 of these divide their lives between the oceans and freshwater, and the rest are fully marine; the majority of these species are associated with the seafloor rather than living higher in the water column. Invertebrate diversity is higher and, unlike the vertebrates, invertebrates are typically studied in terms of communities rather than by species or higher taxonomic units. From functional and logistical perspectives, these communities are further divided by habitat, as this is how they are targeted during sampling: those associated with the sea ice, those within the water column (plankton) and those living on the seafloor (benthos). Of the functional groups of organisms considered thus far, with the exception of birds and mammals, species discovery continues within them all, especially now with the use of molecular tools for species identification (Gradinger et al., 2010). The Arctic is a mosaic of habitats across all its marine realms, within which information is very unequally collected between regions and over time. Habitat complexity, combined with the logistical challenges of sampling in the Arctic, and the generally limited interest in surveying the perceived low-diversity invertebrate fauna of the Arctic, have greatly impaired our ability to construct precise regional and temporal understanding of marine invertebrate diversity. Although changes have been noted in population size, abundance, growth, condition and behaviour of several marine mammals and fish, few changes for planktonic and benthic systems are documented. No comprehensive monitoring activities are conducted in the Central Arctic Ocean; oceanographic and ecosystem sampling has been occurring largely on an opportunistic basis, as part of national (e.g. Canadian) and international research programmes. Some efforts have been directed at developing directed systematic programmes for data collection, sometimes relying on community-based collaborations, but these are recent developments and baseline data from the Canadian Arctic Ocean are very limited in their spatial and temporal extent.

Examples of these collaborations are the Arctic Council CBMP (Circumpolar Biodiversity Monitoring Program), the Russian-American Long-term Census of the Arctic (2004-12), which involves a partnership of several United States and Russian Federation institutions to create a baseline dataset in the Pacific gateway area of the Bering Strait and southern Chukchi Sea, and other bi-national collaborations (e.g. United States and Canada, Russian Federation and Norway). Recently, the United States Bureau of Ocean Energy Management (BOEM) initiated programmes in Arctic waters to provide estimates of abundance and species composition of marine fishes and invertebrates, as well as information on the macro- and microzooplankton communities and their oceanographic environment. In 2008, the oil and gas industry began new biological assessment programmes (Chukchi Sea Offshore Monitoring in Drilling Area: Chemical and Benthos) in the Chukchi Sea in response to the sale of leases for new offshore prospect areas (Fautin et al., 2010).

3.2 Northern Hemisphere: focus on Canada

Canada borders the Pacific, Arctic, and Atlantic Oceans; and its coastline of 243,791 km (16.2 per cent of the global coastline) exceeds all of Europe combined. At 2,687,667 km^2, Canada's territorial sea covers 14.3 per cent of the global total. The Census of Marine Life program (Archambault et al., 2010), in collaboration with the Canadian Healthy Oceans Network (CHONe) (Snelgrove et al., 2012), conducted an assessment of the status of knowledge on marine biodiversity in Canada's oceans which included four biogeographical provinces: the Canadian Arctic (including the sub-Arctic Hudson Bay System), Eastern Canada, St. Lawrence estuary and Gulf, and Western Canada (Pacific coast). The taxonomic assessment encompassed the status of knowledge on microbes, phytoplankton, macroalgae, zooplankton, benthic infauna, fishes, and marine mammals resulting in an estimate of between 15,988 and 61,148 taxa (including microbes), a number that is most likely to be an underestimate, as many poorly sampled taxa and regions still exist. This assessment noted that significant data gaps exist, that larger species (e.g., mammals and fishes) are better known than small species (e.g., microbes), and that knowledge of diversity is inversely related to both water depth and geographical remoteness. Thus, even for well-known groups, such as fishes, deep-water and Arctic environments continue to yield new species. The Eastern Canada waters are the best-sampled province of Canada.

The Census of Marine Life was very active in Canada and significantly helped advance knowledge of marine biodiversity. The Barcode of Life developed barcodes for many Canadian species, and the Future of Marine Animal Populations (FMAP) programme provided many new insights on trends in fisheries, global patterns in biodiversity, and movements of animals in the oceans. The Pacific Ocean Shelf Tracking (POST) project provided new insight into movements of Pacific salmon species, sturgeon, and other species along the North Pacific coastline. The Canadian Healthy Oceans Network (CHONe) (Snelgrove et al., 2012), a five-year national research programme to establish biodiversity baselines in poorly sampled areas, grew out of the Census of Marine Life. Several other Census projects sampled in Canadian waters: include the Arctic Ocean Diversity (ArcOD) project, and the Natural Geography of Inshore Areas (NaGISA), and the Gulf of Maine Area (GoMA) project. The latter project assembled species lists for that region and worked closely with the Canadian scientists of the Ocean Biogeographic Information System (OBIS) program, which assembled extensive datasets produced by Fisheries and Oceans Canada over several decades.

Current monitoring programmes, largely by the Department of Fisheries and Oceans Canada (DFO), the lead agency responsible for monitoring Canada's three oceans (Atlantic, Pacific and Arctic) and freshwater habitats, will further improve knowledge of Canadian oceans. Many of DFO's monitoring activities were initiated to address operational requirements dealing with commercial exploitation of marine and freshwater populations, but over time many have evolved to provide assessments of the state of local ecosystems in the context of a consistent national approach. A general assessment of aquatic monitoring in Canada, conducted in 2005-2006 (Chadwick, 2006), provides an overview of the diversity of activities carried out by DFO and other agencies.

Most programmes that contribute to biodiversity assessments derive data from: (1) broad-scale regional multispecies bottom trawl surveys that provide information on the distribution and abundance of fish and invertebrate species, (2) oceanographic surveys that collect observations of phytoplankton and zooplankton abundance and taxonomic composition, and (3) single-species surveys that can yield knowledge for non-target species caught or observed during data collection. Most surveys are carried out by DFO, but in several instances, most importantly in the Pacific region, partner organizations contribute significantly, such as assessments of coral and sponge diversity distributions aided by academic and by non-governmental organizations (NGO) activities. Coverage varies greatly among aquatic environments.

Monitoring of the Atlantic and Pacific continental shelves and slopes is fairly extensive and generally conducted annually for focus areas. On the Pacific coast, Canada has maintained one of the longest – in duration – datasets that exist on ocean conditions, phytoplankton and zooplankton through the Line P/Station Papa programme. For other species groups, such as marine mammals, groundfish and salmon on the west coast, monitoring remains a DFO focus. Differences in the extent and intensity of survey activities in specific ecosystems within these two ocean areas will affect our ability to detect changes in biodiversity. For example, the coastal sea near Vancouver and the Fraser River salmon runs are the focus of sustained monitoring for many species; however, little information on the biodiversity of the northern west coast fjord exists. Furthermore, protocols for data collection, taxonomic resolution and expertise, and quality assurance vary greatly among survey types and location, which are also likely to significantly affect the estimation of Canadian marine biodiversity, particularly with respect to the occurrence of rare or difficult-to-identify species, including invasive species.

In all of Canada's oceans, information sources on habitat structure, invasive species, food web structure and interactions, species at risk, and any effects of cumulative anthropogenic impacts are limited. There are few systematic efforts to assess ecosystem health, particularly in nearshore and coastal areas, and data pertaining to pelagic species other than plankton are restricted in scope and coverage. Finally, almost all marine observations are collected from ships, yet the number of sea days declined by half between 1995 and 2005, while costs have doubled (Chadwick, 2006).

3.3   North Atlantic: The East Coast of the United States

The marine biodiversity of the United States is extensively documented; however, even the most complete taxonomic inventories are based on records scattered in space and time. The best-known taxa are those of commercial importance or large body sizes. Best-known areas are the shore and shallow waters. Measures of biodiversity other than species diversity, such as ecosystem and genetic diversity, are poorly documented. In the North-east Continental Shelf region, scientific sampling of coastal intertidal and shallow subtidal organisms extends back to the mid-1800s. Off-shore, early assessments in the late 1800s and early 1900s include those conducted by the *Fish Hawk*, the *Albatross*, and by Henry Bryant Bigelow.

In the last decade, the Gulf of Maine Area Program of the Census of Marine Life assessed this ecoregion, plus the southern and western Scotian Shelf, the continental slope to 2,000 m, and the western New England Seamounts. In the South-east Continental Shelf region, assessments began during the United States colonial period (seventeenth and eighteenth centuries). Early offshore studies focused on finding exploitable fish populations. In the late 1800s, exploratory surveys were aimed primarily at bottom-living organisms. Since the mid-twentieth century, the United States National Oceanic and Atmospheric Administration (NOAA) and its predecessor agencies (e.g., the Bureau of Commercial Fisheries) have explored habitats and their natural resources off the coast of the south-eastern United States. Beginning in the 1950s, several ships conducted exploratory fishing surveys using trawl nets; they found concentrations of snappers, groupers, and other economically valuable fishes, along with other significant fishery resources (drums, flatfishes, mullets, herrings, shrimps). Additional surveys using dredges, grabs, and other benthic samplers collected invertebrates and new species.

Valuable fish surveys have been carried out by the NOAA Marine Resources Monitoring, Assessment and Prediction (MARMAP) and Southeast Area Monitoring and Assessment Program (SEAMAP) monitoring programmes. Significant regional invertebrate surveys of the South Atlantic Bight (SAB) were conducted under the auspices of the Bureau of Land Management (BLM) and the Minerals Management Service (MMS). From the 1970s until now, surveys of the continental shelf and slope off North Carolina and in the tropical western North Atlantic have been made by the Duke University Marine Laboratory (DUML) and the Rosenstiel School of Marine and Atmospheric Sciences (RSMAS) of the University of Miami, respectively. The RSMAS collections and archives (Marine Invertebrate Museum: http://rsmas.miami.edu/divs/mbf/invert-museum.html) document the biodiversity of the Atlantic and Gulf of Mexico's tropical and deep-sea species and include material from the Straits of Florida and the Florida Keys National Marine Sanctuary. Marine resource agencies of the individual states have also conducted faunal and fishery surveys within state waters, particularly within estuaries (Fautin et al., 2010).

3.4   North Atlantic: The Gulf of Mexico

The most recent survey of the Gulf of Mexico's biodiversity appeared in book form (Felder and Camp, 2009), and as an open-access online database for utilization by anyone, as well as for updating and expansion by taxonomists (see GulfBase at www.gulfbase.org/biogomx; Moretzsohn et al., 2011). Over 15,400 species are listed in the database, with full biological and zoogeographical information for each species.

Historically, environmental studies or assessments on the Gulf of Mexico's biota can be divided into four different periods: (1) Exploratory Period (1850-1939), (2) Local Coastal Study Period (1940-1959), (3) Multidisciplinary Investigation and Synthesis Period (1960-2009/2010), and (4) Ecosystem Focus Period (2009/2010-present). The initial period involved the exploratory work of early oceanographically equipped ships, such as the *Blake* and the *Albatross*, and coastal expeditions from northeastern United States universities and institution. During the second period, over a dozen marine laboratories were established around the shores of the Gulf of Mexico, and scientists at these facilities expanded our biodiversity knowledge in those early locations in the United States, Mexico, and Cuba. Recently, a dedicated issue of the journal Gulf of Mexico Science (Volume 28, 2010) mapped the current 35 laboratories around the Gulf and presented a history of 21 of them; many are still instrumental in biodiversity assessments. Important fisheries vessels, such as the *Alaska* and the *Oregon I* and *II*, also operated in this second period and expanded our knowledge of faunal distributions in the region. In addition, although not comprehensive, the first biotic inventory of species in the Gulf of Mexico was published by Galtsoff (1954).

The third period involved large-scale, multidisciplinary investigations and synthesis projects in selected regions, primarily in the United States and Mexico. In the United States during the early to mid-1960s, the Hourglass Cruises were among the first large-scale projects, focused on the biota of the West Florida Shelf, and funded by the state of Florida. The United States Department of the Interior, Bureau of Land Management, Minerals Management Service, and now the Bureau of Ocean Energy Management, funded decades of environmental studies, including biotic surveys, mainly related to the oil and gas industry and its impact in the northern Gulf of Mexico. These studies first focused on the continental shelves, but as the oil and gas industry began exploring and producing in deeper and deeper water down the continental slope, studies focused on that area and out onto the adjacent abyssal plain. Those reports can be found at: http://www.data.boem.gov/.

The fourth period marks its beginning with the publication of the comprehensive inventory of all Gulf of Mexico species (Felder and Camp, 2009; an affiliate Census of Marine Life project) and the Deepwater Horizon blowout and oil spill in 2010, including the establishment of the Gulf of Mexico Research Initiative funded by BP with 500 million United States dollars to study the Gulf and its ecosystems over the next 10 years (at 50 million dollars per year) (See Appendix 1-Gulf of Mexico: selection of the major assessments or surveys - the Gulf of Mexico).

Gaps in knowledge include selected taxa and selected geographic areas or depths within the Gulf of Mexico. Similar to most well-studied areas, the larger taxa are well known, but smaller ones, particularly meiofauna and microbiota (viruses, microbes, fungi, benthic nematodes and harpacticoid copepods, etc.), gelatinous plankton and other soft-bodied invertebrates (that often do not preserve well in non-targeted sampling), as well as parasitic groups, are little known. Biomass, ecology, trophic interactions and diseases are poorly known for most species. Geographically, still not well known are the abyssal plain in the deepest part of the Gulf, and selected areas in the southern Gulf, such as off the northern coast of Tamaulipas and the very southern coast of Veracruz off the San Andres Tuxtlas Mountains.

3.5   North Atlantic: The Caribbean

Historically, knowledge of marine biodiversity for the Caribbean islands has resulted from inclusions within larger marine surveys and assessments funded by foreign institutions. For example, the Universities of Havana and Harvard (1938 to 1939) carried out a joint marine expedition which was the first such baseline information for the Cuban archipelago.

Additionally, some territories have benefited from visiting research vessels (e.g., the 1969 RV *John Elliott Pillsbury* expedition to the Lesser Antilles). More recently, local institutions dedicated to marine research have been established in several islands, such as: The Oceanology Institute (Cuba), the Institute of Marine Affairs (Trinidad) and The Discovery Bay Marine Lab (Jamaica). The Association of Marine Laboratories of the Caribbean (AMLC) (http://www.amlc-carib.org/) is an umbrella organization (with over 30 institutions) which has been promoting collaborations in marine sciences since 1968. Other organizations supporting research initiatives in the region are the International Oceanographic Commission-Caribe (IOCARIBE), the Nature Conservancy (TNC)), and several universities (e.g., the University of the West Indies island campuses of Mona in Jamaica, Cavehill in Barbados, and St. Augustine in Trinidad and Tobago).

To date not a single comprehensive marine assessment has detailed the Caribbean island territories, although several projects have targeted certain ecosystems and taxonomic groups. One of the most successful research programmes to date (1993 to present) is the Caribbean Coastal Marine Productivity Program (CARICOMP), which covers islands throughout the Wider Caribbean (e.g., Barbados, Dominican Republic, Jamaica, Puerto Rico (Mona Island) and Trinidad and Tobago) and has over 30 participating institutions. The project datasets include: percentage coral cover, sea urchin density, gorgonian density, seagrasses (growth, biomass and leaf area), mangrove forest structure and productivity, sea water temperature, salinity and clarity, daily maximum and minimum air temperature, and rainfall.

Furthermore, the Atlantic and Gulf Rapid Reef Assessment (AGRRA) Program is an international collaboration of scientists and managers aimed at determining the regional condition of reefs in the Western Atlantic and Gulf of Mexico and includes some Caribbean territories. Additionally, the Northern Caribbean and Atlantic node of the Global Coral Reef Monitoring Network (GCRMN) monitors coral reefs and their status in the Bahamas, Bermuda (United Kingdom), Cayman Islands (United Kingdom), Cuba, the Dominican Republic, Haiti, Jamaica and the Turks and Caicos Islands (United Kingdom). The Centre for Marine Sciences (CMS) at the University of the West Indies (UWI), at Mona, Jamaica, is the repository for these three databases (CARICOMP, AGRRA, GCRMN). Reef Check is another coral programme and is active in: Anguilla (United Kingdom), Antigua and Barbuda, Bahamas, Barbados, Belize, Dominica, Grenada, Haiti, Jamaica, Montserrat (United Kingdom), St. Kitts and Nevis, St. Lucia, St. Vincent and the Grenadines, and Trinidad and Tobago.

The Caribbean Sea Ecosystem Assessment (CARSEA) was a sub-global assessment (2005 to 2008) out of the global Millennium Ecosystem Assessment (MEA, 2001 to 2005) which made a major contribution to Caribbean biodiversity knowledge, and provided analytical status reports, including trends in some populations such as fish and coral reefs. The Census of Marine Life resulted in a detailed review of the known marine biodiversity of several Caribbean islands, including Bermuda (United Kingdom), , Cuba, the Dominican Republic, Jamaica and Puerto Rico (United States) and of the marine biodiversity along the Caribbean coasts of Colombia, Costa Rica, Mexico and Venezuela (Miloslavich and Klein, 2005). The nearshore (NaGISA) project followed shortly thereafter, with biodiversity surveys in Colombia, Cuba, Trinidad and Tobago, and Venezuela. This assessment examined patterns of biodiversity at both global and local scales on rocky shores and seagrass beds and made a major contribution to marine biodiversity (Miloslavich et al., 2010).

Overall, within the larger assessments (CARICOMP, CARSEA, CoML), certain ecosystems (mangroves, coral reefs, seagrass beds and rocky shores) have been studied in detail and certain marine taxonomic groups (marine mammals, turtles, seabirds, fish, corals, sponges) have also been comprehensively reviewed/assessed by local researchers, scientists and post-graduate students. Macrobenthic organisms for both Trinidad and Tobago and Jamaica, and plankton (for Jamaica) have been well documented. These important baseline data were improved by later surveys by the Institute of Marine Affairs in Jamaica (IMA) during the period 1980 to 1992.

Despite these efforts, there are still many gaps in our knowledge of Caribbean biodiversity (e.g. offshore and deep regions, small sized taxonomic groups) (Miloslavich et al., 2010). To fill these gaps and build regional capacity, almost all Caribbean countries have strengthened their environmental institutions (e.g., Coastal Zone Management Institute-Belize; Coastal Zone Management Authority-Barbados; Darwin Initiative by the Smithsonian Tropical Research Institute) and administrative capacities (e.g., the Environmental Management Authority-Trinidad and Tobago), to integrate environmental considerations into physical planning. Another initiative that has improved capacity building and aimed towards maintaining functional and structural integrity and biodiversity in this region is the Caribbean Large Marine Ecosystem Project (CLME). While not a monitoring programme, the CLME Program has one pilot project on Reef Biodiversity and Reef Fisheries which implemented the ecosystem-based approach for the conservation and effective management of coral reef ecosystems and associated resources.

3.6 North Atlantic: Oceans around Europe

There is early evidence of European marine biota assessments from the 3rd century B.C., but formal scientific studies did not begin until the 18th century in the Mediterranean region and early 19th century throughout the remainder of Europe (Coll et al. 2010, Ojaveer et al., 2010). For several taxa these early works provide historical baselines against which to compare contemporary biodiversity data.

In the European Atlantic, as mentioned earlier in the plankton section, the longest time-series sets exist for planktonic organisms. Regarding benthic organisms, in shelf seas, such as the North and Baltic Seas, benthic survey and monitoring programmes have been in place. One of the priority research areas, directly linked to provision of management advice, in European seas has been commercial fish and fisheries. The related surveys include egg and larval fish surveys, young fish surveys, experimental bottom trawl surveys and, more recent hydro-acoustic surveys. Some of these data-sets date back to before the 1950s. However, information on non-commercial fish is scarce and incomparable and should be considered as a major drawback in drawing conclusions about the spatio-temporal patterns and dynamics of fish communities. A representative overview of the status and trends of non-indigenous species in European waters is assembled in the database "Information system on aquatic non-indigenous and cryptogenic species in Europe", AquaNIS.

In the North-East Atlantic, the Convention for the Protection of the Marine Environment of the North-East Atlantic (OSPAR, formerly the Oslo-Paris Convention) established in 1972, is aimed towards the conservation of the marine environment and its resources. As such, OSPAR has pioneered ways of ensuring monitoring and assessment of the quality status of the seas, including the implementation of a Biodiversity and Ecosystem Strategy under the coordination of the Joint Assessment and Monitoring Program (JAMP).

3.7 North Pacific: focus on the West Coast of the United States

3.7.1 The Gulf of Alaska

There have been many scientific expeditions to the Gulf of Alaska over the years since early times and a historical review of scientific exploration of the North Pacific Ocean from 1500 to 2000 is available. Early explorations were carried out mostly for mapping and species identification (i.e. fishes, birds, and invertebrates). Marine survey expeditions in the late 1800s include the United States steamer *Tuscarora* in the Aleutian Trench, the *Albatross*, and the Harriman Alaska Expedition from Seattle through Prince William Sound, out to the Aleutians, and north along the Russian Federation coast of the Bering Sea. In the 1950s, major expeditions were carried out by NORPAC (North Pacific), the Japanese research vessel *Oshoro Maru*, the University of Washington's *Brown Bear*, the Russian Federation research vessel *Vityaz*, the Bering Sea Commercial Research Expedition, and the Pacific Research Institute of Fisheries and Oceanography (TINRO). Recent survey programmes are funded by MMS and NOAA. The MMS Outer Continental Shelf Environmental Assessment Program (OCSEAP) began in 1974 and is still active. The bottom trawl surveys run by NOAA collect information on fishes and many species from the Bering Sea, the Aleutians and Gulf of Alaska to support fishery management decisions by the North Pacific Fishery Management Council and the United States Secretary of Commerce. Biodiversity information has also been collected by the *Exxon Valdez* Oil Spill Trustee Council in Prince William Sound, continuous plankton recorder surveys across the North Pacific, Seward Line zooplankton collections in the Gulf of Alaska, and Hokkaido University's annual training cruises on the *Oshoro Maru* to the Bering Sea and Strait and, less frequently, to the Chukchi Sea. More recently, the NOAA Office of Ocean Exploration supported cruises (2002-05) to study biodiversity implementing the use of ROVs.

Large-scale research programmes off Alaska (see, e.g., the joint National Science Foundation–North Pacific Research Board Bering Sea study at http://bsierp.nprb.org) contribute to broader knowledge of biodiversity, and continue adding to the many efforts over the past 40 years to enumerate species from the coastal rocky headlands to the deep ocean and even in sea ice. However, no species inventory of all realms exists for any region of Alaska. Important databases containing biodiversity information are listed in Fautin et al. (2010) and efforts are underway to compile data in databases (e.g., Alaska Resources Library and Information Services - ARLIS; the *Exxon Valdez* Oil Spill Trustee Council - EVOSTC).

3.7.2 The California Current ecosystem:

Early surveys in this region began in the late 1700s and 1800s by European explorers, including Cook, La Perouse, Vancouver, and Bodega y Quadra. In the 1800s, United States naval expeditions collected information on fishes and whales. In the 1900s, marine research laboratories were established (e.g., Hopkins Seaside Laboratory in Monterey, California) which today form the Monterey Bay Crescent Ocean Research Consortium. Today, 40-50 marine research facilities operate in the region under the umbrella of the Western Association of Marine Laboratories. At present, many of the available long-term data are a product of fishery management efforts, mostly funded by NOAA. Biodiversity databases of this region are listed in Fautin et al. (2010). Two major assessments in this region are the California Current Ecosystem Long Term Ecological Research (CCE LTER) and the California Cooperative Oceanic Fisheries Investigations programme(CALCOFI), both focused on the pelagic realm. The CalCOFI programme is a 60+ year survey including zooplankton with strong relations to biodiversity and a world-recognized data base allowing analysis of temporal trends (Kang and Ohman, 2014).

3.7.3 Insular Pacific–Hawaiian Large Marine Ecosystem:

Initial surveys of the Hawaiian Islands began in the early 1800s by French, Russian, and United States expeditions. The first plankton samples were taken by the *Challenger* in mid-1875, while major collections from Hawaii were initiated by the *Albatross* Expedition in the early 1900s. Between 1923 and 1924, four trips were made with the *Tanager* to survey 13 Hawaiian Islands, Johnston Atoll and Wake Island. Results from the *Tanager* expedition were published in Marine Zoology of Tropical Central Pacific, and included crustaceans, echinoderms, polychaetes, and foraminiferans. Between July and September 1930, an expedition led by P.S. Galtsoff to Pearl and Hermes, surveyed the abundance of pearl oysters for potential commercial use, and also noted the corals, algae, sponges, molluscs, crustaceans, and echinoderms.

Since these early cruises, conducting inventories of the biota of Hawaii has largely been the responsibility of the Bishop Museum, which at present has been designated the Hawaii Biological Survey (HBS). Surveys have occurred in targeted sites in the main Hawaiian islands, such as Kaneohe Bay and Pearl Harbor on the island of Oahu, and waters around the island of Kahoolawe.

Since 1995, surveys have also covered Midway Atoll, French Frigate Shoals, and Johnston Atoll. Electronic datasets for Hawaiian marine biodiversity include: http://hbs.bishopmuseum.org/ (Hawaii Biological Survey); http://cramp.wcc.hawaii.edu/ (Reef Assessment and Monitoring Program); http://www.nbii.gov/portal/community/Communities/Geographic_Perspectives/Pacific_Basin/ (National Biological Information Infrastructure (NBII), Pacific Basin Information Node); and http://www.nbii.gov/portal/community/Communities/Habitats/Marine/Marine_Data_ (OBIS-USA)/. Intensive biological inventories have been carried out on fishes, stony corals, crustaceans, and molluscs (Fautin et al., 2010).

3.8 North Pacific: focus on Japan

In Japan, nationwide censuses of biodiversity of coastal areas, such as tidal flats, coral reefs, seagrass and algal beds, were conducted by the Ministry of the Environment, and showed long-term decline of these important habitats during the 1970s-1990s. However, the survey frequency was insufficient to identify the causal mechanisms of changes in relation to various environmental factors. Since 2002, the Ministry started a new type of monitoring programme, called "Monitoring Sites 1000" which aims to monitor the 1000 most important ecosystems in Japan over the whole 21st century. In this programme, ca. 50 coastal sites, including tidal flats, rocky intertidal shores, seagrass beds, algal beds and coral reefs are being monitored annually over the long term. These data will be utilized for various purposes, such as the prediction of coastal ecosystem response to global climate changes and other more local factors, as well as the impact assessment of the catastrophic disturbance by the 2011 earthquake and tsunami.

However, the number of sites is too small to set out in detail the changes in the coastal areas of the entire Japanese coast. In the meantime, local prefectural governments, fisheries agencies and certain NGOs have been conducting assessments of local coastal habitats of their areas, although the systems for sharing the information gathered are not well established at present. Certain ongoing network activities, such as the Japan Biodiversity Observation Network (JBON) and the Japan Long-term Ecological Research Network (JaLTER), are expected to collect this scattered information for use in developing integrated analyses of coastal ecosystem changes (Fujikura et al., 2010) (see Appendix 1-Japan for a list of assessments).

3.9 North Pacific: focus on China

The marine biological investigations in China started late. Until the early twentieth century, only limited areas had been explored and scattered taxonomic groups had been collected and researched. Additionally, due to the lack of special research institutes and taxonomists, many precious samples were lost.

From 1919 to 1949, some independent investigations and research on marine biological work in China were conducted. This period saw the real beginning of marine biological research in the country; the first qualitative benthic trawling investigation was launched in this time. But research conditions were very precarious and no research vessels for marine or fisheries science existed; therefore, surveys were relatively limited. During this period, research mainly focused on the coastal areas of Qingdao, Yantai, Xiamen, Beidaihe and Hainan Island.

Since the establishment of the People's Republic of China in 1949, many marine research institutions were set up gradually (e.g. Chinese Academy of Sciences - Institute of Oceanology (IOCAS) and South China Sea Institute of Oceanology (SCSIOCAS), State Oceanographic Administration (SOA), Ocean University of China (OUC), Chinese Academy of Fishery Sciences, etc.). Comprehensive oceanographic surveys were carried out from the 1950s to the 1980s. The National Comprehensive Oceanographic Survey (1958-1960; the First National Marine Census) was the first large-scale national comprehensive marine survey with participation of over 60 organizations and more than 600 researchers, which covered most coastal areas of the China seas north to the Taiwan Strait and most parts of the northern South China Sea. The biological investigation of this survey included assessments of plankton, benthos and nekton. More than 200,000 biological specimens were collected.

Other large-scale comprehensive marine surveys include the National Coastal Zone and Beach Resources Comprehensive Survey (1981-1987) and the First National Island Resources Comprehensive Survey (1988-1996). These two surveys covered over 50,000 km^2, and involved microbial, planktonic, benthic and nektonic community investigations. These surveys investigated the coastal and island natural environments from China, and comprehensively evaluated the quantity, quality and distribution of biological resources. By the late 1980s, most of the waters under the jurisdiction of China had been investigated, and the diversity, distribution and utilization of the main marine biological species were roughly identified.

From the 1990s to date, large-scale comprehensive marine surveys include: the Continental Shelf Environment and Living Resources Survey (1997-2000), the National Offshore Comprehensive Oceanographic Survey and Evaluation (2004-2010), also referred to as the Second National Marine Census, and the ongoing Second National Island Resources Comprehensive Survey. These surveys were very intensive and thorough, providing supplemental data to the earlier efforts.

In the past 20 years, more regional investigations, including assessments in Bohai Gulf, Liaodong Bay, Jiaozhou Bay, Changjiang Estuary, Dayawan Bay, Quanzhou Bay, Hainan Island, and some islands in the South China Sea have been performed, with continued study in several regions. More studies were focused on particularly diverse habitats, such as coral reefs, mangrove forests and seagrass beds. Oceanographic exploration is reaching further into areas adjacent to China's seas, including in the West Pacific Ocean, Indian Ocean, even the North and South Poles, as well as cold seeps and seamounts in the deep sea of South China Sea.

Although significant advances have been made in marine biodiversity research in China since the 1950s, much insufficiency still remains. First, the marine biological specimen collection and biodiversity research is considered inadequate, especially from coral reefs, the deep sea and other special habitats. Second, the current investigations are considered as lacking systematic and thorough data publication. Third, the phylogenetic and biogeographic studies on marine living organisms are insufficient. Last, supervision and conservation are weak, and many species are critically endangered (Liu, 2013) (see Appendix 1-China, for a list of assessments).

3.10   Indian Ocean: focus on India

The two major institutions concerned with surveys and inventories of the fauna and flora in India are the Zoological Survey of India and the Botanical Survey of India, along with other research organizations, such as the Central Marine Fisheries Research Institute and the National Institute of Oceanography.

The published literature on coastal and marine biodiversity of India comprises an inventory indicating that 17,795 species of faunal and floral communities were reported from seas around India (see Appendix 1-Species diversity India). The taxonomy of many of the minor groups, particularly invertebrates (especially sponges, octocorals, ctenophores, tunicates, polychaetes and other worms, as well as small size invertebrates) remain a challenge to specialists; as a result these taxa continue to be inadequately known from Indian seas. However, considerable knowledge on the taxonomy of groups, such as seaweeds, seagrasses, mangroves, scleractinian corals, crustaceans, molluscs, echinoderms, fishes, reptiles and marine mammals, is available in India.

Most of the marine biodiversity data come from surveys that sample up to 200 meters. There are large data gaps for smaller taxa and for large parts of the shelf and deep sea ecosystems, including seamounts (Wafar et al., 2011). The data provided in this paper warrant continued taxonomic research on the least-studied and unknown groups, in light of current threats to marine biodiversity. The full extent of biodiversity in any of the world's oceans may never be known, and the rate at which our understanding is increasing (Keesing and Irvine, 2005) is likely to be lowest in Indian seas. The impacts of climate change will alter coastal marine ecosystems, affecting the range of species and their ecology at a rate faster than it is possible to record their presence and abundance (Keesing and Irvine, 2005). In conclusion, it is evident that comprehensive taxonomic coverage of the marine biota of the entire region remains a monumental task, beyond the capacity of existing local taxonomic expertise.

Thus, to gain an appreciable knowledge on the patterns of diversity in the region, it will be necessary to identify indicator species to assess responses to unpredictable climate change. It would be quite appropriate to plan systematic studies rather than continue the present system of haphazard and opportunistic description of new species as and when they are discovered.

Within the largest Indian Ocean basin, the International Indian Ocean Expedition (IIOE) was held during years 1962-1965. This expedition was one of the greatest international, interdisciplinary oceanographic coordinated research efforts to explore the Indian Ocean in almost all disciplines in the marine sciences. The culmination of IIOE led to birth of the National Institute of Oceanography at Goa, the first regional institute for oceanographic research. At present, the plan for a second International Indian Ocean Expedition (IIOE2) has been drafted by the Science Council for Oceanic Research (SCOR) and will include more biological aspects than the first, particularly in marine biodiversity. The need for these expeditions as well as other continued studies are of overall importance in a region recognized by having growing concerns on food security, biodiversity loss, coastal erosion and pollution, along with a pressing need of conservation for tourism and sustainable fisheries.

3.11   Sub-Saharan Africa

Along the coastline of Sub-Saharan Africa, states of knowledge of marine biodiversity vary dramatically between the east, southern and west coasts of Africa. The biota of the east coast is moderately known. A general field guide to marine life in the region exists and two reviews have attempted to tabulate and assess states of knowledge of regional marine biodiversity. However, this listing is far from complete. Some well-known taxa, such as reptiles, birds and mammals, are simply omitted from the tabulation. Other larger and/or more economically valuable taxa, such as seaweeds, flowering plants, fishes, corals, larger molluscs and crustaceans, etc., are probably fairly accurately represented.

However, many smaller and difficult-to-identify taxa are not included in the lists at all (for example, Nematoda, Copepoda and Ostracoda) or are likely to be severely under-represented, and probably less than half of the actual numbers of marine species present in the region have been described. Notable regional differences in sampling effort are found: Kenya, United Republic of Tanzania and southern Mozambique are the best-sampled regions, and northern Mozambique and especially Somalia are the least studied. In all regions, sampling effort declines rapidly with depth and distance from the coast; the deeper continental slope and abyssal habitats are almost completely unexplored. Regional taxonomic capacity is very limited and adequate marine collections in regional museums are lacking.

The marine biota of South Africa is by far the best known on the continent. The region has a relatively strong history of marine taxonomic research and a reasonably comprehensive and well-curated museum collection network, totalling some 291,000 marine records. South Africa has more than a dozen institutions with a strong focus in marine science (e.g. South African Institute of Aquatic Biodiversity or SAIAB, formerly the JLB Smith Institute of Ichthyology), with the largest concentration of marine scientists found in the Cape Town region.

Several regional guides to marine life, such as Branch et al. (2010), list more specialized taxonomic monographs. The regional data centre, AfrOBIS, houses some 3.2 million records of more than 23,000 species. These are derived from the wider African region, although the vast majority of data points originate from within South Africa. The total number of recorded marine species stands at 12,914 and these are tabulated by taxonomic group by Griffiths et al. (2010), who also list taxonomic resources and experts for the region. Many groups, particularly of smaller invertebrates, still remain poorly studied, however. In terms of regional coverage, shallow waters have been relatively well sampled, but sampling effort declined dramatically with increasing depth: 99 per cent of all samples have been taken in depths shallower than 1000 m. The 75 per cent of the EEZ that lies deeper than 1,000 m thus remains extremely poorly explored and is a priority for future research.

The marine biota of West Africa is poorly known. No regional marine guide exists and no comparative analyses of regional marine biodiversity have been compiled. Some reports purport to list the biodiversity for various individual countries in the region, but these are clearly superficial and fail to adequately reflect the diversity of smaller taxa. For example, in the Namibian EEZ, only 1,053 species are documented, of which more than half are fishes. This amounts to less than 10 per cent of the total known from South Africa, where fish comprise less than 20 per cent of the recorded taxa, indicating that the Namibian estimate is strongly biased towards larger, more conspicuous taxa. Similar biases are evident in other national estimates, which appear to radically underestimate smaller, less conspicuous components of the biota and to concentrate on fishes and other 'target species'. This entire region probably remains amongst the least explored of coastal marine areas and a pressing need remains for taxonomic study of most invertebrate groups in the region. As with other regions sampling effort in waters deeper than 1,000 m is particularly lacking (see Appendix 1-Africa for a summary of assessments).

3.12 South Pacific: focus on Australia and New Zealand

3.12.1 Australia

Although Australia has the world's third largest EEZ, extending more than 5,000 km from the tropics (9°S) to temperate latitudes (47°S), it has a comparatively small marine survey capacity. At the same time, Australia has been very active in progressing marine conservation planning (called Marine Bioregional Planning), including the identification of a network of representative marine reserves covering 36.4 per cent of the EEZ. The need to support marine bioregional planning encouraged researchers to recover, validate and make accessible older surveys going back to the 1950s. Some of the most widespread data are associated with 4,000 exploratory demersal trawls by Russian Federation fishing vessels for the period 1965-78. Cooperation with scientists in Vladivostok, in the Russian Federation, and in Australia enabled data for earlier surveys to be made accessible and, where necessary to be aligned with modern taxonomy.

Aboriginal people accumulated much knowledge of Australia's flora, fauna, and ecological systems, including those of its "sea country," over the last 40,000 to 60,000 years, but much of this knowledge and understanding remains cryptic (Butler et al., 2010). European scientific study began with the first scientifically staffed voyages of discovery, notably those of James Cook in 1770, Nicolas Baudin in 1801–03, and Matthew Flinders in 1802. Charles Darwin visited Australia in the *Beagle* in 1836. The voyage of HMS *Challenger*, 1872–76, included Australian samples in its global investigation of the deep sea, and its reports are a basis of many disciplines. Soon after the British established the colony of New South Wales in 1788, scientific societies and natural history museums entered an active period of research. Discovery in the sea was more difficult and more limited than on land, but there was much activity during the twentieth century.

The taxonomy and descriptive ecology of organisms on accessible shores were an early focus, which has developed into a strong tradition of experimental ecology on seashores and in shallow water, as well as a determined effort to produce identification guides (Butler et al., 2010). The study of plankton and of benthopelagic coupling is less well developed than the study of benthos in Australia. Publications on phytoplankton have been available in Australia since the 1930s, but species lists are available only for limited locations. Research on zooplankton ecology has increased recently and several transects are now surveyed regularly with the Continuous Plankton Recorder. Mesopelagic organisms are being assessed on several cross-Tasman transects from commercial vessels using standardized mid-water acoustic survey techniques supported by periodic mid-water trawls. These two standardized approaches are part of the Integrated Marine Observing System (IMOS; www.imos.org.au).

The diversity of sources from which Australian marine biodiversity data are obtained means that there are few repeat surveys – typically each survey has set out to answer a particular research question with scant regard to long-term comparison. A notable exception is the Australian Institute of Marine Science (AIMS) Long-Term Monitoring Program; it recently analyzed 2,258 standardized surveys from 214 different reefs between 1985 and 2012 and showed that coral cover had declined from 28.0 per cent to 13.8 per cent (0.53 per cent y-1). This programme, together with the much newer IMOS, are the only long-term sustained monitoring programmes in Australian waters, although individual researchers have conducted repeat surveys using standardized sampling techniques for individual research projects, or have collated a variety of historical data sources to answer particular questions. Australian scientists recognize the need to develop longer time-series of survey data to support national State of the Environment reporting and to measure the effectiveness of the marine reserve network. This will require increased capabilities and capacity for biological sampling, which need to be brought to a similar level of standardization, replication, sustainability, interpretation and communication, as has been achieved by physical oceanographers.

Beginning in the first half of the twentieth century, energetic research was targeted at fisheries by Australian state agencies and by CSIRO's Division of Fisheries and its predecessors (Mawson et al., 1988). Although searching for commercial prospects, this work collected many non-commercial fish and invertebrates that were lodged in museums throughout the country, including the Australian National Fish Collection at CSIRO Marine and Atmospheric Research (CMAR). These fish collections have recently provided the most comprehensive and useful biological dataset for bioregionalization of Australian waters. In the 1960s, a period of intensive environmental research began, targeting in particular, bays, estuaries, and continental shelf near major capital cities (Wilson, 1996).

More recent work has explored deeper waters, with interests in exploration, the conservation of biodiversity and research on sustainable fisheries, more recently as part of the Australian Government's National Environmental Research Program Marine Biodiversity Hub (and predecessors), set up to provide the scientific information to support government policy and decision-making. Thus, museums are building important collections of Australian specimens from depths as great as 2,000 m and, in restricted parts of the shelf and slope, quite comprehensive faunal collections. An important component of the taxonomists' work, besides describing the 30-50 per cent of species that are new to science found on each survey, has been to provide regionally consistent descriptions of species so that broader bioregional patterns can be established. Although this has been available for fish species for many years, supported by genetic barcodes, it has only recently been possible for some invertebrate taxa.

With the declaration of Australia's Commonwealth Marine Reserve network (http://www.environment.gov.au/marinereserves/), survey emphasis is shifting from discovery to monitoring (or establishing the first quantitative baseline). Non-destructive sampling approaches, including autonomous underwater vehicles (e.g., www.imos.org.au/auv) and possibly genetic approaches, will be important additions to what will remain Australia's most prevalent deepwater activity – commercial fisheries which will continue outside the marine reserve network and inside the network in multiple-use areas, collecting data from their fishing operations and additionally, through cooperation with scientists, to routinely collect scientific information. In shallower waters it is likely that standardized citizen science will become increasingly valued.

3.12.2　New Zealand

The New Zealand's EEZ is one of the largest in the world. Despite important exploration efforts begun more than 200 years ago by James Cook followed by Louis Duperry and Dumont D'Urvillefor, Charles Darwin, the Challenger, and continued at present, much of this region remains unexplored biologically, especially at depths beyond 2,000 m. The major oceanographic data repository is the National Institute of Water and Atmospheric Research (NIWA), which is also data manager and custodian for fisheries research data owned by the Ministry of Fisheries. Museum collections in New Zealand hold more than 800,000 registered lots representing several million specimens. During the past decade, 220 taxonomic specialists (85 marine) from 18 countries engaged in the review of New Zealand's entire biodiversity, which ended in a major three-volume publication (Gordon, 2009). Current marine biodiversity in New Zealand surpasses the 17,000 species, and a list of all described New Zealand marine Animalia is available through OBIS (Gordon et al., 2010).

Multiple surveys (2000-2008) have been commissioned by the Ministry of Agriculture and Forestry Biosecurity New Zealand (MAFBNZ) in ports to detect alien species which have generated baseline information of species composition in these areas. At present, marine research (including marine biodiversity assessments) has a significant momentum in New Zealand with funding from the Ministry of Fisheries, the Foundation of Research, Science and Technology, the Ministry of Agriculture and Forestry, Biosecurity New Zealand, and the Universities Performance Based Research Fund. The Ocean Survey 20/20 programme(administered by Land Information New Zealand), could perhaps be noted as one of the most significant biodiversity assessments carrying out biodiversity sampling and habitat mapping in the New Zealand EEZ and Ross Sea/Southern Ocean on a yearly basis. Large areas have been surveyed on the Chatham Rise, Challenger Plateau (down to about 1,200 m), the Ross Sea and Southern Ocean (down to about 3,500 m), and currently in a large area of the northeastern North Island shelf out to 200 m. Data from many of these surveys is still being processed (Gordon et al., 2010).

3.13　Oceans around South America

The first studies of the South American coastal biota were carried out during a series of expeditions by European and North American researchers in the late 1700s and the first half of the 1800s, with naturalists Alcyde d'Orbigny, Alexander Von Humboldt, Aimé Bonpland, and Charles Darwin, among others. In the late 1800s, several other important oceanographic expeditions, including the HMS *Challenger*, collected samples along the coasts of Ecuador, Peru, Chile, Argentina, Uruguay, and Brazil. In the 1900s, the Deutsche Südpolar Expeditions in 1901–03, the Swedish Lund University expedition to Chile in 1948–49, the Royal Society Expedition to Southern Chile, the Soviet Antarctic Expedition in 1955–58, and the Calypso campaigns in 1961–62 were among the most significant European expeditions to South America. Other important campaigns during the second half of the twentieth century which increased the knowledge of marine biodiversity and strengthened the local research capacities were carried out by the R/V *Academik Knipovich* (1967), the R/V *Almirante Saldanha* (1966), the R/V *Atlantis II*, (1971), the R/V *El Austral* (1966–67), the R/V *Vema* (1962), and the R/V *Walther Herwig* (1966–71). At present, the oceanographic vessel *Polarstern* from the Alfred Wegener Institute (Germany) has been carrying out exploration voyages to the southern regions of the continent and the Southern Ocean for more than 20 years.

In the northern latitudes of the continent, the Tropical Eastern Pacific (TEP) Biogeographic Region has a rich history of oceanographic and biological explorations dating back to the voyage of Charles Darwin to the Galápagos Islands aboard the HMS *Beagle* in 1835 and the Eastern Pacific Expedition of the United States National Museum of Natural History in 1904 aboard the United States Fish Commission steamer *Albatross*. A series of research cruises and expeditions organized by North American institutions in the first half of the twentieth century contributed greatly to the discovery and knowledge of the marine fauna and flora existing in the rich area between the low-tide mark and 200 m of depth in the Panama Bight, including Panama, Colombia, and Ecuador (e.g., the *Saint*

George to Gorgona Island in 1927, the Allan Hancock cruises aboard the *Velero III* and *IV* vessels (1931-1941), the Askoy Expedition of the American Museum of Natural History in 1941). Taxonomic and ecological studies have been carried out in the last three decades in Costa Rica, Panama, Colombia, and Ecuador, mostly in the Gulf of Nicoya, the Bay of Panama, the Pearl Islands, the Bay of Buenaventura, Gorgona Island, and the Gulf of Guayaquil.

Important collections or libraries of regional marine fauna are maintained by the Los Angeles County Museum, the Scripps Institution of Oceanography at La Jolla, California, the California Academy of Sciences in San Francisco, and the Smithsonian Tropical Research Institute (STRI) in Panama City.

In the Tropical Western Atlantic (TWA), the natural history of Guyana (formerly British Guiana) was described by early explorers Sir Walter Raleigh (circa 1600) and Charles Waterton (early 1800s), who reported his discoveries in the book "Waterton's Wanderings in South America". In French Guiana, the first studies were carried out after World War II, for fish inventories and later, in the 1950s, on benthic (mostly shrimps) and demersal continental shelf fauna, from 15 to 100 m depth. The Venezuelan Atlantic Front was until recently almost completely unexplored, and the little information available concerned commercially valuable species of fish and shrimp.

In the southern part of the continent, the local and regional academic community also had important historical representatives and in the 1900s, research on coastal biodiversity received a strong stimulus due to the immigration of many European scientists who contributed to knowledge and capacity-building mainly through their involvement in local universities and natural science museums. Although a few research institutions were established in the region early in the twentieth century, such as the Smithsonian Tropical Research Institute (STRI) in Panama (1923), the most important stimulus to regional, autochthonous marine science was given by the establishment of several marine research institutions, mostly in the 1950s and 1960s. These institutions changed the way that marine science was done by incorporating time series of the environmental variables and their effect on biodiversity into the traditional taxonomic studies.

In the 1960s, the Food and Agriculture Organization of the United Nations began to develop projects giving an impetus to fisheries, especially in the southwest Pacific, an upwelling zone of extraordinary productivity that was responsible for 20 per cent of the world's fisheries by the end of that decade. In the 1980s and 1990s, centres for marine biodiversity research were created along the coasts of several countries, especially Brazil, Argentina, and Chile. The natural history museums in South America have been fundamental to preserving the regional marine biodiversity patrimony, both in collections and in the literature, and are considered to be taxonomically indispensable.

Some of the most relevant museums are the Museo de La Plata and the Museo Argentino de Ciencias Naturales (Argentina), the Museo de Historia Natural (Quinta Normal) in Chile, the Museo Dámaso Larrañaga and the Museo de Historia Natural in Uruguay, and the Museo de Boa Vista (Brazil). Other important collections are held at research institutions such as the STRI in Panama, the Instituto del Mar del Perú (IMARPE) in Peru, the Instituto de Investigaciones Marinas y Costeras (INVEMAR) in Colombia, and at universities.

Today, South America benefits greatly from regional cooperation. One example of cooperation was the Census of Marine Life that incorporated the region into several of its field projects (e.g., Shore Areas, Antarctic Life, Continental Margins, Marine Microbes (ICoMM), and the Mid-Atlantic Ridge Ecosystem (MAR-ECO) projects), which all contributed greatly to increasing the knowledge of marine biodiversity in the region. South America also has contributed nearly 300,000 records to OBIS from almost 7,000 species through its regional node.

At present, some of the main marine biodiversity assessments carried out in the region are:

(a) SARCE: South American Research Group in Coastal Ecosystems (regional); since 2010. Aimed to study biodiversity and ecosystem function in the intertidal zone of rocky shores: http://sarce.cbm.usb.ve/;
(b) Pampa Azul: South Atlantic (Argentina, approved in 2014). Aimed to carry out research for conservation purposes along with technology development and outreach;
(c) SIMAC: Sistema Nacional de Monitoreo de Arrecifes Coralinos en Colombia; since 1998. Yearly sampling to monitor state of coral reefs. Taxa monitored include corals, macroalgae, invertebrates, and fish;
(d) IMARPE (Instituto del Mar del Perú) and Universidad Nacional Mayor de San Marcos have initiated at least four projects characterizing marine biodiversity in several areas along the Peruvian coasts since 2009/2010 focused on benthic groups.
(e) The Colombian National Authority for Aquaculture and Fisheries (AUNAP) in conjunction with INVEMAR carry out scientific research programmes to evaluate the Colombian potential to take advantage of new marine fishery resources such as tuna, dolphin fish, billfish, snappers, groupers and other fish of high commercial importance.

The objective of these programmes is to establish the current status of these resources in order to take management measures to promote the extraction of unconventional resources and discourage fishing pressure on those resources that are over-exploited. These studies provide not only information on highly important commercial species, but also a characterization on the status of marine biodiversity in the exclusive economic zone of Colombia.

3.14 The Southern Ocean

Whilst the economic exploitation of Antarctica's marine resources dates back to the 18th century, scientific research on its marine ecosystems only began in the mid-19th century. The HMS *Challenger*, the *Belgica* and the *Discovery* Investigations were among the first to undertake systematic sampling of the marine biology in the region. Taxonomic studies from these early expeditions provide the foundations of modern taxonomy in the Southern Ocean. Advances in technology, such as SCUBA diving, ice-capable research vessels and underwater imagery from remotely operated vehicles, have heralded a new era for marine ecological work in polar regions. Together with the recognition by the Scientific Committee on Antarctic Research (SCAR) and some national agencies of the importance of fundamental taxonomy, the rate of discovery and description of new species in the Southern Ocean has increased significantly.

In the framework of the Census of Marine Life a Decade of Discovery, life in the world's oceans has been investigated and questions about the known, the unknown and the unknowable have kept marine researchers busy. Many resources were made available for future research within the CoML community, but also for the public and policy-makers. One of the flagship projects was the five-year Census of Antarctic Marine Life (CAML), which investigated the distribution and abundance of Antarctica's marine biodiversity, how it is affected by climate change, and how change will alter the nature of the ecosystem services currently provided by the Southern Ocean for the benefit of mankind. In this framework and within the International Polar Year 2007-2009, 19 research voyages were coordinated by CAML, involving more than 400 biologists from over 30 nations.

The CAML community explored the unknown bathyal and abyssal Southern Ocean (SO) and many shallow sites. Within the project about 16,000 SO taxa were identified and included in a database of Antarctic Marine Life; see the SCAR-Marine Biodiversity Information Network (www.scarmarbin.be). The CAML projects barcoded more than 3,000 species, a SO Plankton Atlas was established, life underneath the collapsed Larsen A and B ice shelves was studied and many scientists worked on the biodiversity, biogeography and conservation of various marine taxa.

Moreover, more than 700 species new to science were discovered (Brandt et al., 2007) and new and unknown habitats were explored, e.g., the SO deep sea, and the Amundsen Sea. The lasting legacy of CAML is a benchmark, a system (or database) for monitoring change in the SO. Another major legacy of the CAML project and the SCAR Marine Biodiversity Network is the Biogeographic Atlas of the Southern Ocean (De Broyer and Koubbi, 2014) which compiles in more than 80 chapters an extensive review of the state of knowledge of the distributional patterns of the major benthic and pelagic taxa and of the key communities in the SO within an ecological and evolutionary framework. The Atlas relies on vastly improved datasets, and on insights provided by innovative molecular and phylogeographic approaches, and new methods of analysis, visualisation, modelling and prediction of biogeographic distributions. A dynamic online version of the Biogeographic Atlas will be hosted on www.biodiversity.aq.

The development of molecular techniques is a technical advance which promises to revolutionize work on the diversity and biogeography of Antarctic marine biota. CAML supported these efforts through its DNA Barcoding program. This technology is rapidly evolving and becoming ever more sophisticated. In particular such work is starting to uncover a wealth of cryptic species within what were once regarded as single, widely distributed species. Not only does this work increase the known species richness of the SO, but it also changes biogeographic patterns (typically reducing the range size or depth range) and hence affects

The boundaries and names shown and the designations used on this map do not imply official endorsement or acceptance by the United Nations.

Figure 35.5 | The total number of all marine sample sites and species found within each cell of a 3° of latitude by 3° of longitude grid, from distribution records in SCAR-MarBIN. The red line indicates the mean position of the Antarctic Polar Front, which defines the maximum northern extent of this study. A. Sample sites. B. Species richness (total number of all marine species recorded in that cell). Taken from Griffiths, 2010.

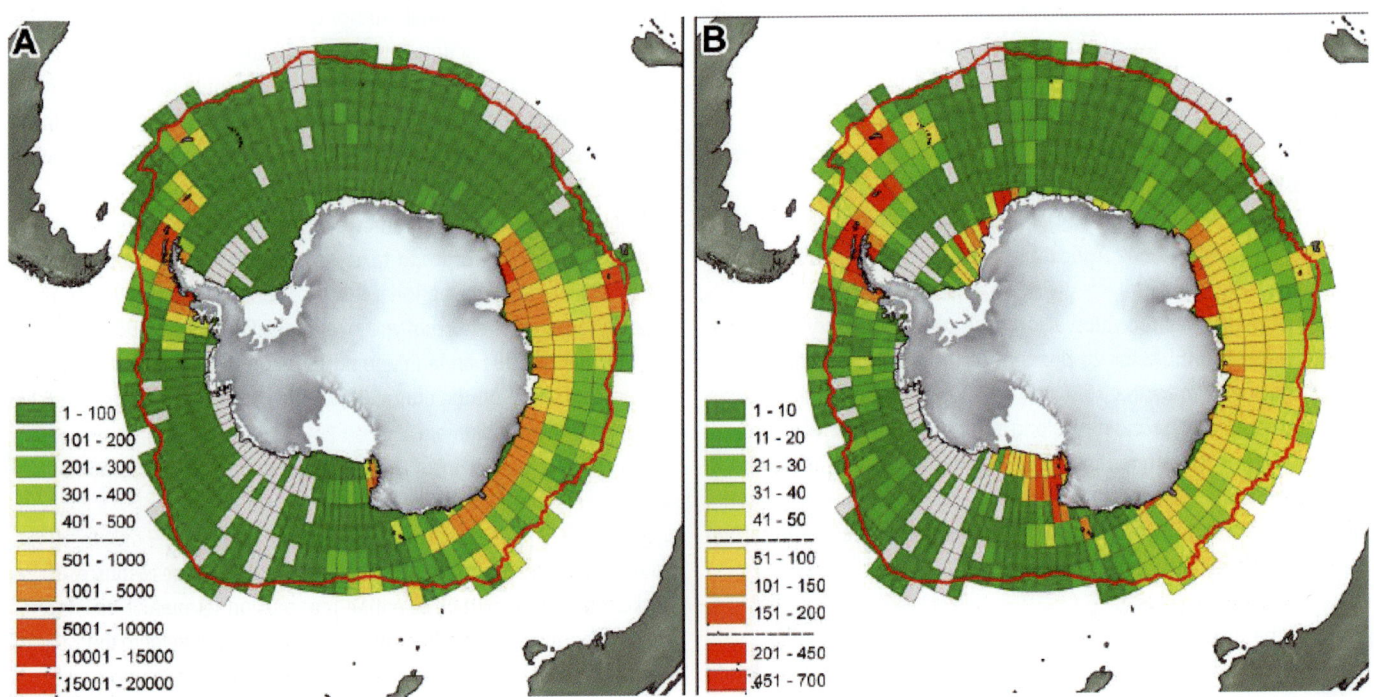

Chapter 35 — Extent of Assessment of Marine Biological Diversity

Despite the recent outstanding progress, some problems, gaps in knowledge and questions remained unaddressed (Griffiths et al., 2011). Besides obvious geographic and sampling gaps which still exist in the SO (East Antarctic, the Amundsen and Bellingshausen Seas and the SO deep sea, Figure 35.5), the extent of our knowledge of the biology, distribution, zoogeography, and evolution of Antarctic species is size-dependent.

The smaller the species are (nano-, meiobenthos, <1mm), the less is known about them. This also includes a lack of information on species' life histories and their diets, as our knowledge of the SO food web is mainly based on the diets of large pelagic predators. It is largely unknown what bottom dwellers feed on. This might be due to the fact that scientific effort and sampling of the benthos has predominantly concentrated on the continental shelves. Besides this bias in sampling depths, additional knowledge gaps are due to bias in the sampling gear used, with different gear considered as being either quantitative or qualitative (corers vs. trawled gear). Different working groups and expeditions have used different mesh sizes (not all scientists use fine mesh-sized gear and sieves), protocols and fixation methods.

In many marine areas (especially in the deep sea) more than 50 per cent of all species are rare and occur in samples as singletons. The fact that species occurrences are unevenly distributed – depending on their evolution and the availability of food sources – makes it difficult to understand the phenomena of patchiness and/or rarity. Very little is known at the community level about the potential effects of ongoing environmental change in the region. Although some shallow-water species have been physiologically investigated (and physiological adaptations are known for certain single species), community-scale effects of stressors, such as temperature rise, ocean acidification, increased frequency of iceberg scouring, etc., are very little known completely unknown (See Appendix 1-Southern Ocean for a summary of assessments).

4 Status of Knowledge of Marine Biodiversity: A Synthesis United the Ocean Biogeographic Information System (OBIS)

4.1 Taxonomic completeness

Appeltans et al. (2012) estimated at about 226,000 the number of eukaryotic marine species described. More importantly, these authors report that more species were described in the past decade (about 20,000) than in any previous one and that the number of authors describing new species has been increasing at a faster rate than the number of new species described in the past six decades, demonstrating that progress has been made globally. Despite this, between one-third and two-thirds of marine species may be undescribed, representing a major gap. Costello et al. (2010), based on the regional reviews of the state of knowledge of marine biodiversity compiled worldwide by the Census of Marine Life, provided a global perspective on what is known and what

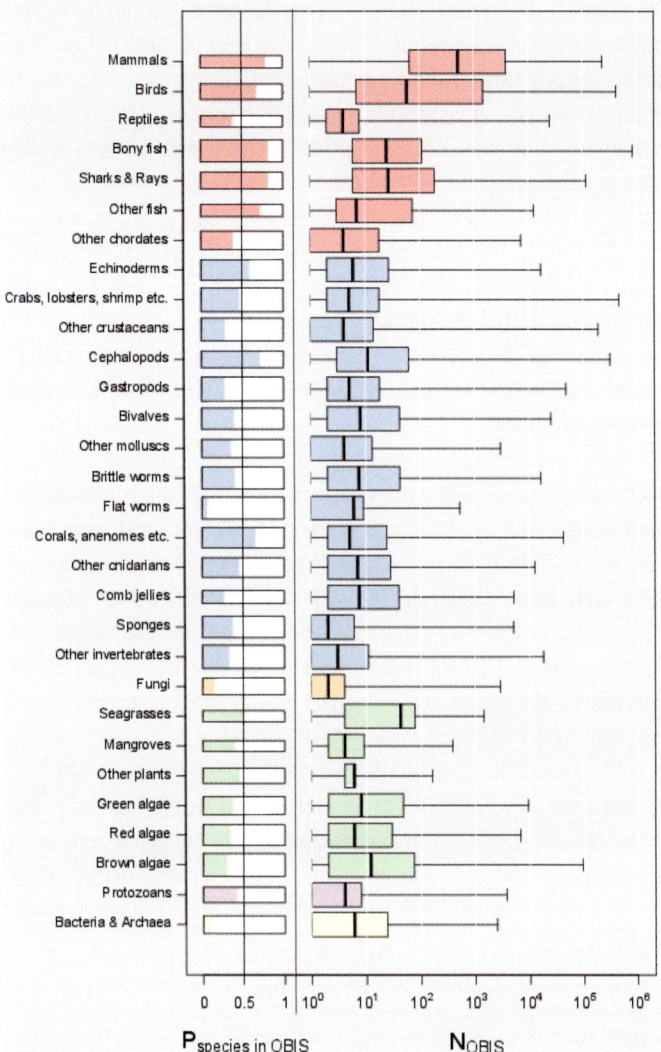

Figure 35.6 | Summary of the current global availability of biogeographic knowledge across all major marine taxa, using Ocean Biogeographic Information System data. The left-hand panel shows, for each taxonomic group, the proportion of all known species within that group which have at least one distribution record in OBIS (P species in OBIS), with the solid vertical line indicating data available for 50 per cent of species. The thickness of each bar is scaled to the number of described species in each group, according to the World Register of Marine Species (WoRMS; WoRMS Editorial Board, 2014). The right-hand panel shows for each group the number of records across all species occurring in OBIS (N OBIS). The solid bar is the median, the coloured box shows the interquartile range, and the lines extend to the minimum and maximum number of records for each group. Colours indicate: red (vertebrates), blue (invertebrates), orange (fungi), green (plant and algae), purple (protozoans), yellow (bacteria and archaea).

our interpretation of the evolutionary history of the fauna. Along with these techniques, recent advances in satellite and aerial imagery will also become important for mapping and visualization and will help improve knowledge of marine ecosystems in the Southern Ocean. The Commission for the Conservation of Antarctic Marine Living Resources (CCAMLR) is regarded as a model for regional cooperation and maintains scientific research programmes (including ecosystem monitoring) to address risks to commercially exploited fish stocks in the SO using an ecosystem-based approach.

the major scientific gaps are. They concluded that although there have been significant surveys and research efforts over the years, many habitats had been poorly sampled, especially in the deep sea, and several species-rich taxonomic groups, especially of smaller organisms, were still poorly studied. The best-known groups, which together comprise more than 50 per cent of total known biodiversity across regions, are crustaceans, molluscs, and fishes. However, knowledge of marine biodiversity is not only related to surveys but also to the availability of local and regional taxonomic expertise, and to commercial value (e.g. fish and crustaceans). Here we examine the current global availability of biogeographic knowledge across all major marine taxa, using OBIS (OBIS; IOC of UNESCO, 2014) data.

Overall, the figure includes data for 228,935 accepted marine species across all taxonomic kingdoms (Figure 35.6). Forty per cent (90,921) of these species, including at least one representative from each major taxonomic group, contribute to the total of 28,369,304 OBIS distribution records. This figure shows that very few groups have more than 50 per cent of their species represented in OBIS (mammals, birds, bony fish, sharks and rays, and other fishes among the vertebrates and echinoderms, cephalopods, corals and anemones among the invertebrates). For example, over 80 per cent (13.7K) of the 16.7K known species of bony fish have a record in OBIS; on average these 13.7K species are known from between 10 and 100 distribution records (median = 25). The maximum number of records in OBIS is 849,179 and corresponds to the Atlantic cod *Gadhus morhua*. The typical species occurring in OBIS has just 6 distribution records, and 20 per cent of them (18,181 species) are represented by only a single record. Nonetheless, 27 of the 30 groups considered here include at least one species with >1,000 distribution records, with 17 and 7 groups, respectively, including species with >10,000 and >100,000 records.

5 Final remarks

Marine biodiversity assessments are very variable among taxonomic groups and among ecosystems. Best assessed are groups such as fish, sea mammals, sea birds, turtles, and plankton, and ecosystems such as coral reefs. However, assessments are mostly limited in time, as very few have long term series data (as, for example, the CPR -Continuous Plankton Recorder has), and are limited by geographic range and taxonomic representation. Regarding taxonomic representation, for example, among fish efforts are mostly focused on commercial species (stock assessments) and top predators.

Among large vertebrates, efforts are focused on "iconic" and/or under-threat large species such as whales and turtles. Regarding geographic

Figure 35.7 | This figure provides a visual representation of our knowledge measured as number of observations for species (7A), sampling (7B), and records (7C) for the different taxonomic groups comparing coastal and continental shelf environments versus open ocean and deep sea waters for the seven ocean basins. In general, it is clear from Figure 7A that fishes, along with crustaceans and molluscs, are the most diverse groups in all ocean basins. Figures 7B and C show that the North Atlantic is the best-known ocean basin for all groups. (www.iobis.org). These figures also demonstrate that, for each of the ocean basins, knowledge is significantly higher in the coastal and continental shelf environments in comparison to the open ocean and deep sea environments which reflects the same situation exposed by Costello et al. (2010), four years later despite important efforts in advancing deep sea research. To analyse geographic completeness, we show an estimate of the number of species using the Chao index for the different seas within the seven ocean basins using the OBIS database. It is evident from this graph that the best sampled areas have been in the northern hemisphere, and that the southern hemisphere, with the exception of the Southern Ocean, has been poorly sampled (Figure 8).

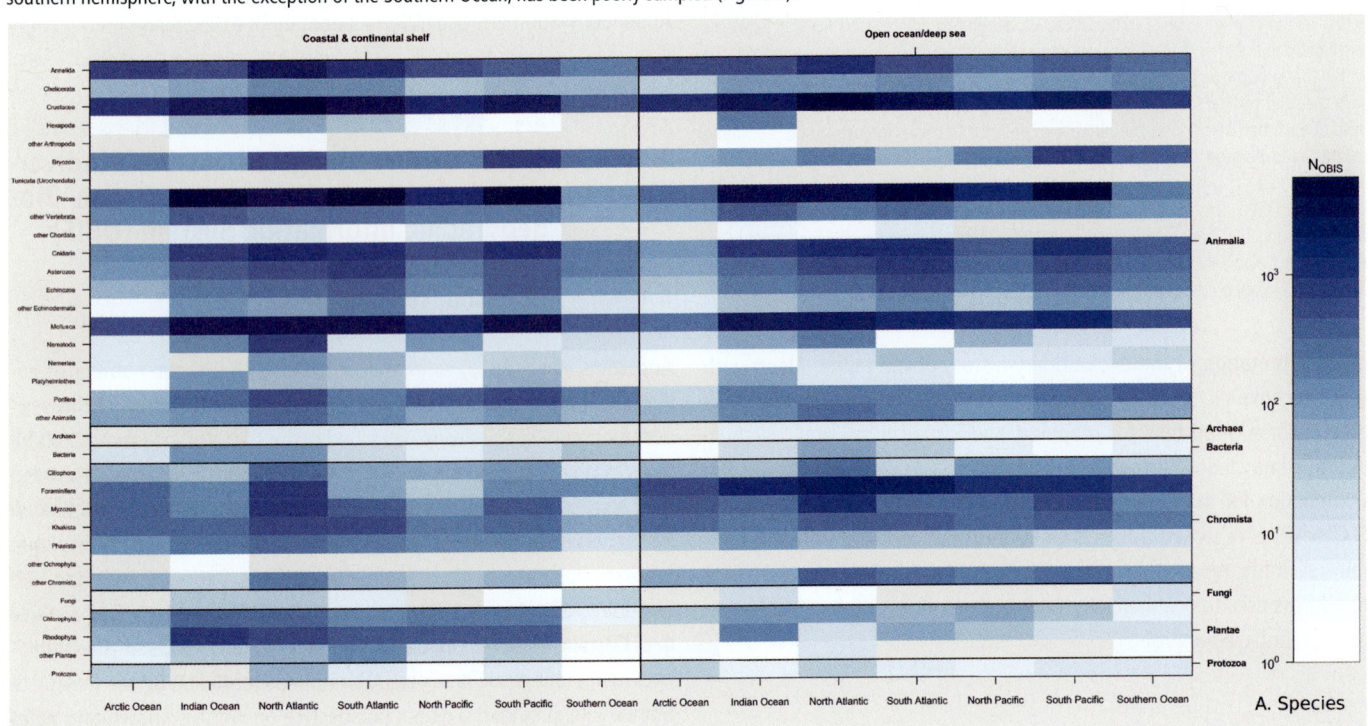

Chapter 35   Extent of Assessment of Marine Biological Diversity

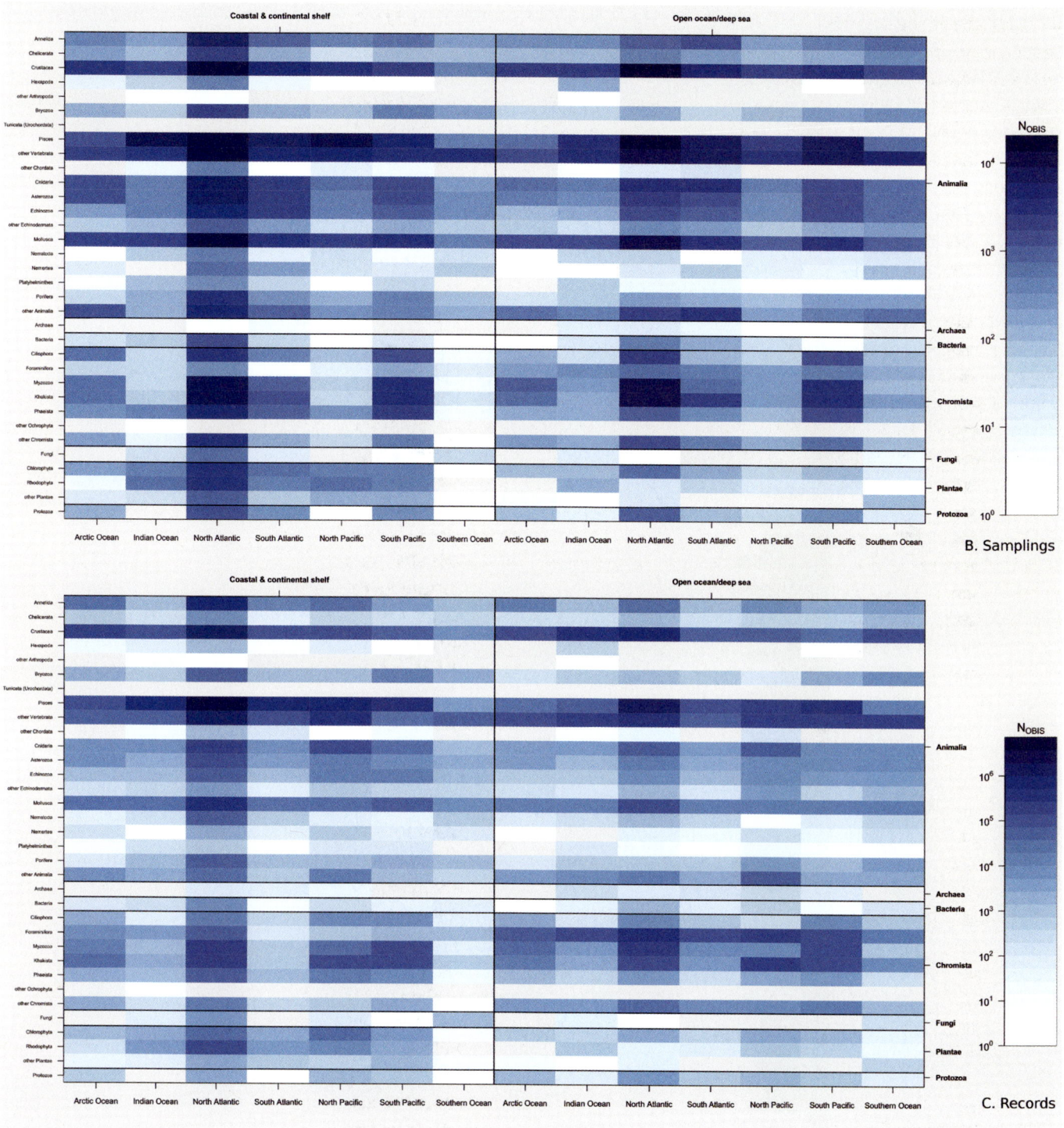

B. Samplings

C. Records

range, there is a considerable amount of information on coastal shelves and slopes along developed nations (e.g. Europe, United States, Canada, Australia, Japan, South Africa), however, even in these regions, knowledge is patchy in time (very few sustained long term efforts) and space (concentrated in particular areas of those coasts). The Arctic and Southern Ocean have received considerable attention (again the "charismatic" reason), but due to habitat complexity and logistical challenges, knowledge is fragmented, with some areas very poorly known. A generalized problem common to developed and developing countries, is that there is much unpublished data (at least not available through open access databases).

In addition, the ecosystem-approach type of assessment leading to an integrated management strategy is very recent, and still not widely used. Coral reefs may be the pioneer ecosystems in which this approach has been used, as monitoring programmes measure live cover, abundance and biomass in addition to biodiversity. This approach is also extending to other shallow water communities such as rocky shores through the integration of data and the creation of international networks. In the

549

Extent of Assessment of Marine Biological Diversity

Chapter 35

Figure 35.8 | Estimate of the number of species, using the Chao index, for the different seas within the seven ocean basins using the OBIS database (www.iobis.org).

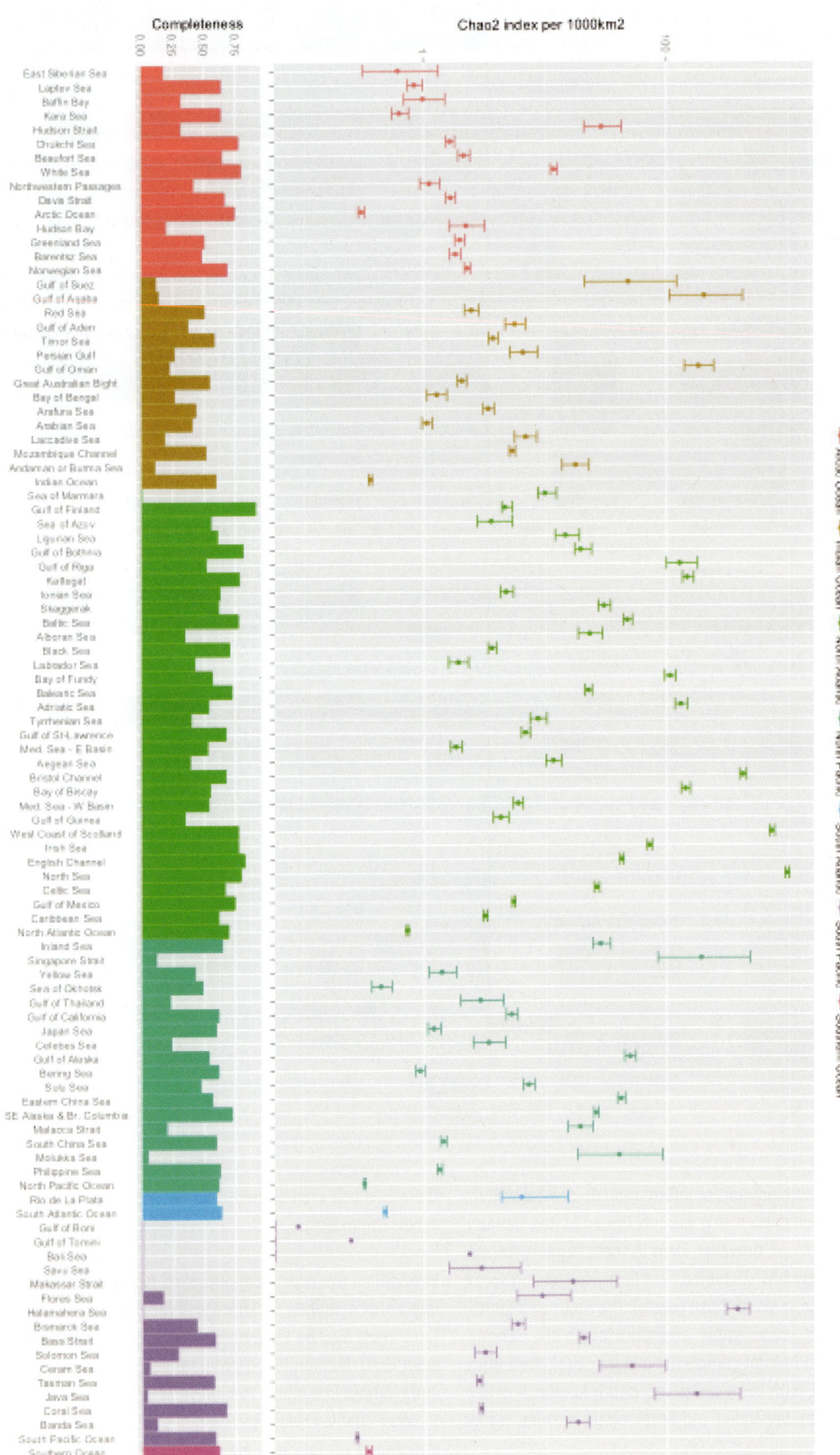

deep sea, seamounts seem to be the best assessed ecosystems, again maybe due to their potential economic value for fisheries or other extractive harvests such as minerals, as well as their potential to support significant biodiversity. This creates the urge to understand what they have in terms of living resources so that they can be managed properly before serious exploitation begins. On geologically active ecosystems such as vents and seeps, no assessments have been carried out, and information about these is very recent, very patchy, and very scarce.

We continue to stress the importance of taxonomy, systematics, and studies of biodiversity to advance our knowledge of ecology, ecosystem-based management, and understanding/valuation of ecosystem services. These are especially needed with increasing extinction rates, continued anthropogenic pressures on biodiversity, and the consequences of human-induced climate change. In this sense, biogeographic information is of fundamental importance for discovering marine biodiversity hotspots, detecting and understanding impacts of environmental changes, predicting future distributions, monitoring biodiversity, or supporting conservation and sustainable management strategies. The major challenges and needs for obtaining a more comprehensive overview of global marine biodiversity are the need to: (1) invest in taxonomy and capacity-building; (2) standardize methodologies to ensure proper comparisons; (3) increase sampling effort, exploring new habitats, and identifying and mapping biodiversity hotspots; (4) make historical and new data increasingly more accessible through open access data portals such as OBIS; (5) quantify ecosystem services and the impact of loss of biodiversity on these goods and services in different marine habitats and ecosystems across regions, and analyze how cumulative and synergistic anthropogenic impacts may affect these services; and (6) continue to enhance the importance of biodiversity in marine management policy decisions.

References

Abercrombie, D.L., Clarke, S.C., and Shivji, M.S. (2005). Global-scale Genetic Identification of Hammerhead Sharks: Application to Assessment of the International Fin Trade and Law Enforcement. *Conservation Genetics* 6: 775–788.

Allain, V., Kerandel, J.-A., Andrefouet, S., Magron, F., Clark, M.R., Kirby, D.S., and Muller-Karger, F.E. (2008). Enhanced Seamount Location Database for the Western and Central Pacific Ocean: Screening and Cross-checking of 20 Existing Datasets. *Deep-Sea Research* 55(8): 1035–1047.

Appeltans, W., Ahyong, S.T., Anderson, G., Angel, M.V., Artois, T., Bailly, N., Bamber, R., Barber, A., Bartsch, I., Berta, A., Blazewicz-Paszkowycz, M., Bock, P., Boxshall, G., Boyko, c.B., Nunes Brandão, S., Bray, R.A., Bruce, N.L., Cairns, S.D., Chan, T.-Y., Cheng, L., Collins, A.G., Cribb, T., Curini-Galletti, M., Dahdouh-Guebas, F., Davie, P.J.F., Dawson, M.N., De Clerck, O. (2012). The Magnitude of Global Marine Species Diversity, *Current Biology*, http://dx.doi.org/10.1016/j.cub.2012.09.036

Aranda, M., de Bruyn, P., and Murua, H. (2010). A Report Review of the Tuna RFMOs: CCSBT, IATTC, IOTC, ICCAT and WCPFC. EU FP7 Project No. 212188 TXOTX, *Deliverable* 2.2, 171 pp.

Archambault, P., Snelgrove, P.V.R., Fisher, J.A.D., Gagnon, J.-M., Garbary, D.J., Harvey, M., Kenchington, E.L., Lesage, V., Levesque, M., Lovejoy, C., Mackas, D.L., McKindsey, C.W., Nelson, J.R., Pepin, P., Piché, L., and Poulin, M. (2010). From Sea to Sea: Canada's Three Oceans of Biodiversity. *PLoS ONE* 5(8): e12182. doi:10.1371/journal.pone.0012182.

Beaulieu, S. E., Baker, E.T., German, C.R., and Maffei, A. (2013). An Authoritative Global Database for Active Submarine Hydrothermal Vent Fields. *Geochemistry, Geophysics, Geosystems* 14(11): 4892–4905.

Block, B.A., Jonsen, I.D., Jorgensen, S.J., Winship, A.J., Shaffer, S.A., Bograd, S.J., Hazen, E.L., Foley, D.G., Breed, G.A., Harrison, A.L., Ganong, J.E., Swithenbank, A., Castleton, M., Dewar, H., Mate, B.R., Shillinger, G.L., Schaefer, K.M., Benson, S.R., Weise, M.J., Henry, R.W., and Costa, D.P. (2011). Tracking Apex Marine Predator Movements in a Dynamic Ocean. *Nature* 475: 86–90.

Branch, G.M., Griffiths, C.L., Branch, M.L., and Beckley, L.E. (2010). *Two Oceans: A Guide to the Marine Life of Southern Africa*. Struik Publishers, Cape Town. 456 pp.

Brandt, A., Gooday, A.J., Brix, S.B., Brökeland, W., Cedhagen, T., Choudhury, M., Cornelius, N., Danis, B., De Mesel, I., Diaz, R.J., Gillan, D.C., Ebbe, B., Howe, J., Janussen, D., Kaiser, S., Linse, K., Malyutina, M., Brandao, S., Pawlowski, J., and Raupach, M. (2007). The Southern Ocean Deep Sea: First Insights into Biodiversity and Biogeography. *Nature* 447: 307–311.

Burke, L., Reytar, K., Spalding, M., Perry, A. (2011). *Reefs at Risk Revisited*. World Resources Institute.

Butler, A.J., Rees, T., Beesley, P., and Bax, N.J. (2010). Marine Biodiversity in the Australian Region. *PLoS ONE* 5(8): e11831. doi:10.1371/journal.pone.0011831.

CAFF (2013). Arctic Biodiversity Assessment. Status and trends in Arctic biodiversity. Conservation of Arctic Flora and Fauna, Akureyri (http://www.arcticbiodiversity.is/the-report/chapters).

Cairns, S.D. (2007). Deep-water corals: an overview with special reference to diversity and distribution of deep-water scleractinian corals. Bulletin of Marine Science, 81(3): 311-322.

Chadwick, M. (2006). Aquatic monitoring in Canada: A Report from the DFO Science Monitoring Implementation Team. DFO Canadian Science Advisory Secretariat, Proceedings Series 2006/003, iv+48 p.

Clark, M.R.; Rowden, A.A.; Schlacher, T.; Williams, A.; Consalvey, M.; Stocks, K.I.; Rogers, A.D.; O'Hara, T.D.; White, M.; Shank, T.M.; Hall-Spencer, J. (2010) The ecology of seamounts: structure, function, and human impacts. *Annual Review of Marine Science* 2: 253–278.

Coll, M., Piroddi, C., Steenbeek, J., Kaschner, K., Lasram F.B.R., Aguzzi, J., Ballesteros, E., Bianchi, C.N., Corbera, J., Dailianis, T., Danovaro, R., Estrada, M., Froglia, C., Galil, B.S., Gasol, J.M., Gertwage, R., Gil, J., Guilhaumon, F., Kesner-Reyes, K., Kitsos, M.-S., Koukouras, A., Lampadariou, N., Laxamana, E., de la Cuadra, C.M.L.-F., Lotze, H.K., Martin, D., Mouillot, D., Oro, D., Raicevich, S., Rius-Barile, J., Saiz-Salinas, J.I., Vincente, C.S., Somot, S., Templado, J., Turon, X., Vafidis, D., Villanueva, R., and Voultsiadou, E. (2010). The Biodiversity of the Mediterranean Sea: Estimates, Patterns, and Threats. *PLoS ONE* 5(8): e11842. doi:10.1371/journal.pone.0011842.

Collette, B.B., Carpenter, K.E., Polidoro, B.A., Juan-Jorda, M.J., Boustany, A., Die, D.J., Elfes, C., Fox, W., Graves, J., Harrison, L.R., McManus, R., Minte-Vera, C.V., Nelson, R., Restrepo, V., Schratwiser, J., Sun, C.-L., Amorim, A., Brick Peres, M., Canales, C., Cardenas, G., Chang, S.-K., Chiang, W.-C., de Oliveira Leite Jr., N., Harwell, H., Lessa, R., Fredou, F.L., Oxenford, H.A., Serra, R., Shao, K.-T., Sumalia, R., Wang, S.-P., Watson, R., and Yáñez, E. (2011). High Value and Long Life—Double Jeopardy for Tunas and Billfishes. *Science* 333: 291–292.

Costello, M.J., Coll, M., Danovaro, R., Halpin, P., Ojaveer, H., Miloslavich, P. (2010) A Census of Marine Biodiversity Knowledge, Resources, and Future Challenges. *PLoS ONE* 5(8): e12110. doi:10.1371/journal.pone.0012110.

Croxall, J.P., Butchart, S.H.M., Lascelles, B., Stattersfield, A.J., Sullivan, B., Symes, A., Taylor, P. (2012). Seabird conservation status, threats and priority actions: a global assessment. *Bird Conservation International* 22: 1-34.

Dulvy, N.K., Fowler, S.L., Musick, J.A., Cavanagh, R.D., Kyne, P.M., Harrison, L.R., Carlson, J.K., Davisdson, L.N.K., Fordham, S., Francis, M.P., Pollock, C.M., Simpfendorfer, C.A., Burgess, G.H., Carpenter, K.E., Compagno, L.V.J., Ebert, D.A., Gibson, C., Heupel, M.R., Livingstone, S.R., Sanciangco, J.C., Stevens, J.D., Valenti, S., and White, W.T. (2014). Extinction Risk and Conservation of the World's Sharks and Rays. *eLIFE* 3:e00590.

Edwards, M., Beaugrand, G., Hays, G.C., Koslow, J.A., andRichardson, A.J. (2010). Multi-decadal Oceanic Ecological Datasets and their Application in Marine Policy and Management. *Trends in Ecology and Evolution* 25: 602–610.

Edwards, M., Helaouet, P., Johns, D.G., Batten, S., Beaugrand, G., Chiba, S., Flavell, M., Head, E., Hosie, G., Richardson, A.J., Takahashi, K., Verheye, H.M., Ward, P., and Wootton, M. (2012). *Global Marine Ecological Status Report: Results from the Global CPR Survey 2010/2011*. SAHFOS Technical Report 9: 1–40. Plymouth, U.K.

Fautin, D., Dalton, P., Incze, L.S., Leong, J.-A.C., Pautzke, C., Rosenberg, A., Sandifer, P., Sedberry, G., Tunnell Jr., J.W., Abbot, I., Brainard, R.E., Brodeur, M., Eldredge, L.G., Feldman, M., Moretzsohn, F., Vroom, P.S., Wainstein, M., and Wolff, N. (2010). An Overview of Marine Biodiversity in United States Waters. *PLoS ONE* 5(8): e11914. doi:10.1371/journal.pone.0011914.

Felder, D.L., and Camp, D.K. (eds.). (2009). *Gulf of Mexico Origin, Waters, and Biota*. Vol. 1. Biodiversity. Texas A&M University Press, College Station, Texas. 1393pp.

Galtsoff, P. (ed.). (1954). Gulf of Mexico: Its Origin, Waters, and Marine Life. *Fishery Bulletin of the Fish and Wildlife Service*, Vol. 55(89). Washington, DC. 604 p.

Fujikura, K., Lindsay, D., Kitazato, H., Nishida, S., and Shirayama, Y. (2010). Marine Biodiversity in Japanese Waters. *PLoS ONE* 5(8): e11836. doi:10.1371/journal.pone.0011836.

German, C.R., Ramirez-Llodra, E., Baker, M.C., and Tyler, P.A. (2011). Deep-water Chemosynthetic Ecosystem Research during the Census of Marine Life Decade and Beyond: A Proposed Deep-ocean Road Map. *PLoS ONE* 6(8): e23259. doi:10.1371/journal.pone.0023259.

Gordon, D.P., ed. (2009) New Zealand inventory of biodiversity. Volume 1. Kingdom Animalia: Radiata, Lophotrochozoa, Deuterostomia. Christchurch: Canterbury University Press. pp 568+16.

Gordon, D.P., Beaumont, J., MacDiarmid, A., Robertson, D.A., Ahyong, S.T. (2010) Marine Biodiversity of Aotearoa New Zealand. *PLoS ONE* 5(8): e10905.doi:10.1371/journal.pone.0010905

Gradinger, R., Bluhm, B.A., Hopcroft, R.R., Gebruk, A., Kosobokova, K.N., Sirenko, B., and Wesławski, J.M. (2010). Marine Life in the Arctic. In: McIntyre, A.D. (ed.), *Life in the World's Oceans*. Blackwell Publishing Ltd., New York, pp. 183–202.

Griffiths, C.L., Robinson, T.B., Lange, L., and Mead, A. (2010). Marine Biodiversity in South Africa: an Evaluation of Current States of Knowledge. *PLoS ONE* 5(8): e123008. doi: 10.1371/journal.pone.0012008.

Griffiths, H.J. (2010). Antarctic Marine Biodiversity – What Do We Know About the Distribution of Life in the Southern Ocean? *PLoS ONE* 5(8): e11683. doi:10.1371/journal.pone.0011683.

Griffiths, H.J., Danis, B., and Clarke, A. (2011). Quantifying Antarctic Marine Biodiversity: The SCAR-MarBIN Data Portal. *Deep Sea Research Part II: Topical Studies in Oceanography* 58(1), 18–29.

Hällfors, H., Backer, H., Markku Leppänen, J. M., Hällfors, S., Hällfors, G., Kuosa, H. (2013). The northern Baltic Sea phytoplankton communities in 1903-1911 and 1993-2005: a comparison of historical and modern species data. *Hydrobiologia*, 707: 109-133.

Halpin, P.N., Read, A.J., Fujioka, E., Best, B.D., Donnelly, B., Hazen, L.J., Kot, C., Urian, K., LaBrecque, E., Dimatteo, A., Cleary, J., Good, C., Crowder, L.B., and Hyrenbach, K.D. (2009). OBIS-SEAMAP The World Data Center for Marine Mammal, Sea Bird, and Sea Turtle Distributions. *Oceanography* 22:104–115.

Hodgson, G. (1999). A global assessment of human effects on coral reefs.. *Marine Pollution Bulletin* 38, 345.

IOC of UNESCO (2014). The Ocean Biogeographic Information System. Web: http://www.iobis.org. Accessed 2014-05-06.

Jackson, J.B.C., Donovan M., Cramer K., Lam V. (2014). *Status and Trends of Caribbean Coral Reefs: 1970-2012*. GCRMN/ICRI/UNEP/IUCN. Pp. 245.

Kang, Y.S., Ohman, M.D. (2014). Comparison of long-term trends of zooplankton from two marine ecosystems across the North Pacific: Northeastern Asian marginal sea and Southern California current system. *California Cooperative Oceanic Fisheries Investigations Reports*. 55:169-182.

Kaschner, K., Quick, N., Jewell, R., Williams, R., Harris, C.M. (2012). Global Coverage of Cetacean Line-transect Surveys: Status quo, Gaps and Future Challenges. *PLoS ONE* 7(9): e44075. doi: 10.1371/journal.pone.0044075.

Keesing, J. and Irvine, T. (2005). Coastal Biodiversity in the Indian Ocean: the Known, the Unknown and the Unknowable. *Indian Journal of Marine Sciences* 34(1): 11–26.

Knowlton, N., Brainard, R. E., Fisher, R., Moews, M., Plaisance, L., Caley, M. J. (2010) Coral Reef Biodiversity. In: *McIntyre, A. (Ed.) Life in the World's Ocean: Diversity, distribution, and abundance*. Wiley-Blackwell, UK, p. 65-78.

Kvile, K.Ø., Taranto, G.H., Pitcher, T.J., and Morato, T. (2014). A Global Assessment of Seamount Ecosystems Knowledge. *Biological Conservation* 173: 108–120.

Liu, J.Y. (2013). Status of Marine Biodiversity of the China Seas. *PLoS ONE* 8(1): e50719. doi:10.1371/journal.pone.0050719.

Mawson, V., Tranter, D.J., Pearce, A.F. (eds.). (1988). *CSIRO at Sea: 50 Years of Marine Science*. Hobart, Tasmania: CSIRO Marine Laboratories. 216 pp.

Miloslavich, P., Díaz, J.M., Klein, E., Alvarado, J.J., Díaz, C., Gobin, J., Escobar-Briones, E., Cruz-Motta, J.J., Weil, E., Cortés, J., Bastidas, A.C., Robertson, R., Zapata, F., Martin, A., Castillo, J., Kazandjian, A., and Ortiz, M. (2010). Marine Biodiversity in the Caribbean: Regional Estimates and Distribution Patterns. *PLoS ONE* 5(8): e11916. doi:10.1371/journal.pone.0011916.

Miloslavich, P., Klein, E., Díaz, J.M., Hernández, C.E., Bigatti, G., Campos, L., Artigas, F., Castillo, J., Penchaszadeh, P.E., Neill, P.E., Carranza, A., Retana, M.V., Díaz de Astarloa, J.M., Lewis, M., Yorio, P., Piriz, M.L., Rodríuez, D., Yoneshigue-Valentin, Y., Gamboa, L., and Martín, A. (2011). Marine Biodiversity in the Atlantic and Pacific Coasts of South America: Knowledge and Gaps. *PLoS ONE* 6(1): e14631. doi:10.1371/journal.pone.0014631.

Moretzsohn, F., Brenner, J., Michaud, P., Tunnell, J.W., and Shirley, T. (2011). Biodiversity of the Gulf of Mexico Database (BioGoMx). Version 1.0. Harte Research Institute for Gulf of Mexico Studies (HRI), Texas A&M University-Corpus Christi (TAMUCC), Corpus Christi, Texas. Available at http://gulfbase.org/biogomx/.

National Research Council (NRC) (2010). *Assessment of Sea-Turtle Status and Trends: Integrating Demography and Abundance*. National Academies Press, Washington, DC.

O'Dor, R., Miloslavich, P., Yarincik, K. (2010). *Marine Biodiversity and Biogeography – Regional Comparisons of Global Issues, an Introduction*. PLoS ONE 5(8): e11871. doi:10.1371/journal.pone.0011871.

Ojaveer, H., Jaanus, A., MacKenzie, B.R., Martin, G., Olenin, S., Radziejewska, R., Telesh, I., Zettler, M.L., and Zaiko, A. (2010). *Status of Biodiversity in the Baltic Sea*. PLoS ONE 5(9): e12467. doi:10.1371/journal.pone.0012467.

Reid, P.C., Edwards, M. and Johns, D.G.,(2008). *Trans-Arctic invasion in modern times*. Science, 322: 528-529.

Reid, P.C., Johns, D.G., Edwards, M., Starr, M., Poulin, M. & Snoeijs, P. (2007). A biological consequence of reducing Arctic ice cover: the arrival of the Pacific diatom *Neodenticula seminae* in the North Atlantic for the first time in 800, 000 years. *Global Change Biology*, 13: 1910-1921.

Roberts, Callum M., McClean, C.J., Veron, J.E.N., Hawkins, J.P., Allen, G.R., McAllister, D.E., Mittermeier, C.G., Schueler, F.W., Spalding, M., Wells, F., Vynne, C., Werner. T.B. (2002). Marine Biodiversity Hotspots and Conservation Priorities for Tropical Reefs. *Science* 295:1280-84. DOI: 10.1126/science.1067728.

Robertson, H.A., Dowding, J.E., Elliott, G.P., Hitchmough, R.A., Miskelly, C.M., O'Donnell, C.F.J., Powlesland, R.G., Sagar, P.M., Scofield, R.P., Taylor, G.A. (2013). Conservation status of New Zealand birds, (2012). New Zealand Threat Classification Series 4, Department of Conservation, Wellington. 22 p.

Seuss, E. (2010). Marine Cold Seeps. In: Timmis, K. (ed.), *Handbook of Hydrocarbon and Lipid Microbiology*: Berlin, Springer-Verlag: 187–203.

Snelgrove, P.V.R., Archambault, P., Juniper, S.K., Lawton, P., Metaxas, A., Pepin, P., Rice, J.C., and Tunnicliffe, V. (2012). The Canadian Healthy Oceans Network (CHONe): An Academic-Government Partnership to Develop Scientific Guidelines in Support of Conservation and Sustainable Usage of Canada's Marine Biodiversity. *Fisheries* 37: 296–304.

Veron, J.E.N. (2000). *Corals of the World, Vol. 1,2,3*. Australian Institute of marine Science, Townsville, Australia, 1382 p.

Wafar, M., Venkataraman, K., Ingole, B., Ajmal Khan, S., and Loka Bharathi, P., (2011). State of Knowledge of Coastal and Marine Biodiversity of Indian Ocean Countries. *PLoS ONE* 6(1): e14613. doi: 10.1371/journal.pone.0014613.

Wallace, B.P., DiMatteo, A.D., Hurley, B.J., Finkbeiner, E.M., Bolten, A.B., Chaloupka, M.Y., Hutchinson, B.J., Abreu-Grobois, F.A., Amorocho, D., Bjorndal, K.A., Bourjea, J., Bowen, B.W., Dueños, R.B., Casale, P., Choudhury, B.C., Costa, A., Dutton, P.H., Fallabrino, A., Girard, A., Girondot, M., Godfrey, M.H., Hamann, M., López-Mendilaharsu, M., Marcovaldi, M.A., Mortimer, J.A., Musick, J.A., Nel, R., Pilcher, N.J.,

Seminoff, J.A., Troëng, S., Witherington, B., and Mast, R.B. (2010). Regional Management Units for Marine Turtles: A Novel Framework for Prioritizing Conservation and Research Across Multiple Scales. *PLoS ONE* 5(12): e15465. doi:10.1371/journal.pone.0015465.

Wallace, B.P., DiMatteo, A.D., Bolten, A.B., Chaloupka, M.Y., Hutchinson, B.J., et al. (2011). Global Conservation Priorities for Marine Turtles. *PLoS ONE* 6(9): e24510. doi:10.1371/journal.pone.0024510.

Wilkinson, C. (1999, 2000, 2002, 2004 and 2008). *Status of coral reefs of the World*. AIMS/ICRI.

Wilson, J.R (ed.). (1996). *Sydney Deepwater Outfalls Environmental Monitoring Program: Final Report Series*. Volumes 1–7. Chatswood, NSW: Australian Water and Coastal Studies Pty Ltd. For the Environment Protection Authority.

WoRMS Editorial Board (2014). World Register of Marine Species. Available at http://www.marinespecies.org at VLIZ. Accessed 2014-05-06.

36 Overview of Marine Biological Diversity

36A North Atlantic Ocean

Writing team:
Jake Rice (Convenor, Lead Member and Editor of Part VI), Christos Arvanitidis, Laura Boicenco, Panagiotis Kasapidis, Robin Mahon, Thomas Malone, William Montevecchi, Marta Coll Monton, Fabio Moretzsohn, Patrick Ouellet, Hazel Oxenford, Tim Smith, John Wes Tunnell, Jan Vanaverbeke and Saskia Van Gaever (Co-Lead Member)

Chapter 36A

1 Introduction

The North Atlantic is characterized by relatively wide continental shelves, particularly in its northerly portions, with steep slopes to the abyssal plain[1]. The width of the shelf decreases towards the south, with typical boundary current systems, characterized by strong seasonal upwelling, off the Iberian Peninsula and northwest Africa. Two chains of volcanic islands, the Azores and the Canaries, are located in the east central North Atlantic, and a large number of islands of volcanic origin, many with associated warm water coral reefs, are found in the southwest portion of the North Atlantic. In the far north of the region is the world's largest island, Greenland, primarily of Precambrian origin, whereas Iceland and the Faroe Islands are of more recent volcanic origin. All have rugged coastlines with rich faunas.

The biota of the North Atlantic is strongly influenced by both the warm Gulf Stream flowing north-eastward from the Gulf of Mexico and the Caribbean to northwest Europe, and the cold, fresh Labrador Current flowing south from the Canadian Archipelago and Greenland to the northeast coast of the United States. Major oceanographic and associated biotic regime shifts have been documented in the North Atlantic, but not with the frequency or scale of the North Pacific.

Around the coasts of the North Atlantic are a number of semi-enclosed seas. These seas have distinct oceanographic and bathymetric regimes, and ecosystems with many characteristics determined by local-scale processes and pressures. Hence each of these semi-enclosed seas, including the Black Sea, Mediterranean Sea, Baltic Sea (and similar coastal estuaries of the United States), the Caribbean Sea, and the Gulf of Mexico, receive some individual consideration in this assessment. Within the North Atlantic, there are several habitat types of special importance for biodiversity, such as seagrass beds and cold- and warm-water corals. Since these are important where they are found on the globe, they are treated in an integrated manner, respectively, in chapters 47, 42-43, rather than separately.

Coastal areas of the Northeast Atlantic have been settled in and used for several millennia. Commercial fisheries, both coastal and, as technology developed, offshore, have exploited fish and shellfish resources for centuries as well (Garcia et al., 2014), with periods of widespread overfishing in the twentieth century. Industrialization developed first in northwest Europe and eastern North America, and land-based pollution and coastal infrastructure have been significant pressures on coastal biodiversity of the North Atlantic for nearly two centuries. Large urban centres developed on the coasts of the North Atlantic at the time of industrialization, and below the boreal latitudes most of the North Atlantic coastal areas have been altered by various combinations of urban or municipal development, industry, agriculture, and tourism. Although some tens of kilometres of coasts and a few hundreds of square kilometres of coastal seabed are now protected in various ways, almost all biotic communities have been altered by centuries of pressures from human uses.

2 Coastal and Shelf Holoplankton

2.1 Status[2]

2.1.1 Phytoplankton

Diatoms and dinoflagellates account for most phytoplankton species (> 2 μm) in coastal and shelf waters of the North Atlantic (Tables 36A.1 and 36A.2)[3], with diatom blooms typically peaking during spring and dinoflagellate blooms during summer (cf., McQuatters-Gollop et al., 2007). The ubiquitous prokaryotic species *Synechoccocus* spp. (< 2 μm) also peaks in abundance (> 107 cells liter-1) during summer (cf., Wang et al., 2011). On a decadal time scale (1960-2009), dinoflagellate species richness has increased while abundance has decreased relative to diatoms in the North-East North Atlantic and North Sea, a trend that has been attributed to the combined effect of increases in sea surface temperature (SST) and wind shear during the summer (Hinder et al., 2012).

2.1.2 Mesozooplankton (200 - 2000 μm)

Calanoid copepods dominate the holoplankton in coastal and shelf waters throughout the North Atalntic (Table 36A.3). Many of these species are cosmopolitan, e.g., > 60 per cent of the species described from the Caribbean Sea and Gulf of Mexico are also found in the North-East Atlantic (Park, 1970). A decadal scale (1958 – 2005) progressive increase in the abundance of warm-temperate calanoid species (e.g., *Calanus helgolandicus, Centropages typicus*) and a decline of cold-temperate calanoid species (e.g., *Calanus finmarchicus, Euchaeta norvegica*) has been documented in the North-East North Atlantic (Beaugrand et al., 2002, 2009; Chust et al., 2014). Coincident with this trend, the mean size of copepods decreased as their species diversity increased and SST increased (Beaugrand et al., 2010).

2.2 Long-Term (multi-decadal) Trends and Pressures in Holoplankton

2.2.1 Regime Shifts: Overfishing and Climate Change

The primary pressures responsible for regime shifts in shelf ecosystems are overfishing and climate-driven changes (hydro-climate pressures including Arctic ice melting, ocean warming, and mode variability) in the marine environment (Steele, 2004; Edwards et al., 2006; Kane, 2011).

---

1 Biodiversity of the abyssal plain and mid-Atlantic Ridge that divides the eastern and western Atlantic is dealt with in Chapter 36F.

2 Sampling for species identification employs different techniques (e.g., water samples, net samples, continuous plankton recorder) and, therefore, are not comparable from region to region in terms of species richness.

3 Note that the lists of abundant taxa in Tables 36A.1 and 36A.2 are not meant to be comprehensive. For example, a phytoplankton check list for the Baltic Sea describes over 1,500 species (Hällfors, 2004).

Table 36A.1 | Abundant phytoplankton species of selected taxa (based on abundance and number of genera represented) in coastal and shelf waters of the western North Atlantic (* produce mucilage and foam, ** potentially toxic species).

Domain	Location	Division	No.	Abundant Species
Coastal & Shelf	Gulf of Maine[1]	Heterokontophyta (Diatoms)	386	Chaetoceros spp., Navicula spp., Nitzschia spp., Rhizosolenia hebetata, Coscinodiscus spp., Pleurosigma spp., Thalassiosira spp., Gyrosigma spp., Phaeoceros spp.
		Alveolata (Dinoflagellates)	151	Peridinium spp., Alexandrium spp.**, Amphidinium spp., Ceratium spp.
		Haptophyta	31	Chrysochromulina spp.**, Diacronema spp., Emiliania huxleyi, Pavlova spp., Prymnesium spp.
		Cyanophyta	22	Synechococcus spp.
		Chlorophyta	20	Halosphaera viridia, Micromonas pusilla, Ostreococcus sp., Pycnococcus provasolii, Tetraselmis spp.
		TOTAL	665	
	Gulf of Maine & New York Bight[2]	Heterokontophyta	274	Leptocylindrus danicus, Skeletonema costatum, Asterionella glaccialis, Pseudo-nitzschia pungens**, Rhizosolenia delicatula
		Alveolata	332	Procentrum micans**, P. minimum
		Haptophyta	19	Emiliana huxleyi
		Cyanophyta	12	Nostoc commune
		Chlorophyta	13	Nannochloris atomus
		TOTAL	678	
Coastal	New York Bight[3]	Heterokontophyta	-	Skeletonema costatum, Asterionella japonica, Leptocylindrus danicus, Thahsionema nitzschioides, Chaetoceros debilis
		Alveolata	-	Gymnodinium splendens, Prorocentrum micans**, P. triangulatum
		Chlorophyta	-	Nannochloris atomus
Shelf	New York Bight[3]	Heterokontophyta	-	Rhizosolenia abta, R. faeroense, Chaetoceros socialis, Cylindrotheca closterium
		Alveolata	-	Ceratium tripos, C. macroceros, C. furca, Peridinium depressum
Coastal & Shelf	Gulf of Mexico[4]	Heterokontophyta	152	Guinarda spp., Hemiaulus senensis, Leptocilyndrus danicus, Thalassionema spp., Cylindrotheca closterium, Pseudo-nitzschia delicatissima
		Alveolata	124	Ceratium spp., Dinophysis caudate**, Gyrodinium fusiforme, Scrippsiella trochoidea
		Cyanophyta	18	Trichodesmium spp.
		TOTAL	306	

1   Li et al., 2011
2   Marshall and Cohn, 1982
3   Malone, 1977
4   Merino-Virgilio et al., 2013

Synchronous, system-wide regime shifts in plankton communities were initiated during the late 1980s and early 1990s in the Baltic Sea, North Sea, Scotian Shelf and Gulf of Maine (Reid et al., 2001; Edwards et al., 2002; Alheit et al., 2005; Record et al., 2010; Kane, 2011; Möllmann, 2011). In each case, synergies between trophic cascades triggered by overfishing and changes in hydro-climate were the primary pressures with overfishing reducing resiliency and hydro-climate forcing initiating the regime shift (Drinkwater, 2005; Beaugrand et al., 2008; Fogarty et al., 2008; Hilborn and Litzinger, 2009; Möllmann, 2011).

Gulf of Maine – During 1961-2008, the annual cycle of species richness and abundance was characterized by a seasonal peak during spring when diatoms dominated (Kane, 2011). The most abundant taxa were *Thalassiosira* spp., *Rhizosolenia hebetate, Phaeoceros* spp., *Thalassiothrix longissima*, and *Thalassionema nitzschioides*. On a decadal time scale, Kane (2011) documented three consecutive multi-year periods of varying species richness and abundance: below average (1961 – 1989), above average (1990 – 2001), and average (2002 – 2008). Decadal changes were more pronounced for diatoms than dinoflagellates (Möllmann, 2011), and the most striking feature in the time-series was the persistent positive anomaly of the 1990s.

Zooplankton species richness and abundance also increased sharply during the 1990s as the abundance of smaller copepod species increased and larger species declined (Pershing et al., 2005; Kane, 2007; Record et al., 2010). Increases in zooplankton and phytoplankton stocks were also reported during this period over the Newfoundland and Scotian shelves as Arctic species originating from the Gulf of St. Lawrence and the Labrador Current (e.g., *Calanus glacialis, Calanus hyperboreus*) became more abundant and warmer water species (e.g., *Calanus finmarchicus, Centropages typicus, Metridia lucenss, Temora stylifera*) became less abundant (Head and Sameoto, 2007; Möllmann, 2011). This interdecadal zooplankton biodiversity signal was significantly correlated with phytoplankton biomass (Record et al., 2010).

North Sea – As indicated by rapid changes in plankton biomass and species diversity, a regime shift was initiated between 1983 and 1988, apparently as a consequence of increases in SST, a positive phase of the North Atlantic Oscillation (NAO), and increases in advection from the

Table 36A.2 | Abundant phytoplankton species of selected taxa (based on abundance and number of genera represented) in coastal and shelf waters of the eastern North Atlantic (* produce mucilage and foam, ** potentially toxic species).

Domain	Location	Division	No.	Abundant Species
Coastal & Shelf	NE Atlantic[1]	Heterokontophyta (Diatoms)	59	*Bacteriastrum* spp., *Chaetoceros* spp., *Cylindrotheca closterium**, *Guinardia delicatula*, *Odontella aurita*, *Proboscia alata*, *Pseudo-nitzschia* spp.**, *Rhizosolenia* spp., *Skeletonema costatum*, *Thallassionema nitzschioides*
		Alveolata (Dinoflagellates)	48	*Ceratium furca*, *C. fusus*, *C. horridum*, *C. lineatum*, *C. longipes*, *C. macroceros*, *C. tripos*, *Dinophysis* spp.**, *Gonyaulax* spp.**, *Noctiluca scintillans*, *Prorocentrum* spp., *Protoperidinium* spp.
		Haptophyta	1	*Phaeocystis pouchetii**
		TOTAL	170	
Coastal	German Bight[2]	Heterokontophyta (Diatoms)	109	*Chaetoceros curvisetus*, *Chaetoceros* spp., *Coscinodiscus* spp., *Coscinodiscus wailesii**, *Guinardia flaccida*, *Odontella sinensis*, *Pseudo-nitzschia pungens***, *Rhizosolenia imbricate*, *Rhizosolenia styliformis*
		Alveolata (Dinoflagellates)	26	*Noctiluca scintillans*, *Ceratium tripos*, *Ceratium fusus*, *Ceratium longipes*, *Gyrodinium spirale*, *Protoperidinium depressum*
		Haptophyta	3	*Emiliana huxleyi*, *Phaeocystis sp.**
		TOTAL	292	
Coastal	English Channel[3]	Heterokontophyta (Diatoms)	131	*Guinardia* spp., *Phaeocystis globosa*, *Paralia sulcata*, *Pseudo-nitzschia* spp**., *Chaetoceros* spp., *Thalassiosira* spp.
		Alveolata (Dinoflagellates)	28	*Prorocentrum* spp.**
		Haptophyta		*Phaeocystis globosa**
		TOTAL	178	
Coastal	Iberian Peninsula[4]	Heterokontophyta (Diatoms)	68	*Chaetoceros* spp., *Leptocylindrus danicus*, *Pseudo-nitzschia delicatissima***, *Guinardia* spp., *Rhizosolenia fragile*, *Thalassiosira* spp., *Nitzschia longissima*
		Alveolata (Dinoflagellates)	69	*Prorocentrum micans***, *Amphidinium curvatum*, *Dinophysis* spp.** *Ceratium lineatum*, *Gymnodinium sp.***, *Scrippsiella trochoidea*
		TOTAL	161	

1 Barnard et al., 2004; Edwards et al., 2006
2 Wasmund et al., 2012
3 Guilloux et al., 2013
4 Rodrígueza et al., 2003; Not et al., 2007; Ospina-Alvarez et al., 2014

North-East North Atlantic (Beaugrand, 2004; McQuatters-Gollop et al., 2007). Mean phytoplankton chlorophyll levels peaked in 1989 (Reid et al., 1998), and the new regime (1990 - 2003) maintained 13 per cent and 21 per cent higher chlorophyll concentrations in open and coastal waters, respectively (McQuatters-Gollop et al., 2007). The regime shift was also marked by increases in the abundance and diversity of dinoflagellates relative to diatoms, decreases in the abundance of *Ceratium* spp. (e.g., *C. furca*, *C. fusus*, *C. horridum*, *C. tripos*, *C. lineatum*), increases in diversity and abundance of warm water calanoid copepod species (e.g., *Rhincalanus nasutus*, *Eucalanus crassus*, *Centropages typicus*, *Candacia armata*, *Calanus helgolandicus*), decreases in cold water species (e.g., *Heterorhabdus norvegicus*, *Euchaeta norvegica*, *Calanus finmarchicus*), and increases in the frequency of jellyfish outbreaks (most notably the hydrozoan *Aglantha digitale* and the scyphozoan *Pelagia noctiluca*) (Attrill et al., 2007; Beaugrand et al., 2009, 2010; Edwards et al., 2009; Richardson et al., 2009; Licandro et al., 2010).

2.2.2 Toxic Phytoplankton Blooms: Invasions, Eutrophication and Climate Change

The frequency of toxic phytoplankton blooms has increased over the last three decades in coastal waters of both the western and eastern North Atlantic (Anderson et al., 2012). Toxic taxa include dinoflagellates (*Alexandrium* spp., *Gymnodinium catenatum*, *Karenia mikimotoi*, *Karenia brevis*, *Dinophysis* spp., *Protoperidinium crassipes*, *Prorocentrum* spp.), diatoms (*Pseudo-nitzschia* spp.), and microflagellates (*Chrysochromulina polylepis*, *Chattonella* spp., *Fabrocapsa japonica*).4 Increases in toxic events associated with these species have been attributed to more frequent and comprehensive observations, the dispersal of invasive toxic species via the ballast water of ships, coastal eutrophication, and climate change (Skjodal and Dundas, 1991; Hallegraeff and Bolch, 1992; Belgrano et al., 1999; Anderson et al., 2002; Sellner et al., 2003; Glibert et al., 2005; Dale et al., 2006; Edwards et al., 2006; Moore et al., 2008; Fu et al., 2012). The risk of harmful phytoplankton blooms in the future has increased due to synergy between climate-driven changes and anthropogenic nutrient inputs (Hallegraeff, 2010).

2.2.3 Invasive Species: Ballast Water, Aquaculture and Climate Change

Large numbers of non-indigenous species have been introduced to coastal marine and estuarine ecosystems, largely due to transoceanic shipping (ballast water) and aquaculture (cf., Reise et al., 1999; Gollasch et al., 2009). Many of these species become invasive. Four ecologically significant invasions have been unequivocally documented (Birnbaum,

4 http:www.marbef.org/wiki/OSPAR_eutrophication_assessment; http://www.whoi.edu/redtide/species/by-syndrome

Table 36A.3 | Abundant mesozooplankton species for selected taxa (based on abundance, number of genera represented and their importance as indicators of climate-driven changes in hydro-climate) in coastal and shelf regions of the North Atlantic (NE – British Isles, Baltic and North Seas; SE - Bay of Biscay, Iberian coast, west Africa; North – Labrador and Norwegian Seas, Greenland and Iceland; NW – New York Bight, Gulf of Maine, Newfoundland and Scotian Shelves; SW – Caribbean Sea, Gulf of Mexico, South Atlantic Bight).

Location	Taxa	No.	Abundant Species
NE North Atlantic[1,2]	Calanoida	286	*Acartia spp., Calanus finmarchicus, Calanus helgolandicus, Centropages spp., Clausocalanus spp., Eurytemora affinis, Metridia lucens, Paracalanus spp., Para-Pseudocalanus spp., Pseudocalanus spp., Temora longicornis,*
	Cyclopoida	1	*Oithona spp.*
	Cladocera	3	*Evadne spp., Podon spp., Penilia avirostris*
	Thecosomomata	4	*Limacina spp.*
	Copepod Total	381	
SE North Atlantic[1,2,3]	Calanoida	580	*Acartia spp., Calanoides carinatus, Calanus helgolandicus, Candacia armata, Centropages typicus, Clausocalanus spp., Ctenocalanus vanus, Euchaeta hebes, Metridia lucens, Paraeuchaeta gracilis, Para-Pseudocalanus spp., Pseudocalanus spp., Temora stylifera, Undeuchaeta spp., Lucicutia flavicornis, Nannocalanus minor, Paracalanus parvus, Ctenocalanus vanus, Neocalanus gracilis, Rhincalanus cornutus, Eucalanus subtenuis*
	Cyclopoida	4	*Oithona spp.*
	Cladocera	7	*Evadne spp., Podon spp*
	Thecosomomata	4	*Limacina inflata, Limacina trochiformis, Creseis acicula*
	Copepod Total	747	
North North Atlantic[1,2]	Calanoida	188	*Acartia spp., Calanus finmarchicus, Heterorhabdus norvegicus, Paraeuchaeta norvegica*
	Cyclopoida	-	*Oithona spp.*
	Cladocera	-	*Evadne spp.*
	Thecosomomata	-	*Limacina spp.*
	Copepod Total	204	
NW North Atlantic[1,2,4]	Calanoida	204	*Acartia spp., Calanus finmarchicus, Calanus glacialis, Calanus hyperboreus, Centropages spp., Labidocera aestiva, Metridia lucens, Para-Pseudocalanus spp., Pseudocalanus spp., Clausocalanus arcuicornis, Paracalanus spp., Temora longicornis, Tortanus discaudatus*
	Cyclopoida	-	*Oithona spp.*
	Cladocera	-	*Evadne spp., Podon spp., Penilia spp.*
	Thecosomomata	-	*Limacina retroversa, Limacina spp.*
	Copepod Total	261	
SW North Atlantic[2,5]	Calanoida	553	*Acartia spp., Calanus tenuicornis, Centropages spp., Clausocalanus spp., Corycaeus spp., Eucalanus spp., Euchaeta spp., Haloptilus longicornis, Labidocera aestiva, Lucicutia flavicornis, Paracalanus crassirostris, Paracalanus parvus, Parvocalanus spp., Pleuromamma gracilis, Nannocalanus minor, Temora spp.*
	Cyclopoida		*Oncaea venusta, Corycaeus amazonicus, Oithona brevicornis, Oithona nana, Oithona plumifera, Oithona simplex*
	Cladocera		*Evadne sp.*
	Thecosomomata		*Limacina trochiformis*
	Copepod Total	715	

1 Barnard et al., 2004
2 Van Ginderdeuren, 2012; Laakmann et al., 2013; Razouls et al., 2014
3 Thiede, 1975; Poulet et al., 1996; Valdés et al., 2007; Albaine and Irigoien, 2007, Hernández-León et al., 2007
4 Malone, 1977, Johns et al., 2001; Durbin et al., 2003; Runge and Jones, 2012
5 Grice, 1960; Park, 1970, 1975; Cummings, 1983; Elliott et al., 2012

2006; Javidpour et al., 2006; Reid et al., 2007; Riisgård, 2007). The nuisance diatom (produces mucilage), *Coscinodiscus wailesii*, was introduced in the 1970s. The increase in abundance and geographic expansion of this species from the English Channel into the North Sea over the last three decades is a prototypical example of a planktonic species invasion. Likewise, by the 1990s the cladoceron *Cercopagis pengoi* and comb jelly *Mnemiopsis leidyi*, both voracious planktivores with the potential to disrupt trophic dynamics, had spread from the Gulf of Riga into the Baltic Sea and Gulf of Finland. These species appear to have been introduced via ballast water. The Pacific diatom *Neodenticula seminae* was transported into the Labrador Sea via the Canadian Arctic Archipelago in the late 1990s. It has since spread south to Georges Bank and further east, south of Iceland. The geographic expansion of this species portends of more trans-Arctic invasions from the Pacific Ocean as climate-driven Arctic ice melt continues.

2.2.4 Jellyfish: Overfishing, Eutrophication and Climate Change

Jellyfish blooms have increased in frequency in the northeast North Atlantic since 2002, e.g., *Pelagia noctiluca*, *Aglantha digitale*, and *Physalia physalis* outbreaks in the North-East North Atlantic (Attrill et al., 2007;

Doyle et al., 2007; Richardson et al., 2009; Licandro et al., 2010), and mounting evidence suggests that these outbreaks may lead to trophic shifts from fish to jellyfish and other gelatinous zooplankton as the dominant consumers (Richardson et al., 2009). Primary pressures leading to a more gelatinous state include overfishing (Lynam et al., 2006; Bakun and Weeks, 2006; Roux et al., 2013), coastal eutrophication and hypoxia (Purcell et al., 2001; Condon et al., 2001), and climate-driven ocean warming and mode variability, e.g., the NAO (Purcell et al., 2001; Atrill et al., 2007; Gibbons and Richardson, 2008; Doyle et al., 2007), although decline in predation by sea turtles has also been implicated as a factor in the outbreaks.

2.2.5 Calcifying Plankton: Climate Change

A multi-decadal time series (1960-2009) for the North-East North Atlantic region shows that changes in the abundance and distribution of foraminifera (*Globerigina spp.*), coccolithophores (*Emiliania huxleyi*), pteropods (*Clione limacine*, *Limacina helicina*), non-pteropod molluscs and echinoderm larvae were positively correlated with decadal changes in annual SST and the Atlantic Multidecadal Oscillation and negatively correlated with pH, i.e., abundance increased as pH decreased (Beaugrand et al., 2013). Beare et al. (2013) found no statistical relationship between the abundance of calcifying plankton in the North Sea and pH, although *Globerigina* spp, *Emiliania huxleyi* and echinoderm larvae increased during the period, and pteropods and bivalve larvae decreased. Thus, although acidification may become a serious threat to calcifying plankton, observations to date suggest that the primary driver of calcifying plankton abundance has been ocean warming.

3 Benthos

Most studies dealing with "benthos" in the North Atlantic focussed on macrobenthos (i.e., metazoan animals living in the seafloor and retained on a sieve of 500-1000 μm). Studies on meiobenthos are more scattered, and often targeted towards the dominant meiobenthic taxon: the nematodes.

Much of the available information has been compiled at the beginning of this century, in the framework of large-scale projects including the Census of Marine Life (CoML, 2000-2010), and the European Network of Excellence MarBEF (2004-2009). MarBEF was highly successful in compiling data on coastal benthos of the North-East Atlantic, and CoML captured many data of the deeper parts of the Atlantic in general.

3.1 Coastal Benthos

Within MarBEF, a large database (MacroBen), consisting of more than 460,000 distribution records on the distribution of 7,203 taxa from almost 23,000 stations was compiled (Gage et al., 2004; Vanden Berghe et al., 2009). Data were collected between 1972-2005 in the North-East Atlantic and adjacent semi-enclosed seas (North Sea, Baltic Sea, Mediterranean Sea) in a depth range of 0-450 m. Highest sampling density was in the North Sea and North-East Atlantic. As the analysis of this entire database reduces the potential problem of a disproportionally large effect of site-specific features on large-scale patterns (Renaud et al., 2009), and as the analyses of this database took into account methodological issues related to sampling and treatment of samples, the findings from these analyses are taken to reflect the general trends in macrobenthic communities in the North Atlantic. A first macro-ecological analysis showed that most macrobenthic communities follow the same right-skewed frequency distribution known from terrestrial ecosystems, revealing that most species are rare, and only few species have a wide distribution (Webb et al., 2009). Except for polychaetes, macrofaunal communities on a geographical scale of tens of km^2 or less are not random subsets of a species list at a wider geographical scale: species belonging to the same community tend to be more closely related to each other than would be expected when communities were randomly assembled from a regional species pool (Somerfield et al. 2009). Hence, community composition for most taxa is determined by regional processes, whereas random assembly, followed by local environmental and ecological processes, is more important for polychaete communities.

After removal of the confounding effect of depth, sampling effort and the low diversity in the Baltic, Renaud et al. (2009) demonstrated a modest increase in diversity with increased latitude. Much stronger effects were observed for the diversity-depth gradient where a unimodal trend with water depth (within the 0-450 m depth range) was observed: species richness peaked at 100-150 m depth, and maximum values for the expected number of species in a sample with 50 individuals (ES(50)) were observed at 200-350 m (Escaravage et al., 2009). Diversity was negatively related to the fraction of primary production reaching the seafloor, corresponding with the decreasing part of the unimodal productivity-diversity curve (Escaravage et al., 2009).

3.2 Offshore Benthos

General patterns on the offshore benthos from the eastern and western part of the North Atlantic, here defined as the macrobenthos from continental slopes and the deep sea, have been investigated in relation to both latitudinal and depth-related patterns. However, data are scarce in comparison with the more shallow areas, and trends have been deduced from a limited number of studies, focusing on a limited number of taxa. A poleward decrease in diversity was observed for deep-water molluscs and crustaceans (Rex et al., 1993) and cumaceans (Gage et al., 2004). Trends for the cumaceans were stronger in the eastern part of the basin. Data seem to be too scarce at the moment to make statements about the mechanisms behind the patterns (Narayanaswamy et al., 2010). Trends in the relationship between diversity and depth differ between sites for the same taxon: bivalve diversity (measured as ES(50)) is a significant unimodal function of depth, with diversity peaking at mid-bathyal depths in the western part of the North Atlantic, whereas the diversity-depth relationship can be described as a significant, linear function of depth (Brault et al., 2013) in the eastern part. The absence of the unimodal relationship between depth and diversity has been described for other taxa in other areas as well (Narayanaswamy et al., 2010). However, "depth" itself is not the explanatory variable, as it co-varies with a variety of environmental characteristics and not always in the same manner (Narayanaswamy et al., 2010). Strong differences exist in the flux of organic carbon to the seafloor between the eastern and the western part of the Atlantic at depths > 3800m, where the flux of organic carbon is 56 per cent higher in the eastern basin, which is probably reflected in the macrofaunal densities and the feeding mode composition of bivalves (Brault et al., 2013). Although the flux of organic matter is indeed important, processes regulating diversity at local, regional and global scales in the deep sea are multivariate, and smaller-scale processes are hierarchically embedded in larger-scale processes and tend to occur at faster rates (Levin et al., 2001). Processes at the local scale involve biological interactions (competition, facilitation, and predation), patch type characteristics (biogenic structures, nutrient concentrations, topography) and disturbance and recruitment. All of these are hierarchically embedded in environmental gradients at the regional scale, dispersal, metapopulation dynamics and gradients in habitat heterogeneity. Processes at the global scale include, amongst others, speciation/extinction, large-scale disturbances and large-scale environmental gradients.

Apart from patterns in diversity, the relationship between biomass and depth received considerable attention as well. The decrease in total biomass with increasing depth has been observed at both sites of the Atlantic Ocean (i.e., Heip et al., 2001; Rex et al., 2006), and is explained by a decreased flux of organic matter when depth increases (Johnson et al., 2007). A global-scale analysis (Wei et al., 2010) showed that this decrease in total biomass is related to a decrease in individual size of the organisms.

3.3   Dominant Pressures

Many human activities have been documented to have impacts on benthic communities (Rice et al., 2010). Effects of mobile bottom-contacting fishing gear on coastal and shelf benthic communities have been documented essentially everywhere that such gear has been used. However, the nature of those impacts and their duration have been shown in many reviews to depend on the type of substrate and frequency of trawling (Collie et al., 1997; FAO, 2009; Hiddink et al., 2006; Kenchington et al., 2007; National Research Council, 2001), with the longest-lasting impacts on hard-bodied biogenic structures, such as corals and glass sponges (see chapters 42 and 43). Recovery of benthic communities following cessation of bottom trawling has also been documented in several studies (Grizzle et al., 2009; Kaiser et al., 2006). Similar effects have been documented for other physical disturbances, such as aggregate extraction (Barrio Froján et al., 2011), with moderate recovery rates from local extraction events (Boyd et al., 2005), although such disturbances are usually concentrated in coastal areas (ICES, 2009).

Coastal benthic communities are also documented to be affected by pollution from land-based and coastal sources (see Chapter 20), such as nutrient runoff from the land, and shoreline alteration for human recreation and infrastucture. These impacts are nearly universal where pollution and nutrient inputs occur, and where coastlines are urbanized or adapted for tourism. However, their nature depends on the type, intensity, and duration of the pollution or nutrient input and extent of alteration, although persistent pressures of this type can greatly alter the species composition and biomass of benthos directly and infirectly through processes such as hypoxia (Borja et al., 2008; Gagné et al., 2006; HELCOM, 2009a; Middelburg and Levin, 2009). These effects can be specific enough that benthic community composition and/or productivity is often used as an indicator of ecosystem stress from pollutants or nutrient inputs (Borja et al., 2009; Quintaneiro et al., 2006; Solimini et al., 2006).

Climate change, including multi-year climate variability, is another major and growing pressure on the benthos. Not only are there effects of coastal warming, but the dependency of deep-sea benthos on the export of organic matter from the ocean surface seems to be at the basis of possible large-scale changes in biomass distribution under future climate scenarios (Jones et al., 2013). Surface ocean warming can result in increased stratification and a less efficient nutrient supply for primary production, leading to a projected decrease in upper ocean biomass (Joos et al., 1999; Steinacher et al., 2010), and a subsequent decrease in organic matter flux to the open ocean seafloor communities (McClain et al., 2012). Jones et al. (2012) modelled particulate organic carbon (POC) fluxes to the seafloor and the resulting macrofaunal biomass distribution under future climate scenarios (RCP 8.5 and RCP 4.5). Under the more severe RCP 8.5 scenario, global macrofaunal biomass is predicted to decrease by 3.771 per cent as a result of decreased organic matter fluxes. Both the greatest negative (-49.7 per cent) and positive (+36.79 per cent) change are predicted for the North Atlantic, with positive changes mainly located on the western side and major negative changes to the east. These shifts can result in range changes of species and facilitate colonization by invasive species in certain areas (Thatje, 2005). The reduction in total biomass of macrofauna will coincide with a size-shift toward smaller organisms (Jones et al., 2012, 2013), which has important biological consequences, including increasing respiration rates and reduction in overall biomass production efficiency (Brown et al., 2004; McClain et al., 2012; Smith et al., 2008) and a reduced energy transfer to higher trophic levels (Brown et al., 2004). Ocean acidification is an increasing threat to coastal marine benthos. Species having calcareous exoskeletons may have particularly high vulnerability to acidification, but it has been found to affect many species of benthos (Andersson et al., 2011). It can be concluded that climate change, through its effect on primary production, is likely to have an important impact on the deep-water benthos, as well as coastal and shelf benthic communities.

4   Fish Communities

Much has been written about the status, trends and drivers of North Atlantic fish communities. These fish communities have been exploited for centuries, and have experienced additional pressures from coastal development, land-based inputs to the seas, and climate variation and change. Much effort has been devoted to teasing apart the influences of exploitation and environmental conditions as drivers of change in fish populations and communities, but definitive answers are elusive. Although long time-series of survey data and commercial catches exist, most effort has gone into assessing the status of commercially exploited populations.

Studies of fish biodiversity in the wider sense are much less numerous, and usually restricted to portions of the North Atlantic only. Thus general patterns have to be assembled from multiple studies, with many differences among them. Consolidating results of multiple studies as a basis for inferences on status, trends and pressures on fish communities has risks, because of the need to account for differences in the catchability of various species (Fraser et al., 2008), the underrepresentation of smaller fish in surveys (Cook and Bundy, 2012), the large sampling effort needed to document fish diversity (Greenstreet and Piet, 2008), and the dependence of inferred patterns of community change on the spatial scale of the analyses (Gaertner et al., 2007). Hence there will be many exceptions to generalizations and many gaps in knowledge. Some species groups, such as tuna and other large pelagics, and elasmobranchs, are addressed in separate chapters of this Assessment, and should be reviewed there. Likewise the parts of this Chapter on the several semi-enclosed seas have information on fish communities at those scales, and should also be consulted. The rest of this section looks at general patterns of near-coastal and shelf fish communities more generally.

4.1 Coastal Fish Communities

The need to piece together emergent messages from numerous separate studies is particularly true for near-coastal fish communities, where time series are usually much shorter and of localized spatial coverage. Most studies devote more attention to explaining variation among coastal fish community properties relative to features of the physical and chemical habitats, including temperature, salinity, oxygen and nutrient levels, clarity of and pollutants in the water column, and to depth, sediment types, benthic communities, contaminant levels, oxygen levels, and disturbance regimes of the seafloor. All of these factors have been shown to influence fish community composition and structure in at least some coastal areas around the North Atlantic.

A few generalizations can be drawn from the diversity of results, but many are common sense. Studies have documented that when any of a large number of environmental factors are altered substantially from values historically characteristic of a coastal area, fish communities are highly likely to be altered as well. Large effects have been documented for factors like oxygen depletion (Oguz and Gilbert, 2007; Stramma et al., 2010), eutrophication (Buchheister et al., 2013; Olsen et al., 2012), sedimentation (França et al., 2012; Jordan et al., 2010; Kopp et al., 2013), and contaminants (Bergek et al., 2012; McKinley et al., 2011; Pato et al., 2008). Emergent macroalgae are an important determinant of coastal fish community diversity and fish abundance, and correspondingly many studies have documented that altering aquatic vegetation can have large effects on fish community status (Pihl et al., 2007; Waycott et al., 2009; York et al., 2012).

However, effects are often situation-specific, and multiple factors interact. For example, coastal flow regimes can dominate over nutrient loading, possibly through ensuring reoxygenation of water and sediments (Kotta et al., 2009; Reiss et al., 2010). Scales of wave energy can also influence how strongly local perturbations of habitat conditions are reflected in changes in fish communities (Jordaan, 2011). The local scale at which fish community structure is determined and variation is documented (e.g., Bonaca and Lipej, 2005) can be amplified, because many drivers of change in coastal fish communities are either both local in scale, such as coastal infrastructure development (e.g., Bulleri et al., 2012) or episodic, such as major oil spills (e.g., Mendelssohn et al., 2012).

Unfortunately the local scales at which coastal fish communities are structured, and where many impacts are experienced, means that it is not possible to present a quantitative accounting of trends in coastal fish communities on regional or North Atlantic scales. Long-term or large-scale studies do document that effects of major oceanographic drivers, such as warming or cooling trends, can be seen in coastal fish communities, documenting that large-scale as well as local factors affect community status (Henderson et al., 2011; Hurst et al., 2004). Species composition generally will react to such large-scale drivers at greater rates than more integrative community metrics (Bui et al., 2010; Hurst et al., 2004).

Given the intensification of use of coastal areas for aquaculture (Chapter 12), infrastructure (several chapters in part V), and land-based inputs to coastal areas (Chapter 20), it is likely that overall the status of coastal fish communities around the North Atlantic has been altered, and in areas with high human use and large habitat changes, the alterations could be large, with a reduction in species diversity and simplification of community structure (Lotze et al., 2006; Waycott et al., 2009). In addition, invasive species can have a large impact in coastal fish communities, and in cases such as the lionfish invasion of Caribbean and coastal southern northern American waters, may spread rapidly from multiple points of initial establishment, seriously disrupting native fish communities (Whitfield et al., 2007; Muñoz et al., 2011). Such changes would, in turn, have consequences much wider than the local scale of the impacts, given the important role of coastal systems as nursery habitats (Beck et al., 2001; Persson et al., 2012).

Consequently, even if the overall trends in coastal fish communities cannot be quantified on the scale of the North Atlantic, the impacts of many pressures on these communities have been documented, as have the effects of larger-scale oceanographic and climatic drivers. With the increase in intensity of human activities causing many of these pressures (Sections IV and V) and a background of a changing ocean climate, there is ample justification for attention to the conservation of these systems. The evidence also indicates that appropriate management regimes need to be designed and implemented on local scales, to accommodate local communities and pressures, even if the overarching policies are developed at larger scales (sensu FAO Ecosystem Approach to Fishing (Staples and Funge-Smith, 2009)).

4.2 Shelf Fish Communities

A few studies have reconstructed fish communities and their variation over centuries into the past, albeit usually for just a few selected species and using catch records, sediment layers, or middens for local areas. These studies have consistently shown major changes in the composition of the fish community over the full time series, sometimes in regime-like ways. Likely impacts of overfishing were already evident early in the second half of the previous millennium (Mackenzie et al., 2007; Poulsen et al., 2007), but changes to the fish community associated with warmer and cooler periods of the North Atlantic are documented for the last several centuries (Enghof et al., 2007).

The current status of shelf fish stocks is best evaluated by the assessments done by the major fisheries management authorities around the North Atlantic. When data are sufficient, assessments provide estimates of fishing mortality and biomass, and interpret these relative to sustainability benchmarks. The biomass benchmarks reflecting that a stock is not overfished vary among jurisdictions and often are based on data series that do not extend back to a time when the stocks were unexploited (Lotze and Worm, 2009; Greenstreet et al., 2012). Nevertheless, Table 36A.4 presents the evaluations for most of the major assessment jurisdictions. The general messages are clear: many stocks are overfished and/or experiencing current overfishing, based on their current status relative to their management benchmarks, and the status of a number of other exploited stocks is not known. However, that only reflects part of the picture. For a large fraction of these stocks, the severe overfishing occurred in the 1990s and 2000s, and their status is improving. The improvement is consistently attributed to reductions in fishing effort (ICES and NOAA websites).

Of course the status of exploited stocks is only part of the fish diversity of the North Atlantic shelf systems. Many studies have analysed trends in the properties of fish communities, but these studies have varied greatly in the time intervals used, the parts of the North Atlantic examined, the metrics of community status quantified, and the species included in the metrics. Given that the results of community analyses are scale-dependent (e.g., Gaertner et al., 2007), metrics are often partially redundant but not interchangeable (e.g., Greenstreet et al., 2012), and both fishing pressure and environmental conditions have changed substantially over the past several decades (ter Hofstede et al., 2012), only a few broad generalizations can be drawn from the diversity of results reported (Table 36A.5).

Based on Table 36A.5 and consistent with other overviews (ICES annual reports, NOAA annual reports), it is without question that in nearly every area of the North Atlantic examined, even moderate fishing pressure has been associated with a decrease in the proportion and absolute number of large fish in the community. Everywhere that heavy fishing was reported, not only does the size composition of the community continue to be truncated, but dominance usually declines as the most common species are reduced in abundance and species of lower or no commercial importance increase in at least relative and often absolute abundance. Whether this changes the diversity metrics depends on case-specific properties of the fish community (how dominant were the most dominant species) and fishing pressure (how intense, how sustained, and how selective).

It is also without question that for every time series of even moderate (decadal) length, effects sought of changing oceanographic conditions on fish community have been documented (e,g, Perry et al., 2005; Lucey and Nye, 2010; Pinsky et al., 2013) . Warming is usually associated with increases in richness and diversity metrics, as the pool of warm-water species that can move into an area from the south during warm periods is almost always larger than the pool of cold-water species that can move in from the north during cold periods. Occasionally one of these environmentally sensitive species becomes very abundant (e.g., the Snake Pipefish outbreak in the North Sea around 2005, Harris et al., 2007), affecting diversity and evenness metrics. However, a number of studies report a negligible or even no trend in community metrics over moderate periods of varying environmental conditions, yet report large changes in the species composition of the community underlying the aggregate metrics. This highlights the strong buffering capacity that is increasingly being argued, where functional redundancy (Schindler et al., 2010; Widemann et al., 2012) gives resilience to fish communities, even if the species composition is changing substantially and without strong structuring processes (Rice et al., 2012).

In a few cases, time series of several decades are available. In all these reports the effects of both changes in fishing effort and in oceanographic conditions are apparent. These can be seen in individual species (e.g., herring, Harma et al., 2012; Larsson et al., 2010; cod, Drinkwater, 2010; Eero et al., 2011; Lilly et al., 2013; sole, Horwood, 2010) and such metrics of community as can be assembled from samples over long time periods (Foch et al., 2014; Greenstreet et al., 1999; ter Hofstede and Rijnsdorp, 2011; Shackell et al., 2012).

Bringing together the results of studies that look at how environmental drivers and fisheries have affected North Atlantic fish communities, the key messages include:

Table 36A.4 | Tabulation of conclusions of assessment authorities on stock status. Each authority has its own standards for benchmarking status. Where quantitative reference points are not estimated, a stock was counted as "healthy or cautious" if abundance was reported as average or high, or as increasing if below average. Stocks reported as depleted or low and declining were counted as negative status. F = status relative to fishing mortality reference points; B = status relative to biomass reference points (,)

Authority	US-NMFS	Canada-DFO	NAFO	ICES
Positive or Cautious Status	195 (B) 290 (F)	72 Healthy 31 Cautious	5	60
Negative Status	38 overfished 27 overfishing	17	5	42
Unknown	247	35	4	75

Table 36A.5 | Tabulation of a number of primary publications documenting status and trends in fish community metrics for areas in the North Atlantic. A large number of community metrics were used, and have been grouped into several categories: "size" includes metrics of body size; "diversity" includes any of the typical indices of species diversity; "richness" and "evenness" include numbers of species recorded and how numbers were distributed among species; "dominance" includes measures of how much the abundance of the few most common species in a community comprised of all the individuals in the community; "N" includes measures of total abundance of individuals in a community, "B" includes measures of community biomass; "slope" and "intercept" are parameters of community size spectra; "species composition" are diverse ways of reporting how the representation of particular species in the community changed. For reporting trends in the metric, "+" means an overall upward trend, "-" means an overall downward trend and "nt" means no overall trend, although there could be substantial interannual variation in the metric.

Location	Indicators and trends	Interval	Comments	Reference
Georges Bank	Size +, diversity+, Biomass/area+	1960s - 2009	Decline in F and effort	Collie et al. 2013
North Sea	Size +, diversity+, B+	1960s - 2009	Decline in F and effort	Collie et al. 2013
Ionian Sea	N+, Richness+, Size+	1998-2008	Decline in F and effort	Tsagarakis et al. 2011
Galicia	Diversity, Richness, N, B	1980-1991	Species composition changed more than indices	Farina et al. 1997
Multiple	Slope +, intercept *	Various; 2 decades+	Increases in F	Bianchi et al. 2000
Scotian Shelf	Slope+, intercept+, diversity nt	1970-1997	Increases in F	"
North Sea	Slope+, intercept+, richness+	1972 - 1998	Increase F and expansion of southern species	"
Portugal	Slope nt, intercept nt, evenness nt	1982 – 1998 with gaps	Large variation in species composition w/o trend	"
NW North Sea	Diversity -, dominance +	1920 – 1990s (with gaps)	Large changes in non-commercial species as commercial ones decreased	Greenstreet et al. 1999
Canary Islands	Diversity nt, richness nt	1990s	Large changes w/o trend in species composition	Uiblein et al. 1996
Baltic and Kattegat	Richness +, N nt	1990 - 2008	Difference in connectance not a factor	Hiddink and Coleby 2012
West of Scotland	Richness -	1997-2008	F stayed high	ter Hofstede et al. 2010
North Sea	Richness+	"	Southern species incursions	"
Celtic Sea	Richness+	"	"	"
Iceland	Richness +, diversity nt	1996-2007	Warming	Steffansdottir et al. 2010
Iceland	Species Composition	1970s-2010s	Warming	Valdimarsson et al. 2012
Dogger Bank	Diversity-, dominance +	1991 - 2010	Warming, and common species increased	Sonnewald and Turkay 2012
NE Shelf	Richness+, Diversity+	1980 - 2008	Abrupt regime like change. Species makeup changed most	Simpson et al. 2011
North Sea	Richness+ (eggs & larvae)	1958 - 2005	Warm species entering	Beaugrand et al. 2008
Barents Sea	Functional diversity nt	2004 - 2009	Many changes in species comp.	Wiedmann et al. 2014
Scotian Shelf	B-, Size-, Evenness-	1970 - 2006	Period of heavy fishing	Shackell et al. 2012
W of Scotland	Richness nt, diversity nt	1980s – 2000s	Fishing stable but high	Campbell et al. 2011
North Sea	Slope+, Intercept+, Lmax-	1972 -2000s	Large species becoming rarer. Greater in high F areas	Daan et al. 2005
Scotian Shelf	Evenness+. Dominance -	1970 - 2000	F high and increasing	Shackell and Frank 2003
Medit. and NE Atlantic	Length nt, N nt	1997 - 2007	Large variability but no persistent trends	Rochet et al. 2010
Portuguese shelf	Richness nt, Diversity nt	1989 - 1999	High spatial patchiness, no overall trends	Sousa et al. 2006
NW Atlantic	Diversity nt	1970s – 2000s	Latitudinal trend but no time trend	Fisher et al. 2008
NW Atlantic	N-, MaxAge-, Size-	1978 various	F high and not declining	Hutchings and Baum
Medit.	Abundance-, Richness-	1948-2005	Just sharks	Ferretti et al. 2013

(a) Essentially every shelf fish community in the North Atlantic has been altered by decades to centuries of fishing. For many areas, excessive fishing persisted long enough for target species to be depleted to states where recovery has been slow, and whole communities have had their diversity reduced, with size metrics showing the greatest effects at community scales. However, in the twenty-first century, fishing effort has been reduced in most parts of the North Atlantic shelves, particularly where stocks and communities were most stressed, and there is evidence of recovery in most of these areas, albeit at different rates for different species, with some species having recovered to target levels.

(b) Where data have been examined, every shelf fish community has had its species composition change as oceanographic conditions have changed. Responses to warming and cooling trends seem to be most prevalent, but these also have been looked for most often. Regime-like changes in fish community composition have been documented often, but they are not universal. Aggregate community metrics have often changed much less than the abundance of the species contributing to them.

(c) On case-by-case examples it is often hard to definitively untangle the effects of fishing and of environment on fish communities although some at least partial successes are being reported (Bell et al., 2014). However, unless fishing is kept at a sustainable level, community-scale effects of depletion of target species are highly likely, and may reduce resilience to environmental drivers (Shackell et al., 2012).

5 Seabirds[5]

5.1 North Atlantic Overview

Overall, populations of breeding seabirds in the North Atlantic appear to be decreasing. This contention is the outcome of an integration of negative trends in both the North Atlantic Fisheries Organization (NAFO) and the International Council for the Exploration of the Sea (ICES) Regions. Most of the uncertainty about the Iceland population centres on estimates of the very abundant auk species which drive the overall population patterns. Further resolution of these estimates is essential.

Trends are considered for all species and for diving and surface-feeding taxa, which often have different sensitivities to climatic and anthropogenic environmental changes. Considerations of marine bird biodiversity are swamped by these most abundant species, although some aspects of species trends and community changes are addressed in the appended regional accounts.

The overall picture here indicates that surface-feeders (storm-petrels, gulls, terns) drive the negative NAFO trend, and diving auks (Dovekie *Alle alle*, Thick-billed Murre *Uria lomvia*, Common Murre *Uria aalge*) in Iceland drive the negative ICES trend, with the ICES decrease being six times greater than that reported for NAFO.

Within regional trends considerable variation is observed (Table 36A.6), with different areas exhibiting increasing trends (E Baffin Island, Newfoundland/Labrador, E Canada + US, Faroes) or decreasing trends (W Greenland, Gulf of St. Lawrence, Caribbean, E Greenland, Iceland, Norwegian and North Seas).

5.2 NAFO Area

The negative trend in the NAFO Region is driven by surface-feeding species (gulls, terns, petrels) that are decreasing in eastern Canada (Cotter et al., 2012) and in the Caribbean (Bradley and Norton, 2009). The decline is also driven by an inferred decreasing trend in a diving planktivore (Dovekie) in Western Greenland based on North American Christmas Bird Counts (BirdLife International, 2014). Otherwise, divers are increasing in all regions, with the exception of the Caribbean, where a small population of Brown Pelicans (*Pelecanus occidentalis*) is declining (Bradley and Norton, 2009). Decreasing trends in surface-feeders and

increasing trends in diving species are associated with fisheries closures in eastern Canada and the concurrent cessation of discards and gill-net removals (Bicknell et al., 2013; Regular et al., 2013). Surface-feeders are vulnerable to sea-surface temperature perturbations (Schreiber and Schreiber, 1984) and long-line fishing (Zydelis et al., 2009). Some ocean regions, notably the Gulf of Mexico, are data-deficient.

5.3 ICES Area

The decreasing trend in the ICES Regions is overridden by the uncertain negative Icelandic estimates. Positive trends are reported for the Faroes Island (Denmark)/Western United Kingdom and for the Barents Sea (which is excluded from consideration as it is in the Arctic rather than in the North Atlantic region).

Decreasing trends in auks in the Norwegian Sea (Anker-Nilsen et al., 2007) are associated with warming ocean trends and the consumption of forage prey by warm-water predatory fishes (e.g., Atlantic mackerel) moving into the region (T. Anker-Nilsen, pers. comm.).

6 Marine mammals

Many marine mammals primarily inhabit the margins of the North Atlantic Ocean, especially the continental shelf and within the many

Table 36A.6 | Population trends of breeding seabird taxa in the North Atlantic Ocean. Divers include shearwaters, gannets, cormorants, shags, pelicans, auks; surface-feeders include fulmars, storm-petrels, frigatebirds, skuas, jaegers, gulls, terns.

REGION	ALL SPECIES	DIVERS	SURFACE-FEEDERS
NORTH ATLANTIC	Decrease	Decrease	Decrease
NAFO	Decrease	Increase	Decrease
E Baffin Is	Increase	Increase	?
W Greenland	Decrease	Decrease	No Change?
Newfoundland/Labrador	No Change	Increase	Decrease
Gulf St. Lawrence	Decrease	Increase	Decrease
E Canada/United States	Increase	Increase	Increase
Gulf of Mexico	?	?	?
Caribbean	Decrease	Decrease	Decrease
ICES	Decrease	Decrease	Increase
E Greenland/Iceland	Decrease	Decrease	Increase
Barents Sea[a]	Increase	Increase	Decrease
Norwegian Sea	Decrease	Decrease	Increase
Faroes Island (Denmark), Shetland, Western United Kingdom	Increase	Increase	Decrease
N Sea/English Channel	Decrease	No Change?	Decrease
Baltic Sea, Skagerrak, Kattegat	?	?	?
France, Iberia, Azores	?	?	?

---

[5] Information in this subchapter is based on the appended spreadsheets of regional trends, numbers and sources, and can be modified as gaps are filled and new information obtained. Trends in breeding seabird populations from the 1970s/80s through the 2000s are reported where possible, although more often only the most recent decade[s] is available.

semi-enclosed regional seas. Other species primarily occupy the North Atlantic gyre, bounded by clockwise flowing currents most famously defined by the Gulf Stream in the north and the Canary Current in the east (Figure 36A.1). Many of these latter species also utilize habitats in areas further north and south of the gyre, at least seasonally.

The gyre species as identified here include many historically subjected to whaling, including slower whales such as right whales, humpback whales and sperm whales, all targeted by open boat whalers through the nineteenth century. Faster whales, such as blue whales and fin whales, were targeted beginning in the late nineteenth century. Several medium-sized whales were also subject to whaling, with some continuing to be so down to the present, including long-finned pilot whales (targeted, for example, for a millennium in a drive fishery in the Faeroe Islands), the northern bottlenose whale (targeted from ships from the mid-nineteenth to the mid-twentieth century), and minke whales (targeted from ships by twentieth- and twenty-first-century whalers). The effects of whaling range from slight for species such as long-finned pilot whales (Taylor et al., 2008) to near extinction for right whales (Reeves et al., 2007). In recent decades, species such as humpback whales that migrate across the North Atlantic gyre from breeding grounds near islands in the Caribbean have recovered from earlier effects of whaling. However, those humpbacks breeding near the Cabo Verde islands apparently have not (Reilly et al., 2008).

Other gyre cetaceans include short-finned pilot whales, killer whales, pygmy killer whales, various other bottlenose whales and common dolphins. Generally these have not been subject to intense whaling.

In addition to the effects of historical and ongoing whaling, cetaceans in the North Atlantic are subject to various forms of disturbance (see Chapter 37). For example, disturbance and sometimes injury of individual animals by noise, including sound generated from military operations and from seismic operations has been demonstrated for some species of beaked whales (Cox et al., 2006; Whitehead, 2013). Similarly, mortality from ship strikes and entanglement in fishing gear have been demonstrated for humpback whales and right whales in the North Atlantic, and, especially for the small population of right whales that survived whaling, such mortality can be significant (Laist et al., 2001).

Harbour porpoise and common dolphin occupy continental shelf regions in the North Atlantic, and also occur further north. These have been subject to entanglement in fishing gear in many areas, especially bottom-

Figure 36A.1 | Currents defining the North Atlantic gyre. From http://earth.usc.edu/~stott/Catalina/Oceans.html

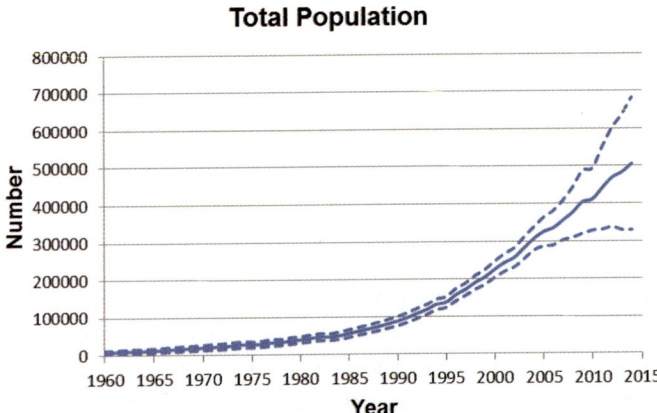

Figure 36A.2 | Estimated recent population trends of Northwest Atlantic gray seals. Taken from Hammill et al., 2014. Estimated based on surveys and population model; dashed lines represent 95% confidence interval of estimates.

tending gill nets (NOAA, 2014). Pinniped species, such as harbour seals, gray seals and harp seals, generally occur further north, but also occupy northern continental shelf regions around New England in the west and the United Kingdom of Great Britain and Northern Ireland in the east. They are dependent on haul-out areas, beaches and ice cover, and have often been thought to compete with fishermen for prey. Gray and harp seals have been subject to predator control programmes and, especially in the western North Atlantic, commercial harvests. However, both have increased in abundance in recent decades to relatively high levels. For example, in the northwest North Atlantic, gray seals have increased, reaching roughly half a million animals in 2014 (Figure 36A.2).

7   Specific areas of the North Atlantic

The predominance of semi-enclosed seas with characteristic biota around the North Atlantic, particularly the more southern and central portions of the region, and the concentration of human pressures around these seas, result in many important trends in biodiversity being observed most clearly at the scale of these seas. Hence this chapter includes brief summaries of the main patterns of and pressures on biodiversity for a number of these regional seas.

7.1   Black Sea

The Black Sea is a very deep inland sea with an area of 432,000 km². The thin upper layer of marine water (up to 150 m) supports the unique biological life in the Black Sea ecosystem. The deeper and more dense water layers are saturated with hydrogen sulphide, that over thousands years, accumulated from decaying organic matter in the Black Sea. Due to the unique geomorphological structure and specific hydrochemical conditions, specific organisms, basically on the level of protozoa, bacteria, and some multi-cellular invertebrates inhabit the deep-sea waters. Knowledge about biological forms of life in the deep waters of the Black Sea is very limited. The disturbance of the natural balance between the two layers could trigger irreversible damage to the people and ecosystem of the Black Sea[6].

The recently published evidence raises the number of species, including supra-specific taxa, inhabiting the Black Sea to 5,000 (Gomoiu, 2012).

The distribution diagram of different physiological types of species from the Black Sea fauna shows the coexistence of four categories of species, according to a salinity gradient: (1) marine species, (2) freshwater species, (3) brackish water species, and (4) Ponto-Caspian relic species (Skolka and Gomoiu, 2004). The Black Sea biota consist of 80 per cent of Atlantic-Mediterranean origin species, and 10.4 per cent and 9.6 per cent of species of freshwater and Ponto-Caspian origin, respectively (Shiganova and Ozturk, 2010). The eastern sector is one of the biologically richest regions on Earth and is recognized as a biodiversity hotspot, along with other parts of the Caucasus Biodiversity Hotspot Region (Kazanci et al., 2011).

Genetic studies confirm the recent origin of many Black Sea marine taxa from the Mediterranean. The majority of these taxa most probably entered the Black Sea through the Marmara Sea and the Straits linking the Black Sea and the Marmara Sea after the last glacial maximum, when a connection between the Mediterranean and the Black Sea was re-established (Ryan et al., 1997). For this reason, these Black Sea populations are genetically similar to the Mediterranean ones, although in some cases they have already diverged, implying reduced genetic connectivity (e.g., Durand et al., 2013). There are also cases of Black Sea taxa, such as the copepod *Calanus euxinus*, to which the species status has been attributed, although they have only recently diverged (Papadopoulos et al., 2005). On the other hand, genetic studies have confirmed the ancient origin of the Ponto-Caspian species and have even revealed an additional diversity in the form of cryptic species (Audzijonyte et al., 2006).

The migration of marine species from the Mediterranean is hampered by a number of ecological barriers: (1) low salinity and ionic composition and the difference in the thermal conditions, (2) the presence of hydrogen sulphide in the Black Sea bottom areas, and (3) the lack of tides (Skolka, 2004; Gomoiu, 2004).

Our current knowledge on the biodiversity of certain taxa in the region shows a well-defined zoogeocline from the Marmara Sea and Bosphorus Strait to the inner parts of the region (Azov Sea), depicted both as a pattern in overall species composition and species (or taxa) numbers. As a general trend, species numbers decrease along with the decrease in salinity towards the inner parts of the basin. The trend is homologous to that seen in the benthic invertebrate inventories of all the major European semi-enclosed regional seas. Salinity and food availability appear to be the dominant abiotic factors correlated, though weakly, with the various patterns deriving from the taxonomic/zoogeographic categories (Surugiu et al., 2010).

The invasion of the basin by alien species began in the Middle Ages, with the bivalve *Teredo navalis*, as the first one recorded. The anthropogenic disturbance gradually increased and reached an unprecedented amplitude, becoming evident during the 1950s, concurrently with the penetration of the Indo-Pacific predator gastropod *Rapana thomasiana*, which has severely reduced stocks of native oysters - *Ostrea sublamelossa* (Gomoiu, 1998; Skolka, 1998).

The invasive species may affect not only the ecosystem but also various sectors of the economy, with devastating effects for some of these sectors. This is the case of the American comb jelly, *Mnemiopsis leidyi*, accidentally introduced into the Black Sea by ship ballast water in the early 1980s. The introduced comb jelly nearly led to the collapse of pelagic fish populations (over 26 commercial Black Sea fish stocks), and finally caused a major shift in the marine ecosystem. Only after the penetration of a new warm-water ctenophore, *Beroe ovata*, ten years later, did the *M. leidyi* population diminish, allowing the ecosystem to recover its entire trophic web.

Another particularly important case of accidental immigration is that of toxic microscopic algae, whose outbreaks produce the so-called toxic blooms. The biotoxins produced inside the algal cells (i.e., domoic, okaidic, yessatoxine or azaspiracids acids), can have toxic effects on the other taxa and even on humans. The vast majority of these species, such as *Noctiluca scintillans*, also can reduce or deplete the oxygen concentration in the water column and sediments, leading to hypoxia or anoxia.

During recent decades, the temperature increased both in the surface mixed and in the cold intermediate water layers. This has been shown to be another factor accelerating the establishment of more thermophilic species populations, promoting their northward expansion from the Mediterranean (Shiganova and Ozturk, 2010).

In the conditions present during periods after major eutrophication outbreaks have returned towards more typical states, two antagonistic and synergistic processes have taken place: (a) the penetration of some opportunistic species, and (b) the disappearance of some economically valuable native species. In 1999, the first Black Sea Red Book was issued, which includes 160 endangered and rare plant and animal species. Sturgeons are the most endangered species, along with those that inhabit shallow coastal waters (turbot, sharks), seals, shrimp and oyster species.

The same critical status is also attributed to most of the coastal margin neritic and open sea habitats, and to 13 out of 37 benthic habitats. Among the habitats at risk are included the neritic water column, coastal lagoons, estuaries/deltas and wetlands/saltmarshes (Goriup, P., 2009).

In the Black Sea Region, almost a quarter of the habitat types listed in the European Union (EU) Habitats Directive can be found. Many are located in the intertidal zone and are consequently heavily influenced by the presence of salt water and continuous wave action. They include

---

6 http://www.blacksea-commission.org/_geography.asp

extensive areas of mud and sand flats, salt meadows and marshes, and long stretches of white sandy beaches (Sundseth and Barova, 2009).

Coastal forests are also well represented, especially on the low-lying hills in the south of Bulgaria and within the Danube Delta. They include a variety of rare habitat types listed in the Habitats Directive, such as the Western Pontic beech forests, as well as floodplain forests, alluvial and mixed riparian forests, all of which are important roosting and resting habitats for bats and birds (Sundseth and Barova, 2009).

7.2 Mediterranean Sea

The Mediterranean Sea is a marine biodiversity hot spot. Approximately 17,000 marine species have been reported from the Mediterranean Sea (Coll et al., 2010). Of these, at least 26 per cent are prokaryotic (Bacteria and Archaea) and eukaryotic (Protists) marine microbes. Phytoplankton includes more than 1,500 species. Macrophytes include approximately 850 species. Among microzooplankton, the foraminifera are the main group with more than 600 species, and about 100 species are commonly present in Mediterranean waters (Dolan, 2000). However, it is within the Animalia group that there is published evidence for the majority of the species so far reported (~11,500) with the greatest contribution coming from the Crustacea (13.2 per cent) and the Mollusca (12.4 per cent), followed by the Annelida (6.6 per cent), the Plathyhelminthes (5.9 per cent), the Cnidaria (4.5 per cent), the subphylum Vertebrata (4.1 per cent), the Porifera (4.0 per cent), the Bryozoa (2.3 per cent), the subphylum Tunicata (1.3 per cent), and the Echinodermata (0.9 per cent). With regard to the Vertebrata, there are ~650 marine species of fish, of which approximately 80 are elasmobranchs and the rest are mainly from the Actinopterygii class (86 per cent). Nine species of marine mammals (five belong to the dolphins, Delphinidae, and one each to the Ziphiidae, Physeteridae, Balaenopteridae, and Phocidae (seals)) and three species of sea turtles (the green *Chelonia mydas*, the loggerhead *Caretta caretta* and the leatherback *Dermochelys coriacea*) are regularly recorded in the Mediterranean Sea. A total of 15 species of seabirds frequently occur in the Mediterranean Sea (Coll et al., 2010).

However, estimates of marine diversity are still incomplete as yet-undescribed species will be added in the future (Coll et al., 2010; Danovaro et al., 2010). In many cases, several cryptic species, mainly invertebrates, have been revealed through molecular approaches (Calvo et al., 2009), thus increasing the number of reported species. Moreover, diversity for microbes is substantially underestimated, and the deep-sea areas and portions of the southern and eastern regions are still poorly known (Coll et al., 2010). The next generation of sequencing data are already producing a wealth of unique sequences, in unprecedented rates, many of which are new operational taxonomic units (OTUs). Tens of thousands of OTUs are produced in the context of minimal monitoring projects (Pavloudi et al., in press), a fact which will soon alter the numbers so far reported. In addition, the invasion of alien species is a crucial factor that will continue to change the biodiversity of the Mediterranean (Zenetos et al., 2010), mainly in its eastern basin, where invading species can spread rapidly northwards and westwards due to the warming of the Mediterranean Sea (Lejeusne et al., 2010).

Genetic diversity, especially the presence of genetically distinct populations within a species, is another important component of biodiversity. Although the lack of strong physical barriers in the marine environment and the high dispersal ability of many marine taxa tend to diminish genetic differentiation, several marine species within the Mediterranean exhibit a strong genetic structure as a result of their life history traits and of the complex geography and hydrography of the Mediterranean. Diverse taxa, from small pelagic fish, like anchovy (Magoulas et al., 2006), to molluscs (Cordero et al., 2014) have been shown to consist of genetically distinct populations, with low connectivity, which calls for more local-scale management and conservation actions, especially for commercially exploited or vulnerable species. This differentiation often occurs across well-known biogeographic barriers, like the Almeria-Oran front (Paternello et al., 2007) and the Siculo-Tunisian strait (Mejri et al., 2009).

Spatial patterns of species diversity show a general gradient characterized by the decreasing number of species from the west to the east (Arvanitidis et al., 2002, 2009; Coll et al., 2010). For certain taxa such as polychaetes, indications of habitat diversification, such as the average island distance from the nearest coast, number of islands and island surface area, have been reported to be best correlated with their multivariate community patterns (Surugiu et al., 2010). The decrease in species richness gradient has also been attributed to a synergy of variables, such as food availability, salinity (Surugiu et al., 2010) and current knowledge gaps with regard to the biota over large sectors, such as the northern African and eastern coasts (Coll et al., 2010; Coll et al., 2012). Lower productivity rates (oligotrophism) are related to the significantly lower size of the species in the eastern basin, a phenomenon known as "Levantine nanism" (dwarfism) (Por, 1989). Biodiversity is also generally higher in coastal areas and continental shelves, and with some exceptions it decreases with depth (Coll et al., 2010; Danovaro et al., 2010). However, fish biodiversity components, measured as species richness of total, endemic and threatened coastal fish assemblages, as well as their functional and phylogenetic diversity, have been mapped and described as spatially mismatched between regions of the Mediterranean Sea (Mouillot et al., 2011).

The Mediterranean Sea is also diverse in terms of habitats and ecosystem types, due to its unique biogeography (Bianchi et al., 2012). Although empirical data are insufficient to have a full representation of habitat types (Danovaro et al., 2010; Levin et al., in press) and are only fully available for some coastal habitats (Giakoumi et al., 2013), a series of surrogates or modelling techniques are used to characterize marine habitats in the whole Mediterranean basin (Micheli et al., 2013; Martin et al., 2014).

Temporal trends have indicated that overexploitation of some fish and macro-invertebrates and habitat loss have been the main human drivers of historical changes in biodiversity (Coll et al., 2010; Lotze et al., 2011;

Coll et al., 2012). At present, habitat loss and degradation, followed by fishing, climate change, pollution, eutrophication, and the establishment of invasive species, are the most important factors that affect most of the taxonomic groups and habitats (Claudet and Fraschetti, 2010; Coll et al., 2010; Abdul Malak et al., 2011: Lotze et al., 2011; Bianchi et al., 2012; Coll et al., 2012; Micheli et al., 2013). All these impacts are expected to grow in importance in the future, especially climate change and habitat degradation.

7.3   Baltic Sea

The Baltic Sea is a small sea on a global scale, but as one of the world's largest and most isolated bodies of brackish water, it is ecologically unique. Eutrophication, caused by nutrient pollution, is a major concern in most areas of the Baltic Sea. The biodiversity status was classified as being unfavourable in most of the Baltic Sea, as only the Bothnia Sea and some coastal areas in the Bothnian Bay were classified as having an acceptable biodiversity status. The results indicate that changes in biodiversity are not restricted to individual species or habitats; the structure of the ecosystem has also been severely disturbed (Helcom, 2010).

Baltic Sea biodiversity and human pressures on it have been summarized for all components except bacteria in Helcom (2009a) integrated thematic assessment. Alongside the general deterioration of the Baltic Sea biodiversity positive signs were also found for grey seals and some fish and bird species. A recent expert evaluation of endangered species in the Baltic Sea by Helcom (2013) shows the risk for extinction among a number of plant and animal species still existing.

The Baltic Sea is characterized by large areas (ca 30 per cent) that are less than 25 m deep, interspersed by a number of deeper basins with a maximum depth of 459 m. The western and northern parts of the Baltic have rocky seabeds and extended archipelagos; the seafloor in the central, southern, and eastern parts consists mostly of sandy or muddy sediment (ICES, 2008b).

The Baltic Sea phytoplankton community is a diverse mixture of microscopic algae representing several taxonomic groups, with more than 1,700 species recorded. The species composition of the phytoplankton depends on local nutrients and salinity levels and changes gradually from the southwest to the northeast. Primary production exhibits large seasonal and interannual variability (Helcom, 2002). In the southern and western parts, the spring bloom is dominated by diatoms, and by dinoflagellates in the central and northern parts (Helcom, 2002, 2009a).

Cyanobacteria are a natural component of the phytoplankton community in most parts of the Baltic Sea area. They usually dominate in summer in the coastal and open areas of most sub-basins of the Baltic Sea, with the exception of the Belt Sea and the Kattegat. Cyanobacterial blooms in the Baltic Proper are typically formed by the diazotrophic species Aphanizomenon flos-aquae, Anabaena spp. and Nodularia spumigena that can fix molecular nitrogen. N. spumigena blooms are potentially toxic, whereas no toxic blooms of A. flos-aquae have been recorded in the Baltic Sea. The blooms of N_2-fixing cyanobacteria as such do not necessarily indicate strengthened eutrophication (Helcom, 2009b).

The zooplankton of the Baltic Sea is dominated by calanoid copepods and cladocerans. The species composition is influenced by the salinity gradient. Generally marine species (e.g., *Pseudocalanus* spp.) prevail in the southern more saline part, while brackish species (e.g., *Eurytemora affinis* and *Bosmina longispina maritima*) dominate in the northern areas (ICES, 2008b). The latitudinal distribution of marine macrozoobenthos in the Baltic Sea is limited by the gradient of decreasing salinity towards the north. The decreasing salinity reduces macrozoobenthic diversity, affecting both the structure and function of benthic communities. In addition, the distribution of benthic communities is driven by strong vertical gradients. Generally, the more species-rich and abundant communities in shallow-water habitats (with higher habitat diversity) differ from the deep-water communities, which are dominated by only a few species (Helcom, 2009a).

The composition of the benthos depends both on the sediment type and salinity level, with suspension-feeding mussels important on hard substrata and deposit feeders and burrowing forms dominating on soft bottoms. The species richness of the zoobenthos is generally poor and declines from the southwest towards the north due to the drop in salinity. However, species-poor areas and low benthos biomass are also found in the deep basins in the central Baltic due to the low oxygen content of the bottom water (ICES, 2008b).

The distribution of the roughly 100 fish species inhabiting the Baltic is largely governed by salinity levels. Marine species (some 70 species) dominate in the Baltic proper, and fresh-water species (some 30–40 species) occur in coastal areas and in the innermost parts (Nellen and Thiel, 1996, cited in Helcom, 2002). Cod, herring, and sprat comprise the large majority of the fish community in both biomass and numbers. Commercially important marine species are sprat, herring, cod, various flatfish, and salmon. Sea trout and eel, once abundant, are now in very low populations. Sturgeon was a very important component of local exploited fish fauna for centuries, especially in the southern Baltic. Currently, sturgeon is a red-listed fish in the Baltic Sea and a reintroduction programme has been initiated (Helcom, 2009a). The seabirds in the Baltic Sea comprise pelagic species like divers, gulls, and auks, and benthic-feeding species like dabbling ducks, sea ducks, mergansers, and coots (ICES, 2003). The Baltic Sea is more important for wintering (ca 10 million) than for breeding (ca 0.5 million) seabirds and sea ducks. The common eider exploits marine waters throughout the annual cycle, but ranges from being highly migratory (e.g., in Finland) to being more sedentary (e.g., in Denmark).

The marine mammals in the Baltic consist of gray (*Halichoerus grypus*), ringed (*Phoca hispida*), and harbour seals (*Phoca vitulina*), and a small population of harbour porpoise (*Phocoena phocoena*). Seals and harbour porpoise were much more abundant in the early 1900s than they are today (Elmgren, 1989; Harding and Härkönen, 1999) where their fish consumption may have been an important regulating factor for the abundance of fish (MacKenzie et al., 2002). Baltic seal populations – harbour seals, gray seals, and ringed seals – are generally increasing. The recent abundance of the harbour porpoise in the Baltic Proper is low (Helcom, 2009a).

7.4 North Sea

The North Sea is a large semi-enclosed sea on the continental shelf of northwest Europe, formed by flooding in the Holocene period. The sea is shallow, deepening towards the north. The seabed is predominantly sandy, but muddy in deeper parts and in southern coastal areas with extensive river influence.

The strong coupling between benthic and pelagic communities in the shallow parts of the North Sea makes it one of the most productive marine areas in the world, with a wide range of plankton, fish, seabirds and benthic communities.

The most commonly found zooplankton genus in the North Sea is of the copepod *Calanus*. Hays et al. (2005) observed between 1960 and 2003 a clear decrease in the abundance of *Calanus finmarchicus*, and an increase in *C. helgolandicus*, with a marked overall decrease in both species combined. Beaugrand et al. (2002) also found a decrease in the abundance of cold water and Arctic zooplankton species and an increase in warmer water ones.

The 50-m, 100-m, and 200-m depth contours broadly define the boundaries between the main benthic communities in the North Sea (Künitzer et al., 1992; Callaway et al., 2002). Bottom temperature, sediment type, and trawling intensity have been identified as the main environmental variables affecting local community structure. Epifaunal communities are dominated by free-living species in the south and by sessile species in the north.

Throughout the year, the pelagic fish component is dominated by the herring *Clupea harengus*. The mackerel *Scomber scombrus* and the horse mackerel *Trachurus trachurus* are mainly present in summer, when they enter the area from the south and from the northwest. Dominant gadoid species are the cod *Gadus morhua*, the haddock *Melanogrammus aeglefinus*, the whiting *Merlangius merlangus*, and the saithe *Pollachius virens*; the main flatfish species are the common dab *Limanda limanda*, the plaice *Pleuronectes platessa*, the long rough dab *Hippoglossoides platessoides*, the lemon sole *Microstomus kitt*, and the sole *Solea vulgaris*. The major forage fish species are the sandeel *Ammodytes marinus*, the Norway pout *Trisopterus esmarki*, and the sprat *Sprattus sprattus*, but juvenile herring and gadoids also represent an important part of the forage stock. However, large annual variations in species composition occur as a consequence of natural fluctuations in the recruitment success of individual species. Fish species richness is highest around the edges of the North Sea (particularly along the coast of Scotland, in the Southern Bight, and in the Kattegat) and lowest in the central North Sea (ICES, 2008).

Certain highly migratory species that historically were fairly common in the North Sea have become very rare (e.g., tunas and the halibut *Hippoglossus hippoglossus*). The stocks of most elasmobranchs are at low levels (ICES, 2008). The spurdog (*Squalus acanthias*) was the most common shark species, but is now considered to be depleted to approximately 5 per cent of its virgin biomass in the whole Northeast Atlantic (Hammond and Ellis, 2005). Decades of intensive fishing have been shown to have altered both the species (ICES, 2008; Piet et al., 2009; Greenstreet et al., 2010) and size composition (Daan et al., 2005) in the North Sea, with greatest effects where fishing has been most intense. There is some evidence that these effects are being reversed since fishing pressure was reduced in the late 2000s, but community metrics are still far from their values observed prior to the 1970s (Greenstreet et al., 2011).

About 2.5 million pairs of seabirds, belonging to some 28 species (ICES, 2008) breed around the coasts of the North Sea. Although most species breed in dense colonies along the coast, they make very different use of the marine ecosystem. During the breeding season, some species depend on local feeding conditions within tens of kilometres around their colony, whereas others may cover several hundreds of kilometres during their foraging trips. Outside the breeding season, some species stay quite close to their breeding grounds, and others migrate across the North Sea or elsewhere, even as far as the Antarctic. Feeding habits also diverge. Auks and cormorants dive from the surface, gannets and terns use plunge diving, and gulls feed mostly from the surface. A few (especially skuas) are kleptoparasites (Dunnet et al., 1990). Their food resources vary accordingly, ranging from plankton to small schooling fish and discards. Because of all these differences, seabirds do not represent a single homogeneous group that responds to fisheries in a uniform way. A few species profit directly from human consumption fisheries, either discards or offal, e.g., fulmars and gulls. Current seasonal distributions, status, and trends of these species have shown an increasing trend over the last century. Auks and cormorants are now protected in some areas (e.g., the southern North Sea and Kattegat). Gull numbers have been controlled in many areas. Fulmars may have benefited from the expansion in fishing. Skuas may have profited directly from the increase in population size of seabirds in general. On a shorter time scale, 12 out of 28 species show an increasing trend during the last decade and four a decreasing trend; four appear to be stable and for another four the situation is unknown (ICES, 2008).

Many cetacean and pinniped species have been observed within the North Sea, but most of these must be considered vagrants and only a few are resident representatives of the North Sea ecosystem. Harbour *Phoca vitulina* and gray *Halichoerus grypus* seals have undergone large population changes over the past century. Populations of harbour seals along the continental coast reached an all-time low in the 1970s. Subsequently, these populations increased steadily at an annual rate of 4 per cent, with two major interruptions in 1988 and 2002, when the populations were hit by outbreaks of the phocine distemper virus (ICES, 2008).

Although several cetacean species visit the North Sea, the dominant species are minke whales *Balaenoptera acutorostrata*, harbour porpoises *Phocoena phocoena*, and whitebeaked dolphins *Lagenorhynchus albirostris*. Preliminary abundance estimates from a survey conducted in 2005 indicate the *status quo* for all these species. Harbour porpoises, however, have shifted their focal distribution from the northern part of the North Sea to the southern part. The reasons for this southward shift of harbour porpoise distribution are unknown; however, a change in distribution and availability of prey species is considered the most likely explanation, although other explanations are possible (ICES, 2008).

A number of sand banks across the North Sea qualify for protection under the European Union Habitats Directive, mainly along the UK coast, the eastern Channel, the approaches to the Skagerrak, and the Dogger Bank. Extensive biogenic reefs of *Lophelia* have recently been mapped along the Norwegian coastline in the eastern Skagerrak, and *Sabellaria* reefs have been reported in the south, although their distribution and extent are not known. Gravel deposits also qualify for protection, but comprehensive maps at a total North Sea scale are not readily available (ICES, 2008).

7.5 Gulf of St. Lawrence

The Gulf of St. Lawrence (hereafter the Gulf) is a relatively small (236,000 km^2) sea located along the southern Canadian Atlantic in the Northwest Atlantic connected to the ocean by the shallow (60 m) and narrow (17 km) Strait of Belle Isle in the northeast, where dense Labrador Shelf water enters the Gulf and the deep (480 m) and wide (104 km) Cabot Strait in the south (Figure 36A.2bis). The Gulf receives large quantities of freshwater from the St. Lawrence River system, more than half of the freshwater inputs from the east coast of North America, which gives the Gulf an estuarine-like circulation. The Gulf has a complex geomorphology, with a broad shallow shelf (<100 m) in the south and a northern region characterized by narrow coastal shelves and deep (>300 m) channels. The environment is influenced by climate variability in the Northwest Atlantic. The seasonality is characterized by severe winters with low temperatures and ice covering a large surface from December to April, and by warm surface water in summer. The sea surface temperature differences between winter (less than 0°C) and summer (close to 20°C in the south) are among the largest recorded around the Atlantic. The Gulf's multilevel complexity provides the conditions for a highly diverse fauna, with a mixture of boreal, temperate and sub-tropical species, and productive biological communities (Benoît et al., 2012).

Global climate change, including ocean warming affects the Gulf as indicated by a clear decreasing trend in winter ice coverage and volume and increasing sea surface temperatures in recent years (DFO, 2013). The characteristics of the deep layer are influenced by the quality and relative proportions of the deep Atlantic water masses from which it originated. Low levels of dissolved oxygen (and associated low pH levels) are found in the deep channels and some areas (e. g., the head of the Laurentian Channel) have been hypoxic (oxygen concentration < 20 per cent) since the mid-1980s (Benoît et al., 2012). Appropriate oxygen and pH levels are important conditions for the health of the organisms; the observed levels have been attributed in part to local biological oxygen consumption in the deep layer, but also to changes in the source water masses from the North Atlantic. About 1 M people live in periphery of the Gulf and depend (82 per cent of regional employment, 79 per cent of regional GDP) directly on marine activities (DFO, 2013). The St. Lawrence River upstream and other tributaries from the urban and industrial zones are sources of contaminants that, combined with human activities in the coastal zone (e.g., agriculture/aquaculture, habitat destruction/modification, nutrient loading), can have disturbing effects on the ecosystem. The Gulf is also an important maritime route and the heavy commercial shipping connecting the Atlantic (and other oceans of the world) with the Great Lakes industrialized region in mid-North America represent another pressure (e.g., noise, potentially invasive species from ballast water) on the ecosystem. Human activity may have directly contributed to the establishment of 20 non-indigenous (alien) aquatic species in the Gulf; half of these since 1994 (SOTO, 2012). For decades, if not centuries, the principal human activity in the Gulf has been fishing, both commercial and recreational. That may change in the near future, however, with the development of oil and gas exploitation projects.

The timing of the phytoplankton bloom is affected by the timing of the winter ice retreat and water stratification. The bloom may have been earlier in the late 1990s than today, but no clear trend in timing and intensity is detected (Dufour et al., 2010). However, systematic monitoring of the lower level of the ecosystem (phytoplankton, zooplankton) is recent, less than 20 years, for the Gulf, and temporal trends are difficult to detect from the important interannual variability. Phytoplankton community structures vary regionally but, generally, diatoms (the group of phytoplankton typically associated with intense spring blooms in productive northern seas) and small-celled planktonic organisms (e.g., dinoflagellates) are found in equal proportions (Bérard-Therriault et al., 1999). Recently, decreasing diatom/dinoflagellate and flagellate ratios have been observed due to an increase in dinoflagellates and flagellates that may reflect increasing stratification, temperature and nutrient loading in the system (Dufour et al., 2010). Ecosystems dominated by dinoflagellates and flagellates are less productive. The Crustacea are the

Figure 36A.2bis | Bathymetric chart of the Gulf of St. Lawrence in eastern Canada. Cabot Strait in the southeast and the Strait of Belle Isle in the northeast connect the Gulf to the continental shelf regions of the Northwest Atlantic. (Prepared by Marie-Noëlle Bourassa, DFO, Canada)

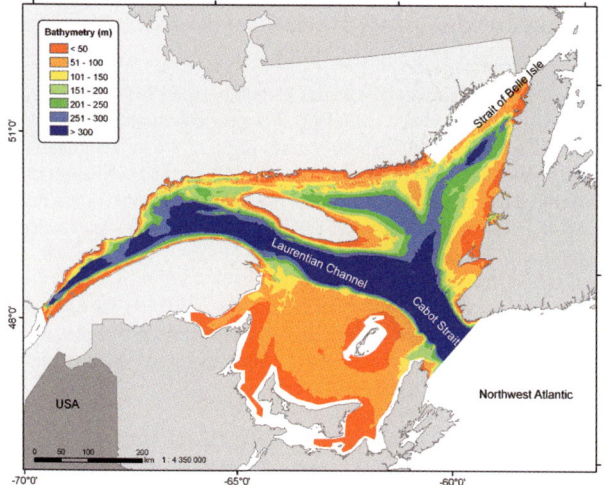

Table 36A.7 | Estimates of indigenous species richness and pressures on them in the Chesapeake Bay estuarine system (sans meiofauna, bacteria and microzooplankton). Of the phytoplankton, diatoms accounted for 46%, chlorophytes 19%, dinoflagellates 13%, cyanobacteria 9% and toxic species 2%. Of the benthic (soft bottom) macrofauna, arthropods accounted for 37%, annelids 25%, and mollusks 25%. (Data sources: Orris, 1980; Musick et al., 1985; Brownlee and Jacobs, 1987; Birdsong et al., 1989; Birdsong and Buchanan, 1993; Wagner, 1999; Wagner and Austin, 1999; Nizinski, 2003; and Marshall et al., 2005)

Trophic Level	Category	Number of Species	Major Anthropogenic Pressures
Primary Producers	Marsh grasses	19	Coastal development, Sea level rise, Invasive species
	Submerged vascular plants	15	Nutrient loading, Invasive species
	Macroalgae	25	Nutrient loading, Invasive species
	Phytoplankton	1453	Nutrient loading, Harvest of pelagic & benthic filter feeders
Consumers	Zooplankton (> 200 μm)	400	Nutrient loading, Harvest of pelagic filter feeders & predators, Ocean warming
	Fin fish	348	Fishing, Habitat loss, Invasive species, Ocean warming
	Benthic macrofauna	696	Fishing, Seasonal hypoxia, Ocean warming
	Waterfowl	49	Fishing, Habitat loss
TOTAL		3005	

most important group (60 per cent) in the zooplankton, and the Copepoda account for 70 per cent of all species (Brunel et al., 1998). There are close associations between the abundance of some small copepod species, the young stages of larger species and feeding by young fish. In the Gulf, all copepods seem to have increased around 2005, but the abundance of the key species *Calanus finmarchicus* has been below average since 2009 (DFO, 2013). Although there are important uncertainties about total biomass and trend, high concentrations of krill (Euphausiids) are found at specific sites and large numbers of blue whales from the Northwest Atlantic population migrate to the Gulf to feed (Gagné et al., 2013). At least 12 species of whales migrate to the Gulf every year, which makes the Gulf (and the Estuary) one of the best whale-watching sites in the World. In addition, the head of the Laurentian Channel in the Estuary is the refuge of an endangered Beluga whale population. Despite conservation efforts (control on contaminants, noise, and a protected areas project), the Beluga whale population shows no sign of increase; at present it is, at best, stable.

The total number of fish and invertebrates species is approximate, due in part to the poor sampling of the shallow or coastal zones. The current inventory lists some 2200 marine invertebrate species (Brunel et al., 1998). An analysis of scientific bottom trawl survey records indicates that ~130 species of fish may be present in the northern Gulf alone (Dutil et al., 2006, 2009). The total fish biomass is dominated by a small number of large species, most of commercial interest; hence many species are relatively rare or occasional visitors. Nonetheless, the importance of the habitat is shown by at least 48 species of ichthyoplankton (eggs and larval fish) recorded in the Gulf. As for the Canadian shelf regions, dramatic shifts to both the northern and southern Gulf ecosystems occurred in the late 1980s, particularly in response to fishing and, to a lesser extent, to changes in environmental conditions. These shifts include changes in species abundance and/or biomass and food web structure and functioning (Savenkoff et al., 2008) (Figure 36A.3). The ecosystems that were dominated by large demersal fish predators (e.g., Atlantic cod, redfish and white hake) are today dominated by small-bodied forage species and invertebrates. The biomass of shrimps (close to 20 species) has increased since the early 1980s and at an accelerated rate beginning in the early 1990s. Despite a 15-year moratorium on harvesting, the ecosystem structure has not returned to its previous state. In the southern Gulf, the natural mortality of cod remained high, causing a decline in their abundance. Evidence is mounting that predation by top predators (e.g., seals, whose abundance has been increasing) is the cause of this mortality (Benoît et al., 2012).

Fishing can affect population structure, entire communities, and the habitat itself. The only important bottom trawl use remaining in the northern Gulf is by the northern shrimp fishery. The fishery is well regulated and the use of a separator grate is mandatory to reduce the catch of large fishes. Nonetheless, an analysis in 2012 showed small specimens from 69 taxa (the majority in low numbers) as remaining in the catch (Savard, 2013).

7.6 Chesapeake and other coastal estuaries and bays of the United States

The Chesapeake Bay estuarine system[7] supports more than 3,000 species of plants and animals (Table 36A.7). A subset of species has been identified as being *ecologically valuable*[8] (Table 36A.8), based on their importance in (1) regulating the flow of carbon through the food web, (2) providing habitat, and/or (3) supporting ecosystem services (Baird and Ulanowicz, 1989; Costanza et al., 1997; Jordán, 2001; Martínez et al., 2007; Koch et al., 2009). Marshall et al. (2005) documented 1454 phytoplankton species in the Bay during 1984-2004; diatoms were

---

[7] Chesapeake Bay (including tidal waters of its tributaries) is the largest estuary in the U.S. (~ 320 km long, 11,700 km²). It is a partially stratified, coastal plain estuary (drowned river valley) with a mean depth of 8.4 m and three salinity zones: oligohaline, 0 – 10 psu; mesohaline, 11 – 18 psu; and polyhaline, 19 – 36 psu (Schubel and Pritchard, 1987). Eleven major rivers flow into the Bay through six states and a drainage basin of 172,000 km². River flows are typically highest during spring and lowest during summer. The climate is moderate with mean water temperatures ranging from a winter low of ~ 5°C to a summer high of ~25°C. With a large ratio of watershed to estuarine area (14:1), the Bay is closely connected to the landscape. With a relatively long residence time of water (~ 6 months), the Bay is susceptible to impacts from land-based inputs of nutrients and toxic contaminants (Kemp et al., 2005).

[8] The Ecologically Valuable Species Workgroup of the Living Resources Subcommittee for the Chesapeake Bay Program (http://nepis.epa.gov/Exe/ZyPDF.cgi/P1001WSJ.PDF?Dockey=P1001WSJ.PDF).

Table 36A.8 | Representatives of ecologically important species drawn from Heck and Orth (1980), Orth et al. (1987), Baird and Ulanowicz (1989), Sellner and Marshall (1993), Birdsong and Buchanana (1993), Houde (1993), Dauer et al. (1993), Newell and Breitburg (1993), McConaugha and Rebach (1993), Jorde et al. (1993), Stevenson and Pendleton (1993), Jordán (2001), Buchanan et al. (2005), Orth et al. (2006), and Chambers et al. (2008). (*Iconic species, **Toxic species).

Category	Species
Tidal marsh grasses	Pontederia cordata, Zizania aquatica, Scripus olneyii, Spartina cynosuroides, Spartina alterniflora, S. patens
Submerged vascular plants	Vallisneria americana, Stuckenia pectinata, Potamogeton perfoliatus, Ruppia maritime, Zostera marina
Macroalgae	Ulva lactuca, Agardhiella spp., Enteromorpha spp., Cladophora spp.
Phytoplankton	Diatoms – Ceratulina pelagica, Rhizosolenia fragilissima, Leptocylindrus minimus, Skeletonema costatum, Asterionella glacialis; Dinoflagellates – Gymnodinium spp., Ceratium lineatum, Prorocentrum minimum**, Dinophysis acuminata*; Cyanobacteria – Microcystis aeruginosa**
Zooplankton	Bosmina longirostris, Leptodora kindtii, Acartia tonsa, Eurytemora affinis, Nemopsis bachei, Mnemiopsis leidyi, Chrysaora quinquecirrha
Benthic infauna	Mya arenaria, Macoma balthica, M. mitchelli, Nereis succinea
Benthic epifauna	Crassostrea virginica*, Callinectes sapidus*
Forage fish	Alosa pseudoharengus, A. aestivalis, A. sapidissima, A. mediocris, Anchoa mitchilli, Brevoortia tyrannus, Gobiosoma bosci, Menidia menidia, Fundulus heteroclitus, Cyprinodon variegates, Gambusia holbrooki
Intermediate predators	Micropogonias undulatus, Trinectes maculatus, Leiostomus xanthurus*, Morone americana*, Ictalurus punctatus
Top predators	Pomatomus saltatrix, Cynoscion regalis, Paralichthys dentatus, Morone saxatilis*
Waterfowl	Aythya americana, Aythya valisineria, Anas rubripes, A. americana, Phalacrocorax auritus, Ardea herodias, Branta canadensis*, Cygnus columbianus, Haliaeetus leucocephalus
Invasive species	Marsh grass – Phragmites australis, Lythrum salicaria, Trapa natans; Benthic macrofauna – Dreissena polymorpha; Fin Fish – Ictalurus furcatus, Pylodictis olivaris; Other –Cygnus olor, Myocastor coypus
Threatened & endangered species	Sturgeon – Acipenser oxyrinchus, A. brevirostrum, Mussels – Alasmidonta heterodon, Turtles – Caretta caretta, Lepidochelys kempi

the most abundant taxon (Table 36A.6). The ratio of planktonic centric diatom species to benthic pennate diatom species increased from ~ 1 prior to European settlement to ~ 5 today (Cooper and Brush, 1991), a trend that coincided with a decrease in species richness and an increase in phytoplankton biomass due to increases in anthropogenic nutrient loading and a rapid decline in the abundance of filter feeders (oysters) during the twentieth century (Newell, 1988; Cooper and Brush, 1991; Marshall et al., 2003; Kemp et al., 2005). Projected increases in water temperature and winter-spring precipitation associated with climate change are likely to enhance these trends and promote the growth of toxic dinoflagellates (Pyke et al., 2008).

Species richness is correlated with fisheries productivity (Worm et al., 2006). Historically important fisheries in the Bay included striped bass (*Morone saxatilis*), Atlantic sturgeon (*Acipenser oxyrinchus*), American shad (*Alosa sapidissima*), Atlantic menhaden (*Brevoortia tyrannus*), blueback herring (*Alosa aestivalis*), alewife (*Alosa pseudoharengus*), soft-shelled clam (*Mya arenaria*), eastern oyster (*Crassostrea virginica*), and blue crab (*Callinectes sapidus*). Of these, landings of sturgeon,[9] shad,[10] soft-shelled clams[11] and eastern oysters (Rothschild et al., 1984; Wilberg et al., 2011) experienced dramatic declines (> 98 per cent) during the twentieth century due to overfishing and habitat loss. Herring[12] and menhaden[13] landings have declined by ~80 per cent.

Habitat loss has been expressed primarily in terms of increases in the spatial extent of summer hypoxia and decreases in the spatial extent of tidal marshes, submerged vascular plant beds, and oyster reefs. Oyster reefs, submerged vascular plant beds and tidal marshes support high species diversity (Heck and Orth, 1980; Orth et al., 1985; Newell, 1988; Chambers et al., 1999; Coen et al., 1999; Jackson et al., 2001; Wyda et al., 2002; USACE, 2009; Philine et al., 2012). During the course of the twentieth century, the spatial extent of these habitats declined significantly: oyster reefs by 92 per cent (USACE, 2009; Wilberg et al., 2011), submerged vascular plants by 65 per cent (Kemp et al., 1983, 2005; Orth and Moore, 1983, 1984) and marshes by 60 per cent.[14]

Sea level rise is expected to result in even greater losses of marshes, putting hundreds of species of fish, invertebrates and birds at risk (Titus and Strange, 2008), and estuarine acidification poses a significant threat to oyster restoration efforts in the Bay (USACE, 2009; Waldbusser et al., 2011; Sanford et al., 2014). Loss of these habitats exacerbates the impacts of overfishing and is one of the main pressures on species richness,

---

[9] http://www.dnr.state.md.us/fisheries/fishfacts/atlanticsturgeon.asp

[10] http://www.fws.gov/chesapeakebay/SHAD.HTM

[11] http://www.dnr.state.md.us/irc/docs/00000260_04.pdf

[12] http://www.dnr.state.md.us/fisheries/fishfacts/herring.asp

[13] http://www.asmfc.org/species/atlantic-menhaden

[14] http://chesapeakebay.noaa.gov/wetlands

North Atlantic Ocean

Chapter 36A

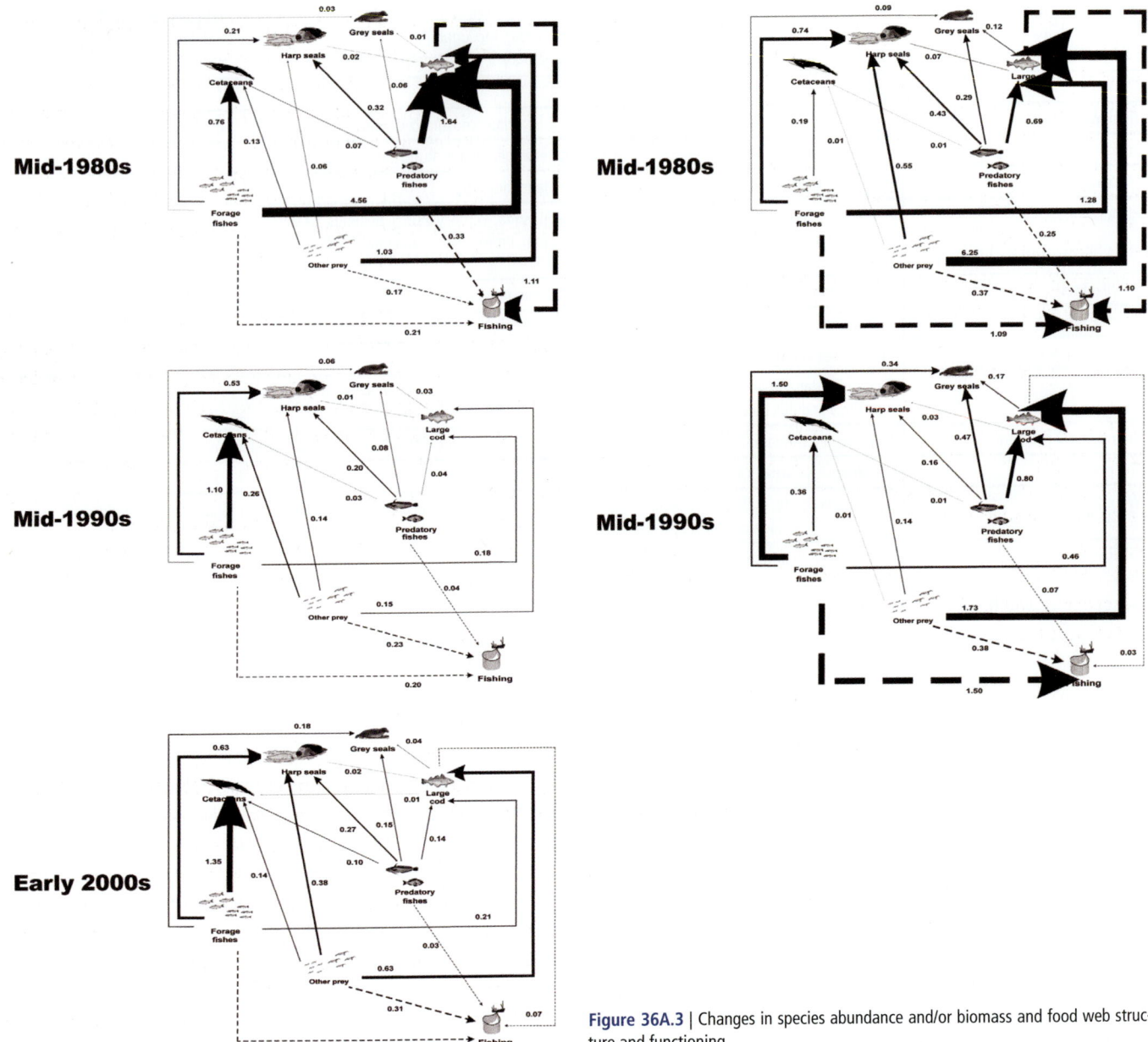

Figure 36A.3 | Changes in species abundance and/or biomass and food web structure and functioning

often leading to species extirpation (Orth and Moore, 1983, 1984; Ruiz et al., 1993; Duarte et al., 2008; Heck et al., 2008; Keppel et al., 2012).

Recurring deep-water hypoxia (Cooper and Brush, 1991; Malone, 1991)15 represents a major loss of pelagic and benthic habitat during a critical period for reproduction and growth of benthic macrofauna and fish, resulting in declines in their abundance (Llansó, 1992; Ruiz et al., 1993; Baird et al., 2004; Kemp et al., 2005; Buchheister et al., 2013) and the threat of extirpation of commercially valuable species that have been overfished, e.g., *Acipenser oxyrinchus* (Secor et al., 2000) and *C. virginica* (Wilberg et al., 2011). Projected increases in climate-driven water temperature and winter-spring precipitation over the twenty-first century may increase the pressure of summer hypoxia on species richness (Pyke et al., 2008), an impact that may be exacerbated by the direct effects of rising water temperatures: e.g., *M. arenaria* is near its southern distribution limit and may be extirpated if summer temperatures approach and remain near 32°C, and temperate fish species such as white perch (*Morone americana*), striped bass (*Morone saxatilis*), and summer flounder (*Paralichthys dentatus*) may experience similar fates (Najjar et al., 2010).

More than 170 known or possible non-native species have invaded the Bay.[16] Of these, at least eight are potentially major threats to species richness in the Chesapeake Bay estuarine system (Table 36A.7).[17]

---

15 http://mddnr.chesapeakebay.net/eyesonthebay/documents/DeadZoneStatus_Summer2013.pdf

16 http://invasions.si.edu/nemesis/chesapeake.html

17 http://www.mdsg.umd.edu/topics/aquatic-invasive-species/aquatic-invasive-species

Given the collective importance of vegetated habitats and oyster reefs as refugia for a broad diversity of species and the impacts of fishing, seasonal hypoxia and invasive species, it is highly likely that the species diversity of the Chesapeake Bay estuarine system significantly declined during the 20th C. Continued declines in habitat extent, combined with the impacts of seasonal hypoxia and climate-driven sea level rise, estuarine warming, and acidification, portend increases in the rate of extirpations and declines in species diversity during the twenty-first century. Given similarities in pressures, all of the estuaries of the Virginian Province[18] (Hale et al., 2002) are likely to exhibit similar trends in their capacity to support species diversity.

7.7 Caribbean Sea

The Caribbean is the most biologically diverse area of the Atlantic Ocean, hosting approximately 10 per cent of the world's coral reefs, including the Mesoamerican Barrier Reef System; extensive coastal mangroves and shallow banks with seagrass communities; as well as sandy beaches, rocky shores and many bays, lagoons and brackish estuaries. The Caribbean also has open-ocean and lesser-known deep-sea environments, and has been listed as a global-scale hotspot of marine biodiversity (Roberts et al., 2002).

The Caribbean Sea receives primarily oligotrophic, high-salinity North Atlantic water from the North Equatorial Current, but it also receives South Atlantic water entrained in North Brazil Current rings which transport water from the Amazon into the Caribbean basin via the Guiana Current (Cowen et al., 2003). The persistent through-flow of the warm Caribbean Current is modulated by a highly complex and variable pattern of mesoscale eddies (Lin et al., 2012) and upwelling along the South American coastline. Two significant South American rivers, the Orinoco and the Magdalena, also discharge directly into the southern Caribbean. The considerable spatial heterogeneity of physical environments and habitats across the Caribbean Sea influences the distribution, population connectivity and biodiversity of marine organisms found there. Several significant barriers to gene flow in Caribbean reef populations have been recognized (Cowen et al., 2006). This has led to relatively high levels of endemism. Miloslavich et al. (2010) estimate a value of 25.6 per cent regional endemism across 21 of 78 marine taxa examined in the Caribbean, with values ranging from 45 per cent for fish, 26 per cent for molluscs and 2 per cent for copepods. They also summarize the diversity, distribution and key threats to marine biota in the Caribbean and conclude that the 12,046 species currently reported is a gross underestimation, considering that the marine biota is far from well known in this area.

The significant drivers of declines in Caribbean marine biodiversity are overexploitation and environmental degradation. These are being exacerbated by external drivers, including climate variability and change, and alien species invasions. Iconic Caribbean mega-vertebrates have suffered from historical overexploitation (Jackson, 1997), including the now extinct West Indian monk seal (*Monachus tropicalis*); the endangered Caribbean manatee (*Trichechus manatus manatus*); the North Atlantic humpback whale (*Megaptera novaeangliae*); and marine turtles, of which five species are found here, all endangered. The number of fishery stocks that are fully exploited or overexploited has grown over the last few decades and total landings have declined significantly since the late 1980s, driven by increasing market demand, inadequate fisheries management, and exacerbated by habitat degradation (Salas et al., 2011; Sea Around Us Project, www.seaaroundus.org).

The abundance of reef fishes has decreased region-wide (Paddack et al., 2009). Most fishable reef species are fully exploited or overexploited; those most vulnerable to fishing are now rare (Roberts, 2012). Notable is the Nassau grouper (*Epinephelus striatus*), once of great commercial importance and now endangered (Sadovy, 1999). There is concern regarding the decline of key functional groups on reefs, especially herbivores, such as parrotfishes, that are vital to reef resilience (Mumby et al., 2006). Large highly migratory pelagic species, such as the billfishes (swordfish and marlins), have suffered significant population declines from fishing by foreign fleets operating in the Atlantic.

Caribbean coral reefs are considered globally unique (UNEP, 2005). Overexploitation and deterioration of coastal water quality (primarily due to high nutrient, sediment and bacterial loads, and toxins from domestic, agricultural and industrial activities in coastal watersheds) have altered reef communities and resilience, leading to region-wide decreases in live coral cover (Gardner et al., 2003; Jackson et al., 2014) and reef structural complexity (Alvarez-Filip et al., 2009) over the last three decades. Concomitant increases in disease epidemics (Rogers, 2009) and in macro-algae (Bruno et al., 2009) have resulted in ecosystem shifts from coral-dominated to algal-dominated reefs (Hughes et al., 2007). The once dominant *Acropora palmata* was severely reduced by coral disease from the late 1970s through the 1980s, and this genus is now listed as endangered in the United States Caribbean (http://www.nmfs.noaa.gov/pr/laws/esa/). The 1982-1984 mass die-off of the *Diadema antillarum* sea urchins left Caribbean reefs without a keystone herbivore (Jackson et al., 2014). Pioneering coral species, such as *Porites astreoides*, are becoming more prevalent (Green et al., 2008). The degradation is exacerbated by global climate change resulting in warming causing mass coral bleaching and associated coral mortality (Eakin et al., 2010), physical destruction from more intensive storms (Gardner et al., 2005; Wilkinson and Souter, 2008) and the threat of ocean acidification. Caribbean reef biodiversity is being further affected by the alien invasive Pacific red lionfish (*Pterios volitans*), which has spread across the Caribbean in the last decade (Schofield, 2010).

Mangroves and their associated biodiversity occur throughout the insular and continental Caribbean coastlines (Bossi and Cintron, 1990). The Caribbean has nine mangrove tree species (Polidoro et al., 2010), but is reported to host the world's richest mangrove-associated invertebrate fauna (Ellison and Farnsworth, 1996). Mangrove area has declined by about 1 per cent annually over the last three decades, representing the second highest loss rate globally (FAO, 2007). *Pelliciera rhizopho-*

---

18 Chesapeake Bay, Delaware Bay, Hudson-Raritan system, and Long Island Sound

rae, endemic to Central America, is now listed as vulnerable (Polidoro et al., 2010). Mangrove declines are driven by exploitation (of timber); deteriorating water quality (especially petroleum pollution); and coastal development (aquaculture ponds, marinas, reclamation for coastal construction and agriculture); and climate change (Ellison and Farnsworth, 1996; Polidoro et al., 2010).

Seagrass communities occur throughout the Caribbean and support a high diversity of epiphytic and other species (van Tussenbroek et al., 2010). Seven native seagrass species are known from the region; two (*Halophila engelmanni* and *H. baillonii*) are considered to be near-threatened and vulnerable, respectively. A recently introduced species, *Halophila stipulacea*, is spreading rapidly through the Lesser Antilles (Willette et al., 2014). CARICOMP found that most study sites showed a decline in seagrass health between 1993 and 2007 (van Tussenbroek et al., 2014).

Caribbean seabirds comprise 25 breeding species, of which seven are regionally endemic species or subspecies: the abundant laughing gull subspecies *Larus atricilla atricilla*; the rare and declining white tailed tropic bird subspecies *Phaethon lepturus catesbyii*; the near threatened Audubon's shearwater *Puffinus lherminieri lherminieri*; the endangered brown pelican subspecies *Pelicanus occidentalis occidentalis*; the critically endangered black capped petrel *Pterodroma hasitata* and black noddy subspecies *Anous minutus minutes*; and the probably extinct Jamaica petrel *Pterodroma caribbaea* (Lee and Makin, 2012). Most of the 25 species including both surface feeders and divers are declining, threatened by human disturbance and nest predation by introduced species; pollution of ocean waters; and fishery by-catch impacts (Schreiber and Lee, 2000; Lee and Mackin, 2012).

Caribbean economies are the most tourism-dependent in the world (CLME, 2011). Declining marine biodiversity will have enormous social and economic consequences, through loss of goods and critical ecosystem services. Caribbean-wide degradation of coral reef, mangrove and seagrass ecosystems, ecosystems that are fundamental to the Caribbean tourism product and at the core of the region's ability to cope with climate change sea level rise, will mean annual losses amounting to billions of United States dollars (CARSEA, 2007; Burke et al., 2011).

7.8 Gulf of Mexico

The Gulf of Mexico is a semi-enclosed sea and one of the most economically and ecologically productive bodies of water in the world (Tunnell, 2009). The Gulf is connected to the Caribbean Sea through the Yucatan Channel between the Yucatan Peninsula and Cuba, where warm, tropical water flows into the Gulf and forms the Loop Current, the dominant Gulf current, and then exits via the Florida Straits between Florida and Cuba into the Atlantic Ocean, where it forms the Gulf Stream, one of the world's strongest and most important currents.

As a large receiving basin, the Gulf of Mexico receives extensive drainage from five countries (the United States, Canada, Mexico, Guatemala, and Cuba). The Mississippi River dominates this drainage, which includes over two-thirds of the U.S. watershed in the north, and the Grijava-Usumacinta River System dominates in the south. Thirty-three major rivers and 207 estuaries and lagoons are found along the Gulf coastline (Darnell and Defenbaugh, 1990; Tunnell, 2009).

Biologically, the shallow waters of the northern Gulf are warm-temperate (Carolinian Province) and those in the southern Gulf are tropical (Caribbean Province) (Briggs, 1974). Oyster reefs and salt marshes are the dominant estuarine habitat type in northern, low-salinity estuaries, and seagrass beds are common in clearer, more saline bays. In the tropical southern Gulf, mangroves line bay and lagoon shorelines; some oyster reefs, salt marshes, and seagrasses are distributed in similar salinity conditions as in the northern Gulf. Offshore, coral reefs are common in the Florida Keys, Cuba, and the southern Gulf off the state of Veracruz and on the Campeche Bank (Tunnell et al., 2007; other topographic highs or hard bottoms are sporadic on the normally smooth, soft substratum of the continental shelves (Rezak and Edwards, 1972; Rezak et al., 1985). Unique, recently discovered, and highly diverse habitats in deeper Gulf waters include chemosynthetic communities and deepwater coral communities (*Lophelia* reefs) (CSA International Inc., 2007; Brooks et al., 2008; Cordes et al., 2008).

Regarding the biodiversity of the Gulf of Mexico, the Harte Research Institute for Gulf of Mexico Studies at Texas A&M University-Corpus Christi recently led a multi-year, multi-international effort (Biodiversity of the Gulf of Mexico Project) involving 140 taxonomists from 80 institutions in 15 countries to prepare a comprehensive list of all species (Felder and Camp, 2009). This list of 15,419 species with range, distribution, depth, habitat-biology, and updated taxonomy was subsequently added to GulfBase in 2011 at http://www.gulfbase.org/biogomx/biosearch.php, where it is openly accessible and completely searchable by any topic or species. The database has since been used in two major papers comparing biodiversity of other United States regions (Fautin et al., 2010) and four other global case studies in marine biodiversity (Ellis et al., 2011).

Several Gulf of Mexico iconic or well-known species are of historical, social, and economic importance (Davis et al., 2000). The West Indian monk seal (*Monachus tropicalis*) was probably the first large animal to become extinct because of human activity in the Gulf and Caribbean region. It was last seen on the Campeche Bank islands in the Gulf in 1948 and in the Caribbean in the early 1950s (Wursig et al., 2000). Other species that have become endangered include the Kemp's Ridley sea turtle (*Lepidochelys kempii*), brown pelican (*Pelecanus occidentalis*), and whooping crane (*Grus americana*). Restoration programs for each of these have increased their populations in recent decades. West Indian manatees (*Trichechus manatus*) are greatly reduced, and they only exist now in certain drainage areas along the west coast of Florida. The largest commercial fishery by weight in the Gulf is for menhaden (*Brevoortia patronus*), and the penaeid shrimp fishery is the largest by value (the white shrimp *Litopenaeus setiferus*, the pink shrimp *Farfantepenaeus duorarum*, and the brown shrimp *Farfantepenaeus aztecus*).

Predominant commercial estuarine shellfish in the northern Gulf include the eastern oyster (*Crassostres virginica*) and blue crab (*Callinectes sapidus*) (Nelson, 1992; Patillo et al., 1997). In the tropical southern Gulf, spiny lobster (*Panulirus argus*) and queen conch (*Eustrombus gigas*) are taken. However, these are now commercially extinct in many areas and are taken only by recreational fishers, sometimes under strict regulations (Tunnell et al., 2007).

The Gulf-wide bottlenose dolphin (*Tursiops truncatus*) is probably the single most recognizable Gulf species by the public, as it is abundant in coastal bays and estuaries, as well as offshore in the northern Gulf (Wursig et al., 2000).

Gulf-wide biodiversity patterns cannot be completely explained, for lack of complete information, although we do know that the Gulf of Mexico exhibits great habitat complexity that probably supports high levels of biodiversity due to both endemic and cosmopolitan species (Rabalais et al., 1999). Linkage to the Caribbean Sea with large-scale circulation provides the southern and eastern Gulf with a distinct Caribbean biota. However, strong regional endemism appears to exist, as demonstrated in large-scale studies across the entire northern Gulf (Rabalais et al., 1999; Harper, 1991; Carney et al., 1993). Eventual analysis of databases from the Biodiversity of the Gulf of Mexico Project on GulfBase will provide considerable insight into the spatial distribution of species. Of the 15,419 species found, 1,511 (10 per cent) are endemic to the Gulf of Mexico and 341 (2 per cent) are non-indigenous (Felder and Camp, 2009). The most diverse taxa include crustaceans (2,579 species), mollusks (2,455), and vertebrates (1,975), and the least diverse include kinorhynchs (2 species), entoprocts (2), priapulids (1), hemichordates (5), and cephalochordates (5). In addition, other taxa are known to exist in the Gulf of Mexico (placozoans, orthonectids, loriciferans, and pogonophorans), but representatives have not yet been identified (Felder and Camp, 2009; Fautin et al., 2010).

A recent ecosystem status report for the Gulf of Mexico, utilizing the DPSER (Drivers, Pressures, States, Ecosystem Services, Responses) conceptual modelling framework, gives a high-level overview of the state of the Gulf (Karnauskas et al., 2013). Major, large-scale climatic drivers include the Atlantic Multidecadal Oscillation, Atlantic Warm Pool, sea surface temperature, Loop Current, and geostrophic transport in the Yucatan Channel and Florida Current. Long-term trends or changes in these drivers in turn cause fluctuations or changes in selected pressures, such as hurricanes or hypoxic zones. Other pressures include contamination by pollution (e.g., mercury, cadmium), oil and gas exploration and production (including major oil spills, such as Ixtoc I in 1979 and Deepwater Horizon in 2010), bacterial water quality problems, and habitat destruction, mainly caused by coastal development. Harmful algal blooms (HABs), such as red tide and brown tide, are well documented in the Gulf, as are some invasive species (Tunnell, 2009; Fautin et al., 2010; Karnauskas et al., 2013).

The recent Deepwater Horizon oil spill prompted a study of the ecosystem services of the Gulf of Mexico by a leadership committee of the United States National Research Council (NRC). This comprehensive report utilizes the ecosystem services approach and frames for the first time the goods and services provided by the Gulf for an economically and ecologically healthy ecosystem (NRC, 2013).

8 Factors of Sustainability

The biodiversity of the North Atlantic has supported harvesting and trade by bordering cultures for millennia. Pressures from human uses both diversified and intensified with industrialization and associated coastal development already more than two centuries ago. Every form of direct use of North Atlantic biodiversity and every indirect effect of human activities on coastal populations and habitats have been unsustainable in at least some times and places. Some of these impacts, such as the depletion of populations of the great whales by overharvesting (section 36A.6) will take centuries to recover, even with effective policies and high compliance.

Efforts towards sustainability have been greatly aided by coordinated international efforts to provide scientific and technical information on the status and trends in biodiversity, and threats to sustainable uses. The Quality Status reports (QSRs) from the OSPAR Commission at the start of each of the past three decades have proven invaluable in assessing status and trends in many marine environmental indicators and the biodiversity they represent, and guiding policies and management measures to address poor or declining marine environmental quality in the northeast Atlantic (OSPAR, 2010, and earlier QSRs). Other examples of such efforts are the coordinated processes carried out within the framework of the Convention on Biological Diversity for identifying Ecologically and Biologically Significant Areas in the Northwest Atlantic (CBD, 2014a), and Mediterrean (CBD, 2014b), and the on-going process in the Northeast Atlantic (ICES, 2013a), and the processes to identify Vulnerable Marine Ecosystems in the NAFO (NAFO, 2013) and North-East Atlantic Fisheries Commission (NEAFC) (FAO, 2015).

Although there are well-documented examples in the sections above of cases where habitats, lower-trophic-level productivity, benthic communities, fish communities, or seabirds or marine mammal populations were severely altered by pressures from a specific activity, such as overfishing, pollution, nutrient loading, physical disturbance, or transplanted species, many biodiversity impacts, particularly at larger scales, are the result of cumulative and interactive effects of multiple pressures from multiple drivers. It has repeatedly proven difficult to disentangle the effects of the individual pressures, impeding the ability to address the individual causes (Fu et al., 2012; Blanchard et al., 2005; ter Hofstede and Rijnsdorp, 2011). Particularly given that the North Atlantic is surrounded by many of the best marine research centres in the world, has many of the longest and most systematic data sets, and has an international science organization, the International Council for Exploration of the Seas (ICES), that has functioned for over a century to promote and coordinate scientific and technical cooperation among many of the circum-North

Atlantic countries with the highest science capacities, this inability to consistently disentangle causation of unsustainable uses of, and impacts on, marine biodiversity may seem initially discouraging.

On the other hand, well-documented examples also exist of the benefits that can follow from actions to address past unsustainable practices. Many of the fish stocks depleted by overfishing in both the Northeast and Northwest Atlantic have shown increasing trends in abundance and recovery of range when unsustainable levels of fishing effort have been reduced (Table 36A.4). Efforts to control pollution and nutrient inputs, driven by the EU Water Framework Directive and the United States Environmental Protection Act, have led to reduction in these pressures and in many cases to the commencement of the recovery of benthic communities (EEA, 2012). Coastal habitat restoration activities have also shown clear benefits in improved environmental quality and biodiversity measures in many coastal areas around the North Atlantic (Pendleton, 2010). All of these improvements have come with at least short-term costs, which are sometimes large, such as displaced or reduced fishing opportunity (see Part IV), the costs of pollution control and nutrient management in coastal areas and watersheds (costs summarized in the chapters of Part V), and the direct costs of habitat restoration, which may run to the millions of dollars for restoration projects of even moderate scale (Diefenderfer et al., 2011; Kroeger and Guennel, 2014).

In summary, the North Atlantic presents examples of both the extent to which unsustainable actions can adversely affect biodiversity and the benefits that can accrue from policies and programmes that are well developed, adequately resourced, and effectively implemented. The best examples of effective policies and programmes have been designed to address the dominant pressures from the twentieth century: overharvesting of living marine resources and pollution and excessive nutrient inputs from coastal and land-based sources. In the twenty-first century, additional pressures are growing, particularly climate change, invasive species (both responding to changing environmental conditions and transported by shipping), and in many areas, particularly at lower latitudes, ocean-based tourism. Lessons learned from dealing successfully with the earlier pressures, if applied proactively, may help safeguard biodiversity from unsustainable impacts, and result in healthy ecosystems producing many ecosystem services of value to the circum-Atlantic human populations.

References

Abdul Malak, D., Livingstone, S.R., Pollard, D., Polidoro, B.A., Cuttelod, A., Bariche, M., Bilecenoglu, M., Carpenter, K.E., Collette, B.B., Francour, P., Goren, M., Kara, M.H., Massutí, E., Papaconstantinou, C., Tunesi, L. (2011). *Overview of the Conservation Status of the Marine Fishes of the Mediterranean Sea.* IUCN, vii + 61pp., Gland, Switzerland and Malaga, Spain.

Albaina, A. and Irigoien, X. (2007). Fine scale zooplankton distribution in the Bay of Biscay in spring 2004. *Journal of Plankton Research*, vol. 29, No. 10, pp. 851–870.

Alheit, J., Möllmann, C., Dutz, J., Kornilovs, G., Loewe, P., Mohrholz, V., Wasmund, N. (2005). Synchronous ecological regime shifts in the central Baltic and the North Sea in the late 1980s. *ICES Journal of Marine Science,* vol. 62, pp. 1205–1215.

Alvarez-Filip, L., Dulvy, N.K., Gill, J.A., Côté, I.M, Watkinson, A.R. (2009). Flattening of Caribbean coral reefs: region-wide declines in architectural complexity. *Proceedings of the Royal Society B-Biological Sciences*, vol. 276, No. 1669, pp. 3019–3025. Available from http://rspb.royalsocietypublishing.org/content/early/2009/06/05/rspb.2009.0339.full.pdf+html.

Andersson, A.J., Mackenzie, F.T., Gattuso, J.-P. 2011. Effects of ocean acidification on benthic processes, organisms, and ecosystems. Ocean Acidification. (Gattuso J, Hansson L, Eds.).: xix,326p, Oxford England; New York: Oxford University Press

Anderson, D.M., Glibert, P.M., Burkholder, J.M. (2002). Harmful algal blooms and eutrophication: Nutrient sources, composition, and consequences. *Estuaries*, vol. 25, No. 4b, pp. 704–26.

Anderson, D.M., Cembella, A.D., Hallegraeff, G.M. (2012). Progress in understanding harmful algal blooms: paradigm shifts and new technologies for research, monitoring, and management. *Annual Review of Marine Science*, vol. 4, pp. 143–176.

Anker-Nilsen T., Høyland, T., Barrett, R., Lorentsen, S.-H., Strøm, H. (2007). Dramatic breeding failures among Norwegian seabirds. *Seabird Group Newsletter*, vol. 106, pp. 7–8.

Arvanitidis, C., Bellan, G., Drakopoulos, P., Valavanis, V., Dounas, C., Koukouras, A., Eleftheriou, A. (2002). Seascape biodiversity patterns along the Mediterranean and the Black Sea: lessons from the biogeography of benthic polychaetes. *Marine Ecology Progress Series*, vol. 244, pp. 139–152.

Arvanitidis, C., Somerfield, P.J., Rumohr, H., Faulwetter, S., Valavanis, V., Vasileiadou, A., Chatzigeorgiou, G., Vanden Berghe, E., Vanaverbeke, J., Labrune, C., Grémare, A., Zettler, M.L., Kedra, M., Wlodarska-Kowalczuk, M., Aleffi, I.F., Amouroux, J.M., Anisimova, N., Bachelet, G., Büntzow, M., Cochrane, S.J., Costello, M.J., Craeymeersch, J., Dahle, S., Degraer, S., Denisenko, S., Dounas, C., Duineveld, G., Emblow, C., Escavarage, V., Fabri, M.C., Fleischer, D., Gray, J.S., Heip, C.H.R., Herrmann, M., Hummel, H., Janas, U., Karakassis, I., Kendall, M.A., Kingston, P., Kotwicki, L., Laudien, J., Mackie, A.S.Y., Nevrova, E.L., Occhipinti-Ambrogi, A., Oliver, P.G., Olsgard, F., Palerud, R., Petrov, A., Rachor, E., Revkov, N.K., Rose, A., Sardá, R., Sistermans, W.C.H., Speybroeck, J., Van Hoey, G., Vincx, M., Whomersley, P., Willems, W., Zenetos, A. (2009). Biological geography of the European seas: results from the MacroBen database. *Marine Ecology Progress Series*, vol. 382, pp. 265–278.

Attrill, M.J., Wright, J., Edwards, M. (2007). Climate related increases in jellyfish frequency suggest a more gelatinous future for the North Sea. *Limnology and Oceanography*, vol. 52, No. 1, pp. 480–485.

Audzijonyte, A., Daneliya, M.E., Väinölä, R. (2006). Comparative phylogeography of Ponto-Caspian mysid crustaceans: isolation and exchange among dynamic inland sea basins. *Molecular Ecology*, vol. 15, No. 10, pp. 2969–2984.

Baird, D., and Ulanowicz, R.E. (1989). The seasonal dynamics of the Chesapeake Bay ecosystem. *Ecological Monographs,* vol. 59, No. 4, pp. 329–364.

Baird, D., Christian, R.R., Peterson, C.H., Johnson, G.A. (2004). Consequences of hypoxia on estuarine ecosystem function: Energy diversion from consumers to microbes. *Ecological Applications*, vol. 14, No. 3, pp. 805–822.

Bakun, A., and Weeks, S.J. (2006). Adverse feedback sequences in exploited marine systems: are deliberate interruptive actions warranted? *Fish and Fisheries,* vol. 7, No. 4, pp. 316–333.

Barnard, R., Batten, S.D., Beaugrand, G., Buckland, C., Conway, D.V.P., Edwards, M., Finlayson, J., Gregory, L.W., Halliday, N.C., John, A.W.G., Johns, D.G., Johnson, A.D., Jonas, T.D., Lindley, J.A., Nyman, J., Pritchard, P., Reid, P.C., Richardson, A.J., Saxby, R.E., Sidey, J., Smith, M.A., Stevens, D.P., Taylor, C.M., Tranter, P.R.G., Walne, A.W., Wootton, M., Wotton, C.O.M., Wright, J.C. (2004). Continuous plankton records: Plankton atlas of the North Atlantic Ocean (1958–1999). II. Biogeographical charts, *Marine Ecology Progress Series*, Supplement, pp. 11–75.

Barrio Froján, C.R.S., Cooper, K.M., Bremner, J., Defew, E.C., Wan Hussin, W.M.R., Paterson, D.M. (2011). Assessing the recovery of functional diversity after sustained sediment screening at an aggregate dredging site in the North Sea. *Estuarine, Coastal and Shelf Science*, vol. 92, No. 3, pp. 358–366.

Beaugrand, G. (2004). The North Sea regime shift: Evidence, causes, mechanisms and consequences. *Progress in Oceanography*, vol. 60, No. 2–4, pp. 245–262.

Beaugrand, G., Edwards, M., Brander, K., Luczak, C., Ibañez, F. (2008). Causes and projections of abrupt climate-driven ecosystem shifts in the North Atlantic. *Ecology Letters*, vol. 11, No. 11, pp. 1157–1168.

Beaugrand, G., Edwards, M., Legendre, L. (2010). Marine biodiversity, ecosystem functioning, and carbon cycles. *Proceedings of the National Academy of Sciences of the United States of America*, vol. 107, No. 22, pp. 10120–10124.

Beaugrand, G., Luczak, C., Edwards, M. (2009). Rapid biogeographical plankton shifts in the North Atlantic Ocean. *Global Change Biology*, vol. 15, No. 7, pp. 1790–1803.

Beaugrand, G., McQuatters-Gollop, A., Edwards, M., Goberville, E. (2013). Long-term responses of North Atlantic calcifying plankton to climate change. *Nature Climate Change*, vol. 3, pp. 263–267.

Beaugrand, G., Reid, P.C., Ibañez, F., Lindley, J.A., Edwards, M. (2002). Reorganization of North Atlantic marine copepod biodiversity and climate. *Science*, vol. 296, No. 5573, pp. 1692–1694.

Beare, D., McQuatters-Gollop, A., van der Hammen, T., Machiels, M., Teoh, S.J., Hall-Spencer, J.M. (2013). Long-term trends in calcifying plankton and pH in the North Sea. *PLoS ONE* 8(5): e61175. doi:10.1371/journal.pone.0061175.

Beck, M.W., Heck, Jr., K.L., Able, K.W., Childers, D.L., Eggleston, D.B., Gillanders, B.M., Halpern, B., Hays, C.G., Hoshino, K., Minello, T.J., Orth, R.J., Sheridan, P.F., Weinstein, M.P. (2001). The identification, conservation, and management of estuarine and marine nurseries for fish and invertebrates. *BioScience*, vol. 51, No. 8, pp. 633–641.

Belgrano, A., Lindahl, O., Hernroth, B. (1999). North Atlantic Oscillation primary productivity and toxic phytoplankton in the Gullmar Fjord, Sweden (1985–1996). *Proceedings of the Royal Society B-Biological Sciences*, vol. 266, No. 1418, pp. 425–430. Available from http://www.ncbi.nlm.nih.gov/pmc/articles/PMC1689790/pdf/YA35QW7PWX7MH8DC_266_425.pdf.

Bell, R.J., Richardson, D.E., Hare, J.A., Lynch, P.D., & Fratantoni, P.S. (2014). Disentangling the effects of climate, abundance, and size on the distribution of marine fish: an example based on four stocks from the Northeast US shelf. *ICES Journal of Marine Science: Journal du Conseil*, fsu217.

Benoît, H. P., Gagné, J.A., Savenkoff, C., Ouellet, P., Bourassa M.-N. (eds.) (2012).

State-of the-Ocean Report for the Gulf of St. Lawrence Integrated Management (GOSLIM) Area. *Canadian Manuscript Report of Fisheries and Aquatic Sciences*, vol. 2986, viii + 73 pp.

Bérard-Therriault, L., Poulin, M., Bossé, L. (1999). Guide d'identification du phytoplancton marin de l'estuaire et du golfe du Saint-Laurent incluant également certains protozoaires, *Canadian Special Publication of Fisheries and Aquatic Sciences* No. 128, 387pp.

Bergek, S., Ma, Q., Vetemaa, M., Franzén, F., Appelberg, M. (2012). From individuals to populations: impacts of environmental pollution on natural eelpout populations. *Ecotoxicology and Environmental Safety*, vol. 79, pp.1–12.

Bianchi, C.N, and Morri, C. (2000). Marine Biodiversity of the Mediterranean Sea: Situation, Problems and Prospects for Future Research. *Marine Pollution Bulletin*, vol. 40, No. 5, pp. 367–376.

Bianchi, C.N., Morri, C., Chiantore, M., Montefalcone, M., Parravicini, V., Rovere, A. (2012). Mediterranean Sea biodiversity between the legacy from the past and a future of change. In: Stambler, N. (ed.), *Life in the Mediterranean Sea: A Look at Habitat Changes*. Nova Science Publishers, New York, pp. 1-55.

Bianchi, G., Gislason, H., Graham, K., Hill, L., Jin, X., Koranteng, K., Manickchand-Heileman, S., Payá, I., Sainsbury, K., Sanchez, F., Zwanenburg, K. (2000). Impact of fishing on size composition and diversity of demersal fish communities. *ICES Journal of Marine Science*, vol. 57, No. 3, pp. 558–571.

Bicknell, A.W.J., Oro, D., Camphuysen, K.C.J., and Voitier, S.C. (2013). Potential consequences of discard reform for seabird communities. *Journal of Applied Ecology*, vol. 50, No. 3, pp. 649–658.

BirdLife International (2014). Species factsheet: Little Auk *Alle alle*. Downloaded from http://www.birdlife.org. Accessed 07 Sept. 2014.

Birdsong, R.S., Marshall, H.G., Alden, R.W., and Ewing, R.M. (1989). *Chesapeake Bay Plankton Monitoring Program*, Final Report, 1987–1988, Old Dominion University Research Foundation, Norfolk, Virginia.

Birdsong, R.S., and Buchanan, C. (1993). Zooplankton, In: The Ecologically Valuable Species Workgroup of the Living Resources Subcommittee, *Chesapeake Bay Strategy for the Restoration and Protection of Ecologically Valuable Species*, The Marlyand Development of Natural Resources Tidewater Administration Chesapeake Bay Research and Monitoring Division for the Chesapeake Bay Program, United States, pp. 45–50.

Birnbaum, C. (2006). Online Database of the North European and Baltic Network on Invasive Alien Species, NOBANIS (www.nobanis.org).

Blanchard, J.L., Dulvy, N.K., Jennings, S., Ellis, J.R., Pinnegar, J.K., Tidd, A., Kell, L.T. (2005). Do climate and fishing influence size-based indicators of Celtic Sea fish community structure? *ICES Journal of Marine Science*, vol. 62, No. 3, pp. 405–411.

Bonaca, M.O., and Lipej, L. (2005). Factors affecting habitat occupancy of fish assemblage in the Gulf of Trieste (Northern Adriatic Sea). *Marine Ecology*, vol. 26, No. 1, pp. 42–53.

Borja, A., Bricker, S.B., Dauer, D.M., Demetriades, N.T., Ferreira, J.G., Forbes, A.T., Hutchings, P., Jia, X., Kenchington, R., Marques, J.C., Zhu, C. (2008). Overview of integrative tools and methods in assessing ecological integrity in estuarine and coastal systems worldwide. *Marine Pollution Bulletin*, vol. 56, pp. 1519–1537.

Borja, A., Muxika I., Rodrıguez J.G. (2009). Paradigmatic responses of marine benthic communities to different anthropogenic pressures, using M-AMBI, within the European Water Framework Directive. *Marine Ecology*, vol. 30, No. 2, pp. 214–227.

Bossi, R.H., and Cintrón, G. (2010). *Mangroves of the Wider Caribbean: Toward Sustainable Management*, UNEP, Nairobi, Kenya, 30pp.

Boyd, S.E., Limpenny, D.S., Rees, H.L., Cooper, K.M. (2005). The effects of marine sand and gravel extraction on the macrobenthos at a commercial dredging site (results 6 years post-dredging). *ICES Journal of Marine Science*, vol. 62, No. 2, pp. 145–162.

Bradley, P.E., and Norton, R.L. (Eds.) (2009). *An Inventory of Breeding Seabirds of the Caribbean*, University Press of Florida, United States.

Brault, S., Stuart, C.T., Wagstaff, M.C., McClain, C.R., Allen, J.A., Rex, M.A. (2013). Contrasting patterns of α-and β-diversity in deep-sea bivalves of the eastern and western North Atlantic. *Deep Sea Research Part II: Topical Studies in Oceanography*, vol. 92, pp. 157–164.

Brown, J.H., Gillooly, J.F., Allen, A.P., Savage, V.M., West, G.B. (2004). Toward a metabolic theory of ecology. *Ecology*, vol. 85, 1771–1789.

Brownlee, D.C., and Jacobs, F. (1987). Mesozooplankton and microzooplankton in the Chesapeake Bay. In: Majumdar, S.K., Hall, L.W., Austin, A.M. (eds.), *Contaminant Problems and Management of Living Chesapeake Bay Resources*. Pennsylvania Academy of Sciences, Eston, PA, United States, pp. 218–269.

Brunel, P., Bossé, L., Lamarche, G, (1998). *Catalogue of the Marine Invertebrates of the Estuary and Gulf of Saint Lawrence*, Canadian Special Publication of Fisheries and Aquatic Sciences, NRC Press, Canada, pp. 1–405.

Bruno, J.F., Sweatman, H., Precht, W.F., Selig, E.R., Schutte, V.G. (2009). Assessing evidence of phase shifts from coral to macroalgal dominance on coral reefs. *Ecology*, vol. 90, 1478–1484.

Buchanan, C., Lacouture, R., Marshall, H., Olson, M., Johnson, J. (2005). Phytoplankton reference communities for Chesapeake Bay and its tidal tributaries. *Estuaries*, Vol. 28, No. 1, pp. 138–159.

Buchheister, A., Bonzek, C.F., Gartland, J., Latour, R.J. (2013). Patterns and drivers of the demersal fish community of Chesapeake Bay. *Marine Ecology Progress Series*, vol. 481, pp. 161–180.

Bui, A.O.V., Ouellet, P., Castonguay, M., Brêthes, J.-C. (2010). Ichthyoplankton community structure in the northwest Gulf of St. Lawrence (Canada): Past and present. *Marine Ecology Progress Series*, vol. 412, pp. 189–205.

Bulleri, F., and Chapman, M.G. (2010). The introduction of coastal infrastructure as a driver of change in marine environments, *Journal of Applied Ecology*, vol. 47, No. 1, pp. 26–35.

Burke, L., Reytar, K., Spalding, M., Perry, A. (2011). *Reefs at Risk Revisited*. World Resources Institute, Washington, DC.

Callaway, R., Alsvåg, J., de Boois, I., Cotter, J., Ford, A., Hinz, H., Jennings, S., Kröncke, I., Lancaster, J., Piet, G., Prince, P., Ehrich, S. (2002). Diversity and community structure of epibenthic invertebrates and fish in the North Sea. *ICES Journal of Marine Science*, vol. 59, No. 6, pp. 1199–1214.

Calvo, N., Templado, J., Oliverio, M., Machordom, A. (2009). Hidden Mediterranean biodiversity: molecular evidence for a cryptic species complex within the reef building vermetid gastropod *Dendropoma petraeum* (Mollusca: Caenogastropoda). *Biological Journal of the Linnean Society*, vol. 96, No. 4, pp. 898–912.

Campbell, N., Neat, F., Burns, F., Kunzlik, P. (2011). Species richness, taxonomic diversity, and taxonomic distinctness of the deep-water demersal fish community on the Northeast Atlantic continental slope (ICES Subdivision VIa). *ICES Journal of Marine Science*, vol. 68, pp. 365–376.

CARSEA, (2007). Agard, J.B.R., Cropper, A. (eds). Caribbean Sea Ecosystem Assessment (CARSEA), *A sub-global component of the Millennium Ecosystem Assessment prepared by the Caribbean Sea Ecosystem Assessment Team*. Caribbean Marine Studies, Special Edition, 85pp.

CBD 2014s Report of the North-West Atlantic Regional Workshop To Facilitate the Description of Ecologically Or Biologically Significant Marine Areas. UNEP/CBD/

EBSA/WS/2014/2/4; 120 pp.

CBD 2014b Report of the Mediterranean Regional Workshop to Facilitate the Description of Ecologically or Biologically Significant Marine Areas. UNEP/CBD/EBSA/WS/2014/3/4

Chambers, R.M., Havens, K.J., Killeen, S., Berman, M. (2008). Common reed *Phragmites australis* occurrence and adjacent land use along estuarine shoreline in Chesapeake Bay. *Wetlands*, vol. 28, No. 4, pp. 1097–1103.

Chambers, R. M., Meyerson, L.A., Saltonstall, K. (1999). Expansion of *Phragmites australis* into tidal wetlands of North America. *Aquatic Botany*, vol. 64, No. 3–4, pp. 261–274

Chust, G., Castellani, C., Licandro, P., Ibaibarriaga, L., Sagarminaga, Y., Irigoien, X. (2014). Are *Calanus* spp. shifting poleward in the North Atlantic? A habitat modelling approach. *ICES Journal of Marine Science*, vol. 71, pp. 241–253.

Claudet, J., and Fraschetti, S. (2010). Human-driven impacts on marine habitats: A regional meta-analysis in the Mediterranean Sea. *Biological Conservation*, vol. 143, No. 9, pp. 2195–2206.

CLME, (2011). *Caribbean Large Marine Ecosystem Regional Transboundary Diagnostic Analysis*, 148 pp. Available from www.clmeproject.org.

Coen, L.D., Luckenbach, M.W., Breitburg, D.L. (1999). The role of oyster reefs as essential fish habitat: a review of current knowledge and some new perspectives. *American Fisheries Society*, Symposium 34, pp. 303–307.

Coll, M., Piroddi, C., Steenbeek, J., Kaschner, K., Ben Rais Lasram, F., Aguzzi, J., Ballesteros, E., Nike Bianchi, C., Corbera, J., Dailianis, T., Danovaro, R., Estrada, M., Froglia, C., Galil, B.S., Gasol, J.M., Gertwagen, R., Gil, J., Guilhaumon, F., Kesner-Reyes, K., Kitsos, M.-S., Koukouras, A., Lampadariou, N., Laxamana, E., López-Fé de la Cuadra, C.M., Lotze, H.K., Martin, D., Mouillot, D., Oro, D., Raicevich, S., Rius-Barile J., Saiz-Salinas, J.I., San Vicente, C., Somot, S., Templado, J., Turon, X., Vafidis, D., Villanueva, R., Voultsiadou, E. (2010). The biodiversity of the Mediterranean Sea: Estimates, patterns and threats. *PLoS ONE* 5:doi:10.1371.

Coll, M., Piroddi, C., Albouy, C., Ben Rais Lasram, F., Cheung, W., Christensen, V., Karpouzi, V., Le Loc, F., Mouillot, D., Paleczny, M., Palomares, M.L., Steenbeek, J., Trujillo, P., Watson, R., Pauly, D. (2012). The Mediterranean Sea under siege: spatial overlap between marine biodiversity, cumulative threats and marine reserves. *Global Ecology and Biogeography*, vol. 21, pp. 465–480.

Collie, J., Rochet, M.-J., and Bell, M. (2013). Rebuilding fish communities: the ghost of fisheries past and the virtue of patience. *Ecological Applications*, vol. 23, pp. 374–391.

Collie, J.S., Escanero, G.A., Valentine, P.C. (1997). Effects of bottom fishing on the benthic megafauna of Georges Bank. *Marine Ecology Progress Series*, vol. 155, pp. 159–172.

Condon, R., Decker, M.B., Purcell, J.E. (2001). Effects of low dissolved oxygen on survival and asexual reproduction of scyphozoan polyps (*Chrysaora quinquecirrha*). *Hydrobiologia*, vol. 451, pp. 89–95.

Cook, A.M., and Bundy, A. (2012). Use of fishes as sampling tools for understanding biodiversity and ecosystem functioning in the ocean. *Marine Ecology Progress Series*, vol. 454, pp. 1–18.

Cooper, S. R., and Brush, G.S. (1991). Long-term history of Chesapeake Bay anoxia. *Science*, vol. 254, pp. 992–996.

Cordero, D., Peña, J., Saavedra, C. (2014). Phylogeographic analysis of introns and mitochondrial DNA in the clam *Ruditapes decussatus* uncovers the effects of Pleistocene glaciations and endogenous barriers to gene flow. *Molecular Phylogenetics and Evolution*, vol. 71, pp. 274–287.

Costanza, R., d'Arge, R., de Groot, R., Farberk, S., Grasso, M., Hannon, B., Limburg, K., Naeem, S., O'Neill, R.V., Paruelo, J., Raskin, R.G., Suttonk, P., van den Belt, M. (1997). The value of the world's ecosystem services and natural capital. *Nature*, vol. 387, pp. 253–260.

Cotter, R.C., Rail, J.-F., Boyne, A.W., Robertson, G.J., Weseloh, D.V.C., Chaulk, D.G. (2012). *Population status, distribution, and trends of gulls and kittiwakes breeding in eastern Canada 1998–2007*. Environment Canada, Gatineau, QC, Canada, 93pp.

Cowen, R.K., Paris, C.B., Srinivasan, A. (2006). Scaling connectivity in marine populations. *Science*, vol. 311, No. 5760, pp. 522–527.

Cowen, R.K., Sponaugle, S., Paris, C.B., Lwiza, K.M.M., Fortuna, J.L., Dorsey, S. (2003). Impact of North Brazil Current rings on local circulation and coral reef recruitment to Barbados, West Indies, In: Goni, G.J., and Rizzoli, P.M. (eds.), *Interhemispheric Water Exchange in the Atlantic Ocean*. Elsevier, Amsterdam, pp. 443–462.

Cox, T.M., Ragen, T.J., Read, A.J., Vos, E., Baird, R.W., Balcomb, K., Barlow, J., Caldwell, J., Cranford, T., Crum, L., D'Amico, A., D'Spain, G., Fernandez, A., Finneran, J.J., Gentry, R.L., Gerth, W., Gulland, F., Hildebrand, J., Houser, D., Hullar, T., Jepson, P.D., Ketten, D.R., MacLeod, C.D., Miller, P., Moore, S., Mountain, D.C., Palka, D., Ponganis, P., Rommel, S., Rowles, T.K., Taylor, B., Tyack, P.L., Wartzok, D., Gisiner, R.C., Mead, J.G., Benner, L. (2006). Understanding the impacts of anthropogenic sound on beaked whales. *Journal of Cetacean Research and Management*, vol. 7, pp. 177–187.

Cummings, J.A. (1983). Habitat dimensions of calanoid copepods in the western Gulf of Mexico. *Journal of Marine Research*, vol. 42, No. 1, pp. 163–188.

Daan, N., Gislason, H., Pope, J. G., Rice, J. C. (2005). Changes in the North Sea fish community: evidence of indirect effects of fishing? *ICES Journal of Marine Science*, vol. 62, No.2, pp. 177–188.

Dale, B., Edwards, M., Reid, P.C. (2006). Climate change and harmful algal blooms. In: Graneli, E., and Turner, J. (eds.), Ecology of Harmful Algae. *Ecological Studies*, vol. 189, Springer-Verlag: Berlin, pp. 367–378.

Danovaro, R., Company, B.J., Corinaldesi, C., D'Onghia, G., Galil, B.S., Gambi, C., Gooday, A., Lampadariou, N., Luna, G., Morigi, C., Olu, K., Polymenakou, P., Ramirez-Llodra, E., Sabbatini, A., Sardà, F., Sibuet, M., Tselepides, A. (2010). Deep-sea biodiversity in the Mediterranean Sea: the known, the unknown and the unknowable. *PLoS ONE* 5:e11832.

Dauer, D.M., Gerritsen, J., Ranasinghe, J.A. (1993). Benthos. In: The Ecologically Valuable Species Workgroup of the Living Resources Subcommittee, *Chesapeake Bay Strategy for the Restoration and Protection of Ecologically Valuable Species*, The Marlyand Development of Natural Resources Tidewater Administration Chesapeake Bay Research and Monitoring Division for the Chesapeake Bay Program, US, pp. 55–60.

Diefenderfer, H.L., Thom, R.M., Johnson, G.E., Skalski, J.R., Vogt, K.A., Ebberts, B.D., Roegner, G.C., Dawley, E.M. (2011). A levels-of-evidence approach for assessing cumulative ecosystem response to estuary and river restoration programs. *Ecological Restoration*, vol. 29, No. 1–2, pp. 111–132.

DFO (2013). *Oceanographic conditions in the Atlantic zone in 2012*. Canadian Science Advisory Secretariat Science Advisory Report 2013/057.

DFO (2013). In : Gagné, J.A., Ouellet, P., Savenkoff, C., Galbraith, P.S., Bui, A.O.V., Bourassa, M.-N. (eds.), *Rapport intégré de l'initiative de recherche écosystémique (IRÉ) de la région du Québec pour le projet : les espèces fourragères responsables de la présence des rorquals dans l'estuaire maritime du Saint-Laurent*. Secr. Can. de Consult. Sci. du MPO. Doc. de rech. 2013/086, vi + 181 pp.

DFO (2011). *Synopsis of the Social, Economic, and Cultural Overview of the Gulf of St. Lawrence*. OHSAR Pub. Ser. Rep. NL Region, No. 0005: vi + 32pp.

Dolan J. R., 2000. Tinitinnid ciliate diversity in the Mediterranean Sea: Longitudinal patterns related to water column structure in late spring-early summer. *Aquatic Microbial Ecology* 22: 69-78.

Doyle, T., Houghton, J.D.R., Buckley, S.M., Hays, G.C., Davenport, J. (2007). The broad-scale distribution of five jellyfish species across a temperate coastal environment. *Hydrobiologia*, vol. 579, No. 1, pp. 29–39.

Drinkwater, K. (2009). Comparison of the response of Atlantic cod (*Gadus morhua*) in the high-latitude regions of the North Atlantic during the warm periods of the 1920s–1960s and the 1990s–2000s. *Deep-Sea Research II: Topical Studies in Oceanography*, vol. 56, No. 21–22, pp. 2087–2096.

Drinkwater, K.F. (2005). The response of Atlantic cod (*Gadus morhua*) to future climate change. ICES *Journal of Marine Science* 62, 1327–1337.

Duarte, C.M., Borum, J., Short, F.T., Walker, D.I. (2008). Seagrass ecosystems: Their global status and prospects. In: Polunin, N.V.C. (ed.), *Aquatic Ecosystems*. Cambridge: Cambridge University Press, pp. 281–306.

Dufour, R., Benoît, H., Castonguay, M., Chassé, J., Devine, L., Galbraith, P., Harvey, M., Larouche, P., Lessard, S., Petrie, B., Savard, L., Savenkoff, C., St-Amand, L., Starr, M. (2010). *Ecosystem status and trends report: Estuary and Gulf of St. Lawrence ecozone*. Canadian Science Advisory Secretariat. Research Document 2010/030.

Dunnet, G.M., Furness, R.W., Tasker, M.L., Becker, P.H. (1990). Seabird ecology in the North Sea. *Netherlands Journal for Sea Research*, vol. 26, No. 2–4, pp. 387–425.

Durand, D., Blel, H., Shen, K.N., Koutrakis, E.T., Guinand, B. (2013). Population genetic structure of *Mugil cephalus* in the Mediterranean and Black Seas: a single mitochondrial clade and many nuclear barriers. *Marine Ecology Progress Series*, vol. 474, pp. 243–261.

Durbin, E.G., Campbell, R.G., Casas, M.C., Ohman, M.D., Niehoff, B., Runge, J., Wagner, M. (2003). Interannual variation in phytoplankton blooms and zooplankton productivity and abundance in the Gulf of Maine during winter. *Marine Ecology Progress Series*, vol. 254, pp. 81–100.

Dutil, J.-D., Miller, R., Nozères, C., Bernier, B., Bernier, D., Gascon, D. (2006). Révision des identifications de poissons faites lors des relevés scientifiques annuels d'évaluation de l'abondance des poissons de fond et de la crevette nordique dans l'estuaire et le nord du golfe du Saint-Laurent. *Rapport manuscrit Canadien des Sciences Halieutiques et Aquatiques*, vol. 2760: x + 87pp.

Dutil, J.-D., Nozères, C., Scallon-Chouinard, P.-M., Van Guelpen, L., Bernier, D., Proulx, S., Miller, R., Savenkoff, C. (2009). Poissons connus et méconnus des fonds marins du Saint-Laurent. *Le Naturaliste Canadien*, vol. 133, No. 2, pp. 70–82.

Eakin, C.M., Morgan, J.A., Heron, S.F., Smith, T.B., Liu, G., Alvarez-Filip, L., Baca, B., Bartels, E., Bastidas, C., Bouchon, C., Brandt, M., Bruckner, A.W., Bunkley-Williams, L., Cameron, A., Causey, B.D., Chiappone, M., Christensen, T.R.L., Crabbe, M.J. C., Day, O, de la Guardia, E., Díaz-Pulido, G., DiResta, D., Gil-Agudelo, D.L., Gilliam, D.S., Ginsburg, B.N., Gore, S., Guzmán, H.M., Hendee, J.C., Hernández-Delgado, E.A., Husain, E., Jeffrey, C.F.G., Jones, R.J., Jordán-Dahlgren, E., Kaufman, L.S., Kline, D.I., Kramer, P.A., Lang, J.C., Lirman, D., Mallela, J., Manfrino, C., Maréchal, J.-P., Marks, K., Mihaly, J., Miller, W.J., Mueller, E.M., Muller, E.M., Toro, C.A.O., Oxenford, H.A., Ponce-Taylor, D., Quinn, N., Ritchie, K.B., Rodríguez, S., Ramírez, A.R., Romano, S., Samhouri, J.F., Sánchez, J.A., Schmahl, G.P., Shank, B.V., Skirving, W.J., Steiner, S.C.C., Villamizar, E., Walsh, S.M., Walter, C., Weil, E., Williams, E.H., Roberson, K.W., Yusuf, Y.(2010). Caribbean corals in crisis: record thermal stress, bleaching, and mortality in 2005. *PLoS ONE* 5: 9pp.

Edwards, M., Beaugrand, G., John, A.W.G., Johns, D.G., Licandro, P., McQuatters-Gollop, A., Reid, P.C. (2009). *Ecological Status Report: results from the CPR survey 2007/2008*. SAHFOS Technical Report, vol. 6, pp. 1–12. Plymouth, U.K. ISSN 1744-0750.

EEA (2012). *European waters – assessment of status and pressures*. EEA Report 8/2012.

Edwards, M., Beaugrand, G., Reid, P.C., Rowden, A.A., Jones, M.B. (2002). Ocean climate anomalies and the ecology of the North Sea. *Marine Ecology Progress Series*, vol. 239, pp. 1–10.

Edwards, M., Johns, D.G., Leterme, S.C., Svendsen, E., Richardson, A.J. (2006). Regional Climate Change and Harmful Algal Blooms in the Northeast Atlantic. *Limnology and Oceanography*, vol. 51, No. 2, pp. 820–829.

Eero, M., MacKenzie, B.R., Köster, F.W., Gislason, H. (2011). Multidecadal responses of a cod (*Gadus morhua*) population to human-induced trophic changes, fishing, and climate. *Ecological Applications*, vol. 21, No. 1, pp. 214–226.

Elliott, D.T., Pierson, J.J., Roman, M.R. (2012). Relationship between environmental conditions and zooplankton community structure during summer hypoxia in the northern Gulf of Mexico. *Journal of Plankton Research*, vol. 34, No. 7, pp. 602–613.

Ellison, A.M, and Farnsworth, E.J. (1996). Anthropogenic disturbance of Caribbean mangrove ecosystems: Past impacts, present trends, and future predictions. *Biotropica*, vol. 28, No. 4a, pp. 549–565.

Elmgren, R. (1989). Man's impact on the ecosystem of the Baltic Sea: Energy flows today and at the turn of the century. *Ambio*, vol. 18, pp. 326–332.

Enghoff, I.B., MacKenzie, B.R., Nielsen, E.E. (2007). The Danish fish fauna during the warm Atlantic period (ca. 7000–3900 BCE): Forerunner of future changes? *Fisheries Research*, vol. 87, No. 2–3, pp. 167–180.

Escaravage, V., Herman, P., Merckx, B., Bodarska-Kowalczuk, M., Amouroux, J., Degraer, S., Grémare, A., Heip, C., Hummel, H., Karakassis, I. (2009). Distribution patterns of macrofaunal species diversity in subtidal soft sediments: biodiversity-productivity relationships from the MacroBen database. *Marine Ecology Progress Series*, vol. 382, pp. 253–264.

FAO (2007). *The World's Mangroves 1980-2005*. FAO Forestry Paper 153, 77pp.

FAO (2009). *Report of the Technical Consultation on International Guidelines for the Management of Deep-sea Fisheries in the High Seas*. Rome, 4–8 February and 25–29 August 2008. Available from ftp://ftp.fao.org/docrep/fao/011/i0605t/i0605t00.pdf.

FAO (2013). NEAFC VME fact sheets 2015. Altair. [online]. Rome. Updated 6 February 2015. [Cited 20 April 2015]. http://www.fao.org/fishery/

Fisher, J.A.D, Frank, K.T., Leggett, W.D. (2010). Global variation in marine fish body size and its role in biodiversity–ecosystem functioning. *Marine Ecology Progress Series*, vol. 405, pp. 1–13.

Fisher, J.A.D, Frank, K.T., Petrie, B., Leggett, W.C., Shackell, N.L. (2008). Temporal dynamics within a contemporary latitudinal diversity gradient. *Ecology Letters*, vol. 11, No. 9, pp. 883–897.

Fock, H.O., Kloppmann, M.H.P, Probst, W.N. (2014). An early footprint of fisheries: Changes for a demersal fish assemblage in the German Bight from 1902–1932 to 1991–2009. *Journal of Sea Research*, vol. 85, pp. 325–335.

Fogarty, M., Incze, L., Hayhoe, K., Mountain, D., Manning, J. (2008). Potential climate change impacts on Atlantic cod (Gadus morhua) off the northeastern USA. *Mitigation and Adaptation Strategies for Global Change*, vol. 13, pp. 453–466.

França, S., Vasconcelos, R., Fonseca, V., Tanner, S., Reis-Santos, P., Costa, M., Cabral, H. (2012). Predicting fish community properties within estuaries: Influence of habitat type and other environmental features. *Estuarine, Coastal and Shelf Science*, vol. 107, pp. 22–31.

Frank, K.T., Petrie, B., Shackell, N.L., Leggett, W.C. (2005). Trophic cascades in a formerly cod-dominated ecosystem. *Science*, vol. 308, No. 5728, pp. 1621–1623.

Fraser, H.M., Greenstreet, S.P.R., Fryer, R.J., Piet, G.J. (2008). Mapping spatial variation in demersal fish species diversity and composition in the North Sea: accounting for species- and size-related catchability in survey trawls. *ICES Journal of Marine Science*, vol. 65, No. 4, pp. 531–538.

Fu, F.X., Tatters, A.O., Hutchins, D.A. (2012). Global change and the future of harmful algal blooms in the ocean. *Marine Ecology Progress Series*, vol. 470, pp. 207–233.

Fu, C., Gaichas, S., Link, J.S., Bundy, A., Boldt, J.L., Cook, A.M., Gamble, R., Utne, K.R., Liu, H., Friedland, K.D. (2012). Relative importance of fisheries, trophodynamic and environmental drivers in a series of marine ecosystems. *Marine Ecology Progress Series*, vol. 459, pp. 169–184.

Gaertner, J.-C., Bertrand, J.A., Relini, G., Papaconstantinou, C., Mazouni, N., de Sola, L.G., Durbec, J.P., Jukic-Peladic, S. (2007). Spatial pattern in species richness of demersal fish assemblages on the continental shelf of the northern Mediterranean Sea: a multiscale analysis. *Marine Ecology Progress Series*, vol. 341, pp. 191–203.

Gage, J., Lambshead, P., Bishop, J.D., Stuart, C., Jones, N.S. (2004). Large-scale biodiversity pattern of Cumacea (Peracarida: Crustacea) in the deep Atlantic. *Marine Ecology Progress Series*, vol. 277, pp. 181–196.

Gagné, F., Blaise, C., Pellerin, J., Pelletier, E., Strand, J. (2006). Health status of *Mya arenaria* bivalves collected from contaminated sites in Canada (Saguenay Fjord) and Denmark (Odense Fjord) during their reproductive period. *Ecotoxicology and Environmental Safety*, vol. 64, No. 3, pp. 348–361.

Gardner, T.A., Côté, I.M., Gill, J.A., Grant, A., Watkinson, A.R. (2003). Long-term region-wide declines in Caribbean corals. *Science*, vol. 301, No. 5635, pp. 958–960.

Gardner, T.A., Côté, I.M., Gill, J.A., Grant, A., Watkinson, A.R. (2005). Hurricanes and Caribbean coral reefs: Impacts, recovery, patterns and role in long-term decline. *Ecology*, vol. 86, No. 1, pp. 174–184.

Giakoumi, S., Sini, M., Gerovasileiou, V., Mazor, T., Beher, J., Possingham, H.P., Abdulla, A., Çinar, M.E., Dendrinos, P., Gucu, A.C., Karamanlidis, A.A., Rodic, P., Panayotidis, P., Taskin, E., Jaklin, A., Voultsiadou, E., Webster, C., Zenetos, A., Katsanevakis, S. (2013). Ecoregion-based conservation planning in the Mediterranean: Dealing with large-scale heterogeneity. *PLoS ONE* 8:e76449.

Gibbons, M.J., and Richardson, A.J. (2008). Patterns of pelagic cnidarian abundance in the North Atlantic. *Hydrobiologia*, vol. 616, pp. 51–65.

Glibert, P., Seitzinger, S., Heil, C.A., Burkholder, J.M., Parrow, M.W., Codispoti, L.A., Kelly, V. (2005). The role of eutrophication in the coastal proliferation of harmful algal blooms: new perspectives and new approaches. *Oceanography*, vol. 18, pp. 198–209.

Gollasch, S., Haydar, D., Minchin, D., Wolff, W.J., Reise, K. (2009). Introduced aquatic species of the North Sea coasts and adjacent brackish waters. In: G. Rilov, J.A. Crooks (eds.), *Biological Invasions in Marine Ecosystems*. Ecological Studies 204, Springer-Verlag, pp. 507–528.

Gomoiu, M.-T. (2012). *Biodiversity in the Black Sea*, Available from http://www.ibiol.ro/man/2012/Diaspora/04_Gomoiu_Biodiversitatea%20Marii%20Negre.pdf.

Gomoiu, M.-T., and Skolka, M. (1998). *Increase of biodiversity by immigration - new species for the Romanian fauna.* An.Univ.OVIDIUS Constanta, Seria Biologie-Ecologie, vol. II, pp. 181-202.

Goriup, P. (2009), *Guidelines for the Establishment of Marine Protected Areas in the Black Sea*, the Environmental Collaboration for the Black Sea Project, p. 13. Available from http://www.enpi-info.eu/files/publications/Guidelines%20on%20Black%20Sea%20MPAs%20Mar09.pdf.

Green, D.H., Edmunds, P.J., Carpenter, R.C. (2008). Increasing relative abundance of Porites astreoides on Caribbean reefs mediated by an overall decline in coral cover. *Marine Ecology Progress Series*, vol. 359, pp. 1–10.

Greenstreet, S.P.R., Fraser, H.M., Cotter, J., Pinnegar, J. (2010). Assessment of the "State" of the demersal fish communities in OSPAR Regions II, III, IV, and V. *OSPAR Commission Quality Status Report Assessment of the Environmental Impact of Fishing* No. 465/2009.

Greenstreet, S.P.R., Fraser, H.M., Rogers, S.I., Trenkel, V.M., Simpson, S.D., Pinnegar, J.K. (2012). Redundancy in metrics describing the composition, structure, and functioning of the North Sea demersal fish community. *ICES Journal of Marine Science*, vol. 69, pp. 8–22.

Greenstreet, S.P.R., Rogers, S.I., Rice, J.C., Piet, G.J., Guirey, E.J., Fraser, H.M., Fryer, R. J. (2011). Development of the EcoQO for the North Sea fish community. *ICES Journal of Marine Science*, vol. 68, pp. 1–11.

Greenstreet, S.P.R., Spence, F.E., McMillan, J.A. (1999). Fishing effects in northeast Atlantic shelf seas: patterns in fishing effort, diversity and community structure. V. Changes in structure of the North Sea groundfish species assemblage between 1925 and 1996. *Fisheries Research*, vol. 40, pp. 153–183.

Grice, G.D. (1960). Calanoid and cyclopoid copepods collected from the Florida Gulf coast and Florida Keys in 1954 and 1955. *Bulletin of Marine Science*, vol. 10, No. 2, pp. 217–226.

Grizzle, R.E., Ward, L.G., Mayer, L.A., Malik, M.A., Cooper, A.B. (2009). Effects of a large fishing closure on benthic communities in the western Gulf of Maine: recovery from the effects of gillnets and otter trawls. *Fishery Bulletin*, vol. 10, pp. 308–317.

Guilloux, L., Rigaut-Jalabert, F., Jouenne, F., Ristori, S., Viprey, M., Not, F., Vaulot, D., Simon, N. (2013). An annotated checklist of Marine Phytoplankton taxa at the SOMLIT-Astan time series off Roscoff (Western Channel, France): data collected from 2000 to 2010. *Cahiers de Biologie Marine*, vol. 54, pp. 247–256.

Hale, S.S., Hughes, M.M., Strobel, C.J., Buffum, H.W., Copeland, J.L., Paul, J.F. (2002). Coastal ecological data from the Virginian Biogeographic Province, 1990–1993. *Ecology*, vol. 83, No. 10, p. 2942.

Hallegraeff, G.M. (2010). Ocean climate change, phytoplankton community responses, and harmful algal blooms: a formidable predictive challenge. *Journal of Phycology*, vol. 46, No. 2, pp. 220–235.

Hallegraeff, G.M., and Bolch, C.J. (1992). Transport of diatom and dinoflagellate resting spores via ship's ballast water: implications for plankton biogeography and aquaculture. *Journal of Plankton Research*, vol. 14, No. 8, pp. 1067–1084.

Hällfors, G. (2004). Checklist of Baltic Sea phytoplankton species (including some heterotrophic protistan groups). *Baltic Sea Environment Proceedings* No. 95, Helsinki Commission, Baltic Marine Environment Protection Commssion, 208pp.

Hammill, M. O., den Heyer, C.E. and Bowen, W.D. 2014. Grey Seal Population Trends in Canadian Waters, 1960-2014. DFO Can. Sci. Advis. Sec. Res. Doc. 2014/037. iv + 44 p.

Hammond, T. R., and Ellis, J. R. (2005). Bayesian Assessment of North-east Atlantic Spurdog Using a Stock Production Model, with Prior for Intrinsic Population Growth Rate Set by Demographic Methods. *Journal of Northwest Atlantic Fishery Science*, vol. 35, pp. 299-308

Harding, K. C., and Härkönen, T. J. (1999). Development in the Baltic grey seal (*Halichoerus grypus*) and ringed seal (*Phoca hispida*) populations during the 20th century. *Ambio*, vol. 28, pp. 619–627.

Harma, C., Brophya, D., Mintoa, C., Clarke, M. (2012). The rise and fall of autumn-spawning herring (*Clupea harengus* L.) in the Celtic Sea between 1959 and 2009: Temporal trends in spawning component diversity. *Fisheries Research*, vol.

121–122, pp. 31–42.

Harris, M.P., Beare, D., Toresen, R., Nøttestad. L., Kloppmann, M., Dörner, H., Peach, K., Rushton, D.R.A., Foster-Smith, J., Wanless, S. (2007). A major increase in snake pipefish (*Entelurus aequoreus*) in northern European seas since 2003: potential implications for seabird breeding success. *Marine Biology*, vol. 151, No. 3, pp. 973–983.

Hays, G.C., Richardson, A.J., Robinson, C. (2005). Climate change and marine plankton. *TRENDS in Ecology and Evolution*, vol. 20, No. 6, pp. 337–344.

Head, E.J.H., and Sameoto, D.D. (2007). Inter-decadal variability in zooplankton and phytoplankton abundance on the Newfoundland and Scotian shelves. *Deep Sea Research II*, vol. 54, No. 23–26, pp. 2686–2701.

Heck, K.L. and Orth, R.J. (1980). Structural components of eelgrass (*Zostera marina*) meadows in the lower Chesapeake Bay-decapod crustacea. *Estuaries*, vol. 3, pp. 289–295.

Heck, K.L., Carruthers, T.J.B., Duarte, C.M., Hughes, A.R., Kendrick, G., Orth, R.J., Williams, S.W. (2008). Trophic transfers from seagrass meadows subsidize diverse marine and terrestrial consumers, *Ecosystems*, vol. 11, pp. 1198–1210.

Heip, C., Duineveld, G., Flach, E., Graf, G., Helder, W., Herman, P., Lavaleye, M., Middelburg, J., Pfannkuche, O., Soetaert, K. (2001). The role of the benthic biota in sedimentary metabolism and sediment-water exchange processes in the Goban Spur area (NE Atlantic). *Deep Sea Research Part II: Topical Studies in Oceanography*, vol. 48, pp. 3223–3243.

HELCOM (2002). Environment of the Baltic Sea area 1994–1998. *Baltic Sea Environment Proceedings* No. 82 B., 215pp,, Helsinki, Finland. Available from www.helcom.fi.

HELCOM (2009a). Biodiversity in the Baltic Sea -- "An integrated thematic assessment on biodiversity and nature conservation in the Baltic Sea." Baltic Sea Environment Proceedings No. 116B. Available from www.helcom.fi.

HELCOM (2009b). Eutrophication in the Baltic Sea – An integrated thematic assessment of the effects of nutrient enrichment and eutrophication in the Baltic Sea region. *Baltic Sea Environment Proceedings* No. 115 B., p. 152. Available from www.helcom.fi.

HELCOM (2010). Ecosystem Health of the Baltic Sea 2003–2007: HELCOM Initial Holistic Assessment. *Baltic Sea Environment Proceedings* No. 122. Available from www.helcom.fi.

HELCOM (2013). HELCOM Red List of Baltic Sea species in danger of becoming extinct. Baltic Sea Environment Proceedings 140. Available from www.helcom.fi.

Henderson, P.A., Seaby, R.M.H., Somes, J.R. (2011). Community level response to climate change: The long-term study of the fish and crustacean community of the Bristol Channel. *Journal of Experimental Marine Biology and Ecology*, vol. 400, pp. 78–89.

Hernández-León, S., Gómez, M., Arístegui, J. (2007). Mesozooplankton in the Canary Current System: The coastal–ocean transition zone. *Progress in Oceanography*, vol. 74, No. 2–3, pp. 397–421.

Hiddink, J., Jennings, S., Kaiser, M., Queirós, A., Duplisea, D., Piet, G. (2006). Cumulative impacts of seabed trawl disturbance on benthic biomass, production, and species richness in different habitats. *Canadian Journal of Fisheries and Aquatic Sciences*, vol. 63, pp. 721–736.

Hiddink, J.G., and Coleby, C. (2012). What is the effect of climate change on marine fish biodiversity in an area of low connectivity, the Baltic Sea? *Global Ecology and Biogeography*, vol. 21 (2012), pp. 637–646.

Hilborn, R., and Litzinger, E. (2009). Causes of decline and potential for recovery of Atlantic cod populations. *The Open Fish Science Journal*, vol. 2, pp. 32–38.

Hinder, S.L., Hays, G.C., Edwards, M., Roberts, E.C., Walne, A.W., Gravenor, M.B. (2012). Changes in marine dinoflagellate and diatom abundance under climate change. *Nature Climate Change*, vol. 2, pp. 271–275.

Horwood, J. (2010). Marine ecosystem management: Fish abundance and size under exploitation. *Marine Policy*, vol. 34, pp. 1203–1206.

Houde, E.D. (1993). Forage fishes, inThe Ecologically Valuable Species Workgroup of the Living Resources Subcommittee, *Chesapeake Bay Strategy for the Restoration and Protection of Ecologically Valuable Species*, The Marlyand Development of Natural Resources Tidewater Administration Chesapeake Bay Research and Monitoring Division for the Chesapeake Bay Program, US, pp. 51–54.

Hughes, T.P., Rodrigues, M.J., Bellwood, D.R., Ceccarelli, D., Hoegh-Guldberg, O., McCook, L., Moltschaniwskyj, N., Pratchett, M.S., Steneck, R.S., Willis, B. (2007). Phase shifts, herbivory and the resilience of coral reefs to climate change. *Current Biology*, vol. 17, No. 4, pp. 360–365.

Hurst, T.P., McKown, K.A., Conover, D.O. (2004). Interannual and long-term variation in the nearshore fish community of the mesohaline Hudson River estuary. *Estuaries*, vol. 27, pp. 659–669.

Hutchings, J.A., and Baum, J.K. (2005). Measuring marine fish biodiversity: temporal changes in abundance, life history and demography. *Philosophical Transactions of the Royal Society B*, vol. 360, No. 1454: doi: 10.1098/rstb.2004.1586.

ICES (2008). Report of the ICES Advisory Committee 2008. *ICES Advice*, 2008. Book 6, 326pp.

ICES (2008b). Report of the ICES Advisory Committee, 2008. *ICES Advice*, 2008. Book 8, 133pp.

ICES (2009). Effects of extraction of marine sediments on the marine environment 1998-2004. *ICES Cooperative Research Report*, vol. 297, 181pp.

ICES. 2013a. Report of the Workshop to Review and Advise on EBSA Proposed Areas (WKEBSA), 27 - 21 May 2013, ICES HQ, Copenhagen, Denmark. ICES CM 2013/ACOM:70. 127 pp.

ICES. 2013b. Report of the Workshop to Review and Advise on EBSA Proposed Areas (WKEBSA), 27 - 21 May 2013, ICES HQ, Copenhagen, Denmark. ICES CM 2013/ACOM:70. 127 pp.

Jackson, J.B.C. (1997). Reefs since Columbus. *Coral Reefs*, vol. 16, Suppl: S23–S32.

Jackson, J.B.C., Donovan, M.K., Cramer, K.L., Lam, V. (eds.) (2014). *Status and trends of Caribbean coral reefs: 1970–2012*, Global Coral Reef Monitoring Network, IUCN, Gland, Swizerland; 304pp..

Jackson, E.L., Rowden, A.A., Attrill, M.J., Bossey, S.J., Jones, M.B. (2001). The importance of seagrass beds as a habitat for fishery species. *Oceanography and Marine Biology: An Annual Review*, vol. 39, pp. 269–303.

Javidpour, J., Sommer, U., Shiganova, T. (2006). First record of *Mnemiopsis leidyi* A. Agassiz 1865 in the Baltic Sea. *Aquatic Invasions*, vol. 1, No. 4, pp. 299–302.

Johannesen, E., Høines, Å.S., Dolgov, A.V., Fossheim, M. (2012). *Demersal Fish Assemblages and Spatial Diversity Patterns in the Arctic-Atlantic Transition Zone in the Barents Sea*. PLoS ONE 7: e34924. doi:10.1371/journal.pone.0034924.

Johns, D.G., Edwards, M., Batten, S.D. (2001). Arctic boreal plankton species in the Northwest Atlantic, *Canadian Journal of Fisheries and Aquatic Sciences*, vol. 58, No. 11, pp. 2121–2124.

Johnson, N.A., Campbell, J.W., Moore, T.S., Rex, M.A., Etter, R.J., McClain, C.R., Dowell, M.D. (2007). The relationship between the standing stock of deep-sea macrobenthos and surface production in the western North Atlantic. *Deep Sea Research Part I: Oceanographic Research Papers*, vol. 54, pp. 1350–1360.

Jones, D.O., Yool, A., Wei, C.L., Henson, S.A., Ruhl, H.A., Watson, R.A., Gehlen, M. (2013). Global reductions in seafloor biomass in response to climate change.

Global Change Biology, vol. 20, No. 6: doi: 10.1111/gcb.12480.

Joos, F., Plattner, G.-K., Stocker, T.F., Marchal, O., Schmittner, A. (1999). Global warming and marine carbon cycle feedbacks on future atmospheric CO2. *Science*, vol. 284, pp. 464–467.

Jordán, F. (2001). Seasonal changes in the positional importance of components in the trophic flow network of the Chesapeake Bay. *Journal of Marine Systems*, vol. 27, pp. 289–300.

Jordan, S.J., Lewis, M.A., Harwell, L.M., Goodman, L.R. (2010). Summer fish communities in northern Gulf of Mexico estuaries: Indices of ecological condition. *Ecological Indicators*, vol. 10, pp. 504–515.

Jordaan, A. (2010). Fish assemblages spatially structure along a multi-scale wave energy gradient. *Environmental Biology of Fishes*, vol. 87, pp. 13–24.

Jorde, D.G., Haramis, G.M., Forsell, D.J. (1993). Waterbirds, In The Ecologically Valuable Species Workgroup of the Living Resources Subcommittee, *Chesapeake Bay Strategy for the Restoration and Protection of Ecologically Valuable Species*, The Marlyand Development of Natural Resources Tidewater Administration Chesapeake Bay Research and Monitoring Division for the Chesapeake Bay Program, US, pp. 69–76.

Kaiser, M.J., Clarke, K.R., Hinz, H., Austen, M.C.V., Somerfield, P.J., Karakassis, I. (2006). Global analysis of response and recovery of benthic biota to fishing, *Marine Ecology Progress Series*, vol. 311, pp. 1–14.

Kane, J. (2007). Zooplankton abundance trends on Georges Bank, 1977–2004. *ICES Journal of Marine Science*, vol. 64, pp. 909–919.

Kane, J. (2011). Multiyear variability of phytoplankton abundance in the Gulf of Maine. *ICES Journal of Marine Science*, vol. 68, No. 9, pp. 1833–1841.

Karnauskas, M., Schirripa, M. J., Kelble, C. R., Cook, G. S., Craig, J. K. (eds.) 2013. Ecosystem status report for the Gulf of Mexico. NOAA Technical Memorandum NMFS-SEFSC-653, 52 p.

Kazanci, N., Öz, B., Turkmen, G., Başören, Ö.E. (2011). Contributions to aquatic fauna of a Biodiversity Hotspot in Eastern Blacksea Region of Turkey with records from runningwater interstitial fauna. *Review of Hydrobiology*, vol. 4, No. 2, pp. 131–138.

Kemp, W.M., Boynton, W., Adolf, J., Boesch, D., Boicourt, W., Brush, G., Cornwell, J., Fisher, T., Glibert, P., Hagy, J., Harding, L., Houde, E., Kimmel, D., Miller, W.D., Newell, R.I.E., Roman, M., Smith, E., Stevenson, J.C. (2005). Eutrophication of Chesapeake Bay: Historical trends and ecological interactions. *Marine Ecology Progress Series*, vol. 303, pp. 1–29.

Kemp, W.M., Boynton, W.R., Stevenson, J.C., Twilley, R.R., Means, J.C. (1983). The decline of submerged vascular plants in upper Chesapeake Bay: summary of results concerning possible causes. *Marine Technology Society Journal*, vol. 17, No. 2, pp. 78–89.

Kenchington, E., Kenchington, T.J., Henry, L.-A., Fuller, S.D., Gonzalez, P. (2007). Multidecadal changes in the megabenthos of the Bay of Fundy: The effects of fishing. *Journal of Sea Research*, vol. 58, pp. 220–240.

Keppel, G., van Niel, K.P., Wardell-Johnson, G.W., Yates, C.J., Byrne, M., Mucina, L., Schut, A.G.T., Hopper, S.D., Franklin, S.E. (2012). Refugia: identifying and understanding safe havens for biodiversity under climate change. *Global Ecology and Biogeography*, vol. 21, pp. 393–404.

Koch, E.W., Barbier, E.B., Silliman, B.R., Reed, D.J., Perillo, G.M.E., Hacker, S. D., Granek, E. F., Primavera, J. H., Muthiga, N., Polasky, S., Halpern, B. S., Kennedy, C. J., Kappel, C. V., Wolanski, E. (2009). Non-linearity in ecosystem services: temporal and spatial variability in coastal protection. *Frontiers in Ecology and the Environment*, vol. 7, No. 1, pp. 29–37.

Kopp, D., Le Bris, H., Grimaud, L., Nérot, C., Brind'Amour, A. (2013). Spatial analysis of the trophic interactions between two juvenile fish species and their preys along a coastal–estuarine gradient. *Journal of Sea Research*, vol. 81, pp. 40–48.

Kotta, J., Kotta, I., Simm, M., Põllupüü, M. (2009). Separate and interactive effects of eutrophication and climate variables on the ecosystem elements of the Gulf of Riga. *Estuarine, Coastal and Shelf Science*, vol. 84, No. 4, pp. 509–518.

Kroeger, T., and Guannel, G. (2014). Fishery enhancement, coastal protection and water quality services provided by two restored Gulf of Mexico oyster reefs, In: K. Ninan (ed.), *Valuing Ecosystem Services-Methodological Issues and Case Studies*. Edward Elgar Press, London. In press.

Künitzer, A., Basford, D., Craeymeersch, J.A., Dewarumez, J.M., Dorjes, J., Duineveld, G.C.A., Eleftheriou, A., Heip, C., Herman, P., Kingston, P., Niermann, U., Rachor, E., Rumohr, H., de Wilde, P.A.J. (1992). The benthic infauna of the North Sea: species distribution and assemblages. *ICES Journal of Marine Science*, vol. 49, pp. 127–143.

Laakmann, S., Gerdts, G., Erler, R., Knebelsmerger, T., Arbizu P.M.I., Raupach, M.J. (2013). Comparison of molecular species identification for North Sea calanoid copepods (Crustacea) using proteome fingerprints and DNA sequences. *Molecular Ecology Resources*, vol. 13, No. 5: doi: 10.1111/1755-0998.12139.

Laist, D.W., Knowlton, A.E., Mead, J.G., Collet, A.S., Podesta, M. (2001). Collisions between ships and whales. *Marine Mammal Science*, vol. 17, pp. 35–75.

Lambert, Y. (2011). Environmental and fishing limitations to the rebuilding of the northern Gulf of St. Lawrence cod stock (*Gadus morhua*). *Canadian Journal of Fisheries and Aquatic Sciences*, vol. 68, No. 4, pp. 618–631.

Larsson, L.C., Laikre, L., Andre, L., Dahlgren, T.G., Ryman, N. (2010). Temporally stable genetic structure of heavily exploited Atlantic herring (*Clupea harengus*) in Swedish waters. *Heredity*, vol. 104, pp. 40–51.

Lee, D.S., and Mackin, W.A. (2012). *West Indian breeding seabird atlas*. Available from http://www.wicbirds.net/species.html.

Lejeusne, C., Chevaldonné, P., Pergent-Martini, C., Boudouresque, C., Pérez, T. (2010). Climate change effects on a miniature ocean: the highly diverse, highly impacted Mediterranean Sea. *Trends in Ecology and Evolution*, vol. 25, pp. 250–260.

Levin, L.A., Etter, R.J., Rex, M.A., Gooday, A.J., Smith, C.R., Pineda, J., Stuart, C.T., Hessler, R.R., Pawson, D. (2001). Environmental influences on regional deep-sea species diversity. *Annual Review of Ecology and Systematics*, vol. 32, pp. 51–93.

Levin, N., Coll, M., Fraschetti, S., Gal, G., Giakoumi, S., Göke, C., Heymans, J.J., Katsanavakis, S., Mazor, T., Öztürk, B., Rilov, G., Gajewski, J., Steenbeek, J., Kark, S. (2014). Review of biodiversity data requirements for systematic conservation planning in the Mediterranean Sea. *Marine Ecology Progress Series.* In press.

Li, W.K.W., Andersen, R.A., Gifford, D.J., Incze, L.S., Martin, J.L., Pilskaln, C.H., Rooney-Varga, J.N., Sieracki, M.E., Wilson, W.H., Wolff, N.H. (2011). Planktonic microbes in the Gulf of Maine area, *PLoS ONE* 6(6): e20981.

Licandro, P., Conway, D.V.P., Yahia, M.N.D., de Puelles, M.L.F., Gasparini, S., Hecq, J.H., Tranter, P., Kirby, R.R. (2010). A blooming jellyfish in the northeast Atlantic and Mediterranean. *Biology Letters*, vol. 6, No. 5, pp. 688–691.

Lilly, G.R., Nakken, O., Brattey, J. (2013). A review of the contributions of fisheries and climate variability to contrasting dynamics in two Arcto-boreal Atlantic cod (Gadus morhua) stocks: Persistent high productivity in the Barents Sea and collapse on the Newfoundland and Labrador Shelf. *Progress in Oceanography*, vol. 114, pp. 106–125.

Lin, Y., Sheng, J., Greatbatch, R.J. (2012). A numerical study of the circulation and monthly-to-seasonal variability in the Caribbean Sea: the role of Caribbean eddies. *Ocean Dynamics*, vol. 62, pp. 193–211.

Llansó, R.J. (1992). Effects of hypoxia on estuarine benthos: The lower Rappahannock River (Chesapeake Bay), a case study. *Estuarine, Coastal, and Shelf Science*, vol. 35, pp. 491–515.

Lotze, H.K., Coll, M., Dunne, J. (2011). Historical changes in marine resources, food-web structure and ecosystem functioning in the Adriatic Sea. *Ecosystems*, vol. 14, pp. 198–222.

Lotze, H.K., Lenihan, H.S., Bourque, B.J., Bradbury, R.H., Cooke, R.G., Kay, M.C., Kidwell, S.M., Kirby, M.X., Peterson, C.H., Jackson, J.B.C.(2006). Depletion, Degradation, and Recovery Potential of Estuaries and Coastal Seas, *Science*, vol. 312, No. 5781, pp. 1806–1809.

Lotze, H.K., and Worm, B. (2009). Historical baselines for large marine animals. *Trends in Ecology and Evolution*, vol. 24, pp. 254–262.

Lucey, S. M., & Nye, J.A. (2010). Shifting species assemblages in the northeast US continental shelf large marine ecosystem. *Marine Ecology Progress Series*, 415, 23-33.

Lynam, C.P., Gibbons, M.J., Axelsen, B.E., Sparks, C.A.J., Coetzee, J., Heywood, B.G., Brierley, A.S. (2006). Jellyfish overtake fish in a heavily fished ecosystem. *Current Biology*, vol. 16, No. 13, pR492–R493.

MacKenzie, B.R., Alheit, J., Conley, D.J., Holm, P., Kinze, C.C. (2002). Ecological hypothesis for a historical reconstruction of upper trophic level biomass in the Baltic Sea and Skagerrak. *Canadian Journal of Fisheries and Aquatic Sciences*, vol. 59, No. 1, pp. 173–190.

MacKenzie, B.R., Bager, M., Ojaveer, H., Awebro, K., Heino, U., Holm, P., Must, A. (2007). Multi-decadal scale variability in the eastern Baltic cod fishery 1550-1860 — Evidence and causes. *Fisheries Research*, vol. 87, pp. 106–119.

Magoulas, A., Castilho, R., Caetano, S. (2006). Mitochondrial DNA reveals a mosaic pattern of phylogeographical structure in Mediterranean populations of anchovy (*Engraulis encrasicolus*). *Molecular Phylogenetics and Evolution*, vol. 39, pp. 734–746.

Malone, T.C. (1977). *Plankton Systematics and Distributions*. MESA New York Bight Atlas Monograph 13, New York Sea Grant Institute, Albany, New York.

Malone, T.C. (1991). River flow, phytoplankton production and oxygen depletion in Chesapeake Bay. In: Tyson, R.V. and T.H. Pearson (eds.), *Modern and Ancient Continental Shelf Anoxia*. Geological Society Publication No. 58, pp. 83-94.

Marshall, H.G., and Cohn, M.S. (1982). *Seasonal Phytoplankton Assemblages in Northeastern Coastal Waters of the United States*. NOAA Technical Memorandum NMFS-F/NEC-15, 31pp.

Marshall, H.G., Nesius, K.K., Lane, M. (2003). Long-term phytoplankton trends and related water quality trends in the lower Chesapeake Bay, Virginia, US. *Environmental Monitoring and Assessment*, vol. 81, No. 1–3, pp. 349–360.

Marshall, H.G., Burchardt, L., Lacouture, R. (2005). A review of phytoplankton composition within Chesapeake Bay and its tidal estuaries. *Journal of Plankton Research*, vol. 27, No. 11, pp.1083–1102.

Martin, C.S., Giannoulaki, M., De Leo, F., Scardi, M., Salomidi, M., Knitweiss, L., Pace, M.L., Garofalo, G., Gristina, M., Ballesteros, E., Bavestrello, G., Belluscio, A., Cebrian, E., Gerakaris, V., Pergent, G., Pergent-Martini, C., Schembri, P.J., Terribile, K., Rizzo, L., Ben Souissi, J., Bonacorsi, M., Guarnieri, G., Krzelj, M., Macic, V., Punzo, E., Valavanis, V., Fraschetti, S. (2014). Coralligenous and maërl habitats: predictive modelling to identify their spatial distributions across the Mediterranean Sea. *Scientific Reports,* Article no. 5073. 4:10.1038/srep05073.

Martínez, M.L., Intralawan, A., Vázquez, G., Pérez-Maqueo, O., Sutton, P., Landgrave, R. (2007). The coasts of our world: Ecological, economic and social importance. *Ecological Economics*, vol. 63, pp. 254–272.

Mastrototaro, F., Ocaña, O., Zingone, A., Gambi, M.C., Streftaris, N. (2010). Alien species in the Mediterranean Sea by 2010. A contribution to the application of European Union's Marine Strategy Framework Directive (MSFD). Part I. Spatial distribution. *Mediterranean Marine Science*, vol. 11, No. 2, pp. 381–493.

McConaugha, J.R., and Rebach, S. (1993). Crustaceans, In The Ecologically Valuable Species Workgroup of the Living Resources Subcommittee, *Chesapeake Bay Strategy for the Restoration and Protection of Ecologically Valuable Species*, The Marlyand Development of Natural Resources Tidewater Administration Chesapeake Bay Research and Monitoring Division for the Chesapeake Bay Program, US, pp. 65–68.

McClain, C.R., Allen, A.P., Tittensor, D.P., Rex, M.A. (2012). Energetics of life on the deep seafloor. *Proceedings of the National Academy of Sciences*, vol. 109, pp. 15366–15371.

McKinley, A.C., Miskiewicz, A., Taylor, M.D., Johnston, E.L. (2011). Strong links between metal contamination, habitat modification and estuarine larval fish. *Environmental Pollution*, vol. 159, pp. 1499–1509.

McQuatters-Gollop, A., Raitsos, D.E., Edwards, M., Pradhan, Y., Mee, L.D., Lavender, S.J., Attrill, M.J. (2007). A long-term chlorophyll data set reveals regime shift in North Sea phytoplankton biomass unconnected to nutrient trends. *Limnology and Oceanography*, vol. 52, No. 2, pp. 635–648.

Mejri, R., Lo Brutto, S., Hassine, O.K., Arculeo, M. (2009). A study on *Pomatoschistus tortonesei* Miller 1968 (Perciformes, Gobiidae) reveals the Siculo-Tunisian Strait (STS) as a breakpoint to gene flow in the Mediterranean basin. *Molecular Phylogenetics and Evolution*, vol. 53, No. 2, pp. 596–601.

Mendelssohn, I.A., Andersen, G.L., Baltz, D.M., Caffey, A.H., Carman, K.R., Fleeger, J.W., Joye, S.B., Lin, Q., Maltby, E., Overton, E.B., Rozas, L.P. (2012). Oil impacts on coastal wetlands: Implications for the Mississippi River Delta ecosystem after the *Deepwater Horizon* oil spill. *BioScience*, vol. 62, No. 6, pp. 562–574.

Merino-Virgilio, F. del C., Okolodkov, Y.B., Aguilar-Trujillo, A.C., Herrera-Silvera, J.A. (2013). Phytoplankton of the northern coastal and shelf waters of the Yucatan Peninsula, southeastern Gulf of Mexico. *Journal of Species Lists and Distribution*, vol. 9, No. 4, pp. 771–779.

Micheli, F., Halpern, B.S., Walbridge, S., Ciriaco, S., Ferretti, F., Fraschetti, S., Lewison, R., Nykjaer, L., Rosenberg, A.A. (2013). Cumulative human impacts on Mediterranean and Black Sea marine ecosystems: Assessing current pressures and opportunities. *PLoS ONE* 8:e79889.

Middelburg, J.J., and Levin, L.A. (2009). Coastal hypoxia and sediment biogeochemistry. *Biogeosciences Discussion*, vol. 6, pp. 3655–3706.

Miloslavich, P., Díaz, J.M., Klein, E., Alvarado, J.J., Día, C., Gobin, J., Escobar-Briones, E., Cruz-Motta, J.J., Weil, E., Cortés, J., Bastidas, A.C., Robertson, R., Zapata, F., Martin, A., Castillo, J., Kazandjian, A., Ortiz, M. (2010). Marine biodiversity in the Caribbean: regional estimates and distribution patterns. *PLoS ONE* 5(8): e11916. doi:10.1371/journal.pone.0011916.

Möllmann, C. (ed.) (2011). *Regime shifts in marine ecosystems: How overfishing can provoke sudden ecosystem changes*. European Parliament, Directorate General for Internal Policies, Policy Department B: Structural and Cohesion Policies, Fisheries, 170pp.

Moore, S.K., Trainer, V.L., Mantua, N.J., Parker, M.S., Laws, E.A., Backer, L.C., Fleming, L.E. (2008). Impacts of climate variability and future climate change on harmful algal blooms and human health. *Environmental Health*, vol. 7 (Suppl 2): S4, 12 pp.

Mouillot, D., Albouy, C., Guilhaumon, F., Ben Rais Lasram, F., Coll, M., DeVictor, V., Douzery, E., Meynard, C., Pauly, D., Troussellier, M., Velez, L., Watson, R., Mouquet, N. (2011). Protected and threatened components of fish biodiversity in the Mediterranean Sea. *Current Biology*, vol. 21, pp. 1044–1050.

Mumby, P.J., Dahlgren, C.P., Harborne, A.R., Kappel, C.V., Micheli, F., Brumbaugh, D.R., Holmes, K.E., Mendes, J.M., Broad, K., Sanchirico, J.N., Buch, K., Box, S., Stoffle, R.W., Gill, A.B. (2006). Fishing, trophic cascades, and the process of grazing on coral reefs. *Science*, vol. 311, No. 5757, pp. 98–101.

Muñoz, R. C., Currin, C. A., & Whitfield, P. E. (2011). Diet of invasive lionfish on hard bottom reefs of the Southeast USA: insights from stomach. Marine Ecology Progress Series, 432, 181-193.

Musick, J.A., Colvocoresses, J.A., Foell, E.J. (1985). Seasonality and the distribution, availability, and composition of fish assemblages in Chesapeake Bay, In: A.Yáñez-Arancibia (ed.), *Fish Community Ecology in Estuaries and Coastal Lagoons: Towards an Ecosystem Integration*. Universidad Nacional Autónoma de Mexico Press, Mexico City, Mexico, pp. 451–474.

NAFO 2013. Report of the Scientific Council June Meeting 2013. NAFO SCS Doc 13/17. Serial No 6208. 246 pp Halifax, Canada

Najjar, R.G., Pyke, C.R., Adams, M.B., Breitburg, D., Hershner, C., Kemp, M., Howarth, R., Mulholland, M.R., Paolisso, M., Secor, D., Sellner, K., Wardrop, D., Wood, R. (2010). Potential climate-change impacts on the Chesapeake Bay. *Estuarine, Coastal and Shelf Science*, vol. 86, pp. 1–20.

Narayanaswamy, B.E., Renaud, P.E., Duineveld, G.C., Berge, J., Lavaleye, M.S., Reiss, H., Brattegard, T. (2010). Biodiversity trends along the western European margin. *PLoS One* 5, e14295.

National Research Council (NRC) (2001). *Effects of trawling and dredging on seafloor habitat*. National Academy Press, Washington, DC.

Nellen, W., and Thiel, R. (1996). Fische, In: Rheinheimer, G. (ed.), *Meereskunde der Ostsee*. Berlin, Heidelberg, New York (Springer), pp. 190–196.

Newell, R.I.E. (1988). Ecological changes in Chesapeake Bay: Are they the result of overharvesting the American oyster, *Crassostrea virginica*? In: Lynch, M.P., and Krome, E.C., (eds.), *Understanding the estuary: advances in Chesapeake Bay research*. Chesapeake Research Consortium Publ. 129, Gloucester Point, VA, pp. 536–546.

Newell, R.I.E., and Breitburg, D. (1993). Oyster reefs, In: The Ecologically Valuable Species Workgroup of the Living Resources Subcommittee, *Chesapeake Bay Strategy for the Restoration and Protection of Ecologically Valuable Species*, The Maryland Development of Natural Resources Tidewater Administration Chesapeake Bay Research and Monitoring Division for the Chesapeake Bay Program, US, pp. 61–64.

Nizinski, M.S. (2003). Annotated checklist of decapod crustaceans of Atlantic coastal and continental shelf waters of the United States. *Proceedings of the Biological Society of Washington*, vol. 116, No. 1, pp. 96–157.

NOAA (2014). *Stock assessment reports*. Available from http://www.nmfs.noaa.gov/pr/sars/. Accessed 01 May 2014.

Not, F., Zapata, M., Pazos, Y., Campaña, E., Doval, M., Rodríguez, F. (2007). Size-fractionated phytoplankton diversity in the NW Iberian coast: a combination of microscopic, pigment and molecular analyses. *Aquatic Microbial Ecology*, vol. 49, pp. 255–265.

Oguz, T., and Gilbert, D. (2007). Abrupt transitions of the top-down controlled Black Sea pelagic ecosystem during 1960–2000: Evidence for regime-shifts under strong fishery exploitation and nutrient enrichment modulated by climate-induced variations. *Deep-Sea Research I*, vol. 54, pp. 220–242.

Olsen, E.M., Carlson, S.M., Gjøsæter, J., Stenseth, N.C. (2009). Nine decades of decreasing phenotypic variability in Atlantic cod. *Ecology Letters*, vol. 12, pp. 622–631.

Olsson, J., Bergström, L., Gårdmark, A. (2012). Abiotic drivers of coastal fish community change during four decades in the Baltic Sea. *ICES Journal of Marine Science*, vol. 69, pp. 961–970.

Orris, P.K. (1980). A revised species list and commentary on the macroalgae of the Chesapeake Bay in Maryland. *Estuaries*, vol. 3, No. 3, pp. 200–206.

Orth, R.J., and Moore, K.A. (1983). Chesapeake Bay: An unprecedented decline in submerged aquatic vegetation. *Science*, vol. 222, pp. 51–53.

Orth, R.J., and Moore, K.A. (1984). Distribution and abundance of submerged aquatic vegetation in Chesapeake Bay: An historical perspective. *Estuaries*, vol. 7, pp. 531–540.

Orth, R.J., Batiuk, R.A., Nowak, J.F. (1995). *Trends in the Distribution, Abundance, and Habitat quality of Submerged Aquatic Vegetation in Chesapeake Bay and its Tidal Tributaries: 1971-1991*. Chesapeake Bay Program, Annapolis, MD, CBP/TRS 137/95. EPA 903-R-95-009. 187pp.

Orth, R.J., Carruthers, T.J.B., Dennison, W.C., Duarte, C.M., Fourqurean, J.W., Heck, K.L., Hughes, A.R., Kendrick, G.A., Kenworthy, W.J., Olyarnik, S., Short, F.T., Waycott, M., Williams, S.L. (2006). A global crisis for seagrass ecosystems. *BioScience*, vol. 56, pp. 987–996.

Orth, R.J., Simons, J., Capelli, J., Carter, V., Frisch, A., Hindman, L., Hodges, S., Moore, K., Rybicki, N. (l987). *Distribution of submerged aquatic vegetation in the Chesapeake Bay and tributaries and Chincoteague Bay—l986*. Final Report, U.S.E.P.A., 191pp.

OSAPR 2010. OSPAR Quality Status Report 2010 (including regional assessments). Available at http://qsr2010.ospar.org/en/ch01.html.

Ospina-Alvarez, N., Varela, M., Doval, M.D., Gómez-Gesteira, M., Cervantes-Duarte, R., Prego, R. (2014). Outside the paradigm of upwelling rias in NW Iberian Peninsula: Biogeochemical and phytoplankton patterns of a non-upwelling ria. *Estuarine, Coastal and Shelf Science*, vol. 138, pp. 1–13.

Paddack, M.J., Reynolds, J.D., Aguilar, C., Appeldoorn, R.S., Beets, J., Burkett, E.W., Chittaro, P.M., Clarke, K., Rene, E., Fonseca, A.C., Forrester, G.E., Friedlander, A.M., García-Sais, J., González-Sansón, G., Jordan, L.K.B., mcClellan, D.B., Miller, M.W., Molloy, P.P., Mumby, P.J., Nagelkerken, I., Nemeth, M., Navas-Camacho, R, Pitt, J., Polunin, N.V.C., Reyes-Nivia, M.C., Robertson, D.R., Rodríguez-Ramírez, A., Salas, E., Smith, S.R., Spieler, R.E., Steele, m.A., Williams, I.D., Wormald, C.L., Watkinson, A.R., Côté, I.M. (2009). Recent region-wide declines in Caribbean reef fish abundance. *Current Biology*, vol. 19, pp. 590–595.

Park, T.S. (1970). Calanoid copepods from the Caribbean Sea and Gulf of Mexico. 2. New species and new records from plankton samples. *Bulletin of Marine Science*, vol. 20, No. 2, pp. 472–546.

Park, T.S. (1975). Calanoid copepods of the family Euchaetidae from the Gulf of Mexico and Western Caribbean Sea. *Smithsonian Contribution to Zoology*, No. 196, 26pp.

Patarnello, T., Volckaert, F., Castilho, R. (2007). Pillars of Hercules: is the Atlantic-Mediterranean transition a phylogeographical break? *Molecular Ecology*, vol. 16, pp. 4426–4444.

Pato, P., Válega, M., Pereira, E., Vale, C., Duarte, A.C. (2008). Inputs from a Mercury-contaminated lagoon: Impact on the nearshore waters of the Atlantic Ocean. *Journal of Coastal Research*, vol. 24, pp. 28–38.

Pavloudi, C., Oulas, A., Vasileiadou, K., Sarropoulou, E., Kotoulas, G., Arvanitidis, C.V. (in press). Bacterial biodiversity in Mediterranean lagoons: providing metagenetics data for ecological description and environmental impact assessment. *Journal of Sea Research*.

Pendleton, L. (2010). Measuring and Monitoring the Economic Effects of Habitat Restoration: A Summary of a NOAA Blue Ribbon Panel, Available from http://www.era.noaa.gov/pdfs/NOAA%20RAE%20BRP%20Estuary%20Economics_FINAL.pdf.

Permanent Secretariat, (2009). Strategic Action Plan for the Environmental Protection and Rehabilitation of the Black Sea. Available from http://www.blackseacommission.org/_bssap2009.asp.

Perry, A. L., Low, P. J., Ellis, J. R., & Reynolds, J. D. (2005). Climate change and distribution shhifts in marine fishes. science, 308(5730), 1912-1915.

Persson, A., Ljungberg, P., Andersson, M., Götzman, E. and Nilsson, P.A. (2012). Foraging performance of juvenile Atlantic cod Gadus morhua and profitability of coastal habitats. *Marine Ecology Progress Series*, vol. 456, pp. 245–253.

Philine, S.E., Ermgassen, Z., Spalding, M.D., Blake, B., Coen, L.D., Dumbauld, B., Geiger, S., Grabowski, J.H., Grizzle, R., Luckenbach, M., McGraw, K., Rodney, W., Ruesink, J.L., Powers, S.P., Brumbaugh, R. (2012). Historical ecology with real numbers: past and present extent and biomass of an imperiled estuarine habitat. *Proceedings of the Royal Society B-Biological Sciences*, vol. 279, No. 1742, pp. 3393–3400. doi:10.1098/rspb.2012.0313.

Piet, G.J., Jansen, H.M., Rochet, M.-J. (2008). Evaluating potential indicators for an ecosystem approach to fishery management in European waters. *ICES Journal of Marine Science*, vol. 65, pp. 1449–1455.

Pihl, L., Baden, S., Kautsky, N., Rönnback, P., Söderqvist, T., Troell, M., Wennhage, H. (2006). Shift in fish assemblage structure due to loss of seagrass *Zostera marina* habitats in Sweden. *Estuarine, Coastal and Shelf Science*, vol. 67, No. 1–2, pp. 123–132.

Pinsky, M. L., Worm, B., Fogarty, M. J., Sarmiento, J. L., & Levin, S. A. (2013). Marine taxa track local climate velocities. Science, 341(6151), 1239-1242.

Polidoro, B.A., Carpenter, K.E., Collins, L., Duke, N.C., Ellison, A.M., Ellison, J.C., Farnsworth, E.J., Fernando, E.S., Kathiresan, K., Koedam, N.E., Livingstone, S.R., Miyagi, T., Moore, G.E., Nam, V.N., Ong, J.E., Primavera, J.H., Salmo III, S.G., Sanciangco, J.C., Sukardjo, S., Wang, Y., Yong, J.W.H. (2010). The loss of species: mangrove extinction risk and geographic areas of global concern. *PLoS ONE* 5: e10095. doi:10.1371/journal.pone.0010095.

Por, F. (1989). *The Legacy of Tethys: An Aquatic Biogeography of the Levant*. Kluwer Academic Publishers, Dordrecht, 225 pp.

Poulet, S.A., Laabir, M., Chaudron, Y. (1996). Characteristic features of zooplankton in the Bay of Biscay. *Scientia Marina*, vol. 60 (Supl. 2), pp. 79–95.

Poulsen, B., Holm, P., MacKenzie, B.R. (2007). A long-term (1667–1860) perspective on impacts of fishing and environmental variability on fisheries for herring, eel, and whitefish in the Limfjord, Denmark. *Fisheries Research*, vol. 87, pp. 181–195.

Purcell, J. E., Breitburg, D.L., Decker, M.B., Graham, W.M., Youngbluth, M.J., Rastoff, K.A. (2001). Pelagic cnidarians and ctenophores in low dissolved oxygen environments: a review, In Coastal Hypoxia: Consequences for Living Resources and Ecosystems. Rabalais, N.N., and Turner, R.E., eds. *Coastal and Estuarine Studies*, vol. 58, pp. 77–100, American Geophysical Union, Washington, DC.

Pyke, C.R., Najjar, R.G., Adams, M.B., Breitburg, D., Kemp, M., Hershner, C., Howarth, R., Mulholland, M., Paolisso, M., Secor, D., Sellner, K., Wardrop, D., Wood, R. (2008). *Climate Change and the Chesapeake Bay: State-of-the-Science Review and Recommendations*. A Report from the Chesapeake Bay Program Science and Technical Advisory Committee (STAC), Annapolis, MD.

Quintaneiro, C., Monteiro, M., Pastorinho, R., Soares, A.M.V.M., Nogueira, A.J.A., Morgado, F., Guilhermino, L. (2006). Environmental pollution and natural populations: A biomarkers case study from the Iberian Atlantic coast. *Marine Pollution Bulletin*, vol. 52, No. 11, pp. 1406–1413.

Razouls, C., de Bovée, F., Kouwenberg, J., Desreumaux, N. (2014). *Diversity and Geographic Distribution of Marine Planktonic Copepods*. Observatoire Océanologique de Banyuls Université Pierre et Marie Curie (Paris VI) - CNRS/INSU, available at http://copepodes.obs-banyuls.fr/en/index.php.

Record, N.R., Pershing, A.J., Jossi, J.W. (2010). Biodiversity as a dynamic variable in the Gulf of Maine continuous plankton recorder transect. *Journal of Plankton Research*, vol. 32, pp. 1675–1684.

Reeves, R.R., Smith, T., Josephson, E. (2007). Near-annihilation of a species: Right whaling in the North Atlantic, In: Kraus, S.D., and Rolland, R.M. (eds.), *The urban whale: North Atlantic right whales at the crossroads*. Harvard University Press, Cambridge, MA.

Reid, P., de Borges, M., Svendsen, E. (2001). A regime shift in the North Sea circa 1988 linked to changes in the North Sea mackerel fishery. *Fisheries Research*, vol. 50, No. 1–2, pp. 163–171.

Reid, P.C., Johns, D.G., Edwards, M., Starr, M., Poulin, M., Snoeijs, P. (2007). A biological consequence of reducing Arctic ice cover: arrival of the Pacific diatom *Neodenticula semiae* in the North Atlantic for the first time in 800,000 years. *Global Change Biology*, vol. 13, pp. 1910–1921.

Reid, P.M., Edwards, H.H., Warner, A. (1998). Phytoplankton change in the North Atlantic. *Nature*, vol. 391, p. 546.

Reilly, S.B., Bannister, J.L., Best, P.B., Brown, M., Brownell Jr., R.L., Butterworth, D.S., Clapham, P.J., Cooke, J., Donovan, G.P., Urbán, J., Zerbini, A.N. (2008). *Megaptera novaeangliae*, In: IUCN 2013, *IUCN Red List of Threatened Species*. Version 2013.2, available at www.iucnredlist.org. Downloaded on 01 May 2014.

Reise, K., Gollasch, S., Wolff, W.J. (1999). Introduced marine species of the North Sea coasts. *Helgoländer Meeresuntersuchungen*, vol. 52, pp. 219–234.

Reiss, H., Degraer, S., Duineveld, G.C.A, Kröncke, I., Aldridge, J., Craeymeersch, J.A., Eggleton, J.D., Hillewaert, H., Lavaleye, M.S.S., Moll, A., Pohlmann, T., Rachor, E., Robertson, M., Berghe, E.V., Hoey, G.V., Rees, H.L. (2010). Spatial patterns of infauna, epifauna, and demersal fish communities in the North Sea. *ICES Journal of Marine Science*, vol. 67, No. 2, pp. 278–293.

Regular, P.M., Montevecchi, W.A., Hedd, A., Robertson, G.J., Wilhelm, S.I. (2013). Canadian fishery closures provide a large scale test of gillnet bycatch on seabird populations. *Biology Letters*, vol. 9, No. 4. doi:10.1098/rsbl.2013.0088

Renaud, P., Webb, T., Bjørgesæter, A., Karakassis, I., Kedra, M., Kendall, M., Labrune, C., Lampadariou, N., Somerfield, P., Wlodarska-Kowalczuk, M. (2009). Continental-scale patterns in benthic invertebrate diversity: insights from the MacroBen database. *Marine Ecology Progress Series*, vol. 382, pp. 239–252.

Rex, M.A., Etter, R.J., Morris, J.S., Crouse, J., McClain, C.R., Johnson, N.A., Stuart, C.T., Deming, J.W., Thies, R., Avery, R. (2006). Global bathymetric patterns of standing stock and body size in the deep-sea benthos. *Marine Ecology Progress Series*, vol. 317, pp. 1–8.

Rex, M.A., Stuart, C.T., Hessler, R.R., Allen, J.A., Sanders, H.L., Wilson, G.D. (1993). Global-scale latitudinal patterns of species diversity in the deep-sea benthos. *Nature*, vol. 365, pp. 636–639.

Richardson. A.J., Bakun, A., Hays, G.C., Gibbons, M.J. (2009). The jellyfish joyride: causes, consequences and management responses to a more gelatinous future. *Trends in Ecology and Evolution*, vol. 24, No. 6, pp. 312–22.

Rice, J., Arvanitidis, C., Borja, A., Frid, C., Hiddink, J., Krause, J., Lorance, P., Ragnarsson, S.Á., Sköld, M., Trabucco, B. (2010). Marine Strategy Framework Directive – Task Group 6 Report Seafloor integrity. *JRC Scientific and Technical Reports*, 81 pp.

Riisgård, H.U., Bøttiger, L., Madsen, C.V., Purcell, J.E. (2007). Invasive ctenophore *Mnemiopsis leidyi* in Limfjorden (Denmark) in late summer 2007 - assessment of abundance and predation effects. *Aquatic Invasions*, vol. 2, No. 4, pp. 395–401.

Roberts, C.M., McClean, C.J., Veron, J.E.N., Hawkins, J.P., Allen, G.R., McAllister, D.E., Mittermeier, C.G., Schueler, F.W., Spalding, M., Wells, F., Vynne, C., Werner, T.B. (2002). Marine biodiversity hotspots and conservation priorities for tropical reefs, *Science*, vol. 295, No. 5558, pp. 1280–1284.

Roberts, C. (2012). *The ocean of life: the fate of man and the sea*. Viking, New York, 416pp.

Rochet, M.-J., Trenkel, V.M., Carpentier, A., Coppin, F., De Sola, L.G., Léauté, J.-P., Mahé, J.-C., Maiorano, P., Mannini, A., Murenu, M., Piet, G., Politou, C.-Y., Reale, B., Spedicato, M.-T., Tserpes, G., Bertrand, J.A. (2010). Do changes in environmental and fishing pressures impact marine communities? An empirical assessment. *Journal of Applied Ecology*, vol. 47, No. 4, pp. 741–750.

Rodríguez, F., Pazos, Y., Maneiro, J., Zapata, M. (2003). Temporal variation in phytoplankton assemblages and pigment composition at a fixed station of the Ría of Pontevedra (NW Spain). *Estuarine, Coastal and Shelf Science*, vol. 58, pp. 499–515.

Rogers, C. (2009). Coral bleaching and disease should not be underestimated as causes of Caribbean coral reef decline. *Proceedings of the Royal Society B-Biological Sciences*, vol. 276, pp. 197–198.

Rothschild, B.J., Ault, J.S., Goulletquer, P., Hérel, M. (1994). Decline of the Chesapeake Bay oyster population: a century of habitat destruction and overfishing. *Marine Ecology Progress Series*, vol. 111, pp. 29–39.

Roux, J.-P., van der Lingen, C.D., Gibbons, M.J., Moroff, N.E., Shannon, L.J., Smith, A.D.M., Cury, P.M. (2013). Jellyfication of marine ecosystems as a likely consequence of overfishing small pelagic fishes: Lessons from the Benguela. *Bulletin of Marine Science*, vol. 89, No. 1, pp. 249–284.

Ruiz, G.M., Hines, A.H., Posey, M.H. (1993). Shallow water as a refuge habitat for fish and crustaceans in non-vegetated estuaries. An example from the Chesapeake Bay. *Marine Ecology Progress Series*, vol. 99, pp. 1–16.

Runge, J.A., and Jones, R.J. (2012). Results of a collaborative project to observe coastal zooplankton and ichthyoplankton abundance and diversity in the Western Gulf of Maine: 2003–2008. *American Fisheries Society Symposium* 79, 16pp.

Ruppert, J.L.W., Fortin, M.-J., Rose, G.A., Devillers, R. (2010). Environmental mediation of Atlantic cod on fish community composition: an application of multivariate regression tree analysis to exploited marine ecosystems. *Marine Ecology Progress Series*, vol. 411, pp. 189–201.

Ryan, W.B.F., Pitman III, W.C., Major, C.O., Shimkus, K., Moskalenko, V., Jones, G.A., Dimitrov, P., Gorur, N., Sakinc, M., Yuce, H. (1997). An abrupt drowning of the Black Sea shelf. *Marine Geology*, vol. 138, No. 1–2, pp. 119–126.

Sadovy, Y. (1999). The case of the disappearing grouper: *Epinephelus striatus*, the Nassau grouper, in the Caribbean and Western Atlantic. *Proceedings of the Gulf and Caribbean Fisheries Institute*, vol. 45, pp. 5–22.

Salas, S., Chuenpagdee, R., Charles, A., Seijo, J.C. (2011). *Coastal fisheries of Latin America and the Caribbean*. FAO Fisheries and Aquaculture Technical Paper No. 544, 430pp.

Sanford, E., Gaylord, B., Hettinger, A., Lenz, E.A., Meyer, K., Hill, T.M. (2014). Ocean acidification increases the vulnerability of native oysters to predation by invasive snails. *Proceedings of the Royal Society B-Biological Sciences*, vol. 281. No. 1778. DOI: 10.1098/rspb.2013.2681.

Savenkoff, C., Morissette, L., Castonguay, M., Swain, D.P., Hammill, M.O., Chabot, D., Hanson, J.M. (2008). Interactions between marine mammals and fisheries: Implication for cod recovery. In: Chen, C., and Guo, J. (eds.), *Ecosystem Ecology Research Trends*, Nova Science Publishers, Inc., Hauppauge, NY, US.

Savard, L. (2013). *Update of bycatch in the Estuary and Gulf of St. Lawrence Northern shrimp fishery in 2012*. Canadian Science Advisory Secretariat (CSAS), Research Document 2013/010.

Schofield, P.J. (2010). Update on geographic spread of invasive lionfishes (*Pterois volitans* [Linnaeus, 1758] and *P. miles* [Bennet, 1828]) in the Western North Atlantic Ocean, Caribbean Sea and Gulf of Mexico. *Aquatic Invasions*, vol. 5 (suppl. 1), S117–S122.

Schreiber, R., and Schreiber, E.A. (1984). Central Pacific Seabirds and the El Niño Southern Ocean Oscillation: 1892-1983 Perspectives. *Science*, vol. 225, No. 4663, pp. 713–716.

Schreiber, E.A., and Lee, D.S. (eds.) (2000). *Status and conservation of West Indian seabirds*. Society of Caribbean Ornithology, Special Publication 1, Ruston, Louisiana, 10pp.

Secor, D.H., Niklitschek, E., Stevenson, J., Gunderson, T., Minkkinnen, S., Richardson, B., Florence, B., Mangold, M., Skjeveland, J., Henderson-Arzapalo, A. (2000). Dispersal and growth of yearling Atlantic sturgeon, *Acipenser oxyrinchus* released into Chesapeake Bay. *Fishery Bulletin*, vol. 98, No. 4, pp. 800–810.

Sellner, K., and Marshall, H.G. (1993). Phytoplankton, In: The Ecologically Valuable Species Workgroup of the Living Resources Subcommittee, *Chesapeake Bay Strategy for the Restoration and Protection of Ecologically Valuable Species*, The Marlyand Development of Natural Resources Tidewater Administration Chesapeake Bay Research and Monitoring Division for the Chesapeake Bay Program, US, pp. 35–44.

Sellner, K.G., Doucette, G.J., Kirkpatrick, G.J. (2003). Harmful algal blooms: Causes, impacts and detection. *Journal of Industrial Microbiology and Biotechnology*, vol. 30, No. 7, pp. 383–406.

Shackell, N.L., and Frank, K.T. (2003). Marine fish diversity on the Scotian Shelf, Canada. *Aquatic Conservation: Marine and Freshwater Ecosystems*, vol. 13, No. 4, pp. 305–321.

Shackell, N.L., Fisher, J.A.D., Frank, K.T., Lawton, P. (2012). Spatial scale of similarity as an indicator of meta-community stability in exploited marine systems. *Ecological Applications*, vol. 22, pp. 336–348.

Shackell, N. L., Bundy, A., Nye, J. A., and Link, J. S. 2012. Common large-scale responses to climate and fishing across Northwest Atlantic ecosystems. ICES Journal of Marine Science, 59:151-162.

Shiganova, T., and Ozturk, B. (2010). Northward movement of species. Trend on increasing Mediterranean species arrival into the Black Sea, In: Commission on the Protection of the Black Sea against Pollution Permanent Secretariat, *Diagnostic Report to guide improvements to the regular reporting process on the state of the Black Sea Environment*.

Simpson, S.D., Jennings, S., Johnson, M.P., Blanchard, J.L., Schön, P.-J., Sims, D.W., Genner, M.J. (2011). Continental shelf-wide response of a fish assemblage to rapid warming of the sea, *Current Biology*, vol. 21, No. 18, pp. 1565–1570.

Skjoldal, H.R., and Dundas, I. (1991). The *Chrysochromulina polylepis* bloom in the Skagerrak and the Kattegat in May–June 1988: Environmental conditions, possible causes and effects. *ICES Cooperative Research Report*, vol. 175, pp. 1–59.

Skolka, M., and Gomoiu, M.-T. (2004). *Specii invazive în Marea Neagră: Impactul ecologic al pătrunderii de noi specii în ecosistemele acvatice*, Ovidius University Press, Constanța, 180pp.

Smith, C.R., De Leo, F.C., Bernardino, A.F., Sweetman, A.K., Arbizu, P.M. (2008). Abyssal food limitation, ecosystem structure and climate change. *Trends in Ecology and Evolution*, vol. 23, pp. 518–528.

Solimini, A.G., Cardoso, A.C., Heiskanen, A.S. (2006). *Indicators and Methods for the Ecological Status Assessment under the Water Framework Directive: Linkages between Chemical and Biological Quality of Surface Waters*. Institute for Environment and Sustainability, Joint Research Center, European Communities, Ispra, 262pp.

Somerfield, P.J., Arvanitidis, C., Faulwetter, S., Chatzigeorgiou, G., Vasileiadou, A., Amouroux, J., Anisimova, N., Cochrane, S., Craeymeersch, J., Dahle, S. (2009). Assessing evidence for random assembly of marine benthic communities from regional species pools. *Marine Ecology Progress Series*, vol. 382, pp. 279–286.

Sonnewald, M., and Turkay, M. (2012). Environmental influence on the bottom and near-bottom megafauna communities of the Dogger Bank: a long-term survey. *Helgoland Marine Research*, vol. 66, No. 4, pp. 503–511.

SOTO (2012). *Canada's State of the Oceans Report*, 2012. Fisheries and Oceans Canada Ottawa, Ontario K1A 0E6, DFO/2012-1818.

Sousa, P., Azevedo, M., Gomes, M.C. (2006). Species-richness patterns in space, depth, and time (1989-1999) of the Portuguese fauna sampled by bottom trawl. *Aquatic Living Resources proofs*, vol. 19, pp. 93–103.

Staples, D. & Funge-Smith, S. (2009) Ecosystem approach to fisheries and aquaculture: Implementing the FAO Code of Conduct for Responsible Fisheries. FAO Regional Office for Asia and the Pacific, Bangkok, Thailand. RAP Publication 2009/11, 48 pp.

Steele, J.H. (2004). Regime shifts in the ocean: reconciling observations and theory. *Progress in Oceanography*, vol. 60, pp. 135–141.

Stefansdottir, L., Solmundsson, J., Marteinsdottir, G., Kristinsson, K., Jonasson, J.P. (2010). Groundfish species diversity and assemblage structure in Icelandic waters during recent years of warming, *Fisheries Oceanography*, vol. 19, No. 1, pp. 42–62.

Steinacher, M., Joos, F., Frölicher, T., Bopp, L., Cadule, P., Cocco, V., Doney, S.C., Gehlen, M., Lindsay, K., Moore, J.K. (2010). Projected 21st century decrease in marine productivity: a multi-model analysis. *Biogeosciences*, vol. 7, pp. 979–1005. doi:10.5194/bg-7-979-2010.

Stevenson, J.C., and Pendleton, E. (1993). Wetlands, In: The Ecologically Valuable Species Workgroup of the Living Resources Subcommittee, *Chesapeake Bay Strategy for the Restoration and Protection of Ecologically Valuable Species*, The Marlyand Development of Natural Resources Tidewater Administration Chesapeake Bay Research and Monitoring Division for the Chesapeake Bay Program, US, pp. 83–88.

Stramma, L., Schmidtko, S., Levin, L.A., Johnson, G.C. (2010). Ocean oxygen minima expansions and their biological impacts. *Deep-Sea Research*, vol. 57, pp. 587–595.

Sundseth, K., and Barova, S. (2009). *The Black Sea Region*. European Communities, Luexmbourg.

Surugiu, V., Revkov, N., Todorova, V., Papageorgiou, N., Valavanis, V., Arvanitidis, C. (2010). Spatial patterns of biodiversity in the Black Sea: An assessment using benthic polychaetes. *Estuarine, Coastal and Shelf Science*, vol. 88, pp. 165-174.

Taylor, B.L., Baird, R., Barlow, J., Dawson, S.M., Ford, J., Mead, J.G., Notarbartolo di Sciara, G., Wade, P., Pitman, R.L. (2008). *Globicephala melas*, In: IUCN 2013, *IUCN Red List of Threatened Species*. Version 2013.2, available at www.iucnredlist.org. Downloaded on 01 May 2014.

ter Hofstede, R., Hiddink, J.G., Rijnsdorp, A.D. (2010). Regional warming changes fish species richness in the eastern North Atlantic Ocean. *Marine Ecology Progress Series*, vol. 414, pp. 1–9.

ter Hofstede, R., and Rijnsdorp, A.D. (2011). Comparing demersal fish assemblages between periods of contrasting climate and fishing pressure. *ICES Journal of Marine Science: Journal du Conseil*, vol. 68, No. 6, pp. 1189–1198.

Thatje, S. (2005). The future fate of the Antarctic marine biota? *Trends in Ecology and Evolution*, vol. 20, pp. 418–419.

Thiede, J. (1975). *Relative abundance of pteropod species of the plankton pump samples in the surface water of the eastern North Atlantic Ocean* (Table 4),doi:10.1594/PANGAEA.510774.

Supplement to: Thiede, J. (1975). *Shell- and skeleton-producing plankton and nekton in the eastern North Atlantic Ocean*. Meteor Forschungsergebnisse, Deutsche Forschungsgemeinschaft, Reihe C Geologie und Geophysik, Gebrüder Bornträger, Berlin, Stuttgart, C20, pp. 33–79.

Titus, J.G., and Strange, E.M. (eds.) (2008). *Background Documents Supporting Climate Change Science Program Synthesis and Assessment Product 4.1: Coastal Elevations and Sensitivity to Sea Level Rise*. EPA 430R07004, U.S. EPA, Washington, DC.

Tsagarakis, K., Mytilineou, C., Haralabous, J., Lorance, P., Politou, C.-Y., Dokos, J. (2013). Mesoscale spatio-temporal dynamics of demersal assemblages of the Eastern Ionian Sea in relationship with natural and fisheries factors, *Aquatic Living Resources*, vol. 26, pp. 381–397.

UNEP (2005). *Caribbean environment outlook: special edition for the Mauritius International Meeting for the 10-year review of the Barbados Programme of Action for the sustainable development of small island developing states* (ed. S. Heileman). United Nations Environment Programme, Nairobi, Kenya, 114pp.

USACE (2009). *Final environmental assessment and finding of no significant impact: Chesapeake Bay oyster restoration using alternate substrate*. Dept. of the Army, Corp of Engineers, Baltimore, MD.

Valdés, L., López-Urrutia, A., Cabal, J., Alvarez-Ossorio, M., Bode, A., Miranda, A., Cabanas, M., Huskin, I., Anadón, R., Alvarez-Marqués, F., Llope, M., Rodríguez, N. (2007). A decade of sampling in the Bay of Biscay: What are the zooplankton time series telling us? *Progress in Oceanography*, vol. 74, pp. 98–114.

Valdimarsson, H., Astthorsson, O.S., Palsson, J. (2012). Hydrographic variability in Icelandic waters during recent decades and related changes in distribution of some fish species. *ICES Journal of Marine Science*, vol. 69, pp. 816–825.

Vanden Berghe, E., Claus, S., Appeltans, W., Faulwetter, S., Arvanitidis, C., Somerfield, P.J., Aleffi, I.F., Amouroux, J.M., Anisimova, N., Bachelet, G., Cochrane, S.J., Costello, M.J., Craeymeersch, J., Dahle, S., Degraer, S., Denisenko, S., Dounas, C., Duineveld, G., Emblow, C., Escaravage, V., Fabri, M.C., Fleischer, D., Gremare, A., Herrmann, M., Hummel, H., Karakassis, I., Kedra, M., Kendall, M.A., Kingston, P., Kotwicki, L., Labrune, C., Laudien, J., Nevrova, E.L., Occhipinti-Ambrogi, A., Olsgard, F., Palerud, R., Petrov, A., Rachor, E., Revkov, N., Rumohr, H., Sarda, R., Sistermans, W.C.H., Speybroeck, J., Janas, U., Van Hoey, G., Vincx, M., Whomersley, P., Willems, W., Wlodarska-Kowalczuk, M., Zenetos, A., Zettler, M.L., Heip, C.H.R. (2009). MacroBen integrated database on benthic invertebrates of European continental shelves: a tool for large-scale analysis across Europe. *Marine Ecology Progress Series*, vol. 382, pp. 225–238.

Van Ginderdeuren, K., Fiers, F., de Backer, A., Vincx, M., Hostens, K. (2012). Updating the zooplankton species list for the Belgian part of the North Sea. *Belgian Journal of Zoology*, vol. 142, No. 1, pp. 3–22.

van Tussenbroek, B.I., Cortés, J., Collin, R., Fonseca, A.C., Gayle, P.M.H., Guzmán, H.M., Jácome, G.E., Juman, R., Koltes, K.H., Oxenford, H.A., Rodríguez-Ramirez, A., Samper-Villarreal, J., Smith, S.R., Tschirky, J.J., Weil, E. (2014). Caribbean-wide, long-term study of seagrass beds reveals local variations, shifts in community structure and occasional collapse. *PLoS ONE* 9(3): e90600. doi:10.1371/journal.

pone.0090600.

van Tussenbroek, B.I., Santos, M.G.B., Wong, J.G.R., van Dijk, J.K., Waycott, M. (2010). *A guide to the tropical seagrasses of the Western Atlantic*. Universidad Nacional Autónoma de Mexico, 79pp.

Wagner, M.C. (1999). Expression of the estuarine species minimum in littoral fish assemblages of the lower Chesapeake Bay tributaries. *Estuaries*, vol. 22, pp. 304–312.

Wagner, C.M., and Austin, H.M. (1999). Correspondence between environmental gradients and summer littoral fish assemblages in low salinity reaches of the Chesapeake Bay, USA. *Marine Ecology Progress Series*, vol. 177, pp. 197–212.

Waldbusser, G.G., Voigt, E.P., Bergschneider, H., Green, M.A., Newell, R.I.E. (2011). Biocalcification in the eastern oyster (*Crassostrea virginica*) in relation to long-term trends in Chesapeake Bay pH. *Estuaries and Coasts*, vol. 34, pp. 221–231.

Wang, K., Wommack, K.E.W., Chen, F. (2011). Abundance and distribution of *Synechococcus* spp. and Cyanophages in the Chesapeake Bay. *Applied and Environmental Microbiology*, vol. 77, No. 21, pp. 7459–7488.

Wasmund, N., Postel, L., Zettler, M.L. (2012). Biologische Bedingungen in der deutschen ausschließlichen Wirtschaftszone der Nordsee im Jahre 2011. *Meereswissenschaftliche Berichte*, Warnemünde, vol. 90, pp. 1–103.

Waycott, M., Duarte, C.M., Carruthers, T.J.B., Orth, R.J., Dennison, W.C., Olyarnik, S., Calladine, A., Fourqurean, J.W., Heck, Jr., K.L., Hughes, A.R., Kendrick, G.A., Kenworthy, W.J., Short, F.T., Williams, S.L. (2009). Accelerating loss of seagrass across the globe threatens coastal ecosystems. *Proceedings of the National Academy of Sciences of the United States of America*, vol. 106, pp. 12377–12381.

Webb, T.J., Aleffi, I.F., Amouroux, J.M., Bachelet, G., Degraer, S., Dounas, C., Fleischer, D., Gremare, A., Herrmann, M., Hummel, H., Karakassis, I., Kedra, M., Kendall, M.A., Kotwicki, L., Labrune, C., Nevrova, E.L., Occhipinti-Ambrogi, A., Petrov, A., Revkov, N.K., Sarda, R., Simboura, N., Speybroeck, J., Van Hoey, G., Vincx, M., Whomersley, P., Willems, W., Wlodarska-Kowalczuk, M. (2009). Macroecology of the European soft sediment benthos: insights from the MacroBen database. *Marine Ecology Progress Series*, vol. 382, pp. 287–296.

Whitehead, H., 2013. Trends in cetacean abundance in the Gully submarine canyon, 1988–2011, highlight a 21% per year increase in Sowerby's beaked whales (*Mesoplodon bidens*). *Canadian Journal of Zoology*, 2013, 91(3): 141-148, 10.1139/cjz-2012-0293

Whitfield, P.E., Hare, J.A., David, A.W., Harter, S.L., Munoz, R.C., & Addison, C.M. (2007). Abundance estimates of the Indo-Pacific lionfish Pterois volitans/miles complex in the Western North Atlantic. Biological Invasions, 9(1), 53-64.

Wilberg, M.J., Livings, M.E., Barkman, J.S., Morris, B.T., Robinson, J.M. (2011). Overfishing, disease, habitat loss, and potential extirpation of oysters in upper Chesapeake Bay. *Marine Ecology Progress Series*, vol. 436, pp. 131–144.

Wilkinson, C., and Souter, D. (eds.) (2008). *Status of Caribbean coral reefs after bleaching and hurricanes in 2005*. Global Coral Reef Monitoring Network, and Reef and Rainforest Research Centre, Townsville, Australia, 152pp.

Willette, D.A., Chalifour, J., Dolfi Debrot, A.O., Engel, S., Miller, J., Oxenford, H.A., Short, F.T., Steiner, S.C.C., Védie, F. (2014). Continued expansion of the trans-Atlantic invasive marine angiosperm *Halophila stipulacea* in the Eastern Caribbean. *Aquatic Botany*, vol. 112, pp. 98–102.

Woodland, R.J., Secor, D.H., Fabrizio, M.C., Wilberg, M.J. (2012). Comparing the nursery role of inner continental shelf and estuarine habitats for temperate marine fishes. *Estuarine, Coastal and Shelf Science*, vol. 99, pp. 61–73.

Worm, B., Barbier, E.B., Beaumont, N., Duffy, J.E., Folke, C., Halpern, B.S., Jackson, J.B.C., Lotze, H.K., Micheli, F., Palumbi, S.R., Sala, E., Selkoe, K.A., Stachowicz, J.J., Watson, R. (2006). Impacts of biodiversity loss on ocean ecosystem services. *Science*, vol. 314, No. 5800, pp. 787–790.

Worm, B., Hilborn, R., Baum, J.K., Branch, T.A., Collie, J.S., Costello, C., Fogarty, M.J., Fulton, E.A., Hutchings, J.A., Jennings, S., Jensen, O.P., Lotze, H.K., Mace, P.M., McClanahan, T.R., Minto, C., Palumbi, S.R., Parma, A.M., Ricard, D., Rosenberg, A.A., Watson, R., Zeller, D. (2009). Rebuilding Global Fisheries. *Science*, vol. 325, No. 5940, pp. 578–584.

Wyda, J.C., Deegan, L.A., Hughes, J.E., Weaver, M.J. (2002). The response of fishes to submerged aquatic vegetation complexity in two ecoregions of the mid-Atlantic Bight: Buzzards Bay and Chesapeake Bay. *Estuaries*, vol. 25, pp. 86–100.

York, P.H., Kelaher, B.P., Booth, D.J., Bishop, M.J. (2012). Trophic responses to nutrient enrichment in a temperate seagrass food chain. *Marine Ecology Progress Series*, vol. 449, pp. 291–296.

Zenetos, A., Gofas, S., Verlaque, M., Cinar, M.E., Raso, G., Bianchi, C.N., Morri, C., Azzurro, E., Bilecenoglu, M., Froglia, C., Siokou, I., Violanti, D., Sfriso, A., San Martín, G., Giangrande, A., Katağan, T., Ballesteros, E., Ramos-Esplá, A., Mastrototaro, F., Ocaña, O., Zingone, A., Gambi, M.C., Streftaris, N. (2010). Alien species in the Mediterranean Sea by 2010. A contribution to the application of European Union's Marine Strategy Framework Directive (MSFD). Part I. Spatial distribution. *Mediterranean Marine Science*, vol. 11, pp. 381–493.

Zydelis, R., Bellebaum, J., Osterblom, H., Vetemaa, M., Schirmeister, B., Stipniece, A., Dagys, M., van Eerden, M., Garthe, S. (2009). Bycatch in gillnet fisheries – An overlooked threat to waterbird populations. *Biological Conservation*, vol. 142, No. 7, pp. 1269–1281.

Also:

Centre for Marine Biodiversity (CMB) http://www.marinebiodiversity.ca/cmb.

Extra References for Gulf of Mexico:

Literature cited – all references originally provided were NOT listed in manuscript

Briggs, J. C. (1974). *Marine Zoogeography*. McGraw-Hill, New York. 480 pp.

Brooks, J. M., Fisher, C., Roberts, H., Bernard, B., McDonald, I., Carney, R., Joye, S., Cordes, E., Wolff, G., Goehring, E. (2008). *Investigations of Chemosynthetic Communities on the Lower Continental Slope of the Gulf of Mexico. Interim Report 1*. U.S. Dept. of the Interior, Minerals Management Service, Gulf of Mexico OCS Region, New Orleans, Louisiana. OCS Study MMS 2008-009. 332 pp.

Carney, R. S. (1993). *Review and Reexamination of OCS Spatial-temporal Variability as Determined by MMS Studies in the Gulf of Mexico. OCS Study MMS 93-0041*. New Orleans, Louisiana: U.S. Department of the Interior, Minerals Management Service, Gulf of Mexico OCS Regional Office, New Orleans, Louisiana.

Cordes, E., McGinley, E., Podowski, M.P., Becker, E.L., Lessard-Pilon, S., Viada, S.T., Fisher, C.R. (2008). Coral communities of the deep Gulf of Mexico. *Deep-Sea Research I*, vol. 55, pp. 777–787.

CSA International, Inc. (2007). *Characterization of Northern Gulf of Mexico Deepwater Hard Bottom Communities with Emphasis on Lophelia Coral*. U.S. Department of the Interior, Minerals Management Service, Gulf of Mexico OCS Region, New Orleans, Louisiana. OCS Study MMS 2007-044. 169 pp. + app.

Darnell, R.M., Defenbaugh, R.E. (1990). Gulf of Mexico: Environmental Overview and History of Environmental Research. *American Zoologist*, vol. 30, pp. 3-6.

Davis, R.W., Evans, W.E., Würsig, B. (eds.). (2000). *Cetaceans, Sea Turtles and Seabirds in the Northern Gulf of Mexico: Distribution, Abundance and Habitat Associations*. Vol. II. Technical Report. Prepared by Texas A&M University at Galveston and the National Marine Fisheries Service. U.S. Department of the Interior, Geological Survey, Biological Resources Division, USGS/BRD/CR-1999–0006 and Minerals Management Service, Gulf of Mexico OCS Region, New Orleans, Louisiana, OCS Study MMS 2000–003. 346pp.

Ellis S.L., Incze, L.S., Lawton, P., Ojaveer, H., MacKenzie, B.R., Pitcher, C.R., Shirley, T.C., Eero, M., Tunnell, Jr., J.W., Doherty, P.J., Zeller, B.M. (2011). Four regional marine biodiversity studies: Approaches and contributions to ecosystem-based management. *PLoS ONE*, vol. 6, No. 4: e18997. doi:10.1371/journal.pone.0018997.

Fautin, D., Dalton, P., Incze, L.S., Leong, J.A.C., Pautzke, C., Rosenberg, A., Sandifier, P., Sedberry, G., Tunnell, Jr., J.W., Abbott, I., Brainard, R.E., Brodeur, M., Eldredge, L.G., Feldman, M., Moretzsohn, F., Vroom, P.S., Wainstein, M., Wolff, N. (2010). An overview of marine biodiversity in United States waters. *PLoS ONE*, vol. 5, No. 8: e11914. doi:10.1371/journal.pone.0011914.

Felder, D.L., Camp, D.K. (eds.) (2009). *Gulf of Mexico—Origin, Waters, and Biota*. Vol. 1. Biodiversity. Texas A&M University Press, College Station, Texas. 1393 pp.

Harper, Jr., D.E. (1991). Macrofauna and macroepifauna. In: Brooks, J. M. (ed) *Mississippi-Alabama Continental Shelf Ecosystem Study Data summary and synthesis*. Vol. II. Technical. U.S. Department of the Interior, Minerals Management Service, Gulf of Mexico OCS Region.

36B South Atlantic Ocean

Contributors:
Enrique Marschoff (Convenor and Lead Member), Javier Calcagno, Beatrice Padovani Ferreira (Co-Lead Member), Flavia Lucena-Frédou, Monica Muelbert, Angel Perez, Andrea Raya Rey, Laura Schejter and Alexander Turra

1 Introduction

In this chapter we refer to the area of the Atlantic Ocean south of the Equator and north of the Polar Front (Antarctic Convergence). The main topographical feature in the South Atlantic is the Mid-Atlantic Ridge which runs between Africa and South America from approximately 58° South to Iceland in the north. A rift valley is associated with the Ridge. The Ridge is of volcanic origin and the development of transverse ridges creates a number of basins: the Argentine, Brazil, Guinea, Angola and Cape Basins.

The Atlantic coast of South America is influenced by three major rivers, Orinoco, Amazon and La Plata, that discharge large amounts of freshwater and sediment into the Atlantic Ocean. The Amazon discharges about one-fifth of the world's total freshwater runoff into the Atlantic (Curtin, 1986) and it is transported offshore up to 500 km seaward (Lentz, 1995). The heavy sediment discharge ($2.9 \cdot 10^8$ tons year)$^{-1}$ is not deposited over the outer shelf, but is carried by the North Brazil Current to Guyana's shelf, where it forms extensive mud deposits (Gratiot et al., 2008). The continental shelf is wider along its West Coast, both in the north at the Amazon (≈300 km) and in southern Argentina, where it reaches up to 600 kilometres (Miloslavich et al., 2011). The shelf is narrower along the East Coast of the Atlantic and also along the east coast of Brazil, where riverine muds give way to calcareous deposits and the shelf in some areas reaches a minimum of 8 km width (Miloslavich et al., 2011).

The continental slope is cut by deep canyons connecting shelf and deep waters. High benthic richness was reported at the head of the submarine canyons, and about half of the species are shared with the shelf-break community (Bertolino et al., 2007; Schejter et al., 2014b). The ~7500 km of the Brazil coasts comprise a combination of freshwater, estuarine and marine ecosystems, with diverse but poorly known habitats in its northern part and with sandy beaches, mangrove forests, rocky shores, lagoons and coral reefs to the south (Miloslavich et al., 2011). Uruguay's coasts are dominated by sandy beaches; a narrow rocky portion has high biodiversity (Calliari et al., 2003). The coasts of Argentina are mostly sandy beaches, with some rocky formations located mainly at Mar del Plata, Peninsula Valdes and Tierra del Fuego; pebble beaches are common in Patagonia. The coasts of South Africa are part sandy beaches, rocks and rocks mixed with sand on the upper shore and a wave-cut rocky platform (Bally et al., 1984).

South Atlantic waters are characterized by the counterclockwise central subtropical gyre of surface and intermediate waters running close to South America and South Africa, with more complex currents developing on the coasts of both continents (Campos et al., 1995; McDonagh and King, 2005). The gyre is approximately 4,500 km in diameter and lies between the equator and 40° S (Piontkovski et al., 2003). The eastern (African) branch transfers warm water to the northern hemisphere and the western (South American) branch carries cold water from the north, both as part of the global circulation known as the Atlantic Meridional Overturning Circulation. The gyre is closed to the south by the subantarctic branch of the Antarctic Circumpolar Current (West Wind Drift). The gyre constitutes a biogeographic province with distinct physical and biological properties relative to adjacent regions (Longhurst, 1998).

The currents, closely linked to the topography, result in a series of surface fronts with varying depth expressions associated with the different ridges which effectively limit exchange of water among the different basins. The Benguela Current runs northward off the African continent carrying cold water towards the equator, with a coastal branch close to the continent (Griffiths et al., 2010). Its counterpart near South America is the Brazil current, which meets the northward current derived from the Antarctic Circumpolar Current (West Wind Drift) running close to the Patagonian shelf (see chapter 36H, Figures 1 to 3).

The offshore waters in the South Atlantic Gyre occupy more than half of the South Atlantic Ocean; the Gyre is characterized by low concentrations of nutrients, and low phytoplankton and zooplankton abundances. Along the South American continental shelf, circulation patterns are modified by topography, upwelling and continental runoff (Gonzalez-Silvera et al., 2004). The South Subtropical Convergence is the intersection point of low-macronutrient subtropical gyre waters and high-macronutrient Antarctic Circumpolar Current waters and therefore represents one of the most dynamic nutrient environments in the oceans (Ito et al., 2005).

On the African shelf, as in the South American coasts, the Agulhas retroversion, and the Benguela and the Angola Currents are responsible for upwelling processes and enhanced primary production (Lutjeharms and Ballegooyen, 1988). The different water masses resulting from these processes offer an ample variety of habitats for pelagic biodiversity. The Congo and, on the western side, the Amazon and the Plata plumes can also seasonally contribute to the enhancement of plankton, thus becoming one of the most productive marine areas of the Atlantic Ocean, which is essentially due to the presence of a great variety of frontal productive systems (Acha et al., 2004; da Cunha and Buitenhuis, 2013).

On the western coasts and shelves, the plumes of the Amazon and of the De la Plata Rivers extend up to 1000 km into the ocean, modifying coastal waters. The De la Plata River plume impinges on coastal and shelf waters of Argentina, Uruguay and Brazil (Muelbert et al., 2008). Lateral mixing results in a water mass typical of the De la Plata plume waters (Piola et al., 2008) with high nutrient concentrations and primary productivity. The Amazon plume influence is felt on the northwest coast of Brazil and modifies the vertical structure of the Equatorial West Ocean (Hu et al., 2004).

Modelling and observed trends predict that the low-productivity subtropical gyre will expand as a result of climate warming. Such an increase will affect not only the plankton communities, but also fisheries on both sides of the South Atlantic Ocean. Since 1998, the gyre has been expanding at average rates between 0.8 per cent/yr and 4.3 per cent/yr. The rate of expansion is greater during the winter (Polovina et al., 2008; Henson et al., 2010).

2 Plankton and Nekton

Water-mass properties and movements are the main factors behind plankton distribution in the South Atlantic. The quantitative distribution of pelagic life in the South Atlantic parallels those found in the other oceans: a large area of poor central waters bound to the north and south by richer equatorial and subpolar bands, respectively, with the biologically richest sectors found in the coastal regions, especially along Africa (Boltovskoy et al., 1999). These conditions are exacerbated at the inshore component of the Benguela Current, which exhibits strong a wind-driven upwelling with a periodicity of 5 to 10 days; more intense in summer (Shannon and Nelson, 1996). Much of the organic matter associated with this high productivity sinks onto the relatively wide continental shelf, where decay results in the reduction of dissolved oxygen in bottom waters. Periodically, these low-oxygen conditions extend close inshore, sometimes reaching the shoreline itself and resulting in mass mortalities of fish, rock lobster, and other invertebrates (Griffiths et al., 2010).

In general, knowledge of the plankton taxonomy and ecology is relatively scarce in the South Atlantic, in particular the oligotrophic waters of the South Atlantic gyre are very little known; little sampling has been conducted there (Piontkovski et al., 2003). Table 36B.1 presents a summary of the relevant parameters for different regions of the South Atlantic Ocean.

Primary production values in the central gyre range around >0.1-0.2 g C m-2 d^{-1}, with phytoplankton concentrations below 103 cells l^{-1}. In the vicinity of the equator, biological richness is enhanced by equatorial divergence and by seaward advection of nutrient- and biomass-rich Benguela upwelling waters (Boltovskoy et al., 1999). A 30-fold difference in mean surface chlorophyll and a 100-fold difference in phytoplankton biomass were found between the centre of the gyre and the Benguela Current waters (Piontkovski et al. 2003). The high productivity region extends to the north along the African coast. In the Southwestern Atlantic, primary production and chlorophyll-a measurements show a number of areas of enhanced phytoplankton output. Diatoms play an important role in this biomass build-up (Olguin et al. 2006).

Diversity increases in oligotrophic waters; the abundance in the Antarctic Circumpolar Current decreases from West to East, whereas the diversity index increases. In the Benguela Current, the number of species is similar to that in the Antarctic Circumpolar Current, but the diversity index is lower; the taxonomic composition of the South Equatorial Current and the central gyre are similar (Greze, 1984). South of the Subtropical Convergence, the primary limiting nutrient is Fe, whereas to the north the phytoplankton standing crop seems to be limited by macronutrients (Browning et al., 2014).

In general, the taxonomic study of plankton species in the South Atlantic is poor to average, approximately 2500 zooplankton species have been identified in the South Atlantic and it is expected to find 300 more (Boltovskoy et al., 2003). Copepods, in terms of numbers and biomass, are the main component of zooplankton; within this group the largest proportion is made up of the smaller copepods (less than 0.3 mm). Euphausiids and amphipods are important components, especially in neritic zones (Thompson et al., 2013).

Squid are important components of the South Atlantic marine ecosystem, for ecological and socioeconomic reasons. They show high predation rates and contribute substantially to the flux of energy and nutrients to higher trophic levels (Rosas-Luis et al., 2013). The Argentine shortfin squid (*Illex argentinus*) is a common species on the western shelf; mainly feeding on amphipods and euphausiids (Ivanovic, 2010), it sustains an important fishery by trawlers and jigging vessels, and *Doryteuthis gahi* is also targeted by the commercial fishery with substantial annual catches (Arkhipkin et al. 2013). On the African coast, the chokka squid (*Loligo vulgaris reynaudii*) is closely linked to the Agulhas ecosystem; its catches and biomass are highly variable (Roberts, 2005) and, in the south Brazil ecosystem, *Loligo plei* is an important link between

Table 36B.1 | Major structural-functional characteristics of the South Atlantic (0–100 m layer). Data presentation adapted from Greze (1984) and Boltovskoy et al. (1999) in Piontkovski et al., 2003.

Parameters	Benguela Current	South Equatorial Current	Brazil Current	West Wind Drift	Central Gyre
Extension of current (km)	4000	4000	4800	6400	
Current velocity (cm s^{-1})	50–150	40–70	150	50–65	
Primary production (mgC m^{-2} d^{-1})					
(1) Koblentz-Mishke (1977)	250–>500	150–250	<100–150	100–250	<100
(2) Greze (1984)	1000–5000	175–480	117–547	285–515	95–201
(3) Longhurst et al. (1995)	880	360–430	830	330–370	210
Chl a (surface) (mg m–3)	3.0 ± 1.5	0.15 ± 0.05	0.30 ± 0.10	0.22 ± 0.10	0.09 ± 0.04
C:Chl a (0–100 m) (mg mg–1)	35	70	45	91	85
Phytoplankton total biomass (mgC m^{-3})	105	15	14	18	13
>5-µm phytoplankton (mgC m^{-3})	103	3	2.9	1.2	1
Species of >5-µm phytoplankton	110	264	155	70	233
Phytoplankton diversity (a)	13	49	46	27	56
Zooplankton biomass (mgC m^{-3})	3.5	3.4	1.6	1.1	1.3
Copepod species (number)	176	215	280	161	300
Copepod diversity H' (bit ind^{-1})	3.1	3.9	4.4	2.5	4.5

pelagic and demersal energy pathways (Gasalla et al. 2010), supporting small-scale fisheries around São Sebastião Island (Postuma and Gasalla, 2010). Other squid species, such as *D. gahi, Onykia ingens* and *Histioteuthis atlantica*, are important components of the ecosystem on the outer Patagonian shelf.

3 Benthos

In general, the development of the benthic communities is mainly linked to the availability of food (primary producers and nutrients), closely related with the development of the seasonal and permanent frontal systems, as well as upwelling processes. Benthic habitats are variable in the South Atlantic, with unique and highly diverse ecosystems (Miloslavich et al., 2011), such as kelp forests (Rozzi et al., 2012) and huge rhodolith beds (Amado-Filho et al., 2012). The services derived from benthic habitats (e.g., Copertino, 2011) support several human activities, such as fisheries (Salas et al., 2011) and tourism. Distribution patterns of the different taxa, benthic communities and assemblages mainly obey oceanographic conditions; the majority of them are distributed according to biogeographical regions (Kröncke et al., 2013), although patterns are not always clear because of the fragmented and unequal sampling effort along the total extent of the South Atlantic.

To date, benthos studies have been based on basic sampling methodologies, were usually limited to descriptive results, and population and ecological features were mostly studied in communities dominated by commercial species (mussels, scallops, oysters). There is still a lack of knowledge in coastal and mid-shelf waters, whereas benthic communities beyond the shelf-break are only poorly known.

The available information on the fauna inhabiting the deep basins is scarce: ophiuroids and surface deposit feeders are dominant in the Cape Basin, sponges, sipunculids and fish in the Angola Basin, asteroids, crustaceans and fish in the Eastern Guinea Basin, and sipunculids in the Western Guinea Basin. The content of chlorophyll in sediments is consistent with primary production and flux rate of organic matter in the three basins of the south-east Atlantic. The structure and function of the three basin communities correlate with the amount of seafood reaching the seafloor (Wei et al., 2010).

Along the South American coast, many research programmes have been developed that focus on individual, but mainly coastal, communities and species. Several adopted an integrated approach to the study of benthos in deep waters, such as the REVIZEE Programme (Programme for the Evaluation of the Sustainability Potential of Living Resources in the Exclusive Economic Zone), which so far is the most concerted effort to increase knowledge (see Lana et al., 1996, for a baseline of the REVIZEE) on the benthic diversity on the continental shelf and slope, recording more than 1000 taxa in 322 samples. More recently, the Pampa Azul initiative focused on the interdisciplinary study of the marine environment in the South Atlantic (www.mincyt.gob.ar/accion/pampa-azul-9926). As part of the environmental requirements of the licensing policies, the offshore oil industry produced an important amount of data on benthos that are not yet available for scientific research purposes.

In coastal areas the benthos is better known; several research groups work along the South Atlantic, although with distinct hotspots of effort. Most of the information regarding benthic biodiversity, richness and distribution patterns must be obtained from publications devoted to a single taxonomic group or from local ecological studies.

Although there are areas with no information on benthic habitats and diversity, there are places where studies are concentrated. One example is the Aracá Bay, Southeastern Brazil, where 733 benthic species were historically recorded (Amaral et al., 2010) and where, in a recent and continued study, more than 1,000 species have already been recorded in this area. In the SW Atlantic Ocean, 134 echinoderm species have been identified from the five classes (Brogger et al., 2013; Souto et al., 2014), about 360 benthic molluscs (bivalves+gastropods) (Zelaya, 2014 and pers. comm.), of which 27 are of present or potential commercial interest (Roux et al., 2010), 102 crustacean decapod species, with five of commercial interest (Boschi, 2010), at least 212 amphipod species (López Gappa et al., 2006), at least 218 sponge species (López Gappa and Landoni, 2005; Schejter et al., 2006; Bertolino et al., 2007; Goodwin et al., 2011), 246 bryozoan species (López Gappa, 2000), 88 hydrozoan species (Souto et al., 2010) and at least other 27 cnidarian species, including corals (Zamponi, 2008), at least 70 polychaete species (Bremec et al., 2010) and 79 ascidians, including the records of exotic species (Tatián et al., 2013). Many other minor groups (e.g., brachyopods, nemertina, sipunculida, echiurida, other molluscs, etc.) contribute to the total richness of the benthic realm.

Studies on the deep benthic macro- and mega-fauna in the South Atlantic have concentrated on the South American and African continental margins; the deep central areas have remained one of the least studied areas of the world ocean (Perez et al., 2012). Much of the diversity data in the southern Mid-Atlantic Ridge, for example, still derive from large-scale expeditions conducted in the late nineteenth century, such as the *HMS Challenger* expedition, which recorded over 80 species of echinoderms, polychaetes and bryozoans (Murray, 1895), and surveys conducted by the former USSR in the second half of the twentieth century, which focused mostly on seamounts and trenches (e.g., Malyutina, 2004). Fishing surveys conducted on seamounts of the Walvis Ridge provided further descriptions of crustacean (McPherson, 1984) and scleractinian coral faunas (Zibrowius and Gili, 1990), including several new species and the extension of geographic distribution ranges of species from the North Atlantic and Southern Oceans.

More recently, efforts to increase knowledge on deep benthic fauna in the South Atlantic have derived from global initiatives such as the Census of Marine Life (e.g., German et al., 2011; Perez et al., 2012) on the Mid-Atlantic Ridge and Walvis Ridge. Vent sites 3° – 7° south of the Equator were found to contain the mussel *Bathymodiolus puteoserpentis*, the vesicomyd clam *Abyssogena southwardae*, and the alvinocarid shrimp *Rimicaris exoculata*, also common in North Atlantic vent sites. These records imply that the Equatorial Fracture Zone may not be a significant barrier to dispersal of North and South Mid-Atlantic ridge fauna (German et al., 2011). Nearly 190 benthic species records were obtained in non-chemosynthetic environments of the Mid-Atlantic Ridge and Walvis Ridge, with particularly increased diversity found on the Romanche Fracture Zone (Perez et al., 2012). Among these records new species of Hemichordates, amphipods and caridean shrimp were recently described (Cardoso and Fransen, 2012; Holland et al., 2013; Serejo, 2014). Findings such as the ones described will tend to escalate as the deep areas of the South Atlantic are more and better sampled in the future.

National efforts to increase the knowledge on marine biodiversity are taking place in recent years. The Long-Term Ecological Studies Programme(PELD, in Portuguese), funded by the Brazilian National Science Foundation (CNPq), and the SISBIOTA Programme(National Biodiversity System), funded by national and state science-funding agencies, are examples of structured initiatives to produce relevant information on benthic habitats. Several groups are producing temporal series of benthic data to enable the understanding of the impacts of local and global changes. The Brazilian Network for Monitoring Benthic Coastal Habitats (ReBentos; rebentos.org) is a strategy to aggregate and support this kind of study, linking the scientific efforts to public policies related to marine conservation, such as the National System of Marine Protected Areas, the National Plans for Adaptation to Climate Changes, and the National Action Plans for Coral Reefs and Mangroves. Several research institutions operate along the coastline. A number of regions meeting the criteria set for Ecologically or Biologically Significant Marine Areas (EBSAs) of ecological and socioeconomic importance have been identified in the South Western Atlantic (Falabella et al., 2013); they contain mussel beds, reproductive areas for mammals and birds (Península de Valdés), high primary productivity (shelf-break frontal system), oceanic biodiversity, including corals and sponges (Namuncurá – Burdwood Marine Protected Area), king crab, mussel beds and corals (Beagle and Isla de los Estados). Along the southern South American Atlantic coast, the rocky intertidal community is dominated by the bivalves *Brachidontes rodriguezi* (north to 38ºS, warm temperate waters), *Perumytilus purpuratus* (=*Brachidontes purpuratus*, ca. 42-44ºS, cold temperate waters) or *Mytilus chilensis* (south to 47ºS, cold waters) forming dense beds (Olivier et al., 1966; Penchaszadeh, 1973; Zaixso and Pastor, 1977; Zaixso et al., 1978; Sánchez and Zaixso, 1995; Adami et al., 2004; Bazterrica et al., 2007; Hidalgo et al., 2007; Kelaher et al., 2007; Liuzzi and López Gappa 2008).

On the eastern coast, the UNEP-CBD Regional Workshop (Anon., 2013) recognized the biological and ecological importance at the regional level of the Subtropical Convergence Zone, the Walvis Ridge and the Mid-Atlantic Ridge, and at the subregional level the Guinea-Canary Currents convergence, the migratory corridor along the Guinea Current, the seamounts facing the Congo Basin, the Congo Basin and adjacent canyons' marine area, the Guinea-Benguela Currents convergence zone, and the equatorial production zone. This latter zone stretches along both sides of the Equator to the convergence of the Guinea-Canary Currents; the area was described for its high productivity. It is also a breeding ground and migration area for tuna and related species, as well as of marine mammals. Overall 45 areas of interest were identified as requiring further research in the fields of oceanography, geomorphology, ecology and taxonomy.

The benthic communities are subject to different natural and anthropogenic disturbances, depending on the area. Overfishing, trawling, chemical contamination in harbours and coastal areas, changes in the habitats due to the introduction of alien species, oil prospective and extractive activities are the main activities influencing benthic communities. Succession and stability in rocky intertidal communities are subject to artisanal shellfish gathering, which in some areas occurs so intensely that it may cause the local disappearance of these communities.

Global analyses of environmental impacts reveal that the South Atlantic is experiencing diffuse but increasing impacts that affect mainly the coastal areas (Halpern et al., 2008). Although the Central South Atlantic is characterized as subjected to a "low impact," most of the region (about 70 per cent) exhibits indicators of higher impacts. "High" and "Very High" affected areas are spread along the coastal zones and concentrated close to the most urbanized and/or populated centres, such as Southeastern Brazil and the Gulf of Guinea. This scenario was also evident based on the Ocean Health Index (Halpern et al., 2012), which indicates the sustainable use of marine and coastal ecosystem services, with the Southeastern Atlantic and Gulf of Guinea presenting the worst performances. Several drivers are responsible for this scenario, which is in essence a consequence of public policy implementation gaps associated with evident indices of poverty. One element that directly influences the benthic habitats is the nutrient and sewage input in coastal areas. Diaz and Rosemberg (2008) reported the occurrence of several dead zones (areas with no oxygen to support life), along the Southwestern Atlantic Coast and the Gulf of Guinea.

Additionally, climate change and global warming are also acting in the South Atlantic Ocean. However, concerted efforts to understand the effects of global environmental changes on the South Atlantic lag behind other regions worldwide, leaving society ill-prepared to cope with future changes (Turra et al., 2013). In fact, the paucity of time-series data in the southern hemisphere is especially acute in developing countries (Rosenzweig et al., 2008).

Bottom fishing has represented a variable source of threats to benthic communities in deep areas of the South Atlantic. The development of deep water fishing in slope areas off southern Brazil may have produced significant impacts on benthic organisms as reported by Perez and Wahrlich (2005) and Perez et al. (2013). Further south, off the Patagonian Shelf, benthic communities along the shelf-break front are dominated by scallop beds (Bogazzi et al., 2005) that have been studied and exploited since 1996, when the Patagonian scallop fishery started (Lasta and Bremec, 1998; Bremec and Lasta, 2002). However, interactions of bottom trawling operations and vulnerable benthic communities were found to be generally low (Portela et al., 2012; Schejter et al., 2014a).

A limited number of vessels operate on seamounts of the Walvis Ridge aiming at bottom resources, most notably orange roughy and alfonsino. Although a general paucity of data exists regarding ecosystem impacts of these operations (Bensch et al., 2008; Rogers and Gianni, 2010), catch reports made by different countries throughout the 1980s and 1990s suggest that some seamounts may have been heavily fished by bottom gear, producing an uncertain impact on benthic organisms, including scleractinian corals and sponges (Clark et al., 2007). Countries such as Spain and Namibia have made efforts to describe these communities in different deep fishing areas in the South Atlantic high seas and to identify those considered to be "vulnerable marine ecosystems" (VMEs), whose protection would be a priority in the process of managing deep sea fisheries worldwide (FAO, 2009; Durán Muñoz et al., 2012). In 2006, the Southeast Atlantic Fishery Organization (SEAFO) precautionarily adopted the closure of ten seamount areas for bottom fishing, which were reviewed in 2010 and some reopened (Weaver et al., 2011). Currently, a total of eleven areas are closed in order to protect VMEs (SEAFO Conservation Measure 29/14 Annex 2).

Offshore extraction of oil and gas may cause potential harm to deep benthic communities mostly in association with the production of waste deposits and discharges of chemical pollutants (Davies et al., 2007). Deep areas of Brazilian and West African margins have large reserves, which will be increasingly exploited in the next decade. Nearly 80 per cent of all oil produced annually in Brazil derives from deep oil fields in Campos Basin, off southeastern Brazil, where important efforts to describe benthic diversity and define environmental baselines have been undertaken (Lavrado and Brasil, 2010). It is critical that these efforts expand in the upcoming years, as an even greater offshore oil extraction activity is expected to develop in the large pre-salt reserves recently discovered in the Santos Basin (Abreu, 2013).

Deep mineral deposits have been prospected in the South Atlantic "Area," particularly in association with abyssal plains (polymetallic nodules), the Mid-Atlantic Ridge (seafloor massive sulphides) and seamounts (cobalt-rich crusts) (Hein et al., 2013). In 2013-14 a first plan for exploration of cobalt-rich crusts in a large seamount area named Rio Grande Rise was proposed by Brazil and approved by the International Seabed Authority (ISA, 2014). This contract will aim, in a 15-year programme, to characterize virtually undescribed benthic communities and produce an environmental baseline and monitoring plan for the claimed area. This programme follows ISA regulations for contractors, which include assessing potential disturbances of benthic habitats and organisms caused by crust-removing activities on the seafloor. Although most claims for mineral exploration are currently concentrated in the Pacific Ocean, it may be expected that interest in other areas, such as the South Atlantic, will also develop as knowledge of the South Atlantic increases (Perez et al., 2012).

One promising tool to reconcile benthic conservation and industrial development is offered by the implementation of Marine Protected Areas (MPA) (Marone et al., 2010; Turra et al., 2013). Studies and efforts to increase the number of MPAs are ongoing, such as the survey of priority areas for marine conservation in Brazil (MMA, 2007). In fact, the pressure of particular stakeholders (as shrimp farmers) reduced the protection of mangroves and wetlands along the Brazilian coast (Rovai et al., 2012). In addition, the recent findings of high oil and gas reserves in the Pre-Salt layer off the Southern-Southeastern Brazilian coast, as well as the potential of mining activities, raise awareness about the conservation of the fragile deep-sea benthic habitats.

Due to growing maritime traffic, the record of exotic species is expected to increase. In the Southwest Atlantic, a survey of exotic species in coastal and shelf areas of Uruguay and Argentina revealed that 31 species were introduced and 46 were cryptogenic. Coastal ecosystems between La Plata and Patagonia have been modified. Only exposed sandy beaches appear to be free from the pervasive ecological impact of invasion by exotic species. Poor knowledge of the regional biota makes it difficult to track invasions (Orensanz et al., 2002). Alien species (more than 40 reported for Argentina) severely modified native habitats and may cause loss of biodiversity. The most significant examples are: the Japanese alga *Undaria pinnificata* that highly transformed the benthic structure found in gulfs and bays in Patagonia (Dellatorre et al., 2012), the introduced barnacles that greatly modified hard substrates in harbours and surrounding areas, the polychaete *Ficopomatus enigmaticus* that built reefs in the coastal lagoon Mar Chiquita (Buenos Aires) (Schwindt et al., 2004), the alien oyster *Crassostrea gigas* that transformed and colonized coastal and intertidal areas in Buenos Aires and in Patagonia (Castaños et al., 2009; Giberto et al., 2012), the *Rapana venosa* with an increasing population (Giberto et al., 2006), a voracious gastropod that has already caused huge financial losses in other countries in scallop, mussel and oyster culture and natural populations. In Brazil, a recent report indicated the occurrence of 58 exotic species, with 9 considered invasive (MMA, 2009), one in the phytobenthos, the green alga *Caulerpa scalpelliformis* var. *denticulate*, and six in the zoobenthos: the cnidarian anthozoans *Tubastraea coccinea* and *Tubastraea tagusensis*, the mollusc bivalves *Isognomon bicolor* and *Myoforceps aristatus*, the crustacean decapod *Charybdis hellerii* and the ascidian *Styela plicata*.

Due to continental sources of materials (sediment) and pollutants, including chemicals, nutrients and solid wastes, as well as the strong and widespread impact of human occupation, fishing, mining and oil industry activities, benthic coastal habitats are in danger in the South Atlantic and deserve proper attention from governments and society.

4   Fish

4.1   Status

The area of direct influence of the Amazon and Tocantins rivers is highly heterogeneous in terms of the dynamics of sedimentary deposition and freshwater discharge. This determines the characteristics of its fauna (Coelho, 1980; Camargo and Isaac, 2001) and flora (Prost and Rabelo, 1996), including species richness and distribution patterns (Giarrizo and Krumme, 2008). This area supports high fish biodiversity. Camargo and

Isaac (2003) estimated that >300 fish species inhabit this area from 23 orders and 86 families, with a high degree of diversification mainly of the families Sciaenidae and Ariidae, but also of Rajiformes, Pleuronectiformes and Tetradontiformes. Many fish species of these families were also reported in French Guyana, Suriname and Guyana (Lowe McConnell, 1962; Planquette et al., 1996; Le Bail et al., 2000, Keith et al., 2000) and Venezuela (Cervigon, 1996). Souza and Fonseca (2008), who also included information on the shelf and shelf break, identified more than 500 species from 106 families. A river-ocean gradient in the distribution of different species reflects their capacity to tolerate varying levels of salinity. Seasonal changes occur in the composition of the fish community, with predominance of freshwater species during the rainy season and marine species during the dry season (Camargo and Isaac, 2001). The high productivity of the area offers a high potential for fishery activities due to the numerous rivers and estuaries that empty into the Atlantic Ocean, forming a complex aquatic environment with high biological productivity.

Further south along the east Brazil LME, fringing and barrier reefs occur along the coast and over the shelf, harbouring diverse reef fish communities (Maida and Ferreira, 1997; Floeter et al., 2001). Offshore are located a major oceanic plateau, the Ceara Rise, and the Fernando de Noronha Ridge, with a chain of seamounts and the only atoll in the South Atlantic Ocean, Atol das Rocas. The Southwestern Atlantic region (SWA; including Brazilian oceanic islands and Argentina) has an impoverished reef fish fauna in relation to the Northwestern Atlantic and Caribbean, with only over half (471) of the reef species richness and 25 per cent of endemic species distinguishing the 'Brazilian Province' (Rocha, 2003; Floeter et al., 2007).

The Amazon freshwater and sediment outflow is a strong (albeit permeable) barrier to shallow-water reef fish and other organisms, and it is probably responsible for most of the endemism found in Brazilian coastal habitats (Rocha, 2003). Where the continental shelf is narrower and unusually steep, reef formations on the shelf-edge zone and slope down to 500 m depth support important multi-species fisheries, harbour critical habitats for the life cycle of many reef fish species, including spawning aggregation sites (Olavo et al., 2011), serve as a faunal corridor that extends beyond the Amazon mouth area (Collette and Rutzler, 1977), including the hump of Brazil and connect cold habitats in southern Brazil and the Caribbean (Olavo et al., 2011). Snapper dominates the demersal fisheries in the region (Frédou et al., 2006). Further south the shelf widens and the Abrolhos Bank forms a physical barrier to the Brazil Current, hence upwelling and land conditions create even more diversity, especially for the reef fauna.

The South Brazil Shelf Large Marine Ecosystem (LME) extends roughly over the entire continental shelf off southeastern South America. The shelf waters result from the mixing of several water types: coastal, sub-Antarctic, sub-tropical and mixed waters (Bisbal, 1995). The Rio de la Plata represents the greatest freshwater inflow to the region, discharging on average 22,000 m3/s (with an annual fluctuation of 22 per cent). About 185 species of fish have been identified in South Brazil and around 540 in Argentine and Uruguayan shelf areas (Miloslavich et al., 2011). Two of the most commercially important finfish species exploited from this system are common hake (*Merluccius hubbsi*) and Patagonian hake (*Merluccius australis*) (Bisbal, 1995). Many Sciaenid species are also important as a fishery resource. In this region more than 55 per cent of the stocks are overexploited (MMA, 2006).

At the mouth of the Río de la Plata, the fish fauna comprise 53 marine species (Nión, 1998). The most abundant species undergo migrations related to changes in hydrographic conditions. In the coastal zone fishes from the family Sciaenidae are dominant to a depth of 50 m (Calliari et al., 2003). South of the Rio de la Plata the patagonian shelf is much wider and highly productive with great fish diversity sustaining fisheries on more than 50 species. Common hake (*M. Hubbsi*), patagonian grenadier (*Macruronus magellanicus*), whitemouth croaker (*Micropogonias furneri*), sardinella (*Sardinella brasiliensis* and *S. aurita*), southern blue whiting (*Micromesistius australis*), anchovy (*Engraulis anchoita*), prawns (*Pleoticus muelleri*) and several species of skates and rays (http://www.fao.org/fishery/statistics/en).

Along the eastern margin of the South Atlantic, fisheries resources are highly productive, supported by the upwelling resulting from the Guinea and Canary Currents along the coast. Currently fish stocks in the area are already overexploited by the foreign distant-water fleets fishing in the Exclusive Economic Zone of West African countries (Alder and Sumaila 2004; Atta-Mills et al., 2004) under bilateral agreements with the European Union (Alder and Sumaila, 2004).

4.2   Pressures and Trends

Landings of fisheries in Brazil far exceeded sustainable target levels in the main portion of this area, according to the results from fisheries assessments from a multi-year Brazilian research programme called REVIZEE; the majority of stocks were either fully (23 per cent) or overexploited (33 per cent) and little room remained for expansion into new fisheries (MMA, 2006). The decline of landings of many demersal stocks has been reported in the Southern areas of Brazil [Perez et al., 2003].

Brazilian marine fishes were regionally assessed according to the IUCN Red List criteria. For 151 marine Chondrichthyes species, 39 per cent were categorized as threatened, mainly due to intense and unmanaged fisheries (Peres et al., 2012), and 35 per cent marine teleost species were also assessed as threatened.

Some species in the south west Atlantic have experienced local collapse (e.g., *Cynoscion guatucupa*, a migratory species) at Bahía Blanca from the increasing fishing pressure exerted by the industrial fishing fleet operating in open waters (López Cazorla et al., 2014). Information on the status of the exploited species can be found in www.inidep.edu.ar/pesquerias/ppales-pesquerias.

Global climate change affects fish and fisheries. The effects range from increased oxygen consumption rates in fishes, to changes in foraging and migrational patterns in polar seas, to fish-community changes in bleached tropical coral reefs. Projections of future conditions portend further impacts on the distribution and abundance of fishes associated with relatively small temperature changes (Roessig et al., 2004). The information on the effects of climate change and long-term studies to assess those effects in the Southwest Atlantic are scarce or non-existent. Schroeder and Castello (2010) modelled effects of climate change scenarios on Patos Lagoon estuarine-dependent resources, notably pink shrimp, white-mouth croaker and grey mullet. ENSO cycles and climate changes may increase the limnic and decrease the saline influence in the estuary. This scenario may affect the biology and dynamics of estuarine-dependent species and their fisheries, because temperature influences metabolism, which affects the growth of individuals. The natural mortality of larvae may increase due to metabolic stress, although increased growth rates may also reduce the period during which the young are vulnerable to predation. A decrease in the maximum size of the species is also expected, as well as a shift in biomass peaks and the effect of fisheries. West Africa is considered one of the regions most vulnerable to the impact of climate change on fisheries because of the threats to livelihoods and well-being of communities depending on fisheries. The model's estimates indicate a 21 per cent drop in landed value and a 50 per cent loss in fisheries-related jobs (Lam et al., 2012).

5 Marine Mammals

5.1 Status

Marine mammals comprise a diverse group of taxonomically and ecologically distinct aquatic vertebrates that inhabit pelagic to coastal and estuarine waters, from the photic zone to deep ocean canyons (Berta et al., 2006). Most marine mammals are top predators that have spatial and temporal separation of feeding and nursing areas; they must maximize the energy obtained during foraging events and thus are considered good indicators of environmental conditions and health because they depend on the ocean for food and survival (Block et al., 2002; Hooker and Boyd, 2003; Fedak, 2013). Along the coast of South America, several species of marine mammals are found at different (partial or full) stages of their life cycles (Miloslavich et al., 2011 and references herein).

Worldwide, 129 species of marine mammals are described, of which 60 have also been reported for the South Atlantic Ocean (Perrin et al., 2009). Among resident, frequent and occasional visitors, the South Atlantic Ocean is home to approximately 20 species of the Order Carnivora (Suborder Pinnipedia and Family Mustelidae), and ~45 species of the Order Cetacea. Three Mysticete families (seven species of baleen whales), five Odontocete families (27 species of toothed whales), two Pinniped families (10 species), two Mustelidae species and one Sirenid family (a manatee) were reported for Patagonian and Brazilian coastal waters. Throughout South America, we find marine mammals that are endemic or limited in distribution (La Plata River dolphin, Austral dolphin, Commerson dolphin and manatees), and others with wider distribution that depend on coastal areas of the region for important stages of their life cycles.

Some baleen whales, such as the southern right whale and the humpback whale, breed in waters off Santa Catarina, Brazil (28°S), the north Patagonian gulfs (34°S), or in the Abrolhos Bank (17°S), Northeast Brazil, and on the coast of Southwest Africa. The only representative of the manatees in the SAO, *Trichechus manatus*, occurs discontinuously along coastal waters of Northeast Brazil (Alagoas -9°S to Amapá 0°) where it is under serious threat (Luna et al., 2010). Manatees (*Trichechus* spp.), that are commonly found in mangrove areas in the North and Northeast regions and along the Amazon River Basin, were hunted in the past for their meat and skin and were at risk of extinction, but they are currently protected by the Brazilian Government. Humpback whales (*Megaptera novaengliae*) frequent the southern tip of South America, the Beagle Channel and South Africa. Blue whales (*Balaenoptera musculus*) are only seen sporadically along northern Argentine and South African coasts. Southern right whales, *Eubalaena asutralis*, inhabit the north Patagonian gulfs, one of the most important breeding grounds for the species, and are also regularly seen along the coasts of Uruguay and Brazil. Three coastal dolphins are endemic to the region: Peale's dolphin *Lagenorhynchus australis*, Commerson's dolphin *Cephalorhynchus commersonii* and the La Plata or Franciscana dolphin *Pontoporia blainvillei* (Bastida and Rodriguez, 2010). The orca *Orcinus orca* presents smaller populations with a characteristic predatory behaviour in north Patagonia (Lewis and Campagna, 2008) and Isla de los Estados (Raya Rey, pers. obs.). Given advances in technology, the at-sea movements of several species have been revealed, from the pelagic southern elephant seals (Campagna et al., 1999; Campagna et al., 2006) and the more coastal South American sea lion (Campagna et al., 2001) to the endemic and coastal La Plata River dolphin (Bordino et al., 2008) and breeding and post-breeding humpback whales (Horton et al., 2011).

Resident species also include the South American sea lion *Otaria flavescens*, widely distributed all along the southern coast of South America, including Isla de los Estados and Falkland Islands (Malvinas), South American fur seals *Arctocephalus australis* and southern elephant seals *Mirounga leonina*, with the only land-breeding colony along the coasts of Península Valdés and Punta Ninfa (Chubut, Argentina) believed to be the same stock as the seals breeding at the Falkland Islands (Malvinas) (Lewis and Campagna, 2008). On the African coast between the resident species are found the Cape fur seal *Arctocephalus pusillus pusillus*.

Among the cetaceans that visit and live within the Southwest Atlantic Ocean, five species are considered "vulnerable" or "endangered" worldwide. Among these are the blue whale, the humpback whale, the sperm whale *Physeter macrocephalus*, the La Plata River dolphin *Pontoporia blainvillei* and the manatee *Trichechus manatus*. Most of the Odontoceti species are considered to be data-deficient (Lewis and Campagna, 2008). Two of the three mustelid species, *Lontra provocax* and *L. felina*, that inhabit the southern region, are considered endangered.

5.2 Pressures

Some coastal species are threatened by anthropogenic activities, such as pollution, fishing and fisheries by-catch, tourism activities, coastal development and habitat destruction. Pelagic species are also threatened by increasing traffic of ocean vessels (boats and ships), seismic prospection, fishing, and oil and gas activities.

The indirect effects of fisheries have also been observed in seal populations. For the South American sea lion it has been shown that the level of harvested squid and hake could have a negative impact on seal populations (Koen Alonso and Yodzis, 2005). In the Benguela ecosystem, the interaction of seals (Cape fur seal *Arctocephalus pusillus pusillus*) and fisheries have been also described (Yodzis, 1998).

Artisanal fisheries and entanglement pose a major threat for small cetacean populations, in particular the endangered La Plata dolphin (Praderi et al., 1989; Pérez Macri and Crespo, 1989; Secchi et al., 1997). Although not in big numbers and without a clear impact on the population, some other species caught in fishing nets are the Commerson's dolphin (Crespo et al., 1994; Crespo et al., 1997; Crespo et al., 2000; Schiavini and Raya Rey, 2001; Dans et al., 2003), dusky dolphin, common dolphin and seals (Dans et al., 1997; Crespo et al., 2000; Dans et al., 2003). Elephant seals are also known to become entangled in squid fishing gear (Campagna et al., 2007), as well as many sea lions dying every year with plastic rings around their necks, although this fact has not been quantified and therefore its effect on the population is not known.

Recently the Southern right whale populations have suffered from kelp gull (*Larus dominicanus*) attacks. This problem started during the 1970s as a consequence of gull population growth due to an increase in food supply of human origin (fisheries and home garbage). Although wounds inflicted by kelp gulls do not pose a real threat for the population, gull attacks impose a change in behaviour by the target whale as it increases its attention to the source of the attacks in 24 per cent of cases, to the detriment of other "natural" forms of behaviour (Rowntree et al., 1998; Rowntree et al., 2001; Bertellotti and Perez Martinez, 2008). On average, 27 southern right whales, mostly pups, die annually in Península Valdés; saw a record of 83 dead whales from unknown causes. Although pathogens are not believed to be a major threat to biodiversity, the growth of human activities and climate change could promote their expansion (Uhart et al., 2008).

5.3 Trends

Recent studies have provided insights into marine biodiversity in South America, specifically regarding the Patagonian Shelf (Uruguay and Argentina) and the Brazilian Continental Shelf (Miloslavich et al., 2011). Even though Miloslavich and colleagues' assessment represented a major breakthrough in our knowledge of marine biodiversity in South America, their focus was on *macroalgae, cnidarians, molluscs, crustaceans, echinoderms, and fish*. The assessment represented an initiative by the Census of Marine Life (CoML) to promote a thorough background check into the information on marine biodiversity produced and accessed only locally in South America and to make it available worldwide via a marine diversity database (OBIS). The knowledge generated by marine research in South America has been limited, given the poor access to oceanographic vessels, isolation between researchers, and the lack of coordination between scientific programmes (Ogden et al., 2004). The lack of dedicated efforts to examine the trends in biodiversity and species richness for the region has represented a major bottleneck for the development of efficient conservation and management measures. Whereas governmental agencies (e.g., the National Oceanic and Atmospheric Administration (NOAA) and the National Marine Fisheries Service (NMFS) of the United States and the Department of Fisheries and Oceans of Canada (DFO)) in the North Atlantic are focused on performing marine mammal assessments to help establish their status and improve their conservation, south of the Equator most assessments are a result of isolated research initiatives with local or regional scope, given the costs associated with this type of assessment.

The South American sea lion is the most abundant marine mammal in the region, with several breeding grounds along the coast from Uruguay to Tierra del Fuego, Staten Island and Falkland Islands (Malvinas) (Reyes et al., 1999; Dans et al., 2004; Schiavini et al., 2004). Most of its populations have been recovering over the last decades from the devastating exploitation suffered between the 17th and 20th centuries (Bastida and Rodriguez, 2010). Nevertheless, whereas colonies in north Patagonia are growing at a 5.4 per cent (Dans et al., 2004) annual rate, the Uruguay populations are decreasing by 4 per cent annually (Páez, 2005). The South American fur seal, with smaller populations, is also recovering, with a 1-2 per cent annual increase rate in the Uruguay populations (Bastida and Rodriguez, 2010). Southern elephant seal populations also increased at an annual rate of 3 per cent from the 1980s to 2000, and have remained stable in numbers since then (Lewis and Campagna, 2002; Lewis and Campagna, 2008).

Data on cetacean population numbers and trends are scarce for the region. Southern right whales show a 7 per cent annual increase rate in Península Valdés (Cooke et al., 2001). Population numbers are known for some of the coastal species (Brownell et al., 1998; Lescrauwaet, 1997; Schiavini et al., 1999; Pedraza, 2008) with no trends available, although given the mortality rate and the narrow distributional range, the La Plata dolphin populations are thought to be decreasing (see threats). No population estimates are known for the Falkland Islands' (Malvinas') resident populations (Otley et al., 2008).

The Brazilian Ministry of Environment has established MPAs and implemented action plans for wildlife, particularly marine mammals (Barreto et al., 2010; Di Benedetto et al., 2010; Luna et al., 2010; Campos et al., 2011). In Uruguay and Argentina, MPAs are increasing, but the implementation of specific action plans for marine mammals has not yet been achieved. In Argentina, marine and coastal resources (Yorio et al., 1998; Sapoznikow et al., 2008), have been under protection in about 59 coastal and marine protected areas, which include marine organisms, such as seabirds and marine mammals, among their main conservation targets (GEF, 2013) . In Uruguay, the process of establishing MPAs is incipient, but a National System of Protected Areas is responsible for this process, and three coastal areas are currently being considered (Santa Lucía, Cabo Polonio, and Cerro Verde), as well as a proposal for a network of

MPAs (Defeo et al., 2009). Recently, ecosystem-based fishery management and MPAs are emerging as promising tools to conserve marine environments, in view of declining fisheries indicators in the South Western Atlantic Ocean (Mugetti et al., 2004; Milessi et al., 2005; Defeo et al., 2009). The overall aim is to ensure ecosystem resilience and adaptation to a changing environment while maintaining ecosystem processes and a sustainable use of marine resources. Thus it is important to focus not only on vulnerable species, such as the coastal manatees, sea lions or breeding baleen whales, but also on vulnerable areas, or areas of ecological significance for many species of marine mammals as well as also other groups, such as birds, turtles and fish (Mittermeier et al., 2011).

6 Seabirds

6.1 Status

Tropical waters are relatively poor in seabirds as a result of low productivity. About 130 coastal and marine species are found north of the Río de la Plata. The larger part of these birds comes from the northern hemisphere between September and May and from the southern seas between May and August to reproduce in areas such as Atol das Rocas, which are crucial for the maintenance of their populations (Miloslavich et al. 2011).

South of the Río de la Plata, shelves are high-productivity areas which maintain a great diversity of seabirds. This ecosystem not only harbours many birds that come each summer to breed, but also thousands of seabirds forage within its waters (Yorio et al., 1999; Croxall and Wood, 2002; Favero and Silva, 2005). Pelagic and coastal waters are home to about 50 species that belong to the orders Procellariiformes, Sphenisciformes, and the families Stercorariidae, Sterniidae, Lariidae and Phalacrocoraciidae.

The main breeding sites in the South West for these populations are concentrated in three areas: (a) Península Valdés and adjacent coasts; (b) Tierra del Fuego and adjoining areas; and (c) the Falkland Islands (Malvinas) (Croxall et al., 1984; Strange, 1992; Woods and Woods, 1997; Yorio et al., 1999). Three species (Magellanic penguin *Spheniscus magellanicus*, southern rockhopper penguin *Eudyptes chrysocome* and black-browed albatross *Thalassarche melanophris*) have over half their world population in the area (Boersma et al., 2013; Pütz et al., 2013) and two others (gentoo penguin *Pygoscelis papua* and thin-billed prion *Pachyptila belcheri*) probably have more than one-quarter of their world population in the region (Croxall and Wood, 2002).

Seabird diversity and abundance have long been studied by ship surveys (e.g., Cooke and Mills, 1972; Jehl, 1974; Veit, 1995; Orgeira, 2001a; Orgeira, 2001b). With advances in technology (satellite tracks, global positioning system devices and geolocators), the origin, sex, age and status of birds using the area can be established and quantified (Falabella et al., 2009). Therefore, it is known that the area is intensively used by a wide range of species: from pelagic flying birds (e.g., Jouventin and Weimerskirch, 1990; Weimerskirch et al., 1997; Prince et al., 1998; Berrow et al., 2000; González-Solís et al., 2002; Quintana and Dell'Arcriprete, 2002; Trathan and Croxall, 2004; Masello et al., 2010) to penguins (Stokes et al., 1998; Stokes and Boersma; 1999; Pütz et al., 2002; Pütz et al., 2007; Wilson et al., 2007; Raya Rey et al., 2007; Sala et al., 2014; Rattcliffe et al., 2014) and coastal birds (Suárez and Yorio, 2005; Suárez et al., 2012). The waters are not only used by resident species but also by seabirds that breed in distant colonies: wandering albatrosses *Diomedea exulans* from the South Georgias Islands extensively use the Patagonian shelf, two fulmarine petrels, cape petrel *Daption capense* and Antarctic fulmar *Fulmarus glacialoides*, which breed on the Antarctic Peninsula and Continent, are very common visitors over the Patagonian shelf (Orgeira, 2001a). Wilson's storm petrel *Oceanites oceanicus*, a trans-equatorial migrant, with a large Antarctic but small Falkland Islands (Malvinas) breeding population is also common in the region (Orgeira, 2001a). Although only present in small numbers, seabirds breeding at Tristan da Cunha and Gough (mainly soft-plumaged petrel *Pterodroma mollis*, Atlantic petrel *P. incerta*, Kerguelen petrel *Lugensa brevirostris* and great shearwater *Puffinus gravis*) visit the southern part of the Patagonian Shelf (Orgeira, 2001a). In addition, substantial numbers of Tristan albatross *Diomedea dabbenena*, Atlantic yellow-nosed albatross *Thalassarche chlororhynchos* and spectacled petrel occur in similar areas of northern Argentina, Uruguay and southern Brazil (Olmos et al., 2000). Finally, northern royal albatross *Diomedea sanfordi*, endemic to New Zealand, is known to forage along the Patagonian Shelf (Nicholls et al., 2005).

Of the seabirds which breed in the region, two species (southern rockhopper penguin and white-chinned petrel *Procellaria aequinoctialis*), qualify under the IUCN criteria for globally threatened status; magellanic penguin, black-browed albatross and gentoo penguin are regarded as Near Threatened. Also seven of the non-resident visitor species qualify for the globally threatened status (BirdLife International, 2014). Seabirds in the region are considered in the International Plans of Action presented by Argentina, Brazil, Chile, South Africa and Uruguay.

South African waters are of prime importance for conserving seabirds because the Benguela upwelling system and the Agulhas Bank provide rich foraging opportunities for a wide diversity of seabirds (Petersen et al., 2009). The Benguela Ecosystem harbours 15 seabird species, of which 10 are endemic to South Africa. Species belong to the families Spheniscidae, Hydrobatidae, Pelecanidae, Sulidae and Haemotopodidae with one species each, three species of the Laridae family and four species in the Phalacrocoracidae and Sternidae families (Kemper et al., 2007). Seabirds from the Benguela Ecosystem are highly threatened; in particular, the African Penguin *Spheniscus demersus* and Cape Cormorant *Phalacrocorax capensis* are now considered endangered following ongoing decreases (Birdlife, 2014). The Southeast Atlantic Ocean also has numerous islands (Ascension, St. Helena, Inaccessible, Tristan and Nightingale) rich in seabird species (Cuthbert 2004; Birdlife, 2014). Most of these islands are home to the endangered northern rockhopper penguin *Eudyptes moseleyi* (Birdlife, 2014), and 15 albatrosses and petrels foraging within these waters qualify under the IUCN criteria for globally threatened status, such as the Tristan albatross which is critically endangered (Abrams, 1983; Abrams, 1985; Ryan and Moloney, 1988; Nel and Taylor, 2002; Wanless et al., 2009; Petersen et al., 2009; Birdlife, 2014).

6.2 Trends

Population trends of resident and non-resident seabirds that forage in the Southwest Atlantic Ocean present different trajectories over the years, with some species showing opposite trends at different locations. Some of the large Procellariiformes species have declined over the past decades (i.e., wandering albatross) and this trend continues (Poncet et al., 2006), but others, such as the southern giant petrel, are recovering at least at some colonies (Reid and Huin, 2005; Quintana et al., 2006; Wolfaardt, 2012). Small petrels' trends are not well known in the area (Otley et al., 2008). Among penguins, southern rockhopper penguins have experienced dramatic declines between 1930 and 2005 (Pütz et al., 2003), although this trend seems to have reverted within the Falkland Islands (Malvinas) (Baylis et al., 2013), whereas the population at Staten Island has reduced its numbers during the last decade (Raya Rey et al., 2014). Gentoo and Magellanic penguins present different trends, depending on the colony. Gentoos in the Falkland Islands (Malvinas) presented a 42 per cent decrease, which was attributed to a paralytic shellfish poisoning in 2002, with a later increase of 95 per cent since 2005 (Pistorius et al., 2010); a recent study showed interannual fluctuations without a clear trend (Baylis et al., 2012), but in the meantime the small population of Tierra del Fuego has increased (Raya Rey et al., 2014). Magellanic penguin population trends are variable: some of the bigger colonies are decreasing, but at the same time new colonies are being established in northern Patagonia (Boersma, 2008; Boersma et al., 2013); the population in the Falkland Islands (Malvinas) does not present a clear trend, whereas populations in Tierra del Fuego are increasing (Raya Rey et al., 2014).

Long-term population trends for coastal birds, such as gulls, cormorants, skuas and terns, are scarce and limited for certain regions. Cormorants present opposite trends depending on the species and site, but in general numbers have been stable with some populations slightly increasing or decreasing (Frere et al., 2005; Raya Rey et al., 2014). Kelp gulls take advantage of human garbage, and their populations are increasing all along the Patagonian coast (Lisnizer et al., 2011; Raya Rey et al., 2014). Terns and skuas are the least studied of the species, with small populations and frequent variations in colony locations with unknown trends (Yorio, 2005; Otley et al., 2008).

Southeast Atlantic seabird population trends contrast between species, but several have experienced severe decreases during the last decades, such as the African penguin, the Cape Gannet *Morus capensis* and Cape cormorant (Kemper et al., 2007). Some gull populations are increasing, which is largely attributable to the provision of additional food sources from human activities (Crawford, 1997) and the cessation of population control measures (mainly the destruction of eggs and chicks) at most breeding localities (Hockey et al., 2005). The decrease in Northern rockhopper penguins is evident from population estimates in the Tristan da Cunha group and Gough Island, which indicate a decline of more than 50 per cent (Cuthbert et al., 2009). The Tristan albatross also decreased severely at 3 per cent per year, and the sooty albatross *Phoebetria fusca*, Atlantic yellow-nosed albatross *Thalassarche chlororhynchos* and southern giant petrels *Macronectes giganteus* remained stable during the last decade (Cuthbert et al., 2014).

6.3 Pressures

Direct and indirect discharge of chemical pollutants, industrial and expanded city pollution, bycatch, entanglement, climate change and alien species pose severe threats for seabird populations both at sea and at their colonies in the Southwest Atlantic Ocean. Oil pollution in Argentine inshore waters is of major concern and kills thousands of Magellanic penguins annually (Gandini et al., 1994; Garcia-Borboroglu et al., 2006; Garcia-Borboroglu et al., 2010). Negative consequences of garbage disposal have also been documented in the region (Copello and Quintana, 2003; Otley and Ingham, 2003).

Bycatch (incidental mortality) of seabirds in fishing gear has been a foremost conservation issue, due to the large number of albatrosses and petrels killed by longline fishing vessels (Croxall, 1998; Neves and Olmos, 1998; Olmos et al., 2000; Prince et al., 1998; Schiavini et al., 1998; Stagi et al., 1998). Seabird mortalities decreased by one order of magnitude towards the end of the decade (2001-2010), not due to lower bycatch rates but rather to a drop in the number of hooks set per year (Favero et al., 2013). Black-browed albatrosses, white-chinned petrels, southern giant petrels and southern royal albatrosses are the most common species interacting with trawlers. The total annual mortality of these birds associated with the trawl fleet under investigation was roughly estimated to be from several hundred to over a thousand albatrosses (Favero et al., 2011). Entanglement of penguins in trawl nets is considerable, and other inshore feeding species are doubtless at risk (although this risk is in some cases minor) in various other net fisheries (Gandini et al., 1999; González-Zevallos et al., 2007; González- Zevallos and Yorio, 2011; Seco Pon et al., 2012; Seco Pon et al., 2013).

Seabirds foraging and breeding in the Southeast Atlantic are subject to many threats, such as: human disturbance of breeding colonies; destruction of breeding habitats by development (du Toit et al., 2003); predation by domestic cats and mice (Wanless et al., 2009); egg and chick predation by Kelp Gulls and Great White Pelicans (Crawford, 1997; du Toit et al., 2003); competition with commercial fisheries for food (du Toit et al., 2003). Longline fisheries pose a serious threat in particular for the Tristan albatross, and also for albatrosses and petrels in general (Baker et al., 2007). Introduced mice species at Gough Island are known to affect albatrosses and petrels (Cuthbert et al., 2013). Food supplies for northern rockhopper penguin may be affected by squid fisheries, climate change and shifts in marine food webs (Cunningham and Moors 1994; Guinard et al., 1998; Hilton et al., 2006). African penguins have dramatically decreased during the last decade, which is related to prey abundance (Crawford et al., 2011).

References

Abrams, R.W. (1983). Pelagic seabirds and trawl-fisheries in the southern Benguela Current region. *Marine Ecology Progress* Series 11: 151–156.

Abrams, R.W. (1985). Pelagic seabird community structure in the southern Benguela region: changes in response to man's activities? *Biological Conservation* 32: 33–49.

Abreu, F.V. (2013). O Pré-sal Brasileiro e a legislação do novo marco regulatório: Uma avaliação geoeconômica dos recursos energéticos do pré-sal. *Revista de Geologia*, 26(1): 7-16.

Acha, E.M., Mianzan, H.W., Guerrero, R.A., Favero, M., Bava, J., (2004). Marine fronts at the continental shelves of austral South America. Physical and ecological processes. *Journal of Marine Systems* 44, 83-105.

Adami M.L., Tablado A., López Gappa J.J. (2004). Spatial and temporal variability in intertidal assemblages dominated by the mussel Brachidontes rodriguezii (d'Orbigny, 1846). *Hydrobiologia* 520:49-59.

Alder J, Sumaila U.R. (2004). Western Africa: a fish basket of Europe past and present. *Journal of Environment and Development* 13:156–178.

Amado-Filho, G.M., Moura R.L., Bastos A.C. et al. (2012). Rhodolith beds are major CaCO3 bio-factories in the tropical South West Atlantic. *PLoS ONE*, 7, e35171.

Amaral, A., Zacagnini, C., Migotto, Á.E.; Turra, A., Schaeffer-Noveli, Y., (2010). Araçá biodiversidade, impactos e ameaças. *Biota Neotropica (Edição em Português. On-line)*, v. 10, p. 219-264.

Anon. (2013). *Report of the South-Eastern Atlantic Regional Workshop* UNEP/CBD/RW/EBSA/SEA/1/4.

Arkhipkin, A.I., Hatfield, E., & Rodhouse, P.G. (2013). *Doryteuthis gahi*, Patagonian long-finned squid. *Advances In Squid Biology, Ecology And Fisheries*, 123.

Atta-Mills J, Alder J, Sumaila U.R. (2004). The decline of a regional fishing nation: the case of Ghana and West Africa. *Natural Resources Forum* 28: 13–21.

Baker, B.G., Double, M.C., Gales, R., Tuck, G.N., Abbott, C.L., Ryan, P.G., Petersen, S.L., Robertson, C.J.R., and Alderman, R. (2007). A global assessment of the impact of fisheries-related mortality on shy and white-capped albatrosses: conservation implications. *Biological Conservation* 137, no. 3: 319-333.

Bally, R., Mc Quaid, C.D., Brown, A.C. (1984). Shores of mixed sand and rock: an unexplored ecosystem. *S A J Sci* 80: 500–503.

Barreto, A.S, Rocha Campos, C.C., Rosas, F.C.W., Silva Jr., J.M., Dalla Rosa, L. Flores, P.A.C., da Silva, V.M.F. (2010). Plano Nacional para a Conservação dos Mamíferos Aquáticos (Pequenos Cetáceos). *Instituto Chico Mendes de Conservação da Biodiversidade*, ICMBio,132 pp. Brasília. ISBN: 978856184235-2

Bastida, R., Rodríguez, D. (2010). *Mamíferos marinos de Patagonia y Antártida*. Vazquez Mazzini, Buenos Aires.

Baylis A.M., Zuur A.F., Brickle P., Pistorius P.A. (2012). Climate as a driver of population variability in breeding gentoo penguins *Pygoscelis papua* at the Falkland Islands. *Ibis* 154:30-41

Baylis, A.M.M., Wolfaardt, A.C., Crofts, S., Pistorius, P.A., Ratcliffe, N. (2013). Increasing trend in the number of Southern Rockhopper Penguins (*Eudyptes c. chrysocome*) breeding at the Falkland Islands. *Polar Biology* 36:1-12

Bazterrica, M. Cielo; Brian R. Silliman; Fernando J. Hidalgo; Caitlin M. Crain; Mark D. Bertness. (2007). Limpet grazing on a physically stressful Patagonian rocky shore. *Journal of Experimental Marine Biology and Ecology* 353 (1): 22-34.

Bensch, A., M. Gianni, D. Gréboval, J.S. Sanders, A. Hjort. (2008). Worldwide review of bottom fisheries in the high seas. *FAO Fisheries and Aquaculture Technical Paper*. No. 522. Rome, FAO. 145p.

Berrow, S.D., Wood, A.G., Prince, P.A. (2000). Foraging location and range of White chinned Petrels Procellaria aequinoctialis breeding in the South Atlantic. *Journal of Avian Biology* 31:303-311.

Berta, A., Sumich, J.L., and Kovacs, K. (2006). *Marine Mammals: Evolutionary Biology*. 2nd ed. Elsevier, 547 pp.

Bertellotti, M.I., Pérez Martínez, D. (2008). "Gaviotas, ballenas y humanos en conflicto" en Estado de Conservación del Mar Patagónico y Áreas de Influencia. [online]. Puerto Madryn, publicación del Foro, available in: http://www.marpatagonico.org

Bertolino, M., Schejter, L., Calcinai, B., Cerrano, C., Bremec, C., (2007). Sponges from a submarine canyon of the Argentine Sea, In: Custódio, M.R., Hajdu, E., Lôbo-Hajdu, G., Muricy, G. (Eds.) *Porifera Research: Biodiversity, Innovation, Sustainability*. Museu Nacional, Rio de Janeiro, pp. 189-201.

BirdLife International (2014). *IUCN Red List for birds*. Downloaded from http://www.birdlife.org on 04/07/2014.

Bisbal, G.A. (1995). The Southeast South American shelf large marine ecosystem: Evolution and components. *Marine Policy*, 19(1), 21-38.

Block, B.A., Costa, D.P., Boehlert, G.B. and Kochevar, R.E. (2002). Revealing pelagic habitat use: the tagging of Pacific pelagics program. *Oceanologica Acta* 25:5, 255-266

Boersma, P.D. (2008). Penguins as marine sentinels. *BioScience* 58:597-607.

Boersma, D., Garcia-Borboroglu, P., Frere, E., Kane, O., Pozzi, L., Pütz, K., Raya Rey, A., Rebstock, G.A., Simeone, A., Smith, J., Yorio, P., Van Buren, A. (2013). Magellanic Penguins. In *PENGUINS: Natural History and Conservation*. (García Borboroglu, P.G. and Boersma, P.D. eds.) University of Washington Press, Seattle U.S.A. 233-26.

Bogazzi, E., Baldoni, A., Rivas, A., Martos, P., Reta, R., Orensanz, J.M., Lasta, M., Dell'Arciprete, P., Werner, F. (2005). Spatial correspondence between areas of concentration of Patagonian scallop (*Zygochlamys patagonica*) and frontal systems in the southwestern Atlantic. *Fisheries Oceanography* 14, 359-376.

Boltovskoy, D., Gibbons, M., Hutchings, L., and Binet, D. (1999) General biological features of the South Atlantic. In *D. Boltovskoy (ed.) South Atlantic Zooplankton*, Backhuys Publishers, Leiden, The Netherlands 1999.

Boltovskoy, D., Correa, N., and Boltovskoy, A. (2003). Marine zooplanktonic diversity: a view from the South Atlantic. *Oceanologica Acta* 25: 271–278.

Bordino, P., Wells, R.S., Stamper, M.A. (2008). Satellite tracking of Franciscana Dolphins Pontoporia blainvillei in Argentina: preliminary information on ranging, diving and social patterns. In *IWC Scientific Committee Meeting*, June, Santiago, Chile.

Boschi E.E., (2010). Crustáceos decápodos. In: *M.B. Cousseau (Ed.). Peces, crustáceos, y moluscos registrados en el sector del Atlántico Sudoccidental comprendido entre 34°S y 55°S, con indicación de las especies de interés pesquero. INIDEP Serie Informe Técnico*, 5, Mar del Plata, 65-78.

Bremec, C.S., Lasta, M.L., (2002). Epibenthic assemblage associated with scallop (*Zygochlamys patagonica*) beds in the Argentine shelf. *Bulletin of Marine Science* 70, 89-105.

Bremec, C., Souto, V., Genzano, G., (2010). Polychaete assemblages in SW Atlantic: results of "Shinkai Maru" IV, V, X and XI (1978-1979) cruises in Patagonia and Buenos Aires. *Anales Instituto Patagonia (Chile)* 38, 47-57.

Brogger, M.I., Gil, D.G., Rubilar, T., Martínez, M.I., Díaz de Vivar, M.E., Escolar, M., Epherra, L., Pérez, A.F., Tablado, A., (2013). Echinoderms from Argentina: Biodiversity, distribution and current state of knowledge. In: *Alvarado, J.J., Solís-Marín, F.A. (Eds.) Echinoderm Research and Diversity in Latin America*. Springer-Verlag,

Berlin pp. 359-400.

Brownell Jr. R.L., Crespo E.A., Donahue M. (1998). Peale's Dolphin Lagenorhynchus australis. En *Handbook of Marine Mammals, Volume 6: The Second Book of Dolphins and the Porpoises*, Ed. S. Ridgway and R. Harrison, pp 105-120.

Browning,T.J., Bouman, H.A., Moore, C.M., Schlosser, C., Tarran, G.A., Woodward, E.M.S., and Henderson, G.M. (2014). Nutrient regimes control phytoplankton ecophysiology in the South Atlantic. *Biogeosciences*, 11, 463–479

Calliari, D., Defeo, O., Cervetto, G., Gómez, M., Giménez, L., Scarabino, F., Brazeiro, A., and Norbis, W. (2003). Marine Life of Uruguay: Critical Update and Priorities for Future Research. *Gayana* 67(2): 341-370.

Camargo, M. and Isaac, V.J. (2001). Os peixes estuarinos da região norte do Brasil: lista de espécies e considerações sobre sua distribuição geográfica, *Zoologia*.

Camargo, M. and Isaac, V. (2003). Ictiofauna estuarina. Pp. 105-142. In: Fernandes, M.E.B. (Ed.). *Os manguezais da costa norte brasileira*. Fundação Rio Bacanga, 142 p.

Campagna, C., Falabella, V., Lewis, M. (2007). Entanglement of southern elephant seals in squid fishing gear. *Marine Mammal Science* 23: 414-418.

Campagna, C., Fedak, M.A., McConnell, B.J. (1999). Post-breeding distribution and diving behavior of adult male southern elephant seals from Patagonia. *Journal of Mammalogy* 1341-1352.

Campagna, C., Piola, A.R., Rosa Marin, M., Lewis, M., Fernández, T. (2006). Southern elephant seal trajectories, fronts and eddies in the Brazil/Malvinas Confluence. *Deep Sea Research Part I: Oceanographic Research Papers* 53: 1907-1924.

Campagna, C., Werner, R., Karesh, W., Marín, M. R., Koontz, F., Cook, R., Koontz, C. (2001). Movements and location at sea of South American sea lions (*Otaria flavescens*). *Journal of Zoology* 255: 205-220.

Campos, E.J.D., Gonçalves, J.E., and Ikeda, Y. (1995). Water mass characteristics and geostrophic circulation in the South Brazil Bight: Summer of 1991, J. *Geophys. Res.*, 100(C9), 18537–18550, doi:10.1029/95JC01724.

Campos, C.C.R., Moreno, I.B., Rocha, J.M., Palazzo Jr., J.T., Groch, K.R., Oliveira, L.R., Goncalves, L., Engel, M., Marcondes, M.C., Muelbert, M.M.C., Ott, P.H., Silva, V.M.F. (2011). *Plano de Ação Nacional para Conservação dos Mamíferos Aquáticos – grandes cetáceos e pinípedes, versao* III. 2011.156 pp. Brasilia :ISBN 978856184216-1

Cardoso, I. A., Fransen, C.H.J.M. (2012). A new species of the deepwater shrimp genus Leontocaris (Hippolytidae: Caridea) from the South Mid-Atlantic Ridge. *J. Mar. Biol.* Assoc. U. K., v. 92, n. 5, p. 1083-1088.

Castaños, C., Pascual, M. and Pérez Camacho, A. (2009). Reproductive Biology of the Nonnative Oyster, *Crassostrea gigas* (Thunberg, 1793), as a Key Factor for Its Successful Spread Along the Rocky Shores of Northern Patagonia. Argentina. *Journal of Shellfish Research* 28(4):837-847.

Cervigon, F. (1996). *Los peces marinos de Venezuela* (Spanish). Fundación Cientifica Los Roques, Caracas (Venezuela), 2. ed., 4 v.

Clark, M.R., Vinnichenko, V.I., Gordon, J.D., Beck-Bulat, G.Z., Kukharev, N.N., Kakora, A.F. (2007). Large-scale distant-water trawl fisheries on seamounts. In: *Pitcher, T.J., Morato, T., Hart, P.J.B., Clark, M.R., Haggan, N., Santos, R.S. (Eds.), Seamounts: ecology, fisheries & conservation*. Blackwell Publishing, Oxford, pp. 361–399.

Coelho, P.A., Porto, M.R., Koening, M.L. (1980). Biogeografia e bionomia dos crustáceos do litoral equatorial do Brasil. *Trabalhos Oceanográficos da Universidade Federal de Pernambuco* 15: 7-138.

Collette, B.B., Rutzler K. (1977). *Reef fishes over sponge bottoms off the mouth of the Amazon River*. Proceedings of the 3rd International Coral Reef Symposium, Miami, Florida, U.S.A. pp. 305-310.

Cooke, F., Mills, E.L. (1972). Summer distribution of pelagic birds off the coast of Argentina. *Ibis* 114:245-251.

Cooke, J.G., Rowntree, V.J., Payne, R. (2001). Estimates of demographic parameters for southern right whales (*Eubalaena australis*) observed off Península Valdés, Argentina. J. Cetacean Res. *Manag*. 125-132.

Copello, S., Quintana, F. (2003). Marine debris ingestion by Southern Giant Petrels and its potential relationships with fisheries in the Southern Atlantic Ocean. *Marine Pollution Bulletin* 46:1513-1515.

Copertino, M.S. (2011). Add coastal vegetation to the climate critical list. *Nature*, 473, 255.

Crawford, R.J.M. (1997). In: Harrison J.A., Allan D.G., Underhill L.G., Herremans M., Tree A.J., Parker V., Brown C.J. (eds). Kelp Gull *Larus dominicanus*. *The atlas of southern African birds*. Vol. 1: Nonpasserines. BirdLife South Africa, Johannesburg: 462–463.

Crawford, R., Altwegg., B.J. Barham., P.J. Barham., J.M. Durant., B.M. Dyer., D., Geldenhuys, A.B. Makhado, L. Pichegru, P.G. Ryan, L.G. Underhill, L. Upfold, J. Visagie, L.J. Waller, and P.A. Whittington. (2011). Collapse of South Africa's penguins in the early 21st century *African Journal of Marine Science* 33(1): 139–156.

Crespo, E.A., Corcuera, J., Lopez Cazorla, A. (1994). Interactions between marine mammals and fisheries in some fishing areas of the coast of Argentina. *International Whaling Commission. Special Issue* 15:283- 290.

Crespo, E.A., Koen Alonso M., Dans S.L., García N.A., Pedraza S.N., Coscarella M.A., González R. (2000). Incidental Catch Of Dolphins In Mid-Water Trawls For Southern Anchovy Off Patagonia. *Journal of Cetacean Research and Management* 2:11-16.

Crespo, E. A., Pedraza S.N., Dans S.L., Koen Alonso M., Reyes L.M., Garcia N.A., Coscarella M., Schiavini A.C.M. (1997). Direct and Indirect Effects of the Highseas Fisheries on the Marine Mammal Populations in the northern and central Patagonian Coast. J. Northwest Atlantic Fish. *Sci,* 22:189-207.

Croxall, J.P., McInnes, S.J., Prince, P.A. (1984). The status and conservation of seabirds at the Falkland Islands. Status and conservation of the world's seabirds. *ICBP Technical Publication* 271-292.

Croxall, J.P., Wood, A.G. (2002). The importance of the Patagonian Shelf for top predator species breeding at South Georgia. Aquatic Conservation: *Marine and Freshwater Ecosystems* 12:101-118.

Croxall, J.P. (1998). Research and conservation: a future for albatrosses? Pp. 267–288 in: Robertson G. and Gales R. (eds.). *Albatross biology and conservation*. Surrey Beatty and Sons, Chipping Norton.

Cunningham, D.M., and Moors, P.J. (1994). The decline of rockhopper penguins Eudyptes chrysocome at Campbell Island, Southern Ocean and the influence of rising sea temperatures. *Emu*, 94(1), 27-36.

Curtin, T.B., (1986). Physical observations in the plume region of the Amazon River during peak discharge--III. *Currents. Cont. Shelf Res*., 6: 73-86.

Cuthbert, R. (2004). Breeding biology and population estimate of the Atlantic Petrel, *Pterodroma incerta*, and other burrowing petrels at Gough Island, South Atlantic Ocean. *Emu* 104: 221–228.

Cuthbert, R., Cooper, J., Burle, M.H., Glass, C.J., Glass, J.P., Glass, S., Glass, T., Hilton, G.M., Sommer E.S., Wanless, R.M., and Ryan, P.G. (2009). Population trends and conservation status of the Northern Rockhopper Penguin *Eudyptes moseleyi* at Tristan da Cunha and Gough Island. *Bird Conservation International*, 19(01), 109-120.

Cuthbert, R.J., Louw., H., Lurling, J. Parker, G., Rexer-Huber, K., Sommer, E., Visser, P. and Ryan, P.G. (2013). Low burrow occupancy and breeding success of burrowing petrels at Gough Island: a consequence of mouse predation. *Bird Conservation International*, pp. 1:12.

Cuthbert, R.J., Cooper, J. and Ryan, P.G. (2014). Population trends and breeding success of albatrosses and giant petrels at Gough Island in the face of at-sea and on-land threats. *Antarctic Science* 26(2), 163–171.

da Cunha, L.C. and Buitenhuis, E.T. (2013). Riverine influence on the tropical Atlantic Ocean biogeochemistry. *Biogeosciences*, 10, 6357–6373.

Dans S.L., Crespo, E.A., Garcia, N.A., Reyes, L.M., Pedraza, S.N., Koen Alonso, M. (1997). Incidental Mortality of Patagonian Dusky Dolphins. In Mid-Water Trawling: Retrospective Effects From The Early 80's. *Report of The International Whaling Commission* 47:699-704.

Dans S.L., Crespo, E.A., Pedraza, S.N., Koen Alonso, M. (2004). Recovery of the South American sea lion population in northern Patagonia. *Canadian J. Fisheries and Aquatic Science*, 61:1681-1690.

Dans S.L., Koen Alonso, M., Pedraza, S.N., Crespo, E.A. (2003). Incidental catch of dolphins in trawling fisheries off Patagonia, Argentina: can populations persist? *Ecological Applications* 13:754-762.

Davies, A.J., Roberts, J.M., Hall-Spencer, J. (2007). Preserving deep-sea natural heritage: Emerging issues in offshore conservation and management. *Biological Conservation*, 138:299-312.

Defeo, O., Horta, Carranza, A., Lercari, D., Álava, A. Gómez, J., Martínez, G., Lozoya, J.P., & Celentano, E. (2009). *Hacia un manejo ecosistémico de pesquerías*. Montevideo: Facultad de Ciencias-DINARA. 122 pp.

Dellatorre, F.G., Amoroso, R. and Barón, P.J., (2012). *El alga exótica Undaria pinnatifida en Argentina: Biología, distribución y potenciales impactos*. Editorial Académica Española, 60pp.

Di Beneditto, A.P.M., Rocha Campos, C.C., Danilewicz, D.S., Secchi, E.R., Moreno, I.B., Hassel, L.B., Tavares, M., Ott, P.H., Siciliano, S. Souza, S.P., & Alves, V.C. (2010). *Plano Nacional para a Conservação do pequeno cétaceo TONINHA - ICMBio-MMA. Plano de ação nacional para a conservação do pequeno cetáceo Toninha: Pontoporia blainvillei*. Instituto Chico Mendes de Conservação da Biodiversidade, ICMBio, 2010. 76 pp. Brasília. ISBN: 978856184217-8.

Diaz, R.J. & Rosemberg, R., (2008). Spreading dead zones and consequences for marine Ecosystems. *Science*, 321(5891): 926-929.

Durán Muñoz, P., Sayago-Gil, M., Murillo, F.J., Del Río, J.L., López-Abellán, L.J., Sacau, M., Sarralde, R., (2012). Actions taken by fishing Nations towards identification and protection of vulnerable marine ecosystems in the high seas: The Spanish case (Atlantic Ocean). *Marine Policy* 36, 536-543.

du Toit M, Boere G.C., Cooper J., de Villiers, M.S., Kemper J., Lenten, B, Simmons, R.E., Underhill, L.G., Whittington, P.A. (eds.) (2003). *Conservation Assessment and Management Plan for southern African coastal seabirds*. Cape Town: Avian Demography Unit and Apple Valley: Conservation Breeding Specialist Group.

Falabella, V., Campagna, C., Croxall, J. (eds.) (2009). *Atlas del Mar Patagónico. Especies y espacios*. Buenos Aires, Wildlife Conservation Society y BirdLife International.

Falabella, V., Campagna, C., Caille, G., Krapovickas, S., Moreno, D., Michelson, A., Piola, A., Schejter, L. y Zelaya, D., (2013). Banco Burdwood: Contribuciones para el establecimiento de una línea de base y plan de manejo de la futura Área Marina Protegida. *Informe Técnico*, 51pp.

FAO (2009). International guidelines for the management of deep-sea fisheries in the high-seas. Rome/ FAO, 73p.

Favero, M., Blanco, G., Copello, et al. (2013). Seabird bycatch in the Argentinean demersal longline fishery, 2001–2010. *Endanger Species Research* 19:187-199.

Favero, M., Blanco, G., García, et al. (2011). Seabird mortality associated with ice trawlers in the Patagonian shelf: effect of discards on the occurrence of interactions with fishing gear. *Animal Conservation* 14:131-139.

Favero, M., Silva Rodríguez, M.P. (2005). Estado actual y conservación de aves pelágicas que utilizan la plataforma continental argentina como área de alimentación. *Hornero* 20:95-110.

Fedak, M.A. (2013). The impact of animal platforms on polar ocean observation. *Deep Sea Research Part II: Topical Studies in Oceanography* 88-89, 7-13.

Floeter, S.R., Guimarães, R.Z.P., Rocha, L.A., Ferreira, C.E.L., Rangel, C.A., Gasparini, J.L. (2001). Geographic variation in reef-fish assemblages along the Brazilian coast. *Global Ecol. Biogeogr*. 10: 423– 433.

Floeter, S.R., Krohling, W., Gasparini, J.L., Ferreira, C.E., & Zalmon, I.R. (2007). Reef fish community structure on coastal islands of the southeastern Brazil: the influence of exposure and benthic cover. *Environmental Biology of Fishes*, 78(2), 147-160.

Floeter, S.R., Halpern, B.S., Ferreira, C.E.L. (2011). Effects of fishing and protection on Brazilian reef fishes. *Biological Conservation*, v. 21, p. 199-209.

Frédou, T., Ferreira, B.P., & Letourneur, Y. (2006). A univariate and multivariate study of reef fisheries off northeastern Brazil. *ICES Journal of Marine Science: Journal du Conseil*, 63(5), 883-896.

Frere, E., Quintana, F., Gandini, P. (2005). Cormoranes de la costa patagónica: estado poblacional, ecología y conservación. *Hornero* 20:35-52.

Gandini, P.A., Frere, E., Pettovello, A.D., Cedrola, P.V. (1999). Interaction between Magellanic penguins and shrimp fisheries in Patagonia, Argentina. *Condor* 783-789.

Gandini, P., Boersma, P.D., Frere, E., Gandini, M., Holik, T., Lichtschein, V. (1994). Magellanic penguins (*Spheniscus magellanicus*) affected by chronic petroleum pollution along coast of Chubut, Argentina. *The auk* 20-27.

García-Borboroglu, P., Boersma, P.D., Ruoppolo, V., et al. (2006). Chronic oil pollution harms Magellanic penguins in the Southwest Atlantic. *Marine Pollution Bulletin* 52:193-198.

García-Borboroglu, P., Boersma, P. D., Ruoppolo, V., et al. (2010). Magellanic penguin mortality in 2008 along the SW Atlantic coast. *Marine pollution Bulletin* 60:1652-1657.

Gasalla, M.A., Rodrigues, A.R. and Postuma, F.A. (2010). The trophic role of the squid *Loligo plei* as a keystone species in the South Brazil Bight ecosystem. *ICES Journal of Marine Science,* 67.

GEF (2013) Governance Strengthening for the Management and Protection of Coastal- Marine Biodiversity in Key Ecological Areas and the Implementation of the Ecosystem Approach to Fisheries(EAF). GEF 5112

German, C.R., Ramirez-Llodra, E., Baker, M.C., Tyler, P.A. and the ChEss Scientific Committee. (2011). Deep-water chemosynthetic ecosystem research during the Census of Marine Life decade and beyond: A proposed deep-ocean road map. *PLoS ONE* 6(8):e23259.

Giarrizo, T., Krumme, U. (2008). Heterogeneity in intertidal fish fauna assemblages along the world's longest mangrove area in Northern Brazil. *J Fish Biol* 72: 773-779.

Giberto, D., Bremec, C.S., Schejter L., Schiariti, A., Mianzan, H., Acha, E.M., (2006). The invasive rapa whelk *Rapana venosa* (Valenciennes 1846): status and potential ecological impacts in the Río de la Plata estuary, Argentina-Uruguay. *Journal of Shellfish Research* 25, 919-924.

Giberto D.A., Bremec, C.S., Schejter, l., Escolar, M., Souto, V., Schiariti, A., and Dos Santos, E.P. (2012). La ostra del pacífico *Crassostrea gigas* (Thunberg 1793) en la provincia de Buenos Aires: reclutamientos naturales en Bahía Samborombón.

Revista de Investigación y Desarrollo Pesquero *INIDEP* 21: 21-30.

Gonzalez-Silvera, A., Santamaria-del-Angela, E., Garcia, V.M.T., Garcia, C.A.E., Millán-Nuñeza, R. and Muller-Kargerd, F. (2004). Biogeographical regions of the tropical and subtropical Atlantic Ocean off South America: classification based on pigment (CZCS) and chlorophyll-a (SeaWiFS) variability. *Continental Shelf Research* 24: 983–1000.

González-Solís, J., Croxall, J., Briggs, D. (2002). Activity patterns of giant petrels, Macronectes spp., using different foraging strategies. Marine Biology 140:197-204.

González-Zevallos, D., Yorio, P. (2011. Consumption of discards and interactions between Black-browed Albatrosses (*Thalassarche melanophrys*) and Kelp Gulls (*Larus dominicanus*) at trawl fisheries in Golfo San Jorge, Argentina. *Journal of Ornithology* 152:827-838.

González-Zevallos, D., Yorio, P., Caille, G. (2007). Seabird mortality at trawler warp cables and a proposed mitigation measure: A case of study in Golfo San Jorge, Patagonia, Argentina. Biological Conservation 136:108-116.

Goodwin, C., Jones, J., Neely, K., Brickle, P., (2011). Sponge biodiversity of the Jason Islands and Stanley, Falkland Islands with descriptions of twelve new species. *Journal of the Marine Biological Association of the United Kingdom* 91, 275-301.

Gratiot N., Anthony E.J., Gardel A., Gaucherel C., Proisy C. & Wells J.T. (2008). Significant contribution of the 18.6 year tidal cycle to regional coastal changes. *Nature Geoscience*, 169-172.

Greze, V.N. (Ed.). (1984). *The bioproductive system of the macroscale oceanic gyre*. Naukova Dumka, Kiev 264 p (in Russian).

Griffiths, C.L.; Robinson, T. B.; Lange, L.; Mead, A. (2010). Marine Biodiversity in South Africa: An Evaluation of Current States of Knowledge. *PLoS ONE* 2010.

Guinard, E., Weimerskirch, H., and Jouventin, P. (1998). Population changes and demography of the northern Rockhopper Penguin on Amsterdam and Saint Paul Islands. *Colonial Waterbirds*, 222-228.

Halpern, B.S., Longo, C., Gardy, D., Mcleod, K., Samhouri, J.F., Katonas, S.K., Kleisner, K., Lester, S.E., O'leary, J., Ranelletti, M., Rosemberg, A.A., Scarborough, C., Seligs, E.R., Best, B.D., Brumbaugh, D.R., Chapin, F.S., Crowder, L.B., Daly, K.L., Doney, S.C., Elfes, C., Fogarty, M.J., Gainess, S.D., Jacobsens, K.I., Karrer, L.B., Leslie, H.M., Neeley, E., Paylu, D., Polasky, S., Ris, B., Martin, K., Stone, G.S., Sumaula, R.U., Zeller, D. (2008). A global map of human impact on marine ecosystems. *Science*, 319, 948-952.

Halpern, B.S.; Longo, C.; Hardy, D.; Mcleod, K.L.; Samhouri, J.F.; Katona, S.K.; Kleisner, K.; Lester, S.E.; O'leary, J.; Ranelletti, M.; Rosenberg, A.A.; Scarborough, C.; Selig, E. R.; Best, B. D.; Brumbaugh, D.R.; Chapin, F.S.; Crowder, L.B.; Daly, K.L.; Doney, S. C.; Elfes, C.; Fogarty, M.J.; Gaines, S.D.; Jacobsen, K.I.; Karrer, L.B.; Leslie, H.M.; Neeley, E.; Pauly, D.; Polasky, S.; Ris, B.; Martin, K. S.; Stone, G.S.; Sumaila, U.R.; Zeller, D. (2012). An index to assess the health and benefits of the global ocean. *Nature*, 488: 615-120.

Hein, R.H., Mizell, K., Koschinsky, A., Conrad, T.A. (2013). Deep-ocean mineral deposits as a source of critical metals for high- and green-technology applications: comparisons with land-based resources. *Ore Geology Reviews*, 51: 1-14.

Henson, S.A., Sarmiento, J.L., Dunne, J.P., Bopp, L., Lima, I.D., Doney, S.C., John, J., and Beaulieu, C. (2010). Detection of anthropogenic climate change in satellite records of ocean chlorophyll and productivity. *Biogeosciences*, 7, 621-640.

Hidalgo, F.J., Silliman, B.R., Bazterrica, M.C., Bertness, M.D. (2007). Predation on the rocky shores of Patagonia, Argentina. *Estuaries and Coasts* 30:886–894.

Hilton, G.M., Thompson, D.R., Sagar, P.M., Cuthbert, R.J., Cherel, Y., and Bury, S.J. (2006). A stable isotopic investigation into the causes of decline in a sub-Antarctic predator, the rockhopper penguin *Eudyptes chrysocome*. *Global Change Biology*, 12(4), 611-625.

Hockey, P.A.R.,Dean, W.R.J. and Ryan, P.G. (2005). *Roberts' birds of southern Africa* (7th ed.). Cape Town, South Africa: The Trustees of the John Voelcker Bird Book Fund.

Holland, N.D., Osborn, K.J., Gebruk, A.V., Rogacheva, A. (2013). Rediscovery and augmented description of the HMS Challenger acorn worm (Hemichordata, Enteropneusta), *Glandiceps abyssicola*, in the Equatorial Atlantic abyss. *J. Mar. Biol. Ass. UK*, 93(8): 2197-2205.

Hooker, S.H. and Boyd, I.A. (2003). Salinity sensors on seals: use of marine predators to carry CTD data loggers. *Deep Sea Research Part I: Oceanographic Research Papers* 50:7, 927-939.

Horton, T.W., Holdaway, R.N., Zerbini, A.N., Hauser, N., Garrigue, C., Andriolo, A., and Clapham, P.J. (2011). Straight as an arrow: humpback whales swim constant course tracks during long-distance migration. *Biology letters*, rsbl20110279.

Hu, C., Montgomery, E.T., Schmitt, R.W. and Muller-Karger, F.E. (2004). The dispersal of the Amazon and Orinoco River water in the tropical Atlantic and Caribbean Sea: Observation from space and SPALACE floats. In: *Deep Sea Research Part II: Topical Studies in Oceanography*.[S.l.]. 1151–1171.

ISA (2014). Report and recommendations of the Legal and Technical Commission to the Council of the International Seabed Authority relating to an application for the approval of a plan of work for exploration for cobalt-rich ferromanganese crusts by Companhia de Pesquisa de Recursos Minerais. *ISBA/ 20/ C/ 17*.

Ito, T., Parekh, P., Dutkiewicz, S., and Follows, M.J. (2005). The Antarctic Circumpolar Productivity Belt. *Geophysical Research Letters*, 32, L13604, doi:10.1029/2005gl023021, 2005.

Ivanovic, M. (2010). Alimentación del calamar *Illex argentinus* en la región patagónica durante el verano de los años 2006, 2007 y 2008. *Rev. Invest.Desarr. Pesq*. 20:51-63.

Jehl, J. R. (1974). *The distribution and ecology of marine birds over the continental shelf of Argentina in winter*. San Diego Society of Natural History.

Jouventin, P., Weimerskirch, H. (1990). Satellite tracking of wandering albatrosses. *Nature* 746-748.

Keith, P., Bail, P.Y.L., Planquette, P. (2000). *Atlas des poissons d'eau douce de Guyane (tome 2, fascicule I)*. Paris: Publications scientifiques du Muséum national d'Histoire naturelle. 286 p.

Kemper, J., Underhill, L.G., Crawford, R.J.M., Kirkman, S.P. (2007). Revision of the conservation status of seabirds and seals breeding in the Benguela ecosystem. In: *Kirkman SP (ed.), Final report of the BCLME (Benguela Current Large Marine Ecosystem) project on top predators as biological indicators of ecosystem change in the BCLME*. Cape Town: Avian Demography Unit. pp 697–704.

Koen-Alonso, M., Yodzis, P. (2005). Multispecies modelling of some components of the northern and central Patagonia marine community, Argentina. *Canadian Journal of Fisheries and Aquatic Sciences* 62:1490-1512.

Kröncke, I., Reiss, H. and Türkay, M. (2013). Macro and megafauna communities in three deep basins of the South-East Atlantic. *Deep Sea Research Part I: Oceanographic Research Papers* Volume 81, Pages 25–35.

La Mesa, M., Riginella, E., Melli, V., Bartolini, F., & Mazzoldi, C. Biological traits of a sub-Antarctic nototheniid, Patagonotothen ramsayi, from the Burdwood Bank. *Polar Biology*, 1-9.

Lam, V.W.Y., Cheung, W.W.L., Swartz, W. and Sumaila, U.R. (2012). Climate change impacts on fisheries in West Africa: implications for economic, food and nutritional security. *African Journal of Marine Science*, 34:1, 103-117.

Lana, P.C., Camargo, M.G., Brogim, R.A., Isaac, V.J. *O bentos da costa brasileira. Aval-*

iação crítica e levantamento bibliográfico. Ministério do Meio Ambiente, dos Recursos Hídricos e da Amazônia Legal/ Comissão Interministerial Para Os Recursos do Mar/Fundação de Estudos do Mar, Rio de Janeiro, 431 Pp.. Rio de Janeiro: MMA/CIRM/FEEMA, (1996). 431p Lasta, M.L. and Bremec, C.S., 1998. *Zygochlamys patagonica* in the Argentine sea: a new scallop fishery. *Journal of Shellfish Research* 17, 103-111.

Laptikhovsky, V., Arkhipkin, A., and Brickle, P. (2013). From small bycatch to main commercial species: Explosion of stocks of rock cod Patagonotothen ramsayi (Regan) in the Southwest Atlantic. *Fisheries Research*, *147*, 399-403.

Lavrado, H.P. and Brasil, A.C.S. (2010). *Biodiversidade da região oceânica profunda da Bacia de Campos*: Megafauna e Ictiofauna Demersal. Rio de Janeiro: SAG Serv. 376p.

Le Bail P.Y., Keith P., and Planquette P. (2000). Atlas des Poissons d'Eau douce de Guyane. Tome 2, fascicule II: Siluriformes. 307 p. *Patrimoines Nat.*, 43(2). Paris: MNHN/SPN.

Lentz, S.J. (1995). The Amazon River plume during AMASSEDS: subtidal current variability and the importance of wind forcing. *Journal of Geophysical Research*, 100, 2377–2390.

Lescrauwaet, A.K. (1997). Notes on the behaviour and ecology of the Peale's dolphin, *Lagenorhynchus australis*, in the Strait of Magellan, Chile. *International Whaling Commission* 47:747-755.

Lewis, M., Campagna, C. (2002). Los elefantes marinos de Península Valdés. *Ciencia Hoy*, 12: 12-22.

Lewis, M., Campagna, C. (2008). *"Mamíferos marinos" en Estado de Conservación del Mar Patagónico y Áreas de Influencia*. [online]. Puerto Madryn, publicación del Foro, available in: http://www.marpatagonico.org.

Lisnizer, N., Garcia-Borboroglu, P., Yorio, P. (2011). Spatial and temporal variation in population trends of Kelp Gulls in northern Patagonia, *Argentina. Emu* 111:259-267.

Liuzzi, M.G., López Gappa, J. (2008). Macrofaunal assemblages associated with coralline turf: species turnover and changes in structure at different spatial scales. *Mar Ecol Prog Ser*.Vol. 363: 147–156.

Longhurst, A.R., (1998). *Ecological Geography of the Sea*. Academic Press, San Diego.

Longhurst, A., Sathyendranath, S., Platt, T., Caverhill, C. (1995). An estimate of global primary production in the ocean from satellite radiometer data, *Journal of Plankton research*, 17(6): 1245-1271.

Lopez Cazorla, A., Molina, J.M., Ruarte, C. (2014). The artisanal fishery of Cynoscion guatucupa in Argentina: Exploring the possible causes of the collapse in Bahía Blanca estuary. *Journal of Sea Research*, Volume 88, Pages 29-35.

López Gappa J., (2000). Species richness of marine Bryozoa in the continental shelf and slope off Argentina (south-west Atlantic). *Diversity and Distributions* 6: 15-27.

López Gappa J., Alonso G.M. and Landoni N.A., (2006). Biodiversity of benthic Amphipoda (Crustacea: Peracarida) in the Southwest Atlantic between 35ºS and 56ºS. *Zootaxa* 1342: 1–66.

López Gappa, J., Landoni, N.A., (2005). Biodiversity of Porifera in the Southwest Atlantic between 35 S and 56 S. *Revista del Museo Argentino de Ciencias Naturales* 7, 191-219.

Lowe Mc-Connell, R.H. (1962). The fishes of the British Guiana continental shelf, Atlantic coast of South America, with notes on their natural history. *Zoological Journal of the Linnean Society* 44, 669-700.

Luna, F.O., da Silva, V.M.F. Andrade, M.C.M., Marques, C.C., Normande, I.C., Veloso T.M.G. & Severo, M.M. (2010). *Plano Nacional para a Conservação dos Sirênios*. Instituto Chico Mendes de Conservação da Biodiversidade (ICMBio), 80 pp. Brasília. ISBN: 978856184221-5.

Lutjeharms J.R.E., van Ballegooyen R.C. (1988). The retroflection of the Agulhas Current. *J Physical Oceanog* 18: 1570–1583.

Maida, M., Ferreira, B.P. (1997). Coral reefs of Brazil: an overview. In: *Proceedings of the 8th International Coral Reef Symposium*, Panama, 1996

Malyutina, M.V. (2004). Russian deep-sea investigations of Antarctic fauna. *Deep Sea Research Part II* 51:1551–1570.

Marone, E., Lana, P.C., Andriguetto, J.M., Seixas, C.S., Turra, A., Knoppers, B.A. . Coastal Ecosystems and Human Well-Being. The case of MAFU Brazil and a program in progress with India and South Africa. (2010). Cátedra UNESCO sobre Desarrollo Sostenible y Educación *Ambiental de la UPV/EHU*, v. 4, p. 113-125.

Masello, J.F., Mundry, R., Poisbleau, M., Demongin, L., Voigt, C.C., Wikelski, M., and Quillfeldt, P. (2010). Diving seabirds share foraging space and time within and among species. *Ecosphere*, 1(6), art19.

McDonagh, E.L., and King, B.A. (2005). Oceanic fluxes in the South Atlantic. *Journal of physical oceanography*, *35*(1), 109-122.

McPherson, E. (1984). Crustáceos decápodos del Banco Valdivia (Atlántico Sudoriental). Resultados de Expediciones Científicas (*Suplemento de Investigaciones Pesqueras*), 12:39–105.

Milessi, A.C., Arancibia, H., Neira, D., Defeo, O. (2005). The mean trophic level of Uruguayan landings during the period 1990–2001. *Fish Res* 74: 223–231.

Miloslavich, P., Klein, E, Díaz, J.M., Hernandez, C.E., Bigatti,G., Campos, L., Artigas, F., Castillo, J., Penchaszadeh, P., Neill, P., Carranza, A., Retana, M., Díaz de Astarloa, J.M., Lewis, M., Yorio, P., Piriz, M., Rodriguez, G., Yoneshigue-Valentin, Y., Gamboa, L., Martín, A. (2011). Marine Biodiversity in the Atlantic and Pacific Coasts of South America: Knowledge and Gaps. *PLOS ONE*, Vol. 6, pp. 1 – 44.

Mittermeier R.A., Turner W.R., Larsen F.W., Brooks T.M., Gascon C. (2011). Global Biodiversity Conservation: The Critical Role of Hotspots. In*: Zachos F.E., Habel J.C., editors. Biodiversity Hotspots*. Berlin Heidelberg: Springer-Verlag.

MMA. (2006). *Programa REVIZEE - Relatório Executivo. 1ed.Brasília*: Ministério do Meio Ambiente.

MMA, (2007). Áreas Prioritárias para Conservação, Uso Sustentável e Repartição de Benefícios da Biodiversidade Brasileira: Atualização - Portaria MMA nº9, de 23 de janeiro de 2007. / Ministério do Meio Ambiente, Secretaria de Biodiversidade e Florestas. – Brasília: MMA, p. : il. color. ; 29 cm. (*Série Biodiversidade*, 31).

MMA, (2009). *Informe sobre as espécies exóticas invasoras marinhas no Brasil* / Ministério do Meio Ambiente; Rubens M. Lopes/IO-USP... [et al.], Editor. – Brasília:

Muelbert, J.H., Acha, M., Mianzan, H., Guerrero, R., Reta, R., Braga, E.S., Garcia, V., Berasategui, A., Gomez-Erache, M. and Ramírez, F. (2008). Biological, physical and chemical properties at the Subtropical Shelf Front Zone in the SW Atlantic Continental Shelf. *Continental Shelf Research* 28 1662–1673.

Mugetti, A.C., Calcagno, A.T., Brieva, C.A., Giangiobbe, M.S., Pagani, A., and Gonzalez, S. (2004). Aquatic habitat modifications in La Plata River Basin, Patagonia and associated marine areas. *Ambio* 33: 78–87.

Murray, J. (1895). *A Summary of the Scientific Results. Report of Scientific Results of the Voyage of H.M.S. Challenger During the Years 1873–76*. 1607 pp.

Nel, D.C., Taylor, F.E. (2002). *Globally threatened seabirds at risk from longline fishing: international conservation responsibilities*. Cape Town: BirdLife International Seabird Conservation Programme.

Neves, T., Olmos, F. (1998). Albatross mortality in fisheries off the coast of Brazil. Albatross Biology and Conservation. Pp 214-219.in: *Robertson G. & Gales R. (eds.) Albatross biology and conservation*. Surrey Beatty and Sons, Chipping Norton.

Nicholls, D.G., Robertson, C.J.R., and Naef-Daenzer, B. (2005). Evaluating distribution modelling using kernel functions for northern royal albatrosses (Diomedea sanfordi) at sea off South America. *Notornis*, 52(4), 223.

Nión, H. (1998). Peces del Río de la Plata y algunos aspectos de su ecología. Capítulo 6. pp: 169-190. In: *Wells, P.G. & G.R. Daborn (Eds.) El Río de la Plata. Una revisión Ambiental*. Dalhousie University, Halifax, Nova Scotia, Canada. 256 pp.

Ogden, J., Podestá, G., Zingone, A., Wiebe, W.J., Myers, R.A. (2004). Las ciencias del mar en la Argentina. *Ciencia Hoy* 13: 23–46.

Olavo, G., Costa, P.A., Martins, A.S., and Ferreira, B.P. (2011). Shelf-edge reefs as priority areas for conservation of reef fish diversity in the tropical Atlantic. Aquatic Conservation: *Marine and Freshwater Ecosystems*, 21(2), 199-209.

Olguin, H.F., Boltovskoy, D., Lange, C.B. and Brandini, F. (2006). Distribution of spring phytoplankton (mainly diatoms) in the upper 50 m of the Southwestern Atlantic Ocean (30–61°S). *J. Plankton Res*. 28(12):1107-1128.

Olivier, S.R., de Paternoster, I.K., Bastida, R. (1966). Estudios biocenóticos en las costas de Chubut (Argentina) I. Zonación biocenológica de Puerto Pardelas (Golfo Nuevo). *Boletín Instituto de Biología Marina* 10:5-71.

Olmos, F., Bastos, G.C.C., da Silva Neves, T. (2000). Estimating seabird bycatch in Brazil. *Marine Ornithology* 28(2).

Orensanz, J.M.L., Schwindt, E., Pastorino, G., Bortolus, A., Casas, G., Darrigran, G., Elías, R., Lopez Gappa J.J., Obenat, S. Pascual, M., Penchaszadeh, P., Piriz, M.L., Scarabino, F., Spivak, E.D. and Vallarino, E.A. (2002). No longer the pristine confines of the world ocean: a survey of exotic marine species in the southwestern Atlantic. *Biological Invasions*, 4(1-2), 115-143.

Orgeira, J.L. (2001a). Distribución espacial de densidades de aves marinas en la plataforma continental argentina y Océano Atlántico Sur. *Ornitología Neotropical* 12:45-55.

Orgeira, J.L. (2001b). Nuevos registros del Petrel Atlántico (Pterodroma incerta) en el océano Atlántico Sur y Antártida. *Ornitología Neotropical* 12:165-171.

Otley, H., Munro, G., Clausen, A., Ingham, B. (2008). *Falkland Islands State of the Environment Report 2008*. Falkland Islands Government and Falklands Conservation, Stanley.

Otley, H., Ingham, R. (2003). Marine debris surveys at Volunteer beach, Falkland Islands, during the summer of 2001-2002. *Marine Pollution Bulletin* 46:1534-39.

Pedraza, S.N. (2008). *Ecología poblacional de la tonina overa Cephalorhynchus commersonii (Lacépède, 1804) en el litoral patagónico*. Tesis Doctoral, Universidad de Buenos Aires, Buenos Aires, Argentina, 213 pp.

Penchaszadeh P.E. (1973). Ecología de la comunidad del mejillín (*Brachydontes rodriguezii* d'Orb.) en el mediolitoral rocoso de Mar del Plata (Argentina): el proceso de recolonización. *Physis* 32:51-64.

Peres M.B., Barreto R., Lessa R., Vooren C., Charvet P., et al. (2012). *Heavy fishing puts Brazilian sharks and rays in great trouble*. Abstract. 6th World Fisheries Congress, 7–11 May 2012, Edinburgh, Scotland. p.21.

Perez, J.A.A., Wahrlich, R. (2005). A bycatch assessment of the gillnet monkfish *Lophius gastrophysus* fishery off southern Brazil. *Fisheries Research* 72, 81–95.

Perez, J.A.A., Pezzuto, P.R., Lucato, S.H.B., Vale, W.G. (2003). Frota de arrasto de Santa Catarina. In: *Cergole M.C., Rossi-Wongtscowski C.L.D.B., editors. Dinamica das frotas pesqueiras – Analise das principais pescarias comerciais do Sudeste-Sul do Brasil. Avaliação do potencial sustentavel de recursos vivos na Zona Economica Exclusiva, Programa REVIZEE*, Score Sul. Evoluir, São Paulo, pp. 117–183.

Perez, J.A.A., dos Santos Alves, E., Clark, M.R., Aksel Bergstad, O., Gebruk, A. Azevedo Cardoso, I. and Rogacheva, A. (2012). Patterns of life on the southern Mid-Atlantic Ridge: Compiling what is known and addressing future research. *Oceanography* 25(4):16–31.

Perez, J.A.A., Pereira, B.N., Pereira, D.A., Schroeder, R. (2013). Composition nd diversity patterns of megafauna discards in the deep-water shrimp trawl fishery off Brazil. *Journal of Fish Biology* 83, 804-825.

Pérez Macri, G., Crespo, A. (1989). Survey of the franciscana, Pontoporia blainvillei, along the Argentine coast, with a preliminary evaluation of Pontoporia blainvillei (foto Pablo Bordino). mortality in coastal fisheries.. In *Biology and Conservation of the River Dolphins (W. F. Perrin, R. L. Brownell Jr., K. Zhou and J. Liu, eds.)*. Pages 57-63. Occasional Papers of the IUCN Species Survival Commission (SSC) 3.

Perrin, W.F., Wursig, B. and Thewissen, J.G.M. (2009). *Encyclopedia of Marine Mammals*. 2nd ed. Academic Press, San Diego. 1316 pp.

Petersen, S.L., Honig, M.B., Ryan, P.G. and Underhill, L.G. (2009). Seabird bycatch in the pelagic longline fishery off southern Africa. African. *Journal of Marine Science*, 31, 191–204.

Planquette, P., Keith, P., Le Bail, P.Y., (1996). Atlas des Poissons d'Eau douce de Guyane (tome 1). *Collection Patrimoines Nat*., 22. 429 p. Paris: IEGB-MNHN, INRA, CSP, Min. Environ

Piola, A.R., Matano, R.P., Palma, E.D., Möller, O.O. Jr., and Campos, E.J.D. (2008). The influence of the Plata River discharge on the western South Atlantic shelf. *Geophysical Research Letters*, Vol. 32,

Piontkovski S.A., Landry, M.R., Finenko, Z.Z., Kovalev, A.V., Williams, R., Gallienne, C.P., Mishonov, A.V., Skryabin, V.A., Tokarev, Y.N., Nikolsky, V.N. (2003). Plankton communities of the South Atlantic anticyclonic gyre. *Oceanologica Acta* 26 (2003) 255–268.

Pistorius, P., Huin, N., Crofts, S. (2010). Population change and resilience in Gentoo Penguins (*Pygoscelis papua*) at the Falkland Islands. *Marine Ornithology* 38: 49-53.

Polovina, J.J., Howell, E.A., and Abecassis, M. (2008). Ocean's least productive waters are expanding. *Geophysical Research Letters*, vol. 35, L03618, doi:10.1029/2007GL031745.

Poncet, S., Robertson, G., Phillips, R.A., Lawton, K., Phalan, B., Trathan, P.N., Croxall, J.P. (2006). Status and distribution of wandering, black-browed and grey-headed albatrosses breeding at South Georgia. *Polar Biology* 29:772-781.

Portela, J., Acosta, J., Cristobo, J., Muñoz, A., Parra, S., Ibarrola, T., Del Río, J.L., Vilela, R., Ríos, P., Blanco, R., Almón, B., Tel, E., Besada, V., Viñas, L., Polonio, V., Barba, M., Marín, P., (2012). Management Strategies to Limit the Impact of Bottom Trawling on VMEs in the High Seas of the SW Atlantic, In: Cruzado, A. (Ed.) *Marine Ecosystems*., pp. 199-228.

Postuma, F.A. and Gasalla, M.A. (2010). On the relationship between squid and the environment: artisanal jigging for *Loligo plei* at São Sebastião Island (24°S), southeastern Brazil. *ICES Journal of Marine Science*, 67(7): 1353-1362.

Praderi R., Pinedo, M.C., Crespo, E.A. (1989). Conservation and management of Pontoporia blainvillei in Uruguay, Brazil and Argentina. In *Biology and Conservation of the River Dolphins (W.F. Perrin, R.L. Brownell Jr., K. Zhou and J. Liu, eds.)*. Pages 52-56. Occasional Papers of the IUCN Species Survival Commission (SSC) 3.

Prince, P.A., Croxall, J.P., Trathan, P.N., Wood, A.G. (1998). The pelagic distribution of South Georgia albatrosses and their relationship with fisheries. Pp. 137–167 in: *Robertson G. & Gales R. (eds) Albatross biology and conservation*. Surrey Beatty and Sons, Chipping Norton.

Prost M.T.R.C., Rabelo B.V. (1996). Variabilidade fito-espacial de manguezais litorâneos e dinâmica costeira: exemplos da Guiana Francesa, Amapá e Pará. Bol Mus Para E Goeldi 8: 101-121.Souza and Fonseca. 2008.

Pütz, K., Raya Rey, A., Otley, H. (2013). Southern Rockhopper Penguin. En *PENGUINS:*

Pütz, K., Ingham, R.J., Smith, J.G., Lüthi, B.H. (2002). Winter dispersal of rockhopper penguins *Eudyptes chrysocome* from the Falkland Islands and its implications for conservation. *Marine Ecology Progress Series* 240:273-284.

Pütz, K., Schiavini, A., Rey, A.R., Lüthi, B.H. (2007). Winter migration of magellanic penguins (*Spheniscus magellanicus*) from the southernmost distributional range. Marine Biology 152:1227-1235.

Pütz, K., Clausen, A.P., Huin, N., Croxall, J.P. (2003). Re-evaluation of historical Rockhopper Penguin population data in the Falkland Islands. *Waterbirds* 26:169-175.

Quintana, F., Punta, G., Copello, S., Yorio, P. (2006). Population status and trends of Southern Giant Petrels (*Macronectes giganteus*) breeding in North Patagonia, Argentina. *Polar Biology* 30:53-59.

Quintana, F., Dell'Arciprete, P.O. (2002). Foraging grounds of southern giant petrels (*Macronectes giganteus*) on the Patagonian shelf. *Polar Biology* 25:159-161.

Raya Rey, A., Liljesthröm, M., Saenz Samaniego, R., Schiavini, A. (2014). Species-specific population trends detected for penguins, gulls and cormorants over 20 years in sub-Antarctic Fuegian Archipelago Polar Biology Online DOI 10.1007/s00300-014-1526-6.

Raya Rey, A., Trathan, P., Pütz, K., & Schiavini, A. (2007). Effect of oceanographic conditions on the winter movements of rockhopper penguins *Eudyptes chrysocome chrysocome* from Staten Island, Argentina. *Marine Ecology Progress* Series 330:285-295.

Reid, T., Huin, N. (2005). Census of the Giant-petrel population of the Falkland Islands. *Falklands Conservation Newsletter*: 1-2.

Reyes, L.M., Crespo, E.A., and Szapkievich, V. (1999). Distribution and population size of the southern sea lion (Otaria flavescens) in central and southern Chubut, Patagonia, Argentina. *Marine Mammal Science*, 15(2), 478-493.

Rocha, L.A. (2003). Patterns of distribution and processes of speciation in Brazilian reef fishes. *J. Biogeogr*. 30: 1161–1171.

Rogers, A.D., Gianni, M. (2010). The implementation of UNGA resolutions 61/105 and 64/72 in the Management of Deep-sea fisheries on the High Seas. *Report prepared for the Deep Sea Conservation Coalition. International Programme on the State of the Ocean*, London, United Kingdon, 97p.

Roessig, J.R., Christa M. Woodley, Joseph J. Cech Jr., Lara J. Hansen. Effects of global climate change on marine and estuarine fishes and fisheries. *Reviews in Fish Biology and Fisheries* (2004), Volume 14, Issue 2, pp 251-275.

Rosenzweig C., Karoly D., Vicarelli M. et al. (2008). Attributing physical and biological impacts to anthropogenic climate change. *Nature*, 453, 353–357.

Roux A., Bremec C., Schejter L. and Lasta M., (2010). Gasterópodos y Bivalvos. In: M.B. Cousseau (Ed.). Peces, crustáceos, y moluscos registrados en el sector del Atlántico Sudoccidental comprendido entre 34ºS y 55ºS, con indicación de las especies de interés pesquero. *INIDEP Serie Informe Técnico, 5, Mar del Plata*, 79-112.

Roberts, M.J. (2005). Chokka squid (Loligo vulgaris reynaudii) abundance linked to changes in South Africa's Agulhas Bank ecosystem during spawning and the early life cycle. *ICES Journal of Marine Science*, 62: 33-55

Roessig, J.R., Christa M. Woodley, Joseph J. Cech Jr., Lara J. Hansen. Effects of global climate change on marine and estuarine fishes and fisheries. *Reviews in Fish Biology and Fisheries* (2004) Volume 14, Issue 2, pp 251-275

Rosas-Luis, R., Sánchez, P., Portela, J.M., del Rio, J.L. (2013). Feeding habits and trophic interactions of *Doryteuthis gahi*, *Illex argentinus* and *Onykia ingens* in the marine ecosystem off the Patagonian Shelf. *Fisheries Research* 152: 37– 44.

Rovai, A.S., Menghini, R.P., Schaeffer-Novelli, Y., Molero, G.C., Coelho Jr., C., (2012). Protecting Brazil's coastal wetlands. *Science*, 335: 1571-1572.

Rowntree, V., Mc Guiness, P., Marshall, K., Payne, R., Sironi, M., Seger, J. (1998). Increased harassment of Right Whales (*Eubalaena australis*) by kelp gulls (*Larus dominicanus*) at Peninsula Valdés, Argentina. *Marine Mammal Science*, 99: 115.

Rowntree, V., Payne, R., Schell, D.M. (2001). Changing partterns of hábitat use by southern Right Whales (Eubalaena australis) on their nursery ground at Península Valdés, Argentina, and in their long-ragne movement. *International Whaling Commission*. Special Issue 2:133- 143.

Rozzi, R., Armesto, J.J., Gutierrez, J.R. et al. (2012). Integrating ecology and environmental ethics: Earth stewardship in the southern end of the Americas. *BioScience*, 62, 226–236.

Ryan P.G., Moloney C.L. (1988). Effect of trawling on bird and seal distributions in the southern Benguela region. *Marine Ecology Progress* Series 45: 1–11.

Sala, J.E., Wilson, R.P., Frere, E., Quintana, F. (2014). Flexible foraging for finding fish: variable diving patterns in Magellanic penguins Spheniscus magellanicus from different colonies. *Journal of Ornithology*, 1-17.

Salas, S., Chuenpagdee, R., Charles, A., Seijo, J.C. (eds) (2011). Coastal fisheries of Latin America and the Caribbean. *FAO Fisheries and Aquaculture Technical Paper*. No. 544. FAO, Rome.

Sánchez, V., Zaixso, H.E. (1995). Secuencias de recolonización mesolitoral en una costa rocosa del Golfo San José (Chubut, Argentina). Naturalia Patagónica, Ciencias Biológicas, 3:57-83.

Sapoznikow, A., Giaccardi, M., Tagliorette, A. (2008). "Indicadores: Cobertura de Áreas Costeras y Marinas Protegidas" en Estado de Conservación del Mar Patagónico y Areas de Influencia. Puerto Madryn: Publicación del Foro. Available: http://www.marpatagonico.org.

Seco Pon, J.P., Copello, S., Moretinni, A., et al. (2013). Seabird and marine-mammal attendance and by-catch in semi-industrial trawl fisheries in near-shore waters of northern Argentina. Marine and Freshwater Research 64: 237-248.

Seco Pon, J.P., García, G., Copello, et al. (2012). Seabird and marine mammal attendance in the Chub mackerel Scomber japonicus semi-industrial Argentinian purse seine fishery. Ocean & Coastal Management 64:56-66.

Schejter, L., Calcinai, B., Cerrano, C., Bertolino, M., Pansini, M., Giberto, D., Bremec, C., (2006). Porifera from the Argentine Sea: Diversity in Patagonian scallop beds. *Italian Journal of Zoology* 73, 373-385.

Schejter L., Escolar M., Marecos A. and Bremec C., (2014a). . Asociaciones faunísticas en las unidades de manejo del recurso "vieira patagónica" en el frente de talud durante el período 1998-2009. (Technical Report) *Informe de Investigación INIDEP* Nº14, 29pp.

Schejter, L., López Gappa, J. and Bremec, C., (2014b). . Epibiotic relationships on Zygochlamys patagonica (Mollusca, Bivalvia, Pectinidae) increase biodiversity in a submarine canyon in Argentina. *Deep Sea Research* II 104: 252-258.

Schiavini, A., Frere, E., Gandini, P., García, N. and Crespo, E. (1998). Albatross–fisheries interactions in Patagonian shelf waters, pp. 208–213, In G. *Robertson and R. Gales, eds. Albatross biology and conservation*. Australia: Surrey Beatty and Sons.

Schiavini, A., Raya Rey, A. (2001). Aves y Mamíferos en Tierra del Fuego. Estado de situación, interacción con actividades humanas y recomendaciones para su manejo. Informe preparado bajo contrato con el Proyecto Consolidación e Implementación del Plan de Manejo de la Zona Costera Patagónica. Proyecto ARG/97/G31 GEF/ PNUD/MREIC.

Schiavini, A.C.M., Pedraza, S.N., Crespo, E.A., Gonzalez, R., Dans, S.L. (1999). The abundance of dusky dolphins (Lagenorhynchus obscurus) off north and central Patagonia, Argentina, in spring and a comparison with incidental catch in fisheries. Results from a pilot survey in spring 1995. *Marine Mammal Science*, 15:828-840.

Schiavini, A.C.M., Crespo, E.A., & Szapkievich, V. (2004). Status of the population of South American sea lion (*Otaria flavescens* Shaw, 1800) in southern Argentina. *Mammalian Biology-Zeitschrift für Säugetierkunde*, 69(2), 108-118.

Schwindt E., De Francesco, C.G. and Iribarne O.O., (2004). Individual and reef growth of the invasive reef-building polychaete *Ficopomatus enigmaticus* in a southwestern Atlantic coastal lagoon. *Journal of the Marine Biological Association of the UK* 84(05): 987 - 993.

Serejo, C.S. (2014). A new species of Stilipedidae (Amphipoda: Senticaudata) from the south Mid-Atlantic Ridge. *Zootaxa* 3852 (1): 133-140.

Shannon, L.V., Nelson, G. (1996). The Benguela: Large-scale features and process and system variability. In: *Wefer G, Berger WH, Siedler G, Webb DJ, editors. The South Atlantic: Present and past circulation*. Telos: Springer-Verlag. 644 p.

Secchi, E.R., Zerbini, A.N., Bassoi, M., Dalla Rosa, L., Moller, L.M., Roccha-Campos, C.C. (1997). Mortality of franciscanas, Pontoporia blainvillei, in coastal gillnetting in southern Brazil: 1994-1995. *Report International Whaling Commission*, 47:653-658.

Souto, V., Escolar, M., Genzano, G. and Bremec, C., (2014). Species richness and distribution patterns of echinoderms in the southwestern Atlantic Ocean (34-56°S). *Scientia Marina* doi: http://dx.doi.org/10.3989/scimar.03882.26B.

Souto, V., Genzano, G., Bremec, C., (2010). Patrones de distribución y riqueza específica de invertebrados bentónicos en el Mar Argentino: Hydrozoa como caso de estudio. IX Reunión Argentína de Cladística y Biogeografía, November 15-17 2010, La Plata. *Book of ABstracts*: 88.

Souza, R.F.C., Fonseca, A.F. (2008). Síntese do Conhecimento sobre a pesca e a biodiversidade das espécies de peixes marinhos e estuarinos da costa norte do Brasil. Coleção Síntese do Conhecimento sobre a Margem Equatorial. *Amazônica*. v.3, p. 1-29.

Stagi, A., Vaz-Ferreira, R., Marin, Y., Joseph, L. (1998). The conservation of albatrosses in Uruguayan waters, pp. 220–224. In: *Robertson G.& Gales R. (eds). Albatross biology and conservation*. Surrey Beatty and sons, Chipping Norton.

Stokes, D.L., Boersma, P.D. (1999). Where breeding Magellanic penguins Spheniscus magellanicus forage: satellite telemetry results and their implications for penguin conservation. *Marine Ornithology* 27:59-65.

Stokes, D.L., Boersma, P.D., Davis, L.S. (1998). Satellite tracking of Magellanic Penguin migration. Condor 376-381.

Strange, I.J. (1992). *Wildlife of the Falkland Islands: And South Georgia*. Harper Collins.

Suárez, N., Retana, M.V., Yorio, P. (2012). Spatial patterns in the use of foraging areas and its relationship with prey resources in the threatened Olrog's Gull (Larus atlanticus). *Journal of Ornithology* 153:861-871.

Suárez, N., Yorio, P. (2005). Foraging patterns of breeding dolphin gulls Larus scoresbii at Punta Tombo, Argentina. *Ibis*, 147: 544-551.

Tatián, M., Alurralde, G., Lagger, C., Maggioni, T., Schwindt, E., Taverna, A. and Varela, M.M., (2013). Present knowledge on ascidian biodiversity at the SW Atlantic (Argentine Sea) with emphasis in invasive species. 7th. Tunicate Meeting, July 22-26 of 2013, Naples, Italy. *Book of Abstracts*: 54-55.

Thompson, G.A., Dinofrio, E.O. and Alder, V.A. (2013). Structure, abundance and biomass size spectra of copepods and other zooplankton communities in upper waters of the Southwestern Atlantic Ocean during summer. *J. Plankton Res*. 35 (3) 610-629.

Trathan, P.N., Croxall, J.P. (2004). Marine predators at South Georgia: an overview of recent bio-logging studies. Memoirs of National Institute of Polar Research 58:118-132.

Turra, A., Cróquer, A., Carranza, A., Mansilla, A., Areces, A. J., Werlinger, C., Martínez-Bayón, C., Nassar, A., Plastino, E., Schwindt, E., Scarabino, F., Chow, F., Figueroa Berchez, F., Hall-Spencer, J.M., Soto, L.A., Buckeridge, M.S., Copertino, M.S., De Széchy, M.T., Menezes, Ghilardi-Lopes, N. P., Horta, P., Coutinho, R., Fraschetti, S., Leão, Z.M.A.N. (2013). Global environmental changes: setting priorities for Latin American coastal habitats. *Global Change Biology*, v. 19, p. 1965-1969.

Uhart, M., Karesh, W., Cook, R. (2008). "¿Es el Mar Patagónico un ecosistema saludable?" en Estado de Conservación del Mar Patagónico y Áreas de Influencia. [online]. Puerto Madryn, publicación del Foro, available in: http://www.marpatagonico.org.

Veit, R.R. (1995). Pelagic communities of seabirds in the South Atlantic Ocean. *Ibis*, 137(1), 1-10.

Wanless, R.M., Ryan, P.G., Altwegg, R., Angel, A., Cooper, J. Cuthbert, R. and Hilton, G.M. (2009). From both sides: dire demographic consequences of carnivorous mice and longlining for the Critically Endangered Tristan Albatrosses on Gough Island. *Biol. Conserv*. 142: 1710–1718.

Weaver, P.P.E., Benn, A., Arana, P.M., Ardron, J.A., Bailey, D.M., Baker, K., Billett, D.S.M., Clark, M.R., Davies, A.J., Durán Muñoz, P., Fuller, S.D., Gianni, M., Grehan, A.J., Guinotte, J., Kenny, A., Koslow, J.A., Morato, T., Penney, A.J., Perez, J.A.A., Priede, I.G., Rogers, A.D., Santos, R.S., Watling, L. (2011). The impact of deep-sea fisheries and implementation of the UNGA Resolutions 61/105 and 64/72. *Report of an international scientific workshop*, National Oceanography Centre, Southampton, 45 pp

Wei, C., Rowe, G.T., Escobar-Briones, E., Boetius, A., Soltwedel, T., Caley, M.J., Soliman, Y., Huettmann, F., Qu, F., Yu, Z. and others. (2010). Global patterns and predictions of seafloor biomass using random forests. *PloS ONE* e15323.

Weimerskirch, H., Mougey, T., Hindermeyer, X. (1997). Foraging and provisioning strategies of black-browed albatrosses in relation to the requirements of the chick: natural variation and experimental study. *Behavioral Ecology* 8: 635-643.

Wilson, R.P., Liebsch, N., Davies, et al. (2007). All at sea with animal tracks; methodological and analytical solutions for the resolution of movement. Deep Sea Research Part II: *Topical Studies in Oceanography* 54: 193-210.

Wolfaardt, A. (2012). *An assessment of the population trends and conservation status of black-browed albatrosses in the Falkland Islands. ACAP UK South Atlantic Overseas Territories*, Joint Nature Conservation Committee, UK.

Woods, W., Woods, A. (1997). *Atlas of breeding birds of the Falkland Islands*. Redwood Books, Trowbridg.

Yodzis, P. (1998). Local trophodynamics and the interaction of marine mammals and fisheries in the Benguela ecosystem. *Journal of Animal Ecology*. 67: 635-658

Yorio, P., Tagliorette, A., Harris, G., Giaccardi, M. (1998). Áreas protegidas costeras de la Patagonia: síntesis de información, diagnosis sobre su estado actual de protección y recomendaciones preliminares. *Fundación Patagonia Natural*. Puerto Madryn. 75 pp.

Yorio, P. (2005). Estado poblacional y de conservación de gaviotines y escúas que se reproducen en el litoral marítimo argentino. *El hornero* 20: 75-93.

Yorio, P., Frere, E., Gandini, P., Conway, W. (1999). Status and conservation of seabirds breeding in Argentina. Bird Conservation International 9: 299-314.

Zaixso, H.E., Boraso, A.L., Lopez Gappa, J.J. (1978). Observaciones sobre el mesolitoral rocoso de la zona de Ushuaia (Tierra del Fuego, Argentina). *Ecosur* 5:119-130.

Zaixso, H.E. and C.T. Pastor. (1977). Observaciones sobre la ecología de los mitílidos de la ría Deseado. I. Distribución y análisis biocenótico. *ECOSUR*, 4(7): 1-46.

Zamponi, M., (2008). La corriente de Malvinas: ¿una vía de dispersión para cnidarios bentónicos de aguas frías? *Revista Real Academia Galega de Ciencias* XXVII, 183-203.

Zelaya, D.G., (2014). Marine Bivalves from the Argentine coast and shelf: a reassessment of species diversity and faunistic affinities. Mollusca. México, June 22-27 of 2014. *Book of Abstracts*: 239-240.

Zibrowius, H., and Gili, J.M. (1990). Deep-water Scleractinia (Cnidaria: Anthozoa) from Namibia, South Africa and Walvis Ridge, Southeastern Atlantic. *Scientia Marina* 54(1):19–46.

36C North Pacific Ocean

Contributors:
Thomas Therriault (Convenor), Chul Park (Lead Member) and Jake Rice (Co-Lead Member and Editor of Part VI)

Chapter 36C

North Pacific Ocean

1 Introduction

The Pacific is the largest division of the World Ocean, at over 165 million km2, extending from the Arctic Ocean in the north to the Southern Ocean in the south (Figure 36C.1). Along the western margin are several seas. The Strait of Malacca joins the Pacific and the Indian Oceans to the west, and the Drake Passage and the Strait of Magellan link the Pacific with the Atlantic Ocean to the east. To the north, the Bering Strait connects the Pacific with the Arctic Ocean (International Hydrographic Organization, 1953). The Pacific Ocean is further subdivided into the North Pacific and South Pacific; the equator represents the dividing line. The North Pacific includes the deepest (and, until recently, the least explored) place on Earth, the Mariana Trench, which extends to almost 11 km below the ocean's surface, although the average depth of the North Pacific is much less, at approximately 4.3 km. Thus, the North Pacific encompasses a wide variety of ecosystems, ranging from tropical to arctic/sub-arctic with a wide diversity of species and habitats. Further, the volcanism that creates the "rim of fire" around the Pacific has resulted in unique undersea features, such as hydrothermal vents (including the Endeavor Hydrothermal Vents) and seamount chains (including the Hawaiian-Emperor Seamount Chain). Both create unique habitats that further enhance biodiversity in the North Pacific. The continental shelves around the North Pacific tend to be very narrow with highly variable productivity, with the exception of the continental shelf of the Bering Sea, which is one of the largest and most productive in the World Ocean (Miles et al., 1982). Further influencing productivity and biological diversity in the North Pacific is a series of large-scale oceanic currents on both sides of the basin, especially the Kuroshio and Oyashio Currents on the western side and the Alaska and California Currents on the eastern side. Also, the North Pacific Transition Zone (NPTZ) is an oceanographic feature of special importance to the biology of many species in the North Pacific Ocean. This 9,000 km wide upper water column oceanographic feature is bounded by thermohaline fronts thereby establishing a highly productive habitat that aggregates prey resources, attracts a number of pelagic predators, and serves as a migratory corridor. Ocean climate indices, such as the Pacific Decadal Oscillation (PDO), reflect spatial and temporal variability observed in the North Pacific (Mantua and Hare, 2002). For example, the PDO tends to indicate that a cool eastern North Pacific is associated with a warmer central and western North Pacific and *vice versa*, thereby contributing to spatial and temporal variability in ecosystem productivity and shifting patterns of biological diversity. The density of human habitation around the North Pacific is more concentrated in southern latitudes and on the western side of the basin. This in turn influences the anthropogenic stressors affecting biodiversity and productivity.

The boundaries and names shown and the designations used on this map do not imply official endorsement or acceptance by the United Nations.

Figure 36C.1 | Sources: Bathymetry extracted from the GEBCO Digital Atlas (GDA): IOC, IHO and BODC, 2003. Centenary Edition of the GEBCO Digital Atlas, published on CD-ROM on behalf of the Intergovernmental Oceanographic Commission and the International Hydrographic Organization as part of the General Bathymetric Chart of the Oceans, British Oceanographic Data Centre, Liverpool, U.K. More information at http://www.gebco.net/data_and_products/gebco_digital_atlas/ Ocean and Sea names extracted from ESRI, DeLorme, HERE, GEBCO, NOAA, National Geographic, Geonames.org, and other contributors More information at http://www.arcgis.com/home/item.html?id=0fd0c5b7a647404d8934516aa997e6d9. With the kind assistance of the FAO.

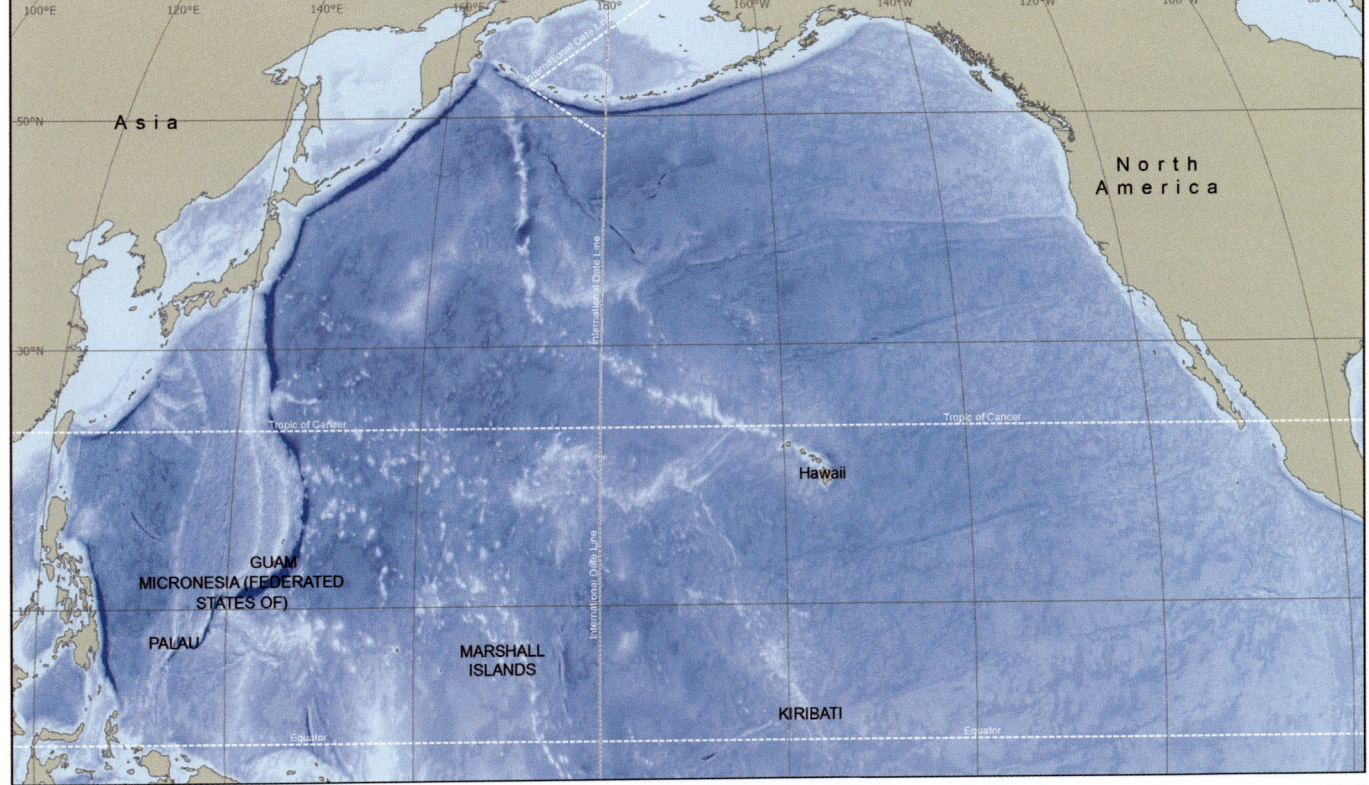

2 Coastal Areas of the North Pacific

Like other oceanic basins, the coastal areas of the North Pacific encompass a wide variety of complex habitat patches, each with different levels and types of biological diversity. Spalding et al. (2007) identify at least 50 ecoregions around the North Pacific, based in part on their relatively homogenized biological diversity and differentiation from adjacent areas, but status and trend information for biodiversity is not available even at this intermediate spatial scale. Limited information is derived from localized, smaller-scale studies conducted for specific habitat patches (e.g., coral reefs, estuaries, etc.) or fish stocks, but synthesis at the basin scale remains a critical gap for coastal areas of the North Pacific. For example, Japan has established a programme to track community-structure changes at 1,000 monitoring sites (both terrestrial and marine) and many countries around the North Pacific conduct stock assessments for major commercial species, but higher-level synthesis remains a gap. Furthermore, coastal systems are under different pressures in different parts of the basin, which will only complicate higher-level synthesis of status and trends.

2.1 Biodiversity status and trends

2.1.1 Primary producers

Climatic variability continues to increase in the North Pacific Ocean, especially in the eastern part of the basin, where both extreme warm and cool events have occurred in the Gulf of Alaska, the California Current and equatorial waters in recent years (Sydeman et al., 2013). At finer spatial scales, eddies and current meanders are important determinants of ecosystem productivity. For example, in the Gulf of Alaska region, eddies influence nutrients, phytoplankton, and even higher trophic levels (Ream et al., 2005). In the California Current, chlorophyll concentrations have increased (Kahru et al., 2009), but this has resulted in a shift to a community more dominated by dinoflagellates, at least in Monterey Bay, that has resulted in significant ecosystem changes, including impacts at higher trophic levels. In the Kuroshio Current region, the species-composition time-series is limited, hence it is not possible to identify trends in biomass, but the dominant taxa have been highly variable with an obvious diatom spike in 2004, possibly due to the meandering of this current (Sugisaki et al., 2010; Figure 36C.2). In general, large-scale, taxonomically diverse time series for phytoplankton are lacking.

2.1.2 Zooplankton communities

One of the most significant biological changes in the North Pacific is the explosion of gelatinous macrozooplankton in the western portion of the basin, especially the Yellow Sea, where medium to large jellyfish have become overly abundant in recent years and have resulted in increased reports of impacts (Purcell et al., 2007; Figure 36C.3). This increase in jellyfish has had unforeseen biological (e.g., effects on productivity and diversity) and economic consequences (e.g., effects on fisheries, industry, and tourism) with resulting impacts to ecosystem and human services.

Studying the California Current system Chelton et al. (1982) showed a strong correlation between zooplankton biomass anomalies and temperature anomalies. Thus, it is not surprising that recent changes between warm and cool periods in the eastern North Pacific coincided with large-scale changes in zooplankton community composition and abundance. Cool periods favour northern copepod species that tend to be larger and energy rich, making them good prey items while warm periods favour southern copepod species that tend to be smaller and energy poor making them less suitable prey (McKinnell et al., 2010; Figure 36C.4). Anomalously strong upwelling further influences the zooplankton community composition and abundance in the California Current system. On the western side of the North Pacific, the hydrography of the Kuroshio Current acts to differentiate zooplankton biomass and diversity between the onshore and offshore sides and main stream of this current. Further, copepod biomass varies interannually with different seasonal peaks but the overall trend remains relatively constant (Sugisaki et al., 2010). Large-scale, taxonomically diverse time series

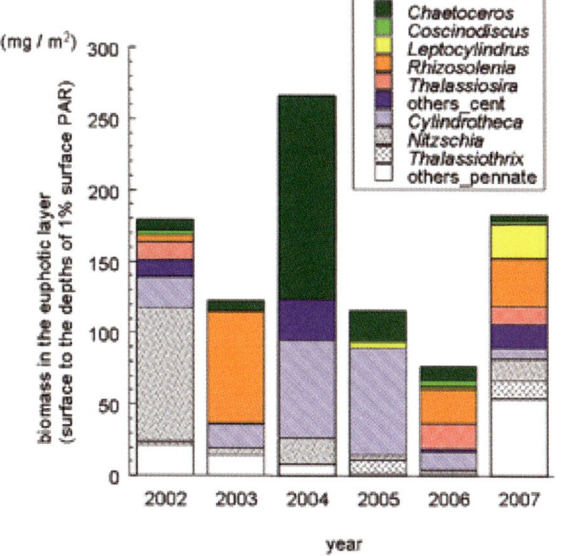

Figure 36C.2 | Composition of diatoms in the euphotic zone at Station B03 (34°N 138°E) in May (from Sugisaki et al., 2010).

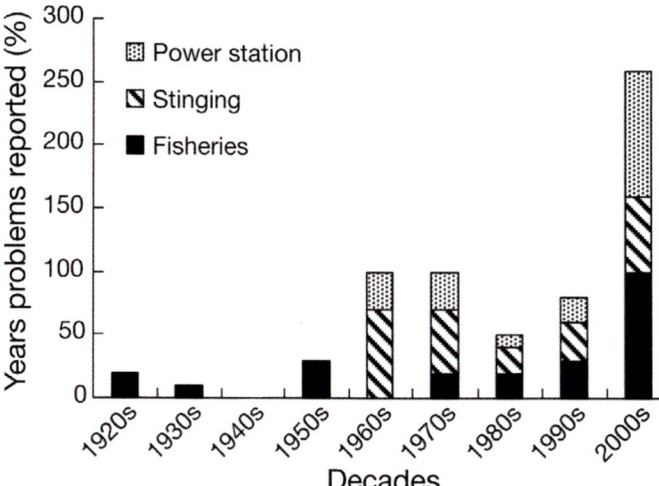

Figure 36C.3 | Percentage of years in each decade with reports of human problems with jellyfish in Japan (from Purcell et al., 2007).

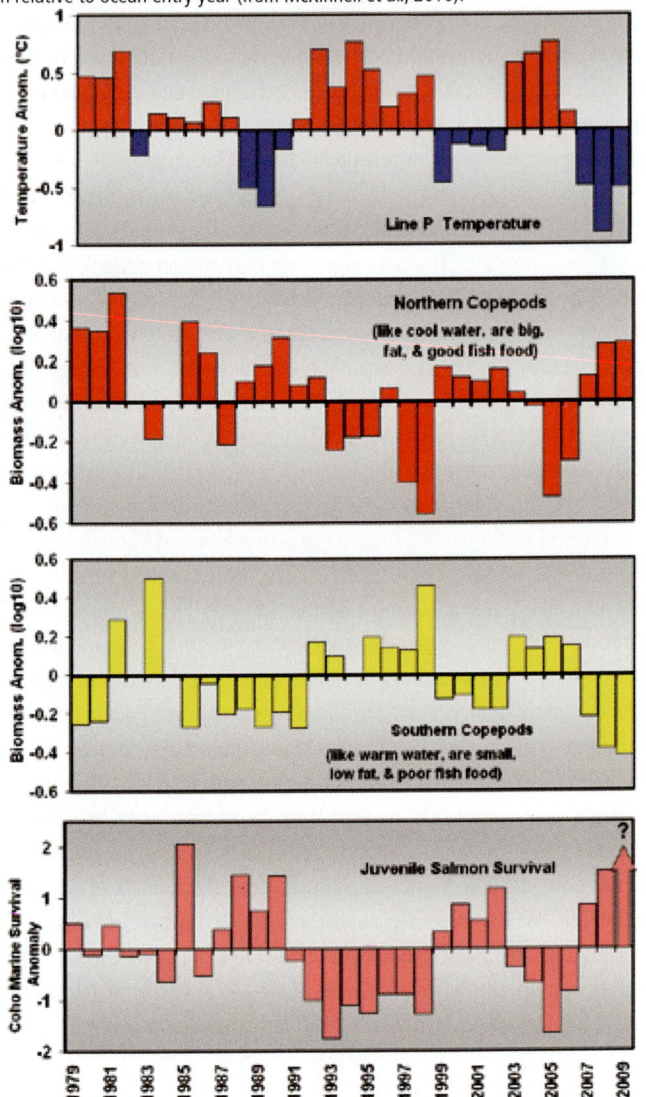

Figure 36C.4 | Northeast Pacific anomaly time series for upper ocean temperature, biomass of "Northern" and "Southern" copepods, and marine survival of coho salmon relative to ocean entry year (from McKinnell et al., 2010).

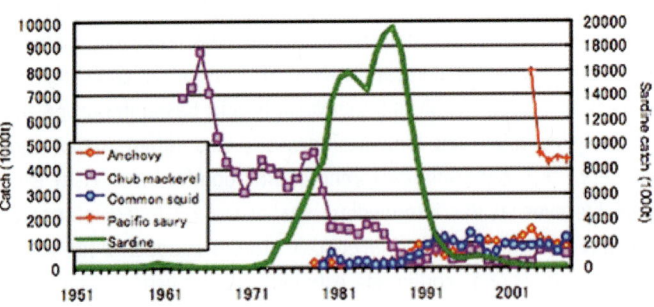

Figure 36C.5 | Biomass of sardine, anchovy, chub mackerel, Pacific saury and common squid (winter spawning stock) along the Pacific coast of Japan (from Chiba et al., 2010).

areas of the North Pacific. However, Kodama et al. (2010) and Kodama and Horiguchi (2011) document periods of defaunation in Tokyo Bay for macrobenthic and megabenthic communities, suggesting there have been decreases in benthic community diversity at least at local scales around the North Pacific (beyond the scope of this Assessment).

2.1.4 Higher trophic levels

McKinnell et al. (2010) provide the only intra-basin comparison of changes in key fish and invertebrate stocks between 1990-2002 and 2003-2008. In this study many taxa in the Sea of Okhotsk and Oyashio regions increased and many taxa in the California Current, Yellow Sea, and East China Sea decreased (Table 36C.1). In addition to changes in abundance, distributional shifts occurred, related at least in part to changing ocean conditions; these shifts can have ecological and economic consequences on ocean services (e.g., Mueter and Litzow, 2008).

In the eastern North Pacific, a mid-water trawl survey for the California Current system provides evidence that the forage fish community of this ecosystem tends to alternate between a less productive warm community and a more productive cool community in response to widely recognized regime shifts in oceanic conditions (NOAA's Southwest Fisheries Science Center (SWFSC) in Bograd et al., 2010). Similarly, on the western side of the basin in the Kuroshio-Oyashio system, where a strong latitudinal gradient in annual productivity (Pope et al., 1994) exists, evidence of decadal-scale changes in fish communities or "species replacements"

are lacking for other important zooplankton species (e.g., arrow worms, pteropods, salps, krill).

2.1.3 Benthic communities

Although cold, deep-water corals and sponges have received some attention in recent years (and some have been afforded special protection at regional or local scales), our understanding of the diversity and distribution of these organisms at larger spatial scales is very incomplete, making inferences about status of and trends in diversity impossible. Given their very slow growth rates and long regeneration times, they are particularly sensitive to disturbances, such as bottom-contact fishing gear, harvesting, natural resource exploration and extraction, submarine cable/pipelines, climate change, ocean acidification, and invasive species (Hourigan et al., 2007). Corals and sponges are not the only benthic taxa but no large-scale synoptic information was identified on status and trends in the diversity of other benthic communities in the coastal

Figure 36C.6 | Average marine survival of up to 45 coho salmon (O. kisutch) stocks in the northern California Current region by year of ocean entry (from Bograd et al., 2010).

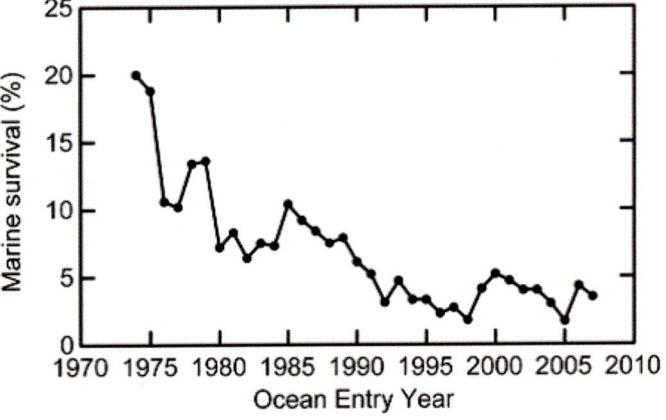

Chapter 36C

North Pacific Ocean

linked to regime shifts have been observed (Chiba et al., 2010; Figure 36C.5).

Pacific salmon are an economically and culturally important species in the North Pacific. Marine survival for over 40 coho salmon stocks has decreased substantially in the California Current system since the early 1970s, due at least in part to poor marine survival (Bograd et al., 2010; Figure 36C.6); extremely low survival corresponds to the 2005 smolt entry year; this trend is also detected in marine birds (see 2a (v) below). Similarly, masu salmon (*Oncorhynchus masou*) in Japan have experienced significant declines in returns over the same period (Chiba et al., 2010). Additional higher trophic level species in the North Pacific include a variety of fish and invertebrate species, including: small pelagic species (e.g., anchovy, sardine, saury, mackerel, squid); large pelagic species (e.g., tuna, shark, billfish, ray); benthopelagic species (e.g., rockfish, croaker, cod); and demersal species (e.g., pollock, flatfish, crab), some of which may have experienced population declines at regional or sub-regional scales.

For Pacific salmon spawning in Canadian waters, the Canadian Department of Fisheries and Oceans has provided outlooks since 2002. In the most recent iteration, 91 stocks were assessed and an outlook provided for 84, of which 28 were linked to a conservation concern, despite 21 units showing improvement, compared to 9 that have worsened since

Table 36C.1 | Interregional comparison of levels in biomass or abundance indices of fishes and invertebrates since 2003 compared to 1990-2002. Colour codes are: blue (increase), red (decrease), orange (change <|10%|), grey (not relevant to the region), and white (no data). The symbol © indicates that the evaluation for that taxon/region is based on catch data. In some regions, flatfish data were not reported by species, so any trends that are indicated apply only to the aggregate of flatfish species caught in that region, and not necessarily to the individual species listed in the column headers (from McKinnell et al., 2010).

Table 36C.2 | Status of commercial fishery stocks in the North Pacific.

Stock	Status	Region	Source
Bigeye tuna	Overfishing	Pacific/Western Pacific	NOAA (2013)
Pacific bluefin tuna	Overfishing	Pacific/Western Pacific	NOAA (2013)
Striped marlin (Central Western Pacific)	Overfishing	Western Pacific	NOAA (2013)
Blue king crab (Pribilof Islands)	Overfished	North Pacific	NOAA (2013)
Canary rockfish	Overfished	Pacific	NOAA (2013)
Pacific ocean perch	Overfished	Pacific	NOAA (2013)
Yelloweye rockfish	Overfished	Pacific	NOAA (2013)
Striped marlin (Central Western Pacific)	Overfished	Western Pacific	NOAA (2013)
Seamount groundfish complex (Hancock Seamount)	Overfished	Western Pacific	NOAA (2013)
Pacific bluefin tuna (Pacific)	Overfished	Pacific and Western Pacific	NOAA (2013)

the previous period (DFO, 2014). In the United States, the National Oceanic and Atmospheric Administration (NOAA) reports on the status of 480 managed stocks and stock complexes, including rockfishes, flatfishes, and gadoids, relative to fishing mortality and biomass reference points. Several of these stocks from the Pacific Ocean have been identified as overfished, but all domestic stocks are rebuilding and one stock (Sacramento River Fall Chinook) was recently removed from this list (NOAA, 2013). No domestic stocks in the Pacific currently are experiencing overfishing, although several stocks of highly migratory species are under international management and two species (Pacific bluefin tuna and striped marlin) were added to the list of overfished stocks in 2013 (NOAA, 2013).

2.1.5 Other biota

At least eleven species of marine birds and nine species of marine mammals designated as being at risk by the IUCN are found in the North Pacific; overall, it does appear that populations are either stable or increasing. Only planktivorous auklets in the Sea of Okhotsk appear to be the exception for marine birds (McKinnell et al., 2010; Table 36C.3); Steller sea lions and harbour seals in the central and western Aleutian Islands, northern fur seals from the Pribilof Islands, and potentially harbour seals in Prince William Sound, Alaska, are the exceptions for marine mammals (McKinnell et al., 2010; Table 36C.4) have experienced population declines. Additional species that are critically endangered in the North Pacific include the vaquita (*Phocoena sinus*) which is on the verge of extinction: only 241 animals were estimated in 2008 (Gerrodette et al., 2011) and Hawaiian monk seals (*Monachus schauinslandi*). Additional marine birds and mammals may be considered at-risk at regional or subregional scales that are beyond the scope of this Assessment.

In the California Current, increased variability has resulted in significant responses at higher trophic levels, including marine birds and mammals, and the cumulative effects of human-mediated stressors on marine predators can be difficult to unravel (Maxwell et al., 2013). For example, Cassin's auklets experienced an almost complete breeding failure in 2005-2006, due to changes in upwelling phenology that affected eu-

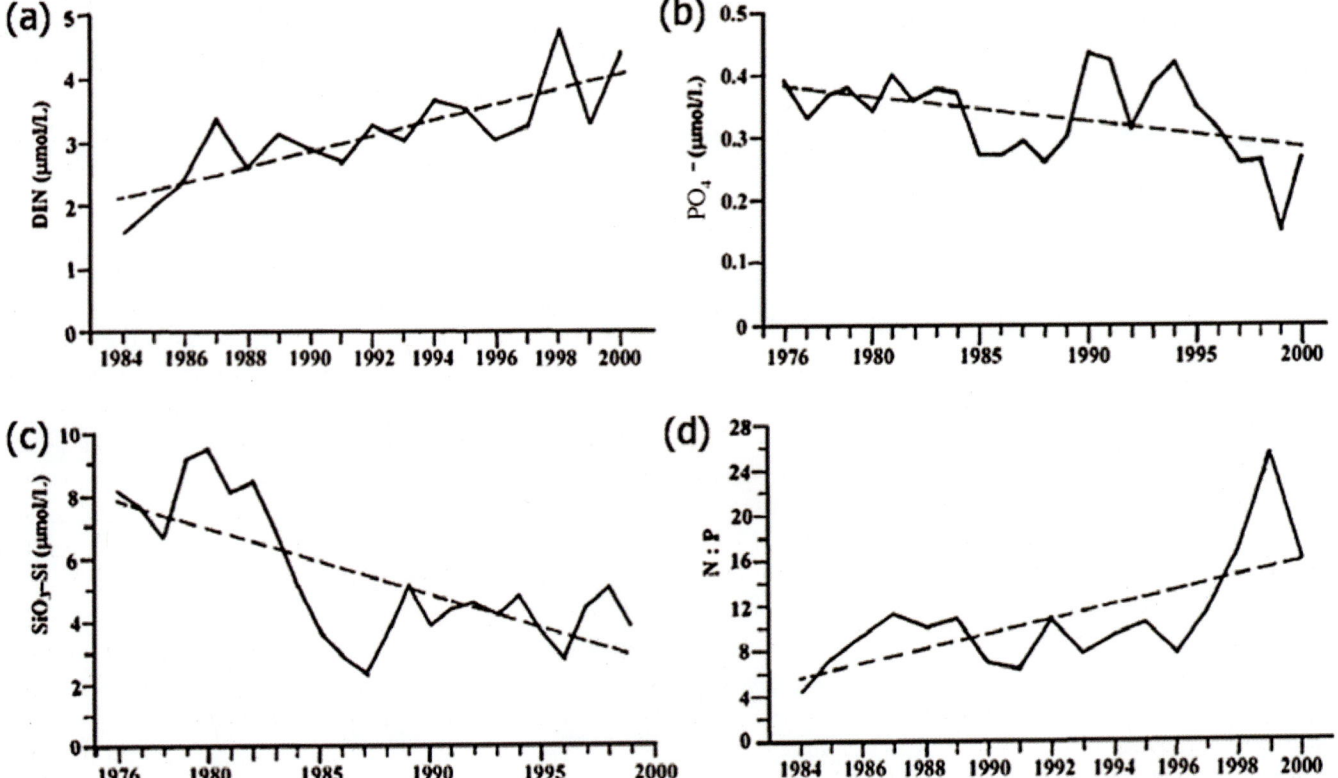

Figure 36C.7 | Long-term trend in the nutrients along a transect across 36°N in the Yellow Sea. (a) Dissolved Inorganic Nitrate (DIN), (b) phosphates, (c) silicates, (d) N:P ratio (from Yoo et al., 2010, and modified from Lin et al., 2005).

Table 36C.3 | Trends in the numbers or productivity of planktivorous species of marine birds and baleen whales. [CA= California, USA; BC=British Columbia, Canada; PRBO= Point Reyes Bird Observatory in California, K = carrying capacity] (from McKinnell et al., 2010)

Location	Species	Metric	Dates used	Trend	Reference
California Current					
Farallon Is., CA	Cassin's auklet	Population trend	1998 - 2008	No trend	PRBO - pers. comm.
Farallon Is., CA	Cassin's auklet	Productivity	2002 - 2008	Down	PRBO - pers. comm.
California & Oregon	Blue whale	Population trend	1991 - 2008	Up <3% y^{-1}	Calambokidis 2009
California, Oregon & Washington	Blue whale	Population trend	2001 - 2005	No trend	Caretta et al. 2009
California, Oregon & Washington	Fin whale	Population trend	2001 - 2005	No trend	Caretta et al. 2009
California & Oregon	Humpback whale	Population trend	1990 - 2008	Up 7.5% y^{-1}	Calambokidis 2009
California, Oregon & Washington	Humpback whale	Population trend	1999 - 2003	Up	Caretta et al. 2009
British Columbia and Southeast Gulf of Alaska					
Triangle Is., BC	Cassin's auklet	Population trend	1999 - 2009	No trend	Hipfner, pers. comm.
Triangle Is., BC	Cassin's auklet	Productivity	1998 - 2006	No trend	Hipfner, pers. comm.
British Columbia	Humpback whale	Population trend		Up 4.1%	Ford et al. 2009
Northern and western Gulf of Alaska					
Northern Gulf of Alaska	Humpback whale	Population count	1987 - 2003	Up 6.6% y^{-1}	Allen & Angliss 2009
Shumagin-Kodiak areas	Fin whale	Population count	1987 - 2003	Up 4.8% y^{-1}	Allen & Angliss 2009
Sea of Okhotsk					
Talan Island	Crested auklet	Population count	1989 vs 2008	Down	Andreev et al., In Press
Talan Island	Ancient murrelet	Population count	1989 vs 2008	Down	Andreev et al., In Press
Talan Island	Parakeet auklet	Population count	1989 vs 2008	Down	Andreev et al., In Press
Western North Pacific					
Asia stock	Humpback whale	Population count	1991-93 vs. 2004-06	Probably Up	Allen & Angliss 2009

phausiid prey populations (Sydeman and Thompson, 2010). Also, the California sea lion (*Zalophus californianus*), where the number of pups produced at the Channel Island reference site has shown a quadratic increase since the mid-1970s (Bograd et al., 2010), has experienced a recent decline in abundance and poor pup health. The spotted seals of the Yellow Sea also have decreased precipitously since the 1960s, due to overharvesting and habitat destruction; this has resulted in local extirpation and some rookeries support fewer than 150 individuals.

2.2 Major pressures in the coastal area and major groups affected by the pressures

In addition to global climate change impacts, including ocean acidification, there are a large number of coastal pressures affecting the North Pacific, similar to other coastal marine ecosystems, due largely to the diverse human-mediated activities in these environments. These include, but are not limited to: habitat loss; over-exploitation and fishing impacts; shipping; energy development/exploration; aquaculture; pollution (both direct and indirect), eutrophication and resulting impacts (pathogenic bacteria, harmful algal blooms; hyp/anoxia); species introductions/invasions; watershed alteration and physical alterations of coasts; tourism; and marine litter. None have been quantified at the scale of the North Pacific, but some regional patterns can be highlighted. Studies such as by Halpern et al. (2008) demonstrate that coastal area can be severely affected by human activities, including those in the western North Pacific. Furthermore, the Yellow and East China Seas area is one of the most densely populated areas of the world; approximately 600 million people inhabit this area, resulting in immense anthropogenic stressors on this coastal system. Urbanization in Asia is not unique and other coastal areas of the North Pacific also have experienced increased urbanization and an increase in a wide variety of ecosystem stressors.

Runoff from the Fraser and Columbia Rivers in the California Current region, the Amur River in the Sea of Okhotsk, the Changjiang River in the East China Sea, and the Pearl River and Mekong River in the South China Sea all play important roles in driving coastal processes and resulting ecosystem services. The Changjiang River is the world's third-longest river; its watershed of approximately 1.8 million km^2 encompasses about one-third of China's population and 70 per cent of its agricultural production. The widespread use of fertilizers for agricultural production has resulted in increased nutrient discharge to the coastal environment, causing increased eutrophication since the early 1970s. As a result, in the Yellow Sea, nitrogen:phosphorus and nitrogen:silicon ratios have been increasing basin-wide for decades (Yoo et al., 2010; Figure 36C.7). This in turn has resulted in an increase in the frequency and intensity of harmful algal bloom events and a shift in the phytoplankton community from diatoms to dinoflagellates that have affected ecosystem services and increased the severity of hypoxic events in the estuary (Yoo et al., 2010). A related anthropogenic activity that could significantly alter riverine discharges is large-scale water diversion projects that would result

Table 36C.4a | Piscivorous species in the North Pacific [PRBO= Point Reyes Bird Observatory in California, USA; K= carrying capacity; CA= California, USA; WA= Washington State, USA; BC= British Columbia, Canada; GOA= Gulf of Alaska; DFO= Canadian Department of Fisheries & Oceans] (from McKinnell et al., 2010)

Location	Species	Metric	Dates used	Trend	Reference
California, Oregon & Washington	California sea lion	Population trend	2000 - 2006	No trend	Caretta et al. 2009
San Miguel Is., CA	Northern fur seal	Population trend	1998 - 2005	Up	Caretta et al. 2009
San Miguel Is., CA	Northern fur seal	Pup production	1972-76 vs 2002-06	Up - interrupted by El Niño	Olesiuk 2009
Channel Islands, CA	California sea lion	Population trend	2004 - 2008	Up	Bograd et al. 2010
Channel Islands, CA	Northern elephant seal	Population trend	2000 - 2005	Up	Caretta et al. 2009
Farallon Is. CA	Common murre	Population trend	1998 - 2008	Up	PRBO – pers. comm.
Farallon Is., CA	Common murre	Productivity	2002 - 2008	No trend	PRBO – pers. comm.
Farallon Is, CA	Rhinoceros auklet	Population trend	1998 - 2008	Unknown	PRBO – pers. comm.
Farallon Is., CA	Rhinoceros auklet	Productivity	2002 - 2008	No trend	PRBO – pers. comm.
Farallon Is. CA.	California sea lion	Population trend	1998 - 2008	No trend	PRBO – pers. comm.
Farallon Is., CA	Northern fur seal	Population trend	1998 - 2008	Up	PRBO – pers. comm.
Farallon Is., CA	Northern elephant seal	Population trend	1998 - 2008	No trend	PRBO – pers. comm.
Central California	Steller sea lion	Non-pup count	1996 - 2004	No trend	Caretta et al. 2009
Northern California & Oregon	Steller sea lion	Non-pup count	1996 - 2002	No trend, at K	Caretta et al. 2009
California	Harbour seal	Population trend	1995 - 2004	No trend, at K	Caretta et al. 2009
Oregon & Washington	Harbour seal	Population trend	1995 - 2004	No trend, at K	Caretta et al. 2009
Tatoosh Is., WA	Common murre	Productivity	1998 - 2008	Up	Parrish, pers. comm.
Triangle Is., B.C.	Rhinoceros auklet	Population trend	1999 - 2009	Up ?	Hipfner, pers. comm.
Triangle Is., B.C.	Rhinoceros auklet	Productivity	1998 - 2007	Up ?	Hipfner, pers. comm.
British Columbia	Steller sea lion	Pup count	1980s - 2006	Up 7.9% y^{-1}	DFO, 2008
British Columbia	Steller sea lion	Non-pup count	1998 - 2002	Up	Allen & Angliss 2009
St. Lazaria Is., E GOA	Rhinoceros auklet	Population trend	1994 - 2006	Up	Slater, pers. comm.
St. Lazaria Is., E GOA	Rhinoceros auklet	Population trend	1998 - 2006	Up 5% y^{-1}	Dragoo, pers. comm.
St. Lazaria Is., E.GOA	Rhinoceros auklet	Productivity	2002 - 2006	Up ?	Dragoo, pers. comm.
St. Lazaria Is., E GOA	Unid. murre	Population trend	1998 - 2006	No trend	Dragoo. pers. comm.
St. Lazaria Is., E GOA	Unid. murre	Population trend	1994 - 2006	Down	Slater & Byrd 2009
St. Lazaria Is., E GOA	Unid. murre	Population trend	2001 - 2006	No trend	Slater & Byrd 2009
Southeast Alaska	Steller sea lion	Pup counts	1996 - 2009	Up 5.0% y^{-1}	DeMaster, 2009
Southeast Alaska	Harbour seal	Population trend	1990s - 2002	Variable no trend	Allen & Angliss 2009
Eastern GOA	Steller sea lion	Pup count	2001 - 2009	No trend	DeMaster 2009
Central GOA	Steller sea lion	Pup count	1994 - 2009	Down 0.6% y^{-1}	DeMaster 2009
Middleton Is., GOA	Unid. murre	Population count	1998 - 2007	Down	Hatch, pers. comm.
Middleton Is., GOA	Rhinoceros auklet	Population count	1998 - 2007	Up	Hatch, pers. comm.
Middleton Is., GOA	Black-legged kittiwake	Population count	1998 - 2007	Down	Hatch, pers. comm.
Western GOA	Steller sea lion	Pup count	1998 - 2009	Up 2.6% y^{-1}	DeMaster 2009
Prince William Sound	Harbour seal	Population trend	1984 - 1997	Down	Allen & Angliss 2009
Kodiak Region, GOA	Harbour seal	Population trend	1993 - 2001	Up 6.6% y^{-1}	Allen & Angliss 2009
Semidi Is, W GOA	Black-legged kittiwake	Population trend	1998 - 2007	No trend	Dragoo, pers. comm.
Semidi Is., W GOA	Common murre	Population trend	1999 - 2007	No trend	Dragoo, pers. comm.

in less discharge to coastal environments around the North Pacific. For example, much of the flow of the Columbia and Fraser Rivers is used for agricultural production that can result in less discharge reaching ocean in some years. This can result in reduced nutrient inputs, which in turn lowers productivity, and that reduction adversely affects the diversity that depends on it. As increased climate variability intersects with growing human populations and increased irrigation demands in coastal environments, reduced river discharges could have profound impacts on coastal productivity and biodiversity.

The introduction of non-indigenous species continues to result in economic and ecological consequences, including negative impacts on native biodiversity (Sala et al., 2000). In the first synoptic study of non-indigenous species in the North Pacific, Lee II and Reusser (2012)

Table 36C.4b | Piscivorous species in the eastern Bering Sea and Aleutian Islands (from McKinnell et al., 2010)

Location	Species	Metric	Dates Used	Trend	Reference
St. Paul Is., E Bering	Black-legged kittiwake	Population trend	1999 - 2008	No trend	Dragoo, pers. comm.
St. Paul Is., E. Bering	Common murre	Population trend	1999 - 2008	No trend	Dragoo, pers. comm.
St. George Is., E. Bering	Black-legged kittiwake	Population trend	1999 - 2008	No trend	Dragoo, pers. comm.
St. George Is., E. Bering	Common murre	Population trend	1999 - 2008	No trend	Dragoo, pers. comm.
Pribilof Is., Bering Sea	Northern fur seal	Pup count	1972 - 76 vs. 2002 - 06	Down 2.7% y^{-1}	Olesiuk 2009
St. Paul Is., Pribilofs	Northern fur seal	Pup count	1998 - 2006	Down 6.1% y^{-1}	Allen & Angliss 2009
St. George Is., Pribilofs	Northern fur seal	Pup count	1998 - 2006	Down 3.4% y^{-1}	Allen & Angliss 2009
Bogoslof Is., Bering Sea	Northern fur seal	Population trend	1972 - 76 vs. 2002 - 06	Rapid growth	Olesiuk 2009
Bogoslof Is., Bering Sea	Northern fur seal	Pup count	2005 - 2007	Up	Allen & Angliss 2009
Bering Sea	Harbour seal	Population trend	1980s - 1990s	Probably down	Allen & Angliss 2009
Aiktak Is., Eastern Aleutian Islands	Unidenitified murre	Population trend	1998 - 2007	No trend	Dragoo, pers. comm.
Eastern Aleutian Islands	Steller sea lion	Pup count	1998 - 2009	Up 4.2% y^{-1}	DeMaster 2009
Eastern Aleutian Islands	Harbour seal	Population trend	1977 - 82 vs. 1999	Down 45%	Allen & Angliss 2009
Koniuji Is., C. Aleutian Islands	Black-legged kittiwake	Population trend	1998 - 2007	No trend	Dragoo, pers. comm.
Koniuji Is., C. Aleutian Islands	Unidentified murre	Population trend	2001 - 2007	No trend	Dragoo, pers. comm.
Ulak Is., C. Aleutians	Unidentified murre	Population trend	1998 - 2008	Up 6.2% y^{-1}	Dragoo, pers. comm.
C. Aleutian Islands	Harbour seal	Population trend	1977 - 82 vs. 1999	Down 66%	Allen & Angliss 2009
Buldir Is., W. Aleutians	Black-legged kittiwake	Population trend	1998 - 2007	No trend	Dragoo, pers. comm.
Bering Sea Stock	Harbour seal	Population trend	1980s - 1990s	Probably down	Allen & Angliss 2009
Western Aleutian Islands	Harbour seal	Population trend	1977 - 82 vs. 1999	Down 86%	Allen & Angliss 2009
Aleutian Islands	Steller sea lion	Pup count	1994 - 2009	Down 1.6% y^{-1}	DeMaster 2009
Western Aleutian Islands	Steller sea lion	Pup count	1997 - 2008	Down 10.4% y^{-1}	DeMaster 2009

Table 36C.4c | Piscivorous species in the western Pacific, including western Bering Sea, Sea of Okhotsk, Oyashio, and Yellow Sea (from McKinnell et al., 2010).

Location	Species	Metric	Dates Used	Trend	Reference
E. Kamchatka	Steller sea lion	Non-pup count	2001 – 2008	No trend	Burkanov et al. 2009
Commander Islands	Steller sea lion	Non-pup count	2000 - 2008	No trend	Burkanov et al. 2009
Commander Islands	Northern fur seal	Pup production	1972 - 76 vs. 2002 - 06	No trend	Olesiuk 2009
Kuril Islands	Northern fur seal	Pup production	1972 - 76 vs. 2002 - 06	Up, 3% y^{-1}	Olesiuk 2009
Robben Is., Okhotsk	Northern fur seal	Pup production	1972 - 76 vs. 2002 - 06	No trend	Olesiuk 2009
Kuril Islands	Steller sea lion	Non-pup count	2000 – 2007	Up	Burkanov et al. 2009
N. Okhotsk	Steller sea lion	Non-pup count	1996 – 2006	Up	Burkanov et al. 2009
Sakhalin Island, Okhotsk	Steller sea lion	Non-pup count	2000 – 2009	Up	Burkanov et al. 2009
Talan Is., Okhotsk	Horned puffin	Population count	1989 vs 2008	Up	Andreev et al., In Press
Talan Is., Okhotsk	Black-legged kittiwake	Population count	1989 vs 2008	Up	Andreev et al., In Press
Talan Is., Okhotsk	Unidentified murre	Population count	1989 vs 2008	No trend	Andreev et al., In Press
Teuri Is., W. Hokkaido	Japanese cormorant	Nest count	2007 - 2008	No trend	Watanuki, pers. comm.
Teuri Is., W. Hokkaido	Rhinoceros auklet	Population count	1985 -1997	Up	Watanuki, pers. comm.

identified 746 species that were present in, but not native to, at least one ecoregion in the North Pacific. Of these, 32 per cent were native elsewhere in the North Pacific, 48 per cent were native to regions outside the North Pacific, and 20 per cent were cryptogenic (of unknown origin). Furthermore, the Hawaiian and Northeastern Pacific regions had considerably more introduced species than the Northwestern Pacific (Lee II and Reusser, 2012; Figure 36C.8). Given the continued increase in global trade, it is expected that the number of species that will be introduced to new environments also will increase. Combined with the high species richness and density of non-indigenous species already reported for many regions that have significantly altered population, community, and ecosystem processes (Ruiz et al., 1997), additional, cumulative consequences of these invasions should be expected.

Other major stressors in the North Pacific include hypoxia, habitat destruction, pollution, and overfishing. However, none of these have been quantified at the scale of the North Pacific. Ocean acidification has dramatically impacted some calcifying organisms such as pteropods

North Pacific Ocean — Chapter 36C

Figure 36C.8 | Number of non-indigenous species in central/northern North Pacific marine ecoregions (from Lee II and Reusser, 2012).

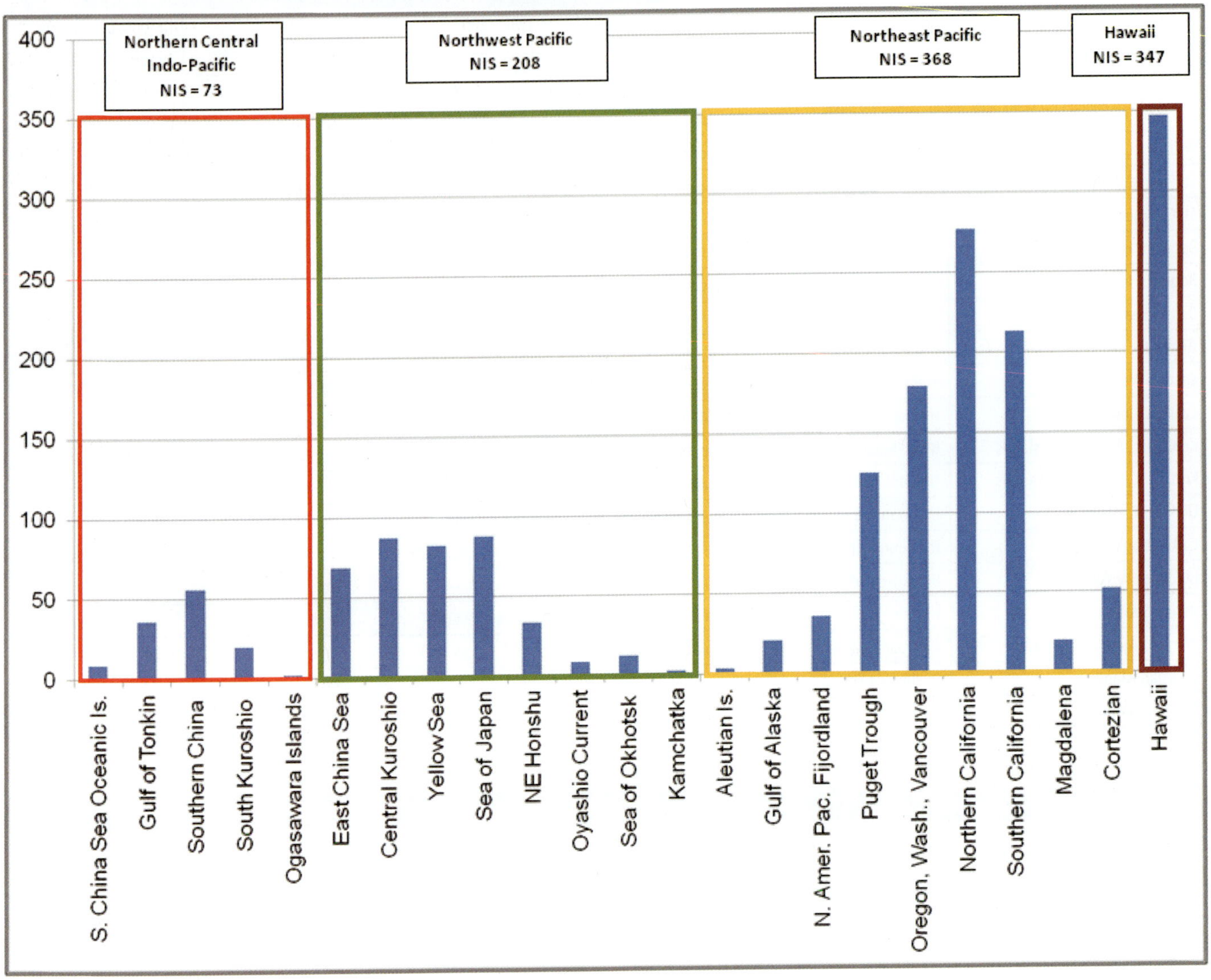

(Orr et al., 2005). Hypoxia has not only increased in the Yellow Sea, but continues to be a major pressure on coastal ecosystems in the eastern North Pacific, including off Oregon, with lethal consequences for benthic species (Grantham et al., 2004) and the western North Pacific, including Tokyo Bay where there has been a reduction in nutrient recycling (Kodama and Horiguchi, 2011). Further, the shoaling of this continental hypoxic zone has reduced habitat for several species, including some commercially important ones, which could alter ecosystem services. Habitat destruction is the leading cause of biodiversity loss (Sala et al., 2000). Many forms of habitat destruction and/or degradation are occurring around the North Pacific, including shoreline hardening/development and land creation, but quantifying the amount of habitat lost or impaired at the scale of the North Pacific remains a gap. Overfishing continues to be a major pressure in some coastal areas of the North Pacific. For example, in the Yellow Sea, overfishing has contributed to trophic cascades, resulting in fishing down the food chain. In addition, Hutchings (2000) has shown that most stocks are very slow to recover from overfishing; this has consequences for ecosystem services and can adversely affect biodiversity. However, it should be noted that some managed fisheries systems in the North Pacific are doing well (Hilborn et al., 2005).

The boundaries and names shown and the designations used on this map do not imply official endorsement or acceptance by the United Nations.

Figure 36C.9 | Ratio of mean chlorophyll a between 1998-2002 (denominator) and 2003-2007 periods. White colour indicates minimal change between the two periods (ratios = 0.9-1.1) (from McKinnell et al., 2010).

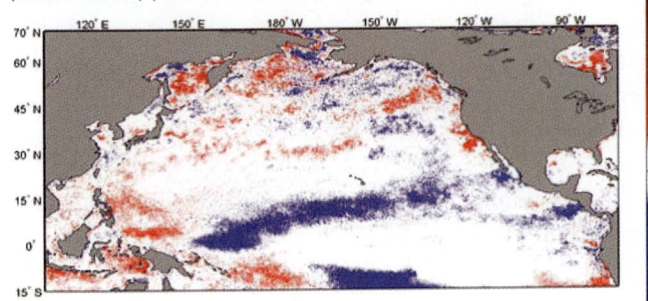

Chapter 36C

North Pacific Ocean

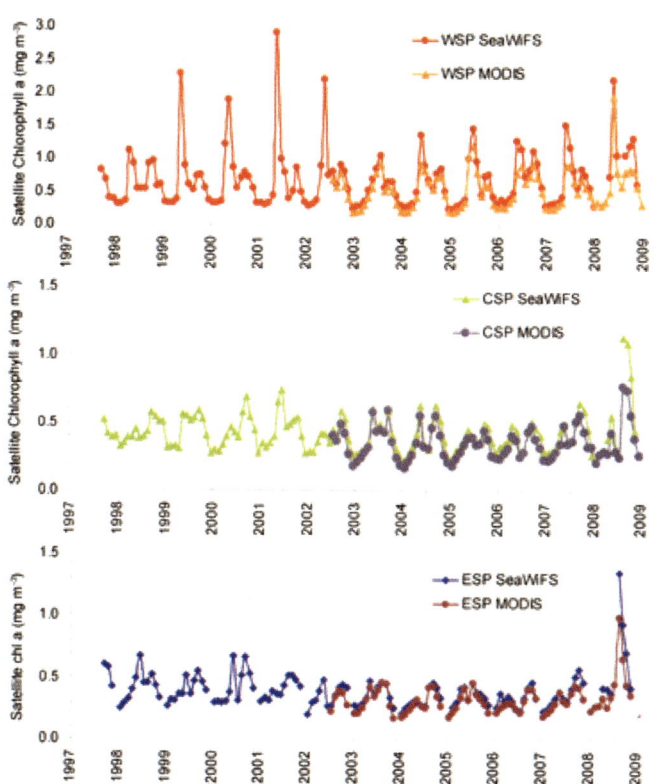

Figure 36C.10 | Chlorophyll a concentration for three regions (western subarctic Pacific (WSP), 155-172°E, 45-53°N; eastern subarctic Pacific (ESP), 140-155°W, 49.5-57°N; and central subarctic Pacific (CSP), 45-51°N 160-180°W) estimated from ocean colour-sensing satellites (from Batten et al., 2010).

2.3 Major ecosystem services being affected by the pressures

2.3.1 Ecosystem services being lost

Although it is expected that ecosystem services being lost in coastal areas of the North Pacific would be consistent with those affected by these pressures globally, these data are lacking at the scale of the North Pacific. Worm et al. (2006) showed that reduced biodiversity increased the rate of resource collapse and decreased recovery potential, stability, and water quality; in contrast, restoration of biodiversity increased productivity and decreased variability. Furthermore, Francis et al. (1998) document the ecological consequences of major species re-distribution in the northeast Pacific following the major regime shift in 1977. Species invasions (and extinctions) also reorganize coastal ecosystems; Hooper et al. (2005) show that this has altered ecosystem goods and services in many well-documented cases and that most are difficult, expensive, or impossible to reverse.

The Pacific Ocean has the largest pool of low-oxygen water in the global ocean and in recent decades this pool has been expanding, with reduced oxygen concentrations observed both on the western and eastern sides of the basin (Ono et al., 2001; Emerson et al., 2004; Bograd et al., 2008). Generally, global climate models predict that global warming will lead to deoxygenation of the deep ocean because warmer surface waters will hold less oxygen and will be more stratified, resulting in less ventilation of the deep ocean (Sarmiento et al., 1998; Keeling et al., 2010). This will adversely affect benthic and pelagic ecosystems (Levin et al., 2009; Stramma et al., 2010; Koslow et al. 2011). Koslow et al. (2011) showed decreased mid-water oxygen concentrations were correlated with the decline of 24 mid-water fish taxa from eight families. Extended to larger scales, this could have significant adverse ecological and biogeochemical effects.

2.3.2 Human services being lost

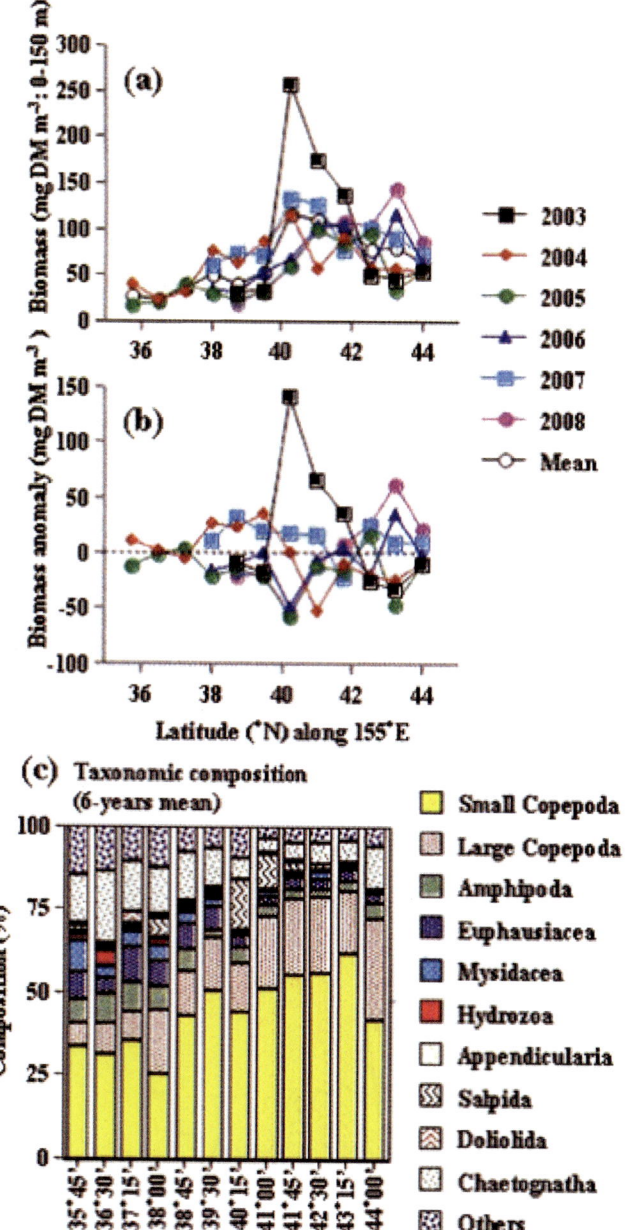

Figure 36C.11 | (a) Zooplankton dry biomass at 35°45'N-44°00'N along 155°E in the western North Pacific between 10-20 May, 2003-2008. (b) biomass anomalies compared to a 6-year mean, (c) and the taxonomic composition (from Batten et al., 2010).

Changes to marine biodiversity in coastal systems, and hence in ecosystem structure and function, can result from both direct impacts (e.g., exploitation, pollution, species invasion, and habitat destruction), or indirect impacts, via climate change and related perturbations of ocean biogeochemistry (e.g., acidification, hypoxia); these can have severe consequences on human services. Although no basin-scale studies quantifying these impacts were found, smaller-scale examples can highlight what might be expected. Myers and Worm (2003) demonstrated a dramatic decline in large predatory fish globally while Ovetz (2007) showed how industrial longline fishing has extensive negative economic and social consequences for coastal communities, especially those heavily reliant on fish protein, as biomass/species changes resulted in cascading effects, some of which were not predicted.

Also, Schroeder and Love (2002), who compared rockfish assemblages among three differently fished areas, showed large differences in fish density, size structure, and species composition. Only the protected area had both higher density and larger fish and greater species composition. Finally, Jackson et al. (2001) highlight how overfishing that precedes other forms of human disturbance to coastal ecosystems (pollution, degradation of water quality, and anthropogenic climate change) has resulted in ecological extinctions, especially of large vertebrate predators, with significant ecological and economic consequences.

In the western North Pacific, increased jellyfish blooms have had both direct and indirect negative impacts on human services. These blooms have reduced tourism, affected fishing and aquaculture and increased industrial costs, by, e.g., clogging the cooling-water intake screens of power plants (e.g., Purcell et al., 2007). Furthermore, blooms of gelatinous zooplankton have indirect effects on fisheries by feeding on zooplankton and ichthyoplankton; thus they are predators on and potential competitors with fish. Similarly, Cooley et al. (2009) showed that ocean acidification could affect a range of ecosystem services, such as fishery/aquaculture harvests, coastal protection, tourism, cultural identity, and ecosystem support, by adversely affecting calcifying marine organisms; they also showed that these impacts are expected to be greater in developing countries. Thus, more research is needed to understand how pressures are affecting services at the basin scale.

Coastal eutrophication is a growing concern, especially in the western North Pacific, where increases in nutrient loading have been linked with the development of large biomass blooms and harmful algal blooms, resulting in anoxia and toxic/harmful impacts on fisheries, aquaculture, ecosystems, human health and recreation (see Anderson et al., 2002; McKinnell and Dagg, 2010).

3 Oceanic Areas of the North Pacific

Unlike the coastal realm, where Spalding et al. (2007) proposed marine ecoregions, the oceanic realm of the North Pacific does not have a similar delineation scheme. Longhurst (2007) identifies zones in the North Pacific, but these are delineated more on the basis of oceanographic conditions and have relatively little weighting based on differences in biological diversity, species composition and productivity. It is probable that biological diversity in the oceanic North Pacific is also patchy, but the nature and scale of this patchiness needs to be determined.

3.1 Biodiversity Status and Trends

3.1.1 Primary producers

Using SeaWiFS satellite data, it is possible to detect large-scale changes in chlorophyll concentrations between 1998-2002 and 2003-2007. During this period, average chlorophyll decreased in parts of the eastern North Pacific (with the exception of the California Current region) and increased in the western North Pacific (McKinnell et al., 2010; Figure 36C.9). Also of note is the significant decline in average chlorophyll across the entire tropical/subtropical zone from Indonesia to Baja California, Mexico. However, at finer spatial and temporal scales, interannual variations in the location, timing, and magnitude of surface chlorophyll levels can be considerable around the North Pacific (Yoo et al., 2008). The general trend towards increased sea surface temperatures has resulted in an expansion of the low surface-chlorophyll extent in the subtropical North Pacific (Polovina et al., 2008). The areal increase in these low-chlorophyll waters of the central North Pacific from 1998 to 2006 is about 2 per cent per year (Batten et al., 2010); much of this expansion is eastward. The expansion of low surface-chlorophyll waters is consistent with increased vertical stratification due to ocean warming; this situation also has been identified in the South Pacific, North Atlantic, and South Atlantic (Polovina et al., 2008). Two satellite-derived time series exist for chlorophyll estimates (SeaWiFS and MODIS) for three major domains in the North Pacific (Batten et al., 2010; Figure 36C.10). Both the central subarctic Pacific (CSP) and eastern subarctic Pacific (ESP) are completely oceanic; the western subarctic Pacific (WSP) does infringe upon the coastal environment and thus is subject to potential biases in spring-bloom characteristics (both concentration and seasonal variability were greatest for this domain). The spike in 2008 is obvious

Figure 36C.12 | Total annual catch (tons reported) of North Pacific albacore tuna by all nations, 1966-2007 (from Batten et al., 2010, with data taken from ISC, 2008).

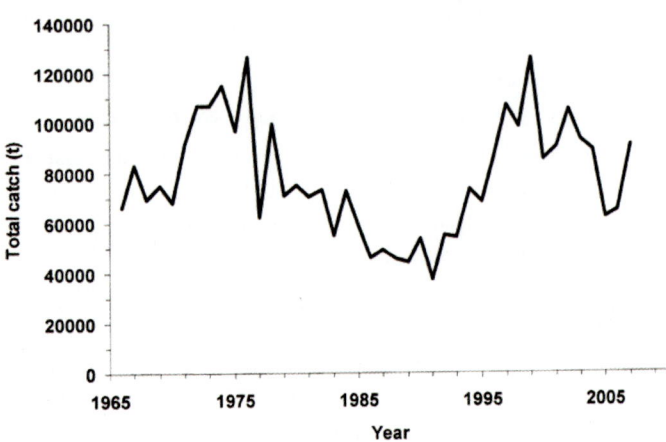

in both the ESP and CSP series. As noted above, the NPTZ is a key feature of the oceanic realm of the North Pacific and the transition-zone chlorophyll front (TZCF), which indicates a strong meridional gradient in surface chlorophyll at the boundary of the subarctic and subtropical gyre, migrates from south to north over 1,000 km annually (Polovina et al., 2001). Ocean productivity estimates derived from models and satellite observations (Behrenfeld and Falkowski, 1997) indicate high annual average phytoplankton production throughout the NPTZ, in particular in the west, related to the Kuroshio Extension region. Surface chlorophyll concentrations in the subtropical gyre are usually <0.15 mg/m³, whereas in the subarctic gyre and NPTZ they can be >0.25 mg/m³. Further, the expansion of the subtropical central gyre appears to have resulted in a change in primary productivity from a nitrate-limited, diatom-dominated phytoplankton community to one that is dominated by the N_2-fixing *Prochlorococcus* (Karl, 1999; Karl et al., 2001).

3.1.2 Zooplankton communities

The NPTZ supports higher secondary productivity with respect to zooplankton biomass (McKinnell and Dagg, 2010) relative to other areas of the North Pacific, but no single comprehensive index exists for mesozooplankton time-series trends in the North Pacific. Efforts at sub-basin scales are focused on the Alaska Gyre (Continuous Plankton Recorder (CPR) Survey, Line P), near Hawaii (Hawaii Ocean Time Series), and in the western Pacific along 155oE (Hokkaido University T/S *Oshoro Maru*). The Weathership surveys provided a zooplankton time series within the Alaska Gyre along Line P off the West Coast of Vancouver Island from 1956-1980 (see Fulton, 1983) but only intermittent sampling was conducted opportunistically, until more regular surveys were initiated in 1997 (Mackas and Galbraith, 2002). In addition, a CPR survey has provided additional zooplankton productivity and diversity measures since 2000 (Batten et al., 2010). Although zooplankton taxa are similar between nearshore and offshore stations, the dominance hierarchies differ between the shelf margin (Mackas et al., 2001) and the oceanic zone, (Mackas and Galbraith, 2002) where "subarctic oceanic" copepod species dominate. These species have distributions that span the Pacific basin north of the subarctic front and their interannual variability has been attributed both to temperature variability and increased transport by the North Pacific Current; this could result in re-distribution of other marine species (Batten et al., 2010; Figure 36C.11). Zooplankton time-series information also exists for station ALOHA, due to survey efforts by the Hawaii Ocean Time Series programme showed an increase in biomass between 1994 and 2004, after which it either stabilized or decreased slightly (Sheridan and Landry, 2004; Batten et al., 2010). The sampling along the 155oE transect provides an opportunity to look at productivity and diversity along a north:south gradient that encompasses the subarctic front, transition domain, subarctic boundary and subtropical current system. Batten et al. (2010) demonstrate that zooplankton biomass tends to be higher in the transition domain, and during the unprecedented warm year of 2008, this transition domain extended north into the historical zone of the subarctic front. Furthermore, taxonomic composition differs along this latitudinal gradient, with small and large copepods most prevalent at the subarctic front and transition domain; amphipods, euphausids, chaetognaths, and copepods are represented in the subarctic boundary and subtropical current system.

3.1.3 Benthic communities

Perhaps with the exception of the limited snap-shot surveys of hydrothermal vent and seamount chain communities, no large-scale synoptic information is available on status of and trends in the diversity of benthic communities in the oceanic realm of the North Pacific. Much of this area is extremely deep, making survey efforts virtually impossible until recently, leaving much to be explored and characterized. The limited studies of vent communities in the North Pacific suggest high levels of endemism and diversity, especially within microbial communities that have different physiologies/metabolisms and thermal and salinity tolerances (e.g., Hedrick et al., 1992; Tunnicliffe et al., 1993). Furthermore, for these unique systems, the chemosynthesis that forms the basis of these deep-water food webs is critically important (Zhou et al., 2009). Stone and Shotwell (2007) have identified at least 140 coral species associated with seamounts in Alaska, representing at least six major taxonomic groups.

3.1.4 Higher trophic levels

Taxonomic diversity in both the eastern and western divisions of the North Pacific contains a mix of subtropical, temperate, subarctic and arctic species. The eastern North Pacific shows a gradient in diversity from east to west (Mueter and Norcross, 2002), with most fish biomass (and exploited stocks) on the continental shelf or coastal nearshore areas. Relatively little is known about the demersal species in the oceanic realm due to the great depths. Some exploitation of species associated with seamounts occurs in the Gulf of Alaska and along the west coast of North America, for species such as sablefish (*Anaplopoma fimbria*), but within exclusive economic zones (EEZs), most seamounts have some level of protection due to restrictions on bottom-contact fishing gear and seamounts have been identified as Ecologically and Biologically Significant Areas (CBD, 2014). Furthermore, limited surveys/data mean that no time series are available.

The NPTZ once supported large-scale squid (*Ommastrephes bartrami*) driftnet fisheries, until a United Nations General Assembly ban on this gear was imposed in 1992 (see resolution 46/215) (PICES, 2004). Now the NPTZ supports the pelagic longline fishery based in Hawaii, with many vessels targeting tunas (including albacore, *Thunnus alalunga*), billfish, and squid. Albacore tuna is an economically important and widely distributed species in the North Pacific. Reported catches for albacore have been variable over time, but peaked in 1976 and 1997, and were rather depressed until the early 1990s, in part due to overfishing and below average recruitment (Cox et al., 2002; Batten et al., 2010; Figure 36C.12). The catch of other tuna species, especially skipjack and yellowfin, has increased substantially since the 1950s (Sibert et al., 2006). However, Sibert et al. (2006) found that, although biomass was lower than that predicted in the absence of fishing (and perhaps higher than management targets), a reduction in the proportion of large fish, and the decreased trophic level of the catch suggest fisheries impacts on these top-level predators.

3.1.5 Other biota

As noted in 2(a) above, at least eleven species of marine birds and nine species of marine mammals designated as being at risk by the IUCN are found in the North Pacific. Although it appears that some populations are either stable or increasing, significant threats remain. No additional entirely oceanic organisms are known to be designated as being at risk for the North Pacific.

3.2 Major pressures in the oceanic area and major groups affected by the pressures

It appears that the open oceanic area of the North Pacific is significantly less affected than coastal areas, where multiple point-source stressors are routinely encountered. However, the large-scale stressors that are affecting this oceanic area will require substantial international efforts to mitigate: specifically the climate impacts that have resulted in changes to both the physical and biogeochemical properties of the ocean. For example, global climate change is altering temperature, salinity, mixed-layer depth, and pCO_2 (acidification) in the open ocean. It has been demonstrated that marine organisms have shifted their distributions in response to changing marine conditions and this can result in local extinctions or incursions (Cheung et al., 2009). Furthermore, Cheung et al. (2009) suggest that under certain climate change scenarios up to 60 per cent of current biomass could be affected, and disrupt existing ecosystem services.

In addition to climate change, overfishing, illegal, unreported and unregulated fishing, and commercial shipping are major pressures in the oceanic North Pacific. Other major pressures in the oceanic areas of the North Pacific include ocean dumping and increased UV-B radiation (Gray, 1997). However, time series are unavailable for these oceanic stressors at the scale of the North Pacific and require more study.

A rather unique but more localized feature of the North Pacific that could be affecting the open-ocean environment is the Pacific garbage patch. Day and Shaw (1987) have shown that the amount of plastic material in the oceans has increased over historical levels, and as oceanic eddies represent favourable locations for accumulation of floating debris like plastic, it should not be surprising that Moore et al. (2001) found the highest concentrations of plastic recorded in the Pacific within the North Pacific Subtropical Gyre. Also, mesopelagic fish species such as myctophids, have been shown to ingest microplastics in both the eastern (Davison and Asch, 2011) and western (Van Noord, 2013) parts of the basin. Thus, there are potential implications of this microplastic on a variety of organisms and on ecosystem structure and function although the specific effects are less clear.

3.3 Major ecosystem services being affected by the pressures

3.3.1 Ecosystem services being lost

It has been shown that zooplankton on both sides of the North Pacific respond strongly to regime shifts and hence should be expected to respond similarly to climate change. Thus, if the results of the study by Beaugrand et al. (2010) for the North Atlantic translate to the reorganization of the planktonic ecosystem towards smaller organism dominance that affects carbon flows, there could be adverse effects. As in coastal ecosystems, oceanic fishing operations can cause different types of ecological impacts, including bycatch of non-target species, habitat damage, mortality caused by lost or discarded gear, pollution, generation of marine debris, etc. Understanding the specific ecosystem services lost due to these activities will require more study.

3.3.2 Human services being lost

As in the coastal areas, changes to marine biodiversity (and hence to ecosystem structure and function) in oceanic systems also can result from direct impacts (e.g., fishing, pollution, and habitat destruction), and indirect impacts via climate change and related perturbations of ocean biogeochemistry (e.g., acidification) and these can have severe consequences on human services. As with the coastal areas, no basin-scale studies quantifying these impacts have been found and the lack of smaller-scale examples from oceanic areas of the North Pacific suggest that significant gaps exist. Although some studies have suggested fisheries targeting top predators have resulted in fewer large fish available for fishermen (Myers and Worm, 2003), other impacts on human services are less clear, suggesting that additional research is needed.

Historically, plastic debris was seen as a major concern for organisms that became entangled in or were ingesting it, especially marine mammals, seabirds, turtles, and fish (Laist, 1987). More recently, Boerger et al. (2010) showed that 35 per cent of the planktivorous fish sampled in the North Pacific Gyre (Garbage Patch) had ingested plastics; with this increased recognition of microplastics in the marine environment, the potential services being lost could be greater than initially thought, suggesting that more research is warranted.

4 Specific Areas of the North Pacific

The Bering Sea is a semi-enclosed subarctic sea that connects the North Pacific and Arctic Oceans and is bounded to the north by the Bering Strait and to the south by the Aleutian Archipelago. The deep central basin of this sea is bordered by a western shelf that extends from the Gulf of Anadyr along the Kamchatka Peninsula and a very broad eastern shelf extending from Alaska to the Russian Federation. Sea ice is important for determining the extent of the cold pool and regulates the timing of the spring bloom that has important cascading effects on ecosystem productivity and biological diversity (Stabeno and Hunt, 2002; Stabeno et al., 2007). Similarly, sea ice is a critical component of the structure and function of the Sea of Okhotsk. As with the Bering Sea, the Sea of Okhotsk has a relatively large shelf zone covering approximately 40 per cent of the basin (Udintsev, 1957). Predictions of potential impacts of global climate change are expected to be more severe at higher latitudes, suggesting native biodiversity could be adversely affected.

The South China Sea is a tropical system that includes diverse habitats such as mangrove forests, seagrass beds, and coral reefs. It lies within the Tropic of Cancer and has an area of 3.5 million km^2, of which 30 per cent is relatively deep sea with an average depth of about 1,400 m. The unique feature of this sea includes the effect of a monsoonal climate and complex surface-current patterns (Huang et al., 2010 and references therein). The complex surface-current system greatly influences the structure of the marine ecosystem, which is a mixture of tropical and subtropical communities. In addition, the Pearl and Mekong rivers discharge a huge amount of nutrients into the South China Sea. These characteristics support very diverse fauna and flora, with over 2,300 fish species (Caihua et al., 2008), 58 cephalopod species, and many other invertebrates (Jia et al., 2004). However, the dramatic expansion of fishing effort and improved fishing technology (Pang and Pauly, 2001) resulted in over-exploitation of fisheries resources here (Cheung and Sadovy, 2004).

Two additional semi-enclosed seas around the North Pacific deserve attention, due to increased human population growth and anthropogenic stress (e.g., urbanization, pollution, fisheries, and invasive species). On the eastern side of the basin is the Salish Sea; this relatively large estuarine system extends from the Strait of Georgia/Desolation Sound in the north to Puget Sound in the south and the Strait of Juan de Fuca to the west. With over seven million people living in the basin, the number of anthropogenic stressors is large with many having the potential to adversely affect biodiversity and ecosystem structure and function. Similarly, on the western side of the basin is the Seto Inland Sea, which serves as an important transportation link between the Pacific Ocean and the adjacent sea and between industrial centres around Japan. Many unique species call the Seto Inland Sea home, but anthropogenic impacts have been severe. For example, increased frequency of red tide (HAB) events and jellyfish blooms, possibly due to changes in nutrients in recent years, have resulted in significant losses to fisheries and aquaculture production.

A number of major rivers terminate in the North Pacific, but the ones that empty into semi-enclosed basins can result in unique attributes there. For example, the Colorado River, that discharges into the upper portion of the Gulf of California, results in biophysical features and oceanographic characteristics (strong tidal mixing, significant freshwater influx) that has resulted in a high level of endemism, such as the vaquita (a critically endangered porpoise; Gerrodette et al., 2011). In addition, other marine megafauna, such as the totoaba (*Totoaba mcdonaldi*) and the curvina golfina (*Cynoscion othonopterus*) have disjointed distributions in the upper Gulf of California.

The Emperor Seamount Chain and Hawaiian Ridge extends over 3,000 km from the Aleutian Trench to the Hawaiian Islands and seamounts outside of the United States EEZ were identified as meeting the criteria for Ecologically and Biologically Significant Areas (CBD, 2014). Hart and Pearson (2011) identified 49 fish species associated with this seamount chain and commercial fisheries targeting North Pacific armorhead (*Pseudopentaceros wheeleri*) and Splendid alfonsin (*Beryx splendens*) have operated since the late 1960s. Further, Japanese surveys have identified a variety of coral species inhabiting this chain including; Gorgonaceans (8 families, 24 genera), Alcyonaceans (6 families, 7 genera), Antipatharians (4 families, 5 genera) and Scleractinians (6 families, 16 genera). Also, the more productive surface waters provide good foraging environments for a variety of seabird species, including albatrosses.

Other specific areas in the North Pacific include the Mariana Trench and deep-sea hydrothermal vents. The Mariana Trench is unique in being the deepest location known on Earth. Relatively few studies exist; most characterize or describe the unique bacterial communities inhabiting this environment. Globally, hydrothermal vents are relatively rare and a unique geological feature associated with the spreading of tectonic plates. These sites support chemosynthetically driven ecosystems that support a diverse array of unique organisms (see Chapter 45).

5   Special conservation status issues

5.1   Taxonomic groups

Corals are often identified as a taxonomic group requiring special conservation consideration. In the North Pacific, approximately 30 per cent of the world's coral reefs are located in Southeast Asia; Wilkinson et al. (1993) suggest that more than half are already destroyed and being destroyed by sedimentation, overexploitation (including by dynamite and chemicals), and pollution. In addition to these warm-water corals, a growing number of cold-water corals and sponges also should be considered (e.g., Stone and Shotwell, 2007). Both warm- and cold-water corals are covered in more detail in Chapters 43 and 42, respectively.

Pacific salmon (*Oncorhynchus* species) are ecologically, commercially, and culturally important around the North Pacific. As anadromous species they require both freshwater and marine habitats for their continued survival and productivity but increased human activities have reduced productive habitat (in both freshwater and marine environments) and resulted in a number of additional stressors. The unique homing nature of salmon has resulted in a high degree of stock differentiation. For example, Slaney et al. (1996) identified 9,662 anadromous salmon stocks in British Columbia and the Yukon, including 866 Chinook, 1,625 Chum, 2,594 Coho, 2,169 Pink, 917 Sockeye, 867 Steelhead and 612 sea-run Cutthroat trout stocks. Maintaining genetic diversity will be important for maintaining productive Pacific salmon stocks; a goal of Canada's Wild Salmon Policy (DFO, 2005).

Other taxonomic groups often identified for special conservation status include many large or apex predators, such as tunas, sharks, billfish, and sea turtles, including Loggerhead (*Caretta caretta*) and Olive Ridley (*Lepidochelys olivacea*) sea turtles often found associated with the NPTZ (Polovina et al., 2004), because of their vulnerability to overexploitation and their role in ecosystem structure and function. Each of these is considered in more detail in Chapters 37-41.

5.2 Habitats

Much remains to be discovered with respect to biodiversity in the North Pacific, but additional conservation measures could be considered for several habitats, including hydrothermal vents, seamounts, large river deltas, kelp forests, mangroves, and coastal lagoons. In general, seamounts are often highly productive ecosystems that can support high biodiversity (Pitcher et al., 2007; Chapter 51), especially where their summit reaches into the euphotic zone and can be utilized by pelagic species, including marine birds and mammals. However, they can be susceptible to overfishing (see Douglas, 2011). Closer to the equator, coastal habitats, such as mangroves and coastal lagoons, are important habitats supporting relatively higher levels of biodiversity where degradation and/or complete destruction are significant concerns (see Chapters 48 and 49).

6 Factors for sustainability

It is clear that the maintenance of biodiversity contributes to ecosystem stability and sustainability and, like the other world oceans, the North Pacific is not unique in being under a barrage of anthropogenic stressors, that threaten the biodiversity and the sustainability it provides both for ecosystem services and human well-being. However, these stressors are not uniformly distributed across the North Pacific, with many more stressors noted for coastal ecosystems relative to the oceanic North Pacific. Furthermore, as research expands into the realm of how multiple stressors interact to affect biodiversity and ecosystem structure, function, and productivity, evidence is mounting that ecosystems are responding in complex, non-linear, non-additive - but cumulative - ways. Understanding and managing human activities to maintain or enhance biodiversity will make a substantial contribution to ecosystem sustainability globally (e.g., Hughes et al., 2005) and in the North Pacific specifically. However, as human populations, many of which are dependent on coastal or oceanic ecosystems for their existence, continue to expand around the North Pacific, there will be challenges.

References

Anderson, D.M., Glibert, P.M., Burkholder, J.M. (2002). Harmful algal blooms and eutrophication: nutrient sources, composition, and consequences. *Estuaries* 25, 704-726.

Batten, S., Chen, X., Fling, E.N., Freeland, H.J., Holmes, J., Howell, E., Ichii, T., Kaeriyama, M., Landry, M., Lunsford, C., Mackas, D.L., Mate, B., Matsuda, K., McKinnell, S.M., Miller, L., Morgan, K., Pena, A., Polovina, J.J., Robert, M., Seki, M.P., Sydeman, W.J., Thompson, S.A., Whitney, F.A., Woodworth, P., Yamaguchi, A. (2010). Status and trends of the North Pacific oceanic region, 2003-2008, pp. 56-105 In McKinnell, S.M. and Dagg, M.J. (eds.).*Marine Ecosystems of the North Pacific Ocean, 2003-2008*. PICES Special Publication 4, 393 p.

Beaugrand, G., Edwards, M., and Legendre, L. (2010). Marine biodiversity, ecosystem functioning, and carbon cycles. *Proceedings of the National Academy of Sciences of the United States of America* 107, 10120-10124.

Behrenfeld, M.J., and Falkowski, P.G. (1997). Photosynthetic rates derived from satellite-based chlorophyll concentration. *Limnology and Oceanography* 42, 1–20.

Boerger, C.M., Lattin, G.L., Moore, S.L., and Moore, C.J. (2010). Plastic ingestion by planktivorous fishes in the North Pacific Central Gyre. *Marine Pollution Bulletin* 60, 2275-2278.

Bograd, S.J., Castro, C.G., Di Lorenzo, E., Palacios, D.M., Bailey, H., Gilly, W., and Chavez, F.P. (2008). Oxygen declines and the shoaling of the hypoxic boundary in the California Current. *Geophysical Research Letters* 35, L12607.

Bograd, S.J., Sydeman, W.J., Barlow, J., Booth, A., Brodeur, R.D., Calambokidis, J., Chavez, F., Crawford, W.R., Di Lorenzo, E., Durazo, R., Emmett, R., Field, J., Gaxiola-Castro, G., Gilly, W., Goericke, R., Hildebrand, J., Irvine, J.E., Kahru, M., Koslow, J.A., Lavaniegos, B., Lowry, M., Mackas, D.L., Manzano-Sarabia, M., McKinnell, S.M., Mitchell, B.G., Munger, L., Perry, R.I., Peterson, W.T., Ralston, S., Schweigert, J., Suntsov, A., Tanasichuk, R., Thomas, A.C., Whitney, F. (2010). Status and trends of the California Current region, 2003-2008, pp. 106-141 In McKinnell, S.M. and Dagg, M.J. (eds.).*Marine Ecosystems of the North Pacific Ocean, 2003-2008*. PICES Special Publication 4, 393 p.

Caihua, M.A., Kui, Y., Meizhao, Z., Fengqi, L., and Dagang, C. (2008). A preliminary study on the diversity of fish species and marine fish faunas of the South China Sea. *Oceanic and Coastal Sea Research* 7(2), 210-214.

CBD (Convention on Biological Diversity) (2014). Report of the North Pacific regional workshop to facilitate the description of ecologically or biologically significant marine areas. UNEP/CBD/RW/EBSA/NP/1/4. 187 pp.

Chelton, D.A., Bernal, P.A., and McGowan, J.R. (1982). Large-scale interannual physical and biological interaction in the California Current. *Journal of Marine Research* 40, 1095-1125.

Cheung, W.W.L., Lam, V.W.Y., Sarmiento, J.L., Kearney, K., Watson, R., and Pauly, D. (2009). Projecting global marine biodiversity impacts under climate change scenarios. *Fish and Fisheries* 10, 235-251.

Cheung, W.W.L., and Sadovy, Y. (2004). Retrospective evaluation of data-limited fisheries: a case from Hong Kong. *Reviews in Fish Bilogy and Fisheries* 14, 181-206.

Chiba, S., Hirawake, T., Ishizaki, S., Ito, S., Kamiya, H., Kaeriyma, M., Kuwata, A., Midorikawa, T., Minobe, S., Okamoto, S., Okazaki, Y., Ono, T., Saito, H., Saitoh, S., Sasano, D., Tadokoro, K., Takahashi, K., Takatani, Y., Watanabe, Y., Watanabe, Y.W., Watanuki, Y., Yamamura, O., Yamashita, N., and Yatsu, A. (2010). Status and trends of the Oyashio region, 2003-2008, pp. 300-329 In McKinnell, S.M. and Dagg, M.J. (eds.).*Marine Ecosystems of the North Pacific Ocean, 2003-2008*. PICES Special Publication 4, 393 p.

Cooley, S.R., Kite-Powel, H.L., and Doney, S.C. (2009). Ocean acidification's potential to alter global marine ecosystem services. *Oceanography* 22, 172-181.

Cox, S.P., Martell, S.J.D., Walters, C.J., Essington, T.E., Kitchell, J.F., Boggs, C.H., and Kaplan, I. (2002). Reconstructing ecosystem dynamics in the central Pacific Ocean, 1952-1998: I. Estimating population biomass and recruitment of tunas and billfishes. *Canadian Journal of Fisheries and Aquatic Sciences* 59, 1724-1735.

Davison, P., and Asch, R.G. (2011). Plastic ingestion by mesopelagic fishes in the North Pacific Subtropical Gyre. *Marine Ecology Progress Series* 432, 173–180.

Day, R.H., and Shaw, D.G. (1987). Patterns in the abundance of pelagic plastic and tar in the North Pacific Ocean, 1976-1985. *Marine Pollution Bulletin* 18, 311-316.

DFO (Department of Fisheries and Oceans Canada) (2005). Canada's Policy for Conservation of Wild Pacific Salmon. 57 p.

DFO (Department of Fisheries and Oceans Canada) (2014). *Preliminary 2014 Salmon Outlook*. Accessed at: http://www.pac.dfo-mpo.gc.ca/fm-gp/species-especes/salmon-saumon/outlook-perspective/salmon_outlook-perspective_saumon-2014-eng.html (Jun 20, 2014)

Douglas, D.A. (2011). The Oregon Shore-Based Cobb Seamount Fishery, 1991-2003: Catch Summaries and Biological Observations. Oregon Department of Fish and Wildlife, *Information Reports* Number 2011-03.

Emerson, S., Watanabe, Y.W., Ono, T., Mecking, S. (2004). Temporal Trends in Apparent Oxygen Utilization in the Upper Pycnocline of the North Pacific: 1980-2000. *Journal of Oceanography 60*, 139-147.

Francis, R.C., Hare, S.R., Hollowed, A.B., and Wooster, W.S. (1998). Effects of interdecadal climate variability on the oceanic ecosystems of the NE Pacific. *Fisheries Oceanography* 7, 1-21.

Fulton, J. (1983). Seasonal and annual variations of net zooplankton at Ocean Station "P", 1956-1980. *Canadian Data Report of Fisheries and Aquatic Sciences* 374, 65p.

Gerrodette, T., Taylor, B., Swift, R., Jaramillo, A., and Rojas-Bracho, L. (2011). A combined visual and acoustic estimate of 2008 abundance, and change in abundance since 1997, for the vaquita, *Phocoena sinus*. *Marine Mammal Science* 27, E79-E100.

Grantham, B.A., Chan, F., Nielsen, K.J., Fox, D.S., Barth, J.A., Huyer, A., Lubchenco, J., and Menge, B.A. (2004). Upwelling-driven nearshore hypoxia signals ecosystem and oceanographic changes in the northeast Pacific. *Nature* 429, 749-754.

Gray, J.S. (1997). Marine biodiversity: patterns, threats and conservation needs. *Biodiversity and Conservation* 6, 153-175.

Halpern, B.S., Walbridge, S., Selkoe, K.A., Kappel, C.V., Micheli, F., D'Agrosa, C., Bruno, J.F., Casey, K.S., Ebert, C., Fox, H.E., Fujita, R., Heinemann, D., Lenihan, H.S., Madin, E.M.P., Perry, M.T., Selig, E.R., Spalding, M., Steneck, R., and Watson, R. (2008). A global map of human impact on marine ecosystems. *Science* 319, 948-952.

Hart, P.J.B., and Pearson, E. (2011). An application of the theory of island biogeography to fish speciation on seamounts. *Marine Ecology Progress Series* 430, 281-288.

Hilborn, R., Parrish, J.K., and Litle, K. (2005). Fishing rights or fishing wrongs? *Reviews in Fish Biology and Fisheries* 15, 191-199.

Hedrick D.B., Pledger, R.D., White, D.C., and Baross, J.A. (1992). In situ microbial ecology of hydrothermal vent sediments. *FEMS Microbiology Letters*101, 1-10.

Hooper, D.U., Chapin III, F.S., Ewel, J.J., Hector, A., Inchausti, P., Lavorel, S., Lawton, J.H., Lodge, D.M., Loreau, M., Naeem, S., Schmid, B., Setälä, H., Symstad, A.J., Vandermeer, J., and Wardle, D.A. (2005). Effects of biodiversity on ecosystem functioning: a consensus of current knowledge. *Ecological Monographs* 75, 3-35.

Hourigan, T.F., Lumsden, S.E., Dorr, G., Bruckner, A.W., Brooke, S., and Stone, R.P. (2007). State of deep coral ecosystems of the United States: introduction and national overview. In: Lumsden, S.E., Hourigan, T.F., Bruckner, A.W., and Dorr, G. (eds). The State of Deep Coral Ecosystems of the United States. NOAA Technical Memorandum CRCP-3. Silver Spring MD 365 pp.

Huang, B., Cheung, W., Lam, V.W.Y., Palomares, M.L.D., Sorongon, P.M.E, and Pauly, D. (2010). Toward an account of the biodiversity in Chinese shelf waters: The role of SeaLifeBase and FishBase. In Palomares, M.L.D.and Pauly, D. (eds.). *Marine biodiversity in Southeast Asian and Adjacent Sea.* Fisheries Centre Research Report 18(3), 2-14.

Hughes, T.P., Bellwood, D.R., Folke, C., Steneck, R.S., and Wilson, J. (2005). New paradigms for supporting the resilience of marine ecosystems. Trends in Ecology and Evolution 20, 380-386.

Hutchings, J.A. (2000). Collapse and recovery of marine fishes. *Nature* 406, 882-885.

International Hydrographic Organization. (1953). *Limits of the Oceans and Seas* (Special Publication 28), 3rd Edition.

ISC (International Scientific Committee for Tuna and Tuna-like Species in the North Pacific Ocean). (2008). Report of the Eighth Meeting of the International Scientific Committee for Tuna and Tuna-like Species in the North Pacific Ocean. Plenary Session, 22-27 July 2008, Takamatsu, Japan. 47p. Available at: http://isc.ac.affrc.go.jp/isc8/ISC8rep.html

Jackson, J.B.C., Kirby, M.X, Berger, W.H., Bjorndal, K.A., Botsford, L.W., Bourque, B.J., Bradbury, R.H., Cooke, R., Erlandson, J., Estes, J.A., Hughes, T.P., Kidwell, S., Lange, C.B., Lenihan, H.S., Pandolfi, J.M., Peterson, C.H., Steneck, R.S., Tegner, M.J., and Warner, R.R. (2001). Historical overfishing and the recent collapse of coastal ecosystems. *Science* 293, 629-637.

Jia, X., Li, Z., Li, C. Qiu, Y., and Gan, J. (2004). *The ecosystem and fisheries resources in the commercial zone and the continental shelf of the South China Sea.* Science Press, Beijing. 647 p. (in Chinese).

Kahru, M., Kudela, R., Manzano-Sarabia, M., and Mitchell, B.G. (2009). Trends in primary production in the California Current detected with satellite data. *Journal of Geophysical Research* 114, 1978-2012.

Karl, D.M. (1999). A sea of change: biogeochemical variability in the North Pacific subtropical gyre. *Ecosystems* 2, 181-214.

Karl, D.M., Bidigare, R.R., and Letelier, R.M. (2001). Long-term changes in plankton community structure and productivity in the North Pacific Subtropical Gyre: the domain shift hypothesis. *Deep-Sea Research II* 48, 1449-1470.

Keeling, R.F., Kortzinger, A., and Gruber, N. (2010). Ocean deoxygenation in a warming world. *Annual Reviews in Marine Science* 2, 199-229.

Kodama, K., Oyama, M., Lee, J.-h, Kume, G., Yamaguchi, A., Shibata, Y., Shiraishi, H., Morita, M., Shimizu, M., and Horiguchi, T. (2010). Drastic and synchronous changes in megabenthic community structure concurrent with environmental variations in a eutrophic coastal bay. *Progress in Oceanography* 87, 157-167.

Kodama, K., and Horiguchi, T. (2011). Effects of hypoxia on benthic organisms in Tokyo Bay, Japan: A review. *Marine Pollution Bulletin* 63, 215-220.

Koslow, J.A., Goericke, R., Lara-Lopez, A., and Watson, W. (2011). Impact of declining intermediate-water oxygen on deepwater fishes in the California Current. *Marine Ecology Progress Series* 436, 207-218.

Laist, D.W. (1987). Overview of the biological effects of lost and discarded plastic debris in the marine environment. *Marine Pollution Bulletin* 18, 319-326.

Lee II, H., and Reusser, D.A. (2012). Atlas of nonindigenous marine and estuarine species in the North Pacific. Office of Research and Development, National Health and Environmental Effects Research Laboratory, EPA/600/R/12/631.

Levin, L., Ekau, W., Gooday, A.J., Jorissen, F., Middelburg, J.J., Naqvi, S.W.A., Neira, C., Rabalais, N.N., and Zhang, J. (2009). Effects of natural and human-induced hypoxia on coastal benthos. *Biogeosciences* 6, 2063-2098.

Lin, C., Ning, X. Su, J., Lin, Y., and Xu, B. (2005). Environmental changes and the responses of the ecosystems of the Yellow Sea during 1976–2000. *Journal of Marine Systems* 55, 223-234.

Longhurst, A.R. (2007). *Ecological geography of the sea--2nd ed.* Amsterdam; Boston, MA Elsevier Academic Press. 542 p.

Mackas, D.L., and Galbraith, M. (2002). Zooplankton community composition along the inner portion of Line P during the 1997-98 El Nino event. *Progress in Oceanography* 54, 423-437.

Mackas, D.L., Thomson, R.E., Galbraith, M. (2001). Changes in the zooplankton community of the British Columbia continental margin, 1985-1999, and their covariation with oceanographic conditions. *Canadian Journal of Fisheries and Aquatic Sciences* 58, 685-702.

Mantua, N.J., and Hare, S.R. (2002). The Pacific Decadal Oscillation. *Journal of Oceanography* 58, 35-44.

Maxwell, S.M., Hazen, E.L., Bograd, S.J., Halpern, B.S., Breed, G.A., Nickel, B., Teutschel, N.M., Crowder, L.B., Benson, S., Dutton, P.H., Bailey, H., Kappes, M.A., Kuhn, C.A., Weise, M.J., Mate, B., Shaffer, S.A., Hassrick, J.L., Henry, R.W., Irvine, L., McDonald, B.I., Robinson, P.W., Block, B.A., and Costa, D.P. (2013). Cumulative human impacts on marine predators. *Nature Communications* 4, 2866.

McKinnell, S.M. and Dagg, M.J. (eds.) (2010). *Marine Ecosystems of the North Pacific Ocean, 2003-2008.* PICES Special Publication 4, 393 p.

McKinnell, S.M., Batten, S., Bograd, S.J., Boldt, J.L., Bond, N., Chiba, S., Dagg, M.J., Foreman, M.G.G., Hunt Jr., G.L., Irvine, J.R., Katugin, O.N., Lobanov, V., Mackas, D.L., Mundy, P., Radchenko, V., Ro, Y.J., Sugisaki, H., Whitney, F.A., Yatsu, A., Yoo, S. (2010). Status and trends of the North Pacific Ocean, 2003-2008, pp. 1-55 In McKinnell, S.M. and Dagg, M.J. (eds.). *Marine Ecosystems of the North Pacific Ocean, 2003-2008.* PICES Special Publication 4, 393 p.

Miles, E., Gibbs, S., Fluharty, D., Dawson, C., and Teeter, D. (eds). (1982). *The Management of Marine Regions: The North Pacific.* University of California Press.

Moore, C.J., Moore, S.L., Leecaster, M.K., and Weisberg, S.B. (2001). A comparison of plastic and plankton in the North Pacific Central Gyre. *Marine Pollution Bulletin* 42, 1297-1300.

Mueter, F.J., and Litzow, M.A. (2008). Sea ice retreat alters the biogeography of the Bering Sea continental shelf. *Ecological Applications* 18, 309-320.

Mueter, F.J., and Norcross, B.L. (2002). Spatial and temporal patterns in the demersal fish community of the shelf and upper slope regions of the Gulf of Alaska. *Fishery Bulletin* 100, 559-581.

Myers, R.A., and Worm, B. (2003). Rapid worldwide depletion of predatory fish communities. *Nature* 423, 280-283.

NOAA (2013). Status of Stocks 2013 Annual Report to Congress on the Status of U.S. Fisheries. 8 pp.

NOAA (2014). Status of Stocks 2014 Annual Report to Congress on the Status of U.S. Fisheries. 8p.

Ono, T., Midorikawa, T., Watanabe, Y.W., Tadokoro, K., and Saino, T. (2001). Temporal increases of phosphate and apparent oxygen utilization in the subsurface waters of western subarctic Pacific from 1968 to 1998. *Geophysical Research Letters* 28, 3285-3288.

Orr, J.C., Fabry, V.J., Aumont, O., Bopp, L., Doney, S.C., Feely, R.A., Gnanadesikan, A., Gruber, N., Ishida, A., Joos, F., Key, R.M., Lindsay, K., Maier-Reimer, E., Matear, R., Monfray, P., Mouchet, A., Najjar, R.G., Plattner, G.-K., Rodgers, K.B., Sabine, C.L.,

Sarmiento, J.L., Schlitzer, R., Slater, R.D., Totterdell, I.J., Weirig, M.-F., Yamanaka, Y., and Yool, A. (2005). Anthropogenic ocean acidification over the twenty-first century and its impact on calcifying organisms. *Nature* 437, 681-686.

Ovetz, R. (2007). The bottom line: An investigation of the economic, cultural and social costs of high seas industrial longline fishing in the Pacific and the benefits of conservation. *Marine Policy* 31, 217-228.

Pang, L., and Pauly, D. (2001). Chinese marine capture fisheries from 1950 to the late 1990s: the hopes, the plans and the data. In Watson, R. Pang, L., and Pauly, D. (eds.), The Marine Fisheries of China: Development and Reported Catches. *Fisheries Centre Research Report* 9(2), 1-27.

PICES (2004). *Marine ecosystems of the North Pacific*. PICES Special Publication 1, 280p.

Pitcher, T.J., Morato, T., Hart, P.J.B., Clark, M.R., Haggan, N., and Santos, R.S. (eds.) (2007). *Seamounts, Fisheries and Conservation*. Blackwell Publishing, 527 pp.

Polovina, J., Howell, E., Kobayashi, D., and Seki, M. (2001). The transition zone chlorophyll front, a dynamic global feature defining migration and forage habitat for marine resources. *Progress in Oceanography* 49, 469-483.

Polovina, J.J, Balazs, G.H., Howell, E.A., Parker, D.M., Seki, M.P., and Dutton, P.H. (2004). Forage and migration habitat of loggerhead (*Caretta caretta*) and olive ridley (*Lepidochelys olivacea*) sea turtles in the central North Pacific Ocean. *Fisheries Oceanography* 13, 36-51.

Polovina, J.J., E.A. Howell, and Abecassis, M. (2008). Ocean's least productive waters are expanding. *Geophysical Research Letters* 35, 5.

Pope, J.G., Shepherd, J.G., and Webb, J. (1994). Successful surf-riding on size spectra: the secret of survival in the sea. *Philosophical Transactions of the Royal Society of London, B* 343, 41-49.

Purcell, J.E., Uye, S., and Lo, W.-T. (2007). Anthropogenic causes of jellyfish blooms and their direct consequences for humans: a review. *Marine Ecology Progress Series* 350, 153-174.

Ream, J.A., Sterling, J.T., and Loughlin, T.R. (2005). Oceanographic features related to northern fur seal migratory movements. *Deep-Sea Research II* 52, 823-843.

Ruiz, G.M., Carlton, J.T., Grosholz, E.D., and Hines, A.H. (1997). Global invasions of marine and estuarine habitats by non-indigenous species: mechanisms, extent, and consequences. *American Zoologist* 37, 621-632.

Sala, O.E., Chapin, S.F., Armesto, J.J., Berlow, E., Bloomfield, J., Dirzo, R., Huber-Sanwald, E., Huenneke, L.F., Jackson, R.B., Kinzig, A., Leemans, R., Lodge, D.M., Mooney, H.A., Oesterheld, M., Poff, N.L., Sykes, M.T., Walker, B.H., Walker, M. & Wall, D.H. (2000). Global biodiversity scenarios for the year 2100. *Science*, 287, 1770–1774.

Sarmiento, J.L., Hughes, T.M.C., Stouffer, R.J., and Manabe, S. (1998). Simulated response of the ocean carbon cycle to anthropogenic climate warming. *Nature* 393, 245-249.

Schroeder, D., and Love, M.S. (2002). Recreational fishing and marine fish populations in California. *CalCOFI Report* 43, 182-190.

Sheridan, C.C., and Landry, M.R. (2004). A 9-year increasing trend in mesozooplankton biomass at the Hawaii Ocean Time-series Station ALOHA. *ICES Journal of Marine Science* 61, 457-463.

Sibert, J., Hampton, J., Klieber, P., and Maunder, M. (2006). Biomass, size, and trophic status of top predators in the Pacific Ocean. *Science* 314, 1773-1776.

Slaney, T.L., Hyatt, K.D., Northcote, T.G., and Fielden, R.J. (1996). Status of anadromous salmon and trout in British Columbia and Yukon. *Fisheries* 21, 20-35.

Spalding, M.D., Fox, H.E., Allen, G.R., Davidson, N., Ferdaña, Z.A., Finlayson, M., Halpern, B.S., Jorge, M.A., Lombana, A., Lourie, S.A., Martin, K.D., McManus, E., Molnar, J., Recchia, C.A., Robertson, J. (2007). Marine ecoregions of the world: a bioregionalization of coast and shelf areas. *BioScience* 57, 573-583.

Stabeno, P.J., and Hunt, G.L. (2002). Overview of the inner front and southeast Bering Sea carrying capacity programs. *Deep-Sea Research II* 49, 6157-6168.

Stabeno, P.J., Bond, N.A., and Salo, S.A. (2007). On the recent warming of the southeastern Bering Sea shelf. *Deep-Sea Research II* 54, 2599-2618.

Stone, R.P., and Shotwell, S.K. (2007). State of deep coral ecosystems in the Alaska Region: Gulf of Alaska, Bering Sea and the Aleutian Islands. Pp 65-108. In: Lumsden, S.E., Hourigan, T.F., Bruckner, A.W., Dorr, G. (eds.). *The state of Deep Coral Ecosystems of the United States*. NOAA Technical Memorandum CRCP-3. Silver Spring MD 365 pp.

Stramma, L., Schmidt, S., Levin, L.A., and Johnson, G.C. (2010). Ocean oxygen minima expansions and their biological impacts. *Deep-Sea Research I* 210, 587-595.

Sugisaki, H., Nonaka, M., Ishizaki, S., Hidaka, K., Kameda, T., Hirota, Y., Oozeki, Y., Kubota, H., Takasuka, A. (2010). Status and trends of the Kuroshio region, 2003-2008, pp. 330-359 In McKinnell, S.M. and Dagg, M.J. (eds.). *Marine Ecosystems of the North Pacific Ocean, 2003-2008*. PICES Special Publication 4, 393 p.

Sydeman, W.J., and Thompson, S.A. (2010). The California Current Integrated Ecosystem Assessment, Module II: Trends and Variability in System State. Report to NOAA-NMFS-ERD. Farallon Institute for Advanced Ecosystem Research, Petaluma, CA.

Sydeman, W.J., Santora, J.A., Thompson, S.A., Marinovic, B., and DiLorenzo, E. (2013). Increasing variance in the North Pacific climate relates to unprecedented ecosystem variability off California. *Global Change Biology* 19, 1662-1675.

Tunnicliffe V., Desbruyeres D., Jollivet D., Laubier L. (1993). Systematic and ecological characteristics of *Paralvinella sulfincola* Desbruyères and Laubier, a new polychaete (family Alvinellidae) from northeast Pacific hydrothermal vents. *Canadian Journal of Zoology* 71, 286-297.

Udintsev, G.V. (1957). Bottom relief of the Sea of Okhotsk. *Trudy IOAN USSR* 22, 3-76. (in Russian).

Van Noord, J.E. (2013). Diet of five species of the family Myctophidae caught off the Mariana Islands. *Ichthyological Research* 60, 89-92.

Worm, B., Barbier, E.B., Beaumont, N., Duffy, J.E., Folke, C., Halpern, B.S., Jackson, J.B.C., Lotze, H.K., Micheli, F., Palumbi, S.R., Sala, E., Selkoe, K.A., Stachowicz, J.J., Watson, R. (2006). Impacts of biodiversity loss on ocean ecosystem services. *Science* 314, 787-790.

Wilkinson, C.R., Chou, L.M., Gomez, E., Ridzwaan, A.R., Soekano, S. and Sudra, S. (1993). *Global Aspects of Corals; Health, Hazards and History*. University of Miami.

Yoo, S., Batchelder, H.B., Peterson, W.T., and Sydeman, W.J. (2008). Seasonal, interannual and event scale variation in North Pacific ecosystems. *Progress in Oceanography* 77, 155-181.

Yoo, S., An, Y.-R., Bae, S., Choi, S., Ishizaka, J., Kang, Y.-S., Kim, Z.G., Lee, C., Lee, J.B., Li, R., Park, J., Wang, Z., Wen, Q., Yang, E. J., Yeh, S.-W., Yeon, I., Yoon, W.-D., Zhang, C.-I., Zhang, X., Zhu, M. (2010). Status and trends in the Yellow Sea and East China Sea region, pp. 360-393 In McKinnell, S.M. and Dagg, M.J. (eds.) (2010). *Marine Ecosystems of the North Pacific Ocean, 2003-2008*. PICES Special Publication 4, 393 p.

Zhou, H., Li, J., Peng, X., Meng, J., Wang, F., and Ai, Y. (2009). Microbial diversity of a sulfide black smoker in main endeavour hydrothermal vent field, Juan de Fuca Ridge. *The Journal of Microbiology* 47, 235-247.

36D South Pacific Ocean

Contributors:
Nic Bax (Convenor), Patricio Bernal (Lead Member), Marilú Bouchon Corrales, Martin Cryer, Karen Evans, Günter Försterra, Carlos F. Gaymer, Vreni Häussermann and Jake Rice (Co-Lead Member and Editor of Part VI)

Chapter 36D

South Pacific Ocean

1 Introduction

The Pacific Ocean is the Earth's largest ocean, covering one-third of the world's surface. This huge expanse of ocean supports the most extensive and diverse coral reefs in the world (Burke et al., 2011), the largest commercial fishery (FAO, 2014), the most and deepest oceanic trenches (General Bathymetric Chart of the Oceans, available at www.gebco.net), the largest upwelling system (Spalding et al., 2012), the healthiest and, in some cases, largest remaining populations of many globally rare and threatened species, including marine mammals, seabirds and marine reptiles (Tittensor et al., 2010).

The South Pacific Ocean surrounds and is bordered by 23 countries and territories (for the purpose of this chapter, countries west of Papua New Guinea are not considered to be part of the South Pacific), which range in size from small atolls (e.g., Nauru) to continents (South America, Australia). Associated populations of each of the countries and territories range from less than 10,000 (Tokelau, Nauru, Tuvalu) to nearly 30.5 million (Peru; Population Estimates and Projections, World Bank Group, accessed at http://data.worldbank.org/data-catalog/population-projection-tables, August 2014). Most of the tropical and sub-tropical western and central South Pacific Ocean is contained within exclusive economic zones (EEZs), whereas vast expanses of temperate waters are associated with high seas areas (Figure 36D.1). The eastern and western extremes of the ocean basin contain two major boundary currents:

the poleward-flowing East Australian Current (EAC), which runs along Australia's North-West shelf in the west (Ridgway and Dunn, 2003) and the northward-flowing Humboldt Current, which runs along South America's continental shelf in the east (Montecino and Lange, 2009). The dominant shallow water ecosystems of the region are tropical coral reef and lagoon systems and mangrove communities in the sub-tropics and tropics and temperate rocky reefs and kelp beds in temperate zones. Other marine communities across tropical, sub-tropical and temperate zones include rocky intertidals, mudflats, seagrass beds, estuaries and salt marshes in inshore areas and seamount, hydrothermal vents and trenches in offshore zones. Five Large Marine Ecosystems (www.lme.edc.uri.edu) have been defined across the South Pacific Ocean, including the Humboldt Current, the northeast Australian shelf, east-central Australian shelf, southeast Australian shelf and New Zealand shelf.

Physical processes of the basin play an important role in driving shelf and coastal marine processes and climate across the region. Northern parts of the South Pacific Ocean are dominated by a basin-scale sub-tropical gyre, whose northern branch forms the South Equatorial Current (SEC; Figure 36D.2; Reid, 1997). The SEC is predominantly driven by prevailing easterly trade winds and as water moves from the east to the west, a thick layer of warm water (>29°C), the Western Warm Pool (WWP) is formed west of ~170°E (Picaut et al., 1996). As the westward-flowing SEC encounters islands and land masses, it splits into several currents and jets, some of which, particularly the New Guinea Coastal

The boundaries and names shown and the designations used on this map do not imply official endorsement or acceptance by the United Nations.

Figure 36D.1 | The South Pacific Ocean. Sources: Bathymetry extracted from the GEBCO Digital Atlas (GDA): IOC, IHO and BODC, 2003. Centenary Edition of the GEBCO Digital Atlas, published on CD-ROM on behalf of the Intergovernmental Oceanographic Commission and the International Hydrographic Organization as part of the General Bathymetric Chart of the Oceans, British Oceanographic Data Centre, Liverpool, U.K. More information at http://www.gebco.net/data_and_products/gebco_digital_atlas/. Ocean and Sea names extracted from ESRI, DeLorme, HERE, GEBCO, NOAA, National Geographic, Geonames.org, and other contributors. More information at http://www.arcgis.com/home/item.html?id=0fd0c5b7a647404d8934516aa997e6d9. With the kind assistance of the FAO.

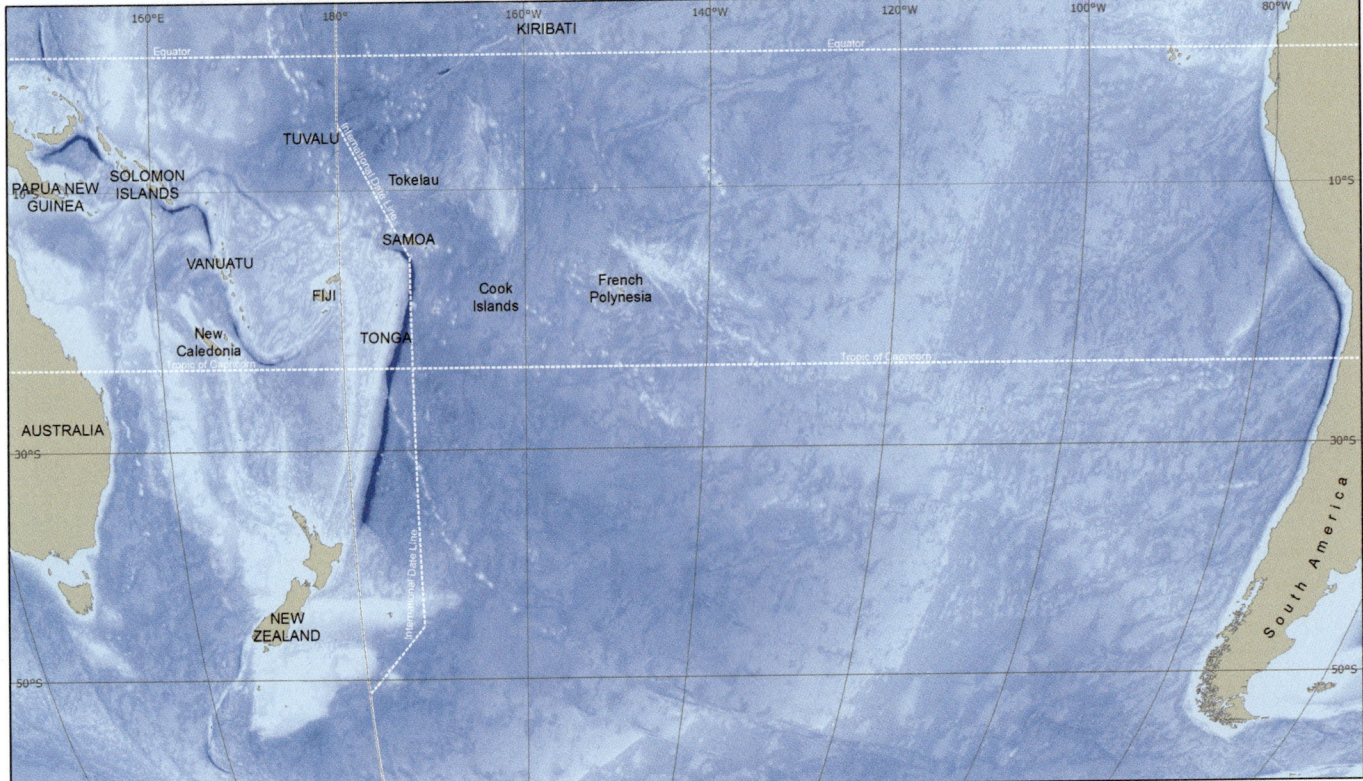

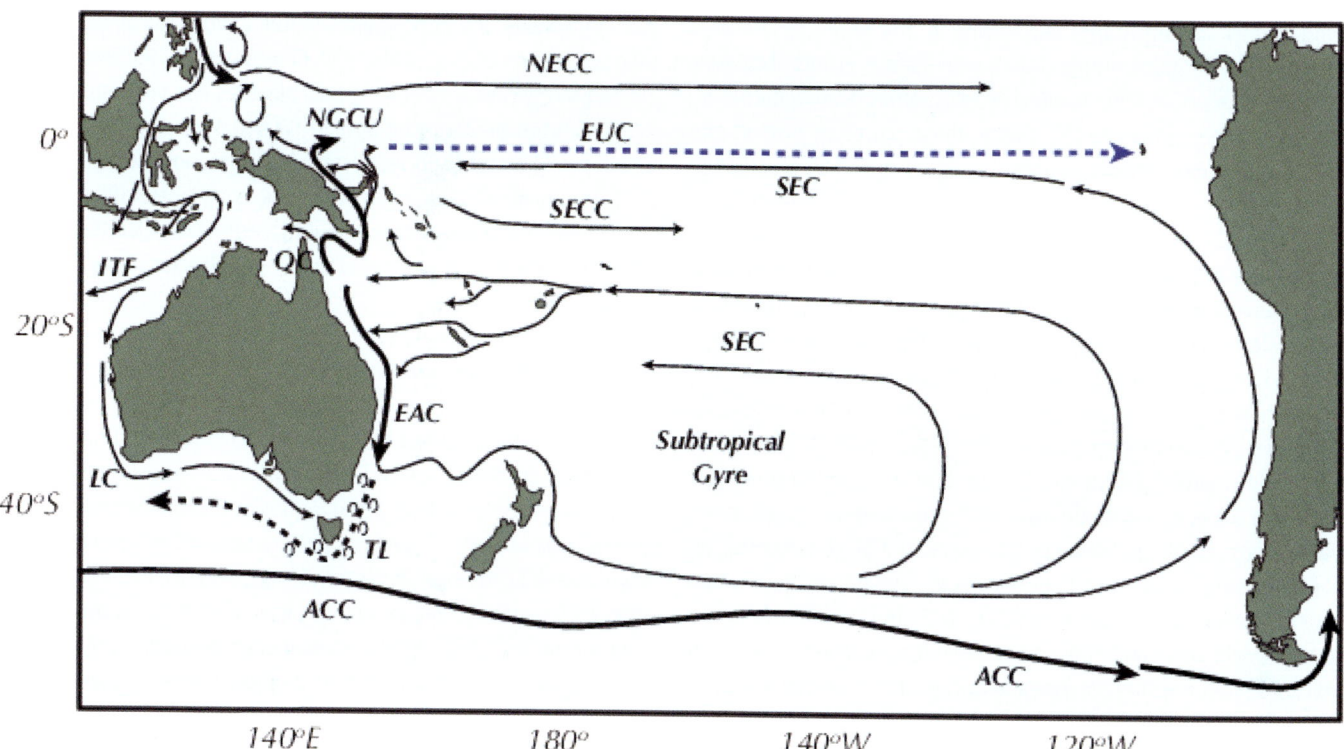

Figure 36D.2 | Major currents of the South Pacific Ocean. NECC: North Equatorial Counter-Current; NGCU: New Guinea Coastal Under-current; EUC: Equatorial Under-Current; SEC: South Equatorial Current; SECC: South Equatorial Counter-Current; ITF: Indonesian Through-Flow; QC: Queensland Current; EAC: East Australian Current; LC: Leeuwin Current; ACC: Antarctic Circumpolar Current.

Under-current (NGCU), contribute to the Equatorial Under-Current (EUC; Figure 36D.2). The EUC contributes significantly to equatorial thermocline waters and is thought to modulate the El Niño-Southern Oscillation (ENSO; Grenier et al., 2011). The EUC is also the primary source of iron in the photic layer of the region, and variability in the EUC drives regional biological productivity (Ryan et al., 2006). Once the jets meet the land mass of Australia, they form the poleward-flowing western boundary current, the East Australian Current (EAC).

As the EAC flows south along the Australia's North-West shelf, eddies separate from the main body of the EAC, forming a region of upwelling and downwelling. Outflow from the EAC forms a band of zonal eastward flow, the Tasman Front. The band of zonal eastward flow associated with the Tasman Front contributes to the East Auckland Current and eventually to the subtropical gyre moving northward and contributing to the SEC.

The eastern boundary of the gyre combines waters from the trans-Pacific West Wind Drift (WWD) with a northward-flowing arm of the Antarctic Circumpolar Current. At its arrival on the South American coast, the WWD diverges into the poleward-flowing Cape Horn Current and the northward-flowing Humboldt or Peru-Chile Current (PCC) which flows along western South America (Strub et al., 1998). Similarly to the EAC, the PCC is characterized by significant mesoscale variability in the form of fronts, eddies and filaments (Hill et al., 1998; Montecino and Lange, 2009), which intensify closer to the coast. The Humboldt Current system is highly productive due to the combined effect of advection of nutrient-rich waters from the south and upwelling. Cool, nutrient-rich waters are brought to the surface north of ~40°S as a result of coastal upwelling driven by winds and the impingement of the subtropical gyre along the coast (Morales et al., 1996; Montecino and Lange, 2009). Upwelling occurs seasonally across the ~30°S-40°S region (Strub et al., 1998; Shaffer et al., 1999), whereas in Northern Chile and off the Peruvian coast, upwelling is permanent (Hill et al., 1998; Vasquez et al., 1998). As the EUC encounters the Galápagos Islands it splits: one arm forms an undercurrent that reaches South America near the equator and becomes the poleward-flowing Gunther or Peru-Chile Under-Current, which flows beneath the PCC across the slope and outer shelf. The other arm flows to the southeast of the Galápagos Islands and forms the poleward-flowing Peru-Chile Counter-Current which divides the PCC into two branches: a coastal and an oceanic branch (Strub et al., 1998).

The physical dynamics of the region vary markedly with ENSO: during La Niña, stronger trade winds increase the intensity of the SEC, pushing the WWP west, and upwelling and productivity in the Pacific Equatorial Divergence (PEQD) increase. During El Niño, trade winds weaken, the SEC weakens, allowing the WWP to extend east and upwelling and productivity in the PEQD decrease (Ganachaud et al., 2011). Shifts in the intensity of the SEC have flow-on effects for both basin-scale circulation and shelf systems at the basin edges where shifts result in weakening/strengthening of the boundary currents.

Interaction of the easterly trade winds and ocean currents with island topography modifies the flow of water downwind of the islands, creating counter-currents, eddies and upwelling. This results in enhanced mixing of deeper nutrient-rich waters with surface waters, increasing biological production and enriching coastal waters (Ganachaud et al., 2011). For many South Pacific islands, these processes support rich coastal ecosystems in regions which would otherwise be regarded as oligotrophic.

Coordinated assessments of the state of the environment (including the marine environment) have been undertaken by several countries over the last decade, including: Australia (State of the Environment Committee, 2011), French Polynesia (Gabrie et al., 2007), Kiribati (Ministry of Environment Lands and Agricultural Development, 2004), New Zealand (Ministry for the Environment, 2007), Palau (Sakuma, 2004), Peru (World Bank, 2006), Samoa (Ministry of Natural Resources and Environment, 2013), the Solomon Islands (Ministry of Environment Conservation and Meteorology, 2008) and Vanuatu (Mourgues, 2005). A number of regional assessments have also been undertaken: The State of the Environment in Asia and Pacific 2005 (UNESCAP, 2005), the Pacific Environment Outlook (McIntytre, 2005), the Global International Waters Assessment (UNEP, 2006a), including the Global Assessment and Synthesis Reports from the Millennium Ecosystem Assessment (www.unep.org/maweb/en/Global; UNEP 2006b), the Pacific Ocean Synthesis (Center for Ocean Solutions, 2009), the UNEP Large Marine Ecosystems Report (Sherman and Hempel, 2009), the Global Biodiversity Outlook (Secretariat of the Convention on Biological Diversity, 2010), the Global Environment Outlook (UNEP, 2012) and the Pacific Environment and Climate Change Outlook (SPREP, 2012). This chapter summarizes the available assessments and current knowledge from peer-reviewed literature on the status of, immediate and long-term concerns for, and threats to the coastal and shelf marine ecosystems of the South Pacific Ocean.

2 Status and trends of biodiversity

Across the South Pacific Ocean, the most reliable time-series of the status of biodiversity across the region from which trends can be derived are largely limited to high-level indicators, including some oceanographic parameters (e.g., sea-surface temperatures, sea level) and industrial commercial fisheries (e.g., tuna, anchoveta). Indicators of pressures and impacts are similarly limited to high-level indicators of population and socio-economic measures. Long-term monitoring initiatives (e.g., those spanning multiple decades) are sparse and are largely limited to within-country monitoring of a few indicators associated with specific objectives. The TAO/TRITON array, which consists of approximately 70 moorings deployed across the tropical Pacific Ocean to collect primarily physical and meteorological data, can be considered the most extensive ocean observation system currently functional in the South Pacific Ocean. For a short period in the late 1990s, biological and chemical sensors (i.e., continuous pCO_2 analyzers, three biospherical irradiance meters, nitrate analyzers, and PAR sensors) were added to several buoys. This enabled continuous monitoring of biological productivity during deployment, improving understanding of biophysical coupling from inter-annual (ENSO events) to intra-seasonal (tropical instability waves) time scales (e.g., Chavez et al., 1998; Chavez et al., 1999; Strutton et al., 2001). More recently, several regional alliances and programmes under the Global Ocean Observing System (GOOS), including the Australian Integrated Marine Observing System (www.imos.org.au; see Lynch et al., 2014), Pacific Islands Global Ocean Observing System (PI-GOOS) and GOOS Regional Alliance for the South-East Pacific region (GRASP) are expanding physical and biological monitoring of ecosystems across the South Pacific Ocean.

2.1 Primary producers

High inter-annual variability in surface chlorophyll concentrations throughout the South Pacific Ocean tends to be associated with the eastern boundary upwelling of the Humboldt Current system, restricted regions east of New Zealand, around islands and in coastal margins where variability in local dynamics is high (Dandonneau et al., 2004). Although information on the assemblages of plankton is available for most coastal and shelf regions across the South Pacific Ocean, data on seasonal and inter-annual variability or longer-term trends are sparse.

In the western equatorial region, during the northwest monsoon, an area of upwelling develops along the coast of Papua New Guinea (Ueki et al., 2003), bringing nutrient-rich waters to the surface and resulting in increased concentrations of surface chlorophyll which are evident in satellite imagery (Messié and Radenac, 2006). This pool of nutrient- and chlorophyll-rich waters advects eastward during westerly wind events with concentrations of phytoplankton rapidly declining as the oligotrophic waters of the WWP are reached. This decline in concentration is thought to be associated with low nitrate concentrations in surface waters resulting from a stratified salinity layer at the base of the WWP that creates a barrier layer to nutrients (Messié and Radenac, 2006). Shifts in the nitrocline depth, which allow mixing of surface waters with deep nutrient-rich waters associated with eastward expansion of the WWP during La Niña events, contribute to positive primary production anomalies observed in the western equatorial Pacific Ocean (Radenac et al., 2001; Turk et al., 2001; Messié et al., 2006). This is further enhanced by changes in NGCU circulation which enhances iron transport from the shelf and upper slope of Papua New Guinea to the EUC (Ryan et al., 2006). Diatoms in the region have been observed to increase their concentration fourfold as a result of this increased nutrient input (Rousseaux and Gregg, 2012).

In situ sampling across the shelf region of north-eastern Papua New Guinea has recorded a phytoplankton community during the austral summer dominated by nanoeucaryotes and *Prochlorococcus* (Everitt et al., 1990; Higgins et al., 2006). Variability in phytoplankton community assemblages has been observed to be high with a gradient in concentrations from the coast (high) to waters further offshore (low). High nitrogen-fixation rates have also been observed in coastal regions of north-east Papua New Guinea, associated with nanoplankton cyanobacteria and *Trichodesmium* spp. (Bonnet et al., 2009). Across the western

Figure 36D.3 | A coccolithophorid bloom in the coastal waters of north-east Tasmania, Australia. Photo taken in October 2004. Image courtesy of CSIRO, Australia.

equatorial region, nitrogen fixation is dominated by fractions of less than 10 mm associated with unicellular photosynthetic diazotrophs.

Further to the south in the western Pacific Ocean (south of ~23°S) and in the region of New Caledonia, seasonal enrichment of surface chlorophyll concentrations during the austral winter months has been observed and is associated with surface cooling and vertical mixing (Dandonneau and Gohin, 1984; Menkes et al., 2015). Assemblages are dominated by the cyanobacteria *Prochlorococcus*, with lower concentrations of *Synechococcus* and picoeukaryotes (Dandonneau et al., 2004; Menkes et al., 2015). Around the New Caledonian and Vanuatu archipelagos, blooms of *Trichodesmium* spp. are often reported during the austral summer months and have been associated with increased nutrient inputs from islands as a result of seasonal patterns in rainfall (Rodier and Le Borgne, 2008; Le Borgne et al., 2011), although direct linkages between *Trichodesmium* blooms and seasonal land-based nutrient input are not clear (Peter Thompson, CSIRO, pers. comm., 21 August 2014). In contrast to the productivity observed around high islands, such as Papua New Guinea, New Caledonia and Vanuatu, productivity around low islands and coral atolls is rarely enhanced. This is because in general these islands and atolls release very few sediments and nutrients into coastal and shelf regions (Le Borgne et al., 2011).

Overall, a distinct latitudinal gradient in phytoplankton is observed in eastern Australian coastal and shelf waters: higher concentrations of the picoplankton *Prochlorococcus* and *Synechococcus* are found in the north, which gradually decline to the south (Thompson et al., 2011). Tropical shelf waters are typified by phytoplankton communities similar to those observed in the oligotrophic waters of the northern Coral Sea and those around New Caledonia and southern parts of the Coral Sea.

The Great Barrier Reef lagoon supports a high diversity of nanoplankton and picoplankton species, which demonstrate a seasonal progression in community structure and biomass across the austral summer months (chlorophyll *a* concentrations have been observed to increase by up to 50 per cent). In outer regions of the lagoon this is associated with intrusion of nutrient-enriched Coral Sea water (Furnas and Mitchell, 1986; Brodie et al., 2007) and in inner regions associated with sediment-laden river plumes (Revelante et al., 1982). Surface chlorophyll concentrations are frequently, although not always, higher in lagoon regions of the Great Barrier Reef than in adjacent shelf regions (Furnas et al., 1990). Assessments of surface chlorophyll *a* concentrations throughout the Great Barrier Reef lagoon suggest that relatively short (5 – 8 years) time-series may provide spurious estimates of longer-term trends, given the high variability in multi-year patterns (Brodie et al., 2007).

In general, phytoplankton assemblages in the EAC are diatom-dominated in inshore regions; flagellates dominate offshore regions (Young et al., 2011). Assemblages associated with the mesoscale features of the EAC are highly variable with distinct spatial separation of phytoplankton species observed across individual eddy systems (Jeffrey and Hallegraeff, 1980; Jeffrey and Hallegraeff, 1987). Further south, across the temperate neritic province, episodic phytoplankton blooms driven by seasonal intrusions of nitrate-rich water into the euphotic zone occur (Hallegraeff and Jeffrey, 1993; Bax et al., 2001). These seasonal blooms can include diatoms such as *Pseudonitzschia*, which is responsible for amnesic shellfish poisoning (Hallegraeff, 1994), *Thalassiosira partheneia* (Bax et al., 2001), and also coccolithophorids (e.g., Figure 36D.3).

Waters around the north of New Zealand typically demonstrate similar seasonal patterns in chlorophyll concentrations to those observed in the broader Tasman Sea: higher concentrations in the austral spring and autumn and lowest in the winter. Phytoplankton maximas in north-eastern New Zealand shelf regions have been associated with diatom blooms with community succession to dinoflagellates, nanoflagellates and picophytoplankton as blooms decline (Chang et al., 2003). Along the west coast, diatoms are most abundant close to shore, phytoflagellates most abundant seaward of shore areas and dinoflagellates most abundant in areas further offshore (Chang, 1983). In the High-Nitrate-Low-Chlorophyll (HNLC) subantarctic waters southeast of New Zealand, episodic elevated chlorophyll events have been observed (Boyd et al., 2004) with phytoplankton assemblages dominated by cyanobacteria (Bradford-Grieve et al., 1997) and diatoms (Boyd et al., 1999).

Declines in austral spring bloom biomass and growth rates of chlorophyll *a* along the south-eastern Australian region have been suggested to be associated with a long-term decrease in dissolved silicate concentrations. This decrease is thought to be driven by increased intensity in the EAC (Thompson et al., 2009) associated with decadal climate variability (see Section 3). Range expansion of some species has also been reported (Hallegraeff, 2010). The drivers of these expansions have not been established and may be associated with eutrophication, ballast water translocation or oceanographic changes associated with a changing climate (see also Section 3).

Further east in the waters of the PCC, marine ecosystem dynamics are driven by intra-seasonal, annual and inter-annual changes in the upwelling systems that typify the region (Alheit and Niquen, 2004; Montecino and Lange, 2009). Productivity is highest in inshore areas of high upwelling. Coastal upwelling regions off Peru are mainly composed of early successional stages of small diatoms (5-30 nm) with high reproduction rates, whereas in later successional stages, they are characterized by larger species (Tarazona et al., 2003). Small phytoplankton, including nano- and picoplankton, have been reported to account for a large proportion (> 60 per cent) of the primary production and chlorophyll a concentration in the coastal waters between 18oS and 30oS. (Escribano et al., 2003).

South of ~30°S, where upwelling is seasonal, coastal surface chlorophyll a concentrations demonstrate a maximum in the austral summer. Further offshore, concentrations are out of phase with upwelling events, demonstrating a winter maximum potentially associated with offshore advection of productivity away from coastal regions by eddy systems (Morales et al., 2007). Assemblages are diatom- and silico-flagellate-dominated; similar assemblages are observed offshore associated with coastally derived filaments and eddy systems. Assemblages demonstrate little variability throughout the year. The northern Coquimbo upwelling system, despite being an area of persistent upwelling, demonstrates lower production than the more southern Antofagasta and Concepción Shelf areas where production values have been reported to be some of the highest in the ocean (Daneri et al., 2000). Abundances vary considerably, both temporally and spatially, in response to high variability in upwelling systems and associated coastally derived filaments and eddy systems (Daneri et al., 2000; Morales et al., 2007). Such variability makes describing trends in productivity associated with climate variability (such as ENSO) difficult over shorter time scales (Daneri et al., 2000).

2.2 Zooplankton communities

Within coastal mangrove and seagrass systems in the tropics and subtropics, the composition of zooplankton communities is principally controlled by diel changes in tidal flows and seasonal changes in salinities, influenced by the seasonal outflow of freshwater from estuaries or seasonal changes in rainfall in coastal lagoon systems (Grindley, 1984; Duggan et al., 2008). Estimates of the biomass of zooplankton in inshore and estuarine systems tend to be much higher than adjacent coastal regions – tropical Australian mangrove and seagrass habitats have been recorded as having a zooplankton biomass an order of magnitude higher than adjacent embayments (Robertson et al., 1988). Similar spatial gradients in abundance have been recorded elsewhere in the subtropical western Pacific Ocean (Kluge, 1992; Champalbert, 1993; Le Borgne et al., 1997; Carassou et al., 2010).

Within coral reef systems, abundances of zooplankton can vary in relation to the structure of reef/lagoon systems. Enclosed and semi-enclosed atolls (e.g., the Tuamoto archipelago in French Polynesia) have higher densities of zooplankton species than lagoon systems that are more open and have regular exchange of waters with adjacent open oceans (e.g., Uvea lagoon in New Caledonia and the Great Barrier Reef lagoon in Australia; Le Borgne et al., 1997; Niquil et al., 1998).

Composition of zooplankton in tropical and subtropical coastal regions can vary on scales of metres; assemblages in mangrove forests, mangrove streams, coral mounds and sandy floor areas vary distinctly (Jacoby and Greenwood, 1988; Robertson et al., 1988). Mangrove systems have been observed to contain higher amounts of meroplankton species than seagrass habitats, which are more similar to habitats further offshore in bay and lagoon areas (Robertson et al., 1988). Reef systems have been observed to comprise a mix of resident, swarming and demersal species, most of which associate with benthic communities during the day, ascending into the water column during the night. The resident demersal component of zooplankton communities on reef systems can be 5-20 times more abundant than pelagic communities (Alldredge and King, 1977; Roman et al., 1990). In the western Pacific Ocean, cyclopoid copepods, such as *Oithona* spp., and calanoid copepods, such as *Parvocalanus* spp., have been recorded as being numerically the most abundant taxa across all habitats (Robertson and Howard, 1978; Saisho, 1985; Jacoby and Greenwood, 1988; Robertson et al., 1988; Roman et al., 1990; Kluge, 1992; Robertson and Blaber, 1993; Le Borgne et al., 1997; McKinnon and Klumpp, 1998; Duggan et al., 2008). Abundances of zooplankton in the tropical coastal waters of northern Peru are some of the highest recorded in the eastern Pacific Ocean and are associated with high primary production resulting from coastal upwelling across the region (Fernández-Álamo and Färber-Lorda, 2006).

Zooplankton communities in tropical and subtropical habitats demonstrate high variability across tidal to seasonal time scales (Jacoby and Greenwood, 1988; Hamner et al., 2007). The biomass of most communities tends to be highest in the austral summer in the western Pacific Ocean and is associated with peaks in primary productivity (see Section 2.1) and seasonal spawning of invertebrate and vertebrate plankton predators. Not all species demonstrate peaks in abundance during the austral summer; some species demonstrate a peak in the austral winter (Jacoby and Greenwood, 1988). In the eastern Pacific Ocean, increased zooplankton in coastal upwelling regions occurs after winter wind-driven production with peaks in October–December (Fernández-Álamo and Färber-Lorda, 2006).

Inter-annually, zooplankton community structure and abundances in tropical and sub-tropical habitats vary in association with rainfall, associated terrestrial run-off and turbulence, water temperatures, adjacent offshore upwelling in the western Pacific Ocean (Robertson et al., 1988; McKinnon and Thorrold, 1993) and inter-annual variability in upwelling in the eastern Pacific Ocean associated with ENSO. Although clear patterns associated with ENSO cycles are not yet established (Alamo and Bouchon, 1987; Fernández-Álamo and Färber-Lorda, 2006), zooplankton abundances have been observed to be higher during "cold" decades, particularly in the 1960s, than more recent "warm" decades and have been associated with regime shifts in fish communities in the region (see Section 2.4; Figure 36D.4; Ayón et al., 2004). Similar shifts in the size distribution of zooplankton have also been observed in the eastern Pa-

Figure 36D.4 | Spatial and temporal variability in zooplankton biomass in the tropical and subtropical coastal waters of the eastern South Pacific Ocean. Reproduced from Ayón et al., 2004.

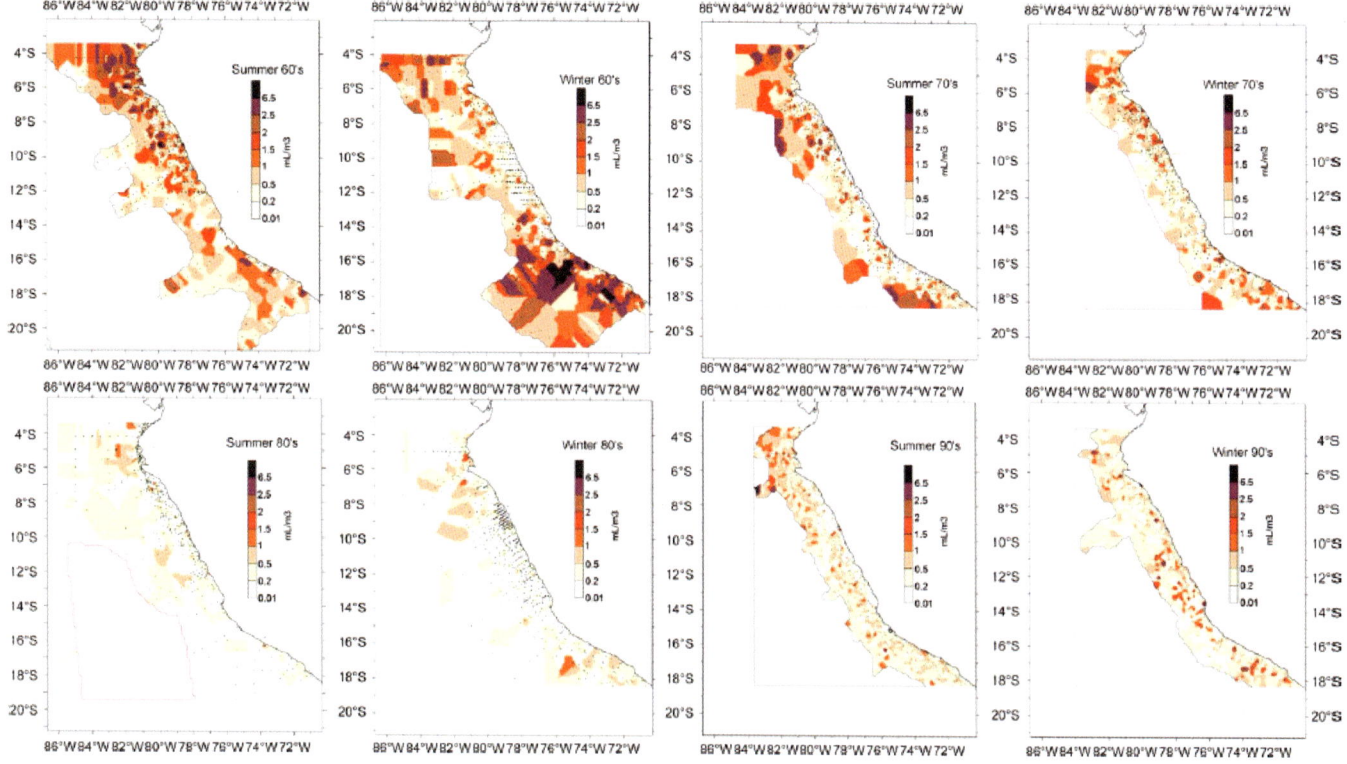

cific Ocean with smaller zooplankton dominating during warmer, lower upwelling conditions (Ayón et al., 2011).

Zooplankton communities in temperate waters of the South Pacific Ocean, similarly to those in the tropics and sub-tropics, are dominated by copepod species (Tranter, 1962; Bradford, 1972; Escribano et al., 2007). Swarming gelatinous species, such as salps, can also dominate; increased abundance of salps is associated with declines in copepod abundances (Tranter, 1962). In the western South Pacific Ocean, assemblages and abundances are spatially and temporally highly variable, reflecting high variability in the physical features and primary productivity of the EAC. Eddy systems associated with the EAC can contain distinct abundances and assemblages of taxa in comparison to adjacent waters and other eddies, which evolve, becoming less distinct as the eddy ages (Griffiths and Brandt, 1983; McWilliam and Phillips, 1983; Tranter et al., 1983; Young, 1989). Further south, the shelf waters off south-eastern Australia are dominated by the euphausid *Nyctiphanes australis;* peaks in abundance occur in the austral autumn months in association with seasonal upwelling onto shelf regions and resulting peaks in primary production (Young et al., 1996). The temperate zooplankton communities of east New Zealand waters are also typified by *N. australis*, gelatinous zooplankton and copepod species (Jillet, 1971; Jillet, 1976; Bradford, 1972). Peaks in abundance vary between species, but most species have been observed to demonstrate peaks in abundance in the late austral winter/early spring (Bradford, 1972). Shifts in the composition of zooplankton communities from predominantly cold water species to more warm water species have been reported from the shelf waters off south-eastern Australia and have been associated with large shifts in regional oceanography (Johnson et al., 2011).

In the eastern South Pacific Ocean, temperate zooplankton communities are dominated by copepods, euphausids and gelatinous zooplankton (Escribano et al., 2007). Community composition in coastal and shelf regions off Peru tends to correspond with large oceanographic features throughout the region; coastal upwelled waters are comprised of copepod species and meroplanktonic larvae. Waters further offshore are largely comprised of large holopankters, such as euphausids, copepods, siphonophores and chaetognaths (Tarazona et al., 2003; Fernández-Álamo and Färber-Lorda, 2006; Ayón et al., 2011). In the coastal regions of central Chile, where assemblages are influenced by subantarctic waters, copepods dominate offshore shelf regions, whereas euphausids dominate the fjord region (Escribano et al., 2003). Abundances of zooplankton in the region are both strongly and positively associated with the vertical distribution of the oxygen minima zone. Abundances of gelatinous zooplankton have been observed to peak in the austral winter and spring: copepods demonstrate a peak in the austral spring and summer in association with seasonal upwelling conditions. Euphausids, dominated by the endemic *Euphausia mucronata*, demonstrate little seasonal variability in abundances (Gonzalez and Marin, 1998; Escribano et al., 2007). Little inter-annual variability in species assemblages and abundances has been observed in the region, although shifts in species assemblages of zooplankton in coastal Chilean waters have been associated with ENSO phase. Species alternate between copepods and euphausids during upwelling events associated with La Niña and cyclopoid copepods during warmer events associated with El Niño (Hidalgo and Escribano, 2001).

2.3 Benthic communities

Benthic communities across the South Pacific Ocean occupy a diverse range of habitats, including estuaries, mangroves, rocky intertidals, seagrass beds, kelp forests, soft bottoms (ranging from sandy to muddy), coral reefs and rocky reefs, and form one of the richest assemblages of species in the marine environment (Snelgrove, 1999). Many species are subject to recreational, artisanal and commercial fisheries. These include bivalves (e.g., giant clams, scallops) echinoderms (e.g., sea cucumbers, sea urchins) and gastropods (e.g., *Trochus*, abalone). Benthic fish and macro-invertebrates are discussed in Section 2.4.

Given the diversity of benthic community habitats across the South Pacific Ocean, few have been sampled comprehensively. Most assessments have focused on classification and documentation of the sediments or plants forming the basis of a habitat type or on individual species, rather than on benthic community assemblages (e.g., Kennelly, 1987; Heap et al., 2005; Fisher et al, 2011; Waycott et al, 2009). Because of a lack of assessments, surveys often discover previously undescribed species that may only be documented once in a survey (Snelgrove, 1999; Williams et al., 2010). Establishing trends in benthic communities in most regions is at present difficult, in large part due to the sparse data available across all habitat types and the high variability in faunal assemblages associated with each habitat (e.g., Waycott et al., 2005).

Coral reef communities are one of the better documented benthic communities throughout the South Pacific Ocean. Although 75 per cent of the world's coral reefs are found in the Indo-Pacific region, few long-term trends have been documented for these communities (Bruno and Selig, 2007). Densities of species in coral reef communities exhibit considerable variability spatially across multiple scales and reflect larval dispersion, recruitment success, competition for substrate and local environmental conditions. Community structure within reef communities tends to follow a nutrient gradient and is also associated with proximity to inner lagoon or outer open-ocean regions.

Coral reef assemblages attenuate gradually in diversity with increasing distance from the Coral Triangle; 60 genera of reef-building (hermatypic) corals have been reported at 9°S in comparison to 18 genera at 30°S (Wells, 1955; Veron, et al. 1974; Bellwood and Hughes, 2001). Eastern South Pacific coral reefs are low in diversity; however, most species within these communities are unique (Glynn and Ault, 2000). In areas where water temperatures seasonally drop to below 18°C, hermatypic corals become sparse: coral reef communities give way to rocky reef and soft substrate communities (Veron, 1974). Factors suggested to be associated with attenuation of coral reef communities include water temperature, aragonite saturation, light availability, larval dispersal and recruitment, and competition with other species, including macroalgae (Harriott and Banks, 2002).

Overall, coral communities across the South Pacific Ocean are considered to be generally healthy (Burke et al., 2011), although coral cover is gradually declining (Bruno and Selig, 2007). Exposure and resilience to disturbance vary depending on reef type and location; major drivers of changes in species densities, assemblages and spatial distributions are associated with natural phenomena such as storms and cyclones, outbreaks of natural predators (e.g., crown-of-thorns starfish) and seasonal local coral bleaching events (Chin et al., 2011; Hoegh-Guldberg et al., 2011). Coral reef communities in the eastern South Pacific Ocean recorded the earliest mass bleaching and mortality of any region; widespread losses are linked to high ocean temperature anomalies that occur during El Niño events (Glynn and Ault, 2000; Burke et al., 2011).

Figure 36D.5 | Estimates of hard coral cover across the Great Barrier Reef 1986 – 2012, based on data collected from 214 reefs. Dashed lines represent the standard error. Taken from Great Barrier Reef Marine Park Authority (2014) (modified from De'ath et al. (2012).

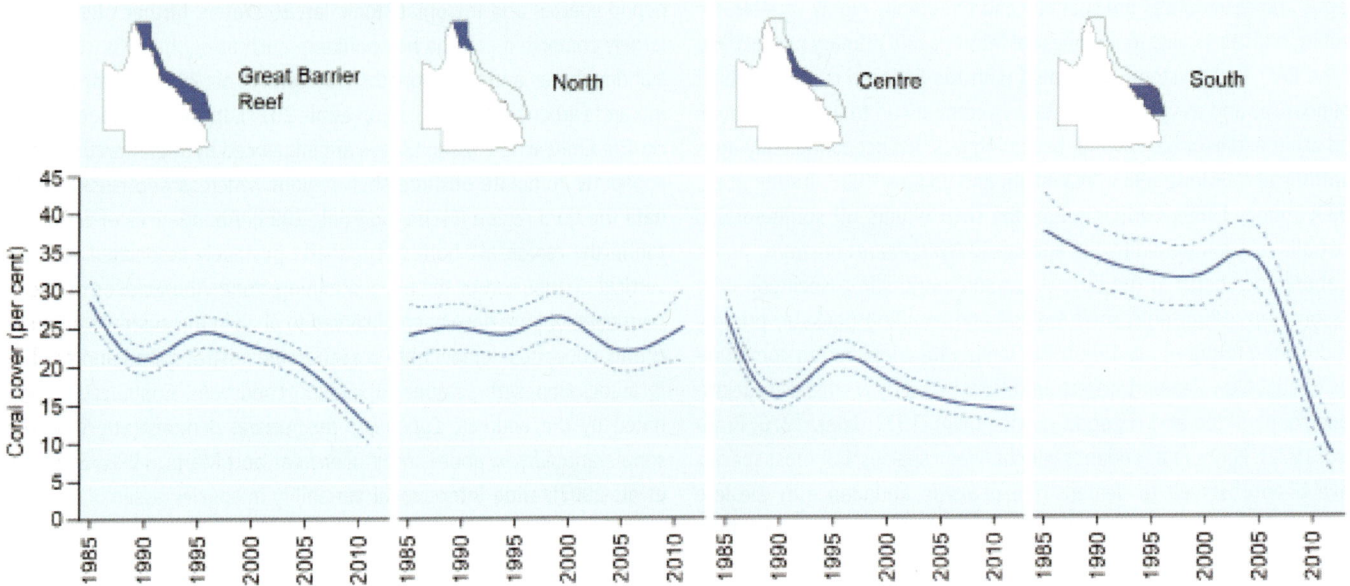

Coral cover from reefs in the Indo-Pacific region has been declining relatively uniformly since records began in the 1960s and 1970s; yearly estimates of loss average 1 – 2 per cent over the last 20 years (Bruno and Selig, 2007). Hard coral cover on the Great Barrier Reef has declined from 28.0 per cent in 1985 to 13.8 per cent in 2012; the rate of decline is increasing in recent years (De'ath et al., 2012; Figure 36D.5). Declines have been most severe on reefs south of 20°S, particularly since 2006, and on areas of inshore reefs which have been documented to have declined by 34 per cent since 2005 (GBRMPA, 2014). Tropical cyclones, coral predation by the crown-of-thorns starfish and coral bleaching have accounted for 48 per cent, 42 per cent and 10 per cent, respectively, of declines observed; elevated loads of nutrients, sediments and pollutants via terrestrial run-off affect reef resilience and the potential for recovery (De'ath et al., 2012; Wiedenmann et al., 2013).

Local declines in species densities, assemblages and spatial distributions are increasingly being observed, particularly in areas close to population centres where over-fishing, pollution from terrestrial run-off and sewage, and damage from coastal developments are occurring (see Section 3). Bleaching events are becoming more widespread, increasing in intensity and frequency as surface waters of the South Pacific Ocean warm with long-term climate change (Burke et al., 2011; Chin et al., 2011; Hoegh-Guldberg et al., 2011). Recovery from bleaching events is possible if local factors which affect reef systems, such as coastal run-off and overfishing, are minimized (Figure 36D.6); however, as marine conditions alter with climate change, the ability of coral reefs to recover from bleaching events is expected to decline (see section 3.1; De'ath et al., 2012).

Figure 36D.6 | Variation in coral, turf and macroalgal cover at the Tiahura outer reef sector, French Polynesia, 1991 – 2006, in relation to coral bleaching and cyclone events. Dotted lines indicate the standard deviation. Taken from Adjeroud et al., 2009; with permission of Springer.

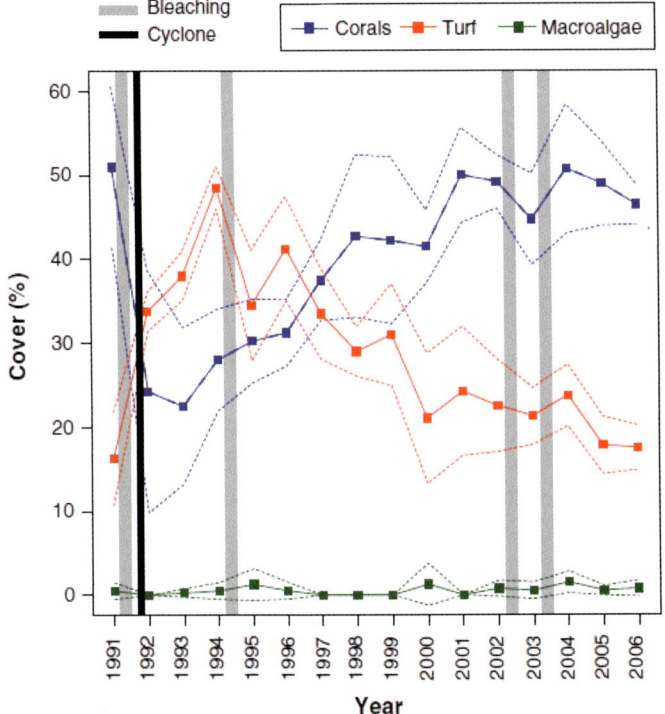

Monitoring of approximately 30 intertidal seagrass meadows along the central and southern coast of the Great Barrier Reef suggests that their overall abundance, reproductive effort and nutrient status have all declined. Although shallow subtidal seagrass meadows are less extensively monitored, many sites also show signs of declines in abundance. Causes for these declines include cyclones and poor water quality (GBRMPA, 2014).

Across temperate regions, benthic communities of rocky intertidal and rocky reef habitats are reasonably well documented, although studies are often local and only include assessments of a few sites. In the western South Pacific Ocean, subtidal coastlines are typified by patches of common kelp, ascidians and crustose coralline algae barrens, which are often associated with the sea urchin *Centrostephanus rodgersii* (Connell and Irving, 2008). Other herbivores common in these habitats include limpets and topshells (Underwood et al., 1991). Community assemblages vary with depth; sponges, ascidians and red algae are more abundant in deeper sheltered areas (Underwood et al., 1991). In protected areas where removal of urchin predators is restricted and predator populations are provided with the opportunity to increase, sea urchin barrens demonstrate a decline and macroalgal forests are more extensive (Babcock et al., 1999; Shears and Babcock, 2003; Barrett et al., 2009). This suggests that removal of urchin predators (e.g., via fishing) can have widespread effects on community structure in subtidal habitats (Connell and Irving, 2008). Declines in the density of giant kelp beds and increased densities of sea urchins and urchin barrens have also been associated with increased southward larval advection of urchin larvae following shifts in large oceanographic features throughout the southeastern Australian region (Johnson et al., 2011). Commercial and recreational fishing practices have been associated with declines in benthic communities. Oyster reef communities in south-eastern Australia have been largely destroyed by fishing and mining practices: over 60 per cent are considered functionally extinct, and of the remaining, 90 per cent of the original area of the reef is lost (Beck et al., 2011). The flow-on impacts of such reductions include reduced habitat and food sources for other species and reduced water-filtering capacity, resulting in a reduction in overall water quality (Beck et al., 2011).

Estuarine and coastal soft-sediment benthic communities have been routinely monitored, mostly by councils and unitary authorities, at over 70 sites in New Zealand for up to 25 years (see http://geodata.govt.nz/geonetwork_memp). Time series established demonstrate a wide variety of trends and cycles, and it is difficult to draw overall conclusions. For example, Mahurangi Harbour, which has a predominantly rural catchment, has demonstrated a declining trend in species sensitive to increased sediment loading (Halliday et al., 2013), whereas Manukau Harbour, adjacent to New Zealand's largest city, has demonstrated no declining trends that might indicate that the habitat is becoming degraded (Greenfield et al., 2013). New Zealand is developing a Marine Environmental Monitoring Programme to coordinate and report on the diverse monitoring being conducted.

In temperate regions of the eastern South Pacific Ocean, rocky intertidal communities are dominated by mussel beds, corticated algae and herbivorous gastropods in shallow regions and kelps, crustose algae, chitons and fissurellids in deeper regions (Broitman et al., 2001). Carnivorous gastropods and crustaceans also dominate communities. Latitudinal gradients occur in some species; mussel and crustose algae densities decrease with decreasing latitudes, whereas ephemeral algae increase with decreasing latitudes (Broitman et al., 2001). Further south, recent inventory studies conducted in the relatively easily accessible areas of the Chilean fjord region have documented that over 30 per cent of the sampled specimens represent new species; 10 per cent of these are in benthic communities (Häussermann and Försterra, 2009). Many of the species documented exhibited unexpected distribution patterns: for example, species thought to be limited to the Peruvian faunal province were discovered far to the south; others, presumed to be subantarctic, were present in northern Chilean fjords, and deep water species were found in shallow depths of the fjords (e.g., Försterra et al., 2005). The species richness of several benthic taxa has been observed to decrease with increasing latitude and then increase again south of 40-45°S. This is possibly associated either with an increase in the presence of Antarctic fauna or an increase in habitat with the broadening of the continental shelf south of 40°S (Escribano et al., 2003).

Temporal changes in the density and assemblage of benthic communities have been observed at two long-term sites in the Chilean fjord region. Within the Comau Fjord, mussel banks, meadows of sea whips, sea anemones, large calcified bryozoans and the rare gorgonian *Swiftia* n. sp. have reduced by at least 50 per cent and in some cases, completely disappeared. Associated decapod species have also experienced declines in abundance (Häussermann et al., 2013). Declines in biodiversity in this fjord are thought to be associated with increased mussel harvesting in the area, eutrophication, increased organic sedimentation and the extensive use of chemicals in aquaculture operations (Häussermann et al., 2013; Forsterra et al., 2014; Mayr et al., 2014).

Further detail on estuaries, mangroves, kelp forests, seagrass and corals can be found respectively, in chapters 44, 48, 47 and 42-43.

2.4 Fish and macro-invertebrates

Fish and macro-invertebrates occurring in coastal and shelf regions of the South Pacific Ocean range from highly resident species (e.g., cardinal fishes, Apogonidae; wrasses, Labridae), species that move relatively small distances, but utilize multiple habitats during their lifespan (e.g., penaeid prawns; yellowfin bream, *Acanthopagrus australis*), pelagic species that roam shelf waters extensively (e.g., Australian salmon or kahawai *Arripis* spp., white sharks, *Carcharodon carcharias*), to highly migratory pelagic species that utilize shelf regions periodically or seasonally (bigeye tuna, *Thunnus obesus*; southern bluefin tuna, *T. maccoyii*). A few species are anadromous (e.g., shorthead and pouch lampreys, Mordaciidae) and some are catadromous (e.g., barramundi, *Lates calcarifer*; short-finned eel, *Anguilla australis*).

Time series of indicators of populations are largely limited to species that are the focus of recreational/sport, subsistence and commercial fisheries and are subject to varying degrees of management (Bates et al., 2014). Subsistence fishing tends to be more important in rural areas throughout the South Pacific Ocean and is much larger than commercial fishing in these areas (Dalzell et al., 1996; Kulbicki et al., 1997). In the Pacific Islands region, where fish consumption in some countries is at least twice the level needed to supply 50 per cent of the recommended protein requirements, 60 – 90 per cent of fish consumed is caught by subsistence fishers (Bell et al., 2009). Across much of the rest of the South Pacific Ocean, and in the shelf areas of the Pacific Island region, commercial fishing is much larger than subsistence or recreational fishing.

Small-scale subsistence and commercial reef fisheries across the Pacific Island region and coastal commercial fisheries in Ecuador and northern Peru can catch up to 200 – 300 species (Dalzell et al., 1996; Heileman et al., 2009; FAO, 2010). The primary fish families important to Pacific Island communities are the Acanthuridae (surgeonfish), Scaridae (parrot fish), Lutjanidae (snapper), Lethrinidae (emperor fish) and Balistidae (triggerfish; Dalzell et al., 1996; Pratchett et al., 2011). A number of fish species are also commercially fished for the live reef food and the aquarium trade. Macroinvertebrates of importance include sea cucumbers (22 species are commercially caught across the region for the production of bêche-de-mer), trochus (*Trochus* spp.), green snail (*Turbo marmoratus*), giant clams (*Tridacna* spp.), cephalopods (primarily cuttlefish and octopus) and crustaceans (penaeid prawns, crabs, lobsters). Commercial pearl oyster aquaculture operations occur in the eastern (Melanesian) part of the Pacific Islands region; two-thirds of all production occurs in French Polynesia (Dalzell et al., 1996). Species of importance in the eastern South Pacific include silverside (*Odontesthes regia*), flathead grey mullet (*Mugil cephalus*), lorna drum (*Sciaena deliciosa*), Peruvian scallop (*Argopecten purpuratus*), and the cephalopods *Loligo gahi* and *Dosidicus gigas* (FAO, 2010).

Reef fish abundances are influenced by the extent and condition of coral cover, and can vary two- to ten-fold over time, largely in association with loss and subsequent recovery of coral reef habitat following storm events (Halford et al., 2004; Kulbicki et al., 2007; Wilson et al., 2008; Brewer et al., 2009). Macro-carnivores, micro-herbivores and plankton feeders show some of the highest variability (Kulbicki, 1997).

A large-scale assessment of coral reef fish and invertebrate communities in 17 Pacific Island countries and territories found that across 63 sites, less than one-third of the sites had resources that were in good condition; most were in average/low or poor condition (Pinca et al., 2009). Herbivores and smaller fish were more abundant in reefs of below average condition, whereas reefs in good condition had higher biomasses of carnivores and greater numbers of larger fish (Pinca et al., 2009). More recently an assessment of the status of reef fish assemblages on fished reefs estimated that reef fish assemblages around Papua New Guinea were at a point indicating fisheries collapse (Mac Neil et al., 2015). Declines have also been observed in sea cucumber species, giant clam and *Trochus* spp. Sea cucumber fisheries in Palau, Papua New Guinea, Samoa, Solomon Islands, Tonga and Vanuatu have been closed due to overfishing (Purcell et al., 2013). Some species of giant clam have been declared extinct in a number of countries and all giant clam species are now listed under Appendix II of the Convention of International Trade in Endangered Species1 (CITES), which covers species that may become threatened if their trade is not effectively regulated (Pinca et al., 2009). Formal monitoring and regulation of coastal fishery resources is largely lacking throughout the region, particularly for subsistence fisheries resources (Dalzell et al., 1996; Gillett, 2010). Landings that are reported are considered to be underestimates; reconstructed catches from coral fisheries in American Samoa have been estimated to be 17-fold greater than those officially reported (Zeller et al., 2006).

Large components of tropical reef systems along eastern Australia are managed as part of the multiple use Great Barrier Reef Marine Park, some of which is open to commercial, recreational and traditional fishing for a range of fish (for food and the live aquarium trade), crustaceans (penaeid prawns, crabs, lobsters), sea cucumbers, trochus, beach or sand worms and live coral. Monitoring of 214 fish populations in the Great Barrier Reef Marine Park since the 1990s indicate high inter-annual variability (Figure 36D.7); population declines are associated with a declining coral cover, particularly in southern parts of the park, where coastal development is greatest (GBRMPA, 2014). Overall abundance and size of commercially caught species have declined when compared to historical abundance and size; the fishery for black teatfish sea cucumber (*Holothuria whitmaei*) was closed in 1999 (GBRMPA, 2014). Illegal fishing is known to occur in areas closed to fishing and is likely to have contributed to overall declines in fish populations. Recreational fishing catches throughout the marine park are largely unmonitored; however, in 2010 an estimated 700,000 recreational fishers throughout Queensland caught over 13 million fish, approximately half of which were returned to the water (Taylor et al., 2012). Recreational catches appear to have been declining over the last decade.

Temperate fish and macro-invertebrates are also affected by the physical and biological attributes of temperate habitats (De Martini and Roberts, 1990; Curley et al., 2003; Anderson and Millar, 2004). Spatial variability in assemblages has been associated with type of reef habitat (e.g., urchin barren, kelp forest, sponge habitat), benthic topography and depth (Williams and Bax, 2001; Curley et al., 2003; Hill et al., 2014). Long-term changes in temperate fish assemblages have been observed in eastern Australia, associated with fishing, introduced alien species and ongoing changes to the marine environment as a result of climate change and coastal development (Last et al., 2011; State of the Environment Committee, 2011; Bates et al., 2014).

Coastal waters of the tropical eastern Pacific are some of the least explored in the region (Cruz et al., 2003; Zapata and Robertson, 2007); approximately 70 per cent of fish are endemic to the region. The unique oceanographic conditions and heterogeneity of the coastal regions of Chile have resulted in high levels of endemism in many invertebrate groups (Griffiths et al., 2009; Miloslavich et al., 2011). Endemism is also high in the waters of small oceanic islands in the eastern South Pacific Ocean; approximately 77 per cent of the fish at Easter Island, 73 per cent at Salas y Gómez, 72 per cent at Desventuradas and 99 per cent at the Juan Fernández Archipelago are endemic (National Geographic/Oceana/Armada de Chile, 2011, Friedlander et al., 2013; National Geographic/Oceana, 2013). Most of the oceanic islands of the eastern South Pacific are thought to have relatively healthy biomasses of fish and macro-invertebrates, with the exception of Easter Island, where fisheries have been operating for over 800 years (Hunt and Lipo 2011). Within the last three decades, a dramatic decrease in the marine resources of Easter Island has been observed; this is largely associated with overexploitation, increasing tourist numbers with associated increases in demand for resources, illegal industrial fishing and lack of surveillance and enforcement procedures (Gaymer et al., 2013).

Shallow reef habitats of the Galápagos archipelago are reported to have undergone major transformation as a result of the severe 1982/1983 El

Figure 36D.7 | Time series of the abundance of some coral reef fish species in the Great Barrier Reef Marine Park 1991 – 2003. Taken from Great Barrier Reef Marine Park Authority (2014) and adapted from Australian Institute of Marine Science Long-term Monitoring Program (2008 and 2014).

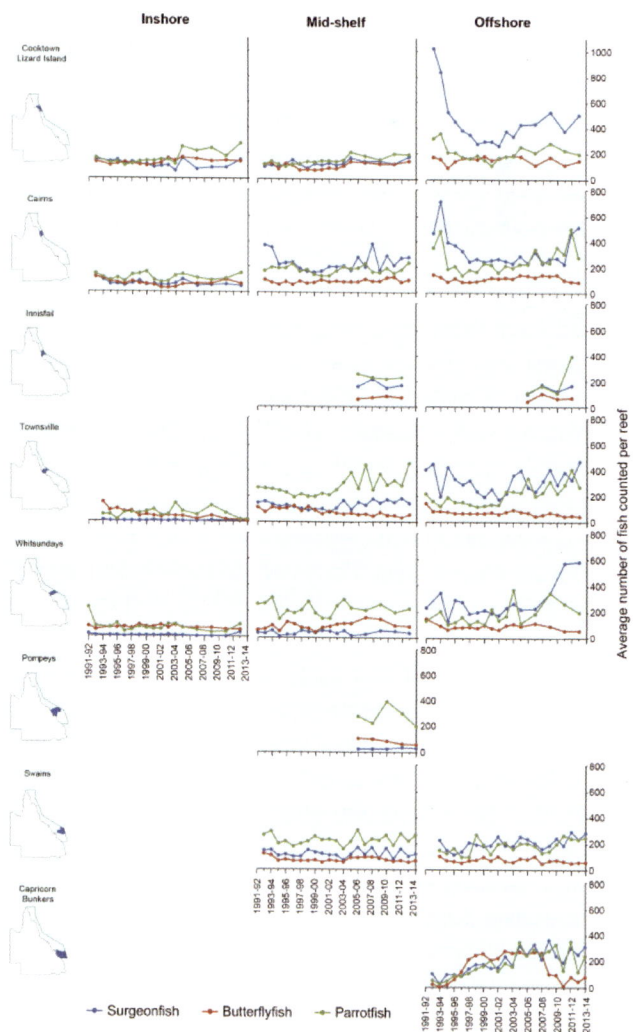

---

1 United Nations, *Treaty Series*, vol. 993, No. 14537.

Niño warming event, resulting in local and regional decline in biodiversity, including a number of identified extinctions (Edgar et al., 2010). Artisanal fishing for lobster and fish species is thought to have magnified the impacts of the El Niño event; the grouper *Mycteroperca olfax* is characterized as functionally extinct in the central Galápagos region (Ruttenberg, 2001; Okey et al., 2004). Commercial fishing within the Galápagos Islands reserve has been largely banned from the area, except for artisanal fishing, which has been allowed in the reserve since 1994. The region has been subject to extensive illegal fishing for sharks, sea cucumber and a range of fish in the region and a lack of controls on or enforcement of management measures for artisanal fishing and a lack of credible assessment of stocks have resulted in over-exploitation (Hearn, 2008; Castrejón et al., 2013).

Coastal regions of Ecuador and northern Peru have been the site of extensive shrimp (penaeid prawn) mariculture operations since the 1960s. The establishment of these operations has been responsible for extensive reduction in mangrove forests and associated fish and invertebrate populations (Bailey, 1988; Primavera, 1997).

A diverse range of fish and macro-invertebrate species are targeted by fisheries in the coastal waters of Chile, including molluscs, gastropods, echinoderms, cephalopods and fish. Fisheries catches in general were low prior to the 1980s, reflecting low levels of effort and predominantly local consumption of catches. During the 1980s, export markets grew and catches increased substantially. Catches of many invertebrate species subsequently declined and have remained at low levels since (Thiel et al., 2007). Many species targeted by coastal fisheries demonstrate fluctuations in abundances linked to El Niño/Southern Oscillation (ENSO) (e.g., Wolff, 1987).

A wide range of fish and macro-invertebrate species are the focus of commercial and recreational fisheries in Australia and New Zealand; most coastal and shelf stocks are considered to be sustainably fished (e.g., MPI, 2013; Andre et al., 2014). For example, of 93 stocks managed at the national level in Australia, 77 (83 per cent) were considered as not subject to overfishing, four were considered to be subject to overfishing and 12 were considered as uncertain (Woodhams et al., 2013). In New Zealand, 99 of 114 (87 per cent) were considered as not subject to overfishing. The proportion of assessed stocks experiencing overfishing declined from 25 per cent in 2009 to 13 per cent in 2013 (MPI, 2013). Inshore stocks are the least assessed in New Zealand waters, particularly inshore fish species. Species of ongoing concern tend to be long-lived, slow-maturing species which have been subject to a number of decades of fishing, such as southern bluefin tuna, orange roughy (*Hoplostethus atlanticus*) and school shark (*Galeorhinus galeous*). Species that are caught in lesser amounts and non-targeted species are not routinely assessed and as a result the status of populations is largely unknown (Woodhams et al., 2013; MPI, 2013). In Australian waters, the grey nurse shark (*Carcharias taurus*) is listed as critically endangered, the white shark (*Carcharodon carcharias*), black rockcod (*Epinephelus daemelii*), and the whale shark (*Rhincodon typus*) are listed as vulnerable, and orange roughy, gemfish (*Rexea solandri*), southern bluefin tuna and several shark species, including school shark, have been listed as conservation-dependent under the Environment Protection and Biodiversity Conservation Act 1999.

Recreational fisheries are subject to variable levels of assessment and monitoring and most are regulated only via size and bag limits. In some areas (e.g., New South Wales, Australia) and for some species recreational fishing licenses are required. In Australia, recreational catches of many highly sought-after species are thought to be larger than commercial catches (State of the Environment Committee, 2011). Catches of snapper (*Chrysophrys auratus*), which comprise New Zealand's largest recreational fishery from the north-east coast of the North Island, were 3,750 t in 2011/12, similar to commercial landings (Hartill et al., 2013). The first reliable and comprehensive assessment of recreational harvest across all stocks in New Zealand was completed in 2013 (Hartill et al., 2013). Populations of commercially exploited fish and macro-invertebrates in the temperate coastal regions of New Zealand are considered mainly to be in good condition (Mace et al., 2014).

Offshore fisheries resources in tropical, sub-tropical and temperate shelf regions of the western and central South Pacific Ocean largely consist of commercial operations targeting tuna and billfish species, as well as a number of other large pelagic species, such as mahi mahi (*Coryphaena hippurus*), rainbow runner (*Elegatis bipinnulata*), wahoo (*Acanthocybium solandri*) and Spanish mackerel (*Scomberomorus commerson*), and small pelagic species, such as flying fish, pilchards and sardines. Several shark species are either directly targeted or caught as by-catch. Across the Pacific Island region, pelagic fish species are estimated to contribute on average up to 28 per cent of coastal fisheries production (range 10 – 70 per cent; Pratchett et al., 2011). This proportion is likely to increase as increasing populations lead to further exploitation of pelagic species by coastal populations (Bell et al., 2009).

All of the tuna and billfish species, that comprise the majority of commercial catches in the Western and Central Pacific Fisheries Commission (WCPFC) area, except bigeye tuna, are not considered to be in an overfished state, although recent catches of skipjack tuna (*Katsuwonus pelamis*) and yellowfin tuna (*T. albacares*) are at levels that marginally exceed maximum sustainable yield and catches of swordfish (*Xiphias gladius*) are at levels that exceed maximum sustainable yield (Hoyle et al., 2012; Davies et al., 2013; Davies et al., 2014; Rice et al., 2014). In the Inter-American Tropical Tuna Commission (IATTC) area in the eastern Pacific Ocean, all of the main tuna and billfish species are not considered to be in an overfished state, although some uncertainty exists as to whether current catches of skipjack tuna are at levels that exceed maximum sustainable yield (IATTC, 2012; IATTC, 2014). Catches of species other than the main tuna and billfish species that comprise the majority of commercial catches are largely unmonitored and so the status of populations and ongoing sustainability of resources is unknown. Recent assessments of shark species caught in substantial numbers throughout the WCPFC and IATTC areas have indicated that fishing mortalities are well above those considered to be sustainable (e.g., Rice and Harley, 2013; IATTC, 2013). Current catches of most small pelagic species throughout the Pacific Island region are considered to be sustainable (Blaber, 1990; Pratchett et al., 2011).

Pelagic fisheries in the eastern South Pacific have been responsible for approximately 10 – 20 per cent of the world landings over the last 50 years (Chavez, 2008; Fréon et al., 2008) and are dominated by the anchoveta fishery, which primarily targets anchovy (*Engraulis ringens*) and sardine (*Sardinops sagax*). Jack mackerel (*Trachurus murphy*), chub mackerel (*Scomber japonicus*) and, in the southern Peru-Chile region, a herring-like sardine (*Strangomera bentincki*) also sustain important pelagic fisheries. Fisheries across the region also target the common hake (*Merluccius gayi*), swordfish, tunas, cephalopods, primarily *D. gigas,* and a number of crustaceans. The Chilean fjord region is an important region for catches of gadiform fishes, such as Patagonian grenadier (*Macruronus magellanicus*) and the southern hake (*Merluccius australis*; FAO, 2011). Variability in the population size and distributions of anchovies and sardines in the waters of the eastern South Pacific Ocean and linkages with environmental variability have been well documented (e.g., Arntz and Tarazona, 1990; Ayón et al., 2004; Bertrand et al., 2004; Cubillos et al., 2007) and are now recognised to be associated with short-term dynamics associated with ENSO and longer-term dynamics associated with decadal climate variability (Alheit and Nuiquen, 2004; Fréon et al., 2008). The two northern anchovy stocks appear to not be overfished, whereas the southern stock is considered to be depleted and estimated to be at around 8 per cent of virgin stock biomass (Fish Source, www.fishsource.com, accessed 17 August 2014). Currently no recovery plan is in place for this fishery. Fisheries for sardine are based on four stocks: northern, central and southern Peru and northern Chile. Stocks between Peru and Chile are assessed independently and management measures are not co-ordinated between the two countries.

For many other species subject to commercial catches, information on important biological parameters (larval ecology, spawning, movements) is lacking and consequently, little is known about the effects of inter-annual variability in oceanographic conditions on their population dynamics (Thiel et al., 2007). Artisanal fishing, which comprises a significant proportion of the total fishing effort in the region, is largely unreported and is unregulated across the region. Most fisheries throughout the region, other than those mentioned above, are considered to be overexploited (De Young, 2007).

Further detail on sharks and other elasmobranchs and tuna and billfish can be found in chapters 40 and 41.

2.5   Other Biota

The South Pacific Ocean contains a diverse assemblage of marine mammals, seabirds and marine reptiles, most of which have been subject to some level of direct and indirect harvesting (SPREP, 2012). The region contains the northernmost populations of penguins and fur seals, both of which breed on the Galápagos Islands, breeding populations of six of the seven species of sea turtles and populations of the only sirenian in the family Dugonidae. Although harvesting of many of these species has been either banned or limited in many countries across the region and protection measures have been put in place for some species in some countries, many species interact with commercial fishing operations (e.g., Table 36D.1) and for some, this is a substantial source of mortality (Waugh et al., 2012; Reeves et al., 2013; Richard and Abraham, 2013; Wallace et al., 2013; Lewison et al., 2014). Populations of species throughout the region demonstrate varying trends as a result.

Overall, global populations of sea turtles are considered to have declined (Wallace et al., 2011) with those that occur in the South Pacific Ocean demonstrating varying trends. Two distinct genetic stocks of green turtles (*Chelonia mydas*) occur within the Great Barrier Reef Marine Park which experience different pressures and demonstrate differing population trajectories. The southern stock has demonstrated a consistent increase in population size, whereas the northern stock may be in the early stage of decline (GRMPA, 2014). The loggerhead turtle (*Caretta caretta*) population is increasing after substantial decline, whereas the hawksbill turtle (*Eretmochelys imbricata*) population has declined and the flatback turtle (*Natator depressus*) population has remained stable (GRMPA, 2014). Little is known of the current population status of leatherback turtles (*Dermochelys coriacea*), which are known to nest in Papua New Guinea, the Solomon Islands and Vanuatu in the western South Pacific Ocean, but regionally they are considered to be in decline (Dutton et al., 2007). In the eastern Pacific Ocean, leatherback, green and hawksbill turtles nest along the coast in Ecuador, with vagrant nesting sites occurring in Peru. Large numbers are caught in small-scale fisheries off the coasts of Ecuador and Peru (Alfaro-Shigueto et al., 2011).

Table 36D.1 | Fisheries interactions with species of conservation concern 2006 – 2012. Reproduced from Great Barrier Reef Marine Park Authority (2014) using data from Queensland Department of Agriculture, Fisheries and Forestry (2013).

SPECIES	COMMERCIAL FISHERY GEAR TYPE			
	Otter trawl	Net	Line	Pot
Green sawfish	•			
Narrow sawfish	•	•		
Leafy seadragon	•			
Unspecified seahorse	•			
Unspecified seasnake	•	•	•	
Estuarine crocodile		•		
Unspecified crocodile		•		•
Flatback turtle	•			
Green turtle	•	•		
Hawksbill turtle	•	•		
Loggerhead turtle	•	•		
Unspecified marine turtle	•	•	•	
Seabird: gannets and boobies	•			
Dugong		•		
Offshore bottlenose dolphin		•	•	
Humpback whale		•	•	
Minke whale			•	

International protection and management of saltwater crocodile populations in Australia and Papua New Guinea after periods of commercial harvesting have resulted in increases in populations to pre-exploitation levels (Thorbjarnarson, 1999; Tisdell and Swarna Nantha, 2005). Few data on the population abundances of sea snake species are available, despite substantial numbers being caught in fishing operations (e.g., Milton 2001; Wassenberg et al., 2001). What data have been collected indicate population declines (Goiran and Shine, 2013).

Most large cetaceans occurring in coastal and shelf regions throughout the South Pacific Ocean are seasonal visitors, spending large periods of time in offshore regions (e.g., Birtles et al., 2002; Hucke-Gaete et al., 2004; Olavarría et al., 2007). Of those species that historically have been subject to widespread commercial harvesting, some populations have been documented to be increasing, while the status of others in the South Pacific Ocean region is still uncertain (e.g., Baker and Clapham, 2004; Branch et al., 2007; Magera et al., 2013). Smaller coastal cetacean species, although largely lacking in population data, are considered to demonstrate varying trends: some are relatively stable and others are decreasing (e.g., Gerrodette and Forcada, 2005; Parra et al., 2006; Currey et al., 2009). Dugong (*Dugong dugon*) populations across the western South Pacific are considered to be declining, although estimates of abundance are lacking for most countries (Marsh et al., 1995; Marsh et al., 1999; Marsh et al., 2002). Fisheries for the species throughout the region are considered to be unsustainable (Marsh et al., 1997; Garrigue et al., 2008). Most pinniped populations, although substantially reduced due to commercial exploitation during the 1800s, are considered in some areas to be increasing at varying rates in temperate regions, whereas in others they may be decreasing (e.g., Kirkwood et al., 2010; Robertson and Chilvers, 2011). Seabirds throughout the region demonstrate varying trends; most species that forage in offshore regions are considered to be decreasing (e.g., Majluf et al. 2002; Baker and Wise, 2005).

Further details on marine mammals, marine reptiles and seabirds can be found in chapters 37-39.

3 Major pressures

3.1 Climate change and oceanographic drivers

Changes to ocean environments occurring as a result of long-term changes to the global climate are likely to be highly variable across the South Pacific Ocean. Ocean temperatures have risen across most of the South Pacific Ocean over the last century and are expected to continue to rise into the future (IPCC 2014). Inter-comparison of climate models used to explore future changes to the global climate under emission scenarios (see Taylor et al., 2012) has identified numerous biases in ocean parameters, both within and across models. These biases are particularly evident in the tropical Pacific Ocean and are associated with difficulties in simulating sea-surface temperatures, precipitation and salinity (Sen Gupta et al., 2009; Ganachaud et al., 2011). Use of a multi-model mean derived from models used in inter-comparisons considerably reduces these biases, although certain regions still retain sizeable biases, indicating systematic biases across models (Sen Gupta et al., 2009). In particular, the eastern tropical Pacific cold tongue is placed too far west and the South Pacific Convergence Zone is too elongated towards the east, resulting in biases in precipitation and ocean surface salinity, which has implications for projections of climate relating to a number of Pacific Islands. Along the Chilean shelf edge, problems with the representation of local atmospheric processes and upwelling lead to biases in cloud formation and radiative heat transfer, with flow-on impacts on ocean salinity (Randall et al., 2007; Sen Gupta et al., 2009; Brown et al., 2013Ganachaud et al. 2013). The resolution at which most climate models are run does not take into account processes occurring in the near-coastal ocean, so pressures and associated projections derived from climate models are extrapolated from observations made offshore (Rhein et al., 2013). This is particularly problematic for projections relating to islands in the South Pacific Ocean and also for mesoscale and submesoscale processes that are important for delivering nutrients to the photic zone (Ganachaud et al., 2011). Bearing in mind the biases and the resolutions of current models, a summary of observed and projected changes to the South Pacific Ocean are presented here.

Projections of surface temperatures are robust at a large scale and suggest a warming rate of the surface ocean during the 21st century that is approximately seven times that observed in the 20th century (Sen Gupta et al., 2015). Within the South Pacific Ocean, the western tropical Pacific Ocean is projected to warm and the region associated with the EAC and its extensions is projected to undergo enhanced warming (Cravatte et al., 2009; Ganachaud et al. 2013.). Intensification of south-easterly trade winds in the eastern South Pacific Ocean region will result in weakened warming (Sen Gupta et al., 2015).

Observations of surface salinity within the subtropical gyre in the South Pacific Ocean have demonstrated an increasing trend, particularly in the east, whereas in the equatorial WWP, surface waters have freshened (Durak and Wijffels, 2010). Model projections suggest an increase in rainfall across tropical latitudes in association with increased evaporation and enhanced convection, which will have implications for ocean salinity in these regions. The area of the WWP in the tropical western Pacific Ocean is projected to continue to freshen and the area of relatively fresher water is projected to move east (Cravatte et al., 2009; Ganachaud et al. 2013). At mid-latitudes, rainfall is projected to decrease, particularly over the central and eastern Pacific Ocean, which will result in increasing surface salinity (Ganachaud et al., 2011).

Changes in wind stress forcing over the past two decades has resulted in changes in circulation in the South Pacific Ocean (Rhein et al., 2013). The southern limb of the South Pacific subtropical gyre and the subtropical cells have intensified in response to intensification of Southern Hemisphere westerlies. In addition, the gyre and cells have moved poleward (Roemmich et al., 2007; Rhein et al., 2013). This has also resulted in a southward expansion of the EAC into the Tasman Sea (Ridgway, 2007; Hill et al., 2008). These wind-driven changes are most likely due to interannual to decadal variability (i.e., intensification of the Southern Annular Mode); time series are currently not substantial enough to determine longer-term trends (Ganachaud et al., 2011; Rhein et al., 2013). Interactions between large scale oceanographic and atmospheric processes and island topography are expected to have local effects on the waters surrounding islands in the South Pacific Ocean; however, local projections of confidence are scarce (Ganachaud et al., 2011).

The combined effect of changes to surface temperature and salinity will result in changes to the stability of the water column and the level of stratification. The level of stratification of the water column affects the potential for vertical exchange of ocean properties, such as oxygen or nutrients, which has flow-on effects for primary productivity (Ganachaud et al. 2011; 2013). Surface warming of the tropical South Pacific, in combination with freshening in the WWP, have resulted in an increase in stratification over the upper 200 m (Cravatte et al., 2009). Stratification is projected to continue being most pronounced in the WWP and the South Pacific Convergence Zone (Ganachaud et al., 2011; 2013). In conjunction, the annual maximum depth of the mixed layer is projected to shoal across most of the tropical South Pacific Ocean (Ganachaud et al. 2013). Although the mixed layer depth is expected to shoal, in the eastern South Pacific Ocean this is not expected to affect primary production. This is because nitrate concentrations due to upwelling processes are still likely to exceed levels at which the supply of iron currently limits phytoplankton growth (Le Borgne et al., 2011). Within the western South Pacific Ocean, use of high-resolution ocean models has suggested that projected increased mixing due to changes in currents (which are not fully resolved in lower-resolution models), will result in increased subsurface primary production. This is expected to result in close to no change in overall net primary production throughout the region (Matear et al., 2015).

Oxygen concentrations in the tropical South Pacific Ocean thermocline have decreased over the past 50 years in association with changes in oxygen solubility (resulting from warming), ventilation and circulation. This has resulted in a major westward expansion of oxygen minimum waters in the eastern Pacific Ocean (Stramma et al., 2008). Recent observations for the period 1976–2000 have shown that dissolved oxygen concentrations have declined at a faster rate in the coastal ocean than in the open ocean and have also occurred at a faster rate than in the period 1951 – 1975 (Gilbert et al., 2010). Projected changes to surface temperatures and stratification are likely to result in a decreased transfer of oxygen from the atmosphere, resulting in lower concentration of oxygen in waters above the thermocline across the tropics (Ganachaud et al., 2011). Existing oxygen minimum waters in the eastern South Pacific Ocean are projected to intensify (Ganachaud et al., 2011), although uncertainty in model projections limits projections associated with the evolution of oxygen concentrations in and around oxygen minimum zones (Ciais et al., 2013). Outside the tropics, trends in oxygen concentrations are less obvious (Keeling et al., 2010), but it is expected that warming of the ocean will result in declines in dissolved oxygen in the ocean interior (Rhein et al., 2013). In coastal regions, because hypoxia is largely driven by eutrophication and is therefore controlled by the flow of nutrients from terrestrial origins, any increase in nutrient run-off associated with increasing agriculture or industrialization of coastal regions will also result in increasing coastal water deoxygenation (Rabalais et al., 2010; Ciais et al., 2013; see also section 3.2).

Observations of carbon concentration in the ocean demonstrate considerable variability associated with seasonal, interannual (associated with ENSO) and decadal (associated with the Pacific Decadal Oscillation) changes in wind and circulation (Rhein et al., 2013). Taking into account this variability, trends in surface ocean carbon dioxide have increased, resulting in a decrease in surface pH. This decrease varies regionally: the subtropical South Pacific Ocean demonstrates the smallest reduction in pH (Rhein et al., 2013). Continued increased storage of carbon dioxide in the ocean will result in further decreases in the pH of the ocean; surface ocean pH is projected to decrease by 0.06 – 0.32 depending on the emission scenario used in projections (Ciais et al., 2013). Generally, projected changes to pH are greatest at the ocean surface; surface waters are projected to become seasonally corrosive to aragonite at higher latitudes in one to three decades. In the subtropics, however, the greatest changes to pH are projected to occur at 200 – 300 m where lower carbonate buffering capacity results in lower pH, although carbon dioxide concentration might be similar to that at the surface (Orr, 2011). The horizon separating shallower waters supersaturated with aragonite from deeper under-saturated waters will shoal, resulting in a decline in the global volume of ocean with supersaturated waters (Steinacher et al., 2009). In areas of freshwater input (e.g., around river mouths), reduction in pH and the aragonite saturation state will be exacerbated (Ciais et al., 2013). Overall, projected decreases in pH will be greater at higher latitudes than at lower latitudes (Le Borgne et al., 2011).

Taking into account inter-annual fluctuations associated with ENSO, time series of global sea level measurements demonstrate that mean sea level has risen at a rate of 1.7 mm yr^{-1} over the last century in association with ocean warming and redistribution of water between continents, ice sheets and the ocean (Church and White, 2011; Church et al., 2013). This rate has increased over the last two decades (to a mean rate of 3.2 mm yr^{-1}), but it is unclear whether or not this reflects decadal variability or an increase in the long-term trend (Church et al., 2013). In the western Pacific Ocean, sea level has risen up to three times the rate of global sea level over the last two decades, largely in association with intensified trade winds which may be related to the Pacific Decadal Oscillation (Merrifield et al., 2012). Increases in mean sea level have resulted in an increase in sea-level extremes. Short-term drivers of sea level (e.g., tides) are not projected to change substantially, whereas longer-term drivers (e.g., ice melt, thermal expansion of the ocean) are projected to continue. Over the next century the rate of global mean sea-level rise is expected to increase to 4.4 – 7.4 mm yr^{-1} depending on the emission scenario used, noting that the rate of regional sea level rise can differ from the global average by more than 100 per cent as a result of climate variability (Church et al., 2013). In the South Pacific Ocean, coastlines are projected to experience an increase in sea level from approximately 0.3 m to over 0.8 m by 2100 depending on the emission scenario used and noting that projections of land-ice melt have large uncertainties.

These uncertainties result in considerable variability in projected patterns of sea-level change between climate models (Church et al., 2013).

At present little evidence exists of any trend or long-term change in tropical or extra-tropical storm frequency or intensity in the South Pacific Ocean (Rhein et al., 2013). Increases in observed sea-level extremes have primarily been associated with an increase in mean sea level rather than the level of storminess (Church et al., 2013; Rhein et al., 2013). Across the South Pacific Ocean, the monsoon area is projected to expand over the central and eastern tropical Pacific and the strength of monsoon systems is projected to increase. Monsoon seasons are also likely to lengthen and so precipitation throughout tropical regions is projected to increase (Christensen et al., 2013). However, because the South Pacific Convergence Zone is projected to move to the northeast, precipitation over many South Pacific islands is projected to decrease. Projections of tropical cyclones suggest that although the global frequency of cyclones is likely to remain the same, their intensity is likely to increase and a poleward shift in storm tracks is likely, particularly in the Southern Hemisphere. Regional projections are not yet well quantified; many climate models fail to simulate observed temporal and spatial variations in tropical cyclone frequency (Walsh et al., 2012). As a result, projections of the frequency and intensity of cyclones at the level of ocean basins are highly uncertain and confidence in projections is low (Christensen et al., 2013).

The potential impacts of changes to the physical and chemical structure of the South Pacific Ocean and on the biodiversity of the region will depend on the capacity of organisms to adapt to these changes over the time scales at which they are occurring. As waters warm, some species are expected to alter their distribution and already evidence exists that some species have extended their distributions poleward in line with warming trends in the South Pacific Ocean (Sorte et al., 2010; Last et al., 2011). The introduction of new species into regions via expansion of their distribution has the potential to alter marine communities and it is likely that at least some marine communities will undergo major changes to their community structure (Hughes et al., 2003). Conversely, other species may demonstrate range contraction as range edges become thermally unsuitable and the time scales at which changes are occurring exceed the adaptive ability of species (e.g., Smale and Wernberg, 2013). For example, it is likely that the latitudinal and bathymetric range of kelp communities will become restricted. Although other species might replace these climatically sensitive species, reductions in kelp production will have important consequences for the communities that rely on them and other near-shore habitats that depend on the export of kelp detritus (Harley et al., 2006).

Bleaching events as a result of thermal stress induced by higher ocean temperatures, in combination with a reduced ability of corals to calcify due to ocean acidification, are expected to result in steep declines in coral cover across the South Pacific Ocean over the next decades, even under good management (Hughes et al., 2003; Bell et al., 2013). Already evidence exists that corals within the Great Barrier Reef are calcifying at lower rates than those prior to 1990 (De'ath et al., 2009). This will have flow-on effects for benthic organisms and fish and macro-invertebrate populations associated with reef communities. The differing abilities of coral species to migrate in response to climate change and their genetic ability to adapt to warmer waters will, however, result in changes to community structure beyond the immediate effect of selective mortality caused by severe bleaching (Hughes et al., 2003).

Altered temperatures may decouple population processes of taxonomic groups that are reliant on the population processes of (and) other group(s). For example, the breeding processes of many marine species are timed to coincide with peaks in forage-species populations, whose timing is often driven by temperature. If the timing of the two is altered so that they no longer match, this will likely affect population recruitment (e.g., Philippart et al., 2003).

Because phytoplankton have differing sensitivities to carbon dioxide concentrations and utilise carbon in differing ways, changes in carbon dioxide concentrations will not only change the activity of individual phytoplankton species, but will also tend to favour some species over others. Increasing ocean carbon dioxide is therefore likely to result in shifts in phytoplankton community structure, which will in turn influence the community structure of higher trophic organisms reliant on phytoplankton for food (Hays et al., 2005). Furthermore, phytoplankton and zooplankton species that depend on current saturation levels of aragonite to build robust shells and skeletons (e.g., coccolithophorids, pteropod molluscs, gastropods) are expected to be most affected by ocean acidification. Reduced capacity to build shells and skeletons will make such organisms more fragile and vulnerable to predation and, in some cases, may result in the disappearance of these organisms from food webs (e.g., Coleman et al., 2014). This is likely to have unpredictable and cascading effects on marine food webs. Higher ocean carbon dioxide concentrations may have physiological impacts on a range of species, altering the metabolism and growth rates of some species and affecting the sensory systems of fish (Poloczanska et al., 2007; Munday et al., 2009; Munday et al., 2010; Appelhans et al., 2014).

Altered precipitation and increased storm intensity will affect the dynamics of coastal marine ecosystems through fluctuations in wave height and intensity, salinity, turbidity and nutrients. In regions where precipitation is expected to decrease, such as many Pacific Islands, these ecosystems will experience higher salinity environments, whereas those in regions where precipitation is expected to increase, such as eastern Australia, will experience fresher environments. Mangrove, seagrass and coral reef communities will be particularly prone to these changes (see Fabricus, 2005; Harley et al., 2006; Polaczanska et al., 2007).

3.2 Social and economic drivers

The South Pacific Ocean is a highly diverse region, featuring considerable variation in the social, economic, cultural and infrastructural composition of the countries and territories located within its bounds. Although climate change is considered to be one of the largest threats to marine environments over the long term, management of social and economic stressors on marine environments can be considered to be the most significant challenge over the short term (Bell et al., 2009; Center for Ocean Solutions, 2009; Brander et al., 2010). Coastal habitats have increasingly come under pressure as human populations grow. Pacific Island regions have been increasing at >3 per cent in the last two decades (Figure 36D.8); urban areas are growing at twice the national growth rate (SPREP, 2012). The economic performance by countries throughout the South Pacific Ocean varies, and in some Pacific Island countries poor economic performance has resulted in *per capita* incomes stagnating (McIntyre, 2005). As a result of poor economic performance and growing inequalities, poverty is a growing problem in some countries. The majority of Pacific Island countries have relatively limited opportunities for development and are highly dependent on overseas development assistance (McIntyre 2005). Agriculture and fisheries are the mainstay of many of the economies of South Pacific Ocean countries, and support both subsistence livelihoods and commercial production. Logging and mining are significant in countries such as Australia, Chile, Ecuador, Fiji, Nauru, New Caledonia(France), Papua New Guinea, Peru and the Solomon Islands (Observatory of Economic Complexity, www.atlas.media.mit.edu, accessed 25 August 2014). Tourism is an important economic sector throughout the region and is growing in importance in the Pacific Islands (SPREP, 2012).

Major pressures on coastal and shelf environments associated with social and economic drivers can be grouped into three broad categories: (i) habitat loss or conversion as a result of coastal development, destructive fisheries, deforestation and extraction of resources; (ii) habitat degradation as a result of various forms of pollution, increased salinization of estuarine areas and introduction of alien species; and (iii) overfishing and exploitation as a result of increasing demand at local, regional and global scales, poor fisheries management and a breakdown of tradition-

al regulation systems (Table 36D.2; UNEP, 2006a; UNEP, 2006b; Center for Ocean Solutions, 2009; UNEP, 2012). Many of these pressures have risen indirectly from larger changes to global populations, economies, industry and technologies.

Nearshore development associated with urbanization, growing populations and tourism replaces vegetated landscapes with hard surfaces and converts marine habitats into new land (e.g., Maragos, 1993; Table 36D.2). Modification of shorelines alters currents and sediment delivery, often inducing erosion and receding beaches. Increased development is often also coupled with increasing land-based pollution (e.g., Ministry of Natural Resources and Environment, 2013; Table 36D.2). The extent of land-use planning varies across the South Pacific Ocean, resulting in varying management of habitat conversion, construction activities and pollution. Many of the Pacific Islands are charting a new path from subsistence and traditional management systems to market-based economies (Center for Ocean Solutions, 2009). In many regions this has led to the slow breakdown of traditional land- and marine tenure systems, resulting in unregulated development and exploitation of coastal regions (Table 36D.2; see also section 2.4 in regard to overexploitation of coastal fisheries). The unique natural environments of many islands in the South Pacific Ocean and the desire to experience these environments can end up contributing to the reason these environments are under threat. Unregulated coastal tourism development can result in the destruction of highly regarded environments (Table 36D.2); the Galápagos Islands World Heritage Site was placed on the List of World Heritage in Danger in 2007 largely as a result of unregulated tourism development and overexploitation of marine resources (see section 5.1).

Poor management of watersheds often leads to degradation of estuaries and coastal environments (Table 36D.2). Agricultural and grazing prac-

Figure 36D.8 | Observed and projected increase in Pacific Island populations. Taken from SPREP (2012).

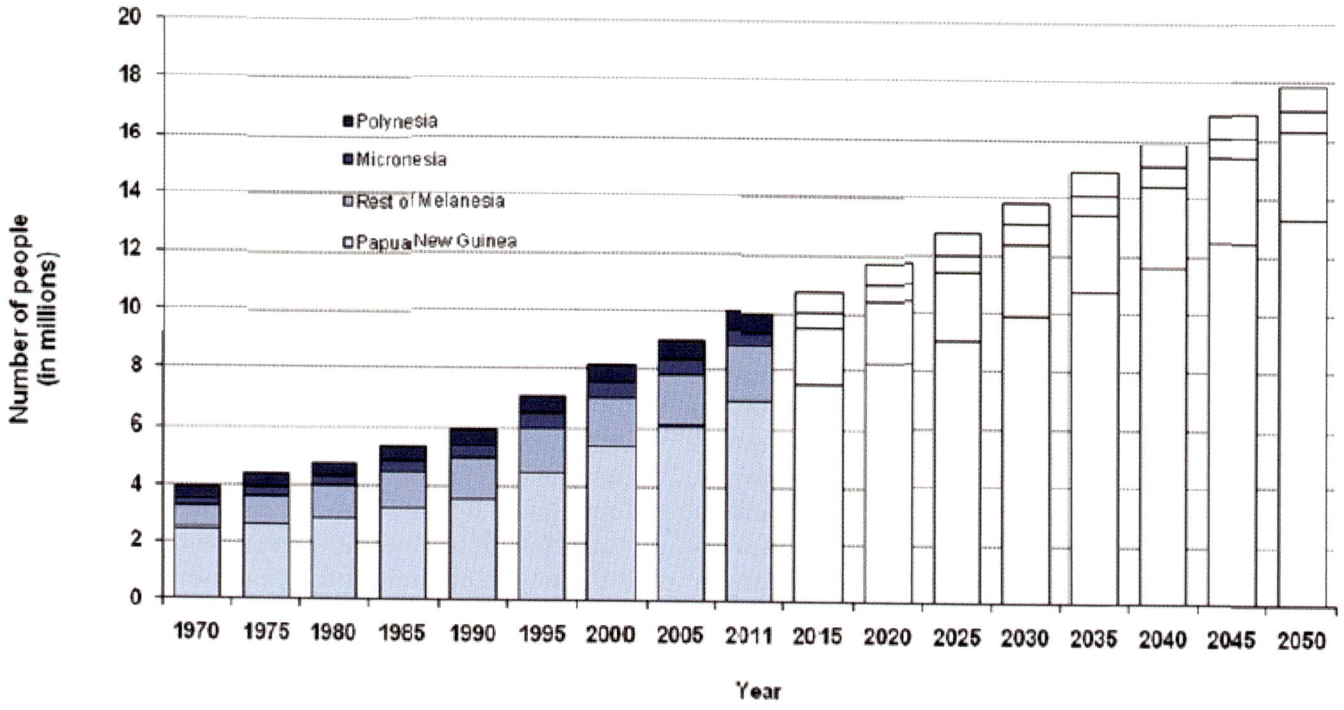

Table 36D.2 | Social and economic drivers of change in coastal and shelf ecosystems of the South Pacific Ocean. Modified from UNEP (2006b).

DIRECT DRIVERS	INDIRECT DRIVERS
Habitat loss or conversion	
Coastal development (ports, urbanization, tourism-related development, industrial development, civil engineering works)	Population growth; transport and energy demands; poor urban planning and industrial development policy; tourism demand; environmental refugees and internal migration
Destructive fishing practices (dynamite, cyanide, bottom trawling)	Shift to market economies; on-going demand for live food fish, aquarium species; increasing competition associated with diminishing resources
Coastal deforestation	Lack of alternative materials; increasing competition associated with diminishing resources; global commons perceptions
Mining (coral, sand, minerals, dredging)	Lack of alternative materials; global commons perceptions
Aquaculture-related habitat conversion	International demand for luxury items (including new markets); regional demand for food; demand for fishmeal in aquaculture and agriculture; decline in wild stocks or decreased access to fisheries (or inability to compete with larger-scale fisheries)
Habitat degradation	
Eutrophication from land-based sources (agricultural waste, sewage, fertilizers)	Population growth; urbanization; lack of infrastructure (stormwater, sewage systems); lack of sewage treatment; unregulated agricultural development and management; loss of natural catchments (wetlands, etc.)
Pollution: toxins and pathogens from land-based sources	Increasing pesticide and fertiliser use; lack of regulations associated with use; lack of awareness of impacts; unregulated industries
Pollution: dumping and dredge spoil	Lack of alternative disposal methods; decreasing terrestrial options; increasing regulation and enforcement of terrestrial disposal; lack of awareness of impacts
Pollution: shipping-related	Increased ship-based trade; substandard shipping, pollution and violation of marine safety regulations; flags of convenience
Increased salinization of estuaries due to reduced freshwater flows	Increased and unregulated agricultural development; increased demand for electricity and water
Introduction of alien species	Lack of regulations on the discharge of ballast; increased aquaculture-related escapes; lack of agreements and policies on deliberate introductions
Overexploitation	
Directed take of low-value species at volumes exceeding sustainable levels	Population growth; demand for subsistence and markets; globalization of trade networks, increased demand for aquaculture feed, industrialization of fisheries; improved fishing technologies; poor management and enforcement; breakdown of traditional regulation systems; introduction/maintenance of subsidies
Directed take of high-value species for luxury markets at volumes exceeding sustainable levels	Demand for speciality foods, medicines, aquarium fish, globalization of trade networks, lack of awareness of or concern about impacts
Directed take of commercial species; decreasing availability for subsistence and artisanal use	Population growth, globalization of trade networks, industrialization of fisheries; improved fishing technologies; poor management and enforcement; breakdown of traditional social systems; introduction/maintenance of subsidies
Incidental take or by-catch	Poor management and enforcement; lack of awareness of or concern about impacts

tices that destroy natural riparian habitats can result in floods and burial of natural estuarine and coastal habitats under silt and enriched sediment (e.g., Fabricus, 2005). Interruption of natural water flow via extraction for agriculture or power restricts water and nutrient flow into estuarine environments, reducing flushing and dilution of pollution (fertilizers, pesticides, sewage, debris, chemicals, and stormwater run-off), causing siltation and, in extreme cases, closure of estuary mouths, and increasing the salinization and toxicity of estuary areas. Agricultural practices often result in excessive nutrient loading of estuarine and coastal environments, causing these areas to become eutrophied, resulting in algal blooms and dead zones. Land-based sedimentation, combined with nutrient inputs, is a major water-quality threat to many of the coastal environments of the western and central South Pacific (e.g., Maragos and Cook, 1995; Hughes et al., 2003; Orth et al., 2006; Center for Ocean Solutions, 2009; GBRMPA, 2014). Higher nutrient concentrations associated with run-off from coastal urbanized areas have been documented to drive shifts in phytoplankton community composition and abundance (Jacquet et al., 2006).

Port development, such as infilling, dredging, channelling, and installation of harbour works including seawalls and groins, often results in alterations to estuaries and embayments (Table 36D.2). Alterations to soft bottom habitats in these areas often create conditions for new assemblages of species, and facilitate range expansions of invasive species (Ruiz and Crooks, 2001). Furthermore, the movement of ships and other transport vehicles into these areas from around the globe has enabled the spread of many marine species (Table 36D.2). Introduction of invasive species facilitated by shipping (via fouling, boring, nestling into the hull, anchor chain, and ballast water) has been reported extensively across the South Pacific Ocean; alien species are reported from most countries and territories in the region (Carlton, 1987; Bax et al., 2003; Hewitt et al., 2004; Ministry of the Environment, 2004; Sakuma, 2004; Mourgues, 2005; Gabrie et al., 2007; Ministry of Environment Conservation and Meteorology, 2008; Ministry of Natural Resources and Environment, 2013). In the south-east region of Australia, invasive species such as starfish, sea urchins, plankton, algae, molluscs, crustaceans and worms have had major impacts on coastal marine environments. Port Phillip Bay, the site of the Port of Melbourne, has been described as one of the most invaded marine ecosystems in the Southern Hemisphere: more than 150 alien species are reported from the embayment (Bax et al., 2003). Another site of equal note is the Derwent River estuary in Tasmania (State of the Environment Committee, 2011). In New Zealand, invasive species have been detected in virtually all coastal habitat types (Hewitt et al., 2004).

Coastal aquaculture operations, although bringing important socio-economic benefits to countries, can result in modification of coastlines and benthic habitats and pollution of coastal habitats (Table 36D.2). Shrimp and salmonid aquaculture in the coastal regions of Ecuador, Peru and Chile contributes significantly to the economies of each country; Chile is one of the main producers of salmonids in the world (De Young, 2007). After lengthy periods of sustained growth, aquaculture operations in Ecuador have resulted in the destruction of large tracts of mangrove forest and coastal wetlands (Bailey, 1988; Martinez-Porchas and Martinez-Cordova, 2012). Operations in Chile have caused significant loss of benthic biodiversity and local changes in the physical and chemical properties of sediments have occurred in areas with salmonid farms (Buschmann et al., 2006). Pulses in dinoflagellate densities have increased and it is suggested that escaped farmed fish may have an impact on native species, although their survival in the wild appears low. In addition, the abundance of omnivorous diving and carrion-feeding marine birds in areas of aquaculture operations has increased two - fivefold (Buschmann et al., 2006).

4   Major ecosystem services

Coastal and shelf ecosystems provide a diverse range of services to marine and terrestrial environments and benefits to human society (Table 36D.4). Globally, coastal and shelf marine habitats are estimated to provide over 14 trillion United States dollars' worth of ecosystem goods (e.g., food and raw materials) and services (e.g., disturbance regulation and nutrient cycling) per year (Costanza et al., 1997). Valuable natural resources, such as fisheries, oil, deep sea mineral deposits and pharmaceutical constituents, are abundant throughout coastal, shelf and offshore waters of the South Pacific Ocean. Inshore regions provide coastal protection and artisanal fisheries, aquaculture, and tourism provide significant income for local communities (SPREP, 2012). The natural environment of coastal and inshore regions is an integral part of the culture, tradition, history and way of life for many communities. These resources are therefore essential to the livelihoods of communities throughout the South Pacific Ocean, as well as being desirable for the global community. On-going use of coastal ecosystems and associated declines in the health of these ecosystems have flow-on effects on the benefits and ecosystem services these provide to the environment and to the communities that rely on them.

4.1   Services to ecosystem being lost

Loss of coastal ecosystem biodiversity has been identified as affecting three primary ecosystem services: provision of nursery habitats, filtering and detoxification services and maintenance of trophic stability (Worm et al., 2006). Estuaries, salt marshes, mangroves, lagoons, seagrass meadows and kelp forests serve as nurseries for many marine species, provide interconnectivity of habitats for the life stages of some species and provide essential food resources across multiple trophic levels (Figure 36D.9; Robertson and Blaber, 1993; Dayton et al., 1998; Orth et al., 2006; UNEP, 2006b). Mangroves, seagrass meadows and coral reefs provide protective services for the coastline, binding sediments and dissipating wave action (Moberg and Folke, 1999). Mangroves, via their ability to trap water, control the chemistry of estuarine water and the flow rates of mangrove creeks, both of which are important for water-column biota survival and dispersal (Robertson and Blaber, 1993). Mangrove forests and seagrass meadows are both an atmospheric carbon dioxide sink and an essential source of oceanic carbon, providing an essential supply of organic matter in marine environments (Suchanek et al., 1985; Duarte et al., 2005; Duke et al., 2007). Coral reefs are nitrogen fixers in otherwise nutrient-poor environments (Sorokin, 1993) and the release of excess nitrogen by coral reef systems is important for the productivity of adjacent communities (Sorokin, 1990). Reduction of these communities imperils dependent fauna with their complex habitat linkages, and endangers physical benefits like the buffering of seagrass beds and coral reefs by mangroves against the impacts of river-borne siltation (Duke et al., 2007) and protection by coral reefs against the impacts of currents, waves and storms (Moberg and Folke, 1999).

Many coastal marine habitats contain species that regulate ecosystem processes and functions through grazing and predation (Moberg and Folke, 1999). These processes operate across all trophic levels and disruption at any one trophic level can have flow-on effects across other trophic levels. For example, reduction in herbivorous and predatory reef fish in coral reef communities, as a result of overfishing in the South Pacific Ocean, has been found to result in alterations to community structure. Alterations include increases in coral-eating starfish densities, leading to a decline in reef-building corals and an increase in non-reef-building species, such as filamentous algae (Hughes et al., 2003; Dulvy et al., 2004). Once algae become abundant, coral recovery is suppressed unless herbivores return to reduce algal cover, and corals can then recruit. Recent research on coral communities in Fiji has demonstrated, however, that chemical cues emitted by algae in degraded reefs repulse coral recruits, resulting in coral juveniles actively avoiding recruiting to these areas (Dixson et al., 2014). Declines in coral cover have flow-on effects for ecosystem processes, such as reef building, primary and sec-

Table 36D.4 | Examples of the services coastal and shelf ecosystems provide. Taken from UNEP (2006b).

SERVICE	ESTUARIES/ MARSHES	MANGROVES	LAGOONS/SALT PONDS	INTERTIDAL	KELP	ROCKY REEFS	SEAGRASS	CORAL REEFS	SHELVES
Services to ecosystems									
Biodiversity	•	•	•	•	•	•	•	•	•
Biological regulation	•	•	•	•	•	•	•	•	
Hydrological balance	•		•						
Atmospheric and climate regulation	•	•	•	•	•	•	•	•	•
Biochemical	•	•	•	•	•	•	•	•	•
Nutrient cycling and fertility	•	•	•	•	•	•	•	•	•
Flood/storm protection	•	•	•	•		•	•	•	
Erosion control	•	•	•					•	
Services to humans									
Food	•	•	•	•	•	•	•	•	•
Fibre, timber, fuel	•	•	•						
Waste processing	•	•					•	•	
Atmospheric and climate regulation	•	•	•	•	•	•	•	•	•
Flood/storm protection	•	•	•	•		•	•	•	
Erosion control	•	•	•	•		•	•	•	
Culture and amenity	•	•	•	•	•	•	•	•	•
Recreational	•	•	•	•	•	•	•	•	•
Aesthetics	•	•	•	•	•	•	•	•	•
Education and research	•	•	•	•	•	•	•	•	•
Medicines, other resources	•	•	•		•			•	
Human disease control	•	•	•			•	•	•	

require careful management to ensure that the socio-economic benefits from these resources are maintained sustainably (Bell et al., 2013).

Many costal ecosystems provide communities with materials essential for construction and fuel. Mangroves provide timber, fibre and fuel, coral reefs provide lime and other building materials, and sand mining occurs in many coastal regions across the South Pacific Ocean. Shelf regions provide oil and gas and various other minerals. Over-exploitation and reduction of such finite resources will require identification of alternatives and adaptive strategies to ensure transfer to alternative economies. Logging has been identified as the most pressing issue facing the Solomon Islands: current rates are unsustainably high. Maintenance of unsustainably high rates of logging will result in serious impacts on the country's economy when the revenue stream collapses and on the population when building resources are no longer available and watersheds deteriorate (Ministry of Environment Conservation and Meteorology, 2008). Subsequent run-off of sediments from cleared areas will have further impacts on coastal reef environments and associated food resources. Construction of causeways on South Tarawa, Kiribati Islands, which block the migration pathways of several species of fish that are the focus of subsistence fisheries, has been associated with the collapse of their populations. Coastal erosion as a result of infrastructure development, overcrowding and overexploitation of the physical resources of the coastal zone of South Tarawa has resulted in the loss of houses, roads and agricultural land (Ministry of Environment, Lands and Agricultural Development, 2004).

5 Areas of special conservation significance and associated issues of the South Pacific

ondary production, which then in turn affect higher trophic levels, and reduce ecosystem functioning.

4.2 Services to humans being lost

Many of the development goals of Pacific Island countries and territories are intricately linked to marine ecosystems and the benefits provided by them. Nowhere else in the world do so many countries depend on marine resources, and in particular, fishery resources, for economic development, food security, government revenue and livelihoods (Bell et al., 2013). Loss of coral reef habitat will affect coral reef fisheries, the majority of which are already considered to be either fully or overexploited. Anticipated human population growth will place increasing pressures on these resources and it has been estimated that an additional 196,000 km² of coral reef habitat will be required to sustain current levels of fishing (Newton et al., 2007). Projections of further reductions in coral reefs as a result of climate change are more than likely to have flow-on effects on coral reef fisheries and, as a result, reliance on pelagic resources for protein is expected to increase throughout Pacific Island countries and territories (Bell et al., 2009). Increased reliance on pelagic resources will

5.1 World Heritage Sites

Two of the largest World Heritage sites are in the South Pacific Ocean – the Phoenix Islands Protected Area and the Great Barrier Reef. Whereas the Phoenix Islands Protected Area is comprised of largely oceanic, deep water ecosystems, the Great Barrier Reef is entirely shelf-based. Other World Heritage sites located in the South Pacific Ocean with protected marine components include the Lord Howe Island Group in Australia, East Rennell in the Solomon Islands, the lagoons of New Caledonia and the Galápagos Islands in Ecuador.

The Great Barrier Reef is the world's largest coral reef system (34 million hectares), extending 2,000 kilometres along the eastern Australian coast. It comprises over 2,500 individual reefs and 900 islands. Declared in 1981, it was one of the first World Heritage sites. It is home to over 400 types of coral and is one of the richest areas in the world for animal biodiversity. The diversity of species and habitats, and their interconnectivity, make the Great Barrier Reef one of the richest and most complex natural ecosystems on earth. Key threats affecting the site include

Figure 36D.9 | Conceptual depiction of major mechanisms of seagrass and related key ecosystem services loss for (a) tropical and (b) temperate seagrass ecosystems. Taken from Orth et al., 2006.

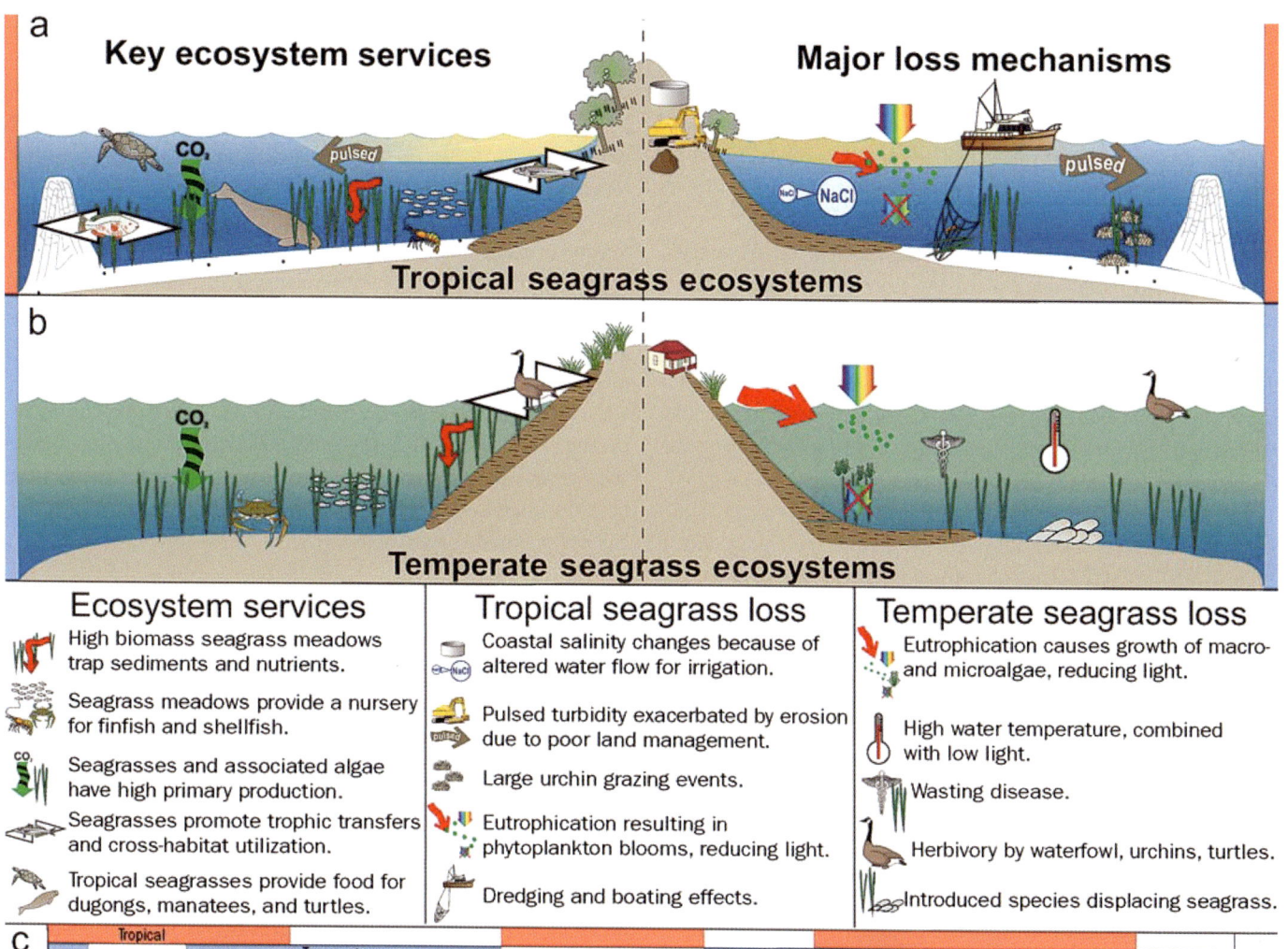

coastal development, development of ports and liquefied natural gas facilities, extreme weather events, grounding of ships, water quality and oil and gas (UNESCO, 2014a). The 2014 Great Barrier Reef Outlook Report (GBRMPA, 2014) concludes that: "Even with the recent management initiatives to reduce threats and improve resilience, the overall outlook for the Great Barrier Reef is poor, has worsened since 2009 and is expected to further deteriorate in the future." Serious declines in the condition of the Great Barrier Reef, including coral recruitment and reef building across large parts of the reef, have been observed and the report concludes that a 'business as usual' approach to managing the reef is not an option (GBRMPA, 2014; UNESCO 2014a).

The Phoenix Islands Protected Area (PIPA) was declared in 2010; at 40.8 million hectares, it is the world's largest World Heritage Site. It consists of eight oceanic coral atolls, most of which are uninhabited, two submerged reefs and fourteen identified seamounts. Its isolation and low population density have helped the area remain comparatively undisturbed and it provides important habitat for migratory and pelagic/planktonic species. It is an important breeding area for marine and seabird species and is considered a sentinel of the impacts of climate change on coral reef health (Anon., 2009). Key threats affecting the site include illegal fishing and overfishing by licensed and unlicensed vessels and degradation of seamounts. A phased zoning scheme has been proposed to ensure the site's long-term conservation. The first phase has been implemented by designating approximately 3.1 per cent of the total area of the site as a "no take" zone. As of 1 January 2015, the closure of the exclusive economic zone of the PIPA to fishing came into effect. Implementation of the second phase will involve increasing no-take areas to 25 per cent of the site and reducing offshore fishing effort for tunas. However, this implementation relies on the establishment of a trust fund which will only become operative once its capital reaches a level which will compensate the Kiribati government for losses in distant-water fishing nation license fees associated with the reduction in fishing effort. Currently, no timelines are set for reducing fishing effort (UNESCO, 2012).

The Lord Howe Island Group, declared a World Heritage site in 1982, spans 146,300 hectares and contains the world's most southerly true coral reef. The small land area within the site provides an important breeding ground for many seabirds and its marine system demonstrates a rare example of a transition zone between algal and coral reefs. Marine assemblages consist of cohabiting tropical and temperate species and endemism is high. Key threats affecting the site include invasive plants and animals, climate change, tourism and fishing (Anon., 2003).

The East Rennell World Heritage Site, declared in 1998, comprises the southern third of the world's largest raised coral atoll, Rennell Island, the southernmost island in the Solomon Islands, whose marine area extends three nautical miles to sea. Coastal waters around the island provide important habitat for migratory and pelagic/planktonic species and it is an important site for speciation processes, especially with respect to bird species. Key threats affecting the site include logging, invasive species, overexploitation of marine resources, climate change and management of the site. These threats have resulted in the site being placed on the List of World Heritage in Danger. A state-of-conservation assessment for the removal of the site from the List of World Heritage in Danger is currently underway (UNESCO 2014b).

The lagoons of New Caledonia were declared a World Heritage Site in 2008; at 1.57 million hectares, they comprise the third-largest reef system in the world. Reef systems within the site contain the most diverse concentration of reef structures in the world, ranging from barrier offshore reefs and coral islands to near-shore reticulate reefs. It contains intact ecosystems with top predators and a large diversity and abundance of large fish (Anon., 2008). Key threats affecting the site include mining, fishing and aquaculture, tourism and climate change. One of the main management issues for the site is a lack of capacity and resources for some of the existing co-management committees tasked with managing the site in enforcing fisheries and water-quality regulations and responding to incursions (UNESCO, 2011).

The Galápagos Islands were designated a World Heritage Site in 1978. They are renowned for their unique species and inspiration for the theory of evolution by natural selection proposed by Charles Darwin in the mid-1800s. The archipelago of 19 islands lies at the confluence of three ocean currents and is highly influenced by ENSO, generating one of the richest and most diverse marine ecosystems in the world. The direct dependence on the marine environment by much of the island's wildlife (e.g., seabirds, marine iguanas, sea lions) intricately links terrestrial and marine environments in the site. Lack of management of commercial, sport and illegal fishing, leading to overfishing of the marine environment, a lack of quarantine measures enabling alien species invasions, and unsustainable and uncontrolled tourism development contributed to the islands being placed on the List of World Heritage in Danger in 2007. Following strengthened quarantine, fishing and tourism management and governance of the islands, the site was removed from this list in 2010. Key threats affecting the site include changes in the identity, social cohesion and nature of the local population and community, illegal activities, tourism, visitors and recreation and the related infrastructure and management activities, systems and plans (UNESCO, 2014c).

5.2 Large Marine Ecosystems and Ecologically and Biologically Significant Areas

The South Pacific Ocean contains five Large Marine Ecosystems (LMEs), three along the eastern coastline of Australia (the north-east Australian shelf/Great Barrier Reef, the east-central Australian shelf and the south-east Australian shelf), one on the New Zealand shelf and one incorporating the Humboldt Current. The definition of these areas is based on four ecological criteria: (i) bathymetry; (ii) hydrography; (iii) productivity; and (iv) trophic relationships and definitions. These criteria provide a framework to focus on marine science, policy, law, economics and governance on a common strategy for assessing managing, recovering and sustaining marine resources and their environments (Sherman and Alexander, 1986). The approach uses five modules to measure and provide indicators of changing states within the ecosystem of each LME, including productivity, fish and fisheries, pollution and ecosystem health, socio-economics and governance. Because a lot of these factors have been discussed in previous sections of this chapter, details of each LME will not be provided again here.

The Strategic Plan for Biodiversity 2011-2020 developed under the Convention on Biological Diversity[2], provides a framework for reducing biodiversity loss and maintaining ecosystem services. It is centred around 20 targets, the Aichi Biodiversity Targets, organized under five strategic goals and the identification of marine areas in need of protection and within which the targets can be best focused known as Ecologically or Biologically Significant Marine Areas (EBSAs). Identification of these areas is based on seven scientific criteria, including (i) uniqueness or rarity; (ii) special importance for life-history stages of species; (ii) importance for threatened, endangered or declining species and/or habitats; (iv) vulnerability, fragility, sensitivity, or slow recovery; (v) biological productivity; (vi) biological diversity; and (vii) naturalness (Secades et al., 2014). To date, 26 EBSAs have been identified from the western South Pacific Ocean and 13 identified from the eastern South Pacific Ocean (Table 36D.3).

6 Factors for sustainability

The conservation and sustainable use of marine ecosystems is a goal articulated under a number of national and international policies and the development plans of countries in the South Pacific Ocean. It is becoming evident that the extent to which marine ecosystems can absorb recurring natural and anthropogenic perturbations and continue to regenerate without continued degradation will require improvements to current resource management (Hughes et al., 2005). Furthermore, current resource management and supporting marine policy will need to incorporate multi-scale ecological and social information in order to sustain delivery of ecosystem services and benefits. With this in mind, coastal and ocean managers confront a growing diversity of challenges

---

[2] United Nations, *Treaty Series*, vol. 1760, No. 30619

Table 36D.3 | Ecologically and Biologically Significant Areas (EBSAs) identified by the Convention on Biological Diversity in the South Pacific Ocean.

1.	Phoenix Islands	14.	Vatu-i-ra/Lomaiciti	27.	Equatorial high-productivity zone (east)
2.	Ua Puakaoa Seamounts	15.	South Tasman Sea	28.	Galápagos archipelago and western extension
3.	Seamounts of West Norfolk Ridge	16.	Equatorial high-productivity zone (west)	29.	Carnegie Ridge – Equatorial Front
4.	Remetau Group: south-west Caroline Islands and northern New Guinea	17.	Central Louisville Seamount chain	30.	Gulf of Guayaquil
5.	Kadavu and the southern Lau region	18.	Western South Pacific high aragonite saturation state zone	31.	Humboldt Current upwelling system in Peru
6.	Kermadec-Tonga-Louisville junction	19.	Clipperton fracture zone petrel foraging areas	32.	Permanent upwelling cores and important seabird areas of the Humboldt Current in Peru
7.	Monowai Seamount	20.	Northern Lord Howe Ridge petrel foraging area	33.	Northern Chile Humboldt Current upwelling system
8.	New Britain Trench region	21.	Northern New Zealand/South Fiji basin	34.	Central Chile Humboldt Current upwelling system
9.	New Hebrides Trench region	22.	Taveuni and Ringgold Islands	35.	Southern Chile Humboldt Current upwelling system
10.	Rarotonga outer reef slopes	23.	Manihiki Plateau	36.	Salas y Gómez and Nazca Ridge
11.	Samoan archipelago	24.	Niue Island and Beveridge Reef	37.	Juan Fernandez Ridge seamounts
12.	Suwarrow National Park	25.	Palau southwest	38.	West Wind Drift Convergence
13.	South of Tuvalu/Wallis and Futuna/north of Fiji Plateau	26.	Tongan archipelago	39.	Grey petrel's feeding area in the South East Pacific Rise

in balancing environmental and socio-economic needs throughout the South Pacific Ocean.

6.1 Ecosystem-based management and integrated coastal zone management

Ecosystem-based management (EBM) approaches are broadly accepted as cornerstones to effective marine conservation and resource management (Levin et al., 2009). Ideally frameworks for EBM should consider multiple external influences, value ecosystem services, integrate natural and social science into decision-making, be adaptive, identify and strive to balance diverse environmental and socioeconomic objectives, and make trade-offs transparent. Integrated coastal zone management (ICZM) can be used within an EBM framework to address the ecological and human complexity of interconnected systems. Development of ICZM, in principle, should incorporate an integrated, adaptive approach for coastal management that addresses all aspects of the coastal and neighbouring ocean zone, including land–coastal interactions, climate change, geographical and political boundaries, in an effort to achieve long-term sustainable use and reduce conflicts. It requires the careful balancing of a wide range of ecological, social, cultural, governance, and economic concerns. Although some examples exist of the implementation of ICZM throughout the South Pacific Ocean (e.g., see National Resource Management Ministerial Council, 2006; Department of Conservation, 2010; section 6.2), for many countries, comprehensive coastal management remains a challenge.

Within the Pacific Islands region, a Framework for a Pacific Oceanscape has been developed and endorsed by 23 countries and territories within the region. This framework, finalised in 2012, draws on the Pacific Islands Regional Oceans Policy and has been designed to address six strategic priorities associated with marine resource conservation, habitat protection and fisheries management (Pratt and Govan, 2010). These priorities will be met via the development of terrestrial and marine protected areas, identification of risks and mitigation strategies for climate change and the provision of research and leadership capacity development throughout the region (Pratt and Govan, 2010).

Uptake of EBM approaches to resources and, in particular, to commercial fisheries has also been slow and although such an approach may have been adopted at a policy level, practical implementation is largely lacking (Garcia et al., 2003; Smith et al., 2007). Traditional management of fisheries, which is still conducted by most national and international management agencies throughout the South Pacific Ocean, concentrates on individual fish populations strictly in demographic terms, i.e., accounting for the input of individuals as population growth or immigration and the output in terms of natural and fishing mortality. Fish populations, however, are also affected by variability in external factors, such as predator and prey abundances and variability in their biophysical environment. At the same time, changes in the abundance of populations will affect all the surrounding ecosystems of which fishes are part. Federal, state and territorial fisheries management agencies in Australia have adopted ecosystem-based fisheries management as the approach to future management (Smith et al., 2007). Tools to facilitate this approach have largely been developed and implemented for most Commonwealth fisheries and are in various stages of development for state and territory fisheries.

6.2 Marine management areas

The establishment of representative systems and networks of marine management areas is regarded internationally and nationally as one of the most effective mechanisms for protecting biodiversity and a tool for resource sustainability. Protected areas, including national parks, managed resource protected areas, locally managed marine areas, marine reserves, protected seascapes, and habitat management areas, occur in varying degrees in the coastal and offshore regions of countries and territories throughout the South Pacific Ocean (see also section 5.1). Co-ordinated networks of protected or managed areas providing for protection of ecosystems representative of regions are largely lacking, with the exception of Peru's Guano Islands, Islets and Capes National Reserve System (RNSIIPG), and enforcement is an issue for many marine management areas. Australia's National Representative System of Marine Protected Areas (NRSMPA) has been developed for Australian marine waters, but is currently under review and yet to be implemented. Community-based management areas throughout the Pacific Islands and territories have shown some level of success, largely because those that benefit from sustainable resource use are those directly involved in managing those resources. However, managers of community-based management areas are often not equipped to ensure that management is effective. Across many communities, knowledge about the long-term effects of current use of marine resources, sustainability issues, and the requirements for management, research, and monitoring is poor. The need to strengthen education has been identified by a number of countries. Frameworks for the identification and implementation of marine protected areas at regional scales are also being developed under the Convention on Biological Diversity[3] (see section 5.2) and the Framework for a Pacific Oceanscape (see section 6.1). Recent research has demonstrated that in areas where full protection of marine regions is untenable because of dependence of communities on marine resources, even simple forms of fisheries restrictions can have substantive positive effects on functional groups (MacNeil et al. 2015).

6.3 Integration of climate change adaptation and mitigation into marine policy, planning and management

Over the long term, one the largest threats to coastal and marine systems within the South Pacific Ocean is climate change. Responding to the environmental and socio-economic consequences of climate change in order to maintain ecosystem services requires coordinated and integrated efforts in incorporating adaptation and mitigation options into marine policy, planning and management. International efforts at coordinated adaptation and mitigation planning have occurred, largely through the United Nations Framework Convention on Climate Change, either via National Adaptation Programmes of Action (NAPAs) in the case of least developed countries, or National Communications for Annex I countries (see www.unfcc.int). At present, however, examples of the implementation of climate change adaptation actions are limited, even though acceptance is widespread of the need for adaptation and for significant investments in adaptation planning. Factors affecting implementation include local adaptive capacity, inabilities and inefficiencies in the application of existing resources, and limited institutional support and integration, particularly between and across governments (Christensen et al., 2007; Noble et al., 2014). In order to overcome this, adaptation assessments may need to link more directly to particular decisions and tailor information to local contexts to facilitate the decision-making process (Noble et al., 2014). Key components in the integration of adaptation and mitigation options should include (i) stakeholder participation in decision making; (ii) capacity development; (iii) communication, education and public awareness; (iv) development of alternative income-generating activities; (v) monitoring; (vi) addressing uncertainty and (vii) analysis of trade-offs (UNEP, 2006b). Without ensuring that adaptation options are integrated into coastal zone management, it is likely that ecosystem services will not be maintained into the future (Bell et al., 2013).

7 Acknowledgements

We thank Alex Sen Gupta for providing Figure 36D.1. Neville Barrett, Camille Mellin and Peter Thompson are thanked for providing useful comments on the chapter.

---

3   United Nations, *Treaty Series*, vol. 1760, No. 30619.

References

Adjeroud, M., Michonneau, F., Edmunds, P.J., Chancerelle, Y., Lison de Loma, T., Penin, L., Thibaut, L., Vidal-Dupiol, J., Salvat, B., Galzin, R. (2009). Recurrent disturbances, recovery trajectories, and resilience of coral assemblages on a South Central Pacific reef. *Coral Reefs* 28, 775-780.

Alamo, A., Bouchon, M. (1987). Changes in the food and feeding of the sardine (*Sardinops sagax sagax*) during the years 1980–1984 off the Peruvian coast. *Journal of Geophysical Research* 92, 14411-14415.

Alfaro-Shigueto, J., Mangel, J.C., Bernedo, F., Dutton, P.H., Seminoff, J.A., Godley B.J. (2011). Small-scale fisheries of Peru: a major sink for marine turtles in the Pacific. *Journal of Applied Ecology* 48, 1432-1440.

Alheit, J., Niquen, M. (2004). Regime shifts in the Humboldt Current ecosystem. *Progress in Oceanography* 60, 201-222.

Alldredge, A. and King, J. (1977). Distribution, abundance, and substrate preferences of demersal reef zooplankton at Lizard Island Lagoon, Great Barrier Reef. *Marine Biology* 41, 317-333.

Anderson, M.J. and Millar, R.B. (2004). Spatial variation and effects of habitat on temperate reef fish assemblages in northeastern New Zealand. Journal of Experimental Marine *Biology and Ecology* 305, 191-221.

André, J., Lyle, J., Hartmann, K. (2014). *Tasmanian Scalefish fishery assessment 2010/12*. Institute for Marine and Antarctic Studies, University of Tasmania.

Anon. (2003). *State of Conservation of the World Heritage Properties in the Asia-Pacific Region. Australia. The Lord Howe Island Group*. United Nations Educational, Scientific and Cultural Organization, Paris.

Anon. (2008). *Les lagons de Nouvelle-Calédonie diversité récifale et écosystèmes associés. Dossier de présentation en vue l'inscription sur la liste du Patrimoine Mondial de l'UNESCO au titre d'un bien naturel*. Haut-Commissariat de la République en Nouvelle Calédonie, Nouméa.

Anon. (2009). *Phoenix Islands Protected Area Kiribati. Nomination for a World Heritage Site*. Ministry of Environment, Lands and Agricultural Development, Tarawa.

Appelhans, Y.S., Thomsen, J., Opitz, S., Pansch, C., Melzner, F., Wahl, M. (2014). Juvenile sea stars exposed to acidification decrease feeding and growth with no acclimation potential. *Marine Ecology Progress Series* 509, 227-230.

Arntz, W.E. and Tarazona, J. (1990). Effect of El Niño on benthos, fish and fisheries off the South American Pacific coast, in: Glynn, P.W. (ed.) Global ecological consequences of the 1982-83 El Niño-Southern Oscillation. *Elsevier*, pp 323-360.

Ayón, P., Purca, S., Guevara-Carrasco, R. (2004). Zooplankton volume trends off Peru between 1964 and 2001. *ICES Journal of Marine Science* 61, 478-484.

Ayón, P., Swartzman, G., Espinoza, P., Bertrand, A. (2011). Long-term changes in zooplankton size and distribution in the Peruvian Humboldt Current System: conditions favouring sardine or anchovy. *Marine Ecology Progress Series* 422, 211-222.

Babcock, R.C., Kelly, S., Shears, N.T., Walker, J.W., Willis, T.J. (1999). Changes in community structure in temperate marine reserves. *Marine Ecology Progress Series* 189, 125-134.

Bailey, C. (1988). The social consequences of tropical shrimp mariculture development. *Ocean and Shoreline Management* 11, 31-44.

Baker, B.G., Wise, B.S. (2005). The impact of pelagic longline fishing on the flesh-footed shearwater *Puffinus carneipes* in Eastern Australia. *Biological conservation* 126, 306-316.

Baker, C.S., Clapham, P.J. (2004). Modelling the past and future of whales and whaling. *Trends in Ecology and Evolution* 19, 365-371.

Barrett, N.S., Buxton, C.D., Edgar, G.J. (2009). Changes in invertebrate and macroalgal populations in Tasmanian marine reserves in the decade following protection. *Journal of Experimental Marine Biology and Ecology* 370, 104-119.

Bates, A.E., Pecl, G.T., Frusher, S., Hobday, A.J., Wernberg, T., Smale, D.A., Sunday, J.M., Hill, N.A., Dulvy, N.K., Colwell, R.K., Holbrook, N.J., Fulton, E.A., Slawinski, D., Feng, M., Edgar, G.J., Radford, B.T., Thompson, P.A., Watson, R.A. (2014). Defining and observing stages of climate-mediated shifts in marine systems. *Global Environmental Change* 26, 27-38.

Bax, N.J., Burford, M., Clementson, L., Davenport, S. (2001). Phytoplankton blooms and production sources on the south-east Australian continental shelf. *Marine and Freshwater Research* 52, 451-462.

Bax, N., Williamson, A., Aguero, M., Gonzalez, E., Geeves, W. (2003). Marine invasive alien species: a threat to global biodiversity. *Marine Policy* 27, 313-323.

Beck, M.W., Brumbaugh, R. D., Airoldi, L., Carranza, A., Coen, L.D., Crawford, C., Defeo, O., Edgar, G.J., Hancock, B., Kay, M.C., Lenihan, H.S., Luckenback, M.W., Toropova, C.L., Zhang, G., Guo, X. (2011). Oyster reefs at risk and recommendations for conservation, restoration, and management. *BioScience* 61, 107-116.

Bell, J.D., Ganachaud, A., Gehrke, P.C., Griffiths, S.P., Hobday A.J., Hoegh-Guldberg, O., Johnson, J.E., Le Borgne, R., Lehodey, P., Lough, J.M., Matear, R.J., Pickering, T.D., Pratchett, M.S., Sen Gupta, A., Inna Senina, I., Waycott, M. (2013). Mixed responses of tropical Pacific fisheries and aquaculture to climate change. *Nature Climate Change* 3, 591-599.

Bell, J.D., Kronen, M., Vunisea, A., Nash, W.J., Keeble, G., Demmke, A., Pontifex, S., Andréfouët, S. (2009). Planning the use of fish for food security in the Pacific. *Marine Policy* 33, 64-76.

Bellwood, D.R. and Hughes, T.P. (2001). Regional-scale assembly rules and biodiversity of coral reefs. *Science* 292, 1532-1535.

Bertrand, A., Segura, M., Gutiérrez, M., Vásquez, L. (2004). From small-scale habitat loopholes to decadal cycles: a habitat-based hypothesis explaining fluctuations in pelagic fish populations off Peru. *Fish and Fisheries* 5, 296-316.

Birtles, R.A., Arnould, P.W., Dunstan, A. (2002). Commercial swim programs with dwarf minke whales on the northern Great Barrier Reef, Australia: some characteristics of the encounters with management implications. *Australian Mammalogy* 24: 23-38.

Blaber, S.J.M. (1990). Workshop summary, in: Blaber, S.J.M., Copland, J.W. (eds.). *Tuna baitfish in the Indo-Pacific region: proceedings of a workshop, Honiara, Solomon Islands*, 11-13 December 1989. ACIAR Proceedings No. 30. Australian Centre for International Agricultural Research, Canberra.

Bonnet, S., Biegala, I.C., Dutrieux, P., Slemons, L.O., Capone, D.G. (2009). Nitrogen fixation in the western equatorial Pacific: rates, diazotrophic cyanobacterial size class distribution, and biogeochemical significance. *Global Biogeochemical Cycles* 23, GB3012.

Boyd, P., LaRoche, J., Gall, M., Frew, R., McKay, M.L. (1999). Role of iron, light, and silicate in controlling algal biomass in subantarctic waters SE of New Zealand. *Journal of Geophysical Research: Oceans* 104, 13395-13408.

Boyd, P.W., McTainsh, G., Sherlock, V., Richardson, K., Nichol, S., Ellwood, M., Frew, R. (2004). Episodic enhancement of phytoplankton stocks in New Zealand subantarctic waters: contribution of atmospheric and oceanic iron supply. *Global Biogeochemical Cycles* 18, GB1029.

Bradford, J.M. (1972). Systematics and ecology of New Zealand central east coast plankton sampled at Kaikoura. *New Zealand Oceanographic Institute Memoir No 54, New Zealand Department of Scientific and Industrial Research Bulletin 207*. New Zealand Department of Scientific and Industrial Research, Wellington.

Bradford-Grieve, J.M., Chang, F.H., Gall, M., Pickmere, S., Richards, F. (1997). Size-fractionated phytoplankton standing stocks and primary production during austral winter and spring 1993 in the Subtropical Convergence region near New Zealand. *New Zealand Journal of Marine and Freshwater Research* 31, 201-224.

Branch, T.A., Stafford, K.M., Palacios, D.M., Allison, C., Bannister, J.L., Burton, C.L.K., Cabrera, E., Carlson, C.A., Vernazzani, B.G., Gill, P.C., Hucke-Gaete, R., Jenner, K.C.S., Jenner, M.N.M., Matsuoka, K., Mikhalev, Y.A., Miyashita, T., Morrice, M.G., Nishiwaki, S., Sturrock, V.J., Tormosov, D., Anderson, .RC., Baker, A.N., Best, P.B., Borsa, P., Brownell, R.L., Childerhouse, S., Findlay, K.P., Gerrodette, T., Ilangakoon, A.D., Joergensen, M., Kahn, B., Ljungblad, D.K., Maughan, B., McCauley, R.D., McKay, S., Norris, T.F., Oman Whale and Dolphin Research Group, Rankin, S., Samaran, F., Thiele, D., Van Waerebeek, K., Warneke, R.M. (2007). Past and present distribution, densities and movements of blue whales Balaenoptera musculus in the Southern Hemisphere and northern Indian Ocean. *Mammal Review*, 37: 116-175.

Brander, K. (2010). Impacts of climate change on fisheries. *Journal of Marine Systems* 79, 389-402.

Brewer, T.D., Cinner, J.E., Green, A., Pandolfi, J.M. (2009). Thresholds and multiple scale interaction of environment, resource use and market proximity on reef fishery resources in the Solomon Islands. *Biological Conservation* 142, 1797-1807.

Brodie, J., De'Ath, G., Devlin, M., Furnas, M., Wright, M. (2007). Spatial and temporal patterns of near-surface chlorophyll a in the Great Barrier Reef lagoon. *Marine and Freshwater Research* 58, 342-353.

Broitman, B.R., Navarrete, S.A., Smith, F., Gaines, S.D. (2001). Geographic variation of southeastern Pacific intertidal communities. *Marine Ecology Progress Series* 224, 21-34.

Brown, J.N., Sen Gupta, A., Brown, J.R., Muir, L.C., Risbey, J.S., Whetton, P., Zhang, X., Ganachaud, A., Murphy, B., Wijffels, S.E. (2013). Implications of CMIP3 model biases and uncertainties for climate projections in the western tropical Pacific. *Climatic Change* 119, 147-161.

Bruno, J.F., Selig, E.R. (2007). Regional decline of coral cover in the Indo-Pacific: timing, extent, and subregional comparisons. *PLoS One* 2, e711.

Burke, L., Reytar, K., Spalding, M., Perry, A. (2011). *Reefs at risk revisited*. World Resources Institute, Washington D.C.

Buschmann, A.H., Riquelme, V.A., Hernández-González, M.C., Varela, D., Jiménez, J.E., Henríquez, L.A., Vergara, P.A., Guíñez, R., Filún, L. (2006). A review of the impacts of salmon farming on marine coastal ecosystems in the southeast Pacific. *ICES Journal of Marine Science* 63, 1338-1345.

Carassou, L., Le Borgne, R., Rolland, E., Ponton, D. (2010). Spatial and temporal distribution of zooplankton related to the environmental conditions in the coral reef lagoon of New Caledonia, Southwest Pacific. *Marine Pollution Bulletin* 61, 367-374.

Carlton, J.T. (1987). Patterns of transoceanic marine biological invasions in the Pacific Ocean. *Bulletin of Marine Science* 41, 452-465.

Castrejón, M. and Charles, A. (2013). Improving fisheries co-management through ecosystem-based spatial management: the Galapagos Marine Reserve. *Marine Policy*, 38: 235-245.

Center for Ocean Solutions. (2009). *Pacific Ocean synthesis. Scientific literature review of coastal and ocean threats, impacts and solutions*. The Woods Center for the Environment, Stanford University, California.

Chaigneau, A., Pizarro, O. (2005). Mean surface circulation and mesoscale turbulent flow characteristics in the eastern South Pacific from satellite tracked drifters. *Geophysical Research Letters* 32, L08605.

Champalbert, G. (1993). Plankton inhabiting the surface layer of the southern and southwestern lagoon of New Caledonia. *Marine Biology* 115, 223-228.

Chang, F. (1983). Winter phytoplankton and microzooplankton populations off the coast of Westland, New Zealand, 1979. *New Zealand Journal of Marine and Freshwater Research* 17, 279-304.

Chang, F.H., Zeldis, J., Gall, M., Hall, J. (2003). Seasonal and spatial variation of phytoplankton assemblages, biomass and cell size from spring to summer across the north-eastern New Zealand continental shelf. *Journal of Plankton Research* 25, 737-758.

Chavez, F.P. (2008). The northern Humboldt Current System: brief history, present status and a view towards the future. *Progress in Oceanography* 79, 95-105.

Chavez, F.P., Strutton, P.G., McPhaden, M.J. (1998). Biological-physical coupling in the central equatorial Pacific during the onset of the 1997-98 El Niño. *Geophysical Research Letters* 25, 3543-3546.

Chavez, F.P., Strutton, P.G., Friederich, G.E., Feely, R.A., Feldman, G.C., Foley, D.G., McPhaden M.J. (1999). Biological and chemical response of the equatorial Pacific Ocean to the 1997-98 El Niño. *Science* 286, 2126-2131.

Chin, A., Lison, T., Reytar, K., Planes, S., Gerhardt, K., Clua, E., Burke, L., Wilkinson, C. (2011). *Status of coral reefs of the Pacific and outlook: 2011*. Global Coral Reef Monitoring Network.

Christensen, J.H., Hewitson, B., Busuioc, A., Chen, A., Gao, X., Held, I., Jones, R., Kolli, R.K., Kwon, W.-T., Laprise, R., Magaña Rueda, V., Mearns, L., Menéndez, C.G., Räisänen, J., Rinke, A., Sarr, A., Whetton, P. (2007). Regional climate projections, In: Solomon, S., Qin, D., Manning, M., Chen, Z., Marquis, M., Averyt, K. B., Tignor, M., Miller, H. L. (eds.). *Climate Change 2007: The Physical Science Basis. Contribution of Working Group I to the Fourth Assessment Report of the Intergovernmental Panel on Climate Change*. Cambridge University Press, Cambridge, pp. 847-940.

Christensen, J.H., Krishna Kumar, K., Aldrian, E., An, S.-I., Cavalcanti, I.F.A., de Castro, M., Dong, W., Goswami, P. Hall, A., Kanyanga, J.K., Kitoh, A., Kossin, J., Lau, N.-C., Renwick, J., Stephenson, D.B., Xie S.-P., Zhou, T. (2013). Climate phenomena and their relevance for future regional climate change, In: *Stocker, T.F., Qin, D., Plattner, G.-K., Tignor, M., Allen, S.K., Boschung, J., Nauels, A., Xia, Y., Bex, V., Midgley, P.M. (eds.). Climate Change 2013: The Physical Science Basis. Contribution of Working Group I to the Fifth Assessment Report of the Intergovernmental Panel on Climate Change*. Cambridge University Press, Cambridge, pp 1217-1308.

Church, J.A., Clark, P.U., Cazenave, A., Gregory, J.M., Jevrejeva, S., Levermann, A., Merrifield, M.A., Milne, G.A., Nerem, R.S., Nunn, P.D., Payne, A.J., Pfeffer, W.T., Stammer, D., Unnikrishnan, A.S. (2013). Sea Level Change, In: *Stocker, T.F., Qin, D., Plattner, G.-K., Tignor, M., Allen, S.K., Boschung, J., Nauels, A., Xia, Y., Bex, V., Midgley, P.M. (eds.). Climate Change 2013: The Physical Science Basis. Contribution of Working Group I to the Fifth Assessment Report of the Intergovernmental Panel on Climate Change*. Cambridge University Press, Cambridge, pp 1137-1216.

Church, J.A., White, N.J. (2011). Sea-level rise from the late 19th to the early 21st century. *Surveys in Geophysics* 32, 585-602.

Ciais, P., Sabine, C., Bala, G., Bopp, L., Brovkin, V., Canadell, J., Chhabra, A., DeFries, R., Galloway, J., Heimann, M., Jones, C., Le Quéré, C., Myneni, R.B., Piao, S., Thornton, P. (2013). Carbon and Other Biogeochemical Cycles, In: *Stocker, T.F., D. Qin, G.-K. Plattner, M. Tignor, S.K. Allen, J. Boschung, A. Nauels, Y. Xia, V. Bex and P.M. Midgley (eds.). Climate Change 2013: The Physical Science Basis. Contribution of Working Group I to the Fifth Assessment Report of the Intergovernmental Panel on Climate Change*. Cambridge University Press, Cambridge, pp 465-570.

Coleman, D.W., Byrne, M., Davis, A.R. (2014). Molluscs on acid: gastropod shell repair and strength in acidifying oceans. *Marine Ecology Progress Series* 509, 203-211.

Connell, S.D., Irving, A.D. (2008). Integrating ecology with biogeography using landscape characteristics: a case study of subtidal habitat across continental Australia. *Journal of Biogeography* 35, 1608-1621.

Costanza, R., d'Arge, R., de Groot, R., Farber, S., Grasso, M., Hannon, B., Limburg, K., Naeem, S., O'Neill, R.V., Paruelo, J., Raskin, R.G., Sutton, P., van den Belt, M. (1997). The value of the world's ecosystem services and natural capital. *Nature* 387, 253-260

Cravatte, S., Delcroix, T., Zhang, D., McPhaden, M., Leloup, J. (2009). Observed freshening and warming of the western Pacific warm pool. *Climate Dynamics* 33, 565-589.

Cruz, M., Gabor, N., Mora, E., Jiménez, R., Mair, J. (2003). The known and unknown about marine biodiversity in Ecuador (continental and insular). *Gayana (Concepción)* 67, 232-260.

Cubillos, L.A., Ruiz, P., Claramunt, G., Gacitúa, S., Núñez, S., Castro, L.R. (2007). Spawning, daily egg production, and spawning stock biomass estimation for common sardine (*Strangomera bentincki*) and anchovy (*Engraulis ringens*) off central southern Chile in 2002. *Fisheries Research* 86, 228-240.

Curley, B.G., Kingsford, M.J., Gillanders, B.M. (2003). Spatial and habitat-related patterns of temperate reef fish assemblages: implications for the design of Marine Protected Areas. *Marine and Freshwater Research* 53, 1197-1210.

Currey, R.J.C., Dawson, S.M., Slooten, E. (2009). An approach for regional threat assessment under IUCN Red List criteria that is robust to uncertainty: the Fjordland bottlenose dolphins are critically endangered. *Biological Conservation* 142, 1570-1579.

Dalzell, P., Adams, T.J.H., Polunin, N.V.C. (1996). Coastal fisheries in the Pacific Islands. *Oceanography and Marine Biology: an Annual Review* 34, 395-531.

Dandonneau, Y., Deschamps, P.-Y., Nicolas, J.-M., Loiosel, H., Blanchot, J., Montel, Y., Thieuleux, F., Bécu, G. (2004). Seasonal and interannual variability of ocean color and composition of phytoplankton communities in the North Atlantic, equatorial Pacific and South Pacific. *Deep Sea Research* II 51, 303-318.

Dandonneau, Y., Gohin, F. (1984). Meridional and seasonal variations of the sea surface chlorophyll concentration in the southwestern tropical Pacific (14–32° S, 160–175° E). *Deep Sea Research* 31, 1377-1393.

Daneri, G., Dellarossa, V., Quiñones, R., Jacob, B., Montero, P., Ulloa, O. (2000). Primary production and community respiration in the Humboldt Current System off Chile and associated oceanic areas. *Marine Ecology Progress Series* 197, 41-49.

Davies, N., Harley, S., Hampton, J., McKechnie, S. (2014). Stock assessment of yellowfin tuna in the western and central Pacific Ocean. *Working paper WCPFC-SC10-2014/SA-WP-04 presented to the Western and Central Pacific Fisheries Commission Scientific Committee tenth regular session, 6 – 14 August 2014*, Majuro, Republic of the Marshall Islands.

Davies, N., Pilling, G., Harley, S., Hampton, J. (2013). Stock assessment of swordfish (Xiphias gladius) in the southwest Pacific Ocean. *Working paper WCPFC-SC9-2013/SA-WP-05 presented to the Western and Central Pacific Fisheries Commission Scientific Committee ninth regular session, 6 – 14 August 2013*, Pohnpei, Federated States of Micronesia.

Dayton, P.K., Tegner, M.J., Edwards, P.B., Riser, K.L. (1998). Sliding baselines, ghosts, and reduced expectations in kelp forest communities. *Ecological Applications* 8, 309-322.

De'ath, G., Fabricius, K.E., Sweatman, H., Puotinen, M. (2012). The 27-year decline of coral cover on the Great Barrier Reef and its causes. *Proceedings of the National Academy of Sciences* 109, 17995-17999.

De'ath, G., Lough, J.M., Fabricius, K.E. (2009). Declining coral calcification on the Great Barrier Reef. *Science* 323, 116-119.

De Martini, E.E., Roberts, D.A. (1990). Effects of giant kelp (Macrocystis) on the density and abundance of fishes in a cobble-bottom kelp forest. *Bulletin of Marine Science* 46, 287-300.

Department of Conservation. (2010). *New Zealand coastal policy statement 2010*. Department of Conservation, Wellington.

De Young C. (2007). Review of the state of world marine capture fisheries management: Pacific Ocean. *FAO Fisheries Technical Paper 488/1*. Food and Agriculture Organisation, Rome.

Dixson, D.L., Abrego, D., Hay, M.E. (2014). Chemically mediated behaviour of recruiting corals and fishes: a tipping point that may limit reef recovery. *Science* 345, 892-897.

Duarte, C.M., Middelburg, J., Caraco, N. (2005). Major role of marine vegetation on the oceanic carbon cycle. *Biogeosciences* 2, 1-8.

Duggan, S., McKinnon, A.D., Carleton, J.H. (2008). Zooplankton in an Australian tropical estuary. *Estuaries and Coasts* 31, 455-467.

Duke, N.C., Meynecke, J.-O., Dittmann, S., Ellison, A.M., Anger, K., Berger, U., Cannicci, S., Diele, K., Ewel, K.C., Field, C.D., Koedam, N., Lee, S.Y., Marchand, C., Nordhaus, I., Dahdouh-Guebas. (2007). A world without mangroves? *Science* 317, 41-42.

Dulvy, N.K., Freckleton, R.P., Polunin, N.V.C. (2004). Coral reef cascades and the indirect effects of predator removal by exploitation. *Ecology Letters* 7, 410-416.

Durack, P.J., Wjiffles, S.E. (2010). Fifty-year trends in global ocean salinities and their relationship to broad-scale warming. *Journal of Climate* 23, 4342-4362.

Dutton, P.H., Hitipeuw, C., Zein, M., Benson, S.R., Petro, G., Pita, J., Rei, V., Ambio, L, Bakarbessy, J. (2007). Status and genetic structure of nesting populations of leatherback turtles (*Dermochelys coriacea*) in the western Pacific. *Chelonian Conservation and Biology* 6, 47-53.

Edgar, G.J., Banks, S.A., Brandt, M., Bustamante, R.H., Chiriboga, A., Earle, S.A., Garske, L.E., Glynn, P.W., Grove, J.S., Henderson, S., Hickman, C.P., Miller, K.A., Rivera, F., Wellington, G.M. (2010). El Niño, grazers and fisheries interact to greatly elevate extinction risk for Galapagos marine species. *Global Change Biology* 16, 2876-2890.

Escribano, R., Fernández, M., Aranis, A. (2003). Physical-chemical processes and patterns of diversity of the Chilean eastern boundary pelagic and benthic marine ecosystems: an overview. *Gayana* 67, 190–205.

Escribano, R., Hidalgo, P., González, H., Giesecke, R., Riquelme-Bugueño, R., Manríquez, K. (2007). Seasonal and inter-annual variation of mesozooplankton in the coastal upwelling zone off central-southern Chile. *Progress in Oceanography* 75, 470-485.

Everitt, D.A., Wright, S.W., Volkman J.K., Thomas, D.P., Lindstrom, E.J. (1990). Phytoplankton community compositions in the western equatorial Pacific determined from chlorophyll and carotenoid pigment distributions. *Deep Sea Research* 37, 975-997.

Fabricus, K.E. (2005). Effects of terrestrial runoff on the ecology of corals and coral reefs: review and synthesis. *Marine Pollution Bulletin* 50, 125-146.

FAO. (2010). *National fisheries sector overview Peru. Fishery and Aquaculture Country Profile FID/CP/PER*. Food and Agriculture Organisation of the United Nations, Rome.

FAO. (2011). *Review of the state of world marine fisheries resources*. FAO Fisheries and Aquaculture Technical Paper 569. Food and Agriculture Organisation of the United Nations, Rome.

FAO. (2014). *The state of world fisheries and aquaculture*. Food and Agriculture Organisation of the United Nations, Rome.

Fernández-Álamo, M.A., Färber-Lorda, J. (2006). Zooplankton and the oceanography of the eastern tropical Pacific: a review. *Progress in Oceanography* 69, 318-359.

Fisher, R., Knowlton, N., Brainard, R.E., Caley, M.J. (2011). Differences among major taxa in the extent of ecological knowledge across four major ecosystems. *Plos One* 6, e26556.

Försterra, G., Beuck, L., Häussermann, V., Freiwald, A. (2005). Shallow water *Desmophyllum dianthus* (Scleractinia) from Chile: characteristics of the biocenoses, the bioeroding community, heterotrophic interactions and (palaeo)-bathymetrical implications, in *Freiwald, A., Roberts, J.M. (eds.). Cold-water corals and ecosystems*. Springer, Berlin, pp. 937-977.

Försterra, G., Häussermann, V., Laudien, J., Jantzen, C., Sellanes, J., Muñoz, P. (2014). Mass die off of the cold-water coral *Desmophyllum dianthus* in the Chilean Patagonian fjord region. *Bulletin of Marine Science* 90, 895-899.

Fréon, P., Bouchon, M., Mullon, C., García, C., Ñiquen, M. (2008). Interdecadal variability of anchoveta abundance and overcapacity of the fishery in Peru. *Progress in Oceanography* 79, 401-412.

Friedlander, A.M., Ballesteros, E., Beets, J., Berkenpas, E., Gaymer, C.F., Gorny, M., Sala, E. (2013). Effects of isolation and fishing on the marine ecosystems of Easter and Salas y Gómez Islands, Chile. *Aquatic Conservation: Marine and Freshwater Ecosystems* 23, 515-531.

Furnas, M.J., Mitchell, A.W. (1986). Phytoplankton dynamics in the central Great Barrier Reef—I. Seasonal changes in biomass and community structure and their relation to intrusive activity. *Continental Shelf Research* 6, 363-384.

Furnas, M.J., Mitchell, A.W., Gilmartin, M., Revelante, N. (1990). Phytoplankton biomass and primary production in semi-enclosed reef lagoons of the central Great Barrier Reef, Australia. *Coral Reefs* 9, 1-10.

Gabrie, C., You, H, Farget, P. (2007). L'état de l'environnement en Polynésie Française 2006. Ministère de l'Environnement.

Ganachaud, A., Sen Gupta, A., Brown, J.N., Evans, K., Maes, C., Muir, L.C. and Graham, F.S. (2013). Projected changes in the tropical Pacific Ocean of importance to tuna fisheries. *Climatic Change* 119, 163-179.

Ganachaud, A.S., Sen Gupta, A., Orr, J.C., Wijffels, S.E., Ridgway, K.R., Hemer, M.A., Maes, C., Steinberg, C.R., Tribollet, A.D., Qiu, B., Kruger, J.C. (2011). Observed and expected changes to the tropical Pacific Ocean, in: *Bell. J.D., Johnson, J.E., Hobday, A.J. (eds). Vulnerability of tropical Pacific fisheries and aquaculture to climate change*. Secretariat of the Pacific Community, Nouméa, pp 101–187.

Garcia, S.M., Zerbi, A., Aliaume, C., Do Chi, T., Lasserre, G. (2003). The ecosystem approach to fisheries. Issues, terminology, principles, institutional foundations, implementation and outlook. *FAO Fisheries Technical Paper 443*. Food and Agriculture Organisation of the United Nations, Rome.

Garrigue, C., Patenaude, N., Marsh, H. (2008). Distribution and abundance of the dugong in New Caledonia, southwest Pacific. *Marine Mammal Science* 24, 81-90.

Gaymer, C.F., Tapia, C., Acuña, E., Aburto, J., Cárcamo, P.F., Bodini, A., Stotz, W. (2013). Base de conocimiento y construcción de capacidades para el uso sustentable de los ecosistemas y recursos marinos de la ecorregión de Isla de Pascua. *Informe Final Proyecto SUBPESCA Licitación No 4728-33-LE12*.

GBRMPA. (2014). *Great Barrier Reef outlook report 2014*. Great Barrier Reef Marine Park Authority, Townsville.

Gerrodette, T. Forcada, J. (2005). Non-recovery of two spotted and spinner dolphin populations in the eastern tropical Pacific Ocean. *Marine Ecology Progress Series* 291, 1-21.

Gilbert, D., Rabalais, N. N., Diaz, R. J., Zhang, J. (2010). Evidence for greater oxygen decline rates in the coastal ocean than in the open ocean. *Biogeosciences* 7, 2283-2296.

Gillett, R. (2010). Marine fishery resources of the Pacific Islands. Food and Agriculture Organization of the United Nations, Rome.

Glynn, P.W., Ault, J.S. (2000). A biogeographic analysis and review of the far eastern Pacific coral reef region. *Coral Reefs* 19, 1-23.

Goiran, C., Shine, R. (2013). Decline in sea snake abundance on a protected coral reef system in the New Caledonian Lagoon. *Coral reefs* 32, 281-284.

Gonzalez, A., Marín, V.H. (1998). Distribution and life cycle of *Calanus chilensis* and *Centropages brachiatus* (Copepoda) in Chilean coastal waters: a GIS approach. *Marine Ecology Progress Series* 165, 109-117.

Greenfield, B., Hewitt, J., Hailes, S. (2013). *Manukau Harbour ecological monitoring programme: report on data collected up until February 2013*. Auckland Council technical report, TR2013/027. National Institute of Water and Atmospheric Research, Auckland.

Grenier, M., Cravatte, S., Blanke, B., Menkes, C., Koch-Larrouy, A., Durand, F., Melet, A., Jeandel, C. (2011). From the western boundary currents to the Pacific Equatorial Undercurrent: modeled pathways and water mass evolutions. *Journal of Geophysical Research: Oceans* 116, C12044.

Griffiths, F.B., Brandt, S.B. (1983). Mesopelagic Crustacea in and around a warm-core eddy in the Tasman Sea off eastern Australia. *Marine and Freshwater Research* 34, 609-623.

Griffiths, H.J., Barnes, D.K.A., Linse, K. (2009). Towards a generalized biogeography of the Southern Ocean benthos. *Journal of Biogeography* 36, 162-177.

Grindley, J.R. (1984). The zooplankton of mangrove estuaries, in: Por, F.D., Dor, I. (Eds.), *Hydrobiology of the Mangal*. Dr. W. Junk Publishers, The Hague, pp. 79-88.

Halford, A., Cheal, A.J., Ryan, D., Williams, D.McB. (2004). Resilience to large-scale disturbance in coral and fish assemblages on the Great Barrier Reef. *Ecology* 85, 1892-1905.

Hallegraeff, G.M. (1994). Species of the diatom genus *Pseudonitzschia* in Australian waters. *Botanica Marina* 37, 397-411.

Hallegraeff, G.M. (2010). Ocean climate change, phytoplankton community responses and harmful algal blooms: a formidable predictive challenge. *Journal of Phycology* 46, 220-235.

Hallegraeff, G.M., Jeffrey, S.W. (1993). Annually recurrent diatom blooms in spring along the New South Wales coast of Australia. *Australian Journal of Marine and Freshwater Research* 44, 325-34.

Halliday, J., Edhouse, S., Lohrer, D., Thrush, S., Cummings, V. (2013). *Mahurangi Estuary ecological monitoring programme: report on data collected from July 1994 to January 2013*. Auckland Council technical report, TR2013/038. National Institute for Water and Atmospheric Research, Auckland.

Hamner, W.M., Colin, P.L., Hamner, P.P. (2007). Export-import dynamics of zooplankton on a coral reef in Palau. *Marine Ecology Progress Series* 334, 83-92.

Harley, C.D.G., Hughes, A.R., Hultgren, K.M., Miner, B.G., Sorte, C.J.B., Thornber, C.S., Rodriguez, L.F., Tomanek, L., Williams, S.L. (2006). The impacts of climate change in coastal marine systems. *Ecology Letters* 9, 228-241.

Harriott, V.J., Banks, S.A. (2002). Latitudinal variation in coral communities in eastern Australia: a qualitative biophysical model of factors regulating coral reefs. *Coral Reefs* 21, 83-94.

Hartill, B., Bian, R., Rush, N., Armiger, H. (2013). *Aerial-access recreational harvest estimates for snapper, kahawai, red gurnard, tarakihi and trevally in FMA1 in 2011–12*. New Zealand Fisheries Assessment Report 2013/70. 44 p. Ministry for Primary Industries, Wellington.

Hearn, Alex. (2008). The rocky path to sustainable fisheries management and conservation in the Galápagos Marine Reserve. *Ocean & Coastal Management* 51 : 567-574.

Häussermann, V., Försterra, G. (eds.). (2009). *Marine Benthic Fauna of Chilean Patagonia*. Nature in Focus, Santiago, Chile.

Häussermann, V., Försterra, G., Melzer, R.R., Meyer, R. (2013). Gradual changes of benthic biodiversity in Comau Fjord, Chilean Patagonia – lateral observations over a decade of taxonomic research. *Spixiana* 36, 161-288

Hays, G.C., Richardson, A.J., Robinson, C. (2005). Climate change and marine plankton. *Trends in Ecology and Evolution* 20, 337-344.

Heap A.D., Harris P.T., Hinde A., Woods M. *(*2005). Benthic marine bioregionalisation of Australia's exclusive economic zone. *Report to the National Oceans Office on the development of a national benthic marine bioregionalisation in support of regional marine planning*. Geoscience Australia, Canberra.

Heileman, S., Guevara, R., Chavez, F., Bertrand, A., Soldi, H. (2009). Xvii-56 Humboldt Current LME, in *Sherman, K., Hempel, G. (eds.). The UNEP Large Marine Ecosystem Report, A perspective on changing conditions in LMEs of the world's regional seas. UNEP Regional Seas Report and Studies No. 182*. United Nations Environment Programme, Nairobi, pp 749-762.

Hewitt, C.L., Willing, J., Bauckham, A., Cassidy, A.M., Cox, C.M.S. Jones, L., Wotton, D.M. *(2004)*. New Zealand marine biosecurity: delivering outcomes in a fluid environment. New Zealand. *Journal of Marine and Freshwater Research* 38, 429-438.

Hidalgo, P., Escribano, R. (2001). Succession of pelagic copepod species in coastal waters off northern Chile: the influence of the 1997–98 El Niño. *Hydrobiologia* 453, 153-160.

Higgins, H.W., Mackey, D.J., Clementson, L. (2006). Phytoplankton distribution in the Bismarck Sea north of Papua New Guinea: the effect of the Sepik River outflow. *Deep Sea Research* I 53, 1845-1863.

Hill, A.E., Hickey, B.M., Shillington, F.A., Strub, P.T., Brink, K.H., Barton, E.D., Thomas, A.C. (1998). Eastern ocean boundaries, *in: Robinson, A.R., Brink, K.H. (Eds.), The Sea*. J. Wiley and Sons, Inc, New York, pp 29-68.

Hill, N.A., Lucieer, V., Barrett, N.S., Anderson, T.J., Williams, S.B. (2014). Filling the gaps: Predicting the distribution of temperate reef biota using high resolution biological and acoustic data. *Estuarine, Coastal and Shelf Science* 147, 137-147.

Hoegh-Guldberg, O., Andréfouët, S., Fabricus, K.E., Diaz-Pulido, G., Lough, J.M., Marshall, P.A., Pratchett, M.S. (2011). Pages 251–296 in *J.D. Bell, J.E. Johnson and A.J. Hobday (eds.). Vulnerability of tropical Pacific fisheries and aquaculture to climate change*. Secretariat of the Pacific Community, Nouméa.

Hoyle, S., Hampton, J., Davies, N. (2012). *Stock assessment of albacore tuna in the south Pacific Ocean*. Working paper WCPFC-SC8-2012/SA-WP-04-REV1 presented to the Western and Central Pacific Fisheries Commission Scientific Committee eighth regular session, 7 – 15 August 2012, Busan, Republic of Korea.

Hucke-Gaete, R., Osman, L.P., Moreno, C.A., Findlay, K.P., Ljungblad, D.K. (2004). Discovery of a blue whale feeding and nursery ground in southern Chile. *Proceedings of the Royal Society* B 271, S170-S173.

Hughes, T.P., Baird, A.H., Bellwood, D.R., Card, M., Connolly, S.R., Folke, C., Grosberg, R., Hoegh-Guldberg, O., Jackson, J.B.C., Kleypas, J., Lough, J.M., Marshall, P., Nyström, M., Palumbi, S.R., Pandolfi, J.M., Rosen, B., Roughgarden, J. (2003). Climate change, human impacts and the resilience of coral reefs. *Science* 301, 929-933.

Hughes, T.P., Bellwood, D.R., Folke, C., Steneck, R.S., Wilson, J. (2005). New paradigms for supporting the resilience of marine ecosystems. *Trends in Ecology and Evolution* 20, 380-386.

Hunt, T., Lipo, C. (2011). *Unravelling the mystery of Easter Island. The Statues that Walked*. Free Press/Simon and Schuster, New York.

IATTC. (2012). Meeting report. Inter-American Tropical Tuna Commission Scientific Meeting La Jolla, California, 15 – 18 May 2012.

IATTC. (2013). Meeting report. Inter-American Tropical Tuna Commission Scientific Meeting La Jolla, California 29 April – 3 May 2013.

IATTC. (2014). Meeting report. Inter-American Tropical Tuna Commission Scientific Meeting La Jolla, California, 12 – 16 May 2014.

IPCC. (2014). Summary for Policymakers, in: Edenhofer, O., Pichs-Madruga, R., Sokona, Y., Farahani, E., Kadner, S., Seyboth, K., Adler, A., Baum, I., Brunner, S., Eickemeier, P., Kriemann, B., Savolainen, J., Schlömer, S., von Stechow, C., Zwickel, T., Minx, J.C. (eds.). *Climate Change 2014, Mitigation of Climate Change. Contribution of Working Group III to the Fifth Assessment Report of the Intergovernmental Panel on Climate Change*. Cambridge University Press, Cambridge, pp 1-31.

Jacoby, C., Greenwood, J.G. (1988). Spatial, temporal, and behavioral patterns in emergence of zooplankton in the lagoon of Heron Reef, Great Barrier Reef, Australia. *Marine Biology* 97, 309-328.

Jacquet, S., Delesalle, B., Torréton, J.-P., Blanchot, J. (2006). Response of phytoplankton communities to increased anthropogenic influences (southwestern lagoon, New Caledonia). *Marine Ecology Progress Series* 320: 65-78.

Jeffrey S.W., Hallegraeff, G.M. (1980). Studies of phytoplankton species and photosynthetic pigments in a warm core eddy of the East Australian Current. *Marine Ecology Progress Series* 3, 285-294.

Jeffrey S.W., Hallegraeff, G.M. (1987). Phytoplankton pigments, species and light climate in a complex warm core eddy of the East Australian Current. *Deep Sea Research*, vol. 34.

Jillett, J.B. (1971). Zooplankton and hydrology of Hauraki Gulf. *New Zealand Department of Scientific and Industrial Research Bulletin 204*. New Zealand Department of Scientific and Industrial Research, Wellington.

Jillett, J.B. (1976). Zooplankton associations off Otago Peninsula, south-eastern New Zealand, related to different water masses. *New Zealand Journal of Marine and Freshwater Research* 10, 543-557.

Johnson, C.R., Banks, S.C., Barrett, N.S., Cazassus, F., Dunstan, P.K., Edgar, G.J., Frusher, S.D., Gardner, C., Haddon, M., Helidoniotis, F., Hill, K.L., Holbrook, N.J., Hosie, G.W., Last, P.R., Ling, S.D., Melbourne-Thomas, J., Miller, K., Pecl, G.T., Richardson, A.J. Ridgway, K.R., Rintoul, S.R., Ritz, D.A., Ross, D.J., Sanderson, J.C., Shepherd, S.A., Slotwinski, A., Swadling, K.A., Taw, N. (2011). Climate change cascades: Shifts in oceanography, species' ranges and subtidal marine community dynamics in eastern Tasmania. *Journal of Experimental Marine Biology and Ecology* 400, 17-32.

Keeling, R. F., Körtzinger, A., Gruber, N. (2010). Ocean deoxygenation in a warming world. *Annual Reviews in Marine Science* 2, 199-229.

Kennelly, S.J. (1987). Physical disturbances in an Australian kelp. *Marine Ecology Progress Series* 40, 145-153.

Kirkwood, R., Pemberton, D., Gales, R., Hoskins, A.J., Mitchell, T., Shaughnessy, P.D., Arnould, J.P.Y. (2010). Continued population recovery by Australian fur seals. *Marine and Freshwater Research* 61, 695-701.

Kluge, K. (1992). Seasonal Abundances of Zooplankton in Pala Lagoon. *DMWR Biological Report Series No. 36*. Department of Marine and Wildlife Resources, Pago Pago.

Kulbicki M., (1997). Bilan de 10 ans de recherche (1985-1995) par l'ORSTOM sur la structure des communautés des poissons lagonaires et récifaux en Nouvelle-Calédonie. *Cybium 21 Suppl.*, 47-79

Kulbicki, M., Sarramégna, S., Letourneur, Y., Wantiez, L., Galzin, R., Mou-Tham, G.,

Chauvet, C., Thollot, P. (2007). Opening of an MPA to fishing: natural variations in the structure of a coral reef fish assemblage obscure changes due to fishing. *Journal of Experimental Marine Biology and Ecology* 353, 145-163.

Last, P.R., White, W.T., Gledhill, D.C., Hobday, A.J., Brown, R., Edgar, G.J., Pecl, G. (2011). Long-term shifts in abundance and distribution of a temperate fish fauna: a response to climate change and fishing practices. *Global Ecology and Biogeography* 20, 58-72.

Le Borgne, R., Allain, V., Griffiths, S.P., Matear, R.J., McKinnon, A.D., Richardson, A.J., Young, J.W. (2011). Vulnerability of open ocean food webs in the tropical Pacific to climate change, in: *Bell, J.D., Johnson, J.E., Hobday, A.J. (eds.). Vulnerability of tropical Pacific fisheries and aquaculture to climate change.* Secretariat of the Pacific Community, Noumea, pp 189–249.

Le Borgne, R., Rodier, M., Le Bouteiller, A., Kulbicki, M. (1997). Plankton biomass and production in an open atoll lagoon: Uvea, New Caledonia. *Journal of Experimental Marine Biology and Ecology* 212, 187-210.

Levin, P.S., Fogarty, M.J., Murawski, S.A., Fluharty, D. (2009). Integrated ecosystem assessments: developing the scientific basis for ecosystem-based management of the ocean. *Plos Biology* 7, e1000014.

Lewison, R.L., Browder, L.B., Wallace, B.P., Moore, J.E., Cox, T., Zydelis, R., McDonald, S., DiMatteo, A., Dunn, D.C., Kot, C.Y., Bjorkland, R., Kelez, S., Soykan, C., Stewart, K.R., Sims, M., Boustany, A., Read, A.J., Halpin, P., Nichols, W.J., Safina, C. (2014). Global patterns of marine mammal, seabird, and sea turtle by-catch reveal taxa-specific and cumulative megafauna hotspots. *Proceedings of the National Academy of Sciences* 111, 5271-5276.

Lynch, T.P., Morello, E.B., Evans, K., Richardson, A.J., Rochester, W., Steinberg, C.R., Roughan, M., Thompson, P., Middleton, J.F., Feng, M., Sherrington, R., Brado, V., Tilbrook, B., Ridgway, K., Allen, S., Doherty, P., Hill, K., Moltmann, T.C. (2014). IMOS National Reference Stations: a continental scaled physical, chemical and biological coastal observing system. *PloS One*, doi: 10.1371/journal.pone.0113652.

Mace, P.M., Sullivan, K.J., Cryer, M. (2014). The evolution of New Zealand's fisheries science and management systems under ITQs. *ICES Journal of Marine Science* 71, 204-215.

MacNeil, M.A., Graham, N.A.J., Cinner, J.E., Wilson, S.K., Williams, I.D., Maina, J., Newman, S., Friedlander, A.M., Jupiter, S., Polumim, N.V.C., McClanahan, T.R. (2015). Recovery potential of the world's coral reef fishes. *Nature*, doi:10.1038/nature14358.

Magera, A.M., Mills Flemming, J.E., Kaschner, K., Christensen, L.B., Lotze, H.K. (2013). Recovery trends in marine mammal populations. *Plos One* 8, e77908.

Majluf, P., Babcock, E.A., Riveros, J.C., Schreiber, M.A., Alderete, W. (2002). Catch and by-catch of sea birds and marine mammals in the small-scale fishery of Punta San Juan, Peru. *Conservation Biology* 16, 1333-1343.

Maragos, J.E. (1993). Impact of coastal construction on coral reefs in the U.S.-affiliated Pacific islands. *Coastal Management* 21, 235-269.

Maragos, J.E., Cook, C.W. (1995). The 1991-1992 rapid ecological assessment of Palau's coral reefs. *Coral Reefs* 14: 237-252.

Marsh, H., Eros, C., Corkeron, P., Breen, B. (1999). A conservation strategy for dugongs: implications of Australian research. *Marine and Freshwater Research* 50, 979-990.

Marsh, H., Harris, A.N.M., Lawler, I.R. (1997). The sustainability of the indigenous dugong fishery in Torres Strait, Australia/Papua New Guinea. *Conservation Biology* 11, 1375-1386.

Marsh, H., Penrose, H., Eros, C., Hugues, J. (2002). Dugong status report and action plans for countries and territories. Early warning and assessment report series UNEP/DEWA/RS.02-1. United Nations Environment Programme, Nairobi.

Marsh, H., Rathbun, G.B., O'Shea, T.J., Preen, A.R. (1995). Can dugongs survive in Palau? *Biological Conservation* 72, 85-89.

Martinez-Porchas, M., Martinez-Cordova, L.R. (2012). World aquaculture: environmental impacts and troubleshooting alternatives. *The Scientific World Journal* 2012, 389623.

Matear, R.J., Chamberlain, M.A., Sun, C., Feng, M. (2015). Climate change projection for the western tropical Pacific Ocean using a high resolution ocean model: implications for tuna fisheries. *Deep Sea Research II* 113, 22-46.

Mayr, C., Rebolledo, L., Schulte, K., Schuster, A., Zolitschka, B., Försterra, G., Häussermann, V. (2014). Responses of nitrogen and carbon deposition rates in Comau Fjord (42°S, southern Chile) to natural and anthropogenic impacts during the last century. *Continental Shelf Research* 78, 29-38.

McIntyre, M. (2005). Pacific environment outlook. Special edition for the Mauritius international meeting for the 10-yr review of the Barbados programme of action for the sustainable development of small island developing states. South Pacific Regional Environment Programme and United Nations Environment Programme.

McKinnon, A.D., Klumpp, D.W. (1998). Mangrove zooplankton of North Queensland, Australia II. Copepod egg production and diet. *Hydrobiologia* 362, 145-160.

McKinnon, A.D., Thorrold, S.R. (1993). Zooplankton community structure and copepod egg production in coastal waters of the central Great Barrier Reef lagoon. *Journal of Plankton Research* 15, 1387-1411.

McWilliam, P.S., Phillips, B.F. (1983). Phyllosoma larvae and other crustacean macrozooplankton associated with eddy J, a warm-core eddy off south-eastern Australia. *Australian Journal of Marine and Freshwater Research* 34, 653-663.

Menkes, C., Allain, V, Rodier, M., Gallois, F., Lebourges-Dhaussy, A., Hunt, B.P.V., Smeti, H., Pagano, M., Josse, E., Daroux, A., Lehodey, P., Senina, I., Kestenare, E., Lorrain, A., Nicol, S. (2015). Seasonal oceanography from physics to micronekton in the south-west Pacific. *Deep Sea Research II* 113, 125-144.

Merrifield, M.A., Thompson, P.R., Lander, M. (2012). Multidecadal sea level anomalies and trends in the western tropical Pacific. *Geophysical Research Letters* 39, L13602.

Messié, M., Radenac, M.-H. (2006). Seasonal variability of the surface chlorophyll in the western tropical Pacific from SeaWiFS data. *Deep Sea Research I* 53, 1581-1600.

Messié, M., Radenac, M.-H., Lefévre, J., Marchiesiello, P. (2006). Chlorophyll bloom in the western Pacific at the end of the 1997–1998 El Niño: the role of the Kiribati Islands. *Geophysical Research Letters: Oceans* 33, L14601.

Miloslavich, P., Klein, E., Díaz, J.M., Hernández, C.E., Bigatti, G., Campos, L., Artigas, F., Castillo, J., Penchszadeh, P.E., Neill, P.E., Carranza, A., Retana, M.V., Díaz de Astarloa, J.M., Lewis, M., Yorio, P., Piriz, M., Rodríguez, D., Yoneshigue-Valentin, Y., Gamboa, L., Martín, A. (2011). Marine biodiversity in the Atlantic and Pacific coasts of South America: knowledge and gaps. *Plos One* 6, e14631.

Milton, D.A. (2001). Assessing the susceptibility to fishing of populations of rare trawl by-catch: sea snakes caught by Australia's Northern Prawn Fishery. *Biological Conservation* 101, 281-290.

Ministry of Environment Conservation and Meteorology. (2008). Solomon Islands state of environment report 2008. Ministry of Environment Conservation and Meteorology.

Ministry of Environment Lands and Agricultural Development. (2004). State of the environment report 2000-2002. Government of the Republic of Kiribati.

Ministry of Natural Resources and Environment. (2013). Samoa's state of the environment (SOE) report 2013. Government of Samoa.

Ministry for the Environment. (2007). Environment New Zealand 2007. Ministry for the Environment, Wellington.

Moberg, F., Folke, C. (1999). Ecological goods and services of coral reef ecosystems. *Ecological Economics* 29, 215-233.

Montecino, V., Lange, C.B. (2009). The Humboldt Current System: Ecosystem components and processes, fisheries, and sediment studies. *Progress in Oceanography* 83, 65-79.

Morales, C.E., Blanco, J.L., Braun, M., Reyes, H., Silva, N. (1996). Chlorophyll-a distribution and associated oceanographic conditions in the upwelling region off northern Chile during the winter and spring 1993. *Deep Sea Research I* 43, 267-289.

Morales, C.E., González, H.E., Hormazabal, S.E., Yuras, G., Letelier, J., Castro, L.R. (2007). The distribution of chlorophyll-a and dominant planktonic components in the coastal transition zone off Concepción, central Chile, during different oceanographic conditions. *Progress in Oceanography* 75, 452-469.

Mourgues, A. (2005). Vanuatu environment profile. Available at www.sprep.org/Vanuatu/country-reports.

MPI. (2013). The status of New Zealand's fisheries 2013. Ministry of Primary Industries. Available at http://fs.fish.govt.nz/Page.aspx?pk=16&tk=478.

Munday, P.L., Dixson, D.L., Donelson, J.M., Jones, G.P., Pratchett, M.S., Devitsina, G.V., Døving, K.B. (2009). Ocean acidification impairs olfactory discrimination and homing ability of a marine fish. *Proceedings of the National Academy of Sciences* 106, 1848-1852.

Munday, P.L., Dixson, D.L., McCormick, M.I., Meekan, M., Ferrari, M.C.O., Chivers, D.P. (2010). Replenishment of fish populations is threatened by ocean acidification. *Proceedings of the National Academy of Sciences* 107, 12930-12934.

Murphy, R.J., Pinkerton, M.H., Richardson, K.M., Bradford-Grieve, J.M., Boyd, P.W. (2001). Phytoplankton distributions around New Zealand derived from SeaWiFS remotely-sensed ocean colour data. *New Zealand Journal of Marine and Freshwater Research* 35, 343-362.

National Geographic, Oceana. (2013). Islas desventuradas: Biodiversidad Marina y Propuesta de Conservación. Informe de la Expedición "Pristine Seas".

National Geographic, Oceana, Armada de Chile. (2011). Expedición a Isla de Pascua y Salas y Gómez. Febrero-Marzo 2011. Informe Científico.

Natural Resource Management Ministerial Council. (2006). National cooperative approach to integrated coastal zone management. Framework and implementation plan. Australian Government Department of the Environment and Heritage, Canberra.

Newton, K., Côté, I.M., Pilling, G.M., Jennings, S., Dulvy, N.K. (2007). Current and future sustainability of island coral reef fisheries. *Current Biology* 17, 655-658.

Niquil, N., Jackson, G.A., Legendre, L., Delesalle, B. (1998). Inverse model analysis of the planktonic food web of Takapoto Atoll (French Polynesia). *Marine Ecology Progress Series* 165, 17-29.

Noble, I.R., Huq, S., Anokhin, Y.A., Carmin, J., Goudou, D., Lansigan, F.P., Osman-Elasha, B., Villamizar, A. (2014). Adaptation needs and options, in: Field, C.B., Barros, V.R., Dokken, D.J., Mach, K.J., Mastrandrea, M.D., Bilir, T.E., Chatterjee, M., Ebi, K.L., Estrada, Y.O., Genova, R.C., Girma, B., Kissel, E.S., Levy, A.N., MacCracken, S., Mastrandrea, P.R., White, L.L. (eds.). *Climate Change 2014: Impacts, Adaptation, and Vulnerability. Part A: Global and Sectoral Aspects. Contribution of Working Group II to the Fifth Assessment Report of the Intergovernmental Panel on Climate Change*. Cambridge University Press, Cambridge.

Okey, T.A., Banks, S., Birn, A.F., Bustamante, R.H., Calvopiña, M., Edgar, G.J., Espinoza, E., Fariña, J.M., Garske, L.E., Reck, G.K., Salazar, S., Shepherd, S., Toral-Granda, V., Wallem, P. (2004). A trophic model of a Galápagos subtidal rocky reef for evaluating fisheries and conservation strategies. *Ecological Modelling* 172, 383-401.

Olavarría, C., Baker, C.S., Garrigue, C., Poole, M., Hauser, N., Caballero, S., Flórez-González, L., Brasseur, M., Bennister, J., Capella, J., Clapham, P., Dodemont, R., Donoghue, M., Jenner, C., Jenner, M.-N., Moro, D., Oremus, M., Paton, D., Rosenbaum, H., Russell, K. (2007). Population structure of South Pacific humpback whales and the origin of the eastern Polynesian breeding grounds. *Marine Ecology Progress Series* 330, 257-268.

Orr, J.C. (2011). Recent and future changes in ocean carbonate chemistry, in: *Gattyso J.-P., Hansson, L. (eds.). Ocean acidification*. Oxford University Press, Oxford, pp: 41-66.

Orth, R.J., Carruthers, T.J.B., Dennison, W.C., Duarte, C.M., Fourqurean, J.W., Heck, K.L., Hughes, A.R., Kendrick, G.A., Kenworthy, W.J., Olyarnik, S., Short, F.T., Waycott, M., Williams, S.L. (2006). A global crisis for seagrass ecosystems. *BioScience* 56, 987-996.

Parra, G.J., Corkeron, P.J., Marsh, H. (2006). Population sizes, site fidelity and residence patterns of Australian snubfin and Indo-Pacific humpback dolphins: implications for conservation. *Biological conservation* 129, 167-180.

Philippart, C.J.M., van Aken, H.M., Beukema, J.J., Bos, O.G., Cadée, G.C., Dekker, R. (2003). Climate-related changes in recruitment of the bivalve Macoma balthica. *Limnology and Oceanography* 48, 2171-2185.

Picaut, J., Ioualalen, M., Menkes, C., Delcroix, T., McPhaden, M.J. (1996). Mechanism of the Zonal Displacements of the Pacific Warm Pool: Implications for ENSO. *Science* 274, 1486-1489.

Pinca, S., Kronen, M., Friedman, K., Magron, F., Chapman, L., Tardy, E., Pakoa, K., Awira, R., Boblin, P., Lasi, F. (2009). *Pacific Regional Oceanic and Coastal Fisheries Development Programme Regional Assessment Report: profiles and results from survey work at 63 sites across 17 Pacific Island Countries and Territories*. Secretariat of the Pacific Community, Nouméa.

Poloczanska, E.S., Babcock, R.C., Butler, A, Hobday, A.J., Hoegh-Guldberg, O., Kunz, T.J., Matear, R. Milton, D.A., Okey, T.A., Richardson, A.J. (2007). Climate change and Australian marine life. *Oceanography and Marine Biology: an Annual Review* 45: 407-478.

Pratchett M.R., Munday, P.L., Graham, N.A.J., Kronen, M., Pinca, S., Friedman, K., Brewer, T.D., Bell, J.D., Wilson, S.K., Cinner, J.E., Kinch, J.P., Lawton, R.J., Williams, A.J., Chapman, L.J., Magron, F. Webb, A. (2011). Vulnerability of coastal fisheries in the tropical Pacific to climate change, in: *Bell, J.D., Johnson, J.E., Hobday, A.J. (eds.). Vulnerability of tropical Pacific fisheries and aquaculture to climate change*. Secretariat of the Pacific Community, Nouméa, pp 493-576.

Primavera, J.H. (1997). Socio-economic impacts of shrimp culture. *Aquaculture Research* 28, 815-827.

Purcell, S.W., Mercier, A., Conand, C., Hamel, J.-F., Toral-Granda, M.V., Lovatelli, A., Uthicke, S. (2013). Sea cucumber fisheries: global analysis of stocks, management measures and drivers of overfishing. *Fish and Fisheries* 14, 34-59.

Rabalais, N.N., Diaz, R.J., Levin, L.A., Turner, R.E., Gilbert, D., Zhang, J. (2010). Dynamics and distribution of natural and human-caused hypoxia. *Biogeosciences* 7, 585-619.

Radenac, M.-H., Menkes, C., Vialard, J., Moulin, C., Dandonneau, Y., Delcroix, T., Dupouy, C., Stoens, A., and Deschamps, P.-Y. (2001). Modeled and observed impacts of the 1997-1998 El Niño on nitrate and new production in the equatorial Pacific. *Journal of Geophysical Research* 106, 26879-26898.

Randall, D.A., Wood, R.A., Bony, S., Colman, R., Fichefet, T., Fyfe, J., Kattsov, V., Pitman, A., Shukla, J., Srinivasan, J., Stouffer, R.J., Sumi A., Taylor, K.E. (2007). Climate models and their evaluation, in: *Solomon, S., Qin, D., Manning, M., Chen, Z., Marquis, M., Averyt, K.B., Tignor M., Miller, H.L. (eds.). Climate Change 2007: The Physical Science Basis. Contribution of Working Group I to the Fourth Assessment Report of the Intergovernmental Panel on Climate Change*. Cambridge University Press, Cambridge, pp 589-662.

Reeves, R.R., McClellan, K., Werner, T.B. (2013). Marine mammal by-catch in gillnet and other entangling net fisheries, 1990 to 2011. *Endangered Species Research* 20, 71-97.

Reid, J.L. (1997). On the total geostrophic circulation of the Pacific Ocean: flow patterns, tracers, and transports. *Progress in Oceanography* 39, 263-352.

Revelante, N., Williams, W.T., Bunt, J.S. (1982). Temporal and spatial distribution of diatoms, dinoflagellates and Trichodesmium in waters of the Great Barrier Reef. *Journal of Experimental Marine Biology and Ecology* 63, 27-45.

Rhein, M., Rintoul, S.R., Aoki, S., Campos, E., Chambers, D., Feely, R.A., Gulev, S., Johnson, G.C., Josey, S.A., Kostianoy, A., Mauritzen, C., Roemmich, D., Talley, L.D., Wang, F. (2013). Observations: Ocean, in: *Stocker, T.F., Qin, D., Plattner, G.-K., Tignor, M., Allen, S.K., Boschung, J., Nauels, A., Xia, Y., Bex, V., Midgley, P.M. (eds.). Climate Change 2013: The Physical Science Basis. Contribution of Working Group I to the Fifth Assessment Report of the Intergovernmental Panel on Climate Change*. Cambridge University Press, Cambridge, pp 255-315.

Rice, J., Harley, S. (2012). *Stock assessment of oceanic white tip sharks in the western and central Pacific Ocean*. Working paper WCPFC-SC8-2012/SA-WP-03 presented to the Western and Central Pacific Fisheries Commission Scientific Committee ninth regular session, 7-15 August 2012, Busan, Republic of Korea.

Rice, J., Harley, S. (2013). *Updated stock assessment of silky shark in the western and central Pacific Ocean*. Working paper WCPFC-SC9-2013/SA-WP-06 presented to the Western and Central Pacific Fisheries Commission Scientific Committee eighth regular session, 6-14 August 2013, Pohnpei, Federated States of Micronesia.

Rice, J., Harley, S., Davies, N., Hampton, J. (2014). *Stock assessment of skipjack tuna in the western and central Pacific Ocean*. Working paper WCPFC-SC10-2014/SA-WP-05 presented to the Western and Central Pacific Fisheries Commission Scientific Committee tenth regular session, 6-14 August 2014, Majuro, Republic of the Marshall Islands.

Ridgway, K.R. (2007). Seasonal circulation around Tasmania: an interface between eastern and western boundary dynamics. *Journal of Geophysical Research* 112, C10016.

Ridgway, K.R., Dunn, J.R. (2003). Mesoscale structure of the mean East Australian Current System and its relationship with topography. *Progress in Oceanography* 56, 189-222.

Robertson, A.I., Blaber, S.J.M. 1993. Plankton, epibenthos and fish communities. *Coastal and Estuarine Studies* 41, 173-224.

Robertson, B.C., Chilvers, B.L. (2011). The population decline of the New Zealand sea lion Phocarctos hookeri: a review of possible causes. *Mammal Review* 41, 253-275.

Robertson, A.I., Dixon, P., Daniel, P.A. (1988). Zooplankton dynamics in mangrove and other nearshore habitats in tropical Australia. *Marine Ecology Progress Series* 43, 139-150.

Robertson, A.I., Howard, R.K. (1978). Diel trophic interactions between vertically-migrating zooplankton and their fish predators in an eelgrass community. *Marine Biology* 48, 207-213.

Rodier, M., Le Borgne, R. (2008). Population dynamics and environmental conditions affecting *Trichodesmium* spp. (filamentous cyanobacteria) blooms in the south-west lagoon of New Caledonia. *Journal of Experimental Marine Biology and Ecology* 358, 20-32.

Roemmich, D., Gilson, J., Davis, R., Sutton, P., Wijffels, S., Riser, S. (2007). Decadal spinup of the South Pacific subtropical gyre. *Journal of Physical Oceanography* 37, 162-173.

Roman M.R., Furnas, M.J., Mullin, M.M. (1990). Zooplankton abundance and grazing at Davies Reef, Great Barrier Reef, Australia. *Marine Biology* 105, 73 – 82.

Rousseaux, C.S., Gregg, W.W. (2012). Climate variability and phytoplankton composition in the Pacific Ocean. *Journal of Geophysical Research: Oceans* 117, C10006.

Ruiz, G.M. Crooks, J.A. (2001). Biological invasions of marine ecosystems: patterns, effects, and management, in: *Bendell-Yound, L., Gallagher, P. (eds.). Waters in Peril*. Kluwer Academic Publications, Berlin, pp. 1-17.

Ruttenberg, B.I. (2001). Effects of artisanal fishing on marine communities in the Galapagos Islands. *Conservation Biology* 15, 1691-1699.

Ryan, J.P., Ueki, I., Chao, Y., Zhang, H., Polito, P.S., Chavez, F.P. (2006). Western Pacific modulation of large phytoplankton blooms in the central and eastern equatorial Pacific. *Journal of Geophysical Research: Biogeosciences* 111, G02013.

Saisho, T. (1985). Notes on the plankton community in the habitat of nautilus off the southeast coast of Viti Levu, Fiji. *Kagoshima University Research Center of the South Pacific Occasional Papers* 4, 80-83.

Sakuma, B. (2004). *Status of the environment in the Republic of Palau*. Palau Conservation Society.

Secades, C., O'Connor, B., Brown, C., Walpole, M. (2014). Earth observation for biodiversity monitoring: a review of current approaches and future opportunities for tracking progress towards the Aichi Biodiversity Targets. *CBD Technical Series 72*. Secretariat of the Convention on Biological Diversity, Montreal.

Secretariat of the Convention on Biological Diversity. (2010). *Global Biodiversity Outlook 3*. Secretariat of the Convention on Biological Diversity, Montreal.

Sen Gupta, A., Brown, J.N., Jourdain, N.C., van Sebille, E., Ganachaud, A., Vergés, A. (2015). Episodic and non-uniform shifts of thermal habitats in a warming ocean. *Deep Sea Research* II 113, 59-72.

Sen Gupta, A., Santoso, A., Taschetto, A.S., Ummenhofer, C.C., Trevena, J., England, M.H., (2009). Projected changes to the Southern Hemisphere ocean and sea ice in the IPCC AR4 climate models. *Journal of Climate* 22, 3047-3078.

Shaffer, G., Hormazabal, S, Pizarro O., Salinas, S. (1999). Seasonal and interannual variability of currents and temperature off central Chile. *Journal of Geophysical Research* 104, 29951-29961.

Shears, N.T., Babcock, R.C. (2003). Continuing trophic cascade effects after 25 years of no-take marine reserve protection. *Marine Ecology. Progress Series* 246, 1-16.

Sherman, K., Alexander, L.M. (editors). (1986). *Variability and management of large marine ecosystems*. American Association for the Advancement of Science Selected Symposia 99, Westview Press, Boulder.

Sherman, K., Hempel, G. (editors). (2009). *The UNEP Large Marine Ecosystem Report: a perspective on changing conditions in LMEs of the world's Regional Seas*. United Nations Environment Programme, Nairobi.

Smale, D.A., Wernberg, T. (2013). Extreme climatic event drives range contraction of a habitat-forming species. *Proceedings of the Royal Society* B 280, 20122829.

Smith, A.D.M., Fulton, E.J., Hobday, A.J., Smith, D.C., Shoulder, P. (2007). Scientific tools to support the practical implementation of ecosystem-based fisheries management. *ICES Journal of Marine Science* 64, 633-639.

Snelgrove, P.V.R. (1999). Getting to the bottom of marine biodiversity: sedimentary habitats. *BioScience* 49, 129-138.

Sorokin, Y. I. (1993) (editor). Coral Reef Ecology. *Ecological Studies 102*. Springer

Verlag, Berlin.

Sorokin, Y. I. (1990). Aspects of trophic relations, productivity and energy balance in reef ecosystems, in: Dubinsky, Z. (ed.). *Ecosystems of the World 25: Coral Reefs*. Elsevier, New York, pp. 401-410.

Sorte, C.J.B., Williams, S.L., Carlton, J.T. (2010). Marine range shifts and species introductions: comparative spread rates and community impacts. *Global Ecology and Biogeography* 19, 303-316.

Spalding, M.D., Agostini, V.N., Rice, J., Grant, S.M. (2012). Pelagic provinces of the world: A biogeographic classification of the world's surface pelagic waters. *Ocean and Coastal Management* 60, 19-30.

SPREP. (2012). *Pacific environment and climate change outlook*. Secretariat of the Pacific Regional Environment Programme, Apia.

State of the Environment Committee. (2011). *Australia state of the environment 2011*. Independent report to the Australian Government Minister for Sustainability, Environment, Water, Population and Communities. Department of Sustainability, Environment, Water, Population and Communities, Canberra.

Steinacher, M., Joos, F., Frölicher, T.L., Plattner, G.-K., Doney, S.C. (2009). Imminent ocean acidification in the Arctic projected with the NCAR global coupled carbon cycle-climate model. *Biogeosciences* 6, 515-533.

Stramma, L., Johnson, G.C., Sprintall, J., Mohrholz, V., (2008). Expanding oxygen-minimum zones in the tropical oceans. *Science* 320, 655-658.

Strub, P.T., Mesías, J., Montecino, V., Rutllant, J., Salinas, S. (1998). Coastal ocean circulation off western South America, in: Robinson, A.R., Brink, K.H. (Eds.), *The Sea*. J. Wiley & Sons, Inc., New York, pp 273-313.

Strutton, P.G., Ryan, J.P., Chavez, F.P. (2001). Enhanced chlorophyll associated with tropical instability waves in the equatorial Pacific. *Geophysical Research Letters* 28, 2005-2008.

Suchanek, T.H., Williams, S.W., Ogden, J.C., Hubbard, D.K., Gill, I.P. (1985). Utilization of shallow-water seagrass detritus by Caribbean deep-sea macrofauna: δ 13C evidence. *Deep Sea Research* 32, 2201-2214.

Tarazona, J., Gutiérrez, D., Paredes, C., Indacochea, A. (2003). Overview and challenges of marine biodiversity research in Peru. *Gayana* 67, 206-231

Taylor, S., Webley, J., McInnes, K. (2012). *2010 statewide recreational fishing survey*. Queensland Department of Agriculture, Fisheries and Forestry, Brisbane.

Thiel, M., Macaya, E.C., Acuna, E., Arntz, W.E., Bastias, H., Brokordt, K., Camus, P.A., Castilla, J.C., Castro, L.R., Cortés, M., Dumont, C.P., Escribano, R., Fernandez, M., Gajardo, J.A., Gaymer, C.F., Gomez, I., González, A.E., González, H.E., Haye, P.A., Illanes, J.-E., Iriarte, J.L., Lancellotti, D.A., Luna-Jorquera, G., Luxoro, C., Manriquez, P.H., Perez, E., Marín, V., Muñoz, P., Navarrete, S.A., Perez, E., Poulin, E., Sellanes, J., Sepúlveda, H.H., Stotz, W., Tala, F., Thomas, A., Vargas, C.A., Vasquez, J.A., Alonso Vega, J.M. (2007). The Humboldt Current System of northern and central Chile: oceanographic processes, ecological interactions and socioeconomic feedback. *Oceanography and Marine Biology* 45, 195-344.

Thompson, P.A., Baird, M.E., Ingleton, T., Doblin, M.A. (2009). Long-term changes in temperate Australian coastal waters: implications for phytoplankton. *Marine Ecology Progress Series* 394, 1-19.

Thompson, P.A., Bonham, P., Waite, A.M., Clementson, L.A., Cherukuru, N., Hassler, C., Doblin, M.A. (2011). Contrasting oceanographic conditions and phytoplankton communities on the east and west coasts of Australia. *Deep Sea Research II* 58, 645-663.

Thorbjarnarson, J. (1999). Crocodile tears and skins: international trade, economic constraints, and limits to the sustainable use of crocodilians. *Conservation Biology* 13, 465-470.

Tisdell, C., Swarna Nantha, H. (2005). Management, conservation and farming of saltwater crocodiles: an Australian case study of sustainable commercial use. *Working paper no. 126. Working papers on economics, ecology and the environment*. University of Queensland, St. Lucia.

Tittensor, D.P., Mora, C., Jetz, W., Lotze, H.K., Ricard, D., Vanden Berghe, E., Worm, B. (2010). Global patterns and predictors of marine biodiversity across taxa. *Nature* 466, 1098-1101.

Tranter, D.J. (1962). Zooplankton abundance in Australasian waters. *Australian Journal of Marine and Freshwater Research* 13, 106-142.

Tranter, D. J., Leech, G.S., Airey, D. (1983). Edge enrichment in an ocean eddy. *Australian Journal of Marine and Freshwater Research* 34, 665-80.

Turk, D., McPhaden, M.J., Busalacchi, A.J., Lewis, M.R. (2001). Remotely sensed biological production in the equatorial Pacific. *Science* 293, 471-474.

Ueki, I., Kashino, Y., Kuroda, Y. (2003). Observation of current variations off the New Guinea coast including the 1997–1998 El Niño period and their relationship with Sverdrup transport. *Journal of Geophysical Research: Oceans* 108, 3243.

Underwood, A.J., Kingsford, M.J., Andrew, N.L. (1991). Patterns in shallow subtidal marine assemblages along the coast of New South Wales. *Australian Journal of Ecology* 16, 231-249.

UNEP. (2006a). *Challenges to international waters – regional assessments in a global perspective*. United Nations Environment Programme, Nairobi.

UNEP. (2006b). *Marine and coastal ecosystems and human wellbeing: a synthesis report based on the findings of the Millenium Ecosystem Assessment*. United Nations Environment Programme, Nairobi.

UNEP. (2012). Global *Environment Outlook 5. Environment for the future we want*. United Nations Environment Programme, Nairobi.

UNESCAP. (2005). *The state of the environment in Asia and the Pacific*. United Nations Economic and Social Commission for Asia and the Pacific, Bangkok.

UNESCO. (2011). *Lagoons of New Caledonia: Reef Diversity and Associated Ecosystems*. United Nations Educational, Scientific and Cultural Organization, Paris.

UNESCO. (2012). State *of Conservation (SOC) Phoenix Islands Protected Area*. United Nations Educational, Scientific and Cultural Organization, Paris.

UNESCO. (2014a). State *of Conservation (SOC) Great Barrier Reef*. United Nations Educational, Scientific and Cultural Organization, Paris.

UNESCO 2014b). *State of Conservation (SOC) East Rennell*. United Nations Educational, Scientific and Cultural Organization, Paris.

UNESCO. (2014c). State *of Conservation (SOC) Galápagos Islands*. United Nations Educational, Scientific and Cultural Organization, Paris.

Vasquez, J.A., Camus, P.A., Ojeda, P. (1998). Diversidad, estructura y funcionamiento de ecosistemas costeros rocosos del norte de Chile. *Revista Chilena de Historia Natural*, 71, 479-499.

Veron, J.E.N., How, R.A., Done, T.J., Zell, L.D., Dodkin, M.J., O'Farrell, A.F. (1974). Corals of the Solitary Islands, central New South Wales. *Australian Journal of Marine and Freshwater Research* 25, 193-208.

Wallace, B.P., Kot, C.Y., DiMatteo, A.D., Lee, T., Crowder, L., Lewison, R.L. (2013). Impacts of fisheries by-catch on marine turtle populations worldwide: toward conservation and research priorities. *Ecosphere* 4, 40.

Wallace, B.P., DiMatteo, A.D., Bolten, A.B., Chaloupka, M.Y., Hutchinson, B.J., Abreu-Grobois, F.A., Mortimer, J.A., Seminoff, J.A., Amorocho, D., Bjorndal. K.A., Bourjea, J., Bowen, B.W., Briseño Dueñas, R., Casale, P., Choudhury, B.C., Costa, A., Dutton, P.H., Fallabrino, A., Finkbeiner, A.M., Girard, A., Girondot, M., Hamann, M., Hurley, B.J., López-Mendilaharsu, M., Marcovaldi, M.A., Musick, J.A., Nel, R., Pilcher, N.J., Troëng, S., Witherington, B., Mast, R.B. (2011). Global conservation priorities for marine turtles. *Plos One* 6, e24510.

Walsh, K., McInnes, K., McBride, J. (2012). Climate change impacts on tropical cyclones and extreme sea levels in the South Pacific - a regional assessment. *Global Planetary Change* 80–81, 149–164.

Wassenberg, T.J., Milton, D.A., Burridge, C.Y. (2001). Survival rates of sea snakes caught by demersal trawlers in northern and eastern Australia. *Biological Conservation* 100, 271-280.

Waugh, S.M., Filippi, D.P., Kirby, D.S., Abraham, E., Walker, N. (2012). Ecological risk assessment for seabird interactions in Western and Central Pacific longline fisheries. *Marine Policy* 36, 933-946.

Waycott, M., Duarte, C.M., Carruthers, T.J.B., Orth, R.J., Dennison, W.C., Olyarnik, S., Calladine, A., Fourqurean, J.W., Heck, K.L., Hughes, A.R., Kendrick, G.A.M. Kenworthy, W.J., Short, F.T., Williams, S.L. (2009). Accelerating loss of seagrasses across the globe threatens coastal ecosystems. *Proceedings of the National Academy of Sciences* 106, 12377-12381.

Waycott, M., Longstaff, B.J., Mellors, J. (2005). Seagrass population dynamics and water quality in the Great Barrier Reef region: A review and future research directions. *Marine Pollution Bulletin* 51, 343-350.

Wells, J.W. (1955). *A survey of the distribution of reef coral genera in the Great Barrier Reef region.* Government Printer, South Africa.

Wiedenmann, J., D'Angelo, C., Smith, E.G., Hunt, A.N., Legiret, F.-E., Postle, A.D., Achterberg, E.P. (2013). Nutrient enrichment can increase the susceptibility of reef corals to bleaching. *Nature Climate Change* 3, 160-164.

Williams, A., Althaus, F., Dunstan, P.K., Poore, G.C., Bax, N.J., Kloser, R.J., McEnnulty, F.R. (2010). Scales of habitat heterogeneity and megabenthos biodiversity on an extensive Australian continental margin (100–1100 m depths). *Marine Ecology* 31, 222-236

Williams, A., Bax, N. (2001). Delineating fish-habitat associations for spatially-based management: an example from the south-eastern Australian continental shelf. *Marine and Freshwater Research* 52, 513-536.

Wilson, S.K., Fisher, R, Pratchett, M.S., Graham, N.A.J., Dulvy, N.K., Turner, R.A., Cakacaka, A., Polunin, N.V.C, Rushton, S.P. (2008). Exploitation and habitat degradation as agents of change within coral reef fish communities. *Global Change Biology* 14, 2796-2809.

Woodhams, J., Vieira, S, Stobutzki, I. (eds.). (2013). *Fishery status reports 2012*. Australian Bureau of Agricultural and Resource Economics and Sciences, Canberra.

World Bank. (2006). *Republic of Peru environmental sustainability: a key to poverty reduction in Peru. Country environmental analysis. Volume 2: full report*. Environmentally and Socially Sustainable Development Department, Latin America and Caribbean Region, World Bank.

Wolff, M. (1987). Population dynamics of the Peruvian scallop Argopecten purpuratus during the El Niño phenomenon of 1983. *Canadian Journal of Fisheries and Aquatic Science*s 44, 1684-1691.

Worm, B., Barbier, E.B., Beaumont, N., Duffy, J.E., Folke, C., Halpern, B.S., Jackson, J.B.C., Lotze, H.K., Micheli, F., Palumbi, S.R., Sala, E., Selkoe, K.A., Stachowicz, J.J., Watson, R. (2006). Impacts of Biodiversity Loss on Ocean Ecosystem Services. *Science* 314, 787-790.

Young, J.W. (1989). The distribution of hyperiid amphipods (Crustacea:Peracarida) in relation to warm-core eddy J in the Tasman Sea. *Journal of Plankton Research* 11, 711-728.

Young, J.W., Bradford, R.W., Lamb, T.D., Lyne, V.D. (1996). Biomass of zooplankton and micronekton in the southern bluefin tuna fishing grounds off eastern Tasmania, Australia. *Marine Ecology Progress Series* 138, 1-14.

Young, J.W., Hobday, A.J., Campbell, R.A., Kloser, R.J., Bonham, P.I., Clementson, L.A., Lansdell, M.J. (2011). The biological oceanography of the East Australian Current and surrounding waters in relation to tuna and billfish catches off eastern Australia. *Deep Sea Research II* 58, 720-733.

Zapata, F.A., Robertson, D.R. (2007). How many species of shore fishes are there in the Tropical Eastern Pacific? *Journal of Biogeography* 34, 38-51.

Zeller, D., Booth, S., Craig, P., Pauly, D. (2006). Reconstruction of coral reef fisheries catches in American Samoa, 1950–2002. *Coral Reefs* 25, 144-152.

# 36E Indian Ocean

Group of Experts:
Renison Ruwa (Lead Member) and Jake Rice (Co-Lead Member and Editor of Part VI)[1]

---

[1] The members of the Group of Experts would like to thank Cosmas Munga, Melchzedeck Osore, and Nina Wambiji for their substantive input to this chapter.

Chapter 36E

Indian Ocean

1 Introduction

The Indian Ocean is the third largest ocean in the world. It is mostly surrounded by a rim of developing countries and island States, one of which is the fourth largest island in the world, Madagascar. The Indian Ocean is bound by Asia to the north, by Africa to the west, Australia to the east and Antarctica to the south. It has two major seas, the Red Sea between the Arabian Peninsula and Africa, and the Arabian Sea to the west of India; and the largest bay, the Bay of Bengal, to the east of India. Following the FAO statistical fishing areas, the Indian Ocean is divided into two major parts: the Western Indian Ocean (WIO) and Eastern Indian Ocean (EIO) (FAO, 1990-2015).

In terms of the oceanographic physical environment of the Indian Ocean, the major epipelagic atmospheric and ocean currents in relation to other global features are as depicted in Figure 36E.1. The detailed seasonal characteristics of the reversing wind systems of the monsoon are shown in Figure 36E.2. The system is important in the distribution of global heat, salinity and biogeochemical cycling of carbon and inorganic elements (Wajih et al., 2006). There are basically two monsoonal seasons, but it is common to have a third inter-monsoonal season: North East Monsoon (NEM), from February to May; South West Monsoon (SWM) from June to October and an Inter-Monsoon Season (IMS) from November to January.

It is noted that:

(a) From a wide geographical perspective, most of the major ocean area is under-sampled with regard to both coastal and oceanic environments. The oceanic areas are particularly unsampled and therefore the biological diversity is still incompletely described for most ecosystems;

(b) In terms of human scientific capacity, there is an extreme lack of taxonomists and therefore most of the species are still undescribed or are simply unknown;

(c) Much of the area has largely been studied using satellite technology, so observations are based on remote sensing and therefore driving forces at species and community level are relatively vague or unknown; there is a need to undertake ground truth sampling to support satellite data;

(d) Most studies are based on isolated collections in localized areas and are not continuous, making it difficult to discern possible trends;

(e) Coastal and offshore ocean sampling are rarely synchronized in space and time, increasing data gaps in data collection;

(f) At regional scales most of the sampling methods are not standardized making the data difficult to compare and a weak basis for describing status and trends or creating baselines for benchmarking. There is a need to form regional multidisciplinary research teams to address these needs. Such teams could create the necessary synergy to share research capacity in terms of both human skills and infrastructure, standardize research methodologies, synchronize sampling programmes and plans, establish sampling stations for continuous sampling and data generation for long-term research data requirements, and create databases.

The boundaries and names shown and the designations used on this map do not imply official endorsement or acceptance by the United Nations.
Figure 36E.1 | Map to show the epipelagic water masses and current patterns in the Indian Ocean in relation to other global circulations in the world oceans (Source: Pierrot-Bults and Angel 2013).

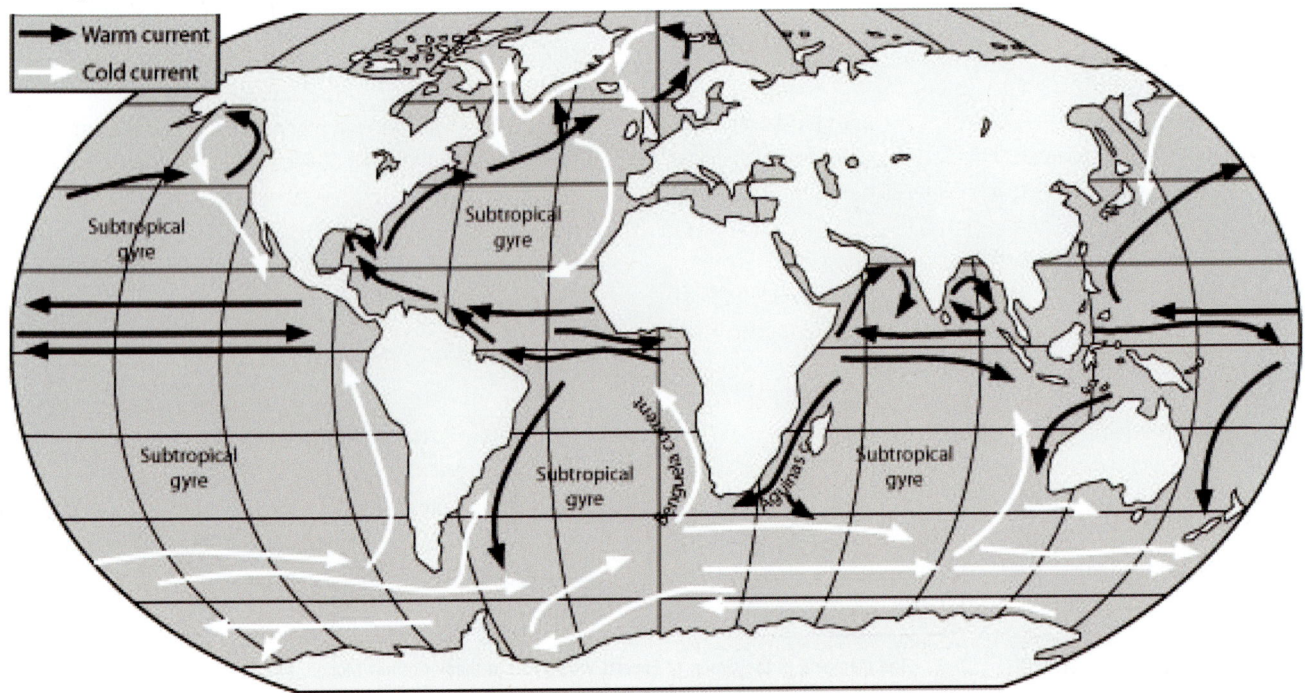

Chapter 36E

Indian Ocean

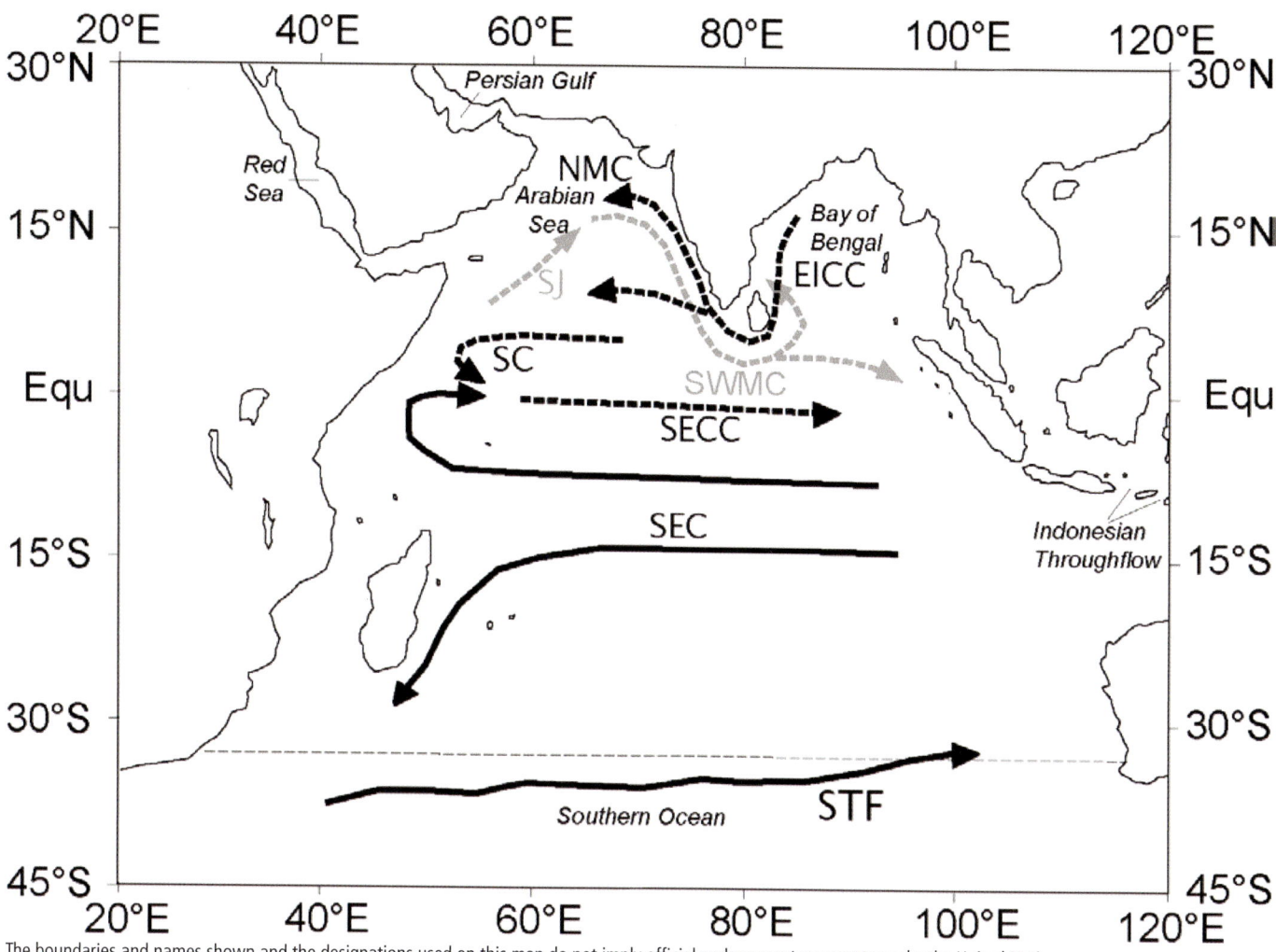

Figure 36E.2 | Major features of the surface circulation in the Indian Ocean [after Schott and McCreary, 2001; Wajih et al., 2006]. The SEC (South Equatorial Current), SECC (South Equatorial CounterCurrent), and STF (Subtropical Front) are present throughout the year. Surface currents during the Northeast Monsoon include the NMC (Northeast Monsoon Current), SC (Somali Current), and EICC (East Indian Counter Current). Surface currents during the Southwest Monsoon include the SWMC (Southwest Monsoon Current) and SJ (Somali Jet). (Source: Bates et al., 2006).

2   Indian Ocean Biodiversity

The Indian Ocean covers about 30 per cent of the total global ocean area and being predominantly a tropical ocean, accounts for a significant part of tropical coastal biodiversity and deep-sea oceanic biodiversity in various marine ecosystems. It accounts for 30 per cent of the total global coral reef cover, 40,000km^2 mangrove cover, besides supporting various types of biodiversity found in its various ecosystems (Table 36E.1). There has been progress in addressing marine and coastal biodiversity since the major surveys undertaken in the first International Indian Ocean Expedition (IIOE) (1960-1965) about 50 years ago (http://www.incois.gov.in/portal/iioe/aboutus.jsp). The present review of Indian Ocean biodiversity will address the long-term status, trends and research gaps in relation to:

(a) Marine fisheries including tuna, focusing on their exploitation and species diversity over wide geographic coverage;
(b) Threatened megafauna species, particularly: marine mammals, marine reptiles and seabirds, focusing on describing the status and trends including their associated drivers and general abundances and what dominant taxa groups exist;
(c) Description of phytoplankton production, zooplankton and benthos structures focusing on their abundance and diversity, including the drivers of change and possible effects of climate change; identifying hot spots for primary production in both coastal and deep sea over various time and geographical scales and major influences of seasonality.

3   Fish Biodiversity

This section mostly presents information on marine capture fisheries, as reported by the FAO and the Indian Ocean Tuna Commission (IOTC).

3.1   Marine Finfish

Table 36E.1 | Types and area cover of marine ecosystems in the Indian Ocean (Source: Wafar et al., 2011)

ECOSYSTEM	Area (in million km²)
Open ocean	
Oligotrophic	19.6
Transitional	23.8
Equatorial divergence	18.9
Coastal	
Upwelling zones	7.9
Other neritic waters	5.3
Other	
Coral reefs	0.2
Mangroves	0.04
Sandy and rocky beaches	0.004
Estuaries	-
Hypersaline water bodies/lagoons	<0.005

The contribution of coastal and marine capture fisheries (finfish, shellfish and molluscs) from the Indian Ocean (average of 11.01 million tons annually) to the global landings is third after the Pacific Ocean (average of 48.3 million tons annually) and the Atlantic Ocean (average of 11.03 million tons annually) based on the 2003, 2011 and 2012 FAO estimates (FAO, 2014). This chapter describes the coastal and marine fisheries finfish production excluding tuna in the Indian Ocean, focusing on the status and trends in exploited species, long-term species surveys and different kinds of diversity indices over the FAO statistical areas of the EIO and the WIO. These areas have recorded increasing overall catch trends since 1950 (Figure 36E.3) however, incidences of reduced catches have been reported in inshore areas. This increase in catches may be due to expansion of fishing to new areas or species, and the improved recording of fish landing statistics over time. The EIO and WIO together contributed 28 per cent of the total global marine catches of finfish, shellfish and molluscs in 2011 (FAO, 2014).

The WIO shows a similar scenario, in which the largest catches are made up of the category "marine fishes nei" followed by the small pelagic "In-

Figure 36E.3 | Long term trends in total finfish landings excluding tuna in the EIO, WIO and overall Indian Ocean (data source: FishstatJ – FAO Fishery and Aquaculture Global Statistics).

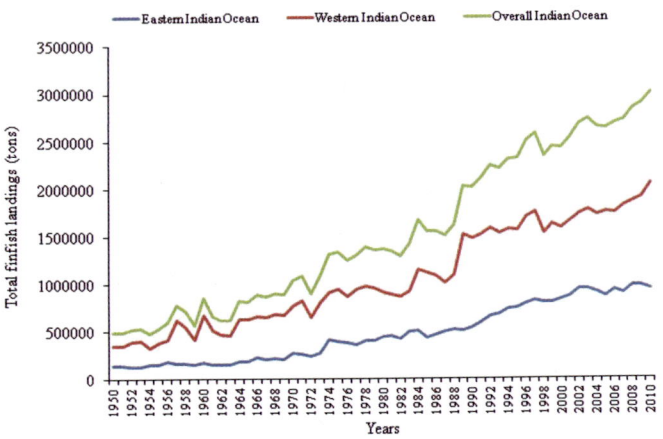

Figure 36E.4 | Top twenty highest landed finfish species except tuna from the Eastern Indian Ocean based on total catches from 1950-2010 data in Australia and India (data source: FishstatJ – FAO Fishery and Aquaculture Global Statistics).

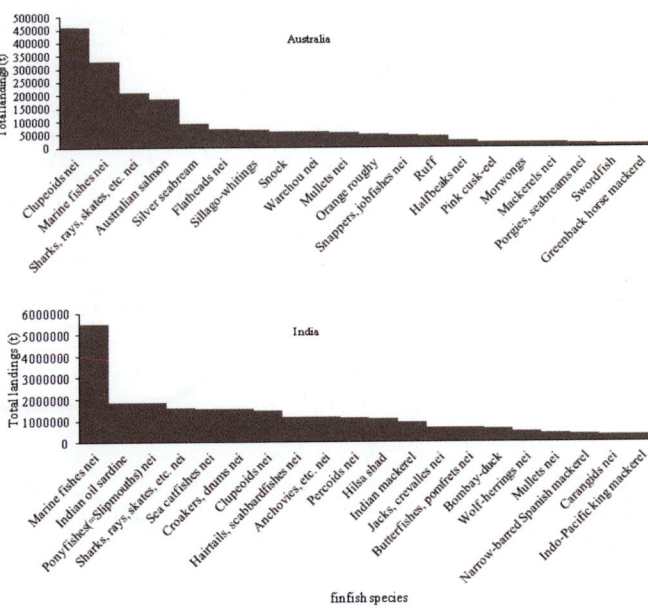

dian oil sardine", ponyfishes, and sharks, rays and skates. Total landings in the WIO reached a peak of 4.5 million tons in 2006, but then declined slightly, with 4.2 million tons in 2011 (FAO, 2014). A recent assessment has shown that the narrow-barred Spanish mackerel (Scomberomorus commerson) is overfished, and this species is among the 20 most highly

Figure 36E.5 | Long-term trends in total landings of (a) sharks, rays and skates in EIO and WIO, and (b) narrow-barred Spanish mackerel in the WIO.

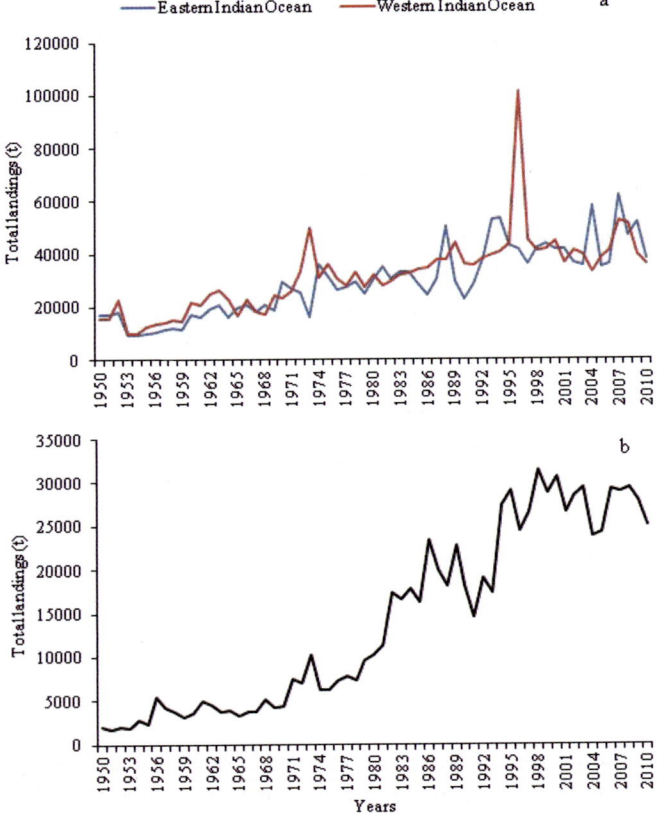

landed (FAO, 2014; Figure. 6) in the WIO. Long term catch data in the Indian Ocean, especially the WIO, are often not detailed enough for stock assessment and species composition purposes, a situation aggravated by the lack of adequate resources to conduct scientific studies, monitoring and enforcement (McClanahan and Mangi, 2004). However, the Southwest Indian Ocean Fisheries Commission (SWIOFC) conducted stock assessments for 140 species in 2010 based on best available data and information (FAO, 2014). Overall, 75 per cent of fish stocks were estimated to be fully fished or under-fished, and 25 per cent fished at unsustainable levels. There are many other species in the Indian Ocean where the level of exploitation is unknown or is extremely difficult to determine. Long-term trend analysis by individual fish taxa indicates that catches of sharks, rays and skates together started to decline or level off from the mid-1990s in both the EIO and WIO (Figure 36E.5a). In the late 1990s, a similar trend is observed with the narrow-barred Spanish mackerel in the WIO (Figure 36E.5b).

Fish species diversity studies in the Indian Ocean, especially in the WIO are biased to coral reef areas. Fish diversity in relation to coral reefs in the region covering about 200 sites situated in Kenya, Madagascar, Maldives, Mauritius, Mayotte (France), Mozambique, Reunion (France), Seychelles, South Africa and the United Republic of Tanzania was studied (McClanahan et al., 2011). This study found that the region from southern Kenya to northern Mozambique across to northern eastern Madagascar and the Mascarene Islands and the Mozambique-South Africa border are areas with moderate to high fish diversity. The WIO fish fauna is one of the richest marine fish faunas in the world, with some 3,200 species or about 20 per cent of the world marine fish fauna. Despite considerable effort by ichthyologists over the past two centuries, the taxonomy of WIO fishes is ongoing. Of the 329 new marine species described between the years 2002 and 2012, 140 were from the WIO (http://www.saiab.ac.za/coastal-fishes-of-the-western-indianocean.htm).

Long-term fish species surveys are scanty in the Indian Ocean. The South African line fishery however, has been monitored since the late 1900's. This fishery is multispecies targeting over 200 species with about 50 being economically important. Due to concerns of overfishing, management measures were first introduced in the 1940s. Stock assessment of the line fishery has been based on both fishery dependent and independent data, as well as data from marine protected areas. Since 1985, the South African line fishery has been one of the largest spatially referenced marine line fishery data sets. After the introduction of management measures, monitoring results have indicated that, generally, the over-exploited line fish stocks are now slowly recovering except for *Polysteganus undulosus* which has remained significantly reduced (SWIOFC, 2012).

This chapter has identified key gaps in relation to the Indian Ocean fisheries, as follows:

- Total catch statistics data is mostly still poor in terms of temporal and spatial coverage, and catches are in many cases estimates of actual catches. This is attributed to lack of human and financial capacity as well as remoteness of some of the fish landing sites;
- Lack of a comprehensive species' composition data. To date, the largest proportion of catches is categorized as "unidentified". This is attributed mainly to the inadequate knowledge in fish taxonomy in the region;
- Long-term research surveys in the region are rare due to lack of professional expertise, infrastructure and financial capacity. Most of the research surveys are short term and sporadic depending on availability of donor funding. The International Indian Ocean Expedition, if regularly implemented could be the best source of long-term research survey data.
- The impacts of fishing gears on target fisheries, by-catches and habitats are hardly studied, and bottom contacting fishing gears are used indiscriminately in the region, resulting in biodiversity losses.

3.2 Tuna Species

Tuna and tuna-like species form the most important resources of the offshore pelagic fishery. In the Indian Ocean, both the EIO and the WIO, at least seven different tuna species, including tuna-like species, have been reported in the landing statistics. The four main commercially fished tuna species in the Indian Ocean are: albacore (*Thunnus alalunga*), skipjack tuna (*Katsuwonus pelamis*), yellowfin tuna (*T. albacares*) and bigeye tuna (*T. obesus*). The other species are: frigate and bullet tunas (*Auxis* sp.), kawakawa (*Euthynnus affinis*), southern bluefin tuna (*Thunnus maccoyii*), and tuna-like species. Since 2010, after three years (2007–09) during which piracy negatively affected fishing in the WIO, tuna catches have recovered. During the 2007-2009 period, total tuna catches decreased by 30 per cent as piracy deterred fishing operations (FAO, 2014). Among the 23 major fish species in the global marine capture fisheries, skipjack tuna ranked third with increasing landings of 2.2 million tons in 2003, 2.6 million tons in 2011, and 2.8 million tons in 2012 (FAO, 2014). The yellowfin tuna was ranked eighth, however with variable landings of 1.5, 1.2 and 1.4 million tons in 2003, 2011 and 2012 respectively.

Figure 36E.6 | Long term trends in total tuna and tuna-like landings in the EIO, WIO and overall Indian Ocean (data source: FishstatJ – FAO Fishery and Aquaculture Global Statistics).

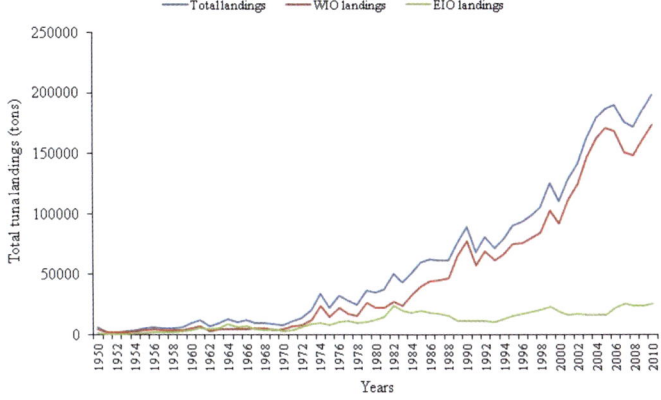

Indian Ocean   Chapter 36E

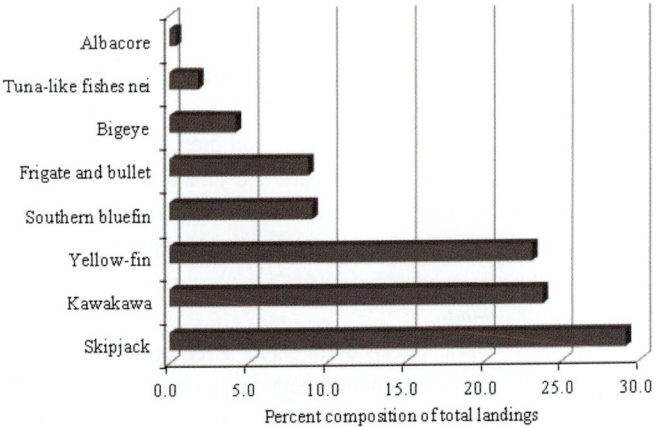

Figure 36E.7 | Species percent composition of total landings of tuna and tuna-like from the Indian Ocean based on total catch data from 1950-2010 (data source: FishstatJ – FAO Fishery and Aquaculture Global Statistics).

In the last 5 decades, total landings of tuna and tuna-like species in the Indian Ocean have been increasing (Figure 36E.6). This is especially evident in the WIO region, whose global contribution in total tuna landings is 30 per cent. The increasing trend of total tuna and tuna-like landings in the EIO region is not pronounced as landings have remained just about 20 000 tons annually for a long time between 1982 and 2010. During this period, a total of 7 tuna species and tuna-like species were recorded in the Indian Ocean (Figure 36E.7). The contribution of tuna and tuna-like landings for the WIO region in the last 5 decades came from India, Kenya, Madagascar, Mauritius, Mayotte (France), Mozambique, Seychelles, South Africa and the United Republic of Tanzania. On the other hand, landings for the EIO during the same period were reported from Australia, India, Madagascar and Seychelles.

The landing statistics of tuna and tuna-like species in the last 5 decades in the Indian Ocean show a great variation in terms of species percent composition of total landings (Figure 36E.7). In this period, skipjacks, kawakawa and yellowfin tuna contributed the highest percent composition of 29 per cent, 24 per cent and 23 per cent respectively. The lowest percent composition was made up of albacore (0.4 per cent), tuna-like fishes nei (2 per cent), and bigeye tuna (4 per cent). The species frigate and bullet tunas, and southern bluefin tuna contributed intermediate percent composition of about 9 per cent each.

Recent stock assessment estimates from the Indian Ocean Tuna Commission (IOTC) indicate that yellowfin, bigeye tuna stocks, skipjack and albacore are not overfished and not subject to overfishing (IOTC, 2014). Estimates of the total and spawning stock biomasses show a marked decrease over the last decade, accelerated in recent years by the high catches of 2003-2006 (Figure 36E.7). The Spawning Stock Biomass was estimated to be 57 per cent for the skipjack tuna, 38 per cent for the yellowfin tuna, 40 per cent for the bigeye tuna and 57 per cent for the albacore of the unfished level. However current fishing mortality has not exceeded the Maximum Sustainable Yield (MSY) level for these species (IOTC, 2014).

The estimated catches for the skipjack tuna was 455,000 tons in 2009 and 428,000 tons in 2010 with an average catch of 500,000 tons between 2005 and 2010 being lower than the median value of the estimated Maximum Sustainable Yield. IOTC recommended that catches should not exceed 500,000 tons. The MSY for yellowfin tuna for the whole Indian Ocean should not exceed 300,000 tons, while the MSY estimates for bigeye tuna is estimated at 102,664 tons. Based on available data, the major challenge in the region at large is declining spawning stock biomass and the possibility of recruitment overfishing.

3.3   Research gaps

There is a need to research the impacts of the target fish catches and fishing gear on non-target fish or bycatch, food chains cycles and overall on species biodiversity, especially focusing on various taxa over long-term periods in order to also account for climate change effects.

4   Plankton Diversity

The contribution of the Indian Ocean plankton data into the World Ocean Database (WOD) is still very minimal. Similarly, except for India and South Africa and to some extent Indonesia and Pakistan, very little research is undertaken by the countries of the Indian Ocean region. The national contribution of plankton data work in the WOD09 by the countries bordering the Indian Ocean is less than 1.5 per cent Likewise, among the major international oceanographic projects that contribute to the plankton data, only a minimum has involved the Indian Ocean.

4.1   Phytoplankton

Marine phytoplankton are an essential component in marine life as they play a fundamental role in the biodiversity and bio-productivity of the marine ecosystem. They are mainly microscopic plants that float passively throughout the pelagic zone, pushed by the dominant ocean current. They also play a crucial role in the food chains and food webs, as phytoplankton represent the primary producers of organic matter and zooplankton are the link between the phytoplankton and higher trophic levels. In addition, plankton play a crucial role in the biogeochemical cycle of numerous chemical elements in the ocean.

In a balanced ecosystem, phytoplankton provide food for a wide range of sea creatures including whales, shrimp, snails and jellyfish. During unusually high availability of nutrients, phytoplankton may grow out of control and form harmful algal blooms (HABs). These blooms can produce extremely toxic compounds that have harmful effects on fish, shellfish, mammals, birds, and even humans. The Intergovernmental Oceanographic Commission of the United Nations Educational, Scientific and Cultural Organization (IOC-UNESCO) has supported effort towards the detection, identification and management of HABs in the WIO region but the data are still inadequate to generate trends.

Chapter 36E

Indian Ocean

The boundaries and names shown and the designations used on this map do not imply official endorsement or acceptance by the United Nations.

Figure 36E.8 | Boundaries for the 12 major oceanographic basins (Source: Gregg et al., 2003).

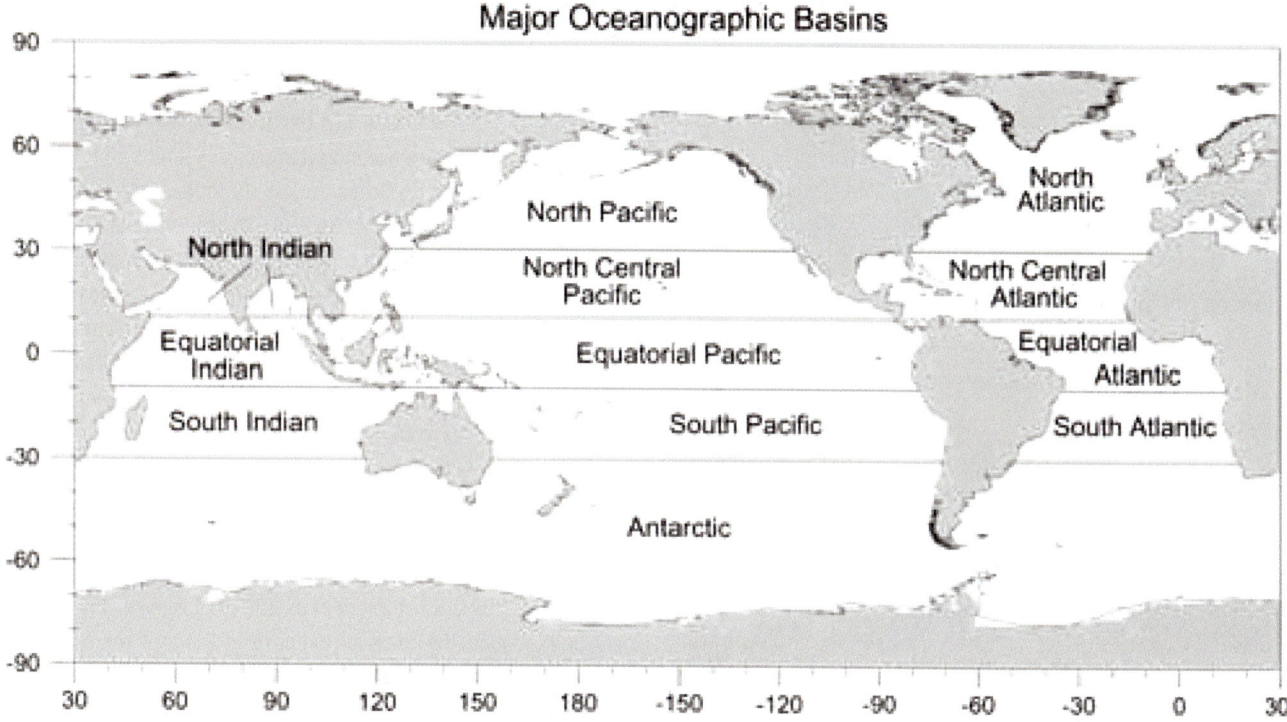

Figure 36E.9 | Differences between SeaWiFS (1997–2002) and CZCS (1979–1986) in the 12 major oceanographic basins. Differences are expressed as SeaWiFS-CZCS. Top left: Annual primary production (Pg C y_1). An asterisk indicates the difference is statistically significant at $P < 0.05$. Top right: SST (degrees C). Bottom left: iron deposition (%). Bottom right: mean scalar wind stress (%) (Source: Gregg et al., 2003).

675

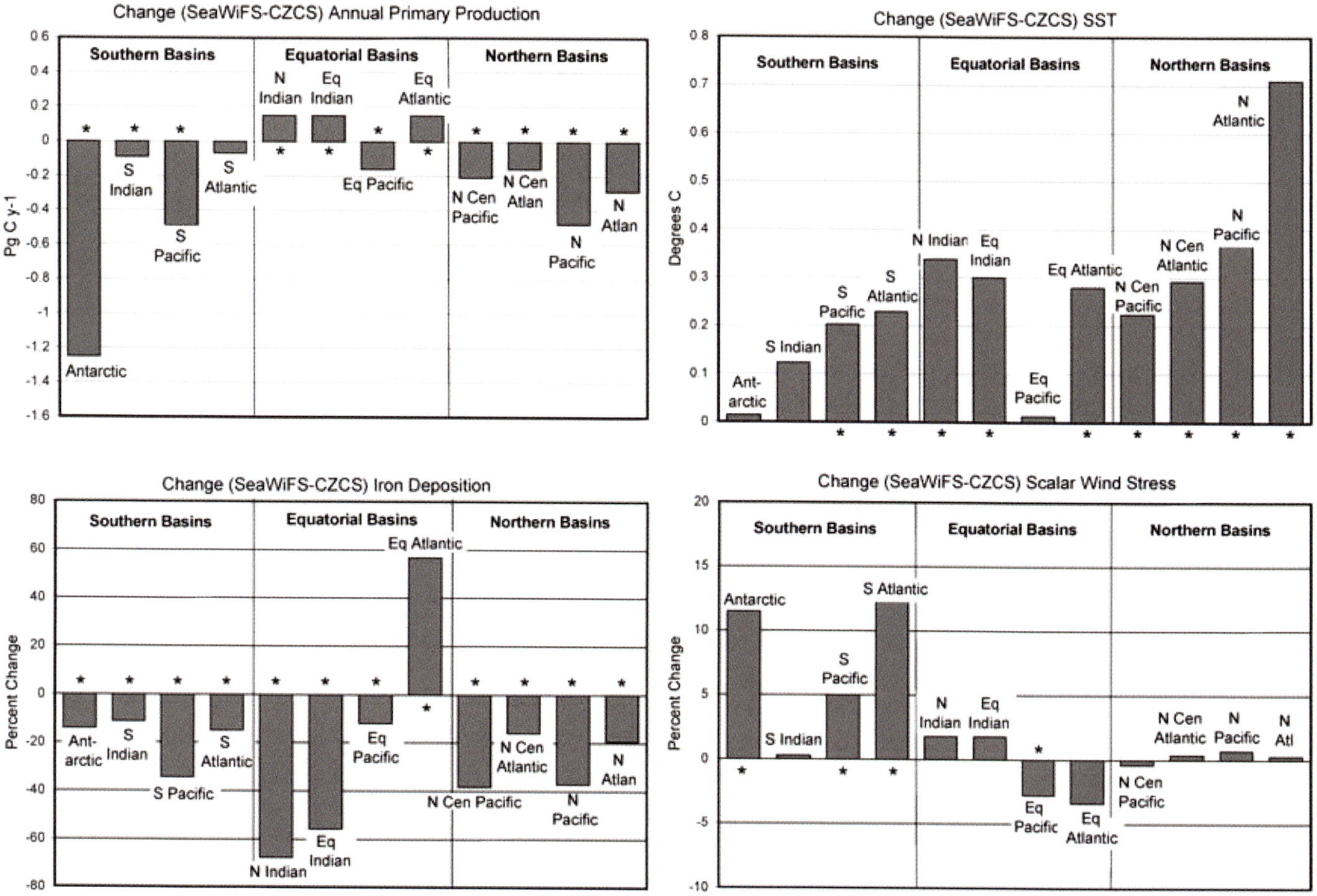

Figure 36E.10 | Primary production distributions for the SeaWiFS era (1997-mid-2002), the CZCS era (1979-mid-1986) 327 and the difference. Units are gCm-2y-1. White indicates missing data (Source: Gregg et al., 2003).

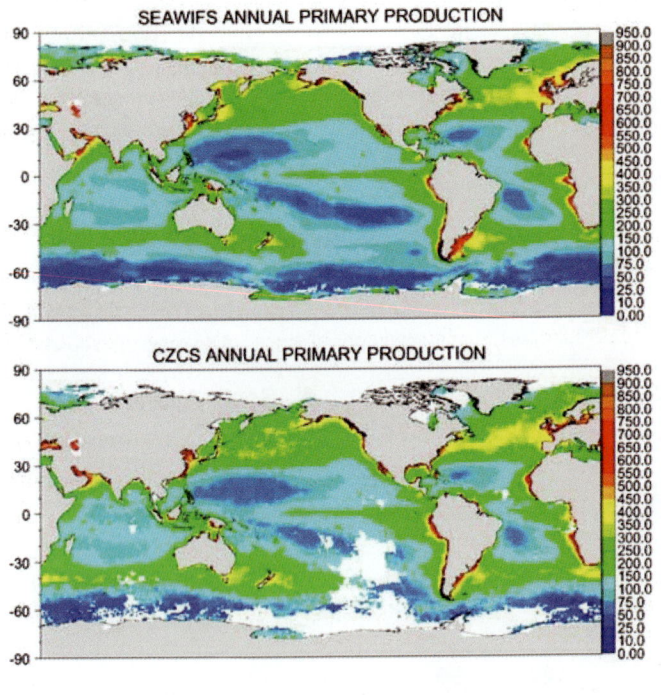

Results from the Tyro Expedition in the WIO under the Netherlands Indian Ocean Programme (NIOP, 1991-1995) established that the seasonal change in monsoon regime affect the nitrogen nutrition of marine phytoplankton conspicuously (Wafar et al., 2011). Seasonal variation of phytoplankton abundance and diversity is a common occurrence in many bays and creeks of the Indian Ocean. This is driven by the interchanging monsoon regime almost every half year as well as by water quality as result of land based activities.

Research undertaken by India in its west coast, in the Arabian Sea, and east coast, in the Bay of Bengal, has demonstrated strong monsoonal seasonality using satellite and remote sensing technology and sampling cruises (Moharana and Patra, 2013) to measure chlorophyll and analyze patterns of distribution over time. The same technology has also been used to study the global impact of climate change on in primary production in oceans using chlorophyll concentrations (Gregg et al., 2003)

Climate change effects on primary production in the oceans have been a major subject of study globally. Using satellite technology to study chlorophyll records over 22 years, from 1980, in 12 major oceanographic basins (Figure 36E.8), it has been established that the global annual ocean primary production has declined by 6 per cent since 1980. The question is whether the Indian Ocean has behaved similarly. The Indian Ocean is divided in three major oceanographic basins which are: North Indian Ocean, Equatorial Indian Ocean and South Indian Ocean. On the basis of the differences between the SeaWiFS (1997-2002) and CZCS (1979-1986), the analysis which not only involved primary production but also environmental parameters including surface sea temperatures, nutrients and wind stress in the 12 major oceanographic basins (Figure 36E.9 and Figure 36E.10), is summarized in Table 36E.2. It is clear from this analysis that primary production increases in the North and Equatorial Indian Ocean but decreases in the South Indian Ocean basin, which is at higher latitudes and close to the Antarctica basin, where also the primary production decreases. It was further noted that the highest increase occurs in the western portion of the Arabian Sea, in the west coast of India, and in the North Indian oceanographic basin, which experiences upwelling (Gregg et al., 2003). Chaturvedi et al., (2013) observing the chlorophyll behaviour around the coast of India, recorded that there was higher chlorophyll from December to March in the Arabian Sea whereas in the Bay of Bengal, the peak occurs in February to March.

4.2 Zooplankton

Implementation of the Indian Ocean chapter of the Census of Marine Life Programme (IOCoML) inaugurated in 2003 has vastly increased the knowledge of marine biodiversity of the Indian Ocean countries. Among the major achievement of IO-CoML is the discovery of more than 40 new

Table 36E.2 | Percent change in ocean primary production (SeaWiFS-CZCS) by basin and surface area of the basins (10km-2) where data from both SeaWiFS and CZCS were sampled (Source: Gregg et al., 2003

OCEAN BASIN	AREA	CHANGE
N. Atlantic	1.83	- 6.7%
N. Pacific	2.32	- 9.3%
N. Central Atlantic	1.53	-7.0%
N. Central Pacific	3.22	-5,8%
N. Indian	**0.46**	**13.6%**
E. Atlantic	1.15	6.9%
E. Pacific	3.72	-3.8%
Eq. Indian	**1.37**	**8.8%**
S. Atlantic	1.20	-3.8%
S. Pacific	2.69	-14.0%
S. Indian	**1.77**	**-4.2%**
Antarctic	8.28	-10,4%
Global	29.73	-6.3%

zooplankton species including from groups of mysids, chaetognaths and sponges.

There are two types of zooplankton differentiated by aspects of their life cycles. There are categories of zooplankton that partly live as zooplankton, for example as larvae, and grow to sub-adults and then adults and leave the plankton community as they grow. Others live as zooplankton throughout their lives. Those that live partly as zooplankton are referred to as meroplankton, e.g. fish larvae, whereas those that live in planktonic form throughout their lives are known as holoplankton, e.g. copepods.

The biophysical factors that affect phytoplankton similarly affect zooplankton and their peaks tend to be in rhythm with a small time lag of 1-1.5 months because zooplankton depend on phytoplankton for food (Fabian et al., 2005). For vertebrate animals like fish the peak abundances are highly influenced by taxonomic category, since some taxa which are r-selected can produce eggs profusely throughout their short lives whereas k-selected tend to have fewer larvae which take long to mature and may not be in rhythm with other taxonomic groups. To discern these patterns, long-term research to cover various biophysical factors associated with seasonality, climate change, predation, pollution and eutrophication are essential and have to be multidisciplinary. These types of coordinated research protocols are lacking in the region, making it difficult to describe zooplankton status and trends.

Most studies are one-off events at short time intervals and can only be taken to represent season samples. Only in rare cases does sampling cover longer periods of the continuous cycles, although in most cases the abundance and species composition of the zooplankton vary considerably across seasons (Fazel et al., 2013). The situation is compounded by the lack of taxonomic expertise as well lack of taxonomic descriptions of many of the major zooplankton in the area (Chesalina et al., 2013). Since samples are rarely continuously collected for long-term monitoring, and most samples are collected to represent a part of a season, therefore the variances are high, making quantification of trends very difficult. This emphasizes the need to have permanent sampling platforms in the region for long-term continuous monitoring, preferably daily sampling to account for diel cycles also.

One of the clearest scenarios shown in the field relating to nutrient levels is the comparison of performance of production in the Arabian Sea and the Bay of Bengal. It is clearly observed that the production of zooplankton and phytoplankton shows a higher standing crop in the west coast of India, in the Arabian Sea, than in the east coast of India, in the Bay of Bengal. The difference is attributed to upwelling in the Arabian Sea whereas the Bay of Bengal depends only on nutrient inputs from the major rivers (Moharana and Patra 2013; Dorgham 2013).

4.3   Research gaps

The gaps identified requiring research in both phytoplankton and zooplankton are similar and therefore most efficiently addressed together as a plankton group, with emphasis appropriately made for either of the two groups when necessary. For regional and global comparisons of data and information to be effective the sampling methodologies need to be standardized, including equipment, and the time of cruises synchronized since plankton are highly affected by atmospheric processes and ocean currents which vary regionally and globally in time and space. There is therefore a need to establish the frequency and regular nature of sampling, number of sampling stations and their location, number of samples to be taken, methods of collection (net and/or water samples, depth of collection (surface and/or other defined and agreed depth as per sampling protocol)). Without such a coordinated research protocol approach the studies will remain fragmented and consequently the data and information will be difficult to use for comparative studies, or to provide baselines for documenting trends. The following specific needs should be addressed:

(a) Plan synchronized regular, multidisciplinary and comprehensive study for both phytoplankton and zooplankton and have a comprehensive database for the Indian Ocean;
(b) Identify exotic plankton carried in various areas by ship ballast;
(c) Collect detailed information about harmful algal blooms;
(d) Establish fixed stations for continuous regular plankton studies for time series analysis;
(e) Establish coastal fixed stations in water masses covering special benthic communities in critical habitats namely: Corals, seagrass, mangroves and intertidal zones;
(f) Undertake continuous special studies on dinoflagellates cysts to establish potential of harmful algal species; and
(g) Establish satellite networks for regional studies of primary production in relation to various environmental parameters in order to relate to climate change.

5   Benthos

Marine benthic organisms are organisms that live on or are associated with the seabed. Their usual mode of effective dispersal is through their planktonic larval or immature stages, transported by ocean currents. Key benthic habitats include: coastal water bottoms, mangrove habitats, coral reefs are benthic hot spots, seagrass beds, intertidal zones, deep water continental shelf and slopes or in the depth of ocean trenches. Since the initial IIOE in 1959-1965, there have been various other research expeditions to the northern part of the WIO that have yielded a substantial amount of results about the benthos. These have especially been conducted by the former USSR and also by the United States, the United Kingdom, Germany, France and the Netherlands. This region still lacks the capacity to support and execute research offshore. In fact the only States bordering the Indian Ocean that have demonstrated capability to conduct oceanographic research are Australia, India, Pakistan and South Africa. These countries have continued to conduct research even after the IIOE ended, especially in their own territorial waters.

As documented in Chapter 34, species diversity gradients exist across latitudes and from coastal waters to deep or oceanic water, such that there is an increase in species diversity from high to low latitudes and a decrease from shallow coastal waters to deep oceanic waters (Gray 1997, Chapter 34). However, little data and information exist in the Indian Ocean to describe these scenarios and establish how effectively this pattern could be applied for conservation purposes especially when interlinked with within – habitat or alpha diversity (Fisher et al., 1943; Whittaker, 1967) and between habitat or beta biodiversity (Whittaker, 1975, 1977) and at a larger regional scale seascape gamma diversity (Ray 1991). Due to scanty studies and many undescribed benthic species in the Indian Ocean, the reliability of species diversity indices for the Indian Ocean benthos is similarly poor due to the likely underestimation of species numbers (richness underestimated) and relative abundance of species, particularly less common species (evenness not accurately measured) (Wafar et al., 2011).

Further challenges are due to threats to benthic biodiversity occurring at different intensities with time and space. The coastal and oceanic benthic habitats and species diversity are threatened by habitat loss and perturbations or alteration especially due to mining, sediment deposits, fishing with bottom-contacting gears, and dumping of solid wastes. Other threats are due to climate change, overexploitation of benthic species which affect their feeding cycles and ecological balances. There is a tendency for opportunistic species succession resulting from ecological population imbalances and pollution, especially in shallow coastal habitats where eutrophication and toxicity from algal blooms may lead to loss of species biodiversity. Sand and coral mining, including dynamite fishing which blasts coral reefs, can lead to complete loss of habitat, with restoration often difficult, if possible at all, and costly. There is evidence of various degrees of continuous benthic habitat loss especially coral, seagrass and mangrove in various parts of the Indian Ocean (Gray 1997; Wafar et al., 2011).

Due to taxonomic limitations, the major taxa encountered in cruises have also been limited to common names or groups but identified to species levels, even for species that have been described. Consequently, it is difficult to provide even snapshots of species richness or species diversity of any given space and time in the region, even for relatively well sampled areas like the Bay of Bengal and the Arabian Sea. The results from the latter areas have contributed to demonstration of the existence of species gradients with depths. However the need for multidisciplinary approaches to describe the drivers of these gradients in an ecosystem context cannot be overemphasized. Based on meiofauna and macrofauna occurrence across shallow to deep sea, from 20m to 6km water depth, both groups showed a similar gradient pattern that the density and species variety of the meiofauna and macrofauna decreased significantly with increasing water depth. The taxon-specific depth affinities have also formed the basis for macrobenthic organisms to be selected as suitable indicator organisms for environmental stress, taking advantage of their sedentary habits, so changes in population abundance and species composition reflect changes in habitat quality (Dauer and Corner, 1980). There is also a positive significant correlation between chlorophyll concentration and macrobenthic density (Pavithran et al., 2009). In the Arabian Sea, the surface water chlorophyll a is higher due to upwelling than that of Bay of Bengal (Prasanna et al., 2002; and Madhuparatap et al., 2003). Therefore the macrobenthic density is also higher in the Arabian Sea than in Bay of Bengal (Parulekar et al., 1982; Mahapatro et al., 2011). This highlights the need for bringing interdisciplinary ecological relationships into focus through ecosystem approach rather than dealing with single factors in isolation.

5.1   Research Gaps

- There is a lack of, or there are relatively few, quantitative data on local species extirpations in the Indian Ocean region.
- Predatory gastropod snails are fished or collected to be used as souvenirs in various places and since they play key roles in controlling prey population their local extirpation can lead to major changes in biodiversity and this needs to be studied.
- Overexploitation of other benthic species is occurring and there is need to properly evaluate the status and trends in exploited benthos.
- Particularly where benthos are being over-exploited or suffering high stress from coastal inputs, there is a need to get better information about the rate and magnitude of loss of species and the implication of these losses for ecosystem processes, including trophodynamics.
- Coastal benthic critical habitats e.g. corals, seagrass and mangrove, are threatened by various anthropogenic activities e.g. overfishing, mining, dynamiting, pollution, beach seining and trawling. There is need to study the effect of these on latitudinal and longitudinal loss of biodiversity and species diversity focusing also on endemic species.
- There is a need to have coordinated rapid regional assessments of benthos to provide a baseline report on the status of species diversity in the various benthic habitats in the Indian Ocean.
- Taxonomic experts are generally lacking for the various numerous benthic taxa groups and there is need to train and create a critical number of taxonomists to develop accurate identification and complete databases for benthic species diversity.
- For deep-sea or ocean diversity, there is a need for scientific assessment of little-known benthic ocean species biodiversity, particularly because it is increasingly being threatened by various anthropogenic activities, especially ocean dumping, mining and climate change.
- There is an urgent need to undertake regional scale assessments and long-term monitoring of habitat loss and species loss, record and document it using Geographical Information Systems (GIS). The assessment methodology should be robust enough to discern effects due to climate change and seasonality that may cause habitat loss or disturbance that may cause species loss, abundance and population density.
- There is a need for monitoring marine litter disturbance of benthic habitats, especially intertidal zones due to dumping of solid wastes.
- There is a need for undertaking long-term regional assessments of root causes or causal chain analyses of habitat loss and degradation to combat loss of biodiversity.

6 Megafauna

The megafauna, namely marine mammals (cetaceans and the sirenians), seabirds, and reptiles (sea turtles and sea snakes) may be described characteristically as large species with low fecundity and productivity, slow growth, and late age at maturity. Such biological characteristics have important implications for their sustainability in fisheries, especially as by-catch because they can sustain only very low rates of mortality. Moreover, they typically depend on a healthy, stable environment and generally have limited capacity to sustain and recover from depleted populations, such as result from heavy fishing pressure.

The megafauna in marine ecosystems play a significant role in the structure and functions of the ecosystems and in the economic sector, especially in tourism. The representative groups in consideration are as follows: (a) marine mammals, (b) marine reptiles, and (c) seabirds. These groups have characteristic species whose lives are interconnected with maritime zones, coastal and shelf waters and deep sea oceanic habitats as grazers or primary consumers and predators or secondary consumers in the ecological food chain cycles. In these predator-prey relationships these megafauna groups play an important ecological role in regulation of marine biodiversity, species richness and environment quality. However, as much as they play their important roles, they face various challenges that threaten their lives and ability to play their roles efficiently and effectively. The major threats are primarily anthropogenic pressures, especially habitat loss and degradation, overexploitation or unsustainable exploitation, pollution and climate change, whose root causes are a major concern (Wafar et al., 2011; Chaturvedi et al., 2013; Bellard et al., 2014).

6.1 Marine Mammals

The entire Indian Ocean region has 31 species of marine mammals found in the pelagic shelf and near-shore waters (De Boer et al., 2003). Though well surveyed, with further surveys the number of species is likely to increase. The North Western Indian Ocean region has 25 species, the North Eastern Indian Ocean 28 species, the South Eastern Indian Ocean 30 species and the South Western Indian Ocean has 25 species (De Boer et al., 2003). Whales are known to be highly migratory over long distances, across the hemisphere, to their nesting and feeding grounds.

The marine mammals are threatened by various human activities especially fishing activities using gill nets, seine nets, beach seines and drift nets, in which they are caught as by-catch; habitat degradation and loss, as well as pollution, including marine debris. These threats lead to the destruction of their breeding and feeding grounds. A further threat is deliberate hunting for food. Being slow to grow and reach maturity, coupled with having low fecundity, overexploitation leads to population destruction and collapse, leading to extinction (Kiszka et al., 2008).

6.2 Research Gaps

(a) Although the taxonomy of marine mammals is fairly well known, there is a need to train and equip local scientists and equip local institutions to effectively collect and archive quality data in suitable databases, to enable accurate analysis of status and trend of stock status of marine mammals across various time scales in various geographical locations, in a regional way due to their migratory behaviours.
(b) The existing research studies are patchy and one-off event types and their data and information make it difficult to standardize and to generate trends. Therefore there is a need for coordinated long-term monitoring, using standardized multidisciplinary methodologies which will allow to document status and trends in time and space, including climate change impacts.
(c) For conservation of whales, it is important to undertake analysis of the biology and ecology of the various species of whales to protect and avoid polluting their habitats.
(d) Undertake regular by-catch research including the fishing gears that are used so that the impacts of fishing on whales are understood and appropriate mitigating measures can be undertaken.
(e) Genetic studies are also needed to be able to understand the interconnectedness and dispersal nature of the similar species over the wider geographical ranges that are encountered in the region.

6.3 Sea Turtles

Sea turtles are herbivorous or sometimes invertebrate-eating reptiles that play a very important role in marine ecosystems in the maintenance of healthy seagrass beds and coral reefs and assist in avoiding trophic cascades. As part of their life is spent on land to breed, sea turtles also play an important role in nutrient cycling from land to water and vice-versa (through their faeces).

In the coral ecosystems, sea turtles also forage on sponges which are known to compete aggressively with corals and can reduce coral growth. Where sponges can colonize aggressively and grow, they can limit the growth of corals and modify the overall structure of a given coral reef ecosystem. Sea turtle predation on sponges can prevent the expansion of sponges, and thus protect coral reefs. In terms of pelagic food webs, sea turtles predate on jelly fish and especially leatherback sea turtles which may be the dominant predator on jellyfish. Decline in this key predator may lead to jelly-fish population explosion, possibly leading to gradual replacement of once abundant fish species (Lynam et al., 2006, Purcell, J.E. et al., 2007).

The Taxonomy of sea turtles in the Indian Ocean is well known and they are represented by only a few species which migrate extensively to various oceans. According to WWF (2012), the sea-turtle taxa are:

Leatherbacks (*Dermochelys coriacea*)

Green sea turtles (*Chelonia mydas*)

Hawksbills (*Eretmochelys imbricatea*)

Olive ridleys (*Lepidochelys olivecea*)

Loggerhead (*Caretta caretta*)

According to the Convention on International Trade in Endangered Species of Wild Fauna and Flora[1] (CITES) and IUCN all the species of sea turtles in the Indian Ocean require protection from anthropogenically induced activities. Specifically, CITES lists all sea turtle species under its Appendix I meaning that commercial international trade in the species is prohibited (CITES, 2015). The explanation of CITES listing is as follows:

- Appendix I lists species that are the most endangered among which are those threatened with extinction and CITES prohibits international trade in specimens of these species except when the purpose of the import is not commercial.
- Appendix II lists species that are not necessarily now threatened with extinction but that may become so unless trade is closely controlled.
- Appendix III is a list of species included at the request of a Party that already regulates trade in the species and that needs the cooperation of other countries to prevent unsustainable or illegal exploitation.

The IUCN Red List of Threatened Species (IUCN Red List) classified the status of leatherback turtles globally as 'Vulnerable' (IUCN, 2015). Specifically, the Southwest Indian Ocean subpopulations are 'Critically Endangered', while the Northeast Indian Ocean subpopulations are 'Data Deficient'. Green turtles and loggerhead turtles are listed in the IUCN Red List as globally 'Endangered', while Olive Ridley turtles are globally 'Vulnerable' and Hawksbill turtles are 'Critically Endangered'. Further threats to sea turtles are also discussed in chapter 39.

6.4   Research Gaps on Sea Turtles

- There is a need to undertake research on bycatch intensity by taxa and gear type for effective conservation management;
- Methods of assessment of bycatch should be standardized so that the data and information can be comparable using GIS and reduce high variances especially at regional levels;
- There is a need for long-term multidisciplinary research study to understand the life histories and ecology of the various species of sea-turtles in the various regions, how they relate to seasonality and climate change;
- There is a need to deploy observers in industrial fishing vessels to collect all the essential information for management of bycatch;
- Nesting and feeding grounds at national and regional levels need to be mapped;
- The genetic connectivity of the various taxa groups needs to be known in order to understand the nature of regional connectivity of the sea turtles.

6.5   Seabirds

Seabirds are characterized by their nature and behaviour to live partly in a terrestrial environment and partly in marine littoral, pelagic and oceanic habitats. Essentially they exploit the terrestrial environment for reproduction strategies and the benthic coastal marine and pelagic oceanic environment for foraging or feeding, playing the role of a fisher like human beings (Burger 1988, chapter 38).

Seabirds, in terms of their species numbers, contribute significantly to the overall marine species biodiversity and they can play a significant role in the predator – prey relationships in the marine food webs. Global patterns and trends in seabird biodiversity, and threats to the populations, are discussed in chapter 38.

Information exists on the seabird taxa in the Indian Ocean. These families are mainly *Diomedeidae* (albatrosses), *Procellariidae* (petrels) *Hydrobatidae* (storm-petrels), *Pelecanoididae* (diving-petrels), *Phaethonidae* (tropic birds), *Sulidae* (gannets and boobies), *Fregatidae* (frigatebirds), *Stercorariidae* (skuas) *Phalacrocoracidae* (cormorants) and *Laridae* (gulls and terns). Most of these taxa are migrant seabirds from European and Asian regions and their most preferred habitats are the estuaries of large rivers along the African continent. There are many species of seabirds, but there is a low degree of endemism (Wanless, 2012). There are still very few studies on seabirds and these are mostly patchy. Long-term studies to analyze status and trends are lacking. There is a need to focus these studies on measuring the impacts of climate change and bycatch due to industrial commercial fishing, and their effects on biodiversity changes in the coastal, offshore and deep or open ocean ecosystems. Seabirds in the tropical Indian Ocean region are primarily migratory birds from breeding grounds in the Arctic, Antarctic and temperate regions that come wintering in the tropics, after travelling tens of thousands kilometres. Chapter 38 discusses the major global threats to seabirds, including bycatch mortality in fisheries, habitat degradation and loss, over-exploitation of their food supply, bioaccumulation of pollutants and toxins, and sea-level rise. Because the seabirds use the Indian Ocean for only part of their annual cycle, it is very difficult to distinguish the impacts of pressures in the Indian Ocean from other pressures on the same populations. Combined with the few targeted studies of seabirds in the Indian Ocean, particularly offshore, it is hard to evaluate status and trends of most seabirds in the Indian Ocean and many research gaps exist.

6.6   Research Gaps on Seabirds

There is a need to address the following research gaps:

---

[1]   United Nations, *Treaty Section*, vol. 993, No. 14537.

- Lack of comprehensive taxonomic knowledge of species of seabirds at national and regional levels in the various countries of the Indian Ocean region;
- Lack of data and information of the bioaccumulation of toxic substances in seabirds arising from marine food chains and their impacts on species biodiversity;
- Lack of comprehensive long-term biophysical impacts due to habitat degradation and loss including sea-level rise impacts on seabirds migration and their biodiversity at taxa levels;
- Lack of comprehensive understanding of long-term bycatch impacts due to various fishing methods and gears on seabirds at taxa levels and different habitats ranging from coastal to open sea at various depths where the seabirds forage;
- Need to create GIS maps for various migratory routes of the seabirds encountered in the region and their hotspots at taxa levels;
- Need to undertake coordinated research using harmonized methodologies for improved quality of data and information and facilitate comparisons.

7 Conclusion

Compared to other world oceans, the Indian Ocean biodiversity is relatively still scarcely known in terms of the taxonomic composition of the species found therein and their geographical distribution, except for some continental shelf areas around the Indian Subcontinent, Southwestern Australia and Southern Africa. However, deep sea species remain the most unknown, hence the need for concerted research efforts to understand their ecological roles in the diverse ecosystems of the Indian Ocean where they may be encountered. It is known that fish taxa and their geographical distribution are comparatively fairly well known and documented but from a fisheries activity perspective, which also generates ecological bycatch concerns. Since fish do not live in isolation and are part of the ecological systems, their role in relation to other biological taxa including their interaction with the environment can be fully understood if comprehensive integrated ecological studies are undertaken across seasons and time scales.

References

Bates, N.R., Christine, P. and Sabine, C.L. (2006). Ocean carbon cycling in the Indian Ocean: 2. Estimates of net community production. *Global Biogeochemical Cycles*, Vol. 20, GB3021, doi: 10.1029/2005GB 002492, pp 1-14.

Bellard, C., Leclerc, C. and Courchamp, F. (2014). Impact of Sea level rise on the 10 insular biodiversity hotspots. *Global Ecology and Biogeography* 23:203-212.

Burger, J. (1988). *Seabirds and other marine vertebrates competition, predation and interactions*, 312pp Columbia UP.

Chaturvedi, N., Shah, M. and Jasrai, Y. (2013). Is there impact of climate change on biological productivity in the Indian Ocean. *Indian Journal of Geo-Marine Sciences* Vol.42(1) pp 5057.

Chesalina, T., Al-Kharusi, A. Al-Aisry, A., Al-Abri, N., Al-Mukhaini, E., Al-Maawali, A. and

Al Hasani, L. (2013). Study of diversity and abundance of fish larvae in the South-Western Part of the Sea of Oman 2011-2012. *Journal of Biology, Agriculture and Healthcare* Vol.3 No.1 22224-3208.

CITES (2015). Appendix I, II and III Classification of the Convention on International Trade in Endangered Species. http://www.cites.org/eng/app/index.php Accessed 10 May 2015.

Dauer D.M. and Corner, W.G. (1980). Effect of moderate sewage input on benthic polychaetespopulation. *Estuarine, Coastal and Shelf Science*. 10:335-362.

De Boer, R., Baldwin, C.L.K., Burton, E.L., Eyre, K.C.S., Jenner, M.N.M., Jenners, S.G., Keith, K.A., McCabes, E.C.M., Parsons, V.M., Peddemors, H.C., Rosenbaum, R., Rudolph and Simmonds, M.P. (2003). Cetaceans in the Indian Ocean Sanctuary: A review. A WDCS (Whale and Dolphin Conservation Society) *Science Report*: 1-52.

Dorgham, M.M. (2013). Plankton research in the ROPME Sea Area, achievements and gaps *International Journal of Environmental Research* 7(3): 767-778.

FAO (1990-2015). CWP Handbook of Fishery Statistical Standards. Section H: fishing areas for statistical purposes. CWP Data Collection. In: *FAO Fisheries and Aquaculture Department* [online]. Rome. Updated 19 February 2015. [Cited 10 December 2015]. http://www.fao.org/fishery/cwp/handbook/h/en.

FAO. (2014). *The State of World Fisheries and Aquaculture: Opportunities and challenges*. Rome. 243 pp.

FAO. (2011). Review of the state of world marine fishery resources. FAO Fisheries and aquaculture. *Technical Paper* No. 569. Rome, FAO. 2011. 334 pp.

Fabian, H., Koppelmann, R. and Weikert, H. (2005). Full-depths zooplankton composition at two deep sites in the Western and Central Arabian Sea. *Indian Journal of Marine Sciences* Vol.34(2) June 2005, pp. 174-187.

Fazel, N., Savari, A., Nabavi, S.M.B. and Zare, R. (2013). Seasonal variation of zooplankton abundance, composition and biomass in the Chabahar Bay, Oman Sea. *International Journal of Aquatic Biology* 1 (6): 294-305.

Fisher, R.A., Corbet, A.S. and Williams, C.B. (1943). The relationship between the number of species and the number of individuals in a random sample of an animal population. *Journal of Animal Ecology* 12:42-58.

Gray, J.S. (1997). Marine biodiversity: patterns, threats and conservation needs. *Biodiversity and Conservation* 6; 153-175.

Gregg, W.W., M.E. Conkright, P. Ginoux and J.E. O'Reilly (2003). Ocean Primary production and climate: Global decoded changes. *Geophysical Research letters* Vol, 30, No.15,1809, doi 1029/2003 GL 016889 p 1-3.

IOTC (Indian Ocean Tuna Commission) (2011). Report of the Seventh Session of IOTC Working Party on Ecosystems and Bycatch. 99 pp.

IOTC (2014) Skipjack tuna: http://iotc.org/sites/default/files/documents/science/species_summaries/english/Skipjack%20tuna.pdf;Yellowfin tuna: http://iotc.org/sites/default/files/documents/science/species_summaries/english/Yellowfin%20tuna.pdf; Bigeye tuna: http://iotc.org/sites/default/files/documents/science/species_summaries/english/Bigeye%20tuna.pdf (Accessed: May 10, 2015)

IUCN (2015). *The IUCN Red List of Threatened Species*. http://www.iucnredlist.org/ Accessed 10 May 2015.

Kiszka, J., Muir, C., Poonian, C., Cox, T.M., Amir, O.A., Bourjea, J., Razafindrakoto, Y., Wambiji, N. and Bristol, N. (2008). Marine Mammal bycatch in the Southwest Indian Ocean: Review and need for a comprehensive status assessment Western Indian Ocean. *Journal of Marine Sciences* No.2, pp. 119-136.

Lynam, C.P., Gibbons, M.J., Axelsen, B.E., Sparks, C.A.J., Coetzee, J., Heywood, B.G., Brierley, A.S. (2006). Jellyfish overtake fish in a heavily fished ecosystem. *Current Biology* 16(13): R492-R493.

Madhuparatap, M., Gauns, M., Ramaiah, N., Prasanna Kumar, S., Muraleedharan, P.M., DeSousa, S.N., Sardessai, S., Muraleedharan, S. (2003). Biogeochemistry of the Bay of Bengal during summer monsoon 2001. *Deep Sea Research* (II) 50:881-896.

Mahapatro, D., Panigrahy, R.C., Naik, S., Pati, S.K. and Samal, R.N. (2011). Macrobenthos of the Shelf Zone off Dhamara estuary, Bay of Bengal. *Journal of Oceanography and Marine Science* Vol.2(2), pp 32-42.

Moharana, P. and Patra, A.K. (2013). Spatial distribution and seasonal abundance of plankton population of Bay of Bengal at Digha sea-shore in West Bengal. *Indian Journal of Scientific Research* 4(2): 93-97.

McClanahan, T.R., Mangi, S.C. (2004). Gear-based management of a tropical artisanal fishery based on species selectivity and capture size. *Fisheries Management and Ecology* 11: 51 – 60.

McClanahan, T.R., Maina, J.M., Muthiga, N.A. (2011). Associations between climate stress and coral reef diversity in the Western Indian Ocean. *Global Change Biology*, doi:10.1111/j.1365-2486.2011.02395.x.

Parulekar, A.H., Harkantra, S.N. and Ansari, Z.A. (1982). Benthic production and assessment of dermersal fishery resource of Indian Seas, *Indian Journal of Marine Sciences* 11:107-114.

Pavithran S., Ingole B.S., Nanajkar, M., Raghukumar. C., Nath, B.N., Valsangkar, A.B. (2009). Composition of macrobenthos from the Central Indian Ocean Basin. *Journal of Earth System Science* 118 No.6 pp. 689-700.

Pierrot-Butts, A.C. and Angel, M.V. (2013). Pelagic biodiversity and biogeography of the oceans. *Biology International* Vol.51:9-35.

Prasama Kumar. S., Muraleedharan, P.M., Prasad, T.G., Gauns, M., Ramaiah, N. De-Souza, S.N., Sardesai, S., Madhupratap, M. (2002). Why is the Bay of Bengal less productive in summer monsoon compared to the Arabian Sea? *Geophysical Research Letters*, 29(4): 88-1-88-4. DOI: 10.1029/2002GL016013.

Purcell, J.E., Shin-ichi, U., Wen-Tseng, L. (2007). Anthropogenic causes of jellyfish blooms and their direct consequences for humans: A review. *Marine Ecology Progress Series* 350:153-174.

Ray, G.C. (1991). Coastal zone biodiversity patterns. *Bioscience* 41:490-498.

Schott, F., and McCreary, J.P. (2001). The monsoonal circulation of the Indian. *Progress in Oceanography*, 51, 1-123.

SWIOFC (2012). *Proceedings of the Fifth Session of the Scientific Committee of the South West Indian Ocean Fisheries Commission (SWIOFC)*, 27th February to 1st March 2012, 15 Orange Hotel, Cape Town, South Africa.

Wafar, M., Venkataraman, K., Ingo, B., Ajmal Khan, S., LokaBharathis, P. (2011). State of knowledge of coastal and marine biodiversity of Indian Ocean countries. *PLoS*

ONE 6(1): E14613. Doi10.1371/journal.pone.0014613.

Wajih, S., Naqvi, A., Narvekar, P.V. and Desa, E. (2006). Coastal biogeochemical processes in the North Indian Ocean, In *The Sea, vol.14A, The Global Coastal Ocean: Interdisciplinary Regional Studies and Syntheses*, edited by A.R. Robinson and K.Brink, pp.723-780, John Wiley, Hoboken, N.J.

Wanless, R.M. (2012). Seabirds of the Western Indian Ocean: A review of the status, distribution and interaction with fisheries in the South West Indian Ocean. In: Rudy Van der Elst (ed.), *Mainstreaming biodiversity in fisheries management: A retrospective analysis of existing data on vulnerable organisms in the South West Indian Ocean. South West Indian Ocean Project (SWIOFP) report.*

Whittaker, R.H. (1977). Evolution of species diversity in land communities. *Evolutionary Biology*.10:1-67

Whittaker, R.H. (1975). *Communities and ecosystems*. Macmillan 2nd ed. New York.

Whittaker, R.H. (1967). Gradient analysis of vegetation. *Biological Reviews* 42:207-264.

WWF (2012) *Global Marine Turtle Strategy 2012-2020*. Gland, Switzerland, 52p.

36F Open Ocean Deep Sea

Contributors:
Jeroen Ingels (Convenor), Patricio Bernal (Lead Member), Malcolm R. Clark, Katsunori Fujikura, Andrew R. Gates, Daniel O. B. Jones, Lisa A. Levin, Bhavani E. Narayanaswamy, Jose Angel A. Perez, Imants G. Priede, Ashley A. Rowden, Henry A. Ruhl, Ricardo Santos, Craig R. Smith, Paul Snelgrove, Thomas Soltwedel, Tracey Sutton, Andrew K. Sweetman, Saskia Van Gaever (Co-Lead Member), Michael Vecchione, Moriaki Yasuhara and Linda Amaral Zettler

1. Introduction to the open ocean deep sea

The deep sea comprises the seafloor, water column and biota therein below a specified depth contour. There are differences in views among experts and agencies regarding the appropriate depth to delineate the "deep sea". This chapter uses a 200 metre depth contour as a starting point, so that the "deep sea" represents 63 per cent of the Earth's surface area and about 98.5 per cent of Earth's habitat volume (96.5 per cent of which is pelagic). However, much of the information presented in this chapter focuses on biodiversity of waters substantially deeper than 200 m. Many of the other regional divisions of Chapter 36 include treatments of shelf and slope biodiversity in continental-shelf and slope areas deeper than 200 m. Moreover Chapters 42 and 45 on cold water corals and vents and seeps, respectively, and 51 on canyons, seamounts and other specialized morphological habitat types address aspects of areas in greater detail. The estimates of global biodiversity of the deep sea in this chapter do include all biodiversity in waters and the seafloor below 200 m. However, in the other sections of this chapter redundancy with the other regional chapters is avoided, so that biodiversity of shelf, slope, reef, vents, and specialized habitats is assessed in the respective regional or thematic chapters.

This truly vast deep-sea realm constitutes the largest source of species and ecosystem diversity on Earth, with great potential for mineral, energy, and living resources (e.g., Koslow, 2007). Despite major technological advances and increased deep-sea exploration in the past few decades (Danovaro et al., 2014), a remarkably small portion of the deep sea has been investigated in detail (Ramirez-Llodra et al., 2010), particularly in terms of time-series research (Glover et al., 2010). For the pelagic areas much less than 0.0001 per cent of the over 1.3 billion km3 of deep water has been studied. The inevitable result is weaker characterization of deep-sea biodiversity compared to the shelf, slope and terrestrial realms. Correspondingly this also means that continued scientific and surveying efforts may potentially change our current understanding of deep-sea biodiversity. There is strong evidence that the richness and diversity of organisms in the deep sea exceeds all other known biomes from the metazoan to the microbial realms (Rex and Etter, 2010; Zinger et al., 2011) and supports the diverse ecosystem processes and functions necessary for the Earth's natural systems to function (Thurber et al., 2014). Moreover, the extensive species, genetic, enzymatic, metabolic, and biogeochemical diversity hosted by the deep ocean also holds the potential for new pharmaceutical and industrial applications. With up to millions of estimated deep-sea species (cf. Chapter 34; CoML, 2010; Grassle and Maciolek, 1992), although the true number of species may be less, (Appeltans et al., 2012, Costello et al., 2013; Mora et al., 2013a), it would take many generations to document deep-sea diversity in its entirety. In fact, this may not even be possible given the huge taxonomic effort required (Mora et al., 2013a) and the rate of species extinctions (Pimm et al., 1995). Nor is it necessary to have fully quantified deep-sea biodiversity to commence identification of risks and opportunities, and design of programmes for its conservation and sustainable use, even if new knowledge is later acquired that enables such programmes to be improved.

Over the years, deep-sea ecologists have posited several theories to explain high deep-sea biodiversity; many highlight aspects of habitat heterogeneity and the extended time scales at which the deep sea is thought to operate (e.g. Levin and Dayton, 2009; Rex and Etter, 2010; Snelgrove and Smith, 2002). Most experts agree that the presence of different habitats, along with temporal variation, critically support deep-sea diversity; for instance, geomorphological structures such as canyons, seamounts (Figure 36F.1; cf. Chapter 51), hydrothermal vents and methane seeps (cf. Chapter 45), as well as biotic structures, such as cold-water coral reefs (Figure 36F.1; Chapter 42), and whale falls sustain unique assemblages of organisms, diversifying the deep-sea species pool (Reed et al., 2013). At the same time, however, many deep-sea species are widely distributed (e.g., Havermans et al., 2013; Ingels et al., 2006; Pawlowski et al., 2007), although new genetic tools already suggest many species are less cosmopolitan than was previously thought. Small-scale heterogeneity further enhances diversity, through the provision of phytodetrital patches, biogenic structures such as sponges and xenophyophores, organic food falls, pits, and hillocks, (Buhl-Mortensen et al., 2010). Anthropogenic structures such as deep-water oil rigs and shipwrecks harbouring highly diverse faunal assemblages reflect deep-sea faunal responses to smaller-scale habitat heterogeneity (Church et al., 2009; Taylor et al., 2014; Friedlander et al., 2014).

Deep-sea ecosystems are crucial for global functioning; e.g., remineralization of organic matter in the deep sea regenerates nutrients that help fuel the oceanic primary production that accounts for about half of atmospheric oxygen production. Whilst coastal and shallow-water processes and functions produce services within tangible time scales and local and regional spatial scales, the deep-sea processes and ecosystem functions that occur on the scale of microns to meters and time scales up to years often translate to useful services only after centuries of integrated activity (Thurber et al., 2014). Evidence demonstrates, however, that interannual changes in climate can influence deep-sea systems over time scales not fundamentally different from terrestrial habitats. Climatically driven changes in sinking particulate organic matter can alter deep-sea abundance, community structure, diversity and functioning within days to months, depending in part on body size (Ruhl et al., 2008; Ruhl and Smith, 2004) along with temperature-driven interannual diversity changes (Danovaro et al., 2014).

Numerous human activities affect deep-sea ecosystems, goods, and services directly and indirectly now and will do so increasingly in the future (Glover and Smith, 2003; Mengerink et al., 2014; Ramirez-Llodra et al., 2011). These are addressed in various chapters of Parts IV and V of this Assessment, with Chapters 11 (Capture Fisheries), 21 (Offshore Hydrocarbon Industries), 20 (Land-based Inputs), 23 (Other Mining Industries), 25 (Marine debris) and 27 (Tourism) of particular relevance.

2 Benthic realm

2.1 Deep-sea margins

The global continental margins extend for ~150,000 km (Jahnke, 2010) and encompass estuarine, open coast, shelf, canyon, slope, and enclosed-sea ecosystems (Levin and Sibuet, 2012). Deep-sea margins are those areas that lie beyond the shelf break, where the seafloor slopes down to the continental rise at abyssal depths, and encompasses bathyal depths. Numerous canyons and channels incise the continental slope (see Chapter 51), often featuring cold-water coral reefs (Chapter 42) or oxygen minimum zones (OMZs) as distinct habitats along the deep margin. Sediment covers much of the deep continental margin, but with exposed bedrock in areas where topography is too steep for sediment accumulation (e.g., steep canyon walls) or where sediment is washed away (e.g., parts of seamounts). Different faunas inhabit soft- and hard-bottom substrates.

Relative to their area, the margins account for a disproportionately large fraction of global primary production (10-15 per cent), nutrient recycling, carbon burial (>60 per cent of total settling organic carbon), and fisheries production (Muller-Karger et al., 2005). They are also exceptionally dynamic systems with ecosystem structures that can oscillate slowly or shift abruptly, but rarely remain static (Levin et al., 2014).

2.1.1 Status of and trends for biodiversity

In the well-studied North Atlantic, local macrofaunal (300 µm-3 cm) species diversity on the continental slope exceeds that of the adjacent continental shelf, and estimates of bathyal diversity in other parts of the world ocean are comparably high (Rex and Etter, 2010), but local environmental conditions drive regional differences: e.g., the Gulf of Mexico, the Norwegian and Mediterranean Seas (Narayanaswamy et al., 2013), the Eastern. Pacific and the Arabian Sea (Levin et al., 2001). Most researchers agree that habitat heterogeneity on different spatial scales drives high diversity along the margins (Narayanaswamy et al., 2013) and that margins often exhibit upwelling and increased production that enhances biodiversity. Nonetheless, excess food availability can reduce diversity.

Depth-related species diversity gradients in macrofauna often peak unimodally at mid-bathyal depths of about 1500-2000 m (Rex and Etter, 2010), although shallower peaks in diversity have been observed in Arctic waters (Narayanaswamy et al., 2005; 2010; Svavarsson, 1997; Yasuhara et al., 2012b) for bivalves, polychaetes, gastropods and cumaceans (Rex, 1981), as well as for the entire macrofauna (Etter and Mullineaux, 2000; Levin et al., 2001) and some meiofauna (Yasuhara et al., 2012b) (32 µm-1000 µm). Even regions with very low diversity can host highly specialized species (e.g., OMZs) and contribute to overall margin diversity (Gooday et al., 2010). Thus, throughout their depth gradient, continental margin slope areas exhibit the highest macrofaunal diversity and offer a potentially important refuge against future climate change, as mobile organisms could migrate upslope or downslope in search of suitable conditions (Rodriguez-Lazaro and Cronin, 1999; Yasuhara et al., 2008; 2009).

The diversity of meiofauna (32 µm-1,000 µm) exceeds that of the macrofauna and their diversity generally increases with depth; however, groups such as foraminifera and ostracods exhibit unimodal peaks in diversity (Yasuhara et al., 2012b). Meiofaunal diversity may decline or increase with increasing bathyal depths (Narayanaswamy et al., 2013), generally driven by food availability and intensity and regularity of disturbance regimes, as well as by temperature and local environmental conditions (Corliss et al., 2009; Yasuhara et al., 2012a; 2009; 2012b; 2014).

Russian and Scandinavian deep-sea expeditions described peak benthic megafaunal (>3 cm) diversity at mid-bathyal depths as early as the 1950s and 1960s, despite observing much lower megafaunal than meio- and macrofaunal diversity (Vinogradova, 1959). Sponges, cnidarians, crustaceans (decapods and isopods) and echinoderms (echinoids, asteroids, crinoids, holothurians) all display this pattern; however later studies confirmed the pattern for some megafaunal invertebrates, but showed a decline or even increase in diversity with increasing depth for some taxa. Evidence to date suggests lower species richness in deep-sea bacterial communities than in coastal benthic environments, with the caveat that deep-sea environments remain underexplored (Zinger et al., 2011). However, the presence of extreme environments in the deep sea which have high phylogenetic diversity promises a rich source of bacterial diversity and genetic innovation (Sogin et al., 2006).

Several faunal groups also exhibit latitudinal gradients in species diversity (Narayanaswamy et al., 2010; Rex and Etter, 2010; Yasuhara et al., 2009): diversity of crustaceans, molluscs and foraminifera declines poleward (Gage et al., 2004; Rex et al., 2000), whilst others such as nematodes respond to phytodetrital input (Lambshead et al., 2000). Latitudinal gradients have also been identified in bacteria (Fuhrman et al., 2008; Sul et al., 2013) but recent modelling indicates peak bacterial richness in temperate areas in winter (Ladau et al., 2013). The effect of seasons on macro-ecological patterns in the microbial ocean warrants continued investigation to test the mechanisms that underlie latitudinal patterns in different fauna.

Broad-scale depth and latitudinal patterns in benthic diversity are modified regionally by a variety of environmental factors operating at different scales. For example, OMZs strongly affect diversity where they impinge on the seafloor. OMZs typically occur between 200 m and 1000 m, often at major carbon burial sites along the continental margins where high productivity results in high carbon fluxes to the seafloor and low oxygen. The organic-rich sediments of these regions often support mats of large sulphide-oxidizing bacteria (*Thioploca, Beggiatoa, Thiomargarita*), and high-density, low-diversity metazoan assemblages. Protists are also well represented in OMZs such as the Cariaco Basin, where representatives of all major protistan clades occur (Edgcomb et al., 2011). Depressed diversity near OMZs centres favours taxa that can tolerate hypoxia, such as nematodes (Cook et al., 2000; Levin, 2003) and certain annelids and foraminifera (Levin, 2003). Other taxa that cannot tolerate low-oxygen conditions may aggregate at the OMZs fringes where food is often abundant.

2.1.2 Major pressures

Multiple anthropogenic influences affect deep-sea habitats located close to land (e.g., canyons, fjords, upper slopes when continental shelves are very narrow), including organic matter loading (see Chapter 20), mine tailings disposal (Kvassnes and Iversen, 2013; Kvassnes et al., 2009), litter (Pham et al., 2014), bottom trawling (Pusceddu et al., 2014) and overfishing (Clark et al., 2007), enhanced or decreased terrestrial input, oil and gas exploitation (Ramirez-Llodra et al., 2011) and, potentially in future, deep-sea mining (see Chapter 23). Fishing on margins can also have indirect ecological effects at deeper depths (Bailey et al., 2009). These anthropogenic influences can modify deep-margin habitats through physical smothering and disturbance, sediment resuspension, organic loading, and toxic contamination and plume formation, with concomitant losses in biodiversity, declining energy flow back to higher trophic levels, and impacts on physiology from exposure to toxic compounds (e.g., hydrocarbons, polycyclic aromatic hydrocarbons (PAHs), heavy metals) (see Ramirez-Llodra et al., 2011 for review).

2.2 Abyss

2.2.1 Status and trends for biodiversity

The abyss (~3-6 km water depth) encompasses the largest area on Earth. Its vast areas of seafloor plains and rolling hills are generally covered in fine sediments with hard substrates associated with manganese nodules, rock outcrops and topographic highs (e.g. seamounts). The absence of *in situ* primary production in this comparatively stable habitat (apart from scant occurrence of chemosynthesis at hydrothermal vents and cold seeps; cf. Chapter 45) characterize an ecosystem adapted to a limiting and variable rain of particulate detrital material that sinks from euphotic zones. Nonetheless, the abyss supports higher levels of alpha and beta diversity of meiofauna, macrofauna and megafauna than was recognized only decades ago (Rex and Etter, 2010). The prevalence of environmental DNA preserved in the deep sea biases estimates of richness, at least in the microbial domain, adding a challenge to biodiversity study in the abyss using molecular methods (Pawlowski et al., 2011).

Despite poorly known biodiversity patterns at regional to global scales (especially regarding species ranges and connectivity), some regions, such as the abyssal Southern Ocean (Brandt et al., 2007; Griffiths, 2010) and the Pacific equatorial abyss, are likely to represent major reservoirs of biodiversity (Smith et al., 2008).

2.2.2 Major pressures

The food-limited nature of abyssal ecosystems, and reliance on particulate organic carbon (POC) flux from above, suggest that all groups, from microbes to megafauna, will be highly sensitive to changes in phytoplankton productivity and community structure, and especially to changes in the quantity and quality of the export flux (Billett et al., 2010; Ruhl et al., 2008; Ruhl and Smith, 2004; Smith et al., 2008; Smith et al.,2013). Climate warming in some broad areas may increase ocean stratification, reduce primary production, and shift the dominant phytoplankton community structure from diatoms to picoplankton, and reduce export efficiency, driving biotic changes over major regions of the abyss, such as the equatorial Pacific (Smith et al., 2008). However the effects of climate change, including ocean warming, on biodiversity are likely to vary regionally and among species groups in ways that are poorly resolved with current models and knowledge of ecosystem dynamics in the deep sea. In the future, deep sea mining may also become a pressure on abyssal areas of the deep sea, and potential effects are addressed in Chapter 21.

2.3 Hadal

2.3.1 The Hadal zone

The Hadal zone, comprising ocean floor deeper than 6000 m, encompasses 3,437,930 km2, or less than 1 per cent of total ocean area (Harris et al., 2014) and represents 45 per cent of its depth and related gradients. Over 80 separate basins or depressions in the sea floor comprise the hadal zone, dominated by 7 great trenches (>6500 m) around the margins of the Pacific Ocean, five of which extend to over 10 km depth: the Japan-Kuril-Kamchatka, Kermadec, Tonga, Mariana, and Philippine trenches. The Arctic Ocean and Mediterranean Sea lack hadal depths. These trenches are often at the intersection of tectonic plates, exposing them as potential epicentres of severe earthquakes which can directly cause local and catastrophic disturbance to the trench fauna.

2.3.2 Status and trends for biodiversity

Although the hadal zone contains a wide range of macro- and megafaunal taxa (cnidarians, polychaetes, bivalves, gastropods, amphipods, decapods, echiurids, holothurians, asteroids, echinoids, sipunculids, ophiuroids and fishes (Beliaev, 1989; Wolff, 1970), all trenches occur below the Carbonate Compensation Depth (CCD), reducing the numbers of calcified protozoan and metazoan species found there (Jamieson, 2011). Chemosynthetic seep biota, including vesicomyid and thyasirid clams, occur in hadal depths in the Japan Trench; the deepest known methane seeps and associated communities are found at 7,434 m in this area (Fujikura et al., 1999; Watanabe et al., 2010). Cold seep communities also commonly occur in trench areas, such as the Aleutian and Kuril Trenches (Juniper and Sibuet, 1987; Ogawa et al., 1996; Suess et al., 1998). Benthic foraminifera are among the most widespread taxa at hadal depths and include calcareous, large agglutinated, and organic walled species (Beliaev, 1989; Gooday et al., 2008). Abundant metazoan meiofaunal taxa, such as nematodes, at hadal depths (Gambi et al., 2003; Itoh et al., 2011; Kitahashi et al., 2013; Tietjen, 1989; Vanhove et al., 2004) may exceed those found at bathyal depths by 10-fold (Danovaro et al., 2002); small numbers of ostracods, halacarids, cumaceans, kinorhynchs, and meiofaunal-sized bivalves are also found there (Vanhove et al., 2004). Nematode and copepod communities in trenches differ greatly from those found at bathyal and abyssal depths (Gambi et al., 2003; Kitahashi et al., 2013), driven by opportunistic taxa and meiofaunal dwarfism in trench systems (Danovaro et al., 2002; Gambi et al., 2003).

Although not yet well quantified, and the mechanisms remain to be discerned, higher densities of fauna (Jamieson et al., 2009) and respiration have been found at trench axis points than would be expected from a purely vertical rain of POC flux (Glud et al., 2013).The exact number of species in trenches is not known, but the few quantitative studies made so far suggest that diversity is lower compared to diversity at abyssal depths (Grassle, 1989). Reasons for the lower diversity levels are not well understood but the high pressure, relatively high food supply and organic matter accumulation, relatively elevated temperature (due to adiabatic heating), or a combination thereof may attenuate trench diversity.

Sampling to date suggests that hadal basins are populated by a higher proportion of endemic species compared to much shallower waters, species that can survive the extreme hydrostatic pressure and, in some instances, remoteness from surface food supply (Wolff, 1970). Physiological and other evidence suggests that fishes cannot survive at depths greater than 8000 m (Yancey et al., 2014); the deepest hadal fish, the liparids (snail-fish), are unique to each trench system. Decapod crustaceans have been observed only to 8200 m (Gallo et al., in revision).

At depths over 8000 m, scavenging amphipod crustaceans dominate the mobile megafauna, along with potential predators, including penaeid shrimp, princaxelid amphipods and ulmarid jellyfish, as observed in the New Britain Trench and the Sirena Deep (Mariana Trench). Comparison of scavenging and epibenthic/demersal biota suggests that density, diversity, and incidence of demersal (near bottom) lifestyles all increase with greater food supply (Blankenship and Levin, 2007; Blankenship et al., 2006).

Wide separation between trenches in the northern and southern hemispheres and between the different oceans has likely facilitated speciation to result in distinct assemblages of fauna in each hadal basin (Fujii et al., 2013). Some 75 per cent of the species in Pacific Ocean trenches may be endemic to each trench. Despite their remoteness from the surface, many hadal trenches are close to land and receive organic inputs from terrestrial and coastal sources, yielding higher mega-, macro- and meio-faunal densities than expected for greater depths (Danovaro et al., 2003; Danovaro et al., 2002; Jamieson, 2011; Jumars and Hessler, 1976; Vanhove et al., 2004).

2.3.3 Major pressures

The proximity of some trenches to land also increases their vulnerability to human activity in terms of dumping of materials and effluents, as well as from disaster debris, run off from land and pollution from ships. Some of these items, including anthropogenic litter, have been observed down to 7,200 m depth (George and Higgins, 1979). Evidence for the vulnerability of trench fauna is also provided by the levels of the radioisotope ^{134}Cs detected in sediments in the Japan Trench, four months after the Fukushima Dai-ichi nuclear disaster (Oguri et al., 2013).

2.3.4 Knowledge gaps

Trenches are arguably the most difficult deep-sea environments to access and current facilities are very limited worldwide, and consequently knowledge of their biodiversity is particularly incomplete.

In general, biodiversity patterns of non-nematode meiofauna and non-foraminiferal protists are especially poorly known in the deep sea.

Most information about biodiversity in the deep sea is for the predominant soft-substrate habitats. However, hard substrates abound in the deep sea in nearly all settings, and organisms that cannot be seen in a photograph or video image are hard to sample and study quantitatively. Thus knowledge of small-taxon biodiversity is best developed for deep-sea sediments.

Beyond cataloguing diversity, even in those systems we have characterized, almost nothing is known about the ranges of species, connectivity patterns or resilience of assemblages and their sensitivity to climate stressors or direct human disturbance. There is also currently a lack of appropriate tools to adequately evaluate human benefits that are derived from the deep sea (Jobstvogt et al., 2014a; 2014b; Thurber et al., 2014).

3 Pelagic realm

3.1 Status and trends for biodiversity

Between the deep-sea bottom and the sunlit surface waters are the open waters of the deep pelagic or "midwater" environment. This huge volume of water is the least explored environment on our planet (Webb et al., 2010). The deep pelagic realm is very diffuse, with generally low apparent abundances of inhabitants, although recent observations from submersibles indicate that some species may concentrate into narrow depth bands (Herring, 2002).

The major physical characteristics structuring the pelagic ecosystems are depth and pressure, temperature, and the penetration of sunlight. Below the surface zone (or epipelagic, down to about 200 m), the deep layer where sunlight penetrates with insufficient intensity to support primary production, is called the mesopelagic zone. In some geographic areas, microbial degradation of organic matter sinking from the surface zone results in low oxygen concentrations in the mesopelagic, called OMZs (Robinson et al., 2010). This mesopelagic zone is a particularly important habitat for fauna controlling the depth of CO_2 sequestration (Giering et al., 2014).

Below the depth to which sunlight can penetrate (about 1,000 m) is the largest layer of the deep pelagic realm and by far the largest ecosystem on our planet, the bathypelagic region. This comprises almost 75 per cent of the volume of the ocean and is mostly remote from the influence of the bottom and its communities. Temperatures there are usually just a few degrees Celsius above zero. The boundary layer where both physical and biological interactions with the bottom occur is called 'benthopelagic'.

The transitions between the various vertical layers are gradients, not fixed surfaces; hence ecological distinctions among the zones are somewhat blurred across the transitions. Recent surveys have shown a great deal of connectivity between the major pelagic depth zones (Sutton, 2013). The abundance and biomass of organisms generally varies among these layers from a maximum near the surface, decreasing through the mesopelagic, to very low levels in the bathypelagic, increasing somewhat in the benthopelagic (Angel, 1997; Haedrich, 1996). Although abundances are low, because such a huge volume of the ocean is bathypelagic, even species that are rarely encountered may have very large total population numbers (Herring, 2002).

The life cycles of deep-sea animals often involve shifts in vertical distribution among developmental stages. Even more spectacular are the daily vertical migrations of many mesopelagic species (Benoit-Bird and Au, 2006; Hays, 2003). This vertical migration may increase physical mixing of the ocean water and also contributes to a "biological pump" that drives the movement of carbon compounds and nutrients from the surface waters into the deep ocean (Robinson et al., 2010).

Sampling the deep pelagic biome shares the logistical difficulties of other deep-sea sampling, compounded by the extremely large volume and temporal variability of the environment and the widely dispersed populations of its inhabitants. New species continue to be discovered regularly. Whereas scientific information on the composition of mesopelagic assemblages is rapidly improving, very little is known of the structure of the deeper lower bathyal and abyssal pelagic zones.

Possibly because of high mobility and transport by ocean current, the overall diversity of species seems to be less than that found in other ecosystems (Angel, 1997). However, the number of distinct major evolutionary groups (i.e., phyla, classes, etc.) found in the deep pelagic is high.

Studies of microbes and their roles in the deep pelagic ecosystems are just beginning to reveal the great diversity of such organisms. The species richness of deep ocean bacteria surpasses that of the surface open ocean (Zinger et al., 2011).

As is true in other pelagic systems, crustaceans make up a large percentage of the deep zooplankton in both abundance and numbers of species. These crustaceans include numerous and diverse copepods, amphipods, ostracods and other major groups. Some groups, like arrow worms, are almost all pelagic and are important in deep waters. Large gelatinous animals, including comb jellies, jellyfishes, colonial siphonophores, salps and pyrosomes, are extremely important in deep pelagic ecosystems (Robison, 2004).

The strong swimmers of the deep pelagic, the "nekton", include many species of fishes and some sharks, crustaceans (shrimps, krill, and other shrimplike animals), and cephalopods (including squids, "dumbo" and other octopods, and "vampire squids") (Hoving et al., 2014). In terms of global fish abundance, deep pelagic fishes are by far the numerically dominant constituents; the genus *Cyclothone* alone outnumbers all coastal fishes combined and is likely to be the most abundant vertebrate on earth. Furthermore, at an estimated ~1,000 million tons, mesopelagic fishes dominate the world's total fish biomass and constitute a major component of the global carbon cycle. Acoustic surveys now suggest that an accurate figure of mesopelagic fish biomass may be an order of magnitude higher (10,000 - 15,000 million tons; Irigoien et al., 2014; Kaartvedt et al., 2012; Koslow, 2009). When bathypelagic fish biomass is included, deep pelagic fish biomass is likely to be the overwhelming majority of fish biomass on Earth (Sutton, 2013). The deep pelagic fauna is also important prey for mammals (toothed whales and elephant seals) and even birds (emperor penguins) and reptiles (leatherback sea turtles). The amount of deep-sea squids consumed by sperm whales alone annually has been estimated to exceed the total landings of fisheries worldwide (Rodhouse and Nigmatullin, 1996).

Horizontal patterns exist in the global distribution of deep pelagic organisms. However, the faunal boundaries of deep pelagic assemblages are less distinct than those of near-surface or benthic assemblages (Pierrot-Bults and Angel, 2012). Generally, the low-latitude oligotrophic regimes that make up the majority of the global ocean house more species than higher-latitude regimes (Hopkins et al., 1996). Some major oceanic frontal boundaries, such as the polar and subpolar fronts, extend down into deep waters and appear to form biogeographic boundaries, although the distinctness of those boundaries may decrease with increasing depth.

The dark environment also means that production of light by bioluminescence is almost universal among deep pelagic organisms. Some animals produce the light independently, whereas others are symbiotic with luminescent bacteria.

3.2   Major pressures

A fundamental biological characteristic throughout the deep pelagic biome is that little or no primary production occurs and deep pelagic organisms are dependent on food produced elsewhere. Therefore, changes in surface productivity will be reflected in changes in the deep midwater. When midwater animals migrate into the surface waters at night, they are subjected to predation by near-surface species. Shifts in the abundance of those predators will affect the populations of the migrators and, indirectly, the deeper species that interact with the vertical migrators at their deeper daytime depths. Either or both of these effects may be caused by global climate change, fishing pressure and the impact of pollutants in surface waters (Robinson et al., 2010; Robison, 2009).

Climate change will likely increase stratification caused by warming of surface waters and expanded OMZs resulting from the interaction of shifts in productivity with increased stratification. If the so-called conveyor-belt of global circulation weakens, transport of oxygen by the production of deep water will affect the entire deep sea. The biomass of mesopelagic fishes in the California Current, for instance, has declined dramatically during recent decades of reduced midwater oxygen concentrations (Koslow et al., 2011). Furthermore, increases in carbon dioxide resulting in acidification may affect diverse deep pelagic animals, including pteropods (swimming snails) and crustaceans which use calcium carbonate to build their exoskeletons, fishes that need it for internal skeletons, and cephalopods for their balance organs. Acidification also changes how oxygen is transported in the blood of animals and those living in areas of low oxygen concentration may therefore be less capable of survival and reproduction (Rosa and Seibel, 2008).

Few fisheries currently target deep pelagic species, but fisheries do affect the ecosystem. Whaling reduced worldwide populations of sperm whales and pilot whales to a small fraction of historical levels (Roman et al., 2014). Similarly, fisheries for surface predators such as sharks, tunas and billfishes, and on seamounts, reduce predation pressure, particularly on vertical migrators like squids and lantern fishes (Zeidberg and Robison, 2007).

Increasing extraction of deep-sea hydrocarbon resources increases the likelihood of accidental deep release of oil and methane (Mengerink et al., 2014), as well as the deep use of dispersants to minimize apparent effects of such spills at the surface (See Chapter 21).

Deep sea mining and some forms of renewable energy production may also affect the pelagic realm of the deep ocean (Ramirez-Llodra et al., 2011), and potential effects are addressed in Chapters 23 and 22 respectively.

3.3 Knowledge gaps

Any summary of deep pelagic ecosystems emphasizes how little is known, especially relative to coastal systems. Sampling has been intensively conducted in only a few geographic areas, using selective methods, each of which illuminates only a fraction of the biodiversity. Sampling at lower bathyal or abyssal depths has been limited, and virtually nothing is known about pelagic fauna associated with deep trenches. There is also limited knowledge of the performance of conservation and management measures developed for coastal and shelf marine ecosystems when applied in deep ocean systems characterized by large spatial scales and variable but sometimes vertically and/or horizontally high-mobility organisms, and incomplete knowledge of ecosystem structure and processes.

4 Special areas typical for the open ocean deep sea

4.1 Ocean ridges

The Mid-Ocean Ridge system is a continuous single feature on the earth's surface extending ca. 50,000 km around the planet; it defines the axis along which new oceanic crust is generated at tectonic plate boundaries (Heezen, 1969). The ridge sea floor is elevated above the surrounding abyssal plains, reaching the sea surface at mid-ocean islands, such as Iceland, the Azores and Ascension Island in the Atlantic Ocean, Easter Island and Galapagos in the Pacific Ocean. Typically there is a central axial rift valley bounded by ridges on both sides. A series of sediment-covered terraces slope down on the two sides of the ridge axis to the abyssal plains. The global ridge system, including associated island slopes, seamounts and knolls, represents a vast area of mid-ocean habitat at bathyal depths, accessible to fauna normally associated with narrow strips of suitable habitat on the continental slopes. The ocean ridges sub-divide the major ocean basins, but fracture zones at intervals permit movement of deep water and abyssal organisms between the two sides of the ridge.

Much attention has been directed to the importance of Mid-Ocean Ridges as sites of the hydrothermal vents and their unique fauna found close to the geothermally active ridge axis (German et al., 2011). However, the total area of hydrothermal vents is small and the dominant fauna on the mid-ocean ridges is made up of typical bathyal species known from adjacent continental margins (See Chapter 45). The biomass of benthic fauna and demersal fishes on the ridges is generally similar to that found at corresponding depths on the nearest continental slopes (Priede et al., 2013). New species, potentially endemic to mid-ocean ridges, have been discovered (Priede et al., 2012). But these are likely to be found elsewhere as exploration of the deep sea progresses. The island slopes and summits of seamounts associated with ocean ridges are important areas for fisheries; evidence suggests that biodiversity, including large pelagic predators, is enhanced around such features (Morato et al., 2010; Morato et al., 2008). Chapter 51 considers the biodiversity of these mid-ocean ridges, and its threats, in greater detail.

4.2 Polar deep sea

Polar marine ecosystems differ in many ways from other marine ecosystems on the planet (see Chapters 36G and 36H).

4.3 Arctic

Arctic deep-sea areas have generally been poorly studied; although several studies over the past two decades have greatly advanced our knowledge of its marine diversity and deep-sea processes. They indicate that the Arctic deep sea is an oligotrophic area, featuring steep gradients in benthic biomass with increasing depth that are primarily driven by food availability (Bluhm et al., 2005, 2011).

The Arctic deep basins comprise ~50 per cent of the Arctic Ocean seafloor and differ from those of the North Atlantic, as the Arctic Sea is relatively young in age, semi-isolated from the world's oceans, and largely ice-covered. Moreover, the high Arctic experiences more pronounced seasonality in light, and hence in primary production, than lower latitudes.

The history and semi-isolation of the Arctic basin play a major role in its biodiversity patterns (Golikov and Scarlato, 1990). Originally an embayment of the North Pacific, the Arctic deep sea was influenced by Pacific fauna until ~80 million years ago, when the deep-water connection closed (Marincovich Jr. et al., 1990). Exchange with the deep Atlantic began ~40 Ma ago, coinciding with a strong cooling period (Savin et al., 1975). Although some Arctic shelf and deep-sea fauna were removed by Pleistocene glaciations, other shelf fauna in the Atlantic sector of the Arctic found refuge in the deep sea and are considered the ancestral fauna at least for some of the recent Arctic deep-sea fauna (Nesis, 1984). The bottom of the Arctic basin is filled with water originating from the North Atlantic (Rudels et al., 1994); the sediments are primarily silt and clay whilst the ridges and plateaus have a higher sand fraction (Stein et al., 1994). Exceptions include ice-rafted dropstones, enhancing diversity by providing isolated hard substrata and enhanced habitat heterogeneity for benthic fauna (Hasemann et al., 2013; Oschmann, 1990). Considerable inputs of refractory terrestrial organic matter from the large Russian and North American rivers characterize the organic component of sediments along the slopes, and in the basins (Stein and Macdonald, 2004). The only present-day deep-water connection to the Arctic is via the Fram Strait (~2,500m), providing immigrating species access via the high water flux through this gateway. Submarine ridges within the Arctic form physical barriers, but current evidence suggests that these do not form biogeographic barriers (Deubel, 2000; Kosobokova et al., 2011; Vinogradova, 1997).

Bluhm et al. (2011) conservatively estimated the number of benthic invertebrate taxa in the Arctic deep sea to be ~1,125. As in other soft-sediment habitats, foraminiferans and nematodes generally dominate the meiofauna, whereas annelids, crustaceans and bivalves dominate the macrofauna, and echinoderms dominate the megafauna. The degree of endemism at the level of both genera and species is far lower than in the Antarctic, which has a similarly harsh environment. Just over 700 benthic species were catalogued from the central basin a decade ago (Sirenko, 2001). The latitudinal species-diversity gradient has been observed in the Arctic Ocean (Yasuhara et al., 2012b) and the peak of the unimodal species-diversity depth gradient occurs at much shallower depths compared to other oceans (Clarke, 2003; Svavarsson, 1997; Yasuhara et al., 2012b).

The Arctic, is populated by species that have experienced selection pressure for generalism and high vagility (Jansson and Dynesius, 2002), and should have inherent resilience in the face of climate change.

In a warmer future Arctic with less sea ice altered algal abundance and composition will affect zooplankton community structure (Caron and Hutchins, 2012) and subsequently the flux of particulate organic matter to the seafloor (Wohlers et al., 2009), where the changing quantity and quality of this matter will impact benthic communities (Jones et al., 2014; Kortsch et al., 2012).

4.4   Antarctic

The Southern Ocean comprises three major deep ocean basins, i.e., the Pacific, Indian and Atlantic Basins, separated by submarine ridges and the Scotia Arc island chain. Oceanographically, the Southern Ocean is a major driver of global ocean circulation and plays a vital role in interacting with the deep water circulation in each of the major oceans.

Chapter 36H describes the general dynamics of the Southern Ocean, including seasonal changes. The winter sea-ice formation creates cold, dense, salty water that sinks to the seafloor and forms very dense Antarctic Bottom Water (Bullister et al., 2013). This in turn pushes the ocean's nutrient-rich, deep water closer to the surface, generating areas of high primary productivity in Antarctic waters, similar to areas of upwelling elsewhere in the world.

The remote Southern Ocean is home to a diverse and rich community of life that thrives in an environment dominated by glaciations and strong currents (Griffiths, 2010). However, although relatively little is known about the deep-sea fauna, or about the complex interactions between the highly seasonally variable physical environment and the species that inhabit the Southern Ocean, but our knowledge of Southern Ocean deep-sea fauna and biogeography is increasing rapidly (Griffiths, 2010; Kaiser et al., 2013). The range of ecosystems found in each of the marine realms can vary greatly within a small geographic area (e.g. Grange and Smith, 2013), or in other cases remain relatively constant across vast areas of the Southern Ocean. The region also contains many completely un-sampled areas for which nothing is known (e.g., Amundsen Sea, Western Weddell Sea, Eastern Ross Sea). These areas include the majority of the intertidal zone, areas under the floating ice shelves, and the greater benthic part of the deep sea. However, several characteristic features of Southern Ocean ecosystems include circumpolar distributions and eurybathy of many species (Kaiser et al., 2013).

Both pelagic and benthic communities tend to show a high degree of patchiness in both diversity and abundance. The benthic populations show a decrease in biomass with increasing depth (Arntz et al., 1994), with notable differences in areas of disturbance due to anchor ice and icebergs in the shallows (Smale et al., 2008) and in highly productive deep fjord ecosystems (Grange and Smith, 2013). Hard and soft sediments from the region are known to be capable of supporting both extremes of diversity and biomass. In some cases, levels of biomass are far higher than those in equivalent habitats in temperate or tropical regions. A major international study led by Brandt revealed comparably high levels of biodiversity (higher than in the Arctic), thereby challenging suggestions that deep-sea diversity is depressed in the Southern Ocean (Brandt et al., 2007). Understanding of large-scale diversity distributions is improving (Brandt and Ebbe, 2009; Kaiser et al., 2013). For example, depth-diversity gradients of several taxa are known to be unimodal with a shallow peak comparable to those of the Arctic Ocean (Brandt et al., 2007; Brandt and Ebbe, 2009).

Longline fishing continues in the Southern Ocean, where the Commission for the Conservation of Antarctic Marine Living Resources (CCAMLR) has been implementing conservation measures for toothfish, icefish and krill fisheries, and has closed almost all of the regulatory area to bottom trawling since the 1980s (Reid et al., 2010; Hanchet et al., 2015). Climate change, is also a significant potential threat to the Antarctic marine communities (Griffiths, 2010; Smith et al., 2012), for reasons similar to those presented for the Arctic.

4.5 Seamounts

Seamounts are important topographic features of the open ocean. Although they are small in area relative to the vast expanse of the abyssal plains, accounting for <5 per cent of the seafloor (Yesson et al., 2011), three important characteristics distinguish them from the surrounding deep-sea habitat (Figure 36F.1; see Chapter 34). First, they are "islands" of shallow sea floor, and provide a range of depths for different communities. Second, bare rock surfaces can be common, enabling sessile organisms to attach to the rock, in contrast to the majority of the ocean sea floor, which is covered with fine unconsolidated sediments. Third, the physical structure of some seamounts drives the formation of localised hydrographic features and current flows that can keep species and production processes concentrated over the seamount, even increasing the local deep pelagic biomass. They are a sufficiently important part of marine deep-sea biodiversity that seamounts are fully treated in Chapter 51 of this Assessment.

4.6 Organic falls

The decay of large sources of organic matter (e.g., whales, wood, jellyfish) that 'fall' from surface or midwater provide a concentrated source of food on the deep sea floor directly, and indirectly through the decay of the organic matter, can yield hydrogen sulphide and methane. An array of scavenging species (hagfish, amphipods, ophiuroids, and crabs) is adapted to rapidly finding and consuming organic matter on the deep seabed. In addition, lipid-rich whale bones and wood support specialized taxa that have evolutionarily adapted to consume the substrate via symbionts (Smith and Baco, 2003; Smith et al., 2015). At least 30 species of polychaetes in the genus *Osedax* colonize and degrade whale bones, with the aid of heterotrophic symbionts in the group Oceanspirales (Goffredi et al., 2005; Rouse et al., 2009; Smith et al., 2015). *Osedax* and other taxa colonizing whale falls exhibit biogeographic separation, succession during the life of the whale fall (Smith and Baco, 2003; Braby et al., 2007; Glover et al., 2005; Smith et al., 2015), *Adipicola* and other deep-sea mussels also harbour chemoautotrophic endosymbionts and colonize sulphide-rich whale remains (Fujiwara et al., 2007; Thubaut et al., 2013). Similarly, members of the bivalve genus *Xylophaga* colonize and consume wood in the deep sea, with symbionts that aid cellulose degradation and nitrogen fixation. The activities of these 'keystone' species, in conjunction with microbial decay, transform the environment and facilitate colonization by a high diversity of other taxa, for example >100 species thus far found only on deep-sea whale falls (Smith et al., 2015). Human impacts have likely already affected these organic-fall ecosystems. For example, 20th century whaling drastically reduced the flux of whale carcasses to the deep seafloor (Roman et al., 2014; Smith, 2006; Smith et al., 2015).

Numerous areas throughout the world's oceans have experienced large jellyfish population expansions. Although numerous studies have sought to identify the driving forces behind and the impacts of live jellyfish on marine ecosystems (Purcell, 2012; Purcell et al., 2007), very few have focused on the environmental consequences from the deposition of jellyfish carcasses (from natural die-off events). Recently it has become apparent that jellyfish carcasses have very high sinking speeds (1,500 m d-1, Lebrato et al., 2013a; 2013b). Thus, jellyfish blooms may affect seafloor habitats through the sedimentation of jellyfish carcasses (but also of macro-zooplankton, see Smith et al. (2014)), the smothering of extensive areas of seafloor and reducing oxygen flux into seafloor sediments leading to hypoxic/anoxic conditions. Jelly falls may also be actively consumed by typical deep-sea scavengers, enhancing food-flux into deep-sea food webs (Sweetman et al., 2014). Jellyfish falls have so far been observed in the Atlantic, Indian and Pacific oceans (Billett et al., 2006; Lebrato and Jones, 2009; Yamamoto et al., 2008; Lebrato et al., 2013a; 2013b; Sweetman and Chapman, (2011), and are reviewed in Lebrato et al. (2012).

4.7 Methane seeps

Continental margins host a vast array of geomorphic environments associated with methane seepage and other types of seeps. Many support assemblages reliant on chemosynthesis fuelled by methane and sulphide oxidation (Levin and Sibuet, 2012; Sibuet and Olu, 1998). Their specialized biodiversity features are assessed in Chapter 45.

5 Major ecosystem services being affected by the pressures

Despite its apparent remoteness and inhospitability, the deep ocean and seafloor play a crucial role in human social and economic wellbeing through the ecosystem goods and services they provide (Armstrong et al., 2012; Thurber et al., 2014; van den Hove and Moreau, 2007). Whilst some services, such as deep-sea fisheries, oil and gas energy resources, potential CO2 storage, and mineral resources directly benefit humans, other services support the processes that drive deep-sea and global ecosystem functioning. Despite its inaccessibility to most people, the deep sea nonetheless supports important cultural and existence values. The deep sea acts as a sink for anthropogenic CO2, provides habitat, regenerates nutrients, is a site of primary (including chemosynthetic) and secondary biomass production, as well as providing other biodiversity-related functions and services, including those the deep water and benthic assemblages provide (Irigoien et al., 2014).

Ocean warming and acidification associated with climate change already affect the deep sea, reaching abyssal depths in some areas (Østerhus and Gammelsrød, 1999). Ongoing global climatic changes driven by increasing anthropogenic emissions and subsequent biogeochemical changes portend further impacts for all ocean areas, including the deep-sea and open ocean (Mora et al., 2013b). Data from pre-anthropocene times indicates millennial-scale climate variability on deep-sea biodiversity (Cronin and Raymo, 1997; Cronin et al., 1999; Hunt et al., 2005; Wollenburg et al., 2007; Yasuhara and Cronin, 2008; Yasuhara et al., 2012a; 2009), as well as decadal-centennial climate events (Yasuhara et al., 2008; 2014). The potential impacts of climate change on the ocean are addressed in Part II of the Intergovernmental Panel on Climate Change (IPCC) 5th Assessment Report, Working Group II Chapters 6 and 30. Consistent with the mandate of this Assessment, they are only briefly summarized here.

Some impacts of climate change will be direct. For example, altered distributions and health of open-ocean and deep-sea fisheries are expected to result from warming-induced latitudinal or depth shifts (Brander, 2010); deoxygenation will induce habitat compression (Prince and Goodyear, 2006; Stramma et al., 2012; Koslow et al., 2011); and acidification will stress organismal function and thus organismal distribution. Climate change-related stressors are also likely to act in concert, and effects could be cumulative (Rosa and Seibel, 2008). Shifts in bottom-up, competitive, or top-down forcing will produce complex and indirect effects on the services described above. Acidification-slowed growth of carbonate skeletons, delayed development under hypoxic conditions, and increased respiratory demands with declining food availability illustrate how climate change could exacerbate anthropogenic impacts and compromise deep-sea ecosystem structure and function and ultimately benefits to human welfare.

The most important ecosystem service of the deep pelagic region is arguably the "biological pump", in which biological processes, such as the daily vertical migration, package and accelerate the transport of carbon compounds, nutrients, and other materials out of surface waters and into the deep sea. However, the microbial diversity and processes of the deep-pelagic realm are not sufficiently known to predict confidently how the biological pump ecosystem service will respond to perturbations.

6 Deep-sea exploitation

6.1 Deep-sea fisheries

Deep-sea fishing has a long history, but it did not become an important activity until the mid-twentieth century, when technological advancement allowed the construction of large and powerful vessels, and the development of line and trawl gear that could be deployed to continental slope depths. FAO (2009) acknowledges that deep-sea fisheries often exploit species which have relatively slower growth rates, reach sexual maturity later and reproduce at lower rates than shelf and coastal species.

Deep-sea fish species were the basis of major commercial fisheries in the 1970s to early 2000s (Japp and Wilkinson, 2007) but started to decline as aggregations were fished out, and realisation grew about the low productivity, and hence low yields, of these species (Clark, 2001; Sissenwine and Mace, 2007) and impacts of some of these fisheries on seafloor structure and benthos (Clark and Dunn 2012). Globally the main commercial deep-sea fish species at present number about 20, comprising alfonsino, toothfish, redfish, slickheads, cardinalfish, scabbardfish, armourhead, orange roughy, oreos, roundnose and rough-headed grenadiers, blue ling and moras. The current commercial catch of these main deep-sea species is about 150,000 tons, and has been similar over the last five years, although the proportional species mix has changed. The ecosystem effects of these fisheries are discussed in Chapter 11 of this Assessment and in Chapter 51 relative to the seamounts which are centres for many of these fisheries.

6.2 Deep gas and oil reserves

The oil and gas industry has been active in the open ocean since the 1970s. Over 10,000 hydrocarbon wells have been drilled globally; at least 1,000 are routinely drilled in water depths >200 m, and as deep as 2,896 m in the Gulf of Mexico. The scale of the exploration and development of hydrocarbon reserves and then ecosystem effects are discussed in Chapter 21.

6.3 Minerals

Great interest exists in exploiting the deep sea for its various reserves of minerals, which include polymetallic nodules, seafloor massive sulphide (SMS) deposits, mineral-rich sediments and cobalt-rich crusts. Currently no commercial mining projects have started, although several projects are in the exploratory or permitting phase. From those exploratory studies and related research some knowledge of potential ecosystem effects is accumulating.

Experimental studies to assess the potential impact of mining polymetallic nodules in the abyss have indicated that seafloor communities may take many decades before showing signs of recovery from disturbance (Bluhm, 2001; Miljutin et al., 2011), and may never recover if they rely directly on the nodules for habitat.

The recovery of communities at active hydrothermal vents where SMS deposits may be exploited may be relatively rapid, because vent sites undergo natural disturbances which have seen some communities appear to recover from catastrophic volcanic activity within a few years (Tunnicliffe et al., 1997). However, the rates of recovery of benthic communities are likely to vary among sites.

Figure 36F.1 | Deep-sea habitats. Top left: coral garden in the Whittard Canyon, NE Atlantic at approx. 500 metres depth (2010; image courtesy of Jeroen Ingels); top right: A sea anemone, Boloceroides daphneae, on cobalt crust covering a seamount off Hawaii, 1000 metres depth (image courtesy of Chris Kelly, HURL); bottom left: An orange roughy (Hoplostethus atlanticus) aggregation at 890 metres depth near the summit of a small seamount (termed "Morgue") off the east coast of New Zealand (image courtesy of Malcolm Clark); bottom right: A reef-like coverage by stony corals (Solenosmilia variabilis) together with prominent orange brisingid seastars on the summit of a small seamount (termed "Ghoul") feature at 950 metres off the east coast of New Zealand (image courtesy of Malcolm Clark).

Other potential mining activities include exploiting mineral-rich sediments. For example in some deep marine sediments, phosphorite occurs as "nodules" (2 to >150 mm in diameter), in a mud or sand matrix, which can extend beneath the seafloor sediment surface to tens of centimetres depth.

No mining has yet been authorized for such deposits but could result in the removal of large volumes of both the phosphorite nodules and the surrounding soft sediments, together with associated faunal communities and generate large sediment plumes. In addition, cobalt-rich ferromanganese crusts are promising sources of cobalt and rare minerals required to sustain growing human population demands and emerging high and green technologies (Hein et al., 2013). Conditions favouring their formation are found in abrupt topography, especially on the flanks and summits of oceanic seamounts and ridges at depths of 800-2500 m, where the most Cobalt-rich deposits are known to concentrate, in habitats dominated by suspension-feeding sessile organisms (mostly cold-water corals and sponges) and comparatively rich biological communities (Clark, 2013; Clark et al., 2011; Fukushima, 2007; Schlacher et al. 2013). Interest in cobalt-rich crust resources is growing, although mining for cobalt-rich crusts has not yet started, and technological challenges mean it may develop later than for polymetallic nodule or SMS resources. Further information on these mining activities is found in Chapter 23, and the seamount and seep/vent habitats in Chapters 51 and 45, respectively.

7 Special conservation/management issues and sustainability for the future

7.1 Special habitats (VMEs, EBSAs, MPAs) and conservation measures

The United Nations General Assembly has adopted a number of resolutions that called for the identification and protection of vulnerable marine ecosystems (VMEs) from significant adverse impacts of bottom fishing (for example 61/105 of 2006), which has facilitated the development of the 2008 International Guidelines for the Management of Deep-Sea Fisheries in the High Seas (FAO, 2009). The concept and developments of VMEs and their protection is addressed in Chapter 11. Also in the 2000s, in response to the call in the World Summit on Sustainable Development (WSSD) for greater protection of the open ocean, the Conference of Parties to the Convention on Biological Diversity (CBD) developed and adopted criteria for the description of ecologically or biologically significant areas (EBSAs) in open-ocean waters and deep-sea habitats.

The application of the EBSA criteria is a scientific and technical exercise, and areas that are described as meeting the criteria may receive protection through a variety of means, according to the choices of States and competent intergovernmental organizations (decision X/29 of the CBD COP10). Expert reviews have concluded that both approaches can be complementary in achieving effective sustainable management in the deep sea (Rice et al., 2014; Dunn et al., 2014).

7.2 Protection of the marine environment in the Area

With regard to deep-sea mining the International Seabed Authority (ISA), established in 1994, is required to take the necessary measures ensure that the marine environment is protected from harmful effects from activities in the Area under its jurisdiction. Such measures may include assessing potential environmental impacts of deep-sea activities (exploration and possible mining) and setting standards for environmental data collection, establishment of environmental baselines, and monitoring programmes (ISA, 2000, 2007 2013).

7.3 Deep-ocean observatories-ocean networks

Deep-sea observatories are becoming increasingly important in monitoring deep-sea ecosystems and the environmental changes that will affect them. The first long-term and real-time deep-sea observatory was deployed in 1993 at a methane seep site at 1,174 m depth in Sagami Bay, Japan (JAMSTEC, Japan), and is still operating. Several internationally organized projects have been initiated to achieve global integration of deep-sea observatories (e.g., Global Ocean Observing System (GOOS, NSF); FixO3 (Fixed Point Open Ocean Observatories, European Union Framework Programme 7), largely based on existing observing networks (e.g., Porcupine Abyssal Plain in the North Atlantic, (NOC, UK), Hausgarten Site in the transition between the North Atlantic and the Arctic (AWI, Germany), Ocean Network Canada with the Neptune Observatory on Canada's west coast) and aiming at achieving multidisciplinary integration, including physics, climate, biogeochemistry, biodiversity and ecosystems, geophysics with integration across sectors, and economics and sociology. Whilst moving towards a global strategy to obtain maximum efficiency, one of the major goals of deep-sea observatory initiatives is to better understand and predict the effects of climate change on the linked ocean-atmosphere system, and on marine ecosystems, biodiversity and community structure, In terms of biodiversity and ecosystems, several objectives need addressing: exploration and observation; prediction of future biological resources; understanding the functioning of deep-sea ecosystems; and understanding the roles of relationships between ecosystems and the services they provide.

References

Angel, M.V. (1997). Pelagic Biodiversity. In: Ormond, R.F.G., Gage, J.D., and Angel, M.V., editors. *Marine biodiversity: patterns and processes*. Cambridge University Press, New York.

Appeltans et al., The Magnitude of Global Marine Species Diversity, Current Biology (2012), http://dx.doi.org/10.1016/j.cub.2012.09.036

Armstrong, C.W., Foley, N.S., Tinch, R., van den Hove, S. (2012). Services from the deep: Steps towards valuation of deep sea goods and services. *Ecosystem Services* 2, 2-13.

Arntz, W.E., Brey, T., and Gallardo, V.A. (1994). Antarctic zoobenthos. *Oceanographic Marine Biology* 32: 241-304.

Bailey, D.M., Collins, M.A., Gordon, J.D.M., Zuur, A.F., Priede, I.G. (2009). Long-term changes in deep-water fish populations in the northeast Atlantic: a deeper reaching effect of fisheries? *Proceedings of the Royal Society B: Biological Sciences*. DOI: 10.1098/rspb.2009.0098.

Beliaev, G.M. (1989). *Deep-sea ocean trenches and their fauna*. Moscow: Nauka. 385 pp.

Benoit-Bird, K.J., and Au, W.W.L. (2006). Extreme diel horizontal migrations by a tropical nearshore resident micronekton community. *Marine Ecology Progress Series* 319: 1–14.

Billett, D.S.M., Bett, B., Jacobs, C., Rouse, I., Wigham, B. (2006). Mass deposition of jellyfish in the deep Arabian Sea. *Limnology and Oceanography* 51 (5), 2077-2083.

Billett, D.S.M., Bett, B.J., Reid, W.D.K., Boorman, B., and Priede, I.G. (2010). Long-term change in the abyssal NE Atlantic: The 'Amperima Event' revisited. *Deep-Sea Research II* 57: 1406–1417.

Blankenship, L., Yayanos, A., Cadien, D., and Levin, L. (2006). Vertical zonation patterns of scavenging amphipods from the hadal zone of the Tonga and Kermadec trenches. *Deep-Sea Research Part I: Oceanographic Research Papers* 53: 48-61.

Blankenship, L.E., and Levin, L.A. (2007). Extreme food webs: foraging strategies and diets of scavenging amphipods from the ocean's deepest 5 km. *Limnology and Oceanography* 52: 1685-1697.

Bluhm, B.A., Ambrose, W.G., Bergmann, M., Clough, L.M., Gebruk, A.V., Hasemann, C., Iken, K., Klages, M., MacDonald, I.R., Renaud, P.E., Schewe, I., Soltwedel, T., and Wlodarska-Kowalczuk, M. (2011). Diversity of the Arctic deep-sea benthos. *Marine Biodiversity* 41: 87-107.

Bluhm, B.A., MacDonald, I.R., Debenham, C., Iken, K. (2005). Macro- and megabenthic communities in the high Arctic Canada Basin: initial findings. *Polar Biology* 28: 218-231.

Bluhm, H. (2001). Re-establishment of an abyssal megabenthic community after experimental physical disturbance of the seafloor. *Deep Sea Research Part II: Topical Studies in Oceanography* 48(17–18), 3841-3868.

Braby, C.E., Rouse, G.W., Johnson, S.B., Jones, W.J., Vrijenhoek, R.C. (2007). Bathymetric and temporal variation among Osedax boneworms and associated megafauna on whale-falls in Monterey Bay, California. *Deep-Sea Research Part I: Oceanographic Research Papers* 54 (10), 1773-1791.

Brander, K. (2010). Impacts of climate change on fisheries. *Journal of Marine Systems* 79: 389–402.

Brandt, A., De Broyer, C., De Mesel, I., Ellingsen, K.E., Gooday, A.J., Hilbig, B., Linse, K., Thomson, M.R.A., Tyler, P.A. (2007). The biodiversity of the deep Southern Ocean benthos. *Philosophical Transactions of the Royal Society B-Biological Sciences* 362 (1477), 39-66.

Brandt, A., Ebbe, B. (2009). Southern Ocean deep-sea biodiversity-From patterns to processes. *Deep-Sea Research Part II: Topical Studies in Oceanography* 56 (19-20), 1732-1738.

Buhl-Mortensen, L., Vanreusel, A., Gooday, A.J., Levin, L.A., Priede, I.G., Buhl-Mortensen, P., Gheerardyn, H., King, N.J., Raes, M. (2010). Biological structures as a source of habitat heterogeneity and biodiversity on the deep ocean margins. *Marine Ecology* 31, 21-50.

Bullister, J.L., Rhein, M., and Mauritzen, C. (2013). Deepwater Formation. In: Siedler, G., Griffies, S.M., Gould, J., Church, J.A. (eds.) Ocean Circulation and Climate - A 21 Century Perspective. *International Geophysics* 103: 227-253.

Caron, D.A., Hutchins, D.A. (2012). The effects of changing climate on microzooplankton grazing and community structure: drivers, predictions and knowledge gaps. *J. Plankton Res.*, 235-252, doi:10.1093/plankt/fbs091

Church, R.A., Warren, D.J., Irion, J.B. (2009). Analysis of deepwater shipwrecks in the Gulf of Mexico: Artificial reef effect of Six World War II shipwrecks. *Oceanography* 22(2), 50-63.

Clark, M. (2001). Are deepwater fisheries sustainable? - the example of orange roughy (Hoplostethus atlanticus) in New Zealand. *Fisheries Research* 51: 123–135.

Clark, M.R. (2013). Biology associated with Cobalt-rich Ferromanganese crusts. In: Baker, E., Beaudoin, Y. (Eds.), Secretariat of the Pacific Community. *Deep Sea Minerals: Cobalt-rich Ferromanganese Crusts, a physical, biological, environmental, and technical review*. Vol. 1C, SPC.

Clark, M.R., and Dunn, M.R. (2012). Spatial management of deep-sea seamount fisheries: balancing exploitation and habitat conservation. *Environmental Conservation* 39(2): 204-214. Doi:10.1017/S0376892912000021.

Clark, M.R, Vinnichenko, V.I., Gordon, J.D.M., Beck-Bulat, G.Z., Kukharev, N.N., and Kakora, A.F. (2007). Large scale distant water trawl fisheries on seamounts. Chapter 17. In: Pitcher, T.J., Morato, T., Hart, P.J.B., Clark, M.R., Haggan, N., and Santos, R.S., editors. *Seamounts: ecology, fisheries, and conservation*. Blackwell Fisheries and Aquatic Resources Series 12 Blackwell Publishing, Oxford. pp. 361–399.

Clark, M.R., Kelley, C., Baco, A., and Rowden, A. (2011). Fauna of cobalt-rich ferromanganese crust seamounts. *International Seabed Authority Technical Study No. 8*. p. 83.

Clarke, A. (2003). The polar deep seas. In: Tyler, P.A., ed. *Ecosystems of the World. Ecosystems of the Deep Oceans*. Vol. 28, Elsevier, Amsterdam, pp. 239–260.

CoML, (2010). First Census of Marine Life 2010: Highlights of a Decade of Discovery. In: Ausubel, J.H., Crist, D.T., Waggoner, P.E. (Eds.), *Census of Marine Life International Secretariat*. Consortium for Ocean Leadership, Washington, p. 64.

Cook, A.A., Lambshead, P.J.D., Hawkins, L.E., Mitchell, N., Levin, L.A. (2000). Nematode abundance at the oxygen minimum zone in the Arabian Sea. *Deep-Sea Research Part II: Topical Studies in Oceanography* 47 (1-2), 75-85.

Corliss, B.H., Brown, C.W., Sun, X. and Showers, W.J. (2009). Deep-sea benthic diversity linked to seasonality of pelagic productivity. *Deep-Sea Research I* 56: 835–841.

Costello, M.J., May, R.M., Stork, N.E. (2013). Can We Name Earth's Species Before They Go Extinct? *Science* 339(6118), 413-416.

Cronin, T.M. and Raymo, M.E. (1997). Orbital forcing of deep-sea benthic species diversity. *Nature* 385(6617): 624–627.

Cronin, T.M., DeMartino, D.M., Dwyer, G.S., and Rodriguez-Lazaro, J. (1999). Deep-sea ostracode species diversity: response to late Quaternary climate change. *Marine Micropaleontology* 37(3-4): 231–249.

Danovaro, R., Gambi, C. and Della Croce, N. (2002). Meiofauna hotspot in the Atacama Trench, eastern South Pacific Ocean. *Deep-Sea Research, Part I: Oceanographic Research Papers* 49: 843–857.

Danovaro, R., Della Croce, N., Dell'Anno, A., Pusceddu, A. (2003). A depocenter of organic matter at 7800 m depth in the SE Pacific Ocean. *Deep Sea Research Part I: Oceanographic Research Papers* 50 (12), 1411-1420.

Danovaro, R., Snelgrove, P.V.R., Tyler, P. (2014). Challenging the paradigms of deep-sea ecology. *Trends in Ecology & Evolution* 29 (8), 465-475.

Deubel, H. (2000). Structures and nutrition requirements of macrozoobenthic communities in the area of the Lomonosov Ridge in the Arctic Ocean (in German). *Reports on Polar Research* 370: 1-147.

Dunn, D.C., Ardron, J., Bax, N., Bernal, P., Cleary, J., Cresswell, I., Donnelly, B., Dunstan, P., Gjerde, K., Johnson, D., Kaschner, K., Lascelles, B., Rice, J., von Nordheim, H., Wood, L., Halpin, P.N. (2014). The Convention on Biological Diversity's Ecologically or Biologically Significant Areas: Origins, development, and current status. *Marine Policy* 49, 137-145.

Edgcomb, V., Orsi, W., Bunge, J., Jeon, S., Christen, R., Leslin, C., Holder, M., Taylor, G.T., Suarez, P., Varela, R., and Epstein, S. (2011). Protistan microbial observatory in the Cariaco Basin, Caribbean. I. Pyrosequencing vs Sanger insights into species richness. *The ISME Journal* 5(8): 1344-1356.

Etter, R., Mullineaux, L. (2000). Deep-sea communities. In: Bertness, M.D., Gaines, S., Hay, M. (Eds.), *Marine Community Ecology*. Sinauer Associates, Inc., Sunderland, MA, USA, pp. 367-393.

FAO (2009). *International Guidelines for the Management of Deep-sea Fisheries in the High Seas*. Rome, Italy: FAO: 73 pp. Available from: http://www.fao.org/docrep/011/i0816t/i0816t00.htm.

Friedlander, A.M., Ballesteros, E., Fay, M., Sala, E. (2014). Marine Communities on Oil Platforms in Gabon, West Africa: High Biodiversity Oases in a Low Biodiversity Environment. *PLoS ONE* 9(8), e103709.

Fuhrman, J.A., Steele, J.A., Hewson, I., Schwalback, M.S., Brown, M.V., Green, J.L., and Brown, J.H. (2008). A latitudinal diversity gradient in planktonic marine bacteria. *Proceedings of the National Academy of Science USA* 105(22): 7774–7778.

Fujii, T., Kilgallen, N.M., Rowden, A.A., Jamieson, A.J. (2013). Deep-sea amphipod community structure across abyssal to hadal depths in the Peru-Chile and Kermadec trenches. *Marine Ecology Progress Series* 492, 125-138.

Fujikura, K., Kojima, S., Tamaki, K., Maki, Y., Hunt, J., and Okutani, T. (1999). The deepest chemosynthesis-based community yet discovered from the hadal Zone, 7326 m deep, in the Japan Trench. *Marine Ecology Progress Series* 190: 17-26.

Fujiwara, Y., Kawato, M., Yamamoto, T., Yamanaka, T., Sato-Okoshi, W., Noda, C., Tsuchida, S., Komai, T., Cubelio, S.S., Sasaki, T., Jacobsen, K., Kubokawa, K., Fujikura, K., Maruyama, T., Furushima, Y., Okoshi, K., Miyake, H., Miyazaki, M., Nogi, Y., Yatabe, A., Okutani, T. (2007). Three-year investigations into sperm whale-fall ecosystems in Japan. *Marine Ecology* 28 (1), 219-232.

Fukushima, T. (2007). Amounts of megabenthic organisms in areas of manganese nodules, cobalt-rich crusts and polymetallic sulphides occurrences. Proceedings of the International Seabed Authority's (ISA) Workshop, September 2004: Polymetallic Sulphides and Cobalt-Rich Ferromanganese Crust Deposits: Establishment of Environmental Baselines and an Associated Monitoring Programme During Exploration (ed. by ISA), pp. 356–368. International Seabed Authority, Kingston, Jamaica. Available at: (http://www.isa.org.jm/en/ documents/ publications) (accessed 18 October 2013).

Gage, J.D., Lambshead, P.J.D., Bishop, J.D.D., Stuart, C.T., Jones, N.S. (2004) Large-scale biodiversity pattern of Cumacea (Peracarida : Crustacea) in the deep Atlantic. *Marine Ecology-Progress Series* 277, 181-196.

Gallo, N.D., Cameron, J., Hardy, K., Fryer, P., Bartlett, D., and Levin, L.A. Submersible and lander-observed community patterns in the Mariana and New Britain Trenches: Influence of productivity and depth on benthic community structure (in revision, *Deep-Sea Research Part I-Oceanographic Research Papers*).

Gambi, C., Vanreusal, A., and Danovaro, R. (2003). Biodiversity of nematode assemblages from deep-sea sediments of the Atacama Slope and Trench. *Deep Sea Research I Part I: Oceanographic Research Papers* 50: 103-117.

George, R.Y., and Higgins, R.P. (1979). Eutrophic Hadal Benthic Community in the Puerto Rico Trench. *Ambio Special Report, No. 6, The Deep Sea: Ecology and Exploitation*, pp. 51-58.

German, C.R., Ramirez-Llodra, E., Baker, M.C., Tyler, P.A., and the ChEss Scientific Steering Committee (2011). Deep-Water Chemosynthetic Ecosystem Research during the Census of Marine Life Decade and Beyond: A Proposed Deep-Ocean Road Map. *PLoS ONE* 6(8): e23259. Doi:10.1371/journal.pone.0023259.

Giering, S., Sanders, R., Lampitt, R., Anderson, T., Tamburini, C., Boutrif, M., Zubkov, M., Marsay, C., Henson, S., Saw, K., Cook, K., and Mayor, D. (2014). Reconciliation of the carbon budget in the ocean's twilight zone. *Nature* 507: 480-483.

Glover, A.G., and Smith, C.R. (2003). The deep-sea floor ecosystem: current status and prospects of anthropogenic change by the year 2025. *Environmental Conservation* 30: 219–41.

Glover, A.G., Gooday, A.J., Bailey, D.M., Billett, D.S.M., Chevaldonné, P., Colaço, A., Copley, J., Cuvelier, D., Desbruyères, D., Kalogeropoulou, V., Klages, M., Lampadariou, N., Lejeusne, C., Mestre, N.C., Paterson, G.L.J., Perez, T., Ruhl, H., Sarrazin, J., Soltwedel, T., Soto, E.H., Thatje, S., Tselepides, A., Van Gaever, S., and Vanreusel, A. (2010). Temporal change in deep-sea benthic ecosystems: a review of the evidence from recent time-series studies. *Advances in Marine Biology* 58: 1-95.

Glover, A.G., Källström, B., Smith, C.R., Dahlgren, T.G. (2005). World-wide whale worms? A new species of Osedax from the shallow north Atlantic. *Proceedings of the Royal Society B: Biological Sciences* 272 (1581), 2587-2592.

Glud, R.N., Wenzhofer, F., Middelboe, M., Oguri, K., Turnewitsch, R., Canfield, D.E., Kitazato, H. (2013). High rates of microbial carbon turnover in sediments in the deepest oceanic trench on Earth. *Nature Geoscience* 6 (4), 284-288.

Goffredi, S.K., Orphan, V.J., Rouse, G.W., Jahnke, L., Embaye, T., Turk, K., Lee, R., Vrijenhoek, R.C. (2005). Evolutionary innovation: a bone-eating marine symbiosis. *Environmental Microbiology* 7 (9), 1369-1378.

Golikov, A.N., and Scarlato, O.A. (1990). History of the development of the Arctic marine ecosystem and their functional peculiarities. In: Kotlyakov, V.M., and Sokolov, V.E., eds., *Arctic Research: Advances and prospects, Proceedings of the Conference of Arctic and Nordic countries on coordination of research in the Arctic*, Leningrad, December 1988, Moscow, pp. 196-206.

Gooday, A.J., Todo, Y., Uematsu, K., and Kitazato, H. (2008). New organic-walled Foraminifera (Protista) from the ocean's deepest point, the Challenger Deep (western Pacific Ocean). *Zoological Journal of the Linnean Society* 153: 399–423.

Gooday, A.J., Bett, B.J., Escobar, E., Ingole, B., Levin, L.A., Neira, C., Raman, A.V., Sellanes, J. (2010). Habitat heterogeneity and its influence on benthic biodiversity in oxygen minimum zones. *Marine Ecology* 31, 125-147.

Grange, L. J. and Smith, C.R. (2013). Megafaunal Communities in Rapidly Warming Fjords Along the West Antarctic Peninsula: Hotspots of Abundance and Beta Diversity. *PLoS ONE*, 8(11): e77917. doi:10.1371/journal.pone.0077917

Grassle, J.F. (1989). Species diversity in deep-sea communities. *Trends in Ecology and Evolution* 4 (1), 12-15.

Grassle, J.F., Maciolek, N.J. (1992). Deep-sea species richness: Regional and local diversity estimates from quantitative bottom samples. *American Naturalist* 139 (2), 313-341.

Griffiths, H.J. (2010). Antarctic Marine Biodiversity – What Do We Know About the

Distribution of Life in the Southern Ocean? *PLoS ONE* 5(8): e11683.

Haedrich, R.L. (1996). Deep-water fishes: evolution and adaptation in Earth's largest living spaces. *Journal of Fish Biology* 49(Suppl. A): 40–53.

Hanchet, S., Sainsbury, K., Butterworth, D., Darby, C., Bizikov, V., Rune Godø, O., Ichii, T., Kock, K.-H., López Abellán, L., Vacchi, M. (2015). CCAMLR's precautionary approach to management focusing on Ross Sea toothfish fishery. *Antarctic Science* FirstView, 1-8.

Harris, P.T., Macmillan-Lawler, M., Rupp, J., and Baker, E.K. (2014). Geomorphology of the oceans. *Marine Geology* 352: 4–24. Doi:10.1016/j.margeo.2014.01.011.

Hasemann, C., Bergmann, M., Kanzog, C., Lochthofen, N., Sauter, E., Schewe, I., and Soltwedel, T. (2013). Effects of dropstone-induced habitat heterogeneity on Arctic deep-sea benthos with special reference to nematode communities. *Marine Biological Research* 9(3): 276-292.

Havermans, C., Sonet, G., d'Udekem d'Acoz, C., Nagy, Z.T., Martin P., Briz, S., Riehl, T., Agrawal, S., and Held, C. (2013). Genetic and Morphological Divergences in the Cosmopolitan Deep-Sea Amphipod Eurythenes gryllus Reveal a Diverse Abyss and a Bipolar Species. *PLoS ONE* 8(9): e74218. Doi:10.1371/journal.pone.0074218.

Hays, G.C. (2003). A review of the adaptive significance and ecosystem consequences of zooplankton diel vertical migrations. *Hydrobiologia* 503: 163–170.

Heezen, B.C. (1969). The world rift system: an introduction to the symposium. *Technophysics* 8: 269–279.

Hein, J.R., Mizell, K., Koschinsky, A., and Conrad, T.A. (2013). Deep-ocean mineral deposits as a source for critical metals for high- and green technology applications: comparison with land-based resources. *Ore Geology Reviews* 51: 1-14.

Herring, P. (2002). *The biology of the deep ocean*. Oxford University Press, Oxford, UK, 314 pp.

Hopkins, T.L., Sutton, T.T., and Lancraft, T.M. (1996). Trophic structure and predation impact of a low latitude midwater fish community. *Progress in Oceanography* 38: 205-239.

Hoving, H.T., Perez, J.A.A., Bolstad, K.S.R., Braid, H.E., Evans, A.B., Fuchs, D., Judkins, H., Kelly, J.T., Marian, J.E.A.R., Nakajima, R., Piatkowski, U., Reid, A., Vecchione, M., and Xavier, J.C.C. (2014). The Study of Deep-Sea Cephalopods. In: Vidal, E.A.G., editor. *Advances in Marine Biology, Vol. 67*. Oxford, UK.: 235-359.

Hunt, G., Cronin, T.M., and Roy, K. (2005). Species–energy relationship in the deep sea: a test using the Quaternary fossil record. *Ecology Letters* 8: 739–747.

Ingels, J., Vanhove, S., De Mesel, I., and Vanreusel, A. (2006). The biodiversity and biogeography of the free-living nematode genera Desmodora and Desmodorella (family Desmodoridae) at both sides of the Scotia Arc. *Polar Biology* 29(11): 936-949.

Irigoien, X., Klevjer, T.A., Rostad, A., Martinez, U., Boyra, G., Acuña, J.L., Bode, A., Echevarria, F., Gonzales-Gordillo, J.I., Hernandez-León, S., Agusti. S., Aksnes, D.L., Duarte, C.M., Kaardvedt, S. (2014). Large mesopelagic fishes biomass and trophic efficiency in the open ocean. Nature Communications 5:3271.

ISA (2013). *Recommendations for the guidance of contractors for the assessment of the possible environmental impacts arising from exploration for marine minerals in the Area*. ISBA/19/LTC/8.

ISA (2000). *Decision of the Assembly relating to the regulations on prospecting and exploration for polymetallic nodules in the Area*. ISBA/6/A/18.

ISA (2007). *Polymetallic Sulphides and Cobalt-Rich Ferromanganese crusts deposits: Establishment of environmental baselines and an associated monitoring programme during exploration*. Proceedings of the International Seabed Authority Workshop held in Kingston, Jamaica, 6-10 September 2004, 491 pp.

Itoh, M., Kawamura, K., Kitahashi, T., Kojima, S., Katagiri, H., and Shimanaga, M. (2011). Bathymetric patterns of meiofaunal abundance and biomass associated with the Kuriland Ryukyu trenches, western North Pacific Ocean. *Deep-Sea Research I* 58: 86–97.

Jahnke, R.A. (2010). Global Synthesis. In: Liu, K.-K., Atkinson, L., Quinones, R., Talaue-McManus, L. (Eds.) *Carbon and Nutrient Fluxes in Continental Margins*. Springer, pp. 597-615.

Jamieson, A.J. (2011). Ecology of Deep Oceans: Hadal Trenches. In: *Encyclopedia of Life Sciences (ELS)*. John Wiley & Sons, Ltd, Chichester. Doi: 10.1002/9780470015902.a0023606.Jamieson, A.J., Fujii, T., Mayor, D.J., Solan, M., Priede, I.G. (2009). Hadal trenches: the ecology of the deepest places on Earth. *Trends in Ecology and Evolution* 25 (3), 190-197.

Jansson, R., and Dynesius, M. (2002). The fate of clades in a world of recurrent climate change: Milankovitch oscillations and evolution. *Annual Review of Ecological Systematics* 33: 741-777.

Japp, D.W., and Wilkinson, S. (2007). Deep-sea resources and fisheries. Report and documentation of the expert consultation on deep-sea fisheries in the High Seas. *FAO Fisheries Report 838*: pp. 39–59. Rome, Italy: FAO. Available from: ftp://ftp.fao.org/docrep/fao/010/a1341e/a1341e00.pdf.

Jobstvogt, N., Hanley, N., Hynes, S., Kenter, J., Witte, U. (2014a). Twenty thousand sterling under the sea: Estimating the value of protecting deep-sea biodiversity. *Ecological Economics* 97, 10-19.

Jobstvogt, N., Townsend, M., Witte, U., Hanley, N. (2014b). How Can We Identify and Communicate the Ecological Value of Deep-Sea Ecosystem Services? *PLoS ONE* 9 (7), e100646.

Jones, D.O.B., Yool, A., Wei, C.L., Henson, S.A., Ruhl, H.A., Watson, R.A., and Gehlen, M. (2014). Global reductions in seafloor biomass in response to climate change. *Global Change Biology* 20: 1861–1872, Doi: 10.1111/gcb.12480.

Jumars, P.A., and Hessler, R.H. (1976). Hadal community structure: implications from the Aleutian Trench. *Journal of Marine Research* 34: 547–560.

Juniper, S.K., and Sibuet, M. (1987). Cold seep benthic communities in Japan subduction zones: spatial organization, trophic strategies and evidence for temporal evolution. *Marine Ecology Progress Series* 40: 115-126.

Kaartvedt, S., Staby, A., and Aksnes, D. (2012). Efficient trawl avoidance by mesopelagic fishes causes large underestimation of their biomass. *Marine Ecology Progress Series* 456: 1–6.

Kaiser, S., Brandao, S.N., Brix, S., Barnes, D.K.A., Bowden, D.A., Ingels, J., Leese, F., Schiaparelli, S., Arango, C.P., Badhe, R., Bax, N., Blazewicz-Paszkowycz, M., Brandt, A., Brenke, N., Catarino, A.I., David, B., De Ridder, C., Dubois, P., Ellingsen, K.E., Glover, A.G., Griffiths, H.J., Gutt, J., Halanych, K.M., Havermans, C., Held, C., Janussen, D., Lorz, A.N., Pearce, D.A., Pierrat, B., Riehl, T., Rose, A., Sands, C.J., Soler-Membrives, A., Schuller, M., Strugnell, J.M., Vanreusel, A., Veit-Kohler, G., Wilson, N.G., Yasuhara, M. (2013). Patterns, processes and vulnerability of Southern Ocean benthos: a decadal leap in knowledge and understanding. *Marine Biology* 160 (9), 2295-2317.

Kitahashi, T., Kawamura, K., Kojima, S., and Shimanaga, M. (2013). Assemblages gradually change from bathyal to hadal depth: A case study on harpacticoid copepods around the Kuril Trench (north-west Pacific Ocean). *Deep Sea Research I* 74: 39–47.

Kortsch, S., Primicerio, R., Beuchel, F., Renaud, P.E., Rodriguez, J., Lønne, O.J., Gulliksen, B. (2012). Climate-driven regime shifts in Arctic marine benthos. *Proc. Natl. Acad. Sci. USA* 109(35), 14,052-14,057.

Koslow, J.A. (2009). The role of acoustics in ecosystem-based fishery management. *ICES Journal of Marine Science* 66: 966–973.

Koslow, J.A. (2007). The biological environment of cobalt-rich ferromanganese crusts deposits, the potential impact of exploration and mining on this environment, and data required to establish environmental baselines. In: *Polymetallic Sulphides and Cobalt-Rich Ferromanganese crusts deposits: Establishment of environmental baselines and a monitoring program during exploration.* Proceedings of the International Seabed Authority's Workshop held in Kingston, Jamaica, 6-10 September 2004, p: 274-294.

Koslow, J.A., Goericke, R., Lara-Lopez, A., Watson, W. (2011. Impact of declining intermediate-water oxygen on deepwater fishes in the California Current. *Marine Ecology Progress Series* 436, 207-218.

Kosobokova, K.N., Hirche, H.J., and Hopcroft, R.R. (2011). Patterns of zooplankton diversity through the depths of the Arctic's central basin. *Marine Biodiversity* 41: 29-50.

Kvassnes, A.J.S., Iversen, E. (2013). Waste sites from mines in Norwegian Fjords. *Mineralproduksjon* 3, A27-A38.

Kvassnes, A.J.S., Sweetman, A.K., Iversen, E., Skei, J. (2009). Sustainable use and future of submarine tailings placements in the Norwegian extractive Industry. *Securing the Future (Mining, metals and the environments in a sustainable society) and 8th ICARD (International Conference on Acid Rock Drainage).* http://www.proceedings-stfandicard-2009.com/, Skelleftea, Sweden.

Ladau, J., Sharpton, T.J., Finucane, M.M., Jospin, G., Kembel, S.W., O'Dwyer, J., Koeppel, A.F., Green, J.L. and Pollard, K.S. (2013). Global marine bacterial diversity peaks at high latitudes in winter. *The ISME journal* 7(9): 1669-1677.

Lambshead, P.J.D., Tietjen, J., Ferrero, T., Jensen, P. (2000). Latitudinal diversity gradients in the deep sea with special reference to North Atlantic nematodes. *Marine Ecology-Progress Series* 194, 159-167.

Lebrato, M., Jones, D.O.B. (2009). Mass deposition event of Pyrosoma atlanticum carcasses off Ivory Coast (West Africa). *Limnology and Oceanography* 54(4), 1197-1209.

Lebrato, M., Pitt, K., Sweetman, A., Jones, D.B., Cartes, J., Oschlies, A., Condon, R., Molinero, J., Adler, L., Gaillard, C., Lloris, D., Billett, D.M. (2012). Jelly-falls historic and recent observations: a review to drive future research directions. *Hydrobiologia* 690(1), 227-245.

Lebrato, M., Mendes, P., Steinberg, D.K., Cartes, J.E., Jones, B., Birsa, L.M., Benavides, R., Oschlies, A. (2013a). Jelly biomass sinking speed reveals a fast carbon export mechanism. *Limnology and Oceanography* 58 (3), 1113-1122.

Lebrato, M., Molinero, J.-C., Cartes, J.E., Lloris, D., Mélin, F., Beni-Casadella, L. (2013b). Sinking Jelly-Carbon Unveils Potential Environmental Variability along a Continental Margin. *PLoS ONE* 8 (12), e82070.

Levin, L.A. (2003). Oxygen minimum zone benthos: Adaptation and community response to hypoxia. *Oceanography and Marine Biology*, 41, 1-45.

Levin, L.A., Etter, R.J., Rex, M.A., Gooday, A.J., Smith, C.R., Pineda, J., Stuart, C.T., Hessler, R.R., and Pawson, D. (2001). Environmental influences on regional deep-sea species diversity. *Annual Review of Ecological Systematics* 32: 51-93.

Levin, L.A., Liu, K.-K., Emeis, K.-C., Breitburg, D.L., Cloern, J., Deutsch, C., Giani, M., Goffart, A., Hofmann, E.E., Lachkar, Z. (2014). Comparative biogeochemistry–ecosystem–human interactions on dynamic continental margins. *Journal of Marine Systems*.

Levin, L.A., Sibuet, M. (2012). Understanding Continental Margin Biodiversity: A New Imperative. *Annual Review of Marine Science* 4 (1), 79-112.

Levin, L.A. and Dayton, P.K. (2009). Ecological theory and continental margins: where shallow meets deep. *Trends in Ecology and Evolution* 24: 606-617.

Marincovich, L. Jr., Brouwers, E.M., Hopkins, D.M., and McKenna, M.C. (1990). Late Mesozoic and Cenozoic paleogeographic and paleoclimatic history of the Arctic Ocean Basin, based upon shallow-water marine faunas and terrestrial vertebrates. In: Gantz, A., Johnson, L., Sweeny, J.F., editors. *The Arctic Ocean Region. The Geology of North America, vol. L.* Geological Society of America, Boulder, Colorado. pp. 403-426.

Mengerink, K.J., Van Dover, C.L., Ardron, J., Baker, M., Escobar-Briones, E., Gjerde, K., Koslow, J.A., Ramirez-Llodra, E., Lara-Lopez, A., Squires, D., Sutton, T.T., Sweetman, A.K., and Levin, L.A. (2014). A Call for Deep-Ocean Stewardship. *Science* 344: 696-698.

Miljutin, D.M., Miljutina, M.A., Arbizu, P.M., and Galéron, J. (2011). Deep-sea nematode assemblage has not recovered 26 years after experimental mining of polymetallic nodules (Clarion-Clipperton Fracture Zone, Tropical Eastern Pacific). *Deep-Sea Research*. 58(8): 885-897.

Mora, C., Rollo, A., Tittensor, D.P. (2013a). Comment on Can We Name Earth's Species Before They Go Extinct?. *Science*, 341(6143), p. 237. DOI: 10.1126/science.1237254.

Mora, C., Wei, C.-L., Rollo, A., Amaro, T., Baco, A.R., Billett, D., Bopp, L., Chen, Q., Collier, M., Danovaro, R., Gooday, A.J., Grupe, B.M., Halloran, P.R., Ingels, J., Jones, D.O.B., Levin, L.A., Nakano, H., Norling, K., Ramirez-Llodra, E., Rex, M., Ruhl, H.A., Smith, C.R., Sweetman, A.K., Thurber, A.R., Tjiputra, J.F., Usseglio, P., Watling, L., Wu, T., Yasuhara, M. (2013b). Biotic and Human Vulnerability to Projected Changes in Ocean Biogeochemistry over the 21st Century. *PLoS Biol* 11 (10), e1001682.

Morato, T., Hoyle, S.D., Allain, V., and Nicol, S.J. (2010). Seamounts are hotspots of pelagic biodiversity in the open ocean. *Proceedings of the National Academy of Science USA* 107: 9707–9711.

Morato, T., Varkey, D.A., Dâmaso, C., Machete, M., Santos, M., Prieto, R., Santos, R.S., and Pitcher, T.J. (2008). Evidence of a seamount effect on aggregating visitors. *Marine Ecology Progress Series* 357: 23-32. Doi:10.3354/meps07269.

Muller-Karger, F.E., Varela, R., Thunell, R., Luerssen, R., Hu, C., Walsh, J.J. (2005). The importance of continental margins in the global carbon cycle. *Geophysical research letters* 32 (1), L01602.

Narayanaswamy, B.E., Bett, B.J., and Gage, J.D. (2005). Ecology of bathyal polychaete fauna at an Arctic-Atlantic boundary (Faroe-Shetland Channel, North-east Atlantic). *Marine Biology Research* 1: 20-32.

Narayanaswamy, B.E., Renaud, P., Duineveld, G., Berge, J., Lavaleye, M.S.S., Reiss, H. and Brattegard, T. (2010). Biodiversity trends along the western European Margin. *PLoS ONE* 5(12): e14295.

Narayanaswamy, B.E., Coll, M., Danovaro, R., Davidson, K., Ojaveer, H., and Renaud, P.E. (2013). Synthesis of knowledge on marine biodiversity in European Seas: from Census to sustainable management. *PLoS ONE* 8(3): e58909. Doi:10.1371/journal.pone.0058909.

Nesis, K.N. (1984). A hypothesis on the origin of western and eastern Arctic distribution of areas of marine bottom animals. *Soviet Journal of Marine Biology* 9: 235-243.

Ogawa, Y., Fujioka, K., Fujikura, K. and Iwabuchi, Y. (1996). En echelon patterns of Calyptogena colonies in the Japan Trench. *Geology* 24: 807-810.

Oguri, K., Kawamura, K., Sakaguchi, A., Toyofuku, T., Kasaya, T., Murayama, M., Fujikura, K., Glud, R.N., and Kitazato, H. (2013). Hadal disturbance in the Japan Trench induced by the 2011 Tohoku–Oki Earthquake. *Scientific Reports* 3: 1915. Doi: 10.1038/srep01915.

Oschmann, W. (1990). Dropstones - rocky mini-islands in high-latitude pelagic soft substrate environments. *Senkenbergiana Marit* 21: 55-75.

Østerhus, S., and Gammelsrod, T. (1999). The abyss of the Nordic Seas is warming.

Journal of Climate 12: 3297–3304.

Pawlowski, J., Christen, R., Lecroq, B., Bachar, D., Shahbazkia, H.R., Amaral-Zettler, L., and Guillou, L. (2011). Eukaryotic richness in the abyss: insights from pyrotag sequencing. *PLoS One* 6(4): e18169.

Pawlowski, J., Fahrni, J., Lecroq, B., Longet, D., Cornelius, N., Excoffier, L., Cedhagen, T., and Gooday, A.J. (2007). Bipolar gene flow in deep-sea benthic foraminifera. *Molecular Ecology* 16(19): 4089-4096.

Pham, C.K., Ramirez-Llodra, E., Alt, C.H.S., Amaro, T., Bergmann, M., Canals, M., Company, J.B., Davies, J., Duineveld, G., Galgani, F., Howell, K.L., Huvenne, V.A.I., Isidro, E., Jones, D.O.B., Lastras, G., Morato, T., Gomes-Pereira, J.N., Purser, A., Stewart, H., Tojeira, I., Tubau, X., Van Rooij, D., Tyler, P.A. (2014). Marine Litter Distribution and Density in European Seas, from the Shelves to Deep Basins. *PLoS ONE* 9 (4), e95839.

Pierrot-Bults, A., and Angel, M. (2012). Pelagic Biodiversity and Biogeography of the Oceans. *Biology International* 51: 9-35.

Pimm, S.L., Russell, G.J., Gittleman, J.L., Brooks, T.M. (1995). The future of biodiversity. *Science* 269(5222), 347-349.

Priede, I.G., Bergstad, O.A., Miller, P.I., Vecchione, M., Gebruk, A., Falkenhaug, T., Billett, D.S.M., Craig, J., Dale, A.C., Shields, M.A., Tilstone, G.H., Sutton, T.T., Gooday, A.J., Inall, M.E., Jones, D.O.B., Martinze-Vicente, V., Menezes, G.M., Niedzielski, T., Sigurosson, P., Rothe, N. Rogacheva, A., Alt, C.H.S., Brand, T., Abell, R., Brierley, A.S., Cousins, N.J., Crockard, D., Hoelzel, A.R., Hoines, A., Letessier, T.B., Read, J.F., Shimmield, T. Cox, M.J., Galbraith, J.K, Gordon, J.D.M., Horton, T., Neat, F., and Lorance, P. (2013). Does Presence of a Mid-Ocean Ridge Enhance Biomass and Biodiversity? *PLoS ONE* 8(5): e61550. Doi:10.1371/journal.pone.0061550.

Priede, I.G., Osborn, K.J., Gebruk, A.V., Jones, D., Shale, D., Rogacheva, A., and Holland N.D. (2012). Observations on torquaratorid acorn worms (Hemichordata, Enteropneusta) from the North Atlantic with descriptions of a new genus and three new species. *Invertebrate Biology* 131: 244-257. Doi:10.1111/j.1744-7410.2012.00266.x.

Prince, E.D., Goodyear, C.P., (2006). Hypoxia-based habitat compression of tropical pelagic fishes. *Fisheries Oceanography* 15 (6), 451-464.

Purcell, J.E. (2012). Jellyfish and ctenophore blooms coincide with human proliferations and environmental perturbations. *Annual Review of Marine Science* 4, 209-235.

Purcell, J.E., Uye, S.-i., Lo, W.-T. (2007). Anthropogenic causes of jellyfish blooms and their direct consequences for humans: a review. *Marine Ecology - Progress Series* 350, 153.

Pusceddu, A., Bianchelli, S., Martín, J., Puig, P., Palanques, A., Masqué, P., Danovaro, R. (2014). Chronic and intensive bottom trawling impairs deep-sea biodiversity and ecosystem functioning. *Proceedings of the National Academy of Sciences,* 111 (24) 8861-8866. Doi: 10.1073/pnas.1405454111.

Ramirez-Llodra, E., Brandt, A., Danovaro, R., De Mol, B., Escobar, E., German, C.R., Levin, L.A., Martinez Arbizu, P., Menot, L., Buhl-Mortensen, P., Narayanaswamy, B.E., Smith, C.R., Tittensor, D.P., Tyler, P.A., Vanreusel, A., Vecchione, M. (2010). Deep, diverse and definitely different: unique attributes of the world's largest ecosystem. *Biogeosciences* 7 (9), 2851-2899.

Ramirez-Llodra, E., Tyler, P.A., Baker, M.C., Bergstad, O.A., Clark, M.R., Escobar, E., Levin, L.A., Menot, L., Rowden, A.A., Smith, C.R., Van Dover, C.L. (2011). Man and the Last Great Wilderness: Human Impact on the Deep Sea. *PLoS ONE* 6 (8), e22588.

Reed, J.K., Messing, C., Walker, B., Brooke, S., Correa, T., Brouwer, M., and Udouj, T. (2013). Habitat characterization, distribution, and areal extent of deep-sea coral ecosystem habitat off Florida, southeastern United States. *Journal of Caribbean Science* 47: 13-30.

Reid, E., Sullivan, B., Clark, J. (2010). Mitigation of seabird captures during hauling in CCAMLR longline fisheries. *CCAMLR Science* 17, 155-162.

Rex, M.A. (1981). Community structure in the deep-sea benthos. *Annual Review of Ecology and Systematics* 12, 331-353.

Rex, M.A., and Etter, R.H. (2010). *Deep-Sea Biodiversity: Pattern and Scale*. Harvard University Press, Boston, United States, 354 pp.

Rex, M.A., Stuart, C.T., Coyne, G. (2000). Latitudinal gradients of species richness in the deep-sea benthos of the North Atlantic. *Proceedings of the National Academy of Sciences* 97 (8), 4082-4085.

Rice, J., Lee, J., Tandstad, M., 2014. Parallel initiatives. Governance of Marine Fisheries and Biodiversity Conservation. John Wiley & Sons, Ltd., pp. 195-208.

Robinson, C., Steinberg, D.K., Anderson, T.R., Arístegui, J., Carlson, C.A., Frost, J.R., Ghiglione, J.F., Hernández-León, S., Jackson, G.A., Koppelmann, R., Quéguiner, B., Ragueneau, O., Rassoulzadegan, F., Robison, B.H., Tamburinim, C., Tanaka, T., Wishner, K.F., and Zhang, J. (2010). Mesopelagic zone ecology and biogeochemistry – a synthesis. *Deep-Sea Research Part II* 57: 1504-1518.

Robison, B.H. (2004). Deep pelagic biology. *Journal of Experimental Marine Biology and Ecology* 300: 253-272.

Robison, B.H. (2009). Conservation of deep pelagic biodiversity. *Conservation Biology* 23(4): 847-858.

Rodhouse, P.G., and Nigmatullin, C.M. (1996). Role as consumers. *Philosophical Transactions of the Royal Society of London* B 351 (1343), 1003-1022.

Rodriguez-Lazaro, J., Cronin, T.M. (1999). Quaternary glacial and deglacial Ostracoda in the thermocline of the Little Bahama Bank (NW Atlantic): palaeoceanographic implications. *Palaeogeography, Palaeoclimatology, Palaeoecology* 152 (3–4), 339-364.

Rogers, A.D. (2007). Evolution and biodiversity of Antarctic organisms: a molecular perspective. *Philosophical Transactions of the Royal Society B: Biological Sciences* 362 (1488), 2191-2214.

Roman, J., Estes, J., Morissette, L., Smith, C.R., Costa, D., McCarthy, J., Nation, J.B., Nicol, S., Pershing, A., Smetacek, V. (2014). Whales as ecosystem engineers. Frontiers in Ecology and the Environment, doi:10.1890/130220.

Rosa, R., and Seibel, B.A. (2008). Synergistic effects of climate –related variables suggest future physiological impairment in a top oceanic predator. *Proceedings of the National Academy of Science USA* 105(52): 20776–20780.

Rouse, G.W., Wilson, N.G., Goffredi, S.K., Johnson, S.B., Smart, T., Widmer, C., Young, C.M., Vrijenhoek, R.C. (2009). Spawning and development in Osedax boneworms (Siboglinidae, Annelida). *Marine Biology* 156 (3), 395-405.

Rudels, B., Jones, E.P., Anderson, L.G., and Kattner, G. (1994). On the intermediate depth waters of the Arctic Ocean. *Geophysical Monogram* 85: 33-46.

Ruhl, H.A., and Smith, K.L. Jr. (2004). Shifts in deep-sea community structure linked to climate and food supply. *Science* 305: 513-515.

Ruhl, H.A., Ellena, J.A., and Smith, K.L. Jr. (2008). Connections between climate, food limitation, and carbon cycling in abyssal sediment communities: a long time-series perspective. *Proceedings of the National Academy of Science USA* 105: 17006–17011.

Savin, S.M., Douglas, R.C., and Stehli, F.G. (1975). Tertiary marine paleotemperatures. *Geological Society of American Bulletin* 86: 1499-1510.

Schlacher, T.A., Baco, A.R., Rowden, A.A., O'Hara, T.D., Clark, M.R., Kelley, C., and Dower, J.F. (2013). Seamount benthos in a cobalt-rich crust region of the central Pacific: conservation challenges for future seabed mining. *Diversity and Distribu-*

tions 1-12.

Sibuet, M., Olu, K. (1998). Biogeography, biodiversity and fluid dependence of deep-sea cold-seep communities at active and passive margins. *Deep-Sea Research Part II* 45 (1-3), 517-567.

Sirenko, B.I. (2001). List of species of free-living invertebrates of Eurasian Arctic seas and adjacent deep waters. *Explorations of the Fauna of the Seas* 51: 1-129.

Sissenwine, M.P., and Mace, P.M. (2007). Can deep water fisheries be managed sustainably? In: *Report and documentation of the Expert Consultation on Deep-Sea fisheries in the High Seas.* FAO Fisheries Report 838. Rome, Italy: FAO. pp. 61–111.

Smale, D.A., Barnes, D.K.A., Fraser, K.P.P., and Peck, L.S. (2008). Benthic community response to iceberg scouring at an intensely disturbed shallow water site at Adelaide Island Antarctica. *Marine Ecology Progress Series* 355: 85-94.

Smith, C.R. (2006). *Bigger is better: The role of whales as detritus in marine ecosystems*. In: Whales, Whaling and Ocean Ecosystems, Estes, J.A., DeMaster, D.P., Brownell Jr., R.L., Doak, D.F., and Williams, T.M. (eds.). University of California Press, Berkeley, CA, USA, pp. 286 – 301.

Smith, C.R., De Leo, F.C., Bernardino, A.F., Sweetman, A.K., Arbizu, P.M. (2008). Abyssal food limitation, ecosystem structure and climate change. *Trends in Ecology and Evolution* 23 (9), 518-528.

Smith, C.R., Grange, L., Honig, D.L., Naudts,L., Huber, B., Guidi, L. and Domack, E. (2012). A large population of king crabs in Palmer Deep on the West Antarctic Peninsula and potential invasive impacts. *Proceedings of the Royal Society B*, 279: 1017-1026. doi: 10.1098/rspb.2011.1496

Smith, C.R. and Baco, A.R. (2003). The ecology of whale falls at the deep-sea floor. *Oceanography and Marine Biology Annual Review*, 41: 311-354.

Smith, C.R., Glover, A.G., Treude, T., Higgs, N.D. and Amon, D.J. (2015). Whale-fall ecosystems: recent insights into ecology, paleoecology and evolution. *Annual Review of Marine Science*, 96. doi: 10.1146/annurev-marine-010213-135144.

Smith, K.L., Ruhl, H.A., Kahru, M., Huffard, C.L., Sherman, A.D. (2013). Deep ocean communities impacted by changing climate over 24 y in the abyssal northeast Pacific Ocean. *Proceedings of the National Academy of Sciences.* Doi: 10.1073/pnas.1315447110.

Smith, K. L., Sherman, A.D., Huffard, C.L., McGill, P.R., Henthorn, R., Von Thun, S., Ruhl, H.A., Kahru, M., Ohman, M.D. (2014). Large salp bloom export from the upper ocean and benthic community response in the abyssal northeast Pacific: Day to week resolution. *Limnology and Oceanography* 59 (3), 745-757.

Snelgrove, P.V.R., Smith, C.R. (2002). A riot of species in an environmental calm: The paradox of the species-rich deep-sea floor. In: Gibson, R.N., Barnes, M., Atkinson, R.J.A. (Eds.), *Oceanography and Marine Biology*, Vol. 40. Taylor & Francis Ltd, London, pp. 311-342.

Sogin, M.L., Morrison, H.G., Huber, J.A., Welch, D.M., Huse, S.M., Neal, P.R., Arrieta, J.M., Herndl, G.J. (2006). Microbial diversity in the deep sea and the underexplored "rare biosphere". *Proceedings of the National Academy of Sciences USA* 103 (32), 12115-12120.

Stein, R., Grobe, H., and Wahsner, M. (1994). Organic carbon, carbonate, and clay mineral distribution in eastern central Arctic surface sediments. *Marine Geology* 119: 269-285.

Stein, R., MacDonald, R.W., eds (2004). *The organic carbon cycle in the Arctic Ocean*. Springer, Berlin, 363 pp.

Stramma, L., Prince, E.D., Schmidtko, S., Luo, J., Hoolihan, J.P., Visbeck, M., Wallace, D.W., Brandt, P., Körtzinger, A. (2012). Expansion of oxygen minimum zones may reduce available habitat for tropical pelagic fishes. *Nature Climate Change* 2 (1), 33-37.

Suess, E., Bohrrnann, G., von Huene, R., Linke, P., Waiimann, K.W., Larnmers, S., and Sahling, H. (1998). Fluid venting in the eastern Aleutian subduction zone. *Journal of Geophysical Research* 103: 2597-2614.

Sul, W.J., Oliver, T.A., Ducklow, H.W., Amaral-Zettler, L.A, and Sogin, M.L. (2013). Marine bacteria exhibit a bipolar distribution. *Proceedings of the National Academy of Science USA*. Doi:10.1073/pnas.1212424110.

Sutton, T.T. (2013). Vertical ecology of the pelagic ocean: classical patterns and new perspectives. *Journal of Fish Biology* 83: 1508-1527.

Svavarsson, J. (1997). Diversity of isopods (Crustacea): new data from the Arctic and Atlantic Oceans. *Biodiversity and Conservation* 6: 1571-1579.

Sweetman, A.K., Chapman, A. (2011). First observations of jelly-falls at the seafloor in a deep-sea fjord. *Deep Sea Research Part I: Oceanographic Research Papers* 58 (12), 1206-1211.

Sweetman, A.K, Smith, C.R., Dale, T. and Jones, D.O.B. (2014). Rapid scavenging of jellyfish carcasses reveals the importance of gelatinous material to deep-sea food webs. *Proceedings of the Royal Society B*: 281: 20142210

Taylor, J.R., DeVogelaere, A.P., Burton, E.J., Frey, O., Lundsten, L., Kuhnz, L.A., Whaling, P.J., Lovera, C., Buck, K.R., Barry, J.P., 2014. Deep-sea faunal communities associated with a lost intermodal shipping container in the Monterey Bay National Marine Sanctuary, CA. *Marine Pollution Bulletin* 83(1), 92-106.

Thubaut, J., Puillandre, N., Faure, B., Cruaud, C., Samadi, S. (2013). The contrasted evolutionary fates of deep sea chemosynthetic mussels (Bivalvia, Bathymodiolinae). *Ecology and Evolution* 3:4748–66

Thurber, A.R., Sweetman, A.K., Narayanaswamy, B.E., Jones, D.O.B., Ingels, J., Hansman, R.L., (2014). Ecosystem function and services provided by the deep sea. *Biogeosciences* 11 (14), 3941-3963.

Tietjen, J.H. (1989). Ecology of deep-sea nematodes from the Puerto Rico Trench area and Hatteras Plain. *Deep-Sea Research* 36: 1579–1594.

Tunnicliffe, V., Embley, R.W., Holden, J.F., Butterfield, D.A., Massoth, G.J., Juniper, S.K. (1997). Biological colonization of new hydrothermal vents following an eruption on Juan de Fuca Ridge. *Deep Sea Research Part I: Oceanographic Research Papers* 44 (9), 1627-1644.

Van den Hove, S., Moreau, V. (2007). *Deep-Sea Biodiversity and Ecosystems: A scoping report on their socio-economy, management and governance*, UNEP-WCMC Biodiversity Series 28. 88pp.

Vanhove, S., Vermeeren, H., and Vanreusel, A. (2004). Meiofauna towards the South Sandwich Trench (750–6300m), focus on nematodes. *Deep-Sea Research II* 51: 1665–1687.

Vinogradova, N. (1959). The zoogeographical distribution of the deep-water bottom fauna in the abyssal zone of the ocean. *Deep Sea Research* (1953) 5 (2), 205-208. Doi: 10.1016/0146-6313(58)90012-1.

Vinogradova, N.G. (1997). Zoogeography of the abyssal and hadal zones. *Advanced Marine Biology* 32: 326-387.

Watanabe, H., Fujikura, K., Kojima, S., Miyazaki, J.I., and Fujiwara, Y. (2010). Ch. 12 Japan: Vents and seeps in close proximity. In: Kiel, S., editor. *The Vent and Seep Biota: Aspects from Microbes to Ecosystems*. Springer, Dordrecht, Netherlands, pp. 379-402

Webb, T., Vanden Berghe, E., and O'Dor, R. (2010). Biodiversity's Big Wet Secret: The Global Distribution of Marine Biological Records Reveals Chronic Under-Exploration of the Deep Pelagic Ocean. *PLosOne* 5(8): e10223.

Wohlers, J., Engel, A., Zöllner, E., Breithaupt, P., Jürgens, K., Hoppe, H.-G., Sommer, U., Riebesell, U. (2009). Changes in biogenic carbon flow in response to sea surface warming. *Proceedings of the National Academy of Sciences of the United States of America* 106, 7067-7072.

Wolff, T. (1970). The concept of hadal or ultra abyssal fauna. *Deep-Sea Research* 17: 983-1003.

Wollenburg, J.E., Mackensen, A., and Kuhnt, W. (2007). Benthic foraminiferal biodiversity response to a changing Arctic palaeoclimate in the last 24,000 years. *Palaeogeography, Palaeoclimatology, Palaeoecology* 255: 195–222.

Yamamoto, J., Hirose, M., Ohtani, T., Sugimoto, K., Hirase, K., Shimamoto, N., Shimura, T., Honda, N., Fujimori, Y., and Mukai, T. (2008) Transportation of organic matter to the sea floor by carrion falls of the giant jellyfish *Nemopilema nomurai* in the Sea of Japan. *Marine Biology* 153: 311-317.

Yancey, P.H., Gerringera, M.E., Drazen, J.C., Rowdenc, A.A., and Jamieson, A. (2014). Marine fish may be biochemically constrained from inhabiting the deepest ocean depths. *Proceedings of the National Academy of Sciences of the United States of America* 111: 4461–4465.

Yasuhara, M., and Cronin, T.M. (2008). Climatic influences on deep-sea ostracode (Crustacea) diversity for the last three million years. *Ecology* 89(11): S52–S65.

Yasuhara, M., Cronin, T.M., deMenocal, P.B., Okahashi, H., and Linsley, B.K. (2008). Abrupt climate change and collapse of deep-sea ecosystems. *Proceedings of the National Academy of Sciences of the United States of America* 105(5): 1556–1560.

Yasuhara, M., Hunt, G., Cronin, T.M., and Okahashi, H. (2009). Temporal latitudinal-gradient dynamics and tropical instability of deep-sea species diversity. *Proceedings of the National Academy of Sciences of the United States of America* 106(51): 21717–21720.

Yasuhara, M., Hunt, G., Cronin, T.M., Hokanishi, N., Kawahata, H., Tsujimoto, A., and Ishitake, M., (2012a). Climatic forcing of Quaternary deep-sea benthic communities in the North Pacific Ocean. *Paleobiology* 38: 162–179.

Yasuhara, M., Hunt, G., van Dijken, G., Arrigo, K.R., Cronin, T.M., and Wollenburg, J.E. (2012b). Patterns and controlling factors of species diversity in the Arctic Ocean. *Journal of Biogeography* 39: 2081–2088.

Yasuhara, M., Okahashi, H., Cronin, T.M., Rasmussen, T.L., and Hunt, G. (2014). Deep-sea biodiversity response to deglacial and Holocene abrupt climate changes in the North Atlantic Ocean. *Global Ecology and Biogeography*. Doi:10.1111/geb.12178.

Yesson, C., Clark, M.R., Taylor, M., and Rogers, A.D. (2011). The global distribution of seamounts based on 30-second bathymetry data. *Deep Sea Research I*. 58: 442–453. Doi: 10.1016/j.dsr.2011.02.004.

Zeidberg, L.D., and Robison, B.H. (2007). Invasive range expansion by the Humboldt squid, Dosidicus gigas, in the eastern North Pacific. *Proceedings of the National Academy of Sciences of the United States of America* 104, 12948–12950.

Zinger, L., Amaral-Zettler, L.A., Fuhrman, J.A., Horner-Devine, M.C., Huse, S.M., Welch, D.B.M, Martiny, J.B.H., Sogin, M., Boetius, A., and Ramette, A. (2011). Global patterns of bacterial beta-diversity in seafloor and seawater ecosystems. *PLoS ONE* 6(9): e24570.

36G Arctic Ocean

Contributors:
Lis Lindal Jørgensen (Convenor), Philippe Archambault, Claire Armstrong, Andrey Dolgov, Evan Edinger, Tony Gaston, Jon Hildebrand, Dieter Piepenburg, Jake Rice (Lead Member and Editor of Part VI), Walker Smith, Michael Vecchione and Cecilie von Quillfeldt

Referees:
Arne Bjørge, Charles Hannah.

1 Introduction

1.1 State

The Central Arctic Ocean and the marginal seas such as the Chukchi, East Siberian, Laptev, Kara, White, Greenland, Beaufort, and Bering Seas, Baffin Bay and the Canadian Archipelago (Figure 36G.1) are among the least-known basins and bodies of water in the world ocean, because of their remoteness, hostile weather, and the multi-year (i.e., *perennial*) or seasonal ice cover. Even the well-studied Barents and Norwegian Seas are partly ice covered during winter and information during this period is sparse or lacking. The Arctic has warmed at twice the global rate, with sea-ice loss accelerating (Figure 36G.2, ACIA, 2004; Stroeve et al., 2012, Chapter 46 in this report), especially along the coasts of Russia, Alaska, and the Canadian Archipelago (Post et al., 2013). Changes in ice cover, ocean warming, altered salt stratification, alterations in water circulation and fronts, and shifts in advection patterns show that oceans within the Arctic are subjected to significant change, and may face even more change in future (Wassmann, 2011 and references within). The Central Arctic Ocean and the marginal seas are home to a diverse array of algae and animals, some iconic (e.g., polar bear), some obscure, and many yet to be discovered. Physical characteristics of the Arctic, important for structuring biodiversity, include extreme seasonality resulting in short growing seasons and annual to multi-annual ice cover. The Central Arctic Ocean has a deep central basin (>4000 m depth) surrounded by the most extensive shelves of all the world's oceans, and is characterized by extensive (albeit declining) ice cover for much of the year. This offers a vast number of different habitats created by the shape of the seabed, latitude, history of glaciations, proximity to the coastline and rivers, oceanic currents, and both the seabed and the ice as a substrate. Barriers for dispersal, such as the ice plug in the Canadian High Arctic, effectively separate stocks of some marine mammals (Dyke et al., 1996). Polynyas, which are open water areas surrounded by ice, provide important foraging and refuge areas and contribute to Arctic biodiversity. Differences in ice cover, mixing between warm- and cold-water currents, or currents with different nutrient content, create a mosaic of nutrient-poor areas which is reflected in species diversity (ABA, 2014, Figure 36G.3). Despite this heterogeneity, the Arctic is less diverse than lower-latitude areas for several taxa, including mammals and birds, but equal to, or higher than those areas for bottom animals (Renaud et al., 2009; Piepenburg et al., 2011), marine crustaceans and phytoplankton (algae plankton) (Archambault et al., 2010). The marine areas in the Arctic support species of algae, plankton, nekton, fish, benthos, mammals, and birds (see sections below) but also thousands of species of fungi, endoparasites and microorganisms (Figure 36G.3, see also ABA 2014 for more information). Due to mixing of sub-Arctic and Arctic fauna, the biodiversity is high in the vicinity of the Arctic Gateways of the North Atlantic and Pacific Oceans (ABA, 2014). The Red List of the International Union for Conservation of Nature (IUCN) includes 13 Arctic or seasonal mammalian inhabitants and 21 Arctic or Arctic-breeding seabirds as threatened species (IUCN, 2012), and eight targeted fish stocks and five Arctic fish species are evaluated according to the IUCN red list criteria (Christiansen et al., 2014, http://www.iucnredlist.org/technical-documents/categories-and-criteria). Humans in the Arctic lead lives based on traditional hunting, fishing and gathering of marine resources or commercial fishing and other economic and recreational activities. Along the coast and on islands, the marine environment plays a central role in food, housing, settlement patterns, and cultural practices and boundaries.

1.2 Trends and Pressures

Climatically, ecologically, culturally and economically, the Arctic is changing, with implications throughout the region (ABA, 2014). Primary producers, such as sea-ice algae and sub-ice phytoplankton, have lost over 2 million km^2 of Arctic ice since the end of the last century (Figure 36G.2, Kinnard et al., 2011), representing a loss of habitat. The largest changes will take place in the northern sections of today's seasonal ice zones, which will expand and eventually cover the entire Arctic Ocean while the multi-year ice will be declining (Wassmann, 2011). The seasonal timing of the ice-algal bloom, driven by light penetration through thinning sea ice, is critical to reproduction of some zooplankton, and the subsequent algal bloom is critical for the survival of zooplankton offspring (Søreide et al., 2010). The annual zoo- and phytoplanktonic pulses of productivity fuel the Arctic marine food web (Darnis et al., 2012) affecting zooplankton production and the Arctic cod that feed on them (Ji et al., 2013), as well as their seabird and marine mammalian predators (Post et al., 2013). It also affects the underlying benthic communities such as bivalves, crabs, sea urchins, which are in turn key prey for seabottom feeding specialists, such as diving sea ducks, bearded seals, walrus, and gray whales (Grebmeier and Barry, 2007). Vertebrate species are also directly affected and walrus and polar bears are moving their habitats from the diminishing sea ice to land (Fischbach et al., 2009). Arctic warming and sea-ice loss will facilitate invasion by new species, hosts, harmful microorganisms, and diseases (Post et al., 2013).

As sea ice retreats (see also chapter 46) and living commercial resources migrate northward, shipping (AMSA, 2009), fishing, petroleum activities, tourists, and consequently the risks of oil spills, noise, pollution and disturbances follow. These risks are often found where fish and marine mammals are abundant (AMSA IIc, 2013). Some of the largest populations of seabirds in the northern hemisphere are intersected by major shipping routes. Boreal fish stocks may move into unexploited parts of the Arctic, depending on the sensitivity and adaptive capacity of the affected species (Hollowed et al. 2013). New Arctic and sub-Arctic species have recently been reported from the Canadian Beaufort Sea and harvested Atlantic species have moved poleward into Arctic Seas. These patterns likely represent both altered distributions resulting from climate change and previously occurring but unsampled species. (Mueter et al., 2013). As targeted boreal stocks move into as yet unexploited parts of the seas, Arctic fish species turn up as unprecedented by-catch and could be vulnerable to large-scale industrial fisheries (Christiansen et al., 2014, ABA 2014). Bottom-dwelling fisheries harvest near the seabed and they reshape bottom morphology and impoverish, perturb and change the functional composition of benthic communities (Puig et al., 2012). Cold-water coral, sponges and sea pens, which form a more complex habitat, are protected species (Fuller et al. 2008, FAO 2009)

Chapter 36G — Arctic Ocean

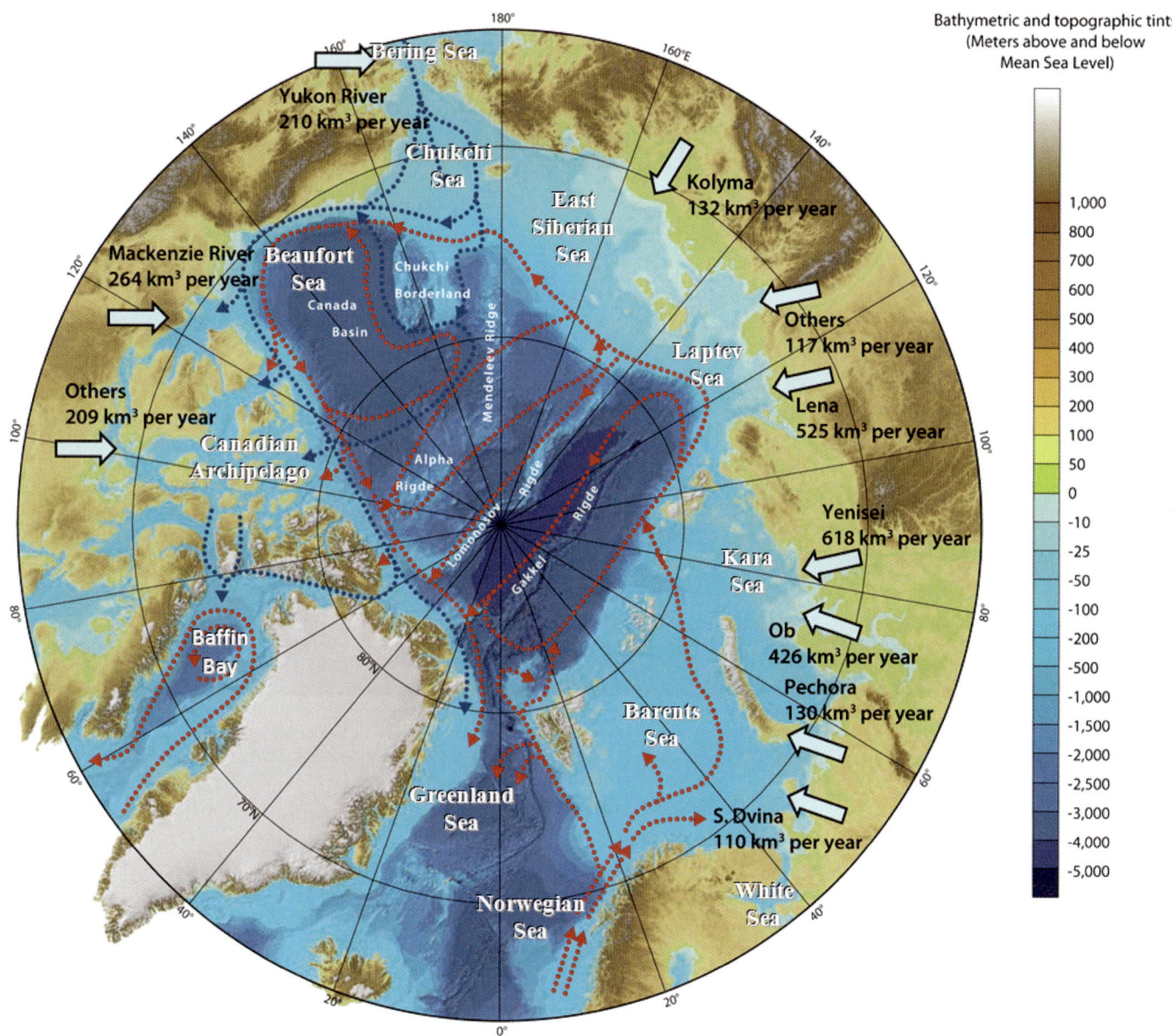

The boundaries and names shown and the designations used on this map do not imply official endorsement or acceptance by the United Nations.

Figure 36G.1 | The deep Central Arctic Ocean and the marginal seas such as the Chukchi, East Siberian, Laptev, Kara, White, Greenland, Beaufort, Barents, Norwegian and Bering Seas, Baffin Bay and the Canadian Archipelago. Blue arrows show freshwater inflow, red arrows water circulation. (adapted from CAFF 2013, Arctic Biodiversity Assessment, figure 14.1).

and areas potentially inhabited by these vulnerable taxonomic groups are mainly found north of 80°N in the Barents Sea, the Greenland Sea, and North of Greenland (Jørgensen et al, 2015; Jørgensen et al 2013; Boertmann and Mosbech 2011; Tendal et al 2013; Klitgaard and Tendal, 2004). Bans on industrial fisheries in the Chukchi and Beaufort Seas are currently in place in the USA (http://alaskafisheries.noaa.gov), along with restrictions on new commercial fishing operations in the Canadian Beaufort (http://news.gc.ca/web/article-en.do?mthd=index&crtr.page=1&nid=894639) and "protected areas" (those regions with reduced or strictly controlled fishing) are debated for the Arctic region (Barry and Price, 2012). Competition for use of marine space might increase in the Arctic, together with increased demand for products from the sea (food, minerals, recreation, etc). Climate-induced changes in the severity of storms and intensity of extreme events might pose challenges to the exploitation of resources in the Arctic. Arctic oil and gas fields provide a substantial part of the world's supply at present, and many fields have yet to be developed. There might be a threat of oil spills and introduction of invasive species (AMAP, 2009). Contaminants are present in organisms at the base of the food web, and they accumulate from one level of the food web (trophic level) to the next (AMAP, 2011). In addition to coastal wave erosion and changes in wildlife movement patterns and cycles, managers also face increases in ocean acidity due to increased CO_2 concentrations. The oceans within the Arctic are especially vulnerable to ocean acidification and Arctic marine ecosystems are highly likely to undergo significant change due to ocean acidification (AMAP 2013). The suite of stressors experienced by species living in the Arctic today is novel, making past periods of climate change an imperfect analogue for the challenges now facing biodiversity in the

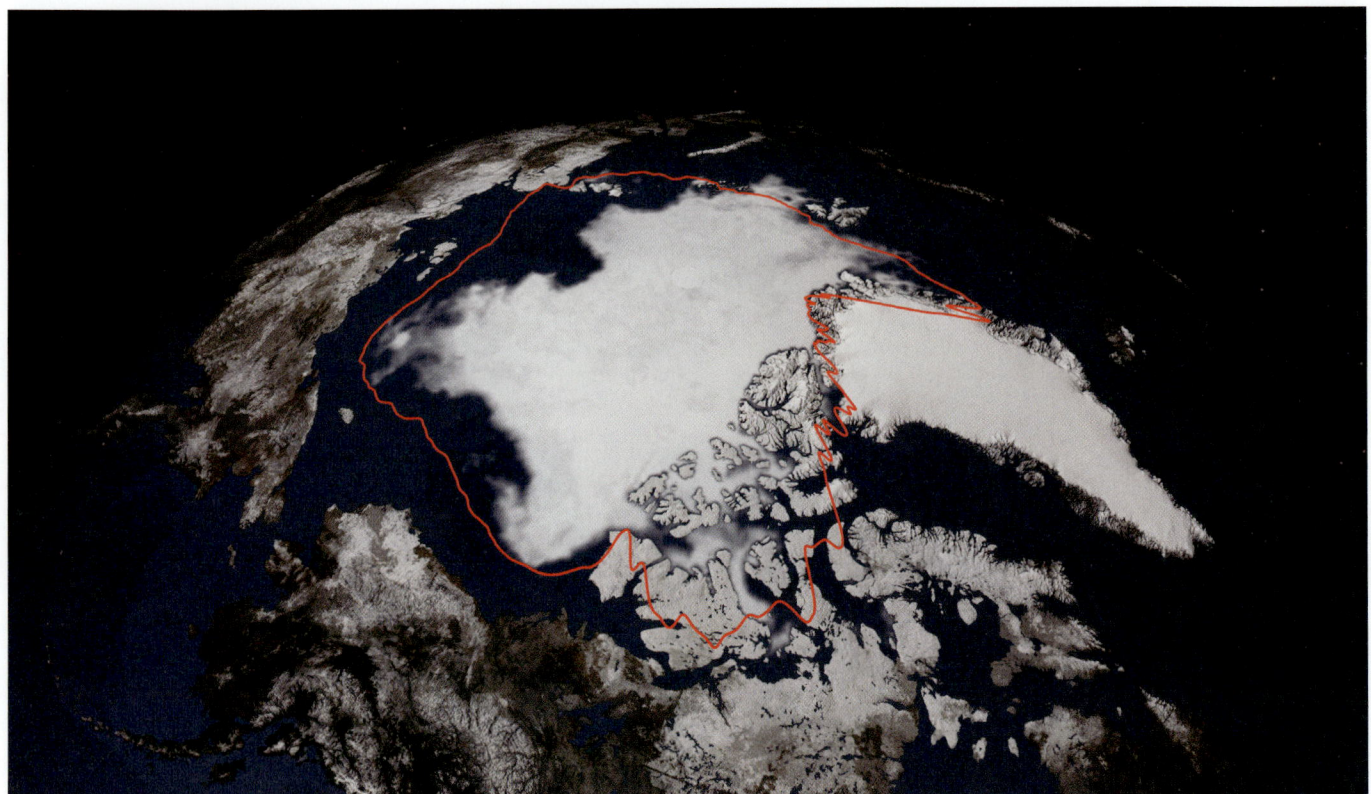

The boundaries and names shown and the designations used on this map do not imply official endorsement or acceptance by the United Nations.
Figure 36G.2 | Sea ice acts as an air conditioner for the planet, reflecting energy from the Sun. On September 17, the Arctic Sea ice reached its minimum extent for 2014 — at 1.94 million square miles (5.02 million square kilometres) the sixth lowest extent of the satellite record. With warmer temperatures and thinner, less resilient ice, the Arctic sea ice is on a downward trend. The red line in the still image indicates the average ice extent over the 30 year period between 1981 and 2011. NASA/Goddard Scientific Visualization Studio, 2014. Printed with permission from NASA's Earth Science News Team patrick.lynch@nasa.gov.

Arctic (ABA 2014). Global climate change threatens to alter the population dynamics of many species for which rates of demographic change and estimates of population size are imprecise or lacking. There is an urgent need to continuously monitor their distribution and occurrence as significant changes occur in the ecosystem. At present, scientists are unable to provide valid answers to questions about safe and sustainable operations. The availability of solid interdisciplinary baseline data is rare but crucial, and it is clear that ongoing and future changes can only be detected through long-term monitoring of key species, communities and processes, providing adequate seasonal coverage in key regions of the Arctic, utilizing new technologies, and making existing historical data accessible to the international research community. Up-to-date knowledge is needed and gaps in knowledge and key mechanisms driving change must be identified in order to secure Arctic biodiversity.

2 Primary producers

2.1 Introduction

Arctic microalgae can be divided by function (e.g., ice algae and phytoplankton). Phytoplankton live suspended in the upper layer of the water column, but ice algae live attached to ice crystals, in the interstitial water between crystals, or associated with the under-surface of the ice (Horner et al., 1988).

2.2 Status

The study of phytoplankton, ice algae and macroalgae of Arctic seas dates back more than one hundred years (e.g., Ehrenberg, 1841; Cleve, 1873; Kjellmann, 1883; Rosenvinge, 1898). Early studies concentrated on diversity and on temporal changes in species composition or distribution relative to oceanographic structure, and were of local or regional character. Poulin et al. (2010) reported 2,016 taxa with 1,874 phytoplankton and 1,027 sympagic (ice algae) taxa in Arctic waters. Daniëls et al. (2013) concluded that few biodiversity assessments of benthic microalgae exist across the Arctic, but estimate ca. 215 seaweed species. Most of the algal species in the Arctic are cold water or temperate species, although some are distributed globally and a few are warm water species (Hasle and Syvertsen, 1996; von Quillfeldt, 1996). The species composition in different Arctic areas is often comparable, which is likely to be due to advection (horizontal transportation) of cells by the currents in the Arctic (Carmack and Swift, 1990; Abelmann, 1992). Differences occur on a smaller scale, often as a result of local environmental conditions (Cota et al., 1991; von Quillfeldt, 2000). Prominent forcing factors on species diversity in the Arctic include the extreme seasonality of light, combined with sea-ice distribution (Bluhm et al., 2011), but the result (increase/decrease) depends on season and locality. However, a

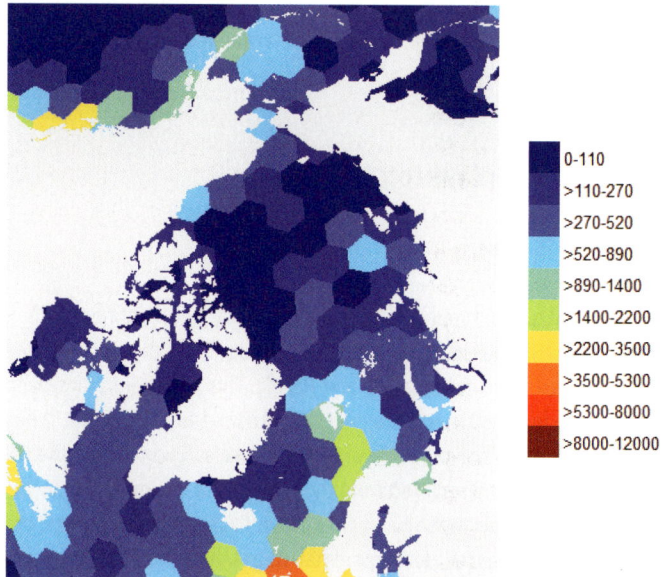

The boundaries and names shown and the designations used on this map do not imply official endorsement or acceptance by the United Nations.
Figure 36G.3 | Pan-Arctic map showing the number of marine species from the OBIS database in a gridded view of hexagonal cells (OBIS, 2015).

suite of environmental variables (e.g., nutrients, light, water stratification, salinity, temperature) determines abundance, biomass and taxonomic composition over time (Poulin et al., 2010). Many species have a wide environmental tolerance (Degerlund and Eilertsen, 2009).

The composition of the phytoplankton varies seasonally (von Quillfeldt, 2000; Lovejoy et al., 2002; Ratkova and Wassmann, 2002; Wassmann et al., 2006; Sukhanova et al., 2009). Most species can be characterized as winter, spring, summer or autumn species, but a few are seasonally independent. Several decades ago, few Arctic areas had been sampled during winter, but the importance of flagellates during winter had been suggested (Schandelmeier and Alexander, 1981; Horner and Schrader, 1982; Rey, 1986). Recently the importance of extremely small (< 20 μm) forms in Arctic waters throughout the year has been confirmed (e.g., Vørs, 1993; Lovejoy and Potvin, 2011; Niemi et al., 2011; Sørensen et al., 2012; Terrado et al., 2012; Kilias et al., 2013).

Ice communities are widespread throughout the Arctic (e.g., Apollonio, 1965; Meguro et al., 1967; Grainger, 1977; Hsiao, 1980; Horner et al.,1988, Horner et al.,1992, Syvertsen 1991), but the different types of communities are characterized by specific species (von Quillfeldt, 1997; Gradinger, 1999; Melnikov et al.. 2002; Zheng et al., 2011). Solitary diatoms (a type of microalga) are common in interstitial communities or sometimes in older ice, whereas the majority of colonial algae, except for *Melosira arctica*, are most common in sub-ice communities of one-year-old ice and in more offshore areas (Dunbar and Acreman, 1980; De Sève and Dunbar, 1990; von Quillfeldt, 1996; von Quillfeldt et al., 2003). Irradiance is the most important factor in determining abundance of ice algae. Snow depth and ice thickness control light in sea ice and thereby algal abundance, as does the ice structure (Gosselin et al., 1997; Robineau et al., 1997; Krembs et al., 2000). Ice algae are distributed throughout the ice during winter and become concentrated at the bottom in spring as a result of brine drainage and active migration of cells through brine channels (Hsiao, 1980; Horner, 1985). Furthermore, a south-north spatial gradient similar to the seasonally dependent gradient in the species composition is often observed. The oldest and most specialized ice community occurs in the far north (Syvertsen, 1991).

Lüning (1990) divided Arctic seaweeds into flora with a distinct vegetation structure; many species are distributed throughout the Arctic, and a few are found only within the Arctic Basin. Macroalgal (multi celled algae attached to the seabed) diversity decreases with increasing latitude and from the Atlantic to the Pacific sector (Pedersen, 2011). Temperature is a primary factor in macroalgal distribution (Lüning, 1990). Wulff et al. (2011) emphasized that macroalgae can be of either Atlantic or Pacific origin, but more macroalgae are of Pacific origin than previously thought. Substratum characteristics are important for the distribution of benthic algae (Zacher et al., 2011.) Along the Russian Arctic coast are areas where a soft substratum prevails and macroalgae are absent (Lüning, 1990). Areas exposed to mechanical effects of sea ice or icebergs will also be devoid of macroalgae (Gutt, 2001; Wulff et al., 2011). The Arctic is also strongly affected by marked changes in surface salinity due to melting of sea ice and freshwater input from rivers. Thus, macroalgae must be able to withstand large variations in salinity over the year. Fricke et al. (2008) described the succession of macroalgal communities in the Arctic and the effect of disturbances on communities of different ages and their changes with depth.

2.3 Trends

Daniëls et al. (2013) comment that it is difficult to estimate trends in Arctic phytoplankton, sea-ice algae and benthic algae due to the rela-

Box 36G.1 | General information on primary producers

Primary producers (algae) in Arctic marine waters are dominated by small, solitary photosynthetic cells containing different types of pigments, and reproducing by the formation of spores and gametes (Daniëls et al., 2013). They consist of numerous heterogeneous and evolutionarily different groups (Adl et al., 2012) and include both single-celled organisms (microalgae) and multicellular organisms (macroalgae). In addition, the prokaryotic Cyanobacteria also occur throughout the ocean. Microalgae occur as solitary cells or form colonies with different shape and structure. The size varies between 0.2 and 200 μm, a few up to 400 m (pico: < 2μm, nano: 2-20 μm, micro: 20-200 μm). Macroalgae are seaweeds that are visible to the naked eye, take a wide range of forms, and range from simple crusts, foliose and filamentous forms with simple branching structures, to more complex forms with highly specialized structures for light capture, reproduction, support, flotation, and attachment (Diaz-Pulido and McCook, 2008).

tively poor knowledge of algal distributions prior to the period of rapid environmental changes. Baseline data are generally lacking, and it is challenging to distinguish between natural variations and changes in assemblages due to anthropogenic modification. The high variability in the number of single-celled algae across the Arctic can be related to sampling effort in time and space, rather than actual differences, and a strong bias towards large cells (Poulin et al., 2010) and sampling in coastal areas has been observed. Knowledge of the biodiversity improved recently as a result of improved sampling techniques, advanced microscopic and molecular methods, electronic databases and gene libraries, and increased international cooperation (Daniëls et al., 2013), as well as increased sampling in the central basins (e.g., Melnikov, 1997; Katsuki et al., 2009; Joo et al., 2012; Tonkes, 2012).

Some surveys indicate that climate-mediated changes appear to be occurring, but geographical differences are also found. For example, less sea ice and an increase in atmospheric low-pressure systems that generate stronger winds (and deeper mixing of the upper ocean), as well as a warming and freshening of the surface layer are likely to favour smaller species (Sakshaug, 2004; Li et al., 2009; Tremblay et al., 2012). However, Terrado et al. (2012) found that some small-celled phytoplankton species were specifically adapted to colder waters, and are likely to be vulnerable to ongoing effects of surface-layer warming. Altered discharge rates of rivers and accompanying changes of composition will also affect the composition of the phytoplankton (Kraberg et al., 2013). Emiliania huxleyi, a prymnesiophyte, has become increasingly important: blooms of this species have occurred in the Atlantic, presumably related to changing climate conditions (Sagen and Dalpadado, 2004; Hegseth and Sundfjord, 2008). The reappearance of the North Pacific planktotic diatom Neodenticula seminae may also be a consequence of regional climate warming (Poulin et al., 2010). Harrison et al. (2013) predict that northward movement of Atlantic waters will replace cold-water phytoplankton with temperate species and shift transition zones farther north. Increased amounts of annual sea ice relative to multi-year ice will influence ice-algal composition (Poulin et al., 2010). Warming could alter benthic algal distribution and favour invasion by temperate species (Campana et al., 2009). Models suggest that some macroalgal species will shift northwards and that the geographic changes will be most pronounced in the southern Arctic and the southern temperate provinces (Jueterbock et al., 2013). Reduced sea-ice cover and retreating glaciers will continue to alter light, salinity, sedimentation and disturbance processes (Campana et al., 2009).

3 Zooplankton

3.1 Status

The zooplankton community structure in coastal and continental shelf waters of the Arctic is largely controlled by proximity to rivers and the areas of influx from the Atlantic and Pacific. This community has been studied in a few restricted areas (e.g., Walkusz et al., 2010), but it has not been comprehensively reviewed. Many species known from the Atlantic and Pacific and reported from the neritic (shallow marine environment extending from mean low water down to 200m depths) Arctic are found only as advanced developmental stages, and therefore probably are non-reproductive expatriates. However, evidence is increasing that some North Atlantic species are reproducing in the polar Arctic.

Kosobokova et al. (2011) recently reviewed what is known about multicellular zooplankton in the central Arctic based largely on depth-stratified net collections from multiple projects during 1975-2007. They reported 174 oceanic species, of which 70 per cent were crustaceans. Although large copepods are very important in this assemblage, including *Calanus* species (Copepod) typical of the North Atlantic, as well as Arctic endemics (prevalent in or limited to a particular region), the abundance of this fauna is strongly dominated by small copepods. In addition to copepods, amphipods are important crustaceans, again including Arctic endemics, as well as Atlantic species; other important taxonomic groups include larvaceans (pelagic tunicates), chaetognaths (arrow worms), and pteropod molluscs (pelagic snails) (Gradinger et al., 2010). Relative to the westerlies and trade-wind regions, polar systems are relatively enriched with copepods and pteropods, and reduced in species with jelly-like bodies (Longhurst, 2007).

Despite "Thorson's Rule" that the proportion of species with planktonic larvae decreases at high latitudes, many benthic invertebrates of the Arctic, North Atlantic, and North Pacific develop through planktonic stages. These species contribute seasonally to the diversity of the endemic assemblage, as well as to the many non-native species carried by currents into the Arctic from the Atlantic and Pacific Oceans, at times reaching abundances similar to those of the holoplankton (Hopcroft et al., 2010).

Vertical stratification of the zooplankton community in the central Arctic basins is strongly influenced by water-mass distribution and advective input of low-salinity surface water and layering of Pacific and Atlantic intrusions at mid-depths. Whereas the surface waters and waters below the surface layer are considered to be "well-characterized" (Gradinger et al., 2010), the large percentage of species in the very deep waters of the bathypelagic (i.e, open waters >1000 m depth) zone that are either new to science or previously unknown from the Arctic indicate that much

---

Box 36G.2 | General information on zooplankton

Plankton are animals drifting in the sea. Many are microscopic, but some (such as jellyfish, medusa and comb jellies) are visible with the naked eye. Multicellular zooplankton, such as the copepods are called the ocean's "grass and grazers". In addition to copepods, amphipods, another group of small crustaceans are important. Larvaceans are solitary, free-swimming tunicates and live in the pelagic zone. They are transparent, planktonic animals with a tail. Chaetognaths (arrow worms) are transparent dart-shaped animals, while pteropod molluscs are pelagic snails.

remains to be learned about the truly deep pelagic fauna. The evidence to date indicates that little difference among the deep basins exists in the bathypelagic (depth generally between 1000-4000 m) zooplankton.

In-situ observations by remotely operated vehicles, as well as net collections, have shown that gelatinous megaplankton can be important in the central Arctic (Raskoff et al., 2010). This assemblage is dominated by "true" jellyfishes (medusae) and other, similar forms. Similarly to the net-collected mesozooplankton (planktonic animals in the size range 0.2-20 mm), the vertical structure of this assemblage is strongly associated with the vertical distribution of water masses. The overall abundance of gelatinous megaplankton, especially medusae, decreased dramatically over shallow slope, ridge and plateau areas relative to that found in the central basin, whereas the abundance of ctenophores (comb jellies), which are typically present in the surface layer, remained high in these shallower areas (Raskoff et al., 2010).

In addition to water-mass dynamics, vertical distribution of zooplankton is strongly linked to the penetration of solar radiation and the availability of food. Although the seasonal irradiance signal is extreme in the Arctic, diel variability exists and may act as a physiological cue for diel vertical migration (Rabindranath et al., 2011). However, the major irradiance patterns in the Arctic are seasonal, with continuous darkness in winter and 24-h photoperiods in summer. In winter, many species reside deeper in the water column for a diapause, or resting, phase in their life cycle. Winter diapause is not universal; some species are reproductively active under the winter ice (Hirche and Kosobokova, 2011). Additional variability in summer results from breakup of sea ice. When present, ice cover strongly limits irradiance penetration and the seasonal melt further controls both the phytoplankton bloom and the release of ice algae, important factors for the timing of zooplankton life-cycle events and their vertical distribution. In deeper water the vertical linkage between phytoplankton and zooplankton becomes progressively weaker (Longhurst and Harrison, 1989).

Kosobokova et al. (2011) categorized 6 per cent of Arctic zooplankton species as ice-associated ("cryopelagic'). Gradinger et al. (2010) listed 39 invertebrate species as being ice-associated, although the division between cryopelagic and cryobenthic (or sympagic) is not clear. The association between the animals and the sea ice can be based on physical substrate or on the food web based on ice algae (Hop et al., 2011). Furthermore, the association may extend throughout the life cycle or just include a portion (e.g., dependence of a larval copepod stage on ice algae).

Where polynyas maintain open-water conditions in areas surrounded by solid ice cover, the zooplankton community is more similar to the open-water community rather than that found under ice. The pelagic food web of the polynya contributes to transfer of resources to the benthos (Deibel and Daly, 2007).

3.2   Trends and pressures

Limited long-term comparisons within the central Arctic indicate that species inventories, other than newly discovered species, seem to be unchanged. Whereas Pacific species on the shelves are probably non-reproducing populations transported from native waters of the North Pacific, at least some species typical of the Atlantic are found on Arctic shelves and seem to be reproducing successfully. Recent evidence (e.g., Kraft et al., 2013) indicates increasing reproductive success of Atlantic species. Therefore although the diversity inventory has not changed, physiological effects related to climate change appear to be shifting functions in the ecosystem. This "Atlantification" could have several possible results affecting the pelagic food web and transfer of energy to the benthos, in addition to the structure of the zooplankton community. Examples of such possible effects include differences in lipid-storage dynamics and timing of zooplankton reproduction relative to blooms of primary producers failing to provide adequate food to Arctic predators of zooplankton and changes in the production of "marine snow" (including faecal pellets, moults, discarded mucus-feeding structures),

Box 36G.3 | General information on the Benthos

The benthos, an important component of the ocean system, is the scientific term for the community of organisms that inhabit the seabed, ranging from the tidal coastal zone to the abyssal depths of the deep sea. The seabed environment includes a great variety of physically diverse habitats that differ from each other in terms of depth (intertidal to abyssal), temperature, light availability, and type of substratum (ranging from hard through soft, muddy bottoms). It encompasses organisms from a wide variety of taxa, sizes, life forms and ecological niches. Benthic animals, seaweeds (phytobenthos, incl. microalgae, macroalgae, seagrass), bacteria and protists (microbenthos), account for 98 per cent of the marine biodiversity in terms of species; the remaining 2 per cent are pelagic. Furthermore, some benthic fauna live in the sediment (endobenthos or infauna), attached to the seafloor (epibenthos or epifauna) or living above it (suprabenthos). Hyperbenthic (or suprabenthic) animals do not live directly on or in the seabed but very close above the seafloor in the near-bottom part of the water column. The benthic fauna are typically classified into size categories. Microbenthic organisms are bacteria and protists smaller than 0.1 mm. Meiobenthos consists of tiny benthic organisms that are less than 0.5 mm but greater than 0.1 mm in size, mostly inhabiting the interstitial space between the sediment grains. By far the best-studied is the macrobenthos, encompassing forms larger than 0.5 mm that are visible to the naked eye, mostly polychaete worms, bivalves, crustaceans, anthozoans, echinoderms, sponges, and ascidians. Finally, the term megabenthos has been operationally defined as including large, often mobile benthic animals, mostly fish and crustaceans that are big enough to be visible in seabed images or to be caught by towed sampling gear.

which is important for transport of surface productivity necessary to feed the deep benthic communities.

3.3 Climate Change and Oceanographic Drivers Affecting Zooplankton.

Climate-induced changes in the timing and extent of sea-ice melt and breakup could have far-reaching effects on zooplankton structure and function within the pelagic food web, including coupling with the benthos and air-breathing vertebrates. The end of dormancy and initiation of feeding for lipid storage to fuel reproduction in the large Arctic copepods is linked to the ice-edge bloom. Because lipid dynamics differ in North Atlantic congeners, the "Atlantification" of the Arctic may be favoured by early and extensive breakup of the ice. The Atlantic species, which do not build up lipid reserves extensively prior to spawning, as do the Arctic endemics, may not provide adequate food for predators.

Increased ultra-violet (UV) radiation may have extensive effects on epipelagic species. This UV radiation can have substantial impacts on all plankton and can be lethal for zooplankton, especially eggs. Acidification can have a variety of effects, especially on species for which calcification is important, either for formation of exoskeletons (e.g., molluscs and crustaceans) or of sensory organs (e.g., otoliths of fish larvae and statoliths of cephalopods). Particular sensitivity can be expected in the abundant pteropod (swimming snail) species *Limacina helicina*, in which the aragonitic shell is even more vulnerable to dissolution than are calcite structures (see ocean acidification in chapter 5). Climate change impacts in the Arctic are expected to be significant and to be expressed earlier than in other oceanic realms due to the modification of the ice cover currently being observed (Stammerjohn et al., 2012).

4 Benthos

4.1 Status

The current knowledge on the biodiversity of the benthic fauna in coastal, shelf and deep-sea regions has been summarized in three papers (Weslawski et al., 2011; Piepenburg et al., 2011; Bluhm et al., 2011a). These large-scale studies were conducted as contributions to the Arctic Ocean Diversity (ArcOD) project (http://www.arcodiv.org; Bluhm et al., 2011b), which in turn was part of the Census of Marine Life (http://www.coml.org; Snelgrove, 2010) and the International Polar Year 2007/2008. It aimed at coordinating research efforts examining the diversity in each of the three major realms (sea ice, water column, and sea floor) of Arctic marine ecosystems to consolidate what is known and fill gaps in our knowledge.

4.2 Coasts

Weslaswski et al. (2011) reviewed the pattern of occurrence and recent changes in the distribution of macrobenthic organisms in fjords and coastal (nearshore) Arctic waters. In addition, likely future changes were hypothesized. The biodiversity patterns observed were demonstrated to differ among regions and habitat types. The North Atlantic Current along Scandinavia to Svalbard and the Bering Strait was shown to be a major area of biotic advection, where larvae and adult invertebrates are transported from the sub-Arctic areas to Arctic areas. There, increased temperature associated with increased advection in recent decades has favoured the immigration of more boreal-sub Arctic species, increasing the local biodiversity when local cold-water species may be suppressed. On the opposite side, in the Canadian Archipelago, the Nares Strait (between Greenland and Ellesmere Island), Lancaster Sound, Barrow Strait and M'Clure Strait are conduits for cold Arctic water flowing to the North Atlantic. Other large coastal areas, such as the Siberian shores, were shown to be little influenced by advected waters.

4.3 Shelf seas (30 to 500 m)

The knowledge of Arctic shelf seas has increased in the past decade, but benthic diversity was investigated at regional scales only. Piepenburg et al. (2011) presented a first pan-Arctic account of the species diversity of the macro- and megabenthic fauna inhabiting Arctic shelves. It was based on an analysis of 25 published and unpublished species-level data sets, together encompassing 14 of the 19 marine Arctic shelf regions and comprising 2,636 species, including 847 Arthropoda, 668 Annelida (669 if we include the new species described by Olivier et al., 2013), 392 Mollusca, 228 Echinodermata, and 501 species of other phyla. Furthermore, gross estimates of the expected species numbers of the major four phyla were computed on a regional scale. Some areas, such as the Canadian Archipelagos, we have not compiled because of the lack of data. Extrapolating to the entire fauna and study area leads to a conservative estimate: 3,900-4,700 macro- and megabenthic species can be expected to occur on the Arctic shelves. These numbers are smaller than analogous estimates for the Antarctic shelf, but the difference is on the order of about two and thus is less pronounced than previously assumed. On a global scale, the Arctic shelves are apparently characterized by intermediate numbers of macro- and megabenthic species. This preliminary pan-Arctic inventory provided an urgently needed assessment of current diversity patterns that will be used by future investigations for evaluating the effects of climate change and anthropogenic activities in the Arctic.

4.4 Central Arctic Ocean

Bluhm et al. (2011a) compiled a benthic species inventory of 1,125 taxa from various sources for the central Arctic deeper than 500 m, and bounded towards the Atlantic by the Fram Strait. An additional 115 taxa were added from the Greenland–Iceland–Norwegian Seas (GIN). The inventory was dominated by taxa of Arthropoda (366), Foraminifera (197), Annelida (194), and Nematoda (140). A large overlap in taxa with Arctic shelf species supported previous findings that part of the deep-sea fauna originates from shelf species. Macrofaunal abundance, meiofaunal abundance and macrofaunal biomass decreased significantly with water depth. Species evenness increased with depth and latitude. No mid-depth peak in species richness was observed. Multivariate analysis of the Eurasian, Amerasian and GIN Seas polychaete occurrences revealed a strong Atlantic influence, the absence of modern Pacific fauna, and the lack of a barrier effect by mid-Arctic ridges. Regional differences are apparently moderate on the species level and minor on the family level, although the analysis was confounded by a lack of methodological standardization and inconsistent taxonomic resolution. Bluhm et al. (2011a) concluded that more consistent methods to observe temporal trends should be used in future efforts to help fill the largest sampling gaps (e.g., eastern Canada Basin, depths >3,000 m, megafauna). This is necessary to be able to adequately address how ocean warming, and the shrinking of the perennial ice cover, will alter deep-sea communities. The findings of Boetius et al. (2013) indicated that the benthic-pelagic coupling is more intense in the Arctic deep sea than expected and suggested strong alteration of this area in the future.

4.5 Trends and pressures

In a recent manuscript, Wassmann et al. (2011) reviewed the evidence reported in the scientific literature as of mid-2009 on whether – and how – climate change has already caused clearly discernible changes in marine Arctic ecosystems. In general, they found that most reports concerned marine mammals, particularly polar bears, and fish, whereas the number of well-documented changes in planktonic and benthic systems was surprisingly low. Quantitative data on abundance and distribution are still generally lacking, and particularly few footprints of climate change have been reported from particularly remote and difficult-to-access regions, such as the wide Siberian shelf and the central Arctic Ocean, due to the limited research effort made in these environments. Wassmann et al. (2011) concluded that despite the alarming nature of climate change and its strong potential effects in the Arctic Ocean, the amount of reliable data on – as well as the research effort evaluating – the impacts of climate change in this region is rather limited. However, during a 30-year period (1980–2010), featuring a gradually increasing seawater temperature and decreasing sea-ice cover in Svalbard, Kortsch et al. (2012) documented rapid and extensive structural changes in the rocky-bottom communities of two Arctic fjords. They observed a reorganization of the benthic communities, led by an abrupt increase in macroalgal cover.

Because data on the effects of climate change on Arctic benthic fauna are limited, it is not yet possible to make sound predictions on trends and pressures on a quantitative level. However, some reports exist on general trends related to climate-change effects. Based on the available evidence of recent and on-going changes in Arctic systems, Wassmann et al. (2011) forecast that the ecological responses to climate change will encompass range shifts and changes in abundance, growth/condition, behaviour/phenology, as well as community/regime shifts, all of which will inevitably have a strong influence on regional and temporal patterns in diversity. In their attempt to predict possible changes in the diversity patterns of coastal benthic fauna in response to climate change, Weslawski et al. (2011) hypothesized that, in areas that are little influenced by advected waters, such as the Siberian shores and the coasts of the Canadian Archipelago, the local Arctic communities are exposed to increasing ocean temperature, decreasing salinity and a reduction in ice cover, with unpredictable effects on biodiversity. On the one hand, benthic species in Arctic fjords are exposed to increasing siltation from glacial meltwater and to decreasing salinities, which together may lead to habitat homogeneization and a subsequent decrease in biodiversity. On the other hand, the innermost basins of Arctic fjords are able to maintain pockets of very cold, dense, saline water and thus may act as refugia for cold-water species.

Furthermore, all the current and anticipated climate-related changes in the Arctic are accompanied by an increase of anthropogenic activities, such as fisheries. These are known to impact marine ecosystems worldwide and have become an important environmental issue (Pauly et al. 1998; Link et al. 2010; Zhou et al. 2010). Particularly bottom trawling is assumed to be one of the most destructive fishing methods, causing severe damage to seafloor structure and benthic communities due to the passage of fishing gears and frequent by-catch (Jones 1992; Tillin et al. 2006; Thurstan et al. 2010). Using trawls is only feasible in largely ice-free areas, which are becoming more abundant. Trawling has recurrently been demonstrated to severely modify benthic communities (Watling and Norse 1998; Collie et al. 2000) and fish habitats (Auster 1998; Kaiser et al.,1999, Collie et al. 2000; Lindholm et al.,2001; Thrush et al., 2002; Moritz et al., 2015), primarily because of the reduction of bottom complexity through the smoothening of sediments and removal of biogenic structures (Collie et al., 1997; Collie et al., 2000; Thrush et al., 1995; Thrush et al., 1998; Hall-Spencer and Moore, 2000). A number of field studies (Auster et al., 1995; Tupper and Boutillier, 1995a; Tupper and Boutillier, 1995b; Kaiser et al., 1999; Lindholm et al., 1999; Anderson and Gregory, 2000; Linehan, 2001; Stoner and Titgen, 2003) have related habitat complexity to survival of juvenile fishes. Cold-water corals, sponges and sea pens form biogenic structures that provide complex habitats for a diverse associated fauna. Although they are protected marine species (Fuller et al., 2008; FAO 2009), they are most vulnerable to the first passing of trawls. Areas protected from bottom trawling due to ice cover, thus remaining pristine and potentially inhabited by these vulnerable taxonomic groups, are mainly found north of 80° N in the Barents Sea, the Greenland Sea, and off northern Greenland (Jørgensen et al, accepted; Jørgensen et al., 2013; Boertmann and Mosbech, 2011; Tendal et al., 2013; Klitgaard and Tendal, 2004). Exploratory trawling fisheries will be carried out in a number of areas of the Canadian Arctic, such as Hudson Strait and northern Hudson Bay. Furthermore, there are shrimp pot fisheries, which are less damaging to benthic habitats, in Baffin Bay. There are currently no commercial fisheries in the Beaufort Sea. The United States has adopted a precautionary approach and placed a moratorium on fishing in the United States EEZ of the Arctic Ocean until further scientific information is available (Wilson and Ormseth, 2009). A similar strategy was recently adopted in the western Canadian Arctic with the signing of the Beaufort Sea Integrated Fisheries Management Framework (http://news.gc.ca/web/article-en.do?nid=894639), which outlines an agreement between the Government of Canada and Inuvialuit to co-manage marine mammal and fish resources in the Canadian Beaufort Sea.

5 Nekton (including demersal and holopelagic vertebrates and invertebrates)

5.1 Status

Arctic fishes include two main groups – typically marine species which are confined to the marine environment, and anadromous species (fish migrating from salt water to spawn in fresh water such as salmonids and coregonids) which occur in freshwater and coastal areas, such as bays, inlets and estuaries, ascending rivers from the sea for breeding. Depending on how widely the Arctic region is defined, total fish diversity ranges from 242 to 633 marine fish species (from 106 families) and 18-49 freshwater species that occur in marine/brackish waters (Chernova, 2011; Mecklenburg et al., 2011; Christiansen et al., 2013). Marine species comprise 88-90 per cent of total fish diversity. Species numbers in the Arctic are rather low, for both marine and freshwater species compared to the total number of fish species globally (approximately 16 and 12 thousand, respectively); 92 per cent of Arctic species are bony fishes; cartilaginous fishes (sharks and skate) comprise only 8 per cent. Most Arctic species are teleost (fishes with bony skeletons) fishes (92 per cent); cartilaginous (having a skeleton composed either entirely or mainly of cartilage) fishes (sharks and skates) comprise only 8 per cent (Lynghammar et al., 2013).

Fish diversity declines from the Arctic gateway regions near the Atlantic and Pacific Oceans, such as the Norwegian and Barents Seas (Atlantic) and Bering and Chukchi Seas (Pacific) to the farthest and most strictly Arctic seas. This diversity gradient is driven primarily by the presence of many boreal species in the Arctic gateway seas; such species cannot reproduce under the consistently colder conditions of the high Arctic. This spatial pattern holds in both the Eurasian and North American shelf seas (Karamushko, 2012; Christensen and Reist, 2013; Coad and Reist, 2004).

From a zoogeographic point of view, only 10.6 per cent of the bony fishes are considered as being strictly Arctic, and able to reproduce in waters below 0°C, whereas 72.2 per cent are boreal or Arctic-boreal species. Demersal fish species prevail in the group of strictly Arctic species (which includes 64 species or 14 per cent of the global marine fish fauna) (Chernova, 2011; Christensen and Reist, 2013).

Species composition and structure of fish communities vary in different depth zones and regions. Coastal brackish areas are usually inhabited by freshwater and anadromous fishes (whitefish, char, etc.). Fjords provide important habitats for fishes in some areas of the Arctic Seas, particularly along steep, bedrock-dominated coasts, such as are found in Greenland, Spitsbergen/Svalbard, Northern Norway, and the eastern parts of the Canadian Arctic Archipelago. Fjord fish faunas tend to be dominated numerically by the cryopelagic (of cold, deep oceanic waters) species (polar cod, Arctic cod), and by anadromous species. Fjord fish faunas include a wide cross-section of Arctic bottom-living (demersal) fishes (Christiansen et al., 2012), including diverse sculpins (Cottidae) and eelpouts (Zoarcidae) on sills and along fjord walls, as found in rocky areas of the continental shelves, and flatfish (Pleuronectidae) on sand and mud bottoms in fjord basins (Haedrich and Gagnon, 1991).

> **Box 36G.4 | General information on the Nekton**
>
> The nekton includes the bony fishes and the cartilaginous fishes (sharks and skate). Demersal fish species live and feed on or near the bottom, while pelagic fish live in open water. Brackish areas are usually inhabited by freshwater and anadromous fishes (whitefish, char, etc.). Cephalopod (a molluscan group consisting of, for example, octopus, squid, and cuttlefish) and shrimp species are also part of the nekton.

Fish communities of shelf seas are composed of common abundant pelagic (herring, capelin, polar cod, etc.) and demersal (gadoids, flatfish, sculpins, eelpouts, etc.) fishes.

Fish species composition in deeper waters in the Arctic Basin, as well as in many parts of the outer shelf regions, remains poorly investigated. Species richness is lower compared to coastal areas and especially shelf seas; the most abundant fishes are cryopelagic (e.g., polar cod, Arctic cod) (Andriashev et al., 1980; Melnikov and Chernova, 2013) or deepwater (e.g., snailfish) (Tsinovsky and Melnikov, 1980) fish. Cryopelagic species, which are ecologically dependent on sea ice, including the circumpolar polar cod *Boreogadus* and the ice cod *Arctogadus*, are important prey species for many larger fish and marine mammals. Although the most abundant species are widely distributed in the Arctic and adjacent waters, the demersal fauna of the Arctic pseudo-abyss (the zone from 200 to 500–1,000 m in different parts of the ocean; characterized by a mixture of fauna) is represented mainly by endemic species (Chernova, 2011).

Commercial fisheries in the Arctic are located mainly in shelf seas where boreal species dominate. The most important areas are the Norwegian and Barents Sea in the Northeast Atlantic, Baffin Bay in the Northwest Atlantic, and in the Bering Sea in Pacific (Christiansen et al., 2014). In total 59 stocks are target species and 60 stocks are by-catch species taken by fisheries in Arctic and sub-Arctic areas. Most of the targeted species (50) are boreal species, six species are Arcto-boreal and three species are Arctic. The dominant families exploited in fisheries are herring (Clupeidae), capelin (Osmeridae), cod (Gadidae), flatfish (Pleuronectidae), and rockfish (called redfish in the Atlantic, Scorpaenidae). Wolf-fish (Anarhichadidae, an endemic Holarctic marine family) and grenadiers (Macrouridae) are important target species in the Eurasian Arctic, but in North America the effects of fisheries on these fish are dominantly through bycatch. High-latitude species in both families are considered to be endangered (Kearley, 2012). The landlocked Atlantic cod (*Gadus morhua*), found in meromictic (stratified lakes that consist of two layers that do not completely mix) Arctic lakes (Hardie and Hutchings, 2011), is considered to be of special concern.

The diversity of invertebrate nekton is much less than that of the fish, although some species are important prey for high-level predators (Gardiner and Dick, 2010). None are cryopelagic or estuarine, although some may be found in fjords. Nesis (2001) considered only seven cephalopod (a molluscan group consisting of, for example, octopus, squid, and cuttlefish) species to be resident in the Arctic. Although Arctic records of many other cephalopod species were added by Gardiner and Dick (2010) and Golikov et al. (2012), the only species that may be Arctic endemics are poorly known benthic octopods in the Bering and Chukchi Seas. Even fewer shrimp species are known from the Arctic and almost all have been reported from the North Atlantic, North Pacific, or both. One species, *Hymenodora glacialis*, is a major component of the pelagic biomass in the deep basins (Auel and Hagen, 2002).

5.2 Trends and pressures

Fisheries affect mainly the traditional marine target fish in the shelf seas located in the narrow Arctic to boreal regions of the Atlantic and Pacific Oceans (e.g., the Barents Sea and Bering Sea), and their effects are generally lower in other seas (e.g., the Laptev Sea and East Siberian Sea). But with further warming, fisheries areas are expected to shift into previously unfished Arctic regions where they will affect strictly Arctic fish communities (Christiansen et al., 2014). In coastal areas a few species, predominantly freshwater and anadromous, are harvested by indigenous peoples, but these catches are generally much lower than those of commercial fisheries. Furthermore, fisheries for anadromous species rarely involve mobile gear, which poses the greatest risk of causing extensive habitat damage in previously unfished areas if used in ways that contact habitat features (e.g., Anderson and Clarke, 2003; Rice et al., 2006).

Under continuous ocean warming conditions, shifts of native species and new appearances of warm-water species may result in changes to fish community structure and subsequently to trophic pathways, depending on the sensitivity and adaptive capacity of the affected species (Hollowed et al. 2013). Higher water temperatures may cause an increase in the abundance and proportion of boreal species in the Arctic community. The deep Central Basin will probably be affected less than the shallower shelf seas of the Arctic, as most abundant boreal species are demersal or neritic (the relatively shallow part of the ocean above the drop-off of the continental shelf, approximately 200 m in depth) and such species are not likely to be found in areas deeper than 800-1000 m (Dolgov and Karsakov, 2011).

Occasional appearances of new species have been observed in the Arctic for decades, but these are apparently becoming more frequent. In 1950s, pink salmon was introduced from the Pacific to the Barents and White Seas (Atlas of Russian freshwater fishes, 2002). Norwegian pollock *Theragra finnmarchica* has been known in the Barents Sea since the 1950s (Christiansen et al., 2005; Privalikhin and Norvillo, 2010), but it is now considered to be a junior synonym of the Pacific walleye pollock *Gadus chalcogrammus* (Ursvik et al., 2007; Byrkjedal et al., 2008), reflecting the possibility of recent connections between Pacific and Atlantic waters across the Arctic. Range expansion of boreal species into the Arctic as ocean temperatures rise, has been observed both in the Eurasian Arctic (Christiansen et al. 2013) and in the North American Arctic, specifically in the Canadian Beaufort Sea (Mueter et al. 2013). If commercial fishing activities expand northward following these species, more bycatch may occur of endemic Arctic fish species that were previously unexploited and relatively unperturbed. In southern Newfoundland and Norwegian fjords, mesopelagic fish, especially the Myctophid (Lanternfish) *Benthosema glaciale* and the sternoptychid (small deep-sea ray-finned fish of the stomiiform family Sternoptychidae) *Maurolicus muelleri*, are important elements. These cold-water fish, in places very abundant on the high seas, are very likely to move from the sub-Arctic to the full Arctic as ice retreats.

Although some high-latitude areas have high planktonic productivity and high fish production, e.g., the Barents Sea and Bering Sea, many strictly Arctic Seas have limited primary productivity due to ice cover, lack of nutrient replenishment, etc. Most Arctic fish, similar to deep-sea fish, have adapted to these low-productivity conditions with life-history characteristics that cause them to be readily overfished, either as directly targeted species or as by-catch (Koslow et al., 2000; Roberts, 2002; Baker et al., 2009). The risk of rapid overexploitation is high for Arctic fish populations, as it already seems to be for Greenland halibut (*Reinhardtius hippoglossoides*) in the Western Atlantic. By contrast, the Barents Sea stock has grown over the last decade, when the fishery was closed completely at first but later reopened at a low intensity.

As with many other groups, evidence exists that invertebrate nekton species are spreading into the Arctic from lower latitudes, especially the North Atlantic. Indeed, Hamilton et al. (2003) reported that fisheries in western Greenland have shifted to the northern shrimp, *Pandalis borealis*, a North Atlantic species also fished in the Barents Sea (Standal, 2003). Furthermore, the presence of foraging schools of ommastrephid squid (Golikov et al., 2012) could indicate an important shift in the pelagic food web of the Arctic.

6 Mammals

6.1 Status

Thirty-five species of marine mammals are known to be present in Arctic waters. Seven of these (narwhal, beluga, bowhead whale, ringed seal, bearded seal, walrus, and polar bear) inhabit the Arctic year-round and are dependent upon sea ice for at least part of the year. Four additional species (spotted seal, ribbon seal, harp seal, and hooded seal) use sea ice for pupping in the winter and spring, but range widely in open waters of the Arctic and sub-Arctic the rest of the year. These eleven species of marine mammals are ice-dependent for at least some of their annual cycle; their reproduction, moulting, resting and/or feeding behaviour are closely linked to the presence of sea ice.

> **Box 36G.5 | General information on the marine mammals**
>
> Seals, together with toothed (killer whales, sperm whales, dolphins) and baleen (bowheads, blue whale) whales are part of the marine mammals.

Another 24 marine mammal species occur in low Arctic waters or seasonally migrate to the Arctic to feed, including four species of pinnipeds, nineteen species of cetaceans, and a carnivore, the sea otter. The northern fur seal and Steller sea lion are found in the Okhotsk and Bering Seas; the gray seal is found in the Atlantic Arctic, and the harbour seal in Arctic waters of both the Atlantic and Pacific. Nineteen species of cetaceans use Arctic waters seasonally, including: the North Pacific right whale and gray whale that are confined to the Pacific Arctic; the North Atlantic right whale in Arctic waters near Greenland; the blue whale, fin whale, sei whale, minke whale, and humpback whale in both Pacific and Atlantic Arctic waters during summer; the sperm whale in low Arctic waters; Baird's beaked whale, Stejneger's beaked whale and Cuvier's beaked whale in the low Arctic waters of the Pacific; and the northern bottlenose whale in the low Arctic waters of the Atlantic. Delphinids that are present in the Arctic during summer include: killer whale, white-beaked dolphin, long-finned pilot whale and Atlantic white-sided dolphin. Dall's porpoise occur in low Arctic waters of the Pacific, and harbour porpoise in low Arctic waters of both the Atlantic and Pacific. These species occur in Arctic waters primarily to feed, based on the high seasonal productivity.

6.2   Trends

There is a history spanning several centuries of commercial whaling and seal hunting in the Arctic. In some cases, over-harvesting has reduced Arctic marine mammal populations to low numbers and contracted their ranges. Two of the three hooded seal populations were subjected to intense commercial hunting over the past two centuries. In the East Greenland Sea a substantial decrease in hooded seal abundance took place between the 1940s and 1980s (ICES, 2008), and recent surveys suggest that a downward trend continues. Regulation of commercial harvests has led to stabilization or recovery of some other marine mammal populations. All bowhead whale populations were severely depleted by commercial whaling, which began in the Atlantic in the 17th century (Ross, 1993). The global bowhead population now appears to be increasing, and the Bering-Chukchi-Beaufort subpopulation has recovered to close to its pre-whaling level. Indigenous harvesting of Arctic marine mammals also has a long history, and indigenous peoples have strong cultural and economic ties to marine mammals. In most cases the subsistence harvest is not a factor affecting marine mammal populations; however, a sharp decline of the Cook Inlet beluga population occurred in the 1990s and is attributed to subsistence overharvesting (Mahoney and Shelden, 2000); and they remain critically endangered.

Assessing Arctic marine mammal populations is challenging because of the difficulty of working in this region and the large seasonal ranges of many of these animals. Documenting changes in the abundance and distribution of marine mammals requires study on long time-scales. For the eleven ice-dependent marine mammal species, population trends are discussed here to illustrate the state of our knowledge. Trends in abundance are unavailable for most beluga sub-populations, but three subpopulations are known to be declining: the Cook Inlet (Hobbs et al., 2012), the eastern Hudson Bay (Gosselin et al., 2009), and the White Sea (Burdin et al., 2009). Although population estimates are available for most narwhal stocks (Heide-Jørgensen et al., 2010; Richard et al., 2010), they are not adequate to establish population trends. The Bering-Chukchi-Beaufort population of bowhead whales has increased since the late 1970s (George et al., 2004), and bowhead whales in West Greenland have increased since 2000 (Wiig et al., 2011), whereas trends in the bowhead subpopulations in the Svalbard-Barents Sea and the Sea of Okhotsk are unknown. Population trends for ringed seals are unknown, yet ringed seal density estimates in western Hudson Bay show an approximate 10-year cycle of fluctuation (Ferguson and Young, 2011). Walrus populations in West Greenland and the North Water have been in steady decline, whereas the population in East Greenland has been increasing (Witting and Born, 2005). Walrus numbers at Svalbard have increased slowly during 1993-2006 (Lydersen et al., 2008). Pacific Walrus populations recovered from a depleted state to historical high levels in the 1980s (Fay et al., 1997). Data are insufficient to estimate trends for spotted seals, ribbon seals and bearded seals. Harp seal birth rates in the White Sea stock have experienced significant declines since 2004 (Chernook and Boltnev, 2008). Recent models (ICES, 2008) revealed that since about 1970, the population of harp seals in East Greenland increased in size from its earlier depleted state. There was a moderate increase in the NW Atlantic hooded seal population between the mid-1980s and 2005 (Hammill and Stenson, 2007), but the NE Atlantic hooded seal population has declined by 85-90 per cent over the last 40-60 years (Øigård et al., 2010). For nineteen polar bear subpopulations, seven are declining, four are stable, one is increasing, and insufficient data are available to determine a trend for seven subpopulations (Obbard et al., 2010). The sea otter is believed to have undergone a population decline exceeding 50 per cent over the past 30 years (Estes et al., 2005).

6.3   Pressures

Reductions in sea ice represent an on-going threat to marine mammals in the Arctic. Recent sea ice declines are well documented (Stroeve et al., 2012), and modelling predicts that the Arctic may be ice-free in summer within three decades (Wang and Overland, 2012). These reductions in sea ice are forcing ice-dependent marine mammals, such as polar bears, seals and walrus, to modify their feeding, reproduction and resting behaviour and locations. Pacific walrus have begun hauling out on land in the summer due to loss of annual Arctic sea ice and the summer retreat of the pack ice beyond the continental shelf (Garlich-Miller et al., 2011). Early sea-ice melt and longer open-water periods cause increased primary production in the Arctic (Arrigo and van Dijken, 2011), but are likely to decrease nutrient fluxes to the seafloor. As a result, walrus, bearded seals, and other marine mammals specializing in benthic feeding may experience reductions in prey availability (Bluhm and Gradinger, 2008).

The Arctic is also experiencing more human maritime activity, primarily related to hydrocarbon and mineral development and the opening of shipping routes. These changes bring risks for marine mammals of direct mortality, displacement from critical habitats, noise disturbance, and increased exposure to hunting. Arctic marine mammals also have high levels of contaminants (Norstrom and Muir, 1994), notably organo-chlorines (an organic compound containing at least one covalently bounded atom of chlorine) and heavy metals, as a result of the presence of these substances in the Arctic food web. Little evidence exists of demographic effects in wild marine mammals, but the need is growing to understand the origins of pollutants, and to coordinate efforts to reduce them at their source.

7 Marine birds

7.1 Status

Arctic waters are host in summer to many millions of marine birds which come to nest. Unlike many animal groups, marine birds are more diverse and abundant in cold seas than they are in warm ones (Gaston, 2004). In the Northern Hemisphere, the highest breeding densities of seabirds occur in Arctic waters (Cairns et al., 2008). Forty-four species of seabirds, ten sea ducks (eiders and scoters) and one marine goose (brant) are listed by Ganter and Gaston (2013) as breeding in the Arctic, of which 23 species of seabirds, seven sea ducks and the brant occur in the high Arctic, with most being endemic to the region. The majority of Arctic marine birds are members of the order Charadriiformes (34 species), including four endemic genera, all containing only one immediately subordinate taxon (little auk, ivory gull, Sabine's gull, Ross's gull). Nineteen species are circumpolar, breeding in Canada, Alaska and over most of the Russian Arctic, whereas 11 occur only in the Atlantic basin (East Canada-Svalbard) and 14 in the Pacific basin (East Siberia-Yukon). Four species of sea ducks, one loon, one gull and one auk are considered to be vulnerable, near-threatened or endangered by Birdlife/IUCN (IUCN, 2012).

Most Arctic marine birds are migrants, occurring in Arctic waters only in summer, and moving to boreal or warmer waters, or in some cases to the Southern Hemisphere, in winter (Newton, 2007). Only two species of gull (Ross's and ivory), two auks (black guillemot and thick-billed murre/Brunnich's guillemot) and the four species of eider occur in Arctic waters throughout the year. The extent of migratory behaviour means that the population sizes and trends of many Arctic marine birds are potentially affected by events on their wintering ranges outside the Arctic (Ganter and Gaston, 2013).

Scoters, jaegers/skuas, some terns and some gulls occur mainly in terrestrial or freshwater habitats while breeding, but make use of marine waters while in passage, and, in the case of scoters, while moulting.

> **Box 36G.6 | General information on the marine birds**
>
> Marine birds are adapted for the marine environment. Most species nest in colonies. Many undertake long annual migrations. Ducks, goose, auks (such as guillemots), loon, gulls, scoters, jaegers/skuas, terns, are all part of the marine birds.

Other marine birds, although feeding at sea, must all visit land to breed, so that all have a presence in the coastal zone in summer. The vast majority feed in shelf waters, although some may also feed away from the continental shelf. The most numerous birds making use of Arctic marine waters in summer are the northern fulmar, black-legged kittiwake, thick-billed murre, and little auk, all of which have world populations centred in the Arctic and numbering more than 10 million individuals (Ganter and Gaston, 2013). All these species make use of pelagic habitats. The number of seabirds making use of the central Arctic Ocean is small, although this is a major post-breeding dispersal area for Ross's Gull (Hjort et al., 1997).

Some Arctic marine bird populations provide valuable subsistence resources in the Arctic. Eiders, or their eggs and down, are harvested throughout the region and they are important for traditional food and lifestyle, not only in many Arctic communities, but also in SE Canada and the Baltic region (Merkel and Barry, 2008). In some countries, especially Iceland, down-feather collection constitutes a significant commercial industry (Bédard et al., 2008). Auks are also harvested by native peoples in Alaska and Canada.

7.2 Trends

Most Arctic seabird populations for which reliable information is available have shown negative trends in recent years. These current trends are superimposed on a situation where several important populations were substantially depressed by anthropogenic mortality, compared with numbers in the first half of the 20th century (Ganter and Gaston, 2013).

Some evidence exists for the recent northward spread of predominantly temperate or low Arctic species: e.g., glaucous-winged gull (Winker et al., 2002) in the Bering Sea, horned puffin in the Beaufort Sea (Moline et al., 2008), great skua in Svalbard (Anker-Nilssen et al., 2000; Krasnov and Lorentsen, 2000) and lesser black-backed gull in Greenland (Boertmann, 2008). At the same time, evidence exists of a retreat for at least one high Arctic species: the range of the ivory gull has contracted in Canada; most colonies in the southern part of the Canadian range are deserted, whereas numbers have remained stable farther north (Environment Canada, 2010). Southern colonies of ivory gull are also decreasing in Greenland (Gilg et al., 2009b).

Black-legged kittiwake, an abundant species throughout circumpolar Arctic waters, has shown significant population declines throughout almost the entire Atlantic sector of the Arctic, especially around the Barents Sea (Barrett et al., 2006), in Iceland (Garðarsson, 2006) and in West Greenland (Labansen et al., 2010). Thick-billed murre populations have shown downward trends over much of their range in the past thirty years. The population of thick-billed murres in central West Greenland is much lower than it was in the early 20th century, as a result of heavy harvesting of adults at colonies (Evans and Kampp, 1991; Kampp et al., 1994) and shows no sign of recovery (K. Kampp and F. Merkel, pers. comm.). Similarly, numbers in Novaya Zemlya are considerably lower than in the early 20th century: down from two million to one million birds (Bakken and Pokrovskaya, 2000). In Svalbard, numbers of thick-billed murres were thought to be stable up to the 1990s, but they have since decreased, especially in the southern part of the archipelago (CAFF Circumpolar Seabird Working Group, unpubl.). In Iceland, numbers of thick-billed murres decreased at 7 per cent per year between 1983 and 1985 and 2005-2008, whereas numbers of common murres and Atlantic puffins decreased between 1999-2005 after modest increases earlier (Garðarsson, 2006; Garðarsson, 2009).

7.3 Pressures

Eider populations declined in the 1980s and 1990s in Alaska, Canada, Greenland and Russia; in some cases because of human disturbances, excessive harvest of eggs and birds, together with severe climatic events (Robertson and Gilchrist, 1998; Suydam et al., 2000; Merkel 2004a). The current trend of common eider populations varies, but at least some populations in Alaska, Canada and Greenland are now recovering with improved harvest management (Chaulk et al., 2005; Gilliland et al., 2009; Merkel, 2010). Breeding populations in the Barents Sea region appear stable (Bustnes and Tertitski, 2000). A recent outbreak (2005-present) of avian cholera in the East Canadian Arctic reversed a population increase and reduced the population of a large colony by 30 per cent in just three years (Buttler, 2009). Bycatch in fisheries gillnets is also a significant problem in some areas (Bustnes and Tertitski, 2000; Merkel, 2004b; Merkel, 2011) and may be a more widespread concern.

Some recent changes in the status of Arctic seabirds have been linked with climate changes, mostly ascribed to causes operating through the food chain (Durant et al., 2004; Durant et al., 2006; Sandvik et al., 2005; Irons et al., 2008), but direct effects have been documented in a few cases: White et al. (2011) showed that expansion of the great cormorant population in central West Greenland may be related to increased sea-surface temperature. Several potential causes of the decline of ivory gulls in Canada have been identified: mortality from hunting of adults in Greenland (Stenhouse et al., 2004), high levels of mercury in eggs (Braune et al., 2006) and changes in ice conditions (Gilchrist et al., 2008; Environment Canada, 2010). In Hudson Bay in recent years a combination of warm summer weather and earlier emergence by mosquitoes caused the death or reproductive failure among thick-billed murres (Gaston et al., 2002). In addition, polar bears, coming ashore earlier than usual, ate many eggs, chicks and adults of murres and common eiders, leading to complete reproductive failure at some colonies (Gaston and Elliott, 2013; Iverson et al., 2014). Such mortality has increased sharply over the past three decades.

Substantial research has been carried out on concentrations and trends of contaminants in Arctic marine birds, especially organohaline compounds and heavy metals (Braune et al., 2001; Helgason et al., 2008; Letcher et al., 2010). Very high levels of mercury have been found in the eggs of ivory gulls from Canada (Braune et al., 2006) and high levels of organohaline compounds in those from Svalbard (Miljeteig et al., 2009). High organohaline concentrations occur also in glaucous gulls from Svalbard (Bustnes et al., 2003; Bustnes et al., 2004), perhaps causing mortality in some cases (Gabrielsen et al., 1995; Sagerup et al., 2009). These species scavenge marine mammal carcasses, which places them high up the food chain and hence they become subjected to an increasing concentration of a substance, such as a toxic chemical, in the tissues of organisms at successively higher levels in a food chain (biomagnifications). As a result of biomagnification, organisms at the top of the food chain generally suffer greater harm from a persistent toxin or pollutant than those at lower levels. The extension of offshore oil and gas exploitation and transport in and through Arctic waters poses a potential threat to all marine birds (Meltofte et al., 2013), especially auks and sea ducks, which are among the birds most vulnerable to mortality from oil spills (Clark, 1984). These two groups, along with cormorants, are also very susceptible to drowning in gill nets (Tasker et al., 2000). If, as expected, a general retreat of Arctic sea ice allows an extension of hydrocarbon exploitation, shipping and fisheries in Arctic waters, then special care will be required to safeguard populations of these birds. Moreover, changes in the timing of the open-water season are affecting the timing of seasonal events in marine ecosystems, and this is affecting the optimal timing of breeding for marine birds, especially in low Arctic areas (Gaston et al., 2009). Changes in the distributions of predators and parasites have also been noted, and these may have important consequences for Arctic seabirds (Gaston and Elliott, 2013). Because of the number of Arctic endemic marine bird taxa, the decline of Arctic marine birds presages a significant loss of global biodiversity.

8 Socioeconomic Aspects

8.1 Biodiversity and ecosystem services in the Arctic

Biodiversity, whether it is functional, genetic or species-based, plays a role in fundamental processes of nature, i.e., so-called ecosystem processes or intermediate ecosystem services[1], which feed into all final ecosystem services, whether these are provisioning, regulating or cultural services. These latter services contribute directly to human wellbeing, and these benefits can often be valued in economic terms.

Although the ecosystem processes/intermediate services of biodiversity may be essential for most final services, their values as such cannot be

---

[1] The Millennium Ecosystem Assessment (2005) coined the expression *Supporting services*, which was later referred to as *Ecosystem processes or Intermediate services* by the UK National Ecosystem Assessment (2011) and others.

added to the value of benefits from final services, as this would imply a double counting. However, it is important to ascertain the significance of biodiversity as an intermediate service in order to ensure that human actions do not limit these services to such a degree that a loss in final services occurs, and that the value of this loss exceeds the value from the human actions that led to them. And despite the remote nature of the Arctic, ecosystem processes related to biodiversity taking place there may provide important services far removed in space and time.

Biodiversity may also be a final service and, for example, it may be included in cultural services, in the sense that humans value biodiversity directly. The Arctic is a sparsely populated part of the world, but indigenous and commercial uses related to Arctic biodiversity are nevertheless present. Furthermore, humans who may never set foot in the Arctic may value the existence of Arctic biodiversity, hence the services in the Arctic may have greater importance than what immediately meets the eye.

In the following we focus on ecosystem services from Arctic biodiversity being affected by climate change pressures, where these involve services to the ecosystem (i.e., ecosystem processes or intermediate services) and services to humans (provisioning, regulating or cultural services).

8.2 Services to ecosystems being affected

Climate change will make species of boreal ecosystems move into what are now sub-Arctic areas, transforming these into boreal ecosystems, whereas sub-Arctic ecosystems will move into more Arctic areas. Consequently, Arctic ecosystems will remain in smaller and possibly more fragmented areas such as cold, dense, saline water basins (inner fjords and the abyss) and may thus act as important refugia for cold-water species.

In a spatial context, Arctic biodiversity will therefore decline. In some areas, Arctic biodiversity will disappear, but species of boreal ecosystems will increasingly move northwards, increasing boreal biodiversity in these areas. However, the absolute biodiversity may increase, decrease or remain unchanged, due to the combination of extinction and immigration. Increased biodiversity may especially be the case in the shallow marginal seas of the Arctic, but also in a presumed interim period, where both Arctic and boreal species co-exist. This may temporally affect ecosystem processes/intermediate services. The biodiversity dynamics depend on a number of factors, such as immigration, extinctions, possible hybridization, competitive pressures and new pathogens/parasites, as well as human pressures (harvesting, bycatch of Arctic species in targeted harvests of boreal species, bioaccumulation of pollution, stress from ship traffic and oil exploitation, harvesting of eggs and birds, ocean acidification).

As the ice cover declines, the Arctic biodiversity comes under pressure, and some ecosystem services may be lost due to smaller and possibly fragmented suitable areas. This is particularly the case for species that have parts of their life cycle/ history strategy dependent on ice (e.g., seals nursing on ice). A loss of ecosystem processes/intermediate services involving failures in reproduction, predator-prey interactions and habitat composition is then likely.

8.3 Services to humans being affected

On current sub-Arctic shelf areas, where boreal species will become more prominent, ecosystem services, such as those related to fisheries, may increase. This may be advantageous for human coastal communities, indigenous and otherwise, by increasing or securing values connected to benefits of cultural and provisioning services from fisheries. Off-shelf areas may not give increased ecosystem services, despite ice-cover decline, due to stratification inhibiting the mixing of the water masses and thereby limiting the nutrients needed for productive ecosystems (Wassmann, 2011). However, great uncertainty remains regarding these future processes.

Ice decline will have consequences for Arctic biodiversity. This is particularly the case for species that spend part of their life cycle on land and part on ice (e.g., polar bears, seals and walrus). These species supply provisioning and cultural services for commercial and indigenous users in the Arctic, and cultural services for people worldwide due to existence values.

IPCC (2014) identifies a number of climatic change effects that are expected to affect directly the way of life of Arctic indigenous peoples. The indirect effects via marine biodiversity change are more uncertain. Yet as mentioned above, both positive effects regarding fisheries and negative effects in relation to marine mammals may be possible. Where, how, what, and when changes may arise are unsure, and point towards significant knowledge gaps with regard to socio-economic consequences of climate change for indigenous peoples.

8.4 Management

The loss or reduction of services from Arctic ecosystems points to the need to protect the remaining Arctic and Arctic ice areas against activities that might reduce biodiversity (pollution, diseases/parasites, physical and vocal stress); e.g., securing protection in relation to activities of exploitation (fish, oil, minerals, tourism) in ice areas, and transport routes through the Arctic and Arctic ice.

The final service losses likewise point to the need for adaptive and ecosystem based management efforts to limit negative effects of existing and potential human use. This involves sustainable management of current use of resources, and restrictions on aggregating anthropogenic effects in relation to vulnerable Arctic ecosystems and species. It is clear that we will discover and develop ecosystem services in the future that we are not aware of today. Option values related to these services, for example from bioprospecting, underline the need to secure ecosystem services for the future.

References

Introduction

Andersson, A.J., Mackenzie, F.T. (2012). Revisiting four scientific debates in ocean acidification research. *Biogeosciences* 9: 893–905.

Andersson, A.J., Gledhill, D. (2013). Ocean acidification and coral reefs: effects on breakdown, dissolution, and net ecosystem calcification. *Annual Reviews of Marine Science* 5, 321–48.

ABA (2014). Arctic Biodiversity Assessment, full Scientific Report. The Conservation of Arctic Flora and Fauna (CAFF). 673 pp. http://www.arcticbiodiversity.is/the-report

ACIA (2004). Impacts of a Warming Arctic. Cambridge University Press. p 140 http://www.amap.no/arctic-climate-impact-assessment-acia

AMAP (2009). Oil and gas activities in the Arctic: effects and potential effects. Arctic Monitoring and Assessment Program, Oslo.

AMAP (2011). Mercury in the Arctic. Arctic Monitoring and Assessment Program, Oslo.

AMSA (2009). Arctic Marine Shipping Assessment 2009 Report. Arctic Council.

AMSA IIc ((2013). AMAP/CAFF/SDWG Identification of Arctic marine areas of heightened ecological and cultural significance: Arctic Marine Shipping Assessment. Oslo. 114 pp.

Archambault, P., Snelgrove, P.V.R., Fisher, J.A.D., Gagnon, J.-M., Garbary, D.J., et al. (2010). From Sea to Sea: Canada's Three Oceans of Biodiversity. *PLoS ONE* 5(8, e12182).

Barry, T., Price, C. (2012). The Arctic Species Trend Index 2011. Key findings from an in-depth look at marine species and development of spatial analysis techniques. CAFF Assessment Series No. 9.

Bluhm, B.A., Gebruk, A.V., Gradinger, R., Hopcroft, R.R., Huettmann, F., Kosobokova, K.N., Sirenko, B.I. and Weslawski, J.M. (2011). Arctic marine biodiversity: An update of species richness and examples of biodiversity change. *Oceanography* 24(3):232–248, http://dx.doi.org/10.5670/ oceanog.2011.75.

Boertmann, D., Mosbech, A., 2011. Eastern Baffin Bay- A strategic environmental impact assessment of hydrocarbon activities. Scientific Report from Danish Centre for Environment and Energy, No 9. Aarhus University.

CAFF (2013). *Arctic Biodiversity Assessment: Status and trends in Arctic biodiversity*. Conservation of Arctic Flora and Fauna, Akureyri. http://arcticlcc.org/assets/resources/ABA2013Science.pdf

Christiansen, J.S., Mecklenburg, C.W., Karamushko, O.V. (2014). Arctic marine fishes and their fisheries in light of global change. *Global Change Biology* 20, 352–359.

Darnis, G., Robert, D., Pomerleau, C. et al. (2012). Current state and trends in Canadian Arctic marine ecosystems: II. Heterotrophic food web, pelagic-benthic coupling, and biodiversity. *Climate Change* 115, 179-205.

Dyke, A.S., Hooper, J., Savelle, J.M. (1996). A history of sea ice in the Canadian Arctic Archipelago based on postglacial remains of the bowhead whale (*Balaena mysticetus*). *Arctic* 49, 235–255.

Fischbach, A.S. et al. (2009). Enumeration of Pacific walrus carcasses on the beaches of the Chukchi Sea in Alaska following a mortality event, September 2009 (USGS, Washington, DC).

Fuller S.D., Murillo Perez, F.J., Wareham V., Kenchington E. (2008). Vulnerable Marine Ecosystems dominated by deep-water corals and sponges in the NAFO Conventional Area. *NAFO Scientific Council Research Document* 08/22, N5524, 24p.

Grebmeier, J.M., Barry, J.P. (2007). Benthic processes in polar polynyas. Smith, W.O. and Barber, D.G. (Eds.) *Polynas: Windows to the World.* 262-290.

IUCN (2013). *The IUCN Red List of Threatened Species. Version 2013.2.* <http://www.iucnredlist.org>. Downloaded on 21 November 2013.

Ji, R.B., Jin, M. B. and Varpe. (2013). Sea ice phenology and timing of primary production pulses in the Arctic Ocean. *Global Change Biology* 19, 734-741.

Jørgensen, L.L., Planque, B., Thangstad, T.H., Certain, G. (2015). Vulnerability of megabenthic species to trawling in the Barents Sea. *ICES Journal of Marine Research*, doi: 10.1093/icesjms/fsv107 .

Jørgensen, O.A., Tendal, O.S., Arboe, N.H. (2013). Preliminary mapping of the distribution of corals observed off West Greenland as inferred from bottom trawl surveys 2010-2012. Scientific Council Meeting, June 2013. NAFO SCR Doc. 13/007, Serial No. N6156

Kelly, B.P., Whiteley, A., Tallmon, D. (2010). The Arctic melting pot. *Nature* 468, 891.

Kinnard, C., Zdanowicz, C.M., Fisher, D.A., Isaksson, E., Vernal, A. de, Thompson, L.G. (2011). Reconstructed changes in Arctic sea ice over the past 1,450 years. *Nature*, 479 (7374), 509-512.

NASA/Goddard Scientific Visualization Studio, (2014). "Arctic sea ice hit its annual minimum on Sept. 17, 2014". Digital image taken from: *NASA News Release 2014 Arctic Sea Ice Minimum Sixth Lowest on Record*. NASA, http://svs.gsfc.nasa.gov/cgi-bin/details.cgi?aid=4215 accessed 21/94/15.

OBIS (2015). Pan-Arctic map showing the number of species in a gridded view of hexagonal cells [Map] (Available: Ocean Biogeographic Information System. Intergovernmental Oceanographic Commission of UNESCO. http://www.iobis.org. Accessed: 2015-04-21).

Piepenburg, D., Archambault, P., Ambrose, W. G. Jr., Blanchard, A., Bluhm, B. A., Carroll, M. L., Conlan, K., Cusson, M., Feder, H. M., Grebmeier, J. M., Jewett, S. C., Lévesque, M., Petryashev, V. V., Sejr, M. K., Sirenko, B., Włodarska-Kowalczuk, M. (2011). Towards a pan-Arctic inventory of the species diversity of the macro- and megabenthic fauna of the Arctic shelf seas. *Marine Biodiversity* 41, 51-70.

Puig, P., Canals, M., Company, J.B., Martín, J., Amblas, D., Lastras, G., Palanques, A., Calafat, A.M. (2012). Ploughing the deep sea floor. *Nature* 489, 286-289.

Post, E., Bhatt, U.S., Bitz, C.M., Brodie, J.F., Fulton, T.L., Hebblewhite, M., Kerby, J., Kutz, S.J., Stirling, I., Walker, D.A. (2013). Ecological consequences of sea-ice decline. *Science* 341, 519-524.

Renaud, P.E., Webb, T.J., Bjørgesæter, A., Karakassis, I. and others. (2009). Continental-scale patterns in benthic invertebrate diversity: insights from the MacroBen database. *Marine Ecology Progress Series* 382, 239–252.

Lynghammar, A., Christiansen, J.S., Mecklenburg, C.W., Karamushko, O.V., Møller, P.R., Gallucci, V.F. (2013). Species richness and distribution of chondrichthyan fishes in the Arctic Ocean and adjacent seas. *Biodiversity*, 14, 57-66.

Stroeve, J.C., Serreze, M.C., Holland, M.M., Kay, J.E., Malanik, J., Barrett, A.P. (2012). The Arctic's rapidly shrinking sea ice cover: a research synthesis. *Climatic Change* 110, 1005–1027.

Søreide, J.E., Leu, E., Berge, J., Graeve, M., Falk-Petersen, S. (2010). Timing of blooms, algal food quality and *Calanus glacialis* reproduction and growth in a changing Arctic. 11, 3154 – 3163.

Tendal, O.S., Jørgensbye, M.I.Ø., Kenchington, E., Yashayev, I., Best, M. (2013) Greenlands first living deep-water coral reef. *ICES Insight* 50:6 pp

Klitgaard, A.B., Tendal, O.S. (2004) Distribution and species composition of mass occurrences of large-sized sponges in the northeast Atlantic. *Progress in oceanography* 61:57-98

Wassmann, P. (2011). Arctic marine ecosystems in an era of rapid climate change. *Progress in Oceanography* 90, 1-4.

Primary producers

Abelmann, A. (1992). Diatom assemblages in Arctic sea ice - indicator for ice drift pathways. *Deep-Sea Research* 39, 525-538.

Adl, S.M., Simpson, A.G.B., Lane, C.E., et al. (2012). The revised classification of Eukaryotes. *Journal of Eukaryotic Microbiology*, 59, 429-493.

Apollonio, S. (1965). Chlorophyll in arctic sea ice. *Arctic*, 18, 118-122.

Bluhm, B.A., Gebruk, A.V., Gradinger, R., Hopcroft, R.R., Huettmann, F., Kosobokova, K.N., Sirenko, B.I., Weslawski, J.M. (2011). Arctic marine biodiversity: An update of species richness and examples of biodiversity change. *Oceanography* 24, 232-240.

Campana, G.L., Zacher, K., Fricke, A., Molis, M., Wulff, A., Quartino, M.L., Wiencke, C. (2009). Drivers of colonization and succession in polar benthic macro- and microalgal communities. *Botanica Marina* 52, 655-667.

Carmack, E.C., Swift, J.H. (1990). Some aspects of the large-scale physical oceanography of the Arctic Ocean influencing biological distribution, in Medlin, K., Priddle, J. (Eds.): Polar marine diatoms. British Antarctic Survey, Cambridge, England, pp. 35-46.

Cleve, P.T. (1873). On diatoms from the Arctic Sea. Bihang till Kungleg Svenska Vetenskaps-Akademiens Handlingar 1, 1-28.

Cota G.F., Legendre L., Gosselin M., Ingram R.G. (1991). Ecology of bottom ice algae: I. Environmental controls and variability. *Journal of Marine Systems* 2, 257-277.

Daniëls, F.J.A., Gillespie, L.J., Poulin, M., Afonina, O.M., Alsos, I.G., Bültmann, H., Ickert-Bond, S., Konstantinova, N.A., Lovejoy, C., Väre, H., Westergaard, K.B. (2013). Chapter 9. Plants, in: Meltofte, H. (ed.) 2013. Arctic Biodiversity Assessment. Status and trends in Arctic biodiversity. Conservation of Arctic Flora and Fauna, Akureyri, pp. 311-353.

Degerlund, M., Eilertsen, H.C. (2009). Main species characteristics of phytoplankton spring blooms in NE Atlantic and Arctic Waters (68-80° N). *Estuaries and coasts* 33, 242-269.

De Sève, M.A., Dunbar, M.J. (1990). Structure and composition of ice algal assemblages from the Gulf of St. Lawrence, Magdalen Islands Area. *Canadian Journal of Fisheries and Aquatic Sciences* 47, 780-788.

Diaz-Pulido, G., McCook, L. (2008), 'Macroalgae (Seaweeds),' in Chin. A. (ed) The State of the Great Barrier Reef On-line, Great Barrier Reef Marine Park Authority, Townsville. http://www.gbrmpa.gov.au/corp_site/info_services/publications/sotr/downloads/SORR_Macroalgae.pdf . Viewed on (25.01.14)

Dunbar, M.J., Acreman, J. (1980). Standing crops and species composition of diatoms in sea ice from Robeson Channel to the Gulf of St. Lawrence. *Ophelia* 19, 61-72.

Ehrenberg, C.G. (1841). Einen Nachtrag zu dem Vortrage über Verbreitung und Einfluß des mikroskopischen Lebens in Süd- und Nord-Amerika. D. Akad. Wiss., Berlin, Monatsber, pp. 202-207.

Fricke, A., Molis, M., Wiencke, C., Valdivia, N., Chapman, A.S. (2008). Natural succession of macroalgal-dominated epibenthic assemblages at different water depths and after transplantation from deep to shallow water on Spitsbergen. *Polar Biology* 31, 1191-1203.

Gosselin, M., Levasseur M., Wheeler P.A., Horner R.A., Booth, B. (1997). New measurements of phytoplankton and ice algae production in the Arctic Ocean. *Deep-Sea Research II* 44, 1623-1644.

Gradinger, R. (1999). Vertical fine structure of the biomass and composition of algal communities in Arctic pack ice. *Marine Biology* 133, 745-754.

Grainger, E.H. (1977). The annual nutrient cycle in sea-ice, in: Dunbar, M.J. (ed.): Polar Oceans. Arctic Institute of North America, Calgary, pp. 285-299.

Gutt, J. (2001). On the direct impact of ice on marine benthic communities, a review. *Polar Biology*, 24, 553-564.

Harrison, W.G., Børsheim K.Y., Li, W.K.W., Maillet, G.L., Pepin, P., Sakshaug, E., Skogen, M., Yeats, P.A. (2013). Phytoplankton production and growth regulation in the Subarctic North Atlantic: A Comparative study of the Labrador Sea-Labrador/Newfoundland shelves and Barents/Norwegian/Greenland seas and shelves. *Progress in Oceanography* 114, 26-45.

Hasle, G.R., Syvertsen, E.E. (1996). Marine diatoms, in: Thomas, C.R. (ed.): Identifying marine diatoms and dinoflagellates. Academic Press, Inc., San Diego, California, pp. 5-385.

Hegseth, E.N., Sundfjord A. (2008). Intrusion and blooming of Atlantic phytoplankton species in the high Arctic. *Journal of Marine Systems* 74, 108-119.

Horner, R. 1985. Ecology of sea ice microalgae, in, Horner, R. (ed.): Sea ice biota, CRC Press, Florida, pp. 83-103.

Horner, R., Ackley, S.F., Dieckmann, G.S., Gulliksen, B., Hoshiai, T., Legendre, L., Melnikov, I. A., Reeburgh, W.S., Spindler, M., Sullivan, C.W. (1992). Ecology of sea ice biota. 1. Habitat, terminology, and methodology. *Polar Biology* 12, 417-427.

Horner, R., Schrader G.C. (1982). Relative contributions of ice algae, phytoplankton and benthic microalgae to primary production in nearshore regions of Beaufort Sea. *Arctic* 35, 485-503.

Horner, R., Syvertsen, E.E., Thomas, D.P., Lange, C. (1988). Proposed terminology and reporting units for sea ice algal assemblages. *Polar Biology* 8, 249-253.

Hsiao, S.I.C. (1980). Community structure and standing stock of sea ice microalgae in the Canadian Arctic. *Arctic* 33, 768–793.

Joo, H.M., Lee, S.H., Jung, S.W., Dahms, H.-U., Lee, J.H. (2012). Latitudal variation of phytoplankton communities in the western Arctic Ocean. Deep-Sea Research II 81-84, 3-17.

Jueterbock, A., Tyberghein, L., Verbruggen, H., Coyer, J.A., Olsen, J.L., Hoarau, G. (2013). Climate change impact on seaweed meadow distribution in the North Atlantic rocky intertidal. *Ecology and Evolution* 3, 1356-1373.

Katsuki, K., Takahashi, K., Onodera, J., Jordan, R.W., Suto, I. (2009). Living diatoms in the vicinity of the North Pole, summer 2004. *Micropalentology* 55, 137-170.

Kilias, E., Wolf, C., Nöthig, E.-M., Peeken, I., Metfies, K. (2013). Protist distribution in the western Fram Strait in summer 2010, based on 454-pyrosequencing of 18S rDNA. *Journal of Phycology* 49, 996-1010.

Kjellmann, F.R. (1883). The algae of the Arctic Sea: a survey of the species together with an exposition of the general characters and development of the flora. Kungleg Svenska Vetenskaps-Akademiens Handlingar 20(5), 1-351, 31 plates.

Kraberg, A.C., Druzhkova, E., Heim, B., Loeder, M.J.G., Wiltshire, K.H. (2013). Phytoplankton community structure in the Lena Delta (Siberia, Russia) in relation to hydrography. *Biogeosciences* 10, 7263-7277.

Krembs C., Gradinger R., Spindler M. (2002). Implication of brine channel geometry and surface area for the interaction of sympagic organisms in Arctic sea ice. *Journal of Experimental Marine Biology and Ecology* 243, 55-80.

Li, W.K.W., McLaughlin, F.A., Loveloy, C., Carmack, E.C. (2009). Smallest algae thrive as the Arctic Ocean freshens. *Science* 326, 539-539.

Lovejoy C., Legendre L., Martineau M.-J., Bâcle J., von Quillfeldt C.H. (2002). Distribution of phytoplankton and other protists in the North Water Polynya (Arctic). Deep-Sea Research II 49, 5027-5047.

Lovejoy, C., Potvin, M. (2011). Microbial eukaryotic distribution in a dynamic Beaufort Sea and the Arctic Ocean. *Journal of Plankton Research* 33, 431-444.

Lüning, K., (1990). *Seaweeds. Their environment, biography, and ecophysiology*. John Wiley & Sons, Inc., New York.

Meguro, H., Ito, K., Fukushima, H. (1967). Ice flora (bottom type): a mechanism of primary production in polar seas and the growth of diatoms in sea ice. *Arctic* 20, 114-133.

Melnikov, I.A. (1997). *The Arctic Sea ice ecosystem*. Gordon and Branch Science Publisher, Amsterdam.

Melnikov, I.A., Kolosova, E.G., Welch, H.E., Zhitina, L.S. (2002). Sea ice biological communities and nutrient dynamics in the Canada Basin of the Arctic Ocean. *Deep Sea Research I*, 49, 1623–1649.

Mueter,F.J., Reist,J.D., Majewski,A.R., Sawatzky,C.D., Christiansen,J.S., Hedges,K.J., et al.(2013). Marine fishes of the Arctic. , In *Arctic Report Card: Update for 2013:Tracking Recent Environmental Changes*. Available online at: http://www.arctic.noaa.gov/reportcard/marine_fish.html

Niemi, A., Michel, C., Hille, K., Poulin, M. (2011). Protist assemblages in winter sea ice: setting the stage for the spring ice algal bloom. *Polar Biology* 34, 1803-1817.

Pedersen, P.M. (2011). Grønlands havalger. Forlaget Epsilon.dk., Denmark.

Poulin, M., Daugbjerg, N., Gradinger, R., Ilyash, L., Ratkova, T., von Quillfeldt, C.H. (2010). The pan-Arctic biodiversity of marine pelagic and sea-ice unicellular eukaryotes: A first-attempt assessment Marine Biodiversity. *Marine Biodiversity* 41, 13-28.

Rat'kova, T.N., Wassmann, P. (2002). Seasonal variation and spatial distribution of phyto- and protozooplankton in the central Barents Sea. *Journal of Marine Systems* 38, 47-75.

Rey, F. (1986). Planteplankton-artssammensetning i Barentshavet i januar 1985, in Hassel, A., Loeng, H. and Skjoldal, H.R. (Eds.): Marinøkologiske undersøkelser i Barentshavet i januar 1985. Report number FO 8604, Havforskningsinstituttet i Bergen, Appendix B, 3 pp.

Robineau B., Legendre L., Kishino M., Kudoh S. (1997). Horizontal heterogeneity of microalgae biomass in the first-year ice of Saroma-Ko Lagoon (Hokaido, Japan). *Journal of Marine Systems* 11, 81-91.

Rosenvinge, L.K. (1898). Deuxième mémoire sur les algues marines du Groenland. Meddelser om Grønland 20, 1-125.

Sagen H., Dalpadado P. (2004). *Emiliania huxleyi*-oppblomstringen i Barentshavet sommeren 2003 observert ved hjelp av satellitt. Fisken og havet, særnummer 2, 96-97.

Sakshaug E. (2004). Primary and secondary production in the Arctic Sea, in Stein, R., Macdonald, R.W. (Eds): *The organic carbon cycle in the Arctic Ocean*. Springer, Berlin, pp. 57-81.

Schandelmeier, L., Alexander, V. (1981). An analysis of the influence of ice on spring phytoplankton population structure in the southeast Bering Sea. *Limnology and Oceanography* 26, 935-943.

Sørensen, N., Daugbjerg, N., Gabrielsen T.M. (2012). Molecular diversity and temporal variation of picoeukaryotes in two Arctic fjords, Svalbard. *Polar Biology* 35, 519-533.

Sukhanova, I.N., Flint, M.V., Pautova, L.A., Stockwell, D.A., Grebmeier, J.M., Sergeeva, V.M. (2009). Phytoplankton of the western Arctic in the spring and summer of 2002: Structure and seasonal changes. *Deep-Sea Research II* 56, 1223-1236.

Syvertsen, E.E. (1991). Ice algae in the Barents Sea: types of assemblages, origin fate and role in the ice-edge phytoplankton bloom. *Polar Research* 10, 277–288.

Terrado, R., Scarcella, K., Thaler, M., Vincent, W.F., Lovejoy, C. (2012). Small phytoplankton in Arctic seas: vulnerability to climate change. *Biodiversity*, doi:10.1080/14888386.2012.704839.

Tonkes, H. (2012). Phytoplankton composition of central Arctic Ocean in summer 2011: with special emphasis on pico- and nanoplankton. Major thesis, Wageningen University.

Tremblay, J.-E., Robert, D., Varela, D., Lovejoy, C., Darnis, G., Nelson, R.J., Sastri, A. (2012). Current state and trends in Canadian Arctic marine ecosystems: I Primary production. *Climate Change*, doi: 10.1007/s10584-012-0496-3.

von Quillfeldt, C.H. (1996). Ice algae and phytoplankton in north Norwegian and arctic waters: species composition, succession and distribution. Ph D Thesis, University of Tromsø.

von Quillfeldt, C.H. (1997). Distribution of diatoms in the Northeast Water Polynya, Greenland. *Journal of Marine Systems* 10, 211-240.

von Quillfeldt, C.H. (2000). Common diatom species in arctic spring blooms: their distribution and abundance. *Botanica Marina* 43, 499-516.

von Quillfeldt, C.H., Ambrose, W.G., Clough, L.M. (2003). High number of diatom species in first year ice from the Chukchi Sea. *Polar Biology* 26, 806-818.

Vørs, N., 1993. Heterotrophic amoebae, flagellates and heliozoan from Arctic marine waters (North West Territories, Canada and West Greenland). *Polar Biology* 13, 113–126.

Wassmann P., Reigstad M., Haug T. Rudels B., Carroll M.L., Hop H., Gabrielsen G.W., Falk-Petersen S., Denisenko S.G., Arashkevich E., Slagstad D., Pavlova O. (2006). Food webs and carbon flux in the Barents Sea. *Progress in Oceanography*. 71, 232-287.

Wulff, A., Iken, K., Quartino, L.M., Al-Handal, A., Wiencke, C., Clayton, M.N. (2011). Biodiversity, biography and zonation of marine benthic micro- and macroalgae in the Arctic and the Antarctic, in: Wiencke, C. (ed.) Biology of polar benthic algae. De Gruyter, Berlin, pp.23-52.

Zacher, K., Rautenberger, R., Hanelt, D., Wulff, A., Wiencke, C. (2011). The abiotic environment of polar marine benthic algae, in: Wiencke, C. (ed.) *Biology of polar benthic algae*. De Gruyter, Berlin, pp. 9-21.

Zheng, S., Wang, G., Zhang, F., Cai, M., He, J. (2011). Dominant diatom species in the Canada Basin in summer 2003, a reported serious melting season. *Polar Record* 47, 244-261.

Zooplankton

Deibel, D., Daly, K.L. (2007). Zooplankton processes in Arctic and Antarctic polynas. *In*: Smith, W.O. and D.G. Barber (Eds.) *Polynas: Windows to the World*. Elsevier. Pp. 271-332.

Gradinger, R., Blum, B.A., Hopcroft, R.R., Gebruk, A.V., Kosobokova, K., Sirenko, B., Weslawski, J.M. (2010). Marine life in the Arctic. *In*: McIntyre, A.D. (ed.) *Life in the World's Oceans*. Blackwell Publishing, Ltd. Pp. 183-202.

Hirche, H.-J., Kosobokova, K.N. (2011). Winter studies of zooplankton in Arctic Seas: the Stofjord (Svalbard) and adjacent ice-covered Barents Sea. *Marine Biology* 158, 2359-2376.

Hop, H., Mundy, C.H., Gosselin, M., Rossnagel, A.L., Barber, D.G. (2011). Zooplankton boom and ice amphipod bust below melting sea ice in the Amundsen Gulf, Arctic Canada. *Polar Biology* 34, 1947-1958.

Hopcroft, R.R., Kosobokova, K.N., Pinchuk, A.I. (2010). Zooplankton community patterns in the Chukchi Sea during summer 2004. Deep-Sea Research II 57, 27-39.

Kosobokova, K.N., Hopcroft, R.R., Hirche, H.-J. (2011). Patterns of zooplankton diversity through the depths of the Arctic's central basins. *Marine Biodiversity* 41, 29-50.

Kraft, A., Nöthig, E.M., Bauerfeind, E., Wildish, D.J., Pohle, G.W., Bathmann, U.V., Beszczynska-Möller, A,. Klages, M. (2013). First evidence of reproductive success in a southern invader indicates possible community shifts among Arctic zooplankton. *Marine Ecology Progress Series* 493, 291-296.

Longhurst, A.R. (2007). Ecological Geography of the Sea. Academic Press, London.

Longhurst, A.R., Harrison, W.G. (1989). The biological pump: profiles of plankton production and consumption in the upper ocean. *Progress in Oceanography* 22,

47-123.

Rabindranath, A., Danse, M., Falk-Petersen, S., Wold, A., Wallace, M.I., Berge, J., Brierly, A.S. (2011). Seasonal and diel vertical migration of zooplankton in the High Arctic during the autumnal midnight sun of 2008. *Marine Biodiversity* 41, 365-382.

Raskoff, K., Hopcroft, R.R., Kosobokova, K.N., Purcell, J.E.,Youngbluth, M. (2010). Jellies under ice: ROV observations from the Arctic 2005 hidden ocean expedition. *Deep-Sea Research II* 57, 111–126.

Stammerjohn, S.E., Massom, R., Rind, D., Martinson, D.G. (2012). Regions of rapid sea ice change: an inter-hemispheric seasonal comparison. *Geophysical Research Letters* 39:L06501, doi:10.1029/2012GL050874.

Walkusz, W., Paulic, J.E., Kwasniewski, S., Williams, W.J., Wong, S., Pabst M.H. (2010). Distribution, diversity and biomass of summer zooplankton from the coastal Canadian Beaufort Sea. *Polar Biology* 33, 321-335.

Benthos

Anderson, J.T. and Gregory, R.S. (2000). Factors regulating survival of northern cod (NAFO 2J3KL) during their first three years of life. *ICES Journal of Marine Science* 57:349-359.

Auster P. (1998). A conceptual model of the impacts of fishing gear on the integrity of fish habitat. *Conservation Biology* 12:1198-1203.

Bluhm, B.A., Ambrose, W.G., Bergmann, M., Clough, L.M., Gebruk, A.V., Hasemann, C., Iken, K., Klages, M., MacDonald, I.R., Renaud, P.E., Schewe, I., Soltwedel, T., Wlodarska-Kowalczuk, M. (2011a). Diversity of the arctic deep-sea benthos. *Marine Biodiversity* 41:87–107. doi: 10.1007/s12526-010-0078-4

Bluhm, B.A., Gradinger, R., Hopcroft, R.R. (2011b). Editorial - Arctic Ocean Diversity: synthesis. *Marine Biodiversity* 41:1–4. doi: 10.1007/s12526-010-0080-x

Boetius, A., Albrecht, S., Bakker, K., Bienhold, C., Felden, J., Fernández-Méndez, M., Hendricks, S., Katlein, C., Lalande, C., Krumpen, T., Nicolaus, M., Peeken, I., Rabe, B., Rogacheva, A., Rybakova, E., Somavilla, R., Wenzhöfer, F., Polarstern, R.V. (2013) Export of Algal Biomass from the Melting Arctic Sea Ice. *Science* 22 Vol. 339 no. 6126 pp. 1430-1432. DOI: 10.1126/science.1231346.

Collie, J., Escanero, G. and Valentine, P.C. (1997). Effects of bottom fishing on the benthic megafauna of Georges Bank. *Marine Ecology Progress* 155: 159-172.

Collie, J.S., Hall, S.J., Kaiser, M.J., Poiner, I.R. (2000). A quantitative analysis of fishing impacts on shelf-sea benthos. *Journal of Animal Ecology* 69: 785-798.

FAO (2009). International Guidelines for the Management of Deep-sea Fisheries in the High Seas. Rome, 73p.

Hall-Spencer, J.M. and Moore, P.G. (2000). Scallop dredging has profound long-term impacts on maerl habitat. ICES *Journal of Marine Science* 57: 1407-1415.

Jones, J.B. (1992). Environmental impact of trawling on the seabed: a review. *New Zealand Journal of Marine and Freshwater Research* 26: 59-67.

Kaiser, M.J., Rogers, S.I., and Ellis, J. (1999). Importance of habitat complexity for demersal fish assemblages. *American Fisheries Society Symposium* 22:212-223.

Kortsch, S., Primicerio, R., Beuchel, F., Renaud, P.E., Rodrigues, J., Jørgen Lønne, O., Gulliksen, B. (2012). Climate-driven regime shifts in Arctic marine benthos. *Proceedings of the National Academy of Sciences* 109:14052–14057, doi: 10.1073/pnas.1207509109.

Lindholm, J.B., Auster, P.J., Ruth, M., and Kaufman, L. (2001). Modeling the effects of fishing and implications for the design of marine protected areas: juvenile fish responses to variations in seafloor habitat. *Conservation Biology* 15:424-437.

Linehan, J.E. (2001). Predation risk of age-0 cod (*Gadus*) relative to depth and substrate in coastal waters. *Journal of Experimental Marine Biology and Ecology* 261:25-44.

Lindholm, J.B., Auster, P.J., and Kaufman, L.S. (1999). Habitat-mediated survivorship of juvenile (0-year) Atlantic cod *Gadus morhua*. *Marine Ecology Progress Series* 180:247-255.

Link, J.S., Yemane, D., Shannon, L.J., Coll, M., Shin, Y.J., Hill, L., Borges, M.F., (2010). Relating marine ecosystem indicators to fishing and environmental drivers: an elucidation of contrasting responses. *ICES Journal of Marine Science* 67: 787-795.

Moritz, C., Gravel, D., Savard, L., McKindsey, C.W., Brêthes, J.-C., Archambault, P. (2015). No more detectable fishing effect on Northern Gulf of St. Lawrence benthic invertebrates. *ICES Journal of Marine Science*, doi: 10.1093/icesjms/fsv124.

Pauly, D., Christensen, V., Dalsgaard, J., Froese, R., Torres, F., (1998). Fishing down marine food webs. *Science* 279: 860-863.

Piepenburg, D., Archambault, P., Ambrose, W.G., Blanchard, A.L., Bluhm, B.A., Carroll, M.L., Conlan, K.E., Cusson, M., Feder, H.M., Grebmeier, J.M., Jewett, S.C., Lévesque, M., Petryashev, V.V., Sejr, M.K., Sirenko, B.I., Wlodarska-Kowalczuk, M. (2011). Towards a pan-Arctic inventory of the species diversity of the macro- and megabenthic fauna of the Arctic shelf seas. *Marine Biodiversity* 41:51–70. doi: 10.1007/s12526-010-0059-7.

Snelgrove, P.V.R. (2010). *Discoveries of the Census of Marine Life: Making Ocean Life Count*. Cambridge University Press.

Stoner, A.W. and Titgen, R.H. (2003). Biological structures and bottom type influence habitat choice made by Alaska flatfishes. *Journal of Experimental Marine Biology and Ecology* 292 : 43-59.

Thrush, S.F., Hewitt, J., Cummings, V.J., and Dayton, P.K. (1995). The impact of habitat disturbance by scallop dredging on marine benthic communities: What can be predicted from the results of experiments? *Marine Ecology Progress Series*. 129: 141-150.

Thrush, S.F., Hewitt, J.E., Cumming, V.J., Dayton, P.K., Cryer, M., Turner, S.J., Funnell, G.A., Budd, R., Milburn, C.J. and Wilkinson, M.R. (1998). Disturbance of the marine benthic habitat by commercial fishing: impacts at the scale of the fishery. *Ecological Applications* 8 866-879.

Thrush, S.F., Schultz, D., Hewitt, J.E. and Talley, D. (2002). Habitat structure in soft-sediment environments and abundance of juvenile snapper *Pagrus auratus*. *Marine Ecology Progress Series*. 245 273-280.

Tillin, H.M., Hiddink, J.G., Jennings, S., Kaiser, M.J. (2006). Chronic bottom trawling alters the functional composition of benthic invertebrate communities on a sea-basin scale. *Marine Ecology Progress Series* 318: 31-45.

Thurstan, R.H., Brockington, S., Roberts, C.M. (2010). The effects of 118 years of industrial fishing on UK bottom trawl fisheries. *Nature Communications* 1: 10.1038/ncomms1013.

Tupper, M. and Boutillier, R.G. (1995a). Size and priority at settlement determine growth and competitive success of newly settled Atlantic cod. *Marine Ecology Progress Series* 118 295-300.

Tupper, M. and Boutillier, R.G. (1995b). Effects of habitat on settlement, growth and postsettlement survival of Atlantic cod (*Gadus mothua*). *Canadian Journal of Fisheries and Aquatic Sciences* 52 1834-1841.

Wassmann, P., Duarte, C., Agustí, S. (2011) Footprints of climate change in the Arctic Marine Ecosystem. *Global Change Biology* 17:1235–1249. doi: 10.1111/j.1365-2486.2010.02311.x.

Watling, L. and Norse, E.A. (1998). Disturbance of the sea bed by mobile fishing gear: a comparison to forest clearcutting. *Conservation Biology* 12: 1180-1197.

Wilson, W.J., Ormseth, O.A. (2009). A new management plan for the Arctic waters of the United States. *Fisheries* 34:555–558.

Zhou S., Smith, A.D.M., Punt, A.E., Richardson, A.J., Gibbs, M., Fulton, E.A., Pascoe, S., Bulman, C., Bayliss, P., Sainsbury, K. (2010). Ecosystem-based fisheries management requires a change to the selective fishing philosophy. *Proceedings of the National Academy of Sciences of the United States of America* 107: 9485-9489.

Nekton (including demersal and holopelagic vertebrates and invertebrates).

Anderson, O.F., Clark, M.R. (2003) Analysis of bycatch in the fishery for orange roughy, *Hoplostethus atlanticus*, on the South Tasman Rise. *Marine & Freshwater Research* 54: 643–652.

Andriashev, A.P., Mukhamediarov, B.F., Pavshtiks, E.A. (1980) On dense concentrations of cryopelagic fishes *Boreogadus saida* and *Arctogadus glacialis* in the near-pole areas of Arctic. In: Vinogradov ME, Melnikov IA (Eds.) *Biology of the Central Arctic Basin*. Moscow, Nauka Publishing, p 196–211. (in Russian).

Atlas of Russian Freshwater fishes. (2002). Reshetnikov, Yu. S. (ed.) Moscow, Nauka publishing. V.1. 379 pp, V.2 253 pp. (in Russian).

Auel, H. and Hagen, W. (2002). Mesozooplankton community structure, abundance and biomass in the central Arctic Ocean. *Marine Biology*. 140:1013-1021.

Baker, K.D., Devine, J.A., Haedrich, R.L. (2009). Deep-sea fishes in Canada's Atlantic: population declines and predicted recovery times. *Environmental Biology of Fishes* 85: 79-88.

Byrkjedal, I., Rees, D.J., Christiansen, J.S., Fevolden, S.-E. (2008). The taxonomic status of *Theragra finnmarchica* Koefoed, 1956 (Teleostei: Gadidae): perspectives from morphological and molecular data. *Journal of Fish Biology* 73:1183–1200.

Chernova, N.V. (2011). Distribution patterns and chorological analysis of fish fauna of the Arctic region. *Journal of Ichthyology* 51(10), p 825-924.

Christiansen, J.S., Reise, J.D. and 33 others (2013). Fishes. In: *Arctic Biodiversity Assessment*, Conservation of Arctic Flora and Fauna (CAFF).

Christiansen, J.S., Fevolden, E., Byrkjedal, I. (2005). The occurrence of *Theragra finnmarchica* Koefoed, 1956 (Teleostei, Gadidae), 1932–2004. *Journal of Fish Biology* 66:1193–1197.

Christiansen, J.S., Mecklenburg, C.W., Karamushko, O.V. (2014) Arctic marine fishes and their fisheries in light of global change. *Global Change Biology* 20, 352–359.

Coad, B.W. and Reist, J.D. (2004). *Annotated list of Arctic marine fishes of Canada*. Canadian Manuscript Report of Fisheries and Aquatic Sciences 2674, iv+112 p.

Dolgov, A.V., Karsakov, A.L. (2011). Species-specific habitat conditions and possible changes in the distribution of fishes in the Barents Sea under climate change. In: T. Haug, A. Dolgov, K. Drevetnyak, I. Røttingen, K. Sunnanå and O. Titov (Eds.) *Climate change and effects on the Barents Sea marine living resources*. 15th Russian-Norwegian Symposium Longyearbyen, 7-8 September 2011.

Gardiner, K. and Dick, T.A. (2010). Arctic cephalopod distributions and their associated predators. *Polar Research* 29:209-227.

Golikov, A.V., Sabirov, R.M., Lubin, P.A. and Jorgensen, L.L. (2012). Changes in distribution and range structure of Arctic cephalopods due to climatic changes of the last decades. *Biodiversity* 14: 28-35.

Haedrich, R.L. and Gagnon, J.-M. (1991). Rock wall fauna in a deep Newfoundland fjord. *Continental Shelf Research* 11: 1199-1208.

Hamilton, L.C., Brown, B. C. and Rasmussen, R.O. (2003). West Greenland's cod-to-shrimp transition: local dimensions of climatic change. *Arctic*. 56:271-282.

Hardie, D.C., and Hutchings, J.A. (2011). The ecology of Atlantic cod (*Gadus morhua*) in Canadian Arctic lakes. *Arctic* 64: 137-150.

Hollowed, A.B., Planque, B. and Loeng, H. (2013). Potential movement of fish and shellfish stocks from the sub-Arctic to the Arctic Ocean. *Fisheries Oceanography*, 22(5), 355-370.

Karamushko, O.V. (2012): Structure of ichthyofauna in the Arctic seas off Russia. Berichte zur Polar- und Meeresforschung. Reports on Polar and Marine Research. *Arctic Marine Biology*, 129-136.

Kearley, W. (2012). *Here's the catch: the fish we harvest from the northwest Atlantic*. Boulder Publications, Portugal Cove-St Phillip's, Newfoundland & Labrador. 263 pp.

Koslow, J.A., Boehlert, G.W., Gordon, J.D.M., Haedrich, R.L., Lorance, P., and Parin, N. (2000). Continental slope and deep-sea fisheries: implications for a fragile ecosystem. *ICES Journal of Marine Science*, 57: 548–557.

Lynghammar, A., J.S. Christiansen, C.W. Mecklenburg, O.V. Karamushko, P.R. Møller, and V.F. Gallucci (2013): Species richness and distribution of chrondrichthyan fishes in the Arctic Ocean and adjacent seas. *Biodiversity*, 14, 57-66.

Mecklenburg, C.W., Møller, P.R. and Steinke, D. (2011): Biodiversity of arctic marine fishes: taxonomy and zoogeography. *Marine Biodiversity*, 41, 109-140.

Melnikov, I.A., Chernova, N.V. (2013) Characteristics of under-ice concentrations of polar cod *Boreogadus saida* (Gadidae) in the Central Arctic basin. *Journal of Ichthyology* 53(1), p 22-30.

Mueter, F.J., Reist, J.D., Majewski, A.R., Swatzky, C.D., Christiansen, J.S., Hedges, K.J., Coad, B.W., Karamushko, O.V., Lauth, R.R., Lynghammar, A., MacPhee, S.A., Mecklenburg, C.W. (2013). *Marine fishes of the Arctic*. Arctic Report Card: update for 2013, tracking recent environmental changes. US NOAA, http://www.arctic.noaa.gov/reportcard/marine_fish.html. Dec 6, 2013.

Nesis, K.N. (2001). West-Arctic and East-Arctic distributional ranges of cephalopods. *Sarsia* 86:1-11.

Privalikhin, A.M., Norvillo, G.V. (2010) On the finding of a rare species—Norwegian pollock *Theragra finnmarchica* Koefoed, 1956 (Gadidae)—in the Barents Sea. *Journal of Ichthyology* 50:143–147.

Rice, J. (2006). Impacts of mobile bottom gears on seafloor habitats, species, and communities: a review and synthesis of selected international reviews. DFO CSAS Research Document 2006/057, Ottawa, Canada, iii+35 p.

Roberts, C.M. (2002). Deep impact: the rising toll of fishing in the deep sea. *Trends in Ecology and Evolution* 5: 242-245.

Standal, D. (2003). Fishing the last frontier—controversies in the regulations of shrimp trawling in the high Arctic. *Marine Policy* 27:375-388.

Tsinovsky, V.D., Melnikov, I.A. (1980) On occurrence of *Liparis koefoedi* (Liparidae, Osteichtyes) in the waters of the Central Arctic Basin. In: Vinogradov, M.E., Melnikov, I.A. (Eds.) *Biology of the Central Arctic Basin*. Moscow, Nauka Publishing, p 211-214 (In Russian).

Ursvik, A., Breines, R., Christiansen, J.S., Fevolden, S.-E., Coucheron, D.H., Johansen, S.D. (2007) A mitogenomic approach to the taxonomy of pollocks: *Theragra chalcogramma* and *T. finnmarchica* represent one single species. *BMC Evolutionary Biology* 7:86.

Marine birds

Anker-Nilssen, T., Bakken, V., Strøm, H., Golovkin, A.N., Bianki, V.V., Tatarinkova, I.P. (2000). *The status of marine birds breeding in the Barents Sea region*. Norsk Polarinstitutt, Norway.

Bakken, V., Pokrovskaya, I.V. (2000). Thick-billed Murre, in: Anker-Nilssen, T., Bakken, V., Strøm, H., Golovkin, A.N., Bianki, V.V., Tatarinkova, I.P. (Eds.), *The status of marine birds breeding in the Barents Sea region*. Norsk Polarinstitutt, Tromsø, Norway, pp. 119-124.

Barrett, R.T., Lorentsen, S.H., Anker-Nilsson, T. (2006). The status of breeding seabirds in mainland Norway. *Atlantic Seabirds* 8, 97-126.

Bédard, J., Nadeau, A., Giroux, J.-F., Savard, J.-P. (2008). *Eiderdown: Characteristics and harvesting procedures*. Société Duvetnor Ltée and Canadian Wildlife Service, Environment Canada, Québec.

Boertmann, D. (2008). The Lesser Black-backed Gull, *Larus fuscus*, in Greenland. *Arctic* 61, 129-133.

Braune, B.M., Donaldson, G.M., Hobson, K.A. (2001). Contaminant residues in seabird eggs from the Canadian Arctic. Part I. Temporal trends 1975-1998. *Environmental Pollution* 114, 39-54.

Braune, B.M., Mallory, M.L., Gilchrist, H.G. (2006). Elevated mercury levels in a declining population of ivory gulls in the Canadian Arctic. *Marine Pollution Bulletin* 52, 978-982.

Bustnes, J.O., Tertitski, G.M. (2000). Common eider *Somateria mollissima*, in: Anker-Nilssen, T., Bakken, V., Strøm, H., Golovkin, A.N., Bianki, V.V., Tatarinkova, I.P. (Eds.), *The status of marine birds breeding in the Barents Sea region*. Norsk Polarinstitutt, Tromsø, Norway, pp. 46-50.

Bustnes, J.O., Erikstad, K.E., Skaare, J.U., Bakken, V., Mehlum, F. (2003). Ecological effects of organochlorine pollutants in the Arctic: a study of the glaucous gull. *Ecological Applications* 13, 504–15.

Bustnes, J.O., Hanssen, S.A., Folstad, I., Erikstad, K.E., Hasselquist, D., Skaare, J.U. (2004). Immune function and organochlorine pollutants in arctic breeding glaucous gulls. *Archives of Environmental Contamination and Toxicology* 47, 530-541.

Buttler, E.I. (2009). *Avian cholera among arctic breeding common eiders: temporal dynamics and the role of handling stress in reproduction and survival*. M.Sc. Thesis, Department of Biology, Carleton University, Carleton.

Cairns, D.K., Gaston, A.J., Heutemann, F. (2008). Endothermy, ectothermy and the global structure of marine vertebrate communities. *Marine Ecology Progress Series* 356, 239-250.

Chaulk, K.G., Robertson G.J., Montevecchi, W.A. (2004). Breeding range update for three seabird species in Labrador. *Northeastern Naturalist* 11, 479-485.

Chaulk, K.G., Robertson, G.J., Collins, B.T., Montevecchi, W.A., Turner, B.C. (2005). Evidence of recent population increases in Common Eiders breeding in Labrador. *Journal of Wildlife Management* 69, 805-809.

Clark, R.B. (1984). Impact of oil pollution on seabirds. *Environmental Pollution, Series A. Ecological and Biological* 33, 1-22.

Durant, J.M., Anker-Nilssen, T., Hjermann, D.O., Stenseth, N.C. (2004). Regime shifts in the breeding of an Atlantic puffin population. *Ecology Letters* 7, 388-394.

Durant, J.M., Anker-Nilssen, T., Stenseth, N.C., 2006. Ocean climate prior to breeding affects the duration of the nestling period in the Atlantic puffin. *Biology Letters* 2: 628-631.

Environment Canada (2010). Recovery Strategy for the Ivory Gull (*Pagophila eburnea*) in Canada [draft]. *Species at Risk Act Recovery Strategy Series*. Environment Canada, Ottawa.

Evans, P.G.H., Kampp, K. (1991). Recent changes in thick-billed murre populations in West Greenland. *Canadian Wildlife Service Occasional Papers* 69, 7-14.

Gabrielsen, G.W., Skaare, J.U., Polder, A., Bakken, V. (1995). Chlorinated hydrocarbons in glaucous gulls (*Larus hyperboreus*) in the southern part of Svalbard. *Science of the Total Environment* 160/161, 337-346.

Ganter, B., Gaston, A.J. (2013). Birds, in: Meltofte, H. (Ed.), *Arctic Biodiversity Assessment. Status and trends in Arctic biodiversity conservation*. Conservation of Arctic Flora and fauna, Akureyri, pp. 142-180.

Garðarsson, A. (2006). *Nýlegar breytingar á fjölda íslenskra bjargfugla* [Recent changes in cliff-breeding seabirds in Iceland]. Bliki 27, 13-22.

Garðarsson, A., Guðmundsson, G.A., Lilliendahl, K., Vigfúsdóttir, F. (2009). *Status of cliff-breeding seabirds in Iceland in 2005-08*. Poster for Seabird group Conference, Brugges, March 2009.

Gaston, A.J. (2004). *Seabirds: a Natural History*. Yale University Press, New Haven.

Gaston, A.J., Elliott, K.H. (2013). Effects of climate-induced changes in parasitism, predation and predator-predator interactions on reproduction and survival of an Arctic marine bird. *Arctic* 66, 43-51.

Gaston, A.J., Gilchrist, H.G., Mallory, M.L. and Smith, P.A. (2009). Changes in seasonal events, peak food availability and consequent breeding adjustment in a marine bird: a case of progressive mismatching. *The Condor* 111: 111-119.

Gaston, A.J., Hipfner, J.M., Campbell, D. (2002). Heat and mosquitoes cause breeding failures and adult mortality in an Arctic-nesting seabird. *Ibis* 144, 185-191.

Gilchrist, H.G., Strøm, H., Gavrilo, M.V., Mosbech, A. (2008). International ivory gull conservation strategy and action plan. *CAFF Technical Report* no. 18.

Gilg, O., Boertmann, D., Merkel, F., Aebischer, A., Sabard, B. (2009b). Status of the endangered ivory gull, *Pagophila eburnea*, in Greenland. *Polar Biology* 32, 1275-1286.

Gilliland, S., Gilchrist, H.G., Rockwell, R.F., Robertson, G.J., Savard, J.-P., Merkel, F.R., Mosbech, A. (2009). Evaluating the sustainability of harvest among Northern Common Eiders in Greenland and Canada. *Wildlife Biology* 15, 24-36.

Helgason, L.B., Barrett, R., Lie, E., Polder, A., Skaare, J.U., Gabrielsen, G.W. (2008). Levels and temporal trends (1983-2003) of persistent organic pollutants (POPs) and mercury (Hg) in seabird eggs from Northern Norway. *Environmental Pollution* 155, 190-198.

Hjort, C., Gudmundsson, G.A., Elander, M. (1997). Ross's Gulls in the Central Arctic Ocean. *Arctic* 50, 289-292.

Irons, D.B., Anker-Nilssen, T., Gaston, A.J., et al. (2008). Magnitude of climate shift determines direction of circumpolar seabird population trends. *Global Change Biology* 14: 1455-1463.

IUCN (2012). IUCN Red List of Threatened Species. www.iucnredlist.org/apps/redlist [accessed 12 February 2014].

Iverson, S.A., Gilchrist, H.G., Smith, P.A., Gaston, A.J., Forbes M.R. (2014). Longer ice-free seasons increase the risk of nest depredation by polar bears for colonial breeding birds in the Canadian Arctic. *Proceedings of the Royal Society B* 281, http://dx.doi.org/10.1098/rspb.2013.3128.

Labansen, A.L., Merkel, F., Boertmann, D., Nyeland, J. (2010). Status of the black-legged kittiwake (Rissa tridactyla) breeding population in Greenland, 2008. *Polar Research* 29, 391-403.

Letcher, R.J., Bustnes, J.O., Dietz, R., Jenssen, B.M., Jørgensen, E.H., Sonne, C. et al. (2010). Exposure and effects assessment of persistent organohalogen contaminants in arctic wildlife and fish. *Science of the Total Environment* 15, 2995-3043.

Kampp, K., Nettleship, D.N., Evans, P.G.H. (1994): Thick-billed Murres of Greenland: status and prospects. In: D.N. Nettleship, J. Burger & M. Gochfeld (eds.). Seabird on islands: threats, case studies and action plans, pp. 133-154. *BirdLife Conservation Series No. 1*.

Krasnov, Y.V., Lorentsen, S.-H. (2000). The great skua Catharacta skua. In: T. Anker-Nilssen, V. Bakken, H. Strøm, A.N. Golovkin, V.V. Bianki & I.P. Tatarinkova (eds.). *The status of marine birds breeding in the Barents Sea region*, pp. 79-81. Norsk Polarinstitutt, Tromsø.

Meltofte, H., Barry, T., Berteau, D., et al. (2013). Synthesis: implications for conservation, in: Meltofte, H. (Ed.), *Arctic Biodiversity Assessment. Status and trends in Arctic biodiversity conservation*. Conservation of Arctic Flora and fauna, Akureyri, pp. 21-65.

Merkel, F.R. (2004a). Evidence of population decline in Common Eiders breeding in western Greenland. *Arctic* 57, 27-36.

Merkel, F.R. (2004b). Impact of hunting and gillnet fishery on wintering eiders in Nuuk, Southwest Greenland. *Waterbirds* 27, 469-479.

Merkel, F.R. (2010). Evidence of recent population recovery in common eiders breeding in western Greenland. *Journal of Wildlife Management* 74, 1869-1874.

Merkel, F.R. (2011). Gillnet bycatch of seabirds in Southwest Greenland, 2003-2008. *Technical Report* No. 85, Pinngortitaleriffik, Greenland Institute of Natural Resources.

Merkel, F.R., Barry, T. (Eds.) (2008). Seabird harvest in the Arctic. Circumpolar Seabird Group (CBird), CAFF *Technical Report* No. 16.

Miljeteig, C., Strøm, H., Gavrilo, M.V., Volkov, A., Jenssen, B.M., Gabrielsen, G.W. (2009). High Levels of Contaminants in Ivory Gull Pagophila eburnea eggs from the Russian and Norwegian Arctic. *Environmental Science and Technology* 43, 5521-5528.

Moline, M.A., Karnovsky, N.J., Brown, Z., Divoky, G.J., Frazer, T.K., Jacoby, C.A. et al. (2008). High latitude changes in ice dynamics and their impact on polar marine ecosystems. *Annals of the New York Academy of Science*. 1134, 267-313.

Newton, I. (2007). *The migration ecology of birds*. Academic Press, London.

Robertson, G.J. & Gilchrist, H.G. (1998). Evidence of population declines among Common Eiders breeding in the Belcher Islands, Northwest Territories. *Arctic* 51, 378-385.

Sagerup, K., Helgason, L.B., Polder, A., Strøm, H., Josefsen, T.D., Skåre, J.U., Gabrielsen, G.W. (2009). Persistent organic pollutants and mercury in dead and dying glaucous gulls (Larus hyperboreus) at Bjørnøya (Svalbard). *Science of the Total Environment* 407, 6009-6016.

Sandvik, H., Erikstad, K.E., Barrett, R.T., Yoccoz, N.G. (2005). The effect of climate on adult survival in five species of North Atlantic seabirds. *Journal of Animal Ecology* 74, 817-831.

Stenhouse, I.J., Robertson, G.J., Gilchrist, H.G. (2004). Recoveries and survival rate of Ivory gulls banded in Nunavut. *Waterbirds* 27, 486-492.

Suydam, R.S., Dickson, D.L., Fadely, J.B. & Quakenbush, L.T. (2000). Population declines of King and Common Eiders of the Beaufort Sea. *The Condor* 102, 219-222.

Tasker, M.L., Camphuysen, C.J., Cooper, J., Garthe, S., Montevecchi, W.A., Blaber, S.J. (2000). The impacts of fishing on marine birds. *ICES Journal of Marine Science* 57, 531-547.

White, C.R., Boertmann, D., Grémillet, D., Butler, P.J., Green, J.A., Martin, G.R. (2011). The relationship between sea surface temperature and population change of Great Cormorants Phalacrocorax carbo breeding near Disko Bay, Greenland. *International Journal of Avian Science*, DOI: 10.1111/j.1474-919X.2010.01068.x

Winker, K., Gibson, D.D., Sowls, A.L., Lawhead, B.E., Martin, P.D., Hoberg, E.P., Causey, D. (2002). The birds of St. Matthew Island, Bering Sea. *Wilson Bulletin* 114, 491-509.

Marine Mammals

Arrigo, K.R. and van Dijken, G.L. (2011). Secular trends in Arctic Ocean net primary production. *Journal of Geophysical Research: Oceans (1978–2012)*, 116(C9).

Bluhm, B.A. and Gradinger, R. (2008). Regional variability in food availability for Arctic marine mammals. *Ecological Applications*, 18(sp2), S77-S96.

Burdin, A., Filatova, O. & Hoyt, E. (2009). Marine mammals of Russia: a guidebook. Kirov, Moscow.

Chernook, V.I. and Boltnev, A.I. (2008). Regular instrumental aerial surveys detect a sharp drop in the birthrates of the harp seal in the White Sea. *Marine Mammals of the Holarctic* 4: 100-104.

Estes, J.A., Tinker, M.T., Doroff, A.M. and Burn, D.M. (2005), Continuing sea otter population declines in the Aleutian Archipelago. *Marine Mammal Science*, 21: 169–172.

Fay, F.H., Eberhardt, L.L., Kelly, B.P., Burns, J.J. and Quakenbush, L.T. (1997). Status of the Pacific walrus population, 1950-1989. *Marine Mammal Science* 13: 537-565.

Ferguson, S.H. and Young, B.G. (2011). Aerial survey estimates of hauled-out ringed seal (Pusa hispida) density in western Hudson Bay, June 2009 and 2010. *Science Advisory Report* 2011/029, Department of Fisheries and Oceans Canada, Ottawa.

Garlich-Miller, J.L., MacCracken, J.G., Snyder, J., Meehan, R., Myers, M.J., Wilder, J.M., Lance, E. and Matz, A. (2011*). Status review of the Pacific walrus (Odobenus rosmarus divergens)*. Marine Mammals Management, United States Fish and Wildlife Service, Anchorage.

George, J.C., Zeh, J., Suydam, R. and Clark, C. (2004). Abundance and population trend (1978-2001) of western Arctic bowhead whales surveyed near Barrow, Alaska. *Marine Mammal Science* 20: 755-773.

Gosselin, J. F., Lesage, V. and Hammill, M.O. (2009). Abundance indices of beluga in James Bay, eastern Hudson Bay and Ungava Bay in 2008. *Research Document* 2009/006. Science Advisory Secretariat, Department of Fisheries and Oceans Canada, Ottawa.

Hammill, M.O. and Stenson, G.B. (2007). Application of the precautionary approach and conservation reference points to the management of Atlantic seals. *ICES Journal of Marine Sciences* 64: 701-706.

Heide-Jørgensen, M.P., Laidre, K.L., Burt, M.L., Borchers, D.L., Hansen, R.G., Rasmussen, M. and Fossette, S. (2010). Abundance of narwhals (Monodon monoceros) in Greenland. *Journal of Mammalogy* 91(5): 1135-1151.

Hobbs, R. C., Sims, C.L. and Shelden, K.E.W. (2012). *Estimated abundance of belugas in Cook Inlet, Alaska, from aerial surveys conducted in June 2012*. NMFS, NMML Unpublished Report. 7 pp.

ICES (2008). Report of the Joint ICES/NAFO Working Group on Harp and Hooded Seals, 27-30 August 2008, Tromsø, Norway. ICES Report CM 2008/ACOM 17, International Council for the Exploration of the Sea (ICES), Copenhagen.

Lydersen, C., Aars, J. and Kovacs, K.M. (2008). Estimating the number of walruses in Svalbard based on aerial surveys and behavioural data from satellite telemetry. *Arctic* 61: 119-128.

Mahoney, Barbara A. and Shelden, Kim E.W. (2000). Harvest History of Belugas, Delphinapterus leucas, in Cook Inlet, Alaska. *Marine Fisheries Review*, 62(3), pp. 124-133.

Norstrom R. J. and Muir, D.C.G. (1994) Chlorinated hydrocarbon contaminants in arctic marine mammals. *The Science of the Total Environment* 154:107-128.

Obbard, M.E., Thiemann, G.W., Peacock, E. and DeBruyn, T.D. (eds.) (2010). Proceedings of the 15th Working Meeting of the IUCN/SSC Polar Bear Specialist Group, 29 June - 3 July 2009, Copenhagen, Denmark. *Occasional Paper* No. 43 of the IUCN Species Survival Commission, IUCN, Gland.

Øigård, T.A., Haug, T. and Nilssen, K.T. (2010). Estimation of pup production of hooded seals and harp seals in the Greenland Sea in 2007: Reducing uncertainty using generalized additive models. *Journal of the Northwest Atlantic Fishery Science*. 42: 103-123.

Richard, P.R., Laake, J.L., Hobbs, R.C., Heide-Jørgensen, M.P., Asselin, N.C. and Cleator H. (2010). Baffin Bay narwhal population distribution and numbers: aerial surveys in the Canadian High Arctic, 2002-2004. *Arctic* 63: 85-99.

Ross, W.G. (1993). Commercial whaling in the North Atlantic sector. pp. 511-61. In: Burns, J.J. Montague, J.J. and Cowles, C.J. (eds.) Special Publication. No. 2. *The*

Bowhead Whale. 1st. Edn. Society of Marine Mammalogy, Lawrence, KS. 787pp.

Stroeve, J.C., Serreze, M.C., Holland, M.M., Kay, J.E., Malanik, J. and Barrett, A.P. (2012). The Arctic's rapidly shrinking sea ice cover: a research synthesis. *Climatic Change* 110: 1005-1027.

Wang, M. and Overland, J.E. (2012). A sea ice free summer Arctic within 30 years: An update from CMIP5 models. *Geophysical Research Letters* 39: L18501. doi:10.1029/2012GK052868

Wiig, Ø., Bachmann, L., Heide-Jørgensen, M.P., Lindqvist, C., Laidre, K.L., Postma, L., Dueck, L., Palsbøll, P.J., Bachmann, L. (2011). Recaptures of genotyped bowhead whales (*Balaena mysticetus*) in eastern Canada and west Greenland. *Endangered Species Research* 14: 235-242.

Witting, L. and Born, E. (2005). An assessment of Greenland walrus populations. *ICES Journal of Marine Sciences* 62: 266-285.

Socioeconomic Aspects

Millennium Ecosystem Assessment (2005). *Ecosystems and Human Well-being: Synthesis*. Washington, DC, Millennium Ecosystem Assessment, Island Press.

UK National Ecosystem Assessment (2011). *The UK National Ecosystem Assessment: Synthesis of the Key Findings*. UNEP-WCMC, Cambridge.

36H Southern Ocean

Contributors:
Viviana Alder, Maurizio Azzaro, Rodrigo Hucke-Gaete, Enrique R. Marschoff (Lead Member), Renzo Mosetti, José Luis Orgeira, Liliana Quartino, Andrea Raya Rey, Laura Schejter and Michael Vecchione

Southern Ocean

Chapter 36H

The Southern Ocean is the common denomination given to the southern extrema of the Indian, Pacific and Atlantic Oceans, extending southwards to the Antarctic Continent. Its main oceanographic feature, the Antarctic Circumpolar Current (ACC), is the world's only global current, flowing eastwards around Antarctica in a closed circulation with its flow unimpeded by continents. The ACC is today the largest ocean current, and the major means of exchange of water between oceans; it is believed to be the cause of the development of Antarctic continental glaciation by reducing meridional heat transport across the Southern Ocean (e.g., Kennett, 1977; Barker et al., 2007). The formation of eddies in the Antarctic Circumpolar Current has a significant role in the distribution of plankton and in the warming observed in the Southern Ocean.

As with the ACC, the westward-flowing Antarctic Coastal Current, or East Wind Drift (EWD), is wind-driven. These two current systems are connected by a series of gyres and retroflections (e.g., gyres in the Prydz Bay region, in the Weddell Sea, in the Bellingshausen Sea) (Figure 36H.1).

The boundaries and names shown and the designations used on this map do not imply official endorsement or acceptance by the United Nations.
Figure 36H.1 | From Turner et al. (eds.), 2009. Schematic map of major currents south of 20°S (F = Front; C = Current; G = Gyre) (Rintoul et al., 2001); showing (i) the Polar Front and Sub-Antarctic Front, which are the major fronts of the Antarctic Circumpolar Current; (ii) other regional currents; (iii) the Weddell and Ross Sea Gyres; and (iv) depths shallower than 3,500m shaded (all from Rintoul et al, 2001). In orange are shown (a) the cyclonic circulation west of the Kerguelen Plateau, (b) the Australian-Antarctic Gyre (south of Australia), (c) the slope current, and the (d) cyclonic circulation in the Bellingshausen Sea, as suggested by recent modelling studies (Wang and Meredith, 2008), and observations – e.g., the eastern Weddell Gyre - Prydz Bay Gyre (Smith et al., 1984), westward flow through Princess Elizabeth Trough (Heywood et al., 1999), and circulation east of Kerguelen Plateau (McCartney and Donohue, 2007).

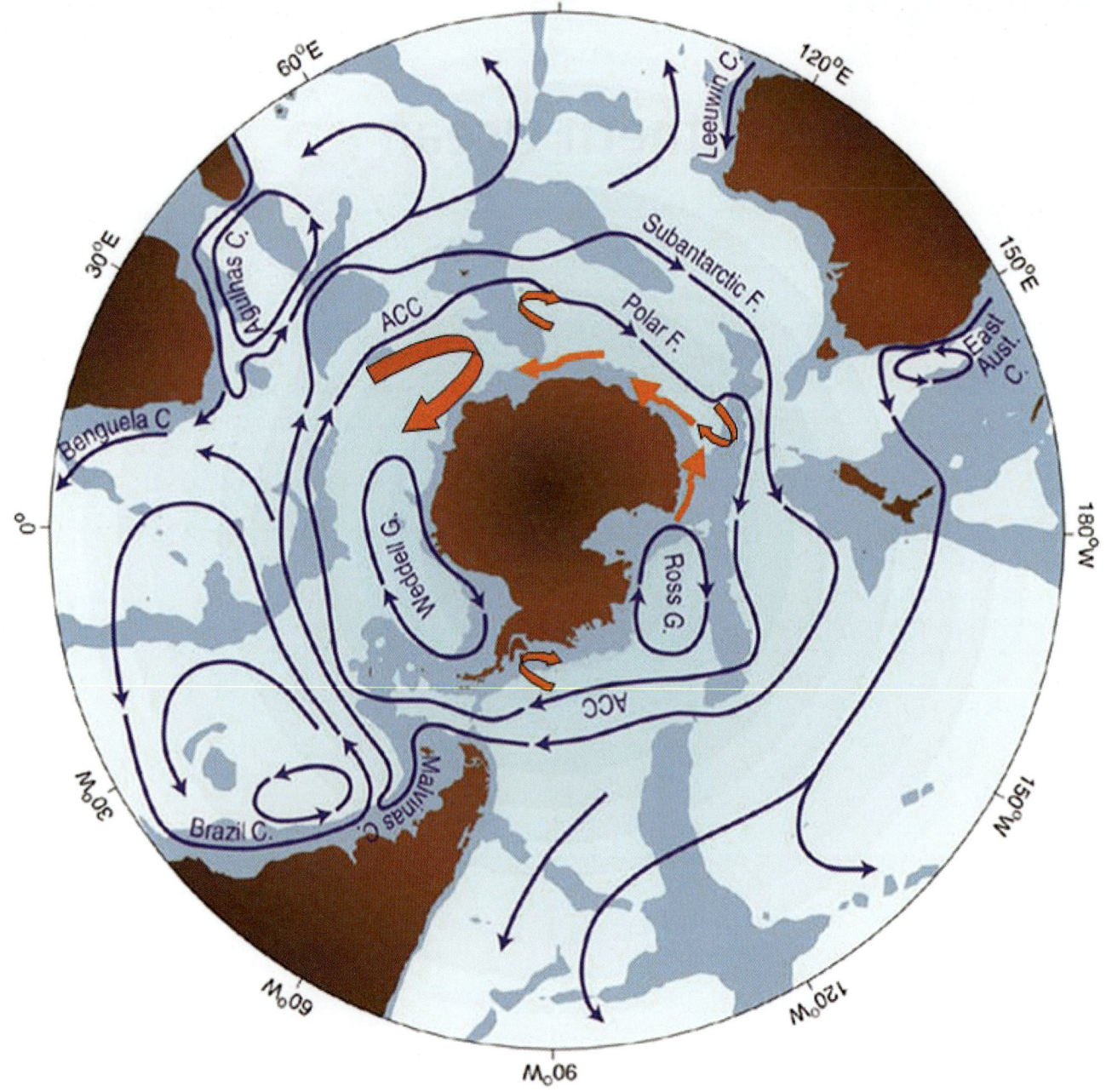

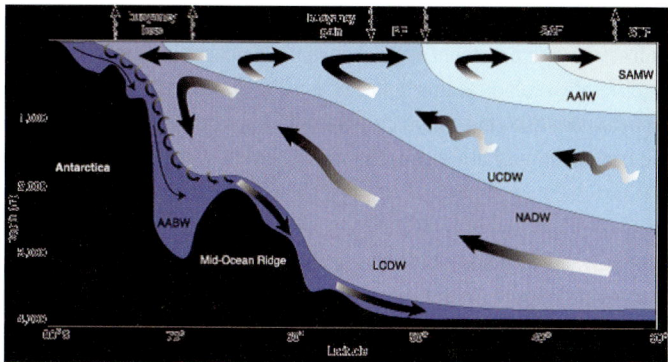

The boundaries and names shown and the designations used on this map do not imply official endorsement or acceptance by the United Nations.

Figure 36H.2 | From Turner et al. (eds.), 2009. South (left) to north (right) section through the overturning circulation in the Southern Ocean. South-flowing products of deep convection in the North Atlantic are converted into upper-layer mode and intermediate waters and deeper bottom waters and returned northward. Marked are the positions of the main fronts (PF – Polar Front; SAF – Sub-Antarctic Front; and STF – Subtropical Front), and water masses (AABW – Antarctic Bottom Water; LCDW and UCDW, Lower and Upper Circumpolar Deep Waters; NADW – North Atlantic Deep Water; AAIW – Antarctic Intermediate Water and SAMW – Sub- Antarctic Mode Water) (from Speer et al., 2000, ©American Meteorological Society. Used with permission.). Note that as well as water moving north to south or vice-versa, it is also generally moving eastward (i.e., towards the observer in the case of this cross-section), except along the coast where coastal currents move water westward (away from the observer).

The circumpolarity of the circulation is the principal factor determining the development of circumpolar frontal zones associated with this system of currents (Orsi et al., 1995). The biogeographical importance of these fronts was recognized practically from the beginning of Antarctic research (Tate Regan, 1914); their approximate positions are shown in Figure 36H.2.

The ACC is usually considered to be the northern border of the Southern Ocean. As the ACC links the ocean basins of the Atlantic, Indian and Pacific Oceans, the waters carried in the ACC contain a mix of waters originating in different parts of the world. Water flows away from the ACC, to the north and to the south, where it becomes a primary source for the Antarctic Bottom Water. In the ACC the three oceans exchange heat, salinity and nutrients, playing an important role in the regulation of temperature and flow of the global conveyor belt. Along its course in the ACC, the water exchanges oxygen and carbon dioxide with the atmosphere while cooling; the resulting dense water sinks and transfers heat and gases into the deep ocean. These exchanges create water masses with different properties and distribution patterns which are responsible for water properties in all the world's oceans (Figure 36H.3); see Turner et al. (eds.) (2009) for further information.

About 50 per cent of the Southern Ocean is covered by ice in winter, decreasing to 10 per cent in summer. Ice cover has important effects both on climate and on the biota (e.g., Ainley et al., 2003). It is a defining structure in polar ecosystems. Antarctic sea ice is inhabited by prokaryotes, protists, algae, crustacea, worms (Schnack-Schiel et al., 1998), fish eggs and larvae (Vacchi et al., 2012), birds and seals (Ainley and DeMaster, 1990).

Overall, the Antarctic sea-ice cover has been increasing in the satellite records from 1978 to 2010 (Parkinson and Cavalieri, 2012; see Chapter 47), but modelling predicts a reduction of 33 per cent by the end of this century. This masks dramatic regional trends; declines in sea ice in the Bellingshausen Sea region have been matched by opposing increases in the Ross Sea (Maksym et al., 2012). Besides the seasonal sea ice, large portions of coastal waters are covered by permanent ice shelves. Ice shelves derive from land ice where glaciers or whole ice sheets flow towards the coastline and over the ocean surface (Trathan et al., 2013). Ice cover defines three biogeographic zones (Tréguer and Jacques, 1992): the northernmost part of the ACC, permanently ice free with high nutrient concentrations but low primary productivity (see below); the region that is covered seasonally, where the movements of the ice margin significantly affect the cycle of primary production and zooplankton aggregations, and the sea below ice shelves where the fauna develop under unique oligotrophic conditions (Gutt et al., 2010). Of particular interest are the regions of contact between the sea-ice cover and the shelf ice, where regions of highly productive open-water areas develop (Comiso and Gordon, 1996; Smith and Comiso, 2008).

The present characteristics of the Antarctic were established at the time of the separation between Antarctica and South America, allowing the unimpeded flow of the ACC (Barker et al., 2007; Turner et al. (eds.), 2009) and the development of the Polar Front. Uncertainty exists with regard to the date of the opening, but it is widely accepted that it occurred about 34 million years ago in the Eocene/Oligocene limit (Barker et al., 2007).

From that time onwards, the oceans south of the Polar Front have been part of a single system comprising the basins of the Atlantic, Indian and

Figure 36H.3 | Model of the global ocean circulation, emphasizing the central role played by the Southern Ocean. NADW = North Atlantic Deep Water; CDW = Circumpolar Deep Water; AABW = Antarctic Bottom Water. Units are in Sverdrups (1 Sv = 106 × m3 of water per second). The two primary overturning cells are the Upper Cell (red and yellow), and the Lower Cell (blue, green, yellow). The bottom water of the Lower Cell (blue) wells up and joins with the southward-flowing deep water (green or yellow), which connects with the upper cell (yellow and red). This demonstrates the global link between Southern Ocean convection and bottom water formation and convective processes in the Northern Hemisphere. From Lumpkin and Speer (2007; ©American Meteorological Society. Used with permission.) in Turner et al. (eds.), 2009.

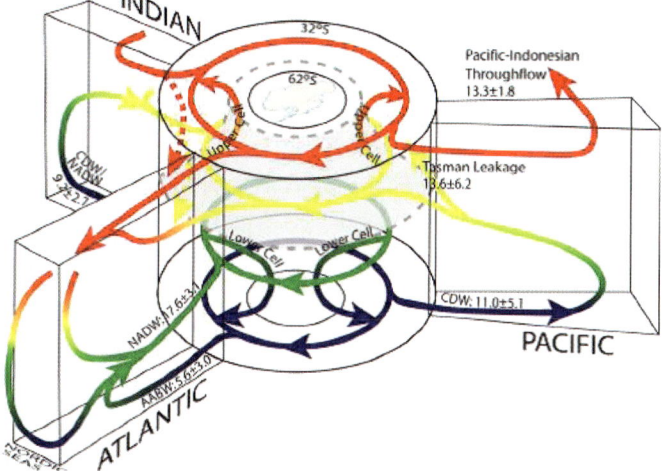

Pacific Oceans and isolated from other shelf areas in the Southern Hemisphere. The Antarctic marine environment experienced a slow transition from warm water conditions to the present cold water system (from 15°C to 1.87°C) (Turner et al. (eds.), 2009). The result of evolution under these conditions is a highly specialized marine biota with high Antarctic endemism and little tolerance for warming (Bilyk and DeVries, 2011).

1 Primary Production

Research on Antarctic primary productivity started around 1840, the age of the pioneering expeditions to Antarctica. For some 100 years, most studies were of a qualitative nature and largely focused on net phytoplankton (>20 μm: diatoms, dinoflagellates, silicoflagellates, etc). The results of these investigations showed that phytoplankton distribution was linked to seasonality and latitude, with a fast and early growth in northern sectors and a southward shift of growth maxima as summer progresses.

A period of change started around 1950 with the development and application of new methods which involved a faster collection of quantitative data associated with the amount of biomass produced per unit of space and time. Such methods enabled estimating, for example, chlorophyll-a concentration as a proxy for actual phytoplankton biomass and primary productivity, and assimilation of dissolved inorganic carbon by phytoplankton as a proxy for the rate of photosynthetic production of organic matter in the euphotic zone. To date, these methods are the most widely used for estimating *in situ* primary productivity in Antarctic ecosystems at all scales, and at a global level as well.

More sophisticated techniques and equipment developed during the 1980s enabled quantifying fragile cells that are difficult to preserve. These improvements represented a substantial progress in the knowledge of newly identified taxa contributing to phytoplankton biomass, such as flagellates, and of some groups' capability of alternating their trophic strategies following the fluctuations of certain environmental variables, such as nutrient and/or prey availability and light. Phytoplanktonic communities comprise at least three main size classes of algae: picoplankton (<2 μm), nanoplankton (2-20 μm) and microplankton (>20 μm). Blooms of microplanktonic and nanoplanktonic algae (e.g., diatoms, dinoflagellates, colonial and flagellated *Phaeocystis* cells) are mostly detected during the summer within the marginal ice zone (e.g., Buma et al., 1992; Olguín and Alder, 2011). However, little knowledge exists about the importance of mixotrophic groups, such as flagellates and dinoflagellates, as food for primary consumers (some of commercial importance) and their contribution to phytoplanktonic biomass (and chlorophyll-a levels) and primary productivity.

Finally, from 1990 on, many investigations on productivity have been largely based on data provided by satellites equipped with colour scanners. At present, ocean color remote sensing is our most effective tool for understanding ocean ecology and biogeochemistry at basin-to-global scales (Figure 36H.4). Many of the algorithms used in satellite data processing and a number of predictive mathematical models employed at different ecological levels (fisheries, CO2 dynamics, etc.) are based on *in situ* measurements which yield differing results depending on the methods used and are a cause of much current debate (Strutton et al., 2012).

At any scale, light and nutrients are the most crucial resources for phytoplankton growth. In addition, diverse physical, chemical and biological variables act as conditioning factors for phytoplankton development and biomass levels. For example, temperature, water column stability, advection, grazing, sinking, bottom topography, offshore distance, etc., usually lead to temporal and spatial variations in primary producers at different scales and also in primary productivity levels.

The overall distribution of phytoplankton biomass and primary production is associated with the position of frontal zones and water circulation resulting from the cyclonic circulation linked to the topography. This general scheme of distribution has been known since the first reports on Antarctic phytoplankton (Hardy and Gunther, 1935; Balech, 1968; El-Sayed, 1968a; 1968b; 1970). High local variability is superimposed on this general pattern as demonstrated by satellite information showing spots of very high chlorophyll concentration within areas of generally low concentration (El-Sayed and Hofmann, 1986).

Knox (2007) reviewed the levels reached by phytoplankton, chlorophyll and primary productivity in distinct Antarctic areas and showed the strong variability associated with different processes and sectors. In the case of Antarctic phytoplankton, variability is generally attributed to (a) extreme seasonal variability in solar radiation, (b) availability of iron (Fe), which is considered as a key limiting factor in the deep and open waters of the Southern Ocean, and (c) the extent, duration, and seasonality of sea ice and glacial discharge, which influence the life cycles of most Antarctic organisms (Ducklow et al., 2013). The annual retreat and melting of sea ice in spring causes the stratification of the upper ocean layer, thus activating the development of important phytoplankton blooms. The magnitude of these blooms is related to the winter extent of ice cover, which acts as a barrier to wind mixing (Ducklow et al., 2012; 2013).

Antarctic continental shelf regions have an annual productivity that ranges from 10 g C m^{-2} to 200 g C m^{-2}; the greatest rates occur in the Ross Sea and the western Antarctic Peninsula, but elevated productivity is found in nearly all coastal polynyas (Catalano et al., 2010; Smith et al., 2010).

Large variability in primary productivity was observed along a twelve-year time series (1995-2006; Palmer Long-Term Ecological Research). The average daily integrated primary productivity varied by an order of magnitude, from 250 C m^{-2} d^{-1} to 1100 mg C m^{-2} d^{-1}, with an average of 745 mg C m^{-2} d^{-1}. A marked onshore–offshore gradient from 1000 C m^{-2} d^{-1} to 100 mg C m^{-2} d^{-1} was found along the shelf with higher production rates inshore. Inter-annual regional variability ranged from 248 C m^{-2} d^{-1} to 1788 mg C m^{-2} d^{-1} (Vernet et al., 2008).

Satellite (SeaWiF) measurements of chlorophyll concentrations in the Southern Ocean from October 1997 through September 1998 reveal: (a) low-mean values (0.3 mg m^2- 0.4 mg m^2); (b) phytoplankton blooms and highest chlorophyll concentrations (>1.0 mg m^2) located in three areas: coastal waters above the continental shelf, the seasonally retreating sea ice, and the vicinity of the major fronts; (c) the SeaWiFS global chlorophyll algorithm works better than the Coastal Zone Color Scanner (CZCS); (d) based on the production model of Behrenfeld and Falkowski (1997), annual primary production south of 50°S was estimated in 2.9 Gt C yr-2 (Moore and Abbott, 2000).

Marine food webs depend on primary productivity, which also contributes to the sequestration of carbon in the oceanic reservoir. In this region the models and observations for the global balance of CO_2 differ most in magnitude (Gruber et al., 2009). This disagreement highlights the importance of obtaining better estimates of all terms in the oceanic carbon budget, including primary productivity (Strutton et al., 2012).

In the seasonally ice-covered region, microalgae grow on sea ice. Their production is greatly exceeded by phytoplankton production in open waters, but their ecological role is significant. They constitute an important environment that provides refuge to cells and spores that will later seed blooms in open water (Lizotte, 2001) and provide food for ice-associated grazers, such as developmental stages of krill.

2   Zooplankton and Nekton

Multicellular animals in the water column are generally divided by size and swimming ability. This distinction is not always clear, however; some active swimmers may be quite small, whereas some large animals may be such poor swimmers that they are little more than drifters. Zooplankton species range from microscopic animals so small that the water is for them a very viscous environment, to large (sometimes very large) but slow-moving gelatinous animals from several evolutionary lineages. In addition to animals that are planktonic throughout their life cycles, zooplankton sometimes include eggs and larvae or, for some species, spawning stages of bottom-living (benthic) animals. The nekton include generally larger animals with swimming ability adequate to overcome movement by currents, e.g., primarily fish, shrimp, and cephalopods. The term nekton also encompasses diving air-breathers, including marine mammals, birds (e.g., penguins) and, in more temperate regions, reptiles; the first two groups are addressed in separate sections of this chapter.

Important for understanding zooplankton and especially for nekton is the distinction between the swimming animals that live on or closely associated with the bottom (i.e., demersal and benthopelagic) and those that spend their lives higher up in the water column and are not dependent on the bottom (pelagic). A further distinction in polar regions involves the species that are dependent, at least at some stage in their life, on sea ice (cryopelagic).

The overlap between the zooplankton and nekton that occurs in the smallest nekton (=micronekton) and large plankton (=megaplankton) is particularly important in the Southern Ocean. A unique characteristic of pelagic marine ecosystems is the alternating dominance as keystone grazers in the food webs by species from two very different evolutionary and ecological groups (Atkinson et al., 2004). These two groups are euphausiids, primarily the Antarctic Krill (*Euphausia superba*), and salps (especially *Salpa thompsoni*). The former are actively swimming shrimplike crustaceans, the adults of which dominate the micronekton community, whereas the latter form "blooms" and colonial chains of gelatinous megaplankton.

Good reviews of current knowledge about zooplankton and nekton in the Southern Ocean have recently been published. These include several chapters in Knox (2007), products of the Census of Antarctic Marine Life (Gutt et al., 2010; DeBroyer et al. (ed.), 2011; Schiaparelli and Hopcroft (eds.), 2011 and papers within that volume), and recent summaries of national research programmes, particularly around the Antarctic Peninsula (e.g., Ducklow et al., 2007; Steinberg et al., 2012) and in the Ross Sea (Faranda et al. (eds.), 2000).

The boundaries and names shown and the designations used on this map do not imply official endorsement or acceptance by the United Nations.
Figure 36H.4 | From Strutton et al. (2012). Austral summer ocean pigment concentrations from the Coastal Zone Colour Scanner (CZCS). The data are available at 4 km and 9 km (this image) spatial resolution. The northern and southern solid lines represent the subtropical front and the polar front, respectively.

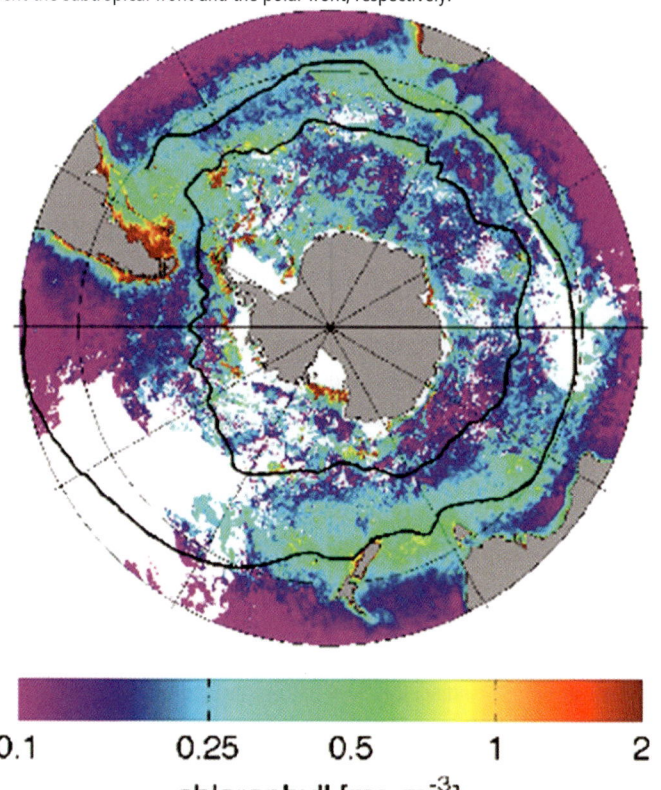

Because of their swarming behaviour, krill form large patches of extremely high abundance and biomass, which are targeted by many Antarctic predators, including fish, squid, birds and mammals (Knox, 2007). The patches are also targeted for harvesting by humans (Nicol et al., 2012). The commercial potential of krill has, in turn, stimulated studies of zooplankton and micronekton assemblages and pelagic ecology more generally around Antarctica, especially near the Peninsula and in the Ross Sea.

Salps, which are megaplankton with jelly-like bodies, are chordates more closely related to vertebrates than to true jellyfish. They have complex life cycles, including both sexual and asexual (budding) reproduction and alternating generations of solitary individuals and chain-like colonial aggregates. Because they are capable of asexual reproduction, population abundance can respond rapidly to favourable conditions, resulting in a "bloom" (i.e., very high abundance).

Because a developmental stage of krill is dependent on ice algae and therefore on the amount of sea-ice habitat during that stage in the krill life cycle, alternation between high abundances of krill and salps is related to the amount of sea ice during the previous winter (Loeb et al., 1997). This alternation has very important implications for food webs in the Southern Ocean, especially pelagic food webs, but for benthic food webs as well. Krill are the preferred prey of many mid-level Antarctic predators (Knox, 2007). Although many of them can also feed on salps, the food quality obtained from salps compared with the energy expended in feeding is much less than for krill. Therefore, krill swarms are very important to maintain population levels for many species in the Southern Ocean.

Other than salps and krill, the numerically dominant group of zooplankton is the copepod crustaceans. Copepods are also the most diverse group of zooplankton, with more than 70 species in the upper 100 m of the water column. Another important crustacean group is the amphipods, which may be free-swimming or associated as predators or commensals with gelatinous megaplankton.

Other important zooplankton groups, sometimes abundant although not diverse, include pteropod mollusks, pelagic polychaete worms, chaetognaths (arrow worms), larvaceans, and ostracods.

Because many Antarctic species develop directly in large eggs rather than as larvae hatching from many small eggs, benthic species that spend their early life history in the water column (meroplankton) make up a less important component of the zooplankton in the Southern Ocean than elsewhere, although larvae may be seasonally abundant in some coastal areas.

Among the nektonic fish, cephalopods, and shrimp, the fish are clearly dominant in diversity, abundance and biomass. Only about 300 fish species are found south of the South Polar Front; of these, slightly over 100 species are considered to be endemic to the Southern Ocean (Knox, 2007). Most of these fish species are closely associated with the bottom (demersal). The common demersal fish are dominated by a unique evolutionary group, the suborder Notothenioidei, referred to as Antarctic blennies or southern cod-icefish.

The few species of shrimp (Caridea, <10 species) are primarily demersal. Three species have been reported locally in high abundances (Gorny et al., 1992). Southern Ocean cephalopods include both pelagic squid and demersal octopods. The former includes a few species each in at least nine families. Although rarely collected in large numbers, squid are important prey for many fish, birds and mammals, indicating that current sampling methods are not representative. Some attain large sizes, including *Mesonychoteuthis hamiltoni*, one of the largest known invertebrates. The octopods include one finned (Cirrrata) species that can be locally very abundant (Vecchione et al., 1998) and a surprising diversity of incirrate species. Whereas the incirrate fauna were long thought to comprise a few variable species, careful examination of both morphology and DNA sequences has shown the species to number at least two dozen, including one very large species, *Megaleledone setebos* (the giant Antarctic octopus). One genus, *Pareledone*, is especially diverse and is an example of both circumpolar distributions and "ring evolution" of cryptic species around the Antarctic continent (Allcock et al., 2011). The incirrate octopods have also been shown to be both an evolutionary source of species for the world's deep oceans (Strugnell et al., 2008) and to have colonized the Southern Ocean from deep oceans elsewhere (Strugnell et al., 2011).

In general, both zooplankton and nekton include some nearshore species that are circumpolar and others for which distribution appears to be regionally limited. Some of the former, however, might include complexes of cryptic species with limited distributions. The regional patterns are strongest closest to shore. For both fully pelagic and demersal animals, a gradient exists from coastal through continental shelf and slope to fully oceanic assemblages. Also, species interactions occur across the Polar Front where some subpolar zooplankton and nekton species intrude into polar waters and the ranges of some Southern Ocean species extend into lower latitudes.

In spite of the extreme polar light regime, pelagic animals of the Southern Ocean undertake vertical migrations as are well known from other oceans. Daily variability in light intensity penetrates into the upper waters and diel vertical migrations by pelagic zooplankton and nekton follow this signal. Of course the seasonal variability in light is extreme, resulting from both the angle of solar incidence and seasonal expansion and contraction of sea ice. This seasonal variability results in extensive changes in the vertical distribution of many pelagic species, usually manifested by different life-history stages occupying different depths in the water (ontogenic vertical migration). In addition to the importance (e.g., role in the food web, development through life-history stages, etc.) of these vertical migrations to the individual species, they are also important in understanding the flow of biomass, nutrients, carbon, etc., between the surface and deeper waters and ultimately to the benthic environment.

Another important association for pelagic animals in the Southern Ocean is with ice, in the form of sea ice and icebergs. As mentioned above, some stages of some species, notably krill, are dependent on the protection of the physical ice structure and on the food that grows on sea ice (Thomas and Dieckmann, 2003). Furthermore, it has recently been demonstrated that drifting icebergs are "hotspots" of pelagic productivity and biomass (Smith et al., 2007; Vernet et al., 2012). As regards prokaryotic assemblages, a very high abundance has been reported for the bottom layer of sea ice (Archer et al., 1996; Delille and Rosiers, 1996) and the platelet ice (Guglielmo et al., 2000; Riaux-Gobin et al., 2000).

3   Microbes

The microbial community plays a pivotal role in the pelagic food web of the Southern Ocean; it controls many processes, including primary production, turnover of biogenic elements, degradation of organic matter and mineralization of xenobiotics and pollutants (Azam et al., 1991; Azzaro et al., 2006; Fuhrman and Azam, 1980; Karl, 1993; Manganelli et al., 2009; Smith et al., 2010; Yakimov et al., 2003). Prokaryotic abundance and activity shift significantly over the annual cycle as sea ice melts and phytoplankton blooms develop (Ducklow et al., 2001; Pearce et al., 2007). Microbial food chains develop even in regions where large euphausiids are abundant. These chains involve small metazoans and predominate in the northern open waters (Atkinson et al., 2012) with multiple trophic levels (copepods, chaetognaths, amphipods, myctophids, fish and birds) in contrast with the classical short chain of diatoms – krill – vertebrates. Marine microbes exhibit a diversity which also depends on the timing, location and sampling method (Pearce 2008; Murray et al., 1998); research devoted to this group is increasing, using genetic and molecular approaches in Antarctic surface (Murray and Grzymski, 2007) and deep waters (Moreira et al., 2004). Studies on diversity of bacterioplankton suggest that the diversity seems to rival that found in other ocean systems, although many polar phylotypes host a distinct biogeographic signal (Pommier et al., 2005). The Archaea (DeLong et al., 1994) have a distinct seasonal cycle in which Marine Group I. the Crenarchaeota, are abundant in late winter in surface Antarctic waters (Murray et al., 1998; Church et al., 2003; Murray and Grzymski, 2007). Southern Ocean environmental genomic studies focusing on identifying organisms and metabolic capabilities of the microbial community are limited (Béja et al., 2002; Grzymski et al., 2006). Grzymski et al. (2012) found that the most noteworthy change in the bacterioplanktonic community in nearshore surface waters of the Antarctic Peninsula was the presence of chemolithoautotrophic organisms in winter and their virtual absence in summer when incident solar irradiance is at a maximum and primary productivity is high. If chemolithoautotrophy is widespread in the Southern Ocean in winter, this process may be a previously unidentified carbon sink. Research trends point to microbial diversity in marine invertebrates (Webster et al., 2004; Webster and Bourne, 2007; Riesenfeld et al., 2008).

4   Benthos

Antarctic shelves are very deep, a process that developed from the glaciation that began with the isolation of the Southern Oceans because of the weight of the continent's massive ice sheet. Due to the isostatic depression of the Antarctic continent by the extant ice sheets, the features of Antarctic continental margins are distinct from those of the rest of the ocean and the continental shelf break occurs at a depth of ca. 1,000 m compared to about 200 m and less elsewhere in the world (Smith et al., 2010). Such evidence implies that (a) the continental margins tend to be narrow and often have deep canyons, (b) essentially no organic matter is derived from continental sources, and (c) the only significant effects of the continent are to provide mineral material to the continental shelves via ice-rafted debris and glacial meltwater to restricted coastal environments. More than 95 per cent of the shelves are at depths outside the reach of photosynthetically active radiation (Turner et al. (eds.), 2009). Some 33 per cent of the continental shelves are covered by the floating ice shelves.

The shelf benthic fauna are dominated by sessile particle feeders with high biomass and low productivity. Many species are long-lived, have low metabolic rates, lack the pelagic larval phase and need longer development time. Benthic communities cover a full range from an extremely high biomass of several kg wet-weight m^{-2} to extremely low biomass, abundances and metabolic processes below ice shelves (Azam et al., 1979).

Knowledge of the benthic fauna has developed from historical surveys, individual national projects and cooperative projects, such as the European Polarstern Study (EPOS), Ecology of the Antarctic Sea Ice Zone (EASIZ), Evolution in the Antarctic (EVOLANTA), Census of Antarctic Marine Life (CAML), Latitudinal Gradient Project (LGP), Food for Benthos on the Antarctic Continental Shelf (FOODBANCS), Antarctic Benthos (BENTANTAR), and the Antarctic benthic DEEP-sea biodiversity: colonization history and recent community patterns (ANDEEP), but regionally many gaps in survey data remain (Griffiths et al., 2009). Some 4,000 species have been described (White, 1984; Arntz et al., 1997; Clarke and Johnston, 2003), of an estimated total macrofauna of more than 17,000 species (Gutt et al., 2004); the deep benthos remains largely unsampled (Brandt et al., 2004).

Close to the shoreline, benthic communities are strongly affected by ice scouring. This phenomenon provokes a continuous recolonization of benthos. Iceberg impacts are catastrophic events eliminating up to 96 per cent of the biomass of the community (Smale et al., 2008), interfering with community development in those areas where ice scouring becomes chronic (Dayton et al., 1974; McCook and Chapman, 1993; Barnes, 1995; Pugh and Davenport, 1997). In essence, long-lived species are selected against by this process which results in widely different community structures among areas with different scouring histories.

Benthic communities are marked by the absence of crabs and sharks and by a limited diversity of skates and finfish; skeleton-breaking predation is limited. Slow-moving invertebrates are present at high trophic levels. These characteristics, together with dense ophiuroid and crinoid populations, resemble the worldwide Palaeozoic faunal assemblages (Aronson and Blake, 2001). Detritivores, feeding on deposited organic material, include the infauna (mainly mollusks) and vagrant deposit feeders such as holothurians (Gutt, 2007).

At regional and local levels, patchiness is high, due to differences in environmental conditions, food supply and disturbances; but at very coarse spatial resolution, benthic assemblages are typical of the Antarctic with circumpolar distributions (Turner et al. (eds.), 2009).

Below the depth scoured by drifting ice, the invertebrate benthic fauna typically comprise a dense community of sessile species (e.g., sponges, ascidians, gorgonians, anemones, corals, bryozoans, crinoids) in a three-dimensional pattern (Arntz et al., 1994, 1997; Gutt, 2000) with an associated mobile fauna (echinoids, pycnogonids, isopods, amphipods, polychaetes, etc.) developing complex relations among different species (Figure 36H.5).

The highly seasonal primary production in the Antarctic results in a seasonal flux of organic material deposited in the sediment. This provides an abundant persistent food supply for detritivores which might be resuspended as a result of water mixing or ice scouring. This was identified as a "food bank" by Mincks et al. (2005) and Mincks and Smith (2007).

Macroalgae are common elements of nearshore hard-substratum communities in Antarctic and Sub Antarctic regions (Wulff et al., 2011). The areas with hard substrata (e.g., rocks and boulders) are particularly suitable for macroalgal colonization (Quartino et al., 2013). Macroalgae occur in a distinct vertical zonation, mainly between the intertidal and the subtidal zone, down to 30 m depth. The lower distribution is related to their capacity to survive under low light conditions. The South Atlantic Ocean is a nutrient-replete system where nutrients rarely become limiting for macroalgae (Zacher et al., 2009).

Macroalgal communities play a key role in the coastal ecosystem. They are important primary producers, constituting food supply for benthic organisms, such as amphipods, gastropods, annelids and fish (Barrera-Oro, 2002), and represent a significant contribution to the particulate and dissolved organic matter for the coastal food web (Iken et al., 2011). Furthermore, macroalgae provide habitat and structural refuges (Barrera-Oro, 2002; Huang et al., 2007). Macroalgal coastal carbon production seems to be an important food source for the benthic Antarctic communities. If not grazed, macroalgae die and decompose, returning particulate organic matter and mineral nutrients to the system (Quartino and Boraso, 2008).

The sublittoral rocky shores are colonized by macroalgae; the deeper macroalgal assemblages are dominated by canopies of large brown algae from the Order Desmarestiales (*Desmarestia anceps*, *D. menziessi* and *Himantothallus grandifolius*), which replace ecologically the role of the Order Laminariales (i.e., the kelp *Macrocystis pyrifera*) in temperate waters (Wulff et al., 2011).

The strong isolation of the benthic seaweed flora of the South Atlantic Ocean has resulted in a high degree of endemism in Antarctica. Thirty-five per cent of all seaweed species are endemic to the Antarctic region. Within the Heterokontophyta (brown and golden algae) 44 per cent of the species are endemic, within the Rhodophyta (red algae) 36 per cent and within the Chlorophyta (green algae) 18 per cent; and the number of endemic species is continuously increasing (Wiencke and Amsler, 2012). The northern distribution of endemic Antarctic species is often limited by the temperature demands for growth. The southern-most location of open water where macroalgae occur is the Ross Sea, Antarctica (Wiencke and Clayton 2002).

Figure 36H.5 | From Mintenbeck et al. 2012. (A) Trematomus cf. nicolai hiding inside a sponge; (B) Pogonophryne sp. on top of a sponge (ANT XXVII-3 in 2011, western Weddell Sea). Photos: ©Tomas Lundälv, University of Gothenburg.

5 Fish

The Antarctic ichthyofauna is small in size and less diverse than might be expected, given the size and age of the Antarctic marine ecosystem (Eastman, 1995). The knowledge of Antarctic fish began in the nineteenth century through zoogeographic and taxonomic descriptions. Fish fauna in the Antarctic is dominated by the Suborder Notothenioidei; approximately 66 per cent of the Antarctic species and 95 per cent in numbers belong to this Suborder. They live from tide pools (genus *Harpagifer*) to great depths (genus *Bathydraco*). Conversely, Antarctic habitats dominate within the Suborder: from about 100 species of notothenioids, 92 are Antarctic, 12 are found in Patagonia, 4 in New Zealand, 2 in Tasmania and 1 from Saint Paul and Amsterdam Islands. In Antarctica, the Notothenioidei are represented by six families: Bovichtidae, Nototheniidae, Harpagiferidae, Artedidraconidae, Bathydraconidae and Channichthyidae (Kock, 1992). The family Chaennichthydae (ice-fish) is exceptional because it has a colorless blood due to lack of haemoglobin (Kock, 2005); only one species lives outside the Antarctic (*Champsocephalus esox*, Calvo et al., 1999).

Notothenioids lack a swim bladder and have antifreeze glycoproteins in their blood (Matschiner et al., 2011). They have developed a wide range of feeding strategies, which allow them to utilize food resources in a variety of habitats (Gröhsler, 1994). This diversification has been supported by a trend towards pelagization of demersal species (Nybelin, 1947), which might be related to the abundance of available food in the water column, such as krill, in zones of the Southern Ocean. Thus, fish are main predators of benthos, zooplankton and nekton in the water column, including krill, copepods, hyperiid amphipods, squid and fish.

Besides the Notothenioids, demersal fish of the families Zoarcidae, Liparidae, Muraenolepidae, Macrouridae, Moridae, Achiropsettidae, etc. are represented with significant numbers of species endemic to the Southern Ocean. Chondrichthyes (sharks and rays) are also found with bottom dwelling (e.g. *Somniosus antarcticus*, *Amblyraja spp.* and *Bathyraja spp.*) and mesopelagic species (e.g. *Lamna nasus*).

The Southern Ocean lacks the epipelagic fish typically found in surface waters of other oceans. The few species of mesopelagic fish, living in the open ocean down to depths of about 1000 m, are members of cosmopolitan families. Typically antarctic is the nothothenid *Pleurogramma antarctica*. Closely related with ice are the species *Trematomus borchgrevinki*,

T. amphitreta, and *Pagothenia brachysoma*. A recent revision of Southern Ocean fish, their diversity and biogeography can be found in Duhamel et al. 2014.

The environmental factors related to fish distribution can only be described in general terms. On the deep Antarctic shelves (500 m deep on average in the Antarctic, against some 200 m deep worldwide) lying in the area of seasonal ice and the islands in the Scotia arc, the fish fauna are dominated by the families Nototheniidae and Channichthydae.

In the high Antarctic, although the biomass and numbers are smaller, the diversity and endemism are the highest (Kock, 1992) (e.g., genera *Trematomus*, *Pleuragramma*, *Aethotaxis* and *Pagothenia*).

Pelagic fish include occasional species like *Lampris spp.*, *Lamna nasus* and *Thunnus maccoyii*, and, in general, species also found in waters north of the Polar Front. About 85 per cent of the shelf fish fauna are endemic to the Antarctic against only 25 per cent of the deep sea fish. The vertical distribution of mesopelagic fish is related to the Antarctic surface water (Lubimova et al., 1983).

High energetic costs are associated with pelagic feeding, which may hamper the development of shark species in the southern basins. On the other hand, the benthos is a seasonally stable resource, but most of the benthic epifauna are not very suitable for utilization by fish (Kock, 1992).

The mesopelagic fish fauna are mainly composed of Myctophidae (the dominant group) and Gonostomatidae (Kozlov, 1995); the distribution is mainly circumpolar and always related to the Antarctic surface water (Lubimova et al., 1983). This water mass drifts northwards and sinks at the Polar Front. *Gymnoscopelus nicholsi*, found in surface waters down to 700 m near the subtropical convergence, reaches more than 2000 m depth; they are prey of other fish, squid, fur seals and penguins (Sabourenkov, 1991). *Pleuragramma antarcticum* is the notothenid present in the mesopelagic over the shelves.

Inshore, the ecological role of demersal fish is more important than that of krill. There, demersal fish are major consumers of benthos and also feed on zooplankton (mainly krill in summer). They are links between lower and upper levels of the food web and are common prey of other fish, birds and seals. Offshore, pelagic fish (e.g., myctophids, *Pleuragramma antarcticum*) play an important role in the energy flow from macrozooplankton to higher trophic levels (Barrera-Oro, 2002). As krill predators, fish play an important role in the Southern Ocean ecosystems (Kock et al., 2012). Notothenids and Channichthyids are relevant predators of krill and myctophids (Kock et al., 2012); the latter also prey on all development stages of krill.

6 Higher-order predators

Many sub-Antarctic species of birds, pinnipeds and cetaceans occur in northern ice-free waters of the Southern Ocean, and move south in the summer as the pack ice recedes (see chapter 36B). Oceanic fronts present sharp discontinuities in the properties of surface water and food availability; it is well known that predators associate with fronts where they find favourable feeding conditions and are critical for the distribution of seabirds and marine mammals (Bost et al., 2009).

Numerous species of seabirds have been recorded in the Southern Ocean; most are vagrant with only 16 of them nesting in the Antarctic continent (Clements, 2000; Woehler et al., 2001; Harris et al., 2011; Coria et al., 2011; Santora and Veit, 2013; Joiris and Dochy, 2013; Ropert-Coudert et al., 2014). Vagrant species forage within the productive Southern Ocean waters during summer and come mainly from sub-Antarctic islands, although some, such as the Arctic tern *Sterna paradisaea*, fly thousands of kilometres from very distant places (Egevang et al., 2010). A community of seabirds with very stable composition is found in the pack ice; it is probably the most unvarying of any seabird assemblage in the Southern Hemisphere (Ribic and Ainley, 1988). Penguins (Adélie and Emperor) are the typical species, together with snow and Antarctic petrels and, in summer, the South Polar skua and Wilson and storm petrels.

Penguins are the dominant component of the seabird communities in the Southern Ocean in terms both of biomass and prey consumption (Croxall and Lishman, 1987). Nine out of the 18 penguin species inhabit the Southern Ocean; their distributions are reflected in their diets and adaptations to the particular environmental conditions found in their respective ranges, as summarized by Ratcliffe and Trathan (2011):

- Emperor penguins (*Aptenodytes forsteri*) are inhabitants of the high Antarctic; this is the only species that breeds on the land-fast ice along the Antarctic coast during winter. When foraging during winter, emperor penguins have to travel to the edge of the fast ice to feed (Wienecke and Robertson 1997; Zimmer et al., 2008).
- King penguins (*Aptenodytes patagonicus*) feed close to the Polar Front in summer, predominantly on myctophids (*Krefftichthys anderssoni* and *Electrona carlsbergi*). In winter the birds move closer to the ice edge.
- Adélie penguins (*Pygoscelis adeliae*) breed on the Antarctic continent and nearby islands, but their breeding season is in summer, roughly from October to March, and their foraging activity is heavily dependent on sea-ice conditions (Ainley, 2002). Their diet is dominated by euphausiid crustaceans and fish (e.g., Coria et al., 1995; Libertelli et al., 2003). Foraging is mainly confined to pack ice, and seasonal variations in the distribution of this ice cause marked seasonal and spatial variations in foraging ranges, migration routes and wintering areas.

Chinstrap penguins (*Pygoscelis antarctica*) have a diet comprised almost entirely of *Euphausia superba*; diet and reproductive success are dependent on ice conditions (Rombolá et al., 2003; 2006). Chinstrap penguins tend to forage in open water and avoid areas of pack ice (Ainley et al., 1992).

Gentoo penguins (*Pygoscelis papua*) have a diet comprised of a wide range of crustacean and fish taxa, with crustaceans typically less important than for other *Pygoscelis* or *Eudyptes* spp. breeding at the same sites.

Royal penguins (*Eudyptes schlegeli*) are found only on Macquarie Island and macaroni penguins (*Eudyptes chrysolophus*) are found at all other localities. Their diet comprises mostly euphausiid crustaceans and myctophid fish throughout their biogeographic range, with small contributions by amphipods and squid.

Among flying seabirds, the families that are best represented in the Antarctic marine avifauna are Procellariiformes, including albatrosses (Diomedeidae), petrels, prions and shearwaters (Procellariidae), storm petrels (Hydrobatidae) and diving petrels (Pelecanoididae). The order Suliformes is represented by cormorants (Phalacrocoracidae) and the order Charadriiformes by skuas (Stercorariidae) and, to a lesser extent, the gulls and terns (Laridae). Most of the Procellariiformes travel hundreds or thousands of kilometres from the colony during the breeding season to feed on patchily distributed resources and they migrate even further during the non-breeding period (Phillips et al., 2008).

Antarctic marine mammals can be defined as those species whose populations rely on the Southern Ocean as a critical habitat for a part or all of their life history, either through the provision of habitat for breeding and/or through the provision of a major food source (Boyd, 2009). The Southern Ocean accounts for about 10 per cent of the world's oceans, but is estimated to support 80 per cent of the world's pinniped biomass (Laws, 1977) and is a critical feeding ground for several cetaceans, particularly the highly migratory baleen whales (Mackintosh, 1965). Many subfamilies and genera are missing in the Southern Ocean. In spite of the species richness of the family Otariidae (sea lions and fur seals) in the South Atlantic, Indian and Pacific Oceans, only Antarctic fur seals (*Arctocephalus gazella*) are found south of the Polar Front in island rookeries and open waters; sometimes they reach the boundary of the pack ice during the austral summer, and some 50 per cent of the population migrates north during winter. This species feeds mainly on krill, with fish and squid found in their diet in proportions that vary with area and season.

True seals (family Phocidae) are represented by five species. The elephant seal (*Mirounga leonina*), also found north of the Polar Front in open waters, is seldom found in the pack-ice area and also migrates to the north in winter. The remaining seal species are more or less associated with the pack ice: the leopard seal (*Hydrurga leptonix*) preys on krill (about half of its diet), seabirds (mainly penguins), other seals and fish. The Weddell seal's (*Leptonychotes weddellii*) diet is practically all fish and a small proportion of krill. The crabeater seal is the most abundant marine mammal in the world (*Lobodon carcinophaga*) and is a pack-ice inhabitant feeding mostly on krill. Finally, the Ross seal (*Ommatophoca rossii*) is very scarce and little is known of its diet.

A main ecological distinction exists between seals: those breeding in shore colonies (fur and elephant seals) and those breeding on the pack ice (leopard, Weddell, crabeater and Ross seals). The difference is a key element in our ability to estimate population sizes: ice-breeding seals can only be studied through large-scale surveys and it is very difficult to sample the same population year after year; shore colonies offer easier conditions (Southwell et al., 2012). Crabeater seal numbers were estimated from the Antarctic Pack Ice Seals (APIS) International Programme at 10 million individuals, albeit with large confidence intervals and this is likely to be an overestimate (Southwell et al., 2012). Populations of the other three species are much smaller.

Cetaceans in the Southern Ocean are represented by six species of baleen whale: blue (*Balaenoptera musculus*), fin (*Balaenoptera physalus*), sei (*Balaenoptera borealis*), humpback (*Megaptera novaeangliae*), Antarctic minke (*Balaenoptera bonaerensis*) and southern right whales (*Eubalaena australis*). Among these, the Antarctic blue whale was depleted close to extinction by the whaling industry (from 239,000 (95 per cent confidence interval; 202,000-311,000) to a low of 360 (150-840) in 1973) (Branch et al., 2004). Current estimates suggest that some populations are recovering, but that others are not (e.g., those seldom sighted in the Antarctic Peninsula region); others, such as the southern right whale and especially the humpback whale, are both increasing in numbers (Branch, 2011).

At least nine species of odontocetes are found: sperm whale (*Physeter macrocephalus*), southern bottlenose (*Hyperoodon planifrons*), Arnoux's (*Berardius arnuxii*), Cuvier's (*Ziphius cavirostris*) and strap-toothed (*Mesoplodon layardii*) beaked whales, long-finned pilot whale (*Globicephala melas*), orca (*Orcinus orca*), hourglass dolphin (*Lagenorhynchus cruciger*) and the spectacled porpoise (*Phocoena dioptrica*)) (Brownell, 1974; Laws, 1977; Jefferson et al., 2008). Female sperm whales do not reach the Southern Ocean, and only large adult males reach the pack ice.

All in all, cetaceans in the Southern Ocean represent a little less than one-fifth of the world's cetacean species in spite of the large diversity of this family (86 species). Those species that sustain a large biomass are related to the direct plankton food chain (diatoms-krill-vertebrates) which has on average one trophic level less and is more efficient in terms of the transfer of energy and mass than those that include squid or fish as intermediate steps (Boyd, 2009).

7   Pressures and Trends

By-catch, habitat loss, introduced species, human disturbance, pollution and climate change pose severe, albeit of different intensity, threats for seabirds at sea and in colonies in the Southern Ocean and along the Antarctic continent (Micol and Jouventin, 2001; Croxall et al., 2002; Weimerskirch et al., 2003; Jenouvrier et al., 2005). Population trends are variable between species and colonies within a species (Woehler et al., 2001). Significant decreases in populations are evident for those species known to be caught on longline fisheries (albatrosses, Southern giant petrel and *Procellaria* spp.: Woehler et al., 2001; Tuck et al., 2003). Penguin population trends vary in terms of degree and direction among species and geographical areas (Forcada et al., 2006; Lynch et al., 2010; Trivelpiece et al., 2011; Coria et al., 2011). Burrowing petrel species are poorly known, in particular their abundance and trends (Woehler et al., 2001).

8   Harvesting of living resources

Early exploration of the Southern Ocean was driven by the potential of harvesting its nekton – first mammals, then finfish, and finally krill. The discovery of islands lying south of the Antarctic Polar Front rapidly led to the initiation of massive sealing expeditions from various nations during the early 1800s. The outcome of these intensive sealing activities was the near-extermination of fur seals in Antarctic and Sub-Antarctic Islands by the mid-19th century. Combined with the hunting of fur seals, a rather less relentless pursuit of elephant seals (*Mirounga leonina*) followed for the production of an oil equivalent to whale oil (Bonner, 1984). In 1812, a new method to process seal skins was introduced in London factories, increasing the value of southern pelts. Despite efforts to regulate the catches, Patagonian seals were depleted below commercial levels by 1825, and the Antarctic exploration and harvesting finally resulted in severe depletion and loss of commercial value of seal colonies by 1840. Around 1870, technological improvements led to the growth of pelagic whaling with the development of faster, steam-powered catching vessels, and whaling in the Southern Ocean entered a new era in the early years of the 20th century, which saw the industrialization of whale exploitation. In 1904, the first shore station was built at Grytviken. In 1912, about 11,000 whales were killed annually to be processed at six Antarctic shore stations, a level deemed to be unsustainable (Suarez, 1927). Over the following decades (1904-1960s), more than 2 million large whales were caught, reducing their populations to less than 35 per cent of their initial numbers and 16 per cent of their original biomass (Laws, 1977; Clapham et al.,1999).

The ensuing extraction of other resources followed the same pattern as in other parts of the world; from the highest trophic levels down the trophic web (Kock, 2007; Ainley and Pauly, 2014). The impacts of the reduction to less than 20 per cent of their original size of several fish stocks by 1980, stocks which are not experiencing significant recovery despite management actions by the Convention for the Conservation of Antarctic Marine Living Resources (CCAMLR) (Ainley and Blight, 2008; Marschoff et al., 2012) are still felt in spite of significant harvest reductions (including fisheries closures in large areas). Species' relationships might be altered by harvesting: for example, evidence exists that the decrease in demersal fish in the fishing area was followed by a long-term increase in populations of benthic octopods (Vecchione et al., 2009).

Although the targets and intensity of harvesting have shifted over the years, the ecosystem effects continue. Removal of large predators, such as seals and especially whales, reduces predation pressure on species in the mid-level of the food web, including fish and squid. Some of the large predators (e.g., baleen whales, fur and crabeater seals) feed directly on krill, whereas others (e.g., toothed whales) feed on krill predators. Therefore, the ecosystem effects of sealing, whaling and fishing are potentially complex and were initially related to the "whale reduction" or "krill surplus" hypothesis (Sladen, 1964), in which the outcome from the dramatic exploitation of whale stocks was a presumed excess of food (krill) which was being redistributed throughout the system. Although sealing has ceased and historical levels of whaling have been reduced (a reduced harvest of baleen whales continues), impacts from the reduced levels of these top-level species arguably still reverberate through the pelagic ecosystem.

The reduction of whaling brought about increased harvesting of fish. Rapid reduction of the target fish stocks was a major reason for the adoption of the CCAMLR, which implemented international fisheries and ecosystem research and resulted in a moratorium on bottom fishing for notothenioids. Kock and Jones (2007) reviewed the fisheries data for the primary fishing area around the Antarctic Peninsula. They found that populations of several fish species declined as a result of the fishery primarily targeting mackerel icefish, *Champsocephalus gunnari*, and marbled notothenia, *Notothenia rossii*. Since the CCAMLR moratorium was implemented in 1989-90 for the South Shetland area, populations of several fish species have recovered, but not of the two main target species. The observed recovery in *Notothenia rossii* in the South Shetland Islands shelf is taking longer than the objective set by CCAMLR of two or three decades and *Gobionotothen gibberifrons* remains at low levels (Barrera-Oro and Marschoff, 2007).

Finfish fishing was conducted by bottom trawling until 1985, when regulation of by-catch of depleted finfish species moved the *Champsocephalus gunnari* fishery to introduce midwater trawls. Since 2008, fisheries where the fishing gear interacts with the bottom (e.g., longlining and demersal trawling) are subject to mitigation measures to protect Vulnerable Marine Ecosystems, for example the positions where substantial amounts of VME indicator species are encountered are closed to fishing (see CCAMLR – Conservation Measures at www.ccamlr.org). Further, marine reserves have been established by France and Australia (Welsford et al. 2011; Falguier and Marteau 2011) and in South Orkney Islands where fishing is restricted.

Antarctic fur seals may have recovered (and may even have become overpopulated on South Orkneys during the late 1990s (Hodgson et al., 1998)) after being severely depleted, but some breeding rookeries have reached carrying capacity well below historical records (Hucke-Gaete et al., 2004). Trends in the ice-breeding seals are difficult to establish (Southwell et al., 2012). In other circumstances it might become impossible to determine the causes behind the observed population trends. For example, it is difficult (if not impossible) to disentangle the effects of climate change, recovery of seals, and variations in krill availability on the population trends observed for pack-ice seals (Trathan et al., 2012). However, recent evidence suggests that climate change might actually be responsible for the declining trend in Antarctic fur seals, where food stress provoked by climate variation has significantly reduced female longevity, juvenile and adult survival, fecundity and pup birth weight, among other symptoms, since 2003, after a three-decade monitoring programme of biometric, life history and genetic aspects (Forcada and Hoffman, 2014).

A decrease in density of krill (*Euphausia superba*) and a correlated increase in salp abundance has been suggested from the analysis of net samples (Atkinson et al., 2004). Krill decrease has also been inferred by stable isotope studies in krill predators (e.g., Huang et al., 2011).

The regulation of Antarctic fisheries under CCAMLR operates in the framework of the Antarctic Treaty System. Since its inception (1980), CCAMLR requires the application of the ecosystem approach, aiming to limit the changes induced by the fisheries to those reversible in two to three decades. Catch limits and *inter alia*, fishing methods and data collection requirements are established by a Comission, based on the assessments and advice provided by the Scientific Committee. To date, no methods of catch allocation among members are in place. Several of the 25 Commission Members and 11 acceding do not participate in harvesting. However, the management developed along 30 years has proved to be effective in the sense that this organization is highly regarded in terms of the achievement of conservation objectives (Cullis-Suzuki and Pauly, 2010).

The krill fishery is the largest in the Southern Ocean. Recent annual catch has exceeded 200,000 tons. The fishery developed on a relatively small scale in the 1970s, but rapidly increased during the 1980s to a peak of >500,000 tons/year (Nicol et al., 2012). This is actually much less than the precautionary catch limit set by CCAMLR at a total of over 8.6 million tons. Therefore, krill are considered to be "underexploited", but the fishery is expanding and management methods to take into account ecosystem considerations are under development (e.g., CCAMLR, 2013, paragraph 5.5; SC-CAMLR, 2013, paragraphs 3.11 to 3.27). A trigger level (a level that cannot be exceeded until more advanced management procedures are in place) of 620,000 tons throughout the main fishing ground is being applied by CCAMLR. However, ecosystem effects of the removal of large numbers of krill remain to be determined, especially when considered in light of climate change.

9   Climate change

In addition to harvesting, the other major pressure on Antarctic biota is the changing climate. The Scientific Committee on Antarctic Research (SCAR) produced a comprehensive Antarctic Climate Impact Assessment (Turner et al. (eds.), 2009). The following discussion is largely based on this report.

For the past 50 years the Antarctic marine ecosystem has been affected by climate change, especially on the western side of the Peninsula, with its warming water and declining sea ice. Westerly winds around the continent have increased by 20 per cent since the 1970s and surface air temperature has increased over the Antarctic Peninsula. Information from ice cores suggest that warming started around 1800. The Antarctic Circumpolar Current temperature increased by approximately 0.5°C between 300 m to 1000 m. Böning et al. (2008) analyzed historical and recent data from drifting buoys, finding that the wind-driven Antarctic Circumpolar Current has not augmented its transport, but reported warming and freshening of the current on a hemispherical scale extending below 1000 m, meaning that transport and meridional overturning are insensitive to changes in wind stress. Although the response of the Antarctic Circumpolar Current and the carbon sink to wind-stress changes is under debate, it has been suggested (Hallberg and Gnanadesikan, 2006; Meredith and Hogg, 2006) that the Antarctic Circumpolar Current's response to an increase in wind is a change in eddy activity rather than a change in transport. Given the importance of the Antarctic Circumpolar Current and its system of eddies in structuring the pelagic ecosystem, the consequences of these changes cannot be foreseen.

Ship observations suggest that the extent of sea ice was greater in the first half of the twentieth century, but satellite measurements from 1979 to 2006 show a positive trend of around 1 per cent per decade. The greatest increase, at around 4.5 per cent per decade, occurred in the Ross Sea; the reduction in sea-ice cover affected the Bellingshausen sea.

The pelagic ecosystem was affected by the consequences of the regional sea-ice reduction. Krill population has not increased after the near-extinction of some whale stocks. Although predation by seals and birds increased, the total bird and seal biomass remains only a fraction of that of the former whale population (Flores et al., 2012). The krill stock, of which 150 million tons were being eaten by whales, would have been an estimated three times larger in the pre-whaling time. Commensurate primary production would be around that estimated for the North Sea, not leaving much for other grazers and copepods. This means that phytoplankton also decreased, but the details of the phenomenon are still unclear.

Sea bird monitoring in the Scotia Sea has shown a significant decline in the abundance of krill predators, such as the cape petrel, *Daption capense*, the southern fulmar, *Fulmarus glacialoides,* and Wilson's storm petrel, *Oceanites oceanicus*; other species with generalist diets have increased their number: the Antarctic Prion, *Pachyptila desolata,* and the Black-browed Albatross, *Thalassarche melanophris* (Orgeira and Montalti, 1998). Other non-Antarctic species, such as the white-chinned petrel, *Procellaria aequinoctialis,* have extended their pelagic ranges further south, (Montalti et al., 1999).

At least a conceptual model of the structure and functioning of the ecosystem is necessary to understand these phenomena. *In-situ* iron fertilization experiments demonstrated that iron, as a micronutrient, may limit phytoplankton growth even in presence of large concentrations of nitrate and phosphate. In the whale feeding grounds, krill stocks were close to, if not at, the carrying capacity of the ecosystem prior to whaling; this is consistent with the frequent observations in the 1930s of krill swarms at the surface, an observation now seldom made from tourist or scientific vessels.

Estimates of krill abundance derived from the analysis of net samples indicate a decline of up to 81 per cent in the krill stock (Atkinson et al., 2004) and an increase in salp populations, suggesting the replacement of krill by salps and of the typical short food chain of diatoms-krill-higher predators by the longer food chain implied in the microbial food webs to which salps are better adapted. The actual dimension of these changes is currently under debate, because of the large difficulties associated with the analysis of the simultaneous effects of whale depletion, sea-ice retreat at one of the most important recruitment sites of krill (the western Antarctic Peninsula), iron-limited phytoplankton growth, and more complex ecological phenomena (Ainley et al., 2007; Nicol et al., 2007). The lower rate of recycling of iron in the microbial planktonic food web when compared to the short diatom-krill-predators "chain of the giants" may also contribute to the reduction in iron. Large predators also contribute to iron recycling while accumulating blubber and excreting nutrients in surface waters; a significant proportion of plankton biomass is degraded below the euphotic zone. Thus the productive "chain of the giants" may have maintained itself via recycling the nutrients at a rate compatible with the growth of phytoplankton.

With the decline in sea ice, more phytoplankon blooms should be supplying food to benthic organisms on the shelf. A resulting increase in phytodetritus on the shelf may cause a decline in suspension feeders adapted to limited food supplies, and to their associated fauna. The positive correlation between the extent and duration of sea-ice cover over krill reproduction and survival (Loeb et al., 1997), the negative trends of sea-ice extent (Stammerjohn et al., 2008) and the overall decrease in krill biomass over the last decades (Siegel and Loeb 1995; Atkinson et al., 2004) would be expected to have profound implications for the Southern Ocean food web and is the most relevant issue affecting krill-dependent fauna particularly. When ice shelves collapse, the changes from a unique ice-shelf-covered ecosystem to a typical Antarctic shelf ecosystem, with high primary production during a short summer, are likely to be among the largest ecosystem changes on the planet, a process that seems to develop faster than was previously thought (Gutt et al., 2013).

Another expected impact of climate change is the change in pH levels, with seawater becoming more acid. It seems likely that the skeletons of planktonic pteropoda and of cold water corals will become thinner. Hatching rates of krill eggs are also demonstrated to be negatively affected by the level of ocean acidification projected for the end of the century and beyond (Kawaguchi et al., 2013). The Southern Ocean is at higher risk from this than other oceans, because it has low saturation levels of $CaCO_3$.

10   Invasive species

The slow rates of growth and endemism of Antarctic species may lead to the establishment of non-indigenous species, probably restricted by their own physiological limits. The incomplete taxonomic knowledge of the Antarctic biota will make it difficult to recognize whether a particular specimen is the result of a natural southern distribution limit or an invasive species. Examples include: the occasional findings of anomuran and brachyuran larvae in the South Shetland Islands (Thatje and Fuentes, 2003); *Euphausia superba* in Chilean fjords; Antarctic diatoms in Tasmania, etc. (Clarke et al., 2005).

There is concern that several factors associated with ocean warming and increased vessel activity (scientific expeditions, tourism, fisheries, etc.) in the Antarctic increase the risk of the introduction of alien species and even pathogens (Kerry and Riddle, 2009). In the crab-eater population of the Antarctic Peninsula, one-third of the population carries antibodies to the canine distemper (Bengtson et al., 1991), attributed to contagion from sled dogs, which were removed from the Antarctic Treaty Area. The introduction of non-native living organisms is banned, except in accordance with a permit.

11 Contamination

At the local level, contaminants from coastal stations are introduced through waste water, dump sites and particulates from the activity of stations and ships. Persistent organic pollutants (POPs) have been found in water, sediments and organisms in the vicinity of several stations (e.g., UNEP, 2002; Bargagli, 2005). Since 1991, the Protocol on Environmental Protection to the Antarctic Treaty[1] has imposed severe restrictions and regulations on disposal and treatment of wastes and emissions from stations and tourism vessels. Thus, locally originated contamination is not expected to become a significant problem.

Global contamination reaches Antarctica through the global circulation of the oceans. Persistent pollutants are transported and biomagnified, these include DDT (dichlorodiphenyltrichloroethane) and other organophospates. While DDT has been little used since the 1970s a possible source of DDT maintaining high levels in penguin populations is glacier ablation (Geisz et al., 2008). Anthropogenic radionuclides stemming from above-ground nuclear bomb testing are also present throughout Antarctica, including evidence of the Chernobyl nuclear accident (Dibb et al., 1990), and have even been used to provide dating controls within long-lived biological systems (Clarke, 2008). Snow samples enabled the reconstruction of lead pollution of Antarctica that started as early as the 1880s, related to non-ferrous metal production activities in South America, South Africa and Australia and coal-powered ships that crossed Cape Horn en route between the Atlantic and Pacific Oceans. Lead pollution declined in the 1920s, correlated with the opening of the Panama Canal in 1914, and decreased from the mid-1980s because of lead-free modern cars. Antarctica is significantly contaminated with other metals, such as Cr, Cu, Zn, Ag, Bi and U, as a consequence of long-distance transport from the surrounding continents.

---

[1] United Nations, *Treaty Series*, No. 5778.

References

Ainley, D.G. and DeMaster, D.P. (1990). The upper trophic levels in polar marine ecosystems. In: Smith, W.O. Jr. (ed.), *Polar Oceanography, Part B: Chemistry, biology, and geology*, pp. 599–630. Academic Press, San Diego, California.

Ainley, D.G., Ribic, C.A. and Fraser, W.R. (1992). Does prey preference affect habitat choice in Antarctic seabirds. *Marine Ecology Progress Series*, 90: 207–221.

Ainley, D.G. (2002). *The Adélie Penguin: Bellwether of Climate Change*. New York: Columbia University Press, 310 pp.

Ainley, D.G., Tynan, C.T., and Stirling, I. (2003). Sea ice: A critical habitat for polar marine mammals and birds. In: Thomas, D. N. and Dieckmann, G. S. (eds.) *Sea Ice: An Introduction to Its Physics, Chemistry, Biology, and Geology*. Oxford, Blackwell Science, 240–266, 2003.

Ainley, D.G., Ballard, G., Ackley, S., Blight, L., Eastman, J.T., Emslie, S.D., Lescroel, A., Olmastroni, S., Townsend, S.E., Tynan, C.T., Wilson, P. and Woehler, E. (2007). Paradigm lost, or is topdown forcing no longer significant in the Antarctic marine ecosystem? *Antarctic Science*, 19, 283-290.

Ainley, D.G. and Blight, L.K. (2008). Ecological repercussions of historical fish extraction from the Southern Ocean. *Fish and Fisheries*, 9, 1–26.

Ainley, D.G. and Pauly, D. (2014). *Fishing down the food web of the Antarctic continental shelf and slope. Polar Record*. pp.1-16. Cambridge. Cambridge University Press. doi:10.1017/S0032247412000757.

Allcock, A.L., Barratt, I., Eleaume, M., Linse, K., Norman, M.D., Smith, P.J., Steinke, D., Stevens, D.W. and Strugnell, J.M. (2011). Cryptic speciation and the circumpolarity debate: A case study on endemic Southern Ocean octopuses using the COI barcode of life. *Deep-Sea Research II* 58:242-249.

Archer, S.D., Leakey, R.J.G., Burkill, P.H., Sleigh, M.A. and Apple, C.J. (1996). Microbial ecology of sea ice at a coastal Antarctic site: community composition, biomass and temporal change. *Marine Ecology Progress Series*, vol. 135, 179-195.

Arntz, W.E., Brey, T. and Gallardo, V.A. (1994). Antarctic zoobenthos, *Oceanography and Marine Biology – An Annual review*, 32, 241-304.

Arntz, W.E., Gutt, J. and Klages, M. (1997). Antarctic marine biodiversity: an overview. In: Battaglia, B. (ed.) *Antarctic communities: species, structure and survival*. Cambridge University Press, 3-14.

Aronson, R.B. and Blake, D.B. (2001). Global Climate Change and The Origin of Modern Benthic Communities in Antarctica. *American Zoologist*, 41:27–39.

Atkinson, A., Siegel, V., Pakhomov, E., Rothery, P. (2004). Long-term decline in krill stock and increase in salps within the Southern Ocean. *Nature*, 432: 100–103.

Atkinson, A., Ward, P., Hunt, B.P.V., Pakhomov, E.A. and Hosie, G.W. (2012). An Overview of Southern Ocean Zooplankton Data: Abundance, Biomass, Feeding and Functional Relationships. *CCAMLR Science*, Vol. 19: 171–218.

Azam, F., Beers, J.R., Campbell, L., Carlucci, A.F., Holm-Hansen, O., Reis, F.M.H. and Karl, D.M. (1979). Occurrence and metabolic activity of organisms under the Ross Ice Shelf, Antarctica, at station J9. *Science*, 203, 451- 453.

Azam, F., Smith, D.C. and Hollibaugh, J.T. (1991). The role of the microbial loop in Antarctic pelagic ecosystems. *Polar Research*, 10: 239–244.

Azzaro, M., La Ferla, R., Azzaro, F. (2006). Microbial respiration in the aphotic zone of the Ross Sea (Antarctica). *Marine chemistry*, 99 (1): 199-209.

Balech, E. (1968). Dinoflagellates. *American Geographical Society; Antarctic Map Folio Series* 10:8-9.

Bargagli, R. (2005). *Antarctic ecosystems: environmental contamination, climate change, and human impact*. Berlin: Springer, 395 pp.

Barker, P.F., Filippelli, G.M., Florindo, F., Martin, E.E. and Scher, H.D. (2007). Onset and Role of the Antarctic Circumpolar Current. *Deep Sea Research Part II: Topical Studies in Oceanography*, Vol. 54, Issues 21–22, 2388–2398.

Barnes, D.K.A. (1995). Sublittoral epifaunal communities at Signy Island, Antarctica. II. Below the ice-foot zone. *Marine Biology* 121:565–572.

Barrera-Oro, E.R. (2002). The role of fish in the Antarctic marine food web: differences between inshore and offshore waters in the southern Scotia Arc and west Antarctic Peninsula. *Antarctic Science*, 14 (4): 293–309.

Barrera-Oro, E.R. and Marschoff, E.R. (2007). Information on the status of fjord Notothenia rossii, *Gobionotothen gibberifrons* and *Notothenia coriiceps* in the lower South Shetland Islands, derived from the 2000–2006 monitoring program at Potter Cove. *CCAMLR Science*, 14, 83–87.

Behrenfeld, M.J. and Falkowski, P.G. (1997). Photosynthetic rates derived from satellite-based chlorophyll concentration. *Limnology and Oceanography*, 42(1): 1-20, DOI: 10.4319/lo.1997.42.1.0001.

Béja, O., Koonin, E.V., Aravind, L., Taylor, L.T., Seitz, H., Stein, J.L., Bensen, D.C., Feldman, R.A., Swanson, R.V., Delong, E.F. (2002). Comparative genomic analysis of archaeal genotypic variants in a single population and in two different oceanic provinces. *Applied Environmental Microbiology*, 68: 335–345.

Bengtson, J.L., Boveng P., Franzén U., Have P., Heide-Jorgensen M.P., Harkonen T.J. (1991). Antibodies to canine distemper virus in Antarctic Seals. *Marine Mammal Science*, 7 (1):85-87.

Bilyk, K.T. and DeVries, A.L. (2011). Heat tolerance and its plasticity in Antarctic fishes. *Comparative Biochemistry and Physiology*, Part A 158, 382–390.

Böning, C.W., Dispert, A., Visbeck, M., Rintoul, S. and Schwarzkopf, F.U. (2008). Response of the Antarctic Circumpolar Current to Recent Climate Change." *Nature Geoscience*, 1: 864-69.

Bonner, W.N. (1984). Conservation and the Antarctic. pp. 821-850 in Laws RM, ed. *Antarctic Ecology*. vol. 2(15) Academic Press.

Bost, C.A., Cotté, C., Bailleul, F., Cherel, Y., Charrassin, J.B., Guinet, C., Ainley, D.G., Weimerskirch, H. (2009). The importance of oceanographic fronts to marine birds and mammals of the southern oceans. *Journal of Marine Systems*, 78: 363–376.

Boyd, I.L. (2009). Antarctic Marine Mammals in: William F. Perrin, Bernd Wursig, J.G.M. Thewissen(eds.) *Encyclopedia of Marine Mammals*, 2nd. Edition. Elsevier Academic Press.

Branch, T. A., Matsuoka, K. and Miyashita, T. (2004). Evidence for increases in Antarctic blue whales based on Bayesian modelling. *Marine Mammal Science*, 20:726-754.

Brandt, A., De Broyer, C., Gooday, A.J., Hilbigd, B., Thomson, M.R.A. (2004). Introduction to ANDEEP (Antarctic benthic DEEP-sea biodiversity: colonization history and recent community patterns)—a tribute to Howard L. Sanders *Deep-Sea Research II*, 51:1457–1465.

Branch, T.A. (2011). Humpback whale abundance south of 60°S from three complete circumpolar sets of surveys. *Journal of Cetacean Research and Management*, 3: 53-69.

Brownell, R.L. Jr. (1974). Small odontocetes of the Antarctic. In: V.C. Bushnell (ed.) *Antarctic Map Folio Series*, folio 18, pp. 13-19. New York: American Geographical Society.

Buma, A.G.J., Gieskes W.W.C, Thomsen, H. (1992). Abundance of cryptophyceae and chlorophyll b-containing organisms in the Weddell-Scotia Confluence area in the spring of 1988". *Polar Biology*. 12(1):43-52.

Calvo, J., Morriconi, E. and Rae, G.A. (1999). Reproductive biology of the icefish *Champsocephalus esox* (Gunther, 1861) (Channichthyidae). *Antarctic Science* I I (2): 140-149.

Catalano, G., G. Budillon, G., La Ferla, R., Povero, P., Ravaioli, M., Saggiomo, V., Accornero, A., Azzaro, M., Carrada, G.C., Giglio, F., Langone, L., Mangoni, O., Misic, C. and Modigh, M. (2010). The Ross Sea. In: K.-K. Liu, L. Atkinson, R. Quiñones, L. Talue-McManus (eds.), *Carbon and Nutrient Fluxex in Continental Margins*. Springer –Verlag, The IGBP Series, 303-318.

CCAMLR (2013). *Report of the Thirty-second Meeting of the Commission (CCAMLR-XXXII)*. CCAMLR, Hobart, Australia.

Clapham, P.J., Young, S.B., Brownell Jr., R. (1999). Baleen whales: conservation issues and the status of the most endangered populations. *Mammal Review*, 29: 35-60.

Clarke, A. and Johnston, N.M. (2003). Antarctic marine benthic diversity, *Oceanography and marine Biology: An Annual Review*, 41, 47-114.

Clarke, A., Barnes, D.K.A., Hodgson, D.A. (2005). How isolated is Antarctica? *Trends in Ecology and Evolution*, 20 (1), 1-3.

Clarke, L.J. 2008. Resilience of the Antarctic moss *Ceratodon purpureus* to the effects of elevated UV-B radiation and climate change. PhD thesis, U. Wollongong; L.S. Peck and T. Brey 1996, *Nature*, 380: 207-208.

Clements, J.F. (2000). *Birds of the World: a Checklist*. Cornell University Press. 880p.

Comiso, J.C. and Gordon, A.L. (1996). Cosmonaut polynya in the Southern Ocean: Structure and variability, *Journal of Geophysical Research-Oceans*, 101, Issue C8, 18297-18313.

Coria, N.R., Spairani, H., Vivequin, S. and Fontana, R. (1995). Diet of Adélie penguins *Pygoscelis adeliae* during the post-hatching period at Esperanza Bay, Antarctica, 1987/88. *Polar Biology*, 15: 415–418.

Coria, N.R., Montalti, D., Rombolá, E.F., Santos, M.M., Garcia Betoño, M.I. & Juares, M.A. (2011). Birds at Laurie Island, South Orkney Islands, Antarctica: breeding species and their distribution. *Marine Ornithology*, 39: 207–213.

Croxall, J.P. and Lishman G.S. (1987). The food and feeding ecology of penguins. In: Croxall JP (ed.) *Seabirds and role in marine ecosystems*. Cambridge University Press, Cambridge, pp. 101– 133.

Croxall, J.P., Trathan, P.N., Murphy, E.J. (2002). Environmental change and Antarctic seabird populations. *Science*, 297(5586), 1510-1514.

Cullis-Suzuki and Pauly, S.D. (2010). Failing the high seas: A global evaluation of regional fisheries management organizations *Marine Policy*, Volume 34, Issue 5, 1036-1042.

Church, M.J., Delong, E.F., Ducklow, H.W., Karner, M.B., Preston, C.M. and Karl, D.M. (2003). Abundance and distribution of planktonic Archaea and Bacteria in the waters west of the Antarctic Peninsula. *Limnology and Oceanography*, 48: 1893–1902.

Dayton, P.K., Robbiliard G.A., Paine R.T., Dayton L.B. (1974). Biological accommodation in the benthic community at McMurdo Sound, Antarctica. *Ecological Monographs*, 44:105–128.

DeBroyer, C., Danis, B. with 64 SCAR-MarBIN Taxonomic (eds.) (2011). How many species in the Southern Ocean? Towards a dynamic inventory of the Antarctic marine species. *Deep-Sea Research II*, 58:5–17.

Delille, D. and Rosiers, C. (1996). Seasonal changes of Antarctic marine bacterioplankton and sea ice bacterial assemblages. *Polar Biology*, 13:463-470.

DeLong, E.F., Wu, K.Y., Prézelin, B.B., Jovine, R.V. (1994). High abundance of Archaea in Antarctic marine picoplankton. *Nature*, 371: 695–697.

Dibb, J., Mayewski, P.A., Buck, C.F. and Drummey, S.M. (1990). Beta radiation from snow, *Nature*, 344 (6270), 25.

Ducklow, H., Carlson, C., Church, M., Kirchman, D., Smith, D. and Steward, G. (2001). The seasonal development of the bacterioplankton bloom in the Ross Sea, Antarctica, 1994–1997. *Deep Sea Research II*, 48: 4199–4221.

Ducklow, H.W., Baker, K., Martinson, D.G., Quetin, L.B., Ross, R.M., Smith, R.C., Stammerjohn, S.E., Vernet, M. and Fraser, W. (2007). Marine pelagic ecosystems: the West Antarctic Peninsula. *Philosophical Transactions of the Royal Society* B 362:67-94.

Ducklow, H., Clarke, A., Dickhut, R., Doney, S.C., Geisz, H., Huang, K., Martinson, D.G., Meredith, M.P., Moeller, H.V., Montes-Hugo, M., Schofield, O., Stammerjohn, S.E., Steinberg, D., Fraser, W. (2012). The Marine System of the Western Antarctic Peninsula, In: A.D. Rogers, N.M. Johnston, E.J. Murphy and A. Clarke (eds.) *Antarctic Ecosystems: An Extreme Environment in a Changing World*. John Wiley & Sons, Ltd, Chichester, UK. doi: 10.1002/9781444347241.ch5.

Ducklow, H.W., Fraser, W.R., Meredith, M.P., Stammerjohn, S.E., Doney, S.C., Martinson, D.G., Sailley, S.F., Schofield, O.M., Steinberg, D.K., Venables, H.J. and Amsler C.D. (2013). West Antarctic Peninsula: An ice-dependent coastal marine ecosystem in transition. *Oceanography*, 26(3):190–203, http://dx.doi.org/10.5670/oceanog.2013.62.

Duhamel, G., Hulley, P.A., Causse, R., Koubbi, P., Vacchi, M., Pruvost, P., Vigetta, S., Irisson, J.O., Mormède, S., Belchier, M., Dettai, A., Detrich, H.W., Gutt, J., Jones, C.D., Kock, K.H., Lopez Abellan, L.J., and Van de Putte, A. (2014). Biogeographic Patterns of Fish in De Broyer, C., Koubbi, P., Griffiths, H.J., Raymond, B., Udekem d'Acoz, C. d', Van de Putte, A.P., Danis, B., David, B., Grant, S., Gutt, J., Held, C., Hosie, G., Huettmann, F., Post, A., Ropert-Coudert, Y. (eds.), 2014. *Biogeographic Atlas of the Southern Ocean*. Scientific Committee on Antarctic Research, Cambridge, Chapter 7.

Eastman, J.T. (1995). The evolution of Antarctic fishes: questions for consideration and avenues for research. *Cybium*, 19, 371–389.

Egevang, C., Stenhouse, I.J., Phillips, R.A., Petersen, A., Fox, J.W., Silk, J.R. (2010). Tracking of Arctic terns Sterna paradisaea reveals longest animal migration. *Proceedings of the National Academy of Sciences*, 107(5), 2078-2081.

El-Sayed, S.Z. (1968a). On the productivity of the Southwest Atlantic Ocean and the waters west of the Antarctic Peninsula. *Biology of the Antarctic Seas III*, vol. 11:15-47.

El-Sayed, S.Z. (1968b). Productivity of antarctic and subantarctic waters. *American Geographical Society*; *Antarctic Map Folio Series* 10:8-9.

El-Sayed, S.Z. (1970). On the productivity of the Southern Ocean. Antarctic Ecology. M.W. Holdgate (ed.). *Academic Press I*, 119-135.

El-Sayed, S.Z. and Hofmann, E. (1986). Drake Passage and Western Scotia Sea (Antarctica). In: Hovis, W.A. (ed.) *Nimbus-7 CZSC coastal zone color scanner imagery for selected coastal regions*. NASA, pp. 97-99.

Faranda, F.M., L. Guglielmo and A. Ianora (eds.) (2000). *Ross Sea Ecology*, Springer-Verlag, Berlin, 1-604.

Flores, H., Atkinson, A., Kawaguchi, S., Krafft, B.A., Milinevsky, G., Nicol, S., Reiss, C., Tarling, G.A., Werner, R., Bravo Rebolledo, E., Cirelli, V., Cuzin-Roudy, J., Fielding, S., Groeneveld, J.J., Haraldsson, M., Lombana, A., Marschoff, E., Meyer, B., Pakhomov, E.A., Rombolá, E., Schmidt, K., Siegel, V., Teschke, M., Tonkes, H., Toullec, J.Y., Trathan, P.N., Tremblay, N., Van de Putte, A.P., van Franeker, J.A., Werner, T. (2012). Impact of climate change on Antarctic krill. *Marine Ecology Progress Series*, 458: 1-19

Falguier, A. and Marteau, C. (2011). The management of the marine reserve of the Terres australes françaises (French Southern Lands). In: Duhamel, G. and Welsford, D.C. (eds.) The Kerguelen Plateau: Marine Ecosystem and Fisheries. Paris: Société française d'ichtyologie, pp. 293-296.

Forcada, J., Trathan, P.N., Reid, K., Murphy, E.J., Croxall, J.P. (2006). Contrasting population changes in sympatric penguin species in association with climate warming.

Global Change Biology, 12(3), 411-423.

Forcada, J. and Hoffman, J.I. (2014). Climate change selects for heterozygosity in a declining fur seal population. *Nature*, 511: 462-465.

Fuhrman, J.A. and Azam, F. (1980). Bacterioplankton secondary production estimates for coastal waters of British Columbia, Antarctica, and California. *Applied Environmental Microbiology*, 39:1085-1095.

Geisz, H.N., Dickhut, R.M., Cochran, M.A., Fraser, W.R. and Ducklow, H.W. (2008) Melting Glaciers: A Probable Source of DDT to the Antarctic Marine Ecosystem. *Environmental Science & Technology*, 2008, 42 (11), pp 3958–3962.DOI: 10.1021/es702919n

Gorny, M., Arntz, W.E., Clarke, A. and Gore, D.J. (1992). Reproductive biology of caridean decapods from the Weddell Sea. *Polar Biology* 12:111-120.

Griffiths, H.J., Barnes, D.K.A. and Linse, K. (2009). Towards a generalized biogeography of the Southern Ocean benthos. *Journal of Biogeography*, 36, 162-177.

Gröhsler, T. (1994). Feeding habits as indicators of ecological niches: Investigations of Antarctic fish conducted near Elephant Island in late autumn/winter 1986. *Archive of Fishery and Marine Research*, 42, 17–34.

Gruber, N., Gloor, M., Mikaloff-Fletcher, S.E., Doney, S.C., Dutkiewicz, S., Follows, M.J., Gerber, M., Jacobson, A.R., Joos, F., Lindsay, K., Menemenlis, D., Mouchet, A., Muller, S.A., Sarmiento, J.L. and Takahashi, T. (2009). Oceanic sources, sinks and transport of atmospheric CO_2. *Global Biogeochemical Cycles*, 23: doi: 10.1029/2008GB003349.

Grzymski, J.J., Carter, B.J., DeLong, E.F., Feldman, R.A., Ghadiri, A., Murray, A.E. (2006). Comparative genomics of DNA fragments from six Antarctic marine planktonic bacteria. *Applied Environmental Microbiology*, 72: 1532–1541.

Grzymski, J.J., Riesenfeld, C.S., Williams, T.J., Dussaq, A.M., Ducklow, H., Erickson, M., Cavicchioli, R. and Murray, A.E. (2012). A metagenomic assessment of winter and summer bacterioplankton from Antarctica Peninsula coastal surface waters. *The ISME Journal*, 6: 1901–1915.

Guglielmo, L., Carrada, G.C., Catalano, G., Dell'Anno, A., Fabiano, F., Lazzara, L., Mangoni, O., Pusceddu, A. and Saggiomo, V. (2000). Structural and functional properties of sympagic communities in the annual sea ice at Terra Nova Bay (Ross Sea, Antarctica). *Polar Biology*, 23, 137–146.

Gutt, J. (2000). Some "driving forces" structuring communities of the sublittoral Antarctic macrobenthos, *Antarctic Science*, 12 (3), 297-313.

Gutt, J., Sirenko, B.I., Smirnov, I.S. and Arntz, W.E. (2004). How many macrobenthic species might inhabit the Antarctic shelf? *Antarctic Science*, 16, 11-16.

Gutt, J. (2007). Antarctic macro-zoobenthic communities: a review and an ecological classification, *Antarctic Science*, 109 (2), 165-182.

Gutt, J., Hosie, G. and Stoddart, M. (2010). Marine life in the Antarctic. In: McIntyre, A.D. (ed.). *Life in the World's Oceans: Diversity, Distribution, and Abundance*. Wiley-Blackwell, Oxford, UK, doi: 10.1002/9781444325508.ch11.

Gutt, J., Cape, M., Dimmler, W., Fillinger, L., Isla, E., Lieb, V., Lundälv, T. and Pulcher, C. (2013). Shifts in Antarctic megabenthic structure after ice-shelf disintegration in the Larsen area east of the Antarctic Peninsula." *Polar Biology*, 36, no. 6:895-906.

Hallberg, R. and Gnanadesikan, A. (2006). The Role of Eddies in Determining the Structure and Response of the Wind-Driven Southern Hemisphere Overturning: Results from the Modeling Eddies in the Southern Ocean (MESO) Project. *Journal of Physical Oceanography*, 36: 2232-2252.

Hardy, A.C. and Gunther, E.R. (1935). The plankton of the South Georgia whaling grounds and adjacent waters, 1926-1927. *Discovery Reports*, vol. XI: 1-456.

Harris, C.M., Carr, R., Lorenz, K. and Jones, S. (2011). *Important Bird Areas in Antarctica: Antarctic Peninsula, South Shetland Islands, South Orkney Islands – Final Report*. Prepared for BirdLife International and the Polar Regions Unit of the UK Foreign & Commonwealth Office. Environmental Research & Assessment Ltd., Cambridge.

Heywood, K.J., Sparrow, M.D., Brown, J., Dickson, R.R. (1999). Frontal structure and Antarctic Bottom Water flow through the Princess Elizabeth Trough, Antarctica, *Deep-Sea Research I*, 46, 1181-1200.

Hodgson, D.A., Johnston, N.M., Caulkett, A.P., Jones, V.J. (1998). Palaeolimnology of Antarctic fur seal *Arctocephalus gazella* populations and implications for Antarctic management. *Biological Conservation*, 83(2): 145-154.

Huang, Y.M., Amsler, M.O., McClintock, J.B., Amsler, C.D., Baker, B.J. (2007). Patterns of gammaridean amphipod abundance and species composition associated with dominant subtidal macroalgae from the western Antarctic Peninsula. *Polar Biology*, 30(11), 1417-1430.

Huang, T., Sun, L., Stark, J., Wang, Y., Cheng, Z., et al. (2011). Relative Changes in Krill Abundance Inferred from Antarctic Fur Seal. *PLoS ONE*, 6(11): e27331. doi:10.1371/journal.pone.0027331.

Hucke-Gaete, R., Osman, L.P., Moreno, C.A. and Torres, D. (2004). Examining natural population growth from near extinction: the case of the Antarctic fur seal at the South Shetlands, Antarctica. *Polar Biology*, 27: 304-311.

Iken, K., Amsler C.D., Amsler, M.O., McClintock J., Baker B.J. (2011). Field studies on deterrent properties of phlorotannins in Antarctic brown algae. In: Wiencke C (Ed.). *Biology of Polar Benthic Algae, Marine and Freshwater Botany*. Berlin/New York: Walter de Gruyter GmbH & Co. K.G.

Jefferson, T., Webber, M., Pitman, R. (2008). *Marine Mammals of the World: A Comprehensive Guide to their Identification*. San Diego, CA: Academic Press.

Jenouvrier, S., Weimerskirch, H., Barbraud, C., Park, Y. H., Cazelles, B. (2005). Evidence of a shift in the cyclicity of Antarctic seabird dynamics linked to climate. *Proceedings of the Royal Society B: Biological Sciences*, 272(1566), 887-895.

Joiris, C.R. and Dochy, O. (2013). A major autumn feeding ground for fin whales, southern fulmars and grey-headed albatrosses around the South Shetland Islands, Antarctica. *Polar Biology*, 36(11), 1649-1658.

Karl, D.M. (1993). Microbial processes in the southern oceans. In: Friedmann E.I. (ed.) *Antarctic Microbiology*, 1-63.

Kawaguchi, S., Ishida, A., King, R., Raymond, B., Waller, N., Constable, A., Nicol, S., Wakita, M., Ishimatsu, A. (2013) Risk maps for Antarctic krill under projected Southern Ocean acidification. *Nature Climate Change* 3:843-847.

Mintenbeck, K., Barrera-Oro, K.R., Brey, T., Jacob, U., Knust, R., Mark, F.C., Moreira, E., Strobel, A., Arntz, W.E. (2012). Impact of Climate Change on Fishes in Complex Antarctic Ecosystems. In: Jacob, U. and Woodward, G. (eds.) *Advances In Ecological Research*, Vol. 46, Burlington: Academic Press, pp. 351-426.

Kennett, J.P. (1977). Cenozoic evolution of Antarctic glaciation, the Circum-Antarctic ocean, and their impact on global paleoceanography. *Journal of Geophysical Research*, 82: 3843-3860.

Kerry, K.R. and Riddle, M. (2009). Health of Antarctic Wildlife: An Introduction. In: Kerry, K.R. and Riddle, M. (eds.). *Health of Antarctic Wildlife*. Springer-Verlag Berlin Heidelberg. 470 pp.

Knox, G. (2007). *The biology of the Southern Ocean*. Cambridge, UK: Cambridge University Press.

Kock, K.-H. (1992). *Antarctic Fish and Fisheries*. Cambridge University Press Cambridge, New York: 359 pp.

Kock, K.-H. (2005). Antarctic icefishes (Channichthyidae): a unique family of fishes. A review, Part 1. *Polar Biology*, 28: 862–895.

Kock, K.-H. (2007). Antarctic Marine Living Resources – exploitation and its management in the Southern Ocean. *Antarctic Science*, 19 (2): 231–238.

Kock, K.-H. and Jones, C.D. (2007). Fish stocks in the southern Scotia Arc region – A review and prospects for future research. *Reviews in Fisheries Science* 13:75-108.

Kock, K.-H., Barrera-Oro, E., Belchier, M., Collins, M.A., Duhamel, G., Hanchet, S., Pshenichnov, L., Welsford, D. and Williams, R. (2012). The role of fish as predators of krill (*Euphausia superba*) and other pelagic resources in the Southern Ocean. *CCAMLR Science*, Vol. 19: 115–169.

Kozlov, A.N. (1995). A Review of the Trophic Role of Mesopelagic Fish of the Family Myctophidae in the Southern Ocean Ecosystem. *CCAMLR Science*, Vol. 2: 71-77.

Laws, R.M. (1977). Seals and whales of the Southern Ocean. *Philosophical Transactions of the Royal Society Biological Sciences, B.* 279: 81-96.

Libertelli, M., Coria, N. and Marateo, G. (2003). Diet of the Adélie penguin during three consecutive chick rearing periods at Laurie Island. *Polish Polar Research*, 24: 133–142.

Lizotte, M.A. (2001). The Contributions of Sea Ice Algae to Antarctic Marine Primary Production. *American Zoologist*, 41:57–73.

Loeb, V., Siegel, V., Holm-Hansen, O., Hewitt, R., Fraser, W., Trivelpiece, W.Z., Trivelpiece, S. (1997). Effects of sea-ice extent and krill or salp dominance on the Antarctic food web. *Nature*, 387: 897-900.

Lubimova, T.G., Shust, K.V., Troyanovski F.M. and Semenov, A.B. (1983). To the ecology of mass species of myctophids from the Antarctic Atlantic. In: *Soviet Committee of Antarctic Research*. The Antarctic. The Committee Report, 22: 99-106.

Lumpkin, R. and Speer, K. (2007). Global ocean meridional overturning, *Journal of Physical Oceanography*, 37, 2550-2562.

Lynch, H.J., Fagan, W.F., Naveen, R. (2010). Population trends and reproductive success at a frequently visited penguin colony on the western Antarctic Peninsula. *Polar Biology*, 33(4), 493-503.

Mackintosh, N.A. (1965). *The stocks of whales*. London: Fishing News (Books) Limited.

Maksym, T., Stammerjohn, S.E., Ackley, S.and Massom, R. (2012). Antarctic sea ice—A polar opposite? *Oceanography* 25(3):140–151.

Manganelli, M., Malfatti, F., Samo, T.J., Mitchell, B.G., Wang, H. and Azam, F. (2009). Major role of microbes in carbon fluxes during austral winter in the Southern Drake Passage. *PLoS ONE*, DOI: 10.1371/journal.pone.0006941.

Marschoff, E.R., Barrera-Oro, E.R., Alescio, N.S., Ainley, D.G. (2012).Slow recovery of previously depleted demersal fish at the South Shetland Islands, 1983–2010. *Fisheries Research*, 125– 126: 206– 213.

Matschiner, M., Hanel, R., Salzburger, W. (2011). On the origin and trigger of the notothenioid adaptive radiation. *PLoS ONE*, 6, e18911.

Mccartney, M.S. and Donohue, K.A. (2007). A deep cyclonic gyre in the Australian-Antarctic Basin, *Progress in Oceanography*, 75, 675-750.

McCook, L.J. and Chapman, A.R.O. (1993). Community succession following massive ice-scour on a rocky intertidal shore: recruitment, competition and predation during early, primary succession. *Marine Biology*, 115:565–575.

Meredith, M.P. and Hogg, A.M. (2006). Circumpolar response of Southern Ocean eddy activity to a change in the Southern Annular Mode, *Geophysical Research Letter*, 33 (16): L16608, doi: 10.1029/2006GL026499.

Micol, T. and Jouventin, P. (2001). Long-term population trends in seven Antarctic seabirds at Pointe Géologie (Terre Adélie), Human impact compared with environmental change. *Polar Biology*, 24(3), 175-185.

Mincks, S.L., Smith, C.R. and Demaster, D.J. (2005). Persistence of labile organic matter and microbioal biomass in Antarctic shelf sediments: evidence of a sediment 'food bank', *Marine Ecology Progress Series*, 300, 3-19.

Mincks, S.L. and Smith, C.R. (2007).Recruitment patterns in Antarctic Peninsula shelf sediments: evidence of decoupling from seasonal phytodetritus pulses. *Polar Biology* 30:587–600.

Montalti, D., Orgeira, J.L. and Di Martino, S. (1999). New records of vagrant birds in the South Atlantic and in Antarctic. *Polish Polar Research*, 20: 4 347-354.

Moore, J.K. and Abbott, M.R. (2000). Phytoplankton chlorophyll distributions and primary production in the Southern Ocean. *Journal of Geophysical Research*, 105(C12), 28709–28722, doi:10.1029/1999JC000043.

Moreira, D., Rodríguez-Valera, F. and López-García, P. (2004). Analysis of a genome fragment of a deep-sea uncultivated Group II euryarchaeote containing 16S rDNA, a spectinomycin-like operon and several energy metabolism genes. *Environmental Microbiology*, 6: 959–969.

Murray, A.E., Preston, C.M., Massana, R., Taylor, L.T., Blakis, A., Wu, K. and Delong, E.F. (1998). Seasonal and spatial variability of bacterial and archaeal assemblages in the coastal waters off Anvers Island, Antarctica, *Applied and Environmental Microbiology*, 64, 2585-2595.

Murray, A.E. and Grzymski, J.J. (2007). Diversity and genomics of Antarctic marine micro-organisms, *Philosophical Transactions of the Royal Society Biological Sciences*, 362, 2259-2271.

Nicol, S., Croxall, J., Trathan, P., Gales, N. and Murphy, E. (2007). Paradigm misplaced? Antarctic marine ecosystems are affected by climate change as well as biological processes and harvesting, *Antarctic Science*, 19, 291-295.

Nicol, S., Foster, J. and Kawaguchi, S. (2012). The fishery for Antarctic krill – recent developments. *Fish and Fisheries*, 13:30-40.

Nybelin, O. (1947). Antarctic fishes. *Scientific Results of the Norwegian Antarctic Expedition*, 26, 1–76.

Olguín, H. and Alder, V.A. (2011). Species composition and biogeography of diatoms in Antarctic and Subantarctic (Argentine shelf) waters (37-76ºS). *Deep-Sea Research II*, vol. 58, 139–152.

Orgeira, J.L. and Montalti, D. (1998). Autumn seabird observations on the South Shetland Islands. *Hornero*, 15 (1): 60-64. Buenos Aires. ISSN 0073-3407.

Orsi, A.H., Whitworth, T. and Nowling, W.D. (1995). On the meridional extent and fronts of the Antarctic Circumpolar Current. *Deep Sea Research*, Series I, 42, 641–673.

Parkinson, C.L. and Cavalieri, D.J. (2012). Antarctic sea ice variability and trends, 1979–2010 *The Cryosphere*, 6, 871–880.

Pearce, I., Davidson, A.T., Bell, E.M. and Wright, S. (2007). Seasonal changes in the concentration and metabolic activity of bacteria and viruses at an Antarctic coastal site. *Aquatic Microbial Ecology*, 47: 11–23.

Pearce, D. (2008). Biodiversity of the bacterioplankton in the surface waters around Southern Thule in the Southern Ocean. *Antarctic Science*, 20, 291-300.

Phillips, R.A., Croxall, J.P., Silk, J.R.D., Briggs, D.R. (2008). Foraging ecology of albatrosses and petrels from South Georgia: two decades of insights from tracking technologies. *Aquatic Conservation: Marine and Freshwater Ecosystems*, 17, S6–S21.

Pommier, T., Pinhassi, J. and Hagström, A. (2005). Biogeographic analysis of ribosomal RNA clusters from marine bacterioplankton. *Aquatic Microbial Ecology*, 41:79-89.

Pugh, P.J.A. and Davenport, J. (1997). Colonisation vs. disturbance: the effects of sustained ice-scouring on intertidal communities. *Journal of Experimental Marine Biology and Ecology*, 210:1–21.

Quartino, M.L. and Boraso, de Zaixso A.L. (2008). Summer macroalgal biomass in Potter Cove, South Shetland Islands, Antarctica: its production and flux to the ecosystem. *Polar Biology*, 31: 281–294.

Quartino, M.L., Deregibus D., Campana G.L., Latorre G.E.J., Momo F.R. (2013). Evidence of macroalgal colonization on newly ice-free areas following glacial retreat in Potter Cove (South Shetland Islands), Antarctica. *PLoS ONE*, 8(3): e58223. doi:10.1371/ journal.pone. 0058223.

Ratcliffe, N. and Trathan, P. (2011). A Review of The Diet And At-Sea Distribution Of Penguins Breeding Within The Camlr Convention Area. *CCAMLR Science*, Vol. 18: 75–114.

Riaux-Gobin, C., Tréguer, P., Poulin, M. and Vétion, G. (2000). Nutrients, algal biomass and communities in land-fast ice and seawater off Adélie Land (Antarctica). *Antarctic Science*, 12:160-171.

Ribic, C.A. and Ainley, D.G. (1988). Constancy of seabird species assemblages: an exploratory look. *Biological Oceanography*, 6, 175-202.

Riesenfeld, C.S., Murray, A.E. and Baker, B.J. (2008). Characterization of the microbial community and polyketide biosynthetic potential in the Palmerolide producing tunicate, *Synoicum adareanum*. *Journal of Natural Products*, 71, 1812-1818.

Rintoul, S.R., Hughes, C.W. and Olbers, D. (2001). The Antarctic Circumpolar Current system. In: G. Siedler, J. Church and J. Gould (eds.), *Ocean circulation and climate; observing and modelling the global ocean*. International Geophysics Series, 77, 271-302, Academic Press.

Rombolá, E., Marschoff, E. and Coria, N. (2003). Comparative study of the effects of the late pack -ice break-off on chinstrap and Adélie penguins' diet and reproductive success at Laurie Island, South Orkney Islands, Antarctica. *Polar Biology*, 26: 41–48.

Rombolá, E., Marschoff, E. and Coria, N. (2006). Interannual study of chinstrap penguin's diet and reproductive success at Laurie Island, South Orkney Islands, Antarctica. *Polar Biology*, 29: 502–509.

Ropert-Coudert, Y., Hindell M.A., Phillips R., Charassin J.B., Trudelle L., Raymond B. (2014). CHAPTER 8. BIOGEOGRAPHIC PATTERNS OF BIRDS AND MAMMALS. In: De Broyer C., Koubbi P., Griffiths H.J., Raymond B., Udekem d'Acoz C. d', et al. (eds.). *Biogeographic Atlas of the Southern Ocean*. Scientific Committee on Antarctic Research, Cambridge, pp. 364-387.

Sabourenkov, E. (1991). *Myctophids in the diet of Antarctic predators*. Selected Scientific Papers, 1991 (SC-CAMLR-SSPI8). CCAMLR, Hobart, Australia: 335-360.

Santora, J.A. and Veit, R.R. (2013). Spatio-temporal persistence of top predator hotspots near the Antarctic Peninsula. *Marine Ecology Progress Series*, 487, 287-304.

SC-CAMLR (2013). *Report of the Thirty-second Meeting of the Scientific Committe (SC-CAMLR-XXXII)*.CCAMLR, Hobart, Australia.

Schiaparelli, S. and Hopcroft, R. (eds.) (2011). Census of Antarctic Marine Life: Diversity and change in the Southern Oceans Ecosystems. *Deep-Sea Res. II* 58(1-2): 1-276.

Schnack-Schiel, S.B., Thomas, D., Dahms, H.-U., Haas, C. and Mizdalski, E. (1998). Copepods in Antarctic sea ice. In M. P. Lizotte and K. R. Arrigo (eds.), Antarctic Sea ice: Biological processes, interactions, and variability. *Antarctic Research Series* 73: 173–182.

Siegel, V. and Loeb, V. (1995). Recruitment of Antarctic krill *Euphausia superba* and possible causes for its variability. *Marine Ecology Progress Series* 123: 45-56.

Sladen, W.J.L. (1964). The distribution of the Adelie and chinstrap penguins. Pp. 359-365 In: Carrick, R., M.W. Holdgate and J. Prevost (eds.) *Antarctic Biology*. Paris: Hermann.

Smale, D.A., Barnes, D.K.A., Fraser, K.P.P., Peck, L.S. (2008). Benthic community response to iceberg scouring at an intensely disturbed shallow water site at Adelaide Island, Antarctica. *Marine Ecology Progress Series* 355: 85–94.

Smith, K.L., Robison, B.H., Helly, J.H., Kaufmann, R.S., Ruhl, H.A. Shaw, T.S., Twinning, B.S. and Vernet, M. (2007). Free-drifting icebergs: Hot spots of chemical and biological enrichment in the Wedell Sea. *Science*, 317:478-482.

Smith, N.R., Dong, Z. Kerry, K.R. and Wright, S. (1984). Water masses and circulation in the region of PrydzBay, *Antarctica, Deep-sea Research*, 31: 1121-1147.

Smith, W.O. and Comiso, J.C. (2008). Influence of sea ice on primary production in the Southern Ocean: A satellite perspective. *Journal of Geophysical Research*, 113, C05S93, doi:10.1029/2007JC004251.

Smith, W.O., Peloquin, J.A. and Karl, D.M. (2010). Antarctic Continental Margins. In: K.-K. Liu, L. Atkinson, R. Quiñones, L. Talue-McManus (eds.), *Carbon and Nutrient Fluxes in Continental Margins*. Springer –Verlag, The IGBP Series, 318-330.

Southwell, C., Bengtson, J., Bester, M., Blix, A.S., Bornemann, H., Boveng, P., Cameron, M., Forcada, J., Laake, J., Nordøy, E., Plötz, J., Rogers, T., Southwell, D., Steinhage, D., Stewart, B.S. and Trathan, P. (2012). A review of data on abundance, trends in abundance, habitat use and diet of ice-breeding seals in the Southern Ocean. *CCAMLR Science*, Vol. 19: 49–74.

Speer, K., Rintoul, S.R. and Sloyan, B. (2000). The diabatic Deacon cell. *Journal of Physical Oceanography*, 30: 3212–3222.

Stammerjohn, S.E., Martinson, D.G., Smith, R.C., Yuan, X., Rind, D. (2008). Trends in Antarctic annual sea ice retreat and advance and their relation to El Niño–Southern Oscillation and Southern Annular Mode variability. *Journal of Geophysical Research*, 113: C03S90.

Steinberg, D.K., Martinson, D.G. and Costa, D.P. (2012). Two decades of pelagic ecology of the western Antarctic Peninsula. *Oceanography*, 25:56-67.

Strugnell, J.M., Rogers, A.D., Prodohl, P.A., Collins, M.A., and Allcock, A.L. (2008). The thermohaline expressway: the Southern Ocean as a centre of origin for deep-sea octopuses. *Cladistics*, 24:108.

Strugnell, J., Cherel, Y., Cooke, I.R., Gleadall, I.G., Hochberg, F.G., Ibáñez, C.M., Jorgensen, E., Laptikhovsky, V.V., Linse, K., Norman, M., Vecchione, M., Voight, J.R. and Allcock, A.L. (2011). The Southern Ocean: source and sink? *Deep Sea Research II*. 58:196-204.

Strutton, P.G., Lovenduski, N.S., Mongin, M., Matear, R. (2012). Quantification of Southern Ocean Phytoplankton Biomass and Primary Productivity via Stellite Observations and Biogeochemical Model, *CCAMLR Science*, 19: 247-265.

Suarez, J.L. (1927). *Rapport au Conseil de la Société des Nations. Exploitation des Richesses de la Mer*. Publications de la Société des Nations V. Questions Juridiques. V.1. 120:125.

Tate Regan, C. (1914). Fish. British Antarctic (Terra Nova) Expedition 1910. *Natural History Report. Zoology*, 1, 125-156.

Thatje, S. and Fuentes, V. (2003). First record of anomuran and brachyuran larvae (Crustacea: Decapoda) from Antarctic waters, *Polar Biology* 26 (4): 279-282.

Thomas, D.N. and Dieckmann, G.S. (2003). *Sea Ice: An Introduction to its Physics, Chemistry, Biology and Geology*. Wiley-Blackwell, 1-416.

Trathan, P.N., Ratcliffe and Masden, E.A. (2012). Ecological drivers of change at South Georgia: the krill surplus, or climate variability. *Ecography*, 35: 983–993.

Trathan, P.N., Grant, S.M., Siegel V. and Kock, K.-H. (2013). Precautionary spatial protection to facilitate the scientific study of habitats and communities under ice shelves in the context of recent, rapid, regional climate change. *CCAMLR Science*, Vol. 20: 139–151.

Tréguer, P. and Jacques, G. (1992). Dynamics of nutrients and phytoplankton, and fluxes of carbon, nitrogen and silicon in the Antarctic Ocean. *Polar Biology*, 12:149-162.

Trivelpiece, W.Z., Hinke, J.T., Miller, A.K., Reiss, C.S., Trivelpiece, S.G., & Watters, G.M.

(2011). Variability in krill biomass links harvesting and climate warming to penguin population changes in Antarctica. *Proceedings of the National Academy of Sciences*, 108(18), 7625-7628.

Tuck, G.N., Polacheck, T. and Bulman, C.M. (2003). Spatio-temporal trends of longline fishing effort in the Southern Ocean and implications for seabird bycatch. *Biological Conservation*, 114(1), 1-27.

Turner, J., Bindschadler, R.A., Convey, P., di Prisco, G., Fahrbach, E., Gutt, J., Hodgson, D., Mayewski, P. and Summerhayes, C. (eds.) (2009). *Antarctic Climate Change and the Environment.* Scientific Committee on Antarctic Research, Cambridge, 526 p.

UNEP (2002). *Regionally Based Assessment of Persistent Toxic Substances.* Antarctica Regional Report, Global Evironoment Facitiy.

Vacchi, M., DeVries, A.L., Evans, C.W., Bottaro, M., Ghigliotti, L., Cutroneo, L. and Pisano, E. (2012). A nursery area for the Antarctic silverfish *Pleuragramma antarcticum* at Terra Nova Bay (Ross Sea): first estimate of distribution and abundance of eggs and larvae under the seasonal sea-ice. *Polar Biology*, 35, 1573-1585.

Vecchione, M., Piatkowski, U. and Allcock, A.L. (1998). Biology of the cirrate octopod Grimpoteuthis glacialis (Cephalopoda; Opisthoteuthidae) in the South Shetland Islands, Antarctica. *South African Journal of Marine Science* 20:421-428.

Vecchione, M., Piatkowski, U., Allcock, A.L., Jorgensen, E. and Barratt, I. (2009). Persistent elevated abundance of octopods in an overfished Antarctic area. pp. 197-203 In: Krupnik, I. et al. (eds.) *Smithsonian at the Poles: Contributions to International Polar Year Science.* Smithsonian Institution Scholarly Press, Washington, DC.

Vernet, M., Martinson, D., Iannuzzi, R., Stammerjohn, S., Kozlowski, W., Sines, K., Smith, R.C. and Garibotti, I. (2008). Primary production within the sea-ice zone west of the Antarctic Peninsula: I—Sea ice, summer mixed layer, and irradiance. *Deep Sea Research Part II*, 55:2,068–2,085, http://dx.doi.org/10.1016/j.dsr2.2008.05.021.

Vernet, M., Smith, K.L., Cefarelli, A.O., Helly, J.J., Kaufmann, R.S., Lin, H., Long, D.G., Murray, A.E., Robison, B.H., Ruhl, H.A., Shaw, T.J., Sherman, A.D., Sprintall, J., Stephenson, G.R., Stuart, K.M. and Twinning B.S. (2012). Islands of ice: Influence of free-drifting icebergs on pelagic marine ecosystems. *Oceanography*, 25:38-39.

Wang, Z. and Meredith, M.P. (2008). Density-driven Southern Hemisphere subpolar gyres in coupled climate models, *Geophysical Research Letters*, 35(14) 5, pp. 10.1029/2008GL034344.

Webster, N.S., Smith, L.D., Heyward, A.J., Watts, J.E., Webb, R.I, Blackall, L.L., Negri, A.P. (2004). Metamorphosis of a scleractinian coral in response to microbial biofilms. *Applied and Environmental Microbiology* 70 (2): 1213-1221.

Webster, N.S. and Bourne, D. (2007). Bacterial community structure associated with the Antarctic soft coral, Alcyonium antarcticum, Fems. *Microbiol Ecology*, 59, 81-94.

Weimerskirch, H., Inchausti, P., Guinet, C., Barbraud, C. (2003). Trends in bird and seal populations as indicators of a system shift in the Southern Ocean. *Antarctic Science*, 15(2), 249-256.

Welsford, D.C., Constable, A.J. and Nowara, G.B. (2011). The Heard Island and McDonald Islands Marine Reserve and Conservation Zone – A model for Southern Ocean Marine Reserves? In: Duhamel, G. and Welsford, D.C. (eds.) The Kerguelen Plateau: Marine Ecosystem and Fisheries. Paris: Société française d'ichtyologie, pp. 297-304.

White, M.G. (1984). Marine benthos. In: Laws, R.M. (ed.) *Antarctic Ecology*. vol. 2. Academic Press, London, 421-461.

Wienecke, B.C. and Robertson, G. (1997). Foraging space of emperor penguins Aptenodytes forsteri in Antarctic shelf waters in winter. *Marine Ecology-Progress Series* 159, 249–263.

Wiencke, B.C. and Clayton, M.N. (2002). Antarctic Seaweeds, In: J. W. Wägele and J. Sieg (eds.) *Synopses of the Antarctic Benthos*. Ruggell: Gantner ; Königstein: Koeltz Scientific Books., 239 p.

Wiencke, B.C. and Amsler, C.D. (2012). Seaweeds and their communities in polar regions. In: Wiencke, C. and K. Bischof (eds.) *Seaweed Biology*, pp.265-291, Springer, Berlin Heidelberg.

Woehler, E.J., Cooper, J., Croxall, J.P., Fraser, W.R., Kooyman, G.L., Miller, G.D., Nel, D.C., Patterson, D.L., Peter, H.-U., Ribic, C.A., Salwicka, K., Trivelpiece, W.Z., Weimerskirch, H. (2001). *A statistical assessment of the status and trends of Antarctic and Subantarctic seabirds*. SCAR.

Wulff, A., Iken, K., Quartino, M.L., Al-Handal, A., Wiencke, C. , Clayton M.N. (2011). Biodiversity, biogeography and zonation of marine benthic micro- and macroalgae in the Arctic and Antarctic. (Capítulo 3, 23-52). In: Wiencke, C. (ed.) *Biology of Polar Benthic Algae*. De Gruyter, Berlin. pp. 23-52.

Yakimov, M. M., Giuliano, L., Gentile, G., Crisafi, E., Chernikova, T.N., Abraham, W.-R., Lünsdorf, H., Timmis K.N. and Golyshin P.N. (2003). Oleispira antarctica gen. nov., sp. nov., a novel hydrocarbonoclastic marine bacterium isolated from Antarctic coastal sea water. *International Journal of Systematic and Evolutionary Microbiology*, 53: 779–785.

Zacher, K., Rautenberger, R., Hanelt, D., Wulff, A., Wiencke, C. (2009). The abiotic environment of polar marine benthic algae. *Botanica Marina*, 52(6), 483-490.

Zimmer, I., Wilson, R.P., Gilbert, C., Beaulieu, M., Ancel, A., Plötz, J. (2008). Foraging movements of emperor penguins at Pointe Géologie, Antarctica. *Polar Biology*, 31, 229–243.

B

Marine Ecosystems, Species and Habitats Scientifically Identified as Threatened, Declining or Otherwise in need of Special Attention or Protection

Marine Species

37 Marine Mammals

Contributors:
Tim D. Smith (Convenor), John Bannister, Ellen Hines, Randall Reeves, Jake Rice (Lead Member and Editor of Part VI), Lorenzo Rojas-Bracho and Peter Shaughnessy

1 Introduction

Marine mammals occupy a wide range of marine and some freshwater habitats around the world. They have been used by humans for millennia for food and to obtain other products. Marine mammals consist of cetaceans (whales, dolphins and porpoises), pinnipeds (seals, sea lions and walruses), sirenians (dugongs and manatees), mustelids (sea otters and marine otters) and the polar bear – 130 or more species in total, including several that range into fresh water and some that exclusively occupy rivers and lakes. Human interactions, both direct and indirect, have negatively affected most marine mammal species at least to some degree. Historically, industrial harvesting greatly reduced the abundance of many populations. Although the intensity of such exploitation has declined in recent decades for many species, humans continue to use certain species of marine mammals in some places for food, skins, fur, ivory and increasingly as bait for fisheries. In addition, human activities continue to reduce the availability and quality of marine mammal habitats and cause substantial numbers of marine mammals to die incidentally as a result of entanglement or entrapment in fishing gear and from being struck by vessels. Mitigation of ongoing and future threats to marine mammals from human activities requires improved knowledge of those human activities and of the animals' ecology, behaviour and habitat use.

1.1 Changes in Biological Diversity

Globally, certain species of marine mammals have become extinct over the last several centuries, i.e., Steller's sea cow, the Japanese sea lion, the Caribbean monk seal. The Yangtze River dolphin (baiji) is also likely to have become extinct although there has been one unconfirmed sighting in 2005. In addition, many populations have been reduced to remnant status, such that they no longer play a significant role in the ecosystem, i.e., they are functionally extinct. For example, right whales have essentially disappeared from the eastern North Atlantic, and walruses are gone from their former strongholds in south-eastern Canada (Gulf of St. Lawrence and Sable Island). The loss of marine mammal diversity due to actual and functional extinctions has had significant effects on marine ecosystems at varying scales from local to ocean-basin-wide (Estes et al., 2006).

1.2 Magnitude of changes

Besides the above-mentioned extinctions and extirpations, the numerical abundance (and biomass) of many other marine mammal populations has been greatly reduced. All of the commercially valuable great whales were depleted by whaling. A good example is the world's largest animal, the blue whale. Within the 20th century more than 360,000 Antarctic blue whales were taken, leaving a remnant population of only hundreds of animals and a current abundance of roughly 5 per cent of original numbers (Branch et al., 2007). Some populations of virtually all types of marine mammals have been depleted to low levels; for example, southern and northern hemisphere populations of fur seals and elephant seals were all depleted by harvesting for either their fur or their oil (Busch, 1985). Similarly, many populations of sirenians and sea otters have been reduced to small fractions of historic levels.

Although the abundance of almost all populations of marine mammals has been reduced by human activities, a number of them have recovered since they were protected from deliberate exploitation. Eastern Pacific gray whales, northern elephant seals, eastern Steller sea lions, humpback whales and some populations of right whales provide some of the most clear-cut examples. Certain populations of those species, however, have not recovered and remain at small fractions of historic levels (e.g., western Pacific gray whales, Arabian Sea humpback whales, western Steller sea lions, North Atlantic right whales, and southern right whales in some areas).

2 Population trends or conservation status

2.1 Aggregated at global scale

Because of the great diversity of marine mammal species and their habitats, it is difficult to characterize their conservation status and population trends in the aggregate or at a global scale. Many of the marine species suffered major declines over the past several hundred years as a result of commercial hunting. Massive changes in human demography and economy have unquestionably affected the environmental carrying capacity, particularly in coastal regions, and as a result less suitable habitat (including forage base) is available to support marine mammal populations. These changes make 'full' recovery unfeasible for some species.

2.2 Major taxonomic and/or geographic subdivisions

2.2.1 Large whales

Commercial exploitation began as long ago as the 10th century, and by the late 19th century intensive whaling had caused severe depletion, and even near-extinction, of some species and populations. Industrial mechanized whaling in the 20th century led to further major declines. Some populations of large whales have been recovering in recent decades: for example, humpback whales globally, blue whales in some regions and southern hemisphere right whales when treated as a single group (Figure 37.1, from IWC, 2013a). At the same time, many populations have failed to recover to anywhere near their original abundance. For example, right whales are effectively extirpated from the eastern North Atlantic, and are only barely surviving in the eastern North Pacific the eastern South Pacific, and around New Zealand (Jackson et al., 2011).

2.2.2 Pelagic dolphins and porpoises

Pelagic (off-shore) dolphins are generally less susceptible to human interactions than many marine mammals because they are relatively small, of little commercial importance, wide-ranging (some species are

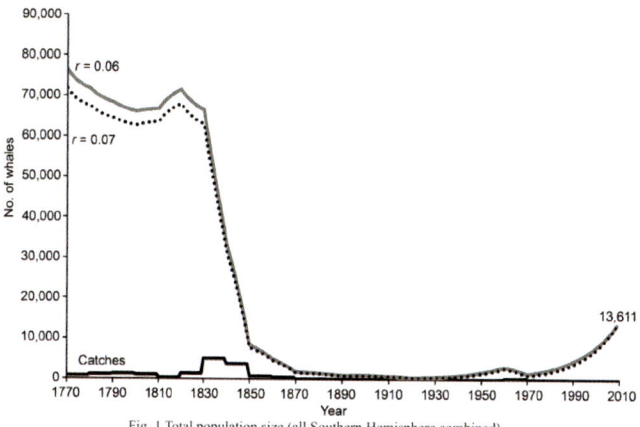

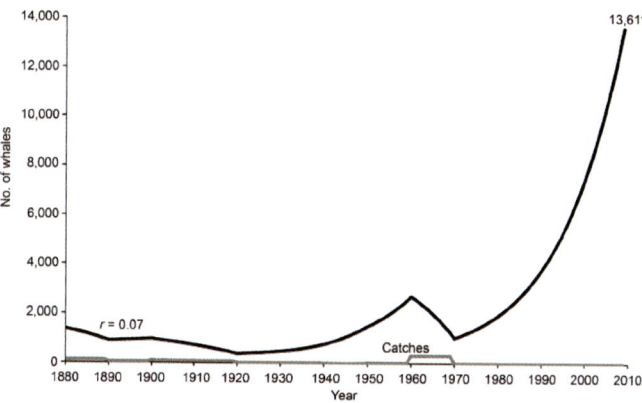

Figure 37.1 | Catches (solid dark line) and estimated population size for southern right whales from 1770 to 2010, assuming a maximum annual net rate of increase of 6 per cent (grey line) and 7 percent (dotted line), culminating in an estimated population size of 13,600 in 2010 (IWC, 2013a).

also extremely numerous) and live far from most human activities. There are potential interactions with military activity, and clear interactions with fishing gear in the eastern tropical Pacific. Several dolphin species in that region have been significantly affected by humans because they have symbiotic relationships with other pelagic animals that are of commercial interest, notably yellowfin tuna. Purse seine fishermen learned to chase and encircle these dolphins to facilitate the capture of tuna, and in the past large numbers of dolphins drowned in the nets. This mortality greatly reduced the abundance of several dolphin species in the latter half of the 20th century. Due to international efforts, fishing methods have changed and the by-catch has been reduced significantly (Hedley, 2010, NMFS, undated).

2.2.3 Coastal and Estuarine dolphins and porpoises

Coastal dolphins and porpoises are hunted for food and bait by a number of States. Furthermore, they often die accidentally in fishing gear in coastal commercial and artisanal fisheries. This has contributed to declines in abundance, at times to critical levels. For example, vaquitas in the Gulf of California (Mexico) declined by 57 per cent from 1997-2008, and continue to decline (Gerrodette et al., 2011). In New Zealand, the Māui dolphin (a subspecies of Hector's dolphin) has declined to very low

levels (Currey et al., 2012; Hamner et al., 2012). In both cases, there have been management interventions aimed at minimizing direct mortalities and entanglements. However, under current levels of direct mortalities and entanglements, these trends are expected to continue, with both the vaquita and the Māui dolphin at risk of becoming extinct within the next few decades (CIRVA 2014).

Several species of small cetaceans live in estuaries and in large rivers, often in close proximity to humans, and are threatened by artisanal fisheries, pollution, and coastal development. Two species of South American river dolphins are relatively widespread and at least locally common in Amazonia and Orinoquia: the tucuxi and the boto (Trujillo et al., 2010). The endemic blind river dolphins of the South Asian subcontinent still inhabit large portions of the Indus, Ganges, Brahmaputra and Karnaphuli rivers. Although they are locally common in a few places, their numbers and ranges have declined markedly over the past century. Three relict populations of 100 or fewer Irrawaddy dolphins persist in the Ayeyarwady (Irrawaddy) River of Myanmar, the Mahakam River of eastern Borneo, Indonesia, and the Mekong River of Cambodia and Laos. All three populations are at high risk of extirpation from a variety of human activities (Kreb et al., 2010). The baiji (*Lipotes vexillifer*), endemic to the Yangtze River, (China), is likely to have become extinct in the early 21st century (Turvey et al., 2007), and the Yangtze population of narrow-ridged finless porpoise (*Neophocaena phocaenoides asiaeorientalis*) is declining rapidly, probably due to mortality in fishing gear, vessel strikes and habitat degradation (Wang et al., 2013).

2.2.4 Pinnipeds

Populations of many species of commercially exploited pinnipeds were greatly reduced in abundance from the 18th to the 20th century. They were targeted mainly for the oil from their insulating blubber and for their fur. Some species, such as the Caribbean monk seal and the Japanese sea lion, were driven to extinction, and others, such as the Mediterranean monk seal, as well as fur seals in many areas, were reduced to very low numbers. The hooded seal in the Greenland Sea has been hunted down to 10-15 per cent of its original population. However, some regional and ocean-basin-wide pinniped populations have been recovering, such as the harp seal in Canada and grey and common seals in the United Kingdom of Great Britain and Northern Ireland. In some cases they have been recolonizing habitat where they had long been absent, for example, the New Zealand fur seal has expanded into former habitats in New Zealand and Australia (Shaughnessy et al., 2014). In other cases populations have increased to historical levels, such as the Antarctic krill-eating seals (Kovacs et al., 2012).

2.2.5 Sirenians

Manatees and dugongs are the only totally aquatic herbivorous mammals, feeding primarily on submerged vegetation. Sirenians have a global range in about 90 tropical and subtropical countries. They have been and continue to be hunted, trapped, and netted for food, and human activities of many kinds have modified or otherwise degraded their habitat. Overall, populations have been greatly reduced in terms of both abundance and range. One species discovered in 1741, Steller's sea cow, was hunted to extinction in less than a quarter of a century after its discovery. The present status varies regionally (Marsh et al. 2014, Hines et al. 2014).

2.2.6 Mustelids

Two species of mustelids are truly marine: the sea otter (*Enhydra lutris*), along the Pacific coast of North America and East Asia, and the marine otter (*Lontra felina*), along the Pacific coast of South America. Both species are very coastal and they have been extensively exploited for their pelts. The sea otter has recovered substantially in many areas of its original distribution, whereas the marine otter has not, as it continued to be harvested through much of the 20th century. Its contracted range is fragmented due to various kinds of human interaction (IUCN, 2015).

2.2.7 Polar bears

Polar bears are endemic to high latitudes of the northern hemisphere. They have a circumpolar distribution and depend on both sea ice (for hunting pinnipeds and for denning and reproduction in some areas) and land (for hunting hauled-out pinnipeds such as walruses, as well as for denning and reproduction). Most populations have been subjected to extensive killing, at least historically, mainly for meat, hides, and sport, and to protect human life and property. Commercial hunting is now prohibited, although substantial numbers of bears are killed every year (legally) by aboriginal people in Alaska (United States of America), Canada and Greenland (Denmark). The main long-term and range-wide threat to polar bears is the projected loss of sea ice habitat associated with climate change. Sea ice provides essential breeding habitat for ringed seals, the principal prey of polar bears, and the bears rely on ice as a hunting platform to gain access to the seals. Limited access to food leads to mobilization of fat reserves and the release of hazardous substances stored in fat, which are transported by the blood circulation to vital organs such as liver and brain. High levels of contaminants have been associated with negative health effects (e.g. enzyme activation, hormonal disturbance and weakened immune systems) in polar bears in Svalbard (Norway), Greenland (Denmark), and Hudson Bay, Canada (Atkinsson et al 1996).

2.3 Special Conservation Status Issues

2.3.1 Sources of conservation information

International Whaling Commission: The conservation issues associated with large whales and, to a lesser extent, small cetaceans are addressed in the reports of the Scientific Committee of the International Whaling Commission. Besides directed hunting, these reports increasingly have addressed other topics of concern, including habitat degradation, ship strikes, climate change, fisheries by-catch, noise disturbance, and ecological interactions. The status of many of the great whales is summarized at www.iwc.int/status#species.

International Union for Conservation of Nature (IUCN): The IUCN Red List is an authoritative and regularly updated source of conservation information for all species and subspecies of marine mammals and for many threatened subpopulations. This information is easily accessible (www.redlist.org). The authority for listing determinations resides within the relevant Specialist Groups (SG) of the IUCN Species Survival Commission: the Pinniped SG, Cetacean SG, Sirenian SG, Otter SG and Polar Bear SG.

National and regional sources: In addition to the international sources of conservation information, some nations undertake research in support of management. Information obtained from such research is often available online. Major sources include Australia, Canada, New Zealand and the United States. The United States National Marine Fisheries Service and Fish and Wildlife Service provide updated Stock Assessment Reports on marine mammals under their respective jurisdictions (www.nmfs.noaa.gov/pr/sars; www.fws.gov/alaska/fisheries/mmm/stock/stock.htm). The United Sates Marine Mammal Commission also provides much information on marine mammal science and conservation (www.mmc.gov). In Canada, the Department of Fisheries and Oceans publishes information on population and ecosystem status (www.isdm-gdsi.gc.ca/csas-sccs). In addition, the Committee on the Status of Endangered Wildlife in Canada posts status reports on all marine mammal species of concern in Canada (www.sararegistry.gc.ca). Australia provides information on cetaceans, pinnipeds and dugongs that occur in areas within national jurisdiction (www.environment.gov.au/topics/marine/marine-species). New Zealand undertakes its own assessments (Hitchmough, 2010). In addition, a regional body in the North Atlantic, the North Atlantic Marine Mammal Commission, provides information on marine mammals in that region (www.nammco.no).

2.3.2 Potential value of management methods

Regulation of directed and indirect takes: The most obvious approach to management of human activities that result in the killing of marine mammals is to control directed (i.e. deliberate) takes. Numerous examples demonstrate the effectiveness of this approach, such as the strong recoveries of some severely depleted populations following protection from whaling, sealing, or other forms of deliberate exploitation. The status of marine mammal populations, especially in relation to the effects of directed takes, is usually measured in terms of the size of populations relative to historical levels. The likelihood of population recovery after takes are controlled depends on the survival of a viable nucleus of individuals in the population at the time of protection. Furthermore, suitable habitat, including food resources, needs to exist to support recovery.

In some cases, especially when a population has been reduced to low levels, it is also necessary to manage sources of incidental (i.e. unintentional) mortality, injury, and disturbance due to other human activities. This can include requiring the use of less wasteful fishing practices and less by-catch-prone fishing gear and the regulation of ship traffic to reduce the incidence of vessel strikes. Techniques to reduce by-catch include the use of acoustic deterrents, gear modifications, time-area closures, and gear switching, for example, from gillnets to hook-and-line or traps/pots (i.e., Knowlton and Kraus, 2001; Read et al., 2006). Some of these approaches have been the subject of considerable research (www.bycatch.org). Other anthropogenic threats may also need to be addressed, including reduction of noise from anthropogenic sources, and reduction of contamination by toxic substances (including oil) and of marine debris (including discarded or derelict fishing gear, and plastics of all sorts) (see chapter 25. The effects of tourism, for example whale and pinniped watching, in modifying behaviour through close approach, particularly to breeding and nursery areas, may also need to be addressed (Higham et al., 2014). In any event, successful marine mammal management requires selecting and implementing suitable methods of enforcement.

Marine Protected Areas: Protecting marine mammals in specific areas (e.g., feeding, breeding, and resting areas) can sometimes be effective for addressing certain threats (Gormley et al., 2012). Understanding the life history of the species, the degree of localization of the threats and the needs and interests of local human populations is critical to the design of area management. Effective protection depends on devising plans that consider local community needs, as well as establishing effective control and monitoring. Furthermore, marine mammals tend to be highly mobile, and some species migrate across multiple ecosystems and even entire ocean basins, so protection may be needed in more than one area. There is considerable interest in spatial management to protect and conserve marine mammals through the establishment of parks, reserves and sanctuaries (e.g., Reeves, 2000; Marsh and Morales-Vela, 2012; and see www.icmmpa.org).

3 Key pressures linked to trends

3.1 Direct Removals

There are various types of directed takes, including commercial harvests, scientific sampling, captive display, and subsistence. Some of these are under management while others are unauthorized and unmanaged. The nature of direct removals varies among major marine mammal groups.

There is a very long-standing, but relatively small (811 whales over four centuries), capture of northern bottle-nosed whales in the Faeroe Islands (Denmark), recorded from at least the sixteenth century (Bloch et al., 1996). Coastal dolphins continue to be taken not only in the Solomon Islands (Oremus et al., 2013), but also in a large-scale drive fishery in Japan. Dolphins are hunted with harpoons in various countries for human consumption or for bait, especially in shark fisheries (IWC, 2013b). The consumption of by-catch as well as deliberately taken cetaceans and sirenians is a growing concern in Africa, Asia, and some parts of Latin America (Clapham and Van Waerebeek, 2007). In the Arctic, narwhals and belugas continue to be hunted by aboriginal people in Canada, Greenland (Denmark), the Russian Federation and the United States. Marine mammals are also harassed and sometimes deliberately killed around aquaculture facilities in many parts of the world.

Whales are taken by several States for commercial, aboriginal and scientific purposes. The nature and regulation of these takes are described by the International Whaling Commission (www.iwc.int). Cetaceans occasionally strand on the shore in large groups: so-called mass strandings. The cause of this behaviour is poorly understood, but it has occurred for millions of years (Pyenson et al., 2014). Unless a population has been depleted by other causes, mass strandings are unlikely to be a major threat in themselves.

Pinnipeds – including walruses and numerous seal species – are hunted in large parts of the Arctic and sub-Arctic, primarily for subsistence. A few commercial hunts for pinnipeds continue, including those for Cape fur seals in Namibia, harp and hooded seals in Canada, and harp seals in Norway and the Russian Federation. Because pinnipeds often interact directly with fishing operations, sometimes fishermen and aquaculturists exert pressure to limit or reduce pinniped populations by commercial hunting or culling.

For manatees and dugongs, commercial hunting no longer occurs, but harvesting for meat and for dugong tusks continues in many areas. Although often illegal, there is little enforcement in most areas. The largest aboriginal hunts of dugongs persist in Australia and the western Pacific (Dobbs et al., 2012). In Australia, both Aboriginal and Torres Strait Islander people continue their traditional subsistence hunting of dugongs, but commercial hunting of dugong is prohibited.

3.2 Fisheries Interactions

While the effects of fisheries on marine mammals from entanglement and by-catch are well known (see section 2.3.2.1), the effects of marine mammals on fish populations and the effects of fisheries on the prey base of marine mammals are less clear. The diet of most marine mammals includes fish and the possibility is often raised that some marine mammals are competing with fisheries or impeding the recovery of depleted fish stocks. Although some cetaceans, such as sperm whales, killer whales, pilot whales, and false killer whales, are known to depredate fishing operations, the significance of such depredation on fish populations and fish catches is not always clear.

Similarly, the effects of pinniped predation on valuable fish populations remain uncertain. For example, in the north-eastern Pacific sea lions feed on small populations of endangered salmon during spawning migration in rivers, especially in connection with dams (NMFS 1999). On the other hand, the significance of some other interactions is less clear (NMFS, 1999). Cape fur seals prey on two commercially fished species of hake off the west coast of South Africa (Punt et al., 1995) and grey seals prey on cod populations off the east coast of Canada (Fisheries and Oceans Canada, 2010). Population modelling studies of these fishery interactions suggest that seal culls are likely to have small if any effects on fish catch rates. Indeed, the effect of Cape fur seal predation may be overshadowed by predation of one hake species on the other. This suggests the need for further ecosystem-level research to clarify complex foraging relationships and interactions before conclusions are drawn, as discussed in section 4.1. (Yodzis, 2001; Gerber et al., 2009; Morisette et al., 2010). The biological and economic significance of marine mammal interactions varies with the fisheries and fish species involved. The areas of conflict are geographically limited and fishery interactions do not appear to be a global problem (Kaschner and Pauly, 2004).

3.3 Habitat Alterations

3.3.1 Disturbance

Many human activities generate underwater noise, including ship movements, military exercises involving the use of sonar or explosives, offshore oil and gas exploration and pile driving associated with construction of renewable energy facilities. Potential effects of such noise on marine mammals include direct acoustic injury, interference with foraging and communication, and food-web disruption. Disturbances are more likely to be significant for populations that have also been affected by other factors, such as harvesting or bycatch. Among the effects of disturbance that have been confirmed are the tendency of some whales to modify their movement patterns to avoid fixed-point noise sources, as for bowhead whales near an offshore oil production facility (McDonald et al., 2012), and the mortality of deep-diving cetaceans exposed to naval sonar (Cox et al., 2006; Fahlman et al., 2014; Ketten, 2014).

Offshore oil and gas development creates unique problems related to oil spills, whether at drill sites or during shipping. For depleted species, the additional mortality from ship strikes have been shown to be significant and vessel speed controls are demonstrably effective at reducing such strikes, as for right whales in the western North Atlantic (Laist et al., 2001; Laist et al., 2014).

3.3.2 Coastal and riverine development

Development in coastal and freshwater areas affects marine mammals in a variety of ways. Residential and urban development can make pinniped haul-out habitat inaccessible or hazardous. However, some species have shown remarkable adaptability to human-modified beach environments. For example, in the United States, California sea lions regularly haul out on piers and even on moored yachts in San Francisco. Although monk seals are extremely sensitive to disturbance in some areas, in Hawaii they sometimes rest and even nurse pups on crowded bathing beaches. In such cases, the animals seem capable of co-existing with the human presence as long as they are not molested.

Coastal habitats used by whales and dolphins make them vulnerable to increasingly intensive aquaculture operations of many types (Hucke-Gaete et al., 2013). A well-studied case is salmon farming in southern Chile, which operates in concentrated areas using open-cage net pens, moorings and anchoring, external supplementary feeding (rich in nutrients) and a significant quantity of chemical products (antimicrobials and pesticides) (Buschmann et al., 1996). However, there are differences globally in fish-farm design and operation, so general conclusions are difficult.

The construction of dams, barrages, and other structures in rivers and estuaries has led to fragmentation of dolphin and manatee populations in Asia and South America, making such populations more vulnerable to various threat factors, including entrapment in canals and mortality in flood-control gates. Runoff from agricultural fields, livestock feedlots, factories, and city streets contributes to chemical and biological contamination of freshwater, estuarine and coastal food webs, with often uncertain but likely negative effects on marine mammal health. For example, sea otter deaths in California (United States), have been linked to protozoan parasites known to breed in domestic cats (Johnson et al., 2009) and toxoplasmosis has been identified in Hector's dolphins (Roe et al. 2013).

3.3.3 Climate change

Climate change, both natural and human-induced, has the potential to affect the spatial distribution, reproductive success, foraging, and health of marine mammals (Leaper et al., 2006; Burek et al., 2008). The direction of such effects, negative or positive, is likely to be variable, with some species suffering from the loss of habitat and others able to take advantage of new habitat. MacLeod (2009) predicted that the ranges of most cetacean species (88 per cent) would be affected by changes in water temperature resulting from global climate change. This author predicted that the effects would be unfavourable for about half (47 per cent) of cetacean species. Little is known about the ability of most marine mammals to adapt to rapid environmental change. For example ice seals and polar bears, which are dependent on sea ice (Ferguson, et al. 2006), may be especially vulnerable to predicted climate change effects on ice habitat. Foraging habitat of right whales, which are dependent on small zooplankton, may change with increasing water temperature (Torres et al., 2013). Other species with more generalized diets and the ability to thrive in multiple types of habitat, such as bottlenose dolphins, may be more resilient (e.g., Heide-Jørgensen, 2009; Salvadeo et al., 2010). The overall effects of sea-level rise have been studied for northern elephant seals (Funayama et al., 2012) and Hawaiian monk seals (Baker, Littman and Johnston, 2006).

4 Major ecosystem services provided by marine mammals

4.1 Services to the ecosystem

Marine mammals can affect their ecosystems in several ways. Some species, such as sea otters, dugongs and walruses, structure their foraging habitat (e.g. Estes and Duggins, 1995). Depletion of these animals can result in major habitat changes; for example, kelp beds thrive when sea urchins are suppressed by sea otter predation. Other species, such as killer whales and leopard seals, play key roles as high-order predators, and their absence can affect prey resources of other marine mammal, bird or fish populations (Estes et al., 1998; Williams et al., 2004). Additionally, some species, such as sperm whales and blue whales, may have a large effect on nutrient recycling, with nutrient transport from deep ocean feeding areas to the surface (Lavery et al., 2010; Lavery et al., 2014). These ecosystem-level effects are understood for only a few species, but they can be critical for maintaining diverse and productive ecosystems (Bowen, 1997; Roman and McCarthy, 2010).

4.2   Direct services to humans including economic and livelihood services

The economic value of products obtained from marine mammals – meat, oil, ivory, fur, and many others – has been large, and this has contributed to these animals' extreme depletion and in a few instances led to their extinction (Steller's sea cow, Caribbean monk seal). Many groups of people continue to benefit from hunting marine mammals, including in some instances from selling products in international markets (e.g., narwhal and walrus ivory and seal skins). Aboriginal people in the Arctic and sub-Arctic continue to consume products from cetaceans and pinnipeds on a regular basis. Local people in Amazonia, northern Australia, and West Africa continue to harvest sirenians for food. In Nunavut and the Northwest Territories in Canada, regulated polar bear sport hunts provide income to Inuit who serve as guides and are required to use dog teams and sleds to pursue the animals.

In contrast, many people benefit from non-consumptive or low-consumptive uses of marine mammals, especially through whale-, dolphin-, and seal-watching tourism (see chapter 27). In addition, many people enjoy seeing marine mammals in the wild on their own. However, such activities can negatively affect small localized populations, for example bottlenose dolphins in Shark Bay, Australia (Bejder et al. 2006). Additionally, public display of captive marine mammals can make people more aware and appreciative of them, but it is extremely controversial, in part because the capture of marine mammals from the wild for display in captivity could threaten small wild populations (Fisher and Reeves, 2005).

5   Conservation responses and factors for sustainability

Like most large animals, marine mammals have limited capacity to reproduce and increase their numbers. Therefore, factors that can result in either low recruitment (e.g. impairment of reproduction by chemical contaminants such as organochlorines in food webs; (Dierauf and Gulland, 2001) or human-induced mortality rates higher than replacement (e.g. hunting or by-catch) need to be addressed to achieve conservation goals. Harmful algal blooms, ocean acidification, and expansion of hypoxia zones are among the most intractable factors affecting marine mammal populations. Conservation requires understanding of the organisms and their habitat requirements, and a balancing of human needs and desires with the natural productivity and the carrying capacity of the environment. The identification of key limiting factors is a first step toward developing management measures that can help populations to recover.

For the most part, population recoveries are regarded as successes, although in some cases they have led to unanticipated conflicts. One such conflict is with fishermen who have become accustomed to low levels of marine mammal abundance. For example, the recovery of numbers and range of sea otters has increased competition with fisheries for high-value molluscs in Alaska (United States) and the eastern North Pacific. Similarly, as mentioned above, sea lions prey on endangered fish, such as salmon and sturgeon, and recovering grey seals in Europe have had negative effects on seabird nesting habitat. Growing pinniped populations have also led to increased interactions with recreational fishers, vessel owners, and marina managers.

The conservation of marine mammals, like conservation more generally, should be understood as a dynamic and continuing process. Consequences of management actions need to be anticipated and unforeseen consequences addressed as they arise. Especially in the case of animals like marine mammals, that are widely distributed and rarely occur within the jurisdiction of only one State, multilateral approaches are essential. For example, the global ban on large-scale pelagic drift net fishing on the high seas imposed by the United Nations in 1994 was a major step in limiting the by-catch of several marine mammal (and seabird) species that were especially vulnerable to entanglement. Other international instruments, such as Convention on the International Trade in Endangered Species (CITES) and the International Convention for the Regulation of Whaling, have helped limit the damage caused by over-exploitation of the great whales.

With a broad understanding of the many aspects of both conservation and use of marine mammals and their roles in marine ecosystems, it should be possible to address at least some of the issues arising from human interactions.

References

Atkinson, S.N., Nelson, R.A., Ramsay, M.A., (1996). Changes in the body composition of fasting polar bears (Ursus maritimus): The effect of relative fatness on protein conservation. *Physiological Zoology* 69, 304-316.

Baker, J. D., Littnan, C. L., Johnston, D.W. (2006). Potential effects of sea level rise on the terrestrial habitats of endangered and endemic megafauna in the Northwestern Hawaiian Islands. *Endangered Species Research* 4:1-10.

Bejder, L., Samuels, A., Whitehead, H., Gales, N., Mann, J., Connor, R.C., Heithaus, M.R., Watson-Capps, J., Flaherty, C. and Krutzen, M. (2006). Decline in Relative Abundance of Bottlenose Dolphins Exposed to Long-Term Disturbance. *Conservation Biology*, 20: 1791–1798.

Bloch, D., Desportes, G., Zachariassen, M. and Christensen, I. (1996). The northern bottlenose whale in the Faroe Islands, 1584-1993, *Journal of Zoology*, 239/1: 123–140.

Bowen, W.D. (1997). Role of Marine Mammals in Aquatic Ecosystems. *Marine Ecology Progress Series* Vol 158:267-274.

Branch, T.A., Stafford, K.M., Palacios, D.M., Allison, C., Bannister, J.L. and many others (2007). Past and present distribution, densities and movements of blue whales Balaenoptera musculus in the southern hemisphere and northern Indian Ocean. *Mammal Review* 37: 116-175.

Burek, K.A., Gulland, M.D. and O'Hara, T.M. (2008). Effects of climate change on Arctic marine mammal health. *Ecological Applications* 18:S126–S134.

Busch, B.C. (1985). *The War against the Seals: A History of the North American Seal Fishery*. McGill University Press.

Buschmann, A.H., López, D.A., Medina A. (1996). A review of the environmental effects and alternative production strategies of marine aquaculture in Chile. *Aquacultural Engineering* 15: 397–421.

CIRVA (2014). Report of the fifth meeting of the 'Comité Internacional para la Recuperación de la Vaquita.' www.worldwildlife.org/publications/report-5th-meeting-of-the-international-committee-for-the-recovery-of-the-vaquita-cirva.

Clapham, P. and Van Waerebeek, K. (2007). Bush-meat and bycatch: the sum of the parts. *Molecular Ecology* (2007) 16, 2607–2609 doi: 10.1111/j.1365-294X.2007.03378.x

Cox, T.M., Ragen, T.J., Read, A.J., Vos, E., Baird, R.W., and many others. (2006). Understanding the impacts of anthropogenic sound on beaked whales. *Journal of Cetacean Research and Management* 7:177-187.

Currey, R.J.C., Boren, L.J., Sharp, B.R. and Peterson, D. (2012). *A risk assessment of threats to Maui's dolphins*. Ministry for Primary Industries and Department of Conservation, Wellington. 51 pp.

Dierauf, L. and Gulland, F.M.D (eds.). (2001). *CRC Handbook of Marine Mammal Medicine: Health, Disease, and Rehabilitation*. CRC Press: Boca Raton.

Dobbs, K., Lawler, I. and Kwan, D. (2012). Dugongs in Australia and the Pacific. In: Hines, E., Reynolds, J., Mignucci-Giannoni, A, Aragones, L.V., and M. Marmontel (eds.). 2012. *Sirenian Conservation: Issues and Strategies in Developing Countries*. The University Press of Florida. pp. 99-105.

Estes, J.A. and Duggins, D.O. (1995). Sea otters and kelp forests in Alaska: generality and variation in a community ecology paradigm. *Ecological Monographs* 65:75-100.

Estes, J.A., Tinker, M.T., Williams, T.M. and Doak, D.F. (1998). Killer whale predation on sea otters linking oceanic and nearshore ecosystems. *Science* 282: 473-476.

Estes J.A., DeMaster, D.P., Doak, D.F., Williams, T.M., Brownell, R.L. Jr (Eds.). (2006). *Whales, whaling and ocean ecosystems*. Berkeley: University of California Press.

Fahlman, A, Tyack, P.L., Miller, P.J.O. and Kvadsheim, P.H. (2014). How man-made interference might cause gas bubble emboli in deep diving whales. *Frontiers in Physiology* 5:13.

Ferguson, S.H., Stirling, I. and McLoughlin, P. (2005). Climate change and ringed seal (Phoca hispida) recruitment in western Hudson Bay. *Marine Mammal Science*, 21: 121-135.

Fisher, S.J. and Reeves, R.R. (2005). The global trade in live cetaceans: implications for conservation. *Journal of International Wildlife Law and Policy* 8(4): 315-340.

Fisheries and Oceans Canada (2010). Impacts of grey seals on fish populations in eastern Canada; summary. Science Advisory Report 2010/071. Available at www.dfo-mpo.gc.ca/csas-sccs/Publications/SAR-AS/2010/2010_071-eng.html. Consulted 23 December 2013.

Funayama, K., Hines, E., Davis, J., and Allen, S. (2012). Effects of sea-level rise on northern elephant seal breeding habitat at Point Reyes Peninsula, California. *Aquatic Conservation: Marine and Freshwater Ecosystems* 23: 233-245.

Gerber, L.R., Morissette, L., Kaschner, K., and D. Pauly. (2009). Should whales be culled to increase fishery yield? *Science* 323: 880-881.

Gerrodette, T., Taylor, B.L., Swift, R., Rankin, S., Jaramillo-Legorreta, A.M. and Rojas-Bracho. L. (2011). A combined visual and acoustic estimate of 2008 abundance, and change in abundance since 1997, for the vaquita, Phocoena sinus. *Marine Mammal Science* 27(2): E79–E100.

Gormley, A.M., Slooten, E., Dawson, S., Barker, R.J., Rayment, W., du Fresne, S. and Brager, S. (2012). First evidence that marine protected areas can work for marine mammals. *Journal of Applied Ecology* 2012, 49, 474–480.

Hamner, R.M., Oremus, M., Stanley, M., Brown, P., Constantine, R., and Baker, C.S. (2012). *Estimating the abundance and effective population size of Maui's dolphins using microsatellite genotypes in 2010–11, with retrospective matching to 2001–07*. Department of Conservation, Auckland, 44pp.

Hedley, C. (2010). The 1998 Agreement on the International Dolphin Conservation Program: Recent Developments in the Tuna-Dolphin Controversy in the Eastern Pacific Ocean. *Ocean Development & International Law* 32:1.

Heide-Jørgensen, M.P., Iversen, M., Hjort Nielsen, M., Lockyer, C., Stern, H. and Ribergaard, M.H. (2011). Harbour porpoises respond to climate change. *Ecology and Evolution* 1(4): 579–585. doi: 10.1002/ece3.51.

Higham, J., Bejder, L. and Williams, R. (eds.). (2014). *Whale-watching: Sustainable Tourism and Ecological Management*. Cambridge University Press, Cambridge, UK, 387 pages.

Hines, E., Ponnampalam, L., Jamal, F., Jackson-Ricketts, J., and Whitty, T. (2014). *Report of the 3rd Workshop on the Biology and Conservation of Cetaceans and Dugongs of South-East Asia*. UNEP/CMS Secretariat, Bonn, Germany.

Hitchmough, R. (2010). Conservation status of New Zealand marine mammals (suborders Cetacea and Pinnipedia), 2009, *New Zealand Journal of Marine and Freshwater Research* (http://dx.doi.org/10.1080/00288330.2010.482970).

Hucke-Gaete, R., Haro, D., Torres-Florez, J. P., Montecinos, Y., Viddi, F.A., Bedriñana, L., and Ruiz, J. (2013). A historical feeding ground for humpback whales in the Eastern South Pacific revisited: the case of northern Patagonia, Chile. *Aquatic Conservation: Marine & Freshwater Ecosystems* 23: 858–867. DOI: 10.1002/aqc.2343.

IUCN (2015). International Union for the Conservation of Nature, *Red List under the species Enhydra lutris and Lontra felina* (http://www.iucnredlist.org/details/7750/0 and /12303/0 accessed 12 July 2015).

IWC (2013a). Report of the IWC Workshop on the Assessment of Southern Right Whales. *Journal of Cetacean Research and Management* 14 (Supplement): 439-462.

IWC (2013b). Report of the Scientific Committee for 2013. *Journal of Cetacean Research and Management* 14 (Supplement):59.

Jackson, J.A., Carroll, E, Smith, T.D., Patenaude N. and Baker, C.S. (2011). Taking stock – the historical demography of the New Zealand right whale (the Tohora) 1820-2008. Final Report, Taking Stock Project. Ministry of Fisheries, New Zealand, 35 pp.

Johnson, C.K., Tinker, M.T., Estes, J.A., Conrad, P.A., Staedler, M.S., Miller, M.A., Jessup, D.A. and Mazet, J.A. (2009). Prey choice and habitat use drive sea otter pathogen exposure in a resource-limited coastal system. *Proceedings of the National Academy of Sciences of the USA* 106(7): 2242-2247.

Kaschner, K. and Pauly, D. (2004). *Competition between Marine Mammals and Fisheries: Food for thought*. Humane Society, Washington DC, 28 pp.

Ketten, D.R. (2014). Sonars and strandings: are beaked whales the aquatic acoustic canary? *Acoustics Today* 10 (3): 46-56.

Knowlton, A.R. and Kraus, S.D. (2001). Mortality and serious injury of northern right whales (Eubalaena glacialis) in the western North Atlantic Ocean. *Journal of Cetacean Research and Management* 2: 193-201.

Kovacs, K.M., Aguilar, A., Aurioles Gamboa, D., Burkanov, V., Campagna, C., Gales, N., Gelatt, T. and Goldsworthy, S. (2012). Global threats to pinnipeds. *Marine Mammal Science* 28(2): 414-436.

Kreb, D., Reeves, R.R., Thomas, P.O., Braulik, G. and Smith, B.D. (Eds.). (2010). *Establishing protected areas for Asian freshwater cetaceans: Freshwater cetaceans as flagship species for integrated river conservation management*, Samarinda, 19-24 October 2009. Final Workshop Report, Yayasan Konservasi RASI, Samarinda, Indonesia.

Laist, D.W., Knowlton, A.E., Mead, J.G., Collet A.S. and Podesta, M. (2001). Collisions between ships and whales. *Marine Mammal Science* 17:35-75.

Laist, D.W., Knowlton, A.E., and Pendleton, D. (2014). Effectiveness of mandatory vessel speed limits for protecting North Atlantic right whales. *Endangered Species Research* 23:133-147.

Lavery, T.J., Roudnew, B., Gill, Semout, J., Seuront, L., Johnson, G., Mitchell, J.G. and Smetacek, V. (2010). Iron defecation by sperm whales stimulates carbon export in the Southern Ocean. *Proceedings of the Royal Society of London B* ,vol. 277 no. 1699: 3527-3531.

Lavery, Trish J., Roudnew, B., Seymour, J., Mitchell, J.G., Smetacek, V. and Nicol, S. (2014). Whales sustain fisheries: Blue whales stimulate primary production in the Southern Ocean. *Marine Mammal Science* 30(3): 888-904.

Leaper, R., Cooke, J., Trathan, P., Reid, K., Rowntree, V. and Payne R., (2006). Global climate change drives southern right whale (Eubalaena australis) population dynamics. *Biology Letters* 2: 289-292.

McDonald, T.L., Richardson, W.J., Greene, C.R. Jr., Blackwell, S.B., Nations, C.S., Nielson, R.M. and Streever, B. (2012). Detecting changes in the distribution of calling bowhead whales exposed to fluctuating anthropogenic sounds. *Journal of Cetacean Research and Management* 12: 91-106.

MacLeod, C.D. (2009). Global climate change, range changes and potential implications for the conservation of marine cetaceans: a review and synthesis. *Endangered Species Research* 7: 125–136, doi: 10.3354/esr00197.

Marsh, H., O'Shea, T.J. and Reynolds, J.E. III, (2012). *Ecology and conservation of the sirenia*. Cambridge University Press.

Marsh, H. and Morales-Vela B. (2012). Guidelines for developing protected areas for sirenians. In Hines, E., Reynolds, J., Mignucci-Giannoni, A, Aragones, L.V., and M. Marmontel (eds.) *Sirenian Conservation: Issues and Strategies in Developing Countries*. The University Press of Florida.

Morissette, L., Kaschner, K., and Gerber, L.R. (2010). 'Whales eat fish'? Demystifying the myth in the Caribbean marine ecosystem. *Fish and Fisheries* 11:388-204.

NMFS (National Marine Fisheries Service), (1999). *Report to Congress: Impacts of California sea lions and Pacific harbor seals on salmonids and West Coast ecosystems*. USDOC/NOAA.NMFS. NMFS (National Marine Fisheries Service). (undated). ETP Cetacean Assessment, https://swfsc.noaa.gov/textblock.aspx?Division=PRD&ParentMenuId=228&id=1408), accessed 19 April 2015.

Oremus, M., Leqata, J. and Baker, C.S. (2013). The resumption of traditional drive-hunts of dolphins in the Solomon Islands in early 2013. IWC Scientific Committee Meeting Document. SC/65a/SM08.

Punt, A.E. and Butterworth, D.S. (1995). The effects of future consumption by the Cape fur seal on catches and catch rates of the Cape hakes. 4. Modelling the biological interaction between Cape fur seals Arctocephalus pusillus pusillus and the Cape hakes Merluccius capensis and M. paradoxus. *South African Journal of Marine Science* 16: 255-285.

Pyenson N.D. et al. (2014). Repeated mass strandings of Miocene marine mammals from Atacama Region of Chile point to sudden death at sea. *Proceedings of the Royal Society of London B* 281:20133316. http://dx.doi.org/10.1098/rspb.2013.3316.

Read A.J., Drinke, P. and Northridge, S. (2006). Bycatch of Marine Mammals in U.S. and Global Fisheries. *Conservation Biology* 20 (1): 163–169.

Reeves, R.R. (2000). The Value of Sanctuaries, Parks, and Reserves (Protected Areas) as Tools for Conserving Marine Mammals. Final report for MMC contract T74465385. 50 pp. Available from the Marine Mammal Commission, Bethesda, Maryland.

Roe, W.D., Howe, L., Baker, E.J., Burrows, L. and Hunter, S.A. (2013). An atypical genotype of Toxoplasma gondii as a cause of mortality in Hector's dolphin (Cephalorhynchus hectori). *Veterinary Parasitology* 192: 67-74.

Roman, J. and McCarthy, J.J. (2010). The whale pump: marine mammals enhance primary productivity in a coastal basin. *PLoS ONE* 5, e13255.

Salvadeo, C.J., Lluch-Belda, D., Gómez-Gallardo, D.A., Urbán-Ramírez J. and MacLeod, C.D. (2010). Climate change and a poleward shift in the distribution of the Pacific white-sided dolphin in the northeastern Pacific. *Endang Species Research*11: 13–19, doi: 10.3354/esr00252.

Shaughnessy, P.D., Goldsworthy, S.D., and Mackay, A.I. (2014). Status and trends in abundance of New Zealand fur seal populations in South Australia. South Australian Research and Development Institute (Aquatic Sciences), Adelaide. SARDI Publication No. F2014/000338-1. SARDI *Research Report Series No. 781*. 33 pp.

Torres, L.G., Smith, T.D. Sutton, P., MacDiarmid, A., Bannister, J. and Miyashita, T. (2013). From exploitation to conservation: habitat models using whaling data predict distribution patterns and threat exposure of an endangered whale. *Diversity and Distributions* 19: 1138-1152.

Trujillo, F., Crespo, E., Van Damme, P.A. and J.S. Usma (eds). (2010). *The Action Plan for South American River Dolphins 2010 – 2020*. WWF, Fundación Omacha, WDS, WDCS, Solamac. Bogotá, D.C., Colombia. 249 pp. Available at: www.iucn-csg.org/index.php/downloads/.

Turvey, S.T., Pitman, R.L., Taylor, B.L., Barlow, J., Akamatsu, T. and others. (2007). First human-caused extinction of a cetacean species? *Biology Letters* 3:537-540.

Wang, D., Turvey, S.T., Zhao, X. & Mei, Z. (2013). Neophocaena asiaeorientalis ssp. asiaeorientalis. In: IUCN 2013. IUCN Red List of Threatened Species. Version 2013.2. <www.iucnredlist.org>. Downloaded on 09 February 2014.

Williams, T.M., Estes, J.A., Doak, D.F. and Springer, A.M. (2004). Killer appetites: assessing the role of predators in ecological communities. *Ecology* 85:3373–3384.

www.dx.doi.org/10.1890/03-0696

Yodzis, P. (2001). Must top predators be culled for the sake of fisheries? *Trends in Ecology & Evolution* 16: 78-84.

38 Seabirds

Contributors:
Ben Lascelles (Convenor), Jake Rice (Lead Member and Editor of Part VI), Mayumi Sato, Marguerite Tarzia and Ross McLeod Wanless[1]

---

1   The writing team thanks Esteban Frere for his substantial contribution to this chapter.

Seabirds Chapter 38

1 Introduction

Seabirds are the most threatened bird group and their status has deteriorated faster over recent decades. Globally 28 per cent are threatened (5 per cent are in the highest category of Critically Endangered) and a further 10 per cent are Near Threatened. Of particular concern are those species whose small range or population is combined with decline (64 species). Pelagic species are disproportionately represented in comparison with coastal species; those listed under the Agreement on the Conservation of Albatross and Petrels[1] have fared worst of all.

Declines have been caused by ten primary pressures. At sea these include: incidental bycatch (in longline, gillnet and trawl fisheries); pollution (oil spills, marine debris)); overfishing; energy production and mining. On land, invasive alien species, problematic native species (e.g. those that have become super-abundant), human disturbance, infrastructural, commercial and residential development, hunting and trapping have driven declines. Climate change and severe weather affect seabirds on land and at sea.

Given their imperilled conservation status, many seabirds have been highlighted for special conservation status and action under a range of international, regional and national agreements and mechanisms. Data on distribution, abundance, behaviour and pressures can be used to inform the design of effective management regimes for seabirds. Management decisions can be guided by: (1) where the key areas are, (2) when these areas are used, (3) what variables explain seabird presence in a given area, (4) the threat status of species in a given area, (5) what pressures may be adversely affecting the species, associated habitats and processes, (6) what management actions are needed to address these threats, and (7) how any management intervention can best be monitored to assess its effectiveness.

Seabirds provide many ecosystems services and their role as potential indicators of marine conditions is widely acknowledged. Many studies use aspects of seabird biology and ecology, especially productivity and population trends, to infer relationships with and/or effects on and/or correlate with aspects of the marine environment, particularly food availability.

2 Population trends or conservation status

2.1 Aggregated at global scale

Croxall *et al.* (2012) reviewed 346 seabird species and found that overall, seabirds are more threatened than other comparable groups of birds and their status has deteriorated faster over recent decades. In terms of the categories used in the International Union for the Conservation of Nature (IUCN) Red List, globally 97 species (28 per cent) are threatened,

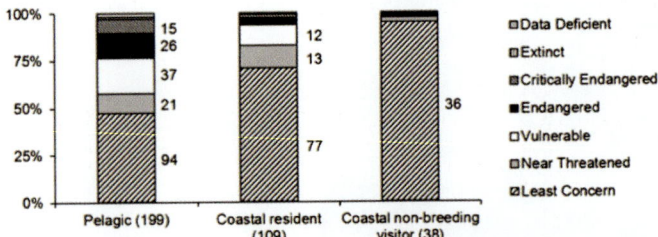

Figure 38.1 | Proportion of species in each IUCN Red List category for pelagic species, coastal residents and coastal non-breeding visitors. Figures give number of species (for totals >5). Source: Croxall et al., 2012.

with 17 species (5 per cent) in the highest category of Critically Endangered and a further 10 per cent Near Threatened. Only four species, all storm petrels, are regarded as Data Deficient; three species are considered Extinct, and two other species are Possibly Extinct. Of the 132 threatened and Near Threatened seabird species 70 (53 per cent) qualify by virtue of their very small population and/or range. 66 species (50 per cent) qualify by virtue of having undergone rapid population decline. Of particular concern are those with both small range and/or population as well as having undergone decline (64 species; 48 per cent); this includes six species of penguins, 17 of gadfly petrels and eight of cormorants. Pelagic species are disproportionately represented in all categories in comparison with coastal species (Figure 38.1). 57 species (17 per cent) are increasing; for many, such as the 17 gull species, this is doubtless due to their abilities to exploit close links with human activities.

A broader, but less sensitive, measure of overall trends is provided by the Red List Index (Butchart *et al.*, 2004; 2007), which measures trends in extinction risk (based on the movement of species through IUCN Red List categories owing to genuine improvement or deterioration in status) and is virtually the only trend indicator currently available for seabirds on a worldwide and/or regional basis. It shows (Figure 38.2) that, over the last 20 years, seabirds have had a substantially poorer conservation status than non-seabirds and that they have deteriorated

Figure 38.2 | Red List Index of species survival for all bird species (n=9,853 non-Data Deficient species extant in 1988), all seabirds (n=339) and ACAP (Agreement on Conservation of Albatross and Petrels)-listed species (n=29). Values for the latter are projected to 2012 based on data from the 2012 IUCN Red List to be published later this year. RLI values relate to the proportion of species expected to remain extant in the near future without additional conservation action. An RLI value of 1.0 equates to all species being categorized as of Least Concern, and hence that none are expected to become extinct in the near future. An RLI value of zero indicates that all species have become Extinct. See Butchart et al 2004 for further explanation. Source: BirdLife International 2012.

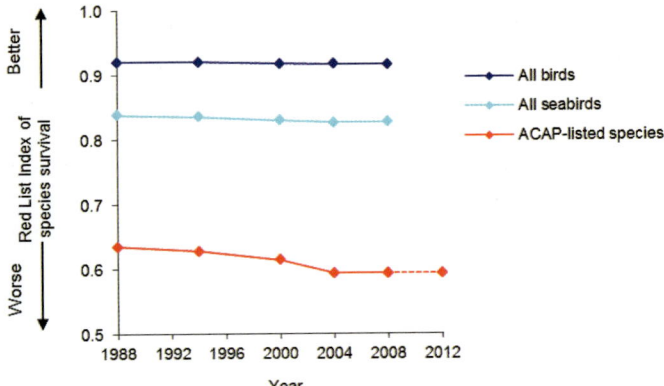

---
[1] United Nations, *Treaty Series*, vol. 2258, No. 40228.

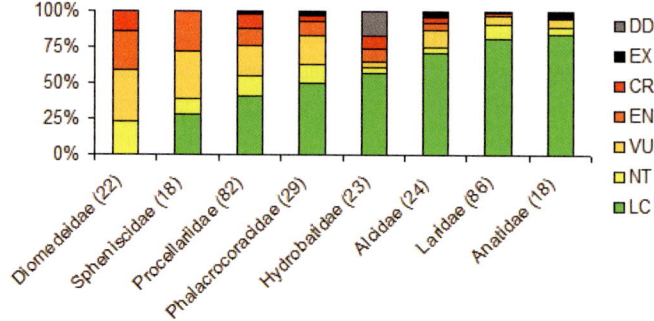

Figure 38.3 | Percentage of species in each IUCN Red List category for the major seabird families. Figures give number of species. Source: Croxall et al., 2012.

faster over this period. Pelagic species are more threatened and have deteriorated faster than coastal species, and this difference is particularly pronounced for the albatrosses and large petrels that are covered by the 2004 Agreement on the Conservation of Albatross and Petrels ([ACAP] BirdLife International, 2012).

Reviewing the pattern taxonomically (Figure 38.3) reveals that, of the main families (which together account for 87 per cent of species), the most threatened are the albatrosses/petrels (Diomedeidae/Procellariiformes and penguins (Sphenisciformes).Together these (represent nearly one half (43 per cent) of all seabirds and contain many pelagic species. Conservation of Diomedeidae benefits considerably from ACAP. Within Procellariiformes the genera *Pterodroma* and *Pseudobulweria* are the next most threatened and a special internet forum has recently been established to promote priority conservation action for them: Gadfly Petrel Conservation Group; www.gadflypetrel.ning.com.

2.2 Special conservation status issues

Given their imperilled conservation status, many seabirds have been highlighted for special conservation status and action under a range of international, regional and national agreements and mechanisms. However, because seabirds are highly mobile and migrate, they are exposed to vagaries of differing levels of protection across international (and non-governmental) regions. Those agreements and mechanisms currently most actively undertaking work include ACAP (30 species), EU Birds Directive (all seabirds in the EU), the Convention for the Protection of the Marine Environment of the North-East Atlantic[2] (OSPAR Convention) (9 species), the Agreement on the Conservation of African-Eurasian Migratory Waterbirds[3] (82 species), East Asian-Australasian Flyway Partnership (39 species), the Convention for the Protection of the Mediterranean Sea Against Pollution[4] (Barcelona Convention) (14 species), Convention on the Conservation of Migratory Species of Wild Animals[5] (CMS; 20 seabird species are listed on Annex I; 50 on Annex II), the Convention on the conservation of European wildlife and natural habitats[6] (Bern Convention) (over 30 species), Helsinki Commission (HELCOM; 11 species), the Convention on the Protection of the Black Sea Against Pollution[7] (Bucharest Convention) (2 species), Commission for the Conservation of Antarctic Marine Living Resources (CCAMLR; 7 species), Convention for Arctic Flora and Fauna (3 species), Migratory Bird Treaty Act (139 species), North American Agreement on Environmental Cooperation (1 species), Trilateral Committee for Wildlife and Ecosystem Conservation and Management (1 species), and the Convention on International Trade in Endangered Species of Wild Fauna and Flora[8] (CITES) (6 species). Other agreements that have this remit but are not yet active include the Nairobi Convention for the Protection, Management and Development of the Marine and Coastal Environment of the Eastern African Region[9] (Nairobi Convention) (47 species), the Regional Convention for the Conservation of the Red Sea and Gulf of Aden Environment[10] (Jeddah Convention) (lists not yet provided by contracting parties), the Convention for Cooperation in the Protection, Management and Development of the Marine and Coastal Environment of the Atlantic Coast of the West, Central and Southern Africa Region[11] (Abidjan Convention) (considering adding a species list), and the Convention for the Protection and Development of the Marine Environment in the Wider Caribbean Region (WCR)[12] (Cartagena Convention) (5 species). In addition to the above agreements, Regional Fisheries Management Organisations (RFMOs) have also begun to adopt strategies that address incidental seabird bycatch. Level of regulation varies across RFMOs but includes combinations of the use of one or more bycatch mitigation measures in certain areas, data collection through observer programmes and use of monitoring, surveillance and compliance measures.

3 Key pressures linked to trends

The majority of seabirds are highly migratory species that require a variety of marine and terrestrial habitats during different seasons and life stages (Lascelles *et al*, 2014). Many seabirds are long-lived and slow reproducing. These characteristics make them particularly vulnerable to a wide range of pressures, where even quite small increases in mortality can lead to significant population declines. In addition, many seabirds have highly specialised diets, being reliant on just a few prey species, the abundance and distribution of which can alter dramatically in response to abrupt environmental changes.

---

2   United Nations, *Treaty Series*, vol. 2354, No. 42279.

3   Ibid., vol. 2365, No. 42632.

4   Ibid., vol. 1102, No. 16908.

5   Ibid., vol. 1651, No. 28395.

6   Ibid., vol. 1284, No. 21159.

7   Ibid., vol. 1764, No. 30674.

8   Ibid., vol. 993, No. 14537.

9   http://www.unep.org/NairobiConvention/The_Convention/index.asp

10 http://www.persga.org/Documents/Doc_62_20090211112825.pdf

11 http://abidjanconvention.org/index.php?option=com_content&view=article&id=100&Itemid=200&lang=en

12 United Nations, *Treaty Series*, vol. 1506, No. 25974.

Croxall et al. (2012) found that globally, of the top 10 pressures on threatened seabirds (Figure 38.4), invasive species typically acting at the breeding site potentially affect 73 species (75 per cent) of all threatened seabird species and nearly twice as many as any other single threat, although in some cases the threat is of a potential future impact. The remaining pressures are fairly evenly divided between: (a) those acting mainly at the breeding site, namely problematic native species (e.g. those that have become superabundant - 31 species, 32 per cent), human disturbance (26 species, 27 per cent), infrastructural, commercial, and residential development (14 species, 14 per cent) and (b) those acting mainly at sea in relation to foraging, moulting or migration areas/aggregations, namely, bycatch in longline, gillnet and trawl fisheries (40 species, 41 per cent), pollution (30 species, 31 per cent), overfishing and/or inappropriate spatial management of fisheries (10 species, 10 per cent). Hunting and trapping (23 species, 24 per cent) and energy production and mining (10 species, 10 per cent) affect both domains, the former more at breeding sites, the latter more in relation to foraging areas, flight paths and flyways. Climate change and severe weather (39 species, 40 per cent), as currently assessed, largely reflect adverse weather and flooding at breeding sites. However, the impact of sea level rise is clearly an important driver of change that is increasingly affecting seabirds in many ways, albeit mainly in the medium to long term (i.e., at time frames mostly outside those of relevance to IUCN Red List criteria). The relative importance of threats is largely similar when only those of high impact are considered, although bycatch becomes almost as significant as the effects of invasive alien species (Croxall et al., 2012).

Commercial fisheries are the most serious at-sea pressure facing the world's seabirds, affecting both adult and juvenile birds. Despite data gaps, each year incidental bycatch in longline fisheries is estimated to kill 160,000-320,000 seabirds from 70 species, although there is evidence of substantially reduced bycatch in some key fisheries where the

Figure 38.4 | Threats to threatened (a) seabirds (n=346 species); (b) pelagic seabirds (n=197 species); (c) coastal seabirds (n=146 species). Source: Croxall et al., 2012.

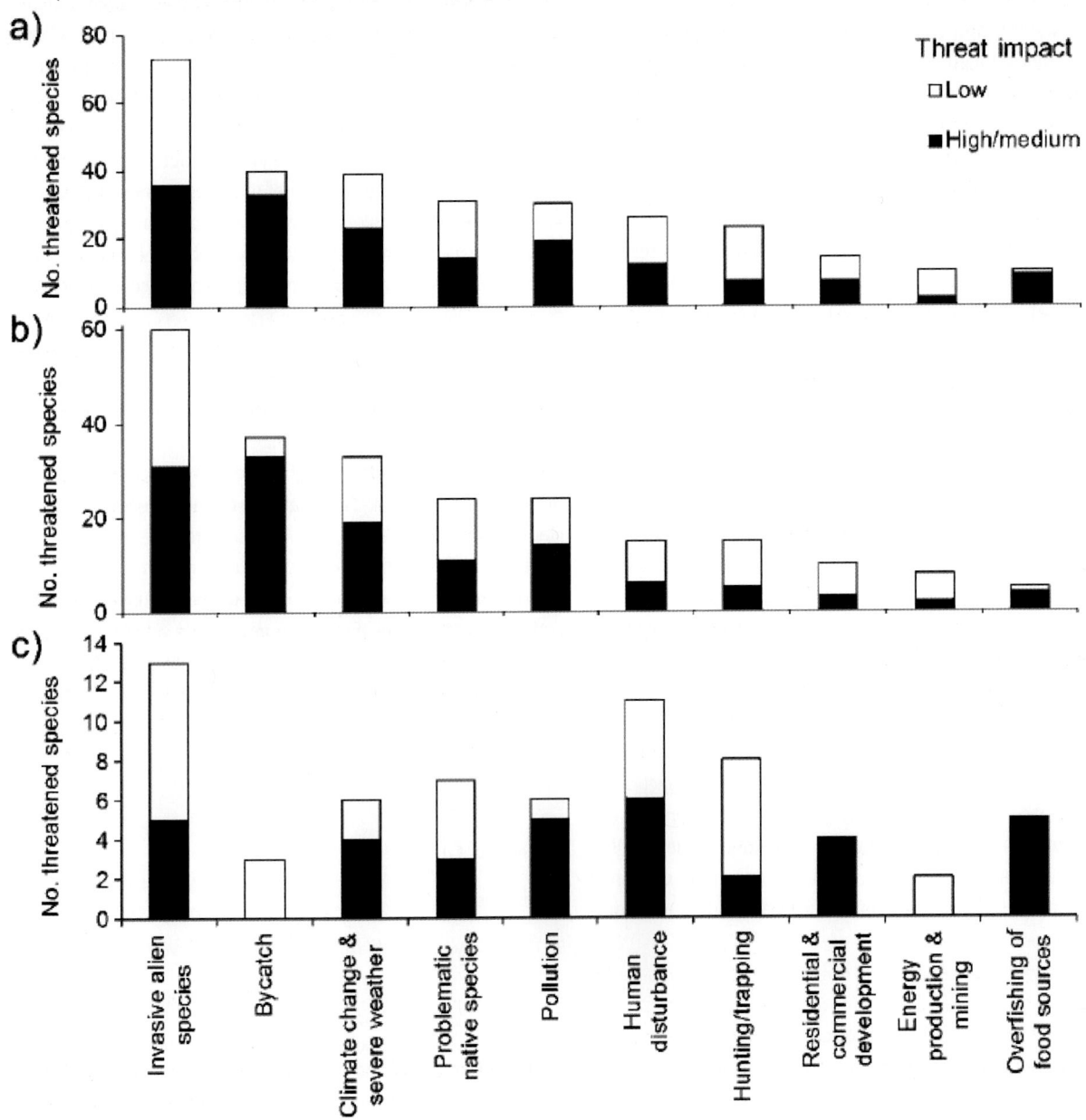

pressure has been managed (Anderson et al., 2011). Several papers have reviewed seabird bycatch rates in both demersal (bottom) and pelagic (upper water column) longline fisheries in various regions (e.g., Brothers, 1991; Dunn and Steel, 2001; BirdLife International, 2007; Steven et al., 2007; Bugoni et al., 2008; Rivera et al., 2008; Waugh et al., 2008; Kirby et al., 2009, Waugh et al., 2012), and two assessments have been made on a global scale (Nel and Taylor, 2003; Anderson et al., 2011). The fleets identified as having the highest levels of seabird bycatch include the Spanish hake fleet in the Gran Sol area, the Japanese pelagic tuna fleet in the North Pacific, the Namibian hake fleet and the Nordic demersal fleets (Anderson et al., 2011). The impacts of illegal, unreported, and unregulated fishing (IUU) on seabirds have been estimated in the thousands of individuals each year south of 30° S but are inherently difficult to assess here and elsewhere (Anderson et al., 2011).

Since 1992 a global moratorium has been imposed on the use of all large-scale pelagic drift-net fishing on the high seas, including enclosed and semi-enclosed seas (General Assembly resolution 46/215). Gillnet fisheries (both set and drift nets) are, however, still permitted to operate within a State's Exclusive Economic Zone (EEZ). Although many data gaps remain, hampering assessment, a review of existing data shows that gillnets are responsible for the incidental capture of large numbers of birds, sharks and marine mammals (e.g., Northridge, 1991; Hall, 1998; Tasker et al., 2000; Johnson et al., 2005; Rogan and Mackey, 2007; Žydelis et al., 2013). Amongst birds, the pursuit-diving species, such as divers (loons), grebes, seaducks, auks and cormorants, are the most vulnerable to entanglement (Piatt and Nettleship, 1987; Žydelis et al., 2009). The most recent global review estimated incidental bycatch in gillnet fisheries at 400,000 seabirds from 150 coastal and diving species each year (Žydelis et al., 2013). The highest bycatch has been reported in the Northwest Pacific, Iceland and the Baltic Sea (Žydelis et al., 2013).

Although seabird bycatch in long-line fishing has been known since the 1980s, the threat posed by trawl fisheries has also become apparent in recent years (Bartle, 1991; Weimerskirch et al., 2000; Sullivan et al., 2006). No global review of the impact of trawl fishing on seabirds has been undertaken, but a number of regional and national levels studies highlight the significance of the problem (Gonzalez-Zevallos et al., 2007; Petersen et al., 2008; Yorio et al., 2010) with tens of thousands of 40 larger species of seabird thought to be killed each year. Trawling can also alter benthic habitats which may have indirect impacts on seabirds via the effect this has on forage fish species (see, for example, Chapter 36A).

Fisheries may compete with seabirds for their prey items, and overfishing of both forage species and predatory species that help aggregate food sources for seabirds have been cited as a reason for the decline in several species (e.g., Becker and Beissinger, 2006; Camphuysen, 2005). Cury et al. (2011) assessed prey abundance and breeding success for 14 bird species within the Atlantic, Pacific, and Southern Oceans and found that when less than one third of the maximum prey biomass was available to seabirds, their productivity was adversely affected.

Climate change and severe weather driven by habitat shifts and alterations, storms and flooding, and temperature extremes are already affecting some seabird species. Species' sensitivity and adaptive capacity depend on a suite of taxon-specific biological and ecological traits; as well as the degree to which they are exposed to changes in climate (Foden et al. 2013). Known negative impacts may include loss of habitat, decreased marine productivity causing shifts in location of prey, and shifts in range and migration routes due to changes in winds, ocean currents and sea surface temperature (e.g. Forcada and Trathan, 2009; Hazen et al., 2012; Sydeman et al., 2012).

Pollution in various forms is a widespread problem adversely affecting many seabirds. Oil spills, from both offshore facilities and shipping tankers, can cause mortalities that lead to population-level impacts, particularly when they occur within the most sensitive sites. Single spills have been recorded as killing up to a quarter of a million birds (García et al., 2003) and causing the loss of 7 per cent of regional populations of certain species (Piatt and Ford, 1996). Since its advent, plastic in the form of solid waste materials has become ubiquitous in all oceans of the world and entanglement and ingestion of this material by seabirds is now a widespread problem, affecting at least 100 species (Laist, 1997, Provencher et al. 2014).

Attraction to artificial sources of light has been recorded in at least 21 species of Procellariiformes, as well as in several other seabird groups, and has a detrimental effect on some globally threatened populations (Reed et al., 1985), notably shearwater (e.g. Day et al. 2003) and Pterodroma (e.g. Ainley et al. 1997; Le Corre et al. 2002;) species around their breeding colonies. Light-induced seabird collisions at sea, either with fishing vessels (such as those emitting light to catch squid) or with marine oil platforms, are difficult to quantify, occurring episodically particularly in low-visibility conditions and probably exacerbated by seabirds' attraction to bright lights and flares (Ronconi et al 2014). However, up to tens of thousands of seabirds have been observed in a single collision event (Montevecchi, 2006).

Impacts from shipping may include water and air pollution, disturbance, and collision. The level of impact may increase in the future as ship traffic increases, particularly in sensitive areas such as the Arctic, where key seabird habitats and potential shipping routes may overlap, and further exacerbate impacts from predicted climate change (Humphries and Huettman 2014).

4 Major ecosystem services provided by the species group and impacts of pressures on provision of these services

4.1 Services to ecosystems

The role of seabirds as potential indicators of marine conditions is widely acknowledged (e.g., Boyd et al., 2006; Piatt et al., 2007; Parsons et al.,

2008). Many studies use aspects of seabird biology and ecology, especially productivity and population trends, to infer relationships with and/ or effects on and/or correlate with aspects of the marine environment, particularly food availability.

Seabirds play a key role in nutrient cycling via the shaping of the plant community in their terrestrial and coastal breeding habitat. Seabirds transport allochthonous nutrients (i.e., fixed nitrogen, phosphorus, and trace elements), mainly via their guano, to seabird colonies (i.e., cross-ecosystem subsidies). They also shape plant communities in their breeding habitat by creating physical disturbance, dispersing seeds, and bioturbating the soil with their burrowing (Ellis, 2005; Bancroft et al., 2005). These functions provided by seabirds increase productivity and diversity in terrestrial and coastal ecosystems surrounding seabird colonies (Powell et al., 1991; Bosman et al., 1986; Brimble et al., 2009).

4.2 Direct services to humans including economic and livelihood services

Seabirds contribute several provisioning (e.g., protein, guano) services, play an important cultural role in many countries (e.g., for the Maori of New Zealand and the Tsimshian of Alaska), and feature in Greek, Hawaiian and Christian mythology. Seabird breeding colonies are increasingly used as a means to generate tourism income.

Seabird guano has excellent properties as a natural fertilizer enriching both terrestrial (Havik et al. 2014) and marine (Gagnon et al. 2013) environments. It consists of nitrogen-rich ammonium oxalate and urate, phosphates, as well as some earth salts and impurities. It typically contains 8 to 16 per cent nitrogen (the majority of which is uric acid), 8 to 12 per cent equivalent phosphoric acid, and 2 to 3 per cent equivalent potash. Archaeological evidence suggests that Andean peoples have collected seabird guano for well over 1,500 years (Collar et al., 2007). A harvest boom in the nineteenth century, called the "white gold rush", saw tens of thousands of workers extracting guano from the Peruvian seabird breeding islands and loading thousands of tons onto each ship. Other harvested guano islands were located in the Caribbean, atolls in the Central Pacific, and off the coast of Namibia, South Africa, Oman, Patagonia, and Baja California (Skaggs, 1994). This unsustainable harvest resulted in massive deposits of guano, in some cases more than 50 m deep, being severely depleted. Many areas of the industry collapsed, although some Peruvian islands are still managed for guano on a rotational system (Méndez, 1987).

Harvesting of seabird adults, chicks, eggs, and feathers have been important activities for some coastal communities for many centuries, but have also driven seabird declines. Bones were used to make fishing hooks and musical instruments and to engrave tattoos; feathers featured prominently in the millinery trade and are still used in some countries for local arts and handicrafts, e.g., to make cloaks and hair adornments (Spennemann, 1998). The meat and eggs still form key sources of protein. Harvest methods have changed over time to include more efficient tools, making the seabirds more exposed to excessive harvesting. Declines in a number of species have been attributed to over-exploitation. Harvesting quotas exist in some areas, such as in the Seychelles (limited to 20 per cent of Sooty Tern eggs each year), New Zealand (limited to 13 per cent of Sooty Shearwater chicks each year (Newman et al., 2009), and the United Kingdom (limited to 2,000 Northern Gannet each year). However, unregulated harvesting is a substantial problem in the entire Arctic region (2 million adults and countless eggs of several species of Alcidae are taken each year (Merkel, 2008)), the Tuamotus and the Marquesas (egg collection), Peru (Waved Albatross and Humboldt Penguin), Madagascar (egg collection), Jamaica (egg collection (Haynes, 1987)) and Indonesia.

For centuries fishers have used seabirds as a visual guide to locate fishing areas. They remain important for artisanal operations (such as in Hawaii, Comoros, Madagascar and Tanzania), which search for flocks of seabirds in order to find fish. Without seabirds, these livelihoods (e.g., catching small skipjack and juvenile yellow-fin tuna) could disappear or be substantially adversely affected.

Viewing seabirds is an increasingly popular pastime for many tourists; many spectacular breeding colonies are accessible to visitors and revenues generated contribute substantially to local economies (Steven et al., 2013). For example, in Australia, the Phillip Island Little Penguin colony receives half a million visitors a year, spending 35 million Australian dollars (Marsden Jacob Associates, 2008). A single African Penguin colony in South Africa generates United States dollars 2 million/yr in tourist revenue (Lewis et al., 2012). In New Zealand, nature-based tourism relying primarily on the Yellow-eyed Penguin returned 100 million dollars annually to the Dunedin economy, hence a single breeding pair could be worth 60,000 dollars/yr (Tisdell, 2008). The Royal Society for the Protection of Birds (RSPB) estimated that four of its seabird reserves in the UK (one each in England, Northern Ireland, Scotland and Wales) together generated around 1.5million dollars/yr for the local economies (RSPB 2010). Tourism in the Galapagos is thought to generate over 62 million dollars each year; seabirds are a prime reason for visiting. Pelagic trips to view seabirds at sea have also become popular, particularly in Europe, North America and the Southern Ocean. The value of these trips has not been quantified to any degree, but is likely to be significant; for example, 80,000 dollars was spent on a single pelagic trip off South Africa (Turpie and Ryan, 1999).

5 Conservation responses and factors for sustainability

Data on seabird distribution, abundance, behaviour and pressures can be used to inform the design of effective management regimes (Lascelles et al 2012). Management decisions can be guided by: (1) where the key areas are, (2) when these areas are used, (3) what variables explain seabird presence in a given area, (4) the threat status of species in a given area, (5) what pressures may be adversely affecting the species, associated habitats and processes, (6) what management actions are needed to address these threats, and (7) how any management intervention can best be monitored to assess its effectiveness.

Depending on the species, the priority actions needed may involve: (a) formal and effective protection of the most important sites. For site protection to be effective, it should ensure that areas are large enough to capture critical behaviour (such as key breeding sites, the marine areas around them used for maintenance and more distant feeding and aggregation sites), consider temporal and spatial variations, and have adequate regulation to minimise effects of any pressures. Where national, regional and global networks of Marine Protected Areas (MPAs) are being developed, inclusion of key sites in those networks would contribute substantially to the necessary site protection; (b) removal or control of invasive, and especially predatory, alien species from areas used for seabird breeding, feeding and/or aggregation, as part of habitat and species recovery initiatives; and (c) reduction of bycatch to levels that do not pose a threat of species decline. For many uncommon species or species of low productivity, this likely can only be achieved when bycatch is reduced to near zero. Other, more generic actions, such as education and awareness-raising and accompanying stakeholder involvement, are also high priorities, as are some more species-specific activities, such as harvest management, species reintroductions and species recovery. Although it is relatively straightforward to derive these generic recommendations for conservation action, it can be costly and difficult to implement them effectively and at a sufficient scale to make a difference to the conservation status of seabird species. However, progress has been achieved in recent years in terms of the three highest priority actions, but despite these successes, problems will continue without further action.

Where simple seabird mitigation measures have been implemented, there is evidence of substantially reduced bycatch in some key fisheries where the pressure has been managed (e.g. Anderson *et al.*, 2011), including a greater than 95 per cent reduction in some areas (Maree *et al.*, 2014). The main tuna RFMOs now have voluntary or binding regulations in place that require the use of a combination of mitigation techniques in different geographies, though their effectiveness may be hampered by a lack of monitoring and/or enforcement.

Key sites for seabirds have begun to be protected in several countries, primarily covering selected breeding sites on land, though marine designation for seabirds has also advanced, with new MPAs in Europe, the Antarctic and the Americas in recent years. Where eradications and/or controls of invasive alien species have been undertaken, recoveries of seabird populations have been rapid and dramatic (e.g Pitman *et al.*, 2005), and a great number of larger islands are now being tackled. Translocations of some species to new locations have also proved an effective conservation strategy for several species (e.g. Carlile *et al.*, 2003; Madeiros 2004).

Actions that are implementing an ecosystem approach to capture fisheries management are discussed in Part IV of this assessment; many of those measures, including better management of selectivity of fishing gear and including ecosystem feeding requirements in setting fishery harvest limits, will contribute to improving the conservation status of seabirds if implemented effectively.

References

Ainley, D.G., Podolsky, R., DeForest, L. and Spencer, G. (1997) New insights into the status of the Hawaiian petrel on Kauai. Colon. *Waterbirds* 20: 24–30.

Anderson, O.R.J., Small, C.J., Croxall, J.P., Dunn, E.K., Sullivan, B.J., Yates, O., and Black, A. (2011). Global seabird bycatch in longline fisheries. *Endangered Species Research* 14: 91–106.

Bancroft, W.J., Roberts, J.D., and Garkaklis, M.J. (2005). Burrowing seabirds drive decreased diversity and structural complexity, and increased productivity in insular-vegetation communities. *Australian Journal of Botany* 53: 231–241.

Bartle, J.A. (1991). Incidental capture of seabirds in the New Zealand Subantarctic Squid trawl fishery, 1990. *Bird Conservation International* 1: 351–359.

Becker, B.H., and Beissinger, S.R. (2006). Centennial decline in the trophic level of an endangered seabird after fisheries decline. *Conservation Biology* 20: 470–479.

BirdLife International (2007). Distribution of albatrosses and petrels in the WCPFC Convention Area and overlap with WCPFC longline fishing effort. Paper submitted to WCPFC Scientific Committee Third Regular Session 13–24 August 2007, Honolulu, United States of America. WCPFC-SC3-EB SWG/IP-17.

BirdLife International (2012). The Red List Index for species covered by the Agreement on the Conservation of Albatrosses and Petrels. MoP4 Inf 03 Agenda Item 7.5. Fourth Meeting of the Parties Lima, Peru, 23 – 27 April 2012.

Bosman, A.L., Du Toit, J.T., Hockey, P.A.R., and Branch, G.M. (1986). A field experiment demonstrating the influence of seabird guano on intertidal primary production. *Estuarine, Coastal and Shelf Science* 23(3): 283-294.

Boyd I.L., Wanless, S., and Camphuysen, K., editors (2006). *Top predators in marine ecosystems: their role in monitoring and management*. Cambridge, UK: Cambridge University Press.

Brimble, S.K., Blais, J.M., Kimpe, L.E., Mallory, M.L., Keatley, B.E., Douglas, M.S.V., and Smol, J.P. (2009). Bioenrichment of trace elements in a series of ponds near a northern fulmar (Fulmarus glacialis) colony at Cape Vera, Devon Island. *Canadian Journal of Fisheries and Aquatic Science* 66: 949–958.

Brothers, N.P. (1991). Albatross mortality and associated bait loss in the Japanese longline fishery in the Southern Ocean. *Biological Conservation* 55: 255–268.

Bugoni, L., Neves, T.S., Leite, N.O. Jr., and Carvalho, D. (2008). Potential bycatch of seabirds and turtles in hook and- line fisheries of the Itaipava Fleet, Brazil. *Fish Research* 90: 217–224.

Butchart, S.H.M., Stattersfield, A.J., Bennun, L.A., Shutes, S.M., Akçakaya, H.R., Baillie, J.E.M., Stuart, S.N., Hilton-Taylor, C., Mace, G.M. (2004). Measuring global trends in the status of biodiversity: Red List Indices for birds. *PLoS Biology* 2: 2294–2304.

Butchart, S.H.M., Akçakaya, H.R., Chanson, J., Baillie, J.E.M., Collen, B., Quader, S., Turner, W.R., Amin, R., Stuart, S.N., Hilton-Taylor, C., and Mace, G.M. (2007). Improvements to the Red List Index. *PLoS One* 2(1): e140. doi:10.1371/journal.pone.0000140.

Camphuysen, C.J. (2005). Understanding marine foodweb processes: an ecosystem approach to sustainable sandeel fisheries in the North Sea. *IMPRESS Final Report* Project# Q5RS-2000-30864.

Carlile, N., Priddel, D., Zino, F., Natavidad, C., Wingate, D.B. (2003). A review of four successful recovery programmes for threatened sub-tropical petrels. *Marine Ornithology* 31: 185-192.

Collar, N.J., Long, A.J., Robles-Gil, P., and Rojo, J. (2007). *Birds and people: bonds in a timeless journey*. Mexico City: CEMEX.

Croxall, J.P., Butchart, S.H.M., Lascelles, B., Stattersfield, A.J., Sullivan, B., Symes, A., Taylor, P. (2012). Seabird conservation status, threats and priority actions: a global assessment. *Bird Conservation International* 22: 1-34.

Cury, P.M., Boyd, I.L., Bonhommeau, S., Anker-Nilssen, T., Crawford, R.J.M., Furness, R.W., Mills, J.A., Murphy, E.J., Österblom, H., Paleczny, M., Piatt, J.F., Roux, J.P., Shannon, L., and Sydeman, W.J. (2011). Global Seabird Response to Forage Fish Depletion—One-Third for the Birds. *Science* 334(6063): 1703-1706.

Day, R.H., Cooper, B.A. and Telfer, T.C. (2003). Decline of Townsend's (Newell's) shearwater (Puffinus auricularis newelli) on Kauai, Hawaii. *Auk* 120: 669–679.

Dunn, E., and Steel, C. (2001). The impact of longline fishing on seabirds in the northeast Atlantic: recommendations for reducing mortality. *NOF Rapportserie Rep. No. 5*, The Royal Society for the Protection of Birds (RSPB), Sandy, United Kingdom.

Ellis, J.C. (2005). Marine birds on land: a review of plant biomass, species richness, and community composition in seabird colonies. *Plant Ecology* 181: 227–241.

Foden, W.B., Butchart, S.H.M., Stuart, S.N., Vié, J.C., Akçakaya, H.R., Angulo, A., DeVantier, L.M., Gutsche, A., Turak, E., Cao, L., et al. (2013). Identifying the world's most climate change vulnerable species: a systematic trait-based assessment of all birds, amphibians and corals. *PLoS ONE* 8: e65427.

Forcada, J., Trathan, P.N. (2009). Penguin responses to climate change in the southern Ocean. *Global Change Biology* 15: 1618–1630.

Gagnon, K., Rothäusler, E., Syrjänen, A., Yli-Renko, M., VJormalainen, V. (2013). Seabird Guano Fertilizes Baltic Sea Littoral Food Webs. *PLoS ONE* DOI:10.1371/journal.pone.0061284.

García, L., Viada, C., Moreno-Opo, R., Carboneras, C., Alcade, A., and Gonzalez, F. (2003). *Impacto de la marea negra del "Prestige" sobre las aves marinas*. Madrid: SEO/BirdLife

Gonzalez-Zevallos, D., Yorio, P., and Caille, G. (2007). Seabird mortality at trawler warp cables and a proposed mitigation measure: a case of study in Golfo San Jorge, Patagonia, Argentina. *Biological Conservation* 136: 108–116.

Hall, M.A. (1998). An ecological view of the tuna-dolphin problem: impacts and trade-offs. *Reviews in Fish Biology & Fisheries* 8: 1–34.

Havik, G., Catenazzi, A., Holmgren, M. (2014). Seabird Nutrient Subsidies Benefit Non-Nitrogen Fixing Trees and Alter Species Composition in South American Coastal Dry Forests. PlosOne DOI: 10.1371/journal.pone.0086381

Haynes, A.M. (1987). Human exploitation of seabirds in Jamaica. *Biological Conservation* 41: 99-124.

Hazen, E.L., Jorgensen S., Rykaczewski, R.R., Bograd, S.J., Foley, D.G., Jonsen, I.D., Shaffer, S.A., Dunne, J.P., Costa, D.P., et al. (2012). Predicted habitat shifts of Pacific top predators in a changing climate. *Nature Climate Change* 3: 234–238.

Humphries, G.R.W., Huettmann, F. (2014). Putting models to a good use: a rapid assessment of Arctic seabird biodiversity indicates potential conflicts with shipping lanes and human activity. *Diversity and Distributions* 02/2014; 20(4). DOI:10.1111/ddi.12177

Johnson, A., Salvador, G., Kenney, J., Robbins, J., Landry, S., Clapham, P., and Kraus, S. (2005). Fishing gear involved in entanglements of right and humpback whales. *Marine Mammals Science* 21: 635–645.

Kirby, D., Waugh, S., and Filippi, D. (2009). *Spatial risk indicators for seabird interactions with longline fisheries in the western and central Pacific*. Western and Central Pacific Fisheries Commission-SC5-2009/EB-WP-06.

Laist, D.W. (1997). Impacts of marine debris: entanglement of marine life in marine debris including a comprehensive list of species with entanglement and ingestion records. In Coe, J.M., and Rogers, D.B., editors. *Marine Debris- Sources, Impacts, and Solutions*. New York: Springer-Verlag. pp. 99–139.

Lascelles B., Notarbartolo Di Sciara, G., Agardy, T., Cuttelod, A., Eckert, S., Glowka, L., Hoyt, E., Llewellyn, F., Louzao, M., Ridoux, V., and Tetley, M.J., (2014) Migratory marine species: their status, threats and conservation management needs, *Aquatic Conservation: Marine and Freshwater Ecosystems*, 24, pages 111–127. doi: 10.1002/aqc.2512.

Lascelles et al. (2012). From Hotspots to Site Protection: Identifying Marine Protected Areas for Seabirds around the Globe, *Biological Conservation* 165: 5-14.

Le Corre, M., Ollivierb, A., Ribes, S., Jouventin, P. (2002). Light-induced mortality of petrels: a 4-year study from Reunion Island (Indian Ocean). *Biological Conservation* 105: 93–102.

Lewis, S.E.F., Turpie, J.K., and Ryan, P.G. (2012). Are African penguins worth saving? The ecotourism value of the Boulders Beach colony. *African Journal of Marine Science* 34(4): 497-504.

Madeiros, J. (2004). The 2004 translocation of Cahow chicks to Nonsuch Island. B*ermuda Audubon Society 50thAnniversary Report*: 21–23.

Maree, B.A., Wanless, R.M., Fairweather, T.P., Sullivan, B.J. and Yates, O. (2014), Significant reductions in mortality of threatened seabirds in a South African trawl fishery. *Animal Conservation*, 17: 520–529. doi: 10.1111/acv.12126.

Marsden Jacob Associates (2008). *The potential impacts of climate change on the Phillip Island Little Penguin colony - regional economic impacts*.

Méndez, C. (1987). *Los trabajadores guaneros del Perú, 1840–1879*. Lima: Universidad Nacional Mayor de San Marcos.

Merkel, F., and Barry, T., (eds). (2008). *Seabird harvest in the Arctic*. CAFF International Secretariat, Circumpolar Seabird Group (CBird), CAFF Technical Report No. 16.

Montevecchi, W.A. (2006). Influences of artificial light on marine birds. In Rich, C. and Longcore, T., (eds). *Ecological consequences of artificial night lighting*. Washington, D.C.: Island Press.

Nel, D.C., and Taylor, F.E. (2003). *Globally threatened seabirds at risk from longline fishing: international conservation responsibilities*. BirdLife International Seabird Conservation Programme, BirdLife South Africa, Cape Town.

Newman, J., Scott, D., Bragg, C., McKechnie, S., Moller, H., and Fletcher, D. (2009). Estimating regional population size and annual harvest intensity of the sooty shearwater in New Zealand. *New Zealand Journal of Zoology*, 36: 307–323.

Northridge, S. (1991). Driftnet fisheries and their impacts on non-target species: a worldwide review. Rome: FAO (*FAO Fisheries Technical Paper* No. 320).

Parsons, M., Mitchell, I., Butler, A., Ratcliffe, N., Frederiksen, M., Foste, S., and Reid, J.B. (2008). Seabirds as indicators of the marine environment. *ICES Journal of Marine Science* 65: 1520– 1526.

Petersen, S.L., Nel, D.C., Ryan, P.G., and Underhill, L.G. (2008). Understanding and mitigating vulnerable bycatch in southern African trawl and longline fisheries. *WWF South Africa Report Series*, 2008/Marine/002.

Piatt, J.F., and Nettleship, D.N. (1987). Incidental catch of marine birds and mammals in fishing nets off Newfoundland, Canada. *Marine Pollution Bulletin* 18:344–349.

Piatt, J.F., Ford, R.G. (1996). How many seabirds were killed by the Exxon Valdez oil spill? In Rice, S.D., Spies, R.B., Wolfe, D.A., and Wright, B.A., editors. Proceedings of the Exxon Valdez oil spill symposium. American Fisheries Society Symposium 18. pp. 712-720.

Piatt, J.F., Sydeman, W.J., and Wiese, F. (2007). Introduction: a modern role for seabirds as indicators. *Marine Ecological Progress Series* 352: 199–204.

Pitman, R.L., Balance, L.T. and Bost, C. (2005). Clipperton Island: Pig sty, rat hole and booby prize. *Marine Ornithology* 33: 193–194.

Powell, G.V.N., Fourquerean J.W., Kenworthy, W.J., and Zieman, J.C. (1991). Bird colonies cause seagrass enrichment in a subtropical estuary: Observational and experimental evidence. *Estuarine, Coastal and Shelf Science* 32(6): 567–579.

Provencher et al. 2014. Prevalence of marine debris in marine birds from the North Atlantic. Marine Pollution Bulletin 84:411-417.

Reed, J.R., Sincock, J.L., and Hailman, J.P. (1985). Light attraction in endangered procellariiform birds: reduction by shielding upward radiation. *The Auk, vol.* 102: 377–383.

Rivera, K.S., Henry, R.W. III, Shaffer, S.A., LeBoeuf, N., and VanFossen, L. (2008). Seabirds and fisheries in the IATTC area: an update. 9th Meeting of the Inter-American Tropical Tuna Commission (IATTC) Working Group on Stock Assessment, 12–16 May 2008, La Jolla, CA. IATTCSARM- 9-11a.

Rogan, E., and Mackey, M. (2007). Megafauna bycatch in drift nets for albacore tuna (Thunnus alalunga) in the NE Atlantic. *Fisheries Research* 86: 6–14.

Ronconi et al 2014. Bird interactions with offshore oil and gas platforms: Review of impacts and monitoring techniques. Journal of Environ mental Management 147:34-45.

RSPB (Royal Society for the Protection of Birds) (2010). *The local value of seabirds: Estimating spending by visitors to RSPB coastal reserves and associated local economic impact attributable to seabirds*. The RSPB, Sandy, United Kingdom.

Skaggs, J. (1994). *The Great Guano Rush: Entrepreneurs and American Overseas Expansion*. New York: St. Martin's. ISBN 0312103166.

Spennemann, D.H.R. (1998). Excessive exploitation of Central Pacific seabird populations at the turn of the 20th Century. *Marine Ornithology* 26: 49–57.

Steven, R., Castley, J.G., and Buckley, R. (2013). Tourism revenue as a conservation tool for threatened birds in protected areas. *PLoS ONE* 8(5): e62598. doi:10.1371/journal.pone.0062598.

Sullivan, B.J., Reid, T.A., and Bugoni, L. (2006). Seabird mortality on factory trawlers in the Falkland Islands and beyond. *Biological Conservation* 131: 495–504.

Sydeman, W.J., Thompson, S.A., Kitaysky, A. (2012). Seabirds and climate change: roadmap for the future. *Marine Ecology Progress Series* 454: 107–117.

Tasker, M.L., Camphuysen, C.J., Cooper, J., Garthe, S., Montevecchi, W.A., and Blaber, S.J.M. (2000). The impacts of fishing on marine birds. *ICES Journal of Marine Science* 57: 531–547.

Tisdell, C. (2008). Wildlife conservation and the value of New Zealand's Otago peninsula: economic impacts and other considerations Working Paper No. 149. *Economics, Ecology and the Environment*. University of Queensland. ISSN 1327-8231.

Turpie, J., and Ryan, P. (1999). What are birders worth/ the value of birding in South Africa. *Africa Birds & Birding*. pp. 64-68.

Waugh, S.M., Baker, G., Gales, R., and Croxall, J.P. (2008). CCAMLR process of risk assessment to minimise the effects of longline fishing mortality on seabirds. *Marine Policy* 32(3): 442–454.

Waugh, S.M., Filippi, D.P., Kirby, D.S., Abraham, E., and Walker, N. (2012). Ecological Risk Assessment for seabird interactions in Western and Central Pacific longline fisheries. *Marine Policy* 36: 933-946. doi.org/10.1016/j.marpol.2011.11.005.

Weimerskirch, H., Capdeville, D., and Duhamel, G. (2000). Factors affecting the number and mortality of seabirds attending trawlers and long-liners in the Kerguelen area. *Polar Biology* 23: 236–249.

Yorio, P., Quintana, F., Dell'arciprete, P., and Gonzalez-zevallos, D. (2010). Spatial overlap between foraging seabirds and trawl fisheries: implications for the effectiveness of a marine protected area at Golfo San Jorge, Argentina. *Bird Conservation International* 20: 320–334.

Žydelis, R., Bellebaum, J., Österblom, H., Vetemaa, M., Schirmeister, B., Stipniece, A., Dagys, M., van Eerden, M., and Garthe, S. (2009). Bycatch in gillnet fisheries—

an overlooked threat to waterbird populations. *Biological Conservation* 142: 1269–1281.

Žydelis, R., Small, C., French, G. (2013). The incidental catch of seabirds in gillnet fisheries: A global review. *Biological Conservation* 162: 76–88.

39 Marine Reptiles

Contributors:
Bryan P. Wallace (Convenor), Peter H. Dutton, Maria Angela Marcovaldi, Vimoksalehi Lukoschek and Jake Rice (Lead Member and Editor of Part VI)

1 Assessment Frameworks

Although several other frameworks assess marine turtle status at global and sub-global scales, in this chapter we focus on results from the International Union for the Conservation of Nature (IUCN) Red List assessments and the IUCN Marine Turtle Specialist Group's conservation priorities portfolio (Wallace et al., 2011) because these are the most comprehensive and widely recognized assessment frameworks at present. For a comprehensive summary of other assessment frameworks for marine turtles, please see Chapter 35. In this chapter, we provide an overview of the two above-mentioned IUCN assessments with regard to marine turtles, and we also present available information on the conservation status of sea snakes and marine iguanas.

2 Status Assessments

2.1 IUCN Red List

The primary global assessment framework for marine turtle species is the IUCN *Red List of Threatened Species*TM (www.iucnredlist.org). The universally applicable criteria and guidelines of the Red List make it the most widely used and accepted framework for assessing the conservation status of species worldwide.

The IUCN Marine Turtle Specialist Group (MTSG), one of the IUCN/Species Survival Commission's specialist groups, is responsible for conducting regular Red List assessments of each marine turtle species on a global scale. However, because marine turtle population traits and trajectories can vary geographically, the global extinction risk assessment framework represented by the Red List does not adequately assess the conservation status of spatially and biologically distinct marine turtle populations (see Seminoff and Shanker, 2008 for review).

2.2 Subpopulation or regional assessments

To address the challenges presented by the mismatched scales of global Red List assessments and regional/population-level variation in status, the MTSG developed an alternative assessment framework and a new approach to Red List assessments that better characterize variation in status and trends of individual populations (Wallace et al., 2010; Wallace et al., 2011; see next section). This new approach centres on assessing marine turtle subpopulations, as well as the global population (i.e., species), using Red List guidelines, which results in official Red List categories for subpopulations in addition to the single global listing. This working group first developed regional management units (RMUs) (i.e., spatially explicit population segments defined by biogeographical data of marine turtle species) as the framework for defining biologically meaningful population segments for assessments (Wallace et al., 2010). RMUs are functionally equivalent to IUCN subpopulations, thus providing the appropriate demographic unit for Red List assessments. Next, the group developed a flexible yet robust framework for assessing population viability and degree of threats that could be applied to any subpopulation in any region (Wallace et al., 2011). Population viability criteria included abundance, recent and long-term trends, rookery vulnerability, as well as genetic diversity, and threats included by-catch (i.e., incidental capture in fishing gear), human consumption of turtles or turtle products, coastal development, pollution and pathogens, and climate change. The final product was a "conservation priorities portfolio" for all subpopulations globally. It includes identification of critical data needs, as well as risk and threats criteria by subpopulation, and reflects the wide variety of conservation objectives held by different stakeholders, depending on institutional or regional priorities.

3 Conservation Status of Marine Reptiles

3.1 Marine Turtles

Currently, global Red List categories for marine turtle species are: Vulnerable (leatherback, *Dermochelys coriacea*; olive ridley, *Lepidochelys olivacea*), Endangered (loggerhead, *Caretta caretta*; green turtle, *Chelonia mydas*), Critically Endangered (Kemp's ridley, *Lepidochelys kempii*; hawksbill, *Eretmochelys imbricata*), and DataDeficient (flatback, *Natator depressus*).

However, as mentioned above, the MTSG is actively appraising Red List assessments to include all subpopulations, as well as the global listing for each marine turtle species. In 2013, the MTSG completed the first complete suite of subpopulation assessments—in addition to the global listing—for any marine turtle species (Wallace et al., 2013a). The updated Red List assessments for leatherback turtles changed the global status for this species from Critically Endangered to Vulnerable—due to new data becoming available and to one large and increasing subpopulation (Northwest Atlantic Ocean)—and added new listings for each of the seven leatherback subpopulations, which ranged from Critically Endangered (East Pacific Ocean; West Pacific Ocean; Southwest Atlantic Ocean; Southwest Indian Ocean) to Least Concern (Northwest Atlantic Ocean) to Data-Deficient (Southeast Atlantic Ocean; Northeast Indian Ocean) (Wallace et al., 2013a). Updated global and subpopulation assessments are expected to be completed in 2016-2018.

3.2 MTSG's conservation priorities portfolio

Marine turtle Red List assessments have been and will continue to be informed by the MTSG's conservation priorities portfolio (Wallace et al., 2011), the results of which are presented briefly here.

Average values of population risk and threats criteria across marine turtle subpopulations assessed by Wallace et al. (2011) are presented in Table 39.1. Globally, long-term population trends are declining on average across marine turtle subpopulations, but are stable or perhaps even increasing in recent years (Table 39.1). In general, population viability criteria tend to cluster around moderate values across subpopulations.

Table 39.1 | Average values of population risk and threats criteria across marine turtle subpopulations. Scores range from 1 (high abundance, increasing trends, high diversity, low threats) to 3 (low abundance, declining trends, low diversity, high threats).

RISK SCORES					
	population size	recent trend	long-term trend	rookery vulnerability	genetic diversity
mean	1.95	1.81	2.47	1.72	1.90
No. subpop'ns scored	58	43	38	57	58
THREATS SCORES					
	fisheries by-catch	human consumption	coastal development	pollution and pathogens	climate change
mean	2.21	2.08	1.93	1.70	2.20
No. subpop'ns scored	56	57	53	25	20

At ocean-basin scales (i.e., Atlantic Ocean and Mediterranean Sea, Indian Ocean, Pacific Ocean), subpopulations in the Pacific Ocean had the highest average risk (i.e., population viability) score, whereas subpopulations in the Atlantic Ocean (as well as in the Mediterranean Sea) had the highest average risk and threats score (Table 39.2). Indian Ocean subpopulations had the highest average data uncertainty scores for both risk and threats (Table 39.1), as well as the most populations assessed as "critical data needs" (Table 39.3).

One-third of all marine turtle subpopulations were assessed as "high risk-high threats"—i.e. low, declining abundance and low diversity simultaneously under high threats—which could be considered as the world's most endangered populations (Wallace et al., 2011). Between 20 and 30 per cent of subpopulations in each ocean basin were "high risk-high threats" (Table 39.3). More than half of *E. imbricata* subpopulations and roughly 40 per cent of *C. caretta* and *D. coriacea* subpopulations were categorized as High Risk-High Threats (Figure 39.1).

One-fifth of marine turtle subpopulations globally were categorized as "low risk-low threats"—i.e., high and stable or increasing abundance, high diversity while experiencing low to moderate threats—a pattern that was reflected at the ocean-basin scale as well (Table 39.2). These included five *C. mydas* subpopulations, three *E. imbricata* subpopulations, two *D. coriacea* subpopulations, and one each for *C. caretta* and *L. olivacea* (Fig. 1).

These results illustrate both the large degree of variation and level of uncertainty in the conservation status of marine turtles within and among species and regions, as well as the importance of flexible assessment frameworks capable of reflecting these sources of variation.

Table 39.2 | Average risk and threats scores (and accompanying data uncertainty indices) of subpopulations that occur in each ocean basin.

ocean basin	average risk score	average risk score data uncertainty	average threats score	average threats score data uncertainty
Atlantic/Med (n=19)	1.81	0.26	2.16	0.35
Indian (n=18)	1.92	0.78	2.08	0.68
Pacific (n=21)	2.03	0.32	1.96	0.48

3.3 Sea snakes

Elapid sea snakes comprise two evolutionary lineages: live-bearing true sea snakes (at least 63 species) and egg-laying amphibious sea kraits (genus *Laticauda* - 8 species). True sea snakes are further divided into two monophyletic groups, the *Aipysurus* group (> 10 species in two genera, predominantly associated with coral reefs) and the *Hydrophis* group (> 50 species in ten nominal genera, mostly associated with inter-reefal habitats) (Lukoschek and Keogh, 2006). Marine elapids are found throughout the Indian and Pacific Oceans, but do not occur in the Atlantic Ocean, Mediterranean or Caribbean Seas. Highest species richness occurs in Southeast Asia and northern Australia (Elfes et al., 2013). Marine snakes are poorly studied: new species continue to be described, and revisions to taxonomic status and geographic ranges are not uncommon, resulting in changes in the numbers of recognized species and complicating assessments of their conservation status.

In 2009, the first Red List global marine assessment of extinction risk was conducted for 67 of the 71 elapid sea snake species recognized at the time (Elfes et al., 2013). Six species were classified in one of the threatened categories (Critically Endangered, Endangered or Vulnerable) and four species were classified as Near Threatened. The three most threatened species were *Aipysurus* congeners, two of which were Critically Endangered (*A. apraefrontalis* and *A. foliosquama*) and one Endangered (*A. fuscus*). At the time of the Red List Assessments, these three species were regarded as being endemic to a small number of reefs in the Timor

Table 39.3 | Categories in which RMUs occurred in each basin (including critical data needs RMUs). Categories: HR-HT=High Risk-High Threats; HR-LT=High Risk-Low Threats; LR-LT=Low Risk-Low Threats; LR-HT=Low Risk-High Threats. * One RMU (C. mydas, northeast Indian Ocean) was scored critical data needs only.

ocean basin	critical data needs	HR-HT	HR-LT	LR-LT	LR-HT	Total
Atlantic/Med (n=19)	1	5	2	3	9	19
Indian (n=18) *	8	6	3	4	4	17*
Pacific (n=21)	3	8	4	5	4	21
Total	12	21	9	12	15	57*

Sea, where they had undergone catastrophic population declines since the mid-1990s (Lukoschek et al., 2013). However, recent sightings of at least one of these three species on coastal reefs in Western Australia suggest that further research is needed to confirm their true geographic ranges (Lukoschek et al., 2013). Of the eight species of *Laticauda*, two were classified as Vulnerable and three as Near Threatened (Elfes et al., 2013). Both Vulnerable species of *Laticauda* were small-range endemics (*L. crockeri* restricted to Lake Te-Nggano in the Solomon Islands; *L. schistorhyncha* to Niue), as were two of the three Near Threatened species (*L. frontalis* occurring only in Vanuatu and the Loyalty Islands; *L. guineai* restricted to Southern New Guinea). The third Near Threatened species, *L. semifasciata*, had undergone significant historical declines in the Philippines due to harvest for skin and food. *Hydrophis semperi* (endemic to Lake Taal in the Philippines, was classified as Vulnerable, and *Hydrophis pacificus* (endemic to North-east Australian waters) was classified Near Threatened. Of the remaining 57 species, 34 were classified as of Least Concern and 23 as Data-Deficient (Elfes et al., 2013). Several species classified as Data-Deficient are known only from a few museum specimens collected many years ago and may not be valid species. At the same time, some species listed as Data-Deficient may, in fact, be threatened and clarification of threat status for Data-Deficient species is needed (Elfes et al., 2013).

3.4 Marine iguanas

Marine iguanas (*Amblyrhynchus cristatus*) are the world's only marine lizard species, and are endemic to the Galápagos Islands (Ecuador). Ten subpopulations occur on separate islands within the archipelago, but the status of most of these subpopulations is unknown. Marine iguanas occupy rocky coastal areas and intertidal areas, and forage on marine algae in nearshore waters (Nelson et al. 2004). Although abundance estimates are unavailable for seven of the subpopulations, abundance estimates of three subpopulations range between 1,000-2,000 individuals (Rabida Island), 4,000-10,000 (Marchena Island), and 15,000-30,000 (Santa Fe Island) (Nelson et al. 2004). Due to their restricted distribution and area of occupancy, marine iguanas are classified as Vulnerable according to the IUCN Red List (Nelson et al. 2004).

4 Threats to Marine Reptiles Globally

4.1 Marine Turtles

Dutton and Squires (2011) highlight the need for a holistic conservation approach that addresses all sources of mortality and deals with the trans-boundary nature of these multiple threats. Decades of overharvest of eggs on nesting beaches have driven historic declines of some breeding populations, rendering them more vulnerable to impacts from fisheries by-catch and other threats. According to Wallace et al. (2011), fisheries by-catch was scored as the highest threat across marine turtle subpopulations, followed by human consumption and coastal development (Table 39.1). Climate change was scored as Data-Deficient in two-thirds of all RMUs, whereas pollution and pathogens were scored as Data-Deficient in more than half of all RMUs (Table 39.1).

A recent global assessment of fisheries by-catch impacts documented the Mediterranean Sea, Northwest and Southwest Atlantic, and East Pacific Oceans as regions with particularly high by-catch threats to marine turtle subpopulations (Wallace et al., 2013b). This assessment also highlighted the disproportionately large impact that by-catch in small-scale fisheries in coastal areas can have on marine turtle populations. Efforts to reduce turtle by-catch have included changes in gear configuration and/or fishing method, time-area closures, and enforcement of by-catch quotas, but by-catch reduction has only been successful when tailored to local environmental factors and characteristics of fishing gear and methods (Lewison et al., 2013). At a global scale, the FAO has adopted guidelines to reduce sea turtle mortality in fishing operations and encourages States to adopt and implement sea turtle by-catch reduction measures according to the those guidelines. Human consumption of marine turtles and turtle products has occurred as traditional and subsistence use, as well as commercially, around the world for centuries. The full magnitude of the effects of this human consumption on marine turtle populations has not been quantified, but unsustainable rates of consumption have contributed to declines in abundance in several places (e.g., *C. mydas*, *D. coriacea*, *L. olivacea* in the East Pacific Ocean, Abreu-Grobois et al., 2008; Seminoff and Wallace, 2012; *E. imbricata* in the Wider Caribbean, Southeast Asia, West Pacific; Mortimer and Donnelly, 2008). Consumption of turtles and turtle products has been reduced in recent decades due to top-down enforcement of national and international regulations against trade and use of turtle products (e.g., Convention on International Trade in Endangered Species of Wild Fauna and Flora (CITES), national endangered species laws), but both legal and illegal turtle harvest continues in many countries (Humber et al., 2014).

Although climate change has been suggested as a major potential threat to marine turtles globally—e.g., possible skewing of sex ratios (which are controlled by temperature), habitat alteration related to increased frequency and severity of storms affecting nesting beaches, among other effects (Hamann et al. 2013)—specific impacts have not been quantified widely to date (Wallace et al., 2011). Increased beach sand and air temperatures and decreased precipitation might negatively affect hatchling production from nesting beaches, and fluctuating oceanographic conditions might alter migratory routes and foraging areas (Hawkes et al., 2009). More quantitative analyses of potential impacts to marine turtles related to climate change are warranted.

4.2 Sea Snakes

Sea snakes are a diverse group of meso-predators with varying habitat and prey requirements that range on a spectrum from being generalists to highly specialised. Some species of true sea snakes occur predominantly in inter-tidal and estuarine habitats, others are restricted to coral reefs, and others occur in reefal, inter-reefal and estuarine habitats. Egg-laying amphibious sea kraits require intact coral reefs for feeding, as well as intertidal and terrestrial sites for nesting and resting. In terms of

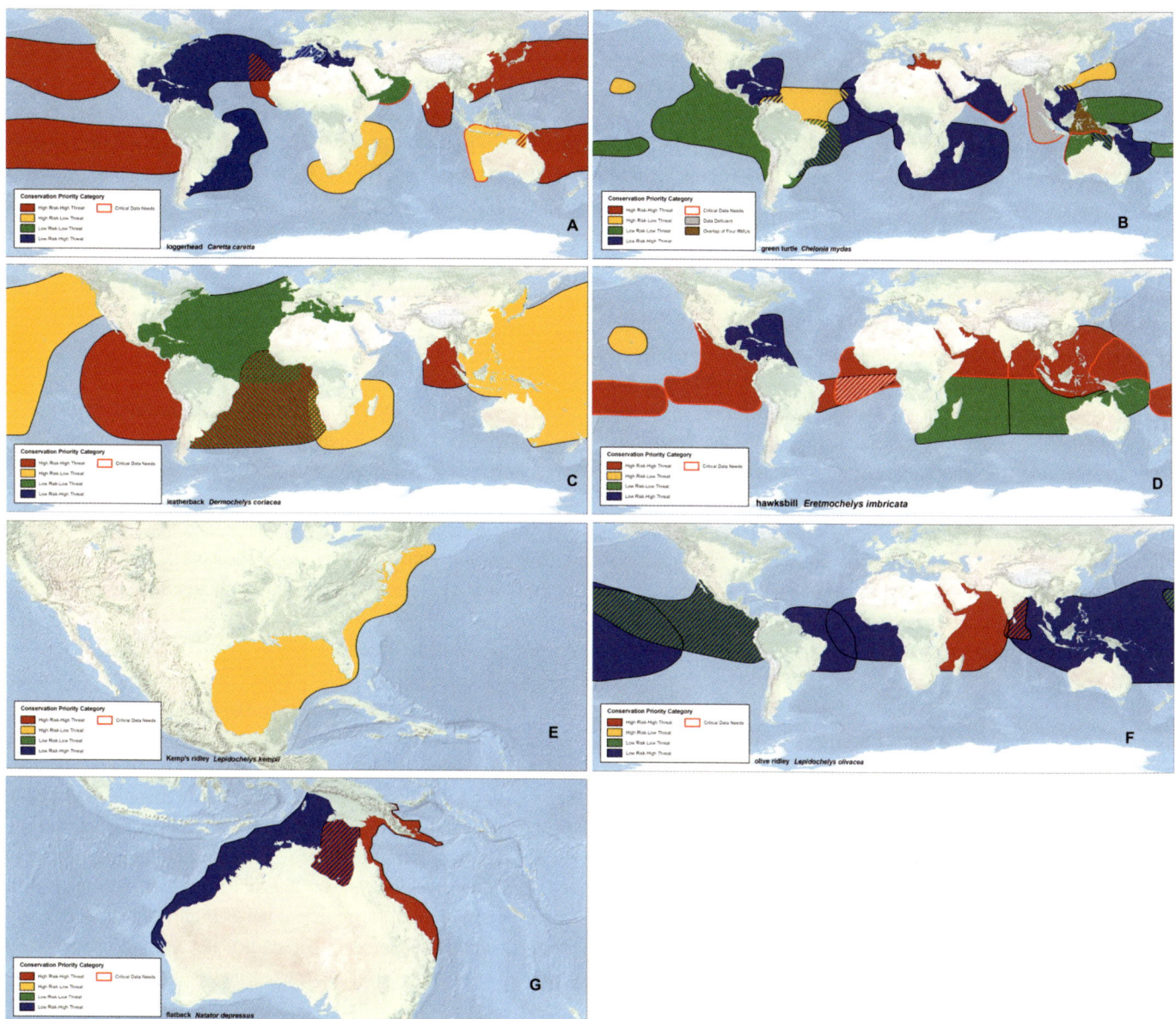

The boundaries and names shown and the designations used on this map do not imply official endorsement or acceptance by the United Nations.

Figure 39.1 | Conservation status of marine turtles: Four conservation priority categories are displayed: (red) high risk – high threat, (yellow) high risk – low threat, (green) low risk – low threat, (blue) low risk – high threat. Panels: (A) loggerheads (*Caretta caretta*), (B) green turtles (*Chelonia mydas*), (C) leatherbacks (*Dermochelys coriacea*), (D) hawksbills (*Eretmochelys imbricata*), (E) Kemp's ridleys (*Lepidochelys kempii*), (F) olive ridleys (*Lepidochelys olivacea*), (G) flatbacks (*Natator depressus*). Subpopulations were classified as having critical data needs (outlined in red) if the data uncertainty indices for both risk and threats ≥1 (denoting high uncertainty). Hatched areas represent spatial overlaps between subpopulations. The brown area in panel B highlights an overlap of four subpopulations, and the grey area in panel B represents the *C. mydas* Northeast Indian Ocean subpopulation, which had excessive data-deficient scores and was not included in overall calculations and categorization. Figure from Wallace et al. (2011) PLoS ONE 6(9): e24510. doi:10.1371/journal.pone.0024510.

diet, generalist species feed on a variety of small fish, eels, squid, and crustaceans, whereas dietary specialists, such as *Emydocephalus* spp., exclusively forage on eggs of small reef fish, and most sea kraits forage exclusively on eels. Range extents also vary enormously, with some species having extensive ranges (Persian Gulf to Australia), and others being restricted to a single island or inland lake, or a handful of coral reefs. The differing ecologies, diets and geographic ranges mean that potential threatening processes vary among species and among geographically disparate populations of the same species.

Globally sea snakes are taken as by-catch, particularly in trawl fisheries in inter-reefal and/or estuarine habitats. Most information about the nature and extent of sea snake by-catch comes from northern and eastern Australia, and indicates that species composition and abundance vary spatially, temporally and between fisheries (Courtney et al., 2009). For example, trawl fisheries on Queensland's east coast catch > 100,000 sea snakes from 12 species annually, of which approximately 25 per cent die; however, 59 per cent of all sea snake catches and ~85 per cent of deaths occur in just one fishery, due to the spatial overlap of habitats between the red-spot king prawns, *Melicertus longistylus*, being harvested and reef-associated sea snakes (Courtney et al., 2009). Nonetheless, risk assessments for Australia's Northern Trawl Fishery indicated that no sea snake species was at risk under the existing fishing effort (Milton et al., 2008). While the use of by-catch reduction devices (BRDs), which are placed the regulation 120 meshes from the codend drawstring, did not reduce sea snake by-catch (Milton et al., 2008), the use of some BRDs placed closer to the drawstring (<70 meshes) has been shown to reduce the number of snakes taken by 40-85 per cent without significant prawn loss (Milton et al., 2008).

In Southeast Asia, many reptile species are heavily harvested for the commercial food, medicine and leather trades; however, very limited information exists about the extent to which marine snakes are targeted and about potential impacts (Auliya, 2011). To some extent, this lack of information probably reflects the fact that to date no sea snake species has been CITES-listed. One anecdotal account of a tannery in West Malaysia indicates that over 6,000 spine-bellied sea snakes (*Lapemis curtus*) were harvested per month (Auliya, 2011), suggesting that the impact might be high if this account is representative of other locations. Nonetheless, *L. curtus* has a large geographic range, is a voracious generalist predator (feeding on a variety of small fish, eels, squid, crustaceans) and typically occurs in large numbers in many habitat types, so it may be able to sustain heavy harvests (Auliya, 2011).

The three most threatened sea snake species are endemic to coral reefs in the Timor Sea, including Ashmore Reef, a renowned sea snake biodiversity hotspot. Species diversity at Ashmore Reef has declined from at least nine species in 1973 and 1994 to just two species in 2010 (Lukoschek et al., 2013) and abundances have declined > 90 per cent from the estimated standing stock of > 40,000 snakes in the mid-1990s (Guinea and Whiting, 2005; Lukoschek et al., 2013). In addition to the three threatened species from the genus *Aipysurus*, two species that disappeared (*Aipysurus duboisii*, endemic to Australasia, and *Emydocephalus annulatus*, also in the *Aipysurus* group), typically occur on coral reefs, suggesting that their declines might be due to loss or degradation of reef habitats. Reef-associated sea snakes shelter and forage under ledges and within the reef matrix, where they might be affected by reductions in coral cover, diversity and habitat complexity following coral bleaching events. A mass bleaching event in 2003 caused widespread coral mortality at Ashmore Reef; however, the most pronounced sea snake declines occurred between the mid-1990s and 2002 (Lukoschek et al., 2013), preceding the 2003 coral loss. The cause of these declines is unknown (Lukoschek et al., 2013). Widespread bleaching associated with the 1998 El Niño event affected many Australian reefs, including Scott Reef in the Timor Sea, but Ashmore Reef experienced minimal coral loss in 1998 (Lukoschek et al., 2013). Moreover, two additional species that disappeared from Ashmore Reef (*Hydrophis coggeri* and *Acalyptophis peroni*) were predominantly associated with soft-sediment habitats. Illegal harvesting on Timor Sea reefs targets invertebrates and sharks, but there is no evidence that sea snakes have ever been taken (Lukoschek et al., 2013). Moreover, Ashmore Reef was declared a National Nature Reserve (IUCN Category 1a) in 1983 and a National Parks or Customs presence, maintained for much of the year since 1986, has limited illegal fishing at Ashmore Reef (Lukoschek et al., 2013). Similar declines of *Aipysurus* group species have occurred on protected reefs in New Caledonia (Goiran and Shine, 2013) and the southern Great Barrier Reef (Lukoschek et al., 2007a). Possible reasons for these apparently enigmatic declines of sea snakes include reproductive failure due to the sub-lethal or lethal effects of increased sea surface temperatures, disease, and pollution; however, compared with other marine vertebrates, limited research has been conducted quantifying the extent to which these processes affect sea snakes. There has been no research into the effects of ocean acidification on sea snakes.

Sea snakes tend to have highly patchy or aggregated distributions throughout their ranges. Genetics research on species from the *Aipysurus* group (Lukoschek et al., 2007b; Lukoschek et al., 2008; Lukoschek and Shine, 2012) suggests that dispersal (gene flow) between geographically disparate populations is limited and that local population declines or extinctions are unlikely to be reversed by dispersal over ecological time-scales relevant for conservation (Lukoschek et al., 2013).

4.3 Marine Iguanas

Periods of extremely high water temperatures and poor nutrient availability associated with El Niño events cause declines in food resources available to marine iguanas; dramatic (60-90 per cent) population declines related to El Niño have been documented (Vitousek et al. 2007). Introduced predators could also negatively affect marine iguana populations on some islands (Nelson et al. 2004). Increased stress responses and related changes in immune function have been documented in marine iguanas subject to consistent presence of tourists, which could pose a significant sub-lethal threat, particularly when compounded by periods of low resource availability (French et al. 2010).

5 Assessment and Conservation Needs

In general, an urgent need remains for enhanced monitoring and reporting of marine reptile population status and trends, as well as of threats to marine reptiles globally. For example, insufficient information was available to assess recent and long-term trends for roughly 25-30 per cent of all subpopulations, and threats such as climate change also remain poorly quantified (Wallace et al., 2011). Significant efforts to quantify fundamental marine reptile demographic rates and processes (NRC, 2010) are still required to improve assessments of marine reptile status at global, regional, and local scales. Understanding biogeographical factors that influence the biology and ecology of marine reptiles, as well as the anthropogenic pressures on marine reptile species and populations, will improve status assessments and inform conservation strategies.

References

Abreu-Grobois FA, Plotkin PT, (assessors) (2007) IUCN Red List Status Assessment of the olive ridley sea turtle (*Lepidochelys olivacea*) IUCN/SSC-Marine Turtle Specialist Group. 39 p.

Auliya M (2011) *Lapemis curtus* (SERPENTES: ELAPIDAE) harvested in West Malaysia. *IUCN/SSC Sea Snake Specialist Group Newsletter*, 1, 6-8

Courtney AJ, Schemel BL, Wallace R, Campbell MJ, Mayer DG, Young B. (2009) Reducing the impact of Queensland's trawl fisheries on protected sea snakes, pp. 1-123. Queensland Primary Industries and Fisheries, Brisbane.

Dutton, P.H. and Squires, D. (2011). A Holistic Strategy for Pacific Sea Turtle Conservation, in Dutton, P.H., Squires, D. and Mahfuzuddin, A., (Eds), *Conservation and Sustainable Management of Sea Turtles in the Pacific Ocean*, University of Hawaii Press, 481pp.

Elfes C, Livingstone SR, Lane A, Lukoschek V, Sanders KL, Courtney AJ, et al. (2013). Fascinating and forgotten: the conservation status of the world's sea snakes. *Herpetological Conservation and Biology*, 8:37-52.

French, S.S., DeNardo, D.F., Greives, T.J., Strand, C.R., and Demas, G.E. (2010). Human disturbance alters endocrine and immune responses in the Galapagos marine iguana (*Amblyrhynchus cristatus*). *Hormones and Behavior* 58: 792-798.

Goiran C, Shine R (2013) Decline in sea snake abundance on a protected coral reef system in the New Caledonian Lagoon. *Coral Reefs*, 32, 281-284.

Guinea ML, Whiting SD (2005) Insights into the distribution and abundance of sea snakes at Ashmore Reef. *The Beagle*, Supplement 1, 199-205.

Hamann, M. et al. (2013). Climate change and marine turtles. Pp 353-397. in Wyneken, J. et al eds. *The Biology of Sea Turtles* Volume III, CRC Press, Boca Raton, FL

Hawkes, LA, Broderick, AC, Godfrey, MH, and Godley, BJ (2009) Climate change and marine turtles. *Endangered Species Research*, 7: 137-154.

Humber, F, Godley, BJ, Broderick, AC (2014) So excellent a fishe: a global overview of legal marine turtle fisheries. *Diversity and Distributions* 20(5): 579-590. DOI: 10.1111/ddi.12183.

Lewison RL, Wallace BP, Alfaro-Shigueto J, Mangel J, Maxwell S, Hazen E. (2013) Fisheries by-catch of marine turtles: lessons learned from decades of research and conservation. In: J. Wyneken, J.A. Musick (eds). *The Biology of Sea Turtles*, Vol 3. CRC Press, Boca Raton, FL. pp 329-352.

Lukoschek V, Keogh JS. (2006) Molecular phylogeny of sea snakes reveals a rapidly diverged adaptive radiation. *Biological Journal of the Linnean Society*, 89: 523-39.

Lukoschek V, Heatwole H, Grech A, Burns G, Marsh H (2007a) Distribution of two species of sea snakes, *Aipysurus laevis* and *Emydocephalus annulatus*, in the southern Great Barrier Reef: metapopulation dynamics, marine protected areas and conservation. *Coral Reefs*, 26, 291-307.

Lukoschek V, Waycott M, Marsh H (2007b) Phylogeographic structure of the olive sea snake, *Aipysurus laevis* (Hydrophiinae) indicates recent Pleistocene range expansion but low contemporary gene flow. *Molecular Ecology*, 16, 3406-3422.

Lukoschek V, Waycott M, Keogh JS (2008) Relative information content of polymorphic microsatellites and mitochondrial DNA for inferring dispersal and population genetic structure in the olive sea snake, *Aipysurus laevis*. *Molecular Ecology*, 17, 3062-3077.

Lukoschek V, Shine R (2012) Sea snakes rarely venture far from home. *Ecology and Evolution*, 2, 1113-1121.

Lukoschek V, Beger M, Ceccarelli DM, Richards Z, Pratchett MS. (2013) Enigmatic declines of Australia's sea snakes from a biodiversity hotspot. Biological Conservation, 166:191-202.

Mortimer JA, Donnelly M, (assessors) (2008) Marine Turtle Specialist Group 2007 IUCN Red List Status Assessment, Hawksbill Turtle (*Eretmochelys imbricata*), 121 pages.

National Research Council (NRC) (2010) *Assessment of Sea-Turtle Status and Trends: Integrating Demography and Abundance*. National Academies Press, Washington, DC.

Nelson, K., Snell., and Wikelski, M. (2004). *Amblyrhynchus cristatus*. In: *The IUCN Red List of Threatened Species. Version 2014*.3. www.redlist.org. Downloaded on 15 April 2015.

Seminoff J., Shanker K. (2008) Marine turtles and IUCN Red Listing: A review of the process, the pitfalls, and novel assessment approaches. *Journal of Experimental Marine Biology and Ecology*, 356:52-68

Seminoff, J.A., Wallace, B.P. (2012) *Sea Turtles of the Eastern Pacific: Advances in Research and Conservation*. University of Arizona Press.

Vitousek, M.N., Rubenstein, D.R., and Wikelski, M. (2007). The evolution of foraging behavior in the Galapagos marine iguana: natural and sexual selection on body size drives ecological, morphological, and behavioral specialization. In: *Lizard Ecology: The Evolutionary Consequences of Foraging Mode*. S.M. Reilly, L.D. McBrayer, and D.P. Miles (eds.). Cambridge University Press.

Wallace B.P., DiMatteo A.D., Hurley B.J., Finkbeiner E.M., Bolten B.A., et al. (2010) Regional Management Units for marine turtles: A novel framework for prioritizing conservation and research across multiple scales. *PLoS ONE* 5(12): e15465. doi:10.1371/journal.pone.0015465.

Wallace B.P., DiMatteo A.D., Bolten A.B., Chaloupka M.Y., Hutchinson B.J. (2011) Global conservation priorities for marine turtles. *PLoS ONE* 6(9): e24510. doi:10.1371/journal.pone.0024510.

Wallace, B.P., Tiwari, M. and Girondot, M. (2013a). *Dermochelys coriacea*. In: IUCN 2013. IUCN Red List of Threatened Species. Version 2013.2. <www.iucnredlist.org>. Downloaded on 30 April 2014. <www.iucnredlist.org>. Downloaded on 30 April 2014.

Wallace, B. P., Kot, C. Y., DiMatteo, A. D., Lee, T., Crowder, L. B., and Lewison, R. L.. (2013b). Impacts of fisheries by-catch on marine turtle populations worldwide: toward conservation and research priorities. *Ecosphere* 4(3):40. doi.org/10.1890/ES12-00388.1

40 Sharks and Other Elasmobranchs

Contributors:
Steven E. Campana, Francesco Ferretti and Andrew Rosenberg (Lead Member)

Sharks and Other Elasmobranchs Chapter 40

Sharks and rays are among the most endangered group of marine animals and include many species for which there is little information on abundance and distribution. There are no global abundance trends for elasmobranchs as a group, and very few robust regional trend indicators. Population-level stock assessments, which provide the most reliable index of abundance, are available for only about 10 per cent of 1,088 chondrichthyan species (FAO 2012; Worm et al., 2013; Dulvy et al., 2014; Cortés et al. 2012). Almost all of these assessments report a depleted and/or over-exploited population. In light of the scarcity of time series of absolute abundance indicators, the conservation status of elasmobranchs as a group is most commonly based on trends in reported landings, trajectories of standardized catch rates or indices of current status.

1 Global catches and trends

Global landings of sharks, rays and chimaeras (chondrichthyans) as reported to the Food and Agriculture Organization of the United Nations (FAO) have increased steadily since the 1950s, peaking at about 888,000 mt in 2000 before declining (Figure 40.1). In 2012, landings were 14 per cent lower than in 2000. The increasing trend in global catches reflected a combination of fisheries expansions into previously-unexploited regions, changes in the species composition of catches, and changes in the way countries reported landed catch for sharks and rays (e.g. changes in the taxonomic resolution of the reported landings; Ferretti et al. 2010). Historically, many species of sharks and rays had low commercial value, and were not regularly recorded in fisheries statistics. Since the 1980s, sharks became an alternative resource for some fisheries as many fish stocks collapsed and demand for shark fins in Asian markets strongly increased. The resulting increased fishing pressure on elasmobranchs and reports of severely depleted populations attracted the attention of management agencies, which increasingly reported shark statistics.

Although it has been assumed that the recent decline in reported landings may reflect better management for the species (FAO 2010), a recent analysis of the FAO chondrichthyan landings disaggregated by countries (Davidson et al. 2015) evaluated the importance of direct and indirect indices of fishing exploitation and measures of fisheries management performance, revealing that the decline is more closely related to fishing pressure and population declines (Eriksson and Clarke 2015, Davidson et al. 2015). Yet reported landings continue to be a gross underestimation of actual catches (Dulvy et al. 2014). Catches estimated from the volume of the shark-fin trade suggest global shark catches are on the order of 1.7 million mt in recent years (Clarke et al., 2006). More recently, Worm et al. (2013), using various assumptions about reporting rate, discarding and post-release mortality, derived a similar global estimate of 1.41 million mt in 2010, which is twice the figure reported from FAO statistics. Not included in that value would be any post-release mortality of discarded, unfinned catch. Thus, actual trends in shark catch and landings are unknown. Similar reporting issues confound shark catch statistics reported by other regional fisheries management organizations (RFMOs), such as is the case with blue sharks reported by International Commission for the Conservation of Atlantic Tunas (Campana et al., 2006).

Regional differences in the status of shark and ray populations might reflect different histories of fishing exploitation (Figure 40.1). The North-West Pacific, North-East Atlantic and Mediterranean were the areas where industrial fisheries began before the 1950s. These three areas, in decreasing order, recorded the highest initial catches per unit shelf area. The North Atlantic, Mediterranean, around Australia and New Zealand, and the North Pacific are the regions with the longest history of intensive fishing, but are also among the best-monitored sectors of the world's oceans, with stock assessments available for some species. The analysis of International Union for the Conservation of Nature (IUCN) Red List species indicated that these regions (as a group) were somewhat less likely to have a higher proportion of threatened species (<20 per cent) than were the less populated Indian, central Pacific and south and central Atlantic Oceans (>20 per cent) (Dulvy et al., 2014). Unassessed fisheries in the central Pacific and central and southern Atlantic were also likeliest to be characterized by low relative biomass on the basis of time series of catch and fisheries development, and species life histories traits (Costello et al., 2012); this is indicative of overfishing.

1.1 Conservation status

A comprehensive analysis of 1,041 chondrichthyan species on the IUCN Red List (www.redlist.org) reported that 17 per cent of the species were considered threatened with extinction (Critically Endangered, Endangered or Vulnerable, Dulvy et al., 2014). Moreover, just 241 species (23 per cent of the total) were considered to be safe from extinction threats (categorized as Least Concern according to IUCN criteria), which is the lowest fraction of safe species among all vertebrate groups studied by IUCN to date. Assessed shark and ray species with large body sizes were considered to be in the most danger, especially those living in shallow waters that were accessible to fisheries. Almost half of the examined species (47 per cent) were considered data-deficient, meaning that their

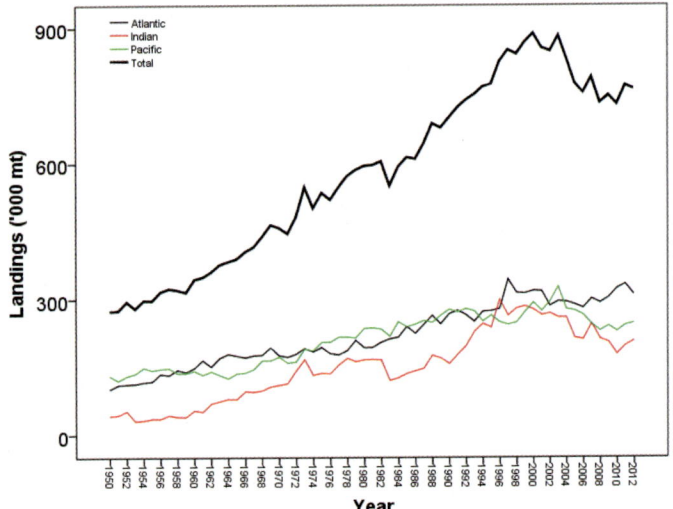

Figure 40.1 | Trends in reported landings of elasmobranchs as reported to FAO. The trend line for the Atlantic includes the Mediterranean.

conservation could not be assessed for lack of adequate data on abundance and distribution. However, in an independent study, Costello et al. (2012) used a multivariate regression analysis of assessed finfish species to estimate the current population biomass relative to maximum sustainable yield (MSY) in 1,793 unassessed marine fisheries around the world. Chondrichthyans had the lowest relative biomass values (40 per cent of the MSY (optimal) level) of any fish species or group; this was considered indicative of considerable over-exploitation (Costello et al., 2012).

2 Drivers of shark decline

2.1 Sharks are intrinsically vulnerable

Most sharks and rays are characterized as having low productivity associated with low fecundity, a slow growth rate, and a late age at sexual maturation (Musick, 1999). These life history characteristics are more similar to those of marine mammals than of the more productive bony fishes (Myers and Worm, 2005), which make them particularly susceptible to fishing pressure (Walker, 1998). Deep-sea sharks appear to be particularly susceptible, due to their very low productivity (Clarke et al., 2003; Forrest and Walters, 2009).

However, the similarity of their environmental preferences to several commercially valuable teleost species can also increase the likelihood of their capture in some fisheries. For example, blue shark catch rates and seasonal distributions tend to co-vary with those of swordfish, making the blue sharks difficult to avoid in a swordfish fishery (Bigelow et al., 1999). Similarly, many skates and rays are captured in bottom trawling for groundfishes such as flounders, and subsequently discarded dead (Enever et al., 2009; Damalas and Vassilopoulou 2011; Graham et al., 2001).

2.2 Fishing

Mortality due to fishing is almost entirely responsible for the world-wide declines in shark and ray abundance. Although directed shark fishing is still practised in some countries, a much larger proportion of overall shark mortality is associated with by-catch in non-shark fisheries (Lewison et al., 2004).

2.2.1 By-catch

Sharks have typically been exploited as a by-catch of commercial fisheries targeting more valuable bony fishes, especially tuna and billfish (ICCAT, 2005) and in trawl fisheries exploiting groundfishes and shrimps (Shepherd and Myers, 2005). In many countries, shark by-catch is partially or primarily retained for the fin and/or food trade. But even where living sharks are released at sea because they are considered unwanted catch, post-release mortality rates can exceed 18 per cent for some species (Campana et al., 2009; Musyl et al., 2011). In the North-West Atlantic, blue shark by-catch from an international pelagic longline fleet outnumbers the target swordfish catch by about 3:1, resulting in an annual post-release blue shark mortality of ~20,000 mt (Campana et al., 2009). Similar calculations of capture and post-release mortality of released sharks, using conservative mortality estimates for all shark species, suggest total shark mortalities of non-landed sharks of about 34,000 mt per year (Worm et al., 2013). Discarded skate and ray by-catch of bottom trawl fisheries are ubiquitous and poorly documented, but appear to be responsible for steep declines in abundance, and even risk of extinction, in some areas (Shepherd and Myers, 2005; McPhie and Campana, 2009).

Elasmobranch species living in the high seas appear to be particularly susceptible to undocumented and/or illegal catches.

2.2.2 Historical shark fisheries

At local scales, targeted shark fisheries have developed in multiple regions of the world. In the Mediterranean Sea, for example, fishing for sharks constituted an important off-season activity for fishing communities relying on harvests of tuna or small pelagic species, such as sardines and anchovies (Ferretti et al., 2008). In the Adriatic Sea, at the beginning of the last century, there were elasmobranch fisheries for angel sharks, skates and dogfishes; some have persisted into recent times (Ferretti et al., 2013; Costantini et al., 2000). In Monterey Bay, United States, between the 1930s and the 1940s, large numbers of basking sharks stimulated the development of a directed fishery used for the production of liver oil and other pharmaceutical products (Castro et al., 1999). Similar fisheries developed in the northeast Atlantic for basking and porbeagle sharks (Fowler et al., 2004; Sims, 2008). Directed fisheries for the meat of porbeagle shark and/or spiny dogfish still persist in both the North-West Atlantic and the North-East Pacific Oceans (Campana et al., 2008; Rago and Sosebee, 2009).

2.2.3 Shark fishing for fins

In recent decades, an increasing demand for shark fins from the Asian market stimulated the conversion of many industrial fisheries from bony fishes to sharks (Amorim et al., 1998; Aires-da-Silva et al., 2008). For countries in central America and in southeastern Asia, shark finning has become an important source of income (Dell'Apa et al., 2014).

The commercial trade in shark fins has been a primary driver of shark mortality. With prices of up to 2,000 United States dollars per kg, and a total estimated market value of about 350 million dollars, the fin trade is a strong motivator for retaining shark by-catch (Worm et al., 2013). The fin trade (which also includes fins of landed sharks) has been linked to a median annual estimate of 38 (CI: 26 – 73) million sharks landed, resulting in fishing mortality rates which are unsustainable for some species (Clarke et al., 2006, 2013).

Some elasmobranch species are also the target of a lucrative international trade of body parts, above and beyond that of shark fins, leading to unsustainable mortality rates for some species. The jaws and teeth of white sharks have been sold for as much as 50,000 US dollars (Fergusson et al. 2009), resulting in the intentional killing of many accidentally caught sharks that might otherwise be released. Sawfishes are critically endangered in most parts of the world, in part because of the high value of their toothed rostra (Harrison and Dulvy, 2014, Dulvy et al. 2014).

2.2.4 Recreational fisheries

Recreational shark fisheries provide an economic value to local communities greatly exceeding the value of commercial fisheries for the same species (Babcock, 2008). Although some recreational shark fisheries have been converted to no-kill, catch-and-release fisheries, recreational shark fishing remains a significant source of fishing mortality in regions such as the southeastern United States. Shark tournaments (a specific type of recreational shark fishing) have been conducted for decades in several countries (Campana et al., 2006; Pradervand et al., 2007), but even shark tournaments have converted to catch-and-release in some regions, with rewards given for sharks that are tagged and released (NOAA, 2013).

2.3 Habitat destruction

Population declines have been linked to habitat destruction in some regions and for some species, but the linkage is often indirect. Habitat destruction is considered an issue for elasmobranch species living in estuarine or mangrove habitats and in demersal communities exploited by trawl fisheries. For example, in the Adriatic Sea, elasmobranch catch rates in an aggregate of five trawl surveys declined by 94 per cent between 1948 and 2005. Exploitation history and spatial gradients in trawl fishing pressure, one of the most destructive forms of fishing in use (Walting and Norse, 1998), explained most of the declining patterns in abundance and diversity (Ferretti et al., 2013).

2.4 Pollution

Persistent bioaccumulation of toxins and heavy metals have been documented in sharks feeding at high trophic levels, at concentrations which can be toxic to human consumers, but their effect on the host shark remains unclear (Storelli and Marcotrigiano, 2001; Mull et al., 2012).

3 Ecosystem effects of shark depletion

3.1 Community changes through predator or competitor release

Sharks are very abundant and diverse in unperturbed ecosystems (Nadon et al., 2012; Ferretti et al., 2010). However, because of their slow population productivity low levels of fishing mortality may rapidly deplete these communities, with consequent pervasive effects on the structure and functioning of marine ecosystems. The overfishing of sharks can trigger community changes because of changing interspecific interactions among shark species and between sharks and other marine animals. The overfishing of large sharks triggered range expansions of more prolific broad-ranging competitors in coastal and offshore areas (Baum and Myers, 2004; Dudley and Simpfendorfer, 2006; Myers et al., 2007), and increases in small elasmobranchs released from shark predation (van der Elst, 1979; Myers et al., 2007; Ferretti et al., 2010). Sharks are often the sole consumers of small meso-predators, such as small sharks and rays, and of other long-lived marine organisms like turtles, tuna, billfish, and marine mammals, especially during their juvenile stages. Hence when large predatory sharks are removed, a rapid increase in the numbers of dogfishes, skates and rays or the recovery of historically depleted megafauna populations has been observed in some areas (Ferretti et al., 2010).

Changes in community structure due to shark depletion can also have indirect effects on community structure. Green turtles and dugongs affected by the presence of tiger sharks influence the distribution and species composition of seagrass beds through foraging and excavation (Heithaus et al., 2008; Preen, 1995; Aragones, 2000). In the North Atlantic, decades of overfishing on large sharks along the United States east coast coincided with a generalized increase in small shark and ray populations, and a substantial increase in one of these species (the cownose ray) adversely affected the abundance of bay scallops (Myers et al., 2007). Similarly, 50 years of shark netting along the KwaZulu Natal shore in South Africa triggered a trophic cascade involving smaller sharks and bony fishes (Ferretti et al., 2010).

3.2 Effect on stability

In addition to inducing trophic cascades, overfishing of sharks can make communities more prone to perturbations through a reduction in omnivory (Bascompte et al., 2005). In the Caribbean ecosystem, the observed change of many coral reefs from coral- to seaweed-dominated reefs was attributed to the depletion of shark populations and a consequent increase in fish consumers, which ultimately depressed herbivore density. Analyses of long-term time series of cost per unit effort (CPUE) in the Mediterranean Sea demonstrated that the removal of large predatory sharks from coastal ecosystems destabilized the community by reducing resistance, resilience and reactivity (Britten et al., 2014).

4 Shark management

Historically, shark species have been a low management priority for RFMOs and national management bodies (Ferretti et al. 2010; Dulvy et al., 2014). However, this trend is changing as shark and ray species are increasingly representing a larger proportion of protected species relative to other fishes. Regional fisheries management organizations such as the Northwest Atlantic Fisheries Organization (NAFO) and ICCAT now require all countries to report all shark catches, while the Commission for the Conservation of Antarctic Marine Living Resources (CCAMLR) prohibits directed fishing on any shark species, other than for scientific research. The Bahamas, Maldives and Palau have enacted legislation to prohibit shark fishing within areas under national jurisdiction (Dell'Apa et al., 2014; Techera and Klein, 2014), and the great white shark has become a protected species in all four countries where it is abundant (Australia, New Zealand, South Africa and the United States).

Although sustainable shark fisheries are theoretically possible, most industrial fisheries targeting elasmobranch resources have been characterized by a "boom and bust" trajectory of landings, culminating with a strong depletion of exploited populations (Castro et al., 1999; Campana et al. 2008). A few such fisheries that are apparently sustainable are now in place, but they require more conservative benchmarks and perhaps a higher level of enforcement (Walker, 1998; Gedamke et al., 2007).

4.1 Seafood certification

Certification of the sustainability of a fishery (e.g., the Marine Stewardship Council) is intended to provide an indirect economic incentive towards ensuring that the fished population is not threatened, and certification has been granted for a small number of shark fisheries, such as northwest Atlantic spiny dogfish. However, such certification is usually only possible for directed fisheries; in by-catch fisheries (such as is the case for most elasmobranchs), alternate conservation and/or recovery actions would have to be taken.

4.2 By-catch mitigation options

Reduced by-catch of sharks is usually the preferred option, since it results in both reduced shark mortality and reduced loss of fishing gear and bait (and therefore increased profits) by fishermen.

4.3 Spatial or seasonal closures

In principle, by-catch can be reduced by restricting access to "by-catch hotspots" through spatial or seasonal closures, although this approach is complicated by the similar habitat preferences of the target species and the shark by-catch. To this point, there is still little evidence of the effectiveness of large sanctuaries for sharks (mainly because of the absence of empirical data), although analyses of shark abundance and distribution along spatial gradients suggest that these might be effective management options (Ferretti et al., 2013).

Closure of shark mating and pupping grounds to fishing increases the protection of sensitive life-history stages (i.e., Campana et al., 2008). By-catch can also be reduced through modifications to fishing gear; for example, the introduction of the circle hook has reduced shark hooking mortality relative to the traditional J hook (Kaplan et al., 2007). However, other attempts to reduce shark catchability through use of rare earth metals and electrical fields have largely been disappointing (Godin et al., 2013).

4.4 Catch and release

Recreational shark fishing is a relatively small source of fishing mortality, despite its public visibility (Campana et al., 2006). Nevertheless, the introduction of catch-and-release fishing tournaments has reduced the mortality of some species.

4.5 Better monitoring

Ensuring the survival of a shark population is very challenging if the status of the population is unknown. Improved reporting of all catches and discards, the introduction (or expansion) of scientific observer programmes on commercial fishing vessels, and the inclusion of dead discards and estimates of post-release mortality rates in stock assessments, would all lead to improved assessments of population status, and thus simplify recovery efforts.

4.6 Fin trade bans and restrictions on finning

The fin trade has been one of the primary drivers of global shark mortality. Bans on fin sales have been adopted by some cities and in some states of the United States on the presumption that sales would decline in the absence of a legal market. Customer education in some Asian markets is also reducing the demand for wedding soup, and thus fin sales (Eilperin, 2011). Fisheries regulations requiring that the entire shark carcass be landed, and not just the fins, would also reduce shark mortality, as boat capacity is much more limited by the presence of entire sharks than by the much smaller fins. In some countries there is a fin-to-carcass ratio regulation which requires fishers to land no more than a given percentage of fin weight relative to total landings (Davidson et al., 2015).

4.7 Implementation of international policies

In response to the perception that many of the world's elasmobranch species are severely depleted, several international organizations have moved to actively conserve some shark and ray species. The Food and Agriculture Organization of the United Nations (FAO) released an *International Plan of Action for the Conservation and Management of Sharks* urging immediate action to better document and conserve shark and ray species (FAO 1998). The Convention on the Conservation of Migratory Species of Wild Animals[1] (CMS) has listed eight shark species for international conservation and protection (CMS 2014; http://www.cms.int/en/species). Finally, the Shark Specialist Group of the International Union for Conservation of Nature (IUCN) provides information and guidance to governments and non-governmental organizations associated with the conservation of threatened shark species and populations. The SSG released their report on the Global Status of Oceanic Pelagic Sharks and Rays in 2009. As a final step of protection, the international Convention on International Trade in Endangered Species of Wild Fauna and

---

[1] United Nations *Treaty Series*, vol. 1651, No. 28395.

Flora[2] (CITES) attempts to protect endangered species through international trade regulations, such as restrictions on import and export. To this point, CITES has listed 18 shark and ray species under their Appendices I and II trade restrictions (CITES 2014; http://checklist.cites.org), which will remain in place until it can be demonstrated that the population is being managed sustainably. CITES trade restrictions appear to have tangible effects on the trade of listed shark species, and thus reduce the demand (Wells and Barzdo, 1991). However, it is yet to be seen if CITES listings can be implemented in time to protect species, which have already reached the brink of extinction (e.g., sawfish).

5   Ecotourism

Ecotourism in the form of shark diving has become a burgeoning industry generating millions of dollars for local economies worldwide (Musick and Bonfil, 2005; Gallagher and Hammerschlag, 2011). One estimate suggests that shark ecotourism currently generates more than 314 million US dollars per year and supports about 10,000 jobs. Projections suggest that this figure could double in the next 20 years and thus surpass the landed value of global shark fisheries (Cisneros-Montemayor et al., 2013). Indeed, in terms of individual value, sharks in some localities may be worth more alive than if landed and marketed. In the Maldives, it has been estimated that an individual free-swimming grey reef shark is worth 33,500 dollars per year compared to 32 dollars for the same individual sold dead by local fishermen. In the Bahamas, shark diving generates annual revenues of 78 million dollars (Gallagher and Hammerschlag, 2011). In the Maldives (where shark fishing has been banned), ecotourism contributed >30 per cent of the Maldivian GDP (Gallagher and Hammerschlag, 2011).

---

2   United Nations *Treaty Series*, vol. 993, No. 14537.

References

Aires-da-Silva, A., Hoey, J., and Gallucci, V. (2008). A historical index of abundance for the blue shark (*Prionace glauca*) in the western North Atlantic. *Fisheries Research* 92(1): 41–52.

Amorim, A., Arfelli, C., and Fagundes, L. (1998). Pelagic elasmobranchs caught by longliners off southern Brazil during 1974-97: an overview. *Marine and Freshwater Research* 49: 621–632.

Aragones, L.V. (2000). A review of the role of the green turtle in tropical seagrass ecosystems. In: Pilcher, N., and Ismail, G. (eds.), *Sea Turtles of the Indo-Pacific: Re- search, Management and Conservation*. Academic Press Ltd, London, UK, pp. 69–85.

Babcock, E.A. (2008). Recreational Fishing for Pelagic Sharks Worldwide. In: Camhi, M.D., Pikitch, E.K., and Babcock, E.A. (eds.), Sharks of the Open Ocean: Biology, Fisheries and Conservation. Blackwell Publishing, Oxford, UK, pp.193–204.

Bascompte, J., Melián, C.J., and Sala, E. (2005). Interaction strength combinations and the overfishing of a marine food web. *Proceedings of the National Academy of Sciences of the United States of America* 102, 5443–5447.

Baum, J.K. and Myers, R.A. 2004. Shifting baselines and the decline of pelagic sharks in the Gulf of Mexico. *Ecology Letters 7*, 135-145.

Bigelow, K.A., Boggs, C.H., and He, X. (1999). Environmental effects on swordfish and blue shark catch rates in the US North Pacific longline fishery. *Fisheries Ocean* 8:178–198.

Britten, G.L., Dowd, M., Minto, C., Ferretti, F., Boero, F. and Lotze, H.K. (2014). Predator decline leads to decreased stability in a coastal fish community. *Ecology Letters*, 17, 1518-15-25.

Campana, S., Joyce, W., Marks, L., Hurley, P., Natanson, L.J., Kohler, N.E., Jensen, C.F., and Myklevoll, S. (2008). The rise and fall (again) of the porbeagle shark population in the Northwest Atlantic. In: Camhi, M.D., Pikitch, E.K., and Babcock, E.A. (eds.), Sharks of the Open Ocean: Biology, Fisheries and Conservation. Blackwell Publishing, Oxford, UK, pp. 445–461.

Campana, S.E., Joyce, W., and Manning, M.J. (2009). By-catch and discard mortality in commercially caught blue sharks *Prionace glauca* assessed using archival satellite pop-up tags. *Marine Ecology Progress Series* 387: 241–253.

Campana, S.E., Marks, L., Joyce, W., and Kohler, N.E. (2006). Effects of recreational and commercial fishing on blue sharks (*Prionace glauca*) in Atlantic Canada, with inferences on the North Atlantic population. *Canadian Journal of Fisheries and Aquatic Sciences* 63(3): 670–682.

Castro, J., Woodley, C.M., and Brudek, R. (1999). A Preliminary Evaluation of the Status of Shark Species. FAO.

Cisneros-Montemayor, A.M., Barnes-Mauthe, M., Al-Abdulrazzak, D., Navarro-Holm, E., and Sumaila, U.R. (2013). Global economic value of shark ecotourism: implications for conservation. Oryx 47(3): 381–388.

CITES (Convention on International Trade in Endangered Species of wild Fauna and Flora (2014). Appendices I, II and III. https://cites.org/sites/default/files/eng/app/2015/E-Appendices-2015-02-05.pdf (accessed in 2014).

Clarke, M.W.; Kelly, C.J.; Connolly, P.L., and Molloy, J.P. (2003). A life history approach to the assessment and management of deepwater fisheries in the Northeast Atlantic. *Journal of Northwest Atlantic Fishery Science* 31:401-411.

Clarke, S.C., McAllister, M.K., Milner-Gulland, E.J., Kirkwood, G.P., Michielsens, C.G.J., Agnew, D.J., Pikitch, E.K., Nakano, H., and Shivji, M.S. (2006). Global estimates of shark catches using trade records from commercial markets. *Ecology Letters* 9(10): 1115–1126.

Clarke, S.C.; Harley, S.J.; Hoyle, S.D. and Rice, J.S. (2013). Population Trends in Pacific Oceanic Sharks and the Utility of Regulations on Shark Finning. *Conservation Biology* 27, 197-209.

Costello, C., Ovando, D., Hilborn, R., Gaines, S.D., Deschenes, O., and Lester, S.E. (2012). Status and solutions for the world's unassessed fisheries. *Science* 338(6106): 517–520.

Costantini, M., Bernardini, M., Cordone, P., Giuliani, P.G., and Orel, G. (2000). Osservazioni sulla pesca, la biologia riproduttiva ed alimentare di Mustelus mustelus (Chondrichtyes, Triakidae) in Alto Adriatico. Biologia Marina Mediterranea 7(1): 427–432.

Cortés, E.; Brooks, E.N. and Gedamke, T. (2012). Population dynamics, demography, and stock assessment. *Biology of Sharks and Their Relatives*, CRC Press, 453-86

Damalas D. and Vassilopoulou, V.C. (2011). By-catch and discards in the demersal trawl fishery of the central Aegean Sea (Eastern Mediterranean) *Fisheries Research, 108*, 142-152

Davidson, L.N.K, Krawchuck, M.A. and Dulvy, N.K. (2015). Why have global shark and ray landings declined: improved management or overfishing? *Fish and Fisheries*.

Dell'Apa, A., Smith, M.C., and Kaneshiro-Pineiro, M.Y. (2014). The Influence of Culture on the International Management of Shark Finning. Environmental management, 54(2): 1–11.

Dudley, S. and Simpfendorfer, C. 2006. Population status of 14 shark species caught in the protective gillnets off KwaZulu-Natal beaches, South Africa, 1978-2003. *Marine and Freshwater Research, 57*, 225

Dulvy, N.K. et al. (2014). Extinction risk and conservation of the world's sharks and rays. eLife 2014;3:e00590. DOI: 10.7554/eLife.00590 .

Dulvy, N.K. et al. (2014). Ghosts of the coast: global extinction risk and conservation of sawfishes *Aquatic Conservation: Marine and Freshwater Ecosystems*. DOI: 10.1002/aqc.2525.

Eilperin, J. (2011). *Demon fish*. Pantheon Books, New York. 295 p.

Enever, R., Catchpole, T.L., Ellis, J.R., and Grant, A. (2009). The survival of skates (Rajidae) caught by demersal trawlers fishing in UK waters. *Fisheries Research* 97: 72–76.

Eriksson, H., Clarke, S. (2015). Chinese market responses to overexploitation of sharks and sea cucumbers. *Biological Conservation* 184: 163-173.

FAO (Food and Agriculture Organization of the United Nations). (1998) International Plan of Action for the Conservation and Management of Sharks. FAO Document FI:CSS/98/3, Rome. Italy.

FAO (Food and Agriculture Organization of the United Nations). (2010). The state of the World Fisheries and Aquaculture. Food and Agriculture Organization of the United Nations, Rome, Italy, 218 pp.

FAO (Food and Agriculture Organization of the United Nations). (2012) The state of world fisheries and aquaculture. FAO, Rome, Italy. 209pp.

Fergusson, I., Compagno, L.J.V. and Marks, M. (2009). Carcharodon carcharias. The IUCN Red List of Threatened Species. Version 2014.3. <www.iucnredlist.org>. Downloaded on 08 May 2015.

Ferretti, F., Myers, R.A., Serena, F., and Lotze, H.K. (2008). Loss of large predatory sharks from the Mediterranean Sea. Conservation Biology 22(4): 952–964.

Ferretti, F., Worm, B., Britten, G.L., Heithaus, M.R., and Lotze, H.K. (2010). Patterns and ecosystem consequences of shark declines in the ocean. *Ecology Letters* 13(8): 1055–1071.

Ferretti, F., Osio, G.C., Jenkins, C.J., Rosenberg, A.A., and Lotze, H.K. (2013). Long-term change in a meso-predator community in response to prolonged and heterogeneous human impact. Scientific Reports 3, Article number 1057. doi:10.1038/

srep01057.

Forrest, R.E. and Walters, C.J. (2009) Estimating thresholds to optimal harvest rate for long-lived, low-fecundity sharks accounting for selectivity and density dependence in recruitment. *Canadian Journal of Fisheries and Aquatic Sciences* 66:2062-2080.

Fowler, S., Raymakers, C. and Grimm, U. (2004). Trade in and Conservation of two Shark Species, Porbeagle (Lamna nasus) and Spiny Dogfish (Squalus acanthias). BfN - Skripten 118.

Gallagher, A.J., and Hammerschlag, N. (2011). Global shark currency: the distribution, frequency, and economic value of shark ecotourism. Current Issues in Tourism 14(8): 797–812.

Gedamke, T., Hoenig, J.M., Musick, J.A., DuPaul, W.D., and Gruber, S.H. (2007). Using demographic models to determine intrinsic rate of increase and sustainable fishing for elasmobranchs: pitfalls, advances, and applications. *North American Journal of Fisheries Management* 27: 605–618.

Godin, A.C., Wimmer, T., Wang, J.H., and Worm, B. (2013). No effect from rare-earth metal deterrent on shark by-catch in a commercial pelagic longline trial. *Fisheries Research* 143: 131–135.

Graham, K.J., Andrew, N.L. and Hodgson, K.E. (2001). Changes in the relative abundance of sharks and rays on Australian south east fishery trawl grounds after twenty years of fishing. *Marine and Freshwater Research* 52:549-561.

Harrison, L.R., Dulvy, N.K. (2014). *Sawfish: A Global Strategy for Conservation*. International Union for the Conservation of Nature Species Survival Commission's Shark Specialist Group, Vancouver, Canada. 112 pp.

Heithaus, M.R., Frid, A., Wirsing, A.J., and Worm, B. (2008). Predicting ecological consequences of marine top predator declines. *Trends in Ecology and Evolution* 4, 202–210.

ICCAT (2005). Report of the 2004 inter-sessional meeting of the ICCAT Subcommittee on by-catches: shark stock assessment. *Collective Volume of Scientific Papers* ICCAT 57: 1–46.

Kaplan, I.C., Cox, S.P., and Kitchell, J.F. (2007). Circle hooks for Pacific longliners: not a panacea for marlin and shark by-catch, but part of the solution. *Transactions of the American Fisheries Society* 136: 392–401.

Lewison, R.L., Crowder, L.B., Read, A.J., and Freeman, S.A. (2004). Understanding impacts of fisheries by-catch on marine megafauna. *Trends in Ecology and Evolution* 19: 598–604.

McPhie, R.P. and Campana, S.E. (2009). Reproductive characteristics and population decline of four species of skate (Rajidae) off the eastern coast of Canada. *Journal of Fisheries Biology* 75:223-246.

Mull, C.G., Blasius, M. E., O'Sullivan, J.B. and Lowe, C.G. (2012). Heavy metals, trace elements, and organochlorine contaminants in muscle and liver tissue of juvenile White Sharks, Carcharodon carcharias, from the Southern California Bight *Global perspectives on the biology and life history of the White Shark*. CRC Press, Boca Raton, Florida, 59-75.

Musick, J.A. (1999). Ecology and conservation of long-lived marine animals. *American Fisheries Society Symposium* 23: 1–10.

Musick, J.A. and Bonfil, R. (2005). *Management Techniques for Elasmobranch Fisheries*. Chapter14: Shark utilization. Food and Agriculture Organization, pp. 323–336.

Musyl, M.K., Brill, R.W., Curran, D.S., Fragoso, N.M., McNaughton, L.M., Nielsen, A., Kikkawa, B.S., and Moyes, C.D. (2011). Postrelease survival, vertical and horizontal movements, and thermal habitats of five species of pelagic sharks in the central Pacific Ocean. *Fishery Bulletin* 109(4): 341–368.

Myers, R.A. and Worm, B. (2005). Extinction, survival or recovery of large predatory fishes. *Philosophical Transactions of the Royal Society B:* 360, 13–20.

Myers, R.A., Baum, J.K., Shepherd, T., Powers, S.P. and Peterson, C.H. (2007). Cascading Effects of the Loss of Apex Predatory Sharks from a Coastal Ocean *Science*, 315, 1846-1850.

Nadon, M., Baum, J., Williams, I., McPherson, J., Zgliczynsky, B., Richards, B., Schroeder, R. and Brainard, R. (2012). Re-Creating Missing Population Baselines for Pacific Reef Sharks *Conservation Biology, 26*, 493-503.

NOAA (2013). http://www.nmfs.noaa.gov/stories/2013/08/best_fishing_practices_sharks.html.

Pradervand, P., Mann, B.Q., and Bellis, M.F. (2007). Long-term trends in the competitive shore fishery along the KwaZulu-Natal coast, South Africa. *African Zoology* 42(2): 216–236.

Preen, A. (1995). Impacts of dugong foraging on seagrass habitats: observational and experimental evidence for cultivation grazing. *Marine Ecology Progress Series* 124: 201–213.

Rago, P.J., and Sosebee, K.A. (2009). The agony of recovery: scientific challenges of spiny dogfish recovery programs. In: Gallucci, V.F., McFarlane, G.A., and Bargmann, G.G. (eds.), Biology and management of dogfish sharks. American Fisheries Society, Bethesda, MD, pp. 343–372.

Shepherd, T.D., and Myers, R.A. (2005). Direct and indirect fishery effects on small coastal elasmobranchs in the northern Gulf of Mexico. Ecology Letters 8(10): 1095–1104.

Sims, D. (2008). Sieving a living: a review of the biology, ecology and conservation status of the plankton-feeding basking shark Cetorhinus maximus. *Advances in Marine Biology* 54: 171–220.

Storelli, M.M., and Marcotrigiano, G.O. (2001). Persistent organochlorine residues and toxic evaluation of polychlorinated biphenyls in sharks from the Mediterranean Sea (Italy). Marine Pollution Bulletin 42(12): 1323–1329.

Techera, E. and Klein, N. (2014). *Sharks: Conservation, Governance and Management* Taylor & Francis.

Van der Elst, R.P. (1979). A proliferation of small sharks in the shore-based Natal sport fishery *Environmental Biology of Fishes, 4*, 349-362.

Walker, T.I. (1998). Can shark resources be harvested sustainably? A question revisited with a review of shark fisheries. *Marine and Freshwater Research* 49(7):553–572.

Watling, L. and Norse, E. A. (1998). Disturbance of the seabed by mobile fishing gear: a comparison to forest clear cutting *Conservation Biology*, 12, 1180-1197

Wells, S.M., and Barzdo, J.G. (1991). International trade in marine species: Is CITES a useful control mechanism? Coastal Management 19(1):135–154.

Worm, B., Davis, B., Kettemer, L., Ward-Paige, C.A., Chapman, D., Heithaus, M.R., Kessel, S.T., and Gruber, S.H. (2013). Global catches, exploitation rates, and rebuilding options for sharks. *Marine Policy* 40: 194–204.

Chapter 41

Tunas and Billfishes

Contributors:
Victor Restrepo (Convenor), Bruce B. Collette, Flávia Lucena Frédou, Maria José Juan-Jordá and Andrew Rosenberg (Lead Member)

1 Introduction

Tunas and billfishes are epipelagic marine fishes that live primarily in the upper 200 metres of the ocean and are widely distributed throughout the tropical, subtropical and temperate waters of the world's oceans (Collette and Nauen 1983; Nakamura 1985). Tunas (Tribe Thunnini, family Scombridae) include five genera (*Thunnus, Katsuwonus, Euthynnus, Auxis* and *Allothunnus*) with 15 species altogether (Collette et al., 2001). Seven of the 15 species of tunas are commonly known as "principal market tunas" due to their economic importance in the global markets (Majkowski 2007). These include albacore (*Thunnus alalunga*), bigeye tuna (*T. obesus*), Atlantic bluefin tuna (*T. thynnus*), Pacific bluefin tuna (*T. orientalis*), southern bluefin tuna (*T. maccoyii*), yellowfin tuna (*T. albacares*) and skipjack tuna (*Katsuwonus pelamis*). The principal market tunas have extensive oceanic distributions and are highly migratory. They sustain diverse fisheries worldwide, from highly industrialized commercial fisheries, to small and medium scale artisanal fisheries, and also lucrative recreational fisheries. The non-principal market tuna species including longtail tuna (*Thunnus tonggol*), blackfin tuna (*Thunnus atlanticus*), kawakawa (*Euthynnus affinis*), little tunny (*E .alletteratus*), black skipjack (*E. lineatus*), bullet tuna (*Auxis rochei*), frigate tuna (*A. thazard*) and slender tuna (*Allothunnus fallai*) have in general more coastal distributions, except for the slender tuna which is found worldwide in the Southern Ocean. These species also sustain important small to medium industrial and artisanal fisheries throughout their distributions (Collette and Nauen, 1983; Majkowski, 2007).

Billfishes are highly migratory fishes that live also primarily in the upper 200 metres of the ocean and have widespread oceanic distributions. They are distinguished by their elongate spears or swords on their snouts. Some billfish species are targeted by commercial and recreational fisheries world-wide, but generally billfish species are caught as a by-product of the tuna fisheries (Kitchell et al., 2006). Billfishes include ten species in two families (Xiphiidae and Istiophoridae); the monotypic Xiphiidae (swordfish, *Xiphias gladius*) and Istiophoridae containing five genera and nine species: blue marlin (*Makaira nigricans*), sailfish (*Istiophorus platypterus*), black marlin (*Istiompax indica*), striped marlin (*Kajikia audax*), white marlin (*Kajikia albida*), and four spearfishes, shortbill spearfish (*Tetrapturus angustirostris*), roundscale spearfish (*Tetrapturus georgii*), longbill spearfish (*Tetrapturus pfluegeri*), and Mediterranean spearfish (*Tetrapturus belone*) (Collette et al., 2006).

Due to the highly migratory nature, widespread distributions, and global economic importance of tunas and billfishes, five Regional Fisheries Management Organizations (RFMOs) are in charge of their management and conservation (hereinafter referred to as tuna RFMOs). The five tuna RFMOs are the International Commission for the Conservation of Atlantic Tunas (ICCAT, Atlantic Ocean), the Indian Ocean Tuna Commission (IOTC, Indian Ocean), the Inter-American Tropical Tuna Commission (IATTC, Eastern Pacific Ocean), the Western and Central Pacific Fisheries Commission (WCPFC, Western Pacific Ocean), and the Commission for the Conservation of Southern Bluefin Tuna (CCSBT, Southern Ocean).

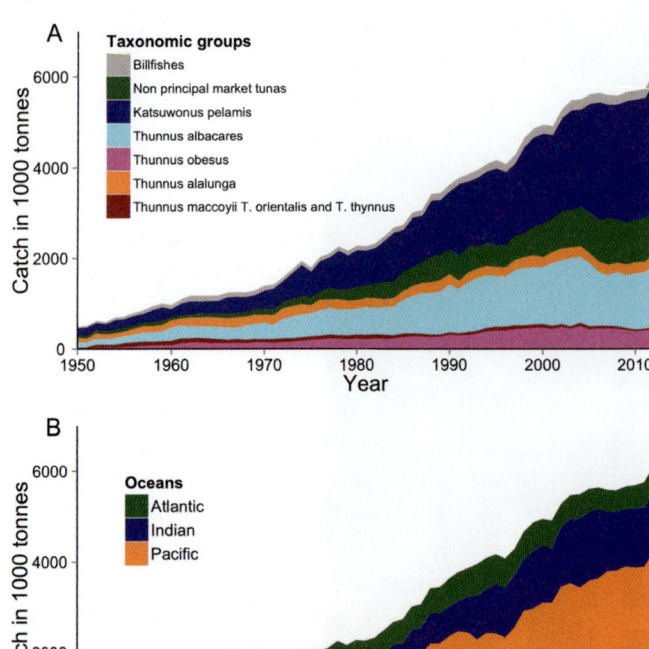

Figure 41.1 | Global catch trends of tuna and billfish species (FAO, 2014). (A) Global aggregated temporal trends of catches by major taxonomic groups. (B) Global aggregated temporal trends of catches by oceans.

2 Population trends or conservation status

2.1 Aggregated at global scale

Annual catches of tunas and billfishes have risen continuously since the 1950s, reaching at least 6 million tons in 2012 (Figure 41.1A). In 2012, the total catches of tunas and billfish species combined contributed up to 9.3 per cent of the annual total marine fish catch (FAO, 2014). Although the global increase in catches of all marine fishes reached a peak at the end of the 1980s and has since then stabilized, tuna and billfish catches have not reached a plateau yet. However, a plateau will likely be reached in the short term as many of the world's most important tuna and billfish fisheries are considered fully exploited now with limited room for sustainable growth (Miyake et al., 2010; Juan-Jordá et al., 2011; ISSF, 2013a). The current exploitation status of principal-market tuna and billfish populations is summarized according to the latest fisheries stock assessments and biological reference points[1] carried out by the five tuna RFMOs. Currently the tuna RFMOS have formally assessed a total of 44 stocks (13 species) of tuna and billfish species, including 23 principal market tuna stocks (7 species) and 21 billfish stocks (6 species) (Appendix 1). Hereinafter, the term "population" is used instead of "stock". Each tuna RFMO has its own convention objectives ranging from ensuring the long term conservation and sustainable use of tuna

---

[1] Definitions of the term "reference points" are available at the FAO Term Portal (http://www.fao.org/faoterm/en/) and in ISSF (2013b).

and tuna-like species to, in some cases, ensuring the optimum utilization of stocks.[2] Scientific advisory groups or science providers within these tuna RFMOs routinely carry out stock assessments and estimate two common standard biological reference points, B/B_{MSY} and F/F_{MSY}, which are used to determine the current exploitation status of the populations. B/B_{MSY} is the ratio of the current biomass (often measured only for the spawning fraction of the population) relative to the biomass that would provide the maximum sustainable yield (MSY). A population whose biomass has fallen below B_{MSY} (i.e., $B/B_{MSY} < 1$) is considered to be "overfished" with regards to this target. F/F_{MSY} is the ratio of current fishing mortality relative to the fishing mortality rate that produces MSY. Overfishing is occurring for a population whose fishing mortality exceeds F_{MSY} (i.e., $F/F_{MSY} > 1$).

According to the most recent fisheries stock assessments (2010-2014, Appendix 1), 51.2 per cent of the tuna and billfish populations are not overfished and are not experiencing overfishing (21 populations), 14.6 per cent of populations are not overfished but are experiencing overfishing (6 populations), 22 per cent of populations are overfished and are experiencing overfishing (9 populations), and 12.2 per cent of populations are overfished but are not experiencing overfishing anymore (5 populations) (Figure 41.2A). However, the total catches and abundance differ markedly among tuna and billfish species and populations, around 3 orders of magnitude between the population with the smallest catches (eastern Pacific sailfish ~300 tons/annually) and the population with the largest catches (western and central Pacific skipjack ~1,700,000 tons annually). When accounting for their relative contributions to their total global catches, a different global picture of the status of these species emerges (Figure 41.2B). In terms of their relative contributions to the total catches, 86.2 per cent of the global catch of tuna and billfish comes from healthy populations, for which the biomass is not overfished and whose populations are not experiencing overfishing, 4.5 per cent of the catch comes from populations that are not overfished but are experiencing overfishing, 1.4 per cent of the catch comes from populations that are overfished and are experiencing overfishing, and 8 per cent of the catch comes from populations that are overfished and are not experiencing overfishing anymore (Figure 41.2B). This distinct pattern of global exploitation status is mostly driven in part by the fact that tropical skipjack and yellowfin tuna populations contribute 68 per cent of the global tuna catches and their populations are largely at healthy levels and not experiencing overfishing. In contrast, most of the populations that are overfished and experiencing overfishing are mostly temperate bluefin tuna and billfish populations, whose combined catches make up a relatively small fraction of the total catch.

Although the current exploitation status for the principal market tunas is relatively well known globally, knowledge of the exploitation status for the non-principal market tuna and billfish populations and species is fragmentary and uncertain. Currently, all the populations for all seven species of principal market tunas are formally assessed on a regular basis (every 2-4 years depending on the population) by the scientific staff or scientific committees in the five tuna RFMOs, and have management and conservation measures in place. Not all billfish populations and species have been formally assessed yet, therefore the global picture of their current exploitation status may be biased towards the most commercially productive and resilient species of billfish. Furthermore, tuna RFMOs have not yet conducted formal fisheries stock assessments or adopted management and conservation measures for any of the eight non-principal market tuna species. Therefore their current exploitation status is unknown or highly uncertain throughout their neritic distributions. There are some exceptions and some species of non-principal market tunas have been assessed locally by national government fisheries agencies or recently by IOTC. For the South Atlantic Ocean off the coast of Brazil, *Thunnus atlanticus* was assessed in the year 2000, concluding the population was as at healthy levels and not experiencing overfishing (Freire, 2009). In the Indian Ocean, *Thunnus tonggol* was assessed for the first time in the year 2013 and 2014 by the IOTC Working Party on neritic tunas. The assessments concluded that *Thunnus tonggol* was likely subject to overfishing in recent years while not being in an overfished state (IOTC-SC17, 2014). Therefore the exploitation status for the majority of non-principal market tuna populations and species is mostly unknown throughout their ranges, despite the importance of their commercial fisheries for many coastal fishing communities in many developed and developing countries around the world.

2.2 Four major taxonomic and/or geographic subdivisions

Since the 1950s, principal market tunas have made up the majority of the global catches of tunas and billfish combined (Figure 41.1A). In 2012, the catch of principal market tunas accounted for 80 per cent of the total catch, the catch of non-principal market tunas accounted for 16 per cent and billfish catch accounted for 4 per cent. Among principal market tunas, skipjack tuna and yellowfin tuna make up 46 per cent and 22 per cent of the global catch in 2012, followed by bigeye tuna (7 per cent), albacore tuna (4 per cent) and the three bluefin tuna species (1 per cent). The increasing trend in the total catch of principal market tunas is mainly due to the increase in catches in tropical tuna species since the 1950s until today, a trend driven by skipjack tuna, followed by yellowfin tuna and then bigeye tuna. By contrast, temperate principal market tuna species, including albacore tuna and the three bluefin tuna species, show an increasing trend in catch up to the 1970s, and since then the trend has stabilized or shown a decrease. Over two-thirds of the world's tunas and billfishes catches currently come from the Pacific Ocean (69 per cent), 22 per cent come from the Indian Ocean and 9 per

---

2 See Agreement for the Establishment of the Indian Ocean Tuna Commission (United Nations, *Treaty Series*, vol. 1927, No 32888); Convention between the United States of America and the Republic of Costa Rica for the establishment of an Inter-American Tropical Tuna Commission (United Nations, *Treaty Series*, vol. 80, No. 1041); Convention for the Conservation of Southern Bluefin Tuna (United Nations, *Treaty Series*, vol. 1819, No. 31155); Convention for the strengthening of the Inter-American Tropical Tuna Commission established by the 1949 Convention between the United States of America and the Republic of Costa Rica; Convention on the Conservation and Management of Highly Migratory Fish Stocks in the Western and Central Pacific Ocean (United Nations, *Treaty Series*, vol. 2275, No. 40532); International Convention for the Conservation of Atlantic Tunas (United Nations, *Treaty Series*, vol. 673, No. 9587).

Tunas and Billfishes

Figure 41.2 | Global exploitation status of principal market tuna and billfish species according to the latest fisheries stock assessments conducted by tuna RFMOs. (A) Proportion of populations by exploitation status. (B) Relative contribution of the total catch by exploitation status. (C) Exploitation status by major taxonomic groups and oceans.

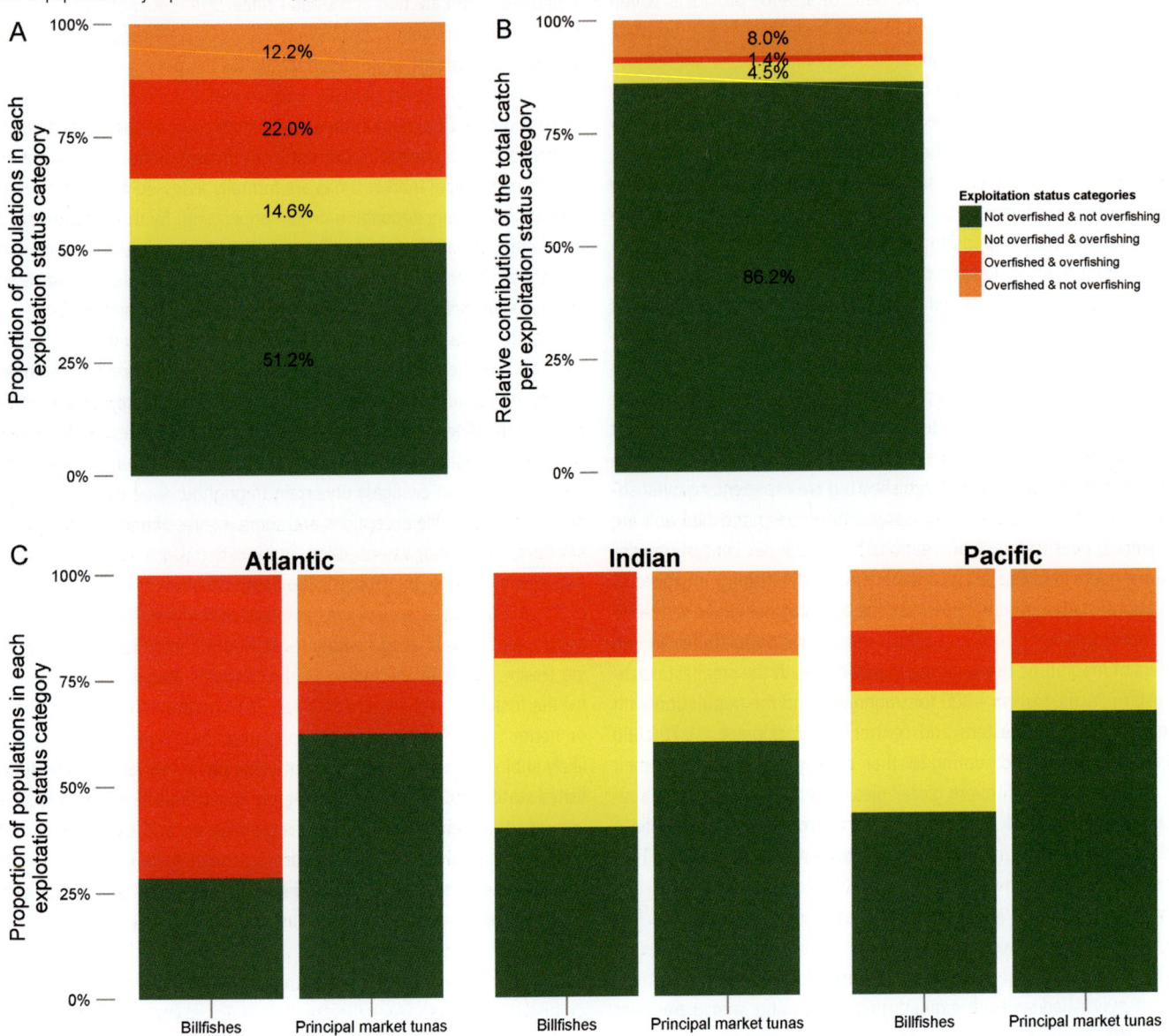

cent from the Atlantic Ocean (Figure 41.1B). Although catches in the Atlantic Ocean have increased only until the early 1980s and since then have declined slightly, in the Pacific and Indian Oceans catches have increased continuously since the 1950s.

Among the non-principal market tuna species, frigate and bullet tunas combined (*Auxis rochei* and *A. thazard*) make up 40 per cent of the catch and kawakawa (*Euthynnus affinis*) makes up 33 per cent of the catch. Among billfishes, swordfish (*Xiphias gladius*) makes up 51 per cent of the catch and Atlantic blue marlin (*Makaira nigricans*) makes up 17 per cent of the catch. Global catches for non-principal market tunas and billfish have also shown a continuous increase since the 1950s, accelerating in the 1980s, a result that is likely to be derived from better reporting of the catch for these species. However, it is generally agreed that catch estimates for non-principal market tunas and billfish have been and still are underestimated as the majority of these species are caught by small scale fisheries or as a by-catch[3] of principal market tuna fisheries. Small-scale coastal fisheries targeting both principal market tunas and the smaller non-principal market tunas are poorly reported. Similarly, billfish catches, of which the majority come from industrial tuna fisheries as bycatch, have also been commonly poorly reported and monitored (Miyake et al., 2010).

According to the latest tuna RFMO fisheries stock assessments (Appendix 1), the global picture of the exploitation status of tunas and billfishes indicates that principal market tuna populations are relatively better managed than billfish populations (Figure 41.2C). Although 37 per cent of billfish populations (7 of 19 populations) are currently overfished and experiencing overfishing, 9 per cent of the principal market tunas (2

---

3 Definitions of the term "by-catch" are available at the FAO Term Portal and in Gilman et al. (2014).

of 22 populations) are considered to be overfished and experiencing overfishing. The majority of principal market tunas are at healthy levels with 64 per cent of the populations not overfished and not experiencing overfishing, and 18 per cent of the populations, although overfished, are no longer experiencing overfishing and therefore are on the path to recovery, if fishing mortality continues to be controlled. The exploitation status of tunas and billfishes also differs among the three major oceans (Figure 41.2C). In the Atlantic Ocean, the status of only 47 per cent of the populations is currently healthy (not overfished and not experiencing overfishing), in the Indian Ocean the status of half of the populations (50 per cent) is healthy, and in the Pacific Ocean over half of the populations (~56 per cent) is currently healthy and within sustainable levels.

When accounting for the relative contributions of their catches, principal market tuna populations provide the majority of the catches from healthy populations when compared with billfish species. Although 87 per cent of the total catches of principal market tunas come from healthy populations (not overfished and not experiencing overfishing) and only 0.9 per cent come from unhealthy populations (overfished and experiencing overfishing), 60.8 per cent of the total catches of billfish populations come from healthy populations and 16.1 per cent come from unhealthy populations. Healthy populations of skipjack in every ocean make up a large portion of the total tuna catches, whereas healthy swordfish populations make up the largest portion of the total billfish catches. As previously mentioned, the exploitation status remains unknown for some billfish species and populations, and therefore this global picture might be biased towards the most commercially data-rich billfish species. Among the three oceans, the large majority of tuna and billfish catches in the Indian and Pacific Oceans come from healthy populations (92.3 and 87.5 per cent, respectively), and in the Atlantic Ocean 66.4 per cent of the catches come from healthy populations. In the Atlantic, currently 7.9 per cent of the tuna and billfish catches come from unhealthy populations (overfished and experiencing overfishing) and 25.7 per cent of catches come from overfished populations for which overfishing is no longer occurring and therefore might be on their path to recovery.

3 Special conservation status issues (CITES, national listing or priority for Marine Protected Area)

In 2011, the International Union for Conservation of Nature (IUCN) assessed for the first time the global conservation status for all tuna and billfish species using the IUCN criteria (Collette et al. 2011). The IUCN conservation assessments provide a complementary tool to existing fisheries stock assessment for setting conservation priorities at the global levels for this group of species and a platform for identifying species with long-term sustainability issues. The IUCN assessments utilize the IUCN Red List Criteria and all the available species information, including their global distribution, ecology, landing trends, biomass trends (mostly derived from fisheries stock assessments), and impacts of major threats, in order to classify species into the IUCN Red List categories. These categories range from Least Concern and Near Threatened, to the three threatened categories (Vulnerable, Endangered, Critically Endangered), providing a species ranking in terms of their relative risk of global extinction and conservation needs.

There is also a Data Deficient category where species with insufficient information to be evaluated are placed. Nonetheless, the information used to categorize tuna and billfishes in the IUCN Red List categories vary greatly among species; whereas some species such as *Allothunnus fallai* and *Tetrapturus angustirostris* are data poor due to scarce and incomplete landing and biological data against which to apply the IUCN criteria, some species such as the majority of the principal market tuna species are data rich with relatively extensive and highly detailed biological studies and fisheries stock assessment models for multiple populations throughout their distribution, which makes applying the IUCN criteria relatively easy. The IUCN Red List evaluation for the 25 species of tunas and billfishes resulted in 48 per cent (12 of 25 spp.) of species being listed under the Least Concern category, 12 per cent of tunas and billfishes listed in the Near Threatened category (*Thunnus alalunga*, *T. albacares* and *Kajikia audax*), and 24 per cent (6 spp.) had declined sufficiently in biomass to trigger listing under the Threatened categories. *Thunnus maccoyii* is listed as Critically Endangered, *T. thynnus* is Endangered, and *T. obesus*, *T. orientalis*, *Kajikia albida* and *Makaira nigricans* are Vulnerable. Lastly, 16 per cent (4 spp.) of tunas and billfish were listed as Data Deficient (Collette et al., 2011; Collette et al., 2014).

It should be noted that the IUCN criteria sometimes conflict with fisheries management objectives (Davies and Baum, 2012). For example, a species whose abundance declines from the unfished level by one-half in a given period of time may be classified in a threatened category in the IUCN Red List, but it might be well managed (i.e., not overfished and not experiencing overfishing) from an RFMO perspective. Conversely, a species that remains severely overfished for a period of time may be of grave concern to an RFMO but not classified in a threatened category in the IUCN Red List. Nevertheless, from a global conservation perspective, the latest IUCN assessments and derived conservation status were largely consistent with the current knowledge about the exploitation status of tuna and billfish populations derived from the RFMO fisheries stock assessments, in that the three longer-lived bluefin tuna species with geographically restricted spawning sites are more vulnerable to overfishing and are in need of more stringent management and conservation measures than the shorter-lived and more resilient tropical tuna species such as skipjack tuna for which spawning occurs in multiple broad spawning grounds throughout the tropics (Reglero et al., 2014).

More importantly, the IUCN conservation assessments are a useful tool particularly for those tuna and billfish species which are commercially exploited but lack formal fisheries population assessment, whose exploitation status is unknown and highly uncertain, and which do not have any management and conservation measures in place to ensure their long-term sustainability. According to the latest IUCN assessments, the following four IUCN Data Deficient species, *Thunnus tonggol*, *Tetrapturus angustirostris*, *Tetrapturus georgii* and *Istiompax indica*, should be the focus of future management and conservation efforts to ensure that their current fishing exploitation, and absence of fishery population assessments and management plans, do not jeopardize the long-term sustainability of these species.

A proposal to list Atlantic bluefin tuna on the Convention on International Trade in Endangered Species of Wild Fauna and Flora (CITES) was introduced at the fifteenth meeting of the Conference of the Parties in March 2010 by Monaco and supported by the United States and several European countries, but the proposal failed. There have also been several attempts to list several tuna and billfish species under National Listings. In the United States, the Center for Biological Diversity petitioned the United States Department of Commerce National Marine Fisheries Service to list both the eastern and western Atlantic populations of the bluefin tuna as endangered under the United States Endangered Species Act (ESA) (Center for Biological Diversity, 2010) but the Department rejected the petition, although declaring the Atlantic bluefin to be "a species of interest" after a review (Atlantic Bluefin Tuna Status Review Team, 2011). A petition from the Biodiversity Legal Foundation requesting listing of the Atlantic white marlin (*Kajikia albida*) as a threatened or endangered species under the ESA was found to be not warranted at that time by the National Marine Fisheries Service (White Marlin Biological Review Team, 2007). In Canada, the Committee on the Status of Endangered Wildlife in Canada (COSEWIC) determined in 2011 that the western Atlantic bluefin tuna was Endangered (Maguire, 2012). In Brazil, a number of specialists, coordinated by the Brazilian Ministry of Environment (MMA) through the Instituto Chico Mendes de Conservação da Biodiversidade (ICMBIO), evaluated the risk of extinction of marine species following IUCN Red list of Threatened Species. Most species have been listed in the same category as the global list, however *Xiphias gladius* was categorized as Near Threatened, *Makaira nigricans* as Endangered, *Thunnus alalunga* and *T. albacares* as Least Concern and *Auxis rochei* and *A. thazard* as Data Deficient, differently from the global list (ICMBIO. In press).

Marine protected areas (MPAs), or time-area closures, a term mostly used by fisheries managers, are one of the many tools available to fishery managers to reduce fishing mortality or redistribute fishing effort to protect a segment of a fish population (e.g. spawning adults) or vulnerable fish habitats, among many other applications. Marine protected areas or time-area closures can vary from complete prohibition on fishing or other forms of exploitation "no-take zone" to a continuum of spatial, temporal and user restrictions allowing numerous options and applications. Currently, the role of pelagic MPAs or time-area closures for the conservation and management of tunas and billfishes is a major topic of discussion, given their high mobility, their wide distributions, the dynamic physical nature of pelagic habitats, as well as the small number of empirical and theoretical studies showing their effectiveness (Davies et al., 2012; Dueri and Maury 2012). In the last decade, tuna RFMOs have tested and implemented several types of time-area closures, always in combination with other tools to control catch and effort, to reduce fishing effort and reduce by-catch of non-target species (Sibert et al. 2012). Past experiences indicate that time-area closures, if used alone, might be ineffective and inefficient to manage tuna species (ISSF, 2012). However, if time-area closures are used in combination with other fishery management tools, closures could have substantial benefits when the objectives are clearly defined and their implementation is accompanied by close evaluation, monitoring and enforcement (ISSF, 2012).

The future success of pelagic MPAs or time-area closures as a fisheries, conservation and management tool for tuna and billfish species relies on more theoretical modelling and long-term empirical studies to compare and contrast their effectiveness with other fishery management tools and in combination with these tools.

4 Key pressures linked to trends

Commercial fishing has been identified as the primary pressure driving tuna and billfish population declines and causing the overexploitation of some populations (Collette et al., 2011). Over the last century, industrial fisheries targeting tuna and billfish species have sequentially expanded from coastal areas to the high seas and now their fisheries cover the majority of the world's oceans (Miyake et al., 2004). Globally tuna and billfish catches and fishing effort have risen consistently since the 1950s and may not have yet reached a plateau (Juan-Jordá et al., 2011). Currently the global demand for tuna meat is still increasing, and fishing capacity, with the construction of new fishing vessels, especially purse seiners, and improved technology, is also increasing (Justel-Rubio and Restrepo, 2014). As concluded by Allen (2010), managing capacity and eliminating overcapacity where it exists has been identified as one of the major challenges jeopardizing the long-term sustainability of tuna and billfish species.

Climate change is another potential pressure that needs to be accounted for in the biology, economics and management of tuna and billfish species (McIlgorm, 2010). It is projected that by increasing ocean water temperatures, and altering oceanic circulation patterns and the vertical stratification of the water column, climate change will lead to a decrease in primary productivity in the tropics and a likely increase in higher latitudes (Intergovernmental Panel on Climate Change (IPCC, 2007)). Climate change, and the resultant increase in ocean temperatures, is also increasing the extension of areas with hypoxic waters and oxygen-depleted dead zones (Altieri and Gedan, 2015).

The extension of deep hypoxic bodies of water limits the distribution of tunas and billfishes by compressing their preferred habitat into a narrow surface layer, making this species more vulnerable to over-exploitation by surface gears (Prince and Goodyear, 2006). Thus, climate change might have an effect on tuna and billfish species by changing their physiologies, temporal and spatial horizontal and vertical distributions and abundances within the water column. A growing number of studies are evaluating the current and future impacts of climate change on the physiology, distribution, abundance and reproductive and feeding migrations of these species (Dufour et al., 2010; McIlgorm, 2010; Muhling et al., 2011; Bell et al., 2013; Dueri et al., 2014).

A study modelling the impacts of climate change on skipjack tuna in the tropical world's oceans suggests that the spatial distribution and abundance of skipjack tuna may change substantially with current suitable tropical habitats deteriorating and habitat suitability improving at higher latitudes (Dueri et al., 2014). In the Western and Central Pacific, another modelling study evaluated the effect of climate change on the food webs, habitat and main fish resources of the region, and found that distribution of skipjack tuna, the major tuna resource of the area, may move further east across the region. This eastward movement of skipjack tuna could benefit some nations by increasing their access to tuna resources and adversely affect other nations which would lose access to optimum tuna fishing grounds (Bell et al., 2013).

In the Atlantic Ocean, it has been documented that each year North Atlantic albacore tuna and East Atlantic bluefin tuna have arrived progressively earlier in the Bay of Biscay area, a major feeding ground, indicating that these species may be progressively adapting the timing of their feeding migrations and latitudinal distributions in response to climate change (Dufour et al., 2010). Another modelling study has also suggested that climate change might alter the temporal and spatial spawning and migratory activity of the West Atlantic bluefin tuna in the Gulf of Mexico with subsequent effects on population sizes and fisheries (Muhling et al. 2011). The impacts of climate change on tuna and billfish species are raising increasing concerns and need to be further understood, in order that governments and tuna RFMOs can respond rapidly to climate change by developing mitigation and adaptation programs.

5 Major ecosystem services provided by the species group and impacts of pressures on provision of these services

5.1 Ecosystem services

The impacts of fishing on the abundance of fishes and food web dynamics can have consequences on the structure, functioning and resilience of marine ecosystems (Heithaus et al., 2008; Baum and Worm, 2009). Consequently, population declines in tuna and billfish species and changes in their food web dynamics may be impairing the ocean's capacity to generate basic ecosystem processes which are vital to enable the maintenance and delivery of other ecosystem services benefiting human health, welfare and economic activities. To what extent widespread declines in tuna and billfish populations have altered the capacity of the ocean to support vital ecosystem processes, functions and services by altering species interactions and food web dynamics is poorly known (Kitchell et al., 2006; Hunsicker, 2012; IATTC, 2014a).

Tuna and billfish species are large predatory fishes, acting as apex and mesopredators and occupying high trophic levels in the marine food web; their removal could have ecological consequences for predator-prey interactions through trophic cascading effects (Baum and Worm, 2009). To fully understand the effects of removing tunas and billfishes from marine ecosystems, and their value in maintaining key ecosystem processes and services, requires better understanding of their unique role as predators and prey, and their interactions and dynamics using modelling and empirical approaches. This requires the collection of accurate information on trophic links and biomass flows through the food web in open marine ecosystems and accounting for environmental forcing (IATTC, 2014a).

To date, tuna RFMOs have conducted limited research and have a limited track record for incorporating food-web and ecosystem considerations into the management of tuna and billfish fisheries because traditionally their management has focused on achieving MSY for each of their targeted species individually. Consequently, according to de Bruyn et al., (2013), tuna RFMOs have implemented limited conservation measures to address the wide ecological effects of fishing . However, in the last decade tuna RFMOs, and especially IATTC and WCPFC, have increased their research activities to ensure that ecosystem considerations are part of their agendas (IATTC 2014a). These actions have mostly focused on monitoring, quantifying and mitigating incidental by-catch, increasing the coverage of the observing programmes and modifying fishing gear technology (Gilman et al., 2014; IATTC, 2014a).

5.2 Direct services to humans including economic and livelihood services

Tuna and billfish species provide a wide variety of direct ecosystem services to humans by supporting food production and creating vital coastal livelihoods, economies and recreational opportunities such as sport fisheries (Gilman et al., 2014). At present more than 80 countries have tuna fisheries, thousands of tuna fishing vessels operate in all the oceans, and tuna fishery capacity is still growing in the Indian and Pacific Oceans (ISSF, 2010). The popularity of tuna meat has increased remarkably around the globe and now tuna meat is considered to be a relatively low-cost source of protein, which is traded as a global "commodity" product (i.e. high volume, low value, low margins) (Hamilton et al., 2011). The canning and sashimi industries are the major players in the global trade of tuna, particularly focused on the principal market tuna species.

At the other extreme, in some regions of the world tuna and billfish species still contribute substantially to the subsistence of many fishing communities by providing the great majority of dietary animal protein (Bell et al., 2009). The global economic activity that tuna fisheries can generate directly and indirectly is remarkable. Every year at least 2.5 million tons of the global tuna catch is destined for the canning industry and globally around 256 million cases are consumed (3.2 million tons whole round equivalent), valued at 7.5 billion United States dollars (Hamilton et al., 2011). Therefore, ensuring the long-term sustainability of the world's tuna and billfish fisheries is intrinsically linked with providing food security, vital livelihoods and economic benefits in many regions of the world.

The dependency on healthy and sustainable tuna populations and the direct ecosystem services they provide is particularly strong for countries in the tropical western and central Pacific Ocean which is the most important tuna fishing area in the world. The tuna catch in the West Pacific Ocean is greater than that of the Atlantic, Indian and East Pacific Oceans combined (Miyake et al., 2010). Countries in the tropical west Pacific Ocean depend heavily on tuna resources for their nutrition, food security, economic development, employment, government revenue, livelihoods, culture and recreation (Gillett et al., 2001; Gillett, 2004; Gillett, 2009). Pacific States and territories in the west Pacific Ocean derive a large share of their taxes (up to 40 per cent) and Gross Domestic Product (up to 20 per cent) from fishing licenses sold to distant-water fishing nations and fish processors (Gillett, 2009; Bell et al., 2013).

Tuna and billfish also provide valuable recreational services; these fishes are considered to be valuable sportfishes, which gives them an important status in recreational fisheries in many regions of the world. Although the global picture of the recreational catch, effort and economic data for this industry is very fragmentary or unknown, for those countries with better records, the aggregate impact of the recreational tuna and billfish industry in terms of revenue and employment can be substantial for the local economies. For example, the total annual aggregate value of the recreational billfish industry in Costa Rica, Mexico, the United States Atlantic coast and Puerto Rico (United States) combined ranges between 203 and 340 million United States dollars, creating vital economic development, employment and recreation in the region (Ditton and Stoll, 2003).

6 Conservation responses and factors for sustainability

Tuna RFMOs face several challenges to ensure the long-term sustainability of tunas and billfishes and associated ecosystems within their Convention areas. Some of the main challenges have been considered to be:

(a) the existing overcapacity of fishing fleets;
(b) the equitable allocation of fishing rights among fishing nations;
(c) the possible implementation of the precautionary approach[4] and ecosystem approach;
(d) the monitoring of by-catch of vulnerable species; and
(e) the adequacy of financial resources to eliminate illegal, unreported and unregulated fishing and implement effective Monitoring, Control and Surveillance.

Tuna RFMOs have increasingly adopted a series of conservation and management measures to specifically address each of these challenges although their success and implementation have been mixed, and more time is needed to fully evaluate their success.

Tuna RFMOs control the amount of fishing of each stock through a variety of tools including catch limits, time-area closures and other input or output controls. Nevertheless, management of fleet capacity remains an issue of special concern, especially in the long term, because it tends to increase pressure on resources and management. The open access nature of fisheries, particularly in the high seas, has led to overcapacity of fleets in every tuna RFMO convention area (Allen, 2010; Miyake et al., 2010). Once overcapacity develops, it is difficult to reduce it because the fishing industry will continue operating as long as profits exceed costs (ISSF, 2010b). The IATTC has adopted a closed vessel registry for its purse seine fleet, a first and key step in managing overcapacity. However, overcapacity in the region remains well above the target (IATTC, 2014b). ICCAT, IOTC and WCPFC also have measures to limit capacity for some of their fisheries, but the problem of overcapacity has not been addressed in the RFMOs as a whole. It has been proposed that the establishment of exclusive rights to fish can be a formula to prevent overfishing, reduce overcapacity, achieve maximum economic benefits and sustainability in tuna and billfish fisheries, but its application is currently being debated (Allen, 2010; ISSF, 2010b; Squires et al., 2013). Ultimately, the global nature of tuna and billfish fisheries and industries might need cooperation among tuna RFMOs to manage fleet capacity successfully.

The equitable allocation of fishing rights is another challenge, given that allocating fishing access or catch quotas among the different member countries continues to be one of the most contentious matters in the RFMOs decision-making progress, impeding other more timely relevant conservation and management measures from moving forward, according to the International Seafood Sustainability Foundation (ISSF, 2013c). Nowadays, tuna RFMO allocation negotiations occur in a decision-making climate that is basically consensus-driven, which can result in overall catch levels being higher than scientifically-recommended levels. Identifying solutions requires recognizing the complexity and heterogeneity of tuna fisheries and the diverse objectives of RFMO member countries (ISSF, 2011).

Endorsing the precautionary and ecosystem approach requires the adoption of harvest control rules including limit and target reference points for tunas and billfishes and associated species, a long-standing recommendation of several international FAO Agreements and Guidelines over the past 15 years (Caddy and Mahon, 1995) and part of the Agreement for the Implementation of the Provisions of the United Nations Convention on the Law of the Sea of 10 December 1982 relating to the Conservation and Management of Straddling Fish Stocks and Highly Migratory Fish Stocks.[5] This is also part of the more modern RFMO Conventions, such as the WCPFC and IATTC. The CCSBT has adopted a

---

4 Definitions of the term "precautionary approach" are available at the FAO Term Portal and in ISSF (2013b).

5 United Nations, *Treaty Series*, vol. 2167, No. 37924.

formal management procedure[6] for deciding on Total Allowable Catch levels to rebuild the southern bluefin tuna population to 20 per cent of the unfished abundance level by 2035. The other RFMOs have not adopted such formulaic approaches to decision-making, but all are making progress in adopting population-specific limit and target reference points and discussing the use of harvest control rules. The adoption of harvest control rules and limit and target reference points is also a common requirement of several eco-label certifications, such as the Marine Stewardship Council Management Program.

The fifth aforementioned challenge reflects the paucity of knowledge about the impacts of tuna and billfish fisheries on other less productive species such as sharks, on species interactions and food web dynamics, and on the greater marine ecosystems (Dulvy et al., 2008; Gerrodette et al., 2012; de Bruyn et al. 2013; Gilman et al., 2014; IATTC, 2014a). One issue of concern is the widespread use of Fish Aggregating Devices (FADs) by industrial purse seine tuna fisheries and its potential impacts on tuna populations (especially on very small bigeye), higher levels of bycatch relative to setting nets on free-swimming schools, and possible ecosystem impacts (Dagorn et al., 2012; Fonteneau et al., 2013).

RFMOs have increasingly adopted several measures to monitor and regulate the use of FADs, and to increase data reporting requirements specific to FADs. Moreover, new research initiatives have also been emerging that aim to identify best practices in FAD fishing, as well as modification of gears, and new technology to reduce the catch of non-target species by FAD fisheries. For example, IATTC, IOTC and ICCAT have adopted measures to require a transition to non-entangling FADs that would reduce unobserved mortality of sharks and other species. Pelagic longline tuna and swordfish fisheries have higher levels of bycatch of sensitive species such as sharks, turtles and seabirds (Gilman, 2011). In addition, mitigation measures in longline fisheries targeting tunas and swordfishes have been developed and adopted by the RFMOs to reduce the by-catch of species like sea birds and sea turtles, although their successful implementation and effectiveness in reducing by-catch levels vary greatly among tuna RFMOs (Small, 2005; Gilman, 2011).

The last challenge encompasses the difficulty of eliminating illegal, unreported and unregulated fishing (IUU) and implementing effective monitoring, control and surveillance (MCS) measures in a context of insufficient financial resources (ISSF, 2013c). Effective MCS is required to successfully implement any conservation and management measure in place and combat IUU fishing. MCS measures can be very diverse, from operating transparent catch documentation schemes, implementing effective at-sea observer programs, requiring vessels to acquire unique vessel identifiers, maintaining comprehensive IUU vessel lists, and operating regular reports of transshipments. The extent to which tuna RFMOs have successfully adopted MCS measures varies greatly (ISSF, 2013c). The compliance mechanisms used by the different tuna RFMOs vary considerably (Koehler, 2013). The identification of best practices, successful measures and incentives to promote best practices is a first step forward, which would require global collaboration among all tuna RFMOs.

---

6   See CCSBT (2011).

References

Allen, R. (2010). International management of tuna fisheries: Arrangements, challenges and a way forward. *FAO Fisheries and Aquaculture Technical Paper*. No. 536. Rome, FAO. pp. 47.

Altieri, A.H. and Gedan, K.B. (2015). Climate change and dead zones. *Global Change Biology*, 21: 1395–1406.

Atlantic Bluefin Tuna Status Review Team (2011). Status *Review Report of Atlantic bluefin tuna (Thunnus thynnus)*. Report to National Marine Fisheries Service, Northeast Regional Office. March 22, 2011. 104 pp.

Baum, J.K. and Worm, B. (2009). Cascading top-down effects of changing oceanic predator abundances. *The Journal of Animal Ecology*, 78: 699-714.

Bell, J. D., Ganachaud, A., Gehrke, P. C., Griffiths, S. P., Hobday, A. J., Hoegh-guldberg, O., Johnson, J. E., Borgne, R. L., Lehodey, P., Lough, J. M., Matear, R. J., Pickering, T. D., Pratchett, M. S., Gupta, A. S. and Senina, I., (2013). Mixed responses of tropical Pacific fisheries and aquaculture to climate change. *Nature Climate Change*, 3: 591-599.

Bell, J. D., Kronen, M., Vunisea, A., Nash, W. J., Keeble, G., Demmke, A., Pontifex, S. and Andréfouët, S. (2009). Planning the use of fish for food security in the Pacific. *Marine Policy*, 33: 64-76.

Caddy, J.F. and Mahon, R. (1995). Reference points for fisheries management. *FAO Fisheries Technical Paper*. No. 347. Rome, FAO. pp. 83.

CCSBT (2011). Report of the Eighteenth Annual Meeting of the Commission, 10-13 October 2011, Bali, Indonesia.

Center for Biological Diversity (2010). *Petition to list the Atlantic bluefin tuna (Thunnus thynnus) as endangered under the United States Endangered Species Act*. 44 pp.

Collette, B.B., Reeb, C., Block, B.A. (2001). Systematics of the tunas and mackerels (Scombridae). In: *Tuna: Physiology, ecology, and evolution*, BA Block and ED Stevens, eds. Academic Press, San Diego, pp. 1-33.

Collette, B.B., Carpenter, K.E., Polidoro, B.A., Juan-Jordá, M.J., Boustany, A., Die, D.J., Elfes, C., Fox, W., Graves, J., Harrison, L.R., McManus, R., Minte-Vera, C.V., Nelson, R., Restrepo, V., Schratwieser, J., Sun, C.-L., Amorim, A., Brick Peres, M.B., Canales, C., Cardenas, G., Chang, S.-K., Chiang, W.-C., de Oliveira Leite Jr., N., Harwell, H., Lessa, R., Fredou, F.L., Oxenford, H.A., Serra, R., Shao, K.-T., Sumaila, R., Wang, S.-P., Watson, R. and Yáñez, E. (2011). High value and long life - Double jeopardy for tunas and billfishes. *Science*, 333: 291-292.

Collette, B.B., McDowell, J.R. and Graves, J.E. (2006). Phylogeny of Recent billfishes (Xiphioidei). *Bulletin of Marine Science*, 79: 455-468.

Collette, B.B. and Nauen, C.E. (1983). FAO Species Catalogue. Vol. 2. Scombrids of the world: an annotated and illustrated catalogue of tunas, mackerels, bonitos and related species known to date. *FAO Fisheries Synopsis*. No 125. Rome, FAO. pp. 137.

Collette, B., Fox, W., Juan Jorda, M., Nelson, R., Pollard, D., Suzuki, N. and Teo, S. 2014. Thunnus orientalis. The IUCN Red List of Threatened Species. Version 2014.3. <www.iucnredlist.org>. Downloaded on 25 March 2015.

Dagorn, L., Holland, K.N., Restrepo, V. and Moreno, G. (2012). Is it good or bad to fish with FADs? What are the real impacts of the use of drifting FADs on pelagic marine ecosystems? *Fish and Fisheries*, 14: 391–415.

Davies, T.D. and Baum J.K. (2012). Extinction Risk and Overfishing: Reconciling Conservation and Fisheries Perspectives on the Status of Marine Fishes. *Scientific Reports* 2:1-9.

Davies, T.K., Martin, S., Mees, C., Chassot, E. and Kaplan, D.M. (2012). A review of the conservation benefits of marine protected areas for pelagic species associated with fisheries. *ISSF Technical Report 2012-02*. International Seafood Sustainability Foundation, McLean, Virginia, USA.

de Bruyn, P., Murua, H. and Aranda, M. (2013). The precautionary approach to fisheries management: How this is taken into account by tuna regional fisheries management organisations (RFMOs). *Marine Policy*, 38: 397-406.

Ditton, R.B. and Stoll, J.R. (2003). Social and economic perspective on recreational billfish fisheries. *Marine and Freshwater Research*, 54: 545-554.

Dueri, S., Bopp, L. and Maury, O. (2014). Projecting the impacts of climate change on skipjack tuna abundance and spatial distribution. *Global Change Biology*, 20: 742–753.

Dueri, S. and Maury, O. (2012). Modelling the effect of marine protected areas on the population of skipjack tuna in the Indian Ocean. *Aquatic Living Resources*, 26: 79-84. doi:10.1051/alr/2012031.

Dufour, F., Arrizabalaga, H., Irigoien, X. and Santiago, J. (2010). Climate impacts on albacore and bluefin tunas migrations phenology and spatial distribution. *Progress in Oceanography*, 86: 283–290.

Dulvy, N. K., Baum, J. K., Clarke, S., Compagno, L., Cortés, E., Domingo, A., Fordham, S., Fowler, S., Francis, M. and Gibson, C. (2008). You can swim but you can't hide: the global status and conservation of oceanic pelagic sharks and rays. *Aquatic Conservation: Marine and Freshwater Ecosystems*, 482: 459-482.

FAO (2014). Fisheries Global Information System (FAO-FIGIS) In: *FAO Fisheries and Aquaculture Department* [online]. Rome. <www.fao.org/fishery/figis/en>. Downloaded on May 2014.

Fonteneau, A., Chassot, E. and Bodin, N. (2013). Global spatio-temporal patterns in tropical tuna purse seine fisheries on drifting fish aggregating devices (DFADs): Taking a historical perspective to inform current challenges. *Aquatic Living Resources*, 26: 37–48.

Freire, K. (2009). Dinâmica de Populações e Avaliação dos Estoques dos Recursos Pesqueiros do Nordeste do Brasil; Espécies Pelágicas- Thunnus atlanticus In: Lessa, RPT, Bezerra Júnior, L. Nóbrega, MF. Dinâmica de Populações e Avaliação dos Estoques dos Recursos Pesqueiros do Nordeste do Brasil.1 ed.Fortaleza : Martins & Cordeiro LTDA, v.1, p. 76-89.

Gerrodette, T., Olson, R., Reilly, S., Watters, G. and Perrin, W. F. (2012). Ecological metrics of biomass removed by three methods of purse-seine fishing for tunas in the eastern tropical Pacific Ocean. *Conservation Biology*, 26: 248–256.

Gillett, R. (2004). *Tuna for tomorrow? Some of the science behind an important fishery in the Pacific Islands*. Asian Development Bank and Secretariat of the Pacific Community.

Gillett, R. (2009). *Fisheries in the economies of the Pacific island countries and territories*. Mandaluyong City, Philippines: Asian Development Bank, 2009.

Gillett, R., Mccoy, M., Rodwell, L. and Tamate, J. (2001). *Tuna: A Key Economic Resource in the Pacific Islands - A Report Prepared for the Asian Development Bank and the Forum Fisheries Agency*. Pacific Studies Series. pp 95.

Gilman, E. (2011). Bycatch governance and best practice mitigation technology in global tuna fisheries. *Marine Policy* 35: 590-609.

Gilman, E., Passfield, K. and Nakamura, K. (2014). Performance of regional fisheries management organizations: ecosystem-based governance of bycatch and discards. *Fish and Fisheries* 15:327–351.

Hamilton, M. J., Antony, L., McCoy, M., Havice, E. and Campling, L. (2011). Market and industry dynamics in the global tuna supply chain. Pacific Islands Forum Fisheries Agency (FFA), Honiara, Salomon Islands.

Heithaus, M.R., Frid, A., Wirsing, A.J. and Worm, B. (2008). Predicting ecological consequences of marine top predator declines. *Trends in Ecology and Evolution*, 23: 202–210.

Hunsicker, M.E., Olson, R., Essington, T., Maunder, M., Duffy, L. and Kitchell, J.F. (2012). Potential for top-down control on tropical tunas based on size structure of predator–prey interactions. *Marine Ecology Progress Series*, 445: 263–277.

IATTC (2014a). *Ecosystem considerations*. Document SAC-05-13. Inter-American Tuna Commission. Scientific Advisory Committee Fifth Meeting, La Jolla, California, USA, 12-16 May 2014.

IATTC (2014b). *Utilization of Vessel Capacity Under Resolutions C-02-03, C-12-06, and C-12-08*. Fifteenth Meeting of the Permanent Working Group on Fleet Capacity, Lima, Peru, 12-13 July 2014.

ICMBIO. In press. *Diagnóstico do risco de extinção das espécies da fauna brasileira*.

IOTC-SC17. 2014. Report of the Seventeenth Session of the IOTC Scientific Committee. Seychelles, 8–12 December 2014. IOTC–2014–SC17–R[E]: 357 pp.

Intergovernmental Panel on Climate Change (IPCC) (2007). Intergovernmental Panel on Climate Change (IPCC), 2007. Summary for policymakers. Cambridge University Press, Cambridge, UK and New York, USA.

ISSF (2010). Status of the world fisheries for tuna. Section A -Introduction. International Seafood Sustainability Foundation, Washington, D.C., USA.

ISSF (2010b). Bellagio Framework for Sustainable Tuna Fisheries: Capacity controls, rights-based management, and effective MCS. International Seafood Sustainability Foundation, Washington, D.C., USA.

ISSF (2011). The Cordoba Conference on the Allocation of Property Rights in Global Tuna Fisheries. International Seafood Sustainability Foundation, Washington, D.C., USA.

ISSF (2012). Report of the 2012 ISSF Workshop: Review of spatial closures to manage tuna fisheries. *ISSF Technical Report 2012-08*. International Seafood Sustainability Foundation, Washington, D.C., USA.

ISSF (2013a). ISSF Tuna Stock Status Update, 2013(3): Status of the world fisheries for tuna. *ISSF Technical Report 2013-04B*. International Seafood Sustainability Foundation, Washington, D.C., USA.

ISSF (2013b). *ISSF Stock Assessment Workshop: Harvest Control Rules and Reference Points for Tuna RFMOs*, San Diego, California, USA, March 6-8, 2013, International Seafood Sustainability Foundation, Washington, D.C., USA.

ISSF (2013c). Status of the world fisheries for tuna: management of tuna stocks and fisheries. *ISSF Technical Report 2013-05*. International Seafood sustainably Foundation, Washington, D.C., USA.

Juan-Jordá, M.J., Mosqueira, I., Cooper, A.B. and Dulvy, N.K. (2011). Global population trajectories of tunas and their relatives. *Proceedings of the National Academy of Science*, USA 51: 20650-20655.

Justel-Rubio, A. and Restrepo, V.R., (2014). *A Snapshot of the Large-Scale Tropical Tuna Purse Seine Fishing Fleets at the Beginning of 2014*. ISSF Technical Report 2014-07. International Seafood Sustainability Foundation, McLean, Virginia, USA.

Kitchell, J.F., Martell, S. J.D., Walters, C.J., Jensen, O. P., Kaplan, I.C., Watters, J., Essington, T.E. and Boggs, C.H. (2006). Billfishes in an ecosystem context. *Bulletin of Marine Science*, 79: 669-682.

Koehler, H. (2013). Promoting Compliance in Tuna RFMOS: A Comprehensive Baseline Survey of the Current Mechanics of Reviewing, Assessing and Addressing Compliance with RFMO Obligations and Measures. *ISSF Technical Report 2013-02*. International Seafood Sustainability Foundation.

Maguire J.-J. (2012). Bluefin tuna (*Thunnus thynnus*) in Atlantic Canadian waters: biology, status, recovery potential, and measures for mitigation. DFO Can. Sci. Advis. Sec. Res. Doc. 2012/002, 28 pp.

Majkowski, J. (2007). Global fishery resources of tuna and tuna-like species. *FAO Fisheries Technical Paper*. No. 483. Rome, FAO. pp. 1-54.

McIlgorm, A. (2010). Economic impacts of climate change on sustainable tuna and billfish management: Insights from the Western Pacific. *Progress in Oceanography*, 86: 187–191.

Miyake, M.P., Guillotreau, P., Sun, C., Ishimura, G. (2010). Recent developments in the tuna industry: Stocks, fisheries, management, processing, trade and markets. *FAO Fisheries and Aquaculture Technical Paper*. No. 536. Rome, FAO. pp. 1-125.

Miyake, M.P., Miyabe, N., Nakano, H. (2004). Historical trends of tuna catches in the world. *FAO Fisheries Technical Paper*. No. 467. Rome, FAO. pp. 74.

Muhling, B.A., Lee, S-K., Lamkin J.T. and Liu Y. (2011). Predicting the effects of climate change on bluefin tuna (*Thunnus thynnus*) spawning habitat in the Gulf of Mexico. *ICES Journal of Marine Science*, 68: 1051–1062.

Nakamura I. (1985). FAO Species Catalogue. Vol. 5. *Billfishes of the world: an annotated and illustrated catalogue of marlins, sailfishes, spearfishes, and swordfishes known to date*. FAO Fisheries Synopsis. No 125. Rome, FAO. 65. pp

Prince, E.D. and Goodyear, C.P. (2006). Hypoxia-based habitat compression of tropical pelagic fishes. *Fisheries Oceanography*, 15: 451–464.

Reglero, P., Tittensor, D.P. Álvarez-Berastegui, D., Aparicio-González, D. and Worm, B. (2014). Worldwide distributions of tuna larvae: revisiting hypotheses on environmental requirements for spawning habitats. *Marine Ecology Progress Series* 591:207-224.

Sibert, J., Senina, I., Lehodey, P. and Hampton, J. (2012). Shifting from marine reserves to maritime zoning for conservation of Pacific bigeye tuna (*Thunnus obesus*). *Proceedings of the National Academy of Science*, USA, 109: 18221–18225.

Small, C.J. (2005). *Regional Fisheries Management Organizations: their duties and performance in reducing bycatch of albatrosses and other species*. Cambridge, UK: BirdLife International.

Squires, D., Allen, R. and Restrepo, V. (2013). Right-based management in international tuna fisheries. *FAO Fisheries and Aquaculture Techinical Paper*. No. 571, Rome, FAO pp. 79.

White Marlin Biological Review Team. (2007). *Atlantic white marlin status review. Report to the National Marine Fisheries Service*, Southeast Regional Office, Dec. 10. 88 pp.

II Marine Ecosystems and Habitats

Cold-Water Corals

Contributors:
Erik Cordes (Convenor), Sophie Arnaud-Haond, Odd-Aksel Bergstad, Patricio Bernal (Lead Member), Ana Paula da Costa Falcão, Andre Freiwald and J. Murray Roberts

1 Inventory and Ecosystem Functions

Globally viewed, cold-water corals cover a wide range of depths (39 - 2000 m) and latitude (70°N – 60°S). In this Chapter, we will focus on the corals found below 200 m, the average depth below which photosynthesis does not occur, to avoid overlap with other chapters. The term "corals" refers to a diverse group of species in the Phylum Cnidaria, including the scleractinian hard corals, octocorals including the sea fans and soft corals, antipatharian black corals, and stylasterid lace corals. Although the majority of the species-level diversity of scleractinians is in the solitary corals (Cairns, 2007), some of the scleractinian corals may form extensive reef structures, occasionally accumulating into large carbonate mounds, or bioherms. Many of the ecological patterns discussed in this chapter are derived from the study of these structures, simply because they have been the focus of the most extensive research in this developing field. However, other types of cold-water corals can also form highly significant structural habitat and these are also discussed. The most representative cold-water, framework-building, scleractinian corals are *Enallopsammia rostrata*, *Goniocorella dumosa*, *Lophelia pertusa* (Figure 42.1) *Madrepora oculata*, *Oculina varicosa* and *Solenosmilia variabilis* (Roberts et al., 2006). The most common and widespread of the large, structure-forming octocorals are found in the genera *Corallium*, *Isidella*, *Paragorgia*, *Paramuricea*, and *Primnoa* (Watling et al., 2011) (Figure 42.2).

Cold-water corals (CWC) most commonly occur in continental slope settings, on deep shelves and along the flanks of oceanic banks and seamounts. The majority of CWC occur between the depths of 200 to 1000 m, with the bathymetric ranges becoming shallower towards the poles (Roberts et al., 2009). However, there are numerous, dense coral gardens (primarily octocorals and black corals) found on the slopes

Figure 42.1 | Examples of dense cold-water Lophelia pertusa reef frameworks, including provision of fish habitat. (a) and (b) from 400-500 m depth in the Viosca Knolls region of the Gulf of Mexico. (c) and (d) from 600 to 800 m depth on the Logachev coral carbonate mounds on the Rockall Bank in the Northeast Atlantic.
All photos are property of the contributors to this chapter and should also be attributed to: (a) and (b): Ecosystem Impacts of Oil and Gas Inputs to the Gulf (ECOGIG), a consortium funded by the Gulf of Mexico Research Initiative (GoMRI), and the Ocean Exploration Trust; (c) and (d): Roberts, J.M., Changing Oceans Expedition 2012, funded by UK Ocean Acidification programme (NERC, DECC, Defra).

Figure 42.2 | Octocoral gardens from different depths within the Gulf of Mexico. (a): A 2m tall Leiopathes glaberrima black coral colony from 200 m depth. (b): A diverse community of Stichopathes sp. black corals, keratoisid bamboo corals, and other octocorals from 500 m depth. (c): Large, habitat-forming Paramuricea sp. colonies from 1000 m depth. (d): A diverse community of octocorals including Iridogorgia sp., keratoisid bamboo corals, Paramuricea biscaya, and Corallium sp. from 2000 m depth.
All photos are property of the contributors to this chapter, and should also be attributed to: (a) and (b): The Lophelia II project, funded by the United States Bureau of Ocean Energy Management (US BOEM) and the National Oceanic and Atmospheric Administration, Office of Ocean Exploration and Research (NOAA OER); (c) and (d): Ecosystem Impacts of Oil and Gas Inputs to the Gulf (ECOGIG), a consortium funded by the Gulf of Mexico Research Initiative (GoMRI), and the Ocean Exploration Trust.

of seamounts and the base of the continental slope to over 3000 m, and some soft corals and sea pens are found on soft substrata down to abyssal depths (Yesson et al., 2012). The shallowest occurrences of typically deep-water species are in high latitudes associated with the rocky slopes and sills of fjords (*L. pertusa* off of Norway at 37 m depth, Wilson, 1979) or narrow passes between islands (the octocorals *Paragorgia arborea* and *Plumarella* spp. at 27 m depth in Alaska, (Stone, 2006)). Continental slopes exhibit a variety of specific topographic irregularities that provide suitable substrate for cold-water coral larvae to settle. In many parts of the world ocean, the shelf edge is incised by gullies and submarine canyons (Harris and Whiteway, 2011; Harris et al., 2014). Some prominent examples are located at the canyon-rich slope of the Gulf of Lion off the coast of France (Fabri et al., 2014), the Bay of Biscay under the national jurisdiction of France and Spain (De Mol et al., 2011; Sánchez et al., 2014), the Gully off the coast of Nova Scotia (Mortensen and Buhl-Mortensen, 2005), and the canyons off the eastern United States (Watling and Auster, 2005; Brooke and Ross, 2014). Narrow straits between land-masses may also provide suitable substrate, such as the Straits of Florida (Correa et al., 2012), Gibraltar (De Mol et al., 2012), Sicily (Freiwald et al., 2009), and the Yucatan (Hebbeln et al., 2014). Open-slope CWC mounds are known from the large reefs off the Norwegian coast (Mortensen et al., 2001; Buhl-Mortensen et al., 2014), the Northeast Atlantic along the Rockall and Porcupine Banks (Van der Land et al., 2014), the Southeast coast of the United States (Stetson et al., 1962; Reed et al., 2006), the Gulf of Mexico (Reed et al., 2006; Cordes et al., 2008), Southwestern Atlantic Ocean (Viana et al., 1998; Sumida et al., 2004; Pires, 2007; Carranza et al., 2012), and off Mauritania (Colman et al., 2005). These mounds are not randomly distributed over the slope but show a strong affinity with distinct water mass boundaries passing along the slope (Mienis et al., 2007; Arantes et al, 2009; White and Dorschel, 2010). Open-slope coral gardens appear to be common along most of the continental margins of the world (Figure 42.3, Yesson et al., 2012). Oceanic seamounts represent another important cold-water coral-rich environment (see Chapter 51), such as the Tasmanian seamounts off South Australia (Thresher et al., 2011), the seamount speckled Chatham Rise off the coast of New Zealand (Tracey et al., 2011), seamounts of the central Pacific (Rogers et al., 2007), and seamounts of the Mid-Atlantic Ridge system (Mortensen et al., 2008). A compilation of framework-forming cold-water coral occurrences is displayed in Figure 42.4 based on the UNEP-WCMC database (Freiwald et al., 2005) and more recent findings. The current information on deep-water octocorals suggests that they are ubiquitous along continental margins and seamounts on hard substrata, as well as occasionally on soft-bottom in the case of the sea pens and a few species of bamboo corals. A combination of octocorals collections and observations along with a predictive habitat suitability model is displayed in Figure 42.3 (Yesson et al., 2012).

Cold-water corals have been known since the first descriptions in the 18th century and the first deep-water research expeditions of the 19th century (Roberts et al., 2006). The presence of large reef structures in deep water was not broadly appreciated by the scientific community until the first submersibles were available in the late 20th century (Cairns, 2007). Using these new tools, a more complete set of distribution records and characterization of the habitat requirements of CWC were developed. Based on these recent data, the use of habitat modelling has led to the discovery of numerous cold-water coral sites and habitats. As an example, scleractinians were discovered on steep submarine cliffs after modelling (Huvenne et al., 2011) and field observation in the Mediterranean (Naumann et al., 2013) and the Bay of Biscay (De Mol et al., 2011; Reveillaud et al., 2008). Similarly, an extensive screening of newly available mapping and visualization technology in the Mediterranean revealed additional and more extensive coral formations than anticipated hitherto (Freiwald et al., 2009). Habitat modelling has thus far mostly been applied to a few of the most common species at a global and regional scale (Rengstorf et al., 2013; Yesson et al., 2012) at a coarse spatial resolution (Ross and Howell, 2013). However, models are now being applied at finer resolution levels in order to guide surveys with the visual tools of remotely-operated and manned submersibles (Georgian et al., 2014). Additional fine-grained and broad-scale habitat modelling, specifically incorporating the best available taxonomic identifications (Henry and Roberts, 2014) is still needed to discover additional habitats, and to forecast the fate of CWC facing both direct (fisheries) and indirect (environmental) impacts (Guinotte et al., 2006; Clark and Tittensor, 2010).

Cold-water coral reefs, mounds, and gardens support a highly diverse community, comprising faunal biomass that is orders of magnitude above that of the surrounding seafloor (Mortensen et al., 1995; Henry and Roberts, 2007; Cordes et al., 2008; Roberts et al., 2008; Rowden et al., 2010). In addition to this tightly-associated community, cold-water corals may also serve as important spawning, nursery, breeding and feeding areas for a multitude of fishes and invertebrates (Koslow et al., 2001; Fossa et al., 2002; Husebo et al., 2002; Colman et al., 2005; Stone, 2006; Ross and Quattrini, 2009; Baillon et al., 2012; Henry et al., 2013), and habitat for transient diel vertical migrators (Davies et al., 2010). The ability to construct massive calcium carbonate frameworks, which makes both shallow and deep-water coral reefs unique, provides an important biogeochemical function in both the carbonate system (Doney et al., 2009) and in calcium balance (Moberg and Folke, 1999). CWC skeletons also provide an information function (*sensu* de Groot

The boundaries and names shown and the designations used on this map do not imply official endorsement or acceptance by the United Nations.

Figure 42.3 | Global octocoral distribution. Direct observations and collections are noted by "x" while the shading represents the habitat suitability probability for the presence of one order (lighter orange) or all nine orders (darker orange) Adapted from Yesson et al., 2012.

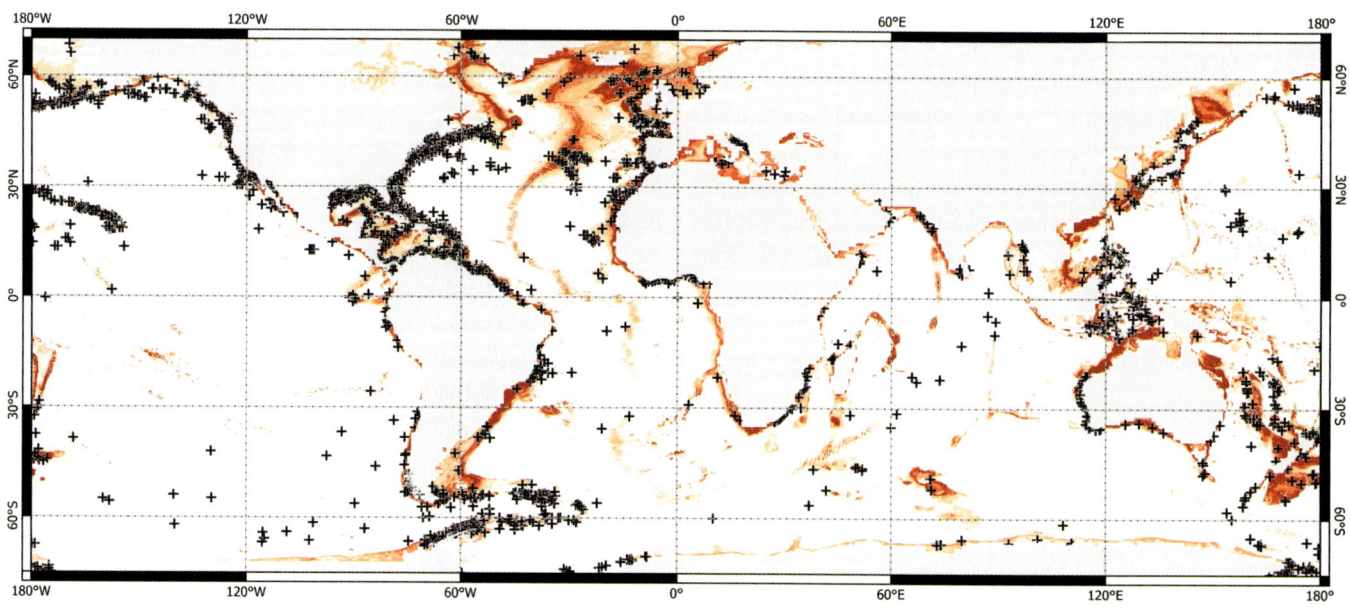

+ Observations   □ 1 Suborder predicted   ■ 7 Suborders predicted

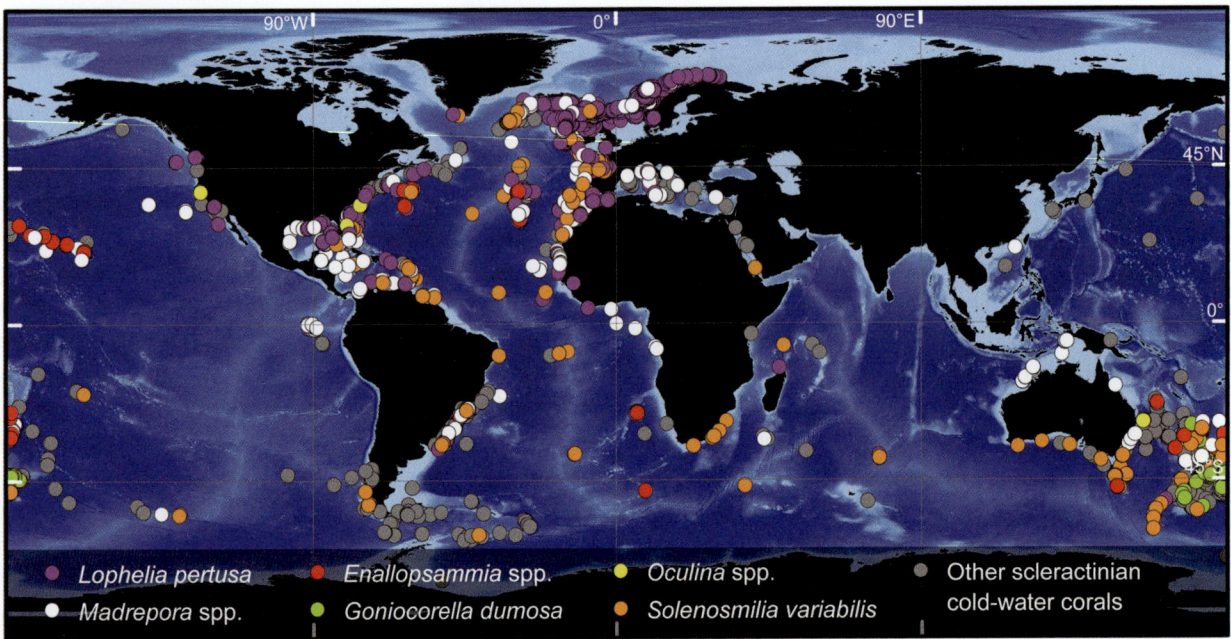

Figure 42.4 | Global distribution of the major framework-forming cold-water corals. Source: Freiwald et al., 2005, and more recent published data, n = 7213 entries

et al., 2002) through their archiving of paleoclimate signals (Adkins et al., 1998; Williams et al., 2006). Besides this, CWC ecosystems possess an inherent aesthetic value (*sensu* de Groot et al., 2002) demonstrated through countless films, photographs, and paintings of reefs or reef organisms.

Cold-water corals and the communities they support rely on surface productivity as their primary source of nutrition; either through the slow, relatively steady deposition of particulate organic carbon (POC) in the form of marine snow, which may be enhanced by hydrographic mechanisms (e.g. Davies et al., 2009; Kiriakoulakis et al., 2007), or through more active transport of carbon provided by vertical migrators (Mienis et al., 2012). However, *L. pertusa* has been shown to incorporate everything from dissolved organic carbon (DOC) to POC to algal biomass to small zooplankton (van Oevelen et al., 2009). As in shallow-water systems, corals and sponges of the deep reefs recycle these nutrients and form both the structural and trophic foundation of the ecosystem. In addition to these ties from shallow to deep water, the transport of nutrients from deep to shallow water is accomplished both by the diel vertical migrations of plankton and small fishes (Davies et al., 2010) as well as by periodic down- and upwelling that can occur near some of the reefs (Mienis et al., 2007; Davies et al., 2009). Although the mechanisms for deep-to-shallow water transport are well established, the input of deep-water secondary productivity to shallow ecosystems remains unquantified.

2 Features and Trends

All geological structures mentioned share some environmental factors that facilitate coral settlement and subsequent growth: provision of current-swept hard substrate, and often topographically-guided hydrodynamic settings. It has been suggested that corals are preferably confined to narrow seawater density (Sigma-theta) envelopes (Dullo et al., 2008) in which along-slope larval dispersal propagation may be facilitated. Survival and growth may be most closely associated with specific hydrodynamic settings including tidal-driven internal-wave fronts hitting continental slopes and seamounts (Mienis et al., 2007; Henry et al., 2014), specific up- and downwelling currents affecting the summits of shallow-water seamounts (Ramirez-Llodra et al., 2010), and tidal-driven downwelling phenomena on inner shelf settings (Davies et al., 2009; Findlay et al., 2013). These hydrographic transfer processes tend to concentrate or prolong the retention time of nutrients and food that sustain the metabolic demands of the suspension-feeding community.

Another perspective on the occurrence of coral habitat is a combined biogeophysical and hydrochemical analysis of the ambient seawater, a very recent endeavour in the still young research history of cold-water coral systems (e.g., Findlay et al., 2014; Flögel et al., 2014; Henry et al., 2014; Lunden et al., 2013). These forms of data along with species presence data were incorporated into global habitat suitability study by Davies and Guinotte (2011) that was conducted on the six major cold-water framework-building corals (*Enallopsammia rostrata*, *Goniocorella dumosa*, *Lophelia pertusa*, *Madrepora oculata*, *Oculina varicosa* and *Solenosmilia variabilis*) using the Maximum Entropy modelling approach (MAXENT). This approach uses species-presence data, global bathymetry 30-arc second grids (1 km2 resolution) and incorporates environmental data from several global databases. Viewed on such a global scale, these corals generally thrive in waters that: (1) are supersaturated with respect to aragonite, (2) occur shallower than 1500 m water depth, (3) contain dissolved oxygen concentrations of >4 ml l-1, (4) have a salinity range between 34 and 37 ppt, and (5) show a temperature range between 5 and 10°C. Laboratory experiments have confirmed many of these ranges, with *L. pertusa* being the most commonly studied species. Mediterranean *L. pertusa* and *M. oculata* colonies survived and grew at 12oC for three weeks, with *M. oculata* showing a greater sensitivity to high temperature (Naumann et al., 2014). Gulf of Mexico *L. pertusa* colonies survived and grew at up to 12oC, but died when exposed to 14oC for 8 days (Lunden et al., 2014). Studies of *L. pertusa* from west of Scotland, United Kingdom of Great Britain and Northern Ireland, demonstrated that this species can maintain respiratory independence and even survive periods of reduced oxygen (Dodds et al., 2007).

However, some remarkable outliers to these trends exist in the Red Sea and the Gulf of Mexico. The Red Sea represents the warmest and most saline deep-sea basin on Earth, with temperatures >20°C throughout the water column and salinity in excess of 40 ppt. Recent findings of typically deep-dwelling corals in these habitats shed new light on the persistence of corals in deep waters (Roder et al., 2013; Qurban et al., 2014). Although none of the coral species found in the Red Sea are among the most common globally (see above for list), limited framework growth is recorded mainly by *Eguchipsammia fistula* under food- and oxygen deprived conditions (1.02 – 2.04 ml l-1). Coral survival under such extreme environmental conditions may follow the strategy of metabolic depression (*sensu* Guppy and Withers, 1999), including depressed aerobic respiration and calcification rates. However, the high temperatures in combination with high aragonite saturation values of 3.44-3.61 in the Red Sea may facilitate calcification under these otherwise adverse conditions (Roder et al., 2013). The cold-water coral communities in the northern Gulf of Mexico belong to the most intensively studied sites in waters of the United States (e.g., Cordes et al., 2008). The major framework-constructor is *L. pertusa* and most environmental variables (i.e., temperature, salinity and aragonite saturation state) reflect the ranges known from Atlantic *Lophelia* sites (Davies et al., 2010; Lunden et al., 2013). However, dissolved oxygen values appear to be low, 2.7–2.8 ml l−1 are typically observed (Davies et al., 2010) and values as low as 1.5 ml l−1 have been recorded adjacent to coral mounds (Georgian et al., 2014). Coral nubbins from these Gulf of Mexico populations survived and grew in the lab at oxygen levels as low as 2.9 ml l-1, but eight-day incubations at lower oxygen concentrations (1.5 ml l-1) caused complete mortality, suggesting that these conditions are short-lived *in situ* (Lunden et al., 2014). Similarly, low oxygenation levels were found in the newly discovered *Lophelia-Enallopsammia* coral mounds in the Campeche Bank coral mound province, in the southern Gulf of Mexico (Hebbeln et al., 2014). It is possible that the low oxygen concentrations of the Gulf of Mexico result in lower growth rates observed for *L. pertusa* on natural (Brooke and Young, 2009) and man-made substrata (Larcom et al., 2014), although this remains to be examined empirically.

There have been numerous recent advances in our knowledge of the oceanographic variables describing coral habitat in the deep sea. However, knowledge gaps still remain when up-scaling from local to regional to global scales. Furthermore, limited capacity to carry out long-term *in situ* measurements with benthic landers and cabled observatories persists. This knowledge is of utmost importance to understand the consequences of already perceptible environmental change, such as ocean acidification, spread of oxygen minimum zones, and rising temperatures, on deep-sea ecosystems.

3 Major Pressures Linked to the Trends

Numerous anthropogenic threats to cold-water coral communities exist, the most significant of which include fisheries, hydrocarbon exploration and extraction, and mining, as well as global ocean change including warming and acidification. An improved understanding of the function of cold-water corals as habitat, feeding grounds and nurseries for many fishes including certain deep-sea fisheries targets has emerged along with concerns as to the impact of fisheries on these ecosystems (Costello et al., 2005; Grehan et al., 2005; Stone, 2006; Hourigan, 2009; Maynou and Cartes, 2012). Physical impacts from both trawl fisheries and long-lining, now being conducted as deep as 1500-2000 m, are likely to be significant anywhere deep-water fisheries are active, but have been well-demonstrated in the North Atlantic and Norwegian Seas (Roberts et al., 2000; Fossa et al., 2002; Hall-Spencer et al., 2002, Reed, 2002), on the Australian seamounts (Koslow et al., 2001), off the coast of New Zealand (Probert et al., 1997, Clark and Rowden, 2009), and Southwestern Atlantic slope (Kitahara, 2009). Trawl fisheries have the most severe impacts, by removal of large volumes of organisms and of cold-water coral framework from the seafloor and the concomitant destruction of the habitat, but long-lining impacts have also been observed (Heifetz et al., 2009). Recovery times from these types of disturbance are likely to require settlement and regrowth of the corals, which based on radiometric dating of cold-water coral species, can require decades to centuries (Andrews et al., 2002; Prouty et al., 2014) or in the case of the black corals, could require millennia (Roark et al., 2009). Direct evidence of recovery times is consistent with these estimates, indicating that there was no apparent recovery 5-10 years after the closure of seamount fisheries on the Tasmanian seamounts (Althaus et al., 2009). These impacts have also been the most recognized in terms of management efforts, thus far (see below).

Installation of oil and gas offshore facilities and drilling activities (see Chapter 21) have a great potential to impact cold-water coral communities. The potential impact should be higher in areas where much of the available substrate is from authigenic carbonates related to natural oil and gas seepage, such as the Gulf of Mexico (Cordes et al., 2008), some locations on the Norwegian margin (Hovland, 2005), and the New Zealand margin (Baco et al., 2010). Most of the typical impacts would be from infrastructure installation and the deposition of drill tailings that can include high concentrations in barium, among other potential toxins (Continental Shelf Associates, 2006). These impacts are typically confined to a few hundred metres, but can have been shown to extend over 2 kilometres in some cases (Continental Shelf Associates, 2006). The most glaring example of oil and gas industry impacts in the deep sea is the Deepwater Horizon disaster in 2010 in the Gulf of Mexico. Material conclusively linked to the spill was discovered on octocoral colonies (primarily *Paramuricea biscaya*) approximately 11 km away from the site of the drilling rig (White et al., 2012a). These colonies suffered tissue loss and many have continued to decline in health since the spill (Hsing et al., 2013). Subsequent surveys detected at least two additional sites, extending the impacts to 26 km from the site of the well, and from 1,370 m to 1,950 m water depth (Fisher et al., 2014). One of the primary lessons learned from this tragic incident is that there is an urgent need for improved baseline surveys in deep waters prior to industrial activity. Offshore energy industry activity in the form of wind and wave energy is also increasing (see Chapter 22), and physical structure placed on the seafloor, including pipelines and cables, could have an impact on cold-water corals if the appropriate surveys are not completed prior to installation.

Mining activities have increased in the deep sea in recent years. This activity has mainly focused on massive seafloor sulphide deposits near hydrothermal vents, cobalt-rich crusts on seamounts, and also on polymetallic nodules on the abyssal plain (Ramirez-Llodra et al., 2011). These forms of mining would involve removal of a large area of the seafloor surface, and complete removal of any associated communities, along with the generation of large sediment and tailing plumes that may impact filter feeding communities at a distance from the mining activity (Ramirez-Llodra et al., 2011). On the seamounts of the Kermadec Arc, some which have already been leased for mining, cold-water coral communities consisted of scleractinian, schizopathid, stylasterid, primnoid, and isidid corals primarily associated with inactive areas away from hydrothermal venting (Boschen et al., 2015). Deep-sea corals are often found on the hard substrata in inactive vent fields, and may be subject to significant impacts from their removal due to their long life spans and low recruitment rates.

Global climate change is affecting every community type on Earth, and its effects are already being felt in the deep sea. Ocean warming has been recorded in numerous deep-water habitats, but is particularly significant in marginal seas, which are home to many of the world's cold-water coral reefs. In particular, there is evidence that the Mediterranean has warmed by at least 0.1°C between 1950 and 2000 (Rixen et al., 2005), and this change has been shown to impact the deep-sea communities there (Danovaro et al., 2004). Cold-water corals are highly sensitive to warming waters because of their upper thermal limits, and the temperature excursions around this general upward trend are likely to be much higher.

Ocean acidification is another pervasive threat (see Chapter 5). Continued additions of CO_2 into the atmosphere exacerbate the problem as the oceans absorb approximately 26 per cent of the CO_2 from the atmosphere (Le Quere et al., 2009). Because the carbonate saturation state in seawater is temperature-dependent, it is much lower in cold waters and therefore cold-water corals lie much closer to the saturation horizon (the depth below which the saturation state is below 1 and carbonate minerals will dissolve) than shallow-water corals. As ocean acidification proceeds, the saturation horizon will become shallower, thus exposing more cold-water corals to undersaturated conditions (Guinotte et al., 2006). Solitary corals of the South Pacific are already facing saturation states below 1 (Thresher et al., 2011), and small reef frameworks constructed by *Solenosmilia variabilis* grow in periodically undersaturated waters on Northeast Atlantic (Henry and Roberts 2014; Henry et al., 2014) and New Zealand seamounts (Bostock et al., 2015). The *Lophelia* reefs of the Gulf of Mexico lie very close to the saturation horizon, at a minimum saturation state of approximately 1.2 (Lunden et al., 2013). Since these recent studies represent the baseline for the deep-water carbonate system, the extent to which anthropogenic CO_2 contributes to these low values remains unclear.

Other possible effects of global climate change include deoxygenation and changes in sea-surface productivity. Declines in oxygen availability are primarily linked to increasing water temperature, but also to synergistic effects of pollution and agricultural runoff, which are most significant in shallow water. However, because some cold-water corals live at oxygen-minimum zone depths (Davies et al., 2010; Georgian et al., 2014), even small changes in oxygen concentration could be significant. Because cold-water corals live below the photic zone and rely for their nutrition on primary productivity transferred from the surface waters to depth, changes in surface productivity could have significant negative impacts. In particular, the increased stratification of surface waters above the thermocline will lead to decreased productivity in high latitude spring-bloom and upwelling ecosystems (Falkowski et al., 1998). This includes the North Atlantic, where the most extensive examples of the known cold-water coral reefs exist.

Through *in situ* habitat characterization as well as by experimental approaches, it has become clear that acidification and the expansion of oxygen minimum zones, together with rising temperatures, will affect the average metabolism and physiology of most scleractinians (Gori et al., 2013; Lartaud et al., 2014; McCulloch et al., 2012; Naumann et al., 2013). However, whether such changes will result in range shifts, massive extinctions (as suggested by Tittensor et al., 2010), or if species possess the resources to cope with variations through phenotypic plasticity or adaptive genetic changes, is still largely unknown. The solitary coral *Desmophyllum dianthus* and colonial scleractinian *Dendrophyllia cornigera* have shown resistance to high temperature in aquaria (Naumann et al., 2013). The *L. pertusa* colonies from the North Atlantic and Mediterranean have shown the ability to acclimatize to ocean acidification in long-term experiments (Form and Riebesell, 2012; Maier et al., 2012). In other experiments, certain genotypes of *L. pertusa* from the Gulf of Mexico were able to calcify at saturation states as low as 1.0, suggesting a possible genetic basis to their sensitivity to ocean acidification (Lunden et al., 2014). However, to date no long-term studies combining acidification with temperature stress have been produced and long-term effects on bare skeletal structure are unknown. In addition, some cold-water coral species seem to be resilient to some of these processes, and may hold some of the answers for coral survival in future global climate-change scenarios. Regardless, the projected shoaling of the aragonite saturation horizon (Orr et al., 2005) threatens the future integrity of deep-water scleractinian reef structures world-wide (Guinotte et al., 2006).

The ability of these organisms to keep up with the pace of ocean change and disperse into a new environment or to recolonize depleted areas depends on the capacity for mid- or long-distance dispersal. This capacity has been demonstrated for *L. pertusa* by isotope reconstruction and genetic analysis (Henry et al., 2014), supporting the hypothesis of a post-glacial recolonization of the Atlantic by refugees in the Mediterranean (De Mol et al., 2002; De Mol et al., 2005; Frank et al., 2009).

Overall, *L. pertusa* shows a pattern of relative homogeneity within regions (e.g. the North Atlantic), and modest but significant differentiation among regions, both for the Western Atlantic (e.g. Gulf of Mexico vs. Southeast United States vs. North Atlantic; Morrison et al., 2011), as well as along Eastern Atlantic margins from the Bay of Biscay to Iceland for both *L. pertusa* and *M. oculata* (Becheler, 2013). Previous studies on the Eastern Atlantic margin had shown less extensive connectivity, possibly reflecting the peculiar position of fjord populations in Sweden and Norway (Le Goff-Vitry et al., 2004). Preliminary studies on *D. dianthus* suggest a lack of barrier to large-scale dispersal (Addamo et al., 2012), although bathymetric barriers to gene flow are evident (Miller et al., 2011). Bathymetric barriers to dispersal are also apparent in the phylogenetic community structure of deep-water octocoral assemblages in the Gulf of Mexico (Quattrini et al., 2014).

Finally, the distribution of genetic polymorphism among populations of octocorallians and antipatharians across seamounts of the Pacific spanning 1700 km also showed no evidence for strong endemism, supporting the ability for large scale dispersal of the species studied (Thoma et al., 2009). It is only recently that the embryonic and larval biology of *Lophelia pertusa* has been described (Brooke and Järnegren, 2013). The settlement and benthic juvenile stages have not been observed. Knowledge on the possible effects of ocean acidification on coral reproduction so far comes from tropical corals but it is reasonable to believe that there are many similarities (Albright, 2011).

Altogether, the present state of knowledge of genetic connectivity of deep-water corals suggests that the potential exists for some species to disperse and colonize across large distances in response to major environmental changes, and some species have a more limited dispersal capacity. However, more studies need to be conducted at a finer spatial scale using specific genetic markers (e.g. Dahl et al., 2012) to improve the understanding of the impact of environmental changes on connectivity and persistence at the local scale. These different degrees of differentiation among and within ocean basins indicate the need for regional-scale conservation strategies.

4   Implications for Services to Ecosystems and Humanity

Impacts on cold-water corals and the structures they form would have significant implications for the functioning of the surrounding deep sea and wider oceanic ecosystems. The linkages from shallow to deep water, and back again, implicate cold-water corals as key components of the broader oceanic ecosystem. The physical structures created by cold-water corals support fisheries through the direct provision of habitat, refuge, or nursery grounds, which is likely to lead to increases in commercially significant fish populations. These effects are most pronounced where cold-water corals are known to be highly abundant, such as on the North Atlantic, North Pacific, and Australian and New Zealand seamounts.

The ecosystem services provided go beyond the direct provision of substrate and shelter (see review by Foley et al., 2010). The complex habitat formed by cold-water corals increases the heterogeneity of the continental margin, promoting higher diversity (Cordes et al., 2010). As in other ecosystems (e.g. Tilman et al., 1997), increased diversity mostly promotes higher levels of ecosystem function, including carbon cycling. This specific ecosystem service may be important in relatively oligotrophic regions such as the Gulf of Mexico and the Mediterranean where cold-water corals-enhanced nutrient cycling and remineralization would generate nutrients that may be transported back to the surface. Recent findings from reefs off Norway demonstrated their significant role in carbon cycling, raising additional concerns as to the impact of their disappearance on global biochemical cycles (White et al., 2012b).

Cold-water corals also hold genetic resources that may provide services to humanity, either directly or through their function as biodiversity hotspots in the deep sea (Arrieta et al., 2010). Taxa such as cnidarians, sponges, and molluscs have been shown to harbour the highest abundance of natural marine products of interest for biotechnology development (Molinski et al., 2009; Rocha et al., 2011). As an example, the anti-AIDS drug AZT was developed from an extract of a sponge from a shallow Caribbean reef (de la Calle, 2009). At least half, and likely far more, of the diversity of corals and sponges lies in deep, cold waters (Cairns, 2007; Hogg et al., 2010), and therefore, these understudied and often unknown species have the highest potential for new discoveries. With this potential comes a management concern, especially as some of the potential genetic resources (see also chapter 29) harboured within the genomes of cold-water corals and sponges lie in areas beyond national jurisdiction (Bruckner, 2002; de la Calle, 2009).

5   Conservation Responses

Raised awareness of the susceptibility of cold-water coral communities to impacts of human activities in recent decades has resulted in national and international actions to protect cold-water corals and facilitate recovery of coral areas adversely affected in the past. In some areas where significant damage was documented, e.g. along the continental shelf off Norway (Fossa et al., 2002) and on seamounts in Australia and New Zealand (Koslow et al., 2001), national legislation was introduced and specific management measures were implemented. A growing number of protected areas and fisheries closures in areas within national jurisdiction in the Atlantic and North Pacific have followed, and in some countries, e.g. Norway, it is illegal to deliberately fish in coral areas even if the area is not formally closed as a protected area.

Since the mid-2000s a series of United Nations General Assembly (UNGA) resolutions (e.g. 61/105, 64/72, 66/68) on sustainable fisheries have called for a number of measures, including the implementation of the International Guidelines for the Management of Deep-Sea Fisheries in the High Seas (FAO, 2009), and action to avoid significant adverse impacts of fisheries on vulnerable marine ecosystems,[1] including e.g. cold water corals.[2] These resolutions focus in particular on areas beyond national jurisdiction. Some of the protective efforts, including fisheries closures, within EEZs predate the UNGA resolutions, but the resolutions stimulated further action. Such actions run in parallel with efforts to create networks of marine protected areas in areas within national jurisdiction, partly motivated by the need to protect corals.

In response to the measures called for by the General Assembly, seamounts and continental slope habitats with a documented or assumed coral presence have now been set aside as marine reserves or fisheries closures by competent authorities. These areas are protected partly through area-based management tools (for example, by Australia, New Zealand and the United States within areas under their respective national jurisdictions) and partly by regional fisheries management organizations and arrangements (RFMO/As) in the high seas of the North and South-eastern Atlantic. In the north-eastern Atlantic, substantial areas have been protected within national jurisdictions of European Union Member States, as well as of Iceland and Norway. Beyond areas of national jurisdiction, RFMOs/As with competence to regulate bottom fisheries (for example, the Northwest Atlantic Fisheries Organization (www.nafo.int) and the North East Atlantic Fisheries Commission (www.neafc.org)) have closed a range of seamounts and seabed areas to bottom fishing. Within their regulatory areas, these RFMOs have also restricted fishing to a limited agreed set of sub-areas outside the "existing fishing areas", and have created strict rules and impact assessment requirements for these sub-areas. These measures are intended not only to protect known areas with significant concentrations of cold-water coral, but also essentially to reduce the incentive for exploratory bottom fishing outside existing fishing areas. Similar rules apply in the southeast Atlantic high seas implemented by the South East Atlantic Fisheries Organization (SEAFO, www.seafo.org) which closed selected ridge sections and seamounts to fishing, and restricted fisheries to certain subareas. In the Mediterranean, the General Fisheries Commission for the Mediterranean (GFCM, www.gfcm.org) implemented fisheries restriction zones in specific coral sites.

The Commission for the Conservation of Antarctic Marine Living Resources (CCAMLR, http://www.ccamlr.org/) banned bottom trawl fishing within the CCAMLR Convention area. Bottom fishing regulations and area closures aim to facilitate responsible fisheries and to prevent adverse impacts on bottom-associated vulnerable marine communities as defined by FAO (2009). Marine protected areas in this area are being considered but only one MPA has been established thus far.

Currently little information exists to assess the impacts on target or bycatch species by deep-sea fishing on seamounts in the Indian Ocean. The Southern Indian Ocean Deepsea Fishers Association declared a number of seamounts in the Southern Indian Ocean as voluntary areas closed to fishing. The entry into force in 2012 of the Southern Indian Ocean Fisheries Agreement[3] (SIOFA), a new regional fisheries management arrangement for the region, may lead to better documentation and regulation of seamount fisheries.

In the North Pacific, the United States designated Habitat Areas of Particular Concern (HAPCs) that contain Essential Fish Habitat (EFH) and closed subareas of the shelf and upper slope from California to the Aleutian Islands to bottom trawling. Additional areas of *L. pertusa* habitat have recently been designated as HAPCs off the southeast coast of the United States. Canada also has a strategy to develop and implement further measures. In areas beyond national jurisdiction in the North and South Pacific, respectively, States which participated in the negotiations to establish the North Pacific Fisheries Commission (NPFC) and the South Pacific Fisheries Management Organization (SPRFMO) introduced measures similar to those adopted by the Atlantic RFMOs.

Within areas under national jurisdiction of the United States in the Gulf of Mexico, mitigation areas are established around mapped seafloor seismic anomalies that often coincide with hardgrounds that may support cold-water coral communities. Although these measures may prevent most direct impacts from infrastructure, the persistent threat of deep-water fishing, accidental loss of gear, and catastrophic oil spills remains a concern.

A continued challenge is to assess the effectiveness of current and new protective measures and to develop management in areas that need greater attention, such as those for which no RFMOs exist. The fisheries sector is often perceived as representing the major threat to cold-water corals, but a growing challenge is to avoid adverse impacts from other industries moving into areas containing known coral habitats, e.g. mining, oil and gas industries, and renewable energy industries operating under different management regimes.

---

1 The International Guidelines for the Management of Deep-Sea Fisheries in the High Seas describe vulnerable marine ecosystems and list characteristics to be used as criteria in the identification of such ecosystems.

2 The Annex to the Guidelines refers to "certain coldwater corals" as part of examples of species groups, communities and habitat forming species that are documented or considered sensitive and potentially vulnerable to deep-sea fisheries in the high seas, and which may contribute to forming vulnerable marine ecosystems.

3 United Nations, *Treaty Series*, vol. 2835, No. 49647.

References

Addamo, A.M., Reimer, J.D., Taviani, M., Freiwald, A., and Machordom, A. (2012). *Desmophyllum dianthus* (Esper, 1794) in the Scleractinian Phylogeny and Its Intraspecific Diversity. *Plos One* 7, e50215.

Adkins, J.F., Cheng, H., Boyle, E.A., Druffel, E.R.M., and Edwards, R.L. (1998). Deep-sea coral evidence for rapid change in ventilation of the deep North Atlantic 15,400 year ago. *Science* 280, 725–728.

Albright, R. (2011). Reviewing the effects of ocean acidification on sexual reproduction and early life history stages of reef-building corals. *Journal of Marine Biology* (2011), ID 473615. doi:10.1155/2011/473615.

Althaus, F., Williams, A., Schlacher, T.A., Kloser, R.J., Green, M.A., Barker, B.A., Bax, N.J., Brodie, P. and Schlacher-Hoenlinger, M.A. (2009). Impacts of bottom trawling on deep-coral ecosystems of seamounts are long-lasting. *Marine Ecology Progress Series* 397, 279-294.

Andrews, A.H., Cordes, E.E., Mahoney, M.M., Munk, K., Coale, K.H., Cailliet, G.M., Heifetz, J. (2002). Age, growth and radiometric age validation of a deep-sea, habitat-forming gorgonian (*Primnoa resedaeformis*) from the Gulf of Alaska. *Hydrobiologia* 471, 101-110.

Arantes, R.C.M., Castro, C.B., Pires, D.O., and Seoane, J.C.S. (2009). Depth and water mass zonation and species associations of cold-water octocoral and stony coral communities in the southwestern Atlantic. *Marine Ecology Progress Series* 397, 71-79.

Arrieta, J., Arnaud-Haond, S., and Duarte, C.M. (2010). What lies underneath: Conserving the Ocean's Genetic Resources. *Proceedings of the National Academy of Sciences* 107, 18318-18324.

Baco, A. R., Rowden, A. A., Levin, L. A., Smith, C. R., and Bowden, D. A. (2010). Initial characterization of cold seep faunal communities on the New Zealand Hikurangi margin. *Marine Geology*, 272(1), 251-259.

Baillon, S., Hamel, J.F., Wareham, V.E., and Mercier, A. (2012). Deep cold-water corals as nurseries for fish larvae. *Frontiers in Ecology and the Environment*; doi:10.1890/120022.

Becheler, R. (2013). Feedbacks between genetic diversity and demographic stability in clonal organisms, Ifremer, Département Environnement Profond. IUEM: Institut Universitaire Européen de la Mer, Brest.

Boschen, R.E., Rowden, A.A., Clark, M.R., Barton, S.J., Pallentin, A., and Gardner, J.P.A. (2015). Megabenthic asssemblage structure on three New Zealand seamounts: implications for seafloor massive sulfide mining. *Marine Ecology Progress Series* 523, 1-14.

Bostock, H. C., Tracey, D. M., Currie, K. I., Dunbar, G. B., Handler, M. R., Fletcher, S. E. M., Smith, A.M., and Williams, M. J. (2015). The carbonate mineralogy and distribution of habitat-forming deep-sea corals in the southwest Pacific region. *Deep Sea Research Part I: Oceanographic Research Papers*, 100, 88-104.

Brooke, S. and Young, C.M. (2009). In situ measurement of survival and growth of *Lophelia pertusa* in the northern Gulf of Mexico. *Marine Ecology Progress Series* 397, 153-161.

Brooke, S. and Järnegren, J. (2013) Reproductive periodicity of the deep-water scleractinian coral, Lophelia pertusa from the Trondheim Fjord, Norway. *Marine Biology* 160:139-153.

Brooke, S. and Ross, S.W. (2014). First observations of the cold-water coral *Lophelia pertusa* in mid-Atlantic canyons of the USA. *Deep-Sea Research II* 104, 245-251.

Bruckner, A.W.(2002). Life-Saving Products from Coral Reefs Issues in *Science and Technology* online.

Buhl-Mortensen, L., Olafsdottir, S.H., Buhl-Mortensen, P., Burgos, J.M., and Ragnarsson, S.A. (2014). Distribution of nine cold-water coral species (Scleractinia and Gorgonacea) in the cold temperate North Atlantic in light of bathymetry and hydrography. *Hydrobiologia*. DOI: 10.1007/s10750-014-2116-x.

Cairns, S. (2007). Deep-water corals: an overview with special reference to diversity and distribution of deep-water Scleractinia. *Bulletin of Marine Science* 81, 311-322.

Carranza, A., Recio, A.M., Kitahara, M., Scarabino, F., Ortega, L., López, G., Franco-Fraguas, P., De Mello, C., Acosta, J., Fontan, A. (2012). Deep-water coral reefs from the Uruguayan outer shelf and slope. *Marine Biodiversity* 42, 411–414.

Clark, M.R. and A.A. Rowden (2009). Effect of deepwater trawling on the macro-invertebrate assemblages of seamounts on the Chatham Rise, New Zealand. *Deep Sea Research I* 56, 1540-1544.

Clark, M. R. and Tittensor, D. P. (2010). An index to assess the risk to stony corals from bottom trawling on seamounts. *Marine Ecology* 31, 200-211.

Colman, J.G., Gordon, D.M., Lane, A.P., Forde, M.J., and Fitzpatrick, J.J. (2005). Carbonate mounds off Mauritania, Northwest Africa: status of deep-water corals and implications for management of fishing and oil exploration activities. In: Freiwald A., Roberts, J.M. (eds.) *Cold-water corals and ecosystems*. Springer, Heidelberg, pp 417-441.

Continental Shelf Associates, Inc. (2006). *Effects of Oil and Gas Exploration and Development at Selected Continental Slope Sites in the Gulf of Mexico*. Volume I: Executive Summary. U.S. Department of the Interior, Minerals Management Service, Gulf of Mexico OCS Region, New Orleans, LA. OCS Study MMS 2006-044. 45 pp.

Cordes, E.E., McGinley, M.P., Podowski, E.L., Becker, E.L., Lessard-Pilon, S., Viada, S.T., and Fisher, C.R. (2008). Coral communities of the deep Gulf of Mexico. *Deep-Sea Research I* 55, 777-787.

Cordes, E.E., Cunha, M.M,, Galeron, J., Mora, C., Olu-Le Roy, K., Sibuet, M., Van Gaever, S., Vanreusel, A., and Levin, L.(2010). The influence of geological, geochemical, and biogenic habitat heterogeneity on seep biodiversity. *Marine Ecology* 31: 51-65.

Correa, T.B.S., Eberli, G.P., Grasmueck, M., Reed, J.K., and Correa, A.M.S. (2012). Genesis and morphology of cold-water coral ridges in a unidirectional current regime. *Marine Geology* 326-328, 14-27.

Costello, M.J., McCrea, M., Freiwald, A., Lundälv, T., Jonsson, L., Bett, B.J., van Weering, T.C.E., de Haas, H., Roberts, J.M., and Allen, D. (2005). Role of cold-water *Lophelia pertusa* coral reefs as fish habitat in the NE Atlantic. In: Freiwald, A., Roberts, J.M. (eds.) *Cold-water Corals and Ecosystems*. Berlin, Germany, Springer, 771-805, 1243 pp.

Dahl, M.P., Pereyra, R.T., Lundalv, T., and Andre, C. (2012). Fine-scale spatial genetic structure and clonal distribution of the cold-water coral *Lophelia pertusa*. *Coral Reefs* 31, 1135–1148.

Danovaro, R., Dell'Anno, A., and Pusceddu, A. (2004). Biodiversity response to climate change in a warm deep sea. *Ecology Letters*, 7(9), 821-828.

Davies, A.J., Duineveld, G.C.A., Lavaleye, M.S.S., Bergman, M.J.N., Van Haren, H. and Roberts, J.M. (2009). Downwelling and deep-water bottom currents as food supply mechanisms to the cold-water coral *Lophelia pertusa* (Scleractinia) at the Mingulay Reef complex. *Limnology and Oceanography* 54, 620-629.

Davies, A.J., Duineveld, G.C.A., van Weering, T.C.E., Mienis, F., Quattrini, A.M., Seim, H.E., Bane, J.M. and Ross, S.W. (2010). Short-term environmental variability in cold-water coral habitat at Viosca Knoll, Gulf of Mexico. *Deep-Sea Research I* 57, 199-212.

Davies, A.J. and Guinotte, J.M. (2011). Global habitat suitability for framework-forming cold-water corals. *PLos ONE* 6: e18483.

de Groot, R.S., Wilson, M.A., Roelof, M.J. and Boumans, R.M.J. (2002). A typology for the classification, description and valuation of ecosystem functions, goods and services. *Ecological Economics* 41 393–408.

de la Calle, F. (2009). Marine Genetic Resources: A Source of New Drugs - The Experience of the Biotechnology Sector. *International Journal of Marine and Coastal Law* 12, 209-220.

De Mol, B., Van Rensbergen, P., Pillen, S., Van Herreweghe, K., Van Rooij, D., McDonnell, A., Huvenne, V., Ivanov, M., Swennen, R. and Henriet, J.P. (2002). Large deep-water coral banks in the Porcupine Basin, southwest of Ireland. *Marine Geology* 188, 193-231.

De Mol, B., Henriet, J.P. and Canals, M. (2005). Development of coral banks in Porcupine Seabight: do they have Mediterranean ancestors? In: Freiwald A. and Roberts J.M., (eds). *Cold-Water Corals and Ecosystems*. Erlangen Earth Conference Series, Springer. pp 515-533.

De Mol, B., Amblas, D., Alvarez, G., Busquets, P., Calafat, A., Canals, M., Duran, R., Lavoie, C., Acosta, J. and Munoz, A. (2012). Cold-water coral distribution in an erosional environment: the Strait of Gibraltar Gateway. In: Harris PT, Baker EK (eds.) *Seafloor Geomorphology as Benthic Habitat*. Elsevier, Amsterdam, pp 635-643.

De Mol, L., Van Rooij, D., Pirlet, H., Greinert, J., Frank, N., Quemmerais, F. and Henriet, J.P. (2011). Cold-water coral habitats in the Penmarc'h and Guilvinec canyons (Bay of Biscay): Deep-water versus shallow-water settings. *Marine Geology* 282, 40-52.

Dodds, L.A., Roberts, J.M., Taylor, A.C., Marubini, F. (2007). Metabolic tolerance of the cold-water coral *Lophelia pertusa* (Scleractinia) to temperature and dissolved oxygen change. *Journal of Experimental Marine Biology and Ecology* 349, 205-214.

Doney, S. C., Fabry, V. J., Feely R.A., J. Kleypas J.A. (2009). Ocean acidification: The other CO_2 problem. *Annual Review of Marine Science* 1, 169-192.

Dullo, W.C., Flögel, S., Rüggeberg, A. (2008). Cold-water coral growth in relation to the hydrography of the Celtic and Nordic European continental margin. *Marine Ecology Progress Series* 371, 165-176.

Fabri, M.C., Pedel, L., Beuck, L., Galgani, F., Hebbeln, D., Freiwald, A. (2014). Megafauna of vulnerable marine ecosystems in French Mediterranean submarine canyons: Spatial distribution and anthropogenic impacts. *Deep-Sea Research* II 104, 184-207.

Falkowski, P. G., Barber, R. T., & Smetacek, V. (1998). Biogeochemical controls and feedbacks on ocean primary production. *Science*, 281(5374), 200-206.

Findlay, H.S., Wicks, L., Navas, J.M., Hennige, S., Huvenne, V., Woodward E.M.S., Roberts J.M. (2013). Tidal downwelling and implications for the carbon biogeochemistry of cold-water corals in relation to future ocean acidification and warming. *Global Change Biology* 19, 2708-2719.

Findlay, H.S., Hennige, S.J., Wicks, L.C., Navas, J.M., Woodward, E.M.S., Roberts, J.M. (2014). Fine-scale nutrient and carbonate system dynamics around cold-water coral reefs in the northeast Atlantic. *Nature Scientific Reports* 4: 3671.

Fisher, C.R., Hsing, P.Y., Kaiser, C., Yoerger, D., Roberts, H.H., Shedd, W., Cordes, E.E., Shank, T.S., Berlet, S.P., Saunders, M., Larcom, E.A., Brooks, J. (2014). Footprint of Deepwater Horizon blowout impact to deep-water coral communities. *Proceedings of the National Academy of Sciences* 111, 11744-11749.

Foley, N.S., van Rensburg, T.M., Armstrong, C.W. (2010). The Ecological and Economic Value of Deep Water Corals. (Review paper) *Ocean and Coastal Management* 53, 313-326.

Food and Agriculture Organization of the United Nations. (2009). International Guidelines for the Management of Deep-sea Fisheries in the High Seas. Rome, FAO. 2009. 73 pp.

Flögel, S., Dullo, W.C., Pfannkuche, O., Kiriakoulakis, K., Rüggeberg, A. (2014). Geochemical and physical constraints for the occurrence of living cold-water corals. *Deep-Sea Research* II 99, 19-26.

Form, A.U., Riebesell, U. (2012). Acclimation to ocean acidification during long-term $CO2$ exposure in the cold-water coral *Lophelia pertusa*. *Global Change Biology* 18, 843-853.

Fossa, J.H., Mortensen, P.B., Furevik, D.M. (2002). The deep-water coral *Lophelia pertusa* in Norwegian waters: distribution and fishery impacts. *Hydrobiologia* 471, 1-12.

Frank, N., Ricard, E., Lutringer-Paquet, A., van der Land, C., Colin, C., Blamart, D., Foubert, A., Van Rooij, D., Henriet, J.-P., de Haas, H., van Weering, T. (2009). The Holocene occurrence of cold water corals in the NE Atlantic: Implications for coral carbonate mound evolution. *Marine Geology* 266, 129-142.

Freiwald, A., Rogers, A., Hall-Spencer, J. (2005). *Global distribution of cold-water corals* (version 2). Cambridge (UK): UNEP World Conservation Monitoring Centre.

Freiwald, A., Beuck, L., Rueggeberg, A., Taviani, M., Hebbeln, D. (2009). The white coral community in the Central Mediterranean Sea Revealed by ROV Surveys. *Oceanography* 22, 58-74.

Georgian, S.E., Shedd, W., Cordes, E.E. (2014). High resolution ecological niche modelling of the cold-water coral *Lophelia pertusa* in the Gulf of Mexico. *Marine Ecology Progress Series* 506, 145-161.

Gori, A., Orejas, C., Madurell, T., Bramanti, L., Martins, M., Quintanilla, E., Marti-Puig, P., Lo Iacono, C., Puig, P., Requena, S., Greenacre, M., Gili, J.M. (2013). Bathymetrical distribution and size structure of cold-water coral populations in the Cap de Creus and Lacaze-Duthiers canyons (northwestern Mediterranean). *Biogeosciences* 10, 2049-2060.

Grehan, A.J., Unnithan, V., Roy, K.O.L., Opderbecke, J. (2005). Fishing impacts on Irish deepwater coral reefs: Making a case for coral conservation. In: *Benthic Habitats and the Effects of Fishing* (eds. Barnes BW, Thomas JP), pp. 819-832.

Guinotte, J. M., Orr, J., Cairns, S., Freiwald, A., Morgan, L., George, R. (2006). Will human-induced changes in seawater chemistry alter the distribution of deep-sea scleractinian corals? *Frontiers in Ecology and the Environment*, 4, 141-146.

Guppy, M., Withers, P. (1999). Metabolic depression in animals: physiological perspectives and biochemical generalizations. *Biological Reviews of the Cambridge Philosophical Society* 74: 1-40.

Hall–Spencer, J., Allain, V., Jan Helge Fosså, J.H. (2002).Trawling damage to Northeast Atlantic ancient coral reefs. *Proceedings of the Royal Society of London B* 269, 507-51.

Harris, P.T., MacMillan-Lawler, M., Rupp, J., Baker, E.K. (2014). Geomorphology of the oceans. *Marine Geology* 352, 4-24.

Harris, P.T., Whiteway, T. (2011). Global distribution of large submarine canyons: geomorphic differences between active and passive continental margins. *Marine Geology* 285, 69–86.

Hebbeln, D., Wienberg, C., Wintersteller, P., Freiwald, A., Becker, M., Beuck, L., Dullo, C., Eberli, G.P., Glogowski, S., Matos, L., Forster, N., Reyes-Bonilla, H., Taviani, M. (2014). Environmental forcing of the Campeche cold-water coral province, southern Gulf of Mexico. *Biogeosciences* 11, 1799-1815.

Heifetz, J., R.P. Stone, and S.K. Shotwell (2009). Damage and disturbance to coral and sponge habitat of the Aleutian Archipelago. *Marine Ecology Progress Series* 397, 295-303.

Henry, L.A., Roberts, J.M. (2007). Biodiversity and ecological composition of macrobenthos on cold-water coral mounds and adjacent off-mound habitat in the bathyal Porcupine Seabight, NE Atlantic. *Deep-Sea Research* I 54, 654–672.

Henry, L.A., Navas J.M., Hennige, S.J., Wicks, L., Vad, J., Roberts, J.M. (2013). Cold-water coral reef habitats benefit recreationally valuable sharks. *Biological Conservation* 161, 67-70.

Henry, L.A., Frank, N., Hebbeln, D., Wienberg, C., Robinson, L., van de Flierdt, T., Dahl, M., Douarin, M., Morrison, C.L., Lopez Correa, M., Rogers, A.D., Ruckelshausen, M., Roberts, J.M. (2014). Global ocean conveyor lowers extinction risk in the deep sea. *Deep-Sea Research* Part I-Oceanographic Research Papers 88, 8-16.

Hogg, M.M., O.S. Tendal, K.W. Conway, S.A. Pomponi, R.W.M. Van Soest, J. Gutt, M. Krautter, J.M. Roberts (2010). Deep-sea sponge grounds: Reservoirs of biodiversity. *UNEP-WCMC Biodiversity Series* No. 32. UNEP-WCMC, Cambridge, UK.

Hourigan, T.F. (2009). Managing fishery impacts on deep-water coral ecosystems of the USA: emerging best practices. *Marine Ecology Progress Series* 397, 333-340.

Hovland, M. (2005). Pockmark-associated coral reefs at the Kristin field off Mid-Norway. In: *Cold-Water Corals and Ecosystems*, Erlangen Earth Conference Series, pp 623-632.

Hsing, P.Y., Fu, B., Larcom, E.A., Berlet, S.P., Shank, T.M., Govindarajan, A.F., Lukasiewicz, A.J., Dixon, P.M., Fisher, C.R. (2013). Evidence of lasting impact of the Deepwater Horizon oil spill on a deep Gulf of Mexico coral community. *Elementa*. 1: 000012.

Husebo, A., Nottestad, L., Fossa, J.H., Furevik, D.M., Jorgensen, S.B. (2002). Distribution and abundance of fish in deep-sea coral habitats. *Hydrobiologia* 471, 91-99.

Huvenne, V.A., Tyler, P.A., Masson, D.G., Fisher, E.H., Hauton, C., Huhnerbach, V., Le Bas, T.P., Wolff, G.A. (2011). A picture on the wall: innovative mapping reveals cold-water coral refuge in submarine canyon. *PloS one* 6, e28755.

Kiriakoulakis, K., Freiwald, A., Fisher, E. and Wolff, G.A. (2007). Organic matter quality and supply to deep-water coral/mound systems of the NW European continental margin. *International Journal of Earth Sciences*, 96, 159-170.

Kitahara, M.V. (2009). A pesca demersal de profundidade eos bancos de corais azooxantelados do sul do Brasil. *Biota Neotropica* 9, 35-43.

Koslow J. A., Gowlett-Holmes, K., Lowry, J. K., O'Haram T., Poore, G. C. B., Williams, A. (2001). Seamount benthic macrofauna off southern Tasmania: community structure and impacts of trawling. *Marine Ecology Progress Series* 213, 111-125.

Larcom, E.A., McKeana, D.L., Brooks, J.M., Fisher, C.R. (2014). Growth rates, densities, and distribution of *Lophelia pertusa* on artificial structures in the Gulf of Mexico. *Deep-Sea Research* I 85, 101-109.

Lartaud, F., Pareige, S., de Rafelis, M., Feuillassier, L., Bideau, M., Peru, E., De la Vega, E., Nedoncelle, K., Romans, P., Le Bris, N. (2014). Temporal changes in the growth of two Mediterranean cold-water coral species, in situ and in aquaria. *Deep-Sea Research Part II-Topical Studies in Oceanography* 99, 64-70.

Le Goff-Vitry, M.C., Rogers, A.D., Baglow, D. (2004). A deep-sea slant on the molecular phylogeny of the Scleractinia. *Molecular Phylogenetics and Evolution* 30, 167-177.

Le Quere, C., Raupach, M.R., Canadell, J.G., et al. (2009). Trends in the sources and sinks of carbon dioxide. *Nature Geoscience* 2:831-836.

Lunden, J.J., Georgian, S.E., Cordes, E.E. (2013). Aragonite saturation states at cold-water coral reefs structured by Lophelia pertusa in the northern Gulf of Mexico. *Limnology and Oceanography* 58, 354-362.

Lunden, J.J., McNicholl, C.G., Sears, C.R., Morrison, C.L., Cordes, E.E. (2014). Sensitivity of the deep-sea coral *Lophelia pertusa* to global climate change and ocean acidification varies by individual genotype in the Gulf of Mexico. *Frontiers in Marine Science*, vol. 1: Article 78.

Maier, C., Watremez, P., Taviani, M., Weinbauer, M.G., J. P. Gattuso, J.P. (2012). Calcification rates and the effect of ocean acidification on Mediterranean cold-water corals. *Proceedings of the Royal Society of London* B 279, 1716-1723.

Maynou, F., Cartes, J.E. (2012). Effects of trawling on fish and invertebrates from deep-sea coral fades of Isidella elongata in the western Mediterranean. *Journal of the Marine Biological Association of the United Kingdom* 92, 1501-1507.

McCulloch, M., Trotter, J., Montagna, P., Falter, J., Dunbar, R., Freiwald, A., Foersterra, N., Lopez Correa, M., Maier, C., Ruggeberg, A., Taviani, M. (2012). Resilience of cold-water scleractinian corals to ocean acidification: Boron isotopic systematics of pH and saturation state up-regulation. *Geochimica et Cosmochimica Acta* 87, 21-34.

Mienis, F., de Stigter, H., White, M., Duineveld, G.C.A., de Haas, H., van Weering, T. (2007). Hydrodynamic controls on cold-water coral growth and carbonate-mound development at the SW and SE Rockall Trough Margin, NE Atlantic Ocean. *Deep-Sea Research* I 54, 1655-1674.

Mienis, F., Duineveld, G.C.A., Davies, A.J., Ross, S.W., Seim, H., Bane, J., van Weering, T.C.E. (2012). The influence of near-bed hydrodynamic conditions on cold-water corals in the Viosca Knoll area, Gulf of Mexico. *Deep-Sea Research* I 60: 32-45.

Miller, K.J., Rowden, A.A., Williams, A., Häussermann, V. (2011). Out of their depth? Isolated deep populations of the cosmopolitan coral *Desmophyllum dianthus* may be highly vulnerable to environmental change. *Plos One* 6, e19004.

Moberg, F., Folke, C. (1999). Ecological goods and services of coral reef ecosystems. *Ecological Economics* 29, 215–233.

Molinski, T.F., Dalisay, D.S., Lievens, S.L., Saludes, J.P. (2009). Drug development from marine natural products. *Nature Reviews Drug Discovery* 8, 69-85.

Morrison, C.L., Ross, S.W., Nizinski, M.S., Brooke, S., Jaernegren, J., Waller, R.G., Johnson, R.L., King, T.L. (2011). Genetic discontinuity among regional populations of *Lophelia pertusa* in the North Atlantic Ocean. *Conservation Genetics* 12, 713-729.

Mortensen, P.B., Hovland, M., Brattegard, T., Farestveit, R. (1995). Deep water bioherms of the scleractinian coral *Lophelia pertusa* (L.) at 641N on the Norwegian shelf: structure and associated megafauna. *Sarsia* 80, 145–158.

Mortensen, P.B., M.T. Hovland, J.H. Fossä & D.M. Furevik (2001). Distribution, abundance and size of *Lophelia pertusa* coral reefs in mid-Norway in relation to seabed characteristics. *Journal of the Marine Biological Association of the UK* 81, 581-597.

Mortensen, P.B., Buhl-Mortensen, L. (2005). Deep-water corals and their habitats in The Gully, a submarine canyon off Atlantic Canada. In: Freiwald, A., Roberts, J.M. (eds.) *Cold-water corals and ecosystems*. Springer, Heidelberg. pp247-277.

Mortensen, P.B., Buhl-Mortensen, L., Gebruk, A.V., Krylova, E.M. (2008). Occurrence of deep-water corals on the Mid-Atlantic Ridge based on MAR-ECO data. *Deep-Sea Research* II 55, 142-152.

Naumann, M.S., Orejas, C., Ferrier-Pages, C. (2013). High thermal tolerance of two Mediterranean cold-water coral species maintained in aquaria. *Coral Reefs* 32, 749-754.

Naumann, M.S., Orejas, C., Ferrier-Pagès, C. (2014). Species-specific physiological response by the cold-water corals *Lophelia pertusa* and *Madrepora oculata* to variations within their natural temperature range. *Deep Sea Research* Part II 99, 36–41.

Orr, J.C., Fabry, V.J., Aumont, O., Bopp, L., Doney, S.C., Feely, R.A., Gnanadesikan, A., Gruber, N., Ishida, A., Joos, F., Key, R.M., Lindsay, K., Maier-Reimer, E., Matear, R., Monfray, P., Mouchet, A., Najjar, R.G., Plattner, G.K., Rodgers, K.B., Sabine, C.L., Sarmiento, J.L., Schlitzer, R., Slater, R.D., Totterdell, I.J., Weirig, M. F., Yamanaka,

Y. and Yool, A. (2005). Anthropogenic ocean acidification over the twenty-first century and its impact on calcifying organisms. *Nature*, 437, 681-686.

Pires, D.O. (2007). The azooxanthellate coral fauna of Brazil. In: George, R.Y. and S.D. Cairns, (eds.). *Conservation and adaptive management of seamount and deep-sea coral ecosystems*. Rosenstiel School of Marine and Atmospheric Science, University of Miami. Pp 265-272.

Probert, K., Knight, D.G.M., Grove, S.L. (1997). Benthic invertebrate bycatch from a deep-water trawl fishery, Chatham Rise, New Zealand. *Aquatic Conservation: Marine and Freshwater Ecosystems* 27-40.

Prouty, N.G., Roark, E.B., Koenig, A., Demopoulos, A.W., Batista, F.C., Kocar, B.D., Selby, D., McCarthy, M.D., Mienis, F. (2014). Deep-sea coral record of human impact on watershed quality in the Mississippi River Basin. *Global Biogeochemical Cycles* 28, 29-43.

Quattrini, A.M., Etnoyer, P.J., Doughty, C.L., English, L., Falco, R., Remon, N., Rittinghouse, M., Cordes, E.E. (2014). A phylogenetic approach to octocoral community structure in the deep Gulf of Mexico. *Deep-Sea Research* II. 99, 92-102.

Qurban, M.A., Krishnakumar, P.K., Joydas, T.V., Manikandan, K.P., Ashraf, T.T.M., Quadri, S.I., Wafar, M., Qasem, A., Cairns, S.D. (2014). In-situ observation of deep water corals in the northern Red Sea waters of Saudi Arabia. *Deep-Sea Research* I 89, 35-43.

Ramirez-Llodra, E., Brandt, A., Danovaro, R., De Mol, B., Escobar, E., German, C.R., Levin, L.A., Martínez-Arbízu, P., Menot, L., Buhl-Mortensen, P., Narayanaswamy, B.E., Smith, C.R., Tittensor, D.P., Tyler, P.A., Vanreusel, A., Vecchione, M. (2010). Deep, diverse and definitely different: unique attributes of the world's largest ecosystem. *Biogeosciences* 7, 2851-2899.

Ramirez-Llodra, E., Tyler, P.A., Baker, M.C., Bergstad, O.A., Clark, M.R., Escobar, E., Levin, L.A., Menot, L., Rowden, A.A., Smith, C.R., Van Dover, C.L. (2011). Man and the last great wilderness: human impact on the deep sea. *PlosONE* 6(8), e22588. Doi: 10.1271/journal.pone.022588.

Reed, J.K. (2002). Deep-water Oculina coral reefs of Florida: biology, impacts, and management. *Hydrobiologia* 471, 43-55.

Reed, J.K., Weaver, D.C., Pomponi, S.A. (2006). Habitat and fauna of deep-water *Lophelia pertusa* coral reefs off the southeastern US: Blake Plateau, Straits of Florida, and Gulf of Mexico. *Bulletin of Marine Science* 78, 343–375.

Rengstorf, A.M., Yesson, C., Brown, C., Grehan, A.J. (2013). High-resolution habitat suitability modelling can improve conservation of vulnerable marine ecosystems in the deep sea. *Journal of Biogeography* 40, 1702-1714.

Reveillaud, J., Freiwald, A., Van Rooij, D., Le Guilloux, E., Altuna, A., Foubert, A., Vanreusel, A., Olu-Le Roy, K., Henriet, J.-P. (2008). The distribution of scleractinian corals in the Bay of Biscay, NE Atlantic. *Facies* 54, 317-331.

Rixen, M., Beckers, J.M., Levitus, S., Antonov, J., Boyer, T., Maillard, C., Fichaut, M., Balopoulos, E., Iona, S., Dooley, H., Garcia, M.J., Manca, B., Giorgetti, A., Manzella, g., Mikhailov, N., Pinardi, N., Zavatarelli, M. (2005). The Western Mediterranean deep water: A proxy for climate change. *Geophysical Research Letters* 32, L12608.

Roark, E.B., Guilderson, T.P., Dunbar, R.B., Fallon, S.J., Mucciarone, D.A. (2009). Extreme longevity in proteinaceous deep-sea corals. *Proceedings of the National Academy of Sciences* 106(13), 5204-5208.

Roberts, M., Harvey, S.M., Lamont, P.A., Gage, J. D. (2000). Humphery Seabed photography, environmental assessment and evidence for deep-water trawling on the continental margin west of the Hebrides. *Hydrobiologia* 441, 173-183.

Roberts, J.M., Wheeler, A.J., Freiwald, A. (2006). Reefs of the deep: the biology and geology of cold-water coral ecosystems. *Science* 312, 543-547.

Roberts, J.M., Henry, L.A., Long, D., Hartley, J.P. (2008). Cold-water coral reef frameworks, megafaunal communities and evidence for coral carbonate mounds on the Hatton Bank, north east Atlantic. *Facies* 54, 297-316.

Roberts, J.M., Wheeler, A.J., Freiwald, A., Cairns, S.D. (2009). *Cold-water Corals: The Biology and Geology of Deep-sea Coral Habitats*. Cambridge: Cambridge University Press.

Rocha, J., Peixe, L., Gomes, N.C.M., Calado, R. (2011). Cnidarians as a Source of New Marine Bioactive Compounds—An Overview of the Last Decade and Future Steps for Bioprospecting. *Marine Drugs* 9, 1860-1886.

Roder, C., Berumen, M.L., Bouwmeester, J., Papathanassiou, E., Al-Suwailem, A., Voolstra, C.R. (2013). First biological measurements of deep-sea corals from the Red Sea. *Scientific Reports* 3, 2801.

Rogers, A. D., Baco, A., Griffiths, H., Hart, T., & Hall-Spencer, J. M. (2007). Corals on seamounts. In: *Seamounts: ecology, fisheries and conservation*, 141-69.

Ross, S.W., Quattrini, A.M. (2009). Deep-sea reef fish assemblage patterns on the Blake Plateau (Western North Atlantic Ocean). *Marine Ecology-an Evolutionary Perspective* 30, 74-92.

Ross, R.E., Howell, K.L. (2013). Use of predictive habitat modelling to assess the distribution and extent of the current protection of 'listed' deep-sea habitats. *Diversity and Distributions* 19, 433-445.

Rowden, A.A., Schlacher, T.A., Williams, A., Clark, M.R., Stewart, R., Althaus, F., Bowden, D.A., Consalvey, M., Robinson, W., Dowdney, J. (2010). A test of the seamount oasis hypothesis: seamounts support higher epibenthic megafaunal biomass than adjacent slopes. *Marine Ecology* 31, 95-106.

Sánchez, F., González-Pola, C., Druet, M., García-Alegre, A., Acosta, J., Cristobo, J., Parra, S., Ríos, P., Altuna, Á., Gómez-Ballesteros, M., Muñoz-Recio, A., Rivera, J., Díaz del Río, G., (2014). Habitat characterization of deep-water coral reefs in La Gaviera Canyon (Avilés Canyon System, Cantabrian Sea). *Deep-Sea Research* II 106, 118-140.

Stetson, T.R., Squires, D.F., Pratt, R.M. (1962). Coral banks occurring in deep water on the Blake Plateau. *American Museum Novitates* 2114, 1–39.

Stone, R.P. (2006). Coral habitat in the Aleutian Islands of Alaska: depth distribution, fine-scale species associations, and fisheries interactions. *Coral Reefs* 25, 229-238.

Sumida, P.Y.G., Yoshinagaa, M.Y., Madureirab, L.A.S.P., Hovland, M. (2004). Seabed pockmarks associated with deepwater corals off SE Brazilian continental slope, Santos Basin. *Marine Geology* 207, 159–167.

Thoma, J.N., Pante, E., Brugler, M.R., France, S.C. (2009) Deep-sea octocorals and antipatharians show no evidence of seamount-scale endemism in the NW Atlantic. *Marine Ecology Progress Series* 397, 25-35.

Thresher, R.E., Tilbrook, B., Fallon, S., Wilson, N.C., Adkins, J. (2011). Effects of chronic low carbonate saturation levels on the distribution, growth and skeletal chemistry of deep-sea corals and other seamount megabenthos. *Marine Ecology Progress Series* 442, 87-99.

Tilman, D., Lehman, C.L., Thomson, K.T. (1997). Plant diversity and ecosystem productivity: Theoretical considerations. *Proceedings of the National Academy of Sciences* US 94, 1857-1861.

Tittensor, D.P., Baco, A.R., Hall-Spencer, J.M., Orr, J.C., Rogers, A.D. (2010). Seamounts as refugia from ocean acidification for cold-water stony corals. *Marine Ecology-an Evolutionary Perspective* 31, 212-225.

Tracey, D.M., Rowden, A.A., Mackay, K.A., Compton, T. (2011). Habitat-forming cold-water corals show affinity for seamounts in the New Zealand region. *Marine Ecology Progress Series* 430, 1-22.

van der Land, C., Eisele, M., Mienis, F., De Haas, H., Hebbeln, D., Reijmer, J.J.G., Van Weering, T.C.E. (2014). Carbonate mound development in contrasting settings on the Irish margin. *Deep-Sea Research* II 99, 297-326.

van Oevelen, D., Duineveld, G., Lavaleye, M., Mienis, F., Soetaert, K., Heip, C.H.R. (2009). The cold-water coral community as a hot spot for carbon cycling on continental margins: A food-web analysis from Rockall Bank (northeast Atlantic). *Limnology and Oceanography* 54(6), 1829-1844.

Viana, A.R., Faugères, J.C., Kowsmann, R.O., Lima, J.A.M., Caddah, L.F.G., Rizzo, J.G. (1998). Hydrology, morphology and sedimentology of the Campos continental margin, offshore Brazil. *Sedimentary Geology* 115, 133-157.

Watling, L. and P.J. Auster (2005). Distribution of deepwater alcyonacea off the northeast coast of the United States. p. 279-296. In: A. Freiwald and J.M. Roberts (eds.) *Cold-water Corals and Ecosystems*, Springer-Verlag, Berlin Heidelberg.

Watling, L., France, S.C., Pante, E. & Simpson, A. (2011). Biology of deep-water octocorals. *Advances in Marine Biology*, 60, 41–122 http://dx.doi.org/10.1016/B978-0-12-385529-9.00002-0.

White, H.K., Hsing, P.Y., Cho, W., Shank, T.M., Cordes, E.E., Quattrini, A.M., Nelson, R.K., Camilli, R., Demopoulos, A., German, C.R., Brooks, J.M., Roberts, H.H., Shedd, W., Reddy, C.M., Fisher, C.R. (2012)a. Impact of the Deepwater Horizon oil spill on a deep-water coral community in the Gulf of Mexico. *Proceedings of the National Academy of Sciences* 109, 20303-20308.

White, M., Dorschel, B. (2010). The importance of the permanent thermocline to the cold water coral carbonate mound distribution in the NE Atlantic. *Earth and Planetary Science Letters* 296, 395-402.

White, M., Wolff, G.A., Lundalv, T., *et al.* (2012)b. Cold-water coral ecosystem (Tisler Reef, Norwegian Shelf) may be a hotspot for carbon cycling. *Marine Ecology Progress Series* 465, 11-23.

Williams, B., Risk, M.J., Ross, S.W., Sulak, K.J. (2006). Deepwater Antipatharians: proxies of environmental change. *Geology* 34, 773–776.

Wilson, J.B. (1979). The distribution of the coral *Lophelia pertusa* (L.) [*L. prolifera* (Pallas)] in the north-east Atlantic. *Journal of the Marine Biological Association of the United Kingdom* 59, 149-164.

Yesson, C., Taylor, M.L., Tittensor, D.P., Davies, A.J., Guinotte, J., Baco, A., Black, J., Hall-Spencer, J.M., Rogers, A.D. (2012). Global habitat suitability of cold-water octocorals. *Journal of Biogeography* 39, 1278-1292.

43. Tropical and Sub-Tropical Coral Reefs

Writing team:
Clive Wilkinson (Convenor), Angelique Brathwaite, C. Mark Eakin, Beatrice Padovani Ferreira (Lead Member), Peter Harris (Co-Lead Member), Ronaldo Francini-Filho, Bernard Salvat and Nicole Webster

1 Introduction

Many activities and businesses are judged on three criteria, the triple bottom line: economic evaluation; social responsibility; and environmental conservation. Coral reefs make major contributions towards "people, planet, profit"; they are economically beneficial to many countries, especially small island developing States (SIDS), in the provision of food, materials and income from tourism and fisheries; coastal and island societies are often largely or nearly completely dependent on adjacent coral reefs, with cultures developed around those reefs; and reefs contain the largest reservoirs of biodiversity in the world. Moreover, these reefs constitute a very special ecosystem, forming a link between humans on the land and the ocean around them.

Of the 193 Member States of the United Nations, 79 States (41 per cent) have coral reefs in their maritime zones, including a large number of SIDS. These reefs are estimated to cover 249,713 km^2 (Burke et al., 2011a) to 284,300 km^2 (Spalding et al., 2001), with an additional 600,000 km^2 of sandy lagoons. Reefs and nearby seagrass and mangrove ecosystems are of major importance for 275 million people who depend on associated fisheries as their major source of animal protein (UNSG, 2011) and play a role in social stability, especially within a subsistence economy which is often declining in sustainability. Of these 79 States, more than 30 SIDS have coral reefs that provide the major source of food, coastal protection, and a limited amount of rock and sand; and valuable income from tourism; the continual provision of these ecosystem services is dependent on actions focused on sustaining and conserving healthy, productive coral reef ecosystems.

Coral reefs around the world have been in a state of continual decline over the past 100 years, and especially over the past 50 years. The Global Coral Reef Monitoring Network, which has reported since 1998 in the "Status of Coral Reefs of the World" series assessed that approximately 19 per cent of the world's coral reefs were severely damaged with no immediate prospects of recovery, and 35 per cent of the remaining coral reefs were under imminent risk of degradation from direct human pressures (assessment by the Global Coral Reef Monitoring Network; Wilkinson, 2008; with 372 contributing authors from 96 States and territories). Similar estimates of large-scale degradation have been reported both before and since (Burke et al., 2002; Burke and Maidens, 2004; Bruno and Selig, 2007; Bellwood et al., 2004; Obura et al., 2008). A more recent study by the World Resources Institute in the "Reefs at Risk Revisited" report (Burke et al., 2011a) calculated that more than 60 per cent of the world's coral reefs are under immediate threat. Indeed the latest Intergovernmental Panel on Climate Change (IPCC (2014)) report suggests that "coral reefs are one of the most vulnerable ecosystem on Earth" and will be functionally extinct by 2050, without adaptation (worst case scenario), or by 2100 with biological adaptation of the whole ecosystem. Presently the level of threats varies considerably in different geographical regions; reefs of the Pacific Ocean are least threatened, but those throughout Asia and the wider Caribbean and Atlantic regions are under greater threats.

Coral reefs developed throughout millions of years under a wide range of "natural" stresses, such as storms, variations in sea level, volcanic and tectonic plate activity. However recent anthropogenic stresses are overwhelming the natural reef resistance/resilience and recovery mechanisms, resulting in major losses and declines in the reefs and their biological resources in many regions. The major threats are: overfishing and destructive fishing practices; pollution and increased sedimentation; habitat destruction; increases in diseases and predation; and especially impacts of climate change and ocean acidification (OA). This chapter highlights the threats to the world's coral reefs, lists their current status and reports conservation actions that so far have been successful to ensure that reefs continue to provide ecosystem services to several billion people around the world.

Coastal protection and reef fisheries are of utmost socioeconomic importance for coastal communities; and reefs constitute the basis of many cultures. In addition, they are a source of rock and sand aggregate for construction but frequently such exploitation is unsustainable. The economic value of reefs,, only as a source of raw materials, has been estimated at 28 United States dollars per hectare (Costanza et al., 2014; see also Chapter 7). Reefs underpin the reef-based tourism industry and harbour biodiversity as natural capital.

1.1 Cultural

Since humans began to inhabit coral reef areas, they developed strong cultural links with this ecosystem, both with the habitats and also with many species. Such cultural themes associated with reef ecosystems developed through popular beliefs and the ecosystem services essential to their livelihoods. More importantly for many people, the coral reefs constituted and sustained the land on which they lived. Some of the human communities in South East Asia, which had settled near coral reef waters, migrated outwards during the Holocene and progressively colonized islands throughout the Pacific Ocean. Many Pacific communities developed strong cultural affinities towards the reefs and many of these remain active and recognized by local and national governments.

1.2 Coastal protection

Reefs and mangrove forests provide coastal protection for land resources and human infrastructure, especially where large areas of shallow reef flats are adjacent to the shore and reefs have a distinct crest. This is a continual service, which is especially important during storms and

Table 43.1 | Annual net global benefits from coral reef-related ecosystem services in dollars assessed in 2010, with two important States included for emphasis. Values are expressed in millions of United States dollars as net benefits, including costs (from Burke et al., 2011b).

Region & Total	Tourism	Reef Fisheries	Shoreline Protection
Global 29 000	11 500	6 800	10 700
Indonesia 2 014	127	1 500	387
Philippines 1 283	133	750	400

cyclones. This service also includes some attenuation of tsunami waves, as was the case during the 2004 Indian Ocean tsunami (Wilkinson et al., 2006). Coastal protection provided by coral reefs is valued at 10.7 billion dollars (Table 43.1), which can be considered as a natural alternative to the cost of building seawalls along coasts that are otherwise protected from ocean swell and storm waves by offshore barrier reef systems.

1.3 Fisheries and food

About 275 million people worldwide depend directly on ecosystem services provided by coral reefs and associated ecosystems (Newton et al., 2007; Cinner et al., 2008). This is particularly crucial for SIDS and coastal developing countries (Burke et al., 2011a; see also Chapter 15). Estimates of the value of all goods, services, and livelihoods associated with coral reefs (including tourism, fisheries and protection) exceed 30 billion dollars (Cesar et al., 2003). Fisheries in the tropics feed millions of people (Whittingham et al., 2003); but the importance of reefs extends far beyond economically measurable values, as the identity of many coastal peoples is linked to reefs through their socio-cultural practices (Johannes, 1981; Cinner et al., 2008; Kittinger et al., 2012).

1.4 Rock and sand

Coral reefs produce large amounts of exploitable sand and rock (Reid et al., 2005), which are valuable for many coastal communities, especially those living on coral islands with no other sources of these materials. The use of coral blocks taken off the reef for building construction was sustainable when human populations were lower. However, with increasing demand from growing populations, the practice became unsustainable in some areas, and excessive harvesting of coral rock and sand exposed shorelines to increased erosion, resulting in damage to adjacent communities. The reef flat around the main island of the Maldives, Malé, was so seriously mined over centuries that the shoreline protection was virtually lost, such that in 1987 storm waves penetrated throughout the city causing massive saltwater damage, including contamination of the groundwater system. Replacement concrete tetrapod seawalls cost more than 10 million dollars per km in the 1990s; the cost would be much higher now (Talbot and Wilkinson, 2001). Such problems create economic dilemmas for governments, as it may be cheaper to mine fringing reefs and sand flats, rather than take the material from land or remote coral structures. This will be exacerbated with climate change-related sea-level rise. Mining also occurs at deeper areas. Large-scale mining projects are predicted for eastern Brazil to explore one of the largest rhodolith beds (i.e., nodules of calcareous coralline algae) in the world (Amado-Filho et al., 2012), aimed at extracting micronutrients and correcting soil acidity for sugar cane plantations.

1.5 Recreation and tourism

Reef-related tourism generates 11.5 billion dollars per year in revenue for the global economy (Table 43.1). Tourism and recreation in Australia's Great Barrier Reef alone sustain 69,000 jobs and are valued at either 4.4 billion dollars per year (Deloitte Access Economics, 2013) or 11.5 to 15.5 billion dollars (Stoeckl et al. 2014) depending on the methods employed. Reefs contribute about 1 billion dollars per year to the economy of Hawaii, United States of America (Bishop et al., 2011). In 2000-2001, the artificial and natural reefs off southeast Florida supported almost 28 million person-days of recreational diving, fishing and viewing activities. These activities generated about 4.4 billion dollars in local sales, almost 2 billion dollars in local income, and sustained 70,400 full and part-time jobs (Johns, et al., 2001). In Belize, coral reef- and mangrove-associated tourism contributed an estimated 150 - 196 million dollars to the national economy in 2007. Belize is an example of many small developing countries where tourists provide a large proportion of foreign currency earnings. Reef-based tourism is especially sensitive to reef condition, and thus the sector is particularly vulnerable to degradation (Cooper et al., 2008).

1.6 Biodiversity

Coral reefs are the largest reservoirs of biodiversity on earth: they host 32 of 34 recognised phyla and approximately one-third of all marine biodiversity (Spalding et al., 2001; Groombridge and Jenkins, 2002; Roberts et al., 2002; Bouchet, 2006). The centre of global coral reef biodiversity is the "Coral Triangle" (CT), including eastern Indonesia and Malaysia, the Philippines, Timor-Leste, Papua New Guinea and the Solomon Islands (Figure 43.1). There are more than 550 species of hard corals in the CT area; the diversity decreases away from this focus to the West and to the

The boundaries and names shown and the designations used on this map do not imply official endorsement or acceptance by the United Nations.

Figure 43.1 | The diversity of hard coral species is greatest within Southeast Asia and the West Pacific; declining diversity radiates out from this area, which is called the Coral Triangle. Much lower diversity of corals is found in the Atlantic and wider Caribbean (from Veron et al., (2015).

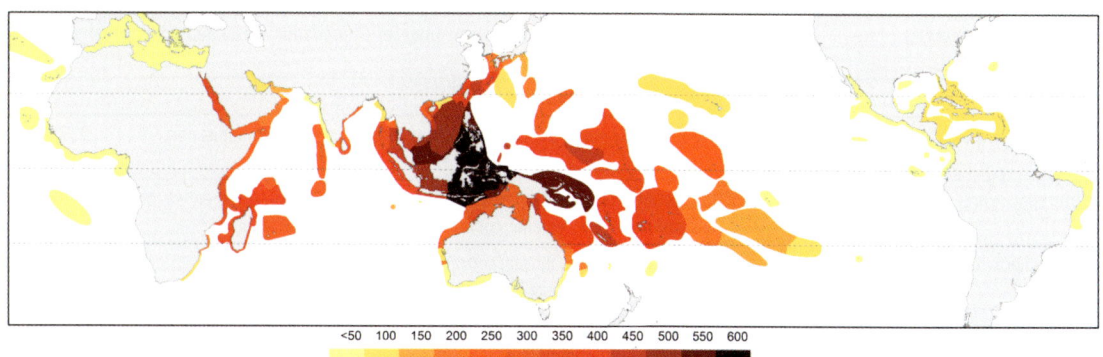

East, such that less than half this number of species are found in French Polynesia (France), the Hawaiian Islands (United States), and the East African coast. Reef biodiversity in the Caribbean and Atlantic region is also lower; only 65 different coral species are recorded on all these reefs.

2 Major threats

Modern coral reefs have developed since the end of the last ice age (the Pleistocene) when global sea level rose approximately 120 m to just above current levels about 6,500 years ago (Woodroffe and Webster, 2014). Coral reef growth has continued throughout this period, especially during relatively stable sea level (the Holocene); until recently the major stressors were local natural damage, e.g., storms, earthquakes, extreme low tides.

The current serious and further deteriorating status of coral reefs around the world is directly due to damaging stresses that arose during the Anthropocene (Bradbury and Seymour, 2009; Hoegh-Guldberg, 2014); effectively since the mid-18th century, and particularly since 1950, when human pressures ramped up to destructive levels. Assessments of coral reefs cited above and anecdotal reports (Sale and Szmant, 2012) indicate that most reefs were largely "pristine" until direct and indirect human pressures and the advent of "new technology" started affecting many reefs, commencing in the 1970s. This "new technology" permitted far more extensive resource exploitation over far greater areas and to greater depths. This technology (discussed below) includes monofilament lines and nets, and boats with motors. Problems with catchment management, in the face of deforestation for agricultural purposes, have also affected coral reefs, especially coastal reefs off Africa, Australia and South America (Wilkinson and Brodie, 2011).

The degradation of many coral reefs around the world is both directly and incidentally due to increasing anthropogenic pressures arising from increasing population pressures on reefs and their resources, especially through increased economic capacity to use these resources. The major threats include extractive activities, pollution, sedimentation, physical destruction, and the effects of anthropogenic climate change. Such stressors often interact synergistically with natural stressors, such as

Table 43.2 | Natural and anthropogenic stresses divided into three direct damage categories and one group of organizational factors [summarised from Wilkinson and Salvat, 2012].

1. Natural factors	not readily amenable to conservation measures
i. Catastrophic geological: earthquake, tsunami, volcano, meteors	Potential for rare, but major local damage to coral reefs, especially in Indonesia and South-West Pacific (Papua New Guinea, Solomon Islands, Vanuatu)
ii. Meteorological and climatic: tropical storms, floods, droughts, extremes of heat and cold	Severe storms smash coral reefs or bury them under sediments following floods. Temperature extremes cause coral bleaching and death.
iii. Extreme low tides	Exposes coral reefs leading to widespread mortality e.g., Red Sea
2. Direct human pressures	**major target for conservation measures**
i. Exploitation: overfishing, bomb fishing and trawler damage (exacerbated by global market pressures)	Harvesting of fishes and invertebrates beyond sustainable yields, includes damaging practices (bomb, cyanide fishing); boat scour and anchor damage to reefs
ii. Sedimentation increases: logging, farming, development	Excess sediment and mud on coral reefs from poor land use, deforestation, dredging; reduces photosynthesis; and associated with disease;
iii. Nutrient and chemical pollution	Organic and inorganic chemicals in sediments, untreated sewage, agriculture, animal husbandry and industry wastes; includes complex organics and heavy metals. Turbidity reduces light, promotes growth of competing algae on corals. Herbicides kill algae associated with coral reefs.
iv. Development of coastal areas	Removal or burial of coral reefs for urban, industrial, transport and tourism developments (e.g., airports); mining reef rock and sand beyond sustainable limits
3. Global change threats	**need major global focus; local conservation can assist by increasing reef resilience and raising awareness;**
i. Elevated sea-surface temperatures	Bleaching in corals, i.e., loss of photosynthetic zooxanthellae either temporary or lethal; stimulates algal blooms on reefs; increases disease susceptibility; reduces larval survival.
ii. Increased storms, wider climatic fluctuations	Stronger storms will smash or bury coral reefs; increased rain increases sediment flows; can reduce thermal stress locally.
iii. Rising CO2 dissolved in seawater with increasing ocean acidification	Increased CO_2 in seawater increases acidity, which decreases calcification in corals and other organisms and reef cementation and increases erosion (including bioerosion); higher CO_2 may increase algal productivity;
iv. Diseases, plagues and invasive species	Intensity and frequency of coral diseases and plagues of predators correlated with global climate change, especially higher temperatures.
4. Governance, awareness, political will	**major target for conservation measures**
i. Rising poverty, increasing populations, alienation from land and sea	More poor, dispossessed people use coral resources for subsistence and habitation.
ii. Poor management capacity and lack of resources	Few trained personnel for coastal management, raising awareness, enforcement and monitoring; lack of funds and logistics for conservation, e.g., smaller countries.
iii. Poor political will and poor oceans governance	Political ignorance, indifference, inertia; corruption and low transparency in governance at global and regional levels all impede decision-making and waste resources.
iv. Uncoordinated global and regional conservation arrangements	Inadequate coordination among multilateral environmental agreements and international donors results in overlapping meeting and reporting requirements which exhaust conservation capacity in smaller countries.

Table 43.3 | A numerical compilation of anthropogenic threats to coral reefs summarized in the graphics in the first map of Burke et al. (2011a), shows that threats are greatest in Southeast and East Asia, with almost 50 percent of reefs at High to Very High threat levels, whereas threats in the wider Pacific and around Australia are much less. Predicted climate change damage, however, will affect all reefs in the world in the next two to three decades. Methodological details are in Burke et al., 2011a.

Region	Low %	Medium %	High %	Very High %
Southeast and East Asia	6	47	28	20
Indian Ocean	34	32	21	13
Caribbean and Atlantic	25	44	18	13
Middle East	35	44	13	8
Pacific	52	28	15	5
Australia	86	13	1	0
World – all areas	39	34	17	10

storms (Table 43.2). Carpenter and 38 other authors (Carpenter et al., 2008) have estimated that 33 per cent of all reef-building corals could become extinct due to damage from local threats combined with climate change impacts.

2.1 Overfishing

The major traditional use for coral reefs is extractive exploitation of tropical fisheries resources. For many centuries, these resources, particularly fishes and also turtles, algae, molluscs, crustaceans and echinoids, served as the major animal protein source for many coastal and island communities throughout all oceans. These resources are socially and economically important in sustaining livelihoods of traditional coastal communities, especially through ensuring their food security. However the rate and ease of exploitation has increased, such that in many areas it has reached unsustainable levels and is seriously damaging the ecological integrity of coral reefs. The rate and ease of exploitation has increased in recent decades with the introduction of aluminium boats and motors, monofilament lines and nets, metal hooks, dive masks and spear-guns (now frequently used with underwater lights to catch sleeping fish at night) and use of compressed air (SCUBA and hookah gear). Habitat-damaging practices, such as use of explosives, cyanide or other poisons, also pose a serious threat (Johannes and Riepen, 1995). External markets have driven the increase in the exploitation rate and extension, especially in Asia, to support the tourist demand (see live reef-fish trade below) and also in the wider Caribbean and South America for fresh reef seafood and for export of conch and lobster to the United States. Rapid economic growth throughout Asia has stimulated the lucrative live reef food-fish trade, which is expanding rapidly, with reef fish taken largely through the use of cyanide and other destructive practices. This trade particularly targets large attractive edible fish, such that one species, the humphead wrasse (*Cheilinus undulatus*) is now listed on the International Union for Conservation of Nature (IUCN) Red List as "Endangered" and several groupers, particularly larger species, are listed as "Near Threatened" (Sadovy et al., 2013). This trade is so valuable that industrial-scale fishing across the Indo-West Pacific targets mass fish-spawning aggregations (Sadovy and Domeier, 2005). Reef fish spawning aggregations have also been drastically reduced across the Caribbean by artisanal fishers. A notable example is the Nassau grouper (*Epinephelus striatus*), once of great commercial importance and now listed as "Endangered" and commercially extinct across much of its range in the Caribbean (Sadovy, 1999). More than a quarter of global records of fish aggregations show a declining trend in numbers of fish aggregating, and 4 per cent are documented as having disappeared entirely (Status Report - Worlds Fish Aggregations 2014; Russell et al., 2014).

Another particularly destructive form of industrial scale fishing is via muro ami (drive net) fishing, observed to operate predominantly from the southern Philippines (Jennings and Polunin, 1996). This practice has been banned from many areas; however, illegal fishing with this method still occurs.

A detailed assessment by the Secretariat for the Pacific Community of current and predicted coral reef fisheries resources in 49 island States reported that catch rates in 55 per cent of them are unsustainable and unlikely to be able to provide food security into the future (Figure 43.2; Bell et al., 2011b in Bell et al., 2011a). The human population of Oceania has increased fourfold since the middle of the last century. A large proportion of this population is still based on a subsistence economy. The extra fish stocks required will have to come from pelagic species, such as tuna, or through aquaculture, as reef fisheries are declining alarmingly due to over-exploitation, especially through the use of "modern" technology (Figure from Bell et al., 2011).

Data from more than 300 coral reefs in the wider Caribbean show a three to six per cent decline in total fish populations per year over a 50-year period (Paddack et al., 2009). This is in parallel to the decline in mean coral cover from 50 per cent to 10 per cent over a 25-year period (Gardner et al., 2003), and a major loss of reef structural complexity over a 30-year period (Jackson et al., 2014). Large-bodied herbivorous fish from the family Scaridae (parrotfish) play key ecological roles and favour coral health and abundance by controlling overgrowth by algae (Bellwood et al., 2004). A recent large-scale synthesis of peer-reviewed and unpublished data indicates that overfishing of coral reef herbivorous

Figure 43.2 | Current and predicted rates of population (upper diagrams), and the fish stocks needed for food security (lower diagrams) in urban (dark colour) and more remote (pale colour) areas of Melanesia, Micronesia and Polynesia between 2010 and 2030. Note that the scale bar for Melanesia is 10 times larger than the other regions (source: SPC and Bell et al., 2011b in Bell et al., 2011a).

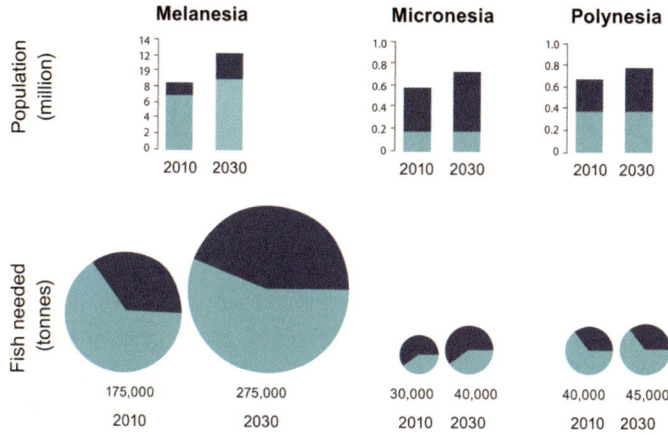

fish is a worldwide problem that deserves urgent attention (Edwards et al., 2013).

Similar declines in reef fish stocks in the Indian Ocean due to over-exploitation are documented (McClanahan et al., 2008; McClanahan et al., 2011), matching reports from the Pacific (Dalzell and Adams, 1996; Zeller et al., 2006).

Many coral reef fishes periodically and predictably aggregate to spawn, making them vulnerable to fishing. The problem for management of fishing on these aggregations is particularly challenging because little is known about aggregating fish behaviour and the impacts of fishing, although clear evidence exists of serious declines in several species. Although information on the level of management and monitoring is limited, it appears that 35 per cent have some form of management in place such as marine protected areas, seasonal protection from fishing and/or sale, or fisheries harvest controls, and about 25 per cent have some form of monitoring, such as fish counts (Russell et al., 2014). Multiple management measures are needed for those species, however it is clear that whenever uncontrolled exploitation continues it may lead to major depletions for both fish populations and fisheries and livelihoods they support. (De Michelson et al., 2008; Russell et al., 2014)

2.2   Pollution and sedimentation

Water quality (including elevated nutrient, sediment and contaminant concentrations) is a significant environmental driver for the health of coral reefs. Coral reefs are threatened by a wide range of chemical pollution pressures that are likely to increase with further industrial development and land use (see chapter 20 for more detail). Trace metal contaminants are accumulating in fish, with a clear link to coastal contamination from mining in New Caledonia (France), while contamination by persistent organic pollutants (POPs) occurs across the whole lagoon region (Briand et al., 2014). Millions of tons of dust are transported in the atmosphere each year from Africa and Asia to the Caribbean. This is a significant input source of trace metals, organic contaminants and potential microbial pathogens in the reef ecosystem which is likely to adversely affect the health of corals (Garrison et al., 2003).

Excess nutrients result in poor water quality and eutrophication. Reefs exposed to poor water quality show significant increases in macroalgal cover and reduced coral richness and recruitment (De'ath and Fabricius, 2010; Fabricius et al., 2012; Vega Thurber et al., 2014). In the mid-1990s, global models of coral reef pollution estimated that 22 per cent of all reefs were classified as being at high (12 per cent) or medium (10 per cent) risk from pollution and soil erosion (Bryant et al., 1998). On the Great Barrier Reef (GBR), central and southern rivers are reported to deliver five- to nine-fold higher nutrient and sediment loads compared with pre-European settlement, largely due to changes in land-use practices, including land clearing, fertilization and urbanization (Kroon et al., 2012). Flood events that deliver high nutrient and sediment loads via river runoff are now directly affecting up to 15 per cent of GBR reefs (De'ath and Fabricius, 2010; Kroon et al., 2012).

Pressures related to elevated sediments include sedimentation, total suspended solids and light attenuation. All of these can damage coral reef species via smothering, shading and blocking of the filter-feeding systems. Specific assessments of sediment stress have been experimentally examined in only 10 per cent of all known reef-building corals; these studies indicate sediment thresholds and also identify response and adaptation mechanisms that corals employ to cope with excess sediments (Erftemeijer et al., 2012). Reduced coral recruitment success and reef overgrowth by microalgae are significant effects of increased sedimentation on coral reefs. In addition, chronic effects from increased sediment loads include reduced reef calcification, shallower photosynthetic compensation points, changes in the community structure of corals, and reduced species richness. This decreased diversity and increased simplification of reef ecosystems with increasing sediment exposure may compromise their ability to maintain critical ecosystem functions (Fabricius, 2005). The impacts of dredging on coral reefs are primarily linked to the intensity, duration and frequency of exposure to increased total suspended solids and sedimentation (Erftemeijer et al., 2012) and whether the sediments include particulate organic matter or dissolved inorganic nutrients (Fabricius, 2005). Total suspended sediment thresholds reported for coral reef systems range from <10 mg L^{-1} to >100 mg L^{-1} while the maximum sedimentation rates tolerated by corals range from <10 mg cm^{-2} d^{-1} to >400 mg cm^{-2} d^{-1} (Erftemeijer et al., 2012).

Pesticides including herbicides have been widely studied in tropical systems. Most pesticides have no natural sources; concentrations detected in the nearshore lagoon of the GBR are positively correlated with low salinity associated with river runoff. The composition and concentration of pesticides entering the marine environment typically mirror agricultural use in the catchments adjacent to the GBR (Kennedy et al., 2012; Lewis et al., 2009) and on reefs of French Polynesia (France) (Salvat et al., 2012).

Herbicides that inhibit photosystem II in plants are highly persistent in marine environments and are regularly detected in coral reef systems (Schaffelke et al., 2013); with concentrations of herbicides periodically exceeding regulatory guidelines for the GBR during flood plume events (Lewis et al., 2012). These concentrations are known to deleteriously affect corals (Jones and Kerswell, 2003; Negri et al., 2005), microalgae (Bengtson Nash et al., 2005; Magnusson et al., 2008), crustose coralline algae (Negri et al., 2011), foraminifera (van Dam et al., 2012), and seagrass (Haynes et al., 2000; Gao et al., 2011).

The sensitivity of a coral reef to poor water quality largely depends on the pre-existing health of the ecosystem, overall reef resilience and the baseline conditions that the reef normally experiences. For example, the proportion of reefs at risk is highest in countries and entities with widespread land clearing (Burke et al., 2002). It is important that recent research shows that reducing runoff of nutrients, sediments and pesticides from the land will at least partially offset increasing stress and deleterious effects from climate factors for coral reefs (Schaffelke et al., 2013).

2.3 Diseases and predators

Coral disease is reported as one of the most prominent drivers of recent coral reef declines (Aronson and Precht, 2006; Bruckner and Hill, 2009; Rogers, 2009; Sokolow, 2009; Weil and Cróquer, 2009). In particular, the Caribbean has been designated as a "coral disease hotspot" due to the rapid spread, high prevalence, and virulence of diseases associated with corals and other reef organisms (Harvell et al., 2002; Weil et al., 2002). Although the Caribbean is home to only eight per cent of the world's coral reefs, approximately 66 per cent of all coral diseases are found across 38 Caribbean States and territories (Green and Bruckner, 2000). Disease is also reported as the major factor behind a 25-year decline in Caribbean coral reefs, with mean coral cover declining from 50 per cent to only 10 per cent across the entire Caribbean region (Gardner et al., 2003). Disease has also reshaped the community structure of many Caribbean reefs over the last few decades, including: (i) the virtual elimination of Acroporid corals by White Band Disease in the 1980s (Gladfelter, 1982; Ritchie and Smith, 1998; Aronson and Precht, 2001; Bythell et al. 2001; Kline and Vollmer, 2011); (ii) the loss of many *Acropora palmata* by White Pox in the late 1990s (Sutherland et al., 2011); and (iii) the mass mortality of the keystone grazer species, *Diadema antillarum*, by an unidentified disease in the early 1980s (Hughes et al., 1985). The loss of this sea urchin, coupled with declines in herbivorous fish strongly contributed to overgrowth of reefs by macroalgae (Lessios, 1984; Hunte and Younglao, 1988; Hughes, 1994; Jackson et al., 2001), believed to be a contributing factor to coral disease (Nugues et al., 2004).

Coral diseases have now been documented within all major reef systems and ocean basins (Ruiz-Moreno et al., 2012). Indo-Pacific coral reefs are home to 75 per cent of the world's coral reefs and at least 10 identified coral diseases (~30 per cent of known coral diseases; Willis et al., 2004). It is unclear whether coral disease will have the same impact on Indo-Pacific reefs as it has in the Caribbean due to fundamental differences in their coral reef communities (Wilson et al., 2014). A higher level of diversity and functional redundancy in herbivorous fishes and coral communities, slower macroalgal growth, and less dependence on fragmentation as a reproductive mode, may protect Indo-Pacific reefs from dramatic phase-shifts (Roff and Mumby, 2012).

The recent and rapid increase in disease occurrence worldwide is correlated with increasing environmental stressors that have local and global impacts e.g., elevated seawater temperatures, nutrient enrichment, sedimentation, and fish farming (Sutherland et al., 2004; Sato et al., 2009; Pollock et al., 2014; Vega Thurber et al., 2014; Randall and vanWoesik, 2015). Research is only now starting to determine how diseases are contracted and/or spread from one colony to another. Changes in the coral-associated microbial community and subsequent disease severity are often correlated with bleaching stress in warm summer or winter months (Willis et al., 2004; Bourne et al., 2008; McClanahan et al., 2009; Heron et al., 2010). A significant relationship was shown between the frequency of warm temperature anomalies and white syndrome outbreaks during six years across 48 reefs in the GBR (Bruno et al., 2007). White syndrome was described as either an additional emergent disease, or a group of diseases, among Pacific reef-building corals. Proliferation of disease during the hotter months or during mild winters may be correlated with a greater virulence of coral pathogens at higher temperatures (Miller et al., 2006; Harvell et al., 2007; Heron et al., 2010). Additional anthropogenic factors that are considered to influence disease events include nutrient enrichment from fertilizers (Bruno et al., 2003), sewage pollution (Sutherland et al., 2011), fish farming (Garren et al., 2009), and increased macroalgal abundance as a result of overfishing and disease outbreaks (Nugues et al., 2004).

Crown-of-thorns starfish (COTS; *Acanthaster planci*) were not considered a major problem until the last 40 years or so. Population outbreaks in the 1970s devastated large parts of the GBR and similar outbreaks were reported on other reefs of the Indo-Pacific (COTS do not occur in the wider Caribbean). These outbreaks subsided and most reefs recovered their previous coral cover. However, repeated damaging outbreaks have occurred since, such that COTS are reported as the major destructive factors on reefs in French Polynesia (France) (Adjeroud et al., 2009; Kayal et al., 2012), Fiji, Japan, and parts of the Red Sea (Wilkinson, 2008); and COTS contributed 42 per cent of the recent damage to the GBR, alongside storms (48 per cent) and climate change-related damage (10 per cent) (Great Barrier Reef Marine Park Authority 2014; De'ath et al., 2012).

Previous population outbreaks of COTS are reported from the Red Sea around Egypt, in Kenya and the United Republic of Tanzania, and in Southeast and East Asia, especially in China, Japan and the Philippines, and in the Pacific in Fiji, French Polynesia (France), Guam (United States) and Majuro Atoll (Marshall Islands). In the past, these plagues caused massive losses (often in the vicinity of 90 per cent) of the living coral cover (Wilkinson, 2002). Similar outbreaks are reported of the coral-eating mollusc (*Drupella cornus*) on reefs in western Australia and southern China. After apparently abating, major outbreaks have occurred simultaneously with mass coral bleaching in 2005 and 2006.

Four widely supported but not mutually exclusive theories to explain COTS outbreaks are: (a) fluctuations in COTS populations are a natural phenomenon; (b) removal of natural predators (such as large molluscs and some fishes) of the COTS has allowed populations to expand; (c) human-induced increases in the nutrients flowing to the sea have resulted in an increase in planktonic food for larvae of the COTS which leads to an increase in the number of adult starfish causing outbreaks (Fabricius et al., 2010); and (d) increased COTS larval survival as ocean temperatures increase (Uthicke et al., 2014).

2.4 Natural stresses (cyclones, tsunami)

Although many reefs lie outside the zone of frequent tropical cyclones and hurricanes (approximately between 7°N and 7°S latitude), storms regularly damage coral reefs outside this latitudinal zone (Figure 43.3). Storm damage is exacerbated by storm surge and both reduce the ability of coral reefs to return to their mean pre-disturbance state or condition by slowing coral recruitment, growth, and reducing fitness (Nyström et

Tracks and Intensity of All Tropical Storms

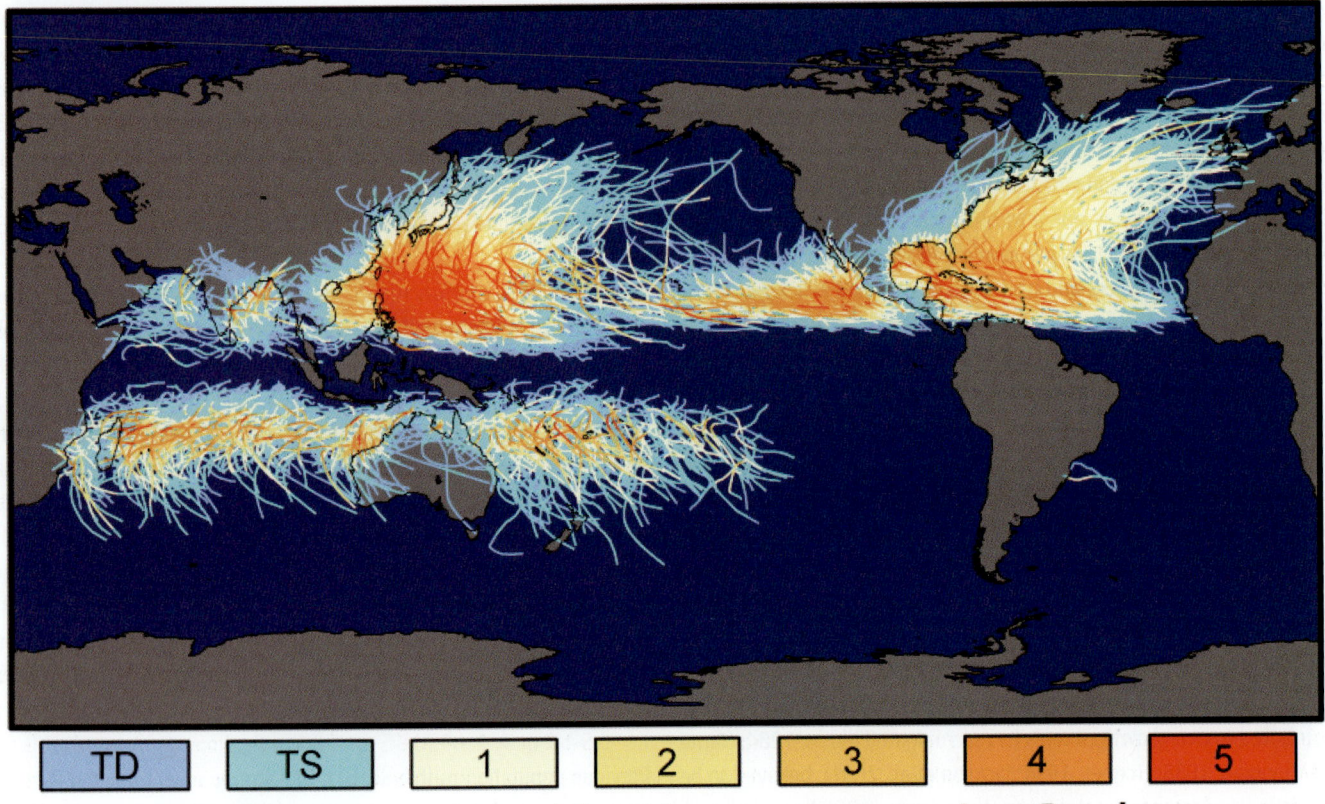

Figure 43.3 | These plots of tropical cyclones (and typhoons) over the past 100 years illustrate that damaging storms are rare within a band between 7o North and South of the Equator, such that a large proportion of the high biodiversity reefs in Indo-Pacific are rarely damaged by damaging storms (courtesy of NASA, USA, 2008). There are predictions that under increasing climate change, the damaging strength of cyclones will increase with more category 4 and 5 storms, but the number of storms may not change (Wilkinson and Souter, 2008).

al., 2000). The combination of tropical storms with other stressors has caused successive and substantial losses of corals worldwide (Harmelin-Vivien, 1994; Done, 1992; Miller et al., 2002; Fabricius et al., 2008; Williams et al., 2008a). However, tropical storms also benefit reefs when the storms are sufficiently distant to not inflict damage, but close enough to cool waters through enhanced wave-induced vertical mixing and to reduce bleaching risk (Szmant and Miller, 2005; Manzello et al., 2007; Carrigan and Puotinen, 2014). A recent modelling study predicted that Caribbean coral reefs with intact herbivore fish and urchin populations would likely maintain their community structure and function under any expected level of tropical cyclone activity, as long as other stressors, such as local pollution and thermal bleaching, are minimal (Edwards et al., 2010).

2.5 Climate change effects and predictions

The most recent report of the IPCC (2014) stated that "Coral reefs are one of the most vulnerable marine ecosystems (*high confidence*) and more than half of the world's reefs are under medium or high risk of degradation". The effects of anthropogenic climate change on coral reefs include: (a) thermal stress causing coral bleaching; (b) storm damage to reefs; (c) sea-level rise; and (d) acidification causing reduced coral accretion and increased erosion.

2.5.1 Thermal stress and coral bleaching

Coral bleaching was a relatively unknown phenomenon until the early 1980s, when a series of local bleaching events occurred principally in the eastern tropical Pacific and wider Caribbean regions, but was also noticed in the Indo-Pacific. Coral bleaching refers to the expulsion of symbiotic algae, the zooxanthellae, in response to stress. Corals can withstand mild to moderate bleaching but severe, prolonged or repeated bleaching can lead to colony mortality. Corals' physiological processes are optimized to the warmest temperatures they normally experience, so an increase of only 1 -2°C above the normal local seasonal maximum can induce bleaching (Fitt and Warner, 1995). Although most coral species are susceptible to bleaching, thermal tolerance varies amongst taxa and along geographic gradients (Marshall and Baird, 2000; McClanahan et al., 2007). Bleaching is best predicted by using an index of accumulated thermal stress above a locally established threshold (Atwood et al., 1992; Eakin et al., 2009). Many heat-stressed and/or bleached corals subsequently die from coral diseases (reviewed in Burge et al., 2014).

The strong El Niño - La Niña events of 1998 brought a global focus on coral bleaching when approximately 16 per cent of the world's coral reefs in almost all tropical ocean basins were massively damaged and lost most of their corals (Wilkinson, 2000). Rising temperatures have accelerated bleaching and mass mortality during the past 25 years (Brown, 1997a; Eakin et al., 2009), when coral bleaching was documented throughout various parts of the world (Eakin et al., 2009; Eakin et al., 2010; Wilkinson and Souter, 2008; Williams and Bunkley-Williams, 1990). A global analysis of threats to coral reefs shows that this widespread threat has significantly damaged most coral reefs around the world (Burke et al., 2011a).

Although some recovery occurred in the Caribbean from the 1987 (Fitt et al., 1993) and the 1995 bleaching events, bleaching in 1998 and 2005 caused high coral mortality at many reefs with little evidence of recovery (Eakin et al., 2010; Goreau et al., 2000; Wilkinson and Souter, 2008). The subsequent strongest recovery was on reefs that were highly protected from anthropogenic pressure. This led to recognition of the importance of maintaining resilience in coral reef ecosystems (Nyström et al., 2000; Hughes et al., 2007; Anthony et al., 2014). An example of reef resilience was observed on the remote Scott Reef off western Australia, when the reef was severely damaged during the 1997-98 El Niño. However, the herbivore fish population grew rapidly to control algal overgrowth, allowing many new coral recruits to restore most of the lost coral cover after 12 years (Gilmour et al., 2013). Additionally, certain factors such as reef depth and structural complexity were shown to increase reef resilience after the 1998 bleaching in the Seychelles (Graham et al., 2015).

A comparison of the recent and accelerating thermal stress events with the slow recovery rate of most reefs (Baker et al., 2008), suggests the temperature increase has exceeded the balance between event recurrence and recovery rate. It appears that some coral species are less sensitive to short-term temperature anomalies than others, although there are significant geographic variations (McClanahan et al, 2007) and some corals may have already adapted or acclimatised to warming (Guest et al., 2012), albeit not quickly enough to prevent major losses (Logan et al., 2013). Some heritable epigenetic adaptation to frequent heat stress may occur in some species of lagoonal corals (Palumbi et al., 2014; Eakin, 2014). However, adaptation potential may be limited in species where larval survival has been shown to decline at high temperatures (Randall and Szmant, 2009a; 2009b).

Climate models are able to predict the potential consequences of future warming on corals, including the future frequency of thermal events exceeding the bleaching threshold for a given area (map 3.3 in Burke et al., 2011a). In the absence of adaptation, there are predictions that many of the world's coral reefs will experience annual bleaching by mid-century (Donner et al., 2005; Donner, 2009; Logan et al., 2013; van Hooidonk et al., 2013a).

2.5.2 Storm damage to reefs

One consequence of global climate change will be an increase in the frequency of more damaging Category 4 and 5 tropical cyclones; however the number of tropical storms is not predicted to increase. Such intense Category 4 and 5 tropical cyclones (hurricanes) will significantly damage coral reefs and the communities that depend upon them in the wider Caribbean, where evidence is already available (Salvat and Wilkinson, 2011); whereas in other regions, the evidence is less clear (IPCC, 2013).

Corals have withstood and recovered from tropical cyclones for millennia; a seriously damaged reef will normally recover in 15 to 20 years, provided there are no other disturbances during that period (Salvat and Wilkinson, 2011). However, in the last 100 years the combination of natural and anthropogenic stresses (bleaching, sedimentation, eutrophication, ocean acidification) has reduced the ability of many coral reefs to recover from storm damage by slowing coral recruitment and growth, and reducing fitness (Nyström et al., 2000).

2.5.3 Sea-level rise

If CO_2 emissions continue to increase at current rates (exceeding Representative Concentration Pathways RCP 8.5), sea level is predicted to rise 0.5-1.0 m by 2100 (IPCC, 2013), and the impacts on coral reefs will vary depending on local conditions. Corals may be able to colonise reef flats as sea levels rise, and oceanic reefs will not be adversely affected but may benefit from new space to grow upwards. Rates of reef growth at many sites kept up with rising sea levels after the last ice age (about 20 mm yr^{-1} Dullo, 2005, Montaggioni et al., 2005; and up to 40 mm yr^{-1} Camoin et al., 2012) but reefs are now are accreting more slowly (Perry et al., 2013). However, reefs adjacent to coasts may be affected by increased wave action in lagoons, and flooding of polluted coastal plains will increase erosion of coastal sediments (Adey et al., 1977; Lighty et al., 1978), increase sediment transport (Hopley and Kinsey, 1988), and increase turbidity (Storlazzi et al., 2011). This will reduce the ability of corals and reefs to keep up with rising sea level. Simultaneously, increasing ocean acidification will decrease coral reef accretion. Finally, as sea level rises, some coastal systems may undergo landward retreat, and others will experience coastal squeeze as eroding shorelines approach hard, immobile, structures. These may be either natural or man-made; the latter are increasing by coastal hardening to protect human infrastructure. Coastal squeeze may shrink habitats, affecting the survivability of a variety of organisms (Jackson and McIlvenny, 2011).

2.5.4 Ocean acidification

The first detailed prediction of the potential for increasing ocean acidification to damage coral reefs was made in 1992 at the 7th International Coral Reef Symposium (Buddemeier 1993). Experimental studies confirmed these predictions of damage to coral calcification in the 1990s (Gattuso et al., 1998; Gattuso et al., 1999). The IPCC (2014) report determined that under medium- to high-emission scenarios (RCP4.5, 6.0 and 8.5), ocean acidification poses substantial risks to coral reefs through its effects on the physiology, behaviour, and population dynamics of individual species from phytoplankton to animals (*medium* to *high confidence*, IPCC, 2014). Also the lowering of pH will favour the dissolution of the calcareous matrix of coral reefs. These effects will be additive or synergistic with damage from rising sea-surface temperatures. Further experiments with increased concentrations of CO_2 in seawater have shown decreased calcification rates by corals and other calcium carbonate-secreting organisms (Barker and Elderfield, 2002; Doney et al., 2009; Riebesell et al., 2000; see also Chapter 7). A doubling of current atmospheric CO_2 concentrations reduced calcification by 11 per cent to 37 per cent in many corals (Langdon et al., 2003; Marubini et al., 2003; Langdon and Atkinson, 2005). However, some corals show either limited or no response when provided with elevated nutrients (Holcomb et al., 2010; Chauvin et al., 2011). This suggests that nutrient-enriched corals may use more dissolved inorganic carbon to maintain calcification rates.

Ocean acidification also reduces calcification and skeletal growth in post-settlement and juvenile corals (Albright et al., 2008; Albright et al., 2010; Cohen et al., 2009; Kurihara, 2008; Suwa et al., 2010). Fertilization success during spawning and subsequent settlement of *Acropora palmata* were significantly reduced at increased CO_2 levels (Albright et al., 2010); and larvae of *Acropora digitifera* showed reduced metabolism and suppressed metamorphosis (Nakamura et al., 2011). No effect was observed in *Porites astreoides* larvae (Albright et al., 2008).

Reefs found in naturally acidic waters are poorly cemented, unstable, and fragile (Manzello et al., 2008) and show rapid rates of bioerosion (Eakin, 1996; 2001; Glynn, 1988; Reaka-Kudla et al., 1996). Similarly, in "natural experiments" where coral is reduced or absent around volcanic seeps of CO_2 near Papua New Guinea (Fabricius et al., 2011) and Italy (Rudofo-Metalpa et al., 2011), coral calcification is reduced and species composition changes along the pH gradient. Bioerosion by filamentous eroding algae (Tribollet et al., 2009) and boring sponges (Fang et al. 2013; Wisshak et al., 2012) are enhanced under acidified conditions. Other experiments show there may be declines in the growth of crustose coralline algae (Jokiel et al., 2008; Kuffner et al., 2007).

3 Social and economic considerations.

Economic valuation of coral reefs is a relatively recent process (Cesar, 1996; Cesar et al., 2003) to demonstrate the importance of reef ecosystem services and encourage greater conservation efforts. However, there is a potential critical error in that high-value, short-term economic gains that result from development activities can occur at the expense of longer-term benefits. Economic valuation provides more complete information on the economic consequences of decisions that lead to degradation and loss of natural resources, as well as the short- and long-term costs and benefits of environmental protection. Many studies have assessed the value of ecosystem services provided by coral reefs, at local to global scales. The focus is predominantly on tourism and reef-related fisheries; because these are widely studied and direct-use data are more readily available. It is more difficult to estimate indirect-use values, such as shoreline protection, and most difficult with controversial methods to estimate non-use values, such as cultural, biodiversity and heritage values. The annual net global benefits from coral reefs have been estimated at 29 billion dollars (11.5 billion dollars tourism; 6.8 billion dollars fisheries; 10.7 billion dollars shoreline protection) (Burke et al., 2011a). This emphasizes that tourism and fisheries are especially important as direct money earners for coral-reef communities and their countries. But such an evaluation is for current values and does not take into account all future consequences of changes, such as cultural aspects, community livelihoods, and social and political stability in coral reef communities and their countries, which, if disrupted, will result in other cascading damage. A specific example is the reported value of the GBR to the Australian economy. The total estimated value varies between 4.4 and 15.5 billion dollars, comprising 84 per cent for tourism, 4.6 per cent for other recreational activities, 2.6 per cent for fisheries, and 1.5 per cent for scientific research and management with employment estimated at 69,000 people (Deloitte, 2013; Stoeckl et al. 2014)).

4 Management and conservation.

Calls for increased protection of the marine environment from many organizations and in conventions have specifically addressed the need to protect coral reefs. This includes developing and facilitating the use of diverse approaches and tools, including the ecosystem approach, the elimination of destructive fishing practices, the establishment of marine protected areas (MPAs) consistent with international law and based on scientific information, and the establishment of representative networks and time/area closures. Among the Aichi Biodiversity Targets adopted at the 10th Meeting of the Conference of the Parties to the Convention on Biological Diversity in 2010 was Target 10: "By 2015, the multiple anthropogenic pressures on coral reefs, and other vulnerable ecosystems impacted by climate change or ocean acidification are minimized, so as to maintain their integrity and functioning". The United Nations General Assembly has supported these calls (amongst others) in "The Future We Want" (resolution 66/288) with specific mentions in paragraph 177 and subsequent paragraphs on SIDS.

According to the World Resources Institute, an estimated 2,679 MPAs that coincide with coral reef areas exist worldwide, encompassing approximately 27 per cent of the world's coral reefs (Burke et al., 2011a). Nevertheless global protection of coral reefs is considered by Burke et al. (2011a) to provide effective protection for only 6 per cent of coral reefs, due to shortcomings in planning, management and enforcement of regulations. The benefits of MPAs for achieving targets of conservation of coral reef areas, however, have been reported widely in the scientific literature, in particular when extractive activities are not allowed, as in the no-take areas or marine reserves (Lubchenco et al., 2003; Halpern 2003).

The designs of MPAs range from small units to networks of no-take areas (NTMRs) and large scale marine protected areas (LSMPAs). The first major MPA was the Great Barrier Reef Marine Park in 1975 with 20,679 km² of coral reefs. It now has established no-take areas that protect 33.5 per cent of coral reefs (6,928 km²), to form a network of no-take areas.

Since 2004, ten LSMPAs were established in the Pacific and Indian Oceans in areas within national jurisdiction, and two-thirds of them were declared as marine reserves representing more than 80 per cent of the worldwide MPA coverage (Leenhardt et al., 2013; Table 43.3).

Those areas face major logistical and economic challenges of implementing, managing and monitoring (Leehardt et al., 2103).

Emslie et al. (2015) showed that expanding NTMR networks had clear benefits for fishery target, but not non-target, species. During the study, a cyclone caused widespread degradation, but target species biomass was retained within NTMRs, with greater recovery potential for adjacent areas.

MPAs, even with no-take management, cannot be assured of full protection to reefs. Reefs inside MPAs may still be affected by pollution and sedimentation. In these cases, catchment management has been shown to be effective in promoting reef recovery (many examples in Wilkinson and Brodie, 2011).

Another mechanism targeted at conserving vulnerable species, including those on coral reefs, has been through the Convention on International Trade of Endangered Species and Wild Fauna and Flora (CITES), listing them in Appendices II and III (http://www.cites.org/).

5 Integrated assessment of the status of the habitat.

Reefs in Southeast Asia, the Caribbean and along the East coast of Africa are the most threatened, and this is correlated with high levels of human exploitation of, and dependence on, coral reef resources. In the wider Caribbean, live coral cover has declined by 80 per cent between 1976 and 2001 (Gardner et al., 2003). Further declines following mass coral bleaching linked to climate change occurred in 2005 (Wilkinson and Souter, 2008; Eakin et al., 2010). According to Burke et al. (2011a), coral reefs around Australia were less degraded, although a year later De'ath et al. (2012) reported a loss of 50 per cent of initial coral cover occurred over the 1985-2012 period on the GBR, especially for the central and southern sections where more anthropogenic disturbances occur. In the Central Pacific, far from continents and with low human pressure, reefs are much less threatened and are in better condition and more resilient to natural destructive effects (Salvat et al., 2008; Burke et al., 2011a; Chin et al., 2011). On a regional basis, and based mainly on material from Wilkinson (2008), the condition of reefs is summarized as follows:

5.1 Indian Ocean

During the first half of 1998, the most severe El Niño event ever recorded resulted in the loss of more than 90 per cent of live coral cover throughout large parts of the Indian Ocean. Damage was particularly severe in the Maldives, Chagos Archipelago, Seychelles and Kenya. Prior to 1998, reefs adjacent to large human populations along the coast of East Africa, India and Sri Lanka had already suffered serious damage from excessive and destructive fishing, nutrient pollution, increased sediment input from land and direct development over the reefs, including coral mining.

Reefs on remote islands and in the Red Sea were generally in good health prior to 1998. Since 1998, coral recovery has been minimal in the Persian Gulf and Gulf of Oman, with recovery often reversed by more bleaching. Throughout the Arabian Peninsula region, massive coastal development and dredging to create oil industrial sites and residential and tourist complexes has occurred. Many reefs in the Red Sea continue to be healthy, although COTS (crown-of-thorns starfish) have caused damage, and expanding tourism in the Northern Red Sea is accelerating some coral losses.

Table 43.3 | Large Marine Protected Areas that have been established to include significant areas of coral reefs.

Name of Marine Protected Area	Country	Date	Area km2
Pacific Remote Islands Marine National Monument	United States	2014	2,025,380
Le Parc Naturel de la Mer de Corail (Natural Park of the Coral Sea) (New Caledonia)	France	2014	1,291,000
Cook Islands Marine Park	Cook Islands	2012	1,065,000
Coral Sea Commonwealth Marine Reserve	Australia	2011	989,842
Kermadec Benthic Protection Area	New Zealand	2007	620,500
Chagos Marine Protected Area	United Kingdom[1]	2010	545,000
Phoenix Islands Protected Area	Kiribati	2008	408,250
Papahānaumokuākea (Northwestern Hawaiian Islands)	United States	2006	362,100
Great Barrier Reef Marine Park	Australia	1975	344,400

[1] In its award of 18 March 2015 in the matter of the Chagos Marine Protected Area Arbitration (Mauritius v. United Kingdom), the Arbitral Tribunal established under Annex VII to the United Nations Convention on the Law of the Sea, found, inter alia, that, as a result of undertakings given by the United Kingdom in 1965 and repeated thereafter, Mauritius holds legally binding rights (i) to fish in the waters surrounding the Chagos Archipelago, (ii) to the eventual return of the Chagos Archipelago to Mauritius when no longer needed for defence purposes, and (iii) to the preservation of the benefit of any minerals or oil discovered in or near the Chagos Archipelago pending its eventual return. The Tribunal held that in declaring the Marine Protected Area, the United Kingdom failed to give due regard to these rights and had breached its obligations under the United Nations Convention on the Law of the Sea.

Along the coastline of Eastern Africa, a mix of reef recovery and reef degradation is observed as management efforts are directed at controlling the effects of rapidly growing populations and at involving local communities in coastal management. All States are increasing their networks of marine protected areas and States are improving management capacity and legislation.

Reefs of the southwestern islands in the Indian Ocean continue to recover after devastation in 1998. Some reefs of the Seychelles and Comoros have regained about half or more of their lost coral cover but recovery has been poor on reefs damaged by human activities. Recovery rates in the Seychelles varied, in part, due to factors that have now been shown to increase reef resilience – depth and structural complexity (Graham et al., 2015).

The reef decline in South Asia continues, as large human populations further impact coral reefs, adding to the damage that occurred in 1998. Recovery has been observed in the reefs of the western Maldives, Chagos Archipelago, the Lakshadweep Islands (India) and off northwest Sri Lanka, with seemingly locally extinct corals making major recoveries, e.g., some reefs have gone from less than five per cent coral cover to 70 per cent in 10 years. The 2004 Indian Ocean earthquake and tsunami caused significant reef damage at some sites, but many are recovering. In Sri Lanka, bleaching was reported in 2010, fisheries continue to be the biggest chronic impact, and pollution has increased tremendously in the coastal waters of Colombo. Although fisheries management areas have been declared, lack of enforcement is still hindering effectiveness.

5.2 Southeast and Northeast Asia

The reef areas of Southeast Asia contain the highest concentration of biodiversity and also the largest concentrations of human populations. Overfishing, increasing sedimentation and urban and industrial pollution from rapid economic development are accelerating reef degradation and more than 50 per cent of the region's mangroves have been lost.

Coral reefs in Northeast Asia have shown an overall decline since 2004; most reefs are coming under significant levels of human pressures, as well as bleaching and COTS stress. In China, coastal development and overfishing has destroyed 80 per cent of coral cover over the past 30 years (Hughes et al., 2013). A few reefs with high coral cover remain, such as Dongsha Atoll between Taiwan Province of China and the mainland of China. Increased coral reef monitoring and research, including the establishment of a regional database, is occurring in Japan; Hong Kong, China; Taiwan Province of China; and Hainan Island (China), and the region is stimulating more awareness and cooperation by having held the Asia Pacific Coral Reef Symposium in 2006, 2010 and 2014. Awareness of the need for coral reef conservation is rising rapidly in most countries.

5.3 Australia and Papua New Guinea

Australian reefs continue to be relatively stable due to several management measures. Since 2004, no major bleaching events have occurred, although two significant cyclones have resulted in major damage to some reefs. Particular features are the effective partnerships between coral reef science and management. The future outlook for the GBR is regarded as poor, especially in the southern half of the area, where anthropogenic stresses are strongest. Climate-change impacts are considered to be the greatest long-term threat to the whole GBR system (GBRMPA 2014).

In Papua New Guinea, capacity-building for reef management is being conducted via large NGOs working with local communities. Papua New Guinea still has vast areas of healthy and biologically diverse coral reefs, but human pressures are increasing.

5.4 Wider Pacific

The coral reefs of the Pacific remain the most healthy and intact, compared to reefs elsewhere. Many of these reefs grow on seamounts in deep oceans far removed from land-sourced pollution. Moreover, the human populations are not concentrated as they are in Asia and the Caribbean. In the broader Micronesian region, reefs are recovering well after major coral bleaching in 1998, when, coral mortality was as high as 90 per cent on many reefs around Palau. The Federated States of Micronesia, Marshall Islands, Palau, Guam (United States) and Northern Mariana Islands (United States) seek to conserve 30 per cent of their marine resources by 2020 through the designation of more protected areas (www.themicronesiachallenge.org/).

Climate-related coral bleaching continues to be the greatest threat to the reefs of the southwestern Pacific; human impacts, although growing, are not (yet) resulting in major reef loss on large scales. The University of the South Pacific and the CRISP (Coral Reef Initiatives for the South Pacific) programme (www.crisponline.net) focused on building more capacity for monitoring and conservation, with the Locally Managed Marine Area network developed in Fiji leading the way in the establishment of community-managed MPAs. It is noted that periodically harvested reserves (modelled on the traditional Qoliqoli or rahui system of management) have significantly higher target fish biomass than other fished areas. Outbreaks of COTS have re-appeared in Fiji, starting in the Mamanucas (2006-10), then moving to the Coral Coast and Beqa (2009-12). Currently active outbreaks exist in Taveuni and the lower Lomai Viti Islands. Large reef areas of New Caledonia (France) have gained World Heritage listing in recognition of the large extent and high biodiversity content of the reefs and adjacent ecosystems.

Climate change impacts, tropical cyclones and COTS have also caused major reef damage in the Southeast Pacific (Polynesia). The reefs have remained relatively stable since the 1998 bleaching event, although COTS are still present in some sites, especially in French Polynesia (France). Reef awareness and conservation activities have gradually increased. Many coral reefs surround uninhabited islands; climate-change bleaching and ocean acidification are at present the only major future threats.

Thus, many Pacific reefs are considered to be ideal targets for the creation of "reservoir" protected areas to conserve species threatened with over-exploitation or other human stresses. Kiribati has recognized this with the declaration of the Phoenix Islands Protected Area (PIPA), which is also a World Heritage site.

The United States Pacific islands are regarded as globally important reservoirs of virtually pristine coral reefs. Thus the Northwestern Hawaiian Islands were declared to be the Papahānaumokuākea Marine National Monument and in 2014 more islands were included in the enormous Pacific Remote Islands Marine National Monument. Management is increasing around the main Hawaiian Islands, but overfishing and sediment pollution continue as major threats. The depletion of aquarium species is being addressed through the establishment of industry-recognised MPAs.

Warm water corals are limited to the northern region of New Zealand with the situation in Kermadec Ridge being unique with warm- and cold-water corals present. The warm-water (hermatypic) zooxanthellate stony corals are at or near their southernmost limit at shallow depths around the various Kermadec Islands, with *Pocillopora* and *Tubinaria* genera prevalent. Of the 17 hermatypic species, 16 are found on the Australian Great Barrier Reef; but these corals do not form coral reefs (Brook 1999). Ahermatypic corals without zooxanthellae occur in deeper waters along the ridge, including black, gorgonian, scleractinian, and stylasterid corals.

5.5 The Wider Caribbean

These reefs suffered massive losses from coral diseases since the mid-1980s and more recently during the major climate-related events of 2005, when all regions of the Wider Caribbean were affected by record coral bleaching and tropical cyclone (hurricane) damage.

Reefs of the United States Caribbean are the focus of increased scientific and conservation efforts and results are variable: some improvements but also major coral reef losses are observed. The reefs immediately adjacent to the Florida protected areas (Florida Keys National Marine Sanctuary) are showing minimal recovery, if any, as pollution and excessive tourism threats impede many years of management efforts. More remote reefs, like the Tortugas and Flower Garden Banks, are healthier, but Puerto Rico and the United States Virgin Islands are threatened by overfishing, pollution from the land, and these threats are all compounded by coral bleaching and disease.

Reefs in the Northern Caribbean and Western Atlantic were also severely damaged in 2005 including those under strong conservation efforts. A wide disparity exists in the economic status of the States and territories in the region. Some wealthier territories, such as Bermuda (United Kingdom) and the Cayman Islands (United Kingdom), are applying considerable reef management programmes. Some encouraging signs of coral recovery after major losses in the 1980s and 1990s are found, especially around Jamaica, but unusually frequent and intense tropical cyclones are affecting reef recovery. A ban on using fish traps has been followed by significant increases in fish populations, accompanied by coral cover increases, especially in Bermuda (United Kingdom).

The 2005 coral bleaching event caused major damage in the Lesser Antilles, where coral cover was reduced by about 50 per cent on many reefs. Recovery has been slow or non-existent in reefs under high human pressures. Algal cover has increased and coral diseases have been particularly prevalent since 2005. Most of these small islands depend heavily on their coral reefs for tourism income and fisheries, and this awareness is increasing calls for reef conservation, such as the Caribbean Challenge (http://www.caribbeanchallengeinitiative.org/), as well as local initiatives. Reefs of the Netherlands Antilles harbour some of the highest coral cover seen throughout the wider Caribbean.

Reef status along the Mesoamerican Barrier Reef and Central America has similarly declined, after a long series of losses that started in the 1980s. Bleaching and especially tropical cyclones in 2005 caused considerable destruction around Cozumel (Mexico). The trend is for decreasing coral cover, averaging around 11 per cent since 2004, and some reefs have lost more than 50 per cent coral cover. Major programmes have considerably raised capacity and improved management of MPAs, but sedimentation and overfishing continue to impede reef recovery. While fisheries regulations like the 2009 ban on the take of parrotfish have helped, MPAs in Belize have not been adequately managed, such that the Belize Barrier Reef Reserve System was listed as World Heritage in Danger in 2014.

The main drivers of coral decline in the Southern Tropical Americas are pollution, sedimentation and overfishing. Coastal reefs have been historically affected by sedimentation due to deforestation of the Atlantic forests (Macedo and Maida, 2011). Coral bleaching associated with the El Niño phenomenon is affecting both coastal and oceanic systems, in varying degrees of intensity (Ferreira et al., 2013; Kelmo and Attrill, 2013). Coastal reefs in the region are particularly affected by pollution and sedimentation (Bruce et al., 2012; Silva et al., 2013). Overfishing of key large-bodied herbivorous fish is a worrying trend (Francini-Filho and Moura, 2008; Ferreira et al., 2012). An important threat to coral reefs in Brazil is the invasion and rapid spread of the sun coral *Tubastraea coccinea* and *Tubastraea tagusensis* (Silva et al., 2014). Diseases were first recorded in 2005, and now represent an increasing threat (Francini-Filho et al., 2008). Comparison with reports from earlier surveys indicate dramatic declines in costal reefs during the last 50 years (Ferreira and Maida, 2006), with signs of stability in coral cover (Francini-Filho et al., 2013) or disturbance followed by recovery (Kelmo and Attrill, 2013) in the last two decades. Recent trends include the Brazilian Coral Reef National Action Plan and regulation of fisheries over reef fish species considered as threatened (MMA, 2014).

6 Gaps in scientific knowledge

One long-lasting difficulty with monitoring the state of marine ecosystems is the lack of long-standing databases. Although coral reefs have been monitored for decades within countries in many parts of the world, in other regions monitoring is more recent, or interrupted, or collected with a wide range of methods that preclude standardization. Coral reefs are iconic ecosystems and around the world national governments and voluntary organizations have been engaged in coral reef monitoring. The International Coral Reef Initiative (ICRI)[1] has specifically assisted many countries with assessment and monitoring of their coral reefs by supporting the Global Coral Reef Monitoring Network. Other networks for monitoring, awareness and protection are also organized by NGOs; the largest is Reef Check, operating in 90 countries since 1998 (www.reefcheck.org).

A study published by Wilson et al. (2010) canvassed the opinions of 33 experts to identify crucial knowledge gaps in current understanding of climate-change impacts on coral reef fishes. Out of 153 gaps reported by the experts, 42 per cent related to habitat associations and community dynamics of fish, reflecting the established effects and immediate concerns pertinent to climate change-induced coral loss and habitat degradation (i.e., how does coral mortality influence the capacity of a wide range of fish populations to persist?).

Existing maps of the spatial distribution of coral reefs largely are based on satellite images and aerial photographs. Submerged coral reefs (also known as mesophotic coral reefs) that occur below a water depth of around 30 m cannot easily be detected using satellites or aerial photography, even in clear waters. Consequently, their spatial distribution and even their existence are unknown in most reef provinces. For this reason, deeper reefs have been underestimated in analyses of the available area of coral habitat and are not included in assessments for conservation measures, despite recent evidence that these areas may be significant (Locker et al., 2010; Bridge et al., 2013). A recent study suggests that the area of submerged reefs in the GBR may be equal to that of near-surface reefs (Harris et al., 2013). Understanding the extent of submerged reefs is therefore important, because they can support large and diverse coral communities (Bridge et al., 2012) and hence may provide vital refugia for corals and associated species from a range of environmental disturbances (Riegl and Piller, 2003; Bongaerts et al., 2010).

The scientific consensus is that threats associated with climate change (bleaching, ocean acidification, stronger storms etc.) pose the greatest threat to the medium- to long-term existence of coral reefs around the world. What is unknown is whether reefs can and will respond to these threats with greater resilience. Reefs contain very high biodiversity and have progressed through major climate change events in the geological past; how will they be able to respond in the next decades to rapid climate changes? There are early indications that some corals can adapt to warmer temperatures and grow in more acidic water, but it is predicted that many corals and other reef organisms do not have that capacity. The adaptation potentials of coral reef organisms are areas for more targeted research which will significantly increase our ability to reliably predict how reefs will fare into the future.

7 Final remarks

There are strong economic, cultural, biodiversity and natural-heritage reasons to conserve tropical and sub-tropical coral reefs and to ensure that their goods and ecosystem services continue to be provided to user communities and the world at large. There are three levels at which these pressures come together and emphasise the knowledge and capacity-building gaps in this field:

7.1 At the level of the local community

Coral reefs will not be able to continue to provide the goods and ecosystem services on which local communities have relied for generations, unless:

(a) The fishing techniques that are adopted are focused on maintaining a sustainable fishery, and destructive fishing practices (such as dynamiting) cease;
(b) Populations of breeding fish and invertebrates, including spawning aggregations, are conserved;
(c) The pollution of coastal waters by harmful substances (heavy metals and persistent organic pollutants) is prevented, and amounts of inputs of sediment and nutrients are kept at levels that do not damage the reefs (see chapter 20);
(d) Any development in coastal areas is kept to levels and forms that are consistent with the continued health of the reefs.

Without the active involvement of the coastal communities that have the necessary knowledge and skills, there are likely to be serious difficulties in achieving these goals.

7.2 At the national level

In many of the countries that are the guardians of tropical and sub-tropical coral reefs, there are significant gaps in the knowledge and skills needed for the relevant authorities to play their part in sustaining the reefs. In particular, where marine protected areas are an appropriate method of delivering some of the goals, there are gaps at both national and local levels in capacities for the scientific identification of such areas, for the development of management plans for them, and in enforcing the regulations that may be required.

7.3 At regional and supra-regional levels

---

[1] The Global Coral Reef Monitoring Network (http://www.icriforum.org/gcrmn) is assisted by the International Coral Reef Initiative (ICRI), an informal partnership between 34 States and a range of organizations, governmental and non-governmental, that also establishes committees to deal with several coral reef conservation- and management-related issues.

The conservation of specific local areas can frequently only be achieved as part of a network of such areas, since the ocean is a dynamic ecosystem and biota are commonly mobile in their early life-stages. Given the interactions among many forms of human activity in the ocean and between them and local ecosystems, management methods that do not take account of those interactions will be ineffective in delivering a sustainable future for tropical and sub-tropical reefs. Integrated management methods on a large scale can only be achieved where there is a widespread social understanding and knowledge of the pressures (such as climate change, acidification, fisheries, seabed mining (see Van Dover et al., 2012; Boschen et al., 2013), pollution and coastal development), the scales on which they operate and their interactions. All this implies that, without efforts to promote understanding of the ocean and without cooperation at the appropriate national, regional and (in some cases) global level between the relevant regulatory authorities, the pressures described in this chapter will persistently undermine the continued delivery by tropical and sub-tropical coral reefs of the goods and ecosystem services on which local communities, countries and the world have been relying.

References

Adjeroud, M., Michonneau, F., Edmunds, P.J., Chancerelle, Y., Lison de Loma, T., Penin, L., Thibaut, L., Vidal-Dupiol, J., Salvat, B., Galzin, R. (2009). Recurrent disturbances, recovery trajectories and resilience of coral assemblages on a South Pacific reef. *Coral Reefs* 28: 775-780.

Amado-Filho, G.M., Moura, R.L., Bastos, A.C., Salgado, L.T., Sumida, P.Y., Guth, A.Z., Francini-Filho R.B., Pereira-Filho, G.H., Abrantes, D.P., Brasileiro, P.S., Bahia, R.G., Leal, R.N., Kaufman, L., Kleypas, J.A., Farina, M. and Thompson, F.L. (2012). Rhodolith beds are major $CaCO_3$ bio-factories in the tropical South West Atlantic. *PloS ONE*, 7(4), e35171.

Anthony, K.R.N., Marshall, P.A., Abdullah, A., Beeden, R., Bergh, C., Black, R., Eakin, C.M., Game, E.T., Gooch, M., Graham, N.A.J., Green, A., Heron, S.F., van Hooidonk, R., Knowland, C., Mangubhai, S., Marshall, N., Maynard, J.A., McGinnity, P., McLeod, E., Mumby, P.J., Nyström, M., Obura, D., Oliver, J., Possingham, H.P., Pressey, R.L., Rowlands, G.P., Tamelander, J., Wachenfeld, D. and Wear, S. (2014). Operationalising resilience for adaptive coral reef management under global environmental change. *Global Change Biology*. doi:10.1111/gbc.1270.

Aronson, R. and Precht, W. (2001). White-Band Disease and the changing face of Caribbean coral reefs. In: Porter, J.W. (ed.) The ecology and etiology of newly emerging marine diseases. *Hydrobiologia* 460: 25-38.

Bellwood, D.R., Hughes, T.P., Folke, C. and Nystrom M., (2004). Confronting the coral reef crisis. *Nature*, 429: 827-833.

Bengtson Nash, S.M., McMahon, K., Eaglesham, G., Muller, J.F. (2005). Application of a novel phytotoxicity assay for the detection of herbicides in Hervey Bay and the Great Sandy Straits. *Marine Pollution Bulletin 51*, 351-360.

Bishop, R.C., Chapman, D.J., Kanninen, B.J., Krosnick, J.A., Leeworthy, B. and Meade, N.F. (2011). Total Economic Value for Protecting and Restoring Hawaiian Coral Reef Ecosystems: Final Report. Silver Spring, MD: NOAA Office of National Marine Sanctuaries, Office of Response and Restoration, and Coral Reef Conservation Program. NOAA Technical Memorandum CRCP 16. 406 pp.

Bongaerts, P., Ridgway, T., Sampayo, E., Hoegh-Guldberg, O. (2010). Assessing the 'deep reef refugia' hypothesis: focus on Caribbean reefs. *Coral Reefs*, 29: 309–327.

Bouchet, P. (2006). The magnitude of marine biodiversity. In: *The Exploration of Marine Biodiversity, Scientific and Technological challenges*, Duarte, C.M., ed., Fundación BBVA, 33-64.

Bradbury, R.H., Seymour, R.M. (2009). Coral reef science and the new commons. *Coral Reefs*, 28:831–837.

Briand, M.J., Letourneur, Y., Bonnet, X., Wafo, E., Fauvel, T., Brischoux, F., Guillou, G., Bustamante, P. (2014). Spatial variability of metallic and organic contamination of anguilliform fish in New Caledonia. *Environmenal Science and Pollution Research 21*, 4576-4591.

Bridge, T., Fabricius, K., Bongaerts, P., Wallace, C., Muir, P., Done, T., and Webster, J. (2012). Diversity of Scleractinia and Octocorallia in the mesophotic zone of the Great Barrier Reef, Australia. *Coral Reefs*, 31: 179–189.

Bridge, T.C.L., Hughes, T.P., Guinotte, J.M. and Bongaerts, P. (2013). Call to protect all coral reefs. *Nature Climate Change* 3, 528-530.

Brook, F.J. (1999). The coastal scleractinian coral fauna of the Kermadec Islands, southwestern Pacific Ocean. *Journal of the Royal Society of New Zealand*, 29: 4, 435-460.

Bruce, T., Meirelles, P.M., Garcia, G., Paranhos, R., Rezende, C.E., de Moura, R.L., Francini-Filho R.B., Coni E.O.C., Vasconcelos A.T., Amado-Filho G., Hatay M., Schmieder R., Edwards R., Dinsdale E. and Thompson, F.L. (2012). Abrolhos Bank reef health evaluated by means of water quality, microbial diversity, benthic cover, and fish biomass data. *PloS one*, 7(6), e36687.

Bruckner A. and Hill, R. (2009). Ten years of change to coral communities off Mona and Desecheo Islands, Puerto Rico, from disease and bleaching. *Diseases of Aquatic Organisms* 87:19–31.

Bruno, J., Petes, L., Harvell, C. and Hettinger, A. (2003). Nutrient enrichment can increase the severity of coral diseases. *Ecology Letters* 6(12):1056-1061.

Bruno, J.F. and Selig, E.R. (2007). Regional decline of coral cover in the indo-pacific: timing, extent, and subregional comparisons. *PLoS ONE*, 2 (8), e711 on www.plosone.org.

Bryant, D., Burke, L., McManus, J. and Spalding, M. (1998). *Reefs at Risk: A map-bsed indicator of threats to the world's coral reefs.* (World Resources Institute, Washington, DC, International Center for Living Aquatic Resources Management, UNEP World Conservation Monitoring Centre and United Nations Environment Programme), pp. 1-60.

Buddemeier, R.W. (1993). Corals, climate and conservation. Plenary Address - Proc 7th *International Coral Reef Symposium* 1: 3-10.

Burke, L., Selig, E. and Spalding, M. (2002). *Reefs at Risk in Southeast Asia.* World Resources Institute, Washington, DC 2002: pp. 72.

Burke, L. and Maidens, J. (2004). *Reefs at Risk in the Caribbean.* World Resources Institute, Washington D.C. pp. 80.

Burke, L., Reytar, K., Spalding, M. and Perry. A. (2011a). *Reefs at Risk Revisited.* World Resources Institute, Washington, DC: pp. 114. http://www.wri.org/sites/default/files/pdf/reefs_at_risk_revisited.pdf

Burke, L., Reytar, K., Spalding, M. and Perry. A. (2011b). *Reefs at Risk Revisited in the Coral Triangle.* World Resources Institute, Washington, DC: pp. 73.

Burge, C.A., Eakin, C.M., Friedman, C.S., Froelich, B., Hershberger, P.K., Hofmann, E.E., Petes, L.E., Prager, K.C., Weil, E., Willis, B.L., Ford, S.E. and Harvell, C.D. (2014). Climate Change Influences on Marine Infectious Diseases: Implications for Management and Society, *Annual Review of Marine Science* 6:249–77.

Cairns, S.D. (2012). The Marine Fauna of New Zealand: New Zealand Primnoidae (Anthozoa: Alcyonacea). Part 1. Genera Narella, Narelloides, Metanarella, Calyptrophora, and Helicoprimnoa. NIWA Biodiversity Memoir 126: 71 p.

Carrigan, A.D. and Puotinen, M. (2014). Tropical cyclone cooling combats region-wide coral bleaching. *Global Change Biology* doi: 10.1111/gcb.12541.

Carpenter, K.E., Abrar, M., Aeby, G., and 36 other authors (2008). One-third of reef-building corals face elevated extinction risk from climate change and local impacts. *Science* 321: 560-563.

CBD (1992). Convention on Biological Diversity, United Nations, *Treaty Series*, vol. 1760, p. 79.

Cesar, H. (1996). *Economic analysis of Indonesian coral reefs.* World Bank Environment Department, Washington DC, USA., p. 103.

Cesar, H., Burke, L. and Pet-Soede, L. (2003). *The economics of worldwide coral reef degradation.* Cesar Environmental Economics Consulting and WWF-Netherlands, Arnhem and Zeist, the Netherlands. [online] URL: http://pdf.wri.org/ cesardegradationreport100203.pdf.

Chauvin, A., Denis, V. and Cuet, P. (2011). Is the response of coral calcification to seawater acidification related to nutrient loading? *Coral Reefs* 30:911–923 DOI 10.1007/s00338-011-0786-7.

Cinner, J.E., McClanahan, T.R., Graham, N.A.J., Daw, T.M., Maina, J., Stead, S.M., Wamukota, A., Brown, K. and Bodin, O. (2012). Vulnerability of coastal communities to key impacts of climate change on coral reef fisheries. *Global Environmen-

tal Change, 22, 12-20.

Cooper, E., Burke, L. and Bood, N. (2008). *Coastal Capital: Economic Contribution of Coral Reefs and Mangroves to Belize*. Washington DC: World Resources Institute.

Costanza, R., de Groot, R., Sutton, P., van der Ploeg, S., Anderson, S.J., Kubiszewski, I., Farber, S. and Turner, R.K. (2014). Changes in the global value of ecosystem services. *Global Environmental Change, Volume 26,* May 2014, Pages 152-158.

De'ath, G. and Fabricius, K. (2010). Water quality as a regional driver of coral biodiversity and macroalgae on the Great Barrier Reef. *Ecological Applications 20*, 840-850.

De'ath, G., Fabricius, K.E., Sweatman, H. and Puotinen, M. (2012). The 27–year decline of coral cover on the Great Barrier Reef and its causes. *PNAS*, 109: no. 44, p.17995–17999.

Deloitte Access Economics (2013). *Economic contribution of the Great Barrier Reef*. Great Barrier Reef Marine Park Authority, Townsville. http://www.gbrmpa.gov.au/__data/assets/pdf_file/0006/66417/Economic-contribution-of-the-Great-Barrier-Reef-2013.pdf

De Mitcheson, Y.S., Cornish, A., Domeier, M., Colin, P.L., Russell, M. and Lindeman, K.C. (2008). A global baseline for spawning aggregations of reef fishes. *Conservation Biology*, 22(5), 1233-1244.

Done T.J. (1992). Effects of tropical cyclone waves on ecological and geomorphological structures on the great barrier reef. *Continental Shelf Research*, 12, 859.

Donner, S.D. (2009). Coping with Commitment: Projected Thermal Stress on Coral Reefs under Different Future Scenarios. *PLoS ONE* 4(6): e5712. DOI: 10.1371/journal.pone.0005712.

Donner, S.D., Skirving, W.J., Little, C.M., Oppenheimer, M. and Hoegh-Guldberg, O. (2005). Global assessment of coral bleaching and required rates of adaptation under climate change. *Global Change Biology*, 11: 2251-2265.

Eakin, C.M. (2014). Lamarck was partially right - and that is good for corals. *Science* 344, 798; DOI: 10.1126/science.1254136.

Edwards, C.B., Friedlander, A.M., Green, A.G., Hardt, M.J., Sala, E., Sweatman, H.P. and Smith, J.E. et al (2014). Global assessment of the status of coral reef herbivorous fishes: evidence for fishing effects. *Proceedings of the Royal Society B: Biological Sciences*, 281(1774), 20131835.

Erftemeijer, P.L., Riegl, B., Hoeksema, B.W. and Todd, P.A. (2012). Environmental impacts of dredging and other sediment disturbances on corals: a review. *Marine Pollution Bulletin* 64, 1737-1765.

Fabricius, K. (2005). Effects of terrestrial runoff on the ecology of corals and coral reefs: review and synthesis. *Marine Pollution Bulletin* 40, 125–146.

Fabricius, K.E., De'ath, G., Puotinen, M.L., Done, T., Cooper, T.F. and Burgess, S.C. (2008). Disturbance gradients on inshore and offshore coral reefs caused by a severe tropical cyclone. *Limnology and Oceanography*, 53, 690–704.

Fabricius, K.E., Okaji, K. and De'ath, G. (2010). Three lines of evidence to link outbreaks of the crown-of-thorns seastar Acanthaster planci to the release of larval food limitation. *Coral Reefs* 29, 593-605.

Fabricius, K.E., Langdon, C., Uthicke, S., Humphrey, C., Noonan, S., De'ath, G., Okazaki, R., Muehllehner, N., Glas, M.S. and Lough, J.M. (2011). Losers and winners in coral reefs acclimatized to elevated carbon dioxide concentrations. *Nature Climate Change* 1, 165–169 (2011) doi:10.1038/nclimate1122.

Fabricius, K.E., Cooper, T.F., Humphrey, C., Uthicke, S., De'ath, G., Davidson, J., LeGrand, H., Thompson, A. and Schaffelke, B. (2012). A bioindicator system for water quality on inshore coral reefs of the Great Barrier Reef. *Marine Pollution Bulletin* 65, 320-332.

Fang, J.K.H., Mello-Athayde, M.A., Schönberg, C.H.L., Kline, D.I., Hoegh-Guldberg, O. and Dove, S. (2013). Sponge biomass and bioerosion rates increase under ocean warming and acidification. *Global Change Biology*, 19, 3581-3591. Doi: 10.1111/gcb.12334.

Ferreira, B.P., Floeter S.R., Rocha, L.A., Ferreira, C.E.L., Francini-Filho, R.B., Moura, R.L., Gaspar, A.L. and Feitosa, C. (2012). *Scarus trispinosus*. In: IUCN Red List of Threatened Species. Version 2014.2.

Ferreira, B.P., Costa, M.B.S.F., Coxey, M.S., Gaspar, A.L.B., Veleda, D. and Araujo, M. (2013). The effects of sea surface temperature anomalies on oceanic coral reef systems in the southwestern tropical Atlantic. *Coral reefs*, 32, 441-454.

Ferreira, B.P. and Maida, M. (2006). Monitoring Brazilian Coral Reefs: status and perspectives. *Biodiversity Series* 18, Ministry of Environment, Brasília, Brazil.

Flores. F., Collier, C.J., Mercurio, P. and Negri, A.P. (2013) Phytotoxicity of four photosystem II herbicides to tropical seagrasses. *PLoS ONE* 8: e75798.

Francini-Filho, R.B. and de Moura, R.L. (2008). Dynamics of fish assemblages on coral reefs subjected to different management regimes in the Abrolhos Bank, eastern Brazil. *Aquatic Conservation: Marine and Freshwater Ecosystems*, 18(7), 1166-1179.

Francini-Filho, R.B., Moura, R.L., Thompson, F.L., Reis, R.M., Kaufman, L., Kikuchi, R.K. and Leão, Z.M. (2008). Diseases leading to accelerated decline of reef corals in the largest South Atlantic reef complex (Abrolhos Bank, eastern Brazil). *Marine Pollution Bulletin*, 56(5), 1008-1014.

Francini-Filho, R.B., Coni, E.O., Meirelles, P.M., Amado-Filho, G.M., Thompson, F.L., Pereira-Filho, G.H., Bastos A.C., Abrantes, D.P., Ferreira, C.M., Gibran, F.Z., Güth, A.Z., Sumida, P.Y.G., Oliveira, N.L., Kaufman, L., Minte-Vera, C.V. and Moura, R.L. (2013). Dynamics of coral reef benthic assemblages of the Abrolhos Bank, Eastern Brazil: inferences on natural and anthropogenic drivers. *PloS ONE*, 8(1), e54260.

Gabrié, C., Duflos, M., Dupre, C., Chenet, A. and Clua, E. (2011). *Conservation, management, and development of coral reefs in the Pacific: building of results of six years of research, collaboration and education.* Secretariat of the Pacific Community, Noumea. 166 pp. ISBN: 978-982-00-0507-5.

Gattuso, J.P., Frankignoulle, M., Bourge, I., Romaine, S. and Buddemeier, R.W. (1998). Effect of calcium carbonate saturation on coral calcification. *Global Planetary Change* 18: 37-46.

Gattuso, J.-P., Allemand, D. and Frankignoulle, M. (1999). Photosynthesis and calcification at cellular, organismal and community levels in coral reefs: a review on interactions and control by carbonatechemistry. *American Zoologist* 39:160-183.

Gao, Y., Fang, J., Zhang, J., Ren, L., Mao, Y., Li, B., Zhang, M., Liu, D. and Du, M. (2011). The impact of the herbicide atrazine on growth and photosynthesis of seagrass, *Zostera marina*, seedlings. *Marine Pollution Bulletin* 62, 1628-1631.

Gardner, T.A., Côté, I.M., Gill, J.A., Grant, A. and Watkinson, A.R., (2003). Long-term region-wide declines in Caribbean corals. *Science*, 301, 958-960.

Garrison, V.H., Shinn, E.A., Foreman, W.T., Griffin, D.W., Holmes, C.W., Kellogg, C.A., Majewski, M.S., Richardson, L.L., Ritchie, K.B. and Smith, G.W. (2003). African and Asian dust: from desert doils to coral reefs. *Bioscience 53*, 469-480.

GBRMPA (Great Barrier Reef Marine Park Authority) (2014). *The Great Barrier Reef Outlook report* (2014). Great Barrier Reef Marine Park Authority, Townsville Australia http://elibrary.gbrmpa.gov.au/jspui/handle/11017/2856.

Gilmour, J., Smith, L.D., Heyward, A.J., Baird, A.H. and Pratchett, M.S. (2013). Recovery of an isolated coral reef system following severe disturbance. *Science* 340: 69-71.

Gladfelter, W. (1982). White-Band Disease in *Acropora palmate* - Implications for the structure and growth of shallow reefs. *Bulletin of Marine Science* 32: 639–643.

Graham, N.A.J., Jennings, S., MacNeil, M.A., Mouillot, D. and Wilson, S.K. (2015). Predicting climate-driven regime shifts versus rebound potential in coral reefs.

Nature 518, 94–97.

Green, E. and Bruckner, A. (2000). The significance of coral disease epizootiology for coral reef conservation. *Biological Conservation* 96: 347-361.

Groombridge, B. and Jenkins, M. (2002). *World Atlas of Biodiversity*. California University Press, Berkley.

Grottoli, A.G., Warner, M.E., Levas, S.J., Aschaffenburg, M., Schoepf, V., McGinley, M., Baumann, J. and Matsui, Y. (2014). The cumulative impact of annual coral bleaching turns some coral species winners into losers. *Global Change Biology* 10.1111/gcb.12658. http://onlinelibrary.wiley.com/doi/10.1111/gcb.12658/abstract

Guest J.R., Baird A.H., Maynard J.A., Muttaqin E., Edwards A.J., Campbell S.J., Yendall K., Affendi Y.A. and Chou L.M. (2012). Contrasting patterns of coral bleaching susceptibility in 2010 suggest an adaptative response to thermal stress. *Plus One*, 7, 3: 1-8.

Harmelin-Vivien, M.L. (1994). The effects of storms and cyclones on coral reefs: A Review *J Coast Res Spec Issue* 12:211–231.

Harvell D., Jordan-Dahlgren, E., Merkel, S., Rosenberg, E., Raymundo, L., Smith, G., Weil, E. and Willis, B. (2007). Coral Disease, Environmental Drivers, and the Balance between Coral and Microbial Associates. *Oceanography*, Vol.20, No.1 prepared by the Coral Disease Working Group of the Global Environmental Facility Coral Reef Targeted Research Programme.

Haynes, D., Ralph, P., Prange, J. and Dennison, W.C. (2000). The impact of the herbicide diuron on photosynthesis in three species of tropical seagrass. *Marine Pollution Bulletin 41*, 288-293.

Harris, P.T., Bridge, T.C.L., Beaman, R., Webster, J., Nichol, S. and Brooke, B. (2013). Submerged banks in the Great Barrier Reef, Australia, greatly increase available coral reef habitat. *ICES Journal of Marine Science* 70, 284-293.

Heron, S.F., Willis, B.L., Skirving, W.J., Eakin, M.C., Page, C.A. and Miller, I.R. (2010). Summer hot snaps and winter conditions: modelling white syndrome outbreaks on Great Barrier Reef Corals. *PloS ONE*. doi:10.1371/journal.pone.0012210

Hoegh-Guldberg, O., Mumby, P.J., Hooten, A.J., Steneck, R.S., Greenfield, P., Gomez, E., Harvell, C.D., Sale, P.F., Edwards, A.J., Caldeira, K., Knowlton, N., Eakin, C.M., Iglesias-Prieto, R., Muthiga, N., Bradbury, R.H., Dubi, A. and Hatziolos, M.E. (2007). Coral reefs under rapid climate change and ocean acidification. *Science* 318:1737-1742. http://dx.doi. org/10.1126/science.1152509.

Hoegh-Guldberg, O. (2014). Coral reefs in the anthropocene: persistence or the end of the line? *Geological Society Special Publication, 395* 1: 167-183. doi:10.1144/SP395.17

Hughes, T. (1994). Catastrophes, phase-shifts, and large-scale degradation of a Caribbean coral reef. *Science* 265, 1547-1551.

Hughes, T.P., Huang, H., Young, M. (2013). The Wicked Problem of China's Disappearing Coral Reefs. *Conservation Biology* 27, 261–269.

Hughes T., Keller, B, Jackson, J. and Boyle, M. (1985). Mass mortality of the echinoid *Diadema antillarum Philippi* in Jamaica. *Bulletin of Marine Science* 36: 377-384.

Hughes, T.P., Rodrigues, M.J., Bellwood, D.R., Ceccarelli, D., Hoegh-Guldberg, O., McCook, L., Moltschaniwskyj, N., Pratchett, M.S., Steneck, R.S., Willis, B. (2007). Phase shifts, herbivory, and the resilience of coral reefs to climate change. *Current Biology* 17, 360–365, 2007.

Hughes, T.P., Day, J.C. and Jon Brodie, J. (2015). Securing the future of the Great Barrier Reef. *Nature Climate Change*. doi:10.1038/nclimate2604.

IPCC, (2013). Summary for Policymakers. In: *Climate Change 2013: The Physical Science Basis. Contribution of Working Group I to the Fifth Assessment Report of the Intergovernmental Panel on Climate Change* [Stocker, T.F., D. Qin, G.-K. Plattner, M. Tignor, S.K. Allen, J. Boschung, A. Nauels, Y. Xia, V. Bex P.M. Midgley (eds.)]. Cambridge University Press, Cambridge, United Kingdom and New York, NY, USA.

IPCC, (2014). *Climate Change 2014: Impacts, Adaptation, and Vulnerability. Contribution of Working Group II to the Fifth Assessment Report of the Intergovernmental Panel on Climate Change* [Field, C.B., V.R. Barros, D.J. Dokken, K.J. Mach, M.D. Mastrandrea, T.E. Bilir, M. Chatterjee, K.L. Ebi, Y.O. Estrada, R.C. Genova, B. Girma, E.S. Kissel, A.N. Levy, S. MacCracken P.R. Mastrandrea, and L.L. White (eds.)]. Cambridge University Press, Cambridge, United Kingdom and New York, NY, USA.

Jackson, J.B.C., Kirby, M.X., Berger, W.H., Bjorndal, K.A., Botsford, L.W., Bourque, B.J., Bradbury, R.H., Cooke, R., Erlandson, J., Estes, J.A. et al. (2001). Historical overfishing and the recent collapse of coastal ecosystems. *Science* 293, 629-637.

Jackson, J.B.C., Donovan, M.K., Cramer, K.L. and Lam, V.V. (eds.) (2014). *Status and Trends of Caribbean Coral Reefs: 1970-2012*. Global Coral Reef Monitoring Network, IUCN, Gland, Switzerland.

Jennings, S. and Polunin, N.V.C. (1996). Impacts of fishing on tropical reef ecosystems. *Ambio* 25: 44–49.

Johannes, R.E. (1981). *Words of the lagoon: fishing and marine lore in the Palau District of Micronesia*. University of California Press, Berkeley, California, USA.

Johannes, R.E., and Riepen, M., (1995). *Environmental, economic, and social implications of the live fish trade in Asia and the Western Pacific*. The Nature Conservancy, Hawaii. pp. 82.

Johns, G.M., Leeworthy, V.R., Bell, F.W. and Bonn, M.A. (2001). Socioeconomic Study of Reefs in Southeast Florida. Hazen and Sawyer, Final report for Broward, Palm Beach, Miami-Dade and Monroe Counties, Florida Fish and Wildlife Conservation Commission and National Oceanic and Atmospheric Administration.

Jones, R.J. and Kerswell, A.P. (2003). Phytotoxicology of photosystem II., PSII. herbicides to coral. *Marine Ecology Progress Series 251*, 153-167.

Kayal M., Vercelloni J., Lison de Loma T., Bosserelle P., Chancerelle Y., Geoffroy S., Stievenart C., Michonneau F., Penin L., Planes S. and Adjeroud M., (2012). Predator crown-of-thorns starfish (*Acanthaster planci*) outbreak, mass mortality of corals, and cascading effects on reef fish and benthic communities. *PLoSOne*, DOI: 10.1371/journal.pone.0047363.

Kelmo, F. and Attrill, M.J. (2013). Severe impact and subsequent recovery of a coral assemblage following the 1997–8 El Niño event: a 17-year study from Bahia, Brazil. *PloS one*, 8(5), e65073.

Kennedy, K., Schroeder, T., Shaw, M., Haynes, D., Lewis, S., Bentley, C., Paxman, C., Carter, S., Brando, V.E., Bartkow, M., et al. (2012). Long term monitoring of photosystem II herbicides - Correlation with remotely sensed freshwater extent to monitor changes in the quality of water entering the Great Barrier Reef, Australia. *Marine Pollution Bulletin 65*, 292-305.

Kittinger, J.N., Finkbeiner, E.M., Glazier, E.W. and Crowder, L.B. (2012). Human dimensions of coral reef social-ecological systems. *Ecology and Society* 17(4): 17.

Kline D.S. Vollmer. (2011). *White Band Disease (type I) of Endangered Caribbean Acroporid Corals is caused by Pathogenic Bacteria*. Nature: Scientific Reports 1:doi:10.1038/srep00007.

Kroon, F.J., Kuhnert, P.M., Henderson, B.L., Wilkinson, S.N., Kinsey-Henderson, A., Abbott, B., Brodie, J.E. and Turner, R.D.R. (2012). River loads of suspended solids, nitrogen, phosphorus and herbicides delivered to the Great Barrier Reef lagoon. *Marine Pollution Bulletin* 65, 167-181.

Leenhardt, P., Cazalet, B., Salvat, B., Claudet, J., Feral, F (2013). The rise of large-scale marine protected areas: Conservation or geopolitics? *Ocean and Coastal Management*, 85: 112-118.

Lewis, S.E., Brodie, J.E., Bainbridge, Z.T., Rohde, K.W., Davis, A.M., Masters, B.L., Maughan, M., Devlin, M.J., Mueller, J.F. and Schaffelke, B. (2009). Herbicides: A new threat to the Great Barrier Reef. *Environmental Pollution* 157, 2470-2484.

Lewis, S.E., Schaffelke, B., Shaw, M., Bainbridge, Z.T., Rohde, K.W., Kennedy, K., Davis, A.M., Masters, B.L., Devlin, M.J., Mueller, J.F., et al. (2012). Assessing the additive risks of PSII herbicide exposure to the Great Barrier Reef. *Marine Pollution Bulletin* 65, 280-291.

Locker, S., Armstrong, R., Battista, T., Rooney, J., Sherman, C., and Zawada, D. (2010). Geomorphology of mesophotic coral ecosystems: current perspectives on morphology, distribution, and mapping strategies. *Coral Reefs*, 29: 329–345.

Logan CA. Dunne J.P., Eakin C.M. and Donner S.D. (2013). Incorporating adaptive responses into future projections of coral bleaching. *Global Change Biology*, vol. 20, doi: 10.1111/gcb.12390.

Magnusson, M., Heimann, K. and Negri, A.P. (2008). Comparative effects of herbicides on photosynthesis and growth of tropical estuarine microalgae. *Marine Pollution Bulletin 56*, 1545-1552.

Manzello, D.P., Brandt M., Smith T.B., Lirman D., Hendee J.C. and Nemeth R.S. (2007). Hurricanes benefit bleached corals. *Proceedings of the National Academy of Sciences* 104:12035-12039.

McClanahan, T.R., Ateweberhan, M., Graham, N.A.J., Wilson, S.K., Ruiz Sebastián, C., Guillaume, M.M.M. and Bruggemann, J.H. (2007). Western Indian Ocean coral communities: bleaching responses and susceptibility to extinction. *Marine Ecology Progress Series*, 337: 1-13.

McClanahan, T.R., Hicks, C.C and Darling, E.S. (2008). Malthusian overfishing and efforts to overcome it on Kenyan coral reefs. *Ecological Applications*, 18: 1516-1529.

McClanahan, T.R., Weil, E. and Maina, J. (2009). Strong relationship between coral bleaching and growth anomalies in massive *Porites*. *Global Change Biology*, 15: 1804-1816.

McClanahan, T.R., Graham, N.A.J., MacNeil, M.A., Muthiga, N.A., Cinner, J.E., Bruggemann, J.H. and Wilson, S.K. (2011). Critical thresholds and tangible targets for ecosystem-based management of coral reef fisheries. *Proceedings of the National Academy of Sciences of the United States of America*, 108: 17230–17233.

Miller J., Waara, R., Muller, E. and Rogers, C. (2006). Coral bleaching and disease combine to cause extensive mortality on reefs of the US Virgin Islands. *Coral Reefs* 25:418

Miller, M., Bourque A. and Bohnsack J. (2002). An analysis of the loss of acroporid corals at Looe Key, Florida, USA: 1983-2000. *Coral Reefs* 21:179- 182.

Muller E., Rogers, C., Spitzack, A. and van Woesik, R. (2008). Bleaching increases the likelihood of disease on *Acropora palmata* (Lamarck) at Hawksnest Bay, St. John, US Virgin Islands; *Coral Reefs* 27:191-195.

Nagoya (2010), Nagoya Protocol on Access to Genetic Resources and the Fair and Equitable Sharing of Benefits Arising from their Utilization to the Convention on Biological Diversity, UNEP/CBD/COP/DEC/X/1.

Negri, A., Vollhardt, C., Humphrey, C., Heyward, A., Jones, R., Eaglesham, G. and Fabricius, K. (2005). Effects of the herbicide diuron on the early life history stages of coral. *Marine Pollution Bulletin 51*, 370.

Negri, A.P., Flores, F., Röthig, T., Uthicke, S. (2011). Herbicides increase the vulnerability of corals to rising sea surface temperature. *Limnology Oceanography 56*, 471-485.

Newton, K., Cote, I.M., Pilling, G.M., Jennings, S. and Dulvy N.K. (2007). Current and future sustainability of island coral reef fisheries. *Current Biology* 17, 655–658.

Nugues M., G. Smith, R. van Hooidonk, M. Seabra R. Bak. (2004). Algal contact as a trigger for coral disease. *Ecology Letters* 7:919–923.

Nyström, M., Folke, C. and Moberg, F. (2000). Coral reef disturbance and resilience in a human-dominated environment. *Trends in Ecology and Evolution* vol. 15, no. 10 October 2000.

Obura, D.O., Tamelander, J. and Linden, O., (Eds.) (2008). *Ten years after bleaching – facing the consequences of climate change in the Indian Ocean*. CORDIO Status Report. CORDIO (Coastal Oceans Research and Development, Indian Ocean)/Sida-SAREC. Mombasa. http://www.cordioea.org. 493 pp.

Paddack, M.J., Reynolds, J.D., Aguilar, C., Appeldoorn R.S., Jim Beets, J., and 30 others, (2009). Recent region-wide declines in Caribbean reef fish abundance. *Current Biology*, 19: 1–6.

Palumbi S.R. et al. Mechanisms of reef coral resistance to future climate change. *Science* 344, 895 (2014); DOI: 10.1126/science.1251336.

Randall, C.J. and Szmant, A.M. (2009a). Elevated temperature affects development, survivorship, and settlement of the elkhorn coral, *Acropora palmata* (Lamarck 1816). Biological Bulletin 217: 269–282.

Randall, C.J. and Szmant, A.M. (2009b). Elevated temperature reduces survivorship and settlement of the larvae of the Caribbean scleractinian coral, *Favia fragum* (Esper). *Coral Reefs* 28: 537-545.

Randall, C. J. and van Woesik, R. (2015). Contemporary white-band disease in Caribbean corals driven by climate change. *Nature Climate Change* 5, 375–379 (2015).

Reid, W.V., Mooney, H.A., Cropper, A., Capistrano, D., Carpenter, S.R., Chopra, K., Dasgupta, P., Dietz, T., et al., (2005). *Millennium Ecosystem Assessment Synthesis Report. Report of the Millennium Ecosystem Assessment*; pp. 219 www.millenniumassessment.org.

Riegl, B. and Piller, W.E. (2003). Possible refugia for reefs in times of environmental stress. *International Journal of Earth Sciences*, 92:520–531.

Roberts, C. (2007). *The Unnatural History of the Sea,* Island Press, Washington D.C. pp. 435.

Rogers C. (2009). Coral bleaching and disease should not be underestimated as causes of Caribbean coral reef decline. doi:10.1098/rspb.2008.0606. *Proceedings of the Royal Society*. vol. 276 no. 1655 197-198.

Russ, G.R., Cheal, A.J., Dolman, A.M., Emslie M.J., Evans R.D., Miller I., Sweatman, H. and Williamson, D.H. (2008). Rapid increase in fish numbers follows creation of world's largest marine reserve network. *Current Biology*, 18: R514-515.

Russell, M.W., Sadovy de Mitcheson, Y., Erisman, B.E., Hamilton, R.J., Luckhurst, B.E. and Nemeth, R.S. (2014). *Status Report – World's Fish Aggregations 2014. Science and Conservation of Fish Aggregations*, California, USA. International Coral Reef Initiative.

Sadovy, Y. and Domeier, M., (2005). Are aggregation-fisheries sustainable? Reef fish fisheries as a case study. *Coral Reefs*. 24, 254-262.

Sadovy de Mitcheson, Y., Craig, M.T., Bertoncini, A.A., Carpenter, K.E., Cheung, W.W.L., Choat, J.H., Cornish, A.S., Fennessy, S.T., Ferreira, B.P., Heemstra, P.C., Liu, M., Myers, R.F., Pollard, D.A., Rhodes, K.L., Rocha, L.A., Russell, B.C., Samoilys, Melita A. and Sanciangco, J. (2013). Fishing groupers towards extinction: a global assessment of threats and extinction risks in a billion dollar fishery. *Fish and Fisheries*, 14(2): 119–136.

Sale, P.F. and Szmant, A.M., (eds.). (2012). *Reef Reminiscences: Ratcheting back the shifted baselines concerning what reefs used to be*. United Nations University Institute for Water, Environment and Health, Hamilton, ON, Canada, 35 pp.

Salvat, B. and Wilkinson, C. (2011). Cyclones and Climate Change in the South Pacific. *Revue d'Ecologie* (Terre Vie), vol. 66.

Salvat, B., Roche, H., Berny, P. and Ramade, F. (2012). Recherches sur la contamination par les pesticides d'organismes marins des réseaux trophiques récifaux de Polynésie française. *Revue d'Ecologie* (Terre et Vie) 67: 129-148.

Schaffelke, B., Anthony, K., Blake, J., Brodie, J., Collier, C., Devlin, M., Fabricius, K., Martin, K., McKenzie, L., Negri, A., et al. (2013). Marine and coastal ecosystem impacts. In *Synthesis of evidence to support the reef water quality scientific consensus statement 2013.*

Selig, E.R. and Bruno, J.F., (2010). A Global Analysis of the Effectiveness of Marine Protected Areas in Preventing Coral Loss. *PLoS ONE* 5.

Selig E.R., Harvell C.D., Bruno J.F., Willis B.L., Page C.A., et al. (2006). Analyzing the relationship between ocean temperature anomalies and coral disease outbreaks at broad spatial scales. In: Phinney J, Hoegh-Guldberg O, Kleypas J, Skirving W, Strong A, (eds.) *Coral reefs and climate change: science and management.* Washington, DC: *American Geophysical Union.* pp. 111–128.

Silva, A.S., Leão, Z.M.A.N., Kikuchi, R.K.P., Costa, A.B. and Souza, J.R.B. (2013). Sedimentation in the coastal reefs of Abrolhos over the last decades. *Continental Shelf Research*, 70: 159-167.

Silva, A.G.D., Paula, A.F.D., Fleury, B.G. and Creed, J.C. (2014). Eleven years of range expansion of two invasive corals (*Tubastraea coccinea* and *Tubastraea tagusensis*) through the southwest Atlantic (Brazil). *Estuarine, Coastal and Shelf Science*, 141, 9-16.

Smith, R., Middlebrook, R., Turner, R., Huggins, R., Vardy, S. and Warne, M. (2012). Large-scale pesticide monitoring across Great Barrier Reef catchments - Paddock to Reef Integrated Monitoring, Modelling and Reporting Program. *Marine Pollution Bulletin* 65, 117-127.

Spalding, M.D., C. Ravilious and E.P. Green. (2001). United Nations Environment Programme, World Conservation Monitoring Centre. *World Atlas of Coral Reefs*. University of California Press: Berkeley. 416 pp.

Stoeckl, N., Farr, M., Larson, S., Adams, V.M., Kubiszewski, I., Esparon, M. and Costanza, R. (2014). A new approach to the problem of overlapping values: A case study in Australia's Great Barrier Reef. *Ecosystem Services*, 10: 61-78.

Storlazzi C.D., Elias, E., Field, M.E. and Presto M.K. (2011). Numerical modeling of the impact of sea-level rise on fringing coral reef hydrodynamics and sediment transport. *Coral Reefs* 30:83–96 DOI 10.1007/s00338-011-0723-9.

Sutherland K., Shaban, S., Joyner, J., Porter, J. and Lipp, E. (2011). Human pathogen shown to cause disease in the threatened elkhorn coral *Acropora palmata. PLoS ONE* 6(8): e23468. doi:10.1371/journal.pone.0023468.

Szmant A.M. and Miller, M.W. (2005). Settlement preferences and post-settlement mortality of laboratory cultured and settled larvae of the Caribbean hermatypic corals *Montastraea faveolata* and *Acropora palmata* in the Florida Keys, USA. In: *Proceedings 5th International Coral Reef Symp*osium, Vol. 4, p. 295-300, Tahiti.

Talbot, F. and Wilkinson, C. (2001). *Coral reefs, mangroves and seagrasses: a sourcebook for managers*. Australian Institute of Marine Science, Townsville, 193 pp.

Thurber R., Burkepile, D., Correa, A., Thurber, A., Shantz, A. et al. (2012). Macroalgae Decrease Growth and Alter Microbial Community Structure of the Reef-Building Coral, *Porites astreoides. PLoS ONE* 7(9): e44246. doi:10.1371/journal.pone.0044246.

Tribollet, A., Godinot, C., Atkinson, M. and Langdon, C. (2009). Effects of elevated pCO2 on dissolution of coral carbonates by microbial euendoliths. *Global Biogeochemical Cycles*, 23(3), GB3008, doi: 10.1029/2008GB003286.

Uthicke, S., Logan, M., Liddy, M., Francis, D., Hardy, N. and Lamare, M. (2014) Climate change as an unexpected co-factor promoting coral eating seastar (*Acanthaster planci*) outbreaks. *Scientific Reports* 5:8402, DOI: 10.1038/srep08402.

Van Ael, E., Covaci, A., Blust, R. and Bervoets, L. (2012). Persistent organic pollutants in the Scheldt estuary: environmental distribution and bioaccumulation. *Environmental International* 48, 17-27.

van Dam, J.W., Negri, A.P., Mueller, J.F. and Uthicke, S. (2012). Symbiont-specific responses in foraminifera to the herbicide diuron. *Marine Pollution Bulletin* 65, 373-383.

Van Hooidonk R., Maynard J.A., Manzello D. and Planes S. (2013). Opposite latitudinal gradients in projected ocean acidification and bleaching impacts on coral reefs. *Global Change Biology*, 20(1): 103-112. doi: 10.1111/gcb.12394.

Vega Thurber, R.L., Burkepile, D.E., Fuchs, C., Shantz, A.A., McMinds, R., Zaneveld, J.R., (2014). Chronic nutrient enrichment increases prevalence and severity of coral disease and bleaching. *Global Change Biology* 20, 544-554.

Veron, J., Stafford-Smith, M., DeVantier, L. and Emre Turak, E., (2015). Overview of distribution patterns of zooxanthellate Scleractinia. *Frontiers in Marine Science*; 1; Art 81; 1-19

doi: 10.3389/fmars.2014.00081.

Weil E., Urreiztieta, I. and Garzón-Ferreira, J. 2002. Geographic variability in the incidence of coral and octocoral diseases in the wider Caribbean. *Proceedings 9th International Coral Reef Symposium.*, Bali Indonesia 2:1231-1237.

Weil, E. and Cróquer, A. (2009). Local and geographic variability in distribution and prevalence of coral and octocoral diseases in the Caribbean I: Community-Level Analysis. *Diseases of Aquatic Organisms.* 83:195-208.

Wilkinson, C.R. (1998). *Status of Coral Reefs of the World: 1998*. Australian Institute of Marine Science, Townsville, 194 pp.

Wilkinson, C.R. (2000). *Status of Coral Reefs of the World: 2000*. Australian Institute of Marine Science, Townsville, 363 pp.

Wilkinson, C.R. (2002). *Status of Coral Reefs of the World: 2002*. Australian Institute of Marine Science, Townsville, 378 pp.

Wilkinson, C.R. (2004). *Status of Coral Reefs of the World: 2004*. Australian Institute of Marine Science, Townsville, Volume 1; 301 pp.

Wilkinson, C.R. (2008). *Status of Coral Reefs of the World: 2008*. Global Coral Reef Monitoring Network and Reef and Rainforest Research Centre, Townsville, 298 pp.

Wilkinson, C., Souter, D. and Goldberg, J. (2006). *Status of Coral Reefs in Tsunami Affected Countries: 2005*. Australian Institute of Marine Science and Global Coral Reef Monitoring Network, Townsville and 158 pp.

Wilkinson, C. and Brodie, J. (2011). *Catchment Management and Coral Reef Conservation: a practical guide for coastal resource managers to reduce damage from catchment areas based on best practice case studies*. Global Coral Reef Monitoring Network and Reef and Rainforest Research.

Wilkinson, C. and Souter, D. (2008). Status of Caribbean coral reefs after bleaching and hurricanes in 2005. Global Coral Reef Monitoring Network, and Reef and Rainforest Research Centre Townsville, pp. 148.

Wilkinson, C. and Salvat, B. (2012). Coastal resource degradation in the tropics: does the tragedy of the commons apply for coral reefs, mangrove forests and seagrass beds? *Marine Pollution Bulletin*, 64: 1096-1105.

Williams, D.E., Miller, M.W. and Kramer, K.L. (2008). Recruitment failure in Florida Keys *Acropora palmata*, a threatened Caribbean coral. *Coral Reefs* 27:697-705.

Willis B.L., Page C.A. and Dinsdale E.A. (2004). Coral disease on the Great Barrier Reef. In: *Coral Health and Disease,* edited by E. Rosenberg, Y. Loya, pp.69-104, Springer-Verlag, Berlin.

Wilson, S.K., Adjeroud, M., Bellwood, D.R., Berumen, M.L., Booth, D., Bozec, Y-M, Chabanet, P., Cheal, A., Cinner, J., Depczynski, M., Feary, D.A., Gagliano, M., Graham, N.A.J, Halford, A.R., Halpern, B.S., Harborne, A.R., Hoey, A.S., Holbrook, S.J., Jones, G.P., Kulbiki, M., Letourneur, Y., De Loma, T.L., McClanahan, T., McCormick, M.I., Meekan, M.G., Mumby, P.J., Munday, P.L., Ohman, M.C., Pratchett, M.S., Riegl, B.,

Sano, M., Schmitt, R.J. and Syms, C. (2010). Crucial knowledge gaps in current understanding of climate change impacts on coral reef fishes. *Journal of Experimental Biology* 213(6): 894-900.

Wood L.J., Fish L., Laughren J. and Pauly D. (2008). Assessing progress towards global marine protection targets: shortfalls in information and action. *Oryx*, 42(3), 1–12.

Woodroffe, C.D. and Webster, J.M., (2014). Coral Reefs and Sea-Level Change. *Marine Geology* 352, 248–267.

44 Estuaries and Deltas

Contributors:
Peter Harris (Convenor and Lead Member), Kedong Yin, Kawser Ahmed, Patricio Bernal (Co-Lead Member), Margarita Caso, Beatrice Ferreira (Co-Lead Member), Regina Folorunsho, John Machiwa, José Muelbert, Pablo Muniz, Jake Rice (Co-Lead Member and Editor of Part VI) and Claudia Câmara Vale

1 Introduction.

Estuaries and deltas are amongst the most heavily populated areas of the world (about 60 per cent of the world's population live along estuaries and the coast) making them the most perturbed parts of the world ocean (Kennish, 2002; Small and Cohen, 2004). Of the 32 largest cities in the world, 22 are located beside estuaries. They are adversely affected by invasive species, sedimentation (from soil erosion caused by deforestation, overgrazing, and other poor farming practices), overfishing, drainage and filling of wetlands, eutrophication due to excessive nutrients from fertilizer, sewage and animal (including aquaculture) wastes, pollutants including heavy metals (see Chapter 20), polychlorinated biphenyls, radionuclides and hydrocarbons from sewage inputs and diking or damming for flood control or water diversion. Estuaries and deltas provide protected harbours used as ports that are associated with introduced marine pests. They are foci of human attention, attracting potentially incompatible uses by society such as heavy industry, urbanization and recreation; they are affected by global sea-level rise and climate change (Crossland et al., 2005). Estuaries and deltas "form a major transition zone with steep gradients in energy and physicochemical properties at the interface between land and sea" (Jennerjahn and Mitchell, 2013).

More than 50 per cent of large river systems are affected by dams, based on a global synthesis on river fragmentation and flow regulation (Nilsson et al., 2005), with obvious consequences for the estuaries and deltas at their coastal termini. The mean age of river water at river mouths has increased from about two weeks to over one month on a global scale and to more than one year in extreme cases (Vörösmarty et al., 2003). Over the last few centuries, the global annual sediment flux into the coastal zone has increased by 2.3×10^9 tons due to human-induced soil erosion and decreased by 3.7×10^9 tons due to retention in reservoirs, the net effect being a reduction of sediment input by 1.4×10^9 tons (Syvitski et al., 2005). A major environmental consequence of river sediment starvation is erosion of the coast and attendant loss of habitat.

2 Major threatening processes.

Processes affecting the health and condition of estuaries and deltas may be classified into three broad categories that can interact:

(a) "Short-term" pressures associated with the near-term effects of human expansion (e.g., coastal development, land-based inputs of nutrients, over-fishing, aquaculture, and maritime operations);
(b) "Medium- to long-term" pressures associated with anthropogenic climate change (e.g., sea-level rise, increases in atmospheric heat and CO_2 fluxes into the oceans, a strengthening global hydrological cycle, and the increasing magnitude of tropical cyclones); and
(c) Extreme natural events.

A list of processes and impacts is given in Table 1 (see Chapter 44 Appendix).

3 Social and economic considerations.

Estuaries are tourist attractions and provide a centrepiece for development (a harbour view). Estuaries and deltas provide natural harbours that are used for transport and industry as the ideal location of major port facilities. They have ecological importance to a diverse biota, including economic importance to commercial and subsistence fisheries. People value estuaries for recreation, scientific knowledge, education, aesthetics, and traditional practices. Boating, fishing, swimming, surfing, and bird watching are just a few of the numerous recreational activities people enjoy in estuaries and deltas. Their unique habitats make them valuable laboratories for scientists and students. Considering the sum of human activities that depend upon the existence of estuaries and deltas and their ecosystem services (e.g., Barbier et al., 2011), their total economic value to society is vast (Costanza et al., 1997; Costanza et al., 2014). Costanza et al. (1997) estimated their value at approximately 4.1 trillion United States dollars (equal to 6.1 trillion dollars in 2014 dollars).

Some indications of the social and economic value of functioning estuarine and deltaic ecosystems can be found from examples where human activities have impaired such functions. Economic losses due to anthropogenic changes in river discharge are one example. The down-stream consequences of dam-building are often not fully considered when the decision is taken to build a dam on a river system. The economic losses from reduced fisheries landings, due to the reduction in nutrients entering the Indian Ocean at the Sofala Bank fishery (Arthurton, 2002), following alteration to the Zambezi River freshwater flows, has been estimated at between 10 and 20 million dollars (Turpie, 2006). In an extreme case, the Colorado River, prior to the completion of the Hoover Dam in 1935, delivered a combination of nutrient-rich water and silt to the historic Delta, comprised over 2.5 million acres of wetlands, habitat for an estimated 400 species of plants and wildlife and home to some 20,000 Cocopah Indians (Glenn et al., 2001). All of the freshwater discharge was impounded behind dams by 1963; the wetlands dried up, affecting many dependent species. In 2014, an experimental release of 130 million m^3 of water allowed the restoration of the Colorado Delta to begin, although it will take many years to restore even part of the original wetland area (Witze, 2014).

Some of the first and most severe impacts of climate change will come through greater storm surges caused by a combination of higher sea levels and stronger storms in some areas. In the absence of storm surge, a 20-80 cm rise in mean sea level will place 7 – 300 million additional people at risk of being flooded each year (Geneva Reports, 2009, No. 2, 138 pp. www.genevaassociation.org). Increases in storm surge will increase these numbers substantially. The Organization for Economic Cooperation and Development (OECD) estimates that, in the absence of adaptation, the population in 136 major port cities exposed to storm surges could increase from 40 million in 2005 to ~150 million in the 2070s, with exposed assets rising from 3,000 billion dollars to 35,000 billion dollars (Nicholls et al., 2008). By 2050, sea-level rise in the Ganges-Brahmaputra Delta could directly affect more than three million people and Bangladesh could lose nearly one-quarter of the land area it had in 1989 by the end of this century, in a worst-case scenario (Ericson et al., 2005). As a proportion of GDP, economic losses from flooding are much higher for developing countries than for developed countries (Ramcharan, 2007). Financial losses from weather events are currently doubling every 12 years at an annual rate of 6 per cent (UNEP, 2006). In the Sacramento Delta in San Francisco Bay, California, United States, global sea-level rise places about 500,000 acres of agricultural lands in the inner Delta at significant risk of flooding in the first half of the 21st century. Total losses for the wider area—including multiplier effects—could reach 1,800 jobs per year, 130 million dollars in value added, and nearly 14 million dollars in state and local tax receipts (Medellín-Azuara et al., 2012). These examples provide some context for the potential impacts of water abstraction and global warming and sea-level rise on ecosystem services upon which estuarine- and deltaic-based societies and economies depend.

4   Management and conservation.

Healthy estuaries and deltas maintain water quality that benefits both people and marine life. They provide a natural buffer between the land and ocean, absorbing floodwaters and storm surges. Estuaries and deltas help maintain biodiversity by providing a diverse range of unique habitats, including mangrove forests, salt marshes, mud flats and seagrass beds, which are critical for the survival of many species. Many species of commercially important fish and shellfish use estuarine and deltaic habitats as nurseries to spawn and allow juveniles to grow. Maintaining such ecosystem services is commonly declared as a management goal and is the focus of conservation efforts.

In considering the management of estuaries and deltas, the question of the number of estuaries and deltas on earth arises, given that an inventory of any asset is a prerequisite to its management. The number of estuaries and deltas, in turn, is dependent upon scale and definition of what constitutes an estuary or delta. In their estimate of river sediment discharge based on a 30 minute (55.56 km) grid, Syvitski et al. (2005) identified 4,464 river basins > 100 km² in area that are not covered by ice sheets of the Antarctica, Greenland and portions of the Canadian Archipelago and have a positive discharge to the ocean/sea. Given that every estuary or delta is associated with a river that discharges into the ocean/sea, and noting this size limit on catchment area, therefore about 4,464 estuaries and deltas are found on earth.

A search on the IUCN Protected Areas database (http://protectedplanet.net) for "estuaries" yielded 275 results, of which 156 are in Europe (including 107 in the UK alone), 79 are in the Americas (including 53 in the USA), 19 are in Oceania, 11 in Asia and 10 in Africa. A similar search conducted for "deltas" found 210 results, of which 127 are in Europe (including 35 in Greece), 51 are in the Americas, 17 in Asia, 12 in Africa and three in Oceania. In terms of level of protection, only five out of 275 estuaries and 12 out of 210 deltas are in IUCN category Ia (Strict Nature Reserve) or Ib (Wilderness Area), with over 50 per cent in categories IV (Habitat/Species Management Area) and V (Protected Landscape/ Seascape). However, these figures may not capture all estuaries or deltas under protection since in the exact word "estuary" or "delta" must be contained in the place name for the search to recognize the location as containing as estuary or delta; so a place that is named as a "bay" or other term would not be counted. Furthermore, the protection of the marine habitat may not be effective if the catchment itself is not well managed. Nevertheless, the figures give some broad indication of the level of protection afforded to estuaries and deltas on earth.

5   Integrated assessment of the status of the habitat.

In order to produce a global, integrated assessment of estuary and delta condition, a literature search was carried out for papers and reports that have provided an assessment on estuarine and coastal habitats. Studies that reported on the condition of individual estuaries or groups of estuaries within a broad area were included. Where possible, the results given in the reports were converted into a report card score on a scale of 1 to 4 (Very Good, Good, Poor, Very Poor) and the date of assessment recorded (the criteria used to identify the condition category are given in Appendix, online only). In addition, a trend for overall condition was extracted (declining, stable or improving) and the timeframe over which the trend was observed was recorded. The raw data are recorded in a table (Appendix).

Based on published assessments for 103 areas, the global condition of estuaries and deltas (Figure 44.1) is Poor overall (mean score of 2.07 out of 4). The published assessments gave a Very Poor rating in 31 areas, a Poor rating in 32 areas, Good in 31 areas and a Very Good rating in only eight areas (Table 2 in Appendix). These results are biased by the fact that many studies are carried out in affected areas and hence the scores are skewed (i.e., the overall "Poor" rating is influenced by the many studies that are conducted on affected systems). On the other hand, many of the available assessments are based on only a few measured variables (typically related to water quality or fisheries) and they do not give an overall (integrated) picture of the health and condition of estuarine ecosystems. This factor can influence the outcome of a non-integrated assessment for systems in which the impact is not measured by the parameters used.

For example, one of the six Very Good ratings (Table 2 in Appendix) was assigned by UKTAG (2008) for Estuaries and Lochs in Scotland based on the winter mean of dissolved inorganic nitrogen over a six-year period (2001-2006). However, the ecology of at least one of these Scottish lochs (the Firth of Clyde) has been described by Thurstan and Roberts

Estuaries and Deltas

The boundaries and names shown and the designations used on this map do not imply official endorsement or acceptance by the United Nations.
Figure 44.1 | Estuarine and deltaic condition assessments based on reports for 100 regions (listed in Appendix).

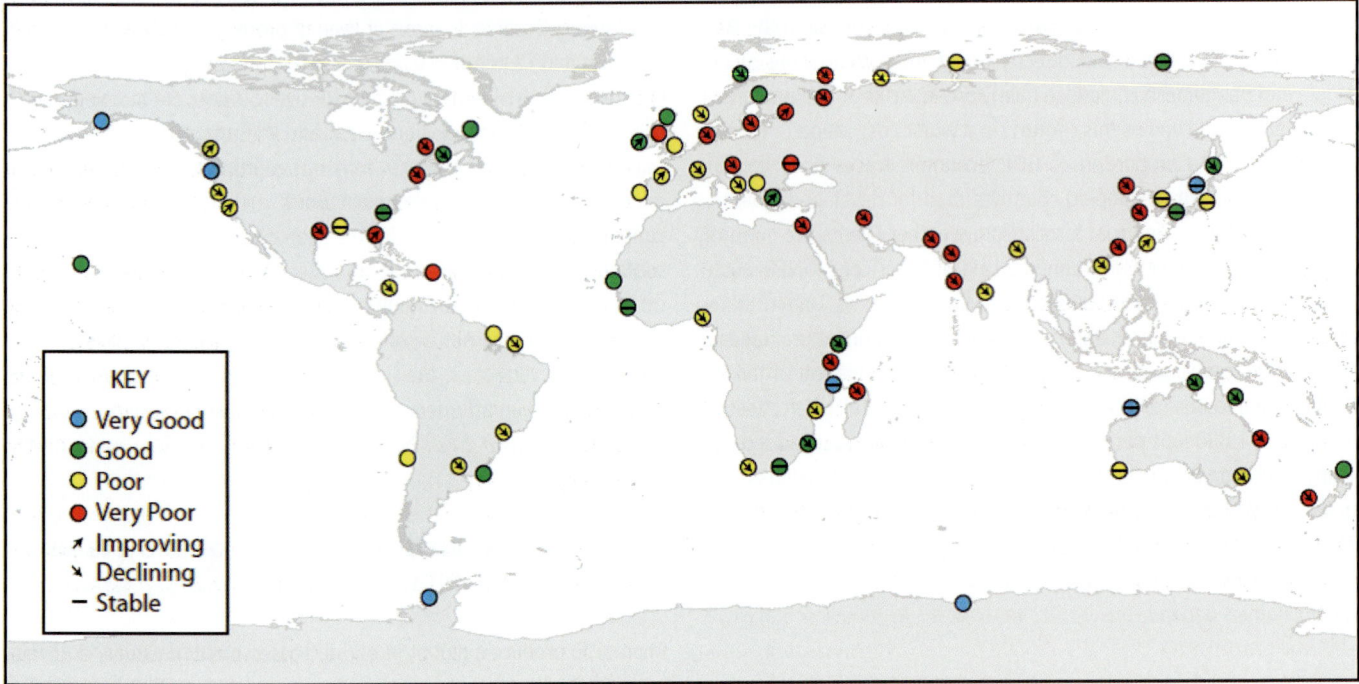

(2010) as "a marine ecosystem nearing the endpoint of overfishing, a time when no species remain that are capable of sustaining commercial catches". Hence, whereas the water quality in this estuary may be rated as very good, the ecosystem has been significantly affected by overfishing to the extent that an integrated assessment would likely give a rating of Very Poor for this estuary. Such cases serve to elevate the global score of "Poor" such that it is unrealistically positive.

Seventy-five studies reported a trend in terms of improving, stable or declining condition (Table 2 in Appendix). Out of those 75 studies, 46 (62 per cent) reported that conditions are declining, 19 (24 per cent) reported conditions were stable and ten (14 per cent) reported an improvement. On no continent does the number of estuaries showing an improving condition exceed the number of assessments of declining condition. Europe has the greatest number of studies that reported improving conditions (five), but only one area was reported to be in a "very good" condition; Africa, Australia and the South Pacific had no studies where conditions were improving. Asia (Japan) Australia and Africa each had one area where the condition was assessed as very good and stable.

6 Gaps in scientific knowledge

Out of the 101 areas assessed, only some are the subject of integrated assessments that include multiple aspects of estuarine environment, including habitats, catchment management, species, ecological processes, physical and chemical processes and socioeconomic aspects. Very few (about 10) areas had assessments that included all aspects of estuarine environments, to provide "fully integrated" assessments. There are 41 areas where assessments included at least three different aspects, producing partially integrated assessments. Another 25 areas had assessments concerned only with some aspect of estuarine water or sediment quality. Thus a critical gap in scientific knowledge is the availability of fully integrated environmental assessments for estuaries and deltas.

Out of the many possible aspects of the environment that could be assessed, water quality and biological aspects are most common, whereas socioeconomic aspects are assessed the least often, which is thus a knowledge gap. One other aspect of condition assessment is the trend (improving, stable or declining) that was assessed in 74 out of 103 areas. The assessment of trends is a critical piece of information for decision-makers, but which is missing in about 26 per cent of assessments. Furthermore, the time interval over which the trend is measured varies between studies, from one year to other arbitrary periods of human impact (as much as a century or longer). Thus the comparison of trends is confounded by differences in the time spans they relate to; international agreement on standards for reporting condition trends is needed to overcome this problem.

References

Barbier, E.B., Hacker, S.D., Kennedy, C., Koch, E.W., Stier, A.C. and Silliman, B.R. (2011). *The value of estuarine and coastal ecosystem services*. Ecological Monographs 81, 169–193.Costanza, R., d'Arge, R., de Groot, R., Farber, S., Grasso, M., Hannon, B., Limburg, K., Naeem, S., O'Neill, R.V., Paruelo, J., Raskin, R.G., Sutton, P., van den Belt, M., (1997). The value of the world's ecosystem services and natural capital. *Nature* 387, 253-260.

Costanza, R., de Groot, R., Sutton, P., van der Ploeg, S., Anderson, S., Kubiszewski, I., Farber, S. and Turner, R. (2014). Changes in the global value of ecosystem services. *Global Environmental Change* 26: 152-158.

Crossland, C., Baird, D., Ducrotoy, J.-P., Lindeboom, H., Buddemeier, R., Dennison, W., Maxwell, B., Smith, S. and Swaney, D. (2005). The Coastal Zone - A Domain of Global Interactions, in: Crossland, C., Kremer, H., Lindeboom, H., Marshall Crossland, J., Le Tissier, M.A. (Eds.), *Coastal Fluxes in the Anthropocene*. Springer, Berlin, pp. 1-37.

Ericson, J.P., Vorosmarty, C.J., Dingman, S.L., Ward, L.G. and Meybeck, M. (2005). Effective sea-level rise and deltas: Causes of change and human dimension implications. *Global Planetary Change* 50, 63-82.

Glenn, E.P., Zamora-Arroyo, F., Nagler, P.L., Briggs, M., Shaw, W. and Flessa, K. (2001). Ecology and conservation biology of the Colorado River Delta, Mexico. *Journal of Arid Environments* 49, 5-15.

Jennerjahn, T.C. and Mitchell, S.B. (2013). Pressures, stresses, shocks and trends in estuarine ecosystems: an introduction and synthesis. *Estuarine, Coastal and Shelf Science* 130, 1-8.

Kennish, M.J. (2002). Environmental threats and environmental future of estuaries. *Environmental Conservation* 29, 78-107.

Medellín-Azuara, J., Hanak, E., Howitt, R. and Lund, J. (2012). *Transitions for the Delta Economy*. Public Policy Institute of California, San Francisco. http://www.ppic.org/content/pubs/report/R_112EHR.pdf

Nicholls, R.J., Hanson, S., Herweijer, C., Patmore, N., Hallegatte, S., Corfee-Morlot, J., Chateau, J. and Muir-Wood, R. (2008). *Ranking Port Cities with High Exposure and Vulnerability to Climate Extremes*. OECD Publishing, OECD Environment Working Papers, 1.

Nicholls, R.J., Wong, P.P., Burkett, V.R., Codignotto, J.O., Hay, J.E., McLean, R.F., Ragoonaden, S. and Woodroffe, C.D. (2007). Coastal systems and low-lying areas. Climate Change 2007: Impacts, Adaptation and Vulnerability. Contribution of Working Group II, in: Parry, M.L., Canziani, O.F., Palutikof, J.P., van der Linden, P.J., Hanson, C.E. (Eds.). *Fourth Assessment Report of the Intergovernmental Panel on Climate Change*, Cambridge University Press, Cambridge, UK, pp. 315-356.

Nilin, J., Moreira, L.B., Aguiar, J.E., Marins, R., Moledo de Souza Abessa, D., Monteiro da Cruz Lotufo, T. and Costa-Lotufo, L.c.V. (2013). Sediment quality assessment in a tropical estuary: The case of Ceará River, Northeastern Brazil. *Marine Environmental Research* 91, 89-96.

Nova Scotia, (2009). *Coastal Water Quality: The 2009 State Of Nova Scotia's Coast Report*. http://www.novascotia.ca/coast/documents/state-of-the-coast/WEB_CWQ.pdf

Ramcharan, R. (2007). Does the exchange rate regime matter for real shocks? Evidence from windstorms and earthquakes. *Journal of International Economics* 73, 31-47.

Ramesh, R., Purvaja, R., Lakshmi, A., Newton, A., Kremer, H.H., Weichselgartner, J. (2009). South Asia Basins: LOICZ *Global Change Assessment and Synthesis of River Catchment: Coastal Sea Interaction and Human Dimensions, Land-Ocean Interactions in the Coastal Zone*, IGBP/IHDP Core Project, LOICZ Research & Studies No. 32. GKSS Research Center, Geesthacht, p. 121.

Richardson, C.J., Hussain, N.A. (2006). Restoring the Garden of Eden: An Ecological Assessment of the Marshes of Iraq. *Bioscience* 56, 477-489.

San Francisco Estuary Partnership (2011). *The State of San Francisco Bay 2011* http://www.bay.org/assets/The%20State%20of%20San%20Francisco%20Bay,%202011.pdf

Seitzinger, S.P., Kroeze, C., Bouwman, A.F., Caraco, N., Dentener, F., Styles, R.V. (2002). Global patterns of dissolved inorganic and particulate nitrogen inputs to coastal systems: Recent conditions and future projections. *Estuaries* 25, 640-655.

Sigmon, C.L.T., Caton, L., Coffeen, G. and Miller, S. (2006). *Coastal Environmental Monitoring and Assessment Program. The Condition of Oregon's Estuaries in 1999, a Statistical Summary*. Oregon Department of Environmental Quality, Laboratory Division, p. 131.

Simboura, N. and Zenetos, A., Pancucci-Papadopoulou, M.A. (2014). Benthic community indicators over a long period of monitoring (2000 - 2012) of the Saronikos Gulf, Greece, Eastern Mediterranean. *Environmental Monitoring and Assessment*, 1-13.

Small, C. and Cohen, J.E. (2004). Continental Physiography, Climate, and the Global Distribution of Human Population1. *Current Anthropology* 45 (2).

SOA (2010). *Bulletin of Marine Environmental Status of China for the year of 2010*, State Oceanic Administration of the People's Republic of China, web site. http://www.soa.gov.cn/zwgk/hygb/zghyhjzlgb/201211/t20121107_5527.html

Syvitski, J.P.M., Vörösmarty, C.J., Kettner, A.J. and Green, P. (2005). Impact of humans on the flux of terrestrial sediment to the global coastal ocean. *Science* 308, 376-380.

Thurstan, R.H. and Roberts, C.M. (2010). Ecological Meltdown in the Firth of Clyde, Scotland: Two Centuries of Change in a Coastal Marine Ecosystem. *PLoS ONE* 5, e11767. doi:10.1371/journal.pone.0011767

Toyama Prefecture, (2009). *Status of water pollution, FY2007* (in Japanese). http://www.pref.toyama.jp/cms_sec/1706/kj00007252-006-01.html

Toyama Prefecture, (2014). *Status of water pollution, FY2012* (in Japanese). http://www.pref.toyama.jp/cms_sec/1706/kj00007252-011-01.html

Turpie, J.K. (2004). *South African National Spatial Biodiversity Assessment 2004: Technical Report. Volume 3: Estuary Component*. Pretoria: South African National Biodiversity Institute. http://www.bcb.uwc.ac.za/pssa/articles/includes/NSBA%20Vol%203%20Estuary%20Component%20Draft%20Oct%2004.pdf

UKTAG, (2008). UK Technical Advisory Group on the Water Framework Directive, *UK Environmental Standards and Conditions (Phase 2)*. UK Water Framework Directive, p. 84.

UNEP, (2006). UNEP Finance Initiative – "Adaptation and vulnerability to climate change: the role of the finance sector" CEO Briefing. UNEP, Geneva.

UNEP, (2011). *Environmental Assessment of Ogoniland*. UNEP Report Job No.: DEP/1337/GE, DJ Environmental, UK.

Vörösmarty, C.J., Meybeck, M., Fekete, B.Z., Sharma, K., Green, P. and Syvitski, J.P.M. (2003). Anthropogenic sediment retention: major global impact from registered river impoundments. *Global and Planetary Change* 39, 169-190.

Witze, A. (2014). Water returns to arid Colorado River delta. *Nature* 507, 286-287.

Appendix. List of condition assessments

List of condition assessments (ranked as very good, good, poor, very poor) for estuaries and deltas in different regions, and the trend in condition (over time interval specified). Year of assessment, the number of estuaries or deltas included and references are indicated for each region. Indicative grading statements used to define the condition are as follows:

Very Good = Estuarine habitats are essentially structurally and functionally intact and able to support all dependent species. Only a few, if any, species populations have declined as a result of human activities or declining environmental conditions. There are no significant changes in physical-chemical-ecological processes or ecosystem services as a result of human activities.

Good = There is some habitat loss or alteration in some small areas, leading to minimal degradation but no persistent substantial effects on populations of dependent species. Populations of a number of significant species but no species groups have declined significantly as a result of human activities or declining environmental conditions. There are some significant changes in physical-chemical-ecological processes as a result of human activities in some areas, but these are not to the extent that they are significantly affecting ecosystem functions or services.

Poor = Habitat loss or alteration has occurred in a number of areas, leading to persistent substantial effects on populations of some dependent species. Populations of many species or some species groups have declined significantly as a result of human activities or declining environmental conditions. There are substantial changes in physical-chemical-ecological processes as a result of human activities, and these are significantly affecting ecosystem functions and services in some areas.

Very Poor = There is widespread habitat loss or alteration, leading to persistent substantial effects on many populations of dependent species. Populations of a large number of species or species groups have declined significantly as a result of human activities or declining environmental conditions. There are substantial changes in physical-chemical-ecological processes across a wide area of the region as a result of human activities, and ecosystem function and services are seriously affected in much of the region.

Key	
	Integrated assessment, including multiple indicators or indices of habitat, biota, water quality and socioeconomics
	Partly integrated assessment based on several (3 or more) indicators and/or multiple studies
	Assessments based on one or only a few (<3) indicators, mainly water quality

Name of Region	Number of estuaries and deltas	Condition	Trend	References	Parameters assessed and other notes
Australia and South Pacific					
Australia, SW Coast	20	Very Poor (2011)	Stable (2006-11)	Department of Environment (2011) Harris and Heap (2003)	Integrated assessment
Australia, NW Coast	162	Very Good (2011)	Stable (2006-11)	Department of Environment (2011) Harris and Heap (2003)	Integrated assessment
Australia, North Coast	164	Good (2011)	Declining (2006-11)	Department of Environment (2011) Harris and Heap (2003)	Integrated assessment
Australia, Moreton Bay	1	Very Poor (2006)	Declining (historic)	Lotze et al. (2006)	Integrated assessment
Australia, NE Coast	199	Good (2011)	Declining (2006-11)	Department of Environment (2011) Harris and Heap (2003)	Integrated assessment
Australia, SE Coast	172	Poor (2011)	Declining (2006-11)	Department of Environment (2011) Harris and Heap (2003)	Integrated assessment
New Zealand, Waikato Region	5	Good (2013)		Waikato Regional Council (2013)	Water quality (dissolved oxygen, pH, turbidity, total ammonia, nitrate, total phosphorus and chlorophyll a)
New Zealand, New River	1	Very Poor (2011)	Declining (2007-2011)	Environment Southland (2011)	Water quality

Chapter 44 — Estuaries and Deltas

North America					
Canada, Nova Scotia	?	Good (2009)	Declining (1985-2000)	Nova Scotia (2009)	Water quality
Canada, Gulf of St Lawrence and Bay of Fundy	1	Very Poor (2006)	Declining (historic)	Lotze et al. (2006)	Integrated assessment
Canada, Salish Sea	>4	Poor (2011)		Barrie et al. (2012)	Benthic habitat mapping – impacted by fishing, pollution, catchment disturbance
Canada, Gilbert Bay, Southern Labrador	1	Good (2011)		Copeland et al. (2012)	Benthic habitat mapping – impacted by scallop dredging
United States, Massachusetts Bay, Delaware Bay, Chesapeake Bay, Pamlico Sound	4	Very Poor (2006)	Declining (historic)	Lotze et al. (2006)	Integrated assessment
United States, NE Coast	12	Very Poor (2000-02)	Declining (2000-02)	EPA (2007); Bricker et al (2007)	Based on five indicators of ecological condition: water quality index (including dissolved oxygen, chlorophyll a, nitrogen, phosphorus, and water clarity), sediment quality index (including sediment toxicity, sediment contaminants, and sediment total organic carbon [TOC]), benthic index, coastal habitat index, and a fish tissue contaminants index.
United States, Hudson River Estuary	1	Very Poor (2012)	Stable (2005-2010)	New York and New Jersey Harbor and Estuary Program (2012)	Assessment of marine habitats, fish, birds and pollution impacts.
United States, SE Coast	2	Good (2000-02)	Stable (2000-02)	EPA (2007); Bricker et al (2007)	Based on five indicators of ecological condition…
United States, Gulf Coast	7	Poor (2000-02)	Stable (2000-02)	EPA (2007); Bricker et al (2007)	Based on five indicators of ecological condition…
United States, W Coast	6	Poor (1999-03)	Improving (1999-03)	EPA (2007); Bricker et al (2007)	Based on five indicators of ecological condition…
United States, Glacier Bay, Alaska	1	Good (2011)		Cochrane et al. (2012)	Benthic habitat mapping – area impacted by fishing and tourism
United States, Alaska		Very Good (1999-03)		EPA (2007)	Based on five indicators of ecological condition…
United States, Hawaii		Good (1999-03)		EPA (2007)	Based on five indicators of ecological condition…
United States, San Francisco Bay	1	Poor (2011)	Declining (2011)	Lotze et al. (2006); San Francisco Bay Partnership (2011)	Integrated assessment (2003) plus water quality (2008)
United States, Oregon	30	Very Good (1999)		Sigmon et al. (2006)	General habitat condition; Water quality; Pollutant exposure and Benthic condition
United States, Puerto Rico	6	Very Poor (2002)		EPA (2007)	Based on five indicators of ecological condition…
United States, Florida Everglades	1	Very Poor (1900-2010)	Improving (1986-2006)	Entry and Gottlieb (2014)	Water quality
Central and South America					
Caribbean river basins	>4	Poor (2002)	Declining	Kjerfve et al. (2002)	Integrated LOICZ assessment covering Caroni River (Trinidad), Kingston Harbour (Jamaica), Parque Nacional Morrocoy (Venezuela) and Magdelena River (Costa Rica)
Uruguay, Río de la Plata, estuarine environment Montevideo Bay (including Montevideo Harbour)	1	Poor to very poor (1997-2010)	Improving	Muniz et al., (2002, 2004, 2005, 2006, 2011, 2012) García-Rodríguez et al. (2010) Burone et al. (2006, 2011) Danulat et al (2002) Gómez –Erache et al. (2001)	Biotic indices based on benthic communities; coastal eutrophication; heavy metals and PAH in sediments

Estuaries and Deltas   Chapter 44

Uruguay, Río de la Plata estuarine environment (seaward), Montevideo Coastal zone	1	Poor to good (1997-2012)	Declining	Muniz et al., (2002, 2004, 2005, 2006, 2011) Burone et al. (2006) Venturini et al. (2004, 2012, in press) Muniz & Venturini (2011) Gómez-Erache et al (2001) García-Rodríguez et al. (2011, 2014)	Biotic indices based on benthic communities; coastal eutrophication; contaminants in water, sediment and biota
Río de la Plata estuarine environment	1	Poor (1980-2000)	Declining	Nagy et al. (2000, 2002) FREPLATA 2005	Eutrophication
Argentina, North Coast of Río de la Plata	1	Good (2005-2010)		Gómez et al .(2012)	Biotic index indicative of eutrophication and organic pollution
East coast of Uruguay (coastal lagoons and sub-estuaries)	6	Good (2007-2008)		Defeo et al. (2009) Muniz et al. (2012) Conde & Rodríguez-Gallego (2002)	Biotic indices based on benthic communities
Brazil, Ceará River	3	Poor (2006-2007)	Declining (2006-2007)	Nilan et al (2013)	Toxicity bioassays and metal distribution
Brazil, Santos-São Vicente Estuary	1	Poor (2007)	Declining (2007)	Buruaem et al. (2013)	Acute toxicity of whole sediment and chronic toxicity of liquid phases, grain size, organic matter, organic carbon, nitrogen, phosphorus, trace metals, polycyclic aromatic hydrocarbons, linear alkylbenzenes and butyltins; benthic community descriptors.
Brazil, Pará River (Amazon estuary)	1	Poor (2009)		Viana et al. (2012)	Multimetric indices of ecosystem integrity: Abundance Biomass Comparation (ABC); Biological Health Index; Estuarine Fish Community, Transitional Fish Classification and Estuarine Biotic Integrity Indexes
Chile, Lenga Estuary	1	Poor		Moscoso et al (2006); Díaz-Jaramillo et al (2013)	Benthic Macroinfauna; oxidative stress responses, including glutathione-S-transferase (GST) activity, total antioxidant capacity (ACAP) and lipid peroxidation levels (TBARS) in estuarine crabs
Chile, Tabul-Raqui Estuary	2	Good		Díaz-Jaramillo et al (2013)	Oxidative stress responses, including glutathione-S-transferase (GST) activity, total antioxidant capacity (ACAP) and lipid peroxidation levels (TBARS) in estuarine crabs.
Asia					
Russian Federation, North Dvina River Estuary	1	Very Poor (1990-2006)	Declining (2006)	Gordeev et al. (2006)	LOICZ – DPSIR approach based mainly on water quality (heavy metals and hydrocarbons, acidification and radionuclide contamination)
Russian Federation, Small rivers of the Kola Peninsula	>5	Very Poor (1990-2006)	Declining (2006)	Gordeev et al. (2006)	LOICZ – DPSIR approach based mainly on water quality
Russian Federation, Pechora Estuary	1	Poor (1990-2006)	Declining (2006)	Gordeev et al. (2006)	LOICZ – DPSIR approach based mainly on water quality
Russian Federation, Ob River	1	Poor (1990-2006)	Stable (2006)	Gordeev et al. (2006)	LOICZ – DPSIR approach based mainly on water quality
Russian Federation, Yenisey River	1	Very Poor (1990-2006)	Stable (2006)	Gordeev et al. (2006)	LOICZ – DPSIR approach based mainly on water quality
Russian Federation, Lena Delta	1	Good (1990-2006)	Stable (2006)	Gordeev et al. (2006)	LOICZ – DPSIR approach based mainly on water quality
Bangladesh, Ganges – Brahmaputra Delta	1	Poor (2005)	Declining	Ramesh et al. (2009); Ahmed et al., (2010, 2011)	LOICZ – DPSIR approach – integrated assessment. Heavy metals in water, sediment and fish from Buriganga River channel; Heavy metal concentrations in macrobenthic fauna from Sundarbans mangrove forest.

India and Sri Lanka, Peninsular rivers	4	Poor (2005)	Declining	Ramesh et al. (2009)	LOICZ – DPSIR approach – integrated assessment
India, small western rivers flowing into the Arabian Sea	?	Very Poor (2005)	Declining	Ramesh et al. (2009)	LOICZ – DPSIR approach – integrated assessment
India and Pakistan, Indus River	1	Very Poor (2005)	Declining	Ramesh et al. (2009)	LOICZ – DPSIR approach – integrated assessment
India, Sabarmati River	1	Very Poor (2014)	Declining (2009-2014)	Haldar et al (2014)	Turbidity, dissolved oxygen, BOD, phenol, and petroleum hydrocarbons, phytoplankton and total and selective bacterial count.
South China Sea estuaries and deltas		Poor (2012)	Declining (2005-2012)	Ward (2012)	Integrated assessment
China, Changjiang (Yangtze) River estuary	1	Very Poor (2003)	Declining (1999-2010)	Xiao et al. (2007); Wang (2007); Liu et al (2013)	Eutrophication - ASSETS and AMBI index methods
China, Jiaozhou Bay	1	Very Poor (2006)	Declining (1980-2005)	Dang et al. (2010); Sun and Sun (2008)	Eutrophication; Index based on macrobenthic community, phytoplankton community, sediments, water quality.
China, Pearl River Estuary	1	Very Poor (2009)	Declining (1980-2009)	Chen et al. (2013)	Ecosystem health index based on biodiversity, water and sediment quality
China, Huanghe (Yellow) River Delta	1	Very Poor (2008)	Declining (2008)	Zhu et al (2003); Fan and Huang (2008)	Water quality; dissolved inorganic nitrogen
Taiwan, Dapeng Bay	1	Poor	Improving (2003-2009)	Hung et al. (2013)	Eutrophication
Iraq, Shat al Arab Waterway	1	Very Poor (2012)	Declining	Richardson and Hussain (2006); Mohamed et al. (2012)	Plant and fish communities and production, changes in water quality, and specific populations of rare and endangered species
Republic of Korea, Gwangyang Bay	1	Good (2012)	Stable (2010-2012)	Kim et al., (2008); Lee et al., (2010); KIOST (2013)	Organochlorine pesticides; Water quality, the carbon isotope ratio of particulate organic matter and sediment, and the nutrients limiting phytoplankton growth; marine ecosystem health index (MEHI) based on water quality, sediment quality, plankton, and benthos
Republic of Korea, Jinhae Bay	1	Poor (2012)	Stable (2010-2012)	Lim et al., (2012); KIOST (2013)	Sediment core records of C, N, CaCO3, trace metals; MEHI
Japan, Mutsu Bay	1	Good (2012)	Declining (2004-12)	Environment Management Bureau, Ministry of the Environment (2005-13)	Environmental status evaluated by Dr. K. Kohata from Japanese Environmental Quality Standards (for Living Environment)
Japan, Toyama Bay	1	Very Good (2012)	Stable (2003-12)	Toyama Prefecture (2009, 2014)	Environmental status evaluated by Dr. K. Kohata from Japanese Environmental Quality Standards
Japan, Tokyo Bay	1	Poor (2012)	Stable (2003-12)	Environment Management Bureau, Ministry of the Environment (2013)	Environmental status evaluated by Dr. K. Kohata from Japanese Environmental Quality Standards
Japan, Ise Bay	1	Poor (2012)	Improving (2003-12)	Environment Management Bureau, Ministry of the Environment (2013)	Environmental status evaluated by Dr. K. Kohata from Japanese Environmental Quality Standards
Japan, Osaka Bay	1	Poor (2012)	Stable (2003-12)	Environment Management Bureau, Ministry of the Environment (2013)	Environmental status evaluated by Dr. K. Kohata from Japanese Environmental Quality Standards
Japan, Ariake Sea and Shimabara Bay	1	Good (2012)	Stable (2003-12)	Environment Management Bureau, Ministry of the Environment (2013)	Environmental status evaluated by Dr. K. Kohata from Japanese Environmental Quality Standards
Japan, Yatsushiro Sea	1	Good (2012)	Stable (2003-12)	Environment Management Bureau, Ministry of the Environment (2013)	Environmental status evaluated by Dr. K. Kohata from Japanese Environmental Quality Standards

Estuaries and Deltas — Chapter 44

Africa					
South Africa, Cool Temperate	23	Poor (1993-99)	Declining (2004)	Turpie (2004); Harrison and Whitfield (2006); Van Niekerk et al (2013)	Integrated assessment
South Africa, Warm Temperate	104	Good (1993-99)	Stable (2004)	Turpie (2004); Harrison and Whitfield (2006)	Integrated assessment
South Africa, Subtropical	62	Good (1993-99)	Declining (2004)	Turpie (2004); Harrison and Whitfield (2006)	Integrated assessment
Guinea Current LME		Poor (2013)	Declining (2008-2013)	Guinea Current Report some.grida.no	Integrated assessment
Niger Delta Nigeria	21	Poor (2011)	Declining 2000-present	Awosika et al (1993); Folorunsho et al. (1994); Awosika and Folorunsho (in press); Folorunsho and Awosika (in press); UNEP (2011)	Petroleum pollution
The Gambia, Gambia Estuary	1	Good (2006)		Simier et al. (2006)	Fish assemblages
Sierra Leone	>1	Good (2014)	Stable (2014)	http://some.grida.no/sierra-leone/1-habitat.aspx	Integrated assessment
Ghana, Iture Estuary	1	Very Poor (2006)		Fianko et al (2007)	Cd, Zn, Se and Pb in water samples
Egypt, Nile Delta	1	Very Poor (2010)	Declining (1984-2010)	El-Asmar and Al-Olayan (2013); Gu et al. (2013)	Satellite image analysis of coastal change; heavy metal pollution
Kenya, Tana and Athi-Sabaki estuaries	2	Good (2009)	Declining	UNEP/Nairobi Convention Secretariat and WIOMSA (2009)	Nutrient levels in the Tana and Athi-Sabaki estuaries are high; the threshold chemical contamination levels of both Tana and Athi-Sabaki rivers have not been attained
Tanzania, Pangani Estuary	1	Very Poor (2009)	Declining	UNEP/Nairobi Convention Secretariat and WIOMSA (2009)	Poor water quality, stream morphology and aquatic life (low DO and eutrophication)
Tanzania, Rufiji Estuary	1	Good (2009)	Stable	UNEP/Nairobi Convention Secretariat and WIOMSA (2009)	DDT and nutrient flows from agricultural activities – local effects only to date.
Mozambique, Ruvuma River Estuary	1	Very Good (2009)	Stable	UNEP/Nairobi Convention Secretariat and WIOMSA (2009)	Best preserved mangrove forests along the coastline
Mozambique, Zambezi River Delta	1	Poor (2009)	Declining	UNEP/Nairobi Convention Secretariat and WIOMSA (2009)	Changed river flow caused loss of fisheries; pollution from sewage and industrial waste
Mozambique, Pungwe River Estuary	1	Good (2009)	Declining	UNEP/Nairobi Convention Secretariat and WIOMSA (2009)	Water abstraction has lead to reduced sediment loads and habitat loss.
Mozambique, Limpopo River Delta	1	Poor (2009)	Declining	UNEP/Nairobi Convention Secretariat and WIOMSA (2009)	Increasing salinity; discharge of untreated or partially treated domestic and industrial effluents; declining of river flows due to escalating demands for water; and discharge of untreated loads from upstream mining activities.
Madagascar, Betsiboka Estuary	1	Very Poor (2009)	Declining	UNEP/Nairobi Convention Secretariat and WIOMSA (2009)	Eutrophication, chemical pollution due to mining, port activities, effluent from oil refinery, harmful algal blooms, loss of mangrove and coral reef habitat, overfishing
Europe					
United Kingdom, England	43	Poor (1993-99)		UKTAG (2008)	Dissolved inorganic nitrogen
United Kingdom, Scotland	51	Very Good (1993-99)		UKTAG (2008)	Dissolved inorganic nitrogen

United Kingdom, Scotland, Firth of Clyde	1	Very Poor	Declining	Thurstan and Roberts (2010)	Fisheries data (species, population, catch statistics)
United Kingdom, Wales	17	Good (1993-99)		UKTAG (2008)	Dissolved inorganic nitrogen
United Kingdom, Bristol Channel	1	Poor (2011)		James et al. (2012)	Benthic habitat mapping
United Kingdom, Northern Ireland	10	Good (1993-99)		UKTAG (2008)	Dissolved inorganic nitrogen
Ireland	67	Good (2001-2005)	Improving (1995-2005)	Environmental Protection Agency (Ireland), (2006); Borja et al (2012)	Eutrophication
Adriatic Sea	1	Very Poor (2006)	Declining	Lotze et al. (2006)	Integrated assessment
Wadden Sea	1	Very Poor (2006)	Declining	Lotze et al. (2006) Dankers et al. (2012)	Integrated assessment
Baltic Sea	1	Very Poor (2006)	Declining	Lotze et al. (2006) Ezhova et al. (2012)	Integrated assessment
Finland, Kvarken Archipelago	1	Good (2010)		Kotilainen et al., (2012)	Benthic habitat mapping
Russian Federation, Neva Bay, Gulf of Finland	1	Very Poor (2005)	Improving (2000-2005)	Balushkina (2009)	Water quality, species diversity of zoobenthos
Spain, Basque Country	18	Good (1995-2003)	Improving (1995-2003)	Muxika et al. (2007); Borja et al (2012)	Physico-chemical, chemical, hydromorphological, and biological (phytoplankton, macroalgae, macroinvertebrates, and fishes) elements
Portugal, Mondego estuary	1	Good (2000-2001)		Chainho et al. (2007)	Benthic invertebrate communities
France, Mediterranean lagoons	7	Poor (2011)	Declining 2000-2011	Ifremer (2011); Borja et al (2012)	Eutrophication, including physicochemical elements in water and sediment, phytoplankton, macroalgae, and macroinvertebrates
Italy, Italian estuaries	22	Poor (2004)	Improving (2004)	Giordani et al. (2005)	LOICZ – water quality, DIN levels
Greece, Amvrakikos Gulf	1	Poor (1996-98)		Tsangaris et al. (2010)	Combination of bioenergetics and biochemical biomarkers in mussels
Greece, Saronikos Gulf	1	Good (2000-2012)	Improving (2000-2012)	Simboura et al (2013)	Benthic communities (BENTIX index)
Romania, Danube Delta	1	Very Poor (2001-2007)	Stable (2001-2007)	Török et al (2008)	Eutrophication
Norway, Southern fjords on the Skagerrak coast		Poor (2014)	Declining	www.environment.no /Topics/Marine-areas/ Coastal-waters/	Eutrophication problem areas, sugar kelp forests have almost disappeared
Norway, Nothern fjords		Good (2014)	Declining	www.environment.no /Topics/Marine-areas/ Coastal-waters/	
Antarctica					
Prydz Bay fjords	1	Very Good (2010)		O'Brien et al. (2012)	Benthic habitat mapping
Antarctic Peninsula, Fjords		Very Good (2010)		Grange and Smith (2013)	Benthic megafaunal abundance, community structure, and species diversity

References related to the table of condition assessments

Ahmed M.K., Islam, M., Shahidul Islam, Md., Rezaul Haque, Md., Shafiur Rahman and Monirul Islam, Md. (2010). Heavy Metals in Water, Sediment and Some Pelagic and Benthic Fishes of Buriganga River, Bangladesh. *International Journal of Environmental Research*. Vol. 4. No. 2, 321-332.

Ahmed, M.K., Mehedi, M.Y., Rezaul Haque and Mandal, P. (2011). Heavy Metals in Benthic Fauna of the Sundarbans Reserved Forest. *Journal of Environmental Monitoring and Assessment*. DOI 10.1007/s10661-010-1651-9.

Awosika, L.F., Ojo, O, Ajayi T.A. et al., (1993). *Implications of climate changes and sea level rise on the Niger Delta, Nigeria Phase 1*. A report for UNEP Nairobi.

Awosika L.F. and Folorunsho, R. (2014). Estuarine and ocean circulation dynamics in the Niger Delta, Nigeria: Implications for oil spill and pollution management. In *Estuaries of the World: Addressing the land/sea interactions challenges in the coastal zone of West Africa*. Pub Springer, 77-86.

Balushkina, E.V. (2009). Assessment of the Neva Estuary ecosystem state on the basis of structural characteristics of benthic animal communities in 1994 to 2005. *Inland Water Biology* 2, 355-363.

Barrie, J.V., Greene, H.G., Conway, K.W., Picard, K., (2012). Ch. 44: Inland Tidal Sea of the Northeastern Pacific, in: Harris, P.T., Baker, E.K. (Eds.), *Seafloor geomorphology as benthic habitat: GeoHAB Atlas of seafloor geomorphic features and benthic habitats*. Elsevier, Amsterdam, pp. 623-634.

Borja, A., Basset, A., Bricker, S., Dauvin, J., Elliot, M., Harrison, T., Marques, J., Weisberg, S., West, R. (2012). Classifying ecological quality and integrity of estuaries, in: Wolanski, E., McLusky, D.S. (Eds.), *Treatise on Estuarine and Coastal Science*. Academic Press, Waltham, pp. 125-162.

Bricker, S., Longstaff, B., Dennison, W.C., Jones, A., Boicourt, K., Wicks, C., Woerner, J. (2007). *Effects of Nutrient Enrichment in the Nation's Estuaries: A Decade of Change*. Series No. 26. National Centers for Coastal Ocean Science. National Oceanic and Atmospheric Administration, Silver Spring, MD, p. 328.

Burone, L., Venturini, N., Sprechmann P., Valente P., Muniz, P. (2006). Foraminiferal responses to polluted sediments in Montevideo coastal zone, Uruguay. *Marine Pollution Bulletin*, v.: 52, p.: 61 – 73.

Burone, L., Muniz, P. et al. (2011). Evolución paleoambiental de la Bahía de Montevideo (Uruguay) – bases para el establecimiento de un modelo ambiental. Libro: *El Holoceno en la zona costera del Uruguay*. p.: 197 - 227, Organizadores: F. García-Rodríguez Editorial: Tradinco, Montevideo. ISSN/ISBN: 9789974007574.

Buruaem, L.M., de Castro, I.B., Hortellani, M.A., Taniguchi, S., Fillmann, G., Sasaki, S.T., Petti, M.A.V., Sarkis, J.E.S., Bícego, M.C., Maranho, L.A., Davanso, M.B., Nonato, E.F., Cesar, A., Costa-Lotufo, L.V., Abessa, D.M.S. (2013). Integrated quality assessment of sediments from harbour areas in Santos-São Vicente Estuarine System, Southern Brazil. *Estuarine, Coastal and Shelf* Science 130, 179-189.

Chainho, P., Costa, J.L., Chaves, M.L., Dauer, D.M., Costa, M.J. (2007). Influence of seasonal variability in benthic invertebrate community structure on the use of biotic indices to assess the ecological status of a Portuguese estuary. *Marine Pollution Bulletin* 54, 1586–1597.

Chen, X., Gao, H., Yao, X., Chen, Z., Fang, H., Ye, S. (2013). Ecosystem health assessment in the Pearl River Estuary of China by considering ecosystem coordination. *PLoS ONE* 8, doi:10.1371/journal.pone.0070547.

Cochrane, G.R., Trusel, L., Harney, J., Etherington, L. (2012). Ch. 18: Habitats and benthos of an evolving fjord, Glacier Bay, Alaska. , in: Harris, P.T., Baker, E.K. (Eds.), *Seafloor geomorphology as benthic habitat: GeoHAB Atlas of seafloor geomorphic features and benthic habitats*. Elsevier, Amsterdam, pp. 299-308.

Conde, D., Rodríguez-Gallego, L. (2002). Problemática ambiental y gestión de las lagunas costeras atlánticas de Uruguay. In: Domínguez, A., Prieto, R. (Eds.), *Perfil Ambiental* 2002. NORDAN, Montevideo, pp. 149–166.

Dang, H., Chen, R., Wang, L., Guo, L., Chen, P., Tang, Z., Tian, F., Li, S., Klotz, M.G. (2010). Environmental factors shape sediment anammox bacterial communities in hypernutrified Jiaozhou Bay, China. *Applied Environmental Microbiology* 76, 7036–7047.

Dankers, N., van Duin, W., Baptist, M., Dijkman, E., Cremer, J. (2012). Ch. 11: The Wadden Sea in the Netherlands: Ecotopes in a World Heritage barrier island system. , in: Harris, P.T., Baker, E.K. (Eds.), *Seafloor geomorphology as benthic habitat: GeoHAB Atlas of seafloor geomorphic features and benthic habitats*. Elsevier, Amsterdam, pp. 213-226.

Danulat, E., Muniz, P., García-Alonso, J., Yannicelli, B. (2002). First assessment of the highly contaminated harbour of Montevideo, Uruguay. *Marine Pollution Bulletin*, v.: 44, p.: 554 – 565.

Deinet, S., McRae, L., De Palma, A., Manley, R., Loh, J., Collen, B. (2010). *The Living Planet Index for Global Estuarine Systems: Technical Report*. Indicators and Assessments Unit, Institute of Zoology, Zoological Society of London, U.K. and WWF International, Gland, Switzerland, p. 49.

Defeo, O., Horta, S., Carranza, A., Lercari, D., de Álava, A., Gómez, J., Martínez, G., Lozoya, J.P., Celentano, E. (2009). *Hacia un Manejo Ecosistémico de Pesquerías. Áreas Marinas Potegidas en Uruguay*. Facultad de Ciencias-DINARA, Montevideo.

Department of Environment (2011). *Australia State of the Environment 2011: Marine Environment*. Canberra: Australian Department of Sustainability, Environment, Water, Population and Communities on behalf of the State of the Environment 2011 Committee. http://environment.gov.au/soe.

Díaz-Jaramillo, M., Socowsky, R., Pardo, L.M., Monserrat, J.M., Barra, R. (2013). Biochemical responses and physiological status in the crab Hemigrapsus crenulatus (Crustacea, Varunidae) from high anthropogenically-impacted estuary (Lenga, south-central Chile). *Marine Environmental Research* 83, 73-81.

Dupra, V., Smith, S.V., David, L.T., Waldron, H., Marshall Crossland, J.I., Crossland, C.J. (2002). *Estuarine systems of Africa: carbon, nitrogen and phosphorus fluxes*., LOICZ Reports and Studies. Land Ocean Interactions in the Coastal Zone (LOICZ), Texel, p. 82.

Dürr, H.H., Laruelle, G.G., Kempen, C.M.v., Slomp, C.P., Meybeck, M., Middelkoop, H. (2011). Worldwide typology of nearshore coastal systems: defining the estuarine filter of river inputs to the oceans. *Estuaries and Coasts* 34, 441-458.

El-Asmar, H.M., Al-Olayan, H.A. (2013). Environmental impact assessment and change detection of the coastal desert along the central Nile Delta coast, Egypt. *International Journal of Remote Sensing Applications* 3, 1-12.

Entry, J., Gottlieb, A. (2014). The impact of stormwater treatment areas and agricultural best management practices on water quality in the Everglades Protection Area. *Environmental Monitoring and Assessment* 186, 1023-1037.

Environment Management Bureau, Ministry of the Environment (2005-2013). *Result of water quality measurement in Public Water Body*, FY2004-2012 (in Japanese). http://www.env.go.jp/water/suiiki/index.html

Environmental Protection Agency (Ireland), (2006). *Water Quality in Ireland 2005. Key Indicators of the Aquatic Environment*. EPA, Wexford, 23 pp.

Environment Southland (2011). *Estuary health, New River Estuary 2010-2011*, New Zealand. http://www.es.govt.nz/media/16133/nre-web.pdf

EPA, (2007). *National Estuary Program Coastal Condition Report*. United States Environmental Protection Agency, Washington DC. http://www.epa.gov/owow/

oceans/nepccr/index.html

Ezhova, E., Dorokhov, D., Sivkov, V., Zhamoida, V., Ryabchuk, D., Kocheshkova, O. (2012). Ch. 43: Benthic habitats and benthic communities in South-Eastern Baltic Sea, Russian sector. , in: Harris, P.T., Baker, E.K. (Eds.), *Seafloor geomorphology as benthic habitat: GeoHAB Atlas of seafloor geomorphic features and benthic habitats*. Elsevier, Amsterdam, pp. 613-622.

Fan, H., Huang, H. (2008). Response of coastal marine eco-environment to river fluxes into the sea: A case study of the Huanghe (Yellow) River mouth and adjacent waters. *Marine Environmental Research* 65, 378-387.

Fianko, J.R., Osae, S., Adomako, D., Adotey, D.K., Serfor-Armah, Y. (2007). Assessment of Heavy Metal Pollution of the Iture Estuary in the Central Region of Ghana. *Environmental Monitoring and Assessment* 131, 467-473.

Folorunsho, R., Awosika, L.F. and Dublin-Green, C.O. (1994). An assessment of river imputs into the Gulf of Guinea shelf. In *Proc. International symposium on the results of the first IOCEA cruise in the Gulf of Guinea, 17-20 May 1994*. p.163-172.

Folorunsho, R., Awosika, L.F., in press. Morphological Characteristics of the Bonny and Cross River (Calabar) Estuaries in Nigeria: Implications for Navigation and Environmental Hazards. In *Estuaries of the World - Addressing the land/sea interactions challenges in the coastal zone of West Africa*. Pub Springer.

FREPLATA, (2004). *Análisis Diagnóstico Transfronterizo del Río de la Plata y su Frente Marítimo*. Documento Técnico. Proyecto Protección Ambiental del Rió de la Plata y su Frente Marítimo. Proyecto PNUD/ GEF/RLA/99/G31.

García-Rodríguez, F., Brugnoli, E., Muniz, P., Venturini, N., Burone, L., Hutton, M., Rodríguez, M., Pita, A., Kandratavicius, N., Pérez, L., Verocai, J. (2013). Warm phase ENSO events modulate the continental freshwater input and the trophic state of sediments in a large South American estuary. *Marine and Freshwater Research*, v.: 65 1, p.: 1 – 11.

Garcia-Rodriguez, F., Del Puerto, L., Venturini, N., Pita, A., Brugnoli, E., Burone, L., Muniz, P. (2011) Diatoms, proteins and carbohydrates content as proxies for coastal eutrophication in Montevideo, Rio de la Plata, Uruguay. *Brazilian Journal of Oceanography*, v.: 59 4, p.: 293 - 310, 2011.

García-Rodríguez, F., Hutton, M., Brugnoli, E., Venturini, N., Del Puerto, L., Inda, H., Bracco, R., Burone, L., Muniz, P. (2010) Assessing the effect of natural variability and human impacts on the environmental quality of a coastal metropolitan area (Montevideo, Bay, Uruguay). *Pan-American Journal of Aquatic Sciences*, v.: 5 1, p.: 90 – 99.

Giordani, G., Viaroli, P., Swaney, D.P., Murray, C.N., Zaldívar, J.M., Crossland, J.I.M. (2005). *Nutrient fluxes in transitional zones of the Italian coast*, LOICZ Reports & Studies No. 28. Land-Ocean Interactions in the Coastal Zone Core Project of the IGBP and the IHDP, Texel, the Netherlands, p. 157.

Gómez, N., Licursi, M., Bauer De, Ambrosio, E.S., Rodríguez-Capítulo, A. (2012). Assessment of biotic integrity of the coastal freshwater tidal zone of a temperate estuary of South America through multiple indicators. Estuaries and Coasts, DOI: 10.1007/s12237-012-9528-5

Gómez-Erache, M., Vizziano, D., Muniz, P. and Nagy, G.J. (2001). The Health of the Río de la Plata system: Northern Coast, Uruguay. In '*Opportunity and Challenges for Protecting, Restoring and Enhancing Coastal habitats in the Bay of Fundy. Proceedings of the 4th Bay of Fundy Science Worshops, Saint John, New Brunswick*. Environment Canada, Atlantic Region. Occasional Report N8 17'. (Eds T. Chopin and P. G. Wells.) pp. 17–35. (Environment Canada: Darmouth, Nova Scotia.)

Gordeev, V.V., Andreeva, E.N., Lisitzin, A.P., Kremer, H.H., Salomons, W., Crossland, J.I.M. (2006). *Russian Arctic Basins*, LOICZ Reports & Studies No. 29. Land-Ocean Interactions in the Coastal Zone Core Project of the IGBP and the IHDP, Geesthacht, Germany, p. 95.

Grange, L.J., Smith, C.R., (2013). Megafaunal Communities in Rapidly Warming Fjords along the West Antarctic Peninsula: Hotspots of Abundance and Beta Diversity. *PLoS ONE* 8, e77917.

Gu, J., Salem, A., Chen, Z. (2013). Lagoons of the Nile delta, Egypt, heavy metal sink: With a special reference to the Yangtze estuary of China. *Estuarine, Coastal and Shelf Science* 117, 282-292.

Haldar, S., Mandal, S., Thorat, R.B., Goel, S., Baxi, K., Parmer, N., Patel, V., Basha, S., Mody, K.H. (2014). Water pollution of Sabarmati River,Äîa Harbinger to potential disaster. *Environmental Monitoring and Assessment* 186, 2231-2242.

Harris, P.T., Heap, A.D. (2003). Environmental management of coastal depositional environments: inferences from an Australian geomorphic database. *Ocean and Coastal Management*, 46, 457-478.

Harrison, T., Whitfield, A. (2006). Application of a multimetric fish index to assess the environmental condition of South African estuaries. *Estuaries and Coasts* 29, 1108-1120.

Hung, J.J., Huang, W.C., Yu, C.S. (2013). Environmental and biogeochemical changes following a decade's reclamation in the Dapeng (Tapong) Bay, southwestern Taiwan. *Estuarine, Coastal and Shelf Science* 130, 9-20.

Ifremer, (2008). *Réseau de Suivi Lagunaire du Languedoc-Roussillon: Bilan des résultats 2007*. Rapport RSL-08/2008, 363 pp. http://rsl.cepralmar.com/bulletin.html (accessed March, 2014).

James, J.W.C., Mackie, A.S.Y., Rees, E.I.S., Darbyshire, T. (2012). Ch. 12: Sand wave field: The OBel Sands, Bristol Channel, U.K. , in: Harris, P.T., Baker, E.K. (Eds.), *Seafloor geomorphology as benthic habitat: GeoHAB Atlas of seafloor geomorphic features and benthic habitats*. Elsevier, Amsterdam, pp. 227-240.

Kim, Y.-S., Eun, H., Cho, H.-S., Kim, K.-S., Sakamoto, T., Watanabe, E., Baba, K., Katase, T. (2008). Organochlorine Pesticides in the Sediment Core of Gwangyang Bay, South Korea. *Archives of Environmental Contamination and Toxicology* 54, 386-394.

KIOST, (2013). *Development of Marine Ecosystem Health Index (MEHI) in the special management areas of the South Sea*. Korea Institute of Ocean Science & Technology, Ansan, Gyeonggi-do, South Korea, p. 13.

Kjerfve, B., Wiebe, W.J., Kremer, H.H., Salomons, W., Crossland, J.I.M., Morcom, N., Harvey, N. (2002). *Caribbean Basins: LOICZ Global Change Assessment and Synthesis of River Catchment/Island-Coastal Sea Interactions and Human Dimensions; with a desktop study of Oceania Basins*., LOICZ Reports & Studies No. 27. LOICZ IPO, Texel, The Netherlands, p. 174.

Kotilainen, A.T., Kaskela, A.M., Bäck, S., Leinikki, J. (2012). Ch. 17: Submarine De Geer moraines in the Kvarken Archipelago, the Baltic Sea. , in: Harris, P.T., Baker, E.K. (Eds.), *Seafloor geomorphology as benthic habitat: GeoHAB Atlas of seafloor geomorphic features and benthic habitats*. Elsevier, Amsterdam, pp. 289-298.

Lee, Y.S., Kang, C.-K. (2010). Causes of COD increases in Gwangyang Bay, South Korea. *Journal of Environmental Monitoring* 12, 1537-1546.

Lim, D., Jung, H.S., Kim, K.T., Shin, H.H., Jung, S.W. (2012). Sedimentary records of metal contamination and eutrophication in Jinhae-Masan Bay, Korea. *Marine Pollution Bulletin* 64, 2542-2548.

Liu, L., Li, B., Lin, K., Cai, W., Wang, Q. (2013). Assessing benthic ecological status in coastal area near Changjiang River estuary using AMBI and M-AMBI. *Chinese Journal of Oceanology and Limnology*, 1-16.

Lotze, H.K., Lenihan, H.S., Bourque, B.J., Bradbury, R.H., Cooke, R.G., Kay, M.C., Kidwell, S.M., Kirby, M.X., Peterson, C.H., Jackson, J.B.C. (2006). Depletion, degradation, and recovery potential of estuaries and coastal seas. *Science*, 312, 1806-1809.

Mohamed, A.-R.M., Resen, A.K., Taher, M.M. (2012). Longitudinal patterns of fish community structure in the Shatt Al-Arab River, Iraq. Basrah *Journal of Science* 30, 65-86.

Moscoso, J., Rudolph, A., Sepúlveda, R.D., Suárez, C. (2006). Effect of Temporary Closure of the Mouth of an Estuary on the Benthic Macroinfauna: Lenga-Chile, A Case Study. *Bulletin of Environmental Contamination and Toxicology* 77, 484-491.

Muniz, P.; Venturini, N.; Martínez, A. (2002). Physico-chemical characteristics and pollutants of the benthic environment in the Montevideo coastal zone, Uruguay. *Marine Pollution Bulletin*, v.: 44, p.: 962 – 968.

Muniz, P., Danulat, E.; Yannicelli, B. García-Alonso, J., Medina, G., Bícego, M. C., (2004). Assessment of contamination by heavy metals and petroleum hydrocarbons in sediments of Montevideo harbour (Uruguay). *Environment International*, v.: 29, p.: 1019 – 1028.

Muniz, P., Venturini N., Pires-Vanin, Ams, Tommasi Lr, Borja A. (2005). Testing the applicability of a marine biotic index (AMBI) to assessing the ecological quality of soft-bottom benthic communities, in the South America Atlantic region. *Marine Pollution Bulletin*, v.: 50, p.: 624 – 637.

Muniz, P., Venturini, N. (2011). Environmental conditions in the coast of Montevideo, Uruguay: historical aspects, present status and perspectives of habitat degradation and uses. Libro: *Encyclopedia of Environmental Health*. v.: 5 , 1, p.: 590 - 601, Elsevier B.V , Amsterdam.

Muniz, P., Venturini, N., Hutton M; Kandratavicius N., Pita A., Brugnoli E., Burone, L., Garcia-Rodriguez F. (2011). Ecosystem health of Montevideo coastal zone: a multi approach using some different benthic indicators to improve a ten-year-ago assessment. *Journal of Sea Research*, v.: 65, p.: 38 – 50.

Muniz, P., Hutton M., Kandratavicius N., Lanfranconi A., Brugnoli E., Venturini N., Gimenez L. (2012). Performance of biotic indices in naturally stressed estuarine environments on the Southwestern Atlantic coast (Uruguay): a multiple scale approach. *Ecological Indicators*, v.: 19, p.: 89 – 97.

Muxika, I., Borja, A., Bald, J. (2007). Using historical data, expert judgement and multivariate analysis in assessing reference conditions and benthic ecological status, according to the European Water Framework Directive. *Marine Pollution Bulletin* 55, 16-29.

Nagy, G. J., Gomez-Erache, M., Lopez, C. H., and Perdomo, A. C. (2002). Distribution patterns of nutrients and symptoms of eutrophication in the Rio de la Plata River Estuary System. *Hydrobiologia* 475/476, 125–139. doi:10.1023/A:1020300906000.

Nagy, G.J., Gómez-Erache, M., Martínez, C.M. and Perdomo, A.C. (2000). "Nutrient over-enrichment surveillance: quintennial time-scale modeling and monitoring in the Rio de la Plata coastal system (Uruguay–Argentina)". Nutrient Over-Enrichment Symposium, Poster session, National Academy of Sciences, Washington D.C., U.S.A., October 2000.

New York and New Jersey Harbor and Estuary Program ((2012)) *State of the estuary 2012*. http://www.harborestuary.org/pdf/StateOfTheEstuary2012/Factsheet_English.pdf

Van Niekerk, L., Adams, J.B., Bate, G.C., Forbes, A.T., Forbes, N.T., Huizinga, P., Lamberth, S.J., MacKay, C.F., Petersen, C., Taljaard, S., Weerts, S.P., Whitfield, A.K., Wooldridge, T.H. (2013). Country-wide assessment of estuary health: An approach for integrating pressures and ecosystem response in a data limited environment. *Estuarine, Coastal and Shelf Science* 130, 239-251.

Venturini, N., Volpedo, A., Muniz, P. (in press). Contamination in the Río de la Plata and its Maritime Front: water, sediment and biota. In: *Environmental assessment and planning of a trans-boundary fluvio-marine ecosystem: the Río de la Plata and its maritime front*. A. Brazeiro; A. Volpedo; M. Gómez-Erache & C. Lasta (Eds.), Springer-Verlag.

Venturini, N., Muniz, P., Rodríguez, M. (2004). Macrobenthic subtidal communities in relation to sediment pollution: the phylum-level meta-analysis approach in a south-eastern coastal region of South America. *Marine Biology*, v.: 144, p.: 119 – 126.

Venturini, N., Pita, A., Brugnoli, E., García-Rodríguez, F., Burone, L., Kandratavicius, N., Hutton, M., Muniz, P. (2012). Benthic trophic status of sediments in a metropolitan area (Rio de la Plata estuary): linkages with natural and human pressures. *Estuarine Coastal and Shelf Science*, v.: 112, p.: 139 – 152.

Waikato Regional Council, (2013). *Estuarine water quality report*. http://www.waikatoregion.govt.nz/Environment/Environmental-information/Environmental-indicators/Coasts/Coastal-water-quality/Estuarine-water-quality-report/

Wang, B. (2007). Assessment of trophic status in Changjiang (Yangtze) River estuary. *Chinese Journal of Oceanology and Limnology* 25, 261-269.

Ward, T.J. (2012). *Workshop Report: Regional Scientific and Technical Capacity Building Workshop on the World Ocean Assessment (Regular Process)*. UNEP/COBSEA, Bangkok, Thailand, 60 pp.

Zhu, Z., Cai, X., Giordano, M., Molden, D., Hong, S., Zhang, H., Lian, Y., Li, H., Zhang, X., Zhang, X., Xue, Y. (2003). Yellow river comprehensive assessment: Basin features and issues, *Working Paper* 57. International Water Management Institute, Colombo, Sri Lanka, p. 31.

45 Hydrothermal Vents and Cold Seeps

Contributors:
Nadine Le Bris (Convenor), Sophie Arnaud-Haond, Stace Beaulieu, Erik Cordes, Ana Hilario, Alex Rogers, Saskia van de Gaever (Lead Member), Hiromi Watanabe

Commentators:
Françoise Gaill, Wonchoel Lee, Ricardo Serrão-Santos

The chapter contains some material (identified by a footnote) originally prepared for Chapter 36F (Open Ocean Deep Sea). The contributors to that chapter were Jeroen Ingels (Convenor), Patricio Bernal (Lead Member), Malcolm R. Clark, Katsunori Fujikura, Andrew R. Gates, Daniel O. B. Jones, Lisa A. Levin, Bhavani E. Narayanaswamy, Jose Angel A. Perez, Imants G. Priede, Ashley A. Rowden, Henry A. Ruhl, Ricardo Santos, Craig R. Smith, Paul Snelgrove, Thomas Soltwedel, Tracey Sutton, Andrew K. Sweetman, Saskia Van Gaever (Co-Lead Member), Michael Vecchione, Moriaki Yasuhara and Linda Amaral Zettler.

Hydrothermal Vents and Cold Seeps

Chapter 45

1 Inventory

Hydrothermal vents and cold seeps constitute energy hotspots on the seafloor that sustain some of the most unusual ecosystems on Earth. Occurring in diverse geological settings, these environments share high concentrations of reduced chemicals (e.g., methane, sulphide, hydrogen, iron II) that drive primary production by chemosynthetic microbes (Orcutt et al. 2011). Their biota are characterized by a high level of endemism with common specific lineages at the family, genus and even species level, as well as the prevalence of symbioses between invertebrates and bacteria (Dubilier et al., 2008; Kiel, 2009).

Hydrothermal vents are located at mid-ocean ridges, volcanic arcs and back-arc spreading centres or on volcanic hotspots (e.g., Hawaiian archipelago), where magmatic heat sources drive the hydrothermal circulation. Venting systems can also be located well away from spreading centres, where they are driven by exothermic, mineral-fluid reactions (Kelley, 2005) or remanent lithospheric heat (Wheat et al., 2004). Of the 521 vent fields known (as of 2009), 245 are visually confirmed, the other being inferred active by other cues such as tracer anomalies (e.g. temperature, particles, dissolved manganese or methane) in the water column (Beaulieu et al., 2013) (Figure 45.1).

Sediment-hosted seeps occur at both passive continental margins and subduction zones, where they are often supported by subsurface hydrocarbon reservoirs. The migration of hydrocarbon-rich seep fluids is driven by a variety of geophysical processes, such as plate subduction, salt diapirism, gravity compression or the dissociation of methane hydrates. The systematic survey of continental margins has revealed an increasing number of cold seeps worldwide (Foucher et al., 2009; Talukder, 2012). However, no recent global inventory of cold seeps is available.

Both vent and seep ecosystems are made up of a mosaic of habitats covering wide ranges of potential physico-chemical constraints for organisms (e.g., in temperature, salinity, pH, and oxygen, CO_2, hydrogen sulphide, ammonia and other inorganic volatiles, hydrocarbon and metal contents) (Fisher et al., 2007; Levin and Sibuet, 2012; Takai and Nakamura, 2010). Some regions (e.g., Mariana Arc or Costa Rica margin) host both types of ecosystems, forming a continuum of habitats that supports species with affinities for vents or seeps (Watanabe et al., 2010; Levin et al., 2012). Habitats indirectly related to hydrothermal venting include inactive sulphide deposits and hydrothermal sediments (German and Von Damm, 2004). Similarly, cold-water corals growing on the carbonate precipitated from the microbial oxidation of methane are among the seep-related habitats, although they typically occur long after seepage activity has ceased (Cordes et al., 2008; Wheeler and Stadnitskaya, 2011).

2 Features of trends in extent or quality

Chemosynthetic ecosystems in the deep sea were first discovered 40 years ago using towed camera systems and manned submersibles; hydrothermal vents in 1977 for diffuse vents on the Galapagos Spreading Center (Corliss et al. 1979), and in 1979 for black smokers on the East Pacific Rise (Spiess et al., 1980), and cold seeps at the base of the Florida escarpment in the Gulf of Mexico in 1984 (Paull et al., 1984). Compared to other deep-sea settings, the exploration of vent and seep

The boundaries and names shown and the designations used on this map do not imply official endorsement or acceptance by the United Nations.

Figure 45.1 | Global map from InterRidge database (http://vents-data.interridge.org/maps) displaying visually confirmed and inferred hydrothermal vents fields. Credits: Beaulieu, S., Joyce, K., Cook, J. and Soule, S.A. Woods Hole Oceanographic Institution (2015); funding from Woods Hole Oceanographic Institution, U.S. National Science Foundation #1202977, and InterRidge. Data sources: InterRidge Vents Database, Version 2.1, release date 8 November 2011; University of Texas PLATES Project plate boundary shapefiles.

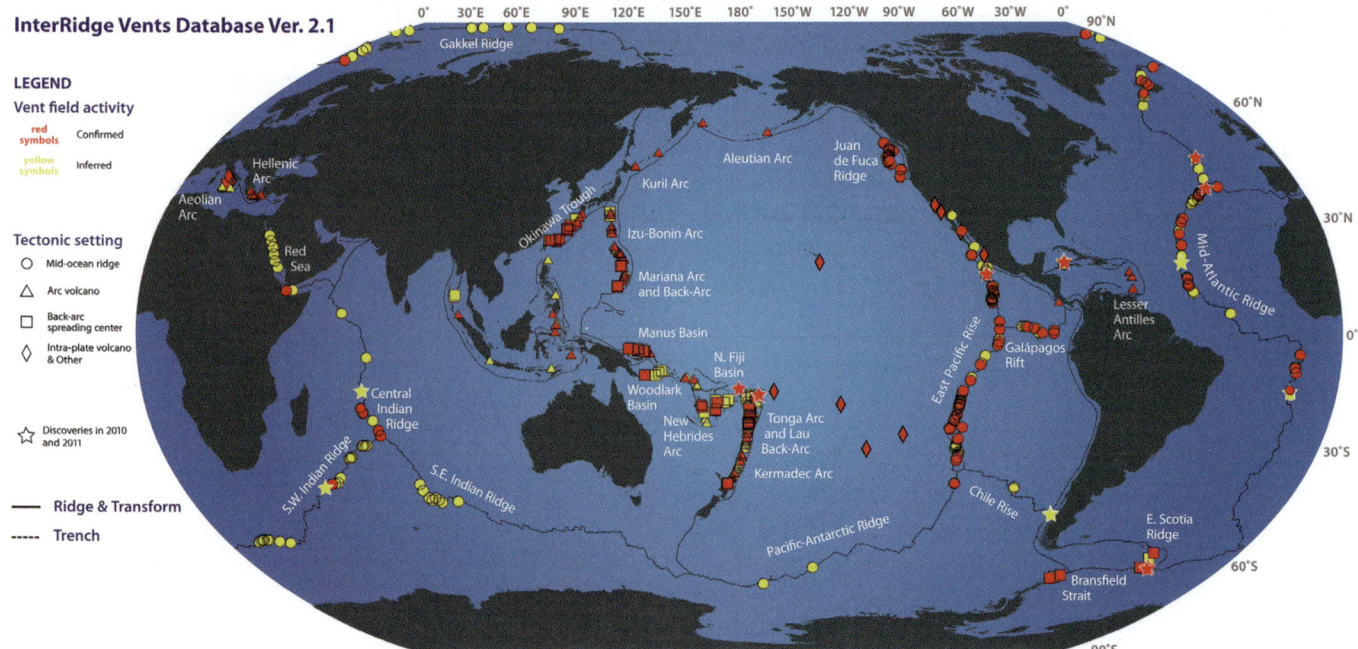

habitats is thus recent (Ramirez-Llodra et al., 2011). In the last decade, high-resolution seafloor mapping technologies using remotely operated vehicles (ROVs) and autonomous underwater vehicles (AUVs) have yet enhanced the capacity to explore the deep seabed.

Since the last global compilation (Baker and German, 2004), the known number of active hydrothermal vent fields has almost doubled, with an increasing proportion of new discoveries being in arc and back-arc settings, as a result of increasing exploration efforts (Beaulieu et al., 2013). These exploration efforts have emphasized the highly heterogeneous and patchy distribution of habitats associated with diffuse and focused-flow vents, hosting diverse microbial and faunal communities.

A large spatial and temporal variability of vent fluid temperature and chemical properties over mid-ocean ridges and arc and back-arc settings has been described, in relation to different geological substrate, hydrothermal activity, volcanic or tectonic instability (e.g. eruptions) (German and Von Damm, 2004; Charlou et al., 2010). This variability generates strong environmental constraints on chemosynthetic primary producers (Amend et al., 2011; Le Bris and Duperron, 2010; Takai and Nakamura, 2010) and dominant fauna including vent-endemic species of tubeworms, mussels, gastropods, clams, shrimp or crabs (Desbruyères et al., 2001; Fisher et al., 2007; Watanabe et al., 2010).

Common features are shared among vent and seep ecosystems. At seeps, microbial consortia oxidizing methane and their end-product (e.g., sulphide) sustain abundant microbial populations exhibiting diverse metabolic pathways. These microbes produce large amounts of organic matter, fuelling high invertebrate biomass. Many forms of symbiosis between chemosynthetic bacteria and host-invertebrates have adapted specifically to the energy-rich environmental conditions of vent and seep environments (Dubilier et al., 2008).

Within methane seeps and other types of seeps, different seepage intensities create distinct habitats dominated by chemosynthetic bacterial mats, and endemic and non-endemic species of tubeworms, mussels, gastropods, clams, shrimp or crabs, and supporting numerous associated invertebrate species. Faunal biodiversity is generally low within each seep habitat (Levin, 2005) but the vast array of geomorphic and biogenic habitats, each with highly adapted species, contributes significantly to beta diversity in the deep-sea (Cordes et al., 2010). Extensive trophic niche partitioning of microbes by heterotrophs also contributes to biodiversity at seeps (Levin et al., 2013). Methane seep sediments share many species with surrounding margin sediments (Levin et al., 2010), and numerous families and genera with hydrothermal vents and organic falls (Bernardino et al., 2012)[1].

Despite recent global efforts, biological inventories are still largely incomplete. At the end of the Census of Marine Life programme (CoML, Crist et al., 2010), the ChEssBase dedicated to chemosynthetic ecosystems reported 700 hydrothermal vent species and 600 species from cold seeps (Ramirez-Llodra and Blanco, 2005; German et al., 2011). Around 200 new species were reported between 2002-2010 (i.e., 25 species/year), most of them belonging to mega and macrofauna. Much remains to be described, particularly in the meiofauna. Currently, vent fauna constitute between 7 and 11 biogeographic provinces, including new discoveries in the Arctic and Southern Oceans (Bachraty et al., 2009; Moalic et al., 2012; Rogers et al., 2012). Each newly identified vent or seep contains a diversity of unidentified species and these biogeographic patterns should be considered as preliminary (German et al., 2011). Furthermore, new types of chemosynthetic ecosystems are still being discovered, such as serpentinite-hosted ecosystems found on continental margins, ridges and trenches (Ohara et al., 2012; Kelley, 2005).

Few vent and seep areas have repeated observations over more than ten years from which temporal trends can be described (Glover et al., 2010). Recolonization of habitats impacted by volcanic eruptions was repeatedly documented on the 9°50′N vent field of the fast-spreading East Pacific Rise (Shank et al., 1998) and over different locations of the intermediate-spreading Juan de Fuca Ridge (Tunnicliffe et al., 1997). The resilience capacity of vent communities, however, cannot be generalized from the re-establishment of microbial communities and few dominant fauna species adapted to these highly unstable systems. Even on those areas, persistent effects on larvae patterns have been documented over years after an eruption suggesting long-lasting impacts on community recovery (Mills et al., 2013). Volcanic activity is furthermore much less frequent at slow or ultra-slow spreading ridges (e.g. Mid-Atlantic Ridge), resulting in a much lower frequency of natural perturbations, and in the absence of knowledge about the potential response of their specific communities to major disturbance. Succession at cold seeps, including later stages of deep-water coral colonization, may proceed over centuries to millennia with slow-growing and long-lived species that should be considered particularly vulnerable to disturbance (Cordes et al., 2009).

Life histories of key species and their links with resource and habitat variability have just started to be described (Ramirez-Llodra et al., 2010). Important biodiversity components supporting ecosystem functions also remain under-studied. In particular, a much lower number of studies dealt with meiofaunal organisms (< 1mm) from vent and seep sites than with macrofauna (Vanreusel et al., 2010). Following the massive molecular inventories allowed by New Generation Sequencing, we are just now getting glimpses into the diversity of microbes in both vent and seep environments. That diversity appears higher by orders of magnitude than those hitherto revealed with classical sequencing and cloning tools. Insufficient knowledge of the drivers of ecosystem and community dynamics at vent and seeps makes anticipating any trends in their ecological status problematic or even elusive in a context of multiple pressures.

---

1 Text originally prepared for Chapter 36F (Open Ocean Deep Sea)

3 Major pressures linked to the trends

The deep sea is being seen as a new frontier for hydrocarbon and mineral resource extraction, as a response to increasing demand for raw materials for emerging high-technology industries and worldwide urbanization. As a consequence, vent and seep ecosystems, so far preserved from direct impacts of human activities, are confronted with increasing pressures (Ramirez-Llodra et al., 2011; Santos et al., 2012).

Offshore oil extraction increasingly occurs in waters as deep as 3000 m and exploration for oil and gas now predominantly occurs in deep water (> 450m) or ultra-deep water (> 1500m depth), where typical seep ecosystems are found. Seafloor installations can directly affect cold seep communities in their impact area, if visual surveys and Environmental Impact Assessments (EIAs) are not completed prior to drilling. In addition, an increasing threat exists of large-scale impacts from accidental spills, such as the 2010 *Deepwater Horizon* blowout in the Gulf of Mexico, which was the largest accidental release of oil into the ocean in human history (McNutt et al., 2012) with a significant impact on surrounding deep-seabed habitats (Montagna et al., 2013; Fisher et al., 2014).

Further pressures on cold seep communities may arise from the combined effects of increasing demand for energy and technological progress in the exploitation of new types of energy resources. This type of development is shown by the world's first marine methane hydrate production test in the Nankai Trough in 2013. Sequestration of CO_2 in deep-sea sedimentary disposal sites and igneous rocks (Goldberg et al., 2008) should also be considered a potential threat specific to these communities (IPCC, 2005).

The increased demand for metals is promoting deep-sea mineral resource exploration both within Exclusive Economic Zones (EEZs) and in the Area (as defined in the United Nations Convention on the Law of the Sea), raising the issue of potential impacts on vent ecosystems (Van Dover, 2012). In 2011, the granting of a mining lease to exploit sulphide minerals for gold, copper and zinc in the EEZ of Papua New Guinea will shortly turn the deep-sea mining industry into a reality. Additionally, in the last five years, the International Seabed Authority has granted two new exploration permits for polymetallic sulphide deposits and two others are about to be signed for sites on the Atlantic and Indian mid-ocean ridges (http://www.isa.org.jm/en/scientific/exploration/contractors).

Significant threats are anticipated on the largely unknown communities associated with active hydrothermal deposits and the typical vent communities that can occur in close proximity to these areas (ISA, 2011). Inactive areas, which no longer had any detectable fluid venting with temperature anomaly as defined in the InterRidge Vents Database, have been mostly described so far in the vicinity of active areas with typical vent communities. These active areas where venting fluid is warmer than ambient, are inclusive of low-temperature diffuse flow and will require systematic exploration surveys and dedicated impact studies. Communities from inactive areas furthermore still need to be described.

It is, for example, unclear whether they encompass species assemblages closely related to deep seabed areas out of any hydrothermal influence, or whether they host specific fauna adapted to the local metal-rich substrate or to the proximity of highly productive chemosynthetic ecosystem at local to regional scale. Furthermore, despite the absence of high temperature associated with black smokers, some of these inferred 'inactive' areas may display diffuse flow vents, that are much more difficult to detect from water column surveys.

Indirect pressures on vent and seep ecosystems resulting from global anthropogenic forcing, including pollution and climate change, are not well constrained. These systems are less sensitive to changes in photosynthetic primary production than other deep-sea ecosystems, but potential threats also exist. Changes in water-mass circulation could affect larval dispersal, potentially reducing the capacity for species' populations to maintain themselves across fragmented habitats (Adams et al., 2011). The extension of hypoxia or anoxia on continental margins and in semi-enclosed seas could also profoundly alter the functioning of these ecosystems because of the high oxygen demand of chemosynthetic activity (Childress and Girguis, 2011). Warming is already affecting the deep ocean waters, especially at high latitudes (e.g., Arctic) and in enclosed seas (e.g., Mediterranean) hosting vent and seep ecosystems (Glover et al., 2010). Cold seep ecosystems could be affected, through direct impacts on the activity of fauna and the microbial consortia or major disturbances, such as land-slides and gas extrusion caused by hydrate destabilization.

These ecosystems occupy fairly small areas of the seabed (typically km-scale) and may be more vulnerable to common deep-sea pressures such as deep-sea fishing or waste dumping. Deep-sea fishing on seamounts flanks and margins down to at least 1500m depth are part of the existing pressure on cold seep, even though rarely documented so far (Ramirez-Llodra et al., 2011). This pressure is potentially exerted on vent communities occurring at those depths on mid-ocean ridge flanks or volcanic arc and back-arc seamount chains. Even activities such as scientific research or bioprospecting can pose a threat to the integrity of these unique communities and their endemic species (Baker et al., 2010). Impacts of ecotourism on vent environments should also be accounted, since this is a growing activity.

4 Implications for services to ecosystems and humanity

Chemosynthetic communities are functionally distinct from other marine communities, with capacity to form very high biomasses relative to other deep-sea ecosystems, though many questions remain open about their distribution, diversity, functioning and environmental features that limit the ability to estimate associated ecosystem services (Armstrong et al., 2012).

Nevertheless, deep-sea vents and seeps represent one of the most physically and chemically diverse biomes on Earth and have a strong potential for discovery of new species of eukaryotes and prokaryotes (Takai and Nakamura, 2011). Their specialized phyla are adapted to a range of environmental constraints. Archaea that live at extremes in pressure, temperature and pH are particularly attractive to industrial sectors (UNU-IAS, 2005). The hydrothermal vent and cold seep animals have evolved traits that allow them to not only tolerate extreme environmental conditions, but in some cases to accumulate and transport chemicals toxic to most other marine species (Childress and Fisher, 1992; Le Bris and Gaill, 2007).

This makes these ecosystems a vast genomic repository of unique value to screen for highly specific metabolic pathways and processes. The vent and seep biota thus constitute a unique pool of potential for the provision of new biomaterials, medicines and genetic resources that has already led to a number of patents (Gjerde, 2006; Arrieta et al., 2010; Thornburg et al., 2010). This great potential value to humankind is accounted for in the public awareness of potential threats and acceptability of deep-sea conservation programmes (Jobstvogt et al., 2014).

Chemosynthetic ecosystems are linked with adjacent deep-sea ecosystems through dispersing larvae and juveniles, and through the export of local productivity to mobile fauna and surrounding deep-sea corals and other filter-feeding communities, but the quantitative importance of their chemosynthetic production at the regional scale still remains to be appraised. At the global scale, a significant role of seep ecosystems is recognized in the regulation of methane fluxes, oxygen consumption and carbon storage from anaerobic methane oxidation by microbial consortia in sediments (Boetius and Wenzhöfer, 2013). Recent evidence shows that hydrothermal vent plumes sustain microbial communities with potential connections to zooplankton communities and biogeochemical fluxes in the deep ocean (Dick et al., 2013). The biological stabilization of metal (e.g., iron, copper) from hydrothermal vents under dissolved or colloidal organic complexes for long-range export in the water column has been documented recently (Wu et al., 2011; Hawkes et al., 2013). Recent assessments of these iron sources indicate their significance for deep-water budgets at oceanic scales and underscore the possibility for fertilizing surface waters through vertical mixing in particular regional settings (Tagliabue et al., 2010) and supporting long-range organic carbon transport to abyssal oceanic areas (German et al., 2015).

Because of their unique biodiversity and ecological functions in the Earth's biosphere, their geophysically-driven primary production sustained by chemosynthesis, their significance in global element cycles (i.e., iron), and their potential for natural products, vent and seep areas hold important (yet largely unknown) implications for services to ecosystems and humanity As such, they will benefit from protection from adverse impacts caused by human activities. Furthermore, beyond the requirement to maintain biodiversity for future generations, cultural ecosystem services such as generation of scientific knowledge and inspiration for citizens to learn about the natural world and for new generations to enter scientific careers, and tourism, should also be recognized in an assessment of their economic value (Jobstvogt et al., 2014).

Table 45.1 | Summary of vent and seep ecosystems protected to date under national or international law (Santos et al., 2012; Calado et al., 2011; ISA, 2011; USFWS, 2012; NTL 2009-G40 ; New Zealand ENMS circular 2007; Gouvernement de Nouvelle Calédonie)

Ocean region	Name of site	Type of chemosynthetic ecosystem	Depth & location	Legal framework
North East Pacific	Endeavour hydrothermal vents MPA	Five vent fields including black smokers	2250m depth, 250km SW of Vancouver Island in Canadian EEZ.	Protected under the Canadian Government's Ocean Act.
North East Pacific	Guaymas Basin Hydrothermal Vents Sanctuary	Hydrothermal vents located in a sedimented seabed.	Gulf of California, depth of ~2500m, Within Mexican EEZ.	Protected under Mexican State Law.
North East Pacific	Eastern Pacific Rise Hydrothermal Vents Sanctuary	Hydrothermal vents located on the East Pacific Rise	East Pacific Rise, depth of ~2800m, in Mexican EEZ.	Protected under Mexican State Law.
North West Pacific	Mariana Trench National Monument	Hydrothermal vents, CO2 vents, sulphur lake.	Located around three northernmost Mariana Islands & Mariana Trench 10m - 1650m depth.	Protected under US Law following Presidential Proclamation.
South West Pacific	Several deep-sea benthic protection areas	Hydrothermal vents	Northern to mid-Kermadec arc	New Zealand
West Pacific	Parc naturel de la mer de Corail (nature park of the Coral Sea)	Hydrothermal vent and coold seeps (suspected)	Up to 7919 m, encompassing the whole French EEZ around New Caledonia	Protected under New Caledonia Government
Gulf of Mexico	Numerous individual sites hosting 'high-densities benthic communities'	Hydrocarbon seeps and associated deep-sea corals	400 - 3300 m, in US EEZ	US Legal Framework: Bureau of Ocean Energy Management Notice to Lessees
North Atlantic	The Azores Hydrothermal Vent MPA	Seven hydrothermal vent fields including Lucky Strike, Menez Gwen, Rainbow and Banco Dom João de Castro. Except for Rainbow they are all Natura 2000 SAC (special areas of conservation under the EU habitats directive)	Amongst or to the south west of Azores Islands, N. Atlantic. 40m - 2300m depth.	Protected under Portuguese national & EU Habitats Directive, Rainbow is the first protected vent site located outside of an EEZ. It is included in the Azores Marine Park.

5   Conservation responses

Action to protect vents and seeps has taken place at national and international levels through the development of informal or voluntary protection plans or codes of conduct and formal protection measures under State or international law. An example of informal measures is the adoption by the scientific community of the InterRidge Statement of Commitment to Responsible Research Practices (Devey et al., 2007). The marine mining industry has also produced the International Marine Minerals Society Code for Environmental Management of Marine Mining (IMMS, 2011), which outlines principles and best practice for use by industry, regulatory agencies, scientists and other interested parties (Boschen et al., 2013). The OSPAR Commission recommended strengthening the protection of hydrothermal vents/fields occurring on oceanic ridges as a threatened and/or declining habitat in order to recover the habitat, to improve its status and ensure its effective conservation in Region V of the OSPAR maritime area (OSPAR, 2014).

Formal protection measures for hydrothermal vent ecosystems have been undertaken mainly within the EEZs of States (Table 45.1). The Rainbow hydrothermal vent field was proposed to be included in the Azores Marine Park by the Portuguese Government at an OSPAR meeting considering these questions, even though it lies outside the EEZ (Ribeiro, 2010; Calado et al., 2011). Portugal proceeded with this area as a Marine Protected Area on the understanding that the area is located on its extended continental shelf. It is also notable that some areas protected from bottom fishing also contain chemosynthetic ecosystems (e.g., on several southern hemisphere ridges), although this protection does not apply to other activities, such as mining.

The Strategic Plan for conservation of Biodiversity 2011-2020 adopted by the Conference of the Parties at the Convention on Biological Diversity (CBD) established a target stating that 10 per cent of marine areas are conserved through systems of protected areas and other effective area-based conservation measures (Aichi Biodiversity Target 11[2]).

In decision IX/20, the Conference of the Parties adopted the scientific criteria for identifying Ecologically or Biologically Significant Marine Areas (EBSAs) in need of protection in areas beyond national jurisdiction, and the scientific guidance for designing representative networks of marine protected areas. Because of their unique biodiversity, ecological properties and potential services, vent and seep areas meet the scientific and technical criteria defined for EBSA (Clark et al., 2014; Dunn et al., 2014; CBD scientific criteria for ecologically or biologically significant areas annex I, decision IX/20). As emphasized by decision X/29 of the Conference of the Parties, the identification of EBSAs and the selection of conservation and management measures is a matter for States and competent intergovernmental organizations, in accordance with international law, including the United Nations Convention on the Law of the Sea (CBD COP decision X/29, para. 26, 2010).

Scientists have called for the development of a cohesive network of such protected areas in which management of marine mining activities would be extremely risk averse, and often mining would be prohibited (Boschen et al.. 2013; Van Dover et al., 2012). It is important to note that, in the context of vents and seeps, natural variability is acknowledged to underlie many of the changes that are happening. Knowledge gaps concerning the ecological dynamics and responses to combined pressures, therefore, currently make it difficult to devise effective conservation measures. In any case, implementation of such measures would require actions at the national, regional and (in some cases) global level to be coordinated with each other.

At present, in the absence of any formal framework for general coordination, voluntary cooperation among the International Seabed Authority (ISA) and RFMOs is taking place. Without further efforts to promote cooperation between the relevant sectoral regulatory authorities and to close gaps in knowledge, both the effectiveness of on-going conservation measures and the development of more wide-ranging protection for vents and seeps are likely to be put at risk.

---

2   Aichi Biodiversity Target 11 states "By 2020, at least 17 per cent of terrestrial and inland water, and 10 per cent of coastal and marine areas, especially areas of particular importance for biodiversity and ecosystem services, are conserved through effectively and equitably managed, ecologically representative and well connected systems of protected areas and other effective area-based conservation measures, and integrated into the wider landscapes and seascapes".

References

Adams, D.K., McGillicuddy, D.J., Zamudio, L., Thurnherr, A.M., Liang, X., Rouxel, O., German, C.R., Mullineaux, L.S. (2011). Surface-Generated Mesoscale Eddies Transport Deep-Sea Products from Hydrothermal Vents. *Science* 332, 580–583. doi:10.1126/science.1201066.

Amend, J.P., Mccollom T.M., Hentscher M., Bach W. (2011). Catabolic and anabolic energy for chemolithoautotrophs in deep-sea hydrothermal systems hosted in different rock types. *Geochimica et Cosmochimica Acta* 75 5736–5748.

Arrieta, J.M., Arnaud-Haond, S., Duarte, C.M., (2010). What lies underneath: Conserving the oceans' genetic resources. *Proceedings of the National Academy of Sciences of the United States of America* 107, 18318-18324.

Armstrong, C.W., Foley, N.S., Tinch, R., van den Hove, S. (2012). Services from the deep: Steps towards valuation of deep sea goods and services. *Ecosystem Services* 2, 2–13.

Bachraty, C., Legendre, P., Desbruyères, D. (2009). Biogeographic relationships among deep-sea hydrothermal vent faunas at global scale. *Deep Sea Research Part II: Oceanography Research Papers* 56, 1371–1378. doi:10.1016/j.dsr.2009.01.009.

Baker, E.T. and German, C.R. (2004). On the global distribution of hydrothermal vent fields. In *Mid-Ocean Ridges: Hydrothermal interactions between the lithosphere and oceans*. Geophysical Monograph Series, Vol. 148, C.R. German, J. Lin, and L.M. Parson (eds.), AGU, 245-266.

Baker, M.C., Ramirez-Llodra, E.Z., Tyler, P.A., German, C.R., Boetius, A., Cordes, E.E., Dubilier, N., Fisher, C.R., Levin, L.A., Metaxas, A., Rowden, A.A., Santos, R.S., Shank, T.M., Van Dover, C.L., Young, C.M., Warén, A. (2010). Biogeography, Ecology, and Vulnerability of Chemosynthetic Ecosystems in the Deep Sea, in: McIntyre, A.D. (Ed.), *Life in the World's Oceans*. Wiley-Blackwell, Oxford, UK, pp. 161–182.

Beaulieu, S.E., Baker, E.T., German, C.R., Maffei, A. (2013). An authoritative global database for active submarine hydrothermal vent fields: Global vent database. *Geochemistry Geophysics Geosystems* 14, 4892–4905. doi:10.1002/2013GC004998

Beaulieu, S., Joyce, K., Cook, J. and Soule, S.A. (2015). Woods Hole Oceanographic Institution.

Bernardino, A.F., Levin, L.A., Thurber, A.R., Smith, C.R. (2012). Comparative composition, diversity and trophic ecology of sediment macrofauna at vents, seeps and organic falls. *PLoS ONE* 7 (4), e33515.

Boetius, A., Wenzhöfer, F. (2013). Seafloor oxygen consumption fuelled by methane from cold seeps. *Nature Geoscience* 6, 725–734. doi:10.1038/ngeo1926.

Boschen, R.E., Rowden, A.A., Clark, M.R., Gardner, J.P.A. (2013). Mining of deep-sea seafloor massive sulfides: A review of the deposits, their benthic communities, impacts from mining, regulatory frameworks and management strategies. *Ocean and Coastal Management* 84, 54–67. doi:10.1016/j.ocecoaman.2013.07.005.

CBD scientific criteria for ecologically or biologically significant areas (EBSAs) annex I, decision IX/20 (http://www.cbd.int/ebsa).

CBD COP decision X/29 (2010). Marine and coastal biodiversity, para. 26.

Calado, H., Ng, K., Lopes, C., Paramio, L. (2011). Introducing a legal management instrument for offshore marine protected areas in the Azores—The Azores Marine Park. *Environmental Science and Policy* 14: 1175-1187.

Charlou, J.L., Donval, J.P., Konn, C., Ondreas, H., Fouquet, Y., Jean-Baptiste P. and Fourre E. (2010). High production and fluxes of H2 and CH4 and evidence of abiotic hydrocarbon synthesis by serpentinization in ultramafic-hosted hydrothermal systems on the Mid Atlantic Ridge. In *Diversity of hydrothermal systems on slow-spreading ocean ridges*, P.A.Rona et al. (Ed.) ,eds.): pp.265-296, AGU Monograph Series, Washington.

ChESSbase (CoML). *Ocean Biogeographic Information System Publication*. http://www.gbif.org/dataset/83b90d28-f762-11e1-a439-00145eb45e9a on 2015-05-06

Childress, J.J. and Fisher, C.R. (1992). The biology of hydrothermal vent animals: Physiology, biochemistry and autotrophic symbioses. *Oceanography and Marine Biology* 30: 337-441.

Childress, J.J. and Girguis, P.R. (2011). The metabolic demands of endosymbiotic chemoautotrophic metabolism on host physiological capacities. *Journal of Experimental Biology*, 214: 312-325. doi:10.1242/jeb.049023.

Clark, M.R., Rowden, A.A., Schlacher, T.A., Guinotte, J., Dunstan, P.K., Williams, A., O'Hara, T.D., Watling, L., Niklitschek, E., Tsuchida, S. (2014). Identifying Ecologically or Biologically Significant Areas (EBSA): A systematic method and its application to seamounts in the South Pacific Ocean. *Ocean and Coastal Management* 91, 65–79. doi:10.1016/j.ocecoaman.2014.01.016.

Cordes, E.E., McGinley, M.P., Podowski, E.L., Becker, E.L., Lessard-Pilon, S., Viada, S.T., Fisher, C.R. (2008). Coral communities of the deep Gulf of Mexico. *Deep Sea Research Part I: Oceanographic Research Papers* 55, 777–787. doi:10.1016/j.dsr.2008.03.005.

Cordes, E.E., Bergquist, D.C., Fisher, C.R. (2009). Macro-Ecology of Gulf of Mexico Cold Seeps. *Annual review of Marine Science*. 1, 143–168.

Cordes, E.E., Cunha, M.R., Galéron, J., Mora, C., Olu-Le Roy, K., Sibuet, M., Van Gaever, S., Vanreusel, A., Levin, L.A. (2010). The influence of geological, geochemical, and biogenic habitat heterogeneity on seep biodiversity. *Marine Ecology* 31, 51-65.

Corliss, J.B., Dymond, J., Gordon, L.I., Edmond, J.M., von Herzen, R.P., Ballard, R.D., Green, K., Williams, D., Bainbridge, A., Crane, K., van Andel, T.H. (1979). Submarine Thermal Springs on the Galápagos Rift, (1979). 203 *Science* 1073-1083.

Crist, D.T., Ausubel, J., Wagoner, P.E. (2010). *First census of marine life 2010: highlights of a decade of discovery*. a publication of the Census of Marine Life, Washington, D.C.

Desbruyères, D., Biscoito, M., Caprais, J.-C., Colaço, A., Comtet, T., Crassous, P., Fouquet, Y., Khripounoff, A., Le Bris, N., Olu, K., Riso, R., Sarradin, P., Segonzac, M. and Vangriesheim, A. (2001). Variations in deep-sea hydrothermal vent communities on the Mid-Atlantic Ridge near the Azores plateau. *Deep Sea Research Part I* 48:1,325–1,346.

Devey, C.W., Fisher, C.R., Scott, S., (2007). Responsible science at hydrothermal vents. *Oceanography* 20: 162-171.

Dick, G.J., Anantharaman, K., Baker, B.J., Li, M., Reed, D.C., Sheik, C.S., (2013). The microbiology of deep-sea hydrothermal vent plumes: ecological and biogeographic linkages to seafloor and water column habitats. *Frontiers in Microbiology* 4. doi:10.3389/fmicb.2013.00124.

Dubilier, N., Bergin, C., Lott, C., (2008). Symbiotic diversity in marine animals: the art of harnessing chemosynthesis. *Nature Reviews Microbiology* 6, 725–740.

Dunn, D.C., Ardron, J., Bax, N., Bernal, P., Cleary, J., Cresswell, I., Donnelly, B., Dunstan, P., Gjerde, K., Johnson, D., Kaschner, K., Lascelles, B., Rice, J., von Nordheim, H., Wood, L., Halpin, P.N., (2014). The Convention on Biological Diversity's Ecologically or Biologically Significant Areas: Origins, development, and current status. *Marine Policy*. doi:10.1016/j.marpol.2013.12.002.

Fisher, C.R., Takai, K. and Le Bris, N. (2007). Hydrothermal Vent ecosystem. *Oceanography* 20 (1) 14-23.

Fisher, C.R., Demopoulos, A.W.J., Cordes, E.E., Baums, I.B., White, H.K., Bourque, J.R., (2014). Coral communities as indicators of ecosystem-level impacts resulting from the *Deepwater Horizon* oil spill. *Bioscience*. 64: 796-807. doi: 10.1093/biosci/biu129

Foucher, J.-P., Westbrook, G.K., Boetius, A., Ceramicola, S., Dupré, S., Mascle, J., Mienert, J., Pfannkuche, O., Pierre, C., Praeg, D. (2009).) Structure and Drivers of Cold Seep Ecosystems. *Oceanography*, 22(1) :): 92-109. Doi: 10.5670/oceanog.2009.11.

Gjerde, K.M., International Union for Conservation of Nature and Natural Resources, United Nations Environment Programme, (2006). *Ecosystems and biodiversity in deep waters and high seas*, UNEP regional seas report and studies. UNEP; IUCN, Nairobi, Kenya : [Switzerland]. ISBN: 92-807-2734-6.

German, C.R. and Von Damm, K.L. (2004). Hydrothermal Processes. in *The oceans and marine geochemistry: Treatise on geochemistry*, vol. 6, ed. H. Elderfield. Elsevier, Amsterdam; Heidelberg.

German, C.R., Ramirez-Llodra, E., Baker, M.C., Tyler, P.A. and the ChEss Scientific Steering Committee. (2011). Deep-Water Chemosynthetic Ecosystem Research during the Census of Marine Life Decade and Beyond: A Proposed Deep-Ocean Road Map. *PLoS ONE* 6: e23259. doi:10.1371/journal.pone.0023259

German, C.R., Legendre, L.L., Sander, S.G., Niquil, N., Luther, G.W., Bharati, L., Han, X., Le Bris, N. (2015). Hydrothermal Fe cycling and deep ocean organic carbon scavenging: Model-based evidence for significant POC supply to seafloor sediments. *Earth and Planetary Science Letters* 419, 143–153.

Goldberg, D.S., Takahashi, T. and Slagle, A.L. (2008). Carbon dioxide sequestration in deep-sea basalt. *Proceedings of the National Academy of Sciences*, 105(29), 9920-9925

Gjerde, K.M. (2006). International Union for Conservation of Nature and Natural Resources, United Nations Environment Programme. *Ecosystems and biodiversity in deep waters and high seas*, UNEP regional seas report and studies. UNEP; IUCN, Nairobi, Kenya : [Switzerland].

Glover, A.G, Gooday, A.J, Bailey, D.M, Billett, D.S.M., Chevaldonné, P.A., Colaco, A., Copley, J., Cuvelier, D., Desbruyères D., Kalogeropoulou, K.V., Klages, M., Lampadariou, N., Lejeusne, C., Mestre, N. C., Paterson, G.L.J., Perez, T., Ruhl ,H., Sarrazin, J., Soltwedel T., Soto, E.H., Thatje, S., Tselepides, A., Van Gaever, S., and Vanreusel, A. (2010). *Advances in Marine Biology*, Vol. 58, p 1-79.

Gouvernement de Nouvelle Calédonie. *Proposition pour un Parc Naturel de la Mer de Corail*. http://www.affmar.gouv.nc/portal/page/portal/affmar/librairie/fichiers/26354258.PDF (27-04-15)

Hawkes, J.A., Connelly, D.P., Gledhill, M., Achterberg, E.P. (2013). The stabilisation and transportation of dissolved iron from high temperature hydrothermal vent systems. *Earth and Planetary Science Letters* 375, 280–290.

IMMS (2011). *International Marine Minerals Society, Code for Environmental Management of Marine Mining*. Revised version. http://www.immsoc.org/IMMS_downloads/2011_SEPT_16_IMMS_Code.pdf.

IPCC (2005) . Caldeira, K. and Akai, M. Ocean Storage. In: Bert Metz, Ogunlade Davidson, Heleen de Coninck, Manuela Loos and Leo Meyer (Eds.) *Special Report on Carbon Dioxide Capture and Storage*. Cambridge University Press, UK. pp 431.

ISA (2011). *Environmental management of deep-sea chemosynthetic ecosystems: justification of and considerations for a spatially-based approach*. International Seabed Authority Technical Report No. 9. ISA, Kingston, Jamaica, 78 pp.

Jobstvogt, N., Hanley, N., Hynes, S., Kenter, J., Witte, U., (2014). Twenty thousand sterling under the sea: Estimating the value of protecting deep-sea biodiversity. *Ecological Economics* 97, 10–19. doi:10.1016/j.ecolecon.2013.10.019.

Kiel, S. (ed.) (2009). The Vent and Seep Biota: Aspects from Microbes to Ecosystems, *Topics in Geobiology* 33, 1 DOI 10.1007/978-90-481-9572-5_1.

Kelley, D.S. (2005). A Serpentinite-Hosted Ecosystem: The Lost City Hydrothermal Field. *Science* 307, 1428–1434. doi:10.1126/science.1102556.

Le Bris, N. and Gaill, F. (2007). How does the annelid *Alvinella pompejana* deal with an extreme hydrothermal environment? *Reviews in Environmental Science and Biotechnology* 6, 197–221. doi:10.1007/s11157-006-9112-1.

Le Bris, N. and Duperron, S. (2010). In: Chemosynthetic communities and biogeochemical energy pathways along the Mid-Atlantic Ridge: The case of *Bathymodiolus azoricus*. American Geophysical Union, Washington D.C.. 188: 9. doi: 10.1029/2008GM000712.

Levin, L.A., (2005). Ecology of cold seep sediments: interactions of fauna with flow, chemistry and microbes. *Oceanography and Marine Biology* 43, 1-46.

Levin, L.A., Sibuet, M., Gooday, A.J., Smith, C.R., Vanreusel, A. (2010). The roles of habitat heterogeneity in generating and maintaining biodiversity on continental margins: an introduction. *Marine Ecology* 31 (1), 1-5.

Levin, L.A., Sibuet, M. (2012). Understanding Continental Margin Biodiversity: A New Imperative. *Annual Review of Marine Science* 4, 79–112. doi:10.1146/annurev-marine-120709-142714.

Levin, L.A., Orphan, V.J., Rouse, G.W., Rathburn, A.E., Ussler, W., Cook, G.S., Goffredi, S.K., Perez, E.M., Waren, A., Grupe, B.M., Chadwick, G., Strickrott, B., (2012). A hydrothermal seep on the Costa Rica margin: middle ground in a continuum of reducing ecosystems. *Proceedings of the Royal Society of London B: Biological Sciences* 279, 2580–2588. doi:10.1098/rspb.2012.0205.

Levin, L.A., Ziebis, W., Mendoza, G., Bertics, V.J., Washington, T., Gonzalez, J., Thurber, A.R., Ebbe, B., Lee, R.W. (2013). Ecological release and niche partitioning under stress: Lessons from dorvilleid polychaetes in sulfidic sediments at methane seeps. *Deep Sea Research Part II: Topical Studies in Oceanography* 92, 214-233.

McNutt, M.K., Rich, C., Crone, T.J., Guthrie, G.A., Hsieh, P.A., Ryerson, T.B., Savas, O. and Shaffer, F. (2012). Review of flow rate estimates of the *Deepwater Horizon* oil spill. *Proceedings of the National Academy of Sciences* 109 (50) 20260-20267.

Mills, S., Mullineaux, L., Beaulieu, S., Adams, D. (2013). Persistent effects of disturbance on larval patterns in the plankton after an eruption on the East Pacific Rise. *Marine Ecology Progress Series*. 491, 67–76. doi:10.3354/meps10463.

Ministry of Fisheries and Department of Conservation. (2008). *Marine Protected Areas: Classification, Protection Standard and Implementation Guidelines*. Ministry of Fisheries and Department of Conservation, Wellington, New Zealand. 54 p. http://www.fish.govt.nz/ennz/Environmental/Seabed+Protection+and+Research/Benthic+Protection+Areas.htm 27/04/15.

Moalic, Y., Desbruyeres, D., Duarte, C.M., Rozenfeld, A.F., Bachraty, C., Arnaud-Haond, S. (2012). Biogeography Revisited with Network Theory: Retracing the History of Hydrothermal Vent Communities. *Systematic Biology* 61, 127-137.

Montagna et al., P.A., Baguley J.G, Cooksey C., Hartwell I., Hyde L.J., Hyland, J.L., Kalke R.D., Kracker, L.M., Reuscher, M., Rhodes, A.C.E. (2013). Deep-Sea Benthic Footprint of the Deepwater Horizon Blowout. *PLoS ONE* 8(8): e70540, doi:10.1371/journal.pone.0070540.

New Zealand ENMS circular (2007). Electronic Net Monitoring Systems – Circular Issued Under Authority of the Fisheries (Benthic Protection Areas) Regulations 2007 (No. F419).

NTL 2009-G40, Deepwater Benthic Communities NTL implemented January 27, 2010. http://www.bsee.gov/Regulations-and-Guidance/Notices-to-Lessees/2009/09-G40/ (27-04-15)

Ohara, Y., Reagan, M.K., Fujikura, K., Watanabe, H., Michibayashi, K., Ishii, T., Stern, R.J., Pujana, I., Martinez, F., Girard, G., Ribeiro, J., Brounce, M., Komori, N., Kino, M. (2012). A serpentinite-hosted ecosystem in the Southern Mariana Forearc. *Proceedings of the National Academy of Sciences* 109, 2831–2835. doi:10.1073/pnas.1112005109.

Orcutt, B.N., Sylvan, J.B., Knab, N.J., Edwards, K.J. (2011). Microbial Ecology of the Dark Ocean above, at, and below the Seafloor. *Microbiology and Molecular Biology Reviews* 75, 361–422. doi:10.1128/MMBR.00039-10

OSPAR (2010). *Background Document for Oceanic ridges with hydrothermal vents/fields*. Biodiversity Series (Publication No. 490/2010): 17pp. OSPAR Commission. ISBN 978-1-907390-31-9.

OSPAR (2014). OSPAR 14/21/1 Annex 16. OSPAR Recommendation 2014/11 on furthering the protection and conservation of hydrothermal vents/fields occurring on oceanic ridges in Region V of the OSPAR maritime area. 6pp. OSPAR Commission.

Paull, C.K. et al., (1984). Biological Communities at the Florida Escarpment Resemble Hydrothermal Vent Taxa, 226 *Science* 965-967

Ramirez-Llodra, E., Freitga, K., Blanco, M. and Baker, C. (2005). ChEssBase: an online information system on biodiversity and biogeography of deep-sea fauna from chemosynthetic ecosystems. Version 2. World Wide Web electronic publications,

Ramirez-Llodra, E., Brandt, A., Danovaro, R., De Mol, B., Escobar, E., German, C.R., Levin, L.A., Martinez Arbizu, P., Menot, L., Buhl-Mortensen, P., Narayanaswamy, B.E., Smith, C.R., Tittensor, D.P., Tyler, P.A., Vanreusel, A., Vecchione, M. (2010). Deep, diverse and definitely different: unique attributes of the world's largest ecosystem. *Biogeosciences* 7, 2851–2899. doi:10.5194/bg-7-2851-2010.

Ramirez-Llodra, E., Tyler, P.A., Baker, M.C., Bergstad, O.A., Clark, M.R., Escobar, E., Levin, L.A., Menot, L., Rowden, A.A., Smith, C.R., Van Dover, C.L., (2011). Man and the Last Great Wilderness: Human Impact on the Deep Sea. *PLoS ONE* 6, e22588. doi:10.1371/journal.pone.0022588.

Report of the Ad Hoc Open-ended Informal Working Group to study issues relating to the conservation and sustainable use of marine biological diversity beyond areas of national jurisdiction and Co-Chairs' summary of discussions. (2013). A/68/399. http://www.un.org/ga/search/view_doc.asp?symbol=A/68/399.

Rogers, A.D., Tyler, P.A., Connelly, D.P., Copley, J.T., James, R., Larter, R.D., Linse, K., Mills, R.A., Naveira Garabato, A., Pancost, R.D., Pearce, D.A., Polunin, N.V.C., German, C.R., Shank, T., Boersch-Supan, P.H., Alker, B.J., Aquilina, A., Bennett, S.A., Clarke, A. Dinley, R.J.J., Graham, A.G.C., Green, D.R.H., Hawkes, J.A., Hepburn, L., Hilario, A., Huvenne, V.A.I., Marsh, L., Ramirez-Llodra, E.,. Reid, W.D.K, Roterman, C.N., Sweeting, C.J., Thatje, S., Zwirglmaier, K. (2012). The Discovery of New Deep-Sea Hydrothermal Vent Communities in the Southern Ocean and Implications for Biogeography. *PLoS Biology* 10(1): e1001234.

Ribeiro, M.C. (2010). The "Rainbow": The First National Marine Protected Area Proposed Under the High Seas. *International Journal of Marine and Coastal Law* 25, 183–207. doi:10.1163/157180910X12665776638669.

Santos, R.S., Morato, T. and Barriga, F.J.A.S. (2012). Increasing Pressure at the Bottom of the Ocean [Chapter 5]: 69-81 [doi: 10.1007/978-94-007-1321-5_5]. In: A. Mendonça, A. Cunha & R. Chakrabarti (Eds.). *Natural Resources, Sustainability and Humanity: A Comprehensive View*. Springer: xvi+199pp.

Shank, T.M., Fornari, D.J., Von Damm, K.L., Lilley M.D., Haymon, R.M. and Lutz, R.A., (1998). Temporal and spatial patterns of biological community development at the nascent deep-sea hydrothermal vents (9°50'N, East Pacific Rise). *Deep-Sea Research II*, 45, 465-515.

Spiess, F.N., MacDonald, K.C., Atwater, T. et al., (1980). East Pacific Rise - hot springs and geophysical experiments. *Science*, 207, 1421-1433.

Tagliabue, A., Bopp, L., Dutay, J.-C., Bowie, A.R., Chever, F., Jean-Baptiste, P., Bucciarelli, E., Lannuzel, D., Remenyi, T., Sarthou, G., Aumont, O., Gehlen, M., Jeandel, C. (2010). Hydrothermal contribution to the oceanic dissolved iron inventory. *Nature Geoscience* 3, 252–256. doi:10.1038/ngeo818.

Takai, K., Nakamura, K. (2010). Compositional, Physiological and Metabolic Variability in Microbial Communities Associated with Geochemically Diverse, Deep-Sea Hydrothermal Vent Fluids, in: Barton, L.L., Mandl, M., Loy, A. (Eds.), *Geomicrobiology: Molecular and Environmental Perspective*. Springer Netherlands, Dordrecht, pp. 251–283.

Takai, K., Nakamura, K. (2011). Archaeal diversity and community development in deep-sea hydrothermal vents. *Current Opinion in Microbiology* 14, 282–291. doi:10.1016/j.mib.2011.04.013.

Talukder, A.R. (2012). Review of submarine cold seep plumbing systems: leakage to seepage and venting: Seeps plumbing system. *Terra Nova* 24, 255–272.

Thornburg, C.C., Zabriskie, T.M. and McPhail, K.L. (2010). Deep-Sea Hydrothermal Vents: Potential Hot Spots for Natural Products Discovery? *Journal of Natural Products*, 73(3), 489-499.

Thomson, R.E., Gordon, R.L. and Dolling, A.G. (1991). An intense acoustic scattering layer at the top of a mid-ocean ridge hydrothermal plume. *Journal of Geophysical Research* 36:4839-4844. dx.doi.org/10.1029/90JC02692.

Tunnicliffe, V., Embley, R.W., Holden, J.F., Butterfield, D.A., Massoth, G.J., and Juniper, S.K., (1997). Biological colonization of new hydrothermal vents following an eruption on Juan de Fuca Ridge. *Deep Sea Research Part I: Oceanographic Research Papers*, 44(9), 1627-1644.

UNEP/CBD/SBSTTA/13/INF/14 (2013). Report of the expert workshop on ecological criteria and biogeographic classification systems for marine areas in needs of protection.

UNU-IAS Report (2005). *Bioprospecting of Genetic Resources in the Deep Seabed: Scientific, Legal and Policy Aspects*. Tokyo, UNU/IAS, 76 pages.

USFWS (2012). *Marianas Trench Marine National Monument Factsheet*. Available at: http://www.fws.gov/marianastrenchmarinemonument/.

Van Dover, C.L., Smith, C.R., Ardron, J. Dunn, D., Gjerde, K., Levin, L., Smith, S., The Dinard Workshop Contributors. (2012). Designating networks of chemosynthetic ecosystem reserves in the deep sea. *Marine Policy* 36: 378-381.

Van Dover, C., Aronson, J., Pendleton, L., Smith, S., Arnaud-Haond, S., Moreno-Mateos, D., Barbier, E., Billett, D., Bowers, K., Danovaro, R., Edwards, A., Kellert, S., Morato, T., Pollard, E., Rogers, A., Warner, R. (2014). Ecological restoration in the deep sea: desiderata. *Marine Policy* 44: 98-106.

Vanreusel, A., De Groote, A., Gollner, S., Bright, M., (2010). Ecology and Biogeography of Free-Living Nematodes Associated with Chemosynthetic Environments in the Deep Sea: A Review. *PLoS ONE* 5, e12449. doi:10.1371/journal.pone.0012449.

Watanabe, H., Fujikura, K., Kojima, S., Miyazaki, J.-I., Fujiwara, Y. (2010). Japan: Vents and Seeps in Close Proximity. In: Kiel, S. (Ed.). *The Vent and Seep Biota*. Springer Netherlands, Dordrecht, pp. 379–401.

Wheat, C.G., Mottl, M.J., Fisher, A.T., Kadko, D., Davis, E.E., and Baker, E., (2004). Heat flow through a basaltic outcrop on a sedimented young ridge flank. *Geochemistry Geophysics Geosystems*, doi:10.1029/2004GC000700.

Wu, J., Wells, M.L., Rember, R., (2011). Dissolved iron anomaly in the deep tropical–subtropical Pacific: Evidence for long-range transport of hydrothermal iron. *Geochimica et Cosmochimica Acta* 75, 460–468.

Wheeler, A.J. & Stadnitskaya, A. (2011). Benthic deep-sea carbonates: reefs and seeps. In: Heiko Hüneke & Thierry Mulder (eds). *Deep-Sea Sediments*. Amsterdam: Elsevier.

46 High-Latitude Ice and the Biodiversity Dependent on it

Contributors:
Jake Rice and Enrique Marschoff (Co-Lead members)

High-Latitude Ice and the Biodiversity Dependent on it

1 Description of the ice systems and their biodiversity

1.1 Annual ice and multi-year ice

The high-latitude ocean areas are ice-covered for much or all of the year. Multi-year ice and annual ice have different physical and chemical properties that make them also differ in terms of their ecological communities. The multi-year ice, in particular, is globally unique, and supports unique communities of ice algae and species of larger invertebrates, fish, birds and mammals wholly or largely dependent on the multi-year ice and on multi-year and annual ice margins. Sea ice is a technical term that refers to floating ice formed by the freezing of seawater. This chapter uses "high-latitude ice" as a more generic term for a variety of critically important high-latitude marine habitats, which include ice shelves, pack ice, sea ice, and the highly mobile ice edge. These forms of high-latitude ice complement and modify other types of habitats, including extensive shallow ocean shelves and towering coastal cliffs (CAFF, 2013. Meltofte, 2013).

The Arctic Ocean is unique in that it contains a deep ocean basin, which until recently was almost completely covered in multi-year ice (chapter 36H). No other area in the world has such an ice-dominated deep ocean. The Southern Ocean is unique in that it contains both icebergs (floating freshwater ice) calved from glaciers and ice shelves and sea ice. Also there is no limit to the northwards extension of the winter pack ice. Thus seasonal variations are much larger in the Antarctic (Ropelewski 1983). Three biogeographic regions have been recognised in the Antarctic, defined by differences in ice cover (Treguer and Jaques, 1992): ice-free, seasonally or permanently covered by high-latitude ice.

1.2 Biodiversity associated with the ice

The high-latitude ice-covered ecosystems host globally significant arrays of biodiversity, and the size and nature of these ecosystems make them of critical importance to the biological, chemical and physical balance of the globe (ACIA 2005). Biodiversity in these systems presents remarkable adaptations to survive both extreme cold and highly variable climatic conditions. Iconic ice-adapted species such as polar bear, narwhal, walrus, seals, penguins and seabirds have adapted to different ice conditions, including extreme examples such as the emperor penguin living among thousands of lesser known species that are adapted to greater or lesser degrees to exploit the habitats created by high-latitude ice (Meltofte, 2013; De Broyer et al. (eds.) 2014).

1.2.1 Primary production and lower trophic level communities

These high-latitude seas are relatively low in biological productivity, and ice algal communities, unique to these latitudes, play a particularly important role in system dynamics. Ice algae are estimated to contribute to more than 50 per cent of the primary production in the permanently ice covered central Arctic (Gosselin et al. 1997, Sakshaug 2004). Ice algae can be divided into communities on the surface, interior and bottom of the ice (Horner et al. 1992). In addition to microalgae, bacteria are an important component of the ice-algal community, but many other groups of organisms (e.g. archaea, fungi, ciliates, kinetoplastids, choanoflagellates, amoebae, heliozoans, foraminiferans and some protists that belong to no known group) also occur in ice communities (Lizotte 2003). Poulin et al. (2010) reported a total of 1027 sympagic taxa in the Arctic Ocean. Many of the dominant ice algae are diatoms that sink and are eaten by different benthic organisms or broken down by bacteria (Boetius et al., 2013), thus creating a link between ice and bottom ecosystems. In the Southern Ocean, the distribution of primary productivity is associated with frontal zones, areas of broken sea ice, and with the divergences linked to the bottom topography, with high horizontal variability at the local scale while the vertical distribution is more regular. Chlorophyll concentration is practically nil below 250 m with maxima around 50 m. This general pattern is highly modified in coastal areas (El-Sayed, 1970).

The primary productivity in the Antarctic is much lower than might be expected given the nutrient concentrations observed. Early in Antarctic research the factors regulating the distribution of primary producers have been discussed (Hart, 1934). In coastal waters nutrients might reach very high values (El-Sayed 1985, Holm-Hansen 1985) and phytoplankton blooms have been observed to deplete these high concentrations (Nelson and Smith 1986, Bienatti et al. 1977).

The marginal ice zone (MIZ), at and near the ice edge, is a highly productive area for phytoplankton (Sakshaug & Holm-Hansen, 1984, Sakshaug & Skjoldal 1989). Stable water masses due to ice melt coupled with high nutrient availability and light result in an intense phytoplankton bloom. As water masses become stratified due to surface heating, nutrient flow from below is inhibited. Consequently, the bloom in marginal ice areas starts earlier than in adjacent areas never experiencing high-latitude ice. The bloom follows the ice edge as it retreats in the spring. This "spring bloom" can occur in autumn in the areas of maximum ice retreat (Falk-Petersen et al. 2008). The ice-edge bloom is likely to weaken with time over the season (Wassmann et al. 2006).

Arctic planktonic herbivores, such as *Calanus hyperboreus*, are able to utilize the vast area of the Arctic Ocean and to feed and store lipids for over-wintering until the sun disappears in October (Falk-Petersen et al. 2008). In the Antarctic the same pattern of seasonal feeding expended in reproductive processes and lipid storage is followed by a suite of herbivores (De Broyer et al. (eds.) 2014) such as euphausiids (i.e. *Euphausia superba, Thysanoessa macrura*), copepods and salps (i.e. *Calanus simillimus, C.propinquus, C. acutus, Salpa thompsoni*).

Around the annual ice, in general there are steep gradients in temperature, salinity, light and nutrient concentrations creating different habitats throughout the ice, the 0.2 m on the lower ice surface having the most favourable conditions for growth among the interior communities (Arrigo 2003). However, with respect to biomass and contribution to primary production, the sub-ice community is the most important in annual ice. In addition there are seasonal trends and inter-annual variations in species composition, biomass and production as a result of several factors, including light, age and origin of the ice (e.g., distance to land and water depth). Thus, there is a high spatial heterogeneity when larger areas are considered.

Sea-ice algae start their growth ahead of phytoplankton. An extended growth season in the Arctic areas forms ice algal communities that are grazed actively by both ice fauna and zooplankton and may be an important component of the diet of some species during the winter. Ice algae contribute 4–26 percent of total primary production in seasonally ice-covered waters (Gosselin et al. 1997, Sakshaug 2004). *Apherusa glacial* is probably the most numerous amphipod species in the central Arctic Ocean. *Onisimus glacialis* may be common in some areas. In the Antarctic sea ice the calanoid copepods are dominant while larvae of *E. superba* benefit from ice for overwintering; to date no species fully dependent on high-latitude ice has been identified (Arndt & Swadling, 2006).

1.2.2   Macrofauna

The ice structure and surfaces include a number of larger invertebrates that also are believed to live their entire life connected to the multi-year ice (e.g. nematode worms, rotifers and other small soft-bodied animals within the ice and amphipods on the underside), Some of these dominate the biomass of macroinvertebrates (Arndt & Swadling 2006) and are the important food items for high-latitude fish. Antarctic euphausiid larvae spend the winter in close association with icebergs or sea ice. The permanent pack-ice zone of the Southern Ocean represents the habitat for a highly confined community of seabirds, the most unvarying of any seabird assemblage in the Southern Hemisphere (Ribic and Ainley, 1988). It is composed of Adélie and Emperor penguins, Snow and Antarctic petrels, with the addition during the summer of South Polar skua and Wilson's storm-petrel.

In the Arctic, Polar bears *Ursus maritimus* are highly dependent on high-latitude ice and are therefore particularly vulnerable to changes in ice extent, duration and thickness. Three ice-associated cetacean species also reside year-round in the Arctic, mostly connected to the marginal ice zone, including the Bowhead whale (*Balaena mysticetus*) that is assessed as highly endangered in part of its range. The reproduction of some Antarctic seal species is also linked to extent of ice (e.g. *Lobodon carcinophagus, Leptonychotes weddellii*) while others are temporarily associated to ice for rest and refuge. Penguin species also use icebergs or sea ice during foraging trips and follow the ice border in winter. Permanent ice shelves strongly modify the habitats and the sea bottom fauna below the shelves (Gutt et al. 2013, Lipps et al., 1977; Lipps et al., 1979; Post et al., 2014).

2   Changes to these systems and their biodiversity

2.1   Changes in the ice structures

These high-latitude ecosystems are undergoing change at a more rapid rate than other places on the globe, threatening the existence of ecosystems such as multi-year high-latitude ice. In the past 100 years, average Arctic temperatures have increased at twice the average global rate (IPCC 2007). Recent changes in Arctic and Antarctic sea-ice cover, driven by climate change including rising temperatures and winds (Stammerjohn et al., 2012), have affected the timing of ice break-up in spring and freeze-up in autumn, as well as the extent and type of ice present in different areas at specific dates. Overall, multi-year ice is rapidly being replaced by first-year ice. The extent of Arctic ice is shrinking in all seasons, but especially in the summer. In some regions of the Antarctic ice shelves are rapidly disappearing, while the maximum extension of winter ice appears to increase (Turner et al., 2009, pg. 130).

In the Arctic Ocean multi-year ice changed from covering more than two-thirds of the Arctic Ocean to less than one-third in less than a decade. For instance, multi-year ice now occupies only part of the deep areas of the Arctic Ocean beyond areas within the national jurisdiction of Canada and is projected to be virtually ice-free in summer within 30 years, with multi-year ice persisting mainly between islands of Canada and in the narrow straits between Canada and Greenland, Denmark (Meltofte, 2013). Similar projections of a largely ice-free Arctic Ocean in summer have been made from the Arctic-Pacific interface as well (Overland and Wand 2013). The multi-year ice that remains is also much younger than previously as the oldest multi-year ice classes have declined more than other classes (AMAP 2011), and even if conditions changed to allow the return of the lost/declined ice cover, it would take many years to return to the state of just a few decades ago.

2.2   Changes in biodiversity

A change in timing and duration of the ice edge bloom increases the probability of a "mismatch" in productivity, which may have severe consequences for zooplankton that are dependent on this bloom today, with potential cascading effects throughout the ecosystem. However, the timing of ice formation and melt also influences the distribution and intensity of the primary production in the water column. Such primary production is likely to increase in areas with less ice but may then become limited by nutrient availability, including trace nutrients such as iron.

Boreal species of algae, invertebrates, fish, mammals (Kaschner et al., 2011) and birds are expanding into these higher latitudes, while some ice-adapted species are losing habitat along the edges of their ranges. Changes are too rapid for evolutionary adaptation, so species with inborn capacity to adjust their physiology or behaviour will fare better. Species with limited distribution, specialized feeding or breeding requirements, and/or high reliance on high-latitude ice for part of their life cycle are particularly vulnerable (Meltofte, 2013. In the Antarctic, seal and penguin species dependent on ice distribution seem to be likely to respond to changes in extreme events, as had happened in some years of anomalous El Niño – Southern oscillation events. Significant declines in ice – even at the regional or local scales – may lead to the replacement of antarctic by subantarctic species (Turner et al., 2009, pg. xxv).

Krill plays a central role in the trophic structure of Antarctic ecosystems. Its abundance and distribution depend on the coupling of reproductive events and oceanic circulation. It is not clear to what extent its population declined and which are the factors involved (Ainley et al., 2005; 2007). The decrease in krill abundance and the increase in salps abundance are thought to be related with changes in ice cover (Loeb et al., 1997).

There are indications that populations of *Pleuragramma antarcticum*, a key fish species of the trophic web, and whose reproduction is closely associated to high-latitude ice, declined at some localities, to be replaced by myctophids, a new food item for predators (Turner et al. 2009, pg. 360).

3 Implications and risks

Reduced high-latitude ice, especially a shift towards less multi-year ice, will affect the species composition in these waters. With decreasing ice cover, the effects on the ice fauna will be strongest at the edges of the multi-year high latitude ice. Seasonal/annual ice has to be colonized every year, as opposed to multiyear ice. In addition, multi-year ice has ice specialists that do not occur in younger ice (von Quillfeldt et al. 2009). The previously very low biological production of the deep basins may also change in this region as light, temperature and storminess increase and currents shift. In addition, wind-driven mixing of the ocean is more efficient over open water and over the thinner, more-mobile, seasonal ice than over multi-year ice, with the potential to also increase productivity.

Due to low reproductive rates and long lifetime, it has been predicted that the polar bears will not be able to adapt to the current fast warming of the Arctic and become extirpated from most of their range within the next 100 years (Gorbunov and Belikov, 2008; Schliebe et al., 2008). Other Arctic ice dependent species such as ringed seal (Kovacs et al. 2008), possibly narwhal, Ross gull (Blomquist & Elander, 1981, Hjort et

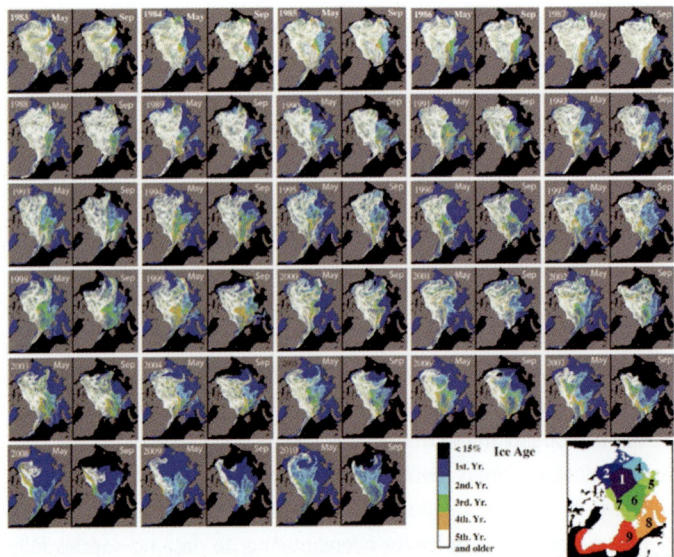

Figure 46.1 | Multi-year Arctic sea-ice 1983 – 2010 (taken from Maslanik et al. 2011).

al., 1997) and ivory gull (Gilg et al., 2010) are also expected to decrease as high-latitude ice, especially multi-year ice, decreases.

The reduction in ice cover in the Arctic is creating the potential for increased utilization of natural resources, including fish stocks, including in the central portion of the Arctic Ocean beyond national jurisdiction (Lin et al. 2012, Arctic Nations 2013). Among non-renewable natural resources, the Arctic is estimated to contain a quarter of the world's remaining oil and gas reserves, the development of which is expected to increase. Already, 10 per cent of the world's oil and 25 percent of the world's natural gas is produced in the Arctic, with the majority coming from the Russian Federation (AMAP 2007, see also chapter 21 of this Assessment).

In the Antarctic, the sea-ice cover is predicted to decrease by 33 per cent in this century (Turner et al., 2009, pg. 384) as well as coastal ice shelves. This would imply a significant stress for marine organisms but no species might be singled out as candidate for extinction in this period. Signy Island and some sites at the West Antarctic Peninsula have witnessed an explosion of the fur-seal numbers that may be related to decreased ice cover resulting in increasing areas available for resting and moulting, but which may also be related to population increases; the growing seal population in Signy Island has had deleterious impacts on the local terrestrial vegetation (Turner et al., 2009, pg. 360).

References

Ainley, D.G., Clarke E.D., Arrigo, K., Fraser, W.R., Kato, A., Barton, K.J., and Wilson, P.R. (2005). Decadal-scale changes in the climate and biota of the Pacific sector of the Southern Ocean, 1950s to the 1990s. *Antarctic Science* 17: 171-182.

Ainley, D.G., Ballard, G., Ackley, S., Blight, L., Eastman, J.T.,Emslie, S.D., Lescroel, A., Olmastroni, S., Townsend, S.E., Tynan, C.T., Wilson, P., and Woehler, E. (2007). Paradigm lost, or is topdown forcing no longer significant in the Antarctic marine ecosystem? *Antarctic Science* 19: 283-290.

AMAP (2007). *Oil and Gas Assessment (OGA)*. Arctic Monitoring and Assessment Programme (AMAP), Oslo.

AMAP (2011). *Snow, Water, Ice and Permafrost in the Arctic (SWIPA); climate change and the cryosphere*. Arctic Monitoring and Assessment Programme (AMAP), Oslo.

ACIA (2005). *Arctic Climate Impact Assessment*. Cambridge University Press, Cambridge, UK, 1042 pp.

Arctic Nations (2013). Arctic nations Arctic Ocean fisheries report. Tromso, October 28-31.

Arndt, C.E. and Swadling, K.M. (2006). Crustacea in Arctic and Antarctic sea ice: distribution, diet and life history strategies. *Advances in Marine Biology* 51: 197-315.

Arrigo, K.R. (2003). Primary production in sea ice. In: Thomas, D.N., and Dieckmann, G.S., editors. *Sea ice. An introduction to its physics, chemistry, biology and geology*, 143-183. Oxford: Blackwell Publishing.

Bienatti, N.L., Comes, R.A. and Spiedo, C. (1977). Primary production in Antarctic waters: seasonal variation and production in fertilized samples during the summer cycle. In: Dunbar, M.J., (ed.). *Polar Oceans*, 377-389. Calgary Arctic Institute of North America.

Blomquist, S., and Elander, M. (1981). Sabine's gull *(Xema sabini)*, Ross's gull *(Rhodostethia rosea)* and ivory gull (Pagophila eburnea). Gulls in the Arctic: A review. *Arctic* 34: 122–132.

Boetius, A., Albrect, S., Bakker, K., Bienhold, C., Felden, J., Fernandez-Mendez, M., Hendricks, S., Katlein, C., Lalande, C., Krumpen, T. Nicolaus, M., Peeken, I., Rabe, B., Rogacheva, A., Rybakova, E. Somavilla, R., Wenzhofer, F. (2013). Export of algal biomass from the melting Arctic sea ice. *Science* 126: 1430-1432.

CAFF (2013). *Life Linked to Ice: A guide to sea-ice-associated biodiversity in this time of rapid change*. CAFF Assessment Series No. 10. Conservation of Arctic Flora, Akureyri.

De Broyer C., Koubbi P., Griffiths H.J., Raymond B., Udekem d'Acoz C. d', Van de Putte A.P., Danis B., David B., Grant S., Gutt J., Held C., Hosie G., Huettmann F., Post A., Ropert-Coudert Y. (eds.), (2014). *Biogeographic Atlas of the Southern Ocean*. Scientific Committee on Antarctic Research, Cambridge.

El-Sayed, S.Z. (1970). On the productivity of the Southern Ocean. In: Holdgate, M.W., ed. *Antarctic Ecology*, 119-135. Academic Press.

El-Sayed, S.Z. (1985). Plankton of the Antarctic Seas. In: Bonner, N., and Walton, D., eds. *Key Environments: Antarctica*. Pergamon Press, Oxford, pp. 135-153.

Falk-Petersen, S., Leu, E., Berge, J., Kwasniewski, S., Nygård, H., Røstad, A., Keskinen, E., Thormar, J., von Quillfeldt, C., Wold, A., and Gulliksen, B. (2008). Vertical migration in high Arctic waters during autumn 2004. *Deep-Sea Research II* 55: 2275-2284.

Gilg, O., Strøm, H., Aebischer, A., Gavrilo, M.V., Volkov, A.E., Miljeteig, C., Sabard, S. 2010. Post-breeding movements of northeast Atlantic ivory gull *Pagophila eburnea* populations. *Journal of Avian Biology* 41 (5): 532–542, doi: 10.1111/j.1600-048X.2010.05125.x

Gorbunov, Yu. A. and Belikov, S.E. (2008). Observations of marine mammals and polar bear in the Arctic Basin. *Marine mammals of the Holarctic. Collection of scientific papers after the fifth International Conference*, Odessa, Ukraine, October 14-18, 220-222. Odessa.

Gutt, J., Barnes, D.K.A., Lockhart, S.J. and van de Putte, A. (2013). Antarctic macrobenthic communities: A compilation of circumpolar information. *Nature Conservation* 4: 1-13.

Gosselin, M., Levasseur, M., Wheeler, P.A., Horner, R.A., and Booth, B.C. (1997). New measurements of phytoplankton and ice algal production in the Arctic Ocean. *Deep-Sea Research* 44: 1623-1644.

Hart, T.J. (1934). On the phytoplankton of the southwest Atlantic and the Bellingshausen Sea. *Discovery Reports* 8: 1-268.

Hjort, C., Gudmundsson, G.A., and Elander, M. (1997). Ross's gulls in the Central Arctic Ocean. *Arctic* 50(4): 289-292.

Holm-Hansen, O. (1985). Nutrient cycles in Antarctic Marine Ecosystems. In: Siegfried, W.R., Condy, P.R., and Laws, R.M., editors. *Antarctic nutrient cycles and food webs*, 6-10. Springer Berlin.

Horner, R., Ackley, S.F., Dieckmann, G.S., Gulliksen, B., Hoshiai, T., Melnikov, I.A., Reeburgh, W.S., Spindler, M., and Sullivan, C.W. (1992). Ecology of Sea Ice Biota. 1. Habitat and terminology. *Polar Biology* 12, 417–427.

IPCC (Intergovernmental Panel on Climate Change)(2007). Summary for Policymakers. In: *Climate Change 2007: The Physical Science Basis*. Contribution of Working Group I to the Fourth Assessment Report of the Intergovernmental Panel on Climate Change.

Kaschner, K., Tittensor, D.P., Ready, J., Gerrodette, T., Worm, B. (2011). Current and Future Patterns of Global Marine Mammal Biodiversity. *PLoS ONE* 6(5): e19653. doi:10.1371/journal.pone.0019653.

Kovacs, K., Lowry, L., Härkönen, T. (2008). Pusa hispida. In: IUCN (2011). IUCN Red List of Threatened Species. Version 2011.1. Available at: www.iucnredlist.org. Accessed on 31 August 2011.

Lin, L., Liao, Y., Zhang, J., Zheng, S., Xiang, P. (2012). Composition and distribution of fish species collected during the fourth Chinese National Arctic Research Expedition in 2010. *Polar Biology* 23: 116-127. doi:10.3724/SP.J.1085.2012.00116.

Lipps, J.H., Krebs, W.N., and Temnikow, N.K. (1977). Microbiota under Antarctic ice shelves. *Nature*, 265, 232.

Lipps, J.H., Ronan, T.E. and DeLaca, T.E. (1979). Life below the Ross ice shelf, Antarctica. *Science*, 203(4379), 447-449.

Meltofte, H. (ed.) (2013). *Arctic Biodiversity Assessment: status and trends in biodiversity*. Conservation of Arctic Flora and Fauna, Akureyri.

Post, A.L., Galton-Fenzi, B.K., Riddle, M.J., Herraiz-Borreguero, L., O'Brien, P.E., Hemer, M.A., McMinn, A., Rasch, D. and Craven, M. (2014). Modern sedimentation, circulation and life beneath the Amery Ice Shelf, East Antarctica. *Continental Shelf Research* 74: 77-87.

Lizotte, M.P. (2003). The microbiology of sea ice. In: Thomas, D.N., and Dieckmann, G.S., editors. *Sea ice: An introduction to its physics, chemistry, biology and geology*, 184-210. Oxford: Blackwell Publishing.

Loeb, V., Siegel, V., Holm-Hansen, O., Hewitt, R., Fraser, W., Trivelpiece, W., and Trivelpiece, S. (1997). Effects of sea-ice extent and krill or salp dominance on the Antarctic food web. *Nature* 387: 897-900.

Maslanik, J., Stroeve, J., Fowler, C. and Emery, W. (2011). Distribution and trends in Arctic sea ice age through spring 2011. *Geophysical Research Letters* 38: L13502. doi:10.1029/2011GL047735.

Nelson, D.M. and Smith, W.O. (1986). Phytoplankton dynamics off the western Ross sea ice edge. *Deep Sea Research* 33:1389-1412.

Overland, J.E. and Wang, M. (2013). When will the summer Arctic be nearly sea ice free? Geophysical Research Letters 40(10), doi: 10.1002/grl.50316, 2097–2101.

Poulin, M., Daugbjerg, N., Gradinger, R., Ilyash, L., Ratkova, T., and von Quillfeldt, C.H. (2010). The pan-Arctic biodiversity of marine pelagic and sea-ice unicellular eukaryotes: A first-attempt assessment Marine Biodiversity. *Marine Biodiversity* 41: 13-28.

Ribic, C.A., and Ainley, D.G. (1988). Constancy of seabird species assemblages: an **exploratory** look. *Biological Oceanography* 6: 175-202.

Ropelewski, C.F. (1983). Spatial and temporal variations in Antarctic sea-ice (1973-82). *Journal of Climate and Applied Meteorology* 22: 470-3.

Sakshaug, E., and Holm-Hansen, O. (1984). Factors governing pelagic production in polar oceans. In: Holm-Hansen, O., Bolis, L., and Gilles, R. (eds.). *Marine phytoplankton and productivity*, 1-18. Springer, Berlin.

Sakshaug, E., and Skjoldal, H.R. (1989). Life at the ice edge. *Ambio* 18: 60-67.

Sakshaug, E. (2004). Primary and secondary production in the Arctic Sea. In: Stein, R., and Macdonald, R.W., (eds.). *The organic carbon cycle in the Arctic Ocean*, 57-81. Springer, Berlin.

Schliebe, S., Wiig, Ø., Derocher, A.E., Lunn, N. (2008). Ursus maritimus. In: IUCN 2011. IUCN Red List of Threatened Species. Version 2011.1. www.iucnredlist.org Downloaded on 31 August 2011.

Stammerjohn, S., Massom, R., Rind, D., Martinson, D. (2012). Regions of rapid sea ice change: An inter-hemispheric seasonal comparison *Geophysical Research Letters* 39.

Tréguer, P., and Jacques, G. (1992). Dynamics of nutrients and phytoplankton, and fluxes of carbon, nitrogen and silicon in the Antarctic Ocean. *Polar Biology* 12:149-162.

Turner, J., Bindschadler, R.A., Convey, P., Di Prisco, G., Fahrbach, E., Gutt, J., Hodgson, D.A., Mayewski, P.A., and Summerhayes, C.P. (eds.) (2009). *Antarctic Climate Change and the Environment*. Cambridge, SCAR, 526 pp.

von Quillfeldt, C.H., Hegseth, E.N., Johnsen, G., Sakshaug, E., and Syvertsen, E.E. (2009). Ice algae. In: Sakshaug, E., Johnsen, G., and Kovacs, K., eds. *Ecosystem Barents Sea*. Tapir Academic Press, Trondheim, Norway, 285-302.

Wassmann, P., Reigstad, M., Haug, T., Rudels, B., Carroll, M.L., Hop, H., Gabrielsen, G.W., Falk-Petersen, S., Denisenko, S.G., Arashkevich, E., Slagstad, D., and Pavlova, O. (2006a). Food webs and carbon flux in the Barents Sea. *Progress in Oceanography* 71(2-4), 232-287.

47 Kelp Forests and Seagrass Meadows

Contributors:
John A. West (Convenor), Hilconida Calumpong (Co-Lead Member), Georg Martin (Lead Member) and Saskia van Gaever (Co-Lead Member)

Kelp Forests and Seagrass Meadows   Chapter 47

1 Introduction

Kelp forests and seagrass meadows form shallow benthic marine habitats. Whereas kelp forests are limited to temperate areas (see Figure 47.1), seagrasses are found throughout all climatic zones (den Hartog 1970; Phillips and Meñez 1988), except in the Polar Regions (see Figure 47.2a). Both provide food and habitat to many economically exploited species, have high productivity (see Chapter 6 for values) and thus, play a significant role in ecological balance. Apart from goods (e.g. associated fisheries, food, phycocolloids) produced by these two ecosystems, kelps and seagrasses also provide many ecosystem services such as carbon sequestration and climate regulation (Raven 1997, Thom 1996; Beer and Koch 1996; Fourqurean et al., 2012), nutrient cycling (Fenchel, 1970; Robertson and Mann 1980; Suchanek et al., 1985; Wahbeh and Mahasneh, 1985; Wood et al., 1969), sediment stabilization and shoreline protection (Barbier et al., 2013), habitat and nursery functions (Duggins et al., 1990; Heck et al., 2003), especially for high value organisms such as crabs, shrimps, clams, flounder, spiny lobster (Kikuchi and Peres, 1977; Tegner and Dayton, 2000). The values of these services vary across geographic regions and cultural groups (see Barbier et al., 2011; Costanza et al., 1997; Cullen-Unsworth, 2014).

The kelp forest is characterized by about 30 species of large brown seaweeds belonging to the order Laminariales (Steneck et al., 2002). Together with its associated animals and other seaweeds, it is considered to be among one of the economically important ecosystems, especially for peoples who have traditionally used them for food, chemicals such as alkali and iodine (Robinson, 2011), and fertilizer and animal feed supplements (Stephenson, 1968). Currently, kelps are still harvested mainly for food and as source of the phycocolloid alginate, which has many uses in industry (see Chapter 14). The brown seaweeds, which are composed primarily of kelps, contribute about half of the total world seaweed production from aquaculture of about 6.8 million tons a year (averaged over a 10-year period between 2003-2012; data from FAO).

The kelp forest is structured like any forest, with different species forming layers or tiers, and large canopy species reaching heights to 45 metres (Steneck et al., 2002). Dominant species differ across regions (see Figure 47.1). Although kelps are not considered to be taxonomically diverse, because most genera are composed of only one species, they support economically important fisheries, such as abalone, lobster, and cod (Steneck et al. 2002). Some host marine mammals, such as sea otters, harbour seals and other pinnipeds (Tegner and Dayton, 2000). Species like the rockfish (*Sebastes* spp.) use kelp habitats during some or most of their life histories (Duggins et al., 1990; Eckman et al., 2003).

Seagrasses are a group of about 72 species of flowering plants in six families (Short et al., 2011) adapted to living and reproducing in the marine environment. They are not true grasses but are named for their close morphological resemblance to terrestrial grasses. They form underwater meadows at depths reached primarily by sunlight in the red wavelength part of the spectrum. In addition, they have unusually high light requirements, approaching 25 per cent of incident radiation for some species (Dennison et al., 1993). Tropical seagrasses tend to have deep lower limits as a result of clear water (and hence, greater light penetration) while most temperate seagrasses are limited to considerably shallower depths. Hence, only a few species grow below 20 m of depth, such as some *Halophila* species which have been reported to oc-

The boundaries and names shown and the designations used on this map do not imply official endorsement or acceptance by the United Nations.

Figure 47.1 | Map showing the approximate location of kelp forests and their dominant species. Modified from: Steneck et al., 2002 extracted from http://commons.wikimedia.org/wiki/File:Kelp_forest_distribution_map.png.

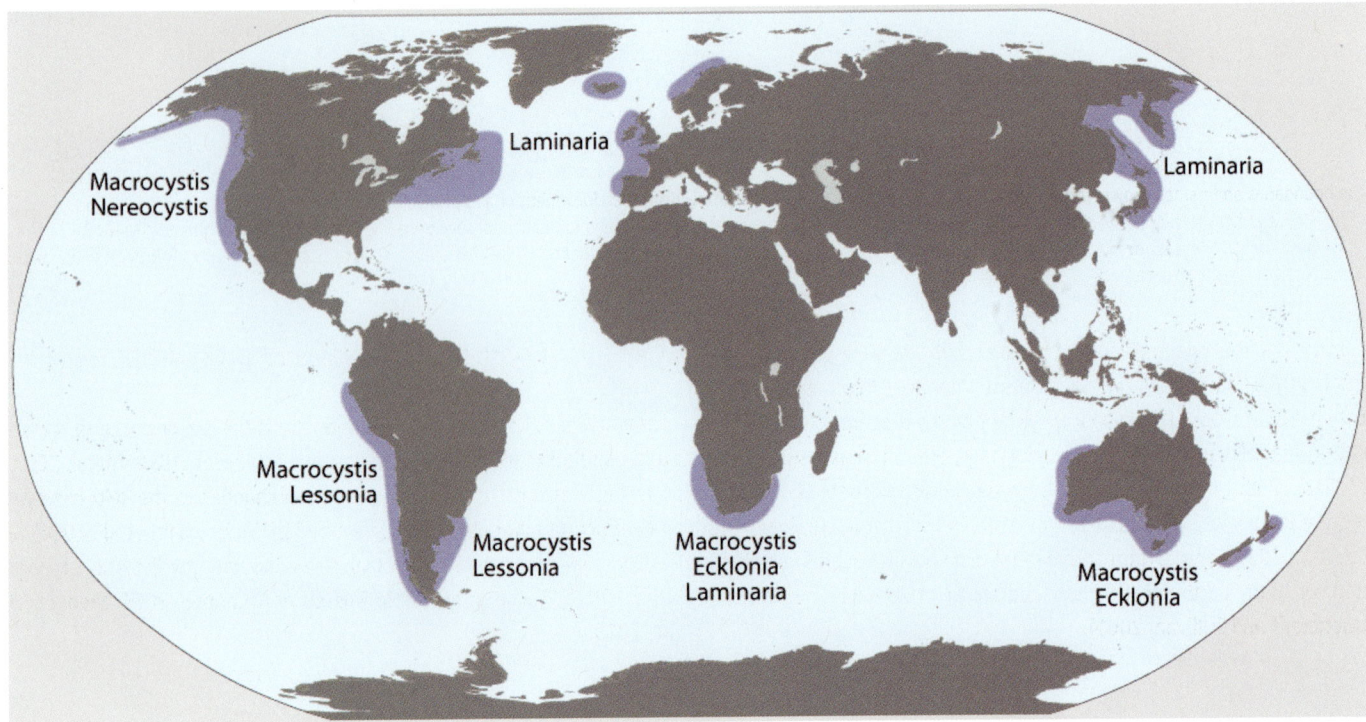

Chapter 47 — Kelp Forests and Seagrass Meadows

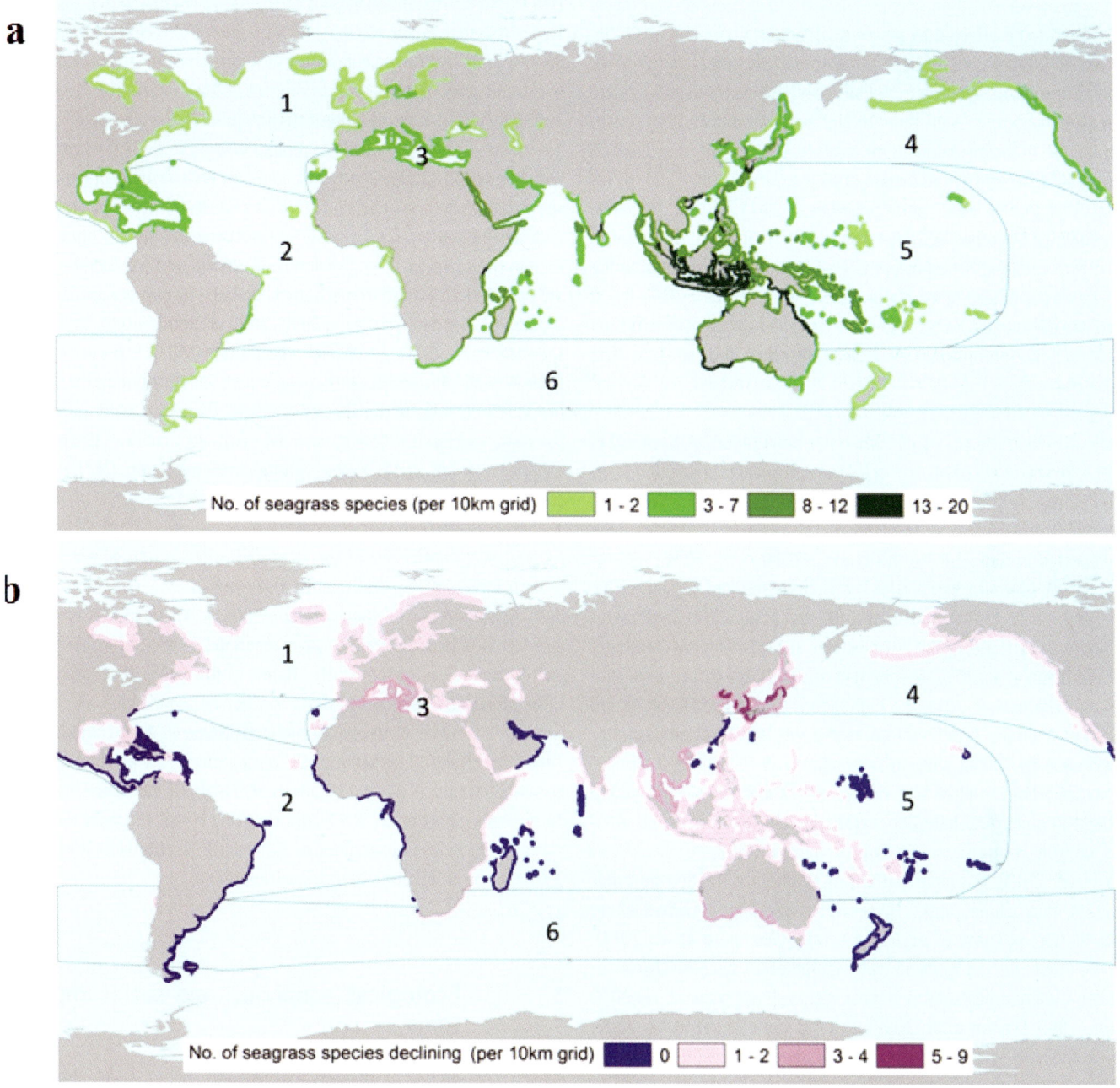

The boundaries and names shown and the designations used on this map do not imply official endorsement or acceptance by the United Nations.

Figure 47.2 | Map showing (a) worldwide distribution and species richness of seagrass meadows; and (b) number of species having declining population trends (sensu IUCN) from Short et al. (2011). Numbers on the map refer to Bioregions. 1-Temperate North Atlantic; 2-Tropical Atlantic; 3- Mediterranean; 4- Temperate North Pacific; 5-Tropical Indo-Pacific; 6-Temperate Southern Oceans.

cur at 40 m (Philipps and Meñez, 1988) and *Posidonia* at 45 m (Pergent et al., 2010). Seagrasses are not presently harvested commercially but they are critical food sources for large herbivores that are specialized for eating seagrass such as manatees, dugongs, green turtles (Philipps and Meñez, 1988) birds, particularly Brant geese (*Branta bernicia*) as they require temperate eelgrass beds as a primary food source (Baldwin and Lovvorn, 1994), and certain commercial fish species such as rabbitfish, and for many other species that feed on the epiphytes and epifauna (Moncreiff and Sullivan, 2001).

2   Population trends and pressures

The harvest of kelps for food and industry is the major pressure on the kelp population worldwide (Vasquez and Santelices, 1990; Millar, 2007; see also Chapter 14). This has resulted in changes in the kelp community structure and habitat well described by McLaughlin et al. (2006) in Chapter 14, and in more recent studies conducted by Estes 2011, Ling et al., 2015; Rocha et al., 2015; Russell and Connell, 2014; Steneck et al., 2013.

Apart from overexploitation, kelp population and distribution worldwide are reported to be affected by a variety of factors. Connell et al. (2008) reported on a wholesale loss of canopy-forming kelp forests (up to 70 per cent) on the Adelaide metropolitan coast of South Australia where urbanization occurred. Overfishing of high value predators often causes explosions in herbivore populations, such as sea urchins, that feed on kelps, resulting in massive reduction of kelp cover and consequently affecting other trophic levels (see Connell et al., 2011; Moy and Christie, 2012; and also Chapter 14). Steneck et al. (2002) reported this threat to kelp beds to be highest roughly in the 40–60° latitude range in both the northern and southern hemispheres. Other mechanisms of kelp forest decline are mechanical damage from destructive fishing gears and boat propellers, pollution, nutrient availability, diseases and parasites, and climatic changes.

Kelp die-off along the coasts of Europe has been reported (Raybaud et al., 2013; Brodie et al., 2014), e.g. in Norway (Moy and Christie, 2012), as well as off the coast of Australia (Smale and Wernberg, 2013; Wernberg et al., 2013). In addition, changes in the distribution of species have been reported in the polar North Atlantic (Müller et al., 2009), and off the coasts of southern England of the United Kingdom (Brodie et al., 2014; Pereira et al., 2011), South Africa (Bolton et al., 2012) and Australia (Connell et al., 2008; Millar 2007; Russell 2011; Smale and Wernberg, 2013; Wernberg et al., 2011; Wernberg et al., 2013) due to increased seawater temperatures. In 2011, the high biodiversity Indian Ocean region of Western Australia experienced a heat wave, which raised seawater temperatures by 2-4°C, causing a significant decline in the canopy-forming brown macroalgae *Scytothalia* and *Ecklonia radiata*, which are important in stabilizing habitats. *Scytothalia* had a 100-km southward retraction from its northernmost limit (Smale and Wernberg, 2013; Wernberg et al., 2013). A similar pattern of southern retraction by other temperate macroalgae caused by seawater warming is evident along the Pacific Ocean coast of eastern Australia (Wernberg et al., 2011). Kelps are most affected by rising water temperature, because sexual reproduction (gamete formation) in most kelps will not occur above 20°C (Dayton, 1985; Dayton et al., 1999). This has been found to be amplified negatively by synergistic interactions between nutrient enrichment and heavy metals, the presence of competitors, low light and increasing temperature and competition with mat-forming seaweeds (Strain et al., 2014).

Seagrass beds are reported to be among the most threatened ecosystems on earth with an estimated disappearance rate of 110 km2 per year since 1980; the rates of decline accelerating from a median of 0.9 per cent per year before 1940 to 7 per cent year - 1 since 1990 (Waycott et al. 2009). According to their assessment, 29 per cent of the known areal extent has disappeared since seagrass areas were initially recorded in 1879. For example, in the Baltic Sea, where only one main seagrass species (*Zostera marina*) exists, seagrass meadows have significantly declined (Boström et al. 2014). In terms of species, Short et al. (2011) reported that 22 of 72 species (31 per cent) of the world's total number of species have declining populations, 29 species (40 per cent) have stable populations, five species (seven per cent) have an increasing population (see Figure 47.2b), and the status of 16 species (22 per cent) is unknown. Two of the species with increasing population (*Halophila stipulacea* and *Zostera japonica*) have been reported to have recently expanded across the Pacific and Atlantic Oceans (Short et al. 2007, Willette and Ambrose 2009). Reported areas of highest decline (80-100 per cent of all species) are in: (a) in China-Korea-Japan region, where the decline is associated with heavy coastal development and extensive coastal reclamation, (b) southeast Asia (one species) due to aquaculture, fisheries and heavy watershed siltation, (c) Australia (three species), and (d) the Mediterranean (four/five species). Declines in Australia and the Mediterranean are primarily attributed to mechanical damage from propellers and ship grounding, degraded water quality, and competition with introduced species such as *Caulerpa* (Williams and Smith 2007). Short et al. (2011, Table 4, p. 1969) rated coastal development as representing the highest threat (93 per cent of the species affected), degraded water quality (53 per cent), mechanical damage (44 per cent), aquaculture (39 per cent), fisheries (38 per cent), excess siltation/sedimentation (36 per cent), competition (7 per cent), and disease (2 per cent).

As for kelps, overfishing of top predators often results in an increase in herbivores, such as sea urchins, that leave barren 'halos' in seagrass beds. Another reported cause of population decline is the "wasting disease" that wiped out the seagrass meadows in the Pacific Northwest and on both sides of the North Atlantic in the 1930s, due to a marine slime mould (*Labyrinthula*) infestation (Rasmussen, 1977); this organism reappeared in New Hampshire and Maine in 1986 (Short et al., 1986). The effects of climate change on seagrasses are just beginning to be studied (Chust et al., 2013; Valle at al., 2014). Of the 72 species, 15 or 24 per cent (Short et al., 2011), are currently classified under the International Union for Conservation of Nature (IUCN) criteria as Threatened (Endangered or Vulnerable) or Near Threatened.

3   Ecological, economic, and social implications

Ecologically, the loss of these two ecosystems will reduce the amount of "blue" carbon stored in submerged marine habitats and thus, increase impacts and changes worldwide on weather patterns, directly putting coastal residents, their livelihoods and food production at risk (see Nelleman et al., 2009; Byrnes, et al., 2011; see also Chapter 6).

Losses of kelp and seagrass beds will affect populations of large marine herbivores, such as manatees, dugongs and green turtles, thus further undermining their already poor conservation status. Short et al. (2011) reported that 115 marine species that live in seagrass beds, including some invertebrates, fishes, sea turtles, and marine mammals, are listed by IUCN as threatened. In addition, reef and mangrove ecosystems biodiversity will be affected by the loss of seagrass habitats since many fish and invertebrate species found in coral reefs and mangroves have been reported to spend their juvenile stages in seagrass beds (Dolar, 1991; Orth, 2006).

Furthermore, the loss of seagrass beds and kelp forests will deprive commercially important fish (such as rabbit fish and cod) and invertebrate species (such as abalone and lobster) of food, habitat and nursery areas, thus undermining their growth and reproductive success and reducing the chances of the stocks either being maintained or being brought back to pre-depletion levels (Orth et al., 2006). This will affect catches of fishers and threaten food security.

Kelp forest losses will reduce the supply of commercially important alginates and fucoidan, thus raising their prices or making them less readily available for new applications (see further Chapter 14). Also, loss of kelp may have an adverse effect on the number of coastal residents whose livelihoods depend on kelp harvesting.

4   Management and conservation responses

Many strategies have been employed in the management of kelps and seagrass meadows. These include: protection through declaration of sanctuaries and protected areas,[1] regulation of harvesting through permitting system for kelps (Leschin-Hoar, 2014), regulation of fishing methods destructive to kelps and seagrasses, such as trawls and seines; transplantation and restoration of seagrass beds (Calumpong and Fonseca, 2001; Fonseca et al., 1998), and systematic monitoring (www.seagrassnet.org).

5   Information and knowledge gaps

The biology and population dynamics of some kelp and seagrass species are still unstudied. Nine of 72 seagrass species are designated by IUCN as Data Deficient due to lack of information about them, while population trends of 16 species remain unknown (Short et al., 2011). Data on relative impacts of anthropogenic factors as well as interactions with climatic changes are lacking (Larkum et al., 2006; Chust et al., 2013; Doney et al., 2009; Duarte, 2002; Grech et al., 2012; Roleda et al., 2012; Valle et al., 2014). Active research is being conducted in the areas of economic valuation of ecosystem services provided by these two ecosystems.

---

[1] See: http://www.westcoast.fisheries.noaa.gov/habitat/habitat_types/kelp_forest_info/kelp_forest_habitat_types.html; http://www.pcouncil.org/habitat-and-communities/habitat/; https://catalog.data.gov/dataset/public-seagrass-compilation-for-west-coast-essential-fish-habitat-efh-environmental-impact-stat; http://www.marinecadastre.gov/news/uses/seagrasses-distribution/.

References

Baldwin, J.R., Lovvorn, J.R. (1994). Expansion of seagrass habitat by the exotic *Zostera japonica*, and its use by babbling ducks and brant in Boundary Bay, British Columbia. *Marine Progress Series* 103: 119-127.

Barbier, E.B., Georgiou, I.Y., Enchelmeyer, B., Reed, D.J. (2013). The Value of Wetlands in Protecting Southeast Louisiana from Hurricane Storm Surges. *PLoS ONE* 8(3): e58715. doi:10.1371/journal.pone.0058715.

Barbier, E.B., Hacker, S.D., Kennedy, C., Koch, E.W., Stier, A.C., Silliman, B.R. (2011). The value of estuarine and coastal ecosystems. *Ecological Monographs* 81(2): 169-193.

Beer, S., Koch, E. (1996). Photosynthesis of seagrasses vs. marine macroalgae in globally changing CO_2 changing environments. *Ecology Progress Series* 141: 199–204.

Bolton, J., Anderson, R., Smit, A., Rothman, M. (2012). South African kelp moving eastwards: the discovery of *Ecklonia maxima* (Osbeck) Papenfuss at De Hoop Nature Reserve on the South Coast of South Africa, *African Journal of Marine Science* 34: 147-151.

Boström, C., Baden, S., Bockelmann, A-C., Dromph, K., Fredriksen, S., Gustafsson, C., Krause-Jensen, D., Möller, T., Nielsen, SL., Olesen, B., Olsen, J., Pihl, L., Rinde, E. (2014). Distribution, structure and function of Nordic eelgrass (*Zostera marina*) ecosystems: implications for coastal management and conservation. *Aquatic Conservation: Marine and Freshwater Ecosystems* 24: 410-434.

Brodie, J., Williamson, C.J., Smale, D.A., Kamenos, N.A., Mieszkowska, N., Santos, R., Cunliffe, M., Steinke, M., Yesson, C., Anderson, K.M., Asnaghi, V., Brownlee, C., Burdett, H.L., Burrows, M.T., Collins, S., Donohue, P.J.C., Harvey, B., Noisette, F., Nunes, J., Ragazzola, F., Raven, J.A., Foggo, A., Schmidt, D.N., Suggett, D., Teichberg, M., Jason M. Hall-Spencer, J.M. (2014). The future of the northeast Atlantic benthic flora in a high CO_2 World. *Ecology and Evolution* 1-12. doi: 10.1002/ece3.1105.

Byrnes, J.E., Reed, D.C., Cardinale, B.J., Cavanaugh, K.C., Holbrook, S.J., Schmitt, R.J., (2011). Climate driven increases in storm frequency simplify kelp forest food webs. *Global Change Biology* 17, 2513-2524.

Calumpong, H.P., Fonseca, M. (2001). Chapter 22: Seagrass transplantation and other seagrass restoration methods. In: Short, F.T., Coles, R.G. (Eds.), *Global Seagrass Research Methods*, Elsevier Science B.V. pp. 426-443.

Chust, G., Albaina, A., Aranburu, A., Borja, Á., Diekmann, O.E., Estonba, A., Franco, J., Garmendia, J.M., Iriondo, M., Muxika, I., Rendo, F., Rodríguez, J.G., Ruiz-Larrañaga, O., Serrão, E.A. and Valle, M. (2013). Connectivity, neutral theories and the assessment of species vulnerability to global change in temperate estuaries. *Estuarine, Coastal and Shelf Science*, 131, 52-63.

Connell, S., Russelll, B., Turner, D., Shepherd, S., Kildea, T., Miller, D., Airoldi, L., Cheshire, A. (2008). Recovering a lost baseline: missing kelp forests from a metropolitan coast. *Marine Ecology Progress Series* 360: 63-72.

Connell, S.D., Russelll, B.D., Irving, A.D. (2011). Can strong consumer and producer effects be reconciled to better forecast catastrophic phase-shifts in marine ecosystems? *Journal of Experimental Marine Biology and Ecology*, 400, 296-301.

Costanza, R., d'Arge, R., de Groot, R., Farber, S., Grasso, M., Hannon, B., Limburg, K., Naeem, S., O'Neill, R.V., Paruelo, J., Raskin, R.G., Sutton, P., van den Belt, M. (1997). The value of the world's ecosystem services and natural capital. *Nature* 387: 253-260.

Cullen-Unsworth, L.C., Nordlund, L.M., Paddock, J., McKenzie, L.J., Unsworth, R.K.F., Baker, S. (2014). Seagrass meadows globally as a coupled social–ecological system: Implications for human wellbeing. *Marine Pollution Bulletin* 83(2): 387-97. doi: 10.1016/j.marpolbul.2013.06.001.

Dayton, P.K. (1985). Ecology of kelp communities. *Annual Review Ecology Systems* 16: 215–245.

Dayton, P.K., Tegner, M.J., Edwards, P.B., Riser, K.L. (1999). Temporal and spatial scales of kelp demography: the role of oceanography climate. *Ecological Monographs* 69: 219–250.

Den Hartog, C. (1970). *The Seagrasses of the World*. Amsterdam, North Holland Publication Co. 275 pp.

Dennison W.C., Orth, R.J., Moore, K.A., Stevenson, J.C., Carter, V., Kollar, S., Bergstrom, P.W., Batiuk, R.A. (1993). Assessing water quality with submersed aquatic vegetation. *BioScience* 43: 86–94.

Dolar, M.L.L. (1991). A survey on the fish and crustacean fauna of the seagrass beds in North Bais Bay, Negros Oriental, Philippines. *Proceedings of the Regional Symposium on Living Resources in Coastal Areas*. Quezon City: University of the Philippines Marine Science Institute, pp. 367-377.

Doney, S.C., Fabry, V.J., Feely, R.A., Kleypas, J.A. (2009). Ocean Acidification: The Other $CO(2)$ Problem. *Annual Review of Marine Science, Annual Reviews,* 169-192.

Duarte, C.M. (2002). The future of seagrass meadows. *Environmental Conservation* 29 (2): 192–206 doi:10.1017/S0376892902000127.

Duggins, D.O., Eckman, J.E., Sewell, A.T. (1990). Ecology of understory kelp environments. II. Effects of kelps on recruitment of benthic organisms. *Journal of Experimental Marine. Biology and Ecology* 143, 27-45.

Eckman, J.E., Duggins, D.O., Siddon, C.E. (2003). Current and wave dynamics in the shallow subtidal: implications to the ecology of understory and surface-canopy kelps. *Marine Ecology-Progress Series* 265, 45-56.

Estes, J.A., Terborgh, J., Brashares, J.S., Power, M.F., Berger, J., Bond, W.J., Carpenter, S.R., Essington, T.F., Holt, R.D., Jackson, J.B.C., Marquis, R.J., Oksanen, L., Oksanen, T., Paine, R.T., Pikitch, E.K., Ripple, W.J., Sandin, S.A., Scheffer, M., Schoener, T.W., Shurin, J.B., Sinclair, A.R.E., Soulé, M.E., Virtanen, R., Wardle, D.A. (2011). Trophic downgrading of planet earth. *Science* 333(6040):301-306. DOI: 10.1126/science.1205106.

Fenchel, T. (1970). Studies on the decomposition of organic detritus derived from the terete grass *Thalassia testudinum*. *Limnology and Oceanography* 15(1): 14-20.

Fonseca, M.S., Kenworthy, W.J., Thayer, G.W. (1998). *Conservation and Restoration of Seagrasses in the United States and Adjacent Waters.* NOAA Coastal Ocean Office, Silver Spring, Maryland Series No. 12 pp. 222.

Fourqurean, J.W., Duarte, C.M., Marbà, N., Holmer, M., Mateo, M.A., Apostolaki, E.T., Kendrick, G.A., Krause-Jensen, D., McGlathery, K.J., Serrano, O. (2012). Seagrass ecosystems as a globally significant carbon stock. *Nature Geoscience* 5: 505–509 doi:10.1038/ngeo1477.

Grech, A., Chartrand-Miller, K., Erftemeijer, P., Fonseca, M., McKenzie, L., Rasheed, M., Taylor, H., Coles, R. (2012). A comparison of threats, vulnerabilities and management approaches in global seagrass bioregions. *Environmental Research Letters* 7.

Heck, K.L., Hays, C., Orth, R.J. (2003). A critical evaluation of the nursery role hypothesis for seagrass meadows. *Marine Ecology Progress Series* 253: 123–136.

Kikuchi, T., Peres, J.P. (1977). *Consumer Ecology in Seagrass Beds.* New York: Marcel Dekker, pp. 153-172.

Leschin-Hoar, C. (2014). Help for Kelp—Seaweed Slashers See Harvesting Cuts Coming. *Scientific American*. http://www.scientificamerican.com/article/help-for-kelp-seaweed-slashers-see-harvesting-cuts-coming/

Ling, S.D., Scheibling, R.E., Rassweiler, A., Johnson, C.R., Shears, N., Connel, S.D.

(2015). Global regime shift dynamics of catastrophic sea urchin overgrazing. *Philosophical Transactions of the Royal Society B: Biological Sciences*, 370(1659), 20130269. DOI: 10.1098/rstb.2013.0269.

Larkum, A., Orth, R.J., Duarte, C., eds. (2006). *Seagrasses: Biology, Ecology and Conservation*. Springer, Dordrecht. 676 pp. DOI 10.1007/978-1-4020-2983-7.

McLaughlin, E., Kelly, J., Birkett, D., Maggs, C., Dring, M. (2006). Assessment of the Effects of Commercial Seaweed Harvesting on Intertidal and Subtidal Ecology in Northern Ireland. *Environment and Heritage Service Research and Development Series*. No. 06/26.

Millar, A.J.K. (2007). *The Flindersian and Peronian Provinces*. In: McCarthy, P., Orchard, A., (Eds.). *Algae of Australia. An Introduction*. CSIRO Publishing, Melbourne, pp. 554-559.

Moncreiff, C.A., Sullivan, M.J. (2001). Trophic importance of epiphytic algae in subtropical seagrass beds: evidence from multiple stable isotope analyses. *Marine Ecology Progress Series*, 215, 93-106.

Moy, F., Christie, H. (2012). Large-scale shift from sugar kelp (*Saccharina latissima*) to ephemeral algae along the south and west coast of Norway. *Marine Biology Research* 8: 309-321.

Müller, R., Laepple, T., Bartsch, I., Wiencke, C. (2009). Impact of oceanic warming on the distribution of seaweeds in polar and cold-temperate waters. *Botanica Marina* 52: 617–638. doi 10.1515/bot.2009.080.

Nelleman, C., Corcoran, E., Duarte, C.M., Valdes, M., DeYoung, C., Fonseca, L., Grimsditch, G., Eds. (2009). *Blue Carbon: A Rapid Response Assessment*. United Nations Environmental Programme. GRID-Arendal.www.grida.no.

Orth, R.J., Carruthers, T.J.B., Dennison, W.C., Duarte, C.M., Fourqurean, J.W., Heck, K.L. Jr., Hughes, A.R., Kendrick, G.A., Kenworthy, W.J., Olyarnik, S. Short, F.T., Waycott, M., Williams, S.L. (2006). A global crisis for seagrass ecosystems. *BioScience* 56(12): 987-996.

Pergent, G., Semroud, R., Djellouli, A., Langar, H., Duarte, C. (2010). *Posidonia oceanica*. The IUCN Red List of Threatened Species. Version 2014.3. <www.iucnredlist.org>. Downloaded on 15 May 2015.

Pereira, T., Engelen, A., Pearson, G., Serrão, E., Destombe, C., Valero, M. (2011). Temperature effects on the microscopic haploid stage development of *Laminaria ochroleuca* and *Saccorhiza polyschides*, kelps with contrasting life histories. *Cahiers de Biologie Marine* 52: 395-403.

Phillips, R., Meñez E.G. (1988). *Seagrasses*. Smithsonian Contributions to the Marine Sciences Number 14, 104 pp.

Rasmussen, E. (1977). The Wasting Disease of Eelgrass (*Zostera marina*) and Its Effects on Environmental Factors and Fauna. In: McRoy, C.P., Helfferich, C. Eds.). *Seagrass Ecosystems: A Scientific Perspective*, Dekker, New York.

Raven, J.A. (1997). Inorganic carbon acquisition by marine autotrophs. In: Callow, J.A. (Ed.), Advances in Botanical Research, Vol 27: Classic Papers. Elsevier, Academic Press Inc., San Diego, p. 85-209.

Raybaud, V., Beaugrand, G., Goberville, E., Delebecq, G., Destombe, C. Valero, M., Davoult, D., Morin, P., Gevaert, F. (2013). Decline in Kelp in West Europe and Climate. *PLoS ONE* 8(6): e66044. doi:10.1371/journal.pone.0066044.

Robertson, A.I., Mann, K.H. (1980). The role of amphipods and isopods in the initial fragmentation of eelgrass detritus in Nova Scotia, Canada. *Marine Biology* 59:63-69.

Robinson, T. (2011). *Connemara - a little Gaelic Kingdom*, Dublin: Penguin, Ireland.

Rocha, J. Yletyinen, J., Biggs, R., Blenckner, T., Peterson, G. (2015). Marine regime shifts: drivers and impacts on ecosystems services. *Philosophical Transactions of the Royal Society B: Biological Sciences*, 370(1659), 20130273. DOI: 10.1098/rstb.2013.0273.

Roleda, M.Y., Morris, J.N., McGraw, C.M., Hurd, C.L. (2012). Ocean acidification and seaweed reproduction: increased CO2 ameliorates the negative effect of lowered pH on meiospore germination in the giant kelp *Macrocystis pyrifera* (Laminariales, Phaeophyceae). *Global Change Biology* 18, 854-864.

Russell, B., Connell, S. (2014). Ecosystem resilience and resistance to climate change. *Global Environmental Change* 1:133-139.

Russell, B., Thomsen, M., Gurgel, F., Bradshaw, C., Poloczanska, E., Connell, S. (2011). Seaweed communities in retreat from Ocean Warming. *Current Biology* 21: 1828-1832.

Short, F.T., Carruthers, T.J.B., Dennison, W.C., Waycott, M. (2007). Global seagrass distribution and diversity: a bioregional model. *Journal of Experimental Marine Biology and Ecology* 350: 3–20.

Short, F.T., Mathieson, A.C., Nelson, J.I., (1986). Recurrence of the eelgrass wasting disease at the border of New Hampshire and Maine, USA. *Marine Ecology Progress Series* 29, 89–92.

Short, F.T., Polidoro, B., Livingstone, S.R., Carpenter, K.E., Bujang, J.S., Calumpong, H.P., Carruthers, T.J.B., Coles, R.G., Bandeira, S., Dennison, W.G., Erftemeijer, P.L.A., Fortes, M.D., Freeman, A.S., Jagtap, T.G., Kamal, A.H.M., Kendrick, G.A., Kenworthy, W.J., La Nafie, Y.A., Nasution, I.M., Prathep, A., Sanciangco, J.C., van Tussenbroek, B., Vergara, S.G., Waycott, M., Zieman, J.C., Orth, R.J. (2011). Extinction risk assessment of the world's seagrass species. *Biological Conservation* 144: 1961–1971.

Smale, D.A., Wernberg, T. (2013). Extreme climatic event drives range contraction of a habitat-forming species. *Proceedings of the Royal Society B Biological Sciences* 280: 20122829.

Steneck, R., Graham, M.H., Bourque, B.J., Corbett D., Erlandson, J.M., Estes, J.A., Tegner, M.J. (2002). Kelp Forest Ecosystems: Biodiversity, Stability, Resilience and Future. *Environmental Conservation* 29 (4): 436–459.

Steneck, R.S., Leland, A., Mcnaught, D.C., Vavrinec, J. (2013). Ecosystem Flips, Locks, and Feedbacks: the Lasting Effects of Fisheries on Maine's Kelp Forest Ecosystem. *Bulletin of Marine Science* 89:31-55. http://dx.doi.org/10.5343/bms.2011.1148.

Stephenson, W.A. (1968). *Seaweed in Agriculture and Horticulture*. Faber and Faber: London, 231 pp.

Strain, E.M.A., Thomson, R.J., Micheli, F., Mancuso, F. (2014). Identifying the interacting roles of stressors in driving the global loss of canopy-forming to mat-forming algae in marine ecosystems. *Global Change Biology*, 20(11): 3300-3312.

Suchanek, T.H., Williams, S.L., Ogden, J.C., Hubbard, D.K., Gill, I.P. (1985). Utilization of shallow-water seagrass detritus by Caribbean deep-sea macrofauna: C13 evidence. *Deep Sea Research* 32(2): 201-214.

Tegner, M.J., Dayton, P.K. (2000). Ecosystem effects of fishing in kelp forest communities. *ICES Journal of Marine Science*, 57: 579–589.

Thom, R.M. (1996). CO2 enrichment effects on eelgrass (*Zostera marina* L.) and bull kelp (*Nereocystis luetkeana* (Mert. P. & R.). *Water, Air, and Soil Pollution* 88: 383-391.

Valle, M., Chust, G., del Campo, A., Wisz, M.S., Olsen, S.M., Garmendia, J.M., Borja, A. (2014). Projecting future distribution of the seagrass *Zostera noltii* under global warming and sea level rise. *Biological Conservation*, 170: 74-85.

Vasquez, J.E., Santelices, B. (1990). Ecological effects of harvesting *Lessonia* (Laminariales, Phaeophyta) in central Chile. *Hydrobiologia* 204/205: 41-47.

Wahbeh, M.I., Mahasneh, M.A. (1985). Some aspects of the decomposition of leaf litter of the seagrass *Halophila stipulacea* from the Gulf of Adaba (Jordan). *Aquatic Botany* 21: 237-244.

Waycott, M., Duarte, C.M., Carruthers, T.J.B., Orth, R.J., Dennison, W.C., Calladine, A., Fourqurean W.J., Heck, K.L.Jr., Hughes, A.R., Kenworthy, W.J., Short, F.T., Williams, S.L., Olyarnik, S., Kendrick, G.A. (2009). Accelerating loss of seagrasses across the globe threatens coastal ecosystems. *PNAS*:106 12377-12381. (www.pnas.org_cgi_doi_10.1073_pnas.0905620106).

Wernberg, T., Russell, B., Thomsen, M., Gurgel, F., Bradshaw, C., Poloczanska, E., Connell, S. (2011). Seaweed communities in retreat from Ocean Warming. *Current Biology* 21: 1828-1832.

Wernberg, T., Smale, D., Tuya, F., Thomsen, M., Langlois, T, de Bettignes, T. Bennett, S., Rousseaux, C. (2013). An extreme climatic event alters marine ecosystem structure in a global biodiversity hotspot. *Letters Nature Climate Change*. 3: 78-82.

Willette, D.A., Ambrose, R.F. (2009). The distribution and expansion of the invasive seagrass *Halophila stipulacea* in Dominica, West Indies with a preliminary report from St. Lucia. *Aquatic Botany* 91: 137-142.

Williams, S.L., Smith, J.E. (2007). Global review of the distribution, taxonomy, and impacts of introduced seaweeds. *Annual Review of Ecology, Evolution, and Systematics* 38: 327-359.

Wood, E.J., Odum, W.E., Zeiman, J.C. (1969). *Influence of Seagrasses on the Productivity of Coastal Lagoons*. Universidad Nacional Autónoma de México: Mexico, pp. 495-502.

Chapter 48

Mangroves

Contributors:
Mona Webber (Convenor), Hilconida Calumpong (Co-Lead Member), Beatrice Padovani Ferreira (Co-Lead Member), Elise Granek, Sean O. Green (Lead Member), Renison Ruwa (Co-Lead Member) and Mário Soares

Mangroves — Chapter 48

1 Definition and significance

Mangroves dominate the intertidal zone of sheltered (muddy) coastlines of tropical, sub-tropical and warm temperate oceans. The word 'mangrove' is used to refer to both a specific vegetation type and the unique habitat (also called tidal forest, swamp, wetland, or mangal) in which it exists (Tomlinson 1986; Saenger, 2003; Duke et al., 2007; Spalding et al., 2010). Mangrove areas often include salt flats, which are mostly observed in arid regions or areas with well-defined dry seasons, and where the frequency of tidal flooding decreases progressively toward the more landward zones of the forest leading to an accumulation of salts. In such mangrove areas, a continuum of features may be observed, which, as described by Woodroffe et al. (1992), may include: (a) mudflats in the zone below mean sea level; (b) mangrove forests in the zone between mean sea level and the level of higher neap tides; and (c) salt flats in the zone above the level of higher neap tides. These transition zones and their tidal positions vary globally as they are dependent on many factors (e.g. climate, topography and hydrology). Mangrove trees, along with other floral inhabitants of the mangrove area, such as shrubs, ferns and palms, are highly adapted with aerial roots, viviparous seeds and salt exclusion/excretion mechanisms (Tomlinson, 1986; Hogarth, 2007), thus coping with periodic immersion and exposure by the tide, fluctuating salinity, low oxygen concentrations in the water and sediments, and sometimes high temperatures (Hogarth, 2007). Mangroves have been used by coastal inhabitants for centuries with the earliest reports from 10,000 - 20,000 years ago (Allen, 1987; Luther and Greenburg, 2009). Mangroves continue to be of tremendous value to humanity through a range of ecosystem services. Several reviews are dedicated to mangrove forests, addressing their global distribution (area covered and biomass), ecology, biology and value/uses (Dittmar et al., 2006; FAO, 2007; Walters et al., 2008; Ellison, 2008; Costanza et al., 2008; Spalding et al., 2010; Giri et al., 2011; Horwitz et al., 2012; McIvor et al., 2012; Hutchinson et al., 2014).

2 Spatial patterns and inventory

Mangrove distribution correlates with air and sea surface temperatures, such that they extend to ~30°N, but to 28°S on the Atlantic coast (Soares et al., 2012), and in the Indo-West Pacific (IWP), to 38°45'S to Australia and New Zealand (Hogarth, 2007). The latitudinal distribution of mangroves is limited by key climate variables such as aridity and frequency of extreme cold weather events (Osland et al., 2013, Saintilan et al., 2014). The distribution and structural development within areas with suitable temperatures is further limited by rainfall or freshwater availability (Osland et al., 2014; Alongi 2015) The area covered by mangroves (between 137,760 and 152,000 km^2) and the number of countries in which they exist (118 to 124) have been the focus of many studies (FAO, 2007; Alongi, 2008; Spalding et al., 2010; Giri et al., 2011). The accuracy of these ranges is affected by the different methods (with varying spatial resolutions) used for area surveys and the exclusion of some countries with small mangrove stands (FAO, 2007; Giri et al., 2011). However, what is more generally accepted is that mangrove coverage is extremely low, accounting for less than 1 per cent of tropical forests and 0.4 per cent of global forest areas (FAO, 2007; Spalding et al., 2010, Van Lavie-

The boundaries and names shown and the designations used on this map do not imply official endorsement or acceptance by the United Nations.

Figure 48.1 | The global range of mangroves is demarcated in red (Giri et al., 2011). Used with permission from UNEP-WCMC.

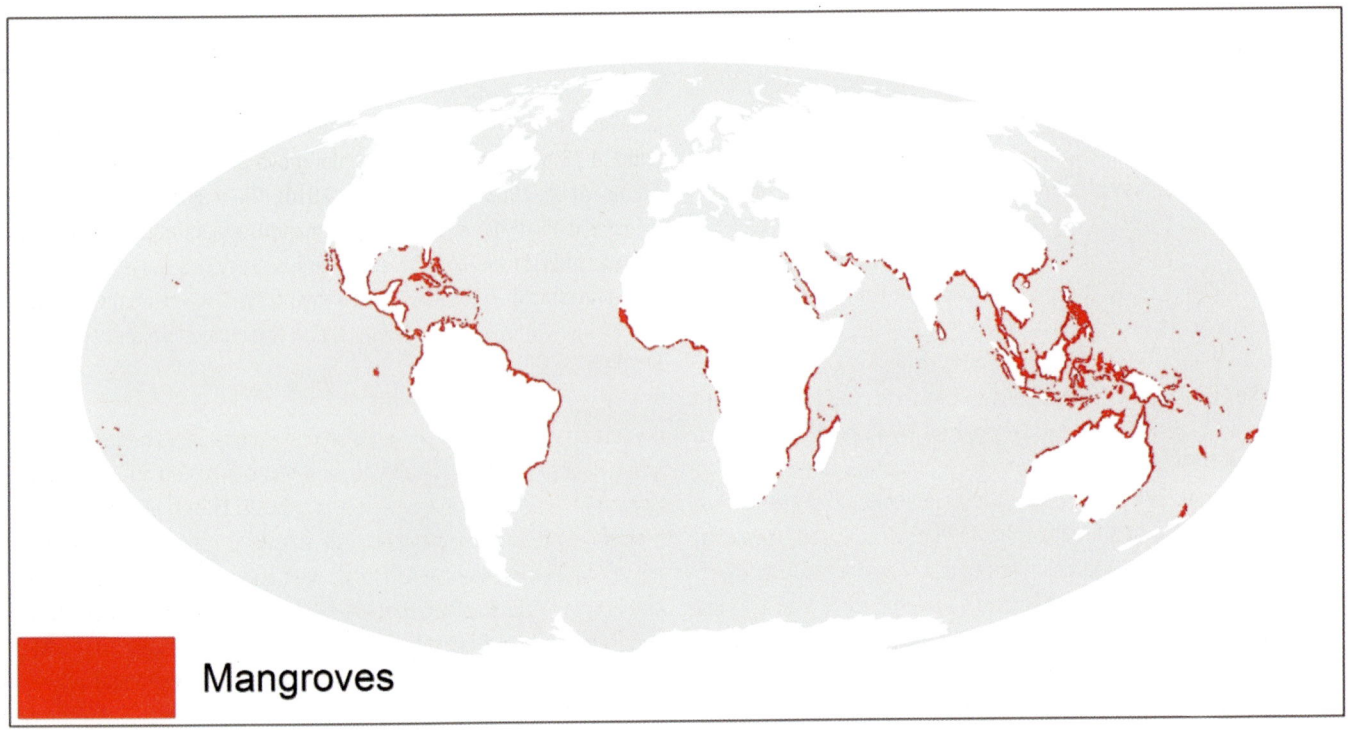

ren et al., 2012). Mangrove area has declined globally over the last 30 years (1980 – 2010), (Polidoro et al., 2010; Donato et al. 2011) and this decline continues in many regions.

Uncertainty also surrounds the number of mangrove species found globally. Spalding et al. (2010) reported 73 mangrove species (inclusive of hybrids), of which 38 were called 'core species', , also called 'foundation species', indicating those which typify mangroves and dominate in most areas (Ellison et al., 2005, Osland et al., 2014. Polidoro et al. (2010) listed a similar number of species (70), which did not include hybrids, but used the criteria of "anatomical and physiological adaptations to saline, hypoxic soils". Thus their list included both 'true' mangroves and mangrove 'associates', classifications by Tomlinson (1986) and Hogarth (2007). Tomlinson (1986) lists the criteria of 'true or strict' mangroves as: (i) occurring in the mangrove environment and not extending into terrestrial communities; (ii) having a major role in the structure of the community; (iii) possessing morphological specialization that adapts them to their environment; (iv) possessing physiological mechanisms for salt exclusion; and (v) having taxonomic isolation from terrestrial relatives, at least at the generic level.

It has been argued that ignoring the distinction between 'true' mangroves and mangrove 'associates' may lead to cryptic ecological degradation, as the latter may include species, such as *Acrostichum aureum*, which can totally replace mangrove trees in some regions, with an accompanying change in mangrove functionality (Dahdouh-Guebas et al., 2005), but without change in areal extent. This idea is controversial and is made more problematic by the inclusion by some of beach grass and scrub vegetation in the category of 'mangrove associates'. It is also difficult to resolve the issue of the exact number of mangrove species recognized worldwide due to taxonomic inconsistencies caused by the use (or not) of the most recent phylogenetic listings. Angiosperm phylogeny listings are constantly updated, most recently with the APGIII (2009). Addressing issues with mangrove taxonomy would enhance our ability to track global species extinctions (Polidoro et al., 2010). Furthermore, it would be useful to base species identification on molecular attributes (not just morphological descriptions), which could address many controversies surrounding use of terms like 'mangrove associates' or 'mangrove hybrids'.

Globally, the IWP and the Atlantic, East Pacific (AEP) have different mangrove species groups (Hogarth, 2007; Spalding et al., 2010). The IWP region has over 90 per cent of species and 57 per cent of global area coverage; the AEP has less than 10 per cent of species and 43 per cent of global area coverage. Fifteen countries account for 75 per cent of global mangrove area (Giri et al., 2011) and these countries are distributed across both regions. Indonesia in the IWP accounts for 22.6 per cent of global mangrove area, and Brazil in the AEP has 8.5 per cent (Spalding et al., 2010). Brazil has the largest continuous mangrove forest (6,516 km2), which lies between Maranhão and Pará in northern Brazil. In the IWP, the Sundarbans, located in India and Bangladesh, extend 85 km inland and cover an area of 6,502 km2 (Spalding et al., 2010). These regions have no true mangrove species in common, except for *Rhizophora mangle/R. samoensis* (Duke and Allen, 2006). *Acrostichum aureum*, which is classified by some as a mangrove 'associate', is also found in both regions. The genera *Rhizophora* and *Avicennia* are unique in having worldwide distribution (Duke et al., 2002).

3 Rate of loss/changes and major pressures

Despite widespread knowledge of their value, mangroves are being lost globally at a mean rate of 1-2 per cent per year (Duke et al., 2007; FAO, 2007), and rates of loss may be as high as 8 per cent per year in some developing countries (Polidoro et al., 2010). Between 20 and 35 per cent of mangroves have been lost since 1980 (FAO, 2007; Polidoro et al., 2010), which is greater than losses of tropical rain forests or coral reefs (Valiela et al., 2001). Spalding et al. (2010) report losses of over 20 per cent in all regions except Australia over a 25-year period (1980-2005). However, their assessments of loss indicate that the global rate of loss has been declining over the last three decades (1.04 per cent in the 1980s; 0.72 per cent in the 1990s and 0.66 per cent in the five year period up to 2005 (Spalding et al., 2010). This could be an indication of increasing resilience of the remaining mangroves or the result of effective conservation and restoration/rehabilitation efforts.

Unfortunately, in some regions, responses to mangrove loss and mitigation remain inadequate, along with the realization that it is more economical to conserve than to restore mangroves (Ramsar Secretariat, 2001; Gilman et al., 2008; Webber et al., 2014). Particular species of mangroves or specific geographic areas have been identified as being more threatened by extinction than others (Polidoro et al., 2011). Although the primary threats to all mangroves are destruction through conversion of mangrove habitat and over-exploitation of resources, pressures that result in loss of area and ecosystem function vary somewhat across regions (Vaiela et al., 2001). Two areas have shown the greatest per cent loss between 1980 and 2005: the Indo-Malay-Philippine Archipelago (IMPA) with 30 per cent reduction, and the Caribbean, with 24-28 per cent reduction in mangrove area (McKee et al., 2007b; Gilman et al., 2008; Polidoro et al., 2010). The major pressure resulting in losses in the IMPA is conversion of mangrove habitat for shrimp aquaculture; while in the Caribbean numerous pressures cause habitat loss, including coastal and urban development, solid waste disposal, extraction of fuel-wood, as well as conversion to aquaculture and agriculture (Polidoro et al., 2010).

Climate change, particularly sea level rise, is considered a threat to mangrove habitat and functionality in all regions (McLeod and Salm, 2006; Gilman et al., 2008; Van Lavieren et al., 2012; Ellison and Zouh, 2012). Mangrove areas most vulnerable to sea level rise are believed to be those of low-relief carbonate islands with a low rate of sediment supply and little available upland space (Schleupner, 2008) as well as those in arid, semi-arid, and dry sub-humid regions (Osland et al., 2014). Mangroves on wet, macrotidal coastlines (>4 m tidal amplitude) with significant riverine inputs, are believed to be least vulnerable (Ellison and

Zouh, 2012). While there are varying opinions on the nature and level effects on mangroves from climate change drivers, it is widely agreed that the vulnerability of mangrove forests is increased by occupation and urbanization of the coastal zone, including the conversion of mangrove area to other land uses (Soares, 2009).

Some of the other effects of climate change (e.g., increased precipitation, temperature and atmospheric CO_2 concentration) may actually increase mangrove productivity (Gilman et al., 2007) and the ability of mangroves to keep pace with sea level rise (Henzel et al. 2006; McKee et al., 2007a; Langley et al., 2009; McKee, 2011; Krauss et al., 2014) because elevated CO_2 increases productivity and biotic controls of soil elevation. Increased temperatures are correlated with mangrove range expansion (Osland et al. 2013), due to the reduction in intensity, duration and frequency of extreme cold-weather events that are expected to support mangrove poleward migration. The genus *Avicenna* has already proliferated at or near their polar limit at the expense of salt marshes (Saintilan et al., 2014). Mangroves may therefore be more resilient to climate change than was previously thought (Alongi, 2007) and certainly the effects will vary greatly depending on local conditions (e.g., geomorphology and shoreline stability). Indeed, the role of mangroves in carbon sequestration and mitigation of climate change effects (Siikamäki et al., 2012) is such that there may be net global economic gains from their protection, especially when all other economic and ecological uses are factored in to the calculation. Mangroves have high rates of atmospheric carbon capture and storage, (Mcleod et al., 2011; Van Lavieren et al., 2012). Their productivity and substantial below- and above-ground biomass, although varying with geomorphology and coastal conditions, can yield sequestration rates of over 174 gCm-2 yr-2 (Alongi, 2012), making them prime targets for not just conservation but active reforestation and restoration. Although mangroves account for a small percentage of the earth's forest cover (Donato et al., 2011; Giri et al., 2011) and hence only 1% of global forest sequestration (Alongi, 2012), they account for 14% of carbon sequestration by the global ocean.

4 Implications for services to the marine ecosystem and humanity

Mangroves provide a suite of regulating, supporting, provisioning and cultural ecosystem services from which humanity benefits (MEA, 2005; Haines-Young and Postchin, 2010; Van Lavieren et al., 2012). Supporting and regulating ecosystem services provided by mangroves include: (i) habitat for a wide range of organisms (Nagelkerken et al., 2000; Granek et al. 2009) including juvenile reef fishes that are essential components of coral reef ecosystems and, in many cases, are important food fish in their own right (Robertson and Duke, 1987; Laegdsgaard and Johnson, 1995; Mumby et al., 2004; Manson et al., 2005); (ii) carbon sequestration (Fujimoto, 2004; Lal, 2005; Donato et al. 2011; Alongi, 2012; 2014); (iii) climate regulation (Mcleod et al., 2011); (iv) shoreline stabilization and coastal protection (Kathiresan and Rajendran, 2005; Wells et al., 2006, 2005; Alongi, 2008; Barbier et al., 2008; Koch et al., 2009), water filtration (Alongi et al., 2003) and pollution regulation (Harbison, 1986; Primavera, 2005; Primavera et al., 2007). Mangroves also provide a suite of provisioning ecosystem services, including: (i) fisheries production (Nagelkerken et al., 2000; Dorenbosch et al., 2004; 2005); (ii) aquaculture production (Minh et al., 2001); (iii) pharmaceutical generation (Goodbody, 2003; Abeysinghe, 2010); (iv) production of timber and fuelwood (the latter being important in the Caribbean and Pacific) (Lugo, 2002; Walters, 2005; Walters et al., 2008). Finally, mangroves provide cultural services that include: (i) recreation and tourism (Bennett and Reynolds, 1992; Thomas et al., 1994; Brohman, 1996); (ii) educational opportunities (Bacon and Alleng, 1992; Field, 1999); (iii) aesthetic and cultural values (e.g., Field, 1999; Ronnback, 1999). The provision of these services is reduced or lost when mangrove habitat is degraded or transformed; this loss of services frequently declines in a non-linear fashion such that beyond a certain threshold (which varies spatially, temporally, and by species), mangroves are no longer able to provide significant coastal protection or fisheries benefits (Barbier et al., 2008; Koch et al., 2009).

Mangrove management is not currently practiced on a global scale. However, there are examples of intensive management of large forests in Asia (Spading et al., 2010). Many such forests are managed for commercial purposes but it would be useful to consider management in light of the tradeoffs among ecosystem services. Because provisioning services are easiest to quantify and assign an economic value, mangroves are frequently managed for one or a few provisioning services at the cost of managing for the full suite of services mangrove ecosystems provide. For example, mangrove ecosystems may be converted to produce aquaculture services; such management can contribute to the decline of other supporting and regulating services, such as pollution regulation and shoreline stabilization. When mangrove management focuses on maximization of one ecosystem service to the detriment of others, some individuals (e.g., aquaculture operators) gain, while others (e.g., coastal residents requiring shoreline protection) often lose. Policies and management of the coastal region that focus on preserving the functional diversity of mangrove ecosystems (multiple services), including the associated salt flats, enhance the possibility of having the highest number of beneficiaries. Whether state management or community-based management will be most effective may be context-dependent and worth consideration (Sudtongkong and Webb, 2008).

5 Conservation responses

The dramatic decline in global mangrove cover (Giri et al., 2011) and the on-going removal of mangrove habitat have led both governmental and non-governmental organizations to take actions to protect mangroves. Worldwide, commercial organizations have exerted, and continue to exert, strong pressures to modify policies that conserve mangroves (Brazil offers one example among many other countries (Glazer, 2004)), yet progress is being made through legislation, new partnerships between governments and local communities, and the REDD+ programme (Reduced Emissions from Deforestation and forest Degradation) in developing countries. Mangrove conservation measures range from traditional approaches including creation of designated areas protected from clearing and legislation restricting or prohibiting clearing, to conservation, education and restoration projects on local, national, regional, or international scales. These often involve local communities and organizations as stewards of mangrove ecosystems and may allow sustainable harvest within the project areas (Lugo et al. 2014).

5.1 Conservation through conventions and protected areas

Multiple international conventions and programs protect mangrove habitats. The Convention on Wetlands of international importance especially as waterfowl habitat[1] (Ramsar convention), an international treaty whereby member countries commit to maintaining the ecological characteristics of their "Wetlands of International Importance", protects mangrove forests at 278 Ramsar mangrove sites in 68 countries (numbers as of 2014). World Heritage sites, UNESCO-designated sites of cultural and natural heritage of outstanding value to humanity, include 26 Sites that protect mangrove habitat within their boundaries and UNESCO Man and the Biosphere Programme sites, many of which include mangrove habitat.

Establishing terrestrial and marine protected areas, including national parks and marine reserves is often used as a management tool to protect mangrove habitat. Examples of national parks that protect mangroves include Mangroves National Park in the Democratic Republic of Congo; Parc Marin de Moheli, Comoros, Kakadu National Park, Australia; Bastimientos Island National Park, Panama; Kiunga Biosphere Reserve, Kenya; Everglades National Park, United States of America; Sirinat National Park, Thailand; Subterranean National Park, Philippines; among others. Despite these efforts, Giri et al. (2011) report that only 6.9 per cent of the world's mangroves fall within existing protected areas networks (IUCN I- Category IV in the IUCN Protected areas management categories).

5.2 Conservation through legislation

In some countries, states, or regions, mangroves are protected through legislation limiting or prohibiting mangrove clearing. Legislation may be national, such as Brazil's Federal Forestry Code (Brazil, 2012), which has been interpreted to prohibit the use of any components of mangrove trees or plants. Other legislation exists at more localized scales, , such as The Mangrove Trimming and Preservation Act enacted in 1996 in the state of Florida, (United States of America), to regulate trimming, disturbance or removal of mangroves in the state.

5.3 Conservation through management, education and restoration projects

The decline in global mangrove cover, combined with the highly recognized ecological and ecosystem services values of mangroves, have given rise to a number of non-governmental organizations engaged in education about and conservation and restoration of mangroves. These include organizations with projects around the world, such as the Mangrove Action Project, Western Indian Ocean (WIO) Mangrove Network, the Mangrove Alliance, and Mangrove Watch, as well as domestic organizations, including Honko, a mangrove conservation and education organization in Madagascar, and the Mangrove Forest Conservation Society of Nigeria, among others. Some countries such as Cuba and Ecuador have invested significant resources and are testing new approaches to mangrove conservation through engagement of local communities in natural resource governance (Gravez et al. 2013; Lugo et al. 2014).

Restoration projects have met with mixed and limited success with many documented efforts resulting in large failure rates in achieving successful mangrove restoration. These failures highlight the importance of considering factors that can doom mangrove restoration including poor site and species selection and failure to utilize advances in the recent science of mangrove restoration (Lewis 2005, Lewis and Brown 2014). For example, the use of biotechnological interventions to produce improved mangrove plantlets (e.g., faster growing plants) could improve the success rate of restoration. It would be useful to have better training at all levels on the concepts and application of mangrove restoration (Lewis and Brown 2014).

5.4 Emerging conservation strategies

The movement to implement "Blue Carbon Solutions" (the carbon sequestered by coastal vegetation, namely mangroves, sea-grasses and salt marsh grasses- McLeod et al., 2011) to reduce atmospheric CO_2 has led to the consideration of tools such as payment for ecosystem services (PES) and REDD-+ schemes to improve conservation outcomes for mangroves (Alongi, 2011; Locatelli et al., 2014). Such approaches may provide novel strategies for mangrove conservation in countries that lack sufficient resources for conservation and management.

Although raising financial resources for whole ecosystem conservation has historically been beneficial, new risks arise from this approach in the emerging paradigm of conservation through commodification of ecosystem functions, such as those related to carbon storage (McAfee, 1999; Igoe and Brockington, 2007; Kosoy and Corbera, 2010; Corbera, 2012). The emerging commodification paradigm, challenges an old ethical and inter-generational argument that nature needs to be managed and protected for the survival of ecosystems and species; it would be useful for mangrove conservation and restoration efforts to consider the risks of trading preservation of ecosystems for their intrinsic value and the emerging paradigm of prioritizing some elements of nature that are economically useful, at the potential cost of other values that are less economically valuable or are useful only to certain groups. In this pro-

---

[1] United Nations, *Treaty Series*, vol. 996, No. 14583.

cess of assigning a monetary value to an ecosystem service, cultural and social values, such as those held by communities that live near and depend directly on the forests and that possess a deep cultural connection with the system, may be strongly devalued. In this way, power asymmetries in the valuation process may further fuel socio-environmental conflicts involving those interested in carbon and the communities interested in the maintenance of the diversity of functions and services, including cultural values, as recently described by Beymer-Farris and Bassett (2012) for the mangrove forests in Tanzania. However, Ecuador's Mangrove Ecosystem Concessions program provides an example of how government agencies can engage local stakeholders by simultaneously providing resource rights and bestowing management responsibilities on those users (Gravez et al. 2013).

As indicated above, although threatened by sea level rise, mangroves have the potential to keep pace with rising sea level if conditions allow them to modify their surface elevation or to adapt through landward migration (Cahoon and Hensel, 2006; Alongi, 2008; Gilman et al., 2008; Soares, 2009, McKee, 2011). It would be useful for mangrove conservation and management efforts to take into account external sediment supply, benthic mats, tree density and root structure, storm impacts, and hydrological factors such as river levels, groundwater inputs and rainfall (McIvor et al., 2012), as well as consider the maintenance and restoration of system resilience (e.g., its capacity to adapt and migrate landward).

6   Capacity building gaps

Capacity building is the process by which individuals, organizations, institutions and societies develop abilities (individually and collectively) to perform functions, solve problems and set and achieve objectives (UNDP, 1997). Capacity building is therefore facilitated through the provision of technical support activities, including coaching, training, specific technical assistance, and resource networking.

Local, regional, national and international initiatives for capacity building in mangrove conservation and sustainable use as a management tool to protect mangrove habitat are widespread around the world, including those led by the United Nations University (UNU), UNESCO's Man and the Biosphere Programme (MAB), Mangrove Action Project (MAP), Mangrove Alliance, Mangroves for the Future (MFF), Mangrove Watch, WIO Mangrove Network and The International Society for Mangrove Ecosystems (ISME). Examples of initiatives specific to different regions include, International Union for the Conservation of Nature - IUCN's Pacific mangrove initiative (PMI), the United Nations Environment Programme's Integrated Coastal Management, with special emphasis on the sustainable management of mangrove forests in Guatemala, Honduras and Nicaragua, the Satoyama Initiative in Benin, and Mangrove Action Project (MPA)-Asia in Thailand, among others.

Although several initiatives are concerned with capacity building, capacity building will be more effective if it is integrated and follows a set of basic assumptions about training and knowledge base. Increased effectiveness can be achieved through: (i) training related to conservation and sustainable use of mangrove forests and their resources; (ii) raising awareness among as many stakeholders as possible (especially policy-makers); (iii) political empowerment of stakeholders; (iv) cooperation within and between governments, institutions, organizations and agencies that are engaged in these activities; (v) identification and development of innovative proposals; (vi) maintaining systems for the reduction and resolution of conflicts; (viii) ensuring that programmes include measures to address threats from climate change and human activities.

Specific ideas for capacity building include: use of standardized methods for mangrove species distribution and area surveys (Manson et al., 2012) and development of capacity in the use of base-maps on digital terrain models. These would display areas where mangroves are mostly at risk from submersion due to sea level rise. Capacity to conduct surveys and geographical information systems (GIS) mapping in all regions would be useful, along with the development of capacity for "climate-smart conservation" (Hansen et al., 2010), which would involve strategies for promoting mangrove adaptation to sea level rise. It would be useful for nations to develop the capacity to better identify and evaluate potential barriers for landward migration in response to sea level rise and have more accurate information regarding the location of landward migration corridors as well as improved strategies for ensuring that these migration corridors are present in the future. It would also be useful to know specifically how other drivers of change (e.g., urbanization, other coastal land uses) may affect the potential for landward migration of mangroves in response to sea level rise.

7   Gaps in scientific knowledge

Comprehensive and comparable data on mangrove species and area distribution from all countries with mangroves would be useful. Lack of information on the current status of mangrove species for the documentation of the various types of mangrove losses in each region has been identified as an important knowledge gap. This could improve a range of conservation and management strategies, along with predictions of habitat loss and species extinctions (Polidoro et al., 2010; Spalding et al., 2010). Other areas of considerations in filling the gaps are: determination of the average changes in the emission or sequestration of greenhouse gases from mangrove forests as a result of human activity; accurate and consistent valuation of mangrove goods and services, and vulnerability mapping.

References

Abeysinghe, P.D. (2010). Antibacterial Activity of some Medicinal Mangroves against Antibiotic Resistant Pathogenic Bacteria. *Indian Journal of Pharmaceutical Sciences* 72: 167-172.

Allen, H.R. (1987). Holocene mangroves and middens in northern Australia and Southeast Asia. *Bulletin of the Indo-Pacific Prehistory Association* 7: 1-16.

Alongi, D.M. (2015). The Impact of Climate Change on Mangrove Forests. *Current Climate Change Reports* 1: 30-39.

Alongi, D.M. (2014). Carbon Cycling and Storage in Mangrove Forests. *Annual Review of Marine Science* 6: 195-219.

Alongi, D.M. (2012). Carbon sequestration in mangrove forests, a review. *Carbon Management* 3: 313-322.

Alongi, D.M. (2011). Carbon payments for mangrove conservation: ecosystem constraints and uncertainties of sequestration potential. *Environmental Science and Policy* 14: 462-470.

Alongi, D.M. (2008). Mangrove Forests: Resilience, protection from tsunamis, and responses to global climate change. *Estuarine Coastal and Shelf Science* 76: 1-13.

Alongi, D.M., Chong, V.C., Dixon, P., Sasekumar, A. and Tirendi, F. (2003). The influence of fish cage culture on pelagic carbon flow and water chemistry in tidally dominated mangrove estuaries of Peninsular Malaysia. *Marine Environmental Research* 55: 313-333.

APGIII (2009). An update of the Angiosperm phylogeny group classification for the orders and families of flowering plants- APGIII. *Botanical Journal of the Linnean Society* 161: 105-121.

Bacon, P.R. and Alleng, G.P. (1992). The management of insular Caribbean mangroves in relation to site location and community type. *The Ecology of Mangrove and Related Ecosystems* 80: 235-241.

Barbier, E.B., Koch, E.W., Silliman, B.R., Hacker, S.D., Wolanski, E., Primavera, J. and Reed, D.J. (2008). Coastal ecosystem-based management with nonlinear ecological functions and values. *Science* 319: 321-323.

Bennett, E.L. and Reynolds, C.J. (1992). The value of a mangrove area in Sarawak. *Biodiversity and Conservation* 2: 359 – 375.

Beymer-Farris, B.A. and Bassett, T.J. (2012). The REDD menace: Resurgent protectionism in Tanzania's mangrove forests. *Global Environmental Change* 22:332-341.

Brasil (2012). *Codigo Florestal Brasileiro*. Lei 12.651 de 25 de maio de 2012.

Brohman, J. (1996). New directions in tourism for third world development. *Annals of Tourism Research* 23: 48–70.

Cahoon, D.R., Hensel, P.F., Spencer, T., Reed, D.J., McKee, K.L. and Saintilan, N. (2006). Coastal wetland vulnerability to relative sea-level rise: wetland elevation trends and process controls. In *Wetlands and natural resource management* (pp. 271-292). Springer Berlin Heidelberg.

Corbera, E. (2012). Problematizing REDD+ as an experiment in payments for ecosystem services. *Current Opinion in Environmental Sustainability* 4: 612–619.

Costanza, R., Pérez-Maqueo, O., Martinez, M., Sutton, P., Anderson, S., and Mulder, K. (2008). The value of coastal wetlands for hurricane protection. *Ambio* 37: 241–248.

Dahdouh-Guebas, F., Jayatissa, L.P., Di Nitto, D., Bosire, J.O., Lo Seen, D. and Koedam, N. (2005). How effective were mangroves as a defence against the recent tsunami? *Current Biology* 15: R443–R447.

Dittmar, T., Hertkorn, N., Kattner, G. and Lara, R. (2006). Mangroves, a major source of dissolved organic carbon to the oceans. *Global Biogeochem Cycles* 20: 1012. DOI:10.1029/2005GB002570.

Donato, D.C., Kauffman, J.B., Murdiyarso, D., Kurnianto, S., Stidham, M., & Kanninen, M. (2011). Mangroves among the most carbon-rich forests in the tropics. *Nature Geoscience* 4: 293-297.

Dorenbosch, M., Van Riel, M.C., Nagelkerken, I., and van der Velde, G. (2004). The relationship of reef fish densities to the proximity of mangrove and seagrass nurseries. *Estuarine Coastal and Shelf Science* 60: 37–48.

Dorenbosch, M., Grol, M.G.G., Christianen, M.J.A. and Nagelkerken, I. (2005). Indo-Pacific seagrass beds and mangroves contribute to fish density and diversity on adjacent coral reefs. *Marine Ecology Progress Series* 302: 63-76.

Duke, N.C., and Allen, J.N. (2006). Rhizophora mangle, R. samoensis, R. racemosa, R. x harrisonii: Atlantic-East Pacific red mangrove. Ver. 3.1. In: Elevitch, C.R., editor. *Species profiles for Pacific Island Agroforestry*. Permanent Agriculture Resources (PAR). Holualoa, Hawaii. Available from: http//www.traditionaltree.org. Accessed: 7 August, 2014.

Duke, N.C., Lo, E.Y. and Sun, M. (2002). Global distribution and genetic discontinuities of mangroves: Emerging patterns in the evolution of Rhizophora. *Trees* 16: 65–79.

Duke, N.C., Meynecke, J., Dittmann, S., Ellison, A., Anger, K.U., Berger, U., Cannicci, S., Diele, K., Ewel, K., Field, C., Koedam, N., Lee, S., Marchand, C., Nordhaus, J., and Dahdouh-Guebas, F. (2007). A world without mangroves. *Science* 317: 41-42.

Ellison, A.M., Bank, M.S., Clinton, B.D., Colburn, E.A., Elliott, K., Ford, C.R., Foster, D.R. (2005). Loss of foundation species: consequences for the structure and dynamics of forested ecosystems. *Frontiers in Ecology and the Environment* 3: 479-486.

Ellison, A.M. (2008). Managing mangroves with benthic biodiversity in mind: moving beyond roving banditry. *Journal of Sea Research* 59: 2–15.

Ellison, J.C. and Zouh, I. (2012). Vulnerability to Climate Change of Mangroves: Assessment from Cameroon. *Central Africa Biology* 1: 617-638. doi:10.3390/biology1030617.

FAO (2007). The World's Mangroves 1980-2005. *FAO Forestry Paper* No. 153. Rome, Forest Resources Division, FAO. pp. 77.

Field, C.D. (1999). Charter for Mangroves. In: Yáñez-Arancibiay, A., and Lara-Domínguez, A.L., editors. *Ecosistemas de Manglar en América Tropical. Instituto de Ecología*. A.C. México, UICN/ORMA, Costa Rica, NOAA/NMFS Silver Spring, MD USA. pp. 380.

Fujimoto, K. (2004). Below-ground carbon sequestration of mangrove forests in the Asia-Pacific region. In: Vannucci, M., editor. *Mangrove Management and Conservation: Present and Future*. pp. 138-146.

Gilman, E., Ellison, J., Duke, N. and Field, C. (2008). Threats to mangroves from climate change and adaptation options: a review. *Aquatic Botany* 89: 237–250.

Giri, C., Ochieng, E., Tieszen, L., Zhu, Z., Singh, A., Loveland, T., Masek, J. and Duke, N. (2011). Status and distribution of mangrove forests of the world using earth observation satellite data. *Global Ecology and Biogeography* 20: 154–159.

Goodbody, I.M. (2003). The Ascidian fauna of Port Royal, Jamaica: 1. Harbor and mangrove dwelling species. *Bulletin of Marine Sciences* 73: 457-476.

Granek, E.F., Compton, J.E., and Phillips, D.L. (2009). Mangrove-exported nutrient incorporation by sessile coral reef invertebrates. *Ecosystems* 12: 462-472.

Gravez, V., Bensted-Smith R, Heylings, P., And Gregoire-Wright, T. (2013). Governance systems for marine protected areas in Ecuador. In. Moksness, E., Dahl, E., Stottrup, J., Eds. *Global challenges in integrated coastal zone management*. John Wiley & Sons, Ltd., Oxford, UK, 145-158.

Haines-Young, R. and Potschin, M. (2010). Proposal for a common international classification of ecosystem goods and services (CICES) for integrated environmental and economic accounting. European Environment Agency.

Hansen, L., Hoffman, J., Drews, C. and Mielbrecht, E. (2010). Designing Climate-Smart Conservation: Guidance and Case Studies. *Conservation Biology* 24: 63-69.

Harbison, P. (1986). Mangrove muds—a sink and a source for trace metals. *Marine Pollution Bulletin* 17: 246-250.

Hogarth, P.J. (2007). *The biology of mangroves and seagrasses* (No. 2nd Edition). Oxford University Press.

Horwitz, P., Finlayson, M. and Weinstein, P. (2012). Healthy wetlands, healthy people: a review of wetlands and human health interactions. *Ramsar Technical Report No. 6*. Secretariat of the Ramsar Convention on Wetlands, Gland, Switzerland, & the World Health Organization, Geneva, Switzerland.

Hutchison, J., Manica, A., Swetnam, R., Balmford, A. and Spalding, M. (2014). Predicting Global Patterns in Mangrove Forest Biomass. *Conservation Letters* 7: 233–240. doi: 10.1111/conl.12060

Igoe, J. and Brockington, D. (2007). Neoliberal conservation: A brief introduction. *Conservation and Society* 5: 432.

Kathiresan, K. and Rajendran, N. (2005). Coastal mangrove forests mitigated tsunami. *Estuarine, Coastal and Shelf Science* 65: 601-606.

Koch, E.W., Barbier, E.D., Silliman, B.R., Reed, D.J., Perillo, G.M.E., Hacker, S.D., Granek, E.F., Primavera, J.H., Muthiga, N., Polasky, S., Halpern, B.S., Kennedy, C.J., Wolanski, E., and Kappel, C.V. (2009). Non-linearity in ecosystem services: temporal and spatial variability in coastal protection. *Frontiers in Ecology and the Environment* 7: 29–37.

Kosoy, N. and Corbera, E. (2010). Payments for ecosystem services as commodity fetishism. *Ecological economics* 69: 1228-1236.

Krauss, K.W., McKee, K.L., Lovelock, C.E., Cahoon, D.R., Saintilan, N., Reef, R. and Chen, L. (2014). How mangrove forests adjust to rising sea level. *New Phytologist* 202: 19-34.

Laegdsgaard, P. and Johnson, C.R. (1995). Mangrove habitats as nurseries: unique assemblages of juvenile fish in subtropical mangroves in eastern Australia. *Marine Ecology Progress Series* 126: 67-81.

Lal, R. (2005). Forest soils and carbon sequestration. *Forest ecology and management* 220: 242-258.

Langley, J.A., McKee, K.L., Cahoon, D.R., Cherry, J.A., & Megonigal, J.P. (2009). Elevated CO2 stimulates marsh elevation gain, counterbalancing sea-level rise. *Proceedings of the National Academy of Sciences* 106: 6182-6186.

Lewis, R.R. (2005). Ecological engineering for successful management and restoration of mangrove forests. *Ecological Engineering* 24: 403-418.

Lewis, R.R and Brown, B. (2014). Ecological mangrove rehabilitation: a field manual for practitioners. Version 3. Mangrove Action Project Indonesia, Blue Forests, Canadian International Development Agency, and OXFAM. 275 p.

Locatelli, T., Binet, T., Kairo, J.G., King, L., Madden, S., Patenaude, G., Upton, C., and Huxham, M. (2014). Turning the Tide: How Blue Carbon and Payments for Ecosystem Services (PES) Might Help Save Mangrove Forests. *AMBIO*. DOI. 10.1007/s13280-014-0530-y.

Lugo, A.E. (2002). Conserving Latin American and Caribbean mangroves: issues and challenges. *Madera y Bosques* 8: 5-25.

Lugo, A. E., Medina, E. and McGinley, K. (2014). Issues and Challenges of Mangrove Conservation in the Anthropocene. Madera y Bosques 20(3):11-38. http://www1.inecol.edu.mx/myb/resumeness/no.esp.2014/myb20esp1138.pdf.

Luther, D.A. and Greenberg, R. (2009). Mangroves: a global perspective on the evolution and conservation of their terrestrial vertebrates. *BioScience* 59: 602-612.

Manson, F.J., Loneragan, N.R., Skilleter, G.A. and Phinn, S.R. (2005). An evaluation of the evidence for linkages between mangroves and fisheries: a synthesis of the literature and identification of research directions. *Oceanography and Marine Biology: an Annual Review* 43: 485-515.

Manson, F.J., Loneragan, N.R., McLeod, M. and Kenyon, R.A. (2012). Assessing techniques for estimating the extent of mangroves: topographic maps, aerial photographs and Landsat TM images. *Marine and Freshwater Research* 52: 787-792.

McAfee, K. (1999). Selling nature to save it? Biodiversity and green developmentalism. *Environment and Planning* 17: 133-154.

McIvor, A., Möller, I., Spencer, T. and Spalding, M. (2012). Reduction of wind and swell waves by mangroves. Natural Coastal Protection Series: Report 1: The Nature Conservancy and Wetlands International. Available from: http://www.wetlands.org/LinkClick.aspx?fileticket=fh56xgzHilg%3D&tabid=56. Accessed: 12 November, 2013.

McKee, K.L. (2011). Biophysical controls on accretion and elevation change in Caribbean mangrove ecosystems. *Estuarine, Coastal and Shelf Science* 91: 475-483.

McKee, K.L., Cahoon, D.R., & Feller, I.C. (2007a). Caribbean mangroves adjust to rising sea level through biotic controls on change in soil elevation. *Global Ecology and Biogeography* 16: 545-556.

McKee, K.L., Rooth, J.E., & Feller, I.C. (2007b). Mangrove recruitment after forest disturbance is facilitated by herbaceous species in the Caribbean. *Ecological Applications* 17: 1678-1693.

Mcleod, E., Chmura, G.L., Bouillon, S. Salm, R., Björk, M., Duarte, C.M., Lovelock, C.E., Schlesinger, W.H. and Silliman, B.R. (2011). A blueprint for blue carbon: toward an improved understanding of the role of vegetated coastal habitats in sequestering CO2. *Frontiers in Ecology and the Environment* 9: 552–560. http://dx.doi.org/10.1890/110004

McLeod, E., and Salm, R.V. (2006). *Managing Mangroves for Resilience to Climate Change*. IUCN, Gland, Switzerland. pp. 64.

Millennium Ecosystem Assessment (2005). *Ecosystems and human well-being*. Washington, D.C., Island Press.

Minh, T.H., Yakupitiyage, A. and Macintosh, D.J. (2001). *Management of the integrated mangrove-aquaculture farming systems in the Mekong delta of Vietnam*. Integrated Tropical Coastal Zone Management, School of Environment, Resources, and Development, Asian Institute of Technology. Available from: mit.biology.au.dk. Accessed: 7 August, 2014.

Mumby, P.J., Edwards, A.J., Arias-Gonzalez, J.E., Lindeman, K.C., Blackwell, P.G., Gall, A., Gorczynska, M.I., Harborne, A.R., Prescod, C.L., Renken, H., Wabnitz, C.C.C. and Lewellyn, G., (2004). Mangroves enhance the biomass of coral reef fish communities in the Caribbean. *Nature* 427: 533-536.

Nagelkerken, I., Van der Velde, G., Gorissen, M.W., Meijer, G.J., Van't Hof, T., and Den Hartog, C. (2000). Importance of mangroves, seagrass beds and the shallow coral reef as a nursery for important coral reef fishes, using a visual census technique. *Estuarine, Coastal and Shelf Science* 51: 31-44.

Osland, M.J., Enwright, N., Day, R.H. and Doyle, T.W. (2013). Winter climate change and coastal wetland foundation species: salt marshes vs. mangrove forests in the southeastern United States. *Global change biology* 19: 1482-1494.

Osland, M.J., Enwright, N., Stagg, C. (2014). Freshwater availability and coastal wetland foundation species: ecological transitions along a rainfall gradient. *Ecology* 95: 2789 – 2802.

Polidoro, B.A., Carpenter, K.E., Collins, L., Duke, N.C., Ellison, A.M., Ellison, J.C., Farnsworth, E.J., Fernando, E.S., Kathiresan, K., Koedam, N.E., Livingstone, S.R., Miyagi, T., Moore, G.E., Nam, V.N., Ong, J.E., Primavera, J.H., Salmo, S.G., Sanciangco, J.C., Sukardjo, S., Wang, Y., and Yong, J.W.H. (2010). The Loss of Species: Mangrove Extinction Risk and Geographic Areas of Global Concern. *PLoS ONE* 5, 1 – 10.

Primavera, J.H. (2005). Mangroves, fishponds and the quest for sustainability. *Science* 310: 57-59.

Primavera, J.H., Altamirano, J.P., Lebata, M.J.H.L., de los Reyes Jr., A.A., and Pitogo, C.L. (2007). Mangroves and shrimp pond culture effluents in Aklan, Panay Is., Central Philippines. *Bulletin of Marine Sciences* 80: 795-804.

Ramsar Secretariat (2001). *Wetland Values and Functions: Climate Change Mitigation*. Gland, Switzerland. Available from: http://www.ramsar.org/cda/ramsar/display/main/main.jsp?zn=ramsar&cp=1-26-253%5E22199_4000_0___. Accessed: 13 August, 2013.

Robertson, A.L. and Duke, N.C. (1987). Mangroves as nursery sites: comparisons of the abundance and species composition of fish and crustaceans in mangroves and other nearshore habitats in tropical Australia. *Marine Biology* 96, 193-205.

Rönnbäck, P. (1999). The ecological basis for economic value of seafood production supported by mangrove ecosystems. *Ecological Economics* 29: 235–252.

Saenger, P. (2003). *Mangrove ecology, siviculture and conservation*. Kluwer Academic Publishers, Dordrecht.

Saintilan, N., Wilson, N., Rogers, K., Rajkaran, A. and Krauss, K.W. (2014) Mangrove expansion and salt marsh decline at mangrove poleward limits. *Global change biology* 20: 147-157.

Sathirathai, S. and Barbier, E.B. (2001). Valuing mangrove conservation in southern Thailand. *Contemporary Economic Policy* 19: 109-122.

Schleupner, C. (2008) Evaluation of coastal squeeze and its consequences for the Caribbean island Martinique. *Ocean & Coastal Management* 51: 383-390.

Siikamäkia, J., Sanchirico, J.N. and Jardine, S.L. (2012). Global economic potential for reducing carbon dioxide emissions from mangrove loss. *PNAS* 109: 14369-14374.

Soares, M.L.G. (2009). A conceptual model for the responses of mangrove forests to sea level rise. *Journal of Coastal Research SI* 56: 267-271.

Soares, M.L.G., Estrada, G.C.D., Fernandez, V. and Tognella, M.M.P. (2012). Southern limit of the Western South Atlantic mangroves: assessment of the potential effects of global warming from a biogeographical perspective. *Estuarine, Coastal and Shelf Science* 101: 44-53.

Spalding, M., Kainuma, M. and Collins, L. (2010). *World Atlas of Mangroves*. ITTO, ISME, FAO, UNEP-WCMC, UNESCO-MAB and UNU-INWEH. Earthscan Publishers Ltd. London.

Sudtongkong, C. and Webb, E.L. (2008). Outcomes of state-vs. community-based mangrove management in southern Thailand. *Ecology and Society* 13: 27.

Thomas, G. and Fernandez, T.V. (1994). Mangrove and tourism: management strategies. *Biodiversity and Conservation* 2: 359-375.

Tomlinson, P.B. (1986). *The botany of mangroves*. Cambridge University Press, NY.

UNDP (1997). *Capacity development resources book*. New York. United Nations Development Programme.

Valiela, I., Bowen, J.L. and York, J.K. (2001). Mangrove Forests: one of the world's threatened major tropical environments. *Bioscience* 51: 807-815.

Van Lavieren, H., Spalding, M., Alongi, D., Kainuma, M., Clüsener-Godt, M. and Adeel, Z. (2012). *Securing the future of mangroves. A Policy Brief.* UNU-INWEH, UNESCO-MAB with ISME, ITTO, FAO, UNEP-WCMC and TNC. pp. 53.

Walters, B.B. (2005). Patterns of local wood use and cutting of Philippine mangrove forests. *Economic Botany* 59: 66-76.

Walters, B.B., Rönnbäck, P., Kovacs, J., Crona, B., Hussain, S., Badola, R., Primavera, J., Barbier, E., and Dahdouh-Guebas, F. (2008). Ethnobiology, socio-economics and management of mangrove forests: a review. *Aquatic Botany* 89: 220–236.

Webber, M., Webber, D. and Trench, C. (2014). Agroecology for sustainable coastal ecosystems: A case for mangrove forest restoration, in: Benkeblia, N. (Ed) *Agroecology, Ecosystems and Sustainability*. CRC Press, Taylor and Francis group, Boca Raton.

Wells, S., Ravilious, C. and Corcoran, E. (2006). *In the front line: shoreline protection and other ecosystem services from mangroves and coral reefs* (No. 24). UNEP/Earthprint.

Woodroffe, C., Robertson, A. and Alongi, D. (1992). Mangrove sediments and geomorphology. In: Robertson, A.I. and Alongi, D.M. (eds.). *Tropical mangrove ecosystems*. American Geophysical Union, Washington, D.C.

49 Salt Marshes

Contributors:
Patricio Bernal (Convenor and Lead Member), K. E. A. Segarra and J. S. Weis

Salt Marshes

1 Inventory

Salt marshes are intertidal, coastal ecosystems that are regularly flooded with salt or brackish water and dominated by salt-tolerant grasses, herbs, and low shrubs. They occur in middle and high latitudes worldwide and are largely replaced by mangroves in the subtropics and tropics (see Chapter 48). They are found on every continent except Antarctica (Figure 49.1). In areas of relatively little sediment delivery, salt marshes are highly organic and often peat-based. In contrast, salt marshes in areas of high sediment delivery, such as sheltered estuaries (see Chapter 44), are often well-developed with inorganic substrates.

2 Features of trends in extent

Salt marshes are among the most productive temperate ecosystems in the world. Contemporary salt marshes developed within the last 8,000 years in low-energy, coastal locations in response to rising sea levels (Milliman and Emery, 1968; Redfield 1967). Their ecology and global importance has been described in classic literature such as Chapman (1960), Ranwell (1972), Doody (2008) and Adam (1990). The physical stresses of salinity and flooding generate zones of salt-tolerant emergent vegetation, including such genera as *Carex*, *Spartina*, *Juncus*, *Salicornia*, *Halimone*, *Puccinellia*, and *Phragmites*. Marsh grasses contribute to the accumulation of organic matter and trapping of inorganic sediment. Salt marsh sustainability is mainly controlled by the relationship between marsh vertical accretion (due to sediment accretion, peat accumulation, belowground decomposition, subsidence) and sea level rise (frequency and duration of tidal flooding - Gagliano et al., 1981; Cahoon and Reed, 1995; Hatton et al., 1983; DeLaune et al., 1983). Other facts that impact their development and structure are tidal, wave and current action, erosion, freshwater influx, nutrient supply, and topography (Mitsch and Gosselink, 1993). Less than 50 per cent of the world's original wetlands remain (Mitsch and Gosselink, 1993) and current loss is estimated at 1-2 per cent per year (Bridgham et al., 2006) making wetlands one of the fastest disappearing ecosystems worldwide. Salt marsh loss coincides with a general historical degradation of estuarine ecosystems (Lotze et al., 2006). Just upstream from salt marshes are brackish marshes with lower salinity regimes, which are also highly productive, subject to the same stresses, and of equally great conservation concern.

3 Major pressures linked to the trends

Over 60 per cent of the globe's population lives on or near the coast, and coastal populations are increasing at twice the average rate (UNEP, 2006a; Nicholls et al., 1999), making coastlines highly vulnerable to human activities. Salt and brackish marshes, formerly viewed as useless wastelands, were filled in for urban or agricultural development. Reclamation of land for agriculture by converting marshland to upland was historically a common practice. Coastal cities worldwide have expanded onto former salt marshes and used marshes for waste disposal sites. Airoldi and Beck (2007) estimate that countries in Europe have lost over 50 per cent of their salt marsh and seagrass areas to coastal development. Estuarine pollution from organic, inorganic, and toxic substances is a worldwide problem. Marshes have been drained, diked, ditched, grazed

The boundaries and names shown and the designations used on this map do not imply official endorsement or acceptance by the United Nations.
Figure 49.1 | Salt Marshes (in orange). Source: UNEP-WCMC, 2015

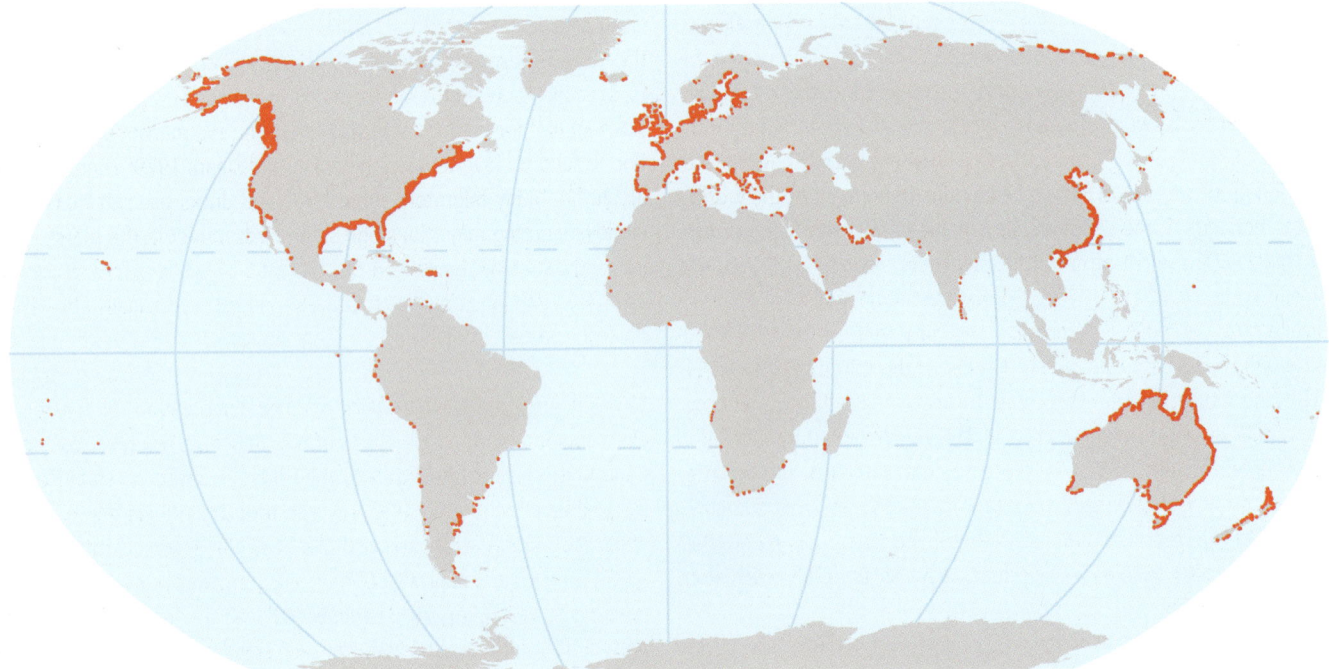

Source: UNEP-WCMC 2015

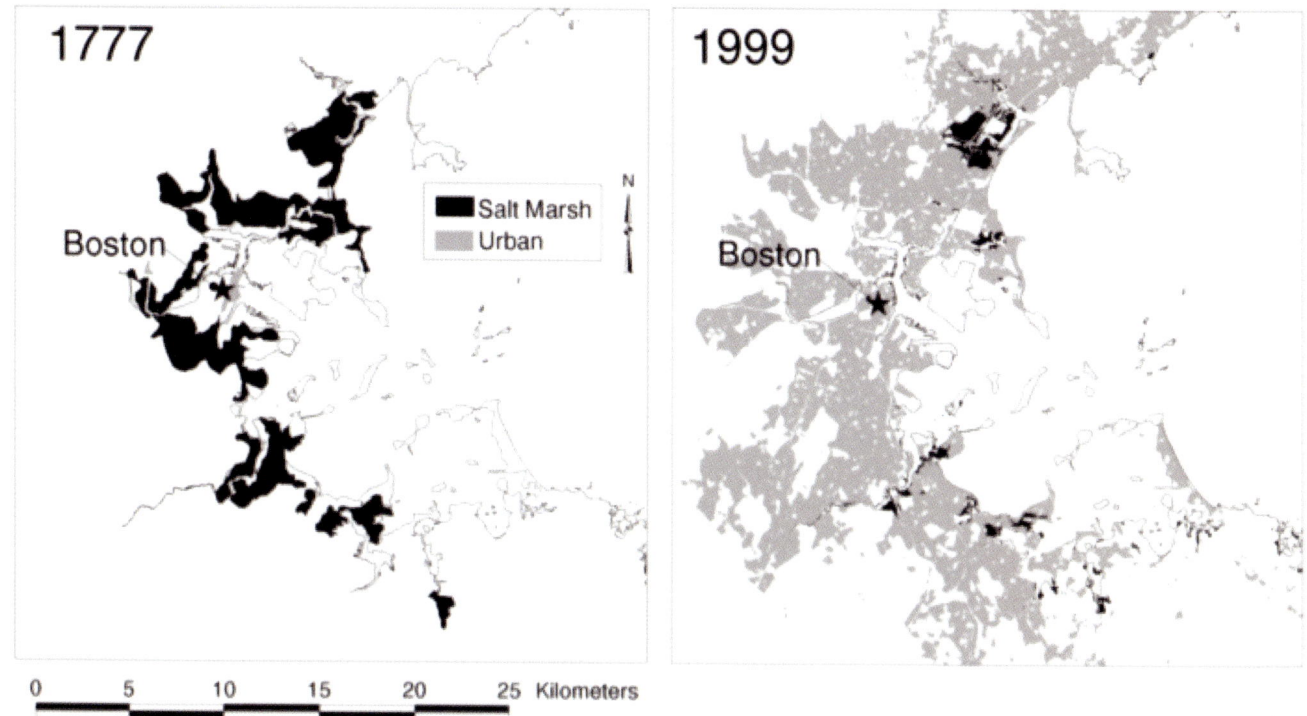

The boundaries and names shown and the designations used on this map do not imply official endorsement or acceptance by the United Nations.
Figure 49.2 | Salt marsh loss in Boston, Massachusetts, United States, between 1777 and 1999 (Bromberg and Bertness, 2005).

and harvested. They have been sprayed for mosquito control, and have been invaded by a range of non-native species that have altered their ecology. As one example, Massachusetts, United States of America, has lost 41 per cent of its salt marshes since the 1770s, with a loss of 81 per cent in Boston (Bromberg and Bertness, 2005; Figure 49.2).

Key threats to salt marshes are land reclamation, coastal development, dredging, sea-level rise (SLR), hydromodification, alteration of processes (e.g. sediment delivery, freshwater input) and eutrophication. Accelerated SLR is the largest climate-related threat to salt marshes. The Intergovernmental Panel on Climate Change predicts with medium confidence a SLR of 0.26-0.98 m by 2100 (Church et al., 2013). Nicholls et al. (1999) predict that 1 m SLR will eliminate 46 per cent of the world's coastal wetlands. Some salt marshes can keep pace with SLR, but others, especially those cut off from their sediment delivery via levees and seawalls cannot (Day et al., 1995). Nutrient pollution, which destabilizes below-ground biomass and increase decomposition, is likely to exacerbate salt marsh loss due to SLR (Turner et al., 2009; Deegan et al., 2012). Subsidence, which contributes to relative SLR in some regions, is an additional stressor. The impact of SLR will depend upon accretion and subsidence rates and other processes that influence the marshes ability to grow vertically and/or to migrate inland. "Coastal squeeze" describes the limitation of marshes to extend landward due to boundaries (Pethick, 2001) such as paved areas, seawalls, and bulkheads from coastal development.

4   Implications for services to ecosystems and humanity

Salt marshes play a large role in the aquatic food web and the cycling of nutrients in coastal waters. They serve as critical habitat for various life stages of coastal fisheries that account for a large percentage of the world's fish catch (UNEP, 2006b). Over half of the commercial fish species of the East coast of the United States utilize salt marshes at some time of their lives (Beck et al., 2001). In addition to providing habitat for juvenile fishes, crabs, and shrimps, marshes support populations of some small forage fishes, which come up on the marsh surface at high tide to feed on invertebrates (Shenker and Dean, 1979; Zimmerman et al., 2000). Many migratory shore birds and ducks use salt marshes as stopovers during migrations and some birds winter in the marsh. Wading birds, such as egrets and herons, feed in salt marshes during the summer. Continued marsh loss could therefore dramatically alter estuarine food webs.

SLR is increasing the vulnerability of coastal populations to coastal erosion, flooding, and storms (IPCC, 2007). Salt marshes serve as natural barriers to these coastal hazards. They serve as shoreline stabilizers because they attenuate wave energy and help prevent erosion (Costanza et al., 2008, Gedan et al., 2011, Moller et al., 2014; see Chapter 26). They also slow and store floodwaters, reducing storm impacts on coastal communities (Cobell et al., 2013). While wetlands do not provide complete protection against coastal hazards, even small salt marshes can provide significant shoreline protection (Gedan et al., 2011). Their preservation and restoration may significantly decrease the economic impact and human losses of extreme events such as hurricanes and tsunamis (Gedan et al., 2011).

Salt marshes remove sediment, nutrients, microbes, and contaminants from runoff and riverine discharge (Gedan et al., 2009), acting as sponges absorbing much of the runoff after major storms and reducing flooding. They sequester pollutants from the water that drains down from the land, protecting nearby estuarine areas and coastal waters from harmful effects. They play a major role in the global carbon cycle and represent a major portion of the terrestrial biological carbon pool. They store excess carbon in their sediments, preventing it from re-entering the Earth's atmosphere and contributing to global warming. Salt marshes are thus an important component of the world's "blue carbon" (McLeod et al., 2011) and currently are being incorporated into global carbon markets. Chmura et al. (2003) estimated that tidal wetlands sequester 10 times the amount of carbon sequestered by peatlands. Salt marshes also provide excellent tourism, education, and recreation services, as well as research opportunities.

It is clear that salt marshes provide enormous benefits to society in the form of "ecosystem services". In this regard, coastal wetlands (which include salt marshes) are among the highest valued coastal ecosystems (Costanza et al., 2014). The serious reduction in salt marsh area reduces their capacity to provide these critical ecosystem services (Gedan et al., 2009; Craft et al., 2009).

5 Conservation Responses and Conclusions

As society has become aware of the environmental and economic values of salt marshes, efforts have commenced to slow their loss and even to restore degraded marshes. These are mostly local initiatives. Concerned individuals and dedicated groups both within and outside government are mobilizing to stop and even reverse the trends. Restoration may involve reconnecting areas to the estuary by excavating channels that had filled in, relying on the tidal flow to allow the marsh to restore itself. Other restoration projects involve removing unwanted invasive vegetation, changing the marsh elevation, and planting the desired species. Monitoring of such projects would need to be done for years after restoration to see if methods are successful or need modification, and to learn how much time it takes before the restored marsh acquires the biodiversity and ecosystem function of a natural marsh (Craft et al., 1999; Zedler and Lindig-Cisneros, 2000; Rozas et al., 2005). Restoration of coastal marshes is now also included among strategies for climate adaptation planning (Arkema et al., 2013, Barbier, 2014) and mitigating greenhouse gas emissions (Olander et al., 2012), highlighting the multiple benefits that may be derived from salt marsh conservation.

Some international legal instruments and policy frameworks, such as the Convention on Wetlands of International Importance, especially as Waterfowl Habitat[1] (Ramsar Convention), the Convention on Biological Diversity[2], and Agenda 21 adopted by the 1992 United Nations Conference on Environment and Development, promote the conservation and wise use of wetlands and support economic valuation to support conservation. Economic valuation can be used to evaluate and compare development uses vis-à-vis conservation uses. Although some estimates have been made (Costanza et al., 1997, Minello et al., 2012), placing a monetary amount on these services is difficult and controversial. Many benefits are non-monetary, which makes comparisons difficult in decision-making (Barbier et al., 2011). Improving the assessment and valuation of salt marsh services could assist current conservation methods.

These important vegetated, intertidal habitats and the ecosystem services they provide, such as fisheries, sequestration of pollutants, and protection from flooding and storm surge, are under threat due to natural and anthropogenic forces. Efforts would be needed worldwide to preserve the remaining salt marshes and restore some of those that have been destroyed or impaired.

---

[1] United Nations, *Treaty Series*, vol. 996, No. 14583.

[2] United Nations, *Treaty Series*, vol. 2226, No. 30619.

References

Adam, P. (1990). *Saltmarsh Ecology*. Cambridge University Press 461 pp.

Airoldi, L. and Beck, M.W. (2007). Loss, status, and trends for coastal marine habitats of Europe. *Oceanography and Marine Biology: An Annual Review* 45: 345-405.

Arkema, K.K., Guannel, G., Verutes, G., Wood, S.A., Guerry, A., Ruckelshaus, M., Kareiva, P., Lacayo, M. and Silver, J.M. (2013). Coastal habitats shield people and property from sea level rise and storms. *Nature Climate Change* 3: 913-918.

Barbier, E.B., Hacker, S.D., Kennedy, C., Koch, E.W., Stier, A.C., Silliman, B.R. (2011). The value of estuarine and coastal ecosystem services. *Ecological Monographs* 81: 169-193.

Barbier, E.B. (2014). A global strategy for protecting vulnerable coastal populations. *Science* 345: 1250-1251.

Beck, M.W. et al. (2001). The identification, conservation, and management of estuarine and marine nurseries for fish and invertebrates. *BioScience* 51: 633-641.

Bridgham, S.D., Megonigal, J.P., Keller, J.K., Bliss, N.B., and Trettin, C. (2006). The carbon balance of North American wetlands. *Wetlands* 26: 889–916.

Bromberg, K.D. and Bertness, M.D. (2005). Reconstructing New England salt marsh losses using historical maps. *Estuaries* 28: 823-832.

Cahoon, D.R. and Reed, D.R. (1995). Relationships among marsh surface topography, hydroperiod, and soil accretion in a deteriorating Louisiana salt marsh. *Journal of Coastal Research* 11(2): 357-369.

Chapman, V.J. (1960). *Salt Marshes and Salt Deserts of the World* Leonard Hill (Books) Ltd. 392 pp.

Chmura, G.L., Anisfeld, S.C., Cahoon, D.R., Lynch, J.C. (2003). Global carbon sequestration in tidal, saline wetland soils. *Global Biogeochemical Cycles* 17(4): 1-11.

Church, J.A., et al. (2013). Sea Level Change. In: Climate Change 2013: The Physical Science Basis. Contribution of Working Group I to the Fifth Assessment Report of the Intergovernmental Panel on Climate Change [Stocker, T.F., et al., eds.]. Cambridge University Press, Cambridge, United Kingdom and New York, NY, USA.

Cobell, Z., Zhao, H., Roberts, H.J., Clark, F.R. and Zou, S. (2013). Surge and wave modeling for the Louisiana 2013 Master Plan. *Journal of Coastal Research* 67: 88-108.

Costanza, R., d'Arge, R., de Groot, R., Farber, S., Grasso, M., Hannon, B., Limburg, K., Naeem, S., Neill, R.V.O., Paruelo, J., Raskin, R.G., Sutton, P. and van den Belt, M. (1997). The value of the world's ecosystem services and natural capital. *Nature* 387: 253-260.

Costanza, R., Perez-Maqueo, O., Martinez, M.L., Sutton, P., Anderson, S.J. and Mulder, K. (2008). The value of coastal wetlands for hurricane protection. *Ambio* 37: 241-248.

Craft, C.B., Reader, J., Sacco, J.N. and Broome, S.W. (1999). Twenty-five years of ecosystem development of constructed *Spartina Alterniflora* (Loisel) marshes. *Ecological Applications* 9: 1405-1419.

Craft, C., Clough, J., Ehman, J., Joye, S., Park, R., Pennings, S., Guo, H. and Machmuller, M. (2008). Forecasting the effects of accelerated sea-level rise on tidal marsh ecosystem services. *Frontiers in Ecology and the Environment* 7: 73-78.

Day Jr., J.W., Pont, D., Hensel, P.F. and Ibañez, C. (1995). Impacts of sea-level rise on deltas in the Gulf of Mexico and the Mediterranean: The importance of pulsing events to sustainability. *Estuaries* 18(4): 636-647.

Deegan, L.A., Johnson, D.S., Warren, R.S., Peterson, B.J., Fleeger, J.W., Fagherazzi, S. and Wollheim, W.M. (2012). Coastal eutrophication as a driver of salt marsh loss. *Nature* 490: 388-392.

DeLaune, R.D., Baumann, R.H., Gosselink. J.G. (1983). Relationships among vertical accretion, coastal submergence, and erosion in a Louisiana Gulf Coast Marsh. *Journal of Sedimentary Research* 53(1): 147-157.

Doody, J.P. (2008). *Saltmarsh Conservation, Management, and Restoration*. Springer 219 pp.

Gagliano, S.M.K., Meyer-Arendt, J. and Wicker, K.M. (1981). Land loss in the Mississippi River deltaic plain. *Transactions – Gulf Coast Association of Geological Societies* 31: 295-300.

Gedan, K. B., Silliman, B.R. and Bertness, M.D. (2009). Centuries of human-driven change in salt marsh ecosystems. *Annual Review of Marine Science* 1: 117-141.

Gedan, K.B., Kirwan, M.L., Wolanski, E., Barbier, E.B. and Silliman, B.R. (2011). The present and future role of coastal wetland vegetation in protecting shorelines: answering recent challenges to the paradigm. *Climatic Change* 106: 7-29.

Hatton, R.S., DeLaune, R.D. and Patrick, Jr. W.H. (1983). Sedimentation, accretion, and subsidence in marshes of Barataria Basin, Louisiana. *Limnology and Oceanography* (3): 494-502.

IPCC (ed.) (2007). Climate change 2007: the physical science basis. Cambridge University Press, New York.

Lotze, H.K., Lenihan, H.S. Bourque, B.J., Bradbury, R.H., Cooke, R.G., Kay, M.C., Kidwell, S.M., Kirby, M.X., Peterson, C.H. and Jackson, J.B. (2006). Depletion, degradation, and recovery potential of estuaries and coastal seas. *Science* 312: 1806-1809.

McLeod, E., Chmura, G.L., Bouillon, S., Salm, R., Bjark, M., Duarte, C.M., Lovelock, C.E., Schlesinger, W.H. and Silliman, B.R. (2011). A blueprint for blue carbon: Toward an improved understanding of the role of vegetated coastal habitats in sequestering CO_2. *Frontiers in Ecology and the Environment* 9: 552-560.

Milliman, J.D. and Emery, K.O. (1968). Sea levels during the past 35,000 years. *Science* 162: 1121-1123.

Minello, T.J., Rozas, L.P., Caldwell, P.A. and Liese, C. (2012). A comparison of salt marsh construction costs with the value of exported shrimp production. *Wetlands* 32: 791-799.

Mitsch, W.J. and Gosselink, J.G. (2007). *Wetlands*, 4th Ed.

Moller, I., Kudella, M., Ruprecht, F., Spencer, T., Paul, M., van Wesenbeeck, R., Wolters, G., Jensen, K., Bouma, T.J., Miranda-Lange, M. and Schimmels, S. (2014). Wave attenuation over coastal salt marshes under storm surge conditions. *Nature Geoscience* 7: 727-731.

Nellemann, C., Corcoran, E., Duarte, C. M., Valdés, L., De Young, C., Fonseca, L., Grimsditch, G. (Eds.). (2009). *Blue Carbon. A Rapid Response Assessment*. United Nations Environment Programme, GRID-Arendal, www.grida.no

Nicholls, R.J., Hoozemans, F.M.J., Marchand, M. (1999). Increasing flood risk and wetland losses due to global sea-level rise: regional and global analyses. *Global Environmental Change* 9: S69-87.

Olander, L.P., Cooley D.M. and Galik, C.S. (2012). The potential role for management of U.S. public lands in greenhouse gas mitigation and climate policy. *Environmental Management* 49: 522-533.

Pethick, J. (2001). Coastal management and sea-level rise. *Catena* 42: 307-322.

Ranwell, D.S. (1972). Ecology *of Salt Marshes and Sand Dunes.* Chapman and Hall 200 pp.

Reed, D.J. (1995). The response of coastal marshes to sea-level rise: survival or submergence? *Earth Surface Processes and Landforms* 20: 39-48.

Redfield, A.C. (1972). Development of a New England salt marsh. *Ecological Monographs* 42: 201-237.

Rozas, L.P., Calswell, P. and Minello, T.J. (2005). The fishery value of salt marsh restoration projects. *Journal of Coastal Research* 40: 37-50.

Shenker, J.M. and Dean, J.M. (1979). The utilization of an interintertidal salt marsh creek by larval and juvenile fishes: Abundance, diversity and temporal variation. *Estuaries* 2(3): 154-163.

Turner, R.E., Howes, B.L., Teal, J.M., Milan, C.S., Swenson, E.M. and Goehringer-Tonerb, D.D. (2009). Salt marshes and eutrophication: an unsustainable outcome. *Limnology and Oceanography* 54: 1634–1642.

UNEP [United Nations Environment Programme]. (2006a). *Our precious coasts.* Nellemann, C. and Corcoran, E. (Eds). GRID Arendal Norway.

UNEP (2006b). Marine and coastal ecosystems and human wellbeing: A synthesis report based on the findings of the Millennium Ecosystem Assessment. UNEP, Nairobi, Kenya. Worm, B. et al. 2006. Impacts of biodiversity loss on ocean ecosystem services. *Science* 314: 787-790.

UNEP-WCMC (2015). Global distribution of saltmarsh (ver. 2.0). Unpublished dataset. Cambridge (UK): UNEP World Conservation Monitoring Centre.

Zedler, J.B. and Lindig-Cisneros, R. (2000). Functional equivalency of restored and natural salt marshes. In: *Concepts and Controversies in Tidal Marsh Ecology*, eds. M.P. Weinstein and D.A. Kreeger. Kluwer Acad. Publishers, Dordrecht, the Netherlands, pp 565-582.

Zimmerman, R.J., Minello, T.J. and Rozas, L.P. (2000). Salt marsh linkages to productivity of penaeid shrimps and blue crabs in the northern Gulf of Mexico. In *Concepts and controversies in tidal marsh ecology*, eds. M. P. Weinstein and D. A. Kreeger, 293-314. Dordrecht, the Netherlands, Kluwer Academic Publishers.

Sargasso Sea

Contributors:
D. Freestone (Convenor), Lorna Inniss (Co-Lead Member), D. d'A. Laffoley, K. Morrison, Jake Rice (Lead Member and Editor of Part VI), H.S.J. Roe and T. Trott

1 Inventory

The Sargasso Sea is a fundamentally important area of the open ocean within the North Atlantic Sub-Tropical Gyre, bounded on all sides by clockwise rotating currents (Laffoley et al., 2011).

Named after its iconic *Sargassum* seaweed, the Sargasso Sea's importance derives from the interdependent mix of its physical oceanography, its ecosystems, and its role in global-scale ocean and earth system processes. It is a place of legend, with a distinct pelagic ecosystem based upon two species of floating *Sargassum*, the world's only macroalgae that spend their whole life-cycle in the water column (holopelagic), which hosts a rich and diverse community, including ten endemic species. *Sargassum* mats are home to >145 invertebrate species and >127 species of fish; the mats act as important spawning, nursery and feeding areas for fish, turtles and seabirds. In deeper water, the Sargasso Sea is the only known spawning area for both the European and American Eels (*Anguilla anguilla*, *A. rostrata*). Porbeagle Sharks (*Lamna nasus*) migrate from Canada to the Sargasso Sea, where they are suspected of pupping in deep water; several other shark species undertake similar migrations and may be using the area as nursery areas. Thirty species of whales occur in the Sargasso Sea and Humpback Whales (*Megaptera novaeangliae*) make regular migrations through the area en route from the Caribbean to the northern North Atlantic. Many other species, including several tuna spp., turtles, rays and swordfish, migrate through the Sargasso Sea: it is truly an ecological crossroads in the Atlantic Ocean, linking its own distinct ecosystem with Africa, the Americas, the Caribbean and Europe. Seamounts and volcanic banks rise up from the sea floor and host diverse and fragile communities of invertebrates and fish, including endemic species and others that are currently undescribed. Many of the species that occur in the Sargasso Sea are endangered or threatened and are listed on the IUCN Red List, and/or in the appendices of the Convention on International Trade in Endangered Species of Wild Fauna and Flora[1] (CITES) or in the annexes of the 1990 Caribbean Protocol Concerning Specially Protected Areas and Wildlife[2] to the Convention for the Protection and Development of the Marine Environment in the Wider Caribbean Region[3] (SPAW) (see Laffoley et al., 2011). Laffoley et al. (2011) present a summary of the scientific case for the protection and management of the Sargasso Sea which maps the area and describes its status, its importance and the threats to its continued existence.

2 Trends

The Sargasso Sea is a globally important area for ocean research and monitoring, hosting the world's longest continuous open-ocean time series of ocean measurements (i.e., Hydrostation S and the associated Bermuda Atlantic Time-series Study (BATS) arrays) which make it possible to observe trends and changes over time (Figure 50.1). These include significant warming of the surface ocean, an increase in salinity in the upper 300 m, and a decrease in surface pH (Figure 50.2). These data are critical for our understanding of global processes and the role of the Sargasso Sea in these processes. The annual net primary production in the Sargasso Sea is surprisingly high, due largely to picoplankton, and as such the area plays a key role in the sequestration of carbon in the global ocean. Changes are occurring in both phytoplankton biomass and primary production in the northern Sargasso Sea, and the possible connections between such changes and any resulting effects in the ecosystem with global climate change is an active area of research (Laffoley et al., 2011).

3 Pressures

In addition to climate change, the Sargasso Sea is threatened by other human activities. Overfishing and the side effects of fishing (e.g., by-catch, lost gear) affect pelagic species e.g., blue marlin and western bluefin tuna are estimated to be overexploited; benthic trawling on seamounts has severely reduced stocks of alfonsino and destroyed benthic communities. Ship-related impacts may include pollution from discharges, introduction of alien species through ballast water, underwater noise, collisions with whales, and physical damage to *Sargassum* mats (Laffoley et al., 2011, 2015 in prep). Surface pollutants, including plastics, accumulate in the central Sargasso Sea, because the encircling currents trap water for periods of 50 years or more. Plastics and debris concentrate in *Sargassum* mats and in frontal zones where animals also concentrate to feed. Other potential pressures include the continuing commercial interest in harvesting *Sargassum*, the impact of submarine cables, and seabed mining (see Laffoley et al., 2011).

4 Ecosystem Services

The economic importance of the Sargasso Sea is derived from direct exploitation, via fisheries and tourism, and indirect benefits from ecosystem services. Pendleton et al. (2015), Sumaila et al. (2013), and Laffoley et al. (2011) provide varying estimates of the values of pelagic fisheries, eel fisheries in Canada, Europe and the United States of America that depend upon eels that spawn in the Sargasso Sea, recreational fishing, reef-associated tourism, and whale and turtle watching. Sumaila et al. (2013) also provide estimates of the indirect-use values for the Sargasso Sea associated with the open ocean, coral reefs, coastal systems and coastal wetlands. The accuracy of many of these estimates is questionable, but all values are large and emphasize the economic importance of the Sargasso Sea and the need to conserve and restore the ecosystem.

---

[1] United Nations, *Treaty Series*, vol. 993, No. 14537.

[2] United Nations, *Treaty Series*, vol. 1506, No. 25974.

[3] United Nations, *Treaty Series*, vol. 1506, No. 25974.

Chapter 50 — Sargasso Sea

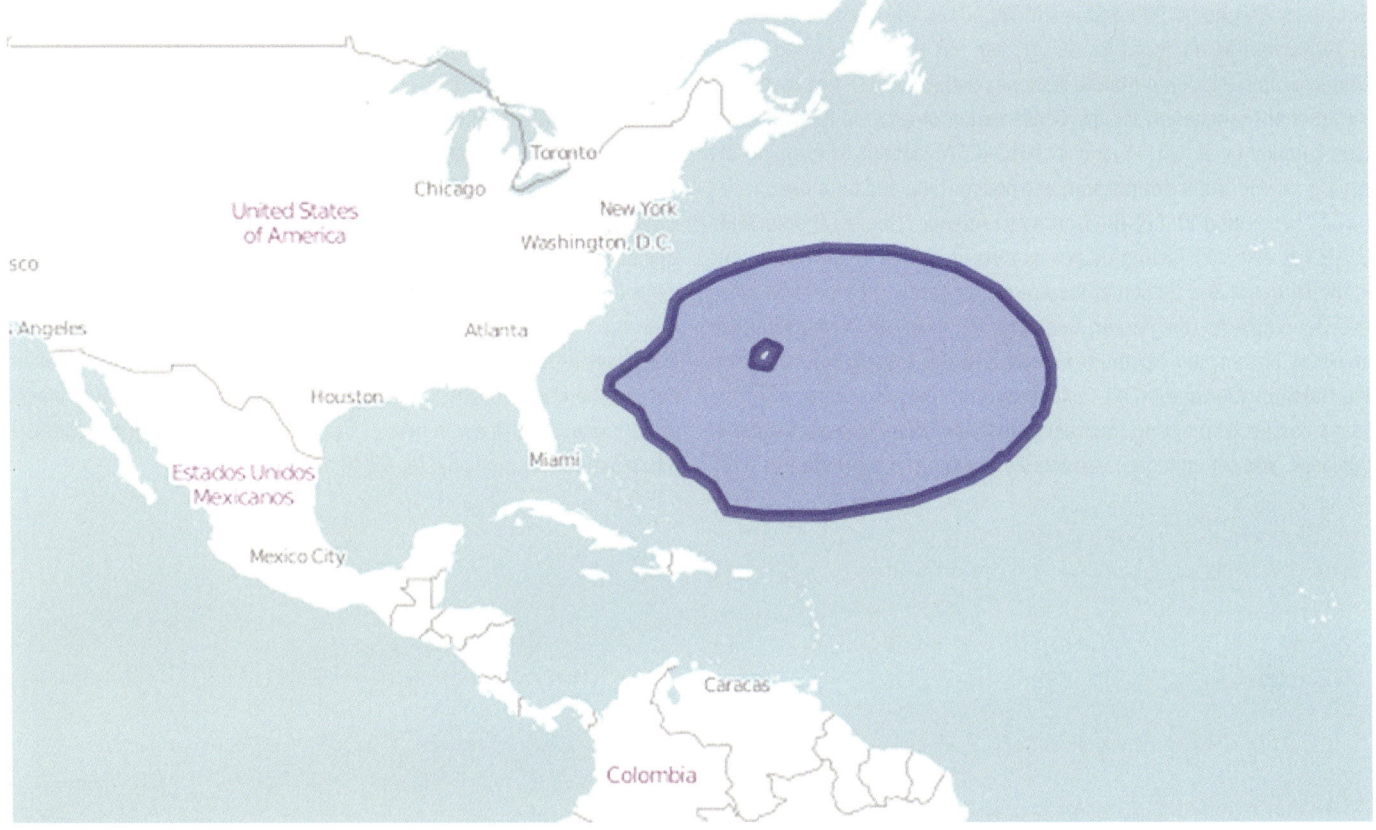

The boundaries and names shown and the designations used on this map do not imply official endorsement or acceptance by the United Nations.
Figure 50.1 | Source: https://chm.cbd.int/database/record?documentID=200098.

Although *Sargassum* generally is considered a unique feature of the Sargasso Sea, in reality mats of *Sargassum* have occasionally been seen in many places in the mid-North Atlantic and even to wash up on island beaches from time to time. However, in 2011 large mats of *Sargassum* appeared on beaches in many Caribbean areas, the coast of Brazil and even the coast of West Africa. Similar mass strandings occurred in 2014 and are continuing in 2015. The source of the *Sargassum* is not the Sargasso Sea but the north equatorial recirculation region (NERR) south of the Sargasso Sea between the north equatorial current and the equator. The causes of these mass blooms and strandings are uncertain but may include nutrient availability from the Amazon and Orinoco Rivers, warmer surface temperatures and changes in circulation associated with climate change (Franks et al 2011-2015 in http://www.usm.edu/gcrl/sargassum/index.php, Johnson et al 2012, Smetacek and Zingone 2013)). The impact of these mass strandings on local economies is severe, affecting tourism and recreation, as the mats are difficult to dispose of and are unsightly and smelly as they decompose (see chapter 27).

5 Conservation Responses

The importance of the Sargasso Sea is now recognised internationally. In October 2012, the Sargasso Sea was accepted by the Conference of Parties to the Convention on Biological Diversity as meeting the criteria for an ecologically and biologically significant area (EBSA) (see https://chm.cbd.int/database/record?documentID=200098). Also in 2012, the Bermuda Government declared the Bermuda exclusive economic zone (EEZ) to be a marine mammal sanctuary and signed a Sister Sanctuary Agreement with the United States' Stellwagen Bank National Marine

Figure 50.2 | Time-series plots of temperature and salinity anomaly at 300 m (STMW) for Hydrostation 'S' 1955-2011. Anomaly computed by subtracting the long-term mean for this depth. Red line shows a one-year central running mean and the observed data are shown as blue dots. Long-term trends for temperature and salinity are determined as 0.009 0C year-1(p>0.01) and 0.002 year-1(p<0.01), respectively. Source: Lomas, et al (2011).

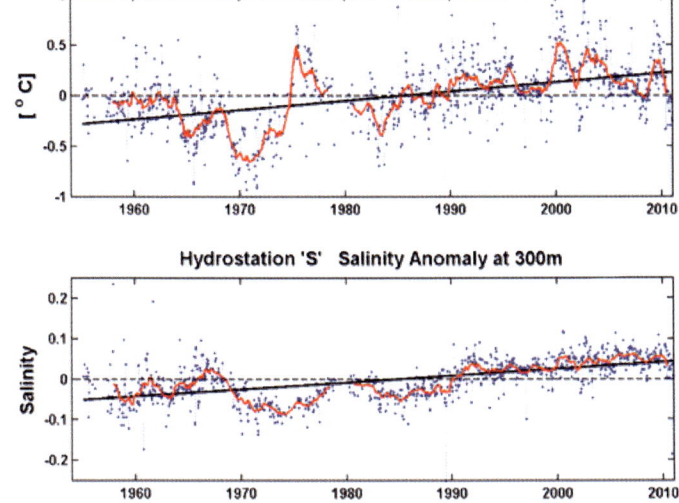

Sanctuary (Bermuda/United States MOA, 2012). The overall importance of *Sargassum* as a habitat for pelagic fish has been recognised by the United States (National Marine Fisheries Service 2003) and by the International Commission for the Conservation of Atlantic Tunas (ICCAT) (see Laffoley et al., 2011), and in 2012 ICCAT agreed to examine the ecological importance of the Sargasso Sea for tuna and tuna-like species (ICCAT Resolution 12-12). The Northwest Atlantic Fisheries Organization (NAFO) is also considering proposals to protect further the seamounts in the Sargasso Sea section of their regulatory area. On 11 March 2014, five Governments (the Azores, Bermuda, Monaco, the United Kingdom of Great Britain and Northern Ireland and the United States) signed the Hamilton Declaration on Collaboration for the Conservation of the Sargasso Sea, committing themselves to collaborate on conservation in this area, and set up a Sargasso Sea Commission to facilitate this work (Freestone and Morrison, 2014). The Sargasso Sea Commission is working with the Convention on the Conservation of Migratory Species of Wild Animals (CMS 1979) regarding conservation of the European eel (*Anguilla anguilla*) and in November 2014 the CMS Conference of the Parties added it to Appendix II as a "having a conservation status which would significantly benefit from international co-operation ...". (UNEP/CMS 2014). It is also in discussions with the International Maritime Organization (IMO) and the Bermudian shipping authorities concerning ways of mitigating shipping risks and it has recently opened a dialogue with the cable-laying industry to develop best environmental practices (SSC Newsletter, 2015). The United Nations General Assembly has taken note of the efforts of the Sargasso Sea Alliance, led by the Government of Bermuda, to raise awareness of the ecological significance of the Sargasso Sea (Resolutions 67/78, 68/70 and 69/245).

References

Angel, M.V. (2011). *The Pelagic Ocean Assemblages of the Sargasso Sea Around Bermuda*. Sargasso Sea Alliance Science Report Series, No. 1, 25pp.

Ardron, J., Halpin, P., Roberts, J., Cleary, J., Moffitt, M. and Donnelly, J. (2011). *Where is the Sargasso Sea? A Report Submitted to the Sargasso Sea Alliance*. Duke University Marine Geospatial Ecology Lab & Marine Conservation Institute. Sargasso Sea Alliance Science Report Series, No. 2, 24pp.

Bermuda/USA MOA. (2014). Memorandum of Understanding between the United States of America – Department of Commerce – National Oceanic And Atmospheric Administration – National Ocean Service – Office of Marine Sanctuaries - and the Government of Bermuda – Ministry of the Environment, Planning and Infrastructure Strategy - to Collaborate on International Protection, Conservation and Management of the Humpbacked Whale. (http://stellwagen.noaa.gov/sister/pdfs/bermuda_moa12.pdf accessed 10 July 2015).

Freestone D. and Morrison K.K. (2014). The Signing of the Hamilton Declaration on Collaboration for the Conservation of the Sargasso Sea: A New Paradigm for High Seas Conservation? (2014) 29 *International Journal of Marine and Coastal Law* 345–362; Text of the Declaration at 355-362.

Gollock, M. (2011). *European eel briefing note for Sargasso Sea Alliance*. Sargasso Sea Alliance Science Report Series, No. 3, 11pp.

Hallett, J. (2011). *The Importance of the Sargasso Sea and the Offshore Waters of the Bermudian Exclusive Economic Zone to Bermuda and its People*. Sargasso Sea Alliance Science Report Series, No. 4, 18pp.

ICCAT (2013). Resolution 12-12; Resolution by ICCAT on the Sargasso Sea. ICCAT Compendium Management Recommendations and Resolutions Adopted by ICCAT for the Conservation of Atlantic Tunas and Tuna-like Species, p. 280, available at http://www.iccat.int/Documents/Recs/ACT_COMP_2013_ENG.pdf.

Johnson, D.R., Ko, D.R., Franks, J.S., Mareno, P, and Snachez-Rubio, G. (2012). *The Sargassum Invasion of the Eastern Caribbean and Dynamics of the Equatorial North Atlantic*. Proceedings of the 65th Gulf and Caribbean Fisheries Institute 65:102-103.

Laffoley, D.d'A., Gerde, K., and Roe, H.S.J. (2015). *A Strategic Assessment of the Risks Posed by Shipping to The Sargasso Sea And Evidence of Impacts*. Sargasso Sea Alliance commissioned paper, 26pp.

Laffoley, D.d'A., Roe, H.S.J., Angel, M.V., Ardron, J., Bates, N.R, Boyd, L.L., Brooke, S., Buck, K.N., Carlson, C.A., Causey, B., Conte, M.H., Christiansen, S., Cleary, J., Donnelly, J., Earle, S.A., Edwards, R., Gjerde, K.M., Giovannoni, S.J., Gulick, S., Gollock, M., Hallet, J., Halpin, P., Hanel, R., Hemphill, A., Johnson, R.J., Knap, A.H., Lomas, M.W., McKenna, S.A., Miller, M.J., Miller, P.I., Ming, F.W., Moffitt, R., Nelson, N.B., Parson, L., Peters, A.J., Pitt, J., Rouja, P., Roberts, J., Roberts, J., Seigel, D.A., Siuda, A., Steinberg, D.K., Stevenson, A., Sumaila, V.R., Swartz, W., Trott, T.M., and Vats, V. (2011). *The protection and management of the Sargasso Sea: The golden floating rainforest of the Atlantic Ocean. Summary Science and Supporting Evidence Case*. Sargasso Sea Alliance, 44pp. ISBN#-978-0-9847520-0-3 available at http://www.sargassoalliance.org/case-for-protection.

Lomas, M.W., Bates, N.R., Buck, K.N. and Knap, A.H. (eds) (2011a). *Oceanography of the Sargasso Sea: Overview of Scientific Studies*. Sargasso Sea Alliance Science Report Series, No. 5, 64pp.

Lomas, M.W., Bates, N.R., Buck, K.N. and Knap, A.H. (2011b). *Notes on "Microbial Productivity of the Sargasso Sea and How it Compares to Elsewhere," and "The Role of the Sargasso Sea in Carbon Sequestration – Better than Carbon Neutral?"* Sargasso Sea Alliance Report Series, No. 6, 10pp.

Miller, M.J., and Hanel, R. (2011). *The Sargasso Sea Subtropical Gyre: The Spawning and Larval Development Area of Both Freshwater and Marine Eels*. Sargasso Sea Alliance Report Series, No. 7, 20pp.

National Marine Fisheries Service (NMFS). (2003). "Fisheries of the Caribbean, Gulf of Mexico and South Atlantic; pelagic Sargassum habitat of the south Atlantic region (Final Rule)" *Federal Register* 68:192 p 57375. http://www.safmc.net/Portals/6/Library/FMP/Sargassum/SargFMPFinalrule.

Parson, L. and Edwards, R. (2011). *The Geology of the Sargasso Sea Alliance Study Area, Potential Non-Living Marine Resources and an Overview of the Current Territorial Claims and Coastal States Interests*. Sargasso Sea Alliance Science Report Series, No. 8, 17pp.

Pendleton, L., Krowicki, F., Strosser, P. and Hallett-Murdoch, J. (2015). *Assessing the Economic Contribution of Marine and Coastal Ecosystem Services in the Sargasso Sea*. NI R 14-05. Durham, NC: Duke University.

Roberts, J. (2011). *Maritime Traffic in the Sargasso Sea: An Analysis of International Shipping Activities and their Potential Environmental Impacts*. Report to IUCN/Sargasso Sea Alliance Legal Working Group by Coastal & Ocean Management, Hampshire, UK. Sargasso Sea Alliance Science Report Series, No. 9, 45pp.

Siuda, A.N.S. (2011). *Summary of Sea Education Association Long-term Sargasso Sea Surface Net Data*. Sargasso Sea Alliance Science Report Series, No. 10, 18pp.

Smetacek, V. and Zingone, A. (2013) Green and golden seaweed tides on the rise. *Nature* 504, 84-88. doi:10.1038/Nature 12860.

SSC Newsletter (2015). Sargasso Sea Commission Newsletter, March 2015. (http://archive.constantcontact.com/fs169/1109154724045/archive/1120341878232.html accessed 10 July, 2015).

Stevenson, A. (2011). *Humpback Whale Research Project*, Bermuda. Sargasso Sea Alliance Science Report Series, No. 11, 11pp.

Sumaila, U.R., Vats, V. and Swartz, W. (2013). *Values from the Resources of the Sargasso Sea*. Sargasso Sea Alliance Science Report Series, No. 12, 24pp.

UNEP/CMS (2014). UNEP/CMS/COP11/Doc.24.1.18.Rev.1.

Chapter 51

Biological Communities on Seamounts and Other Submarine Features Potentially Threatened by Disturbance

Contributors:
J. Anthony Koslow (Convenor), Peter Auster, Odd Aksel Bergstad, J. Murray Roberts, Alex Rogers, Michael Vecchione and Patricio Bernal, Peter Harris and Jake Rice (Editor of Part VI) (Co-Lead Members)

1 Physical, chemical, and ecological characteristics

1.1 Seamounts

Seamounts are predominantly submerged volcanoes, mostly extinct, rising hundreds to thousands of metres above the surrounding seafloor. Some also arise through tectonic uplift. The conventional geological definition includes only features greater than 1000 m in height, with the term "knoll" often used to refer to features 100 – 1000 m in height (Yesson et al., 2011). However, seamounts and knolls do not appear to differ much ecologically, and human activity, such as fishing, focuses on both. We therefore include here all such features with heights > 100 m.

Only 6.5 per cent of the deep seafloor has been mapped, so the global number of seamounts must be estimated, usually from a combination of satellite altimetry and multibeam data as well as extrapolation based on size-frequency relationships of seamounts for smaller features. Estimates have varied widely as a result of differences in methodologies as well as changes in the resolution of data. Yesson et al. (2011) identified 33,452 seamount and guyot features > 1000 m in height and 138,412 knolls (100 – 1000 m), whereas Harris et al. (2014) identified 10,234 seamount and guyot features, based on a stricter definition that restricted seamounts to conical forms. Estimates of total abundance range to >100,000 seamounts and to 25 million for features > 100 m in height (Smith 1991; Wessel et al., 2010). At least half are in the Pacific, with progressively fewer in the Atlantic, Indian, Southern, and Arctic Oceans. Identified seamounts cover approximately 4.7 per cent of the ocean floor, with identified knolls covering an additional 16.3 per cent, in total an area approximately the size of Africa and Asia combined, about three-fold larger than all continental shelf areas in the world's oceans (Etnoyer et al., 2010; Yesson et al., 2011).

Seamounts can influence local ocean circulation, amplifying and rectifying flows, including tidal currents, particularly near seamount summits, enhancing vertical mixing, and creating retention cells known as Taylor columns or cones over some seamounts. These effects depend on many factors, including the size (height and diameter) of the seamount relative to the water depth, its latitude, and the character of the flow around the seamount (White et al., 2007).

Where flows are sufficiently vigorous, they provide a sufficient flow of organic matter to support suspension feeding organisms, such as corals and sponges. Such currents also winnow away the sediment, providing hard substrate necessary for most suspension feeders to settle and attach. Depending on depth and current regime, the seamount benthos may be dominated by an invertebrate fauna typical of the surrounding sediment-covered slope or abyssal plain or a more specialized fauna adapted to high-energy, hard substrate-dominated deep-water environments.

Seamounts that rise to mesopelagic depths or shallower ($\leq$ ~1000 m) often have an associated fish fauna adapted to feed on the elevated flux of micronekton and zooplankton, as well as vertical migrators intercepted by the seamounts during their downward diel migrations (Koslow, 1997; Clark et al., 2010b). More than 70 fish taxa have been commercially exploited around seamounts (Rogers 1994), although the number of species that are found only or principally on seamounts is closer to 13 – 16 (Clark et al., 2007, Watson et al., 2007). Some, such as pelagic armorhead, orange roughy, alfonsino, oreos, and others, are found in substantial aggregations around seamounts, making them efficient targets for fisheries.

Many seamounts are rugged, topographically complex environments, difficult to sample with conventional gear, such as nets. Wilson and Kaufmann's (1987) review reported only 596 invertebrate species recorded from seamounts to that date, with 72 per cent of these species from studies of only five seamounts. Seamount studies ramped up in the 1990s, initially based on concerns about the impacts of deep-water trawl fishing for orange roughy and oreos on seamounts. Based on more intensive and comprehensive sampling, Richer de Forges et al. (2000) reported more than 850 species associated with cold-water coral and sponge communities on seamounts in the Tasman and Coral Seas of the Southwest Pacific, with potentially high levels of endemism. Seamount studies have since been carried out worldwide, significantly stimulated by collaborative efforts such as the Census of Marine Life CenSeam project. A recent review found 1,222 fish species recorded from 184 seamounts (Kvile et al., 2013), approximately doubling the number of seamounts investigated and the number of species recorded from this environment since Wilson and Kaufmann's (1987) review. Overall species richness on seamounts cannot yet be estimated, with < 1 per cent of seamounts sampled and only a few percent of those intensively studied. The number of species recorded from seamounts continues to increase roughly in proportion to sampling effort, with no evidence yet of levelling off (Richer de Forges et al., 2000; Stocks and Hart, 2007; Castelin et al., 2011). Due to the limited sampling, the proportion of species endemic to this habitat is controversial. Some seamounts appear to represent biodiversity hotspots (Samadi et al., 2006; McClain et al., 2009), but variability is extensive, and other studies conclude that species richness is comparable on seamounts and nearby slope habitats (Consalvey et al., 2010; Howell et al., 2010; Castelin et al., 2011).

Biogeographically, seamount faunas generally appear related to the faunas of adjacent basins or continental slopes (Wilson et al., 1985; Parin et al., 1997; Mironov et al., 2006). For seamount fishes, which have been the best studied group of organisms, biogeographical patterns appear to follow the distribution of dominant water masses, such as Antarctic Intermediate Water and North Atlantic Deep Water (Koslow et al., 1994; Clark et al., 2010a and b). Whereas the dominant genera and families of deep-sea demersal and midwater fishes tend to have global distributions, the dominant fish species on seamounts in different ocean basins are often from entirely different genera, families, and even orders. This indicates that seamount-associated fishes in different ocean basins were reproductively isolated and evolved independently. Their similar morphologies and adaptation to the seamount environment is a striking example of convergent evolution (Koslow, 1996).

Chapter 51   Biological Communities on Seamounts and Other Submarine Features Potentially Threatened by Disturbance

Seamounts are the source of significant ecosystem services. In addition to their biodiversity, seamounts often host substantial aggregations of fishes, which have been subject to commercial fisheries. These include species for which seamounts are their primary environment as well as a larger number for which seamounts account for a smaller proportion of their global catch. Annual landings of primary seamount species have fluctuated around 100,000 t since the 1990s, dominated by oreosomatids and orange roughy (Clark et al., 2007; Watson et al., 2007).

Seamounts, along with ridges and plateaus, host ferromanganese crusts that contain cobalt, nickel, and rare earth elements used in high-tech industries and which may have commercial potential, although they are not presently exploited (Hein et al., 2010).

1.2   Ridges and plateaus

One of the more prominent features of the global ocean is the 75,000 km long network of mid-ocean ridges, defined as "the linked major mid-oceanic mountain systems of global extent" (IHO, 2008), essentially created at plate boundary spreading zones, where new crust is being formed as tectonic plates move apart. The most prominent is the Mid-Atlantic Ridge that runs down the middle of the Atlantic from the Arctic to the Southern Ocean, where it connects to the more complex system of ridges in the Indian Ocean and the Pacific Ocean. Mid-ocean ridges in the South Pacific comprise 1.87 million km^2, the largest area in a single ocean (Harris et al., 2014). Ridge features may include islands (e.g. the Azores archipelago (Portugal) and Iceland) and seamounts. Harris et al (2014) distinguish mid-ocean ridges from other ridge features that are isolated, elongate, narrow, steep-sided and at least 1000 m in height. These may overlap with plateaus and seamounts. Hydrothermal vents, which are treated in Chapter 45, are often associated with mid-ocean spreading ridges (Baker and German, 2013).

Studies of ridge features have increased significantly in recent years. Notable among them are studies of the Nazca and Sala y Gomez ridges in the Southeast Pacific (e.g. Parin et al., 1997; Pakhorukov, 2005), the MAR-ECO and ECOMAR studies of the Mid-Atlantic Ridge in the North Atlantic (e.g. special issues of Deep-Sea Research II 55 (1). (2008). and 98 Part B (2013), Bergstad et al., 2008; Vecchione et al., 2010; Priede et al., 2013), as well as studies of the Mid-Atlantic Ridge in the South Atlantic (Perez et al., 2012).

Ridges typically contain seamounts and sedimented slopes; not surprisingly, similarities in the abundance, diversity, and species composition of ridge habitats are found with both seamounts and continental margins (Priede et al., 2013). Priede et al. (2013) also noted that a deep ridge system, such as the mid-Atlantic Ridge (mostly deeper than 1000 m), does not appear to significantly enhance oceanic productivity, although it greatly extends the area of available lower bathyal habitat.

Plateaus and banks are geologically not as well defined or as extensive as ridges, but comprise relatively less steep and comparatively shallow features separated from continental shelves by deep channels. In terms of shape and size, plateaus are wider and much larger than seamounts. Harris et al. (2014) mapped 184 plateaus in the world's oceans, comprising 5.1 per cent of the ocean area. They figure most prominently in the Indian and South Pacific Oceans; Challenger, Campbell, and Kerguelen Plateaus and Chatham Rise around New Zealand are the largest such features. Despite being mostly deeper than 200 m, plateaus may be recognised as oceanic shallows or banks because they are disconnected from continental shelves and coastal waters.

Plateaus share many diversity characteristics and faunas with nearby continental shelves and slopes, and ecosystem services from plateaus are also similar to those of shelves. Most plateaus are nearer to land and are considered richer in terms of harvestable resources than oceanic ridges. Both shallow and deep fisheries on plateaus and banks are therefore relatively substantial. Indeed, most of the deep-water fisheries being conducted at present are either on upper continental slopes or on slopes of plateaus (Koslow et al., 2000; Watson et al., 2007). In addition to fishing, mining and hydrocarbon exploration/extraction are emerging activities on plateaus and ridges (Rona 2003, Ramirez-Llodra et al., 2011).

1.3   Submarine canyons

Submarine canyons are defined as "steep-walled sinuous valleys with V-shaped cross sections, axes sloping outward as continuously as river-cut land canyons, and relief comparable to even the largest land canyons" (Shephard, 1963). Recent estimates of their number and extent vary widely, depending on mapping criteria, from 448 to about 9500 canyons in the global ocean with a total extent ranging from 25,000 to 389,505 km (Ramirez-Llodra et al., 2010; De Leo et al., 2010; Harris and Whiteway, 2011; Harris et al., 2014). Because they cut across the continental shelf and slope, canyons are the deep-sea environment closest to human occupancy, making them convenient to study but also rendering them vulnerable to human stressors.

Canyons have been recognized as distinct topographic features for approximately 150 years (Dana 1863 in Ramirez-Llodra et al., 2010). However, their rough topography compared with nearby slope areas has made studying them difficult. Technological developments of the past few decades, along with international programmes like the Census of Marine Life Continental Margins, or Continental Margin Ecosystems (COMARGE) programme (Menot et al., 2010), and the long-term programme of the Monterey Bay Aquarium Research Institute (MBARI) in Monterey Canyon have led to a renaissance in canyon studies (e.g. Huvenne and Davies, 2013). Whereas most canyons globally have received little or no scientific attention from any discipline, some individual canyons (e.g., Monterey – western North America, "The Gully" – eastern North America, Kaikura – New Zealand, Nazaré – western Europe) have been studied by multidisciplinary teams.

With steep walls and depositional environments along their axes, canyons are geologically complex, including hard substrates and soft sediments, depending on the slope of the walls. Because canyons cut into the ancient sediments of continental margins, many have hydrocarbon seeps and their associated specialized chemosynthetic communities. They also exhibit complex hydrography, intersecting and diverting along-slope and along-shelf currents. The steep bottom topography can intensify these flows by topographic channelling and constriction. Density-driven flows result in episodic cascading down-canyon, transporting shallow waters into the deep sea along with associated material. The intensified currents can result in higher physical disturbance of the benthos relative to nearby slope areas.

Canyons concentrate both biogenic and anthropogenic material along their deep axes and can transport these materials either onto the shelf or into the deep abyssal environment, depending on local flow conditions. Such material includes organic matter produced in the overlying photic zone, as well as pollutants and other anthropogenic material, either inadvertently discharged or deliberately disposed of. The concentration of sinking surface productivity enriches benthic communities along the canyon axes and can result in high biomass. Flow may similarly enhance recruitment of early-life-history stages of both sessile and mobile fauna by local topographic concentration of eggs and/or larvae in particular areas (Vetter et al., 2010).

The hard substrates and particulate transport in canyons can support abundant, diverse sessile suspension-feeding communities (hard and soft corals, sponges) and associated fauna, whereas areas of sediment accumulation support communities of deposit feeders, scavengers and their predators (De Leo et al., 2010). Canyons are also hotspots of pelagic activity, supporting feeding concentrations of air-breathing marine vertebrates (mammals, birds, turtles), including many protected species. Canyons are also targeted by longline, trap and trawl fisheries. Some of these fisheries can damage or destroy both hard- and soft-substrate benthic communities, which may include very long-lived and slow-growing species. As with other bathyal habitats, canyons will be affected by climate change because of changes in circulation, stratification, primary productivity, expansion of oxygen-minimum zones (OMZs), and acidification. Canyons may also serve as conduits to the deep sea for pollution (including trash) or sediment mobilized by mining or bottom trawling (Ramiriez-Llodra et al., 2011).

1.4   Trenches

Trenches are defined as "long, narrow, characteristically very deep and asymmetrical depressions(s) of the seafloor, with relatively steep sides" (IHO, 2008). In addition to featuring steep terrain with typically hard substrates, often narrow terraces and the deepest ocean depths, the trenches have flat floors with accumulated fine sediments. Trenches are formed as oceanic plates collide with continental plates; the heavier oceanic plates are subducted, creating a trench. Trenches are generally narrow (<40 km wide), V-shaped in cross-section and are found near and parallel to island-arc systems and continental land masses (Jamieson et al., 2010). High current velocities (10-32 cm/sec) have been recorded near the bottom in trenches, as have collapsed walls with massive sediment slides and inferred turbidity currents (Ramirez-Llodra et al., 2010). These processes are responsible for the accumulation of sediments and organic matter in the axes of the trenches, similar to what occurs in canyons.

Harris et al. (2014) mapped 56 trenches in the world ocean. Trenches comprise a total area of about 2 million km^2, less than 1 per cent of the total ocean area. About 80 per cent of all trenches, by area, are found in the North and South Pacific Oceans. The seafloor of most trenches is at hadal depths (>6000 m) but some trenches, such as the Hellenic Trench in the Mediterranean, are shallower (Ramirez-Llodra et al., 2010).

Trenches are probably the most poorly-known deep-sea habitat, because of the cost and difficulty of sampling at such depths. The first biological samples from the trench environment were obtained during the 1948 Swedish *Albatross* Expedition, which was soon followed by the Danish *Galathea* and Soviet *Vityaz* Expeditions, which sampled several trench environments. Bruun (1956) first defined the hadal zone (depths > 6000 m) as distinct from the abyss, based on a marked transition in species composition, presumably because of the need to adapt to increased pressure. Hadal depths are also below the carbonate and opal compensation depths. Some animal groups with carbonate or siliceous skeletons are therefore excluded (non-holothurian echinoderms and non-actinarian Cnidaria) or are characterized by decreased skeletal strength compared to their shallower relatives (gastropods and bivalves). As a result of the discontinuities between trenches and apparent difficulties of dispersion, high levels of endemism exist both within the hadal environment as a whole (56 per cent as estimated by Belyaev (1989)) and within individual trench systems at the species level (Vinogradova, 1997; Blankenship-Williams and Levin, 2009). The hadal environment is often considered to have low biodiversity, but Blankenship-Williams and Levin (2009) note that trenches contain diverse habitats (e.g. cold seeps and hydrothermal vents, steep walls) that have been particularly poorly sampled.

A few broad generalities can summarize what little is known about trench fauna. Foraminifera are common even at the very deepest extremes. Patterns of metazoan diversity vary among communities based on substrate and food sources: fine sediments on ledges and the trench axis support infauna, such as macrofaunal polychaetes and meiofaunal nematodes, as well as deposit feeding epifauna dominated by holothurians; hard substrates of the walls are characterized by non-calcified sessile fauna like anemones and their mobile benthic associates (e.g., amphipods); communities near hydrothermal vents and cold seeps are dominated by metazoans dependent on symbiotic chemosynthetic microbes. An assemblage of mobile scavengers is dominated by amphipods, some remarkably large (Jamieson et al., 2013), and fishes such as liparids and macrourids (Jamieson et al., 2009; Fujii et al., 2010).

Because of their extreme depth, trenches have not been subject to commercial activity, such as fishing, mining, or energy extraction. However, they have been subject to dumping, such as of pharmaceuticals in the Puerto Rico trench (Ramirez-Llodra et al., 2011).

2 Documented anthropogenic impacts on the deep ocean including their history (as appropriate) on a regional basis.

2.1 Fishing

A few deep-water artisanal hook and line fisheries around islands and seamounts maintained steady landings with few environmental impacts for decades to centuries (e.g. oilfish in the South Pacific and black scabbardfish around Madeira (Koslow 2007; Silva and Pinho, 2007). Modern large-scale fisheries on seamounts, ridges, and other features with abrupt topographies were initiated after World War II, fostered by technological developments and distant-water industrial fishing. Gillnets, longlines, and both pelagic and bottom trawls are the primary gears (Gianni, 2004; Clark et al., 2007; Bensch et al., 2009). Bottom trawls have had greatest impact, affecting both targeted and non-targeted species including associated benthic communities (Koslow et al., 2001; Clark and Koslow, 2007; Clark and Rowden, 2009). These fisheries have occurred in all oceans except the Arctic.

The first Pacific seamount-associated fisheries were for pelagic species, such as albacore, that aggregated over seamounts in the North Pacific. From 1967 to 1989, a demersal seamount trawl fishery targeting aggregations of pelagic armorhead on the Emperor Seamount chain landed about 800,000 tons of armorhead along with about 80,000 t of alfonsino (Clark et al., 2007). These stocks were depleted and have still not recovered (NPFC, 2014).

Two species of red coral (Corallium spp.) were also depleted sequentially from the Emperor seamounts by a tangle-net fishery between 1965 and 1990 (Clark and Koslow 2007; Koslow, 2007).

Since the mid-1970s, trawl fisheries expanded to seamounts and plateaus in the South Pacific, predominantly for orange roughy and oreos, but also for alfonsino, black cardinalfish, and other species (Clark et al., 2007). A series of these stocks underwent boom-and-bust cycles, mostly in the space of 5 – 10 years and many have not recovered (Clark et al., 2007). Both pelagic and demersal fisheries also occurred on seamounts and ridges in the southeast Pacific, the East Pacific Rise, the Nazca and Sala-y-Gomez Ridges, and the Chilean Rise. Catches were not large nor were they sustainable (Clark et al., 2007).

In the Southern Ocean, seamounts were fished for nototheniids between 1974 and 1991. In the 1990s, the ridges, plateaus, and seamounts around remote sub-Antarctic islands came to be heavily fished for Patagonian toothfish with trawls and longlines. Initially much illegal, unreported and unregulated (IUU) fishing occurred but has declined significantly since 1996 (Agnew et al., 2009).

Large-scale industrial deep-water fisheries in the North Atlantic date to the development of redfish fisheries in the 1950s using both midwater and demersal trawls over the mid-Atlantic Ridge and on some plateaus. Redfish catches peaked at almost 400,000 tons in the 1950s and have declined considerably but several continue to support some harvest (Koslow et al., 2000; ICES, 2013). Fisheries for roundnose grenadier and Greenland halibut first developed on the upper continental slopes of the Northwest Atlantic in the late 1960s, peaking at over 80,000 tons in 1971 and then rapidly declined and moved to the mid-Atlantic Ridge and Rockall-Hatton Bank in 1973 (Troyanovsky and Lisovsky, 1995; Clark et al., 2007). Several other species have been exploited from the seamounts and ridges of the North Atlantic, including alfonsino, orange roughy, deep-water sharks, ling, blue ling, black cardinalfish, tusk, deep-water crabs and shrimp, and others (Clark et al., 2007; Bensch et al., 2009). From the latter half of the 1990s onwards, declines in catch per unit effort (CPUE) indicated that most targeted North Atlantic deep-water fisheries were overfished and some severely depleted (Koslow et al., 2000; Large et al., 2003; Large and Bergstad, 2005; Devine et al., 2006; Bensch et al., 2009; Rogers and Gianni, 2010).

In 2003, the International Council for the Exploration of the Sea (ICES) deemed that most deep-water stocks "were probably outside safe biological limits." In the last decade fisheries on seamounts and ridges in the Atlantic have declined significantly due to a combination of declining fish populations, significantly altered socioeconomic conditions, and increased regulation by national governments and Regional Fisheries Management Organizations (RFMOs).

Retrospective analyses based on research vessel surveys also indicated that several target and non-target species in the Northwest Atlantic had declined by >90 per cent between 1978 and 1994 (Devine et al., 2006). Many fishery restrictions were implemented and recent survey data indicate biomass levels have stabilized in most surveyed areas (Neat and Burns, 2010). However, recovery for deep-water species with low productive capacity will probably take decades or longer (Baker et al. 2009; Neat and Burns, 2010).

Seamount fisheries in the South Atlantic have been undertaken at a smaller scale than in the North Atlantic. However, there have been seamount fisheries targeting orange roughy, alfonsino, cardinal fish, armorhead, Patagonian toothfish and deep-sea red crab, and some continue to the present day (Rogers and Gianni, 2010; www.seafo.org).

Exploratory trawl fishing on seamounts in the Indian Ocean began in the 1970s targeting shallow-water redbait and rubyfish on the Southwest Indian Ocean Ridge, the Mozambique Ridge and the Madagascar Ridge (Romanov, 2003; Clark et al., 2007) and continued into the mid-1980s. In the late 1990s, trawlers working on the Southwest Indian Ocean Ridge targeted deep-water species, such as orange roughy, black cardinalfish, pelagic armorhead, oreosomatids and alfonsino (Clark et al., 2007), but the fishery rapidly collapsed (Gianni, 2004). Fishing has shifted to the many ridges, seamounts and plateaus targeting a variety of species of deep-sea fish and crustaceans (Clark et al., 2007; Bensch et al., 2009; SWIOFC, 2009).

Overall, deep-water demersal fisheries over the continental slope, ridges, seamounts, and plateaus have landed between 800,000 and

1,000,000 t per annum from the mid-1960s to 1990s (Koslow et al., 2000) and annual landings on the order of 100,000 t since about 1990 (Clark et al., 2007; Watson et al., 2007). The vast majority of seamount-associated demersal fisheries have proven unsustainable, undergoing a boom-and-bust cycle that has usually lasted less than 10 years. Many of the stocks have vulnerable life histories and are small, remote, and difficult to monitor and manage effectively (Koslow, 2007). However, during the last 10-15 years many States and intergovernmental organizations have recognized the need for enhanced management action to protect vulnerable marine species and habitats to facilitate the recovery of depleted stocks.

2.2 Fishing impacts on seamount benthic habitats

Energetic seamount habitats that support substantial fish aggregations also often host diverse, productive benthic habitats dominated by corals, sponges, and associated fauna (Rogers et al., 2007; Samadi et al., 2007). Demersal trawling on seamounts generally removes benthic habitat as by-catch along with the target species or destroys it converting reefs and other structure-forming species to rubble. See chapter 11 for more details on the nature of these impacts. The United Nations General Assembly has been looking into the impact of bottom trawling (e.g. resolutions 61/105, 64/72 and 66/68),[1] although no global assessment has been carried out on the extent of benthic impacts. The documented widespread extent of deep-water trawl fisheries has led to pervasive concern for the conservation of fragile benthic habitats. Moreover, on seamounts where trawling has been discontinued, little regeneration is observed even after five to 10 years (Althaus et al., 2009; Williams et al., 2010) and recovery may require centuries to millennia. Because of the close correspondence between the productivity and diversity of seamount benthic habitats, there are likely few diverse seamount habitats within the vertical range of bottom-trawl fishing that remain pristine, and many have been denuded, their coral and sponge habitats reduced to rubble.

Examples of measures taken by regional fisheries management organizations and arrangements (RFMO/As) to avoid or mitigate fishing impacts on benthic habitats include:

- Both the North Atlantic Fisheries Organization (NAFO) and North East Atlantic Fisheries Commission (NEAFC) in the North Atlantic set quotas for deep-sea stocks based on scientific assessments, and have identified and closed to fishing areas that meet the Food and Agricultural Organization of the United Nations (FAO) criteria for vulnerable marine ecosystems.
- The Southeast Atlantic Fisheries Organization (SEAFO) has closed selected ridge sections and seamounts to fishing, restricted fisheries to certain subareas, and introduced catch quotas (TACs) for the fishes and deep-water crab targeted on seamounts.
- States which participated in the negotiations for the establishment of the North Pacific Fisheries Commission have established interim measures for fisheries management and are working towards stock assessments to modify fishing effort to sustainable levels (Rogers & Gianni, 2010; NPFC, 2014).
- The South Pacific Regional Fisheries Management Organization has called for interim conservation measures, including freezing of the fishing footprint and catch based on historical patterns of fishing which have been implemented by some States (SPRFMO, 2014). Efforts are underway by some member States to map vulnerable marine ecosystems and to assess fisheries data in order to estimate stock biomass and sustainable levels of exploitation (SPRFMO, 2014).
- The Commission for the Conservation of Antarctic Marine Living Resources banned bottom trawl fishing; has restricted remaining fishing opportunities to previously licensed fishing areas or exploratory areas and species specific catch quotas (or total allowable catches (TACs)); and is implementing spatial measures to prevent adverse impacts on bottom-associated vulnerable marine communities.
- The Southern Indian Ocean Deep-sea Fishers Association declared a number of seamounts in the Southern Indian Ocean as voluntary closed areas to fishing although levels of compliance amongst non-members are unknown. With the ratification in 2012 of the Southern Indian Ocean Fisheries Agreement (SIOFA), a new regional fisheries management arrangement for the region is expected to lead to better data collection and regulation of seamount fisheries.

While these actions are progressive, their effectiveness in ensuring the sustainability of exploitation of populations or recovery of vulnerable species and communities is not yet clear (see sections 4-6 below). Indeed whether full closures will result in recovery of vulnerable communities to a former state or a shift to some less desirable community state remains uncertain given current knowledge.

2.3 Pollution

The deep sea was once considered as being too remote from the point sources of industrial pollution for pollution to be a significant issue. However, key contaminants of concern, including mercury and many halogenated hydrocarbons (e.g., DDT, PCBs, and many other pesticides, herbicides, and industrial chemicals) are volatile and enter the ocean predominantly through the atmosphere. These are discussed in Chapter 20. As noted there, concentrations of persistent organic pollutants in deep-sea-dwelling fish may be an order of magnitude higher than in surface-dwelling fish, and the deep sea has been described as one of the ultimate global sinks for such contaminants. Butyl tin, an antifoulant that causes imposex in mollusks, is reported in elevated concentrations in deep-sea organisms, particularly in the vicinity of shipping lanes (Takahashi et al., 1997), and microplastics are now widely reported in deep-sea sediments (van Cauwenberghe et al., 2013).

---

[1] Reports of the Secretary-General on this issue have been issued as A/61/154, A/64/305 and A/66/307.

2.4 Climate change, including acidification and deoxygenation

Predicted shoaling in the depth of calcium carbonate saturation horizons will expose large areas of seamount, ridge, plateau and slope habitat to undersaturated waters (Guinotte et al., 2006). Recent reviews and meta-analyses of the impacts of ocean acidification summarize the present understanding of its effects on cold-water corals (e.g. Wicks and Roberts, 2012), although to date no experimental studies have focused on seamount species. Studies have highlighted the ability of live cold-water corals to maintain calcification at reduced pH (Maier et al., 2009; Form and Riebesell 2012) but synergistic effects with increasing temperature and longer-term effects on resource allocation and reproduction remain unknown. It is becoming clear that deep-water ecosystems may experience more natural variability in carbonate chemistry than was previously supposed (Findlay et al., 2013; Findlay et al., 2014) and that calcareous species can persist even in under-saturated conditions on Tasmanian seamounts (Thresher et al., 2011). However, undersaturated waters will be corrosive to dead coral skeletons that provide structural habitat for many other species, a factor potentially explaining the limited scleractinian coral reef framework on the Hebrides Terrace Seamount (Henry et al., 2014). Increased carbon dioxide and reduced pH may also directly affect marine organisms' physiology, growth, and behaviour (Wicks and Roberts, 2012). It is thus necessary to understand their ecosystem-level impacts, such as the effects of acidification on bioerosion of deep-water corals (Wisshak et al., 2012).

Global climate models predict that oxygen concentrations will decline in the deep ocean due to decreased ventilation (a warmer ocean will be a more stratified ocean) and decreased oxygen solubility at warmer temperatures (Sarmiento et al., 1998; Matear and Hirst, 2003; Shaffer et al., 2009). Over the past 20 years, oxygen concentrations have declined in regions around the North Pacific Ocean and the tropical Indian, Atlantic and Pacific Oceans which have pronounced OMZs, with concomitant horizontal and vertical expansion of these OMZs (Whitney et al. 2007, Bograd et al., 2008; Stramma et al., 2008; Keeling et al., 2010). Benthic communities are dramatically affected where OMZs impinge on seamounts, ridges or continental margins, with greatly reduced biomass and biodiversity (Wishner et al., 1990; Levin 2003; Stramma et al., 2010). Deoxygenation may also affect deepwater benthic organisms indirectly through habitat loss and declining food availability. Midwater fishes, the primary food of many deepwater squid and fish species, including orange roughy, declined ~60 per cent during recent periods of low-oxygen availability in the California Current (Koslow et al., 2011). Palaeoceanographic studies have pointed to the significance of perturbations in oxygen concentration in controlling deep coral occurrence in the Eastern Mediterranean (Fink et al. 2012) and on seamounts (Thiagarajan et al., 2013). Most major marine mass extinction events in the geological past are associated with anoxia and acidification (Harnik et al., 2012).

2.5 Mining

There is the possibility of future mining of cobalt-rich ferromanganese crusts on the bare volcanic rock of seamounts, ridges and plateaus found particularly on seamounts within the exclusive economic zones of island States in the western equatorial Pacific (Rona, 2003; see Chapter 23). Significant differences have been found in the communities inhabiting cobalt-crust-hosting seamounts in the northern Pacific and seamounts outside of the cobalt-rich zone (Schlacher et al., 2013). These differences are not related to species richness but more to the relative abundance of species and community composition in the cobalt-crust rich areas versus non-cobalt rich areas. A high level of heterogeneity amongst the seamounts in terms of their biological communities within the region was also found. Thus, although it is suggested that mining operations will severely affect a small percentage of available seamount area (Hein et al., 2010), predicting the impacts of such activities will be complicated and precautionary spatial management of mining activities based on scientific information will be required.

2.6 Dumping

Although in the past dumping has been a significant issue in the deep sea (Thiel, 2003), it has not been a major issue for most of the habitats treated here, except trenches. Pharmaceutical dumping was permitted in the Puerto Rico Trench from 1973-78, with some 378,000 tons dumped. Impact was noted on the microbial community and invertebrates, and the dumping was halted in early 1980s (reviewed in Ramirez-Llodra et al., 2010). Dumping of radioactive waste in trenches was banned in the 1990s (see Chapter 24).

3 Social and economic considerations, including capacity-building needs.

To date, deep-water fisheries comprise the primary documented direct contribution of seamounts, ridges, canyons, and plateaus to human social and economic wellbeing. Estimating this contribution is challenging, and it is certain that other ecosystem services are provided by these ecosystems that have not been specifically identified or in any way valued.

Deep-water fisheries exploiting resources associated with seamounts, ridges, and plateaus are a relatively minor component of global fisheries, comprising about 1 per cent of total annual landings (Koslow et al., 2000; Gianni, 2004; Clark et al., 2007; Watson et al., 2007; Bensch et al., 2009; Sumaila et al., 2010). High-seas bottom-trawl fisheries are carried out predominantly by a few developed countries of Asia, Oceania and Europe: Australia, Denmark (Faroe Islands), Estonia, France, Iceland, Japan, Latvia, Lithuania, New Zealand, Republic of Korea, Russian Federation, Spain and Ukraine (Gianni 2004; Sumaila et al., 2010). These fisheries account for about 2 per cent of the total landings and about 3 per cent of the landed value for these countries' fisheries. Total subsidies (fuel and non-fuel) for the high-seas bottom-trawl fisheries are estimated to be about 25 per cent of their value and substantially more than their net profit (Sumaila et al., 2010). Gianni (2004) estimated that high-seas bottom trawling occupied the equivalent of 100-200 vessels full time out of a total of 3.1 million fishing vessels worldwide.

Deep-water trawling affects the sedentary species on the continental shelf of countries, which can extend beyond 200 nautical miles. Many of these are developing countries which may need capacity-building in this regard (Gianni 2004). Fisheries associated with seamounts and other submarine topographic features are exceptionally difficult to manage sustainably. Capacity-building is desirable in the areas of stock assessment and sustainable management where such fisheries occur around developing States. This may require investment in international infrastructure (e.g. fisheries research vessels) that is often unavailable to developing States as well as investment at a national level (e.g. fisheries research laboratories and scientists, fisheries ministries). An example of the former is the Nansen Programme that has operated around the Africa Coast and in the Indian Ocean for several decades.

If mining of seamounts proceeds, multi-sectoral management will be needed, in particular the need to balance mining and fishery interests with those of conservation. Within areas under national jurisdiction, the agencies that manage mineral resources generally have no authority to manage the exploitation of living marine resources. The same is true of the International Seabed Authority with regard to the "Area".

4   Management and conservation of the habitat and its resources.

Several scientific reviews of deep-water fisheries over seamounts, ridges, and other abrupt topographies in deep waters have called attention to serious deficiencies in their management and conservation (Koslow et al., 2000; Koslow, 2007 (Chapter 10); Clark et al., 2007; Clark 2009; Rogers and Gianni 2010; Norse et al., 2012). Key contributing factors include the life-history characteristics of many exploited and bycatch species: extreme longevity, late maturity, slow growth, and infrequent recruitment events. These characteristics lead to low productivity and high vulnerability to over-exploitation. Low productivity in itself promotes unsustainable harvest practices based on economic incentives to liquidate a relatively unprofitable resource (Clark, 1973). In addition, many of the stocks are small and aggregated predominantly over isolated features, such as seamounts, making the populations highly vulnerable to serial depletion. The fisheries are carried out predominantly with bottom trawls, which are severely destructive to the fishes' associated benthic habitats, which are often dominated by structure-forming taxa such as corals and sponges.

Deep-sea fisheries within the EEZs of coastal States are in a varied state including depleted, overfished and sustainably fished at the present time (e.g. ICES, 2014). Low-productivity deep-sea species have tended to fall into the former two categories. Managament of deep-water fisheries on the high seas, where many of the stocks occur, is more complicated and difficult. To address this situation, the United Nations General Assembly adopted resolutions 61/105, 64/72 and 66/68 in 2006, 2009 and 2011, respectively, calling on RFMOs and States to manage high-seas deep-water fisheries sustainably through the application of the precautionary approach to fisheries management, and in 2008 the FAO Guidelines for the management of deep-sea fisheries on the high seas were adopted (FAO, 2009; Norse et al., 2012). These resolutions of the General Assembly recommended that impact assessments be carried out prior to the development of new fisheries and steps taken, such as setting aside reserves, or eliminating damaging forms of fishing from sensitive areas, to ensure the conservation of vulnerable habitat. This approach should be adopted prior to mining development as well.

Because of deep-water fishing impacts on sensitive benthic habitats, many States and RFMOs have set aside portions of such habitats as marine reserves or bottom-fishing closures in the Atlantic and North and South Pacific Oceans (e.g. NAFO, NEAFC and SEAFO; Clark and Dunn, 2012). In general, many of the areas that were protected or are planned to be protected through area-based management tools by States and by RFMOs are located in areas remote from where commercial activities occur or are expected to fail to protect those species, communities and habitats most threatened (Devillers et al., 2014). Given the difficulties of surveying deep-water habitats, predictive habitat models may prove useful to identify areas that might be designated for protective measures to meet conservation goals (Taranto et al., 2012; Yesson et al., 2012) as recently put forward in SPRFMO (SPRFMO, 2014).

However, Rogers and Gianni (2010) have provided examples of inadequate implementation of the FAO Guidelines. The need to review the efficacy of conservation measures has been underlined (e.g., the details of move-on rules: Auster et al., 2011). In 2013 and 2014 several RFMOs revised their relevant management measures. Reports on actions taken by States and RFMOs in response to United Nations General Assembly resolutions 64/72 and 66/68 are due to be reviewed by the General Assembly in 2016. We believe this is likely to be the case for many deep-water marine reserves, although this issue has not been specifically addressed for the deep sea.

5   Integrated assessment of the status of the habitat. Cross-cutting and emergent conclusions.

The development of deep-water fisheries, particularly those carried out across wide areas with bottom trawls over seamounts, ridges, canyons and plateaus at upper bathyal depths, have been one of the most transformative human impacts affecting such areas of the global ocean in the latter half of the twentieth century (Ramirez-Llodra et al., 2011). Impacts appear to have been greatest on shallow seamounts, because the depth, biological and physical conditions (e.g. accelerated current flows) support fish resources that can easily be targeted. However, a continuum of physical conditions and biological communities is found on these types of features. All of these communities have been subject to deep-water fisheries and their impacts.

The vast majority of deep-water fisheries have been carried out unsustainably, or at least without satisfactory assessments of impacts and sustainability. This has led to the serial depletion of dozens of stocks from about a dozen species commercially harvested from these habitats. Severe impacts have been reported for by-catch species, including other fishes and benthic invertebrates from the diverse coral and sponge communities found on these communities. The extent of benthic impacts has been described for local fishing grounds but has not been assessed globally; however, if the impacts of these regional studies are generalized, we can extrapolate that fishing, and in particular deep-water trawling, has caused severe, widespread, long-term destruction of these environments globally. The time scale for recovery of deep-water reef habitats is unknown but has been estimated to be on the order of centuries to millennia. Although progress has been made toward sustainable management and conservation of fish stocks and associated diverse, vulnerable benthic communities, numerous studies show that progress to date has not been adequate, with fisheries often closed or limited only after severe depletion has already occurred.

Extractive industries, such as mining and oil and gas development, are generally required to carry out baseline monitoring and assess their environmental impacts prior to development. The General Assembly has called upon States to strengthen the procedures for carrying out assessments on the impacts of bottom fishing activities on vulnerable marine ecosystems and to make the assessments publicly available; not all are currently available. Deep-water fishing has until recently been permitted to proceed in areas of highly diverse and vulnerable ecosystems without consideration of environmental impacts (despite the known, highly destructive impacts of these activities). There is an urgent need, even at this late date, for baseline monitoring of seamount habitats in regions of ongoing and potential fishing activity, and in areas set aside for protection. It is critical that relevant scientific expertise is engaged to develop representative networks of marine protected areas, which may then require the development of conservation measures, such as banning of destructive fishing methods. Reviews of existing conservation measures could address whether conservation and management objectives are met. Devillers et al. (2014) consider that high-seas MPAs must address conservation needs and not be merely designated in areas remote from commercial interests.

The impacts of pollution and climate change, including deoxygenation and acidification, remain poorly understood but are potentially severe. There is an urgent need for research to examine how these factors may potentially influence deep-water benthic communities.

Mining of cobalt-rich ferromanganese crusts from these habitats has been mooted but remains uncertain. Mining would remove all benthic organisms where crust is removed and potentially affect a larger area through sediment mobilization. The environmental impact of mining operations would need to be considered carefully, considering the impacts on regional fisheries and benthic communities, and consideration given to setting aside areas for conservation.

6   Gaps in scientific knowledge.

Since 2000, scientific interest in the ecology of seamounts, ridges, and other sensitive submarine benthic habitats has burgeoned. The development of deep-water fisheries and the need to understand and manage these fisheries and their environmental impacts stimulated this interest, with support from international programmes like the Census of Marine Life. However, these habitats are vast, as well as difficult and expensive to study and the research to date has been largely exploratory. Even after more than a decade of scientific activity, it is apparent that these habitats still remain relatively poorly known. Only a few hundred of the $10^5 - 10^7$ seamounts have been sampled, and the rate of discovery of new species still has not levelled off.

Predictive habitat models have recently been developed to indicate where such habitats are likely to occur, but no one has attempted to assess their present status or the global impact of deep-water trawling, and it is doubtful that the data exist to do so. These models remain largely untested; it is urgently necessary to ground-truth them. Furthermore, knowledge of deep pelagic ecosystems is especially poor (Webb et al., 2010) and the ecological interactions between the geological features considered here and the overlying water column comprise a substantial gap.

Deep-sea ecosystems associated with seamounts, ridges, and other topographic features are now and will increasingly be subjected to multiple stressors from habitat disturbance, pollutants, climate change, acidification and deoxygenation. Studies to date on these impacts have been limited and considered in isolation. The scientific understanding of how these stressors may interact to affect marine ecosystems remains particularly poorly developed. For example, the widespread destruction of deep-water benthic communities due to trawling has presumably reduced their ecological and evolutionary resilience as a result of reduced reproductive potential and loss of genetic diversity and ecological connectivity. The synergistic influence of these factors is unknown at present.

Although it is heartening that some seamounts, ridges and other sensitive marine habitats are being protected by fishing closures, Marine Protected Areas and other actions, little scientific understanding of the efficacy of actions implemented to date and few studies to assess this exist. The connectivity between these habitats remains largely unknown, as are the factors that influence colonization, species succession, resilience and variability. Comparative studies of seamount, canyon, and continental margin habitats seem to indicate that many species are shared (but see Richer de Forges et al., 2000); however, community structure differs markedly and the factors influencing such differences remain unknown (McClain et al., 2009). Our starting point in attempting to understand and manage these habitats is, to paraphrase Socrates, that we know almost nothing.

References

Agnew, D.J., Pearce, J., Pramod, G., Peatman, T., Watson, R., Beddington, J.R., Pitcher, T.J. (2009). Estimating the worldwide extent of illegal fishing. *PLoS ONE* 4(2): e4570.

Althaus, F., Williams, A., Schlacher, T.A., Kloser, R.J., Green, M.A., Barker, B.A., Bax, N.J., Brodie, P., Schlacher-Hoenlinger, M.A. (2009). Impacts of bottom trawling on deep-coral ecosystems of seamounts are long-lasting. *Marine Ecology Progress Series* 397: 279-294.

Angel, M.V. (1982). Ocean trench conservation. *The Environmentalist* 2 (suppl 1): 1-17.

Auster, P.J., Gjerde, K., Heupel, E., Watling, L., Grehan, A., Rogers, A.D. (2011). Definition and detection of vulnerable marine ecosystems on the high seas: problems with the "move-on" rule. *ICES Journal of Marine Science* 68: 254–264.

Baker, E.T. and German, C.R. (2013). On the global distribution of hydrothermal vent fields. In: C.R. German, J. Lin & L.M. Parson (Eds.). *Mid-Ocean Ridges*. American Geophysical Union: 245-266.

Baker, K.D., Devine, J.A., Haedrich, R.L. (2009). Deep-sea fishes in Canada's Atlantic: population declines and predicted recovery times. *Environmental biology of fishes* 85(1): 79-88.

Belyaev, G.M. (1989). *Deep-Sea Ocean Trenches and Their Fauna*. Naika Publishing House, Moscow: 385 pp.

Bensch, A., Gianni, M., Gréboval, D., Sanders, J.S., Hjort, A. (2009). Worldwide review of bottom fisheries in the high seas. *FAO Fisheries and Aquaculture Technical Paper* 522. FAO, Rome: 145.

Bergstad, O.A., Falkenhaug, T., Astthorsson, O.S., Byrkjedal, I., Gebruk, A.V., Piatkowski, U., Priede, I.G., Santos, R.S., Vecchione, M., Lorance, P. (2008). Towards improved understanding of the diversity and abundance patterns of the mid-ocean ridge macro-and megafauna. *Deep Sea Research Part II: Topical Studies in Oceanography* 55(1): 1-5.

Blankenship-Williams, L.E. and Levin, L.A. (2009). Living Deep: a synopsis of hadal trench ecology. *Marine Technology Society Journal* 43(5): 137-143.

Bograd, S.J., Castro, C.G., Di Lorenzo, E., Palacios, D.M., Bailey, H., Gilly, W., Chavez, F.P. (2008). Oxygen declines and the shoaling of the hypoxic boundary in the California Current. *Geophysical Research Letters* 35: L12607.

Bruun, A.F. (1956). The abyssal fauna: its ecology, distribution and origin. *Nature* 177: 1105-1108.

Castelin, M., Puillandre, N., Lozouet, P., Sysoev, A., de Forges, B.R., Samadi, S. (2011). Molluskan species richness and endemism on New Caledonian seamounts: Are they enhanced compared to adjacent slopes? *Deep Sea Research Part I: Oceanographic Research Papers* 58(6): 637-646.

Clark, C.W. (1973). The economics of overexploitation. *Science* 181: 630-634.

Clark, M.R. (2009). Deep-sea seamount fisheries: a review of global status and future prospects. *Latin American Journal of Aquatic Research* 37(3): 501-512.

Clark, M.R., Althaus, F., Williams, A., Niklitschek, E., Menezes, G.M., Hareide, N.-R., Sutton, P., O'Donnell, C. (2010a). Are deep-sea demersal fish assemblages globally homogenous? Insights from seamounts. *Marine Ecology* 31: 39-51.

Clark, M.R., Dunn, M.R. (2012). Spatial management of deep-sea seamount fisheries: balancing exploitation and habitat conservation. *Environmental Conservation* 39(2): 204-214.

Clark, M.R. and Koslow, J.A. (2007). Impacts of fisheries on seamounts. In: Pitcher, T.J., Morato, T., Hart, P.J.B. et al (eds.). *Seamounts: Ecology, Conservation and Fisheries*. Blackwell, Oxford, pp. 413-441.

Clark, M.R. and Rowden, A.A. (2009). Effect of deepwater trawling on the macroinvertebrate assemblages of seamounts on the Chatham Rise, New Zealand. *Deep Sea Res. Pt. I* 56: 1540–1554.

Clark, M.R., Rowden, A.A., Schlacher, T., Williams, A., Consalvey, M., Stocks, K.I., Rogers, A.D., O'Hara, T.D., White, M., Shank, T.M., Hall-Spencer, J.M. (2010b). The ecology of seamounts: structure, function, and human impacts. *Annual Review of Marine Science* 2: 253-278.

Clark, M.R., Vinnichenko, V.I, Gordon, J.D.M., Beck-Bulat, G.Z., Kukharev, N.N., Kakora, A.F. (2007). Large-scale distant-water trawl fisheries on seamounts. In: T.J. Pitcher, T.J., Morato, T., Hart, P.J.B. et al (Eds.). *Seamounts: Ecology, Fisheries & Conservation*. Blackwell, Oxford: 361-399.

Consalvey, M., Clark, M.R., Rowden, A.R., Stocks, K.I. (2010). Life on seamounts. In: A.D. McIntyre (Ed). *Life in the world's oceans: diversity, distribution, and abundance*. Wiley-Blackwell, Blackwell Publishing Ltd, UK: 123-138.

Dana, J.D. (1863). *Manual of Geology*, Philadelphia: 798 pp.

De Leo, F.C., Smith, C.R., Rowden, A.A., Bowden, D.A., Clark, M.R. (2010). Submarine canyons: hotspots of benthic biomass and productivity in the deep sea. *Proceedings of the Royal Society B: Biological Sciences* 277(1695): 2783-2792.

Devillers, R., Pressey, R.L., Grech, A., Kittinger, J.N., Edgar, G. J., Ward, T., Watson, R. (2014). Reinventing residual reserves in the sea: are we favouring ease of establishment over need for protection? *Aquatic Conservation: Marine and Freshwater Ecosystems*. DOI: 10.1002/aqc.

Devine, J.A., Baker, K.D., Haedrich, R.L. (2006). Deep-sea fishes qualify as endangered. *Nature* 439: 29.

Etnoyer, P.J., Wood, J., Shirley, T.C. (2010). How large is the seamount biome? *Oceanography* 23: 206-209.

FAO (2009). International Guidelines for the Management of Deep-sea Fisheries in the High Seas. Directives internationales sur la gestion de la pêche profonde en haute mer. Directrices Internacionales para la Ordenación de las Pesquerías de Aguas Profundas en Alta Mar. FAO, Rome: 73 pp.

Findlay, H.S., Hennige, S.J., Wicks, L.C., Navas, J.M., Woodward, E.M.S., Roberts, J.M. (2014). Fine-scale nutrient and carbonate system dynamics around cold-water coral reefs in the northeast Atlantic. *Nature Scientific Reports* 4: 3671.

Findlay, H.S., Wicks, L., Navas, J.M., Hennige, S., Huvenne, V., Woodward, E.M.S., Roberts, J.M. (2013). Tidal downwelling and implications for the carbon biogeochemistry of cold-water corals in relation to future ocean acidification and warming. *Global Change Biology* 19: 2708-2719.

Fink, H.G., Wienberg, C., Hebbeln, D., McGregor, H.V., Schmiedl, G., Taviani, M., Freiwald A. (2012). Oxygen control on Holocene cold-water coral development in the eastern Mediterranean Sea. *Deep-Sea Research Part I Oceanographic Research Papers* 62: 89-96.

Form, A.U., Riebesell, U. (2012). Acclimation to ocean acidification during long-term CO_2 exposure in the cold-water coral *Lophelia pertusa*. *Global Change Biology* 18: 843-853.

Fujii, T., Jamieson, A.J., Solan, M., Bagley, P.M., Priede, I.G. (2010). A large aggregation of liparids at 7703 meters and a reappraisal of the abundance and diversity of hadal fish. *BioScience* 60(7): 506-515.

Gianni, M. (2004). High seas bottom fisheries and their impact on the biodiversity of vulnerable deep-sea ecosystems: summary findings. IUCN, Gland, Switzerland: 83 pp.

Guinotte, J., Orr, J., Cairns, S., Freiwald, A., Morgan, L., George, R. (2006). Will human-induced changes in seawater chemistry alter the distribution of deep-sea scleractinian corals? *Frontiers in Ecology and the Environment* 4(3): 141-146.

Harnik, P.G., Lotze, H.K., Anderson, S.C., Finkel, Z.V., Finnegan, S., Lindberg, D.R., Liow, L.H., Lockwood, R., McClain, C.R., McGuire, J.L., O'Dea, A., Pandolfi, J.M., Simpson, C., Tittensor, D.P. (2012). Extinctions in ancient and modern seas. *Trends in Ecology and Evolution* 27: 608–617.

Harris, P.T., Macmillan-Lawler, M., Rupp, J., Baker, E.K. (2014). Geomorphology of the oceans. *Marine Geology*.

Harris, P.T. and Whiteway, T. (2011). Global distribution of large submarine canyons: geomorphic differences between active and passive continental margins. *Marine Geology* 285: 69–86.

Hein, J.R., Conrad, T.A., Staudigel, H. (2010). Seamount mineral deposits, a source of rare-metals for high technology industries. *Oceanography* 23: 184-189.

Henry, L.-A., Vad, J., Findlay, H.S., Murillo, J., Milligan, R., Roberts, J.M. (2014). Environmental variability and biodiversity of megabenthos on the Hebrides Terrace Seamount (Northeast Atlantic). *Nature Scientific Reports* 4: 5589.

Howell, K.L., Mowles, S.L., Foggo, A. (2010). Mounting evidence: near-slope seamounts are faunally indistinct from an adjacent bank. *Marine Ecology* 31: 52-62.

Huvenne, V.A.I., Davies, J.S. (2013). DSRII Special Issue: Towards a new and integrated approach to submarine canyon research. *Deep Sea Research Part II: Topical Studies in Oceanography*.

ICES (2013). Report of the Working Group on Biology and Assessment of Deep-sea Fisheries Resources ICES: 963 pp.

IHO (2008). Standardization of Undersea Feature Names: Guidelines Proposal for Terminology, Monaco: 32 pp.

Jamieson, A.J., Fujii, T., Mayor, D.J., Solan, M., Priede, I.G. (2010). Hadal trenches: the ecology of the deepest places on Earth. *Trends in Ecology & Evolution* 25(3): 190-197.

Jamieson, A.J., Fujii, T., Solan, M., Matsumoto, A.K., Bagley, P.M., Priede, I.G. (2009). Liparid and macrourid fishes of the hadal zone: in situ observations of activity and feeding behaviour. *Proceedings of the Royal Society B: Biological Sciences* 276(1659): 1037-1045.

Jamieson, A.J., Lacey, N.C., Lörz, A.N., Rowden, A.A., Piertney, S.B. (2013). The supergiant amphipod *Alicella gigantea* (Crustacea: Alicellidae). from hadal depths in the Kermadec Trench, SW Pacific Ocean. *Deep Sea Research Part II: Topical Studies in Oceanography* 92: 107-113.

Keeling, R.F., Kortzinger, A., Gruber, N. (2010). Ocean deoxygenation in a warming world. *Annual Review of Marine Science* 2: 199–229.

Koslow, J.A. (1996). Energetic and life-history patterns of deep-sea benthic, benthopelagic and seamount-associated fish. *Journal of Fish Biology* 49 Supplement A: 54-74.

Koslow, J.A. (1997). Seamounts and the ecology of deep-sea fisheries. *American Scientist* 85, 168-176.

Koslow, J.A., Boehlert, G.W., Gordon, J.D.M., Haedrich, R.L., Lorance, P., Parin, N. (2000). The impact of fishing on continental slope and deep-sea ecosystems. *ICES Journal of Marine Science* 57: 548-557.

Koslow, J.A., Bulman, C.M., Lyle, J.M. (1994). The mid-slope demersal fish community off southeastern Australia. *Deep-Sea Research* 41: 113-141.

Koslow, J.A., Goericke, R., Lara-Lopez, A., Watson, W. (2011). Impact of declining intermediate-water oxygen on deepwater fishes in the California Current. *Marine Ecology Progress Series* 436: 207–218.

Koslow, J.A., Gowlett-Holmes, K., Lowry, J., O'Hara, T., Poore, G., Williams, A. (2001). The seamount benthic macrofauna off southern Tasmania: community structure and impacts of trawling. *Marine Ecology Progress Series* 213: 111-125.

Koslow, T. (2007). *The Silent Deep*. University of Chicago Press, Chicago: 270 pp.

Kvile, K.Ø., Taranto, G.H., Pitcher, T.J., Morato, T. (2013). A global assessment of seamount knowledge and paradigms. *Biological Conservation*.

Large, P.A. and Bergstad, O.A. (2005). Deepwater fish resources in the northeast Atlantic: fisheries, state of knowledge on biology and ecology, and recent developments in stock assessment and management. In: R. Shotton (Ed) *FAO Fisheries Proceedings*. FAO, Rome: 149-161.

Large, P.A., Hammer, C., Bergstad, O.A., Gordon, J.D.M., Lorance, P. (2003). Deepwater fisheries of the Northeast Atlantic: II. Assessment and management approaches. *Journal of Northwest Atlantic Fishery Science* 31: 151-163.

Levin, L.A. (2003). Oxygen minimum zone benthos: adaptation and community response to hypoxia. *Oceanography and Marine Biology* 41: 1-45.

Maier, C., Hegeman, J., Weinbauer, M.G., Gattuso, J.P. (2009). Calcification of the cold-water coral *Lophelia pertusa*, under ambient and reduced pH. *Biogeosciences* 6: 1671-1680.

Matear, R.J., Hirst, A.C. (2003). Long-term changes in dissolved oxygen concentrations in the ocean caused by protracted global warming. *Global Biogeochemical Cycles* 17(4): 1125.

McClain, C.R., Lundsten, L., Ream, M., Barry, J., DeVogelaere, A. (2009). Endemicity, biogeography, composition, and community structure on a Northeast Pacific seamount. *PLOS Biology* 4. (1): e4141.

Menot, L., Sibuet, M., Carney, R.S., Levin, L.A., Rowe, G.T., Billett, D.S.M., Poore, G., Kitazato, H., Vanreusel, A., Galeron, J., Lavrado, H.P., Sellanes, J., Ingole, B., Krylova, E. (2010). New perceptions of continental margin biodiversity. In: A.D. McIntyre (Ed). *Life in the World's Oceans*. Blackwell Publishing Ltd: 79-101.

Mironov, A.N., Gebruk, A.V., Southward, A.J. (2006). *Biogeography of the North Atlantic seamounts*. KMK Scientific Press, Moscow: 196 pp.

Neat, F. and Burns, F. (2010). Stable abundance, but changing size structure in grenadier fishes (Macrouridae) over a decade (1998–2008) in which deepwater fisheries became regulated. *Deep-Sea Research, Part I* 57: 434-440.

Norse, E.A., Brooke, S., Cheung, W.W.L., Clark, M.R., Ekeland, I., Froese, R., Gjerde, K.M., Haedrich, R.L., Heppell, S.S., Morato, T., Morgan, L.E., Pauly, D., Sumaila, R., Watson, R. (2012). Sustainability of deep-sea fisheries. *Marine Policy* 36(2): 307-320.

NPFC (2014). Record of the 6th Session of the Preparatory Conference for the North Pacific Fisheries Commission. SWG12/WP4/J: Biological synopsis of the North Pacific armorhead, *Pseudopentaceros wheeleri* Hardy, 1983: current status of knowledge and information gap.

Pakhorukov, N.P. (2005). Behavior and distribution of bottom and near-bottom fish on the Emperor Seamount Chain (the Pacific Ocean). *Journal of Ichthyology* 45: 103-110.

Parin, N.V., Mironov, A.N., Nesis, K.N. (1997). Biology of the Nazca and Sala y Gomez submarine ridges, an outpost of the Indo-West Pacific fauna in the Eastern Pacific Ocean: composition and distribution of the fauna, its communities and history. *Advances in Marine Biology* 32: 145-242.

Perez, J.A.A., dosSantos Alves, E., Clark, M.R., Bergstad, O.A., Gebruk, A., Cardoso, I.A., Rogacheva, A. (2012). Patterns of life on the South Atlantic mid-oceanic ridge: compiling what is known and addressing future research. *Oceanography* 25(4): 16-31.

Priede, I.G., Bergstad, O.A., Miller, P.I., Vecchione, M., Gebruk, A., Falkenhaug, T., Billett, D.S.M., Craig, J., Dale, A.C., Shields, M.A. (2013). Does presence of a mid-ocean ridge enhance biomass and biodiversity? *PLoS ONE* 8(5): e61550.

Ramirez-Llodra, E., Brandt, A., Danovaro, R., De Mol, B., Escobar, E., German, C.R., Levin, L.A., Arbizu, P.M., Menot, L., Buhl-Mortensen, P., Narayanaswamy, B.E.,

Smith, C.R., Tittensor, D.P., Tyler, P.A., Vanreusel, A., Vecchione, M. (2010). Deep, diverse and definitely different: unique attributes of the world's largest ecosystem. *Biogeosciences* 7(9): 2851-2899.

Ramirez-Llodra, E., Tyler, P.A., Baker, M.C., Bergstad, O.A., Clark, M.R., Escobar, E., Levin, L.A., Menot, L., Rowden, A.A., Smith, C.R., Van Dover, C.L. (2011). Man and the last great wilderness: human impact on the deep sea. *PLoS ONE* 6(8): e22588.

Richer de Forges, B., Koslow, J.A., Poore, G.C.B. (2000). Diversity and endemism of the benthic seamount fauna in the southwest Pacific. *Nature* 405: 944-947.

Rogers, A.D. (1994). The biology of seamounts. *Advances in Marine Biology* 30: 305-350.

Rogers, A.D., Baco, A., Griffiths, H., Hart, T., Hall-Spencer, J.M. (2007). Corals on seamounts. In: T.J. Pitcher, T. Morato, P.J.B. Hartet al (Eds.). *Seamounts: ecology, fisheries and conservation*. Blackwell, Oxford: 141-169.

Rogers, A.D. and Gianni, M. (2010). The Implementation of UNGA Resolutions 61/105 and 64/72 in the Management of Deep-Sea Fisheries on the High Seas. Deep-Sea Conservation Coalition, International Programme on State of the Ocean, London: 97 pp.

Romanov, E.V. (2003). Summary and review of Soviet and Ukrainian scientific and commercial fishing operations on the deepwater ridges of the Southern Indian Ocean. *FAO Fisheries Circular* FAO: 84 pp.

Rona, P.A. (2003). Resources of the sea floor. *Science* 299: 673-674.

Samadi, S., Bottan, L., Macpherson, E., Richer de Forges, B., Boisselier, M.-C. (2006). Seamount endemism questioned by the geographic distribution and population genetic structure of marine invertebrates. *Marine Biology* 149: 1463-1475.

Samadi, S., Schlacher, T.A., Richer de Forges, B. (2007). Seamount benthos. In: T.J. Pitcher, T. Morato, P.J. Hartet al (Eds.). *Seamounts: Ecology, Fisheries & Conservation*. Blackwell, Oxford: 119-140.

Sarmiento, J.L., Hughes, T.M.C., Stouffer, R.J., Manabe, S. (1998). Simulated response of the ocean carbon cycle to anthropogenic climate warming. *Nature* 393: 245-249.

Schlacher, T.A., Baco, A.R., Rowden, A.A., O'Hara, T.D., Clark, M.R., Kelley, C., Dower, J.F. (2013). Seamount benthos in a cobalt-rich crust region of the central Pacific: conservation challenges for future seabed mining. *Diversity and Distributions*: DOI: 10.1111/ddi.12142.

Shaffer, G., Olsen, S.M. Pedersen, O.P. (2009). Long-term ocean oxygen depletion in response to carbon dioxide emissions from fossil fuels. *Nature Geoscience* 2: 105-109.

Shepard, F.P. (1963). *Submarine Geology*. Harper & Row, New York, 557 pp.

Silva, H.M., Pinho, M.R. (2007). Small-scale fishing on seamounts. In: T.J. Pitcher, Morato, T., Hartet, P.J.B., al (Eds.). *Seamounts: Ecology, Fisheries & Conservation*. Blackwell Publishing, Oxford: 335-360.

Smith, D.K. (1991). Seamount abundances and size distributions, and their geographic variations. *Reviews in Aquatic Sciences* 5(3-4): 197-210.

Stocks, K.I., Hart, P.J. (2007). Biogeography and diversity of seamounts. In: T.J. Pitcher, Morato, T., Hartet, P.J.B. al (Eds.). *Seamounts: Ecology, Fisheries & Conservation*. Blackwell, Oxford, pp. 255-281.

Stramma, L., Johnson, G.C., Sprintall, J., Mohrholz, V. (2008). Expanding oxygen-minimum zones in the tropical oceans. *Science* (Washington D C). 320: 655-658.

Stramma, L., Schmidtko, S., Levin, L.A., Johnson, G.C. (2010). Ocean oxygen minima expansions and their biological impacts. *Deep Sea Research I*. 57(4): 587-595.

Sumaila, U.R., Khan, A., Teh, L., Watson, R., Tyedmers, P., Pauly, D. (2010). Subsidies to high seas bottom trawl fleets and the sustainability of deep-sea demersal fish stocks. *Marine Policy* 34(3): 495-497.

SWIOFC (2009). Report of the Third Session of the Scientific Committee. South West Indian Ocean Fisheries Commission, Maputo, Mozambique: 85pp.

Takahashi, S., Tanabe, S., Kubodera, T. (1997). Butyltin residues in deep-sea organisms collected from Sugara Bay, Japan. Environmental Science and Technology 31, 3103-3109.

Taranto, G.H., Kvile, K.Ø., Pitcher, T.J., Morato, T. (2012). An ecosystem evaluation framework for global seamount conservation and management. *PLoS ONE* 7(8): e42950.

Thiagarajan, N., Gerlach, D., Roberts, M.L., Burke, A., McNichol, A., Jenkins, W.J., Subhas, A.V., Thresher, R.E., Adkins, J.F. (2013). Movement of deep-sea coral populations on climatic timescales. *Paleoceanography* 28: 227-236.

Thiel, H. (2003). Anthropogenic impacts on the deep sea. In: Tyler, P.A. (Ed) *Ecosystems of the Deep Oceans*. Elsevier, Amsterdam: 427-471.

Thresher, R.E., Tilbrook, B., Fallon, S., Wilson, N.C., Adkins, J. (2011). Effects of chronic low carbonate saturation levels on the distribution, growth and skeletal chemistry of deep-sea corals and other seamount megabenthos. *Marine Ecology Progress Series* 442: 87-99.

Troyanovsky, F.M., Lisovsky, S.F. (1995). Russian (USSR) fisheries research in deep waters (below 500 m) in the North Atlantic. In: A.G. Hopper (Ed) *Deep-Water Fisheries of the North Atlantic Oceanic Slope*. Kluwer Academic Publishers, Dordrecht, Netherlands: 357-365.

Van Cauwenberghe, L., Vanreusel, A., Mees, J., Janssen, C.R. (2013). Microplastic pollution in deep-sea sediments. *Environmental Pollution* 182: 495-499.

Vecchione, M., Bergstad, O.A., Byrkjedal, I., Falkenhaug, T., Gebruk, A.V., Godø, O.R., Gislason, A., Heino, M., Høines, Å.S., Menezes, G.M., Piatkowski, U., Priede, I.G., Skov, H., Søiland, H., Sutton, T., Wenneck, T. (2010). Biodiversity patterns and processes on the mid-Atlantic Ridge. In: McIntyre, A.D. (Ed) *Life in the World's Oceans*. Blackwell Publishing Ltd: 361 pp.

Vetter, E.W., Smith, C.R., De Leo, F.C. (2010). Hawaiian hotspots: enhanced megafaunal abundance and diversity in submarine canyons on the oceanic islands of Hawaii. *Marine Ecology* 31(1): 183-199.

Vinogradova, N.G. (1997). Zoogeography of the abyssal and hadal zones. *Advances in Marine Biology* 32: 326-387.

Watson, R., Kitchingman, A., Cheung, W.W.L. (2007). Catches from world seamount fisheries. In: T.J. Pitcher, T. Morato, P.J.B. Hartet al (Eds.). *Seamounts: Ecology, Fisheries & Conservation*. Blackwell, Oxford: 400-412.

Webb, T.J., Berghe, E.V., O'Dor, R. (2010). Biodiversity's big wet secret: the global distribution of marine biological records reveals chronic under-exploration of the deep pelagic ocean. *PLoS ONE* 5(8): e10223.

Wessel, P., Sandwell, D.T., Kim, S.-S. (2010). The global seamount census. *Oceanography* 23: 24-33.

White, M., Bashmachnikov, I., Aristegui, J., Martins, A. (2007). Physical processes and seamount productivity. In: T.J. Pitcher, Morato, T., Hartet, P.J.B., al (Eds.). *Seamounts: Ecology, Fisheries & Conservation*. Blackwell Publishing, Oxford, UK: 65-84.

Whitney, F.A., Freeland, H.J., Robert, M. (2007). Decreasing oxygen levels in the interior waters of the subarctic Pacific. *Progress In Oceanography* 75: 179-199.

Wicks, L., Roberts, J.M. (2012). Benthic invertebrates in a high CO2 world. *Oceanography & Marine Biology: An Annual Review* 50: 127-188.

Williams, A., Schlacher, T.A., Rowden, A.A., Althaus, F., Clark, M.R., Bowden, D.A., Stewart, R., Bax, N.J., Consalvey, M., Kloser, R.J. (2010). Seamount megabenthic assemblages fail to recover from trawling impacts. *Marine Ecology* 31: 183-199.

Wilson, R., Smith, K.L., Rosenblatt, R.H. (1985). Megafauna associated with bathyal seamounts in the central North Pacific Ocean. *Deep-Sea Research* 23: 1243-1254.

Wilson, R.R., Kaufmann, R.S. (1987). Seamount biota and biogeography. In: B.H. Keating, P. Fryer, R. Batiza & G.W. Boehlert (Eds.). *Seamounts, Islands and Atolls*. American Geophysical Union, Washington: 355-377.

Wishner, K., Levin, L., Gowing, M., Mullineaux, L. (1990). Involvement of the oxygen minimum in benthic zonation on a deep seamount. *Nature* 346: 57-59.

Wisshak, M., Schönberg, C.H.L., Form, A., Freiwald, A. (2012). Ocean acidification accelerates reef bioerosion. *PLoS ONE* 7(9): e45124.

Yesson, C., Clark, M.R., Taylor, M.L., Rogers, A.D. (2011). The global distribution of seamounts based on 30 arc seconds bathymetry data. *Deep Sea Research Part I* 58: 442–453.

Yesson, C., Taylor, M.L., Tittensor, D.P., Davies, A.J., Guinotte, J., Baco, A., Black, J., Hall-Spencer, J.M., Rogers, A.D. (2012). Global habitat suitability of cold-water octocorals. *Journal of Biogeography* 39(7): 1278-1292.

C Environmental, economic and/or social aspects of the conservation of marine species and habitats and capacity-building needs

52 Synthesis of Part VI: Marine Biological Diversity and Habitats

Group of Experts:
Jake Rice (Lead Member and Editor of Part VI)

Chapter 52

Synthesis of Part VI: Marine Biological Diversity and Habitats

1 Biodiversity itself

Biodiversity has natural patterns globally, at all levels from phytoplankton to top predators, including fish, marine reptiles, seabirds, and marine mammals. Main factors that underlie these patterns include depth and proximity to coastline, latitude, habitat complexity and primary productivity, temperature and substrate (Chapter 34). These patterns occur on many scales from meters to full ocean basins; the mosaic structure of seafloor benthic biodiversity is often particularly strong. Some types of species are particularly widespread and/or have specialized life history characteristics, making them even more vulnerable to threats and pressures than most other species. They receive special attention in both characterizing the factors that determine their patterns of distribution and in assessing their trends and the associated pressures.

Chapter 35 highlights that although many such patterns are well documented, the ocean's diversity of species, communities and habitats is far from completely sampled. As research continues, new species, new patterns of distribution, and new relationships between components of biodiversity and natural and anthropogenic drivers are being discovered. The incompleteness of our knowledge of biodiversity and the factors that affect it means that decision-making about potential impacts will be subject to high uncertainty, and the application of precaution is appropriate. Nevertheless, as documented below, a central message from Part VI is that detrimental trends in biodiversity on many scales can be at least mitigated, and sometimes eliminated, even when knowledge is incomplete, if the available knowledge is enough to use in choosing appropriate measures and the capacity for implementation of the measures is available.

1.1 Biodiversity hotspots

Although nearly all parts of the ocean support marine life, biodiversity hotspots exist where the number of species and the abundance and/or concentration of biota are consistently high relative to adjacent areas. Some are sub-regional, like the coral triangle in the Pacific (Chapter 36D.2.3) and coral reefs in the Caribbean (Chapter 43), cold-water corals in the Mediterranean Sea (Chapter 36A) the deep seas (Chapter 36F) and the Sargasso Sea (Chapter 51). Some are more local and associated with specific physical conditions, such as biodiversity-rich habitat types. This Assessment has several chapters on these types of special habitats, such as hydrothermal vents (Chapter 45), cold-water (Chapter 42) and warm-water (chapter 43) corals, seamounts and related deep-sea habitats (chapter 51), and the sea-ice zone (Chapter 46), that highlight some of the main factors making an area richer in biodiversity than adjacent areas. Key drivers of biodiversity are complex three-dimensional physical structures that create a diversity of physical habitats (e.g., Chapters 42, 43 on corals, 44 on hydrothermal vents, 51 on seamounts), dynamic oceanographic conditions causing higher bottom-up productivity (e.g., the North Pacific Transition Zone discussed in Chapter 36C, the eastern boundary currents discussed in Chapter 36B.1 [Benguela Current]

and 36D.1 [Humboldt Current], and the ice front of the Southern Ocean (Chapter 36H.2.1), as well as the special seasonality of production in the Southern Ocean benthic communities (Chapter 36H2.4)), effects of land-based inputs extending far out to sea (36B1 [Congo, Plata and Amazon Rivers]) and special vegetation features creating unique and productive habitats nearshore (e.g., kelp forests, Chapter 47; mangroves, Chapter 48; salt marshes, Chapter 49; and offshore (Sargasso Sea, Chapter 50).

2 Bridge to trends and impacts

These habitat-based hotpots are of double concern for the following reasons. As hotspots they support high absolute and relative levels of biodiversity, unique species adapted to their special features, and often serve as centres for essential life history processes of species with wider distributions (e.g., lagoons and spawning beaches for sea turtles, Chapter 39.4.1; upwelling and similar high-productivity centres for foraging seabirds, Chapter 38.3; haul-out sites for pinnipeds, Chapter 37.3.3.2); mangroves (Chapter 48); seagrasses (Chapter 47); estuaries (Chapter 44); and cold-water corals (Chapter 42): all harbour juvenile fish which are important for fisheries in adjacent areas.

Sometimes because of the special physical features that contribute to high biodiversity, and sometimes because of the concentration of biodiversity itself, these hotspots are often magnets for human activities; therefore, many societies and industries are most active in the areas that are also biodiversity hotspots. As on land, humanity has found the greatest social and economic benefits in the places in the ocean that are highly productive and structurally complex. For example, of the 32 largest cities in the world, 22 are located in estuaries (Chapter 44), mangroves and coral reefs support small-scale (artisanal) fisheries in developing countries (Chapter 36.D.2.3, Chapters 43, 48), and commercial fishing targets fish aggregations over seamounts (Chapter 51).

These hotspots are also recognized in the scientific and technical information on classes of special habitats, such as Ecologically or Biologically Significant Areas [EBSAs] (CBD) and Vulnerable Marine Ecosystems [VMEs] (FAO), and similar classes of special habitats. These are available to policy-makers and managers in shipping, seabed mining and other sectors. The tendency for biodiversity hotspots to attract human uses and become socio-economic hotspots, with a disproportionate representation of ports and coastal infrastructure (Chapter 18), other coastal land used (Chapter 20), fishing (Chapter 11) and aquaculture (Chapter 12) is one of the major challenges to conservation and sustainable use of marine biodiversity.

3 Trends in biodiversity for species and groups of species

Superimposed on these patterns at all scales are temporal trends. Biodiversity is not static, hence both random variation and multi-year trends would occur without anthropogenic pressures (for example, Chapter 36C, Figures 8,9,10, which show substantial variation in chlorophyll and zooplankton well offshore of the main influences of land-based inputs, and Chapter 36F, Table 36F.1A; Chapter 36D.2.1, 36G.2, 36H.2.1, 36H2.2, showing substantial variation in bottom-up productivity in the open ocean and high-latitude seas where anthropogenic nutrient inputs are not large enough to be major drivers of basin-scale trends).

Human uses of the ocean have imposed much greater temporal trends on all biodiversity components. This Assessment found evidence of these temporal trends due to human drivers in every regional assessment and for all components of biodiversity, with some emergent patterns. They are summarized below.

3.1 Phytoplankton and zooplankton

Natural regime shifts have changed baseline bottom-up productivity to some extent, and the species composition of the phytoplankton and zooplankton to a greater extent (e.g., Chapter 36C, Figures 1, 3, 4) on the change in plankton community composition. Changes in species composition of lower trophic levels have broader ecological consequences, because such changes have been found to affect pathways of energy flow to higher levels, affecting species of fish, reptiles, birds, and mammals (e.g., Chapter 36A.7 [Gulf of St. Lawrence], Chapter 36G, Figure 36G.1, both showing changes in food-web structure; Chapter 36A.3, 36D.2.1, 36D.2.2, and 36H.2.2, all showing changes in animal community composition in response to productivity drivers).

In coastal areas, human pressures on bottom-up processes were documented in all the divisions of the ocean described in chapter 36. Scales can be local to, occasionally, that of full semi-enclosed seas; the largest effects are documented where human populations are most dense (Chapter 36A.7 [Mediterranean, Baltic and North Seas], Chapter 36C, Figure 36C.6), but local effects are even seen in high-latitude seas (Chapter 36G [Trends]; 36H.1).

Many documented cases were found where high levels of contaminants or land-based nutrient runoff dramatically reduced diversity of species (Chapter 36C, Figure 36C.1; Chapter 36E.2; see also Chapter 20) or diminished or sometimes eliminated diversity due to hypoxia (see Chapter 36C.2(c) on hypoxia; Chapters 20, 44).

Many documented cases were also found where adoption of appropriate policies to address sources, along with funding for monitoring, correcting problems at source, and when necessary clean-up of affected areas, has reversed these trends and achieved good environmental quality (examples in Chapter 36A.7 [North Sea, Baltic Sea, Chesapeake Bay]).

Further out on shelves and in the open ocean, anthropogenically driven trends in lower trophic levels were less conclusively documented *except for climate change* (examples in Chapter 36C, Figures 36C-8, 9; 36D.2.1, 36G). Documentation of effects seems to emerge more strongly as monitoring continues. When direct impacts of pressures were found, causes generally were due to changes in temperature, water masses (currents, upwelling) and seasonality (examples in Chapters 36A.2, 36B.2, Table 36B.1; 36C.1, 36D.2.1, 36E.2) and acidification (Chapter 6.5.3).

A second pervasive pressure on phytoplankton and zooplankton in all parts of the ocean, from coastal embayments to open-ocean areas, are alien invasive species. Ocean physical transport processes and incidental transfer by highly migratory species have always resulted in the possibility that new species would be introduced into an area, become established, and alter energy flows and community structure. As climate change affects temperature and salinity conditions in the ocean, species also may respond with changes in range; the relatively cold-adapted species in a community withdraws towards the poles and species relatively more adapted to warmer conditions expand their range towards the poles. Both types of range changes again can affect the patterns of productivity and community relationships in the areas experiencing changes in species composition of lower trophic levels (e.g., Chapter 36D.2.2). Active, albeit unintentional, transport of species with shipping (Chapter 17) and occasionally tourism can lead to invasions of species across basins and sometimes even greater distances. It is a largely academic argument whether changes associated with natural transport processes and climate-related changes in physical ocean conditions are "invasions", but the impacts on system structure and dynamics can range from negligible to dramatic.

Cases are documented in all regions of alien species becoming established in new areas, and a portion of such invasions have caused almost complete restructuring of the plankton communities at scales from bays to semi-enclosed seas, with consequences for the biodiversity of all higher trophic levels (Chapter 36A.7 [Black Sea]; 36C, Figure 36C.7).

3.2 Benthos

Temporal trends have been less widely documented because of the more local scale of patterns in seafloor biota. Quantifying trends requires expensive and local sampling, which has been undertaken for the most part only in rich countries, in restricted sites close to coasts, and often where problems are already thought to exist, usually due to human pressures.

In cases where appropriate monitoring has occurred, trends in benthos are commonly associated with human pressures. Causes include direct removals for harvesting (Chapter 11), indirect impacts due to fishing gear and aggregate extraction (Chapter 36A.7 [North Sea]; 36D.2.3; Chapters 11, 42, 43 [corals], 44 [estuaries], 51 [seamounts]), and indirect effects due to pollution, sedimentation, etc. For example, loss of coral cover has been linked to catchment disturbance (Chapter 36D.3; Chapter 43), and species loss due to pollution is widespread in many estuaries (Chapter 44). Salt marshes have been drained, diked, ditched, grazed, sprayed for mosquito control, and invaded by a range of non-native species that have altered their ecology (Chapter 50). Many examples were found of high pollution, etc., altering benthic communities extensively and changing both species composition and biomass/productivity (Chapter 36A.7, 36B.4 [hydrocarbons]; 36C.2b; Chapters 20, 44). Trends in benthic populations or communities are often used as indicators for effects monitoring, because some benthos are sensitive to specific pressures and have high local patchiness of occurrence in specific response to those particular pressures.

This Assessment also contains many documented cases where adoption of appropriate policies to address sources, along with funding for monitoring, reducing the threat at source, and when necessary taking actions to remediate or restore damaged populations, communities or habitats, have reversed these trends and achieved good environmental quality (Chapter 36A.4.b, 36B.4.3, 36D). For example, coral-reef fish populations have been shown to recover within MPAs after they have been declared (Chapter 43) and management of shrimp aquaculture that prohibits clearing of mangroves and replanting of new forest has resulted in an improved condition of that habitat (Chapters 12, 48). Climate change also affects benthic biodiversity, but documentation and understanding of pathways and consequences are at an early stage (Chapter 36A.3, 36G.3).

For offshore benthos, the overwhelming pressure is the impacts of fishing gears. Trends were documented in all regions, and the commonality of these trends has led to the occasional characterization of all mobile bottom gear as a destructive fishing practice. Many types of seafloor habitats and benthic communities, particularly those comprised of soft bodied and leathery species, do show recovery from bottom trawling when the pressure is released, although just as with the fishery communities that are being exploited, full recovery may requires years to decades. During periods of disturbance and recovery the *relative* species composition is changed, as long-lived species are reduced in abundance and dominance. However, as long as recovery can commence rapidly and is secure, such perturbations are sustainable and the habitats are considered to have resilience (Chapter 36A.3, 36B.4, 36C.3.b; Chapter 11). However, some special types of habitats and benthic communities are not resilient. Pressures, causing changes to seabed structure or increased mortality of species that are more hard-bodied and that create habitat diversity through burrowing or creating three-dimensional structures, may cause large and lasting trends in the benthic community. Productivity can be reduced and recovery, if feasible at all, could take many decades to centuries (e.g., cold-water coral communities, especially on seamounts; Chapters 42, 43, 51). In such cases, spatial management to prevent impacts is the only effective option to mitigate these trends and allow recovery to commence. Some policies that protect these highly vulnerable to sensitive benthic habitats are in place for the high seas and many national jurisdictions. For example, some States and intergovernmental entities have adopted measures for the protection of seamounts and other deep water habitats within EBSAs, VMEs and MPAs, as discussed in Chapters 42 and 51. But this has not been done in most parts of the ocean, since the task of identifying such areas of particular importance to biodiversity is incomplete in some parts of the ocean. In addition, the necessary scientific and technical information is sometimes not available to the relevant States and intergovernmental organizations.

As with the plankton in the water column, invasions of alien species pose a risk of altering benthic biodiversity on scales from local and coastal to seas or large stretches of coastlines. The same processes of natural transport of reproductive propagules, range changes in response to climate-related changes in ocean conditions, and accidental transport with shipping or tourism have all been documented, with resultant major changes in benthic and occasionally pelagic community structure at scales at least of bays of hundreds of kilometres of coastline documented in all regions where sampling is adequate to detect such effects (Chapter 36A.3, 36B.4, 36C.3.b). In addition, a few cases are recorded of intentional introduction of larger invertebrates to develop new harvesting opportunities, with subsequent expansion of the species well beyond the area of introduction (such as Kamchatka crab in the Barents Sea Chapter 36A).

The shipping industry is actively seeking to improve practices and reduce risk of transferring species to new areas, and cost-effective risk-management practices are available (Chapters 17, 27). Detection of new benthic species requires intensive and often costly monitoring, for which capacity is limited in many areas. Once alien species are established, their elimination and remediation of the impacts have proven to be very difficult, costly, and rarely feasible.

3.3 Fish and pelagic macro-invertebrates

As with the other species groups, fish communities have always varied in abundance over time, sometimes by orders of magnitude, especially for small pelagic species in areas with variable oceanographic conditions (examples in Chapter 11, Chapter 36A.4, 36B, 36C, Figure 36C.4; 36D [salmon]; 36D.2.4). In several ocean basins changes in major portions of fish and invertebrate communities are well documented, and these are often related to corresponding changes in the physical ocean (Chapter 36A.4, 36C.3.a.iv, 36G.4).

Range changes of fish and macro-invertebrates in response to naturally changing ocean conditions are also documented in all regions (examples in Chapter 36A.4.4, 36C.3, 36G.4, 36H.2.3). The responses of fish populations and communities to climate change have been a particular priority for mid- and high-latitude parts of the ocean, with documented effects on productivity, timing of life history processes (e.g., Chapter 36H.5), and community structure in essentially all regions, with magnitudes of effects varying both with the life history of the species and the magnitudes and patterns of change in the oceanographic conditions (examples in Chapter 36.A.4, 36C, Figure 36C.4; 36D.2.4).

Another type of documented trends in ranges of fish and invertebrate species are invasions of non-native species (example in Chapter 36C,

Figure 36C.7) almost certainly associated with shipping. Some of the invasions by large pelagic invertebrates, such as comb-jellies, have completely changed the fish community on the scale of bays, and of the entire Black Sea (Chapter 36A.7, 36C.2). Although the magnitude of the disruptions from such invasions may diminish over time due to both natural ecosystem processes and management interventions, it has not been possible to eliminate or reverse such changes quickly, if at all, and costs have been high both in terms of costs to try to control the invading species, and in foregone benefits from the disrupted fish community (e.g., Black Sea). Prevention of introductions is by far the most logical and cost-effective option, and is receiving attention from the shipping industry. Again, however, resources are needed to implement and ensure adherence to best practices.

Trends in fish populations are linked to contaminants, pollution, and particularly habitat degradation due to land-based sources. However, population-scale effects have been restricted to nearshore areas or semi-enclosed seas where contaminant, pollution and/or sediment levels are high and water quality is degraded, with many fish populations and communities particularly susceptible to reduced oxygen levels in the water due to both climate change and increased nutrient enrichment (Chapter 36A.7 [Gulf of St. Lawrence, Chesapeake Bay, Gulf of Mexico]; 36B.2, 36C and F). However, the concern exists that long before population-scale impacts of contaminants may be apparent, fish may accumulate levels of contaminants in their flesh that pose health risks for consumers (Chapters 10, 15). In addition, it was noted earlier that some specialized habitats that are hotspots for fish and invertebrate biodiversity are also particularly attractive for other human uses. Downward trends in fish populations associated with such habitat losses are documented in many coastal areas (Chapter 36A.7 [all cases]; 36F).

Regardless of the cause of habitat loss or degradation, fish populations and communities have been documented as recovering when effective remediation measures have been taken (Chapter 36.A.4, 36C, Figure 36C.11; Chapter 41). Again, however, costs of remediation have often been high, time lags long, and prevention of loss or degradation is usually the more cost-effective option, with less uncertain outcomes than remediation initiatives.

Exceeding all of these other causes of trends in fish populations and communities are the effects of fishing. Fishing necessarily changes the total abundance and size/age composition of the exploited populations, with effects increasing as bycatch rates increase and as fishing becomes more intense and more selective of only particular species and sizes. The search for levels and methods of fishing that have sustainable impacts has gone on for over a century (Chapters 10, 11). Nevertheless, overfishing has not been eliminated, and downward trends in exploited populations, sometimes to depleted levels, can be found in all regions (Chapter 36A.4, 36B.4, 36D.2.4). Estimates of the economic cost of such depletions are available (Chapters 11, 15), but ecosystem costs from the biodiversity impacts of overfishing exist as well. If genetic diversity of populations is depleted, resilience to naturally varying environmental conditions is reduced (Chapter 34; Chapter 36A.4). Also as the abundance of large fish in a community is reduced through fishing at levels that allow few fish to live long enough to reach their full potential size, any top-down structuring of community dynamics through predation is weakened, again weakening the resilience of the community to any other perturbation (Chapter 11 [ecosystem effects]; Chapter 36D.2.4). The properties of harvesting strategies that would keep effects of fishing sustainable for the exploited species and communities are generally known, and many examples show that when fisheries are managed with sustainable practices, populations can rebuild to and subsequently remain at healthy levels, although varying in response to natural perturbations (Chapter 11; Chapter 36.A.4, 36D.2.4, although recovery may take decades; e.g., 36H.4). However, management authorities must adopt sustainable policies and practices that take biodiversity considerations into account, implement them consistently, monitor population status and fishery performance effectively, and ensure compliance through a combination of stewardship, surveillance, and enforcement that is appropriate to the fishery and community of human users (e.g., Chapter 36A.6, 36H.4).

3.4 Marine Reptiles, Seabirds, and Marine Mammals

Many of the species with the greatest declines in abundance are in these groups of top predators (marine mammals, marine reptiles and seabirds); some species of all three higher taxa are assessed as at risk of extinction by IUCN (Chapters 37, 38, 39). In cases where overharvesting was a contributing factor, some of these declines have lasted for over a century (Chapters 37, 39; Chapter 36H.4). All the factors considered for the above groupings of biodiversity are also implicated in the trends in marine reptiles, seabirds and mammals.

Seabirds and marine reptiles have been particularly affected by habitat degradation, e.g., where terrestrial breeding sites were converted to intensive use by coastal industries or tourism (Chapter 39), or, particularly for seabirds, where new predators were introduced on previously isolated breeding sites (Chapter 38). Body burdens of contaminants have been implicated in reduced breeding success of several populations of pinnipeds and smaller cetaceans (Chapter 37; Chapter 20), and in a few cases this pressure alone may be sufficient to pose a risk of extinction to small populations (Chapter 36A.5; Chapter 37).

Bycatches in fishing gear are well-documented threats to populations and species in all three groups of top predators. Most types of fishing gear have been documented to pose potential threats to specific populations or species, including: mobile trawl gear for turtles and sea snakes; longlines for seabirds and turtles; suspended nets for seabirds, small cetaceans, and pinnipeds; and entanglements of whales with lines connected to traps and pots used in fishing (Chapters 37, 38, 39). Practices which mitigate these risks have been proven to be effective for many types of gear, through changes in gear design (e.g., excluder devices), fishing practices (e.g., surface deployment of longlines), and other methods. However, implementation of mitigation techniques often requires training in their use and is specific to the species, fisheries, and areas where the fishing occurs. Hence additional measures, such as periodic and area closures in areas of high bycatches, or closures of fisheries when allowable bycatch numbers are exceeded, are often applied, with or without gear-based measures (Chapters 38, 39). Downward trends in marine reptile, seabird, or marine mammal populations due to bycatch impacts can be stopped and population increases facilitated by the appropriate combination of these mitigation measures, but require expert study of the nature of the bycatch problem and evaluation of the potential effectiveness of alternative mixes of measures, monitoring of the fishery and the populations suffering the bycatch mortality, often requiring expenditures of capital, time and training in acquiring, adapting, and learning to use the tools, and appropriate surveillance and enforcement.

Not all harvesting-related mortality of marine mammals, seabirds, and marine reptiles is due to bycatches in fisheries. Directed take of marine mammals reduced many whale populations to one or a few per cent of historical populations, and recovery of these species has often been extremely slow, even after harvests were largely eliminated (Chapter 37). Directed harvests of seabirds were historically common in many areas, and are still a practice in some places that still depend on subsistence hunting and fishing (Chapters 36A.5, 36B.8, 36G.6; Chapter 38). Directed harvest of sea turtles was intensive until the early 20th century and depleted many populations, but is now prohibited in most jurisdictions, although again recovery has been slow, usually due to other pressures (Chapter 39).

Aside from directed take and bycatches, fisheries can also affect marine reptiles, seabirds and marine mammals through trophodynamic pathways (examples in Chapter 36B.8). Fisheries on small pelagic stocks have been implicated in depleting the food supply of seabirds, particularly when feeding chicks in breeding colonies (Chapter 38), and cases are documented where the discarding of fishing waste at sea has promoted large increases in populations of scavenging seabirds that in turn displace other seabirds from breeding colonies (Chapter 38). Spatial management measures and adjustments to overall harvesting levels have been successful in dealing with the first type of impact (Chapter 38), and the increasing policy dialogue about managing discard practices in fisheries (Chapter 11; Chapter 36A.5) is at least considering the implications for seabird communities.

Climate change has the potential to affect trends in all three of the types of top marine predators. Studies are being conducted on how changing ocean conditions may affect breeding and resting sites (e.g., sea-level rise and turtle and mammal breeding beaches), loss of ice cover affecting high-latitude seabirds, polar bears, cetaceans and pinnipeds, range changes of species from all groups as temperature and salinity patterns change, and indirect food web effects (Chapter 36.G.5.5, 36H and 38). Given the long life expectancies of most of these types of species, population impacts of climate drivers may only show up gradually, but may be hard to reverse. Moreover the impacts may be non-linear once they start to be manifest, with possibly steep "tipping points", further increasing the policy and management challenges. Some interest exists in spatial management tools as a way to partially mitigate impacts of climate on these populations, but work to test and, as appropriate, implement such measures is in the early stages (Chapters 38, 39).

For all of these species groups, measures to mitigate any of the anthropogenic and climate drivers of decreasing trends in populations have opportunity costs from displacing or refraining from conducting an activity providing social and economic benefits and direct costs of deploying more expensive fishing gear, patrolling of breeding sites of turtles and mammals, etc. (Chapters 37, 39). However, the high regard in which many societies hold these types of species (Chapter 8) also provides opportunities for increasing public awareness of all marine biodiversity concerns, and building public and industry support for taking appropriate actions to address trends and ensure practices are sustainable.

4   Trends in biodiversity for habitats

In cases where habitat features make an area a hotspot for marine biodiversity, the potential for negative trends in biodiversity is greater for several different types of reasons:

- Special habitat types can host uncommon or rare species requiring the special features of these habitats for some aspects of their life histories. Their inherent rarity and high ecological specialization can make such species particularly vulnerable to impacts from human activities and changes in environmental conditions. This Assessment found examples of such vulnerable species in most habitat types examined: for example, the general importance of seagrasses for juvenile fish as well as dugongs and human activities that affect seagrass and/or water quality (Chapter 47), and salmon sensitivity to human impacts on coastal areas (Chapter 36C.5).
- Just by being rich in biodiversity, particularly high interdependencies can exist among the species in these specialized habitats. Perturbations of even a few key species in these ecosystems can affect many other species with which they have ecological relationships, spreading and sometimes amplifying the initial perturbations to produce much greater consequences for the biodiversity as a whole. Again, such vulnerabilities of the biotic community to perturbations of even a few components were found in many of the special habitats assessed in the WOA. For example, overfishing of herbivorous reef fish has resulted in decline of coral cover because of overgrowth of algae (Chapter 43).
- Areas that are biodiversity hotspots often are associated with specialized structural features of the seafloor and/or the water column for which many species use those particular areas for some or all life history processes. If those specialized structural features are disturbed intentionally or collaterally by human activities, their ability to serve those functions for all the species depending or attracted to them is reduced, again with the potential for widespread detrimental effects on biodiversity. Some of the specialized habitats examined showed many examples of declines in biodiversity due to such alterations or elimination of the physical attributes of the habitats.

For example, bottom trawling in many areas and cases of oil and gas development in some places have removed large areas of cold water corals and its associated biodiversity (Chapters 42, 51).

- By having specialized physical characteristics and by supporting many different kinds of marine life, the specialized habitats can be particularly vulnerable to some kinds of human uses (e.g., Chapter 36D.2.3). Some are activities that focus on the special physical features, such as use of high local productivity for aquaculture (for example, conversion of mangrove or other estuarine habitats; Chapters 44, 48) and/or on the high concentration of biodiversity, such as fishing or tourism (for example, coral reefs; Chapter 43). Others are activities that may not intentionally focus on the hotspot, but on the physical features and processes that result in the areas supporting high levels of biodiversity; these also result in the areas being especially exposed to collateral impacts from other human activities, such as the types of biodiversity hotspots that are areas of high land-based inputs or that support higher concentrations of coastal residents than other areas. Examples include estuaries (Chapter 44), mangroves (Chapter 48) and salt marshes (Chapter 49).

This Assessment not only found widespread evidence that specialized habitats commonly show particularly strong negative trends in components of biodiversity, it also found global spatial patterns in the types of vulnerabilities of various different types of specialized habitats.

A few of the specialized habitats were either offshore and/or deep-sea habitats (seamounts, hydrothermal vents and seeps, cold-water corals, the Sargasso Sea, high-latitude ice), where human populations and industrial activities have historically not concentrated. These are highly specialized habitats requiring particularly high adaptation by the biological community to their specialized conditions. However, in providing unique features, such as the enhanced biological productivity of seamounts and the ice edge, high three-dimensional structure of the corals, etc., the areas become biodiversity hotspots relative to adjacent areas that are not as productive. In these habitats, recovery from physical damage to the specialized habitat features and/or depletion of the biological populations is often extremely slow and uncertain, just because of the harshness of the background conditions in adjacent areas, and/or the particularly high specialization of the species to these special environments, and/or the complexity of the specialized habitat itself. Climate change poses particular threats to some of these types of special habitats, because ice melting and thermal stress are altering the special habitat features themselves (e.g., reduction in the area of Arctic sea-ice habitat (Chapter 46) caused by global warming). Also coral bleaching is widespread and has affected the quality of coral reef habitat in all areas of the oceans (Chapter 36D2.3; Chapter 43).

Fishing is the primary activity likely to be attracted to the biota at these special habitats, particularly in the deep sea. Both the complex structure of the habitats and the highly specialized fish populations may be highly vulnerable to physical damage and/or exploitation, and the slow recovery potential of much of the associated biodiversity is a particular risk (for example, cold-water coral communities on, e.g., seamounts; Chapters 42, 43). Many human activities, such as shipping lanes, undersea cables, etc., can be designed to avoid these special areas. However, as the capacity to exploit the physical resources of the deep sea increases, extractive uses may be attracted to exactly these specialized habitats (for example, species found associated with deep-sea hydrothermal vents and cold seeps; Chapter 45). The potential for increased shipping and tourism in the high latitudes is also an increasing threat to biodiversity.

The other specialized habitats are generally associated with nearshore and coastal areas (kelps and seagrasses, mangroves, salt marshes, estuaries and deltas), although low-latitude/warm-water corals may extend out onto continental shelves for many kilometres. A much wider range of human activities poses potential threats to these habitats. Many have the high vulnerability to land-based run-off discussed above (e.g., estuarine habitats), high attractiveness of their biota to directed uses, such as fishing and tourism (e.g., coral reefs) and adjacent coastal development, and of their physical features to extractive uses (for example, carbonate mining of reef rock; Chapter 43) or intentional alteration (e.g., conversion of mangrove habitat for aquaculture; Chapter 48).

The productivity of these more coastal areas is often higher, such that recovery from some types of perturbations of the biodiversity may be more rapid and secure than in off-shore, deep-water habitats (Chapter 36H). On the other hand, these specialized habitats and their biodiversity are likely to be exposed to a myriad of pressures from multiple human activities. That diversity of pressures means that the cumulative impacts on these habitats and communities can be high, even when individual pressures may be managed, and the effectiveness of management measures applied to one pressure may be affected by the nature and intensity of other pressures, e.g., salt marshes (Chapter 49). As a result, protection of these specialized habitats and their biodiversity may require complex and coordinated planning and implementation, and it may be hard to motivate single industries to bear the costs of mitigation measures if the pressures from other uses are likely to continue (illustrations from all of Parts V and VI).

In summary, this Assessment finds that specialized habitats that are also biodiversity hotspots face a potential triple threat from higher vulnerability to perturbations, higher attractiveness to many human uses, and higher challenges to recovery from perturbations if they occur. Correspondingly, evidence of degradation of habitats and communities from one, two, or all three of these factors is widespread for all habitat types and in all regions where they are found. Nevertheless, many cases exist where the threats have been adequately managed, the habitats and their biodiversity protected, and at least some recovery from past perturbations has been recorded. For example, 24 estuarine case studies reported that management has resulted in improving estuarine health (Chapter 44) and declaration of MPAs has prompted the recovery of ecosystem and fish populations in coral reefs (Chapter 43). Although the benefits of coordinated planning and management of pressures cannot be overemphasized, targeted and proactive measures can have high payoffs if even one pressure is reduced effectively (e.g., seasonal closures of fisheries to protect spawning aggregations (Chapter 11);

stopping mangrove habitat conversion for aquaculture (Chapter 48)). It is important, however, to match the management tool to the particular needs of the specialized habitat; for example, limits on catches or effort in a fishery may keep a fishery sustainable at the population level, but spatial tools may need to be added if the fishery can concentrate on biodiversity hotspots causing local depletions of species serving key functions in the biotic community (illustrations from Part IV, Chapters 36, 42-51]. However, progress and even success is possible with the proper suite of measures, and the capacity to implement them.

Methods used in the assessment of the status and condition of species and habitats have undergone a rapid transformation in recent years with the advent of predictive habitat modelling (PHM). This approach has revolutionized the study of many habitats and their associated biota. For example, cold-water coral communities were virtually unknown prior to the discovery of extensive bioherms off Norway in the early 1980s, but the application of PHM enabled the discovery of a completely new habitat for scleractinians on steep submarine cliffs in 2011 and was confirmed almost immediately through field observation in the Mediterranean and Bay of Biscay (Chapter 42).

In summary, the diversity of the world's oceans is rich and dynamic, but it has been incompletely quantified, and even descriptions and inventories of marine biodiversity are incomplete for the open ocean, the deep sea and many shelf and coastal areas. Despite our incomplete knowledge, however, trends in measures of biodiversity at the scales of populations, species, communities and habitats are found almost everywhere, demonstrating that our information is sufficient to look for trends, and often to inform development of policies and management measures to address drivers of the trends. Natural processes play some role in these trends, and occasionally can be a prominent driver. However, in the majority of cases, anthropogenic drivers are the major influence on changes in biodiversity.

The nature of the human activity generating the pressure will strongly influence the scale of impacts in space and time. Some pressures, such as climate change, inherently operate at large (global) scales, others, such as fisheries, may operate at the scale of individual fisheries, but fisheries are widespread and have adversely affected biodiversity in most parts of the ocean. Other pressures are inherently more local in both operation and occurrence, such as hydrocarbon extraction and seabed mining, but they can be intensive pressures where they do occur. Moreover, it is common for biodiversity on local to basin-wide scales to be exposed to cumulative effects of multiple pressures interacting in ways that are usually poorly understood.

Certain types of species, such as marine mammals, seabirds, marine turtles, large sharks and fragile benthic taxa, such as corals and sponges, and certain types of habitats, including coral reefs, hydrothermal vents, estuaries, mangroves, and others, are both particularly sensitive to pressures from many types of human activities and attract human uses in large part because of their biodiversity characteristics. These are often the components of biodiversity showing the strongest declines over time, and thus pose particularly great conservation concern.

Notwithstanding the widespread negative trends in biodiversity and the number and ubiquity of anthropogenic pressures associated with those trends for all components of biodiversity and all types of pressures, positive examples of eliminating or mitigating pressures and reversing the unsustainable trends exist. The likelihood of success in managing threats and protecting and recovering biodiversity that has been affected increases with better knowledge of biodiversity and the pressures in the area, adoption of policies appropriate to the context, and the improvement of capacity to implement the policies effectively.

Chapter 53
Capacity-Building Needs in Relation to the Status of Species and Habitats

Contributors:
Renison Ruwa (Convenor and Lead Member), Amanuel Yoanes Ajawin, Sean O. Green, Lorna Inniss, Osman Keh Kamara, Jake Rice (Editor of Part VI) and Alan Simcock (Co-Lead Members)

Chapter 53

Capacity-Building Needs in Relation to the Status of Species and Habitats

1 Introduction

Knowledge of the status of species and habitats forms a fundamental basis for understanding biodiversity at all scales (Chapter 34) and ecosystem functions and services (Part III and Millennium Ecosystem Assessment, 2005). This facilitates identifying the capacity-building needs for appropriate interventions that will enhance and promote sustainability. This creates a need for knowledge of marine biological diversity and habitats from a marine ecosystem approach, and of how biodiversity varies in relation to various levels of anthropogenic perturbations. Gaps in scientific knowledge, technological advances, human skills and infrastructure for the conservation of marine biodiversity and habitats are crucial. Part VI addresses these issues focusing on the major oceans in relation to marine ecosystems, habitats and major species groups that are emerging from our assessments as potentially threatened, declining or needing special attention. All these categories need a variety of capacity building, technical skills, technology and infrastructure to address their trends. To facilitate this capacity building, we undertook the identification of knowledge gaps mainly from the Part VI Chapters, and the capacity building needed to address socio-economic issues for human well-being. Chapter 32 and this Chapter both address capacity-building needs. However, whereas the identification of needs in Chapter 32 was based on outcomes of regional workshops and the Chapters of Part V, this chapter is based on all the authored chapters, which also include identification of gaps from literature reviews on the oceans.

To address the objective of the Regular Process to ensure that capacity building and technology transfer are done through promoting cooperation, not only North to South but also South to South cooperation (UNGA, 2010; UNGA/AHWGW, 2009; UNGA/AHWGW, 2010), the synthesis is done in geographical areas following the oceans and the major regional seas addressed in Chapter 36. Further capacity needs were identified in relation to the knowledge gaps from the chapters focused on the overall status of the major groups of species and habitats, including the socio-economic aspects of their conservation.

The marine species groups that were given special attention or protection are: marine mammals, seabirds, marine reptiles, sharks, tuna and billfish. These were dealt with globally without specifically linking them to particular oceans. The same general analysis was followed for specific marine ecosystems and habitats addressed in Chapters 42-51, including cold-water corals, warm-water corals, estuaries and deltas, open-ocean deep-sea biomass, hydrothermal vents and cold seeps, high-latitude ice, kelp forests and seagrass, mangroves, salt marshes, Sargasso Sea, seamounts and other submarine geological features potentially threatened by disturbances.

There are already many international initiatives to build capacities (both in terms of skills and of equipment) to meet many of the capacity-building gaps identified in this Assessment. One example among many is the programme of the Food and Agriculture Organization of the United Nations, supported by Norway, using the Research Vessel Dr. Fridtjof Nansen. However, on the information available, it is impossible to say what gaps currently exist in arrangements to build these capacities: conclusions on where the capacity-building gaps exist could only be reached on the basis of a survey, country by country, of the capacity-building arrangements that currently exist and how suitable they are for each country's needs. This applies more generally, but is particularly important in relation to capacity-building in relation to marine bodiversity. The initial inventory of capacity-building arrangements[1] compiled by the Division for Ocean Affairs and Law of the Sea as part of the Regular Process would provide some initial information on which to base such a review, but it would take much more detailed study than has been possible in the first cycle of the Regular Process to match this with the needs of each country.

2 Outcomes based on regional workshops on capacity-building needs

The following regional workshops were held: South-West Pacific region (UNGA, 2013a), Wider Caribbean region (UNGA, 2013b), Eastern and South-Eastern Asian Seas (UNGA, 2012a), South-East Pacific region (UNGA, 2011), the joint North Atlantic, Baltic Sea, Mediterranean and Black Sea region (UNGA, 2012b), the Western Indian Ocean (UNGA, 2013c) South Atlantic Ocean (UNGA, 2013d) and Northern Indian Ocean (UNGA, 2014). From the regional synthesis based on the outcomes of the regional workshops, it appeared that some needs were regionally cross-cutting and some were directly relevant to Chapter 53. The following were more specific to species and habitat relationships across the regions:

(a) Taxonomy and genetics
(b) Bio-physical/chemical research on the ocean environment
(c) Socio-economics of oceanic natural resources focusing on biodiversity and habitats
(d) Skills in integrated assessments, including modelling
(e) Infrastructure with relevant supportive technology, especially in research vessels and laboratories to support multidisciplinary research
(f) Geographical Information System mapping skills.

3 Outcomes based on chapters focusing on knowledge gaps to inform capacity-building needs

3.1 Overview of marine biological biodiversity

The global biological diversity patterns are described in relation to key taxa and habitats and to the identification of key environmental and anthropogenic drivers. The gradients in marine biodiversity are assessed using a taxonomic framework of well-known key groups of organisms

---

1 See A/67/87, Annex V.

(for example, marine mammals; turtles; finfish; plankton (phytoplankton and zooplankton), and seabirds in Chapters 34-36. In addition, a habitat framework was used when the taxonomic identity of the species was of secondary importance to the type of community or conditions in which they occurred, the species and habitat framework focusing on marine ecosystems, species and habitats (Chapters 37-51, Section B).

3.2 Overall status of marine biological diversity in the oceans and knowledge gaps

The Atlantic and Pacific Oceans are relatively more studied than the Indian Ocean, which is the third-largest ocean and almost entirely surrounded by developing countries. By contrast, both the Atlantic and Pacific Oceans are mostly surrounded by developed countries or economies in rapid transition. However the North Atlantic and the North Pacific Oceans are comparatively better studied than the South Atlantic and South Pacific Oceans.

In terms of identifying the global diversity patterns, the gradients in marine biodiversity of the North Atlantic and the North Pacific are assessed primarily in terms of taxonomic frameworks, whereas for the South Atlantic and South Pacific the taxonomic framework is used when possible, but is often augmented by the habitat frameworks in areas surrounded by developed countries and developing countries, respectively. For areas surrounding the Indian Ocean, where many knowledge gaps are found, the gradients in marine biodiversity are assessed primarily in terms of habitat frameworks. As regards the Polar waters, the Antarctic has been more studied than the Arctic, but it is necessary to increase scientific efforts for the Arctic and Antarctic due to their uniqueness.

3.3 Deep-sea environment

Shallow coastal waters are comparatively better researched than the deep sea because of their greater accessibility. It is necessary to build the essential capacity, including deep-sea platforms to provide relevant research and technical skills at regional and global levels to address the following problems:

- Despite technological advances and a sharp increase in deep-sea exploration in the past few decades, a remarkably small portion of the deep sea has been investigated in detail. There are therefore large gaps in what we know about the deep sea.
- Although the species which are specifically considered in this Assessment are vertebrates, it is important to improve the knowledge base about invertebrates, microbes and viruses.
- Deep-sea biodiversity is very poorly characterized compared to the shallow-water and terrestrial realms. Without better characterization of deep-sea biodiversity, its protection will be hampered.
- The deep ocean has many species, with genetic, enzymatic, metabolic and biogeochemical properties which may hold potential for major new pharmaceutical and industrial applications. Without better knowledge of these species and their properties, important opportunities may be missed.
- The deep oceans are estimated to have up to millions of species. Because conservation and sustainable use of biodiversity is improved when the species are known and their biological characteristics inventoried, much effort and time will be required to describe them.
- The deep seas are threatened by ongoing global climatic changes due to increasing anthropogenic emissions and resulting biogeochemical changes. The impacts of climate drivers on the deep sea biota and the magnitude of the drivers in the deep sea need to be better documented.
- The deep oceans may be threatened by, e.g., oil and gas exploitation, mining for metals, fishing practices (both destructive fishing techniques and an excessive scale of fishing) and pollution. More measurement is needed of the scale of these pressures and their potential impacts.
- Perhaps the most important knowledge gap is the knowledge of the effectiveness of alternative management options when applied in such a vast, dynamic space, much of which is beyond national jurisdiction, to reduce the impact of man-made stressors.
- The design of protected areas based on geographic definitions must necessarily account for the fluxes through the system as well as the movement of the inhabitants.
- Deep-sea observatories are becoming increasingly important in monitoring deep-sea ecosystems and the environmental changes that will affect them. These observatories aim at addressing important societal issues, such as climate change adaptation, ecosystem conservation and sustainable resource management. Tackling these issues, along with efficient and clear stakeholder communication, is particularly important for the deep sea, which remains largely unexplored, yet affects the lives and livelihoods of the global population directly or indirectly. Technological advances in recent years offer the ability to continuously monitor the ocean in time and space; in particular, the development of *in-situ* sensors, autonomous vehicles, and cyber-infrastructure, including telecommunications and networking. If these technologies are applied more widely in the world's oceans they would add to the capacity to monitor the deep sea and feed the obtained information into science-policy interfaces and marine management and policy.

4 Specific data or knowledge gaps identified in the Assessments by key marine species or habitats

4.1 Marine species

4.1.1 Marine mammals

Data are obtained mostly from ship-board observations and use of satellite telemetry. The latter has improved offshore data acquisition, because most of the data are taken within the Exclusive Economic Zones (EEZs). USA, European and Antarctic waters are the best assessed waters. The largest knowledge gaps occur in Indian Ocean waters. Only by continuing to monitor and assess the marine mammals in EEZs and putting more research effort into the Areas Beyond National Jurisdiction (ABNJ) can sufficient data be obtained to document trends and inform decision-making.

4.1.2 Seabirds

Birdlife International, the IUCN Red List authority for birds, has the most authoritative global database on seabirds. At regional levels, Europe and North America are most thoroughly assessed; many knowledge gaps remain in the developing world.

- Important knowledge gaps exist in studies of seabird migrations, some of which cross continents, or are inter-continental, because these routes are not well known. Other gaps that cannot be filled without additional capacity include improving understanding and increasing data available on seabird coastal habitats; seabird by-catch; vulnerability to pollution (especially oil, garbage in dumpsites, marine litter and plastics); disturbances of coastal and deep-sea habitats; adequacy of habitat protection; whether and what kind of marine protected areas (MPAs) may address this gap globally; their role in ecosystem, socio-economic and livelihood services; the effectiveness of alternative conservation elements for taking the migratory habits of seabirds into account, and other factors for sustainability in protected areas.

4.1.3 Sea turtles

With respect to sea turtles, the issues where gaps in knowledge and capacity-building are involved include:

- Assessments spearheaded by IUCN's Red List of threatened species and the global listing for vulnerable species exist. However, marine turtle population traits and trajectories can vary geographically and the listing criteria could only be applied effectively if there were a better characterization of the status and trends of individual populations and if the information was used to establish categories for regional sub-populations in addition to the single overall global listings.
- Gaps in knowledge of risks due to effects of climate change still remain a challenge because of insufficient data for analysis of long-term trends. Improved conservation of sea turtles could result from an increase in regional assessments for sea turtles due to their migratory nature. Monitoring and reporting criteria would also perform for effectively if they were augmented by information on the status and trends of population sizes, as well as global threats to the sea turtles.
- Data needs are critical for data-poor regions, especially Africa, the Indian Ocean and South East Asia.

Increased capacity to address these gaps at regional and global levels would allow more effective conservation of marine reptiles. Such efforts would benefit from cooperative regional and global partnerships, because sea turtles are migratory and transboundary.

4.1.4 Sharks and other elasmobranchs

In relation to sharks and other elasmobranchs, there is inadequate capacity in many countries and most regions to address the following issues:

- Lack of or deficient monitoring data make it difficult to assess the status of many sharks. The most data-deficient areas are: Western Central Atlantic Ocean, Eastern Central Atlantic Ocean, the Wider Caribbean Sea, South West Indian Ocean and the eastern and southeastern Asian Seas.
- In addition to obtaining data from fisheries, surveys and catch landings increasing the capacity to use emerging technologies, such as satellite tags, acoustic tracking, digital underwater photography, and sophisticated photo identification systems would facilitate population and distribution estimates in defined geographic locations.
- Although the recent decline in reported landings is consistent with declining abundance due to overfishing, any interpretation should consider that reported landings are almost certainly a gross underestimation of actual catches. To ascertain actual trends in shark catch and landings, which are likely to be even worse than expected, would require increasing the management priority of sharks by regional fisheries management organizations (RFMOs) and national management bodies. Better independent catch and bycatch monitoring data are needed to know the effectiveness of conservation measures taken by RFMOs, noting that destructive fishing is still increasing in regions such as the Indian, central Pacific and south and central Atlantic Oceans.
- Mortality due to fishing, both directly and as bycatch, is almost entirely responsible for the worldwide declines in shark and ray abundance. However, knowledge of survival of living sharks released at sea is limited.
- Persistent bioaccumulation of toxins and heavy metals has been documented in sharks feeding at high trophic levels. Levels which can be toxic to human consumers have been reached in some areas, but their effect on the host shark remains unknown. The global extent and specificity in occurrence of various contaminant burdens are unknown. These knowledge gaps would have to be filled before the population-level threat of toxins and heavy metals could be evaluated effectively.
- Elasmobranchs (sharks and rays) play an important role in the marine ecosystem food chains as top predators; they contribute to maintaining balances in species numbers and biomass abundances. This function is, however, not very clear at local and sometimes regional scales; its overall global manifestation is not well known either as the role of temporal variability is poorly understood. These knowledge gaps would have to be filled in order to place shark conservation in the context of ecosystem functioning.
- A key challenge is to secure ongoing assessment activities, particularly the continuance of research surveys, and to expand assessment

activity to encompass not only the largest, most charismatic species, but also the lesser-known species which are often more threatened, particularly the rays and shark-like rays, and the 90 obligate and euryhaline freshwater species. Geographically, greater attention needs to be paid to Central and South America, Africa, and Southeast Asia.

4.1.5 Tuna and Billfishes

These fish are an important part of the global capture fisheries sector. Billfishes are heavily fished and have therefore attracted the attention of IUCN; some species are listed as vulnerable. Capacity-building gaps exist in addressing the following gaps:

- Assessments are done by RFMOs using fisheries stock assessment methodology and capacity is inadequate in many parts of the world to employ this methodology and to establish research infrastructure with the necessary technology, including satellite tracking facilities, to facilitate the required studies. Lack of this capacity hinders conservation and management of these species. A global paucity of data exists on the population status of these species. Only with additional stock assessments would it be possible to identify and protect early enough many species possibly threatened by overfishing for effective conservation measures to be taken. This can only be done effectively if it is approached at both regional and global levels.
- Although the current exploitation status for the principal market tunas is relatively well known globally, knowledge on the exploitation status for the non-tuna billfish stocks and species is fragmentary and uncertain. Furthermore, tuna RFMOs have not yet conducted formal fisheries stock assessment evaluations or adopted management and conservation measures for any of the eight non-principal market tuna species. Therefore their current exploitation status is unknown or highly uncertain throughout their distribution range, and can only be filled by additional capacity to assess their status.
- It is generally agreed that catch estimates for non-principal market tunas and billfishes have been and still are underestimated, as the majority of these species are caught by small-scale fisheries or as a bycatch of principal market tuna fisheries. Therefore effective assessmetns of these species requires improved catch reporting from small-scale coastal fisheries targeting both principal market tunas and the smaller non-principal market tunas. Similarly, billfish catches, which generally come from industrial tuna fisheries as bycatch, have also been commonly poorly reported and monitored.
- Climate change is another potential pressure that needs to be taken into account in the assessment of the biology, economics and management of tuna and billfish species. Climate change might have an effect on tuna and billfish species by changing their physiology, temporal and spatial distribution and abundance, but these possible relationships can only be known with much more study.
- To what extent the widespread declines in tuna and billfish populations have altered the capacity of the ocean to support vital ecosystem processes, functions and services by reducing their abundances and altering species interactions and food web dynamics is poorly known.

- Incorporating ecosystem considerations into the management of tunas and billfish fisheries would help to move their assessments into an ecosystem context.
- The main challenges to conservation responses and factors for sustainability are: (1) reduction in the existing overcapacity of fishing fleets; and (2) adoption of protocols that ensure implementation of effective Monitoring, Control and Surveillance (MSC) techniques.
- A further challenge is the paucity of knowledge of the impacts of tuna and billfish fisheries on other less productive species such as sharks, on species interactions and food web dynamics, and on the greater marine ecosystems.

4.2 Marine ecosystems and habitats

4.2.1 Cold-water corals

With respect to cold-water corals, the issues where gaps in knowledge and capacity-building are involved include:

- Information on cold-water corals (CWC) in the Indian Ocean region is scanty, even though the region covers an area between latitudes 70^0N - 60^0S, a range where seamount CWC are known to occur.
- Technology and skills for discovering CWC are still lacking in some regions, especially the developing world. Additional fine-grained and broad-scale habitat modelling are still needed to discover additional habitats, and to forecast the fate of CWC facing both direct (fisheries) and indirect (environmental) impacts.
- It is necessary to increase knowledge of the characteristic geological structures and environmental factors facilitating CWC settlement and growth. The current list includes provision of hard, current-swept substrate, and often topographically guided hydrodynamic settings. All need to be identified and mapped. The skills needed include knowledge of combined physical, bio-geo-hydro-chemical analytical techniques (e.g., of ambient seawater characteristics and measurement of current velocities.
- Global knowledge is lacking of CWC distribution in terms of their species occurrences and population abundances; this makes it difficult to set up regional cooperation to consider these species.
- Knowledge of how cold-water corals respond to damage inflicted by pollution is limited. Without better knowledge, it will be difficult to design protective regimes and response mechanisms.

4.2.2 Warm-water corals

With respect to warm-water corals, the issues where gaps in knowledge and capacity-building are involved include:

- Damage to warm-water corals may be more serious than currently perceived because submerged reefs below 20m depth cannot be detected using satellite technology. Submerged reefs cover large areas and understanding the extent of submerged reefs is therefore important.

- GIS mapping of coral reefs is necessary to understand their spatial distribution, especially in shallow water areas where the worst affected reefs are found.
- Corals show trends that justify measures to protect them from anthropogenic impacts. Such protection can be enhanced by spatial management tools, including the creation of MPAs. Globally, only six percent of warm-water reefs are contained in marine reserves. Establishment of more spatial management measures including MPAs would address this concern and aid in reducing anthropogenic impacts, and also assist in meeting other challenges.
- Monitoring sites and the flow of information on coral ecosystems (and in some cases other marine habitats such as mangrove and seagrasses beds) have been reduced in some cases. This will not help to improve the little that is known on the status of their ecological interaction with the changing pressures.
- Restoration and enhancement of capacity for monitoring would be required to allow status and trends of these habitats to be assessed effectively.
- Where warm-water corals are damaged by cumulative impacts, measures which address the full range of the pressures will be the most effective response. This includes pressures from tourism (see Chapter 27).
- Corals provide important cultural values. Indigenous people in some developed countries have been granted rights to access and benefit sharing of genetic resources and traditional knowledge. This recognition acknowledges the importance of these cultural aspects that link human populations and reefs. Capacity building for indigenous access and benefit sharing would be beneficial to the well-being of these peoples.

It would be extremely useful to build capacity for studying and managing coral reefs, at national, regional and global levels, to provide the right skills and infrastructure to address the issues identified and continue to enable coral reefs to provide goods and services that contribute to socioeconomic well-being and the health of the planet as a whole.

4.2.3 Estuaries and Deltas

With respect to estuaries and deltas, the issues where gaps in knowledge and capacity-building are involved include:

- A paucity of knowledge exists about the threats due to human activities, global climate change and extreme natural events.
- Globally very few integrated assessments encompass multiple aspects of estuarine environments, i.e., that include habitats, species, ecological processes, biophysical and socio-economic aspects.
- It would be extremely useful for the better conservation of estuaries and deltas to develop and apply the capacity to address these issues, including incorporating hydrological modelling into coastal modelling and forecasting efforts, in order to link better with the land-coast interface where these important habitats are located.

4.2.4 Hydrothermal vents and cold seeps

With respect to hydrothermal vents and cold seeps, the issues where gaps in knowledge and capacity-building are involved include:

- The survey and research activities have mostly been undertaken in the Pacific (especially in the northeast and northwest Pacific) and Atlantic Oceans (especially the north Atlantic). Very few have been conducted in the Indian Ocean, and those few have mostly been carried out in international waters. Therefore a better global picture of trends would require survey and research efforts to be expanded.
- Increasing knowledge of vents and seeps would only be possible if essential capacity to address all these gaps were built. For developing countries, this would need to be greatly increased, because the capacity is at best low, and usually almost non-existent, in many countries.

4.2.5 High-Latitude Ice

With respect to high-latitude ice, the issues where gaps in knowledge and capacity-building are involved include:

- The ecology of the Arctic and Antarctic regions is still little known due to the challenges their unique environments pose to human beings. This has necessitated the use of special skills and technology to undertake the essential research to understand the effects of the emerging threats of climate change, not only in these regions, but also how these effects would consequently affect wider geographical regions. Capacity to apply these skills and technologies would have to be increased to obtain the full benefit of their potential;
- The ability to manage the effects of sea-level rise caused by melting of polar ice is still a challenge. It is causing considerable social and economic losses along continental coasts and is threatening property and life on entire islands. This is due to the loss of habitats and consequently of biodiversity on which humans depend for their well-being. The costs of economic losses and level of human suffering are not fully quantified, and augmenting this knowledge is necessary to perform threat assessments of these factors;
- Further challenges stem from the inadequate understanding of the polar ecosystems; these are under increasing pressure caused by anthropogenic activities in the form of commercial exploitation of polar natural resources, which include oil and gas. With little ecological understanding of these ecosystems and therefore inadequate mitigation measures, a concern is growing as to how to deal with the looming complex environmental degradation and the need to identify and implement mitigation measures. These possible threats can only be assessed and managed if our ecological understanding is improved through expanded research and monitoring.

4.2.6 Kelp Forests and Seagrass Meadows

With respect to kelp forests and seagrass meadows, the issues where gaps in knowledge and capacity-building are involved include:

- The rate of loss of species is very high due to encroachments on these ecosystems and their proximity to coasts and consequently to human activities. The gravity and extent of these losses vary regionally and have yet to be determined in most areas. However, the causes are commonly due to coastal urbanization and industrialization, and conversion of some areas to build recreation facilities and harbours which involve heavy dredging. However, these pressures are rarely well quantified at local scales. Effective conservation and sustainable use of these habitats will require better quantification at local and regional scales.
- The costs of restoring these habitats (in the rare event that restoration is even possible) are high and the requisite restoration technology and skills are yet to be readily available in most regions. Even when restoration efforts are made, it is difficult to attain the original conditions and biodiversity that were present before degradation. Where restoration is desired or necessary for return of ecosystem services, greater study of restoration technologies would be required.
- The multitude and variety of uses of seagrass and kelp habitats (examples: aquaculture, harvesting, recreational and commercial fishing, tourism, etc.) have created conflicts over best management practices within these ecosystems. If these conflicts are going to be managed and best practices applied, improved capacity in integrated management would be necessary to address these conflicts in their early stages.

4.2.7 Mangroves

With respect to mangroves, the issues where gaps in knowledge and capacity-building are involved include:

- Despite considerable regional and global awareness campaigns on the value of mangrove ecosystems, and therefore the need to sustain their integrity so that they can provide their ecosystem services sustainably for the benefit of human well-being and the environment, estimates of increased destruction and loss in mangrove coverage continue to be reported regionally and globally at different levels of exploitation, although the actual data underpinning these estimates are unclear. If these trends are to be reversed, it is essential to document quantitatively, using the best available technological advances in skills, the various types of losses characteristic of each region and the consequences for biodiversity loss or extinction at the relevant taxonomic levels, as well as the ecosystem services that will be lost regionally and globally. This will enable assessment and quantification of the real risks and development of means to mitigate them.
- The ecosystem services provided by salt marshes are largely unknown.
- At regional and global levels, it is still not clear how to distinguish the characteristic biodiversity index of mangrove species taxonomically in a given area because of the ambiguous definition of a mangrove tree or vegetation. With existing technological advances, species identification should be based not only on morphological descriptions but also on their molecular attributes to avoid ambiguous descriptive terms like mangrove associates or hybrids. Use of these technologies in conservation and management will require building capacity for their application.
- Mangrove restoration is still at its early stages of development. It either uses seeds planted directly in the soil of mangrove habitat or seeds that are first nurtured and grown in a nursery before being planted in the mangrove habitats along the shores. These seeds are not improved in any way. If mangrove restoration is to accelerate it would be necessary to promote faster growing mangrove trees, including those improved through the use of biotechnology application and to ensure that the physical and chemical properties of the soils are optimal for their growth and that mangrove pests are eliminated or kept away from the plantations. These activities should involve local communities to enhance their education about and awareness of this ecosystem.
- Conservation and sustainable use of mangroves would benefit from promotion of ecotourism in natural and restored mangrove forests, managed by local communities for income generation; this is expected to instil in them the importance of these ecosystems in supporting their livelihoods without destroying them for unsustainable exploitation.
- To enhance carbon sequestration and at the same time increase their economic income as well as supporting mangrove conservation and enhancing mangrove ecosystem services would require increased carbon credits to local communities that become involved in growing mangrove forests.
- Protection of mangroves will require improved understanding of why naturally occurring bare, salty, and sandy flats occur in mangrove ecosystems, which would also inform the creation of buffer zones in landward areas that will allow mangroves to migrate landward in response to sea-level rise. This is an established practice for integrated coastal zone management.
- Promotion of ecotourism in natural and restored mangrove forests, as well as management by local communities for income generation, will instil in them the importance of these ecosystems for supporting their livelihoods, without destroying the system through unsustainable exploitation.
- Capacity-building needs should be recognized if there is a desire to address acquisition of technological skills to enhance restoration, growth and management of mangrove forests, infrastructure to support development and use of biotechnology techniques to promote faster growing mangroves and to improve soils.

4.2.8 Salt Marshes

With respect to salt marshes, issues where gaps in knowledge and capacity-building are involved include:

- Salt marshes, in both tropical and temperate zones, are one of the fastest disappearing ecosystems worldwide. This is mostly due to anthropogenic activities, yet little is known about them in terms of their ecology and socio-economic contribution to human well-being.

- In the tropics and sub-tropics, the nature of the ecological interaction of salt marshes and mangrove ecosystems where they share a location is largely unknown; one result is the classification of salt marsh vegetation as associate mangrove species. In other words, the ecological role of salt marshes is masked by, or confused with, mangrove vegetation and therefore constitutes a large knowledge gap for both ecosystems.
- The ecological significance of the role of migratory fauna between salt marsh and mangrove vegetation is poorly known.

4.2.9 Sargasso Sea

With respect to the Sargasso Sea, the issues where gaps in knowledge and capacity-building are involved include:

- If the following issues are to be addressed, it is necessary to build techniques, personnel and infrastructure to address them: The Sargasso Sea is a complex habitat characterized by an interdependent mix of its physical oceanography, its ecosystems and its role in the global scale of ocean and earth processes. It is not fully known how these processes operate to produce this unique habitat.
- The Sargasso Sea ecosystem links to ecosystems in Europe, Africa, the Americas and the Caribbean. This provides a unique ecosystem for study to understand how the divergent and convergent ecological functions of these widely spread, but interlinked, geographic ecosystem regimes operate. Targeted research could produce new knowledge of impacts caused by climate change.

4.2.10 Seamounts and other submarine geological features

With respect to submarine geological features, the issues where gaps in knowledge and capacity-building are involved include:

- Seamounts are predominantly submerged volcanoes, generally now extinct, that can rise to a few thousand metres above the surrounding seafloor. The most significant human activities around seamounts so far are fishing and, potentially, mining. To increase the knowledge available to manage activities around these features, it would be necessary to build techniques, personnel and infrastructure to address the following issues: Only about 6.5 per cent of the sea floor is mapped, so the global number of seamounts can only be estimated.
- Globally, overall species richness in seamount ecosystems is poorly known and therefore improving our knowledge of species composition would require undertaking comprehensive studies of the ecology of seamounts, ridges and other sensitive submarine benthic habitats. Appropriate conservation of these ecosystems requires scientific research.
- The interaction of the geological features with the overlying water column is poorly known.
- Impacts of acidification and de-oxygenation on these ecosystems are also unknown, and are not monitored sufficiently to detect impacts: many seamounts already experience low oxygen and low calcium carbonate saturation levels.
- Trawl gear disturbs and destroys benthic fauna and in some seamounts little decolonization is observed, even years after the closure of fishing. The destructive effects of trawl gear on benthic communities are generally incompletely known, but it is possible that these have reduced the ecological resilience and consequently also reduced reproductive potential, and contributed to the loss of genetic diversity and ecological connectivity.
- Capacity for stock assessment and sustainable management, including investment in shared infrastructure (for example, fisheries research vessels), is insufficient and capacity building would improve the possibility that such fisheries could be sustainably managed.
- Mining of seamounts would benefit from multisectoral management, especially for balancing mining and fishery interests. A first step in this direction could be to build the capacities of those involved to participate in the international work on this subject.
- Managing the effects of multiple stressors on seamounts would benefit from expanding both monitoring and research and may require building capacity to address this need. This would include capacity building for personnel and infrastructure, including multidisciplinary research teams, research vessels and laboratories.

References

Millennium Ecosystem Assessment (2005). *Ecosystems and human well-being*. Washington, D.C., Island Press.

UNGA (2010). *Report of the Secretary-General* (A/65/69/Add.1).

UNGA (2011). *Final report of the workshop held under the auspices of the United Nations in support of the regular process for global reporting and assessment of the state of the marine environment, including Socio-economic aspects*. Santiago, Chile, 13-15 September 2011(A/66/587).

UNGA (2012a). *Final report of the Workshop held under the auspices of the United Nations in support of the Regular Process for Global Reporting and Assessment of the state of the Marine Environment, including Socio-economic Aspects*. Sanya, China, 21-23 February 2012 (A/66/799).

UNGA (2012b). *Final Report of the workshop held under the auspices of the United Nations in support of the Regular Process for Global Reporting and Assessment of the State of the Marine Environment, including Socio-economic Aspects*. Brussels, 27 to 29 June 2012 (A/67/679).

UNGA (2013a). *Final Report of the sixth workshop held under the auspices of the United Nations in support of the Regular Process for Global Reporting and Assessment of the State of the Marine Environment including, Socio-economic Aspects*. Brisbane Australia, 25-27 February 2013 (A/67/885).

UNGA (2013b). *Final report of the fourth workshop held under the auspices of the United Nations in support of the Regular Process for Global Reporting and Assessment of the state of the Marine Environment, including Socio-economic Aspects*. Miami, United States of America, 13-15 November 2012 (A/67/687).

UNGA (2013c). *Final report of the fifth workshop held under the auspices of the United Nations in support of the Regular Process for the Global Reporting and Assessment of the State of the Marine Environment, including Socio-economic Aspects*. Maputo, Mozambique, 6 and 7 December 2012 (A/67/896).

UNGA (2013d). *Final report of the fourth workshop held under the auspices of the United Nations in support of the Regular Process for Global Reporting and Assessment of the state of the Marine Environment, including Socio-economic Aspects*. Grand-Bassam, Cote d' Ivoire, 28-30 October 2013 (A/68/766).

UNGA (2014). *Report of the eighth workshop held under the auspices of the United Nations in support of the Regular Process for Global Reporting and Assessment of the State of the Marine Environment, including Socio-economic Aspects*. Chennai, India, 27-29 January 2014 (A/68/812).

UNGA/AHWGW (2009). *Report on the work of the Ad Hoc Working Group of the Whole to recommend a course of action to the General Assembly on the Regular Process for Global Reporting and Assessment of the state of the Marine Environment, including Socio-economic Aspects*. (A/64/347).

UNGA/AHWGW (2010). *Report on the work of the Ad Hoc Working Group of the Whole to recommend a course of action to the General Assembly on the Regular Process for Global Reporting and Assessment of the state of the Marine Environment, including Socio-economic Aspects*. (A65/358).

VII Overall Assessment

54 Overall Assessment of Human Impact on the Oceans

Group of Experts:
Patricio Bernal, Beatrice Padovani Ferreira, Lorna Inniss, Enrique Marschoff, Jake Rice, Andrew Rosenberg and Alan Simcock (Co-Lead Members)

Chapter 54
Overall Assessment of Human Impact on the Oceans

1 Overview of impacts

No part of the ocean has today completely escaped the impact of human pressures, including the most remote areas. One clear example of this is the universal presence of stratospheric fall-out from atmospheric nuclear-weapons testing, but many other pressures on the marine environment are nearly as widespread.

Human pressures impact on the ocean in many and complex ways. They can take effect directly (as when an oil spill kills sea-birds and sessile benthic biota) or indirectly (as when climate change results in changes to the stratification of seawater, with an adverse effect on the nutrient cycle and the production of the plankton on which fish feed). Equally, the effects can be seen both on the natural environment (as when populations of sea turtles are reduced by tourist development on or near their breeding beaches) as well as on human society and economic activities (as when the collapse of a fish stock removes the economic base of coastal communities). Human pressures can also vary widely in their intensity and spread. Sometimes they have a concentrated impact: for example, the annual expansion of a large dead zone in the Gulf of Mexico, resulting from the high level of inputs of nitrogen compounds in the run-off from the Mississippi and other catchments. Sometimes the effects of human pressures have a very widely distributed effect: for example, the diffusion of persistent organic pollutants over the Arctic zone by airborne volatilization (for both examples, see Chapter 20 on land-based inputs) (Halpern, 2008).

1.1 Summarizing the impacts

An analysis of the overall impact of all the human pressures examined in this Assessment has to start by looking at the direct impacts and collateral effects of each pressure and to examine where those impacts and effects are found. However (as argued below), although this is an essential first step, it is not enough. In addition, any review of the effects of human pressures on the marine environment has to look both at the effects on the marine environment and at the consequences for human society and economies. A taxonomy of the main sources of human pressures on the marine environment that need to be considered must include the following (though these are not listed in any order of priority):

(a) Climate change (and ocean acidification, including the resulting changes in salinity, sea-level, ocean heat content and sea-ice coverage, reduction in oxygen content, changes in ultra-violet radiation);
(b) Human-induced mortality and physical disturbance of marine biota (such as capture fisheries, including by-catch, other forms of harvesting, accidental deaths such as through collisions and entanglement in discarded nets, disturbance of critical habitat, including breeding and nursery areas);
(c) Inputs to the ocean (these can be broken down according to the nature of their effects: toxic substances and endocrine disruptors, waterborne pathogens, radioactive substances, plastics, explosives, excessive nutrient loads, hydrocarbons). Remobilization of past inputs also needs to be considered;
(d) Demand for ocean space and alteration, or increase in use, of coasts and seabed (conflicting demands lead to both changes in human use of the ocean and changes to marine habitats);
(e) Underwater noise (from shipping, sonar and seismic surveys);
(f) Interference with migration from structures in the sea or other changes in routes along coasts or between parts of the sea and/or inland waters (for example, wind-farms, causeways, barrages, major canals, coast reinforcement, etc.);
(g) Introduction of non-native species.

It is a matter of debate how any taxonomy should be structured. For example, all inputs might be classed together, since they are all the result of human activities affecting the ocean. However, there are important differences in the ways in which these pressures will affect the littoral, the water column and the benthos. In addition, the way in which these affect the environment and human societies and economies differs significantly. Hazardous substances may have toxic effects (either directly on animals which ingest them or through the food web on animals and humans that eat contaminated fish and seafood), may affect resilience to infections or may affect reproductive success. Waterborne pathogens may affect marine biota, but can be of particular concern when they are likely to affect humans who bathe in the sea or eat seafood. Excessive nutrients may lead to dead zones or cause blooms of algae that generate toxins. Explosives from past wars dumped into the sea may well not affect marine biota, but may kill or maim fishers who bring them up in trawls. Hydrocarbons may kill marine biota directly, but can also be broken down by bacteria and thus enter the food web. The worst effects of some emissions (such as exhaust fumes from ships) may not be the way that they enter the sea, but the way in which they contribute to damage to human health on land through air pollution. No taxonomy of these kinds of pressures, which are operating in very different fields, is likely to be beyond debate. Table 54.1 summarizes the varieties of human pressures on the marine environment, indicating the environmental and the social and economic effects. The categories of pressure aim to bring together the pressures resulting from various human activities that have similar effects, but keep separate some categories which have some effects of a very different nature, even though they may overlap with other categories in creating some effects.

1.2 Environmental effects

This chapter aims to summarize the overall impact of human activities on the ocean. The elements noted in Table 54.1 therefore relate very much to the impact of human activities on the marine environment. As the regional biodiversity assessments in chapter 36 of Part VI of this Assessment show, there are well-documented examples of cases where habitats, lower-trophic-level productivity, benthic communities, fish communities, or seabirds or marine mammal populations have been severely altered by pressures from a specific activity (such as over-fishing, pollution, nutrient loading, physical disturbance, or non-native species). However, many biodiversity impacts, particularly at larger scales, are the result of cumulative and interactive effects of multiple pressures from multiple drivers. It has repeatedly proven difficult to disentangle the effects of the individual pressures. This impedes the ability to address the individual causes.

Even in the Arctic Ocean, where human settlements are relatively few and small, the potentially synergistic effects of multiple stressors come together. And this is against a background of pressures from a changing climate and increasing human maritime activity, primarily related to hydrocarbon and mineral development and to the opening of shipping routes. These changes bring risks of direct mortality, displacement from critical habitats, noise disturbance, and increased exposure to hunting, which are superimposed on high levels of contaminants, notably organochlorines and heavy metals, as a result of the presence of these substances in the Arctic food web.

Likewise, in the open ocean (remote from land-based inputs), shifts in bottom-up forcing (that is, primary productivity) and competitive, or top-down forcing (that is, by large predators) will produce complex and indirect effects on ecosystem services. Stress imposed by lower oxygen, lower pH (that is, higher acidity), or elevated temperature can reduce the resilience of individual species and ecosystems through stressing organism tolerances or shifting community interactions. Where this happens, it retards recovery from disturbance caused by human activities such as oil spills and trawling and (potentially in the future) seabed mining. Acidification-slowed growth of carbonate skeletons, delayed development under hypoxic conditions, and declining food availability illustrate how climate change could exacerbate anthropogenic impacts and compromise deep-sea ecosystem structure and function, and ultimately its benefits to human welfare.

These multiple pressures interact in ways that are poorly understood, but that can amplify the effects expected from each pressure separately. The North Atlantic is comparatively rich in scientific resources. It has many long-term ocean-monitoring programmes and a scientific organization (the International Council for the Exploration of the Sea) that have functioned for over a century to promote and coordinate scientific and technical cooperation among the countries around the North Atlantic. Even here, however, experts are commonly unable to disentangle consistently the causation of unsustainable uses of, and impacts on, marine biodiversity. This may seem initially discouraging. Nevertheless many well-documented examples exist of the benefits that can follow from actions to address past unsustainable practices, even if other perturbations are also occurring in the same area.

Cumulative effects are documented for species groups of the top predators in the ocean, including marine mammals, seabirds, and marine reptiles. Many of these species tend to be highly mobile, and some species migrate across multiple ecosystems and even entire ocean basins, so they can be exposed in their annual cycle to many threats. Direct harvest occurs for some of these species, particularly some pinnipeds (seals and related species), seabirds and sharks, and bycatch in fisheries can cause significant mortality for many species. However, in addition to having to sustain the impacts from these direct deaths, all of these species suffer from varying levels of exposure to land-based pollution sources and increasing levels of noise in the ocean. Land-nesting seabirds, marine turtles and pinnipeds also face habitat disturbance, including invasive predators on isolated breeding islands, disturbance of beaches where eggs are laid, or direct human disturbance from tourism, including ecotourism.

Some global measures have been helpful in addressing specific sources of mortality, such as the global ban on high-seas drift-netting introduced by the United Nations General Assembly in 1994, which was a major step in limiting the bycatch of several marine mammal and seabird species that were especially vulnerable to entanglement. However, for seabirds alone, at least 10 different pressures have been identified that can affect a single population through its annual cycle, with efforts to mitigate one sometimes increasing vulnerability to other pressures. Because of the complexity of these issues, conservation and management must be approached with care and with alertness to the nature of the interactions among the many human interests, the needs of the animals and their role in marine ecosystems.

1.3   Social and economic effects

Many of the human activities that affect the ocean affect not only its environmental condition, but also various social and economic aspects related to the marine environment. Most human activities in and around the ocean are aimed at getting some form of social or economic benefit from the ocean, and Chapter 57 (Overall value of the ocean to humans) attempts to pull together these aspects. Some human activities, in effect, can undermine their own success: capture fisheries and tourism are a good example of this: over-fishing results in keeping harvested species at less than the maximum sustainable yield, while tourism that attracts too many tourists can downgrade the environment that originally attracted them. In addition, many types of human activity may have adverse impacts on the success of other human activities. For example, marine noise from ships may cause the marine mammals to re-locate and thus undermine a previously successful whale-watching activity (see Chapters 27 and 37). The trade-offs among classes of interacting activities need careful consideration – especially as governance arrangements may make it difficult for such trade-offs to be easily considered together. This can happen either because the voices of some of those affected are not easily heard (for example, small-scale fishers) or because the governance arrangements do not address the same areas (for example, long-range aerial or riverine transport of pollutants may start in areas well away from any ocean). Some effects (such as ocean acidification) may only be capable of being addressed at a global scale, but the ecological effects may be much more localized, because of the uneven distribution of the environmental effects. Likewise, the social and economic impacts of such global pressures may be much more unequally distributed than the ecological effects, because of regional differences in uses of the ocean.

Many of the more serious cases of trade-offs of this kind affect food from the sea. As explained in Part IV, overfishing of certain fish stocks is a very clear example of the way in which an activity can undermine its own success in generating economic and social benefits in terms both of food from the sea and of employments and livelihoods. At the same time, excessive inputs of nutrients (among other things, from sewage discharges or agricultural run-off) can lead to dead zones or hypoxic zones, which can seriously affect the recruitment of fish stocks on which both large-scale and small-scale fisheries depend. To these adverse effects on fish stocks can be added further effects such as those from losses of breeding or nursery areas through land reclamation, the effects of hazardous substances on reproductive success and oil pollution from shipping. Since small-scale fisheries are in general less well studied than the larger, more commercial fisheries, the social and economic consequences of these multiple impacts are not easily quantified. Indeed, as noted above, even for larger, more commercial fisheries the overall way in which multiple pressures work together to produce adverse effects is not well understood. Nevertheless, it is clear that some problems are sufficiently well understood that remedial actions can have some success. For example, reductions in the occurrence of liver tumours in fish in Netherlands waters have been linked to decreases in the levels of organic pollutants (OSPAR 2010). On the other hand, improvements in aquaculture techniques have allowed substantially increased production with lower inputs of fishmeal (FAO, 2012).

The changes in marine biodiversity can have knock-on effects on other ecosystem services that humans obtain from the ocean. An illustration of this is the important link between the health of warm-water corals and tourism. Warm-water corals represent a major component of the attractiveness of many tourist resorts in the Caribbean, the Red Sea, the Indian Ocean, south-east Asia and the South Pacific. The competitive position of their resorts would be seriously undermined if the tourists could no longer enjoy the corals. The same applies to other resorts (even in cold-water areas) where one of the attractions is scuba diving to enjoy the marine ecosystems.

The disappearance (or, more commonly, the reduction in numbers) of iconic species can similarly adversely affect traditional practices. For example, native people on the north-east Pacific coast have seen their traditional whale-hunting halted, because of past over-harvesting by others of grey whales (see Chapter 8, Cultural ecosystem services from the ocean). This hunting was an integral part of their cultural heritage, and the affected tribes consider the cultural loss to be very serious. Pollution can have similar effects: for example, the Faeroese authorities are taking measures to control the traditional food obtained in the islands from pilot whales, because of the high levels of pollutants they contain (see Chapter 20, Land-Based inputs to the ocean). Demand for ocean space and alteration of coasts and seabed will lead to destruction of underwater cultural heritage (see Chapter 26 on land/sea physical interaction; and Chapter 27 on tourism and recreation).

2 Information gaps and capacity building gaps

2.1 Information gaps

Taking an overall view of the state of the world's marine environment presents many challenges, because it requires a large number of different sets of data to be brought together. Techniques for doing this are in their infancy, and many difficult problems need to be resolved.

In the first place, as the chapters in Parts III, IV, V and VI of this Assessment demonstrate, there are many gaps in the basic information necessary to build a reliable, world-wide, comprehensive, quantified survey of the state of the ocean. This Assessment shows that a qualitative view can generally be achieved of most aspects of the oceans and that some aspects can, at least in places, be quantified. More quantified information is needed to achieve a robust quantified assessment. The various chapters of Parts III, IV, V and VI of this Assessment identify major information gaps. Most of these will need to be filled before detailed methods of quantification can be developed that will achieve general acceptance.

In pursuing the aim of a more quantified integrated assessment of the ocean, it will therefore be important to try to improve the detailed information available.

2.2 Capacity gaps

At the same time, there is a more general gap in techniques for bringing information on the different aspects of the ocean together to give an overall picture. Various attempts have been made to do this at various levels, both as to the area to be covered and as to the degree of integration sought.

2.3 Ocean Health Index

One of these is the Ocean Health Index (OHI) (OHI, 2014; OHI, 2013; OHI, 2014). This index is mentioned as an illustration of the challenges in preparing even a semi-quantitative, but comprehensive, assessment of the ocean. There is a wide range of expert views of the robustness of this index – and, indeed, of other such indices. At the same time, it should be noted that many of the most important messages drawn from the OHI do correspond to conclusions drawn in this Assessment. Those conclusions have been drawn by other assessments as well.

The OHI is an attempt to produce a comprehensive assessment of the ocean in numerical terms at the highest possible level. Originally covering only coastal waters, it now covers all aspects of the marine environment and all parts of the ocean (220 areas within national jurisdictions and 16 much larger areas beyond national jurisdictions). Its aim is to convert all the information into numerical scores for the status of each of the goals and sub-goals (shown in Table 54.2). Some of these goals have clear gaps: for example, the "Clean Water" goal does not cover

Chapter 54 Overall Assessment of Human Impact on the Oceans

Table 54.1 | Summary of the goals and sub-goals used for the Ocean Health Index

Goal	Sub-goal	Reference point type and brief description of basis
Food Provision	Fisheries	Functional relationship (difference of total landed biomass from estimated maximum sustainable yield)
	Mariculture	Spatial comparison (sustainably harvested yield of mariculture normalised for the area of inshore waters)
Small-scale Fishing Opportunities		Functional relationship (level of demand for small-scale fisheries (estimated from poverty level and degree of regulation of such fisheries)
Natural Products		Temporal comparison (historical benchmark) (level of exports for the area of coral, ornamental fish, fish oil, seaweeds and marine plants, shells, and sponges compared with the highest level achieved, as a substitute for the maximum possible level)
Carbon Storage		Temporal comparison (historical benchmark) (Current area of mangroves, seagrass beds and salt-marshes compared with historical benchmark)
Coastal Protection		Temporal comparison (historical benchmark) (Current area of mangroves, coral reefs, seagrasses, salt marshes, and sea ice compared with historical benchmark and adjusted for the differing protective effects of each)
Coastal Livelihoods & Economies	Livelihoods: jobs and wages	Temporal and spatial comparisons (moving target) (Number of jobs directly and indirectly supported by tourism, commercial fishing, marine mammal watching, aquarium fishing, wave and tidal energy, mariculture, transportation & shipping, ports and harbours, shipbuilding and boatbuilding, compared with average of last five years, and adjusted by the average wage in each sector)
	Economies	Temporal comparison (moving target) (contribution to Gross Domestic Product generated directly or indirectly by the sectors mentioned in the entry of the previous sub-goal, compared with historical benchmark)
Tourism & Recreation		Spatial comparison (Originally based on international tourist arrivals, but since 2013 based on employment in tourism, adjusted by for sustainability in line with the World Economic Forum's Travel and Tourism Competitiveness Index)
Sense of Place	Iconic Species	Known target (Percentage of species in the World-Wide Fund for Nature's lists of Priority Species and Flagship Species for the area that are classed by the International Union for the Conservation of Nature (IUCN) as threatened, weighted by the threat category)
	Lasting Special Places	Established target (The mean of (a) area of coastal marine protected areas as a percentage of an assumed target that 30% of the area within 3 nautical miles of the coast should be protected, and (b) the length of coastline within 1 kilometre of the shore that is protected as a proportion of an assumed target that 30% of such coast should be protected)
Clean Waters		Known target (Geometric mean of (a) number of people in the coastal area without access to enhanced sanitation, rescaled to the global maximum, (b) modelled index of land-based inorganic pollution from urban runoff from impervious surfaces, (c) modelled index of land-based organic pollution from pesticides and (d) modelled index of pollution from shipping and ports)
Biodiversity	Habitats	Temporal comparison (historical benchmark) (average of the assessed conditions of such of the range of mangroves, coral reefs, seagrass beds, salt marshes, sea-ice edge, and sub-tidal soft-bottom habitats as are present in the area; the assessments of conditions are drawn from a variety of wide-ranging assessments of these habitats)
	Species	Known target (Temporal comparison (historical benchmark) (IUCN Global Marine Species Assessment of the extinction risk status of 2,377 species for which distribution maps exist, calculated as the area- and threat-status-weighted average of the number of threatened species within each 0.5° grid cell)

point source discharges. The exercise also derives figures for trends, pressures and resilience to allow forecasting of future status. Given the limitations of the data that are available, various statistical techniques have had to be applied to that data in order to achieve coherent, comprehensive outputs. A detailed study of the efforts involved in developing the Ocean Health Index quickly shows how difficult it is to gather full information.

Having derived numerical scores for the goals and sub-goals, the next step in the OHI process is then to aggregate the indices developed for each goal into a single index figure for the status of each area of sea covered by the exercise, and then into a single figure for the ocean as a whole. It is possible to allow for different weightings between the results for the different goals, based on expert judgement, in order to allow for different views on the balance between preservation and exploitation (OHI, 2013).

The OHI depends crucially on the availability of satisfactory data across many fields, and on the expert judgements made about the weighting to be given to the different fields covered. Much of the necessary data is not available, and estimates of various kinds have to be used instead. The scale of the expert judgements needed means that there is a substantial subjective component in any results.

2.4 Water-quality indexes

At a much less aggregated level, as described in Chapter 20 (Land-based inputs to the ocean), some regional seas organizations and some States have tried to produce a single index of water quality in the parts of the ocean with which they are concerned. Such efforts, too, require judgements on the relative importance of the effects of hazardous substances and of eutrophication problems, and therefore rely to a substantial degree on expert judgement.

2.5 Ecological quality objectives

An alternative approach accepts that there will inevitably be an element of expert judgement involved, and legitimate differences in views on the appropriate weights given to various types of impacts and benefits, and therefore develops measures along a number of axes. There is no

Table 54.2 | Descriptors of Good Environmental Quality for the European Union Marine Strategy Framework Directive

Descriptor	Title	Detail
1	Biodiversity	Biological diversity is maintained. The quality and occurrence of habitats and the distribution and abundance of species are in line with prevailing physiographic, geographic and climatic conditions.
2	Non-indigenous species	Non-indigenous species (NIS) introduced by human activities are at levels that do not adversely alter the ecosystems.
3	Fish and Shellfish stocks	Populations of all commercially exploited fish and shellfish are within safe biological limits, exhibiting a population age and size distribution that is indicative of a healthy stock.
4	Food webs	All elements of the marine food webs, to the extent that they are known, occur at normal abundance and diversity and levels capable of ensuring the long-term abundance of the species and the retention of their full reproductive capacity.
5	Eutrophication	Human-induced eutrophication is minimised, especially adverse effects thereof, such as losses in biodiversity, ecosystem degradation, harmful algae blooms and oxygen deficiency in bottom waters.
6	Benthos	Sea floor integrity is at a level that ensures that the structure and functions of the ecosystems are safeguarded and benthic ecosystems, in particular, are not adversely affected.
7	Hydrography	Permanent alteration of hydrographical conditions does not adversely affect marine ecosystems.
8	Contaminants	Concentrations of contaminants are at levels not giving rise to pollution effects.
9	Fish and seafood quality	Contaminants in fish and other seafood for human consumption do not exceed levels established by Community legislation or other relevant standards.
10	Marine litter	Properties and quantities of marine litter do not cause harm to the coastal and marine environment.
11	Energy introduction	Introduction of energy, including underwater noise, is at levels that do not adversely affect the marine environment.

attempt to convert these various measures into a single quantified measure. Rather, users are left to apply their varying expert judgements on how much importance to attach to each axis, and on how to interpret what the different measures show. One version of this approach, developed by the regional seas organization for the North-East Atlantic, has been to try to find a suitable set of ecological quality objectives (EcoQOs) for an ocean area (OSPAR, 2007). These EcoQOs are derived by considering successively:

(a) What are the important ecosystem components that collectively reflect a high ecological quality?
(b) What are the human impacts on this component and how can they be monitored?
(c) What are the objectives to be achieved, taking into account existing policies?

These EcoQOs may be quite numerous, and no attempt has yet been made to specify what the relation among them should be: the aim is to develop a set of measures that can be used for diagnosing whether there are problems. So far, a pilot project has looked at 11 such EcoQOs for the North Sea (OSPAR, 2007).

2.6 European Union's Marine Strategy Framework Directive

A related approach is being developed for the implementation of the European Union's (EU) Marine Strategy Framework Directive (MSFD) (EU, 2008). As a starting point, this involves each EU coastal Member State assessing the state of its waters against a list of eleven descriptors, shown in Table 54.3. The European Commission has produced a set of criteria and indicators to assist in developing common approaches to making these assessments. An initial assessment should then be made whether assessments show that the waters of the Member States have "good environmental status". Environmental targets, associated indicators and a programme of measures to maintain that state, or to achieve it by 2020, should then be established by 2015. A preliminary report by the European Commission suggests that much work remains to be done to deliver this programme, and agreement on the relative or absolute benchmarks for good environmental status on many of the descriptors has not been reached (EU, 2014).

Unlike the Ocean Health Index, however, these EcoQO and MSFD approaches do not specifically integrate social and economic aspects, although the effects of sustainable uses are taken into account in setting their benchmarks for good environmental status.

2.7 Conclusion on capacity-building gaps

Some attempts have been made to develop ecosystem-based approaches to managing human activities that affect the ocean. Even here, however, much work remains to be done to develop systems for assessing the overall impacts of human activities on the ocean. There thus remains a general need to develop methods for integrated assessments of the marine environment that can deliver an assessment of the marine environment that is not only (1) integrated across environmental, social and economic aspects, (2) integrated across sectors of human activities, and (3) integrated across all the components of the marine environment, but also gives reliable, quantified information about all parts of the world. There is therefore a general need for capacities to develop and implement such assessment methods.

Chapter 54 — Overall Assessment of Human Impact on the Oceans

Table 54.3 | Pressures and Impacts of Human Activities on Environmental and Socioeconomic Aspects of the Marine Environment

No	Pressures from Human Activities[1]	See	Impacts on Environmental Aspects of the Marine Environment	Impacts on Socioeconomic Aspects of the Marine Environment	Mgt[2]
1	Acidification of the ocean (arising from increased CO2 emissions)	Ch 5, Ch 7, Ch 36 A-H, Ch 42, Ch 43, Ch 46	Reduction of reproductive success, recruitment, growth and survival of some species, especially those with (calcareous) exoskeletons (shells etc). Reduced resilience of coral reefs to other stresses. Second-order loss of habitat for other species if coral reefs degrade.	Losses in livelihoods in some small-scale fisheries. Lower production of some commercial fisheries. Loss of competitiveness for tourism dependent on corals. Potential loss of coastal protection services where coral reefs are degraded. Potential costs of reducing CO_2 emissions.	Not yet
2	Changes in sea temperature	Ch 4, Ch 5, Ch 7, Ch 34, Ch 36 A-H, Ch 42-50, Ch 43, Ch 15	Increased sea-surface temperature will probably increase stratification and thus affect nutrient cycling, with effects on productivity. Changes in species distribution and productivities, bottom up ecosystem productivity and community structure. Coral bleaching. Reduction of sea-ice cover in Arctic and Antarctic will impair species dependent on that habitat.	Adverse changes in weather patterns, including increased storms in higher latitudes. Fisheries and aquaculture potential may have to relocate or change preferred species. Changes in high latitude temperature regimes increase access for many industries with the potential for major impacts on Arctic communities.	Not yet
3	Changes in the salinity of seawater (arising from climate change)	Ch 4, Ch 6, Ch 15, Ch 34, Ch 36 A-G	Changes to the thermohaline circulation of the ocean, in some places leading to increased up-welling of nutrients (see also Item 14). Increased likelihood of stratification of seawater, with consequent adverse effects on primary production that supports fish and seabirds.	Potential fundamental changes in availability of fishery resources with implications for food security and other important ecosystem services. Changes in currents may alter the way that ocean moves heat around the planet, with widespread consequences	Not yet
4	Creation of underwater noise (arising from shipping, offshore prospecting, offshore renewable energy installations and tourism and recreation)	Ch 17, Ch 21, Ch 22, Ch 23, Ch 27, Ch 36, Ch 37	Disturbance of fish, macro-invertebrates, and marine mammals. Mortality due to noise rare but disruption of behaviour may have consequences for life history activities including feeding, migration, recruitment and social behaviour.	Potential costs of reducing noise emissions, including potential closure of sensitive areas to certain activities seasonally or permanently, thus limiting economic activity.	Yes
5	Increased demands for marine space for potentially conflicting uses (arising from fisheries, aquaculture, shipping routes, submarine cables and pipelines, offshore hydrocarbon and mining operations, solid waste disposal, tourism)	Ch 11, Ch 12, Ch 14, Ch 18, Ch 19, Ch 21, Ch 22, Ch 23, Ch 24, Ch 26, Ch 27, Ch 48	Depending on the human activity, the ecological functions of natural habitats in the marine space allocated for human use may be altered, degraded, or destroyed (including by removing or smothering marine plants and benthos). Consequent reductions of habitat available for nature. Changes in habitat productivity can alter ecosystems. Disposal of disused offshore installations can create new habitats.	Conflicts among potential uses of a place may arise, causing problems in finding most suitable allocation of space among potential uses, and increases in costs to manage conflicts. Development pressures may favour higher impact uses such as ports or energy production, with negative implications for lower impact uses such as small-scale subsistence fishing, impacting food security. Secondary impacts on harvesting and tourism are possible, if the permitted uses decrease biological productivity or make the area unavailable.	Yes
6	Increased direct mortality of marine animal populations, including those not directly targeted (arising particularly from fisheries, including recreational fisheries)	Ch 11, Ch 15, Ch 17, Ch 27, Ch 36, Ch 37, Ch 38, Ch 39, Ch 40	Decline in populations if the mortality is unsustainable. Alterations in population structures towards ones composed of smaller and younger individuals, with broader impacts on productivity. Potential alterations to ecosystem balance through differential effects on species.	Unsustainable mortality rates imply declines of living marine resources, with implications of decreasing food security, reduced livelihoods in coastal areas, and reduction in recreational enjoyment. The costs of restoring over-exploited resources are generally very high compared to those of preventing overexploitation from occurring.	Yes
7	Increased disturbance of fauna and flora, arising from increased numbers of people in the coastal zone, and increased amounts of shipping	Ch 17, Ch 18, Ch 21, Ch 23, Ch 26, Ch 27, Chs 37 – 44	High levels of the presence of people affect animal behaviour, including breeding, rearing, feeding, and migration. May reduce the carrying capacity of the coastal zone for marine biota.	Need to manage access of people to ecologically significant places can impose costs on development, and limit scale of industries such as eco-tourism	Yes

Overall Assessment of Human Impact on the Oceans — Chapter 54

No	Pressures from Human Activities[1]	See	Impacts on Environmental Aspects of the Marine Environment	Impacts on Socioeconomic Aspects of the Marine Environment	Mgt[2]
8	**Increased ultra-violet radiation** (arising from reductions in ozone layer)	Ch 6	Possible adverse effects on primary production and on fish larvae. Effects on titanium dioxide nanoparticles, creating biocides affecting phytoplankton, and thus potentially the food web.	Potential effects on harvesting if fish stocks are affected.	Yes[3]
9	**Input of explosives and hazardous gases in containers** (from dumping)	Ch 24	Additional source of hazardous substances and seabed smothering: see Items 12 and 17.	Harm to fishers who catch such dumped material in their nets, and to pipeline- and cable-laying in affected areas.	Yes
10	**Input of hydrocarbons** (from land-based sources, offshore installations, pipelines and shipping)	Ch 12 Ch 17 Ch 19 Ch 20 Ch 21 Ch 23 Ch 27 Ch 37 Ch 38 Ch 39	Killing of benthic biota, fish, marine mammals and reptiles and sea birds. Adverse effects on their later reproductive success.	Consequent damage to aquaculture and fisheries. Fouling of beaches and consequent adverse impact on tourism	Yes
11	**Input of nutrients, both airborne and water-borne** (arising from land-based activities, shipping, solid waste disposal).	Ch 6 Ch 12 Ch 17 Ch 20 Ch 24 Ch 25 Ch 27 Ch 36 A-H Ch 43 Ch 44 Ch 48	Coastal eutrophication, leading to dead zones, hypoxic zones and algal blooms (including toxic algal blooms). Shifts of ecosystem regimes. Consequent loss of benthic diversity and adverse effects on fish and shellfish stocks and on seabirds and marine mammals and reptiles. Algal smothering of coral reefs	Adverse effects on human health, especially through shell-fish poisoning and waterborne pathogens. Adverse effects on fisheries and shellfisheries from dead zones and hypoxic areas. Adverse effects on tourism from beaches covered in algae, and loss of competitiveness from reduced marine wildlife (especially where coral reefs are affected) Increased costs of treatment of inputs.	Some
12	**Input of plastics** (from shipping, fishing, offshore installations, poor control of land-based waste disposal, dumping).	Ch 6 Ch 11 Ch 17 Ch 24 Ch 25 Ch 37 Ch 38 Ch 39	Potential effects from breakdown into nanoparticles on food web, through effects on plankton and on filter-feeding species, resulting in changes in productivity. Mortality from ingestion by, and physical entanglement of, fish, marine mammals, reptiles and seabirds. Loss of habitat contaminated with durable debris.	Potential effects on fish and shellfish stocks through changes in the food web. Loss of vulnerable species may impact tourism or cultural needs. Loss of amenity and fouling of beaches. Consequent adverse impacts on tourism. Costs for cleanup of plastics, lost fishing gear etc., are very high.	Yes
13	**Input and transfer of waterborne pathogens** (arising from land-based activities, open-pen aquaculture, shipping and offshore installations).	Ch12 Ch 17 Ch 20 Ch 25 Ch 37 Ch 43	Possible adverse effects on marine fish, bird, turtle and mammal populations due to introduction of diseases. Coral diseases leading to death and impacts in coralline communities.	Damage to human health from the spread of diseases and from contaminated food from the sea.	Yes
14	**Input, or remobilization, of hazardous substances, by both airborne and waterborne routes** (arising from land-based activities, dumping, offshore installations and shipping).	Ch 17 Ch 20 Ch 21 Ch 23 Ch 24 Ch 15	Reduction in reproductive success and in ability to resist disease of marine biota. In extreme cases, killing of marine biota. Bio-accumulation of toxins in organisms that are subsequently harvested.	Damage to human health from contaminated food from the sea. Adverse effects on fisheries and shellfisheries from effects on stocks.	Yes
15	**Interference with aerial migration routes** (from wind-farms)	Ch 22 Ch 38	Potential damage to seabird population from deaths and injuries from collisions with rotors of wind-farms during migration.	Benefits of increasing non-carbon-intense energy sources involve trade-off with risk of increases from a new source of direct mortality.	Yes
16	**Introductions of non-native species or genetic strains** (arising from aquaculture, shipping and recreational boats)	Ch 12 Ch 17 Ch 27 Ch 36 A-H Ch 43	Degrading genetic pools, Reduction in biodiversity. Destruction of existing wild stocks. Potential for disruptions of natural populations and biotic communities.	Interference with fisheries and shellfisheries. Interference with operation of plant. Aquaculture benefits greatly from the use of strains of fish adapted for culture, which are often different genetically from natural populations.	Yes[4]

No	Pressures from Human Activities[1]	See	Impacts on Environmental Aspects of the Marine Environment	Impacts on Socioeconomic Aspects of the Marine Environment	Mgt[2]
17	**Physical alteration of sea-bed habitats** (arising from bottom-fishing, aquaculture, dredging for shipping, ports, submarine cables and pipelines, offshore hydrocarbon industries and mining, coastal defences, land reclamation, solid waste disposal and tourism and recreation).	Ch 11 Ch 12 Ch 18 Ch 19 Ch 21 Ch 22 Ch 23 Ch 24 Ch 27 Ch 36 A-H Chs 42-50	Direct mortality by physical impacts or smothering. Reduction in three-dimensional habitat structure can reduce biodiversity and productivity. Disturbance of sediments can reduce water quality and/or release contaminants, also impacting biotic communities and populations.	Costs of reducing impacts: some activities necessarily require habitat impacts as part of the business (mining, aggregate extraction); other activities result in habitat impacts as a collateral, but sometimes unavoidable, consequence (fishing with mobile bottom-contacting gears).	Some
18	**Sea-based emission of air-polluting substances (nitrogen oxides etc)** (arising from shipping, fishing vessels, offshore hydrocarbon and mining operations).	Ch 17 Ch 20 Ch 21 Ch 23	Additional source of nutrients, and thus of the problems related to them (see Item 14).	Damage to human health from coastal air pollution. Potential costs of controlling emissions.	Yes
19	**Sea-level rise** (arising from climate change).	Ch 4 Ch 7 Ch 26 Ch 36 Ch 43 Ch 47	Changes in coastal habitats. Contaminants from frequent coastal flooding are likely to add to toxics and nutrient pollution. Loss of costal ecosystems such as sea grasses due to increase in turbulence.	Inundation of low-lying States. Inundation of low-lying cities and other areas resulting in loss of property and population displacement. Critical infrastructure built in low lying areas is highly vulnerable (airports, sea ports, highways and train routes). Potential costs of protecting the built environment.	Not yet (Env) Yes (S/E)

1 In alphabetical order, not in any order of importance.
2 Mgt = Management possibilities: "Yes": examples are known of successful management strategies to reduce this pressure generally; "Some": examples are known of successful management strategies to reduce some aspects of this pressure; "Not yet": no such examples are yet known. NOTE – this marking does not allow for measures that ADAPT to changes: for example, the way in which some aquaculture facilities are mitigating some impacts of acidification.
3 There has been some success in reducing the ozone-depleting effects of certain chemicals, with consequent improvements in the UV-filtering effects of the ozone layer.
4 Transfers of foreign species in ships' ballast water can be managed. It is difficult to impose regimes to protect against transfer of species through attachments to the hull, especially on recreational boats.

References

EU (2008). European Union, *Marine Strategy Framework Directive* (2008/56/EC).

EU (2014). European Commission, *Report from the Commission to the Council and the European Parliament: The first phase of implementation of the Marine Strategy Framework Directive (2008/56/EC). The European Commission's assessment and guidance* (COM/2014/097 final).

FAO (2012). *The State of the World Fisheries and Aquaculture 2012*. FAO. Rome.

Halpern, B.S., Walbridge, S., Selkoe, K.A., Kappel, C.V., Micheli, F., D'Agrosa, C., Bruno, J.F., Casey, K.S., Ebert, C., Fox, H.E., Fujita, R., Heinemann, D., Lenihan, H.S., Madin, E.M.P., Perry, M.T., Selig, E.R., Spalding, M., Steneck, R. and Watson, R. (2008). A Global Map of Human Impact on Marine Ecosystems, *Science*, vol. 319, no. 5865.

Halpern, B.S., Longo, C., Hardy, D., McLeod, K.L., Samhouri, J.F., Katona, S.K., Kleisner, K., Lester, S.E., O'Leary, J., Ranelletti, M., Rosenberg, A.A., Scarborough, C., Selig, E.R., Best, B.D., Brumbaugh, D.R., Chapin, F.S., Crowder, L.B., Daly, K.L, Doney, S.C., Elfes, C., Fogarty, M.J., Gaines, S.D., Jacobsen, K.I., Karrer, L.B., Leslie, H.M., Neeley, E., Pauly, D., Polasky, S., Ris, B., St Martin, K., Stone, G.S., Sumaila, U.R., and Zeller, D. (2012). An index to assess the health and benefits of the global ocean, *Nature* 488, together with Supplementary Information available at doi:10.1038/nature11397 (accessed 1 November 2014).

OHI (2013). *Ocean Health Index*, Supplementary Methods, downloaded from http://www.oceanhealthindex.org/About/Methods/ on 1 November 2014.

OHI (2014). *Ocean Health Index, Ocean Health Index, Global Ocean Assessment*, downloaded from http://www.oceanhealthindex.org/About/Methods/ on 1 November 2014.

OSPAR (2007). *OSPAR Commission for the Protection of the Marine Environment of the North-East Atlantic, Ecological Quality Objectives: Working towards a healthy North Sea*, London, United Kingdom. (ISBN: 978-1-905859-57-3).

OSPAR (2010). *OSPAR Commission for the Protection of the Marine Environment of the North-East Atlantic*, Quality Status Report 2010, London, United Kingdom. (ISBN: 978-1-907390-38-8).

55 Overall Value of the Oceans to Humans

Group of Experts:
Patricio Bernal, Beatrice Padovani Ferreira, Lorna Inniss, Enrique Marschoff, Jake Rice, Andrew Rosenberg and Alan Simcock (Co-Lead Members)

Chapter 55
Overall Value of the Oceans to Humans

1 Introduction

The ocean provides countless ecosystem services. The Millennium Ecosystem Assessment describes ecosystem services as "the benefits people obtain from ecosystems" (MEA, 2005). Some of the ecosystem services from the ocean are delivered without human intervention – though they can be affected or disrupted by such intervention. Examples of these are the regulating and supporting ecosystem services (as described in Chapter 3), such as distribution of heat around the planet, the functioning of the hydrological cycle and the absorption of carbon dioxide as part of the carbon cycle. Other ecosystem services are obtained as a result of human activity to acquire the benefits. Most of these are provisioning ecosystem services (as also described in Chapter 3). The obvious example of such acquired ecosystem services is the food provided by capture fisheries, where humans take from the ocean significant amounts of the protein required for human diets. As demonstrated in Part IV and Part V and in Chapter 54, if the human activities are not carefully managed to maintain ecosystem structure and function, the acquisition of such ecosystem services can result in damage to the marine environment and reduction or loss of ecosystem services. Important issues arise for the institutions of ocean governance at global, regional, national and local levels in balancing the benefits of acquiring these services against the disbenefits (referred to by some as detriments) caused by over-exploitation and in preventing or mitigating those disbenefits.

Very different aspects of the concept of value are brought into play by these different kinds of ecosystem services. For most of the ecosystem services obtained from the ocean by human effort, there are global or local markets for the products obtained. Market valuations are therefore possible, although in some cases questions arise whether such a market value captures all the facets of the value to humans. For example, the value of fish in the sea in maintaining biodiversity and ecosystem functions may be more than the market value of the fish if they were caught and sold for consumption. Or, on a lesser scale, the value of their activities to recreational sea anglers may well be more than the market value of the fish (if any) that they catch. For ecosystem services that do not involve human effort to benefit from them, there is no market, and it is a question of producing a valuation in other ways. It is also important to remember that any discussion of value has to take into account the question of who is benefiting (or suffering the disbenefits) – even where they are unaware of what they are benefiting (or suffering).

A further category of ecosystem services of importance to humans are the aesthetic, cultural, religious and spiritual ecosystem services ("cultural ecosystem services" for short). Some of these (such as cultural objects and marine plants and animals that have cultural significance) are on much the same footing as the other material that humans take from the sea, and may well have market value (see Chapter 8). However, even marketed objects of cultural significance have an added dimension that may well not be captured by the market. This is particularly the case where the cultural value lies in communal self-identification through sharing in the activity that wins the cultural objects (such as communal whale-hunting in the North-East Pacific or the Faeroe Islands). Other cultural ecosystem services stand outside any market: for example, the cultural/religious values that are obtained by having access to the sea during rituals, through the existence of special, sacred places, the cultural values that lie in the enjoyment of the seascape or watching the beauty of seabirds, marine mammals or corals and the knowledge that comes from underwater cultural heritage.

2 Quantification

To mention value is almost inevitably to raise the issue of quantification: comparing values and assessing trade-offs require some idea of the relative sizes of what is being compared. There are many ways of measuring the benefit that humans derive from ecosystem services from the marine environment. "Consuming an ecosystem service" can cover all facets of deriving benefit from some aspect or aspects of the marine environment. It can, for example, include the way that, in some countries, houses enjoying a view of the sea can command higher prices than identical houses without such a view. It is not therefore easy to delimit the scope of what are the values that need to be taken into account. This is particularly the case with cultural values.

Looking only at economic valuations of the marine environment, one library among many contains nearly a thousand such valuations, with nearly twice as many valuation estimates (MESP 2014). And economic valuations are not the only forms of valuation that can be made: social and ecological metrics can be equally significant, without necessarily being reduced to a single economic balance-sheet. Among the metrics that can be important for many different groups of stakeholders are:

(a) The net economic value (for example, the net economic benefits that those who enjoy an ecosystem service derive from it (consumer surplus), or the economic value that those who use some component of the marine environment derive from it (producer surplus). This kind of metric can be valuable when the economic services enter directly into commerce;

(b) The gross and net revenues in monetary terms that are gained by those who enjoy an ecosystem service, or use some component of the marine environment. Such metrics focus on the cash flows related to the ecosystem service, and can sometimes be more readily derived as a measure of changes in the enjoyment of an ecosystem service. This kind of metric can be valuable in many contexts of ecosystem services, including those where non-monetary valuation approaches are necessary;

(c) Measures of the numbers of those employed in a human activity and the rewards (in kind or in cash) that they gain. This kind of metric can be valuable when considering livelihoods outside the monetary economy, as well as when labour is engaged for wages;

(d) Numbers of people benefiting from specific forms of ecosystem service, such as coast protection, recreational use of beaches, eating food from the sea or enjoyment of watching wildlife. This kind of metric can be useful in considering the extent to which different groups benefit from the ecosystem service, since it will avoid building differences in economic circumstances between the groups into the valuation;

(e) Direct measurements of the environmental situation (for example, area covered by mangroves, proportion of coral reefs that are in good condition, lengths of beaches with low levels of marine debris, proportion of dead seabirds contaminated with oil). This kind of metric can be useful when considering values where some of the areas are not (or only sparsely) inhabited, and the value lies in the areas' contributions to some other metric of overall global or regional value.

Considering only economic valuations, there are a wide range of techniques that can be used. Table 55.1 sets out some of the main approaches that can be used.

Although such economic valuation methods have been used to varying degrees, and several of them widely and with considerable success, in many cases the results do not achieve general acceptance as a significant factor to be taken into account. The reasons for lack of acceptance can include that:

(a) Decision-makers do not consider that the decision should turn on purely economic factors. They may wish to apply some overriding principle, such as national security or some other long-term goal, which they regard as incommensurable with economic factors;

(b) The techniques may not manage to give an economic value to some policy or other concerns of the decision-makers or of the society involved that is sufficient to carry conviction that a valid value has been calculated;

Table 55.1 | Some economic valuation methods, typical applications, examples, and limitations. Source: adapted from Waite et al, 2014.

VALUATION METHOD	APPROACH	TYPICAL APPLICATIONS	EXAMPLES	LIMITATIONS Some of these may apply to more than one method
Methods using issue-specific data				
Market price	Observe market prices and volumes of trade to analyze the economic activity generated by use of an ecosystem good or service. (Includes economic impact analysis, which examines the impacts of spending related to the good or service, and can also include indirect impacts in related economic sectors, as well as financial analysis, where operating costs are subtracted.)	Coastal goods and services that are traded on markets	Fisheries, tourism, mangrove timber	Market prices can be distorted (for example, by subsidies). Additional data is required to estimate net value added. Under conditions of unsustainable use the value in trade may be higher than the value the ecosystem can provide sustainably. Many ecosystem services are not traded in markets.
Replacement cost	Estimate cost of replacing ecosystem service with man-made service. Requires three conditions be met to be valid: (1) man-made equivalent provides the same level of ecosystem service; (2) man-made equivalent is the least-cost option of providing the service; (3) people would be willing to incur the cost rather than forgo the service.	Ecosystem services that have a man-made equivalent that provides similar benefits	Shoreline protection by reefs and mangroves; water filtration by forests and wetlands	Estimates might not reflect the true value of ecosystem goods and services. The method only seeks equivalence for the subset of services being costed. The co-benefits of other services provided by the ocean feature of concern are not considered. For example, a seawall might effectively protect the shore, but does not provide fish habitat in the way a healthy coral reef does.
Cost of avoided damage	Estimate damage avoided (e.g., from hurricanes or floods) due to ecosystem service	Ecosystem services that provide protection to houses, infrastructure or other assets	Shoreline protection by reefs and mangroves	Difficult to relate damage levels to ecosystem quality.
Production function	Estimate value of ecosystem service as input in production of marketed good	Ecosystem services that provide an input in the production of a marketed good	Commercial fisheries	Technically difficult to determine and model the relationship between ecosystem change and its impact on the provision of the ecosystem service. High data requirements.
Hedonic pricing	Estimate influence of environmental characteristics on price of marketed goods	Environmental characteristics that vary across goods (for example, houses or hotels with a sea view compared with those that do not)	Tourism, shoreline protection	It is possible to value individual units, but much more difficult to generalize from this to broader coverage.

VALUATION METHOD	APPROACH	TYPICAL APPLICATIONS	EXAMPLES	LIMITATIONS Some of these may apply to more than one method
Travel cost	Travel costs people are prepared to incur to access a resource indicate a minimum value	Recreation sites (for example, some marine protected areas)	Tourism	Technically difficult. High data requirements.
Contingent valuation	Ask survey respondents directly for willingness to pay for ecosystem service	Any ecosystem service (most widely used for non-market ecosystem and services)	Tourism	Expensive to implement because of survey costs. Vulnerable to many sources of bias and requires careful survey design.
Choice modelling	Ask survey respondents to trade off ecosystem services to elicit their willingness to pay	Any ecosystem service (most widely used for non-market ecosystem and services)	Tourism	Expensive to implement because of survey costs. Vulnerable to many sources of bias and requires careful survey design.
Methods using data not specific to the particular issue				
Benefits transfer	Value transfer: Use values estimated at other locations ("study sites") Function transfer: Use a value function estimated at another location to predict values	Any ecosystem service	Any ecosystem service	Relies on judgements of what other locations are sufficiently similar. Possible transfer errors if the "study sites" and "policy site" are different.
Meta-analysis	Synthesize results from multiple existing valuation studies, using mathematical methods to estimate a value function. Meta-analysis can be used for benefits transfer.	Any ecosystem service	Any ecosystem service	Requires compilation of multiple studies and power depends on sample size of value estimates. Adequacy of studies may vary. Can lead to a loss of important valuation information during data aggregation process

(c) The margins of error in the techniques are such that the meaning of the results is unclear to users;

(d) The potential users distrust the reliability of the method of valuation. This can sometimes be because the method has not been adequately explained;

(e) The decision-makers may not have an adequate understanding of the techniques or access to the necessary skills.

Furthermore, most of these techniques require detailed data, which in many cases does not exist for them to be applied. However, where they have been applied at local, national and sometimes regional level, the results are interesting, and can be used for a number of purposes.

3 Value of non-marketed ecosystem services

Looking at the ecosystem services that are delivered without human intervention, there are major problems in trying to place a value on them, especially in monetary terms. Some of these ecosystem services (such as the transfer of heat from the equatorial regions towards the poles) are such fundamental and inherent features of the way in which the whole planet operates that it is not possible to imagine the planet without this type of ecosystem service. It is not possible to conceive of the earth with its present populations of plants and animals (including humans) without the ocean. Without the ocean and therefore without the ecosystem services that it provides, the planet would be totally different. We cannot therefore consider scenarios with and without one of these non-marketed ecosystem services, and use the difference between the two scenarios to isolate an absolute value (whether in monetary terms or in some other form) of the benefits conferred by that ecosystem service.

Nevertheless, changes in the way in which the planetary ecosystem services operate can be measured, since it is possible to compare two different situations. The consequences of those changes can be seen to be massive. A good example of this is the El Niño Southern Oscillation (ENSO). This name refers to the way in which the ocean system of the tropical and subtropical Pacific can, in some years, produce a significant warming of the sea off the western coast of North and South America, often greatest off Peru, in the middle of the southern hemisphere summer. The coastal water temperature difference between one year and another, measured on the same day at the same hour, can be as much as 10° Celsius (Glantz, 2001). This produces major changes in weather across not only South and North America (with major increases in rainfall or other changes in Brazil, Peru, and the United States of America, for example), but also in Australia, India, Indonesia, the Philippines, and parts of Africa (see Chapter 5). There are only limited economic studies of the global effects of the variations brought about by the ENSO. Nevertheless, one study concluded that, over the period 1963-1997, the ENSO cycle can explain about 10-20 per cent of the variation in the growth in gross domestic product of the world's seven largest economies and about 20 per cent of real commodity-price movements (Brunner, 2002). A more recent modelling exercise, looking at the period 1979-2013, concluded that Australia, Chile, Indonesia, India, Japan, New Zealand and South Africa faced short-lived falls in economic activity following an El Niño shock, but that the United States may have benefitted from such events (Cashin et al., 2014). The conclusion on the situation in the United States is illuminated by an analysis of the severe El Niño event of 1998/1999. This analysis estimated that, compared with an average year, the 1998/1999 El Niño event led to costs of 4,000 million United States dollars for the United States; however, there were offsetting savings of 19,000 million dollars (in lower expenditure on natural gas and heating oil, increased economic activity, lack of spring flood damages and savings in highway-based and airline transportation) (Changon, 1999). Elsewhere, of course, the costs outweighed the savings. Such estimates of the economic implications of changes in the behaviour of the ocean are, of course, capable of generating endless discussion, but they serve to give an idea of the orders of magnitude of the costs and benefits that such changes can cause. In Part III of this assessment, likely and potential changes in the delivery of non-marketed ecosystem services from the ocean are noted. Such changes would be accompanied by massive economic consequences, but the data to develop sound economic valuations have not yet been assembled.

The distribution of those economic consequences around the globe is likely to follow the distribution of the changes, but it is clear that some of that distribution will have very different effects in different situations. Some States are at risk of finding much, if not all, of their territory lost to the sea as a result of sea-level rise. Elsewhere, some island and coastal communities risk suffering a similar fate, but impacts would be more local – although with consequences (such as population movements) possible far beyond the coastal sites potentially inundated. Where it is possible to safeguard against such losses by improved sea defences, the cost implications can be regarded as the cost of such improvements. Where the whole territory is lost, different considerations of value come into play. Where sites of cultural significance are lost, another dimension is added to the problem. In addition, of course, the capacities to address the costs are not distributed equally around the world.

It has not proved possible to come to conclusions on one important aspect of assessing the marine environment: a quantitative picture of the levels of many of the non-marketed ecosystem services provided by the ocean. There is simply insufficient quantitative information to allow an assessment of the way in which different regions of the globe benefit from these. Nor do current data-collection programmes appear to make robust regional assessments of ocean ecosystem services likely in the near future, especially for the less developed parts of the planet. Calculations can be made on the basis of sweeping assumptions, which allow estimates to be produced, but the assessment would then be an assessment of the assumptions, and not of the situation that is actually present. This is not to say that some valuations cannot be made at a local level where adequate data is available. Such local valuations can be valuable in assessing the marginal trade-offs between options for action in relation to the management of human activities.

4 Value of cultural ecosystem services

If it is difficult to approach the physical value of non-marketed ecosystem services, it is even more difficult to do so in respect of the cultural aspects of ecosystem services. We may be able to rank some cultural ecosystem services in terms of their importance for the cultures that they support. For isolated aspects such as a particular view of a seascape, it may even be possible to produce a monetary valuation, using one of the methods described in Table 55.1. But the more an aspect of an ecosystem services is embedded in a culture, and the more fundamental that aspect is to the culture, the less that kind of approach can work. Putting an explicit value of any kind on a whole cultural system is impossible, since it would involve value judgements for which there is no recognized system. However, the world may sometimes make implicit valuations of such cultural ecosystem services when it allows a cultural ecosystem service to be downgraded or lost as a result of pursuing some other objective.

5 Value of market-related ecosystem services

The discussions in Parts III, IV and V of the various ecosystem services from the marine environment that are linked to markets contain estimates of the values that can be linked to them. It is not, however, sensible to try to compile such estimates to give an overall picture. There are several reasons for this conclusion.

First, there are the problems of the quality of the estimates. If one takes the example of capture fisheries, it appears from Part IV that there is probably under-reporting, including of the scale of activities of small-scale fisheries. Such under-reporting is bound to distort any attempt to bring together estimates of the value of the different fisheries. Moreover, the "value" of fisheries must be viewed through multiple lenses. Estimates suggest that small-scale fisheries contribute about half of global fish catches, and large-scale fisheries the other half. When considering market-price, production-function or even hedonic-pricing valuations, the revenues generated through commerce would indicate large scale fisheries have much higher "value". However, when considering the provision of livelihoods, nutrition and food security to low income, food-deficit parts of the world, then the value of catches contributed by small-scale fishers increases greatly, on account of the importance of affordable fish and employment to populations in developing countries. Any single method of valuation would fail to communicate crucial information about each type of fishery. Even if a view is taken through a single lens – using a single one of these valuation methods – the uncertainty about the statistics in this field mean that overall estimates of the economic value of fisheries, and of the number of people working in them, are not sufficiently well-founded to compare with the figures for, say, seafarers, where much more comprehensive reporting is available.

Second, there are problems of definition. A good example of this is the tourism industries. Estimates exist of the number of people employed in tourism and of the contributions of tourism to Gross Domestic Product, both directly and indirectly (see Chapter 27). At present, however, there is considerable doubt about how consistently definitions are applied to the tourist industries in deriving the data on which those estimates are based (for example, how far back up supply chains the effects are estimated). There is further difficulty in deciding what parts of the values that can be attributed to "tourism" are related to the ocean and coasts directly. Most reports of tourism revenues do not differentiate revenues from tourism directly related to maritime areas from inland tourism. Even where tourism in the coastal zone can be separated from that inland, it may be generated by the direct attractions of the sea and coast (with tourists engaging in ocean-based activities), it may be indirectly linked, (with tourists visiting coastal cities or sites for cultural or historical features that are present because of the links between the places to the sea), or it may not be linked to the sea at all (for example, if a casino simply happens to be at the seaside). Consequently, the value of ocean-related tourism is a matter of inference. Therefore, there are major issues on how far it is appropriate to analyse or aggregate information within this field, or indeed other fields, and how to bring that information together with information from other fields.

Third, there are problems of the availability of data. Tourism is again an example here. For example, the involvement of international companies in the trade gives rise to uncertainty about the levels of value generated in particular areas (see Chapter 27). Shipping is another field where the lack of information is significant. International shipping is the foundation of most global trade. Without it, much economic activity would cease. Some information can be gathered on the earnings of those employed in the industry, and revenues earned in some parts of the world. However, information is not readily available on the overall earnings of the industry, and therefore its share in the world economy (see Chapter 17), nor on the distribution around the globe of direct revenues, profits, and increases in value of trade goods because ocean-based trade is available.

Nevertheless, some States are making efforts to put values on the benefits created from the ocean areas under their jurisdiction. For example, the United Kingdom published in 2010 a first attempt to put monetary values on a range of activities taking place in its waters. The activities covered were offshore oil and gas, maritime transport, telecommunications cables, leisure and recreation, military defence, fisheries, aquaculture, water abstraction, mineral extraction, renewable energy, coastal defence, waste disposal, education, power transmission and storage of gases. For most of these, an estimate was made of the gross value added, but for some it was only possible to estimate the money being invested in the process. The detailed workings of this exercise show the amount of effort needed to achieve even approximate values (DEFRA, 2010).

6   Global distribution

Some observations can, nevertheless, be made about the way in which the values of some of the market-related ecosystem services provided by the ocean are distributed around the world.

6.1   Fish and seafood consumption

Annual per capita consumption of fishery products has grown steadily in developing regions (from 5.2 kg in 1961 to 17.0 kg in 2009) and in low-income food-deficit countries (from 4.9 kg in 1961 to 10.1 kg in 2009) (see Chapters 10 and 11). This total consumption is still considerably lower than in more developed regions, although the gap is narrowing. A sizeable share of fish consumed in developed countries consists of imports, and, owing to a steady demand and declining domestic fishery production (down 10 percent in the period 2000–2010) (see Chapters 10 and 11), their dependence on imports, in particular from developing countries, is projected to grow. Studies have shown that the selling or trading of even a portion of their catch represents as much as a third of the total income of subsistence fishers in some low income countries. Thus an increase in imports of fish by more developed countries from less developed countries has the potential simultaneously to increase wealth in low income communities and to increase inequities in food security and nutrition, unless these considerations are taken into account in global trade arrangements (see Part IV).

Over time, there has been a striking shift in the operation and location of capture fisheries. In the 1950s, capture fisheries were largely undertaken by developed fishing States, and their activities were largely in fishing grounds in the North Atlantic and North Pacific. Since then, developing countries have increased their share, and the distant water fleets of developed countries range further in the sea. It is illuminating to compare the Northern and Southern Hemispheres. Although the two hemispheres do not precisely reflect developed as compared with developing fishing States, the figures are, nonetheless, indicative. In the 1950s, the Southern Hemisphere accounted for no more than 8 per cent of the landed market value of fish. By the 2000s, the Southern Hemisphere's share had risen to 20 per cent (see Part IV).

6.2   Maritime transport

All sectors of maritime transport (cargo trades, passenger and vehicle ferries and cruise ships) are growing in line with the world economy. According to estimates by the United Nations Conference on Trade and Development (UNCTAD), owners from five countries (Greece, Japan, China, Germany and the Republic of Korea) together accounted for 53 per cent of the world tonnage in 2013. Among the top 35 ship-owning countries and territories, 17 are in Asia, 14 in Europe, and 4 in the Americas. It seems likely that profits and losses are broadly proportional to ownership (see Chapter 17).

6.3   Offshore energy businesses

Offshore oil production is predominantly in the Gulf of Mexico (about 60 per cent of the industry is located in the Gulf of Mexico) and the North Sea. The industry accounts for about 1.5 per cent of the United States GDP, 3.5 per cent of the United Kingdom's GDP, 21 per cent of Norway's GDP and 35 per cent of Nigeria's GDP. The large majority of offshore hydrocarbon production is in the hands of international corporations or national companies (often working in partnership with the international companies). This makes the tracking of the distribution of benefits from this sector, other than direct employment in extraction and processing, very difficult (see Chapter 21). Offshore renewable energy production is very much in its infancy, and it will be some time before a clear picture of what will be the long-term future of the industry emerges (see Chapter 22).

6.4 Developments in offshore mining

There is limited information about the value of the offshore mining industry or the number of people employed, but it is unlikely to be significant in comparison to terrestrial mining. For example, in the United Kingdom, which is the world's largest producer of marine aggregates, the industry directly employs approximately 400 people. There seems little doubt that there will eventually be substantial expansion of offshore mining as terrestrial mineral deposits are worked out. In some cases (for example, diamond and tin mining), major international undertakings are involved. In the remaining cases, most offshore mining is within exclusive economic zones (and, indeed, generally close to the shore), and undertaken by relatively local enterprises. Mining in the Area (seabed, ocean floor and subsoil thereof) will be subject to a decision of the International Seabed Authority (see Chapter 23).

6.5 Tourism

Tourism has generally been increasing fairly steadily for the last 40 years (with occasional set-backs or slowing down in times of global recession). In 2012, estimates of international tourism expenditure exceeded 1 trillion dollars for the first time. Total expenditure on tourism – domestic as well as international – is several times that amount. Globally, the direct turnover of tourism was estimated to contribute 2.9 per cent of Gross Domestic Product (GDP) in 2013, rising to 8.9 per cent when the multiplier effect on the rest of the economy is taken into account. The Middle East is the region where tourism plays the smallest part in the economy (6.4 per cent of GDP including the multiplier effect), and the Caribbean the region where it plays the largest part (13.9 per cent including the multiplier effect). In small island and coastal States, coastal tourism is inevitably predominant. Particularly noteworthy is the way in which international tourism is increasing in Asia and the Pacific, both in absolute terms and as a proportion of world tourism, with the implication that pressures from tourism are becoming of significantly more concern in those regions (see Chapter 27).

Tourism is also a significant component of employment. Globally, it is estimated that in 2013, tourism provided 3.3 per cent of employment, looking at the numbers directly employed in tourist industries, and 8.9 per cent when the multiplier effect is taken into account. In the different regions, the proportion of employment supported by tourism is approximately the same as the share of GDP contributed by tourism, although again the proportion of that which is based on the attractions of sea and coast is poorly known (see Chapter 27).

6.6 Use of marine genetic material

The commercial exploitation of marine genetic resources had very modest beginnings in the 20th century. The value of the use of marine genetic material is therefore only just beginning to develop and projections of its potential economic value differ greatly among plausible scenarios for its future development. All scenarios assume that increases in commercial exploitation will be driven primarily by more developed countries, making considerations relating to access and benefit sharing of marine genetic resources an important issue (see Chapter 29).

7 Knowledge and capacity-building gaps

As is apparent from what has been said in this chapter, there are major gaps in the knowledge available for considering the overall value of the ocean to humans. These gaps in knowledge have important implications for the governance of the ocean, because many issues turn on weighing advantages and disadvantages, and how they are distributed among the marine ecosystems and the maritime countries of the planet. Informed consideration of these issues is made the more difficult the less that is known about the values to be put upon those advantages and disadvantages.

So far there is still much debate on methods of valuing the provision of non-marketed ecosystem services. The work of the United Nations Statistics Division on valuing such ecosystem services for the purposes of national accounts may assist in informing the debate on these issues (see Chapter 9 for discussion of this programme). Other initiatives, such as the WAVES (Wealth Accounting and the Valuation of Ecosystem Services) partnership, aimed ensuring that natural resources are mainstreamed in development planning and national economic accounts, may also help in this task (WAVES, 2014).

As has been said, there are also many gaps in the basic information needed for valuation of market-related ecosystem services, in whatever manner this is eventually applied. These gaps are identified in the individual chapters dealing with the various human activities that result in the acquisition of these services, and the need for this information for comparative valuation purposes adds to the case for filling these gaps. Equally, there are gaps in our understanding of the biophysical models linking metrics of the basic features of ecosystems to the production of the goods and services to be valued. Resolving such gaps requires interdisciplinary collaboration.

There is likewise a need for capacity building in most developing countries in the use of valuation techniques and in the collection of the necessary data.

References

Brunner, A.D. (2002). El Niño and World Primary Commodity Prices: Warm Water or Hot Air? *The Review of Economics and Statistics* 84(1), 176-183.

Cashin, P., Mohaddes, K. and Raissi, M. (2014). *Fair Weather or Foul? The Macroeconomic Effects of El Niño*, Cambridge Working Papers in Economics 1418.

Changnon, S.A. (1999). Impacts of 1997—98 El Niño Generated Weather in the United States, *Bulletin of the American Meteorological Society*, Vol 80, 1819-1827.

DEFRA (2010). United Kingdom, Department of Environment, Food and Rural Affairs, Charting Progress II, London, United Kingdom.

Glantz, M.H., (2001). *Currents of Change: Impacts of El Niño and La Niña on Climate and Society*, Cambridge University Press, Cambridge, United Kingdom (ISBN: 0 521 78672 X).

MEA (Millennium Ecosystem Assessment) (2005). (Millennium Ecosystem Assessment). *Ecosystems and Human Well-being: Synthesis*, Island Press, Washington, DC, 2005 (ISBN: 1-59726-040-1).

MESP (2014). *Marine Ecosystems Services Programme.* (http://www.marineecosystemservices.org/home) accessed 19 July 2014.

Waite, R., Burke, L., Gray, E., van Beukering, P., Brander, L., McKenzie, E., Pendleton, L., Schuhmann, P., Tompkins, E. (2014). *Coastal Capital: Ecosystem Valuation for Decision Making in the Caribbean*, World Resources Institute Washington, DC. (www.wri.org/coastal-capital accessed 17 July 2014).

WAVES (2014). The WAVES Partnership (http://www.wavespartnership.org/en/about-us accessed 10 December 2014).

Annexes

ANNEX I

List of Contributors and Commentators

List of Contributors and Commentators — Annex I

Argentina
Viviana Alder
Javier Calcagno
Enrique Marschoff (GoE)[1]
José Luis Orgeira
Alberto Piola
Liliana Quartino
Andrea Raya Rey
Laura Schejter

Australia
Elaine Baker
John Bannister
Nicholas Bax
Marita Bradshaw
Alan J. Butler
Paul J. Durack
Karen Evans
Peter Harris
Vimoksalehi Lukoschek
Joanna Parr
Nigel Preston
Peter Shaughnessy
Marguerite Tarzia
John A. West
Clive Wilkinson
Colin D. Woodroffe

Bangladesh
Md. M. Maruf Hossain
Ahmed Kawser

Barbados
Caroline Bissada
Angelique Brathwaite
John W. Farrell
Lorna Inniss (GoE)
Leonard Nurse
Patrick McConney
Robin Mahon
Hazel Oxenford
Kareem Sabir
Emma Smith

Belgium
Linda Amaral Zettler
Stace Beaulieu
Laura Boicenco
Erik Cordes
Farid Dahdouh-Guebas
Jeroen Engels
Andre Freiwald
Andrew R. Gates
Jeroen Ingels
Daniel O.B. Jones
Panagiotis Kasapidis
Lisa A. Levin
Imants G. Priede
Wouter Rommens
Henry A. Ruhl
Thomas Soltwedel
Tracey Sutton
Saskia Van Gaever (GoE)
Jan Vanaverbeke
Edward Vanden Berghe
Hiromi Watanabe
Moriaki Yasuhara

Benin
Zacharie Sohou

Brazil
Claudia Câmara Vale
Edmo Campos
Anna Paula da Costa Falcão
Fábio Hazin
Flavia Lucena-Frédou
Maria Angela Marcovaldi
Monica Muelbert
José Muelbert
Beatrice Padovani Ferreira (GoE)
Angel Perez
José Angel A. Perez
Marcos Polette
Regina Rodrigues
Mário Soares
Alexander Turra

Canada
Philippe Archambault
Steven E. Campana
Ratana Chuenpagdee
Nicholas Dulvy
Evan Edinger
Tony Gaston
Charles Hannah
Russell Hopcroft
S. Kim Juniper
Ellen Kenchington
Anna Paula Metaxas
William Montevecchi
Gordon Munro
Patrick Ouellet
Pierre Pepin

---

1   GoE – member of the Group of Experts

Annex I — List of Contributors and Commentators

Randall R. Reeves
Jake Rice (GoE)
Mark Shrimpton
Stephen Smith
Paul Snelgrove
Thomas Therriault
Verena Tunnicliffe
Nathan Young

Chile
Rodrigo Hucke-Gaete
Doris Oliva
Patricio Bernal (GoE)
Günter Försterra
Carlos F. Gaymer
Vreni Häussermann

China
Xinzheng Li
Jinhui Wang
Juying Wang (GoE)
Kedong Yin
Yuhui Zhao
Senlin Zheng

Colombia
Julián Reyna Moreno

Ecuador
Jose Santos

Estonia
Georg Martin (GoE)
Henn Ojaveer
Hannes Tõnisson

France
Françoise Gaill
Nicolas Bailly
Sophie Arnaud-Haond
Bernard Salvat
Nadine Le Bris

Germany
Kai Lorenzen
Ralf Ebinghaus
Angelika Brandt
Kristin Kaschner
Dieter Piepenburg

Greece
Aristomenis P. Karageorgis
Christos Arvanitidis

India
Ramalingaran Kirubagaran
C. Raghunathan
Nagappa Ramaiah
V.N. Sanjeevan
Rahul Sharma
K. Venkataraman

Iran (Islamic Republic of)
Peyman Eghtesadi Araghi (GoE)

Italy
Maurizio Azzaro
Giuseppe M.R. Manzella
Renzo Mosetti
Gabriele Procaccini

Jamaica
Sean O. Green (GoE)
Mona Webber

Japan
Katsunori Fujikura
Kunio Kohata (GoE)
Teruhisa Komatsu
Massa Nakaoka
Mayumi Sato
Kazuaki Tadokoro
Fujikura Hiroya Yamano

Kenya
James Kairo
David Obura
Renison Ruwa (GoE)

Madagascar
Arsonina Bera
Clodette Raharimananirina
Eddy Rasolomanana
Jacquis Rasoanaina

Mexico
Margarita Caso
Elva Escobar Briones
Lorenzo Rojas-Bracho

Mozambique
Alberto Mavume

New Zealand
Malcolm R. Clark
Mark Costello
Martin Cryer

Geoffroy Lamarche
Ashley A. Rowden
Craig Stevens
Marjan van den Belt

Nigeria
Babajide Alo
Regina Folorunsho

Norway
Claire Armstrong
Odd Aksel Bergstad
Arne Bjørge
Lars Golmen
Bjorn Einar Grosvik
Lis Lindal Jørgensen
Andrew K. Sweetman
Cecilie von Quillfeldt

Peru
Marilú Bouchon Corrales

Philippines
Angel C. Alcala (GoE)[2]
Hilconida Calumpong (GoE)[3]

Poland
Jan Marcin Węsławski (GoE)

Portugal
Ana Hilario
Nuno Lourenço
Telmo Morato
Ricardo Santos

Qatar
A. Rahman Ali Al-Naama

Republic of Korea
Sung-Ho Kang
Sung Yong Kim
Hong-Yeon Cho
Chul Park (GoE)
Chang Ik Zhang

Russian Federation
Andrey Dolgov

Seychelles
Rolph Antoine Payet (GoE)

Sierra Leone
Osman Keh Kamara (GoE)

South Africa
Ross McLeod Wanless

South Sudan
Amanuel Yoanes Ajawin (GoE)

Spain
Antonio Bode
Marta Coll Monton
Carlos Garcia-Soto
Maria José Juan-Jordá

Sweden
Michael Thorndyke

Trinidad and Tobago
Judith Gobin
Asha Singh

Uganda
Joshua T. Tuhumwire (GoE)

United Kingdom
Maria Clare Baker
Euan Brown
Harry Bryden
D. Freestone
Charles L. Griffiths
Huw J. Griffiths
D. d'A. Laffoley
Ben Lascelles
K. Morrison
Bhavani Narayanaswamy
Chris Reason
J. Murray Roberts
Howard S.J. Roe
Alex Rogers
Alan Simcock (GoE)
Derek Tittensor
Tom Webb

United Republic of Tanzania
John Machiwa

United States of America
Phillip Arkin
Peter Auster
Michael Banks
Ann Bucklin
Deirdre Byrne

---

[2] Member of the Group of Experts from 2011-2013.

[3] Member of the Group of Experts from 2013-2015.

Bernard Coakley
Bruce B. Collette
Amardeep Dhanju
Robert Duce
Peter H. Dutton
C. Mark Eakin
Francesco Ferretti
Elise Granek
Frank R. Hall
Benjamin Halpern
Patrick N. Halpin
Gordon Hamilton
Jon Hildebrand
Ellen Hines
Dan Kamykowski
James C. Kelley
Kim Kiho
J. Anthony Koslow
Eric Leuliette
Rainer Lohmann
Thomas Malone
Jeremy T. Mathis
Susanne Menden-Deuer
John Milliman
Patricia Miloslavich
Fabio Moretzsohn
Douglas Ofiara
Victor Restrepo
Andrew Rosenberg (GoE)
Tatiana Rynearson
Wilford Schmidt
Raymon W. Schmitt
K.E.A. Segarra
Tim D. Smith
Craig R. Smith
Walker Smith
Karen Stocks
John W. Tunnell Jr.
Michael Vecchione
Bryan P. Wallace
J.S. Weis
William Douglas Wilson
Lisan Yu

Uruguay
Pablo Muniz

ANNEX II

Glossary

Abyssal plain

An abyssal plain is an extensive, flat, gently sloping or nearly level region at abyssal depths. *See also* Trenches.

Area

Article 1, paragraph 1, of UNCLOS defines the Area as "the seabed and ocean floor and subsoil thereof, beyond the limits of national jurisdiction." Article 136 of UNCLOS provides that "[t]he Area and its resources are the common heritage of mankind."

Atolls

Atolls are coral islands consisting of a ring-shaped reef, nearly or entirely surrounding a central lagoon. They occur in the warm waters of the tropics and subtropics. These low-lying and vulnerable landforms owe their origin to reef-building corals. The origin of atolls was explained by Charles Darwin as the result of subsidence (sinking) of a volcanic island.

Baseline

The baseline is the line from which the outer limit of a State's territorial sea is measured. The breadth of other maritime zones is also measured from the same line. The United Nations Convention on the Law of the Sea (UNCLOS) sets out several methods for determining the baselines, providing that coastal States may determine baselines by any of these methods (article 15):

Normal baseline: "[e]xcept where otherwise provided in this Convention, the normal baseline for measuring the breadth of the territorial sea is the low-water line along the coast as marked on large-scale charts officially recognized by the coastal State" (article 5). "

- Straight baseline: "In localities where the coastline is deeply indented and cut into, or if there is a fringe of islands along the coast in its immediate vicinity, the method of straight baselines joining appropriate points may be employed in drawing the baseline from which the breadth of the territorial sea is measured" (article 7, paragraph 1). The remaining paragraphs of article 7 establish the criteria to draw straight baselines as follows: "2. Where because of the presence of a delta and other natural conditions the coastline is highly unstable, the appropriate points may be selected along the furthest seaward extent of the low-water line and, notwithstanding subsequent regression of the low-water line, the straight baselines shall remain effective until changed by the coastal State in accordance with this Convention. 3. The drawing of straight baselines must not depart to any appreciable extent from the general direction of the coast, and the sea areas lying within the lines must be sufficiently closely linked to the land domain to be subject to the regime of internal waters. 4. Straight baselines shall not be drawn to and from low-tide elevations, unless lighthouses or similar installations which are permanently above sea level have been built on them or except in instances where the drawing of baselines to and from such elevations has received general international recognition. 5. Where the method of straight baselines is applicable under paragraph 1, account may be taken, in determining particular baselines, of economic interests peculiar to the region concerned, the reality and the importance of which are clearly evidenced by long usage. 6. The system of straight baselines may not be applied by a State in such a manner as to cut off the territorial sea of another State from the high seas or an exclusive economic zone."

- Archipelagic baselines: "[a]n archipelagic State may draw straight archipelagic baselines joining the outermost points of the outermost islands and drying reefs of the archipelago provided that within such baselines are included the main islands and an area in which the ratio of the area of the water to the area of the land, including atolls, is between 1 to 1 and 9 to 1" (article 47). The remaining paragraphs of article 47 establish the criteria to draw archipelagic baselines as follows: "2. The length of such baselines shall not exceed 100 nautical miles, except that up to 3 per cent of the total number of baselines enclosing any archipelago may exceed that length, up to a maximum length of 125 nautical miles. 3. The drawing of such baselines shall not depart to any appreciable extent from the general configuration of the archipelago. 4. Such baselines shall not be drawn to and from low-tide elevations, unless lighthouses or similar installations which are permanently above sea level have been built on them or where a low-tide elevation is situated wholly or partly at a distance not exceeding the breadth of the territorial sea from the nearest island. 5. The system of such baselines shall not be applied by an archipelagic State in such a manner as to cut off from the high seas or the exclusive economic zone the territorial sea of another State. 6. If a part of the archipelagic waters of an archipelagic State lies between two parts of an immediately adjacent neighbouring State, existing rights and all other legitimate interests which the latter State has traditionally exercised in such waters and all rights stipulated by agreement between those States shall continue and be respected. 7. For the purpose of computing the ratio of water to land under paragraph l, land areas may include waters lying within the fringing reefs of islands and atolls, including that part of a steep-sided oceanic plateau which is enclosed or nearly enclosed by a chain of limestone islands and drying reefs lying on the perimeter of the plateau. 8. The baselines drawn in accordance with this article shall be shown on charts of a scale or scales adequate for ascertaining their position. Alternatively, lists of geographical coordinates of points, specifying the geodetic datum, may be substituted. 9. The archipelagic State shall give due publicity to such charts or lists of geographical coordinates and shall deposit a copy of each such chart or list with the Secretary-General of the United Nations."

UNCLOS also establishes the methods to draw the baselines in the presence of reefs (article 6); mouths of rivers (article 9) and bays (article 10), provided that certain conditions are met. In addition, UNCLOS regulates the effect of the outermost permanent harbour works (article 11), roadsteds (article 12) and low-tide elevations (article 13) on baselines.

Beach

A beach is a type of shore and is an area on which the waves break and over which shore debris, such as sand, shingle and pebbles accumulate.

A beach includes backshore and foreshore. The foreshore is often defined as the area covered and uncovered by the tides.

Bioerosion
Bioerosion is an important process in reefs, with bioeroders, such as algae, sponges, polychaete worms, crustaceans, sea urchins, and boring molluscs (e.g., *Lithophaga*), reducing the strength of the framework and producing sediment that infiltrates and accumulates in the porous reef limestone

Biological pump
The biological pump is the process of active biological uptake of CO2 into the biomass and skeletons of plankters (the individuals that collectively form plankton).

Bruun Rule
The Bruun Rule is a simple heuristic that uses the slope of the foreshore and conservation of mass to predict the extent to which sea-level rise will cause erosion and net recession landwards for many beaches.

Canyon
Submarine canyons are defined as "steep-walled sinuous valleys with V-shaped cross sections and axes sloping outward as continuously as river-cut land canyons, with relief comparable to even the largest land canyons".

Clausius-Clapeyron relationship
The Clausius-Clapeyron relationship is the water-holding capacity of the atmosphere, which increases by 7 per cent for every degree Celsius of warming.

Contiguous Zone
According to article 33 of UNCLOS, "[i]n a zone contiguous to its territorial sea, described as the contiguous zone, the coastal State may exercise the control necessary to: (a) prevent infringement of its customs, fiscal, immigration or sanitary laws and regulations within its territory or territorial sea; (b) punish infringement of the above laws and regulations committed within its territory or territorial sea." In addition, according to the same provision "[t]he contiguous zone may not extend beyond 24 nautical miles from the baselines from which the breadth of the territorial sea is measured."

Continental shelf
'Continental shelf' in this Assessment refers (unless stated otherwise) to the geomorphic continental shelf (as shown in Figure 1) and not to the continental shelf as defined in Article 76 of UNCLOS. The geomorphic continental shelf is usually defined in terms of the submarine extension of a continent or island as far as the point where there is a marked discontinuity in the slope and the continental slope begins its fall down to the continental rise or the abyssal plain. UNCLOS provides that "The continental shelf of a coastal State comprises the seabed and subsoil of the submarine areas that extend beyond its territorial sea throughout the natural prolongation of its land territory to the outer edge of the continental margin, or to a distance of 200 nautical miles from the baselines from which the breadth of the territorial sea is measured where the outer edge of the continental margin does not extend up to that distance." This definition is refined in Article 76, paragraphs (4) to (7).

Continental rise
A submarine feature which is that part of the continental margin lying between the continental slope and the deep ocean floor; simply called the rise in the UNCLOS.

Continental slope
The continental slope is that part of the continental margin that lies between the continental shelf and the continental rise floor.

Deep sea
In this Assessment "deep sea" refers to the sea floor of deep-water areas that are beyond (that is, seawards of) the geomorphic continental shelf. It is the benthic zone that lies in deep water (generally >200 metres water depth).

Dead-weight tonnage
Dead weight tonnage is the measure of how much weight a ship can safely carry. It is the aggregate of the weights of cargo, fuel, fresh water, ballast water, provisions, passengers, and crew.

Downwelling and upwelling
Where surface ocean currents move seawater toward coasts, the water is forced to sink, in the process known as coastal downwelling. Coastal upwelling occurs where surface waters are moved away away from the coast: that water is replaced by water that wells up from below. Upwelling and downwelling also occur in the open ocean where winds cause surface waters to move away from a region (leading to upwelling) or to converge toward a region (leading to downwelling).

Ecosystem
The Millennium Ecosystem Assessment defines an ecosystem as "a dynamic complex of plant, animal and micro-organism communities and their non-living environment interacting as a functional unit"

Ecosystem services
The Millennium Ecosystem Assessment (2005) classified ecosystem services as: provisioning services (e.g., food – including food traded in formal markets and subsistence trade and barter - pharmaceutical compounds, building material); regulating services (e.g., climate regulation, moderation of extreme events, waste treatment, erosion protection, maintaining populations of species); supporting services (e.g., nutrient cycling, primary production) and cultural services (e.g., spiritual experience, recreation, information for cognitive development, aesthetics).

El Niño/Southern Oscillation
The name "El Niño/Southern Oscillation" (ENSO) refers to the way in which the ocean system of the tropical and subtropical Pacific can, in some years, produce a significant warming of the sea off the western

coast of North and South America, often greatest off Peru, in the middle of the southern hemisphere summer.

Enclosed and semi-enclosed seas
Article 122 of UNCLOS defines an enclosed or semi-enclosed sea as a "gulf, basin or sea surrounded by two or more States and connected to another sea or the ocean by a narrow outlet or consisting entirely or primarily of the territorial seas and exclusive economic zones of two or more coastal States."

Estuary
An estuary is the tidal mouth of a river, where the seawater is measurably diluted by the fresh water from the river.

Euphotic zone
This is the layer below the ocean's surface that receives enough light for photosynthesis to occur. This is usually taken to be where photosynthetically active radiation [PAR] is more than 1 per cent of the surface intensity.

Exclusive economic zone
UNCLOS provides that "The exclusive economic zone is an area beyond and adjacent to the territorial sea, subject to the specific legal regime established in this Part, under which the rights and jurisdiction of the coastal State and the rights and freedoms of other States are governed by the relevant provisions of this Convention". UNCLOS further provides that "The exclusive economic zone shall not extend beyond 200 nautical miles from the baselines from which the breadth of the territorial sea is measured.", and that "In the exclusive economic zone, the coastal State has: (a) sovereign rights for the purpose of exploring and exploiting, conserving and managing the natural resources, whether living or non-living, of the waters superjacent to the seabed and of the seabed and its subsoil, and with regard to other activities for the economic exploitation and exploration of the zone, such as the production of energy from the water, currents and winds; (b) jurisdiction as provided for in the relevant provisions of this Convention with regard to: (i) the establishment and use of artificial islands, installations and structures (ii) marine scientific research; (iii) the protection and preservation of the marine environment; (c) other rights and duties provided for in this Convention."

Expendable Bathythermograph
An Expendable Bathythermograph (XBT) is a device that, using electronic solid-state transducers, registers and reports temperature and pressure while it free-falls through the water column. Mechanical Bathythermographs (MBTs) are their mechanical predecessors, that were lowered on a wire suspended from a ship, used a metallic thermocouple as transducer.

Fjord and ria areas
Fjord and ria areas are long narrow inlets into the land from the sea (in the case of fjords, between high cliffs), with depth usually diminishing landwards.

Freshwater flux
Freshwater fluxes into the ocean include: direct runoff from continental rivers and lakes; seepage from groundwater; runoff, submarine melting and iceberg calving from ice sheets; melting of sea ice; and direct precipitation, which is mostly rainfall but also includes snowfall.

Gross primary production (GPP)
Gross primary production is the rate at which photosynthetic plants and bacteria use sunlight to convert carbon dioxide (CO_2) and water to the high-energy organic carbon compounds used to fuel growth.

Gyres
Gyres are circular patterns of currents in ocean basins. In the centre of a gyre the seawater moves less than the seawater in the currents around it. There are five major gyres: in the North and South Atlantic Ocean, in the North and South Pacific Ocean and in the Indian Ocean.

Harvest Control Rule
In the context of fisheries management, this rule has been defined as "An agreed rule (algorithm) that describes how harvest is intended to be controlled by management in relation to the state of some indicator of stock status. For example, a harvest control rule can describe the various values of fishing mortality which will be aimed to be achieved at corresponding values of the stock abundance".

High Seas
Article 86 of UNCLOS defines the high seas as "all parts of the sea that are not included in the exclusive economic zone, in the territorial sea or in the internal waters of a State, or in the archipelagic waters of an archipelagic State."

Hydrological cycle
The Earth's hydrological cycle (or water cycle) is the process by which water evaporates from the Earth's surface and is then precipitated as rain, hail or snow. It also includes the processes by which water moves from place to place across land and sea.

Indian Dipole
The Indian Ocean Dipole is an irregular oscillation of sea-surface temperatures: the western part of the equatorial Indian Ocean becomes alternately warmer and then colder than its eastern part. These changes affect the climate of the countries that surround the Indian Ocean.

Isobath
An isobath is an imaginary line representing the horizontal contour of the sea-bed at a given depth.

Large marine ecosystems
Large marine ecosystems (LMEs) are regions of the ocean encompassing coastal areas from river basins and estuaries out to the seaward boundaries of continental shelves and the outer margins of the major current systems. These areas are characterized by their distinct bathymetry, hydrography, productivity and food webs. The term often refers to the set

of 64 LMEs identified by the United Nations Environment Programme's Regional Seas Programme for analytical purposes.

Macronutrients and micronutrients
Macronutrients are elements needed in relatively large quantities for photosynthesis. They include calcium, carbon, nitrogen, magnesium, phosphorus, potassium, silicon and sulphur. Micronutrients are elements needed in lesser quantities and provide the necessary co-factors for metabolism to be carried out. They include iron, copper and zinc.

Mangroves
Mangroves are one of several genera of tropical trees or shrubs which produce prop roots and grow along low-lying coasts into shallow water. The term is also used for the habitats in which mangrove genera grow.

Marine wetlands
Marine wetlands are areas of salt marsh, fen or shallow water, whether natural or artificial, permanent or temporary, with water that is static or flowing. The Convention on wetlands of international importance, especially as waterfowl habitat (Ramsar Convention)[1], includes areas of marine water the depth of which at low tide does not exceed six metres.

Meridional overturning circulation
The meridional overturning circulation is a system of surface and deep currents linking all ocean basins. It transports water (and, with it, heat) salt, nutrients and other substances around the globe, and connects the surface ocean and the atmosphere with the deep sea. *See also* Thermohaline circulation.

Micronutrients *see* macronutrients

Mid-ocean ridges
Mid-ocean ridges are geologically active chains of submarine mountains formed by plate tectonics. In some ocean basins (for example, the Pacific), they are located away from the centre. Most are linked into a 70,000-kilometre long chain encompassing the whole globe.

Natural capital
Natural capital is usually defined as the stocks of natural assets, including rocks, soil, air, water and all living things. Ecosystem services are derived from this natural capital. In the marine context, natural capital contrasts with human-made capital like ports and navigation systems.

Nautical mile
A nautical mile is defined as 1,852 metres (approximately 2,025 yards, or 1.15 miles).

Net primary production (NPP)
Net primary production is Gross Primary Production (GPP) less the respiratory release of CO_2 by photosynthetic organisms, i.e., the net photosynthetic fixation of inorganic carbon into autotrophic biomass.

Ocean acidification
When CO_2 reacts with water, it forms carbonic acid, which then dissociates and produces hydrogen ions. The extra hydrogen ions link with carbonate ions (CO_3^{2-}) to form bicarbonate (HCO_3^-). In this process, the pH and concentrations of carbonate ions (CO_3^{2-}) decrease. As a result, the carbonate mineral saturation states also decrease. The water thus becomes more acid and less basic (alkaline). Due to the increasing acidity, this process in the ocean is commonly referred to as "ocean acidification". (pH is a numeric scale used to specify the acidity or basicity of a water-based solution, calculated as the negative logarithm to base 10 of the activity of the hydrogen ions).

Ocean acidification hotspots
Although the average oceanic pH can vary on interglacial time scales, the changes are usually on the order of ~0.002 units per 100 years; however, the current observed rate of change is ~0.1 units per 100 years, or roughly 50 times faster. Regional factors, such as coastal upwelling, changes in riverine and glacial discharge rates, and sea-ice loss have created "Ocean acidification hotspots" where changes are occurring at even faster rates.

Ocean currents
Ocean currents are continuous, directional movements of seawater generated by forces such as wind, the Coriolis effect from the rotation of the Earth, differences in temperature and salinity of different parts of the ocean and the effects of mixing bodies of seawater with these different qualities. They are distinct from tides, which are caused by the gravitational effects of the moon and the sun. In the upper ocean, the effects of wind and Coriolis effect predominate. In deeper water, the main driver is the thermohaline circulation.

Oxygen minimum zones
Oxygen minimum zones are the places in the ocean where oxygen saturation in the water column is at its lowest. Such zones typically occur at midwater depths (200-1000 m).

Phenology
Phenology is the study of the timing and duration of cyclic and seasonal natural phenomena (e.g., spring phytoplankton blooms, seasonal cycles of zooplankton reproduction), especially in relation to climate and plant and animal life-cycles.

Precautionary Approach
Principle 15 of the 1992 Rio Declaration on Environment and Development[2] provides "In order to protect the environment, the precautionary approach shall be widely applied by States according to their capabilities. Where there are threats of serious or irreversible damage, lack of full scientific certainty shall not be used as a reason for postponing cost-effective measures to prevent environmental degradation". In the context of fisheries management, the Food and Agriculture Organization of the United Nations has defined it as "A set of agreed cost-effective mea-

---

[1] United Nations *Treaty Series*, volume 996-I, No.14583.

[2] A/CONF.151/26.

sures and actions, including future courses of action, which ensures prudent foresight, reduces or avoids risk to the resource, the environment, and the people, to the extent possible, taking explicitly into account existing uncertainties and the potential consequences of being wrong".

Reef
A reef is a solid structure either of rock, or created by accumulations of organisms (often corals) – a "biogenic reef". Reefs rise from the seabed, or at least clearly form a substantial, discrete community or habitat which is very different from the surrounding seabed. The structure of a biogenic reef may be composed almost entirely of the reef-building organism and its tubes or shells, or it may to some degree be composed of sediments, stones and shells bound together by the organisms. Reefs which either reach close to the sea surface or are exposed at low tide are of particular importance as coastal structures.

Reference Points
In the context of fisheries management, "reference points" are defined as "Benchmarks against which the abundance of the stock, the fishing mortality rate or economic and social indicators can be measured in order to determine its status. These reference points can be used as limits or targets, depending on their intended usage."

Runoff
Runoff is the sum of all upstream sources of water, including continental precipitation, fluxes from lakes and aquifers, seasonal snow melt, and melting of mountain glaciers and ice caps.

Salinity
Salinity refers to the level of dissolved salts in seawater. It varies significantly from place to place. On average, it is around 35 grams of salts per litre. In the Red Sea and the Persian Gulf, the most saline major sea areas, it reaches an average 40 grams of salts per litre, because of the high rate of evaporation and the low rate of freshwater inflow.

Salt marshes
Salt marshes are intertidal, coastal ecosystems that are regularly flooded with salt or brackish water and dominated by salt-tolerant grasses, herbs, and low shrubs. They occur in middle and high latitudes worldwide.

Seamounts
Seamounts are predominantly submerged volcanoes, mostly extinct, rising hundreds to thousands of metres above the surrounding seafloor.

Sedimentation
Sedimentation is the consolidation of loose sediments that have accumulated in water or in the atmosphere. The sediments may consist of rock fragments or particles of various sizes (conglomerate, sandstone, shale), the remains or products of animals or plants (certain limestones and coal), the product of chemical action or of evaporation (salt, gypsum, etc.) or a mixture of these materials.

Solubility pump
The solubility pump is the process of physical air-sea flux of atmospheric CO_2 at the ocean surface.

Southern Ocean
The Southern Ocean is defined as all the ocean area south of 60°S.

Steric
Steric refers to density changes in seawater due to changes in heat content and salinity.

Stratification
Ocean stratification occurs when water masses with different properties (such as oxygenation, salinity or temperature) form layers that limit mixing between the water masses.

Territorial sea
Article 2 of UNCLOS provides that "[t]he sovereignty of a coastal State extends, beyond its land territory and internal waters and, in the case of an archipelagic State, its archipelagic waters, to an adjacent belt of sea, described as the territorial sea." Article 3 of UNCLOS provides that "[e]very State has the right to establish the breadth of its territorial sea up to a limit not exceeding 12 nautical miles, measured from baselines determined in accordance with [UNCLOS].

Thermohaline circulation
The thermohaline circulation is a process driven by differences in the density of water water due to temperature ("thermo") and salinity ("haline") in the various parts of the ocean. Currents driven by thermohaline circulation move much slower than surface currents.

Tidal flats
Tidal flats are marine wetlands that are in the zone between high and low tides.

Ton
In this Assessment, the metric ton of 1,000 kilograms is used.

Toxic Algal Blooms
Toxin-producing algae are a diverse group of phytoplankton species with two characteristics in common: (1) they produce toxins which harm people and ecosystems; and (2) their initiation, development and dissipation are governed by species-specific population dynamics and oceanographic conditions.

Trenches
Trenches are long, narrow, characteristically very deep and asymmetrical depressions of the seafloor, with relatively steep sides. See also abyssal plain.

Upper ocean
In this Assessment, the term "upper ocean" is used to describe both the epipelagic zone (between the surface and 200 metres depth) and the

mesopelagic zone (between 200 and 1000 metres depth). The euphotic zone lies within the epipelagic zone (see "euphotic zone").

Water column

The water column is the vertical continuum of water from sea surface to sea-bed.

Water cycle (see Hydrological cycle)

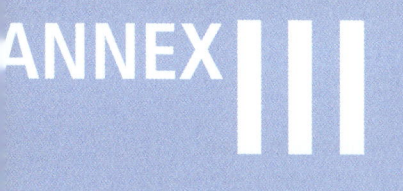

Acronyms

Acronyms

AATAMS	Australian Animal Tracking and Monitoring System
ABA	Arctic Biodiversity Assessment
ABNJ	areas beyond national jurisdiction
ACAP	Agreement on the Conservation of Albatross and Petrels
ACC	Antarctic Circumpolar Current
ACIA	Arctic Climate Impact Assessment
ADCP	Acoustic Doppler Current Profilers
AEWA	Agreement on the Conservation of African-Eurasian Migratory Waterbirds
AEWA	Conservation of African-Eurasian Migratory Waterbirds
AGGRA	Atlantic and Gulf Rapid Reef Assessment
AHWGW	Ad Hoc Working Group of the Whole
AIATSIS	Australian Institute of Aboriginal and Torres Straits Islanders Studies
AIMS	Australian Institute of Marine Science
ALA	Australian Lawyers Alliance
AMAP	Arctic Monitoring and Assessment Program
AMLC	Association of Marine Laboratories of the Caribbean
Anammox	Anaerobic ammonium oxidation
ANAO	Australian National Audit Office
ANDEEP	ANtarctic benthic DEEP-sea biodiversity
ANZECC	Australian and New Zealand Environment and Conservation Council
AoA	Assessment of Assessments
APIS	Antarctic Pack Ice Seals International Programme
ArcOD	Arctic Ocean Diversity project
ARGO	Program of autonomous profiling instruments
AUNAP	Colombian National Authority for Aquaculture and Fisheries
AUV	autonomous underwater vehicle
BAP	Biologically active Phorphorus
BATS	Bermuda Atlantic Time-series Study
BENTANTAR	Antarctic Benthos
BGS	British Geological Survey
BLM	Bureau of Land Management
BMAPA	British Marine Aggregate Producers Association
BOEM	United States Bureau of Ocean Energy Management
BON	biodiversity observation network
BRDs	by-catch reduction devices
CAFF	Conservation of Arctic Flora and Fauna
CALCOFI	California Cooperative Oceanic Fisheries Investigations program
CAML	Census of Antarctic Marine Life
CARICOMP	Caribbean Coastal Marine Productivity Programme
CARSEA	Caribbean Sea Ecosystem Assessment
CBD	Convention on Biological Diversity
CBMP	Arctic Council Circumpolar Biodiversity Monitoring Program
CCAMLR	Commission for the Conservation of Antarctic Marine Living Resources
CCD	Carbonate Compensation Depth
CCE LTER	California Current Ecosystem Long Term Ecological Research
CCSBT	Commission for the Conservation of Southern Bluefin Tuna
CCYIA	China Cruise and Yacht Industry Association
CFCs	chlorofluorocarbons
CHONe	Canadian Healthy Oceans Network
CICES	Common International Classification of Ecosystem Goods and Services
CIRVA	Comité Internacional para la Recuperación de la Vaquita
CITES	Convention on International Trade in Endangered Species of Wild Fauna and Flora
CLIVAR	Climate Variability and Predictability programme
CLME	Caribbean Large Marine Ecosystem Project
CMFRI	Indian Central Marine Fisheries Research Institute
CMS	Convention on the Conservation of Migratory Species of Wild Animals
CNKI	China Knowledge Resource Integrated Database
CNRS	Chize British Antarctic Survey
COBSEA	Coordinating Body on the Seas of East Asia
CoML	Census of Marine Life
CORDIO	Coral Reef Degradation in the Indian Ocean
COSEWIC	Committee on the Status of Endangered Wildlife in Canada
COTS	Crown-of-thorns starfish Acanthaster planci
CPR	Continuous Plankton Recorder
CPUE	catch per unit effort
CSP	central subarctic Pacific
CTD	Conductivity, Temperature and Depth
CTI-CFF	Coral Triangle Initiative on Coral Reefs, Fisheries and Food Security
CTO	Caribbean Tourist Organization
DEFRA	United Kingdom, Department of Environment, Food and Rural Affairs
DFID	United Kingdom, Department for International Development
DFO	Department of Fisheries and Oceans Canada
DfT	United Kingdom, Department for Transport
DPSER	Drivers, Pressures, States, Ecosystem Services, Responses
DPSIR	Driver Pressure State Impact Response
DUML	Duke University Marine Laboratory
DWT	Dead-weight tonnage
EAC	environmental assessment criteria
EASIZ	Ecology of the Antarctic Sea Ice Zone
EBAs	ecosystem-based approaches
EBSAs	Ecologically or Biologically Significant Marine Areas
EBVs	Essential Biodiversity Variables
EcoQOs	ecological quality objectives
ED	electro-dialysis
EDI	electro-de-ionization
EEA	European Environment Agency

Annex III — Acronyms

EEZs	exclusive economic zones
EFH	Essential Fish Habitat
EIA	United States Energy Information Administration
EIA	Environmental Impact Assessments
ENSO	El Niño/Southern Oscillation.
EPOS	European Polarstern Study
ERCCIS	Environment Records Centre for Cornwall
ESCWA	United Nations Economic and Social Commission for Western Asia
ESP	Ecosystem Service Partnership
ESP	eastern subarctic Pacific
EVOLANTA	Evolution in the Antarctic
FESS	Foundation for Environmental Security and Sustainability
FGGE	First GARP Global Experiment
FIPE	Fundação Instituto de Pesquisas Econômicas
FMAP	Future of Marine Animal Populations
FOODBANCS	Food for Benthos on the Antarctic Continental Shelf
GARP	Global Atmosphere Research Programme
GBIF	Global Biodiversity Information Facility
GBRMPA	Great Barrier Reef Marine Park Authority, Australia
GCC	Gulf Cooperation Council
GCN	Global Core Network
GCOS	Global Climate Observing System
GCRMN	Global Coral Reef Monitoring Network
GEMS	Global Environment Monitoring System
GEO BON	Global Biodiversity Observation Network
GEOHAB	Global Ecology and Oceanography of Harmful Algal Blooms
GEOSECS	Geochemical Ocean Sections Study
GESAMP	Joint Group of Experts on the Scientific Aspects of Marine Environmental Protection
GIS	geographic information system
GLOBEC	Global Ocean Ecosystem Dynamics
GLOSS	Global Sea Level Observing System
GMAD	Global Marine Aquarium Database
GODAE	Global Ocean Data Assimilation Experiment
GODP	Global Ocean Drifter Program
GoMA	Gulf of Maine Area project
GoMRI	Gulf of Mexico Research Initiative
GOOS	Global Ocean Observing System
GOSR	Global Ocean Science Report
GPP	Gross primary production
GPS	global positioning systems
GTN-R	Global Terrestrial Network for River Discharge
GTS	WMO's Global Telecommunications System
GWI	Global Water Intelligence
HABs	Harmful Algal Blooms
HAPCs	Habitat Areas of Particular Concern
HBS	Hawaii Biological Survey
HELCOM	Helsinki Commission
HLPE	High Level Panel of Experts on Food Security and Nutrition of the Committee on World Food Security
HNLC	High Nutrient Low Chlorophyll zones
HOTO	Health of the Ocean
IAA	Indo-Australian Arc
IAATO	International Association of Antarctic Tourism Operators
IACS	International Association of Classification Societies
IATTC	Inter-American Tropical Tuna Commission
IBA	International Marine Important Bird Areas
IBAMA	Instituto Brasileiro do Meio Ambiente e dos Recursos Naturais Renováveis
ICCAT	International Commission for the Conservation of Atlantic Tunas
ICES	International Council for the Exploration of the Sea
ICMBIO	Instituto Chico Mendes de Conservação da Biodiversidade
ICRI	International Coral Reef Initiative
ICS	International Chamber of Shipping
ICSICH	Intergovernmental Committee for the Safeguarding of the Intangible Cultural Heritage
ICSU	International Council of Scientific Unions
ICZM	Integrated coastal zone management
IDH	intermediate disturbance hypothesis
IEA	International Energy Agency
IFAW	International Fund for Animal Welfare
IGBP	International Geosphere-Biosphere Programme
IGOSS	IOC/WMO Integrated Global Ocean Services System
IGY	International Geophysical Year (1957-58)
IIOE	International Indian Ocean Expedition
IMA	Institute of Marine Affairs in Jamaica
IMARPE	Instituto del Mar del Peru
IMBER	Integrated Marine Biogeochemistry and Ecosystem Research
IMMS	International Marine Minerals Society
IMO	International Maritime Organization
IMOS	Integrated Marine Observing System
IMR	Norwegian Institute of Marine Research
INVEMAR	Instituto de Investigaciones Marinas y Costeras, Colombia
IOC/UNESCO	Intergovernmental Oceanographic Commission of UNESCO
IOCARIBE	International Oceanographic Commission-Caribe
IOCAS	Chinese Academy of Sciences - Institute of Oceanology
IOCoML	Census of Marine Life Programme
IODP	International Ocean Discovery Programme
IOPCF	International Oil Pollution Compensation Funds
IOSEA	Indian Ocean-Southeast Asia Marine Turtle Memorandum of Understanding
IOTC	Indian Ocean Tuna Commission
IPBES	Intergovernmental Science-Policy Platform on Biodiversity and Ecosystem Services
IPCC	Intergovernmental Panel on Climate Change
IRVSI	University of Delaware, International Research Vessels Schedules and Information

Acronyms

ISA	International Seabed Authority
ISC	International Scientific Committee for Tuna and Tuna-like Species in the North Pacific Ocean
ISME	International Society for Mangrove Ecosystems
ISSF	International Seafood Sustainability Foundation
ITCZ	Intertropical Convergence Zone
ITF	International Transport Workers Federation
ITOPF	International Tanker Owners Federation
IUCN	International Union for the Conservation of Nature
IUU	illegal, unreported and unregulated (fishing)
IWC	International Waterbird Census
IWC	International Whaling Commission
JaLTER	Japan Long-term Ecological Research Network
JBON	Japan Biodiversity Observation Network
JCOMM	WMO-IOC Joint Technical Commission for Oceanography and Marine Meteorology
JGOFS	Joint Global Ocean Flux Study
JNCC	Joint Nature Conservation Committee
JNREG	Joint Norwegian-Russian Expert Group
LC-LP	London Convention and London Protocol
LGP	Latitudinal Gradient Project
LLDCs	Group of Landlocked developing Countries
LMEs	Large Marine Ecosystems
LMR	Living Marine Resources
MAB	UNESCO's Man and the Biosphere Programme
MALSF	Marine Aggregate Levy Sustainability Fund
MAP	Mangrove Action Project
MARAD	United States Department of Transportation, United States Maritime Administration
MAR-ECO	Mid-Atlantic Ridge Ecosystem
MARMAP	Marine Resources Monitoring, Assessment and Prediction
MARPOL	International Convention for the Prevention of Pollution from Ships
MCES	marine and coastal ecosystem services
MCS	monitoring, control and surveillance
MEA	Millennium Ecosystem Assessment
MED	Multiple-Effect-Distillation
MEPC	IMO Marine Environment Protection Committee
MESP	Marine Ecosystems Services Programme.
MFF	Mangrove Alliance, Mangroves for the Future
MGR	marine genetic resources
MIZ	marginal ice zone
MMA	Brazilian Ministry of Environment
MMC	United States Marine Mammal Commission
MMS	Minerals Management Service
MOC	meridional overturning circulation
MPA	Mineral Products Association
MPA	Marine Protected Area
MPMMG	Marine Pollution Monitoring Management Group
MSF	Multi-Stage-Flash
MSFD	Marine Strategy Framework Directive
MSR	Marine Scientific Research
MTSG	Marine Turtle Specialist Group
MTSG	Marine Turtle Specialist Group of IUCN
NAFO	Northwest Atlantic Fisheries Organization
NaGISA	Natural Geography of Inshore Areas
NAMMCO	North Atlantic Marine Mammal Commission
NANI	Net anthropogenic nitrogen input
NAPAs	National Adaptation Programmes of Action
NBSC	National Bureau of Statistic of China
NEAFC	North East Atlantic Fisheries Commission
NEAFC	North East Atlantic Fisheries Commission
NIOP	Netherlands Indian Ocean Programme
NIWA	National Institute of Water and Atmospheric Research, New Zealand
NMFS	USA National Marine Fisheries Service
NMMA	National Marine Manufacturers Association
Norad	Norwegian Agency for Development Cooperation
NOWPAP	North-West Pacific Action Plan
NPFC	North Pacific Fisheries Commission
NPP	Net primary production
NPTZ	North Pacific Transition Zone
NRC	United States National Research Council
NRSMPA	Australia's National Representative System of Marine Protected Areas
NZ EPA	New Zealand Environmental Protection Authority
OA	Ocean Acidification
OBG	Oxford Business Group
OBIS	Ocean Biogeographic Information System
OBIS	Ocean Biodiversity Information System
OBIS-SEAMAP	Ocean Biogeographic Information System Spatial Ecological Analysis of Megavertebrate Populations
OCSEAP	MMS Outer Continental Shelf Environmental Assessment Program Minerals Management Service
OHI	Ocean Health Index
OMZs	oxygen minimum zones
OSPAR	Convention for the Protection of the Marine Environment of the North-East Atlantic, formerly the Oslo-Paris Convention
OUC	Ocean University of China
PAR	photosynthetically active radiation
pCO2	Measurements of the "partial pressure of CO_2"
PDO	Pacific decadal oscillation
PIPA	Phoenix Islands Protected Area
PMI	IUCN's Pacific mangrove initiative
POC	particulate organic carbon
POST	Pacific Ocean Shelf Tracking
PSP	paralytic shellfish poisoning
PW	produced water
PUB	Singapore Public Utilities Board
REDD+	programme - Reduced Emissions from Deforestation and forest Degradation
Redfield Ratio	a C:N:P ratio of 106:16:1 (Redfield et al., 1963)
REMPEC	Regional Marine Pollution Emergency Response Centre for the Mediterranean

RES	relative environmental suitability	SWIOFC	Southwest Indian Ocean Fisheries Commission
RESA	rapid ecosystem service assessment	TAC	total allowable catch
RFMO/As	regional fisheries management organizations and arrangements	TAC	total allowable catches
		TAO	Tropical Atmosphere Ocean project
RFMOs	fisheries management organizations	Tbps	Terabits per second
RMUs	regional management units	TC/OPC	Technical Committee for Ocean Processes and Climate
RNSIIPG	Peru's Guano Islands, Islets and Capes National Reserve System	TEEB	The Economics of Ecosystems and Biodiversity
		TNC	the Nature Conservancy
RO	Reverse osmosis	TOGA	Tropical Ocean-Global Atmosphere Study
ROS	reactive oxygen species	TOPP	Tagging of Pacific Pelagics
ROVs	remotely operated vehicles	TRB	Transportation Research Board
RSMAS	Rosenstiel School of Marine and Atmospheric Sciences of the University of Miami	TRL	Technological Readiness Level
		TSG	ThermoSalinoGraphs
		TZCF	transition-zone chlorophyll front
RSPB	Royal Society for the Protection of Birds	UKHO	United Kingdom Hydrographic Office
RSPB	Royal Society for the Protection of Birds	UKTAG	United Kingdom Technical Advisory Group on the Water Framework Directive
SAB	South Atlantic Bight		
SAFARI	Societal Applications in Fisheries and Aquaculture using Remotely-Sensed Imagery	UNCTAD	United Nations Conference on Trade and Development
		UNEP/GEMS	United Nations Environment Programme, Global Environment Monitoring System
SAIAB	South African Institute of Aquatic Biodiversity		
SARCE	South American Research Group in Coastal Ecosystems	UNESCO	United Nations Educational, Scientific and Cultural Organization
SCAR	Scientific Committee on Antarctic Research		
SCD	Supply Chain Digest	UNGA	United Nations General Assembly
SCSIOCAS	South China Sea Institute of Oceanology	UNSC	United Nations Security Council
SEAFO	Southeast Atlantic Fisheries Organization	UNSCEAR	United Nations Scientific Committee on the Effects of Atomic Radiation
SEAMAP	Southeast Area Monitoring and Assessment Program		
SEEA	System of Environmental-Economic Accounting.	UOHC	upper ocean heat content
SEEF	Seamount Ecosystem Evaluation Framework	USCG	United States Coast Guard
SGIMC	Study Group on Integrated Monitoring of Contaminants and Biological Effects	USGS	United States Geological Survey
		USMA	United States Maritime Administration
SIDS	small island developing States	UWI	University of the West Indies
SIMAC	Sistema Nacional de Monitoreo de Arrecifes Coralinos en Colombia	VC	vapour compression
		VMEs	Vulnerable Marine Ecosystems
SIOFA	Southern Indian Ocean Fisheries Agreement	WAVES	Wealth Accounting and the Valuation of Ecosystem Services
SLR	sea-level rise		
SMOL	Seamounts Online	WCPFC	Western and Central Pacific Fisheries Commission
SOA	State Oceanic Administration of the People's Republic of China	WCPFC	Western and Central Pacific Fisheries Commission
		WCRP	World Climate Research Programme
SOOP	Ship-of-Opportunity Programme	WHO	World Health Organization
SOOPIP	Ship-of-Opportunity Programme and its Implementation Panel	WIDECAST	Wider Caribbean Sea Turtle Conservation Network
		WISE	World Information System on Energy
SOTO	Canada's State of the Oceans Report	WNA	World Nuclear Association
SPAW	Caribbean Protocol Concerning Specially Protected Areas and Wildlife to the Convention for the Protection and Development of the Marine Environment in the Wider Caribbean Region	WOA	World Ocean Assessment
		WOCE	World Ocean Circulation Experiment
		WOD	World Ocean Database
		WOMARS	Worldwide marine radioactivity studies
SPC	Secretariat of the Pacific Community	WRI	World Resources Institute
SPRFMO	South Pacific Fisheries Management Organization	WSP	western subarctic Pacific
SRREN	Special report on Renewable Energy Sources and Climate Change Mitigation	WSSD	World Summit on Sustainable Development
		XBT	Expendable Bathythermograph
SST	Sea-surface temperature	XCTD	expendable disposable conductivity, temperature and depth
SWEC	South-West Economy Centre, University of Plymouth		
SWIOFC	South West Indian Ocean Fisheries Commission		